T. Ishiguro K. Kajimura (Eds.)

Advances in Superconductivity II

Proceedings of the 2nd International Symposium on Superconductivity (ISS '89), November 14-17, 1989, Tsukuba

ISTEC

With 1125 Figures

Springer-Verlag Tokyo Berlin Heidelberg New York
London Paris Hong Kong

Prof. Dr. TAKEHIKO ISHIGURO
Department of Physics
Kyoto University
Kyoto, 606 Japan

Dr. KOJI KAJIMURA
Electrotechnical Laboratory
Tsukuba, 305 Japan

ISBN 978-4-431-68119-9 ISBN 978-4-431-68117-5 (eBook)
DOI 10.1007/978-4-431-68117-5

Softcover reprint of the hardcover 1st edition 1990

Foreword

Since the First International Symposium on Superconductivity (ISS '88) was held in Nagoya, Japan in 1988, significant advances have been achieved in a wide range of high temperature superconductivity research. Although the T_C's of recently discovered oxide superconductors still do not exceed the record high value of 125K reported before that meeting, the enrichment in the variety of materials should prove useful to the investigation of the fundamental mechanism of superconductivity in these exotic materials. The discovery of the *n*-type superconducting oxides proved to oppose the previously held empirical fact that the charge carriers in all oxide superconductors were holes. In addition, optimization of the charge carrier density has been established as a technique to improve the superconducting properties of the previously known oxide materials. Many new experimental and theoretical advances have been made in understanding both the fundamental and the applied aspects of high temperature superconductivity. In this latter area, various new processing techniques have been investigated, and the critical current densities and other significant parameters of both bulk and thin film oxide superconductors are rapidly being improved.

At this exciting stage of research in high temperature superconductivity, it is extremely important to provide an opportunity for researchers from industry, academia, government and other institutions around the world to freely exchange information and thus contribute to the further advancement of research. For this purpose, the International Superconductivity Technology Center (ISTEC) sponsored ISS '89 held in Tsukuba, Japan, the "Science City" where 52 government laboratories and more than 150 laboratories in the private sector are located. Through inter-institutional and international competion and collaboration, we can all carry out a task that is too large for any one individual, institute, company or nation. It is to this goal that we present these proceedings.

SHOJI TANAKA
Vice President, ISTEC

Preface

The Second International Symposium on Superconductivity was held in Tsukuba on November 14 to 17, 1989, under the cosponsorship of the International Superconductivity Technology Center and Ibaraki Prefecture. It was attended by 720 participants including 57 from 14 different countries. One of the characteristics of the Symposium was the intention to offer a forum to the world's scientists, engineers, business administrators and enterpreneurs where they could exchange and discuss their views and findings about superconductivity and its future applications. In particular, many papers from industrial laboratories were presented together with those from universities and national laboratories.

The Symposium covered a wide range of topics from basic reserarch to applications, categorized as physics and chemistry, thin films, bulk materials, wires and tapes, devices and systems. Almost half of the presented papers were classified into the physics and chemistry of high-T_c superconductors. New aspects of the physical properties, properties in newly developed materials, and refined results from qualified materials and new methods were reportred, offering evidence of steady advances being made in basic research. Emphasis was on the flux motion and flux creep, relevant to the limitations in the critical current encountered in these initial stages. To make the materials useful, considerable efforts have been made to develop high quality thin films, refined bulk materials, wires and tapes. Remarkable progress was reported during the Symposium in the improvement of the critical current and zero-resistivity temperature. At the same time, although research and development of the high-T_c superconductors are at their basic stages, views and findings on highly motivated attempts to realize high-perfomance electronic devices and electrical machines were presented during the Symposium.

This volume contains 30 invited papers and 200 refereed contributions presented at the Symposium. We acknowledge the help of the International Advisory Committee and Program Committee in selecting the invited papers and refereeing the contributed papers. We are also most grateful to the many people who worked to organize the Symposium.

Finally we are very grateful to the authors for their cooperation and the care with which they prepared their manuscripts for these Proceedings.

TAKEHIKO ISHIGURO
KOJI KAJIMURA

Organization of ISS '89

Sponsored by:
International Superconductivity Technology Center (ISTEC)

Co-Sponsored by:
Ibaraki Prefecture

Supported by:
Ministry of International Trade and Industry
Ceramic Society of Japan
The Chemical Society of Japan
The Cryogenic Society of Japan
The Federation of Electric Power Companies
The Institute of Electrical Engineers of Japan
The Institute of Electronics Information and Communication Engineers, Japan
The Iron and Steel Institute of Japan
Japan Chemical Industry Association
The Japan Electrical Manufacturers' Association
Japan Electronic Industry Development Association
Japan Fine Ceramics Association
The Japan Institute of Metals
The Japan Iron and Steel Federation
The Japan Key Technology Center
Japan Mining Industry Association
The Japan Society of Applied Physics
The Japanese Electric Wire & Cable Makers' Association
The Magnetic Society of Japan
New Energy and Industrial Technology Development Organization
Physical Society of Japan
Research and Development Association for Future Electron Devices
Research Park Tsukuba Center

Organization Committee

Chairman:

G. HIRAIWA	President, ISTEC

Members:

K. FUEKI	Professor, Science University of Tokyo
H. KASHIWAGI	Director General, Electrotechnical Laboratory
T. KAWAKAMI	Vice President, ISTEC (President, Sumitomo Electric Industries, Ltd.)
K. MITA	Vice President, ISTEC (President, Hitachi Ltd.)
Y. MUTO	Professor, Tohoku University
S. NAKAJIMA	Professor, Tokai University
S. SAITO	Professor Emeritus, Tokyo Institute of Technology
M. SUGIURA	Director-General, AIST, MITI
F. TAKEUCHI	Governor, Ibaraki Prefecture
S. TANAKA	Vice President, ISTEC
S. YAMAMURA	Professor Emeritus, University of Tokyo

International Advisory Committee

B. BATLOGG	AT&T Bell Laboratories
M. R. BEASLEY	Stanford University
H. K. BOWEN	MIT
P. CHAUDHURI	IBM Watson Research Center
C. W. CHU	Houston University
V. J. EMERY	Brookhaven National Laboratory
J. E. EVETTS	University of Cambridge
Ø. FISHER	University of Geneva
D. K. FINNEMORE	AMES Laboratory
F. Y. FRADIN	Argonne National Laboratory
K. FUEKI	Science University of Tokyo
T. H. GEBALLE	Stanford University
D. JÉROME	Universite de Paris-Sud
M. B. MAPLE	University of California
K. A. MÜLLER	IBM Zürich
Y. MUTO	Tohoku University
S. NAKAJIMA	Tokai University
H. R. OTT	ETH Zürich
B. RAVEAU	Universite de Caen
A. W. SLEIGHT	Oregon State University
M. SUENAGA	Brookhaven National Laboratory

Steering Committee

Chairman:

S. TANAKA	ISTEC

Members:

T. HIRAYAMA	The Tokyo Electric Power Co., Inc.
T. INOUE	Nippon Steel Corporation
T. ISHIGURO	Kyoto University
K. KAJIMURA	Electrotechnical Laboratory

Y. KAKUTA	Ibaraki Prefecture
S. KATAOKA	Sharp Corporation
M. KAWASHIMA	Sumitomo Electric Industries, Ltd.
K. KITAZAWA	University of Tokyo
H. OGIWARA	Toshiba Corporation
Y. TAKEDA	Hitachi, Ltd.

Program Committee

Chairman:

T. ISHIGURO	Kyoto University

Members:

J. FUJIE	Railway Technical Research Institute
H. FUKUYAMA	University of Tokyo
Y. FURUTO	The Furukawa Electric Co., Ltd.
S. HASUO	Fujitsu Laboratories Ltd.
H. HAYAKAWA	Nagoya University
O. HORIGAMI	Toshiba Corporation
K. KAJIMURA	Electrotechnical Laboratory
K. KITAZAWA	University of Tokyo
T. KOBAYASHI	Osaka University
N. KOSHIZUKA	SRL, ISTEC
E. MASADA	University of Tokyo
T. MATSUSHITA	Kyushu University
K. SATO	Mitsubishi Electric Industrial Co., Ltd.
Y. SYONO	Tohoku University
T. TANAKA	CRIEPI
A. YAMAJI	NTT
S. YAZU	Sumitomo Electric Industries, Ltd.
R. YOSHIZAKI	University of Tsukuba
K. WASA	Matsushita Electric Industrial Co., Ltd.

General Affairs Committee

Chairman:

K. KAJIMURA	Electrotechnical Laboratory

Members:

K. HARASAWA	The Tokyo Electric Power Co., Inc.
Y. KAWABATA	National Chemical Laboratory for Industry
N. KAWAI	Kobe Steel, Ltd.
T. KITABATA	Ibaraki Prefecture
M. KOYANAGI	Electrotechnical Laboratory
N. MAKI	Hitachi Ltd.
A. MIZUKAMI	Sanyo Electric Co., Ltd.
H. NOMURA	Research Park Tsukuba Center
H. OSANAI	Fujikura, Ltd.
Y. OTEKI	Ibaraki Prefectural Industrial Technology Center
M. TAKISIMA	Tokyo Bureau of International
N. TSUBOUCHI	NEC

Table of Contents

1 Plenary Lectures

2 Material Preparation

2.1 Synthesis and Structure

2.2 Substitution and Doping

2.3 Phase Diagram and Crystal Growth

2.4 Melt Growth

2.5 Wires

3 Physical Properties

3.1 Mechanism of Superconductivity

3.2 Electronic Structures

3.3 Transport and Tunneling Properties

3.4 ESR and Mossbauer Studies

3.5 Optical Properties

3.6 Mechanical Properties

3.7 Magnetic Properties

3.8 Flux Creep and Flux Motion

4 Preparation and Properties of Thin Films

4.1 YBCO and Related Films

4.3 Other Superconducting Films

5 Applications

5.1 Electronic Use

5.2 Energy Systems

1 Plenary Lectures

The Development of Superconductivity Research in Oxides

K. ALEX MÜLLER

IBM Research Division, Zurich Research Laboratory, 8803 Rüschlikon, Switzerland
and University of Zurich, Physics Department, 8001 Zurich, Switzerland

ABSTRACT

Starting with the first observation of superconductivity in an oxide, the history of its development is traced. Basically, and consecutively, three kinds of oxide superconductors have been found: Compounds with normal transition-metal conduction bands, oxides with cations exhibiting charge disproportionation, and finally the cuprates with large coulomb on-site repulsion, U. This discussion will lead over to a characterization of the highest-T_c materials, both concerning their physical properties and application perspectives.

KEY WORDS: superconductivity, oxides

The first oxide in which superconductivity was found is reduced $SrTiO_3$ as reported by Schooley, Hosler and Cohen [1] in 1964. A $T_c \simeq 0.25$ K was observed with only 3×10^{19} electron carriers per cubic centimeter, i.e. at a carrier concentration three orders of magnitude below that of a normal metallic conductor. This phenomenon was clearly outside the accepted BCS picture of electron coupling by shielded phonons. Only recently — 17 years later — was the underlying process understood: because the carrier concentration is so low, the plasma edge is *below* the highest optical phonon branch in $SrTiO_3$. This phonon branch is therefore unshielded and the electron-phonon coupling parameter λ sufficiently large to induce superconductivity. Upon Nb doping of $SrTiO_3$, the electron concentration and the plasma edge increases, the edge passes this phonon branch which becomes shielded, and superconductivity vanishes at $n \simeq 10^{21}$/cc and $T_c \simeq 1.3$ K [2].

Nine months after the $SrTiO_3$ discovery, Matthias' group reported superconductivity in the sodium tungsten bronze Na_xWO_3 with $x \simeq 0.3$ and a $T_c = 0.57$ K [3]. So it was clear that the phenomenon also occurred with rather small electron concentrations — but less so — in another oxide. Interestingly, in the Na_xWO_3 paper no reference to the over half a year older report on $SrTiO_3$ is given. The $SrTiO_3$ crystal showed quite recently another interesting property. Band calculations had yielded two conduction bands 20 meV apart at the center of the Brillouin zone ($\vec{k} = 0$). Thus by increasing the doping, one expected that after the first the second band would begin to fill. Consequently, one- and two-band superconductivity in the *same* material was expected and indeed observed by tunneling in Rüschlikon [4].

In 1965, superconductivity was also reported in TiO and NbO at temperatures of 0.68 and 1.25 K, respectively [5]. However, these results did not meet with great interest because in NbN, with the same NaCl structure, a T_c of 16 K had already been observed back in 1941 [6]. After the superconducting bronzes had been found, the next substantial step forward in the oxides was the observation of a transition in the lithium titanium spinel $LiTi_2O_4$ with $T_c \simeq 11$ K by Johnston *et al.* in 1973 at San Diego [7]. This represented already a quite respectable transition temperature. However, research in the spinel compound was not so intensive, probably because no single crystals became available despite the fact that a T_c of 13.7 K was reached in the mixed phase $Li_{1+x}Ti_{2-y}O_4$ [8].

Transition temperatures of comparable magnitude were reported only two years later in the perovskite $BaPb_{1-x}Bi_xO_3$ system by Sleight *et al.* [9]. Remarkably this quaternary compound did not contain transition metals. Subsequent studies on hot-pressed ceramics [10] and single crystals [11] allowed, in conjunction with expert band calculations [12], the characterization of its properties. The highest T_c was near 11.2 K with $x \simeq 0.25$ and a carrier concentration of 2.4×10^{21} cm^{-1}; correlations lengths were 60−70 Å and the Ginzburg-Landau parameter $\chi \simeq 70-80$. The transition temperature was found to be proportional to the *intensity* of a phonon band at 100 cm^{-1} [13], probably indicative of a substantial Bi(Pb)-O anharmonicity in σ-bonding. More recently, by doping the charge-disproportionated insulator $BaBiO_3$ with K^+ on Ba^{2+} sites, an increase of T_c to 30 K was realized [14]. This is so far the superconductor with the highest T_c that does not contain transition-metal ions. Its isotope effect is substantial, namely $\beta = 0.22$ to 0.4, i.e. nearly fully developed [15], indicating that dynamic atomic motion is responsible for the coupling.

The third class of oxide superconductors was found in 1986 with the Ba^{2+}-doped La_2CuO_4 [16]. Characteristic of this class is the layered structure containing perovskite Cu-O planes. These planes can be stacked singly, doubly, triply or quadruply with either Cu-O chains, planes or single-double planes of other oxides (La, Bi, Tl, etc.) between them. More than a dozen such compounds have been synthesized, with a confirmed maximum of 125 K for $Tl_{1.6}Ca_{1.8}Ba_2Cu_{3.1}O_{10.1}$ [17]. The compounds with the highest T_c's are all *hole* superconductors despite the fact that electron ones have also been found [18]. The coherence length ξ is highly anisotropic, 20 to 30 Å in the planes and 3 to 7 Å perpendicular to them. The Ginzburg-Landau parameter χ is of the order of 100 in plane and perpendicular to it over 1000. This is relevant for applications concerning both critical currents and magnetic fields. The latter become very high in the megagauss range, because $H_{c2} = \phi_2/\pi\xi^2$. The flux pinning observed is atypical. Their carrier concentration is of the order of a few times 10^{21} cm^{-1} [19] substantially lower than that of the earlier high-T_c superconductors, which were all Nb intermetallic alloys or compounds. So far, no other class of high-T_c superconductors has been found – which makes one wonder whether the copper oxides are unique.

In order to make progress, an understanding of the microscopic mechanism is quite relevant. This is reflected in the very large research effort world-wide. Of course, development for use in industry is the other motor. Industrial applications can now progress empirically and do so successfully, based on the experimental findings and expert materials research. These have so far advanced the field [19], and allowed certain microscopic mechanisms to be excluded: on the conservative side those using *harmonic* electron-phonon interactions, on the exotic side the original resonance valence bond and the more recent fractional quantum state theories. Central to all attempts is the insight that a sizeable coulomb repulsion U is operative at the copper and oxygen ions in order to accept further hole carriers on the same site. Thus the ratio U/t, where t stand for the transfer integral, is substantial compared to the ratio Δ/t, where Δ is the energy needed to transfer Cu d-charge onto the neighboring oxygen p orbitals. This is opposite to the process in d-d oxide superconductors with low Hubbard U discovered first [20].

Therefore, one can categorize three classes of known oxide superconductors as follows:
The first ranges from $SrTiO_3$ to $LiTi_2O_4$ and T_c's from 0.2 to 13 K, with $\Delta > U$. In the second class are the valence skippers of the bismuthates with T_c's from 13 to 32 K. The third, with $U \gg \Delta$, comprises compounds with T_c's from 35 K as in La_2CuO_4 to well over 100 K. Regarding their transition temperatures, we therefore note that each of the three categories has its own transition temperatures adjacent to those of next one: the first up to 13 K, valence skippers up to 32 K, and cuprates above them up to 125 K so far. Is there perhaps a fourth category?

REFERENCES

1. Schooley JF, Hosler WR, Cohen ML (1964) Phys. Rev. Lett. 12:474
2. Baratoff A, Binnig G, Bednorz JG, Gervais F, Servoin JL (1982) In: Buckel W, Weber W (Eds) *Superconductivity in d- and f-Band Metals, Proc. IV Conference on Superconductivity in d- and f-Band Metals*, Kernforschungszentrum Karlsruhe GmbH, Karlsruhe, W. Germany. p. 419
3. Raab ChJ, Sweedler AR, Jensen MA, Broadston S, Matthias BT (1964) Phys. Rev. Lett. 13:746
4. Binnig G, Baratoff A, Hoenig HE, Bednorz JG (1980) Phys. Rev. Lett. 45:1352
5. Hulm JK, Jones CK, Mazelsky R, Miller RC, Heim RA, Gibson JW (1965) In: *Low Temperature Physics*, LT-9. Plenum, New York, 1965. Part A, p. 600
6. Aschermann G, Friedrich E, Justi E, Kramer J (1941) Phys. Z. 42:349. A historical NbN sample from January 15, 1943 with $T_c = 16.1$ K is now in the author's possession.

7. Johnston DC, Prakash H, Zachariasen WH, Viswanathan R (1973) Mater. Res. Bull. 8:777

8. Johnston DC (1976) J. Low Temp. Phys. 25:145

9. Sleight AW, Gillson JL, Bierstedt PE (1975) Solid State Commun. 17:27

10. Thanh TD, Koma A, Tanaka S (1980) Appl. Phys. 22:205

11. Batlogg B (1984) Physica 126 B:275

12. Mattheiss LF, Hamann DR (1982) Phys. Rev. B 26:2682

13. Sugai S, Uchida S, Kitazawa K, Tanaka S, Katsui A (1985) Phys. Rev. Lett. 55:426

14. Cava RJ, Batlogg B, Krajewski JJ, Farrow R, Rapp Jr. LW, White AE, Short K, Peck WF, Kometani T (1988) Nature 322:814; Matheiss LF, Gyorgy EM, Johnson Jr DW (1988) Phys. Rev. B 37:3734

15. Hinks, DG, Richards DR, Dabrowski B, Marx DT, Mitchell AW (1988) Nature 335:419; Kondoh S, Sera M, Ando Y, Sato M (1989) Physica C 157:469

16. Bednorz JG, Müller KA (1986) Z. Phys. B 64:189; Bednorz JG, Takashige M, Müller KA (1987) Europhys. Lett. 3:379

17. Beyers RB, Parkin SSP, Lee VY, Nazzal AI, Savoy RJ, Gorman GL, Huang TC, La Placa SJ (1989) p. 228 in Ref. 19

18. Tokura Y, Takagi H, Uchida S (1989) Nature 337:345

19. See the May 1989 issue of the IBM Journal of Research and Development, Vol. 33, No. 3 (1989) with 26 pertinent articles.

20. Zaanen J, Sawatzky GA, Allen JW (1985) Phys. Rev. Lett. 55:418

Present Status of the Theory of High-Temperature Superconductivity

ELIHU ABRAHAMS
Rutgers University, Serin Physics Laboratory, P.O. Box 849, Piscataway, NJ 08855, USA

ABSTRACT

A critical introduction and review is given of theoretical activities connected with the physics of high-temperature superconductors. The topics discussed include: Pairing mechanisms, low-dimensional electron systems with strong correlations, properties of doped and disordered quantum antiferromagnets, model hamiltonians, flux phases, the quantum spin liquid, model wave functions, fractional statistics and superconductivity, theory of anomalous normal state properties.

INTRODUCTION

The wonderful discovery, by Bednorz and Müller [1], of high-temperature superconductivity in perovskite-based cuprates is now three years old. As is well-known, an enormous effort has gone into theoretical work whose goal is to account for the superconductivity in the high-T_c compounds. However, a consensus regarding the mechanism which is responsible for the superconductivity does not exist. Moreover, a systematic theoretical account of the several anomalous normal-state properties has only very recently been attempted [2,3] and is still in a rudimentary state.

By way of introduction to the theoretical issues, let us recall the Bardeen, Cooper and Schrieffer (BCS) [4] theory not only because of its brilliant success in explaining "ordinary" superconductivity but because it is a paradigm for the practice of theoretical condensed matter physics. The BCS theory starts from the complete hamiltonian for the metallic solid and systematically removes the irrelevant degrees of freedom. Even the electron-phonon interaction upon which the theory was based is not treated in detail. The low-energy physics which describes the superconductivity is all contained within the extremely simple BCS "reduced" hamiltonian. The reduced hamiltonian contains only the kinetic energy of the Fermi-liquid quasiparticles which characterize the normal state plus a zero-range interaction between quasiparticles of opposite spin. The reduction of the complicated full hamiltonian to a simple one which contains all the essential physics of a particular problem is of course the standard way to proceed and the BCS theory is a particularly striking example of the success of this approach.

It is therefore natural that a major component of the theoretical activity in high-T_c superconductivity has been a search for a proper model hamiltonian. This search is guided by the structure of the compounds and the experimental phase diagram. Because of the highly anisotropic properties, most people believe that the essential structural elements are the Cu-O planes and the $d_{x^2-y^2}$ and p_σ orbitals of copper and oxygen which appear to make up the valence states responsible for the electronic properties. The phase diagram of lanthanum cuprate doped with strontium or barium is perhaps the simplest one in the high-T_c family. A variety of physical effects are observed, all of which need to be explained: Antiferromagnetism, disordered magnetism, superconductivity, metal-insulator transition, normal state properties, structural instability. Several key questions related to these properties are being emphasized in current theoretical work. These are:

1. What is the nature of the normal metallic state?
2. What is the mechanism for and the characterization of the superconducting state?
3. What is the nature of the disappearance of antiferromagnetism with doping?

The answers to these and other questions may all eventually be given within a single model for the high-T_c compounds.

Because of our experience with conventional metallic oxides and related compounds which in some properties are similar to the cuprate superconductors, we believe that the physics is dominated by Coulomb interactions which may be modeled by strong local repulsions. This tends to inhibit metallic behavior to some extent and puts these materials in a class which can be named [5] "strongly correlated electrons in narrow bands." This is not a new problem and it includes several materials and effects which are a major challenge for contemporary many-body physics:

1. Transition metal oxides.
2. Mott insulators.
3. Magnetism.
4. Metal-insulator transition.
5. Heavy fermion metals.
6. High-T_c superconductivity.

The new feature for the high-T_c problem is the two-dimensional (2D) character of the essential structural elements. For a complete description of the superconductivity, the interplanar coupling must finally be considered [6], but it is believed that the key to understanding the superconducting and normal states is likely to be found in the properties of an appropriate 2D model. With the lesson of BCS in mind, one begins with a 2D hamiltonian of the Cu-O planes containing orbital energies, hybridization terms and Coulomb interactions and tries to simplify it.

MODEL HAMILTONIANS

There is a minimal Hamiltonian [7] which contains essential aspects of the electronic structure of Cu-O superconductors; it is an extended Hubbard model:

$$H = \varepsilon_d \Sigma n_d + \varepsilon_p \Sigma n_p + U_d \Sigma n_{d\uparrow} n_{d\downarrow} + U_p \Sigma n_{p\uparrow} n_{p\downarrow} + \Sigma_{<nn>} [V n_d n_p - t \Sigma_\sigma d_\sigma^+ p_\sigma + h.c.)]. \quad (1)$$

In Eq. (1), the sums are over lattice sites of Cu ions (denoted by d) and O ions (denoted by p). The subscripted sum is over nearest neighbors only. The kinetic energy is represented by the hopping term with the hybridization matrix element t. The intersite Coulomb repulsion V is the leading contribution to the Madelung potential; it affects the charge fluctuation spectrum, produces excitonic effects and tunes, in some regions of parameter space, between spin and charge density wave instabilities and between Mott insulator and charge-transfer insulator. The hamiltonian of Eq. (1) appears to be required for a treatment of pairing superconductivity mediated by the exchange of charge fluctuations [8].

The "simplest" hamiltonian for our problem is the Hubbard model which contains only an on-site repulsion U. It arises from the starting hamiltonian by an approximation developed by Anderson [9] called superexchange. Here it sufficient to consider the Cu-O hybridized valence orbitals on a square lattice and include a strong coulomb repulsion U for electrons in the same orbital:

$$H = -t\Sigma_{<ij>,\sigma} c_{i\sigma}^{+} c_{j\sigma} + U\Sigma_i n_{i\uparrow} n_{i\downarrow}. \qquad (2)$$

This simple hamiltonian does give many of the properties seen in transition metal oxides. For example, at one electron per site ("1/2-filling") and very large U, the ground state is a Mott insulator with a charge gap of order U. This is the situation for the undoped cuprates. Actually, as a consequence of the Pauli principle, there is an attraction between electrons of opposite spin and the ground state is antiferromagnetic (AFM). Therefore, at 1/2-filling, for large U/t, the hamiltonian is even simpler and contains no kinetic energy term:

$$H = J\Sigma_{<ij>} \mathbf{S}_i \cdot \mathbf{S}_j, \qquad J = 4t^2/U, \qquad (3)$$

where $\mathbf{S}_i = c_{i\alpha}^{+} \boldsymbol{\sigma} c_{j\beta}$. Typically, for cuprates, the charge gap U is 60,000 K, while the characteristic spin excitation energy J is of order 1000 K. This separation of charge and spin energy scales has been a key element of much of the thinking about strongly interacting electrons, especially in the high-T_c problem [10]. An alternate view is that the large U/t limit is not appropriate and that the insulating behavior is a consequence of a spin-density wave instability which opens a gap in the electronic spectrum at and near 1/2-filling [11].

When the model is doped away from 1/2-filling, by adding holes, say, the kinetic energy reappears. A popular model to describe the situation in a way which naturally recognizes the separation of charge and spin energy scales is given by the "t-J" hamiltonian [12] in which the strong on-site repulsion represented by U in Eq. (2) does not appear explicitly but is accounted for by the constraint that two electrons may not occupy the same site:

$$H = -t\Sigma_{<ij>,\sigma} c_{i\sigma}^{+} c_{j\sigma} + J\Sigma_{<ij>}[S_i \cdot S_j - n_i n_j/4], \qquad n_i \le 1. \qquad (4)$$

The hamiltonians of Eqs. (2-4) form the basis for several approaches involving magnetic mechanisms (for example, superconductive pairing mediated by the exchange of AFM spin fluctuations), which are motivated by what is perceived as the proximity of the superconducting compositions to an antiferromagnetic state or one with strong long-range AFM-like correlations. The difficult issue in the treatment of hamiltonians is the treatment of the constraint. A solution to the infinite U Hubbard model [Eq. (4) with J=0] in 2D is still a main problem of strong correlation physics.

MODEL QUANTUM STATES

We have seen that the reduction of the degrees of freedom in the hamiltonian, in the case of large Coulomb repulsion, gives rise to a spin-only problem at 1/2-filling. Indeed, the undoped cuprates (pure lanthanum cuprate or $YBa_2Cu_3O_6$) are antiferromagnetic insulators, rather well described by the Heisenberg hamiltonian of Eq. (3). However, the superconductors are not undoped insulators so we may ask how the system behaves as one moves away from the AFM state by doping with holes or otherwise introducing frustrating interactions. One possibility is that a featureless spin liquid, an overall singlet with possible short-range AFM correlations, becomes stable, at sufficiently high doping, as the AFM order melts. This novel 2D quantum spin liquid (QSL) has its origins in Anderson's [10] formulation of the resonating valence bond (RVB) state. The approach here is to discuss the character of the elementary excitations out of the possible QSL ground state. In RVB theory, these are chargeless spin (spinons) and spinless charge (holons) excitations. The spectra and statistics of holons and spinons determine the normal-state and superconducting properties [3,12].

The generalized flux phases are a type of 2D QSL in which local spin and charge current configurations give rise to particular patterns of the phase of the many-body wavefunction. For example, such states may be found in mean-field treatments of the t-J model [13]. It is possible to have a situation, away from 1/2-filling, in which parity P and time reversal T are spontaneously violated. In this case fractional statistics quasiparticles (anyons) may arise [14], a possibility which is a consequence of two-dimensionality [15]. A collection of interacting anyons can exhibit superconducting properties [14,16].

ANOMALOUS NORMAL STATE PROPERTIES

Some time ago [12], P.W. Anderson pointed out that the observed normal-state properties of high-T_c compounds are at variance with a conventional Fermi-liquid picture with well-defined fermionic quasiparticles and must be accounted for by the properties of 2D electrons with strong correlations. The electrical resistivity $\rho(T)$, the thermal conductivity $\kappa(T)$, the optical conductivity $\sigma(\omega)$, the Raman scattering intensity $S(\omega)$, the tunneling conductance as a function of voltage $g(V)$, the nuclear magnetic resonance relaxation rate $T_1^{-1}(T)$, the Hall coefficient $R_H(T)$, and the angle-resolved photoemission spectrum are all anomalous. One is tempted to believe that whatever is responsible for the unusual normal state properties must also be related to the superconductivity.

Recently, the phenomenology of these properties has been addressed in a systematic way [2]. A simple hypothesis about the charge and spin susceptibilities, namely that the energy scale for their low frequency behavior is the temperature rather than the Fermi energy, results in behaviors for $\sigma(\omega)$, $S(\omega)$ and $T_1^{-1}(T)$ which is are reasonable accord with experiment. The hypothesis leads to interesting behavior of the single particle spectrum which determines $\rho(T)$, $g(V)$ and the photoemission spectra. The low-energy single particle excitations have very little quasiparticle character and the usual Fermi-liquid picture breaks down. This has been described as a "marginal" Fermi liquid [2]. The origin of the hypothesized behavior is still under study and the possibility that the same excitations can mediate superconductive pairing has been analysed [17].

Anderson has very recently discussed [3] how the elementary excitations of the 2D RVB state - spinons and holons - give rise to an anomalous, non-Fermi-liquid, density of states (DOS) for the single-particle excitations. Such a DOS leads both to an enhanced Cooper-pair susceptibility in the normal state and to anomalous behaviors of various experimental quantities, in particular photoemission spectra, $\rho(T)$ and $T_1^{-1}(T)$, and which again are in reasonable accord with experiment. Anderson's analysis implies that charge-spin separation (as occurs in 1D) is essential both for the superconductivity and for the normal state properties.

MECHANISMS FOR SUPERCONDUCTIVITY

One view [7] is that the superconductivity arises from pairing mediated by charge polarization excitations and that the fascinating magnetic properties are not a central issue for the superconductivity. This is consistent with the observation of high T_c's in non-cuprate, non-magnetic superconductors, for example $Ba_{1-x}K_xBiO_3$ (T_c = 30K) [18]. The most complete calculation for this mechanism using a generalized random phase approximation (RPA) based on Eq. (1) [8] does give s-wave superconductivity for a narrow range of parameters. That the purely repulsive model of Eq. (1) admits superconductivity in this weak-coupling calculation is due to strong local-field corrections, hence to the particular nature of the electronic structure.

Pairing due to magnetic excitations has also been considered from the weak-coupling side [11]. In this "spin-bag" theory of the Hubbard hamiltonian, U/t is not very large, thus the insulator at 1/2-filling is a spin density wave state which arises because of perfect nesting of the Fermi surface. Away from 1/2-filling, in the metallic phase, strong AFM spin-density wave fluctuations persist and have rather long lifetimes. These are mediating excitations for superconductive pairing. For this argument to work, it is necessary that the correlation length for spin fluctuations be larger than the coherence length of the pairs. This seems rather unlikely as the weight of experimental evidence is against the presence of slow AFM spin fluctuations at compositions which have the highest T_c's [19]. Whether at 1/2-filling one has a Mott insulator (large U) or a spin density wave state (small U) is also controversial. The results of neutron scattering favor large U: For light doping, the effective moment remains roughly constant while the spin correlation length decreases [20].

Superconductivity in the quantum spin liquid has been obtained by various mean-field treatments of the t-J model hamiltonian, Eq. (4). Usually, an anisotropic gap function ("d-wave") is found which appears not to be in accord with experiment. These approaches all suffer from an inadequate treatment of the constraint $n_i \leq 1$ which inhibits double occupancy at a site and therefore have not been convincing. As described above, an interacting anyon gas may describe the excitations of a "chiral" (P and T violating) state of the QSL. In that case, superconductivity can arise and certain anomalous behaviors are predicted [21], including local magnetic fields near charge inhomogeneities and charge associated with vortices in the mixed state. So far, experimental results are negative for all these predictions [22].

Superconductivity arises in the RVB picture via interlayer tunneling which in itself provides the pairing potential [6] for the intralayer pairs which otherwise have no pairing instability, even at T=0. This scenario is consistent with experimental results on the variation of T_c with pressure and with the number of Cu-O planes per unit cell in different compounds.

CONCLUSION

Although there is no consensus as to the correct physics of high-T_c superconductivity, there is general agreement that the phenomenon is driven by an electronic mechanism, rather than by lattice vibrations. It appears likely that the correct description will emerge from a more complete experimental and theoretical understanding of the normal state behavior of charge and spin excitations.

REFERENCES

1. J.G. Bednorz and K.A. Müller (1986) Z. Phys. B64, 189.

2. C.M. Varma, P.B. Littlewood, S. Schmitt-Rink, E. Abrahams, and A.E. Ruckenstein (1989) Phys. Rev. Lett. 63, 1996.

3. P.W. Anderson (1989) submitted to Nature.

4. J. Bardeen, L.N. Cooper and J.R. Schrieffer (1957) Phys. Rev. 108, 1175.

5. J. Hubbard (1963) Proc. Roy. Soc. A276, 238.

6. For example, T. Hsu, J. Wheatley and P.W. Anderson (1988) Nature 333, 121.

7. C.M. Varma, S. Schmitt-Rink and E. Abrahams (1987) Solid State Commun. 62, 681.

8. P.B. Littlewood, C.M. Varma and E. Abrahams (1989) Phys. Rev. Lett. 63, 2602.

9. P.W. Anderson (1959) Phys. Rev. 115, 2.

10. P.W. Anderson (1987) Science 235, 1196.

11. J.R. Schrieffer, X.G. Wen and S. Zhang (1989) Phys. Rev. B39, 11538

12. P. W. Anderson (1988) Frontiers and Borderlines in Many-body Physics; Varenna Lectures. North Holland. Amsterdam.

13. G. Kotliar (1988) Phys. Rev. B37, 3604; I. Affleck and J.B. Marston (1988) Phys. Rev. B37, 3774

14. R. B. Laughlin (1988) Science 242, 525; Y-H Chen, F. Wilczek, E. Witten, and B.I. Halperin (1989) Int. J. Mod. Phys. B3, 1001.

15. F. Wilczek (1982) Phys. Rev. Lett. 49, 957.

16. A.L. Fetter, C.B. Hanna and R.B. Laughlin, (1989) Phys. Rev. B39, 9679.

17. Y. Kuroda and C.M. Varma (1989) submitted to Phys. Rev. Lett.

18. R.J. Cava et al. (1988) Nature 332, 814.

19. R.F. Kiefl, et al (1989) Phys. Rev. Lett. 63, 2136.

20. Y.J. Uemura et al (1987) Phys. Rev. Lett. 59, 1045.

21. B.I. Halperin, J. March-Russell and F. Wilczek (1989) Harvard preprint.

22. R.F. Kiefl, et al (1989) TRIUMF preprint; P.L. Gammel, et al (1989) Bell Laboratories preprint.

Present Status of Experiments in High-T_c Superconductivity

H.R. Ott

Laboratorium für Festkörperphysik, ETH-Hönggerberg, 8093 Zürich, Switzerland

ABSTRACT

Essential experimental results concerning the normal and the superconducting states of oxide-superconductors are summarized. Comparison with other unusual superconductors or metals is made where appropriate.

KEY WORDS:Oxide superconductors, normal state, superconducting state

INTRODUCTION

Time and space restrictions of course do not allow for an in-depth discussion of all experimental results that are available on high-T_c superconductors. It is generally recognized that the oxide materials that are of interest here are not only outstanding with respect to enhanced critical temperatures T_c for superconductivity but also reveal distinct peculiarities with respect to their normal-state behaviour. Furthermore it seems imperative to also investigate materials that are only slightly different in chemical composition but are either magnetically-ordering insulators or else metals that are not superconducting down to very low temperatures. On certain occasions, comparisons with previously observed similar behaviours in other materials like heavy-electron systems or organic conductors and superconductors may be useful. In this sense I shall try to discuss a few key issues by mentioning selected physical properties that seem characteristic for these oxide materials and may therefore give clues for possible successful models describing the behaviour that is experimentally observed.

NORMAL-STATE PROPERTIES

Among the most outstanding properties of the normal state are the temperature dependence of the electrical resistivity $\rho(T)$ and its anisotropy [1]. It is certainly remarkable that for virtually all the new oxide superconductors, a linear-in-T dependence of ρ along the Cu-O planes is observed. In good single-crystalline material, the absolute value of ρ at T_c is a few tens of $\mu\Omega$cm and the slope $\partial\rho/\partial T$ is of order $\mu\Omega$cm/K. Much higher resistivities are measured perpendicular to the Cu-O planes (along the c-direction) and the most extreme anisotropy has been found in Bi-based compounds where ρ_c/ρ_{ab} may be as large as 10^5 as T approaches T_c [2]. For ρ_c, the temperature dependence is less universal from one material to another. In most cases, ρ_c increases with decreasing T, at least near T_c.

With respect to $\rho(T)$ behaviour, we should like to emphasize two cases. The mentioned simple T-dependence of ρ may not be quite adequate to describe the data that were obtained for ρ_{ab} in oxygen-deficient $Ba_2YCu_3O_x$ with $T_c \approx 60$ K [3]. This is shown in fig. 1, where distinct deviations from $\rho \approx T$ behaviour may be seen. On the other hand, the characteristic $\rho(T)$ feature may even survive phase-separation transitions as was inferred in recent work on oxygen-doped $La_2CuO_{4+\varepsilon}$

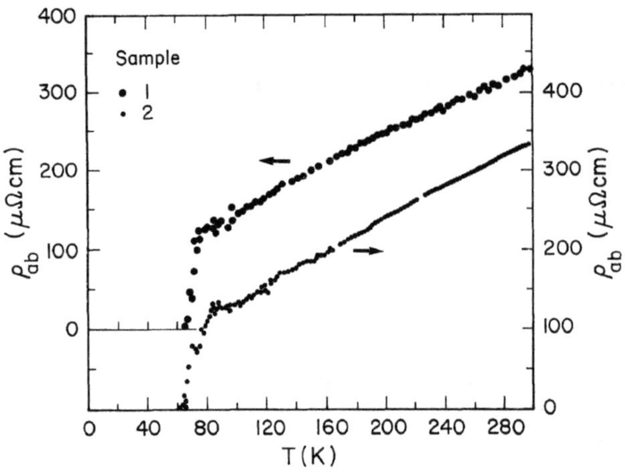

Fig. 1: ρ(T) in (ab) plane for $Ba_2YCu_3O_x$ with $T_c \approx 60$ K (from ref. 3)

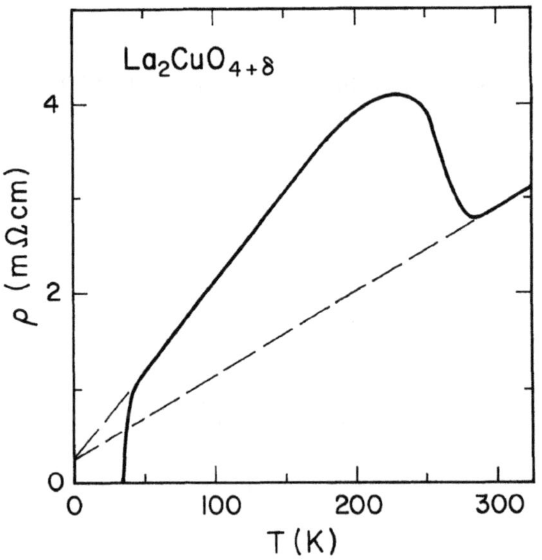

Fig. 2: ρ(T) in (ab) plane of oxygen-doped $La_2CuO_{4+\varepsilon}$ (from ref.4)

[4] and from which the ρ(T) data obtained parallel to the Cu-O planes are shown in fig. 2.

Resistivities varying linearly with T have also been observed in heavy-electron compounds at temperatures before they reach the characteristic Fermi-liquid behaviour at very low temperatures [5]. What is more outstanding in these oxides is probably that this particular ρ(T) behaviour is accompanied by the mentioned considerable anisotropy.

Quite a similar situation is met when analyzing the spin-lattice relaxation times observed in NMR or NQR experiments [6,7] and their temperature dependence. We show a selection of such data for $Ba_2YCu_3O_7$ in fig. 7 [8]. At this point we are concentrating on T_1^{-1} of nuclear spins on various lattice sites in the normal state above T_c. For comparison, $T_1^{-1}(T)$ for pure Cu-metal is also included in the figure as a reference and for demonstrating the Korringa-type behaviour as expected for a conventional metal.

A similar behaviour is obviously only observed for nuclear spins of the O(2,3) atoms in the CuO planes. Here and in other oxide-compound superconductors, the relaxation rates for Cu atoms in these planes are enhanced and tend to deviate distinctly from a strictly linear T-dependence as shown for the Cu(2) atoms in fig. 7 [6,7]. Enhancements and T-dependences of T_1^{-1} similar to these or even more pronounced have also been observed in heavy-electron superconductors like, e.g., in UBe_{13} [9]. The latter compound is known to be a system of highly correlated electrons that have, at the temperatures of investigation in the normal state and at T_c, not yet reached a Fermi-liquid state in the usual sense.Problems regarding the nature of the electronic subsystem in view of high-resolution, angle-resolved photoemission data are discussed by J. Arko in this volume.

OCCURRENCE OF SUPERCONDUCTIVITY

What seems interesting is that also in the Cu-O superconductors a fairly wide span of critical temperatures T_c is observed from a few K for $Bi_2Sr_2CuO_6$ [10] to

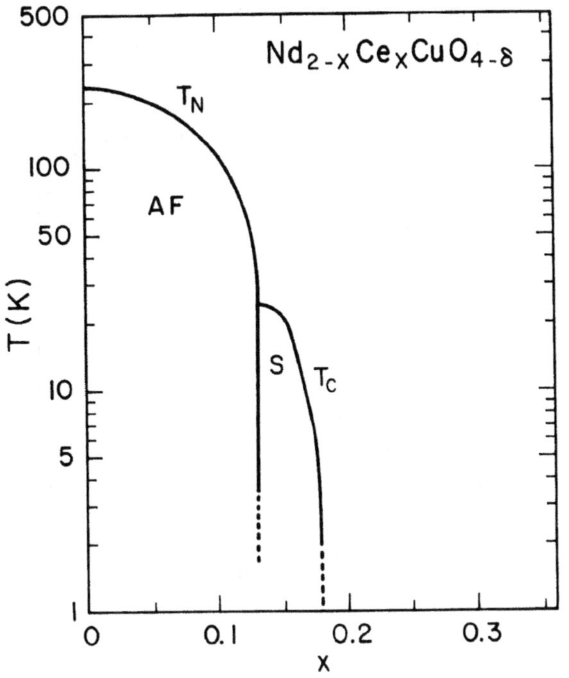

Fig. 3: Ground states for $Nd_{2-x}Ce_xCuO_{4-\delta}$ (from ref. 14).

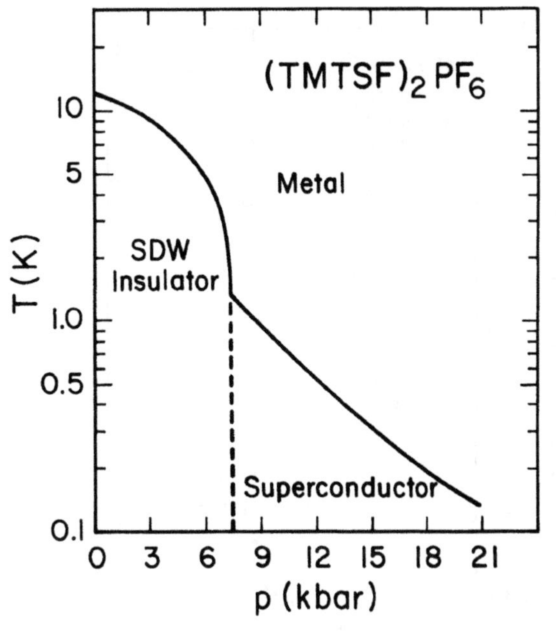

Fig. 4: Ground states for $(TMTSF)_2PF_6$ under external pressure (from ref. 15).

approximately 110 K for Pb-stabilized $Bi_{2-x}Pb_xSr_2Ca_2Cu_3O_{10}$ [11] or 125 K in a Tl-based compound [12]. This is important in view of possible mechanisms for triggering superconductivity in these compounds in the sense that also low T_c's need to be explained.

For most high-T_c superconductors a relatively small change in chemical composition may lead to an antiferromagnetically-ordered insulating ground state below a Néel temperature T_N. Well documented examples are the $La_{2-x}M_xO_4$ (M = Sr, Ba) and the $Ba_2R.E.Cu_3O_{6+x}$ (R.E. = rare earth) compounds where a variation of x leads to phase diagrams with well separated regions of insulating and metallic behaviour and characteristic variations of $T_N(x)$ and $T_c(x)$ [13]. Experimental evidence for coexistence of magnetic order and superconductivity both involving only the Cu-O subsystem at a defined concentration x is not convincing in these cases. Starting from the insulating state, the complete suppression of T_N is followed by a gradual increase of T_c with further increasing x, at least close to the onset concentration x for superconductivity.

Somewhat different seems the situation in $Nd_{2-x}Ce_xCuO_{4-\varepsilon}$ [14]. Increasing values of x suppress T_N and magnetic order is lost rather abruptly at a critical concentration x_c. In the metallic state adjacent to x_c, T_c is apparently at maximum as soon as magnetic order has disappeared. A further increase of x also reduces T_c. This is indicated schematically in fig.3. A similar phase diagram has been obtained for an organic material under external pressure [15] and we illustrate this in fig. 4.

Sensitivity to chemical composition is also apparent in oxide superconductors of the type $Ba_{1-x}M_xBi_{1-y}N_yO_3$ where M and N are alkali metals and Pb or Sb, respectively [16,17]. Instead of a tendency to magnetism, these compounds are close to structural instabilities.

This discussion leads to the obvious question of which parameters are essential for high critical temperatures or what kind of microscopic changes accompany the

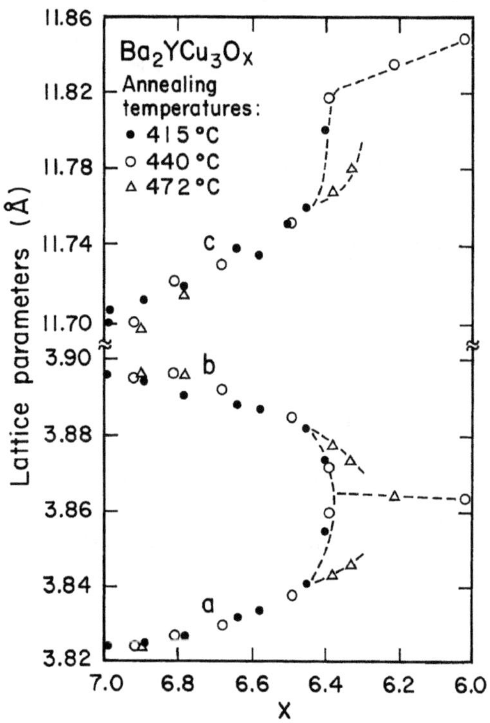

Fig. 5: Variation of lattice parameters for $Ba_2YCu_3O_{6+x}$ with x (from ref. 18).

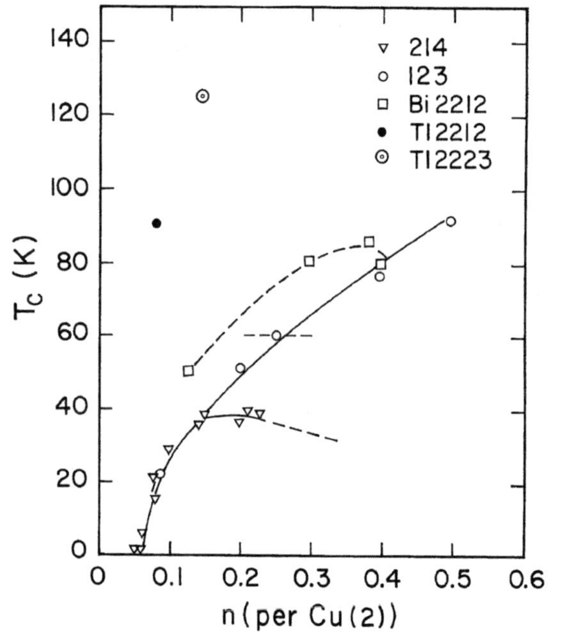

Fig. 6: $T_c(n)$ for various Cu-O superconductors (from ref. 20).

distinct T_c variations. A well documented example is available for $Ba_2YCu_3O_{6+x}$. The variation of x from 0 to 1 induces appreciable changes in the lattice parameters [18] as may be seen in fig. 5. The transition from the insulating to the metallic state is accompanied by a drastic change in the c-parameter and a distortion in the [ab]-plane. The c-anomaly may be traced back to a change in distance between the Cu(2) and the O(1) atoms and is interpreted as being due to a charge transfer from the Cu-O chains to the Cu-O planes. This charge transfer has been evidenced microscopically by inelastic neutron-scattering experiments probing the crystal-electric-field splitting of the 4f-electron ground state in $Ba_2R.R.Cu_3O_{6+x}$. This splitting is influenced by the neighbouring charges surrounding the rare-earth ions and since in this compound series the rare-earth ions are situated between Cu-O planes, the charge variation in the latter between the insulating and the metallic phase has been verified [19].

For some time it was thought that the charge-carrier concentration n determines T_c in the sense that increasing n is causing an increase in T_c up to a certain limit. Figure 6 indicates that such a correlation is not universal in high-T_c materials [20]. Very high values of T_c are obtained in Tl-based materials where n is considerably smaller than that for maximum values of T_c in other compound series. Hence simple rules for high critical temperatures are still lacking but this is not surprising since simple rules that apply for conventional binary systems are worthless for ternary-compound superconductors [21].

THE SUPERCONDUCTING STATE

Beginning with the discovery of high-T_c superconductivity in Cu-oxide materials, speculations about unconventional pairing mechanisms and unusual superconducting

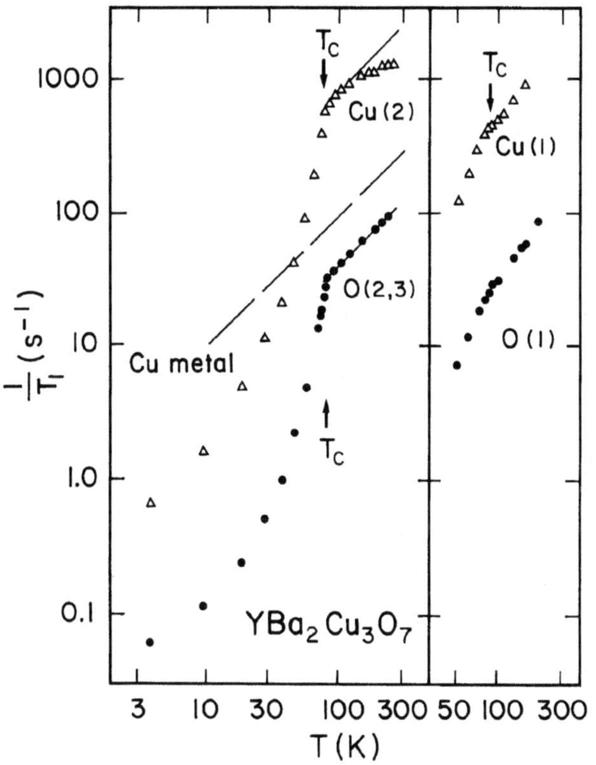

Fig. 7: T_1^{-1} for various nuclear spins in $Ba_2YCu_3O_7$ (from ref. 8).

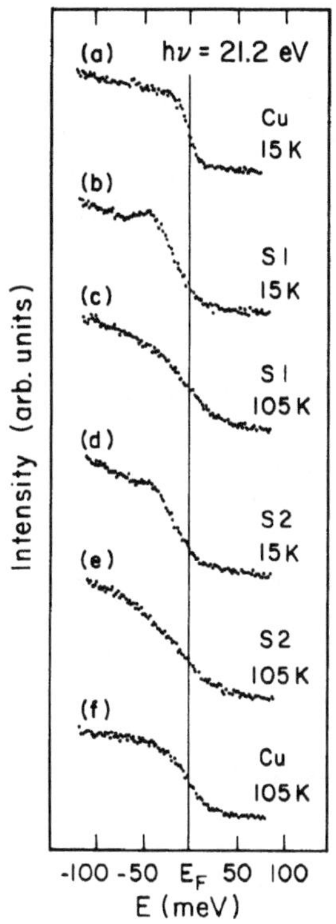

Fig. 8: Photoemission spectra for $Bi_2Sr_2CaCu_2O_8$ close to E_F. S1,2 = sample 1,2 (from ref. 27)

states emerged [22], various experimental data appear to be compatible with conventional interpretations but there are, nevertheless, some puzzles still to be solved.

In none of the numerous NMR or NQR measurements has ever a coherence peak in the spin-lattice relaxation rate been observed below T_c; examples are shown in fig.7 [8], This is, again, also true for heavy-electron superconductors [9]. In these oxides, the drop of T_1^{-1} below T_c is exceedingly fast and not compatible with the usual expectations. The related Knight-shift data, however, are reported to be in agreement with fairly straightforward analyses [23]. Related with the unusual T_1^{-1} behaviour is possibly the observed fast decay of the excess specific heat just below T_c, not compatible with the usual BCS predictions [24].

Somewhat uncertain are magnitudes and structures of gaps in the electronic excitation spectrum below T_c. Experimental evidence for a gap formation has been obtained from NMR and NQR data as those in fig. 7, optical data [25], tunneling experiments [26], and photoemission measurements [27]. An example for results of the latter are shown in fig. 8. What should be noted here is the pile-up of intensity below E_F at $T < T_c$, as is expected upon the gap formation. For $T < T_c$, excess intensity in the gap region is observed. From these and most tunneling data it is inferred that electronic states below the gap energy are still occupied even for $T \ll T_c$. It remains to be seen, whether these findings are simply experimental artefacts or intrinsic features.

All these oxide materials are extreme type-II superconductors, hence implying large penetration depths and very short coherence lengths. This gives rise to peculiar responses to external magnetic fields. These aspects are discussed in more detail in other contributions to this conference.

CONCLUSION

More and more experimental data are reaching a level of considerable reproducibility, thus enhancing the degree of their credibility. Most remaining problems in this respect probably have to be traced back to difficulties in sample preparation and characterization.

REFERENCES

1. Hagen S, Jing TW, Wang ZZ, Horvath J, Ong NP (1988) Phys. Rev.B 37: 7928
2. Martin S, Fiory T, Fleming RM, Schneemeyer LF, Waszczak JV (1988) Phys. Rev. Lett. 60: 2194
3. Brawner DA, Wang ZZ, Ong NP, Phys. Rev. B (in print)
4. Hundley MF, Thompson JD, Cheong SW, Fisk Z, Schirber JE, Phys. Rev. B (in print)
5. Ott HR, Rudigier H, Fisk Z, Willis JO, Stewart GR (1985) Solid State Commun. 53: 235
6. Yasuoka H, Imai T, Shimizu T,(1989) In: Fukuyama H, Maekawa S, Malozemoff AP (eds.) Strong Correlation and Superconductivity. Springer, Berlin, p. 254
7. Kitaoka Y, Ishida K, Fujiwara F, Kondo T, Asayama K, Horvatic M, Berthier Y, Butaud P, Segransan P, Berthier C, Katayama-Yoshida H, Okabe Y, Takahashi T, ref. 6, p. 262
8. Hammel PC, Takigawa M, Heffner RH, Fisk Z, Ott KC, (1989) Phys. Rev. Lett. 63: 1992
9. McLaughlin DE, Tien C, Clark WG, Lan MD, Fisk Z, Smith JL, Ott, HR (1984) Phys. Rev. Lett. 53: 1833
10. Michel C, Hervieu M, Borel MM, Grandin A, Deslandes F, Provost J, Raveau B (1987) Z. Phys. B 68: 421
11. Tanako M (May 23, 1988) Superconductor Week 2, 1
12. Parkin SSP, Lee YY, Engler EM, Nazzal AI, Huang TC, Gorman G, Savoy R, Beyers R (1988) Phys. Rev. Lett. 60: 2539
13. see, e.g. Physica C 153-155, Proc. of HTSC-M^2S Conference (1988) Interlaken, Switzerland
14. Uchida S, Takagi H, Tokura Y, Koshihara N, Arima T, ref. 6, p. 194
15. Greene RL, Chaikin PM (1984)Physica 126 B+C: 431
16. Thanh TD, Koma A, Tanaka S (1980) Appl. Phys. 22: 205
17. Mattheiss LF, Georgy EM, Johnson Jr. DW (1988) Phys. Rev. B 37: 3745
18. Cava RJ, Batlogg B, Rabe KM, Rietman EA, Gallagher PK, Rupp Jr. LW (1988) Physica C 156: 523
19. Allenspach P, Furrer A, Rupp B, Blank H, Physica C (in print)

20. Ong NP, Jing TW, Wang ZZ, Clayhold J, Hagen SJ, Chien TR, ref. 6, p. 204

21. Matthias BT (1981) In: Shenoy GK, Dunlap BD, Fradin FY (eds) Ternary Superconductors. North Holland, Amsterdam, p. 3

22. Anderson PW (1987) Science 235: 1196

23. Durand DJ, Barrett SE, Pennington CH, Slichter CP, Bukowski ED, Friedmann TA, Rice JP, Ginsberg DM, ref. 6, p. 244

24. Schilling A, Ott HR, Hulliger F, Physica C, in print

25. Collins RT, Schlesinger Z, Holtzberg F, Feild C, Koren G, Gupta A, Hinks DG, Mitchell AW, Zheng Y, Dabrowski B, ref. 6, p. 289

26. Gurevitch M, Valles JM, Cucolo AM, Dynes RC, Garno JP, Schneemeyer LF, Waszczak JV (1989) Phys. Rev. Lett. 63: 1008

27. Imer JM, Patthey F, Dardel B, Schneider WD, Baer Y, Petroff Y, Zettl A (1989) Phys. Rev. Lett. 62:336

Scaling Laws for the Irreversible Magnetization of Type-II Superconductors

Y. YESHURUN[1], Y. WOLFUS[1], E.R. YACOBY[1], and I. FELNER[2]

[1] Department of Physics, Bar-Ilan University, Ramat-Gan, 52100, Israel
[2] The Hebrew University of Jerusalem, Racah Institute of Physics, Jerusalem, 91904, Israel

ABSTRACT

A one-parameter scaling of the magnetization curves of three high-temperature ceramic superconductors (Y-Ba-Cu-O, Bi-Sr-Ca-Cu-O and Tl-Ba-Ca-Cu-O), as well as of a powder of a low-temperature superconductor (V_3Si), cause all data points in the irreversible regime to collapse into a single curve $M = H^*f_\pm(H/H^*)$. The scaling field H^* is a function of temperature and f_+ and f_- are the scaling functions for fields H above and below H^*, respectively. On the basis of the available data we conclude that the scaling functions for ceramic samples are "universal". In addition, we demonstrate that the size of the particles, not the sample size, dominate the scaling law. The results are explained in terms of the Bean model.

KEY WORDS: Irreversible magnetization, Scaling laws, Bean model

INTRODUCTION

The magnetization in the Abrikosov mixed state in ideal type-II superconductors (SC) is reversible. In reality, however, most type-II SC are characterized by irreversible magnetic features which result from flux trapping at pinning centers. The magnetization in the irreversible regime is conventionally described by the Bean model(1). This model, and extended versions of it(2-4), have recently been used extensively for the analysis of data for high-temperature superconductors (HTSC).

We have recently discovered a new feature of the magnetization curves of Type-II superconductors: A one-parameter scaling of the magnetization curves in the irreversible regime causes all data points to collapse into a single curve(5-7). The main qualitative observation in the experimental work is that M(H,T) depends on temperature T only through a scaling field variable $H^*(T)$. We have shown(5) that this scaling property is built into the Bean model. In this article we briefly review the experimental data and the analysis within the Bean model. We then discuss the question of universality of the scaling function. We also present new data for a Y-Ba-Cu-O ceramic sample, the purpose of which is to define experimentally the length scale which governs the Bean's equations and the scaling function. Finally, we discuss the similarity of our results with scaling laws of the pinning force and critical currents discovered in conventional type-II superconductors(8-11).

II. SCALING OF MAGNETIZATION CURVES

IIa. Experimental procedure

Data is collected by a direct measurement of the magnetization M vs. field H at a constant temperature T after a zero-field-cooling (zfc) from above the transition temperature T_c. In some cases(5) we have also determined M(H) from measurements of the temperature dependence of the zfc curves. Most measurements have been done on a commercial SHE SQUID susceptometer with the sample oriented in such a way that demagnetization corrections are negligible. Field is ramped in steps from zero and up to 40 kOe. Data points are taken approximately 3 m after each field change. For the Bi-Sr-Ca-Cu-O sample (section IIc below) we have used a 155 PAR vibrating sample magnetometer (VSM) in which the field is ramped up to 15 K Oe at a rate of ≃ 1 kOe/m.

IIb. Y-Ba-Cu-O

The 120-mg ceramic Y-Ba-Cu-O sample - hereafter referred to as YBCO - has a disklike shape with a diameter of 4.95 mm and thickness of 1.35 mm. Grain size is typically 10μm. Sample preparation is described elsewhere [12]. The transition temperature T_c is approximately 91 K. Typical M vs. H isotherms are presented in Fig. 1. Qualitatively all isotherms

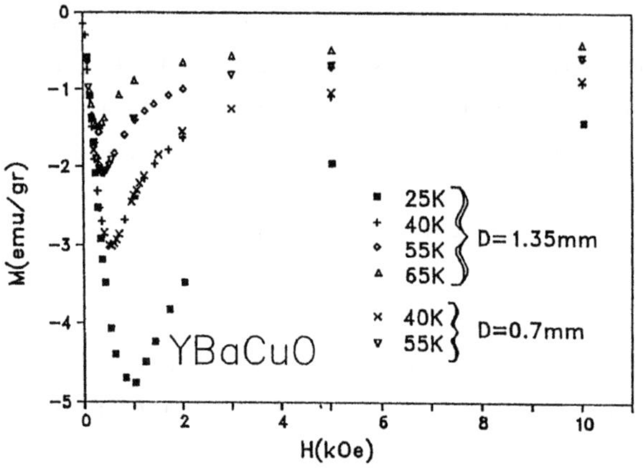

Figure 1: Magnetization curves for YBCO at the indicated temperatures.

look similar but the field H_m for which the magnetization obtains its minimum value M_m is pushed to lower values as temperature is increased. At the same time the absolute value of M_m is reduced. The scaling procedure is straightforward. We scale the field values of each isotherm by H_m and, similarly, we scale the magnetization values by M_m. As a result of this simple scaling procedure, data points for a wide range of temperatures and fields collapse into a single curve, as shown in Fig. 2. The success of this scaling procedure is even more

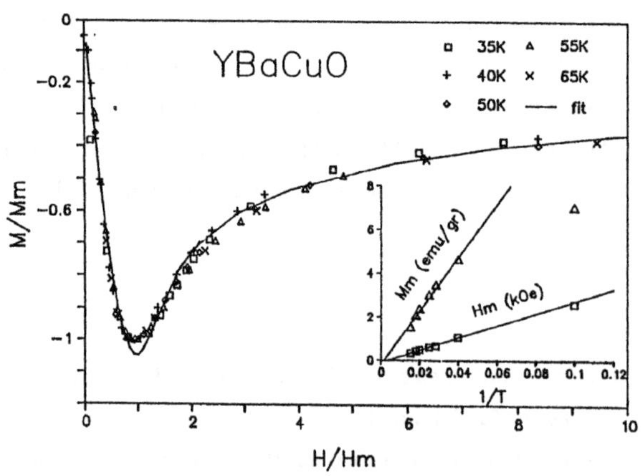

Figure 2: Scaled magnetization curves for YBCO for $H/H_m \leq 10$. Solid line: A fit to Eqs. 1 and 2 with n = 0.45. Inset: The scaling parameters as a function of the inverse temperature.

impressive once we appreciate the implication of the temperature dependence of the two scaling parameters. The inset to Fig. 2 demonstrates that both H_m and M_m scale with the inverse of the temperature. In other words, $M_m \propto H_m$, and hence the scaling procedure is actually a one-parameter scaling.

IIc. Bi-Sr-Ca-Cu-O

The 110-mg disklike Bi-Sr-Ca-Cu-O sample - hereafter referred to as BSCCO - was prepared by a conventional method[13]. For this sample $T_c \simeq 114$ K. The success of the scaling procedure for this material is shown in Ref. 5.

IId. Tl-Ba-Ca-Cu-O

In Ref. 14 we describe preparation of this 285-mg disklike TBCCO sample with $T_c \simeq 113$ K. The scaled magnetization curves are shown in Ref. 5.

IIe. V_3Si

The sample is based on a powder of V_3Si particles of diameter less than 25 μm immersed in a stycast matrix. For this sample $T_c = 16.2$ K. Details of sample preparation are described in Ref. 15. A summary of the magnetization curves for the V_3Si samples at various isotherms is shown in Fig. 3. The scaling features of the magnetic data are apparent in this figure. The inset shows that $M_m \propto H_m$.

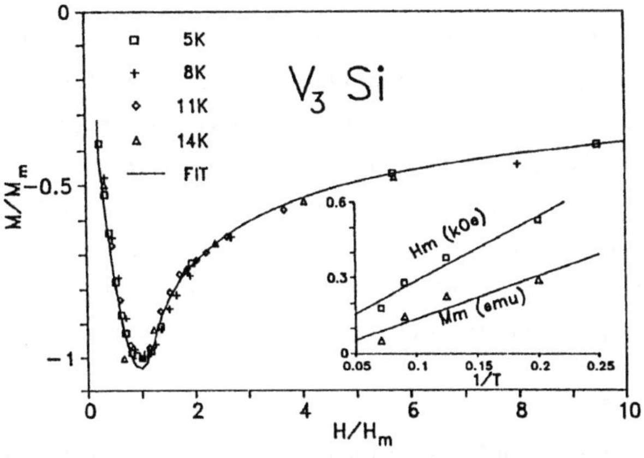

Figure 3: Scaled magnetization curves for V_3Si for $H/H_m \leqslant 10$. Solid line: A fit to Eqs. 1 and 2 with $n = 0.45$. Inset: The scaling parameters as a function of the inverse temperature.

IIf. Universality of scaling functions

Are the scaling functions universal? In Fig. 4 we combine the scaled data for three samples: YBCO, TBCCO and V_3Si, for $H/H_m \leqslant 10$. The result is quite surprising: All isotherms for these three samples collapse into a single curve. At higher scaled fields the three curves disperse. We argue, however, that this dispersion is a result of the paramagnetic background[12] which dominates the magnetic behavior of the TBCCO and V_3Si samples at high fields. (For high enough fields the small diamagnetic signal is of the same order of magnitude as the paramagnetic contribution. For TBCCO this results in positive values for the measured magnetization in the superconducting phase, see Ref. 14).

The scaled data of the magnetization of BSCCO, not shown in Fig. 4, show clear deviations from the scaled data of the other three samples. We tentatively suggest that

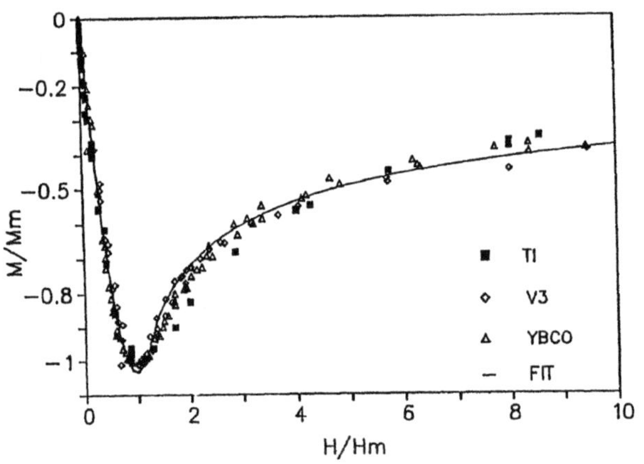

Figure 4: Scaled data for YBCO, TBCCO and V_3Si. Solid line: A fit to Eqs. 1 and 2 with $n = 0.45$.

this deviation is a result of the different ramping and measuring rate for the BSCCO sample, (see section IIa). This assumption is consistent with the fact that the scaled BSCCO data is below (i.e., it is more diamagnetic than) the scaled data for the rest of the samples. It is very plausible that on a VSM data is collected too "early", namely, while fast relaxations are taking place. The slow measurements on the SHE susceptometers define a time-window which is less affected by relaxation.

On the basis of the available data and the discussion above, it is tempting to argue that a universal function may serve to describe the irreversible magnetization of type-II superconductors. This strong conclusion implies, however, that the scaling function for the pinning forces is universal, in apparent contradiction to the conclusions in Refs. 9 and 10. We therefore suggest that the universal behavior exhibited in Fig. 4 is limited to ceramic and powder materials. It is not clear to us whether this universal behavior reflects some intrinsic common mechanism for pinning or that it is a mere reflection of the inevitable averaging over grain sizes and orientations. In an effort to explore the relevance of this last point we are now studying the magnetization curves of YBCO single crystals.

IIg. Effect of length scale

The thickness of the YBCO sample described in Section IIb is $D = 1.35$ mm. We were able to reduce the thickness by approximately a factor of two. Magnetization curves of the reduced sample ($D = 0.7$ mm, weight $\simeq$ 71 mg) are **identical** to the curves obtained for the original sample (see Fig. 1). Similarly, the remanent magnetizations M_{rem} (normalized to the volume of the sample), obtained after switching off a 4 T field, are identical for both the original and the reduced sample. We recall that the critical current in the Bean model is given by[1] $J_c = 30\ M_{rem}/D$. It is therefore apparent that the thickness D is **not** a relevant length scale in this problem. The field is probably large enough to decouple the grains and hence the relevant length scale is the (average) grain size.

III. A SCALING LAW IN THE BEAN MODEL

The scaling features of the irreversible magnetization can be explained in the framework of the Bean model. In its simple version, the model assumes that the critical current J_c is field independent leading to a linear drop of the local magnetic field h with the distance x from the surface. A more realistic model[2] takes $J_c = J_{c1}h^{-n}$, where J_{c1} is the maximum critical current at a given temperature and n is a phenomenological power, typically 0.5-1 in experiments. Such a recent extension of the Bean model yields[2] for a slab of thickness D,

$$\frac{4\pi M}{H^*} = -\left(\frac{H}{H^*}\right) + \frac{n+1}{n+2}\left(\frac{H}{H^*}\right)^{n+2}, \tag{1}$$

$$\frac{4\pi M}{H^*} = -\left(\frac{H}{H^*}\right) + \frac{n+1}{n+2}\left\{\left(\frac{H}{H^*}\right)^{n+2} - \left[\left(\frac{H}{H^*}\right)^{n+1} - 1\right]^{(n+2)/(n+1)}\right\} \tag{2}$$

where $C \equiv (4\pi/10)(n+1)J_{c1}H_{c1}^{n}$ and $H^* \equiv (CD/2 + H_{c1})^{1/(n+1)}$ is the lowest field for which currents flow through the entire volume of the sample. The scaling field in the above equations is H^*, the lowest field for which flux penetrates into the entire volume of the sample. Note that in the Bean model H^* is proportional to, but larger than, H_m. Eqs. 1 and 2 may be summarized as:

$$4\pi M = H^* f_{\pm}(H/H^*), \tag{3}$$

where f_- and f_+ are the scaling functions for $H/H^* \leqslant 1$ and for $H/H^* \geqslant 1$, respectively.

Most type-II superconductors are characterized by irreversible magnetic features which result from flux trapping and pinning centers. The functional dependence of the pinning forces on temperature T and field H determines the irreversible magnetization. This functional form has been a topic of extensive investigations for decades[8-11,16]. Experimentally, it has been established that the pinning forces, and hence the critical current, J_c too, obey a simple scaling law: $J_c = F(T)\cdot G(h)$ where $h = H/H_{c2}$ is the reduced field, H_{c2} is the upper critical field. F and G, the scaling functions, are quite sensitive to microstructure and materials. It is apparent that our results are consistent with the above variable separation. In the absence of such a separation a one parameter scaling would be impossible.

The Bean model also reflects such a variable separation. Algebraically, this is apparent from the fact that H^* depends solely on temperature (even though J_c depends on field, J_{c1}, the maximum critical current at a given temperature, does not). Thus, we may conclude that a variable separation of J_c and a one-parameter scaling of the magnetization are built in the Bean model.

ACKNOWLEDGEMENTS

Important discussions with A.M. Campbell, R.S. Markiewicz, H. Sompolinsky and E. Zeldov are acknowledged. This research is supported in part by the Israel-US Binational Science Foundation, and in part by the Israel Academy for Science and Humanities.

REFERENCES

1. C.P. Bean, Phys. Rev. Lett. 8, 250 (1962) and Rev. Mod. Phys. 36, 31 (1964)
2. Y. Yeshurun, A.P. Malozemoff, F. Holtzberg and T. Dinger, Phys. Rev. B 38, 11828 (1988); Y. Yeshurun, A.P. Malozemoff and F. Holtzberg, J. Appl. Phys. 64, 5797 (1988)
3. G. Ravi-Kumar and P. Chadda, Phys. Rev. B 39, 4704 (1989)
4. L. Ji, R.H. Sohn, G.C. Spalding, C.J. Lobb and M. Tinkham, Phys. Rev.B (to be published)
5. Y. Wolfus, Y. Yeshurun, I. Felner and H. Sompolinsky, Phys. Rev. B 40, 2701 (1989)
6. E.R. Yacoby, Y. Wolfus, Y. Yeshurun, I. Felner and H. Sompolinsky, Proceedings of the M^2HTSC Conference, Stanford (1989). To be published in Physica C.

7. Y. Yeshurun, A.P. Malozemoff, Y. Wolfus, E.R. Yacoby, I. Felner and C.C. Tsuei, Proceedings of the M^2HTSC Conference, Stanford (1989). To be published in Physica C
8. Fietz and Webb, Phys. Rev. 178, 657 (1969)
9. A.M. Campbell and J.E. Evetts, Adv. Phys. 21, 372 (1972)
10. R.G. Hampshire and M.J. Taylor, J. Phys. F 2, 89 (1972)
11. E.J. Kramer, J. Appl. Phys. 44, 1360 (1973)
12. Y. Yeshurun, I. Felner and H. Sompolinsky, Phys. Rev. B 36, 840 (1987)
13. K. Fischer, A. Rojek, S. Thierfeldt, H. Lippert and R.R. Arons, Physica C160, 466 (1989)
14. Y. Wolfus, Y. Yeshurun and I. Felner, Phys. Rev. B 39, 11690 (1989)
15. C.C. Tsuei and J.T.C. Yeh in AIP Conference Proceedings on **Inhomogeneous Superconductors**, Eds. D.U. Gubser, T.L. Francavilla, S.A. Wolf and J.R. Leibowitz, American Physical Institute, New York, p. 67 (1980)
16. J.D. Hettinger et al, Phys. Rev. Lett. 62, 2044 (1989)

Fabrication and Characterization of BSCCO Films

KEIICHI OGAWA, KEIKICHI NAKAMURA, HIROTOSHI HAYAKAWA[1], HISAO HATTORI[2], JUNICHI SATO[3], and SHOZO IKEDA

National Research Institute for Metals, 2-1, Sengen 1-chome, Tsukuba, Ibaraki, 305 Japan

ABSTRACT

A present stage of BSCCO film fabrication techniques is reviewed and metallographic structures of high and low Jc 2223 films are compared. Some difficulties associated with the 2223 film fabrication in an atmosphere of low oxygen partial pressure will be pointed out: 1)oxygen deficiency and 2) slow ordering of chemically similar elements, Sr and Ca. A new vacancy model of defected CuO_2 planes is proposed for the hole source in the BSCCO system and based on this model an explanation is offered for the transformation of the semiconducting 2223 films into superconductors when annealed in an oxidizing atmosphere.

KEY WORDS: 2223 BSCCO films, layer by layer growth, as-deposited superconducting films, oxygen deficiency in CuO_2 planes, chemical ordering of Sr and Ca.

INTRODUCTION

BSCCO films have been fabricated by various techniques such as sputtering [1 − 4], thermal evaporation [5], laser ablation [6], MOCVD [7], LPEG [8], MBE [9] and spray pyrosis [10]. When the fabrication temperature is restricted to be lower than the partial melting temperature, only the first three techniques have been successful so far to fabricate the 2223 film.

We shall first make a brief review on the present stage of the 2223 film fabrication techniques and of physical properties of the resultant films. It has been known that some c-axis oriented 2223 films show an extremely high Jc, e.g. 3.4 MA/cm² at 77.3 K and B = 0 [1], whereas other similarly c-axis oriented films don't [11]. It will be shown that Jc depends critically on whether or not the continuous texture is developed in the c-axis oriented film. Then we shall introduce our study on the layer by layer growth of Bi-O/Sr-Ca-Cu-O films by sputtering [3,12]. Our primary interest lies in the in-situ growth of the superconducting 2223 films preferably with Tc=110K [3]. We shall report effects of three important fabrication factors, i.e. substrate temperature Ts, oxygen incorporation during the cooling cycle and post annealing. Finally, we shall report some physical properties measured on the well characterized films newly available [3,19].

REVIEW ON 2223 FILMS

Table I lists the 2223 films so far successfully fabricated by the low temperature process. Here the low temperature is defined as a temperature lower than the partial melting temperature [13,14]. When the partial melting occurs, the film morphology changes considerably. All the as-deposited films shown in Table I have very low Tc's compared with the expected 110K. These low Tc's apparently arise from the oxygen deficiency in the films as discussed later. Table I also tells us that the post annealing in oxygen atmosphere raises Tc considerably. However, it never reaches 110K. A possible reason for this will be discussed later as well.

2223 FILMS OF HIGH AND LOW Jc's

Cosputtered films predominantly of the 2223 phase can be produced in two different heat cycles [1,15]. Itozaki et al. deposited the 2223 film on the substrate kept at 700℃ and then post annealed it at 885℃ for 1 h. The X-ray diffraction (XRD) pattern showed that the resultant film had a well developed c-axis oriented grains. Its Jc measured at 77K and B=0 was as high as 2×10^6A/cm². A SEM image of the high Jc film is shown in Fig.1(a). Apart from some impurity grains on the surface, the surface appears to be smooth and shiny. The cross-sectional view of the film is also seen in Fig.1(a) and the c-axis oriented grains are connected each other throughout the film. On the other hand, a similarly c-axis oriented 2223

On leave from [1]Yaskawa Electric Mfg. Co Ltd., [2]Sumitomo Electric Ind. and [3]Hitachi Cable Ltd., respectively.

Table I Fabrication, heat treatment and transition temperatures of 2223 films in BSCCO system.

Fabrication Method		As deposited		Post-annealed		Ref.
		Tsub(°C)	Tc zero(K)	Temp.(°C)×Time(hr)	Tc zero(K)	
rf or dc	cosputt.	700	35	885 × 1 *	95	a
magnetron	Bi/SrCu/CaCu/SrCu	650	~10	850 × 5	~80	b
sputtering	Bi/Sr-Ca-Cu-O	665	10-30	810 × 1	82	c
ion beam sputtering	Bi-O/Sr-Ca-Cu-O	605	45	— —	—	d
thermal evaporation	coev.	630	~70	400 × 1	78	e
laser ablation	Bi-O/SrCuOy /CaCuOy/SrCuOy	480	0 *	820 × 5 *	~80 *	f

—— :not specified * : private communication

a) H.Itozaki et. al. : 2nd Workshop on High Temp. Superconducting Electron Devices (1989) Hokkaido ,p293
b) H.Adachi et.al. : Jpn. J. Appl. Phys. 27 (1988) L1883
c) K.Nakamura et.al. : Jpn. J. Appl. Phys. (submitted)
d) J.Fujita et.al. : Appl. Phys. Lett. 54 (1989) 2364
e) T.Sato et.al. : Appl. Phys. Lett. 55 (1989) 702
f) M.Kanai et.al. : Appl. Phys. Lett. 54 (1989) 1802

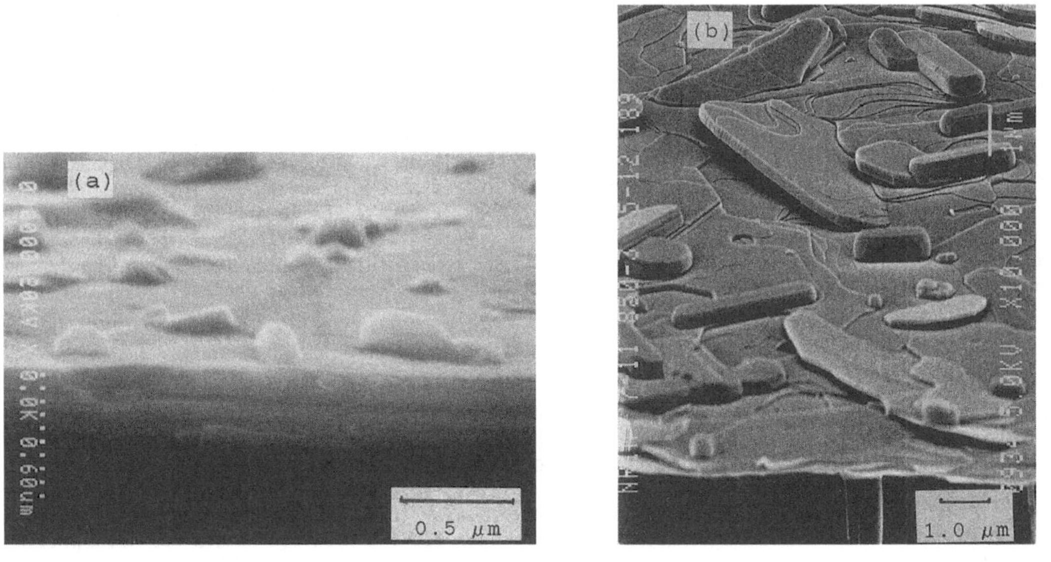

Fig.1 SEM images of 2223 films. (a) high Jc film (after H.Itozaki) and (b) low Jc film.

film can be fabricated from the amorphous state [15] . When the amorphous film is annealed in a Pb vapor atmosphere, the 2223 phase is predominantly formed. As far as the XRD pattern tells, the grains are c-axis oriented better than those of the high Jc film (Fig.1 (a)). The FWHM of the (00ℓ) XRD peaks is by a factor of 2.4 narrower than those of the high Jc film, which implies that the grain size of the present film is significantly larger than the high Jc film. A SEM image of the present film is shown in Fig.1 (b). Some flat grains make an angle of a few degrees with respect to the substrate plane. Because of this morphology, the surface of the present film appears to be frosted. Since the texture is disconnected at those flat grains, the film shows much smaller Jc, e.g. 1000 A /cm² at 77K and B = 0.

Figure 2 a) and b) are schematic drawings of the continuous and discontinuous textures of the high and low Jc films, respectively. When c-axis oriented grains are small, the texture can be easily connected throughout the film (Fig.2 a)). Even though the average misorientation of grains in Fig.1 a) and b) is the same, the texture of Fig.1 b) is forced to be disconnected owing to the large size of some misoriented flat grains. This discontinuous texture reduces Jc by a few orders of magnitude easily.

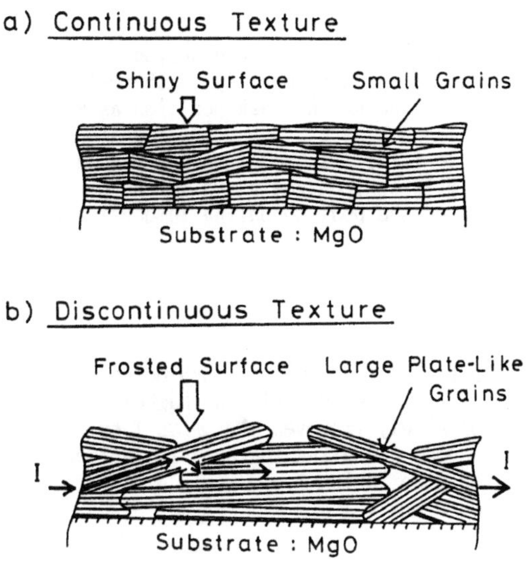

Fig.2 Schematic pictures of c-axis oriented film. a) high Jc film and b) low Jc film. Thin parallel lines drawn in each grain show the CuO_2 plane. Some current path intersecting the CuO_2 plane is shown in b).

ARTIFICIALLY LAYERED Bi-Sr-Ca-Cu OXIDE FILMS

Fabrication Technique [3,12]

Two-target reactive dc and rf magnetron sputtering was employed to fabricate artificially layered BSCCO films. The targets for dc and rf sputtering were metallic Bi and sintered Sr-Ca-Cu-O ceramics with a cationic ratio 2:2:2, respectively. To assist the oxidation of the film during the deposition, Ar-10% O_2 gas was blown to the deposited film through holes drilled on a circular tube to which the gas was fed. The modulated wave length Λ, defined as c/2 was controlled by varying the deposition time of the Sr-Ca-Cu-O layer while keeping that of a Bi-O double layer constant. The constant time was chosen exactly to form a Bi-O double layer. The substrate was a cleaved MgO and its temperature was varied from RT to 700℃. We have successfully fabricated BSCCO films with a designed number of CuO_2 planes per a unit Λ ranging from 1 to 5. Here the substrate temperature was 650℃. The XRD pattern at low angles and at high angles clearly showed that the BSCCO films had designed structures of 22(n-1)n where n=1-5 and that their grains were preferentially c-axis oriend [12] . The SEM image showed that the surface was smooth and without any impurity grain (cf. Fig.1 (a)).

Thermal Stability [3]

Films of 2223 were deposited at three different substrate temperatures, i.e. Ts=600, 645 and 665℃ and subsequently annealed at 760℃ for 1h in air in order to see their thermal stability. XRD patterns of these as-deposited films are shown in Fig.3 (a), (b) and (c). With increasing Ts, (00ℓ) peaks become increasingly narrower, while their peak positions remain constant. When Ts = 665℃, the XRD lines are very sharp, almost comparable to those of the well crystallized 2223 film (Tc zero = 106K) prepared by the high temperature process [15] . Nevertheless, the present as-deposited films show only low Tc zero, typically 0-30K depending on a manner of cooling. The reason for this will be discussed later. XRD patterns of the above films subsequently post annealed at 760℃ for 1h in air are shown in Fig.3 (d), (e) and (f). It is noted that the 2223 film deposited at the lowest Ts, i.e. 600℃ is completely transformed to the 2212

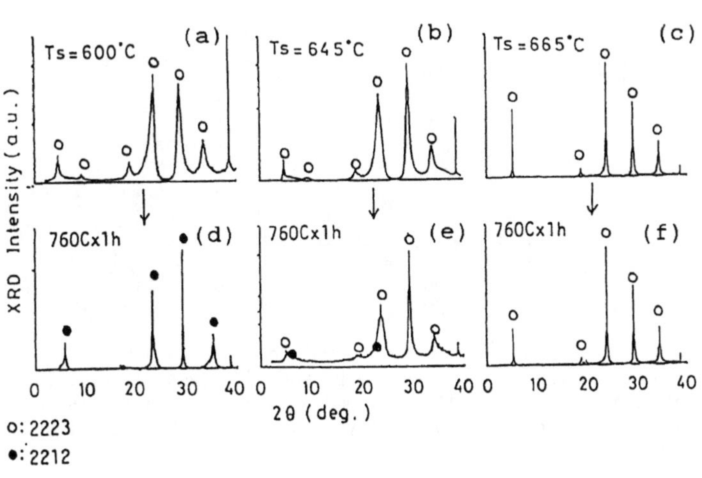

Fig.3 Thermal stability of 2223 films deposited on MgO substrates kept at Ts=600, 645 and 665℃. The upper XRD patterns (CuKα) are for the as-deposited films and the lower ones for the post annealed films at 760℃ for 1 h.

phase (Fig.1(d)) during the post annealing. However, no changes in the XRD pattern are observed for the film deposited at Ts = 665 ℃. In other words, the 2223 film of good crystalline quality is thermally stable at 760 ℃ for 1h. In addition, no changes in the XRD pattern including the peak position as well as the peak intensity imply that atomic rearrangements of heavy elements such as Bi, Sr and Ca didn't occur during the heat treatment. Resistivity vs temperature curves of these films (Ts=665℃) are surprisingly different between prior to and post annealing. The as-deposited film is semiconducting while the post annealed one is supercoducting with Tc zero = 66K and 82K depending on the post annealing temperature, 760 and 810 ℃ for 1h, respectively.

Resistivity vs Temperature Curves

Figure 4 shows resistivity ρ vs temperature T curves of the 2223 films (Ts = 665 ℃), subjected to an improved cooling cycle, i.e. an incorporation of 660 torr oxygen gas into the deposition chamber immediately after the deposition [3]. The above cooling cycle definitely improves the ρ vs T curves, resulting in Tc zero = 35K for the as-deposited film. Although the post annealing improves Tc further (Fig.4), we are still unable to obtain Tc zero as high as 110K (See also Table I).

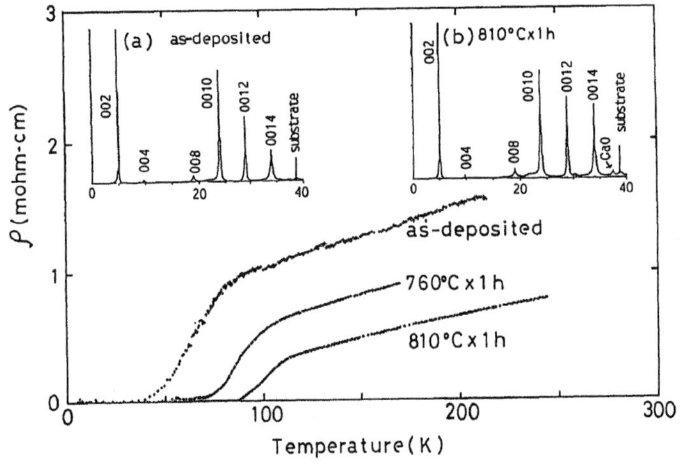

Fig.4 The resistivity vs temperature curves of the 2223 films. The insets show the XRD patterns for the as-deposited and post-annealed 2223 films.

Characterization of Semiconducting and Superconducting 2223 Films

Figure 5 shows cross sectional views of semiconducting and superconducting 2223 as-deposited films. It is surprising that modulated structures are clearly observed in both the pictures. According to a recent analysis of the modulated structure in the BSCCO system based on the X-ray and neutron powder diffration data [16], the modulated structure is a consequence of extra oxygen atoms accomodated in an expanded region of the Bi-O double layer. Clear modulation observed in Fig.5 (a) and (b), therefore, implies that

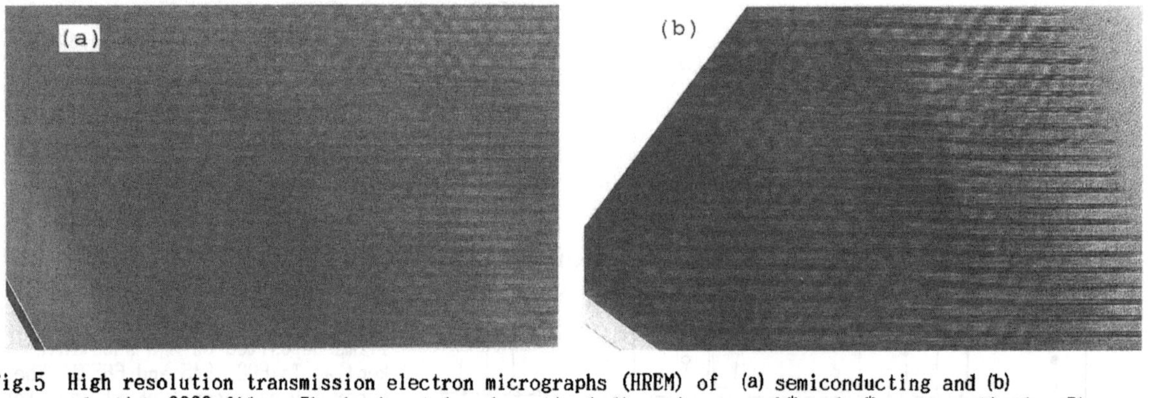

Fig.5 High resolution transmission electron micrographs (HREM) of (a) semiconducting and (b) superconducting 2223 films. The horizontal and vertical directions are b* and c*, respectively. The incident direction of the electron beam is parallel to a*. Well developed modulation is noted in both the pictures (a) and (b). The spacing between the neighboring Bi-O double layers (i.e. dark horizontal double lines) corresponds to c/2 and is equal to 18.5Å.

nearly all the oxygen sites in the Bi-O double layers are occupied by oxygen atoms in both the semiconducting and superconducting 2223 as-deposited films. Supporting evidence for this implication is obtained by the XPS studies of Bi $4f_{7/2}$ electrons in magnetron cosputtered 2223 films [17]. No change in their binding energy is observed before and after the post annealing. The observed binding energy 158.1 eV is close to that of $Bi_2(Sr,Ca)_3Cu_2O_x$ (158.0 eV). When the above as-deposited films were annealed, their Tc zero was raised from 12 to 96K and Cu ion was found to be oxidized increasingly upto $Cu^{+2.2}$. Here Cu $2p_{3/2}$ electron spectra were followed. These observations together with others definitely show that the source of holes in these films is not extra oxygen ions in the Bi-O double layer but is in the CuO_2 plane itself.

SOURCE OF HOLES

A possible source of holes in the BSCCO system generally accepted is schematically shown in Fig.6(1) [18]. When BSCCO films are fabricated at low oxygen partial pressure, oxygen atoms in the Bi-O double layer are deficient, particularly so at the expanded region of the modulated Bi-O double layer [16]. Post annealing in an oxygen atmosphere introduces oxygen atoms into O vacancies in the Bi-O double layer and thereby

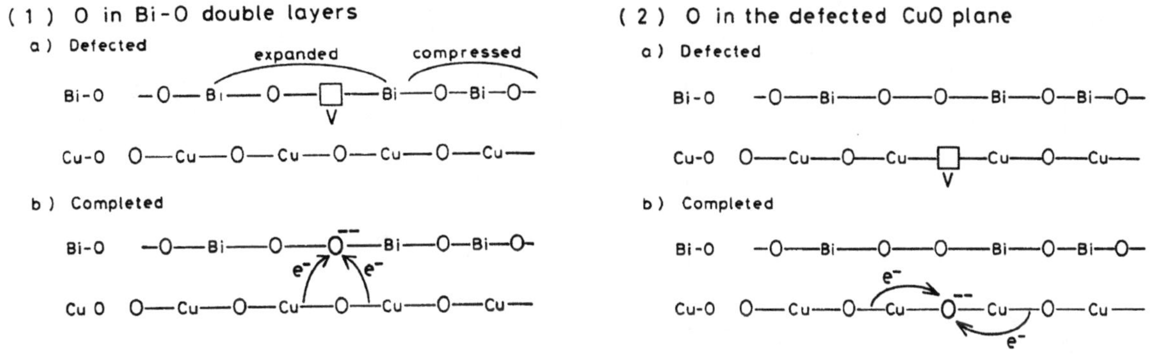

Fig.6 Possible sources of holes in the BSCCO film produced in a reducing atmosphere. Sr layers between the Bi-O and CuO_2 planes are omitted for the sake of clarity.
(1) Incorporation of an oxygen ion into an O vacancy in an expanded portion of the Bi-O layer.
(2) Incorporation of an oxygen ion into an O vacancy in the CuO_2 plane.

creates 2 holes per oxygen atom (Fig.6 (1) b)). The hole generating mechanism described above, however, is not consistent either with the HREM observation (Fig.5) or with the Bi $4f_{7/2}$ and Cu $2p_{3/2}$ core electron spectra [17].

Another more plausible mechanism is shown in Fig.6(2). All the oxygen sites in the Bi-O double layer are already occupied at the as-deposited state. A charge associated with an extra oxygen ion in the Bi-O double layer is balanced by introducing an O vacancy in the CuO_2 plane (Fig.6 (2) a)). This view is chemically plausible because the most difficult element to oxydize is Cu among the four cationic elements involved. It is noted that no holes are generated at this stage. Holes are generated only when oxygen ions are introduced into these O vacancies in the CuO_2 plane. These oxygen atoms are likely to preferentially oxydize the nearby Cu atoms, in agreement with the Cu $2p_{3/2}$ core electron spectroscopy study [17]. The incorporation of oxygen atoms into the O vacancies in the CuO_2 plane, however, is insufficient to cause a sharp superconducting transition at 110K (see Table I and also Fig.4.). All the 2223 films so far fabricated by the low temperature process have tails at a lower temperature end of the resistive transition curve. The observed Tc zero is typically 95K, much lower than 110K expected for the 2223 phase.

ORDERING OF Sr AND Ca

The low temperature fabrication process of the 2223 film so far successfully applied utilizes the vapor deposition process (Table I). Therefore, the atomic arrangement in the film is likely to be far from the equilibrium configuration. The most difficult process is perhaps ordering of Sr and Ca because both the elements are chemically similar. In fact, the XPS studies show that ordering of Sr and Ca is incomplete even in the best annealed film [17]. Therfore, we associate the tail of the resistive transition with this incomplete ordering. This view is in agreement with the observation that the 2223 film shows a sharp transition when heat treated in a partially melted temperature range [15].

SUPERCONDUCTING PROPERTIES OF WELL CHARACTERIZED BSCCO FILMS

One of the advantages of the film fabrication is an easy control over the crystal structures, 22(n-1)n. We were successful to prepare the superconducting 22(n-1)n films with n = 1-5 and to measure Tc. Since the superconducting transition is not sharp in the present films, Tc is defined somewhat arbitrarily as an intersection point of two lines extended from the linear portions of the ρ-T carves in the normal and transition states. Tc's thus determined are plotted in Fig.7 as a function of the number n of the CuO_2 planes intervening between the neighboring Bi-O double layers [19] . Decrease in Tc for $n \geqq 4$ is apparently caused by the deficiency of holes in the CuO_2 planes.

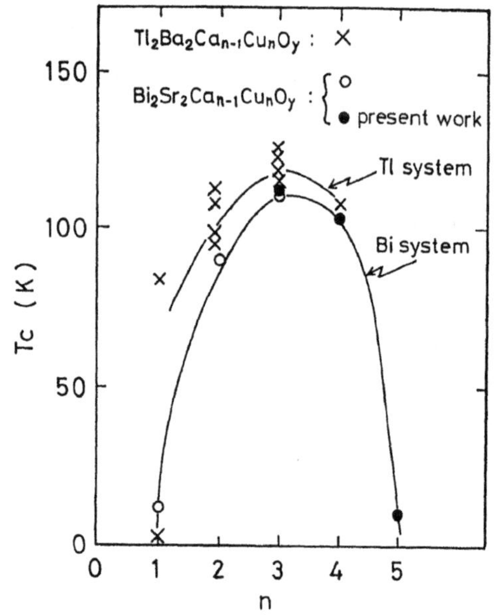

Fig.7 Tc vs number n of the CuO_2 layers intervening between the neighboring Bi-O double layers. A similar plot for the Tl system (after M. Kukuchi et al. (1989) Physica 158:79-82) is also shown.

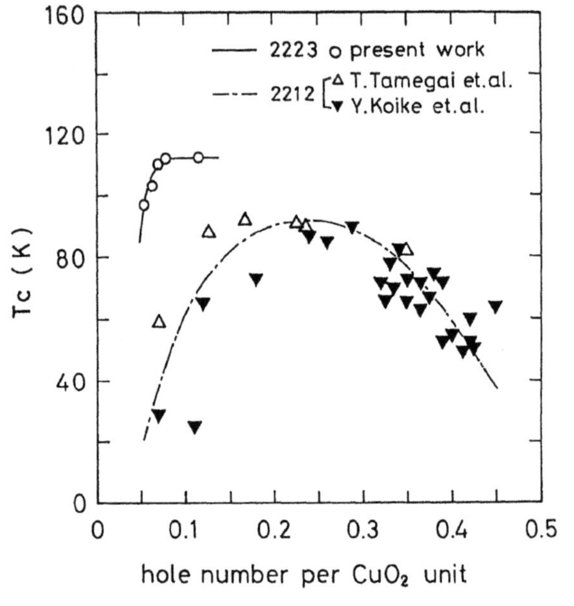

Fig.8 Tc vs the number of holes per CuO_2 unit for 2223 films. A similar plot for the 2212 phase is also shown.

We were also successful to prepare c-axis oriented 2223 films for the first time by annealing the amorphous film in a Pb vapor atmosphere [15] . We measured Tc and the hole number p per CuO_2 for the 2223 films annealed at various oxygen partial pressure. The results are plotted in Fig.8. It is noted that the as-prepared 2223 film corresponding to a maximum p in the plot is located in the middle of a plateau of the Tc vs p curve. This implies that the best Tc of the 2223 film is not sensitive to the oxygen concentration of the film.

SUMMARY

Difficulties associated with the 2223 film fabrication by the low temperature process is two fold: oxygen deficiency in the CuO_2 plane and disordering of Sr and Ca. The former is responsible for the semiconducting and superconducting states of the as-deposited film and the latter for the tail of the resistive transition of the post annealed film. A key factor for high Jc films is a continuous texture of c-axis oriented grains. The layer by layer growth technique is promising to produce a thermally stable film of smooth surface and of a designed structure 22(n-1)n. A new hole generating mechanism in the BSCCO film produced at low oxygen partial pressure is proposed based on the model of O vacancies in the CuO_2 plane.

ACKNOWLEDGMENTS

One of the authors (K.O.) thanks H.Itozaki for informative discussions on the microstructure of the high Jc film and for allowing me to reproduce his unpublished picture (Fig.1(a)). He also thanks A.Yamamoto for helpful communications on extra oxygen atoms in the BiO double layer.

REFERENCES

1. Itozaki H, Tanaka S, Yazu S (1989) Preparation of high Jc superconducting thin film by sputtering. In: 2nd Workshop on High-Temperature Superconducting Electron Devices R&D Association for Future Electron Devices, 7-9 June 1989. Shikabe, Hokkaido, Japan. pp.293-296.

2. Adachi H, Kohiki S, Setsune K, Mitsuyu T, Wasa K (1989) Formation of superconducting Bi-Sr-Ca-Cu-O thin films with controlled c-axis lattice spacings by multitarget sputtering. Jpn.J.Appl. Phys. 27: L 1883-1886.

3. Nakamura K, Sato J, Ogawa K (1990) Formation of thermally stable multilayered BSCCO films with the 2223, 2234 and 2245 structures. Jpn.J. Appl. Phys. (in press).

4. Fujita J, Tatsumi T, Yoshitake T, Igarashi H. (1989) Epitaxial film growth of artificial (Bi-O)/(Sr-Ca-Cu-O) layered structures. Appl. Phys. Lett. 54: 2364-2366.

5. Satoh T, Yoshitake T, Miura S, Fujita J, Kubo Y, Igarashi H (1989) As-grown superconducting Bi-Sr-Ca-Cu-O thin films by coevaporation. Appl. Phys. Lett. 55: 702-704

6. Kanai M, Kawai T, Kawai S, Tabata H (1989) Low-temperature formation of multilayered Bi(Pb)-Sr-Ca-Cu-O thin films by successive deposition using laser ablation. Appl. Phys. Lett. 54: 1802-1804.

7. Yamane H, Kurosawa H, Hirai T, Iwasaki H, Kobayashi N, Muto Y (1988) Formation of bismuth strontium calcium copper oxide superconducting films by chemical vapor deposition. Jpn. J. Appl. Phys. 27: L 1495-1497.

8. Balestrino G, Paoletti A, Paroli P, Romano P (1989) Growth of textured films of $Bi_2Sr_2Ca\ Cu_2\ O_{8+x}$ from KCl solution. Appl. Phys. Lett. 54: 2041-2042.

9. Nakayama Y, Tsukada I, Maeda A, Uchinokura K (1989) Epitaxial growth of Bi-Sr-Ca-Cu-O thin films by molecular beam epitaxy technique with shutter control. Jpn. J. Appl. Phys. 28: L1809-1811.

10. Hsu H M, Yee I, DeLuca J, Hilbert C, Miracky R F, Smith L N (1989) Dense Bi-Sr-Ca-Cu-O superconducting films prepared by spray pyrolysis. Appl. Phys. Lett. 54: 957-959.

11. Hayakawa H (1989) (private communication) .

12. Nakamura K, Sato J, Kaise M, Ogawa K (1989) Synthesis of artificially layered Bi-Sr-Ca-Cu oxide films and their thermal stability. Jpn. J. Appl. Phys. 28: L437-440.

13. Hatano T, Aota K, Ikeda S, Nakamura K, Ogawa K (1988) Growth of the 2223 phase in leaded Bi-Sr-Ca-Cu-O system. Jpn. J. Appl. Phys. 27 : L2055-2058.

14. Aota K, Hattori H, Hatano T, Nakamura K, Ogawa K (1989) Growth of 2223 phase in leaded Bi-Sr-Ca-Cu oxide under reduced oxygen partial pressure. Jpn. J. Appl. Phys. 28 :(in press)

15. Hayakawa H, Kaise M, Nakamura K, Ogawa K (1989) Growth of the 2223 phase in Bi-Sr-Ca-Cu oxide films under a controlled Pb potential. Jpn. J. Appl. Phys. 28 : L 967-969.

16. Yamamoto A, Onoda M, Takayama-Musomachi E, Izumi F, Ishigaki T, Asano H (1989) Rietveld analysis of the modulated structure in the superconducting oxide $Bi_2(Sr,Ca)_3Cu_2O_{8+x}$ Phys. Rev. B (submitted)

17. Kohiki S, Hirochi K, Adachi H, Setsune K, Wasa K (1989) Effect of annealing in oxygen on the structure formation of Bi-Sr-Ca-Cu-O thin films. Phys. Rev. B 39 : 4695-4698.

18. Torardi C C, Parise J B , Subramanian M A, Gopalakrishnan J, Sleight A W (1989) Oxygen nonstoichiometry in copper-oxide based superconductors and related systems: structure of nonsuperconducting $Bi_2Sr_{3-x}Y_xCu_2O_{8+y}$ (x ~0.6-1.0). Physica C 157: 115-123.

19. Hattori H, Nakamura K and Ogawa K (1990) Effects of oxygen-potential-controlled annealing on superconducting properties of $(Bi,Pb)_2Sr_2Ca_2Cu_3O_y$ thin films. Jpn. J. Appl. Phys. (in press).

High-T_c Superconducting Analog Circuits

R.W. Ralston

Lincoln Laboratory, Massachusetts Institute of Technology, 244, Wood St., Lexington, MA 02173-9108, USA

ABSTRACT

Planar transmission line networks incorporating high-transition-temperature (T_c) superconductive thin films can potentially provide signal-distribution, signal-filtering, and signal-processing functions in analog monolithic integrated circuits at higher frequencies, wider bandwidths and larger scales of integration than possible by conventional technology. This paper discusses the materials progress, device concepts and applications prospects for such high-T_c superconducting circuits.

KEY WORDS: microwave circuits, planar transmission-line devices, resonators, filters, analog signal processing

OVERVIEW

Many of the potential applications for the high-T_c superconductive oxides involve the transmission and manipulation of high frequency signals. It is therefore important to develop low-microwave-loss transmission-line structures composed of superconducting oxide films deposited on compatible dielectric substrates. Recent loss measurements of high-T_c striplines are reported. Although the superconductive oxides are not yet to the point of technological utility, requirements for lower-loss films and projections of the impact on high-frequency electronics can be made based on the parameters of prototype devices developed using conventional superconductors. Examples of such devices will be given; these include passive transmission-line structures such as resonators with quality (Q) factors beyond 10^6 and chirp filters with greater than 2.5-GHz bandwidth. Progress toward implementing these devices in high-T_c superconductors is described. Also discussed are more advanced circuits which require switching devices in combination with transmission lines, such as wideband antenna beam-forming networks. This paper concludes with speculation on the long-term applications of superconductive oxide thin films in monolithic microwave circuits.

LOW-LOSS PLANAR TRANSMISSION LINES

For microwave signals, the current flow in both normal conductor and superconductor materials is confined near the surface. However, the behavior of normal conductors and superconductors to microwave fields is fundamentally quite different. The penetration of fields and flow of currents near the surface of a normal conductor decay over a characteristic length, or skin depth δ; this depth scales inversely as the square root of the frequency. In contrast, the characteristic length, or London penetration depth λ, for fields and currents in a superconductor is essentially frequency independent.

Consider the simple two-fluid model of a superconductor. At temperatures below T_c, both normal, unpaired charge carriers and super, paired charge carriers exist, with the population of normal carriers growing and the population of super carriers falling exponentially as the temperature approaches T_c. At zero frequency, the superpairs can fully screen the fields from the normal carriers and current flows with no loss. At microwave frequencies, however, because of the inertia of the pairs, full screening is not achieved and the fields couple to the normal carriers, which in turn lose energy to the lattice. For superconductors in this regime the surface loss R_s, expressed in ohms per square of film area, is expected to scale approximately as the square of frequency [1]. In contrast, in normal conductors R_s scales approximately as the square root of frequency

These loss characteristics are summarized in graphic form in Fig. 1. The expected losses for high-quality Cu and Nb films cooled to 4.2 K are shown, as well as a projection for losses in high-quality $YBa_2Cu_3O_x$ (YBCO) films cooled to 50 K. In each case the superconductor is operated at a reduced temperature T/T_c of ~ 0.5. At this temperature and below, the superpair population is sufficiently large to provide, until the gap frequency is reached, a substantial circuit advantage over normal conductors. The projection for YBCO is based on a simple two-fluid model which assumes isotropic behavior like that of a conventional superconductor and no states in the superconducting gap. The projection, therefore, must be refined. However, recent experimental results at a number of institutions, a few of which are indicated in Fig. 1, indicate that YBCO films can indeed provide a substantial advantage over normal conductors, although not yet as great an advantage as Nb films. These YBCO results will be discussed in a later section of the paper.

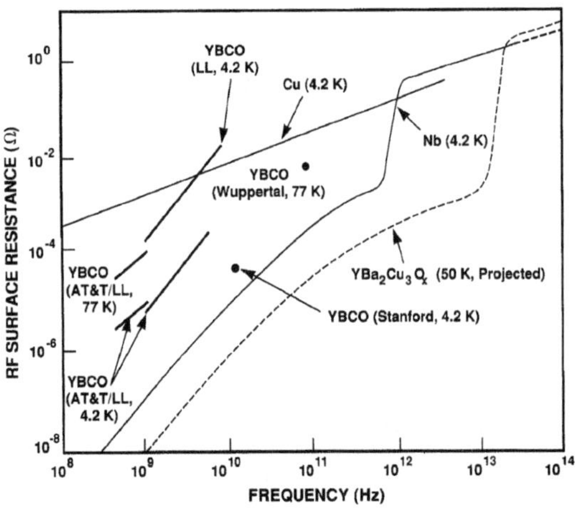

Figure 1. Comparisons of selected microwave surface resistance data with calculated plots for copper, niobium and YBCO.

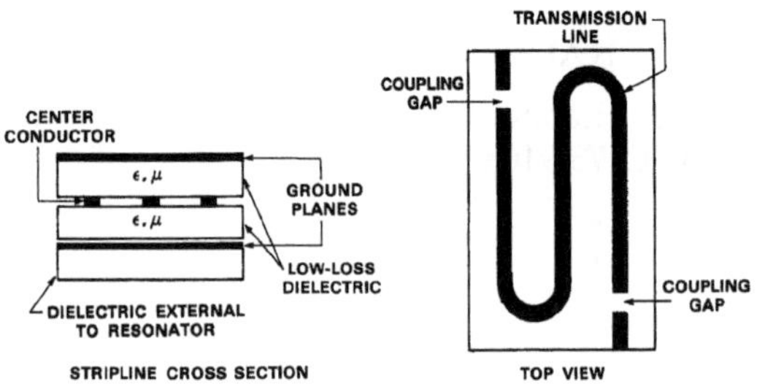

Figure 2. Schematic views of stripline resonator.

Generally, the beneficial impacts of superconductive transmission lines on microwave circuits are as follows. The reduced loss is so substantial as to permit simultaneous increases in density, scale of integration, frequency, bandwidth, and functional performance of planar microwave circuits. An additional, and equally important, feature is that the superconductor has essentially no material dispersion in the frequency regime of interest. Performance will be greatly improved for fully passive circuits, for example, by allowing the replacement of bulky three-dimensional waveguide structures with compact planar stripline circuits. Performance will also be improved for active circuits which, for example, combine semiconductor (or superconductor) switches to facilitate the reception and processing of signals by sensor arrays.

PROTOTYPE CIRCUITS

The use of superconductive transmission lines for analog signal processing was pioneered at MIT Lincoln Laboratory. To illustrate the benefits of superconductivity to microwave circuits, two existing low-T_c devices and an application for each are described.

The first component, a stripline resonator, is shown schematically in Fig. 2. A section of line is capacitively coupled at either end to input and output transmission lines. The circuit will resonate at frequencies for which the isolated line section is an integral number of half-wavelengths long. As fabricated at MIT Lincoln Laboratory, substrates are typically 2.5 x 1.3 cm, and high-quality superconductive films are required over this area. For the surface resistance shown in Fig. 1 for Nb, resonator Qs exceeding 10^6 at 1 GHz have been achieved using substrates of silicon and sapphire [2]. Preliminary results for YBCO will be described later.

The Nb resonators have been used to stabilize feedback oscillators operating near 1 GHz. However, the measured flicker noise of - 100 dBc/Hz at 1 Hz offset for a loaded Q of 2 x 10^5 was excessive [3]. Vibration sensitivity is believed to be one cause and techniques are under development to improve performance.

The second component example is a tapped delay line. A well-developed conventional superconductive technology now exists to produce whole-wafer-scale devices such as that pictured in Fig. 3. The basic concepts are reviewed here; more detail is given elsewhere [4]. Geometrical dispersion is minimized by employing symmetric stripline structures. Niobium thin-film conductors are used on 5-cm-diam. silicon wafers. Two striplines are patterned in close proximity to form a cascaded array of backward-wave couplers. The figure inset shows an enlarged view of one of the coupler structures. The frequency response of the filter is set by the distance between the bends in each of the lines, while the amplitude weighting is determined by the spacing between the coupled lines. The usual response engineered for such a device is a linear frequency-delay characteristic, with either flat or Hamming weighting. Typical parameters are 4-GHz center frequency, 2.6-GHz bandwidth and ~ 40-ns or ~ 90-ns delay. Each 10 ns of delay requires ~ 77 cm of stripline length. With these parameters the superconductive properties are not limiting the performance. Indeed, substantially higher circuit densities and longer delays would be possible if deposited dielectrics of ~ 5-10 μm thickness could be developed with loss tangents < 10^{-6}.

With two of these chirp filters, it is possible to do very rapid spectral analysis by the chirp transform algorithm [4]. The wideband signal to be analyzed is multiplied (in a mixer) by a chirp which has been generated by impulsing one of the chirp filters. For simplicity, consider the input signal to consist of a number of offset frequency tones simultaneously present. The output of the mixer thus consists of multiple, frequency-offset chirps. When these chirps are in turn input to a second filter of matched but opposite chirp slope, then the output of the transform system will appear

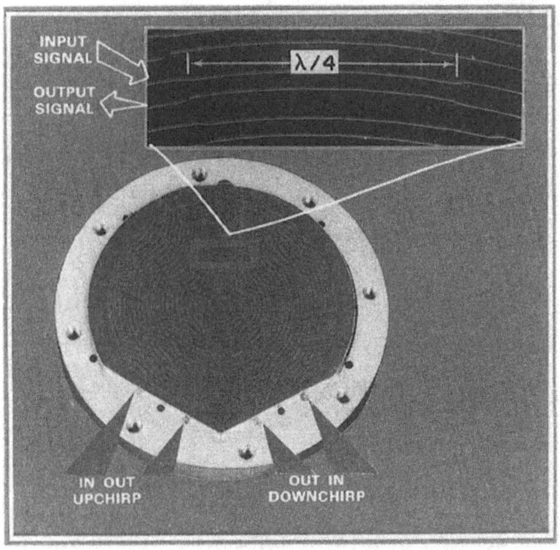

Figure 3. Photograph of 2.6-GHz, 37.5-ns chirp filter fabricated with coupled niobium striplines on a silicon wafer.

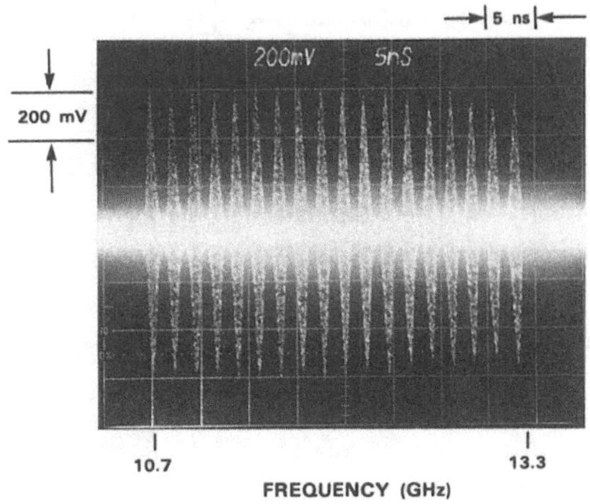

Figure 4. Multiple exposure of sampling-scope output for chirp-transform spectrum analyzer.

as time-offset, compressed pulses. The timing of the individual pulses at the output reflects the frequency of the original tones at the system input; thus frequency has been mapped into time.

A multichannel spectrum analyzer has been demonstrated at MIT Lincoln Laboratory. It is able to handle signals spread over a 2.5-GHz bandwidth in each of four channels [5]. The oscilloscope photograph in Fig. 4 shows the system output for a succession of 18 input tones spaced at 150-MHz intervals. The analysis time is 38 ns, with a frequency resolution of about 40 MHz. Rapid spectral analysis is an important element of a wide variety of signal-processing tasks. The capabilities for speed and bandwidth of analysis are not attainable by any other technology. By comparison, the computational requirements for providing a real-time Fourier analysis digitally by means of the radix-2 Fast Fourier Transform algorithm exceeds 10^{11} operations/s [4].

Unfortunately, the prodigious capability of the analog spectrum analyzer cannot yet be efficiently exploited in systems because sufficiently wideband transient recorders (e.g., flash analog-digital converters) do not exist in convenient, compact form. One of the continuing challenges for superconductive electronics is to develop such devices (in low-T_c and ultimately in high-T_c form).

HIGH-T_c RESONATORS

The stripline resonator structure described in the previous section has been used extensively at MIT Lincoln Laboratory to evaluate the microwave loss characteristics of many oxide superconductor thin film samples [2]. In most cases the bottom Nb ground plane is substituted by the high-T_c sample and R_s is determined as a function of frequency at 4.2 K. This technique requires films over relatively large areas, and is thus technologically more demanding than the more conventional cavity techniques used at higher frequencies by the Wuppertal [6] and other groups to characterize smaller area films. The stripline method tests the films in a configuration very similar to that required for the analog circuit structures which will be essential to so many of the envisioned microwave applications. Recently, AT&T Bell Laboratories has provided YBCO films deposited on $LaGaO_3$ using the co-evaporation followed by post-anneal process [7]. These films had sufficiently low loss to warrant fabrication of all-YBCO resonators. The AT&T/LL results spanning 0.5 - 1 GHz in Fig. 1 indicate that the surface loss in the all-high-T_c device is well below that for Cu even at 77 K.

Additional challenges remain before device-quality material is available. Reproducibility must be improved, substrate twinning and cracking must be reduced, and current-carrying ability must be increased. While the results in Fig. 1 indicate an approximate frequency-squared R_S dependence at very low powers, there are significant field-dependent effects in high-T_c striplines. Preliminary results for YBCO resonators indicate that for increased power, R_S grows proportionate to both frequency and magnetic field [8]. The onset for this behavior occurs substantially below the lower critical field.

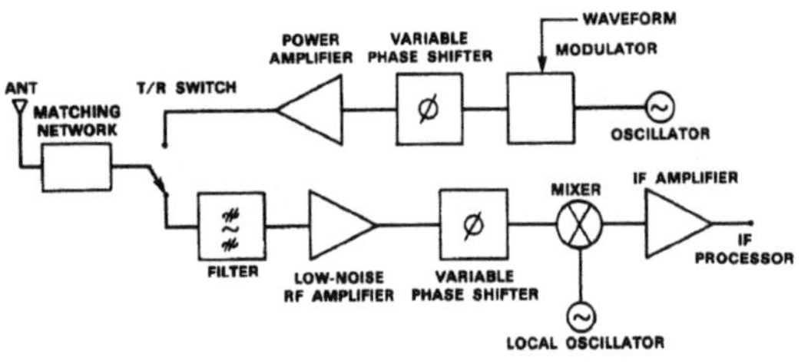

Figure 5. Block diagram of a microwave system.

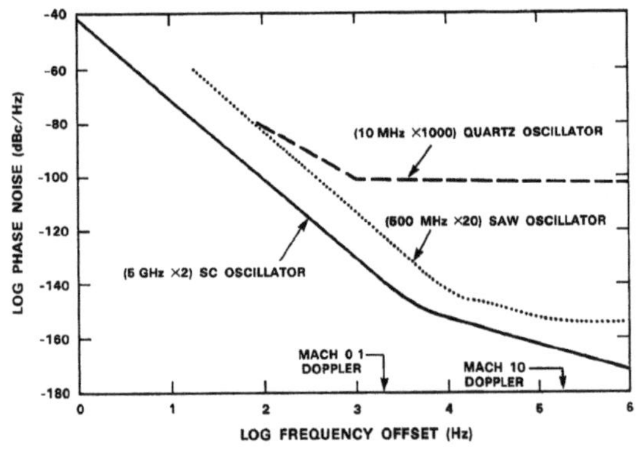

Figure 6. Comparison of single-sideband phase noise for oscillators operating at 10 GHz and using existing acoustic or projected superconductive resonators for stabilization.

PROJECTED HIGH-T_c CIRCUITS

Several superconductive analog circuits and their potential impacts on microwave systems are highlighted. For this purpose it has been assumed that all of the requirements for low-loss superconductors over large area substrates are met at temperatures of $\geq$ 50 K. While no Josephson-effect devices are specified, convenient hybrid interconnection to high-speed semiconductor circuits is essential to most of these components. GaAs seems particularly well matched in analog circuit parameter space to the high-T_c superconductors, and monolithic integration would provide substantial additional benefit. The circuits described provide a context in which to continue the technology development.

Shown in Fig. 5 is a block diagram of a typical transmit/receive microwave system. Low loss and low dispersion in wideband transmission lines would benefit many components in the system, from enabling more efficient power combining and matching within the power amplifier and antenna array for higher output during transmission, to permitting sharper filtering with low insertion loss for greater sensitivity and selectivity during reception. Also, greater stability and lower noise should be achievable in oscillators, and phase shifting of wideband waveforms should be possible for beam steering. Finally, intermediate frequency (IF) signal processors such as the chirp Fourier transformer described previously, working in conjunction with high-speed GaAs-based transient recorders, should permit very efficient demodulation of signals (e.g., frequency-shift keyed information). The capabilities of three components in the microwave system, a low-noise local oscillator, a low-loss filter, and a high-gain electronically steered antenna, are projected in more detail.

A resonator-stabilized oscillator is produced by combining within a feedback loop an amplifier to compensate losses, a phase shifter to adjust the feedback delay to 360°, and a very sharp filter to minimize noise. A small fraction of the loop power is coupled out as the oscillator signal. Figure 6 shows the projected phase noise for a superconducting resonator-stabilized oscillator operating at 5 GHz, frequency doubled to obtained 10-GHz center frequency [3]. The conventional model for noise contributions was used, with flicker noise assumed to be dominated by a GaAs FET amplifier, a resonator Q of 7×10^5 taken and a 0.5-W loop power estimated [5]. Also shown are the phase noise curves which are achieved in today's acoustic technology. Clearly the superconducting oscillator will provide substantially better noise performance provided the flicker noise and current carrying performance of the high-T_c film is improved. Furthermore, the higher resonator frequency reduces the number of multiplier stages required, thereby reducing size and power consumption.

The use of long delay lines of low loss and low dispersion to provide proper phasing of wavefronts arriving over wide beam angles across an antenna aperture of many wavelengths is essential to providing high gain and beam agility for advanced microwave antennas. The block diagram in Fig. 7 shows a segment of such a beam forming and power combining network.

The delays of appropriate lengths must be switched in and out as demanded by the beam direction. At frequencies above a few gigahertz, the losses in anything but waveguide for long runs can be excessive, unless superconductive transmission lines are employed. The weight, performance, and cost advantages of a cooled superconductive feed structure would be substantial for antenna apertures of the order of several hundred elements. A time delay module is shown schematically in Fig. 8.

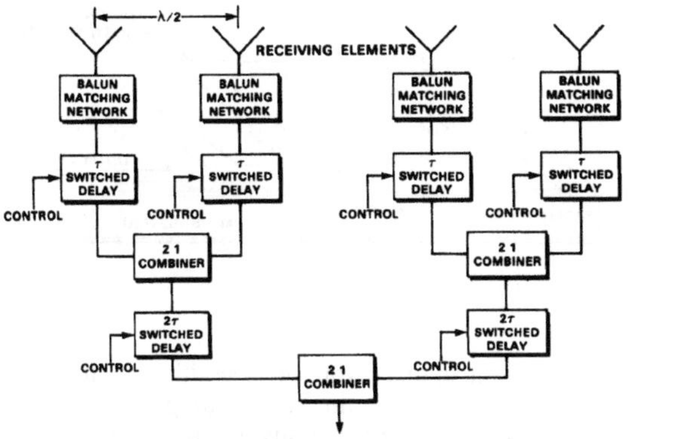

Figure 7. Block diagram of a segment of beam-forming and power-combining network.

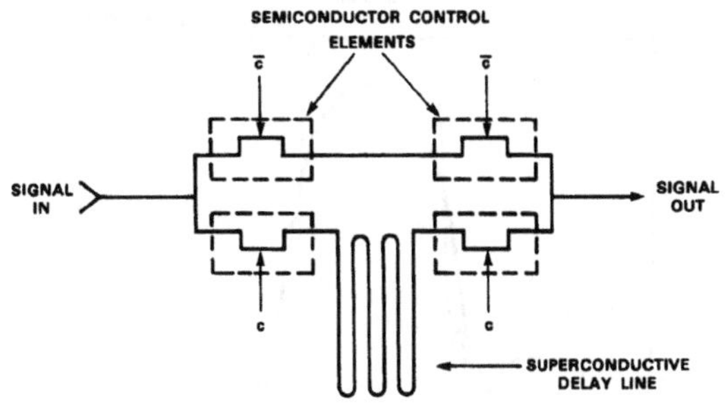

Figure 8. Schematic of a superconductive switched-delay module.

The advantages of a superconductive feed structure would be substantial, as indicated in Fig. 9. This plot clearly shows the effects of distribution loss on antennas operating at 60 GHz. For Cu beyond apertures of about 50 wavelengths, although the spatial beam continues to sharpen, the gain over isotropic drops because of distribution losses. The best reported results for YBCO at 77 K would permit substantial growth of antenna arrays while maintaining efficiency.

Very-low-loss, sharp-skirted filters offer another excellent example of improved circuit performance. The comparison of multipole filters implemented in planar form using normal and superconducting films is shown in Fig. 10. The superconductor has very low loss and thus allows very sophisticated filter functions to be implemented by the use of many sections. This should permit the fabrication of sharp-response filters with essential zero insertion loss in compact form.

Ultimately such filters, as well as the other transmission line components which have been described, might be combined monolithically with GaAs devices. A conceptual layout for a preamplifier is shown in Fig. 11. The lower resistance of the superconductor gate in the transistor would reduce noise figure, as would the zero-insertion-loss, sharp-skirted filter. Furthermore, the reduced dimensions allowed by superconducting lines would permit thinning of the GaAs to achieve both higher circuit densities and better removal of residual heat from active devices.

CONCLUSION

The applications potential for high-T_c superconductors within electronics is extremely promising. This paper has focussed on high frequency analog circuits; equally promising applications exist in low frequency biomedical sensor arrays. There are, however, many material and device challenges to be successfully met if the applications are to be realized. The author expects that the technology development will be best done if conducted within a device context. Application targets provide a focus for the activities of materials scientists, device physicists and circuits engineers. Indeed, the device and circuit structures described in this paper have been adopted at Lincoln Laboratory as instruments by which to measure progress and test the technology.

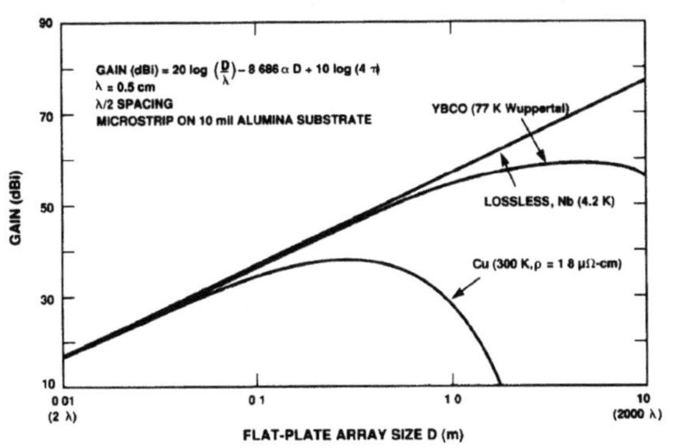

Figure 9. Effect of distribution loss on antenna gain.

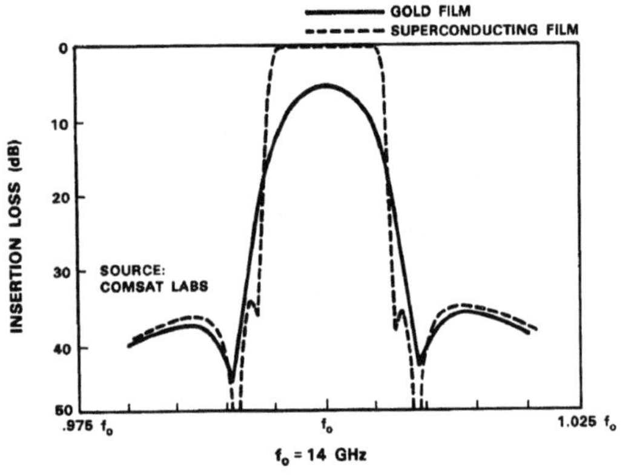

Figure 10. Comparison of calculated microstrip six-pole elliptic-response filters.

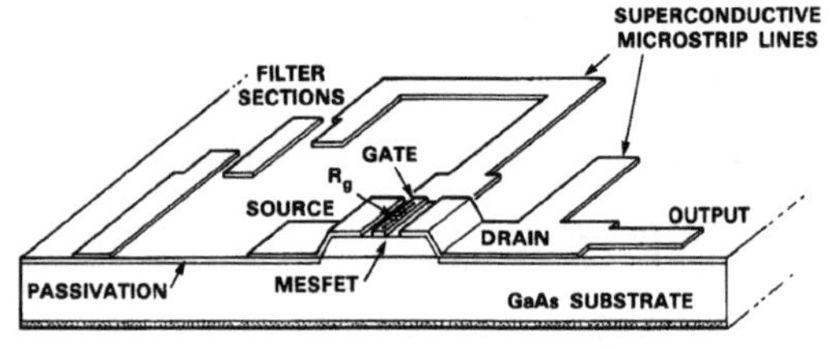

Figure 11. Conceptual layout of monolithic microwave integrated circuit incorporating superconductive transmission lines.

ACKNOWLEDGMENTS

This work was sponsored by the Departments of the Air Force, the Navy and the Army, and by the Defense Advanced Research Projects Agency. The author thanks the many members of the Analog Device Technology Group at MIT Lincoln Laboratory for their contributions. In addition he thanks Dr. Andrew Meulenberg of COMSAT Laboratories for the elliptic filter calculation. Also he thanks Dr. Paul Mankiewich of AT&T Bell Laboratories for YBCO films. The collaborative effort involving YBCO resonators is being conducted within the Consortium for Superconducting Electronics recently formed by MIT, MIT Lincoln Laboratory, IBM Research Division and AT&T Bell Laboratories.

REFERENCES

1. VanDuzer T, Turner CW (1981) Principles of superconductive devices and circuits. Elsevier North Holland, New York

2. Oates DE, Anderson AC (1989) Stripline measurements of surface resistance: relation to HTSC film properties and deposition methods. SPIE symposium series. 10 October 1989 Santa Clara

3. Oates DE, Anderson AC (1989) Superconducting stripline resonators and high-T_c materials. IEEE Trans. Microwave Theory Tech., to be published

4. Ralston RW (1985) High-speed analog signal processing with superconductive circuits. In: Mourou GA, Bloom DM, Lee C-H (eds) Picosecond Electronics and Optoelectronics. Springer-Verlag, Berlin Heidelberg New York Tokyo, pp 228-235 (Electrophysics series, vol 21)

5. Withers RS, Ralston RW (1989) Superconductive analog signal processing devices. Proc IEEE 77: 1247-1263

6. Klein N, Mueller G, Piel H, Roas B, Schultz L, Klein U, Peiniger M (1989) Millimeterwave surface resistance of epitaxially grown $YBa_2Cu_3O_{7-x}$ thin films. Appl Phys Lett 54: 757-759

7. Mankiewich PM, Scofield JH, Skocpol WJ, Howard RE, Dayem AH, Good E (1987) Reproducible techniques for fabrication of thin films of high transition temperature superconductors. Appl Phys Lett 51:1753-1755

8. Oates DE, Mankiewich PM. Private communication

Maglev Trains

SHUNSUKE FUJIWARA
Railway Technical Research Institute, 8-38, Hikari-cho 2-chome, Kokubunji, Tokyo, 185 Japan

ABSTRACT

The superconducting magnets of the maglev vehicle are exposed to various magnetic or mechanical environments, when they are used to generate propulsion, levitation and guidance forces. It is important to know the magnitude of the force distribution on the superconducting coil and of the alternating force. An outline of the test vehicle, its magnets and the forces acting on the superconducting coil is given.

KEY WORDS: maglev, linear motor, levitation, superconducting magnet

INTRODUCTION

Railway Technical Research Institute is engaged in the Maglev development project. At present a test run of the MLU-002 test vehicle is under way on the Miyazaki test track. Recently the plan to construct a longer test track has been finalized, and this means a step forward in the direction of realizing a commercial maglev line. This paper describes the present status, the characteristics of the levitation system and the forces acting on the superconducting magnet.

FEATURES OF MAGLEV VEHICLE

This maglev project is based on the idea that in future a mass transport at superspeed will be required. The superconducting maglev vehicle carries superconducting magnets which serve as field magnets for the linear synchronous motor and electro-dynamic levitation. This system has the following features:

(1) Propulsion and support of the vehicle are carried through non-contact, and they do not depend on adhesion.
(2) As an active guideway is adopted, there is no need to feed propulsive power to the vehicle and the vehicle can be made lightweight.
(3) As the gap between vehicle and guideway is set at about 10 cm, the torelance of the guideway irregularity will be large.
(4) There is no need to control the gap of levitation, but the vehicle can not levitate itself at low speed.

The system which has the above mentioned features includes lightweight and portable superconducting magnets. At present the test vehicle MLU-002 is under test run on the Miyazaki test track. This vehicle has a body of which length and weight are 22m and 17 tons respectively, and has 44 seats(Table 1). There are two bogies and they have each six superconducting magnets. The superconducting magnets are attached vertically to both sides of the bogie, and generate three forces, that is propulsion, levitation and guidance. Guideway has a U-shaped cross section, and on both sides of the vertical surface the combined propulsion and guidance coils are attached, and on the horizontal bottom surface the levitation coils are attached. The propulsion coils are fed a current whose frequency is proportional to vehicle speed, and they generate a propulsive force proportional to the amplitude of the current. When the vehicle runs, currents are induced in the levitation coils, generating a

Table 1 Main features of MLU-002

Vehicle dimensions	
Length, Width, Height	22 x 3.0 x 3.7 m
Mass	17 t
Seating capacity	44
Suspension	
Lift force	196 kN
Effective gap	11 cm
Guidance	
Guidance force	83.3 kN (at 5 cm shift)
Effective gap	More than 15 cm
Propulsion	
Maximum thrust	79.4 kN
Maximum frequency	28 Hz
Maximum speed	420 km/h

Table 2 Superconducting magnet

Coil	
Dimensions	1.7 x 0.5 m
Mass	77 kg
Magnetomotive force	700 kA
Number of turns	1167
Current density	219 A/mm2
Stored energy	550 kJ
Wire	
Copper ratio	1.06
Cross section	1.05 x 2.12 mm
Number of filament	2382
Twist pitch	49 mm

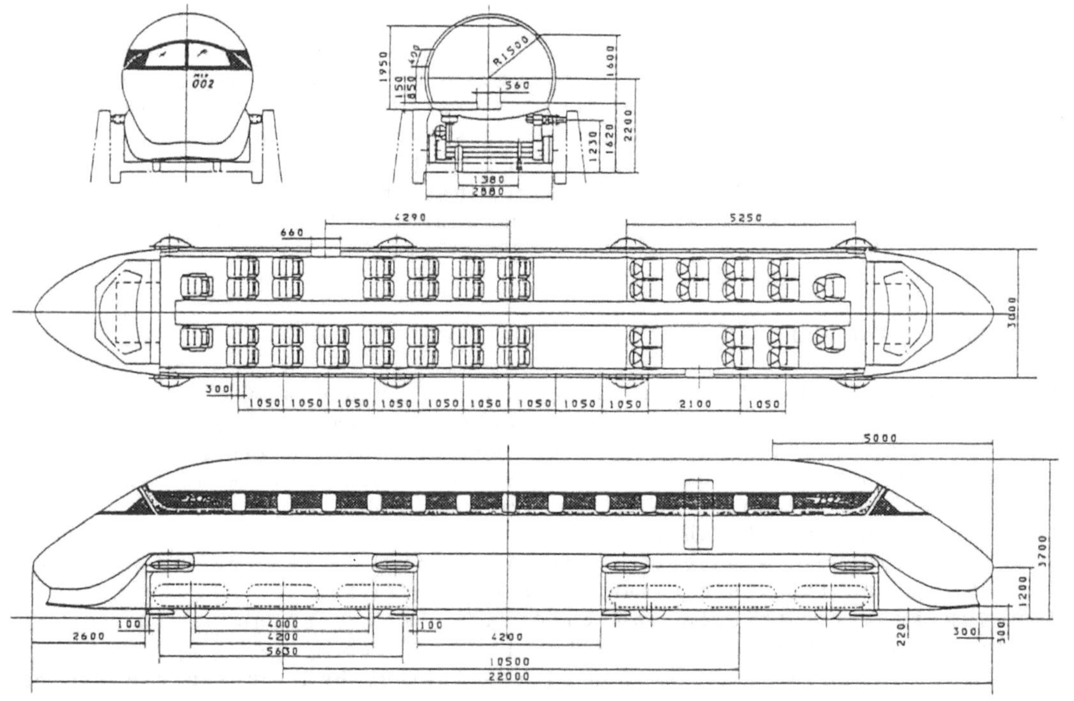

Fig.1 MLU-002

levitation force and the vehicle is levitated at a take off speed.

Superconducting magnet is required to meet following conditions:

(1) Light weight,

(2) Current density in the superconducting coil must be high, because the cross section of the coil must be small to ensure close coupling between on board coil and guideway coil. Then the copper ratio is desirably small,

(3) Cryostat has a minimum dimension from the superconducting coil to the outer surface,

(4) Stationary heat leak must be small,

(5) Levitation, guidance and propulsion forces must be transmitted to the bogie frame and the structure must endure various vibrations,

(6) It is easy to energize or de-energize.

Main features of superconducting coil employed in MLU-002 are listed in Table 2. The length, width and magnetomotive force are 1.7m, 0.5m and 700kA respectively. The wire used is a niobium-titanium alloy with copper ratio about 1. The coil is reinforced by epoxy resin impregnation and housed in the SUS inner vessel, which is fitted by load-bearing members of FRP to the outer vessel of aluminum.

CHARACTERISTICS OF LEVITATION SYSTEM

In the electro-dynamic system it is necessary to induce a current in the guideway levitation coil in order to support the vehicle. Then some energy is dissipated in the levitation coil and it is expressed as a running resistance of the vehicle. The drag-to-lift ratio is an index to goodness of the levitation system. To increase the drag ratio the power loss in the guideway levitation coil must be decreased, and the following two methods have been studied for the purpose:

(1) To decrease resistance of the guidway coil.

(2) To decrease the necessary current in the guideway coil.

The former means an increase in the cross section of the guideway coil and the drag ratio is the proportional to the mass of the material. The latter means an increase in the magnetomotive force of the superconducting magnet or the coupling to be devised between the superconducting magnet and the guideway levitation coil; for instance null flux method is available. Fig.2 shows one of the coil arrangements in which the levitation coils are set on the vertical wall of the guideway[2]. A guideway levitation coil is composed of two unit coils arranged in two rows up and down on the vertical wall, and the two unit coils are connected in reverse direction to each other to make a null flux circuit as shown in fig.3. When the on-board superconducting magnet moves on the center line of the guideway levitation coil, no current is induced in the levitation coil because the mutual inductance is zero. Assuming that the superconducting magnet deviates downward, the induced current flows proportionally to the displacement and then the levitation force is proportional to the displacement. On the other hand the Joule loss in the levitation coil is proportional to the second power of the current; then the drag ratio can be large if the necessary levitation force is obtained with a small displacement. To realize it, the superconducting magnet must have a relatively large magneto-motive force.

In addition the guideway coils on both sides of the wall are connected in reverse direction to each other to make another null flux circuit against the lateral motion as shown in Fig.3. Fig.4(a) shows a current in the coils and the levitation force in the case of no lateral displacement, and Fig.4(b) shows the guidance force in the case of lateral displacement. Fig.5 shows a balanced vertical displacement obtained from Miyazaki test vehicle.

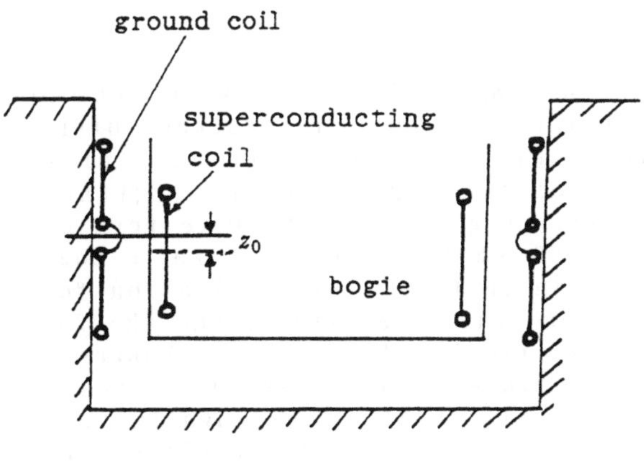

Fig.2 Coil arrangement

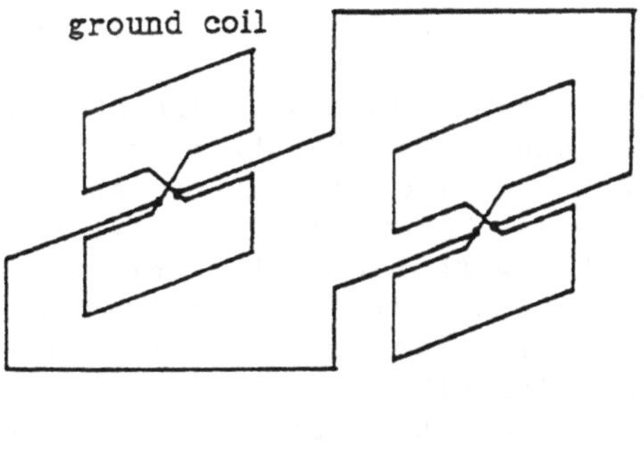

Fig.3 Coil connection

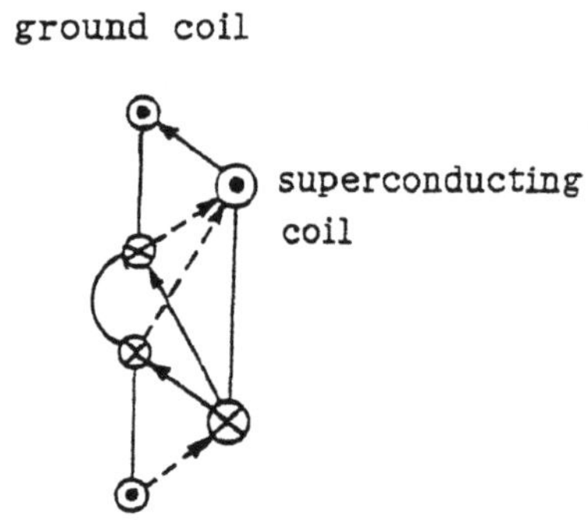

(a) Levitation force

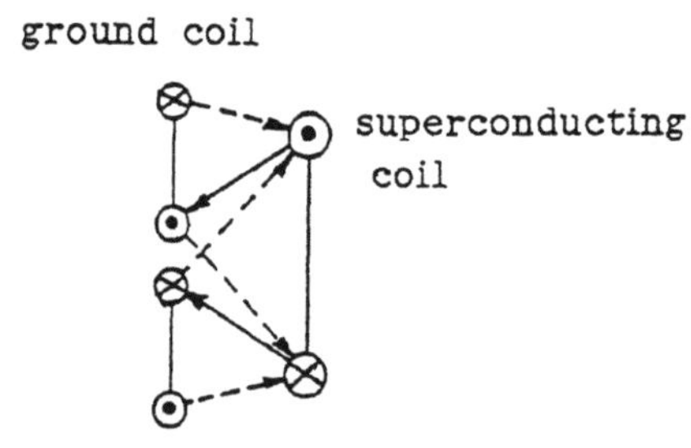

(b) Guidance force

Fig.4 Generation of force

FORCES ACTING ON SUPERCONDUCTING COIL

Force Generated In Propulsion System

Forces generated in the propulsion system are the same as in the ordinary motor, but forces at the coil end are large in comparison with the forces in the ordinary motor. Besides, the structure of the armature coils must be simple and it is hard to eliminate the harmonic magnetic flux. We must design the superconducting magnet taking account of the force distribution or fluctuating forces.

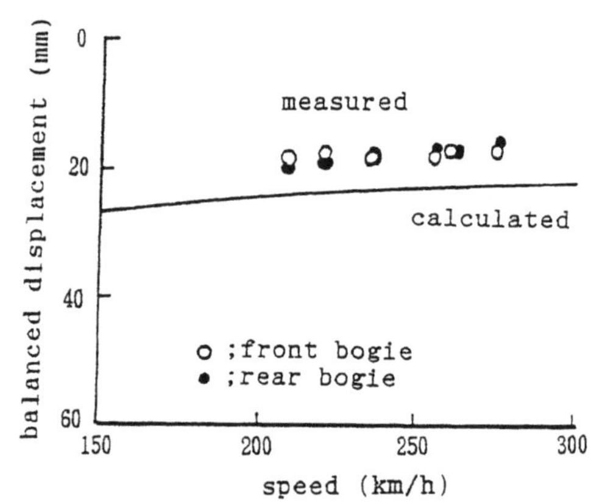

Fig.5 Balanced displacement

Fig.6 shows an example of the force distribution on the superconducting coil in the case that the phase angle of the current coincides with that of induced voltage. Lateral force or vertical force summed up along the coil is zero but its density is almost the same magnitude as the propulsion force density. The superconducting coil is exposed to the alternating magnetic flux density whose frequencies are 3,6,9....times the armature current frequency. The forces generated by these magnetic fields depend on the length of the superconducting coil. Neglecting the shielding effect of the cryostat metals, the dependence of the alternating force on the coil length is calculated as shown in fig.7.

Force Generated In Levitation System

The current in the levitation coil has not a simple sinusoidal wave form and its amplitude and phase angle vary with the vehicle speed. The current differs from the propulsion current in its phase angle, and the directions of the forces differ. The following are examples of the forces in the levitation system with levitation coils on the side wall. Fig.8 shows the wave form of a current flowing in the levitation coil at 500 km/h; it looks like a square wave. The stationary forces distribute as shown in fig.9. The levitation force (fz) distributes almost equally on the upper and the lower sides, and the force in direction of progress distributes on the vertical sides, and it changes the sign on upper half and on lower half of the sides, growing small in the sum along the sides. The two force densities are almost the same magnitude. The frequencies of the alternating magnetic flux density are 6,12,... times those of the armature current. The alternating force depends on the length of the superconducting coil as shown in fig.10.

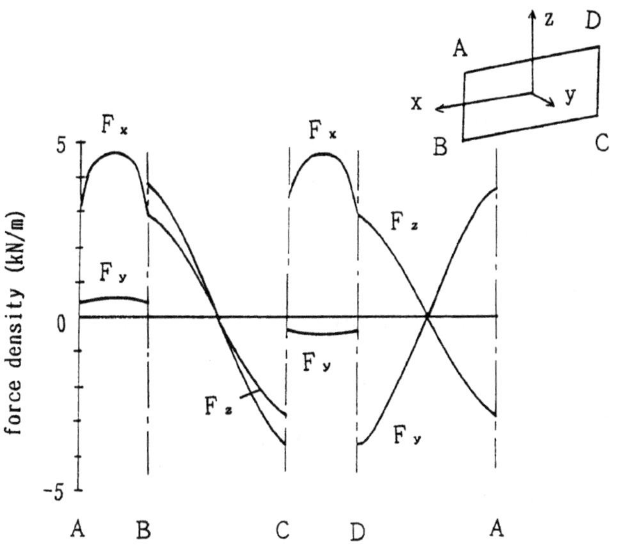

Fig.6 Force distribution on the superconducting coil (LSM)

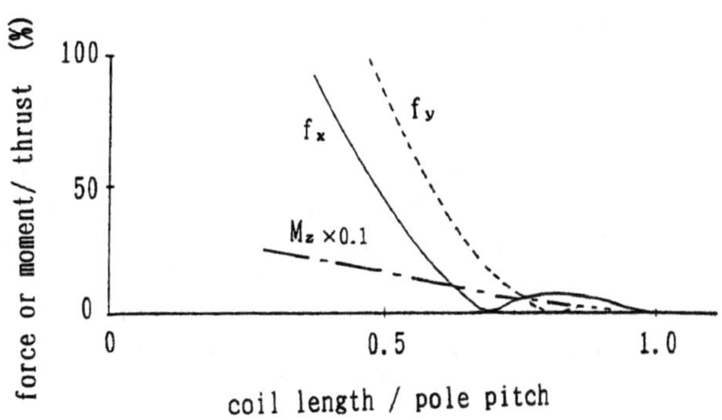

Fig.7 Alternating force versus armature coil length

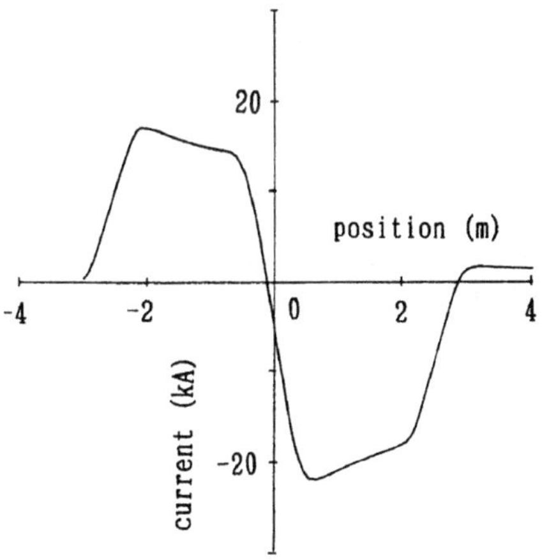

Fig.8 Levitation coil current wave form

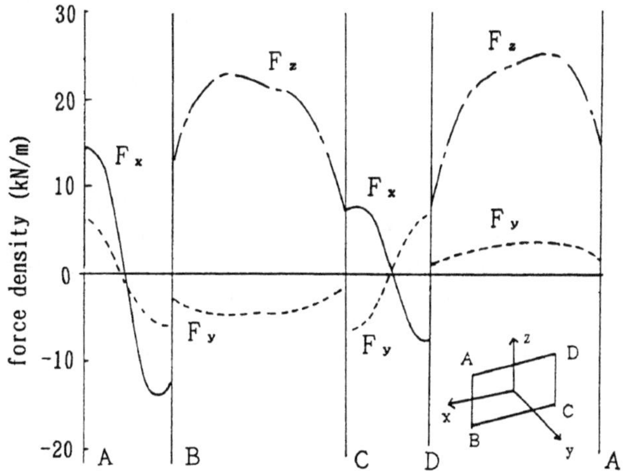

Fig.9 Force distribution on the superconducting coil (levitation)

Other Forces

One of the other forces acting on the superconducting coil that must be considered is that generated in emergency. Should one of the superconducting magnets happen to quench by any chance, the balance in lateral direction will be lost and a large lateral force will be generated. As the above mentioned forces are relatively small, the strength of the coil structure poses no problem, but various modes of vibration must be carefully coped with.

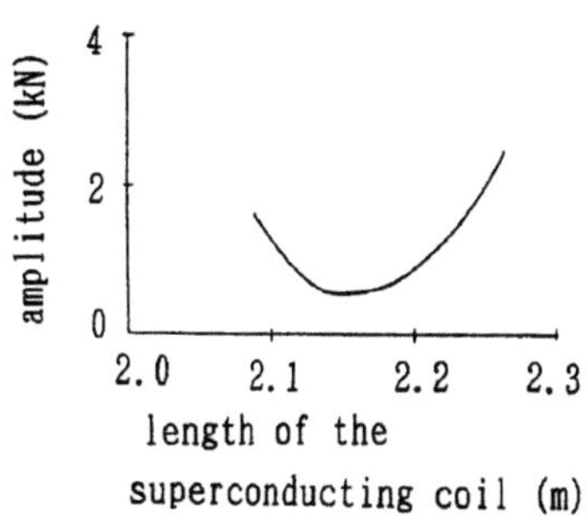

Fig.10 Alternating force

FUTURE PROSPECT

Commercial vehicle will be designed in an articulated type with the superconducting magnets located at the coupling; the passenger room will be isolated from the magnets; and the cross section of the body will be minimized in order to reduce the air drag. The train composed of 14 units will seat about 900 passengers and more than 10,000 passengers can be transported in an hour. A new longer test track is expected to be constructed in order to develop such a commercial system. The track length will be 40-50 km and a 5-unit train will be operated at 500 km/h to study the following items:

(1) Stable running of the vehicle, reliability and durability of various important components in long term high speed running tests.

(2) Standards of various structures, for example, minimum radius of the curve, maximum gradient and cross section of tunnel,

(3) Test and the operation of the control system for plural trains.

(4) Operational cost, operational security and maintenance standard.

A comprehensive development plan is now being drawn up.

CONCLUDING REMARKS

The superconducting magnet is one of the most important components. Many problems have been investigated and practical application has been studied. In future the long term durability will be confirmed on a longer test track. If a high temperature superconductor becomes available, the structure of magnets and refrigeration system will be drastically simplified, and other application of magnetic shielding will be developed. Our attention is focused on the progress in development of the high temperature superconductor.

References

1. Jizo Y, Fujiwara S, Nemoto K (1988) Superconducting magnet of new test vehicle "MLU002" on Japanese EDS system. 10th international conference on magnetically levitated systems, 9-10 June 1988, Hamburg
2. Fujiwara S, Fujimoto T (1989) Characteristics of the combined levitation and guidance system using ground coils on the side wall of the guideway. 11th international conference on magnetically levitated systems and linear drives, 7-11 July 1989, Tokyo

2 Material Preparation

2.1 Synthesis and Structure

Syntheses and Properties of Single CuO_2-Sheet Compounds: Charge-Transfer Gap and Charge-Doping Effect

Y. TOKURA[1], T. ARIMA[1], S. KOSHIHARA[1], T. IDO[2], S. ISHIBASHI[2], H. TAKAGI[2], and S. UCHIDA[2]

[1] Department of Physics, The University of Tokyo, Tokyo, 113 Japan
[2] Engineering Research Institute, The University of Tokyo, Tokyo, 113 Japan

ABSTRACT

Single crystals of parent families of oxide superconductors with single CuO_2-sheets have been synthesized and their optical spectra for charge-transfer and two-magnon excitations have been measured to unravel the puzzling electronic structure. The effect of electron-doping on the charge-transfer gap has been also investigated for $Nd_{2-x}Ce_xCuO_4$ compounds.

KEY WORDS: optical properties, charge-transfer excitation, doping effect

INTRODUCTION

Superconducting copper oxide compounds with high transition temperature (T_c) possess two-dimensional (2D) sheets of corner-linked CuO_4 squares ("CuO_2"-sheets) as a common structural unit. There are three modifications of the CuO_2-sheets, as shown in Fig.1, depending on the presence or absence of apical oxygen(s) positioned above and/or below the Cu atoms, i.e. the 2D sheets of (a) Cu-O squares, (b) pyramids and (c) octahedra. The first-discovered high-T_c compound [1], alkaline earth-substituted La_2CuO_4 (see Fig.2(a)), show the 2D sheets of octahedra, while in most of other copper oxide compounds with higher T_c, such as $YBa_2Cu_3O_7$ and $Bi_2Sr_2CaCu_2O_8$, show adjacent multi-layers of pyramidal CuO_2 sheets. Recent discovery of electron-doping-induced high T_c [2] shows that the CuO_2-sheets alone without apical oxygens (i.e. 2D array of Cu-O squares) also sustain superconducting carriers, as observed in Ce-substituted Nd_2CuO_4 (see Fig.2(d)). The quantitative understanding of electronic structures of various types of CuO_2-sheets and their change upon the carrier-doping are undoubtedly one of the central issues to unravel the mechanism of high T_c. In this report, we investigate the optical properties in single crystals of semiconducting parent compounds of High T_c superconductors with a single CuO_2-sheet per repeated unit. The change in the optical conductivity spectra with doping is also investigated for $Nd_{2-x}Ce_xCuO_4$ with electron-type charge-carriers.

P-TYPE AND N-TYPE BEHAVIOR IN CuO_2 SHEETS

In Fig.2 we show the crystal structures of copper oxide compounds investigated here, which respectively represent typical patterns of 2D Cu-O networks. All the structures show a well-defined network of a single Cu-O sheet within a repeated unit and hence in those compounds there is no

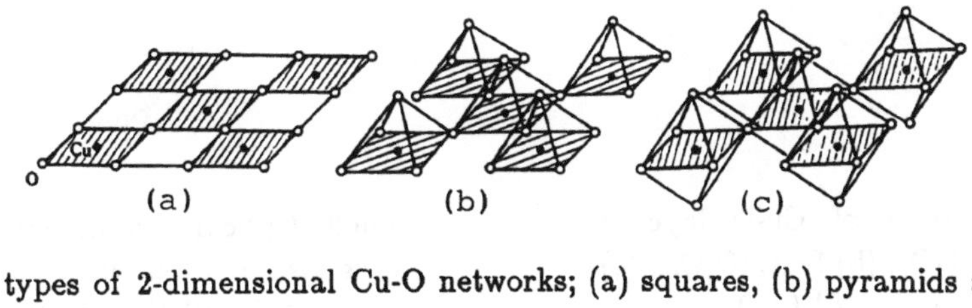

Fig.1. Three types of 2-dimensional Cu-O networks; (a) squares, (b) pyramids and (c) octahedra.

complication due to the combination of two or more types of Cu-O planes as observed, for example, in the Y-Ba-Cu-O systems accompanying the both of pyramidal sheets and chains. This feature is important to obtain clear-cut observation about the Cu-O network dependence of the electronic structures. Single crystals of all these compounds were grown from the melt by using excess CuO or CuO/Bi_2O_3 mixture as flux, detailed procedures of which will be described elsewhere.

Of particular interest are the compounds which can sustain the excess charge in thier CuO_2 sheets; (a) $La_{2-x}Sr_xCuO_4$, (c) $(La,Sm)_{2-x}Sr_xCuO_4$ [3] and (d) $Nd_{2-x}Ce_xCuO_4$. These compounds also represent three typical patterns of CuO network shown in Fig.1. Empirically, it has been established that the octahedral and pyramidal CuO_2 sheets can sustain hole-type carriers (i.e. P-type) and the CuO_2 sheets alone without apical oxygens electron-type carriers (N-type). In other words, the electron type carriers cannot be doped into the octahedral nor pyramidal sheets like Filg.1(b,c), and hole type carriers not into square-type sheets (Fig.1(a)) with no apex. According to preliminary measurements of thermopower, the above empirical rule appears to be valid also for the infinite-layer compound $(Ca,Sr)CuO_2$ shown in Fig.2(e): it shows the N-type behavior though the doping level attained by substitution of Ca with rare-earth elements is quite low. Such a P- or N-type behaviour dependending on the pattern of Cu-O networks implies the importance of electrostatic interaction between the doped carrier and surrounding ionic lattice [4,5].

ELECTRONIC STATES IN INSULATING CuO_2 SHEETS

In Fig.3 are plotted the optical conductivity spectra [6] for six members of Cu-O layered compounds listed in Fig.1 (a)-(f). The conductivity spectra were obtained by the Kramers-Kronig

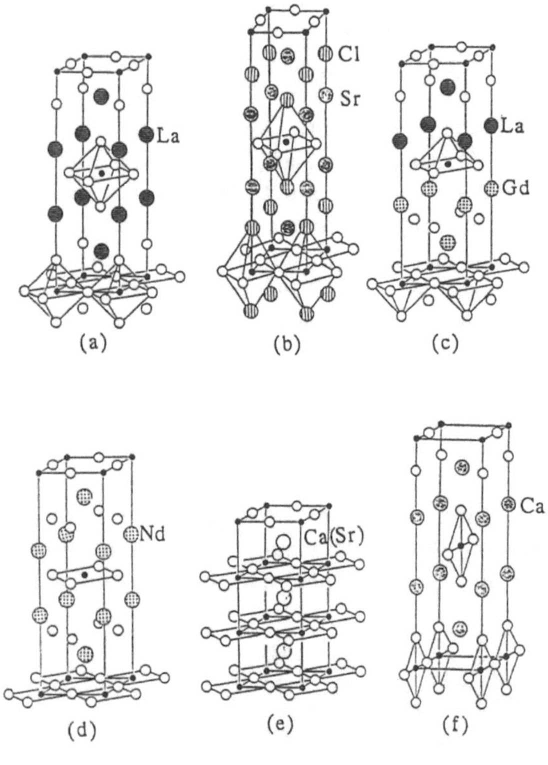

Fig.2. Crystal structures of Cu-O layered compounds; (a) La_2CuO_4, (b) $Sr_2CuO_2Cl_2$, (c) $(La,Gd)_2CuO_4$, (d) Nd_2CuO_4, (e) $(Ca,Sr)CuO_2$, and (f) Ca_2CuO_3.

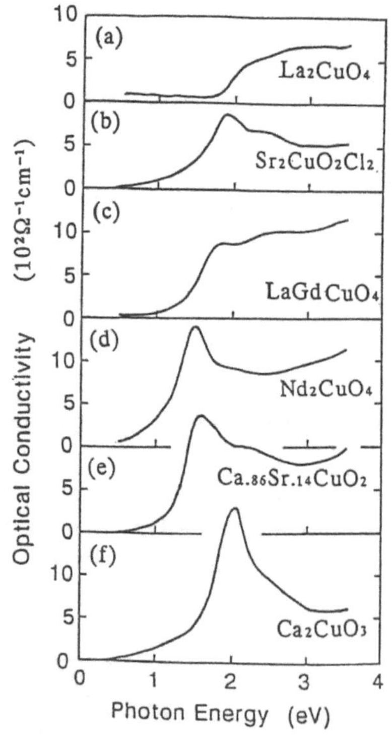

Fig.3. Optical conductivity spectra of single crystals of Cu-O layered compounds listed in Fig.2. In all the spectra the light E-vector is parallel to the basal plane.

transformations of reflectivity data measured on the (001) faces of those single crystals at 290K with the polarization parallel to the Cu-O basal planes. Concerning the reflectivities at photon energies above 6eV, the reflectance data [7] taken with use of the synchrotron radiation source were utilized. In all the compounds investigated here, we have observed the strong optical transitions at energy of 1.5-2.0eV, below which the compounds show no optical active bands except for the optical phonon modes. These optical excitations are strongly polarized along the basal plane of CuO_2 (or parallel to the chain direction in Ca_2CuO_3) and the spectra for the light polarization normal to the sheets show no prominent structure in the photon energy region below 3eV [6]. On these bases, we have assigned these strong absorption bands to the charge-transfer gap excitation mainly from the O-2p to Cu-3d (upper Hubbard) state. The fairly sharp profile of the absorption peaks as observed indicates the excitonic character of the transition rather than the van Hove singularity.

Looking at Fig.3, one may notice the systematic change in the peak position with the number of apical oxygens coordinated around Cu; 2.0eV in octahedral CuO_2 sheets of La_2CuO_4, 1.8eV in pyramidal CuO_2-sheets of $LaGdCuO_4$, and 1.5eV in CuO_2-sheets with no apical oxygen in Nd_2CuO_4 and $(Ca,Sr)CuO_2$. This fact indicates that the relative position of the uppermost filled band (dominatnly of O-2pσ character) and lowest unoccupied band (dominantly of Cu $d_{x^2-y^2}$ character) is quite sensitve to the presence or absence of apical oxygen(s). On the other hand, the oscillator strength of the CT exciton-like absorption increases in going from octaheral to square CuO_2 sheets. The peak energies and oscillator strengths of the CT excitons are summerized in Table 1 for the representative 3 members of single-CuO_2-sheet compounds, La_2CuO_4 (octahedra), $LaGdCuO_4$ (pyramids) and Nd_2CuO_4 (squares). The oscillator strengths were approximately estimated by fitting the optical conductivity spectra of CT excitons with Lorenzian profiles.

In the Raman spectra of these compounds [6], fairly strong bands with the symmetry of A_{1g} and B_{1g} are observed around 3000cm^{-1} (ca.0.4eV). The corrensponding Raman bands in related copper oxide compounds have been assigned to the two-magnon excitation [8]. It was argued [8] that the peak position of the two-magnon Raman band is approximately equal to $2.7J_s$, J_s being the exchange energy between the localized spins on neighboring Cu sites. The observed Raman shifts for the 2-magnon excitations and estimated J-values are also listed in Table 1. In contrast to the systematic change of CT gap energies, the 2-magnon Raman shift appears not to be sensitive to the shape of CuO network.

Table 1 Physical paremeters in CuO_2-sheets.

compounds	La_2CuO_4	$(La,Gd)_2CuO_4$	Nd_2CuO_4
Cu-O network (oxygen coordination)	octahedron (6)	pyramid (5)	square (4)
d_{Cu-O} (CuO_2-sheet)	1.905 A	1.936 A	1.973A
t_{pd}	t_0	$0.937t_0$	$0.869t_0$
Δ (optical gap)	2.0 eV	1.8 eV	1.5 eV
f_{CT}	0.2_8	0.3_1	0.4_3
$(t_{pd}/\Delta)^2$ (arbitrarily normalized)	0.28	0.31	0.38
Raman shift (2-magnon band)	3200 cm^{-1}	2800 cm^{-1}	2900 cm^{-1}
J_s exp.	0.14_7eV	0.12_9eV	0.13_3 eV
J_s cal.*	0.142 eV	0.126 eV	0.133 eV

* assumed U_d=8.0 eV, U_p=4.0 eV, and t_0=0.96 eV.

Based on the observed values for CT and 2-magnon excitation, important features in electronic structures of CuO_2-sheets with various shapes can be discussed. The values of the oscillator strength (f) of CT excitations are quite large (f=0.2-0.5), indicating the strong hybridization between Cu-3d and O-2p orbitals and the strong exciton effect due to the coulombic final state interaction between the excited electron and hole. According to the simplified model for a single CuO_4 cluster, the oscillator strength of the optical CT excitation is approximately proportional to the degree of hybridiazation $(t_{pd}/\Delta)^2$. Here, t_{pd} and Δ stand for the transfer energy and energy difference between the Cu 3d and O 2p orbitals. The transfer energy t_{pd} is known to be proportional to d^{-4}, d being the distance between the neighboring Cu and O within CuO_2-sheet. In Table 1, the change in t_{pd} is shown using the value ($t_{pd}=t_0$) of La_2CuO_4. If we take the observed CT gap energy as an approximation of Δ, we can estimate the relative strength of CT transitions. The caculated results for $(t_{pd}/\Delta)^2$ are also shown in Table 1 in arbitrarily normalized unit. The tendency is in good agreement with the observed results: The degree of hybridization between Cu-d and O-p orbitals is increased as the oxygen coordination around Cu is decreased. This is apparently due to the change in CT gap energy, which is superior to the change in t_{pd}.

In Table 1 are also listed the calculated values of J_s, in which the simple 2-band model is used with the appropriate values of correlation energies for Cu d-electron (U_d=8eV) and O p-electron (U_p=4eV). We used again the observed CT gap as an approximation of Δ and assumed t_0=0.96eV. The agreement with the experimentally derived values is fairly good. Rather small variation in J_s from material to material can be interpreted as a result of cancellation of respective change in t_{pd} and Δ.

EFFECT OF CARRIER-DOPING INTO CuO_2-SHEETS

Up to now, controllable carrier-doping in CuO_2 sheets of single crystals have been capable only in $La_{2-x}Sr_xCuO_4$ and $YBa_2Cu_3O_{6+y}$, by changing the Sr concentration x and oxygen content y, respectively. The latter compound comprises two pyramidal sheets and a CuO_y chain per repeated unit and variation of oxygen content (y) modifies the chain structure, which may cause some complication in the interpretation of optical spectra. In single crystalline films of $La_{2-x}Sr_xCuO_4$, on the other hand, measurements of optical transmittance spectra have been performed by Suzuki [9], who has found doping-induced optical absorption bands in the infrared region and their systematic change with the Sr-concentration (x). (See also the paper by Tajima et al. in the Proceedings.) Here, we present the result for change in optical conductivity spectra observed in single crystals of the electron-doped counterpart $Nd_{2-x}Ce_xCuO_4$.

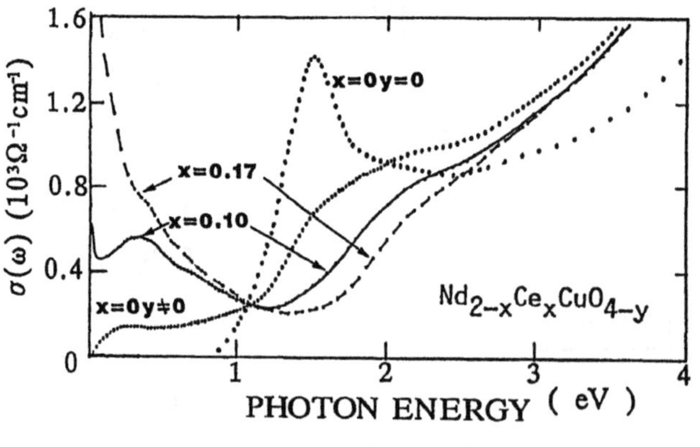

Fig.4. Optical conductivity spectra in single crystals of $Nd_{2-X}Ce_xCuO_{4-y}$.

The E⊥c reflectance spectra have been measured on single crystals of $Nd_{2-x}Ce_xCuO_4$ at 290K. The Ce-composition of each CuO-flux grown single crystal has been determined by the measurement of c-axis length as well as by the ICP method. The transformed optical conductivity spectra, which were obtained by Kramers-Kronig analyses of reflectivity data, are shown in Fig.4. For undoped samples (x=0.0) we also show the spectrum for the intentionally reduced sample which were annealed under the Ar/O_2 gas mixture with the partial O_2-pressure of ca.10^{-4} atm at 900 C. Refering to the cases of polycrystalline samples, the estimated oxygen vacancies is about 1% of the total stoichiometric oxygen content. It shows a blurred feature and the additional broad absorption below the CT gap. This indicates that the reducing procedure in Nd_2CuO_4 introduces the excess electrons which are responsible for the infrared absorption. Upon substitution of Nd with Ce, the introduced electrons in CuO_2 sheets cause the strong infrared absorption below 1.1eV. Remarkably, the isosbetic point is observed at ca.1.1eV with increase of x, indicating that the electron-doping transfers the transition intensity from the CT excitation to the infrared excitations. Effective number of electrons relevant to the transition is given by integration of optical conductivity in the photon enrgy range of the absorption band. Therefore, the existence of isosbetic point with doping indicates that the elctronic states giving rise to the CT excitation and infrared absorption are closely correlated with each other. In other words, the electronic states responsible for the CT gap excitation are modified by doped electrons to the ones relevant to the infrared absorption. This feature is contradictory to the simple picture that the doped carriers are accomodated in the upper-lying rigid band and causes the plasmon-like optical reponse. According to a preliminary estimation, the rate of intensity transfer from CT excitation to infrared band with increasing x ($<$ 0.15) is approximately twice larger in $Nd_{2-x}Ce_xCuO_4$ than that for hole-doping case in $La_{2-x}Sr_xCuO_4$. This may be reflecting the stronger p-d hybridization in Nd_2CuO_4.

REFERENCES

1. J.G.Bedonorz and K.A.Muller, Z. Phys. B **64**, 189 (1986).
2. Y.Tokura, H.Takagi and S.Uchida, Nature **337**, 345 (1989).
3. Y.Tokura et al., Phys.Rev. B **40** 2568 (1989).
4. J.Kondo, J.Phys.Soc.Jpn. **58**, 2884 (1989).
5. J.B.Torrance and R.Metzger, Phys.Rev.Lett.**63**, 1515 (1989).
6. Y.Tokura et al., preprint (submitted for publication).
7. S.Tajima et al., J.Opt.Soc. Am. B6, 475 (1989).
8. K.B.Lyons, P.A.Fleury, J.P.Remeika and T.J.Negran, Phys.Rev. B **37**, 2353 (1988).
9. M.Suzuki, Phys.Rev. B **39**, 2312 (1989).

Structural and Physical Properties of Ca-Substituted $YBa_2Cu_4O_8$

N. Koshizuka, T. Miyatake, K. Yamaguchi, R. Itti, S. Gotoh, M. Kosuge, and S. Tanaka

Superconductivity Research Laboratory, International Superconductivity Technology Center, 10-13 Shinonome 1-chome, Koto-ku, Tokyo, 135 Japan

ABSTRACT

This paper describes the structural and physical properties of $Y_{1-x}Ca_xBa_2Cu_4O_8$ (x=0.0, 0.02, 0.05, 0.1) examined by high resolution transmission electron microscopy(HRTEM), electrical resistivity under magnetic fields, magnetization and photoelectron spectroscopy measurements. The Tc of $YBa_2Cu_4O_8$ is increased from 80 K to 90 K by Ca-substitution for Y. The reason for the increase in Tc is discussed considering the hole concentration in this system. Various superconducting parameters such as upper and lower critical fields are also discussed. The existence of edge structure at Fermi energy is shown in the UPS spectra, which demonstrates no degradiation of the surface state on this compound in vacuum at room temperature.

KEY WORDS : $Y_{1-x}Ca_xBa_2Cu_4O_8$, HRTEM, magnetoresistance, photoelectron spectroscopy, critical field

INTRODUCTION

The $YBa_2Cu_4O_8$ (1-2-4) material is a superconductor with Tc~ 80 K having a double Cu-O chain between two Cu-O planes. This material was first found as a lattice defect in partly decomposed $YBa_2Cu_3O_y$ (1-2-3) powders (1), and then as an ordered defect structure in 1-2-3 films(2, 3). X-ray and transport properties were studied for 1-2-4 thin films by Kapitulnik and Char(4). Bulk synthesis of 1-2-4 was carried out at high oxygen pressure (5, 6) and the space group of Ammm and the orthorhombic lattice constants a=3.871A, b=3.840A and c=27.25A were confirmed by Karpinski et al. from the x-ray and neutron diffraction analysis. Furthermore, it was revealed that, unlike 1-2-3, the oxygen content in 1-2-4 is thermally stable up to 850°C(5), and 1-2-4 has a strong pressure dependence of Tc (dTc/dP=0.55 K/kbar)(7). Neutron diffraction studies of pressure effects on the structure and Tc were reported by Kaldis et al. (8). Afterward, one atmosphere oxygen pressure synthesis of 1-2-4 using reaction rate enhancers(9) and a more straightforward synthesis of $LnBa_2Cu_4O_8$ (Ln=Er, Ho) without using of reaction rate enhancers (10) have been reported. By taking advantage of the thermal stability, NMR studies of 1-2-4 have been made for a wide temperature region up to high temperatures (11). Although the stable oxygen content in 1-2-4 is a great advantage, the Tc of ~ 80 K is too near liq. N_2 temperature for applications. Recently, we reported that the Tc increases from ~ 80 K to ~ 90 K by Ca-substitution for Y in $YBa_2Cu_4O_8$(12). In this paper, we describe the structural and physical properties of $Y_{1-x}Ca_xBa_2Cu_4O_8$ and discuss about the Ca-substitution effects on these properties(12~ 15).

PREPARATION AND STRUCTURAL PROPERTIES

Polycrystalline materials of Ca-substituted 1-2-4 were prepared by solid state reaction with two-step oxygen HIP (hot isostatic pressing) treatments. First, starting materials of Y_2O_3, $Ba(NO_3)_2$, CuO and $CaCO_3$ were calcined at 900°C in flowing oxygen gas and then sintetred at 800°C in oxygen gas atmosphere. The first oxygen HIP treatment was done at 950°C , and the second at 1050°C under oxygen and argon mixed gas pressure. The partial pressure of oxygen was 20 MPa. By using these HIP treatments, we obtained single phase materials without other phases like $YBa_2Cu_3O_y$ in the x-ray diffraction patterns. More details of the sample preparation are described in ref. (12).

A schematic view of the crystal structures of 1-2-4 and 1-2-3 is shown in Fig. 1. As shown here, 1-2-4 has a double Cu-O chain along the b-axis which is different from 1-2-3 with a single chain between two Cu-O planes. The apex positions of the two pyramids sandwhiching a double Cu-O chain are displaced by Cu-O distance in the chain, so the lattice constant c in 1-2-4 is enlarged more than twice of that in 1-2-3. The separation between Cu-O planes in 1-2-4 is larger than that in 1-2-3. Therefore, two-dimensionality of various physical properties is thought to be higher than that in 1-2-3.

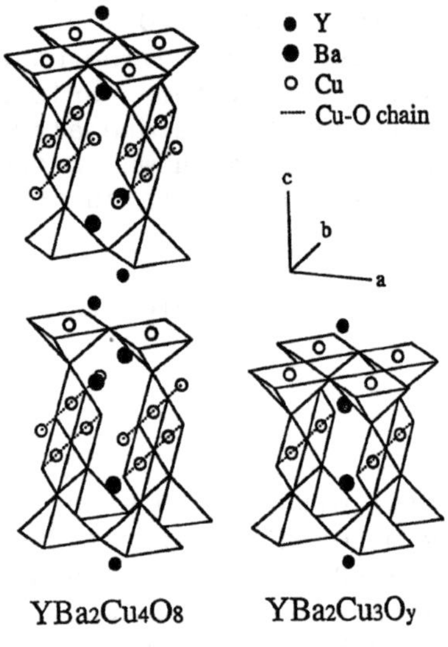

Fig. 1, Schematic view of the crystal structures of 1-2-4 and 1-2-3.

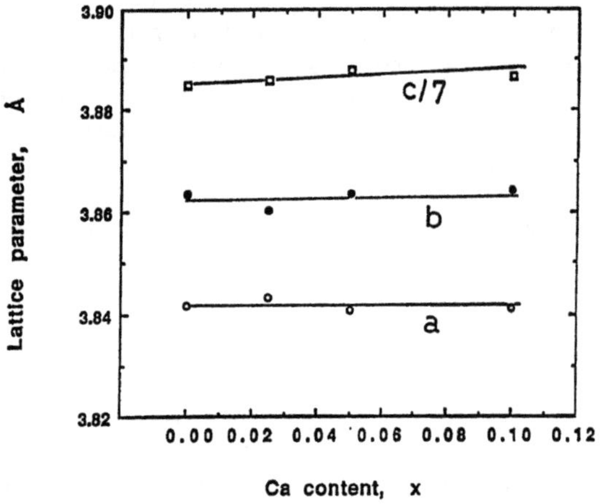

Fig. 2, Lattice parameters a, b and c vs. Ca-content for 1-2-4.

Figure 2 shows Ca-content dependence of lattice parameters. The parameters a and b do not change with increasing Ca content, but we can see the increase in the parameter c. The ionic radius of Ca^{2+} ions is larger than that of Y^{3+} ions (Ca^{2+}:1.12A, Y^{3+}:1.02A). Therefore, the increase in the unit cell length along the c axis means that Y ions are really substituted by Ca ions.

Figure 3 is a high resolution TEM image of Ca-free sintered material taken with the incident electron beam along the a-axis(13). We can see zig-zag arrangements of Cu atoms here. These correspond to double Cu-O chains along the b-axis. We also find a single Cu-O chain along the b-axis. This is a kind of planar defect which is found generally in 1-2-4 materials. We are not sure at present whether these planar defects exist in the whole sample or not. It is possible that these defects may exist locally near the surface region of a grain. As for the Ca-substitution effects on the TEM images, we could not find any significant difference in the images between Ca-free and -substituted samples. And also, we could not find twin structures in the TEM images of ab plane, which is predicted from the absence of orthorhombic-tetragonal phase transition up to high temperatures in 1-2-4 systems.

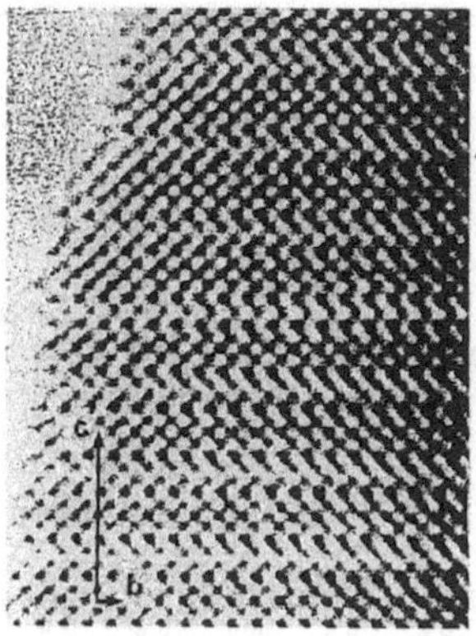

Fig. 3, High resolution TEM image of 1-2-4 with the incident beam along the a-axis(13).

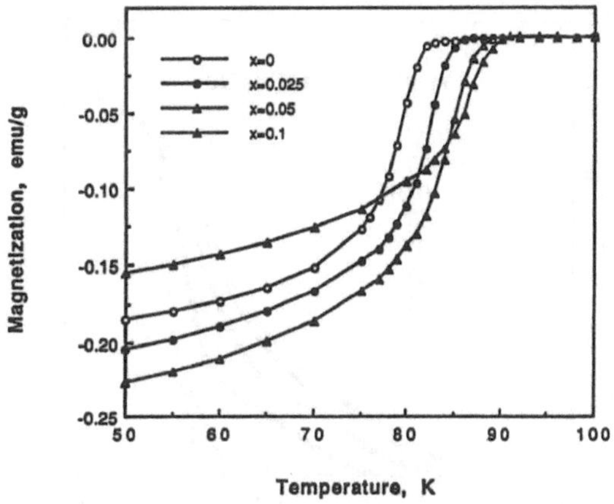

Fig. 4 Magnetization (field-cooled in 50 Oe) vs. temperature for $Y_{1-x}Ca_xBa_2Cu_4O_8$ (12)

ENHANCEMENT OF Tc BY Ca-SUBSTITUTION

According to a report by Miyatake et al. (12), the metallic resistivity decreases and Tc increases from ~80 K to ~90 K with 10 % substitution of Ca for Y in 1-2-4. It is known that the magnitude of resistivity at normal state in 1-2-4 is ~1/4 of that in 1-2-3(4) . This suggests a larger density of states at Fermi level, and it is interesting to compare the transport and optical properties between 1-2-3 and 1-2-4. The increase of Tc by Ca-substitution was confirmed by the measurements of magnetization(12). Figure 4 shows field-cooled magnetization curves for various contents of Ca. The values of Tc derived from these magnetization curves were in good agreement with those from resistivity measurements.

Next, we discuss about the reason why Tc increases by Ca-substitution in 1-2-4 . The average charge or the hole concentration per (Cu-O) unit , p, is thought to be 0.25 for Ca-free 1-2-4 with oxygen content of 8, while, this value p increases to 0.275 by 10 % Ca-substitution. On the other hand, the average charge p is to be 0.333 in 1-2-3 in the case of the oxygen content of 7. Thus a simple explanation for the Tc enhancement may be ascribed to this increase in the average charge p.
According to the diagram of Tc and hole concentration per Cu-O sheet, p_{SH}, in 1-2-3 system by Tokura et al. (16) , Tc increases with hole concentration in a Cu-O sheet and reaches the maximum at about p_{SH}=0.2. In this case, the hole concentration in a Cu-O chain, p_{CH}. becomes 0.6 from the relation $p=(2p_{SH}+p_{CH})/3$
It turned out that the Tc seemed to saturate at 10 % Ca-substitution in 1-2-4. If we can assume that the electronic structure of 1-2-4 is almost same as that of 1-2-3, the maximum Tc may be attained at the same hole concentration per a Cu-O sheet as in 1-2-3, that is , p_{SH}=0.2. More detailed studies of correlation between Tc and hole concentrations in the sheets and chains seem necessary to understand the mechanism of the Tc enhancement.

CRITICAL FIELDS AND SUPERCONDUCTING PARAMETERS

Figure 5 shows temperature dependence of electrical resistivity for 10 % Ca-substituted 1-2-4 under various magnetic fields. The measurements were done by a standard four-probe method using a sintered sample with the size of 6x1.7x 0.2 mm^3. The two-step behaviors of the temperature dependence are characteris-

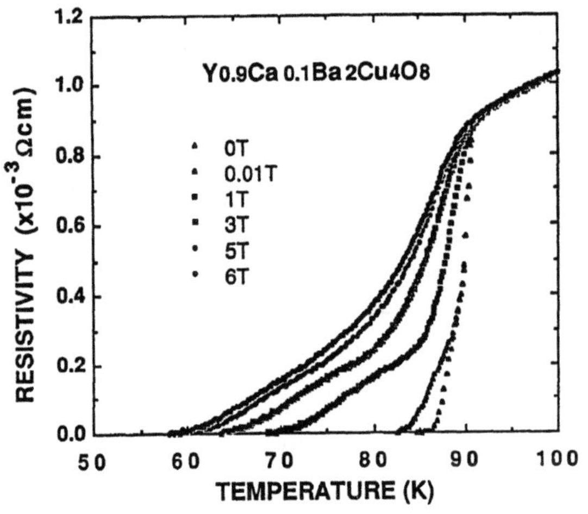

Fig. 5 Electrical resistivity vs. temperature of 10 % Ca-substituted 1-2-4 in various applied fields.

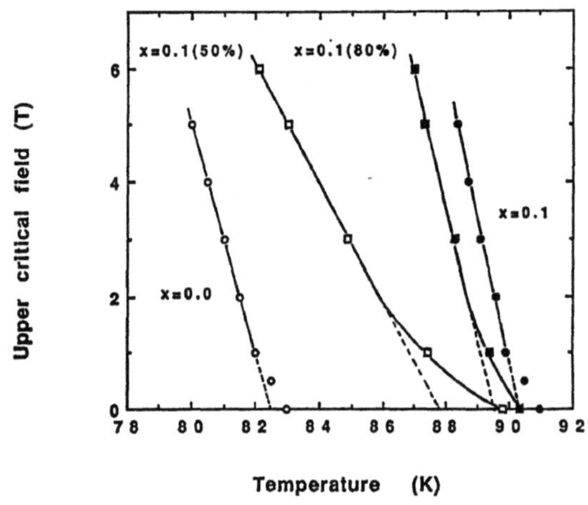

Fig. 6 Upper critical fields H_{c2} vs. temperature for $Y_{1-x}Ca_xBa_2Cu_4O_8$.

tic of sintered materials. The tails at lower temperature side are caused by the transition to the normal state due to an applied magnetic field at weak-links of the grain boundaries.

Temperature dependence of upper critial fields, H_{c2}, are shown in Fig. 6. The outside two curves were obtained from the field-cooled magnetization and the inside two curves were derived from the magnetoresistance curves in Fig. 5. Here, 50 % and 80 % mean the percentages of the transition to the resistivity at normal state. Generally, the resistive transition curves are complicated by an unusual broadening of the transition in a magnetic field. This is caused by field dependence of critical currents in weak-links and dynamical motions of fluxoids such as flux creep and flux flow. In this figure, temperature dependence of the critical field derived from 80% transition points is near the curve from magnetization. This means that the magnetization measurements in an applied field allows a more unique determination of the critical fields by a well defined onset of diamagnetism(17).

Table 1, Various superconducting properties and GL-parameters for 10% Ca-substituted and Ca-free 1-2-4, and pure 1-2-3.

	$YBa_2Cu_4O_8$	$Y_{0.9}Ca_{0.1}Ba_2Cu_4O_8$	$Y_{0.9}Ca_{0.1}Ba_2Cu_4O_8$ b)	$YBa_2Cu_3O_7$ c)
T_c (K)	82.5 a)	90.2 a)	90.3	91.5
$-(dH_{c2}/dT)_{T=T_c}$ (T/K)	2.00	2.77	3.13	2.2
$H_{c1}(0)$ (T)	0.014	0.0185	0.0185 d)	0.101
$H_{c2}(0)$ (T)	114	172	195	141
$H_c(0)$ (T)	0.567	0.796	0.841	1.6
GL κ	142	153	164	42
GL ξ(0) (Å)	17	14	13	14
GL λ(0) (Å)	2400	2100	2100	590

a) T_c is defined as a temperature where the dH_{c2}/dT slope intesects at zero magnetic field.
b) Derived from resisive data.
c) Suenaga et al.
d)Determined by magnetic measurements.

Table 1 gives various superconducting and Ginzburg-Landau parameters of Ca-free and -substituted 1-2-4. Corresponding parameters of 1-2-3 are also given for comparison (18). Here, the lower critical fields, $H_{c1}(T)$, were obtained from the deviation from the linearity in M-H curves and the values of $H_{c1}(0)$ were by extrapolation to 0 K in the temperature dependence. The $H_{c2}(0)$ was obtained from the Werthamer-Helfand-Hohenberg formula, $H_{c2}(0)=-0.69T_c(dH_{c2}/dT)_{Tc}$. Coherent length ξ , flux penetration length λ , and GL-parameter κ ($=\lambda/\xi$) were derived from the $H_{c1}(0)$ and $H_{c2}(0)$ using the following relations: $H_{c2}(0) = \Phi_0/2\pi\xi(0)^2$, where Φ_0 is the flux quantum, $H_{c1}(0)=H_c(0)\ln\kappa/\sqrt{2}\kappa$, and $H_{c2}(0) = \sqrt{2}H_c(0)\kappa$, where $H_c(0)$ is the thermodynamic critical field.

We find that the upper critical field $H_{c2}(0)$ for Ca-substituted 1-2-4 is larger than that for Ca-free 1-2-4, which is ascribed to higher Tc and steep gradient of H_{c2} at Tc in 1-2-4 system. It is also interesting that the lower critical field $H_{c1}(0)$ in 1-2-4 is one order smaller and λ (0) of 1-2-4 is larger than those in 1-2-3. We have no definite answer for these differences. However, these facts might imply that the pinning force in 1-2-4 is smaller than that in 1-2-3. This may be possible because of the absence of twin structures in the 1-2-4 which may act as the pinning centers. Furthermore, we should consider the higher two-dimensionality in 1-2-4 system because of the larger separation between Cu-O planes. It is thought that the higher anisotropy in coherent length results in the reduction of the pinning force. We need to measure the anisotropic properties of these parameters using single crystals.

Fig. 7 UPS spectra for 10% (a), 5% (b) and 0% (c) Ca-substituted 1-2-4, and pure 1-2-3 (d) at room temperature(15).

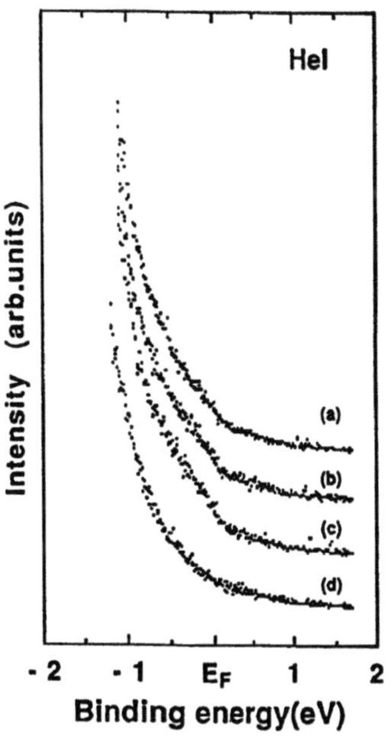

PHOTOELECTRON SPECTROSCOPY

Since the oxygen degradation from surface occurs for 1-2-3 above 20K in vacuum (19), it has been a controversial problem whether an edge structure at the Fermi energy exists or not in the photoelectron spectrum. Figure 7 shows photoelectron spectra (UPS) of 1-2-4 and 1-2-3 sintered materials(15). As we see here, there seems to exist an edge structure at Fermi energy for the 1-2-spectra, but not for the 1-2-3 spectrum. The UPS is very sensitive to the surface state, so we must be careful for analyzing the results especially on sintered materials. If the metallic phase exists as a secondary phase in the grain boundary, it may causes an edge structure at Fermi energy. In our measurements we used high-quality and single phase sintered samples. As a result, we believe that a Fermi-like edge exists at normal state of 1-2-4 , and the oxygen degradation from surface does not occur even in vacuum at room temperature. Synthesis of high-quality single crystals is important for further study of the physical properties of 1-2-4 systems.

Structure and Superconducting Properties of $[(Ln_{1-x}Ln^*_x)_{1/2}(Ba_{1-y}Sr_y)_{1/3}Ce_{1/6}]_8Cu_6O_z$

H. YAMAUCHI, T. WADA, A. ICHINOSE, Y. YAEGASHI, T. KANEKO, S. IKEGAWA, and S. TANAKA

Superconductivity Research Laboratory, International Superconductivity Technology Center, 10-13, Shinonome 1-chome, Koto-ku, Tokyo, 135 Japan

ABSTRACT

A variety of new oxide superconductors that can be represented by the formula, $[(Ln_{1-x}Ln^*_x)_{1/2}(Ba_{1-y}Sr_y)_{1/3}Ce_{1/6}]_8Cu_6O_z$ (Ln, Ln*= lanthanide elements), have been prepared. The crystallographic structures of the oxides were all tetragonal and of the $(Ln^+, Ce)_4(Ln^+, Ba)_4Cu_6O_z$ (Ln^+ = Nd, Sm or Eu) type which had been previously discovered by Sawa et al. As the Sr content, y, increased when Ln=Ln*=Nd, the oxygen content, z, monotonically increased and the superconducting transition temperature, T_c, varied exhibiting a maximum. When z was controlled directly by means of high oxygen pressure sintering techniques, Tc was changed accordingly. Tc's of samples with different combinations of Ln and Ln* and different values of x and y were found to depend on the magnitude of the bond valence sum for a Cu atom located in the bottom plane of the $Cu-O_5$ piramid. Transport and magnitization measurements were carried out to investigate the magnetic field dependence of superconducting properties and to determine the superconductivity material parameters. The Hall coefficients were positive below room temperature and varied yielding a maximum with respect to temperature.

KEY WORDS: crystal structure, high oxygen pressure treatment, magnetic properties, Hall effect

INTRODUCTION

Recently, a new family of superconducting oxides with T_c of 40 K was discovered by Sawa et al.[1] The composition of the superconductors were represented as $(Ln, Ce)_4(Ln, Ba)_4Cu_6O_z$ (Ln:Nd, Sm and Eu) and the crystallographic structure was found to be of a new type and determined as shown in Fig. 1. This structure is perovskite-related[2] and similar to both the triperovskite structure of $YBa_2Cu_3O_y$[3],[4] and that of T'(Nd_2CuO_4) structure.[5] In this structure, there exist two kinds of A sites for lanthanide ions: the smaller ions, i.e. Ce^{4+}, occupy the A_1 sites and the larger ions, i.e. Ba^{2+}, the A_2 sites, while intermediate sized ions, Ln, i.e. Nd^{3+}, Sm^{3+} and Eu^{3+}, can occupy both A_1 and A_2 sites.

We have investigated this family of superconductors. A variety of compounds of new and different compositions were synthesized following the flow diagram shown in Fig. 2. Sawa et al.[1] reported that single-phase samples were obtained in $(Ln_{1-x}Ce_x)_2(Ba_{1-y}Ln_y)_2Cu_3O_{10-\delta}$ only when Ln:Ba:Ce:Cu=6:4:2:9 for Ln=Nd,Sm and Eu. That is, $(Ln_{1/2}Ba_{1/3}Ce_{1/6})_8Cu_6O_z$ (Ln=Nd,Sm,Eu) were of single-phase. First of all (Step 1), possibilities were sought for using La^{3+} (which is larger than Nd^{3+}, Sm^{3+} and Eu^{3+}) together with another lanthanide ion of a smaller size for the Ln site in the original compound.[6] Then (Step 2), the effect of Sr substitution for

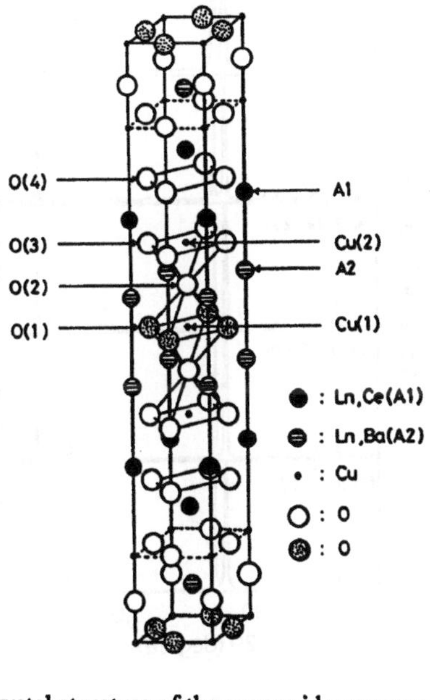

Fig.1 Crystal structure of the new oxide superconductor, $(Eu, Ce)_4(Eu, Ba)_4Cu_6O_z$. [6]

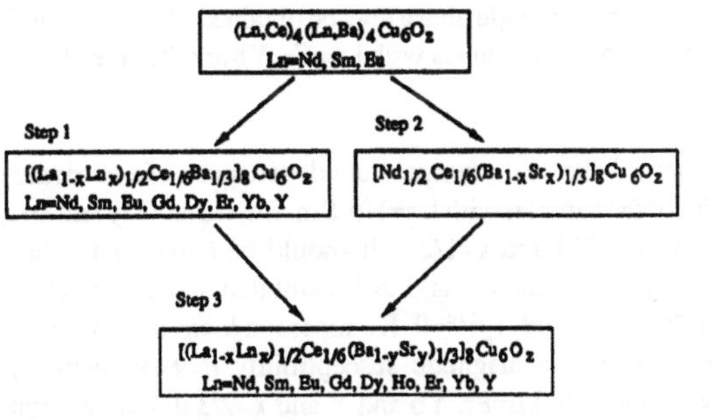

Fig.2 Steps for the development of new materials of $[(Ln_{1-x}Ln^*_x)_{1/2}(Ba_{1-y}Sr_y)_{1/3}Ce_{1/6}]_8Cu_6O_z$.

Ba in the original compound with Ln=Nd was studied.[7] The results of Step1 and Step 2 were combined to obtain compounds with higher transition temperatures than those of the original compounds (step 3).[8]

EXPERIMENTAL

All the samples were prepared by a solid state reaction method. The mixed powder was calcined at 1020 °C for 15 h in O_2 gas flow or O_2 gas of 3 atm. The calcined powder was molded and sintered at 1030 °C for 15 h in O_2 gas flow or O_2 gas of 3 atm, and then held at 600 °C for 5 h and at 400 °C for 20 h. Some of the samples were re-heat-treated at 900 °C for 1 h, and then at 600 °C for 5 h and at 400 °C for 5 h in a high oxygen partial pressure atmosphere of $P(O_2)$=200 atm. The phases present and the lattice constants were determined by powder x-ray diffraction using Cu-Kα radiation. The crystal structures of the single-phase samples were refined by the x-ray Rietveld analysis using RIETAN program.[9] The oxygen contents were analyzed by an inert gas fusion nondispersive IR method.[10] Electrical resistivity was measured by a standard four-probe method. For $(La_{1/6}Gd_{1/3}Ce_{1/6}Ba_{1/6}Sr_{1/6})_8Cu_6O_z$ samples, the temperature dependence of resistivity was measured in magnetic fields and the magnetic properties were measured by a superconducting quantum interference device (SQUID) magnetometer. For both $[Nd_{1/2}Ce_{1/6}(Ba_{0.625}Sr_{0.375})_{1/3}]_8Cu_6O_z$ and $(La_{1/6}Y_{1/3}Ce_{1/6}Ba_{1/6}Sr_{1/6})_8Cu_6O_z$, Hall effect was measured by a specially designed automatic measuring unit using the applied field of 6 T and the applied current of 50 mA.

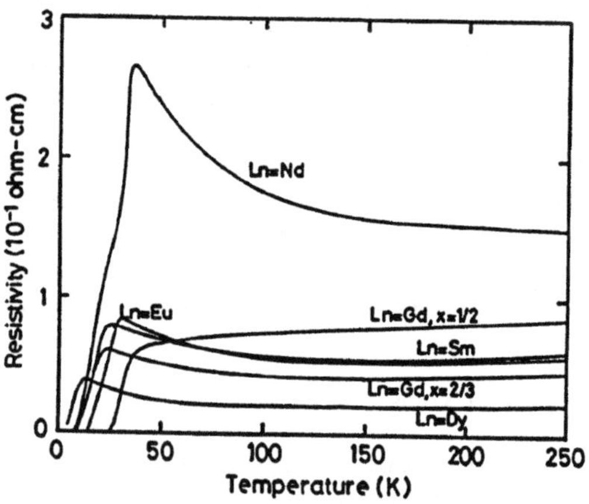

Fig. 3 Temperature dependence of elctrical resistivity of $[(La_{1-x}Ln_x)_{1/2}Ce_{1/6}Ba_{1/3}]_8Cu_6O_z$.

RESULTS AND DISCUSSION

1. $[(La_{1-x}Ln_x)_{1/2}Ce_{1/6}Ba_{1/3}]_8Cu_6O_z$ system

When x=0, i.e. all the original Ln sites were to be occupied only by La, the powder x-ray diffraction data indicated that the strucuture as shown in Fig. 1 did not form. This may be due to the fact that the ionic radius of La^{3+} is significantely larger than those of Nd^{3+}, Sm^{3+} and Eu^{3+}. When x=2/3, although samples with Ln=Nd, Sm, Eu, Gd and Dy were of either single or near single phase, samples with Ln=Er, Yb and Y were of multiple phases. It should be noted that, from a Rietveld analysis for the powder x-ray diffraction data for the sample with Ln=Gd and x=2/3, Gd atoms were found to preferentially occupy the A_1 sites, and La atoms were concluded to occupy the A_2 sites.[11] The following chemical formula is plausible for this compound:$(Gd_{2/3}Ce_{1/3})_4(La_{1/3}Ba_{2/3})_4Cu_6O_z$. Since the A_2 sites are fully occupied with La and Ba in these compounds when x=2/3, it is important for the Ln ions in the A_1 sites to have an ionic radius less than that for La^{3+} and larger than that for Er^{3+}($CN(A_1)$=8). When x=1/2, the excess La ions for the A_2 sites will occupy part of the A_1 sites. Therefore, even with smaller sized ions such as Yb and Y for Ln, samples of single-phase or near single phase may be obtained. Experimental results confirmed that samples with Ln=Er, Yb and Y were of near single phase.

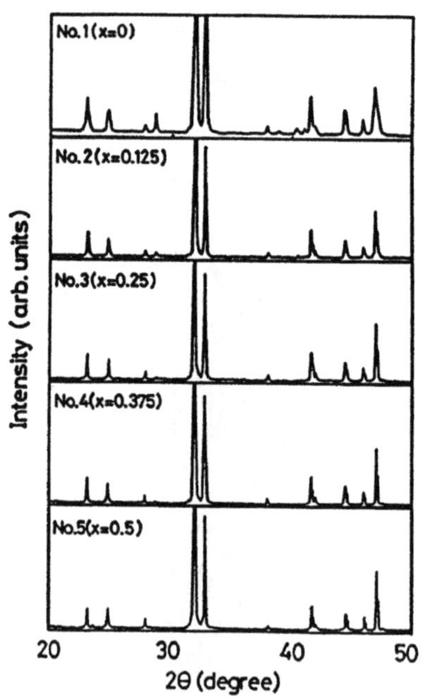

Fig. 4 x-ray diffraction patterns of $[Nd_{1/2}Ce_{1/6}(Ba_{1-x}Sr_x)_{1/3}]_8Cu_6O_z$. [7]

Figure 3 shows the temperature dependence of electrical resistivity of the samples with Ln=Nd, Sm, Eu, Gd and Dy and x=2/3, and with Ln=Gd and x=1/2. It should be noted that although the sample with Ln=Gd and x=0 exhibited superconductivity with T_c^{on}=21 K and $T_c^{R=0}$=9 K, superconductivity was not clearly observed in a magnetic susceptibility measurement.[6] The samples with Ln=Er, Yb and Y and x=2/3 did not exhibit superconductivity. The superconducting transition temperature for the sample with Ln=Gd and x=1/2 was the highest in the $[(La_{1-x}Ln_x)_{1/2}Ce_{1/6}Ba_{1/3}]_8Cu_6O_z$ system. This sample was treated in an oxygen atmosphere of partial pressure of 50 atm. The superconducting transition temperature was then raised from 30 K to 40 K.

2. $[Nd_{1/2}Ce_{1/6}(Ba_{1-x}Sr_x)_{1/3}]_8Cu_6O_z$ system[7]

The crystallographic structure of the near single-phase sample with x=0.5 was refined by an x-ray Rietveld analysis method.[11] It was revealed that the A_1 sites were occupied by Nd and Ce, while the A_2 sites were occupied by Nd, Ba and Sr. Therefore, the following chemical formula is plausible for this compound : $(Nd_{2/3}Ce_{1/3})_4(Nd_{1/3}Ba_{1/3}Sr_{1/3})_4Cu_6O_z$. Figure 4 shows powder x-ray diffraction patterns of compounds in this system. All the samples were of near single phase. The amount of the secondary phase of $BaCeO_3$ in the samples decreased with increasing Sr content. It should be noted that only the sample with x=0.5 was found to contain Nd_2CuO_4 as a minor secondary phase. The powder x-ray diffraction pattern of the sample with x=0.375 was perfectly of single phase. All the peaks in the pattern were indexed for a tetragonal unit cell with the lattice constants of a=3.854 Å and c=28.450 Å. The length of both a- and c-axes decreased with increasing Sr content. This effect of Sr substitution on the lattice constants may be explained by using the ionic radii of Ba^{2+}(1.52 Å) and Sr^{2+}(1.36 Å).[12]

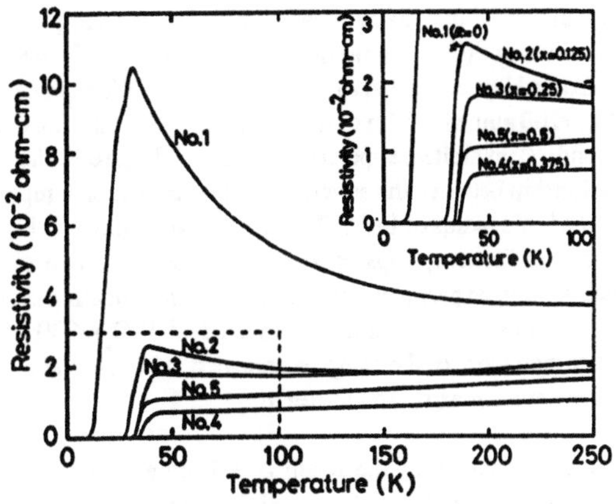

Fig. 5 Temperature dependence of electrical resistivity of $[Nd_{1/2}Ce_{1/6}(Ba_{1-x}Sr_x)_{1/3}]_8Cu_6O_z$. [7]

Figure 5 shows the temperature dependence of electrical resistivity for the samples in this system. The sample without Sr substitution exhibited a temperature dependence of semiconductor type in the normal state and the magnitude of resistivity was higher than those of other samples in this system. As the Sr content increased, the magnitude of electrical resistivity in the normal state first decreased to a minimum at x=0.35 and then increased. The superconducting transition temperature first increased up x=0.35 and then decreased as x increased.

On the contrary, in the $Y(Ba_{1-x}Sr_x)_2Cu_3O_z$ system, the superconducting transition temperature was lowered rather monotonically with increasing Sr content.[13] An important difference between these systems is that the oxygen contents in this system increased as the Sr content increased, while the oxygen content in the $Y(Ba_{1-x}Sr_x)_2Cu_3O_z$ system remained constant even when the Sr content increased. At present, no satisfactory explanations for these two cases of Sr substitution for Ba are known. The appearance of a maximum in the superconducting transition temperature when plotted against the Sr content, x, for the present system may be explained employing the idea of Tokura et al.[14] for the relationship between the superconducting transition temperature and the hole density.

3. $[(La_{1-x}Ln_x)_{1/2}Ce_{1/6}(Ba_{1-y}Sr_y)_{1/3}]_8Cu_6O_z$ system[6],[8]

From a Rietveld analysis for the powder x-ray diffraction data for the sample with Ln=Gd, x=2/3 and y=1/2, Gd atoms were found to preferentially occupy the A_1 sites, and La, Ba and Sr atoms were concluded to occupy the A_2 sites.[11]. Then the chemical formula for this sample may be given by $(Gd_{2/3}Ce_{1/3})_4(La_{1/3}Ba_{1/3}Sr_{1/3})_4Cu_6O_z$. Figure 6 shows x-ray diffraction patterns of the samples with Ln=Eu, Gd, Dy, Ho and Y and y=1/2. These samples were of single phase or near single phase, and the samples with Ln=Nd, Sm, Er, Tm, Yb and Lu and y=1/2 consisted of multiple phases. In this system, $BaCeO_3$ was identified as a minor secondary phase.

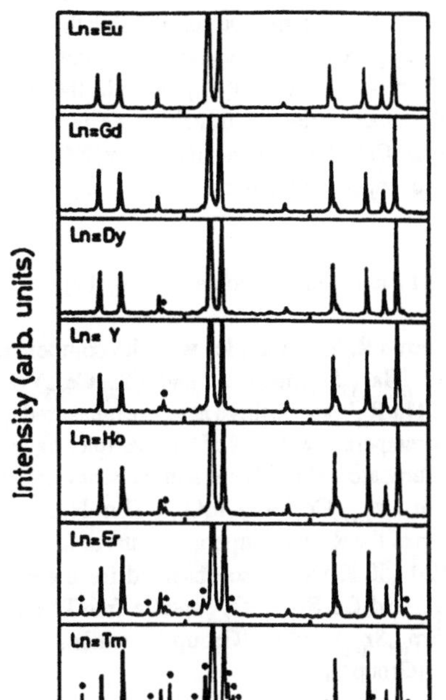

Fig. 6 x-ray diffraction psttems of $[(La_{1-x}Ln_x)_{1/2}Ce_{1/6}(Ba_{1-y}Sr_y)_{1/3}]_8Cu_6O_z$. [8]

The oxygen contents of the single phase or nearly single phase samples with different kinds of Ln atoms heat-treated at $P(O_2)$=3 atm were all nearly equal, being at z=17.8. The oxygen contents of the samples annealed at $P(O_2)$=200 atm were also nearly equal, being at z=18.1.

The lattice constants are plotted against the ionic radius of Ln^{3+} in Fig. 7. The lengths of both a- and c-axes decreased with decreasing ionic radii i.e. in the order of Eu(1.07 Å), Gd(1.05 Å), Dy(1.03 Å), Ho(1.02Å) and Y(1.02Å).[12] The lengths of the c-axis of the samples annealed at $P(O_2)$=200 atm were shorter than those of the samples annealed at $P(O_2)$=3 atm, while the lengths of the a-axis remained constant. A similer tendency has been reported for the oxide superconductors in the $YBa_2Cu_3O_z$ system.[15],[16]

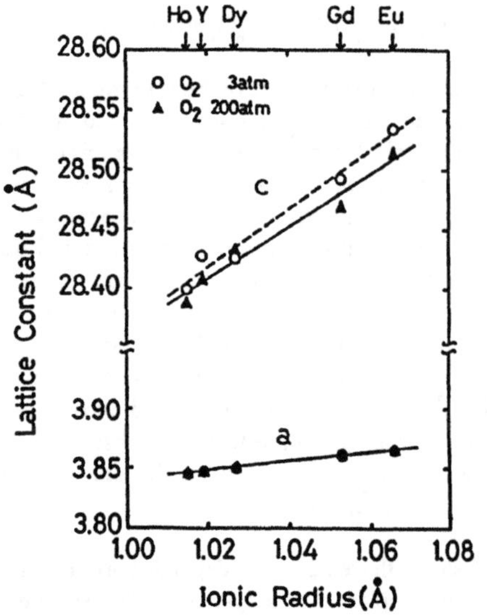

Fig. 7 Lattice constants, a and c, vs. ionic radius of Ln^{3+} for $[(La_{1-x}Ln_x)_{1/2}Ce_{1/6}(Ba_{1-y}Sr_y)_{1/3}]_8Cu_6O_z$. [24]

From electrical resistivity measurements of the samples in this system when $P(O_2)$=3 atm, only the samples with Ln=Nd, Sm, Eu, Gd and Dy, exhibited superconductivity. When annealed at $P(O_2)$=200 atm, all of the samples with Ln=Nd, Sm, Eu, Gd, Dy, Ho and Y, exhibited superconductivity. Figure 8 shows the relationship between the superconducting transition temperature and the ionic radius of Ln^{3+}. Two curves were drawn in Fig.8 for the two different groups of samples prepared at two different levels of oxygen partial pressure. The superconducting transition temperatures for the samples annealed at $P(O_2)$=200 atm are higher than those of the samples annealed at $P(O_2)$=3 atm. Both curves have a maximum at Ln=Gd.

Taking an analogy to the result of a Rietveld analysis for the sample with Ln=Gd x=2/3 and y=1/2, it is plausible that Eu, Gd, Dy, Ho or Y atoms, preferentially occupied the A_1 sites in this structure. It is likely that the ionic radius of the species at the A_1 sites affects the distance between the $Cu(2)O(3)_2$ plane and the O(4) plane. The distance between the $Cu(2)O(3)_2$ plane and the O(4) plane is also likely to affect the hole density on the $Cu(2)O(3)_2$ plane.[17] Therefore, it may be possible that the superconducting transition temperature is dependent on the distance of the $Cu(2)O(3)_2$ plane and the O(4) plane via the relationship[14] between the superconducting transition temperature and the hole density on the $Cu(2)O(3)_2$ plane, if only $Cu(2)O(3)_2$ planes are responsible for superconductivity.

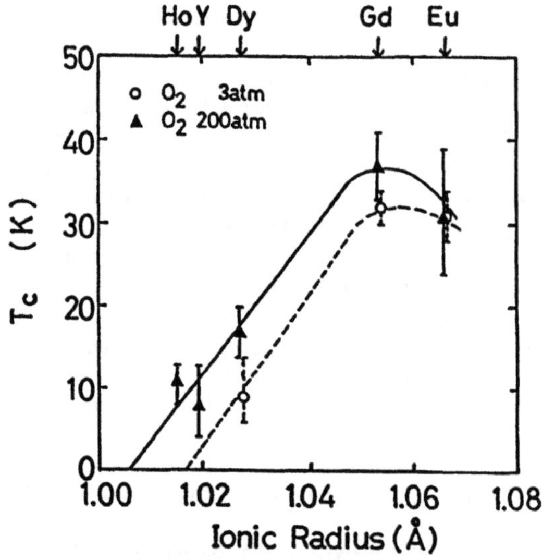

Fig. 8 Relationship between superconducting transition temperature, Tc, and ionic radious of Ln^{3+} for $[(La_{1-x}Ln_x)_{1/2}Ce_{1/6}(Ba_{1-y}Sr_y)_{1/3}]_8Cu_6O_z$. [24]

4. Transport and magnetic properties[18],[19]

Figure 9 shows R_H's of samples with the compositions of $(Nd_{2/3}Ce_{1/3})_4(Nd_{1/3}Ba_{5/12}Sr_{1/4})_4Cu_6O_z$ and $(Y_{2/3}Ce_{1/3})_4(La_{1/3}Ba_{1/3}Sr_{1/3})_4Cu_6O_z$. The R_H's are positive at all the measured temperatures. The R_H-vs-temperature curves of both samples have a broad peak at temperatures around 130 K. A similar behavior to this has been reported for $Ln_{2-x+y}Ce_xBa_{2-y}Cu_3O_{10-\delta}$ (Ln=Nd, Eu)[19]. The broad peaks of the R_H-vs-temperature curves located at temperatures of 120-150 K were also observed for the system of $(Nd, Ce)_2CuO_{4-\delta}$, $(Nd, Ce, Sr)_2CuO_{4-\delta}$ and Ce-free T*-strucutre phase, i.e. $(La_{0.4}Sm_{0.5}Sr_{0.1})_2CuO_{4-\delta}$ [Group 1], while no such peaks were observed for the $YBa_2Cu_3O_z$ system[21] and $La_{2-x}Sr_xCuO_4$ system[22][Group 2].

It may be considered that the difference in the temperature dependence of R_H results from the difference in the crystallgraphic structure.[18] The most striking difference in the crystal structures of Group 1 systems and Group 2 systems is the existance of the fluorite-type layer sandwiched between two CuO_2 planes for compounds of Group 1 systems. The fluorite layer probably affects the electronic states of a neighboring CuO_2 plane or the band structure, and subsequently affect temperature dependence of R_H. This result may correspond to the relation that the superconducting transition temperature for $(Ln_{2/3}Ce_{1/3})_4(La_{1/3}Ba_{1/3}Sr_{1/}$

$_3)_4Cu_6O_z$ is dependent on the distance between the $Cu(2)O(3)_2$ plane and the O(4) plane in the unit cell(Fig.8).

Using a bond valence sum method, effective valence numbers for Cu(1) and Cu(2) in $(Nd_{2/3}Ce_{1/3})_4(Nd_{1/3}Ba_{1/3}Sr_{1/3})_4Cu_6O_z$ were calculated to be 2.50 and 2.31, respectively[8]. If only the $Cu(2)O(3)_2$ planes be responsible for superconductivity, the number of holes for $Cu(2)O(3)_2$ may be considered to be 0.31. On the contrary, using the R_H value at the maximum point in Fig.9, the effective number of holes per $Cu(2)O(3)_2$ is reduced to be 0.10.[18] This may indicate that a certin amount of holes is located or trapped due to a certain mechanism. One of the possibilities for such a mechanism is the fluorite-type layer located neighbouring to the $Cu(2)O(3)_2$ plane.

$H_{c1}(T)$ of the $(Gd_{2/3}Ce_{1/3})_4(La_{1/3}Ba_{1/3}Sr_{1/3})_4Cu_6O_z$ is plotted against temperature in Fig. 10, where, $H_{c1}(T)$ is defined as the field where the magnetism curve, M(H) with respect to the field H, first deviates from linearity. We obtained $H_{c1}(0)$=110 Oe by extrapolating the linear H_{c1}-vs-temperature plot to 0 K. Figure 11 shows the temperature dependences of resistivity in magnetic fields for the estimation of the upper critical field, i.e. $H_{c2}(0)$. We estimated the $H_{c2}(0)$ at 72.7 T for this sample by employing the Werthamer-Helfand-Hohenberg formula for dirty type-II superconductor.[23] The superconductivity material parameters, i.e. GL coherence length, $\xi_{GL}(0)$, penetration depth, $\lambda_{GL}(0)$, GL parameter, κ, thermodynamical critical field, $H_c(0)$, and Sommerfeld parameter, γ, were calculated using $H_{c1}(0)$ and $H_{c2}(0)$. The resultant values for these parameters are summarized in Table 1. The unusually large value for κ may indicates that the compound is an extreme type-II superconductor.

SUMMARY

The conditions for preparing single phase samples of $[(La_{1-x}Ln^*_x)_{1/2}Ce_{1/6}Ba_{1/3}]_8Cu_6O_z$ were investigated. The sizes of ions or average ions which occupy the A_1 sites and the A_2 sites must be within certain ranges in order to obtain single-phase or near single-phase samples. In $[Nd_{1/2}Ce_{1/6}(Ba_{1-x}Sr_x)_{1/3}]_8Cu_6O_z$, the

Table 1 Material parameters for $(Gd_{2/3}Ce_{1/3})_4(La_{1/3}Ba_{1/3}Sr_{1/3})_4Cu_6O_z$ [25]

magnetic T_c (K)	38.0
$[dH_{c2}/dT]_{Tc^*}$ (kOe/k)	26.7
H_{c2} (0) (kOe)	727
H_{c1} (0) (kOe)	0.11
H_c (0) (kOe)	4.08
ξ_{GL} (0) (Å)	21
λ_{GL} (0) (Å)	2680
κ_{GL}	126
γ (mJ/(mol K^2))	49.5

* : determined from 80 % resistivity criterion.

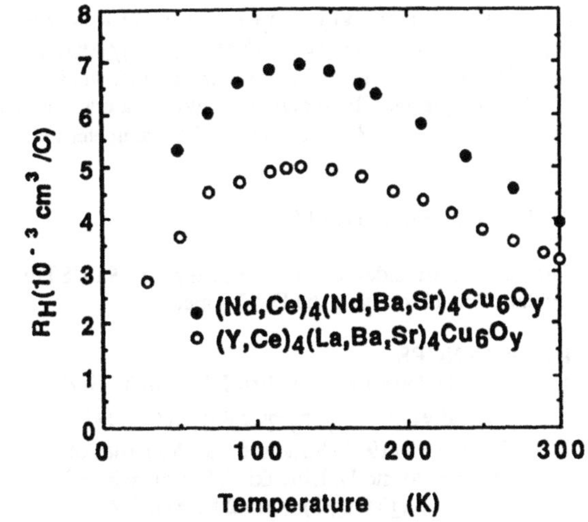

Fig. 9 R_H vs temperature for $(Nd_{2/3}Ce_{1/3})_4(Nd_{1/3}Ba_{5/12}Sr_{1/4})_4Cu_6O_z$ and $(Y_{2/3}Ce_{1/3})_4(La_{1/3}Ba_{1/3}Sr_{1/3})_4Cu_6O_z$.

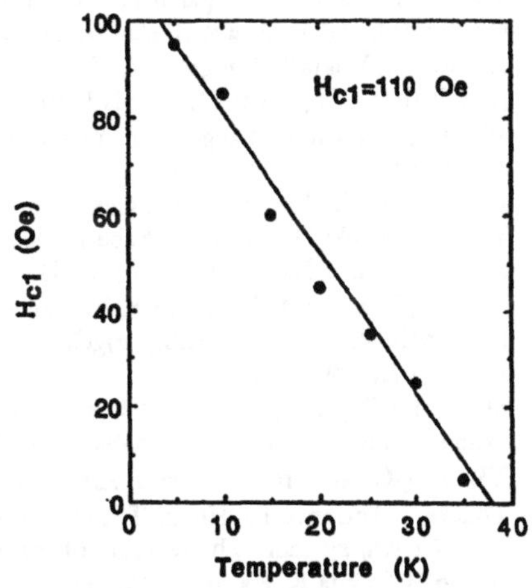

Fig. 10 Temperature dependence of $H_{c1}(T)$. [25]

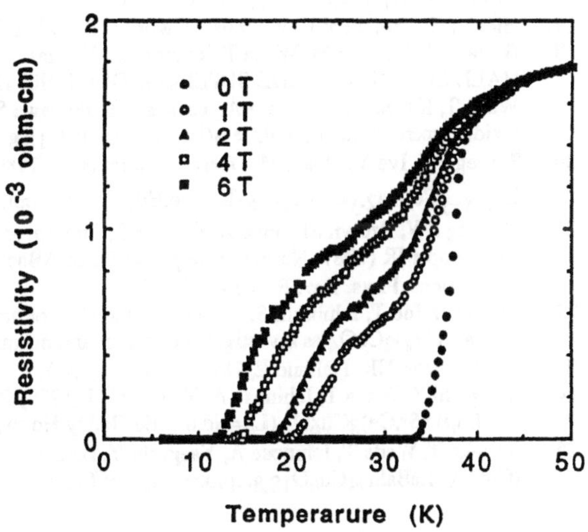

Fig. 11 Temperature dependence of electrical resistivity in mgnetic fields for $(Gd_{2/3}Ce_{1/3})_4(La_{1/3}Ba_{1/3}Sr_{1/3})_4Cu_6O_z$. [25]

highest superconductivity temperature of 39 K was obtaines for x=0.35. It was found that the oxygen content, z, was dependent on the Sr content, x. For $(Ln_{2/3}Ce_{1/3})_4(La_{1/3}Ba_{1/3}Sr_{1/3})_4Cu_6O_z$, T_c was found to depend on the ionic radius of Ln. This fact and the Hall measurements indicated that there might be a certain mechanism which trapeed part of holes the $Cu(2)O(3)_2$ planes. The superconductivity materal parameters were calculated usin the experimantally determined values for $H_{c1}(0)$ and $H_{c2}(0)$. A large value for GL parameter indicated that these superconductors were of an extreme type-2.

ACHNOWLEDGMENTS

The authoes are indebted to Dr. HIdaka od SRL-ISTEC for his discussions of a possible hole trap based on his molecular orbital calculations using the DV-Xa method.

REFERENCES

1. Sawa H, Obara K, Akimitsu J, Matsui Y, Horiuchi S (1989) A New Family of Superconducting Copper Oxides : $(Ln_{1-x}Ce_x)_2(Ba_{1-y}Ln_y)_2Cu_3O_{10-\delta}$ (Ln=Nd,Sm,Eu). J. Phys. Soc. Jpn. 58:2252-2255
2. Galasso FS (1970) Structure and Properties of Inorganic Solids, Pergamon Press, New York
3. Izumi F, Asano H, Ishigaki T, Muromachi ET, Uchida Y, Watanabe N (1987) Crystal Structure of the Orthorhombic Form of $Ba_2YCu_3O_{7-x}$ at 42 K. Jpn. J. Appl. Phys. 26:L1193-L1196
4. Cava RJ, Batlogg B, van Dover RB, Murphy DW, Sunshine S, Siegrist T, Remeika JP, Rietman EA, Zahurak S, Espinosa GP (1987) Bulk Superconductivity at 91 K in Single-Phase Oxygen-Deficient Perovskite $Ba_2YCu_3O_{9-\delta}$. Phys. Rev. Lett. 58:1676-1679
5. Tokura Y, Takagi H, Uchida S (1989) A superconducting copper oxide compound with electrons as the charge carries. Nature 337:345-348
6. Wada T, Ichinose A, Yaegashi Y, Yamauchi H, Tanaka S (in press) Preparation of New Oxide Superconductors, $(La,Gd,Ba,Ce)_8Cu_6O_z$ and $(La,Gd,Ba,Sr,Ce)_8Cu_6O_z$. Phys. Rev. B.
7. Ichinose A, Wada T, Yaegashi Y, Yamauchi H, Tanaka S (1989) Preparation and Superconducting Properties of $[Ln,Ce,(Ba_{1-x}Sr_x)]_8Cu_6O_z$ (Ln=Nd, Sm and Eu). Jpn,J, Appl. Phys. 28:L1765-L1768
8. Wada T, Ichinose A, Yaegashi Y, Yamauchi H, Tanaka S (submitted) Preparation and Properties of superconducting $La_{1/6}Ln_{1/3}Ba_{1/6}Sr_{1/6}Ce_{1/6}]_8Cu_6O_z$ (Ln=Eu,Gd,Dy,Ho and Y). Jpn. J. Appl. Phys.
9. Izumi F (1985) Nippon Kessho Gakkaishi 27:23-31
10. Wada T, Suzuki N, Maeda A, Yabe T, Uchinokura K, Uchida S, Tanaka S (1989) Preparation and properties of superconducting $La_{1+x}Ba_{2-x}Cu_3O_y$ ($0 \leq x \leq 0.5$) ceramics sintered in N_2 gas atmosphre. Phys. Rev. B39:9126-9138
11. Wada T, Ichinose A, Yaegashi Y, Yamauchi H, Tanaka S (1989) Crystal Structure of New Oxide Superconductors, $(Sm,Ba,Ce)_8Cu_6O_z$, $(Nd,Ba,Sr,Ce)_8Cu_6O_z$, $(La,Gd,Ba,Ce)_8Cu_6O_z$ and $(La,Gd,Ba,Sr,Ce)_8Cu_6O_z$. Jpn.J. Appl. Phys. 28:L1779-L1782
12. Shannon RD (1976) Acta Crystallogr. A32:751
13. Wada T, Adachi S, Mihara T, Inaba R (1987) Substitution Effect of Sr for Ba of High-T_c Superconducting $YBa_2Cu_3O_{7-y}$ Ceramics. Jpn. J. Appl. Phys. 26:L706-L708
14. Tokura Y, Torrance JB, Huang TC, Nazzal AI (1988) A broader perspective on the high temperature superconducting $YBa_2Cu_3O_y$-system - The real rool of the oxygen content. Phys. Rev. B38:7156
15. Asano H, Takita K, Katoh H, Akinaga H, Ishigaki T, Nishino M, Imai M, Masuda K (1987) Crystal Structure of the High T_c Superconductor $LnBa_2Cu_3O_{7-\delta}$ (Ln=Sm,Eu and Gd). Jpn. J.Appl. Phys. 26:L1410-L1412
16. Ono A, Ishizawa Y (1987) Preparation and Properties of Three Type of Orthorhombic Superconductor $Ba_2YCu_{3-x}O_{7-y}$. Jpn. J. Appl. 26:L1043-L1045
17. Hidaka H, Yamauchi H:unpublished work at SRL-ISTEC
18. Ikegawa S, Kosuge M, Wada T, Ichinose A, Yaegashi Y, Nakao K, Yamashita T, Sakurai T, Yamauchi H (1989) HALL EFFECT IN OXIDE SUPERCONDUCTORS HAVING FLUORITE-TYPE LAYERS. :proceedings of ISS'89
19. Wada T, Kaneko T, Ichinose A, Yaegashi Y, Ikegawa S, Yamauchi H (in press) Magnetic Properties of 40 K Class Oxide Superconductor, $(Gd,Ce)_4(La,Ba,Sr)_4Cu_6O_{18.8}$. Jpn. J. Appl. Phys.
20. Tamegai T, Iye Y, Ogata M, Obara K, Akimitsu J (1989) Transport Properties of New High-T_c Superconductors $Ln_{2-x+y}Ce_xBa_{2-y}Cu_3O_{10-\delta}$ (Ln=Nd,Eu). Jpn. J. Appl. Phys. 28:L1537-L1540
21. Cheong SW, Brown SE, Fisk Z, Kwok RS, Thompson JD, Zirngiebl E, Gruner G, Peterson DE, Wells GL,Schwarz RB, Cooper JR (1987) Normal-atate properties of $ABa_2Cu_3O_{7-y}$ compounds (A=Y and Gd) : Electron-electron correlations, Phys. Rev. B36:3913-3916
22. Takagi H, Ido T, Ishibashi S, Uota M, Uchida S, Tokura Y (1989) Superconductor-to-nonsuperconductors transition in $(La_{1-x}Sr_x)_2CuO_4$ as investigated transport and magnetic measurements. Phys. Rev B40:2254-2261
23. Werthameter NR, Helfand E, Hohenberg PC (1965)
24. Yaegashi Y, Wada T, Ichinose A, Yanauchi H (1989) PREPARATION AND CHARACTERIZATION OF $(La,Ln,Ba,Sr,Ce)_8Cu_6O_z$ (Ln=Nd,Sm,Eu,Gd,Dy,Ho,Er,Tm,Yb and Y) CERAMICS.:proceedings of ISS'89
25. Kaneko T, Wada T, Ichinose A, Yaegashi Y, Ikegawa S, Yamauchi H (1989) Superconducting properties of $(GdCe)_4(LaBaSr)_4Cu_6O_{18.8}$. :proceedings of ISS'89

Beneficial Effect of Using 211 for the Formation of 123 Superconducting Phase

HYUNGSIK CHUNG, HAE-DOO KIM, JAEWOONG KOH, SUNWHA KIM, and KUK-HYUN HA
Korea Institute of Machinery and Metals, Changwon, Korea

ABSTRACT

Seversal studies demonstrated that slow solidification of partially melted $YBa_2Cu_3O_x$ (123) can produce highly oriented, long grains which yield high critical current density (Jc). The present work shows that the grain morphology of 123 phase formed from melt decomposed phases is very much dependent on the existence of liquid phase during the formation. The 123 phase is considered to grow from Y_2BaCuO_5 (211) and surrounding Ba, Cu-rich phase. Alternatively,a direct utilization of the 211 phase as the base material for 123 formation produces an enhanced development of grain alignments along (001) basal planes. Particular interest is given to the possibility of forming highly oriented 123 thick film on dense 211 substrate.

KEY WORDS : $YBa_2Cu_3O_x$, Y_2BaCuO_5, Grain alignment, 123 Thick film, 211 Substrate.

INTRODUCTION

Following the excitement generated by the discovery of $YBa_2Cu_3O_x$ superconductor, it was soon realized that the current carrying capacity of the material, especially in a bulk form, is too low for any practical applications. The material can yield critical current densities over 10^6 A/cm^2 as demonstrated in epitaxially grown thin films or single crystals but has Jc values of orders of magnitude lower in bulk crystalline form.[1]

It is generally understood that weak coupling between superconducting grains and anisotropic superconductivity of the material are the major cause for the low Jc[2]. Extensive studies have been conducted to increase Jc, and several processing techniques have been reported to yield fairly high Jc.[3-5]

Processes involving solidification of partially melted $YBa_2Cu_3O_x$ were reported to produce highly oriented, long plate type grains which can mininize the weak coupling and anisotropy problem. Critical current densities up to 17,000A/cm^2 and 75,000A/cm^2 were obtained by melt texturing process[6] and a slow solidification process through peritectic transformation[7] respectively. In both processes , Y_2BaCuO_5 solid phase and Ba, Cu-rich phase are supposed to form liquid phase above peritectic temperature of ~1,000 ℃, and solidify through peritectic transformation to yield highly oriented 123 grains.

In this work, we studied the growth of 123 phase from melt decomposed phases in a rapidly solidified ribbon and the possibility of utilizing 211 phase as a substrate to form thick 123 film onto it. Studies involving synthesizing and sintering of the 211 substrate as well as that for the formation of 123 phase are reported in this paper.

GROWTH OF 123 FROM MELT DECOMPOSED PHASES

The formation of 123 phase from melt decomposed phases was studied using rapidly solidified ribbon prepared by melt spinning process. The ribbon consists of rod shaped 211 phase well aligned to spinning direction in Ba, Cu-rich matrix as shown in Fig. 1. The ribbon was heat treated in two routes depicted in Fig. 2 to compare the formation of 123 phase with and without going through peritectic transformation, see pseudobinary phasediagram in Fig. 3. Both of the heat treatment cycles produced 123 phase but with very different microstructures, Fig.4.

The heat treatment (b) which went through peritectic transformation yielded highly oriented rod-type 123 grains whereas the heat treatment (a) produced almost equiaxed type grains. It is conceivable that liquid phase formed above the peritectic temperature promotes mass transport toward 211 phase preferentially to nucleate and grow 123 phase whereas random nucleation and growth can be predominent in a solid-state reaction below the peritectic temperature. The orientation of the 211 phase and 123 phase formed through peritectic transformation may have some relationship. XRD peaks corresponding to (ab0) planses, that is parallel to c-axis, are predominent in the 211 phase whereas (001) or high c-indexed planes, which are perpendicular or near perpendicular to c-axis, are well developed in the 123 phase, Fig. 5.

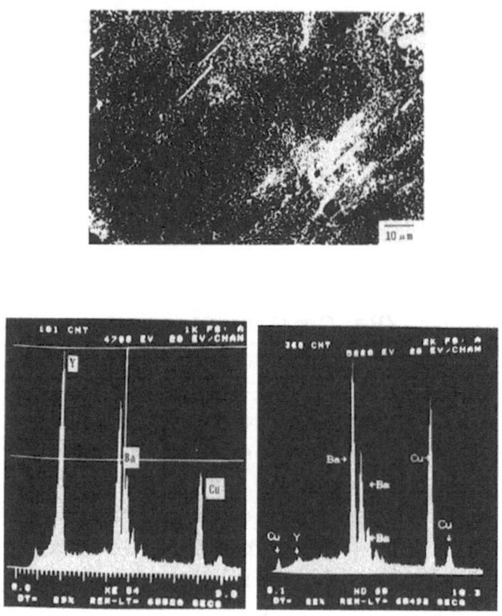

"Fig.1" SEM and EDX of Melt Spun Ribbon
A) 211 Phase B) Matrix

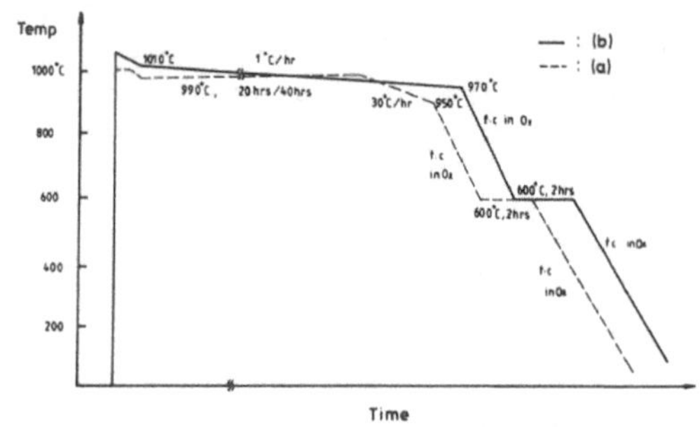

"Fig.2"H.T.Cycle of Melt Spun Ribbon

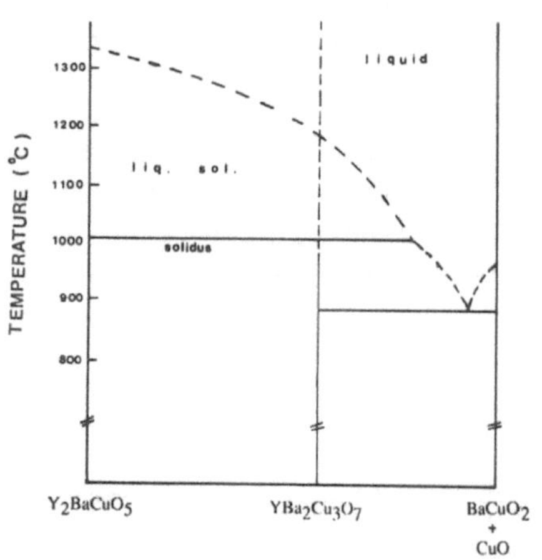

"Fig.3" Binary Phase Diagram of YBCO

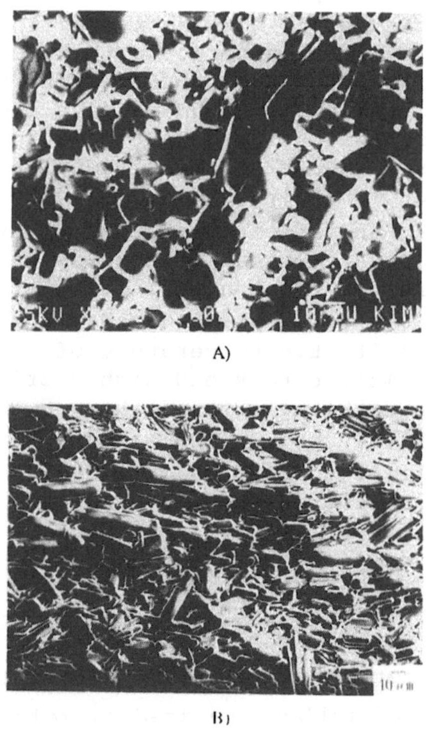

"Fig.4" 123 Structure after H.T.
A) 990°C/4hrs B) 1100-970°C @ 1°C/hr

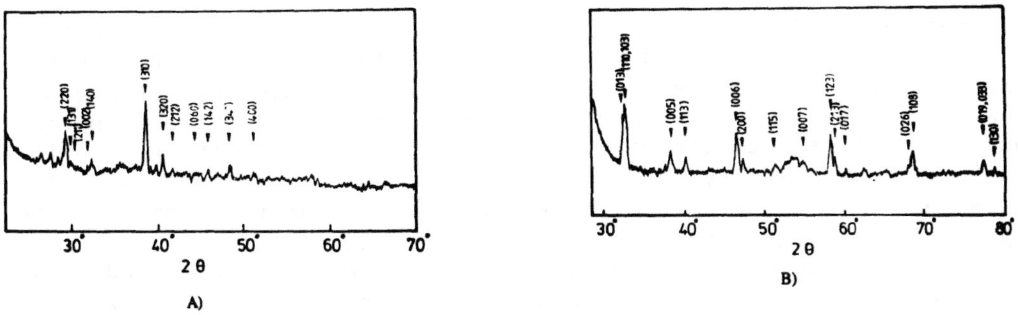

"Fig.5" XRD Pattern of A) 211 Phase in As-Spun Ribbon
B) 123 Phase After H.T. through Peritectic Transformation

FORMATION OF 123 FILM ON 211 SUBSTRATE

The study conducted on a melt spun ribbon led to the idea of utilizing 211 phase as a substrate to form 123 film. Parameters involving the synthesis of 211 powder, sintering 211 substrate, reaction of 211, $BaCuO_2$ and CuO powder mixture and the formation of 123 film were studied.

Preparation of 211 Powder And Substrate

211 powder was synthesized from Y_2O_3, $BaCO_3$, CuO powder mixture by solid state reaction. The powder mixture was calcined from 850 to 1200°C at 50°C interval for 6 hours to determine an optimum condition. DTA-TG analysis was conducted to examine the synthesis condition, and X-ray diffraction pattern was studied after the calcination. Tablets were made from the calcined powder and sintered at temperatures up to 1300°C to examine the stability and sinterability of 211 powder. Figure 6 shows the DTA-TG analysis of the powder mixture. Endothermic peak around 820 °C is probably due to the decomposition of $BaCO_3$ which completes near 1000°C. Small exothermic peak around 905°C is considered for the formation of 211 phase. XRD pattern of calcined powder indicates that 211 phase starts to form at around 900 °C and all the peaks correspond to 211 phase at 1000°C, Fig. 7. Endothermic peak at 1275°C is thought to be the melting point of 211 phase.

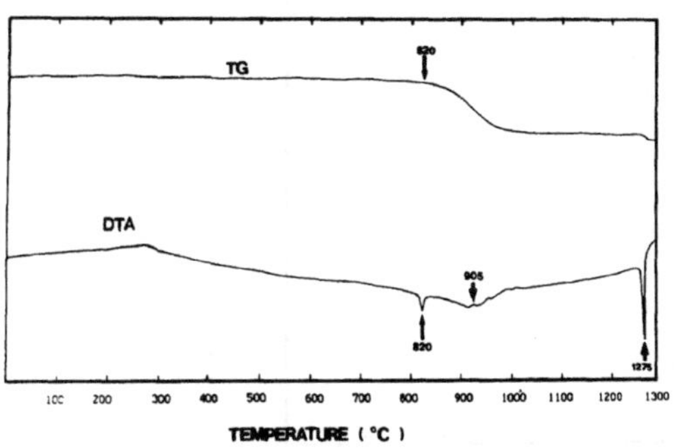

"Fig.6" DTA-TG Analysis During the Calcination of 211 Powder

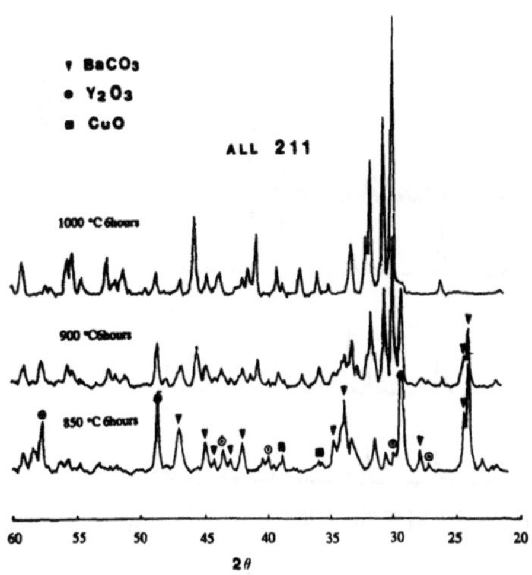

"Fig.7" XRD Pattern After the Calcination of 211 Powder

Unlike 123 Phase which decomposes above peritectic temperature of ~1,000°C, 211 phase is stable at much higher termperatures. Figure 8 shows that there is little change in peak intensities up to 1,250 ℃-3hours sintering. At higher temperatures or longer times, decrease in intensities becoms evident and peaks corresponding to Y_2O_3 start to appear, indicating the decomposition or melting of 211 phase. Sintered density of 211 pellet is affected both by calcination and sintering condition. When sintered at 1,250°C for 6 hours, highest sintered density was obtained with 1,000 °C-6 hours calcined powder, Fig. 9. This is attributable to the incomplete decomposition of $BaCO_3$ below 1,000 °C and large grain growth accompanied at higher calcining temperature. With the powder calcined at 1,000 °C for 6 hours, sintered densities over 5.5g/cc (90% TD) were obtained by sintering at 1,250 °C, Fig. 10. Longer sintering time results in higher density but accompanies a large grain growth and grain rounding, indicating the involvement of liquid phase during the sintering.

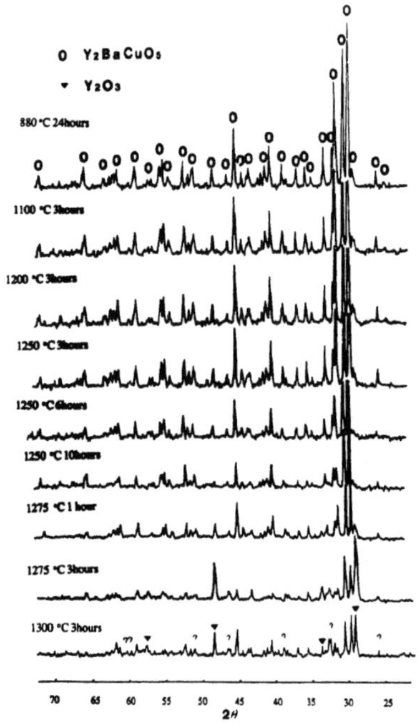

"Fig.8" XRD Pattern of Sintered 211

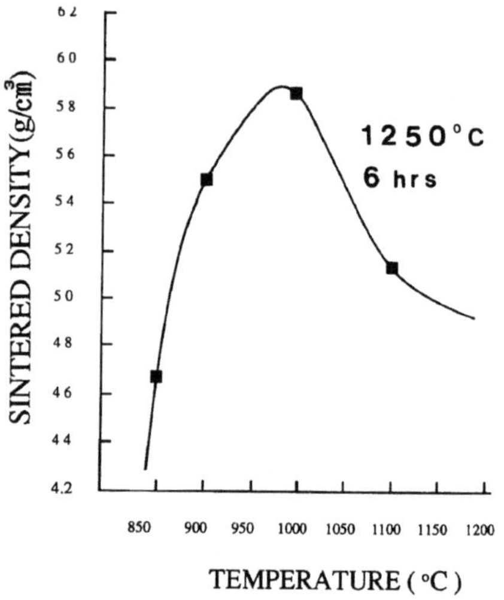

"Fig.9" Calcination Temp. vs. Sintered Density of 211 Pellet

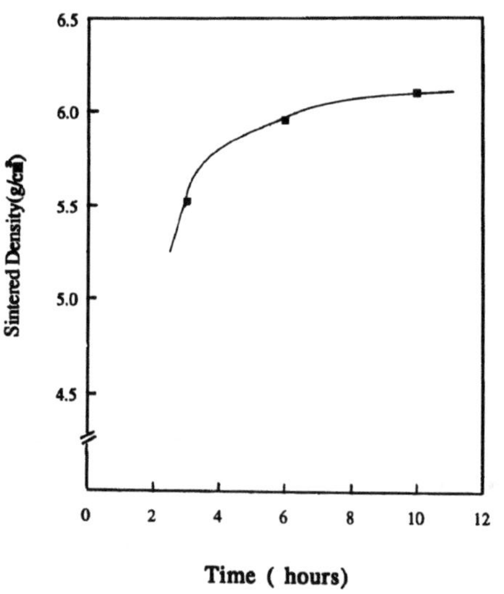

"Fig.10" Sintering Time at 1250°C vs. Sintered Density of 211 Pellet

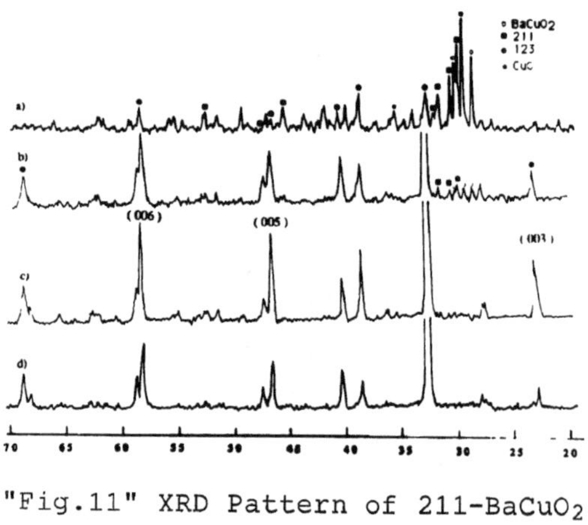

"Fig.11" XRD Pattern of 211-$BaCuO_2$ CuO Powder Compact After Sintering

Synthesis Of 123 Using 211, $BaCuO_2$ And CuO Powder Mixture

The formation of 123 phase from 211, $BaCuO_2$ and CuO was studied by mixing the powder and sintering at 900 °C, 925 °C, 950 °C for 3 hours and 975 °C for 20 to 60 hours. XRD pattern in Fig. 11. shows that 950 °C sintering is required to form 123 phase. It is also noteworthy that this process produces a better development of (001) plane than the usual solid state reaction. Sintering at 975 °C was conducted to study the effect of transient liquid phase for the growth of 123 phase. Unlike peritectic reaction, the transient liquid phase sintering produces long plate type 123 grains of having random orientations, Fig.12. This indicates that the presence of liquid phase is effective to produce long grains but is not sufficient to promote grain alignments, Selective nucleation or sufficient time for aligning 211 phase in mushy state may be necessary for the aligned 123 growth.

Formation Of 123 Film On 211 Substrate

Dense 211 substrate was coated with $3BaCuO_2$+ 2 CuO slurry and sintered at 950 °C for 3 hours. Coating thickness was controlled by changing the ratio of powder content and ethanol carrier in the slurry. Coating thickness after the sintering was measured as 10 - 100μm. XRD pattern was analysed on the coated surface after the sintering. Figure 13 clearly shows that 123 film is formed on 211 substrate with well developed (001) plane aligned on the substrate. The (001) intensities are more pronounced with thinner films applied.

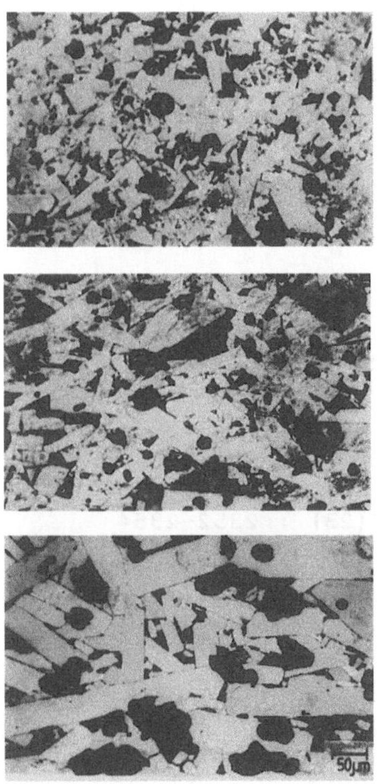

A) B) C)

"Fig.12" Microstructure of 123 Phase Formed By Sintering 211-$BaCuO_2$- CuO Compact
A) 975 °C/20hr B) 975 °C/40hr
C) 975 °C/60hr

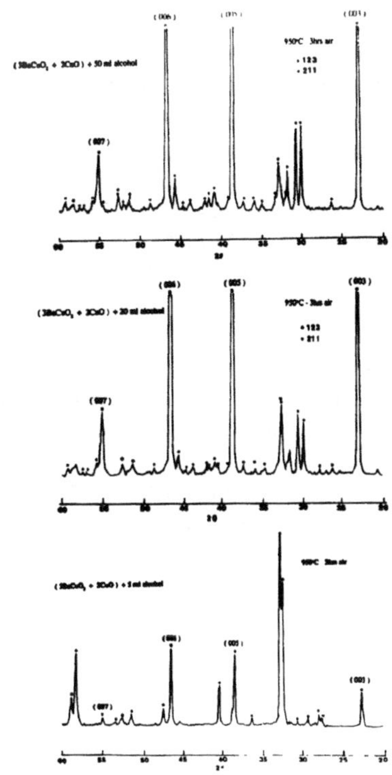

A) B) C)

"Fig.13" XRD Pattern of 123 Film Formed on 211 Substrate

CONCLUSION

The results of this study strongly suggest that using 211 substrate to produce thick 123 superconducting film has some advantages over conventional thick film fabrication. Since 211 substrate can be sintered to higher densities and is stable at higher temperatures than 123 material, it is possible to produce dense 123 film with demensional stability even with the reaction at relatively high temperatures. The reason for the strong directionality of 123 film formed on 211 substrate is not known, but is certainly the another advantage. Better understanding of the kinetics of 123 phase growth from 211 phase is necessary to improve the processing and maximize the advantages.

The authors wish to acknowledge many valuable assistances by collegues in Korea Institute of Machinery and Metals. The present work is supported by Minisitry of Science and Technology in Korean Government.

REFERENCES

1. Tadashi Takenaka, Hideki Noda, Atsuhiko Yoneda, Koichiro Sakata (1988) "Superconducting Properties of grain oriented $YBa_2Cu_3O_x$Ceramics", Jpn.J. Appl. Phys. 27 (7) : L1209-L1212

2. D. Gazit, P.N. Peszkin, R.S. Feigelson (1989) "Preparation of high temperature superconductor-metal wire composites", Mat. Res. Bull. 24 ; 467-474

3. Osamu Kohno, Yoshimitsu Ikeno, Nobuyuki Sadakata, Shinya Aoki, Masaru Sugimoto, and Mikio Nakagawa (1987) " Critical Current Density of Y-Ba-Cu Oxide wires" Jpn. J. Appl. Phys. 26(10) ; 1653-1656

4. G.G. Peterson, B.R. Weinberger, L. Lynds, and H.A. Krasinski (1988) "Improvement of Polycrystalline Y-Ba-Cu-O by the addition of silver", J. Mater. Res. 3(4) ; 605-609

5. Tomoo Takayama, Shojiro Ochiai and Kozo Osamura (1989), "Influence of Microstructure on the critical current density of $Ba_1Y_2Cu_3O_x$superconducting tapes" ; J. Jpn. Inst. Metals, 53 (7) ; 735-743

6. S.Jin, R.C. Sherwood, E.M. Gyorgy, T.H. Tiefel, R.B. van Dover, S. Nakahara, L.F. Schneemeyer, R.A. Fastnacht, and M.E. Davis (1989), "Large magnetic hysteresis in a melt-textured Y-Ba-Cu-O superconductor" Appl. Phys. Lett. 54(6) ; 584-586

7. K.Salama, V. Selvamanickam, L. Gao, K. Sun (1989) "High current density in bulk $YBa_2Cu_3O_x$superconductor". Appl. Phys. Lett. 54(23) ; 2352-2354

Preparation and Characterization of [La, Ln, Ba, Sr, Ce]$_8$Cu$_6$O$_z$ (Ln: Nd, Sm, Eu, Gd, Dy, Ho, Er, Tm, Yb, and Y) Ceramics

Y. YAEGASHI, T. WADA, A. ICHINOSE, and H. YAMAUCHI

Superconductivity Research Laboratory, International Superconductivity Technology Center, 10-13, Shinonome 1-chome, Koto-ku, Tokyo, 135 Japan

ABSTRACT

We have prepared ceramic samples of $[La_{1/6}Ln_{1/3}Ba_{1/6}Sr_{1/6}Ce_{1/6}]_8Cu_6O_z$ (Ln : Nd, Sm, Eu, Gd, Dy, Ho, Er, Tm, Yb and Y). X-ray diffraction patterns indicated that they were of single phase or nearly single phase except for the samples with Ln = Nd, Sm, Er, Tm and Yb. Their crystal structures were of the $(Eu, Ce)_4(Eu, Ba)_4Cu_6O_z$ type. The samples only with Ln = Eu, Gd and Dy when annealed in an oxygen atmosphere of $P(O_2)$ = 3 atm exhibited superconductivity with T_c's in the range of 10-35 K, while the samples with Ln = Eu, Gd, Dy, Ho and Y became superconductors with T_c's in the range of 5-40 K when annealed at $P(O_2)$ = 200 atm. Additionally, we have prepared nearly single-phase $[La_{1/6}Gd_{1/3}Ce_{1/6}(La_xBa_{1-x})_{1/3}]_8Cu_6O_z$ ($0 \leq x \leq 0.5$) ceramics. The structures of the samples with x = 0, 0.1 and 0.5 were all tetragonal and those with x = 0.2, 0.3 and 0.4 were orthorhombic according to the x-ray diffraction patterns. The samples with x = 0 and 0.1 exhibited superconductivity having T_c's in the range of 25-30 K.

KEY WORDS: high-T_c superconductors, high oxygen pressure treatment, rare earth element, tetragonal-orthorhombic phase transition

INTRODUCTION

Recently, a new family of superconducting oxides of $(Ln, Ce)_4(Ln, Ba)_4Cu_6O_z$ (Ln : Nd, Sm and Eu) was discovered by Sawa et al.[1] These superconductors have a crystal structure of new type. In the unit cell, two different types of sites exist for lanthanide ions: "A_1" sites are occupied by both small ions, i.e., Ce^{4+}, and medium-sized ions, i.e., Nd^{3+}, Sm^{3+} and Eu^{3+}, while "A_2" sites are occupied by large ions, i.e., Ba^{2+}, as well as by the medium-sized ions.

Previously, we reported on the syntheses and characterization of samples of the $[Ln_{1/2}(Ba_{1-x}Sr_x)_{1/3}Ce_{1/6}]_8Cu_6O_z$ (Ln : Nd, Sm and Eu) and $[(Gd_{1-x}La_x)_{1/2}(Ba_{1-x}Sr_x)_{1/3}Ce_{1/6}]_8Cu_6O_z$ systems.[2-3] From the Rietveld analysis for the powder x-ray diffraction data for $(La_{1/6}Gd_{1/3}Ba_{1/6}Sr_{1/6}Ce_{1/6})_8Cu_6O_z$, Gd was found to preferentially occupy the A_1 sites, and La and Sr atoms were concluded to occupy the A_2 sites.[4]

In this paper, We studied the crystallographic and superconducting properties of $[La_{1/6}Ln_{1/3}Ba_{1/6}Sr_{1/6}Ce_{1/6}]_8Cu_6O_z$ containing a series of rare earth elements, Nd, Sm, Eu, Gd, Dy, Ho, Er, Tm, Yb and Y for Ln. We also investigated the effect of annealing of the samples at high oxygen partial pressures on the crystallographic and superconducting properties. In addition, we studied the superconducting properties of $[La_{1/6}Gd_{1/3}Ce_{1/6}(La_xBa_{1-x})_{1/3}]_8Cu_6O_z$ (x= 0, 0.1, 0.2, 0.3, 0.4 and 0.5) as the effect of La substitution for Ba in the "A_2" sites.

EXPERIMENTAL

Samples were prepared by a conventional solid-state reaction method. High purity powders of La_2O_3, Nd_2O_3, Sm_2O_3, Eu_2O_3, Gd_2O_3, Dy_2O_3, Ho_2O_3, Er_2O_3, Tm_2O_3, Yb_2O_3, Y_2O_3, CeO_2, $BaCO_3$, $SrCO_3$ and CuO were used as starting materials. Appropriate amounts of powders for each nominal composition of $[La_{1/6}Ln_{1/3}Ba_{1/6}Sr_{1/6}Ce_{1/6}]_8Cu_6O_z$ and $[La_{1/6}Gd_{1/3}Ce_{1/6}(La_xBa_{1-x})_{1/3}]_8Cu_6O_z$

were mixed with ethanol using a ball mill. The mixed powder was calcined at 1020 °C for 15 h in O_2 gas of 3 atm, and then pulverized and pressed into parallelepiped bars. The bars were sintered at 1030 °C for 20 h in O_2 gas of 3 atm and held successively at 600 °C for 5 h and at 400 °C for 20 h. In order to increase oxygen contents, samples of $[La_{1/6}Ln_{1/3}Ba_{1/6}Sr_{1/6}Ce_{1/6}]_8Cu_6O_z$ were re-heat-treated at 900 °C for 1 h, and then at 600 °C for 5 h and at 400 °C for 5 h at a high oxygen partial pressure, $P(O_2)$, of 200 atm.

The lattice constants were determined by powder x-ray diffraction using Cu- $K\alpha$ radiation. For measurements of high resolution, a curved graphite monochromater was placed in the scattering beam path. The oxygen contents were analyzed by an inert gas fusion nondispersive IR method.[5] Electrical resistivity measurements were carried out at temperatures down to the boiling point of liquid He by a conventional dc four-probe method and ac magnetic susceptibility measurements at temperatures down to 20 K by an ac magnetometer.

RESULTS AND DISCUSSIONS

1. $[La_{1/6}Ln_{1/3}Ba_{1/6}Sr_{1/6}Ce_{1/6}]_8Cu_6O_z$

From x-ray diffraction patterns of the $[La_{1/6}Ln_{1/3}Ba_{1/6}Sr_{1/6}Ce_{1/6}]_8Cu_6O_z$ samples, it was found that those with Ln = Eu, Gd, Dy, Ho and Y were of single-phase or near single-phase, and those with Ln = Nd, Sm, Er, Tm and Yb consisted of multi-phases. The crystal structures were all tetragonal and of the $(Eu,Ce)_4(Eu,Ba)_4Cu_6O_z$ type. $BaCeO_3$ was identified as a secondary phase in the samples with Ln = Nd, Sm, Er, Tm and Yb.

The oxygen contents were measured only for single-phase or near single-phase samples. The oxygen contents of the samples with different kinds of Ln elements heat-treated at $P(O_2)$ = 3 atm were all nearly equal: $z \sim 17.8$. The oxygen contents of the samples annealed at $P(O_2)$ = 200 atm were also nearly equal being at $z \sim 18.1$ which is higher than those of samples annealed in an oxygen atmosphere of $P(O_2)$ = 3 atm.

The lattice constants determined are plotted against the ionic radius of Ln^{3+} in Fig. 1. The lengths of both a- and c- axes decreased with decreasing Ln^{3+} ionic radius. Note that ionic radii of the following rare earth elements for the coordination number equal to 8 are given by Shannon [6]: Eu(1.066 Å), Gd(1.053 Å), Dy(1.027 Å), Ho(1.015 Å) and Y(1.019 Å). A parallel tendency has been reported for the lattice constants of $LnBa_2Cu_3O_z$ with respect to the ionic radius of Ln^{3+}.[7] The length of c-axis of a sample annealed at $P(O_2)$ = 200 atm was shorter than that of the sample annealed at $P(O_2)$ = 3 atm. A similar change in the c-axis length in terms of the oxygen content has been reported for $YBa_2Cu_3O_z$.[8] Thus it is likely that the higher the oxygen content, the shorter is the c-axis length of $[La_{1/6}Ln_{1/3}Ba_{1/6}Sr_{1/6}Ce_{1/6}]_8Cu_6O_z$.

From the electrical resistivity measurements of samples of $[La_{1/6}Ln_{1/3}Ba_{1/6}Sr_{1/6}Ce_{1/6}]_8Cu_6O_z$ with Ln = Eu, Gd, Dy, Ho and Y annealed at two different levels of oxygen partial pressure of $P(O_2)$= 3 and 200 atm, only the samples with Ln = Eu, Gd and Dy, annealed at $P(O_2)$ = 3 atm, exhibited superconductivity with T_c's = 10-35 K. The samples with Ln = Eu and Gd also exhibited superconducting transitions at temperatures higher than 20 K according to ac magnetic susceptibility measurements. Those samples with Ln = Eu, Gd, Dy, Ho and Y which were annealed at $P(O_2)$ = 200 atm exhibited superconductivity with T_c's = 5-40 K, while the samples with Ln = Er and Tm had no superconducting transitions down to 4 K.

Figure 2 shows the relationship between T_c and the ionic radius of Ln^{3+}. Two curves were drawn in Fig. 2 for the two different groups of samples prepared at two different levels of oxygen partial pressure. It is shown that the superconducting transition temperatures for the samples annealed at $P(O_2)$ = 200 atm are higher than those of the samples annealed at $P(O_2)$ = 3 atm and that both curves have maximum at Ln = Gd. Similar relationships between T_c and the ionic radius of Ln^{3+} were observed in the $La_{0.8}Ln_{1.0}Sr_{0.2}CuO_y$ systems with Ln = Sm, Eu, Gd, Dy, Tb and Y.[9] In these cases, only samples with Ln = Sm, Eu and Gd exhibited superconductivity. The highest T_c of 20 K was observed for the sample with Ln = Sm.

Taking an analogy to the previously determined crystal structure for $(La_{1/6}Gd_{1/3}Ba_{1/6}Sr_{1/6}Ce_{1/6})_8Cu_6O_z$ [4], one may consider that Ln atoms, that is, Eu, Gd, Dy, Ho or Y atoms, preferentially occupy the A_1 sites in the (Ln, Ce)$_4$(Ln, Ba)$_4$Cu$_6$O$_z$ structure. This crystal structure was shown in Fig.1 in ref[4]. It is likely that the ionic radius of the species at the A_1 sites affects the distance between the $Cu(2)O(3)_2$ plane and the O(4) plane. Therefore, it may be possible that the superconducting transition temperature is dependent on the distance of the $Cu(2)O(3)_2$ plane and the O(4) plane because of the relationship [10] between the superconducting transition temperature and the hole density on the $Cu(2)O(3)_2$ plane.

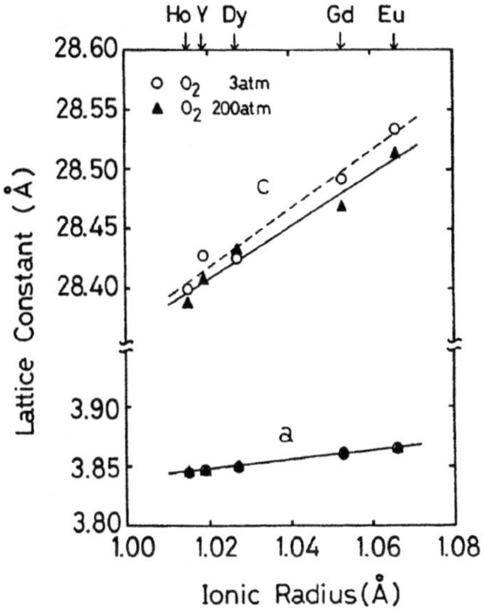

Fig.1. Lattice constants, a and c, vs. ionic radii of Ln^{3+} for $[La_{1/6}Ln_{1/3}Ba_{1/6}Sr_{1/6}Ce_{1/6}]_8Cu_6O_z$ (Ln = Eu, Gd, Dy, Ho and Y) annealed at $P(O_2)$ = 3 and 200 atm.

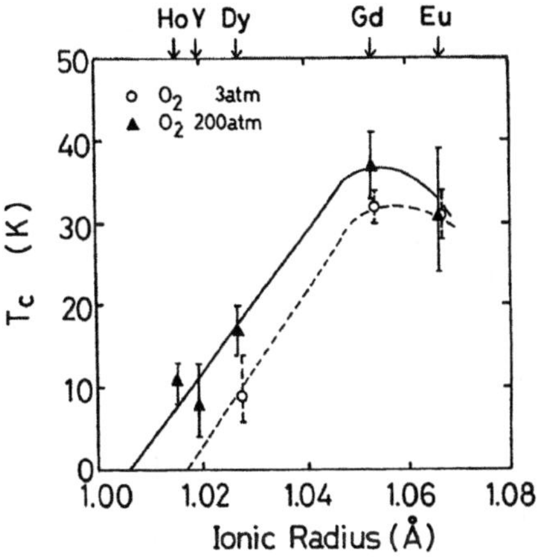

Fig.2. Relationship between superconducting transition temperature, T_c, and ionic radius of Ln^{3+}, for $[La_{1/6}Ln_{1/3}Ba_{1/6}Sr_{1/6}Ce_{1/6}]_8Cu_6O_z$ (Ln = Eu, Gd, Dy, Ho and Y) annealed at $P(O_2)$ = 3 and 200 atm.

2. $[La_{1/6}Gd_{1/3}Ce_{1/6}(La_xBa_{1-x})_{1/3}]_8Cu_6O_z$

X-ray diffraction patterns of $(Gd_{2/3}Ce_{1/3})_4[La_{1/3}(La_xBa_{1-x})_{2/3}]_4Cu_6O_z$ indicated that all the samples were of near single phase. The samples with x = 0, 0.1 and 0.2 contained $BaCeO_3$ as a minor secondary phase. The samples with x = 0, 0.1 and 0.5 were tetragonal, and those with x = 0.2, 0.3 and 0.4 were orthorhombic. Figure 3 shows the relationship between La content, x, and the lattice constants, a, b and c, in $(Gd_{2/3}Ce_{1/3})_4[La_{1/3}(La_xBa_{1-x})_{2/3}]_4Cu_6O_z$. The c-axis linearly decreased from 28.52 to 28.34 Å when x increased from 0 to 0.5. Such a substitution effect on the length of c-axis may be explained by the difference in the ionic radii of Ba^{2+}(1.52 Å) and La^{3+}(1.27 Å).[6] For x =0.2, the a-b splitting occurred and the difference, a-b , became smaller with increasing x. For x = 0.5, the a-b spilliting disappeared. The La content, x, at which a tetragonal-orthorhomobic structure phase transition occurred is in good agreement with the point where superconductivity disappeared.

The temperature dependence of electrical resistivity of $(Gd_{2/3}Ce_{1/3})_4[La_{1/3}(La_xBa_{1-x})_{2/3}]_4Cu_6O_z$ is shown in Fig. 4. The samples with x = 0 and 0.1 were superconductors with T_c's = 30 and 25 K , respectively. In the normal state, these samples were semiconductive. The samples with x = 0.2, 0.3, 0.4 and 0.5 were non-superconductors and exhibited a temperature dependence of a semiconductor type. The magnitude of resistivity became larger with increasing x. The superconductivity in $(Gd_{2/3}Ce_{1/3})_4[La_{1/3}(La_xBa_{1-x})_{2/3}]_4Cu_6O_z$ disappeared even when a small amount of La was added. The transition temperature estimated from the ac magnetic susceptibility measurements was a little lower than the superconductivity onset temperature (T_c^{on}) determined from electrical resistivity measurements.

The oxygen content, z, in $(Gd_{2/3}Ce_{1/3})_4[La_{1/3}(La_xBa_{1-x})_{2/3}]_4Cu_6O_z$ increased from 17.5 to 18.2 with increasing the La content, x. Such a behavior of increasing z with increasing x were not observed in high-quality $La_{1+x}Ba_{2-x}Cu_3O_z$ samples.[5] This is likely due to the fact that the Cu(1)-O between Ba sites in the $La_{1+x}Ba_{2-x}Cu_3O_z$ system has a chain structure, while the Cu(1)-O between Ba sites in $(Gd_{2/3}Ce_{1/3})_4[La_{1/3}(La_xBa_{1-x})_{2/3}]_4Cu_6O_z$ does not form a chain structure and oxygen vacancies are randomly distributed.

The average copper valence which was calculated from the La content, x, and oxygen content, z, was higher than 2.0 for all samples, and slightly increased with increasing x from 0 to 0.2, and decreased with increasing x from 0.3 to 0.5. On the other hand, the superconducting temperature was lowered as x increased from 0 to 0.2 and superconductivity disappeared for the samples with $x \gtrsim 0.2$. This situation may be explained as follows. As x increased, the oxygen content, z, increased.[11] However, since the oxygen atoms introduced by the La doping are likely to occupy half-filled O(1) sites. Then those oxygen atoms may have little influences on the valence of Cu(2) ions which may be responsible for the occurrence of superconductivity. However, the doped La ions may work directly for reducing the valence of Cu(2) ions such that T_c be lowered.

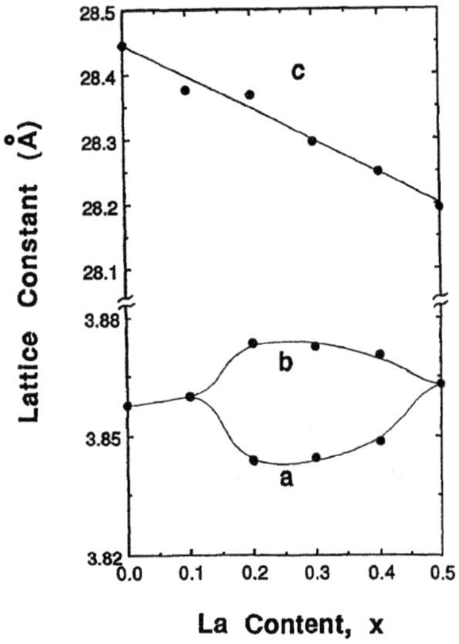

Fig.3. Relationship between La content, x, and lattice constant for $[La_{1/6}Gd_{1/3}Ce_{1/6}(La_xBa_{1-x})_{1/3}]_8Cu_6O_z$.

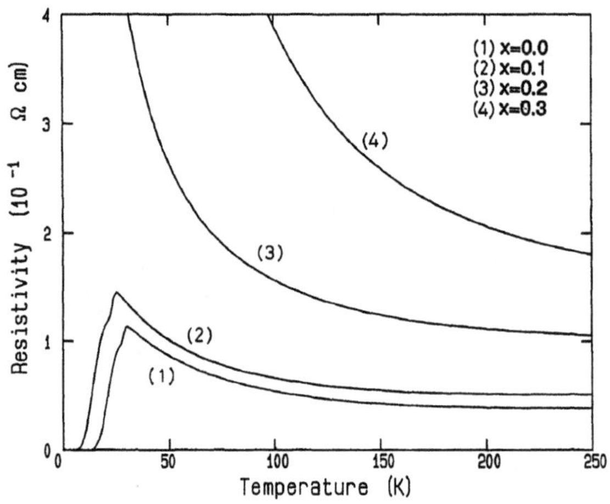

Fig.4. Temperature dependence of electrical resistivity of $[La_{1/6}Gd_{1/3}Ce_{1/6}(La_xBa_{1-x})_{1/3}]_8Cu_6O_z$.

REFERENCES

1. Sawa H, Obara K, Akimitsu J, Matsui Y, Horiuchi S (1989) J.Phys.Soc.Jpn. 58: 2252-2255
2. Ichinose A, Wada T, Yaegashi Y, Yamauchi H, Tanaka S (1989) Jpn.J.Appl. Phys. 28: L1765-L1768.
3. Wada T, Ichinose A, Yaegashi Y, Yamauchi H, Tanaka S (in press) Phys.Rev.B.
4. Wada T, Ichinose A, Yaegashi Y, Yamauchi H, Tanaka S (1989) Jpn.J.Appl. Phys. 28: L1779-L1782.
5. Wada T, Suzuki N, Maeda A, Yabe T, Uchinokura K, Uchida S, Tanaka S (1989) Phys.Rev.B. 39: 9126-9138.
6. Shannon DR (1976) Acta Crystallogr. A32: 751.
7. Asano H, Takita K, Katoh H, Akinaga H, Ishigaki T, Nishino M, Imai M, Masuda K (1987) Jpn.J.Appl.Phys. 26: L1410-L1412.
8. Ono A, Ishizawa Y (1987) Jpn.J.Appl.Phys. 26: L1043-L1045.
9. Muromachi TE, Uchida Y, Kobayashi M, Kato K (1989) Physica C.158: 449-452.
10. Tokura Y, Torrance BJ, Huang CT, Nazzal IA (1988) Phys.Rev.B. 38: 7156.
11. Ichinose A, Wada T, Yaegashi Y, Yamauchi H, Tanaka S: unpublished work at SRL-ISTEC.

High-Resolution Transmission Electron Microscopy of $Y_{1-x}Ca_xBa_2Cu_4O_8$ and Its Defects

K. YAMAGUCHI, T. MIYATAKE, T. TAKATA, S. GOTOH, and N. KOSHIZUKA

Superconductivity Research Laboratory, International Superconductivity Technology Center, 10-13, Shinonome 1-chome, Koto-ku, Tokyo, 135 Japan

ABSTRACT

The micro-structures of $Y_{1-x}Ca_xBa_2Cu_4O_8$ crystals are studied by high-resolution transmission electron microscopy (HRTEM). It is confirmed that no significant difference exists between the crystal structures of Ca-substituted (x=0.1) and Ca-free (x=0) crystals. In both x=0 and x=0.1 materials, there are no twin structures which are observed in $YBa_2Cu_3O_y$. There exist some planar defects with the absence of one chain in the double Cu-O chains. In addition, micro-domains where the directions of the c-axis are different each other are observed in some grains of Ca-substituted crystals.

KEY WORDS: HRTEM, Crystal Defects, $Y_{1-x}Ca_xBa_2Cu_4O_8$

INTRODUCTION

$YBa_2Cu_4O_8$ (1-2-4) is known as a superconductor with $T_c \simeq 80K$ and double Cu-O chains as shown in Fig.1. The crystal structure of this material is examined by X-ray and neutron diffraction [1,2,3] and it is confirmed to be similar to that of $YBa_2Cu_3O_y$ (1-2-3).

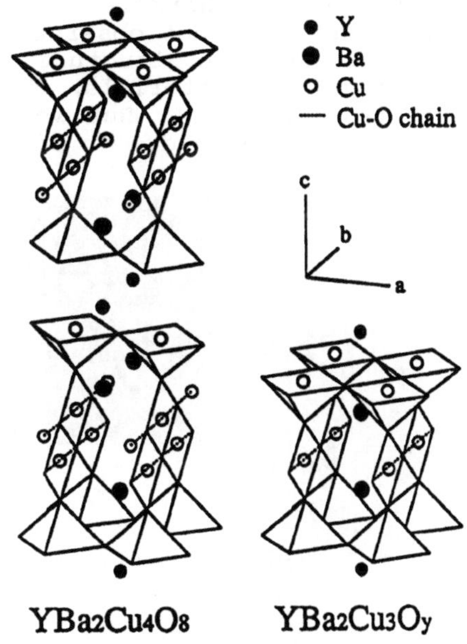

Fig.1 The schematic view of the crystal structures for 1-2-4 and 1-2-3

Ca substitution for Y causes increasing of T_c from 83K to 92K according to Ca content x [4]. In addition, this 1-2-4 material has two features for practical applications. One is excellent stability of the oxygen content and another is no orthorhombic-tetragonal phase transition unlike 1-2-3 material [5]. Powder X-ray diffraction measurements showed that there is no significant difference between substituted and non-substituted materials. So the crystal structure of Ca-substituted material is thought to be same as non-substituted one. Micro-structure of non-substituted 1-2-4 was studied by HRTEM [3,6], and it was shown that there is no twin structure which is always observed in 1-2-3. So, it is of interest to study the micro-structures of 1-2-4 in the context of the research on the pinning mechanism. In this paper, a study of micro-structure of $Y_{1-x}Ca_xBa_2Cu_4O_8$ is reported.

EXPERIMENTAL

Bulk samples of $Y_{1-x}Ca_xBa_2Cu_4O_8$ (x=0-0.1) were synthesized by solid state reaction with oxygen-HIP treatment [4]. It was confirmed by X-ray diffraction that all these samples were of single-phase without other phase like $YBa_2Cu_3O_y$. The Ca contents of the samples were evaluated by inductively coupled plasma atomic emission spectroscopy, and they were in good agreement with the nominal contents. It was confirmed that the Ca contents in every grains were equal to each other by the electron probe micro analysis . The critical temperatures T_c were measured by both resistivity and magnetization measurements, and they were 83K and 92K for x=0 and 0.1, respectively.

The micro-structure of $Y_{1-x}Ca_xBa_2Cu_4O_8$ was examined by HRTEM for Ca contents with x=0 and x=0.1. Samples for HRTEM were prepared by crushing these single-phase samples and dispersed with carbon tetrachloride on holey carbon films. HRTEM observations were done by a transmission electron microscope (JEM-4000EX) having a point resolution of 0.17nm and the operation was performed at 400kV.

RESULTS AND DISCUSSION

Figure 2 is a low magnification TEM image of x=0 material (a), together with the corresponding electron diffraction pattern (b). The diffraction pattern shows that the image was taken with the incident electron beam along the c-axis. From these image and diffraction pattern, it turns out that there is no twin structure in this grain. Although the twin structures are usually observed in $YBa_2Cu_3O_y$ material, they were not observed for more than 10 grains in the non-substituted 1-2-4.

Figure 3 is the TEM image and diffraction pattern of x=0.1 material taken under the same condition as in Fig.2. As the same diffraction patterns are seen in Figs. 2 and 3, the crystal structure is thought to be unchanged by Ca substitution. Furthermore, the twin structure is not seen in Fig.3. The twin structure was not observed for more than 10 grains. Thus, it is concluded that the twin structure observed in $YBa_2Cu_3O_y$ does not exist in Ca substituted 1-2-4 materials. This results agree with a report for non-substituted 1-2-4 [3]. But we can see complicated contrast in Figs.2 and 3, which is regarded as due to dislocations.

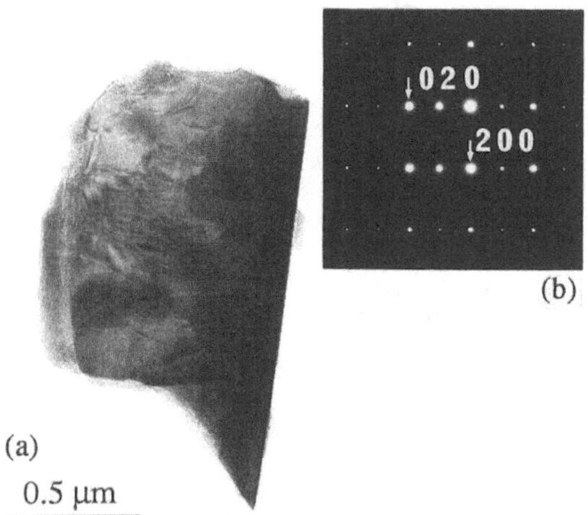

Fig.2 TEM image of x=0 material (a), and the electron diffraction pattern (b), taken with the incident beam along the c-axis.

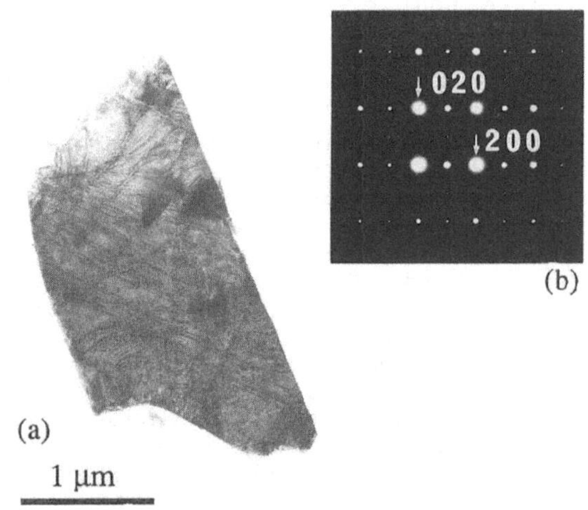

Fig.3 TEM image of x=0.1 material (a), and the electron diffraction pattern (b), taken with the incident beam along the c-axis.

A high-resolution image of x=0 material taken with the incident electron beam along the a-axis is shown in Fig.4-(a), together with the electron diffraction pattern (b). In Fig.4-(a), a planar defect, with the absence of one of the two chains in double Cu-O chains, is observed as marked by arrow [6]. In Fig.4-(c), a HRTEM image of the perfect crystal of 1-2-4 is simulated based on the crystal structure model of Fig.1. From this simulation, the dark spots in the image can be regarded as Ba atoms. Thus, the structure of the planar defects is ascribed to the absence of one Cu-O chain.

Figure 5 is the high-resolution image and the corresponding diffraction pattern of x=0.1 material. This image is similar to the region without the defects in Fig.4. The diffraction patterns in Figs.4 and 5 show good agreement with each other. Thus, the crystal structure of 1-2-4 is considered to be unchanged by Ca substitution for Y between x=0 and 0.1 as mentioned above.

In Fig.6 of the high-resolution image of Ca-substituted 1-2-4, we can find planar defects marked by arrows. The image contrasts of the planar defects for non-substituted and substituted 1-2-4 show a good agreement, so the structures of the planar defects are thought to be the same.

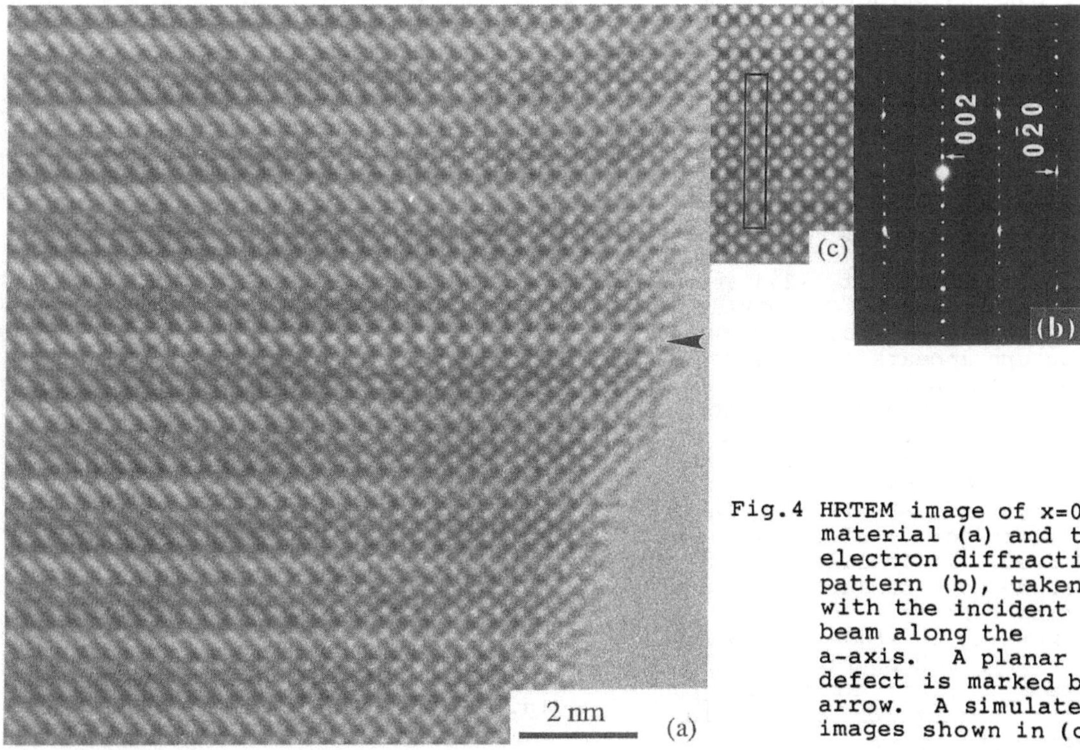

Fig.4 HRTEM image of x=0 material (a) and the electron diffraction pattern (b), taken with the incident beam along the a-axis. A planar defect is marked by arrow. A simulated images shown in (c).

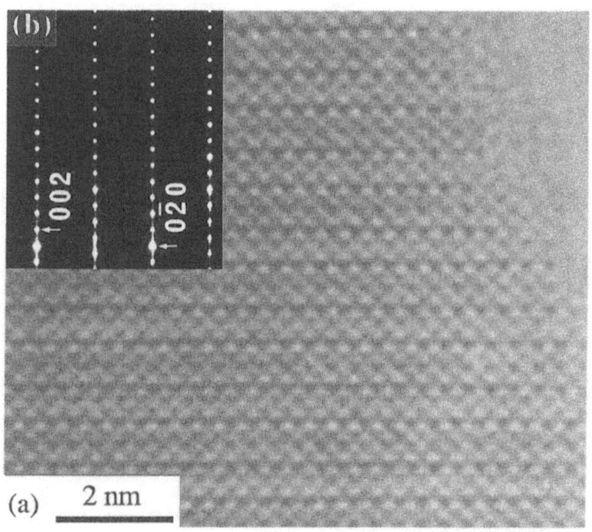

Fig.5 HRTEM image of x=0.1 material of the region without defects (a), and the electron diffraction pattern (b), taken with the incident beam along the a-axis.

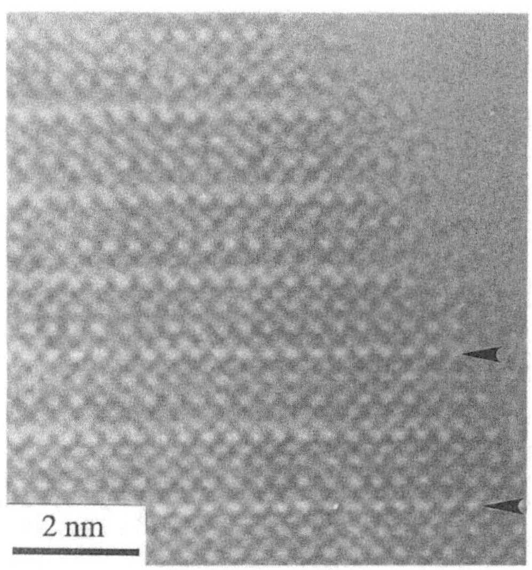

Fig.6 HRTEM image of x=0.1 material with planar defects marked by arrows.

Figure 7 shows a micro-domain structure in x=0.1 material. The c-axes are perpendicular to each other between two domains. From the image at the grain boundaries, the boundaries are thought to be due to Cu-O chains. In 1-2-3, such domain-structures have been reported [7], but the structure of the boundary is not clear. Although we tried to observe for more than 10 grains of non-substituted materials, we could not observe such kinds of micro-domains. Thus, these structures might be specific ones observed in Ca-substituted material.

In conclusion, the crystal- and micro-structures of 1-2-4 seem to be unchanged by Ca-substitution although a micro-domain structure appears in Ca-substituted materials. In 1-2-3 materials, it is thought that the twin structures act as flux pinning centers. Although such twins can not be observed in 1-2-4 materials, planar defects and micro-domain structures are observed. It is likely that these defects can act as flux pinning centers. So, it is interesting to study what kind of defects act as flux pinning centers by using various samples with different Jc. We are in progress to observe the micro-structures of samples prepared by various processing conditions.

Fig.7 HRTEM image of a micro-domain structure in x=0.1 material. White spot lines represent Cu-O chains.

REFERENCES

[1] P.Marsh, R.M.Fleming, M.L.Mandich, A.M.DeSantolo, J.Kwo, M.Hong and L.J.Martinez-Miranda : Nature 334 (1988) 141

[2] P.Fischer, J.Karpinski, E.Kaldis, E.Jilek and S.Rusiecki : Solid State Commun. 69 (1989) 531

[3] E.Kaldis, P.Fischer, A.W.Hewat, E.A.Hewat, J.Karpinski and S.Rusiecki : Physica C 159 (1989) 668

[4] T.Miyatake, S.Gotoh, N.Koshizuka and S.Tanaka : Nature 341 (1989) 41

[5] J.Karpinski, E.Kaldis, E.Jilek, S.Rusiecki and B.Bucher : Nature 336 (1988) 660

[6] K.Yamaguchi, T.Miyatake, T.Takata, S.Gotoh, N.Koshizuka and S.Tanaka : Jpn.J.Appl.Phys. 28 (1989) 1942

[7] Y.Matsui, Y.Kitami, M.Yokoyama, N.Iyi, E.Takayama-Muromachi and S.Takekawa : J.Electron Microsc. 36 (1987) : 246

Synthesis and Crystal Structure of a New Family of Superconductors (Tl, Pb) (R, Sr)$_2$CuO$_5$(R=La, Nd)

T. NAGASHIMA[1], M. WATAHIKI[1], Y. FUKAI[1] T. MOCHIKU[2], and H. ASANO[2]

[1] Faculty of Science and Engineering, Chuo University, Bunkyo-ku, Tokyo, 112 Japan
[2] Institute of Materials Science, University of Tsukuba, Tsukuba, Ibaraki, 305 Japan

ABSTRACT

We have discovered a new family of superconductors (Tl,Pb)(R,Sr)$_2$CuO$_5$(R=La,Nd) with transition temperature of about 40K. The crystal structure of TlLaSrCuO$_5$ was determined by Rietveld analysis of powder X-ray diffraction data. The results indicate Cu-substitution for the Tl site and/or deficiency in the Tl site, and location of La and Sr on the Sr site in the Tl-O monolayer TlSr$_2$CuO$_5$ compound: the compounds have general formulas Tl$_{1-x}$LaSrCu$_{1+x}$O$_5$ and Tl$_{1-x}$LaSrCuO$_5$. In both systems, superconducting properties were improved by Cu-substitution for the Tl site and deficiency of the Tl site.

KEYWORDS : superconductivity, TlLaSrCuO$_5$, (Tl,Pb)(Nd,Sr)$_2$CuO$_5$, powder X-ray diffraction, Rietveld analysis

INTRODUCTION

Recently, superconductivity was observed in the Tl-Sr-Ca-Cu-O system[1][2]. It was found that by doping with Pb and/or rare-earth elements (La, Nd and Y) TlSr$_2$CaCu$_2$O$_7$ is stabilized the crystal structure and its superconducting characteristics are improved[3][4]. Although the general formula of this system are described by TlSr$_2$Ca$_{n-1}$Cu$_n$O$_{3+2n}$ [5][6], it is difficult to synthesize a single-phase of TlSr$_2$CuO$_5$ corresponding to n=1. Recently, superconductivity has been reported for analogous mixed oxides Tl(Ba,La)$_2$CuO$_5$[7] and (Tl,Pb)(R,Sr)$_2$CuO$_5$ (R=Pr, Nd and Sm)[8][9]. Also, we have discovered a new phase of TlLaSrCuO$_5$ with partial substitution of Cu for the Tl site[10][11]. In this paper, the superconductivity and the crystal structure of this new family of superconductors (Tl,Pb)(R,Sr)$_2$CuO$_5$(R=La,Nd) are reviewed.

EXPERIMENTAL

Samples were prepared by solid state reaction of Tl$_2$O$_3$ or Tl$_2$O , La$_2$O$_3$, SrCO$_3$ and CuO powders. Mixed powders with the nominal composition Tl : La : Sr : Cu =1-x : 1 : 1 : 1+x (x=0.0, 0.2, 0.4, 0.6) ,0.7 : 1 : 1 : 1 and 1: 2-x : x : 1 (x=0.0-0.9) were pressed into pellets of 25mm diameter and 1 or 2mm thickness. The samples were calcined and sintered in a covered alumina crucible at 880-900℃ for 5h in air.

RESULTS AND DISCUSSION

Powder X-ray diffraction patterns of Tl$_{1-x}$LaSrCu$_{1+x}$O$_5$ with x=0.0, 0.2, 0.4 and 0.6 are shown in Fig. 1. The samples are almost single phase; most of the diffraction lines can be indexed based on tetragonal unit cells with the space group P4/mmm, although the phase corresponding to the (La,Sr)$_2$CuO$_4$-type structure appears as x increases. Figure 2 shows temperature dependence of the electrical resistivity and magnetic susceptibility for the Tl$_{1-x}$LaSrCu$_{1+x}$O$_5$ system. Judging from the resistance change, the T_c of the sample with x=0.2 is the highest, and volume fraction of the superconducting phase in the sample with x=0.2 obtained

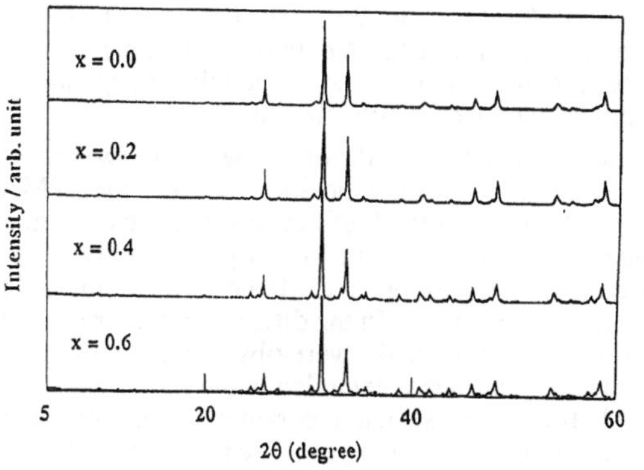

Fig. 1. Powder X-ray diffraction patterns of Tl$_{1-x}$LaSrCu$_{1+x}$O$_5$ with x=0, 0.2, 0.4, 0.6.

from the inductance change is the largest in the $Tl_{1-x}LaSrCu_{1+x}O_5$ system. The T_c determined by the magnetic susceptiblity study is consistent with the results of the electrical resistivity measurement. The EPMA results shown in Table I indicate that samples with x=0.4 and 0.6 have a composition with excess Cu and a deficiency of Tl, which strongly suggests Cu-substitution for the Tl site. The composition of the samples were found to be nearly homogeneous from scanning EPMA measurements.

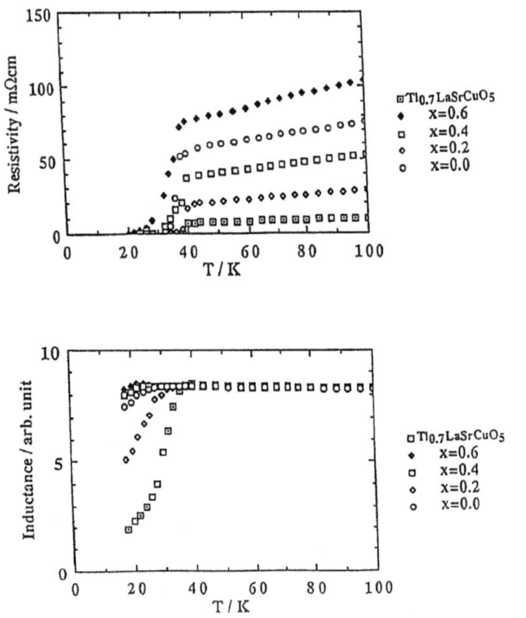

Fig. 2. Temperature dependence of (a) electrical resistivity and (b) magnetic susceptibility of the $(Tl,Cu)(La,Sr)_2CuO_5$ system.

Table I Chemical compositions determined by EPMA and T_c in the $Tl_{1-x}LaSrCu_{1+x}O_5$.

	nominal composition				EPMA result				T_c	
x	Tl	La	Sr	Cu	Tl	La	Sr	Cu	onset	zero
0	1	1	1	1	0.94	1.02	1.06	1.00	38	30
0.2	0.8	1	1	1.2	1.01	0.96	0.97	1.06	42	32
0.4	0.6	1	1	1.4	0.82	1.04	1.08	1.12	40	26
0.6	0.4	1	1	1.6	0.78	1.06	0.99	1.26	38	22

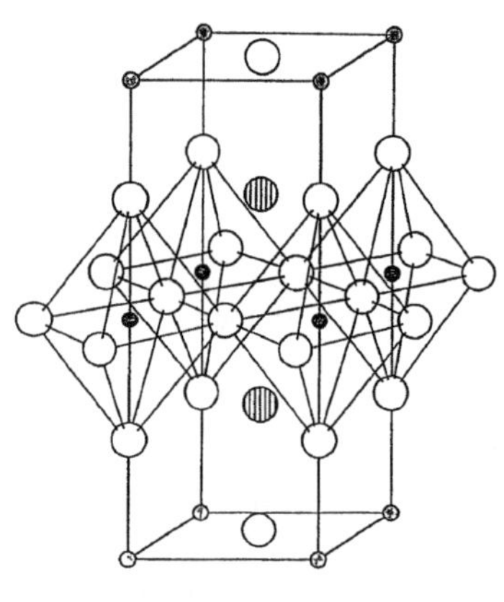

Fig. 3. Crystal structure of $(Tl,Cu)(La,Sr)_2CuO_5$. : (Tl,Cu), : (La,Sr), : Cu, : O.

X-ray diffraction data were analyzed by Rietveld refinement. Initial structure model is based on a $TlBa_2CuO_5$ structure with the space group P4/mmm[12]. The atomic coordinates are

La,Sr	in 2h	: 1/2 1/2 z ,	1/2 1/2 z
Tl,Cu(1)	in 1a	: 0 0 0	
Cu(2)	in 1b	: 0 0 1/2	
O(1)	in 1c	: 1/2 1/2 0	
O(2)	in 2e	: 0 1/2 1/2,	1/2 0 1/2
O(3)	in 2g	: 0 0 z ,	0 0 z.

We assumed from EPMA data that Cu atoms substitute for Tl atoms, and refined the occupation factors (g) under the constraint of $g_{Tl}+g_{Cu}=1$. Occupation factors for the other metal sites were fixed at the nominal composition, because the EPMA study showed that the La/Sr ratio was almost equal to the nominal ratio. Common isotropic thermal parameters were assigned to all the metal atoms and to all the oxygen atoms. In the diffraction pattern, since traces of impurity peaks were observed, intensity data in these regions were excluded in the refinement. Table II lists final structure parameters determined in the present Rietveld analysis. The result of the Rietveld analysis gives g_{Tl} smaller than the EPMA result, which indicates deficiency in the Tl site.

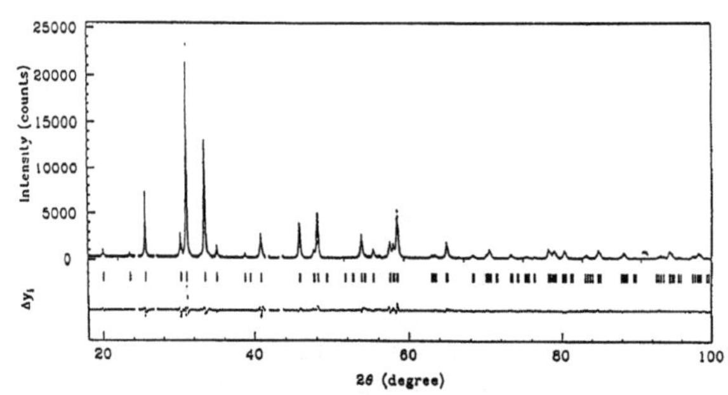

Fig. 4. Rietveld refinement pattern for $Tl_{0.7}LaSrCuO_5$. Crosses show step-scanned intensity data, and the solid line overlying them is the calculated pattern based on the structural model of Fig. 3. Positions of 128 Bragg reflections for $CuK\alpha_1$ and $K\alpha_2$ radiations are indicated by markers, and Δy_i is the difference between observed and calculated intensities.

Next, we prepared $Tl_{0.7}LaSrCuO_5$ which has a deficiency in the Tl site and no Cu substitution. The temperature dependence of the electrical resistivity and magnetic susceptibility is shown in Fig. 2. The result indicates that T_c^{onset} is 40K and T_c^{zero} is 37K. T_c of this sample is higher than that of $Tl_{1-x}LaSrCu_{1+x}O_5$ with the Cu-substitution for the Tl site. The X-ray diffraction data were analyzed by Rietveld refinement in the same manner as in the case of $Tl_{1-x}LaSrCu_{1+x}O_5$ except for disregarding of the possibility of Cu substitution for the Tl site. Figure 4 shows the Rietveld refinement pattern for $Tl_{0.7}LaSrCuO_5$, and final structure parameters obtained are listed in Table I.

Regarding the superconductivity in $Tl(La_{1-x}Sr_x)_2CuO_5$, the temperature dependence of the resistivity and inductance change is shown in Fig.5. The variation of the resistivity just above the critical temperature with composition x is shown in Fig.6. The resistivity decreases linearly with x between x=0.2 and 0.9. X-ray diffraction patterns of these materials are shown in Fig.7. The sample with x nearly equal to 1 has the same structure as $TlLaSrCuO_5$, but the samples with x much smaller than 1 are of multiple phases, containing diffraction lines of the $(La,Sr)_2CuO_4$-type structure. Thus, we consider that the superconductivity of $Tl(La_{1-x}Sr_x)_2CuO_5$ is in fact due to $(La,Sr)_2CuO_4$.

Table II Structure parameters in the $(Tl,Cu)(La,Sr)_2CuO_5$ system. g is the occupation factor and B is the isotropic thermal parameter in Å^2. Numbers in parentheses are standard deviations of the last significant digit. (†) : $Tl_{0.7}LaSrCuO_5$

x	0.0	0.2	0.4	0.6	(†)
a(Å)	3.7796(2)	3.7731(2)	3.7707(3)	3.7714(3)	3.7743(1)
c(Å)	8.8060(7)	8.8268(5)	8.8528(8)	8.8558(9)	8.8308(4)
$z_{La,Sr}$	0.289(1)	0.291(1)	0.290(2)	0.289(2)	0.292(1)
$z_{O(3)}$	0.22(1)	0.22(1)	0.20(1)	0.20(1)	0.23(1)
B_{metal}	1.3(3)	1.2(2)	1.1(4)	1.1(4)	1.2(2)
B_{oxygen}	3.2(19)	2.8(13)	2.0(20)	1.7(20)	1.8(9)
g_{Tl}	1	0.79(3)	0.79(3)	0.68(5)	0.64(6)
R_{wp}(%)	13.6	11.3	14.3	14.8	9.5
R_p(%)	10.1	8.3	11.0	11.3	7.3
R_I(%)	5.9	4.8	6.9	7.9	4.3
R_F(%)	4.0	3.8	5.7	6.5	3.5

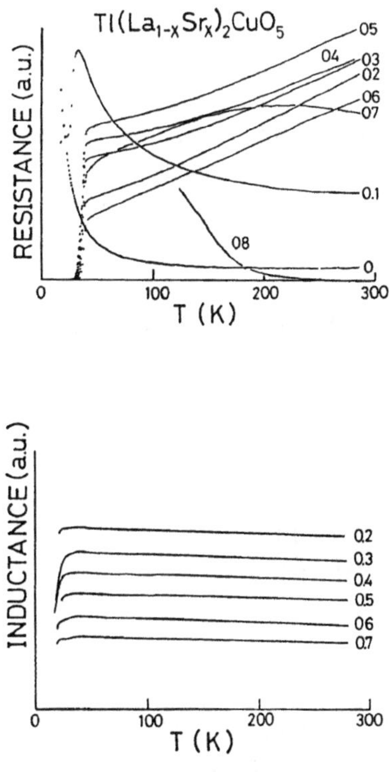

Fig. 5. The temperature dependence of the resistivity and magnetization of $Tl(La_{1-x}Sr_x)_2CuO_5$.

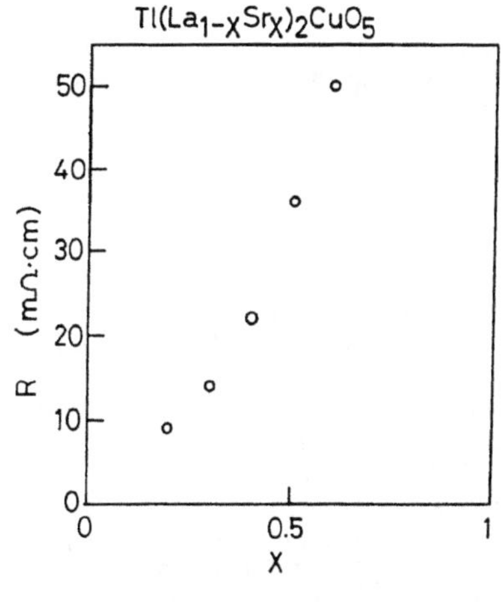

Fig. 6. x dependence of resistivity just above the critical temperature of $Tl(La_{1-x}Sr_x)_2CuO_5$.

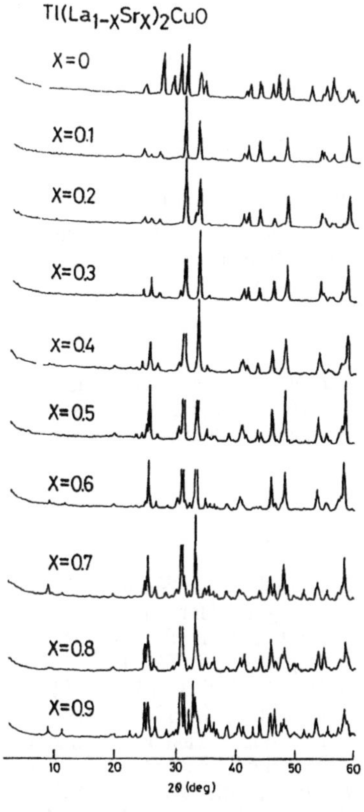

Fig. 7. Powder X-ray diffraction patterns of $Tl(La_{1-x}Sr_x)_2CuO_5$.

We also investigated samples doped with Nd and/or Pb. Figure 8 shows the temperature dependence of the resistivity and magnetization of $Tl_{0.7}Pb_{0.3}LaSrCu_{1.8}O_5$ and $Tl_{0.7}Pb_{0.3}NdSrCu_{1.8}O_5$. Judging from the resistance change, the onset temperature T_c of Pb-doped sample is higher than that of undoped one, although the volume fraction of a superconducting phase to be obtained from the inductance change is larger with the undoped one than that of the doped one. The sample with a nominal composition of $Tl_{0.7}Pb_{0.3}NdSrCu_{1.8}O_5$ was also a superconductor, as shown in Fig.9, while a Pb-undoped sample with a nominal composition $Tl_{0.7}NdSrCu_{1.8}O_5$ was an insulator.

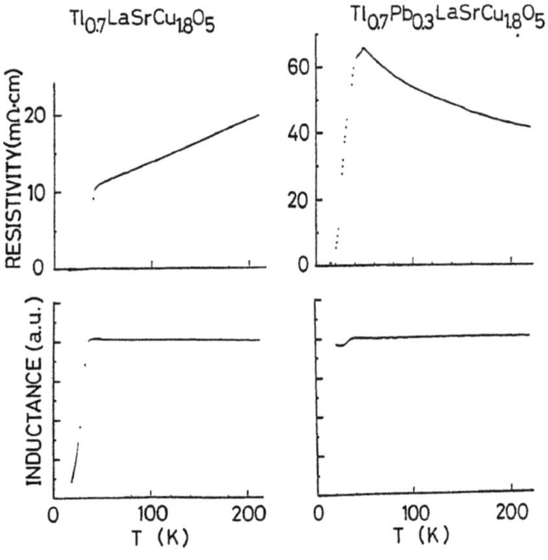

Fig. 8. The temperature dependence of the resistivity and magnetization of $Tl_{0.7}LaSrCu_{1.8}O_5$ and $Tl_{0.7}Pb_{0.3}LaSrCu_{1.8}O_5$.

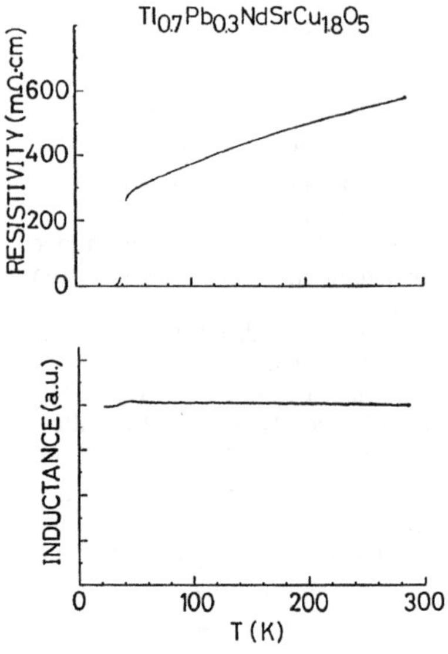

Fig. 9. The temperature dependence of the resistivity and magnetization of $Tl_{0.7}Pb_{0.3}NdSrCu_{1.8}O_5$.

CONCLUSIONS

In summary, we discovered a new family of superconductors $(Tl,Pb)(R,Sr)_2CuO_5$(R=La,Nd) with T_c of 40K. It is suggested that Cu-substitution for the Tl site and/or deficiency of Tl in the Tl site has great influence on the superconducting properties.

Acknowledgments

The authors would like to acknowledge the support of the Chemical Analysis Center, University of Tsukuba for EPMA measurements. This work was partly supported by a Grant-in-Aid for Scientific Research from the Ministry of Education, Science and Culture of Japan.

References

1. Nagashima T, Watanabe K, Saito H, Fukai Y (1988) Jpn. J. Appl. Phys. 27: L1077-L1079.
2. Sheng ZZ, Hermann AM, Vier DC, Schultz S, Oseroff SB, George DJ, Hazen RM (1988) Phys. Rev. B38: 7074-7076.
3. Nagashima T, Watahiki M, Fukai Y (1989) Jpn. J. Appl. Phys. 28: L930-L933.
4. Itoh T, Uchikawa H (1989) Jpn. J. Appl. Phys. 28: L591-L594.
5. Hayri EA, Greeblatt M (1988) Physica C156: 775-780.
6. Matsuda S, Takeuchi S, Soeta A, Suzuki T, Aihara K, Kamo T (1988) Jpn. J. Appl. Phys. 27: 2062-2064.
7. Manako T, Shimakawa Y, Kubo Y, Satoh T, Igarashi H (1989) Physica C158: 143-147.
8. Itoh T, Uchikawa H (1989) Jpn. J. Appl. Phys. 28 : L200-L202.
9. Adachi S, Inoue O, Hirano H, Takahashi Y, Kawashima S (1989) Jpn. J. Appl. Phys. 28: L775-L777.
10. Nagashima T, Watahiki M, Fukai Y, Mochiku T, Asano H: proceedings of Tsukuba Seminor on High T_c Superconductivity, Tsukuba, Ibaraki, June, 1989, pp217-220.
11. Mochiku T, Nagashima T, Watahiki M, Fukai Y, Asano H (1989) Jpn. J. Appl. Phys. 28 No.11 in press.
12. Izumi F (1985) J. Crystallogr. Soc. Jpn.23: 27 [in Japanese].

Superconductivity in the $Bi_2(Ln, Ln^*)_xCa_{3-x}Cu_2O_y$ (Ln, Ln*=Lanthanides and Y) Systems

T. IKEMACHI, T. KAWANO, F. MUNAKATA, A. NOZAKI, and H. YAMAUCHI

Superconductivity Research Laboratory, International Superconductivity Technology Center, 10-13, Shinonome 1-chome, Koto-ku, Tokyo, 135 Japan

ABSTRACT

New superconductors, $Bi_2Nd_xCa_{3-x}Cu_2O_y$ and $Bi_2Pr_xCa_{3-x}Cu_2O_y$ which do not contain Sr, have been successfuly synthesized. The powder x-ray diffraction measurements showed that they were of either single or near single phase having the "2212" structure. *Tc*'s for the samples sintered in O_2 gas flow were about 10K higher than those for the samples sintered in air. It was likely that the increase in the hole concentration due to the introduction of oxygen enhanced the *Tc*. It was also discovered that, a number of compounds of $Bi_2(Ln,Ln^*)_xCa_{3-x}Cu_2O_y$(Ln, Ln*=lanthanides and Y) were superconducting, and the magnitude of *Tc* depended on the average ionic radius of Ln and Ln*. The highest *Tc* in these compounds was obtained at the composition of $Bi_2La_{0.25}Pr_{0.25}Ca_{2.5}Cu_2O_y$: *Tc(onset)* and *Tc(endpoint)* were 65K and 45K, respectively.

KEY WORDS: Sr free Bi cuprate, Bi-(Ln,Ln*)-Ca-Cu-O, O_2 gas flow sintering

INTRODUCTION

Since the discovery of superconductivity in the series of $Bi_2Sr_2Ca_{n-1}Cu_nO_y$[1,2] with n=1, 2 and 3, a number of studies have been reported on these Bi cuprates and other similar cuprates. By the replacement of Tl and Ba for Bi and Sr in $Bi_2Sr_2Ca_{n-1}Cu_nO_y$, $Tl_2Ba_2Ca_{n-1}Cu_nO_y$ were synthesized[3,4]. It is well known that those Tl cuprates have higher *Tc*'s than the corresponding Bi cuprates. The partial substitution of Pb or (Pb,Sb) for Bi was effective to enhance the isolation of the "2223" phase[5]. There are also many studies on the substitutions of a variety of elements for Sr and Ca. The partial substitution of Ln(=Y, Nd, Eu, and Gd) for Ca in $Bi_2Sr_2Ca_{1-x}Ln_xCu_2O_y$ resulted in a decrease of *Tc*. Samples with x>0.5 were antiferromagnetic and insulating[6]. On the contrary, the substitution of Ln(=La, Pr and Nd) for Sr in $Bi_2Sr_{2-x}Ln_xCuO_y$ has been found to raise *Tc* from 8K to 20K[6]. Another series of superconductors having high *Tc*'s of about 80K has been discovered by substituting Ln(=Pr, Nd, Sm, Eu, Gd, Tb, Dy, Ho, Er, Tm and Y) for Sr in $Bi_2Sr_{3-x}Ln_xCu_2O_y$[7,8]. Bi cuprate superconductors not containing Sr were found by Inoue et al[9]. In a sample of the nominal composition of $Bi_2La_{0.5}Ca_{3.5}Cu_3O_y$, *Tc(onset)* and *Tc(endpoint)* were observed around 50K and 21K. The crystallographic structure of this compound was the same as that of $Bi_2Sr_2CaCu_2O_y$(or "2212" phase), but the samples were not of singhe phase.

In this paper, we report on the superconducting properties of novel superconductors, $Bi_2Pr_xCa_{3-x}Cu_2O_y$ and $Bi_2Nd_xCa_{3-x}Cu_2O_y$. The powder x-ray diffraction patterns for the compounds of the both systems were identical to that for a "2212" phase and did not contain peaks due to any secondary phases. Furthermore $Bi_2(Ln,Ln^*)_xCa_{3-x}Cu_2O_y$ with two lanthanide elements exhibited superconductivity. A relationship between *Tc*'s and ionic radii of lanthanide elements was found.

EXPERIMENTAL

Samples were synthesized by a conventional solid-state reaction method. The starting materials for the samples were Bi_2O_3, $CaCO_3$, CuO, La_2O_3, CeO_2, Pr_6O_{11}, Nd_2O_3, Sm_2O_3, Eu_2O_3, Gd_2O_3, Tb_4O_7, Dy_2O_3, Ho_2O_3, Er_2O_3, Tm_2O_3, Yb_2O_3, Lu_2O_3 and Y_2O_3 powders of purity higher than 99.9%. The starting powders weighed to an appropriate ratio were mixed. The mixed powder was pressed into pellets and calcined at 800°C for in 10 hours in air. The pellets were reground and pressed into bars of 3X3X20 mm^3. The bars were sintered at temperature in the range of 830°~840°C for 35 hours in air or in an oxygen atomosphere. (1) The samples sintered in air were cooled to room temperature at a rate of 200°C/hour. (2) The samples sintered in O_2 were cooled to 500°C at a rate of 60°C/hour, and then annealed at 500°C for 10 hours in O_2 and cooled to room temperature at the same time. The electrical resistivity was measurued by a standard four-probe method. Magnetic susceptibility was measured in a magnetic field of H=5 Oe using a SQUID magnetometer(QUANTUM DESIGN: model MPM). The crystal structures of samples were examined by powder x-ray diffraction using CuKα radiation over a 2θ range from 3° to 65°.

RESULTS and DISCUSSION

Fig.1(a) and (b) show the temperature dependences of resistivity for $Bi_2Nd_xCa_{3-x}Cu_2O_y$ and $Bi_2Pr_xCa_{3-x}Cu_2O_y$ samples sintered at 840°C in air. Superconducting transitions are observed for the samples with the composition x, between 0.2 and 0.7. As seen in Fig.1(a) the highest transition temperature for the $Bi_2Nd_xCa_{3-x}Cu_2O_y$ system is obtained at x=0.3, and the onset and endpoint of the transition(*Tc(onset)* and *Tc(endpoint)*) are 60K and 24K, respectively. The composition dependence of superconductivity in $Bi_2Pr_xCa_{3-x}Ca_2O_y$ is similar to the case of Bi-Nd-Ca-Cu-O, and the *Tc(onset)* and *Tc(endpoint)* for the sample with x=0.4 are 60K and 19K, respectively. Additionally we carried out the substitution of other lanthanide elements, *i.e.* Ce, Sm, Eu, Gd, Tb, Dy, Ho, Er, Tm, Yb, Lu and Y, for Ln in $Bi_2Ln_{0.3}Ca_{3.7}Cu_2O_y$ by sintering at 830°C for 35 hours in air, but all the samples showed no superconducting transitions.

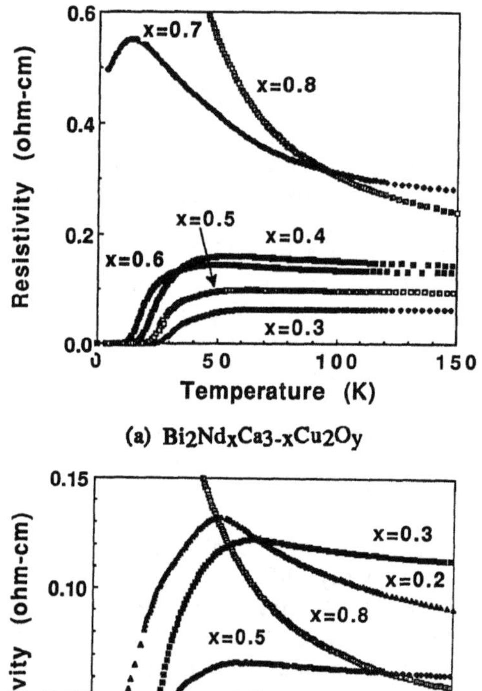

(a) $Bi_2Nd_xCa_{3-x}Cu_2O_y$

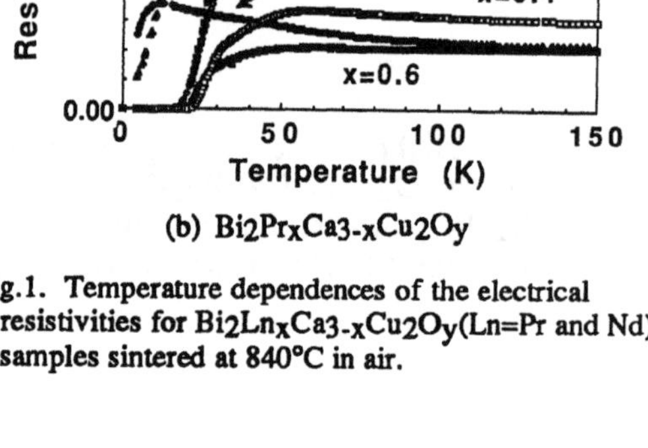

(b) $Bi_2Pr_xCa_{3-x}Cu_2O_y$

Fig.1. Temperature dependences of the electrical resistivities for $Bi_2Ln_xCa_{3-x}Cu_2O_y$(Ln=Pr and Nd) samples sintered at 840°C in air.

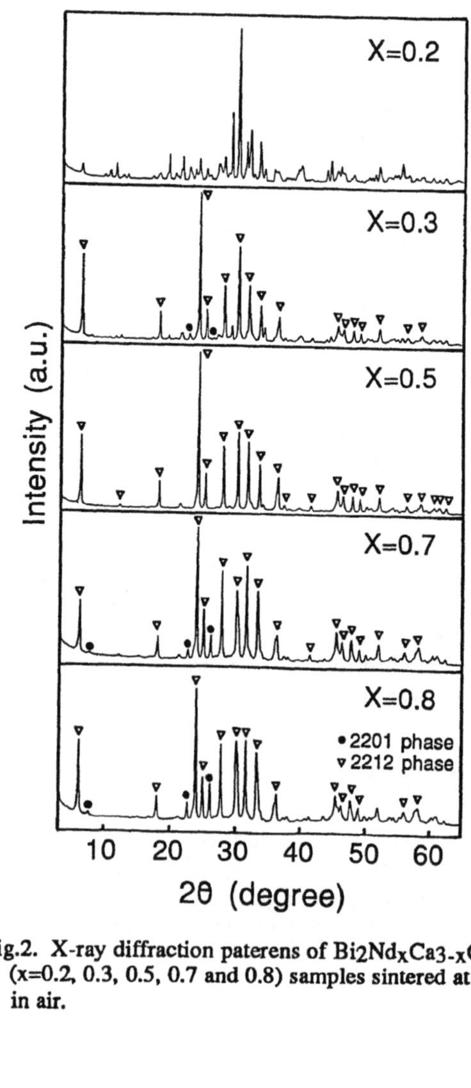

Fig.2. X-ray diffraction paterens of $Bi_2Nd_xCa_{3-x}Cu_2O_y$ (x=0.2, 0.3, 0.5, 0.7 and 0.8) samples sintered at 840°C in air.

Fig.2 shows the x-ray diffraction patterns of $Bi_2Nd_xCa_{3-x}Cu_2O_y$ samples sintered in air. The pattern for the sample with x=0.2 has many peaks, but peaks for the "2212" phase are not clearly seen. The pattern for the sample with x=0.3 includes peaks of both the "2212" and "2201" phases. The sample with x=0.5 exhibits a single phase pattern of the "2212" phase[10]. For the compounds with x=0.7 and 0.8, the existence of the "2201" phase is apparent. The diffraction patterns for $Bi_2Pr_xCa_{3-x}Cu_2O_y$ also depended on the value of x in a manner as the case of $Bi_2Nd_xCa_{3-x}Cu_2O_y$[10].

The effect of oxygen pressure in the preparation process of these superconductors was studied and the results are given in Fig.3 and Table1. Fig.3 shows the temperature dependence of resistivity for $Bi_2Nd_{0.5}Ca_{2.5}Cu_2O_y$ samples sintered both in air and in O_2 gas flow atomosphere. It is observed that the superconducting transition temperature for the sample sintered in O_2 gas flow is about 10K higher than that for the sample sintered in air. More generally, this relation was held for two samples sintered in O_2 gas flow and in air with the same value of x in $Bi_2Pr_xCa_{3-x}Cu_2O_y$ as well as in $Bi_2Nd_xCa_{3-x}Cu_2O_y$. Although *a*-axis remains nearly constant, *c* axis depends on the sintering atomosphare. In comparison with $Bi_2Sr_2CaCu_2O_y$, the *c*-axis becomes shorter. This shrinkage of the *c* axis can be explained by the change in the radius of ions which occupy the "Sr sites".

Table1. Lattice parameters(*a*,*b*,*c*) of $B1_2Nd_{0.5}Ca_{2.5}Cu_2O_y$ and $Bi_2Sr_2CaCu_2O_y$

parameter	$Bi_2Nd_{0.5}Ca_{2.5}Cu_2O_y$		$Bi_2Sr_2CaCu_2O_y$ #)
	sintered in air	sintered in O_2	
a (Å)	5.407	5.403	5.399
b (Å)	5.407	5.403	5.396
c (Å)	29.85	29.87	30.76

#): $Bi_2Sr_2CaCu_2O_y$ was prepared by a conventional solid state reaction method. (calcination: 800°C, 10 hours, in air, 2 times; sintering: 840°C, 70 hours, in air)

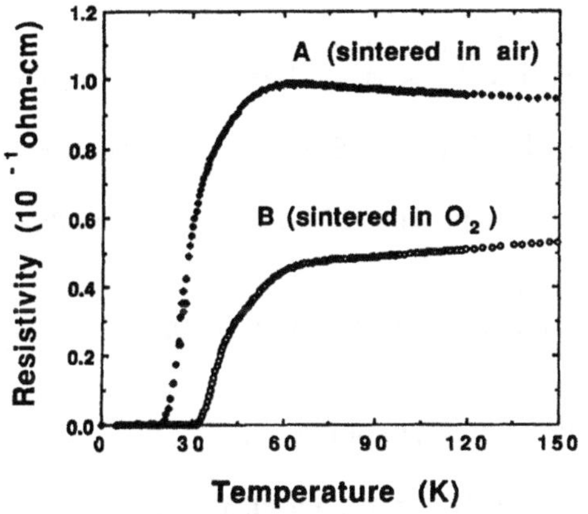

Fig.3. Temperature dependences of the electrical resistivities of $Bi_2Nd_{0.5}Ca_{2.5}Cu_2O_y$, for the sample A sintered in air and for the sample B sintered in air. The sintering temperature for both samples is 840°C.

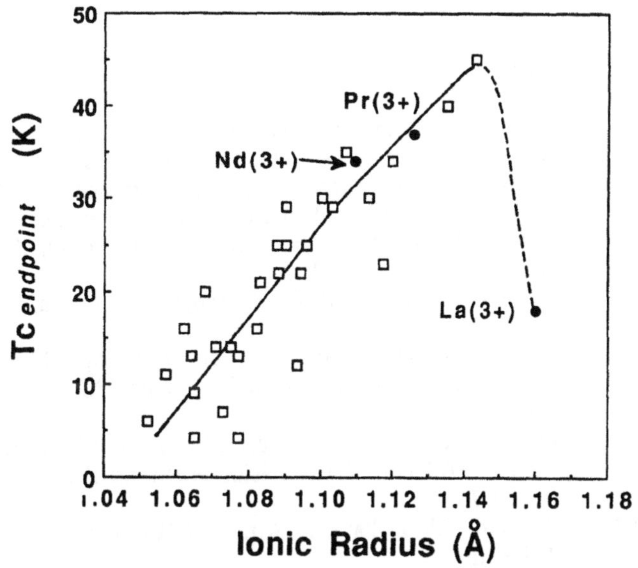

Fig.4. The relationship between Tc and the average ionic radius of Ln and Ln*(Ln, Ln*=lanthanides and Y) in $Bi_2Ln_{0.25}Ln^*_{0.25}Ca_{2.5}Cu_2O_y$ samples sintered at 840°C in O_2. Open square: Bi-(Ln, Ln*)-Ca-Cu-O, solid circle: Bi-Ln-Ca-Cu-O. The data for Bi-La-Ca-Cu-O is according to Ref.[9].

The *Tc(endpoint)*'s of $Bi_2Ln_{0.25}Ln^*_{0.25}Ca_{2.5}Cu_2O_y$ where Ln=La, Pr and Nd, and Ln*=La, Ce, Pr, Nd, Sm, Eu, Gd, Tb, Dy, Ho, Er, Tm, Yb, Lu and Y are plotted against the average ionic radius of Ln and Ln* ions in Fig.4. A number of these compounds are superconducting, but the samples with (Ln, Ln*)=(La, Yb), (La, Lu), (Pr, Yb), (Pr, Lu) and (Nd, Lu) show semiconductor like behavior. In general, as the average ionic radius increase, *Tc(endpoint)* initially increase, has a maximum and then decrease. The size of ions in the "Sr sites" in the "2212" structure may have a certain influence on the Cu-O_2 planes. Among a number of samples of $Bi_2(Ln,Ln^*)_{0.5}Ca_{2.5}Cu_2O_y$ prepared in the present work, $Bi_2La_{0.25}Pr_{0.25}Ca_{2.5}Cu_2O_y$ exhibited the highest superconducting transition.

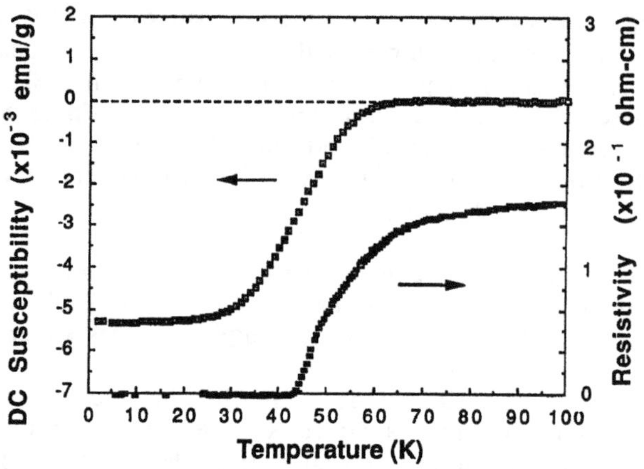

Fig.5. Temperature dependences of the electrical resistivity and the susceptibility for $Bi_2La_{0.25}Pr_{0.25}Ca_{2.5}Cu_2O_y$ sample sintered at 840°C in O_2.

The resistivity and magnetic susceptibility data for the sample are given in Fig.5. The resistive *Tc(onset)* and *Tc(endpoint)* are 65K and 45K, respectively and the magnetic *Tc* is around 60K. The superconducting volume is estimated from the susceptibility data at 40% at 10K.

CONCLUSION

Sr-free oxide superconductors in the Bi-(Ln,Ln*)-Ca-Cu-O systems where Ln=La, Pr and Nd and Ln*=La, Ce, Pr, Nd, Sm, Eu, Gd, Tb, Dy, Ho, Er, Tm and Y were synthesized. These oxides have the "2212" structure, and single phase $Bi_2Nd_{0.5}Ca_{2.5}Cu_2O_y$ samples were successfully obtained. By sintering samples in O_2 gas flow, an increase in *Tc* was observed. This suggested that the hole concentration in $Bi_2(Ln,Ln^*)_xCa_{3-x}Cu_2O_y$ sintered in air was less than that of the same compound sintered in O_2 gas flow. *Tc(endpoint)*'s of these compounds were found to be dependent on the average ionic radius of Ln and Ln* ions. That is, the highest *Tc(endpoint)* of 45K was obtained for the average ionic radius around 1.14Å.

REFERENCES

1. J.Akimitsu, A.Yamazaki, H.Sawa and H.Fujiki: Jpn.J.Appl.Phys.26(1987)L2080.
2. H.Maeda, Y.Tanaka, M.Fukutomi and T.Asano: Jpn.J.Appl.Phys.27(1988)L209.
3. R.M.Harzen, L.W.Finger, R.J.Angel, C.T.Prewit, N.L.Ross, C.G.Hadidiacos, P.J.Heaney, D.R.Veblen Z.Z.Sheng, A.Elali and A.M.Harmann: Phys.Rev.Lett.60(1988)1657.
4. L.Gao, Z.J.Huang, R.L.Meng, P.H.Hor, J.Bechtold, Y.Y.Sun, C.W.Chu, Z.Z.Sheng and A.M.Harmann: Nature 332(1988)623.
5. M.Takano, J.Takada, K.Oda, H.Kitaguchi, Y.Miura, Y.Ikeda, Y.Tomii and H.Mazaki: Jpn.J.Appl.Phys. 27(1988)L1041.
6. N.Fukushima, H.Niu and K.Ando: Jpn.J.Appl.Phys.27(1988)L790.
7. W.Baohofer, Hj.Mattaushu, R.K.Kremer, P.Murugaraj and A.Simon: Phys.Rev.B.39. (1989)7244.
8. T.Den, J.Akimitsu: Jpn.J.Appl.Phys.28(1989)L193.
9. O.Inoue, S.Adachi, Y.Takahashi, H.Hirano and S.Kawashima: Jpn.J.Appl.Phys.28 (1989)L778.
10. T.Kawano, F.Munakata, A.Nozaki, T.Ikemachi, H.Yamauchi and S.Tanaka: Physica C to be published.
11. R.D.Shannon: Acta Cryst.A32(1976)L75.

Superconductivity in Pb-Based Copper Oxides with 1212 Structure

TOSHIHIKO MAEDA, KAZUHIRO SAKUYAMA, SHIN-ICHI KORIYAMA, and HISAO YAMAUCHI[1]

Superconductivity Research Laboratory, International Superconductivity Technology Center, 10-13, Shinonome 1-chome, Koto-ku, Tokyo, 135 Japan

ABSTRACT

Superconducting Pb-based copper oxides with the 1212 structure have been synthesized in an oxidizing atmosphere in contrast to those with the 2213 structure usually synthesized in a reducing atmosphere. Superconductivity of the Pb-based 1212 copper oxides are strongly dependent on the synthetic conditions as well as the chemical compositions. Although the samples of nearly single-phase of the 1212 phase are obtained in a starting composition of $Pb_{0.5}(Sr_{1.5}Ca_{1.33}Y_{0.67})Cu_2O_z$, they are not superconductive at temperatures above 4.2 K. On the other hand, the samples with a starting composition of $PbSr_2Ca_{0.5}Y_{0.5}Cu_2O_z$ are of multi-phase but show a superconductivity transition with an onset temperature of 27 K and a zero resistivity temperature of 16 K.

KEY WORDS: high temperature superconductivity, 1212 structure, Pb-based copper oxides, Pb-Sr-Ca-Y-Cu-O system, oxidizing atomosphere

INTRODUCTION

Since the discovery of superconductivity transition at temperatures above 100 K in Bi-based copper oxides,[1] many attempts have been made to search for new superconducting materials with higher superconductivity transition temperatures (T_C). To date, zero resistivity temperature of 125 K is the highest T_C, which was reported by Parkin et al.[2] for $Tl_2Ba_2Ca_2Cu_3O_z$ which had been found for the first time by Sheng and Hermann.[3] Among the superconducting Tl-based copper oxides, two series of crystallographic structure have been found. One is $Tl_2Ba_2Ca_{n-1}Cu_nO_z$[3] which is isomorphous to $Bi_2Sr_2Ca_{n-1}Cu_nO_z$[4,5] and another is $TlBa_2Ca_{n-1}Cu_nO_z$.[6] Here n represents the number of CuO_2 planes located between two Tl_2O_2 layers in the former structure and between two TlO layers in the latter structure.

Recently, Cava et al.[7] found Bi- and Tl-free new superconducting Pb-based copper oxides of $Pb_2Sr_2(Y,Ca)Cu_3O_z$ (2213 phase) with T_C around 70 K which belongs to the $Pb_2Sr_2Ca_{n-1}Cu_{n+1}O_z$ series.[8] The 2213 phase characteristically contains an oxygen defficient Cu layer between two PbO layers. In the oxygen defficient layer, Cu is supposed to be monovalent, and probably due to the monovalent Cu's contained, the 2213 phase requires a reducing atmosphere (typically, 1 % O_2 in N_2) to be synthesized. More recently, Rouillon et al.[9] reported that Bi- and Tl-free $(Pb_{0.5}Sr_{0.5})Sr_2(Y,Ca)Cu_2O_z$, which was isomorphous to $TlBa_2CaCu_2O_z$ and $(Tl_{0.5}Pb_{0.5})Sr_2CaCu_2O_z$,[10] was synthesized by sintering in an evacuated quartz tube and that the compound was a new superconducting material. The crystallographic structure of these materials coincides with the case of n =2 in $TlBa_2Ca_{n-1}Cu_nO_z$ series (1212 structure). The superconductivity properties of the new compound, however, has not been studied in detail.

In this paper, the preparation of superconducting Pb-Sr-Ca-Y-Cu-O compounds with the 1212 structure and nearly single-phase of the Pb-based 1212 copper oxides are reported.

[1]Leave of absence from Faculty of Engineering, The University of Windsor, Ontario, N9B 3P4, CANADA.

EXPERIMENTAL PROCEDURES

The specimens used in the present study were prepared by a solid state reaction of PbO, CaO, Y_2O_3 and CuO, and Sr_2CuO_3 prepared prior to mixing. These powders were mixed to starting compositions of $Pb(Sr_{1-x-y}Ca_xY_y)_3Cu_2O_z$ (series A) and $Pb_{0.5}(Sr_{1-x-y}Ca_xY_y)_{3.5}Cu_2O_z$ (series B). They were pressed and then sintered at 850 - 1050°C for 10 - 100 h in O_2- or N_2-gas flow. Some specimens were encapsulated in evacuated quartz tubes and sintered as Rouillon et al. employed.[7]

The specimens were characterized by x-ray diffractometry using CuKα radiation, electrical resistivity measurements and ac-susceptibility measurements. Thermogravimetry (TG) and differential thermal analyses (DTA) were performed in O_2 gas flow to simulate the sintering process.

RESULTS AND DISCUSSIONS

Preliminary x-ray powder diffractometry indicated that sintering of a sample in an evacuated quartz tube did not result in a 1212 phase, and that sintering of a sample in N_2 gas flow yielded a 2213 phase. The 1212 phase was obtained by sintering in an oxidizing atmosphere. Figures 1(a) and (b) show the XRD patterns for specimens of series A whose starting compositions were: (a) x = 0.1 and y = 0.3 and (b) x = y = 0.17 being sintered at 1000°C for 10 h in O_2 gas flow. The sintered specimens with composiition (a) were of near single-phase containing a small amount of SrY_2O_4, but showed no superconductivity transition at temperatures down to 4.2 K. Figure 2 and Figure 3 show temperature dependences of electrical resistivity and ac-susceptibility for the sintered specimen with composition (b), respectively. Although this sample show-

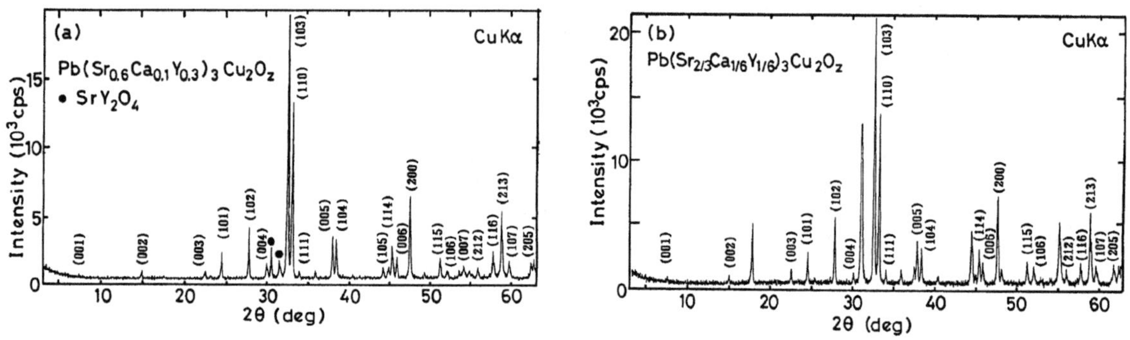

Fig. 1. XRD patterns for the specimens in series A. The starting compositions are: (a) x = 0.1 and y = 0.3, and (b) x = y = 0.17.

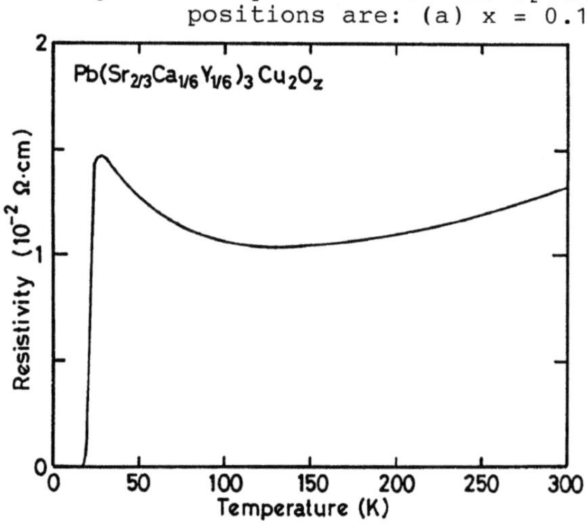

Fig. 2. Temperature dependence of the electrical resistivity for the specimen with a starting composition (b).

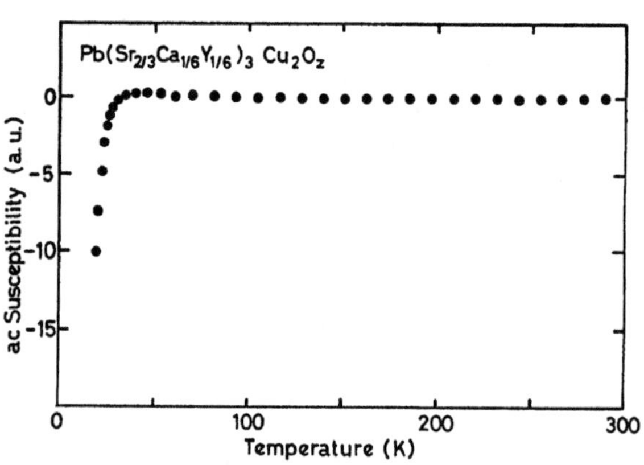

Fig. 3. Temperature dependence of ac-susceptibility for the specimen with a starting composition (b).

ed bulk-superconductivity with an onset temperature of 27 K and zero resistivity temperature of 16 K, it is of multi-phase as shown in Fig. 1 (b). These results for series A suggest that the composition of the superconducting Pb-based 1212 phase must be shifted from the starting composition. Taking an analogy from the case of $(Tl_{0.5}Pb_{0.5})Sr_2CaCu_2O_z$, "Pb sites" in the "PbO layer" in the 1212 structure are suspected to be partially occupied, probably by half,[9] by different atoms. Figure 4 shows results of TG and DTA for the mixed powder of composition (a) in series A. 20 mg of the powder was heated at a rate of 3 deg/min and then kept at 900 or 1000°C for 10 h and finally cooled to room temperature at a rate of 3 deg/min. A remarkable decrease of the weight was observed when the powder was kept at 1000°C, probably due to large volatility of Pb. When it was kept at 900°C, the decrease of the weight was negligibly small. The results of DTA showed that, when sintering was performed at temperatures below 930°C at which a relatively large endothermic peak was observed, the amount of evaporated Pb was supposed to be very small.

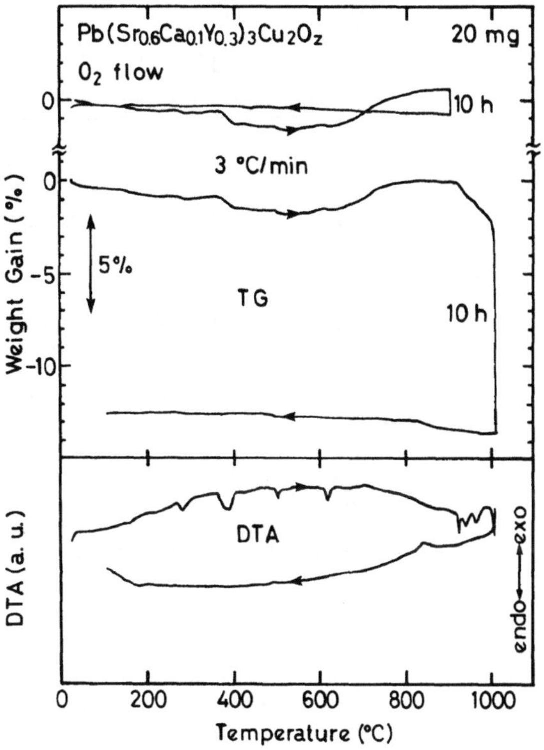

Fig. 4. DTA- and TG-curves for the mixed powder of composition (a).

Figure 5 shows used starting compositions in series B. The mixed powders were pressed, and fired at 1000°C for 1 h and then at 930°C for 30 h in an oxidizing atmosphere. Sintering for a short period of time at 1000°C was quite effective for Sr_2CuO_3 and Y_2O_3 to completely react. Compositions indicated by open circles yielded nearly single-phase samples. Figure 6 shows the x-ray powder diffraction patterns for the specimens with starting compositions of (a) x = 0.38 and y = 0.19, (b) x = y = 0.29 and (c) x = 0.19 and y = 0.38 in series B. The specimen which was the closest to a single 1212 phase was obtained when starting composition was $Pb_{0.5}Sr_{1.5}Ca_{1.33}Y_{0.67}Cu_2O_z$ (x = 0.38 and y = 0.19). Almost all of the x-ray diffraction peaks for this sample were able to be indexed for the tetragonal 1212 crystallographic structure. The minor impurity phase was identified to be CaO using the results of x-ray powder diffractmetry. The intensity of the peaks tended to become weak as x decreased, but relatively strong diffraction peaks due to SrY_2O_4 appeared. In these specimens in series B, superconductivity transition was not observed down to 4.2 K. On the other hand, both specimens with starting compositions of $Pb_{0.5}Sr_{1.8}Ca_{1.13}Y_{0.57}Cu_2O_z$ (x = 0.32 and y = 0.16) and $Pb_{0.5}Sr_2CaY_{0.5}Cu_2O_z$ (x =0.28 and y = 0.14) sintered for 100 h at 900°C in an oxidizing atmosphere were of multiphase containing several impurity phases but showed superconductivity.

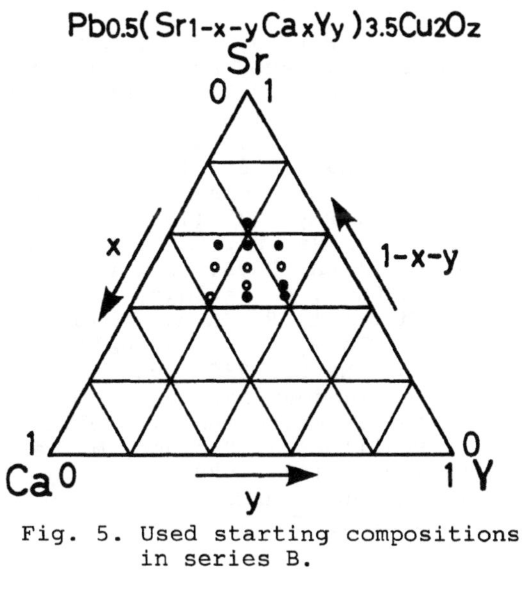

Fig. 5. Used starting compositions in series B.

Superconductivity in the Pb-based 1212 copper oxides was strongly dependent on the synthetic conditions as well as the starting compositions as mentiond so far. The investigation for obtaining Pb-based 1212 compounds with optimum superconducting properties is in progress.

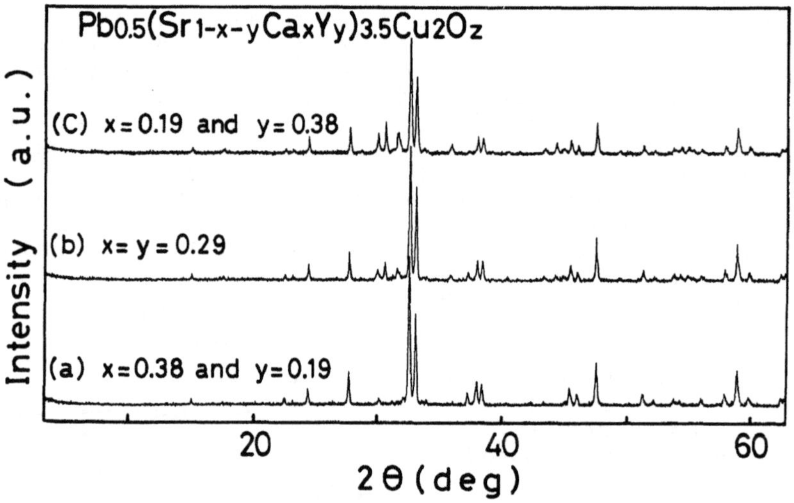

Fig. 6 XRD patterns for the specimens in series B.

CONCLUSION

Pb-based copper oxides of the 1212 structure were synthesized and characterized. The material was prepared in an oxidizing condition in contrast to the Pb-based copper oxides with 2213 structure which were prepared in a reducing condition. Superconductivity of the 1212 materials was strongly dependent on both the chemical compositions and the synthetic conditions, i.e., superconductivity transition of the 1212 material was observed only for samples of multi-phase and samples of nearly single-phase did not exhibit superconductivity at temperatures down to 4.2 K. It was indicated that the compositional range in which single-phase Pb-based copper oxides of the 1212 structure were prepared might be fairly narrow.

ACKNOWLEDGMENT

The authors would like to thank Prof. Shoji Tanaka of SRL-ISTEC for his continuous encouragements. They also thank Dr. Takahiro Wada of SRL-ISTEC for his helpful discussions.

REFERENCES

[1] H. Maeda, Y. Tanaka, M. Fukutomi and T. Asano, Jpn. J. Appl. Phys. **27** (1988) L209.

[2] S. S. P. Parkin, V. Y. Lee, E. M. Enger, A. I. Nazzal, T. C. Huang, G. Gorman, R. Savoy and B. Beyers, Phys. Rev. Lett. **60** (1988) 2539.

[3] Z. Z. Sheng and A. M. Hermann, Nature **332** (1988) 55.

[4] M. A. Subramanian, C. C. Trardi, J. C. Calabrese, J. Gopalakrishnan, K. J. Morrissey, T. R. Askew, R. B. Flippen, U. Chowdhry and A. W. Sleight, Science **239** (1989) 1015.

[5] E. Takayama-Muromachi, Y. Uchida, Y. Matsui, M. Onoda and K. Kato, Jpn. J. Appl. Phys. **27** (1988) L556.

[6] S. S. P. Perkin, V. Y. Lee, A. I. Nazzal, T. R. Savoy, R. Beyers and S. J. La Placa, Phys. Rev. Lett. **61** (1989) 750.

[7] R. J. Cava, B. Batlogg, J. J. Krajewski, L. W. Rupp, L. F. Schneemeyer, T. Siegrist, R. B. vanDover, P. Marsh, W. F. Peck, Jr, P. K. Gallagher, S. H. Glaum, J. H. Marshall, R. C. Farrow, J. V. Waszczak, R. Hull and P. Trevor, Nature **336** (1988) 211.

[8] H. W. Zandbergen, W. T. Fu, J. M. van Ruitenbeek, L. J. de Jongh, G. van Tendeloo and S. Amelinckx, Physica C **159** (1989) 81.

[9] T. Rouillon, J. Provost, M. Herview, D. Groult, C. Michel and B. Raveau, Physica C **159** (1989) 201.

[10] M. A. Subramanian, C. C. Torardi, J. Gopalakrishnan, P. L. Gai, J. C. Calabrese, T. R. Askew, R. B. Flippen and A. W. Sleight, Science **242** (1988) 249.

Rietveld Structure Refinement of Superconducting $YBaSrCu_3O_{7-\delta}$ Using X-ray and Neutron Powder Diffraction Data

E. AKIBA[1], H. HAYAKAWA[1], M. MIZUNO[1], F. IZUMI[2], and H. ASANO[3]

[1] National Chemical Laboratory for Industry, 1-1, Higashi, Tsukuba, Ibaraki, 305 Japan
[2] National Institute for Research in Inorganic Materials, 1-1, Namiki, Tsukuba, Ibaraki, 305 Japan
[3] Institute of Materials Science, University of Tsukuba, Tsukuba, Ibaraki, 305 Japan

ABSTRACT

The structure of high-Tc superconductor $YBaSrCu_3O_{7-\delta}$ which was quenched at various temperatures was investigated by Rietveld method using X-ray powder diffraction data. It was found that tetragonal to orthorhombic transition occurred at quench temperatures between 873K and 823K and the sample quenched at 873K was tetragonal but showed superconductivity (Tc=32K). We adopted a new structural model for Rietveld analysis which contains a usual orthorhombic phase and a tetragonal phase representing the structure around the microtwinning plane. The fitness of refinement using the new model was better than the orthorhombic model for both X-ray and neutron data.

KEY WORDS: Rietveld structure refinement, structure model, $YBaSrCu_3O_{7-\delta}$, tetragonal-orthorhombic transition, microtwinning

INTRODUCTION

Partially Sr substituted $YBa_2Cu_3O_{7-\delta}$ was found to be a superconductor(1-3). We have reported that superconducting transition temperature, Tc, decreased with increase of the amount of Sr substitution except for the Sr/Ba=1 composition(1). It is worth to note that at the composition of $YBaSrCu_3O_{7-\delta}$ (YBSCO) Tc was found to discontinuously increase. It was found that the lattice parameter of the b axis was short and occupancy of O(1)' site became zero as characteristics of the structure at the composition of Sr/Ba=1(1) by Rietveld structural analysis using powder X-ray diffraction data.
Here we report the structure, Tc and oxygen concentration of YBSCO quenched at various temperatures. In addition, we refined the structure by Rietveld method using time-of-flight neutron powder diffraction data in order to obtain precise information on oxygen sites of YBSCO. In Rietveld analysis of a superconducting phase, the orthorhombic(Pmmm) model was usually used. However, the patterns calculated using the model did not fit to some observed peaks(1,2). We adopted a new structure model in which the effect of microtwinning was taken into account for Rietveld analysis. Here, we also report the results of structure refinement using the new model.

EXPERIMENTAL

The YBSCO samples were prepared by solid state reaction from mixtures of Y_2O_3, $BaCO_3$, $SrCO_3$ and CuO in appropriate portion(1). Before quenching at 77K, temperature of the sample was kept at the given temperature for more than 5 hours. All the quenching procedures were carried out under oxygen atmosphere.
X-Ray powder diffraction data was obtained using a Rigaku RAX-01 diffractometer and monochromated CuKα radiation. Program "DBW3.2" by Wiles and Young(4) was used for Rietveld analysis of X-ray data. Neutron powder diffraction patterns were measured using a high resolution time-of-flight neutron diffractometer, HRP, at the KENS pulsed spallation neutron source at the National Laboratory for High Energy Physics (KEK). Rietveld analysis of the neutron data were carried out using "RIETAN" program (5). The oxygen content in the sample was measured by iodometric titration.

RESULTS AND DISCUSSION

Tetragonal to Orthorhombic Transition

Figure 1 shows the relation between Tc and quench temperatures (Tq). Figure 2 and 3 show plots of structural parameters and oxygen concentration vs. quench temperatures, respectively. From Fig.1 YBSCO was a superconductor at Tq below 923K and highest Tc at middle point was found to be 83K (Tq> 673K). However, as shown in Fig. 2 the structure was refined to be tetragonal at Tq> 873K and was orthorhombic Tq< 823K. Therefore, the samples quenched at 773K to 923K were superconductors but had the tetragonal structure. In our knowledge, the tendency that tetragonal to orthorhombic transition temperature differed to that of Tc change has not been observed in YBCO. It is interesting to note that the behavior of Tc to quench temperature is very similar to that of the oxygen content vs. Tq comparing Figs. 1 and 3. If we took iodometric data for oxygen concentration, superconducting transition was not found at the oxygen content less than 6.5 and Tc increased monotonously with increase of the oxygen content up to 6.9. Finally, Tc of 83K was observed at the oxygen content at about 6.95. We also found that the oxygen content in tetragonal YBSCO ($\delta > 0.3$) was higher than that of tetragonal YBCO ($\delta > 0.5$).

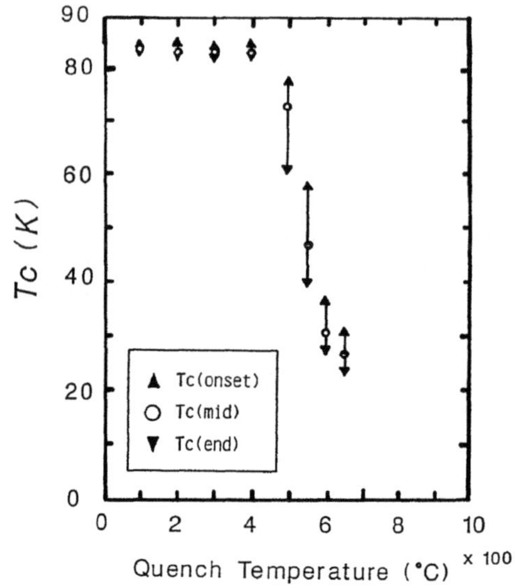

Fig. 1 Relation between Tc and quench temperatures

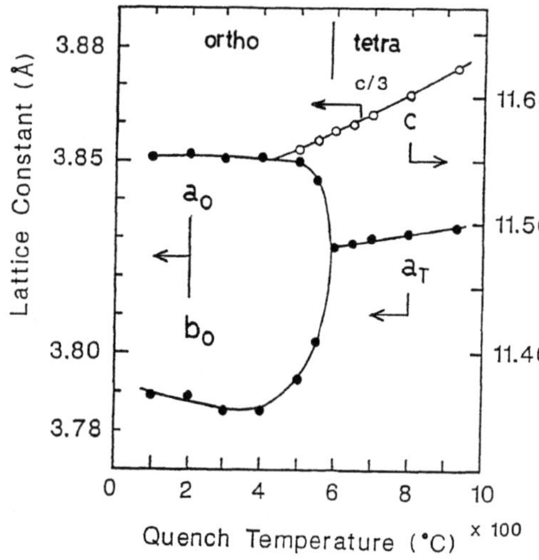

Fig. 2 Relation between lattice parameters and quench temperatures

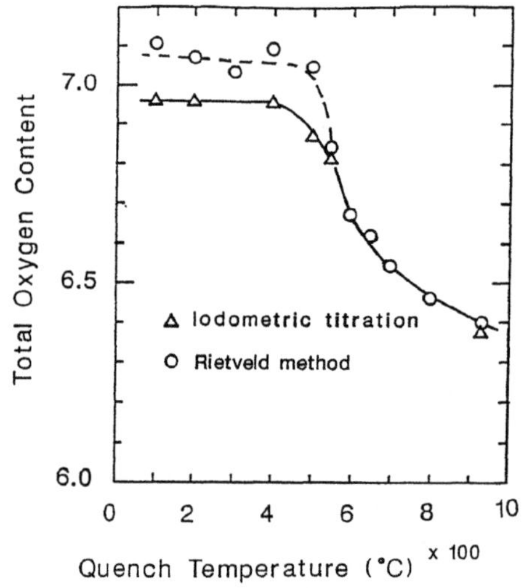

Fig. 3 Relation between oxygen content and quench temperatures

Rietveld Structure Refinement of YBSCO

The structure of superconducting YBSCO was firstly refined in the space group of Pmmm which has been reported in every literature. The profile fitting is shown in Fig.4(a). It was found that one third of the lattice parameter of the c axis was very close to that of the a axis and difference between the lattice parameters of the a and b axes were small compared to that of YBCO. In other words, orthorhombic distortion from the tetragonal structure is smaller than that of YBCO. Therefore, it is very difficult to distinguish between the tetragonal and the orthorhombic structures by looking at the diffraction patterns. To refine structures by Rietveld method was indispensable to obtain structural information of YBSCO.
As shown in Fig. 4(a), calculated diffractions of 200, 006 and 020 at around $2\theta=47°$were not fit to observed one. Microtwinning along (110) plane was observed directly using transmission electron microscope(TEM) by many researchers. The contribution of microtwinning to the powder pattern should be considered but there are only a few reports(6) about such kinds of effects on powder diffraction patterns. We modeled the boundary of microtwinning as schematically shown in Fig. 5. The structure of the microtwin boundary can be described in a tetragonal cell. The relation of lattice parameters of this tetragonal cell to that of the bulk orthorhombic structure is as follow: $a_{tetra}=(a_{ortho}+b_{ortho})/2$, $c_{tetra}=c_{ortho}$. This a very simple model in which the a axis lattice parameter of tetragonal cell is assumed the average of that of a and b axes of the orthorhombic one. Then, the structure model which is introduced here contains two phases, such as the usual orthorhombic phase and the tetragonal phase which expresses the structure of the microtwin boundary. The results of Rietveld refinement using the new model is shown in Fig. 4(b). Obtained structural parameters are listed in Table 1. R-factors were improved from R_{WP}=7.25%(the single phase model) to 5.73%. Fitness around $2\theta=47°$is also improved. The same refinement was carried out using neutron data and better fitness using the new model(R_{WP}= 6.04%) than the single phase model (R_{WP}=7.94%) was obtained. In addition, isotropic thermal parameters were refined for all the sites and much better R-factors were obtained (R_{WP}=5.32%). The detail of the results of neutron work will be published elsewhere. David et al.(6,7) reported a structure model for YBCO to explain hkl-dependent line broadening of diffraction peaks. They assumed that the shape of

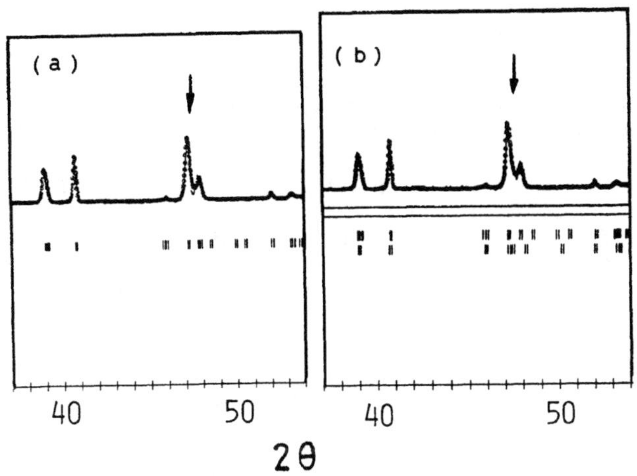

Fig.4 Rietveld pattern fitting diagrams

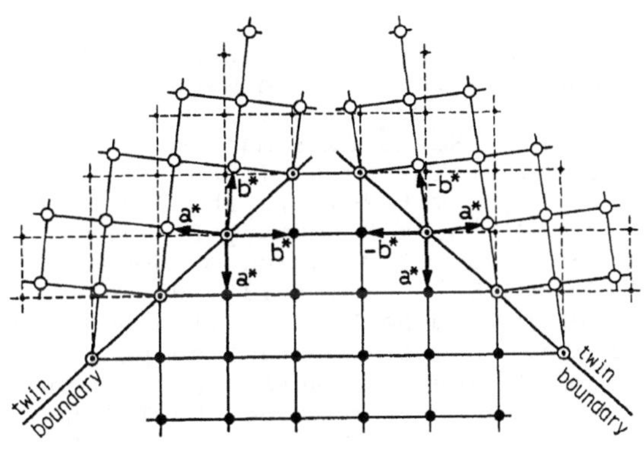

Fig.5 Schematic representation of the microtwin boundary

orthorhombic YBCO grain is a two dimensional sheet with a 110 plane normal. They calculated diffraction peak profile using this sheet model and obtained better fitness than that using the routine manner. However, it was reported that density of the microtwin boundary in YBSCO is larger than in YBCO(2). Therefore, incoherence at the boundary is not neglected and the contribution from the microtwin boundary became great portion of differential in calculated and observed profile obtained by a routine manner. Takeda et al. reported the structure of YBSCO and found that the tetragonal structure appeared again when the oxygen content increased to more than 7.0 while still remaining superconductivity. It indicates that the density of microtwin boundaries increases with increase of the oxygen content and grain of the orthorhombic phase becomes smaller. Therefore, the contribution from the boundary tetragonal symmetry is dominant and the profile is seemed to be tetragonal. This explanation is also supported by another experimental result that orthorhombic distortion is small in the YBSCO lattice. It means that microtwinning occurs much more easily than in YBCO.

Table 1 The Crystallographic data of $YBaSrCu_3O_{7.0}$ using the two phase model

Orthorhombic Pmmm (No.47)
a= 3.8518(2)Å
b= 3.7886(1)Å
c=11.5521(6)Å

Atom	Site	n	x	y	z	B
Ba	2t	0.5	1/2	1/2	0.1833(2)	1.25(6)
Sr	2t	0.5	1/2	1/2	0.1833(2)	1.25(6)
Y	1h	1.0	1/2	1/2	1/2	0.04(8)
Cu(1)	1a	1.0	0	0	0	0.97(14)
Cu(2)	2q	1.0	0	0	0.3543(3)	0.55(7)
O(1)	1b	1.0	1/2	0	0	2.2(9)
O(1)'	1e	0.03(3)	0	1/2	0	2.2(9)
O(2)	2q	1.0	0	0	0.1661(10)	0.2(4)
O(3)	2r	1.0	0	1/2	0.3780(8)	0.2(3)
O(3)'	2s	1.0	1/2	0	0.3764(13)	1.3(5)

Tetragonal P4/mmm (No.123)
a= 3.8227(4)Å
c=11.5521(6)Å

Atom	Site	n	x	y	z	B
Ba	2h	0.5	1/2	1/2	0.1833(2)	1.25(6)
Sr	2h	0.5	1/2	1/2	0.1833(2)	1.25(6)
Y	1d	1.0	1/2	1/2	1/2	0.04(8)
Cu(1)	1a	1.0	0	0	0	0.97(14)
Cu(2)	2g	1.0	0	0	0.3543(3)	0.55(7)
O(1)	2f	0.5	1/2	0	0	2.2(9)
O(2)	2g	1.0	0	0	0.1661(10)	0.2(4)
O(3)	4i	1.0	0	1/2	0.3774(8)	0.2(3)

R_{WP}=5.73%, R_P=4.15%, R_E=2.14%, R_{Bort}=4.76%, R_{Bt}=4.71%

REFERENCES

1. Hayakawa H. Akiba E. Mizuno M. Shin S. Ono S. Ihara H. Ohno E. (in press) The relation between superconductivity and crystal structure on $YBa_{2-x}Sr_xCu_3O_{7-\delta}$ (x=0-1.2). J Ceramic Soc Jpn
2. Takeda Y. Kanno R. Yamamoto O. Takano M. Hiroi Z. Bando Y. Shimada M. Akinaga H. Takita K. (1989) Structure and properties of $YBaSrCu_3O_y$ (y=6.2-7.3). Physica C 157: 358-364
3. Hanic F. Polak M. Gomory F. Plesch G. Horvath I. Lobotka P. Galokova L. (1989) Characterization of doped and substituted high-Tc superconductors Y(Ce)-Ba(Sr)-Cu-O(1:2:3). Br Ceram Trans J 88: 35-40
4. Wiles DB. Young RA. (1981) A new computer program for Rietveld analysis of X-ray powder diffraction patterns. J Appl Cryst 14: 149-151
5. Izumi F. Asano H. Murata H. Watanabe N. (1987) Rietveld analysis of powder patterns obtained by TOF neutron diffraction using cold neutron sources. J Appl Cryst 20: 411-418
6. David WIF. Moze O. Licci F. Bolzoni F. (1988) Neutron powder diffraction studies of microtwinning in the high Tc superconductor, $YBa_2Cu_3O_7$. Solid State Comm 66: 483-485
7. David WIF. Moze O. Licci F. Bolzoni F. Cywinski R. Kilcoyne S. (1989) High resolution neutron powder diffraction studies of microtwinning and stoichiometry variations in the high Tc superconductor, $YBa_2Cu_3O_{7-x}$. Physica B 156 & 157:884-887

Effect of Oxygen HIP on Superconducting Property of High-T_c Oxide Superconductors

K. SHIBUTANI, S. HAYASHI, Y. FUKUMOTO, R. OGAWA, and Y. KAWATE

Superconducting and Cryogenic Technology Center, Kobe Steel, Ltd., 5-5, Takatsukadai, 1-chome, Nishi-ku, Kobe, 673-02 Japan

ABSTRACT

Various synthesis conditions of $Ln_{1-x}Ca_xBa_2Cu_4O_8$ (Ln=Y,Ho, x=0-0.2) system and $SmLa_{1-x}Sr_xCuO_{4-z}$ (x=0.05-0.5) system were investigated by using Oxygen Hot Isostatic Pressing. In $SmLa_{1-x}Sr_xCuO_{4-z}$ system, the mixed phases of T' and T-phase are formed at 950 °C under oxygen partial pressure of 200 atm for 10 hours. The amount of T'-phase increases as the Sr concentration increases. The amount of Meissner signal of $SmLa_{1-x}Sr_xCuO_{4-z}$ system takes the maximum value at x=0.25. The Ca-substituted single 1-2-4 phase can be synthesized at temperatures above 1000 °C and oxygen partial pressure of 200 atm. Critical temperature of $Ln_{1-x}Ca_xBa_2Cu_4O_8$ (Ln=Y,Ho) increases as Ca-substitution increases up to x=0.1.

KEY WORDS: Oxygen HIP, 1-2-4 phase, Ca-substitution, T'-phase, $SmLa_{1-x}Sr_xCuO_{4-z}$

INTRODUCTION

The effects of Oxygen Hot Isostatic Pressing (O_2-HIP) on superconducting property of High-Tc oxide superconductors have been studied from the viewpoints of synthesis of new phases which are stabilized under high oxygen partial pressure (Po_2) and of decreasing oxygen deficiency.[1]
From the viewpoint of formation of new oxide superconductor family, the study of relation between the structure of oxygen coordination around copper and superconducting property have been carried out. Recently, $Nd_{2-x-z}Ce_zSr_xCuO_{4-y}$ with T*-phase structure [2] and $Nd_{2-x}Ce_xCuO_{4-y}$ with T'-phase structure were found to exhibit superconductivity with Tc over 20K.[3] Y.Tokura et al. found a new compound which was partially Sr-substituted LnLaCuO4 (Ln=Sm, Eu and Gd) with T*-phase structure and they have reported that the extensive oxygen annealing made the new compounds metallic or superconducting material.
The superconducting $Y_1Ba_2Cu_4O_8$ (1-2-4) phase has been known to be stable in oxygen content up to high temperatures compared to the $Y_1Ba_2Cu_3O_{7-y}$ (1-2-3) phase.[4] However, the drawback of 1-2-4 phase has been its low critical temperature (Tc) of 80K. Recently, Miyatake et al. discovered the new superconductor $Y_{1-x}Ca_xBa_2Cu_4O_8$.[5] By the Ca partial substitution for Y in 1-2-4 phase, Tc has increased from 80K to 90K. Therefore, the superconductor $Y_{1-x}Ca_xBa_2Cu_4O_8$ has overcome two main drawbacks; instability in oxygen content in 1-2-3 phase and low Tc of 1-2-4 phase.

In this paper, We have investigated what condition is preferable for making the Sr-substituted $LnLaCuO_4$ with T'-phase structure and Ca-substituted 1-2-4 phase superconductor. In addition, We measure magnetization vs. magnetic field hysteresis characteristics for $Y_1Ba_2Cu_4O_8$ and estimate the critical current density by using the critical state model. We measure the relaxation of remanent magnetization for $Y_{0.9}Ca_{0.1}Ba_2Cu_4O_8$ and discuss about the pinning potential of 1-2-4 phase system.

EXPERIMENTAL PROCEDURES

All samples were prepared by the solid state reaction method with oxygen-HIP treatment. In $SmLa_{1-x}Sr_xCuO_{4-z}$ system, Starting materials were 99.9% purity Sm_2O_3, La_2O_3, $SrCO_3$ and CuO. These were mixed at nominal composition x=0.05-0.5 in $SmLa_{1-x}Sr_xCuO_4$ and these powders were calcined in air at 950 ℃ for 10 hours. Then all samples were treated with O_2-HIP technique using mixture gases of O_2 20% and Ar 80% at 950℃ under Po_2=200 atm for 10 hours. In $Y_{1-x}Ca_xBa_2Cu_4O_8$ system, starting materials were 99.9% purity Y_2O_3, Ho_2O_3, $CaCO_3$, $BaCO_3$ and CuO. These were mixed for nominal composition x=0-0.2 in $Ln_{1-x}Ca_xBa_2Cu_4O_8$ and these powders were calcined in air at 880 ℃ for 16 hours. After pre-sintering all samples were treated with O_2-HIP technique. The O_2-HIP treatment conditions were as follows; T=830 ℃, 880 ℃, 930 ℃, 980 ℃, Po_2=200 atm, t=10 hours. Where, t is the holding time at elevated temperature (T) and Po_2 is the oxygen partial pressure. For 1-2-4 system (x=0-0.1), We also tried other method; using oxide for all starting materials without metal carbonate, and well mixed powders were calcined directly with O_2-HIP technique using same gas mixture T=1030 ℃, Po_2=300atm, t=3 hours without pre-sintering. After the HIP treatment, all samples were studied through the measurements of the magnetic susceptibility by using SQUID magnetometer (HSM2000) and measurements of X-ray diffraction patterns for identification of the contained crystalline phases. Other two types of measurements, time dependence of remanent magnetization and hysteresis of magnetization curves were performed by using the vibrating sample magnetometer (VSM) with cryostat.

RESULT AND DISCUSSION

Table 1. shows the comparison of the main peak intensity between T and T' phase, which are contained in the nominal composition of $SmLa_{1-x}Sr_xCuO_4$ (x=0.05-0.5). These samples were synthesized at 950 ℃ under Po_2=200 atm for 10 hours. In result, the amount of T'-phase increases as the Sr concentration increases. This result is understood below. It is known that La_2CuO_4 have the T-phase structure and Sm_2CuO_4 have the T'-phase structure. Sr^{2+} ion can substitute easier for larger La^{3+} ion than for smaller Sm^{3+} ion. So Sr-substitution will make the $La_{2-x}Sr_xCuO_4$ structure (T-structure) which has near lattice parameters of the Sm_2CuO_4 structure (T'-phase). In result, the T'-phase is more stabilized than the T-phase by Sr-substitution.

Figure 1 shows the amount of Meissner signals for $SmLa_{1-x}Sr_xCuO_{4-z}$ system at 4.2K under 100G, as a function of Sr concentration x. The Meissner signals might come from the T-phase. Those behavior is similar to the behavior of $La_{1-x}Sr_xCuO_{4-z}$ system. The amount of meissner signal takes the maximum value at x=0.25, the onset of the critical transition temperature takes the maximum value of 36.9K.

Figures 2-(a), (b), (c) and (d) shows powder X-ray diffraction patterns for $Y_{1-x}Ca_xBa_2Cu_4O_8$ samples with x=0.1 that were O_2-HIP treated 10 hours at 830 ℃, 880 ℃, 930 ℃ and 980 ℃ respectively. In these figures, indices with *-mark correspond to the 1-2-3 phase, and indices with no-mark correspond to that of 1-2-4 phase. It is clear from these figures that as the treatment temperature raises, a major phase changes from the 1-2-3 to the 1-2-4 phase. By the 980 ℃ treatment, the peaks correspond to 1-2-3 phase disappear, though the diffraction pattern correspond to the 1-2-4 phase is rather broad yet. However, all patterns contain some quantity of CuO impurity. Same tendency is observed for Ho-oxide except that the 1-2-4 phase appears at more higher treatment temperature compared with Y-oxide. These results are tabulated in Table 2 for Y-case and Ho-case. The triangle and circle marks show the 1-2-3 phase and the 1-2-4 phase respectively. The superconducting volume fraction shows in %, the onset of Tc shows in K. These measurements were performed after zero field cooling. From this table, it is found that higher treatment temperature is necessary to obtain single 1-2-4 phase structure which has higher Ca-content. According to Morris et al., 1-2-4 phase is stable at high O_2-pressure and at low temperatures.[6] They have reported that Dy (or Y)$Ba_2Cu_4O_8$ was stable at below 1000 ℃ above Po2=100 atm on the phase diagram under thermal equilibrium condition. To synthesize the 1-2-4 phase in short reaction time (below 10 hours), the HIP condition that is above Po_2=200 atm at below 1000 ℃ was employed in the present work. The 1-2-3 phase might be synthesized on the process of pre-sinter at 880 ℃ under 1 atm. So the high treatment temperature is necessary for the decomposition of the 1-2-3 phase and the formation of the 1-2-4 phase.

Figure 3 shows the temperature dependence of Meissner signal for $Y_{1-x}Ca_xBa_2Cu_4O_8$ (x=0 and 0.1, O_2-HIP condition T=980℃, Po_2=200 atm, t=10 hours). From this figure, it is found that Tc increases as Ca-content increases, which is consistent with the result of Miyatake et al. However, the superconducting volume fraction decreases as Ca-concentration increases.
To synthesize the 1-2-4 phase in less reaction time, the other synthesis method was employed. To synthesize the 1-2-4 phase without pre-sinter process, the all starting materials were changed into oxides, Y_2O_3, CaO, BaO and CuO. Well mixed oxide powders were HIP treated directly at 1030 ℃ under 300 atm for 3 hours. After that, the amount of superconducting volume fraction for $YBa_2Cu_4O_8$ shows 43%. The Ca-substituted cases (x=0.05 and 0.1) also show the same level. It is found that the using oxide for all starting materials is effective to synthesize for a few hours.

In Figure 4, magnetization vs. magnetic field hysteresis curves at 4.2K are shown for $Y_1Ba_2Cu_4O_8$ phase powders (this sample was synthesized by using oxide starting materials only and HIP treated at 1030℃ under Po_2=300 atm for 3 hours, and shows 43% of superconducting volume fraction). Diameter of these powders are selected to be between 75 μm and 100 μm. An estimation of a critical current density was attempted for this sample at 0.5T, 1.0T and 1.5T from the M-H curve measurement. When the critical state model is employed, the values of Jc are $1.2x10^4$, $7.8x10^3$ and $6.4x10^3$ A/cm^2 under 0.5T, 1.0T, 1.5T respectivery. Where the length of diameter is employed 80 um. This result suggests that the 1-2-4 phase has the potential for the application at 4.2K.

Figure 4 shows the time dependence of remanent magnetization at 4.2K. According to Anderson's thermal activation model, activation energy is estimated; the almost same value of 0.021 eV for $Y_{0.9}Ca_{0.1}Ba_2Cu_4O_8$ phase at 0.2T, 0.5T and 1T. This value is almost same level of 1-2-3 phase, and slightly larger value than that of Pb-doped Bi (2223) system.

CONCLUSION

The synthesis conditions of Ca-substituted 1-2-4 phase and nominal composition of $SmLa_{1-x}Ca_xCuO_{4-z}$ (0.05-0.5) with T'-phase by using O_2-HIP equipment were clarified.

REFERENCES

1. R.Ogawa, T.Miyatake and K.Shibutani: Proc. of Osaka University Int. symposium on "New Developments in Applied Superconductivity",Osaka(1988)117.
2. H.Sawa, S.Suzuki, M.Watanabe, J.Akimitsu, H.Matsubara, H.Watabe, S.Uchida, K.Kokusho,:H.Asano, F.Izumi and E.Takayama-Muromachi: Nature 337(1989)347.
3. Y.Tokura, H.Takagi and S.Uchida: Nature 337(1989)345.
4. J.Karpinski, E.Kaldis, E.Jilek, S.Rusiecki and B.Bucher:Nature 336(1988)660.
5. T.Miyatake, S.Gotoh, N.Koshizuka and S.Tanaka: Nature 341(1989)41.
6. D.E.Morris, N.G.Asmar, J.H.Nickel, R.L.Sid and J.Y.T.Wei and J.E.Post: Physica C 159(1989)287.

Table 1. Comparison of the main peak intensity between T and T' phase.

Sr concentration	0.05	0.1	0.15	0.2	0.25	0.3	0.4	0.5
T(103)/T'(103)	0.57	0.43	0.38	0.37	0.34	0.32	0.32	0.29

Table 2. Characterization of synthesized phase.

	$Y_{1-x}Ca_xBa_2Cu_4O_8$				$Ho_{1-x}Ca_xBa_2Cu_4O_8$			
X	830℃	880℃	930℃	980℃	830℃	880℃	930℃	980℃
0	△	△+○ 8.3% 80K	○ 19.8% 83K	○ 38.0% 81K	△	△	△+○ 13.3% 77K	△+○ 19.8% 77K
0.05	△	—	—	—	△	△	△+○ 10.1% 84K	△+○ 16.0% 84K
0.1	△	△	△+○ 10.5% 89K	○ 18.5% 89.5K	△	—	△	△
0.2	△	—	△	○ 1.3% 90K	△	—	△	△

Po_2=200 atm t=10 hours △··· 1-2-3 phase ○···1-2-4 phase

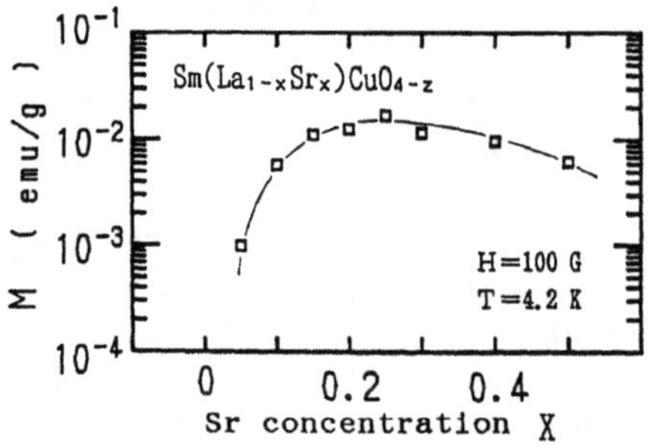

Fig.1 The amount of Meissner signal of $SmLa_{1-x}Sr_xCuO_{4-z}$ system at 4.2K, as a function of Sr concentration.

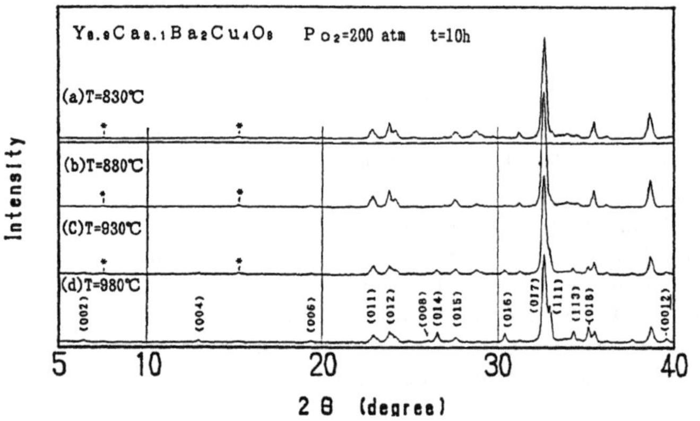

Fig.2 Powder X-ray diffraction patterns for $Y_{0.9}Ca_{0.1}Ba_2Cu_4O_8$ samples, O_2-HIPed at (a) 830°C, (b) 880°C, (c) 930°C, (d) 980°C under Po_2=200 atm for 10 hours. Indexed peaks come from the 1-2-4 phase, and *-marked peaks come from the 1-2-3 phase.

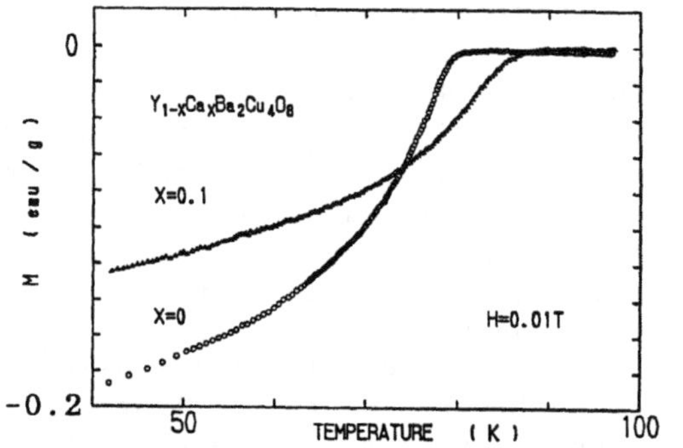

Fig.3 The amount of Meissner signal for $Y_{1-x}Ca_xBa_2Cu_4O_8$ with x=0 and 0.1, as a function of the temperature.

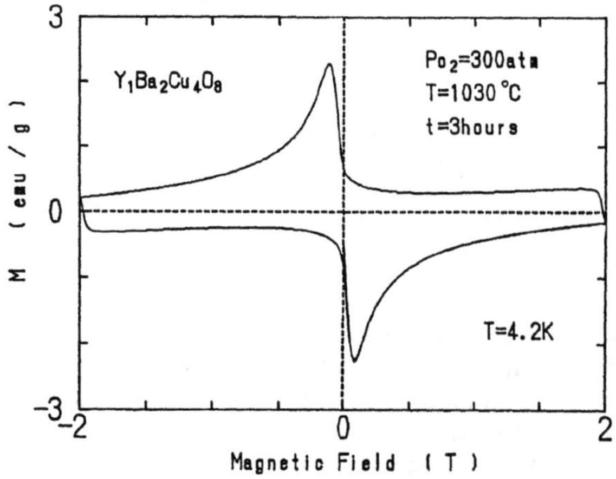

Fig.4 The magnetization vs. magnetic field hysteresis curves for $Y_1Ba_2Cu_4O_8$ sample at 4.2K. This sample was synthesized at 1030°C under Po_2=300 atm for 3 hours.

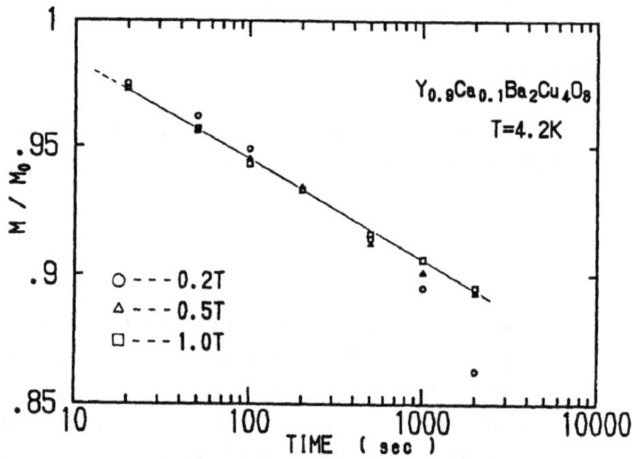

Fig.5 The time dependence of remanent magnetization for $Y_{0.9}Ca_{0.1}Ba_2Cu_4O_8$ at 4.2K.

Determination of Oxygen Content of $YBa_2Cu_3O_{7-y}$ Superconductors

YOKO SUYAMA, MAMORU MATSUMOTO, and TOSHITAKA HAYAKAWA

Japan Fine Ceramics Center, Materials Characterization Division, 4-1, Mutsuno 2-chome, Atsuta-ku, Nagoya, 456 Japan

ABSTRACT

The analytical method to determine the oxygen content precisely was investigated by using standard YBCO samples with different oxygen contents. Besides oxygen, the chemical composition and impurities were also analyzed precisely. It was found that the oxygen content can be determined to four significant figures with high precision by using a fusion-NDIR method. The lattice constants of the samples were refined by Rietveld analysis. An improved relationship between oxygen deficiency and lattice constants of YBCO was presented.

KEY WORDS: $YBa_2Cu_3O_{7-y}$ superconductor, oxygen analysis, chemical composition, oxygen deficiency, lattice constants

INTRODUCTION

The superconducting properties of $YBa_2Cu_3O_{7-y}$ are strongly dependent on the oxygen content. Therefore, the precise quantitative analysis of the oxygen content along with the other constituent elements and then the determination of the whole chemical composition are very important fundamentally to elucidate the properties and the mechanism of superconductors.

In this study, the analytical method to determine the oxygen content precisely was examined by using an oxygen analyzer. The precise oxygen content of several standard samples was successfully determined by controlling the analytical conditions. The relationship between oxygen deficiency and lattice constants of YBCO superconductor was discussed.

EXPERIMENTAL

The standard samples of $YBa_2Cu_3O_{7-y}$ with different oxygen contents were prepared. A mixture of high purity (4N) Y_2O_3, $BaCO_3$, and CuO powders was calcined at 930℃ for 12hrs in air. The calcined powder was ground, pressed into 3mm x 7mm x 20mm plates, and sintered at 950℃ for 24 hrs in O_2, followed by heat treatments to adjust the oxygen content.

Oxygen was analyzed by using an oxygen analyzer: a sample was fused with Sn and Ni in a graphite crucible at 2000 ~ 3000℃ in He gas, oxygen in the sample was extracted as CO by a reaction with C, and the CO was analyzed by NDIR method. The following analytical conditions were investigated: the standard oxides to calibrate the oxygen analyzer, the amount of flux to fuse the sample, and heating temperature and time for extracting oxygen completely from the sample. The major elements such as Y, Ba and Cu were analyzed by wet chemical analyses. Impurities were analyzed with ICP-AES, carbon analyzer and water analyzer.

The superconducting properties, e.g., Tc and Jc were measured by a conventional four-point probe method, and Jc was measured at 77 K.

The crystal structure was examined by X-ray diffraction of CuK_α radiation monochromatized with curved graphite crystal. Intensity data collected at 0.02° step width for 9s over a 2θ range from 18° to 100° were analyzed by a RIETAN program made by Izumi[1].

RESULTS AND DISCUSSION

Electrical Properties

The electrical resistivity of the samples as a function of Temperature is shown in Fig.1. The critical temperatures (onset) were 93.8K, 60.6K and <14K for sample A, B and C, respectively. The critical current density of sample A was $243A/cm^2$ at 77 K.

Crystal Structure and Lattice Constants

The samples A, B and C have the orthorhombic structure. Figure 2 shows the profile fit and difference patterns for sample A, indicating a satisfactory agreement between the calculated and observed intensities. A sufficiently low R factors were obtained: R_{wp}=7.3%, R_p=5.5%, R_E=3.9%, R_I=2.5% and R_F=1.7%. The lattice parameters refined by Rietveld analysis are shown in Table 1.

Chemical Composition and Impurities

The results of quantitative analysis of Y, Ba and Cu are shown in Table 2. The molar ratio of Y : Ba : Cu is approximately 1 : 2.02 : 3.01 in every sample. Trace amount of Fe, Ca, Sr and Si were detected as shown in Table 3. The sum amounts of Fe, Ca, Sr and Si in the samples are about 0.01wt%. The samples also contain C and H_2O as impurities as shown in Table 3. The contents of C are from 0.06 to 0.1wt%, indicating each sample has a different content. The contents of H_2O are from 0.02 to 0.10wt%, indicating the content increases with increasing oxygen deficiency.

As an optimum condition for precise determination of oxygen content, the procedure that about 30mg of sample was heated with 0.8g Ni and 0.5g Sn in a graphite crucible at about 2400℃ for 30s was found. The results of oxygen analysis by using this procedure are shown in Table 2. The oxygen content can be measured to four significant figures with a standard deviation of about 0.07, indicating that this method has high precision. The oxygen contents are 16.38, 15.70 and 15.11wt% for sample A, B and C, respectively.

The various methods such as wet chemical analyses, radioactivation analysis, TG method, and fusion method have been employed for oxygen analysis of YBCO superconductors. Among these methods, TG method has been used in most of previous works. However, TG method is not the absolute determination and has some assumptions, e.g., the weight change is caused by oxygen, and the oxygen content of a reference YBCO is assumed by various way[2$\sim$6]. Therefore, we can not evaluate the accuracy and precision of the results obtained by TG method. On the other hand, the accuracy of the results in the present work will be supported by the precise determination of the whole chemical composition including impurities.

Oxygen Deficiency and Lattice Constants

The relationship between oxygen content and lattice constants is shown in Fig.3 along with several results reported in the past. The a_0 and c_0 increase with increasing oxygen deficiency, and the b_0 decreases with increasing oxygen deficiency. The tendency of the oxygen deficiency dependence of lattice constants is simmilar to that of most previous works. However, there is quantitative difference among all results. It is considered that the relationship obtained in the present work is the improved one, because of the precise determination of oxygen and the refined lattice constants.

CONCLUSIONS

The analytical method of oxygen content of YBCO was investigated. The oxygen content can be determined to four significant figures with high precision by using a fusion-NDIR method. The whole chemical composition of YBCO was determined precisely. The improved relationship between oxygen deficiency and lattice constants of YBCO was presented.

REFERENCES

1. Izumi F, (1985) A Software Package for the Rietveld Analysis of X-ray and Neutron Diffraction Patterns. J. Cryst. Soc. Japan, vol.27, 23. [in Japanese]

2. Kosuge K, Ueda Y, (1988) Preparation of Non-stoichiometric YBCO by Use of High Purity Inert Gases and Their Physical Properties. Funtai oyobi Funmatsuyakin, vol.35, 15

3. Henry JY, Burlet P, Bourret A, Roult G, Bacher P, Jurgens MJGM, Rossat-Mignod J, (1987) ELECTRON MICROSCOPY AND NEUTRON DIFFRACTION STUDIES OF $Ba_2YCu_3O_{7-y}$ COMPOUNDS. Solid State Commun. vol.64, 1037

4. Mckinnon WR, Post ML, Selwyn LS, Pleizier G, Oxygen intercalation in the perovskite superconductor $YBa_2Cu_3O_{6+x}$. Phys. Rev. B, vol.38, 6543

5. Miyamoto H, Miyamoto K, Inamura S, Takahashi Y, (1987) Oxygen deficiency of $Ba_2YCu_3O_{7-\delta}$ and superconductivity. Funtai oyobi Funmatsuyakin, vol.34, 89

6. Ono A, Tanaka T, (1987) Preparation of Single Crystals of the Superconductor $Ba_2YCu_3O_{6.5+x}$. Jpn. J. Appl. Phys. vol.26, L825

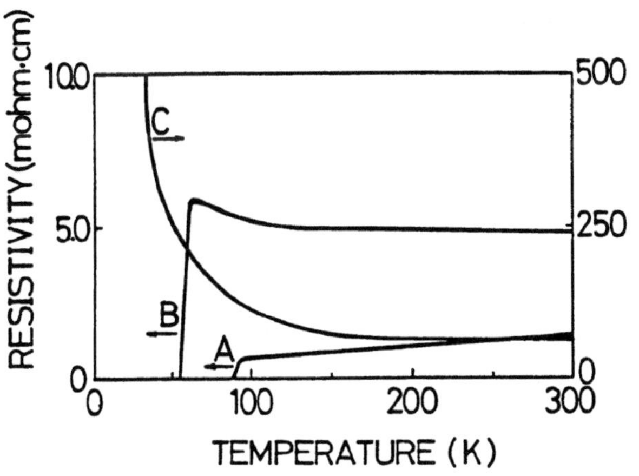

Fig.1 Temperature dependence of electrical resistivity

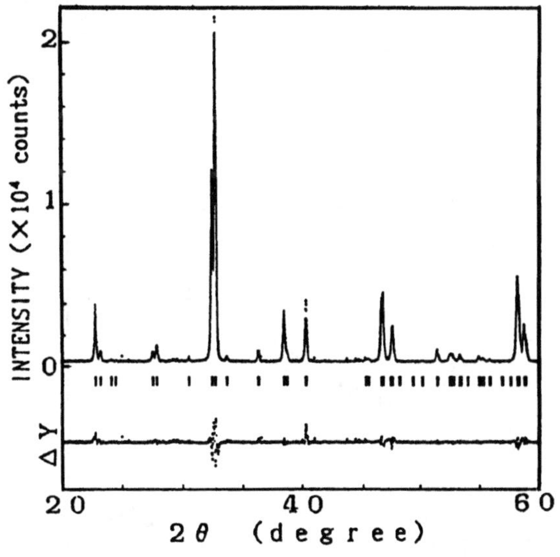

Fig.2 Rietveld refinement patterns for sample A
solid line : calculated
dotted line : observed

Table 1 Lattice Constants and Tc of YBCO

	lattice constant (Å)			Tc (K)
	a_0	b_0	c_0	onset
A	3.8199	3.8833	11.6801	93.8
B	3.8342	3.8806	11.7246	60.6
C	3.8576	3.8653	11.8058	<14

Table 2 Contents of Y, Ba, Cu and O in YBCO samples

Sample \ element	Y	Ba	Cu	O
A	13.26 (1.000)	41.45 (2.024)	28.57 (3.015)	16.38 (6.864)
B	13.38 (1.000)	41.83 (2.024)	28.78 (3.009)	15.70 (6.520)
C	13.50 (1.000)	42.16 (2.022)	29.05 (3.011)	15.11 (6.220)

upper column : wt%, lower column : molar ratio

Table 3 Impurities in YBCO samples

Sample	Fe	Ca	Sr	Si	C	H_2O
A	13	10	63	14	700	200
B	13	8	62	10	1000	400
C	15	18	62	26	600	1000

unit : ppm

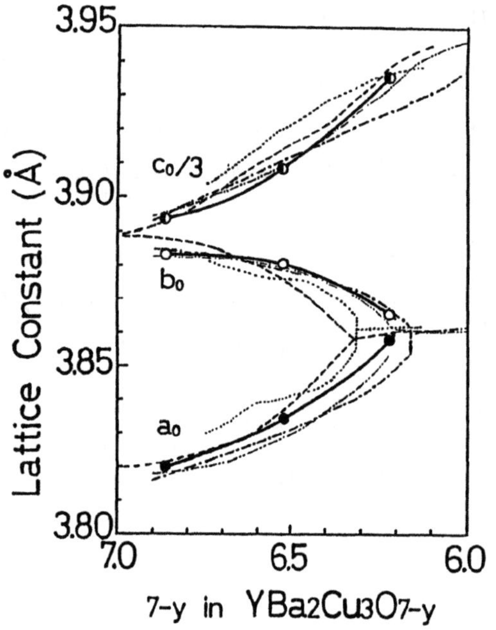

Fig.3 Variation of lattice constants with oxygen deficiency
solid lines : this work
dotted lines : after references

ACKNOWLEDGEMENT

This work was performed under the management of JFCC via ISTEC as a part of the R & D of Basic Technology for Future Industries sponsored by NEDO.

SUMMARY

The experimental results obtained by our investigation on Ca-substituted $YBa_2Cu_4O_8$ are summarized as follows:

1, Tc increases from 80 K to 90 K by Ca-substitution for Y in 1-2-4 systems.
2, Lattice parameters a and b do not change by Ca-substitution, but the parameter c increases.
3, No significant difference is observed in high resolution TEM images between Ca-free and -substituted 1-2-4. It turns out that 1-2-4 crystals have no twin structures, but they have planar defects with the absence of one chain in the double Cu-O chain.
4, Oxygen content is stable up to 800°C in Ca-substituted 1-2-4.
5, Upper critical fields extrapolated to 0 K increases by Ca-substitution.
6. An edge structure is observed at Fermi energy in UPS spectra of 1-2-4 at room temperature.

REFERENCES

1, Zandbergen HW, Gronsky R, Wang K (1988) Nature 331: 596-599
2, Marshall AF, Barton WR, Char K, Kapitulnik A, Oh B, Hammond RH, Laderman SS (1988) Phys.Rev. B 37: 9353
3, Char K, Lee M, Barton RW, Marshall AF, Bozovic I, Hammond RH, Beasley MR, Geballe TH, Kapitulnik A, Laderman SS (1988) Phys.Rev. B 38: 834
4, Kapitulnik A, Char K (1988) Int.Journal of Mod. Phys. B1: 1267-1272
5, Karpinski J, Kaldis E, Jilek E, Rusiecki S, Bucher B (1988) Nature 336: 660-662
6, Morris DE, Nickel JH, Wei JYT, Asmar NG, Scott JS, Scheven UM, Hultgren CT ,Markelz AG, Post JE, Heaney PJ, Veblen DR, Hazaen RM (1989) Phys.Rev. B 39: 7347-7359
7, Bucher B, Karpinski J, Kaldis E, Wachter P (1989) Physica C 157: 478-482
8, Kaldis E, Fischer P, Hewat AW, Hewat EA, Karpinski J, Rusiecki S (1989) Physica C 159: 668-680
9, Cava RJ, Krajewski JJ, Peck Jr. WF, Batlogg B, Rupp Jr. LW, Fleming RM, James ACWP, Marsh P (1989) Nature 338: 328-330
10, Cava RJ, Krajewski JJ, Peck Jr. WF, Batlogg B, Rupp Jr. LW (1989) Physica C 159: 372-374
11, Zimmermann H, Mali M, Brinkmann D, Karpinski J, Kaldis E, Rusiecki S (1989) Physica C 159: 681-688
12, Miyatake T, Gotoh S, Koshizuka N, Tanaka S (1989) Nature 341:41-42
13, Yamaguchi K, Miyatake T, Takata T, Gotoh S, Koshizuka N, Tanaka S (1989) Jpn.J.Appl.Phys. Lett.: to be published
14, Miyatake T, Yamaguchi K, Takata T, Gotoh S, Koshizuka N, Tanaka S (1989) Physica C :
15, Itti R, Miyatake T, Ikeda K, Tajima S, Koshizuka N : to be appeared in Proc. ISS'89 (Tskuba, 1989)
16, Tokura Y, Torrance JB, Huang TC, Nazzal AI (1988) Phys.Rev. B 38: 7156-7159
17, Welp U, Kwok WK, Crabtree GW, Vandervoort KG, Liu JZ (1989) Phys. Rev. Lett. 62: 1908-1911
18, Suenaga M : private communication
19, Arko AJ, List RS, Bartlett RJ, Cheong S-W, Fisk Z, Thompson JD, Olson CG, Yang A-B, Liu R, Gu C, Veal BW, Liu JZ, Paulikas AP, Vandervoot K, Claus H , Campuzano JC, Schirber JE, Shinn ND, (1989) Phys.Rev. B 40: 2268

Pseudo-Elastic Deformation and Environment-Induced Structural Changes in Oxide Superconductors

N. NARITA, K. HIGASHIDA, and S. MISHINA
Department of Metallurgy, Kyoto University, Yoshida-Honmachi, Sakyo-ku, Kyoto, 606 Japan

ABSTRACT

Pseudo-elastic behavior and environmental effect on phase stability were investigated using $YBa_2Cu_3O_y$ (YBCO) and $Bi_{0.7}Pb_{0.3}SrCaCu_{1.8}O_y$ (BPSCCO) ceramics with particular attention to the role of oxygen vacancies. At elevated temperatures, pseudo-elastic deformation due to twinning takes place in the YBCO of orthorhombic phase through the reorientation of oxygen-vacancy arrays. Structural changes during aging in water vapor are markedly promoted in YBCO when excess vacancy content was increased by pre-aging in vacuum. Oxygen vacancies are also introduced in the BPSCCO by heat-treatment under low Po_2 (oxygen partial pressure). Particularly, the BPSCCO of high Tc phase prepared under low Po_2 suffers severe structural changes during aging in water vapor.

KEY WORDS: superconductor, oxygen vacancy, twinning, environmental effect

INTRODUCTION

Behavior of oxygen vacancies often has a strong influence on the structural properties of oxide ceramics [1,2]. In $YBa_2Cu_3O_y$ (YBCO), one-directional arraying of oxygen vacancies leads to the formation of orthorhombic phase. The vacancy arrays must be reoriented by loading at high temperatures to cause twinning through the mutual conversion of a and b axes. Oxygen vacancies may also act as a promoter for the ingress of environmental elements to induce structural changes during aging in some prevalent atmospheres such as water vapor [1,2]. The vacancy content is believed to increase during heat-treatment under low O_2 pressure. Since the treatment under low O_2 pressure promotes the formation of high Tc phase in $Bi_{0.7}Pb_{0.3}SrCaCu_{1.8}O_y$ (BPSCCO) [3], the phase may suffer environment-induced structural changes. The purpose of the present study is to make clear such mechanical and aging effects on structural properties in connection with the role of oxygen vacancies in oxide superconductors.

EXPERIMENTAL

YBCO samples were prepared using the method described previously [4]. BPSCCO samples are prepared from a mixture of Bi_2O_3, Pb_3O_4, $SrCO_3$, $CaCO_3$ and CuO with appropriate molar ratios. The mixture was calcined in air at 1073K for 12h, and then re-ground and pressed into pellets. The pellets were sintered under 1/13 atm of O_2 pressure at 1108K for 72h and were slowly cooled down to room temperature. Compression tests for YBCO samples were made at temperatures between 290 and 723K using strain gauges for the measurement of strains. Structural characterization was made by X-ray measurements using Fe-Kα or Co-Kα radiation. Gravimetric measurements were employed to observe the increase of vacancy content during aging under reduced O_2 pressure. Environment-induced structural changes was examined in an atmosphere of water vapor (1 atm).

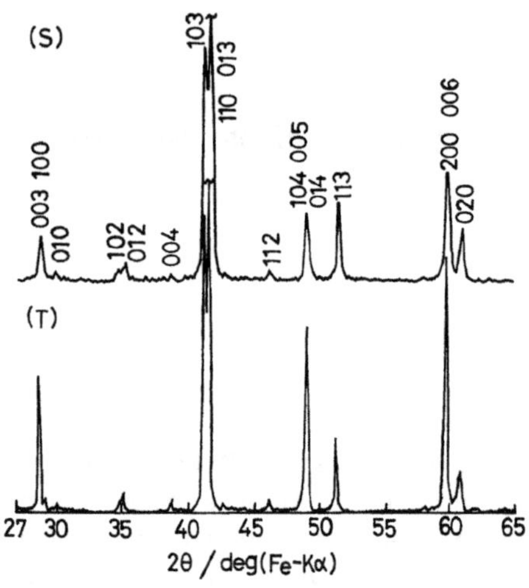

Fig.1 X-ray diffraction patterns for the side and top surfaces of a YBCO sample.

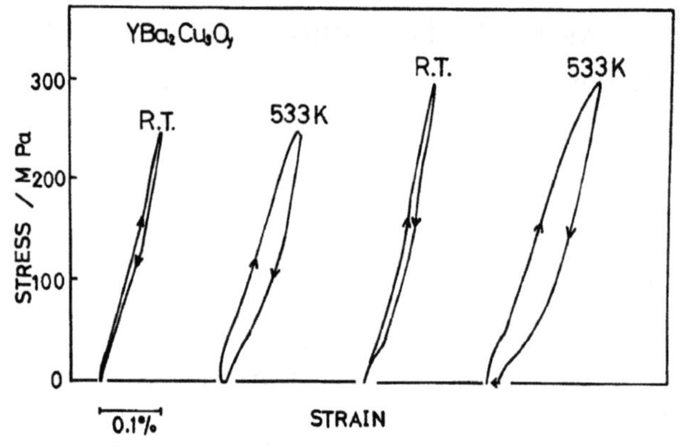

Fig.2 Stress-strain curves for YBCO samples deformed in compression.

RESULTS AND DISCUSSION

Figure 1 shows X-ray patterns obtained from the side (S) and top (T) surfaces of a YBCO sample. Here the T surface is parallel to the surface pressed before sintering and seems to be preferentially oriented parallel to a (001) since the (00n) peaks (n=3,4,5,6) for the T surface are greater than those for the S surface. Typical stress-strain curves for YBCO specimens loaded on the T surface at a strain rate of $7x10^{-7}$/s are shown in Fig.2. Obvious hysteresis of pseudo-elastic strains is seen at 533K, although residual strains are revealed after unloading and are gradually recovered at zero stress level. In samples cooled down after loading, the residual strains were not recovered at room temperature. Since the mobility of oxygen vacancies is increased at elevated temperatures, the pseudo-elastic behavior seems to be caused by the reorientation of vacancy arrays, i.e., twinning or untwinning. The twinning in YBCO may occur by shear deformation along {110} (the left figure in Fig.3), causing the

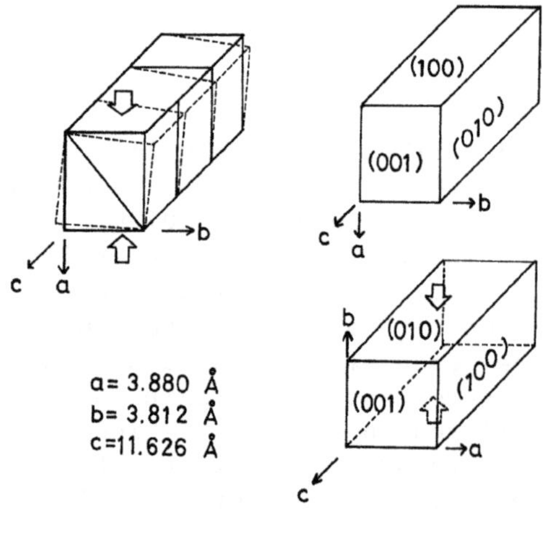

Fig.3 Twinning and the mutual conversion of a and b axes.

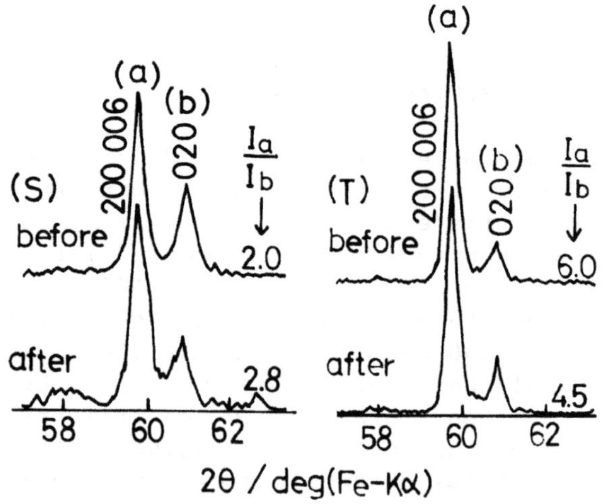

Fig.4 Change of X-ray peaks by loading at 533K (see text).

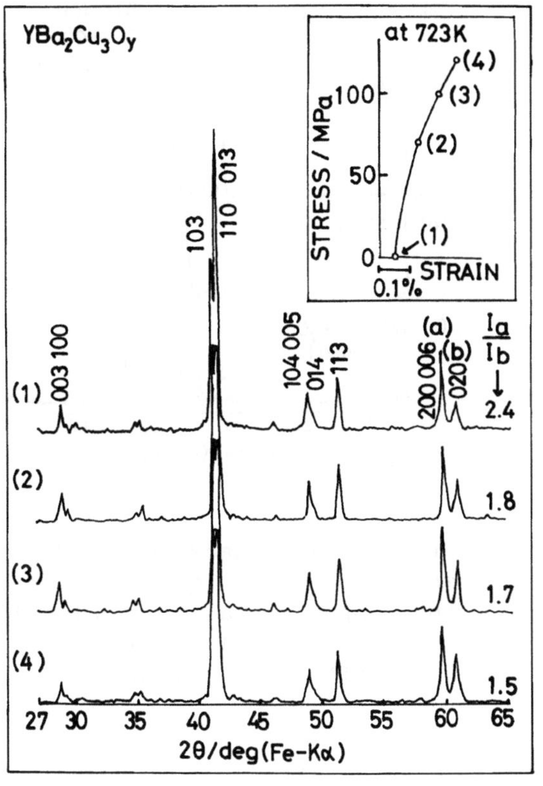

Fig.5 Change of diffraction patterns obtained from a compression surface by loading at 723K.

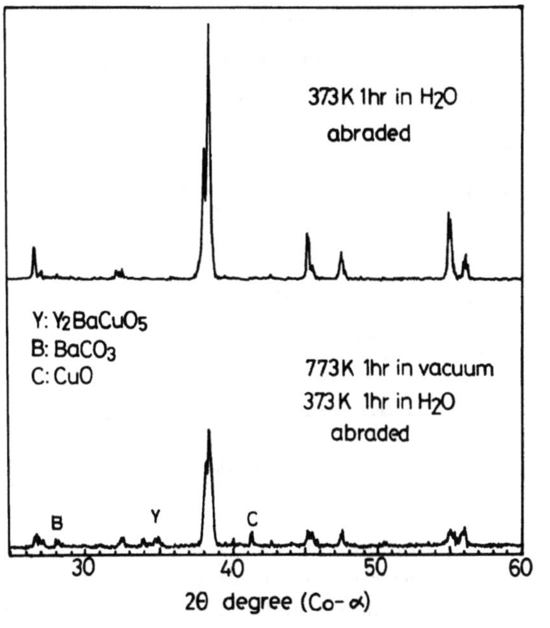

Fig.6 Structural changes by aging in water vapor at 373K for an original YBCO (the upper pattern) and for a YBCO pre-aged in vacuum (2×10^{-3} Pa) at 773K (the lower pattern). The patterns were obtained after the surface layer was removed by 0.2mm.

mutual conversion of a and b axes (the right figures in Fig.3). Therefore, a (100) plane along the T (compression) surface and a (010) along the S surface may convert to (010) and (100) planes by twinning, respectively. Such conversion of the planes is actually seen from the change in the intensity ratio of a (200)+(006) to a (020) peak, I_a/I_b, in Fig.4. By loading up to 300MPa at 533K, the intensity ratio is increased in the S surface and is decreased in the T surface. A similar conversion of the planes was also observed in a case when an external load was applied on a S surface (Fig.5). The X-ray patterns in Fig.5 were obtained from the compression (S) surface of a YBCO sample which was subjected to successive loading at 723K. The decrease of I_a/I_b by loading indicates that the (100) plane along the compression surface converts to the (010) plane with increasing external stress. Further, computer simulation for X-ray patterns by taking account of the multiplicity factors due to preferred orientation showed that the change of peak intensities by loading was mainly ascribed to the mutual conversion of a and b axes. It is, therefore, evident that deformation twinning actually occurs in YBCO samples through the reorientation of vacancy arrays at elevated temperatures.

The content of oxygen vacancies is increased in YBCO samples when they are annealed under low O_2 pressure [2]. In the YBCO annealed to increase vacancy content, structural changes in water vapor are markedly promoted, as is seen from Fig.6. The oxygen vacancy, $V_O^{\cdot\cdot}$, may react with H_2O [1,2], as

$$V_O^{\cdot\cdot} + H_2O \rightarrow O_O^{x} + 2H_i \qquad (1)$$

Here O_O^{x} and H_i denote the oxygen atom occupied at O sites and the interstitial hydrogen ion, respectively. The O_O^{x} and H_i may diffuse inwardly from the surface at high temperatures with the aid of other vacancies and are often recom-

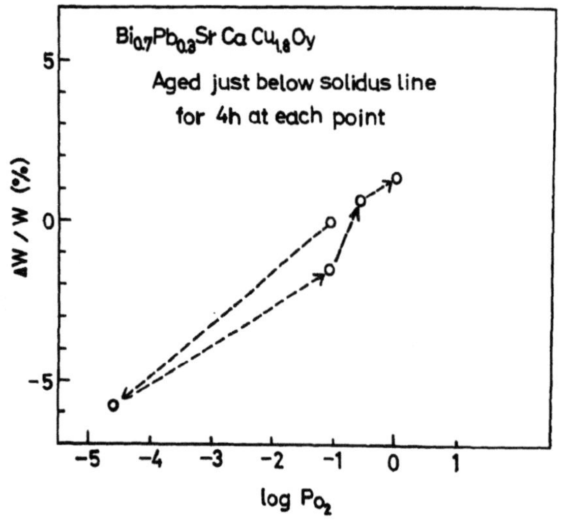

Fig.7 Change in the mass of a BPSCCO sample by successive aging for 4h at each O_2 pressure at temperatures just below solidus line.

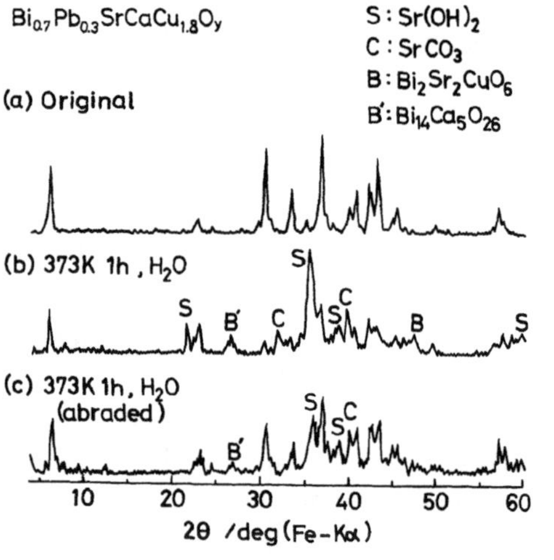

Fig.8 Structural changes in a BPSCCO sample by aging in water vapor; the sample is (a) original,(b) as-aged and (c) as-abraded by 0.2mm after being aged.

bined to form OH^- [1,2]. The OH^- is liable to react with alkaline earth elements so as to form $Ba(OH)_2$ in YBCO, although the $Ba(OH)_2$ reacts with CO_2 in air forming $BaCO_3$ [2]. Thus the phase decomposition can be developed inwardly by the aid of excess vacancies during aging in water vapor. In BPSCCO, excess vacancies may also be introduced during annealing under low O_2 pressure since mass of the samples is decreased (Fig.7). Although the annealing under low O_2 pressure serves to form high-Tc phase in BPSCCO [3], such high-Tc phase really suffers structural changes in water vapor (Fig.8), resulting in the degradation of superconducting properties (Fig.9).

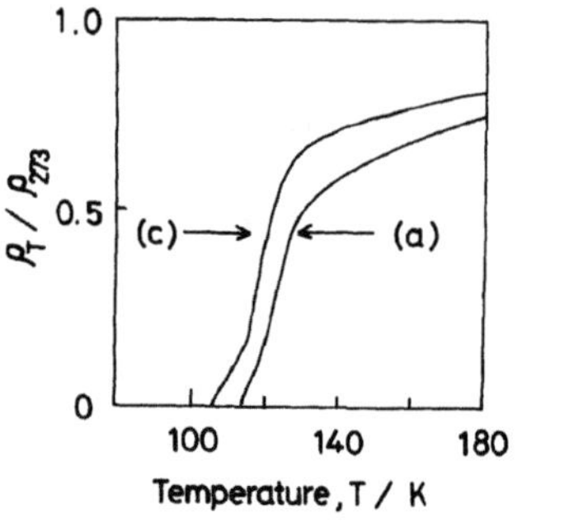

Fig.9 Electrical resistivity for the BPSCCO;(a) original and (c) abraded by 0.2mm after being aged in water vapor for 1h.

In conclusion, the behavior and quantity of oxygen vacancies have a strong influence on structural properties in oxide superconductors. Deformation twinning is induced in YBCO through the reorientation of oxygen-vacancy arrays at elevated temperatures. Structural changes during aging in water vapor are markedly promoted by the introduction of excess oxygen vacancies in BPSCCO as well as those in YBCO.

REFERENCES

1. Narita N (1988) In: Sintering '87. Elsevior Appl.Sci. London. pp 1130-1135

2. Narita N. Leng S. Higashida K (1989) In: Proc.MRS Int.Mtg. on Adv.Mats. Vol 6. Materials Research Society. Pittsburgh. pp 403-408

3. Endo U. Koyama S. Kawai T (1988) Jpn J Appl Phys 27: L1476-L1489

4. Leng S. Narita N. Higashida K (1987) Jpn J Appl Phys 26: L1394-L1397

Hot Deformation and Superconductivity of $YBa_2Cu_3O_{7-x}$ Ceramics

YASUHARU KODAMA and FUMIHIRO WAKAI

Government Industrial Research Institute, Nagoya, 1-1, Hirate-cho, Kita-ku, Nagoya, 462 Japan

ABSTRACT

Compressive deformation of dense polycrystalline $YBa_2Cu_3O_{7-x}$ and that of porous one were studied at elevated temperatures between 850°C and 960°C. The stress exponent measured in plastic deformation of dense material was 2.5. The stress exponent and microstructural observation suggested the intra-granular deformation mechanism. The densification occurred during compressive deformation of porous material till true strain of -0.4, but the crack formation caused by further deformation decreased the apparent density. $T_{c\ on}$ of deformed and annealed sample was 94K due to enhanced grain connectivity by compression. However, zero resistance couldn't be achieved due to the destruction of some superconducting phase.

KEY WORDS : $YBa_2Cu_3O_{7-x}$, compressive deformation, stress exponent, intra-grain mechanism, densification

INTRODUCTION

The studies of oxide high temperature superconductors for application are widely carried out since its discovery [1] [2]. Hot work using deformation such as hot pressing [3], hot isostatic pressing [4], hot forging [5] [6], hot extrusion [7] etc. is one of processes to attain high density, highly grain alignment, and shape forming. The understanding of deformation behavior [8] [9] [10] is important to optimize microstructure concerned to superconductivity through hot works. We studied the deformation and densification behaviors by the compression test of dense polycrystalline $YBa_2Cu_3O_{7-x}$ and porous one at elevated temperature. We also studied effect of deformation on superconductivity.

EXPERIMENT

Two types of samples were used in the compression test; dense materials and porous materials. The dense materials were prepared from commercially available calcined $YBa_2Cu_3O_{7-x}$ powders (Hayashi Chem. Co.). The calcined powders were pressed into column (10mm in diameter and 10mm in height) at 1000kgf/cm^2, and then they were sintered at 960°C for 12 hours in the flowing O_2 atmosphere. The porous materials were cut from $YBa_2Cu_3O_{7-x}$ sputtering target (Mitsubishi Metal Co.) to the size of 5mmx5mmx8mm. The compression tests were performed using Instron-type universal testing machine in air at temperature from 850°C to 960°C with the constant crosshead speed from 0.002 to 0.05 mm/min. To avoid reaction between $YBa_2Cu_3O_{7-x}$ sample and SiC rod, ZrO_2 plates were used as separator. The densities, before and after compression tests, were measured by the Archimedes method. The microstructures were observed by SEM (Scanning Electron Microscope) on the fracture surfaces and by EPMA (Electron Probe Micro Analyzer) on the polished surfaces. Most of samples were quenched after compression tests, but some of them were cooled slowly and then annealed to measure the electrical resistivity. The resistivity was measured by conventional four-probe method.

RESULTS AND DISCUSSION

(1) Dense Materials

The apparent density of the dense materials were 6.0g/cm^3, which were 94% of the theoretical density. True stress-true strain curves are shown in Fig.1. Stress increased proportionally with strain, till true strain of -0.05. The flow stress of the materials deformed at high temperature or at slow cross-head speed was constant after the maximum stress (yield point). But the flow stress of the materials deformed at low temperature or fast cross-head speed decreased with strain. The elastic stress-strain curve was observed till the yield point, then plastic deformation occurred after it. The decrease of stress with increasing strain after yield point was caused by formation of cracks.

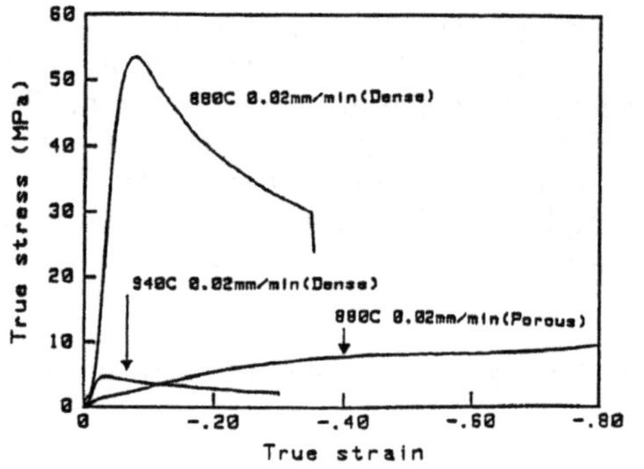

Fig.1 True stress - True strain curve.

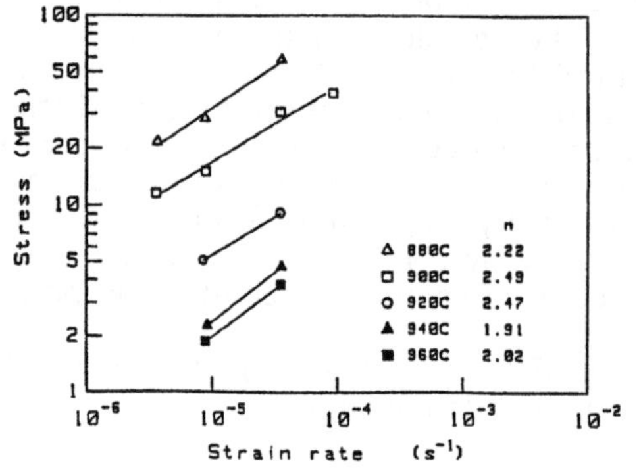

Fig.2 The relationship between stress and strain rate. Vertical axis is maximum stress of true stress - true strain curve. Hrizontal axis is inital strain rate.

The relationship between stress and strain rate is shown in Fig.2. The strain rate of polycrystalline material is generally expressed by the following equation :

$$\dot{\varepsilon}=A\sigma^n\exp(-Q/RT) \qquad (1)$$

where $\dot{\varepsilon}$ is strain rate, A is constant, σ is stress, n is stress exponent, Q is apparent activation energy, R is the gas constant, and T is temperature. The value of stress exponents were calculated to be 2 to 2.5 from Fig.2 and Eq.(1). According to the conventional creep theory, the stress exponent of 1 indicates the diffusional creep, and stress exponent larger than 3 indicates the dislocation creep. The former is characterized by inter-grain mechanism, and the latter by intra-grain mechanism. The relationship between log-stress and reciprocal of temperature is shown in Fig.3. The apparent activation energy calculated from Fig.3 and Eq.(1) is 1080kJ/mol. This value is partially consistent with previous data example Reyes-Morel et al. (n=1.25, Q=1218kJ/mol) [8], Bussod et al. (n=2.5, Q=201kJ/mol) [9], and von Stumberg et al. (n=1, Q=800±100kJ/mol). We suppose that the deviation of n and Q were caused by the deviation of samples.

The composition images of polished surface by EPMA are shown in Fig.4. Figure 4 (a), (b), (c), (d) are as-sintered sample, deformed sample at 880°C, at 920°C and at 960°C, respectively. The as-sintered materials consisted of slightly elongated grains with the size of several μm. CuO existed at grain boundaries. CuO also distributed in grains. Graingrowth was not observed in the sample compressed at 880°C. Grain growth occurred at 920°C, and the length of elongated grain was about 100μm. But, CuO still distributed in grains. Grain growth was observed at 960°C similar to the compressed at 920°C. CuO was not observed in the grains but at triple points. We suppose that $YBa_2Cu_3O_{7-x}$ partially melted at this temperature. The liquid gathered at triple points, and $YBa_2Cu_3O_{7-x}$ and CuO appeared from the liquid. In spite of grain growth, the deformation properties did not change from the data Fig.2 and 3. Then we consider that the deformation can be attribute to the intra-grain mechanism, but we feel that the consideration on co-existent liquid phase is required.

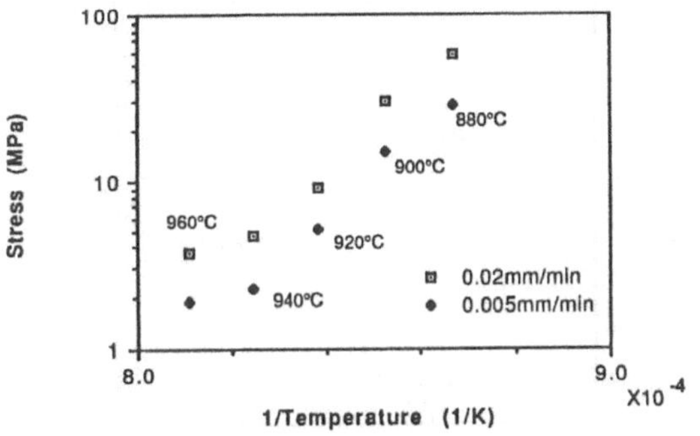

Fig.3 The relationship between log-stress and reciprocal of temperature. Two cross-head speed 0.02mm/min and 0.005mm/min are plotted.

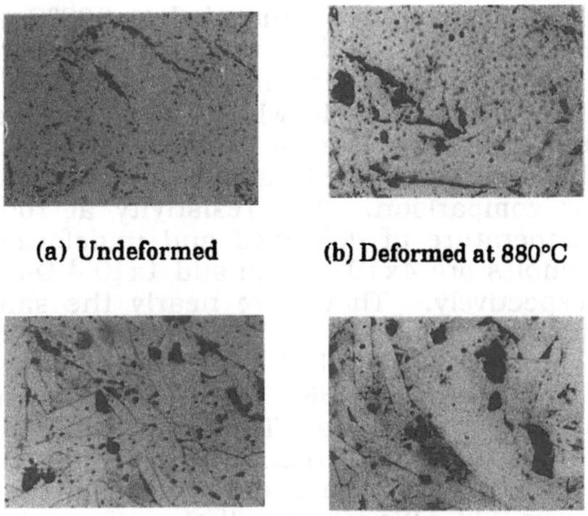

(a) Undeformed (b) Deformed at 880°C

(c) Deformed at 920°C (d) Deformed at 960°C

100μm

Fig.4 Composition image of polished surface of dense materials.

(2) Porous Materials

The apparent density of the porous materials was $4.5 g/cm^3$, which was 70% of the theoretical density. Stress-strain curve of porous material is also shown in Fig.1. The material deformed at relatively low flow stress (1/10 of dense materials) till true strain reached at -0.7, and after that the flow stress increased rapidly with strain. No cracks were observed in the samples with less than -0.6 strain , but some cracks were observed on the side surface of the samples deformed further.

The relation among axial strain ε_z, transverse strain ε_t, shear strain ε_e, and densification strain ε_a is expressed by the following equations,[11]

$$\varepsilon_e = 2/3\,|\varepsilon_z - \varepsilon_t| \qquad (2)$$

$$\varepsilon_a = 2\varepsilon_t + \varepsilon_z \qquad (3)$$

where $\varepsilon_z = \ln(l/l_0)$, $\varepsilon_t = \ln(t/t_0)$, l_0 is initial height, l is height, t_0 is initial width , and t is width. The relationship between shear strain and axial strain and that between densification strain and axial strain are shown in Fig.5. Densification strain was negative when axial strain was less than -0.4 , but when axial strain exceeded -0.6 densification strain became positive. The densification strain at a axial strain was enhanced by elevating temperature.

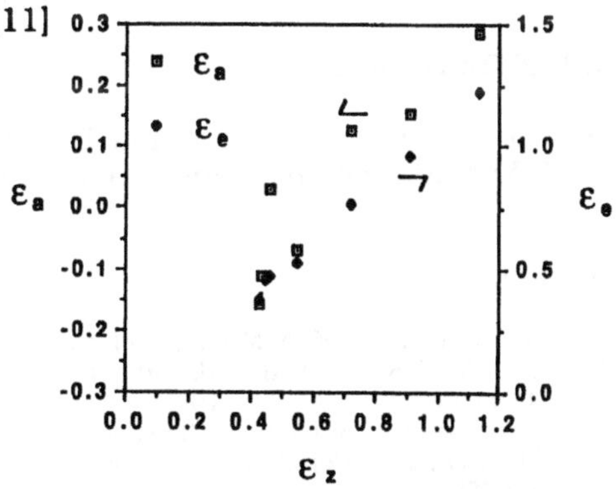

Fig.5 The relationship between shear strain and axial strain and the relationship between densification strain and axial strain

From the observation on the fracture surface by SEM, grain growth occurred above 920°C, but no effect of compression on the grain growth was observed. Grain alignment was not observed in the porous materials as well as in the dense materials.

(3) Effect of deformation on superconductivity

Inside portion of the materials both dense and porous materials which were quenched after compression test behaved semiconductive. The cooling rate was too fast to induce phase transition from tetragonal to orthorhombic. Slowly cooled dense materials were used as samples for measurement of resistivity. Slow cooling was performed as follow, from the test temperature of 880°C to 600°C cooling rate was 5°C/min, from 600°C to 300°C it was 1.5°C/min, and at 300°C samples were taken off from the furnace. Samples were compressed of the constant stress of 70MPa until it was taken off from the furnace. The resistivity of both undeformed and deformed samples slightly decreased at 77K. The resistivity of deformed sample increased semiconductive below 77K. The resistivity of undeformed sample did not become zero until 20K although it decreased with temperature.

Some samples were annealed at 500°C for 24 hours in the flowing O_2 atmosphere. The temperature dependence of the resistivity of the sample which were deformed and annealed is shown in fig. 6. The resistivity of undeformed sample is also shown for comparison. The resistivity at room temperature of deformed and undeformed samples are $4x10^{-2}$ Ω·cm and $1x10^{-2}$ Ω·cm, respectively. They were nearly the same before annealing. Tc on of undeformed sample was 92K, but tailed by 80K, and then Tc zero was 43K. This is caused by insufficient annealing. Tc on of compressed sample was 94K, then the resistivity decreased tailing, and did not reach zero above 20K. We suppose that grain connectivity was enhanced but some superconducting phase was destroyed by the compressive deformation[12].

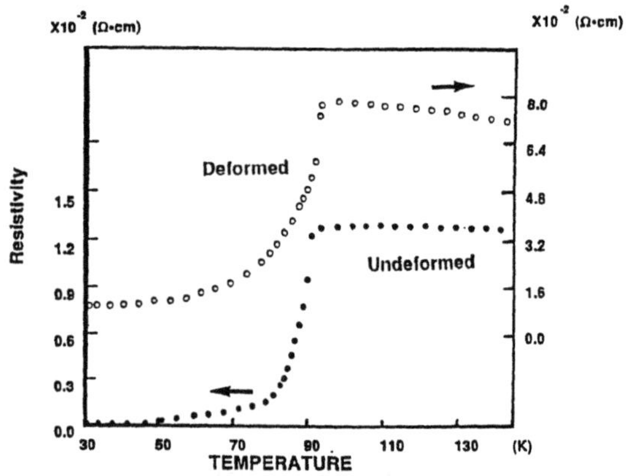

Fig.6 Temperature dependence of resistivity. Tc on of deformed and undeformed sample was 94K and 92K, respectivly. Tc zero of undeformed was 43K but zero resistance could not achieved in deformed sample.

CONCLUSION

Compressive deformation of $YBa_2Cu_3O_{7-x}$ dense and porous materials at elevated temperature was studied. In the dense materials, the stress exponent was 2.5, the apparent activation energy was 1080kJ/mol, and the deformation mechanism was intragrain mechanism. For precise investigation we need fully dense fine polycrystalline. To optimize annealing condition is necessary to evaluate precise effect on superconductivity.

ACKNOWLEDGMENTS

The authors would like to thank E. Sudo and M. Awano for many helpful suggestions, and M. Yamada for instruction to operation of EPMA.

REFERNCES

[1] J. G. Bednorz and K. A. Müller, Z. Phys. B 64, 189 (1986)
[2] M. K. Wu, J. R. Ashburn, C. J. Torng, P. H. Hor, R. L. Meng, L. Gao, Z. J. Huang, Y. Q. Wang, and C. W. Chu, Phys. Rev. Lett. 58, 908 (1987)
[3] R. Yoshizaki, T. Iwazumi, H. Sawada, H. Ikeda, and E. Matsuura, Jpn. J. Appl. Phys. 26, L311 (1987)
[4] J. K. Tien, J. C. Borofka, B. C. Hendrix, T. Caulfield, and S. H. Reichman, Metall. Trans. A, 19A, 1841 (1988)
[5] S. N. Song, Q. Robinson, D. L. Johnson, and J. B. Ketterson, Solid State Commun. 68, 391 (1988)
[6] G. S. Garder, H. M. O'Bryan, and W. W. Rhodes, Appl. Phys. Lett. 52, 1831 (1988)
[7] I-W. Chen, X. Wu, S. J. Keating, C. Y. Keating, P. A. Johnson, and T. Y. Tien, J. Am. Ceram. Soc. 70, C388 (1987)
[8] P. E. Reyes-Morel, X. Wu, and I-W. Chen, in Ceramic Superconductors II, edited by M. F. Yan (American Ceramic Society, Westerville, Oh, 1988) , p. 590
[9] G. Bussod, A. Pechenik, C-T. Chu, and B. Dunn, J. Am. Ceram. Soc. 72, 137 (1989)
[10] A. W. von Stumberg, N. Chen, and K. C. Goretta, J. Appl. Phys. 65, 2079 (1989)
[11] R. Raj, J. Am. Ceram. Soc. 65, C-46 (1982)
[12] B. C. Hendrix, T. Abe, J. C. Borofka, and J. K. Tien, and T. Caulfield, Appl. Phys. Lett. 55, 313 (1989)

Effect of Particle Size on Sintering Behavior of $Bi_2Sr_2CaCu_2O_{8+\delta}$

HITOSHI SAKAI, MANABU YOSHIDA, and HITOSHI YOSHIDA
Materials Research Laboratory, NGK Insulators, Ltd., 2-56, Suda-cho, Mizuho-ku, Nagoya, 467 Japan

ABSTRACT

The effect of particle size of the melt-quenched amorphous starting powder, the composition of which was Bi:Sr:Ca:Cu=2:2:1:2, on the sintering behavior was studied to clarify the changes in microstructure during the partial melting process. The segregation of $(Sr_{0.6}Ca_{0.4})CuO_2$ phase in Bi-rich liquid phase proceeded during the partial melting. The segregation rate increased with decreasing the grain size of the starting powder. It is considered that the segregation rate is related to the melting rate, which can be controlled by the grain size.

KEY WORDS: $Bi_2Sr_2CaCu_2O_y$ superconductor, partial melt, melt-quench, SEM-EPMA, particle size

INTRODUCTION

It is well known that the partial melting process is important to increase the critical current density, Jc, of Bi-contained oxide superconductors because the sintering and the grain growth of superconductors are so slow below melting temperature that the density remains to be low and the grain boundary can rarely be improved in a moderate time. Oka et al.[1,2] studied the the partial melted state of $Bi_2Sr_2CaCu_2O_y$ (2212 phase) and reported that 2212 phase decomposed at 900°C into $Sr_{1-x}Ca_xCuO_2$(x=0.5) and the Bi-rich liquid phase. However, little detail description about the effect of particle size and melting time on the partial melting was made.

It is indicated by Tohge et al.[3] that the amorphous state was easily obtained by rapid quenching within the wide range of the composition in Bi-Sr-Ca-Cu-O system. The melt-quenched amorphous powder is expected to be dense and homogeneous, and may be suitable for the starting powder of partial melting process.

In this study the partial melting process was analyzed in order to clarify the changes in microstructure as a function of partial melting time with emphasis on the grain size of the melt-quenched amorphous starting powder.

EXPERIMENTAL

Starting powder preparation

Bi_2O_3, $SrCO_3$, $CaCO_3$ and CuO powders were mixed so that the cation composition should be Bi:Sr:Ca:Cu=2.0:2.0:1.0:2.0. The mixture was milled with partially stabilized zirconia (PSZ) balls in isopropyl alcohol for 5h. The milled powder was melted at 1100°C for 30min in Pt crucible and then cooled rapidly by using the stainless steel twin-roller. The quenched flaky powder was pulverized in an agate mortar for 30min and milled with PSZ balls in isopropyl alcohol for 1 or 20h. The screening of larger particles than 75μm was done.

The mean particle size and the specific surface area of the two kinds of starting powders were shown in Table 1. The cation composition was estimated to be Bi:Sr:Ca:Cu=1.9:2.0:1.0:2.0 by chemical analysis. The starting powder showed amorphous by X-ray diffractometry as shown in Fig. 1.

Sample preparation and measurement

The starting powder was pressed into the bar with the size of 5x40x1mm. The samples were partially melted on silver plate at 895°C in oxygen atmosphere for various partial melting time up to 180min, then quenched into the liquid nitrogen. The zero melting time indicates that the sample was quenched as soon as the temperature reached 895°C.

The quenched samples were analyzed by the X-ray diffractometer, the scanning electron microscope(SEM) and the electron probe microanalyzer(EPMA). After annealing of the quenched sample at 830°C for 15h in air, the Jc value of the sample was measured at 77K by means of 4 probe technique.

Table 1 Characterization of starting powder

powder	milling time(h)	mean particle size(μm)	specific surface area(m^2/g)
A	1	9.1	0.5
B	20	4.6	2.2

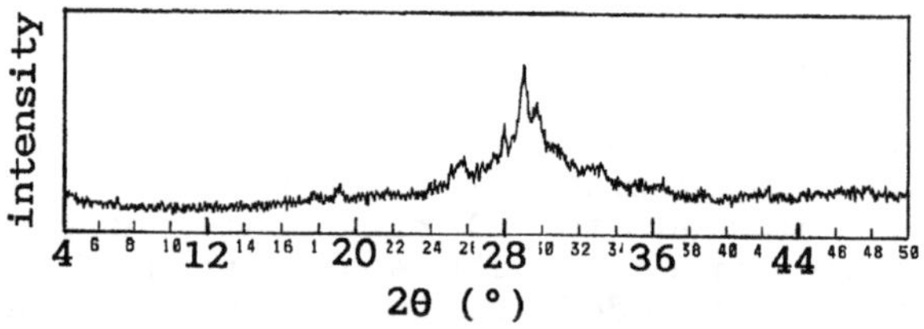

Fig. 1 X-ray diffraction pattern of starting powder

RESULTS

X-ray diffraction

The X-ray diffraction patterns of the quenched samples from powder B(4μm) were shown in Fig. 2. At any melting time, $Bi_2(Sr,Ca)_2CuO_y$(2201 phase) was mainly detected. The relatively small peaks from the unknown phase, which were similar to those from $(Sr_{1.5}Ca_{1.5})Cu_5O_y$ phase reported previously[4], were detected for shorter melting time. As the melting time became longer, $(Sr_{1.5}Ca_{1.5})Cu_5O_y$ phase disappeared and the phase similar to $Sr_{0.5}Ca_{0.5}CuO_2$ phase[2] was formed. The similar results were obtained in the case of the quenched samples from the starting powder A(9μm), except $(Sr_{1.5}Ca_{1.5})Cu_5O_y$ phase remained and little $Sr_{0.5}Ca_{0.5}CuO_2$ phase was detected at longer melting time.

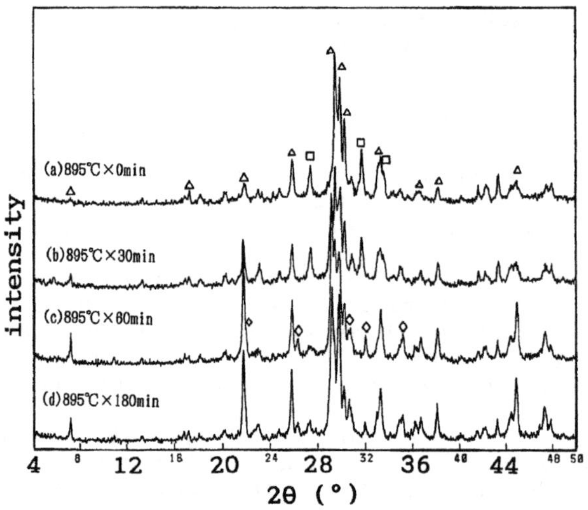

Fig. 2 X-ray diffraction pattern of quenched samples.(powder B)
△ :$Bi_2(Sr,Ca)_2CuO_y$
□ :similar to $(Sr_{1.5}Ca_{1.5})Cu_5O_y$[4]
◇ :similar to $Sr_{0.5}Ca_{0.5}CuO_2$ [2]

microstructure

The microstructure of the quenched sample from powders A and B observed by SEM was shown in Fig. 3. It is clearly seen from Fig.3(a) for powder A that the grain remained small (about 10μm) at 180min. For powder B the microstructure at 30min showed the similar texture to powder A (Fig. 3(b)), while the large columnar grain with the size of 100μm appeared and the sheet-like grain grew remarkable at 180min as seen from Fig.3(c).

Element distribution

The element distributions of quenched samples from powders A and B are shown in Fig. 4. It is seen form Figs.4(b) and (c) that the element distribution was very uniform at 30min for powder B, while the segregation of Bi-poor phase appeared at 180min. The Bi-poor phase identified from the quantitative analysis of EPMA as Bi:Sr:Ca:Cu:O=<0.01:0.63:0.41:1.00:2.12. For powder A as shown in Fig.4(a), the element distribution was uniform up to 60min and the segregation with the small grain was appeared. The composition was analyzed to be Bi:Sr:Ca:Cu:O=0.05:1.95:1.05:5.00:7.65.

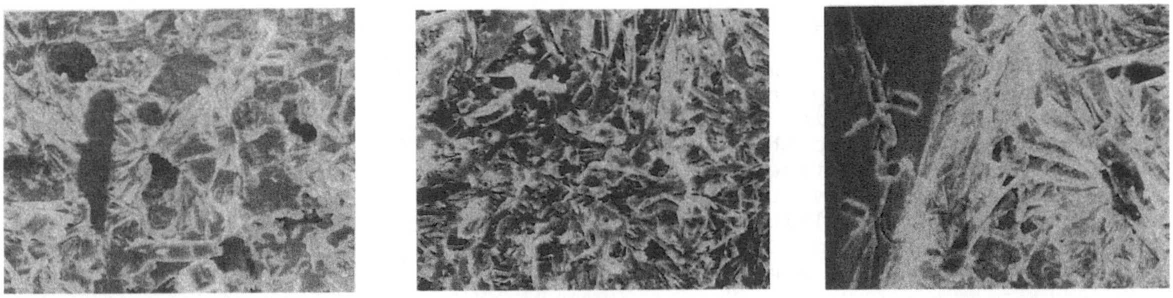

(a)895°Cx180min Powder A (b)895°Cx30min Powder B (c)895°Cx180min Powder B

Fig. 3 Microstructural changes of quenched samples during the partial melting process by SEM. 10μm

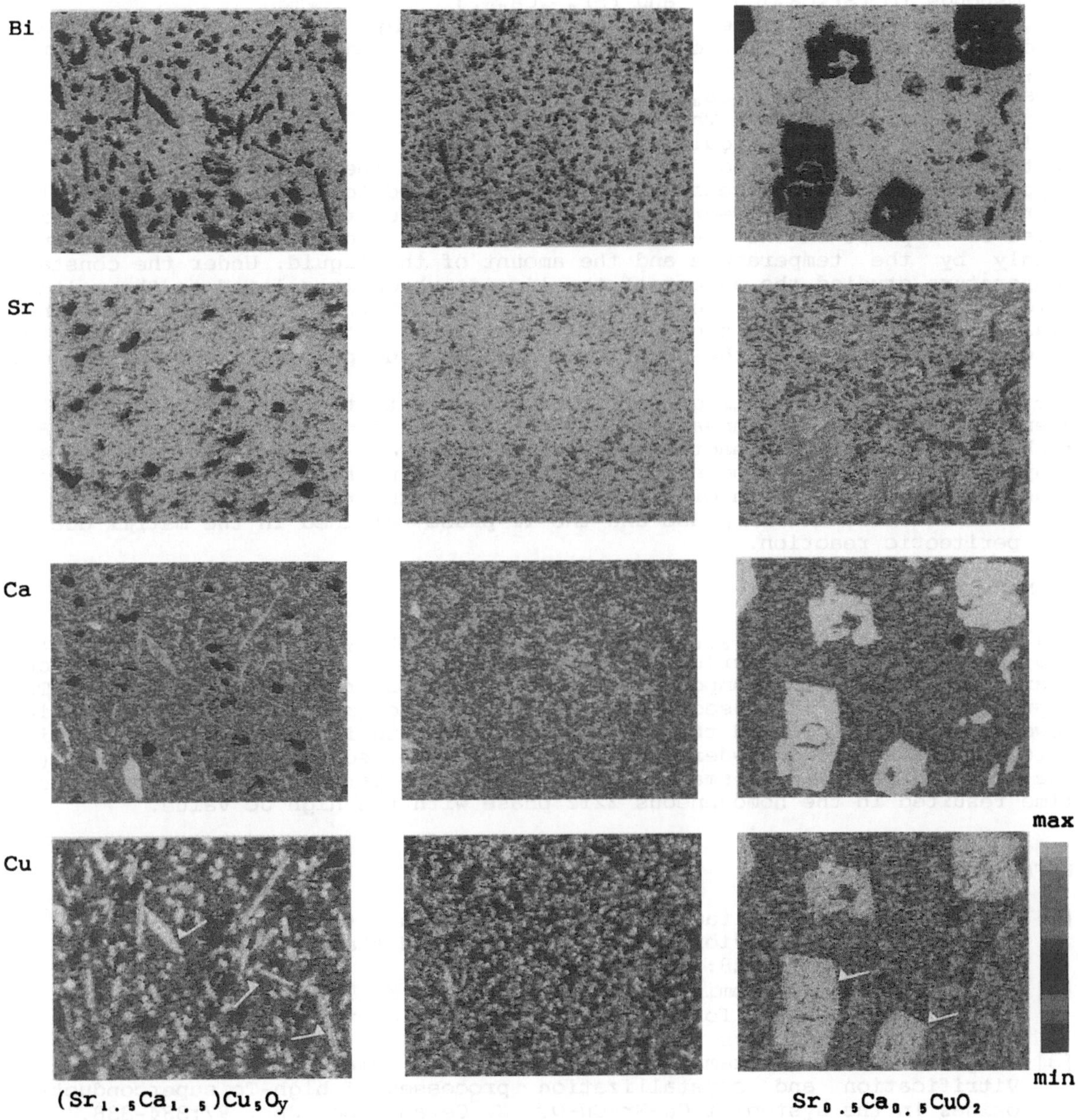

(a)895°Cx180min Powder A (b)895°Cx30min Powder B (c)895°Cx180min Powder B

Fig. 4 Element distributions of quenched samples during the partial melting process by EPMA. 100μm

Jc value after annealing

Jc value of quenched samples from powders A and B after annealing against melting time was plotted as shown in Fig. 5. In both samples from powder A and B, Jc showed the maximum at 30min and 1h, respectively. In the sample from powder A, the Jc value decreased rapidly with increasing the melting time after 30min, while the decrease of Jc more than 1h was relatively small in the sample from powder B.

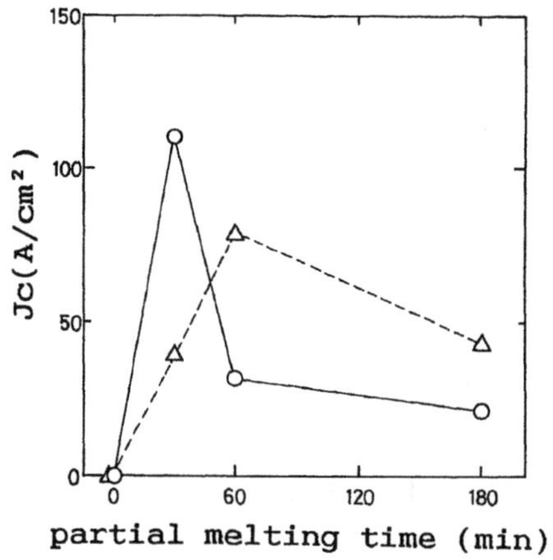

Fig. 5 Jc of annealed samples vs. partial melting time.
△ :Powder A(9μm)
○ :Powder B(4μm)

DISCUSSION

The diffraction patterns detected similar to $(Sr_{1.5}Ca_{1.5})Cu_5O_y$ and $(Sr_{0.5}Ca_{0.5})CuO_2$ phases corresponds to $(Sr_2Ca_1)Cu_5O_y$ and $(Sr_{0.6}Ca_{0.4})CuO_2$, respectively. The amorphous starting powder melted into Bi-rich liquid phase and $(Sr_2Ca_1)Cu_5O_y$ solid phase (35 phase). The 35 phase changed into $(Sr_{0.6}Ca_{0.4})CuO_2$ (11 phase) at longer melting time. The segregation rate of 11 phase in Bi-rich liquid phase during the partial melting depended on the grain size of the starting powder. It is suggested that the segregation rate may be related to the melting rate of the starting powder. The segregation rate, which corresponds to the diffusion speed of 11 phase in the Bi-rich liquid phase, is considered to be determined mainly by the temperature and the amount of the liquid. Under the constant temperature studied,the amount of the liquid, which is related to the melting rate, may affect the segregation. Assuming that the large grain, which has small specific surface area, melted slowly due to the surface reaction, the effect of grain size on the behavior of the melting process can be explained.

The Jc value after annealing was closely related to the microstructure of the quenched sample before annealing. If 11 phase is dispersed finely in the Bi-rich liquid phase, the reaction between 11 phase and liquid phase may be very easy to form the homogeneous reactant, 2212 superconductor phase. On the other hand, If the grain of 11 phase is large, the reaction proceeds only at the surface of the 11 phase and the 11 phase remained in the matrix due to the peritectic reaction.

SUMMARY

(1)The segregation of $(Sr_{0.6}Ca_{0.4})CuO_2$ by way of $(Sr_2Ca_1)Cu_5O_y$in Bi-rich liquid phase proceeded by during the partial melting of rapid-quenched amorphous starting powder, the composition of which was Bi:Sr:Ca:Cu=2:2:1:2. (2) The segregation rate increased with decreasing the grain size of the starting powder. It is considered that the segregation rate is related to the melting rate of the starting powder, which can be controlled by the grain size. (3) The starting powder with small grain size and the partial melting for a short time resulted in the homogeneous 2212 phase with the high Jc value.

REFERENCES

[1] Oka Y, Yamamoto N, Kitaguchi H, Oda K, Takada J (1989) Crystallization behavior and partially melted states in Bi-Sr-Ca-Cu-O. Jpn. J. Appl. Phys. 28: L217-L218.

[2] Oka Y, Yamamoto N, Tomii Y, Kitaguchi H, Oda K, Takada J (1989) Crystalline phases formed in the partially melted states of Bi-Sr-Ca-Cu-O. Jpn. J. Appl. Phys. 28:L801-L803.

[3] Tohge N, Tsuboi S, Akamatsu Y, Tatsumisago M, Minami T (1989) Vitrification and crystallization processes of high-Tc superconducting oxides in the system Bi-Ca-Sr-Cu-O. J. Ceram. Soc. Jpn. 97:334-338.

[4] Oda K, Kitaguchi H, Takada J, Osaka A, Miura Y, Ikeda Y, Takano M, Bando Y, Tomii Y, Oka Y, Yamamoto N, Takeda Y and Mazaki H (1988) Preparation of Bi-Pb-Sr-Ca-Cu-O high-Tc superconductor from coprecipitated oxalates. J. Jpn. Soc. Powder and Powder Metall. 36:959-964.

High Transition Temperature Superconductors, (Tl, Pb, Bi)$Sr_2Ca_2Cu_3O_z$, with Zero Resistance at 120 K

T. WADA, T. KANEKO, H. YAMAUCHI, and S. TANAKA

Superconductivity Research Laboratory, International Superconductivity Technology Center, 10-13, Shinonome 1-chome, Koto-ku, Tokyo, 135 Japan

ABSTRACT

We have synthesized new superconductors of near single-phase with zero resistasnce temperatures at 120 K employing a novel method for sintering. The samples included a nominal composition of $(Tl_{0.64}Pb_{0.2}Bi_{0.16})Sr_2Ca_3Cu_4O_z$. This sample was prepared from the mixture of pre-synthesized "1212" phase (nominal composition of $(Tl_{0.64}Pb_{0.2}Bi_{0.16})Sr_2Ca_2Cu_3O_z$), CaO and CuO powders. The mixed powder was sintered at 920 °C in O_2 gas and post-annealed at 400-600 °C in O_2 gas. The x-ray powder diffraction and high resolution transmission electron microscopy (HRTEM) revealed that the sample was isostructural with the Tl based "1223" phase. The sample exhibited a sharp superconducting transition at T_c^{onset} = 125 K and $T_c^{R=0}$ = 120 K and the superconducting diamagnetic signal at 10 K was larger than 20 % of a full Meissner effect.

KEY WORDS: oxide superconductor, Tl compound, 1223 phase, synthetic process

INTRODUCTION

The discovery of $TlBa_2Ca_{n-1}Cu_nO_{2n+3}$ (n = 1, 2, 3 and 4) was an important addition to the family of superconducting cuprates. It has been demonstrated that the critical temperature (T_c) of $TlBa_2Ca_{n-1}Cu_nO_{2n+3}$ rises with increasing n, i.e. the number of Cu-O layers.[1,2] $TlSr_2Ca_{n-1}Cu_nO_{2n+3}$ with n = 2 and 3 were first synthesized by Matsuda et al.[3] and reported that $TlSr_2CaCu_2O_7$ ("1212" phase for n = 2) had a critical temperature of 75 K and $TlSr_2Ca_2Ca_3O_9$ ("1223" phase for n = 3) 100 K. Subramanian et al.[4] succeeded in raising T_c up to 90 K for the 1212 phase and to 122 K for the 1223 phase by replacing part of Tl by Pb. The highest zero resistance temperature was obtained for $(Tl_{0.5}Pb_{0.5})Sr_2Ca_2Ca_3O_9$ at 115 K. Li and Greenblatt [5] prepared $(Tl,Bi)Sr_2Ca_1Cu_2O_z$ (Tl/Bi ~ 1 and z ~ 7) with T_c ~ 95 K by a solid state reaction using $Sr_2Tl_2O_7$ as a Tl containing precursor. Recently, Inoue et al.[6] and Subramanian et al.[7] reported an increase in T_c^{onset} to around 120 K for $TlSr_2Ca_2Ca_3O_9$ by partially substituting Bi for Tl. However, the zero resistance temperature, $T_c^{R=0}$, of these samples were lower than 115 K. At this stage, it is worthwhile to investigate T_c of $TlSr_2Ca_2Ca_3O_9$ in terms of partial substitution of both Bi and Pb for Tl. In this letter, we report a novel preparation procedure for nearly single- phase $(Tl,Pb,Bi)Sr_2Ca_2Cu_3O_z$ with zero resistsnce temperature at 120 K.

A preliminary investigation was carried out to obtain the optimum composition, x, y and m in $(Tl_{1-x-y}Pb_xBi_y)Sr_2Cu_{2+m}Cu_{3+m}O_z$, at which T_c was highest. It was concluded that the highest T_c was obtained and the phase was stable at the nominal composition with x = 0.2, y = 0.16 for m = 1. Therefore, thoughout this paper, we will deal with samples of the nominal composition of $Tl_{0.64}Pb_{0.20}Bi_{0.16}Sr_{2.00}Ca_{3.00}Cu_{4.00}O_z$ only.

PREPARATION PROCEDURE

We prepared Samples 1 and 2 by two different preparation procedures. Sample 1 was synthesized following the flow of processes given in Fig.1. At Step 1, appropriate amounts of PbO, Bi_2O_3, $SrCO_3$, $CaCO_3$ and CuO powders were mixed well to obtain a uniform mixture of the nominal composition of $Pb_{0.2}Bi_{0.16}Sr_2Ca_2Cu_3O_z$. The mixed powder was pressed into pellets and fired at 850 °C in O_2 gas flow for 10 h to obtain a precursor. Then, the fired precursor was pulverized into a powder. Tl_2O_3 was added to the powder to make a powder mixture of the nominal composition of $Tl_{0.64}Pb_{0.2}Bi_{0.16}Sr_2Ca_2Cu_3O_z$. The mixture was pressed into pellets. The pellets were wrapped with a Au foil and sintered at 920 °C in O_2 gas flow for 20 h (Step 2). The weight loss of pellet during the firing process in Step 2 was measured. The obtained compacts were clashed again and mixed with CaO and CuO powders to yield the nominal composition of $Tl_{0.64-p}Pb_{0.2}Bi_{0.16}Sr_2Ca_3Cu_4O_z$. Here p stands for the amount of Tl lost in Step 2. The powder mixture was pressed into pellet and the pellet was wrapped with a Au foil and sitered at 920 °C in O_2 gas flow for 10 h (Step 3). Finally, the obtained superconducting ceramics was successively annealed at 600 °C for 20 h, at 500 °C for 20 h and at 400 °C for 20 h in O_2 gas flow (Step 4).

Sample 2 was prepared by skipping Step 2 in the process flow shown in Fig. 1. That is, a precursor of the composition of $Pb_{0.2}Bi_{0.16}Sr_2Cu_3Cu_4O_z$ was firstly prepared, and then the powder of the precursor and Tl_2O_3 powder were mixed to the nominal composition of $Tl_{0.64}Pb_{0.2}Bi_{0.16}Sr_2Ca_3Cu_4O_z$. The mixture was sintered and annealed in the same manner as Steps 3 and 4 shown in Fig. 1.

STEP 1 SYNTHESYS OF PRECURSOR
Nominal composition : $Pb_{0.20}Bi_{0.16}Sr_{2.00}Ca_{2.00}Cu_{3.00}O_z$
Starting material : PbO, Bi_2O_3, $SrCO_3$, $CaCO_3$, CuO
Firing : 850 °C in O_2 gas for 10 h

STEP 2 PRECURSOR + Tl_2O_3
Nominal composition : $Tl_{0.64}Pb_{0.20}Bi_{0.16}Sr_{2.00}Ca_{2.00}Cu_{3.00}O_z$
Firing : 920 °C in O_2 gas for 20 h

STEP 3 PRODUCT AT STEP 2 + CaO + CuO
Nominal composition : $Tl_{0.64-p}Pb_{0.20}Bi_{0.16}Sr_{2.00}Ca_{3.00}Cu_{4.00}O_z$
Sintering : 920 °C in O_2 gas for 10 h

STEP 4 ANNEALING in O_2 GAS
Heating : 600 °C for 20 h, 500 °C for 20 h and 400 °C for 20 h

Fig. 1. Flow chart of preparation procedure of the "1223" phase with $T_c^{R=0}$ = 120 K.

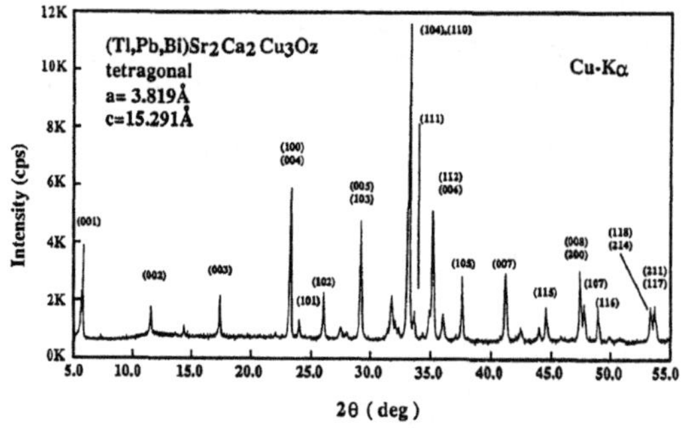

Fig. 2. Powder x-ray diffraction pattern for Sample 1. The peaks are indexed for a tetragonal unit cell with a=3.82 Å and c=15.29 Å.

RESULTS AND DISCUSSION

The identification of the phases present in the samples and the determination of the lattice constants were conducted by powder x-ray diffraction using Cu-Kα radiation. Most reflection peaks in the x-ray diffraction pattern of a sample after firing in Step 2 can be indexed by tetragonal unit cell with a = 3.8 Å and c = 12.1 Å. These values for the lattice constants are in good agreement with those reported for Tl-based 1212 phase.[4] It was likely that $(Tl,Pb,Bi)Sr_2CaCu_2O_7$ (1212 phase) had been formed in Step 2. The x-ray diffraction pattern of the samples after Step 4 is shown in Fig. 2. Most reflection peaks can be indexed employing tetragonal symmetry and the lattice constants of a = 3.819 ± 0.001 Å and c = 15.291 ± 0.002 Å. These values are close to those reported for the $(Tl_{0.5}Pb_{0.5})Sr_2Ca_2Cu_3O_9$ (1223 phase).[4] Thus, (Tl, Pb, Bi)$Sr_2Ca_2Cu_3O_9$ (1223) phase was yielded in Step 3.

The crystallographic structure of thus prepared compound was confirmed to be indeed of a 1223 type by high resolution transmission electron microscopy (HRTEM). The HRTEM experiments were made using an H-9000 ultrahigh resolution-type electron microscope operated at 300 kV. The typical lattice image and corresponding electron diffraction pattern are shown in Fig.3. This was taken with an incident beam parallel to [100]. A projected ideal structure model is inserted in this figure. Tl, Pb or Bi atoms appear as a dark spot and Cu atoms as weak dark spot. The three Cu atoms are piled between the (Tl, Pb, Bi) atoms. This HRTEM micrograph shows that this structure possesses monolayer of (Tl,Bi,Pb)-O and triple layers of Cu-O.

The composition of the sample was analyzed by using an electron probe microanalyzer (EPMA). Three large grains in the ceramic sample were analysed. The compositions of all the three grains were very close. The average composition was $Tl_{0.60}Pb_{0.25}Bi_{0.16}Sr_{2.00}Ca_{1.75}Cu_{3.95}O_z$. The ratio of Tl : Pb : Bi = 0.60 : 0.25 : 0.16 is in good agreement with the nominal atomic ratio of 0.64 : 0.2 : 0.16. This result shows that the amount of Tl lost in Step 2 was small and the sample contained all of Tl, Pb and Bi. The obtained composition may be written as $(Tl,Pb,Bi)_{1.01}Sr_{2.00}Ca_{1.75}Cu_{3.95}$. This may indicate that the main phase in the sample is the 1223 phase although the amount of Cu is close to 4 rather than 3.

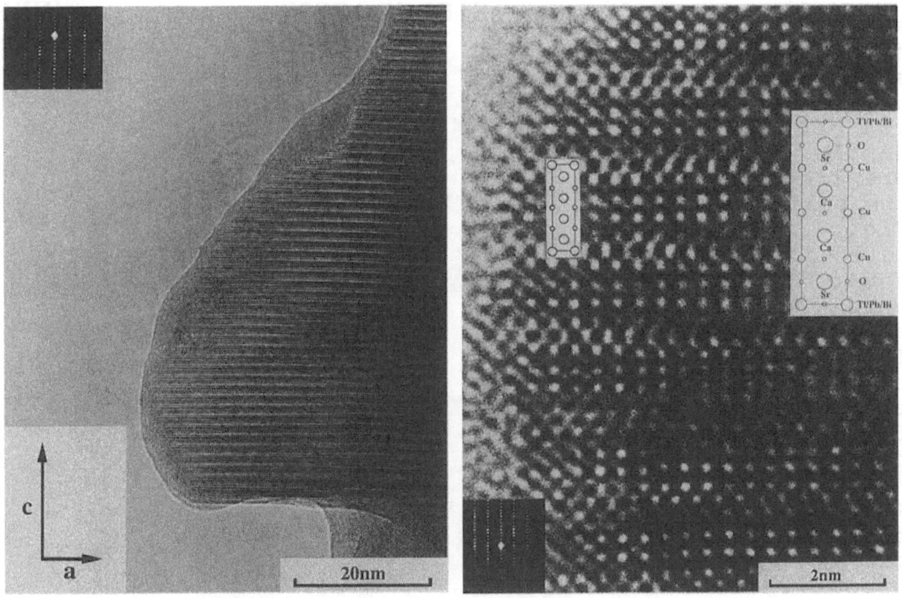

Fig. 3. HRTEM image and corresponding electron diffraction pattern of Sample 1, taking with the incident beam paralell to a axis. A projected structure model is inserted.

The electrical resistivity of the sample was measured by a standard four -probe method. The temperature dependence of electrical resistivity is shown in Fig.4. The resistivity of Sample 1 exhibited a metallic temperature dependence and a sharp superconducting transition at T_c^{onset} = 125 K and $T_c^{R=0}$ = 120 K. This value of 120 K for superconducting zero resistance temperature is 5 K higher than the highest value of 115 K previously reported for $Tc^{R=0}$ of the "1223" phase: $Tc^{R=0}$ = 115 K was reported by Subramanian et al[4] for $(Tl_{0.5}Pb_{0.5})Sr_2Ca_2Cu_3O_9$.

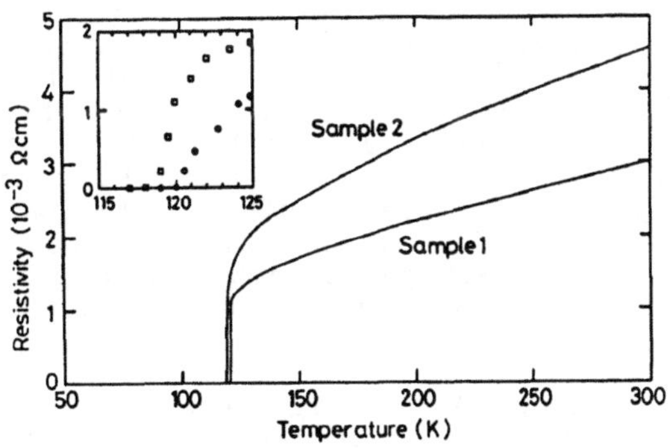

Fig. 4. Temperature dependence of electrical resistivity.

The dc magnetic susceptibility was measured by a superconducting quantum interference device (SQUID) magnetometer, (QUANTUM DESIGN Model MPMS). Figure 5 shows dc magnetic susceptibility in terms of temperature. The magnitude of magnetic field used was 10 Oe. The measurement was made in a field cooling mode. Sample 1 exhibited a sharp diamagnetic (superconducting) transition at 122 K. The figure indicates that the sample had good homogeneity and did not contain phases of the 1212 type whose T_c's were around 90 K. The magnitude of diamagnetic signal at 10 K is larger than 20 % of a full Meissner effect.

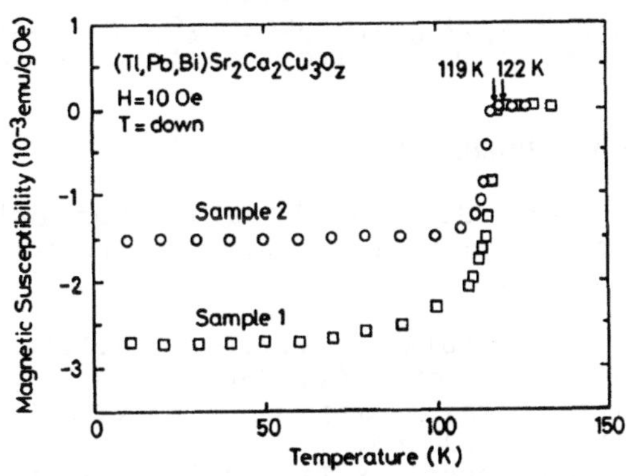

Fig. 5. Temperature dependence of dc magnetic susceptibility.

In order to investigate the advantages of this novel preparation method which is described in Fig. 1, Sample 2 was prepared by skipping Step 2 in the sample preparation procedure. The x-ray diffraction pattern for Sample 2

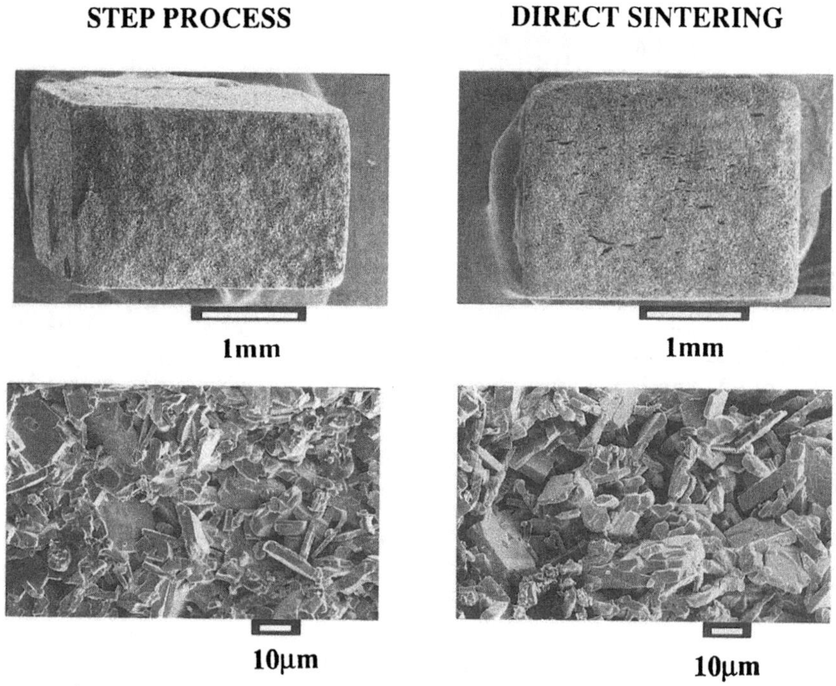

Fig. 6. SEM photographs of fracture surfaces for Samples 1 (left) and 2 (right).

shows that the sample was of near single phase Tl-based 1223 phase as was the case of Sample 1 which was prepared following the process shown in Fig. 1. The lattice constants, a and c, of Sample 2 were very close to those of Sample 1. This result shows that the 1223 phase can be easily formed from the nominal composition of 1234.

The temperature dependence of electrical resistivity for Sample 2 are shown in Fig. 4. The zero resistance temperature ($T_c^{R=0}$) of Sample 2 is 118 K. This value is a little lower than the $T_c^{R=0}$ = 120 K for Sample 1. Figure 5 shows results of dc magnetic susceptibility of Sample 2. This sample exhibited a fairly sharp superconducting transition at 119 K. The magnetic critical temperature was 3 K lower than the T_c^{mag} = 122 K for Sample 1. There are no evidence for the existence of low temperature phase such as the 1212 phase in Sample 2. The magnitude of the diamagnetic signal at 10 K was 12 % for Sample 2 compared with a full Meissner effect. The magnitude of the signal is about half of that for Sample 1. This result may indicates that Sample 1 is more homogeneous in composition than Sample 2. Figure 6 shows fracture surfaces of Samples 1 and 2. These SEM photographs indicated that Sample 1 scarcely included impurity phases and was homogeneous. Actually, Sample 1 was more homogeneous than Sample 2.

In sammary, A high T_c bulk superconductor with the (Tl, Pb, Bi)$Sr_2Ca_2Cu_3O_z$ (1223) phase was prepared employing a novel preparation method. The 1223 phase was easily synthesized in a sample of a nominal composition of 1234. Samples of the 1223 phase with good homogeneity and large Meissner signals were obtained by adding CaO and CuO to a precursor consisting of the 1212 phase. It was found that for the samples of nominal composition of $Tl_{0.64}Pb_{0.2}Bi_{0.16}Sr_2Ca_3Cu_4O_z$ prepared by the procedure described in Fig. 1, superconducting zero resistance temperature exceed 120 K.

ACKOWLEDGEMENTS

We thank S. Kawashima of Matsushita Electric Industrial Co., Ltd. for preparation of samples. We also thank A. Ichinose for his help in preparing this paper.

[1] S.S.P.Parkin, V.Y.Lee, A.I.Nazzal, R.Savoy, R.Beyers and S.J.La Placa: Phys. Rev. Lett. **61**, 750 (1988).
[2] H.Ihara, R.Sugise, K.Hayashi, N.Terada, M.Jo, M.Hirabayashi, A.Negishi, N.Atoda, H.Oyanagi, T.Shimomura and S.Ohashi: Phys. Rev. **B 38**, 11952 (1988).
[3] S.Matsuda, S.Takeuchi, A.Soeta, T.Suzuki, K.Aihara and T.Kamo: Jpn. J. Appl. Phys. **27**, 2062 (1988).
[4] M.A.Subramanian, C.C.Torardi, J.Gopalakrishnan, P.L.Gai, J.C.Calabrese, T.R.Askew, R.B.Flippen and A.W.Sleight: Science **242**, 249 (1988).
[5] S.Li and M.Greenblatt: Physica **C 157**, 365 (1989).
[6] O.Inoue, S.Adachi and S.Kawashima: Jpn. J. Appl. Phys. **28**, L1167 (1989).
[7] M.A.Subramanian, P.L.Gai, A.W.Sleight: Mat. Res. Bull., in press.

The Formation and Stability of Superconducting Phases in Bi-Sr-Ca-Cu-O System

SHIGEKI KOBAYASHI and SHIGETAKA WADA

Toyota Central Research and Development Laboratories, Inc., Nagakute, Aichi, 480-11 Japan

ABSTRACT

The effect of sintering conditions on the formation of superconducting phases has been studied using the sample with a nominal composition of $BiSrCaCu_2O_x$. The preheating at the temperature range of partial melting enhanced the formation of the high Tc phase. Further, quenching to room temperature after the preheating was more effective to increase the content of the high Tc phase.

The phase stability during annealing in various atmospheres was also examined. Oxygen stoichiometry was found to play an important role for controlling Tc's in BSCCO system. Annealing in nitrogen caused to increase the Tc of low Tc phases and decrease the Tc of high Tc phase. The effect was more prominent in the low Tc phase and its Tc could be increased up to 90-100K.

Key Words; two-step sintering, quenching, oxygen stoichiometry, annealing

INTRODUCTION

The Bi-Sr-Ca-Cu-O (BSCCO) system has three different superconducting phases with similar crystallographic structures of c=2.4, 3.0 and 3.7 nm.[1] The problem of this material is the sensitivity of the properties to the processing conditions, especially the sintering temperature. As shown later, a variation of 5-10 ℃ in sintering temperature causes a significant variation in the properties. In this study, the optimization of processing conditions has been explored and the mechanisms for the formation of superconducting phases are discussed. Although the addition of lead is reported to enhance the formation of high Tc phase,[2] undoped (Pb-free) samples were used in this study for the simplicity of the system. At the same time, the effect of annealing after sintering was studied and a significant effect of oxygen content on the Tc's of superconducting phases was found.

EXPERIMENTAL

A powder prepared by a solid-state reaction was used in the major part of this study. In a part of annealing experiment, a commercial coprecipitated powder was also used. The nominal composition of both powders was $BiSrCaCu_2O_x$. The details of preparation conditions were published elsewhere.[3] Sintering was performed in air at 865-900℃ with various heating schedules.

After sintering, some of specimens were annealed in oxygen and nitrogen at 200-800 ℃. For reducing the scatter of experimental data, the same sample was annealed successively for a series of experiment. The samples were evaluated by the AC susceptibility using a self-inductance measuring system. The susceptibility was measured at 1 kHz for the bulk samples during cooling from room temperature to 4.2K.

RESULTS AND DISCUSSION

Optimization of Sintering Conditions

Two types of heating schedules were compared for sintering; a simple pattern with one-step of holding the temperature and one with two-step holding. The first step of the two-step holding was varied over the temperature range of partial melting just below the complete melting and the holding time was 20 min. The temperature of second step was kept at 870 ℃

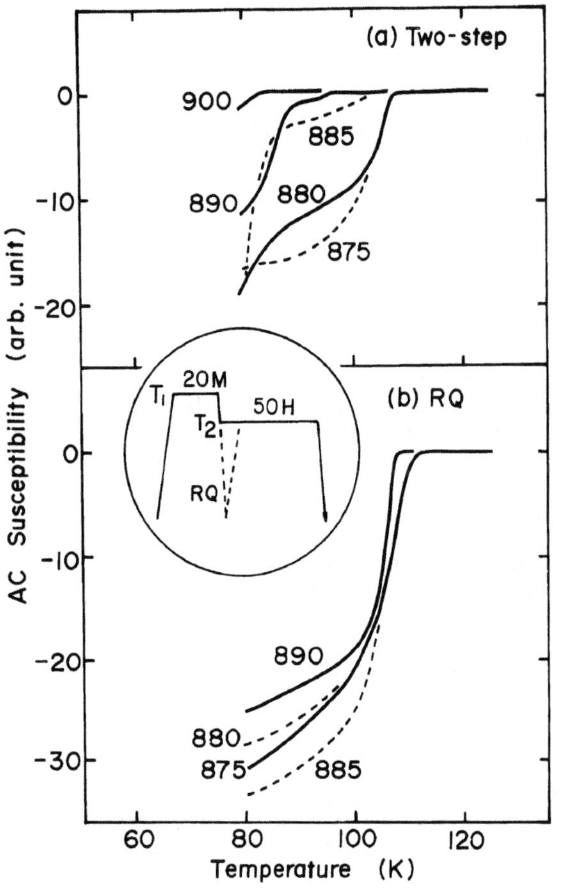

Fig. 1. The variation of Ac susceptibility with the first holding temperature(T_1). the second holding temperature(T_2) is 870℃. (a) The simple two-step pattern, (b) addition of quenching from T_1.

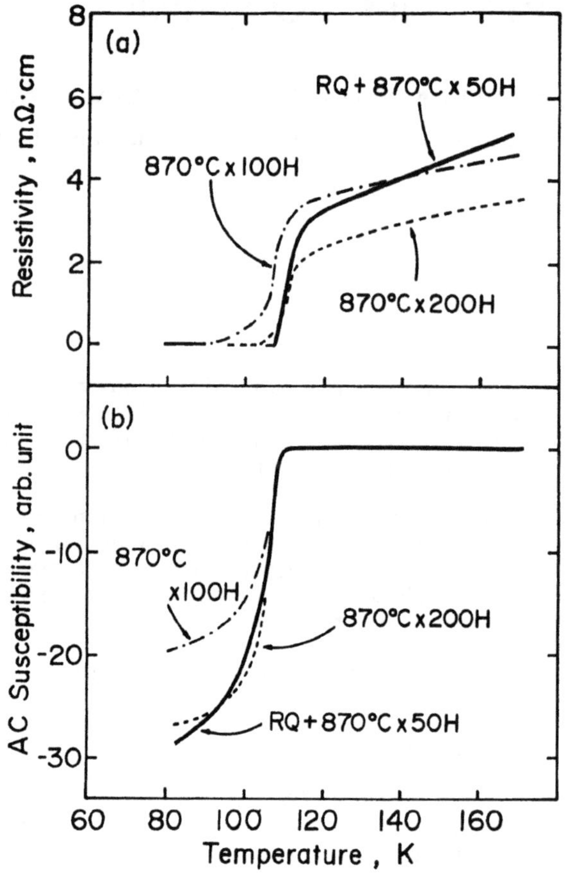

Fig. 2. Comparision of (a) resistivity and (b) AC susceptibility data between the specimens treated by simple two-step patterns and two-step patterns with quenching process.

and the holding time was 50 hrs. Generally, the samples sintered with two-step patterns showed larger signal of AC susceptibility, indicating that the partial melting of specimens was very effective to nucleate superconducting phases. As shown in Fig. 1a, the holding at 875-880 ℃ enhanced the formation of high Tc phase. The amount of the high Tc phase was decreased for the samples sintered at higher temperatures, and at 900 ℃ the specimens finally melted resulting in the rapid deterioration of properties. The optimum temperature for the first step was selected to be 880 ℃. The x-ray diffraction analysis indicated that both the high Tc and low Tc phases were formed and the partial melting enhanced the texture formation. Since the zero resistivity was not obtained above 77K for the samples treated with the optimum two-step pattern, the holding time at 870 ℃ was increased. As shown in Fig. 2, zero resistivity was attained at 90K for 100 hrs, and at 104K for 200 hrs. From the value of susceptibility at 4.2K, the volume of superconducting materials was estimated to be as high as 90% for the sample sintered for 200 hrs.

During the course of experiment, the effective way to shorten the sintering time was accidentally found without losing the high transition temperature. This was the addition of quenching process to the two-step pattern as shown in the inset of Fig. 1. By this quenching, the level of AC susceptibility was increased significantly and the optimum condition could be widened to 875-890 ℃ (Fig. 1b). Although the holding time at 870 ℃ was 50 hrs, the zero resistivity was attained at 107K and the level of AC susceptibility was almost same as that for the 200 hr treatment of simple two-step pattern without quenching (Fig. 2). In the sample treated with the simple two-step pattern, texture formation was noticed and the trend became more prominent with increasing the holding time at 870 ℃. However, the texture formation was less evident in the sample treated with quenching process though the content of high Tc phase was high.

XRD analyses were performed on the as-quenched samples to elucidate the effect of partial melting. For the samples quenched from T≧885℃, the formation of the phase with c=2.4 nm was detected. At T≦ 885 ℃, the low Tc phase was always formed, but the high Tc phase was noticed in the narrow temperature range of 880-885 ℃. It could be concluded from these observations that the enhancement of the high Tc phase formation by preheating was mainly due to the nucleus formation of the high Tc phase in the temperature range of partial melting. At this moment, the reason for the enhancement of high Tc phase formation due to quenching treatment is not clear, but this should be related to the behavior of liquid phase formed by partial melting. The liquid phase might be converted to amorphous phase during quenching and then become nucleus for the high Tc phase on the further heat treatment. The fact that the texture formation was retarded by quenching treatment is consistent with the above idea, but further study is clearly needed to conclude it.

Annealing Effect

The samples sintered in air at 865-875 ℃ were used for annealing experiments. They contained both the high Tc and low Tc phases. Annealing was done in oxygen and nitrogen at 200, 300, 400, 500, 700, 600, 750 ℃, successively using the same specimen. After annealing for 2 hrs, specimens were furnace-cooled with a rate of ~100℃/min. Although detailed behavior slightly depended on the processing conditions, the typical trend can be seen in Fig. 3. The specimen used for the experiments in Fig. 3 was prepared from a coprecipitated powder. The samples prepared from a solid-state reaction powder also showed basically the same behavior. [4]

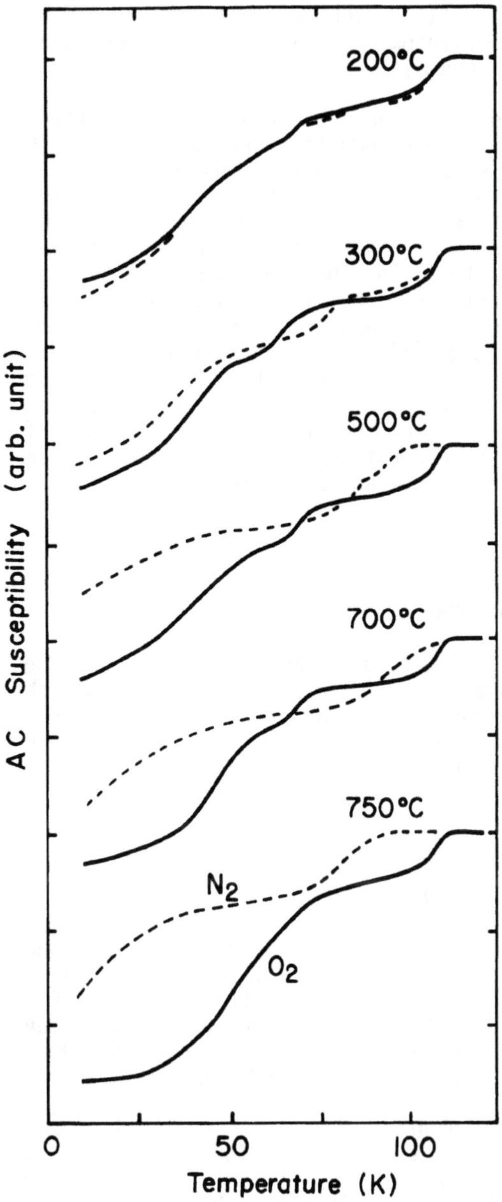

Fig. 3. Variation of AC susceptibility with annealing in oxygen and in nitrogen at 200-750℃.

At 200 ℃, annealing effect was small and at T≧ 300℃ the Tc's of both low Tc and high Tc phases were changed as shown in Figs. 3 and 5. The effect was more prominent in the low Tc phase and at higher temperature. But in the samples annealed in nitrogen at T≧ 700 ℃, the behavior was slightly different for the samples annealed in nitrogen because of reducing reaction. This was clearly visualized by the red color surface of the samples annealed at T≧ 750℃, and at 780 ℃ the superconductivity was completely lost (Fig. 4). The properties of the samples annealed in oxygen was essentially independent on the annealing temperature. Generally, annealing in nitrogen led to higher Tc in low Tc phase and lower Tc in high Tc phase, therefore the difference of the Tc between low Tc and high Tc phases became small. XRD analyses indicated no essential change due to annealing at T< 700 ℃, although the lattice parameters were changed accordingly with the value of Tc. In the sample annealed in nitrogen at T≧ 700 ℃, the intensities of X-ray peaks associated with the high Tc phase decreased. At the same time, the transition for the low Tc phase became obscure on the AC susceptibility curve and Tc could not be determined without any ambiguity. In Fig. 5, the data of high Tc phase at 700 and 750 ℃ in nitrogen are therefore leaving room for discussion, i.e., these data might be for the low Tc phase. In order to make this point clear, further work is needed, especially annealing experiments using specimens containing only low Tc phase (or high Tc phase).

As shown above, annealing did affect the superconducting properties in BSCCO system and the effect was opposite for the low Tc and high Tc phases. These phenomena are considered to be associated with the variation of oxygen content in the specimens. This was confirmed by the reversibility of properties and the corresponding weight change.[4] From the present study, it is now clear that the oxygen stoichiometry also plays an important role to determine the Tc's of superconducting phases in BSCCO system as already known in YBCO.[5]

ACKNOWLEDGEMENTS

The authors wish to acknowledge the members of superconductor research project, especially Dr. T. Hioki, Mr. N. Suzuki and Mr. M. Ohkubo for their interest and helpful discussion.

REFERENCES

1. Takayama-Muromachi E, Uchida Y, Ono A, Izumi F, Onoda M, Matsui Y, Kosuda K, Takekawa S, Kato K (1988) Identification of the superconducting phase in the Bi-Ca-Sr-Cu-O system. Jpn. J. Appl. Phys. 27: L365-L368.

2. Takano M, Takada J, Oda K, Kitaguchi H, Miura Y, Ikeda Y, Tomii Y, Mazaki H (1988) High-Tc phase promoted and stabilized in the Bi,Pb-Sr-Ca-Cu-O system. Jpn. J. Appl. Phys. 27: L1041-L1043.

3. Kobayashi S, Saito Y, Wada S (1989) The effect of sintering conditions on the formation of the high-Tc phase in the Bi-Sr-Ca-Cu-O system. Jpn. J. Appl. Phys. 28: L772-L774

4. Kobayashi S, Wada S (1989) Annealing effect in Bi-Sr-Ca-Cu-O superconductors. Jpn. J. Appl. Phys. in press

5. Takayama-Muromachi E, Uchida Y, Ishii M, Tanaka T, Kato K (1987) High Tc superconductor YBa2Cu3Oy - Oxygen content vs Tc relation. Jpn. J. Appl. Phys. 26: L1156-L1158

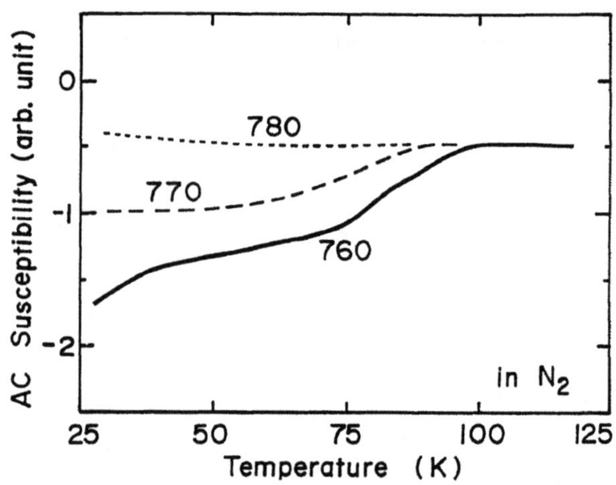

Fig. 4. Variation of AC susceptibility with annealing in nitrogen at 760-780℃. The superconductibity is completely lost because of reducing reaction.

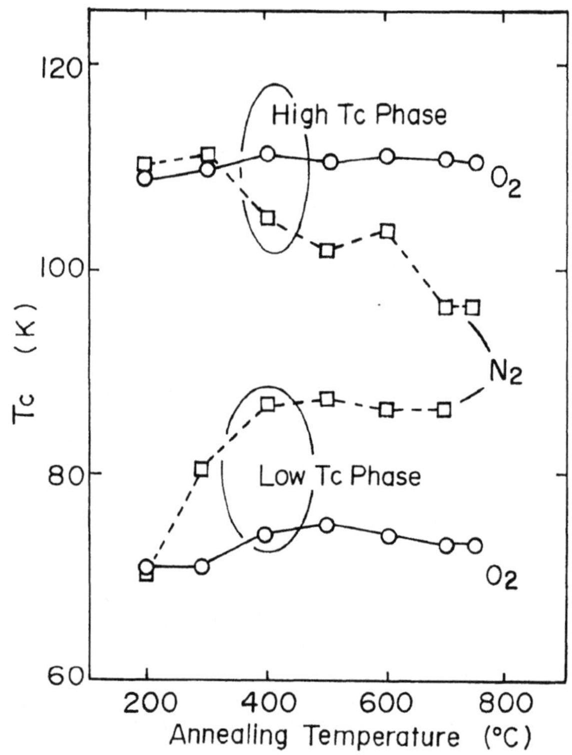

Fig. 5. The effect of annealing in oxygen (solid lines) and in nitrogen (dotted lines) on the Tc's of low Tc and high Tc phases. The annealing effect is more prominent in the low Tc phase.

Growth of High-T_c Phase and Partial Melting in Pb-Doped Bi-Sr-Ca-Cu-O

K. OHBA[1], A. ISHIDA[1], K. IWASAKI[1], H. KUWAJIMA[1], H. NOJI[2], A. KIRIHIGASHI[2], and A. OOTA[2]

[1] Technical Research and Development Laboratories, Topy Industries Ltd., Akemi-cho, Toyohashi, Aichi, 440 Japan

[2] Department of Electrical and Electronic Engineering, Toyohashi University of Technology, Tempaku-cho, Nagoya, 440 Japan

ABSTRACT

Growth of high-T_c phase and partial melting are studied on the oxides with nominal composition $Bi_2Pb_{0.6}Sr_2Ca_2Cu_3O_x$. A single high-$T_c$ phase can be produced from the partially melted phase in a short sintering-period of time in air when the re-press process is introduced. However, the disturbance in current paths at grain boundaries lowers $T_c(\rho=0)$ far below 100K. The prolonged sintering improves this and enhances $T_c(\rho=0)$ above 100K. All the results show the feasibility to single out the high-T_c phase from the partial melt, and suggest the slow reaction to link the grains and to open up current paths.

KEY WORDS: Bi-Pb-Sr-Ca-Cu-O, high-T_c phase, partial melting, current path

INTRODUCTION

Since the discovery of two superconducting transitions at 80 and 110K in Bi-Sr-Ca-Cu-O (BSCCO) system [1], much effort has been made to single out the high-T_c (110K) phase. Sunshine et al [2] and Takano et al [3] have shown that the partial substitution of Pb for Bi in BSCCO promotes and stabilizes the high-T_c phase. The major understanding at the present stage in Pb-doped BSCCO is the efficacy of long sintering-period of time under low oxygen pressure to produce a single high-T_c phase [4], a new mode of one-dimensional structural modulation along the b axis [5] and the location of Pb atoms in the Bi-O layers [6].

The mechanism of the growth of the high-T_c phase in Pb-doped BSCCO has not been solved yet. Hatano et al [7] have pointed out the participation of the partially melted liquid phase in the growth process of the high-T_c phase. We showed that the partial melting enhances the growth rate of the high-T_c phase leading to the volume fraction above 80% within 40h, but it causes a disturbance in current paths and suppresses $T_c(\rho=0)$ far below 100K [8].

The purpose of this study is to make clear the growth process of the high-T_c phase from the partially melted phase in Pb-doped BSCCO prepared through the solid-state reaction in air. The enphases are on the effect of re-press process on the growth of the high-T_c phase and on the formation of transport current paths.

EXPERIMENTAL

The samples were prepared by a solid-state reaction. The nominal composition of the samples was fixed to $Bi_2Pb_{0.6}Sr_2Ca_2Cu_3O_x$, because the growth of the high-T_c phase is enhanced in it [9]. The mixtures of Bi_2O_3, PbO, $SrCO_3$, $CaCO_3$ and CuO were calcined at 800°C for 12h in air. The resultant powders were pulverized, mixed and pressed into a disk shape under the pressure of 1 ton/cm^2. The pellets were sintered at 840°C just below the partial melting temperature $T_{pm}\simeq 842$°C [8], for 20h in air (1st sintering). Subsequently, they were pulverized, mixed and re-pressed into a disk shape under the pressure of 1 ton/cm^2 (re-press process). After that, they were re-sintered in air at the temperature T_s from 835 to 855°C refering to $T_{pm}\simeq 842$°C [8], and for the period of time t_s from 10 to 120h.

RESULTS AND DISCUSSION

The SEM images of fractured surfaces are shown in Fig.1. As can be seen, the surface structures depend on whether or not the sample is subjected to the re-press process. When the sample is sintered at 840°C without the re-press, microcrystals of the high-T_c phase grow up and become coarse during sintering [8]. However, an introduction of the re-press process makes them finer during the subsequent sintering. A prolonged sintering after the re-press does not cause any notable change in the SEM images.

To obtain information on partial melting, the DTA study was made upon heating with a ratio of 10°C/min. Typical curves of DTA are shown in Fig.2. As can be seen, neither the sample under 2nd sintering at T_s = 840°C nor the sample before the re-press shows any evidence for partial melting. They only show an endothermic peak due to the melt. In comparison, the samples under 2nd sintering at T_s =845 and 855°C show the peak due to partial melting [8] besides the melt peak. The value of T_{pm} from the DTA curve turns out to be 842°C independent of t_s, which is in good agreement with our previous study [8]. The DTA results show that the sample goes to the partial melting phase when it is subjected to 2nd sintering at high temperatures $T_s>T_{pm}$ after the re-press.

To examine the effect of the re-press, several samples were subjected to 1st and 2nd sintering without the re-press process. Typical XRD patterns at low angles with Cu K_α radiation for the samples thus prepared are compared with the ones prepared through the re-press process in Fig.3. As can be seen, an introduction of the re-press process enhances the peak intensity of the 002 reflection for the high-T_c phase and suppresses the low-T_c (80K) phase. The efficacy of the re-press process is also convinced in other cases. In Fig.4, XRD patterns with Cu K_α radiation are typically shown for the samples prepared under 2nd sintering at T_s =845 and 840°C for 10h after the re-press. It is evident that the sample under T_s =845°C has a single high-T_c phase and its XRD patterns agree well with the ones reported previously [4]. However, the sample under T_s =840°C has a major high-T_c phase, together with traces

Fig.1. SEM images of fractured surfaces.

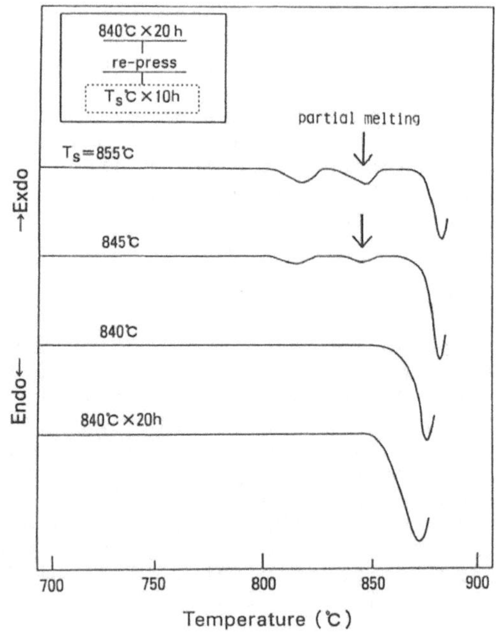

Fig.2. DTA curves upon heating.

of the low-T_c and semiconducting phases. In connection with the DTA results, we can say that a single high-T_c phase can be produced from the partial melt in a very short time (t_s =10h) in air by introducing the re-press process.

The magnetic susceptibility $\chi(T)$ was measured upon cooling (Meissner process). The data under H =10Oe are shown for the samples subjected to 2nd sintering at T_s =845°C, together with the sample before the re-press. As can be seen, the $\chi(T)$ is consistent with the result of XRD. All the samples prepared under 2nd sintering fall on the same curve independent of t_s. The $\chi(T)$ shows a sharp drop at ∿108K and no-two step transition due to the low-T_c phase. In contrast, the sample before the re-press, which is composed of the high-T_c and low-T_c phases in the XRD patterns, shows a marked two-step transition.

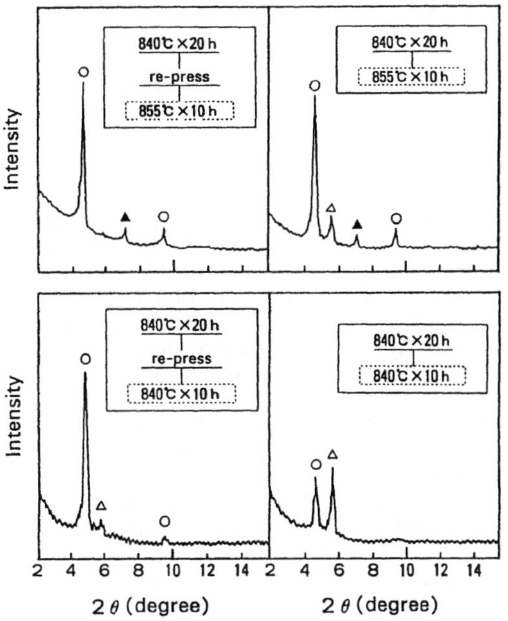

Fig.3. XRD patterns with Cu K_α radiation. O , high-T_c phase; Δ, low-T_c phase; ▲ , semiconducting phase.

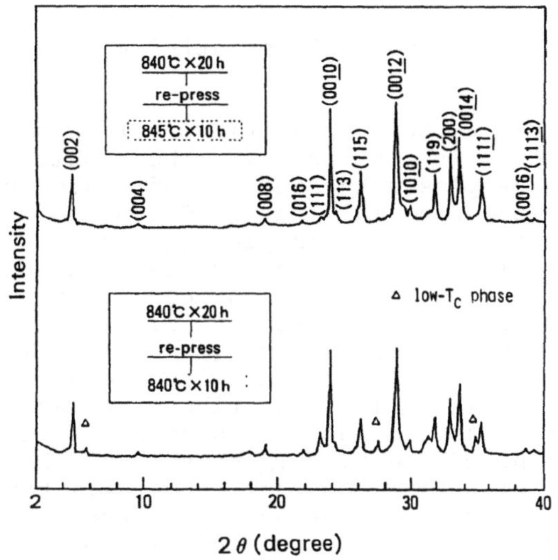

Fig.4. XRD patterns with Cu K_α radiation.

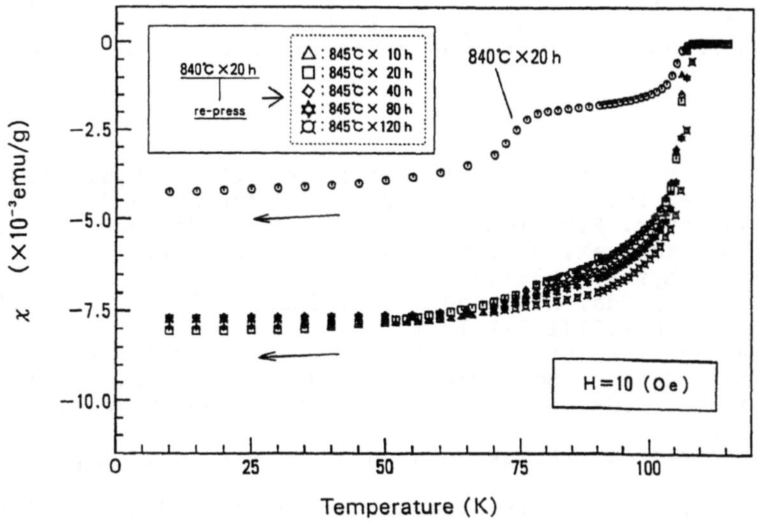

Fig.5. Magnetic susceptibility measured by a SQUID magnetometer under H=10Oe.

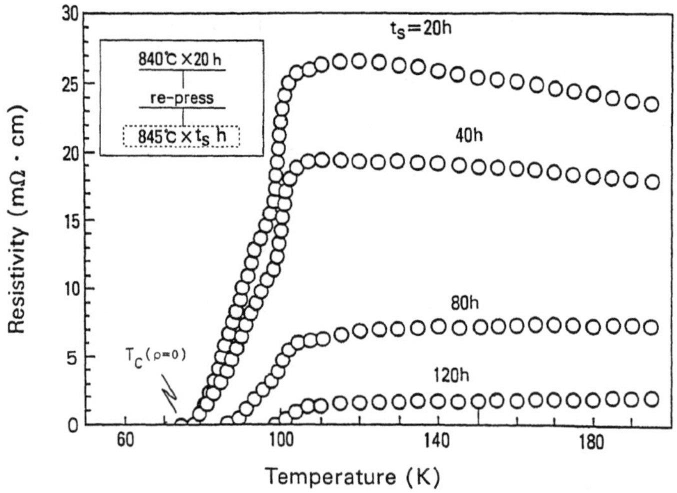

Fig.6. Electrical resistivity as a function of temperature.

In Fig.6, the resistivity ρ(T) is shown for the same samples. The result of ρ(T) is in contrast with the results of XRD (Fig.4) and susceptibility (Fig.5). The samples subjected to 2nd sintering for $t_s \leq 40$h show a semiconducting behavior in normal-state and show a zero-resistivity at T_c(ρ=0) far below 100K. In contrast, the samples subjected to a prolonged sintering of $t_s \geq 80$h show a metallic behavior in normal-state and the value of T_c(ρ=0) is enhanced with increasing t_s. Such an increase of T_c(ρ=0) with t_s is never due to an increase of the fraction v, since the sample has a single high-T_c phase even in a short sintering-time (t_s =10h). Moreover, we have no evidence for new superconducting phases with the same structure as the high-T_c phase. Accordingly, it would be attributed to an improvement for disturbance in current paths at the grain boundaries. Considering the SEM and XRD results, the grains of the high-T_c phase would be rapidly grown at the initial stage of 2nd sintering and then chemically bonded with each other through the intergrain reaction during a prolonged sintering. The slowness of this reaction would require a long sintering period to join the grains and to open up the current paths leading to the improvement of T_c(ρ=0).

CONCLUSIONS

(1) The high-T_c phase can be singled out from the partial melt in a short sintering-time in air when the re-press process is introduced.
(2) The prolonged sintering is indispensable to join the grains and to open up the current paths leading to the enhancement of T_c(ρ=0) above 100K.

REFERENCES

1. H.Maeda, Y.Tanaka, M.Fukutomi and T.Asano; Jpn.J.Appl.Phys.27(1988)L209.
2. S.A.Sunshine, T.Siegrist, L.F.Schneemeyer, D.W.Murphy, R.J.Cava, B.T. Batrogg, R.B.van Dover, R.M.Fleming, S.H.Glarum, S.Nakahara, R.Farrow, J.J.Krajewski, S.M.Zahurak, J.V.Waszczak, J.H.Marshall, P.Marsh, L.W.Rupp, Jr. and W.F.Peck; Phys.Rev.B38(1988)893.
3. M.Takano, J.Takada, K.Oda, H.Kitaguchi, Y.Miura, Y.Ikeda, Y.Tomii and H. Mazaki; Jpn.J.Appl.Phys.27(1988)L1041.
4. U.Endo, S.Koyama and T.Kawai; Jpn.J.Appl.Phys.27(1988)L1476.
5. S.Ikeda, K.Aota, T.Hatano and K.Ogawa; Jpn.J.Appl.Phys.27(1988)L2040.
6. H.Nobumasa, T.Arima, K.Shimizu, Y.Otsuka, Y.Miura and T.Kawai; Jpn.J.Appl. Phys.28(1989)L57.
7. T.Hatano, K.Aota, S.Ikeda, K.Nakamura and K.Ogawa; Jpn.J.Appl.Phys.27(1988) L2055.
8. A.Oota, K.Ohba, A.Ishida, A.Kirihigashi, K.Iwasaki and H.Kuwajima; Jpn.J. Appl.Phys.28(1989)L1171.
9. A.Oota, A.Kirihigashi, Y.Sasaki and K.Ohba; Jpn.J.Appl.Phys.27(1988)L2289.

Formation of a High T_c Phase in Pb-Bi-Sr-Ca-Cu-O Superconductor

S. NISHIKIDA
Sumitomo Metal Industries Ltd., 16, Sunayama, Hasaki, Ibaraki, 314-02 Japan

ABSTRACT

The formation of the 2223 phase (high Tc phase) in Pb-Bi-Sr-Ca-Cu-O superconductor was examined by observing 2223 crystals formed at interface between layers of the 2212 phase (low Tc phase) and Pb-Sr-Ca-Cu-O mixed oxides. At the interface the growth of the 2223 crystals was needle-like and the growth direction was into both the 2212 layer and the mixed oxides layer. The results indicated that the nucleation of the 2223 phase occurred slowly compared to the growth rate of the 2223 crystal. The growth was controlled by the diffusion of Bi atoms to the mixed oxides layer and Ca and Cu atoms to the 2212 layer.

KEY WORDS:high-Tc-superconductor, high-Tc-phase, Bi-Pb-Sr-Ca-Cu-O, formation, nucleation

INTRODUCTION

In the Bi-Sr-Ca-Cu-O superconductors, while the low Tc phase (2212 phase) is easily formed in a wide range of temperature and compositions, the high Tc crystals (2223 phase) is hardly formed as a pure phase. Recently, some effective procedures to increase the amounts of the 2223 phase, for example, Pb addition or sintering under low oxygen partial pressure, were reported. Furthermore, some formation mechanisms of the 2223 phase have been proposed to explain these effects [1][2][3][4]. Most of these mechanisms pointed out that the growth of the 2223 phase was enhanced by the partial melting at the reaction points of the formation of the 2223 phase. These mechanisms, however, did not clarify the crystallographic or compositional relationship between the 2223 phase and the 2212 phase.
In this study, as the first step for understanding the fundamental reactions, the formation of the 2223 phase was examined by observing the crystals formed at the interface between the 2212 phase layer and Pb-Sr-Ca-Cu-O mixed oxides layer.

EXPERIMENTAL

Powders for interface reaction were prepared by a solid state reaction. The composition of the 2212 phase, which was predetermined from the quantitative analysis by electron-probe micro-analysis (EPMA) on samples sintered at temperature as low as not to form the 2223 phase, is shown later in detail. The starting materials were high-purity powders of Bi_2O_3, $SrCO_3$, $CaCO_3$, CuO and PbO. These powders were mixed and calcined at 800°C for 12 hours. The calcination was repeated twice. They were pulverized, mixed, pressed into a disk and finally sintered at 840°C for 24 hours in air. These sintered bodies were pulverized again to use as starting powder for interface reaction. Pb-Sr-Ca-Cu-O mixed oxides powders, of which composition is shown later, were prepared by only calcining at 800°C for 12 hours in air. The powder of the 2212 phase were pressed into a disk and the Pb-Sr-Ca-Cu-O mixed oxides powders were pressed on the surface of the 2212 disk. This layered complex sample was fired at 840°C in air. A cross-section of the interface of this sample was polished and observed by secondary electron microscope (SEM) and EPMA. The cross-sections were also observed by transmission electron microscope (TEM). Specimens for TEM were prepared by ion-thinning method.

RESULTS and DISCUSSION

Preliminary experiments revealed that samples with a composition of Bi1.71 $Pb_{0.43}Sr_{1.71}Ca_{2.14}Cu_3O_y$ formed the 2223 phase by firing at 835°C to 850°C for 100 hours. The calcined powders of the same composition were pressed into a disk and fired at 820°C for 24 hours. The analysis of X-ray powder diffraction showed that there were the 2212 phase, Ca_2PbO_4, Ca_2CuO_3 and CuO in the sintered body. The 2223 phase did not form at all. The chemical composition of the 2212 phase in this sample was A in Table 1 by quantitative analysis by EPMA. Furthermore, the qualitative line analysis by EPMA for Bi atom, shown in Fig.1, revealed that Bi atoms were included only in the 2212 phase and not in the other oxides. Based on these results, nominal composition of the mixture of the other oxides (complex oxides) were determined by subtracting the composition of the 2212 phase from that of the starting composition lead the concentration of Bi atom to be zero. The average composition of the complex oxides was listed in Table 1 as B. These two kinds of oxides, A, the 2212 phase and B,the average composition of complex oxides, were prepared by the methods described in experimental section seperately. Figure 2 shows the X-ray powder diffraction pattern of the complex oxides which was obtained by

Table 1 Chemical compositions of the 2212 phase (A) and the Pb-Sr-Ca-Cu complex oxides. (B) (atomic ratio)

	Bi	Pb	Sr	Ca	Cu
A	1.71	0.3	1.5	1.5	2.0
B	-	0.13	0.21	0.64	1.0

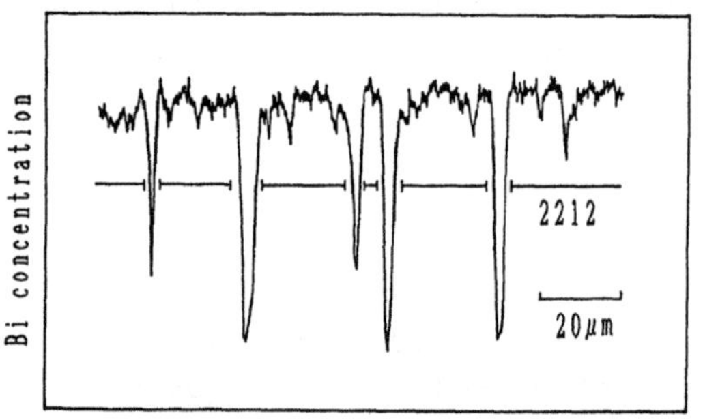

Fig.1 Spectrum of line analysis for Bi in sample sintered at 820°C.

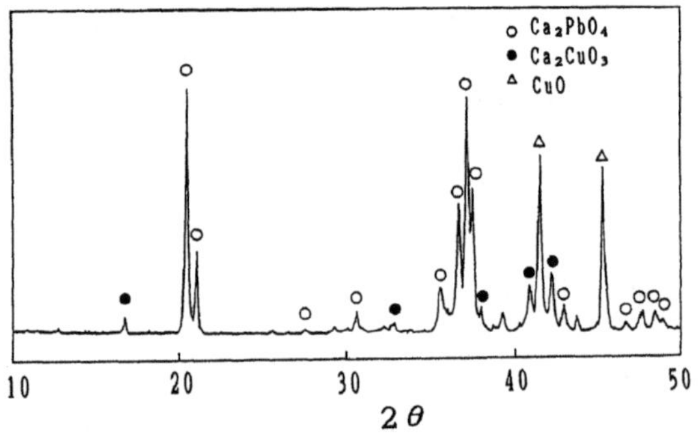

Fig.2 X-ray powder diffraction pattern of the residual oxides. (CoK radiation)

calcining the nominal composition of B in Table 1 at 800°C. The X-ray diffraction pattern reveals that the oxides consist of Ca_2PbO_4, Ca_2CuO_3 and CuO. The calcination of the nominal composition of A resulted in the prevailing the 2212 and a trace of Ca_2PbO_4.

Figure 3 shows SEM photograph of cross section of the layered sample fired at 840°C for 24 hours. Needle like crystals, the 2223, formed at the interface between A and B. As shown in Fig.3, the 2223 crystals did not uniformly cover the interface. This result suggests that in the formation reaction of the 2223 phase, nucleation of the 2223 phase occurs very slowly compared to the growth rate. The growth of the 2223 crystals were directed into both the 2212 disk (A) and the complex oxides disk (B). This means that the crystal growth of the 2223 phase is controlled by the diffusion of Bi atoms in the complex oxides layer and that of Ca and Cu atoms in the 2212 disk, respectively, through the 2223 crystals formed or the interface between the 2223 crystals and the matrix.

Furthermore, by the observation of the cross section, partial melting, as mentioned by some investigators, may occur in the complex oxides layer. In the 2212 layer, however, the partial melting seems not to participate in the reaction, because after the reaction there are only the 2212 phase and the 2223 phase in the 2212 layer and the both phases do not melt at the reaction temperature. Recently, Oota et al.[5] reported that the partial melting enhanced the growth rate of the 2223 phase, but it was not inevitable to form the 2223 phase. Both results consist that partial melting is not the necessary condition of the formation of the 2223 phase.

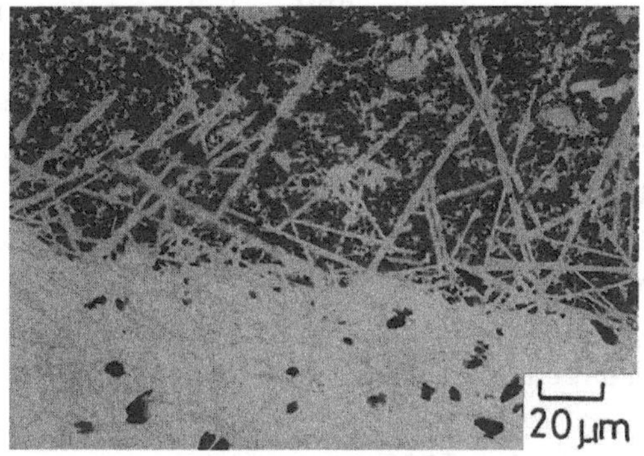

Fig.3 SEM image of the interface between the 2212 phase and the Pb-Sr-Ca-Cu-O complex oxides fired at 840°C for 24 hours.

Fig.4 SEM image of the interface between the 2212 phase and the Pb-Sr-Ca-Cu-O complex oxides fired at 840°C for 80 hours.

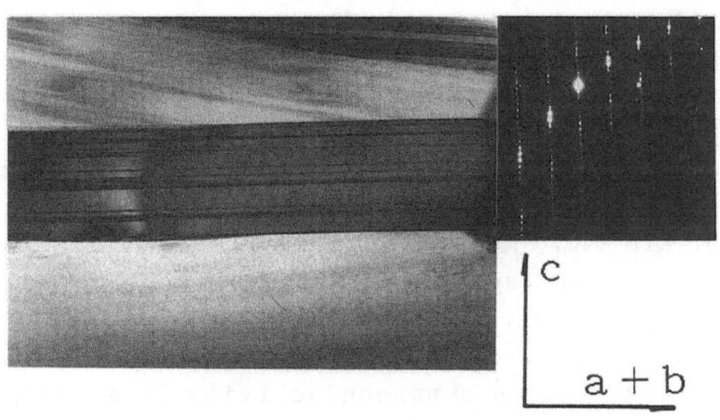

Fig.5 TEM image of the 2223 crystal formed in the 2212 layer and the electron diffraction pattern.

Figure 4 shows a SEM photograph of interface of sample fired at 840°C for 80 hours. Assuming that the growth rate of the 2223 phase obeys the parabolic rate law, the growth rate constants calculated from the length of the 2223 crystals in the complex oxides formed by firing for 24 and 80 hours in Fig.3 and Fig.4, respectively, are about 8x10-10 cm^2/sec. This value agrees with ordinary diffusion constants of cations in oxides. This also indicates that the growth of the 2223 phase is controlled by diffusion of Bi, Ca and Cu atoms. Furthermore, as shown in Fig.4, the thickness of the 2223 crystals also increased. This suggests that Bi, Ca or Cu atoms does not diffuse only to one direction. The growth directions were observed by TEM. In Fig.5, TEM photograph of the 2223 crystals grown in the 2212 layer is shown with the electron diffraction pattern. The electron diffraction patterns indicate the growth direction to be a or b axis.

In order to clarify which oxides in the complex oxides reacts with the 2212 phase to form the 2223 phase, the following individual oxides were used in the reaction with the 2212 phase, separately. The individual oxides, Ca_2CuO_3, Ca_2PbO_4, CuO $(CaSr)_2CuO$ or $(CaSr)_2PbO_4$. however, did not form the 2223 phase. Furthermore, although the combination of Ca_2CuO_3+CuO, Ca_2PbO_4+CuO, $(Ca,Sr)_2CuO$+CuO or (Ca,Sr)PbO_4+CuO were tried to form the 2223 phase, the 2223 did not form. At present only one composition, shown in Table 1 as B, was able to react with the 2212 phase to form the 2223 phase. This result suggests that the three kinds of oxides, CuO, $(Ca,Sr)_2CuO_3$ and $(Ca,Sr)_2PbO_4$ participates in the formation of the 2223 phase, simultaneously.

CONCLUSION

The formation of the 2223 phase was examined by observing the 2223 crystals formed at the interface between the 2212 phase layer and Pb-Sr-Ca-Cu-O mixed oxides layer. In the formation reaction of the 2223 phase, the nucleation occurs very slowly and the growth of the 2223 phase is controlled by the diffusion of Bi atoms in Pb-Sr-Ca-Cu-O mixed layer and that of Ca and Cu atoms in the 2212 layer. The growth directions are a or b axis. The oxides, CuO, $(Ca,Sr)_2CuO_3$ and $(Ca,Sr)_2PbO_4$ participate in the formation of the 2223 phase. The partial melting is not the inevitable condition to form the 2223 phase.

REFERENCES

1. Hatano T. Aota K. Ikeda S. Nakamura K. Ogawa K. (1988) Growth of the 2223 phase in leaded Bi-Sr-Ca-Cu-O system. Jpn. J. Appl. Phys. 27:L2055-L2058
2. Ono A. (1988) Crystallization of 107K superconducting phase and partial welting in Bi-(Pb)-Sr-Ca-Cu-O system. Jpn. J. Appl. Phys. 27:L2276-L2279
3. Kim C.J. Rhee C.K. Lee H.G. Lee C.T. Kang S.J-L. Won D.Y.(1988) The formation of the high-Tc phase in Pb-doped Bi-Sr-Ca-Cu-O system. Jpn. J. Appl. Phys. 28:L45-L48
4. Uzumaki T. Yamanaka K. Kamehara N. Niwa K.(1989) The effect of Ca_2PbO_4 addition on superconductivity in a Bi-Sr-Ca-Cu-O system. Jpn. J. Appl. Phys. 28:L75-L77
5. Oota A. Ohba K. Ishida A. Kirihigashi A. Iwasaki K. Kuwajima H. (1989) Growth process of the (2223) phase in Pb-added Bi-Sr-Ca-Cu-O. Jpn. J. Appl. Phys. 28:L1171-L1174

The Formation Process of the High-T_c Phase in the Bi-Pb-Sr-Ca-Cu-O System

H. ITO[1], Y. IKEDA[1], S. SHIMOMURA[1], Z. HIROI[1], M. TAKANO[1], Y. BANDO[1], J. TAKADA[2], K. ODA[2], T. EGI[2], H. KITAGUCHI[2], and Y. MIURA[2]

[1] Institute for Chemical Research, Kyoto University, Uji, Kyoto, 611 Japan
[2] Department of Applied Chemistry, Faculty of Engineering, Okayama University, Okayama, 700 Japan

ABSTRACT

The formation process of the 2223 (high-T_c) phase in the Bi-Pb-Sr-Ca-Cu-O system has been studied for a nominal composition Bi:Pb:Sr:Ca:Cu=0.9:0.2:1:1:1.6 . From this composition we obtained the high-T_c phase crystallizing homogeneously in the Pb-induced modulation mode as reported previously. The solid phases formed after each firing treatment at and above 795°C were identified, and the melting points of various combinations of these were measured. As a result, the lowest liquid formation temperature was found to be 825°C for a combination of the 2201 phase and Ca_2PbO_4. The liquid formation triggered the formation of the 2223 phase. Formation of the Pb-substituted 2212 phase seems to be the initial step mediated by the liquid formation. The two dimensional size of the plate like particles of the 2223 phase strongly depends upon the thermal history. Possible factors controlling the particle growth are mentioned.

KEY WORDS; Bi-Pb-Sr-Ca-Cu-O system, high-Tc phase, liquid phase, formation process

INTRODUCTION

The finding by Maeda et. al. [1] of high-T_c superconductors with T_c's above liquid nitrogen temperature was followed by a great number of experiments aiming at identification and isolation of their monophasic materials. We have reported that substitution of Pb for Bi facilitates the form of the so-called 2223 phase and have long studied the solid phases and their relations in the Bi-Pb-Sr-Ca-Cu-O system [2]-[7].
Pb substitution has been known to influence both the formation process [8] and the structure modulation [9],[10]. Hatano et. al. [8] pointed out that the 2223 phase precipitated from a liquid phase. On the basis of our systematic study of identifying almost all the phases formed in the above mentioned mother system, we have recently studied the formation process of the 2223 phase in detail. Reported here are some experimental results concerning the identification of the initially formed liquid phase.

EXPERIMENTAL

Samples with a composition Bi:Pb:Sr:Ca:Cu=0.9:0.2:1:1:1.6 were prepared in the following manner. Starting materials Bi_2O_3, PbO, $SrCO_3$, $CaCO_3$, and CuO (all being ≥99.9% pure) were weighted and mixed in agate motors. These were pressed into 10 mm φ and 2mm thick pellets under a pressure of 0.6 t/cm^2. The pellets were calcined at 795°C for 18 h in air. The products were ground, pressed into pellets and fired at various temperatures from 795°C to 850°C for 48 h in air. Quenching to room temperature followed. The samples fired below 850°C were finally treated at 850°C for additional 48 h in air after grinding and pelletization.
Phase identification was carried out by X-ray powder diffraction method (XRD, Cu Kα) and dc resistance was measured with a four terminal method. TG-DTA thermal analysis was also conducted.

RESULTS AND DISCUSSION

Figure 1 shows the XRD patterns measured after the heating at 795°C ~ 850°C for 48 h. The 2212 phase, the 2201 phase, Ca_2PbO_4, and CuO were found to be commonly formed below 820°C . Above 825°C the 2223 phase, the 2212 phase, Ca_2CuO_3 and Ca_2PbO_4 were formed, but the 2201 phase and CuO were not detected . The amount of the 2223 phase was rapidly increased by firing above 825°C at a rate almost proportional to the firing temperature. It was thus suspected that a liquid phase was formed eutectically at this temperature from a certain combination of the preliminarily formed phases and facilitated the 2223 phase formation.

Figure 2 shows the DTA curve for a mixture of the 2201 phase and Ca_2PbO_4 that showed an anomaly at the lowest temperature. Each of the above mentioned four phases formed at 795°C did not melt at this temperature, but the mixture of Ca_2PbO_4 and the 2201 phase was found even by visual inspection to be completed melted by heating at 825°C for 1 h. So, the anomaly in the DTA curve corresponded to the melting.

Figure 3 illustrates the temperature dependence of dc resistance for the samples heated at 825°C, 830°C, and 835°C for 48 h. In comparison with Fig. 1 it is evident that zero resistance (within experimental error of 10^{-6} Ω) was attained even for the samples containing only a very small amount of the 2223 phase. This result has suggested that the 2223 phase tends to be located at grain edges of the preexisting 2212 phases as a result of a liquid-mediated formation reaction. When these samples were subsequently heated at 850°C for 48 h, the 2223 phase dominated the sample volume as can be seen in Fig. 4. Typical ρ-T curves are shown in Fig. 5, where the zero resistance temperature depends on the heating history.

Figure 6 shows the SEM images of the above pair of samples heated at 850°C . Both these consisted of plate-like particles, but the two dimensional size differed remarkably: It is evident

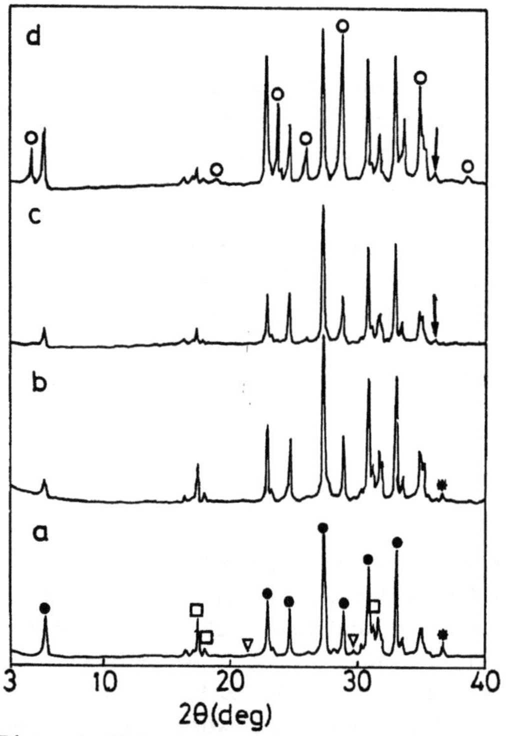

Fig. 1 XRD patterns measured after heating at (a) 795°C , (b) 820°C (c) 825°C , and (d) 840°C for 48h, respectively. The 2212 phase (●), the 2223 phase (○), the 2201 phase (▽), Ca_2CuO_3 (↓), and Ca_2PbO_4 (□) are indicated. The superlattic peak due to modulation (0211) of the 2212 phase is marked with (*).

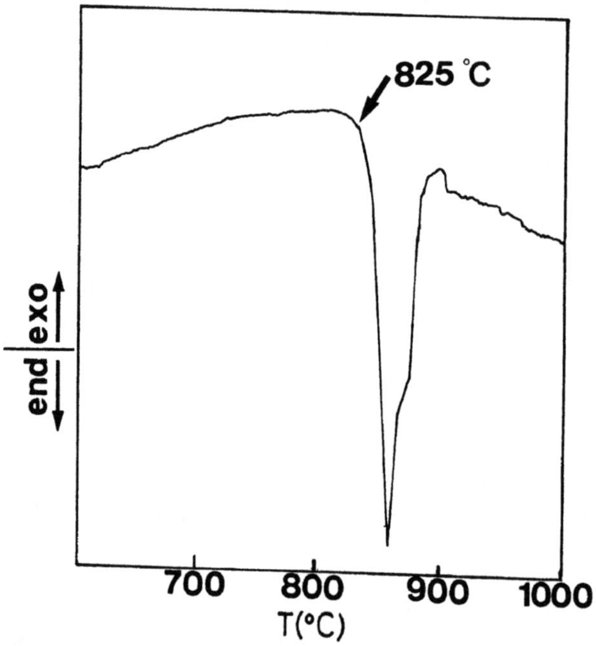

Fig. 2 DTA curve for a combination with the 2201 phase and Ca_2PbO_4.

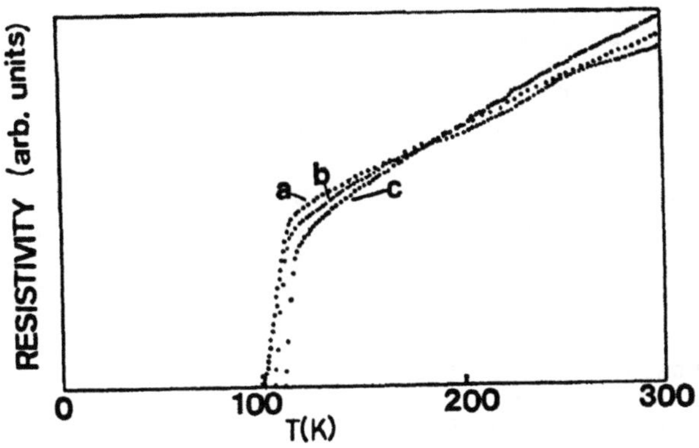

Fig. 3 ρ-T curves of the samples heated at (a) 825°C , (b) 830°C , and (c) 835°C , respectively.

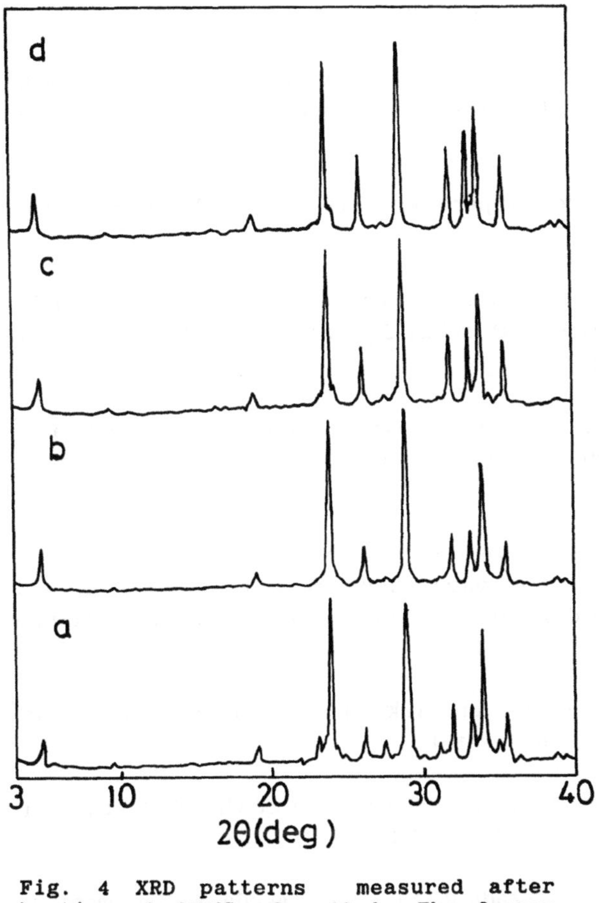

Fig. 4 XRD patterns measured after heating at 850°C for 48 h. The former heating temperature is (a) 795°C , (b) 820°C , (c) 825°C , and (d) 840°C , respectively.

that the two dimensional particle growth rate strongly depends upon thermal history. It is not easy at the present stage to exactly identify the crucial factor controlling the growth rate, but we can mention some possibilities. One is the difference in the composition of the 2212 phase. As can be found by careful inspection of the XRD patterns in Fig. 1, the peak at 2θ=36.78° due to the structure modulation of the bct type is present for the 2212 phase in the sample heated at 795°C but is absent in the sample heated at 825°C . This should be caused by the substitution of Pb for Bi in the 2212 phase in the latter sample[11]. The second is the difference in the composition of the liquid present at 850°C. The presence of Ca_2CuO_3 in the sample heated at 825°C can be seen from the peak marked with an arrow in Fig. 1. According to our experience Ca_2CuO_3 can hardly be formed at this temperature without the help of liquid formation. And so the presence of Ca_2CuO_3 reveals the formation of a liquid phase at 825°C. When the sample is heated subsequently at 850°C, the liquid formed may be deficient at least in Pb and also Ca. Thirdly, if the particles of the 2212 phase formed at and quenched from 825°C are rimmed by the 2223 phase as suggested already and also by some other solidified phases, the successive heat treatment at 850°C may not allow the two dimensional growth of the 2223 phase continued from the previous stage.

In summary we have found that the initial liquid formation occurs from the 2201 phase and Ca_2PbO_4 at 825°C. The formation of the 2223 phase probably occurs on the edges of the particles of the 2212 phase. The two dimensional particle growth rate for the 2223 phase strongly depends upon the thermal history.

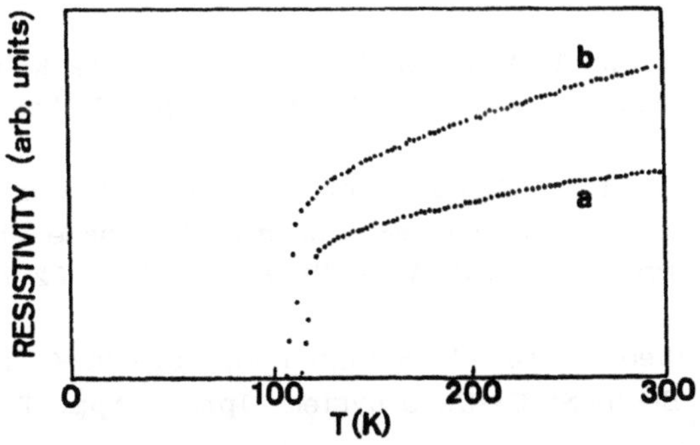

Fig. 5 ρ-T curves measured after the subsequent heating at 850°C for 48h. The former heating temperature is (a) 800°C and (b) 840°C , respectively.

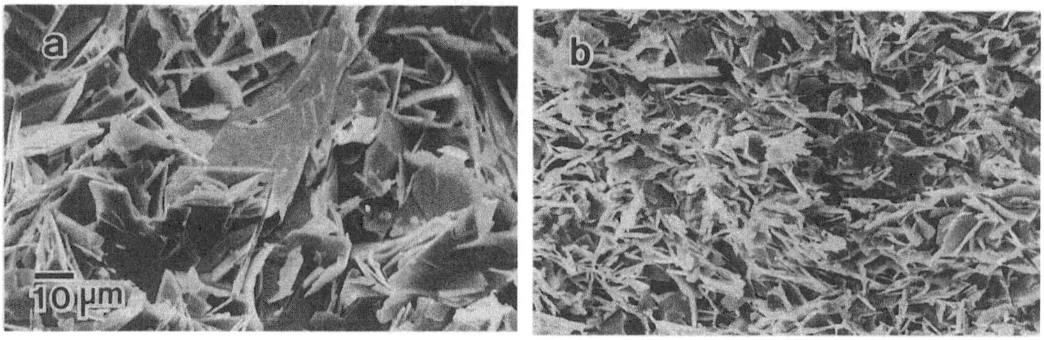

Fig. 6 SEM images obtained after the subsequent heating at 850℃ for 48h. The former heating temperature is (a) 800℃ and (b) 840℃ , respectively.

REFERENCES

1. Maeda H, Tanaka Y, Fukutomi M, and Asano T (1988) A new high-T_c superconductor without rare earth element. Jpn. J. Appl. Phys. 27: L209-L210

2. Takano M, Takada J, Oda K, Kitaguchi H, Miura Y, Ikeda Y, Tomii Y, and Mazaki H (1988) High-Tc phase promoted and stabilized in the Bi,Pb-Sr-Ca-Cu-O system. Jpn. J. Appl. Phys. 27: L1041-L1043

3. Takano M (1989) Studies of high temperature superconductors. A. V. Narlikar, Nova Science Publishers, New York in press

4. Takada J, Ohno M, Kitaguchi H, Oda K, Osaka A, Miura Y, Ikeda Y, Takano and Bando Y (1988) Phase diagrams for the systems Bi_2O_3-PbO-CaO and Bi_2O_3-PbO-SrO at 873K and 1073K in air. Jpn. J. Soc. Powder & Powder Met. 35: 952-958

5. Ikeda Y, Ito H, Hiroi Z, Takano M, Kitaguchi H, Takada J, Oda K, Miura Y, Takeda Y and Mazaki H (1988) The preparation and characterization of Pb-Bi-Sr-Ca-Cu-O high-Tc phase. J. Jpn. Soc. Powder & Powder Met. 35: 965-969

6. Kuniya H, Satoh K, Kitaguchi H, Takada J, Oda K, Osaka A, Miura Y, Ikeda Y, Takano M and Bando Y (1989) Equilibrium phase diagrams for the systems PbO-Cao-SrO, PbO-CaO-CuO and PbO-SrO-CuO at 973K and 1073K in air. proceedings of symposium series. Spring meet. Powder & Powder Met. 23-25 May 1989. Tokyo

7. Ikeda Y, Ito H, Shimomura S, Oue Y, Inaba K, Hiroi Z and Takano M (1989) Phase and their relations in the Bi-Sr-Cu-O system. Physica C 159: 93-104

8. Hatano T, Aota K, Ikeda S, Nakamura K, and Ogawa K (1988) Growth of the 2223 phase in leaded Bi-Sr-Ca-Cu-O system. Jpn. J. Appl. Phys. 27: L2055-L2058

9. Ikeda Y, Takano M, Hiroi Z, Oda K, Kitaguchi H, Takada J, Miura Y, Takeda Y, Yamamoto O, and Mazaki H (1988) The high-T_c phase with a new modulation mode in the Bi,Pb-Sr-Ca-Cu-O system. Jpn. J. Appl. Phys. 27: L2067-L2070

10. Ikeda S, Aota K, Hatano T and Ogawa K (1988) A new mode of modulation observed in the Bi-Pb-Sr-Ca-Cu-O system. Jpn. J. Appl. Phys. 27: L2040-2043

11. Fukushima N, Niu H, Nakamura S, Takeno S, Hayashi M and Ando K (1989) Structural modulation and superconducting properties in $Bi_{2-x}Pb_xSr_2CaCu_2O_{8+d}$ and $Bi_{2-y}Pb_ySr_2YCu_2O_{8+d}$. Physica C 159: 777-783

Preparation of High-T_c Phase in Bi-Sr-Ca-Cu-O System by Means of Addition of Ca_2PbO_4

HAJIME ITO, MASAAKI MATSUI, and MASAO DOYAMA

Department of Materials Science and Engineering, Nagoya University, Nagoya, 464-01 Japan

ABSTRACT

Effect of Ca_2PbO_4 addition has been studied for the preparation of the high T_c superconductor in the Bi-Sr-Ca-Cu-O system. The pure high T_c phase has been obtained by the sintering of the mixture of $Bi_2Sr_2CaCu_2O_{8+\delta}$,Ca_2PbO_4 and CuO at 1123K for 120 hours in air. It is found that the best composition ratio of Ca_2PbO_4 is 0.425 or 0.45 against the 1 mol low T_c phase and 1 mol CuO. The high T_c phase is more homogeneous than that sintered by the conventional solid state reaction.

KEY WORDS: High T_c superconductor ,Bi(Pb)-Sr-Ca-Cu-O system ,High T_c phase , Ca_2PbO_4 ,X-ray diffraction

INTRODUCTION

A new superconducting Bi-Sr-Ca-Cu-O system without rare earth elements was discovered by Maeda et al.[1]. This system mainly contains two superconductuing phases ,a high T_c phase (T_c=110K) and a low T_c phase (T_c=80K). There have been many publications characterizing the crystal structure and the superconducting properties. The low T_c and high T_c phases were found to be $Bi_2Sr_2CaCu_2O_8$ (2212) structure [2] and $Bi_2Sr_2Ca_2Cu_3O_{10}$ (2223) structure [3] ,respectively. The low T_c phase was easily obtained by the conventional sintering method [4] ,while it was considerably difficult to refine the single phase with high T_c. Though the addition of excess Ca [5] or the prolonged sintering at high temperatures [6] was attemped to raise the volume fraction of the high T_c phase ,a lot of amount of the low T_c phase was still remained. Takano et al. [7] reported that the high T_c phase was observed up to 85% in the X-ray diffraction pattern by the partial substitution of Pb for Bi and Endo et al. [8] obtained the high T_c phase completely without the low T_c phase by applying a low oxygen pressure during the solid state reaction of the Bi(Pb)-Sr-Ca-Cu-O system. In this paper ,we propose a method to purify the high T_c phase by means of the sintering of the mixture of $Bi_2Sr_2CaCu_2O_{8+\delta}$ phase ,Ca_2PbO_4 and CuO.

PREPARATION BY THE CONVENTIONAL METHOD

The starting materials of the present samples were high purity powders of Bi_2O_3,PbO_2,$SrCO_3$,$CaCO_3$ and CuO. These powders were well-mixed and pre-calcined at 1073K for 16 hours. We first prepared two samples (sample A and sample B) ,which had the same nominal composition ,Bi:Pb:Sr:Ca:Cu =1.8:0.4:2:2:3.2 ,and were different in the sintering condition. The sample A was sintered at 1128K for 72 hours ,ground into the powder and pressed into the disk. After the procedure was repeated twice ,the disk was finally sintered at 1128K for 96 hours. The total heat treatment was 240 hours at

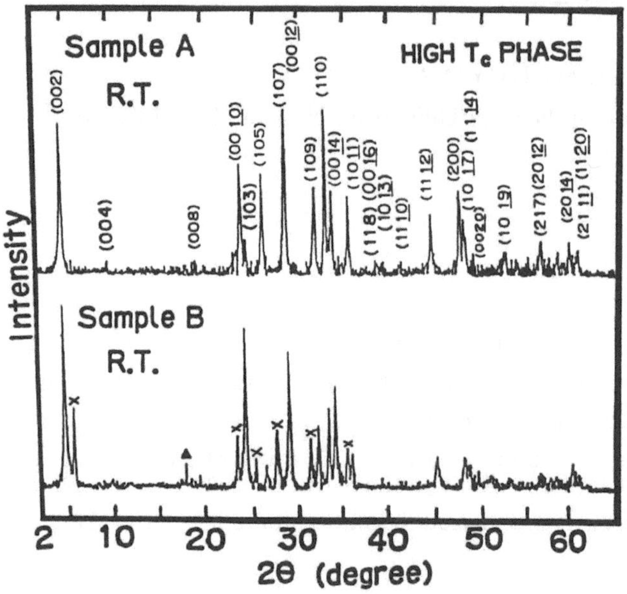

Fig.1 X-ray diffraction patterns at R.T. for sample A and B. "X","▲" indicate the low T_c phase and Ca_2PbO_4 ,respectively.

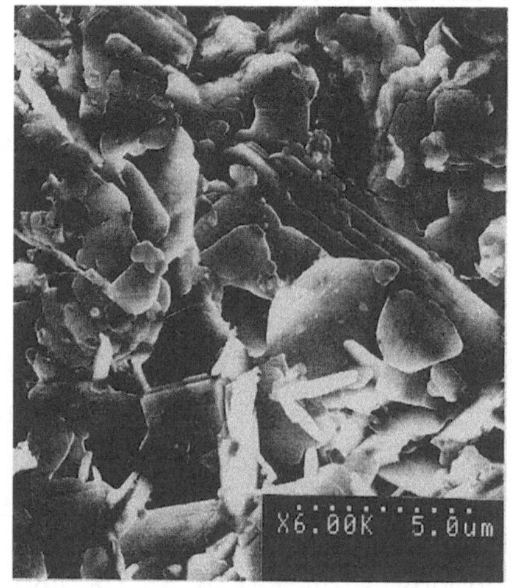

Fig.2 Scanning electron microscope (SEM) photograph of the free surface of sample A.

1128K. The sample B was sintered at 1128K for 240 hours without any additional grinding and pressing.

Figure 1 shows X-ray diffraction patterns of sample A and sample B. There is no indication of the low T_c phase or any other impurities in the diffraction pattern of sample A. All peaks of sample A observed at R.T. could be indexed ,assuming that the high T_c phase has a 2223 structure [4]. Estimated lattice parameters ,a and c ,of the high T_c phase were 3.822 Å and 37.04 Å ,respectively. Those are in agreement with our previous study [4]. On the other hand ,the high T_c ,the low T_c and Ca_2PbO_4 phases were still remained in sample B. The result suggests that the grinding and pressing in the middle of the sample preparation process is effective to purify the high T_c phase. Figure 2 shows the scanning electron microscope (SEM) photograph of the free surface of sample A. The flakes of the high T_c phase are seen in the figure. The ratio of the chemical composition of elements in sample A measured by the energy dispersive X-ray analysis (EDX) was (Bi+Pb):Sr:Ca:Cu=1.79:2.03:1.98:3.20. It should be noted that the value of the Ca ratio is nearly 2 mol ,which is the ideal value of the high T_c phase with 2223 structure.

PREPARATION BY THE ADDITION OF Ca_2PbO_4

We found that Ca_2PbO_4 always appeared in the process of the solid state reaction in such as the heat treatment for sample B. So ,we tried to obtain the pure high T_c phase by means of the addition of Ca_2PbO_4 to the mixture of $Bi_2Sr_{2-y}Ca_{1+y}Cu_2O_{8+\delta}$ and CuO ,where y was varied to obtain exactly 1 mol as the final composition of Ca in the 2212 structure. In the 2212 system ,the region of a solid solution is y=-0.25 to 1.0 [9]. If the Ca composition in the low T_c phase as a starting material is 1 mol and Ca_2PbO_4 supplies 1 mol for Ca ,the high T_c phase will be expected to be formed. According to the

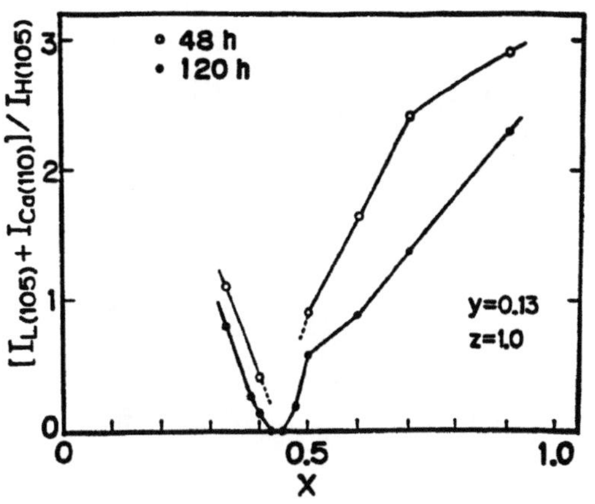

Fig.3 Intensity ratio, $(I_{L(105)}+I_{Ca(110)})/I_{H(105)}$ as a function of the composition of Ca_2PbO_4 with y=0.13 ,z=1.0 for 48 and 120 hours sintering.

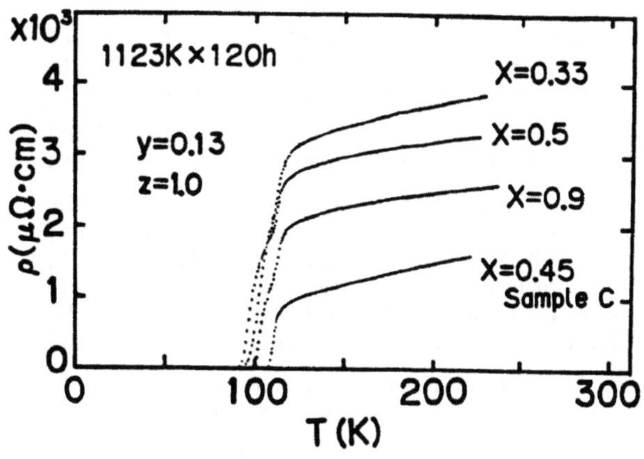

Fig.4 Temperature dependence of the electric resistivity of samples with x=0.33,0.45(sample C),0.5,0.9 sintered at 1123K for 120 hours.

EDX measurement for samples of y=0,0.13,0.25,0.5 in the $Bi_2Sr_{2-y}Ca_{1+y}Cu_2O_{8+\delta}$ system ,it is found that the most appropriate value of y is 0.13. We also synthesized Ca_2PbO_4 by the sintering at 1073K for 72 hours. Changing the composition ratio ,low T_c phase:Ca_2PbO_4:CuO=1:x:z ,the mixture of them were sintered at 1123K.

The amount of the low T_c phase and Ca_2PbO_4 remained in the sample was estimated by the XRD pattern. The ratio between the sum of the intensity of (105) diffraction of the low T_c phase and (110) of Ca_2PbO_4 and (105) of the high T_c phase ,$(I_{L(105)}+I_{Ca(110)})/I_{H(105)}$,was examined. In Fig.3 ,it is shown as a function of the composition of Ca_2PbO_4. As the heating period become longer ,the amount of the low T_c phase and Ca_2PbO_4 decreases by degrees. It is noted that the pure high T_c phase was obtained by sintering for 120 hours for the samples of x=0.425 and 0.45(sample C) when y=0.13 , z=1.0. Sample C was the single phase of the 2223 structure with a=3.821 Å and c=37.10 Å. Fig.4 shows the resistivity of the samples with x=0.33, 0.45 ,0.5 ,0.9 sintered at 1123k for 120 hours. The superconducting on-set temperature and the temperature of zero-resistance of sample C are 115K and 106K,respectively. The resistivity curves of the other samples also have a drop at about 115K but have a tail extended to the zero-resistance state around 90-100K. Static susceptibility for sample C was measured with increasing temperature using SQUID magnetometer,as shown in Fig.5. The susceptibility becomes zero at 108K. The volume fraction of the superconducting phase to the total volume of the

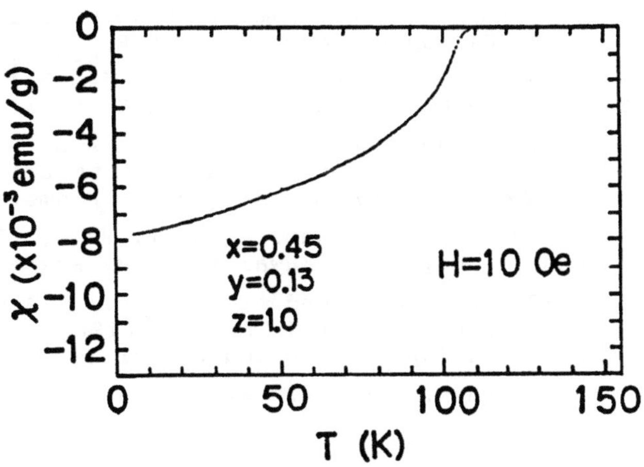

Fig.5 Temperature dependence of the magnetic susceptibility of sample C.

specimen was estimated to be 61.2% at 5K ,assuming the shielding effect. There is no change due to the low T_c phase at around 80K ,suggesting that the specimen had only high T_c phase. Fig.6 shows the SEM photograph of the free surface of sample C. Comparing it with that of sample A (see Fig.2), the sample C is more homogeneous than sample A and the dimension of the flake-like high T_c phase in sample C is larger than that in sample A. Thus,we could obtain the pure high T_c phase by the sintering the low T_c phase,Ca_2PbO_4 and CuO.It is concluded that Ca_2PbO_4 is useful to obtain a homogeneouse sample of the high T_c phase.

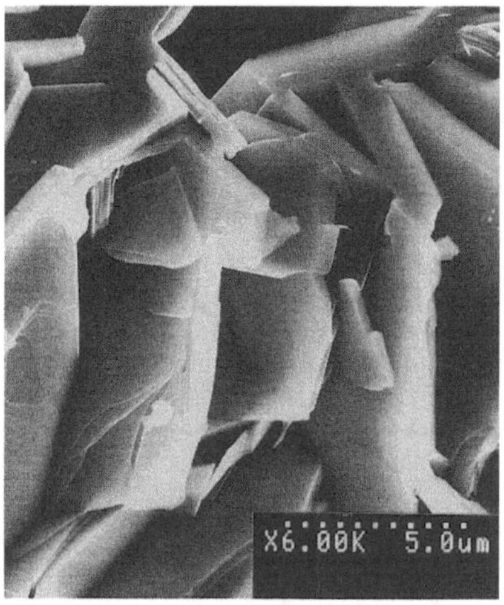

Fig.6 Scanning electron microscope (SEM) photograph of the free surface of sample C.

REFERENCES

[1] Maeda H, Tanaka Y, Fukutomi M, Asano T (1988) A New High-T_c Oxide Superconductor without a Rare Earth Element. Jpn. J. Appl. Phys. 27:L209-210

[2] Tarascon J.M, Le Page Y, Barboux P, Bagley B.G, Greene L.H, McKinnon W.R, Hull G.W, Giroud M, Hwang D.M (1988) Crystal substructure and physical properties of the superconducting phase $Bi_4(Sr,Ca)_6Cu_4O_{16+x}$. Phys. Rev. B37:9382-9389

[3] Tarascon J.M, McKinnon W.R, Barboux P, Hwang D.M, Bagley B.G, Greene L.H, Hull G.W, Le Page Y, Stoffel N, Giroud M (1988) Preparation, structure, and properties of the superconducting compound series $Bi_2Sr_2Ca_{n-1}Cu_nO_y$ with n=1,2, and 3. Phys. Rev. B38:8885-8892

[4] Matsui M, Itoh H, Liu J, Shimizu T, Matsuoka H, Ohmori K, Doyama M (1988) Structural Investigation of High T_c Bi-Sr-Ca-Cu-O. Proceedings of the 1st International Symposium on Superconductivity (ISS'88). 28-31 Aug 1988. pp 891-896

[5] Sumiyama A, Yoshitomi T, Endo H, Tsuchiya J, Kijima N, Mizuno M, Oguri Y (1988) Superconductivity of $Bi_{0.25-y}Sr_{0.25-y}Ca_{2y}Cu_{0.5}O_x$ (y=0.1,0.125, 0.15). Jap. J. Appl. Phys. 27:L542-544

[6] Nobumasa H, Shimizu K, Kitano Y, Kawai T (1988) High T_c Phase of Bi-Sr-Ca-Cu-O Superconductor. Jap. J. Appl. Phys. 27:L846-848

[7] Takano M, Takada J, Oda K, Kitaguchi H, Miura Y, Ikeda Y, Tomii Y, Mazaki H (1988) High-T_c Phase Promoted and Stabilized in the Bi,Pb-Sr-Ca-Cu-O System. Jap. J. Appl. Phys. 27:L1041-1043

[8] Endo U, Koyama S, Kawai T (1988) Preparation of the High-T_c Phase of Bi-Sr-Ca-Cu-O Superconductor. Jap. J. Appl. Phys. 27:L1476-1479

[9] Niu H, Fukushima N, Ando K (1988) Effect of Oxygen Content and Sr/Ca Ratio on Superconducting Properties in $Bi_2Sr_{2-x}Ca_{1+x}Cu_2O_{8+\delta}$. Jap. J. Appl. Phys. 27:L1442-1444

Formation of 2223 Phase and Variation in Composition of Bi(Pb)-Sr-Ca-Cu-O Superconductors

Toranosuke Ashizawa, Shuichiro Shimoda, Minoru Ishihara, Keiji Sumiya, Shozo Yamana, and Hideji Kuwajima

Ibaraki Research Laboratory, Hitachi Chemical Co., Ltd., 1380-1, Tarazaki, Katsuta, Ibaraki, 312 Japan

ABSTRACT

Sintering of $Bi_{1.6}Pb_{0.4}Sr_2Ca_2Cu_{3.6}O_x$ for long time resulted in an increase of the volume fraction of the 2223 phase to 53%. The 2223 phase appears to take the form of flaky grains. The chemical composition of the constituents of the flaky grains are not exact stoichiometric proportions. Substances in the calcined powder start to react around 832℃, then change to (Sr,Ca)-Cu-O and an unknown phase at 862℃. Growth of the 2223 phase is enhanced between these two temperatures.

KEY WORDS: Bi(Pb)-Sr-Ca-Cu-O, 2223 phase, (Sr,Ca)-Cu-O

INTRODUCTION

It has been reported that there are the 2223 phase (Tc 110 K), the 2212 phase (Tc 80 K), and the 2201 phase (Tc<20 K) in Bi-Sr-Ca-Cu-O superconductors [1,2]. It was difficult to produce the 2223 monophase compound, and a number of studies were performed to try to increase the fraction of the 2223 phase[3]. Partial substitution of Pb for Bi is the most efficient method for increasing the yield of the 2223 phase[4,5].

The 2223 phase grows under low oxygen partial pressure, by long period heat treatment, and by choosing an appropriate composition of starting materials[4,5]. The mechanism of formation and growth of the 2223 phase is not clear. It has been proposed that the 2212 phase shows a disproportionation reaction into the 2223 and the 2201 phases [3], and that the 2223 phase grows from the partially melting state of the starting materials[2,6].

In the present study, the growing state of the 2223 phase was investigated by varying the sintering time. Reactions at the temperatures at which the 2223 phase grows were also examined by differential thermal analysis (DTA) and high-temperature X-ray diffraction.

EXPERIMENTAL

Powders of Bi_2O_3, PbO, $SrCO_3$, $CaCO_3$ and CuO in a ratio of Bi:Pb:Sr:Ca:Cu:O= 1.4:0.6:2:2:3.6 were mixed in a ball mill. The powder was calcined in an alumina crucible at 800℃ for 10 hours in air. The calcined material was ground in a ball mill and pressed into pellets 20 mm in diameter and 1.5 mm in thickness. Pellets were sintered at 845℃ for 5 to 200 hours under an atmosphere of O_2:N_2=1:10 and then furnace cooled.

Temperature dependence of ac magnetic susceptibility was determined by measuring the ratio of inductance at 100 Hz of a coil surrounding a specimen and a reference coil (air core). The volume fraction of the 2223 phase was calculated from the difference of ac magnetic susceptibility at the onset temperature and that at 90 K.

Analyses of the formation of the 2223 phase were carried out by Cu Kα radiated X-ray powder diffraction (XRD), scanning electron microscopy (SEM) combined with energy dispersive X-ray spectroscopy (EDAX) and differential thermal analysis (DTA).

RESULTS AND DISCUSSION

Figure 1 shows XRD patterns of sintered specimens. The calcined powder (not sintered) consists of the 2212 phase, Ca_2PbO_4, CuO and $SrCO_3$. The X-ray diffraction intensities from these compounds diminish gradually with increased sintering time. The 2223 phase appears after 10 hours of sintering, and its concentration seems to increase with longer sintering time. The ratios of peak intensities from the 2223 phase and 2212 phases are almost equal after 20 hours of sintering. The specimen sintered for 200 hours appeared to contain only a small amount of the 2212 phase and to be almost all 2223 phase. However the continued existence of the 2212 phase after 200 hours of sintering indicates a slow rate of formation of the 2223 phase and the difficulty of producing 2223 monophase material.

The increase of the volume fraction of the 2223 phase by long time sintering is shown in Fig. 2. Both the results from XRD and measurements of magnetic susceptibility show that higher concentration of the 2223 phase is obtained by long time sintering. The volume fraction of the 2223 phase of the specimen sintered for 200 hours is about 53%. The total volume fraction of all superconductor phases in this sample is about 64% (not shown), calculated from magnetic susceptibility at the onset temperature and at 20 K. X-ray diffraction pattern of the specimen sintered for 200 hours suggests higher fraction of the 2223 phase than that calculated from measurements of magnetic susceptibility. This fact shows that some impurity phases are present even after 200 hours of sintering, and that calculation from XRD may lead to an erroneous estimation of the fraction of the superconductor phases.

Figure 3 shows the surface of a specimen sintered for 100 hours. There are many flaky grains in this sample with a width of about 10 to 30 μm. These grains are common in 2223 phase bulk samples. Chemical composition of the grains identified by numbers in Fig. 3 were analyzed by EDAX (Table 1).

Although the volume fraction of the 2223 phase is estimated to be 40% (Fig. 2), the chemical composition at most positions differs from the exact stoichiometric proportions of 2223 phase material. It seems that some grains contain the 2223 phase plus impurities, and that the variation in composition is caused by intergrowth or substitution. It is considered that formation of the 2223 phase, grain growth and composition variation proceed at the same time. Table 1 shows that a (Sr,Ca)-Cu-O phase exists in this sample (no.2 and 8), although it is not detected by XRD. The main reason for this may be the small concentration of the (Sr,Ca)-Cu-O phase. We suppose that most part of flaky grains have 2223 crystal structure, and that some of them are not superconductors because of their difference in composition from ideal value of the 2223 phase. The existence of these impurity phases is one of the reasons for the low proportion of 2223 phase in Fig.2.

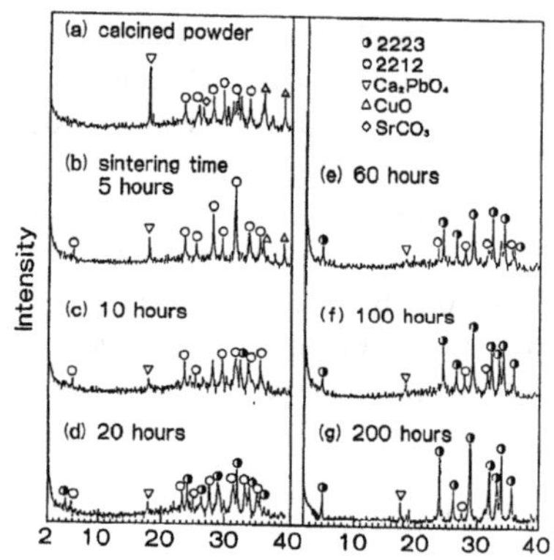

Fig. 1 X-ray diffraction patterns of (a) calcined powder and (b) ~(g) sintered and ground pellets.(sintered at 845°C, O_2 : N_2 =1 : 10)

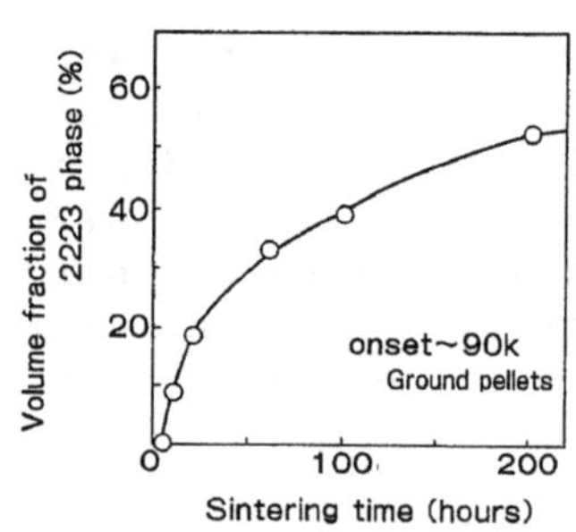

Fig. 2 Volume fraction of 2223 phase vs sintering time. (845°C, O_2 : N_2=1 : 10)

Figure 4 shows DTA curves for the calcined powder, a ground pellet sintered for 20 hours, and a ground pellet sintered for 200 hours. All measurements were carried out under a $1:10=O_2:N_2$ atmosphere with a heating rate of 3.3℃/ min. Every specimen had an endothermal peak between 850℃ and 870℃. The specimen sintered for 200 hours had an unclear endothermal change around 850℃ (slight dip) which seems to agree with an endothermal peak associated with the partial melting[6]. These results show that sintering and formation of the 2223 phase raise the temperature at which the endothermal peak occurs. The calcined powder shows an endothermal change at 845℃ (sintering temperature of our experiments). We consider that this endothermal reaction goes along with formation of the 2223 phase.

A pellet of calcined powder was put into a furnace which was heated to 870℃ (just above the endothermal peak temperature) under a $1:10=O_2:N_2$ atmosphere, kept there for 15 min., then quenched in liquid nitrogen. The surface of this pellet is shown in Fig. 5. This sample consists of rod-like or tabular grains (A) surrounded by an irregular shaped area (B) which seems to be traces of a previously liquid phase. Qualitative analyses by EDAX reveal that grain (A) is (Sr,Ca)-Cu-O, and that surrounding area (B) is Bi-Sr-Ca-Cu-O. It is considered that an endothermal reaction in the calcined powder resulted in formation of solid (Sr,Ca)-Cu-O and liquid Bi-Sr-Ca-Cu-O.

Figures 6 and 7 show high-temperature X-ray diffraction patterns for calcined powder and a pellet sintered for 200 hours, respectively. Samples were kept vertical in a platinum holder and heated in an electric furnace under a $1:10=O_2:N_2$ atmosphere. The 2212 phase, Ca_2PbO_4, CuO, $SrCO_3$ etc. in the calcined powder start to react around 822℃. Above 842℃, the fraction of the 2212 phase decreased, while (Sr,Ca)-Cu-O and an unknown phase formed. The starting materials have almost disappeared at 851℃. The 2212 phase was caused to re-form by cooling the sample. Only 001 diffractions of the 2212 phase can be recognized. This fact shows that the (001) plane of the re-formed 2212 phase aligned parallel to the platinum holder. Our experimental results support the report on high-temperature XRD of Bi-Sr-Ca-Cu-O by Oka et al[7]. (Sr,Ca)-Cu-O and the unknown phase also formed at 862℃ in a pellet sintered for 200 hours and ground. The original compounds disappeared at 869℃, while (Sr,Ca)-Cu-O and the unknown phase still remained. When this sample was cooled from 875℃, the 2212 phase already crystallized at 835℃.

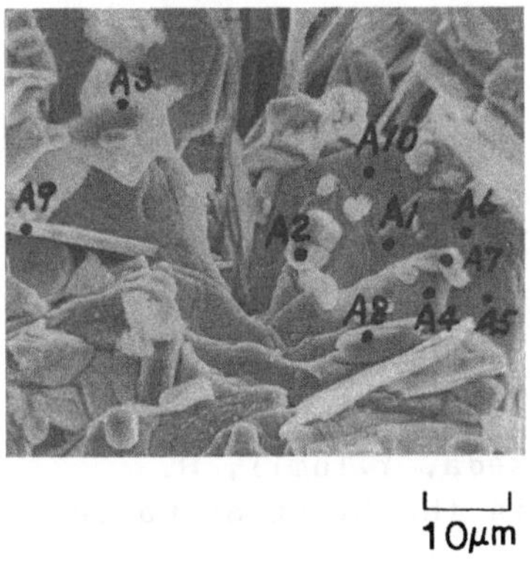

Fig. 3 SEM micrograph of surface of a pellet sintered for 100 hours. Numbers indicate the positions analyzed by EDAX. (sintering 845°C, 100h $O_2:N_2=1:10$)

Table 1 Chemical compositions of typical grains in a pellet sintered for 100 hours. Positions are shown in Fig.3 Analyses were made by means of EDAX.

Position	Atomic ratio				
	Bi	Pb	Sr	Ca	Cu
1	1.50	0.22	1.67	2.33	3.08
2	0	0	0.56	0.68	2.00
3	1.53	0.47	2.02	1.87	3.06
4	1.61	0.39	1.65	2.07	3.46
5	1.64	0.36	1.52	2.43	3.10
6	1.58	0.42	1.65	2.03	3.05
7	1.73	0.27	1.76	1.54	2.91
8	0	0	0.46	0.52	2.00
9	1.74	0.26	1.83	2.07	3.20
10	1.58	0.42	1.61	2.28	2.95

(845°C, $O_2:N_2=1:10$, 100h)

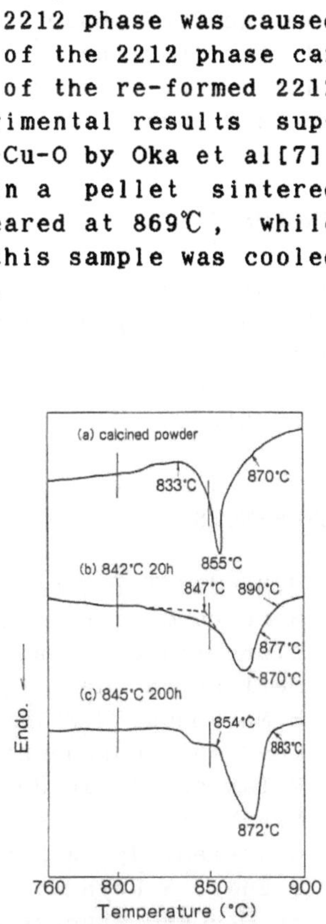

Fig. 4 DTA curves of specimens ($O_2:N_2=1:10$, 3.3°C/min.): (a) calcined powder (2212 phase, Ca_2PbO_4, CuO, $SrCO_3$), (b) sintered for 20 hours and groud pellet (2223 and 2212 phases, Ca_2PbO_4), (c) sintered for 200hours and groud pellet (2223 and 2212 phases, Ca_2PbO_4).

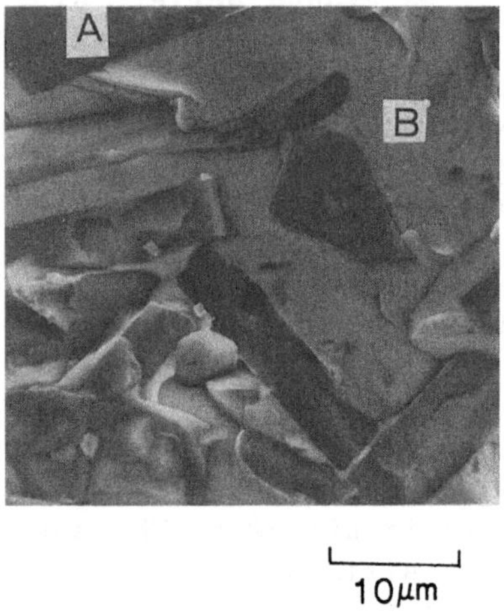

Fig.5 SEM micrograph of surface of pellet sintered at 870°C for 15min and quenched in lquid nitrogen.

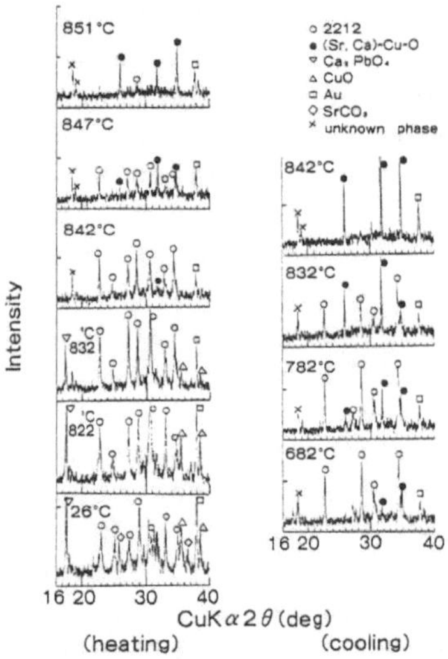

Fig. 6 High temperature X-ray diffraction patterns of calcined powder. ($O_2 : N_2 = 1:10$)

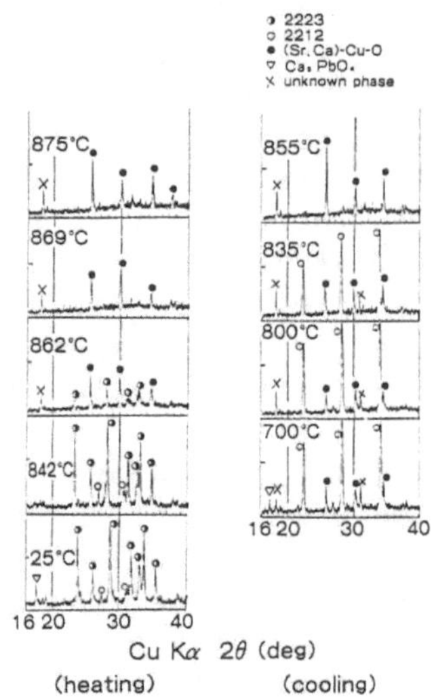

Fig.7 High temperature X-ray diffraction patterns of a pellet sintered for 200hours. ($O_2 : N_2 = 1:10$)

It is reported that formation of the 2223 phase is enhanced between 835℃ and 870℃ [6,7]. Slightly above 832℃, reaction in calcined powder proceeds, and this reaction probably makes a liquid phase. Therefore, growth of the 2223 phase can be accelerated in this partially melting condition. The reaction from the 2223 phase into (Sr,Ca)-Cu-O and unknown phase occurs above 860℃, and the fraction of the 2223 phase decreases at this temperature. It is considered that this reaction limits the temperature range of the formation of the 2223 phase.

CONCLUSIONS

The process of formation of the 2223 phase was investigated. Growth of the 2223 phase proceeds with variation in composition. Reaction in the calcined powder occurs above 832℃, while the 2223 phase changes to (Sr,Ca)-Cu-O and an unknown phase at 862℃. Growth of the 2223 phase is enhanced between these two temperatures.

REFERENCES

1. H.Maeda, T.Tanaka, M.Fukutomi, T.Asano (1988) A new high-Tc oxide Superconductor without a rare earth element. Jpn.J.Appl.Phys.27: L209-L210
2. A.Ono (1988) Crystallizatiion of 107 K superconducting phase and partial melting in the Bi-(Pb)-Sr-Ca-Cu-O system. Jpn.J.Appl.Phys.27: L2276-L2279
3. H.Nobumasa, K.Shimizu, Y.Kitano, T.Kawai (1988) High Tc phase of Bi-Sr-Ca-Cu-O superconductor. Jpn.J.Appl.Phys.27: L846-L848
4. M.Takano, J.Takada, K.Oda, H.Kitaguchi, Y.Miura, Y.Ikeda, Y.Tomii, H. Mazaki (1988) High-Tc phase promoted and stabilized in the Bi,Pb-Sr-Ca-Cu-O system. Jpn.J.Appl.Phys.27: L1041-L1043
5. U.Endo, S.Koyama, T.Kawai (1988) Preparation of the high-Tc phase of Bi-Sr-Ca-Cu-O superconductor. Jpn.J.Appl.Phys.27: L1476-L1479
6. T.Hatano, K.Aota, S.Ikeda, K.Nakamura, K.Ogawa (1988) Growth of the 2223 phase in leaded Bi-Sr-Ca-Cu-O system. Jpn.J.Appl.Phys.27: L2055-L2058
7. Y.Oka, N.Yamamoto, H.Kitaguchi, K.Oda, J.Takada (1989) Crystallization behavior and partially melted states in Bi-Sr-Ca-Cu-O. Jpn.J Appl.Phys.28: L213-L216

Increased Critical Temperature in Bi-Sr-Ca-Cu-O Oxide System

KUMIKO IMAI and HIRONORI MATSUDA

Yokohama R&D Laboratories, The Furukawa Electric Co., Ltd., 4-3, Okano 2-chome, Nishi-ku, Yokohama, 220 Japan

ABSTRACT

The change in critical temperature Tc of $Bi_2Sr_2CaCu_2O_x$ has been investigated. Tc changes with oxygen content which is determined by annealing condition. The critical temperature reaches a maximum value of 95K at an oxygen content about X=8.25.

KEY WORDS: $Bi_2Sr_2CaCu_2O_x$ superconductor, Tc variation, oxygen content, annealing condition

INTRODUCTION

$Bi_2Sr_2CaCu_2O_x$ phase (referred to as the 2212 phase) has been reported that the transition temperature is 85k, however which can be obtained with a shorter sintering time and with a simple process compared to the high Tc phase. It has been reported that Tc of the 2212 phase is changed sensibly by the processing conditions, Sr:Ca ratio or oxygen content.[1-8]
However, these results do not agree among these reports. The disagreement may be due to the samples which have broadly superconducting transition.

We have found a simple process to give a bismuth based bulk superconductor composed mostly of the 2212 phase with a higher Tc of 95K at which it shows a sharp temperature transition.
This report describes the dependence of Tc on annealing conditions and on oxygen content.

EXPERIMENTAL

Preparation of Samples

Bulk samples were prepared by sintering for 15 hours. The materials are Bi_2O_3, $SrCO_3$, $CaCO_3$, CuO mixed to Bi:Sr:Ca:Cu=2:2:1:2, which is the same proportion as that of the stoichiometric low Tc phase.[9-12] The size of samples was about 3mm X 1mm X 20mm. The samples have a critical temperature of 95K by measurement with 1mA current as shown in Fig.1 and have critical current densities at 77K are around 1500 A/cm2, which were measured by transport current with four probe method. Fig.2 shows the X-ray powder diffraction measurement, which revealed that the samples were most structured with 2212 phase,[12] and some other impurities were included.[13,14] The average of the specific gravity was 6.2g/cm^3.

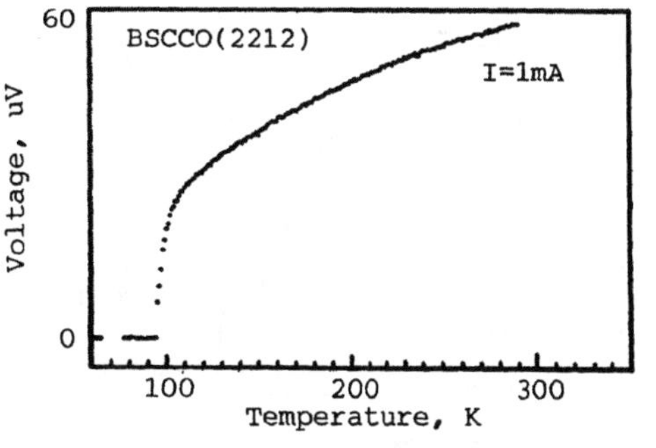

Fig.1 Temperature transition of resistivity of a $Bi_2Sr_2CaCu_2O_X$ bulk sample measured with 1mA current.

Control of Critical Temperature

Annealing temperatures and atmospheres of samples were selected according to required oxygen contents. 2212 phase samples having various oxygen contents were finally obtained by quenching those samples to room temperature.

When samples were treated in oxygen atmosphere, the critical temperature decreased with decrease of annealing temperature.

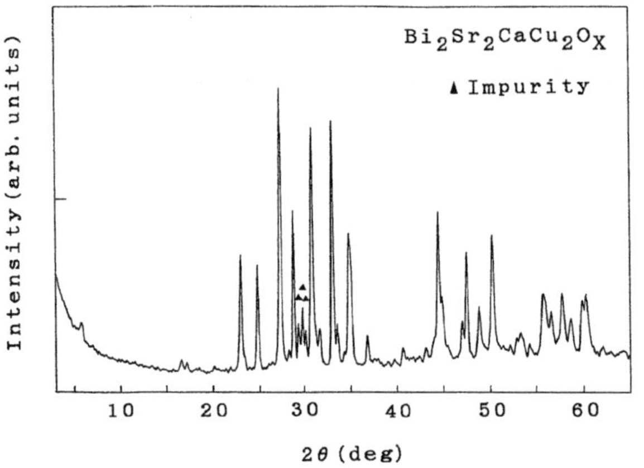

Fig.2 X-ray powder diffraction pattern of a $Bi_2Sr_2CaCu_2O_x$ bulk sample

When samples were treated in oxygen atmosphere, the critical temperature decreased with decrease of annealing temperature.
Fig.3 shows the transition temperatures of the ac susceptibility of these samples. X-ray powder diffraction measurements of all samples showed the same 2212 structure pattern that no significant structural change were not observed. Oxygen contents of these samples increased with decrease of annealing temperature.

When samples having different critical temperature were annealed with an annealing condition in oxygen atmosphere, the samples had an inherent critical temperature in the condition.
The result implies that reversible change occurs in the annealing process in oxygen atmosphere and that critical temperature depends on the oxygen content which was determined by an equilibrium with external oxygen pressure during the annealing process.

On the other hand, when the samples were annealed in nitrogen atmosphere, the critical temperature decreased with annealing temperature and with annealing time. Figure 4 shows the transition temperatures of ac susceptibilities of these samples.

The result implies that irreversible change occurs in the annealing process in nitrogen atmosphere and that decreased critical temperature of samples can not be increased by annealing at different temperature in nitrogen atmosphere.
When the samples are annealed at higher temperature than that shown in Fig.4, the resistivity transition became semiconductive.

Table 1 shows the relation between the annealing conditions of the samples and the corresponding transition temperatures of the susceptibility which were defined by the midpoint temperatures of the transition in Figure 3 and in Figure 4.

Comparing the result with that in oxygen atmosphere, we can suggest that external oxygen pressure during annealing process is a critical factor determining critical temperatures of 2212 bismuth based superconductors.

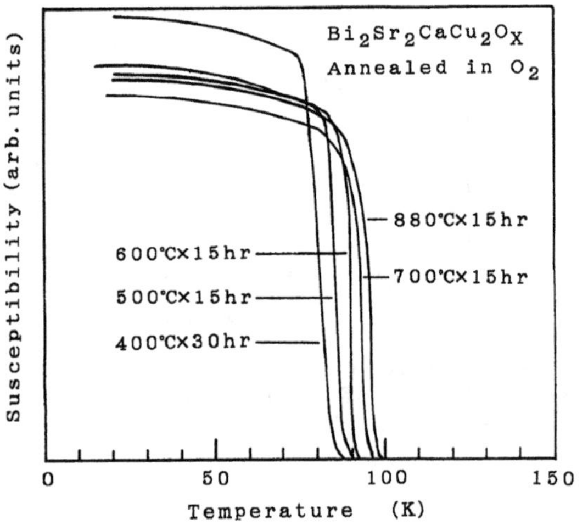

Fig.3 Temperature transition of ac susceptibility of $Bi_2Sr_2CaCu_2O_x$ bulk samples annealed in oxygen atmosphere at 400-880° C. Critical temperature decreases as the annealing temperature decreases.

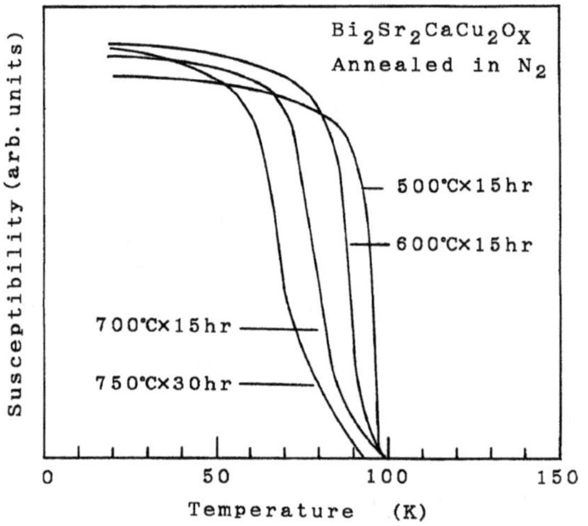

Fig.4 Temperature transition of ac susceptibility of $Bi_2Sr_2CaCu_2O_x$ bulk samples annealed in nitrogen atmosphere at 500-750° C. Critical temperature decreases as the annealing temperature rises and as the annealing time increases.

Table 1
The dependence of critical temperature Tcm on annealing condition Tcm is defined by the midpoint of temperature transition in Fig.3 and Fig.4.

Annealing in O_2		Annealing in N_2	
Annealing Temp.	Tcm,K	Annealing Temp.	Tcm,K
880°C X 15hr	94	750°C X 30hr	68
850°C X 15hr	94	750°C X 15hr	72
700°C X 15hr	91	700°C X 15hr	78
600°C X 15hr	89	600°C X 15hr	87
500°C X 15hr	85	500°C X 15hr	94
500°C X 15hr + 400°C X 30hr	80		

Decision of Oxygen Content

Oxygen contents of the annealed samples were determined being extrapolated by charge valences measured by coulometric titration[15] and weight.

Oxygen stoichiometry in nominal composition $Bi_2Sr_2CaCu_2O_x$ versus the transition temperature are shown in Fig.5. As impurity phase of the sample seems to disturb the measurement of oxygen content, we prepared samples having different content of impurity and these samples were compared by oxygen content. The relation of critical temperature with oxygen content deviation showed similar characteristics among samples with different content of impurity, however oxygen contents deviated depending on the quantity of impurity. Therefore a some constant deviation in oxygen content of 2212 phase may exist owing to the impurity.

When samples were treated in nitrogen atmosphere at more than 700°C, impurity phase decomposed into Cu_2O and significantly disturbed the measurement, so that the data of these samples are not used.

It revealed from Fig.5 that critical temperature of 2212 phase of bismuth based superconductor depends on its oxygen content. Maximum critical temperature of 94K is obtained when the superconductor has a oxygen content, and Tc decreases not only with increase of oxygen content but also with decrease of oxygen content from a critical value. Therefore, we estimates that critical temperature in 2212 phase prepared by us could not exceed 94K by control of oxygen content.

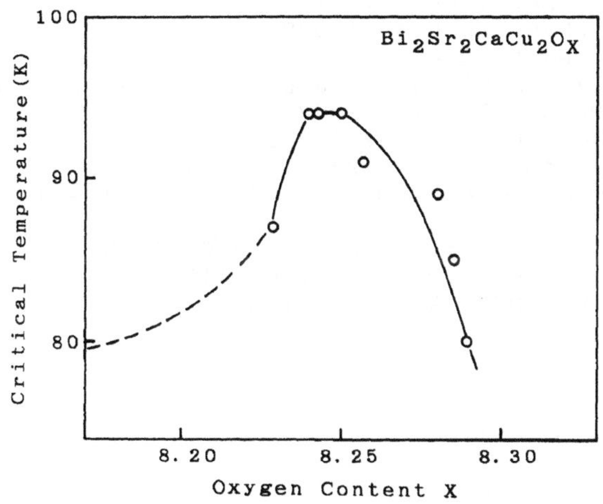

Fig.5 Critical temperature vs. oxygen content X for $Bi_2Sr_2Ca_1Cu_2O_{8+x}$ bulk samples
Critical temperature attains a peak of 95K at an oxygen content about 8.25.

RESULT AND DISCUSSION

In bismuth based system, it has been reported that the fluctuation of Bi-O layer exists along the b-axis.[16,17] The commensurate modulation of periodicity usually exhibits 4 or 5. An extra oxygen is inserted in Bi-O layer every periodicity.[18,19]
It has been reported that the change of oxygen content causes change of binding energy between bismuth and oxygen.[20,21] Oxygen atom seems easily to be inserted or to be eliminated at Bi-O layer. Oxygen content may change from 0.2 to 0.25 in the formula of $Bi_2Sr_2Ca_1Cu_2O_{8+x}$ by periodic modulation of Bi-O layer.
Oxygen content deviation changed by the heat treatment falls in the range of the deviation introduced from the periodic modulation. Therefore it might suggest that oxygen is charged or discharged in Bi-O layer during annealing process and that Tc depends on hall concentration in CuO_2 plane, which is determined by oxygen content in Bi-O layer.
Regarding to 123 structure of ReBaCuO system and $La_{2-X}Sr_XCuO_4$ system, Torrennce, Tokura et al reported that critical temperature depends on hall concentration in CuO_2 layer which is determined by oxygen content or by cation nonstoichiometries, and they also reported that a certain quantity of hall concentration gives maximum critical temperature.[22,23]
In conclusion, we have found that critical temperature of 2212 bismuth system similarly depends on oxygen content and that a certain oxygen content gives the maximum critical temperature as high as 95K.
Critical temperature in 2212 phase shows more steep change with oxygen content than in 123 structure or in lanthanide system.

REFERENCES

1. K.Imai and H.Matsuba, Workshop on High Temp. Superconductivity Huntsville, Alabama, May 23-25, 1989
2. D.E.MOrris, C.T.Hultgren et al. distributed in SC gloval '89
3. T.Ishida, H.Mazaki and T.Sakuma, Jpn.J.Appl.Phys. 27 L1626 (1988)
4. M.Kawasaki, S.Nagata, K.Takeuchi, H.Koinuma, Jpn.J.Appl.Phys. 27 L2227 (1988)
5. M.S. Hybertsen and L.F.Mattheiss, Phys.Rev.Lett. 60 1661 (1988)
6. A.Muto, A.Morimoto and T.Shimizu, Jpn.J.Appl.Phys. 28 L1144 (1989)
7. N.H.Wang, C.M.K.Wang and H.C.I.Kao et al, Jpn.J.Appl.Phys 28 L1505 (1989)
8. H.Niu, N.Fukushima and K.Ando Jpn.J.Appl.Phys. 27 L1442 (1988)
9. F.Herman, R.V.Kasowaski, W.Y.Hsu, Phys.Rev. B 38 204 (1988)
10. J.M.Tarascon, W.R.Mckinnon, P.Barboux, B.G.Bagley, L.H.Green, G.W.Hull, Y.LePage, N.Stoffiel and M.Giroud, Phys.Rev. B 38 8885 (1988)
11. S.A.Sunshine, T.Siegist, L.F.Schneemeyer et al. Phys. Rev. B 38 893 (1988)
12. M.Onoda, A.Yamamoto, E.Takayama-Muromachi and S.Takekawa, Jpn.J.Appl.Phys. 27 L833 (1988)
13. R.M.Hazen et al. Phys. Rev. Lett. 60 1174 (1988)
14. H.Nagano et al, Jpn.J.Appl.Phys. 28 L364 (1989)
15. K.Kurusu and H.Takami accepted to The Analyst
16. Y.Matsui, H.Maeda,Y.Tanaka and S.Hirouchi, Jpn.J.Appl.Phys. 27 L372 (1988)
17. Y.Matsui et al Jpn.J.Appl.Phys. 27 L872 (1988)
18. J.M.Tarascon et al. Phys. Rev. B 39 11587 (1989)
19. S.Sato and T.Tanegai Jpn. J. Appl. Phys. 28 L1620 (1989)
20. P.A.P.Lindberg, P.Soukiassian, Z.-X.Shen, et al. Appl.Phys.Lett. 53 1970 (1989)
21. Z-X.Shen, P.A.D.Lindberg et al. Phys.Rev.B 38 11820 (1988)
22. J.B.Torrance, Y.Tokura, et al. Phys.Rev.Lett. 61 1127 (1988)
23. Y.Tokura, J.B.Torrance, et al. PHys.Rev.B 38 7156 (1988)

HRTEM Observation on Mixed Phases and Interfaces in High-T_c Bi-Based Superconductors

SHIGEO HORIUCHI[1], KAORU SHODA[2], XIAO-JING WU[3], and YOSHIO MATSUI[1]

[1] National Institute for Research in Inorganic Materials, Tsukuba, Ibaraki, 305 Japan
[2] Ube Industries Ltd., Ube, Yamaguchi, 755 Japan
[3] Institute of Physics, Chinese Academy of Science, Beijing, China

For a series of superconductors $Bi_2Sr_2Ca_{n-1}Cu_nO_{2n+4+x}$ (n=1~3), showing the critical temperatures Tc at about 20, 85 and 110K respectively, we have examined the mixing state of the phases, especially the microstructures at interfaces, by high-resolution transmission electron microscopy (HRTEM).[1-4] Here we show the results of observation on two kinds of specimens.

Bi-based superconductors were prepared by doping fluorines.[5] Starting powders of Bi_2O_3, $SrCO_3$, $CaCO_3$, CaF_2 and CuO were heated in air. The mixing ratio of $CaCO_3$ and CaF_2 was changed to control the doping content. On heating at 860°C with the mixing ratio of 9/1 a superconductor with Tc=113K and Tc_{end}=106K was obtained, which can hardly be achieved without fluorines. The grains with the n=2 and 3 phase have grown very large, although it is known from the NMR measurement that most fluorines have been released from the crystal during heating. HRTEM observation reveals a frequent intergrowth of layers. Besides there are an interesting microstructures appearing locally (Fig.1).[6] It is noted that the Bi-O planes are not always in the same level at the center part and, consequently, a local interface is formed. As has been pointed out previously[1] the (020) lattice planes are bended prominently. This causes the formation of the supercell, which are incommensurately modulated. Fig.1(b) schematically representes the lattice bending. The ellipsoids indicate the Bi-concentrated bands. In a normal case they array in the tetragonal symmetry. At the lower right part of the photograph they are in the monoclinic symmetry. We think this type of lattice distortion is essentially important to get high value of Tc.[7] Fig.2 is a simplified expression of the local interface of Fig.1. The Bi-O planes in the both sides are mostly mismatched at the interface but some of Cu-O planes are still linking to each other. This type of interface structure will serve to enhance the critical current density Jc.

Another specimen without F was prepared by quenching from the temperature, at which it has been partially melted. It is consisted of major component of the n=1 phase with a minor component of impurity phases, which are $(Ca,Sr)_3Cu_5O_y$ and Ca_2CuO_3 both with the shape of square column. When the specimen is annealed at 500°C in air, Tc arises at about 75K (Fig.3) and this is due to the formation of small amount of the n=2 phase, which increases the volume fraction on annealing at higher temperature. The n=3 phase becomes the major component on annealing at 880°C. This must be due to the incongruent melting. Fig.4 shows an HRTEM image from the specimen, which has been obtained by the quenching from 900°C. We notice that two plates of the n=1 phase are adjoining making an angle of about 20°. Interesting is that the interface between two plates is not straight but zigzag in a large scale. The contrast disturbance in each plate suggests that there are frequent occurrence of the intergrowth or micro-twinning. Fig.5 is another image from the same specimen. Here we see the interface between an n=1 area and a Ca_2CuO_3 grain. Rather complicated structures in the n=1 area is due to the irradiation damage during ion-thinning. Fig.6(a) is an HRTEM image from the specimen annealed at 500°C for 50hr. Adjoining lamellae are in the relation of micro-twins. This is confirmed by the electron diffraction pattern in (b).

References

1) Y.Matsui, H.Maeda, Y.Tanaka and S.Horiuchi: Jpn.J.Appl.Phys.27(1988)L361 & L372.
2) Y.Matsui, H.Maeda, T.Tanaka, E.Takayama-Muromachi, S.Takekawa and S.Horiuchi: ibid.27(1988)L827.
3) Y.Matsui, S.Takekawa, H.Nozaki, A.Umezono, E.Takayama-Muromachi and S.Horiuchi: ibid.27(1988)L1241.
4) S.Horiuchi, K.Shoda, M.Iwatsuki, Y.Harada and Y.Matsui: ibid.28(1989)L386.
5) S.Horiuchi, K.Shoda, H.Nozaki, Y.Onoda and Y.Matsui: ibid.28(1989)L621.
6) S.Horiuchi, K.Shoda and Y.Matsui: J.Jpn.Ceram.Soc.97(1989)992.
7) S.Horiuchi et al: to be published.

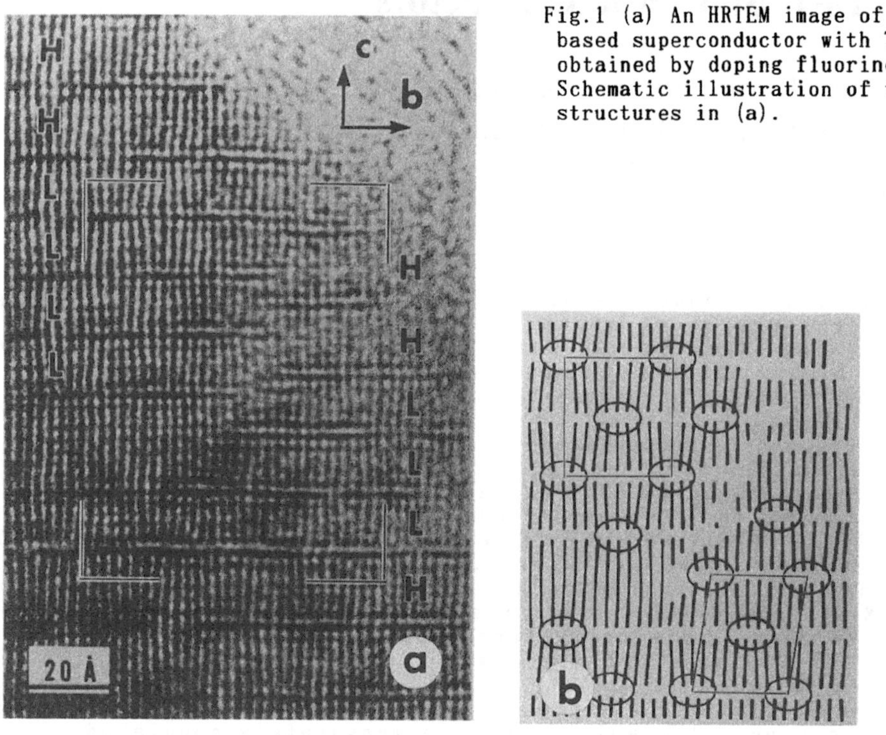

Fig.1 (a) An HRTEM image of a Bi-based superconductor with Tc=113K, obtained by doping fluorines. (b) Schematic illustration of the fine structures in (a).

Fig.2 Schematic illustration of the interface, found in Fig.1. Some of the Cu-O planes are linked at the interface

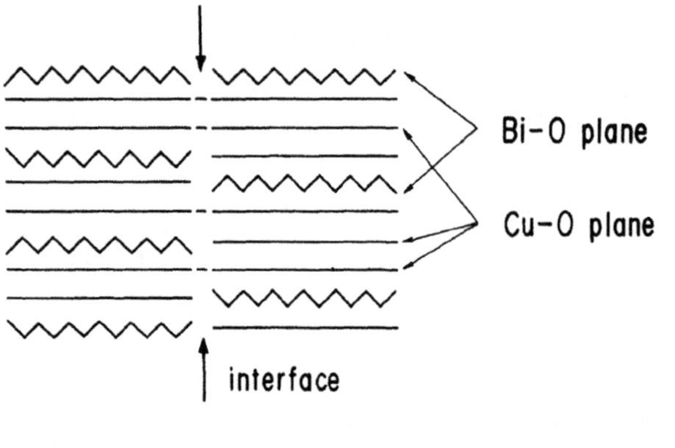

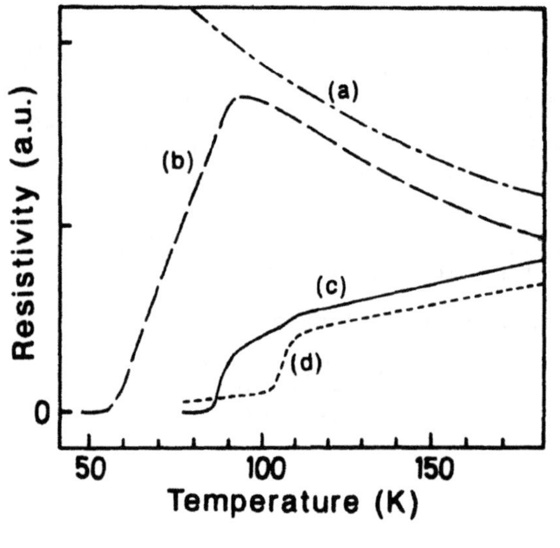

Fig.3 Temperature dependence of the electrical resistivity for the specimens quenched from 900°C (a), annealed at 500°C (b), at 850°C (c) and at 880°C (d) for 50hr.

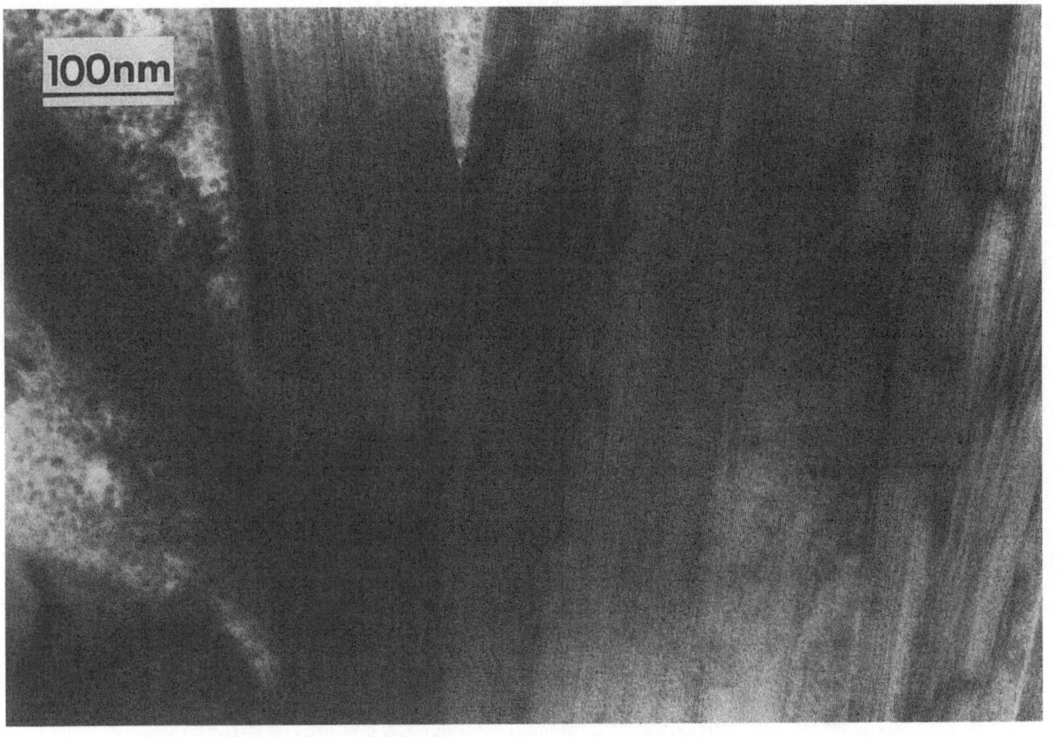

Fig.4 Lamellae of the n=1 phase in the specimen quenched from 900°C. The interface is bended zigzag.

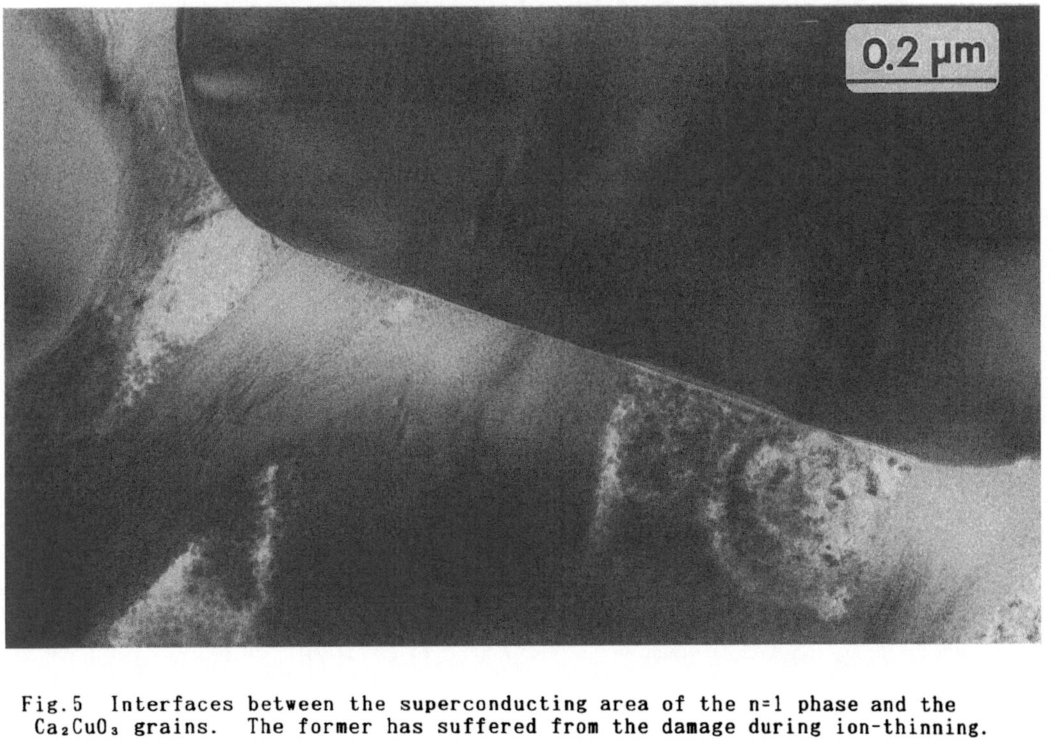

Fig.5 Interfaces between the superconducting area of the n=1 phase and the Ca_2CuO_3 grains. The former has suffered from the damage during ion-thinning.

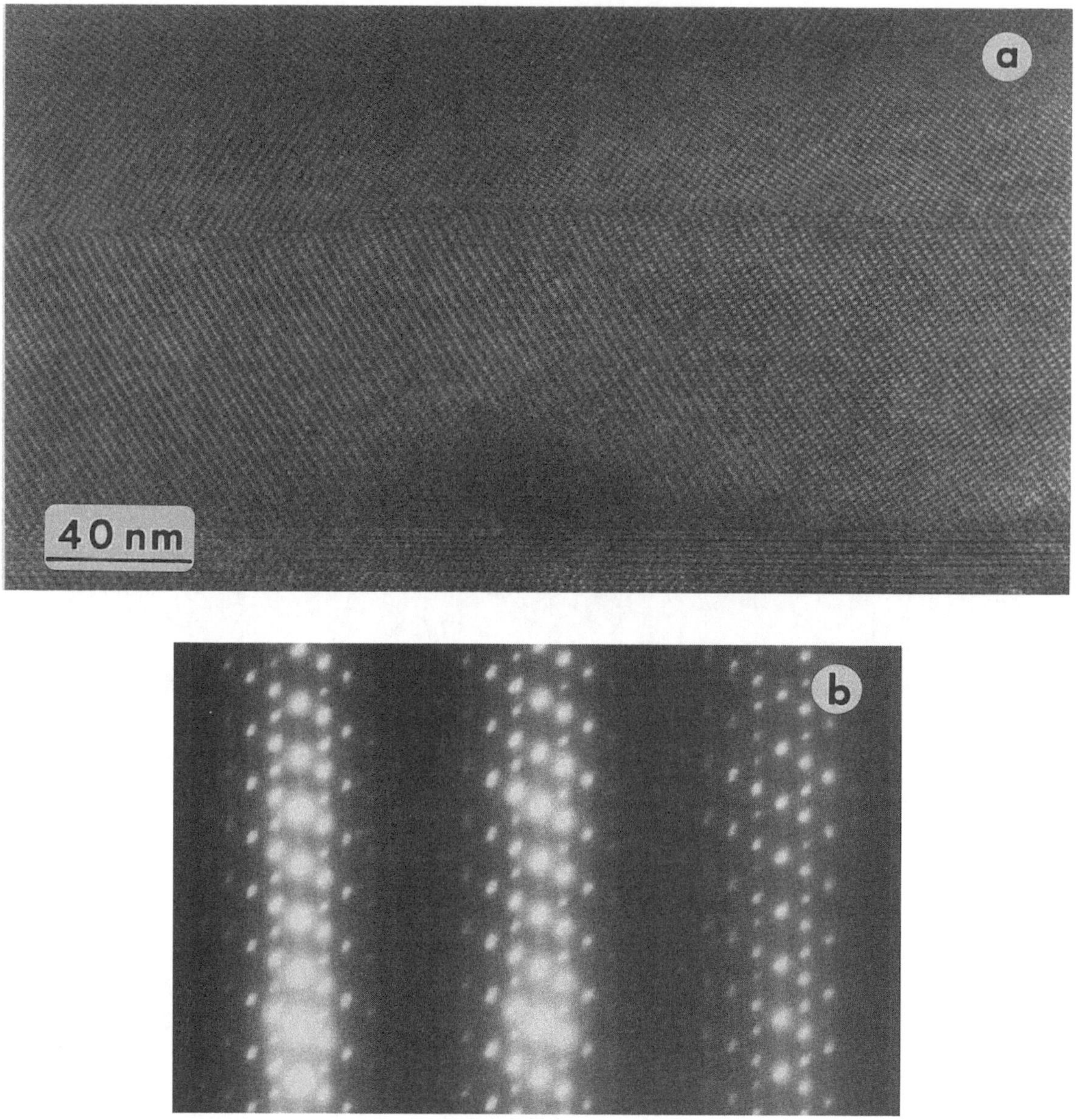

Fig.6 (a) Micro-twins in the specimen quenched from 900°C and annealed at 500°C for 50hr. (b) An electron diffraction pattern corresponding to (a)

Modulated Structures in High-T_c Superconducting Oxides: Bi-Sr-Ca-Y-Cu-O and Tl-Ba-Ca-Cu-O

Y. INOUE[1], M. HASEGAWA[1], M. YAMANAKA[1], and Y. KOYAMA[2]

[1] Central Engineering Laboratories, Nissan Motor Co., Ltd., 1, Natsushima-cho, Yokosuka, 237 Japan
[2] Department of Materials Science and Engineering, Waseda University, Ohkubo, Shinjuku-ku, Tokyo, 169 Japan

ABSTRACT

Features of modulated structures in $Bi_2Sr_2Ca_{1-x}Y_xCu_2O_y$, $Bi_4Sr_3Ca_{3-v}Y_vCu_4O_w$ and $Tl_2Ba_2Ca_2Cu_3O_z$ have been investigated by means of electron microscopy. In $Bi_2Sr_2Ca_{1-x}Y_xCu_2O_y$ and $Bi_4Sr_3Ca_{3-v}Y_v Cu_4O_w$, when the Y content increases, the period of one-dimensional modulated structure along the *b* axis decreases linearly. In their electron diffraction patterns, diffuse streaks along the *a* axis are observed in $v = 2.0$ and $x = 0.8$, and then the splitting of the superlattice spot along the *a* axis is detected in $x = 1.0$. This new structure modulated along both *a* and *b* axes is characterized by the periodic array of wavy stripes along the *b* axis. In $Tl_2Ba_2Ca_2Cu_3O_z$ with T_c of 117 K, thermal diffuse scattering due to a transverse phonon (e // [010]) propagating along the [100] direction was found in electron diffraction patterns. The scattering has an intensity maximum at the middle between two fundamental spots along the <100> directions and its intensity increases with decreasing temperature.

KEYWORDS: superconducting oxide, modulated structure, electron microscopy

INTRODUCTION

Superconductivity in oxides such as Bi-Sr-Ca-Cu-O and Tl-Ba-Ca-Cu-O is one of the most interesting subject in solid-state physics. Both oxides have layered structures with tetragonal symmetry, which is characterized by Cu-O layers. It has been pointed out that T_c depends on the number *n* of Cu-O layers.[1,2] That is, it has been reported that T_c is 75 K for $n = 2$ and 105 K for $n = 3$ in $Bi_2(Sr,Ca)_{n+1}Cu_nO_y$. The highest T_c has been reported to be 125 K for $n = 3$ in $Tl_2(Ba,Ca)_{n+1}Cu_nO_z$. There are two typical compounds, $Bi_2Sr_2CaCu_2O_y$ and $Bi_4Sr_3Ca_3Cu_4O_w$, in $Bi_2(Sr,Ca)_3Cu_2O_y$ system. The difference in the crystal structures is that, in $Bi_4Sr_3Ca_3Cu_4O_w$, Ca occupies not only the Ca site but also a part of the Sr site in $Bi_2Sr_2CaCu_2O_y$. The one-dimensional modulated structure along the *b* axis has been observed in $Bi_2Sr_2CaCu_2O_y$[3] although no modulated structure has been reported in $Bi_4Sr_3Ca_3Cu_4O_w$.
In $Bi_2Sr_2Ca_{1-x}Y_xCu_2O_y$ and $Bi_4Sr_3Ca_{3-v}Y_vCu_4O_w$, an important feature is that the increase in Y content leads to the decrease in T_c, corresponding to the decrease in hole concentration.[4,5] In addition, a semiconductorlike behavior has been observed in $0.6 < x \leq 1.0$ and $1.1 < v \leq 2.0$, and antiferromagnetic ordering appears near $x = 1.0$.[6] Moreover, we have found a new modulated structure in $0.8 < x \leq 1.0$, where the oxides exhibit both the semiconductorlike behavior and antiferromagnetic ordering.
In $Tl_2Ba_2Ca_2Cu_3O_z$, the two-dimensional modulated structure has been observed by other investigators.[7,8] We have found a characteristic thermal diffuse scattering in the vicinity of T_c by means of electron diffraction.
In the present paper, we report the detailed features of the modulated structures in $Bi_4Sr_3Ca_{3-v}Y_vCu_4O_w$ and $Bi_2Sr_2Ca_{1-x}Y_xCu_2O_y$, and describe the features of thermal diffuse scattering found in $Tl_2Ba_2Ca_2Cu_3O_z$.

EXPERIMENTAL

$Bi_2Sr_2Ca_{1-x}Y_xCu_2O_y$, $Bi_4Sr_3Ca_{3-v}Y_vCu_4O_w$ and $Tl_2Ba_2Ca_2Cu_3O_z$ were prepared by the conventional solid-state reaction method. In $Bi_4Sr_3Ca_{3-v}Y_vCu_4O_w$, powder samples were made by mixing powders of

Bi_2O_3, $SrCO_3$, $CaCO_3$, Y_2O_3 and CuO. After calcined at 1123 K for 10 h in air, the mixed powders were ground and pressed into pellets. The pellets were sintered at 1133 K for 30 h in air. The details of the preparation for the others were described elsewhere.[9,10] X-ray-diffraction (XRD) profiles of the Bi-compounds had the same crystal structure which was basically identical to that of $Bi_2(Sr,Ca)_3Cu_2O_y$. Note that XRD profiles of the $Bi_4Sr_3Ca_{3-v}Y_vCu_4O_w$ with $v > 2.0$ exhibit the existence of the impurity phases. The Tl-compound shows a single-phase XRD profile. In order to check the Ca content of the Bi-compounds, the energy dispersive x-ray spectroscopy (EDX) measurement was carried out. The actual content was determined to be 0.00, 0.52, 0.82, and 1.00 for x = 0.00, 0.50, 0.80, and 1.00 in the $Bi_2Sr_2Ca_{1-x}Y_xCu_2O_y$ and 0.00, 1.01, 1.45, and 1.99 for v = 0.00, 1.00, 1.50, and 2.00 in the $Bi_4Sr_3Ca_{3-v}Y_vCu_4O_w$, respectively. The electrical resistance was measured by means of the usual dc four-probe method. $Bi_2Sr_2Ca_{1-x}Y_xCu_2O_y$ with x = 0.00 and 0.52 and $Bi_4Sr_3Ca_{3-v}Y_vCu_4O_w$ with v = 0.00 and 1.01 showed the superconductive behavior, while the others exhibited the semiconductorlike behavior. The zero-resistance temperature of $Tl_2Ba_2Ca_2Cu_3O_z$ was found to be 117 K. The investigation of modulated structures was made by taking not only electron diffraction patterns but also high-resolution electron micrographs, using the Hitachi H-800 equipped with the EDX and JEOL JEM-200CX electron microscope. Specimens for the observation were flakes obtained by crushing the pellets.

RESULTS AND DISCUSSION

Modulated Structure of $Bi_4Sr_3Ca_{3-v}Y_vCu_4O_w$

Figures 1(a)-1(d) are electron diffraction patterns for the four pellets with v = 0.00, 1.01, 1.45, and 1.99. All electron incidences are parallel to the c axis. As have been already observed in $Bi_2Sr_2CaCu_2O_y$,[3] superlattice spots showing the one-dimensional modulated structure along the b axis are seen in all diffraction patterns as well as the fundamental spots. A period of modulation in v = 0.00 is determined to be 4.6b which is slightly smaller than 4.8b in $Bi_2Sr_2CaCu_2O_y$, where b is a lattice parameter of the structure without the modulation. The period of one-dimensional modulated structure along the b axis decreases with increasing the Y content; 4.4b, 4.3b, and 4.2b for v = 1.01, 1.45, and 1.99, respectively. The most important feature in $Bi_4Sr_3Ca_{3-v}Y_vCu_4O_w$ is that the diffuse streak along the a axis is observed in v = 1.99, as indicated by an arrow in Fig. 1(d). The diffuse streak implies that there is a fluctuation of the structure along the a axis. In order to understand the details of the modulated structure, high-resolution electron micrographs were taken in the sample with v = 1.99, shown in Fig. 2. The micrograph shows the periodic array of stripes with the 4.2b period of modulation along the b axis. Some defects were also observed in the micrograph indicated by A in Fig. 2. Further, it is seen that the

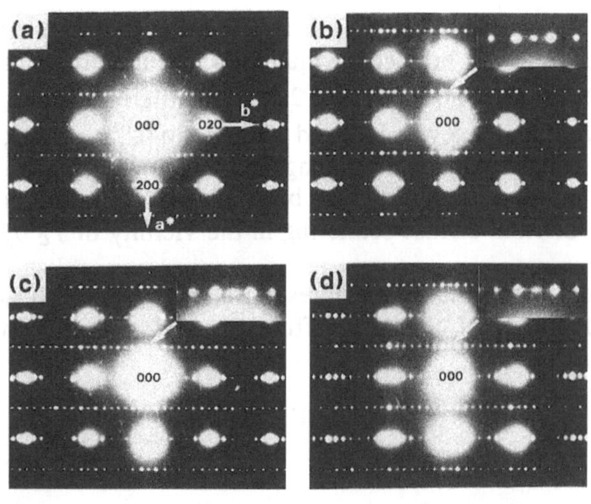

Fig. 1. Electron diffraction patterns of $Bi_4Sr_3Ca_{3-v}Y_vCu_4O_w$ with (a) v = 0.00, (b) 1.01, (c) 1.45, and (d) 1.99. In inserted patterns, superlattice spots around $\bar{1}00$ are shown. Note that fundamental spots are indexed in term of the crystal structure without any modulations.

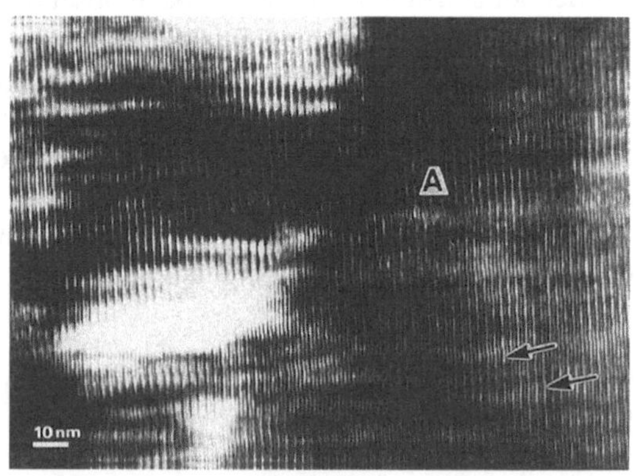

Fig. 2. High-resolution electron micrograph of $Bi_4Sr_3Ca_{3-v}Y_vCu_4O_w$ with v = 1.99, showing the periodic array of the a axis. Electron incidence is parallel to the c axis.

stripes are not partly straight, as indicated by arrows. These defects and wavy stripes make the contrast of narrow bands along the *b* axis. Unfortunately, it is impossible to examine the structure of the compound with $v > 2.0$, because the compounds involves the impurity phases.

Modulated Structure of $Bi_2Sr_2Ca_{1-x}Y_xCu_2O_y$

There is the one-dimensional modulated structure along the *b* axis in $Bi_2Sr_2Ca_{1-x}Y_xCu_2O_y$.[4] The relation between the period of one-dimensional modulated structure along the *b* axis and the Y content, which was obtained in the present experiment, is shown in Fig. 3, together with the relation in $Bi_4Sr_3Ca_{3-v}Y_vCu_4O_w$. The period decreases from 4.8*b* at $x = 0.00$ to 4.0*b* at $x = 0.82$. In this compound, a lock-in transition is found at $x = 0.82$. This result does therefore not coincide with the result obtained previously.[4] A striking feature in $Bi_2Sr_2Ca_{1-x}Y_xCu_2O_y$ is that the the period changes from 4.0*b* to 8.0*b* in the vicinity of $x = 1.00$. In addition, the splitting of the superlattice spots along the *a* axis is observed in the diffraction pattern at $x = 1.00$, shown in Fig. 4(b). The splitting implies that the second modulation takes place along the *a* axis and results in a new modulated structure modulated along the *a* and *b* axes. Figure 4(a) is a high-resolution micrograph in $x = 1.00$ and shows the periodic array of stripes along the *b* axis. The stripes along the *a* axis are wavy ones which can be regarded as a transverse wave. The period of the wave is determined to be about 4.5 nm from the micrograph. Because the wave has the same period as that of modulation along the *a* axis determined from the diffraction pattern, the second modulation is understood to make the periodic array of the wavy stripes.
Let us show that there is a correspondence between the changes in the modulated structures and the superconducting properties. When the Y content increases, the hole concentration decreases.[4,5] This is understood to correspond to the decrease in the period of the modulated structure. In addition, the antiferromagnetic ordering has been found around $x = 1.0$,[6] where appears the new modulated structure. However, the physical origin of the correspondence is not understood at all. Because the mechanism of the superconductivity in the present compounds has not been understood, we believe that it is important to point out the correspondence at present.

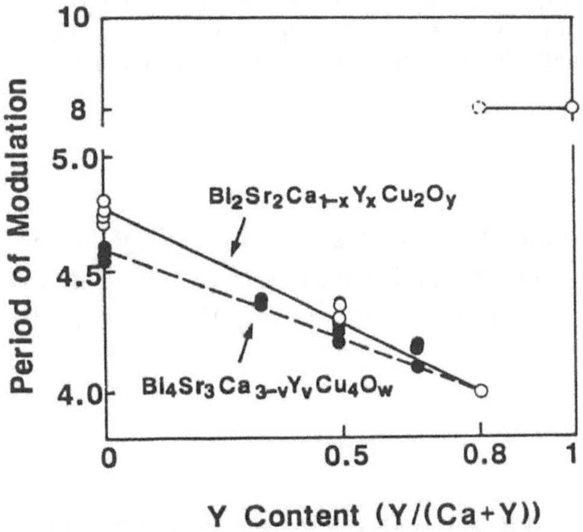

Fig. 3. Periods of one-dimensional modulated structures along the *b* axis of $Bi_2Sr_2Ca_{1-x}Y_xCu_2O_y$ and $Bi_4Sr_3Ca_{3-v}Y_vCu_4O_w$, as a function of the Y content.

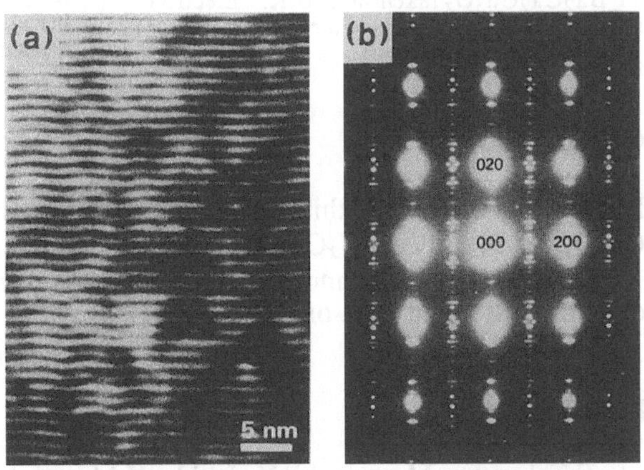

Fig. 4. High-resolution electron micrograph and diffraction pattern of $Bi_2Sr_2Ca_{1-x}Y_xCu_2O_y$ with $x = 1.00$. Electron incidence is parallel to the *c* axis.

Thermal Diffuse Scattering in $Tl_2Ba_2Ca_2Cu_3O_z$

Figure 5 shows an electron diffraction pattern of $Tl_2Ba_2Ca_2Cu_3O_z$ at 100 K. Two kinds of diffuse scatterings, indicated by A and B, are clearly seen in the pattern. Diffuse scattering A appears around a fundamental spot. The same diffuse scattering has been found by other investigators.[7,8] A characteristic feature of the diffuse scattering in the Tl-oxide is diffuse scattering B, which is located between two neighboring fundamental spots. Scattering B was observed as a nonradial diffuse streak and its intensity increases with decreasing temperature. Then the streak is interpreted as thermal diffuse scattering that is ascribed to low-frequency lattice vibrations, and their polarization vectors are perpendicular to directions of the streaks. In addition, because an intensity maximum of the streak is found at the middle between two fundamental spots along the <100> directions, the lattice vibration can be identified as a transverse phonon with q = $\frac{1}{2}$[100], whose polarization vector is parallel to the [010] direction. That is, this phonon seems to be a TA mode at the X point. The intensity at the zone boundary along the [100] direction is plotted against temperature in Fig. 6. It is found that the intensity approximately increases linearly with decreasing temperature in the temperature range between 310 and 110 K. The most striking feature is that the intensity increases remarkably near 100 K. Note that the change in the intensity is reversible during a cycle of cooling and heating and that it is common to all the other nonradial streaks. Based on the theory of thermal diffuse scattering, the intensity of the first-order thermal diffuse scattering is inversely proportional to a square of a phonon frequency. Therefore, the increase in the intensity in cooling represents the softening of the transverse phonon.

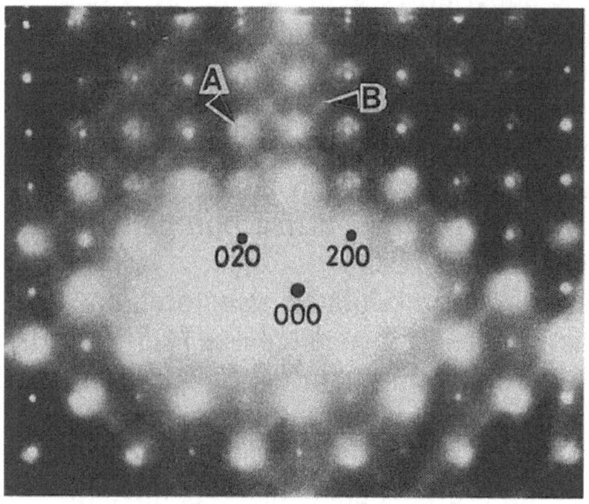

Fig. 5. An electron diffraction pattern of $Tl_2Ba_2Ca_2Cu_3O_z$ taken at 100 K. Electron incidence is almost parallel to the *c* axis.

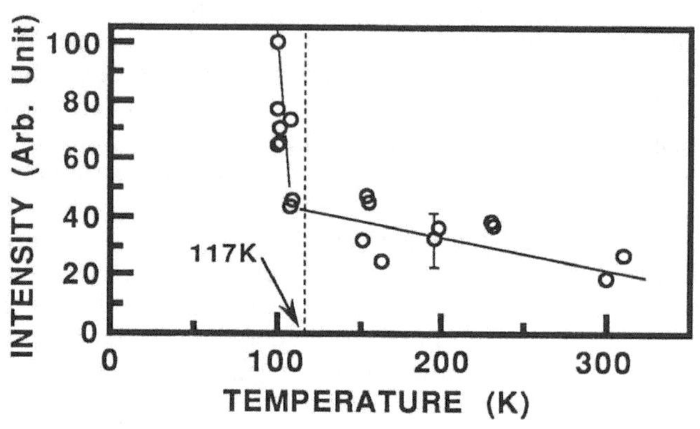

Fig. 6. The diffuse intensity at the middle between two fundamental spots along the [100] direction, as a function of temperature. The error bar is indicated only for one plotted point because the magnitude of the error is the same for all points. The dashed line represents the zero-resistance temperature of 117 K.

CONCLUSION

$Bi_4Sr_3Ca_{3-v}Y_vCu_4O_w$ exhibits the same one-dimensional modulated structure as that in $Bi_2Sr_2Ca_{1-x}Y_x Cu_2O_y$. In $Bi_2Sr_2Ca_{1-x}Y_xCu_2O_y$, the new structure modulated along *a* and *b* axes were observed in x = 1.00. The soft phonon mode due to the transverse phonon (e // [010]) propagating along the [100] direction was found in Tl-oxide.

REFERENCES

1. E.Takayama-Muromachi, Y.Uchida, Y.Matsui, M.Onoda, and K.Kato, Jpn. J. Appl. Phys. **27**, L556 (1988).
2. S.S.Parkin, V.Y.Lee, A.I.Nazzal, R.Savoy, R.Beyers, and S.J.La Placa, Phys. Rev. Lett. **61**, 750 (1988).
3. Y.Matsui, H.Maeda, Y.Tanaka, and S.Horiuchi, Jpn. J. Appl. Phys. **27**, L361 (1988).
4. T.Tamegai, K.Koga, K.Suzuki, M.Ichihara, F.Sadai, and Y.Iye, Jpn. J. Appl. Phys. **28**, L112 (1989).
5. A.Manthiram, and J.B.Goodenough, Appl. Phys. Lett. **53**, 420 (1988).
6. H.Miyatake, T.Ookuma, A.Asamitsu, N.Nishida, T.Tamegai, Y.Iye, R.Yoshizaki, A.Nishiyama, K.Nagamine, Y.Kadono, and J.H.Brewer (unpublished).
7. J.D.Fitz Gerald, R.L.Withers, J.G.Thompson, L.R.Wallenberg, J.S.Anderson, and B.G.Hyde, Phys. Rev. Lett. **60**, 2797 (1988).
8. S.Iijima, T.Ichihashi, Y.Shimakawa, T.Manako, and Y.Kubo, Jpn. J. Appl. Phys. **27**, L1061 (1988).
9. Y.Inoue, Y.Shichi, F.Munakata, M.Yamanaka, and Y.Koyama, Phys. Rev. B **40**, 7307 (1989).
10. Y.Koyama and H.Hoshiya, Phys. Rev. B **39**, 7336 (1989).

Modulated Structures of Bi-Based High-T_c Superconducting Oxides

Y. Hirotsu[1], N. Yamamoto[1], O. Tomioka[1], K. Miyagawa[1], Y. Nakamura[2], Y. Inoue[3], S. Nagakura[1], Y. Iwai[1], and M. Takata[1]

[1] Faculty of Engineering, Nagaoka University of Technology, Nagaoka, 940-21 Japan
[2] Department of Metallurgical Engineering, Tokyo, Institute of Technology, Tokyo, 152 Japan
[3] Central Engineering Laboratories, Nissan Motor Co., Ltd., Yokosuka, 237 Japan

ABSTRACT

Modulated structures in the Bi-Sr-Ca-Cu-O, Bi-Pb-Sr-Ca-Cu-O and Bi-Sr-Ca-Y-Cu-O systems have been investigated by means of high resolution electron microscopy and electron diffraction. Structure models of the atomic displacive modulations are presented for the modulated structures of these oxides.

KEY WORDS: modulated structure, atomic displacive modulation, high resolution electron microscopy, electron diffraction, Bi-based superconductors

INTRODUCTION

A number of structural studies have been made on the low Tc (Tc∼90K) and the high Tc (Tc∼105K) phases in the Bi-Sr-Ca-Cu-O system. Ideal compositions of the low and high Tc phases are $Bi_2Sr_2CaCu_2O_x$ (x∼8;2212 phase) and $Bi_2Sr_2Ca_2Cu_3O_y$ (x∼10;2223 phase) belonging to the polytypoid $Bi_2Sr_2Ca_nCu_{1+n}O_{6+2n}$ structure. The fundamental structures of these phases are tetragonal and basically identical with those of the Tl-based superconductors $Tl_2Ba_2Ca_nCu_{1+n}O_{6+2n}$. In both of these phases, modulated structures with incommensurate periods along their b(or a)-axis have been investigated intensively by high resolution electron microscopy (HREM) and electron diffraction[1,2].
Recently, effects of additive elements on the superconducting properties have been studied in the Bi-based 2212 and 2223 phases. Changes of the modulated structure due to additive elements, however, have not been studied in detail. In the present study, the effect of Pb addition in the 2223 phase and also the effect of replacement of Ca by Y in the 2212 phase on the structure change have been investigated by means of HREM and electron diffraction.

EXPERIMENTAL

Sintered sample with nominal compositions $(Bi_{1-x}Pb_x)_{1.0}Ca_{1.0}Sr_{0.8}Cu_{1.4}O_y$ (x=0, 0.05, 0.12, 0.2 and 0.3) and $Bi_2Sr_2Ca_{1-x}Y_xCu_2O_y$ (x=0, 0.5, 0.8 and 1.0) were prepared. The phase identification of these samples were made by X-ray diffraction. EDX elemental analysis was made by using an analytical 200kV electron microscope (Hitachi). Specimens for electron microscopic observation were prepared by ion-milling and also by crushing techniques. A 200kV electron microscope (JEOL) and a 1 MV electron microscope (Hitachi) were used.

RESULTS AND DISCUSSION

Modulation Modes of 2223 Phase in Bi-Pb-Sr-Ca-Cu-O System

An EDX analysis showed that the composition of each sample was nearly equal to the composition $(Bi_{1-x}Pb_x)_2Sr_2Ca_2Cu_3O_y$. By X-ray diffraction, the samples with x=0 and 0.05 were confirmed to be composed of the 2212 and 2223 phases. In the samples with x larger than 0.12, the 2223 phase was the majority. Selected area electron diffraction patterns showed that the 2223 phase of these samples has two types of modulated structure depending on the content x and that these modulated structures are based on the fundamental tetragonal structure with parameters $a_t=b_t=0.54$ nm and $c_t=3.71$ nm. A selected area [100] diffraction pattern of the 2223 phase with x=0.05 is shown in Fig.1(a), where superlattice reflections from the modulated structure are seen. It was found that the 2223 phase with x=0 and x=0.05 has the same type of modulated structure with the modulation along the b-axis ($b=4.7b_t$) as observed in the 2212 phase of the Bi-Sr-Ca-Cu-O system[1-3]. Hereafter, we call the modulated structure of this type "type-I". In the specimen with x=0.2 and 0.3, regions of the 2223 phase are mostly composed of a new type of modulated structure which gives satellite diffraction spots along the b* direction[4]. We, hereafter, call the modulation type "type-II". A [100] pattern from the "type-II" structure is shown in Fig.1(b). The modulation period measured from the diffraction pattern is about 4.2 nm($b=7.9b_t$). In the specimen with x=0.12, "type-I" and "type-II" structures coexist. In Fig.1(a) and (b), diffraction spots are indexed after commensurate orthorhombic structures with b parameters $5b_t$ and $8b_t$, respectively.

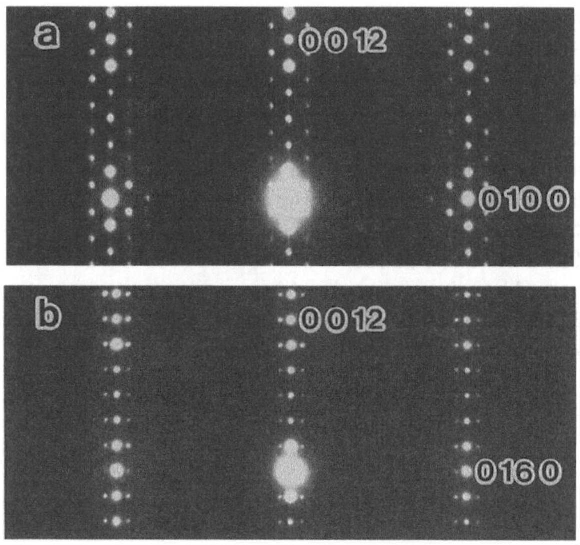

Fig.1 [100] diffraction patterns of the 2223 phase with x=0.05(a) and 0.2(b).

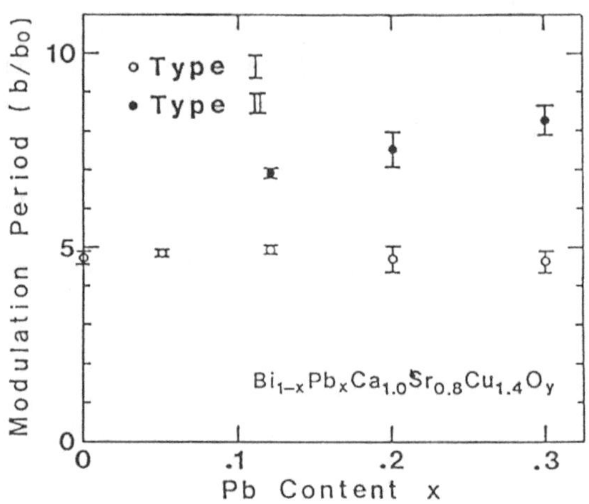

Fig.2 Change of modulation period in the "type-I" and "type-II" structures.

According to the b parameter measurements of the 2223 phase of the present specimens, the modulation period for the "type-I" structure is almost independent of the Pb content x, while that for the "type-II" structure increases with x. The modulation periods (b/b_t) are shown in Fig. 2 as a function of x which were measured from more than ten [100] diffraction patterns for each specimen. These modulation periods are incommensurate in general.
The two types of modulated structure can be distinguished clearly in the high resolution images with beam incidence along [100]. Figure 3(a) shows a [100] image from the "type-I" region in the specimen with x=0.05. A similar periodic lattice distortion is observed in the [100] image of the 2212 phase[1,3]. A [100] high resolution image from the "type-II" structure in the specimen with x=0.2 is shown in Fig.3(b), where the double Bi-layers are imaged as the double wavy chains of dark dots. The period of the sinusoidal displacement wave of the Bi-layers fluctuates from place to place. The atomic displacement of the "type-II" structure is smaller than that of the "type-I" structure.

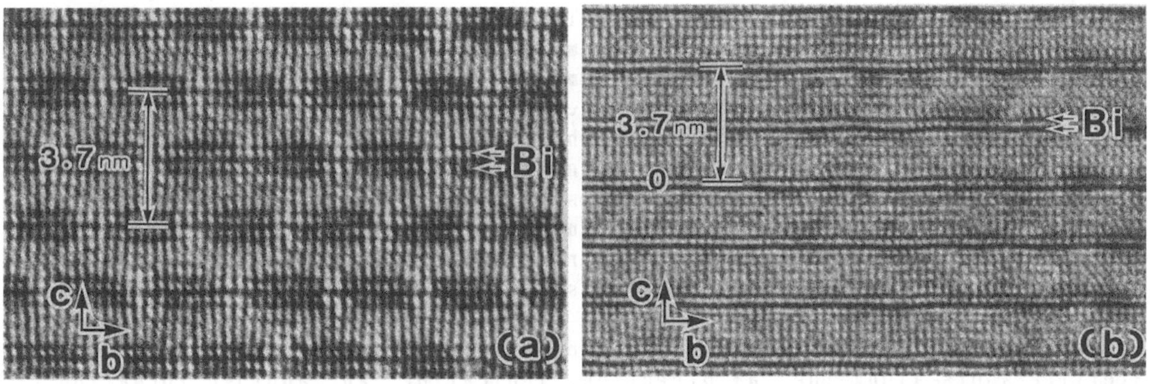

Fig.3 High resolution [100] images for the specimens with x=0.05(a) and 0.2(b)

Modulated Structures of the 2223 Phase in Bi-Pb-Sr-Ca-Cu-O System

Structure analysis of the two types of modulated structure of the 2223 phase in the Pb-doped Bi-Sr-Ca-Cu-O was made by taking advantage of the high resolution structure imaging and the selected area electron diffraction. In these analyses, commensurate periods were assumed and metal atom arrangements are only considered. Along the same way as in our previous analysis of the 2212 phase[3,5], possible structure models were made by taking metal atom displacive waves into consideration.
Figure 4(a) shows the most possible arrangement of Bi atoms in the modulated structure of "type-I" viewed along [001]. The large and small solid circles represent Bi atoms in neighboring double Bi layers, and the large and small open circles also represent Bi atoms in neighboring double Bi layers but separated by c/2 from the former Bi-layers. Metal atom positions in the remaining layers are correlated with those of the Bi atoms. These metal atom displacement waves which are expressed by both transverse and londitudinal waves could explain important zone-axis electron diffraction patterns. The most possible arrangement of Bi atom chains in the "type-II" structure is illustrated in Fig.4(b). The meanings of the solid and open circles are the same as in the above. The considerable displacements of Bi atom chains along [100] and [010] in Fig.4 are thought to be due to the oxygen atom arrangements in or near the double Bi layers. The modulation period of the "type-II" structure longer than that of the "type-I" structure is possibly due to the decrease of the extra oxygen atoms in or near the Bi-layers as a result of divalent Pb ions replacing partially the trivalent Bi ions.

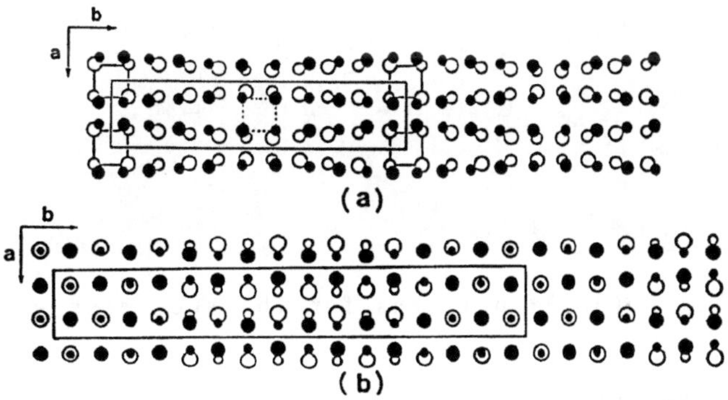

Fig.4 Modulated structure models for the "type-I" (a) and "type-II" (b) structures

Modulated Structures of the 2212 phase in Bi-Sr-Ca-Y-Cu-O system

It has been reported that the oxide is a superconductor up to the composition x=0.5 but changes into an insulating phase by a further replacement and that the modulation period of the 2212 phase changes depending on x [6].

According to the analysis of electron diffraction patterns, the modulation period in the specimen with x=0.5 is incommensurate: $4.3b_t$, while the periods in the specimens with x=0.8 and 1.0 are commensurate:$4.0b_t$ and $8.0b_t$, respectively. The modulated structure type of the specimen with x=0.5 was found to be the same as that of the 2212 phase in the Bi-Sr-Ca-Cu-O system. Figures 5(a), (b) and(c) are [001] diffraction patterns from the specimens with x = 0, 0.8 and 1.0, respectively. In the figures, single arrows indicate forbidden reflections and double arrows indicate double diffractions between the superstructure reflection and the forbidden reflection. The forbidden reflections are thought to be due to growth defects or lateral growth of the layered structure with a growth step of c/2. In [001] diffraction patterns for x=1.0, the superstructure spots are always accompanied by streaks or short period satellite spots along the a* direction, indicating an another formation of long wave modulation period along the [100] direction. Based on the electron diffraction patterns of important zone axes and also on the high resolution images, the commensurate structures with $b=4b_t$(x=0.5) and $8b_t$(x=1.0) were examined. We have come to the conclusion that the modes of atomic displacive modulation are almost the same in these phases when viewed along the [100] direction, but the modes are different from each other when viewed along [010]. Furthermore, it is necessary to introduce a compositional modulation of metal ions for the $4b_t$ and the $8b_t$ structures along their [010] directions. A detailed study on these modulated structures is now in progress.

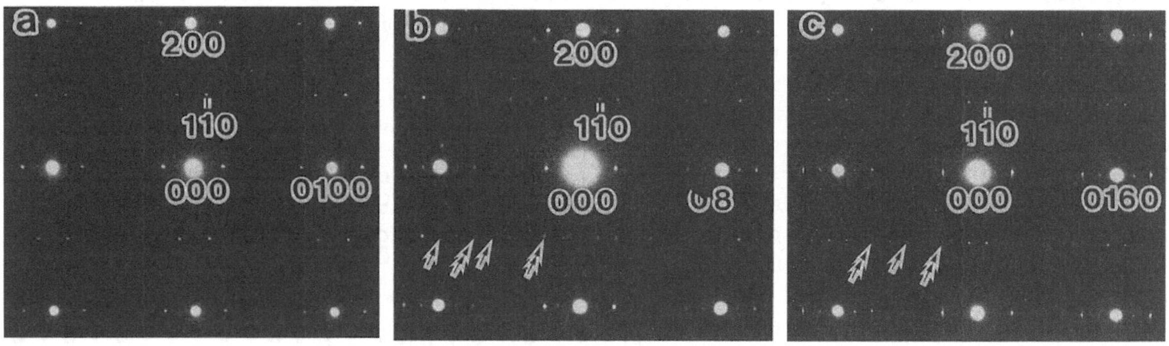

Fig.5 (001)* patterns for the specimens with x=0(a), 0.8(b) and 1.0(c).

REFERENCES

1. Matsui Y, Maeda H, Tanaka Y, Horiuchi S (1988) Possible model of the modulated structure in high-Tc superconductor in a Bi-Sr-Ca-Cu-O system revealed by high-resolution electron microscopy. Jpn. J. Appl. Phys. 27: L372-375
2. Shindo D, Hiraga K, Hirabayashi M, Kikuchi M, Syono Y (1988) Structure analysis of high-Tc superconductor Bi-Ca-Sr-Cu-O by processing of high-resolution electron microscope images. Jpn. J. Appl. Phys. 27: L1018-1021
3. Hirotsu Y, Tomioka O, Ohkubo T, Yamamoto N, Nakamura Y, Nagakura S, Komatsu T, Matsushita K (1988) Modulated strucuture of high Tc superconductor Bi-Ca-Sr-Ca-O studied by high-resolution electron microscopy and electron diffraction. Jpn. J. Appl. Phys. 27: L1869-1872
4. Ikeda S, Aoki K, Hatano T, Ogawa K (1988) A new mode of modulation observed in the Bi-Pb-Sr-Ca-Cu-O system. Jpn. J. Appl.Phys. 27: L2040-2043
5. Yamamoto N, Hirotsu Y, Nakamura Y, Nagakura S (1989) Superspace group analysis of the modulated structure in superconductor Bi-Sr-Ca-Cu-O. Jpn. J. Appl. Phys. 28: L598-601
6. Tamegai T, Koga K, Suzuki K, Ichihara M, Sakai F, Iye Y (1989) Metal-insulator transition in the $Bi_2Sr_2Ca_{1-x}Y_xCu_2O_{8+y}$ system. Jpn. J. Appl. Phys. 28 : L112-115

Preparation of Bi(Pb, Sb)-Sr-Ca-Cu-O System Sperconductors Through Sol-Gel Method

YOSHIO MASUDA[1], TSUYOSHI TATEISHI[2], KAZUYUKI SHIBUTANI[1], RIKURO OGAWA[1], and YOSHIO KAWATE[1]

Superconducting and Cryogenic Technology Center, Kobe Steel Ltd.[1] and Kobelco Research Institute Inc.[2], 5-5, Takatsukadai 1-chome, Nishi-ku, Kobe, 673-02 Japan

ABSTRACT

We prepared superconductive powders in the Bi(Pb,Sb)-Sr-Ca-Cu-O systems through sol-gel method from the acetates. As a result of the examination, it has been revealed that Pb has the effects to lower the partial melting point and to promote the formation and the growth of the superconductive grain, and that in the case of Sb addition, the samples to which only Sb is added can not show the existence of the superconductive phases, but adding small amount of Sb whith Pb, Sb is more effective to produce the high-Tc phase than the case of the addition Pb only.

KEY WORDS: preparation, sol-gel, superconductor, Bi-oxides-superconductors, Pb-Sb

INTRODUCTION

It is known that the Bi-Sr-Ca-Cu-O system superconductors have high-Tc phase (c-axis=3.7nm) having Tc of about 110K and low-Tc phase (c-axis=3.0nm) having about 80K. But it is difficult to obtain single phase composed of high-Tc phase in this system[1]. We have analyzed the phases which are produced when the sol-gel method is applied for synthesise of Bi-Oxide-superconductors. We have also investigated the effects of additions of Pb and Sb on synthesis and propeties of superconductors. The obtained results are discussed in this paper.

EXPERIMENTAL

Acetates of Bi, Sr, Ca, Cu and Pb and metal Sb were used as starting materials. Bi-acetate was dissolved in a mixture solution consisting of acetic acid and hydrogen peroxide. Acetates of Sr, Ca and Cu were added to this Bi-contained solution to get the metal ratio of Bi:Sr:Ca:Cu = 0.96:1.0:1.1:1.6 and dissolved herein. This solution was used as a mother liquor. Pb or/and Sb was added to this solution, and the effects of them on synthesis were examined. As to Pb addition, Pb-acetate was dissolved in water, and then 0.24 mol equivalent quantity of Pb was added to the mother liquor. As to Sb addition, metal Sb was dissolved initially in nitric acid and 0.24 mol equivalent quantity of Sb was added to the mother liquor. A mixture solution containing both Pb and Sb was also prepared and the total amount of Pb and Sb was 0.24 mol equivalent. After adjusting their pH of these solutions, they were heated to 80°C to study how the gels are formed. To study the change from the precursor gel to ceramics the phases which are generated by heat treatment at each transitional temperature were identified by XRD method, and the thermal behavior of the gel was analyzed with TG-DTA technique. The precursor gel was directly calcined in the air for 12 and 100 hrs at 850°C to obtain ceramic powder. Occasionally the powders were annealed at 400°C after heating. For these powders, SEM observation, EDX composition analysis and identification of generated phases by XRD analysis were performed. The superconductivity of them was evaluated also by SQUID magnetmeter(HSM-2000).

RESULTS AND DISCUSSIONS

In the case of the systems to which Sb was not added we could have the transparent gels at pH 5.2 to 5.8. If pH was lower than the specified range, blue sediment appeared but if pH was higher than that range, white sediment appeared. The systems to which Sb was added became muddy white when they were heated and concentrated, so that we could not obtain completely transparent gels. As a result of thermal analysis of these gels, it has been revealed that the systems which gave transparent gels had the exothermic peak at about 230°C and 320°C, accompanied by intensive reduction of weight and that from 400°C to 850°C remarkable thermal change was not observed and weight of samples reduced slowly as temperature rose. Endothermic peak, which seems to indicate fusion, was observed near 853°C in these samples. The Sb-added system had only one exothermic peak (at 250°C) accompanied by intensive reduction of weight at a lower temperature than that of transparent gels and also had endothermic peak at 550°C which is assumed to have been caused by decomposition of $Sr(NO_3)_2$. These exothermic peaks are assumed to have been caused by combustion of organic components.

The powders which were obtained by calcining the dried gels to each transition-al temperatures from gel to ceramics, were examined by XRD analysis. It has been revealed that in the system to which Pb was added the low-Tc phase was more formed already at 800°C than Pb-less systems, and up to 850°C sample was composed almost of low-Tc phase, and that in the systems to which Sb only was added, $Sr(NO_3)_2$ and CuO were formed at 400°C, but $Sr(NO_3)_2$ already disappeared at 600°C and $CaBi_2O_4$ and CuO were formed at 600°C. Whenever heating up to 850°C was performed, superconductive phases could not be formed in this sample. However, when a small amount (0.05 mol each) of Sb was added to Pb contained sample, it seems that Sb promotes the producing reaction of superconductive phases.

Figure 1 shows the result of XRD analysis of the powders obtained by calcining each dried gel at 850°C for 12 hrs. The mother system (a) to which Pb and Sb were not added has almost single phase of the low-Tc phase. The system (b) to which Pb was added gives insignificant peaks of the very low-Tc phase (c-axis = 2.5nm) and Ca_2PbO_4, in addition to low-Tc superconductive phase. The system

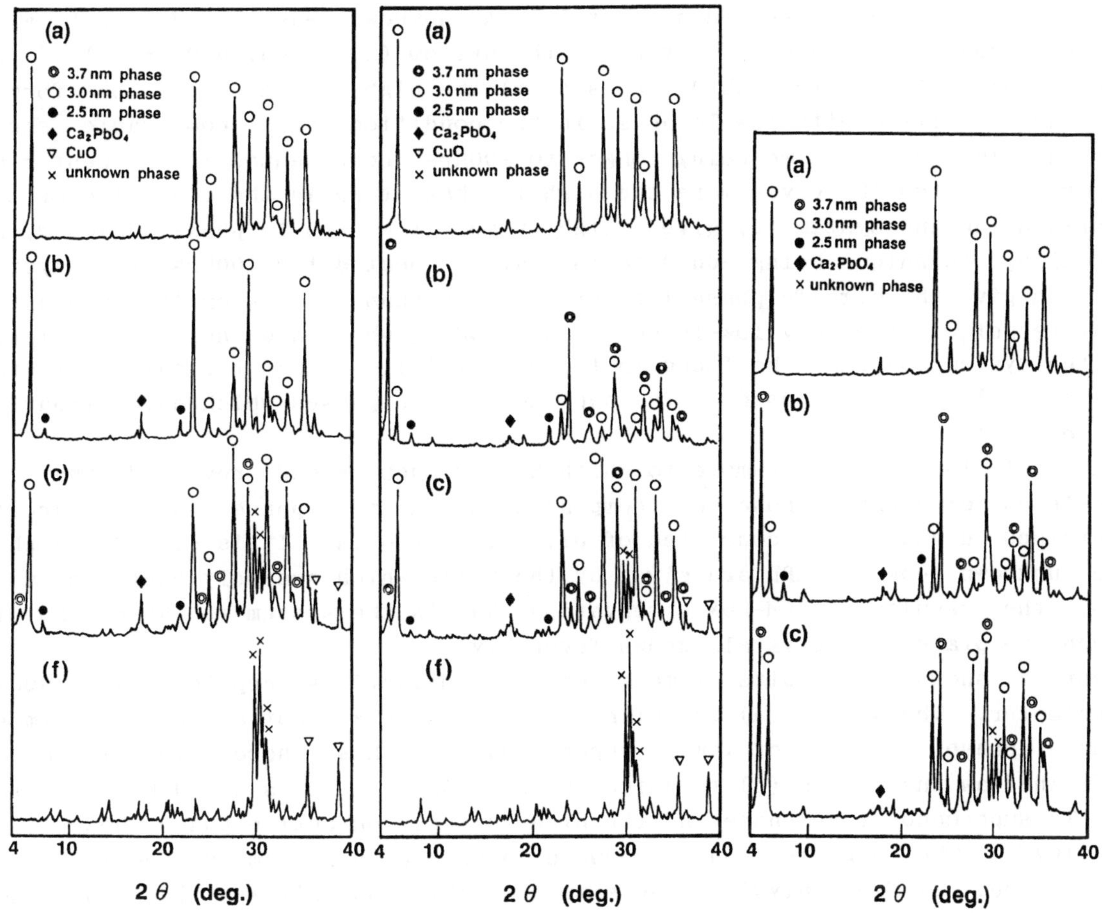

Fig.1 X-ray diffraction patterns of the powders calcined at 850°C for 12hrs (represented left) and 100hrs (represeted center), and the powders annealed at 400°C after being heated at 850°C for 100hrs (represented right).
(a)$Bi_{0.96}Sr_{1.0}Ca_{1.1}Cu_{1.6}O_y$ (b)$Bi_{0.96}Pb_{0.24}Sr_{1.0}Ca_{1.1}Cu_{1.6}O_y$
(c)$Bi_{0.96}Pb_{0.19}Sb_{0.05}Sr_{1.0}Ca_{1.1}Cu_{1.6}O_y$ (f)$Bi_{0.96}Sb_{0.24}Sr_{1.0}Ca_{1.1}Cu_{1.6}O_y$

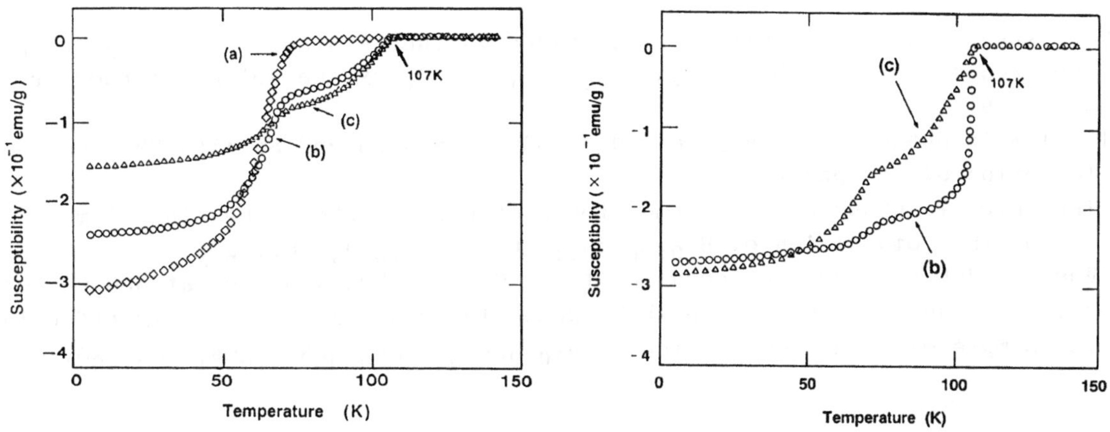

Fig.2 Temperature dependense of the susceptibility of the powders calcined at 850°C for 12hrs (represented left) and the powders annealed at 400°C for 12hrs after being heated at 850°C for 100hrs (represented right).
(a)$Bi_{0.96}Sr_{1.0}Ca_{1.1}Cu_{1.6}O_y$ (b)$Bi_{0.96}Pb_{0.24}Sr_{1.0}Ca_{1.1}Cu_{1.6}O_y$
(c)$Bi_{0.96}Pb_{0.19}Sb_{0.05}Sr_{1.0}Ca_{1.1}Cu_{1.6}O_y$

(f) to which Sb only was added gives CuO and unknown peaks but does not give the superconductive phase. However, in the system (c) to which 0.05 mol of Sb were added with Pb (0.05Sb-0.19Pb system) the high-Tc phase can be formed already a little with low-Tc phase in the condition for a comparable short period. The heating time being longer to 100hrs, represented in the center of figure 1, system (b) gives more the high-Tc phase than in the condition of the 12hr-heating and system (c) gives slightly more the high-Tc phase. However, an annealing technique being added to the samples heated for 100hrs, the system (c) can give the high-Tc phase more effectively than the system (b) and after that technique, the very low-Tc phase and Ca_2PbO_4 which are included a little in the system (b) can't be found in the system (c), so that the sample of the system (c) can be composed of the superconductive phases which have impurity phase little.

EDX analysis of the sample to which Sb was much added revealed that it consisted of a phase whose main component was Cu and a phase which scarcely contained Cu and mainly consisted of Bi, Sb, Sr and Ca. It is considered that the unknown peaks of XRD are given by the phase which scarcely contains Cu and from the results of SEM-EDX observation that in the system to which Pb is added the plate-like crystals grows favorably.

Figure 2 shows the results of measurement of magnetic susceptibility by SQUID magetmeter. The system (a) to which both Pb and Sb are not added has almost single phase composed of low-Tc superconductive phase whose Tc.on is about 75K, whereas the system (c) to which Pb or 0.05Sb-0.19Pb is added has both the low-Tc superconductive phase and the high-Tc superconductive phase whose Tc.on is 107K. The high-Tc phase produced in Pb added system (b) could not be ascertained by XRD analysis. The sample to which Sb only was added did not show transition to superconductor.

It is evident that the sample annealed after sintering contains more high-Tc superconductive phase than the sintered sample without annealing.

CONCLUSIONS

1. The superconductors could be synthesized in the systems of $Bi_{0.96}Sr_{1.0}Ca_{1.1}Cu_{1.6}O_y$ and $Bi_{0.96}(Pb.Sb)_{0.24}Sr_{1.0}Ca_{1.1}Cu_{1.6}O_y$ by the sol-gel method from the acetates.
2. The low-Tc phase is already formed at 800°C, and Pb and Sb with Pb promotes the formation of its phase.
3. The high-Tc phase can be formed for a comparable short heating time at 850°C in the composition of $Bi_{0.96}Pb_{0.19}Sb_{0.05}Sr_{1.0}Ca_{1.1}Cu_{1.6}O_y$.
4. When powder samples are annealed at 400°C after being heated at 850°C for 100hrs, they can have much the high-Tc phase increasingly (especially effective for the sample with the composition of $Bi_{0.96}Pb_{0.19}Sb_{0.05}Sr_{1.0}Ca_{1.1}Cu_{1.6}Oy$).

REFERENCE

[1] H. Maeda et al: Jpn. J. Appl. Phys., 27 (1988) L209

Preparation of Yttrium Barium Cuprate Powder by Sol-Gel Method at Low Temperature

HIROHIKO MURAKAMI, SEIJI YAEGASHI, JUNYA NISHINO, YUH SHIOHARA, and SHOJI TANAKA

Superconductivity Research Laboratory, International Superconductivity Technology Center, 10-13, Shinonome 1-chome, Koto-ku, Tokyo, 135 Japan

ABSTRACT

We have succeeded in forming yttrium-barium-cuprate powder using metal alkoxides at 650℃ by the sol-gel process. The homogeneous solution was synthesized using Y,Ba alkoxides and Cu nitrate. The powders were characterized by X-ray diffraction,TEM,and magnetic susceptibility measurements. The X-ray diffraction pattern of the powders showed a presence of the 123 single phase.The TEM photomicrograph of the powders showed the average agglomerate size to be less than 0.05μm. Magnetic susceptibility measurements showed the powders to be superconducting with a transition at 70K (Tc onset).

KEY WORDS: sol-gel method, ultrafine powder, low-temperature synthetic route, metal alkoxides

INTRODUCTION

The conventional solid-state synthesis of the $YBa_2Cu_3O_7$(123) superconductor typically involves combining yttrium oxide,copper oxide,and barium carbonate,and heating the mixture to about 950℃. In order to improve homogeneity and obtain a higher yield of the superconducting phase, repeatedly grinding and calcination over extensive duration have been carried out. Large powders with poor sinterability have been produced by this method[1].

From the viewpoint of ceramic processing, a sol-gel method is quite attractive with many advantages; including, simplification of fabrication processes, low-temperature synthesis,and improvement in homogeneity and purity of the starting materials. These advantages have been discussed in this field at several workshops[2][3].

There have been enormous efforts to prepare the superconducting materials by many solution routes so-called"sol-gel methods" in the aim to make good use of advantages the method offers. They have been devised to prepare precursors that can be decomposed at relatively mild temperatures to yield finer particles of the 123 phase; however,decomposition temperatures above 800℃ have been required obtaining single phase products[4][5][6][7][8].

The stability of $BaCO_3$, which was formed easily during the pyrolysis process of organic materials, represents a serious impediment to further lowering the 123 synthesis temperature. A complete reaction of this $BaCO_3$ has never been achieved at temperatures lower than 800℃, even with reactive fine particle precursors. Trace amounts of nonreacted $BaCO_3$ at grain boundaries in a sintered compact are known to adversely affect its criticpal current[9].

In this work, the basic experiments were focussed on the selection of metal alkoxides as starting materials and suitable solvents to prepare a chemical homogeneous solution. We report that a highly concentrated mixed alkoxides solution was prepared using $Y(n\text{-}OC_4H_9)_3$, $Ba(OC_2H_5)_2$, and $Cu(NO_3)_2$ as starting materials and that the formation of superconducting powders fired only at 650℃.

EXPERIMENTAL

Figure 1 shows the process for the formation of superconducting powders. $Y(n\text{-}OC_4H_9)_3$ and $Ba(OC_2H_5)_2$, which are soluble in usual organic solvents, were used as starting materials. The use of copper alkoxide as a starting material is limited due to its low solubility in the organic solvents. $Cu(NO_3)_2$ was selected as the source of Cu-compound in this work. Ethanol and xylene were the most suitable solvents among many organic solvents for these starting materials. The starting materials were dissolved in the solvents(ethanol and xylene) mixture with stoichiometric composition of Y/Ba/Cu=1/2/3. The initial concentration of the total metals was 42mmol. The solution was stirred in dry nitrogen and heated at 60℃ for 24hrs. Then, the solution was evaporated with stirring at about 100℃ to obtain amorphous gel powders. The powders were sintered under the vacuum condition($\sim 10^{-1}$Pa) at 650℃ for 10hrs, followed by annealing with flowing O_2 gas at 500℃ for 10hrs. The gel powders were die-pressed and sintered under the same vacuum condition to prepare the specimens for magnetic susceptibility measurements. Then this pellet was annealed with flowing O_2 at 500℃ for longer than 100hrs.

The specimens were characterized by X-ray diffraction and transmission electron microscopy(TEM). The superconducting properties of the pellets were determined by measuring the magnetic susceptibility.

RESULTS AND DISCUSSION

A typical x-ray diffraction pattern is shown in Fig.2. This sample heat-treated at 650℃, which is higher than the crystallization temperature, shows almost all the peaks attributed to a 123 structure. Figure 3 shows the TEM image of 123 powders calcined at 650℃. The resulting powders were shown consisting of agglomerates of about 0.05μm primary particles with some occasionally larger crystallites. These primary particles are lightly sintered together resulting in extensive regions of grain boundary contacts. The properties of the pellets for long annealing (more than 100hrs.) are confirmed by magnetic susceptibility measurement, as shown in Fig.4. Tc onset was approximately 70K for this sample. This low Tc value may be due to the oxygen deficiency resulting from low sintering temperature(650℃).

It has been quite difficult to prepare the concentrated mixed solution of yttrium-barium-copper alkoxides, since most of the copper alkoxides have low solubilities in alcohols. In the present system, we used $Cu(NO_3)_2$ as a source of Cu-compound. The use of $Cu(NO_3)_2$ instead of Cu-alkoxide makes impossible the control of gelation and the completion of metal-oxygen bonding. The low-temperture synthetic route may be regarded as a two-step process. First yielding under vaccum condition, the materials are most likely nonsuperconducting

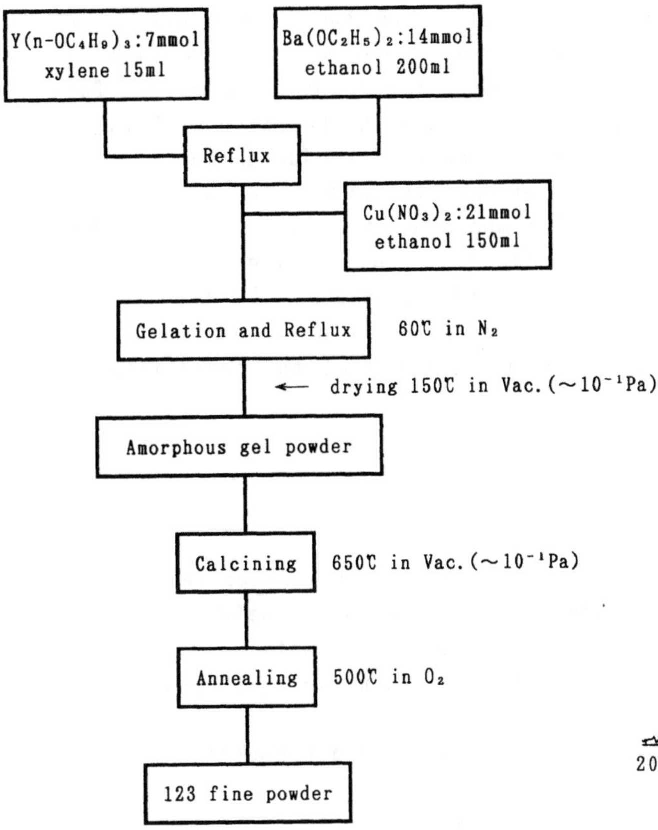

Fig.1 Preparation of 123 single phase powder by sol-gel method at low temperature

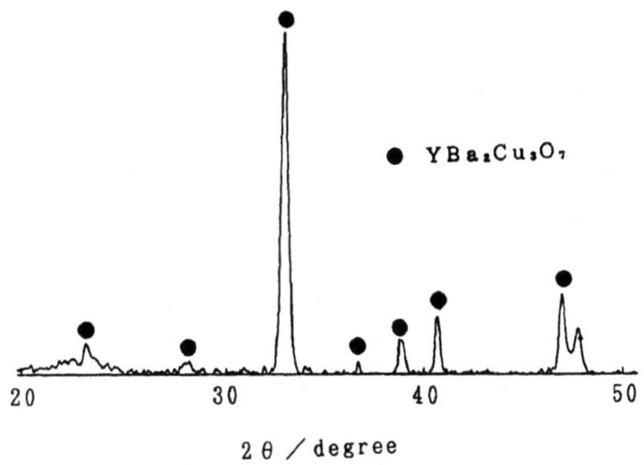

Fig.2 X-ray diffraction pattern of 123 powder prepared at 650℃ for 20h

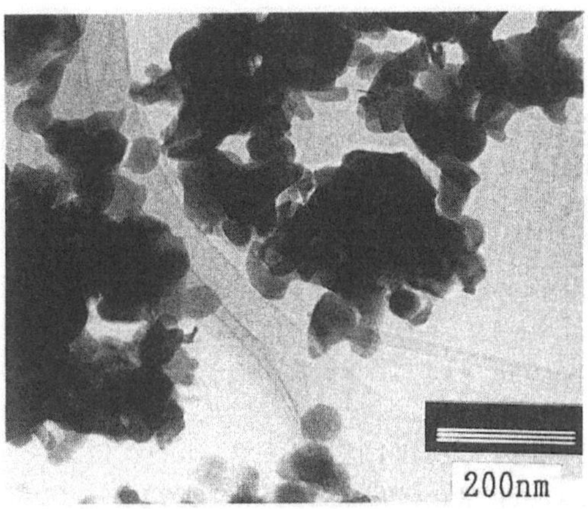

Fig. 3 TEM image of 123 powder prepared at 650℃

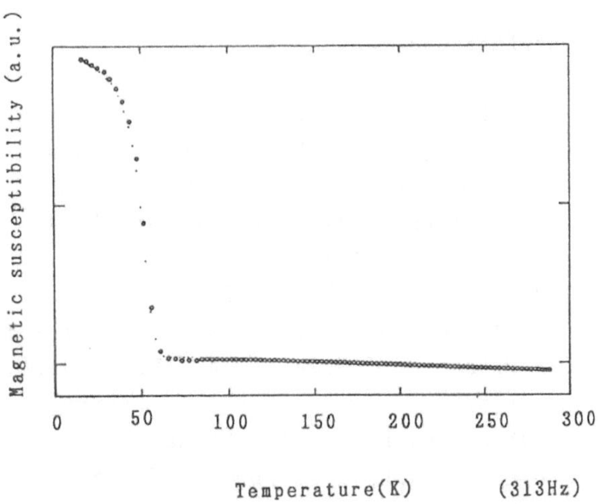

Fig.4 Change of the ac susceptibility for the pelle annealed at 500℃ for 100h

tetragonal 123, transforming with further oxidation to the orthorhombic form during O_2 annealing. The $Cu(NO_3)_2$ brings some benefits to the method. $Cu(NO_3)_2 \cdot xH_2O$ can be easily dissolved in alcohol, resulting in a successfull preparation of the highly concentrated mixed solution with the ratio of Y/Ba/Cu= 1/2/3. Fouthermore, an adequate amount of $(NO_3)^-$ can be introduced, which attack Ba salts. This would produce the nitrates of Ba. In order to avoid $BaCO_3$ formation, the precursors which directly decompose to the oxide are desired. Metal nitrate precursors can serve this purpose[10]. Thus, adding of $(NO_3)^-$ may interruput the formation of troublesome $BaCO_3$ and decompose Ba-salts at low temperature about 550℃[11]. These effects together with the usage of $Y(n\text{-}OC_4H_9)$ and $Ba(OC_2H_5)_2$ made it possible to form ultrafine powders of the consisting 123 single phase at 650℃.

CONCLUSION

The synthesis of the $YBa_2Cu_3O_{7-x}$ superconductor at low temperature was investigated through the sol-gel method, using $Y(n\text{-}OC_4H_9)_3$, $Ba(OC_2H_5)_2$, and $Cu(NO_3)_2$ as starting materials and ethanol and xylene as solvents.

1)Chemically homogeneous ultrafine powders of the single phase $YBa_2Cu_3O_{7-x}$ were successfully prepared at 650℃.

2)The superconducting ultrafine powders showed the average agglomerate size to be 0.05μm.

3)It was confirmed by magnetic susceptibility measurements that the pressed single phase $YBa_2Cu_3O_{7-x}$ pellet showed Tc onset of 70K.

Work supported by R&D of Basic Technology for Future Industries through the New Energy and Technology Development Organization(NEDO).

REFERRENCES

1)H.S.Horowitz, S.J.McLain, A.W.Sleight, J.D.Druliner, P.L.Gai, M.J.VanKavelaar J.L.Wagner,B.D.Biggs and S.J.Poon, SCIENCE, 243,68(1989).

2)Brinker,C.J., Clark,D.E. & Ulrich,D.R.(ed) Better Ceramics through ChemistryⅡ. Elsevir,New York(1984). Mater.Res.Soc.Symposia Proc.vol.73,Mater. Res.Soc.,Pittsburgh(1986).

3)Hench,L.L. & Ulrich,D.R.(ed) Ultrastructure Processing of Ceramics, Glasses and Composites,Wiley,New York,(1984).

4)P.Barboux,J.M.Tarascon,L.H.Greene,G.W.Hull,B.G.Gagley, J.Amppl.Phys.63,2725 (1988).

5)M.J.Cima,R.Chiu,W.E.Rhine, Mat.Res.Soc.Symp.Proc.99,241(1988).

6)G.Kordas,K.Wu,U.S.Brahme,T.A.Friedman,D.M.Ginsgerg,Mat.Lett.5,417(1987) .

7)T.Monde,H.Kozuka and S.Sakka, Chem.Lett. 287(1988).

8)M.Tatsumisago,H.Sato and T.Minami,Chemistry Express,Vol.3,No5,311(1988).

9)B.Dunn, C.T.Chu, L.W.Zhow, J.R.Cooper, G.Gruner, Adv.Ceram.Mat.2,343(1987).

10)B.O.Field and C.J.Hardy, Quarterly Rev.,18,361(1964).

11)Tae-Hyun Sung, K.Takahashi, Y.Ohya, Y.TAKAFGI, Z.Nakagawa, T.Nakamura and Y.Saito, J.Ceram.Soc.Jpn,97,1039(1989).

Preparation of Superconducting Thin Films of $Bi_2Sr_2CaCu_2O_y$ by Metallo-Organic Decomposition Method

SEIJI YAEGASHI, W.C. MOFFATT*, HIROHIKO MURAKAMI, JUN'YA NISHINO, and YUH SHIOHARA

Superconductivity Research Laboratory, Internaitonal Superconductivity Technology Center, 10-13, Shinonome 1-chome, Koto-ku, Tokyo, 135 Japan

ABSTRACT

Metallo-organic decomposition is a nonvacuum technique for thin film fabrication which allows easy alteration of chemical components and no requirement of expensive systems. We report the preparation of ultra-thin film superconductors, which are almost transparent with homogeneous surface by this method using calboxylic acid salts solutions. By optimum coating, pyrolizing, and firing conditions, single phase of 80K superconductor with strongly c-axis aligned parpendicular to the substrate surface was obtained.

KEYWORD: metallo-organic decomposition, superconductor, spin-coating, pyrolysis, thin film

INTRODUCITON

Metallo-organic decomposition which uses solution precursors has a great advantage of mixing each component at molecular level[1]. Superconducting properties of ultra-thin films derived from metallo-organic decomposition is very sensitive to the coating and heat treatment conditions[2][3]. To produce uniform films, coating, drying,and pyrolizing conditons must be carefully controlled. The firing condition is a very important factor to affect superconducting properties, particularly for thin films.

A number of papers appeared reporting metallo-organic decomposition methods and sol-gel methods using solution precursors. Superconducting properties, especially in J_C is inferior to that fabricated by vacuum deposition techniques[2][3][4][5][6][7][8]. The first step to improve superconducting properties by these methods is to produce films with microstructure identical to that by the vacuum techniques.

In this paper, we report the optimization of preparation processes to make good superconducting thin films derived from a metallo-organic decomposition method using naphthenates, 2ethylhexanoates(2EH), and caprylates precursors. Under the optimum condition, prepared superconducting films with the thickness of about 0.3μm are completely oriented to the substrate surface as examined by x-ray diffraction(XRD). Scanning electrom micrograph(SEM) shows the films consist of epitaxially grown like grains to the MgO[100] substrate.

EXPERIMENTAL

Naphthenates solution precursor was prepared by mixing each naphthenate/toluene solution with the Bi:2, Sr:2, Ca:1, Cu:2 molar ratio. 2EH solution precursor was prepared by mixing Bi-, Sr-, Ca 2EH/toluene solutions and Cu-2EH/chloroform solution with the identical ratio to the naphthenates precursor. Capryrates solution precursor was also prepared in the same condition as 2EH. The concentration of each precursor was adjusted to be approximately 4wt%. Viscosity of the precursor solutions was measusred as 2.30, 0.44, and 0.50 mPas for naphthenates, 2EH, and

*Guest researcher (ETL)

caprylates, respectively. Differential thermal analysis(DTA) and thermogravimetric analysis(TG) were performed at a heating rate of 5°C/min in nitrogen atmosphere to compare pyrolizing properties. Measurements of TG and DTA were made on the dried precursor solution at 100°C. Precursor solution was spin coated on a single crystal MgO substrates([100] orientation) at 2000-4000 rpm which resulted in smooth as-coated film surface. Coated films were dried at 100°C in a vacuum oven, pyrolized in N_2 at a heating rate of 2 °C/min from between 200 and 400°C to of 600 °C, then flashed air to complete the pyrolysis process. Coating and pyrolizing were repeated 3-4 times to obtain final film thickness of approximate 0.3μm. Firing was performed at 810-860°C for various durations in air. The films were rapidly cooled by taking out to an ambient atmosphere.

The specimens were characterized by XRD, a.c. susceptibility, optical micrograph, electron probe microanalizer(EPMA), and SEM. d.c. resistivity was measured by a conventional four probe method, using indium soldier contacts. The thickness of films was measured by an optical surface smoothness analyzer and/or mechanical surface smoothness analyzer.

RESULTS AND DISCUSSION

Shown in Figure 1, is a TG result, done in nitrogen atmosphere, of precursors that were naphthenates, 2EH, and caprylates. Three major decomposition steps were observed for naphthenates and 2EH, but two decomposition steps for caprylates. According to the TG results of each component, the steps below 300°C were attributed to the decomposition of Bi- and Cu- salts, and the step around 400°C was due to the decomposition of Sr- and Ca- salts. Difference in these behaviors for naphthenates, 2EH and caprylates reveals the possibility of better uniformity for as-dried films from caprylates precursor.

TG analysis was also performed in oxygen atmosphere. A sudden temperature increase of specimen caused by a exothermic reaction around the temperature of the first step decomposition made it impossible to confirm the precise temperature. Optical observations of the coated films using a gold image furnace showed bloating occured between 200°C and 300°C in air. Accordingly, a careful pyrolizing process is required to produce uniform films without bloating. As-pyrolized products of these films contain Bi_2O_3, $SrCO_3$, $CaCO_3$, and CuO examined by an XRD analysis, but the intensity of the peak assigned to CuO is much stronger for the caprylates precursor than that for the other precursors. This result is consistent with that the solubility of Cu-caprylate is smaller than Cu-naphthenate and Cu-2EH.

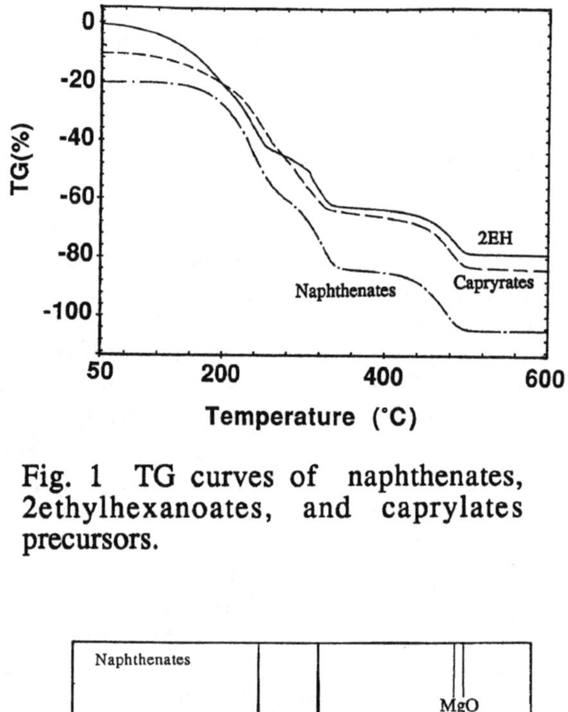

Fig. 1 TG curves of naphthenates, 2ethylhexanoates, and caprylates precursors.

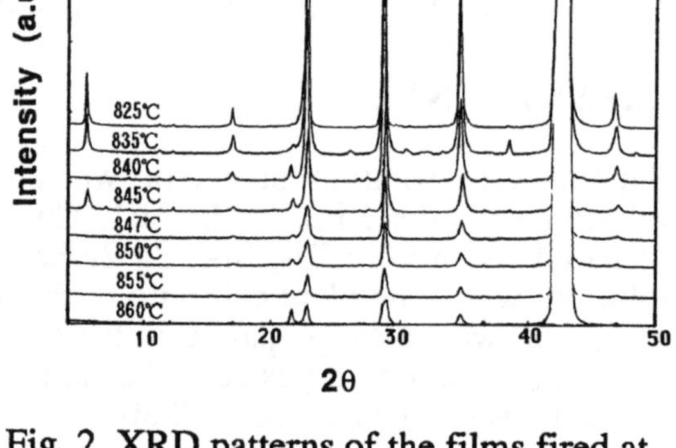

Fig. 2 XRD patterns of the films fired at various temperature for 15 mins.

Figure 2 shows the XRD patterns of the films derived from a naphthenates precursor fired at 810-860 °C for 15mins in air. The films fired around 820-830°C show complete single $Bi_2Sr_2CaCu_2O_y$ phase(2212) with strong c-axis orientation perpendicular to the substrate surface, and the intensity of the peaks tends to decrease as the firing temperature increases. The films fired at 840°C-860°C have another impurity phase of $Bi_2Sr_2CuO_y$(2201). It is noticeable the films fired at below 830°C having smooth surface with gray color but almost transparancy, show the contrast against that fired at over 830°C with rough and not uniform surface. Scanning electron micrographs of the films fired at 825°C and 835°C are shown in Fig. 3.

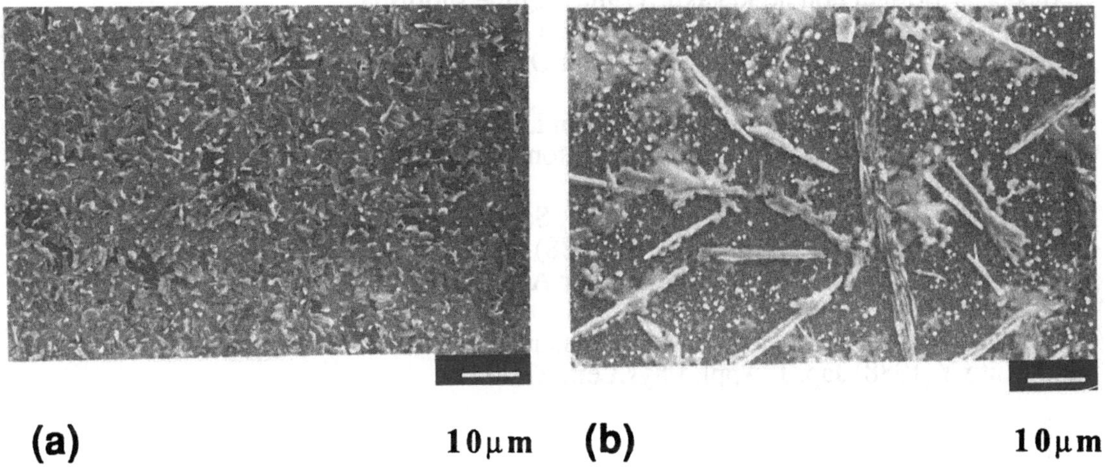

Fig. 3 Scanning electron micrographs of the films fired at : (a) 825°C ; (b) 835°C.

That the film fired at 825°C contains plate-like small grains of 2212 phase. At higher temperatures, larger needle-like grains and small spherical grains were observed. The temperature around 830°C is critical to obtain uniform morphology of the film surface. EPMA result shows these needle-like grains are constituted by (Sr,Ca)CuO_y. Considering with the XRD results, 2212 phase decomposes to yield 2201 phase and (Sr,Ca)CuOy at higher firing temperature. In the case of 2EH and caprylates precursors, similar XRD patterns were recognized. Increasing firing duration does not change XRD patterns drastically, however, crystallography of the film surface tends to be smoother.

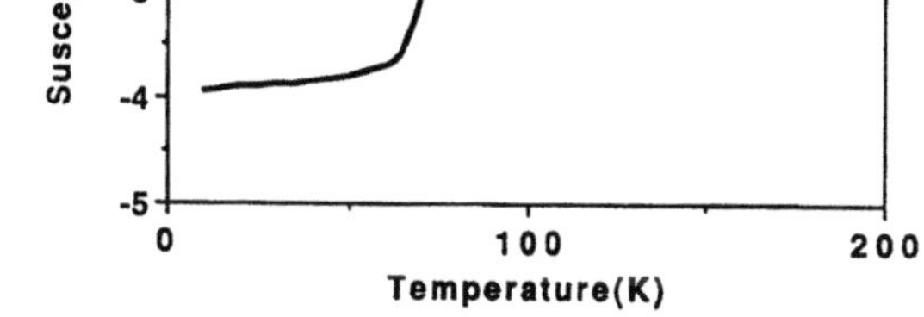

Fig. 4 Magnetic susceptibility of the film fired at 825°C.

The film fired at 825°C for 15hrs has square shape plate-like grains with the length of 2-3μm, indicating the existence of partially epitaxially-grown grains. Fig. 4 shows the temperature dependence of magnetic susceptibility for the film fired at 825°C for 4hrs. A small amount of decrease in susceptibility was detected below 76K, for the film of the thickness about 0.3μm, almost identical to the penetration depth of this superconductor. The result that the film fired for 15mins did not show any susceptibility change suggests that a long time firing improves superconducting properties.

CONCLUSION

We have studied to fabricate pure $Bi_2Sr_2CaCu_2O_y$ superconducting ultra-thin films with the thickness of 0.3μm using metallo-organic precursors by spin-coating on MgO substrates. Uniform film surface was obtained by carefully-controlled pyrolizing process. Preferential orientaion of the films with the c-axis perpendicular to the substrates was achieved by optimizing fabricating conditions.

REFERENCES

1. Sakka S"Sol-gel hou no kagaku" (1988) Agune shofusha
2. Rice CE, van Dover RB, and Fisanick GJ (1987) Appl. Phys. Lett. 51: 1842
3. Nasu H, Kato T, Makita S, Imura T, and Osaka Y(1988) Jap. J. Appl. Phys. Lett.27 :2317
4. Furcone SL and Chiang Y-M, (1988) Appl. Phys. Lett. 52: 2180
5. Kramer SA, Kordas G, McMillan J, Hilton GC, and Harkien DJ (1988)Appl. Phys. Lett. 53: 156
6. Sandstrom RL, Giess EA, Gallagher WJ, Segmuller A, Cooper EI, Chisholm MF, Gupta A, Shinde S, and Laibowits RB(1988) Appl. Phys. Lett. 53:1874
7. Koren G, Gupta A, Giess EA, Segmuller A, and Laibowits RB(1988) Appl. Phys. Lett. 54: 1054
8. Yamane H, Kurasawa H, Iwasaki H, Masumoto H, Hirai T, Kobayashi N, and Muto Y(1988) Jap. J. Appl. Phys.Lett. 27:1275

High Total and High Oxygen Partial Pressure Effects on Bi-Sr-Ca-Cu-O Superconductor During O_2-HIP Sintering

HIROSHI SEINO[1], KOZO ISHIZAKI[1], and MASASUKE TAKATA[2]

Department of Materials Science and Engineering, School of Mechanical Engineering[1], and Department of Electrical Engineering[2], Nagaoka University of Technology, Nagaoka, Niigata, 940-21 Japan

ABSTRACT

The effects of high total and high oxygen partial pressures during O_2-HIPping were studied by using thermodynamic analysis (HIP phase diagrams). A Bi-type superconductor ($Bi_2(Sr_{1-x}Ca_x)_2CuO_{6+\delta}$) was synthesized by O_2-HIP sintering under 100 MPa (P_{O_2} = 5, 10 and 20 MPa) at very high temperature (~1323 K), without any additional treatment for the first time. This paper suggests that high total and high oxygen partial pressures allow to sinter at high temperatures. Consequently the sintering time becomes extremely short to less than 1 hour from about 100 hours of normal sintering.

KEY WORDS: Bi-Sr-Ca-Cu-O, Hot Isostatic Press (HIP), HIP phase diagram, O_2-HIP, Oxygen partial pressure

INTRODUCTION

The Bi-Sr-Ca-Cu-O superconductor can be synthesized by many hours of arduous toil, such as precise sintering temperature control between 1145 and 1150 K[1], sintering under low oxygen partial pressure of 1/130 MPa[2], and mixing of additives like lead[3]. Nevertheless all of them required the sintering time as long as 60 hours or more[1-3].

After obtaining the superconducting phase, however, there is another problem of weak bonding. A few works have been reported to overcome this problem, for example, by hot pressing[4] and by "intermediate (uniaxial) pressing"[5]. These processes have the improvements on the critical current density by densified as well as oriented microstructures[4,5], and are merely additional treatments of the normal sintering of about 100 hours.

This work applies O_2-HIP sintering as high total and high oxygen partial pressures to attain shorter sintering time and consolidate to increase the current density. (A densification processing by a capsule HIPping was reported previously by the authors[6]).

EXPERIMENTAL

The high-T_c phase of the Bi-type superconductor ceramics (i.e. the resistivity drops to zero above 100 K) can be synthesized normally, when they go through partial melting of the samples[7]. Powders of $SrCO_3$ and $CaCO_3$ were calcinated at 1573 K for 5 hours. Then powders of Bi_2O_3, CuO, $SrCO_3$ and $CaCO_3$ were mixed to form a compound with atomic ratio of Bi:Sr:Ca:Cu as 2:2:2:3. The mixed powder was pressed into cylinders (using uniaxial and cold isostatic presses). They were HIPped at various temperatures for 0.25, 0.5 or 2 hours under a high pressure gas mixture of Ar and O_2 (P_{total}=100 MPa). The proportions of O_2 in the gas mixtures were 1, 5, 10 or 20 %.

In the O_2-HIP process, 50MPa of gas mixture was introduced in the pressure vessel at room temperature. Samples were heated to the sintering temperatures between 978K and 1373 K with a heating rate of 400 K/h. The HIPped samples were cut into plates and the resistivity was measured by a dc four-probe method in a cryostat. The crystal structure was examined by XRD.

RESULTS AND DISCUSSIONS

The melting or partial melting temperatures were plotted against P_{O_2} for conventional sintering under 0.1 MPa[2,8] and the present O_2-HIP sintering as in Fig. 1.

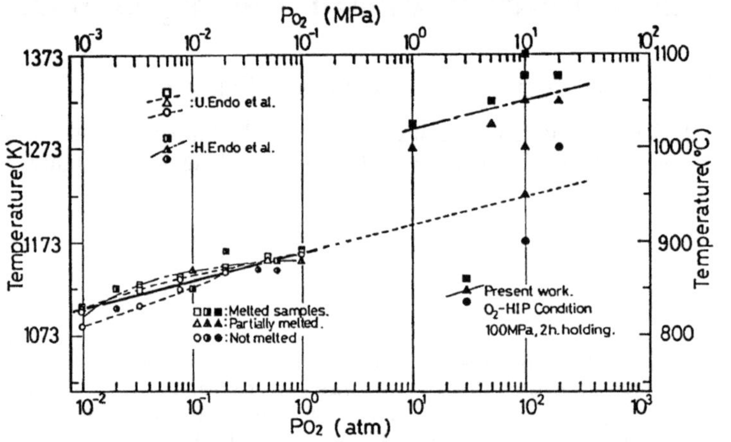

Fig. 1. The relationship between the sintering temperature and PO_2, for both methods (conventional under P_{total}=0.1 MPa and O_2-HIP sintering under P_{total}=100 MPa). The squares, the triangles and the circles indicate respectively the sample conditions of melted, partially melted and not melted. The open and the half open marks correspond respectively to the data of U. Endo et al.[2] and H. Endo et al.[8]. The solid marks correspond to the present work (O_2-HIPping).

The holding time of O_2-HIPping was fixed for 2 hours. The melting temperature rose as oxygen partial pressure (PO_2) augmented. The difference of sintering temperature between the two methods (conventional and O_2-HIP sintering) is due to the influence of the total and the oxygen partial pressures.
The HIP phase diagram on the reaction of high-T_c phase is shown for total pressure (P_{total}) = 0.1 and 100 MPa. The other reaction of 2-2-1-2 and 2-2-0-1 phases must be very close to this line due to similar molar gibbs free energy. The dominant solid gas reactions are almost same for all these phases. HIP phase diagrams were proposed by the authors[9,10]. The phase diagrams consist of three independent variables (P_{total}, PO_2 and temperature: T). The line of P_{total}=0.1 MPa was drawn by using the data of Endo et al.[2,8]. The optimum condition to obtain the high-T_c phase can be estimated from this line, i.e. about 1143 K under PO_2 = 0.02 MPa. The phenomena of the partial melting was considered to be control by the reduction of the oxide component.

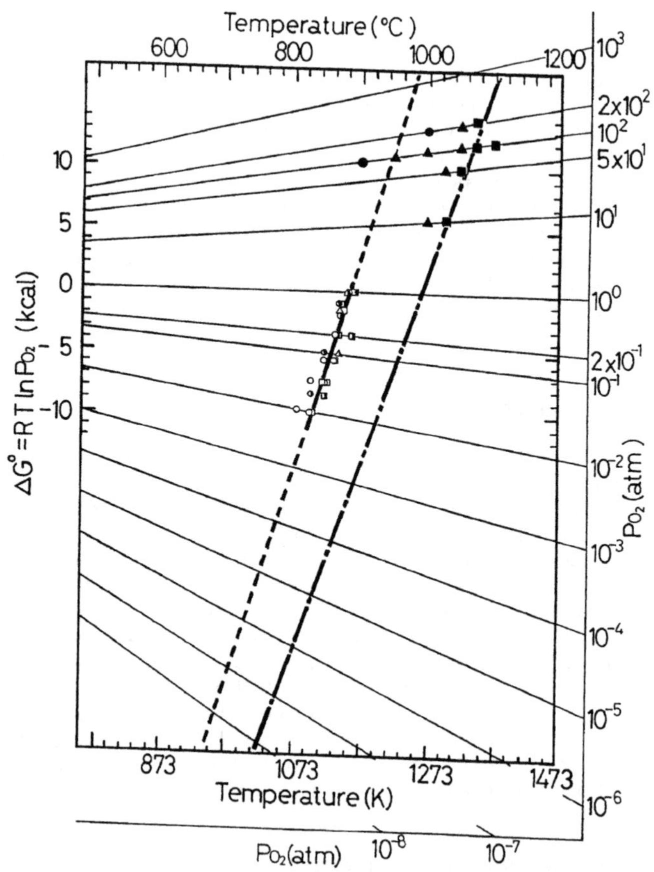

Fig. 2. HIP phase diagram under P_{total}=0.1 and 100 MPa as the formation of high-T_c phase in Bi-Sr-Ca-Cu-O materials. The symbols have identical meanings as in Fig. 1. The broken line with dash indicates the theoretical formation-line of high-T_c phase under P_{total}=100 MPa and the dashed line shows the same formation under P_{total}=0.1 MPa.

The oxidation equilibrium was calculated under P_{total}= 100 MPa for various PO_2[11], and plotted as the P_{total}= 100 MPa line in Fig. 2. The molar gas volume was estimated by an ideal gas, and on the other hand, the molar volumes of the solid product (Bi-Sr-Ca-Cu-O) and the solid reactants (Bi_2O_3, SrO, CaO, CuO or these compounds) were considered negligible.
The data of O_2-HIP sintering are in good agreement with the theoretical formation line under 100 MPa. This fact demonstrates that high total and high oxygen partial pressures, shift the formation line to higher temperatures during O_2-HIP treatment. Hence, one of the O_2-HIPping advantages is the possibility of shortening the sintering time by sifting the partial melting at higher temperatures under high total (P_{total}) and oxygen partial (PO_2) pressures. However, only 2-2-0-1 type superconductor with high T_c of 80 K was obtained by O_2-HIPping, due to very high temperature treatment.

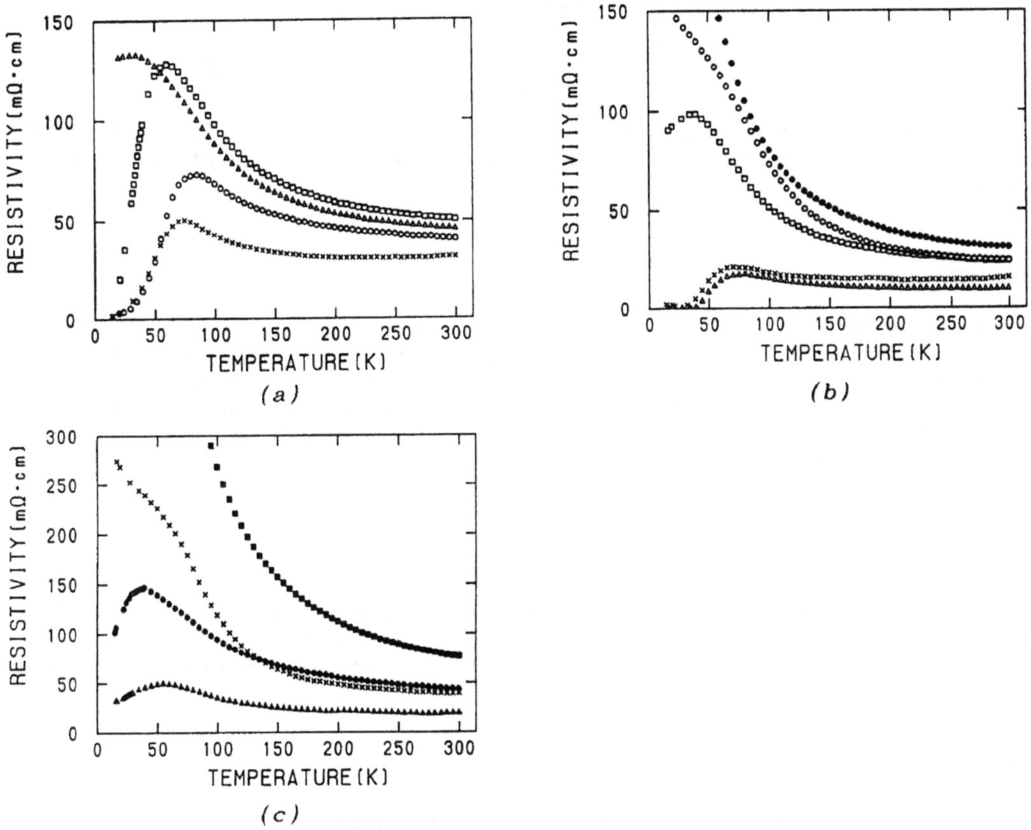

Fig. 3. Temperature dependence of the resistivity for various O_2-HIP treated samples. The total pressure and holding time were fixed to 100 MPa and 0.5 hours. The oxygen partial pressure varied (a) 5 MPa, (b) 10 MPa and (c) 20 MPa. The lines of different marks indicate O_2-HIPping temperatures: Open circles; 1273 K, open triangles; 1283 K, open squares; 1293 K, crosses; 1303 K, closed circles; 1313 K, closed triangles; 1323 K, closed squares; 1333 K.

Figures 3 (a), (b) and (c) indicate the temperature dependence of the resistivity of several samples. They were O_2-HIPped at temperatures between 1273 K and 1333 K for 0.5 hours under P_{total}=100 MPa of various PO_2 gas, (a) PO_2=5 MPa, (b) 10 MPa and (c) 20 MPa. The electric current density for measurement was 20 mA/cm^2.
These specimens behaved like a semiconductor in the resistivity up to $T_{c,onset}$ point, i.e. negative dependency of resistivity on temperature. The superconducting transition of all the samples were broad. However the zero resistances were not found out in these samples.
The density of all the O_2-HIPped samples was about 5000 kg/m^3. The crystal structures of all the samples were similar to that of 2-2-0-1 type superconductor ($Bi_2(Sr_{1-x}Ca_x)_2CuO_{6+\delta}$ ($0 \leqq x \leqq 0.3$) with $T_c \leqq 50$ K[12]), as seen in XRD pattern of Fig. 4. There were also $SrCO_3$ and other compounds like Sr-Ca-Cu-O. The preparation was aimed to attain the 2-2-2-3 phase, however the principal phase obtained was 2-2-0-1 type superconductor and non-reacted

$SrCO_3$. This fact indicates that a considerable amounts of Ca atoms replace Sr sites. An O_2-HIPping at high temperatures above 1273 K permitted to procure three types of $Bi_2(Sr_{1-x}Ca_x)_2CuO_{6+\delta}$ which had the $T_{c,onset}$ of about 80, 60 and 40 K.

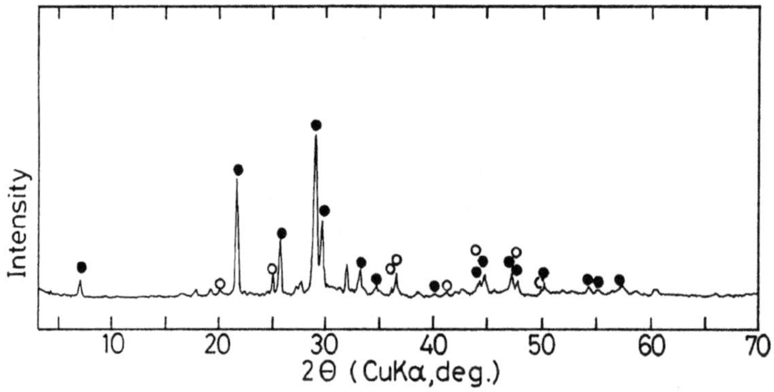

Fig. 4. XRD pattern of Bi-Sr-Ca-Cu-O O_2-HIPped at 1273 K under PO_2 = 5 MPa. Closed and open circles indicate spectra of 2-2-0-1 phase and $SrCO_3$, respectively.

CONCLUSIONS

High total and high oxygen partial pressures of O_2-HIP treatments allow very short time (0.5 hours) and high temperature (~1323 K) synthesis of Bi-Sr-Ca-Cu-O superconductor. $Bi_2(Sr_{1-x}Ca_x)_2CuO_{6+\delta}$ superconductor was obtained by an extremely short time O_2-HIP sintering at the temperature range, from 1273 K to 1323 K, under high total pressure (P_{total}) of 100 MPa (PO_2 = 5, 10 and 20 MPa). O_2-HIP treatment gives three types of $Bi_2(Sr_{1-x}Ca_x)_2CuO_{6+\delta}$ whose $T_{c,onset}$ were 80, 60 and 40 K respectively.

ACKNOWLEDGMENTS

We would like to express our deep thanks to "Monbusho" (Ministry of Education and Culture) as well as to Kobe Steel Co. Ltd. (Kobelco) for providing an excellent opportunity to undertake the present research through a cooperative program.

REFERENCES

1. T. Hatano, K. Aota, S. Ikeda, K. Nakamura and K. Ogawa (1988) Growth of the 2223 phase in leaded Bi-Sr-Ca-Cu-O system. Jpn. J. Appl. Phys. 27: L2055-L2058
2. U. Endo, S. Koyama and T. Kawai (1988) Preparation of the high T_c phase of Bi-Sr-Ca-Cu-O superconductor. ibid. 27: L1476-L1479
3. M. Takano, J. Takada, K. Oda, H. Kitaguchi, Y. Miura, Y. Ikeda, Y. Tomii, and H. Mazaki (1988) High-T_c phase promoted and stabilized in the Bi, Pb-Sr-Ca-Cu-O system. ibid. 27: L1041-L1043
4. N. Murayama, E. Sudo, M. Awano, K. Kani and Y. Torii (1988) Densification and grain-orientation of Bi-Pb-Sr-Ca-Cu-O superconductors by hot-pressing. ibid. 27: L1856-L1858
5. Y. Tanaka, T. Asano, K. Jikihara, M. Fukutomi, J. Machida and H. Maeda (1988) Improvements in the current carrying capacity in high-T_c BiSrCaCuO superconductors. ibid. 27: L1655-L1656
6. H. Seino, K. Ishizaki and M. Takata (1989) HIPped high density Bi-(Pb)-Sr-Ca-Cu-O superconductors produced without any additional treatment. ibid. 28: L78-81
7. Y. Oka, N. Yamamoto, H. Kitaguchi, K. Oda and J. Takada (1989) Crystallization behavior and partially melted states in Bi-Sr-Ca-Cu-O. ibid. 28: L213-L216
8. H. Endo, J. Tsuchiya, N. Kijima, A. Sumiyama, M. Mizuno and Y. Oguri (1988) Thermal stability of the high-Tc superconductor in the Bi-Sr-Ca-Cu-O system. ibid. 27: L1906-L1909
9. K. Ishizaki (in press) Thermodynamics for HIPping of ceramics -proposed HIP phase diagrams-. ASM Int. In Proc. 2nd Int. Conf. hot isostatic pressing. 7-9 June 1989. Gaithersburg, MD, USA.
10. H. Seino, K. Ishizaki and M. Takata (in press) HIPped Bi-(Pb)-Sr-Ca-Cu-O superconductor without any additional treatment and HIPping effects. ibid.
11. K. Watari and K. Ishizaki (1988) Influence of gas pressure on HIP sintered silicon nitride and stability of carbon impurity. J. Ceram. Soc. Jpn., (J. Ceram. Soc. Jpn. Int. Edition) 96: 551-555, (535-540)
12. Z. V. Popovic, C. Thomsen, M. Cardona, R. Liu, G. Stanisic, R. Kremer, W. Konig (1988) Phonon Characterization of $Bi_2(Sr_{1-x}Ca_x)_2CuO_{6+\delta}$ by infrared and raman spectroscopy. Solid State Commun. 66 No.9: pp965-969

Superconductivity of the Tl-Ba-Sr-Ca-Cu System

MASANORI KURODA and MICHIO ARAKI

Government Industrial Research Institute, Chugoku, Agency of Industrial Science and Technology, Ministry of International Trade and Industry, Hiro-Suehiro, Kure, Hiroshima, 737-01 Japan

ABSTRACT

The new superconductivity system is reported in the $Tl_1Ba_1Sr_1Ca_1Cu_2O_x$(11112). The critical temperature (Tc) of this system was determined as 102 K by AC susceptibility measurements. The unit cell derived from XRD measurement was tetragonal with a=3.85A and c=12.3A.

KEY WORDS: high-Tc superconductor, Tl-Ba-Sr-Ca-Cu-O, crystal structure, AC susceptibility

INTRODUCTION

Up to now, four series of high-Tc oxide superconductor have been found. The superconductivity of the Tl-Ba-Cu-O system was found by Sheng and Hermann[1]. Subsequently, a Tl-Ba-Ca-Cu-O system with 125 K superconductivity was reported[2]. In this paper a Tl-Ba-Ca-Cu-O system containing strontium ion is investigated.

EXPERIMENTAL

Samples were prepared by the solid phase reaction method. The intended composition was $Tl_2Ba_1Sr_1Ca_2Cu_3O_x$. Stoichiometric amounts of strontium carbonate, calcium carbonate, barium carbonate, copper(II)oxide were thoroughly ground with a ballmill (Fritsch P-0 type) and calcined in air at 1173 K for 15h. The size of each sample was 10 g. After grinding, the reacted powder was mixed with powder of thallium oxide and heated again in the temperature range from 1073 to 1223 K in air for about 10 to 20 min. The samples were quenched to room temperature at the outside of the furnace. For Tc measurement, the temperature dependence of the AC susceptibility was measured by the computer controlled AC inductance bridge at the temperature range from 30 to 300 K. Temperature was measured using a platinum-cobalt alloy resistance thermometer. The crystal structure of the sample thus obtained was analyzed by XRD using monochromated Cu-Kα radiation.

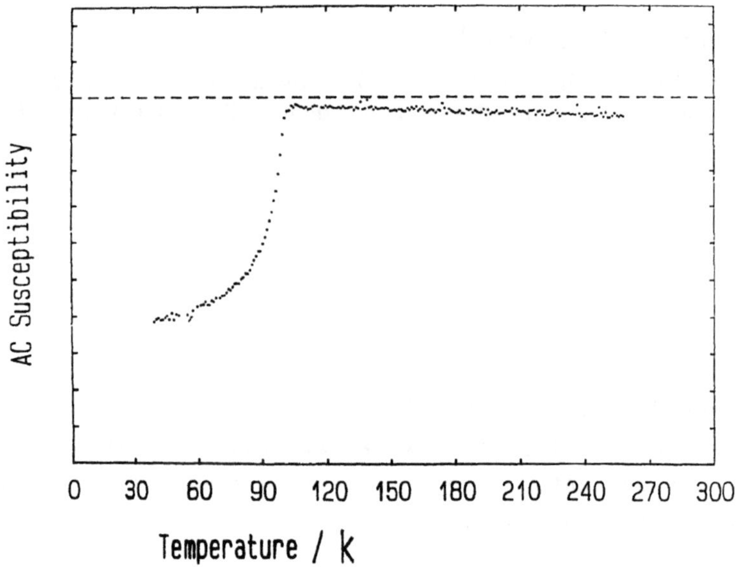

Fig.1 The susceptibility vs. temperature curve of the $Tl_1Ba_1Sr_1Ca_1Cu_2O_x$ superconductor.

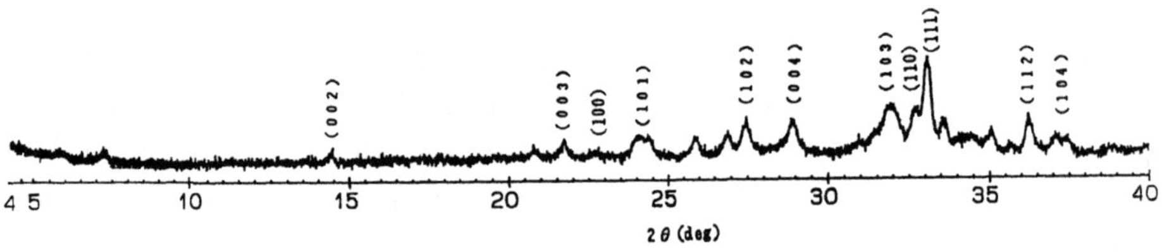

Fig.2-a XRD pattern of the $Tl_1Ba_1Sr_1Ca_1Cu_2O_x$ superconductor

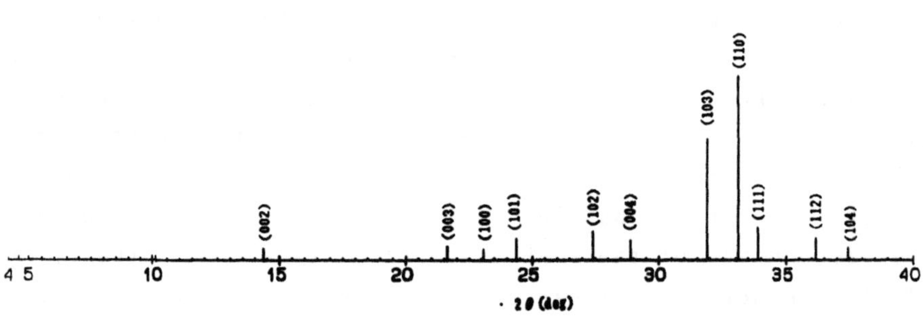

Fig.2-b XRD pattern simulated by Rietveld analysis.

RESULTS and DISCUSSION

A magnetic susceptibility vs. temperature curve of a sample of starting composition of $Tl_2Ba_1Sr_1Ca_2Cu_3O_x$ is shown in Fig. 1. A sharp drop of susceptibility was found at 102 K, and no inflection point was found in the curve. It suggest that the sample contains only one superconducting phase.

The crystal structures of the $Tl_1Ba_1Sr_1Ca_1Cu_2O_x$(11112) were investigated by XRD method. A typical XRD pattern of the sample is shown in Fig.2-a. The lattice parameters were determined by measuring 2θ values of 15 diffraction peaks in the range of 4 to 90 . No peak corresponding to the raw materials was observed. From the X-ray diffraction pattern, it was calculated that the unit cell of this material was tetragonal with a=3.85A and c=12.3A.

The space group of the superconducting material containing a double thallium layer or single thallium layer is I4/mmm or P4/mmm, respectively[3]. From the rule of the absent reflection, it is determined that the material synthesized here belongs to the P4mm space group. Thus, it was found that the sample contained the single thallium layer. This material, $Tl_1Ba_1Sr_1Ca_1Cu_2O_x$ (11112), has a structure comprised of double layer Cu perovskite like slabs separated by monolayer Tl-O units.

The lattice constant of the $Tl_1Ba_2Ca_1Cu_2O_x$(1212) sample is C=12.74A[4]. By comparing this sample to the (1212) sample, it was estimated that the sample had the $Tl_1Ba_1Sr_1Ca_1Cu_2O_x$(11112) composition. The simulated X-ray diffraction pattern by means of Rietveld analysis was shown in Fig2-b. The lattice constants of the (1212) phase were reported a=3.847A and c=12.74A, respectively[4]. As a result of the replacement of one barium atom by a strontium atom, the lattice constant of c-axis of this (11112) phase became slightly shorter than that of the (1212) phase.

In summary, we have found a new superconducting material in one member of the Thallium system superconductor. The composition and Tc was found $Tl_1Ba_1Sr_1Ca_1Cu_2O_x$(11112) and 102 K, respectively. It was confirmed by means of the XRD measurement that the lattice parameter of this new phase was a=3.85A and c=12.3A and the space group was P4mm.

ACKNOWLEDGMENT

We are grateful to F.Izumi of the National Institute for Research in Inorganic Materials for supplying the Rietveld program used in this study.

REfERENCES

1)Z.Z.Sheng and A.M.Hermann:Nature(London) 332 (1988) 138.
2)S.S.P.Parkin, V.Y.Lee, E.M.Engler, A.I.Nazzal, T.C.Huang, G.Gorman, R.Savoy, and R.Beyers; Phys. Rev. Lett. 60 (1988) 2539.
3)T.C.Huang, W.Y.Lee ,V.Y.Lee and R.Karimi: Jpn. J. Appl. Phys. 27 (1988) 1498.
4)P.Haldar, K.Chen, B.Maheswaran, A.Roig-Janiki, N.K.Jaggi, R.S.Markiewicz and B.C.Giesen: Science 241 (1988) 1198.

Crystal Structures of Tl-(Ba, Sr)-Ca-Cu-O Superconductors

ATSUKO SOETA, TAKAAKI SUZUKI, TOMOICHI KAMO, and SHIN-PEI MATSUDA
Hitachi Research Laboratory, Hitachi Ltd., Hitachi, Ibaraki, 319-12 Japan

ABSTRACT

We report on the superconductivity and crystal structures in Tl-(Ba,Sr)-Ca-Cu-O (TBSCC) system. In the case of TBSCC system, the 2223 phase was obtained as a single phase more easily compared with the Tl-Sr-Ca-Cu-O and Tl-Ba-Ca-Cu-O systems. As a result, We achieved a high surperconducting volume fraction, 95% above 100K. The crystallographic transformation from 1223 to 2223 was caused by changing the ratio of Ba to Sr. The crystal structure of $Tl_2Ba_{1.6}Sr_{0.4}Ca_2Cu_3O_{10}$ refined by X-ray powder diffraction data using Rietveld analysis.

KEYWORDS: high-Tc surperconductor, Tl-Sr-Ca-Cu-O, Tl-(Ba,Sr)-Ca-Cu-O, Tl-Ba-Ca-Cu-O, crystal structure

INTRODUCTION

In our previous paper [1], we reported the new superconducting system, Tl-Sr-Ca-Cu-O system (TSCC) with Tc above 100K. The TSCC system had at least two superconducting phases, a high-Tc phase at about 100K and a low-Tc phase at about 75K. Those structures were determined as shown in Figs. 1(a), 1(b) by X-ray and electron diffraction analyses. They were related to the structures of Tl-Ba-Ca-Cu-O (TBCC) and Bi-Sr-Ca-Cu-O (BSCC) systems reported previously [2-4]. They described the six compounds of the form $Tl_mCa_{n-1}Ba_2Cu_nO_{2(n+1)+m}$, where m=1 or 2 and n=1, 2, or 3. Those structures consisted of copper perovskitelike units containing 1, 2 or 3 CuO_2 planes separated by one or two Tl-O bilayers are shown in Fig. 1(c), 1(d). During these experiments, we found that the solid solution between the TSCC and TBCC systems takes place for any ratio of Ba to Sr. The Tl-(Ba,Sr)-Ca-Cu-O system showed unique features compared with the TSCC and TBCC systems. In the case of the TBSCC system, the single phase with high-Tc was obtained more readily, resulting in the higher volume fraction of superconductors. The transformation of crystal structures between Tl-O monolayer and Tl-O bilayers was caused by changing the ratio of Ba to Sr. In this study, we report on the relationship between superconductivity and the crystal structures in the TBSCC system.

EXPERIMENTAL

Samples of TBSCC were prepared as follows. (a) Appropriate amounts of BaO, SrO, CaO, and CuO were thoroughly mixed and calcined at 900°C for 10h. (b) The resultant Ba-Sr-Ca-Cu-O was then mixed with Tl_2O_3 powder and pressed into a pellet was places in a covered alumina crucible and heated in a furnace at 850-900°C for 3h in air. The atomic ratio of Tl:(Ba+Sr):Ca:Cu was 2:2:2:3, corresponding to the high-Tc phase, $Tl_2Ba_2Ca_2Cu_3O_x$ as shown in Fig.1(d). The Ba/Sr ratio was varied at 1/4, 2/3, 1/1, 3/2, and 4/1.
Electrical resistance was measured by the standard four-probe method with indium soldering contacts, and the susceptibility changed by the standard AC method as described previously [5]. X-ray diffraction patterns were obtained using a Rigaku-RAD-1A system with Cu-K_α emission.

RESULTS AND DISCUSSTION

The single-phase samples as determined by X-ray analysis were obtained by sintering at 870°C. A typical susceptibility change in TBSCC with Ba/Sr=4/1 is given in Fig.2; typical susceptibility changes of TBCC and TSCC are also shown for comparison. The susceptibility in TBCC and TSCC samples decreased in two steps, suggesting that each sample contained at least two superconduction phases. This caused off-stoichiometry in samples and as a result, the volume fraction of the superconducting phases became lower in these samples. On the other hand, the Ba-rich sample of TBSCC showed a sharp transition at 114K (90% transition, $\Delta T \leq 4K$), suggesting that the sample consisted of nearly a single phase. The superconduction volume was estimated between 114K and 4.2K to be about 100% of the total volume of the sample, where the AC susceptibility is calibrated by using a known volume of Pb sample.

The relationship between superconductivity and crystal structure was examined by the resistivity study and the powder X-ray diffraction analysis. The typical resistivity transition curves are shown in Fig.3, suggesting that the two samples with different Ba/Sr ratio (4/1 and 1/4) had different T_c 's.

The superconducting transition temperatures (at mid-point) as a function of the Ba/Sr ratio are shown in Fig.4. The T_c decreased gradually from both TBCC and TSCC sides. It should be noted that the value changed discontinuously at Ba/Sr=1.0. Crystal structures were assumed in the analysis of the X-ray diffraction patterns. The typical X-ray diffraction patterns as Ba/Sr=4/1 and 1/4 are illustrated in Fig.5, and the lattice parameters of a unit cell as a function of Ba to Sr ratio are shown in Fig. 6. The lattice parameters were calculated by analyzing the X-ray diffraction patterns. In the Ba-rich side, the crystal structure consisted of three CuO_2 planes sandwiched by Tl-O bilayers as shown in Fig.1(d); in the case of the Sr-rich side, the structure consisted of three CuO_2 planes sandwiched by Tl-O monolayers as shown in Fig.1(b). A crystallographic transformation was observed at Ba/Sr=1.0. It should be noted, however, that the transformation point depends on the heat treatment conditions. The crystallographic transformation may be responsible for the discontinuity in the T_c value at Ba/Sr=1.0. The lattice parameter of the c axis was elongated by Ba substitution of Sr. The changes in the lattice parameters of the a and b axes were relatively small over the entire range of Ba/Sr ratios. This phenomena may be explained by a mutual substitution between Ba ions (lager ionic ratio 1.47Å in 12 coordination) and Sr ions (1.27Å).

To make clear the relationship between crystallographic transformation and the Ba/Sr ratio, we also studied the 1223 compounds. The samples were prepared by the same method as the 2223 compounds, except the atomic ratio of Tl:(Ba+Sr):Ca:Cu.

The lattice parameters of a unit cell as a function of the Ba/Sr ratio are shown in Fig.7. In spite of the different atomic ratio of the starting compounds, 2223 phase was dominantly synthesized in Ba-rich samples. The results suggest that the Ba/Sr ratio may be responsible for the stability of crystal structures. However, it should be noted that the crystallographic transformation took place around Ba/(Sr+Ba)=0.6-0.8 in this case.

We refined the crystal structure of $Tl_2Ba_{1.6}Sr_{0.4}Ca_2Cu_3O_{10}$ sample using the Rietveld analysis computer program RIETAN[6]. From the results of electron diffraction and the tentative analysis of X-ray diffraction patterns, the symmetry of the crystal structure was assumed to be P4/mmm. As the phase includes heavy atoms like Tl and Sr, it is difficult to refine the occupation factors (g) and thermal parameters (B) for the oxygen ions. The occupation factors of the oxygen ions were taken to 1.0 and the thermal parameters were assumed as shown in Table1[8]. The resultant structure parameters and their standard deviations are listed in Table 1. Lattice constants were refined to be a=0.38453(1) nm and c=3.5486(2) nm with space group P4/mmm. R factors were R_{WP} =8.69%, R_P=6.95%, R_E=4.98%, R_I=4.97%, R_F=4.02%. The occupation factors of Ca-sites are considerably larger than 1.0, which suggests that Ca-sites are partially occupied by other metal ions. Judging from the ionic radii and the atomic scattering factor, Tl(III) ion is a likely candidate for the substitution in Ca(II) ion site and perhaps Sr also. On the other hand, the occupation factor of Tl-sites is smaller than 1.0. This suggests as follows. (a) Tl-sites are partially substituted by lighter metal ions. (b) Tl-sites have defects. (c) Tl ions are actually displaced from their ideal positions in order to form a more favorable bonding environment. we have not got the final answer for that so far. The similar phenomena were reported in the case of Tl-Ba-Ca-Cu-O systems[7]. In the case of Ba and Sr sites, the standard deviations are not small enough. It is difficult to discuss the absolute values of the occupation factors. However, the result suggest that Ba and Sr ions may occupy the same sites in this crystal structure.

CONCLUSIONS

We have found that the 2223 phase of TBSCC system was obtained as a single phase more readily compared with the TSCC and TBCC systems. As a result, we achieved the high superconducting volume fraction, 95% above 100K. In the TBSCC system, the Tc decreased gradually from both TBCC and TSCC sides as a function of the Ba/Sr ratio. The Tc value changed discontinuously at Ba/Sr=1.0. By X-ray diffraction analysis, the crystal structure was determined as the 2223 phase in the Ba-rich side, and the 1223 phase in the Sr-rich side. The crystallographic transformation is responsible for the discontinuity of the Tc and the Ba/Sr ratio may change the stability of crystal structures. Rietveld analysis reveal that partial substitutions of Ca-sites and Tl-sites may take place and Ba,Sr ions may occupy the same sites in this system.

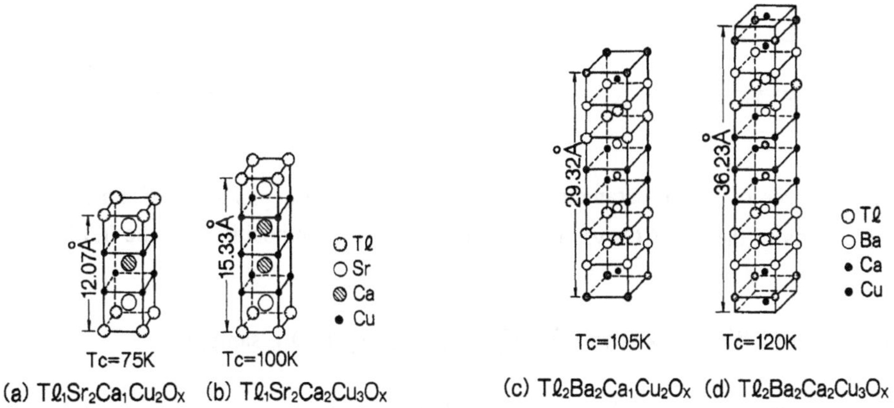

Fig.1. Crystal structures of TBCC and TSCC systems. (a) low-Tc phase, (b) high-Tc phase consisting of Tl-O monolayer and (c) low-Tc phase, (d) high-Tc phase consisting of Tl-O bilayers.

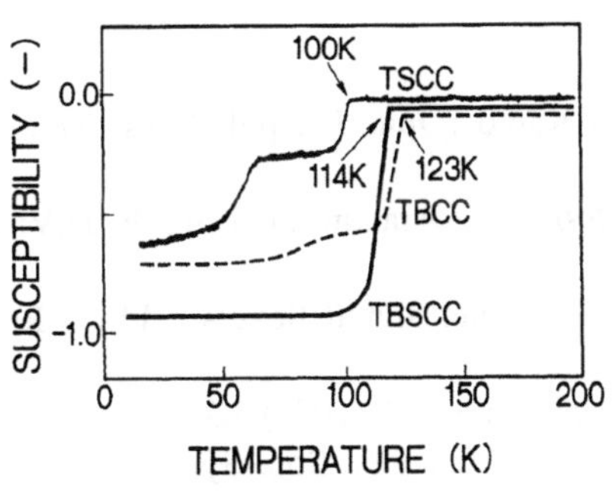

Fig.2. Temperature dependence of susceptibility for TBSCC, TBCC and TSCC systems.

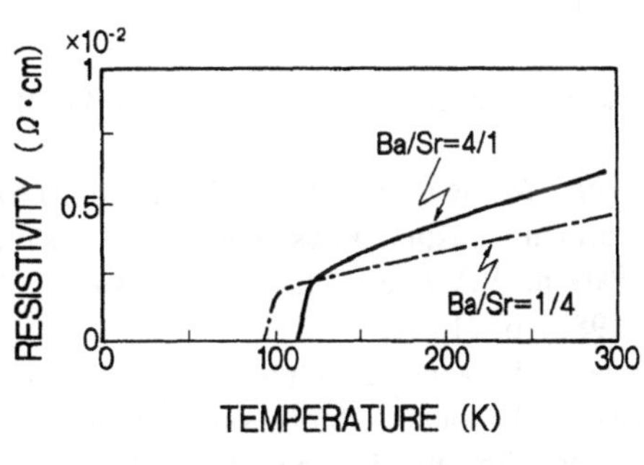

Fig.3. Temperature dependence of resistivity for the TBSCC samples with the Ba/Sr ratios;4/1, 1/4.

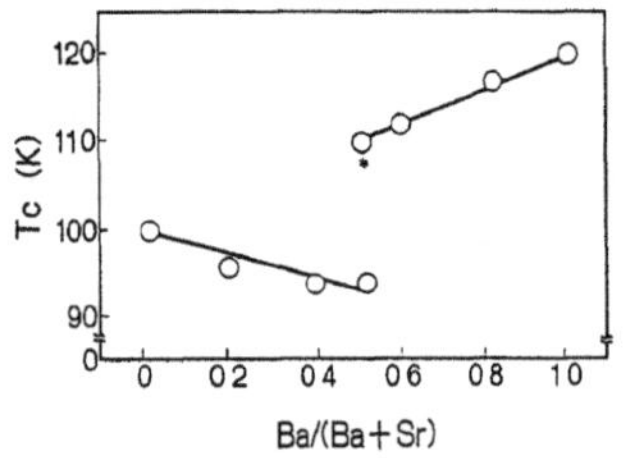

Fig.4. Ba/Sr ratio dependence of Tc values Only the sample marked with * was heated at 850°C for 3h.

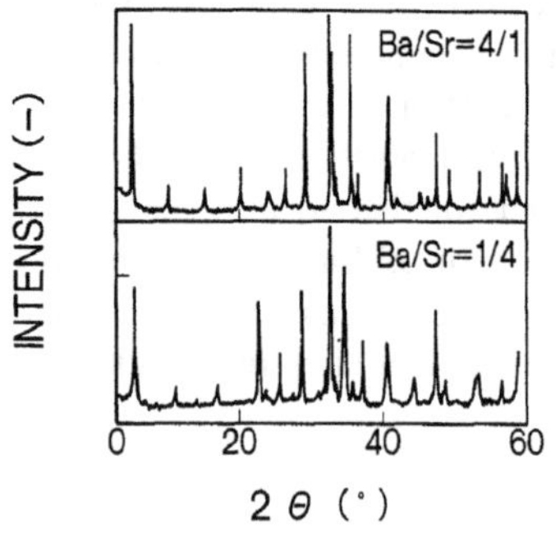

Fig.5. X-ray diffraction patterns

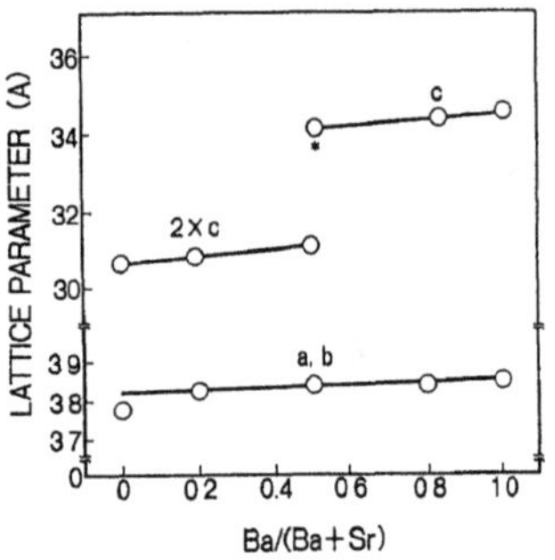

Fig.6. Ba/Sr ratio dependence of lattice parameters for the 2223 compositions in the TBSCC system

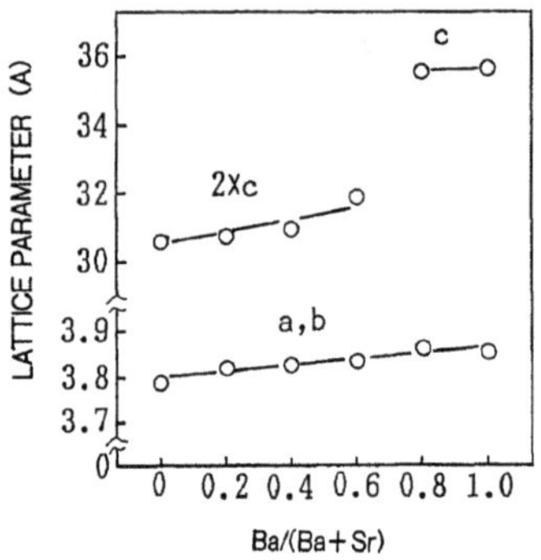

Fig.7. Ba/Sr ratio dependence of lattice parameters for the 1223 compositions in the TBSCC system.

Table 1. Results of Rietveld analysis for $Tl_2Ba_{1.6}Sr_{0.4}Ca_2Cu_3O_{10}$ sample

	X	Y	Z	g	B
Tl	0.0	0.0	0.2202(3)	0.83(3)	0.8
Ba	0.5	0.5	0.1458(3)	0.6(1)	0.5
Sr	0.5	0.5	0.1458(3)	0.4(1)	0.5
Ca	0.5	0.5	0.0471(5)	1.26(9)	0.1
Cu(1)	0.0	0.0	0.0895(7)	1	0.3
Cu(2)	0.0	0.0	0.0	1	0.1
O(1)	0.5	0.5	0.223(3)	1	0.9
O(2)	0.0	0.0	0.165(3)	1	0.9
O(3)	0.0	0.5	0.0873(2)	1	0.8
O(4)	0.5	0.0	0.0	1	0.6

Lattice constants (Å) : a=3.8453(1), c=35.486(2)

g : occupation factor

B : thermal parameter[8]

REFERENCES

1. S.Matsuda, S.Takeuchi, A.Soeta, T.Suzuki, K.Aihara and T.Kamo : Jpn. J. Appl. Phys. 27 (1988) 2062.
2. Z.Z.Sheng, W.Kiehl, J.Bennett, A El Ali, D.Marsh, G.D.Mooney, F.Arammash, J.Smith, D.Viar and A.M.Hermann : Appl. Phys. Lett. 52 (1988) 1738.
3. S.S.P.Parkin, V.Y.Lee, A.I.Nazzal, R.Savoy, T.C.Huang, G.Gorman and R.Beyers : Phys. Rev. B38 (1988) 6531.
4. H.Maeda, T.Tanaka, M.Fukutomi and T.Asano : Jpn. J. Appl. Phys. 27 (1988) L209.
5. S.Matsuda, M.Okada, T.Morimoto, T.Matsumoto and K.Aihara : Mat. Res. Soc. Symp. Proc. (Material Reseach Society, Boston MA, 1988) Vol. 99, P.695.
6. F.Izumi : J. Crystallogr. Soc. Jpn. 27 (1985) 23.
7. M.A.Subramanian, C.C.Torardi, J.Gopalakrishnan, P.L.Gai, J.C.Calabrese, T.R.Askew, R.B.Flippen and A.W.Sleight : Science 24(1988) 249.
8. D.E.Cox, C.C.Torardi, M.A.Subramanian, J.Gopalakrishnan and A.W.Sleight : Phys. Rev. B, 38 (1989) 6624.

Stability of T* and T′ Phases in $Nd_{1.4}Sr_{0.4}Ce_{0.2}CuO_{4-d}$

T. SAKURAI, T. YAMASHITA, S. IKEGAWA, and H. YAMAUCHI

Superconductivity Research Laboratory, International Superconductivity Technology Center, 10-13, Shinonome 1-chome, Koto-ku, Tokyo, 135 Japan

ABSTRACT

The T^* phases in $(Nd,Ce,Sr)_2CuO_{4-d}$ have been found to be superconducting, having holes as charge carriers. Another new-type superconductors have been discovered in the $(Nd,Ce)_2CuO_{4-d}$ having electrons as charge carriers. We have observed that either T^* or T' phase in Nd-Ce-Sr-Cu-O system can be the main and superconducting phase after oxidizing or reducing heat treatment. This T^* ↔ T' phase change can occur reversibly. It is also concluded that the stability of the T^* phase depend on the ionic radii of the constituent elements.

KEY WORDS: phase change, T^* phase, T' phase, oxidizing heat treatment, reducing heat treatment.

INTRODUCTION

New superconductors were discovered by Akimitsu et al [1] in the Nd-Ce-Sr-Cu-O system. It was revealed that these compounds have holes as charge carriers[2] and $Cu\text{-}O_5$ pyramids in the unit cell according to transmission electron microscopic[3] and neutron diffraction analyses[4,5]. The phase having this structure is termed T^* and the structure is shown in Fig. 1(a). Recently, T^* phases were found in other systems, such as, $La(Pr)_{2-x-y}Ln_xSr(Ba,Ca)_yCuO_4$ where Ln denoted a rare earth element[7,8]. Superconductivity was observed in $(La,Eu,Sr)_2CuO_4$, $(La,Gd,Sr)_2CuO_4$ and $(La,Sm,Sr)_2CuO_4$ after heat treatment at a high oxygen pressure[7,8].

In the beginning of 1989, Tokura et al[9] discovered a new family of superconductors having electrons as charge carriers in $(L,Ce)_2CuO_{4-d}$ where L was Nd or Pr. The new superconductors have the Nd_2CuO_4 structure as shown in Fig.1(b) and the phase having this structure is termed T' phase. Later, superconductivity in T' phases was confirmed also in $(Nd,Th)_2CuO_4$[10], $Nd_2CuO_{4-x}F_x$[11], and $(R,Th)_2CuO_{4-d}$ where R was Pr or Sm[12].

In the present study, eight different kinds of samples containing T^* phases as the main phase are prepared and heat treated under different kinds of atmosphere. The stability of T^* phases in these samples will be discussed.

EXPERIMENTAL

Eight different kinds of samples were prepared by a solid state reaction using La_2O_3, CeO_2, Nd_2O_3, Sm_2O_3, Eu_2O_3, Gd_2O_3, Tb_4O_7, Ho_2O_3, $SrCO_3$, $BaCO_3$ and CuO powders. All the powders were of 99.9% or higher purity. Mixed powders with desirable compositions (except Nd-Ce-Sr-Cu-O) were calcined twice at 950 °C for 12 h with an intermediate regrinding. Calcined powders were reground and pressed into rods and sintered at 1050 °C for 12 h. The rods were again reground, pressed into rods and sintered at 1100 °C for 12 h. All of the firing processes were carried out in air. This method was the one firstly employed by Cheong et al[7]. The mixed powders of the Nd-Ce-Sr-Cu-O system were calcined at 950 °C for 10 h and sintered at 1130 °C in air. The samples were then oxygenated at 1100 °C in O_2 gas flow. All the samples were then annealed at 1050 °C in N_2 gas flow and quenched to room temperature (R.T.) in the same atmosphere. The samples were characterized by x-ray diffractometry. The resistivity measurement was carried out using a conventional four-probe technique in the temperature range from R.T. to 4.2 K. A SQUID magnetometer (Quantum Design Model MPMS) was used for measuring magnetic susceptibility. A thermogravimetric and thermal differential analyzer (TG-DTA) (MAC Science) was used for determining the temperature where the $T^* \leftrightarrow T'$ phase change took place.

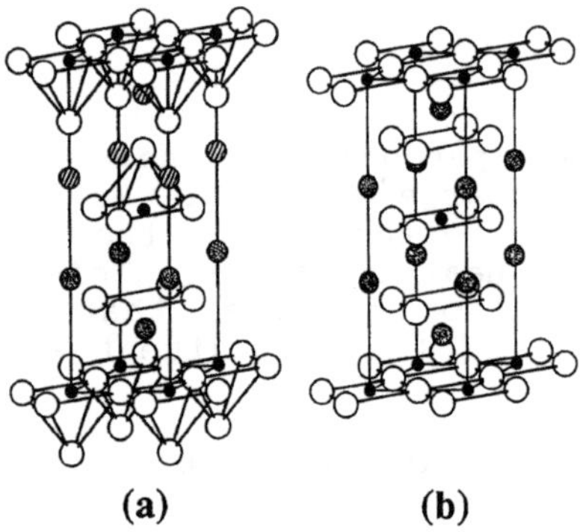

Fig. 1. Crystal structure of (a) T^* phase and (b) T' phase. ⊘ = Sr(Nd)(A2 sites), ⊛ =Nd(Ce)(A1 sites), ● = Cu, ○ = O[6].

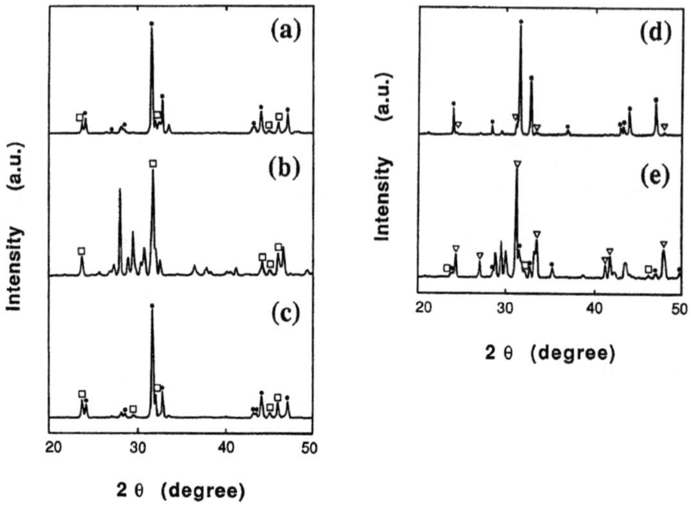

Fig. 2. Powder x-ray diffraction patterns of (a) $Nd_{1.4}Ce_{0.2}Sr_{0.4}CuO_4$ after oxidizing heat treatment, (b) reducing heat treatment, (c) re-oxidizing heat treatment and (d) $La_{1.19}Tb_{0.67}$-$Ba_{0.14}CuO_4$ after oxidizing heat treatment, (e) reducing heat treatment. ● indicates peaks from T^* phase, □ those from T' phase and ▽ those from T phase.

RESULTS AND DISCUSSION

The compositions of the eight samples initially synthesized in air (an oxidizing atmosphere) are listed in Table 1. All the samples contain T^* as the main phase. When annealed in a reducing atmosphere, i.e. annealed at 1050 °C in N_2 gas flow and quenched to R.T. in the same atmosphere, only two samples, i.e. $Nd_{1.4}Ce_{0.2}Sr_{0.4}CuO_4$ and $La_{1.19}Tb_{0.67}Ba_{0.14}CuO_4$, yielded main phases different from those originally contained as shown in Table 1. Figures 2(a) - (e) are powder x-ray diffraction patterns for the two samples before and after the reducing heat treatment. It has been known that the T^* unit cell structure shown in Fig. 1(a) has two different A sites, i.e. A1 and A2. The cordination numbers for an A1 site and an A2 site are 8 and 9, respectively. Izumi et al[4] found that Nd and Sr occupied A1 sites while Nd and Ce A2 sites in the T^* phase of $(Nd,Ce,Sr)_2CuO_4$. Taking this situation into account, the average ionic radius of an A1 and A2 ions can be calculated for all the eight samples using the ionic radii tabulated by Shannon[13]. As shown in Table 2, the average ionic radii are more or less constant for all the eight original samples: 1.07 ± 0.01 Å for an A1 ion and 1.23 ± 0.02 Å for an A2 ion. And also the ratios of the average ionic radii to the radius of Cu^{2+} are almost constant, i.e. 1.77 ± 0.02. It is likely that the difference in the two average ionic radii is essential to obtain T^* phases in the initial samples. After heat treatment in a reducing atmosphere, the T^* phase is still stable for the samples whose chemical formula can be written in the form of $La_{2-x-y}Ln_xSr_yCuO_4$ where Ln is a rare earth element. If the average ionic radii of the elements which are substituted for La and/or Sr in T^* phase are significantly different from those of La^{3+} and Sr^{2+}, the T^* phase becomes unstable after heat treatment in a reducing atmosphere. In fact, the ionic radii of Nd^{3+} ions in an A1 site and in an A2 site are 1.11 and 1.16 Å which are signifficantly smaller than those of La^{3+}, i.e. 1.16 Å for an A1 site and

1.22 Å for an A2 site. On the other hand, the ionic radius of Ba^{2+} is 1.47 Å and this value is significantly larger than that of Sr^{2+}, i.e. 1.31 Å.

The electrical resistivity was measured for all the samples. All the samples showed no superconductivity after heat treatment in air as mentioned by Cheong et al[7]. Only $Nd_{1.4}Ce_{0.2}Sr_{0.4}CuO_4$ samples were heat treated in O_2 gas flow and became superconductive. Since the main phase of this sample was a T^* phase, it was concluded that the T^* phase was responsible for superconductivity[6]. Only two samples of $Nd_{1.4}Ce_{0.2}Sr_{0.4}CuO_4$ and $La_{1.1}Ho_{0.7}Sr_{0.2}CuO_4$ showed superconductivity after being heat treated in a reducing atmosphere. The $Nd_{1.4}Ce_{0.2}Sr_{0.4}CuO_4$ sample showed a $T^* \rightarrow T'$ phase change during heat treatment while $La_{1.1}Ho_{0.7}Sr_{0.2}CuO_4$ did not as shown in Table 1. Because a T' phase becomes superconducting only when the phase is heat treated in a reducing atmosphere[9], the superconducting phase of $Nd_{1.4}Ce_{0.2}Sr_{0.4}CuO_4$ after the reducing heat treatment was concluded to be T'[6]. Thus, the main phase of this sample was superconducting after either type of heat treatment, i.e. oxidizing or reducing. On the other hand, the phases in $La_{1.1}Ho_{0.7}Sr_{0.2}CuO_4$ before and after reducing heat treatment remained unaltered. From SQUID measurements, it was revealed that no bulk superconductivity existed in this sample after annealed in a reducing atmosphere. Thus the reason why this sample showed superconductivity is suspected as follows. The T phase in $La_{1.1}Ho_{0.7}Sr_{0.2}CuO_4$ after heat treated in air was likely to be $La_{2-x}Sr_xCuO_4$ which was probably non-superconducting due to excess holes contained. This phase became superconducting after a reducing heat treatment because this heat treatment removed some oxygens from the phase, i.e. the hole concentration was reduced and became optimum for superconductivity. It was reported that $La_{2-x}Sr_xCuO_{4-y}$ where x = 0.8 and 1 became

Table 1. Phases observed by x-ray diffraction after heat treatment in oxidizing and reducing atmosphere.

	Phases After Heat Treated In			
	Oxidizing Atmosphere		Reducing Atmosphere	
	Main	Others	Main	Others
$La_{1.08}Dy_{0.72}Sr_{0.2}CuO_4$	T^*	T + X	T^*	T + X
$La_{0.9}Eu_{0.9}Sr_{0.2}CuO_4$	T^*	T + T'	T^*	T + T' + X
$La_{0.9}Gd_{0.9}Sr_{0.2}CuO_4$	T^*	T + T'	T^*	T + T' + X
$La_{1.1}Ho_{0.7}Sr_{0.2}CuO_4$	T^*	T	T^*	T + X
$La_{0.8}Sm_1Sr_{0.2}CuO_4$	T^*	T + T'	T^*	T + T'
$La_{1.19}Tb_{0.67}Ba_{0.14}CuO_4$	T^*	T	T	T^* + T'
$La_1Tb_{0.8}Sr_{0.2}CuO_4$	T^*	T	T^*	T + T'
$Nd_{1.4}Ce_{0.2}Sr_{0.4}CuO_4$	T^*	T' + X	T'	X

Table 2. Average ionic radii of ions in A1 and A2 sites(the A1 and A2 sites are defined in Fig.1). And ratio of average ionic radii of A site and Cu^{2+}.

	Average Ionic Radii (A)		Ratio of
	A1 site	A2 site	A site/Cu^{2+}
$La_{1.08}Dy_{0.72}Sr_{0.2}CuO_4$	1.06	1.23	1.77
$La_{0.9}Eu_{0.9}Sr_{0.2}CuO_4$	1.08	1.23	1.78
$La_{0.9}Gd_{0.9}Sr_{0.2}CuO_4$	1.06	1.23	1.77
$La_{1.1}Ho_{0.7}Sr_{0.2}CuO_4$	1.06	1.23	1.76
$La_{0.8}Sm_1Sr_{0.2}CuO_4$	1.08	1.23	1.78
$La_{1.19}Tb_{0.67}Ba_{0.14}CuO_4$	1.08	1.25	1.79
$La_1Tb_{0.8}Sr_{0.2}CuO_4$	1.06	1.23	1.77
$Nd_{1.4}Ce_{0.2}Sr_{0.4}CuO_4$	1.08	1.22	1.77
average	1.07	1.23	1.77

superconducting after heat treatment in a reducing atmosphere[14]. This may be explained using same mechanism proposed in this work.

The $Nd_{1.4}Ce_{0.2}Sr_{0.4}CuO_{4-d}$ samples which were sintered in a reducing atmosphere and confirmed to contain a T' as the main phase were now re-heat treated to testify if the main phase altered to be a T^* phase: the samples were re-heat treated at 1130 °C for 15 h in air and then oxygenated at 1100 °C for 15 h and 500 °C for 10 h in O_2 gas flow. X-ray diffraction analyses revealed that the main phase of the re-oxygenated samples was a T^*, as shown in Fig. 2(c). These samples were superconducting according to resistivity measurement. Thus either T^* phase which was induced by an oxidizing heat treatment or T' phases which was yielded by a reducing heat treatment in $Nd_{1.4}Ce_{0.2}Sr_{0.4}CuO_4$ can be the main phase which is superconducting.

TG-DTA measurements were carried out in N_2 and O_2 gas flow for samples of $Nd_{1.4}Ce_{0.2}Sr_{0.4}CuO_4$ having a T^* and T' as the main phases, respectively. Small endothermic peaks around 930 and 945 °C were respectively observed in the DTA curves on heating. No weight changes were observed on the TG curves at these temperatures. This may suggest that the $T^* \leftrightarrow T'$ phase change occured around 940 °C without oxygen removal as was speculated by Sakurai et al[6].

CONCLUSION

T^* phases were obtaind for 214 compaunds as listed in Table 1 when the difference was certain value in the average ionic radii for ions in the two differnt A sites. Actually, the stability of a T^* phase in a reducing atmosphere depended on the actual radius of ions on the A sites. Either T^* or T' phases in the $Nd_{1.4}Ce_{0.2}Sr_{0.4}CuO_4$ became the main and superconducting phase after oxydizing or reducing heat treatment. $T^* \rightarrow T'$ phase change in the $Nd_{1.4}Ce_{0.2}Sr_{0.4}CuO_4$ was confirmed to be reversible.

ACKNOWLEDGEMENT

We are greatful to M. Kosuge of Superconductivity Research Laboratory for his helpful suggestion for sample preparation.

REFERENCES

1. J. Akimitsu, S. Suzuki, M. Watanabe and H. Sawa, Jpn. J. Appl. Phys. 27, L859-860 (1988).
2. M. Kosuge and K. Kurusu, Jpn. J. Appl. Phys. 28, 810-812 (1989).
3. E. T. Muromachi, Y. Matsui, Y. Uchida, F. Izumi, M. Onoda and K. Kato, Jpn. J. Appl. Phys. 27, L2283-2286 (1988).
4. F. Izumi, E. T. Muromachi, A. Fujimori, T. Kamiyama, H. Asano, J. Akimitsu and H. Sawa, Physica C 158, 440-448 (1989).
5. H. Sawa, S. Suzuki, M. Watanabe, J. Akimitsu, H. Matsubara, H. Watabe, S. Uchida, K. Kokusho, H. Asano, F. Izumi and E. T. Muromachi, Nature 337, 347-348 (1989).
6. T. Sakurai, T. Yamashita, H. Yamauchi and S. Tanaka, Physica C 161, 6-8 (1989).
7. S-W. Cheong, Z. Fisk, J. D. Thompson and R. B. Schwarz, Physica C 159, 407-411 (1989).
8. Y. Tokura, H. Takagi, H. Watabe, H. Matsubara, S. Uchuda, K. Hiraga, T. Oku, T. Mochiku and H. Asano, Phys. Rev. B 40, 2568-2571 (1989).
9. Y. Tokura, H. Takagi and S. Uchida, Nature 337, 345-347 (1989).
10. J. T. Markert and M. B. Maple, Solid State Commun. 70, 145-147, (1989).
11. A. C. W. P. James, S. M. Zahurak and D. W. Murphy, Nature 338, 240-241 (1989).
12. J. T. Market, E. A. Early, T. Bjornholm, S. Ghamaty, B. W. Lee, J. J. Neumeier, R. D. Price, C. L. Seaman and M. B. Maple, Physica C 158, 178 -182 (1989).
13. R. D. Shannon, Acta Cryst. A32, 751 (1976).
14. H. Ihara, M. Hirabayashi, N. Terada, Y. Kimura, K. Senzaki and M. Tokumoto, Jpn. J. Appl. Phys. 26, L463-465 (1987).

Composite-Layered Chalcogenides: A New Candidate for Superconductor

Y. OOSAWA[1], Y. GOTOH[1], and M. ONODA[2]

[1] National Chemical Laboratory for Industry, Higashi, Tsukuba, Ibaraki, 305 Japan
[2] National Institute for Research in Inorganic Materials, Namiki, Tsukuba, Ibaraki, 305 Japan

ABSTRACT

New type of chalcogenides called "composite-layered chalcogenides" with "$BiMX_3$" (M = Ti, V, Nb, Ta; X = S, Se) formula have been prepared and characterized using X-ray powder diffraction, electron diffraction, and electric resistivity measurement. It has been suggested that the structure of the chalcogenides consists of two kinds of layers alternately and incommensurately stacked: two-atom-thick layer of AX and three-atom-thick sandwich of MX_2. They have been prepared from elements by heating in silica tube and are black-grayish microcrystalline powder with luster. Their electric resistivity measured with d. c. four-probe method in the range of 1.7-300 K showed various behaviors: semiconducting, metallic, or superconducting, depending on their compositions.

KEY WORDS: composite-layered chalcogenide, ternary chalcogenide, layer structure, superconductor, mutually-incommensurate

INTRODUCTION

It is well known that chalcogenide superconductors were more important than oxide superconductors for both basic science and application before the discovery of so-called high-Tc superconducting oxides: for example Chevrel phase compound, dichalcogenide, trichalcogenide etc. Because the bonding character of chalcogenide mediates between that of oxide and metal, there is a possibility that chalcogenide offers new superconducting material with merits of oxide and metal at the same time: high Tc, high Jc, tolerability against high magnetic field, etc.

A few ternary chalcogenides with "AMX_3"(A = Pb, Sn; M= Nb, Ta; X = S, Se) formula were reported as superconductor over fifteen years ago[1] but their structures were not reported for long time. Recently, it has become apparent that these teranry chalcogenides have a composite-layered structure similar to that of "$LaCrS_3$"[2]. In "$LaCrS_3$", two-atom-thick layer of LaS with ditorted NaCl structure and three-atom-thick sandwich of CrS_2 with CdI_2 structure are stacked alternately[3]. The sulfide has mutually-incommensurate structure because the ratio of periodic lengths of **b** axes of the LaS and CrS_2 layers is not simple rational. The ideal composition of "$LaCrS_3$" is not strictly $LaCrS_3$ but $(LaS)_{1.2}CrS_2$. This formulation comes from the layered and mutually-incommensurate structure. This is also the case for other "AMX_3" type of composite-layered chalcogenides[2].

We have found another series of composite-layered chalcogenide with "$BiMX_3$"(M = Ti, V, Nb, Ta; X = S, Se) formula and reported preliminarily[4]. We also have found "$PbVS_3$" which belongs to Pb series and have determined its structure[5]. In the present manuscript, we report on "$BiMX_3$": characterization with X-ray powder diffraction, electron diffraction, and electric resistivity measurement using d. c. four-probe method.

EXPERIMENTAL

The composite-layered chalcogenides with "$BiMX_3$" formula were prepared as follows. Starting materials(usually elements with purity above 99.9%) in a stoichiometric amount were mixed together and sealed in a silica tube under vacuum. The tube was placed in an electric furnace, heated at 500 °C, then at 800 °C and cooled to room temperature. Thus the chalcogenides were obtained as black-grayish microcrystalline powder with luster. The X-ray diffraction patterns were taken by the counter-diffractometer technique using Ni-filtered CuKα radiation. Electron diffraction patterns were taken with crushed samples using a 100 kV electron microscope. The temperature dependence of electric resistivity was measured with pellet made of sample powder using a standard d. c. four-probe method in the range of 1.7 K to 300K.

RESULTS AND DISCUSSION

Three series of composite-layered chalcogenide with "AMX_3" formula have been reported up to the present: A = Lanthanides[2, 3, 6]; Pb or Sn[2, 5, 7, 8]; and Bi[2, 4]. All these chalcogenides have similar structures: two-atom-thick AX layer with distorted NaCl structure and three-atom-thick MX_2 sandwich with CdI_2 structure(M is octahedrally coordinated by six X atoms) or trigonal prismatic structure(M is in the center of the prism and coordinated by six X atoms). Figure 1 shows the results of our single crystal structure determination of "$PbVS_3$" as an example[5]. The ideal composition of "$PbVS_3$" is $(PbS)_{1.12}VS_2$. In the sulfide, MX_2 sandwich has a distorted CdI_2 structure.

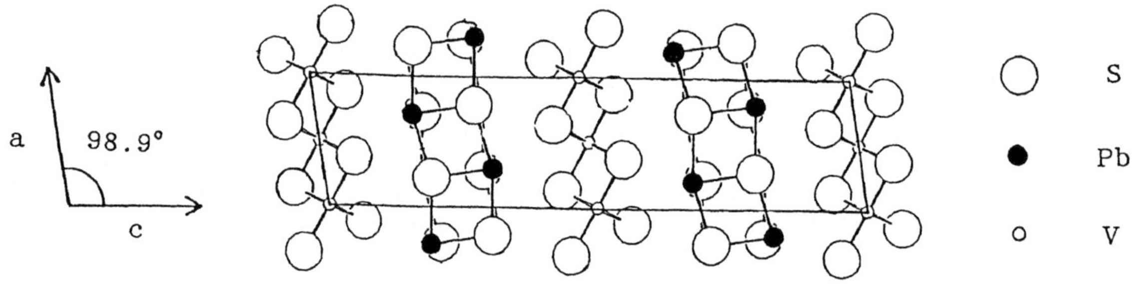

Fig.1. Projection of the structure of "$PbVS_3$" along the b direction.

"$BiMX_3$"(M = Ti, V, Nb, Ta; X = S, Se) were obtained as black-grayish microcrystalline powder with luster[4]. The X-ray diffraction pattern of each chalcogenide obtained using counter diffractometer method consists of several strong peaks and a few weak peaks as shown in Fig. 2. The strong peaks are attributable to a set of parallel planes (0 0 ℓ). This is a reflection of layered structure and extreme preferred orientation. The electron diffraction of "$BiTaS_3$" obtained with an incident beam parallel to the [0 0 1] direction shows a characteristic pattern. The pattern is to be understood as superposition of a pseudo-tetragonal subcell of Bi and S, and pseudo-hexagonal subcell of Ta and S. We adopt orthohexagonal axes $\mathbf{a}^*_{OH}$ and $\mathbf{b}^*_{OH}$ for a pseudo-hexagonal subcell. The two subcells have same periodicity in $\mathbf{a}^*$ axis(a^*_T = a^*_{OH}). They have different periodicities in $\mathbf{b}^*$ axes and are in nearly commensurate relation of $b^*_T/12 \doteq b^*_{OH}/11$. By taking into account the relation found for the periodicities in $\mathbf{a}*$ and $\mathbf{b}*$ axes, the ideal composition of "$BiTaS_3$" would be expressed as $(BiS)_{1.09}TaS_2$. It is expected that a similar situation for the formulation stands also for other "$BiMX_3$" type of chalcogenides.

The electric resistivity of "$BiMX_3$" pellet was measured in a preliminary

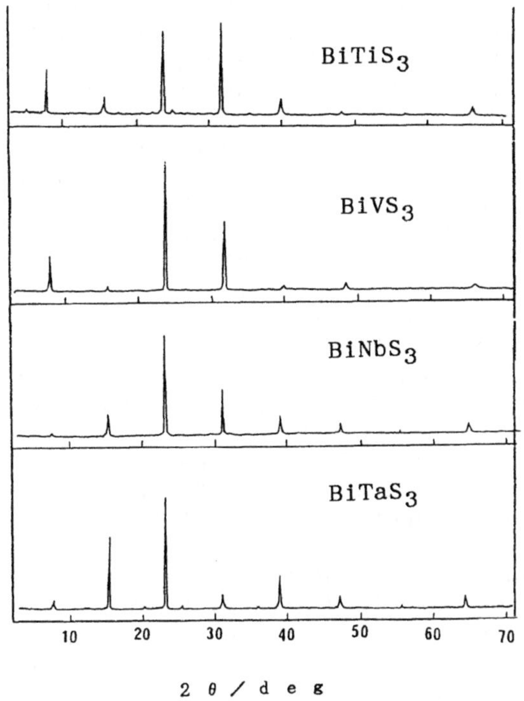

Fig.2. Powder X-ray diffraction pattern of "$BiMS_3$" obtained using counter-difffractometer method.

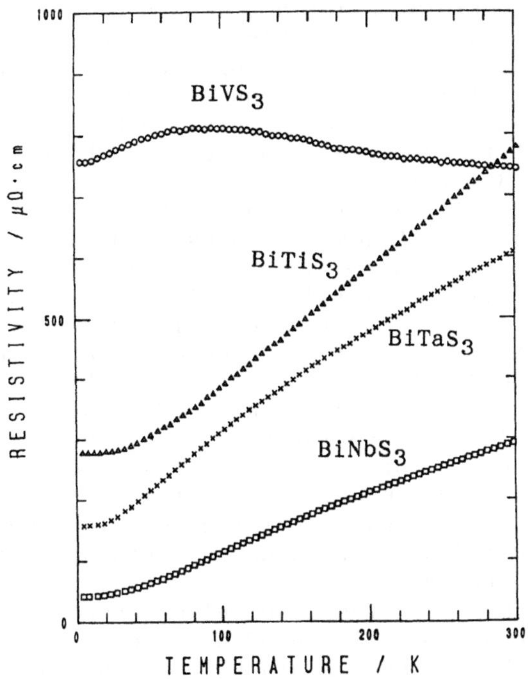

Fig.3. Dependence of electric resistivity on the temperature for "$BiMS_3$".

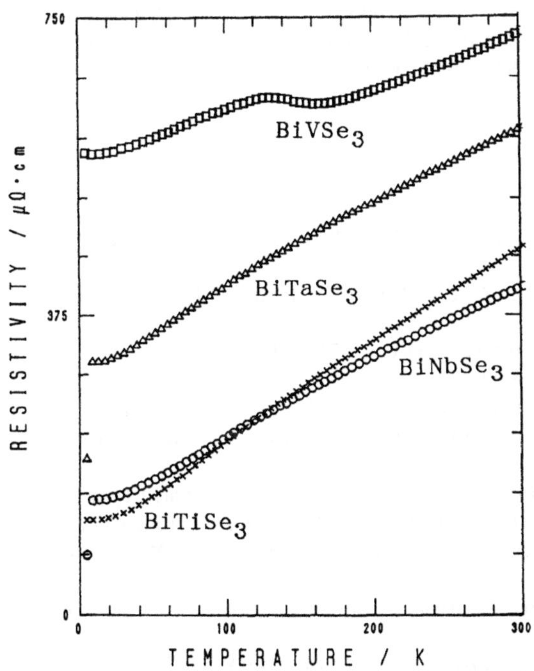

Fig.4. Dependence of electric resistivity on the temperature for "$BiMSe_3$".

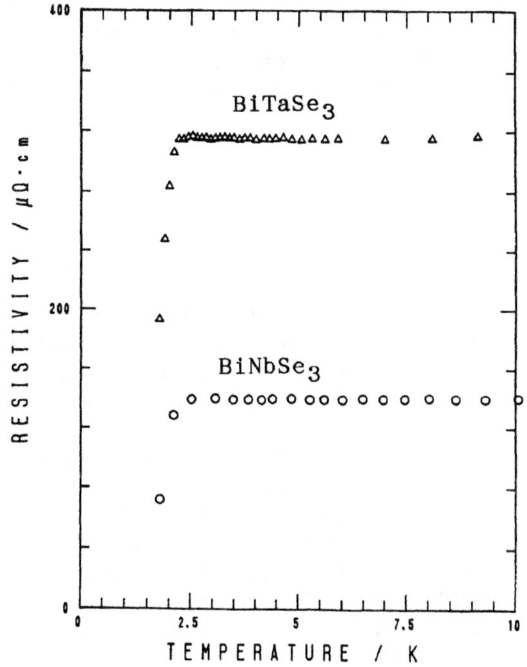

Fig.5. Dependence of electric resistivity on the temperature below 10 K for "$BiNbSe_3$" and "$BiTaSe_3$"

experiment using a standard d. c. four probe method in the range of 1.7 K to 300 K. The stoichiometry used for the preparation of the pellets was strictly 1/1/3 for each chalcogende. Figure 3 shows the results for "$BiMS_3$"(M = Ti, V, Nb, Ta). In the cases of "$BiTiS_3$", "$BiNbS_3$", and "$BiTaS_3$", the electric resistivity showed metallic behavior and gradually decreased with a decrease of temperature. On the other hand, in the case of "$BiVS_3$", the resistivity did not depend on the temperature so much, and there seemed to be some kind of transition below 100 K. No indication of superconductivity was observed for these sulfides. Figures 4 and 5 show the dependence of electric resistivity on the temperature for "$BiMSe_3$"(M = Ti, V, Nb, Ta). The resistivity of "$BiTiSe_3$" showed metallic behavior and decreased with a decrease of temperature. The resistivity of "$BiVSe_3$" showed somewhat unusual behavior like its sulfide analogue. There seemed to be some kind of transition above 100 K. The resistivities of "$BiNbSe_3$" and "$BiTaSe_3$" showed metallic behavior down to around 2.5 K and decreased with a decrease of temperature. And blow 2.5 K, the resistivity decreased drastically as shown in Fig. 5. The drastic decrease of resistivity observed for "$BiNbSe_3$" and "$BiTaSe_3$" seems to be due to superconductivity. It is well-known that certain types of $NbSe_2$ and $TaSe_2$ with trigonal prismatic coordination are superconductors. In "$BiNbSe_3$" and "$BiTaSe_3$", MX_2 layer consists of sandwich of $NbSe_2$ or $TaSe_2$, respectively. Therefore it is suggested that the "superconductivity" of these selenides comes from the diselenide layers. As has been described aleady, however, the ideal composition of these composite-layered chalcogenides is not strictly $BiMX_3$. They should be expressed as $(BiX)_nMX_2$ where n is a number close to 1. On the other hand, the stoichiometry of the elements used for the preparation of the pellets in the present study was $BiMX_3$. Therefore it it expected that some small amount of impurity was included in these pellets. In order to ascertain the superconductivity of these selenides strictly, pure phase has to be prepared at first. Preparation of single crystal is most favorable. The measurement of resistivity in the region lower than 1.7 K is also required. The use of SQUID for the measurement of diamagnetic susceptibility is desirable. Further study is under way.

As has been stated in the introduction, not only oxide but also chalcogenide are promising candidates for medium-to-high Tc superconductor. Search for superconductor not only in oxides but also in chalcogenides seems to be important both for application and basic science.

REFERENCES

1. M. H. van Maaren, Phys. Lett., 40A, 353(1972).
2. G. A. Wiegers et al., Solid State Commun., 70, 409(1989). references therein.
3. (a) K. Kato, I. Kawada, and T. Takahashi, Acta Cryst., B33, 3437(1977).
   (b) K. Kato, Acta Cryst., submitted.
4. (a) Y. Oosawa, Y. Gotoh, and M. Onoda, Chem. Lett., 523(1989).
   (b) Y. Gotoh et al, Chem. Lett., 1559(1989).
5. (a) Y. Gotoh, M. Onoda, M. Goto, and Y. Oosawa, Chem. Lett., 1281(1989).
   (b) Y. Gotoh, M. Goto, K. Kawaguchi, Y. Oosawa, and M. Onoda, Mat. Res. Bull., in the press.
   (c) M. Onoda, K.Kato, Y. Gotoh, and Y. Oosawa, Acta Cryst., submitted.
6. A. Meerschaut, P. Rabu, J. Rouxel, J. Solid State Chem., 78, 35(1989).
7. (a) G. A. Wiegers, A. Meetsma, R. J. Haange, and J. L. de Boer, Mat. Res. Bull., 23, 1551(1988).
   (b) A. Meetsma, G.A. Wiegers, R. J. Haange, and J. L. de Boer, Acta. Cryst. A45, 285(1989).
   (c) S. van Smaalen, J. Phys.: Condens. Matter, 1, 2791(1989).
8. L. Guemas, P. Rabu, A. Meerschaut, and J. Rouxel, Mat. Res. Bull., 23, 1061(1988).

CGS: The Crystal Structure Graphics Display System for Superconducting Materials

S. Ono[1], H. Hayakawa[1], K. Tanabe[1], K. Naito[2], and Y. Imasato[2]

[1] National Chemical Laboratory for Industry, 1-1, Highshi, Tsukuba, 305 Japan
[2] FACOM-HITAC LIMITED, 6-2, Sanbancho, Chiyoda-ku, Tokyo, 102 Japan

INTRODUCTION

Although there are several molecular structure graphics systems which have been developed to display organic molecules or biomolecules, they are not suitable for representing inorganic crystal structures.

Inorganic structures can not be described on the concept of molecules. Bonding character of inorganic crystals is essentially different from that of organic molecules. Ionic or metallic character has no directivity and their structures can basically be represented by packing of spherical atoms in space. The number of bonds is rather large and they are dispersed uniformly in space. Whether a given pair of atoms is bonded or not is not critically important. Instead the coordination state of a given atom is important factor for charaterizing the structure.
New techniques are required in order to specify characteristics of inorganic structures, such as coordination state, chain structure, 2-dimensional network, 3-dimensional network, cluster structure, and layer stacking structure.

A new 3-dimensional crystal structure graphics system, CGS, has been developed in order to make crystal structures including inorganic and organic crystals understood more easily by visualization.

IMPLEMENTATION

The CGS program was initially developed on a host computer, FACOM M780 with a DAIKIN 3-dimensional graphics terminal, COMTEC DS301B, and was written in FORTRAN77 using COMTEC graphic library. Now we have converted it to FACOM S-4 series(UNIX workstation, full-compatible with Sun4 series) and rewritten in C-language using Sun-PHIGS.

The system's specifications for CGS program are below.

- Hardware S-4/260CXP, S-4/330GXP, S-4/370GXP with 24MB main memory, 30MB hard disk.
- Software OS;SunOS 4.x.x, Sun-PHIGS(for S-4/260CXP), PHIGS+(for S-4/330GXP, S-4/370GXP).

DESCRIPTION OF CGS

1. Input data

The input data should be as simple as possible. Although symmetry operations can be derived from the space-group symbols[1], CGS has its own database for symmetry operations which can be operated by the input space-group symbol. The input data file contains

(1) Title,

(2) Space-group symbol,

(3) Unit cell dimensions(a,b,c, α, β, γ),

(4) Number of atoms(N), number of atom species(M), number of bonding parameters(L), etc,

(5) Atom symbol, discriminater, positional parameters, atomic radius, color number, site occupancy factor, etc.(N lines),

(6) Bonding parameters which specify pairs of atoms between which bonds should be created, minimum and maximum interatomic distances limiting the region of bond formation , bond thickness , color (L lines), and

(7) Others.

2. Display models for crystal structures

CGS has three kinds of models for representing crystal structures.

(1) Skeleton model in which only bonds are displayed in color.

(2) Ball & stick model in which spherical atoms in specified size and color and bonds in specified thickness and color are represented.

(3) Skeleton and ball model

In the case of organic molecular crystals, the bonding parameters can be omitted. In that case, CGS has its own database of atomic radii and bonds are automatically created under the condition that the interatomic distance is shorter than the sum of each atomic radius.

Modelling is carried out representing one unit cell. However, multiple cells representation is possible by selection.

Viewpoint adjustment is also available.

3. Functions

Operations can be made with a mouse(S-4's pointing device).

Rotation, translation, extension and reduction can be applied to all models, but real-time movement is only possible in skeleton model.

Undesired atoms or bonds can be deleted by picking with a mouse. Information on interatomic distances and bonding angles is also shown on display by picking related atoms.

A function of 'take out' command is to take out a part of the crystal structure displayed in order to look at details of a part of the structure.

Two methods can be selected by the user.

(1) Rectangular region, or

(2) Spherical region by pointing the central atom and giving a radius.

The former is useful to examine the layer structure, and the latter is especially important to examine the coordination state of a given atom.

Another important function which is under development is building of an imaginary structure by assembling several parts taken out from original structure using 'take out' command. In the case of high Tc oxide superconducting materials, the number of basic layer structures is limited. A number of modified structures is made in which the layer stacking sequences are different. Using the basic structure as parts, many possible new structures can be created in the 3-dimensional modelling.

CONNECTION TO OTHER PROGRAMS

CGS has been built aiming the interfacial tool for the total materials design system. So the interfaces with other useful databases or analytical programs are essential. CGS can use the structural data of the inorganic crystal structure database ICSD which has been developed in Bonn University[2]. CGS is directly connected with the x-ray and neutron powder diffraction pattern fitting program RIETAN (Rietveld analysis program) developed by Izumi, National Institute for Research in Inorganic Materials.

REFERENCES

1. Burzlaff H, Hountas A (1982) J.Appl.Cryst., 15 :464-466

2. Bergerhoff G, Brown I.D, Hundt R, Sievers J (1983) J.Chem.Inf. and Comput.Sci., 23:66-69

3. Izumi F (1985) J.Crystallogr.Soc.Jpn., 27:23-31

2.2 Substitution and Doping

90K Superconductivity in Ca-Substituted $YBa_2Cu_4O_8$

T. MIYATAKE, K. YAMAGUCHI, T. TAKATA, S. GOTOH, N. KOSHIZUKA, and S. TANAKA

Superconductivity Research Laboratory, International Superconductivity Technology Center, 10-13, Shinonome 1-chome, Koto-ku, Tokyo, 135 Japan

ABSTRACT

Superconducting properties of Ca-substituted $YBa_2Cu_4O_8$ are investigated. The transition temperature Tc increases with Ca content. A Tc óf ~90K is achieved for the composition $Y_{0.9}Ca_{0.1}Ba_2Cu_4O_8$. As well as $YBa_2Cu_4O_8$, its Ca-doped compounds shows an excellent thermal stability in oxygen content up to a high temperature (~800°C). The intragranullar critical current densities of $YBa_2Cu_4O_8$ compounds are relatively low, which may be due to the absence of twinning planes.

KEY WORDS : Superconductivity, $YBa_2Cu_4O_8$, Ca-substitution, Transition temperature, Critical current density

1.INTRODUCTION

The superconducting phase $YBa_2Cu_4O_8$ (1-2-4) containing double Cu-O chains, was originaly discovered as lattice defects in the 90K superconductor $YBa_2Cu_3O_7$ (1-2-3) [1], and then produced as an ordered-defect structure in the 1-2-3 thin film [2,3]. Some of the properties of the 1-2-4 phase have been studied in the thin film form. Recently, bulk synthesis of the 1-2-4 phase has been successfully performend by high oxygen pressure annealing method [4,5] and ambient oxygen reaction method using alkali carbonates [6]. Karpinski et al. [4] reported that the 1-2-4 compound had an excellent thermal stability of oxygen content up to a high temperature of ~850°C. This feature may be important for practical applications. However, the lower Tc of 80K in $YBa_2Cu_4O_8$ cannot lead to uses at liquid nitrogen temperature (77K).

In this work, we describe the increase to 90K in critical temperature with Ca substitution for Y in $YBa_2Cu_4O_8$. Furthermore, in order to study a potential of $(Y,Ca)Ba_2Cu_4O_8$ for applications, we evaluate the critical current density through magnetization measurements.

2.EXPERIMENTAL RESULTS

Samples with nominal compositions of $Y_{1-x}Ca_xBa_2Cu_4O_8$ (x=0.0-0.1) were prepared by a solid state reaction method using a high oxygen pressure tecnique. Starting materials of Y2O3, Ba(NO3)2, CuO and CaCO3 with a purity of 99.9% were mixed, and then fired at 900°C in flowing oxygen for 12h. After grinding, the resulting powders were compacted into a rectangular shape at 100MPa, and lightly sintered at 800°C in an oxygen atmosphere. The samples were HIP(Hot Isostatic Pressing)-treated in a gas environment of argon with 20% oxygen at 100MPa twice. First, the HIP treatment at 950°C for 6h was conducted. At this stage , the x-ray diffraction patterns for the samples indicated that the 1-2-4 was the major phase but the broad diffraction peaks suggested a slightly disordered structure. After the second HIP treatment at 1050°C for 3h, high quality polycrystalline 1-2-4 compounds were obtained with no second phase, except for the sample with x=0.1 including a very small amount of CuO as the impurity phase. The lattice parameters of the samples were calculated by a least square fit. The length of the c-axis enlarged with increasing Ca content, but no change of the a- and b-axes were detectable.

Fig.1 shows the temperature dependence of resistivity for $Y_{1-x}Ca_xBa_2Cu_4O_8$. Electrical resistivity measurements were performed by a conventional four-probe technique. As Ca substitution increased, the magnitude of resistivity decreased monotonically. This result indicates that the Ca2+ substitution for Y3+ introduces holes into the 1-2-4 phase. All the samples exhibited sharp superconducting transitions (ΔTc=3~5K). The transition temperature increased with increasing Ca content, x. The sample with x=0.1 revealed the onset temperature of superconductivity at 90.9K and zero-resistance at 87.4K. Values of Tc determined by Meissner signals are in good agreement with the onset temperatures derived from the resistance drop.

Fig.2 shows that the thermogravimetric curves for samples with x=0 and 0.1 are not significantly different, except for a slight reduction of a decomposed temperature coming from Ca-doping. Neither samples decomposes below 800°C. This result shows that the Ca-substituted 1-2-4 , as well as 1-2-4 without Ca, has a high thermal stability in oxygen content up to a high temperature (~800°C). In addition, unlike 1-2-3, the materials are expected to have no orthorhombic-tetragonal phase transformation at an elevated temperature and no twin structures.

We evaluate the critical current density through measurements of the magnetic hysteresis curve. Magnetic hysteresis curves at 5K and the maximum external field of 50kOe for Ca-doped (x=0.1) and undoped samples are shown in Fig.3. Both curves are found to have no significant difference at 5K. A difference ΔM(H) between the field increasing and decreasing magnetization at an external magnetic field H is introduced to estimate the critical current density through the Bean critical-state model [7]. As field increased, the ΔM(H) for the two samples diminished rapidly. Fig.4 shows the magnetic hystreresis curves for the Ca-doped and undoped samples at 77K up to the maximum external field of 5kOe. The ΔM(H) for the Ca-doped sample also decreased rapidly with increasing field. On the other hand, because the temperature of 77K is very close to the Tc (82.5K) of the undoped 1-2-4 sample, the ΔM(H) was not observed.

Based on the critical-state model, the ΔM(H) is expressed as the product of the critical current density and the dimension of the sample for the homogeneous materials. For the inhomogeneous materials such as sintered samples with weak-link networks at grain boundaries, applicability of the critical-state model is not established [8]. However, it is known that, if the sample is ground to powder smaller than the grain size, the ΔM(H) increases with incerasing the powder size [9]. In this case, the critical-state model allows one to determine the critical current density Jc as an intragranular critical current density. Assuming that each particle for powder samples has a circular cross section, we can estimate the Jc by using the following relation: $\Delta M(H)=JcD/30$, where D is the powder diameter. We evaluated the average diameter of the powder samples using a laser diffraction particle size analyzer. The values of the average powder diameter for Ca-doped and undoped samples were 9.04μm and 18.9μm, respectively. Then, the critical current density was calculated employing the above formula. The field dependence of critical current densities at 5K and 77K for the Ca-doped and undoped 1-2-4 compounds are shown in Fig.5. At both 5K and 77K, the Jc of the Ca-doped sample is higher than that of the undoped one. This difference may come from the magnitude of the thermodynamic critical field for each sample, which contribute to the elementary pinning force. Char et al. reported that the Jc of the 1-2-4 thin film yielded about 4x106 A/cm2 at 4.2K and 3 kOe [3]. This value is higher by one order and by factor 5 than those of the undoped and Ca doped samples, respectively.

3. DISCUSSION AND CONCLUSION

Ni et al.[10] measured closed intragrain critical current densities in the sintered 1-2-3 material by using an A.C. inductive method. We will compare the intragranular Jc in the Ca-doped 1-2-4 compound with the intragrain Jc of the sintered 1-2-3 material. Unlike the 1-2-3, the intragranular Jc at 77K of the 1-2-4 with Ca is severely degraded by external magnetic field. The magnitude of the Jc in the 1-2-4 with Ca at 77K is 1.5×10^3 A/cm^2 at 2kOe, which is low by approximately one order compared with the reported values of the 1-2-3 compounds [10]. It is noted that the Jc of the Ca-doped 1-2-4 becomes much smaller than that of the 1-2-3 at higher magnetic fields, although both the Jc of the 1-2-4 and the 1-2-3 materials are almost the same at very low fields near Hc1(T).

The 1-2-4 and the 1-2-3 compounds have a similar crystal structure, except for existance of the additional Cu-O chain in the 1-2-4 and of the twin structures in the 1-2-3. The 1-2-4 materials has no orthorhombic-tetragonal transformation and no twin structures [4]. The addtional Cu-O chain gives strong anisotropy to the crystal structure, changing the cell constant of the c-axis from ~11.6Å in the 1-2-3 to ~27.2Å in the 1-2-4. The anisotropy of the crystal structure might be available as intrinsic pinning centers. On the other hand, Matsushita et al. [11] have predicted that the intragranular critical current densities more than 10^4 A/cm^2 at 77K and 10kOe can be achieved for the 1-2-3 system having pinning centers such as twinning planes. We believe that the additional Cu-O chain gives no serious influence on Jc in the 1-2-4 compounds. As a result, a large degradation of the Jc under magnetic fields may be due to the absence of twin structures in the 1-2-4 compounds.

In conclusion, we demonstrated that the Ca-substitution for Y gave rise to an enhancement of Tc in the $YBa_2Cu_4O_8$ compounds. A 90K superconductor was successfully obtained by Ca substitution by x=0.1 in the $Y_{1-x}Ca_xBa_2Cu_4O_8$. the intragranular critical current densities of $YBa_2Cu_4O_8$ with and without Ca were estimated through the Bean critical-state model. The critical current densities for $YBa_2Cu_4O_8$ with and without Ca are relatively low and sensitive to external magnetic field, which may be due to the absence of twin structures as flux pinning centers.

REFERENCES

1. H.W.Zandbergen, R.Gronski, K.Wang and G.Thomas, Nature 331, 596-599 (1988).
2. A.F.Marshall, R.W.Barton, K.Char, A.Kapitulnik, B.Oh, R.H.Hammond and S.S.Laderman, Phys.Rev. B37, 9353 (1988).
3. K.Char, Mark Lee, R.W.Barton, A.F.Marshall, I.Bozovic, R.H.Hammond, M.R.Beasley, T.H.Geballe and A.Kapitulnik, Phys.Rev. B38, 834 (1988).
4. J.Karpinski, E.Kaldis, E.Jilek, S.Rusiecki and B.Bucher, Nature 336, 660 (1988).
5. D.E.Morris, J.H.Nickel, J.Y.T.Wei, N.G.Asmer, J.S.Scott, U.M.Scheven, C.T.Hultgren, A.G.Markelz, J.E.Post, P.J.Heaney, D.R.Veblen and R.M.Hazen, Phys.Rev. B39, 7347 (1989).
6. R.J.Cava, J.J.Karajewski, W.F.Peck Jr, B.Batlogg, L.W.Rupp Jr, R.M.Fleming, A.C.W.P.James and P.Marsh, Nature 338, 328 (1989).
7. C.P.Bean, Phys.Rev.Lett. 8, 250 (1962).
8. K.Funaki, M.Iwakuma, Y.Sudo, B.Ni, T.Kisu, T.Matsushita, M.Takeo and K.Yamafuji, Jpn.J.Appl.Phys. 26, L1445 (1987).
9. E.Shimizu and D.Itoh, in *Advances in Superconductivity*, edited by K.Kitazawa and T.Ishiguro, (Springer-Verlag, Tokyo, 1989), p451.
10. B.Ni, T.Munakata, T.Matsushita, M.Iwakuma, K.Funaki, M.Takeo and K.Yamafuji, Jpn.J.Appl.Phys. 28, 1658 (1987).
11. T.Matsushita and B.Ni, IEEE Trans.Magn. MAG-25, 2285 (1989).

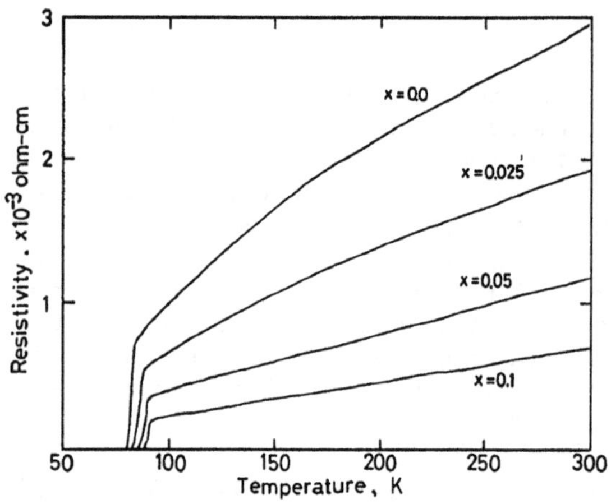

Fig.1. Electrical resistivity of $Y_{1-x}Ca_xBa_2Cu_4O_8$ as a function of temperature.

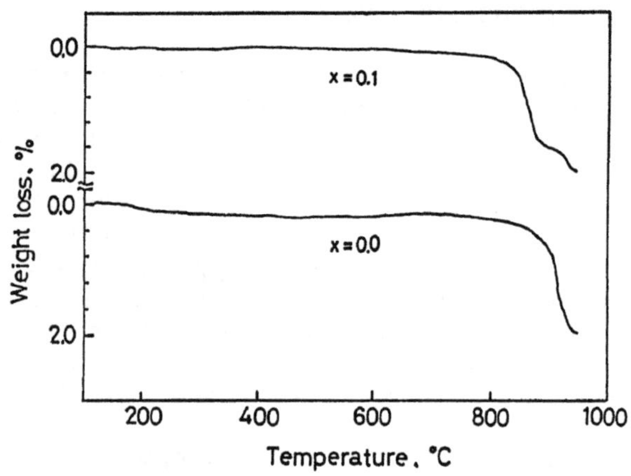

Fig.2. Thermogravimetric curves of $Y_{1-x}Ca_xBa_2Cu_4O_8$. Temperature was increased in air at a rate of 10°C/min.

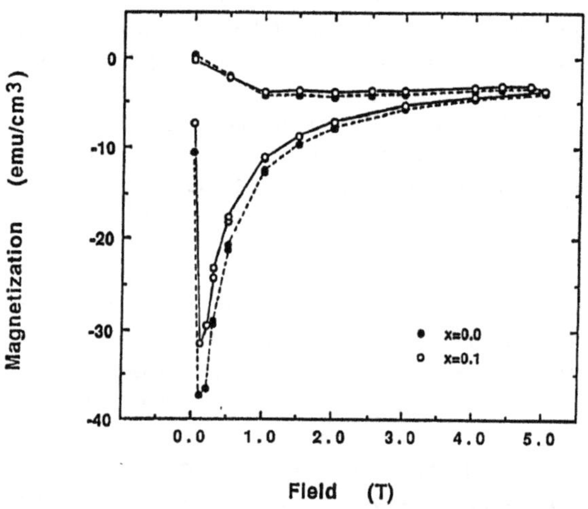

Fig.3. Magnetic hysteresis curves at 5K up to the maximum external field of 50kOe for $Y_{1-x}Ca_xBa_2Cu_4O_8$.

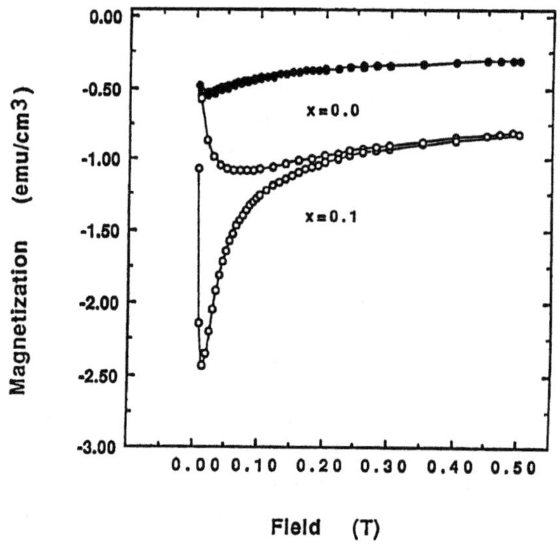

Fig.4. Magnetic hysteresis curves at 77K up to the maximum external field of 5kOe for $Y_{1-x}Ca_xBa_2Cu_4O_8$. Note that the magnetic hysteresis for x=0.0 sample cannot be observed.

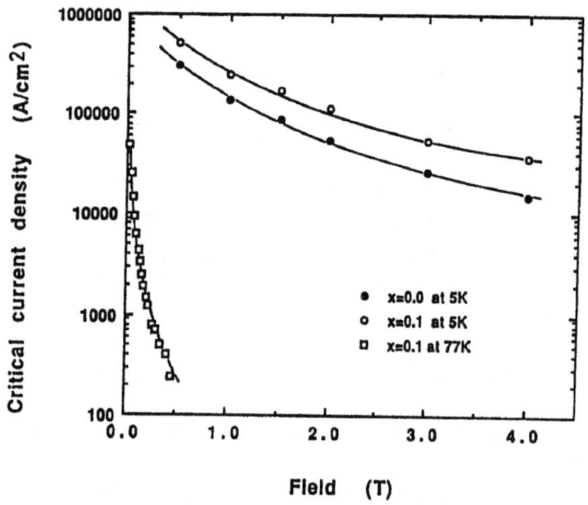

Fig.5. Field dependence of the critical current densities at 5K and 77K for $Y_{1-x}Ca_xBa_2Cu_4O_8$.

Superconductivity in $Bi_2Sr_2Ca_{1-x}Ce_xCu_2O_y$

TAE-SU HAN, AKIHITO SAWA, HIROMOTO UWE, and TUNETARO SAKUDO
Institute of Applied Physics, University of Tsukuba, Tsukuba, Ibaraki, 305 Japan

ABSTRACT

Effect of Ce-ions substitution for Ca ions in the superconducting $Bi_2Sr_2CaCu_2O_y$ has been investigated. Up to the Ce concentration x=0.5, the powder x-ray diffraction exhibits a pattern similar to the non-substituted one. The resistivity measurement shows that Tc slightly increases first with the substitution and then tends to decrease above x=0.05. Further, the system undergoes the metal-insulator transition around x=0.32. The Hall coefficients stay positive and increase with the substitution. These behaviors resemble to those observed for $Bi_2Sr_2Ca_{1-x}$(Y or Tm)$_xCu_2O_y$, provided that $Bi_2Sr_2Ca_{1-x}Ce_xCu_2O_y$ should correspond to $Bi_2Sr_2Ca_{1-2x}$(Y or Tm)$_{2x}Cu_2O_y$ on account of the difference of the formal valences of Ce^{4+} and Y^{3+} or Tm^{3+}.

KEY WORDS: $Bi_2Sr_2CaCu_2O_y$, Hall effect, Ce substitution, Mott-Hubbard insulator

INTRODUCTION

Following the discovery of the Bi-Sr-Ca-Cu-O system by Maeda et al.[1], effect of carrier-number modification on the superconducting properties has been reported by several authors for systems of $Bi_2Sr_2Ca_{1-x}Y_xCu_2O_y$ [2-5] and $Bi_2Sr_2Ca_{1-x}Tm_xCu_2O_y$.[6] The superconducting transition temperature Tc slightly increases at first and then decreases with the Y or Tm substitutions. The increment of Tc with the decreasing hole-number was also found for reducing of $Bi_2Sr_2CaCu_2O_y$[7,8] and the proton doping.[9] On further increasing of the substitution, the system undergoes the metal-insulator transition. In the insulating samples, the μSR experiment[10] or the susceptibility measurement[11] exhibited the magnetic-ordering temperature Tn which increases with the substitution. This behavior that the magnetic ordered phase is adjacent to the superconducting one is common to the systems of $La_{2-x}Sr_xCuO_4$ and $YBa_2Cu_3O_{6+d}$, which gives some hint for the mechanism of the high Tc superconductivity.
In this paper, the substitution effect by Ce^{4+} for Ca^{2+} in $Bi_2Sr_2Ca_{1-x}Ce_xCu_2O_y$ has been studied. Compositional dependence on properties of the system was examined by the measurements of powder x-ray diffraction, electrical resistivity, and Hall-effect. Similar behaviors to those observed for $Bi_2Sr_2Ca_{1-x}Y_xCu_2O_y$ have been found, provided that $Bi_2Sr_2Ca_{1-x}Ce_xCu_2O_y$ should correspond to $Bi_2Sr_2Ca_{1-2x}Y_{2x}Cu_2O_y$ on account of the difference of the formal valences of Ce^{4+} and Y^{3+}.

EXPERIMENTAL

The $Bi_2Sr_2Ca_{1-x}Ce_xCu_2O_y$ compounds were prepared by the solid-solution reaction method. Appropriate amounts of Bi_2O_3, $SrCO_3$, $CaCO_3$, CeO_2 and CuO powders were thoroughly mixed and fired at 830° C for 24 hours in air. The powder was again ground and fired at 880° C for 72 hours in air. The resultant powder was reground and pressed into pellet form, and was sintered at 880° C for 48 hours in air.
Structural properties were studied by powder x-ray diffraction method. The electrical resistivity was measured by the dc four-probe method for a bar-shaped sample cut out from the sintered pellets. Electrodes were prepared by silver paste. Hall effects were

measured under a field of 4 Tesla which was fixed in the persistent mode of a superconducting magnet. The sample thickness was prepared as thin as 0.3-0.5 mm. The samples were rotated by 180° to subtract the contribution from the resistance part and the temperature was controlled within 0.2 K. Here, electrodes were prepared by silver paste after the vacuum evaporation of Au.

RESULTS AND DISCUSSION

Figure 1 shows the x-ray powder diffraction patterns of the $Bi_2Sr_2Ca_{1-x}Ce_xCu_2O_y$ system in the range of 2θ from 26° to 34°, which indicates the same structure as that of $Bi_2Sr_2CaCu_2O_y$ for the Ce-substituted one. For the sample of x=0.5, a small amount of CeO_2 was found. Appreciable splitting of the (200) diffraction line was not found all through the Ce substitutions. Lines derived from the modulation structure of the Bi-O layer[12] disappeared in the Ce-substituted samples. Figure 2 shows the lattice-constant variation with Ce concentration x. With increasing x, the length of c-axis decreases and that of a-axis increase.

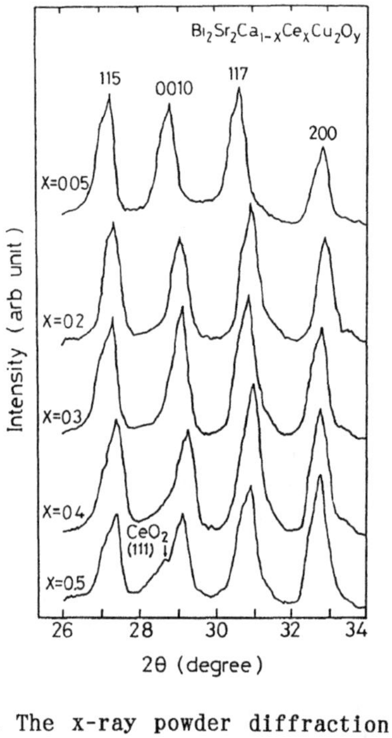

Fig.1 The x-ray powder diffraction patterns of the $Bi_2Sr_2Ca_{1-x}Ce_xCu_2O_y$ system. For the sample of x=0.5, a trace of CeO_2 is found.

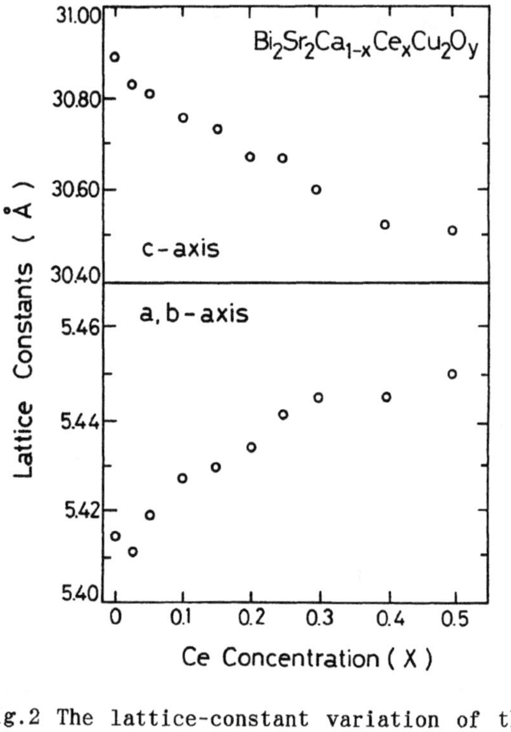

Fig.2 The lattice-constant variation of the $Bi_2Sr_2Ca_{1-x}Ce_xCu_2O_y$ system.

Temperature dependence of the resistivity is shown in Fig.3 as a function of the Ce concentration. With increasing x, the resistive superconducting temperature Tc increases gradually and then takes maximum for the sample of x=0.05. We find that, for the superconducting samples, the resistivity is smaller, the Tc is higher. For the samples with x=0.25-0.3, the resistivity versus temperature curve is no longer metallic but semiconducting. The normal resistivity increases almost by 100 between x=0.05 and 0.4. This result indicates that the metal-semiconductor transition occurs.
Figure 4 shows temperature dependence of the Hall coefficients R_H as a function of the Ce concentration. The sign of the Hall coefficients was positive. In the superconducting range of $x \leqq 0.25$, the Hall coefficient decreases slightly with increasing temperature as in the case of $Bi_2Sr_2Ca_{1-x}Y_xCu_2O_y$ and $YBa_2Cu_3O_7$. With increasing Ce concentration, the Hall coefficient increases.

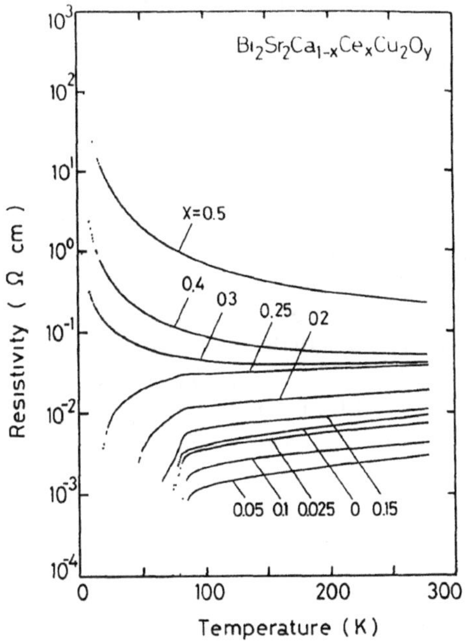

Fig.3 Temperature dependence of the resistivity of the $Bi_2Sr_2Ca_{1-x}Ce_xCu_2O_y$ system.

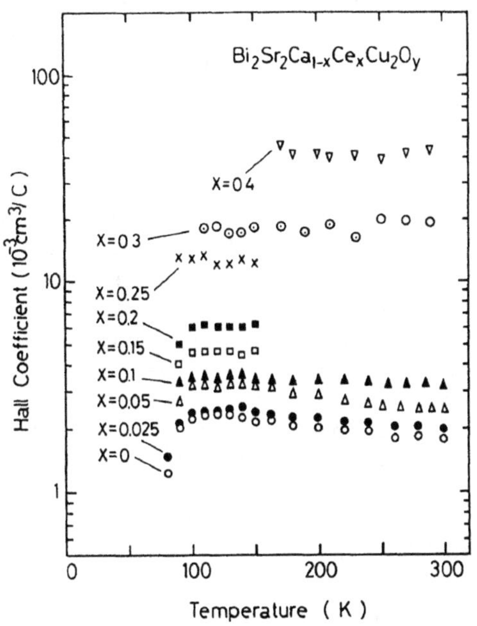

Fig.4 Temperature dependence of the Hall coefficients R_H of the $Bi_2Sr_2Ca_{1-x}$-$Ce_xCu_2O_y$ system.

In Fig.5, Tc and the carrier concentration p_H defined by $1/eR_H$ at 120K are plotted against the Ce concentration. Here, Tc is determined as the midpoint of the resistive transition. The carrier concentration p_H decreases linearly with increasing Ce concentration, and the extrapolation indicates that the carriers are depleted at x_{Ce}=0.31. The steep decrease of p_H indicates the correlation-induced Mott-Hubbard gap in the Cu-O band, as discussed in Ref.6. For the Tm-substituted system, the Tm-concentration per a Cu ion for the carrier-depletion is reported as 0.343, which is close to the x_{Ce} for two Cu ions. Thus, the formal valence of Ce ion in the present system is found to be 4, and the possibility of the formal valence of 3 is excluded. The substituted Ce^{4+} compensates two holes, while Tm^{3+} or Y^{3+} ion does one hole.

For the samples of x>0.35, however, p_H deviates from the linearity. A simple picture for this behavior is that the Ce substitution for x<0.31 compensates the hole carrier on CuO^- but for x>0.31 the excess electron will reside on the Bi-O layer because of Madelung potential.[13] Introduction of excess oxygen ions by the Ce-substitution might not be the case, since the increasing Hall coefficient with the substitution indicates the decrease of the hole concentration.

We find that, with the Ce substitution, Tc increases at first and decreases with further substitution. The normal state conductivity exhibits a concurrent Ce-concentration dependence. However, the carrier concentration deduced from the Hall effect exhibits a monotonous decrease with the Ce substitution.

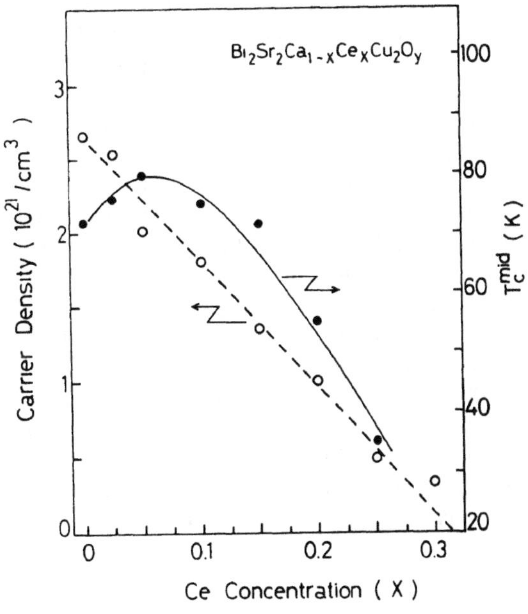

Fig.5 Tc and the carrier concentration p_H defined by $1/eR_H$ at 120K are plotted against the Ce concentration.

Since both the Tc and conductivity reflect the density of states(DOS) at Fermi level, it is concluded that, by the decrease of carrier number in this system, the Fermi level stays once at the optimum high DOS and then moves to the position of the lower DOS.

CONCLUSION

We have studied the solid-solution system of $Bi_2Sr_2Ca_{1-x}Ce_xCu_2O_y$ in which a Ce^{4+} ion substituting for a Ca^{2+} ion compensates two holes. The maximum T_c was obtained around x=0.05. With increasing x, we have observed the Mott-Hubbard-type metal-insulator transition at x=0.31. The hole concentration deduced from the Hall effect decreases linearly with increasing x. Both the Tc and the normal state conductivity exhibit concurrently to increase at first and then decrease with increasing x. These behaviors resemble to those observed for $Bi_2Sr_2Ca_{1-x}$(Y or Tm)$_xCu_2O_y$, provided that $Bi_2Sr_2Ca_{1-x}$-$Ce_xCu_2O_y$ should correspond to $Bi_2Sr_2Ca_{1-2x}$(Y or Tm)$_{2x}Cu_2O_y$ on account of the difference of the formal valences of Ce^{4+} and Y^{3+} or Tm^{3+}.

ACKNOWLEDGMENT

The Hall effect measurement was done at the Cryogenic Center of University of Tsukuba. The authors thank Y.Motoi and T.Iwamatsu for the experimental help.

REFERENCES

1. Maeda H et al (1988) A New High-Tc Oxides Superconductor without a Rare Earth Element. Jpn.J.Appl.Phys. 27:L209-L210.
2. Tarascon JM et al (1988) Crystal Substructure and Physical Properties of Superconducting Phase $Bi_4(Sr,Ca)_6Cu_4O_{16+x}$. Phys.Rev.B37: 9382-9389.
3. Tamegai T et al (1988) Characterization of Non-Superconducting Cuprate $Bi_2Sr_2YCu_2O_{8.5}$. Jpn.J.Appl.Phys.27: L1074-L1076.
4. Fukusima N. Niu H. Ando K. (1988) Electrical and Magnetic Properties in $Bi_2Sr_2Ca_{1-x}Y_xCu_2O_{8+d}$. Jpn.J.Appl.Phys.27:L1432-L1434.
5. Yoshizaki R et al (1988) Superconducting and Magnetic Properties of $Bi_2Sr_2Ca_{1-x}Y_xCu_2O_y$ (0<x<1). Physica C 152: 408-412.
6. Clayhold J et al. (1989) Approaching the Mott-Hubbard Insulator in the 85-K Superconductor $Bi_2(Sr,Ca)_3Cu_2O_{8+d}$ by doping with Tm. Phys. Rev. B39: 7320-7323.
7. Ishida (1988) Anomalous Tc Alteration of Quenched $Bi_2Sr_2Ca_{1-x}Y_xCu_2O_x$. Jpn.J.Appl.Phys. 27: L2327-L2329.
8. Motoi Y. Ikeda Y. Uwe H. Sakudo T.(1989) Enhancement of Tc under Reducing Conditions on $Bi_2Sr_2CaCu_2Ox$. Proc.of M^2S-HTSC, Stanford (in press).
9. Takabatake T et al (1989) Hydrogen Intercalation in Some Superconducting Copper Oxides. Proc.of M^2S-HTSC, Stanford (in press).
10. Nishida N et al (1989) Observation of Antiferromagnetic Ordering in $Bi_2Sr_2YCu_2O_y$ above Room Temperature by the μSR Method. Physica C 156: 625-628.
11. Fujita T. and Tomita T. (1989) Substitution Effect in $Bi_2Sr_2CaCu_2O_{8+d}$. Proc. of M^2S-HTSC, Stanford (in press).
12. Onoda M et al (1988) Assignment of the the Powder x-Ray Diffraction Pattern of superconducting $Bi_2(Sr,Ca)_{3-x}Cu_2O_y$. Jpn.J.Appl.Phys. 27, L833-L836.
13. Kondo J. Asai Y. Nagai S. (1988) The Madelung Energy in Copper-Oxide Based Ceramics. J.Phys.Soc.Jpn. 57: 4334-4342.

Preparative and Structural Studies on Various Substitutions in the Bi-Sr-Ca-Cu-O System

ZHU WENJIE, CHEN FENGXIANG, LI NENG, LI HONGYU, LIN BINGXIONG, and TANG YOUQI

Institute of Physical Chemistry, Peking University, Beijing 100871, China

ABSTRACT

The samples with the norminal compositions of $Bi_{1.8}M_{0.2}Sr_2Ca_2Cu_3O_y$(M=Pb, In, Sb, Nb) and $Bi_{1.7}Pb_{0.2}M_{0.1}Sr_2Ca_2Cu_3O_y$(M=In, Sb, Nb) have been prepared and investigated. Only the statistical substitution of lead for bismuth results in the formation of the 110K phase. In the crystal structures of two homogeneous phases $Bi_{1.8}Pb_{0.2}Sr_2Ca_2Cu_3O_{10}$ and $Bi_{1.7}Pb_{0.2}In_{0.1}Sr_2Ca_2Cu_3O_{10}$, adjacent bismuth layers with one oxygen layer in between display a perfect Aurivillius type structure, and the structural unit $(O_2Cu)Ca(CuO_2)Ca(CuO_2)$ shifts teragonality.

KEY WORDS: crystal structure, substitution, formation mechanism

INTRODUCTION

Chemical substitution in Bi-Sr-Ca-Cu-O system has been reported to improve superconductivit y and eliminate substructure distortion. [1,2] Since the nature of bismuth layers was not well understood, the samples with the norminal compositions of $Bi_{1.8}M_{0.2}Sr_2Ca_2Cu_3O_y$(M=Pb, In, Sb, Nb) and $Bi_{1.7}Pb_{0.2}M_{0.1}Sr_2Ca_2Cu_3O_y$ (M=In, Sb, Nb) (samples A, B, C, D, E, F, G respectively) were prepared, among which two pure phases $Bi_{1.8}Pb_{0.2}Sr_2Ca_2Cu_3O_{10}$ and $Bi_{1.7}Pb_{0.2}In_{0.1}Sr_2Ca_2Cu_3O_{10}$ were obtained to determine whether the bismuth-oxygen layer structure is $(BiO)_2$ or (BiO_2Bi) type and how it affects crystal structure, properties, and formation process.

SYNTHESIS AND CHARACTERIZATION

$Sr_2Ca_2Cu_3O_7$, prepared by firing the mixture of $CaCO_3$, SrO and CuO at 900℃ for 20 hr, was ground with Bi_2O_3 and MxOy (PbO, In_2O_3, Nb_2O_5 respectively) by the composition $Bi_{1.8}M_{0.2}Sr_2Ca_2Cu_3O_y$, pressed into pellets. Then the pellets were calcinated at 870℃ for 20 hr, ground again and fired at 880℃ during 30 hr, and finally quenched in air. The samples $Bi_{1.7}Pb_{0.2}M_{0.1}Sr_2Ca_2Cu_3O_y$(M=In, Sb, Nb) were prepared by the same procedure. Table 1 shows their phase compositions and Tc. Only the substitution of lead causes the formation of 2223 phase.

Table 1. Phase Compositions and Tc

sample	phase	Tc(K)
A	2223	108
B	2212+2201+impurity	72
C	2212+2201+impurity	-
D	2212+2201+impurity	61
E	2223	105
F	2223+2212+impurity	80
G	2223+2212+impurity	86

Figure 1 shows the X-ray diffraction patterns of the homogeneous samples A and E. Electron microprobe analysis carried out on a large number of grains of these two phases revealed metal cation stoichiometries of about $Bi_{1.8}Pb_{0.2}Sr_2Ca_2Cu_3$ and $Bi_{1.7}Pb_{0.2}In_{0.1}Sr_2Ca_2Cu_3$ repectively. The tempereature dependence of resistivity was shown in Fig. 2.

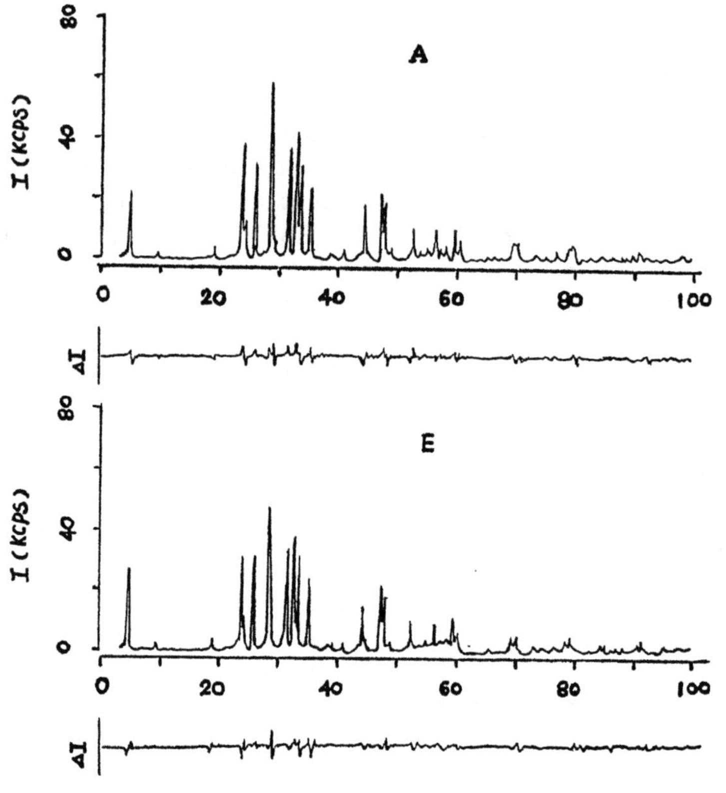

Fig. 1 X-ray powder diffraction patterns for A and E

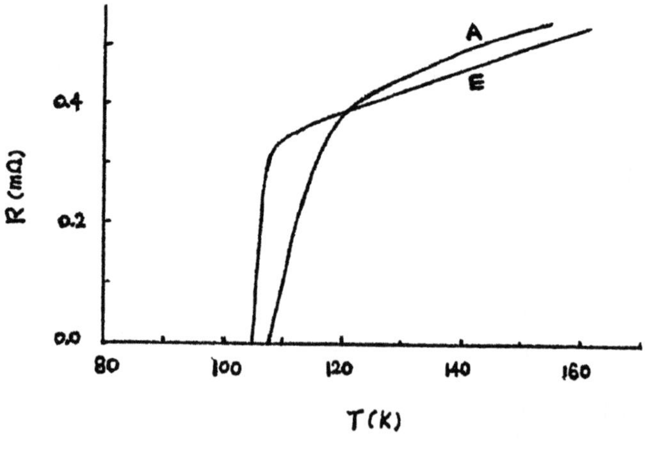

Fig. 2 Resistivity versus temperature

DETERMINATION OF CRYSTAL STRUCTURE

The intensity data were collected on a Rigaku D/max-ra X-ray diffractometer. Operating conditions are: Cu-Kα at 40 Kv and 200 mA; graphite monochromator; $2\theta = 4-100°$; step scan width 0.05°; counting time 10 s.

The diffraction diagram was indexed, which reveals that the unit cell is body centered tetragonal. The lattice parameters are:

	a (Å)	c (Å)
A:	3.813(1)	37.100(15)
E:	3.817(1)	37.040(12)

The unit cell was derived to be a nine decker perovskite structure from the 2212 structure by incorporating an extra double layer $Ca(CuO_2)$ on each Ca layer in 2212. Space group I4/mmm was assigned. The positions for all metal ions were obtained by electron density calculation, while those of oxygen atoms were got from difference Fourier maps and subsequently adjusted on the crystal-chemical basis. The structure parameters were further refined by Rietveld profile analysis. Final goodness of fit achieved is indicated by $R_{|F|} = 0.10$, $R_p = 0.19$ for A and $R_{|F|} = 0.09$, $R_p = 0.17$ for E.

Table 2 gives values of coordinates, thermal parameters B, and atomic occupancy.

Table 2. Structure Parameters for A and E

atom	x	y	z	B(Å)^2	p
X	0	0	0.2086(4)	3.5(5)	1.0
			0.2090(3)	3.8(4)	
Cu(1)	0	0	0	0.4(1)	1.0
				0.4(1)	
Cu(2)	0	0	0.0913(4)	0.5(2)	1.0
			0.0909(4)	0.4(1)	
Sr	1/2	1/2	0.1386(6)	0.8(4)	1.0
			0.1370(5)	0.8(3)	
Ca	1/2	1/2	0.0432(5)	0.7(4)	1.0
			0.0427(6)	0.6(2)	
0(1)	1/2	0	0	2.5(7)	1.0
				2.0(5)	
0(2)	1/2	0	0.0865(8)	2.5(7)	1.0
			0.0862(7)	2.0(5)	
0(3)	0	0	0.1510(10)	2.5(7)	1.0
			0.1504(9)	2.0(5)	
0(4)	1/2	0	1/4	2.5(7)	1.0
				2.0(5)	

X: $(Bi_{0.9}Pb_{0.10})$ for A; $(Bi_{0.85}Pb_{0.10}In_{0.05})$ for E

From 2212 to 2223 phase, c axis increases about 3. 2 Å by incorporating one $Ca(CuO_2)$ slab. Lead and indium statistically substitute bismuth, and thus they bave great thermal factors. The adjacent bismuth layers with one oxygen layer in between display a perfect Aurivillius type structure,[3] rather than the rocksalt type structure.[4] Their distance is 3. 07 Å for A phase and 3. 04 Å for E phase. The bismuth atom is surrounded by five oxygen atoms, forming a square pyramid.

The atomic valences for these two phases were calculated by using the formula $s=(r/r_0)^{-n}$ and the parameters[5] given below:

	Bi	Pb	In	Ca	Sr	Cu
r. (Å)	2. 094	2. 112	1. 902	1. 909	2. 143	1. 718
n	5. 0	5. 5	7. 0	5. 4	7. 0	6. 0

The results in Table 4 indicate that all atoms supporting the skeletons approximately have normal atomic valences, while the strontium atom shows a little more deviation.

Table 4. Atomic Valences for A and E

X	Cu(1)	Cu(2)	Sr	Ca
2. 7	2. 1	2. 3	1. 5	1. 9
2. 7	2. 1	2. 3	1. 6	2. 0

O(1)	O(2)	O(3)	O(4)
2. 0	1. 9	1. 8	1. 8
2. 0	1. 9	1. 8	1. 9

DISCUSSION

The crystal structures in Bi-Sr-Ca-Cu-O series display severe stacking faults along c direction.[6] The resistivity versus temperature curves are various: lower Tc and broad transtion width, which are closely related to calcinating procedure and atmosphere. This phenomenon results from the irregularity of bismuth layers and lower strontium valence, which sensitively change the plane stacking characteristic of (CuO_2) along a, b directions in the structural unit $\{(O_2Cu)Ca\ (CuO_2)\}$ (in 2212 case), and hence make it difficult to incorporate an extra $Ca(CuO_2)$ slab to form the high Tc phase 2223. Since lead and indium have similar ionic size to bismuth, and the lead ions could be accommodated in a square pyramid, which also occurs for bismuth in Aurivillius type materials, the statistical substitutions of lead and indium for bismuth lead bismuth-oxygen layers to alter from the rocksalt-Aurivillius mixing type[7] to an Aurivillius type structure. As a result, the bismuth-oxygen layers stack regularly, an extra Ca (CuO_2) slab incorporates into 2212 phase without distortion, and the structural unit $\{(O_2Cu)\ Ca\ (CuO_2)Ca(CuO_2)\}$ shifts to tetragonal symmetry, which might be the structural origin of the Tc raise of about 20K and narrow transition width.

REFERENCES

1. R. J. cava, B. Batlogg, R. B. Van Dover, R. M. Fleming and S. H. Glarum, Phys. Rev., B38(1988)893.
2. M. Takano, J. Takada, K. Oda, H. Kitaguchi, Y. Miura, Y. Ikeda, Y. Tamii and Majaki, Jpn. J. Appl. Phys., 27(1988)L1041.
3. B. Aurivillius, Arkiv Kemi, 1(1949)499.
4. J. M. Tarascon, Y. Le Page, P. Barboux, B. G. Bagley, L. H. Greene, W. R. Mckinnon, G. W. Hull, M. Giroud and D. M. Hwang, Phys. Rev., B37(1988)9382.
5. I. D. Brown, Acta Cryst. A32(1979) 24; ibid, B41(1985)244.
6. M. Hervieu, C. Michel, B. Domenges, Y. Laligant, A. Lebail, G. Ferey and B. Raveau, Mod. Phys. Lett. B, Vol. 2, No. 1(1988)491.
7. Tang Youqi, Lin Bingxiong et al, Mod. Phys. Lett. B, Vol. 2, No. 2(1988)551.

Effect of Elemental Substitution on the Superconductive Properties in $Bi_2Sr_2Ca_{n-1}Cu_nO_y$

SHIRO KAMBE[1], TERUYUKI MATSUOKA[2], MAKI KAWAI[3], TOMOJI KAWAI[4], and MAKOTO TAKAHASI[2]

[1] F.R.P., The Institute of Physical and Chemical Research, 2-1, Hirosawa, Wako, Saitama, 351-01 Japan
[2] International Christian University, 10-2, Osawa 3-chome, Mitaka, Tokyo, 181 Japan
[3] Research Laboratory of Engineering Materials, Tokyo Institute of Technology, 4259 Nagatsuta, Midori-ku, Yokohama, 227 Japan
[4] The Institute of Industrial and Scientific Research, Osaka University, Mihogaoka, Ibaraki, Osaka, 567 Japan

ABSTRACT

Effects of Pb substitution and post N_2 annealing has been studied in $Bi_2Sr_2Ca_{n-1}Cu_nO_y$. By post annealing in N_2, new structure compound, monoclinic $Bi_{2-2x}Pb_{2x}Sr_2Ca_{n-1}Cu_nO_y$ was found in $0.2 \leq x \leq 0.3$. Tc of monoclinic $Bi_{2-2x}Pb_{2x}Sr_2Ca_{n-1}Cu_nO_y$ was above 97.5K.

KEY WORDS: Pb substitution, post N_2 annealing, monoclinic Bi-Pb-Sr-Ca-Cu-O, hole concentration.

INTRODUCTION

It is well known that Tc's of cuprate superconductors are very sensitive to their hole concentration[1]. In the previous paper, we have shown that the optimum hole concentration per unit cell that gives the highest Tc value in n=3 phase is larger than that in n=2 phase[2]. Recently Fukushima et al.[3] have reported that in n=2 phase, Pb substitution followed by annealing in N_2, removes oxygen from the system. In order to clarify the effect of post N_2 annealing in Pb-substituted Bi system, we studied the structure and the Tc dependence on Pb content before and after the N_2 annealing in n=2 phase. It was found that Pb substitution or low oxygen pressure annealing promotes the structure change into orthorhombic from pseudotetragonal and that with Pb substituted ($x \geq 0.2$) system, post N_2 annealing procedure changes the orthorhombic into the monoclinic structure. Maximum Tc observed for the monoclinic n=2 phase was above 97.5K, which was almost 20K higher than that of orthorhombic one before N_2 annealing.

EXPERIMENTAL

Samples were prepared by dissolving $Bi(NO_3)_3 \cdot 5H_2O$, $Pb(NO_3)_2$, $Sr(NO_3)_2$ and $Ca(NO_3)_2 \cdot 4H_2O$ in water, followed by evaporation and calcination in air at 800℃ for 2 hours. This powder was then reground, pressed into pellets, sintered in 1/18-1/5 atmosphere of O_2 for 40-60 hours at 830℃ and cooled to room temperature. They were post-annealed in N_2 at 750℃ for 8 hours. Four kinds of samples with different composition were

synthesized, whose nominal ratios of Pb/Bi+Pb were (a)0, (b)0.05, (c)0.20 (d)0.30. The composition of Bi, Pb, Sr, Ca and Cu was directly determined by an induction coupled plasma atomic emission spectroscopy, ICP-AES (ICPS-50A, Shimadzu Co.).
Superconductive property was measured by the temperature dependence of resistance and ac magnetic susceptibility. Resistivity measurements were carried out by a standard dc four probe method using indium electrodes attached to the sample by a supersonic solder. Each sample was reground and pelletized prior to the ac susceptibility measurements.
The structure of the compounds were determined by X-ray powder diffraction patterns using CuKα (RAD-IIIB, Rigaku denki). The peaks were assigned according to those indexed by M.Onoda et al[9].

RESULTS AND DISCUSSION

As shown in (a) and (b) of Fig.1, the structure of $Bi_{1.6}Pb_{0.4}Sr_2CaCu_2O_y$ changed drastically with changing the sintering atmosphere. With decreasing oxygen pressure from 1/5 to 1/13, the peak of (2000,0200) split into two peaks. which means that pseudotetragonal changed into orthorhombic. By annealing sample (a) in N_2 at 750℃ for 20 hours, (1130,1150,1170) peaks broaden as shown in (c). Among these broaden peaks, existence of two peaks are clearly observed in (1170). This behavior cannot be explained by pseudotetragonal nor orthorhombic but by monoclinic structure. The peaks are indexed with monoclinic structure in a=5.428Å, b=5.358Å, c=30.77Å and γ=90.9° . The peaks agree well with calculated ones. As shown in Fig.2, monoclinic n=2 phase can be prepared only by post N_2 annealing in the Pb content of $0.2\le x\le 0.3$. The c-

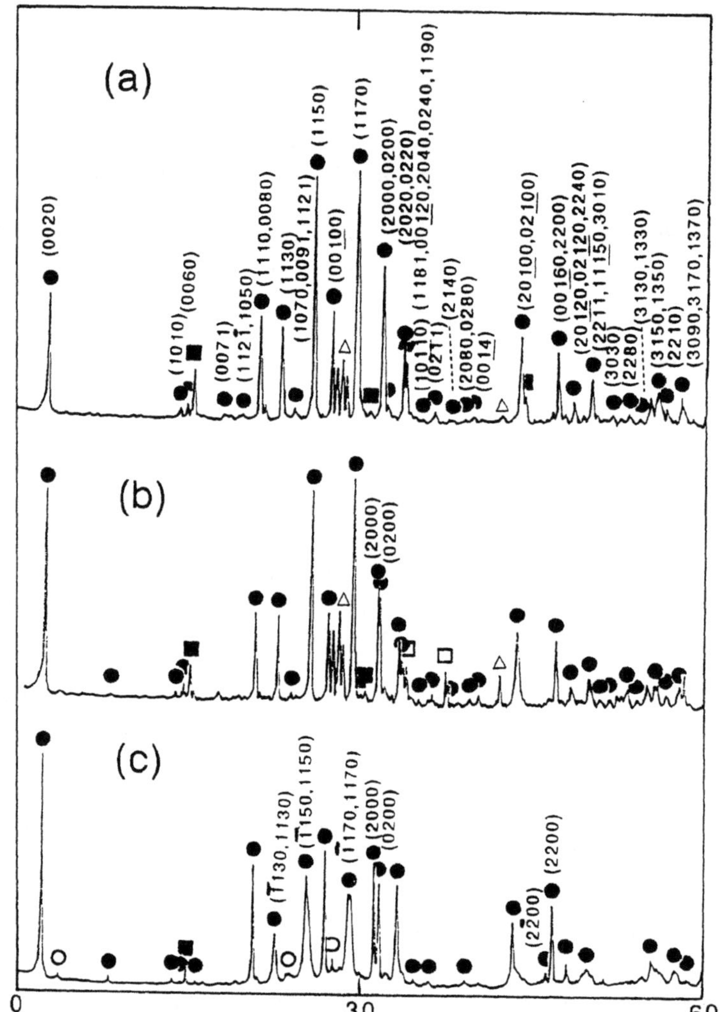

FIg.1 XRD patterns of $Bi_{1.6}Pb_{0.4}Sr_2CaCu_2O_y$

(a) annealed in 1/5 O_2
(b) annealed in 1/13 O_2
(c) annealed in 1/5 O_2 followed by N_2

●:n=2 phase ○:n=1 phase ■:Ca_2PbO_4
□:CuO △:unknown phase[10]

axis length of monoclinic phases were 30.76-30.77Å, which was shorter than that of orthorhombic or pseudotetragonal (30.79-30.85Å). It is already reported that the reduction of n=2 phase leads to the increase of c-axis[4], which is mainly considered to be due to the decrease of extra oxygen in Bi-O layer. The opposite behavior for monoclinic n=2 phase suggests the change of the structure in Bi-O layer. Another interesting feature is the increase of (0020) peak with decreasing oxygen content. More precise analysis is in progress. Tc of monoclinic n=2 phases is no less than 97.5K as shown in Fig.2(a). This temperature is extremely high in comparison with that before N_2 annealing (Fig.2(b)). This is partly caused by the decrease of hole concentration. Tc dependence on the hole concentration in n=2 phase has been studied by several groups[5,6,7,8]. The optimum hole concentration to give the highest Tc in this phase is known to be smaller than that prepared in air. By reducing the system, the value of Tc is expected to increase. Now we are not sure whether the great increase in Tc up to 97.5K can be explained only by the hole

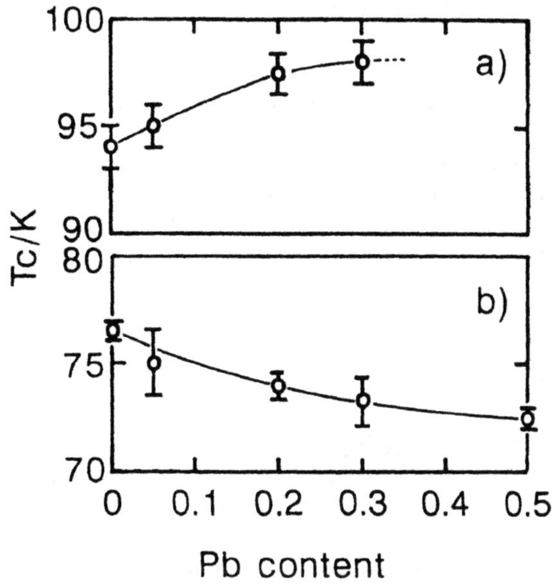

Fig.2 Tc dependence on Pb content

(a) after post N_2 annealing
(b) before post N_2 annealing

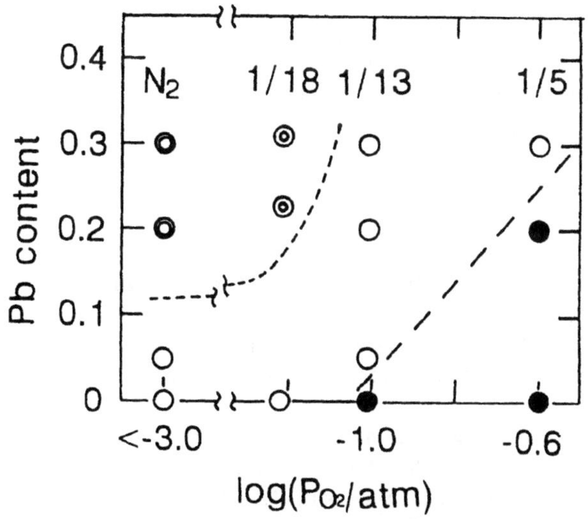

Fig.3 Phase diagram of $Bi_{2-2x}Pb_{2x}Sr_2CaCu_2O_y$

◎ : monoclinic
○ : orthorhombic
● : pseudotetragonal

reduction. Precise measurement on the hole concentration has to be carried out.

In order to elucidate the relation between the crystal structure and preparing condition, phase diagram was produced as shown in Fig.3. Crystal structure strongly depended on both Pb content and the atmosphere during annealing. As Pb content increased or partial pressure of O_2 decreased, pseudotetragonal structure changed into orthorhombic and finally into monoclinic. By annealing in O_2 pressure less than 1/18 atm with $0.2<x<0.3$ of Pb, monoclinic $Bi_{2-2x}Pb_{2x}Sr_2CaCu_2O_y$ can be prepared.

CONCLUSION

Monoclinic $Bi_{2-2x}Pb_{2x}Sr_2CaCu_2O_y$ with $0.2\leq X\leq 0.3$ was synthesized by post N_2 annealing or by low O_2 pressure annealing. In x=0.2, lattice parameters of a, b, c and γ were 5.428Å, 5.358Å, 30.77Å and 90.9°, respectively.

Tc of monoclinic n=2 phase was observed above 97.5K, which was more than 20K higher than that before N_2 annealing.

ACKNOWLEDGMENTS

The authors are grateful to Dr.Y.Takahashi for the ICP-AES measurement, to Dr.Y.Iimura for XRD measurement and to Dr.I.Higashi and Ms.R.Sekine for useful crystallographic discussion about the XRD data. A part of this work was carried out at the Frontier Research Program, the Institute of Physical and Chemical Research.

References

[1]T.Tamegai, A.Watanabe, K.Koga, I.Oguro and Y.Iye: Jpn.J.Appl.Phys., **27** (1988) L1074.

[2]S.Kambe, T.Matsuoka, M.Kawai and M.Takahasi: submitted to Physica C.

[3]N.Fukushima, H.Niu, S.Nakamura, S.Takano, M.Hayashi and K.Ando: PHysica C, **149** (1989) 777.

[4]H.Nagano, R.Liang, Y.Matsunaga, M.Sugiyama, M.Itoh and T.Nakamura: Jpn.J.Appl.Phys., **28** (1989) L364.

[5]T.Tamegai, K.Koga, K.Suzuki, M.Ichihara, F.Sakai and Y.Iye: Jpn.J.Appl.Phys., **28** (1989) L112.

[6]J.Clayhold, S.J.Hagen, N.P.Ong, J.M.Tarascon and P.Barhoux: PHys.Rev., **B39** (1989) 7320.

[7]D.E.Morris, C.T.Hultgren, A.M.Marleltz, J.Y.T.Wei, N.G.Asmar and J.H.Nickel: PHys.Rev.B., **39** (1989) 6612.

[8]A.Manthinam and J.B.Goodenough: Appl.Phys.Lett., **53** (1988) 420.

[9]M.Onoda, A.Yamamoto, E.Takayama-Muromachi and S.Takekawa: Jpn.J.Appl.Phys., **27** (1988) L833.

[10]H.Nagano, R.Liang, Y.Matsunaga, M.Sugiyama, M.Itoh and T.Nakamura: Jpn.J.Appl.Phys., **28** (1989) L364.

Superconductivity of High T_c $TlBa_2Ca_{1-x}Y_xCu_2O_7$ ($0 \leqq x \leqq 1.0$) System

SATORU NAKAJIMA[1], MASAE KIKUCHI[2], NORIO KOBAYASHI[2], HIDEO IWASAKI[2], DAISUKE SHINDO[2], YASUHIKO SYONO[2], and YOSHIO MUTO[2]

[1] Hachioji Research Center, CASIO Computer Co., Ltd., Hachioji, 192 Japan
[2] Institute for Materials Research, Tohoku University, Katahira, Sendai, 980 Japan

ABSTRACT

The effect of Y substitution for Ca in the Tl-Ba-Ca-Y-Cu-O system has been investigated for a wide substitution range. The unit cell dimensions and chemical compositions were determined by X-ray powder diffraction (XPD) analysis and analytical electron microscopy (EDX). Superconducting properties were studied by resistivity and DC susceptibility measurements. T_c of $TlBa_2Ca_{1-x}Y_xCu_2O_7$ increases with increasing x up to x=0.2-0.35, where it turns down and reaches non-superconductor at x=0.5.

KEY WORDS: Tl single layer system, Y substitution for Ca, Cu valence.

INTRODUCTION

The Tl-Ba-Ca-Cu-O system first discovered by Sheng and Hermann [1] has attracted strong attention of researchers because of highest superconducting transition temperature T_c of 120K. This system which has a layered structure shows both double and single Tl layer structures and the number of Cu layers can be varied for 1-4 [2,3] and 2-5 respectively [4-7]. Noteworthy is a marked discrepancy between the electronic states of the Tl double layer and Tl single layer compounds [8,9], although they have similar values of T_c.

The Tl single layer system shows more ionic character of Tl^{3+} ions as revealed by XPS mesurements [9], in contrast to the reduced electronic state in Tl ions in the double Tl layer system due to charge transfer Tl^{3-t}-$(Cu\text{-}O)^p$ [8-10]. The Tl single layer system can be represented as $TlBa_2Ca_{n-1}Cu_nO_{2n+3}$, where n is the number of consecutive Cu-O layers. T_c of single Tl layer compounds increases with increasing number of Cu layer up to n=4, where it turns down [7]. The average Cu valence estimated from the ideal composition varies as 2+1/n with n. The $TlBa_2CaCu_2O_7$ phase shows a superconducting transition at 73K, but the average Cu valence is 2.5^+ which is calculated from

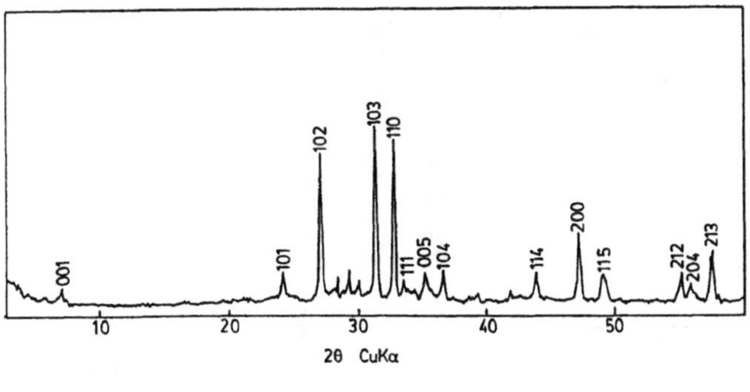

Fig.1 X-ray powder diffraction pattern of $TlBa_2Ca_{0.8}Y_{0.2}Cu_2O_7$. Tetragonal indices are given for the major phase.

stoichiometry, suggesting an over-doping state. Non-superconductor phase for substitution of Ca with Y at x=0.8 and x=1.0 has been reported [11]. Here we report a systematic study on the superconductivity of $TlBa_2Ca_{1-x}Y_xCu_2O_7$ (x=0-1).

EXPERIMENTAL

The Tl single layer compounds can usually be synthesized by long heat treatment for several hours from Tl double layer compounds. We started from 2212 composition to synthesize 1212 phase. The mixture of Tl_2O_3, BaO_2, $BaCuO_2$, CaO, Y_2O_3 and CuO giving a metal composition of $Tl_2Ba_2Ca_{1-x}Y_xCu_2O_8$ was prepared. The mixture was pressed into a pellet form of 10mm diameter and 1mm thickness. The pellet was wrapped with a gold foil and fired at 890°C for 3 hours in a flowing oxygen gas. Reaction products were slowly cooled in a furnace in O_2 atmosphere. The substitution of Ca with Y was confirmed by analytical electron microscopy (JEM 2000FX) equipped with EDX system.

RESULTS AND DISCUSSION

X-ray powder diffraction patterns (XPD) of $TlBa_2Ca_{0.8}Y_{0.2}Cu_2O_7$ measured with CuKα radiation is shown in Fig.1. By XPD measurements, this sample showed a few very weak diffraction lines of $BaCuO_2$. The major diffraction lines of XPD patterns were indexed with the tetragonal indices and the unit cell dimensions were determined from measured d-spacings by the least squares method. Superconducting critical temperature was determined from temperature variation of resistivity and DC susceptibility. The resistivity for a sintered sample was measured by a conventional four probe method with a measuring current of 1mA. Temperature dependences of resistivity for $TlBa_2CaCu_2O_7$ and $TlBa_2Ca_{0.8}Y_{0.2}Cu_2O_7$ are shown in Fig.2. DC susceptibility measurements for the powdered sample were done in a magnetic field of 10 Oe by using a SQUID magnetometer. DC susceptibility for the specimen of $TlBa_2CaCu_2O_7$ and $TlBa_2Ca_{0.8}Y_{0.2}Cu_2O_7$ are shown in Fig.3. Meissner fraction extrapolated at 0K drastically decreases from 46% of the non-doped specimen to 0.2% of the doped one, in spite of a remarkable increase in T_c from 73K to 103K.

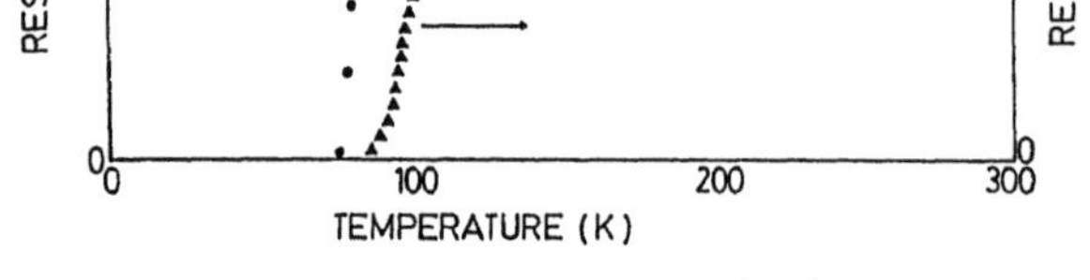

Fig.2 Resistivity vs temperature curve of $TlBa_2CaCu_2O_7$(●) and $TlBa_2Ca_{0.8}Y_{0.2}Cu_2O_7$(▲).

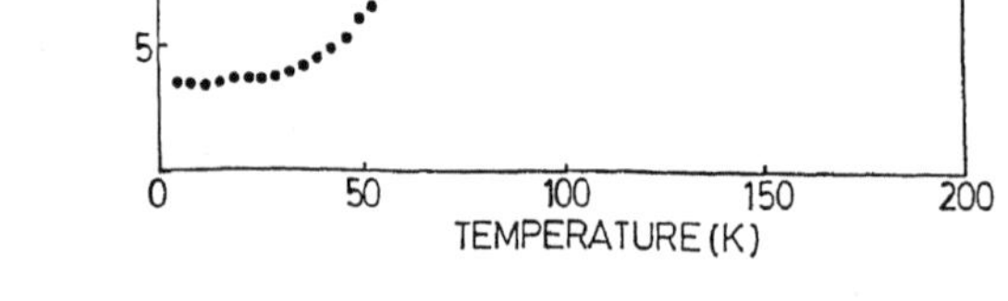

Fig.3 Temperature of DC magnetic susceptibity obtained in the cooling process of $TlBa_2CaCu_2O_7$(●) and $TlBa_2Ca_{0.8}Y_{0.2}Cu_2O_7$(▲).

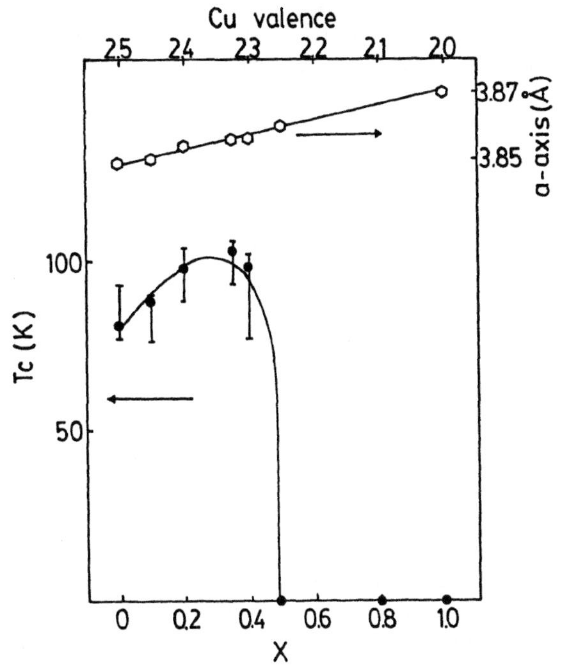

Fig.4 Variation of T_c(●) and unit cell dimension, a(○) with $TlBa_2Ca_{1-x}Y_xCu_2O_7$.

Figure.4 shows transition temperature, T_c, versus Y concentretion substituted for Ca. T_c showed a maximum of 103K around x=0.2-0.35, turned down to 98K at x=0.4, and revealed non-superconductivity down to 4.2K at x=0.5. Unit cell dimension, a, increases monotonously with substitution of Ca with Y. The observed increase in a is consistent with decrease in hole concentration [10]. The average Cu valence, v, of $TlBa_2Ca_{1-x}Y_xCu_2O_7$ is given by v=(5-x)/2, if the ideal stoichiometry is assumed. Decrease in hole concentration with increasing x has actually been observed by EELS measurements [12].

Since the high T_c above 100K was achieved by substitution of Ca with Y, the substituted system probably has pertinent amount of holes i.e. the average Cu valence of 2.3^+ in $YBa_2Cu_3O_7$. T_c of $TlBa_2Ca_{1-x}Y_xCu_2O_7$ took a maximum around x=0.3 which corresponds to the average Cu valence of about 2.35^+ estimated by assuming stoichiometry. The relation between substitution of Ca with Y and T_c suggests an over-doping state in 0<x<0.2 and under-doping state in 0.35<x<0.5. T_c of the Tl single layer system ($TlBa_2Ca_{n-1}Cu_nO_{2n+3}$) showed an inverse correlation with the unit cell dimension, a, which is approximately twice the Cu-O distance within the c-plane. The $TlBa_2Ca_{1-x}Y_xCu_2O_7$ system showed a maximum T_c around x=0.3, while the unit cell dimension, a, increases with increasing x. The sample became non-superconducting at x=0.5, at which unit cell dimension are a=3.860A. We have already reported a correlation between unit cell dimension, a, and T_c [2.3.7]. Upper limit of unit cell dimension, a, for the appearance of superconductivity of the Tl system was empirically known to be a=3.855A by $Tl_2Ba_2CaCu_2O_8$ system [2]. The present results are quite consistent with this correlation between hole concentration and a-dimension for appearance of high T_c.

ACKNOWLEDGEMENTS

The authors are grateful to Professor K.Nagase and Mrs. R.Suzuki, College of General Education, Tohoku University, for warm encouragement and technical support. This work was partly supported by the Grant-in-Aid for Scientific Research on Priority Areas "Mechanism of Superconductivity" given by the Ministry of Education, Science and Culture, Japan.

REFERENCES

1. Sheng ZZ and Hermann AM (1988) Bulk superconductivity at 120K in the Tl-Ca/Ba-Cu-O system. Nature 332:138-139

2. Kikuchi M, Kajitani T, Suzuki T, Nakajima S, Hiraga K, Kobayashi N, Iwasaki H, Syono Y and Muto Y (1989) Preparation and chemical composition of superconducting oxide $Tl_2Ba_2Ca_{n-1}Cu_nO_{2n+4}$ with n=1,2 and 3. Jpn.J.Appl.Phys 28:L382-L385

3. Kikuchi M, Nakajima S, Syono Y, Hiraga K, Oku T, Shindo D, Kobayashi N, Iwasaki H and Muto Y (1989) Preparation of the bulk superconductor $Tl_2Ba_2Ca_3Cu_4O_{12}$. Physica C 158:79-82

4. Parkin SSP, Lee VY, Nazzal AI, Savoy R, Beyers R and La Placa SJ (1988) $Tl_1Ca_{n-1}Ba_2Cu_nO_{2n+3}$ (n=1,2,3): A new class of crystal structures exhibiting volume superconductivity at up to ≈110K. Phys.Rev.Lett 61:750-753

5. Subramanian MA, Parise JB, Calabrese JC, Torardi CC, Gopalakrishnan J and Sleight AW (1988) Crystal stracture of $TlBa_2Ca_2Cu_3O_9$. J.Solid State Chem 77:192-195

6. Liang JK, Zhang YL, Huang JQ, Xie SS, Che GC, Chen XR, Ni YM, Zhen DN and Jia SL (1988) Crystal structures and superconductivity of superconducting phases in the Tl-Ba-Ca-Cu-O system. Physica C 156:616-624

7. Nakajima S, Kikuchi M, Syono Y, Oku T, Shindo D, Kobayashi N, Iwasaki H and Muto Y (1989) Synthesis of bulk high T_c superconductors of $TlBa_2Ca_{n-1}Cu_nO_{2n+3}$ (n=2-5). Physica C 158:471-476

8. Syono Y, Kikuchi M, Nakajima S, Suzuki T, Oku T, Hiraga K, Kobayashi N, Iwasaki H and Muto Y (1989) Structure, composition and superconductivity of high Tc Tl-Ba-Ca-Cu-O system. Proceedings of Symposium, Materials Research Society, Spring Meeting, San Diego, in press.

9. Suzuki T, Nagoshi M, Fukuda Y, Nakajima S, Kikuchi M, Syono Y and Tachiki M (1989) Core level x-ray photoelectron spectroscopy of Tl-Ba-Ca-Cu-O superconductor. Physica C:in press (Proceedings of M^2S-HTSC at Stanford Univ, July)

10. Nakajima S, Kikuchi M, Oku T, Kobayashi N, Suzuki T, Nagase K, Hiraga K, Muto Y and Syono Y (1989) Over-doping of $Tl_2Ba_2CuO_6$ due to charge transfer Tl^{3-t}-$(Cu-O)^p$. Physica C 160:458-460

11. Manako T, Shimakawa Y, Kubo Y, Satoh T and Igarashi H (1988) Non-superconducting $TlBa_2YCu_2O_7$ with a new crystal structure resembling to the superconducting $YBa_2Cu_3O_7$. Physica C 156:315-318

12. Shindo D, Hiraga K, Nakajima S, Kikuchi M, Syono Y, Kobayashi N, Hojou K, Soga T, Furuno S and Otsu H (1989) Oxygen K-edge fine structure of $TlBa_2Ca_{1-x}Y_xCu_2O_7$ studied by electron energy loss spectroscopy. Physica C 159:794-796

Improvement of Water-Resistivity of High-T_c Superconductor by Ag_2O Addition

S. YOSHIZAWA[1], Y. OGAWA[1], K. YAMAMOTO[1], Y. ISHIKAWA[1], and K. SHIOMI[2]

[1] Central Research Laboratory, Dowa Mining Co., Ltd., Tobuki, Hachioji, Tokyo, 192 Japan
[2] Dowa Chemical Co., Ltd., 1781, Nitte, Honjo, Saitama, 367 Japan

ABSTRACT

The deterioration of superconductivity in YBCO system occurs rapidly by moisture adsorption. The addition of Ag_2O into system improves water-resistivity of superconductivity. Appropriate amount of Ag_2O and calcinated YBCO powder were mixed. The mixture was recalcinated, pulverized, pressed into pellets, and then sintered. Immersion of YBCO pellet, which contained no silver, into boiling water for one minute disappeared the superconductivity. By 5wt% and 10wt% addition of Ag_2O to specimens, superconductivity retained even after the thirty minutes and one hour boiling water immersion, respectively. From the results of EPMA and X-ray diffraction pattern, it was suggested that silver deposited in crystallte surface prevents water diffusion into superconducting crystallites.

KEY WORDS: high-Tc superconductor, YBCO system, Ag addition, water-resistivity

INTRODUCTION

The deterioration of superconductivity in YBCO system occurs rapidly by moisture adsorption when it stands for a long time in contact with air or it is used an repeated runs from a low temperature to room temperature where water condenses around superconductor. This deterioration comes from the degradation of YBCO (1,2,3) phase to Y_2BaCuO_5, $BaCO_3$, $Ba_2Cu(OH)_6$, $Y(OH)_3$ and CuO [1,2,3,4,5]. The addition of certain oxides, e.g., In_2O_3 to YBCO system are known to improve the deterioration in moisture [6].

In this report, Ag_2O was chosen as a oxide because there are many papers describing that silver increases the critical current density and improves the mechanical property [7,8,9], and improvement of water-resistivity of YBCO system is studied by addition of Ag_2O which has no effect on the superconductivity. Understanding of the effect by Ag_2O is important for the processing and applications of high Tc superconductors in water containing environments.

EXPERIMENTAL

As shown in Fig.1, specimens were prepared by the solid-state reaction method. Powders of Y_2O_3, $BaCO_3$, and CuO were mixed stoichiometrically at 1:2:3 ratio of Y, Bi, and Cu, respectively. The mixed powder was calcined in air at 950 ℃ for 40 hours. Appropriate amounts of Ag_2O and calcinated YBCO powder were mixed. The mixture was recalcinated, pulverized, pressed into φ 20 x 2 mm thick pellets, and then, sintered in air at 950 ℃ for ten hours.

The water-resistivity of specimens was estimated by the accelerated method; after specimens were immersed into boiling de-ionized water, the residual water was eliminated by drying in an oven at 80 ℃ in vacuo. The electric resistivity was measured by four probe method. X-ray diffraction patterns of specimens were measured to observe the elimination of superconducting phase. In order to observe the existence of element in specimens, the scanning electron micrography (SEM) and the elecron probe micro analysis (EPMA) were used.

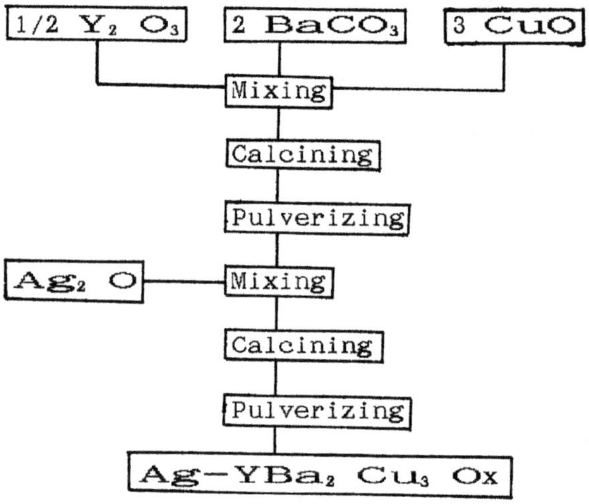

Fig. 1 Preparation flow of Ag-YBCO powder.

RESULTS AND DISCUSSION

Immersion of YBCO pellet, which contains no silver, into boiling water for one minute disappeared the superconductivity. Thus, the electric resistivity increaed monotonously representing the semi-conductive propertity, as the measuring temperature decreased.

Figure 2 shows the X-ray diffraction patterns of YBCO pellet after five minutes of treatment with boiling water. The YBCO (1,2,3) phase disappeared except for the trace quantity of main peak at 33 degree of 2θ. Instead, Y_2BaCuO_5, $Ba(OH)_2$, $Ba(OH)_2(H_2O)_3$, and CuO phases appeared[3,4] even in the inner side of specimen.

By 2wt% addition of Ag_2O to YBCO system, superconductivity was retained after a ten minutes treatment of boiling water, where the YBCO (1,2,3) phase was maintained as shown in Fig. 3(b). Fifteen minutes immersion of pellet deteriorated the superconductivity corresponding to the disappearance of the YBCO (1,2,3) phase in Fig. 3(c).

In the case of 5wt% and 10wt% addition of Ag_2O to specimens, the superconductivity was retained even after 30 minutes and one hour boiling water immersion, respectively. Thus, as shown in Fig. 4, the superconductivity retaining time became longer by the increase of the amount of added Ag_2O. The Tc values of treated specimens were from 92 K to 87 K.

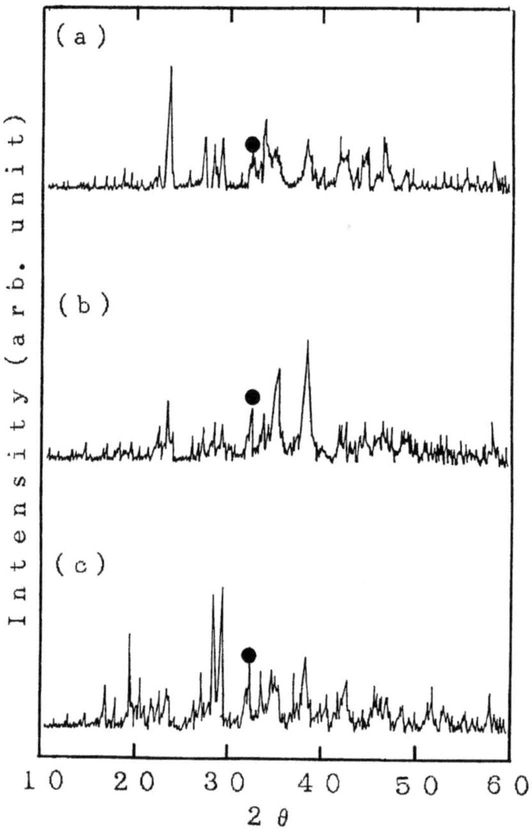

Fig. 2 X-ray diffraction patterns of YBCO pellet immersed in boiling water for five minutes, (a) surface, (b) 0.1 mm deep from the surface, (c) pulverized. ● denotes the main peak of (1,2,3) phase.

For the specimen which 10wt% Ag_2O added to, the superconductivity deteriorated after 70 minutes imersion of boiling water. The X-ray diffraction patterns were shown in Figs. 5(a)(b)(c). YBCO (1,2,3) phase was disappered in the diffraction pattern of surface. However, YBCO (1,2,3) phase was maintained in the diffraction peaks (b) of area 0.1 mm deep from the surface. This observation is definetely difference from the results of non-modifyed YBCO in Fig. 1, where YBCO (1,2,3) phase could not be identified not only in the surface but also inside into the pellet.

The state of added Ag_2O in the YBCO system was studied in terms of X-ray diffraction pattern and EPMA. Two peaks in the diffraction pattern of Ag_2O added YBCO system were observed at 38.1 degree and 44.3 degree of 2 θ which

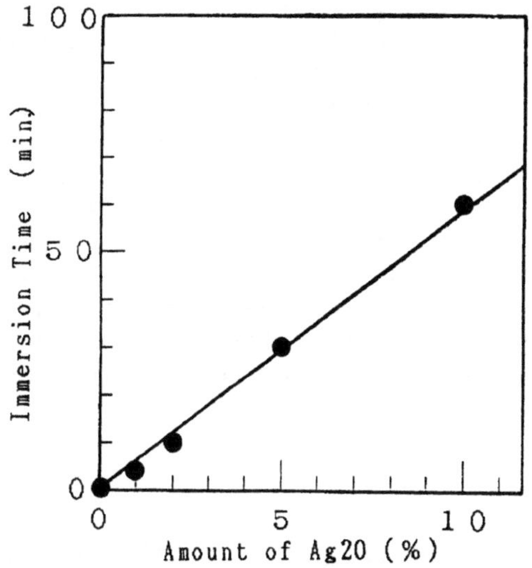

Fig. 4 Elongation of immersion time where superconductivity disapperes by the increase of the amount of added Ag_2O.

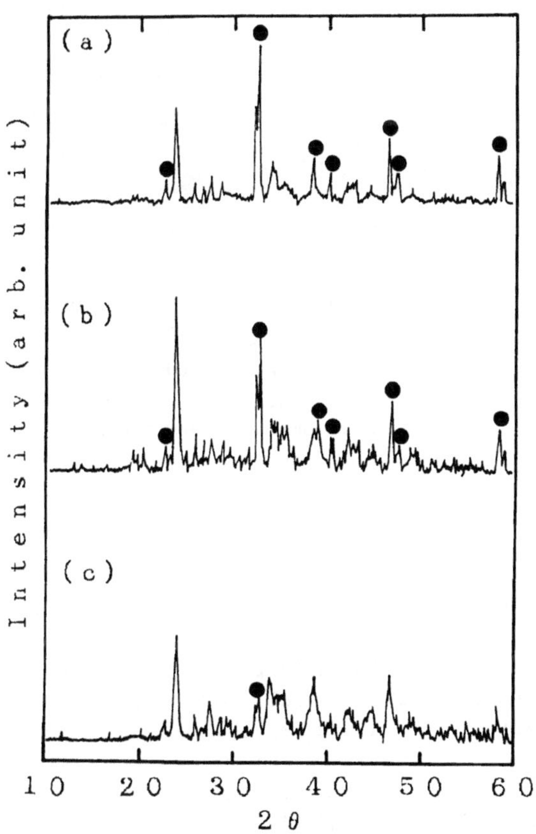

Fig. 3 X-ray diffraction patterns of Ag_2O(2wt%) added YBCO pellet immersed in boiling water. (a)for 5 min. (b)for 10 min. (c)for 15 min.

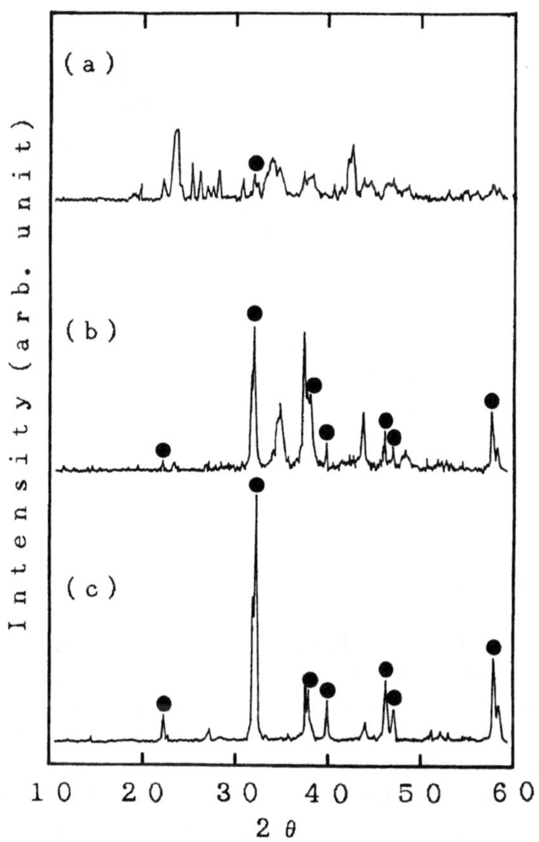

Fig. 5 X-ray diffraction patterns of Ag_2O(10et%) added YBCO pellet immersed in boiling water for 70 minutes, (a)surface, (b) 0.1 mm, (c)0.2 mm deep from the surface. ● denotes the peak of (1,2,3) phase.

were assigned to silver metal. From the results of EPMA of the polished surface in Fig. 6, it was found that solidified silver exists in the crystallite interface. In addition, the superconducting property of Tc, Jc, and Hc was not influenced by adding Ag_2O up to 10wt% discussed in this report.

Figure 7(a) shows the photograph of SEM at a fracture surface of Ag_2O(10wt%) added pellet after 70 minutes immersion in boiling water. The darken upper part exhibits the degraded surface of the pellet. The lower part shows the YBCO (1,2,3) phase. This observation demonstrates that the presence of Ag between the degraded surface and the superconducting phase. It was suggested, therefore, that silver deposited in the crystallite interface prevents water diffusion into superconducting crystallites.

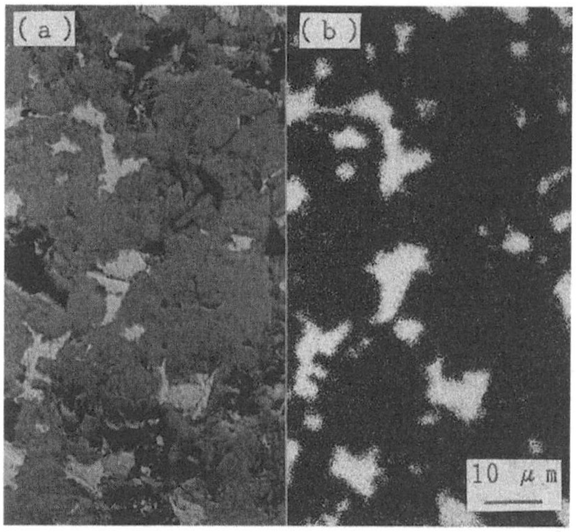

Fig. 6 SEM photograph (a) and EPMA image with Ag (b) of Ag_2O (20wt%) added YBCO polished surface.

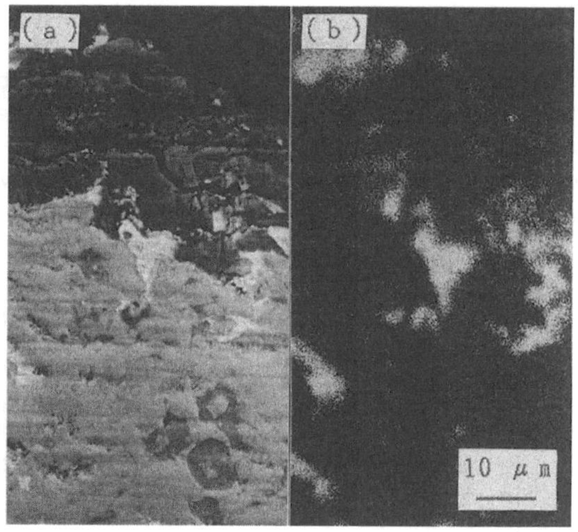

Fig. 7 SEM photograph (a) and EPMA image with Ag (b) at a fracture surface of Ag_2O (10wt%) added YBCO pellet after 70 min. immersion in boiling water.

REFERENCES

1. M.F.Yan, R.L.Barbs, H.M.O' Bryan.Jr., P.k.Gallagher, R.C.Sherwood, S.Jin Appl. Phys. Lett., 51, 532 (1987).
2. B.G.Hyde, J.G.Thompson, R.L.Withers, J.G.FitzGerald, A.M.Stewart, D.J.M. Bevan, J.S.Anderson, J.Bitmead, M.S.Paterson, Nature, 327, 402 (1987).
3. J.M.Rosamilia, B.Miller, L.F.Schneemeyer, J.V.Waszczak, H.M.O' Bryan,Jr., J. Electrochem. Soc., 134, 1863 (1987).
4. K.Komori, H.Kozuka, S.Sakka, J. Mater. Sci., 24, 1889 (1989).
5. L.D.Fitch, V.Burdick, J. Am. Ceram. Soc., 72, 2020 (1989).
6. K.Nagata, T.Iwai, K.Ozaki, T. IEE Jpn., 108-A, 405 (1988).
7. F.Mizuno, H.Masuda, I.Hirabayashi, S.Tanaka, Jpn. J. Appl. Phys., 28, L780 (1989).
8. S.Ochiai, K.Osamura, T. Takayama, Jpn. J. Appl. Phys., 27, L1101 (1988).
9. H.Terada, T.Ido, S.Muto, J. Jpn. Soc. Powder and Powder Metallurgy, 35, 100 (1988).

2.3 Phase Diagram and Crystal Growth

Single Crystal High-T_c Superconductor

YOSHIKAZU HIDAKA
NTT Opto-electronics Laboratories, 162, Tokai, Ibaraki, 319-11 Japan

ABSTRACT

The present stage of High-Tc superconductor single crystal growth is reviewed. The problems for growth of cuprate superconductors are (1) still imperfect knowledge of phase diagrams and difficulty in obtaining equilibrium data (2) inhomogeneous dopant distribution which results from the nature of solid solutions. (3) the contamination effect from the crucible in the slow cooling method. The wide liquidus range both in temperature and concentration in the 2-1-4 system and in the T' phase for electron doped compounds is favorable for growing large single crystals. Taking these facts into consideration from the view point of crystal growth, $La_{2-x}Sr_xCuO_4$ and $Pr_{2-x}Ce_xCuO_4$ are very important systems to be compared.

KEY WORDS: single crystal, phase diagram, solid solution, $La_{2-x}Sr_xCuO_4$, $Pr_{2-x}Ce_xCuO_4$,

INTRODUCTION

Copper oxide superconductors of various kinds have been discovered in recent years. The measured values of the physical properties of these compounds continue to be upgraded by the use of improved samples. Thus at this stage of high-Tc study, it should be stressed that the production of well-characterized high quality materials is fundamental for understanding high temperature superconductivity. In particular, the recent discovery of electron doped copper oxide superconductors[1] gives us a chance to test many high-Tc superconducting mechanisms proposed since the discovery of $La_{2-x}Ba_xCuO_4$(LBCO). Therefore, it is very important to compare the physical properties in both hole and electron-doped systems using high quality single crystals.
Despite the novel two-dimensional magnetic features found in flux grown $La_{2-x}Sr_xCuO_4$(LSCO) crystals, their superconducting properties are still inferior to those of ceramic samples, for example a broad transition at low temperatures. This implies that the detailed study of the evolution of superconductivity in these crystals is difficult. Thus, the growth of large superconducting crystals with a sharp transition is required for further detailed study of the high Tc mechanism. For $YBa_2Cu_3O_{7-y}$(YBCO) and LSCO, crucible contamination even at very low concentrations(<1%) has an effect on physical properties. Pt and Al which substitute for Cu, affect electronic transport in the CuO two-dimensional planes. Consequently, the choice of crucible is very important in the prevention of contamination. In this context, first the present stage of high-Tc single crystal growth is reviewed, then some problems in crystal growth are mentioned, and finally, important conditions that must be satisfied in order to produce high quality single crystals are stated.

PRESENT STAGE OF HIGH-Tc CRYSTAL GROWTH

High-Tc oxide superconductors including the $BaPb_{1-x}Bi_xO_3$(BPB) system are incongruent melting compounds, in which the crystal concentration is different from that of the melt. Therefore, single crystals must be grown from a saturated solution. Consequently, a choice of adequate solvent is very important. Usually, there are two ways of choosing the solvent. One is to use several components of the compound as a self solvent. Another is to use some inorganic-salt like KCl, KOH ... The slow cooling (SC) method, The Top Seeded Solution Growth (TSSG) method, and the Travelling Solvent Floating Zone (TSFZ) method are frequently tried.
Successfully grown single crystals of high-Tc copper oxides, BPB and $Ba_{1-x}K_xBiO_3$(BKB) are listed in Table 1. The slow cooling method from a nonstoichiometric melt is successful for all compounds. The LSCO crystals are separated from the liquid with a platinum net or with a pair of high quality alumina porcelain sticks at high temperature. $Nd_{2-x}Ce_xCuO_4$(NCCO), $Pr_{2-x}Ce_xCuO_4$(PCCO) and YBCO crystals are separated from the aggregate after the liquid is cooled to room temperature[2]. Alkali chloride fluxes used for Bi compound are easily washed from the crystals[3].
The largest single crystals of about 12cc are obtained in the 2-1-4 La-system or in the T' phase system[4].

Recent growth development in the 1-2-3 system yielded thick single crystals of more than 2mm by cooling as slowly as 0.1°C/h[5]. One very important characteristic in these high-Tc copper oxides is growth instability which can only be suppressed by a very slow growth rate under stable conditions. In other higher Tc systems such as Bi or Tl compounds, it is very difficult to obtain a thick homogeneous single crystal because of its tendency to cleave parallel to the basal plane and the fact that a multi-phase mixture easily occurs[6]. Usually only single crystals of less than several 100μm in size are obtained.

Table 1. High Tc oxide single crystals.

Compound	Growth Method	Solvent	Size(mm^3)
$Ba(Pb_{1-x}Bi_x)O_3$	SC, Hydrothermal	$PbO+Bi_2O_3$, KCl	15x15x2
$(Ba_{1-x}K_x)BiO_3$	SC	KOH	1x1x1
$La_{2-x}M_xCuO_4$ (M=Ba, Sr, Ca)	SC, TSSG, TSFZ	La_2O_3 - MO - CuO	50x50x5
$LnBa_2Cu_3O_7$ (Ln=Lanthanides)	SC, TSSG	BaO - Ln_xO_y - CuO	5x5x2
$Bi_2Sr_2CaCu_2O_x$	SC, TSFZ	KCl, Nonstoichiometry	3x8x0.1
$Tl_2Ba_2Ca_{n-1}Cu_nO_{2n+4}$ (n=1-4)	SC	Nonstoichiometry	0.5x0.5x0.1
$La_{2-x}M_xCuO_4$ (M=Ba, Sr, Ca)	SC, TSSG, TSFZ	La_2O_3 - MO - CuO	50x50x5

CRYSTAL GROWTH PROBLEMS IN HIGH Tc SINGLE CRYSTALS

(1) PHASE DIAGRAM

In spite of the great efforts that have been made in the crystal growth field, the knowledge of phase diagrams so far obtained is still imperfect. This is mainly because of the difficulty in Differential Thermal Analysis(DTA) caused by the liquid creeping effect. It makes it difficult to obtain reproducible and reliable equilibrium data, which in turn makes the phase diagram detail ambiguous. Phase diagrams near the liquidus line(or plane) are reported for LSCO, NCCO, YBCO and $Nd_{1+x}Ba_{2-x}Cu_3O_{7-y}$(NBCO). LSCO and NCCO have a wide liquidus line in the temperature and concentration ranges, which facilitates the growth of large single crystals by the slow cooling method as shown in Fig. 1. A wide liquidusplane is observed. While, in YBCO, NBCO, a narrow liquidus range in temperature and concentrationmakes it difficult to obtain large single crystals[7,8]. Then, a very slow cooling rate with a precisely chosen concentration makes it possible to obtain thick YBCO crystals as reported by T. Wolf et al. [5].

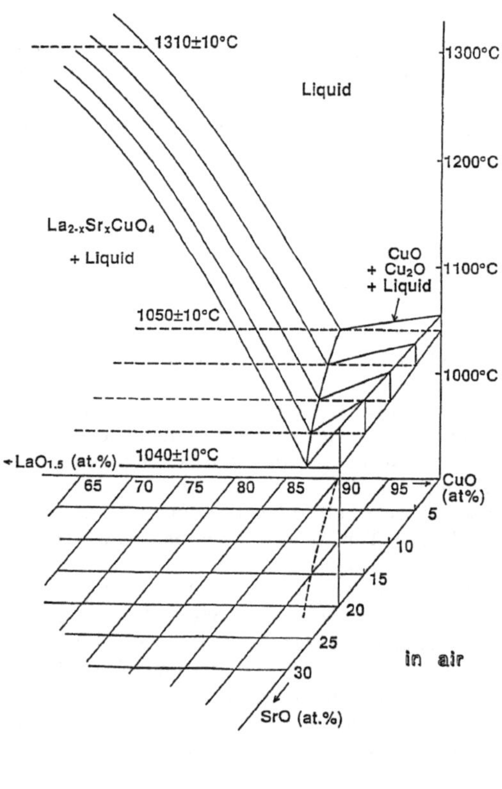

Fig. 1 Phase diagram of LSCO

(2) INHOMOGENEOUS DOPANT DISTRIBUTION

Inhomogeneous dopant distribution is usually observed in doped crystal or solid solution systems, in which deviation must be suppressed as much as possible. EPMA analysis was performed on the a-c plane along the c-axis in LSCO, NCCO and PCCO single crystals as shown in Fig.2. The LSCO single crystal does not exhibit inhomogeneous Sr distribution within an experimental error of 10%. While for the NCCO crystal, largely inhomogeneous Ce distribution was observed in every crystal thicker than 0.5mm. This suggests that large thick single crystals are not useful for physical property measurement. Further, a concentration of about 7% is unexpectedly grown even when the nominal Ce concentration is 2%. This may be closely related to the nature of the solid-solution plane in its phase diagram. On the other hand, the PCCO system does not show such an inhomogeneous Ce distribution though it is in the same electron-doped system. Therefore, at present, LSCO and PCCO large homogeneous single crystals, whose dopant concentration is systematically changed, are very useful for physical property measurements. Figure 3 shows large flux grown LSCO and PCCO single crystals.

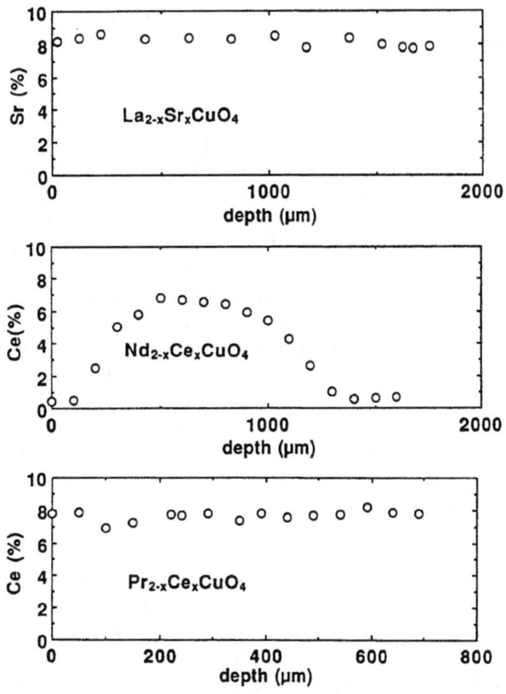

Fig. 2 Depth profile of the dopant(Sr, Ce) along c-axis.

(3) Pt CONTAMINATION EFFECT

Single crystals larger than 10cc can be grown in LSCO and PCCO system by the slow cooling method using a Pt crucible. Cu solvent is known to corrode Platinum. Therefore, the platinum contamination effect in these systems must be clarified.
$La_{1.85}Sr_{0.15}(Cu_{1-x}Pt_x)O_4$ ceramic samples were synthesized using high purity oxide powder La_2O_3(99.99%), $SrCO_3$(99.99%),CuO(99.99%) and Pt(99.9%). The Pt content was changed systematically; 0.002, 0.005, 0.01, 0.015, 0.02, 0.04, 0.06, 0.08, 0.10, 0.15 in 1-x notation. These compounds were calcined at 950°C for 10hr in an oxygen flow. After pulverization, they were again calcined at 1000°C for 10hr in oxygen. Then, after a second pulverization, samples were sintered at a higher temperature from 1070°C - 1120°C for 4hr and cooled to 500°C by 3.5°C/min. The sintering temperatures were;1070°C for 0.002 - 0.015, 1120°C for 0.02 - 0.08, and 1170°C for 0.10, 0.15. After holding the samoles at 500°C for 100hr they were then cooled to room temperature by 1°C/min. These samples were cut into appropriate rectangular shapes for measurement. For Hall measurements, they were ground to about 150μm thick. Neutron diffraction structural analysis and x-ray diffraction analysis were performed to specify the Pt substitution site[9]. The Rietvelt analysis revealed that some Cu is replaced by Pt. The EPMA analysis of single crystals also showed that the atomic ratio of (La, Sr) : (Cu,Pt) = 2 : 1, which also implies that some Cu is substituted by Pt. The c-axis decreases and the a-axis increases as the Pt concentration increases. The Jahn-Teller distortion is released by the Pt replacement.

(a)

(b)

Fig. 3 Non-doped large single crystals of (a) LSCO and (b) PCCO

Resistivity was measured by the four-probe method. Gold was evaporated onto the top surface of the samples and silver wires were attached on them with silver paste. After annealing at 400°C for 30 min., low resistance ohmic contact was obtained. When the Pt contamination is increased, the superconducting transition temperature is lowered and the resistivity rises at the onset critical region. Also the transition width becomes wider than the sharp transition in the non Pt doped sample. Susceptibility was measured by SQUID. The temperature dependence of susceptibility shows that the onset temperature is lowered and the superconducting volume fraction is reduced as Pt is increased. Fig.4 shows the Pt content dependence of the superconducting transition point which is defined at the zero resistivity point. The inset is the volume fraction change as a function of Pt content. The volume fraction in the non-doped sample is defined as 100%. The linear decrease is also observed in Zn or Ga substituted LSCO compounds reported by Gao et al [10,11]. Normal susceptibility was measured as a function of temperature. The susceptibility decreases as the temperature is decreased. The Curie-Weiss behavior was observed in doped samples and temperature dependence becomes steep at low temperatures as the Pt increases. The Curie-Weiss behavior suggests that some amount of local moment is formed as Pt contamination increases.

The Hall coefficient increases as Pt increases, which implies that the hole carrier density decreases.

An Actual LSCO single crystal grown by the slow cooling method contains a Pt contamination of about 1% in the Cu site. This Pt contamination effect may be the reason why the Tc in LSCO crystals grown by the slow cooling method is lower than that in ceramic samples or in TSFZ grown crystals[12,13]. Although the specific heat jump accompanied by the superconducting transition is clearly observed in ceramic samples, they are not observed in flux grown and TSFZ grown single crystals in which a large Meissner volume fraction near 100% is observed. This discrepancy remains to be solved.

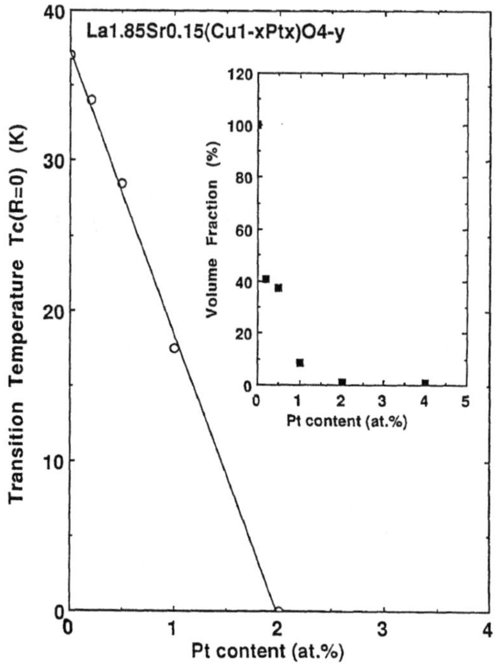

Fig. 4 Pt content dependence of Tc and superconducting volume fraction.

SUMMARY

The growth of high-Tc cuprate single crystals is reviewed emphasizing contamination effects on physical properties especially in flux grown crystals. For further quality improvement in superconducting high-Tc single crystals, these three conditions should be satisfied.

(1) Contamination in the Cu-O plane should be avoided.
The TSFZ method is currently the best method in that a crucible is not used. But, the slow cooling method with either a minimally corrosive crucible or low temperature melting flux is attractive because single crystals larger than 10cc are easily grown.

(2) Homogeneous dopant distribution.
The homogeneity in dopant distribution is closely related to the nature of the compound. Therefore, it is important to choose an appropriate system which does not have such a extreme dopant distribution. In electron-doped systems, PCCO is better than NCCO and in 1-2-3 systems YBCO is better than NBCO.

(3) Specific heat jump should be observed at Tc.
The specific heat jump observed in ceramic samples, should be observed in homogeneous high quality single crystals which are bulk superconductors. The fact that it is not, may suggest that a really good single crystal whose quality is of at least ceramics level has not yet been obtained.

ACKNOWLEDGMENT

The author would like to express his gratitude to N. Yoshimoto, Y. Tajima, M. Matsuda, Y. Endoh, M. Ishikawa, M. Suzuki, T. Murakami and A. Yamaji for useful discussions, and he also thanks T. Inamura and K.Murase for their continuous encouragement in the course of this study.

REFERENCES

1. Y.Tokura, H.Takagi and S.Uchida (1989) A superconducting copper oxide compound with electrons as the charge carriers. Nature 337: 345-348

2. Y.Hidaka,Y.Enomoto, M.Suzuki, M.Oda and T.Murakami(1987) Single crystal growth of $(La_{1-x}A_x)_2CuO_4$ (A=Ba, Sr) and $Ba_2YCu_3O_{7-y}$. J. Cryst. Growth 85: 581-584

3. L.F.Schneemyer, R.B.van Dover, S.H. Glarum, S.A.Sunshine, R.M.Fleming, B.Batlogg, T.Siegrist, J.H.Marshall, J.V.Waszczak & L.W.Rupp (1988) Growth of superconducting single crystals in the Bi-Sr-Ca-Cu-O system from alkali chloride fluxes. Nature 332: 422-424

4. Y.Endoh, M.Matsuda, K.Yamada, K.Kakurai,Y.Hidaka, G.Shirane and R.J.Birgeneau (1989) Two-dimensional spin correlations and successive magnetic phase transitions in Nd_2CuO_4. Phys. Rev. B 40: 7023-7026

5. Th.Wolf, W.Goldacker, B.Obst, G.Roth and R.Flukiger (1989) Growth of thick $YBa_2Cu_3O_{7-y}$ single crystals from Al_2O_3 crucibles. J. Cryst. Growth 96: 1010-1018

6. Y.Hidaka, M.Oda, M.Suzuki, Y.Maeda, Y.Enomoto and T.Murakami(1988) Large Anisotropy of the Upper Critical Magnetic Field in Single Crystal Bi-(Sr,Ca)-Cu-O. Jpn.J.Appl.Phys. 27: L538-L541

7. K.Oka, M.Saito, M.Ito, K.Nakane and H.Unoki (1988) Phase Diagram and Crystal Growth of $RBa_2Cu_3O_{7-y}$ system. Proceedings of ISS'88, 28-31 Aug. 1988 Nagoya: 225-227

8. K.Oka, M.Saito, M.Ito, K.Nakane, K.Murata, Y.Nishihara and H.Unoki (1989) Phase Diagram and Crystal Growth of $NdBa_2Cu_3O_{7-y}$. Jpn.J.Appl.Phys. 28: L219-L221

9. Y.Hidaka, M.Matsuda, H.Asano (unpublished work).

10. M.Z.Cieplak, G.Xiao, A.Bakhshai and C.L.Chien (1989) Superconducting and normal-state properties of $La_{1.85}Sr_{0.15}(Cu_{1-x}Ga_x)O_4$. Phys. Rev. B 39: 4222-4230

11. G.Xiao, A.Bakhshai, M.Z.Cieplak, Z.Tesanovic and C.L.Chien (1989) Correlation between superconductivity and normal-state properties in the $La_{1.85}Sr_{0.15}(Cu_{1-x}Zn_x)O_4$ system. Phys. Rev. B 39: 315-321

12. H.Takagi, T.Ido, S.Ishibashi, M.Uota, S.Uchida and Y.Tokura (1989) Superconductor-to-nonsuperconductor transition in $(La_{1-x}Sr_x)_2CuO_4$ as investigated by transport and magnetic measurements. Phys. Rev. B 40: 2254-2261

13. I.Tanaka and H.Kojima (1989) Superconducting single crystals. Nature 337: 21-22

Phase Equilibria of Bi_2O_3-SrO-CaO-CuO System at 1123K in Air

R.O. SUZUKI, S. KAMBARA, H. TSUCHIDA, K. SHIMIZU, and K. ONO

Department of Metallurgy, Kyoto University, Yoshida-Honmachi, Sakyo-ku, Kyoto, 606 Japan

ABSTRACT

Phase equilibria in Bi-Sr-Ca-Cu-O system was investigated mainly by X-ray powder diffraction measurements. Summarizing the experimental results at 258 compositions, the following results were obtained: 1) The phase equilibria in Bi_2O_3-SrO-CaO, Bi_2O_3-SrO-CuO and SrO-CaO-CuO ternary systems were clarified and a new phase Bi_2SrCaO_y was observed. 2) 8 quaternary phase equilibrias and 9 ternary equilibrias were confirmed in the quasi-quaternary system. 3) No diffraction for "high Tc phase" was detected in these conditions, but "low Tc phase" was at least in equilibrium with 8 compounds and liquid.

KEY WORDS: phase diagram, superconducting oxides, phase equilibria, X-ray diffraction, bismuth

1. INTRODUCTION

It is difficult to produce a high Tc superconducting single phase in Bi-Sr-Ca-Cu-O system, because it may contain a few unexpected phases in preparation. The contamination of the impure phases can have some effects on the superconducting properties. Phase equilibrium studies are therefore essential for determining the phase compatibilities with the other phases.

The purpose of this study is to make clear what phases can coexist with the superconductors. This paper will report the experimental phase diagram in Bi-Sr-Ca-Cu-O system, especially quasi-ternary and quasi-quaternary systems. It is therefore based on the previous reported quasi-binary and ternary phase diagrams[1-6] and on the thermodynamics, especially the phase rule. Since this system contains 5 components, the temperature and oxygen pressure were fixed to be 1123K and air in order to reduce the degree of freedom.

2. EXPERIMENTAL PROCEDURES

A normal sintering process was used to prepare the samples. High purity of Bi_2O_3, $SrCO_3$, $CaCO_3$ and CuO were mixed and calcined. The samples without no bismuth were pre-sintered at 1243 K several times. At the final annealing, they were kept at 1123 K in air. The samples containing a greater amount of Bi were sintered at 1073K before the final annealing. Most of the samples were prepared by mixing the complex oxides formed in advance in order to promote to achieve equilibrium. The sintered samples were characterized by mainly X-ray powder diffraction measurements and electron probe microanalysis (EPMA). When the samples were judged to contain the unreacted regions, they were again ground, pressed into pellets and sintered at 1123K for a few days. We usually repeated these sintering processes 3 to 6 times for a total of 258 compositions. After the final step, most of the samples were cooled in air, but several samples were quenched into water or liquid nitrogen.

3. RESULTS AND DISCUSSION

3.1) Single phases and Bi_2O_3-SrO-CuO System

Fig.1 shows the phases reported in Bi_2O_3-SrO-CaO-CuO system[1-6], by assuming all compounds as the ideal composition about oxygen. This assumption is required to fix the degree of thermodynamic freedom. In this study, the compound and its composition will be given a numerical symbol: $Bi_2Sr_2CaCu_2O_y$, so called 80K phase, will be referred to as (2 2 1 2) in order of Bi, Sr, Ca and Cu. The open circles in Fig.1 show the phases confirmed by our study. A few complex oxides were not confirmed in our experimental conditions, while (2 1 1 0) and (0 1 7 8) were found.

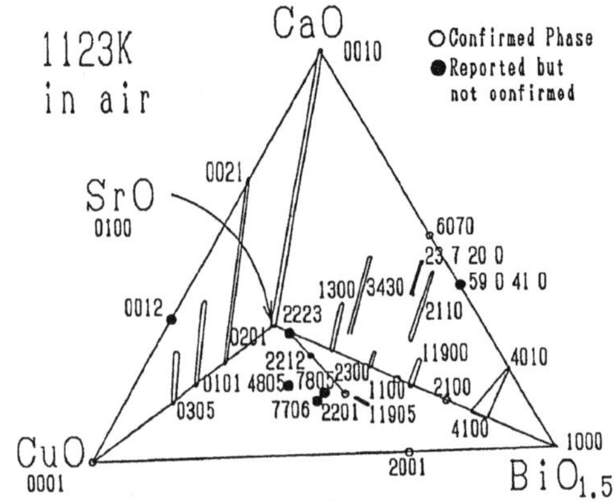

Fig.1 Phases in the quasi-quaternary system.

In the quasi-ternary system of Bi_2O_3-SrO-CuO, 43 composition were studied. Fig.2 shows the X-ray diffraction pattern of the sample with the composition (16 31 0 73), which can be assigned by superimposing three patterns for (2 2 0 1), (0 3 0 5) and CuO. This means these three compounds coexist in equilibrium at the composition (16 31 0 73). From these experimental results, we determined most of phase equilibrias in Fig.3. Since we confirmed the compatibilities between (2 3 0 0) and (0 2 0 1), and among (0 2 0 1), (1 3 0 0) and SrO, we predicted the phase compatibility exists among (0 2 0 1), (2 3 0 0) and (1 3 0 0) by the thermodynamic phase rule. The strontium rich region could be shown only by ternary phase equilibrium triangles, by assuming there are no solid solubilities in these compounds. On the other hand, (11 9 0 5) had a small solubility limit as reported[2]. Corresponding to this solubility, 3 binary phase regions were shown in Fig.3. This work used Takano's formula[6] for (0 3 0 5), while the recent works[8] called it (0 14 0 24) by the structure analysis. Fig.2 disagree with the previous reported phase equilibrias[2,6] only concerning about this phase.

3.2) Bi_2O_3-CaO-CuO and Bi_2O_3-SrO-CaO systems

The liquidus and solidus curves in Fig.3 and 4 are speculated. There are 3 reasons: (A) Since the congruent melting temperature of (2 0 0 1) happened to be just the same temperature as that we studied, the specimens were melted or partially melted in the wide region. (B) The high purity alumina crucible used easily reacted with the bismuth rich samples in the long period. (C) The samples were characterized at room temperature after quenching.

Fig.5 shows (3 4 3 0) makes a wide solid solution as Takano et al. reported [6]. This diffraction pattern was similar as that of (1 3 0 0), but several peaks could not be assigned to those of (3 4 3 0). A new solid solution (2 1 1 0) was found and the diffraction peak positions were not significantly affected by the composition. The analysis of the crystal structure is now in progress. In the bismuth rich region, (4 1 0 0) solid solution stretched toward quasi-ternary system. Because this "beta-solid solution" was reported to be unstable at 1173K[3,5], it might be formed during quenching.

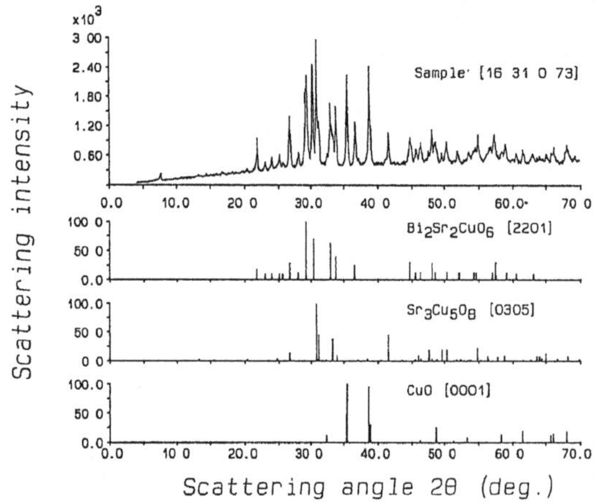

Fig.2 Measured diffraction pattern and those for the related phases.

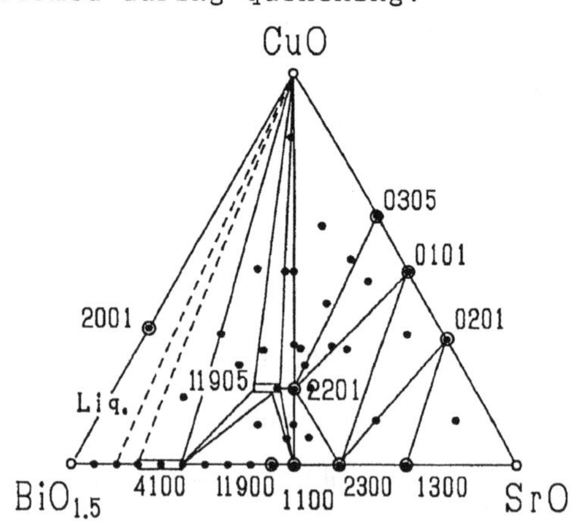

Fig.3 Phase equilibria diagram for for $BiO_{1.5}$-SrO-CuO system.

3.3) SrO-CaO-CuO system

Also in Fig.6(a), we found the substitution between strontium and calcium in a few compounds. Only in Sr rich composition, we could not obtain the correct equilibrium in spite of long annealing and repeated grindings. There are two reasons: (A) The decomposing rate of the starting carbonates was very slow. (B) SrO and CaO are not stable in air even at room temperature. Therefore, we assumed the complete solid solution between calcium oxide and strontium oxide. Two orthorhombic phases, (0 2 0 1) and (0 0 2 1) solved completely at 1123K-1243K. The lattice parameter changed smoothly between two oxides. Based on the lattice parameter analysis, the compositions of the conjugation lines were estimated and shown by dotted lines in Fig.6(a).

A new phase (0 1 7 8) was discovered at 1243K, while this tetragonal phase was not detected at 1123K. By its appearance around 1200K in heating, the phase compatibility is considered to changed between 1123K and 1243K as shown in Fig.6 (a) and (b). Fig.6(b) agreed with the recently reported work[9]. Two possible point groups are proposed as I_{4c2} with a=0.5463nm and c=0.6416nm, and $P_{4/mcc}$ with a=0.3863nm and c=0.6416nm. These structures have essentially the same unit cell as that for (0 14 86 100)[8,9], but ours has double layer of perovskite.

3.4) Bi-Sr-Ca-Cu-O system

By 93 experiments inside the compositional tetrahedron at 1123K in air, the "110K Tc-superconductor" (2 2 2 3) was not detected and only (2 2 1 2) phase existed as shown in Fig.1. Because it is difficult to confirm the coexistence of 4 phases in a specimen only by the X-ray diffraction measurements, by

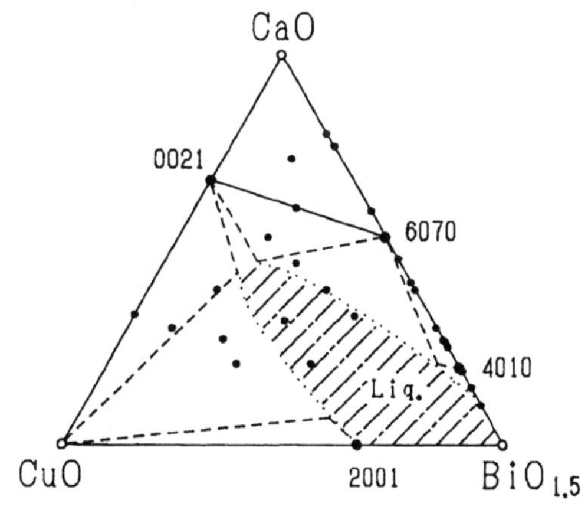

Fig.4 Phase equilibria diagram for $BiO_{1.5}$-CaO-CuO system.

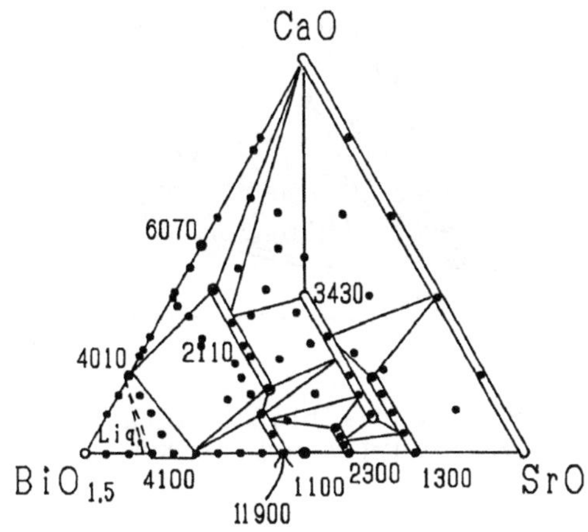

Fig.5 Phase equilibria diagram for $BiO_{1.5}$-SrO-CaO system.

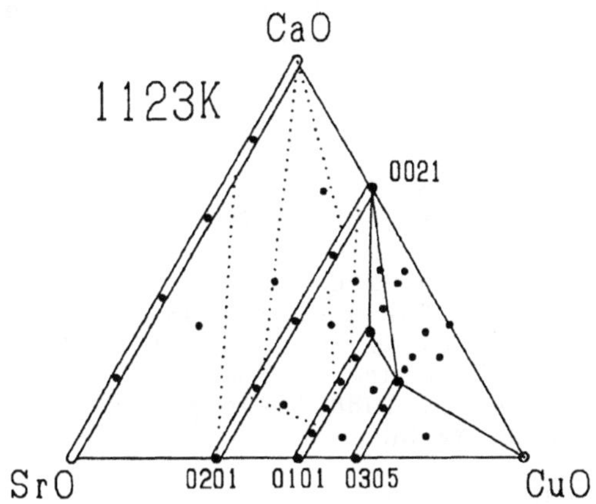

Fig.6(a) Phase equilibria diagram for SrO-CaO-CuO system at 1123K.

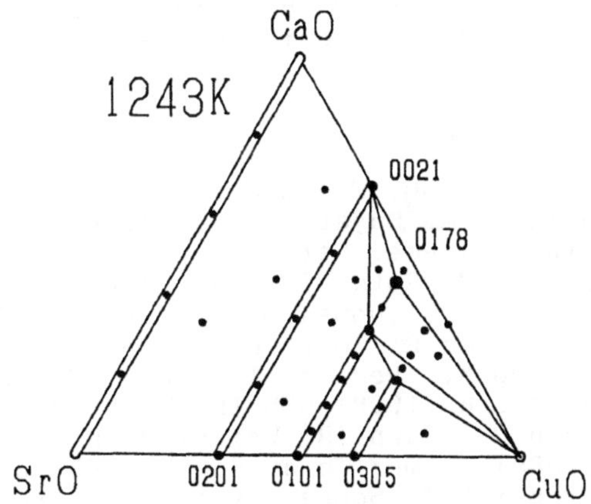

Fig.6(b) Phase equilibria diagram for SrO-CaO-CuO system at 1243K.

the compositional analysis of EPMA we found 8 quaternary phase equilibrias as listed in Table 1, and one of them is shown as a tetrahedron in Fig.7. Fig.7 has still open space near Bi_2O_3-SrO edge and the phase equilibrias related with the liquid could not be determined because of lack of compositional data in the liquid.

Since the quasi-ternary compounds had solid solubilities stretched linearly, ternary phase equilibrias were also shown by the solid bodies in the three dimensional space. If two compounds have no solubility and they are in equilibrium with another solid solution, the compatibility triangles make the tetrahedron(TYPE A) surrounded with 4 flat planes. On the other hand, in the case of a compound and 2 solid solutions, these triangles make the pentahedron(TYPE B) surrounded with 4 planes and a twisted plane. They were detected in the phase diagram and also listed in Table I. The binary equilibrias are shown by the flat or twisted planes in Fig.7.

By EPMA analysis, (3 4 3 0) contained a relatively small amount of copper when it was in equilibrium inside the compositional tetrahedron. Because our analysis neglected this solubility, thermodynamically there should exist extra phase equilibrias concerning with (3 4 3 0), although they could be hardly detected.

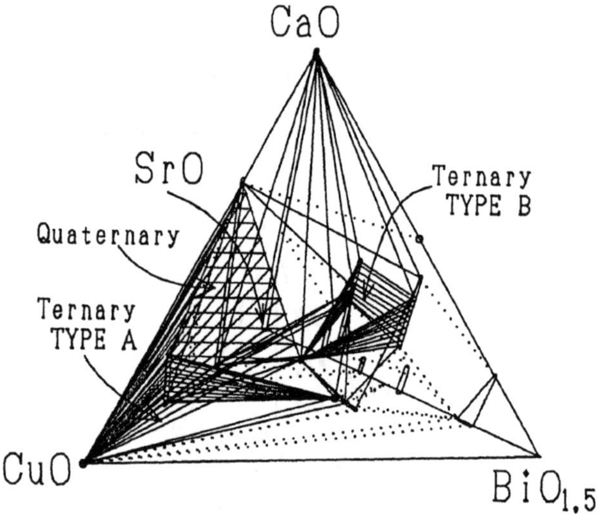

Fig.7 Phase equilibria diagram for the quasi-quaternary system.

Table I Detected phase equilibrias in the quasi-quaternary system.

Quaternary phase equilibria	
	CuO - 033 10 - 2212 - 0021
	CuO - 033 10 - 2212 - 2201
	CuO - 2201 - 2212 - 11 905
	CaO - 3120 - 2212 - 0201
	CaO - 3120 - 2212 - 3340
	2110 - 3430 - 2212 - 11 905
	2201 - 3430 - 2212 - 11 905
	2201 - 3430 - 2212 - 033 10
Ternary phase equilibria	
TYPE A	CuO - 2201 - 0305ss
	CuO - 2212 - 11 905ss
	CaO - 2212 - 2110ss
	CaO - 2212 - 3430ss
	CaO - 2212 - 0201ss
	CaO - 3430 - 0201ss
TYPE B	2212 - 3430ss - 2110ss
	2212 - 2110ss - 11 905ss
	3430 - 0021∼0201ss - 0305ss

4. CONCLUSIONS

Phase equilibria in Bi-Sr-Ca-Cu-O system was studied at 1123 K in air. "Low Tc superconductor" (2 2 1 2) could coexist at least with 8 compounds: CaO, CuO, (3 4 3 0), (2 1 1 0), (1 3 0 0), (11 9 0 5), (0 2 0 1)-(0 0 2 1) and (0 3 0 5). It should be noted (2 2 1 2) was also in equilibrium with the various kinds of liquid, whose compositions could not be determined.

5. REFERENCES

[1] Gadalla AMM, White J (1966) Trans Bri Ceram Soc 65:181-90
[2] Saggio JA, Sujata K,Hahn J, Hwu SJ, Poeppelmeier KR, Mason TO (1989) J Amer Ceram Soc 72:849-53
[3] Levin EM, Roth RS (1964) J Res Nat Bur Stand 64:197-206
[4] Boivin JC, Thomas D, Tridot G (1973) C R Acad Sc Paris C276:1105-7
[5] Guillermo R, Conflant P, Boivin JC, Thomas D(1978) Rev Chem Min 15:153-9
[6] Takano M, Ikeda Y, Takada J, Oda K, Kitaguchi H, Miura Y, Tomii Y, Mazaki H (1988) MRS Int Meeting, Adv Mater. 14 Sep 1988 Tokyo p.1-4
[7] Roth RS, Rawn JR, Whiltler JD, Chiang CK, Wong-Ng WK (1989) J Amer Ceram Soc 72:395-99
[8] Siegrist T, Schneemeyer LF, Sunshine SA, Waszczak JV (1988) Mater Res Bull 23:1429-38
[9] Roth RS, Rawn CJ, Ritter JJ, Burton BP (1989) J Amer Ceram Soc 72:1545-49

Phase Diagram of Bi-Based Superconductors

T. NODA, T. IZUMI, A. NAKAMURA, and Y. SHIOHARA

Superconductivity Research Laboratory, International Superconductivity Technology Center, 10-13, Shinonome 1-chome, Koto-ku, Tokyo, 135 Japan

ABSTRACT

It is necessary to understand phase equilibria in the Bi-system to effectively produce superconductors in this system. Experiments were performed using a quench method together with microscopic study as well as composition analyses by ICP and EPMA. Equilibrium phases at different temperatures were investigated and plotted on a quasi-binary temperature versus composition diagram between the nominal compositions $Bi_2Sr_2Ca_2Cu_3O_x$ (2223, the high Tc phase) and $Bi_2Sr_2CaCu_2O_x$(2212, the low Tc phase). The liquidus temperature of the compound with the 2223 nominal composition found to be above 1100°C, and the primary solid phase formed to be Ca-rich oxide. These results contradict those in previously reported phase diagrams, in which the temperature of the liquidus is about 1000°C. The liquidus lines along the composition axis as well as the primary phase coresponding to these lines as also reported.

KEY WORD:BSCCO, Phase diagram,Liquidus plane

INTRODUCTION

The discovery of superconductivity in Bi-Sr-Ca-Cu-O system has generated a great interest in the field of condensed matter material science. Since the first report was submitted by Michel [1] Maeda et al[2], research has been concentrated in the areas of crystal structures, phase diagrams, superconducting properties, and its synthesis. We have employed a flux method or Czochralski method etc. to obtain the Bi-based supercoductors as grown. When we employ these techniques efficiently, the phase diagram. in hands is required or at least very helpfull, since knowledge from the phase diagram allows us to select starting liquid compositions from which the superconducting phases crystallize. We have studied a portion of the phase diagram including 2223 and 2212 nominal compositions. Liquidus temperatures of 2223 and 2212 nominal compositions were investgated.

EXPERIMENTAL PROCEDURE

Reagent grade starting materials of Bi_2O_3, $SrCO_3$, $CaCO_3$, CuO were used. The nominal compositions of the samples are shown in Table-1. The baches were weight and ground and then mixed with a mortal and pestle for extended time. The baches were heated up at a rate of 250°C/h to the reaction temperature of 800°C and held for 12 hours in the air. After this reaction process, these mixed powders were examined by an X-ray diffraction(XRD) method. If the XRD pattern showed the existence of CaO, the powders were reheated at 880°C for 1 hour. A portion of the phase diagram of $(Bi,Sr)O_{2.5}$-CaO-CuO system has been determined by a quench method using these starting powders. Quenched samples were charcterized by XRD,inductively coupled plasma chemical analysis(ICP), and electron probe micro analysis(EPMA) .

Table-1 Experimental composition

	Bi	Sr	Ca	Cu
B-2223	2.0	2.0	2.0	3.0
B-1	2.0	2.0	1.9	2.9
B-2	2.0	2.0	1.7	2.7
B-3	2.0	2.0	1.5	2.5
B-4	2.0	2.0	1.4	2.4
B-5	2.0	2.0	1.2	2.2
B-6	2.0	2.0	1.1	2.1
B-2212	2.0	2.0	1.0	2.0

Platinum crucibles (5mmØ, 5mmH) were used for containers of the powders. The powders were placed in the crucible. Samples were heated in the air at 1400°C,1200°C,1000°C and 930°C respectively for one hour to achieve an almost equilibrium state, then water quenched together with the crucibles.

Figure-1.shows a method to determine the liquidus line along the composition from CaO to 2203. The melting temperature of CaO reported is 2570°C and small amounts of Bi, Sr and Cu can be soluble in CaO. The composition of the glassy matrix analysed by ICP assummed to be equal to the liquidus composition C_{Ln}(n=1, 2, 3), when the samples were quenched at differnt temperatures Tn(n=1, 2, 3). The lower portion of the samples consisted of only a glassy phase. These glassy phase retained the liquid state (composition) before quenching. The primary crystals dispersed in the glassy matrix near the surface of samples were charcterized by EPMA. The composition of the primary crystal was analyzed and it was $(Ca_{1-x},Sr_x)O$(x=0.05). The liquidus line was drawn through C_{L1},C_{L2} , C_{L3} and interconnecting to the melting temperature of CaO.

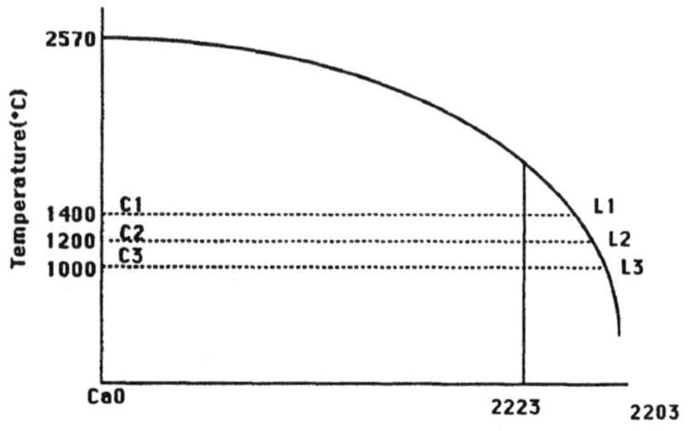

Figur-1 Liquidus line determination from phasediagram

The melting point of nominal composition of 2223 was estimated using this liqudus line. We used similar methods to estimate the melting temperatures for the samples with nominal compositions of B-2,B-3 and B-5. (see Table-1)

RESULTS AND DISCUSSION

Table-2 shows the summarized results of metallographic observations of the samples quenched at few different temperatures. The samples were classified to the following three different types ;

1) glassy phase only
2) single primary phase within a glassy matrix
3)Coexistence of primary and secondary phases

Composition of the primary crystal is $(Ca_x,Sr_{1-x})O$ which is analysed by EPMA. Sr contents increase with increasing temperature. (see Table-3). Primary crystals were observed in the upper region of the sample due to flotation in the liquid. Composition of the secondry crystal is $(Sr_{1-x},Ca_x)CuO_2$. Figure-2 is a $(Bi,Sr)O_{2.5}$-CaO-CuO pseudoternary composition diagram. A straight line is drawn along the primary crystal,2223 and 2203 nominal composition. In the pseudobinary phase diagram, three liquidus temperatures of the primary crystal,2223 and 2203 nominal compositions were included in this line.(see Figure-3) The liquidus line is drawen through the composition of primary crystal,C_{L1},C_{L2} and C_{L3}.The liquidus temperature of the compound with 2223 nominal composition was found to be 1560°C from this liquidus line. Gradient of primary crystal-2223-2203 liquidus line at the 2223 nominal composition is 30°C/%. The liquidus temperatures between 2212 and 2223 were also drawn in a similar way and shown in Figure 5. Gradient of 2223-2212 liquidus line at the 2223 is 13°/%.

A CaO-CuO pseudobinary phase diagram has been reported[3]. The 3-dimensional phase diagram was constructed using the liquidus lines between 2223 and 2203, and between 2223 and 2212,and the CaO-CuO pseudobinary phase diagram. (see Figure-6) This diagram shows several featnes features;

(1) Primary crystals with respected to the liquidus of 2223 and 2212 nominal compositions are (Ca,Sr)O.
(2) Liquidus plane of (Ca,Sr)O exists a wide domain.

(3) Melting points of 2223 and 2212 nominal compositions are 1560°C and about 1100°C respectively.

As we obtain the composition of primary crystal and liquidus plane,next we must search the secondry crystal for melt process.

ACKOWLEDGMENT

This work has been supported by the R&D Basic Technology for Future Industries through the New Energy and Industrial Technology Development.Organizasion(NEDO)

REFERENCE

1.C. Michel, M. Hervieu, M. M.Borel, A. Grandin, F. Deslandes, A. Provost and B. Raveau, Z. Physik B68 (1987)

2 H. Maeda, Y. Tanaka, M. Fukutomi, and T. Asao , Jpn. J. Appl. Phys. 27 (1988) L209

3 A.M.M.Gadalla and J. White and Tranc. Brite. Ceran. Cream. Soc., 65 [4] 185(1966}

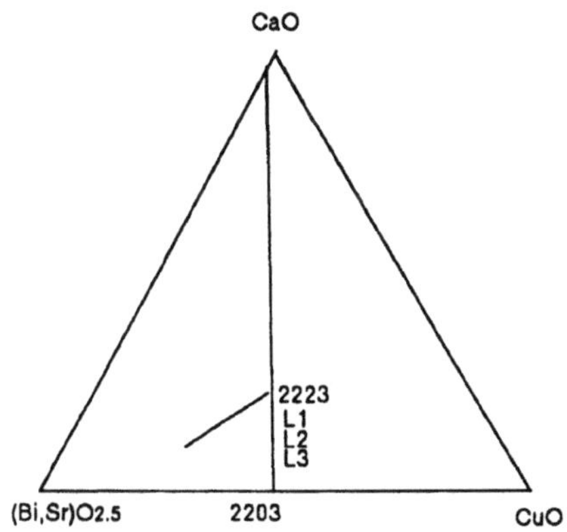

Figure-2 $(Bi,Sr)O_{2.5}$-CaO-CuO pseudoternary phase diagram which is drawen the straight line line on the primary crystal,2223,and 2203.

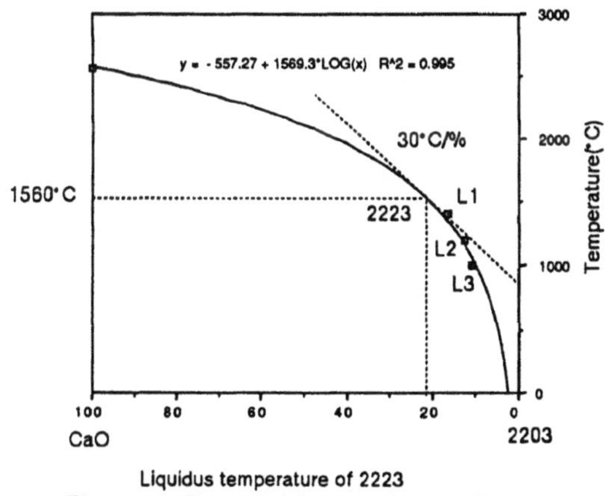

Figure-3 Pseudobinary phase diagram on the primary crystal-2203 straight line.

Table-2 Quench temperature vs pricipitation

Quench temperature vs pricipitation				
	930°C	1000°C	1200°C	1400°C
B-2223	S,P	P	P	P
B-1	S,P	P	P	P
B-2	S,P	P	P	P
B-3	S,P	P	P	P
B-4	S,P	P	P	P
B-5	S,P	P	P	G
B-6	S,P	P	G	G
B-2212	S,P	P	G	G

P:Primary crystal
S:Secondary crystal
G:Glass phase

Table-3 composition of second phase

Temp.(°C)	Sr	Ca
1400	4.8	95.2
1200	3.2	96.8
1000	1.8	98.2

Preparation of Substrates for Superconductive Devices Using Bi-System Low-T_c Single Crystals

K. TAKAHASHI, Y. NAKATANI, M. KAMINO, T. YOKOO, Y. YOSHISATO, and S. NAKANO

Functional Materials Research Center, SANYO Electric Co., Ltd., 18-13, Hashiridani 1-chome, Hirakata, Osaka, 573 Japan

ABSTRACT

Formation of insulating MgO thin layers has been carried out on c-planes of $Bi_2Sr_2CaCu_2O_x$ single crystals as the first attempt at preparing S/I/S structures for the first time.

Single crystals were grown by a self-flux method from Bi_2O_3, $CaCO_3$, $SrCO_3$ and CuO powders with the cation ratios of 2:1:1:2. Temperature dependence of resistivity shows that a sharp superconducting transition occurs at 81K with a transition width of less than 0.1K. MgO films were fabricated on single crystals by an ion beam sputtering method with little diffusion. Preferred orientation along (111) was confirmed in these films.

KEY WORDS: $Bi_2Sr_2CaCu_2O_x$, self flux method, MgO, ion beam sputtering

INTRODUCTION

The discovery of high-Tc superconductors such as Y-Ba-Cu-O[1], Bi-Sr-Ca-Cu-O[2] and Tl-Ba-Ca-Cu-O[3] has a greatly facilitated widespread use of these materials for superconductive devices at liquid-nitrogen temperature.

Most of the superconductive devices need junction structures composed of superconductors (S), insulators (I) and semiconductors (SE). Several types of junction structures, such as S/I/S, S/N/S and S/SE, contribute to the operation of superconductive devices due to their non-linear I-V characteristics and high-frequency response. There are some experimental reports on the fabrication of junction structures such as Y-Ba-Cu-O/Al_2O_x/Nb[4] and Y-Ba-Cu-O/Au/Nb[5] and so on.

However, adequate characteristics of the junctions have not been achieved in spite of tremendous efforts. The main difficulty is that the coherent length of high-Tc oxide superconductors is as short as 2~3 nm at most. It is not easy to prepare reproducibly such a thin and smooth layer with no pinholes between superconductors because of non-homogeneous and irregular interfaces. Therefore, surface morphology of the first superconducting layer, on which the second non-superconducting layer and the third superconducting layer are fabricated, should be precisely controlled for the fabrication of junction structures.

This paper reports on properties of MgO/Bi-Sr-Ca-Cu-O structures for monolithic isolation to obtain good interfaces for superconducting devices. Here bulk single crystals of high-Tc $Bi_2Sr_2CaCu_2O_x$ were used as the lower electrode. The cleavage plane of bulk single crystals are so homogeneous and highly dense that they can also be applied as a substrate for an insulating layer and an upper electrode.

Our methods are characterized by the use of substrates of bulk single crystals as a part of the junction structures, contrary to conventional processes using sputterd films on MgO and $SrTiO_3$ substrates. It is thought that the superconductivity of the substrates themselves will simplify the device structure.

EXPERIMENTAL

The single crystals of high-Tc $Bi_2Sr_2CaCu_2O_x$ were grown by the self-flux method. Mixtures of analytical grade Bi_2O_3, $SrCO_3$, $CaCO_3$ and CuO powders were prepared in cation ratios of 2:1:1:2. The mixtures were ground using a mortar and pestle, placed in alumina crucibles, heated in a furnace at 1000 °C for 24 hours, cooled to 750 °C at a rate of 7 °C/h, and finally cooled to room temperature in the furnace. The reaction product was a solid mass containing single crystals. Single crystals were mechanically isolated by cleavage.

X-ray diffraction analysis using CuKα radiation and energy dispersive X-ray spectroscopy (EDS) were performed to identify the crystallinity and phases, respectively. In addition, crystallization temperature from the molten state was examined by high-temperature X-ray diffraction analysis.

The microstructure was observed using a polarized microscope and scanning electron microscope (SEM). Ion beam sputtering was performed using a single MgO target with the $Bi_2Sr_2CaCu_2O_x$ single-crystal substrate placed 60 mm away from the target. The superconducting transition temperature Tc of a $Bi_2Sr_2CaCu_2O_x$ single-crystal substrate was measured by a standard four-prove resistive method. The quality of the $Bi_2Sr_2CaCu_2O_x$ single-crystal substrate was further examined using AC susceptibility measurements.

RESULTS AND DISCUSSION

1) Preparation of $Bi_2Sr_2CaCu_2O_x$ single crystal substrate

Figure 1 shows X-ray diffraction patterns at 900 °C, 925 °C and 950 °C of Bi_2O_3, $SrCO_3$, $CaCO_3$ and CuO mixtures molten at 1000 °C with a cooling rate of 1 °C/min. $Bi_2Sr_2CuO_x$, that is, the Akimitu and Bi-free ($Sr_{1.5}Ca_{1.5}Cu_5O_x$) phases begin to precipitate at 950 °C when the mixtures are cooled down from 1000 °C. The $Bi_2Sr_2CaCu_2O_x$ phase appears for the first time at 925 °C and peaks of $Bi_2Sr_2CaCu_2O_x$ are more remarkable at 900 °C, reducing both $Bi_2Sr_2CuO_x$ and $Sr_{1.5}Ca_{1.5}Cu_5O_x$ peaks. Therefore, it is considered that the annealing temperature should be over 925 °C in order to obtain $Bi_2Sr_2CaCu_2O_x$ single crystals of good quality.

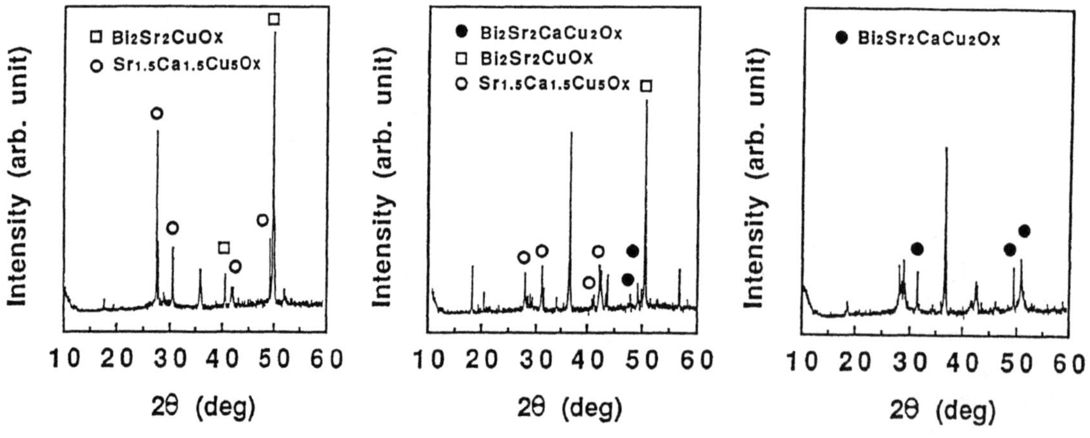

(a) 950 °C (b) 925 °C (c) 900 °C

Fig.1 Crystallization process of $Bi_2Sr_2CaCu_2O_x$ single crystals from molten state

$Bi_2Sr_2CaCu_2O_x$ single crystals were grown using a self-flux method under the conditions already mentioned, and were obtained by cleavage. As shown in Fig. 2, this is a flake with typical dimensions on the order of 0.5x0.3x0.01 mm^3.

The X-ray diffraction pattern of the facets of $Bi_2Sr_2CaCu_2O_x$ single crystals is shown in Fig. 3. This pattern indicates that the crystal is composed of single-phase $Bi_2Sr_2CaCu_2O_x$, with a facet perpendicular to (00*l*).

Fig.2 Single crystal of $Bi_2Sr_2CaCu_2O_x$

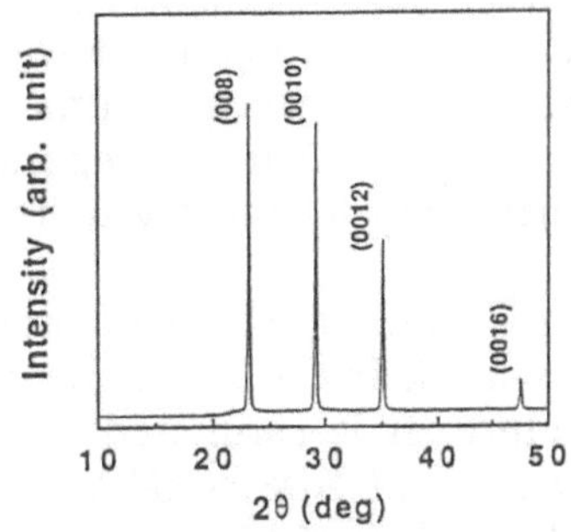

Fig.3 X-ray diffraction pattern for $Bi_2Sr_2CaCu_2O_x$ single crystal

Furthermore, the chemical composition of single crystals was found to be $Bi_2Sr_2CaCu_2O_x$ by using energy dispersive X-ray spectroscopy.

The temperature dependence of resistivity for $Bi_2Sr_2CaCu_2O_x$ single crystals with electrical current perpendicular to the c-axis is shown in Fig.4. A sharp superconducting transition is observed at 81K with a transition width of less than 0.1K.

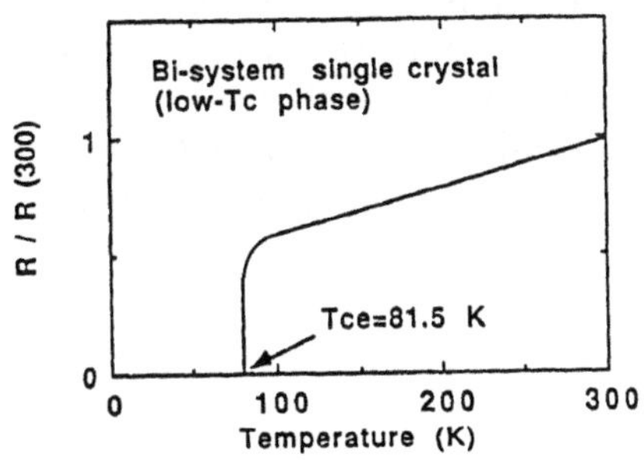

Fig.4 Temperature dependence of resistivity for $Bi_2Sr_2CaCu_2O_x$ single crystal

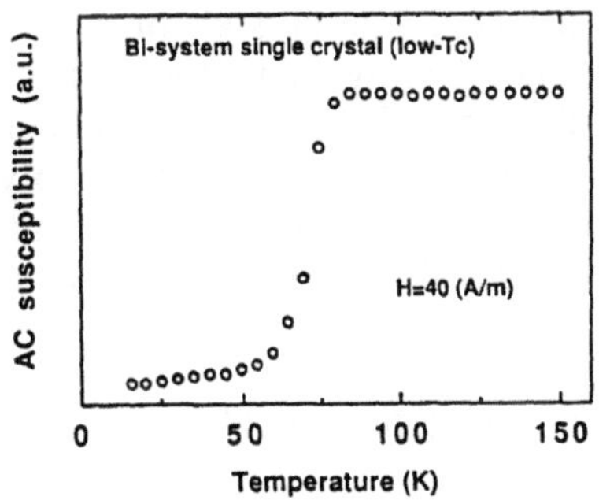

Fig.5 Temperature dependence of magnetization for $Bi_2Sr_2CaCu_2O_x$ single crystal

Figure 5 shows the temperature dependence of magnetization for single crystals in a magnetic field of 40 A/m. An increase in magnetization was observed at 81K with an increases in temperature which is consistent with the resistive Tc.

2) Fabrication of MgO thin films on $Bi_2Sr_2CaCu_2O_x$ substrates

Insulating MgO thin films were fabricated on the substrates of $Bi_2Sr_2CaCu_2O_x$ single crystals by ion beam sputtering under the conditions shown in Table 1. The thickness of these films was about 0.2μm.

Table 1 Ion beam sputtering conditions

Target	MgO (4 inch)
Back pressure	$<10^{-4}$ Pa
Output	120 mA, 1kV
Extractor voltage	150 V
Sputtering pressure	$<10^{-2}$ Pa
Sputtering rate	0.03 nm/sec

Figure 6 shows the X-ray diffraction pattern of the MgO films annealed at 740 °C on the substrates of $Bi_2Sr_2CaCu_2O_x$ single crystals. Preferred orientation with enhanced (111) line was observed in the sputtered film. Furthermore, there were no noticeable peaks besides those found in $Bi_2Sr_2CaCu_2O_x$ substrates and MgO films. Therefore, it is thought that there was not much reaction between films and substrates in spite of high-temperature annealing necessary for the crystallization of MgO.

Figure 7 shows a schematic drawing of the grown $Bi_2Sr_2CaCu_2O_x$ and MgO heterostructure. It is estimated that the lattice parameter of $Bi_2Sr_2CaCu_2O_x$ (00*l*) agrees well with that of MgO (111) within an error margin of as much as 7%. This speculation based on crystal structures is consistent with the results of X-ray measurements.

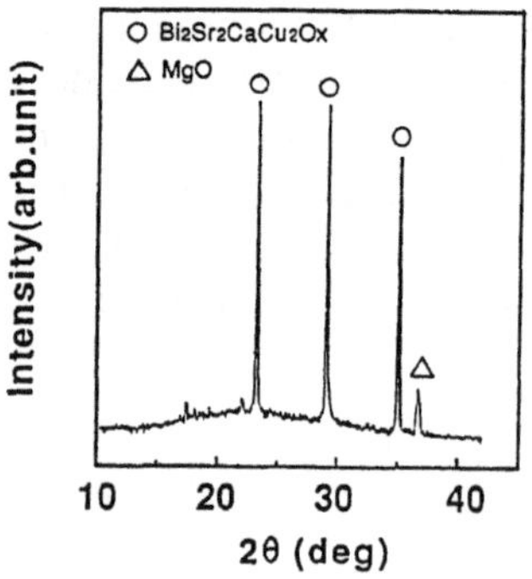

Fig.6 X-ray diffraction pattern for MgO films on $Bi_2Sr_2CaCu_2O_x$ substrates

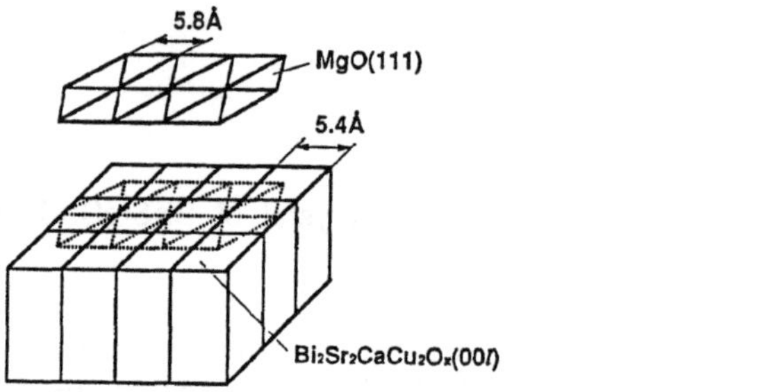

Fig.7 Schematic drawing of the heterostructure of $Bi_2Sr_2CaCu_2O_x$ and MgO

CONCLUSIONS

We have developed a new structure consisting of insulating MgO on a $Bi_2Sr_2CaCu_2O_x$ single crystal for the first time. Single crystals used as substrates show a sharp superconducting transition which occurs at 81K with a transition width of less than 0.1K. MgO films fabricated onto the substrates do not react with substrates and indicate a preferred orientation which is probably controlled to minimize the mismatch in lattice parameters between $Bi_2Sr_2CaCu_2O_x$ and MgO. Therefore, these techniques are considered to be useful for fabricating superconductive devices.

ACKNOWLEDGEMENTS

This work was performed under the management of the R & D Association for Future Electron Devices as a part of the R & D of Basic Technology for Future Industries sponsored by NEDO (New Energy and Industrial Technology Development Organization).

REFERENCES

1. M.K. Wu, J.R. Ashburn, C.J. Trong, P.H. Hor, R.L. Meng, L. Gao, Z.J. Huang, Y.Q. Meng and C.W. Chu: Phys. Rev. Lett., 58,908(1987)
2. H. Maeda, Y. Tanaka, M. Fukutomi and T. Asano: Jpn. J. Appl. Phys., 27,L209(1988)
3. Z.Z. Sheng and A.M. Herman: Nature., 332,55(1988)
4. A. Nakayama, A. Inoue, K. Takeuchi and Y. Okabe: Jpn. J. Appl. Phys.,26,L2055(1987)
5. H. Akoh, F. Shinoki, M. Takahashi and S. Takada., Jpn. J. appl. Phys., 27,L519(1988)

Structure and Superconductivity of $Tl_2Ca_3Ba_2Cu_4Ox$ Single Crystals

T. KOTANI, T. NISHIKAWA, H. TAKEI, and K. TADA

Basic High-Technology Laboratories, Sumitomo Electric Industries, Ltd., 1-3, Shimaya 1-chome, Konohana-ku, Osaka, 554 Japan

ABSTRACT

The crystal structure and superconducting properties of $Tl_2Ca_3Ba_2Cu_4Ox$ single crystals were investigated. Based on the preliminary study of phase relations in the $Tl_2Ca_{n-1}Ba_2Cu_nO_x$ system, 2324-phase single crystals with dimensions of from 0.5 to $1mm^2$ were grown by the flux method. The superconducting onset transition temperatures of grown crystals, determined by DC SQUID magnetometer, were 113 to 115K. According to an X-ray diffraction measurement, the lattice constant of the c-axis was 42.0Å, and the crystal was confirmed to be single phase of 2324. The results of a high-resolution transmission electron microscopic study are also reported.

KEY WORDS: Tl-Ca-Ba-Cu-O system, crystal growth, flux method, transmission electron microscopy

INTRODUCTION

Following the discovery of high-Tc Tl-Ca-Ba-Cu-O superconducting materials with Tc above 100K, many efforts have been made to study the crystal structure and superconducting properties. The Tl-Ca-Ba-Cu-O system contains a wide variety of structures having different superconducting transition temperatures. Much interest has been focused in the relation of the structure to the superconducting properties such as transition temperature. Therefore, the preparation of superconducting phases with many Cu-O layers has been extensively studied. Recently, superconducting phases containing up to 4 Cu-O layers for double Tl-O layered compounds[1,2] and up to 6 Cu-O layers for single Tl-O layered compounds[3,4] have been achieved through a sintering process in the Tl system. But, precise structural analyses and evaluation of superconducting properties should be conducted for the single crystals of distinct structural phases. Previously, we have grown single crystals of $Tl_2Ca_1Ba_2Cu_2Ox$ and $Tl_2Ca_2Ba_2Cu_3Ox$, and examined their structures and superconducting properties[5].

In this work, we grew the $Tl_2Ca_3Ba_2Cu_4Ox$ single crystals with four Cu-O layers in the structure and then studied the superconducting properties by DC SQUID magnetometer, and the micro-structure of the single crystal by means of X-ray diffraction measurement(XRD) and high-resolution transmission electron microscopy(TEM).

EXPERIMENTAL

Raw materials for crystal growth were prepared by mixing the oxide powders of Tl_2O_3, CaO, BaO_2 and CuO in the ratios of Tl : Ca : Ba : Cu = 2 :(2+m): 2 :(3+0.75m) (m=0,1,2,3,4,5,6) thoroughly. Single crystals were grown by the CuO self-flux method. The experimental procedure for the crystal growth is shown below. Raw materials were inserted in sealed Au tubes and then these tubes were placed in a crucible with a tightly covered lid. The materials were rapidly heated to a melting temperature of 925-950°C, and then cooled to 740°C at the rate of 10°C/hour. Thereafter, the crucible was kept at 740°C for 3 hours, followed by further cooling to room temperature. Many plate-like crystals were found in the solidified samples obtained by this heat treatment. Single crystals were picked up from these samples. The solidified samples and single crystals were examined by DC SQUID magnetometer and XRD measurements. High-resolution TEM observations were carried out for a sample of $Tl_2Ca_3Ba_2Cu_4O_x$ single crystal with a 200kV microscope(EM-002B) having a resolution of 1.8Å. Specimens for electron microscopy were prepared by dry grinding, followed by ion beam thinning of (010) plane (a-c plane) of a crystal.

RESULTS AND DISCUSSION

In the composition series studied, the transition temperature of the solidified samples varied with the composition of raw materials as shown in Fig.1. According to the powder XRD measurements of each sample, different superconducting phases were formed in the samples with the different compositions. A 2324-phase was obtained only from the sample with a composition of m=2.

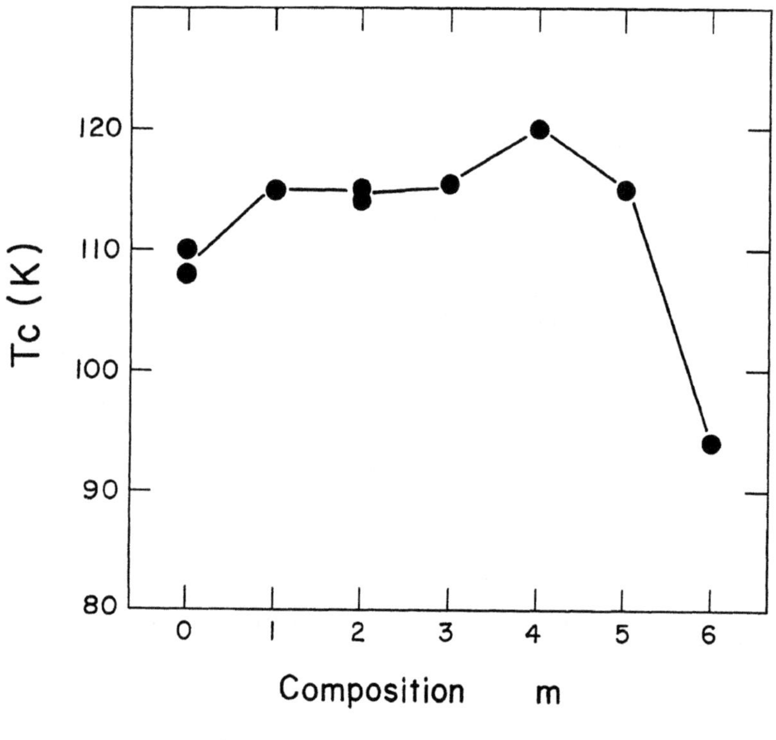

Fig.1. Transition temperature of solidified samples with compositions of Tl : Ca : Ba : Cu = 2 : (2+m) : 2 : (3+0.75m).

Fig.2. A SEM image of the solidified sample of m=2.

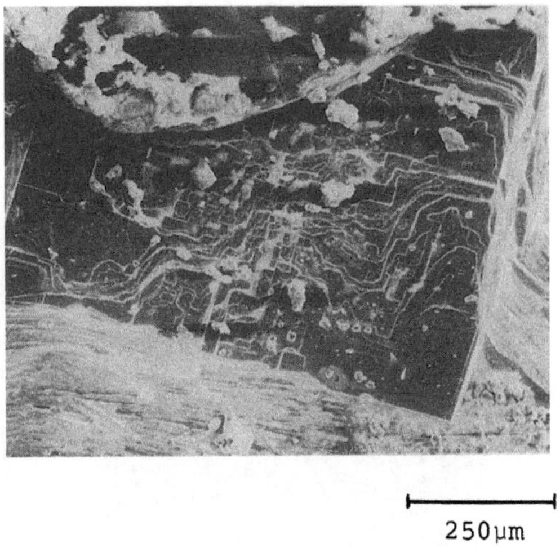

Fig.3. Surface morphology of 2324-phase crystal.

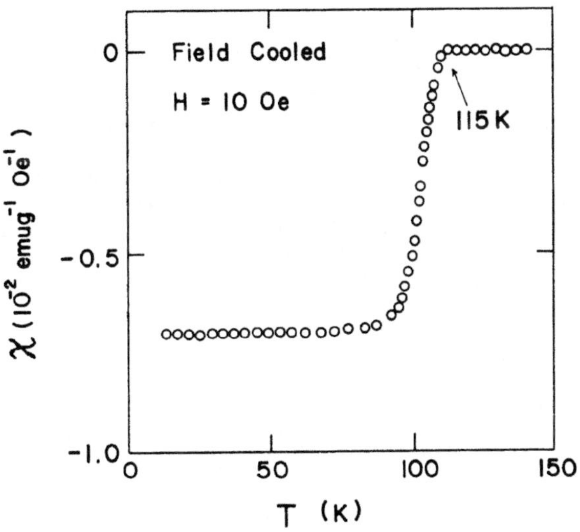

Fig.4. Temperature dependence of DC magnetic susceptibility for a single crystal of 2324-phase.

Figure 2 exhibits SEM image of a solidified sample of m=2 containing many plate-like crystals of 2324-phase. The single crystals were isolated from this solidified sample. The plate size was typically $1\times1mm^2$ wide and about 0.1mm thick. Figure 3 shows the surface morphology of a single crystal. Many growth steps could be observed on the surface. This suggests that the 2324-crystal is primarily crystalized from a liquid phase of m=2 composition. The single crystal was identified by XRD measurement and EDX compositional analyses. Lattice constants of the 2324-phase were determined to be a=b=3.85Å and c=42.0Å. The analyzed composition of a crystal was $Tl_{2.1}Ca_{2.6}Ba_2Cu_{3.8}O_x$. These crystals showed superconducting onset transition temperatures of 113 to 115K by means of DC SQUID magnetometer as shown in Fig.4. These values are slightly lower than that of 2223-phase single crystal.

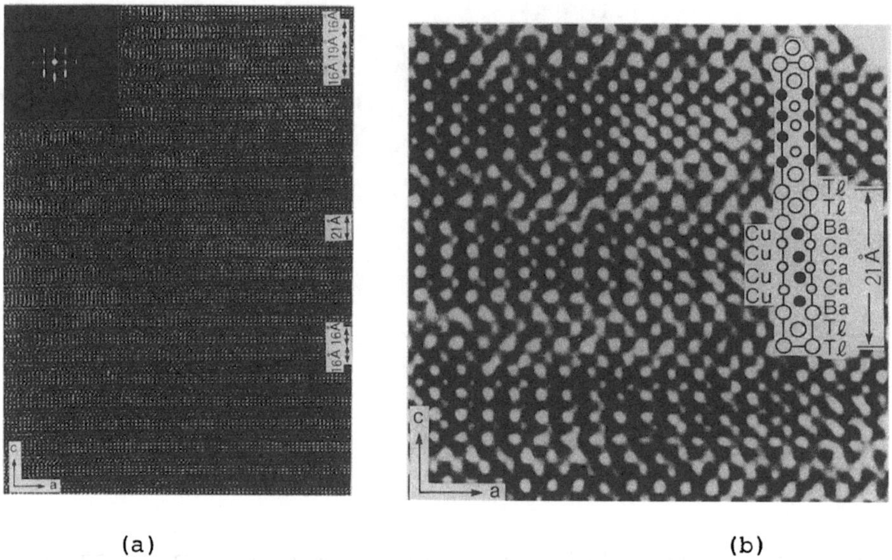

(a) (b)

Fig.5. TEM images of a 2324-phase single crystal.
(a) Low-magnification micrograph, (b) Lattice image of 2324-phase.

High-resolution TEM images of a 2324-phase single crystal were taken with incident electron beam parallel to [010] direction. The low magnification image exhibits layer structure alternating a dark layer with a light one, as shown in Fig. 5 (a). The dark layer is usually considered to be made of heavy metal, Tl or Ba atoms, while the light one is made of Cu or Ca atoms. In this image, the periodicity of stacking is mainly 21Å. However, a few singular stackings with different periodic lengths of 16Å or 19Å were also observed between the normal stacking layers. Figure 5 (b) shows an enlarged micrograph of a region having stacking periodicity of 21Å which is consistent to the half value of c-lattice constant from XRD measurement. This stacking structure have obviously double Tl-O layers and 4 Cu-O layers.

In summary, we grew mm-sized single crystals of $Tl_2Ca_3Ba_2Cu_4O_x$ with superconducting onset transition temperatures of from 113 to 115K. These transition temperatures are slightly lower than that of a $Tl_2Ca_2Ba_2Cu_3O_x$ single crystal having triple Cu-O layers. Furthermore, high-resolution TEM observation of a single crystal confirmed the stacking structure containing double Tl-O layers and 4 Cu-O layers. The stacking periodicity of layer structure is in good agreement with the result of XRD measurement. However, some intergrowth of different stacking structures can be seen within a 2324-phase. Further study of decreasing intergrowths is necessary for precise characterization of these materials.

REFERENCES

1. Z. Z. Sheng and A. M. Hermann (1988) Nature 332:138.
2. S. S. P. Parkin, V. Y. Lee, A. I Nazzal, R. Savoy, T. C. Huang, G. Gorman, and R. Beyers (1988) Bulk superconductivity at 125K in $Tl_2Ca_2Ba_2Cu_3O_x$. Phys. Rev. B38:6531.
3. H. Ihara, R. Sugise, K. Hatashi, M. Terada, M. Jo, N. Hirabayashi, A. Negishi, N. Atoda, H. Oyanagi, T. Shimomura and S. Ohashi (1988) phys. Rev. B38:11952.
4. H. Kusuhara, T. Kotani, H. Takei and K. Tada (to be published).
5. H. Takei, T. Kotani, T. Kaneko and K. Tada (1988) Superconductivity on single crystals of Tl-Ca-Ba-Cu-O system. In: K. Kitazawa and T. Ishiguro (ed) Advance in Superconductivity, Springer-Verlag, Tokyo, p229.

Phase Diagram and Crystal Growth of Nd_2CuO_4 and Pr_2CuO_4 System

KUNIHIKO OKA and HIROMI UNOKI
Electrotechnical Laboratory, 1-4, Umezono 1-chome, Tsukuba, Ibaraki, 305 Japan

ABSTRUCT

Single crystals of $(NdCe)_2CuO_4$ and $(PrCe)_2CuO_4$ were grown by the slow-cooling, travelling-solvent floating-zone (TSFZ) and top-seeded solution growth (TSSG) method. A crystallographic phase diagram of the Nd_2O_3-CuO system was derived by means of differential thermal analysis, quenching technique and X-ray diffraction.

KEY WORDS: superconductivity, phase diagram, crystal growth, $(NdCe)_2CuO_4$ $(PrCe)_2CuO_4$, magnetization

INTRODUCTION

$(NdCe)_2CuO_4$ and $(PrCe)_2CuO_4$ have recently been known to be electron-doped superconductor with T' phase structure (1, 2). Single crystals are necessary to be synthsized, to make clear the mechanism of these new type of superconductive compounds. Therefore, we tried to make up phase diagram of Nd_2O_3-CuO system and to produce the single crystals(3). Here we have made the differential thermal analysis (DTA) measurements in several R_2CuO_4 (R=Pr, Eu, Sm and Gd) compounds. Trials of growing single crystals are also reported to be made by a few groups (4-6).

EXPERIMENTAL PROCESS

The phase diagram of Nd_2O_3-CuO system has been determined using DTA, quenching technique and X-ray diffraction measurements. The DTA experiments were carried out with every 5～10 mol% mixture between Nd_2O_3 and CuO at heating and cooling rate of 20 ℃/min in air. The eutectic and peritectic temperatures at ～1040℃ and ～1240℃, respectively, were roughly estimated by means of DTA experiment. Then, quenching was carried out by heating a few milligrams of the sample in a small platinum sheet, which was put into a muffle furnace at temperatures near 1040 and 1240℃. The furnace was heated stepwise every ten degrees at 15-minute intervals, and the melting was comfirmed by direct observation through the furnace inlet. After heating, the sample was removed from the furnace and quenched in air. The quenched samples were powdered, and phase identification was performed by X-ray powder diffraction. Nd_2CuO_4 melts incongruently at about 1240℃. The phase diagram (3) thus determined is shown in Fig. 1.

The DTA measurments of other R_2CuO_4 compounds were also performed in air at a heating rate of 20 ℃/min in air. All the DTA curves have two endothermic peaks, as shown in Fig. 2. The endothermic peaks at about 1040℃ are due to the eutectic points. The other endothermic peaks cor-

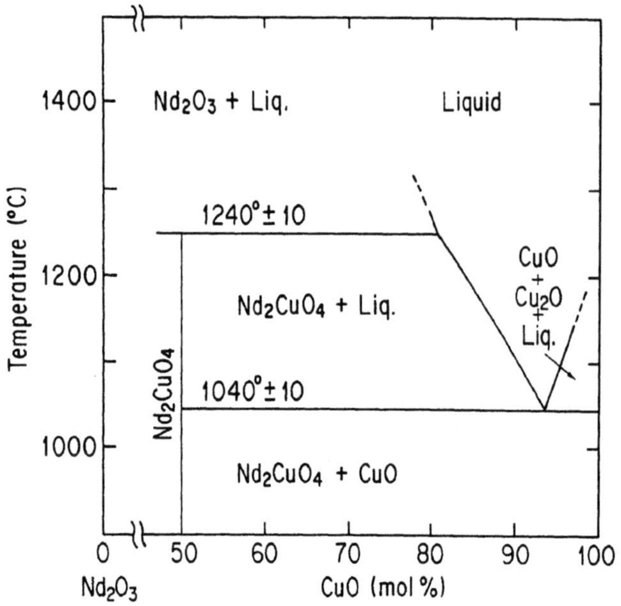

Fig. 1. Phase diagram of the system Nd_2O_3-CuO in air.

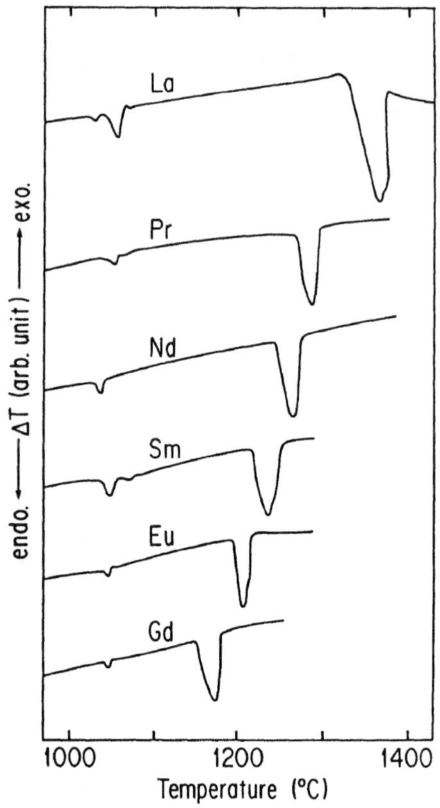

Fig. 2. DTA curves for six R_2CuO_4 compounds in air.

respond to the peritectic points(La:1320℃, Pr:1270℃, Nd:1240℃, Sm:1215℃, Eu:1190℃ and Gd:1150℃). These two separated endothermic peaks imply that, in each phase diagram of the R_2O_3-CuO system, a liquidus line which is more favorable for precipitation of these compounds during cooling could exist. As can be seen in Figs. 1 and 2 and in the phase diagram of La_2O_3-CuO (7) which we reported earlier, the concentration width of the liquidus line increases with increasing ionic radii of R atoms.

RESULTS AND DISCUSSION

Crystals of $(NdCe)_2CuO_4$ and $(PrCe)_2CuO_4$ have been tried to grow in accordance with this phase diagram of the Nd_2O_3-CuO system and the DTA measurements. Nd_2O_3-CuO system suggests that it is possible to grow a single crystal of Nd_2CuO_4 by starting from a CuO-rich solution composed of 82 ~ 93 mol% CuO. The DTA mesurements suggests that the Pr_2CuO_4 crystal should be grown in the same way as the Nd_2CuO_4 crystal could be grown. The starting materials were prepared with Nd_2O_3, Pr_6O_{11}, CeO_2 and CuO of 99.9% purity. The starting materials consisted with $(Nd_{1-x}Ce_x)_2O_3$/CuO and $(Pr_{1-x}Ce_x)_2O_3$/CuO mixtures at the molecular ratio of 15/85 were mixed, which was the identical ratio being used for the solution in the slow-cooling, TSSG and TSFZ methods.

As a first step to grow single crystals, we applied the slow-cooling method. The 100 g of starting material powder to fill the 100 cm^3 platinum crucible was heated at 1260 ℃ in a muffle furnace for 1 h , then cooled slowly at a rate of 15℃/h down to 1040℃; the solid flux was annealed and cooled slowly to room temperature. A crystal grown on the surface of the cruicible by the slow-cooling method is shown in Fig. 3.

Fig. 3. Nd_2CuO_4 single crystal grown by the slow-cooling method.

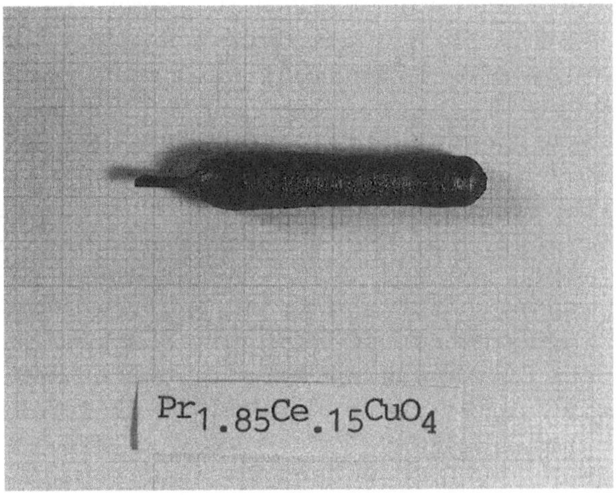

Fig. 4. $Pr_{1.85}Ce_{0.15}CuO_4$ single crystal grown by the travelling-solvent floating-zone method.

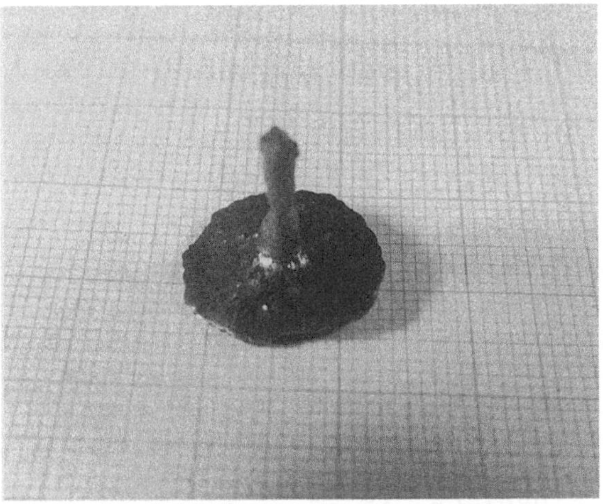

Fig. 5. Nd_2CuO_4 single crystal grown by the top-seeded solution growth method.

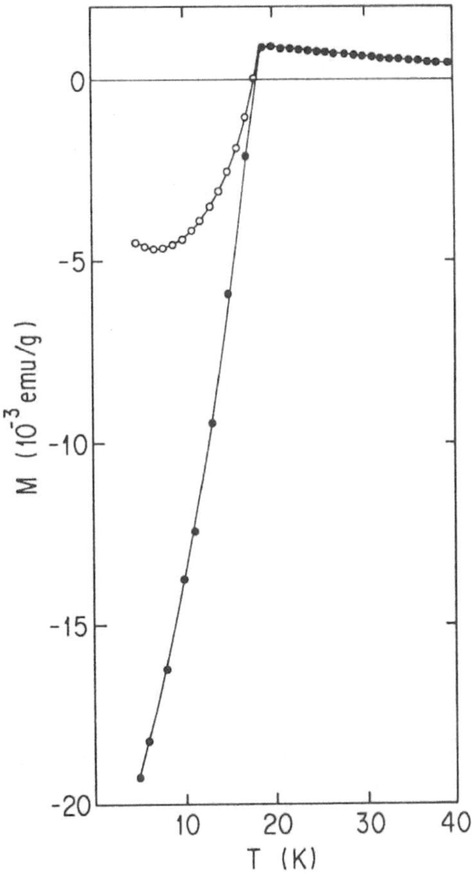

Fig. 6. Temperature dependence of zero-field cooled and field cooled magnetization at 10 Oe in TSFZ-grown $Nd_{1.85}Ce_{0.15}CuO_4$ single crystal

A pure Nd_2CuO_4 crystal is grown in a large plate-like shape. However, attempts to produce Ce-doped $(NdCe)_2CuO_4$ crystals brought us those having only dimensions of 3 mm X 3 mm in C-plate with a thickness of 0.5 mm.

In growing crystals by the TSFZ method we have employed an infrared-radiation-convergence type furnace. The $(Nd_{1-x}Ce_x)_2CuO_4$ and $(Pr_{1-x}Ce_x)_2CuO_4$ starting materials and the solvent materials, $(Nd_{1-x}Ce_x)_2O_3$/CuO and $(Pr_{1-x}Ce_x)_2O_3$/CuO mix-tures at 15/85 ratio, were first calcined at 1000 ℃, then pressed into the rod shape of 8 mm in diameter, and finally sintered at 1220℃ and 1040℃ in air, respectively. Conditions for the typical growth run were as follows. About 1.5～2.0 g of the solvent rod was used. The scanning speed was 0.5 mm/h. The polycrystalline rod and the seed crystal were rotated at 30 rpm in opposite direction to each other. A crystal grown in air by the TSFZ method is show in Fig.4.

For the TSSG method we used a platinum crucible of 50 mm in diameter and 35 mm in depth and an rf-induction heating furnace . The charged material was heated to complete the melting, and then the seed of TSFZ-grown crystal was dipped into the melt, which was cooled at 2～5 ℃/h. Pulling was carried out at a rate of 0.5 mm/h. The seed was rotated at 20～30 rpm. The Nd_2CuO_4 crystal grown by the TSSG method is shown in Fig.5. The crystal has the dimensions of 15 mm in diameter and 3 mm in length and is 2.3 g in weight.

Measurement of the temperature dependence of the zero-field and field cooled magnetization at 10 Oe was performed on the TSFZ single crystal of $Nd_{1.85}Ce_{0.15}CuO_4$, annealed at 1000℃ for 60 hr in the Ar atmosphere, the result is shown in Fig. 6. We found that the crystal is superconductive below ～19 K.

The authors would like to thank Y.Nisihara and T.Katayama for the magnetic susceptibility measurement.

REFERENCES

(1) Tokura Y, Takagi H, Uchida S, (1989) Superconducting copper oxide compound with electrons as the charge carriers. Nature, 337: 345-347

(2) Takagi H, Uchida S, Tokura Y, (1989) Superconductivity produced by electron doping in CuO_2-layered compounds. Phys.Rev.Lett., 62: 1197-1200

(3) Oka K, Unoki H, (1989) Pase diagram and crystal growth of superconductive $(NdCe)_2CuO_4$. Jpn.J.Appl.Phys., 28 : L937-L939

(4) Hidaka Y, Suzuki M, (1989) Growth and anisotropic superconducting properties of $Nd_{2-x}Ce_xCuO_{4-y}$ single crystals. Nature, 338 : 635-637

(5) Tarasacon JM, Wang E, Greene LH, Bagley BG, Hull GW, D'Egidio SM, Miceli PF, Wang ZZ, Jing TW, Clayhold C, Brawner D, Ong NP (1989) Growth, structual, and physical properties of superconducting $Nd_{2-x}Ce_xCuO_4$ crystals. Phys. Rev., B40 : 4494-4502

(6) Cassanho A, Gabbe DR, Jenssen HP, (1989) Growth of single crystals of pure and Ce-doped Nd_2CuO_4. J. Cryst. Growth, 96 : 999-1001

(7) Oka K, Unoki H, (1989) Phase diagram of the La_2O_3-CuO system and crystal growth of $(LaBa)_2CuO_4$. Jpn.J.Appl.Phys.: 26 L1590-L1592

2.4 Melt Growth

Melt Textured Growth: Related Process and Its Characterization

S. JIN
AT&T Bell Laboratories, Murray Hill, NJ 07974, USA

ABSTRACT

Advances toward major technological applications of the bulk, high T_c superconductors have been hindered by two major barriers, i.e., the Josephson weak-links at grain boundaries and the lack of sufficient intragrain flux pinning. It has been demonstrated that the weak link problem can be overcome by extreme alignment of grains such as in Melt-Textured-Growth (MTG) materials. Modified or improved processing by various laboratories has produced further increased critical currents. However, the insufficient flux pinning seems to limit the critical current density in high fields to about $10^4 - 10^5 A/cm^2$ at 77K, which is not satisfactory for many applications. In this paper, processing, microstructure, critical current and flux-pinning behavior of the MTG type superconductors will be described.

KEY WORDS: superconductor, critical current, melt-textured-growth

INTRODUCTION

The recent discovery of high T_c superconductors such as Y-Ba-Cu-O, Bi-Sr-Ca-Cu-O, and Tl-Ba-Ca-Cu-O, has led to unprecedented excitement and intense research effort. For their potential for significant technological applications to be realized, they should carry sufficiently high electrical currents in high magnetic fields.

The presence of Josephson weak links at grain boundaries results in unacceptably low critical current density (J_c) in polycrystalline high T_c superconductors.[1] While the exact nature of the weak links is still not clearly understood, it has been shown that the problem can be overcome by avoiding (or minimizing) grain boundaries in the direction of current flow, for example, by epitaxial growth in thin films[2,3] and melt-textured-growth (MTG) in the bulk Y-Ba-Cu-O superconductor.[4]

In this paper, the processing, microstructure, and critical current behavior of the MTG type superconductors will be discussed.

EXPERIMENTAL

Melt-textured growth processing was carried out by heating sintered bars of $YBa_2Cu_3O_{7-\delta}$ to the partial melt region (~1050°C or above) and continuously cooling in a temperature gradient, as described previously.[4] The resistivity vs temperature, critical current density vs field, ac susceptibility curves, and the optical and SEM micrographs were obtained by standard laboratory procedures.

RESULTS AND DISCUSSIONS

In melt-textured-growth, $YBa_2Cu_3O_{7-\delta}$ (1-2-3 phase) is first decomposed at ~1050°C or above into Y_2BaCuO_x (2-1-1 phase) and Cu, Ba-rich liquid. Upon slow cooling in oxygen atmosphere to below the solidus temperature (~1010°C), the 1-2-3 phase nucleates from the liquid and grows to large parallel plates (packets) due to the preferential crystal growth in the a-b direction. The presence of a temperature gradient during this crystallization period (~1010°C to ~900°C) helps to prevent the undesirable, multiple nucleation of the packets at various locations along the sample length, thus inducing the unindirectional growth of the plates. After the crystallization of the 1-2-3 phase is completed, the material, being very dense, has to be cooled slowly to allow sufficient uptake of oxygen at low temperatures. The MTG sample exhibits very sharp resistivity transition (ΔT (90%-10%) <0.3K) and ac susceptibility transition as compared to sintered samples.

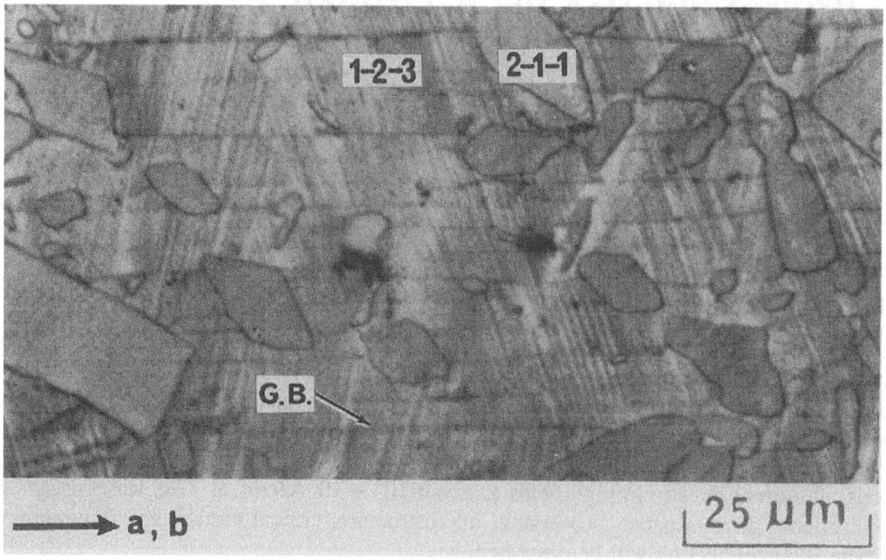

Fig. 1 Optical microstructure of the melt-textured-growth Y-Ba-Cu-O showing a near perfect alignment of the 1-2-3 phase.

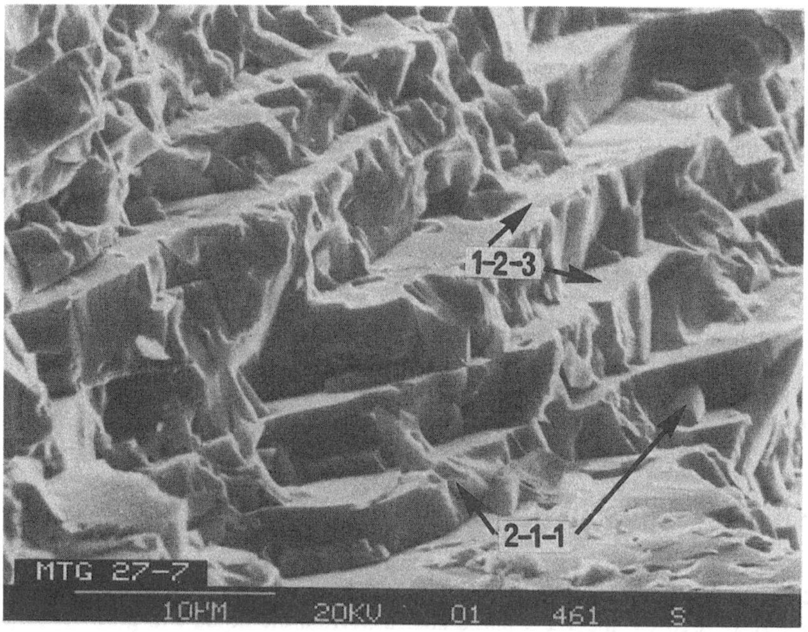

Fig. 2 SEM photograph from the fracture surface of the MTG sample.

Shown in Fig. 1 is an optical microstructure of the MTG material revealing a near-perfect alignment of the 1-2-3 superconductor phase in the a-b direction along the length of the sample. The aligned grain structure is beneficial in resolving three major problems in polycrystalline Y-Ba-Cu-O, i.e., i) weak-link grain boundaries in the paths of current flow, ii) anisotropic thermal contraction and microcracking, and iii) anisotropic superconductivity. The blocky, isolated particles are the 2-1-1 phase, which do not interfere much with the supercurrent flow except for the volume fraction effect. The transport current passed along the sample length (horizontal axis in Fig. 1) does not encounter high-angle grain boundary weak links. Figure 2 is a scanning electron microscope photograph taken from the fracture surface, and shows the plate-like morphology of the 1-2-3 phase.

MTG samples have another unique microstructural feature (in addition to the grain alignment), i.e., very high defect density, as shown by transmission electron microscopy, Fig. 3. Compared to the typical sintered materials, the MTG samples exhibit 2-3 orders of magnitude higher density of dislocations (about $10^9 - 10^{10}$ lines/cm^2), more planar defects, somewhat higher twin density, and dispersed 2-1-1 phase particles (1-20 μm in size). How these various defects influence the critical current behavior of the MTG material is not clearly understood at this time.

As a result of grain alignment in the melt-textured Y-Ba-Cu-O, significantly improved J_c and greatly reduced field dependency of J_c are achieved as shown in Fig. 4. The figure includes recent data obtained by a number of researchers using the MTG type processing, or related or improved melt processing, e.g., by Salama, et al[5] and Murakami, et al.[6] $J_c(H)$ values for melt-textured Y-Ba-Cu-O, in the range of $10^4 - 10^5$ A/cm^2, are improved by orders of magnitude over the values for the random-grained Y-Ba-Cu-O, especially in magnetic fields.

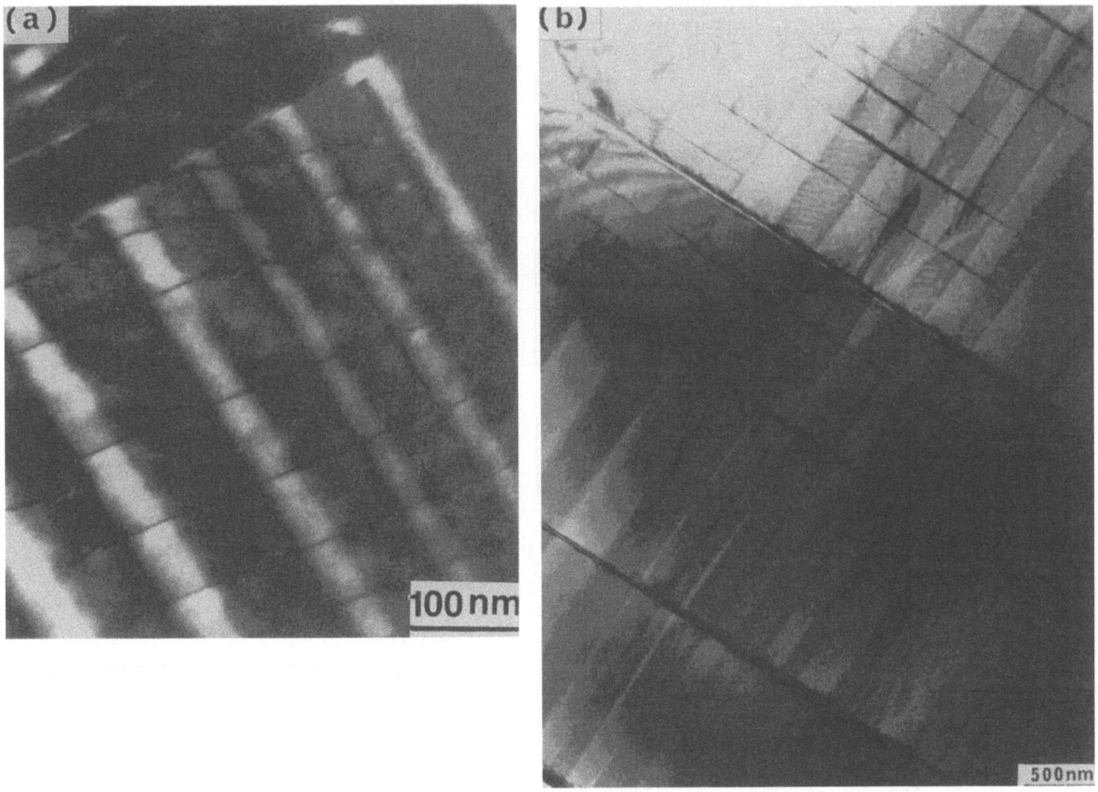

Fig. 3 TEM microstructures of the MTG sample showing (a) dislocations and (b) planar defects.

MTG-type samples show large magnetic hysteresis at 77K as illustrated in Fig. 5. Because of the large positive magnetization after the field is removed, MTG samples exhibit a persistent-current-based, magnet-like behavior as well as a strong suspension phenomenon.[7] From the magnetization loop in Fig. 5, J_c is calculated (using the Bean model and the sample width of 0.4 mm) to be $\sim 10^4$ A/cm^2 at H = 1 Tesla, which agrees roughly with the transport J_c data in Fig. 4.

The time-dependent decay of magnetization in the MTG sample has been studied at 77K and H = 1 Tesla, and the result is shown in Fig. 6. There is an initial rapid decay of M, which slows down considerably after longer time. Approximate 15% decrease in M is observed after 2 minutes, which is attributed to the flux creep in high field.

Many of the potential bulk applications of the high T_c superconductors depend on the capability of the materials to carry sufficiently large currents, i.e., $J_c > \sim 10^5$ A/cm^2 at 77K in H > 1 Tesla. While the weak-link problem in bulk materials can be avoided through grain alignment, the high-field J_c still seems to be limited to $10^4 - 10^5$ A/cm^2 at 77K. Considering that weak-link-free, epitaxial thin films of $YBa_2Cu_3O_{7-\delta}$ are known to exhibit $J_c(H)$ of more than 10^6 A/cm^2, it is most likely to be the low density of flux-pinning sites in the bulk Y-Ba-Cu-O which limits $J_c(H)$ to $10^4 - 10^5$ A/cm^2.

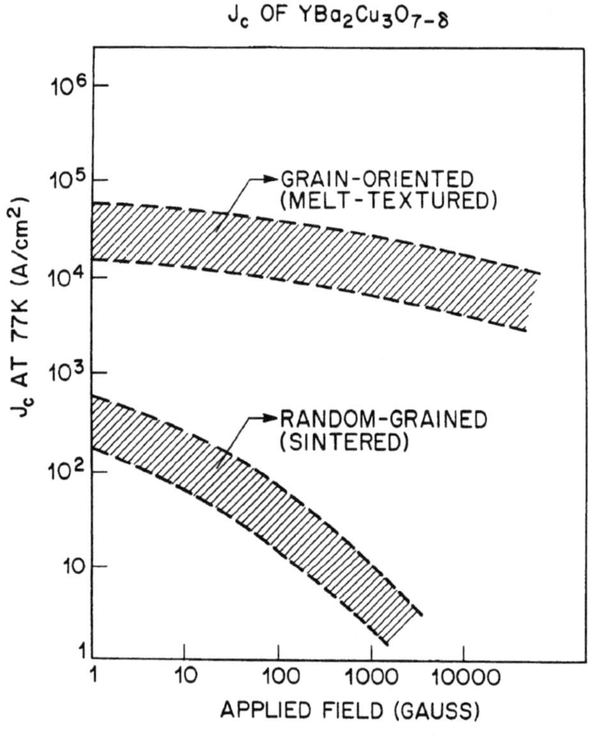

Fig. 4 Transport J_c vs H in bulk Y-Ba-Cu-O.

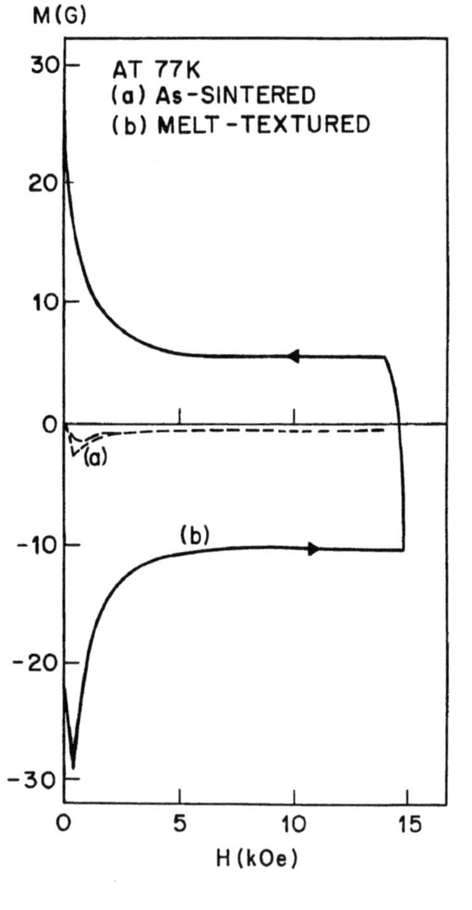

Fig. 5 M-H loop for the MTG sample at 77K.

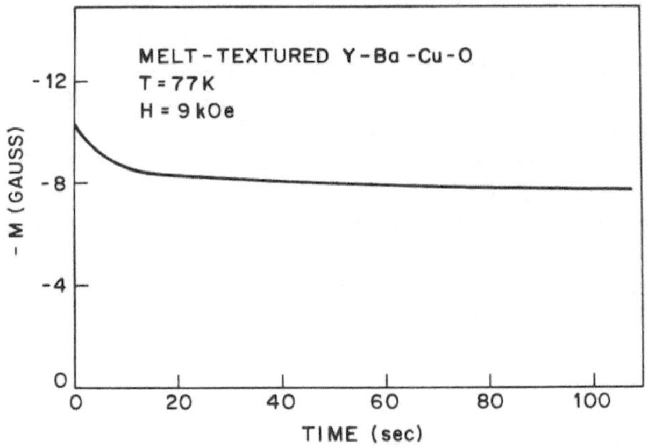

Fig. 6 Time-dependent decay of magnetization in the MTG sample.

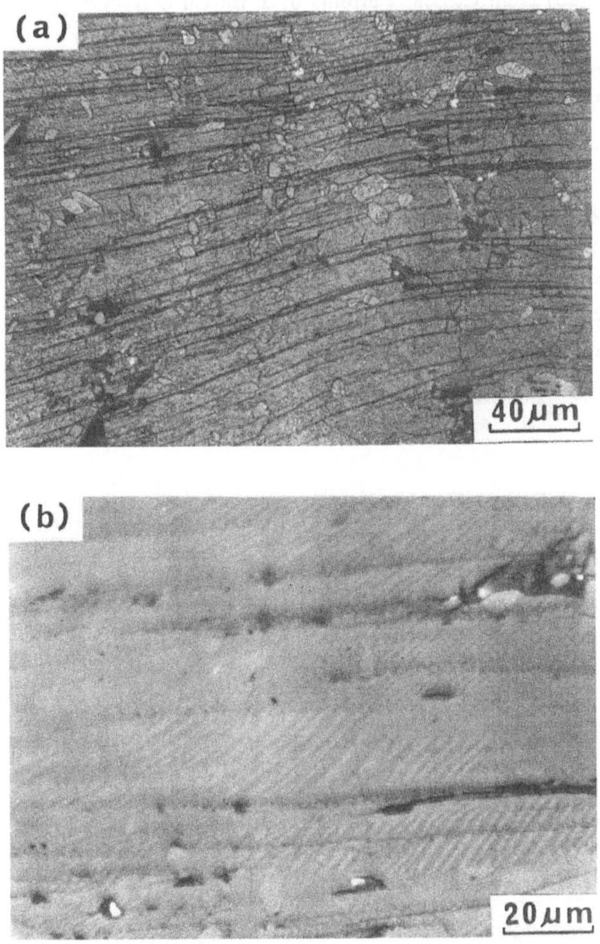

Fig. 7 Evidence of plasticity in some MTG samples revealed by (a) curved low-angle grain boundaries and (b) curved twin traces.

Although a special processing method such as neutron irradiation has proven effective in flux-pinning improvement, it is desirable to have commercially viable processing that will generate a large number of flux-pinning defects. There may be quite a few processing possibilities that have not been fully explored for flux-pinning in high T_c superconductors. One example may be the utilization of high temperature plastic deformation for generation of microscopic defects which could serve as effective pins. The abundance of slip dislocations[8] ([100] or [010] type Burgers vectors and (100), (010) or (001) slip planes) in MTG samples is well documented. Considerable plasticity is observed in some melt-textured Y-Ba-Cu-O as evidenced by curved low-angle grain boundaries (Fig. 7(a)) and bent twin traces (Fig. 7(b)). Other possibilities such as shock wave loading, chemical doping, or phase decomposition and precipitation reactions may also be explored. The creation of efficient flux pinning sites remains as a challenge in the materials science of high T_c superconductors.

REFERENCES

1. Ekin J. W., (1987) Transport Critical Current in Bulk Sintered $YBa_2Cu_3O_x$ and Possibilities for Its Enhancement. Adv. Ceram. Mat. 2: 586-592.

2. Chaudhari P., Koch R. H., Laibowitz R. B., McGuire T. R., Gambino R. J., (1987) Critical Current Measurement in Epitaxial Films of $YBa_2Cu_3O_{7-x}$ Compound. Phy. Rev. Lett. 58: 2684-2686.

3. Dimos D., Chaudhari P., Mannhart J., LeGoues F. K., (1988) Orientation Dependence of Grain Boundary Critical Currents in $YBa_2Cu_3O_{7-\delta}$ Bicrystals. Phys. Rev. Lett. 61: 219-222.

4. Jin S., Tiefel T. H., Sherwood R. C., Davis M. E., van Dover R. B., Kammlott G. W., Fastnacht R. A., Keith H. D., (1988) High Critical Currents in Y-Ba-Cu-O Superconductors. Appl. Phys. Lett. 52: 2074-2076.

5. Salama K., Selvamanickam V., Gao L., Sun K., (1989) High Current Density in Bulk $YBa_2Cu_3O_x$ Superconductor. Appl. Phys. Lett. 54: 2352-2354.

6. Murakami M., Morita M., Doi K., Miyamoto K., (1989) A New Process With the Promise of High J_c in Oxide Superconductors. Jap. J. Appl. Phys. 28: 1189-1194.

7. Jin S., Sherwood R. C., Gyorgy E. M., Tiefel T. H., van Dover R. B., Nakahara S., Schneemeyer L. F., Fastnacht R. A., Davis M. E., (1988) Large Magnetic Hysteresis in a Melt-Textured Y-Ba-Cu-O Superconductor. Appl. Phys. Lett. 54: 584-586.

8. Nakahara S., Jin S., Sherwood R. C., Tiefel T. H., (1988) Analysis of Dislocations in Y-Ba-Cu-O Superconductor. Appl. Phys. Lett. 54: 1926-1928.

Directional Solidification Processing of High T_c Superconducting Oxides

Y. SHIOHARA, M. NAKAGAWA, T. SUGA, K. ISHIGE, T. OYAMA, T. IZUMI, S. NAGAYA, M. MIYAJIMA, I. HIRABAYASHI, and S. TANAKA

Superconductivity Research Laboratory, International Superconductivity Technology Center, 10-13, Sinonome 1-chome, Koto-ku, Tokyo, 135 Japan

ABSTRACT

Unidirectional solidification processing, including Floating Zone Melting, Laser Zone Melting, and Vertical Bridgman methods for making high T_c superconducting oxides has been investigated. The effects of processing parameters on texture, morphology, and structures of unidirectionally grown crystals were investigated. Processing parameters include growth rate (R), temperature gradient (G). Two different crystallization mechanisms were recognized, one is non-equilibrium phase transformation under high GR conditions, and the other is due to local equilibrium phase transformation under low GR conditions. Strongly aligned structures were obtained at high G/R ratios with low GR value. The interrelation between the superconducting characteristics and textured structure produced was also discussed.

KEY WORDS: directional solidification, zone melting, Bridgman method, aligned structure,

INTRODUCTION

Following the discovery of high T_c superconducting oxides [1,2,3], enormous research and development efforts have been stimulated because of the potentially immence technological impact of these novel materials. Preliminary experiments were initiated to determine the feasibility of growing single crystals. Several processes developed to date have yielded large and high quality crystals which have been useful for property studies. Progress toward major application of the bulk high T_c superconductors has been hindered mainly due to the weak coupling between grains, since practical application of these materials require critical current density, J_c, of the order of 10^5~10^6 A/cm^2 It is generally known that J_c is sensitive to microstructure, and the possible sources of reducing J_c are; i) inhomogeneity [4], ii) secondary phases along grain boundaries [5], iii) the presence of porosity due to insufficient densification or microcracks resulting from the anisotropic thermal expansion along different crystallographic directions and phase transformation [6], and iv) anisotropy in conductivity. [7,8]

While the exact nature of the weak-link problem in sintered bulk superconducting oxides is still under further investigation, the potential importance of the melting-solidification processing route was revealed subsequently by the high critical current s demonstrated by Jin *et al.*[9] in textured $YBa_2Cu_3O_{7-X}$ materials produced by "melt-textured growth" (MTG) processing and by Murakami *et al.* [10,11] as well through "quench and melt growth" (QMG) processing. Feigelson *et al.* [12] also reported results for the Bi-Sr-Ca-Cu-O system that clearly demonstrated the feasibility of growing textured fibers of superconducting oxides by the "pedestal" floating zone process. These three processes result in significant improvement of critical currents over samples made by conventional powder processing routes. Brody *et al.* [13] demonstrated the laser heated float zone processing showing great promise for preparing single crystals and/or highly textured ceramics of incongruently melting superconductors. Recently, Levinson *et al.* [14] reported J_c = 450A/cm^2 at 77K in the thick films as processed by laser zone melting with rather slow beam scanning, and improved to 2200 A/cm^2 by post processing anneal.

In the present paper, specifically the following solidification processes are reviewed and discussed from solidification and crystallization standpoints based on our recent experimental studies on Bi-Sr-Ca-Cu-O system: (i) vertical Bridgman unidirectional solidification method, (ii) halogen lamp heated float zone processing, and (iii) laser heated rapid zone melting processing.

UNIDIRECTIONAL SOLIDIFICATION PROCESSES

A variety of different processes are employed to produce alligned (well oriented) polycrystals from melts. These can be grouped in three categories [15] as those in which the entire charge is melted and then solidified from one end, only a small zone of the charge is melted at any one time and then the zone is moved either by moving the heat source with respect to the charge material or vice versa, and a large charge is melted and a small crystal withdrawn slowly from it.
The first category of crystal-growing technique is termed normal freezing. Here, a charge of the charge material is contained within a long crucible of small cross section open at the top. The charge is first melted in a suitable furnace. Next, the crucible is withdrawn slowly from the furnace and the crystal-liquid interface moves until the whole charge is crystallized. This is often termed the *Bridgman method.* In growing single crystals, it is not necessary that entire charge be molten. For some purposes, it is preferable to melt initially only a portion of the charge and move this molten zone through the charge (zone melting). Many types of heat sources are used for zone melting, including induction, resistance, elctron beam, *laser beam*, and *halogen lamp*. Zone melting is done either with or without crucible. The latter type, crucibleless zone melting, or *floating zone melting*, is widely used for reactive and high-melting-point materials. The molten zone is held in place by surface tension forces. Another crystal-growing technique from melts, used widely for nonmetals, is the crystal-pulling, or Czochralski technique. The pedestal growth technique and the flux method somewhat belong to this group. In this case, the charge material is placed in a crucible and melted. A seed crystal is attached to a vertical pull rod, lowered until it touches the melt, allowed to reach thermal equilibrium, and then raised slowly so that crystallization proceeds from the seed crystal. These and other techniques for growth of specific materials have been described[16].

EXPERIMENTAL

Highly aligned polycrystals were produced by unidirectional solidification techniques, including the Bridgman method, the halogen-lamp heated floating zone (FZ) method, and a laser heated zone melting (LZM) method. Precursor samples of $Bi_2Sr_2CaCu_2O_x$ were prepared by mixing of Bi_2O_3, $SrCO_3$, $CaCO_3$ and CuO, grinding, and calcined at 1073 K in air for 12 hrs., then pressed to form a rod (5 mm in diameter) at 2000 atm by Cold Isostatic Pressing (CIP). These precursor samples were used for the Bridgman method and the FZ method. Glassy $Bi_2Sr_2CaCu_2O_x$ precursors for the LZM method were prepared by quenching the molten materials at 1450 K. Details of the experimental procedures have been described elsewhere [17,18,19].

RESULTS AND DISCUSSION

Texture and Morphology

Microstructures of the processed samples were observed by Scanning Electron Microscopy (SEM) and optical microscopy. Figure 1 shows typical microstructure of as grown samples at different growth rates by the FZ method. The sample grown at 1 mm/h composed of highly aligned polycrystals as shown in Fig. 1 (a). The main phases in this specimen are the $Bi_2Sr_2CaCu_2O_x$ (2212) phase and the $Bi_2Sr_2CuO_x$ (2201) phase with a small amount of the $Sr(Ca)CuO_x$ (0101) phase. At the higher growth rate (10mm/h), alignments of the crystals become random and a large amount of the 0101 phase (bright phase) appears. This results suggests that the superconducting 2212 and 2201 phases formed at low growth rates and these phases transforms probably by a peritectic reaction from the liquid and the primary phase of $Sr(Ca)CuO_x$ which requires long transformation time due to involvment of solid diffusion.

Figure 2 shows microstructures of the samples grown by the Bridgman method. In this method, G and R were varied independently while only R could be changed in the FZ method. Microstructures with high G/R ratios consist of the 2212 and 2201 phases with a highly textured manner, as shown in Fig.2(a), while with low G/R ratios grains with randomly oriented were observed, as shown in Fig. 2(b). As was seen in the samples by the FZ method, the primary phase crystals of the $Sr(Ca)CuO_x$ (0101) phase were seen coexisting with the superconducting 2212 and 2201 phases,[Fig.2(c)]. The fractions of this primary phase in the specimen again decreases with decreasing G and/or R.

Microstructures of the samples processed by the LZM method are shown in Figure 3 with the results of phase identifications of the samples by X-ray diffraction (Fig.4). Figure 3(a) shows the samples with a high scanning rate of the YAG laser beam . The XRD pattern indicates noncrystallization, i.e. amorphous formation [Fig. 4(a)]. This structure is similar to that of the precursor materials. The samples processed by the LZM method under certain experimental conditions of laser scanning speed, laser input power, and hot stage temperature to heat the precursor during the process, of 5 mm/s, 0.6 W, and 450°C respectively show about 200 μm width of a processed zone, consisting of three phases including a needle like phase, a dark phase, and a bright phase as shown in Figure 3(b). The X-ray diffraction pattern of this sample, as shown in Fig.4(b), indicates that this sample is composed of the $Bi_2Sr_2CaCu_2O_x$ (2212) phase with a small amount of the Bi_2SrO_x crystals. In addition, the intensity enhancements of the $00l$ peaks suggests that the c-axis is oriented normal to the surface.

Effects of G·R and G/R on Crystallization

In general, unidirectionallity of the crystals strongly depends on the G/R ratio. The higher the G/R ratio, the more highly aligned grain structures were obtained. In other words, with proper control of the growth conditions the phases will be aligned even for high growth rates. The product GR which represents the local cooling rate, generally affects coarsening of the phases, phase selection, and phase transformation. Figure 5 summarizes the effects of G and R on the phase selection as a practical processing window, including the results of the three different processes. In the BSCCO system, the superconducting $Bi_2Sr_2CaCu_2O_x$ (2212) phase can be formed under the two different conditions, high and low GR ranges. In the high GR range, the $Sr(Ca)CuO_x$ (0101) phase, the $Bi_2Sr_2CuO_x$ (2201) phase, the $Bi_2Sr_2CaCu_2O_x$ (2212) phase, and the amorphous phase appears by turns with increasing the GR values.This crystallization mechanism was considered to be non-equilibrium phase transformation, since significant crystallization kinetics was required for high growth rates, and high undercoolings for nucleation as well as for rapid growth were involved. In the low GR range, the 2201 phase, the 2212 phase and the 0101 phase appeared by turns with decreasing the GR values. The crystallization mechanism was considered to be a local equilibrium phase transformation. In this case, appearance of each phase is not necessarily the result from direct crystallization from the liquid. The condition with low GR supplys enough time for diffusion in solid state transformation including a peritectic reaction, in other words, this condition allows a long annealing process included in its entire solidification process. In this range, an increase in volume fraction of the 2212 phase was recognized with decreasing the GR values.

Resistivity Measurements

The transport properties of the samples were measured using a standard four-probe technique. Figures 6 and 7 show the temperature dependence of resistivity of the samples. Figure 6 shows the results of the samples produced with different growth rates by the FZ method. In this method, the temperature gradient at the interface is about 100K/cm which was almost independent of the growth rate. An increase of Tc(R=0) from 75 to 86K was seen with decreasing the growth rates from 5 mm/h to 0.5 mm/h, which indicates the effect of the long annealing. Figure 7 shows the results of the samples grown with different values of GR. The resistivity of the samples with high GR exhibited a semiconducting behavior. The resistivity of the samples with lower GR exhibited a metallic behavior and decreased sharply at around 80K. Finaly, the resistivity of the sample with the GR value of 3.5×10^{-3} K/s reached zero at 80K. Resistivity measurements for the samples produced by the LZM method could not be made due to so small area of the processed zone. Further investigations are required and they are in progress.

ACKNOWLEDGMENTS

The authors express their deep gratitude to the Ministry of International Trade and Industry (MITI) for support of this work in part through the New Energy and Industrial Technology Development Organization (NEDO).

REFERENCES

1. Bednorz JG, Muller KA (1986) Z. Phys. B64: 189
2. Wu MK, Ashburn JR, Torng CJ, Hor PH, Meng RL, Gao L, Huang ZJ, Wang YQ, Chu CW (1987) Phys. Rev. Lett. 58: 908

3. Maeda H, Tanaka Y, Fukutomi M, Asno T (1988) J. Appl. Phys. 27: L209

4. Babcok SE, Kelly TF, Lee PJ, Seuntjens JM, LaVanier LA, Larbalestier DC (1988) Physica C 152: 25

5. Nakahara S, Fisanick GJ, Yan MF, vanDover RB, Boone TJ (1987) Cryst.Growth 85: 639

6. O'Bryan HM, Gallagher PK (1987) Adv. Ceram. Mater. 2: 640

7. Dinger TR, Worthington TK, Gallagher WJ, Sandstrom RL (1987) Phys.Rev.Lett 59: 2687

8. Tozer SW, Kleinsasser AW, Penney T, Kaiser D, Holtzberg F (1987) Phys. Rev.Lett. 59: 1768

9. Jin S, Tiefel RC, Sherwood RC, Davis ME, vanDover RB, Kammlott GW, Fastnacht RA, Keith HD (1988) Appl. Phys. Lett. 52: 2074

10. Murakami M, Matsuda S, Sawano K, Miyamoto K, Hayashi A, Morita M, Doi K, Teshima H, Sugiyama M, Kimura M, Fujinami M, Saga M, Matsuo M, Hamada H (1988) Proceedings of the 1st International Symposium on Superconductivity (ISS'88): 247

11. Murakami M, Morita M, Doi K, Miyamoto K (1989) Japanese Journal of Applied Physics 28, 7: 1189

12. Gazit D, Feigelson RS (1988) Journal of Cryst. Growth 91: 318

13. Brody HD, Haggerty JS, Cima MJ, Flemings MC, Barns RL, Gyorgy EM, Johnson DW, Rhodes WW, Sunder WA, Laudise RA (1989) Journal of Cryst. Growth 96: 225

14.Levinson M, Shah SSP, Wang DY (1989) Appl. Phys. Lett. 55, 16: 1683

15. Fleming MC (1974) Solidification Processing, McGraw-Hill New York: 1

16. Laudice RA (1967) Techniques of Crystal Growth. In: Reiner HS (ed) Crystal Growth, Pergamon Press, New York

17. Izumi T, Oyama T, Shiohara Y (1989) in this proceedings

18. Nakagawa M, Shiohara Y (1989) in this proceedings

19. Ishige K, Suga T, Shiohara Y (1989) in this proceedings

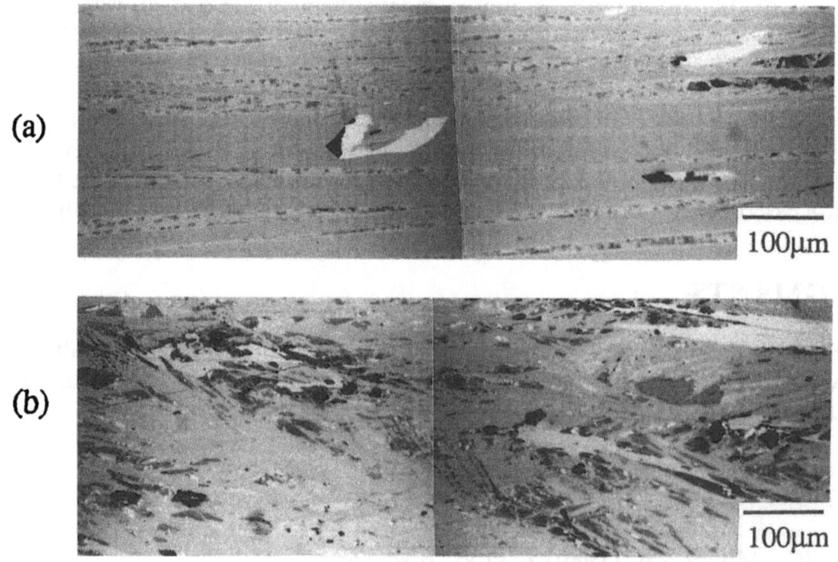

Figure 1 Microstructures of the two samples grown by the FZ method with different growth rates, showing the difference in crystal alignments. Growth rates: (a) 1mm/h, and (b) 10 mm/h.

(a) (b) (c)

Figure 2 Microstructures of the samples grown by the Bridgman method with different G/R ratios, showing the difference in crystal alignments.
G/R: (a) 6.93×10^{10} sK/m^2, and (b) 1.26×10^{10} sK/m^2.
Also, showing unidirectionally grown primary crystals of $Sr(Ca)CuO_x$ (bright phase) in (c).

(a) 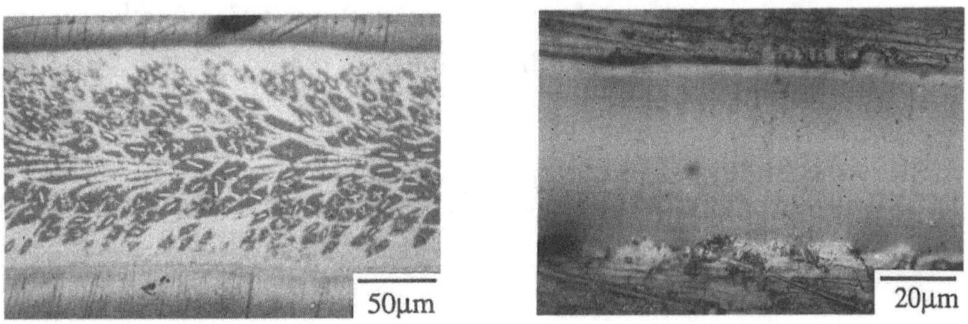 (b)

Figure 3 Microstructures of the samples processed by the LZM method with different GR values.
GR: (a) 1.45×10^4 K/s, and (b) ~ 1.0×10^5 K/s.

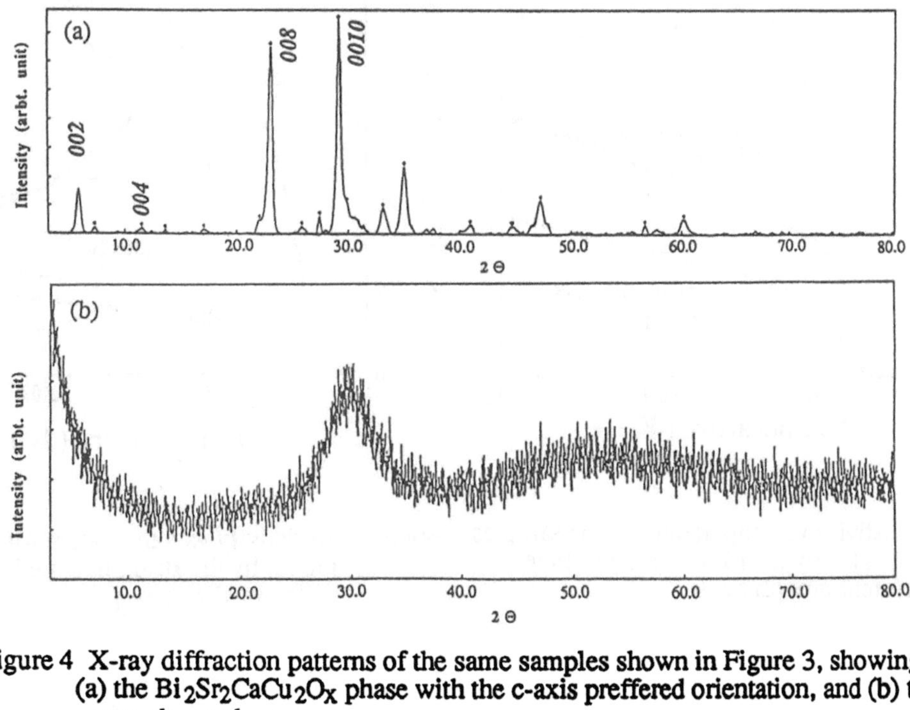

Figure 4 X-ray diffraction patterns of the same samples shown in Figure 3, showing (a) the $Bi_2Sr_2CaCu_2O_x$ phase with the c-axis preffered orientation, and (b) the amorphous phase.

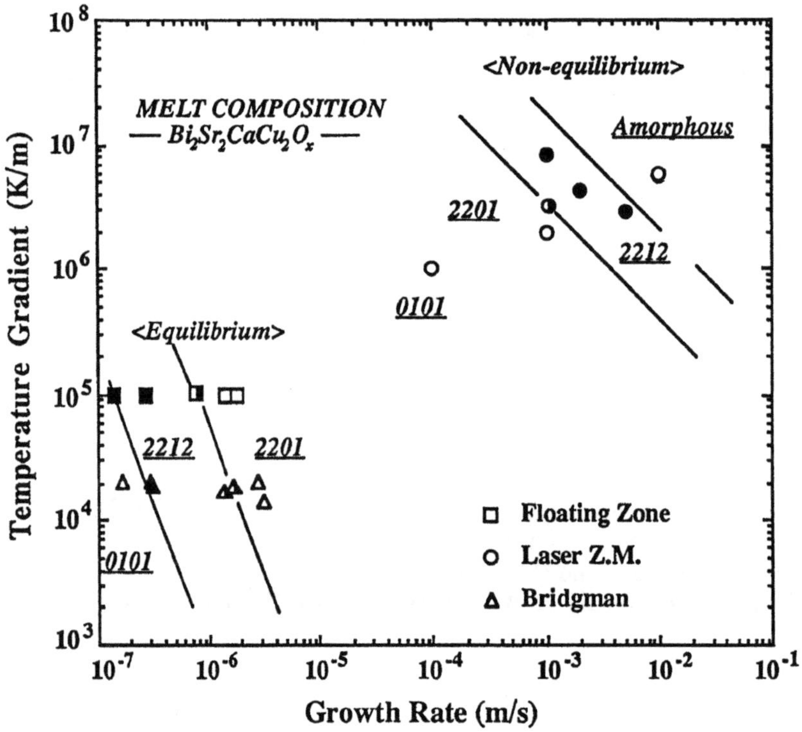

Figure 5 Effects of G and R on the phases appeared in the processed samples, showing two different regions for formation of the $Bi_2Sr_2CaCu_2O_x$ phase. Filled points indicate presence of the phase in the samples.

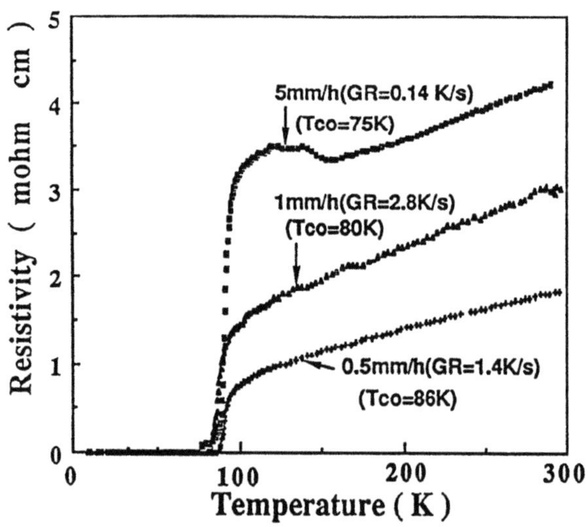

Figure 6 Resistivity vs temperature for the samples processed by the FZ method for three different growth rates.

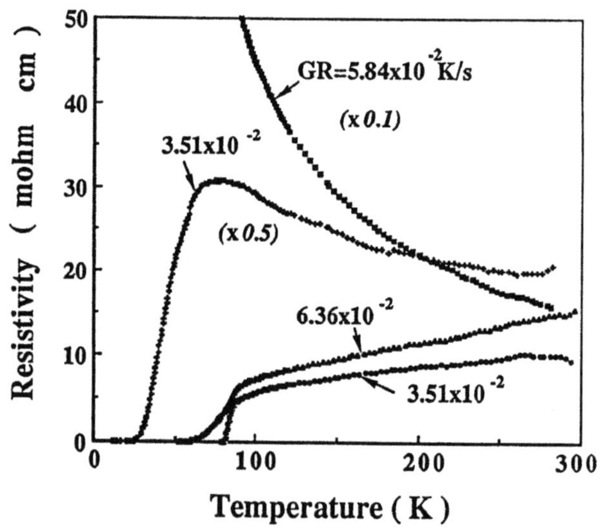

Figure 7 Resistivity vs temperature for the samples grown by the Bridgman method with four different GR values.

Improvement of Grain Boundary Weak-Links in YBCO Tape Shaped Wire Prepared by the Floating Zone Melting Technique

MICHIYA OKADA[1], TOYOTAKA YUASA[1], TOSHIMI MATSUMOTO[1], KATUZO AIHARA[1], MASAHIRO SEIDO[2], and SINPEI MATSUDA[1]

[1] Hitachi Research Laboratory, Hitachi Ltd., Hitachi, Ibaraki, 319-12 Japan
[2] Metal Research Laboratory, Hitachi Cable Ltd., Tsuchiura, Ibaraki, 300 Japan

ABSTRACT

Au-Sheathed Y-Ba-Cu-O(YBCO) tapes were fabricated by the floating zone melting techinique (FZM). A typical microstructure of melt-textured Y-123 policrystallite including fine Y-211 particle was observed. The tapes solidified at an optimized condition yielded Jc=1,000 A/cm^2 at 77K in a magnetic field of 1T. The enhancement of Jc in a magnetic field is suggested to be due to the improvement of weak-links in grain boundaries.

KEYWORDS: unidirectional solidification, superconducting wire, Y-Ba-Cu-O superconductor, critical current density, texture.

INTRODUCTION

Since the discovery of high-Tc superconducting oxides, [1-5] a number of studies have been made on the preparation of wires using the powder method. [6-11] However, the critical current densities Jc for superconducting wires reported so far is in the order of 10-10^2 A/cm^2 at 77K in the presence of magnetic fields. In the application of the superconducting oxides to a power apparatus such as a superconducting magnet, it is first necessary to develop a superconducting wire with a critical current density above 10^4 A/cm^2 in the presence of magnetic fields.
In our previous papers, [6-10] we reported that the drawing-rolling technique [6] is effective to enhance the Jc of Ag-sheathed wires in the absence of magnetic fields. The Jc of the "tape-shaped" wire [6] is enhanced not only by densification, but also by crystallite alignment of the oxide in the Ag sheath through rolling or pressing process, however, it is insufficient to overcome the problem of weak-link. [14, 15] In order to enhance the Jc of wires in magnetic fields, we have to overcome the weak-links in policrystallite. Because intra-grain Jc in bulk superconductor exceeds at least 10^4 A/cm^2 in magnetic fields(>0.1T) without providing any special pinning. [12, 13] In this paper, we applied an unidirectional solidification technique to Au-sheathed high-Tc superconducting wire. The relationship between fabrication conditions and superconducting properties is investigated.

EXPERIMENTAL

The $Y_1Ba_2Cu_3O_7$(Y-123) powder was prepared from Y_2O_3(purity 99.9%), BaO(99.9%), and CuO(99.9%). These powders were mixed and ground with a centrifugal ball mill for 1h, and were calcined at 950°C for 5h in flowing oxygen gas. The resultant compound was then pulverized. The powder was pressed at 1,000 kgf/cm^2 to form pelletes(30mm diameter, 2mm thick). These pellets were sintered under the same conditions as the above mentioned calcination, then ground again. In order to dispers the $Y_2Ba_1Cu_1O_5$(Y-211) during solidification, a small amount of fine Y_2O_3(Size:1-2μm) was added to the powder. This "Y-rich" YBCO powder was then packed into an Au-5%Pd tubing(6mm diameter, 0 .5mm thick, 400mm long). The initial packing density was 2.7g/cm^3. The Composite wire was drawn by using a draw bench with outer diameter of 2.8mm. The wire was, then, cold-rolled into tapes. After cold working, the YBCO core in the tape was unidierectionally solidified. The schematic drawing of experimental facility is shown in Fig. 1. A typical dimension of the sample tape eas 100mm long, 5mm wide and 0.5mm thick. A

pair of bilateral infra-red rays were focused onto a zonal area of the sample tape traveling in the vertical direction, in order to realize a unidirectional solidification of the oxide superconductor core by the floating zone melting technique(FZM). A Ni layer of 20μm thick was caoted on the surface of the tape to increase the absorption ratio of the infra-red ray. The solidification rate of the oxide superconductor core was 3-12 mm/h, and the temperature range was 900-1200°C/h. The thermal gradient was approximately 100-150°C/cm. In this experimental conditions, the unidirectional solidification of oxide superconductor core in metallic sheath was undertaken.
After the solidification, the tape was cut into 30-40mm long, then annealed at 400oC for 100h in a flowing oxygen atomosphere. The critical current density Jc was measured at 77K by the four prove technique. The criterion for the critical current was 1μV/cm. The Jc was also estimated from a magnetization curve acquired by a vibrating sample magnetometer. The microstructure of YBCO core were examined by a SEM and a polarlized optical microscope.

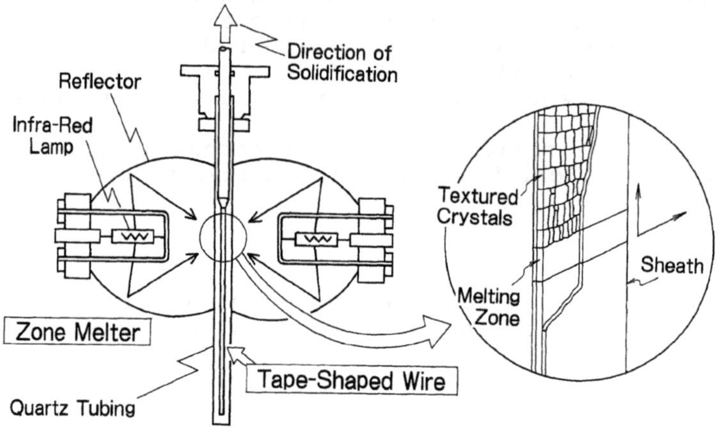

Fig. 1 Schematic picture of the directional solidification (FZM) of supperconductor core in metallic sheath

RESULTS AND DISCUSSION

Figure 2 shows the SEM image of the tape after the solidification (FZM-tape). Large and textured Y-123 crystals were obserbed. The microstructure is quite similar to melt-textured one. [12] The size of the Y-123 grains in the tape was 0.5-5mm, i.e. it is about two orders as large as that of the sintered one. [7] Fine Y-211 particeles with a size of 0.5-1 μm were also observed in the large Y-123 grains. The distribution of Y-211 particles were fairly uniform. The c-plane of Y-123 crystals were almost oriented in the longitudinal direction of the tape as shown in Fig.2a, however, the direction of the c-plane was random in the transversal direction as shown in Fig.2b. Around the grain boundary, the second phase was hardly observed except for the fine Y-211 particles.

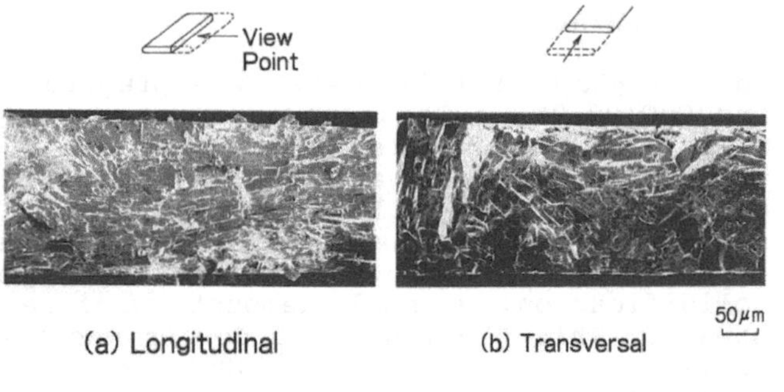

Fig. 2 SEM Images of fractured core cross-section in tape shaped wire

Figure 3 compares the transport Jc's in the magnetic field between the FZM tape and sintered tape. There obserbed no difference in the Jc's of these tapes in the absence of the magnetic field. While the Jc of the sintered tape decreases sharply in the very low magnetic fields, the Jc of the FZM tape still holds 1,000 A/cm^2 even in the magnetic fields of 1T. Figure 4 relates the intra-grain Jc in the magnetic field, estimated from the magetization curves at 77K by the Bean-model. [16] The calculated Jc yields at least 10,000 A/cm^2 at 77K in a magnetic field of 1T for both FZM and sintered tape. The intra-grain Jc of the FZM-tape is slightly higher than that of sintered one, however, there seems to be no significant difference. The fine Y-211 particle of the observed size would make little contributtion to the flux pinning. Therefore, it was assumed that the relatively higher transport Jc of the FZM-tape is mainly governed still by the weak-links in grain boundary. The augumentation of Jc of the FZM tape is thought to be due to the microstructual improvement of grain bounbdary weak-link.

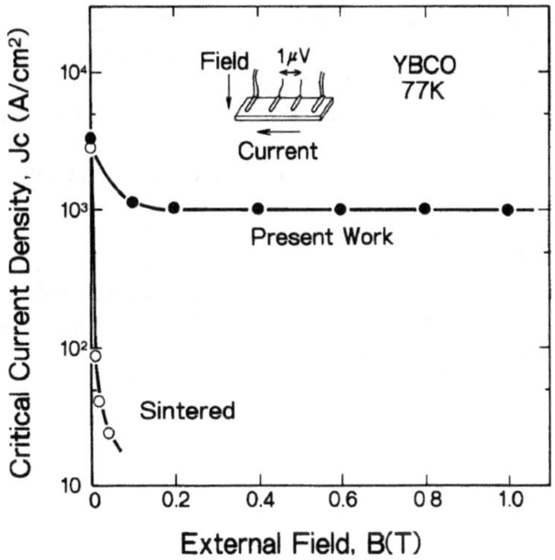

Fig. 3 Magnetic field dependency of critical current density of YBCO tape- shaped wire

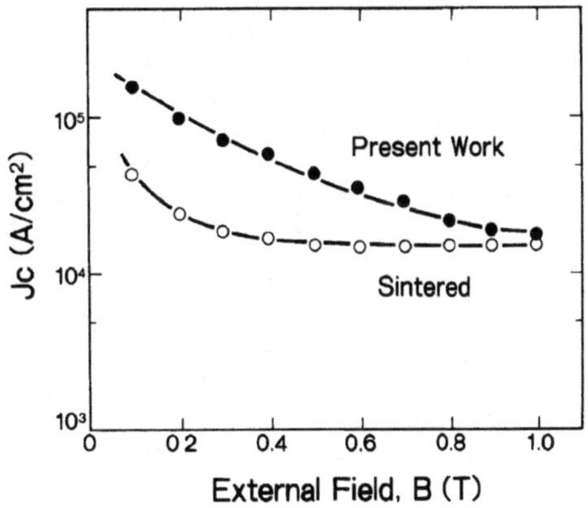

Fig. 4 Magnetic field dependency of intra-grain critical current density of YBCO tape-shaped wire estimated by the magnetization curve.

CONCLUSION

In order to increase the Jc of high-Tc superconducting wire, the unidirectinal solidification technique was applied to Au-sheathed tape-shaped wire using the FZM. The conclusions of this study were summerized as follows:

1. YBCO tape-shaped wire prepared by the floating zone melting technique reached Jc=1,000 A/cm^2 at 77K, 1T.
2. The FZM accelerates crystal grwoth, and is effective to form texture in tape-shaped wire.
3. The enhancement of Jc in magnetic fields is thought to be due to the microstructural modification of garin boundary.

ACKNOWLEDGEMENT

The authors would like to thank Mr. Nabatame and Mr. Takeuchi of Hitach Research Laboratory for their encouragement and many constructive remarks. We also would like to thank Mr. Hosono of Hitachi Cable Ltd. for helpful discussions.

REFERENCES

1. J.G.Bednorz and K.A.Müllar : Z. Phys. B64(1986)189.
2. K. Kshio, K. Kitazawa, S. kanbe, I. Yasuda, N. Sugai, H. Takagi, S. Uchida, F. Fueki and S. Tanaka : Chem. Lett. (1987)429.

3. K. Wu, J. R. Ashburn, C. J. Trong, P. H. Hor, R. L. Memg, L. Gao, Z. J. Hung, Y. Z. Wang and C. W. Chu: Phys. Rev. Lett. 58(1987)908.
4. H. Maeda, Y.Tanaka, M. Fukutomi and T.Asano: Jpn. J. Appl. Phys. 27(1988)L209.
5. Z. Z. Sheng, A. M. Herman, A. El Ali, C. Almasan, J. Estrada and T. Datta: Phys. Rev. Lettl. 60(1988)937.
6. S. Matsuda, M. Okada, T. Morimoto, T. Matsumoto and K. Aihara: High-Temperature Superconductors, ed. M. B. Brodsky, R. C. Dynes, K. Kitazawa and H. L. Tuller(MRS, Pittsburgh, 1988)p695.
7. M.Okada, A. Okayama, T. Morimoto, T. Matsumoto. K. Aihara and S. Matsuda : Jpn. J. Appl. Phys. 27(1988)L185.
8. T.Matsumoto, M. Okada, A. Okayama, T. ,Morimoto, K. Aihara, and S. Matsuda: World Congress on Superconductivity, ed. C. G. Burnham and R. Kane(World Scientific Pub., Singapole, 1988)p321.
9. M. Okada, R. Nishiwaki, T. Kamo, T. Matsumoto, K. Aihara, and M. Seido: Jpn. J. Appl. Phys. 27(1988)L2345.
10. T. Matsumoto, M. Okada, R. Nishiwaki, T. Kamo, K. Aihara, S. Matsuda, M. Seido, K. Ozawa, Y. Morii and S. Funahashi: ISTEC WORKSHOP ON SUPERCONDUCTIVITY, Oiso, Fed. 1-3, 1989, (ISTEC, 1989) p111.
11. Y. Yamada, N. Fukushima, S. Nakayama, H. Yoshino and S. Murase: Jpn. J. Appl. Phys. 26(1987)L865.
12. S. Jin, R. C. Sherwood, R. B. Van Dover, T. H. Tiefel and D. W. Jornson, Jr: Appl. Phys. Lett. 51(1987)203.
13. M. Murakami, S. Matsuda, K. Sawano, K. Miyamoto, A. Hayasi, M. Morita, K. Doi, H. Teshima, M. Sugiyama, M. Kimura, M.Fujinami, M. Saga, M. Matsuo, and H. Hamada: Advances in Superconductivity, (Springer-Verlag, 1988)p247.
14. P. Chaudhari, J. Mannhart, D. Dimos, C. C. Tsuei, J. chi, M. M. Oprysko and M. Scheuermann : Phys. Rev. Lett. 60(1988)1653.
15. D. Dimos, P. Chaudhari, J. Manhart and F. K.LeGoues : Phys. Rev. Lett. 61(1988)219.
16. C. P. Bean : Phys, Rev. Lett. 8(1962)250.

Ag-Doped Bi-Sr-Ca-Cu-O Superconductor Prepared by Floating Zone Method

KAZUO MICHISHITA[1], NORIYUKI SHIMIZU[1], YUTAKA HIGASHIDA[1], HISANORI YOKOYAMA[1], YUMI HAYAMI[1], TSUTOMU TSUNOOKA[1], EIKICHI INUKAI[1], YUKIO KUBO[1], AKIRA SAJI[2], NOBORU KURODA[2], HIROSHI YOSHIDA[2]

[1] Japan Fine Ceramics Center, 4-1, Mutsuno 2-chome, Atsuta-ku, Nagoya, 456 Japan
[2] Chubu Electric Power Co., Inc., 20-1, Kitasekiyama, Odaka-cho, Midori-ku, Nagoya, 459 Japan

ABSTRACT

Ag-doped $Bi_2Sr_2CaCu_2O_y$ bulk samples were prepared by floating zone (FZ) method aiming at enhancement of critical current density (Jc) and lowering of contact resistance at current terminal. It was found that Ag particles tend to be elongated along the growth direction, and be trapped in the oxide grains. Jc value was not significantly affected by Ag-doping up to 10 %, but the contact resistance at current terminal was lowered to about 1/1000 that of undoped sample.

KEY WORDS: Bi-Sr-Ca-Cu-O system, Ag-doping, critical current density, contact resistance, floating zone method

INTRODUCTION

In a recent paper,[1] the authors applied the floating zone (FZ) method to the preparation of a Bi-Sr-Ca-Cu-O superconducting bulk sample, and obtained high critical current density (Jc) exceeding 10^3 A/cm^2. But further enhancement of Jc is necessary for practical use.

On the other hand, there are many reports on Jc enhancement by Ag-doping in sintered samples, in which Ag fills the intergranular space.[2,3,4] These enhancements are considered to be mainly due to improvement of the weak link.

However, there appear to have been no reports on Ag-doping in melt processes including FZ method, in which Ag distribution and the effect on Jc are considered to differ from those in sintered process. Also the contact resistance at current terminal is expected to be lowered by Ag-doping. In this paper, we will describe the results of Ag-doping in the FZ method.

EXPERIMENTAL PROCEDURE

The Ag-doped Bi-Sr-Ca-Cu-O feed rods and seeds for the FZ experiments were prepared in the following manner: the commercially available oxalate coprecipitate with nominal composition of $Bi_2Sr_2CaCu_2O_y$ was calcined and mixed with 0-20 weight % Ag powder (average diameter 2 μm). The mixtures were formed into rod shapes with dimension of 8mmϕ x 60mm under an isostatic pressure of 100 MPa and sintered at around 850 ℃ for 86 hours in air. The FZ experiments were carried out in air at the growth rate of 2 mm/h, the details of which are shown in the literature.[1,5] Grown samples were cut into rectangular bars with typical dimensions of 0.5x0.8x15mm^3 for measuring the critical temperature (Tc) and Jc.

Tc and Jc were measured by the four terminal method. Jc was obtained in liquid nitrogen (77 K) under zero magnetic field using 50 ms wide pulse current. The sample was directly soldered with In metal on the Cu block (50x 50x5mm) current lead; the large heat capacity and high thermal conductivity of the large Cu block were expected to effectively suppress the temperature rise at the current terminal. The contact resistance was measured by the two terminal method below Tc.

The X-ray powder diffraction pattern was obtained using Cu Kα. Ag distribution in samples was observed by means of an optical microscope. The elemental composition of the predominant phase was analyzed by EDX.

RESULTS AND DISCUSSION

Figure 1(a) and 1(b) show Ag distribution observed by an optical microscope in a section parallel to the growth direction in 12.5 % doped sample. Ag grains look white. It is observed that Ag grains are elongated along the growth direction and exist both in (more frequent) and between the oxide grains.

From the X-ray powder diffraction analysis and EDX analysis, it was revealed that Ag grains exist in the metallic state and the predominant superconducting phase is the low-Tc phase. $Bi_2Sr_2CuO_x$[6] and $(Sr,Ca)CuO_2$ exist as a minor phase and the high-Tc phase was not detected.

The melt with nominal composition of $Bi_2Sr_2CaCu_2O_x$ is considered to commence crystallization at about 920 ℃ from differential thermal analysis, whereas Ag crystallizes at around 960 ℃. So Ag is in the solid state when oxide phases crystallizes and Ag grains are considered to be trapped in the oxide grains. On the other hand, as all phases are not completely molten in sintering process, Ag seems to always fill the intergranular space.

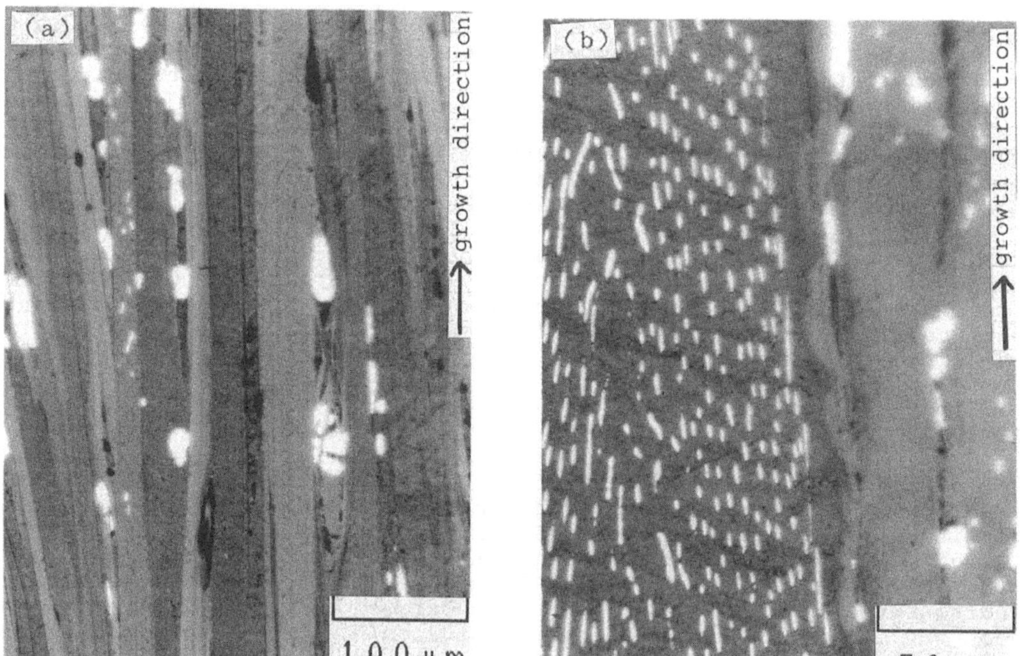

Fig.1 Ag distribution observed by an optical microscope.
(a) Ag fills intergranular space.
(b) Ag particles are trapped in the oxide grains.

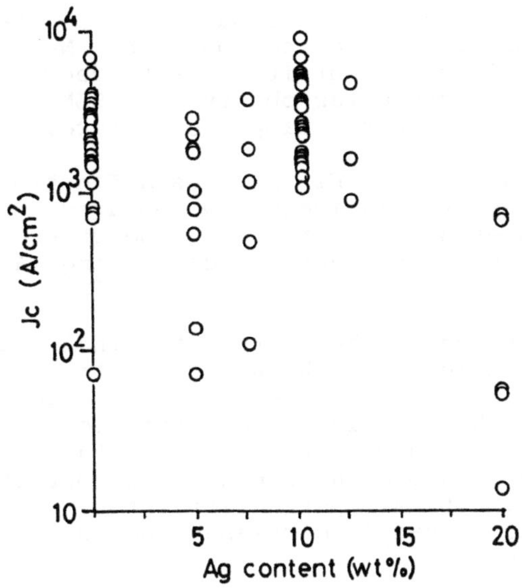

Fig.2 Jc as a function of Ag content.

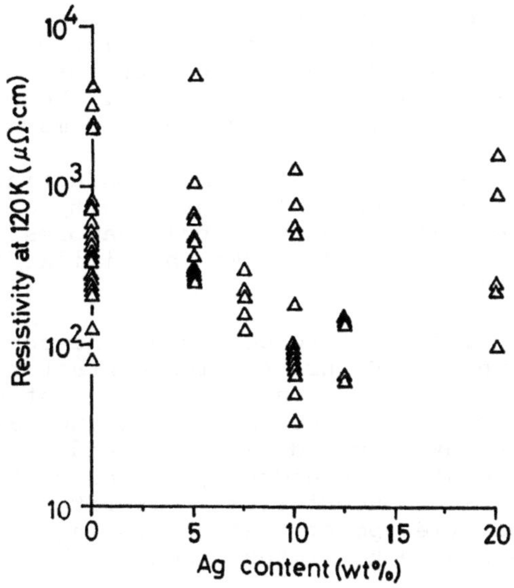

Fig.3 Resistivity at 120 K as a function of Ag content.

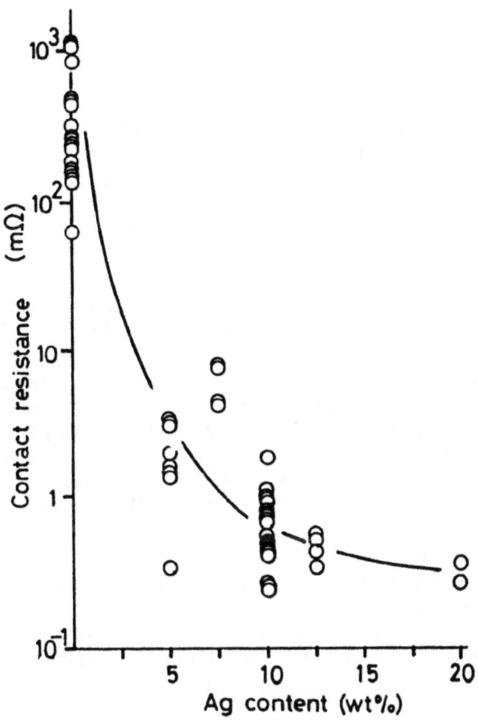

Fig.4 Contact resistance as a function of Ag content.

Figure 2 shows the Jc value as a function of Ag content. Although the deviation of Jc among samples is large, the following features can be seen. Namely, below 10 % Jc value does not change significantly, but it tends to decrease with Ag content over 10 %. Less oriented structure was observed by SEM in 20 % sample; it seems that too much Ag doping spoils the sample structure.

The resistivity at normal conducting state is shown in Figure 3 as a function of Ag content. The resistivity of our samples is relatively low compared to other Bi-Sr-Ca-Cu-O bulk samples. It is interesting that the conductivity (1/resistivity) shows the similar feature with Jc in Figure 2 concerning Ag content.

The effect of Ag doping on the contact resistance at the current terminal is shown in Figure 4. The contact resistance decreases with Ag content; the value at 10 % is about 1/1000 that of undoped sample. The temperature dependence of the contact resistance and the V-I characteristics at the contact were investigated [7]; Ag-doped samples show the ohmic contact with metallike temperature dependence, to the contrary undoped samples have non-ohmic contact with semiconductorlike temperature dependence. It is important from the practical view point that the doping of 10 % Ag to FZ sample decreases the contact resistance to 1/1000 keeping the high Jc value.

CONCLUSIONS

Ag-doped Bi-Sr-Ca-Cu-O bulk samples were prepared by floating zone method. The distribution of Ag particles was observed. Also the effects of the Ag-doping on the Jc characteristics and the contact resistance were investigated. The following results were obtained.

(1)Some of Ag grains are trapped in the oxide grains, not only in the intergranular space.

(2)Jc value is not affected by Ag-doping up to about 10 %.

(3)The contact resistance at current terminal was lowered to about 1/1000 that of undoped sample.

REFERENCES

1.Y.Kubo, K.Michishita, Y.Higashida, M.Mizuno, H.Yokoyama, N.Shimizu, E.Inukai, N.Kuroda, H.Yoshida(1989) Preparation of High Jc Bi-Sr-Ca-Cu-O Bulk Samples by Floating Zone Method. Jpn.J. Appl. Phys. Vo.28, No.4: L606-L608

2.S.Sen, In-Gann Chen, C.H.Chen and D.M.Stefanescu(1989) Fabrication of stable superconductive wires with YBa_2Cu_3Ox/Ag_2O composite core. Appl. Phys. Lett.54(8), 20 February: 766-768

3.B.Dwir, M.Affronte and D.Pavuna(1989) Evidence for enhancement of critical current by intergrain Ag in YBaCuO-Ag ceramics. Appl. Phys. Lett. 55(4)24 July: 399-401

4.S.Jin, R.C.Sherwood, T.H.Tiefel, G.W.Kammlott, R.A.Fastnacht, M.E.Davis and S.M.Zahurak(1988) Superconductivity in the Bi-Sr-Ca-Cu-O compounds with noble metal additions. Appl. Phys. Lett.52(19), 9 May: 1628-1630

5.Y.Kubo, K.Michishita, N.Shimizu, Y.Higashida, H.Yokoyama, Y.Hayami, E.Inukai, A.Saji, N.Kuroda and H.Yoshida(1989) Preparation of a Ag-Doped Bi-Sr-Ca-Cu-O Bulk Sample by the Floating-Zone Method. Jpn. J. Appl. Phys. Vol.28, No.11 (to be published)

6.C.Michel, M.Hervieu, M.M.Borel, A.Grandin, F.Deslandes, J.Provost and B.Raveau(1987) Superconductivity in the Bi-Sr-Cu-O System. Z.Phys.B-Condensed Matter 68: 421-423

7.N.Shimizu, K.Michishita, Y.Higashida, H.Yokoyama, Y.Hayami, Y.Kubo, E.Inukai, A.Saji, N.Kuroda and H.Yoshida(1989) Contact Resistance and V-I Characteristics in a Ag-Doped Bi-Sr-Ca-Cu-O Superconductor. Jpn. J. Appl. Phys. Vol.28, No.11 (to be published)

Effect of Rare Earth Substitution on $REBa_2Cu_3O_x$ Superconductors by Quench and Melt Growth Process

K. SAWANO, M. MORITA, K. DOI, A. HAYASHI, K. KIMURA, K. MIYAMOTO, and S. MATSUDA

R & D Labs-I, Nippon Steel Corporation, 1618, Ida, Nakahara-ku, Kawasaki, 211 Japan

ABSTRACT

Rare earth elements were substituted both completely and partly for Y of $YBa_2Cu_3O_x$ in the Quench and Melt Growth process. In most of the substituted systems, the growth of $REBa_2Cu_3O_x$ (123) was successful. All systems except for Nd-system produced the 123 grains with RE_2BaCuO_5 (211) inclusions which were residues of a peritectic reaction to form 123. Formation temperature of 123 in the Nd-system was about 1060°C, whereas about 900°C in the Yb-system. In the substituted systems, magnetization hysteresis, which corresponds to the critical current density (J_c), did not exceed that in the Y-system. However, since unsatisfactory optimization of the process prevent samples from eliminating weak-links, the J_c is considered to be improved as high as the Y.

KEY WORDS: superconductor, melt process, critical current density, microstructure, rare earth

INTRODUCTION

The critical current density (J_c) is one of the most important properties for superconductors. In oxide superconductors, weak-links such as grain boundaries have been considered to reduce the J_c [1]. Quench and Melt Growth (QMG) process has produced a bulk YBaCuO superconductor with high J_c exceeding 10^4 A/cm^2 even in high magnetic fields such as 1T by eliminating weak-links [2]. This process utilizes a peritectic reaction of Y_2BaCuO_5 (211) and liquid to form $YBa_2Cu_3O_x$ (123) and produces highly oriented 123 with fine dispersion of 211. Because of the nature of the peritectic reaction, the final microstructure is strongly affected by the microstructure before the reaction. Since the J_c is dependent on microstructure, the control of the peritectic reaction is one of the key issues of this process. In 123 system, rare earth (RE) elements substituted for Y are known to exhibit the critical temperature (T_c) as high as Y [3]. On the other hand, the reaction to form 123 is expected to change with types of RE elements. These indicate that there is a possibility to obtain optimum microstructure for the high J_c without changing the T_c. For this reason, substitution experiments for Y were conducted. Microstructure and superconducting properties were examined in comparison with the Y-system.

EXPERIMENTAL PROCEDURE

Samples were prepared by the QMG process. RE was substituted completely and 50% for Y. Starting materials were single phase 123 powders which were synthesized by a conventional solid state reaction process. Off-stoichiometric composition was not tried. Heating pattern of the QMG process is shown in Fig. 1. Melting temperature, partial melting temperature and slow cooling temperature range were adjusted for each RE system. The 123 formed at about 1060°C in the Nd-system, while about 900°C in the Yb-system. In other systems, the formation temperatures were approximately 1000°C which is similar to that in the Y-system. The process was carried out in air. Samples were subsequently annealed in O_2 at 600°C with cooling rate of 50°C/h. Platinum crucibles and two copper plates were used for melting and quenching, respectively. Microstructure was characterized by a polarized light optical microscope, SEM and TEM. Crystalline phases were determined by XRD and EDX. Superconducting properties were measured using a four-probe DC method, AC magnetic susceptibility and a vibrating sample magnetometer (VSM).

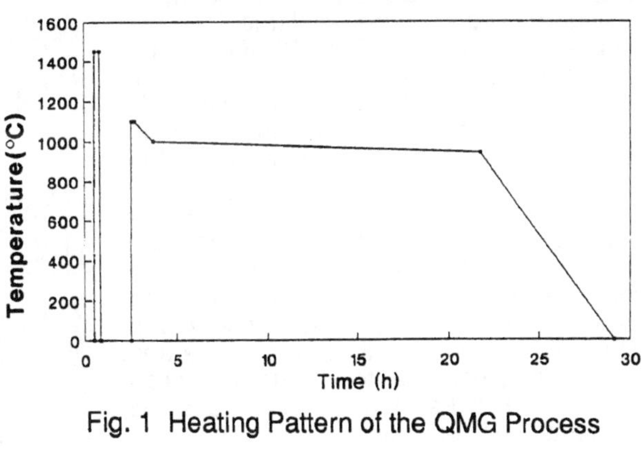

Fig. 1 Heating Pattern of the QMG Process

RESULTS AND DISCUSSION

Melt-Quenched Materials

After melt-quenching from 1450°C in the Y-system, the sample was in plate shape and consisted of spherical Y_2O_3 grains, whose size was about 10 μm, and a solidified liquid phase. From their size and shape, the spherical Y_2O_3 grains are considered to exist as a solid even at 1450°C. It is known that 123 decomposed into 211 and liquid at about 1000°C. The 211 further decomposes to Y_2O_3 and liquid, as indicated by Roth et al [4]. The results agreed with this indication.

Microstructures of melt-quenched materials in some of fully substituted systems are shown in Fig. 2. The Y-system is also shown for comparison. Although some systems had similar microstructure to the Y-, systems such as Sm and Nd included elongated grains which were cylindrical and considered to form during solidification. This indicates that phase diagrams of these systems are different from that of the Y-system. At high temperatures, the liquid phase should contain a considerable amount of RE oxide, which precipitates during solidification. In other words, the liquidus line should be less steep than in the Y-system. In Dy-, Eu- and Ho-system, both spherical and elongated grains were observed, while irregular-shaped grains, sometimes looked hollow, were observed in Gd-system. These grains were oxide of RE elements except for Nd-system in which grains contained both Ba and Cu other than Nd. These grains were initially suspected to be 211. However since attempts to synthesize Nd-211 failed and a similar pattern to $La_6Ba_4Cu_3O_x$ was detected by XRD [5], these grains were considered not to be 211. The EDX analysis supported the XRD result. Microstructures of melt-quenched plate in half-substituted systems are shown in Fig. 3. Unlike fully substituted systems, most systems included spherical RE_2O_3 grains except for Nd-Y system in which grains again contained Ba, Cu, Nd and Y. RE_2O_3 and Y_2O_3 seemed to form complete solid solution according to EDX analysis.

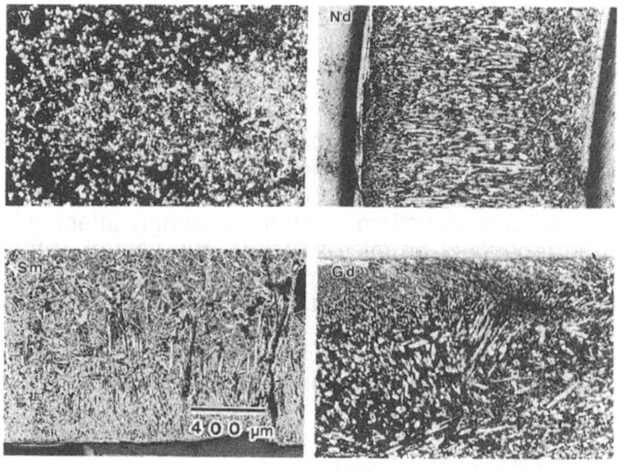

Fig. 2 Microstructure of melt-quenched materials in fully substituted systems.

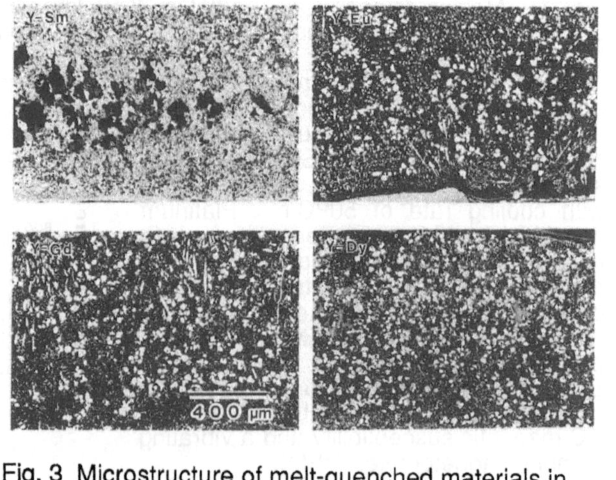

Fig. 3 Microstructure of melt-quenched materials in half-substituted systems.

Microstructure

Microstructures of fully substituted systems after the completion of the QMG process are shown in Fig. 4. Formation of 123 was successful in most systems. However, the size and shape of the 123 were different each other because the optimization of the process for each system was not satisfactory. In most systems, the size of the 123 grains is smaller than that in the Y-system. Cracks, striations and 211 inclusions were observed in the 123 grains. In most systems, the 123 grains seemed to contain more cracks and striations than the Y-system. Although the size of each 211 inclusion was similar, the distribution was less homogeneous. This can be explained by the formation of 211 during partial melting. From the comparison of Fig. 4 with Fig. 2, it is seen that 211 inclusions inherit their distribution from RE_2O_3 grains in the melt-quenched materials. This confirms that the 211 nucleates and grows from RE_2O_3 during partial melting [6]. Since in most systems, RE_2O_3 grains in the melt-quenched materials were larger and more irregular-shaped than those in the Y-system, the distribution of the 211 naturally becomes less homogeneous. This may affect the existence of the cracks and striations because the 211 grains are expected to act as dispersoids in composites and prevent cracks from extending [7]. On the contrary, almost no inclusion was observed in the Nd-system by both optical microscope and TEM. This suggests that the different reaction occurs in the Nd-system. More detailed experiments are in progress.

Microstructures of half-substituted systems are shown in Fig. 5. Cracks and striations again existed more than the Y-. However, the distribution of 211 is rather homogeneous because RE_2O_3 in the melt-quenched materials distributed as spherical grains. Effect of 211 to prevent cracks does not seem clear. Extensive use of 211 such as increasing the amount may be necessary.

Superconducting Properties

All systems exhibited higher T_c than the liquid nitrogen temperature (77K). Since the measurement of J_c by the DC four-probe method was not successful, the J_c's were estimated by the magnetization at 77K. The specimens were slab-shaped with approximate size of 3x5x1 mm^3 and magnetic field (H) was applied in the slab plane. If the critical state is established and the J_c is independent of the H, then the J_c in A/cm^2 can be calculated using $J_c = 20\Delta M/d$, where ΔM is the magnetization hysteresis between increasing and decreasing H in Gauss and d is the slab thickness in cm. Since the establishment of the critical state was not confirmed, only $\Delta M/d$ values at 0.9 T and 77K and the sample thickness were listed in Table 1. Assuming the Bean's critical state[8], the J_c of the Y-system was 1.5×10^4 A/cm^2. All of the substituted systems showed lower $\Delta M/d$ values than the Y-. This is probably due to unsatisfactory optimization of the process. Since the J_c is a strong function of microstructure, the process should be carefully adjusted to each system. Small-grained samples such as Er- showed the small $\Delta M/d$ value which simply indicated that the sample contained a larger amount of weak-links such as grain boundaries. Systems such as Gd-, Dy- and Ho- showed higher $\Delta M/d$ because the 123 grains grew easily in these systems. Since the effort to optimize the process was intensively made in the Yb-system despite the difficulty, the $\Delta M/d$ value became larger compared to the Er-.

In order to understand the weak-link behavior, the sample size dependence of the H in the Gd-system was examined (Fig. 6). In this system, the size of 123 was relatively large. The sample was thinned and measured the ΔM repeatedly. Points are not on a straight line. The $\Delta M/d$ increased 3 times with decreasing thickness and when the thickness was reduced from 0.22 cm to 0.03 cm. The reasons of this phenomenon are considered to be as follows: (1) the critical state is not established, in other words, the specimen is not homogeneous, (2) the J_c is dependent on the H, (3) the sample was too thick for slab geometry. Among these reasons, (1) is considered to be the most significant because the microstructural inhomogeneity is obviously observed in Fig. 4. Dependence of the ΔM on the H can be explained by the reason (2). From this fact, high $\Delta M/d$ value of the Ho-system in Table 1 can partly be explained by small thickness of the sample. The reason why half-substituted systems did not exhibit high $\Delta M/d$ values in spite of microstructural improvement is not understood.

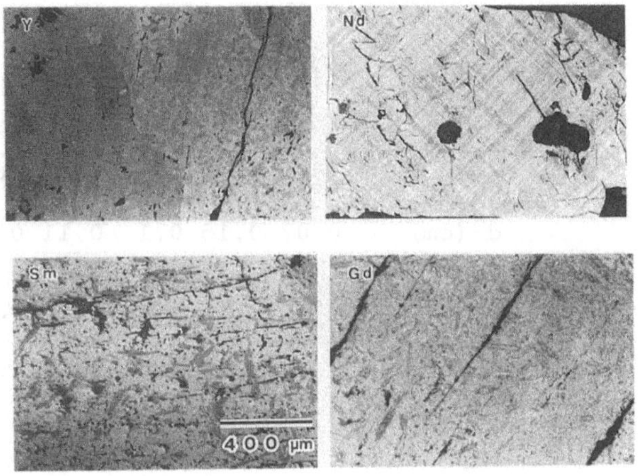

Fig. 4 Final microstructure of fully substituted systems

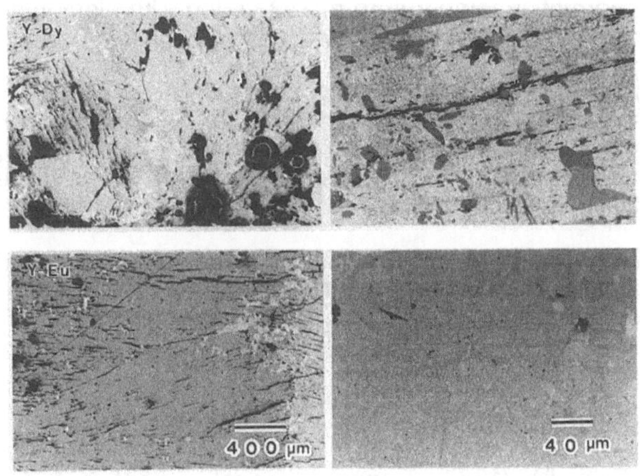

Fig. 5 Final microstructure of half-substituted systems.

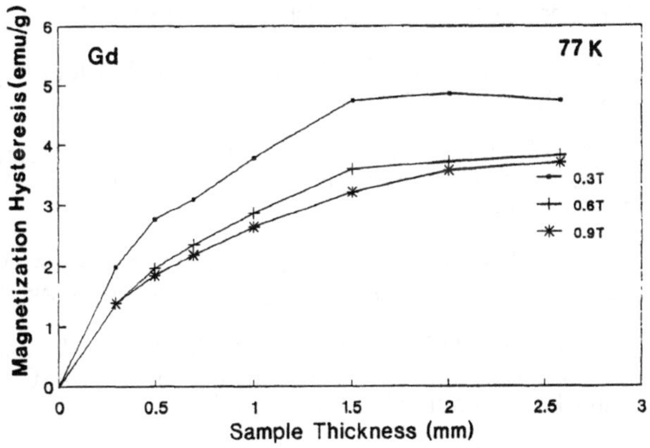

Fig. 6 Sample size dependence of magnetization hysteresis in Gd-system.

From these results, it is expected that the basic performance of substituted systems concerning the J_c is similar to the Y-. Although the Y-system has an advantage of fine, homogeneous 211 dispersion, other systems may be able to have some processing advantages such as low processing temperature in the Yb-system. Further improvement of the process can enhance the advantages of other RE systems. In addition, the absence of inclusions in the Nd-system probably contributes to understand the pinning mechanism.

Table 1 ΔM/d values of RE substituted systems

	Y	Nd	Sm	Eu	Gd	Dy	Ho	Er	Yb
ΔM/d (emu/g.cm)	124	11	13	9	29	89	33	1	34
d (cm)	0.07	0.15	0.1	0.11	0.21	0.03	0.09	0.1	0.09

	Y-Sm	Y-Eu	Y-Gd	Y-Dy	Y-Er
ΔM/d (emu/g.cm)	7	12	16	21	4
d (cm)	0.08	0.09	0.08	0.08	0.12

CONCLUSIONS

(1) In most of the substituted systems, growth of 123 was successful using the QMG process.

(2) Formation temperature of 123 varied with elements. For instance, the Nd-system had the formation temperature about 1060°C, while the Yb-system had about 900°C.

(3) Microstructure, especially distribution of 211 inclusions and resultant cracks and striations in 123 grains, were strongly dependent on the type of rare earth elements. The size and distribution of 211 in the final microstructure directly reflected those of RE_2O_3 after melt quenching. Cracks and striations are strong function of the 211 inclusions which act as dispersoids in composites. In Nd-system, almost no inclusion was observed.

(4) The value of ΔM/d, which is good indication of the J_c, changed because of the microstructural change. Basic superconducting performances of each system are considered to be similar to those of the Y-system. It was not obvious whether the 211 inclusions acted as pinning sites from these experiments.

REFERENCES

1. Ekin JW (1987) Transport critical current in bulk sintered $Y_1Ba_2Cu_3O_x$ and possibilities for its enhancement, Adv Ceram Mater, 2: 586-592

2. Murakami M, Morita M, Doi K, Miyamoto K (1989) A new process with the promise of high J_c in oxide superconductors, Jpn J Appl Phys, 28: 1189-1194

3. Hosoya S, Shamoto S, Onoda M, Sato M (1987) High-T_c superconductivity in new oxide system, Jpn J Appl Phys, 26: L325

4. Roth RS, Rawn CJ, Beech F, Whitler JD, Anderson JO (1988) Phase equilibria in the system Ba-Y-Cu-O-CO_2 in air, In: Yan MF (ed) Research Update, 1988 Ceramic Superconductor II, Am Ceram Soc, Westerville, OH pp13-26

5. Michel J, Er-Rakho L, Raveau B (1981) Les oxydes $La_{4-2x}Ba_{2+2x}Cu_{2-x}O_{10-2x}$: Une structure inedite constituee de groupements CuO_4 carres plans isoles, J Solid State Chem, 39: 161-167

6. Sawano K, Morita M, Miyamoto K, Doi K, Hayashi A, Murakami M, Matsuda S (1989) Effect of synthesis conditions on microstructure of a $YBa_2Cu_3O_x$ superconductor by partial melting process, Seramikkusu Ronbunshi, 97: 1028-33

7. Kimura M, Hayashi A, Morita M, Matsuo M, Sawano K, Matsuda S, (submitted) Microstructural Characterization of $YBa_2Cu_3O_{7-x}$ prepared by Quench and Melt Growth (QMG) process, Proc ISS'89

8. Bean CP (1962) Magnetization of hard superconductors, Phys Rev Lett, 8: 250-3

The Transport Properties of Oxide Superconductors by Melt Process

H. KIKUCHI, K. MATSUMOTO, N. UNO, and M. NAKAJIMA

Yokohama R & D Laboratories, The Furukawa Electric Co., Ltd., 4-3, Okano 2-chome, Nishi-ku, Yokohama, 220 Japan

ABSTRACT

The oriented YBaCuO poly-crystals were prepared by melting and directional solidification. In this specimen, the 211 phases were dispered in the 123 matrix, with a spacing of several micro meters. We measured the magnetic field dependence of the transport critical current density (Jc) by micro four-probe method, and evaluated dominant pinning sites by Jc-B properties based on conventional flux pinning theory. According to the result, the expected site of dominant pinning are normal phases in the grains.

KEY WORDS: high Tc superconductivity, superconductors, flux pinning directional solidification

INTRODUCTION

Since the discovery of high-Tc oxide superconductors, it has been pointed out that the critical current density, Jc, can be increased by many investigations. A number of studies have been recentry proposed[1,2].
We currently made oriented YBaCuO superconductors by melt process.
In this paper, we report the dominant pinning sites by Jc-B properties and conventional flux pinning theory.

EXPERIMENTAL

The raw materials of YBaCuO oxide superconductors were prepared by thorough mixing of appropriate amounts of Y_2O_3, $BaCO_3$, and CuO powders. The resulting mixture was annealed in oxygen atmosphere at 900℃ for 20 hours. After grinding, the powder was pressed at 10(ton), typically of rectangular bar approximately 5x5x50mm in dimensions. These pellets were then sintered at 900℃ in flowing oxygen for 5 hours.
YBaCuO sintered specimens were kept at 1050℃ or above, and were then subjected to low speed movement in a furnace with a temperature gradient (~50℃/cm).
The grown crystal grains were examined by a polarizing microscopy, transmission electron microscopy (TEM), and X-ray diffraction analysis.
The critical current, Ic, of the solidified specimens were measured at 77K using the micro four-probe method. An attempt was made to measure the local Ic of the bulk sample by bonding technique using a gold wire 25μm in diameter.
From these mesurements, we estimated dominant pinning sites using a conventional flux pinning theory.

RESULTS AND DISCUSSION

A microstructure of specimen is shown in Fig.1. The grain sizes are approximately 10 μm in thickness, 10-30 μm in width, and in length 30-40 μm for the shorter ones and 100-200 μm for the longer. In this figure, the second phase and a twinned structure are also shown.
From TEM observation, shown in Fig.2, the spacings of the twin boundaries are 0.1-0.3 μm and the dominant second phase is identified as 211 phase. The sizes of the second phases are 1-3 μm, and a distance between them is also 1-3 μm.
X-ray diffraction analysis (Fig.3) revealed that the ab planes of this specimen spread along the direction of the grain growth.
We measured the magnetization loop at 77K (Magnetic field was applied parallel to the ab plane.). The result is shown in Fig.4. The lower critical field is about 0.01T judgeing from the initial magnetization slope The magnetic hysteresis is large compared with the sintered specimen.
The transport currents were measured up to 10 T. Figure 5 is a sample configuration of micro four-probe method. Figure 6 shows Jc properties as

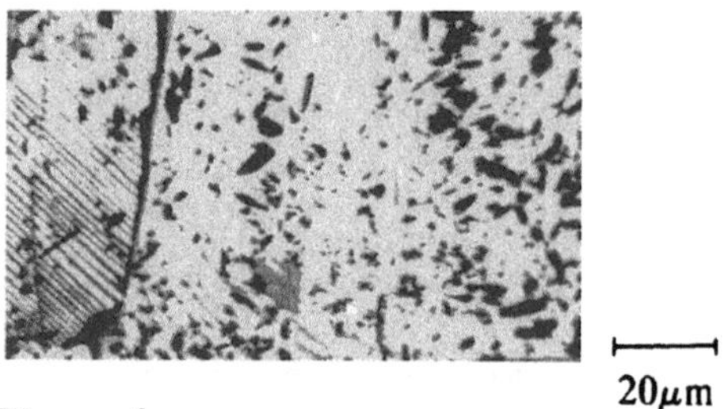

Figure 1
Optical micrograph of microstructure of specimen produced by melting and directional solidification process

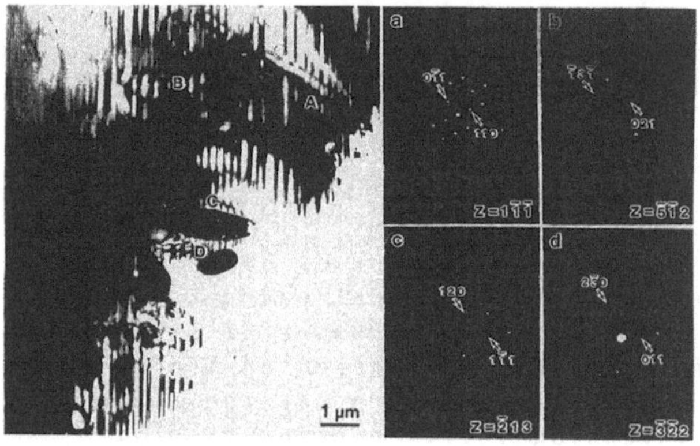

Figure 2
TEM micrograph of specimen (a)-(d) are S.A.D.s taken from Y_2BaCuO_5 (A)-(D) , respectively

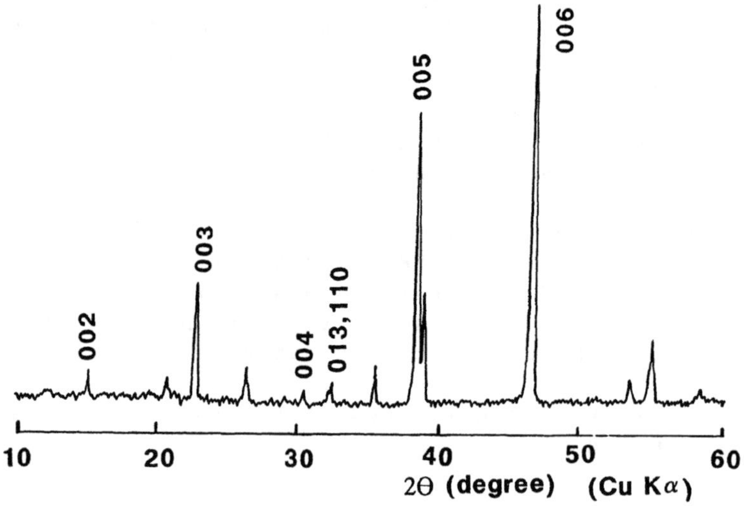

Figure 3
X-ray diffraction pattern of solidified YBaCuO superconductor

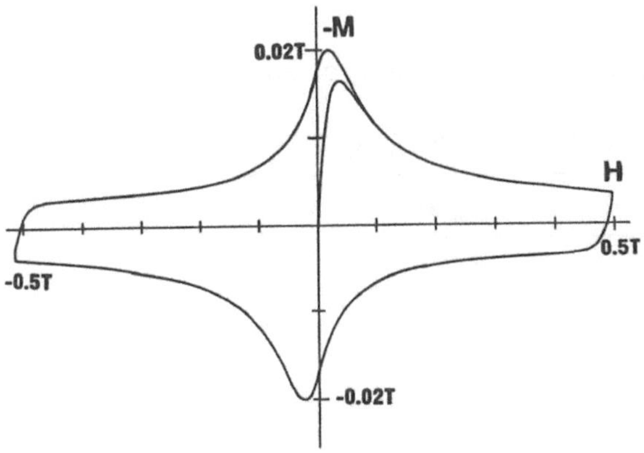

Figure 4
Magnetization hysteresis loop at 77K

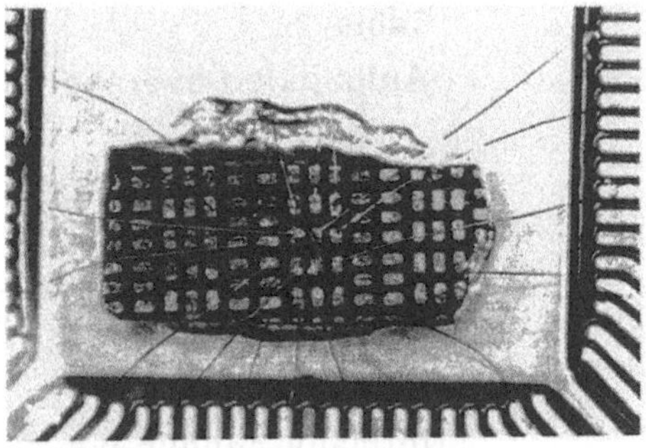

Figure 5
A sample configuration of micro four-probe method

a function of applied magnetic fields, which are in two directions parallel and perpendicular to the ab plane. Transport currents are induced in the ab plane. Dotted line shows Jcm as obtained from magnetization loop in Fig.4. The value of transport Jc in the low field region may be equivalent to the value of Jcm. When magnetic field is parallel to the ab plane, a comparatively large value of Jc can be obtained. The values of 8.5×10^8 Am^{-2} at 0 T and 1.0×10^8 Am^{-2} at 1 T are obtained. Even at medium fields the value obtained are comparatively high; 3.3×10^7 Am^{-2} at 5 T, but at 10 T the value deteriorates precipitously, down to 2×10^6 Am^{-2} or below. When magnetic field is perpendicular to the ab plane , in contrast, values of Jc are low overall; 5×10^7 Am^{-2} at 1 T and 2×10^6 Am^{-2}at 5 T. These specimens were the precipitous degradation of Jc at 1-2 T The pinning force per unit volume, Fp, of superconductors may be calculated

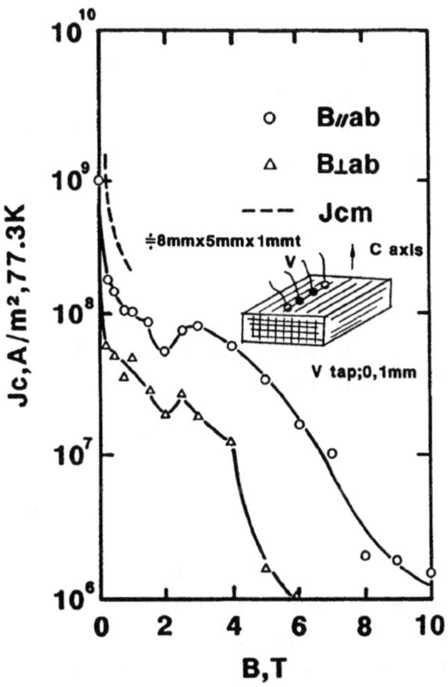

Figure 6
Jc properties of solidified YBaCuO superconductor as a function of applied magnetic field (77K)

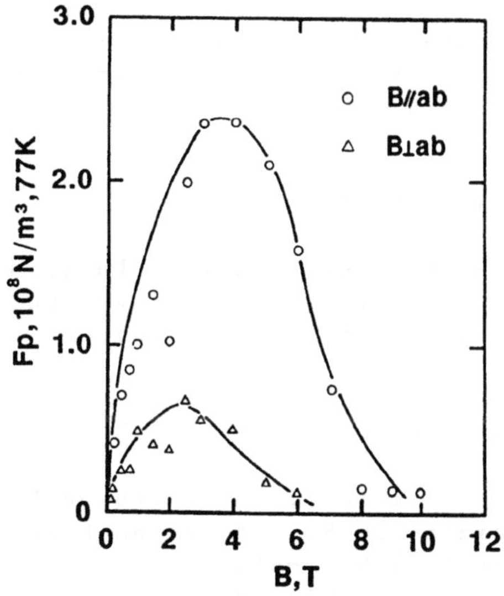

Figure 7
Flux pinning forces plotted as a function of applied magnetic field

Table 1
Anticipated average spacing at various pinning sites

	Average spacing (μm)	
	B// ab	B⊥ab
Grain boundary	2.5	8.7
Twin boundary	—	8.7
Normally conductive phase	2.5	3.5
	B=3.5T Fp=2.4×10^8 Nm^{-3}	B=2.5T Fp=0.6×10^8 Nm^{-3}

by Fp = Jc x B. The results are as shown in Fig. 7. In materials such as Nb_3Sn, in which the elementary pinning force is weak irrespective of whether B is parallel or perpendicular to the ab plane, the curve is of a characteristically saturated type. When B is parallel to the ab plane, the maximum value of Fp is 2.4 $\times10^8$ Nm^{-3}, obtained at 3.5 T, and 0.6 x 10^8 Nm^{-3}, obtained at 2.5 T, when B is perpendicular.
Table 1 shows the expected spacing of various pinning sites. Dtails of the flux pinning theory are described in the other paper[3].
If the maximum value of Fp is not so influenced by weak link and it is determined by bulk pinning force, we think the twin boundary is not an effective pinning site in this specimen. Especially, the normal phases are the best candidates of the pinning site in this specimen by considering the size and the density of them.

CONCLUSIONS

The YBaCuO oxide superconductors made by melt process. We measured the local Ic in the specimen by micro four-probe method. According to that, in this cace, the candidates of dominant pinning site is normal phase in YBaCuO grains. If this estimation is valid, the development of Jc will be expected by controlling the microstructures in the oxide superconductors.

ACKNOWLEDGEMENT

The authors gratefully scknowledge that this work was supported by the group of Tokyo Electric Power Company, Tohoku Electric Power Company, and Hokkaido Electric Power Company.

REFERENCES

1) Jin,s,Tiefel,T.H., Sherwood,R.C., van Dover,R.B., Davis,M.E., Kammlott,G.W. and Fastnacht,R.A. Phys Rev B (1988) 37 7850
2) Murakami,M., matsuda,S., Sawano,K., Miyamoto,K., Hayashi,A., Morita,M., Doi,K., Teshima,H., Sugiyama,M., Kimura,M., Fujinami,M., Saga,M., Matsuo,M., and Hamada,H. Proc Ins Int Symp Superconductivity(1988) 274
3) Matsumoto,K., Kikuchi,H., Uno,N., and Tanaka,Y. Cryogenics

Melt Processing of YBaCuO Oxide Superconductors

H. FUJIMOTO, M. MURAKAMI, S. GOTOH, N. KOSHIZUKA, T. OYAMA, Y. SHIOHARA, and S. TANAKA

Superconductivity Research Laboratory, International Superconductivity Technology Center, 10-13, Shinonome 1-chome, Koto-ku, Tokyo, 135 Japan

ABSTRACT

We report a novel melt process for YBaCuO superconductors that leads to high Jc. Since the superconducting phase in YBaCuO system is produced by a peritectic reaction: $Y_2BaCuO(211)$ + L(liquid) $\rightarrow 2YBa_2Cu_3O_x(123)$, it is essential to finely disperse the 211 phase in liquid to promote the continuous growth of the 123 phase. In order to achieve fine dispersion of 211 phase, we employed pulverizing and mixing process after cooled from partially melted region. The process enables fine dispersion of Y_2O_3 and/or 211 phase in liquid and therefore the continuous growth of the 123 phase, resulting in high Jc.

KEY WORDS: high-Tc superconductivity, $YBa_2Cu_3O_x$, melt process, peritectic reaction, critical current density

INTRODUCTION

Since the discovery of high temperature superconductors with Tc above 77K [1], efforts have been concentrated on searching for processing which yields high critical current density. However until recently critical current density had remained low [2], which had hindered the application of oxide superconductors.

However, recent studies [3,4,5] have shown that high Jc can be obtained in thin films or even in bulk high temperature superconductors. The QMG method(quench and melt growth) which yields large critical current density exceeding 10^4 A/cm^2 at 77 K and 1 T has been reported by Murakami et. al. [5,6,7]. The QMG process leads to suppression of the second phase intrusion, improvement in the connectivity of the superconducting phases and grain alignment, and therefore achievement of high Jc in the YBaCuO system. In this process, the control of precursor microstructure is essential for obtaining well-developed and textured superconducting phase. The QMG process consists of three steps. In the first step, the sample is quickly heated to the Y_2O_3 + L region and quenched, then the sample with finely and homogeneously dispersed Y_2O_3 is obtained. Subsequent reheating to the 211 + L region causes formation of the 211 phase by a peritectic reaction. In the final step, the superconducting phase is produced by another peritectic reaction of the 211 phase. In this process, the role of the 211 phase is important, that is, finely and homogeneously distributed 211 phases cause continuous growth of the superconducting phase leading to the production of high Jc sample. However, as pointed out in Ref.5, it is difficult to obtain homogeneous distribution of fine 211 particles on macroscopic scale due to the difficulty of microstructural control in partial melting region.

In this paper, we report a novel melt process, which enables us to control the 211 phase dispersion by pulverizing a sample after partial melting.

EXPERIMENTAL

Appropriate amounts of mixtures of Y_2O_3,$BaCO_3$ and CuO (99.9% pure) were calcined at 900 °C for 24 hours in flowing oxygen. The powders were melted at 1200 °C - 1400 °C in platinum crucibles and rapidly cooled by using cold copper hammers. Then the melted oxides were pulverized and pressed into pellets. The sample size was 20 mm in diameter and 5 mm in thickness in this experiment. It is possible to form the samples with various sizes by this process. The samples were then heated to 1100 °C for 20 minutes and cooled slowly to room temperature in platinum crucibles under 1 atm air atmosphere. The samples were cooled at the rate of 1 °C/h to 20 °C/h in the temperature range from 1000 °C to 950 °C followed by furnace cooling.

Microstructure was observed with an optical microscope and a scanning electron microscope(SEM). Energy dispersive X-ray analysis(EDX) and X-ray diffraction were used to identify the existing phases. Magnetic properties were measured using a vibrating sample and a SQUID magnetometer.

RESULTS AND DISCUSSION

In order to improve the homogeneity of Y_2O_3 distribution we employed the additional process of pulverizing and mixing the oxides after melted partially at 1200 °C - 1400 °C.

The X-ray diffraction pattern of the pulverized sample after heated at 1350 °C for 15 minutes is shown in Figure 1. The pattern shows the presence of Y_2O_3, $BaCuO_2$ and CuO. Through mixing, these phases are well mixed and homogeneous distribution of Y_2O_3 in liquid-forming phase (a mixture of $BaCuO_2$ and CuO) is realized. After the sample was reheated to 1100 °C and cooled down to 1000 °C, homogeneous distribution of the 211 phase was obtained as shown in the scanning electron micrograph (Figure 2). After the

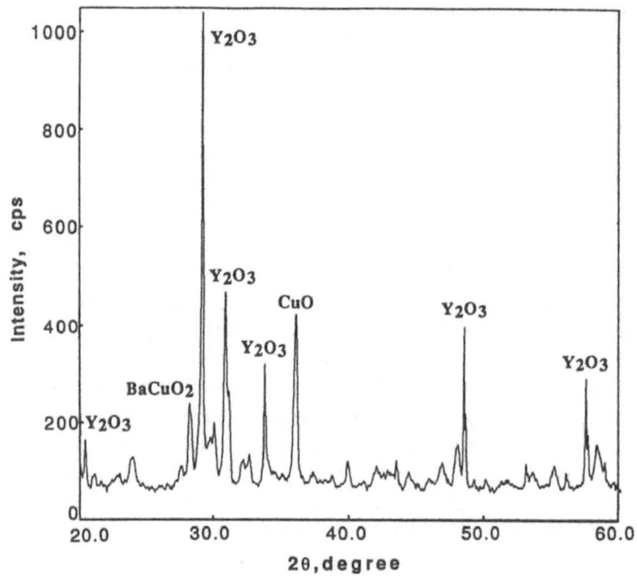

Fig. 1 X-ray diffraction pattern of the powdered sample after melted at 1350 °C.

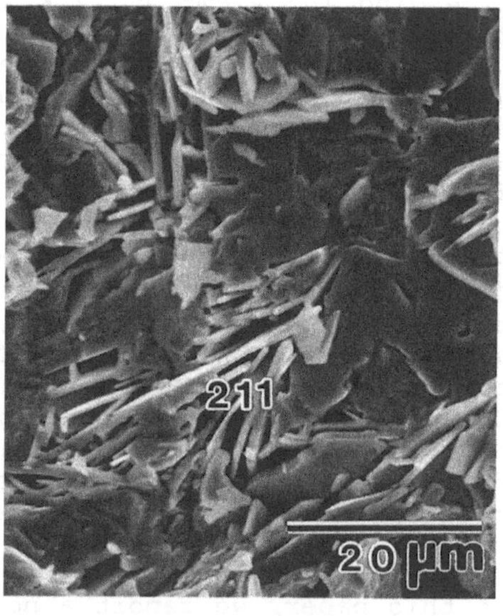

Fig. 2 Scanning electron micrograph for the fracture surface of the sample after reheated to 1100 °C and cooled down to 1000 °C.

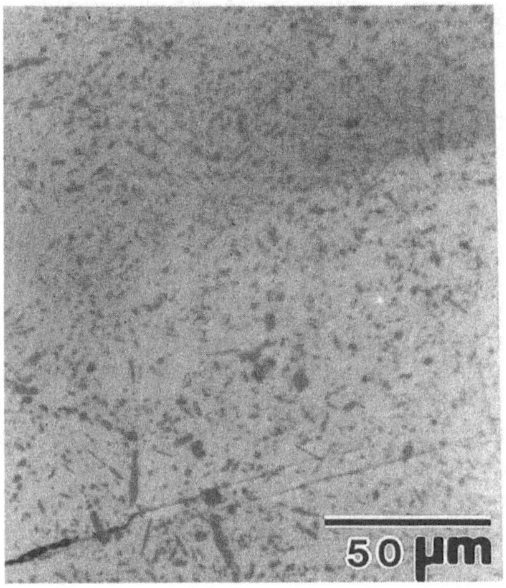

Fig. 3 Optical micrograph of the sample showing the distribution of 211 inclusions.

Fig. 4 Optical micrograph of the sample fabricated using fine powders. Note that no 211 inclusion is present in the sample.

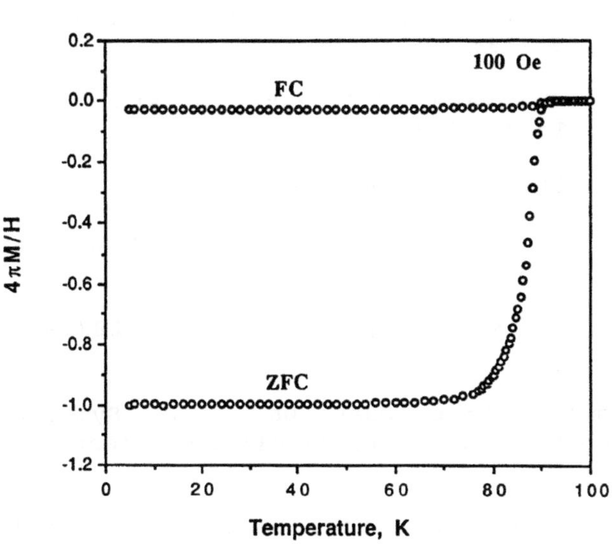

Fig. 5 Temperature dependence of 4πM/H.

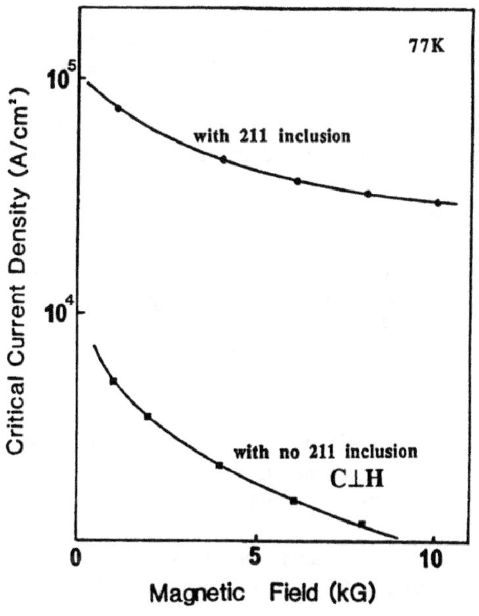

Fig. 6 Magnetic field dependence of Jc for two samples fabricated by the novel melt process.

final stage, the 211 particles, which are represented by dark spots in the optical micrograph (Figure 3), are homogeneously distributed in the 123 superconducting phase.

We have also found that it was possible to eliminate 211 inclusions through refinement and well-mixing of powders. Figure 4 shows the optical micrograph of the sample fabricated using fine powders. No 211 inclusion is observed, indicating that relatively large single crystal can be produced by this method.

Figure 5 shows temperature dependence of magnetic susceptibility in the field-cooled(fc) and zero-field-cooled(zfc) process. Perfect diamagnetism is realized in the zfc with sharp transition, while small magnetization is observed in the fc process. This result shows that the sample has large pinning force [8]. Figure 6 shows magnetic field dependence of Jc for two melt-processed samples at 77K. The figure was obtained by using Jc = 20 $\Delta M/d$ from M-H loops [9]. The sample containing no 211 inclusion exhibits lower Jc value, supporting that the 211 can contribute to flux pinning.

CONCLUSIONS

A novel melt process which is promising for high Jc is presented in this paper. Pulverization of the oxides after partial melting enables homogeneous distribution of Y_2O_3 and consequently fine dispersion of 211 phase in the 123 superconductig phase can be realized, leading to high Jc.

REFERENCES

1. M.K.Wu, J.R.Ashburn, C.J.Torng, P.H.Hor, R.L.Meng, L.Gao, Z.J.Huang, Y.Q.Wang and C.W.Chu: Phys.Rev.Lett. **58** (1987) 908.

2. see,e.g., M.Murakami, M.Morita, K.Sawano, T.Inuzuka, S.Matsuda and H.Kubo: Proc. Sintering 87 (Elsevier, Tokyo, 1989) 1466.

3. S.Tanaka and H.Itozaki: Jpn.J.Appl.Phys. **27**(1988) L662.

4. S.Jin, T.H.Tiefel, R.C.Sherwood, R.B.van Dover, M.E.Davis, G.W.Kammlott and R.A.Fastnacht: Phys.Rev.B **37**(1988) 7850.

5. M.Murakami, M.Morita, K.Doi and K.Miyamoto: Jpn.J.Appl.Phys. **28**(1989) 1189.

6. M.Murakami, M.Morita and K.Miyamoto: Proc. Osaka Univ. Int. Symp. on New Development in Applied Superconductivity, Suita, 1988(World Scientific, Singapore, 1989).

7. M.Murakami and M.Morita: Proc. Mater. Res. Soc. Meeting (Materials Research Society, Tokyo, 1989).

8. M.Murakami et.al.: to be published in Cryogenics (1990).

9. C.P.Bean: Phys.Rev.Lett. **8**(1962) 250.

Highly Textured Superconducting Bi-Sr-Ca-Cu-O Crystals Prepared by the Vertical Bridgman Method

T. IZUMI, T. OYAMA, and Y. SHIOHARA

Superconductivity Research Laboratory, International Superconductivity Technology Center, 10-13, Shinonome 1-chome, Koto-ku, Tokyo, 135 Japan

ABSTRACT

The effects of temperature gradient(G) and growth rate(R) on characteristics of unidirectionally grown crystals by the vertical Bridgman method were investigated. Main phases in all samples were recognized as follows; $(Sr_{0.6}Ca_{0.4})CuO_2$ (primary phase), $Bi_2(Sr,Ca)_3Cu_2O_x$, and $Bi_2(Sr,Ca)_2CuO_x$ (secondary mixture phases). The orientation on all phases were improved with increasing the G/R ratio. A large primary phase was grown and the fraction of $Bi_2(Sr,Ca)_3Cu_2O_x$ in secondary mixture phases were increased with increasing 1/(GR) value.

INTRODUCTION

Since the initial report of superconductivity in the Bi-Sr-Ca-Cu-O system with transition temperature Tc of 80K and 110K by Maeda et.al.[1], enormous research efforts have been focused on understanding of characteristics [2-7] and structure [8-11] of this system. Experimental results using several different unidirectional solidification processes for obtaining crystals with high critical current densities have been reported, including Floating Zone [12], Laser-heated Pedestal Growth[13] and Czochralski methods [14]. However, very few successful runs by the vertical Bridgman technique have been reported. This technique is very useful for not only obtaining aligned crystals but understanding the mechanism of the phase transformation, since temperature gradient(G) and growth rate(R) as process parameters could be controlled independently.

In this paper, the effects of G and R on characteristics of crystals prepared by the vertical Bridgman method were investigated including texture, morphology,grain size, and phases.

EXPERIMENTAL

The experimental apparatus used for crystal growth in this work consists of a vertically mounted tube furnace, resistively heated with programmable time and temperature control, and a quench tank (water or gallium pool). The apparatus is shown schematically in Figure 1. A 50mm-diameter fused alumina tube is used as a furnace liner. The maximum temperature of the furnace for melting is about 1673K, and uniform temperature range is about 12cm long. A 5mm-diameter crucible filled by pressed rod-shaped samples is placed in the center of the liner under air atmosphere. The crucible is drawn downwards through a given temperature gradient (G), which is controlled by the distance between the furnace and the quench tank, at a constant rate (R). These conditions are controllable independently by this method. The values of G and R varied from 5.0×10^3 to 2.3×10^4 K/m and from 0.10 to 30.0 μm/s, respectively.

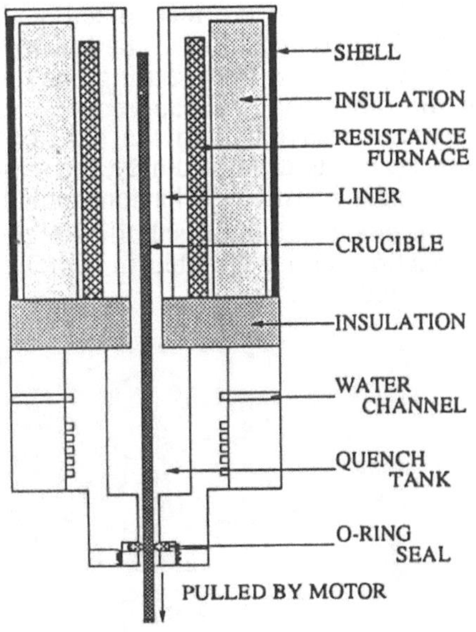

Figure 1 A schematic of the growth apparatus

In this work, samples with starting nominal composition of Bi:Sr:Ca:Cu=2:2:1:2 were used. Powders of Bi_2O_3, $SrCO_3$, $CaCO_3$ and CuO were mixed, ground, calcined at 1073K in air for 12hours. These calcined powder were pressed to form a rod (<5mm diameter) at 2000 atm by Cold Isostatic Pressing (CIP). These rods were placed into a high quality alumina or a magnesia crucible and melted at about 1673K in the furnace. Then the crucible were drawn downwards under several different conditions shown in Table 1. The microstructure and the phase identification of existing crystals in these samples were observed by polarizing microscopy and Electron Probe Microanalyzer (EPMA). The electric characteristics were examined by the standard four-probe method.

Table 1 The conditions of crystal growth

R(m/s) \ G(K/m)	1.30 $(x10^4)$	1.92 $(x10^4)$	2.10 $(x10^4)$
$0.16x10^{-6}$	–	–	◊
$0.31x10^{-6}$	–	◊	◊
$1.67x10^{-6}$	◊	◊	◊
$2.78x10^{-6}$	◊	–	◊

◊ : USED CONDITION

– : UNUSED CONDITION

RESULTS AND DISCUSSION

1 Contamination from crucible materials

Figure 2 shows the temperature dependence of resistivity for samples grown in alumina and magnesia crucibles under the same temperature gradient, G($2.10x10^4$ K/m) at the same growth rate, R(1.67 μm /s). The resistivity of the sample grown in alumina crucible increased with decreasing temperature, and began to decrease at about 60K but did not reach zero by 4.2K. In contrast , the sample prepared in a magnesia crucible exhibited a metallic behavior. The resistivity of this sample began to decrease sharply at about 85K and finally reached zero at 55K.

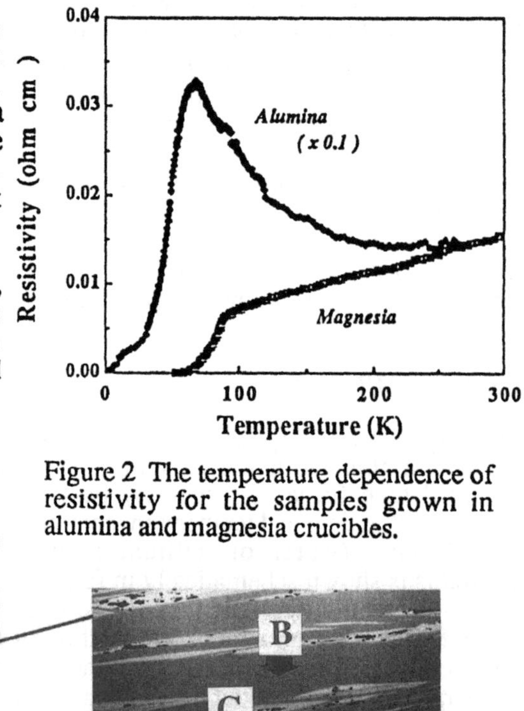

Figure 2 The temperature dependence of resistivity for the samples grown in alumina and magnesia crucibles.

In order to clarify the causes of the difference, the microstructures of these samples were examined. On the sample grown in alumina crucible, aluminum compounds (Al-Sr-Ca-O and Al-Bi-Sr-Ca-O) were recognized at the upper region. The composition in the lower solidified region was shifted from starting one to Cu-rich. On the other hand, no appreciable magnesia compound was found in the case of using a magnesia crucible. Accordingly, the reason of the difference in resistivity behaviors could be due to a composition shift in the solidified region caused by reaction with a crucible in a melted region before solidification.

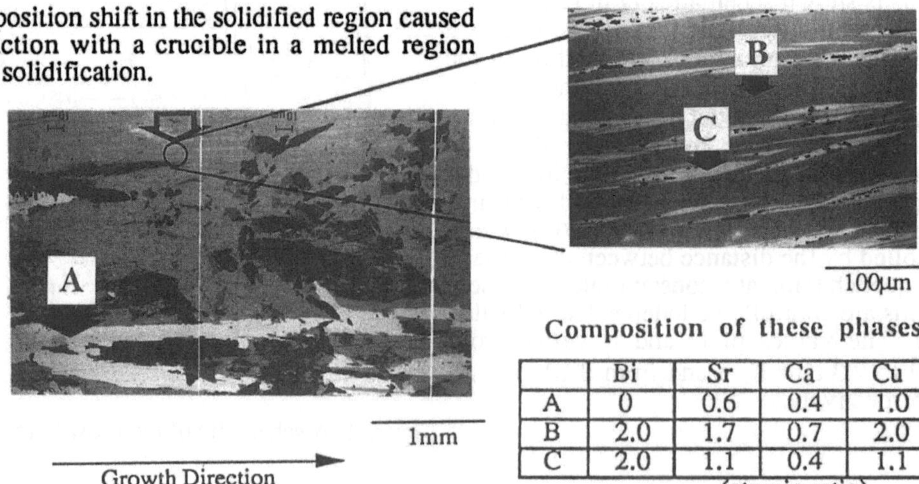

Composition of these phases

	Bi	Sr	Ca	Cu
A	0	0.6	0.4	1.0
B	2.0	1.7	0.7	2.0
C	2.0	1.1	0.4	1.1

(atomic ratio)

Figure 3 The microstructure of the sample grown in a magnesia crucible.

Furthermore, composition analysis of inclusions in the sample grown in the magnesia crucible indicates existence of $(Sr_{0.6}Ca_{0.4})CuO_2$ (primary phase), $Bi_2(Sr,Ca)_3Cu_2O_x$ and $Bi_2(Sr,Ca)_2CuO_x$ (secondary mixture phases) as main phases as shown in Figure 3.

2 Effect of G/R on texture

The distribution curves of angles(q) of the long axis direction of primary crystals tilted from growing direction for different values of G/R are shown in Figure 4. The distribution curve becomes sharper with increasing G/R ratio. Figure 5 shows G/R ratio dependence of aspect ratios of primary phase crystals and microstructures of secondary mixture phases crystals. The aspect ratio increased with increasing of G/R ratio.The grains of secondary mixture phases crystals grew more unidirectionally at higher G/R ratios, while they grew randomly at lower G/R ratios.These results suggest the orientation of both primary and mixture phases crystals depends on the G/R ratio, i.e. supercooled liquid exists in front of the solid/liquid interface which may be due to the constitutional and/or kinetic supercooling.

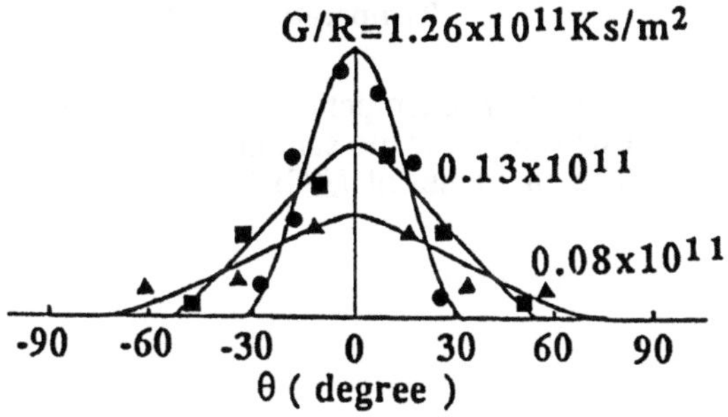

Figure 4 The distribution curves of angles of the long axis direction of primary phase tilted from the growth direction .

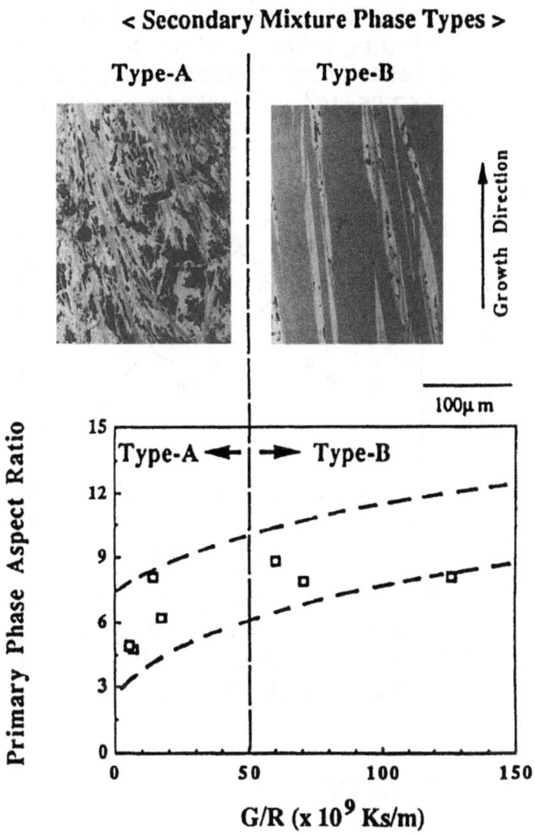

Figure 5 G/R ratio dependence of aspect ratios of primary crystals and two types of microstructure of secondary mixture phases (Superconducting).

3 Effect of 1/(GR) on grain growth (coarsening) and electric characteristics

Figure 6 shows 1/(GR) dependence of the width of the primary crystals grown in the magnesia crucible. The width increases with increasing 1/(GR) value and those relation is as follows ;

$$L = at^n = a[1/(GR)]^n$$

$$n \sim 1/3$$

L : short axis length of the primary phase

a : constant

t : holding times, 1/(GR)

This means that the coarsening mechanism could be somewhat due to the Ostwald ripening mechanism (solute diffusion controlled).

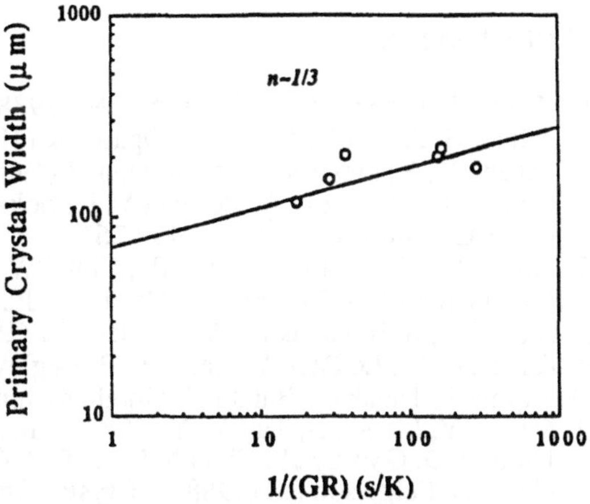

Figure 6 1/(GR) dependence of width of primary crystals.

Figure 7 shows effect of 1/(GR) on fractions of the $Bi_2(Sr,Ca)_3Cu_2O_x$ phase and the $Bi_2(Sr,Ca)_2CuO_x$ phase in the secondary mixture phases estimated by the ratio of X-ray peak intensity ([017] of $Bi_2(Sr,Ca)_3Cu_2O_x$ and [115] of $Bi_2(Sr,Ca)_2CuO_x$). Amounts of $Bi_2(Sr,Ca)_3Cu_2O_x$ phase increased with increasing of 1/(GR) value, while the fraction of the $Bi_2(Sr,Ca)_2CuO_x$ phase decreased. According to this result, the longer reaction (peritectic) time is required for transformation to the superconducting phase ($Bi_2(Sr,Ca)_3Cu_2O_x$).

Figure 8 shows the temperature dependence of resistivity for the samples grown with different values of 1/(GR). The resistivity of the samples grown with lower 1/(GR) values exhibited a semiconducting behavior, on the contrary, the resistivity with higher 1/(GR) values exhibited a metallic behavior and decreased sharply at around 80K. Finally, the resistivity of the sample at 1/(GR) of 2.85×10^2 s/K reached zero at 80K.

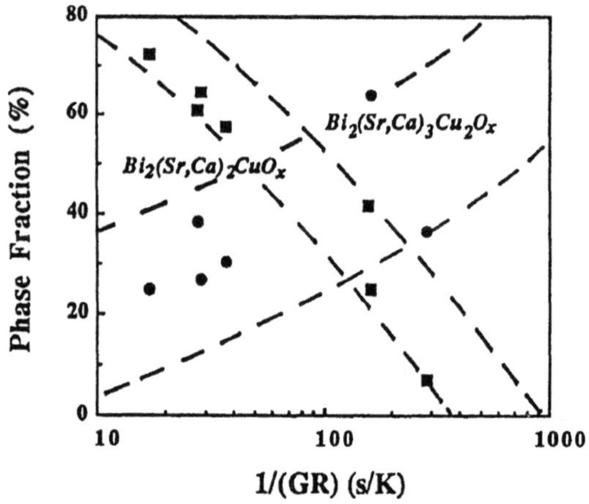

Figure 7 The effect of 1/(GR) on the fractions of the $Bi_2(Sr,Ca)_3Cu_2O_x$ phase and $Bi_2(Sr,Ca)_2CuO_x$ phase in the secondary mixture phases.

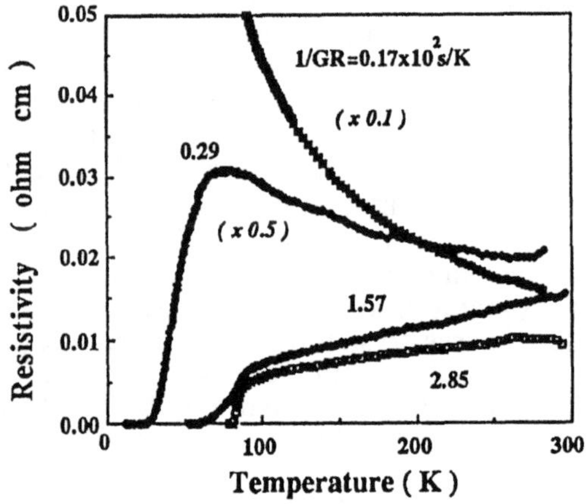

Figure 8 The temperature dependence of resistivities for the samples grown with different values of 1/(GR).

CONCLUSION

Superconducting characteristics was observed in *as-grown* samples produced by the vertical Bridgman method using magnesia crucibles.The effects of temperature gradient (G) and growth rate (R) on characteristics of unidirectionally grown crystals were investigated. The existence of $(Sr_{0.6}Ca_{0.4})CuO_2$ (primary phase), $Bi_2(Sr,Ca)_3Cu_2O_x$ and $Bi_2(Sr,Ca)_2CuO_x$ (secondary mixture phases) were recognized as main phases in the samples. The orientation of both primary and mixture phases depends on G/R ratio. The width of primary crystals depends on the value of 1/(GR). Transformation to the superconducting phase needs longer reaction time, therefore the critical temperature depends on the value of 1/(GR).

This work has been supported by the R&D Basic Technology for Future Industrial Technology Development Organization (NEDO).

REFERENCES

1. Maeda H, Tanaka Y, and Fukutomi M (1988) Jpn.J. Appl. Phys. Lett. 27 : 209
2. Uehara H, Asada Y, Maeda H, Ogawa K (1988) Jpn. J. Appl. Phys. Lett. 27 : 665
3. Togano K, Kumakura H, and Maeda H (1988) Jpn. J. Appl. Phys. Lett.27 : 323
4. Katayama-Yoshida H, Yonezawa Y, Hirooka H (1988) Physica C153 : 425
5. Politis C (1988) Appl. Phys.A45 : 261
6. Hazen RM, Prewitt CT, Angel RJ, Chu CW (1988) Phys. Rev. Lett. 60 : 1174
7. Subramanian MA, Torardi CC, Calabrese JC (1988) Science 239 : 1015
8. Shaw TM, Shivashankar SA, La Placa SJ (1988) Phys. Rev. B37 : 9856
9. Tarascon JM, Le Page Y, Barboux P, Bagley BG (1988) Phys. Rev. B37 : 9382
10. Kijima T, Tanaka J, Bando Y, Onoda M Jpn. (1988) J. Appl. Phys. Lett. 27 : 369
11. Matsui Y, Maeda H, Tanaka Y (1988) Jpn. J. Appl. Phys. Lett. 27 : 361
12. Brody HD, Gyorgy JS, Cima MJ (1988) J. Crystal Growth 96 : 225
13. Gazid D, Feigelson RS (1988) J. Crystal Growth 91 : 318
14. Kurosaka A, Aoyagi M, Tominaga H (1989) Appl. Phys. Lett. 55 : 390

Superconducting High-T_c Oxide/Metal Matrix Composites Produced by Internal Oxidation of Unidirectionally Solidified Ag-Yb-Ba-Cu Alloys

DANIAL R. DIETDERICH, KAZUMASA TOGANO, and HIROAKI KUMAKURA

National Research Institute for Metals, 2-1, Sengen 1-chome, Tsukuba, Ibaraki, 305 Japan

ABSTRACT

Superconducting high-T_c oxide/metal matrix composites were produced by internal oxidation of unidirectionally solidified Ag-Yb-Ba-Cu alloys. The critical temperature of an internally oxidized Ag-23at.%($Yb_1Ba_2Cu_3$) alloy is strongly influenced by the unidirectional solidification processing conditions and the temperature sequence of the subsequent heat treatment. Samples produced with a slow growth rate had higher T_c's than samples produced with a fast growth rate when given the same oxidization heat treatment. This is interpreted as an annealing effect associated with the slow growth rate, since the T_c's of the high growth rate samples were increased by an oxygen-free anneal prior to the oxidization heat treatment. To prevent sample melting, and the resulting loss of the solidification structure, the initial heat treatment temperature should be less than or equal to 650°C. To consistently obtain a material with a zero resistance T_c of about 85K and a narrow transition, ΔT_c less than 5K, the final heat treatment temperature should be at 875°C or above.

KEY WORDS: Yb-Ba-Cu-O, directional solidification, high-T_c oxide, superconducting material, oxide-metal matrix composite

INTRODUCTION

Various fabrication techniques for the high-T_c oxides have been attempted. One of these methods showed that high-T_c oxides could be produced by the oxidation of metallic precursors.[1,2] Since early work on the Y-Ba-Cu-O system showed that the grain boundaries of the "123" (i.e. $ReBa_2Cu_3O_{7-\delta}$) family of high-T_c oxides had weak link character, additions were made to improve the grain boundaries. Good conducting elements, noble metal such as Ag, which are believed not to poison the superconducting phase were added to the oxide in an attempt to remedy the problem. Other factors also made Ag addition attractive. One is the ability to make low resistance electrical contact to the oxide sample and the other is the superior mechanical performance that should result from an oxide/metal composite. Not soon after Ag was added to the 123 metallic precursors producing Ag-Re-Ba-Cu alloys (Re is Y,Eu, and Yb), which when oxidized produced high-T_c superconducting phases.[3,4] However, most of the prior work was performed on alloys with low concentrations of Ag, less that 50at.%, such that, it was only a minor phase in an oxide matrix. Our previous work shows that a Ag-based (Yb-Ba-Cu) alloy with a Ag concentration of 77at.% (80wt.%) alloy could produce a high T_c of 85K.[5] This value is near the highest reported T_c (90K) for the 123 compound of Yb produced by other methods.

EXPERIMENTAL PROCEDURE

An alloy with the nominal composition of Ag-23at.%(0.17Yb-0.33Ba-0.5Cu), i.e. Yb:Ba:Cu=1:2:3, was prepared by arc melting. Cylindrical-shaped specimens with a length of 30-40mm and a diameter of about 8mm were cut from the ingot, placed into a boron-nitrite crucible, and melted at 1000°C in a vertical tube furnace with a flowing Ar atmosphere. Unidirectional solidification was achieved by lowering the crucible through the furnace at a speed of 1.5cm/h or 14cm/h. The temperature gradient at the solid-liquid interface is about 50°C/cm. However, since the temperature profile of the furnace is quadratic the gradient at the solid-liquid interface is a function of the alloy melting point. Small pieces, typically 30mm in length and about 4mm^2 in cross section were cut longitudinally from the UD samples and heat treated in air or flowing oxygen. All of the heat treatments were done in a stepwise manner. Samples heat treated in oxygen received the schedule 650°C for 2d, 750°C for 1d, and 850°C for 1d, followed by 500°C for 1d. Samples treated in air were given the thermal treatment 650°C, 875°C, and 500°C each for 12h. Lastly, all the samples were furnace cooled below 150°C before removal form the furnace. Also, some of the samples grown at the faster rate (14cm/h) were given an oxygen-free anneal prior to the above oxidation heat treatment.

RESULTS AND DISCUSSION

Material Characterization

Since our previous work showed that the Ag-23at.%($YbBa_2Cu_3$) alloy has the best superconducting properties (i.e. T_c and J_c) this paper will focuses on it in an attempt to better understand the source of its superior properties. Figure 1 shows the typical microstructure of this alloy when solidified at 1.5cm/h. It consists of alternating lamella with two distinct microstructures and compositions. Type 1 lamella, the low contrast areas of Fig. 1 denoted by 1 in the legend to the right of the figure, have a composition of Ag-1.7at.%Yb-10.0at.%Ba-4.1at.%Cu (EDS). Judging from the binary phase diagrams of Ag with the other elements one would expect Cu and possibly Yb to form a solid solution at this composition. On the other hand, Ba has negligible solubility in Ag and as a result probably forms a Ag intermetallic compound. The Ag-rich portion of the Ag-Ba phase diagram is not yet will understood, but this compound is expected to have a composition near Ag_4Ba or Ag_5Ba. If one assumes that Ag_5Ba is formed, with all of the Ba is in this compound, then this region is 80-85vol.% Ag_5Ba. Preliminary x-ray diffraction results show that the sample may contain about 80% of Ag_5Ba (JPCS file no. 28-156) and about 10% Ag and 10%Cu. However, several strong reflections have not yet been indexed.

The other lamella, type 2 (denoted by 2 in the legend at the right of fig.1), has four distinct constituents: a region similar to that in the first lamella (region not labelled in lamella 2), plus, Ag-rich regions (labelled), Cu-rich regions (arrows), and fine eutectic-like regions (labelled). Only a few of each region have been identified for clarity. The first region has a composition similar to that of the first lamella except it is slightly more Yb rich, Ag-2.7at%Yb-8.8at.%Ba-3.5at.%Cu. The Ag-rich regions are about 25-50μm in size while the Cu-rich regions are about 5-10μm in size. EDS analyses show that the Ag-rich and Cu-rich regions have the compositions Ag- 7.5at.%Cu and Cu-3.7at.%Ag, respectively. The Cu-rich region may also contain 1-2at.%Yb. The fine eutectic-like region is a mixture of Ag-rich regions and Ag intermetallics. The average composition of this region is Ag-7.0at.%Yb-2.8at.%Ba-8.6at.%Cu.

Figure 2 shows the microstructure of the Ag-23at.%($YbBa_2Cu_3$) alloy with a growth velocity of 15mm/h after internal oxidation. The microstructure has three distinct regions with different oxide morphologies and content. One region, associated with type 1 lamella of fig. 1, has a few small oxide particles distributed in a Ag matrix (top of fig 2). However, a few large Cu-O particles were observed in this region. The second region has a high density of oxide precipitates that may form a continuous network (bottom of fig.2). This region is believed to be associated with the eutectic region of type 2 lamella. The Yb:Ba:Cu proportions of the precipitates in this region is 1:2:3, i.e. the superconducting $YbBa_2Cu_3O_{7-\delta}$ phase. In addition, to the superconducting phase Yb-free granular oxide precipitates with a Ba:Cu proportion of 1:2 is dispersed in this region. The third region is the boundary between the two types of lamella (middle of fig.2). The boundaries have an oxide layer. However, not all the boundaries have an oxide layer.

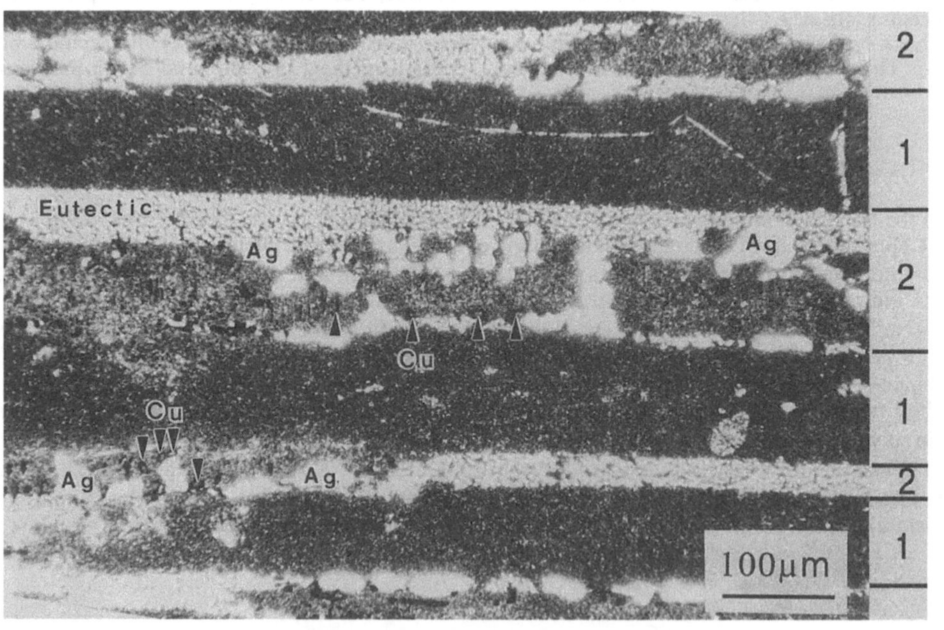

Fig. 1 Ag-23at.% ($Yb_1Ba_2Cu_3$), transverse section after unidirectional solidification at 1.5cm/h. The microstructure consists of alternating lamella with different composition and structures. The scale at the right of the micrograph partitions the lamellar structure into individual lamella. The 1 and 2 of the scale correspond to type 1 and type 2 lamella, respectively.

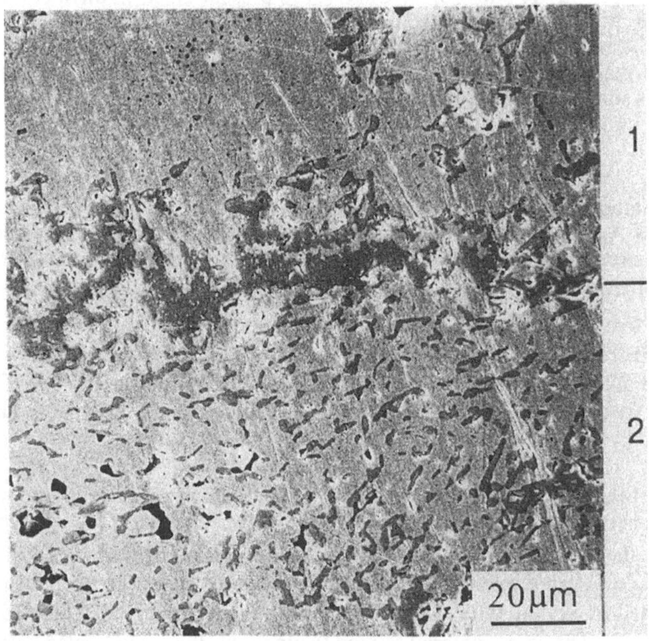

Fig. 2 Scanning electron micrograph (transverse section) of Ag-23at.% ($Yb_1Ba_2Cu_3$) after internal oxidation. The scale to the right of the micrograph is the same as that in Fig. 1. The upper portion of this micrograph coincides with a type 1 lamella and the lower portion with a type 2 lamella.

This is most likely due to the composition variation associated with the different regions of the type 2 lamella. For example, a boundary adjacent to a eutectic region, which have a high Yb concentration, well have an oxide layer. EDS results show that the oxide layers are composed of two oxides with different Yb:Ba:Cu proportions, 1:2:2.5-3 and 2:1:1. This suggests that the layers are composed of both the superconductor $Yb_1Ba_2Cu_3O_{7-\partial}$ and the non-superconductor $Yb_2Ba_1Cu_1O_5$.

The volume fraction of the superconducting phases in the sample is low. Estimates from magnetization curves give 3.5% and 1% for bulk and powder samples, respectively, for a sample heat treated in oxygen and the schedule: 650°C for 2d plus 750°C for 1d, 850°C for 1d, and 500°C for 1d. The larger value obtained for the bulk sample suggests that the superconducting phase is somewhat continuous permitting it to shield normal regions, and thus give a larger volume fraction. The smaller value, 1%, obtained from the powder sample should be nearer the true superconducting phase volume fraction. This value is much smaller than the theoretical maximum. If one assumes that all of the Yb-Ba-Cu reacts to form $YbBa_2Cu_3O_7$ in a Ag matrix a volume fraction of about 30% would be obtained. A substantial increase in the superconducting volume fraction may be achievable by adjusting the UD processing and heat treatment parameters, new variations are currently being considered.

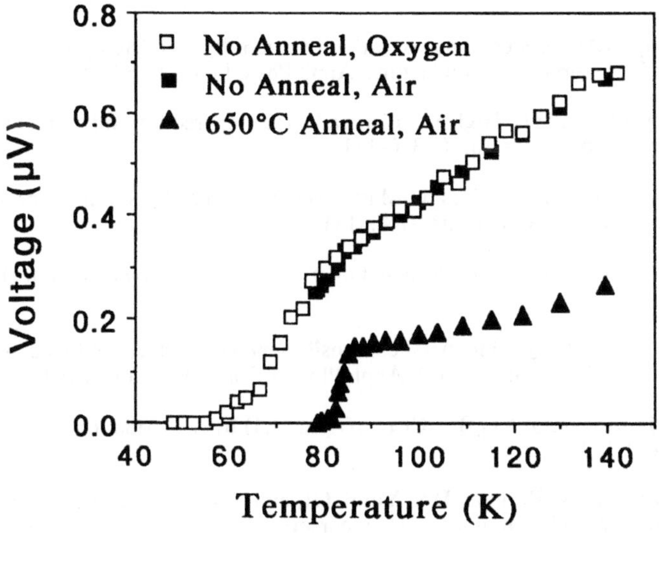

Fig. 3 Voltage vs. temperature dependence of Ag-23at.% ($Yb_1Ba_2Cu_3$) solidified at 14cm/h. Samples with no oxygen-free anneal prior to oxidation have a low T_c regardless of the heat treatment or atmosphere. A sample given a 650°C pre-oxidation anneal produces a high T_c.

To determine whether any of the Cu-rich oxides, $Yb_2Ba_4Cu_7O_x$ and $Yb_1Ba_2Cu_4O_x$, are present electron diffraction analyses were performed. A typical [100] zone axis diffraction pattern from one of the oxides shows that the main reflections are due to $YbBa_2Cu_3O_{7-\delta}$. However, some weak reflections for $YbBa_2Cu_4O_y$ are also present, as well as, streaking along the c* axis. This shows that the material contains many intergrowths, such that, a range of compositions and structures are present. This result is similar to that of Kogure et al. [6]

Processing and Properties

Both oxygen and air were used as heat treatment atmospheres. Heat treating in air is as good as oxygen. Previous results by Kogure et al. for alloys with with 33wt% Ag showed that heat treating in air was superior to an oxygen atmosphere, since less $Yb_2Ba_1Cu_1O_5$ and $BaCuO_2$ are produced.[7] Our results show that the peak heat treatment temperature must be at 875°C or above to consistently obtain a T_c of about 85K and a ΔT_c<5K. Kogure et al. showed that a heat treatment temperature of 900°C produced less $Yb_2Ba_1Cu_1O_5$ and $BaCuO_2$ than one of 800°C.[7] However, a heat treatment directly to 900°C is not good, particularly for directionally solidified alloys, since it melts all of the compounds in the Ag-Yb-Ba-Cu system. Since the melting point of a Ag-23at.%($YbBa_2Cu_3$) alloy is about 675°C (DTA), the initial heat treatment temperature of 650°C was selected. The melting point of this alloy seems to be dominated by the Ag-Ag_9Yb_2 eutectic reaction at 685°C.

Increasing the UD growth velocity from 1.5cm/h to 14cm/h was expected to reduce the lamellar spacing and improve the superconducting properties of the sample. However, this was not the case. The lamellar spacing was reduced by a factor of 5 to 10, from about 50μm to 5-10μm, but the T_c was substantially reduced, to less than 60K (fig. 3). This was the same for two different heat treatment schedules, even though, they had different heat treatment temperatures and times, and atmospheres. One sample heat treated in oxygen with the schedule 650°C for 2d plus 750°C for 1d, 850°C for 1d, and 500°C for 1d has a T_c of 40K (fig. 3). The same material treated in air shows the same treated with temperature to 77K, even thought its heat treatment was 650°C for 12h plus 875°C for 12h plus 500°C for 12h. If the fast growth rate sample is first given an anneal at 650°C (in argon) prior to the oxidization treated its T_c is increased to about 80K (fig. 3). This suggests that a solid state reaction occurs during slow growth or during the 650°C anneal to promote the formation of the superconducting phases. The material seems to be more sensitive to UD processing parameters and heat treatment temperature than heat treatment atmosphere. The microstructure variation between the slow growth and fast growth samples, and the fast growth with and without the pre-oxidation anneal are currently being investigated.

CONCLUSIONS

The critical temperature of an internally oxidized Ag-23at.%($Yb_1Ba_2Cu_3$) alloy is strongly influenced by unidirectional solidification processing conditions and heat treatment temperature. For the same oxidization heat treatment samples produced with a slow growth rate had higher T_c's than samples produced with a fast growth rate. This is interpreted as an annealing effect associated with the slow growth rate, since the T_c's of the high growth rate samples were increased by an oxygen-free anneal prior to the oxidization heat treatment. To prevent sample melting, and the resulting loss of the solidification structure, the initial heat treatment temperature should be less than or equal to 650°C. To consistently obtain a material with a zero resistance T_c of about 85K and a narrow transition, ΔT_c less than 5K, the final heat treatment should be at 875°C or above.

REFERENCES

1. Haldar R, Lu YZ, and Giessen BC (1987) $EuBa_2Cu_3O_x$ produced by oxidation of a rapidly solidified precursor alloy: An alternative areparation method for high T_c ceramic superconductors. Appl. Phys. Lett. 51: 538-54.

2. Matsuzaki K, Inoue A, Kimura H, Aoki K, Masumoto T (1987) High T_c superconductor prepared by oxidation of a liquid-quenched $YbBa_2Cu_3$ alloy foil in air. Japn. J. Appl. Phys 26: L1310-1312.

3. Yurek GJ, Vander Sande JB, Wang W-X, Rudmad DA, Zhang Y, and Matthiesen MM (1987) Synthesis of a superconducting oxide by oxidation of a metallic precursor. Met. Trans. 18A: 1813-1817.

4. Yurek GJ, Vander Sande JB, Rudmad DA, Chiang Y-M (1988) Superconducting microcomposites by oxidation of metallic precursors. J. Metals: 16-18

5. Togano K, Kumakura H, and Dietderich DR (1990) High-T_c oxide/metal composite superconductors produced by oxidation of unidirectionally solidified Ag-Yb-Ba-Cu alloys. (accepted) J. Appl. Physics Comm 67: (not available)

6. Kogure T, Otto A, and Vande Sande JB (1989) Formation of $Yb_nBa_{2n}Cu_{3n+1}O_x$ (n=3,4) by the oxidation of Yb-Ba-Cu-Ag metallic precursors. Physica C 157: 159-163.

7. Kogure T, Zhang Y, Levonmaa R, Kontra R, Wang W-X, Rudman DA, Yurek GJ, and Vander Sande JB (1988) Grain boundary structure of $YbBa_2Cu_3O_{7-\delta}$ formed by oxidation of metallic precursors. Physica C 156: 707-716.

Formation of High-T_c Phase of Bi-Pb-Sr-Ca-Cu-O Oxide Superconductor by Melt Process

A. KUME[1], K. TOMOMATSU[1], O. FUKUDA[1], T. HARA[2], H. ISHII[2], and T. YAMAMOTO[2]

[1] Materials Research Laboratory, Fujikura Ltd., Koto-ku, Tokyo, 135 Japan
[2] Engineering Research Center, Tokyo Electric Power Company, Chofu, Tokyo, 182 Japan

ABSTRACT

High-T_c phase of the Bi-Pb-Sr-Ca-Cu-O superconductor, which shows zero resistance at 108K, has been investigated by melting process with addition of Pb and annealing. The nominal starting composition was $Bi_{2-x}Pb_xSr_2Ca_2Cu_3O_y$ (x= 0, 0.25, 0.5), the calcined powders in Pt vessel were melted in air at 900-950°C for 30 min followed by heat treatment at 860-880°C for 100 hours . The heat treatment enhanced formation of high T_c phase by reaction between the low T_c phase and other phases in the solidified products. The influence on superconductivity by metal element from vessel is also discussed.

KEY WORDS: Bi-Pb-Sr-Ca-Cu-O, high-T_c phase, low-T_c phase, melting process heat treatment

INTRODUCTION

Since the discovery of the superconductor with T_c above 100K in Bi based system[1], much effort has been made to obtain the single phase of the high T_c phase[2]. However, it is not easy to obtain the single high T_c phase because of the coexistance of the low T_c and other phases. An increase of fraction of the high T_c phase is necessary to enhance its superconductivity.[3-5]

Melting process is considered to be preferable for producing pure or directionally grown grain structures. Therefore it is useful for the future application to clarify the nature of the formation of the crystals through melting process.

In this paper, we report the results of the formation of the superconductor by melting process, which contains large fraction of the high T_c phase with the superconducting transition at 108 K. The solidified samples consisted of the low T_c phase and other non-superconducting phases , however, we have succeeded to increase volume ratio of the high T_c phase by following heat treatment near melting temperature.

EXPERIMENTAL

The starting materials of the Bi-Pb-Sr-Ca-Cu-O superconductors were Bi_2O_3, PbO, $SrCO_3$, $CaCO_3$, CuO. The precursors with nominal composition described by $Bi_{2-x}Pb_xSr_2Ca_2CU_3O_y$(x = 0, 0.25, 0.5)were prepared in the following manner: Powders of the starting materials were weighed out in appropriate proportion , mixed well, calcined at 830 °C for 48 h in air and carefully ground in a mortar.

The calcined powder in Pt vessel was heated in air at 900, 925 or 950 °C for melting for 30 min, then cooled to room temperature at the ratio of -100

K/min. After the melting procedure, the solidified samples were heated at 860, 870 or 880 °C each for 100 h in air.

The samples were used for evaluation of superconducting properties. Electric resistivity of the samples was measured by standard four-probe method. X-ray powder diffraction was also performed to analyze the ratio of the high T_c and the low T_c phase by calculating the peak ratio near 5 degrees.

Under preferable condition for the powder in Pt vessel that the nominal starting composition of $Bi_{1.75}Pb_{1.25}Sr_2Ca_2Cu_3O_y$, the melting at 950 °C for 30 min, cooled as well, further heat treatment at 860 °C for 100 hours. The precursors were also melted in Au, Ag or hastelloy vessel, and then heated. As well as the case of Pt vessel, superconducting properties were measured for all the cases.

RESULTS AND DISCUSSION

According to the result of X-ray powder diffraction of the starting composition of $Bi_2Sr_2Ca_2Cu_3O_y$, it was difficult to conjecture wheter the high T_c phase was formed or not. Without Pb, peaks of the low T_c phase were observed while peak at 4.7 degrees of the high T_c phase was not confirmed. But it was recognized that formation of the high T_c phase was possible from results of the T_c measurements shown in figure 1. The T_c curves consisted of two step transition, that is , samples contained not only the low T_c phase but also a little amount of the high T_c phase. The solidified products contained the low T_c phase and other non-superconducting phases, but from figure 1, it was observed that annealing made it possible to convert the low T_c phase into high T_c phase even in the case without Pb.

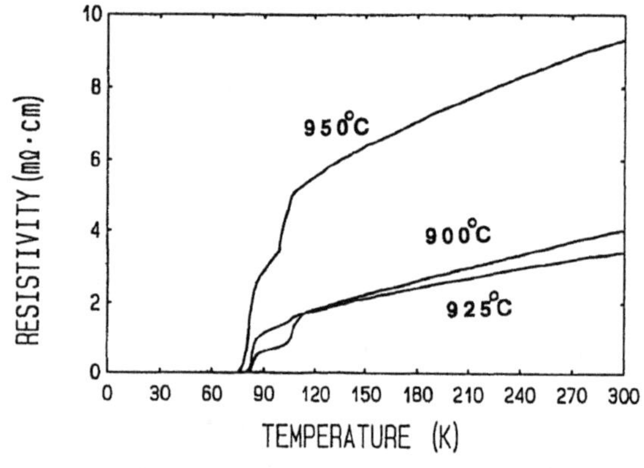

Fig.1 T_c curves for the samples of $Bi_2Sr_2Ca_2Cu_3O_y$ annealed at 880°C for 100 h after melting.

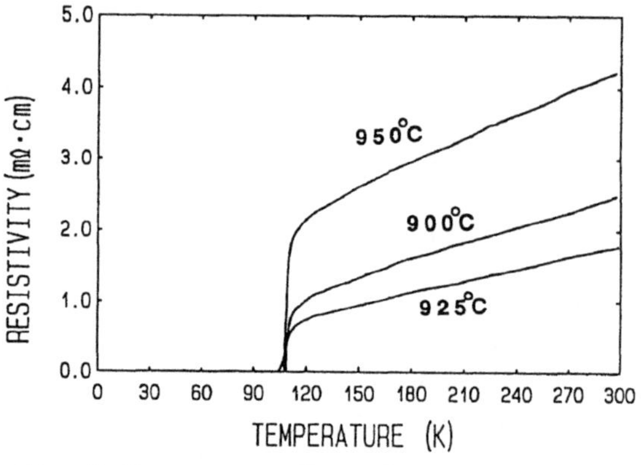

Fig.2 Tc curves for the samples of $Bi_{1.75}Pb_{0.25}Sr_2Ca_2Cu_3O_y$ annealed at 860°C for 100 h after melting.

In the case of Pb addition, influence of the annealing became more clearly. From the results of X-ray powder diffraction, low angle peak at 4.7 degrees of the high T_c phase appeared in most cases, while there was difference in the peak intensity. There were several cases which intensity of the low angle peaks of the high T_c and the low T_c were almost the same level. These results revealed that these samples were consisted of the high T_c phase of more than 50% at the volume ratio. Probably additions of Pb promoted the formation of the high T_c phase. Results of the T_c measurment were shown in figure 2. In the case of $Bi_{1.75}Pb_{0.25}Sr_2Ca_2Cu_3O_y$, temperature of the heat treatment was 860 °C and yielded T_c (R = 0) values were between 105 K and 108 K. All results of T_c measurement were summerized

in Table 1, 2 and 3. From these tables, we confirmed correlation between the melting temperature and additional heat treating temperature. With the respect to quantity of addition of Pb , it seemed that taking the ratio from results of these experiments as Bi : Pb = 1.75 : 0.25 was suitable. To obtain the high T_c phase , it was necessary to take conditions as follows : If we melt powders at the higher temperature, for example 950 ^{o}C, it was preferable to heat samples at the lower temperature, typically 860 ^{o}C.
Through these experiment, solidified products adhered to Pt vessel first, but they separated from the Pt vessel during the followed annealing. It seems to be happened because of the difference of the coefficient of thermal expantion between them. To examine influences of metal element of vessel on the superconductivity, Au, Ag or Hastelloy were used as metal vessel. In the case of Hastelloy, solidified products which were melted at 950 ^{o}C became insulators. Result of X-ray powder diffraction revealed that calcined powders reacted with Hastelloy and the low T_c phase was decomposed into mostly Ca, Bi_2O_3, $SrPbO_3$, PbO. In the case of Au or Ag, samples did not separate from the metal, but the high T_c phase could not be observed according to the X-ray powder diffraction pattern. However, figure 3 shows that formation of the high T_c phase occured in the case of Au vessel. Ag may cause bad influence on the properties of superconductors during the long heat treatment after solidification, resulting difficulty of the formation of the high T_c phase.

Table.1 Effect of melting and annealing temperatures on T_c. ($Bi_2Sr_2Ca_2Cu_3O_y$)

Melting temperature	Annealing temperature 860℃	870℃	880℃
900℃	88K	84K	82K
925℃	83K	79K	81K
950℃	84K	77K	78K

Table.2 Effect of melting and annealing temperatures on T_c. ($Bi_{1.75}Pb_{0.25}Sr_2Ca_2Cu_3O_y$)

Melting Temperature	Annealing temperature 860℃	870℃	880℃
900℃	108K	108K	107K
925℃	108K	106K	107K
950℃	105K	80K	R>0

Table.3 Effect of melting and annealing temperatures on T_c. ($Bi_{1.5}Pb_{0.5}Sr_2Ca_2Cu_3O_y$)

Melting temperature	Annealing temperature 860℃	870℃	880℃
900℃	106K	104K	85K
925℃	107K	101K	84K
950℃	102K	80K	76K

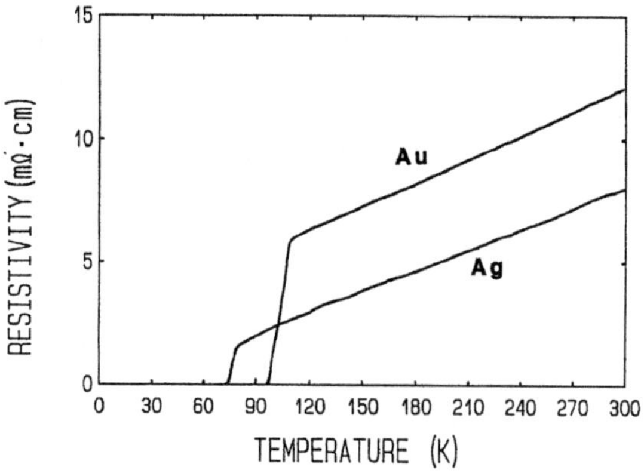

Fig.3 T_c curves prepared in metal vessel annealed at 860°C for 100 h after melting at 950°C for 30 min. ($Bi_{1.75}Pb_{0.25}Sr_2Ca_2Cu_3O_y$)

CONCLUSION

As can be seen from the results in Bi-Pb-Sr-Ca-CU-O, the annealing which was performed after melting was quite effective for the formation of the high T_c phase. With respect to reactivity and adhesion between superconductor and metal crucible, Au was preferable for melting process.

REFERENCES

[1] Maeda, H., Tanaka, Y., Fukutomi, M., and Asano, J., Jpn. J. Appl. Phys. 27, L209 (1988)

[2] Muromachi, E. T., Uchida, Y., Matsui,Y., Onoda, M., and Kato, K., Jpn. J. Appl. Phys. 27,L556 (1988)

[3] Green, S.M., Jiang, C., Mei, Yu., Luo, H.L., and Politis, C., Phys. Rev. 38, L5016(1988)

[4] Takano, M. , Takada, J., Oda, K., Kitaguchi, H., Miura, Y., Ikeda, Y., Momii, Y., and Mazaki,H., Jpn. J. Appl. Phys. 27, L1041(1988)

[5] Ramesh, R., Thomas, G., Green, S.M., Mei,Yu., Jiang, C., and Luo, H. L., Appl. Phy. Lett. 53, L1759(1988)

Preparation of Superconducting Bi-Sr-Ca-Cu-O Rods by Melt-Quenching and Floating Zone Method

K. SAWADA[1], M. HIRAOKA[1], T. SHINTANI[1], T. KOMATSU[2], and K. MATUSITA[2]

[1] Central Research Laboratory, Mitsubishi Cable Industries, Ltd., 8, Nishinocho, Higashimukaijima, Amagasaki, Hyogo, 660 Japan

[2] Department of Chemistry, Nagaoka University of Technology, 1603-1, Kamitomioka, Nagaoka, Niigata, 940-21 Japan

ABSTRACT

Superconducting Bi-Sr-Ca-Cu-O rods were prepared by means of the melt-quenching and floating zone method. The influences of starting composition and growth condition on superconducting properties and microstructures were investigated. It was found that the value of T_c increased and the amount of non-superconducting phase decreased with decreasing growth rate. The low-T_c phase was dominant and the high-T_c phase was not detected in as-grown samples. The most highest value of T_c for the as-grown samples was 80K. The microstructure in the superconducting rods was composed of large crystals with a long pillar shape grown along the growth direction. The c-axis of the low-T_c phase was perpendicular to the growth direction.

KEY WORDS: high-T_c superconductor, Bi-Pb-Sr-Ca-Cu-O system, melt-quenching method, floating zone method

INTRODUCTION

Since the first report on new high-T_c superconducting Bi-Sr-Ca-Cu-O ceramics with different transition temperatures of 105K(high-T_c phase) and 80K(low-T_c phase) by Maeda et al. [1] , many researchers have expended much effort to prepare high-performance superconducting Bi-Sr-Ca-Cu-O ceramics with a large amount of high-T_c phase. Komatsu et al. [2] succeeded in preparing superconducting Bi(Pb)-Sr-Ca-Cu-O ceramics with a $T_{c(zero)}$=100K using the melt quenching method. This method has many advantages such as good forming ability compared with the conventional sintering method. However, since critical current densities in these samples were still low because of porous microstructure, random arrangement of superconducting grains with strong anisotropy and Josephson-like weak links at boundaries between superconducting grains, further work is needed to prepare high-performance superconducting materials. In this paper,the superconducting Bi(Pb)-Sr-Ca-Cu-O rods were prepared by using the melt-quenching and floating zone methods and particularly the orientation of the superconducting crystals was examined.

EXPERIMENTAL

The Bi(Pb)-Sr-Ca-Cu-O feed rods and seeds for floating zone experiments were prepared by the melt-quenching method. The appropriate amounts of commercial powders of guaranteed reagent Bi_2O_3, Pb_3O_4, $SrCO_3$, $CaCO_3$ and CuO were mixed together and calcined at 800 ℃ for 12hr in air. The nominal compositions examined in this study are $Bi_{0.8}Pb_{0.2}SrCaCu_2O_y$ and $Bi_2Sr_2CaCu_2O_y$. The calcined powders were melted in a platinum crucible at 1150-1250 °C for 20min in an electric furnace. The melts pumped up into silica glass tubes with 4.5mm diameter. After the outer glass tubes were removed, the obtained rods were used as the feed rods and seeds. The floating zone apparatus with an infrared convergence-type image furnace with two halogen lamps as a radiation souse was used for present experiments. The feed rods and seed were put coaxially in the center of a silica glass tube. The bottom of the feed rod was melted with the infrared radiation focused with the ellipsoidal reflection mirror. The feed rod was moved downward to contact between molten zone and seed, and the molten zone was traversed along the feed rod. The feed rod and seed were rotated in opposite directions at 25rpm. The growth rates were 1,2 and 5mm/hr.

Measurements of transition temperature T_c and J_c (77K, zero magnetic field) were made by using a four-point probe method. Crystal phases present in the samples were examined by X-ray diffraction (XRD) analyses using Cu-Kα radiation. The observation of microstructure and analysis of compositions of the samples were carried out using scanning electron microscopy (SEM) and energy dispersive X-ray spectroscopy (EDX).

RESULTS AND DISCUSSION

Figure 1 shows the temperature dependence of resistivity for the superconducting $Bi_{0.8}Pb_{0.2}SrCaCu_2O_y$ rods (as-grown) prepared at the growth rates of 1mm/hr and 5mm/hr. The values of T_c for these samples are 80K (1mm/hr) and 64K (5mm/hr). The temperature dependence of resistivity for the superconducting $Bi_2Sr_2CaCu_2O_y$ rods prepared at the growth rates of 2mm/hr and 5mm/hr are shown in Fig.2. The T_c values of these samples are 78K (2mm/hr) and 66K (5mm/hr). These results indicate that the T_c value for the sample grown at the slower growth rate is high. The values of normal-state resistivity of the as-grown rods with a nominal composition $Bi_2Sr_2CaCu_2O_y$ are much smaller than those with a $Bi_{0.8}Pb_{0.2}SrCaCu_2O_y$ composition. The J_c values (77K, zero magnetic field) of these samples were very low.

Figure 3 shows the XRD pattern for the surface parallel to the axial direction in the superconducting $Bi_2Sr_2CaCu_2O_y$ rod prepared at 2mm/hr. The (001) peaks of the low-T_c phase are dominant, indicating that the sample is oriented with the (001) plane of the low-T_c phase parallel to the growth direction. Similar results were obtained for other samples prepared at a series of experiments. No peak corresponding to the high-T_c phase was detected.

Figure 4 is a SEM micrograph of as-grown superconducting $Bi_2Sr_2CaCu_2O_y$ rod prepared at 2mm/hr. It is seen indicate that the microstructure in the sample is composed of large crystals with a long pillar shape grown along the growth direction. From the EDX analyses, it is found that the gray area is the low-T_c phase, the white area is the insulating phase being rich in Bi and the black area is another insulating phase being rich in Ca, Sr or Cu. The amount of non-superconducting phase decreased with decreasing grown rate, indicating that it needs to apply an extremely slow growth rate for the fabrication of the superconducting rods containing no insulating phase.

We also annealed the as-grown sample again at 840 °C for 250hr in air. The T_c value of this annealed sample increased to 100K (J_c=174A/cm^2). This fact indicate that the high-T_c phase was largely formed in the post annealed sample, although the high-T_c phase was not detected in the as-grown sample. Therefore, in the floating zone method with the condition of the growth rate slower than 1mm/hr, it is expected that a large amount of the highly oriented high-T_c phase would be formed even in the as-grown sample.

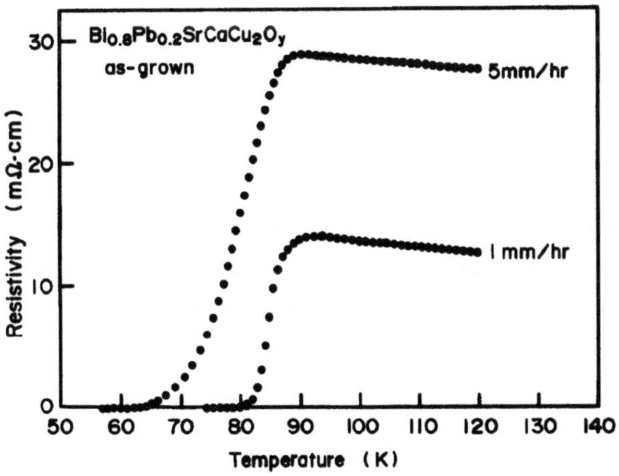

Fig.1 Temperature dependence of resistivity for the superconducting $Bi_{0.8}Pb_{0.2}SrCaCu_2O_y$ rods (as-grown) prepared at the growth rates of 1mm/hr and 5mm/hr.

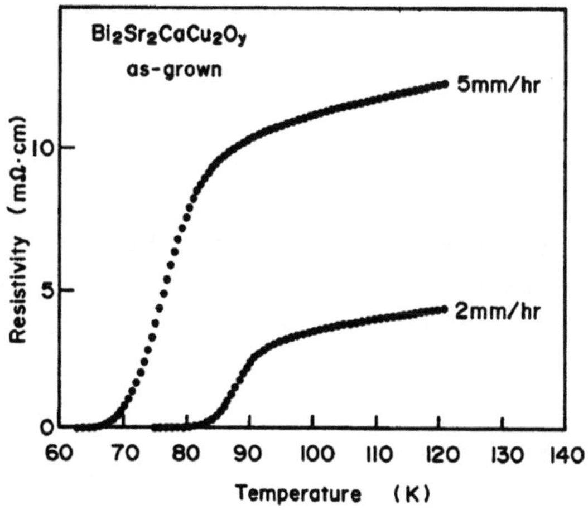

Fig.2 Temperature dependence of resistivity for the superconducting $Bi_2Sr_2CaCu_2O_y$ rods (as-grown) prepared at the growth rates of 2mm/hr and 5mm/hr.

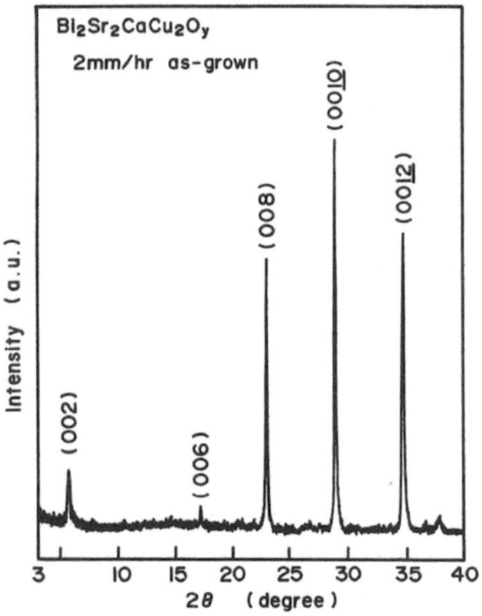

Fig.3 XRD pattern for the surface parallel to the axial direction in the superconducting $Bi_2Sr_2CaCu_2O_y$ rod prepared at 2mm/hr.

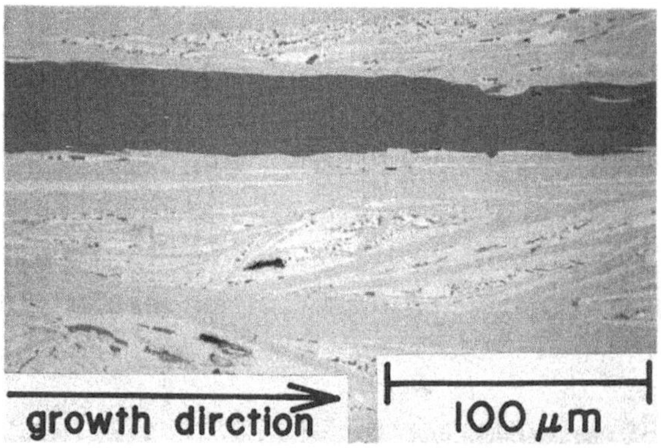

Fig.4 SEM micrograph of as-grown superconducting $Bi_2Sr_2CaCu_2O_y$ rod prepared at 2mm/hr.

CONCLUSION

Superconducting $Bi_{0.8}Pb_{0.2}SrCaCu_2O_y$ and $Bi_2Sr_2CaCu_2O_y$ rods with T_c=80K were prepared by the melt-quenching and floating zone method. The low-T_c phase was dominant in these superconducting rods and orientated unidirectinally.

REFERENCES

1. H. Maeda, T. Tanaka, M. Fukutomi and T. Asano, A New High-T_c Oxide Superconductor without a Rare Earth Element, Jpn. J. Appl. Phys. 27 (1988) L209.

2. T. Komatsu, R. Sato, C. Hirose, K. Matusita and T. Yamashita, Preparation of High-T_c Superconducting Bi-Pb-Sr-Ca-Cu-O Ceramics by the Melt Quenching Method, Jpn. J. Appl. Phys. 27 (1988) L2293.

The Critical Current Density of Y-Rich Y-Ba-Cu-O Superconductor by Half-Melted Process

K. SHIMOHATA, T. USHIJIMA, S. YOKOYAMA, and T. YAMADA

Central Research Laboratory, Mitsubishi Electric Corporation, Tsukaguchi-Honmachi, Amagasaki, 661 Japan

ABSTRACT

We apply half-melted process for the Y-rich Y-Ba-Cu-O systems to improve the critical current density (Jc). The half-melted process improves the Jc without deformation of the bulk Y-Ba-Cu-O. The powder pressed bulk oxide is rapidly heated beyond the melting point and kept a few minutes, and quenched to room temperature. After this process, the sample is annealed and slow cooled in oxgen atomosphere. The Jc of this sample is about 1000A/cm^2 at 0T, 77K. The Jc of the Y-rich Y-Ba-Cu-O systems is more improved than that of the 123 Y-Ba-Cu-O system. The low melting temperature material, $BaCuO_2$, is precipitated in the 123 Y-Ba-Cu-O. But the precipitation of $BaCuO_2$ is suppressed by the excess yttrium of Y-rich Y-Ba-Cu-O.

KEY WORDS : oxide superconductor, critical current density, melt process, stoichiometry

INTRODUCTION

Since high Tc oxides are discovered, many efforts have been denoted to improve the critical current density(Jc)[1-3]. The Jc of the bulk oxide superconductor is much lower than that of thin film or single crystal. In our previous work, we investigated the Jc of the bulk $YBa_2Cu_3O_{7-x}$ by pulse current technique and found the intrinsic critical current density was higher than 10^4 A/cm^2 at 77K, 20T[4]. Some researchers have reported the Jc of bulk oxide superconductors are improved by melt process[5,6]. We must form oxide superconductors into optional shape, when we use that for power applications, for example superconducting magnet or cable. We apply half melted process to improve the Jc of the bulk Y-Ba-Cu-O within the range which hold the pressed shape of the sample.

EXPERIMENTAL

Figure 1 shows the heat treatment procedure of half melt process. The Y-Ba-Cu-O powder is pressed in shape 2.0mm X 0.7mm X 25mm. The pressed bulk Y-Ba-Cu-O are calcined 920℃-5h in the air. The high temperature heat treatment beyond melting temperature is taken within the range which hold the pressed shape. The bulk Y-Ba-Cu-O is rapidly inserted into the pre-heated electric furnace beyond the melting point of $YBa_2Cu_3O_{7-z}$ and kept a few minutes, typically 1200℃-2min, and quenched to room temperature. After this process, that is annealed, typically 900℃-2h, and cooled in oxgen atomosphere. Table 1 shows the composition ratio of the powders. Four kinds of powders of different composition ratio are prepared by mixing calcined $YBa_2Cu_3O_{7-z}$ with Y_2O_3 or $BaCO_3$ powedrs. These composition of sample A,B,C and D corespond to $YBa_2Cu_3O_{7-z}$ in addition 0%, 1%, 3% and 10% of Y_2BaCuO_5 respectively. The Y_2BaCuO_5 is stable above 1050℃. During high temperature heat treatment $YBa_2Cu_3O_{7-z}$ partially decomposes into Y_2BaCuO_5, $BaCuO_2$ and CuO and so on. We expect that the excess Y_2O_3 or $BaCO_3$ compensates the divergence from 123 composition.

$Y_{1+x}Ba_{2+y}Cu_3O_{7-z}$ POWDER
↓
PRESS($P=1t/cm^2$)
↓
CALCINE(920℃-5h, in AIR)
↓
HIGH TEMPERATURE HEAT TREATMENT
(1150-1300℃, 1-5min)
↓
ANNEAL (in O_2gas)
(870-950℃, 0.5-40h)
↓
COOLING (in O_2gas)
(600-400℃, 25℃/h)

Fig.1 Heat treatment procedure of half-melted process.

Table 1 The composision ratio of Y-Ba-Cu-O

Sample	Y	: Ba	: Cu
A	1	: 2	: 3
B	1.017	: 2	: 3
C	1.05	: 2	: 3
D	1.17	: 2.05	: 3

Figure 2 shows the treatment temperature (Tm) dependence of Jc of sample A at 77K, 0T. The treatment time is 1-4min and the annaling condition is 950℃-1h. Unless the high temperature heat treatment is taken, Jc is about $200A/cm^2$ at 77K, 0T. This result suggests that the half-melted process improves the Jc of bulk Y-Ba-Cu-O without deformation. The optimum melting temperature is 1200℃ and treatment time is 2min. Figure 3 shows the annealing time dependence of Jc for the half-melted bulk Y-Ba-Cu-O. The half melting condition is 1200℃-2min and annealing temperature (Ta) is 950℃. The Jc comes to maximum value at annealig time of 1h. As the annealig time comes long, the Jc decreases. Figure 4 shows the SEM (Scanning Electron Microscope) photograph of surface of sample A by half-melted process. Figure 4(1) shows the calcined Y-Ba-Cu-O by the condition of 920℃-5h, the treated sample by 1200℃-2min is shown in fig.4(2), and the annealed samples at 950℃ for 0.5, 1, 2 and 5h are shown in fig.4(3)-(6) respectively. The shape of grain of the calcined bulk Y-Ba-Cu-O is sphereical and the mean diameter is about a few μm. After high temperature heat treatment, the grains are well connected each other. The treated bulk oxide is not superconductor at 77K but amorphous like. The amorphous like oxide is crystallized by annealing at 950℃, but unfortunatly the connected grains are torn to pieces and the connection becomes weak as the annealing time proceeds. Consequently the Jc comes to maximum at annealing time of 1h.

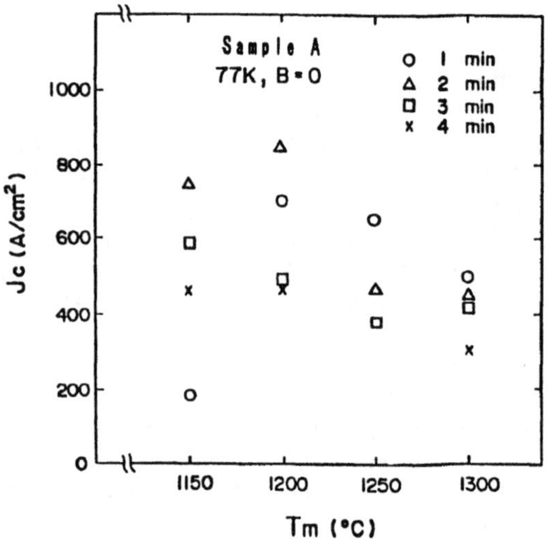

Fig.2 Treatment temperature dependence of Jc at 77K, 0T.

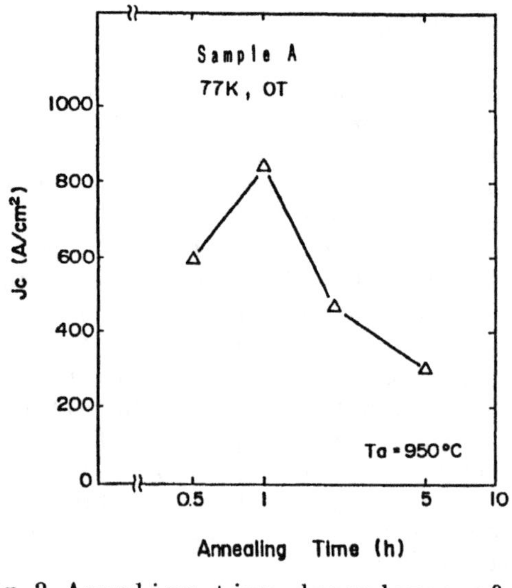

Fig.3 Anneling time dependence of Jc at 77K, 0T.

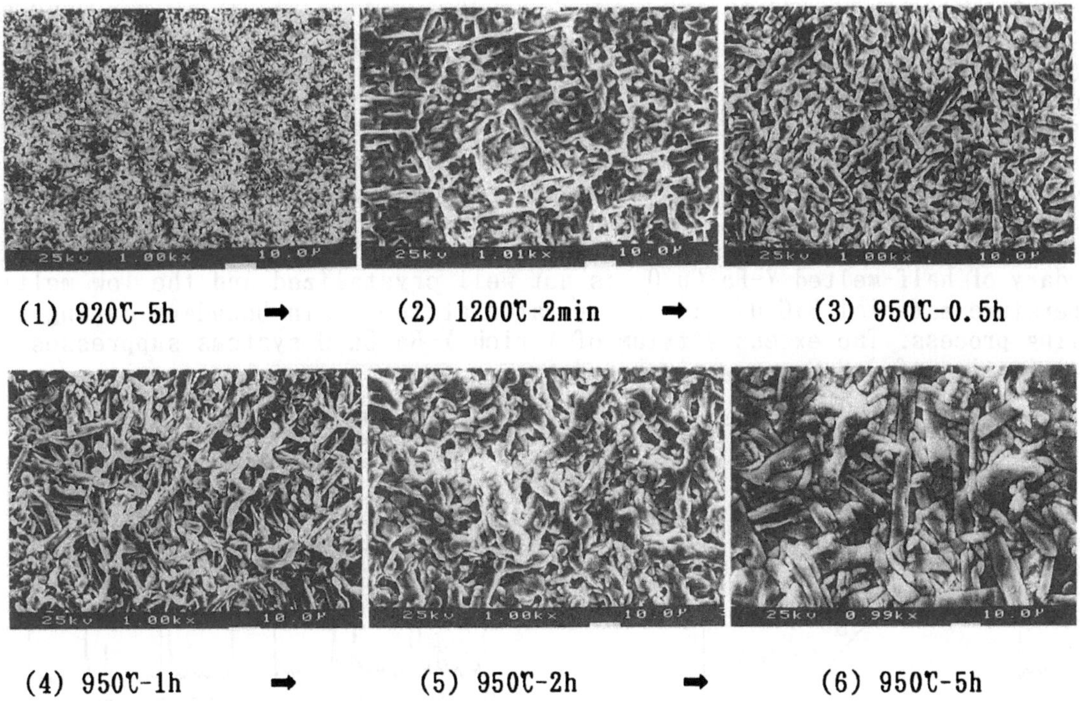

Fig.4 SEM photograph of 123 Y-Ba-Cu-O by half-melted process (1)calcine only (2)after high temperature heat treatment (3)-(6)after annealing

Figure 5 shows the annealing time dependence of Jc at 77K for four samples, A,B,C and D. Figure 5(1) shows the data at zero field, and fig.5(2) shows the data at 0.7T. The high temperature heat treatment condition is 1200℃-2min and the annealing temperature is 900℃. In each sample the Jc at zero field makes no grate difference. In the case of no high temperature heat treatment, the Jc of sample A and B is 300 and 20A/cm^2 respectively. Sample C,D do not reach zero resistance at 77K. Under no external field the Jc of sample which is made by the conventional solid state reaction process is sensitive to the starting composition, but is not sensitive by half-melted process as shown in fig.5(1). The Jc under magnetic field depend on the starting composition, i.e. as the annealing time proceeds, the Jc of sample B,C and D increase but the Jc of sample A decreases.

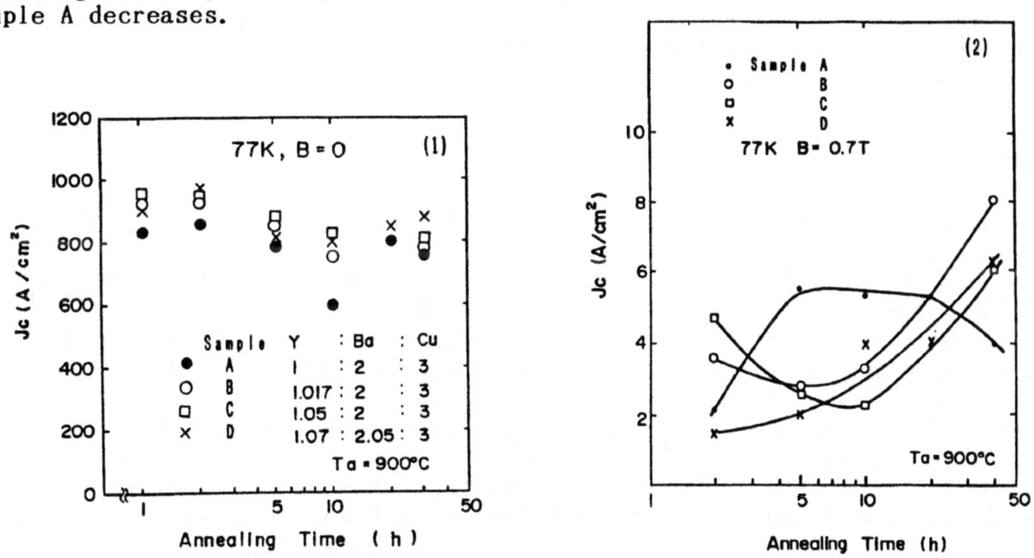

Fig.5 Annealing time dependence of Jc at 77K for different composition ratio of Y-Ba-Cu-O. Fig.5(1) is the data at 0T and fig.5(2) is one at 0.7T.

Figure 6 shows the cooling rate dependence of Jc of sample A and C at 77K, 0T. The treatment condition is 1200℃-2min and the annealing condition is 900℃-10h. As the cooling rate comes slow, the Jc of sample A decreases but the Jc of sample C keeps about 1000A/cm². Figure 7 shows the X-ray diffraction pattern of the bulk oxide surface for the cooling rates of 50, 100 and 400℃/h. Two peeks near 2θ=30°(indicated O) can be seen for sample A and the peek intensity becomes large as the cooling rate becomes slow. The two peeks are identified with $BaCuO_2$. The peeks are suppressd for sample C. It is suggested that the slow coolig make worse the grain boundary condition for sample A. The grain boundary of half-melted Y-Ba-Cu-O is not well crystalized and the low melting temperature material, $BaCuO_2$, is precipitated at the grain boundary during slow cooling process. The excess yttrium of Y-rich Y-Ba-Cu-O systems suppresses the precipitation of $BaCuO_2$ at grain boundary.

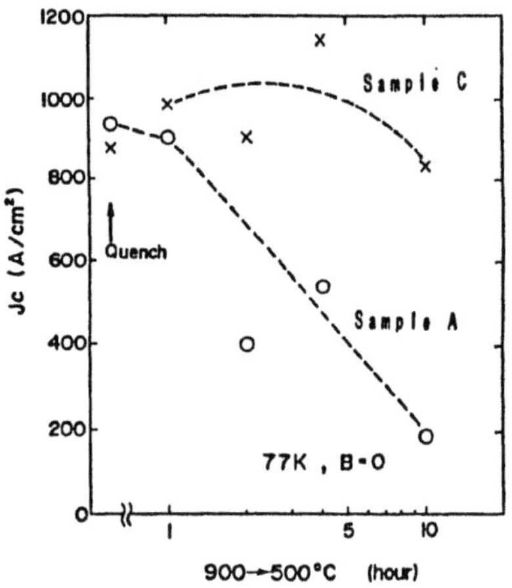

Fig.6 Cooling rate dependence of Jc at 77K, 0T.

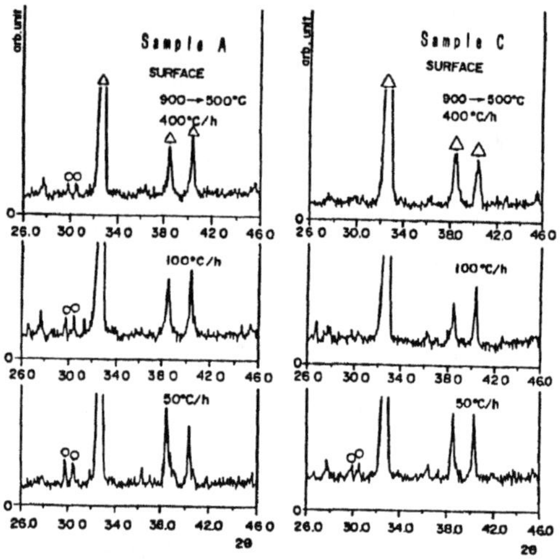

Fig.7 X.D. data for various cooling rate (O:$BaCuO_2$,Δ:$YBa_2Cu_3O_{7-x}$).

CONCLUSION

We apply melt process to the bulk Y-Ba-Cu-O within the range which hold the pressed shape. The Jc about 1000A/cm² at 0T, 77K is obtained. The Jc under external magnetic field depends on the starting composition i,e, as annealing time comes long, the Jc increases for Y-rich composition but decreases for 123 one. The low melting temperature material, $BaCuO_2$, is precipitated for the 123 composition Y-Ba-Cu-O, but the excess yttrium of Y-rich Y-Ba-Cu-O compensate the precipitation at grain boundary.

REFERENCES

[1]M.K.Wu, J.R.Ashburn, C.J.Torng, P.H.Hor, R.L.Meng, L.Gao, Z.J.Wang and C.W.Chu: Phys.Rev.Lett.58,908(1987)
[2]H.Maeda, Y.Tanaka, M.Fukutomi and T.Sano: Jpn.J.Appl.Phys.27,L209(1988)
[3]Z.Z.Sheng and A.M.Hermann: Nature 332,138(1988)
[4]K.Shimohata, Y.Yokoyama, M.Morita, T.Yamada, M.Wakata: International symposium on new developments in applied superconductivity ed.Y.Murakami, Osaka,Japan 17-10 Oct.(1988) P238
[5]S.Jin, T.H.Tiefel, R.C.Sherwood, R.B.van Dover, M.E.Davis, G.W.Kammlott and R.A.Fastnacht, Phys.Rev.B,37,7850(1988)
[6]M.Murakami, S.Matsuda, K.Sawano, K.Miyamoto, A.Hayashi, M.Morita, K.Doi, H.Teshima, M.Sugiyama, M.Kimura, M.Fujinami, M.Saga, M.Matsuo and H.Hamada: Proceedings of the 1st international symposium on superconductivity(ISS'88) ed.Kitazawa and Ishiguro,Nagoya,Japan 28-31 Aug.(1988)P247

Preparation of Bi System Oxide Superconductors by Melt Growth

K. HAYASHI, H. NONOYAMA, N. NAGATA, H. HITOTSUYANAGI, and M. KAWASHIMA
Osaka Research Laboratories, Sumitomo Electric Industries, Ltd., 1-3, Shimaya 1-chome, Konohana-ku, Osaka, 554 Japan

ABSTRACT

BiSrCaCuO superconductors were prepared by several directional-solidification methods: the melt extraction method, the horizontal bridgman method and the laser pedestal growth method. Temperature gradient was an important factor in obtaining highly aligned crystal structures. In the case of the laser pedestal growth method, Tc of 87K and Jc of 3070A/cm² (77.3K, 0T) were obtained. A Tc above 100K was achieved when a Pb added sample by the laser pedestal growth method was annealed.

KEY WORDS : oxide superconductor, directional solidification, crystal growth

INTRODUCTION

Since the discovery of oxide superconductors with high critical temperatures (Tc)[1], many efforts have been made to obtain high Tc superconducting wires with high critical current density (Jc).
It is well-known that melt processed oxide superconductors have high density and highly oriented structure with high Jc.[2][3] It is very easy to prepare the superconducting phase in the Bi-Sr-Ca-Cu-O system, so we tried to apply the several directional solidification methods to this system.
We used the melt extraction method, the horizontal bridgman method and the laser pedestal method as directional solidification. These methods seem to be feasible for obtaining long wire in the future. In the case of the laser pedestal growth method, we also tried the Pb-added Bi-Sr-Ca-Cu-O system.
The relationship between superconducting properties (Tc, Jc) and crystal growth conditions, such as growing speed, was investigated. We will discuss the factors in obtaining highly aligned crystal structures with high Jc, and the process to create the high-Tc phase.

EXPERIMENTAL PROCEDURE AND RESULTS

Bi_2O_3, $SrCO_3$, $CaCO_3$ and CuO were mixed in a ratio of 2:2:1:2 (Bi:Sr:Ca:Cu) and calcined to prepare the raw materials for the melt extraction method and the horizontal bridgman method. Sintered rod of about 4mm in diameter and the same 2212 starting composition was used for the laser pedestal growth method. And in this case, we also chose a 1.4:0.6:2:2:3(Bi:Pb:Sr:Ca:Cu)composition. All the samples after crystal growth were cut into a size of about 1mm×1mm×15mm and annealed in air. The Tc's and Jc's were measured by the four-terminal method. The current flow direction was parallel to the growing direction. The sensitivity of the critical current was 1μV/cm. The micro-structure was observed by optical microscope and EDX analysis.

Melt Extraction Method

Figure 1 shows a schematic illustration of the melt extraction method. Bi-Sr-Ca-Cu oxide was melted in the crucible, pulled through the temperature controlled nozzle by Pt dummy wire and solidified at the outlet of the nozzle. The crucible and nozzle were made of Au-Pd alloy to avoid contamination by impurities.

We succeeded in obtaining a 240mm long sample 2mm in diameter, as shown in Figure 2. The crystal structure was more aligned as the growing speed was decreased. A Tc of 85K and Jc of 173A/cm² (77.3K, 0T) were obtained under the conditions of a 2212 starting composition and 10mm/h growing speed.

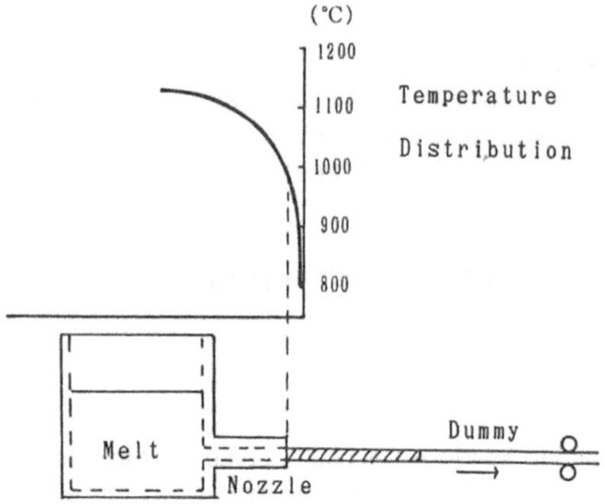

Figure 1. Schematic illustration of melt extraction method.

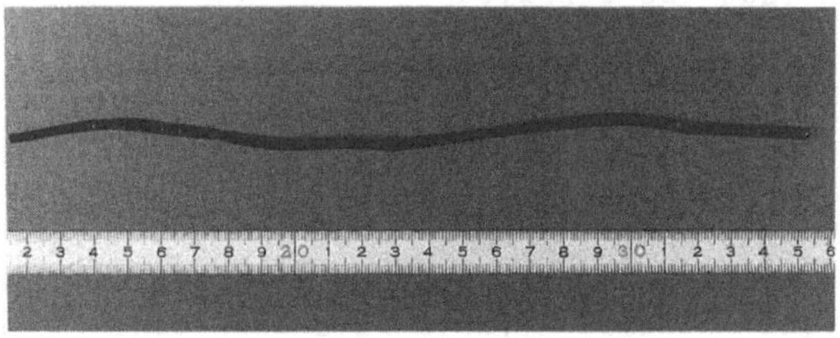

Figure 2. Photograph of obtained sample by melt extraction method.

Horizontal Bridgman Method

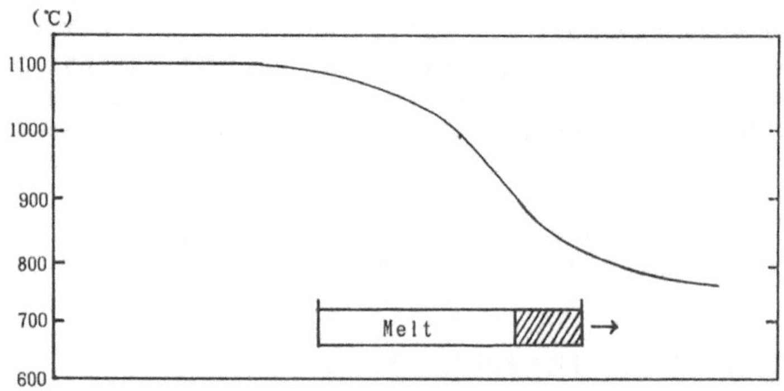

Figure 3. Schematic illustration of horizontal bridgman method

The horizontal bridgman method is famous as the production method of Ⅲ-Ⅴ compound semiconductors. Figure 3 shows the temperature distribution of the furnace used. Bi-Sr-Ca-Cu oxide was melted in a Au-Pd boat and the boat was moved to the low temperature region. Solidification began from the front end of the boat. The crystal structure was aligned as the growing speed was decreased, but the crystal alignment was inferior to the other method. Therefore the highest Jc obtained so far by this method is a low 39A/cm² (77.3K, ,T) under the conditions of a 2212 starting composition and 10mm/h growing speed. In this case, Tc had a relatively low value of 82K.

Laser Pedestal Growth Method

The two beam method of laser pedestal growth was used, as shown in Figure 4. The top of the sintered rod was melted by CO_2 laser and the crystals were grown from the molten zone. The seed crystal was the same sintered rod as the feed

rod. Growing speed and feed speed were controlled individually and the growing crystal and feed rod were rotated inversely in most cases.
When the 2212 starting composition was used, good crystal alignment was observed over the growth length (Figure 5). When the growing speed was over 10mm/h, grown crystals were composed of 2212 (low Tc) phase and Bi-free phase and these phases were aligned in the growing direction. It is well-known that the 2212(low Tc) phase melts incongruently, so when the growing speed was very slow (ex. 1mm/h), primary crystallization phase of the Bi-free phase crystallized first and the composition of melts changed to Bi-rich. After crystallization of the Bi-free phase, only the 2212 phase was obtained. When a much higher growing speed was used, crystal growth succeeded when the growing speed was higher than the feed speed. (i.e. The diameter of growing crystal was smaller than that of the feed rod.) Relatively high Jc's of 1350A/cm^2(77.3K,0T) and 400A/cm^2 (77.3K,0T) were obtained in the 60mm/h and 100mm/h cases, respectively. Tc's had a relatively high value of 87K typically.

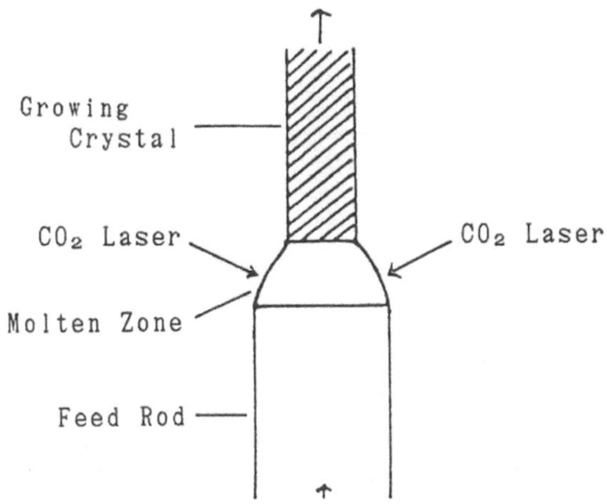

Figure 4. Schematic illustration of laser pedestal growth method.

Figure 5. Longitudinal cross section of 2212 composition.

The Bi system oxide superconductor has a high Tc phase above 100K and Pb stabilizes the high Tc phase, so we tried a Pb addition experiment. The experimental composition was 1.4:0.6:2:2:3 (Bi:Pb:Sr:Ca:Cu). When the growing speed was under 5mm/h, evaporation of Pb easily occurred. The high Tc phase couldn't be obtained directly from the melt. The as-grown structure was composed of an aligned Bi-free phase and a low Tc phase with Pb when the growing speed was 10mm/h. The Bi-free phase and low Tc phase reacted with each other and a high Tc phase was produced at the boundary of these phases. The high Tc phase was not aligned and a Tc of 106K was obtained. A Jc of 187A/cm^2(77.3K, 0T) was obtained in this case.

DISCUSSION

The results of this study are summarized in Table 1.
The laser pedestal growth method was the best way in this study to obtain a highly aligned crystal structure with high Jc. The crystal alignment from melt is related to the growing speed and the temperature gradient at the liquid-solid interface. As the growing speed is slower and the temperature gradient is larger, crystal alignment is easier to obtain. The largest temperature gradient was achieved by the laser pedestal growth method and the smallest by the horizontal bridgman method. Temperature gradients were assumed 1000deg/cm, 200

deg/cm and 50deg/cm for the laser pedestal growth method, the melt extraction method and the horizontal bridgman method, respectively. In the case of the laser pedestal growth method, a large temperature gradient was obtained when the crystal growth diameter was small, so the crystal alignment was maintained at a higher growing speed. A Pb-doped high Tc phase did not appear directly from the melt in this study. A high Tc phase does not seem to coexist with a melt phase. This study is still continuing from this point.

Table 1. Summary of this study

Method	Growing Speed (mm/h)	Jc (A/cm² at 77.3K, 0T)	Tc (K)
Melt Extraction	10	173	85
Horizontal Bridgman	10	39	82
Laser Pedestal Growth (without Pb)	1 60 100	3070 1350 400	(87) typical
Laser Pedestal Growth Bi:Pb:Sr:Ca:Cu =1.4:0.6:2:2:3	10	187	106

ACKNOWLEDGMENT

This work was performed as a part of "R & D on Superconducting Technology for Electric Power Apparatuses" as a subject of Super-GM under the Moonlight Project of Agency of Industrial Science and Technology, MITI, being consigned by New Energy and Industrial Technology Development Organization (NEDO).

REFERENCES

1. J.G. Bednorz and K.A. Muller,"Possible High Tc Superconductivity in the Ba-La-Cu-O system" Z. Phys. B64 (1986):189
2. S. Jin, T.H. Tiefel, R.C. Sherwood, M.E. Davis, R.B. Van Dover, G.W. Kammlott R.A. Fastnacht, and H.D. Keith " High critical current in Y-Ba-Cu-O superconductors " Appl. Phys. Lett. 52 (24) (1988):2074～2073
3. M. Nagata, " Manufacturing Processes and Critical Current Densities of High Tc Oxide Superconductors " 6th US-Japan Workshop on High Field Superconductors, Boulder, Colorado, Feb. 22-24. 1989.

Microstructural Characterization of $YBa_2Cu_3O_{7-x}$ Prepared by the Quench and Melt Growth (QMG) Process

MASAO KIMURA, AKIHIKO HAYASHI, MITSURU MORITA, MUNETSUGU MATSUO, KIYOSHI SAWANO, and SHOICHI MATSUDA

R & D Laboratories-I, Nippon Steel Corporation, 1618, Ida, Nakahara-ku, Kawasaki, 211 Japan

ABSTRACT

The bulk material of $YBa_2Cu_3O_{7-x}$ prepared by the Quench and Melt Growth (QMG) process was characterized from microstructural and crystallographic points of view. Microscopic observation showed that the material was a composite of $YBa_2Cu_3O_{7-x}$ (123) matrix and Y_2BaCuO_y (211) dispersoids with striations and cracks. Because the material was highly oriented, analysis of the crystal structure was carried out by means of a modified method of texture analysis. It was found that the matrix consisted of only a few twinned crystallites of the order of 10^{-2}m. The QMG-processed material was concluded to be a large grain composed of 123 crystallites containing a fine dispersion of 211 phase.

KEY WORDS: oxide superconductor, microstructure, texture analysis, melt process

INTRODUCTION

In oxide superconductors, it is considered that weak-links such as grain boundaries reduce the critical current density (J_c) and that a crystallite itself has a marked anisotropy in particular of superconductivity. Therefore the control of microstructure and crystal orientation is one of the most important factors to produce bulk material with superior properties. We have tried to produce such material by means of a new type of improved melt process: Quench and Melt Growth (QMG) process[1]. In this study, the QMG-processed material is characterized from the microstructural and crystallographic points of view. The material has a highly preferred orientation, and so we have performed a modified texture analysis to study the characteristic structure.

EXPERIMENTAL

Specimens were prepared by the QMG process; $YBa_2Cu_3O_{7-x}$ (123) powders were heated at 1450 °C and the melt was quenched by pressing between copper plates into a disk shape with a thickness of 1mm. The disk was reheated near 1150 °C for 0.5 h and slowly cooled down to room temperature. To produce a larger material, the QMG process was accompanied with the control of starting composition and a uni-directional growth technique:heating a specimen in a temperature gradient. Most of the QMG-processed specimens show a high value of $J_c > 10^8$ A/m^2 at 77K under a magnetic field of 1T.

The microstructure of specimen was characterized by a polarized optical microscope and EPMA. For further information of the structure, an improved method with a texture goniometer was utilized.

Figure 1 shows a schematic diagram of the texture goniometer by Schulz geometry[2]; the intensity of Bragg reflection of a specific crystal plane is measured by varying directions of X-ray incident beam relative to the sample which is tilted(α) and rotated(β). By using this apparatus we can measure the intensity of reflection of planes which are not parallel to the surface of the specimen and investigate the structure analysis of a sample to the order of 10^{-2}m.

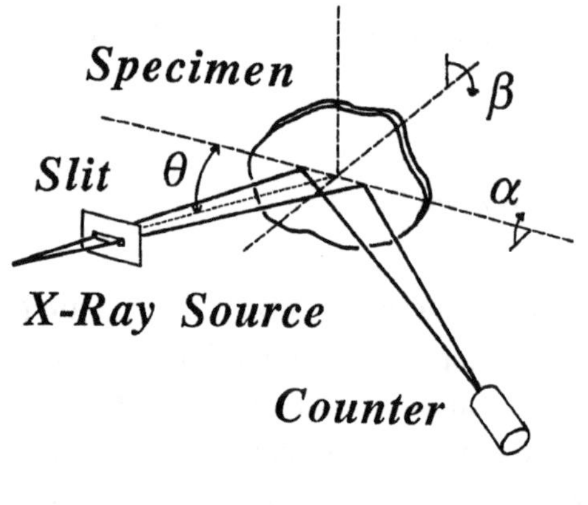

Fig.1 Schematic geometry of the apparatus.

RESULTS AND DISCUSISION

Microstructure

As shown in Fig.2, the QMG-processed specimen, which is composed of grains of the order of 10^2 μm, contains a dispersion of small particles and a parallel alignment of striations and cracks. It was determined by EPMA that the composition was 123 for the matrix and Y_2BaCuO_y (211) for the dispersoid, respectively. An increasing content of dispersoids seems to be effective for a decrease in the density of striations and cracks.

Figure 3 shows the micrograph of a specimen in the size of 10^{-2}m which was prepared by the QMG process with a uni-directional growth technique from a mixture of 211 and 123. The specimen has a fine dispersion of 211 phase and substantially is free of striations,cracks and grain boundaries. It is to be noted that we cannot observe twins with a polarized optical microscope. Therefore, our attention is focused on the the crystal structure of the specimen as the typical microstructure of the QMG-processed material.

Fig.2 Polarized optical micrograph of the specimen prepared by the QMG process.

Fig.3 Polarized optical micrograph of the specimen used in the structure analysis.

Crystal Structure

Figure 4 (lower) shows the diffraction pattern, measured by the conventional 2θ-θ scanning, of the polished surface of specimen. The pattern is composed of an extremely strong peak at 2θ=68° and peaks which can be attributed to polycrystalline 211. The strong peak can be assigned to be that of 123 according to the results of the microstructure analysis and X-ray diffraction study of the pulverized specimen. In other word, this fact suggests that the matrix has a crystallographic preference close to that of a single crystal.

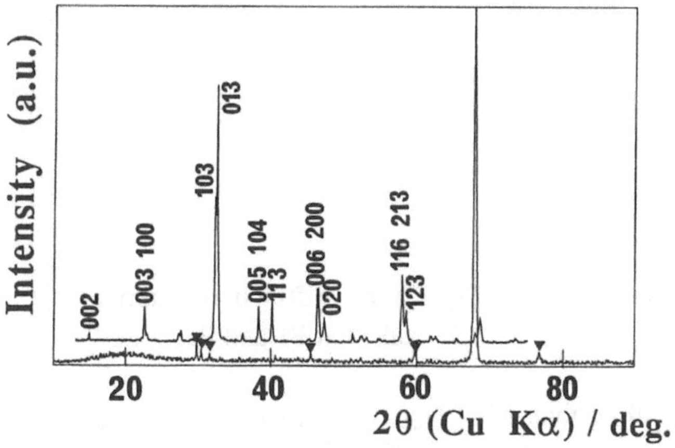

Fig.4 X-ray diffraction patterns of the QMG-processed specimen(lower) and a random polycrystalline(upper).

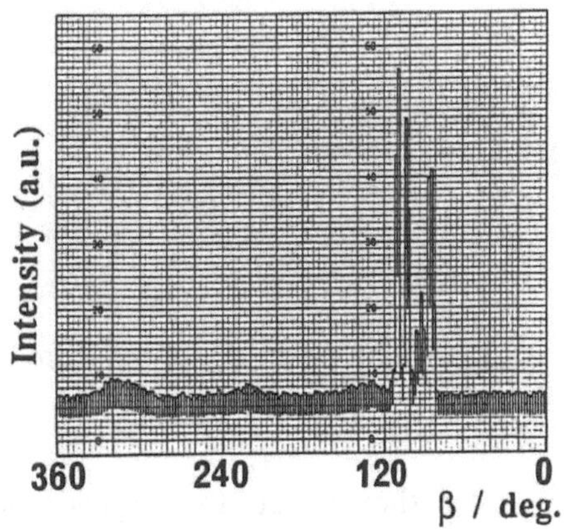

Fig.5 Measured intensity in the range of β=0~360° at 2θ=38.5°, α=82°.

It is most convenient for the analysis using the texture goniometer to measure the distribution of a peak which is composed of one set of {hkl} reflections. This is available in the case of conventional analysis of texture in, for example, steel and other material. Most of the Bragg peaks of 123 involve some sets of {hkl} reflections having nearly the same value of planar spacing, because the lattice constants have the relationship of a>b=c/3. We chose a peak at 2θ=38.5° which is a mixture of {005} and {104} and separated measured intensity into the intensity for each reflection as shown later.

Figure 5 is a measured intensity in the range of β=0~360° at 2θ=38.5°, α=82° and sharp peaks were observed only at β=82,90,102 and 108°. The half-maximum line width of each peak is near 4° suggesting the crystallites are oriented in a direction. In the same way, the intensity was measured in the range of α=20~90° and β=0~360°, and represented on stereographic projection in reference of the specimen's coordinates (Fig. 6). Strong reflections were measured at discrete directions (P1~P5): a cluster of some reflections (P1) and one peak at each direction (P2~P5). This means the existence of a few crystallites which are highly oriented.

Next we identified the peaks at P1~P5 to be {005} or {104} by the 2θ−θ scanning. For example, the diffraction patterns at P1 and P2 are shown in Fig.7. The pattern at P1 involves some peaks corresponding to {001} reflections, though only the peak at 2θ=38.5° is found in the pattern at P2. Therefore, peaks at P1 and P2 were considered to be a {005} and {104} reflection, respectively. All peaks at P1~P5 were identified in the same way as follows; all peaks at P1 are {005}'s and the other peaks at P2~P5 are {104}'s.

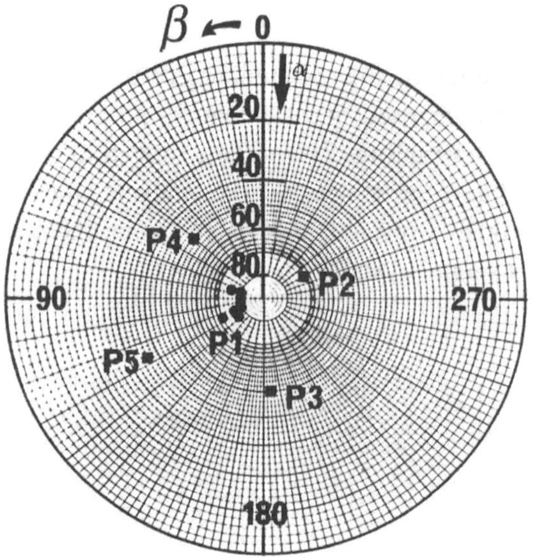

Fig.6 The positions of measured peaks.

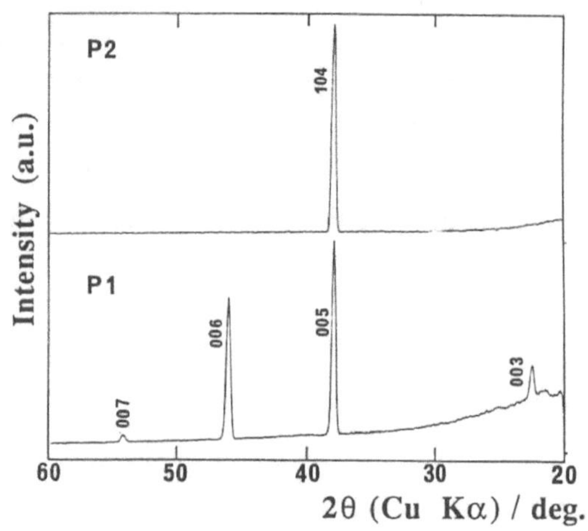

Fig.7 X-ray diffraction patterns taken at P1 and P2.

Figure 6 shows that the angles among directions of P1~P5 have the following relationship. The angles between P1 and P2~P5 are near 36.9° which is close to the angle between {005} and {104} in a crystal. The angles among the directions of P2~P5 are close to calculated angles between {005} and {104} planes in a pair of twinned crystals, which are listed in Table 1.

Table 1 The angle between planes in a pair of twinned crystal. T denotes the twinned crystal.

	104_T	$\bar{1}04$	$\bar{1}04_T$
104	50.7	73.8	50.7
104_T	---	50.7	73.8
$\bar{1}04$	---	---	49.8

This fact suggests that the crystallites are related to each other with boundaries of little difference in angle or that the matrix is a large twinned crystal with some micro-domains. This unique characteristics of the matrix and the fine dispersion of 211 result in a decrease of weak-links such as grain boundaries, cracks and striations and an increase of pinning sites, which may attribute to the high value of J_c.

ACKNOWLEDGMENT

We thank our colleagues at Nippon Steel Corporation for their help in our experiment: K.Miyamoto (sample preparation) and K.Doi (EPMA).

REFERRENCES

1) Murakami M, Morita M, Doi K and Miyamoto K(1989) A new process with the promise of high J_c in oxide superconductors, Jpn J Appl Phys, 28:1189-1194
2) Schulz L G (1949) A direct method of determining preferred orientation of a flat reflection sample using a Geiger counter X-ray spectrometer, J Appl Phys, 20:1030-1033,1033-1036

Microstructure of Bi-Based Superconductor Prepared by Floating Zone Method

M. NAKAGAWA and Y. SHIOHARA

Superconductivity Research Laboratory, International Superconductivity Technology Center, 10-13, Shinonome 1-chome, Koto-ku, Tokyo, 135 Japan

ABSTRACT

Bi-Sr-Ca-Cu-O superconductors were grown by a floating zone method using sintered rods of nominal composition of $Bi_2Sr_2Ca_1Cu_2O$ as precursors. $(Sr,Ca)CuO_2$, $Bi_2Sr_2Ca_1Cu_2O_x$(80K phase), $Bi_2Sr_2Cu_1O_x$(semiconductor phase), and unknown phases were observed in the boules. Volume fraction of $(Sr,Ca)CuO_2$ had a tendency to decrease with decreasing growth rate, while that of the 80K phase increased. Observation of microstructures suggested a peritectic reaction with which liquid and the $(Sr,Ca)CuO_2$ transformed to the 80K phase. The lower the growth rate, the sharper the superconducting transition and the better alignment of the 80K phase crystals was achieved.

KEY WORDS: floating zone method, crystal growth, microstructure, texture, Bi-Sr-Ca-Cu-O superconductor

INTRODUNCTION

High Tc superconducting oxides, such as $YBa_2Cu_3O_{7-x}$[1], $Bi_2Sr_2Ca_1Cu_2O_x$[2], and $Bi_2Sr_2Ca_2Cu_3O_x$[2] are considered to have an extensive merit for practical applications of superconductivity since these compounds have high critical temperatures and could be utilized at liquid nitrogen temperature. Enormous efforts were made to increase the critical current densities of the materials, however they have not yet been successful. Particularly, transport critical current densities measured[3-5] with a standard four-probe method remained far low from the level for a practical use. The low property is attributed to low density of the sintered materials[5-6] and weak links at the grain boundaries[6-7]. Good alignment of the grains[7] is considered to be necessary to enhance the property.
The unidirectional solidification processes[9] are the most promising methods, since elimination of porosities from the boule can be easily attained, and structures and textures can be controlled by selecting appropriate process parameters.
Several applications[10-11] of this method to the Bi-based superconductor systems have been reported, however, crystal growth mechanisms in microstructures of the superconducting phase were not yet made clear. Changes in the microstructures of the as-grown boules grown at different rates were investigated, and the growth mechanism of the $Bi_2Sr_2Ca_1Cu_2O_x$ superconducting phase has been discussed.

EXPERIMENTAL

Sintered rods with nominal composition of $Bi_2Sr_2Ca_1Cu_2O_x$ were prepared as feed and seed materials for the floating zone crystal growth. Calcined $Bi_2Sr_2Ca_1Cu_2O_x$ powders were formed into a rod shape with Cold Isostatic Pressing(CIP) at 2000kg/cm^2. The rods were sintered at 850°C for 12 hours in the ambient atmosphere. Typical diameter of the rods was 7mm.
Crystals were grown using a halogen lamp image furnace, as shown in figure 1. The apparatus consists of a pair of halogen lamps, gold-plated convergence mirrors, and of rotation and pulling system. The growth can be carried out under reducing, oxidizing, or inert atmosphere in ranging from the ambient pressure to 10kg/cm^2. The range of growth rates(R) in this work was 0.5 to 10mm/h(0.14 ~ 2.8mm/s). The boule and the feed rod were rotated at 20rpm in the counter direction. Total power of the lamps was

about 300W. The specimens were quenched in order to observe the solidification interface by means of turning off the lamps during the crystal growth. Temperature distributions on the surface of the specimens during the steady state growth were measured using a JEOL 3210 type infrared pyrometer.

Microstructures of the grown boules were investigated with optical and electron microscopy. Phases observed in the boules were identified by X-ray diffraction and Energy Dispersive X-ray Spectroscopy(EDS). Superconducting transition temperatures (Tc) and critical current densities(Jc) of the as-grown specimens were evaluated with a standard four-probe technique.

RESULTS AND DISCUSSION

Temperature distribution of the specimen along the growth direction was displayed in figure 2. The distributions were measured at different growth rates and direction of rotations, however no significant difference was observed. Temperature gradient(G) in the vicinity of the crystallization interface was found to be about 100K/mm. Optical images of the cross sections of the specimens grown at 10 and 1mm/h are shown in figure 3 a) and b), respectively. Several phases were observed in the boule. Composition of columnar grain(A) was analyzed by the EDS to be typically $Sr_{0.5}Ca_{0.6}Cu_1O_x$. The phase corresponds to $(Sr,Ca)CuO_2$. Average composition of plate-like crystal regions(B) was analyzed to be $Bi_{2.0}Sr_{2.4}Ca_{0.7}Cu_1O_x$. Back scattered electron imagse showed that the region consisted of two phases. They were identified as $Bi_2Sr_2Ca_1Cu_2O_x$ (80K superconducting phase) and $Bi_2Sr_2CuO_x$ (semiconductor phase). Crystals between the plate-like crystals remained unidentified at present. The precipitation is considered to have occurred in the following order. Columnar $(Sr,Ca)CuO_2$ is thought as a primary crystal;

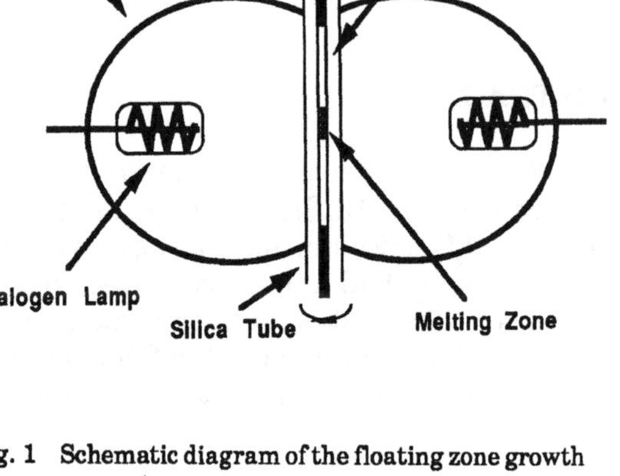

Fig. 1 Schematic diagram of the floating zone growth apparatus.

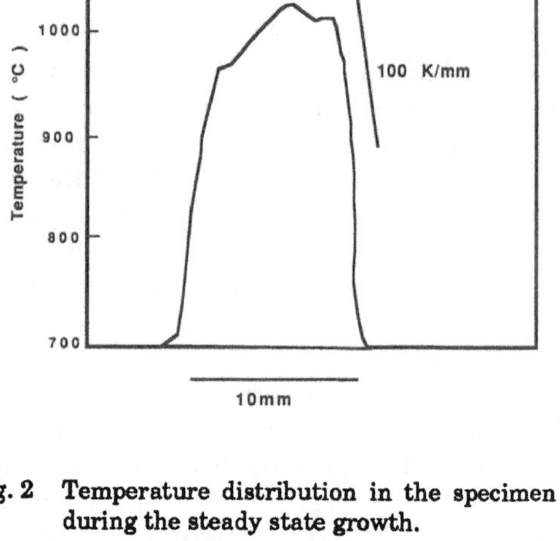

Fig. 2 Temperature distribution in the specimen during the steady state growth.

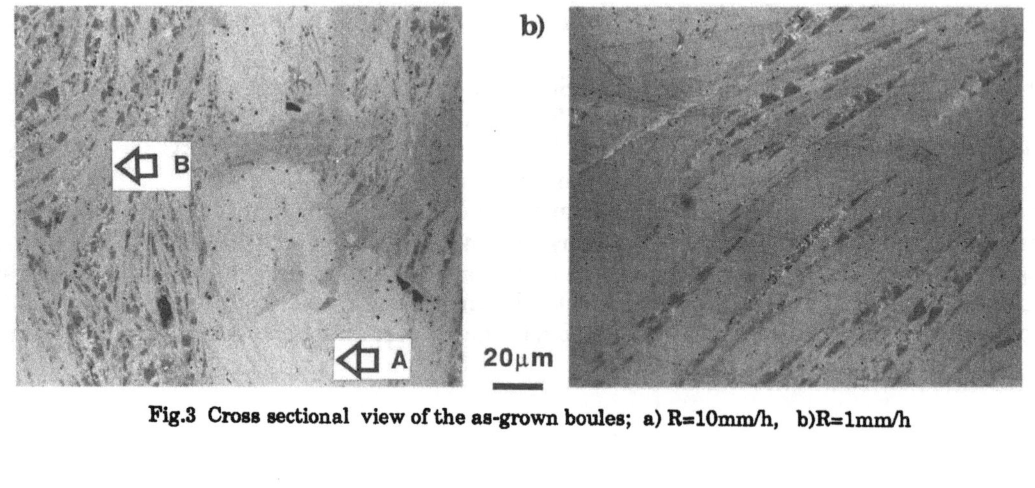

Fig.3 Cross sectional view of the as-grown boules; a) R=10mm/h, b)R=1mm/h

$Bi_2Sr_2Ca_1Cu_2O_x$ and Bi_2Sr_2CuOx grow; liquid left enclosed with the plate-like phases solidified forming the unknown phases. (Sr,Ca)CuO_2 in figure 3 a) was found to be surrounded by $Bi_2Sr_2Ca_1Cu_2O_x$. This fact suggesst a peritectic reaction with which the (Sr,Ca)CuO_2 and liquid form the 80K superconducting phase. In the specimens grown at slower rates had smaller volume fractions of the (Sr,Ca)CuO_2 phase and larger fraction of the $Bi_2Sr_2Ca_1Cu_2O_x$ as shown in figure 3 b). The slower the growth rate (larger 1/GR), the longer duration of the reaction attained and the more the reaction proceeded. This tendency also suggests the reaction with which (Sr,Ca)CuO_2 and liquid transformed into $Bi_2Sr_2Ca_1Cu_2O_x$ phase. These argument is consistent with the work on the phase diagram constructed by Takei[13].

Figure 4 shows an optical micrograph of the longitudinal section of the specimen grown at 1mm/h. $Bi_2Sr_2Ca_1Cu_2O_x$ crystals showed good alignment in the specimen. Results of X-ray diffraction revealed that its c-axis was perpendicular to the growth direction.

Figure 5 shows a temperature dependance of resistivity of the specimens grown at different growth

Fig. 4 Longitudinal view of the as-grown boule R=1mm/h.

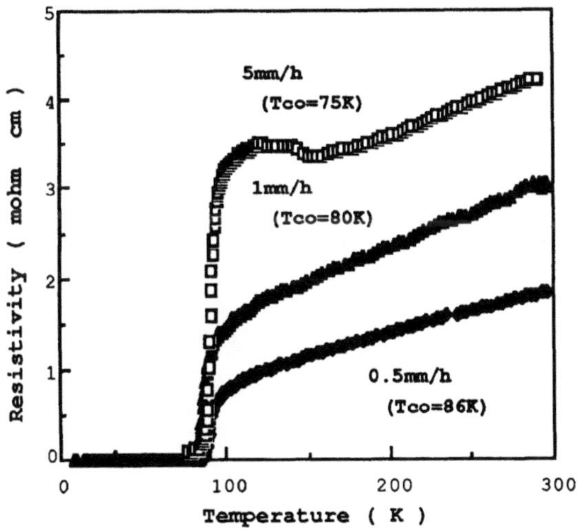

Fig. 5 Temperature dependence of resistivity of the as-grown specimens.

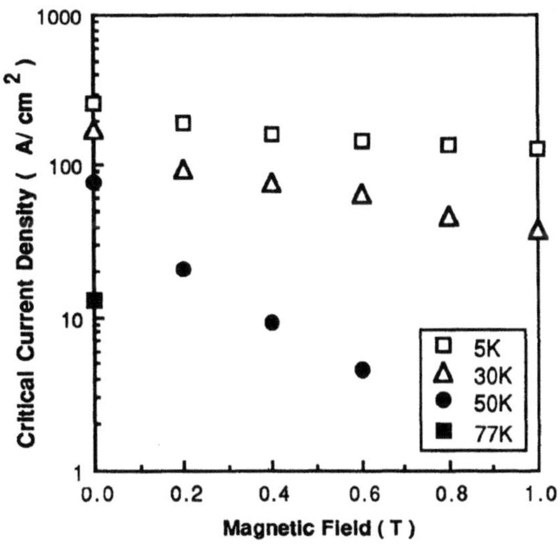

Fig. 6 Magnetic field dependence of Jc of the specimen R=0.5mm/h.

rates. Superconducting onset temperatures were almost identical for these specimens, while sharper transition was achieved in the specimen grown at slow rate. This change is also understood with respect to the time dependent peritectic reaction to form the 80K superconducting phase. Figure 6 shows critical current densities of the specimen grown at 0.5mm/h Magnetic field dependance was strong at higher temperatures and n-value in the power law [13-14] $E \propto J^n$ was as low as 1. These data suggest that the low Jc in the as-grwon specimens may be related with flux creep[14].

CONCLUSION

Bi-Sr-Ca-Cu-O superconductors were grown by a floating zone method using sintered rods of nominal composition of $Bi_2Sr_2Ca_1Cu_2O$ as precursors. $(Sr,Ca)CuO_2$, $Bi_2Sr_2Ca_1Cu_2O_x$, $Bi_2Sr_2Cu_1O_x$, and unknown phases were observed in the boules. Volume fraction of $(Sr,Ca)CuO_2$ had a tendency to decrease with decreasing growth rate, while that of the $Bi_2Sr_2Ca_1Cu_2O_x$ phase increased. Observation of microstructures suggested a peritectic reaction with which liquid and the $(Sr,Ca)CuO_2$ transformed to the $Bi_2Sr_2Ca_1Cu_2O_x$ phase. The lower the growth rate, the sharper the superconducting transition and the better alignment of the 80K phase crystals was achieved.

REFERENCES

[1] C. W. Chu, P. H. Hor, R. L. Meng, L. Gao, Z. J. Haung, and Y. Q. Wang; Phys. Rev. Lett., **58**(1987) 405.

[2] H. Maeda, Y. Tanaka, M. Fukutomi and T. Asano; Jpn. J. Appl. Phys., **27**(1988)L209

[3] H. Kumakura, M. Uehara, M. Suzuki, and K. Moriwaki; Jpn. J. Appl. Phys., **26**(1987)L656

[4] Y. Yamada, N. Fukushima, S. Nakayama, H. Yoshino, and S. Murase; Jpn. J. Appl. Phys., **26**(1987)L865

[5] O. Kohno, Y. Ikeno, N.Sadakata, S. Aoki, M. Sugimoto, and M. Nakagawa; Jpn. J. Appl. Phys., **26**(1987)1653

[6] A.M. Campbell; Jpn. J. Appl. Phys., **Suppl.26-3**(1987)2059

[7] H. Kupfer, I. Apfelstedt, R. Flukiger, C. Keller, R. Meiner-Hirmer, U. Wiech, and T. Wolf; Z. Phys., **B71**(1988)63

[8] D. Dimos; Phys. Rev. Lett., **61**(1988)219

[9] S. Jin, T. H. Tiefel, R. C. Sherwood, M. E. Davis, R. B. van Dover, G. W. Kammlott, R. A. Fastnacht, and H. D. Keith; Appl. Phys. Lett., **52**(1988)2074

[10] D. Gazit and R. S. Feigelson; J. Crystal Growth, **91**(1988)318.

[11] A. Kurosaka, M. Aoyagi, H. tominaga, and O. Fukuda; Appl. Phys. Lett., **55**(1989)390

[12] H. D. Brody, J. S. Haggerty, M. J. Cima, and M. C. Flemings; J. Cryst. Growth, **96**(1989)225

[13] H. Takei; J. Crystal Growth, in press.

[14] J. E. Evetts and B. A. Glowacki; Cryogenics, **28**(1988)641

[15] K. Watanabe, N. Kobayashi, H. Yamane, H. Kurosawa, T. Hirai, H. Kawabe, and Y. Muto; Jpn. J. Appl. Phys., **28**(1989)L1417.

Laser Zone Melting of $Bi_2Sr_2CaCu_2O_x$ Superconductors

K. ISHIGE, T. SUGA, and Y. SHIOHARA

Superconductivity Research Laboratory, International Superconductivity Technology Center, 10-13, Shinonome 1-chome, Koto-ku, Tokyo, 135 Japan

ABSTRACT

Quenched $Bi_2Sr_2CaCu_2O_X$ ceramic precursors were remelted and directionally solidified by a laser surface zone melting method. The X-ray diffraction pattern of the quenched precursor sample shows a glassy characteristic. Samples traversed at different speeds ranging from 0.1 to 200mm/s under continuous-wave laser beams which were focused through a set of lens on their surfaces. After the zone melting, the treated surfaces were characterized by an x-ray diffraction technique. It is shown that even at different scanning rate under comparable solidification conditions, i.e. with same cooling rates, similar phases appeared. The $Bi_2Sr_2CaCu_2O_X$ phase were primarily observed at the laser scanning speeds of 1 to 5 mm/s.

KEY WORDS : high-Tc superconductor, laser, zone melting, Bi-Sr-Ca-Cu-O

INTRODUCTION

Since the discovery of high-Tc superconductors[1,2], many researchers have tried to produce superconductors with high critical current densities by melting and solidification method[3-6]. In a laser zone melting method, a focused beam is scanned across a material to raise the temperatures of the surfaces locally above its melting point. Two potential advantages of this method are as follows. First, the high rate of energy supply enables very high rates of temperature increase to be attained which is followed by a similar rate of cooling due to self-quenching by heat conduction into the bulk material[7-10]. Secondly, extremely high thermal gradients can be attained because the heating effect is concentrated in a very thin and small region at the material surface.

Superconductive phases have been produced at very slow solidification rates less than several tens mm/h, and then the solidification of those methods were restricted to occur under a local equilibrium condition. The rapid solidification process such as laser zone melting is a method which enables to achieve controllable crystallization under the metastable state[11]. Though the material which has a complex crystal structure like $Bi_2Sr_2CaCu_2O_X$ requires extensive local diffusional rearrangements to form the structure during the rapid solidification, it is possible to produce superconductors at high speed and also to prevent segregations which cause the weak links between the grains. Furthermore, in spite of rather rapid solidification, it is possible to form textured microstructures along scanning direction because of the steep thermal gradient.

In this paper, the effect of cooling rate on phases prepared by laser zone melting was investigated.

EXPERIMENTAL PROCEDURE

Glassy $Bi_2Sr_2CaCu_2O_X$ precursors for the laser melting process with thickness of about 1mm were prepared by conventional splat quenching. Commercial powders of Bi_2O_3, $SrCO_3$, $CaCO_3$, and CuO were mixed and calcined at 1100K for 12hrs. Subsequently, the calcined powders were melted in alumina crucibles at about 1450K for 5 mins in a electric furnace. The melts were poured onto a copper plate and pressed quickly.

Figure 1 shows a schematic illustration of laser zone melting apparatus. The samples were placed on an enclosed hot stage held at R.T. or heated up to 803K in air. And the stage traversed under the beam of a YAG laser (CW TEM00 mode, wave length is 1.06μm) so as to move a molten zone. The beam was focused on the sample surface with a set of lens (f=100mm) to supply an approximate point heat source. Laser scanning velocity(scanning speed of samples, v), laser input power(Q) and sample temperature(Tsub.), i.e. hot stage temperature, were mainly varied as experimental parameters. The values of laser scanning rates and laser input powers ranged from 0.1mm/s to 200mm/s and from 0.2W to 10W, respectively. Three temperatures of samples chosen were R.T., 723K and 803K.

Laser processed samples were characterized by optical and scanning electron microscopy and an X-ray diffraction method. Moreover laser beams were scanned across all the surface of samples with constant intervals so as to get the sufficient intensity of X-ray diffraction for characterization.

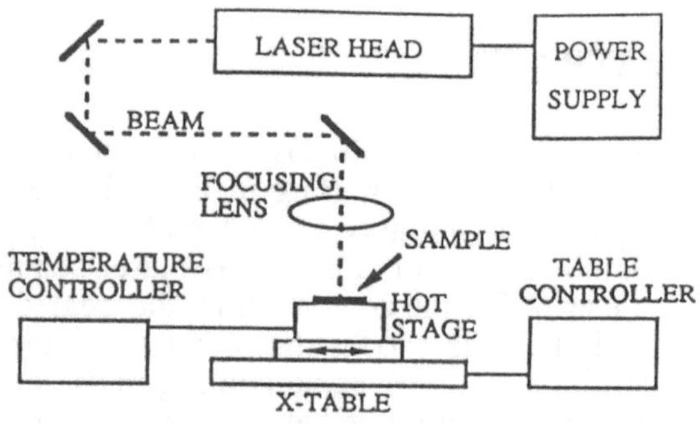

Fig. 1. Schematic of laser zone melting apparatus.

RESULTS AND DISCUSSION

Figure 2 shows a typical optical micrograph of a laser processed surface. Laser scanning speed, laser input power and sample temperature are 5mm/s, 0.6W and 723K respectively. The width of melted zone is about 200μm. Three phases including a needle like phase, a dark phase and a white phase were observed on this surface. The size of the needle-like crystal, which was identified to be (Sr,Ca)CuO$_X$ is typically 10μm long and 3μm in width.

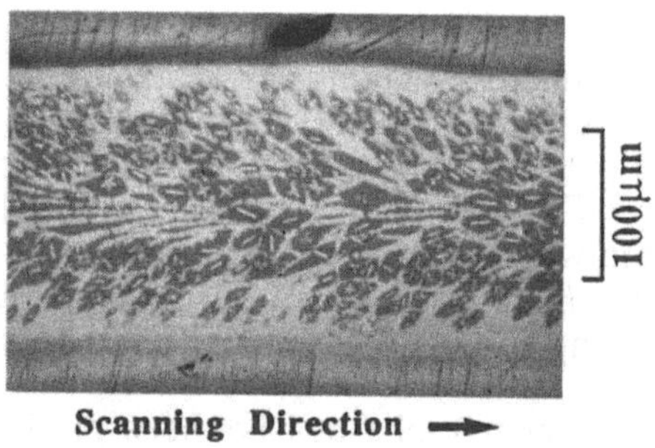

Fig. 2. Optical micrograph of laser processed surface (v=5mm/s, Q=0.6W, Tsub.=723K).

The X-ray diffraction pattern of this surface is shown in figure 3. This pattern indicates that the surface is mainly composed of a $Bi_2Sr_2CaCu_2O_X$ phase(2212) with a small amounts of the $Bi_2Sr_2CuO_X$ phase(2201). In addition, the intensity enhancement of the *(00l)* peak shows that c-axis is oriented normal to the surface.

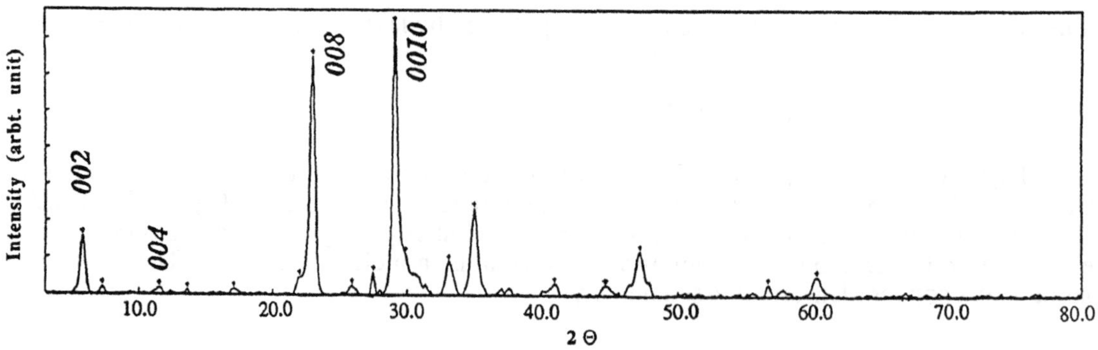

Fig. 3. The X-ray diffraction pattern of laser processed surface (v=5mm/s, Q=0.6W, Tsub.=723K).

Table 1 shows main phases observed in the X-ray diffraction patterns of the samples processed in different experimental conditions including laser scanning speeds, laser input powers and sample temperatures. In this table we were not able to clarify the effect of experimental parameters independently on main phases observed, while cooling rates were adopted as an effective process parameter to arrange the phases observed.

Cooling rates are defined as the product of the temperature gradient at the solid-liquid interface (G) and the solidification rate (R). Scanning velocities were considered to be solidification rates. Temperature gradients at the solid-liquid interface were estimated from the simple calculations of a heat flow equation[12], which is shown below ;

Table 1. Phases observed in the X-ray diffraction patterns of the samples processed in different conditions.

Experimental parameters			Main phases observed by X-ray
Sample temp., Tsub.[K]	Input energy, Q[W]	Scanning velocity, v[mm/s]	
803	1.2	0.1	(Sr,Ca)Cu Oxide
803	0.6	1.0	2201
723	0.5	1.0	2201+2212
803	2.4	10	2201
723	0.2	1.0	2212
723	0.4	2.0	2212
723	0.6	5.0	2212
723	0.4	4.0	2212
723	0.3	10	Amorphous-like
R.T.	1.0	10	Amorphous-like
803	4.8	100	Amorphous

$$T = Tsub. + \frac{Q}{2\pi\kappa} \exp\left(-\frac{v\xi}{2\alpha}\right) \frac{\exp\left(-vr/2\alpha\right)}{r}$$

where $r=\sqrt{x^2+y^2+z^2}$ $\quad \xi = x-vt$

Q : input energy $\quad$ v : scanning velocity

α : thermal diffusivity $\quad$ κ : thermal conductivity

This equation is derivated from several assumptions, such as 1)point heat source 2) steady state, i.e. input energy and scanning velocity are constant. 3) heat diffusivity and heat conductivity are constant. 4) heat loss of radiation is neglected. 5) latent heat of fusion is neglected.

Three dimensional temperature distributions are given by the solutions of the equation. The averages of temperature gradients between 1173K and 1273K along the center line of laser melting zones were adopted as temperature gradients at solid-liquid interfaces.

Figure 4 is the laser process figure, which gives the correlation between main phases and the product of solidification rates and temperature gradients. It is shown that even at different scanning velocities under comparable solidification conditions, i.e. with same cooling rates, similar phases appeared. When cooling rate is above 50×10^3K/s, amorphous or amorphous like phases appear. The cooling rates above 50×10^3K/s are therefore considered to be a sufficient condition of amorphous formations. At slow cooling rate, i.e. 0.1×10^3K/s, enormous needle-like phases, i.e. $(Sr,Ca)CuO_X$, precipitated. Since $(Sr,Ca)CuO_X$ is very stable within the temperature range between 1173K and 1273K, solidification time which is defined as a reciprocal number of a cooling rate, was long enough for $(Sr,Ca)CuO_X$ to grow. The $Bi_2Sr_2CuO_X$ phase precipitated at cooling rates between 2 and 5×10^3K/s. It is thought that macrosegregations occurred with comparatively long solidification times at these conditions.

After all, at the intermediate range of cooling rate between 5 and 20×10^3K/s, The $Bi_2Sr_2CaCu_2O_X$ phases were observed. It is noted that the $Bi_2Sr_2CaCu_2O_X$ phases are produced at high solidification rates of 1 to 5 mm/s.

It is important to get information of morphology of crystallization front. The knowledge of the morphology gives us a great deal of insight on the crystal growth mechanism. To get this information we have therefore contrived an etching method with dilute acetic acid. Figure 5 shows the solidification fronts revealed by this method. The morphology is determined by the heat flow direction and crystal growth kinetics due to supercoolings. In spite of high speed crystal growth, the morphology aligned along scanning direction was achieved in the center of solidified region.

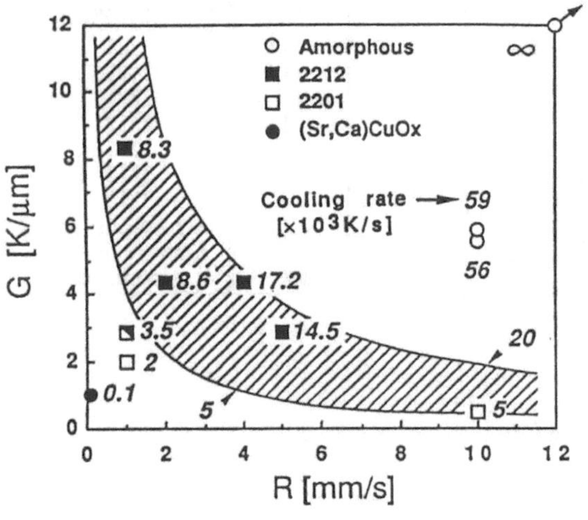

Fig. 4. Effects of G, R and cooling rates on the phases appeared in the samples. The shaded area indicates the range of cooling rates (5 to 20 $\times 10^3$K/s) for presence of the 2212 phase.

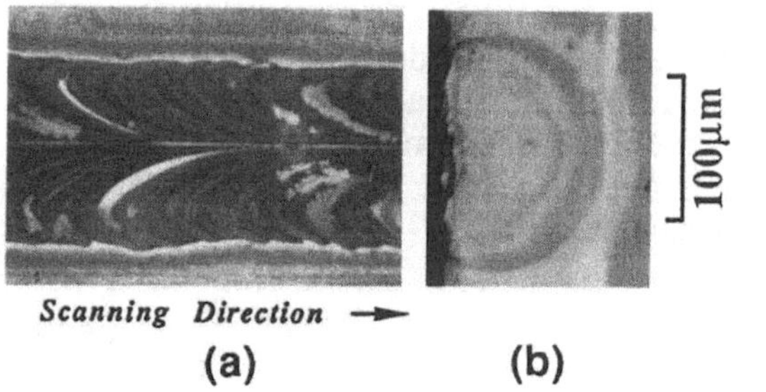

Fig. 5. Microstructure of laser processed sample (v=5mm/s, Q=0.6W, Tsub.=723K). (a) surface, (b) transverse section

CONCLUSION

We achieved the synthesis of $Bi_2Sr_2CaCu_2O_x$ superconductors by the laser zone melting and solidification process. Textured microstructures along laser scanning direction were observed in a treated surface. At the range of cooling rates between 5 and 20×10^3K/s, which was estimated by a heat flow calculation, the $Bi_2Sr_2CaCu_2O_x$ phases were produced.

REFERENCES

1. Bednorz JG, Muller KA (1986) Z. Phys.B64:189
2. Maeda H, Tanaka Y, Fukutomi M (1988) Jpn.J.Appl.Phys.Lett., 27:209
3. Jin S, Tiefel TH, Sherwood RC, Davis ME, van Dover RB, Kammlott GW, Fastnacht RA, Keiyh HD (1988) Appl. Phys. Lett. 52:2074
4. Feigelson RS, Gazit D, Fork DK (1988) J. Crstal Growth, 91:318
5. Kurosaka A, Aoyagi M, Tominaga H, Fukuda O, Osarai H (1989) Appl. Phys. Lett. 55(4), 24:390
6. Levinson M, Shah SSP,Wang DY (1989) Appl.Phys.Lett. 55(16): 1683
7. Anthony TR, Cline HE (1977) J. Appl. Phys., 48:3888
8. Cline HE, Anthony TR (1977) J. Appl. Phys., 48:3895
9. Sepold G (1968) Grundlagen fur das Schweißen mit strahlen hoher Intensitat, Dissertation TU Hannover
10. Bergmann HW, Hunger G, Fritsch H.U (1981) J. of Mat. Science 16:1935
11. Mordike BL, Bergmann HW (1984) Mat. Res. Soc. Symp. Proc., 28:45
12. Rosenthal D (1946) Trans. ASME, 68:849

2.5 Wires

Conductor Design with High-T_c Ceramics: A Review

E.W. COLLINGS

Battelle Memorial Institute, Advanced Materials Group, 505 King Ave., Columbus, OH 43201-2693, USA

ABSTRACT

The most recently available cryophysical property data for high-T_c superconductors are reviewed. Based on the results, the flux-jump- (or intrinsic) stability of high-T_c ceramics at 4, 20, and 80 K are intercompared. Next, a review is presented of the cryostability of rod-like composite superconductors. A simple geometrical model (based on the equal-area principle) is introduced to describe a complete range of cryostabilities. -- from so-called "full cryostability" to the one-dimensional MPZ. Again using recent cryophysical property data, some of the model results are applied to high-T_c ceramics at 4, 20 and 80 K.

KEYWORDS: High-T_c superconductors, thermal conductivity, specific heat, flux-jump stability, cryostability.

INTRODUCTION

The discovery by Chu et al. [1] of superconductivity at 90 K in compounds with the general formula $RBa_2Cu_3O_7$ (RBC, in which R indicates Y or a rare-earth element), followed by those of Maeda et al. [2] and subsequently Herman et al. [3] of superconductivity at higher temperatures in the Bi-Sr-Ca-Cu-O (BSCC) and Tl-Ba-Ca-Cu-O (TBCC) systems, respectively, promoted high-T_c superconductivity (HTSC) from laboratory-curiosity status at 40 K [4] into a technology area with far-ranging practical possibilities. Attention has generally been focussed on high-temperature applications of the new superconductors (SCs). Certainly the possibility of machine and device operation at temperatures of 77 K and above has important technical implications. But along with high T_c, ceramic SCs also posses remarkably high upper critical fields -- typically 200 T for YBC and 400 T for BSCC (for H||ab) at 4.2 K. It is this, coupled with a tendency towards weak flux pinning at elevated temperatures, that has led to the suggestion that ceramic SCs (notably B(Pb)SCC tapes) might find application as high field magnet inserts for 4.2 K [5]. Accordingly it is of interest to explore conductor design not only for 80 K operation (with liquid nitrogen, LN, as coolant) but also for operation at 20 K (liquid hydrogen, LH) and 4 K (in pool-boiling liquid helium, LHe).

A survey of the conductor-design literature has revealed that considerable attention has already been devoted to comparing the properties of HTSCs at about 80 K with those of NbTi at 4.2 K. But, as mentioned above, interest is now turning towards the possible use of HTSCs as *high-field* SCs in LHe (and possibly also in LH). One significant step in support of this approach has been taken by Wipf et al. [6] who have focussed their attention on minimum-propagating-zone (MPZ) cryostability over the temperature range of 4-80 K.

The goal of the present paper is to provide further insights into conductor design for 4-80 K. Using the results of an extensive review of recently acquired cryophysical property data as starting point, calculations are made of a number of representative flux-jump- and cryogenic stability criteria.

SURVEY OF THE HIGH-T_c CONDUCTOR DESIGN LITERATURE

Flux Jump Stability

By now numerous papers have addressed the subject of conductor design with HTSCs. Many of them have compared the properties of the HTSC at 77 or 80 K with those of NbTi at 4 K. But in a few cases the properties of HTSC conductors at lower temperatures (e.g. 20 K [7,8] or the entire 4-80 K range [6]) have been investigated. Usually taking YBC as a model system, several authors have calculated the quench- and flux-jump fields [6,9,10]. Numerous authors have deduced a maximum filament size for adiabatic flux-jump stability in LN [7,9,10,11,12,13] and at lower temperatures -- 20 K [7] and 4-80 K [6]. The intrinsic stability of a HTSC strip in the presence of both thermal conduction and cooling to a cryogen bath has also been recently considered [14]. The dynamic flux-jump stability of a Cu-stabilized HTSC filament at 80 K [7,9,10] and 20 K [7] has also been calculated, the latter being accompanied by a study of the dynamic stability of a composite-tape conductor [8]. Calculations have also been made of the maximum strand diameter for adiabatic self-field stability [8,9,10].

Cryostability

For the construction of very large magnets full cryostability, characterized by high Cu/SC ratios, is usually preferred. Two studies have compared the Cu/SC ratios of fully cryostabilized YBC composites at 77 or 80 K with those of NbTi composites at 4.2 K [9,10,13,15]. As the Cu/SC ratio is reduced, spontaneous recovery from a disturbance is possible only when the normal zone created is less than some critical size (the MPZ [16]). MPZs can be created in the presence of thermal conduction to a cold region (so-called "adiabatic MPZs") or in the presence of that plus cooling to a cryogen bath ("cooled MPZs"). Several authors have calculated the size of the adiabatic MPZ in HTSCs at 77 K [11,12,13,14] or as function of temperature between 4 and 80 K [6], and compared it with the size of a metallic SC adiabatic MPZ at 4.2 K [12,13,14]. The mechanisms of local thermal stability involving *transient* conduction and cooling have also been the subject of a recent study [14].

AC Loss

Issues surrounding AC loss in both low-T_c and high-T_c SCs have been reviewed [9,11]. Particular attention has been paid to the importance of high J_c [9] and small filament diameter [9,11] in the reduction of mixed-state hysteretic loss. Factors which influence eddy-current loss in composite strands have also been addressed [11,12]. Space restriction prevents AC loss from being considered further in this review.

CRYOPHYSICAL PROPERTY DATA

The results of a cryophysical property data survey are summarized in Table I.

Calculations of adiabatic stability properties -- initial flux-jump field, H_{FJ}, quench field, H_Q, adiabatic filament and adiabatic self-field-stable strand diameters, w_{ad} and D_{ad}, respectively -- require as input data the specific heat temperature dependence, C(T), and the normalized reciprocal J_c temperature dependence, $\Delta T_0 \equiv J_c/(-dJ_c/dT)$. We assume that $J_c(T)$ decreases linearly from 10^5 A cm^{-2} at 4 K to zero at a T_c of 90 K; it follows that $\Delta T_0 \equiv (90-T_b)$. The low temperature specific heat of YBC, based on [17,18,19,20], evidently smaller than that of Cu (Eqn.(I-1) [21]), is depicted in Fig. 1. For calculations, we use the C(T) function of Junod et al., Eqn.(I-2) [20]. Inset in Fig. 1 are the specific heats of the BSCC and TBCC materials [22,23] which are clearly greater than that of Cu. From a collection of intermediate-temperature specific heat data for YBC, after [17,24,25,26], an average curve, valid for the temperature range of about 30 to 100 K, has been computed, Eqn.(I-3). Specific heat data are generally reported on a per-mol basis, conversion of which to a per-unit-volume basis requires a knowledge of the density. Density data used in Fig. 1 and Eqns.(I-1) and (I-2) were: ρ(YBC) = 6.37 [27], ρ(BSCC) = 6.5 [28], and ρ(TBCC) = 6.5 g cm^{-3} [29].

Table I. Temperature Dependence of Physical Properties

Quantity	Unit	4 K	20 K	80 K	Ref.(a)
Specific Heat of Cu, C_{Cu}	mJ cm^{-3}	0.820	65.02	1.82×10^3	(b)
Specific Heat of YBC, C_{SC}	mJ cm^{-3}	0.592	45.25	9.74×10^2	
Model CCD (linear to 0 at 90 K), J_c	A cm^{-2}	1.00×10^5	8.14×10^4	1.16×10^4	
Thermal Cond. of Cu (RRR=100, H=0), K_{Cu}	W cm^{-1}K^{-1}	6.27	24.01	5.31	(c)
Thermal Cond. of YBC, K_{SC}	W cm^{-1}K^{-1}	3.92×10^{-3}	3.19×10^{-2}	4.23×10^{-2}	
Elect. Res. of Cu (RRR=100, 7 T), ρ_{Cu}	Ω cm	4.95×10^{-8}	5.03×10^{-8}	2.34×10^{-7}	(d)
Elect. Res. of YBC (normal extrap.), ρ_{SC}	Ω cm	3.27×10^{-4}	3.63×10^{-4}	4.98×10^{-4}	
Model strand filling factor, λ		0.4	0.4	0.4	
Min. pool boiling cooling, Q	W cm^{-2}				
large diam. surface		0.18 (LHe)	0.66 (LH)	0.77 (LN)	
0.4 mm diam. surface		0.18	2.11	1.99	

Temperature Dependences (Actual or Phenomenological)

C_{Cu}(3-30K), J cm^{-3}: $9.7443 \times 10^{-5}.T + 6.7127 \times 10^{-6}.T^3 + 1.4506 \times 10^{-10}.T^5 + 1.4507 \times 10^{-11}.T^7 - 2.3613 \times 10^{-14}.T^9 + 1.2345 \times 10^{-17}.T^{11} - 1.3043 \times 10^{-21}.T^{13}$ (I-1)

C_{SC}(1-17K), J cm^{-3}: $2.6868 \times 10^{-4}/T^2 + 1.0135 \times 10^{-4}.T + 2.5338 \times 10^{-6}.T^3 + 7.1712 \times 10^{-9}.T^5$ (I-2)

C_{SC}(>30-100K), J cm^{-3}: $2.0228 \times 10^{-4}.T^2 - 7.8163 \times 10^{-9}.T^4$ (I-3)

K_{Cu}(1-20K), W cm^{-1}K^{-1}: $1.5448.T + 0.01182.T^2 - 0.001451.T3$ (I-4)

K_{SC}(2-10K), W cm^{-1}K^{-1}: $5.49 \times 10^{-4}.T + 2.7 \times 10^{-5}.T^3$ (I-5)

ρ_{Cu}(1-20 K, 7T), Ω cm: $4.95 \times 10^{-8} + 2.8 \times 10^{-13}.T^2 - 1.2 \times 10^{-13}.T^3 + 1.0 \times 10^{-14}.T^4$ (I-6)

ρ_{SC}(0-250K), μΩ cm: $317.65 + 2.2597.T$ (I-7)

(a) see also text; (b) [21] 3-30 K, [39] 80 K; (c) [38] pp.5.1.1-1 and 2; (d) [38] pp.5.1.1-1 and 7 (0 T), and p.5.1.1-1.1 (10 T).

For the calculation of other flux-jump-related quantities such as the dynamic flux-jump-stable filament diameter, w_{dyn}, as well as various cryogenic-stability (e.g. MPZ) parameters, the thermal conductivity of the SC, K_{SC}, is needed. A function representing $K_{SC}(T)$ between 2 and 10 K was fitted to the data of [30,31,32,33,34,35], Eqn.(I-5). For K_{SC}(20 K) we have taken an average of the 20 K data of [27,30,31,32,33,34,35]; for K_{SC}(80 K) we have averaged the 80 K data of [27,30,31,32,33,34,35,36,37]. Also needed are the electrical and thermal conductivities of Cu which are obtainable from standard sources [38], and for flux-jump-stability calculations, a working value for the strand filling factor, λ = SC vol./total vol. Some estimates of the normal-state resistivity of YBC, Eqn.(I-7) from [19,37], are included in Table I.

For our pool-boiling cryostability calculations minimum cooling rates, Q, for surfaces exposed to liquid cryogens at about 4, 20 and 80 K are needed. Accordingly, from a set of predictive nucleate- and film-boiling correlations for LHe, LH and LN as supplied by Brentari and Smith [40] we have selected, as safe minimum Q-values, data from the feet of the film-boiling traces.

FLUX JUMP STABILITY

Intrinsic Stability

As the applied field acting on a SC is increased beyond some initial value $H_{FJ}(T)$, it is possible for a disturbance cycle to become initiated that leads to a macroscopic ingress of flux -- a flux jump. The first flux motion that occurs is supposed to be insufficient to drive the material normal [41]. According to Swartz and Bean [42]:

$$H_{FJ} = \sqrt{\pi^3 C(T)\Delta T_0} \tag{1}$$

which in the present model is equivalent to $\sqrt{\pi^3 C(T)(90-T_b)}$, where T_b is the bath- or operating temperature. Eventually a field, $H_Q(T)$, is reached whose energy density converted into heat is just sufficient to raise the SC's temperature to T_c. This quench field is given by

$$H_Q^2 = 8\pi\int_T^{T_c} C(T)dT \tag{2}$$

with H in gauss and C(T) in erg cm^{-3}. The temperature dependence of H_{FJ} and H_Q, computed with the aid of Eqns (I-2) and (I-3) are presented in Fig. 2 and Table II. Evidently for temperatures up to 50 K, HTSCs are immune to flux-jump quenching in fields of less than 9 tesla, On the other hand, in LHe, flux-jump instabilities can be expected in fields as low as 0.4 tesla.

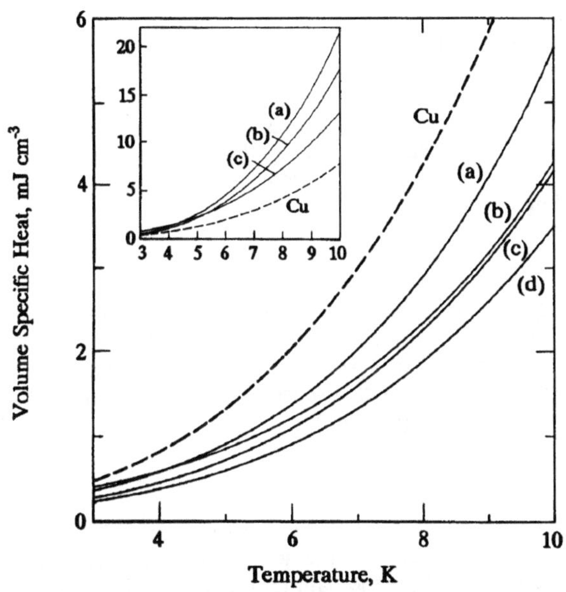

Fig. 1 Low temperature specific heat (LTSH) of YBC after [17] (a), [20] (b), [19] (c), and [18] (d), and of Cu after [21]. Inset are the LTSHs of BSCC after [23] (a) and [22] (b), TBCC after [22] (c) and Cu [21].

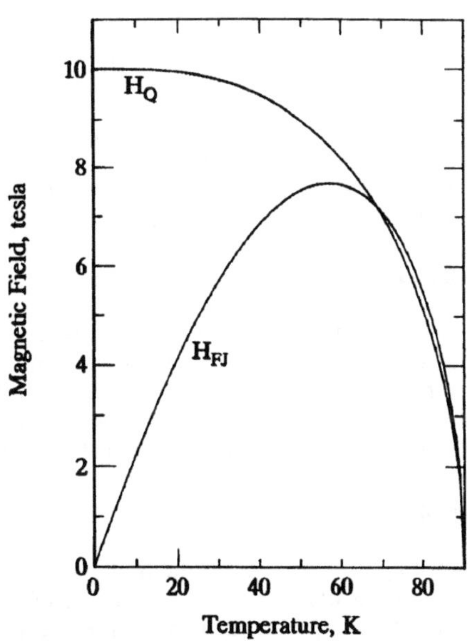

Fig. 2 Initial flux-jump field, $H_{FJ}(T)$, and quench field, $H_Q(T)$, calculated for YBC using Eqns.(1) and (2) and data listed in Table I.

Table II. Intrinsic-Stability- and Strand-Stability Parameters

Parameter	Unit	4 K	20 K	80 K
Initial flux-jump field, H_{FJ}	tesla	0.4	4.2	5.5
Quench field, H_Q	tesla	10.0	9.9	5.1
Adiabatically stable filament diameter, w_{ad}	mm	0.7	6.8	83.0
Dynamically stable filament diameter (zero field), w_{dyn}	mm	3.2	9.9	8.7
Dynamically stable filament diameter (7 tesla), w_{dyn}	mm	1.8	5.7	8.0
Adiabatically self-field stable strand diameter, D_{ad}	mm	1.9	19.0	255.6

Stabilization Against Flux Jumping

Filament Diameter: When flux jumping is to be expected, some form of stabilization is needed. The energy associated with flux ingress can be throttled down by reducing the size of the SC (filamentary subdivision). Two alternative criteria may be applied -- the adiabatic, and the dynamic. The former dominates if the heat capacity is high; otherwise heat conduction to a cryogen bath must be invoked. The maximum filament diameter under the adiabatic criterion (based on a slab calculation, [43] p.134) is

$$w_{ad} = \sqrt{10^9 (3/\pi) C(T) \Delta To} \,/ Jc(T) \qquad (3)$$

The dynamic criterion (based on a cylinder calculation, [43] p.156) is

$$w_{dyn} = 2 \sqrt{8 \left(K_{sc}/\rho_{Cu}\right) \Delta To \, (1-\lambda)/\lambda} \,/ Jc(T) \qquad (4)$$

These quantities, evaluated using the temperature dependence data given in Table I are listed in Table II. There it can be seen that whereas both w_{ad} and w_{dyn} are confined to relatively small values at 4 K, the larger of them is provided by the dynamic criterion. The high specific heat at 80 K inevitably invokes the adiabatic criterion, and leads at the same time to very large flux-jump-stable filament diameters.

Strand Diameter: The self field of the multifilamentary (MF) strand can also induce flux jumping much in the same way that an external field influences the individual filaments. Thus under adiabatic conditions the H_{FJ} of the fully current saturated MF strand is to a first approximation just the full penetration field (cf. Eqn.(3) with (4a) and (4b) of [9]). According to this prescription the maximum adiabatic self-field-stable strand diameter (SFSSD) is given by

$$D_{ad} = \sqrt{10^9 (3/\pi) C_{AV} \Delta To} \,/ (\lambda Jc(T)) \qquad (5)$$

Values of D_{ad} computed for the temperatures 4, 20, and 80 K, are given in Table II. As expected, D_{ad} scales with w_{ad}, and is enormously large at 80 K. At 4 K, where the heat capacities are low, some assistance from conduction and cooling, leading to what then becomes the *dynamic* SFSSD, D_{dyn}, is called for (see [43] p.151). In addition, both D_{ad} and D_{dyn} can be enhanced by operating the strand at a reduced average current density, $\lambda J = i\lambda J_c$. For example, a reduction in i from 0.93 to 0.74 results in a doubling of D.

CRYOSTABILITY

The goal of cryostability is to restore the SC state in an already flux-jump stabilized composite conductor once an electrical (e.g. current overload) or thermomechanical disturbance has taken place. Depending on design variables such as: current density in the SC ($J = iJ_c$), the Cu/SC ratio (R_s), and thermal parameters, conductors may be operated over a wide range of stability states -- from fully cryostable (in the Stekly sense [44]) to metastable (MPZ-stable), as discussed for example by Wipf [16] and reviewed in [45] (Vol.2, pp.306-8).

Full Cryostability

It is well known by now that under the Stekly condition for $J = J_c$-type cryostability, the "heat generation", $G = (\rho/PA)_{Cu} I_c^2$ (full transfer), is balanced by the "cooling", $Q = h(T_c - T_b)$. Here h is the heat-transfer coefficient, P_{Cu} and A_{Cu} are the cooled perimeter and area, respectively, of the Cu, and T_b is the bath temperature. Under full cryostability defined by

$$1 = \frac{G}{Q} = \frac{\rho_{Cu} I_c^2}{(PA)_{Cu} h \Delta T_o} \equiv \alpha \qquad (6)$$

where $\Delta T_0 \equiv T_c - T_b$, it can be shown ([45], Vol.2, p.274) that R_s is related to the (critical) current-carrying capacity according to

$$R_s^2 (R_s + 1) = \frac{\rho_{Cu}^2 J_c^3}{4\pi Q^2} \cdot I_c \qquad (7)$$

The only SC-specific parameters contained in Eqn.(7) are J_c and Q (and the latter only in so far as the choice of SC material permits operation in cryogens other than LHe). The influence of operating temperature on the R_s of a model HTSC-base conductor intended to carry 1000 A is given in Table III. Clearly at 4 K the HTSC may be treated like any other SC with attention being focussed simply on the optimization of J_c. At higher temperatures, R_s decreases, but due more to a reduction in J_c than an increase in Q.

Minimum Propagating Zone (MPZ) Cryostability

The "performance", or $<J_c>_{overall}$, of a conductor can be increased by lowering R_s or increasing I (if possible). A rearrangement of Eqn.(7) indicates that performance can be gauged by a parameter αi defined as

$$\alpha i = \underbrace{\left(\frac{1}{R_s}\right)\left(\frac{\rho}{P}\right)_{Cu}\frac{J_c}{h\Delta T}}_{\text{conductor design}} \cdot \underbrace{I}_{\text{operating current}} \qquad (8)$$

The influence of αi on conductor stability can be demonstrated by means of a series of stability diagrams [16,46]. These are maps of G(T) and Q(T) plotted concurrently against temperature (T), all in reduced coordinates defined by: $t = (T-T_b)/\Delta T_0$, $g(t) = G(T)/h\Delta T_0$, and $q(t) = Q(T)/h\Delta T_0$. Two representative maps are given in Fig. 3. Note that: (i) current sharing in the transition to the normal state commences at $t = 1-i$; (ii) the slope of g(t) is αi; (iii) g(t) intersects q(t) at $t = \alpha i^2$. Note also that an increase in αi, either (a) by increasing i at constant α, or (b) by increasing α at constant i, is accompanied by a decrease in stability as indicated by a decrease in the temperature range needed to bring about g>q.

A conductor can still recover from a thermal excursion sufficiently great to produce g>q provided that some of the heat generated (g) can be removed by thermal conduction. With reference to Figs. 4(a) and 4(b), when conduction is present, stability is always assured for thermal excursions to temperatures less than some t_{max} defined by the equal-area principle: A = A′, and B = B′, respectively [47]. The special Maddock criterion [47,48], Fig. 4(a), provides recovery from a normal zone of unrestricted but finite length. At larger values of αi, the size of a recoverable zone is restricted -- for spontaneous recovery, the zone size must be less than the MPZ.

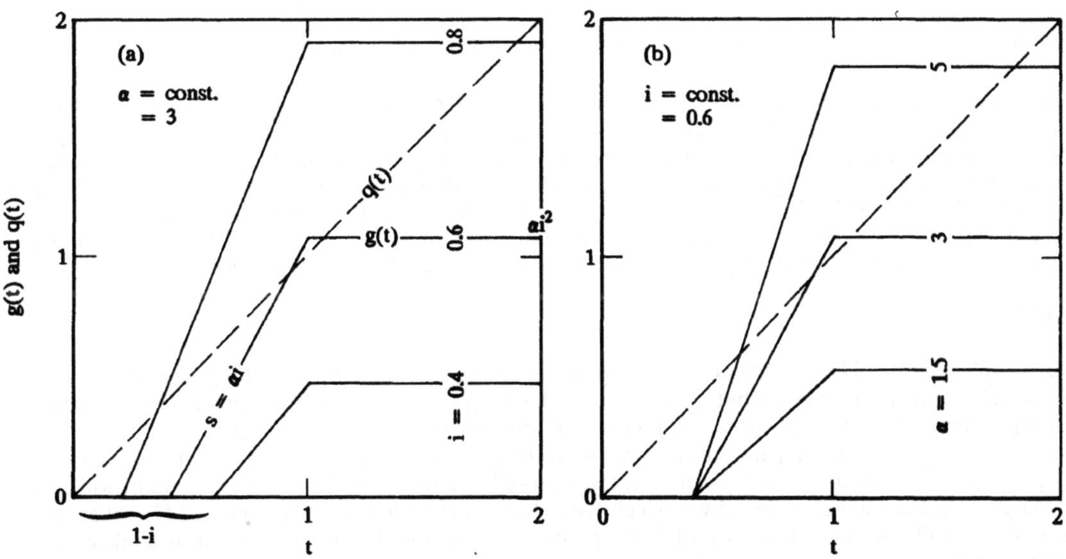

Fig. 3 Stability maps representing conductors with: (a) constant α and increasing i; (b) constant i and increasing α. Note the transition (bottom to top) from ultra cryostable to cooled-MPZ stable (see [45] Vol.2, p.307).

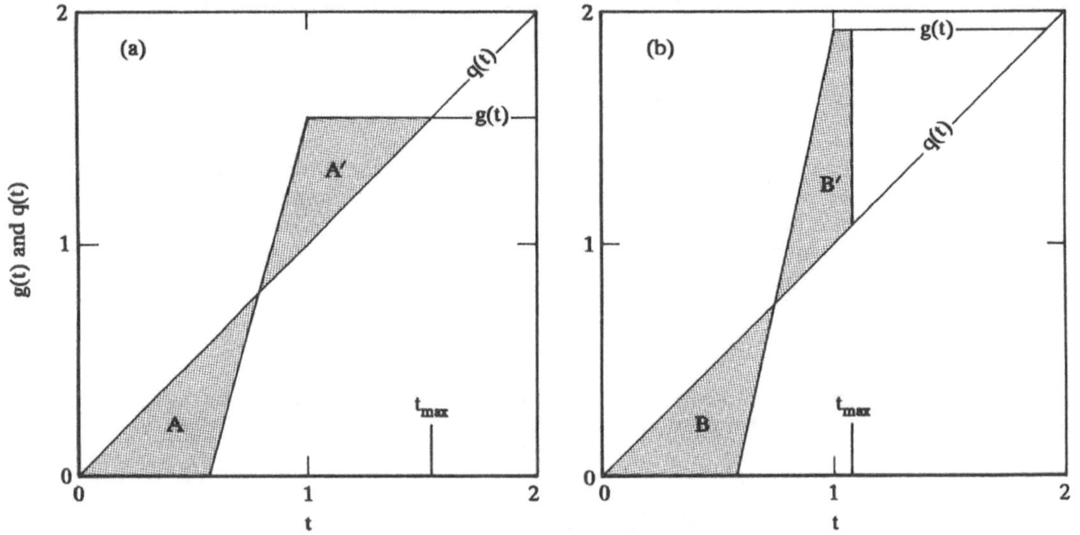

Fig. 4 **Stability maps depicting two examples of the equal-area description of cryostability in the presence of conduction and cooling -- (a) the Maddock-James [48] conservative "cold-end" limit; (b) the higher performance MPZ-cold-end case. See [47] for a unifying discussion.**

Table III. Cryostability Parameters

Full ($J = J_c$) Pool-Boiling Cryostability			4 K	20 K	80 K	
R_s (large-diam. surface)			17.9	5.9_8	1.9_7	
R_s (0.4 mm diam. surface)			17.9	2.6_0	0.9_4	
Adiabatic MPZ Cryostability	λ	R_s	4 K	20 K	80 K	Ref.
MPZ length, cm	0.4	1.5	2.2	4.8	2.8	(a)
	0.1	9	13.3	28.6	16.5	
MPZ velocity, m s^{-1}	0.4	1.5	117	2.65	3.92×10^{-2}	(b)
	0.1	9	107	2.40	3.34×10^{-2}	
MPZ energy (D=2 mm), J	0.4	1.5	2.18×10^{-3}	0.30	0.64	(c)
	0.1	9	1.43×10^{-2}	1.98	4.50	

(a) [43] p. 76; (b) [43] p. 206; (c) calculated using $E_{MPZ} = A_{tot} \cdot length_{MPZ} < C(T) \cdot_{AV} (90-T_b)/2$

Obviously it is not mandatory to operate a conductor under an MPZ criterion. But if MPZ operation is selected, the extent and energy-content of the zone are of interest. Furthermore, if it is aniticipated that zones larger than the MPZ will be created, their rates of expansion should be sufficiently rapid as to quench the entire machine before a destructive hot-spot is formed. A sequence of adiabatic MPZ properties are listed in Table III. Although the zone length, $\sqrt{2}<K>(90-T_b)/(<J_c>^2<\rho>)$, is relatively insensitive to T_b its energy, $MPZ_{vol}<C(T)>(90-T_b)/2$, responds to a thermally increasing C(T). Considerable attention must be paid to the normal-front velocity to ensure that zones larger than the MPZ grow sufficiently rapidly to avert hot-spot formation.

SUMMARY

Large-scale applications of HTSCs will stem not only from their high T_cs, which offer the possibility of operation in LH, LN, as well as refrigerated gases, but also from their high H_{c2}s, especially at low temperatures. Conductor design at high temperatures (compared to 4 K) must take into account the relatively high specific heat, C(T), that materials in general possess at temperatures beyond about 80 K. This is reflected in the adiabatic (C(T)-containing) versions of flux-jump stabilized filament and strand diameter, and MPZ energy -- all of which are relatively large -- and normal-front velocity -- which tends to be small. On the other hand in LHe, when the influences of high C(T) are lost, design with HTSCs proceeds along traditional lines. Thus a dynamic criterion will be favored over an adiabatic one when that option is possible (as in filament and strand design). In fact for all properties containing K(T), by adjusting R_s advantage can be taken of the high thermal conductivity of the stabilizer material. Full cryostabilization is SC-material-independent, excepting insofar as J_c and Q can be thought of as relying on the choice of SC.

ACKNOWLEDGEMENTS

This review was supported in part by Battelle and the US Department of Energy.

REFERENCES

1. M. K. Wu, J. R. Ashburn, C. J. Torng, et al., Phys. Rev. Lett. 58, 908-910 (1987).
2. H. Maeda, Y. Tanaka, M. Fukutomi et al., Japan J. Appl. Phys. 27, L209-210 (1988).
3. Z. Z. Sheng and A. M. Herman, Nature 332, 55-58 (1988).
4. J. G. Bednorz and K. A. Muller, Z. Phys. B 64, 189-193 (1986).
5. H. Sekine, K. Inoue, H. Maeda, et al., Adv. Cryo.. Eng. (Materials), to be published.
6. H. L. Laquer, F. E. Edeskuty, W. V. Hassenzahl, et al., IEEE Trans. Magn. 25, 1516-1519 (1989).
7. T. Ogasawara, Cryogenics 29, 3-5 (1989).
8. T. Ogasawara, Cryogenics 29, 6-9 (1989).
9. E. W. Collings, Cryogenics 28, 724-733 (1988).
10. E. W. Collings, MRS Int'l. Mtg. on Adv. Mats. Vol.6, Materials Research Society, 1989, pp.155-160.
11. D. E. Baynham, in Superconducting Ceramics, Brit. Ceram. Proc. No.40, March 1988, The Inst. of Ceramics, U.K. pp. 283-291.
12. Y. Iwasa, IEEE Trans. Magn. 24, 1211-1214 (1988).
13. D. Ito, Cryog. Eng. Jpn. 22, 383-385 (1987).
14. C. L. Tien, M. I. Flik, and P. E. Phelan, Cryogenics 29, 602-609 (1989).
15. E. W. Collings, Adv. Cryo. Eng. (Materials) 34, 639-646 (1988).
16. S. L. Wipf, "Stability and degradation of superconducting current-carrying devices", Tech. Rept. No. LA-7275, Los Alamos Scientific Laboratory, Dec. 1978.
17. S. von Molnar, A. Torresson, D. Kaiser, et al., Phys. Rev. B (Rapid Comm.) 37, 3762-3765 (1988).
18. T. Sasaki, N. Kobayashi, O. Nakatsu, et al., Physica C 153-155, 1012-1013 (1988).
19. T. Sasaki, O. Nakatsu, N. Kobayashi, et al., Physica C 156, 395-404 (1988).
20. A. Junod, A. Bezinge, D. Eckert, et al., Physica C 152, 495-504 (1988).
21. D. L. Martin, L. L. T. Bradley, W. J. Cazemier et al., Rev. Sci. Instrum. 44, 675-684 (1973).
22. R. A. Fisher, S. Kim, S. E. Lacy, et al., Phys. Rev. B 38, 11942-11945 (1988).
23. T. Sasaki, Y. Muto, T. Shishido, et al., Proc. Int. Conf. M^2S-HTC, Stanford, July 23-28, 1989, to be published.
24. K. Kitazawa, T. Atake, H. Ishii, et al., Japan J. Appl. Phys: Part 2 Letters 26, L748-L750 (1987).
25. M. Lang, T. Lechner, S. Riegel et al., Z. Phys. B - Cond. Matter 69, 459-463 (1988).
26. A. Junod, A. Bezinge, and J. Muller, Physica C 152, 50-64 (1988).
27. J. P. Heremans, D. T. Morelli, G. W. Smith, et al., Phys. Rev. B 37, 1604-1610 (1988).
28. H. Sato, W. Zhu, W. M. Miller, et al., in High Temperature Superconducting Compounds: Processing and Related Properties, ed. by S. H. Whang and A. Das Gupta, TMS, 1989, pp. 479-494.
29. D. S. Ginley, E. L. Venturini, J. F. Kwak, et al., Physica C 152, 217-222 (1988).
30. V. Bayot, F. Delannay, C. Dewitte, et al., Solid State Comm. 63, 983-986 (1987).
31. U. Gottwick, R. Held, G. Sparn, et al., Europhys. Lett. 4, 1183-1188 (1987).
32. C. Uher and A. B. Kaiser, Phys. Rev. B (Rapid Comm.) 36, 5680-5683 (1987).
33. A. Bernasconi, E. Felder, F. Hulliger, et al., Physica C 153-155, 1034-1035 (1988).
34. B. Salce, R. Calemczuk, C. Ayache, et al. Physica C 153-155, 1014-1015 (1988).
35. W. P. Kirk, P. S. Kobiela, R. N. Tsumura, et al., Ferroelectrics 92, 151-157 (1989).
36. G. Sparn, W. Schiebeling, M. Lang, et al., Physica C 153-155, 1010-1011 (1988).
37. K. Noto, W. Watanabe, H. Morita, et al., Cryogenics 29, 648-650 (1989).
38. Metals and Ceramics Information Center, Handbook on Materials for Superconducting Machinery, Nov. 1974 and Supplements, Battelle-Columbus Laboratories, Columbus, OH.
39. Y. S. Touloukian and E. H. Buyco, Specific Heat, Metallic Elements and Alloys: Thermophysical Properties of Matter, Vol. 4, Plenum Press, NY, 1970, p. 51.
40. E. G. Brentari and R. V. Smith, Int. Adv. Cryo. Eng. 10(M-U), 325-341 (1965).
41. R. Hancox and J. A. Catterall, in Cryogenic Fundamentals, Academic Press, 1971, pp.491-571.
42. P. A. Swartz and C. P. Bean, J. Appl. Phys. 39, 4991-4998 (1986).
43. M. N. Wilson, Superconducting Magnets, Clarendon Press, Oxford, 1983.
44. Z. J. J. Stekly, R. Thome, and B. Strauss, J. Appl. Phys. 40, 2238-2245 (1969).
45. E. W. Collings, Applied Superconductivity, Metallurgy, and Physics of Titanium Alloys, Plenum Press, 1986.
46. C. Meuris, in Stability of Superconductors in He I and He II, Int. Inst, Refrig., Paris, 1981, pp. 161- 168.
47. T. Ito and H. Kubota, Cryogenics 29, 621-624 (1989).
48. B. J. Maddock, G. B. James, and W. T. Norris, Cryogenics 9, 261-273 (1969).

Electromagnetic Properties and Structures of BiPbSrCaCuO Superconducting Wires

KEN-ICHI SATO, TAKESHI HIKATA, HIDEHITO MUKAI, TAKATO MASUDA, MUNETSUGU UEYAMA, HAJIME HITOTSUYANAGI, TSUTOMU MITSUI, and MAUMI KAWASHIMA

Osaka Research Laboratory, Sumitomo Electric Industries, Ltd., 1-3, Shimaya 1-chome, Konohana-ku, Osaka, 554 Japan

ABSTRACT

High-Jc silver-sheathed BiPbSrCaCuO superconducting wires were fabricated through the solid reaction method. These wires showed the maximum Jc of 2.5×10^4 A/cm^2 (77.3K) in a zero magnetic field and 1.2×10^4 A/cm^2 at 0.1 Tesla. Key factors for high-Jc wires are: dominant high-Tc phase, highly oriented structures, good bonding at grain boundaries and fine dispersion of non-superconducting phases. When the wire was immersed in liquid helium, maximum Jc of 3.8×10^4 A/cm^2 at 23 Tesla was achieved. 50cm length conductors were fabricated. These conductors were tested and proved to have Ic of practical levels; 170 A at 77.3K and 1000 A at 4.2K. These results show the promising potential of silver-sheathed BiPbSrCaCuO superconducting wires.

KEY WORDS: BiPbSrCaCuO, superconducting wires, alignment, grain boundaries, critical current density

INTRODUCTION

Since the discovery of high-temperature superconductors in 1986[1,2], many experiments have been done to develop superconducting wires using these materials. Among many methods, the metal-sheathed process has advantages such as: (1) conventional techniques for processing metallic superconducting wires are used, so it is possible to obtain long unit lengths and small wire diameters, and the method is suitable for mass production, and (2) the sheath can be used as stabilizer. The critical current density with the metal-sheathed process has been lower than with the film process and melt process. We focused on improving the critical current density through the silver-sheathed process [3-8] using BiPbSrCaCuO superconducting material.[9,10]

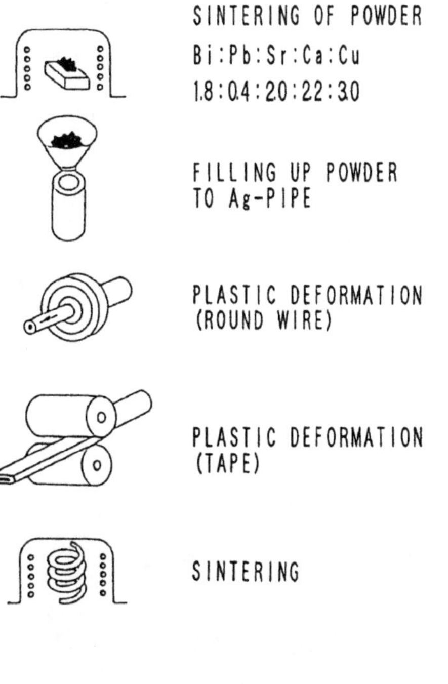

Fig.1. A basic process for fabricating silver-sheathed BiPbSrCaCuO superconducting wires

EXPERIMENTAL

The basic process for fabricating silver-sheathed BiPbSrCaCuO superconducting wires is shown in Fig. 1. A cation ratio of Bi:Pb:Sr:Ca:Cu=1.8:0.4:2:2.2:3 is preferred to 2223 composition for making higher Tc and

Jc. Plastic deformation and sintering are duplicated in the form of flat wires. The first sintering means the formation of the high-Tc phase and the second one means the bonding of the grain boundaries.
A four-probe DC method was used to measure Tc and Jc. Jc criterion was chosen to express the physical meaning properly, using 0.1 μV/cm for Jc-B

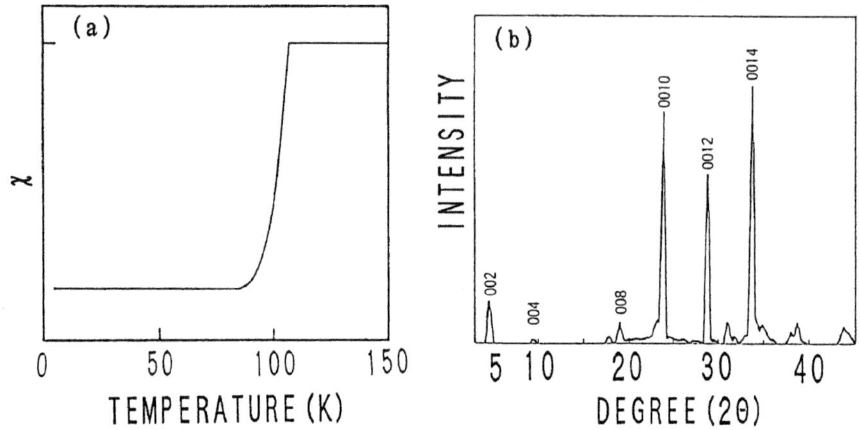

Fig.2. High-Tc phase domination of BiPbSrCaCuO inside silver-sheath
(a) SQUID
(b) XRD

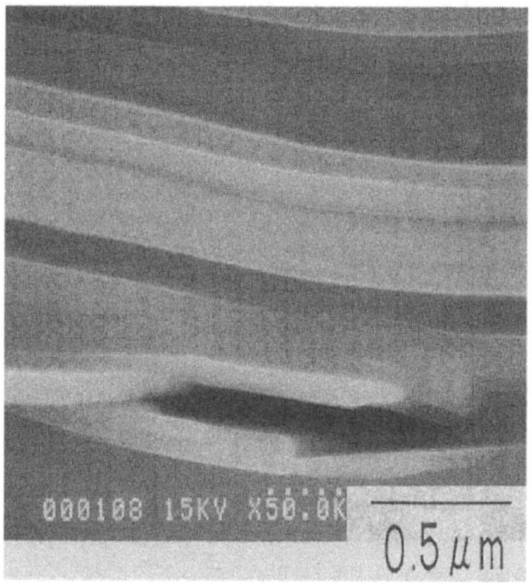

Fig.3. SEM photographs of fracture surface of BiPbSrCaCuO

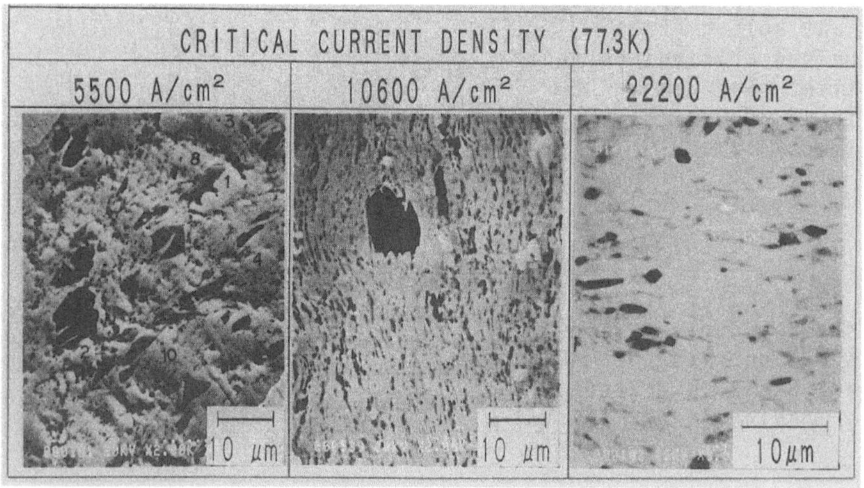

Fig.4. Dispersion of nonsuperconducting phases

characteristics at 77.3K. The magnetic field was applied using The Hybrid Magnet at Tohoku Univ. for measuring Jc-B characteristics up to 23 Tesla at 4.2K.

RESULTS AND DISCUSSION

Through improvement of composition and processing, many factors are revealed as key points for obtaining higher Jc values. These factors are summarized as shown in Figs.2,3 and 4: dominant high-Tc phase, good alignment and good bonding at grain boundaries, followed by fine dispersion of nonsuperconducting phase.

Figure 5 shows Jc-B characteristics of various samples at 77.3K. Improvement of Jc in a zero magnetic field from 1000 A/cm² to 25,000 A/cm² leads to an improvement of the Jc in a magnetic field. The Jc -B properties of the normalized Jc are useful for self-checking the miscalculation of the cross-sectional area of the oxides. The Jc-B results are clearly shown in Fig.6. The maximum Jc at 0.1 Tesla is 12,000 A/cm². When the magnetic field is parallel to the current, Jc-B characteristics are almost the same as when the magnetic field is perpendicular to the current and parallel to the a-b plane as shown in Fig.7. One model proposed to explain this phenomena is shown in Fig.8. This model is supported by the SEM observation. The transport current is forced to flow from thin crystalline platelet to platelet. When the current flows to another platelet, the

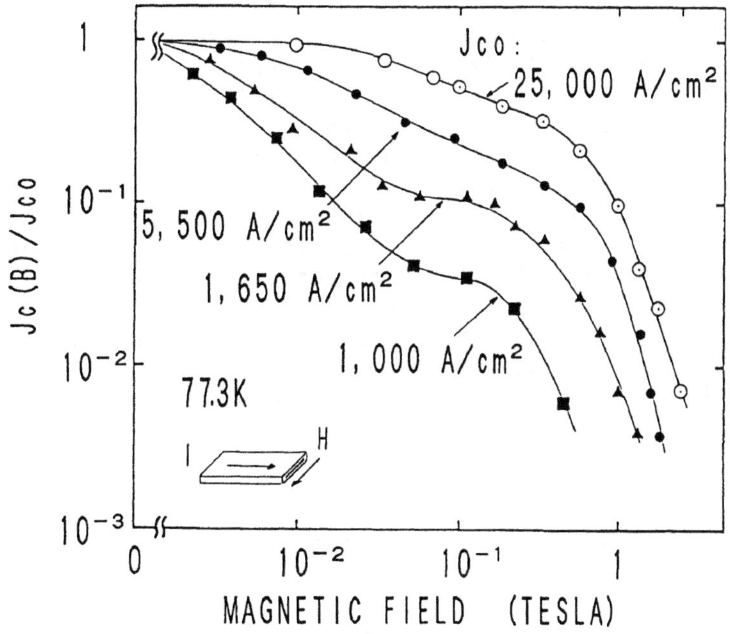

Fig.5. Jc-B characteristics of various samples at 77.3K

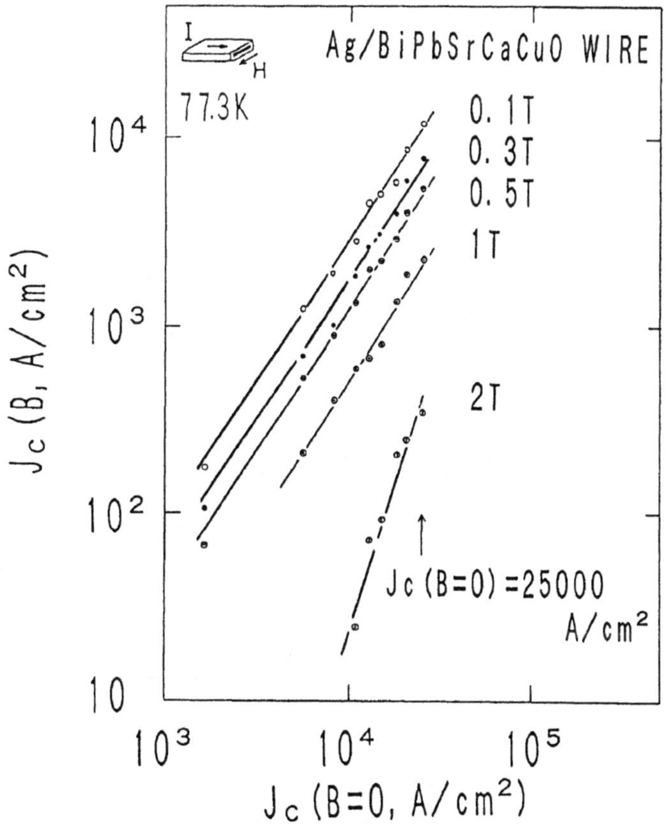

Fig.6. Jc-B improvements by increasing Jc in a zero magnetic field at 77.3K

direction of the current is along the c-axis and the Lorentz force works parallel to the a-b plane in the case of Figs.8(a) and 8(b).
In the case of Fig. 8(c), the magnetic field is perpendicular to the a-b plane and the Lorentz force is parallel to the a-b plane. This model suggests that Jc-B properties are dominated by the Lorentz force parallel to the a-b plane in all directions of applied magnetic field. In the case of Figs. 8(a) and 8(b), the properties of the grain boundaries affected Jc-B values. In the case of Fig. 8(c), the pinning properties of the bulk material affected Jc-B values. We can expect the improvement of grain boundary characteristics to improve Jc-B values when the magnetic field is parallel to the a-b plane. However, when the magnetic field is perpendicular to the a-b plane, Jc-B improvement requires stronger pinning force of bulky state. Figure 9 shows Jc-B characteristics of these superconducting wires at 4.2K. A maximum Jc of 3.8 $\times 10^4$ A/cm ^{2} is obtained at at 23 Tesla using the Hybrid Magnet at Tohoku Univ.

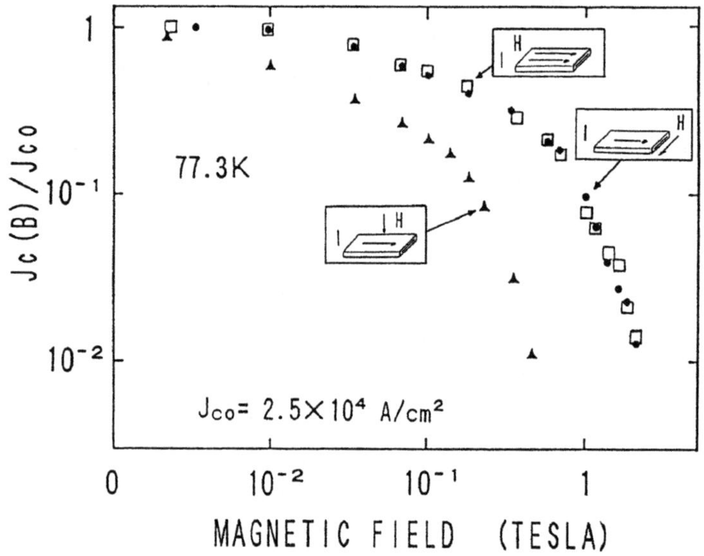

Fig.7. Jc-B characteristics when the magnetic field is applied in three directions for sample of Jc=25,000 A/cm^2

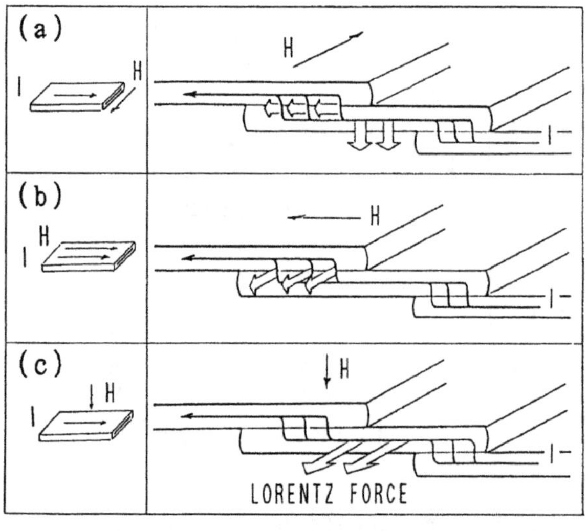

Fig.8. The model of Jc-B characteristics
(a) H⊥I, H//a-b plane
(b) H//I, H//a-b plane
(c) H⊥I, H⊥a-b plane

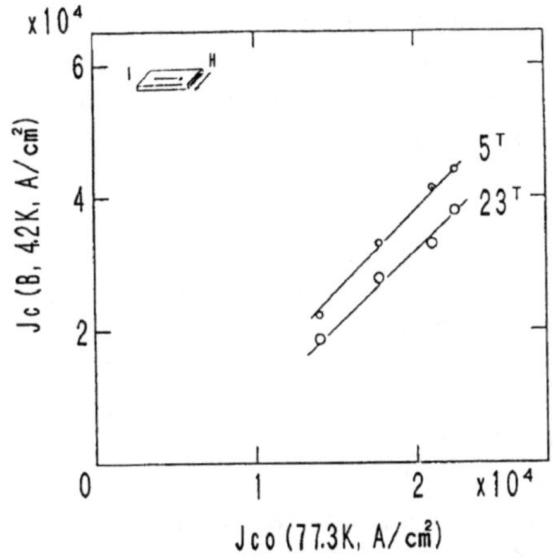

Fig.9. Jc-B characteristics of various samples at 4.2K

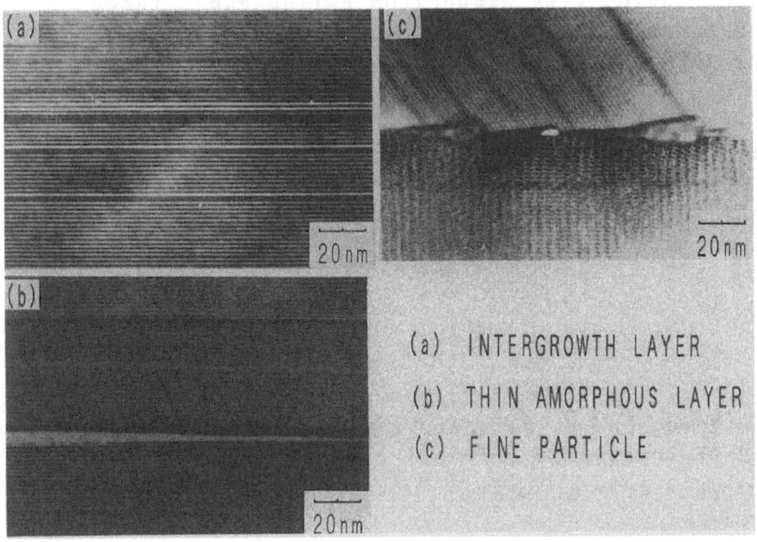

Fig.10. TEM observations of BiPbSrCaCuO inside silver-sheath

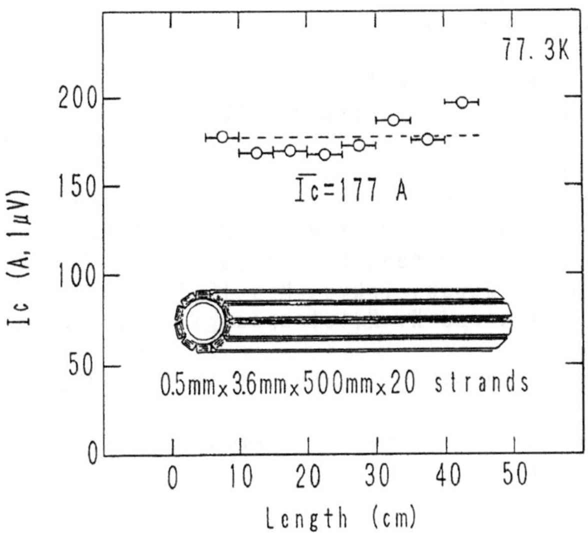

Fig.11. Ic distribution of 50cm length silver-sheathed BiPbSrCaCuO superconductor

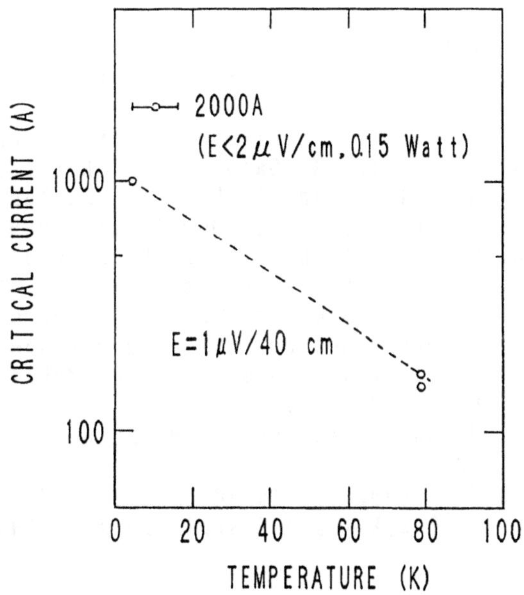

Fig.12. Ic vs temperature of 50cm length silver-sheathed BiPbSrCaCuO superconductors

TEM observations of BiPbSrCaCuO inside silver-sheath are performed after thinning samples through the ion milling method. Intergrowth layers , fine particles and thin amorphous layers are observed. These structures are shown in Figs. 10(a,b,c). They can be expected to act as pinning sites when the Lorentz force is perpendicular to the a-b plane.

Long conductors were made and tested. One was composed of 20 taped wires with thickness of 0.5 mm and length of 500 mm. These wires were bonded with silver pipe during sintering. This conductor was tested in liquid nitrogen and critical current was measured at every 5 cm along the longitudinal direction. The result was shown in Fig. 11. Critical currents were uniform and 150 ampere of Ic was determined with the criterion of 1μV, which means $10^{-12}\Omega\cdot m$.
The other conductor was composed of 30 taped wires with thickness of 0.37 mm and length of 500 mm. These wires were also arranged in decaganal configuration using FRP pipe. As shown in Fig. 12, this conductor was tested and proved to carry critical current of 170 A at 77.3K and 1000 A at 4.2K. When used in temperatures from 4.2K to 77.3K (cooled with evaporated helium gas), these conductors are able to perform as power leads of the superconducting magnets operated at 4.2K and can carry 2000 A generating less than only 2μV/cm.

In summary, we developed high Jc silver-sheathed superconducting wires and demonstrated the possibility of using them at 77.3K in a low magnetic field and at 4.2K in a super-high magnetic field. We also demonstrated long superconductors carrying critical current at practical levels in the temperature range from 77.3K to 4.2K.

ACKNOWLEDGMENT

We greatly appreciate Prof. Y. Muto, Dr. K. Watanabe and their co-workers of HFLSM at Tohoku University for their help in critical current measurements up to 23 Tesla.

REFERENCES

1. J.G.Bendnorz and K.A.Muller: Z. Phys., B64(1986)189.
2. S. Uchida, H. Takagi, and K. Kitazawa, and S. Tanaka: Jpn.J.Appl.Phys.26 (1987) L1.
3. T.Nakahara: Sumitomo Electric Technical Review 134(1989)1.
4. T. Hikata, K. Sato, and H. Hitotsuyanagi: Jpn.Appl.Phys.28(1989)L.82.
5. H. Hitotsuyanagi, K. Sato, S. Takano, and M. Nagata: Proc. of 39th Electronic Components Conf., May 22-24, 1989, Houston, p.2.
6. K. Sato, T. Hikata, H. Mukai and H. Hitotsuyanagi: Proc. of ISTEC Workshop on Superconductivity, February 1-3, 1989, Oiso, (ISTEC, Tokyo, 1989)p.119.
7. K. Sato: Symp. on "Flux Pinning and Critical Current Densities of High-Tc Superconductors." Cryogenic Association of Japan, Tokyo, May 15-16, 1989.
8. T. Hikata, T. Nishikawa, H. Mukai, K. Sato, and H. Hitotsuyanagi: Jpn.J.Appl. Phys.,28(1989)L1204.
9. H. Maeda, Y. Tanaka, M. Fukutomi, and T. Asano: Jpn.J.Appl. Phys.27(1988) L209.
10. M. Takano, J. Takada, K. Oda, H. Kitaguchi, Y. Miura, Y. Ikeda, Y. Tomii, and H. Hazaki: Jpn.J.Appl. Phys.27(1988)L1041.

The Transport Critical Current Property of High T_c Superconducting Wires

NAOKI UNO, NORITSUGU ENOMOTO, HIROYUKI KIKUCHI, KANAME MATSUMOTO, MASANAO MIMURA, and MINORU NAKAJIMA

Yokohama R & D Laboratories, The Furukawa Electric Co., Ltd., 4-3, Okano 2-chome, Nishi-ku, Yokohama, 220 Japan

ABSTRACT

The Ag-sheathed high Tc superconducting wires were prepared by the powder-in-tube method. The transport critical current density(Jc) was measured by the four-probe method with 1μ v/cm criterion. And the magnetic field dependence of transport Jc at 77K and 4.2K were evaluated. In YBCO system, we obtained the Jc of 10^3 A/cm^2 order, but we could not obtain the higher Jc because of the weak link problem. In case of BSCCO system, the Ag-sheathed tapes showed strong grain alignment which could be attained by adjusting the process parameters. So we obtained the maximum Jc of 3.5×10^4 A/cm^2 at 77K and 0T. But this Jc decreased drastically under high magnetic field over 0.1T, because of the weak pinning force and resulting flux creep. At 4.2K, the Jc of BSCCO Ag-sheathed tape under 10T showed 1.7×10^5 A/cm^2.

KEY WORDS : transport critical current, F value, magnetic field dependence, grain alignment, high-Tc superconducting wire

INTRODUCTION

Since the discovery by J.G.Bednorz & K.A.Muller, a variety of high Tc materials have been developed, which has strongly stimulated R&D on the superconductivity application ranging from cryoelectronics to power engineering. The conductor wire is believed as one of the primary materials for the utilization of high Tc superconductivity (HTSC), while over 95% of the current superconductivity market uses materials as wire & cables. Therefor,the wire program is one of the key issues for the commercialization of HTSC. The process development for conductor wire can be devided into the three sequential stages : enhancement of material property, development of composite structure and realization of length productivity, the first of which is primarily under way. As to this stage, supercurrent hurdles of percolation, weak link and pinning could be overcome through the enhancement & qualification of material processing from raw material & precursor to densification & homogenization to texturing and microstructural optimization.

In this report, we would like to show the recent progresses on HTSC wires.

EXPERIMENTAL

The raw materials were prepared by the conventional solid state reaction, in which oxides and carbonates were mixed and calcined at appropriate temperature and then ground to powder. Ag-sheathed wire was prepared by the conventional powder-in-tube method.[1]

The measurement of critical current density(Jc) was carried out by the conventional four-probe method with 1μ v/cm criterion. The magnetic field dependence of transport Jc were measured at 77K and 4.2K. At 4.2K and the magnetic field up to 23T, the Jc-B property was measured by use of a hybrid magnet at Tohoku University.

The samples were subjected to analyses of microscope observation, SEM, TEM and XRD. The c-axis orientation of HTSC was evaluated by using F-value obtained by Xray diffraction as follows;

$$F = (Po - Poo)/(1 - Poo)$$
$$P = \Sigma I(001)/\Sigma I(hkl)$$

where, Poo : X-ray intensity ratio for random sample
Po : X-ray intensity ratio for oriented sample

The super-normal transition of Ag-sheathed wire was evaluated by n-value which was defined by the equation of $E = kI^n$ ($E = 0.2 \sim 1.0\ \mu V/cm$). The activation energy for the flux creep Uo of Ag-sheathed wire was estimated from the magnetic relaxation, dM/d(lnt), using the following equation.[2]

$$dM/d(\ln t) = \Delta MkT/2Uo$$

RESULT AND DISCUSSION

The magnetic field dependence of transport Jc for YBCO Ag-sheathed tape with 0.2mm thickness is shown in Fig.1, where Jc vs magnetic field curve shows the hysteresis. It is explained that this hysteresis is caused of the flux trapping for weak link network in YBCO polycrystals.

Figure 2 shows the influence of the degree of c-axis orientation (F value) to the critical current density for Ag-sheathed BSCCO tapes at 77K and 0T.

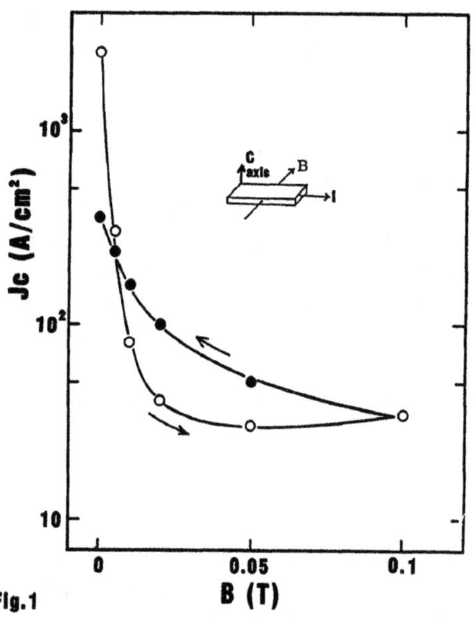

Fig.1 The magnetic field dependence of critical current density for Ag-sheathed YBCO tapes at 77K

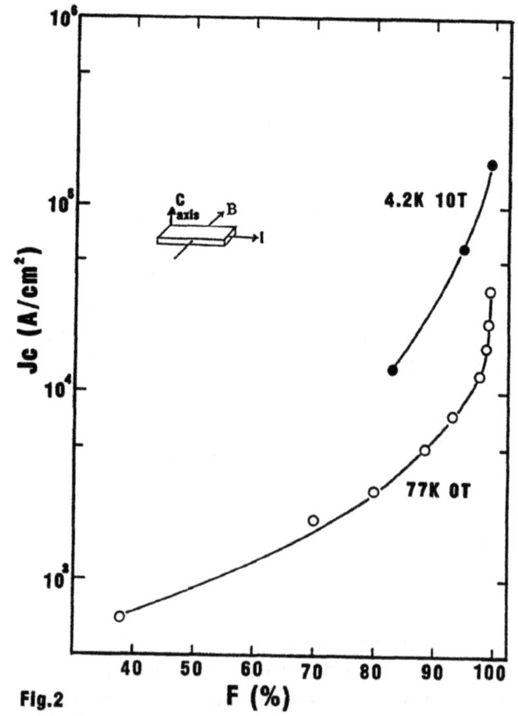

Fig.2 The influence of a degree of c-axis orientation (F) to the critical current density for Ag-sheathed BSCCO tapes at 77K and 4.2K

4.2K and 10T. As increasing the F value, the Jc increased steeply. Over 95% F value, the Jc shows $10^4 A/cm^2$ or higher at 77K 0T. The maximum Jc of Ag-sheathed BSCCO tape with 99% F value was $3.5x10^4 A/cm^2$ at 77K 0T.[3)]

The structural analyses were carried out in order to clear microstructural features of BSCCO wire with 99% F values. Figure 3 shows results on TEM analysis of a cross section of Ag-sheathed BSCCO tape with 99% F. In this figure, (a) is the bright field image with a selector aperture and (b) is the selected area diffraction pattern. The arrow indicates a twist boundary. The incident beam is parallel to [110] of the central grain. The presence of twist boundary shows that the c-axises of upper and lower grain are parallel. Figure 4 shows a high resolution image. In this figure, 't.b.' refers as a twist boundary. In the twist boundary, any impurity is not recognized. The intergrowth is observed in the upper grain. The incident beam is parallel to [110] of the upper grain.

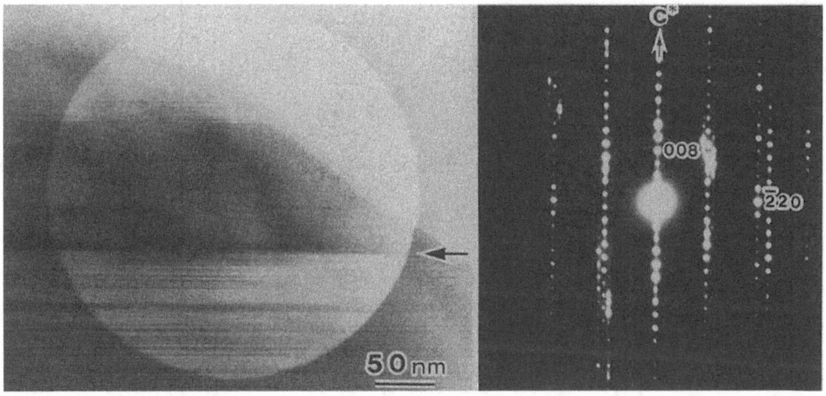

Fig.3 A cross section of a silver sheathed BSCCO tape. (a) B.F.I. with a selector aperture, (b) S.A.D. The arrow indicates a twist boundary. The incident beam is parallel to [110] of the central grain.

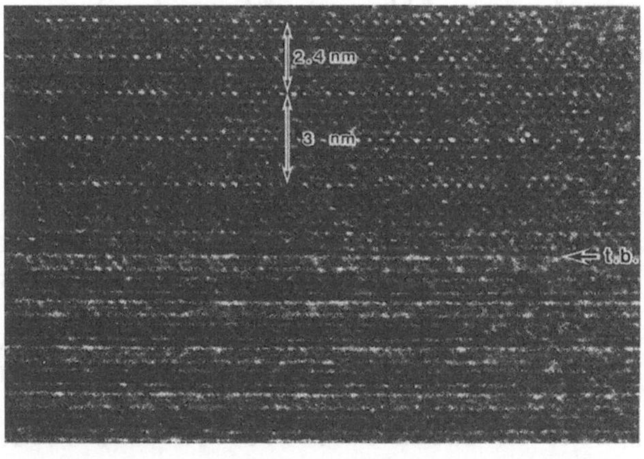

Fig.4 A high resolution image of the cross section of a silver sheathed BSCCO tape. 't.b.' refers as a twist boundary. Intergrowth is observed in the upper grain. The incident beam is parallel to [110] of the upper grain.

It is supposed that the grain alignment of BSCCO tape could be improved by decreasing of twist boundary.

Figure 5 shows the magnetic field dependence for these tapes with various F values at 77K in which the magnetic field was parallel to the tape wide plane. In comparison with YBCO wire Jc-B property was improved remarkably at low magnetic field, because of decreasing of the flux trapping for weak link network in BSCCO wire, which is supposed to be derived out of improved grain alignment with high oriented lamellar structure. However, in Fig.5 this high Jc degradates over 0.1 T field, which is supposedly due to the weakness of intragrain pinning rather than intergrain coupling. The activation energy for the flux creep Uo of YBCO and BSCCO Ag-sheathed tapes are 0.26ev[4)] and 0.087ev respectively, which was estimated at 100 Oe magnetic field This result shows the easy occurrence of flux creep in BSCCO superconductor.

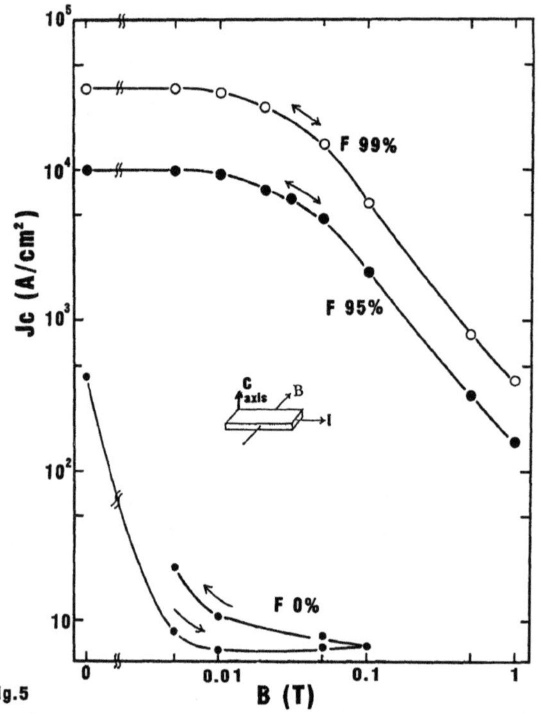

Fig.5 The magnetic field dependence of critical current density at 77K for Ag-sheathed BSCCO tapes with various F values (F : a degree of c-axis orientation)

Figure 6 and 7 show the n-value vs critical current density of YBCO tape wires and BSCCO tapes, respectively. Although the Jc of BSCCO tapes is higher than those of YBCO tape wires, n-values of BSCCO tapes were lower than those of YBCO tape wires. This is relevant to that Jc of BSCCO wire is controlled by flux creep, which was shown at Fig. 5 as above-mentioned.

The BSCCO tapes with different F values were tested at helium temperature of 4.2K. The transport performance under magnetic field at 10T or 23T gives a fairly big encouragement as shown in Fig.8. In this figure, the maximum Jc of $3.2\times10^5 A/cm^2$ at 1T and $1.7\times10^5 A/cm^2$ at 10T were obtained by the sample with 99% F value. When F value decreased down to 95% and 83%, the Jc remained about 2/5 and 1/10 respectively. These Jc-B properties are presented the experimental evidence that HTSC conductor can surpass the conventional superconductors such as $Nb_3(Ge, Al)$ and $PbMo_6S_8$ in terms of Jc and Fp(=JcxB) in magnetic field as high as 20T or more.

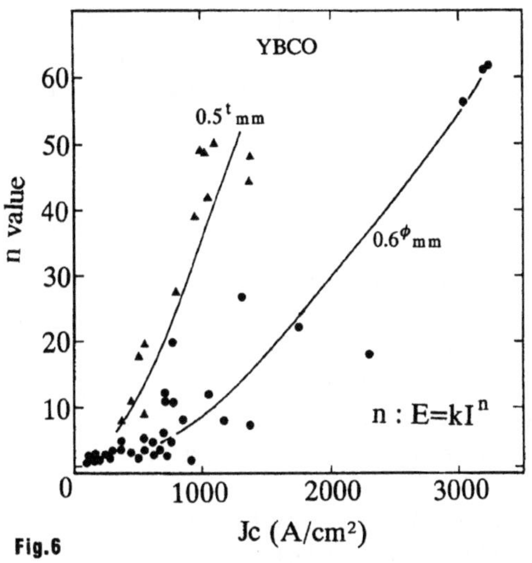

Fig.6
The n-value vs critical current density of Ag-sheathed YBCO tapes and wires

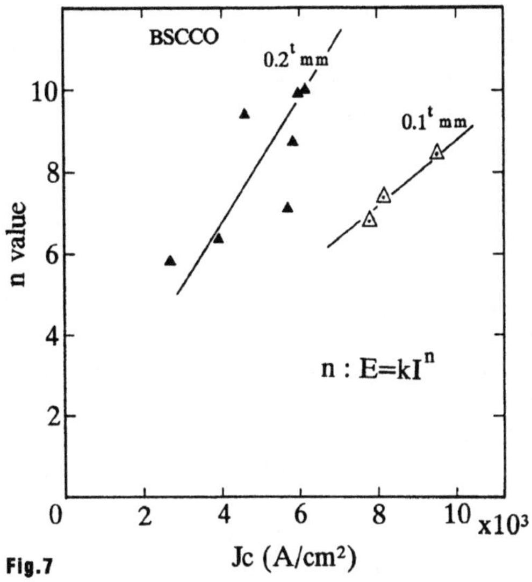

Fig.7
The n-value vs critical current density of Ag-sheathed BSCCO tapes

This is attributable to both the highness of Hc_2 in nature and the coupling strengthened at 4.2K among highly aligned grains.
This result presents a support to a promising applicability of HTSC conductors to high magnetic field technology such as inner coil for conventional superconducting magnet[5].

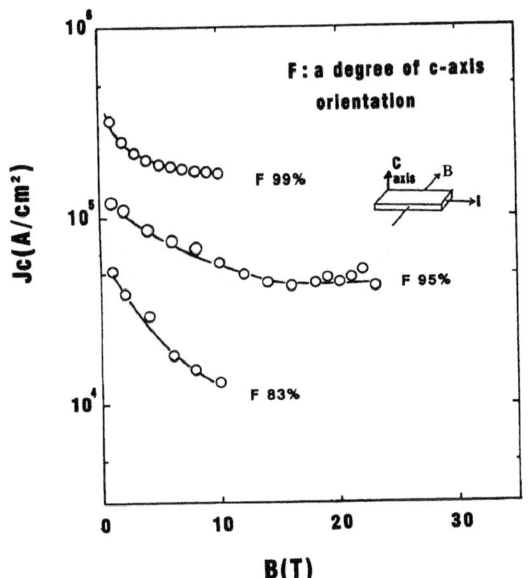

Fig.8
The magnetic field dependence of critical current density for Ag-sheathed BSCCO tapes at 4.2K

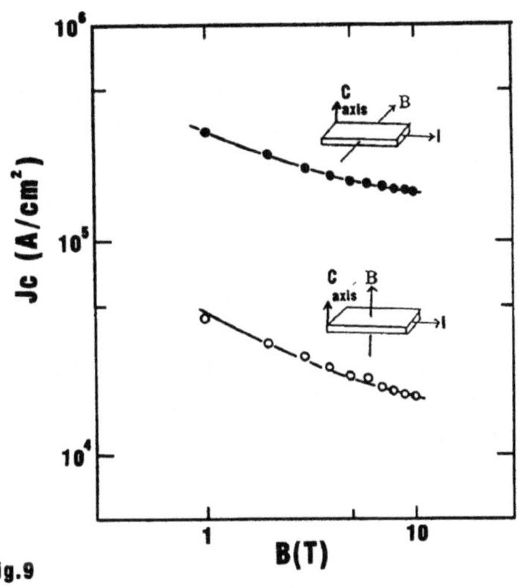

Fig.9
The magnetic field dependence of critical current density, which are two direction, parallel and perpendicular to the wide plane of Ag-sheathed BSCCO tapes at 4.2K

The magnetic field dependence of critical current density at 4.2K is shown Fig.9. In this figure, the Jc-B properties were measured for two direction, parallel and perpendicular to the wide plane of Ag-sheathed BSCCO tapes.

When the magnetic field direction was parallel to the wide plane of tape, the values of Jc were higher about 10 times than the Jc when the magnetic field direction was perpendicular to the wide plane of tape. This result was caused that the BSCCO superconductive oxide had a strong anisotropic property with regard to the magnetic field.

CONCLUSION

The transport critical current density of Ag-sheathed BSCCO tape was strongly and positively related to the degree of c-axis crystal orientation (F) obtained by XRD and that pursuing higher F value up to around 99%, Jc of 3.5×10^4 A/cm^2 and 1.7×10^5 A/cm^2 were achieved at 77K and 0T, and 4.2K and 10T, respectively. This result shows that the grain alignment can supress weak intergranular coupling for the appropriate rolling and heat treatment.

However, the magnetic field dependence of Jc at 77K degradates over 0.1T, which is supposedly due to the weakness of intragrain pinning force.

So we will have to improve the intragrain pinning of BSCCO HTSC.

ACKNOWLEDGEMENT

The authors gratefully acknowledge that this work was supported by Tokyo Electric Power Company, Tohoku Electric Power Company and Hokkaido Electric Power Company.

REFERENCES

1) Y. Tanaka : Proceeding of international Simposium on Superconductivity Aug. 28-31, 1988 Nagoya, Japan.
2) H. Kumakura, K. Togano, E. Yanagisawa, K. Takahashi, M. Nakao and H. Maeda : Jpn. J. Appl. Phys. 28 (1989) L24.
3) N. Enomoto, H. Kikuchi, N. Uno, M. Mimura, M. Nakajima, T. Hara K. Okaniwa and T. Yamamoto : Proceeding of International Simposium on Superconductivity Nov. 14-17, 1989 Tukuba, Ibaragi Japan.
4) T. Hara, H. Hoshino, T. Yamamoto, H. Ishii and M. Nakamura : Proceeding of International Simposium on Superconductivity Nov. 14-17, 1989 Tukuba, Ibaragi Japan.
5) N. Enomoto, H. Kikuchi, N. Uno, M. Ikeda, H. Kumakura, K. Togano and K. Watanabe : Proceeding of 11th International Conference on Magnet Technology 28 Aug.- 1 Sep. 1989 Tukuba, Japan.

Studies on Microstructure-Property Relationship of $YBa_2Cu_3O_{7-x}$ + Ag Composite Wire

SUKHENDU SAMAJDAR and SHYAM K. SAMANTA

Plasticity Laboratory, The University of Michigan, Ann Arbor, MI 48109-2125, USA

ABSTRACT

Microcomposites of YBCO and silver formed through powder processing route have already shown attractive properties, particularly in regard to formability into usable shapes, flexibility (toughness) and zero resistivity (at 77 K) [1]. The material has been characterized, in this effort, by evaluating the microstructure at the various stages of the fabrication route. Effects of different heat treatments on the microstructure and mechanical and electrical properties of the sintered billet and of the extruded as well as rolled wires have been reported in this paper.

KEY WORDS: Formability, flexibility, thermomechanical processing, YBCO-Ag_2O microcomposite, metallization.

The high temperature superconducting ceramic $YBa_2Cu_3O_{7-x}$ is by now a well studied material, albeit metal-ceramic composites using yittrium-barium-copper oxide (YBCO) have not been completely or well characterized. It has already been reported [1] that proper processing of this material can lead to composites of considerable formability. In this brief presentation, the emphasis is on our findings regarding processing-microstructure-property correlations in Ag_2O-YBCO composites prepared through powder metallurgical route followed by thermomechanical processing.

The composition of the composites examined was mainly 50 vol.% Ag_2O + 50 vol.% YBCO. A few composites were also prepared with 70 vol.% and 40 vol.% Ag_2O. The powders are thoroughly mixed and then warm compacted into 3/8" billets which are then sintered [1] at 880°C. The sintered billets can be extruded to 1/8" dia. wire form.

It was realized that uniformity and fineness of particle distribution is an important controlling factor of the formability of the product. Three different sets of sintered billets were prepared following three different powder mixing procedures, viz.

(A) using a homogenizer in alcoholic medium;
(B) mixing by dry grinding in mortar and pestle prior to homogenizing; and
(C) mixing by wet grinding (alcohol) in mortar and pestle prior to homogenizing.

Compression test results shown in Fig.1 for the three different types of specimens, viz.(A), (B) and (C) reveals that although the maximum strength levels are almost the same for all the three, type (A) shows a much enhanced ductility. Comparison with the results of previous workers [2] indicates that these composites have attained superior ductility. It should also be noted that a 50 vol.% Ag_2O powder mix yields a 40 vol.% Ag (approx.) billet upon sintering (assuming complete conversion of Ag_2O into much denser Ag). These results can be correlated very well with the microstructures of the three types of specimens, see Fig. 2. *It is evident, that a uniform and fine mixture of the two phases in the composite aggregate is the key to improved ductility.* This is indeed in conformation with our hypothesis [1] that the metallic phase takes up "most" of the plastic deformation and hence needs to remain as a continuous matrix. However, the origin of the effect of grinding on coarsening the distribution by agglomeration, is not very clear. It is felt that powders of like kind tend to agglomerate because of electrical charge effect, and mechanical grinding seems to aggravate the problem of nonuniform distribution.

Room temperature compression tests were also conducted on warm compacted specimens of 50 vol.% Ag_2O - Figure 3 shows the results. It is interesting to note that with compaction temperature of 175°C the strength is rather low, however for a compaction temperature of 250°C, the compression strength is almost as high as that for sintered specimens. The implication is clear - it is *not* essential to perform prolonged, high temperature sintering to impart considerable strength to the composite billets. An optical micrographic investigation, Fig. 4, of this specimen reveals incipient formation of the metallic silver at the YBCO - Ag_2O interparticle boundaries. It is to be noted that the inherent weakness of "unsintered" powder aggregates essentially reflects - *not* the weakness of the individual powder particles, but the weakness of the bond at the particle boundaries. The preferential formation of nascent silver evidently promises to overcome the problem effectively. It is highlighted that most of the silver germinates preferentially at the boundaries of Ag_2O with YBCO. Although, nucleation of a new phase is more favorable at grain boundaries, the probable influence of oxygen affinity of the YBCO to favor Ag_2O to lose the oxygen preferentially in its vicinity cannot altogether be overruled. In fact, this preferential metallization coupled with associated volume shrinkage of the Ag_2O to Ag transformation is proposed [3] to be *the mechanism of densification of Ag_2O-YBCO composites during sintering*. And in that respect Fig. 5 is claimed to be a metallographic evidence of the *in situ* pressure sintering effect at high temperature. It illustrates that regions in the vicinity of the Ag_2O-YBCO are comparatively denser, and the boundary relatively defect free, while there are some voids away from it. The phenomenological model [3] indeed predicted a superior densification and bonding around the interparticle boundary. And the importance of a nearly defect-free microstructure at the interparticle boundaries cannot be overemphasized in reference to powder processing route of fabrication technology. The void still remaining after sintering are essentially in the metallic silver matrix, and are undoubtedly much less harmful than particle boundary defects. Further, it goes without saying, that minimizing boundary defects will also contribute towards improvement of electrical properties of the composite.

The effect of thermomechanical treatment on the composites was also studied. A nine hour long annealing treatment at 880°C was given to both the sintered billet and the extruded wire of 50 vol.% composition. Although the microstructure of the sintered billet did not change significantly, that of the extruded wires were altered altogether. The resulting microstructure, Fig. 6, shows a continuous interwoven network of the YBCO phase. This microstructure is expected to exhibit particularly good electrical properties. The change in the microstructure is evidently attributed to the deformation processing experienced by the wire (and not by the sintered billet). The implication of such a drastic transformation of the YBCO phase upon thermomechanical treatment is direct - the YBCO ceramic phase is also taking an active part in the deformation of the composite. Transmission electron micrograph [4] illustrating the grain boundary dislocation is a direct evidence for such occurrence. The good welding achieved between the YBCO particles can be deduced from the clear boundary formed with the interfacial dislocation array observed. The dislocation attached to the grain boundary and extending into one of the grains is probably a dislocation generated during deformation.

A short annealing treatment at 550°C for 1 hour, instead of the long anneal, however was found to improve the ductility of the composite. Using this treatment as an intermediate anneal, an extruded wire of 70 vol.% Ag_2O (i.e. 60 vol.% Ag (approx.)) could be successfully section rolled at 450°C to a thinness of 0.75mm diameter.

Recently, it has been possible to extrude a composite of 40 vol.% Ag_2O (i.e. approximately 30 vol.% Ag) successfully into nearly defect-free 1/8" diameter wire. The sintered composite of this composition was subsequently annealed for 18 hours at 880°C followed by an oxygen annealing at 550°C for 12 hours. The specimen shows a sharp resistive transition as shown in Fig. 7. The microstructure corresponding to this treatment shows typical twin domains, in the orthorhombic YBCO grain.

CONCLUSIONS

1. Uniformity and fineness of particle distribution is an important controlling factor of the formability of the Ag_2O-YBCO microcomposite.
2. Powder mixing procedure using a homogenizer in alcoholic medium produces enchanced ductility.
3. Warm compaction promotes the incipient formation of the metallic silver at the YBCO-Ag_2O interparticle boundaries.
4. This preferential metallization coupled with associated volume shrinkage of the Ag_2O to Ag transformation is proposed to be the mechanism of densification of YBCO-Ag_2O microcomposite during sintering.
5. Suitable thermomechanical treatment seemed to achieve a continuous interwoven network of the YBCO phase, promising good superconducting properties.
6. Microcomposite of 40 vol.% Ag_2O-60 vol.% YBCO shows a sharp resistive transition and can be successfully extruded to a defect-free 3mm diameter wire.
7. After interstage annealing, extruded wires can be successfully warm rolled to a thinness of 0.75mm diameter wire.

ACKNOWLEDGEMENTS

The work is supported by The National Science Foundation grants MSM-8718692 and DDM-8911258, and the authors want to express thanks to Professors B.F. von Turkovich, M.F. Devries, R. Komanduri and W. Aung for their support. The experimental support by Mr. W. Durrant, Mr. S. Goel and Mr. John Swanson is also thankfully acknowledged.

REFERENCES

1. Samanta SK. Samajdar S. Durrant WW. Gupta M. (1989) A novel processing technique for fabrication of flexible $YBa_2Cu_3O_{7-x}$ wire. J. Appl. Phys 66: 4532-4534

2. Garland JC. Calabrese JJ. Herbert ST. (1988). A novel metal matrix-high T_c superconducting composite. In: Capone DW. Butler WH. Battlog B. Chu CW. (eds) High-temperature superconductors II. Materials Research Society, Pittsburgh, Pa. pp 319-321

3. Samajdar S. Kumar A. Mallick K. Samanta SK. (1989) (Accepted for Publication) A phenomenological model on the deformation mechanism of $YBa_2Cu_3O_{7-x}$ + Ag_2O composite J. Mat. Sci. Lett.

4. Gouthama. Samanta SK. (submitted) Electron Microscopic studies of interfaces in $YBa_2Cu_3O_{7-x}$ + Ag_2O wire samples. 7th CIMTEC-World Ceramics Congress - High Temperature Superconductors, July 2-5, 1990

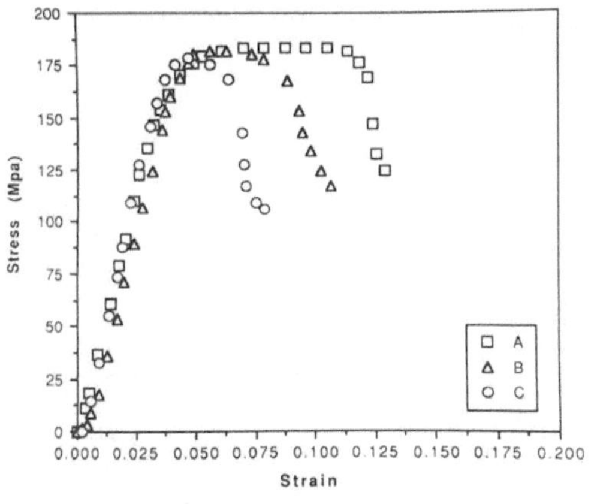

Fig. 1. Compression test results of sintered specimen prepared through different ways - (A), (B), (C) - of mixing.

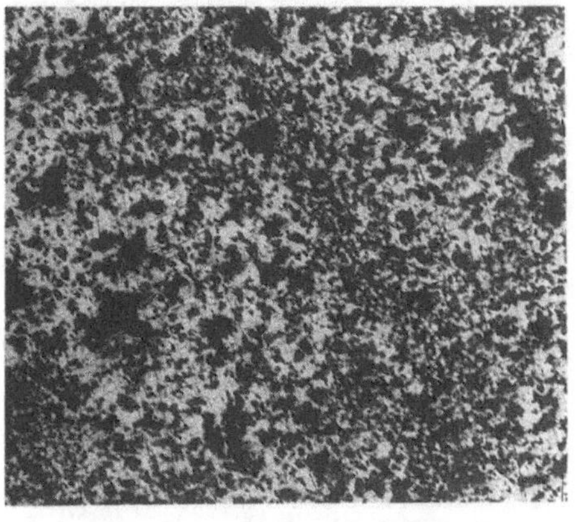

(a)

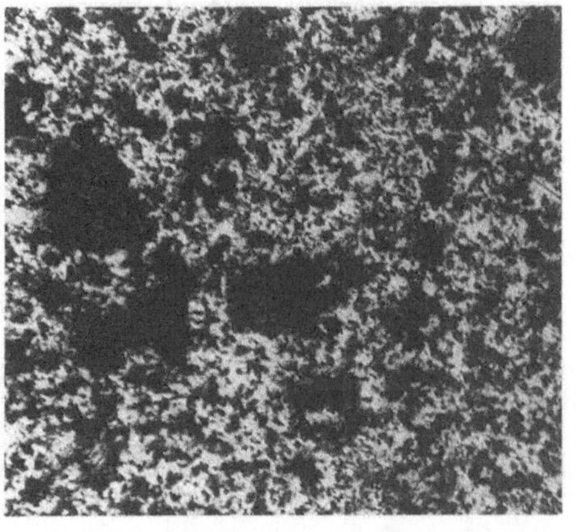

(b)

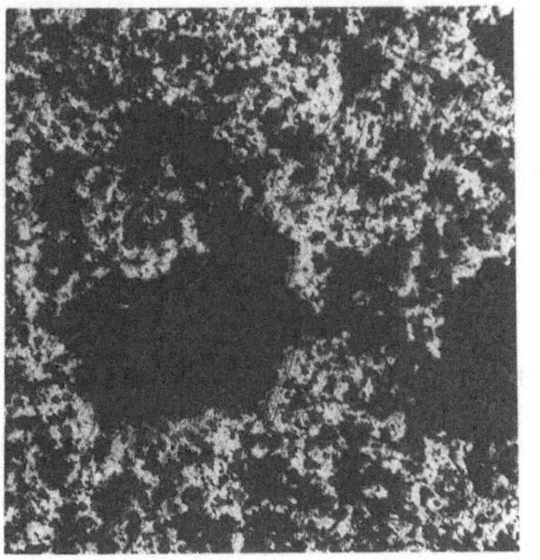

(c)

Fig. 2. Optical micrographs (X100) of sintered specimens of 50 vol.% Ag_2O-YBCO composite subjected to compression testing:
(a) Type (A), (b) Type (B), and (c) Type (C).

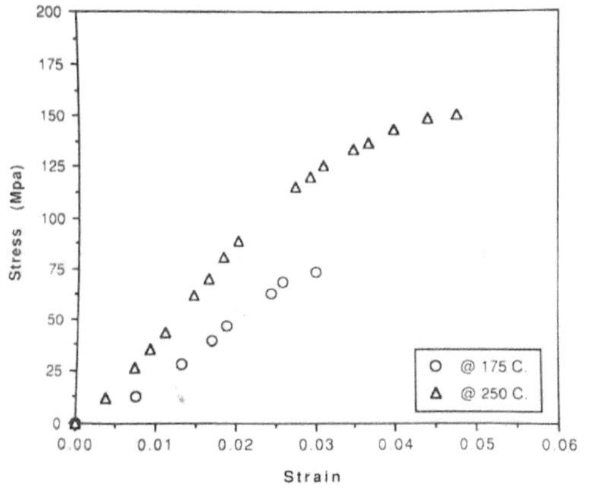

Fig. 3 Compression test results of warm compacted specimens of50 vol.% Ag_2O-YBCO composite.

Fig. 4. Optical micrograph (X1000) of a specimen of 50 vol.% Ag_2O-YBCO composite after being warm compacted at 250°C.

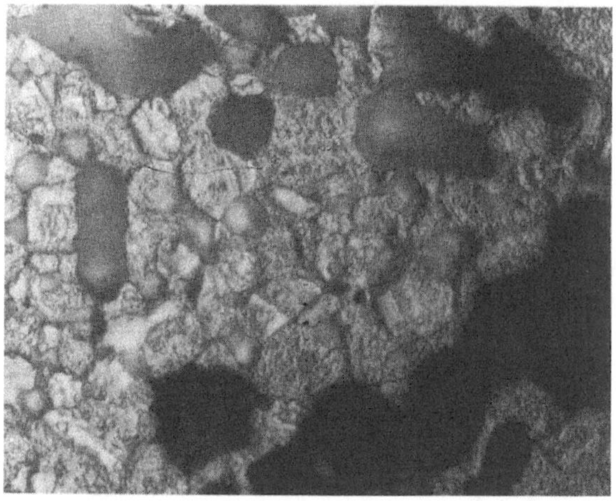

Fig. 5. Optical micrograph (X1000) of a sintered specimen of 50 vol.% Ag_2O-YBCO composite.

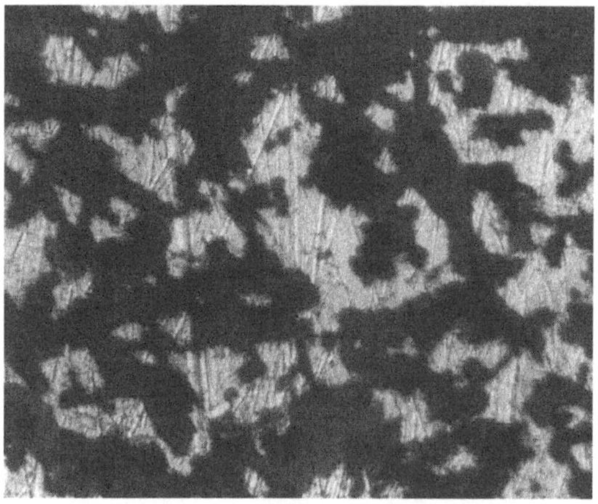

Fig. 6. Optical micrograph (X1000) from an extruded wire of 50 vol.% Ag_2O-YBCO composite, after being subjected to a long annealing treatment.

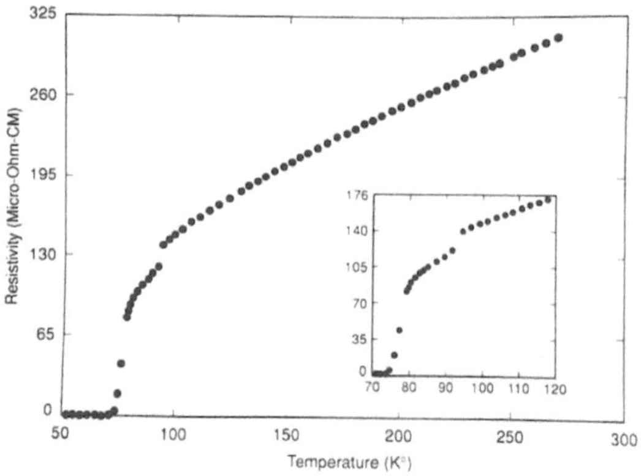

Fig. 7. Resistive transition of a 40 vol.% Ag_2O-YBCO composite.

Mechanical Properties of High-T_c Oxide Composite Superconductors

KOZO OSAMURA, SHOJIRO OCHIAI, and KENJI HAYASHI

Department of Metallurgy, Kyoto University, Sakyo-ku, Kyoto, 606 Japan

ABSTRACT

The mechanical properties of polycrystalline superconductors, YBaCuO and BiPbSrCaCuO, have been investigated. Fracture toughness measured by indentation technique for the polycrystalline YBaCuO was higher than the single crystalline one and the addition of Ag_2O improved toughness. Silver-sheathed YBaCuO and BiPbSrCaCuO wires showed multiple fracture. The cracking stress of the oxide itself distributed mostly in the range between 30 and 100 MPa. It was found that the higher the cracking stress, the higher became the critical current density. A thermal fatigue was observed, when the thermal cycle between 77 K and room temperature was applied to the silver-sheathed wire.

KEYWORDS: mechanical property, multiple fracture, fracture toughness, tensile strength, oxide superconductors

INTRODUCTION

The high T_c oxide superconductors will be used in the magnet technology as a major field of applications. For manufacturing large scale magnets, it is necessary to develop a mass productive process for preparing superconducting wires and tapes. Till now, so extensive studies have been carried out to fabricate composite wire and tape with high critical current density. But mechanical properties have been scarcely paid attention, even though they are very important for practice. The aim of the present work is to make clear the mechanical properties mainly of silver sheathed oxide composites.

EXPERIMENTAL

The silver sheathed oxide wires and tapes were prepared by the powder-in-tube method. Two types of oxides, $YBa_2Cu_3O_{6+x}$ and $Bi_{0.8}Pb_{0.2}Sr_{0.8}CaCu_{1.4}O_y$ were provided. Silver oxide was mixed in some powders. The oxide powder was filled in a silver tube with innner diameter between 2 and 4 mm. The composite was deformed by various cold-working techniques. After swaging or grooved rolling, the reduced wire was cold-rolled or pressed to make tape specimen and heat-treated. In some cases, the heat-treated specimen was again cold-worked. The details were reported elsewhere /1,2/.
The fracture toughness was evaluated by the indentation technique at room temperature. Tensile test was carried out at room temperature and 77K with an Instron type tensile machine at strain rate of $4.2x10^{-4}$/s for a gauge length of 20 mm. The appearance of the oxide core after stressing was observed with a scanning electron microscope by etching away the silver sheath.

RESULTS AND DISCUSSION

A. Fracture Toughness

Fracture toughness and Vickers hardness of $Ba_2YCu_3O_{6+x}$ single crystals were studied by Cook et al/3/. They employed the indentation technique. They reported that the mode I critical stress intensity factor for threshold of fracture in air of the (100) and (001) planes is 0.74 ± 0.2 $MPam^{1/2}$ and Vickers hardness is 8.7 ± 2.7 GPa. On the other hand, Ochiai et al /4/ reported Vickers hardness of 4.5 GPa for the polycrystalline specimen including about 12 volume % voids. The Young's modulus was measured by bending

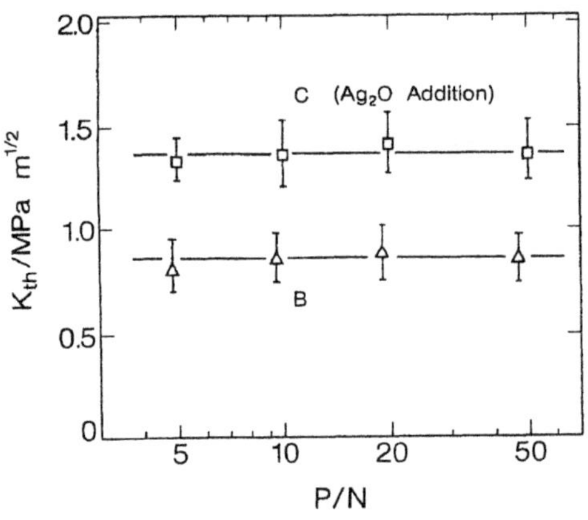

Fig. 1 Stress intensity factor for the threshold of fracture as a function of applied load for two kinds of YBaCuO specimens (C) with or (B) without Ag_2O additive.

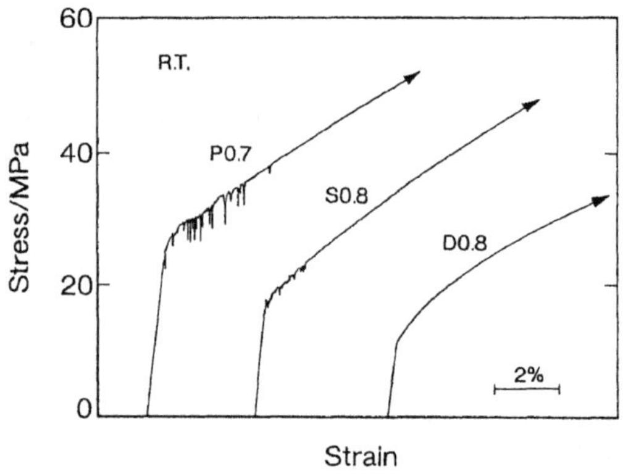

Fig. 2 Stress-strain curve of the Ag sheathed YBaCuO tape and wires. P0.7: tape with 0.7 mm thick by pressing, S0.8: wire with 0.8 mm dia by swaging, D0.8: wire with 0.8 mm dia by drawing.

test by Alford et al/5/. The modulus tended to decrease with increasing voids. By the extrapolation towrds zero porocity, Young's modulus was determined to be 180 GPa.

The experiments of indentation were performed in air on the surface of polished samples with a Vickers diamond pyramid. The mode I stress intensity factor for the threshold of fracture can be evaluated by using the equation,

$$K_{th} = \zeta\ (E/H_v)^{1/2}\ Pc^{-3/2}, \tag{1}$$

where ζ is a material independent constant and put as 0.016, P is the applied load and c is the size of radial crack. Fig. 1 shows the experimental result of the stress intensity factor as a function of applied load for two kinds of the specimens:(C) with or (B) without Ag_2O addition. The K_{th} increased substantially by the addition of Ag_2O. According to Cook et al(1987), the K_{th} of the single crystalline $Ba_2YCu_3O_{6+x}$ in air is 0.74 MPa $m^{1/2}$. Comparing this value with the data shown in Fig.1, the value for sample B is nearly the same, while the value for sample C is higher than that of single crystal in spite of the presence of voids in the polycrystal. Cook et al /3/ suggested that the $Ba_2YCu_3O_{6+x}$ requires considerable toughening in the polycrystalline form to be used in load-bearing applications. The toughening in polycrystalline form is found to be improved by the addition of Ag_2O.

B. Tensile Behaviour Of Silver-Sheathed Oxide Composites.

It is possible to describe silver-sheathed wire and tape as two component composite consisting of ductile metal and brittle ceramic. The tensile property shows a two-step behaviour, because ceramics do not deform plastically. At a first step, both components deform elastically. Beyond the yield point of metallic sheath, the metallic component deforms plastically, but the ceramics

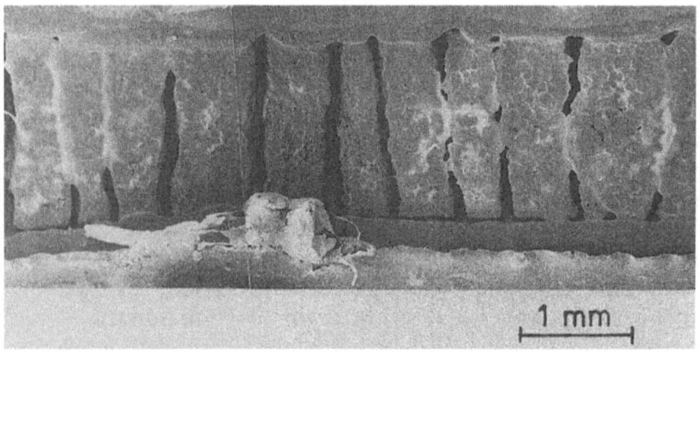

Fig. 3 Multiple fracture in YBaCuO oxide core generated during tensile test.

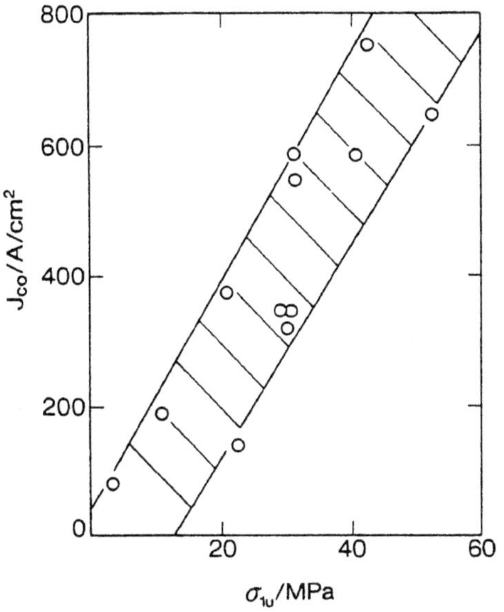

Fig. 4 Relation between the cracking stress of oxide layer and the critical current density for Ag-sheathed YBaCuO wires and tapes.

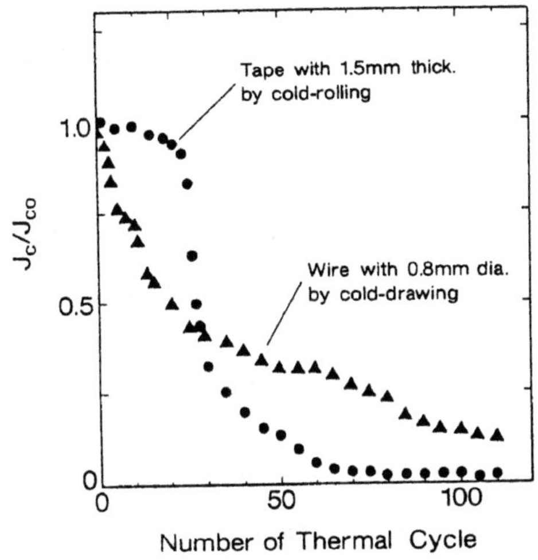

Fig. 5 Change of critical current density as a function of number of thermal cycle between room temperature and 77 K for Ag-sheathed YBaCuO tape prepared by cold-rolling and wire by drawing.

is still elastic. During cooling from the sintering temperature, the internal stress becomes significantly large due to the difference of thermal contraction between two components. As the pure silver has a very samll yield strain of about 0.015%, the silver sheath yielded and plastically deformed at room temperature.

A typical stress-strain curve for the silver sheathed YBaCuO tape is shown in Fig. 2. After an apparent yielding, the drop in stress was observed to be caused by cracking in the oxide. Fig. 3 shows the fracture morphology of the YBaCuO oxide after stressing to make the composite tape fracture as a whole. The oxide fractures nearly perpendicular to tensile axis. As a result of multiple cracking, the length of segments became short. The breakdown of the oxide core in a cross section causes a drop in load bearing capacity of the composite. In this cross-section, however, the metallic sheath work-hardens and then the load bearing capacity at this cross-section rises again. During this process, stress is transferred to the oxide by the shear stress at the core-sheath interface so that the core once fractured fails in another cross-section again. With a repetition of this process, the core fractures into shorter segments.

The first drop in stress in the stress-strain curve is caused by a cracking of the weakest position in the oxide core and it is designated as σ'_c, where the cracking stress of the oxide is expressed as σ_{1u}. The cracking stress of the present oxide was found to distribute in the range up to 100 MPa for the YBaCuO and BiPbSrCaCuO specimens prepared by changing degree of cold working. The room temperature cracking stress was nearly the same as that at 77 K. A strong correlation was found between the critical current density at 77 K and the room temperature cracking stress as shown in Fig. 4. The higher the critical current density corresponds to the higher the cracking stress. The experimental result for the BiPbSrCaCuO specimens is listed in Table 1.

The tensile strength and critical current density of polycrystalline $Ba_2YCu_3O_{6+x}$ oxides are limited in fairly low levels, being attributed to the weak connection of grain boundaries. It is possible to point out a goal for developing the oxide materials from the present data shown in Fig. 4. The critical current is supposed to depend linearly on the cracking stress as $J_{co} = 16\ \sigma_{1u}$. As the elastic properties of both materials Nb_3Sn and $Ba_2YCu_3O_{6+x}$ are nearly the same each other, the strength of $Ba_2YCu_3O_{6+x}$ is expected to be improved to the same level for that of Nb_3Sn, say 2 GPa. When the strength becomes 2 GPa, then the critical current density will reach 32 kA/cm^2. It should be noted that the effort improving the tensile property is strongly connected with improving the critical current density.

Table 1 Correlation between critical current density and cracking stress for BiPbSrCaCuO BiPbSrCaCuO specimens.

Sample No.	J_c at 77 K (A/cm^2)	σ_{1u} (MPa)
S0.8/50-50	1040 - 2520	59 - 84
S1.0D0.6/50-50	1720 - 2290	80
RG/80-80	770 - 1430	81 - 84

C. Thermal Fatigue

As shown in Fig. 5, the silver sheathed tape degrades by the thermal cycle between room temperature and 77 K. The decrease of critical current density is suggested to be caused from the fracture of oxide layer due to thermal fatigue. During the repeat of expansion and contraction, the silver yields every time and plastically deforms. The stress level applied to the oxide layer increases every time due to the plastic deformation of the silver. Beyond a critical level of the stress, the oxide layer will fracture. The cold-rolled specimen showed more homogeneous behaviour for the mechancial property in comparison with that of the cold-drawn specimens.

CONCLUSION

The mechanical properties of polycrystalline oxide superconductors, YBa- CuO and BiPbSrCaCuO, have been investigated. Fracture toughness was measured by means of indentation technique. The polycrystalline YBaCuO showed higher fracture toughness than the single crystalline one and the addition of Ag_2O improved toughness. Silver-sheathed YBaCuO wires showed multiple fracture under applied tensile stress. The cracking stress of the oxide itself was found to be fairly low as 100 MPa at most. There was a correlation between critical current density and cracking stress: the higher the cracking stress, the higher became the critical current density. A thermal fatigue was observed , when the thermal cycle between 77 K and room temperature was applied to the silver-sheathed wire. Its behaviour depended on the history of sample preparation.

Acknowledgements
The authors wish to express their gratitude to Prof. Y.Muto and his coworkers fo HFLSM at Tohoku University for their help in the critical current measurements. This study was made possible by a Scientific Research Grant in-Aid (Project No. 01644520) from the Ministry of Education, Science and Culture of Japan.

References
1. Osamura K. Takayama T. Ochiai S. (1989) Effect of Cold-Working on the Critical Current Density of Ag-Sheathed $Ba_2YCu_3O_{6+x}$ Tapes. Supercond. Sci. Technol. 2: 111-114.
2. Osamura K. Oh S.-S. Ochiai S. (submitted)) Effect of Termomechanical Treatment on the Critical Current Density of Ag-Sheathed B(Pb)SCCO Tapes. Supercond. Sci. Technol.
3. Cook R.F. Dinger T.R. Clarke D.R. (1987) Fracture Toughness Measurements on $YBa_2Cu_3O_x$ Single Crystals. Appl. Phys. Lett. 51: 454-456.
4. Ochiai S. Osamura K. Takayama T. (1988) Fracture Toughness Measurements of $Ba_2YCu_3O_{7-x}$ Superconducting Oxide by Means of Indentation Technique. Jpn.J. Appl. Phys. 27: L1101-L1103.
5. Alford N.M. Birchall J.D. Clegg W.J. Harmer M.A. Kendall K. and Jones H.D. (1988) Physical and Mechnaical Properties of $YBa_2Cu_3O_{7-x}$ Superconductors. J.Mater. Sci. 23 : 761-768.

AC Loss of High-T_c Superconductors

N. Ichiyanagi[1], S. Tanaka[1], T. Hara[2], and K. Okaniwa[2]

[1] The Furukawa Electric Company, Ltd., Ichihara, Chiba, 290 Japan
[2] Tokyo Electric Power Company Inc., Chofu, Tokyo, 182 Japan

ABSTRACT

Applications where high-Tc superconductors hold a good promise to find practical use include high-Tc superconducting power cables. High hopes ride on these new materials for excellent qualities as means of power tranmission, such as high capacity, low voltage and low power loss. As the lower critical field of high-Tc superconductors is considerably smaller than that of metal superconductors, so the surface current density of the former is very low. This poses, however, a technical problem very hard to solve as to how to reduce AC loss. Measurements of AC loss in high-Tc superconducting tapes by the magnetization method have shown that quantitative evaluation in the Bean model provides for satisfactory approximation of AC loss in high-Tc superconducting tapes.

Key words: AC loss, superconducting cable, Bean model, magnetization method

INTRODUCTION

Conventional superconducting cable designs are based on the concept of surface current density. It is expressed as the amperage of current per unit circumferential width of a conductor, a hollow former round which thin superconducting tape is helically wound. In the case of liquid helium cooled superconducting cables, the surface current density is determined so that the conductor surface magnetic self field is held below the lower critical field Hc1 of that superconductor. This places the superconducting tape in a perfect Meissner state so that AC loss in the conductor is held down to an extremely low level. When the field applied to the conductor equals the lower critical field, the surface current density is given by Eq. (1).

$$\mathrm{Is\,(A/cm)} = \frac{5}{2\sqrt{2}\,\pi}\mathrm{Hcl\ (Oe)} \tag{1}$$

$$D = \frac{I}{\pi \cdot \mathrm{Is}} \tag{2}$$

Table 1 shows the critical temperatures and critical fields of main kinds of superconductors. Although high-Tc superconductors show considerable variations in Hcl [1], their Hcl values on the whole are by far smaller than those of metal superconductors [2]. Table 2 shows the surface-current values of main kinds of superconductors in relation to the minimum conductor diameters required to carry a current of 10000 A [2]. According to this table, metal superconductors may have a minimum conductor diameter to carry this current, whereas high-Tc superconductors must have an impracticably large minimum conductor diameter of 1 m or so to do the same job. To reduce the conductor diameter of high-Tc superconducting power cables it is necessary to make the current flow not only in the surface of the

Table 1 Critical Temperature and Field for Main Superconductors

	Superconductors	T_c (K)	Hcl (Oe)	H_{c2} (T)	H_{c2} // / $H_{c2}\perp$
Metal	N_b	9.1	800 -1000	0.19	1
	$N_b\ T_i$	9.8	700 - 900	15	1
	$N_{b3}\ S_n$	18.2	600 - 700	29	1
Oxide	$Y\ B_{a2}\ Cu_3\ Ox$	93	52 - 120	150*	-10
	$B_{i2}\ S_{r2}\ C_{a2}\ C_{u3}\ O_x$	110	6 - 200	220*	-30
	$Tl_2\ B_{a2}\ C_{a2}\ C_{u3}\ O_x$	125	- 150	220*	-40

* Estimated values

Table 2 Surface Current and Minimum Conductor Diameter (for 10000 A)

Superconductor	Surface current (A/cm)	Minimum conductor diameter (cm)
N_b	450 to 563	5.7 to 7.1
$N_{b3}\ S_n$	338 to 394	8.1 to 9.4
$Y\ B_{a2}\ C_{u3}\ O_x$	29 to 67	47.5 to 109.8

conductor but also in its inside. This means that high-Tc superconductors must be used in the range between Hc1 and Hc2. However, if this requirement is met, conductor loss will occur due to field penetration. Consequently, quantitative evaluation of AC loss in high-Tc superconductors is a consideration of great technical importance.

AC loss in high-Tc superconductors occurs in several ways, such as hysteresis loss and viscous-resistance loss in the superconducting layer and coupling loss and eddy-current loss in stabilizers. Considering high-Tc superconducting cables have a maximum self-field of approximately 0.1T. and are to be operated at a frequency of 50 or 60 Hz, it may be presumed that hysteresis loss accounts for the greater part of total AC loss in high-Tc superconducting cables.

EXPERIMENTAL

The magnetization of Bi-Sr-Ca-Cu-O silver-sheathed tape was measured at 77K or 4.2K by the DC magnetization method. Fig. 1 shows the results obtained at 77K and Fig. 2 the results obtained with the same specimen at 4.2K. The hysteresis loss Wh can be calculated from the magnetization curve by Eq. (3).

$$Wh = \oint M dHs \ (J/m^3 \cdot cycle) \tag{3}$$

Hs: surface magnetic field
M: magnetization of superconductor

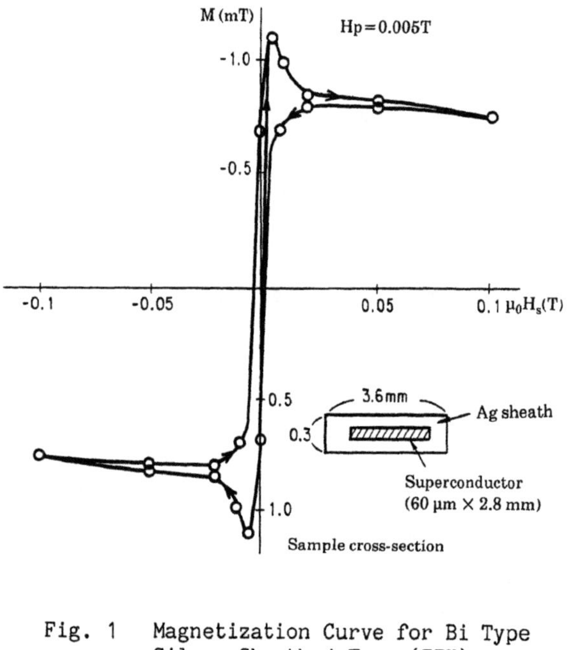

Fig. 1 Magnetization Curve for Bi Type Silver-Sheathed Tape (77K)

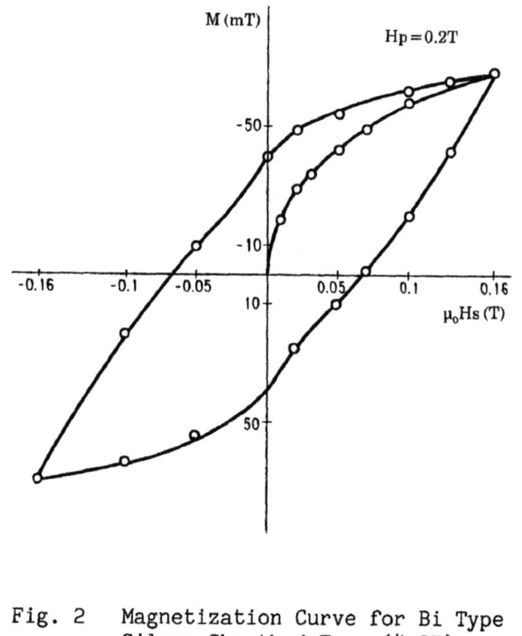

Fig. 2 Magnetization Curve for Bi Type Silver-Sheathed Tape (4.2K)

The Bean model is one of the models which are resorted to when determining the field and current density distribution in a superconductor as required to calculate its hysteresis loss [3]. The hysteresis loss in metal superconductors which is calculated with the aid of this model nearly coincides with the observed values, and it provides the simplest and most convenient tool of calculation to use than any other model. The Bean model holds on the assumption that the current density J is invariably fixed as the critical current density Jc and is not dependent on the field. The J is given by Eq. (4).

$$J = Jc \tag{4}$$

Let the penetration depth from the superconductor surface be x and the surface field be Hs, and the field H shows linear distribution as expressed by Eq. (5), where Hp indicates the magnitude of the surface field which is formed when the quantum flux reaches the center of the superconducting layer. Hp also indicates the value of the field which is externally applied when the absolute value of magnetization is maximized. It can be known from Figs. 1 and 2 that the values of Hp at 77K and 4.2K are 0.005T and 0.2T, respectively.

$$H(x) = Hs - Jc \cdot x \tag{5}$$

$$Hp = Jc \cdot d \tag{6}$$

Let the maximum value of Hs be Hm, the superconducting plate hysteresis loss Wh ($J/m^3\cdot cycle$) in the Bean model is calculated by Eq. (7).

$$\begin{aligned} &Hm \leqq Hp \quad Wh = \frac{2\mu o}{3}\cdot\frac{Hm^3}{Jc\cdot d} \\ &Hm \geqq Hp \quad Wh = 2d\mu oJcHm\left(1-\frac{2}{3}\cdot\frac{Hp}{Hm}\right) \end{aligned} \tag{7}$$

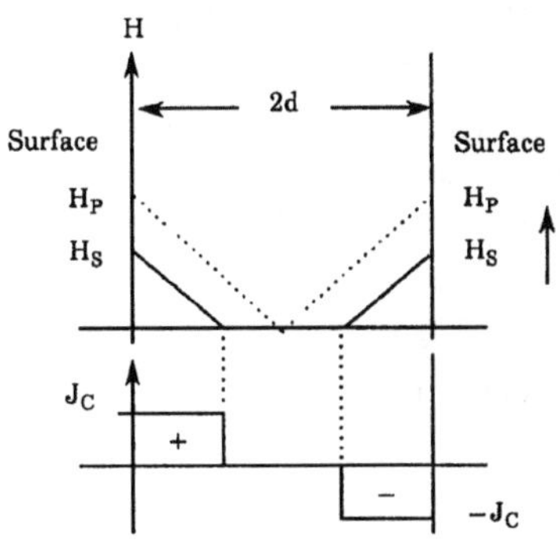

Fig. 3 Field and Current Density Distribution Based on Bean Model (when superconducting infinite plate placed in external field)

Figs. 4 and 5 compare the observed values of hysteresis loss obtained from the magnetization curves in Figs. 1 and 2 with the calculated ones obtained in the Bean model. The values of Jc which were used to obtain the calculated values were derived from Eq. (6).

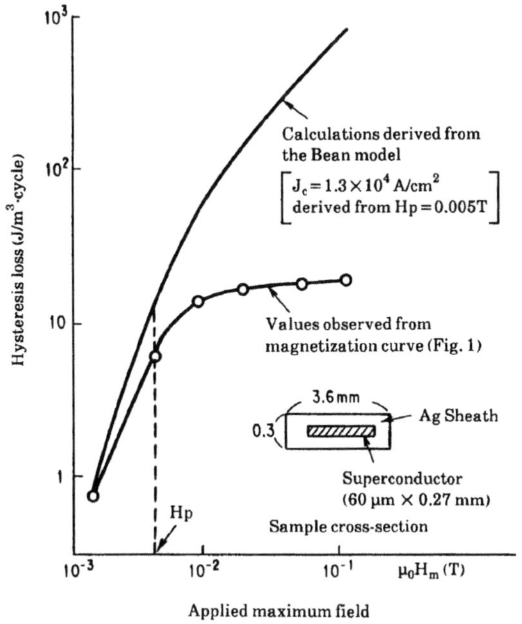

Fig. 4 Hysteresis Loss in Bi Type Silver-Sheathed Tape (77K)

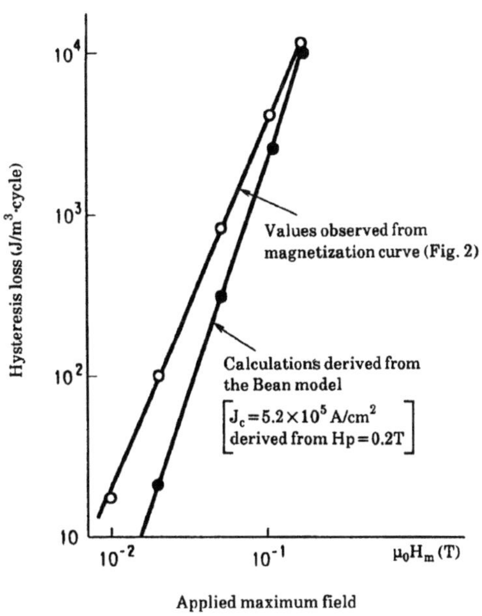

Fig. 5 Hysteresis Loss in Bi Type Silver-Sheathed Tape (4.2K)

As shown in Fig. 4, the observed and calculated values show a similar pattern of field dependency in the range of Hm<Hp, both being in the same order of magnitude, whereas in the range of Hm>Hp the calculated values are 1 to 2 orders of magnitude larger. This difference may be explained by presuming that Jc of the specimens decreased with increasing field with the result that Wh which varies in proportion to Jc also decreased.

Fig. 5, on the other hand, gives the values of applied field and hysteresis loss for the range of Hm<Hp only. This is because Jc of the superconducting tape at 4.2K increased well in excess of the value at 77K. It should be noted, however, that the observed and calculated values are nearly in agreement with each other in the order of magnitude.

If Jc of silver-sheathed superconducting tape at 77K can be further increased substantially, it can be expected that such an approximate coincidence of observed and calculated values as shown in Fig. 5 will be achieved at 77K.
The hysteresis loss which was calculated using the values of Jc obtained by the four-probe method significantly differed from the calculated values. Furthermore, the value of Hp which was calculated using the values of Jc obtained by the four-probe method in Eq. (6), a contradictory result was obtained that the value of Hp was below Hc1.
This contradiction suggests that the shield current flowing in the superconducting grains is dominant over the transport current traversing the grain boundary.
Assuming that only a self-field is applied to the superconducting tape, the relation Hm<Hp invariably holds for such conductors as superconducting tapes are helically wound. It seems therefore reasonable to think that the Bean model finds application in the approximation of hysteresis loss in high-Tc superconductors. The hysteresis loss Wc (W/m) in the entire cylindrical cable conductor is given by Eq. (8) when calculated in the Bean model.

$$Wc = \frac{2\sqrt{2}\mu o}{3} \cdot \frac{I^3}{Jc \cdot Pe^2} \cdot f \tag{8}$$

where, Pe is the conductor perimeter and f, the frequency.

REFERENCES

1. Murayama N. et al. (1988) Remanent moment of high-temperature superconductors. Japan Journal of Applied Pysics. Vol. 27 No. 9: 1629-1630
2. Hara T. et al. (1989) General Annual Meeting of IEEJ. No. 845
3. Bean C.P. (1962) Phys. Rev. Letters 8: 250

Preparation of Ag-Sheathed Tl-Ca-Ba-Cu-O Superconducting Wire

H. TAKEI[1], Y. TORII[1], H. KUGAI[1], T. HIKATA[2], K. SATO[2], H. HITOTUYANAGI[2], and K. TADA[1]

Basic High-Technology Laboratories[1], Osaka Research Laboratories[2], Sumitomo Electric Industries, Ltd., 1-3, Shimaya 1-chome, Konohana-ku, Osaka, 554 Japan

ABSTRACT

In the Tl-Ca-Ba-Cu-O system, Tl-O single layered and double layered structures with general formulas $Tl_1Ca_{n-1}Ba_2Cu_nO_{2n+3}$ (n=1~6) and $Tl_2Ca_{n-1}Ba_2Cu_nO_{2n+4}$ (n=1~4) have been obtained. We fabricated Ag-sheathed Tl-Ca-Ba-Cu-O superconducting wires using powders of Pb-added $Tl_1Ca_3Ba_2Cu_4O_{11}$ and of $Tl_2Ca_2Ba_2Cu_3O_{10}$ phases whose critical temperatures are the highest among the respective structures. The production process of the wires was cold-working and heat treatment. The measurement of transport critical current was carried out in the range of magnetic fields from zero to 2.5T. The critical current density J_c at 77K for the wire of $(Tl,Pb)_1Ca_3Ba_2Cu_4O_{11}$ is 10800A/cm^2, and that value for $Tl_2Ca_2Ba_2Cu_3O_{10}$ wire is 5500A/cm^2. There is a tendency for the wire specimens with homogeneous structure of superconducting phases to have high J_c values. The J_c of the $(Tl,Pb)_1Ca_3Ba_2Cu_4O_{11}$ wire decreases rapidly by 0.1T, while the decrease of J_c above 1T is not so remarkable. The magnetic field dependency of J_c of the $Tl_2Ca_2Ba_2Cu_3O_{10}$ wire shows a two-step decrease in J_c versus B curves; the first step is from zero to 0.1T, and the second step is beyond 1T. The direction of the magnetic field against the plane of the wire does not considerably influence the J_c values of either of the wires, because the anisotropic property of the wire is small.

INTRODUCTION

The technology for manufacturing wire of high-T_c superconducting oxides has been under development for the application of high-Tc superconductors in the fields of power transmission cable, high field magnet and so on. At the present time, it is necessary to improve the current transportation property of the wire up to a practical level. For practical use, a critical current density of above $10^{4\sim5}$A/cm^2 at least is required. While the critical current density at 77K without an external magnetic field for short samples of Ag-sheathed wires using the high-Tc oxides of BiPbCaSrCuO and TlCaSrBaCuO exceeds 10^4A/cm^2, that density with a magnetic field is not sufficiently large(1)(2).

On the other hand, since the discovery of superconductivity in the Tl-Ca-Ba-Cu-O system, a considerable number of superconducting phases have been reported in the Tl-based systems. Single and double Tl-O layered structures with a general formula $Tl_mCa_{n-1}Ba_2Cu_nO_{2n+m+2}$ (m=1,2, n=1~6), where m and n are the number of Tl-O and Cu-O layers respectively, have been obtained in the Tl-Ca-Ba-Cu-O system. It is necessary to choose the most suitable phase among these for the practical application of the wire.

In this paper, we will report on the preparation and superconducting properties of Ag-sheathed wires using high-T_c oxides of $(Tl,Pb)_1Ca_3Ba_2Cu_4O_{11}$ and of $Tl_2Ca_2Ba_2Cu_3O_{10}$ phases.

PRODUCTION PROCESS OF Ag-SHEATHED WIRES

It is fundamentally important to synthesize the homogeneous single phase of the high-T_c oxide for preparing the wire. Some of the authors have found out that a partial substitution of Pb for Tl stabilizes the single Tl-O layered structure of the superconducting phases(3). The relationship between the superconducting transition temperature showing zero resistivity

T_{ci} and a nominal composition of Tl:Pb:Ca:Ba:Cu of (1-x):x:3:1:3, where $0 \leq x \leq 1$, is shown in Fig.1. Superconducting phases with T_{ci} higher than 110K form until the parameter x is 0.7. From the results of X-ray diffraction of the samples, it is known that the double Tl-O layered structure of $Tl_2Ca_2Ba_2Cu_3O_{10}$ phase forms in the sample without Pb (x=0), and even a small substitution of Pb for Tl in the nominal composition makes the double layered structure change to the single layered structure. Almost single phases of $(Tl,Pb)_1Ca_{n-1}Ba_2Cu_nO_{2n+3}$ (n=2~6) are prepared by optimizing the synthetic conditions

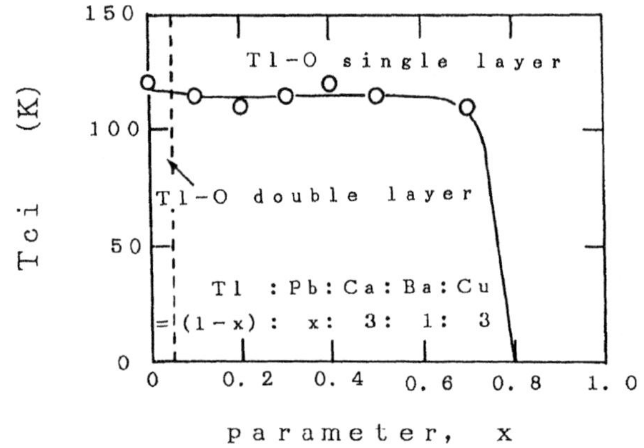

Fig. 1 The relation between critical temperature T_{ci} and parameter x of nominal composition of Tl:Pb:Ca:Ba:Cu =(1-x):x:3:1:3.

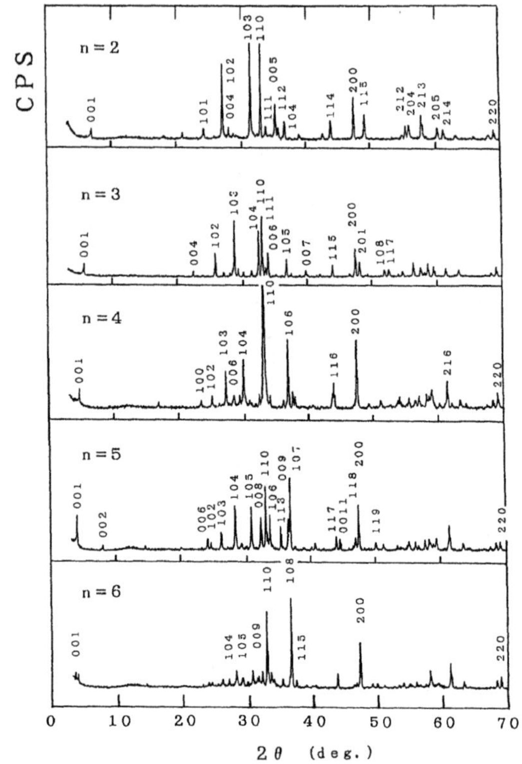

Fig. 2 X-ray diffraction patterns of $(Tl,Pb)_1Ca_{n-1}Ba_2Cu_nO_{2n+3}$ phases(n=2~6)

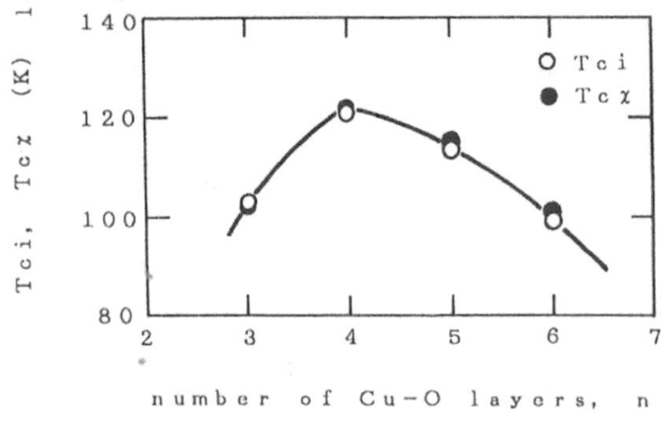

Fig. 3 Critical temperature $T_{c\chi}$, T_{ci} and lattice constants a, c as a function of number of Cu-O layers n.

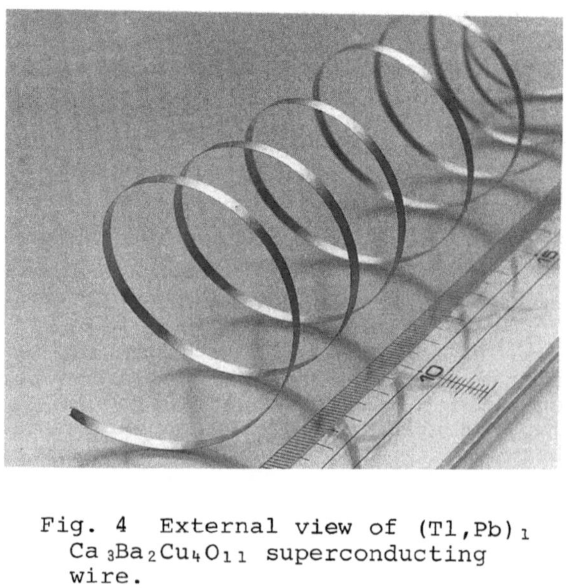

Fig. 4 External view of $(Tl,Pb)_1Ca_3Ba_2Cu_4O_{11}$ superconducting wire.

of the nominal composition and heat treatment. The number of Cu-O layers n in the single Tl-O layered structure tends to increase with a smaller amount of additional Pb atoms and higher sintering temperature.

X-ray diffraction patterns of the superconducting phases $(Tl,Pb)_1Ca_{n-1}Ba_2Cu_nO_{2n+3}$ (n=2~6) with tetragonal symmetry (P4/mmm) are shown in Fig.2. The lattice constants of these phases are shown as a function of the number of Cu-O layers in Fig.3. The superconducting transition temperature showing zero resistivity and that temperature showing diamagnetism $T_{c\chi}$ are also shown in Fig.3. The c-axis lattice constant observed in the present study is in proportion to the number of Cu-O layers n, obaying a equation of c=6.38+3.19n. With an increase in n, both temperatures T_{ci} and $T_{c\chi}$ rise, reaching a maximum value of 121K at n=4, and then fall with further increasing of the Cu-O layers. We adopted the single Tl-O layered structure of $(Tl,Pb)_1Ca_3Ba_2Cu_4O_{11}$ phase and double layered $Tl_2Ca_2Ba_2Cu_3O_{10}$ phase whose critical temperatures are the highest among the respective structures.

The Ag-sheathed wires are fabricated as follows. Appropriate amounts of well-mixed powders of Tl_2O_3, PbO, CaO, BaO_2 and CuO were sintered at around 1140K in oxygen flow for 24hrs, and then ground into powder again. The superconducting powders of $(Tl,Pb)_1Ca_3Ba_2Cu_4O_{11}$ phase and of $Tl_2Ca_2Ba_2Cu_3O_{10}$ phase were put into silver tubes and cold worked into rectangular wires of 3mm in width and 0.15mm in thickness. The wires were finally heat treated at around 1070K. The external view of the $(Tl,Pb)_1Ca_3Ba_2Cu_4O_{11}$ wire is shown in Fig.4.

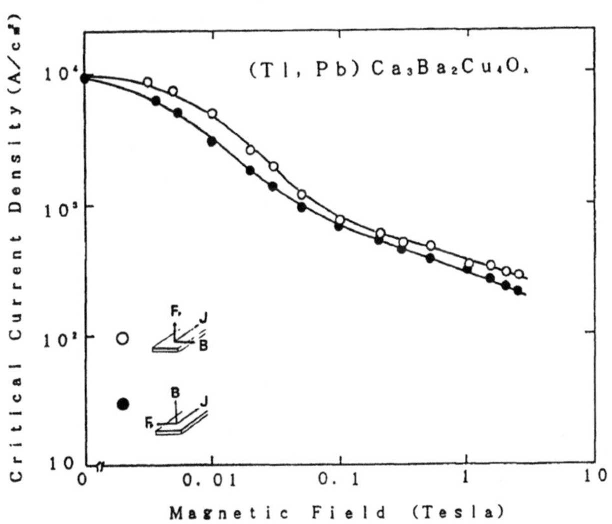

Fig. 5 Magnetic field dependence of critical current density J_c at 77K for $(Tl,Pb)_1Ca_3Ba_2Cu_4O_{11}$ wire.

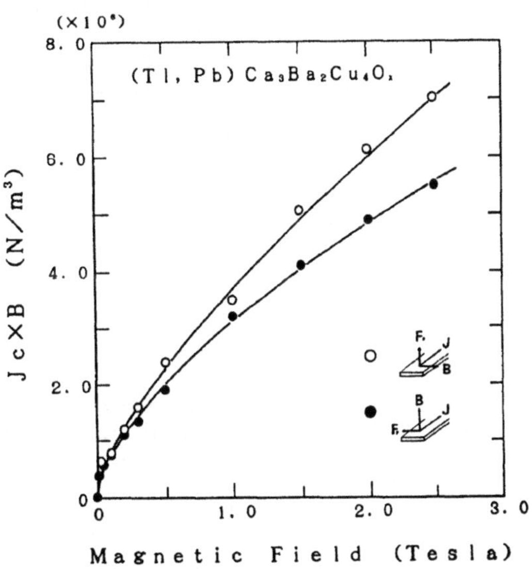

Fig. 6 Magnetic field dependence of pinning force F_p (=J_c×B) at 77K for $(Tl,Pb)_1Ca_3Ba_2Cu_4O_{11}$ wire.

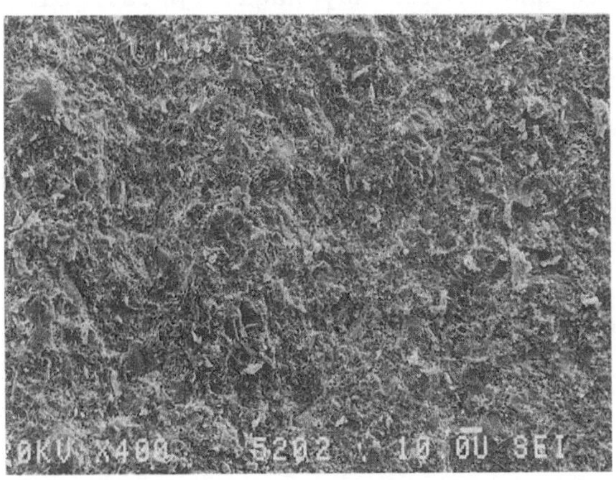

Fig. 7 Scanning electron micrograph of superconducting core of $(Tl,Pb)_1Ca_3Ba_2Cu_4O_{11}$ wire.

SUPERCONDUCTING PROPERTIES OF Ag-SHEATHED WIRES

The magnetic field dependence of the critical current density J_c at 77K for the 10,000A/cm^2 class Ag-sheathed $(Tl,Pb)_1Ca_3Ba_2Cu_4O_{11}$ wire is shown in Fig.5. The measurement of transport critical current was carried out by the four-probe DC method at 77K in the range of magnetic fields from zero to 2.5T. The directions of the external magnetic field applied perpendicularly to the transport current are two cases of parallel and normal to the wide plane of the rectangular wires. The critical current density is defined as current density at which electrical resistivity of $10^{-10}\Omega$cm is generated in the superconductor.

The highest value of J_c without external magnetic field for the $(Tl,Pb)_1Ca_3Ba_2Cu_4O_{11}$ wire reaches 10,800A/cm^2 and that value for the $Tl_2Ca_2Ba_2Cu_3O_{10}$ wire is 5,500A/cm^2. The J_c of the $(Tl,Pb)_1Ca_3Ba_2Cu_4O_{11}$ wire decreases rapidly by 0.1T, while the decrease of J_c above 1T is not so remarkable, as shown in the figure. The magnetic field dependency of J_c for the $Tl_2Ca_2Ba_2Cu_3O_{10}$ wire shows a two step decrease; the first step is from zero to 0.1T, and the second step is beyond 1T. The direction of the magnetic field against the plane of the wire does not considerably influence the J_c values of either of the wires.

The pinning force of the $(Tl,Pb)_1Ca_3Ba_2Cu_4O_{11}$ wire, F_p ($=J_c x B$), is shown in Fig.6 as a function of the magnetic field. The value of F_p increases in proportion to the magnetic field up to 2.5T. The F_p of the $Tl_2Ca_2Ba_2Cu_3O_{10}$ wire increases with an increase in the magnetic field, reaching a maximum value at about 1T, and then decreases with further magnetic increasing. Not only J_c without magnetic field but also the high-field property of the $(Tl,Pb)_1Ca_3Ba_2Cu_4O_{11}$wire is superior to that of the $Tl_2Ca_2Ba_2Cu_3O_{10}$wire. While the pinning sites of the wires are not clear at the present, non-superconductive impurity phases of fine CaCu oxides and the interface of intergrown superconducting phases are thought to be candidates. A scanning electron micrograph of the superconducting core of the $(Tl,Pb)_1Ca_3Ba_2Cu_4O_{11}$ wire after removing the silver sheath is shown in Fig.7. Very fine equiaxised grains are observed at the surface of the core. This isotropic metallurgical structure is thought to be the reason for the small influence of the magnetic field direction on J_c.

CONCLUSION

High J_c Ag-sheathed wires have been prepared using superconducting oxides of $(Tl,Pb)_1Ca_3Ba_2Cu_4O_{11}$ and of $Tl_2Ca_2Ba_2Cu_3O_{10}$. The J_c without external magnetic field for the $(Tl,Pb)_1Ca_3Ba_2Cu_4O_{11}$ wire reaches 10,800A/cm^2 and that value for the $Tl_2Ca_2Ba_2Cu_3O_{10}$ wire is 5,500A/cm^2. The pinning force of the $(Tl,Pb)_1Ca_3Ba_2Cu_4O_{11}$ wire increases in proportion to the magnetic field up to 2.5T at 77K. The direction of the external magnetic field applied perpendicularly to the transport current against the plane of wires does not considerably influence the values of J_c, because the metallurgical structure of the superconducting core of the wires is isotropic.

REFERENCES

(1) T.Hikata, T.Nishikawa, H.Mukai, K.Sato and H.Hitotsuyanagi: Jpn.J.Appl. Phys.,28(1989),L1204.
(2) T.Matsumoto, M.Okada, R.Nisiwaki, T.Kamo, K.Aihara, S.Matsuda, M.Seido, K.Ozawa, Y.Morii and S.Funahashi: Proc.ISTEC Workshop on Superconductivity, Oiso(1989),111.
(3) H.Kusuhara, T.Kotani, H.Takei and K.Tada: to be published in Jpn.J.Appl. Phys.,28,No.10(1989).

Effect of Uniaxial Stress onto the Electromagnetic Properties of Bi-(Pb)-Sr-Ca-Cu-O Superconductors

SEIJI HAYASHI, KAZUYUKI SHIBUTANI, YOSHITO FUKUMOTO, RIKUO OGAWA, and YOSHIO KAWATE

Superconducting and Cryogenic Technology Center, Kobe Steel Ltd., 5-5, Takatsukadai 1-chome, Nishi-ku, Kobe, 673-02 Japan

ABSTRACT

The Bi-(Pb)-Sr-Ca-Cu-O (BPSCCO) superconductors of the rectangular bars without a sheath and tapes with Ag-sheath were prepared by pressing uniaxially. Critical current density (Jc), n-value and pinning potential is measured for these samples. It is elucidated that uniaxial stress is effective for the improvement of Jc and n-value. The larger value of Jc is observed for the sample which contains low-Tc phase partially before sintering process. The existence of the low-Tc phase in sintering process is expected to be useful for the good contact between grains. Typical pinning potential U=0.019eV is obtained at 4.2K in magnetic field H=1T applied parallel to the pressed surface.

KEY WORDS: Bi-Pb-Sr-Ca-Cu-O superconductor, uniaxial stress, critical current density, pinning, weak link

INTRODUCTION

It has been well known that the Bi-Sr-Ca-Cu-O (BSCCO) system had two superconducting phases called low-Tc phase with Tc~85K and high-Tc phase with Tc ~ 110K. Recently, single high-Tc phase has been obtained by partial substitution of Bi by Pb.[1,2] Since pinning potential of these system is very small at 77K, it has been considered to be difficult for practical use at 77K. On the other hand, the upper critical field Hc_2 of this system has been reported to be higher than 100T at 4.2K. In fact, according to K.Heine et al., Jc of $1.5\times10^4 A/cm^2$ could be achieved at 4.2K in magnetic field of 26T.[3] Generally, the sample with higher critical current density showed smaller magnetic field dependence of Jc. Therefore, improvement of Jc is significant for high magnetic field application at 4.2K. It has been reported that uniaxial stress onto the Ag-seathed wire made increase in Jc.[4] We investigate the effect of uniaxial stress and role of low-Tc phase before sintering onto the Jc and n-value.

EXPERIMENTAL

The Ag-seathed wires were fabricated by the conventional powders in tube method, formed into wires by swaging and cold rolling. The BPSCCO tapes were made by applying uniaxial stress to these wires. Two types of powder were packed into a Ag-tube ; one was composed of high-Tc phase only (type-1), and the other was composed of high-Tc and low-Tc phases (type-2). Then, these tubes were swaged and cold rolled to the wires followed by pressing and sintering process. Typical pressing and sintering conditions were as follows; total weight 10ton was applied and then sintered at 835~842°C for 10~70 hours, and this process were repeated several times. The ratio of high-Tc phase to low-Tc phase were checked by SQUID magnetometer (HOXAN HSM-2000) and Jc were measured by conventional four terminal method at after every sintering procedures.

Bulk specimens without seath (type-3) were fablicated by pelletizing the high-Tc single phase BPSCCO powders, pressing uniaxially and sintering at almost the same conditions as that employed for Ag-seathed samples. In order to avoid thermal interference at current lead terminals, samples were formed into narrow in the middle shape. The Jc and I-V characteristics were measured by the four terminal method, then n-value were calculated from these datum. Relaxation of remanent magnetization were measured at 4.2K for these three types of samples.

RESULTS AND DISCUSSION

Figure 1 shows applied stress dependence of Jc and n-value for type-3 samples. In this Figure, horizontal axis corresponds to total weight applied uniaxially to the sample. Since the sample spreads transversely when stress imposed uniaxially, correct strength of the stress cannot be determined. If we roughly estimate the uniaxial stress dividing total weight by pressed surface area, total weight 1.0 ton in Fig.1 corresponds to about 1.4×10^3 kgw/cm^2. The criterion for Jc determination was defined as 1 μV/cm, and n-value was calculated by least-square fitting to the equation; $V=k*(I/Ic)^n$. As seen from this Figure, Jc and n-value increase with applied uniaxial stress. This is considered to be due to the improvement of packing density and grain alignment. This fact is clearly seen in Fig.2(a)-(b), which are the SEM photographs taken on the fractured surfaces of type-3 samples. Figure 2(a) corresponds to the sample without stress, and Fig.2(b) corresponds to 20ton-pressed specimen. However, Jc of type-1 and type-3 samples did not exceed 3000A/cm^2 at 77K and 0T. This is supposed to be due to weak links between grains as mentioned below.

Figure 3 shows that formation of high-Tc phase proceeds with sintering process. As the fraction of high-Tc phase increases, Jc becomes large. Figure 4 shows Jc versus wire diameter relation after high-Tc phase was achieved. Since same weight were applied to these specimens, the horizontal axis is considered to correspond to the degree of stress. As the uniaxial stress increases, the higher Jc-value is achieved, and Jc=9500A/cm^2 (77K,0T) was obtained for type-2 sample with 0.3mm diameter. In these Figures, the criterion for Jc determination was defined also as 1 μV/cm.

Figure 5 shows the example of relaxation of remanent magnetization for type-2 sample at 4.2K. This measurement were performed as follows; when sample was superconducting state, magnetic field was increased from 0T to a certain value, then applied magnetic field was decreased to 0T. From that time, the relaxation of magnetization was measured. This sample show Jc=6710A/cm^2 at T=77K and H=0T. In Fig.5, magnetization decays linearly to the logarism of time, indicating that this decay is caused by the flux creep. According to the Anderson's thermal activation model, $M=M(1-(kT/U)\ln(t/t))$,[5] pinning potential is calculated as listed in Table 1. Pinning potential is larger for the case when flux penetrates parallel to the c-plane.

Figure 6 shows preliminaly result of Ic-H curve at 4.2K for sample which has Jc=2.4×10^4A/cm^2 (4.2K,0T). In this case, magnetic field was applied parallel to the tape surface and perpendicular to the current direction. As seen in this

Figure, Jc of BPSCCO tape specimen is almost kept constant at higher magnetic field of more than 1T. This fact suggests that BPSCCO tape has big potential to practical application in extremely high magnetic field at 4.2K. More detailed mesurement at higher magnetic field for specimen with higher Jc-value is now under investigation.

CONCLUSION

Uniaxial stress is effective for the increase of Jc and n-value. This is due to the improvement of packing density and grain alignment. From the fact that higher Jc-value is obtained for the sample which contains low-Tc phase at sintering process, it is considered that low-Tc phase proceeds good contact between grains, when it is transformed to the high-Tc phase.

REFERENCES

1) S.Green, C.Jiang, Y.Mei, H.L.Luo, C.Politis: Phys. Rev. B 38(1988)5016.
2) M.Takano, J.Takada, K.Oda, H.Kitaguchi, Y.Miura, Y.Ikeda, Y.Tomii and H.Mazaki: Jpn. J. Appl. Phys. 27(1988)L1041.
3) K.Heine, J.Tenbrink, M.Thöner: to be published in Appl. Phys. Lett.
4) T.Hikata, K.Sato and H.Hitotsuyanagi: Jpn. J. Appl. Phys. 28(1988)L82.
5) P.W.Anderson: Phys. Rev. Lett. 9(1962)309.

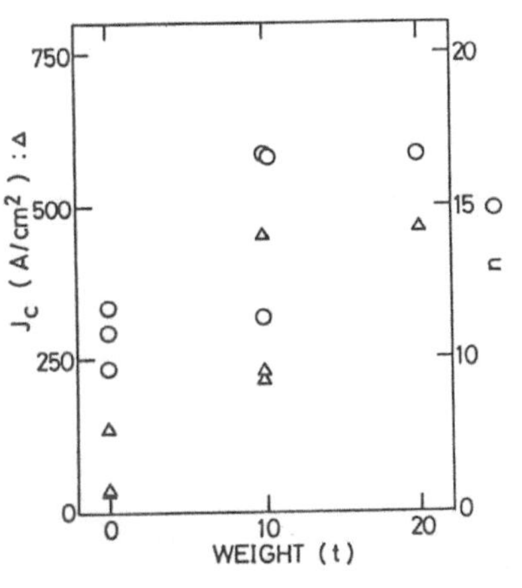

Fig.1 Stress dependence of Jc and n-value measured for samples without sheath.

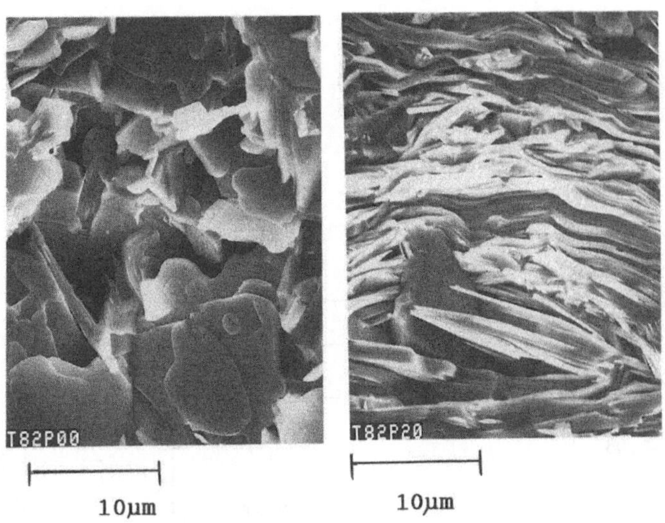

Fig.2(a) Fig.2(b)
The SEM photographs taken on the fractured surface.(a);without stress (b);stress applied

Table 1. Pinning potential at 4.2K ($\times 10^{-2}$eV)

H	0.2T	0.5T	1.0T
H ⊥ tape surface	1.9	1.7	1.6
H // tape surface	4.0	2.3	1.9

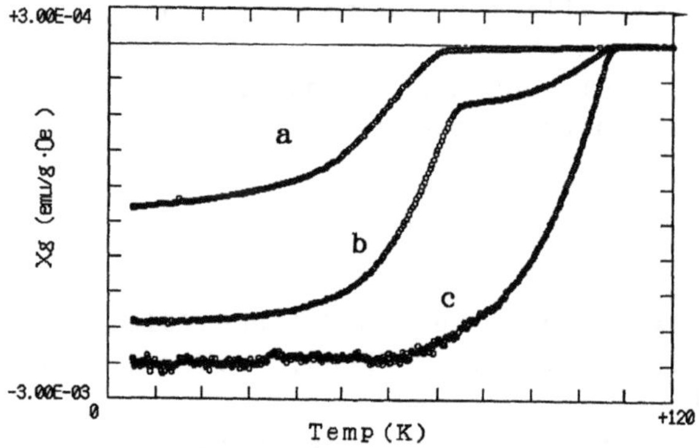

Fig.3 Increase of high-Tc phase by press and sintering process.
a:Magnetization curve of as packed sample.
b:Magnetization curve after first press and sintering. (Jc=930A/cm^2;77K,0T)
c:Magnetization curve after three times repeating of press and sintering process. (Jc=6200A/cm^2;77K,0T)

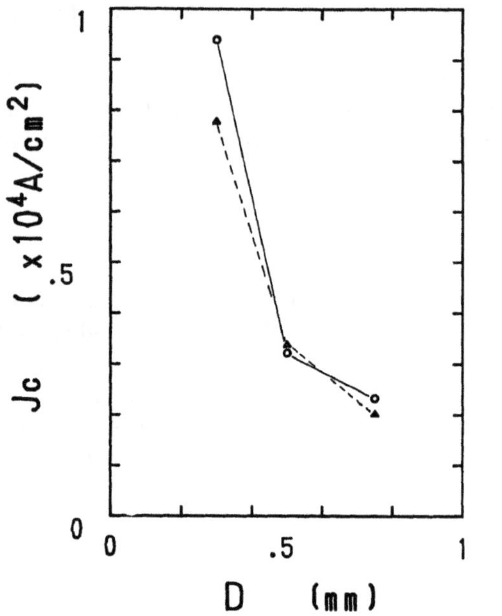

Fig.4 The Jc(77K,0T) versus wire diameter relation. Smaller diameter corresponds to higher uniaxial stress.

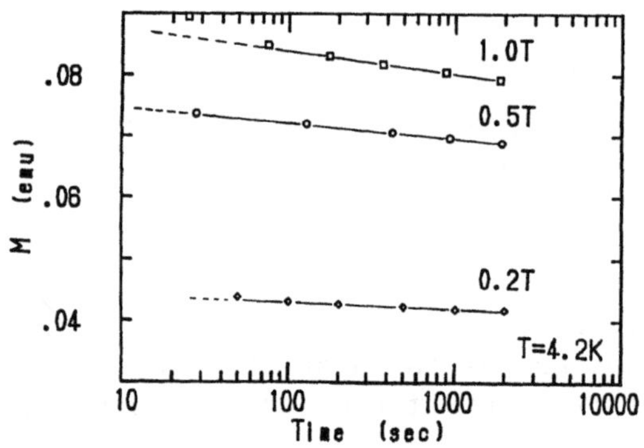

Fig.5 Relaxation of remanent magnetization at 4.2K.

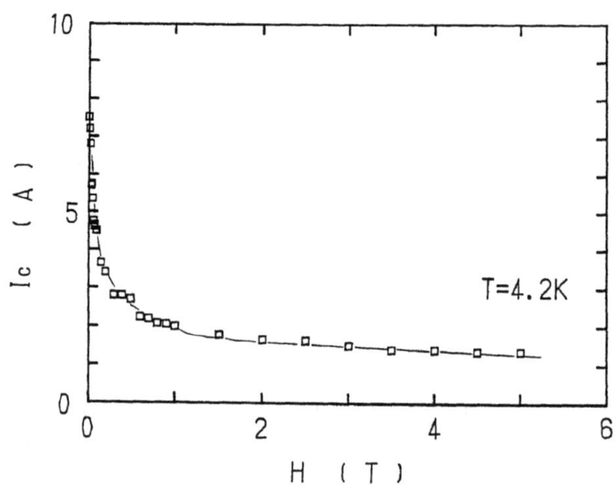

Fig.6 Ic versus H curves measured at 4.2K for BPSCCO tape.

High-T_c Superconducting Bi-Pb-Sr-Ca-Cu-O and Tl-Ca-Ba-Cu-O Filaments Produced by the Suspension Spinning Method

T. Goto and T. Maruyama

Department of Materials Science & Engineering, Nagoya Institute of Technology, Gokiso-cho, Showa-ku, Nagoya, 466 Japan

ABSTRACT

High Tc superconducting Bi-Pb-Sr-Ca-Cu-O and Tl-Ca-Ba-Cu-O filaments were successfully prepared using a combined technique of suspension spinning and densification of the pyrolyzed filaments by pressing and sintering. A zero resistance state was achieved at 104 K and 98 K and the maximum Jc at 77 K, 0 T was 1940 A/cm^2 and 1045 A/cm^2, for Bi and Tl oxide filaments, respectively. The Bi oxide filament consisted of a single high-Tc phase and the grains were oriented with the c-axis perpendicular to the longitudinal direction of the filament. However, the Tl oxide filament was a low-Tc phase.

KEYWORDS: Bi-Pb-Sr-Ca-Cu-O superconducting filament, Tl-Ba-Ca-Cu-O superconducting filament, Suspension spinning, densification, high-Tc phase,

INTRODUCTION

An application of a high-Tc oxide superconductor for the superconducting magnets requires the fabrication of the brittle ceramic materials into tapes or wires with high Jc. We have studied the preparation of the oxide superconducting filament using a suspension spinning method and a high Jc value of more than 1000 A/cm^2 at 77 K, 0 T was attained for $Y_1Ba_2Cu_3O_{7-\delta}$ filament by this method. [1] A Tl oxide superconducting filament was also examined by this method using the precursor of the oxide. [2] However, Tl_2O_3 is chemically instable at the synthesis temperature and the preparation of Tl-Ca-Ba-Cu-O superconducting filament by this method was not easy because of the side reaction with PVA and volatility from large surface area of the filament. The Tl-Ca-Ba-Cu-O filament exhibited the best superconducting properties with an onset temperature as high as 90 K and zero resistivity at 40 K. Recently a high-Tc phase Bi-Pb-Sr-Ca-Cu-O superconducting filament was successfully prepared by using a technique in which suspension spun and pyrolyzed filaments were densified by pressing and sintering. [3] This paper describes Bi-Pb-Sr-Ca-Cu-O and Tl-Ca-Ba-Cu-O superconducting filaments produced by a combined technique of suspension spinning and densification of the pyrolyzed filament by pressing and sintering.

EXPERIMENTAL

Appropriate amounts of Bi_2O_3, PbO, $SrCO_3$, $CaCO_3$ and CuO powders with more than 3 N purity were mixed into the composition of $Bi_{0.86}Pb_{0.24}Sr_1Ca_{1.1}Cu_{1.8}Ox$, calcined at 1073 K for 15 h and pressed into pellets, then sintered at 1123 K for 120 h in air to form the high-Tc phase of the Bi system. The resultant pellet was milled into a fine powder and was suspended in a mixed PVA solution of dimethyl sulfoxide and hexamethylphosphoric triamide. The viscous suspension was extruded as a filament into a precipitating medium of methyl alcohol and coiled on a winding drum. The as-drawn filament of 250 μm in diameter was cold pressed at 200 kg/cm^2 and heated at 773 K for 1h to remove volatile component and was then subjected to sintering under various heating conditions.

$Tl_2Ca_2Ba_2Cu_3Ox$ were prepared from a mixture of Ba-Cu-O, Tl_2O_3 and CaO. Appropriate amounts of Tl_2O_3, CaO and $Ba_2Cu_3O_4$ were completely mixed, ground and pressed into a pellet. Then the pellet was put into a tube furnace, which had been heated at 1133 K and was heated for 600 s in flowing oxygen. The sample was then furnace cooled to room temperature. These procedures were repeated three times. The resultant pellet was milled into a fine powder and the suspension spinning of the powder was made as same as for the Bi system. The as-drawn filament of 250 μm in diameter was heated in an alumina box at 773 K for 1800 s to remove volatile component. The filament was then cold-pressed at 200 kg/cm^2 and put into an alumina crucible with a silver cup on it. The crucible was then put into a furnace kept at various temperatures. After heating for 600 s in flowing oxygen, the filament was subjected to furnace cool. The electrical resistivity (ρ) of the filament heated was measured by a standard four-probe method. Silver paint was used to connect the filament with Ag electrodes of 50 μm in diameter. The specimen temperature was measured using a calibrated chromel-gold + 0.07 % iron thermocouple. The transport Jc measurement was performed at 77 K, 0 T with an criterion of 1 μV/cm.

RESULTS AND DISCUSSION

Bi-Pb-Sr-Ca-Cu-O Filaments

The starting powder of Bi system had a single high-Tc phase but the structure of the filament obtained was dependent on the pyrolyzed condition. The superconducting phase was decomposited if the filament was pyrolyzed at a heating rate more than 0.5 K/s. The pyrolyzed filament had usually mixed structure of high-Tc phase and low-Tc phase and the proportion of the high Tc phase in the filament obtained was dependent on the pyrolyzed condition. A single phase was obtained for the filament cold pressed at 200 kg/cm^2 and heated at 773 K by heating rate of less than 2.5 K/s and hold for 1 h. The pyrolyzed filament was cold pressed and sintered at 1103 K under the oxygen pressure of 1/13 atm in argon gas and was then cooled to room temperature. The Jc of the filament was enhanced by repeatly pressing and sintering as shown in Fig.1. The maximum Jc at 77 K reached 1940 A/cm^2 and zero electrical resistivity was achieved at 104 K. for the filament sintered four times.

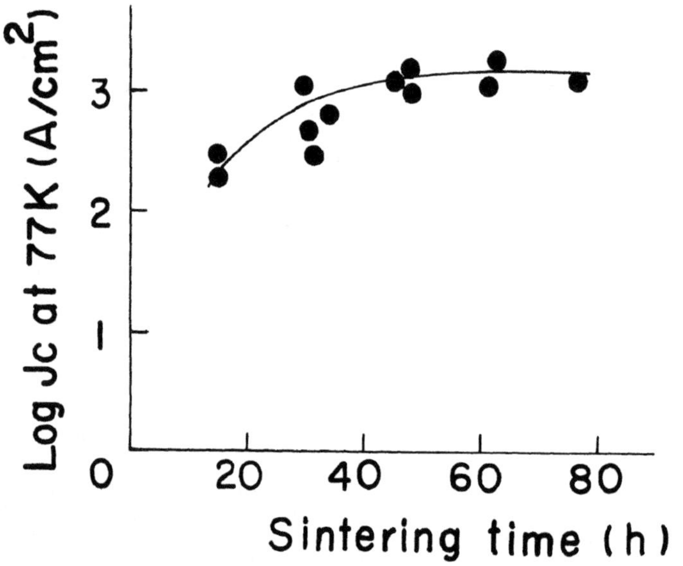

Fig.1 The Jc of the Bi oxide filaments sintered at various times.

X-ray diffraction pattern of the filament with high Jc was examined. X-ray was applied perpendicular to the wide plane in the longitudinal direction. The series of the strong peaks was indexed at (00ℓ) of the high-Tc phase having a tetragonal unit cell with a=b=0.540 nm and c=3.72 nm belonginh to the P4/mmm space group. The grains of the filament were found to be oriented with the c-axis perpendicular to the longitudinal direction.

Tl-Ca-Ba-Cu-O Filaments

The starting powders of the Tl system oxide had a single-low Tc phase and exhibited superconducting at zero resistivity temperature (Tc) of 97 K. The Tl system filament was prepared as same as for the Bi system filament. The pyrolyzed filament was sintered at various temperatures for 600 s after cold press and the electrical resistivity of the filament sintered was measured. Figure 2 shows the temperature dependence of the electrical resistivity for the filaments. The filament exhibited superconductivity at more than 77 K. The Tc, Jc at 77 K, 0 T and ρ value at 150 K of the filaments sintered at various temperatures are listed in Table 1. The Tc increased with increasing the sintering temperature ranging from 1133 K to 1183 K. The filaments sintered at 1183 K for 600 s exhibite high Tc value of 98 K and high Jc value of 1045 A/cm². The densification process was repeated, however, the enhancement of Jc was not detected.

Table 1. Tc, Jc at 77 K and ρ at 150 K of the Tl oxide filaments pressed and then sintered at various temperatures for 600 s after heat treatment at 773 K for 1800 s.

Number	Temperature (K)	Tc (K)	Jc at 77 K (A/cm²)	ρ at 150 K (mΩ · cm)
1	1133	80	0.001	0.84
2	1173	92	42	1.12
3	1183	98	1045	1.08

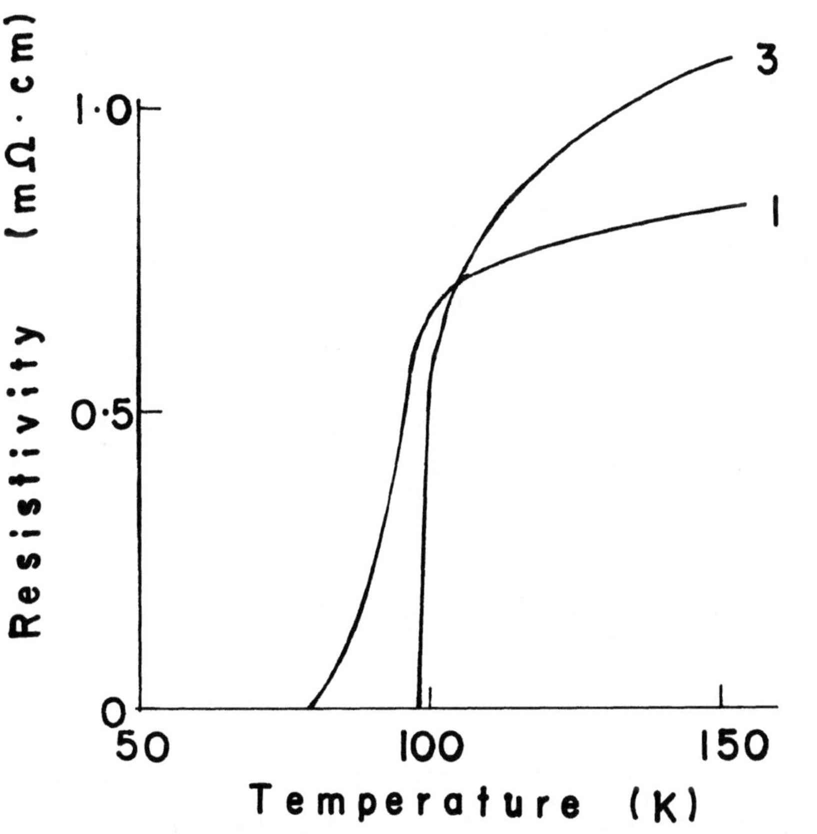

Fig.2. The temperature dependence of the electrical resistivity for the Tl oxide filament sintered at various temperature for 600 s after cold press.
1 sintered at 1133 K 3 sintered at 1183 K

X-ray diffraction pattern of the filament using Cu Kα radiation is shown in Fig.3. The X-ray was applied perpendicular to the wide plane of the longitudinal direction. The series of these strong peaks are indexed at that of the low-Tc phase having a tetragonal unit cell with a=b= 0.385 nm and c=3.561 nm, which is indexed in Fig.3. Slight impurities such as CuO and Ca_2CuO_3 was also observed. Thus the filament produced by this method maintained the structure of starting oxide. The prefered orientation of the filament was not found out.

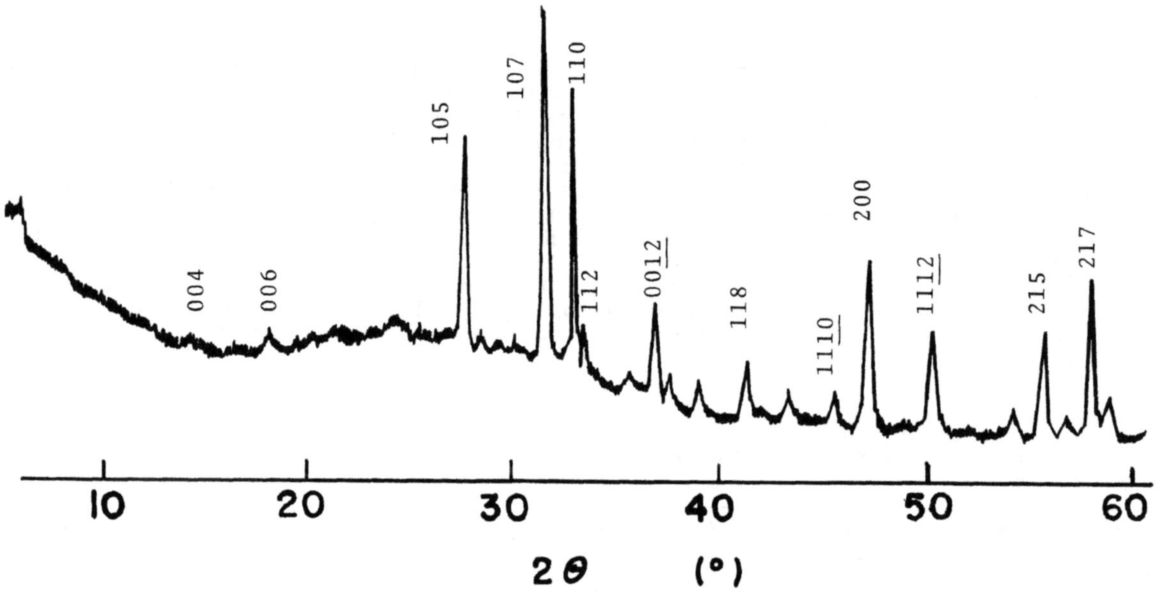

Fig.3. X-ray diffraction pattern of the Tl oxide filaments using Cu Kα radiation. The X-ray was applied perpendicular to the wide plane of the longitudinal direction.

It has frequently been reported that, when the Tl-Ca-Ba-Cu-O compounds are prepared by directly calcining powder oxide, it is often necessary to use a starting mixture with a stoichiometry different from that of the compound to be prepared so as to obtain an appreciable amount of that compound. For example, an oxide mixture with a stoichiometry of $Tl_2Ca_2Ba_2Cu_3$ was often used for preparing the 2122 compound, while a mixture of $Tl_1Ca_3Ba_1Cu_3$ was used for the 2223 compound. [4] Further optimization of the fabrication process for high-Tc phase Tl system filament is now in progress.

In conclusion, Bi-Pb-Sr-Ca-Cu-O and Tl-Ca-Ba-Cu-O filaments having zero resistivity temperature as high as 104 K and 98 K, respectively, were prepared using a combined technique of suspension spinning and densification by pressing and sintering. The highest Jc obtained so far was 1940 A/cm² and 1045 A/cm² at 77 K, 0 T for the Bi and Tl oxide filaments, respectively.

ACKNOWLEGEEMENT

This work was partly supported by a Grant-in-Aid for Special Project Research for the Ministry of Education, Science and Culture. (No.01645511)

REFERENCES

1. Goto T. (to be published) Fine Y-Ba-Cu-O superconducting filaments produced by suspension spinning method. In: Proceeding of M²S-HTSC (Stanford) 23-28 July 1989. to be published in Physica C.

2. Goto T. Inaji H. Takeuch K. Yamada Y. (1989) Preparation of high Tc oxide superconducting filaments by suspension spinning method. In: Kitazawa and Ishiguro (eds) Advances in Superconducting. Spring-Verlag Tokyo. pp 353-358 (Proceedings of ISS' 88)

3. Goto T. (1989) Preparation of high-Tc superconducting Bi-Pb-Sr-Ca-Cu-O filament by the suspension spinning method. Jpn.J.Appl.Phys. 28: L1402-L1404

4. Wu N. Lee S. Yao YD. (1989) New reaction routes for preparing single phase $Tl_2CaBa_2Cu_2O_8$ and $Tl_2Ca_2Ba_2Cu_3O_{10}$ powders from stoichiometric mixture. Jpn.J.Appl.Phys. 28: L-1349-L1351

Transport Property and Microstructure of Silver-Sheathed Bi-Based Superconductors with High Degree of Crystal Orientation

NORITSUGU ENOMOTO[1], HIROYUKI KIKUCHI[1], MASANAONO MIMURA[1], MINORU NAKAJIMA[1], NAOKI UNO[1], TSUKUSHI HARA[2], KIYOSHI OKANIWA[2], and TAKAHIKO YAMAMOTO[2]

[1] Yokohama R & D Labolatories, The Furukawa Electric Co., Ltd., 4-3, Okano 2-chome, Yokohama, 220 Japan

[2] Engineering Research Center, Tokyo Electric Power Co., 4-1, Nishi-Tsutsujigaoka 2-chome, Choufu, Tokyo, 182 Japan

ABSTRACT

Critical current density (Jc) and its dependence on applied magnetic field was improved by alignment and homogenization of the low-Tc phase (90K). It was found that transport Jc of wire specimens was strongly and positively related to degree of c-axis orientation (F) measured by XRD and that pursuing higher F value up to around 99%, Jc of $3.5x10^4 A/cm^2$ was achieved at 77K and 0T. This result shows that grain alignment can supress weak inter-granular coupling.

KEY WORDS: critical current density, Ag-sheathed superconducting tape, $Bi_2Sr_2CaCu_2O_y$, grain alignment

INTRODUCTION

Since the discovery of a new high-Tc superconductor, Bi-Sr-Ca-Cu-O system by Maeda et al. [1], metal-sheathed or powder-in-tube technique has been applied for this material as well as Y-Ba-Cu-O system in course of searching a route for wire conductor in length [2]. This material is easier to obtain laminar structure aligned among grains and showed better Jc-B property in comparison with Y-system. Currently most workers have been concerned with the so-called higher-Tc phase (110K) of $(Bi,Pb)_2Sr_2Ca_2Cu_3O_y$. However, the lower-Tc phase (90K) is thermo-dynamically more stable than the higher-Tc phase, making it easier to form single phase, and therefore higher Jc might be expected along with due to improvement of crystal orientation [3].

In this report, the relationship between critical current density (Jc) and degree of c-axis orientation (F) measured by XRD was studied for Ag-sheathed tape wire of the low-Tc phase. In addition to observation of microstructure, activation energy for flux lines Uo with different F was evaluated.

EXPERIMENTAL

A mixture of Bi_2O_3, $SrCO_3$, $CaCO_3$, and CuO powders in a nominal composition (Bi:Sr:Ca:Cu=2:2:1:2) was calcined at 800°C for 30h and then ground into powder. The powder put into silver tubes, which were swaged and rolled into tapes with thickness 0.5-0.1 mm. The tapes were sintered at 800-900°C for 10 -50h under oxygen atmosphere.
The measurement of critical temperature (Tc) and critical current density (Jc) were carried out by DC susceptivility method (SQUID) and the 4-probe DC method, respectively. X-ray diffraction analysis using Cu-target was performed on the wide surface of each tape specimen after Ag-sheathes were stripped. A microstructure of wire was examined by scanning electron microscopy (SEM) and transmission electron microscopy (TEM). A magnetization at 77K was measured by using a vibrating sample magnetometer (VSM). The degree of c-axis orientation by XRD (F) was calculated by both summation of X-ray intensities, $\Sigma I(001)$ and $\Sigma I(hkl)$ as follows:

$$P=\frac{\Sigma I(001)}{\Sigma I(hkl)}$$

$$F=\frac{P_0 - P_{00}}{1- P_{00}}$$

P_{00}:X-ray intensity ratio of non oriented specimen
P_0 :X-ray intensity ratio of oriented specimen

RESULT AND DISCUSSION

The critical temperature (Tc) of wire specimen was about 93K. The X-ray diffraction pattern for wire specimen showed the series of peaks of the low-Tc phase (hkl). About these wire specimens, the experimental relation between critical current densities (Jc) and degree of c-axis orientation (F) were examined at 0T, 77K. The influence of c-axis orientation ratio (F) to the Jc is shown in Fig.1. The Jc are dependent on the F over the wide region and increased steeply with F. As thickness of the tape is reduced, the Jc per F increased more steeply. The maximum Jc was obtained $3.5x10^4 A/cm^2$ around 99% F.

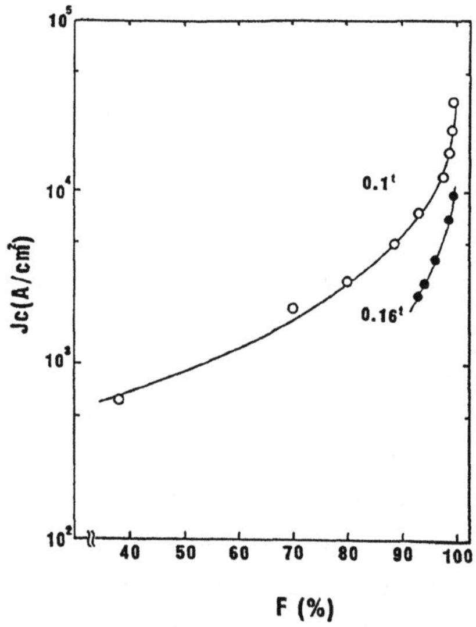

Fig.1 The influence of c-axis orientation ratio(F) to the critical current density for Ag-sheathed $Bi_2Sr_2CaCu_2O_y$ tapes at 77K

at 100 Oe and 77K is calculated as 0.087eV, 0.029eV respectively. This indicate that the activation energy Uo increases about three times in consequence of improvement of weak link at grain boundaries.

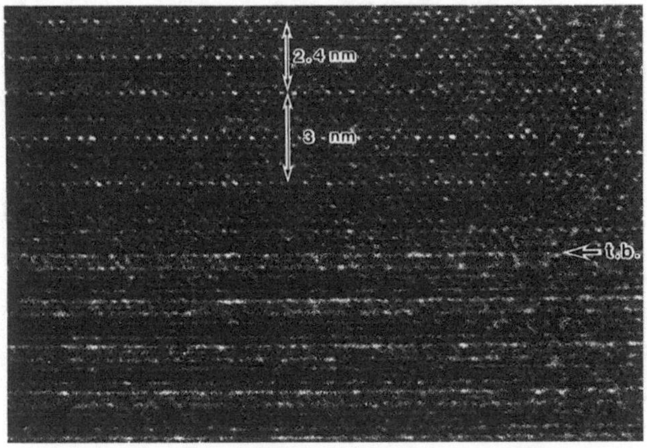

Fig.4 A high resolution image of the cross section of a Ag-sheathed $Bi_2Sr_2CaCu_2O_y$ tape. 't.b.' indicates a twist boundary. The incident beam is parallel to [110] of the upper grain.

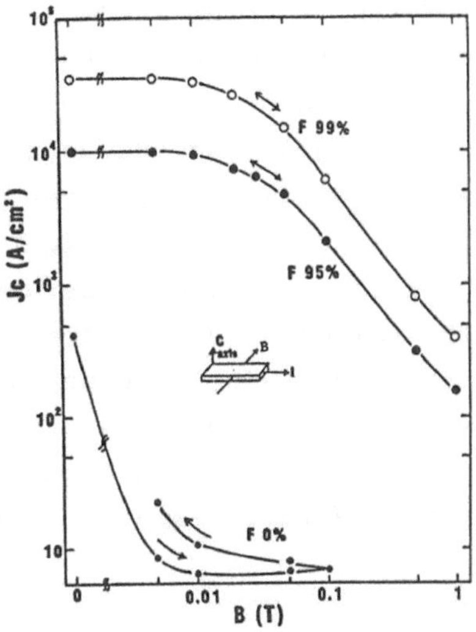

Fig.5 The magnetic field dependence of critical current density for Ag-sheathed $Bi_2Sr_2CaCu_2O_y$ tapes at 77K

CONCLUSION

It was found that the zero field Jc of wire specimen is storongly and positively related to degree of c-axis orientation (F) messured by XRD and that Jc of $3.5x10^4 A/cm^2$ was achieved at 77K and 0T by pursuing higher F value up to around 99%. This result shows that grain alignment can supress weak intergranular coupling and that microscopic measures of structural controlling should be required for the improvement of intagranular property.

REFERENCE

1. H. Maeda, Y. Tanaka, M. Fukutomi and T. Asano:Jpn.J.Appl Phys., Vol.27 (1988) L209.
2. T. Hikata, K. Sato and H. Hitotsuyanagi:Jpn.J.Appl.Phys., Vol.28 (1989) L82
3. N.Enomoto, H. Kikuchi, N. Uno, M. Ikeda, H. Kumakura, K. Togano and K. Watanabe:11th International Conference on Magnet Technology,Tsukuba, Japan, 1989
4. H. Kumakura, K. togano, E. Yanagisawa, K. Takahashi, M. Nakao and H, Maeda:Jpn.J.Appl.Phys., Vol.28 (1989) L24

SEM photograph of wire specimen with 95% F is shown in Fig.2. This indicates that $Bi_2Sr_2CaCu_2O_y$ oxide superconductor is composed of laminar grains well aligned in the same direction. The grain sizes are of the order of 100um in length. An electron diffraction pattern of the cross section of this wire specimen is shown in Fig.3. The incident beam is parallel to [110] of the central grain. In this figure, (a) is bright field image with a selector aperture and (b) is the selected area diffraction pattern. The twist boundary is observed between grains. This indicates that grains of low-Tc phase inside the Ag-sheathed tape consist of c-axis oriented texture perpendicular to the wide plane of wire specimen.

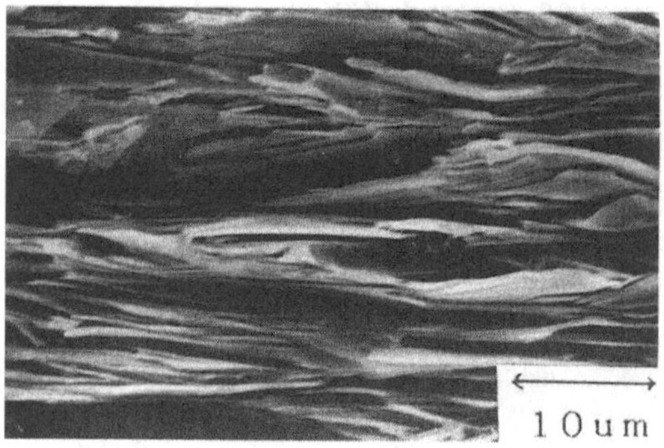

Fig.2 Fracture surface of $Bi_2Sr_2CaCu_2O_y$ oxide superconductor inside Ag-sheathed wire

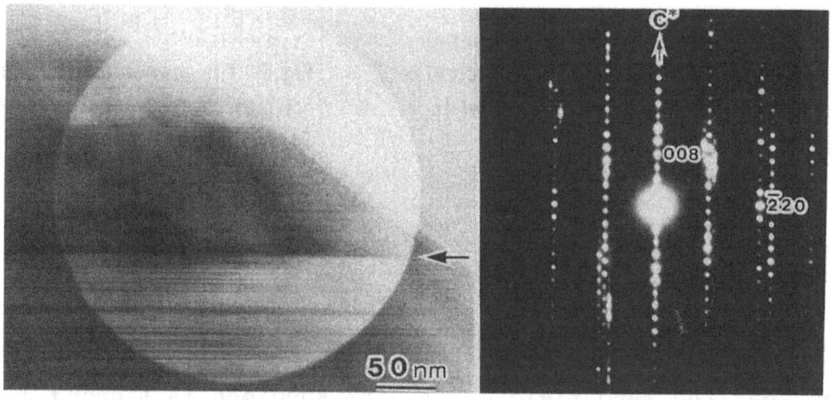

Fig.3 A cross section of a Ag-sheathed $Bi_2Sr_2CaCu_2O_y$ tape. (a)B.F.I. with a selector aperture, (b)S.A.D. The arrow indicates a twist boundary. The incident beam is parallel to [110] of the central grain.

A high resolution image of this specimen is shown in Fig.4. The condition of incident beam is the same mentioned above. Although the intergrowth is observed in the upper grain, any impurity is not recognized in the twist boundary. The grain growth spreads along ab plane and coincides with the current flow direction. The magnetic field dependence of Jc for wire specimen with different F is shown in Fig.5. The magnetic field was applied parallel to the ab plane. The Jc inthe magnetic field was improved greatly when Jc was increased in zero field by pursuing higher F up to 99%. In this case the hysteresis caused mainly by the weak link can not be recognized on Jc-B curve and the Jc could be gradient about $1.5x10^4 A/cm^2$ around 0.05T. However, the Jc decreased steeply beyond 0.1T at 77K. As to the nature of weak coupling, the activation energy was estimated by the magnetic relaxation [4]. The activation energy for flux lines Uo(F=99%) and Uo(F=0%)

2.6 Thick Films and Tapes

Highly Oriented YBaCu-Ceramic Layers Through Reactive Sintering

F. GREUTER, C. SCHÜLER, P. KLUGE-WEISS, and W. PAUL
Asea Brown Boveri, Corporate Research, CH-5405 Baden-Dättwil, Switzerland

A B S T R A C T

Thin layers of $Y_1Ba_2Cu_3O_{7-x}$ with a high degree of c-axis texturing have been prepared by reactive sintering of sub-micron sized powder particles, densely packed on Ag-foils. The processes of reaction, nucleation and growth are briefly discussed. At present the critical current density j_c of the layers is ~5000-8000A/cm^2 (77K, H=0) when measured over a few centimeters length. Locally, j_c might still be higher.

KEY WORDS: superconductors, tape, texture, critical current density.

INTRODUCTION

For sintered bulk samples of the high-T_c compound $Y_1Ba_2Cu_3O_{7-x}$, the current-carrying capacity j_c consistently remains at a low level despite extensive research. The j_c-values in good bulk ceramics converge towards ~1000A/cm^2 (77K, 0T, self-field corresponding to 1x1mm^2 cross section [1]), largely independent on the processing route. This limitation seems to be an inherent property of the polycrystalline material. Generally, the high anisotropy in the crystallo-graphic and electrical properties and the weak link nature of the grain boundaries are thought to be responsible for the low j_c-values [1]. Weak links may be caused by impurity accumulation, deviations from stoichiometry or structural distortions at the grain boundary on a scale comparable to or larger than the very short coherence length ξ. As ξ is at best a few nm, proper control of the microstructure at this level turns out to be very difficult. The relative importance of the anisotropy and of the weak links is hard to disentangle. Both effects might be strongly coupled through the crystallography (and the related chemistry) of the interfaces. It was soon realized that the introduction of a texture in the microstructure is an important, although probably not the only, prerequisite for improving the critical current density j_c.
In the present paper on thin ceramic layers of $Y_1Ba_2Cu_3O_{7-x}$, we report on a pronounced c-axis texturing achieved through a *one-step, reactive sintering process*. By controlling the steps *reaction-nucleation-growth,* a few µm-thick, dense layers can be fabricated on polycrystalline Ag-substrates. We briefly describe the preparation of the ceramic tapes, discuss the oriented nucleation and growth processes and conclude with first j_c-measurements, which show values significantly higher than unoriented bulk materials.

PREPARATION OF CERAMIC LAYERS

Sub-micron sized precursor powders were first prepared by stoichiometric coprecipitation of the oxalates [2]. The amorphous oxalate powder is then decomposed to an intimate mixture of Y_2O_3, $BaCO_3$ and CuO by a heat treatment at

~580°C. Next, a stable organic suspension is prepared by addition of isopropyl alcohol and a small amount (<1 w/o) of a dispersant, followed by a gentle roll-milling to break up most of the agglomerates and by classification to a particle size well below 1 µm. The particles all contain the three elements Y,Ba,Cu with an average concentration ratio of 1:2:3, but show some degree of variation from particle to particle, as verified by STEM/EDX [2]. A well dispersed state for the suspension is essential to the next processing step, which is the formation of a densely packed green layer. Either centrifugal casting or dip-coating were used, with polycrystalline Ag-foils of 50 µm thickness as the substrate. Typical sample dimensions were ~1-2cm^2 or 30-50mm x 5-20mm, respectively. The average layer thickness of the superconductor is controlled by the coverage in the green state and typically was ~2-4µm. Inhomogeneities in the deposited layer thickness, e.g. caused by edge effects, of course translate to local variations of the layer-thickness after sintering. Fig. 1a shows the packing of the powder particles in the green state. A high green density is important for a crack-free sintering of the layer, which is mechanically constrained by the substrate. The high sinter activity of the present powder seems to be favorable for the observed *one-dimensional shrinkage* [3] of the thin layers, which show no lateral contraction. The sintering cycle is schematically shown in Fig. 2. It consists of a one or two step heating to a temperature of ~920-930°C in oxygen, a holding time of ~1-3 hours, followed by cooling and oxygen equilibration. The exact heating schedule is not critical at this stage, except for the top temperature, as we will see below. Silver as a substrate has proved to be unique for the present process, as it is remarkably inert even for few µm thin layers.

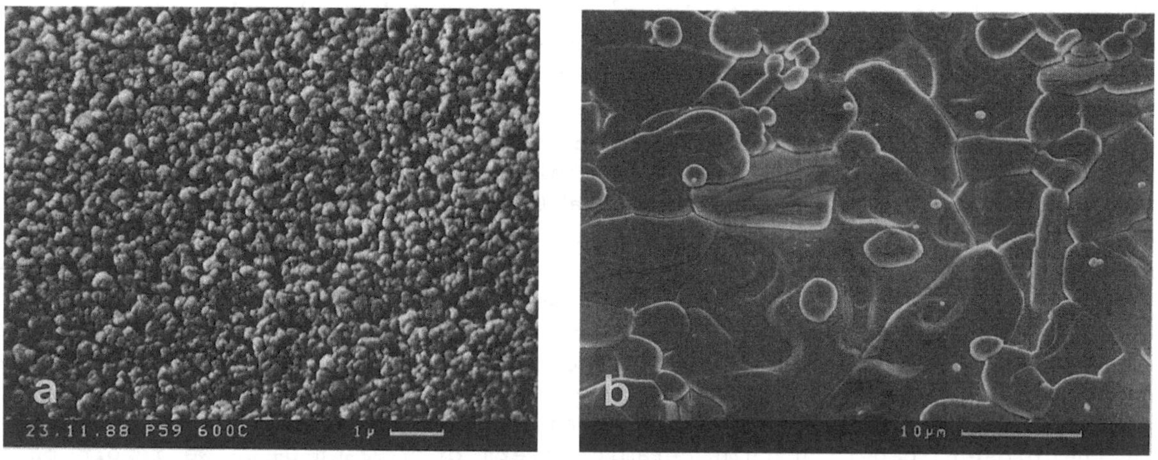

Fig. 1: a) SEM picture of a layer in the green state (centrifugal casting). b) Microstructure of a c-axis textured $Y_1Ba_2Cu_3O_{7-x}$ ceramic layer of ~2µm thickness. Pyramidal growth and a few grains with different orientation are seen.

DEVELOPMENT OF THE MICROSTRUCTURE

Fig. 2 describes the evolution of the microstructure during the sintering process. It can be devided into three sequences: i) reaction of the components Y_2O_3, $BaCO_3$ and CuO to the compound $Y_1Ba_2Cu_3O_{7-x}$, ii) formation of properly oriented nuclei and iii) growth to the final crystal size. For the thin layers, the decomposition of the $BaCO_3$ to BaO starts around ~670°C and is completed already at 850°C, as seen from the TGA-curve in Fig. 2 (heating rate ~150°C/h).

Quenching the sample at 850°C followed by a XRD-analysis shows that the powder has reacted (within XRD limits) to the pseudocubic a≈b≈c/3 phase of YBaCu, reported previously [2,4]. No growth of the particle size can be seen at this stage. Also, the same XRD-spectra are obtained for the top of the layer as well as after "polishing-off" most of the material, leaving only the particles close to the Ag-ceramic interface (= "bottom").
Interrupting the sintering cycle at a slightly higher temperature of 890°C (heating rate ~15°C/h) presents a completely different picture. The onset of grain growth is clearly detected and XRD of the top of the layer shows the transformation to the orthorhombic phase. The XRD-spectrum is characteristic for a *random* orientation of the crystallites probed near the top surface of the layer. However, analyzing the crystal orientation near the Ag-substrate reveals that this part of the layer has already a clear *preferential* alignment of the c-axes perpendicular to the substrate (dominance of (oon) reflexes relative to (013,103,110)). The XRD depth-profiling by sequentially removing parts of the layer is only a qualitative measure for the actual degree of ordering, as a controlled polishing is difficult. The grain size near the interface seems to be somewhat larger than at the top of the layer.

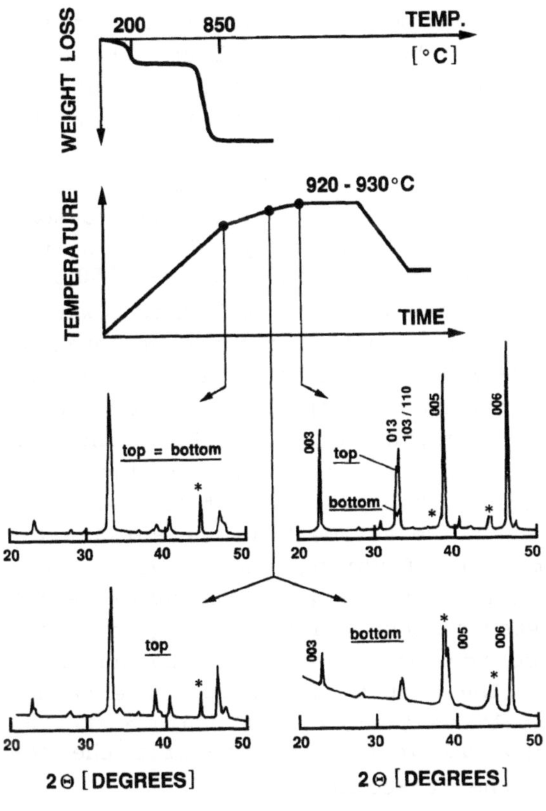

Fig. 2: Weight loss, firing schedule and XRD spectra at different times of the reactive sintering process; top = free surface, bottom = near interface to Ag-substrate, * = Ag.

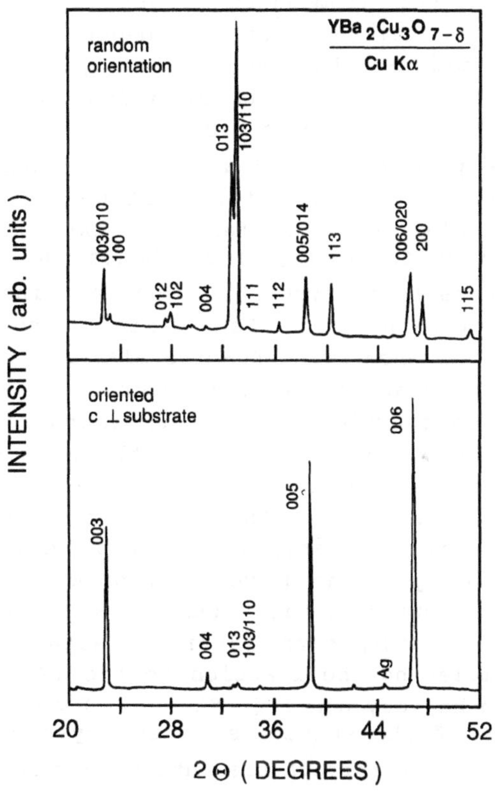

Fig. 3: Comparison of the XRD-spectra of a randomly oriented, polycrystalline bulk sample with a c-axis textured ceramic layer.

At the point where we reach the plateau temperature in Fig. 2, grain growth has further increased, both at the top and the bottom of the layer. Now also the surface shows a texturing, although the degree of orientation is still higher near the substrate than on the free surface. Holding the sintering temperature for a while increases the grain growth and improves the texture. Fig. 3 shows the XRD-diagrams that can be obtained after the complete sintering cycle. If we take the intensity ratio I(006,020)/(103,110) as a measure for the texturing, then enhancements up to a factor of ~ 700 have been obtained in comparison to the random bulk orientation. Good layers are almost completely dense with only a few isolated, µm-sized spots of free Ag-surface. They are brittle but free of cracks, when handled with care, adhere extremely well to the substrate and have a gray-black lustre. In an optical microscope, polished sections of the layers nicely show the orthogonal arrays of twin domains when viewed along the c-axes. Within the layer the a-b-axes are distributed at random. Occasionally some of the grains are found to have a different orientation. The SEM picture in Fig. 1b gives an impression of the morphology of the layers. The crystallites show a pronounced growth of terraces, defined by the a-b planes. This pyramidal growth makes it difficult to determine the effective cross section for the current transport. The easy growth directions for $Y_1Ba_2Cu_3O_{7-x}$ are in the a-b plane, which explains the platelet-like morphology and the appearance of terraces. Reactive sintering seems to promote the platelet-growth, as we have seen for bulk samples, when compared to pellets prepared by grinding and calcination [2]. From Fig. 2 we conclude that the nucleation of the properly aligned grains starts at the interface to the Ag-substrate, from where on the "oriented growth front" moves up through the layer. If the thickness of the green layer is increased beyond several µm, the one-dimensional shrinkage gradually converts to a normal 3-dimensional contraction. Hence, the layer will loose its compactness, surface roughening increases and the surface texturing decreases.

The process of nucleation and growth at the interface to the Ag-substrate is of special interest. At present we can only mention a few observations and add some speculations: the sintering temperature is close to the melting point of silver (~939°C for 1 atm O_2) and hence the Ag-atoms become quite mobile. Above ~850°C Ag starts to recrystallize and the final grain size can be several 100µm. However, the growth of the YBaCu grains is not related to the crystallographic orientation of the underlying Ag-crystals. All of the 123-grains in contact with silver are lying flat on the "macroscopic" Ag-surface, instead of adopting the microscopic reconstruction of the Ag introduced by the thermal etching. Also the a-b orientations of nuclei on the same Ag-crystallite are completely unrelated and show no signs of "epitaxy". The 123-grains ignore the presence of the Ag grain boundaries, which they easily cross during growth. We assume that the surface (or interface) energy is the main driving force for favoring the growth of the properly oriented nuclei. From the observed nucleation and growth behavior we get the impression that these processes are assisted at some point by a liquid phase, which disappears later on either upon cooling or when it is consumed by completing the reaction to the 123-compound. The amount of liquid phase needed can be very small (< XRD-limit) and is difficult to identify. It could either be one of the eutectics in the system Y_2O_3-BaO-CuO or be provided from the Ag-substrate through the interface reaction with the 123-components.

SUPERCONDUCTING PROPERTIES

The measurement of the electrical properties of the thin ceramic layers is complicated by the parallel connection of the well conducting Ag-substrate. However, below T_c the current is almost completely taken up by the superconductor. Now the Ag-substrate is of advantage as it provides a good electrical contact to the ceramic. In the standard four point arrangement, the transition from the normal to the superconducting state is observed at ~90K, with a transition width <3K. The critical current density j_c was measured with dc on ~30mm long strips of ~3-5mm width and a separation of ~10-20mm for the voltage probes. This arrangement samples several hundred grain boundaries in series and a few hundred in parallel to each other. In Fig. 4 we show the I-V characteristic at 77K and zero field. The samples are free standing 50µm Ag tapes obtained by dip-coating. The layers have a *nominal* coating of ~2µm on both sides, as calculated from the coverage in the green state. There are two major sources for overestimating this thickness: i) the coating process on the lab scale is not well optimized and inhomogeneities in the green density exist, mainly towards the ends of the tapes; ii) as the material shows a pyramidal growth, the real contact area between two grains is lower than estimated from the assumption of flat crystals. If we take 1µV/cm as the simplest criterion for the critical current density, then we find $j_c \approx 6\text{-}7000 A/cm^2$ for the *nominal* thickness of the samples in Fig. 4. As discussed above, this represents a rather conservative value for j_c. Based on a cross sectional analysis we estimate that the *effective average* j_c could well be 20-30% higher. Locally, still higher values can be expected [5]. In absolute numbers, the critical current of the tapes in Fig. 4 is ~1A at a width of ~4mm. For a comparison we also show the I-V characteristic of the Ag-foil alone at 77K. It is evident that in the superconducting state almost all of the current is taken up by the ceramic layer.

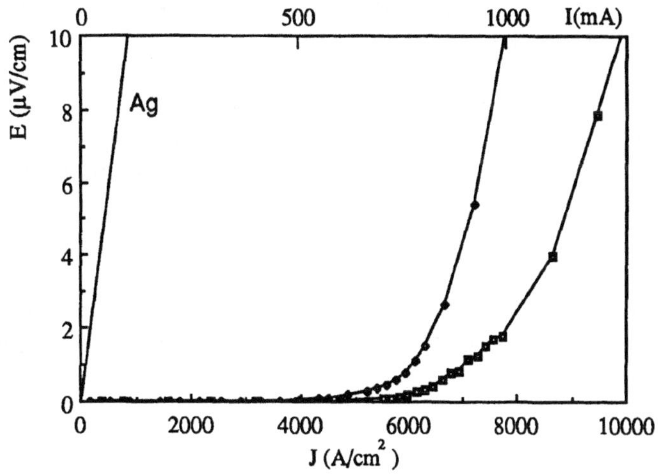

Fig. 4: I-V characteristic of c-axis textured ceramic layers from two different batches (77K, H=0). The current density J is based on the nominal layer thickness and represents a lower limit. For comparison an identical Ag-tape at 77K is shown (versus mA-axis).

The above j_c-values are about a factor of 10 higher than typical values for polycrystalline bulk ceramics. Similar to the bulk material, the current has to cross a huge number of grain boundaries in the ceramic tapes. The major differences, which could explain the higher values, are the texture in the microstructure and the small thickness of the samples: The alignment of the c-axes certainly should be of advantage since it aligns the a-b planes with good superconducting properties and eliminates the crystallographic unfavorable interfaces from the current direction [2,6]. The random in-plane orientation, however, still remains and has a deleterious influence on j_c [5,6]. A j_c-reduction by a factor of ≥30, relative to complete a-b-c-texturing, has to be expected for a random network of c-axis textured material [5]. Likewise deviations in the c-axes alignment of only a few degrees suppress the j_c-values [5]. Deviations of this order of magnitude can well be present for some of the junctions in our layers. Taking these influences from texturing together might bring us from the high values of $\geq 10^6 A/cm^2$ for epitaxial films down to a few

$10^4A/cm^2$ for c-axis textured films. This is not too far from our first data and might indicate an upper limit for such tapes. However, the estimates are very speculative, as the above reduction factors are not only a mere geometrical effect of the grain boundary but also contain a component due to the uncontrolled interface chemistry.

Another point to consider is the geometry of the sample. For a ceramic film with weakly linked grain boundaries, j_c in the self-field limit depends on the inverse layer thickness [1] and could be large simply because of the small thickness. Below a material dependent, minimal thickness this behavior breaks down and j_c becomes independent on the geometry. The data on cold-rolled, Ag-sheathed tapes [7] indicate that this limit lies below 100µm. By assuming reasonable material parameters ($B^* \geq 20G$, $j_o \leq 50kA/cm^2$ [1]) for a weakly linked ceramic layer, we estimate that jc of our 2µm tapes is not limited by the self-field and is a true material property. The magnetic field dependence $j_c(B)$ will show, whether our good samples still are weakly linked.

CONCLUSIONS

Through reactive sintering of $Y_1Ba_2Cu_3O_{7-x}$ ceramic layers on silver substrates, a high degree of c-axis texturing was achieved. The layers are a few µm thick, dense and show a random a-b-orientation within the layer. The oriented nucleation and growth start at the interface to the substrate and move upwards through the layer during sintering. For thicker layers three dimensional sintering and growth habits take over and may limit the maximum layer thickness, unless additional processing steps are found. At present, the first few tapes have a critical current density up to ~8000A/cm^2 (77K, H=0). Further improvement still seems possible through optimizing the different processing steps. The upper limit for the current carrying capacity for such textured ceramic layers, however, still remains an open question.

We thank V. Kottler, R. Loitzl and V. Schmid for their assistance, J. Rhyner for helpful discussions and acknowledge the support by the Swiss National Science Foundation.

REFERENCES

[1] Dersch H, Blatter G (1988) New critical state model for j_c. Phys Rev B38:11391

[2] Schüler C, Kluge-Weiss P, Greuter F (1989) YBaCu powders by oxalate coprecipitation. Proc 2nd Int Conf Ceram Powder Proc Sci and (1988) Prep. and microstructure of high density YBaCu. Physica C 153-155:361

[3] Garino TJ, Bowen HK (1987) Sintering of particle films on a rigid substrate. J Am Ceram Soc 70: C-315

[4] Mauthiram A, Goodenough JB (1987) Synthesis of YBaCu in small particle size. Nature 329: 701

[5] Rhyner J, Blatter G (1988) Limiting path model for j_c in textured YBaCu films. Phys Rev B 40:829

[6] Dimos D, Chaudhari P, Mannhardt J, le Goues FK (1988) Orientation depend. of grain bound. j_c in YBaCu bicrystals. Phys Rev Lett 61: 219, 2476

[7] Okada M, Okayama A, Matsumoto T, Aihara K, Matsuda S, Ozawa K, Morii Y, Funahashi S (1988) Neutron diffraction on preferred orientation YBaCu tape. Japn J Appl Phys 27: L 1715

Contact Resistance and Proximity Effect of the YBCO/Ag Interface

K. MIZUSHIMA, H. KUBOTA, and J. YOSHIDA

Advanced Research Laboratory, Toshiba Research and Development Center, 1, Komukai Toshiba-cho, Saiwai-ku, Kawasaki, 210 Japan

ABSTRACT

Superconducting current through a YBCO/Ag/Pb junction was detected by using a superconducting galvanometer. The critical current was, however, at most 50 mA/mm^2, indicating that Ag was superconducting by proximity mainly to Pb. The nonlinear I-V characteristics observed for deteriorated YBCO stripes as well as for YBCO/AlOx/Ag and YBCO/Ag/PbOx/Pb were attributed to the bulk transport property of the insulator in degraded YBCO films.

KEY WORDS: Proximity effect, super-normal-super junction, superconducting galvanometer

INTRODUCTION

The preparation of planar Josephson junctions between YBCO and conventional superconductors (S) has been hindered by the short coherence length and the vulnerability of the YBCO film surface which is easily deteriorated in air or by contact with most metals. On the other hand, good electric contact to YBCO has been achieved by noble metals (N) with the subsequent heat treatment.[1],[2] Therefore, one of the promising ways for the preparation of a YBCO-S tunnel junction is a YBCO/N/I/S structure, where I is an insulator and the noble metal (N) is superconducting by the proximity effect. Junctions of this type have been reported by two of the present authors and by others.[2],[3] Although the nonlinear I-V characteristic observed for the YBCO/Ag/PbOx/Pb junction[2] was at first attributed to the quasiparticle tunneling between the superconducting Ag and Pb, it has later been revealed that the temperature dependence of the characteristic was difficult to be explained by the above mechanism.

As for the YBCO/N/S junction,[4],[5] although an undetectably low resistance below the critical current Ic and Shapirosteps under microwave irradiation have been reported, the resistance above Ic is much larger than that of a noble metal, indicating that the junction is of a more complicated structure. Furthermore, it is still not clear whether the noble metal is superconducting by proximity to YBCO, S or both.[5]

In this report it will be shown that the nonlinear characteristic which has often been observed for the YBCO/Ag/PbOx/Pb junction[2] can be ascribed to conduction in the insulator which is present on the deteriorated YBCO film surface. An essentially same type of nonlinearity was observed for the YBCO/AlOx/Ag junction, and even in four probe measurements for deteriorated YBCO stripes. The authors have confirmed that supercurrent flows through the junction by using a superconducting galvanometer invented by Clarke[6] for the YBCO/Ag/Pb junction. A critical current Ic up to 50 mA/mm^2 was observed, and the junction resistance R was still below the detection limit, i.e., $R \lesssim 10^{-6}\Omega$ even above Ic. This value of Ic is much smaller than that of the Pb/Ag/Pb junction ($Ic \gtrsim 1A/mm^2$), indicating that Ag is superconducting mainly due to its proximity to Pb.

EXPERIMENTAL

The YBCO thin films used in the present work were prepared on $SrTiO_3$ (110) substrates by multi-target rf magnetron sputtering. Details on film preparation have been published elsewhere.[7] The as-deposited films had a strong (110) orientation with its c-axis along the [001] direction of the substrate. The typical critical temperature of the as-deposited film was around 70K. The AlOx layer of the YBCO/AlOx/Ag junction was formed by oxidizing a 50 Å aluminum film in air for an hour. Deteriorated YBCO stripes were obtained by exposing the stripes to Ar/O_2 plasma (5 m Torr, Ar: O_2= 1:1) at 600°C. The I-V characteristic was measured with the four probe method for stripes on which Ag and In 0.5 mmϕ dot electrodes were formed (Fig. 1). The contact resistance for the In electrode was as large as 10^{-2} Ω cm^2. The existence of the superconducting current through the YBCO/Ag/Pb junction was confirmed using the circuit shown in Fig. 2. Two YBCO/Ag/Pb junctions (Fig. 3) were placed in series with a superconducting galvanometer. The junction area was about 1 mm^2 and the thickness of the Ag layer was varied from 1000 to 4000 Å. If the junction is superconducting, the current from the source is divided between the superconducting wire (path II) and path I composed of the

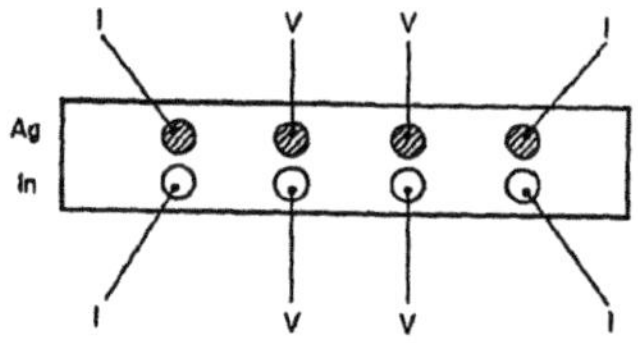

Fig. 1 Electrode configuration for four probe measurement for deteriorated YBCO stripes

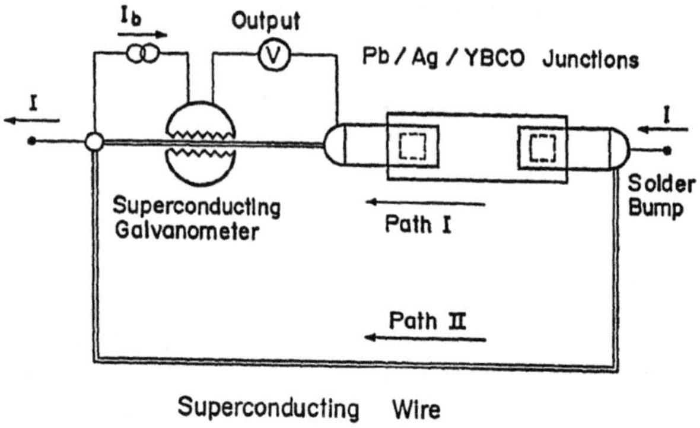

Fig. 2 Circuit used for observation of supercurrent through YBCO/Ag/Pb

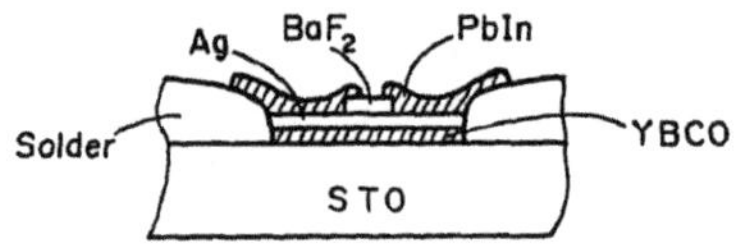

Fig. 3 Double YBCO/Ag/Pb junction

junctions and the galvanometer. On the other hand if the junction is of finite resistance, the current flows only in the superconducting wire.

RESULTS AND DISCUSSION

In the preceding letter [2] the authors reported that a nonlinear I-V characteristic was often observed for the YBCO/Ag/PbOx/Pb junction. To examine further the origin of this nonlinearity, that is, to examine whether it was related to the energy gap of YBCO, the I-V characteristics of YBCO/AlOx/Ag and of deteriorated YBCO stripes were measured. Figure 4 shows the I-V characteristic for the YBCO/AlOx/Ag at 4.2 K, which is very similar to that of the YBCO/Ag/PbOx/Pb. [2] The temperature dependence of the characteristic is shown by logarithmic plot in Fig. 5. The current was proportional to the voltage below 300 μV and was proportional to the square of the voltage in the range from 300 μV to 1 mV. The temperature dependence of the characteristic was very small in these voltage range. The characteristic approached a linear relation again above 1 mV, and the current decreased monotonically with temperature.

The I-V characteristic for the deteriorated YBCO films was measured by four probe measurements. An unexpected result obtained was that the characteristic depended strongly on the electrodes. When the measurement was carried out with the Ag electrodes shown in Fig. 1, the resistance was below the detection limit, i.e., less than 10^{-6} Ω. On the other hand, finite resistance was observed when indium electrodes was used for the voltage probes (Fig. 6 (b)). Furthermore, a strong nonlinearity was observed when the indium electrodes were used for both the current and the voltage probes, as shown in Fig. 9 (a). The reversed I-V characteristic (negative resistance) was often observed.

These results indicate that a filamentary current flowed in the YBCO films, and that high contact resistance electrodes, such as indium electrodes, picked up the voltage difference along a specific current path, as shown in Fig. 7. Measurements with the indium electrodes was carried out for fifteen stripes. The obtained characteristic was classified into three types, as shown in Fig. 8. For curve (a), the current rose in proportion to the square of the voltage, and there existed a point where the derivative changed abruptly. For curve (b), the current rose in proportion to the square of the voltage with a gradual decrease in the slope in the higher voltage region. For curve C, a steep rise of the current with the fourth power of the voltage was observed. Among the fifteen stripes, two were of type (a), another two were of type (c) and the others were of type (b). The voltage region where the nonlinearity was observed was different from sample to sample, ranging from tens of microvolts to tens of millivolts. The temperature dependence of the characteristic was very similar to that of the YBCO/AlOx/Ag junction, that is, a small temperature dependence in the low voltage region and a larger temperature dependence in the high voltage region. For ordinary S/I/S and S/I/N junctions the I-V characteristics below the gap voltage are expressed as

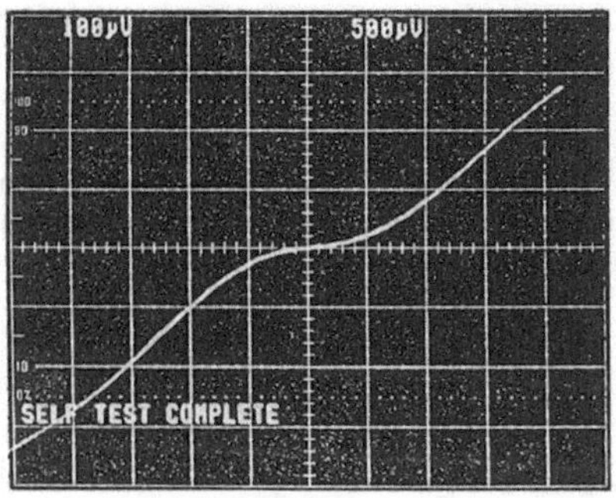

Fig. 4 I-V characteristic for YBCO/AlOx/Ag at 4.2K 500 μV/div, 100 μA/div

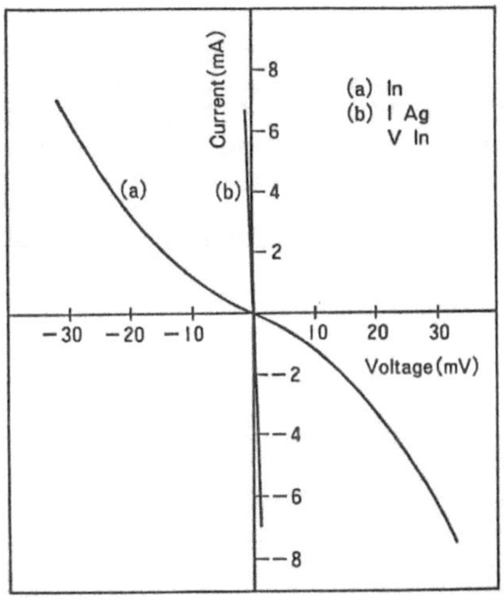

Fig. 6 I-V characteristic for deteriorated YBCO stripes
(a) all probes are indium, (b) current and voltage probes are silver and indium, respectively

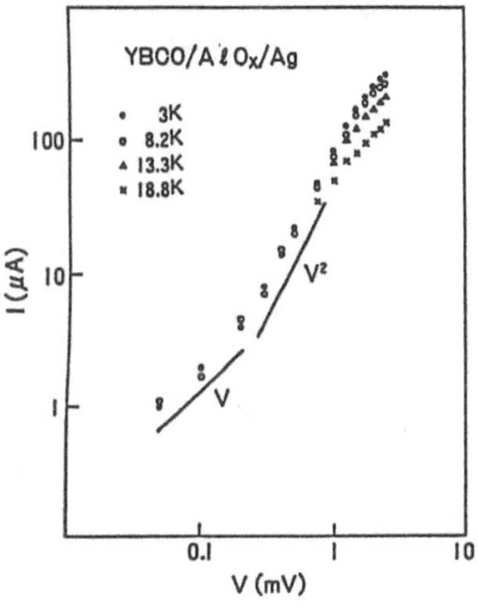

Fig. 5 Temperature dependence of I-V characteristics for YBCO/AlOx/Ag

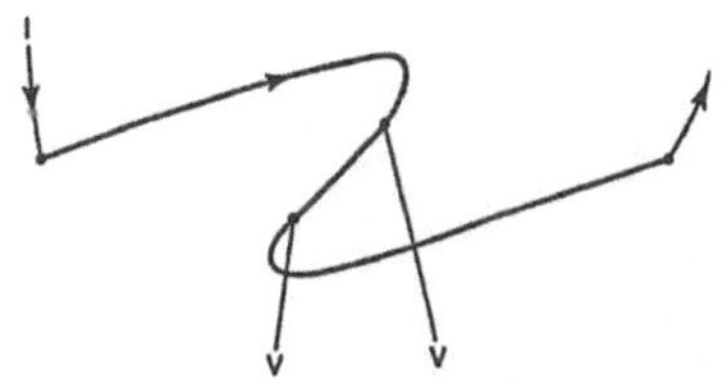

Fig. 7 Filamentary current path in deteriorated YBCO stripe

$$I_{SS} = \frac{2G_N}{e}\, e^{-\frac{\Delta}{kT}} \left(\frac{2\Delta}{eV + 2\Delta}\right)^{\frac{1}{2}} (eV + \Delta)\sinh\left(\frac{ev}{2kT}\right) K_o\left(\frac{eV}{2kT}\right) \tag{1}$$

and

$$I_{SN} = G_N \frac{\Delta}{e}\left(\frac{\pi kT}{2\Delta}\right)^{\frac{1}{2}} e^{\frac{eV-\Delta}{kT}} \tag{2}$$

respectively. The characteristics show a strong temperature dependence below the gap voltage and is almost temperature independent above the gap voltage, which are quite different from those of YBCO/AlOx/Ag and of the deteriorated YBCO stripes. Generally speaking, nonlinearity due to the potential barrier such as Schottky barrier is also expected to depend much on temperature. Another possibility for the strong nonlinearity is the space charge limited current in the insulator. Let us consider a filamentary current path composed of superconducting weak links with an insulator inserted in it (Fig. 9). If the insulator is too thick for electron tunneling, the I-V characteristic is determined by the bulk transport property of the insulator. Space charge limited current in insulator can be expressed as

$$I \propto \varepsilon\mu V^2/L^3 \ , \tag{3}$$

where ε and μ are the dielectric constant and the mobility of the electron, respectively. As the temperature dependence of mobility is expected to be relatively small, eq. (3) explains the V^2 dependence and the small temperature dependence of the current in the low voltage region. The nearly ohmic behavior in the high voltage region can be attributed to the normal resistance of the weak links above the critical current which decreases with temperature. The V^4 dependence which was occasionally observed for the YBCO stripes can be

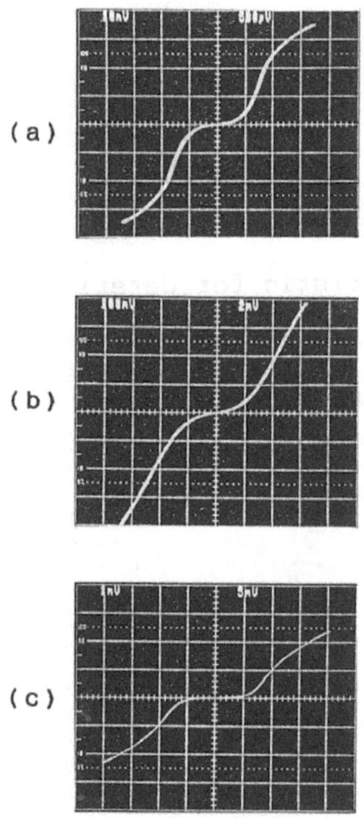

Fig. 8 Three types of nonlinear I-V characteristics observed for deteriorated YBCO stripes

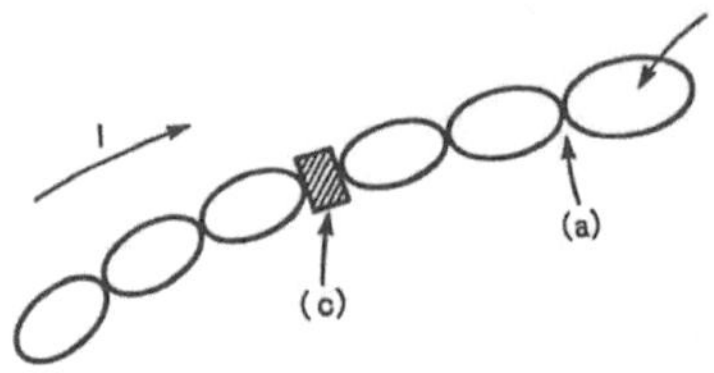

Fig. 9 Filamentary current path in deteriorated YBCO film

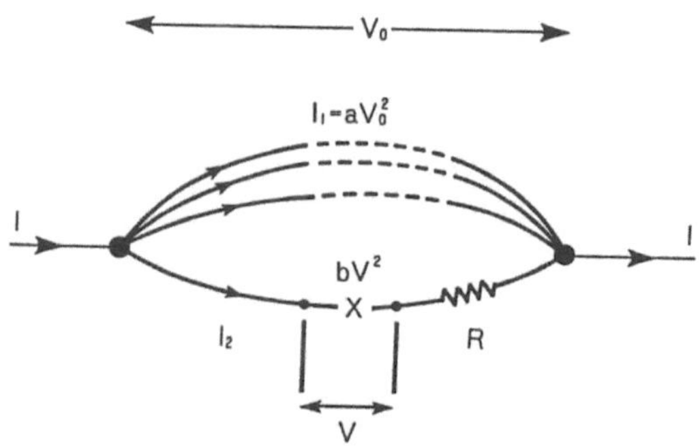

Fig. 10 Model to explain V^4 dependence

explained by the extention of the filamentary current path model shown in Fig. 10. The two current electrodes are connected by many filaments of aV^2 dependence and the voltage electrodes measure the voltage difference on a specific filament of bV^2 dependence. In the case that a certain number of the weak links of the specific filament are in an off state, the resistance R appears on the filament and the observed I-V characteristics is

$$I \sim I_1 \sim ab^2R^2V^4 \tag{4}$$

in the voltage range of $V_o/R << bV_o^2$, where I_1 is the total current of the filaments except for the specific filament, and V and V_o are the voltage difference between the voltage electrodes and between the current electrodes, respectively. The nonlinear I-V characteristics observed for YBCO/Ag/PbOx/Pb and YBCO/AlOx/Ag, as well as for the YBCO stripes, considered to be ascribed to the bulk transport property of the insulator in the filamentary current path in the degraded YBCO films.

It is, therefore, very important to examine in the preparation process of the YBCO/Ag/I/S or the YBCO/Ag/I/N type tunnel junction whether Ag is superconducting by proximity and supercurrent can flow through the YBCO/Ag/S junction. Such an experiment, however, is very difficult when an ordinary measurement circuit configuration is used, because the residual resistance of the Ag layer is very small even if the proximity effect does not work. The authors utilized the superconducting galvanometer invented by Clarke in order to detect the superconducting current through the YBCO/Ag/Pb junction.

Figure 2 shows the measurement circuit configuration. The circuit consisted of two YBCO/Ag/Pb junctions fabricated on a YBCO thin film, (Fig. 3), a superconducting galvanometer, and superconducting interconnections between them. The thickness of the Ag layer was varied between 1000 Å and 4000 Å. When the current was raised from zero to a certain value, the current flowed through both path I and path II in the figure during the current rise time. After the supplied current reached a steady state value, the current through path I gradually fell off in the case that a finite resistance remained at the junctions. This fall off time of the current is given by L/R, where L and R denote the circuit inductance and the residual resistance at the junction, respectively. Thus we can estimate the R value by observing the current decay in path I by the superconducting galvanometer. Since the circuit inductance was estimated to be $\sim 10^{-8}$ H, the detection limit for the resistance value was around 10^{-12} ohm for one hour observation. The detection limit can, in principle, be improved in proportion to the elongation of the observation time. Up to the present, only observations for about two hours has been performed. We have observed no current decay for some junctions through such measurements (Fig. 11 (a)), although the measurements were often disturbed by noises from outside which caused a sudden change in the output voltage of the galvanometer (Fig. 11 (b)). Furthermore, a current redistribution between path I and path II was observed at a certain value of the supplied current. This measurement was carried out by slowly varying the supplied current. One example is shown

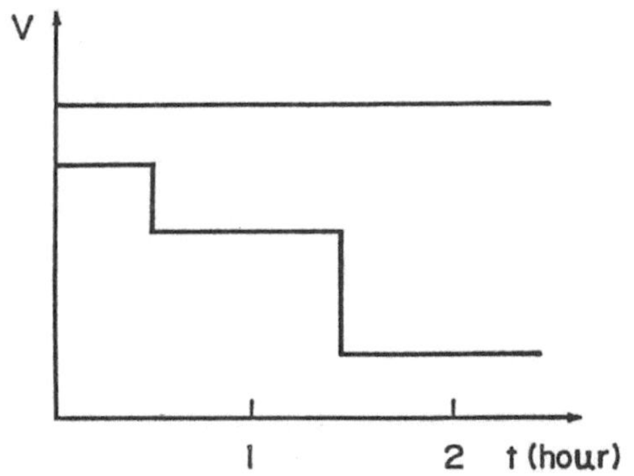

Fig. 11 Variation of output voltage of superconducting galvanometer with time

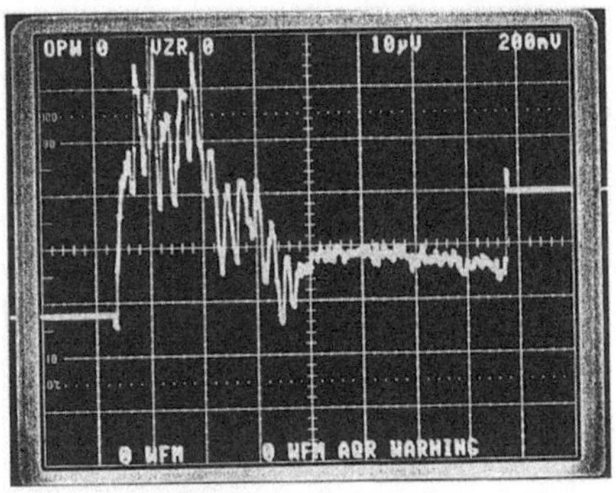

Fig. 12 Variation of output voltage of superconducting galvanometer as a function of supplied current

2mA/div

in Fig. 12, in which the output voltage oscillation of the superconducting galvanometer disappeared around 7 mA supplied current, indicating that the current through path I saturated at this value. Such a behavior can be understood as the existence of a critical current for the junction. These results strongly suggest that a superconducting contact can be formed in the YBCO/Ag/Pb junction. The critical current of the junctions thus determined was at most 50 mA/mm^2 which was much lower than that of a Pb/Ag/Pb junction which exceeded 1 A/mm^2. Furthermore, no systematic dependence of the value of the critical current on the Ag thickness. These results indicate that the superconductivity of Ag is mainly induced by Pb and the critical current of the junction is determined at the YBCO/Ag interface.

CONCLUSIONS

The temperature dependence of the nonlinear I-V characteristics observed for YBCO/Ag/PbOx/Pb and YBCO/AlOx/Ag, as well as for the YBCO stripes, was in contradiction with that of quasiparticle tunneling. The nonlinearity was ascribed to the bulk transport of the insulator in the filamentary current path in degraded YBCO films.

Superconducting current was observed for the YBCO/Ag/Pb junction with a critical current as large as 50 mA/mm2, indicating that Ag is superconducting by proximity to Pb.

ACKNOWLEDGEMENT

A part of this work was supported by NEDO under the management of FED.

REFERENCES

1. Mizushima K. et al. (1988) Appl. Phys. Lett. 52: 1101.
2. Mizushima K. et al. (1988) Jpn. J. Appl. Phys. 27: L1489.
3. Moreland J. et al. (1989) Appl. Phys. Lett. 54: 1477.
4. Akoh H. et al. (1988) Jpn. J. Appl. Phys. 27: L519.
5. Greene et al. (1989) Proc. Intnl. Conf. on Materials and Mechanisms of Superconductivity High-Temperature Superconductivity, Standford, CA.
6. Clarke J. (1965) Phil. Mag. 13: 115.
7. Sagoi M. et al. (1989): Jpn. J. Appl. Phys. 28: L444.

Effect of Cold-Working on the Critical Current Density of Ag-Sheathed Bi(Pb)-Sr-Ca-Cu-O Tapes

SANG-SOO OH, KOZO OSAMURA, and SHOJIRO OCHIAI
Department of Metallurgy, Kyoto University, Sakyo-ku, Kyoto, 606 Japan

ABSTRACT

In order to improve the critical current density (Jc) of Ag-sheathed B(Pb)SCCO tapes, various thermomechanical treatments were applied. The combination of swaging and pressing was most effective at present for obtaining the high critical current densities. Repeating TMT cycle, the densed and preferred oriented granular structure was obtained. This caused Jc to increase, especially the Jc under magnetic fields at 4.2 K.

KEYWORDS : critical current density, thermomechanical treatment, magnetic field dependence, silver sheathed tape, BiPbSrCaCuO.

INTRODUCTION

As Bi based superconducting oxides are expected to be used as a magnet wire, high critical current density is an indispensable property. The Jc of this material as sintered, however, is very low/1/. This might be caused from various factors as weak link at grain boundaries/2/, low packing density and appearence of nonsuperconducting phases. It is known that the powder-in-tube method is a simple and convenient technique to improve the Jc by controlling the microstructure/3,4/. In order to optimize the experimental conditions when Ag sheathed tapes were prepared by thermomechanical treatment after tubing to Ag pipe, several kinds of cold-working processes and heat treatments were applied and their effects on the Jc were investigated.

EXPERIMENTAL

High grade powders of Bi_2O_3,$SrCO_3$,$CaCO_3$,CuO and PbO were mixed in various nominal composition (Bi:Pb:Sr:Ca:Cu=(1-x):x:0.8:1:1.4) and calcined at 1073 K for 86.4 ks and then ground into powder. This powder is called "A" in the present text and filled into Ag pipe. A part of powder A was pelletized and sintered at 1118 K for 353 ks and reground into powder. This is called powder "B". By combining cold-working techniques of swaging(S), rolling(R), drawing(D) and pressing(P), the Ag sheathed pipe was deformed to the tapes with thickness of 0.1 - 0.3 mm and heat treated at various temperatures in air. Pressing and heat treatment were repeated at later stage of this process. Jc measurments were performed at 77 K without magnetic field with criterion of 1μV/cm and 4.2 K under magnetic fields with criterion of 0.1 μV/cm.
The microstructures were observed by optical microscope and scanning electron microscope. X-ray diffraction (XRD) analysis using Cu-target was performed to identify the phases existed.

RESULTS AND DISSCUSION

Fig.1 shows the change of critical current density as a function of Pb content for the tapes processed by SP1HT1P2HT2. Here, SP1HT1P2HT2 indicates the following TMT process; after swaging, the pressing and heat treatment was two times repeated. The mean value of Jc became high in the range of x = 0.15 - 0.25 and reached maximum at x= 0.2. This composition was used for the prepara- tion of tapes.

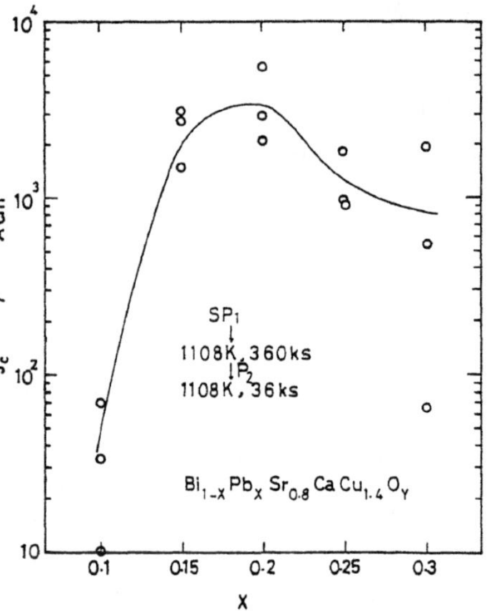

Fig. 1 Change of critical current density at 77 K as a function of amount of Pb additive.

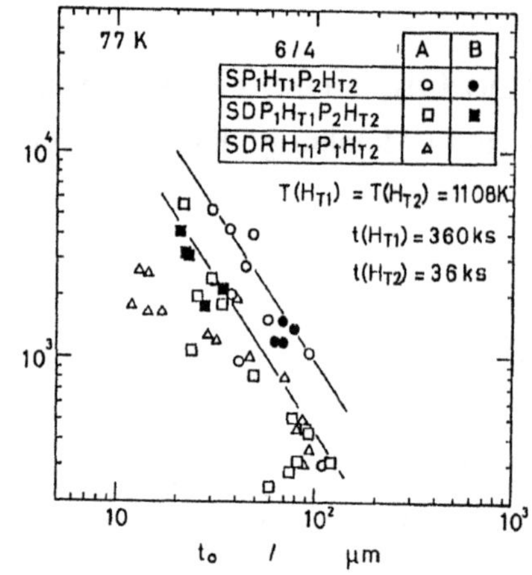

Fig. 2 Change of critical current density at 77 K as a function of oxide layer thickness for the specimens prepared by TMT process. The condition is indicated in the figure.

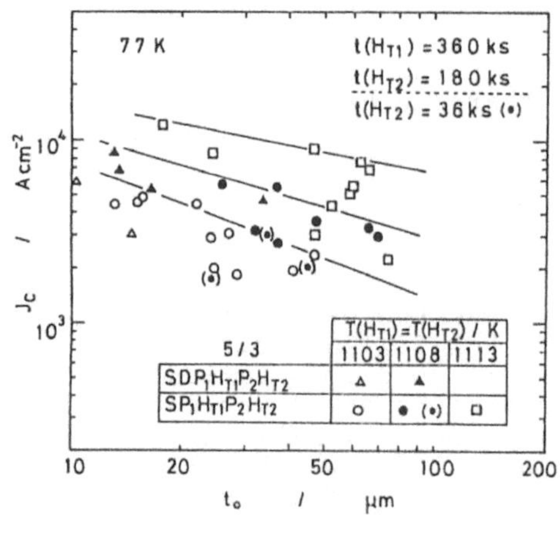

Fig. 3 Change of critical current density at 77 K as a function of oxide layer thickness for the specimens prepared by TMT process indicated in the figure. It should be noted that the J_c increases with increasing heat-treatment temperature. Also the J_c depends on the heat-treatment time.

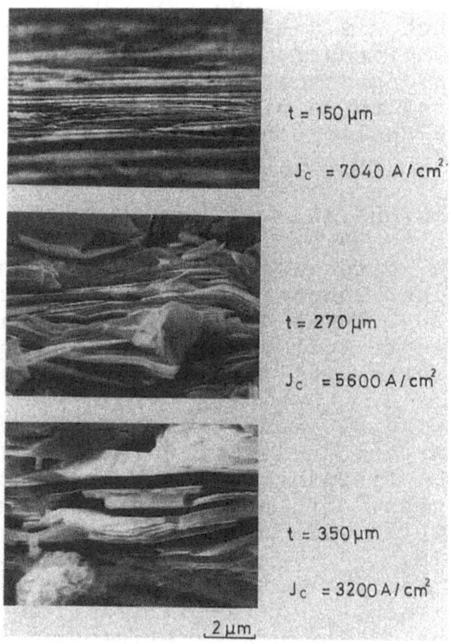

Fig. 4 Fracture surface of the tape specimens with different critical current density as well as oxide layer thickness. The specimen was heavily deformed by cold-pressing.

Fig.2 shows the critical current density as a function of oxide layer thickness for the specimens processed by SP1HT1P2HT2, SDP1HT1P2HT2 or SDRHT1P1HT2. Comparing three different kinds of experimental conditions, SP, SDP and SDR, it is found that the Jc tends to become higher for the tapes prepared by the SP process. From the microstructure observation, it was estimated that the

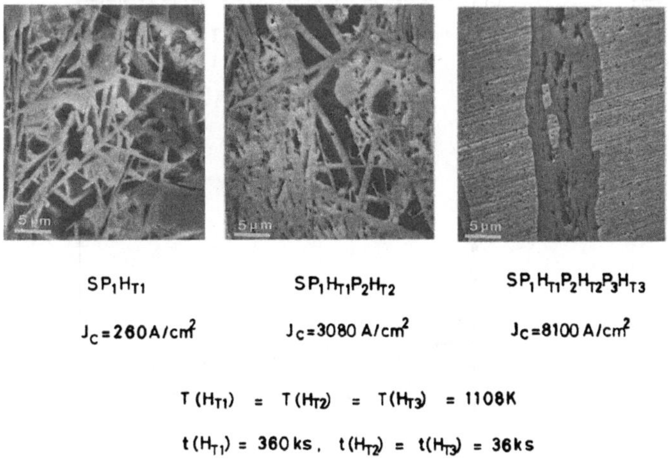

Fig. 5 Microstructure of the tape specimens prepared by different TMT processes.

drawing causes an inhomogeneous deformation, resulting in a degradation of the Jc. Accoding to this fact, the combination of swaging and pressing has been adopted as the cold-working technique hereafter. As shown in Fig.2, there is no difference in the Jc for the tapes prepared by using different powders A and B. By X-ray diffraction analysis, a major phase in the powder A or B was determined to be the low Tc 2212 or high Tc 2223 phase, respectively. On the other hand, the tape specimens prepared from either A or B powder included equally both low Tc and high Tc phases after the final TMT process.
Fig.3 shows the critical current density as a function of oxide layer thickness for the specimens processed by SP1HT1P2HT2 or SDP1HT1P2HT2. It is found that the Jc increased with elevating the heat treatment temperature. Comparing with the result of Fig.2, Jc was improved by the long period-second heat treatment. At present, the Jc of about 12000 A/cm^2 at 77 K was obtained for the specimen heat treated at 1113K.
Fig.4 shows fracture surface for the tapes processed by SP1HT1P2HT2. The pile-uped laminated structure was observed. The degree of orientation of a-b plane with respect to tape surface seems to become higher for the tapes with thinner thickness. Therefore, the increase of Jc was due to the effect of orientation of a-b plane introduced by the combination of cold working and heat treatment.
Fig. 5 shows the microstructures for the specimens processed by SP and repeated heat treatment(HT) and pressing(P). It is observed that superconducting phase distributed randomly and many voids exist among grains for the specimen prepared by one TMT cycle(SP1HT1). The densed and preferred oriented granular structure was observed for the specimens prepared by two or more TMT cycles. It is suggested that Jc increases due to the refinement of microstructure.
Fig.6 shows the magnetic field dependence of critical current density at 4.2 K. The hysteresis effect was still observed for the specimen, which has fairly high Jc. The critical current densities under magnetic fields increased with decreasing the oxide layer thickness, to. Comparing with result from YBCO, the drastic dropping of Jc with repect to magnetic field/5/ was not observed.
Fig.7 shows the effect of applied magnetic field direction with respect to c-axis on the Jc, which became high when the magnetic field was applied to normal with c-axis. According to this fact, the pile-uped lamina-boundary as observed in Fig.4 is suggested to become a dominant pinning center.

CONCLUSION

In the present study, it was made clear that TMT was very useful for preparing the Bi based superconducting tapes. Among various cold working processes, the combination of swaging and pressing was found to be most effective. Repeating TMT cycle, the densed and preferred oriented granular structure was obtained. This improved microstructural factors caused the critical current density to

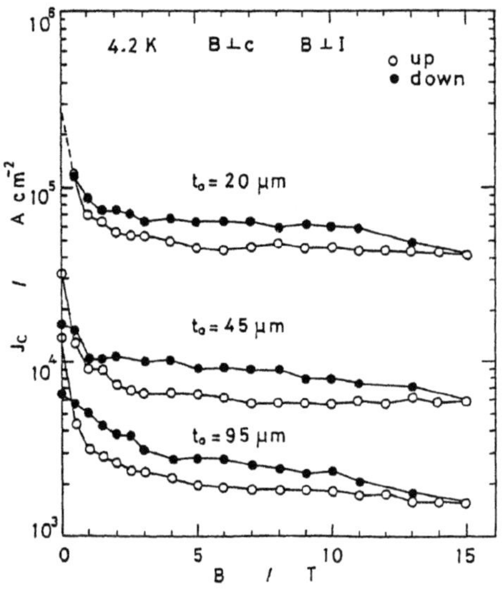

Fig. 6 Magnetic field dependence of critical current density at 4.2 K for the specimens with different layer thickness, where magnetic field was applied parallel to the tape curface and perpendicular to the current flow.

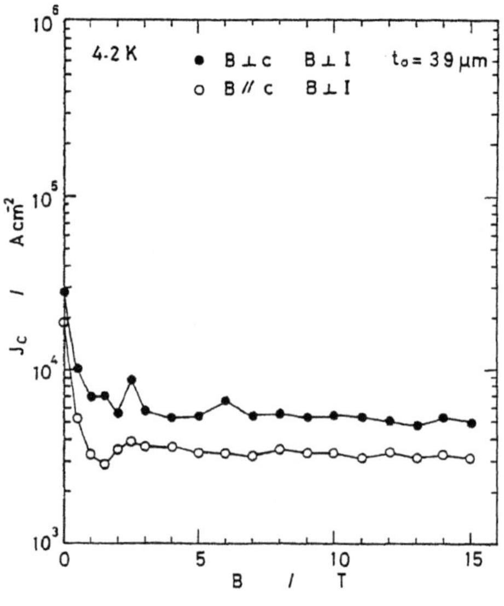

Fig. 7 Magnetic field dependence of critical current density at 4.2 K, where magnetic field applied parallel or perpendicular to the tape surface.

increase. It was confirmed that the critical current density under magnetic fields was improved remarkably for the specimens prepared using the powder-in-tube method.

ACKNOWLEDGEMENTS

The critical current measurments have been partially performed at HFLSM of Tohoku University. This study was made possible by a Scientific Research Grant in-Aid (Proj. No. 01644520) from the Ministry of Education, Science and Culture of Japan.

REFERENCES

1. Umehara M. Asada Y. Maeda H. Ogawa K. (1988) Magnetic properties of $BiSrCaCu_2O_x$ superconductors. Jpn.J.Appl.Phys. 27: L665-L667.
2. Osamura K. Takayama T. Ochiai S. (1989) Weak flux pinning at grain boundaries in Ag sheathed $Ba_2YCu_3O_{6+x}$ tapes. Appl.Phys.Lett. 55(4): 396-398.
3. Hikata T. Sato K. Hitotsuyanagi H. (1989) Ag-sheathed Bi-Pb-Sr-Ca-Cu-O superconducting wires with high critical current density. Jpn.J.Appl.Phys. 28: L82-84.
4. Osamura K. Takayama T. Ochiai S. (1989) Effect of cold-working on the critical current density of Ag-sheathed $Ba_2YCu_3O_{6+x}$ tapes. Supercond. Sci. Technol. 2: 111-114.
5. Takayama T. Ochiai S. Osamura K. (1989) Influence of microstructure on the critical current density of $Ba_2YCu_3O_{6+x}$ superconducting tapes. J. Japan Inst. Metals. 53: 735-743.

New Fabrication Method of Thick Film of Oxide Superconductor by the Combination of Mist Pyrolysis and Collision of Particles to Substrate

M. AWANO and H. TAKAGI

Government Industrial Research Institute, Nagoya, 1-1, Hirate-cho, Kita-ku, Nagoya, 462 Japan

ABSTRACT

Thick film and compact of oxide superconductor were fabricated by the mist pyrolysis method consecutive the collision process of synthesized particles to substrate.
Fine and crystalline Ba-Y-Cu-O and Bi-Pb-Sr-Ca-Cu-O superconducting particles were synthesized directly by the pyrolysis and crystallization of droplets atomized from starting solution of each stoichiometric composition at the optimum condition. Subsequently, produced superconducting powders were carried into vacuum chamber and collided to substrate by the force derived from pressure gap. Obtained thick film or compacted body had dense structure as deposited, and they showed superconductivity after annnealing at suitable temperature condition.

KEY WORDS: Ba-Y-Cu-O, Bi-Pb-Sr-Ca-Cu-O, mist pyrolysis, thick film, collision of particles to substrate

INTRODUCTION

Oxide superconductive material has many problem to be solved for applicationas a superconductor. Improvement of the critical current density (Jc) of oxide superconductor is required for application of this material to wire, coil and so on. For this purpose, sheathed wire and tape [1] or high deposition rate CVD method have been attempted. The other hand, Fine homogeneous superconductive powder are required to obtain the dense homogeneous texture of sintered body [2]. And further improvement of micro-structure of sintered bodies must be studied to improve superconductive and deformation processing properties [3].
For usage of oxide superconductor as bulk, elimination of second phases, improvement of grain linkage and promotion of the crystal orientation or morphology have been requested. And for usage as electronic circuit,for example, fablication processing through heat treatment at lower temperature are required.
In this work, we had established new fablication method of thick film and compacted body (as a kind of bulk) by the combination of a synthetic method of ultrafine, homogeneous and well crystallized superconductive particles (mist pyrolysis method) and formation of dense body as the result of collision and direct deposition on substrate without heat treatment. Current path of superconductivity is not formed by the as-deposited particles, but the formated bulk showed transpert current of superconductivity after annealing at higher temperature.

EXPERIMENTAL

Figure 1 shows the scheme of system of this work. Submicron size particles of Ba-Y-Cu-O and Bi-Pb-Sr-Ca-Cu-O system superconductor were synthesized by the following process; starting solution of each stoichiometric solution of nitrates were atomized by the ultrasonic nebulizer. Produced droplets were separated about their particle size approximately, and carried by carrier gas (oxygen, nitrogen, argon and so on) into reaction zone that was heated at 650 °C - 1000 °C. Passing through the reaction zone, droplets were dried, decomposed and crystallized as a superconductive phase.
Particles were not taken out from system, and they were carried directly

into vacuum chanmber. They were moved speedy through nozzle by the differencial pressure between atmospheric pressure and vacuum in chamber. Dense film or compact were formed on the substrate set near the nozzle. Film shape was varied by movement of the substrate.
Mist-pyrolyzed particles were evaluated about their molphlogy, crystallinity and phase, chemical composition,and superconductivity by means of X-Ray diffraction, SEM, TEM With Energy Dispersive Spectroscopy, and VSM. Charactarization of dense film and compact were carried out for the microstructure (SEM and TEM), crystallinity (XRD) and superconductivity (Tc measurement by four probe method) with as-deposited and annealed film.

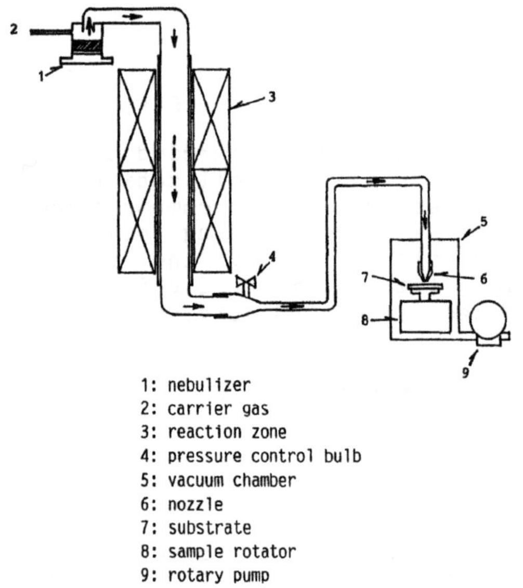

Fig.1: Schematic diagram of mist pyrolysis and collision method.

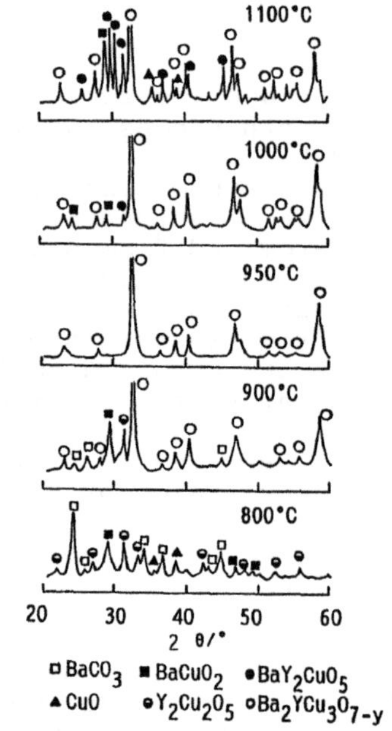

Fig.2: X-Ray diffraction patterns of Ba-Y-Cu-O powders obtained at various reaction temperatures by the mist pyrolysis method.

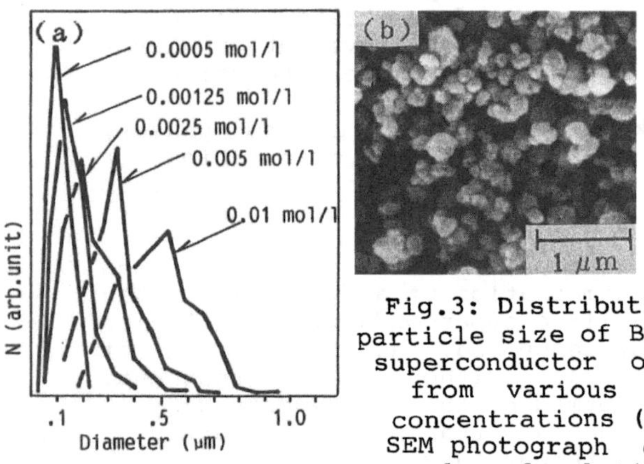

Fig.3: Distribution of particle size of Ba-Y-Cu-O superconductor obtained from various liquid concentrations (a) and SEM photograph of the powder of solution concentration at 0.00125 M (b).

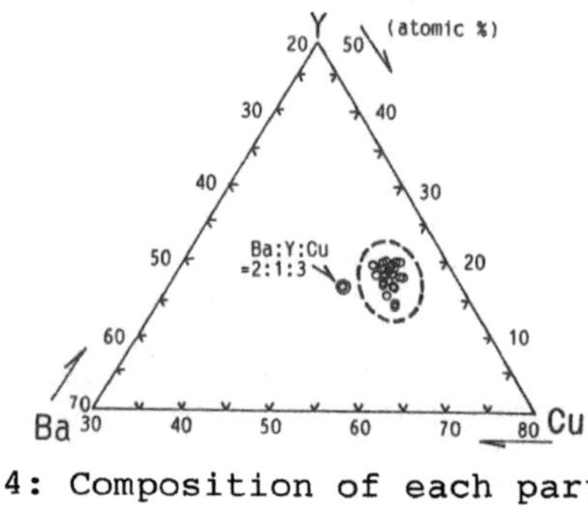

Fig.4: Composition of each particles of Ba-Y-Cu-O superconductor analyzed by EDS method (calculated K-factors were used).

RESULTS AND DISCUSSION

Properties of particles produced by the mist pyrolysis method are effected by the factors of starting solution and reacting conditions.
Figure 2-6 show various properties of particles for Ba-Y-Cu-O superconductor. Crystalline phase and crystallinity of particles depend on the temperature of reaction zone, droplets flow rate and droplets size.
Relation between the temperature of reaction zone and crystalline phase is shown in Fig.2. Superconductive particles are synthesized at 950 °C -1000 °C directly as orthorhombic phase. Particles size are varied by solution concentration. Particle size and size distribution are decreased with the

decreasing of solution concentration (Fig.3). Figure 4 shows EDS analysis for the composition of each particle. This figure reveals the uniformity and stoichiometry of the each particle composition. TEM photographs of the particles show their well crystallized feature. These particles had the phase transition from tetragonal to orthorhombic. The microtwin structure was resulted in this transition [4](Fig.5). Such results consist with their superconductivity (Fig.6) . On the optimum synthetic condition, particles show the Tc onset near 90K and Tc end point less than 80K, and their superconductivity indicated by magnetization-temperature curve effected by crystallinity , orthorhombicity and particle size.
Synthetic condition of reaction temperature, solution concentration and droplets flow rate for Bi-Pb-Sr-Ca-Cu-O supreconducter about nominal composition of so called "high Tc(110K) phase" is shown in Fig.7. Direct synthesizing of high Tc phase by the mist pyrolysis method is impossible at present. X-Ray diffraction patterns (Fig.8) show the crystal phase as "low Tc(80K) phase". Particles shape observed by TEM are rounded or platy figure. Such synthetic condition is similar with the case of nominal composition of low Tc phase.
Conventional fabrication process of thick film, for example, sol-gel method and screen printing, use the precursor of superconductor or organic solvent. So such processes need after or in-situ heat treatment to make superconducting thick film. Above mentioned synthesized particles have advantage that they are already well crystallized and have superconduntivity at as-prepared state. So, if one succceed to bind tightly each particles, resulting polycrystalline bulk have dense structure and superconductivity without any heat treatment except for the weak-link problem.

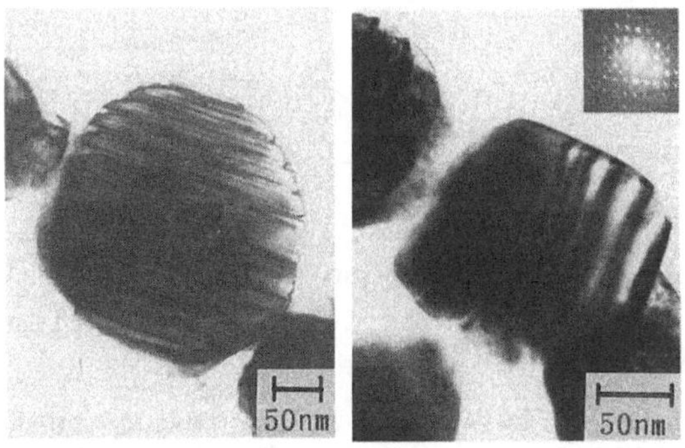

Fig.5: TEM photographs of Ba-Y-Cu-O fine particles produced directly by the mist pyrolysis method.

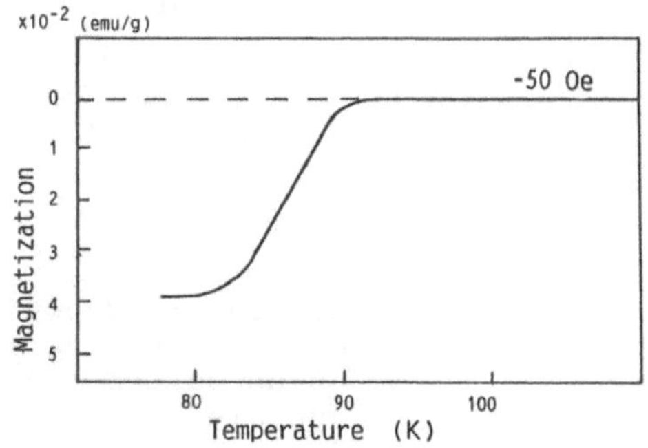

Fig.6: Relation between magnetization and temperature of Ba-Y-Cu-O as-mist pyrolyzed particles.

To obtain dense thick film and compaction body, we attempted to collide the superconducting fine particles to substrate gain high speed in the carrier gas so the particles fixed hardly by inertia force on to substrate or ancient deposited particles. We used a technique of spouting into vacuum chamber through narrow nozzle. Figure 9 shows obtained as-deposited dense film onto MgO and YSZ substrate. Film thickness, wide and length were varied by the control of concentration of starting solution, colliding rate of particles to substrate, nozzle caliber and moving speed of substrate.
The film fixed on the substrate couldn't be removed by tracing with a spatula. X-Ray diffraction pattern of thick film of Ba-Y-Cu-O superconductor on MgO substrate is shown in Fig.10. Nevertheless the film showed orthorhombic phase with random orientation, they couln't show superconductive transport current in the specimen (Fig.11). After heat treatment of the film, it showed the superconductive property in relation of temperature-resistivity curve. This result suggests that as-deposited film has dense structure but each particles have not confirmed linkage each other. So superconducting current could not flow through grain boundary. SEM photographs of the micro structure of annnealed films indicate the grain linkage and recrystallization by annealing.

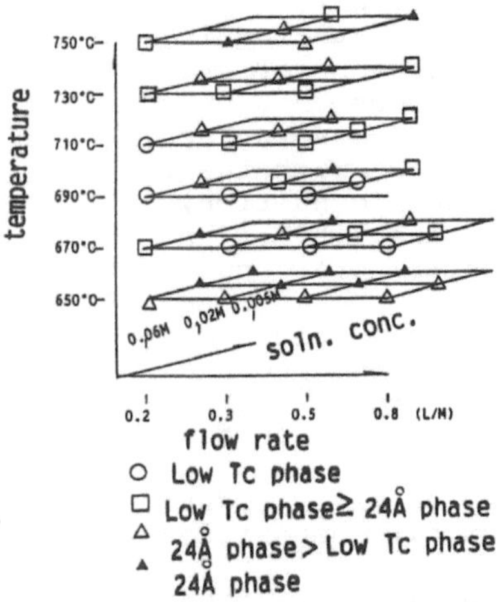

Fig.7: Relationship of the crystalline phases to the pyrolyzing temperature, concentration of starting solution and flow rates of carrier gas.

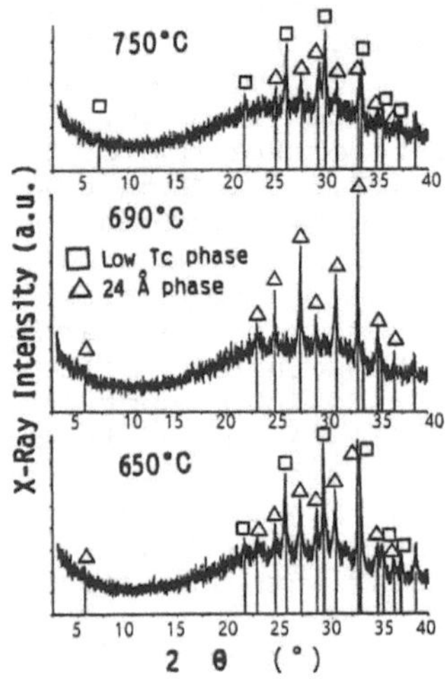

Fig.8: X-Ray diffraction patterns of powders of Bi-Pb-Sr-Ca-Cu-O superconductor obtained at various temperatures by the mist pyrolysis method.

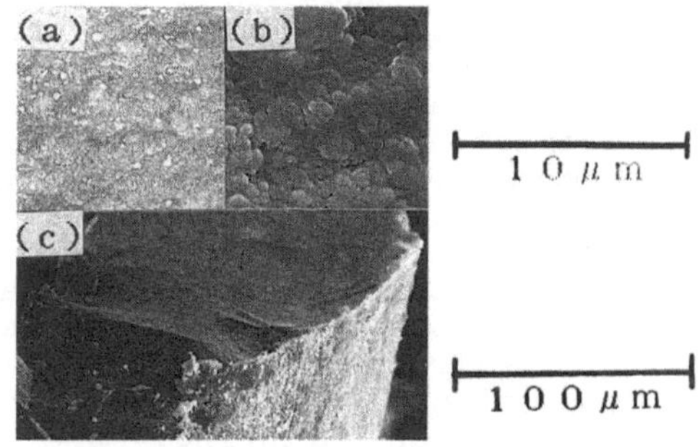

Fig.9: SEM photographs of as-deposited thick film (surface(a) and profile(b)) and compact (c).

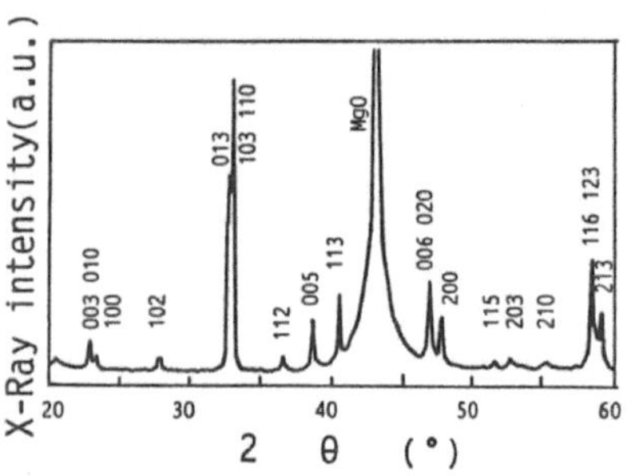

Fig.10: X-Ray diffraction pattern of as-deposited thick film on MgO substrate.

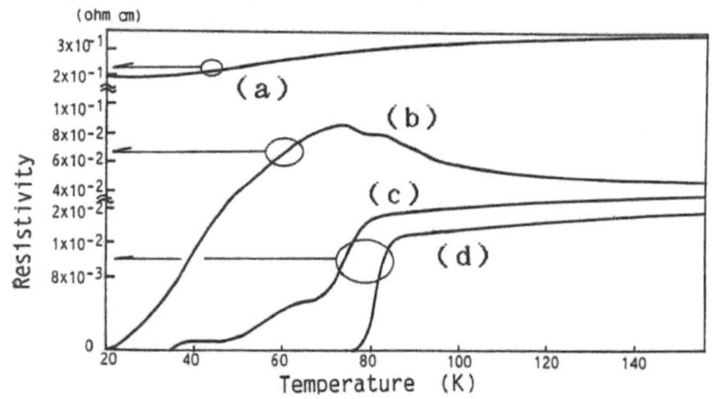

Fig.11: Temperature dependence of the resistivity for the thick film of Ba-Y-Cu-O superconductor;(A)
as-deposited (a)
annealed (b) 950 °C - 3 hours + 500 °C - 12 hours in oxygen flow
(c) 1000 °C - 3 hours + 〃
(d) 1030 °C - 3 hours + 〃

CONCLUSIONS

New fabrication method of thick film and compact was developed by the combination of the mist pyrolysis method and particle deposition by the collision. Dense film and compact bulk body were made without heat treatment on substrate by using of well crystallized superconductive fine particles produced by mist pyrolyzing process. The film showed superconductivity in resistivity-temperature relation only after annealing treatment. Such feature is coincident with grain boundary structure.
Further investigation to make qualified thick film, usage of very fine patricles like a cluster and the study of grain linkage by the heat treatment of substrate at lower temperature through the particles deposition seems to be necessary. This fabrication method will be effective to improvement of flux pinning by microstructure control.

REFERENCES

1. Mimura M. Kumakura H. Togano K. Maeda H. (1989) Improvement of the critical current density in the silver sheathed Bi-Pb-Sr-Ca-Cu-O superconducting tape. Appl.Phys. Lett. 54: 1582-1584
2. Horowits HS. McLain SJ. Sleight AW. Druliner JD. Gai PL. VanKavelaar MJ. Wagner JL. Biggs BD. Poon SJ. (1989) Submicrometer superconducting $YBa_2Cu_3O_{6+x}$ particles made by a low-temperature synthetic route. Science 243: 66-69
3. Wakai F. (1987) Superplasticity of ceramic composites. Bull. Ceram. Soc. Jpn. 22: 844-849
4. Setaka R. Komatsu W. Shibata T. Nakajima M. (1988) Preparation of a single crystalline powder of superconducting $YBa_2Cu_3O_{7-x}$ by the gas phase solidification method. JJAP. 27: L2100-L2102

Fabrication and Critical Current Density of High-T_c Bi-Pb-Sr-Ca-Cu-O Superconducting Thick Films

K. Hoshino[1], H. Takahara[1], and M. Fukutomi[2]

[1] Corporate R & D Center, Mitsui Mining & Smelting Co., Ltd., 1333-2, Haraichi, Ageo, Saitama, 362 Japan

[2] National Research Institute for Metals, Tsukuba Laboratories, 2-1, Sengen 1-chome, Tsukuba, Ibaraki, 305 Japan

ABSTRACT

High-Tc superconducting Bi-Pb-Sr-Ca-Cu-O thick films were successfully prepared on Ag tapes by using combined processes of screen printing, cold rolling and sintering. The critical current density Jc of the tape was very sensitive to the composition and postprinting annealing temperature of the films. The Jc of about 1.2×10^4 A/cm^2 at 77 K in a zero magnetic field was achieved for the film with the compositional ratio of Bi:Pb:Sr:Ca:Cu=1.8:0.4:2:2:3. The magnetic field dependence of Jc was also measured at 4.2 K, where the tape showed the Jc of 1.5×10^4 A/cm^2 in 0.5-1.0 T.

KEY WORDS: oxide superconductor, Bi-Pb-Sr-Ca-Cu-O, screen printing, Ag substrare, critical current density

1. INTRODUCTION

Important progresses as the formation of a nealy perfect single 110 K phase or the improvement of Jc have been made so far in the Bi-Sr-Ca-Cu-O high-Tc oxide superconductor[1-3]. Screen printing is one of the most convenient processes for preparing the thick films of high-Tc superconductors, particularly, which can be used to fabricate the films with a relatively large area. However, the film prepared by this method is usually very porous and the Jc of the film is limited to the order of 10 A/cm^2. we have recently reported that the thick film of the Pb-doped Bi oxide superconductor can be fabricated on a Ag substrate by using the combined processes of screen printing, cold, rolling and sintering[4]. It was found that the cold rolling process drastically improves the packing density of screen printed films as well as the c-axis alignment of the grains. The Jc of those films, however, was still lower than that for the practical applications like superconducting tape or sheet.

The purpose of this investigation is to clarify the relation between the Jc of the thick film on Ag tape and the fabrication conditions. The Jc of the films has been examined as a function of composition, annealing temperature and film thickness.

2. EXPERIMENTAL

The nominal composition ratios of mixed powders for the screen printing paste are listed in Table I. The powders of these compounds were converted into a paste with an acrylic resin. The paste was screen printed on Ag tapes and then subjected to the cold rolling and sintering. Details of the preparation for the tapes are described elsewhere[4]. The superconducting transition temperature Tc(zero) of the film with Ag tape attached was measured using a standard dc four-probe method. The transport critical current Ic of the tape was measured resistively in liquid nitrogen(at 77 K). The measurement of Jc in a magnetic field up to 12 T was carried out in liquid herium(at 4.2K), where the magnetic field was applied perpendicularly to a transport current. The criterion for the determination of Jc was defined as 1 μV/cm in 0 T(at 77 K), and 0.1 μV/cm in a magnetic field(at 4.2 K). Scanning electron microscopy (SEM) and X-ray diffraction analysis were performed on the surface of thick films.

Table 1. Nominal composition ratios of mixed powders.

Composition	Nominal composition				
	Bi :	Pb :	Sr :	Ca :	Cu
(A)	1.84:	0.34:	1.91:	2.03:	3.06
(B)	1.6 :	0.4 :	1.6 :	2.0 :	2.8
(c)	1.6 :	0.4 :	2.0 :	2.0 :	3.0
(D)	1.8 :	0.4 :	2.0 :	2.0 :	3.0

3. RESULTS AND DISCUSSION

The thick films on Ag tapes showed Tc(zero) of around 103-106 K. Table 2 lists the typical values of Jc for the tapes with the different compositions and fabrication conditions. The X-ray diffraction patterns for the tapes with compositions (B) and (D) are shown in Fig. 1. All the films are highly oriented with the c-axis perpendicular to the Ag tape as shown in Fig. 1, where the strong (001) peaks reflected from the 110 K phase were observed. The peak $2\theta = 21.9°$ of the semiconducting phase with a single Cu-O layered structure[5] were observed in tape (B-1). Compositions (A) and (B) have the excess amount of Ca deviated from the stoichiometric composition of 110 K phase (Sr:Ca:Cu=2:2:3). The semiconducting phase appeared in the films with compositions (A) and (B) annealed at high temperatures of 840-850 ℃, which reduced significantly the Jc of the tape as observed in tapes (A-1) and (B-1). The semiconducting phase was observed very little in tape (D-1) even annealed at 850 ℃. Although the so called "gold color phase" of Sr-Ca-Cu-O in the Bi oxide system [6] was found in all the tapes with the compositions

Table 2. Typical values of Jc for the tapes.
Fabrication conditions are also listed.

Tape No.	Composition	Annealing temperature (℃)	Film thickness (μm)	Jc (A/cm^2)
(A-1)	(A)	845(*)845	25	980
(A-2)	(A)	830(*)830	31	2200
(B-1)	(B)	845(*)845	34	830
(C-1)	(C)	840(*)840	27	3100
(D-1)	(D)	850(*)845(*)830	11	12000
(D-2)	(D)	845(*)830	8	8000
(D-3)	(D)	835(*)830	9	4700

(*):Intermediate cold rolling process.

(A), (B), (C) and (D) when annealed at high temperatures, the influence of this phase on the Jc of tapes seems to be less than the semiconducting phase. Since the high temperatre annealing can promote not only the alignment of platelike grains parallel to the Ag tape but also strong bondings between those grains in our films, composition (D) has the advantage in obtaining the higher Jc of films. The highest Jc of 1.2×10^4 A/cm^2 at 77 K was achieved in tape (D-1) when step-annealed from 850 down to 830 ℃ as listed in Table 2. Figure 2 shows the film thickness dependence of the Jc at 77 K. The Jc of tapes tends to increase with decreasing the tape thickness. It is clear that the films with composition (D) have the higher Jc compared with those of other compositions. Figure 3 shows the surface morphology of film (D-1) observed by SEM technique. Platelike grains with a very densified alignment were found on the film surface. The first rolling process after the preheat treatment of 500 ℃ is very effective in improving the packing density of the film. The intermediate rollings between the annealings of 830-850 ℃ promote the c-axis alignment of the platelike grains as well as the growth rate of 110 K phase. Figure 4 shows the typical magnetic field dependence of Jc at 4.2 K for the tapes prepared by the same conditions as tape (D-1). The magnetic field was applied in the paralell (H∥) or perpendicular (H⊥) direction to the tape surface. The magnetic field dependence of Jc in the tape is small at 4.2 K. The Jc was reduced by about one half when the magnetic field was increased from 0.5-1.0 T to 12 T. The Jc depend on the direction of the magnetic field, which may be due to the anisotropy and c-axis alignment of grains in the tape. A rather large hysteresis was observed between the increasing and decreasing magnetic field. This hysteresis may be attributed to the weak coupling between each grain, which was often observed in sintered bulk specimens[7].

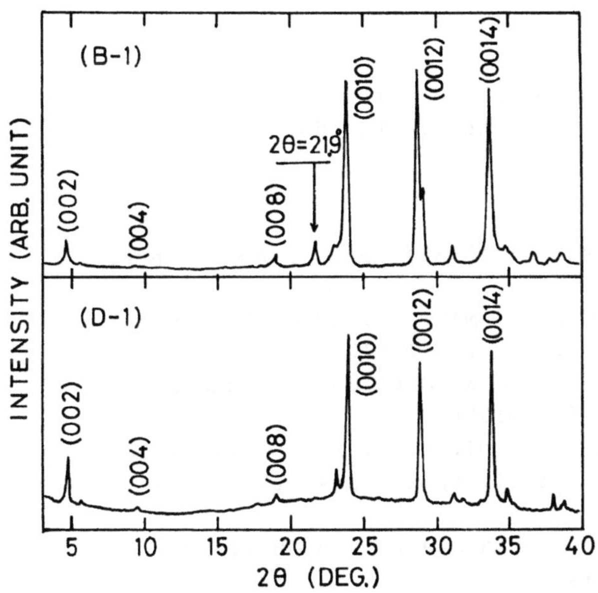

Fig. 1. X-ray diffraction patterns of tapes (B-1) and (D-1).

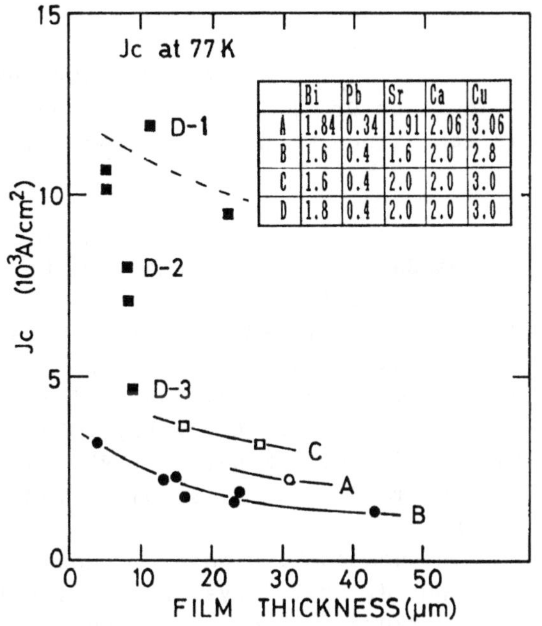

Fig. 2. Film thickness dependence of Jc at 77 K.

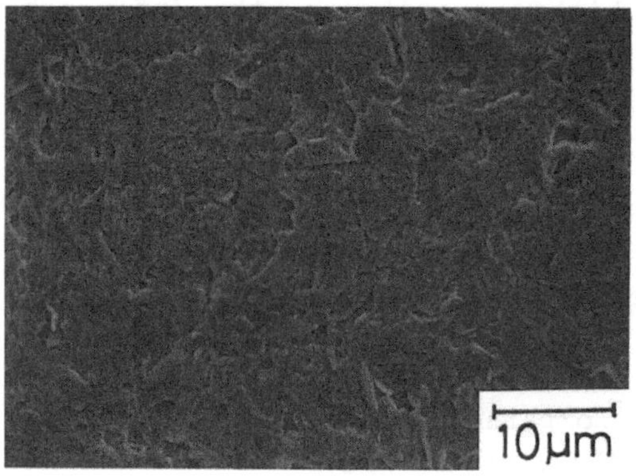

Fig. 3. Surface morphology of tape (D-1).

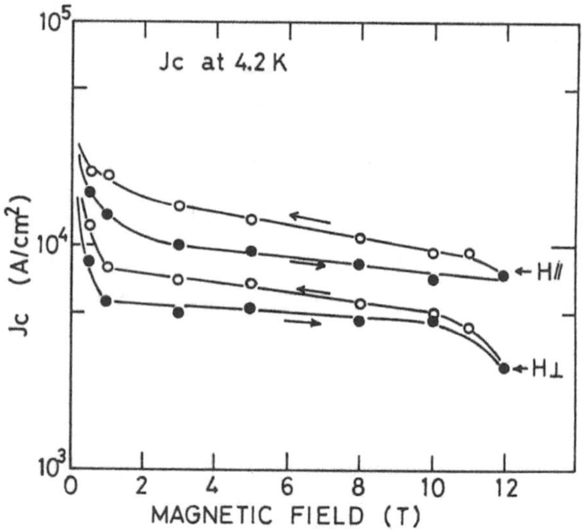

Fig. 4. Magnetic field dependence of Jc at 4.2 K for the tape with about 8000 A/cm^2 at 0 T and 77 K.

In summary, we successfully fabricated superconducting Bi-Pb-Sr-Ca-Cu-O thick film on Ag tape. The highest Jc of $1.2x10^4$ A/cm^2 at 77 K in a zero magnetic field was achieved for the films with compositional ratio of Bi:Pb:Sr:Ca:Cu=1.8:0.4:2:2:3. The magnetic field dependence of Jc was also measured at 4.2 K. The tape showed the Jc of $1.5x10^4$ A/cm^2 in 0.5-1.0 T.

ACKNOWLEDGEMENT

We thank Dr. T. Asano of National Research Institute for metals for his assistance with the Jc measurement in magnetic fields.

REFERENCES

[1] Maeda H. Tanaka Y. Fukutomi M. Asano T (1988) Jpn. J. Appl. Phys. 27: L209-L210.
[2] Takano M. Takada J. Oda K. Kitaguchi H. Miura Y. Ikeda Y. Tomii Y. Mazaki H (1988) Jpn. J. Appl. Phys. 27: L1041-L1043.
[3] Asano T. Tanaka Y. Fukutomi M. Jikihara K. Machida J. Maeda H (1988) Jpn. J. Appl. Phys. 27: L1652-L1654.
[4] Hoshino K. Takahara H (1989) Jpn. J. Appl. Phys. 28: L1214-L1216.
[5] Koyama S. Endo U. Kawai T (1988) Jpn. J. Appl. Phys. 27: L1861-L1863.
[6] Shigematsu K. Higashi I. Hoshino K. Takahara H. Aono M (1989) Jpn. J. Appl. Phys. 28: L1442-L1445.
[7] Kumakura H. Togano K. Maeda H. Yanagisawa E. Morimoto T (1989) Jpn. J. Appl. Phys. 28: L176-L178.

Fabrication and Characteristics of Multi-core Tl-Ba(Sr)-Ca-Cu Oxide Superconducting Tapes

M. SEIDO[1], F. HOSONO[1], T. UMEZAWA[1], A. NOMOTO[1], K. NOMURA[1], and T. MATSUMOTO[2]

[1] Metal Research Laboratory, Hitachi Cable Ltd., Tsuchiura, Ibaraki, 300 Japan
[2] Hitachi Research Laboratory, Hitachi Ltd., Hitachi, Ibaraki, 319-12 Japan

ABSTRACT

The Tl-Ba(Sr)-Ca-Cu oxide superconducting material (TBSCCO) was formed into single- or multi-core (36- or 1332-core) tapes by means of the Ag-sheathed powder method (the rolling or pressing). The maximum critical transport current density Jc of 5800 A/cm^2 was achieved at 77 K and 0 T in a 36 multi-core rolled tape with 0.07 mm thickness. The Ic values of the pressed tapes with single- and 36-core reached 21 and 17 A at 77 K, respectively. It was found that the pressing method improved the Jc of the multi-core tapes in 0-2 T magnetic fields and that there were no indications of difference of Jc values in any direction of the magnetic field.

INTRODUCTION

Since the discovery of the high Tc oxide superconducting materials, $Y_1Ba_2Cu_3O_y$ (YBCO) [1], Bi-Sr-Ca-Cu-O (BSCCO) [2], Tl-Ba-Ca-Cu-O (TBCCO) [3] and Tl-Ba(Sr)-Ca-Cu-O (TBSCCO) [4], various researches have been done to fabricate wires of these new materials for the power apparatus applications; such as, the superconducting magnets with the liquid nitrogen cooling. In many attemped methods to fabricate superconducting wires, the Ag-sheathed powder method [5] is one of the most hopeful techniques. The tape shaped wires fabricated by this method have relatively high critical current densities Jc's, for example, Jc=3-4x10^3 A/cm^2 of the Ag-sheathed YBCO tapes and Jc=10^4 A/cm^2 of the Ag-sheathed single-core TBSCCO tapes at 77 K and 0 T [6]. Multi-core superconducting wires are expected to have some advantages of the power apparatus applications; the higher stability margine and reducing the a.c. losses, even if in the case of oxide superconducting wires.

In the present study Ag-sheathed multi-core TBSCCO tapes were fabricated by the rolling or pressing method, and their superconducting characteristics were investigated in various conditions. The purpose of this research is to investigate the relationship between fabrication conditions and the Jc of Ag-sheathed multi-core TBSCCO tapes.

EXPERIMENTAL

The TBSCCO powder was prepared by sintering the mixture of Tl_2O_3, BaO, SrO, CaO and CuO powder (Tl:Ba:Sr:Ca:Cu=2:1.6:0.4:2:3), in a closed Al_2O_3 crucible at 1073-1223 K. After then the superconductivity of the powder was examined by the a.c.susceptibility change; it revealed that the powder had the Tc (on set) of 118 K and a very high superconducting volume fraction. This oxide superconducting powder was packed into a silver tube, then the tube was reduced by drawing or swaging. The resulted single-core wire was cut into pieces with a certain length. Ag-sheathed wires were assembled into a silver tube. Then this tube was again reduced into a multi-core wire as shown in Fig.1. We employed two methods- the rolling or pressing methods- to form this multi-core wire into a thin multi-core tape. The resulted multi-core tapes were sintered in an oxygen atmosphere at 1070-1170 K in order to sinter oxide particles in the silver sheath. In the pressing method, this sintering process was repeated before and after the pressing.

After the procedure, we measured the superconductivities (Jc and Tc) of the tapes by the four-probe technique. The Jc was also measured in a magnetic field of 0-2 T at nitrogen boiling temperature (77 K). The magnetic fields were applied perpendicular or parallel to the tape surface. Criterion for the critical currents was 1 uV/cm. The microstructures of the tape were observed by SEM and a polarized optical microscope.

RESULTS AND DISCUSSION

Long multi-core superconducting wires and tapes can be formed without any difficulties by the Ag-sheathed powder method. Figure 2 shows the cross sectional views of the Ag-sheathed TBSCCO multi-core wire and tapes of the length of 3 - 10 cm after sintering. The multi-core tapes have a good contact between the TBSCCO and Ag-sheath, and there observed no large voids in the cross sections.

The temperature dependence of the electrical resistivity was measured on the Ag-sheathed TBSCCO multi-core tape. Most of the multi-core tapes presented the critical temperature Tc (zero resistivity) was 107 K.

Figure 3 shows the relation between the tape thickness and the Jc at 77 K under zero magnetic field on the single- and multi-core tapes fabricated by the rolling and pressing methods. In the case of rolled tapes, it was clearly observed that the Jc sharply increased with decreasing the tape thickness less than 0.2 mm. The Jc values of 36-core tapes were higher than those of others. The Jc values of 1332-core tapes were considerably low. The highest Jc values of single-, 36- and 1332 core tapes were 4000, 5800 and 1100 A/cm^2, respectively, at the tape thickness of 0.05 - 0.07 mm. In the case of pressed tapes, relatively high Jc values were attained in the thicker range of the tape thickness, which transported higher critical current Ic than rolled tapes. The highest Ic values of single- and 36-core tapes were respectively 21 and 17 A, where their cross-sections were 0.34x5 and 0.26x6mm^2, respectively.

The Jc values of multi-core tapes depend on uniformity of the oxide superconducting core filaments. Figure 4 shows longitudinal sections of Ag-sheathed multi-core TBSCCO tapes after sintering. The filaments of the 36-core tape had relatively uniform configuration, however those of the 1332-core tape had partially non-uniform area in the macro-scale. The difference in the uniformity of the fillaments caused the difference in the Jc's of the 36- and 1332-core tapes.

The densification of superconducting core effects on the Jc's of the tapes. The densified condition on the superconducting core was examined by the observations of SEM images of fracture surface parallel to the rolled surface of single- and 36-core 0.1 mm thickness tapes. While the single-core

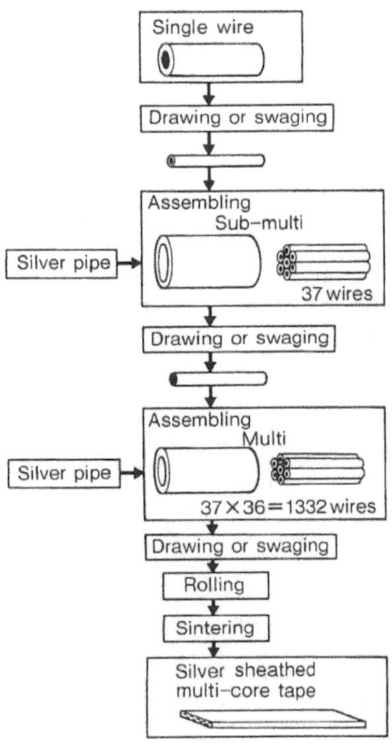

Fig.1 Fabricating process of multi-core tapes

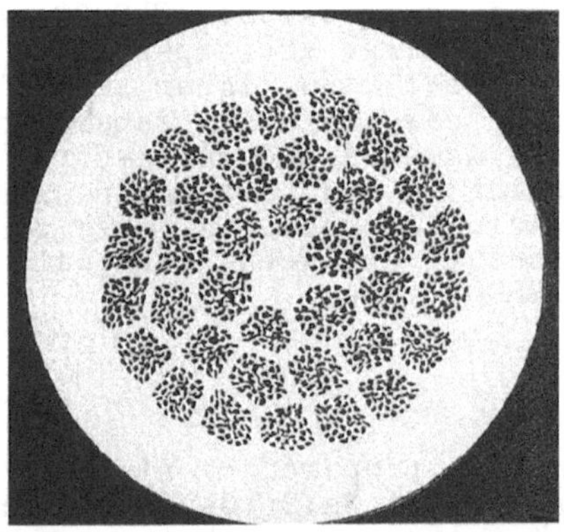

a. 1332-core wire, $\phi 2$

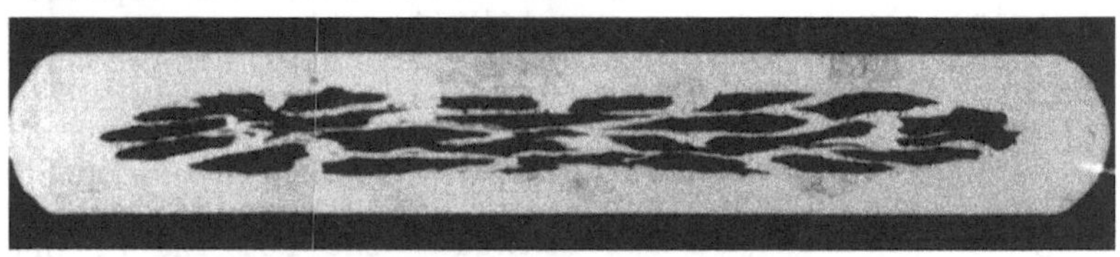

b. 36-core tape

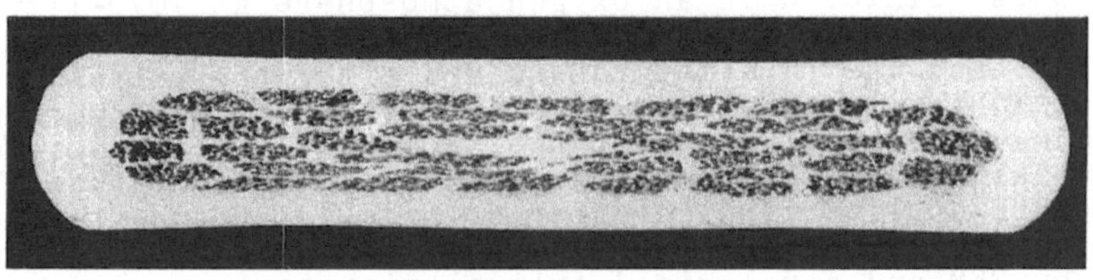

c. 1332-core tape

Fig.2 Cross-sectional view of the Ag-sheathed TBSCCO multi-core wire and tapes

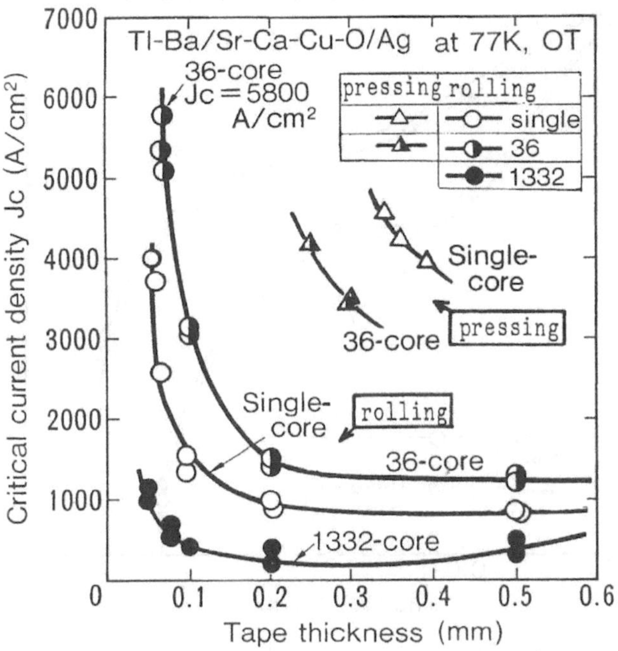

Fig.3 Comparison of critical current densities between single- and multi-core TBSCCO tapes with various thicknesses.

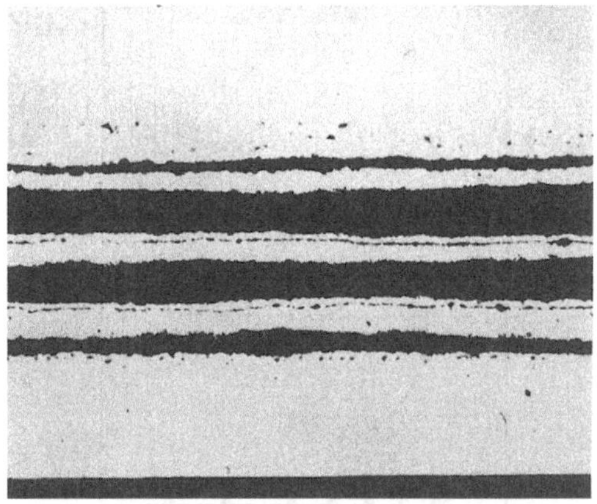

a. 36-core tape (0.43^t)

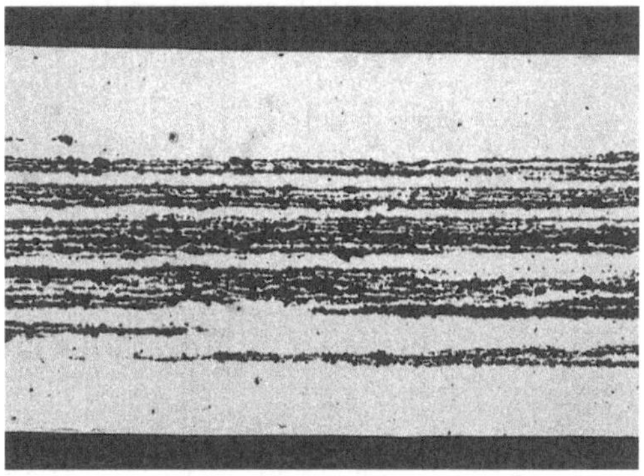

b. 1332-core tape (0.5^t)
Fig.4 Longitudinal views of Ag-sheathed multi-core TBSCCO tapes.

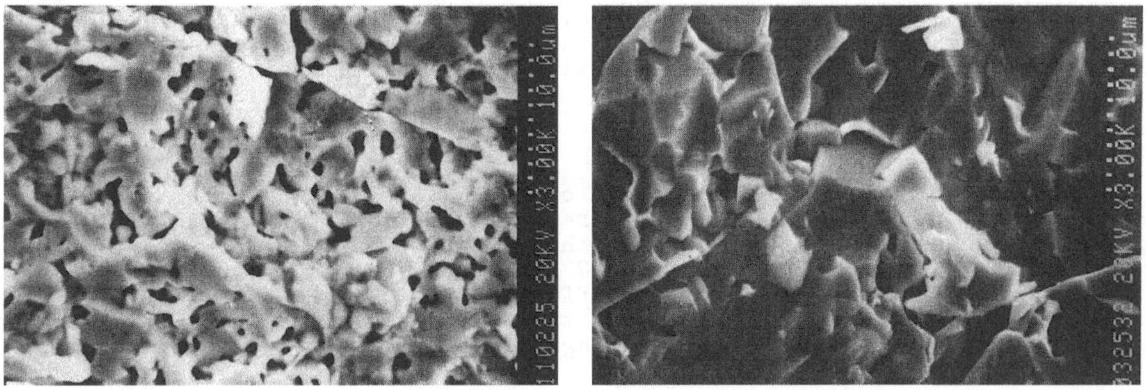

a. single-core tape (0.07^t) b. 36-core tape (0.07^t)
Fig.5 SEM images of TBSCCO fractured core surface parallel to rolled surface

tape has a considerably porous core, the 36-core tape has quite densified cores. The oxide cores of the pressed tapes were more highly densified than those of the rolled tapes. These densification of the oxide cores corresponds to the Jc values of the tapes. Furthermore, superconducting oxide grains were almost randomly aligned.

Figure 6 shows the magnetic field dependence of Jc at 77 K for single-, 36- and 1332-core rolled tapes of 0.05 - 0.07 mm thick. The Jc values of all samples sharply decreased in a low magnetic field, due to the weak-link at the oxide grain boundaries. The Jc values at 0.1 T are about 1/25 - 1/20 times as low as that at 0 T in all kinds of the tapes. In order to obtain a higher Jc wire in a magnetic field, it is necessary to improve the rinks of the superconducting grain boundaries.

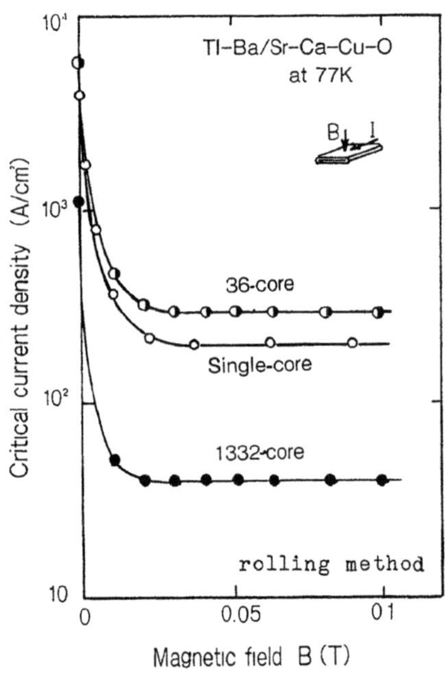

Fig.6 Critical current densities of single- and multi-core TBSCCO tapes in an external magnetic field.

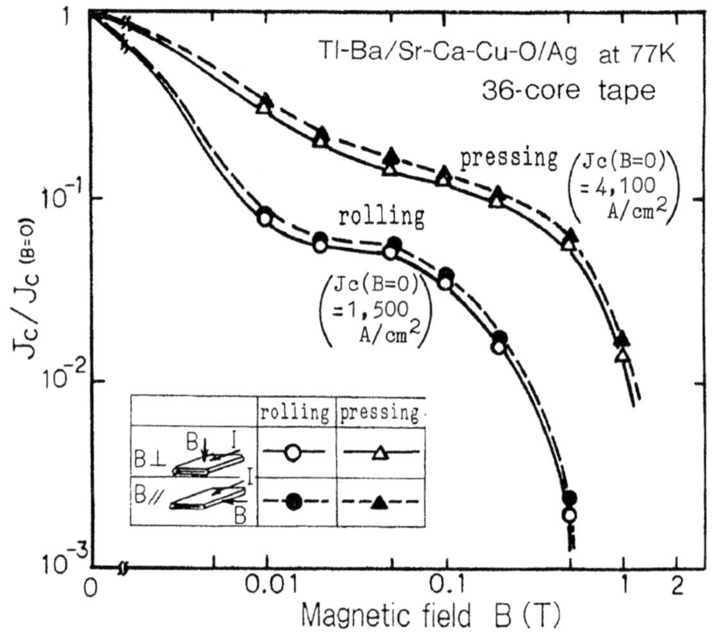

Fig.7 Critical current density ratios of 36-core TBSCCO tapes in 0 - 2 T external magnetic fields.

Figure 7 compares Jc values of rolled tapes and pressed tapes in an external magnetic field. The number of the superconducting cores was 36. It was clearly found that the Jc of the multi-core tapes was much enhanced by pressing in an external magnetic field. In the case of the pressed tape, Jc/Jc(B=0) value was 1/8 at 0.1 T and 1/60 at 1 T. Furthermore, there were no indications of Jc values difference of the tapes in the magnetic fields applied perpendicular or parallel to the tape surface, whether tapes were fabricated by the rolling or pressing. This depends on the random alignment of the oxide superconducting grains in the Ag-sheath as shown in Fig.5.

From these results, it is thought that the Jc enhancement mechanism of the present TBSCCO superconducting tapes is different from that of the BSCCO tapes. In the case of BSCCO tape, the main factor of the high Jc is the C-planes alignment in the same orientation [7]. The main factor of the high Jc of the TBSCCO tape is the densification of the oxide grains in Ag-sheath.

CONCLUSION

We investigated the fabrication of Ag-sheathed TBSCCO multi-core tapes. TBSCCO tapes with 36- and 1332-core could be fabricated by the rolling or pressing method. The Jc of Ag-sheathed TBSCCO tapes with 36- and 1332-core reached 5800 A/cm^2 and 1100 A/cm^2 at 77 K and 0 T, respectively. Furthermore, it was found that the pressing was effective to obtain a high Jc TBSCCO tape in an external magnetic field and that there were no indications of difference of Jc values in any direction of the magnetic field.

REFERENCES

[1] M. K. Wu et al., Phys. Rev. Lett., 58, 908 (1987)
[2] H. Maeda et al., Jpn. J. Appl. Phys., 27, L209 (1988)
[3] Z. Z. Sheng, A. M. Herman et al., Phys. Rev. Lett. 60, 937 (1988)
[4] M. Matsuda et al., Jpn. J. Appl. Phys., 27, 2062 (1988)
[5] M. Seido et al., Proc. 1st Int. Symp. Superconductivity, (1988) p309.
[6] T. Matsumoto et al., ISTEC Workshop on Superconductivity, (1989) p111.
[7] K. Sato et al., proc. ISTEC Workshop on Superconductivity, (1989) p119.

Formation of Y-Ba-Cu-O Thick Films by Plasma Spraying

YOSHIHIDE WADAYAMA, TOSHIMI MATSUMOTO, KATUZO AIHARA, and SINPEI MATSUDA
Hitachi Research Laboratory, Hitachi Ltd., Kuji-cho, Hitachi, 319-12 Japan

ABSTRACT

Chemical composition, pore volume and superconducting properties of plasma sprayed film was investigated in ralation to the spraying conditions. $Y_1Ba_2Cu_3O_{7-y}$ powder was sprayed onto a Ni-Cr alloy substrate. The superconducting film formed was about 100μm thick. The amount of Cu decreases with excessive melting of the powder, related to the powder particle size and power supply. The pore volume could be controlled to less than 5% in area fraction by using fine powder and low atomospheric gas pressure. The Tc and Jc reached values of 91K and 1120A/cm^2(at 77K, OT)respectively for $Y_1Ba_2Cu_3O_{7-y}$ film after post-annealing

KEY WORDS: oxide superconductor, Y-Ba-Cu-O, plasma spraying, thick films

INTRODUCTION

The discovery of Y-Ba-Cu-O superconductors with Tc above 90K has had a great impact on the superconducting fields, and considerable effort has been directed to producing the superconducting wires and films. Plasma spraying, one of the mass production processes, is useful method used to form superconducting thick and large scale films. [1 ~ 3] This paper presents the effect of plasma spraying conditions on chemical composition, pore volume and superconducting properties of the film.

EXPERIMENTAL

Superconducting powders for spraying were prepared by a solid state reaction method. Mixture of high-purity Y_2O_3,$BaCO_3$, CuO powders were calcined twice at 930°C in flowing O_2 for 10h. The powders were sieved into some groups of different particle size. The average particle sizes were 8, 84 and 145μm. Table 1 shows the operating conditions of plasma spraying. Spraying condition such as power supply, gas pressure and powder particle size were varied. The standard condition of this study were underlined in the table.

Table 1 Operating conditions of plasma spraying.
Standard conditions of this study were underlined.

Substrate	Hastelloy X, 100×100×3mm
Source gas	Ar, 50ℓ/min
Current	600, <u>800</u>, 900A
Voltage	34V
Gas pressure	50, 400, <u>760</u> Torr
Particle size of powder	8, <u>84</u>, 145μm
Spraying distance	100, 300mm

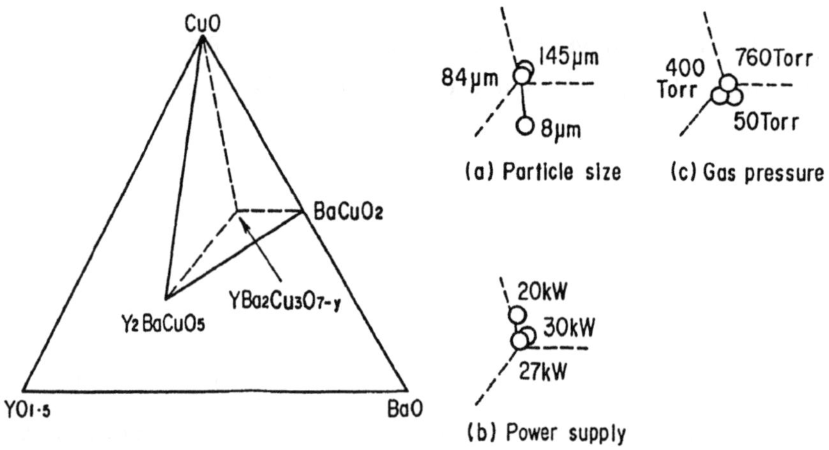

Fig.1 Compositions of plasma sprayed films on the $YO_{1.5}$-BaO-CuO pseudo thernary phase diagrams.

Classified powders were sprayed onto Hastelloy X substrate at room temperature. Hastelloy X substrates were sandblasted by $A\ell_2O_3$ in prior to the plasma spraying. The resulted film is about 100μm thick.

Sprayed films were resistivily measured by a standard dc, four-probe method. Critical temperature, Tc, and critical current density, Jc, were defined by the criteria of 0.1μΩcm and 1μV/cm, respectively. In the Jc measurements, the films were cut into a pattern of about 1mm width in order to reduce the total current Several films prepared by different spraying conditions were characterized by an X-ray diffraction and electron dispersive X-ray spectroscopy(EDX). Their microstructures were also observed by scanning electron microscope(SEM) and optical microscope.

RESULTS AND DISCUSSION

Figure.1 shows the chemical compositions of plasma sprayed films on the $YO_{1.5}$-BaO-CuO pseudo thernaty phase diagrams. The samples for chemical analysis were evaluated by Inductively Coupled Plasma Spectrometer (ICP). The EDX analysis revealed that elements Y, Ba and Cu were uniformly distributed all over the films. The amount of Cu decreased with excessive melting of the powder, itself related to powder particle size. Vaporization of Cu increased with decreasing of powder particle size. Gas pressure was less influential than other parameters.

Fig.2 shows the X-ray diffraction patterns of plasma sprayed films which has been formed by spraying on the substrate substeate at room temperature. Sprayed film was

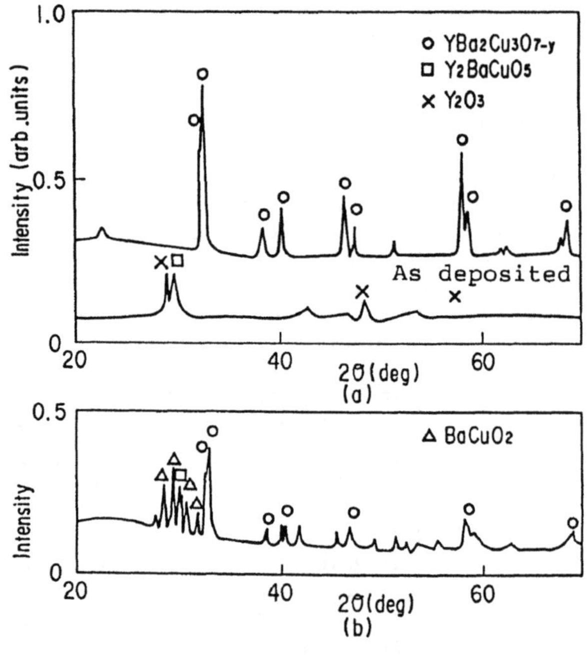

Fig.2 X-ray diffraction patterns of as-deposited and post-annealed films obtained by different particle size. Annealing condition:900°C, 1h 900°C, 1h (a)84μm,(b)8μm

predominantly amorphous with a small amount of Y_2O_3 and Y_2BaCuO_5 under the as-deposited condition, depending on the melt quenching conditions. Crystallization of the film was induced by post-anneling at 900°C for 1h. $Y_1Ba_2Cu_3O_{7-y}$ sigle phase film was obtained under the following condition; powder size:84μm and power supply:27~30kW. The second phases such as $BaCuO_2$ and Y_2BaCuO_5 were observed at the film formed by using fine powder (8μm), due to vaporization of Cu. In order to produce the single phase film, the vaporization of Cu must be suppressed during plasma spraying.

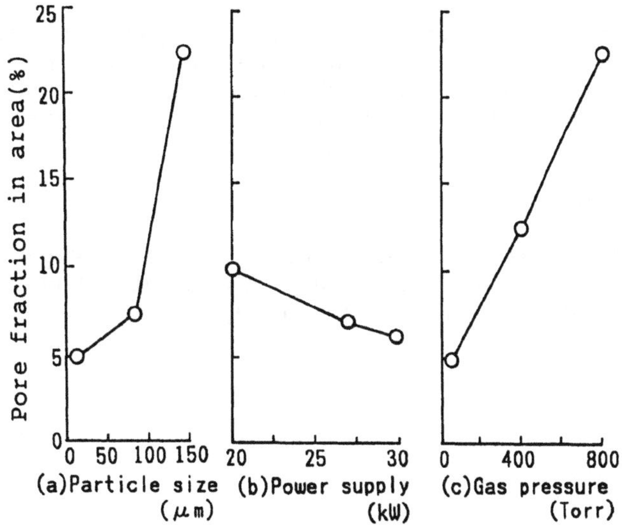

Fig.3 Effect of spraying condition on the porosity of films.

Fig.3 shows the effect of spraying conditions on the pore fraction of sprayed films. The pore fraction in area were affected by the powder particle size and gas pressure. The pore fraction decreased with decreasing of gas pressure, depending on the incease of plasma jet speed. The pore fraction could be controlled less than 5% by using fine powder and low gas pressure. Fine powder, however, results in the excessive vaporization of Cu in the sprayed films. High density films with stoichiometric composition could be obtained by using low gas pressure.

Fig.4 shows the temperature dependence of resistivity of post-annealed films. Sprayed films were annealed at 900°C for 1h in flowing O_2 atmosphere. Superconducting film with Tc of 87K was obtained under the condition;powder size:84μm and power suplly:27kW(a). Resistivity of this film dose not, however, decrease metallically with decrease of temperature. It seem to depend on the excessive deficiency of oxgen atom at the $YBa_2Cu_3O_{7-y}$. After low temperature annealing at 400°C for 100h in flowing oxgen, Tc and Jc of the film were improved to the value of 91K and 1120A/cm^2 (at 77K, OT) respectively.(b)

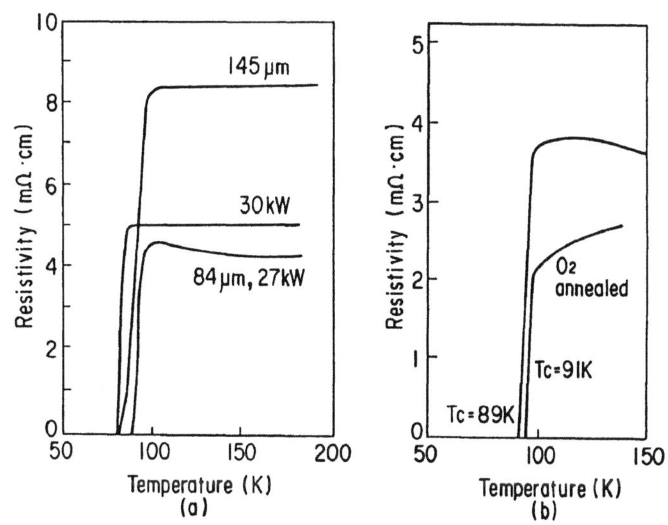

Fig.4 Temperature dependence of resistivity of post-annealed films.(a)900°C,1h (b)900°C, 1h and 400°C,100h

Acknowledgements

This work was performed as a part of "R&D on Superconducting Technology for Electric Power Apparatuses"as a subject of Super-GM under the Moonlight Project of Agency of Industrial Science and Technology, MITI, being consigned by New Energy and Industrial Technology Development Organization(NEDO).

REFERENCE

1. I. Sanakawa et al. Jan. J. Appl. Phys. 27, 6(1988)L1083-L1085
2. Y. Yoshida et al. Jan. J. Appl. Phys. 28, 4(1989)L639-L642
3. N. Mori et al. Jan. J. Appl. Phys. 28, 2(1989)L239-L240

Superconducting Properties and Microstructure of Ag-Sheathed $YBa_2Cu_3O_y$ and Sintered $YBa_2Cu_3O_y$

TSUKUSHI HARA[1], HARUHIKO HOSHINO[2], TAKAHIKO YAMAMOTO[1], HIDEO ISHII[1], and MASAAKI NAKAMURA[1]

[1] Engineering Research Center, Tokyo Electric Power Company, Chofu, Tokyo, 182 Japan
[2] Power and Substation Department, Kandenko, Chofu, Tokyo, 182 Japan

ABSTRACT

Comparative studies of certain properties of Ag-sheathed $YBa_2Cu_3O_y$ wire and sintered $YBa_2Cu_3O_y$ revealed numerous differences between the samples with respect to critical current, superconducting volume fraction, SEM image, magnetic relaxation, and so on. The factors which determine the critical current density and the possibility of higher critical current density are discussed in the light of these data.

KEYWORDS: critical current density, superconducting volume fraction, grain boundary, magnetic relaxation

INTRODUCTION

Superconducting characteristics of Ag-sheathed $YBa_2Cu_3O_y$ (YBCO) wire[1,2] samples were studied and compared with those of YBCO samples that have only been pressed and sintered. The results showed that the critical current density (Jc) of the Ag-sheathed samples was more than five times greater than that of the sintered samples, whereas the superconducting volume fraction of the sintered samples considering the demagnetizing effect was about twice as large as that of the Ag-sheathed samples. The results also revealed that the irreversibility of the Jc with respect to the applied field was much larger in the sintered than in the Ag-sheathed samples. This tendency was common to all the measured samples. In order to further clarify these results, X-ray diffraction patterns, SEM images, and magnetic relaxation were investigated for both types of samples. The factors which determine the Jc and the possibility of higher Jc are discussed on the basis of these experimental results.

BASIC PARAMETERS

Sintered YBCO was obtained using an ordinary ceramic technique with 950°C final heat treatment in O_2 gas flow. To prepare the Ag-sheathed wires, YBCO powder calcined at 920°C was stuffed into a Ag-tube, then swaged and rolled into a tape with a thickness of 0.5 mm and width of 5 mm, and finally heat-treated at 900°C. The Jc of the Ag-sheathed sample measured by the transport method (1 $\mu V/cm^2$) was $1370 A/cm^2$, which was more than five times larger than that of the sintered sample ($195 A/cm^2$) at 77 K. Zero resistivity was achieved at 90.5 K for the former sample, which is almost the same as the corresponding result for the latter, i. e., 91.3 K.

SUPERCONDUCTING VOLUME FRACTION

Figures 1 (a) and 1 (b) show the temperature dependence of the field-cooled (FC) superconducting volume fraction (SVF) of the sintered YBCO and the Ag-sheathed YBCO as estimated from the following relation, taking the demagnetizing effect into account.

$$V\% = -4\pi M_{obs}/VH_a 100\ (1-n). \qquad (1)$$

Here, $V\%$ denotes the calculated SVF, M_{obs} the observed magnetic moment, V the sample volume and H_a the applied magnetic field (10 Oe), respectively. The demagnetizing factors n (using Pb) were investigated in order to obtain as precise a value of SVF as possible; this is described in detail in reference (3). All the measurements were performed with a SQUID susceptometer, model HOXAN HSM-2000. The indexes (A), (B) and (C) given for each curve in Fig. 1 denote the respective directions of the applied magnetic field corresponding to the inserts in the figures. As shown in Fig. 1, the SVF of the sintered sample was about two or more times larger than that of the Ag-sheathed sample with respect to each field direction. The magnitudes of SVF almost coincided in configurations (A) and (B) for both samples, while a far smaller value of SVF was obtained in the case of configuration (C). In configuration

(C), the penetration of magnetic flux at the edge of the sample was attributed to the fact that the local magnetic field at the corners of the rectangular solid sample exceeded the H_{c1} of the YBCO. Therefore, configuration (C) is not appropriate for comparing the SVF of different samples.

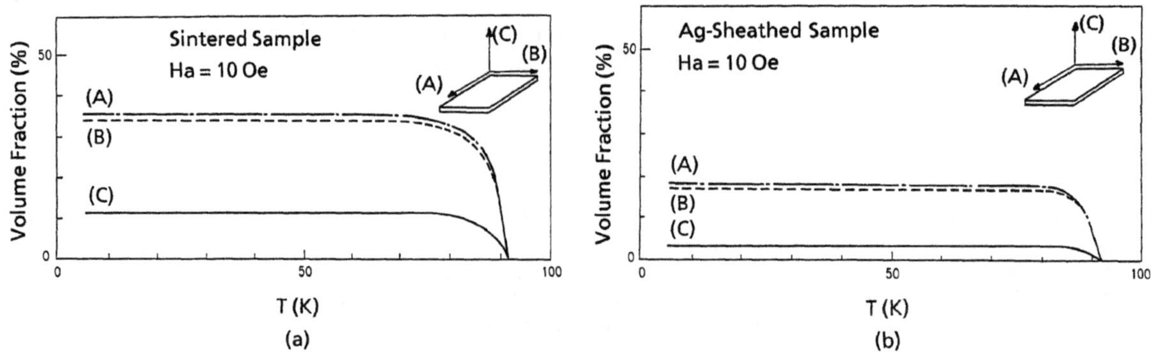

Fig. 1 Temperature dependence of field-cooled (FC) superconducting volume fraction: (a) 4.7 mm×1.4 mm×0.2 mm sintered sample (b) 3.5 mm×2.8 mm×0.14 mm Ag-sheathed sample. (A), (B) and (C) denote the respective directions of the applied field corresponding to the inserts.

The observed values of SVF, however, were far below 100%, indicating that the magnetic field was not perfectly excluded from the interior of the samples. Concerning this point, Malozemoff et al. reported that the magnetic susceptibility measured in the FC process decreased as the applied field was increased for both crystal and ceramic samples of YBCO, and interpreted this result as the effect of flux pinning.[4] A much larger SVF would also have been obtained for our samples by measurements using an applied field weaker than 10 Oe. Presumably one can, at least, compare the relative magnitudes of SVF for different samples using the values obtained in configurations (A) and (B). All of the above results indicated that the proportion of intrinsic superconductive parts was larger for the sintered than for the Ag-sheathed sample. This result seems to be contrary to the higher Jc displayed by the Ag-sheathed sample.

IRREVERSIBILITY OF Jc ON APPLIED MAGNETIC FIELD

The irreversibility of Jc when the magnetic field was decreased (hereafter referred to as "Jc-B") was observed in both the Ag-sheathed and the sintered samples, as shown in Fig. 2. Comparing Figs. 2 (a) and 2 (b), one sees that the irreversibility of Jc in the sintered samples was much larger than that in the Ag-sheathed samples. Evetts et al. reported that irreversibility was strongly affected by grain boundaries.[5] This, combined with the results of the present study, suggests that the grain boundaries have a stronger influence on sintered than on Ag-sheathed wires.

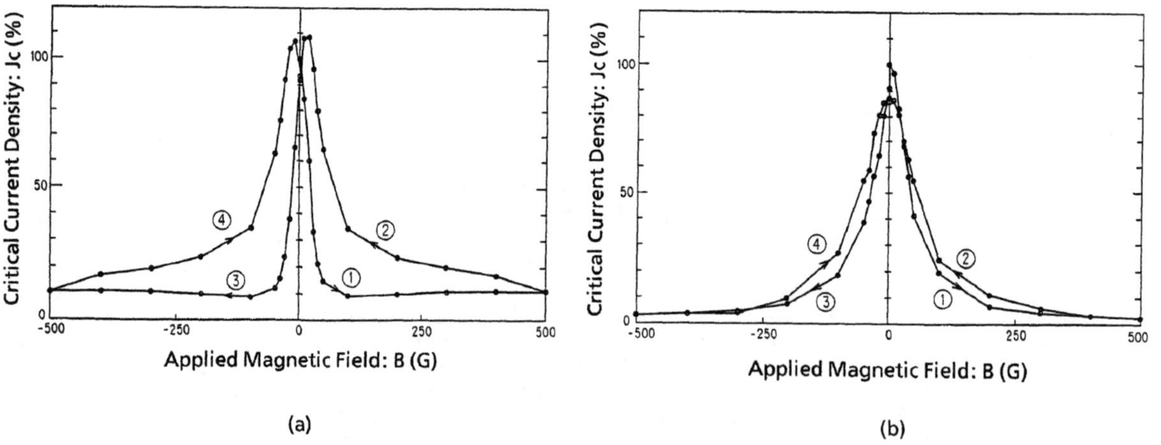

Fig. 2 Irreversibility of critical current density with respect to applied magnetic field at 77 K: (a) sintered sample (b) Ag-sheathed sample.

X-RAY DIFFRACTION PATTERN AND SEM IMAGES

The extent of grain orientation was investigated by X-ray diffraction; however, significant differences between the diffraction patterns and peak intensities of the Ag-sheathed and sintered samples were not observed. Therefore, the main factor responsible for the higher Jc of the Ag-sheathed sample presumably relates to other properties. SEM images were observed in order to investigate the microstructure of the samples directly. SEM photographs are shown in Figs. 3 (a) and 3 (b). Rectangular grains with a size of 10 μm are separated by clearly observed grain boundaries in the sintered sample. The existence of grain boundaries in the sintered sample suggests a strong relationship with the greater irreversibility of the Jc-B characteristics and the low Jc. On the other hand, in the Ag-sheathed sample, the average size of the grains seems to be far smaller than that of the sintered sample and the grains form clusters, with cracks and voids around each cluster. However, coupling between neighboring grains seems to be closer in each cluster, since clear grain boundaries cannot be observed in the polished surface. This situation is even more pronounced in Fig. 3 (c), which is a SEM image in the vicinity of the Ag-sheath. These morphological characteristics are presumably related to the lesser irreversibility of Jc-B characteristics and higher Jc as compared with the sintered samples. One may speculate that the current path in the Ag-sheathed sample allows higher current density than that of the sintered sample, although this path would seem to be of very small cross-section due to the existence of cracks between clusters of grains. These microstructural observations indicate that Jc in the sintered sample is smaller than that in the Ag-sheathed sample due to the many grain boundaries in its path. The cracks and the large voids observed in the Ag-sheathed sample, which are crucial for the small cross-section of the current path and the small superconducting volume fraction, are presumably formed during the heating process because the thermal expansion of Ag is greater than that of YBCO.[6] One may at least expect to achieve higher critical current densities by optimizing the process conditions in order to reduce defects and to obtain a highly oriented grain structure.[6]

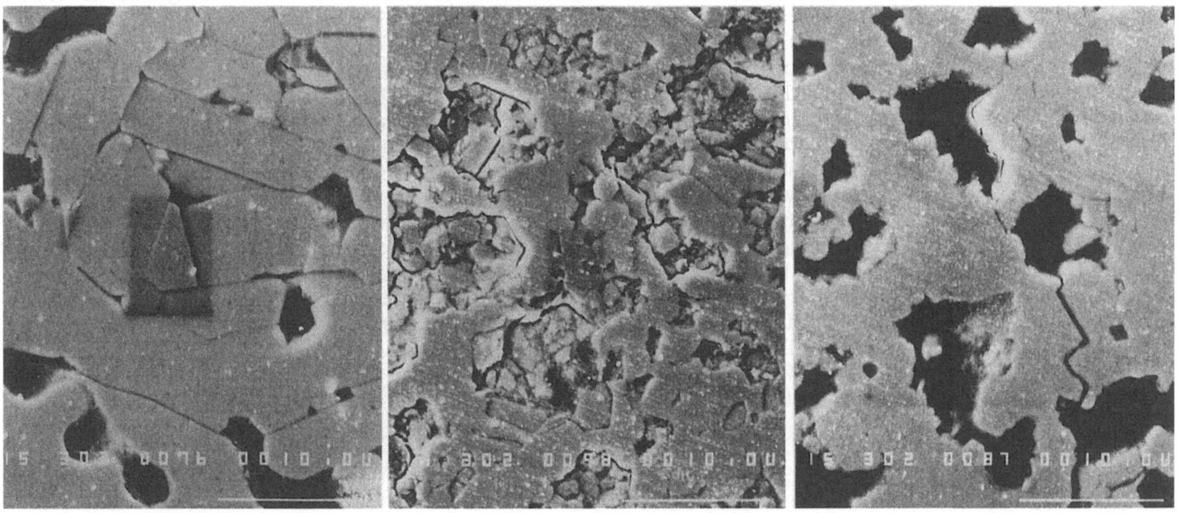

(a) 10 μm (b) 10 μm (c) 10 μm

Fig. 3 Microstructure of samples observed by SEM: (a) sintered sample, (b) Ag-sheathed sample, (c) vicinity of Ag-sheath.

MAGNETIC RELAXATION

Finally, the flux trap was measured and a kind of activation energy U_0 was estimated for both samples. U_0 is calculated from the following relation, which is based on pinning theory:

$$M/M_0=1-(kT/U_0)\ \ln\ (t/t_0) \qquad (2)$$

Here, M_0 and M denote the observed magnetic moment at time zero and time t, respectively, and k is the Boltzmann constant. 50 Oe was applied to the samples before the measurement, and then the magnetic field was switched off. In order to avoid fluctuation due to switched-off-operation, t_0 was set one minute later than the switch-off time. Fig. 4 shows the magnetic relaxation of the sintered and Ag-sheathed samples. U_0 for the Ag-sheathed samples was 0.26 eV, which was a quarter of the U_0 value of the sintered sample.

This result indicates that the effect of the flux trap in the Ag-sheathed sample was smaller than that in the sintered sample, which seems to be consistent with the lesser irreversibility of Jc-B characteristics and the smaller number of grain boundaries in the SEM photos.

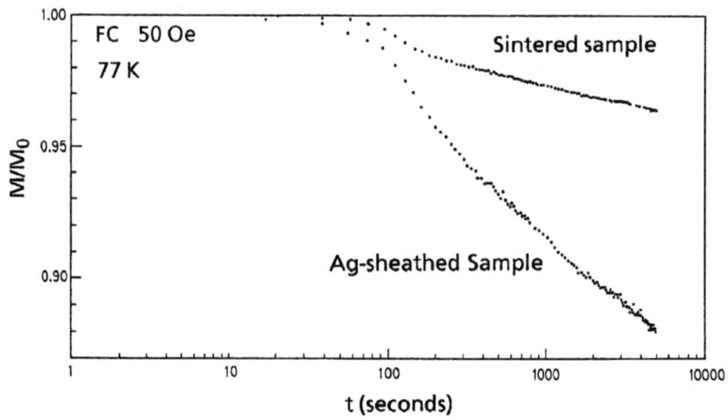

Fig. 4 Magnetic relaxation of Ag-sheathed and Sintered samples

SUMMARY

The results obtained demonstrated that the superconducting volume fraction of the sintered sample was about twice as large as that of the Ag-sheathed sample, whereas the critical current density of the Ag-sheathed sample was more than five times as great as that of the sintered sample. The irreversibility of Jc-B characteristics in the sintered sample was also much greater than that in the Ag-sheathed sample. The X-ray diffraction patterns showed no significant difference between the grain orientations of the two kinds of samples. However, the SEM images revealed larger grain boundaries in the sintered than in the Ag-sheathed sample. The larger grain boundaries in the sintered sample were indicative of the lower Jc, which is generally known. However, it should be pointed out that the larger grain boundaries in the sintered samples observed by SEM may be consistent with the tendency toward greater irreversibility of Jc-B characteristics and smaller magnetic relaxation. Grain boundaries appear to play the role of a kind of flux trap. Hereafter, the roles of grain boundaries in increasing the value of Jc should be clarified in greater detail. Another finding in the present study was the existence of large voids and cracks around clusters which reduced the cross-section of the current path throughout the Ag-sheathed sample. One can expect to achieve higher Jc by reduction of defects and alignment of grains through optimized process conditions and special techniques.

The authors gratefully acknowledge the assistance of Mr. N. Uno and Mr. H. Kikuchi of the Furukawa Electric Co., Ltd., who prepared the samples in accordance with the required specifications.

REFERENCES

1) M. Okada, A. Okayama, T. Morimoto, T. Matsumoto, K. Aihara and S. Matsuda: Jpn. J. Appl. Phys. **27** (1988) L185.
2) T. Hikata, K. Sato, H. Hitotsuyanagi: Jpn. J. Appl. Phys. **28** (1989) L82.
3) H. Ishii, H. Hoshino, T. Hara, M. Nakamura and T. Yamamoto: Jpn. J. Appl. Phys. 28 (1989)
4) A. P. Malozemoff, L. Kurusin-Elbaum, D. C. Cronemeyer, Y. Yeshurun and F. Holtzberg: Phys. Rev. B38 (1988) 6490
5) J.E. Evetts and B. A. Glowacki: Presentation at Symposium on "Critical Currents in High Tc Superconductors", UK (May 1989)
6) T. Takayama, S. Ochiai and K. Osamura: J. Jpn. Inst. Metals **53** (1989) 735. [in Japanese]

Preparation of Superconducting Thick Films of Bi-Pb-Sr-Ca-Cu-O by Gas Deposition of Fine Powder

S. KASHU[1], Y. MATSUZAKI[1], M. KAITO[1], M. TOYOKAWA[2], K. HATANAKA[3], and C. HAYASHI[3]

[1] Vacuum Metallurgical Co., Ltd., 516, Yokota, Sanbu-machi, Chiba, 289-12 Japan
[2] Tohoku Vacuum Metallurgical Co., Ltd., 1-7, Tanosawagashira, Hachinohe, 039-22 Japan
[3] ULVAC JAPAN Ltd., 2500, Hagisono, Chigasaki, 253 Japan

ABSTRACT

Superconducting thick film of BPSCCO is formed on MgO substrate by gas deposition using fine powder of Bi-Pb-Sr-Ca-Cu-O(BPSCCO) having diameter less than 1 μm. The pattern of the deposited film is linear, and the thickness of it can be controlled in the range of 5∿200μm. After heat treatment at 770° C for 10 hr in air atmosphere, Tc(end) showed 105 K and critical current density of 225 A/cm^2 are obtained at 77 K in zero magnetic field. The structure of the surface of thick film by X-ray diffraction pattern, and composition ratio of Bi and Pb in film, were reported. Tape like BPSCCO thick film with width of 10 mm was formed on substrate using similar method. Tape like thick film is expected to utilize as material of superconducting tape.

INTRODUCTION

Elements with superconducting composition were vaporized in gas atmosphere and generated ultra-fine powders were mixed, and thick film was formed by gas deposition method. The superconducting characteristic of this thick film has been investigated[1)] .

For making control of composition certainly and simple in this method, thick film was formed by gas deposition method using fine powder of Bi-Pb-Sr-Ca-Cu-O (BPSCCO) which has composition of superconductivity already, and evaluation of film was conducted. Together with formerly prepared linear film, tape like film with broad width was formed also.

EXPERIMENT

A. Preparation of linear film

Figure 1 shows schematic illustration of arrangment of experiment. Fine powder of BPSCCO which has superconductive composition with average diameter less than 1 μm was contained in mixing container, and open the valve and evacuate film deposition chamber by operating vacuum pump. Then fine particles were wafted in mixing container, and fine particles are carried through transfer tube, and sprayed on substrate from narrow nozzle on the tip of transfer tube.

With moving substrate in the direction of arrow mark, thick film of BPSCCO is formed on substrate by the deposition of fine powder. This method is called gas deposition method.

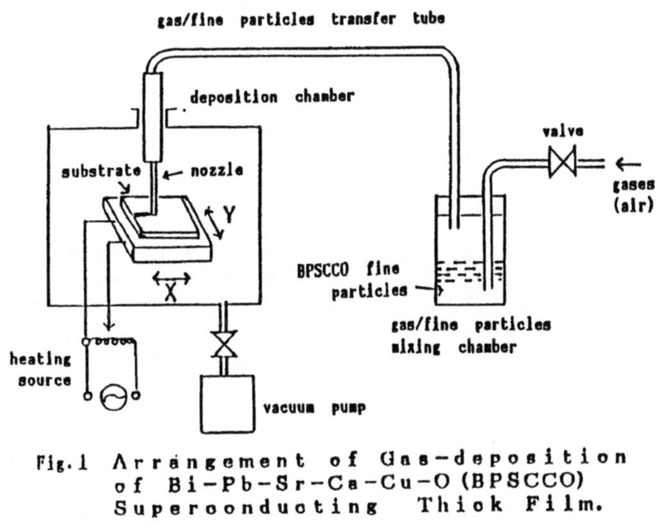

Fig.1 Arrangement of Gas-deposition of Bi-Pb-Sr-Ca-Cu-O (BPSCCO) Superconducting Thick Film.

The material of substrate is MgO and is heated up to 400 °C. The material powder is a mixture of calcined fine powder of BPSCCO, which is shown in Table 1, and fine powder of Pb in ratio of 9:1. The ratio of Bi and Pb is 1:1.

Figure 2 shows as depo. film and cross section of heat treated film which were formed using nozzle with inner diameter of 1 mm. This figure shows that as depo. film is made dense already, because the shrinkage of heat treated film is little. Range of 5∿200 μm is suitable for film thickness.

Table.1 BPSCCO Calcinated Fine Powder

Chemical Composition	$Bi_6Pb_4Sr_{10}Ca_{10}Cu_{15}O_x$
Mean Particle Diameter	less than 1 μm
Specific Surface Area	17.8 m^2/g
Purity	3 N

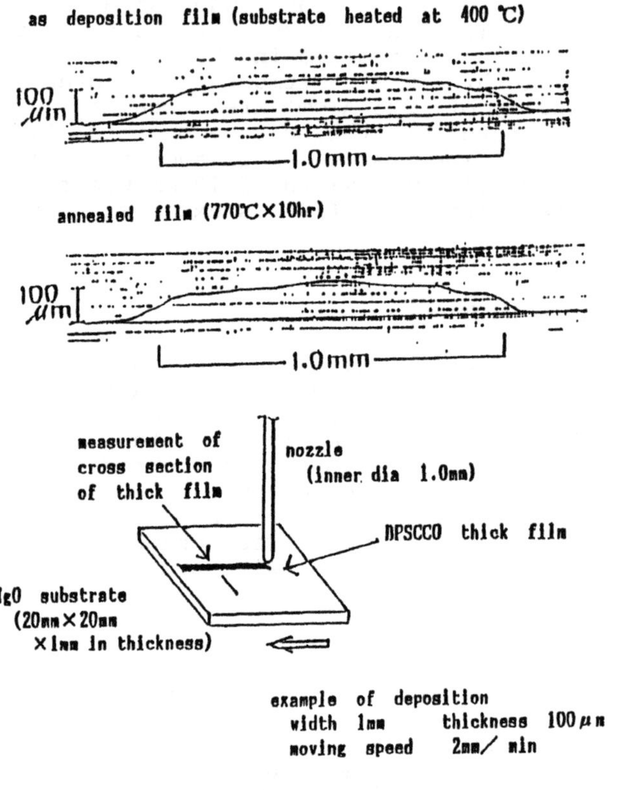

Fig.2 Shape of BPSCCO Thick Film.

B. Heat treatment

As depo. film set in electric furnace, and heated at 730∿790 °C in air and holded for 3∿30 hrs. Both increasing or decreasing rates of temperature are 5 °C/min. This condition of heat treatment is about 100 °C lower in temperature and holding time is as short as one several time compared to the condition of heat treatment which is applied for the preparation of BPSCCO polycrystal specimen.

C. Evaluation of thick film

The structure of thick film was observed by SEM. Further, orientation of crystal, transition temperature of electric resistivity and critical current density and magnetic susceptibility were measured by X-ray diffraction method, four terminal method and induction method, respectively. Distribution of concentrations of Bi and Pb in film were obtained by EPMA analyses.

RESULTS AND DISCUSSION

A. Structure of film

Surface structures of as depo. film and heat treated films which were heated at 730, 750, 770 and 790 °C in air atmosphere for 10 hrs, were observed with SEM. Those SEM photographs are shown in Figure 3. By the rise of temperature of 60 °C from 730 °C to 790 °C, particle shows rapid growth. This phenomenon is supposed attributable to the effect of fine particles.

B. X-ray diffraction

X-ray diffraction patterns of respective thick films which were heat treated with conditions described in former section, are shown in Figure 4. All peaks of film treated at 730 °C are low temperature phase. High temperature phase are indicated by mark H and amount of high temperature phase is most abundant in films which were heat treated at 770 °C.

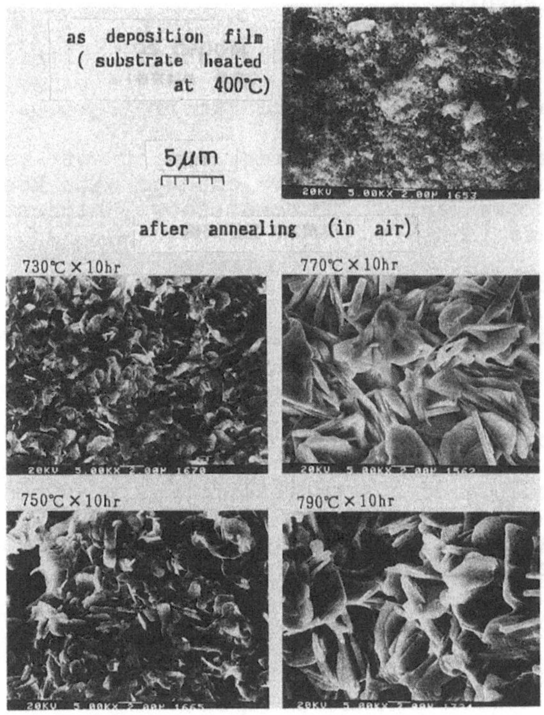

Fig.3 SEM Photograph of the Surface of BPSCCO Thick Films.

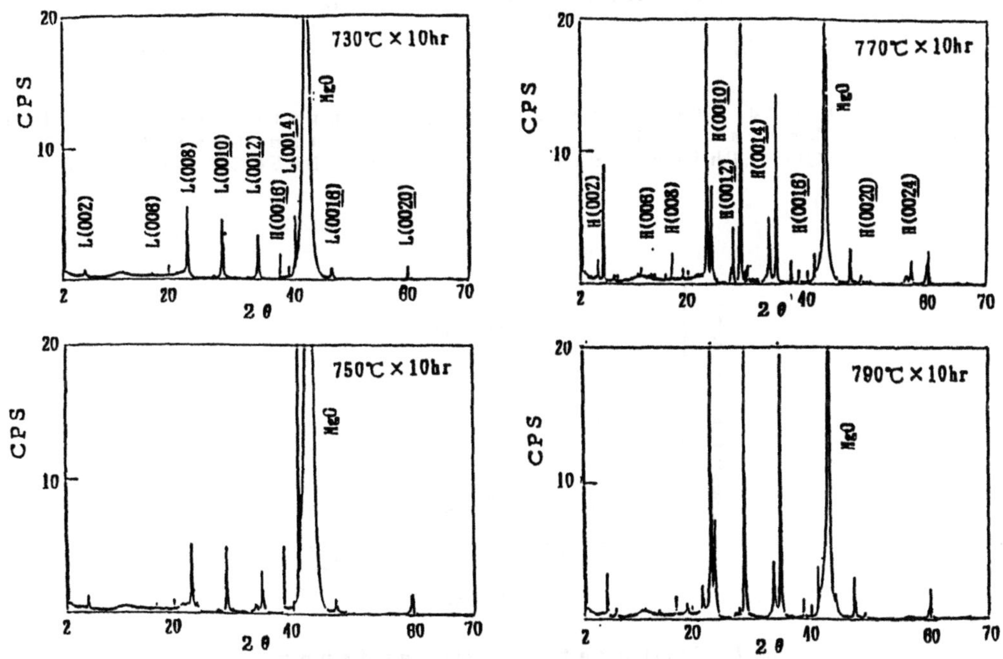

Fig.4 X-ray Diffraction Pattern of BPSCCO Thick Films.

C. Transition temperature

Figure 5 shows resistance temperature characteristic when temperatures of heat treatment were changed. In this temperature range, Tc(on) are 115 K in every case. Film which was heat treated at 770°C shows the highest Tc(end) (105 K).

Figure 6 shows the results of measurement of resistance temperature characteristic when holding times were changed with keeping heat treatment temperature of 770°C constantly. Tc(end) shows highest temperature of 105 K when holded for 10 hrs.

D. Critical current density

Critical current density (Jc) is largest for heat treated film which was heated at 770°C and held for 10 hrs and 1500 A/cm^2 at 4.2 K and 225 A/cm^2 at 77 K were obtained (zero magnetic field for each).

E. Temperature dependence of susceptibility

Temperature dependence of susceptibility was measured by induction method on film which were heat treated at 770°C. As shown in Figure 7, onset temperature in susceptibility measurement fairly agrees with end point of resistance measurement (Fig.5).

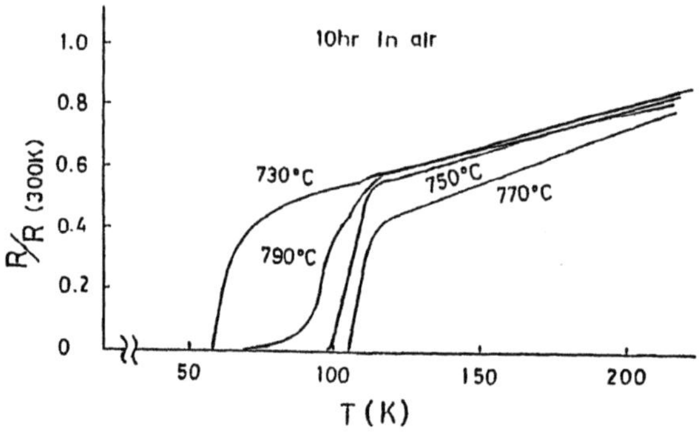

Fig.5 Temperature Dependency of the Resistivity of BPSCCO Thick Films.

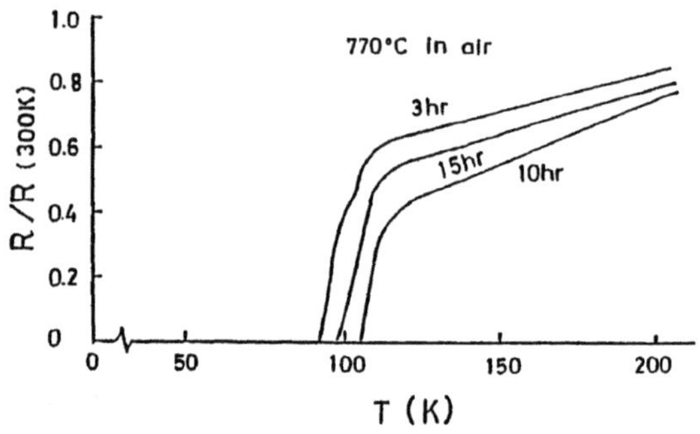

Fig.6 Temperature Dependency of the Resistivity of BPSCCO Thick Films.

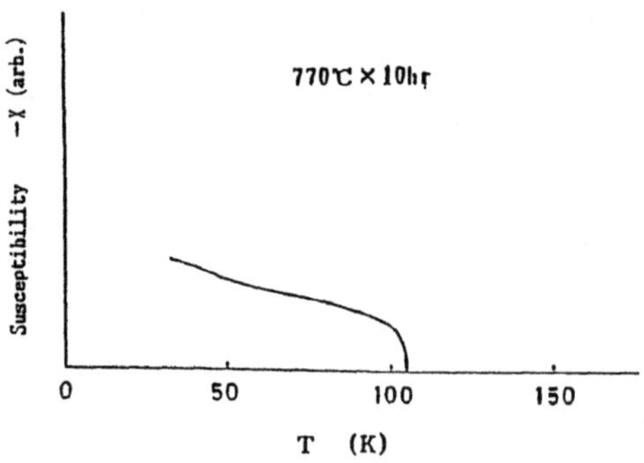

Fig.7 Temperature Dependency of the Susceptibility of BPSCCO Thick Film.

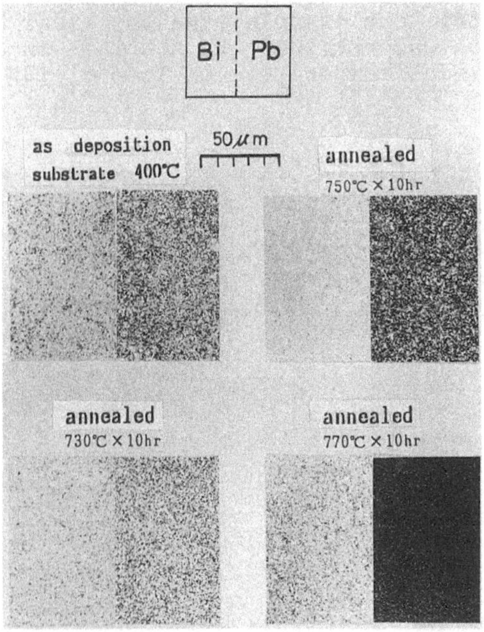

Fig.8 Concentration of Bi and Pb of BPSCCO Thick Films.

F. Compositional ratio of Bi and Pb

For the purpose of making the generation of high temperature phase easy, Pb is mixed for substituting one part of Bi. Ratio of Bi and Pb in material powder in container is 1:1. Compositional ratio of Bi and Pb in film was measured by EPMA plane analysis and compositional ratio is shown in Figire 8. Following the rise of heat treating temperature, decrease of Pb against Bi becomes larger. In film which was heat treated at 770°C, ratio of Pb against Bi is 6%.

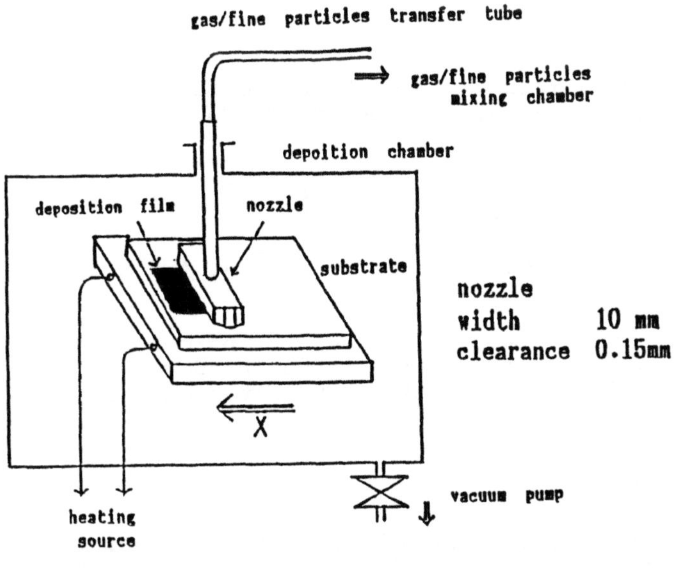

Fig.9 Arrandement of Gas-deposition of BPSCCO Tape-shaped Thick Film.

PREPARATION OF TAPE LIKE THICK FILM WITH BROAD WIDTH

Tape like thick film with width of 10 mm was formed. Film forming chamber, as schematically illustrated in Figure 9 is similar to the method which was shown in Figure 1 in former section, but the nozzle was changed. The spraying port of gas and fine powder is slit shape with 10 mm in width.

Tape like thick film as showed in Figure 10 was formed on substrate (MgO). Upper figure shows cross section of one example, the thickness of which is 75μm. The heat treated film which was obtained by heating in air atmosphere at 770°C for 10 hrs showed critical temperature of Tc(end) 100 K.

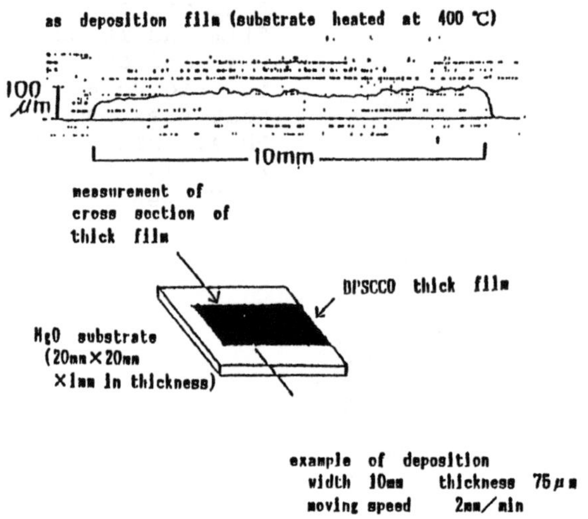

Fig.10 Shape of BPSCCO Tape-shaped Thick Film.

CONCLUSION

a) By Gas Deposition Method which use fine powder with high temperature superconducting composition of BPSCCO, linear and tape like BPSCCO thick film were formed.

b) By heat treatment at 770°C for 10 hrs, thick film showed critical temperature of Tc(end) exceeding 100 K.

c) For Jc of thick film, 1800 A/cm^2 at 4.2 K and 225 A/cm^2 at 77 K were measured.

d) Linear thick film for the use of superconducting wiring material as the method of making pattern without binder, and tape like thick film for the use of superconducting tape material, are expected their development in future.

ACKNOWLEDGEMENT

Cooperations were offered by Miss K. Tsukagoshi for SEM observation, Mr. K. Sekino for X-ray diffraction measurement and Mr. S. Fukui for magnetic susceptibility measurement. This study is forwarded by the support of High Technology Consortium which is promoted by Research Development Corporation of Japan. Authors express their hearty thanks to personnels and organization who offered much assistance to the study.

REFERENCES

[1] K. Hatanaka, M. Kaito, M. Umehara, S. Kashu and C. Hayashi: Proc. 1st Intern. Symp. on Superconductivity (1988) 341.

Effect of the Fabricating Process on the Superconducting Properties of Bi-Pb-Sr-Ca-Cu-O Tapes by the Powder-in-Tube Method

K. YAMAMOTO[1], Y. YAMADA[1,2], S. MURASE[1], and Y. KAMISADA[3]

[1] Toshiba R & D Center, 4-1, Ukishima-cho, Kawasaki-ku, Kawasaki, 210 Japan

[2] Present address: Kernforschungszentrum Karlsruhe GmbH, D7500 Karlsruhe 1, Weberstraße 5, 3640, Federal Republic of Germany

[3] Showa Electric Wire & Cable Co., Ltd., 1-1, Odasakae 2-chome, Kawasaki-ku, Kawasaki, 210 Japan

ABSTRACT

A pseudo phase diagram in the Bi-Pb-Sr-Ca-Cu-O system has been studied by XRD and DTA. A high-Tc phase was formed with a low-Tc phase, Ca_2PbO_4 and CuO. In the powder-in-tube method, this mixed powder was favorable as a calcined powder to obtain higher Jc, because large grain growth and better connectivity occurred due to a partial melt reaction among them. Furthermore, it was found that cold-working improved the flux creep behavior. The origin of the improvement seemed to be microdefects, such as dislocations or stacking faults, observed in the tape specimen by TEM.

KEY WORDS: Ag-sheathed Bi-Pb-Sr-Ca-Cu-O tape, powder-in-tube method, Ca_2PbO_4 microdefects, flux creep

INTRODUCTION

It is well-known that there exists two phases in the Bi system superconducting oxides; a high-Tc phase with Tc = 110 K and a low-Tc phase with Tc = 80 K [1]. Recently, it has been reported that the substitution of Pb for a part of the Bi site enhances the Tc value of the system [2], and thereafter, it has been found that such a substitution promotes the formation of a high-Tc phase [3]. The powder-in-tube method was utilized here in order to make the superconducting oxide in a tape form. The phase diagram is the basis for heat treatment in this method, but few reports exist, especially those suitable for wire or tape fabrication. Few reports exist also concerning the reaction mechanism to form the high-Tc phase and the correlation of the phase of the calcined powder to the critical current density (Jc). Furthermore, the flux creep behavior has been popularly investigated mainly concerning single crystals. However, only Kumakura, et al.[4], has reported on tape specimens in the Bi system. In the present study, a pseudo phase diagram and the reaction mechanism to form a high-Tc phase were investigated, and the effects of both the calcined powder phase on Jc and the cold-working on the flux creep behavior were also studied.

EXPERIMENTAL

First, in order to study the pseudo phase diagram for a fixed composition of $Bi_{1.4}Pb_{0.6}Sr_2Ca_2Cu_3O_x$, co-precipitated oxalate powder was calcined at 850 ^oC for 50 hr in air, and slowly cooled down to room temperature. The calcined powder put into a platinum crucible was heat-treated in a furnace for 15 min - 2 hr at several selected temperatures in the range from 700 to 1000 ^oC. Then, specimens were quenched in liquid nitrogen. X-ray diffraction (XRD) and differential thermal analysis (DTA) were used to determine the phase and melting point, respectively.

Ag-sheathed Bi tapes were fabricated as follows. Co-precipitated oxalate powder of $Bi_{1.72}Pb_{0.34}Sr_{1.83}Ca_{1.97}Cu_{3.13}O_x$ was calcined under two different conditions; (1) at 835 ^oC for 100 hr in Ar+7.7%O_2 for specimens A and B, and (2) at 800 ^oC for 40 hr in air for C and D. These two kinds of calcined powders were packed into silver tubes. All specimens were cold-worked by drawing and rolling, then heat-treated under the condition of 845 ^oC for 50 hr in air. The dimensions of those tape specimens were 4 mm wide and 0.5 mm thick. Specimen B and D were further pressed and heat-treated. The process conditions for fabrication are summarized in Table 1. The critical current

(Ic) of each specimen was measured by the standard four probe method in liquid nitrogen, with the criterion of 1 μV/cm. The XRD was carried out to determine the formed phase and grain orientation. DC magenetization measurements were carried out using a SQUID magnetometer (QUANTUM DESIGN Model MPMS). The relaxation data at a field of 1000 G were taken after zero-field cooling (ZFC). The pinning potential U_0 of the tape was obtained by means of introducing the Bean model to the relaxation of the superconducting current [5]. Transmission electron microscopy (TEM) observation was also carried out.

Table 1 Process conditions and Jc at 77 K , 0 T for tape specimens fabricated in this study

Sample	Calcination condition	Phase of calcined powder	Cold–working *	Jc (A/cm^2)
A	835 °C 100 h in Ar+7.7%O_2	2223 (High–Tc phase)	D+R+HT	120
B			D+R+HT+P+HT	670
C	800 °C 40 h in air	2212 (Low–Tc phase) + Ca_2PbO_4+CuO	D+R+HT	270
D			D+R+HT+P+HT	3000

* D : Drawing R : Rolling P : Pressing
HT : Heat Treatment

RESULTS AND DISCUSSION

Formation Mechanism of The High-Tc Phase

Figure 1 shows the pseudo phase diagram determined by XRD and DTA. It was used only for a fixed composition, but was convenient for heat treatment. The high-Tc phase, $(Bi,Pb)_2Sr_2Ca_2Cu_3O_x$, stably existed in the range of 830 to 860 °C, and it decomposed into Ca_2CuO_3 and the low-Tc phase, $(Bi,Pb)_2Sr_2CaCu_2O_x$, above 860 °C. Besides, the amount of Pb^{2+} or Bi^{3+} decreased by the vaporization of PbO or Bi_2O_3.
On the other hand, a mixed phase of three oxides were found in the lower region below 830 °C; a low-Tc phase, Ca_2PbO_4 and CuO. The authors tried to form a high-Tc phase by adding not only Ca_2PbO_4 but also CuO to the low-Tc phase powder and by annealing at 850 °C for 100 hr in air. Figure 2 shows the XRD pattern of the initial oxides: (a) CuO, (b) Ca_2PbO_4, and (c) a low-Tc phase; and the obtained phase: (d) mostly a high-Tc phase. As a result, it was confirmed that the high-Tc phase was formed from the mixed oxides as follows,

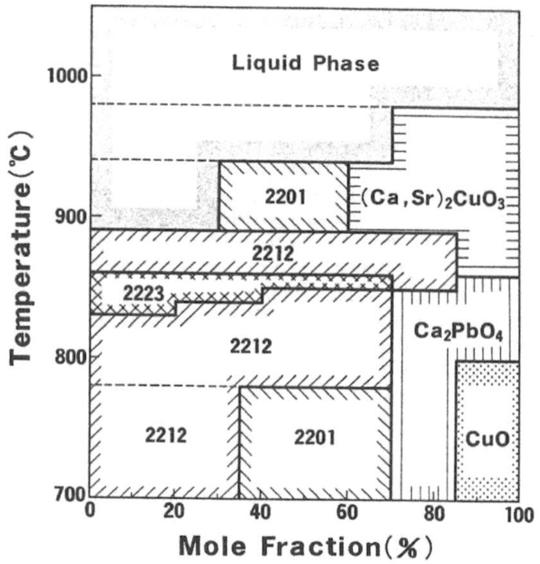

Fig.1 Pseudo phase diagram in $Bi_{1.4}Pb_{0.6}Sr_2Ca_2Cu_3O_x$
2223:High-Tc phase, 2212:Low-Tc phase
2201:Semiconductor phase

$$(Bi,Pb)_2Sr_2CaCu_2O_x + Ca_2PbO_4 + CuO \rightarrow (Bi,Pb)_2Sr_2Ca_2Cu_3O_x \quad (1)$$

Uzumaki, et al., also formed a high-Tc phase by adding Ca_2PbO_4 to the Bi-Sr-Cu-O system. [6]

Phase of Calcined Powder

In order to investigate the effect of calcined powder on Jc, the authors used both a mixed powder of the low-Tc phase, Ca_2PbO_4 and CuO, and a high-Tc phase powder. To make the mixed phase powder of the three, the temperature for the calcination of specimens C and D was set at 800 °C, which was lower than that for the high-Tc phase, as can be seen in Fig.1. The subsequent phase of the calcined powder was confirmed by XRD.

The Jc values measured at 77 K and 0 T are also shown in Table 1 for the tape with these two kinds of calcined powder. Comparing specimen C with A, or, D with B, the Jc's of C and D were several times higher than those of A and B, respectively. The reason is assumed as follows. The crystal grains in the Ag tube are crushed into fine grains by cold-working, such as drawing or rolling. It is necessary for tapes with a higher Jc that the finely crushed crystal grain of the core should be fully re-crystalized and grown to a large size so that they would connect tightly with each other. As described in Ref.7, Ca_2PbO_4 decomposes into CaO and a liquid phase at 822 °C. It is assumed that the liquid phase promotes the growth of high-Tc phase crystal grains and thus obtain better connectivity among them. The phase of the calcined powder affects the Jc of the fabricated wire or tape. The authors concluded that the mixed phase of the three oxides is preferable to the high-Tc one as a calcined powder.

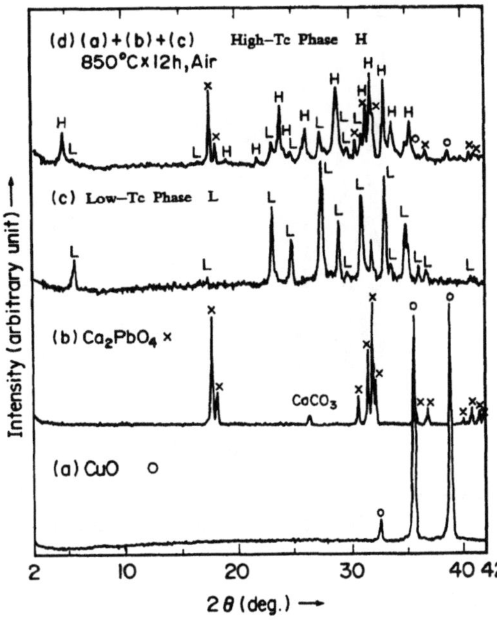

Fig.2 X-ray diffraction pattern of (a)CuO, (b)Ca2PbO4, (c)low-Tc phase and (d) mixed phase of (a)+(b)+(c) sintered at 850 °C for 12 hr in air

Flux Creep and Microstructure of a Cold-worked Tape

The core of a tape is severely deformed by cold-working, such as drawing, rolling or pressing. Kumakura, et al.[4], reported a larger pinning potential U_0 in an oriented BSCCO tape made by the powder method than those of a single crystal measured by Yeshurun, et al.[8]. This means that cold-working improves the flux creep behavior. Figure 3 shows the relaxation rates, dM/d(lnt), measured at an applied field of 1000 G. The obtained values of U_0 calculated from the above rates at 5K were 60 meV for a field perpendicular to the pressed surface, and 180 meV for a parallel field. These values are much larger than those of a low-Tc phase single crystal, 8 meV for a H⊥ 1000 G field by Yeshurun, et al.[8].

TEM photographs for the tape specimen are shown in Fig.4. Many (a) dislocations or (b) stacking faults were observed in the grain. These defects cannot be observed in the single crystal. Such defects seems to be formed by the stress applied in the cold-working process. It is assumed that the microdefects introduced by cold-working act as a pinning center so that they improve the flux creep behavior. And it is also anticipated that they enhance the intragrain Jc of a tape specimen.

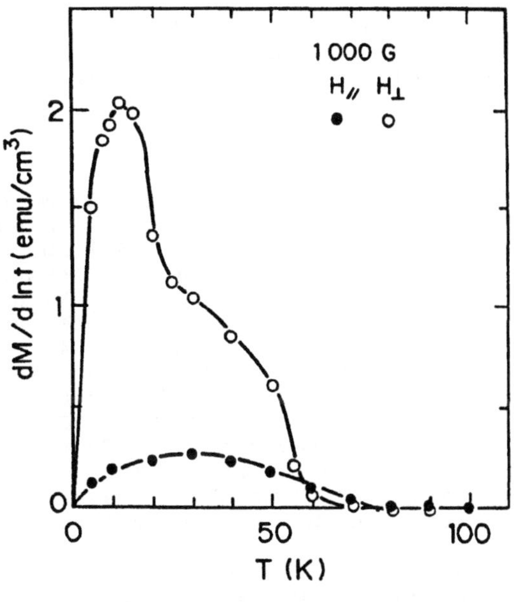

Fig.3 Temperature dependence of logarithmic decay rate dM/d(lnt) for 1000 G field perpendicular and parallel to a rolled surface of a tape

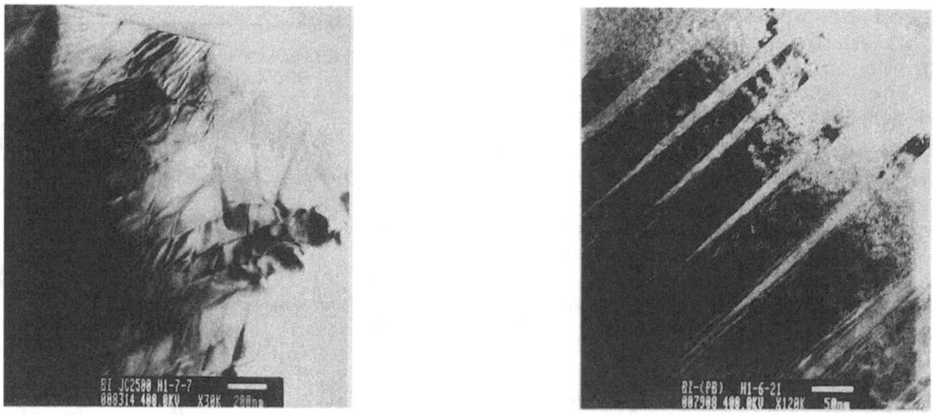

(a) 400nm (b) 100nm

Fig.4 Transmission electron microscopy (TEM) photograph showing some defects in a pressed tape: (a) Dislocations (b) Stacking Faults

CONCLUSIONS

First, a pseudo phase diagram in the Pb-doped Bi system was experimentally determined, and it was confirmed that a high-Tc phase was formed with low-Tc phase, Ca_2PbO_4, and CuO. Second, based on the above results, Ag-sheathed tapes were fabricated, and the relationship between the superconducting properties and the phase of the calcined powder was investigated. As a calcined powder a mixed phase of low-Tc phase, Ca_2PbO_4 and CuO was favorable to obtain a higher Jc. Furthermore, microdefects, such as dislocations or stacking faults were observed in the tape specimen by TEM study. Magnetization measurements revealed that these defects introduced by cold-working enhanced the pinning potential U_0, and improved the flux creep behavior.

ACKNOWLEDGMENTS

The authors are grateful to Mr.Nakamura and Mr.Takeno for TEM observations and also Dr.Horigami for his encouragement. They are also indebted to Mr.F.Umibe for proofreading and correcting the original English manuscript.

REFERENCES

1. Maeda H. Takada Y. Fukutomi M. Asano T.(1988) A New High-Tc Oxide Superconductor without a Rare Earth Element. Jpn J Appl Phys 27:L209-210
2. Yamada Y. Murase S. (1988) Pb Introduction to High-Tc Superconductor Bi-Sr-Ca-Cu-O. Jpn J Appl Phys 27:L996-998
3. Takano M. Takada J. Oda K. Kitaguchi H. Miura Y. Ikeda Y. Tomii Y. Mazaki H. (1988) High-Tc Phase Promoted and Stabilized in the Bi,Pb-Sr-Ca-Cu-O System. Jpn J Appl Phys 27:L1041-1043
4. H.Kumakura, K.Togano, E.Yanagisawa, H.Maeda; Jpn J Appl Phys 28(1989)L185
5. Yamada Y. Murase S. Yamamoto K. Kamisada Y.(1989) Critical current and flux creep in oriented Bi-Pb-Sr-Ca-Cu-O 110 K phase made by powder-in-tube method.: proceedings of Critical Currents in High Tc Superconductors Conference. 24-25 Oct 1989. KFK. Karlsruhe. West Germany
6. Uzumaki T. Yamanaka K. Kamehara N. Niwa K.(1989) The Effect of Ca_2PbO_4 Addition on Superconductivity in a Bi-Sr-Cu-O System. Jpn J Appl Phys 28: L75-77
7. Kuxmann U. Fischer P.(1974) Bietrag zur Kenntnis der Zustandsdiagramme $PbO-Al_2O_3$,PbO-CaO und $PbO-SiO_2$. Erzmetall 27:533-537
8. Yeshurun Y. Malozemoff AP. Worthington TK. Yandrofski RM. Krusin-Elbaum L. Holtzberg FH. Dinger TR. Chandrashekher GV.(1989) Magnetic properties of YBaCuO and BiSrCaCuO crystals: a comparative study of flux creep and irreversibility. Cryogenics 29: 258-262

Concentration of Current to the Surface and Modification by CO_2 Laser for Oxide Superconductor

HAREHIKO NOMURA[1], MAMORU OKUTOMI[1], AKIKAZU KITAGAWA[2], and TOSHITADA ONISHI[1]

[1] Electrotechnical Laboratory, 1-4, Umezono 1-chome, Tsukuba, Ibaraki, 305 Japan
[2] Hitachi-Zosen Technical Research Laboratory, 3-22, Sakurazima 1-chome, Konohana, Osaka, 554 Japan

ABSTRACT

In our study, powder-sintered oxide superconductor have been found difficult to flow superconducting current uniformly to the conductor, but to flow concentrated to the surface of it. Still difficult is a problem that it is poor in mechanical strength. Thus, properties of oxide superconductor are so sensitive to their surface condition that some substantial modifications of the material become very important. Along this line, powder sintered superconductor have been modified by CO_2 laser irradiation. As for the material, $YBa_xSr_{2-x}Cu_3O_{7-y}$ was prepared for improving these surface condition, and investigated its super- conducting and mechanical properties.
We had its current density 10 times and its mechanical strength almost doubled after irradiation. And these processed conductor had showed anistropical electric properties according as the laser traces to current direction.

Key Word. OXIDE SUPERCONDUCTOR, CO_2 LASER, CURRENT DENSITY, SURFACE MODIFICATION

INTRODUCTION

To improve current density in superconductors is nothing but to increase in pinning force. But it has not been sure if pinning distribution in oxide superconductor is following almost to Bean model like metal superconductor. We prepared six samples, same in materials,in length and height but only different in width. The superconducting current did not flow in proportion of cross section, but it has been found only flowing near the surface of the samples (Ref.1). It was therefore why we employed CO_2 laser for the surface modification of the superconductor(Ref.2). We have aimed to improve both in current density and mechanical toughness by this modification.

CURRENT CONCENTRATION

We prepared six samples made of same materials $YBa_2Cu_3O_{7-y}$, same length 25 mm and height 7 mm, but only different in each width 0.2, 0.5, 0.8, 1.5, 2.5 and 5.0 mm respectively. The experiments were curried out for acquiring if current equally flowed into the cross-sectional area of the conductor. The relation between critical current and cross-sectional area of said superconductor showed that critical current was far from the proportion to the cross-sectional area of the conductor. Figure 1 is the critical current data according to each perimeter of the sample. We can learn that superconducting current flow must be biased to the surface of each sample as far as powder sintered superconductor, which have many un-preferable grain boundaries and weak links among their grains through which superconducting current must flow, superconducting order parameter may be changed at each boundary. This makes superconducting current difficult to penetrate into the conductors, but coming near the surface of them. It became therefore clear and important to need the modification of the surface of itself. This is one of the main reason why we have our oxide superconductors irradiated and modified by CO_2 laser.

SAMPLE AND CO_2 LASER IRRADIATION

We prepared Y_2O_3, $BaCO_3$, $SrCO_3$ and CuO for raw powder, calcined at 920 ℃ for 5 hours in air. These process repeated 3 times. The powdered sample was in air at 920 ℃ for 5 hours, and was formed only single $YBa_xSr_{2-x}Cu_3O_{7-y}$ orthorhombic phase (type 1) according as $0.8 \leqq x \leqq 2$. The pellet density was calculated by submersion method around 5.7g/cm^3 (89 % of the ideal value; 6.35g/cm^3). We recognized Meissner effect only in the region of $1.6 \leqq x \leqq 2$. Critical temperature of x = 2 sample was 95 K (on set) and 88 K(off set), the critical current density at 77 K was 150 A/cm^2 by means of four terminal method. To improve surface properties of oxide superconductor, we irradiated CO_2 laser onto the sample. The pellet was set into the carbon holder on the rotating disk which could move X-Y direction. CO_2 Laser beam was irradiated onto the sample on the rotating disk. This disk was sifted at constant velocity 1.6 mm/sec and irradiated for 10 m second (rotating speed of disk : 3000 ~ 750 rpm). We had the workpiece surface always blown with oxygen gas during the irradiation. After the irradiation, sample surface was covered with thin melted film. Laser traces were formed on the surface as shown FIGURE 2. Even by a instant laser irradiation, it was processed from the surface down to 60 ~ 80 μm of its depth. During irradiation of CO_2 laser, there happened instantly both melting and vaporizing of the materials. So we controlled its vaporizing rate and thickness by changing irradiation speed.

IRRADIATED SURFACE

The irradiated pellets according as $0.8 \leqq X \leqq 2.0$ showed perfectly single $YBa_xSr_{2-x}Cu_3O_{7-y}$ phase. Its lattice constant was a=3.866, b=3.816 and C=11.671 A. But under x=0.6 however, both nonsuperconducting and $YBa_xSr_{2-x}Cu_3O_{7-y}$ phases appeared mixedly. X-ray diffraction patterns of the sample surface before and after irradiation are shown in FIGURE 3. These formed compounds were identified as orthorhombic of type 1 and type 2 (lattice parameter is a=3.835, b=3.88 and c=11.744 A, whose length of c axis is longer than type 1 unit cell), Y_2BaCuO_5 and $BaCuO_2$ for non-superconducting phases. The tetragonal system (a=3.87,

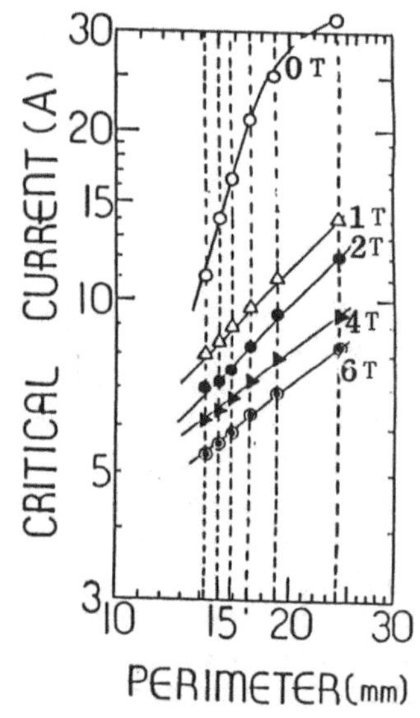

FIGURE 1
Relation between critical current and surface perimeter of each sample.

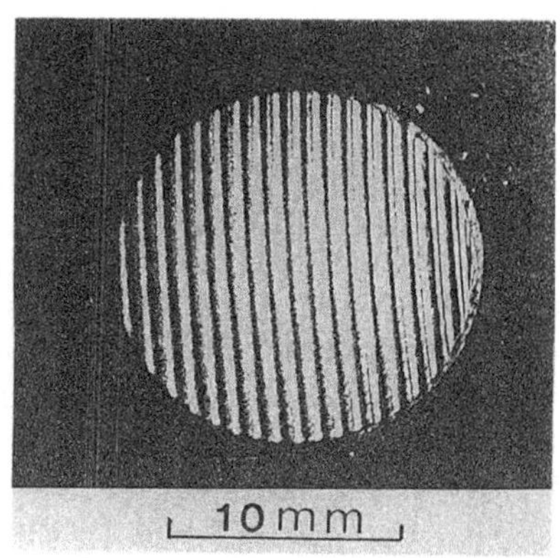

FIGURE 2

Laser irradiated traces on the superconductor.

c=11.63A) was not found. All phases however came bake to the initial orthorhombic system of type 1 or stoichiometry of surface became to 123 superconducting phase by furnaceing at 920 ℃ for 3 hours in air.FIGURE 4 shows the SEM photographs of the fractured surface of initial material and the laser irradiated surface, revealed by etching after polishing. The treated surface looks dense and compacted with crystal grains although many pores existed among grain boundaries in the initial sample. Since oxide superconductor itself is intrisically brittle, the mechanical strength must be improved. It was one of our main reason why we doped Sr for improving mechanical strength of it. The result for example, the surface hardness of $YBa_xSr_{2-x}Cu_3O_{7-y}$ according as for 1.8 ≦ X≦ 2.0 after the irradiation was recorded up to 1400kg/mm^2, being higher than that of ordinary has YBCO (700~ 800 kg/mm^2).

CHARACTERISTICS OF THE SUPERCONDUCTORS

Learning the superconducting characteristics of them, the surface was polished and attached four electrodes to the conductors with silver by acoustic welding. The one pair of electrodes was for putting superconducting current in parallel to the traces swept by laser power, and the other in perpendicular to them. FIGURE 5 shows the characteristics of the relation between temperature and resistivity of said two samples. We learned that the superconducting current was flowed in parallel to the irradiated traces and its resistivity showed metal-like approach to superconducting state. On the contrary, the perpendicular current flow to the traces gave us semiconductor-like approach to superconducting state. Thus, even in the duplicate samples and only different in the direction of current flow, there were two different resistivity approach to superconducting state according as cooling down. This informed us that in the conductor there could happened anistropical recrystallization along the line of laser irradiation. These differences were also acquired when we put large current to each conductor. FIGURE 6 give us current- voltage characteristics of each sample. When current was put to parallel to the traces, weak flow resistivity came out and took up at 17.4 A, but recovered to superconducting state at 7.1 A showing metallic property. But when perpendicular current changing, voltage suddenly came out before 1 A, almost remained constant around 0.3~ 0.15 mV from 3

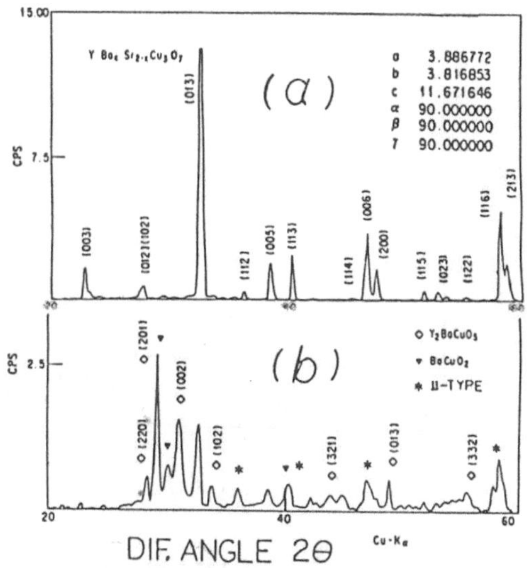

FIGURE 3

X-ray diffraction patterns of the sample surface before (a) and After (b) irradiation.

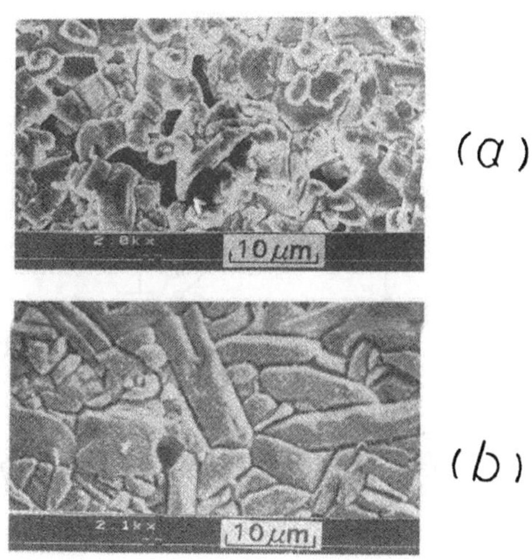

FIGURE 4

SEM photographs of the surface of initial material (a) and the laser irradiated surface revealed by etching after polishing.

A to 16 A and finally around 17 A, it began to run away, and still it had increased yet even a fast decreasing in current. After this, no transport current could flow into the conductor. The superconductor was assumed to change itself to insulating phase by self generated heat.

DISCUSSIONS AND CONCLUSIONS

For making oxide superconductor, powder sintering method is still thought effective, productive and simple. But in this case, superconducting current would come up near the surface of the conductor because of weak links among grain boundaries. We demonstrated here that the CO_2 laser irradiation to superconducting materials was one of more effective method for surface modification. After the irradiation, we have found its current density 4000 A/cm^2 increased by almost 10 times than before the irradiation in consideration of current concentration to the modified thickness. Also did we have its mechanical strength almost doubled after the treatment. We have found that the processed conductor had anistropical electric properties according as the laser traces and current direction. This is thought because that recrystallization had initiated along to the direction through which heat diffused to most, or highest gradient of temperature. It is therefore, more fundamental physics and technique along this method must be needed for getting more controlled, high qualitative and productive oxide superconductors.

REFERENCES

(1) Nomura,H. Okutomi,M. Kitagawa,A. and Onishi,T., Current distribution in Oxide superconductor. Annual meeting report Cryogenic Engineering,Kyushu University Japan. Dec. 1988,B1-20.

(2) Okutomi,M. Nomura,H. Kitagawa,A. and Mitsuhashi,Y., Surface Modification on Superconducting Oxide Y-Ba-Sr-Cu-O system Using CO_2 laser.Creation of High Function Layers on Materials, Joint Symposium between Osaka Univ. and MIT., January 1989, p77.

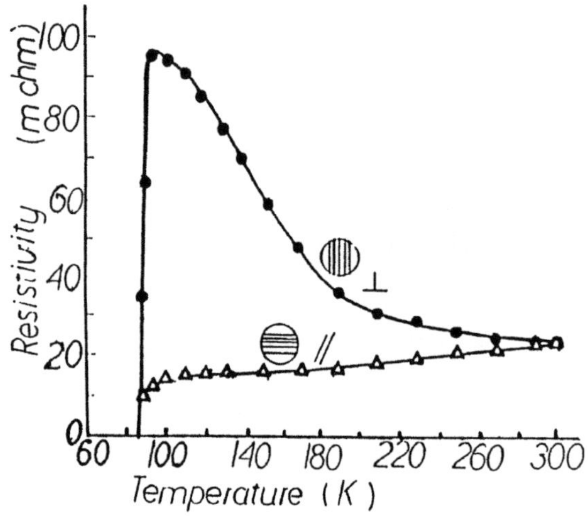

FIGURE 5

The characteristics of resistivities of two samples according as their cool down.

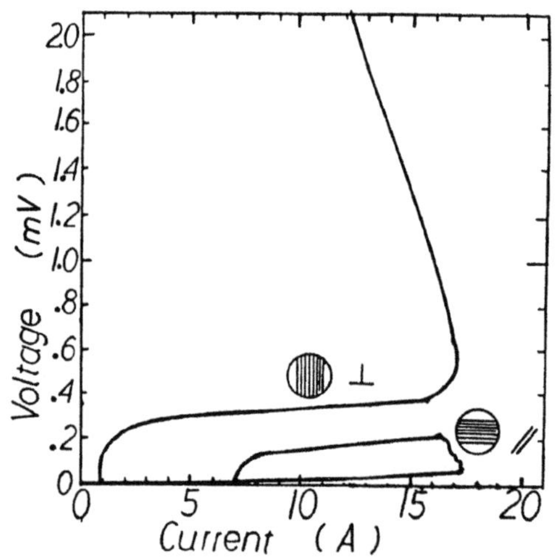

FIGURE 6

Current - Voltage characteristics of each sample, which is only different in the direction of current flow; parallel or perpendicular to the laser traces.

Preparation and Superconducting Properties of Highly Oriented Thick Films of Y-Ba-Cu-O Systems by a Paint-on Method

M. MIYAJIMA[1], S. NAGAYA[1], I. HIRABAYASHI[1], Y. OGAWA[2], Y. MITSUNE[2], Y. ISHIKAWA[2], and S. YOSHIZAWA[2]

[1] ISTEC Superconductivity Research Laboratory, Nagoya Division, 4-1, Mutsuno 2-chome, Atsuta-ku, Nagoya, 456 Japan

[2] Central Research Laboratory, Dowa Mining Co., Ltd., 277-1, Tobuki-cho, Hachioji, Tokyo, 192 Japan

ABSTRACT

The highly oriented films of Y-Ba-Cu-O systems have been prepared on polycrystalline YSZ substrates by a paint-on method. Superconducting properties of these thick films were examined as a function of the particle size of pastes and postprinting annealing conditions. The X-ray diffraction patterns after appropriate heat treatments showed that the highly oriented c-axis was perpendicular to the substrate surface. The orientation was also confirmed by a SEM observation and a magnetization measurement. The typical film of 50 μm thickness showed zero resistance around 90 K. The critical current density of the highly oriented film was 36 and 882 A/cm^2 at 77 and 5 K, respectively. The mechanism of the orientation was composed of two steps.

KEY WORD: High-Tc superconductor, paint-on method, thick film, orientation

INTRODUCTION

Since the discovery of high-temperature superconductors in 1986, there has been widespread interest in producing both thin and thick films of these materials. The reason is that they are required for several applications, such as circuits, devices, tapes and shieldings. The thick-film methods [1-4], e.g. coating and printing are both simple and economical, although a critical current density is lower than that of the films fabricated by thin-film methods. The orientation is one of the significant factors of the critical current density of thick films, because superconductors with a layer type structure have large anisotropies on the electrical conduction paths. Recently, we have succeeded in preparing the highly oriented films of Y-Ba-Cu-O systems on polycrystalline YSZ substrates by a paint-on method. In this paper, we report the relationship between the orientation and the process conditions (the particle size of pastes, the postprinting annealing conditions). Then finally the mechanism of the orientation is discussed.

EXPERIMENTAL

Y-Ba-Cu-O powders were prepared by the ordinary solid state reaction method. Two types of powders with the different average particle sizes were prepared by controling of the grinding time. From the particle size analyses, it was revealed that the average particle sizes of these powders were about 1 and 8 μm. Subsequently, these powders were throughly mixed with an appropriate amount of organic vehicles (ethyl cellulose, telpineol, etc.) using a three-roll mill device to form a paste. The pastes were then painted on flat ceramic substrates using a metal mask and a squeeze. Test patterns (1 mm or 2 mm x 8 mm) were painted on the substrates through a 100 μm stainless-steel mask. Flexible polycrystalline Yttria Stabilized Zirconia (10 mm x10 mm x0.1mm) was used as substrares. After being dried at 120 ℃ for half an hour, the films were sintered at 980-1040 ℃ for 0.1 to 1 hour in flowing oxygen. Both heating and cooling rates were 2 ℃/min, and the oxygen flow rate was 1 ℓ/min.

The crystal structures and the surface morphologies of the films were characterized by an X-ray diffraction spectrometer (Rigaku, RAD-2C) and a scanning electron microscope (JEOL, JSM-840F), respectively. From the X-ray diffraction patterns of the films, the crystalline phases were identified, and the degree of the orientation was estimated. From SEM measurements, the cross sections of the films and the substrates were also observed and the thickness of the films was determined. The compositions of the films and the reacted layer between the films and the substrates were examined by an energy dispersive X-ray microanalysis (JEOL, JED-2000). A critical temperature was determined from the temperature dependence of resistivity and AC susceptibility. A critical current density Jc and its temperature dependence were measured without applying magnetic field. The electrical properties were measured by the standard dc/four-probe method with indium electrodes. Jc was defined by the current density where the generated voltage reached 1 μV from the I-V characteristics of the films. The magnetization was measured by a SQUID magnetometer (QUANTUM DESIGN, MODEL 1802).

RESULTS and DISCUSSION

Figure 1(a) shows the X-ray diffraction patterns of a typical oriented film. For comparison, that of a standard powdered sample is shown in Figure 1(d). The diffraction patterns of the film, in which the intensities of (00ℓ) peaks are larger than that of other peaks, are quite different from that of the standard powdered sample. On the other hand, the diffraction angles of Figure 1(a) are same as those of Figure 1(d), except for some peaks from the substrate (YSZ) and the impurity phase. Y_2BaCuO_5, so-called "green phase" is identified as impurity.

To investigate the relationship between the orientation and the process conditions, the degree of the film orientation was examined as a function of the particle size of pastes and the postprinting annealing conditions. Figure 2(a) and (b) show the annealing temperature and its time dependences of the film orientation for the samples whose average particle size is 8 μm. In this study, we define the degree of the c-axis orientation as the intensity ratio of the (005) peak (2θ=38.5°) to the (110) and (103) peaks (2θ=32.8°). We found that the annealing temperature above 1020 ℃ and the relatively short periods of annealing time enhanced the c-axis orientation. The films with the most c-axis oriented were obtained when the film was annealed at 1025 ℃ for 0.3 hours. The orientation decreased at the higher temperature or for the longer periods than this condition. This tendency is thought to be caused by the interactions between the films and substrates by means of the thermal diffusion during annealing. The films annealed above 1040 ℃ were green in color, and insulating at room temperature.

In order to investigate the reaction at the interface between the films and the YSZ substrates, the cross section was examined by the SEM and EDX. High concentration of barium was found in the substrate near the interface. As Ba ions diffused into the YSZ substrate, the interface layer of several μm was formed, resulted in the segregations of Y and Cu ions within the films. The superconducting films have an excellent adhesion because of this reaction. Tabuchi et al. [4] reported that this reacting product was identified as $BaZrO_3$ by the XRD . Y-rich and Cu-rich regions were predominantly observed near the surface and the interface, respectively. They were Y_2BaCuO_5 and CuO from the EDX analyses. The thickness of the interface layer increased in accordance with raising annealing temperature in the range from 1000 ℃ to 1040 ℃.

The orientation was also confirmed by the SEM observation. The surface morphology and the cross section of the films were observed, as is shown in Figure 3. For the temperature up to 1020 ℃, the grains with the size of 10-20 μm were interconnected loosely. The grain size increases gradually in accordance with raising annealing temperature from 1000 ℃ to 1020 ℃, because of grain growth. Some pores and cracks were observable. The existence of some cracks might be due to the mismatch between the thermal expansion coefficients of the films and the substrates. On the other hand, the films annealed at 1025 ℃ and above show appearence of plate-like feature with some cracks and voids. The plate-like grains having dimensions of a few hundreds μm in plane and several μm in thickness orient parallel to the substrate surface, and stack vertically with each other on the substrates. This large plate-like grains exhibit some evidence for a liquid phase formation during the thermal treatment. This suggests that a melt process is responsible for the formation of the highly oriented films, as reported by other authors [4,5].

Figure 4(a) shows the magnetization curve of the typical film at 5 K under increasing field strength and decreasing field strength between -10 and 10 kOe. Two data were taken with the applied magnetic field parallel and

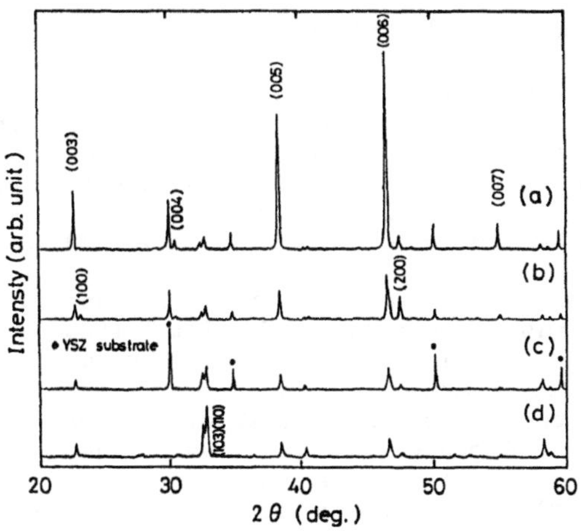

"Fig. 1": X-ray diffraction patterns of the films made from the pastes with the average particle size of (a) 8 μm and (b) 1 μm annealed at 1025 ℃ for 0.2 hrs. and (c) 8 μm dried at 120 ℃ for 0.5 hrs. A standard powdered sample (d) is also shown for comparison.

"Fig. 2": The annealing temperature and its time dependences of the film orientation, the c-axis orientation vs. the annealing temperature (a) and the annealing time (b), the a-axis orientation vs. the annealing temperature (c). Circle and triangle marks correspond to the films made from the pastes with the average particle size of 1 and 8 μm, respectively.

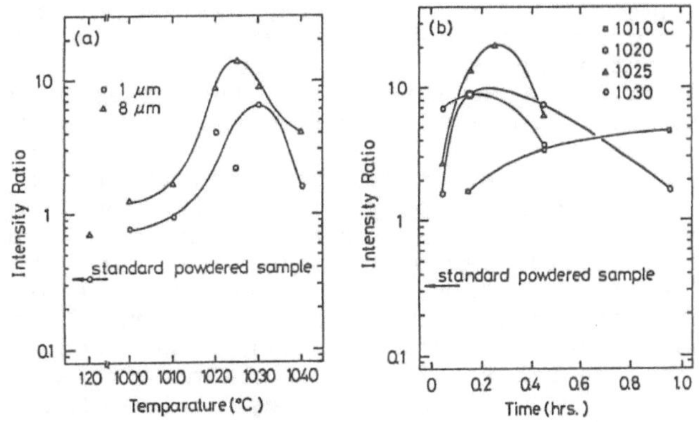

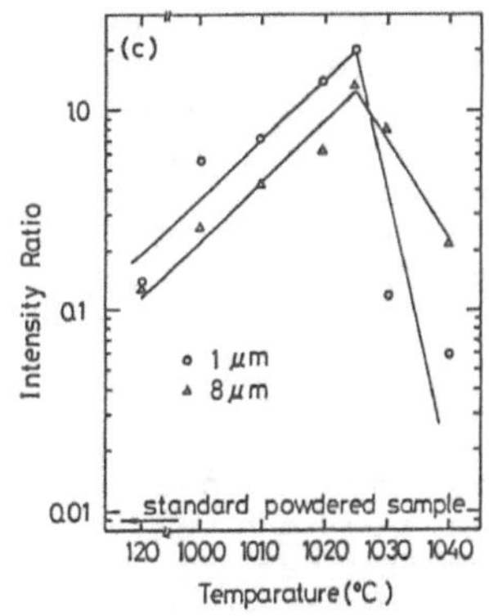

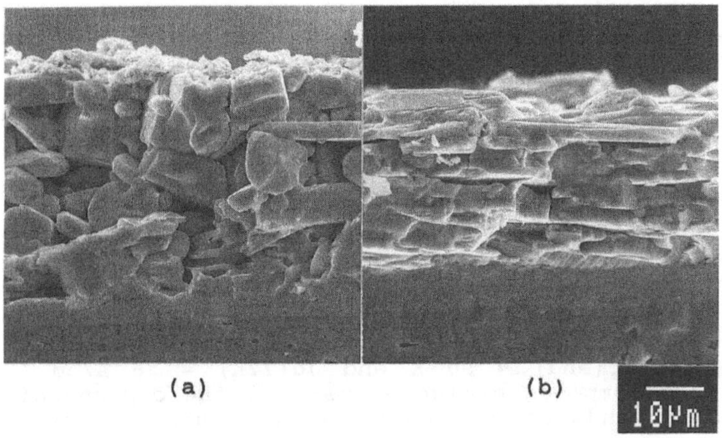

"Fig. 3": Cross section images of the films annealed at (a) 1000 and (b) 1030 ℃ for 0.2hrs. by SEM measurements.

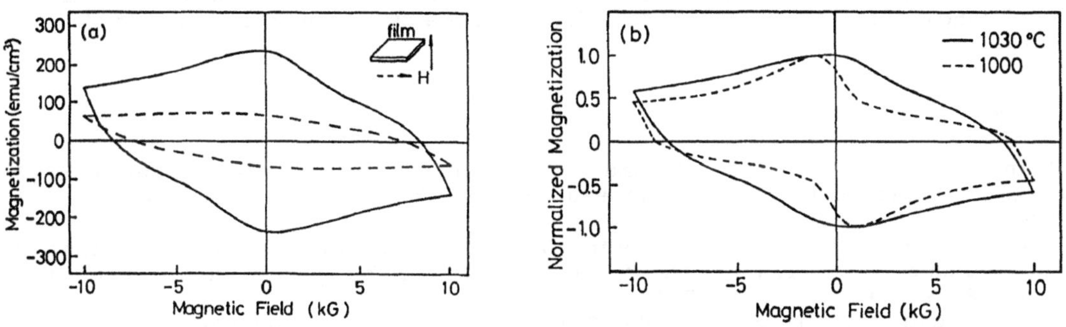

"Fig. 4": The magnetization curve at 5 K of the film annealed at 1030 ℃ for 0.2 hrs. Broken and Solid lines correspond to the data with the applied magnetic field parallel and perpendicular to the film surface, respectively. (b) The normalized magnetization curves of the films annealed at 1030 ℃ (a solid line) and 1000 ℃ (a broken line) for 0.2 hrs. with applied magnetic field perpendicular to the film surface.

perpendicular to the film surface. The anisotropy of the magnetization curve indicates the highly oriented c-axis is perpendicular to the substrate surface. Figure 4(b) shows the normalized magnetization curves of the films annealed at 1000 and 1030 ℃. The normalized magnetization curve swells in accordance with raising annealing temperature. This result indicates that superconducting properties in the magnetic field can be improved by the highly c-axis orientation or the large grain growth. The critical current density at 5 K for the highly oriented films and the moderate oriented films are 883 and 538 A/cm^2, respectively.

The orientation is also dependent on the average particle size of pastes. Figure 1(b) shows the X-ray diffraction patterns of the typical film using the paste with the average particle size of 1 μm. In addition to the c-axis oriented peaks, (h00) peaks are clearly observable, when the films are annealed below 1025 ℃. On the other hand, the films annealed above 1030 ℃ exhibit dominantly the c-axis orientation. Figure 2(c) shows annealing temperature dependence of the a-axis orientation. We define the degree of the a-axis orientation as the intensity ratio of the (200) peak (2θ=47.5°) to the (110) and (103) peaks (2θ=32.8°) The b-axis orientation can not be observed when the c-axis orientation exists, because the diffraction angles of the (010), (020) peaks coincide with those of (003) and (006) peaks, respectively. The magnetization curve of the films annealed below 1020 ℃ was different from that of annealed above 1030 ℃, where the anisotoropy was not observed from the magnetization curve. It was found from the SEM observations, that the films annealed below 1020 ℃ have grains with a rectangular or square shape.

Figure 1(C) shows the X-ray diffraction patterns of the film dried at 120 ℃ for half an hour. The film was made from the paste with the average particle size of 8 μm. Although the film was not annealed at high temperature, some c-axis orientation was observable in comparison with the standard powdered sample. These results suggest that the aspect ratio of the grain is strongly dependent on the orientation. If the average particle size is large, the grains with the large aspect ratio exist. The c-plane is generally thought to be larger than the a or b-plane for such grains. It is natural that a large plane tends to be parallel to the substrate while the pastes are painted on the substrates . On the other hand, if the average particle size is small, the grains tend to be random because of the small aspect ratio.

CONCLUSION

We have succeeded in preparing the highly oriented films of Y-Ba-Cu-O systems on polycrystalline YSZ substrates by a paint-on method. The superconducting properties of these films were Tc(end) = 90 K and Jc(77K) = 36 A/cm^2. The mechanism of the orientation is composed of two steps: (1) the c-plane of the grain with a large aspect ratio tends to be parallel to the substrate surface, (2) the grains with a large aspect ratio grow large in the presence of a liquid phase caused by the interface reaction between the films and the YSZ substrates. It might be possible to control the orientation of thick films on polycrystalline substrates by a paint-on method.

Further study of the mechanism of the orientation and the optimization of the process conditions will lead to thick films with higher critical current densities.

REFERENCES

1. Koinuma H, Hashimoto T, Nakamura T, Kishio K, Kitazawa K, Fueki K.(1987) High Tc Superconducting in Screen Printed Yb-Ba-Cu-O Films. Jpn. J. Appl. Phys. 26: L761-L762
2. Budhani RC, Tzeng SM, Doerr HJ, Bunshah RF. (1987) Synthesis of superconducting films of the Y-Ba-Cu-O system by a screen printing method. Appl. Phys. Lett. 51: 1277-1279
3. Shih I, Qiu CX. (1988) Y-Ba-Cu-O film prepared by a paint-on method. Appl. Phys. Lett. 52: 748-750
4. Tabuchi J, Utsumi K. (1988) Preparation of superconducting Y-Ba-Cu-O thick films with preferred c-axis orientation by a screen-printing method. Appl. Phys. Lett. 53: 606-608
5. Hoshino K, Takahara H, Fukutomi M. (1988) Preparation of Superconducting Bi-Sr-Ca-Cu-O Printed Thick Films on MgO Substrate and Ag Metal Tape. Jpn. J. Appl. Phys. 27: L1297-L1299

A 124 Phase in Y-Ba-Cu-O Film Fabricated by Mocvd

H. Hayashi, T. Sugimoto, K. Kikuchi, S. Yuhya, Y. Yamada, M. Yoshida, K. Sugawara, and Y. Shiohara

Superconductivity Research Laboratory, International Superconductivity Technology Center, 10-13, Shinonome 1-chome, Koto-ku, Tokyo, 135 Japan

ABSTRACT

Y-Ba-Cu-O superconducting thin films were fabricated by metalorganic chemical vapor deposition (MOCVD) using $Y(DPM)_3$, $Ba(DPM)_2$ and $Cu(DPM)_2$ (DPM=2,2,6,6-terамethyl-3,5-heptandionate) as metalorganic (MO) sources. We have succeeded in preparing $YBa_2Cu_3O_y$ thin films. The c-axis is oriented perpendicular to MgO(001) surface and Tc(R=0) is 83K. When MOCVD films were fabricated at susceptor temperature of 800°C and controlling nominal composition within a certain amount, the periodical peaks of 27.247Å in the XRD pattern which may originate from the $YBa_2Cu_4O_y$ phase was detected.

KEY WORD: Y-Ba-Cu-O system, MOCVD, thin film, oxide superconductor, 1-2-4 phase

INTRODUCTION

Since the discovery of the high-Tc oxide superconductor $YBa_2Cu_3O_y$, much efforts have been focused on clarifying physical properties of this material and preparation of thin films. It is well known that the superconductive characteristics of $YBa_2Cu_3O_y$ is sensitive to the oxygen content. At high temperatures, oxygen desorption occurs easily in $YBa_2Cu_3O_y$. Therefore, slow cooling or oxygen annealing is needed to optimize the superconducting property in the 123 system. This property is disadvantageous for an industrial use of the 123 system. On the contrary, the oxygen stoichiometry of the so-called '124' superconductor $YBa_2Cu_4O_y$ is more stable than that of the 123 system. Oxygen contents of the 124 system do not change up to about 850°C, while its crystal structure is stable at high oxygen pressure, therefore bulk 124 superconductors are synthesized at high oxygen pressure [1, 2, 3] or ambient oxygen in the presence of alkali carbonates [4].The value of Tc (80K) is a large demerit for a practical use at liquid nitrogen temperature (77K). Recently, Miyatake et al. reported that Tc of the 124 increased from 80K to 90K by partial substitution of Y by Ca [5].This new superconductor has a good potential for a practical use because of its thermal stability and rather high values of Tc. Thin film formation of the 124 system have been reported by some researchers engaged in reactive magnetron sputtering, electron beam evaporation and laser ablation methods [6, 7]. In these method the 124 phase was synthesized by annealing an amorphous film at appropriate temperature under 1 atm oxygen pressure,and, to our knowledge, there has been no report on as-grown 124 films. Several researchers have succeeded in preparing $YBa_2Cu_3O_y$ films by MOCVD method [8, 9, 10], however no report on the 124 has been seen yet. In this paper, we present our recent study on preparation of the 124 films by MOCVD.

EXPERIMENTAL

Figure 1 shows the MOCVD apparatus used in this study. It consists of a gas-handling system, three vaporizers, a horizontal cold wall type reactor and an evacuation system. Each vaporizer is installed in a separate oven heated independently at appropriate

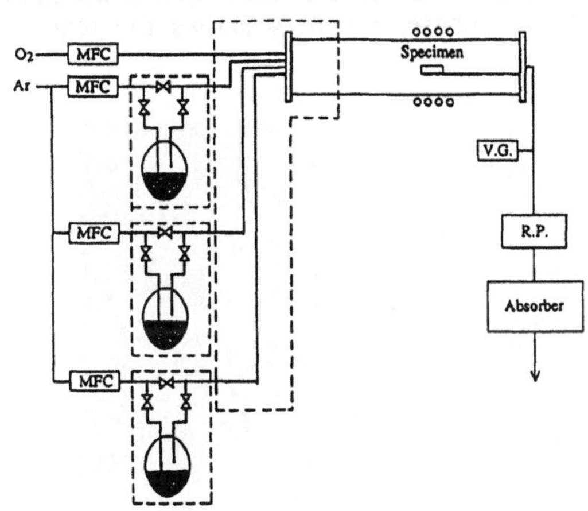

Fig. 1. Schematic diagram of CVD aparatus.

temperature for each gas source (listed in Table 1) and all MO gas lines are also placed into another oven heated at 200°C in order to improve the uniformity of heating and to prevent MO gas from condensation on the inner wall of tubing. Only barium transport tubing is wrapped with heating tapes and maintained at 245°C. Deposition conditions are listed in Table 1. The source materials used are the ß-diketonate metal chelates: bis-(2,2,6,6-tetramethyl-3,5-heptanedionate)-yttrium $Y(DPM)_3$, $Ba(DPM)_2$ and $Cu(DPM)_2$, and temperature of the vaporizers were 100-120°C, 225-235°C, and 100-115°C respectively. Vapors of the chelates were carried independently by Ar gas flow (50-150 sccm) and introduced into a quartz reactor. Oxygen gas was introduced into the reactor separately. The composition ratios of Y, Ba and Cu of the films were controlled by the vaporizer temperature and the carrier gas flow rate. The total gas pressure in the reactor was maintained at 10 Torr during deposition. MgO(001) crystals (10x10x0.5mm) were used as substrates.They were placed on a susceptor heated by radio frequency induction. MOCVD films were fabricated at different susceptor temperatures of 750°C, 800°C, and 850°C. After the deposition, the films were cooled to room temperature in 1 atm oxygen pressure at the rate of 10°C/min. The average compositions of Y, Ba, and Cu were analyzed by inductively coupled plasma chemical analysis (ICP). The structure of the films were analyzed by X-ray diffraction (XRD) using Cu-Kα radiation.The surface morphology was examined by scanning electron microscopy (SEM).And the resistance versus temperature of the sample was mesured in a helium-sealed cryostat. Electric contacts were made to the film by four indium soldered contacts arranged in a standard in line geometry.

Table 1. Deposition condition of Y-Ba-Cu-O films.

Source	Temperature (°C)	Flow (CCM)
Ba(DPM)2	225~235	50~150
Y(DPM)3	100~120	50~150
Cu(DPM)3	100~115	50~150
Oxygen Flow		: 450 CCM
Substrate		: MgO (100)
Temperature of Substrate		: 750~850°C
Gas pressure		: 10 Torr

RESULTS AND DISCUSSION

Figure 2 shows the X-ray diffraction (XRD) pattern of the Y-Ba-Cu-O thin film fabricated on the MgO at 850°C. The composition of the film is Y:Ba:Cu=1.0:1.7:3.0. In the XRD pattern, only (00*l*) peaks of $YBa_2Cu_3O_y$ appeared and other peaks are very weak. According to this result it seems that the c-axis of the film is oriented perpendicular to the MgO(001) surface. Figure 3 shows the surface morphology of this sample. The surface of the film is smooth except existence of some grains which may be Cu-rich phase. Figure 4 shows the resistance

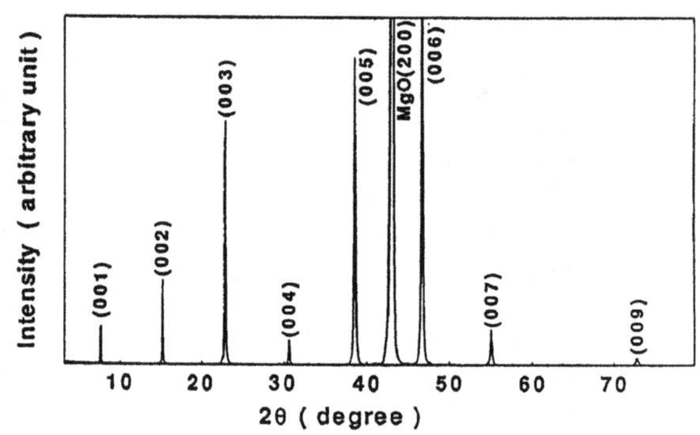

Fig. 2. X-ray diffraction pattern of Y-Ba-Cu-O film deposited on MgO at 850°C.

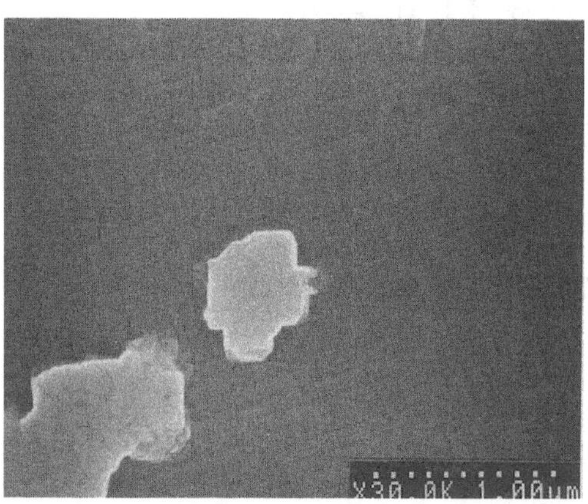

Fig. 3. Scanning electron micrograph of the surface of the Y-Ba-Cu-O film deposited on MgO at 850°C.

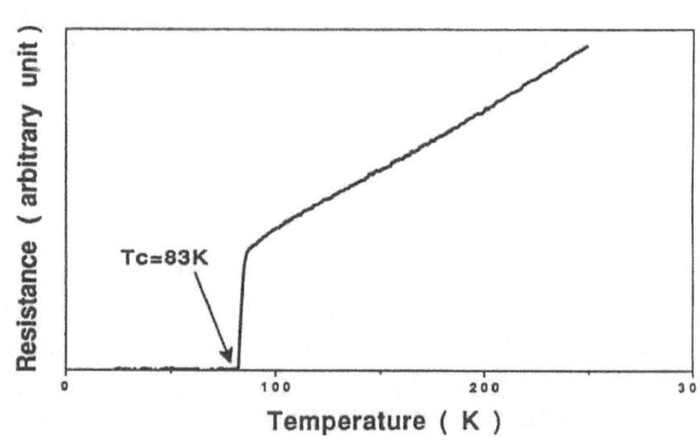

Fig. 4. Resistance versus temperature of the Y-Ba-Cu-O film deposited on MgO at 850°C.

versus temperature of this sample. The resistance clearly decreased at around 86K and reached zero at 83K. From this result we have confirmed that our MOCVD system can prepare the high-Tc superconducting $YBa_2Cu_3O_y$ thin films, in which the c-axis is oriented perpendicular to the substrate surface.

It has been found that the 124 phase decomposes into the 123 phase and the CuO phase at around 850°C under 1 atm oxygen pressure [3]. Considering this fact, we fabricated the MOCVD films at susceptor temperatures of 800°C and 750°C. Figures 5, 6 and 7 show the composition of films fabricated at 850°C, 800°C, and 750°C, respectively. No peak of the 124 phase has been seen in any samples fabricated at 850°C or 750°C (plotted by filled circles in Fig. 5 and Fig. 7). Series of samples, plotted by open circles in figure 6, which were fabricated at 800°C, show interesting traces, which are significantly different from the peaks of the 123 phase, presumably corresponding to the 124 phase. Figure 8 shows the XRD pattern of this sample. In figure 8, there are strong (00ℓ) peaks of $YBa_2Cu_3O_y$, and also there are periodical peaks which are distinctly different from the peaks of the 123 phase. These peaks are assumed to be (00ℓ) peaks of $YBa_2Cu_4O_y$. If these peaks are assumpted to be the (00ℓ) peaks of $YBa_2Cu_4O_y$ the c-axis is estimated as 27.247Å using a least square method, which is close to that 27.240Å in $YBa_2Cu_4O_y$[1]. The reason why no other peak of the YBa_2Cu_4Oy phase was seen is that the c-axis of $YBa_2Cu_4O_y$ is strongly oriented perpendicular to the MgO(001) surface. The 124 peaks did not appear in the films fabricated at 850°C and 750°C because of the following reasons;

i) The temperature of decomposition of the 124 phase into 123 and the CuO phase depends on oxygen pressure. At 1 atm oxygen pressure, e.g. a bulk 124 oxide decomposes at around 850°C, and at oxygen pressure of about 7 torr which is equivalent to the deposition pressure used, the temperature might be lower than 850°C.

ii) The temperature of 750°C is too low to produce the 124 phase, and only a 123 phase appears. These are tentative explanations to the results. In order to fabricate films with the 124 phase, further experiments are needed, in particular, on the oxygen pressure and/or susceptor temperature dependence of the films composition .

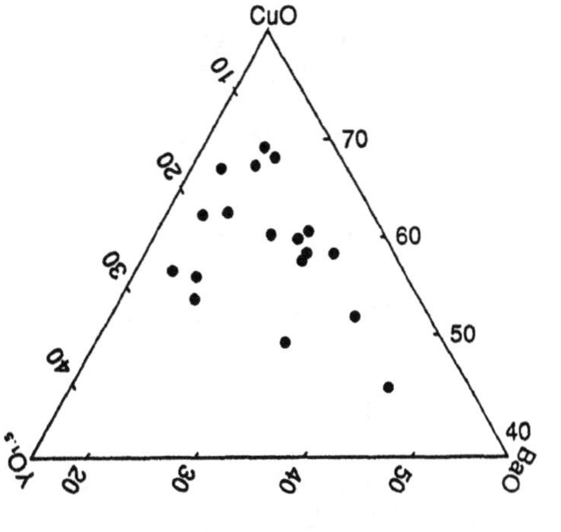

Fig. 5. Composition of films fabricated at 850°C.

Fig. 6. Composition of films fabricated at 800°C, with open circles indicating the compositions of the samples containing the 124 phase.

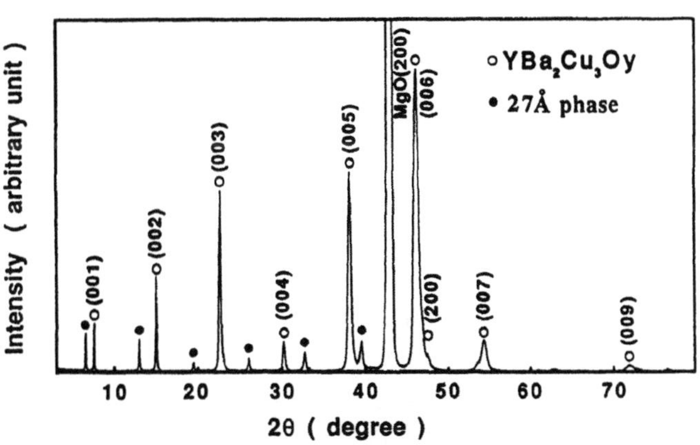

Fig. 8. X-ray diffraction pattern of Y-Ba-Cu-O film deposited on Mgo(100) at 800°C (plotted by open circles in Fig. 6)

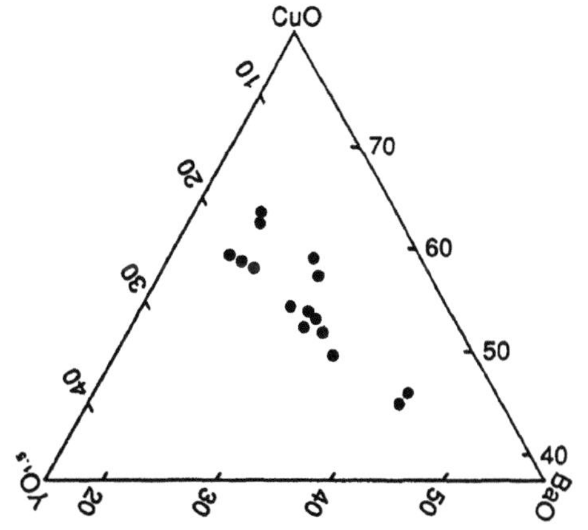

Fig. 7. Composition of films fabricated at 750°C.

CONCLUSIONS

High-Tc superconducting $YBa_2Cu_3O_y$ thin films with orientation of the c-axis perpendicular to the substrate surface have been fabricated by MOCVD. When MOCVD films were fabricated at susceptor temperature of 800°C,we detected a trace of periodical peaks of 27.247Å in XRD patterns which might originate from the $YBa_2Cu_4O_y$ phase.

ACKNOWLEDGEMENT

The authors express their appreciation to Mr. T. Miyatake and Mr. K. Yamaguchi for helpful discussions

REFERENCE

1)'TWO NEW BULK SUPERCONDUCTING PHASE IN THE Y-Ba-Cu-O SYSTEM: $YBa_2Cu_{3.5}O_{7+x}$ (TC≈40K) AND $YBa_2Cu_4O_{8+x}$ (TC≈80K)' J. Karpinski, E. Kaldis, S. Rusiecki, E. Jilek, P. Fischer, P. Bordet, C. Chaillout, J. Chenavas, J. L. Hodeau and M. Marezio, Journal of the Less-Common Metals **150**, 129-137 (1989)

2)'STABILITY OF 124, 123, AND 247 SUPERCONDUCTORS' D. E. Morris, N. G. Asmar, J. H. Nickel, R. L. Sid, J. Y. T. Wei and J. E. Post, Physica C **159**, 287-294 (1989)

3)'Bulk synthesis of the 81-K superconductor $YBa_2Cu_4O_8$ at high oxygen pressure' J.Karpinski, E.Kaldis, E, Jilek, S. Rusiecki and B. Bucher, Nature **336**, 660-662 (1988)

4)'Synthesis of bulk superconducting $YBa_2Cu_4O_8$ at one atmosphere oxygen pressure' R. J. Cava, J. J. Krajewski, W. F. Peck Jr, B. Batlogg, L. W. Rupp Jr, R. M. Fleming, A. C. W. P. James and P. Marsh, Nature **338**, 328-330 (1989)

5)'Tc increased to 90K in $YBa_2Cu_4O_8$ by Ca doping' T. Miyatake, S. Gotoh, N. Koshizuka, and S. Tanaka, Nature **341**, 41-42 (1989)

6) 'Superconducting properties of a 27-Å phase of Ba-Y-Cu-O' M. L. Mandich, A. M. DeSantolo, R. M. Fleming, P. Marsh, S. Nakahara, S. Sunshine, J. Kwo, M. Hong, T. Boone, and T. Y. Kometani, Phys. Rev. B **38**, 5031-5034 (1988)

7)'Properties of Y-Ba-Cu-O thin films with ordered defect structure: $Y_2Ba_4Cu_8O_{20-x}$' K. Char, Mark Lee, R. W. Barton, A. F. Marshall, I. Bozovic, R. H. Hammond, M. R. Beasley, T. H. Geballe, and A. Kapitulnik, Phys. Rev. B **38**, 834-837 (1988)

8)'$Y_1Ba_2Cu_3O_{7-\delta}$ Film Formation by an OM-CVD Method' Hitoshi Abe, Taiji Turuoka and Tomohiro Nakamori, Jpn. J. Appl. Phys. **27**, 1473-1475 (1988)

9)'Tc of c-Axis-Oriented Y-Ba-Cu-O Films Prepared by CVD' Hisanori Yamane, Hideyuki Kurosawa, Hideo Iwasaki, Hiroshi Masumoto, Toshio Hirai, Norio Kobayashi and Yoshio Muto, Jpn. J. Appl. Phys. **27**, 1275-1276 (1988)

10)'Preparation of Y-Ba-Cu-O Superconductig Thin Film by Chemical Vapor Deposition' Kazuhiko Shinohara, Fumio Munakata and Mitugu Yamanaka, Jpn. J. Appl. Phys. **27**, 1683-1685 (1988)

A Thermal Gradient Technique for Accelerated Testing of Tl-HTSC (or, for that Matter, Any Ceramic!)

PETER E.D. MORGAN[2], MICHIYA OKADA[1], TOSHIMI MATSUMOTO[1], and ATSUKO SOETA[1]

[1] Hitachi Research Laboratory of Hitachi Ltd., Hitachi, Ibaraki, 319-12 Japan
[2] On exchange program from: Rockwell International, Thousand Oaks, CA 91360, USA

ABSTRACT

We introduce a thermal gradient technique to accelerate the aquisition of data especially assisting in understanding the mechanisms of sintering and grain-growth interactions in very complex ceramics, specifically here for "$Tl_2Ba_2Ca_2Cu_3O_x$" (Tl-2223) in the form of wires. The introduction of 2%Ag stimulates (de)sintering and grain-growth by a liquid phase solution-reprecipitation reaction commencing as low as 830^{0}C.

KEY WORDS: high temperature superconductor, Tl-2223/Ag, thermal gradient, liquid phase ceramic sintering and grain-growth mechanisms

INTRODUCTION

Ceramics are almost always formed at high temperature; invariably for success, it is necessary arduously to determine the fabrication/temperature response in great detail to optimize the desired properties. Normally, even with rather simple ceramics, e.g. Al_2O_3 and Si_3N_4, many years of tedious research go into sintering/grain growth studies, relating these to the obtained physical properties. With the incomparably more complex ceramic high temperature superconductors (HTSC), ever more onerous experiments are in prospect with even less likelihood of real understanding than has been achieved in simple cases.

To understand the temperature response of a ceramic preform ("green body") it is usual to fire the material at many different temperatures and times attempting to optimize the density/grain-size relationship for the particular application. Additives, to change the rate, or degree, of densification versus the grain size, add more parameters and soon there exists a huge matrix of possible experiments to be carried out. Many different processes occur simultaneously in real ceramic forming, each with its own individual temperature response; it is rarely (if ever!) possible to go from average measurements (such as grain size, shrinkage or density) back to the individual processes which, in integrated combination, produced the changes.

New short-cuts for reducing the number of sintering/densification/grain-growth studies are desperately needed; one attempt was made [1] measuring the electrical properties of a billet of "$Tl_2Ba_2Ca_2Cu_3O_x$" (2223) at temperature while it is sintered; the temperature and rate at which phases, e.g. syntactic intergrowths; "2201"; "2212"; "2223", sequentially form, was monitored in situ. Here we investigate another "short-cut" for reducing this boring enterprise by using a thermal gradient method, which achieves results equivalent to hundreds of firings in one run and although, in this case, we apply the technique to understand the conditions of forming Tl-2223, it is of general application to ceramics, and, indeed, to metals, alloys, composites etc. Hopefully the method can aid in more reliable and reproducible manufacturing. A similar conceptual approach has been applied in the case of mechanical measurements [2], but this approach has not been applied, to our knowledge, in quite this way before.

PROCEDURE AND RESULTS

The method utilizes a furnace designed to establish a computer controlled temperature gradient along the 14cm length of a silver sheathed tape containing 2223/2%Ag composition. There are two sets of high temperature silicon carbide heating elements, the top end of the furnace has a higher temperature set, the bottom a lower. The specimen is placed vertically between them supported in a protective alumina tube; this is radially wrapped in much ceramic wool, so that the heat flow is linear along the length from the hot upper end (T_1) to the cooler lower end (T_2), satisfying the condition for a nearly linear temperature gradient (in this case). The tape, or other ceramic piece, may be previously unfired or could be partially pre-fired, before this technique is used.

A ~1m long silver encased tape, 0.5mm thick and 5mm wide, containing 2223 composition + an addition of 2% Ag powder was prepared by packing the precursor powder in a silver/10% Pd tube, drawing down from 6mm to 2.8mm diameter and then cold rolling [3]. The tape was cut to lengths of 14cm and the ends sealed. Each tape was separately placed in an alumina tube, surrounded by silica wool in a gradient furnace between the two sets of silicon carbide heating elements. Heating up periods, ranging from about 4h to 8h, have been tried, achieving stable T_1 of 920°C and T_2 of 640°C, for a gradient of 2°C/mm. Checks were made for linearity by placing thermocouples in regions of major interest, as becomes apparent later. Constant temperature gradient was maintained for periods of from 1h to 24h.

The specimen is removed from the constant temperature regime, slow cooled or quenched, and is examined in a variety of ways to establish the key temperature dependencies of e.g. crystal structures, density, grain-size, and so forth, in particular it is mounted in epoxy resin, sectioned and polished along the length and examined first by optical microscopy. It is of particular concern to identify, using the appropriately most sensitive techniques, regions where changes occur most rapidly with temperature variation, where sudden changes are seen reflecting changes in mechanisms that may be important in the optimization of efficient and economical manufacturing. At ~830°C, such a sudden change in appearance was readily obvious, Fig.1, for specimens held at constant gradient for 4h and 24h, but was not immediately noticeable in samples held for only 1h.

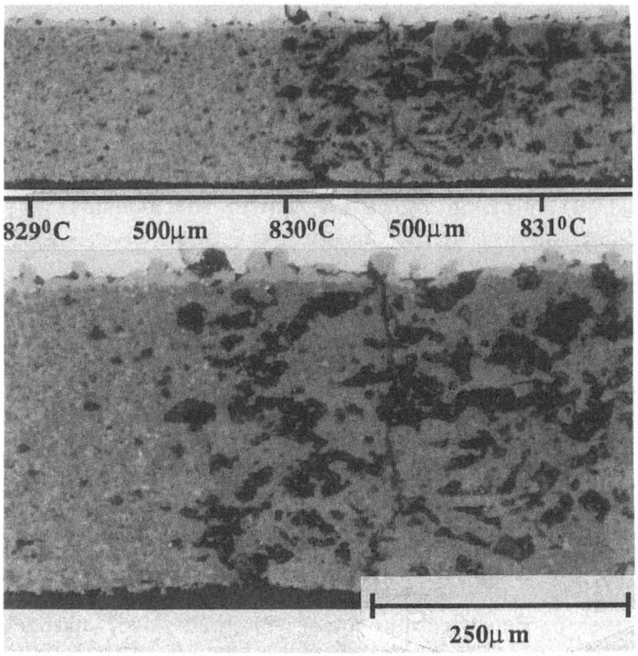

Fig.1. Sudden change region of 24h specimen. Temperature scale is for the upper photograph.

On the lower temperature side, immediately below 830°C, and down to the lowest temperature, little reaction seemed to have occurred, the grains were small and <10μm in size. In contrast, just on the high side of 830°C, much larger, lathe-like grains, of the order of 30-100μm in length were very apparent. Coincident with this change to larger grains, large pores appeared, perhaps consistent with "bloating" caused by the evolution of gases due to previous exposure of the powder to the ambient air. Alternately, the phenomenon may be similar to that found in the Bi system where "desintering" has been noted by several groups [e.g.4].

These changes, occurring within a 1°C range, were most sharp in the 24h case. The 2% of added Ag powder also showed changes in the same temperature regime, this time being most readily apparent in the 1h sample where the other, more uniform features, did not mask the effect. At the lower temperatures the silver particles were several μm in size as originally added. In the immediate vicinity of 830°C, however the Ag was comminuted, as though, perhaps, having been dissolved in a liquid and rapidly reprecipitated. Well beyond the change region, at higher temperatures, the silver reappeared as larger grains, maybe having recrystallized from more abundant liquid farther above a eutectic (or peritectic).

The samples were examined by EDS elemental mapping (to be published elsewhere), which revealed that, below the change, the elements Tl, Ba, Ca and Cu were not uniformly distributed, several phases existed with predominant $Tl_2Ba_2Ca_1Cu_2O_x$ (2212). Above the change, but within <1°C, almost total conversion to 2223 occurred and the elements were consequently quite uniformly distributed. When the grains were large enough, ruling out any interference from the surrounding grains, no Ag presence was detected in the HTSC grains.

If the sudden grain growth is due to a liquid phase, to identify its composition, the regions containing silver were examined closely at ~850°C. Usually these appeared to have been liquid and, on cooling, had precipitated much CuO. We conclude, tentatively, that the liquid responsible is a Ag-CuO rich melt appearing, in this case, at about 830°C. Modern DTA equipment, which uses very tiny amounts of sample, was completely unable to detect the formation of such low levels of liquid. Checking further for traces of liquid, TEM (transmission electron micrography) specimens were then examined. Above the change, regions rich in Ag and Cu were again detected where, on cooling, Ag metal and CuO had precipitated with some Tl rich regions at grain-boundaries and at three grain triple points.

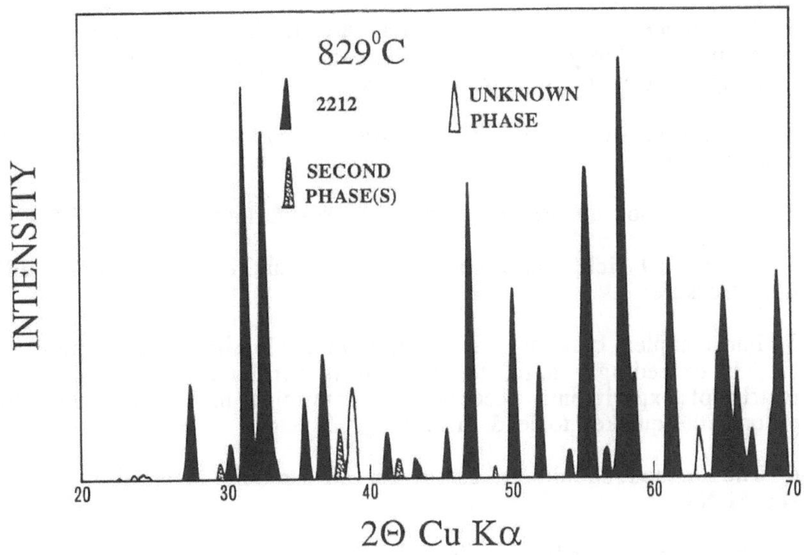

Fig.2. Micro-focus XRD pattern at ~829°C, just below the temperature of sudden change. Much 2212 present.

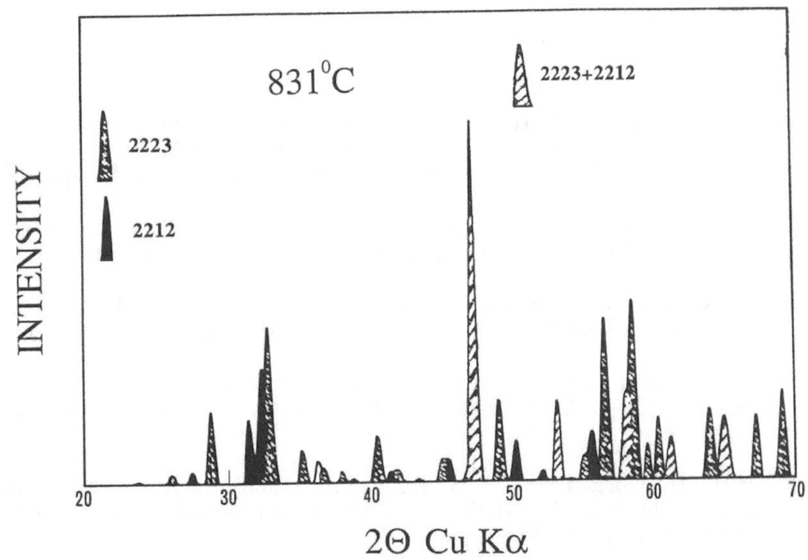

Fig.3. Micro-focus XRD pattern at ~831°C, just above the temperature of sudden change. Much 2223 present.

Microfocus X-ray diffraction (MFXRD), using a 50μm collimated beam (Rigaku, rotating Cu anode, 50KV, 200ma) confirmed mostly 2212 with perhaps ~10% other phases just below the change*, Fig.2; immediately above the 830°C change, within 1°C, almost total conversion to 2223 was confirmed to have taken place in the 24h sample, Fig.3. Lattice parameters, carefully corrected to a Si standard, and calculated from higher angle peaks (70° to 140° 2Θ) were exactly the same as we have seen in earlier work [1] without Ag present, further suggestingthat no Ag had entered the lattice. Chemists would not anticipate Ag replacing Cu, as long as excess Cu is available. Ag more possibly might enter at low levels if there were a deficiency of Cu.

*Puzzling is that, although the original powder contained ~one third 2223, which was still present at the 640°C end, just below the 830°C region, 2223 seemed to have disappeared. Further investigation of this is underway.

CONCLUSIONS

1. The sudden change observations are readily explicable if a eutectic (or peritectic) liquid, appearing suddenly at ~830^{0}C, enhances the reaction by solution-reprecipitation producing rapid grain growth as Tl-2223 forms.

2. A temperature of ~830^{0}C is critically important for the preparation of 2223/2%Ag ceramic composition.

3. Ag does not apparently, enter the crystal lattice of the 2223 structure under our conditions.

4. A Ag-CuO rich liquid implies that the melting temperature will be very sensitive to O partial pressure.

5. For complex ceramics, as here, (and β-aluminas etc.) where the number of possible variables is exceedingly high, the method is particularly useful in cutting down on the size of the matrix of experiments necessary to determine mechanisms and to find useful fabrication conditions: the quicker to lead to high J_c, H_c etc.

6. The method should be useful for all ceramic materials, even "simple" ones, such as Al_2O_3 and Si_3N_4.

7. This thermal gradient technique may provide one of the most sensitive ways of initially detecting eutectic type liquids in ceramics; this will be a useful addition to earlier techniques [5,6].

8. We are well aware that other complicating effects, e.g. thermal diffusion and liquid/vapor transport down the temperature gradient can occur; as we are using a low thermal gradient, these may be minimal. Future work will decide!

REFERENCES

1. J. J. Ratto, J. R. Porter, R. M. Housley and P. E. D. Morgan. Monitoring Sintering/Densification and Crystallization/Grain-Growth in Tl-based High Temperature Superconductors by in-situ Electrical Resistance Mesurements, in press: Jpn. J. Appl. Phys.

2 N. Bhathena, R. G. Hoagland and G. Meyrick (1984) Effects of Particle Distribution on Transformation-Induced Toughening in an MgO-PSZ. J. Am. Ceram. Soc. 67: 799

3. M. Okada, A. Okayama, T. Morimoto, T. Matsumoto, K. Aihara and S. Matsuda (1988) Fabrication of Ag-Sheathed Ba-Y-Cu Oxide Superconductor Tape. Jpn. J. Appl. Phys. 27: L185

4. T. Asano, Y. Tanaka, M. Fukutomi, K. Jikihara and H. Maeda (1989) Properties of Pb-Doped Bi-Sr-Ca-Cu-O Superconductors Prepared by the Intermediate Pressing Process. Jpn. J. Appl. Phys. 28: L595

5. P. E. D. Morgan and M. S. Koutsoutis (1986) Electrical Measurements to Detect Suspected Liquid Phase in the Al_2O_3-$1^m/_0TiO_2$-$0.5^m/_0NaO^1/_2$ and Other Systems. Comm. Am. Ceram. Soc. 69: C254

6. W. W. Ho and P. E. D. Morgan (1987) Dielectric Loss to Detect Liquid Phase in Ceramics at High Temperature. Comm. Am. Ceram. Soc. 70: C209

Characteristics of Superconducting $YBa_2Cu_3O_{7-x}$ Tapes Prepared by Chemical Vapor Deposition

SHIN-YA AOKI, TAICHI YAMAGUCHI, NOBUYUKI SADAKATA, AKIRA KAGAWA, and OSAMU KOHNO

Material Research Laboratory, Fujikura Ltd., 5-1, Kiba 1-chome, Koto-ku, Tokyo, 135 Japan

ABSTRACT

Superconducting $YBa_2Cu_3O_{7-x}$ thin films were prepared by chemical vapor deposition using beta-diketonate chelates of Y, Ba and Cu on single-crystalline $SrTiO_3$ substrates and metal substrates. We obtained superconducting $YBa_2Cu_3O_{7-x}$ thin films oriented c-axis on $SrTiO_3$ (100) substrates at temperature above 840°C. The best film has the zero-resistivity temperature of 92K and the critical current density of $1.0x10^5A/cm^2$. On the other hand, the superconducting films on metal substrates at 750°C have T_c of 84.5K and those with poly crystalline $SrTiO_3$ buffer on metal substrates have T_c of 85K.

KEY WORDS: chemical vapor deposition, high T_c superconducting material, $YBa_2Cu_3O_{7-x}$, thin film, metal substrate,

INTRODUCTION

Since the discovery of the oxide superconductor with high critical temperature in 1986,[1] many techniques have been reported to fabricate the superconducting oxide thin films.[2-4] The use of superconducting oxide for application requires rapid formation of the layer with excellent characteristics in various forms. And high critical current density (J_c) in magnetic field is also necessary as well as high critical temperature(T_c).
Chemical vapor deposition (CVD) method has a great advantage for this viewpoint. Many reports claimed that thin films had three or four orders of magnitude higher J_c than the values of the ordinary sintering process. For example, J_c values of $10^6A/cm^2$ were reported for single crystalline films. It is considered that the thin films have well-aligned grains on the single crystalline substrates.
In this paper, we report the preparation of superconducting $YBa_2Cu_3O_{7-x}$ (YBCO) thin films on single crystalline substrates and metal substrates using beta- diketonate chelates of 2,2,6,6 - tetramethyl -3,5- heptanedione -yttrium ($Y(THD)_3$), -barium ($Ba(THD)_2$) and -copper ($Cu(THD)_2$).

EXPERIMENTAL

The source materials that have sufficient vapor pressure are necessary to realize the CVD method. First, we searched these volatile compounds of Y, Ba and Cu, and not containing halogen elements - especially fluorine, in order to obtaining controllable vapor phases and pure oxide films. At present, we judge that beta-diketone chelates without halogen elements, so-called THD complexes, are the most suitable for CVD of YBCO.
At the beginning of this study, we examined vaporizing characteristics of

these volatile compounds, Y-, Ba- and Cu-THD complexes by thermogravimetry (TG) and differential thermal analysis (DTA). Figure 1 shows TG curves of these complexes under 0.67KPa (5mmHg) air. Next, we calculated the temperature dependence of vaporizing velocity of these complexes and decided vaporizing temperatures of these THD complexes. Figure 2 shows a configuration for CVD used in this experiment. Source materials were set into each vaporizing cylinder and these of Y, Ba and Cu were vaporized at 125-140, 220-250 and 120-135°C, respectively. The source gases from each vaporizer were introduced into the reactor chamber

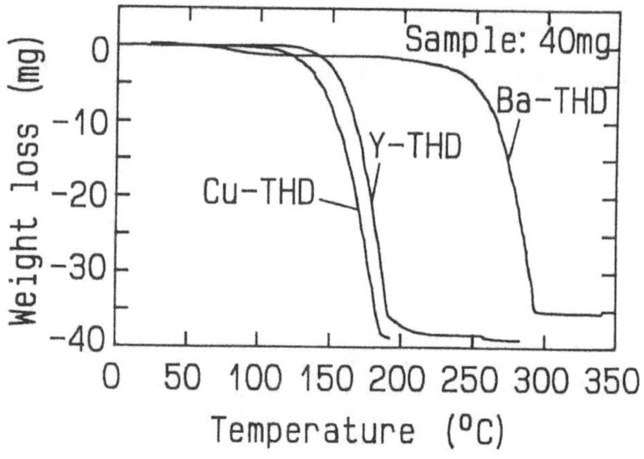

Fig.1 TG curves of $Y(THD)_3$, $Ba(THD)_2$ and $Cu(THD)_2$.

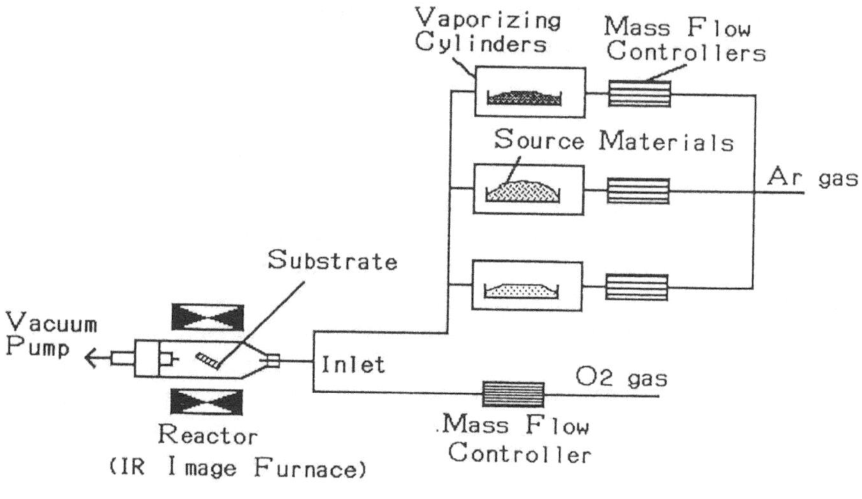

Fig.2 Schematic diagram of a CVD apparatus.

with argon gas used as a carrier gas. Oxygen gas was separately introduced into the same chamber. The mixed source gases were then led to the substrates heated at 750-850°C. The total gas pressure was kept less than 0.67KPa (5mmHg) during the deposition. After the 60 minutes deposition, films were cooled to room temperature at the rate of -300°C/Hr. under 101.3KPa (760mmHg) oxygen atmosphere and subsequently annealed at 500°C for 3 hours under the same condition. The thickness of the films were in the range of 10 to 15 micrometers. In case of the lower vaporizing temperature, the thickness of the films were 0.8 to 2.0 micrometers.

RESULTS AND DISCUSSION

The temperature dependence of resistance was measured by a direct current four terminal method with indium electrodes. Determinant of zero resistance in this experiment was less than 10nV. All films showed an onset superconducting transition at 93K. The best YBCO films on single crystalline

$SrTiO_3$(100) substrates had a critical temperature with zero resistance at 92K.

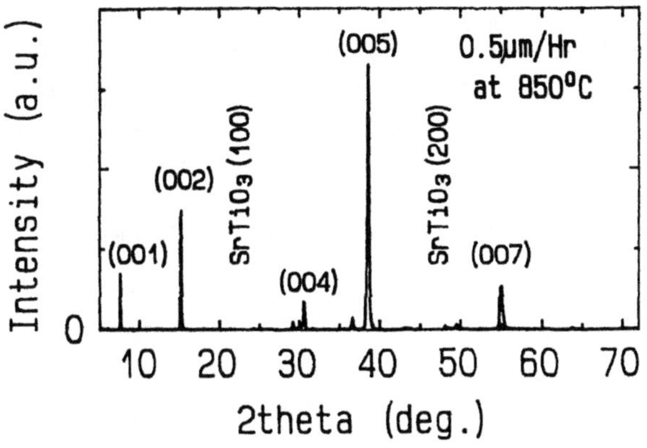

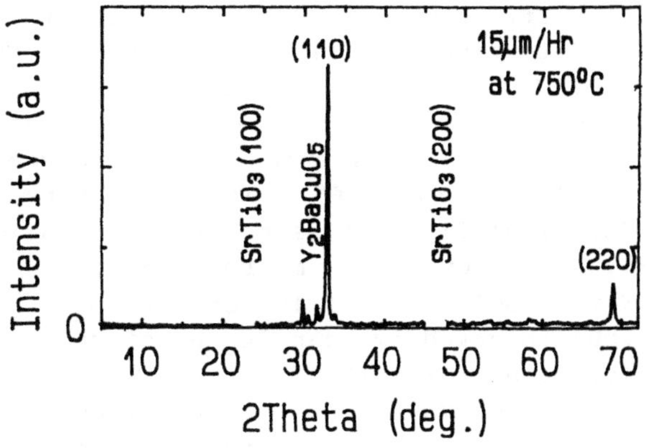

(a) YBCO oriented (00n) (b) YBCO oriented (110)

Fig.3 X-ray diffraction patterns of the YBCO films on $SrTiO_3$(100) single crystals.

Figure 3-(a) shows the X-ray diffraction (XRD) pattern of the YBCO film on single crystalline $SrTiO_3$ (100) by a theta-2theta method using Cu-K alpha radiation. In this case, X-ray was turned off near 23deg. and 46deg. of 2theta to avoid the strong diffraction peaks from single crystalline substrate for protection of the scintillation counter. The sharp diffraction peaks of the YBCO (00n) were observed with weak peaks of the Cu_2O phase. The thickness of this film was about 0.8 micrometers. The XRD patterns of another sample were showed in figure 3-(b). In this figure, the sharp diffraction peaks of the YBCO (110) and (220) were observed with weak peaks of the Y_2BaCuO_5 phase. The thickness of this YBCO (110) film was about 15 micrometers. We can synthesize both YBCO (00n) and (110). This YBCO (110) film was synthesized at more rapid growth rate and at lower temperature of deposition than the YBCO (00n) film.

Critical current at 77K was measured as a function of magnetic field (B) up to 600 mT.

The best YBCO(00n) film, 2.0micrometer in thickness, had a J_c of $1.0x10^5$ A/cm^2 at 77K, 0T.

Figure 4 shows the Jc-B properties of these YBCO films, oriented(00n) and (110) on $SrTiO_3$ (100).Critical current density was reduced using the value of J_c at 0T. The J_c values of the measured YBCO (00n) and (110) were $2.2x10^4 A/cm^2$ and $1.3x 10^3 A/cm^2$, respectively. The dimension of measuring area was 2mm in width and 3mm in length. These J_c values are superior to those of silver sheathed round sintered wire prepared by powder metallurgical technique.[5] The directions of B and J_c were fixed as shown in Figure 4. Within the range of B up

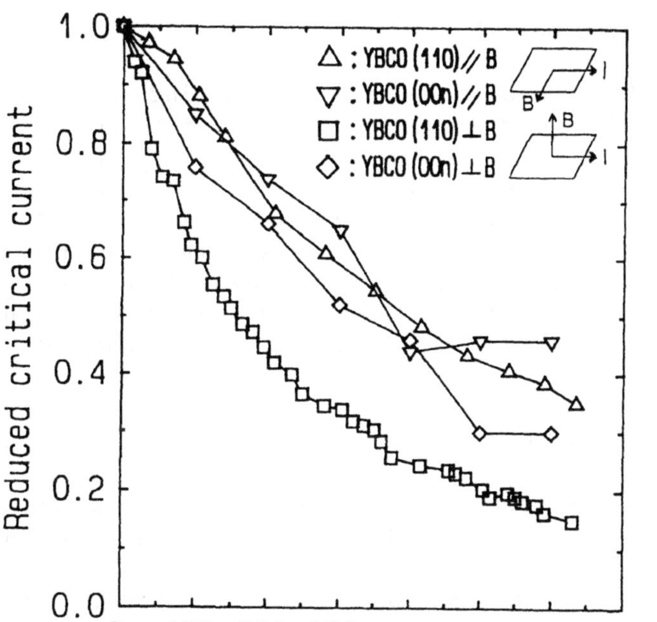

Fig.4 Field dependence of critical current density of the YBCO films on $SrTiO_3$.

to 600mT, J_c of the YBCO (00n) film did not depend on the angle between the direction of B and that of the film plane. But J_c of the YBCO (110) film depends on the angle, volume pinning efficiency perpendicular to the film plane is inferior to that of parallel to the film plane. On the YBCO (110) film, c-axes of the crystals may be random over the film plane direction, so the critical current path must be zigzag. As the critical current is the criterion which produce a Lorentz force strong enough to move the fluxoid off from any pinning centers, in order to raise J_c of YBCO (110) film, it is necessary to align c-axes of the YBCO (110) film.

Figure 5 shows the microstructure of the YBCO (00n) film on $SrTiO_3$ (100) observed by scanning electron microscope (SEM).The film consisted of poly-crystalline pebble-like grains on the substrate. Dense structure was observed and voids were rarely recognized in the film. We tried to synthesize YBCO films on metal substrates, Hastelloy C-276 tapes.These films also showed onset superconducting transitions at 93K. The YBCO films on Hastelloy C-276 had a T_c at 84.5K and those with polycrystalline $SrTiO_3$ buffer on Hastelloy had a T_c at 85K. Both of these YBCO films had J_c values of about $80A/cm^2$ at 0T. We could not measure J_c of these films in the magnetic field. These films were not oriented. But the thickness of these films is about 15micrometers, high T_c films can be fabricated at the rate of about 30 micrometers per hour or more.

Fig.5 Microstructure of the YBCO films oriented (00n) observed by SEM.

CONCLUSIONS

We prepared the superconducting YBCO thin films by CVD on various kinds of substrates, including metal tapes. The YBCO films on single crystalline $SrTiO_3(100)$ substrates have a T_c with zero resistance at 92K and J_c of $1.0x10^5$ A/cm^2 at 77K, 0T. We obtained both YBCO (00n) and (110) aligned films. Within the range of B up to 600mT, J_c of the YBCO (00n) films did not depend on the angle between the direction of B and that of the film plane. The YBCO (00n) films by CVD had higher J_c and superior J_c-B properties than the YBCO (110) films. And we obtained the YBCO films on Hastelloy C-276. Those had a T_c at 85K and a J_c of $80A/cm^2$.

REFERENCES

1. Bednortz, J.G. and Muller, K.A., Z. Phys. B64, 189 (1986).
2. Enomoto, Y., Murakami, T., Suzuki, M. and Moriwaki, K., Jpn. J. Appl. Phys. 26, L1248 (1987).
3. Oh, B., Naito, M., Arnason, S., Rosenthal, P., Barton, R., Beasley, M.R., Geballe, T. H., Hammond, R. H. and Kapitulunik, A., Appl. Phys. Lett. 51, 852 (1987).
4. Yamane, H., Kurosawa, H., Iwasaki, H., Masumoto, H., Hirai,T., Kobayashi,N. and Muto, Y., Jpn. J. Appl. Phys. 27, L1275 (1988).
5. Kohno, O., Ikeno, Y., Sadakata, N., Aoki, S., Sugimoto, M. and Nakagawa,M., Jpn. J. Appl. Phys. 26, 1653 (1987).

3 Physical Properties

3.1 Mechanism of Superconductivity

A Unified Approach to the Description of High-T_c Oxides: Major Normal and Superconducting Parameters

S.A. WOLF[1] and V.Z. KRESIN[2]

[1] Naval Research Laboratory, Washington, DC 20375-5000, USA
[2] Lawrence Berkeley Laboratory, UC Berkeley, Berkeley, CA 94720, USA

Using the Fermi liquid approach, one can evaluate the major normal and superconducting parameters of the cuprate superconductors. The anisotropy of both the electronic and crystal structure and the small values of the both the Fermi energy and the Fermi velocity are the key features of these materials. A complete description of experimentally observed behavior can be constructed using this approach. Also, an unconventional method of determining the strength of the electron-phonon coupling based on an analysis of heat capacity and neutron spectroscopy data has been developed. The results of this analysis lead to the conclusion that the cuprates are described by **strong** coupling, e.g. for $La_{1.85}Sr_{0.15}CuO_4$ the coupling parameter λ is approximately 2. Nevertheless, the observed high T_c values require an additional non-phonon mechanism. In this paper we will summarize many of the results we have obtained and published previously plus present some new results.

In this paper we describe our analysis of the correlations between the normal and superconducting properties of the high transition temperature cuprates. As a result, we have developed a unified approach which allows us to understand many different unusual properties of these materials, including the most important one; the origin of high T_c superconductivity.

Our method is based on a description of these materials in momentum space where the anisotropy of these materials manifest themselves in the topology of the Fermi surface. We predicted the presence of a Fermi surface over two years ago and there are recent experiments that indicate its existence in particular, photoemission and tunneling experiments[3]. We use experimental data that is sensitive to the Fermi surface parameters to reconstruct its features[1-3].

In this paper we we will focus mainly on La-Sr-Cu-O. This material can be treated as a test system, since it is the simplest of the high T_c cuprates yet it contains the essence of the behavior of the whole class of materials.

The Fermi surface for La-Sr-Cu-O is cylindrically shaped and one can obtain expressions for the effective mass and the Fermi energy without specifying the shape of its cross sectional area. Namely, $m^*=3(h^2/\pi)k_B^{-2}d_c\gamma$ and $\varepsilon_F=(\pi^2 k_B^2/3)\eta/\gamma$; here d_c is the interlayer distance, n is the carrier concentration, and γ is the Sommerfeld constant. The effective mass is defined as $m^*=(2\pi)^{-1}\int dl\, v_\perp^{-1}$; the integration is taken over one Fermi curve: $\varepsilon(\vec{k})$, p_z = const. Note that for an isotropic system the density of states N_F and hence, the Sommerfeld constant are proportional to $m^* p_F$ (where p_F is the Fermi momentum which depends on the carrier concentration), whereas for a layered structure $N_F \sim m^* d_c^{-1}$ which enables us to obtain a one to one correspondence between the Sommerfeld constant and the effective mass. The values of the major normal parameters are given in Table I one can also compare the values with those for conventional metals.

TABLE I

Quantity	Conventional Metals	$La_{1.8}Sr_{0.2}Cu\text{-}O_4$
m^*	$1\text{-}15m_e$	$5m_e$
$k_F(cm^{-1})$	10^8	3.5×10^7
$v_F(cm\ sec^{-1})$	$(1\text{-}2) \times 10^8$	8×10^6
$e_F(eV)$	5-10	0.15

We also estimated the value of the Fermi velocity v_F and the Fermi momentum p_F. The small values of E_F and v_F are key features of these cuprates and turn out to be responsible for many of the experimentally observed behaviors[3]. Although the v_F is much smaller than conventional metals the Fermi momentum is also smaller but does not differ drastically from its value in conventional metals. This latter factor is important because it allows a large part of phase space for pairing.

The small value of the Fermi energy leads to anomalous behaviors of the transport properties[2] and critical behavior[4] that have already been observed.

Our results have many implications about the behavior of the collective excitations. For example the quasi-2D low-lying acoustic plasmon branch has a slope of the same order of magnitude as the Fermi velocity. This small slope of the "electronic sound" makes the branch similar to the usual phonon acoustic branch.

SUPERCONDUCTING PROPERTIES, COHERENCE LENGTH, COUPLING STRENGTH PAIRING IN THE QUASI-2D CASE AND H_{C2}

Let us estimate the value of the coherence length ξ_o. Based on the expression $\xi_o=hv_F/\pi\Delta(0)$ and using our previously obtained value of v_F (see Table I), and the value $2\Delta(0) \sim 5k_BT_c$ (see below), we obtain $\xi_o \cong 20Å$ for La-Sr-Cu-O

Evaluation of the strength of one coupling λ responsible for the pairing is a problem of great importance. The value of λ can be obtained from the analysis of heat capacity data. Indeed, near T=OK the values of $\gamma(o)$, as well as effective mass, are renormalized, so that $\gamma(o) \cong \gamma^b$ $(1+\lambda)$, γ^b is the band value. One can show that at T_c the thermal motion leads to the value $\gamma(T_c) \cong \gamma^b$. Therefore, $\lambda \cong (\gamma(o)/\gamma^b)^{-1}$. Using data [8] including the jump in heat capacity, one can obtain $\lambda_{e\text{-}ph} \cong 2.5$ for La-Sr-Cu-O. It means that the cuprates are characterized by a strong electron-phonon coupling; as a result $2\Delta(0)/k_B T_c \cong 5$. Nevertheless, this coupling is not sufficient to give such a high T_c it can be seen from the equation $T_c = 0.25\ \tilde{\Omega}\,[\exp(2/\lambda_{eff})-1]^{-1}$, $\lambda_{eff} = (\lambda-\mu^*)\ [1+2\mu^* + \lambda\mu^* t(\lambda)]^{-1}$, $\lambda = \lambda_{e\text{-}ph}$, $t(\lambda)$ is defined in [9], $\tilde{\Omega}$ is the average phonon frequency; for La-Sr-Cu-O, $\tilde{\Omega} \cong$ 15mev.

Recently, one of the authors and H. Morowitz[7] suggested that a low lying acoustic plasmon branch was responsible for the additional coupling. The efficiency of the coupling is determined by the slope of this branch and the slope is directly proportional to the Fermi velocity. A small v_F implies a small slope and a more efficient coupling so that again the small value of the Fermi velocity is important in understanding the mechanism of the high T_c. Recently this branch has been observed experimentally.

This small value of the coherence length can also be viewed as a direct consequence of the small Fermi energy since it is now clear that the Fermi energy and the energy gap are comparable. This means that almost all the carriers are paired.

Such a picture of pairing is interesting from a different point of view. In 2D, any attractive interaction leads to a bound state (e.g ref. [7]) For conventional superconductors the pairing occurs near the Fermi surface, i.e., the carriers on the Fermi surface form a quasi-2D system in momentum space (this is the essence of the Cooper theorem). The presence of the strongly layered structure in the cuprates is favorable for superconductivity by itself and makes possible the pairing of states distant from the Fermi surface.

The small value of the coherence length is the fundamental reason why the anisotropy of the structure manifests itself in the anisotropy of the superconducting properties. In a conventional superconductor, the coherence length is typically longer than the mean free path so that the properties are averaged over the Fermi surface, but in these new cuprates, the coherence length is typically shorter than the mean free path so that the anisotropies are observed.

For the same reason, one can observe an even stronger effect, ie., multigap structure. The appearance of several energy gaps is due to the presence of the overlapping energy bands. Y-Ba-Cu-O is characterized by quasi-2D and quasi-1D bands; There are now a number of experimental results indicating the presence of two gaps in Y-Ba-Cu-O.

We now have enough information to calculate the upper critical field H_{c2}. Using the expression $H_{c2}(0) = \phi_0/2\pi\xi_{GL}^2$ and the relationship between the BCS and the Ginsberg Landau coherence lengths $\xi_{GL}(0)=a\xi(BCS)$ where a=0.74 for weak coupling and is equal to 0.95 for $\lambda=2$ we find H_{c2} = 88 Tesla. This is a new result and is quite consistent with estimates based on resistive measurements.

To summarize, we have developed a unified approach to the understanding of the new cuprates based on Fermiology. It has been very successful in estimating the major normal and superconducting parameters and describing many of the unusual properties of these very interesting materials. We think that the cuprates are unique because of the presence of the layered conducting structure with small values of v_F and e_F.

References

1. V. Z. Kresin and S. A. Wolf, Novel Superconductivity, S. A. Wolf and V. Z Kresin eds. (Plenum, NY, 1987) p. 287

2. V. Z. Kresin and S. A. Wolf, Sol. State. Comm. 63, 1141 (1987); J. of Super. 1,143, (1988); V. Z. Kresin, G. Deutscher and S. A. Wolf, J. of Super.1, 327, (1988)

3. B. Veal et al., Physica C 156, 269 (1988); A. Arko et al., Phys. Rev. B. (in press); Q. Huang et al (preprint)

4. G. Deutscher, ibid 1, p. 293, Physica C 153-155, 15, (1988)

5. V. Z. Kresin, Phys. Rev. B35,3716, (1987), ibid 1, p.309, V. Z. Kresin and H. Morowitz, ibid 1, p 445, Phys. Rev B37, 7854, (1988); J. Opt. Soc. Am. B6, 490 (1989)

6. V.Z. Kresin and S.A. Wolf, Physica C 158, 76 (1989)

7. L. Landau and E. Lifshitz, Quantum Mechanics,, (Pergamon Press, Oxford, 1977), p. 163

8. N. Phillips et al., ibid 1, p. 739; S. Tanaka, in High Temperature Superconductor, ed. by D. Gubser and M. Schluter, p. 5, MRS, Pittsburgh (1987)

9. V.Z. Kresin, Phys. Lett. A122, 434 (1987)

10. I. Bozovic, Proc. of M²S (Stanford, July 1989) (in press)

Superconducting Transition of 2D Two-Band Systems with Exchange-Like Interaction

KUNIHIKO YAMAJI

Electrotechnical Laboratory, Umezono, Tsukuba, Ibaraki, 305 Japan

ABSTRACT

We show that the exchange-like interaction by which BCS pairs are transferred between two bands can be divergently intensified through a ladder-diagram process in two-dimensional (2D) two-band systems with concentric circular electron and hole Fermi surfaces. In such systems interband polarization function has a peak for wave number equal to the difference of the two Fermi wave numbers, which can lead to the divergence of the ladder-diagram process for important wave numbers, leading to an s-wave superconductivity with T_c attaining to 200 K despite large on-site Coulomb energies. This mechanism is argued to be substantiated in the high-T_c oxides.

KEY WORDS: mechanism of superconductivity, two-band model, exchange-like integral, two-dimensional band, ladder diagram

INTRODUCTION

Remarking the importance of possible multiple-band structure near the Fermi energy in high-T_c oxides, we investigated the possibility that T_c is enhanced by an interband interaction [1] the coupling constant of which is given by the so-called exchange-like integral and which promotes BCS-type pairing by transferring BCS electron pairs between multiple bands. This interaction was found [2] to have its coupling constant of the order of 0.5~1 eV both on the copper and oxygen sites and to be able to enhance T_c. Furthermore, in the case of one-dimensional two-band systems this interaction was shown to be divergently magnified through a process expressed by ladder diagrams in which the exchange-like and interband Coulomb interactions work out this effect together making use of the divergence of the interband polarization function [3]. In this communication the ladder-diagram process is reported to be able to diverge also in two-dimensional (2D) two-band systems.

In 2D systems, although the interband nesting of the Fermi surface is only tangential, the interband polarization function has a cusp-like peak for the tangential nesting wave vector. This can make the ladder-diagram process of the present interest diverge for moderate sets of parameter values, entailing an s-wave superconductivity by itself in the range of high T_c. This mechanism is a new mechanism for high T_c by itself. It can work in enhancing T_c of s-wave superconductivity in concert with the conventional mechanism. It could be substantiated in transition metals, oxides, organics etc. There is a realistic possibility that it plays the major role in the oxide high-T_c superconductors, as will be discussed later. For details see [4].

MODEL AND THEORY

Experimental results [5,6a] suggest that carrier pockets in oxides are located around high-symmetry points in the Brillouin zone. Since we can move off-center pockets around high-symmetry points to the center by formal transformation, we take as our model the following two bands with concentric circular Fermi surfaces:

$$\varepsilon_{1\mathbf{k}} = (k^2 - k_{F1}^2)/2m_1, \qquad \varepsilon_{2\mathbf{k}} = (k_{F2}^2 - k^2)/2m_2 , \tag{1}$$

where the former gives an electron pocket with Fermi wave number k_{F1} and the latter a hole pocket with k_{F2}; $\mathbf{k}$ is a two-dimensional wave vector. We use the convention of units $k_B = \hbar = 1$. Our model Hamiltonian is defined by

$$
\begin{aligned}
H = & \sum_{j=1}^{2} \sum_{\mathbf{k}\sigma} \varepsilon_{j\mathbf{k}} c^{\dagger}_{j\mathbf{k}\sigma} c_{j\mathbf{k}\sigma} + \sum_{j} \frac{U_j}{N} \sum_{\mathbf{k}_1\mathbf{k}_2\mathbf{k}_3\mathbf{k}_4} \sum_{\sigma\sigma'} \delta_{\mathbf{k}_1+\mathbf{k}_2,\mathbf{k}_3+\mathbf{k}_4} c^{\dagger}_{j\mathbf{k}_1\sigma} c^{\dagger}_{j\mathbf{k}_2\sigma'} c_{j\mathbf{k}_3\sigma'} c_{j\mathbf{k}_4\sigma} \\
& + \frac{K}{N} \sum_{\mathbf{k}_1\mathbf{k}_2\mathbf{k}_3\mathbf{k}_4} \delta_{\mathbf{k}_1+\mathbf{k}_2,\mathbf{k}_3+\mathbf{k}_4} (c^{\dagger}_{1\mathbf{k}_1\uparrow} c^{\dagger}_{1\mathbf{k}_2\downarrow} c_{2\mathbf{k}_3\downarrow} c_{2\mathbf{k}_4\uparrow} + \text{H.c.}) \\
& + \frac{U'}{N} \sum_{\mathbf{k}_1\mathbf{k}_2\mathbf{k}_3\mathbf{k}_4} \sum_{\sigma\sigma'} \delta_{\mathbf{k}_1+\mathbf{k}_2,\mathbf{k}_3+\mathbf{k}_4} c^{\dagger}_{1\mathbf{k}_1\sigma} c^{\dagger}_{2\mathbf{k}_2\sigma'} c_{2\mathbf{k}_3\sigma'} c_{1\mathbf{k}_4\sigma} ,
\end{aligned}
\tag{2}
$$

where $c_{j\mathbf{k}\sigma}$ ($c^{\dagger}_{j\mathbf{k}\sigma}$) is the annihilation (creation) operator of the electron in the j-th band with wave vector $\mathbf{k}$ and spin σ; N is the number of the electronic site, U_j is the on-site Coulomb energy for the j-th band, U' is the interband Coulomb energy between the two bands and K is the exchange-like integral between the atomic orbitals composing the two bands; we call the interaction with this coupling constant the exchange-like interaction.

The interband polarization function is defined by

$$
\Pi_{12}(\mathbf{q},\omega_\nu) = -T \sum_{\omega_n} \sum_{\mathbf{k}} G(1,\mathbf{k}+\mathbf{q},\omega_n+\omega_\nu) G(2,\mathbf{k},\omega_n)/N , \tag{3}
$$

where $G(j,\mathbf{k},\omega_n) = (i\omega_n - \varepsilon_{j\mathbf{k}})^{-1}$ is the temperature Green's function for the j-th band with $\omega_n = \pi T(2n+1)$; T is the temperature; $\omega_\nu = 2\pi T\nu$ with ν being an integer. At T = 0, after elementary summation over $\mathbf{k}$ and ω_n eq. (3) is reduced in the whole range of $\mathbf{q}$ to an analytic expression. $\Pi_{12}(\mathbf{q},0)$ is a real function having a cusp-like peak at $q = |k_{F2}-k_{F1}| \equiv k_{21}$. If we displace the Fermi surface of the first band by this wave number, it tangentially touches that of the second band $\varepsilon_{2\mathbf{k}}$. The T-dependence of $\Pi_{12}(q,0)$ can be taken into account as a small correction.

For the irreducible vertex passing a BCS pair from the second band to the first, there appear two divergible processes with the advent of the exchange-like interaction in the diagrammatic perturbation theory in powers of the interaction terms in eq. (2) [3]. The first is expressed in terms of the ladder diagram and the other the bubble diagram. Since the latter is weaker in tendency to divergence, we neglect it in this paper. The irreducible vertex $\Gamma_{12}(-k,k+q)$ represented by the former diagram is summed up to

$$
\Gamma_{12}(-k,k+q) = (K/N)/\mathrm{Re}\{[1-(U'+K)\Pi_{12}(\mathbf{q},\omega_\nu)][1-(U'-K)\Pi^{*}_{12}(\mathbf{q},\omega_\nu)]\}, \tag{4}
$$

where Re{ } means the real part; this vertex links a BCS pair state consisting of $(2,k+q,\uparrow)$ and $(2,-k-q,\downarrow)$ to another pair state of $(1,-k,\uparrow)$ and $(1,k,\downarrow)$; here k and q stand for $(\mathbf{k},\omega_n)$ and $(\mathbf{q},\omega_\nu)$, respectively. Since $\Pi_{12}(\mathbf{q},0)$ has a peak at $q = k_{21}$, Γ_{12} can be greatly enhanced for around this wave number. It can even diverge with decreasing temperature. Since this process promotes the BCS-type superconductivity, transferring the BCS electron pairs between the two bands, especially the pairs whose wave numbers are close to the Fermi surfaces, this divergent enhancement of Γ_{12} must entail superconductivity even with no other help. It should be noted that this process is possible only when K is finite. However, it is remarkable that this instability is facilitated by U'.

The superconducting T_c can be determined as the temperature at which the full scattering vertices of the BCS pair become divergent. This occurs when unity is reached by the maximum eigenvalue of a matrix $\underset{\approx}{A}$ defined by its matrix element $A(jk,j'k') = -T\Gamma_{jj'}(k,k')G(j'k')G(j'k')$, where j is the band index and k denotes $(\mathbf{k}, \omega_n)$; $\Gamma_{jj}(k,k')$ is equal to U_j/N and $\Gamma_{21}(k,k')$ is defined by eq (4). Since there is no special direction concerning the matrix $\underset{\approx}{A}$, the eigenvector corresponding to the maximum eigenvalue is considered to be independent of the direction of $\mathbf{k}$. Then, the problem is reduced to that of a one-dimensional system.

The band energy is conveniently scaled through a parameter $D_2 = 2/m_2a^2$ with a being the lattice constant. This is the band width of the second band if the latter is given by $2t(\cos ak_x + \cos ak_y)$ with eq. (1) being the limiting form at the extrema of this band. This value is fixed by the value of a parameter P defined by $P = (U'+K)\Pi_{12}(k_{21},0)|_{T=0} = (U'+K)F(k_{21},0)|_{T=0}/\pi D_2$, where $F(k_{21},0)|_{T=0}$ is the peak value of the non-dimensional part of $\Pi_{12}(q,0)$. When P = 1, the denominator of Γ_{12} in eq. (4) diverges for $\omega_v = 0$, $q = k_{21}$ and T = 0. Supposing the so-called p_π band as the first band, we equate the lattice constant a to the distance between the nearest neighbor oxygens in the CuO_2 plane, i.e. $a = \sqrt{2}\cdot 1.9$ Å. From the area of the extended Brillouin zone of this band, we choose $k_c = 2\sqrt{\pi}/a$ as the radius of a circle approximating this zone. We choose as a first example $k_{F1} = 0.8/a$, $k_{F2} = 1/a$ and $m_1 = m_2$. For this set of parameter values we get a relatively high value of $F(k_{21},0)|_{T=0} = 3.534$.

In the case of $U_1 = U_2 = U' = U = 1$ eV and K = 1 eV, we get a finite T_c, when $P \geq 0.918$. With increase of P, the value of T_c quickly surpasses 100 K as listed in Table 1(a). T_{div} is the temperature at which Γ_{12} in eq. (4) starts to diverge. The value of D_2 is the band width corresponding to the P value. In the range of $0.918 \leq P < 1$, T_c is finite even though Γ_{12} never diverges. When $P \geq 1.1$, T_c is very high but only marginally higher than T_{div}. In this region superconductivity may be in severe competition with an SDW instability. In the case of U = 5 eV and K = 1 eV, T_c is listed in Table 1(b). Since the divergence of Γ_{12} is facilitated by the larger value of U, the band width D_2 needed for finite T_c is much wider in this case.

Table 1 Obtained values of T_c, T_{div} and D_2 for various values of P and on-site Coulomb energies in the case of K = 1 eV and $m_1 = m_2$.

(a) $U_1 = U_2 = U' = 1$ eV

P	T_c(K)	T_{div}(K)	D_2(eV)
0.918	0.1		2.45
1.0	54	0	2.25
1.1	218	191	2.04
1.2	591	591	1.88
1.3	1047	1046	1.73

(b) $U_1 = U_2 = U' = 5$ eV

P	T_c(K)	T_{div}(K)	D_2(eV)
0.9833	0.1		6.86
1.0	11	0	6.75
1.05	190	166	6.43
1.1	577	576	6.14
1.2	1776	1774	5.62

DISCUSSION

Studies [7] of the spin-fluctuation-mediated superconductivity suggest that the self-energy and vertex corrections may give important influences to the T_c value. However, if the self-energy correction makes the electron mass heavier, it even facilitates superconductivity, in contrast to the case of spin-fluctuations, by increasing the value of $\Pi_{12}(q,\omega_v)$. Optical experiments suggest that the electron mass becomes heavier as T decreases and approaches T_c [8]. This effect may determine the actually realized value of T_c, although $P \sim 1$ does not require extremely narrow bands, as seen in Table 1.

Since the present mechanism gives an s-wave pairing, it concerts with the mechanism mediated by phonons or other bosons in driving superconductivity.

The electronic structure needed for the present mechansim could exist not only in transition metals but also in oxides and organics. Here we discuss whether the high-T_c oxides actually substantiate it or not. The most critical question is if there are two bands crossing the Fermi energy, preferrably, of different carrier types or not. The possibility of the O p_π hole band in the so-called hole-type high-T_c oxides was claimed by Guo et al. [9] and supported by another quantum chemical calculation [10]. This picture is compatible with various photoelectron experiments showing the hole formation in the O $p_{x,y}$ orbitals [11]. It allows a clear interpretation of the positive sign of the Hall coefficient R_H of $La_{2-x}Sr_xCuO_4$ with 0 < x < 0.3 as the dopant Sr pulling out an electron from the O p_π orbital. The change of the sign of R_H with increase of x beyond 0.3 is interpreted as disappearance or diminuition of the O p_π band hole pocket. Accompanying disappearance of superconductivity is in good accord with the present mechanism of supercon-ductivity. Temperature dependence of R_H of all known high-T_c superconductors is not simple, while R_H of non-superconducting oxides is temperature independent [12]. Furthermore, in angle-dependent photoemission data there are clear indications of plural branches cutting the Fermi level in the cases of $Bi_2Sr_2CaCu_2O_8$ [6a] and $YBa_2Cu_3O_7$ [6c]. Positron annihilation results suggest electron and hole pockets in the Bi, La and Y cuprates [5]. Although the interpretations of these experimental results are still controvertial, they justify the try to apply the present mechanism to the oxides.

In actual systems the Fermi surfaces must be deformed from a circle. Therefore, the magnitude of gap parameter promoted by the present mechanism may largely depend on the k-space point on the Fermi surface, since the peak height of $\Pi_{12}(\mathbf{q},0)$, therefore, the vertex function Γ_{12} in eq. (4) depends on the difference of the curvatures of the two Fermi surfaces at the point of tangential nesting. The smaller the difference, the larger $\Pi_{12}(\mathbf{q},0)$, therefore, Γ_{12}. This may be related with the weak enhancement of NMR T_1^{-1} just below T_c due to the existence of small gap regions. The observed large values of ratio $2\Delta_0/k_BT_c$ must be also related to this type of anisotropy.

CONCLUSIONS

(1) In 2D two-band systems the interband polarization function $\Pi_{12}(\mathbf{q},\omega_\nu)$ has a cusp-like peak for frequency $\omega_\nu = 0$ and the wave number corresponding to tangential nesting of the electron and hole Fermi surfaces.
(2) This peak can lead to a real or near divergence of an interband pair transfer process in which BCS pairs are passed between the two bands and which is expressed by ladder diagrams and compiled into a form proportional to $K/[1-(U'+K)\Pi_{12}(\mathbf{q},0)]$, where K is the exchange-like integral and U' is the interband on-site Coulomb energy. This process occurs particularly strongly for BCS pairs composed of electrons near the two Fermi surfaces.
(3) An s-wave superconductivity can occur with T_c on the order of 100 Kelvin with no other pairing mechanism, when $K/[1-(U'+K)\Pi_{12}(\mathbf{q},0)]$ nearly or really diverges, although the present one can concert with the conventional one.
(4) The electronic structure and other conditions necessary for this mechanism appear to exist in oxides and organics as well as in transition metals. From a survey on the features of the high-T_c oxides, we have found many experimental and theoretical indications suggesting the presence of two bands necessary for the present mechanism to be substantiated.

REFERENCES

1 J. Kondo: Progr. Theor. Phys. 29 (1963) 1.
2 K. Yamaji and S. Abe: J. Phys. Soc. Japan 56 (1987) 4237.
3 K. Yamaji: Solid State Commun. 64 (1987) 1157.
4 K. Yamaji: submitted to J. Phys. Soc. Japan.
5 (a) S. Tanigawa, Y. Mizuhara, Y. Hidaka, M. Oda, M. Suzuki and T. Murakami: Proc. MRS 1987 Fall Meeting (Boston); (b) M. Peter, L. Hoffmann and A.A. Manuel: Physica C153-155 (1988) 1724; (c) S. Tanigawa, M. Osawa, S. Kurihara, F. Minami and S. Takekawa: Proc. Annual Meeting of Phys. Soc. Japan (in Japanese), Tokai Univ., March 1989, vol. 3, p. 272.
6 (a) T. Takahashi, H. Matsuyama, H. Katayama-Yoshida, Y. Okabe, S. Hosoya, K. Seki, H. Fujimori, M. Sato and H. Inokuchi: Phys. Rev. B39 (1989) 6636; (b) F. Minami, T. Kimura and S. Takekawa: Phys. Rev. B39 (1989) 4788; (c) Y. Sakisaka, T. Komeda, T. Maruyama, M. Onchi, H. Kato, Y. Aiura, H. Yanashima, T. Terashima, Y. Bando, K. Iijima, K. Yamamoto and K. Hirata: Phys. Rev. B39 (1989) 9080.
7 (a) D. J. Scalapino, E. Loh, Jr. and J.E. Hirsch: Phys. Rev. B35 (1987) 6674; (b) M.T. Béal-Monod, C. Bourbonnais and V.J. Emery: Phys. Rev. B34 (1986) 7716; (c) K. Miyake, S. Schmitt-Rink and C.M. Varma: Phys. Rev. B34 (1986) 6554; (d) H. Shimahara and S. Takada: J. Phys. Soc. Japan 57 (1988) 1044; (e) H. Shimahara: J. Phys. Soc. Japan 58 (1989) 1735.
8 G.A. Thomas, J. Orenstein, D.H. Rapkine, M. Capizzi, A.J. Millis, R.N. Bhatt, L.F. Schneemeyer and J.V. Waszczak: Phys. Rev. Lett. 61 (1988) 1313.
9 Y. Guo, J.-M. Langlois and W.A. Goddard III: Science 239 (1988) 896.
10 Y. Asai: J. Phys. Soc. Japan 58 (1989) 3264.
11 N. Nücker, J. Fink, J.C. Fuggle, P.J. Durham and W.M. Temmerman: Phys. Rev. B37 (1988) 5158.
12 (a) H. Takagi, Y. Tokura and S. Uchida: "Mechanism of High Temperature Superconductivity" (Springer Series in Materials Science vol. 11, ed. H. Kamimura and A. Oshiyama, 1989) p. 238; (b) Y. Tokura: Tsukuba Seminar on High Tc Superconductivity, Univ. of Tsukuba, May-June 1989.

Fermion Confinement Model for High-T_c Superconductivity in a Quasi-Two-Dimensional System

KIMICHIKA FUKUSHIMA[1] and HIKARU SATO[2]

[1] Toshiba Research and Development Center, 4-1, Ukishima-cho, Kawasaki-ku, Kawasaki, 210 Japan
[2] Department of Physics, Hyogo University of Education, Yashiro-cho, Kato-gun, Hyogo, 673-14 Japan

ABSTRACT

Superconducting pairing interaction is investigated for a quasi-two-dimensional system, where charge carriers are confined completely within an xy plane. It is possible to show that the effective electron-electron interaction potential via longitudinal optical (LO) phonons of long wavelength in the second-order perturbation is logarithmic for weak screening because of the momentum conservation law. The calculated values of T_c rise rapidly for the increase in static dielectric constant ε_0 and it is saturated for ε_0 over 100. It is also shown that, for increasing ρ_z, where $\rho = \rho_x\rho_y\rho_z$ represents the density of states for LO phonons, the rise in T_c is monotonic except for the slow increase in the small ρ_z region, when all other parameters are fixed.

KEY WORDS: superconducting mechanism, logarithmic interaction, LO phonon, copper oxides

INTRODUCTION

In spite of many theoretical proposals to elucidate the high-temperature superconductivity for copper oxides, the dominant mechanism has not yet been identified. Angular-resolved photoemission experiments show the existence of Fermi-liquid state near the Fermi level [1], though the Coulomb repulsion on copper sites is rather strong. The measurements by electron-energy loss spectroscopy indicate that charge carriers exist in a two-dimensional (2D) CuO_2 plane [2]. It was also observed that superconducting transition temperature T_c is low, when electron systems are 3D wherein the CuO_2 planes are not isolated from each other [3]. The temperature dependence of magnetic penetration depth measured by muon-spin rotation [4] is consistent with the predictions achieved using the BCS theory. A sharp jump of specific heat at T_c [5] shows that the energy gap is isotropic. These experimental facts indicate the superconducting state has s symmetry. Additionally, large frequency shifts experienced by optical phonons at temperatures below T_c observed by infrared reflection [6] and Raman scattering [7] suggest the important role of optical phonons in superconductivity.

Based on these experimental facts, the authors investigate the superconductivity via longitudinal optical (LO) phonons of long wavelength in a quasi-2D system, where carriers are confined completely within a 2D xy plane. In this system, the carrier scattering is evidently restricted in the xy directions, and the exchanged phonon momentum is oriented in the xy directions because of the momentum conservation law [8,9]. The electron-electron interaction potential in the second-order perturbation for this case is logarithmic for weak screening [8]. The binding energy for the Cooper pair is then increased, which enhances the superconducting T_c. This 2D effect is called fermion confinement in field theory [10]. For synthesizing superconducting materials, it is useful to examine the dependence of T_c on physical parameters.

FERMION CONFINEMENT MODEL

Charge carriers treated here are itinerant in a 2D xy plane and are localized in the z direction. When carriers are itinerant in the z direction, their motion in this direction is coherent over the entire lattice. The z-dependence of the wave function describing these carriers is given by the Bloch function, which is periodic. When carriers are localized around z=0 in the z direction, the electronic correlation is important and the z-dependence of the wave function $\phi(z)$ decays rapidly for increasing value of $|z|$. The localized function $\phi(z)$ does not have the periodicity of the Bloch function. This is discussed in detail by Yamashita and Kurosawa [11]. The wave function for carriers is therefore approximated as $\psi(\vec{x}) = \exp[i(k_x x + k_y y)]\phi(z)$, where $\phi(z)$ is a localized atomic-orbital-like function and is expanded as $\phi(z) = \Sigma_{k_z}\phi_{k_z}\exp(ik_z z)$. Since the total momentum in the z direction is zero, $\phi_{k_z} = \phi_{-k_z}$. For holes, the momentum is measured from $(\pm\pi/a, \pm\pi/a)$ with a being a lattice spacing. Hereafter, the formulations are presented for electrons. To satisfy the momentum conservation law, the field operator for electrons is expanded as $\Psi(\vec{x}) = \Sigma_{\vec{k}} c_{\vec{k}} \exp(i\vec{k}\cdot\vec{x})$ in the second-quantized form in the present context. The Fröhlich interaction is expressed as

$$H' = -i4\pi eF\sum_{\vec{q}}\int d^3x\Psi^+(\vec{x})\,|\vec{q}|^{-1}[b_{\vec{q}}\exp(i\vec{q}\cdot\vec{x}) - b^+_{\vec{q}}\exp(-i\vec{q}\cdot\vec{x})]\Psi(\vec{x}) \quad , \tag{1}$$

in units with $\hbar = 1$, where $b^+_{\vec{q}}$ and $b_{\vec{q}}$ denote the creation and annihilation operators for phonons with momentum $\vec{q}$, respectively. Note that, in Eq. (1), $\delta^3(-\vec{k}' + \vec{k} + \vec{q})$ appears with $\vec{k}$ and $\vec{k}'$ being the initial and final momenta, respectively, which indicates the momentum conservation law. Using $-8\pi F^2/\omega_D = 1/\varepsilon_0 - 1/\varepsilon_\infty$, the second-order electron-electron interaction for Eq. (1) is approximated as

$$H'' = \sum_{\vec{q}}\sum_{\vec{k}\vec{k}'}\left(\frac{1}{\varepsilon_0} - \frac{1}{\varepsilon_\infty}\right)\frac{4\pi e^2}{\vec{q}^2 + k_s^2}\, c^+_{\vec{k}'+\vec{q}} c_{\vec{k}'} c^+_{\vec{k}-\vec{q}} c_{\vec{k}} \quad . \tag{2}$$

Here, ε_0 and ε_∞ are the static and optical dielectric constants, respectively, ω_D is the Debye energy and k_s is a cutoff parameter to remove the infrared divergence.

When $c^+_{\vec{k}-\vec{q}} c_{\vec{k}}$ in H" is operated to the initial state, $|\phi(z)\rangle$ becomes $\Sigma_{k_z}\phi_{k_z}|k_z - q_z\rangle = \exp(-q_z z)|\phi(z)\rangle$. In the summation over q_z with the weight $P(q_z)$, where $P(q_z)$ is the scattering matrix element between the initial and final states, the wave functions with respect to the z direction should not be changed:

$$\frac{1}{2\pi}\int dz\, P(q_z)\exp(-iq_z z)\phi(z) = \phi(z) \quad . \tag{3}$$

Then, $P(q_z)$ must be proportional to $\delta(q_z)$ and $q_z = 0$ is required. When this selection rule is imposed, the electric field is squeezed in the xy direction and the Fourier inverse of the electron-electron interaction potential is proportional to $\rho_z \ln(r)$ for $k_s = 0$, where r is the interelectronic distance and $\rho = \rho_x\rho_y\rho_z$ is the density of states for LO phonons. This means that the fermion confinement occurs, wherein it is difficult to separate the Cooper pair.

When the contribution from the electron-phonon interaction is included into the renormalized effective mass of electrons m*, the following BCS gap equation is no longer a simple weak-coupling one:

$$1 = \int_{-\omega_D}^{\omega_D} d\varepsilon\frac{g}{2\varepsilon}\tanh\frac{\varepsilon}{2T_c} \quad , \tag{4}$$

where $g = -(1/\varepsilon_0 - 1/\varepsilon_\infty)\varepsilon_\infty\mu - \mu^*$ with $\mu^* = \mu/[1 + \mu\ln(\varepsilon_F/\omega_D)]$. Here,

$$\mu = \frac{\xi}{\varepsilon_\infty}\frac{m^*}{k_F}(4\pi e^2\eta) \quad , \tag{5}$$

$$\eta_{k'k} = \frac{2\pi}{[(k^2 + k'^2 + k_s^2)^2 - (2kk')^2]^{1/2}} \quad .$$

The quantities ξ and η are calculated at the Fermi momentum and $\xi = (1/2\pi)^2\rho_z k_F$. In calculating the Coulomb pseudopotential μ^*, the same value is given to ρ_z for photons as for phonons. k_s is set to the inverse of the Thomas-Fermi screening length and is given by $k_s^2 = 4\pi e^2 n/\varepsilon_\infty\varepsilon_F$, where $n = n_x n_y n_z$ is the carrier density and ε_F is the Fermi energy.

RESULTS AND DISCUSSION

Calculations of T_c were carried out for a set of parameters; $n = 4 \times 10^{21}$ cm^{-3}, $\omega_D = 600$ K, $\rho_z = 0.4$ Å^{-1}, $n_z = 0.15$ Å^{-1} and $m^* = 2m_0$, where m_0 is the mass of a free electron. Figure 1 shows the dependence of T_c on the dielectric constants in the quasi-2D system. For $\varepsilon_\infty = 4$ and $\varepsilon_0 = 100$, whose values were determined based on the experimental results [12,13], T_c is below 1 K in 3D, while it is 101 K in the quasi-2D system. This is because the electron-electron interaction is enlarged extremely in the quasi-2D system. It can be seen that T_c increases rapidly for ε_0 up to 50 and it is saturated above $\varepsilon_0 = 100$. ε_0 is between 10 and 30 for usual conducting materials, while the measured value is of the order of 100 for the copper oxides [13]. One of the reasons for the high-T_c observed for copper oxides is therefore the large value for ε_0. The two-dimensionality enhances T_c more for the larger values of $|1/\varepsilon_0 - 1/\varepsilon_\infty|$, which means the stronger electron-phonon interaction. When the electron-phonon interaction is weak, T_c is low even in the quasi-2D system. Figure 2 shows the ρ_z dependence of T_c for $\varepsilon_\infty = 4$ and $\varepsilon_0 = 100$ with other parameters being fixed as mentioned before. T_c rises monotonically up to 101 K with increasing ρ_z in the large ρ_z region. However, T_c increases slowly in the small ρ_z region. This suggests that T_c is raised when density of oxygen atoms is made larger in the z direction. It should be noted that this ρ_z dependence does not directly correspond to the pressure dependence of T_c observed in experiments, since all parameters other than ρ_z change simultaneously in experiments.

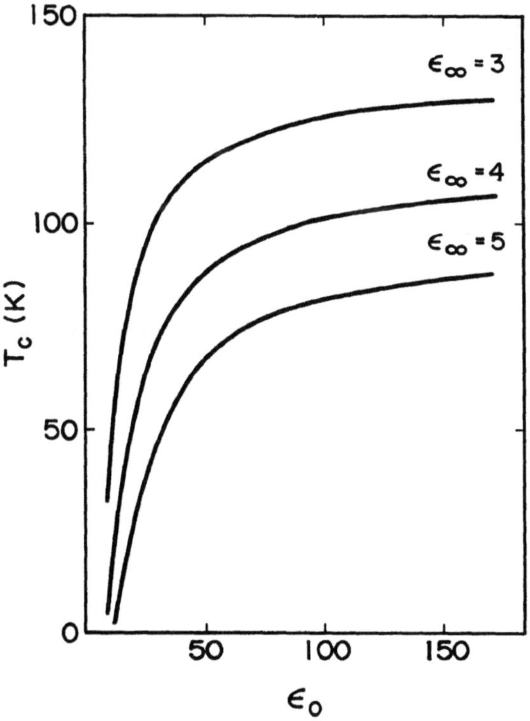

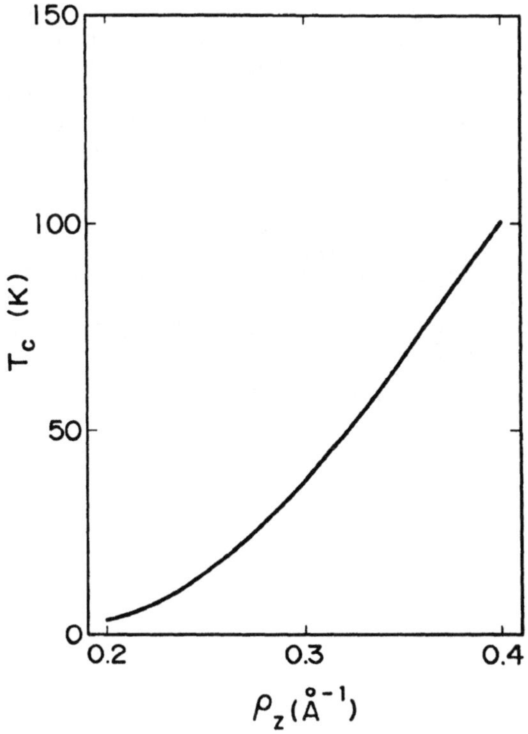

Fig. 1 Dependence of T_c on the dielectric constants. Quantities ε_0 and ε_∞ denote the static and optical dielectric constants, respectively.

Fig. 2 T_c as a function of ρ_z. Density of states for LO phonons is given by $\rho = \rho_x \rho_y \rho_z$.

CONCLUSION

In the present quasi-2D system, the effective electron-electron interaction via LO phonons of long wavelength in the second-order perturbation is proportional to $\rho_z \ln(r)$, which is quite different from the one widely used. This potential leads to the high-T_c superconductivity even within the BCS scheme. It has been found that T_c increases rapidly for ε_0 up to 50, while it is saturated for ε_0 above 100. It has also been shown, for increasing ρ_z, the rise of T_c is monotonic except for the slow increase in the small ρ_z region, with all other parameters being fixed.

ACKNOWLEDGMENT

The authors are grateful to Dr. Y. Gomei for stimulating discussions.

REFERENCES

1. Takahashi T, Matsuyama H, Katayama-Yoshida H, Okabe Y, Hosoya S, Seki K, Fujimoto H, Sato M and Inokuchi H (1989) Phys Rev B39 6636-6639
2. Nücker N, Romberg H, Xi XX, Fink J, Gegenheimer B and Zhao ZX (1989) Phys Rev B39 6619-6629
3. Herman F, Kasowski RV and Hsu WY (1988) Phys Rev B37 2309-2312
4. Harshman DR, Aeppli G, Ansaldo EJ, Batlogg B, Brewer JH, Carolan JF, Cava RJ, Celio M, Chaklader ACD, Hardy WN, Kreitzman SR, Luke GM, Noakes DR and Senba M (1987) Phys Rev B36 2386-2389
5. Inderhees SE, Salamon MB, Goldenfeld N, Rice JP, Pazol BG, Ginsberg DM, Liu JZ and Grabtree GW (1988) Phys Rev Lett 60 1178-1180
6. Thomsen C, Liu R, Wittlin A, Genzel L, Cardona M, König W, Cabañas MV and García E (1988) Solid State Commun 65 219-222
7. Macfarlane RM, Rosen H and Seki H (1987) Solid State Commun 63 831-834
8. Fukushima K and Sato H (1989) phys. status solidi (b) 153 K141-K145
9. Fukushima K and Sato H (1989) In: Saito G and Kagoshima S (ed) Physics and chemistry of organic superconductors. Springer. Berlin (to be published)
10. Kogut JB (1983) Rev Mod Phys 55 775-836
11. Yamashita J and Kurosawa T (1958) J Phys Chem Solids 5 34-43
12. Burns G, Dacol FH, Freitas PP, König W and Plaskett TS (1988) Phys Rev B37 5171-5174
13. Testardi LR, Moulton WG, Mathias H, Ng HK and Rey CM (1988) Phys Rev B37 2324-2325

Ground State Electronic Structure and Mechanism of High-T_c Copper Oxides

YOSHIHIRO ASAI
Electrotechnical Laboratory, Tsukuba, Ibaraki, 305 Japan

ABSTRACT

Ground state electronic structure of high-T_c copper oxides was determined by quantum chemical calculations including the electronic correlation effects. The modified cluster model was used. For the hole doped superconductor, the doped hole partially enters into the $O_{2p\sigma}$ orbital to make a local singlet between the hole on $Cu_{3dx^2-y^2}$ orbital which is in this orbital without doping and also partially enters into the in-plane band conventionally called the $O_{2p\pi}$ band. For the electron doped case, it is plausible that the doped electron enters into Cu_{4s} band. The new model Hamiltonian will be proposed here.

KEY WORDS: quantum chemical calculation, location of supercurrent charge carrier, electronic correlation, Madelung potential, local singlet, in-plane $O_{2p\pi}$ band, Cu_{4s} band

INTRODUCTION

Quantum chemical methods in calculating electronic correlation energy of atoms and molecules have been so successful. It has become possible to calculate the intensity of core level spectroscopy of highly correlated molecules and insulators within *ab initio* theoretical framework. These results not only prove applicability of quantum chemical methods but also provide undeniable assignments of intensity peak, which would be very difficult to obtain without their contributions. Quantum chemical method is an unique first principle method to evaluate the correlation energy of matter quantitatively. Despite the successes of quantum chemical methods in atoms, molecules and insulators, application of these first principle methods in the bulk description of the electronic structure of Mott-Hubbard type and/or charge transfer type solids with periodic boundary condition is not easy. This may be because connected quadruply excited amplitude becomes important in them. Some developments will be awaited to overcome this difficulty.
High-T_c copper oxides are considered to be highly correlated electronic system. Band theory could not describe the correlation gap in the half filled materials. To get some insight of the electronic structure of the copper oxides, quantum chemical methods will play a role. Naive application of the quantum chemical methods on the bulk electronic structure of copper oxides with the periodic boundary condition will encounter the difficulty sooner or later, so another modeling of the bulk materials is necessary. A model employed here is a kind of cluster model.
Effects of Madelung potential comes from outside of the cluster on the electronic structure of the cluster are rigorously incorporated. The charge iteration is performed so that the assumed valences of Cu and O are self-consistent with that estimated from the resultant wavefunction of the cluster. With this model, locations of supercurrent carriers of the hole doped and the electron doped copper oxides are discussed. The theoretical model for the mechanism of superconductivity relevant to the electronic structure obtained is discussed with some comments on the existing theories.

CALCULATIONS

The reader who wants to know the details of the calculation should refer the article[1]. The Madelung potential coming from the outside of cluster is estimated by the summation of 11^3 Evjen's unit cells of point charges. The relative difference of this potential at distinct atomic sites in the cluster coincides with that estimated by using the Ewald's method quite well(maximum error in absolute value is 5.0 meV). We have used the standard basis set for neutral copper and oxygen atoms with additional diffuse p-type single gaussian function on oxygen atoms to describe anion and dianion states of oxygen. Anion state is difficult to describe without additional diffuse function. There is no difficulty in the description of cation state. We adopted $TlBa_2CuO_5$ as the model compound for the hole doped superconductors. It has a hole per unit formula. It has been verified that this compound is a superconductor in contradiction to the earlier observation. This compound has deficiencies of oxygen, but they are not taken into account in the calculation. As the model of the electron doped superconductors, $Nd_{1.5}Ce_{0.5}CuO_4$ was adopted. The clusters CuO_6^{n-}, and $Cu_2O_7^{n-}$ were used for the hole doped and the electron doped compounds. Tl, Ba, Nd, and Ce were assumed to have a nominal valences +3,+2,+3, and +4. The valences of Cu and O and total charge of the cluster (-n) were determined by the self-consistent charge iteration.

For the hole doped superconductor, we first assume that the valences of Cu and O are +3 and -2, respectively. It this case, n is 9. We obtained three states: (i) a hole enters into the $Cu_{3dx^2-y^2}$ - $O_{2p\sigma}$ antibonding orbitals(it means that band picture is valid), (ii) a hole enters into the in-plane $O_{2p\sigma}$ atomic orbitals and makes a local singlet with another hole in $Cu_{3dx^2-y^2}$ atomic orbital which is already in this orbital without doping, (iii) a hole enters into in-plane $O_{2p\pi}$ orbitals(two $O_{2p\pi}$ atomic orbitals in a unit cell forms an antibonding molecular orbital and it is conventionally called $\psi_{2p\pi}{}^*$ in ref.[1] and here it will be called $O_{2p\pi}$, which belongs to the a_{2g} irreducible representation of D_{4h} symmetry) and ferromagnetically couples to a hole in the $Cu_{3dx^2-y^2}$ orbital which is already in this orbital without doping. The lowest state is (iii). The state (ii) is higher than (iii) by 0.44 eV. The state (i) is higher than (ii) by 1.67 eV. The charges of Cu, O(in plane), and O(apical) are +1.8,-1.7, and -2.0, respectively. As the resultant charges are not consistent with those of our first assumption, we should proceed to the next iteration step by assuming the charges of Cu,O(in plane), and O(apical) to be +2.0, -1.5, and -2.0. In this case n is 8. Two holes are doped. Note that our cluster includes oxygen atoms coming from the neighboring unit cell. In this case we obtained two state: (i) a hole enters into the $Cu_{3dx^2-y^2}$ - $O_{2p\sigma}$ antibonding orbital and another hole enters into the $O_{2p\pi}$ orbital defined above, (ii) a hole enters into the in-plane $O_{2p\sigma}$ atomic orbitals and makes a local singlet with a hole in the $Cu_{3dx^2-y^2}$ atomic orbital which is already in this orbital without doping and another hole enters into the $O_{2p\pi}$ orbital. The state (ii) is lower than the state (i) by 1.59 eV. These states are self-consistent in the charge iteration.

For the hole doped superconductors, the doped hole partially enters into the in-plane $O_{2p\sigma}$ atomic orbitals and makes a local singlet with the hole in the $Cu_{3dx^2-y^2}$ atomic orbital, and partially enters into the $O_{2p\pi}$ orbital. A separate estimate by extended Hückel band calculation shows that the in-plane $O_{2p\pi}$ band has 1.3 eV width. The local singlet obtained here was also discussed by Imada[2] using a model Hamiltonian.

For the electron doped superconductor, we assume the charges of Cu and O are +1.5 and -2.0, respectively. In this case, n is 11. We obtained two states: (i) a doped electron enters into $Cu_{3dx^2-y^2}$ -$O_{2p\sigma}$ antibonding band(it means that band picture is valid), (ii) a doped electron enters into the Cu_{4s} band and the antiferromagnetic coupling of the $Cu_{3dx^2-y^2}$ spins remains. The self-consistency of the charge iteration was observed. The state (i) is lower than the state (ii) by 0.55 eV. The bandwidth of Cu_{4s} band is almost 2.0 eV. So, it is very plausible that the band motion of Cu_{4s} electron changes the relative stability of the two state. In this case I have not checked the possibility of the local singlet formation, so the result for the electron superconductors is preliminary.

For the electron doped superconductors, it is plausible that the doped electron enters into the Cu_{4s} band.

The relevant model Hamiltonian for the hole doped superconductors may be

$$
\begin{aligned}
H= & -\sum_{<ij>\sigma} t_{ij}(a^{+}_{di\sigma}a_{pj\sigma} + h.c.) + \varepsilon_d \sum_{i\sigma} n_{di\sigma} + \varepsilon_p \sum_{j\sigma} n_{pj\sigma} \\
& + U_d \sum_{i\sigma} n_{di\sigma} n_{di-\sigma} + U_p \sum_{j\sigma} n_{pj\sigma} n_{pj-\sigma} \\
& - \sum_{<jl>\sigma} t'_{jl}(a^{+}_{\pi j\sigma}a_{\pi l\sigma} + h.c.) + \varepsilon_\pi \sum_{l\sigma} n_{\pi l\sigma} + U_\pi \sum_{l\sigma} n_{\pi l\sigma} n_{\pi l-\sigma} \\
& + V \sum_{j} n_{pj} n_{\pi j} + J_K \sum_{j} \vec{S}_{pj} \vec{S}_{\pi j}
\end{aligned}
\quad (1)
$$

$a_{dj\sigma}$, $a_{pj\sigma}$, and $a_{\pi j\sigma}$ are annihilation operators for $Cu_{3dx^2-y^2}$, $O_{2p\sigma}$, and $O_{2p\pi}$ orbitals for the spin σ, respectively. $n_{i\sigma}$ is the spin specific occupation number. t_{ij}, and t'_{jl} are the hopping integrals between $Cu_{3dx^2-y^2}$ orbital and $O_{2p\sigma}$ orbitals and between $O_{2p\pi}$ and $O_{2p\pi}$ orbitals, respectively. ε and U are the site energy and on-site Coulomb integral. V is the Coulomb integral between $O_{2p\sigma}$ and $O_{2p\pi}$ orbitals. $\vec{S}_{pj}$ and $\vec{S}_{\pi j}$ are the spin operators. J_K is the ferromagnetic coupling integral. The Hamiltonian for the electron doped superconductor may be obtained by replacing the operators for $O_{2p\pi}$ orbitals by Cu_{4s} orbitals, and the V and J_K terms by the corresponding term for the Cu_{4s} orbital and $Cu_{3dx^2-y^2}$ orbital. For the hole doped superconductor where the local singlet between Cu and O is formed, J_K should be zero in the cluster limit. The V term has very important physics which has never been discussed so far.

DISCUSSIONS

Here, we will comment on the other theoretical works on the electronic structure and the mechanism of high-T_c copper oxides.

Quantum chemical methods in calculating the correlation energy in the oxides necessarily needs the cluster model employed here. Reliable quantum chemical calculation should satisfy the following criteria: (a) Do the set of point charges used simulate the functional form of the Madelung potential calculated by well defined Ewald's method?, (b) Is the self-consistency of the point charges and the atomic charges calculated by the resultant wavefunction observed?, (c) Is the anion state of the oxygen site is well described? (if it is not satisfied the orbital energy of HOMO will positive and this mean that the ionized state is more stable), (d) Are all the possible electronic correlations in the cluster are taken into account? There is no published results which satisfies (a) and (b) except [1]. It should be stressed that (a) and (c) is very important. Unreasonable choice of electrostatic potential coming from outside of the cluster leads to an unbelievable result. Neglect of (c) change the energy ordering of Hartree-Fock orbitals and the resultant CI expansion becomes unreliable. We have checked that there is a minimum of the exponent of the added diffuse p-type gaussian orbital. It's value is just we have used in [1]. The difference of the results obtained in [3] and [1] comes form (a),(b) and (c). Guo et al [4] obtained relatively similar result with [1], the difference comes from (a) and (b). Rather large size of cluster was adopted by Martin et al [5]. It becomes more difficult to satisfy (d). I do not think that it is possible to describe low energy excitations of the bulk without introducing a model Hamiltonian. If one adopts a larger size of cluster, it becomes more difficult to satisfy (a) in the whole range of the space that cluster spans. The basis set should be more sophisticated. Band calculations were used to set up model

Hamiltonian for the oxides.[6,7]. The method to incorporate the correlation effect used there is not a logical one, so these should be regarded as complementary works to ours. Phenomenological fittings to high-energy photospectroscopy was also used to set up the model Hamiltonian[8,9]. Essentially, they use only first two lines of eq.(1), and obtained the parameters which simulates main and satellite intensities and peaks. The mechanism that yields the satellite splitting comes only from first two lines of eq.(1). It is dangerous to say that there is no additional terms which do not explicitly contribute the mechanism of formation of the splitting.
There is no theoretical works of the mechanism of superconductivity that involve all of the terms of eq.(1). The $O_{2p\pi}$ band is taken into account in [10,11,12,13], but first two lines of eq.(1) is replaced by Heisenberg Hamiltonian. The last two line of eq.(1) are neglected in [2,12,13,14]. The U_d term in the second line of eq.(1) was underestimated in [15]. A study of the Hamiltonian given by eq.(1) is now in progress, and it will appear elsewhere.

REFERENCES

[1]Y.Asai,J.Phys.Soc.Jpn.,**58**,3264(1989).
[2]M.Imada,J.Phys.Soc.Jpn.,**56**,3793(1987).
[3]M.Eto,R.Saito, and H.Kamimura,Solid State Comm.,**71**,425(1989).
[4]Y.Guo,J.-M.Langlois, and W.A.Goddard III,Science,**239**,896(1988).
[5]R.Martin and P.Saxe,Int.J.Quantum Chem.,**S22**,237(1988).
[6]A.K.McMahan,R.Martin, and S.Sapathy,Phys.Rev.,**B38**,6650(1988).
[7]L.Mattheis and D.R.Hamman,Phys.Rev.,**B40**,2217(1989).
[8]A.Fujimori,Phys.Rev.,**B39**,793(1989).
[9]K.Okada and A.Kotani,J.Phys.Soc.Jpn.,**58**,1095(1989)
[10]K.Hida,J.Phys.Soc.Jpn.,**57**,1544(1988).
[11]Y.Asai,J.Phys.Soc.Jpn.,**57**,3491(1988).
[12]H.Shiba and M.Ogata, *Mechanism of High Temperature Superconductivity,* Eds. by H.Kamimura and A.Oshiyama(Springer,1988) page 44.
[13]M.Imada, *ibid,* page 53.
[14]H.Matsukawa and H.Fukuyama,J.Phys.Soc.Jpn.,**58**,2845(1989).
[15]K.Yamaji,J.Phys.Soc.Jpn., in press.

3.2 Electronic Structures

Normal State Electronic Structure and the Superconducting Energy Gap in HTSC's as Determined from Photoemission Spectroscopy

A.J. Arko[1], R.S. List[1], R.J. Bartlett[1], S.-W. Cheong[1], Z. Fisk[1], J.D. Thompson[1], C.G. Olson[2], A.-B. Yang[2], R. Liu[2], C. Gu[2], B.W. Veal[3], J.Z. Liu[3], A.P. Paulikas[3], K. Vandervoort[3], H. Claus[3], and J.C. Campuzano[3]

[1] Los Alamos National Laboratory, Los Alamos, NM 87545, USA
[2] Ames National Laboratory – USDOE, Iowa State University, Ames, IA 50011, USA
[3] Argonne National Laboratory, Argonne, IL 60439, USA

Photoemission spectroscopy is utilized to determine the electronic structure of high-T_c materials. The observation of dispersive bands at E_F suggests a Fermi surface similar to that obtained from a band calculation. These results, together with a BCS-like energy gap, are not inconsistent with the notion of a Fermi liquid consisting of hybridized p-d bands.

Key words: Photoemission, Band Structure, Fermi Surface, BCS Gap, Correlated Fermi Liquid

I. INTRODUCTION

In spite of intense efforts by a large body of researchers, the nature of the electronic structure of high-T_c materials remains elusive. Early measurements on poorly characterized sintered specimens yielded a variety of confusing and contradictory results which spawned a number of theories[1], almost none of which have been conclusively proven or disproven to this day. The early photoemission measurements[2] on sintered 123- and 214-type samples were a major contributor to the conflicting results, since they yielded spectra that showed no intensity at the Fermi energy, thus fueling speculation that we are possibly dealing with insulators or, at best, semiconductors. At about the same time that the more stable 2212 compounds with their well-defined Fermi edge were discovered, we found that the failure to observe emission at E_F in the 214's and the 123's was due to a surface reconstruction caused either by a near surface oxygen loss, or the formation of an insulating oxide on the surface, or both (The surface reconstruction was actually first observed by Takahashi et al.[3], in the 214's, although they failed to observe a well-defined Fermi edge). Thus at least it was clear that we are dealing with metals, although the verdict is still not in on whether these are conventional metals (i.e., spin 1/2 Fermions) or a more exotic version[4] where spin and charge reside on different particles.

Existing theories can be roughly divided into three categories: 1) Conventional metals, whose properties can be understood from a band picture[5]; 2) Correlated Fermi liquids, whose properties can be derived from some variant of the Hubbard or Anderson model[6]; and 3) Unconventional metals, or non-Fermi liquids[4]. It is difficult to experimentally distinguish between the various approaches, since most theories are still in the evolutionary stages. Within each approach, for example, there is now the prediction of a reasonably sharp momentum cutoff at the Fermi energy and a well-defined Fermi surface, which is often cited, perhaps naively, as the mark of a Fermi liquid. We will try to show that the normal state electronic structure is at least not inconsistent with the notion of a correlated Fermi liquid, and that band calculations can serve as a good starting point in determining the density of states.

In the following sections we will review our most significant data and try to point out the features which can be utilized to distinguish between the various models. We will show that at least in the 123 materials the density of states (DOS) is well described by a band calculation, that in general there exists a Fermi surface resembling a calculated Fermi surface, that the states at E_F consist of hybridized p-d bands, that to within 30 meV the Fermi energy is indistinguishable from that of a Fermi liquid, and finally that a BCS-like superconducting energy gap with its associated pile-up of the DOS at E_F, is clearly distinguishable at $T<<T_c$. All of which would point to a correlated Fermi liquid ground state.

II. EXPERIMENTAL

The research reported here was performed at the Synchrotron Radiation Center in Stoughton, WI, on the Ames-Montana ERG-Seya beamline, as well as on the Minnesota-Argonne-Los Alamos ERG beamline. The angle integrated measurements were done using a CMA electron analyzer with a best overall resolution of 100 meV (usually 200 meV), while the angle-resolved work was done with a VSW ARIES system having an overall best resolution (including beamline) of 30 meV. The single crystals of HTSC materials were cleaved in-situ at T ≈ 20 K to avoid surface reconstruction. A chamber base pressure of 3×10^{-11} Torr insured an adsorbate free surface for a period of at least 24 hrs. As we have shown previously[7-9], it was necessary in the case of 123 materials to make all measurements at T ≈ 20 K since the surface reconstruction rapidly sets in at elevated temperatures. The much more stable 2212's could be heated above T_c (=85 K) without noticeable damage to the surface.

III. RESULTS AND DISCUSSION

A) The Densities Of States

An angle-integrated photoemission spectrum in essence measures the density of states (DOS) of a material, convoluted with an energy-dependent transition cross-section. A number of methods have been developed to calculate the cross-sections for a given material, and a calculation of a photoemission spectrum starting from a DOS is now common procedure[10]. In this section we will compare such theoretical spectra with our results. However, because photoemission is such a surface sensitive measurement, it is necessary first to consider the problem of surface reconstruction, and the need to insure that the measured spectra are representative of the bulk electronic structure.

In Fig. 1 we show angle-integrated spectra of $YBa_2Cu_3O_{6+x}$ for x = 0.9 (spectra A and B) and 0.25 (spectrum C) at hν=50 eV. The energy resolution is 200 meV. Spectrum A was taken at 20K in the a-b plane of a freshly cleaved single crystal. It shows a resolution limited Fermi edge, as well as substantial intensity in the -1.5 to -2 eV region. As the sample is warmed to 300 K (spectrum B) the intensity at E_F , as well as in the -1.5 to -2 eV region, is dramatically reduced. Associated with the loss of intensity at E_F is the growth of the satellite at -9.5 eV which is known to have O-2p character owing to its resonant enhancement at hν = 22 eV. These effects are easily understood if it is postulated that an insulating surface oxide, not related to the 123 structure, has formed on the surface from the 123 structure in the near surface region. In all probability this is some form of Ba-O, since there is reason to believe that the crystal cleaves in the Ba-plane (i.e., Ba core levels show a surface component). At low temperatures this reconstruction is inhibited. The fact that we are not dealing with a simple case of oxygen loss can be deduced from curve C where an oxygen-deficient crystal shows no sign of the -9.5 eV satellite, even while showing no intensity at E_F . The -9.5 eV satellite is thus not related to the 123 structure, and need not be considered as a mystery peak. It is an oxygen satellite from a more localized (insulating, as evidenced by charging effects) surface oxide. We find that the rate of surface reconstruction is a strong function of stoichiometry and surface perfection. Thus the observation of a Fermi edge at room temperature is not inconsistent with our data if the measurements are made on fully oxygenated samples with a non stepped surface.

In Fig. 2a we compare an angle-integrated spectrum for $YBa_2Cu_3O_{6.9}$, taken at 20 K at a photon energy of 70 eV, to a calculated spectrum from Ref. 10. In this figure, the secondary electron background has been subtracted out in the usual fashion in order to better compare with calculations. By shifting the calculated spectrum by 0.5 eV to higher binding energy we get almost exact agreement between band theory and experiment. All the features in the spectrum are resonance. In our 2212 samples the intensity at E_F was generally more than 5% of the valence band maximum so that it was possible to obtain data away from 18 eV where dispersion is much more pronounced. We preferred not to deal with the additional complication of an unknown resonance at 18 eV. With much higher resolution and intensity, there was no doubt about a band crossing. An analysis of the hv-dependence indicated that the bands at E_F consist of about a 30-70 mix 3d and 2p orbitals.

In Fig. 3 we display angle resolved data taken along the [110] direction at a temperature of 20 K, showing only the data within the first 350 meV of E_F. The total dispersion shown in Fig. 3 is approximately half the predicted value (on the order of 400meV), although the Fermi level crossing (at $\theta \approx 20^0$) is at approximately the **k**-value where the Bi-O band is expected to cross E_F. Along the [100]-type directions, the crossing of E_F occurs precisely at the predicted values, so that at least in these directions the calculated Fermi surface has the correct topology.

There are several interesting features of the data in Fig. 3. Fist it will be noticed that the dispersive feature is considerably broader below E_F than at E_F. Indeed, the broadening goes as $(E-E_F)^2$, although because of the large energy scale it is not certain that this can be used as a proof of Fermi liquid behavior. The rapid broadening may, however, be an indication of correlation effects, since the broadening should be related to the effective mass. This is also reflected in the narrower than predicted band width.

A second noticeable feature is the asymmetry of the photoemission peaks due to a relatively large background, which cannot be entirely accounted for on the basis of secondary electrons. Indeed, less than half of the background intensity can be associated with secondaries. It has been suggested that this background is possibly a manifestation of the diffuse background of electronic states in the non-Fermi liquid theories, while the well-defined peaks represent holon or spinon bands. We would point out that while we cannot rule out this possibility, there are other, much more mundane sources of background intensity, such as surface imperfections, which should be considered. This particularly in view of the fact that there is considerable sample dependence to the data. For example, in some crystals of 2212 material, no dispersive features are evident near E_F, in spite of a sharp superconducting transition. Only a featureless background is seen. Under normal circumstances we would interpret this as a bad surface due to a bad cleave. But, in addition, we also wish to point out that in Fig. 3 for $\theta > 22^0$ (i.e., for **k**-values beyond the Fermi level crossing), there is almost no background intensity. This is not inconsistent with the concept of limited surface damage with some residual **k**-dependence in the damaged regions. It would, on the other hand, necessitate one more modification to the non-Fermi liquid theories, in order to include **k**-dependence in the diffuse background of electronic states.

It had been assumed that one could distinguish between a Fermi liquid and a non-Fermi liquid from the sharpness of the Fermi edge, with the Fermi liquid following Fermi-Dirac statistics exactly. However, it is now predicted[16,17] that the differences (if any) may be extremely subtle. In Fig. 4a we show overlayed on the same plot, the Fermi edges of Pt metal and that of $Bi_2Sr_2CaCu_2O_8$ taken at 90 K (i.e., above T_c). To within experimental resolution of 30 meV (i.e., smaller than the broadening due to the Fermi function) there is no difference between them. Any deviation from a Fermi liquid is on a scale smaller than 30 meV, or a temperature equivalent to T_c.

C) The Superconducting Energy Gap

One can see in Fig. 3 from data taken at 20 K, that as the band approaches the Fermi energy, the lineshape of the peak structure changes dramatically. A sharp, resolution limited peak develops at E_F at 20 K (i.e., below T_c), but not above T_c, as seen in Fig. 4a for $\theta = 18^0$. Associated with the peaked structure is a decrease in the Fermi energy, which can be seen in Fig. 4b where Fermi edge overlays for the same samples and the same angles as in Fig. 4a are again shown, but now at 20 K. The most straightforward explanation for the sharp peak and the change in the Fermi energy is reproduced in the calculation, including the bottom of the band at -6 eV. The major disagreements are the 0.5 eV shift, and a sizable reduction in the intensity at E_F, both of which could be a consequence of the effects of electron correlation. Spectra at other photon energies show a similar agreement, thus indicating that the calculation obtains the correct mix of 2p and 3d orbitals. Thus it is strongly suggestive that at least for $YBa_2Cu_3O_{6.9}$, a band calculation correctly captures the essential features of the DOS.

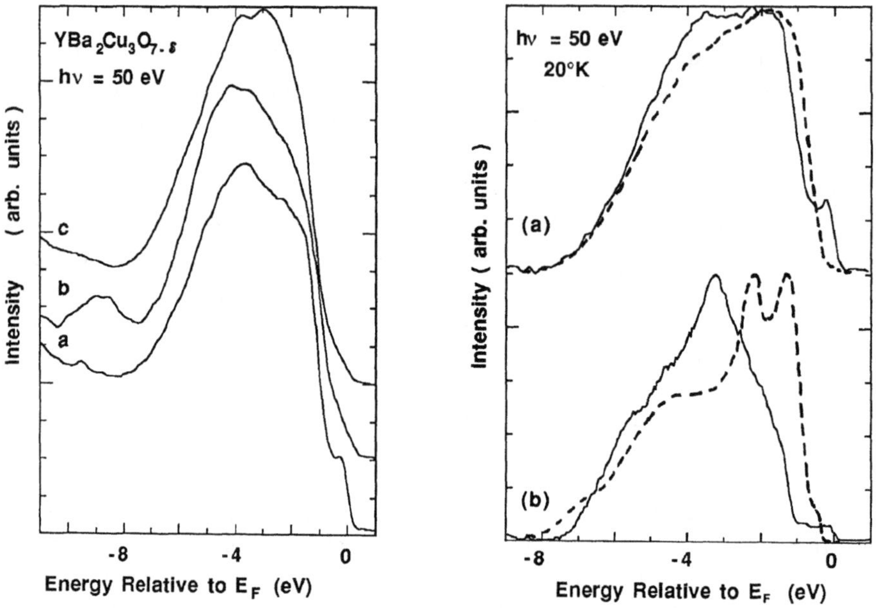

Fig. 1 *Valence band spectra for single crystals of* $YBa_2Cu_3O_{6+x}$ *at* $h\nu = 50$ *eV: a)* $x = 0.9$, $T = 20$ *K; b)* $x = 0.9$, $T = 300$ *K; c)* $x = 0.25$, $T = 20$ *K.*

Fig. 2 *Comparison of valence band spectra (with backgrounds subtracted) to calculated spectra (dotted lines) from Ref. 10 and 11 at* $h\nu = 50$ *eV: a)* $YBa_2Cu_3O_8$ *; b)* $Bi_2Sr_2CaCu_2O_8$.

The same cannot be said for $Bi_2Sr_2CaCu_2O_8$. In Fig. 2b we show a similar comparison of a spectrum to a band calculation[11] as in Fig. 2a. The disagreements are rather obvious. Takahashi has shown that not only the position, but also the curvature of the band at -5.5 eV is completely at odds with band structure results. We would find it unlikely that these disagreements are the result of increased correlation effects in $Bi_2Sr_2CaCu_2O_8$ vs. $YBa_2Cu_3O_{6.9}$. This particularly in view of the fact that the -12 eV Cu satellite, the major indicator of Coulomb correlation, is much smaller in $Bi_2Sr_2CaCu_2O_8$ relative to the valence band maximum than it is in $YBa_2Cu_3O_{6.9}$. It is more likely that the band calculations themselves should be re-examined. In particular, it will be necessary to include relativistic effects of the heavy Bi atom, as well as the effects of the superlattice, before any definitive comparisons can be made. The relativistic effects may be important at E_F as well, where band calculations find a Bi-O band crossing E_F near the M-point in the zone. Indications are that this crossing may not exist (see below).

B) Band Dispersion And The Fermi Edge

Angle-resolved data were first obtained by Takahashi et al. in the 2212 system. They showed a poor correspondence with band calculations. However, the very existence of dispersion and a crossing of the Fermi energy, indicated that a Fermi surface must exist for this system, and that the possibility of a Fermi liquid electronic structure must be considered for the normal state. The data of refs.12 and 13 did not convince all skeptics of the existence of a Fermi surface owing to a low count rate at E_F, and a relatively low resolution. The low count rate necessitated work at 18 eV photon energy in order to enhance the intensity via a presumed Fano resonance. However, because of very strong matrix element effects, the jury is not yet in on whether this is a true Fano that we are observing a BCS-like energy gap opening at E_F, with its associated pile-up of states. Modeling studies show that the spectrum in Fig. 4b (for the 2212 material) can be obtained from the spectrum in Fig. 4a by convoluting the latter with a BCS density of states function, assuming an energy gap $2\Delta = 7\ kT_C$. This gap value, which shows little anisotropy with **k**, is exactly twice the theoretical BCS value, and thus should be an interesting clue in solving the coupling mechanism.

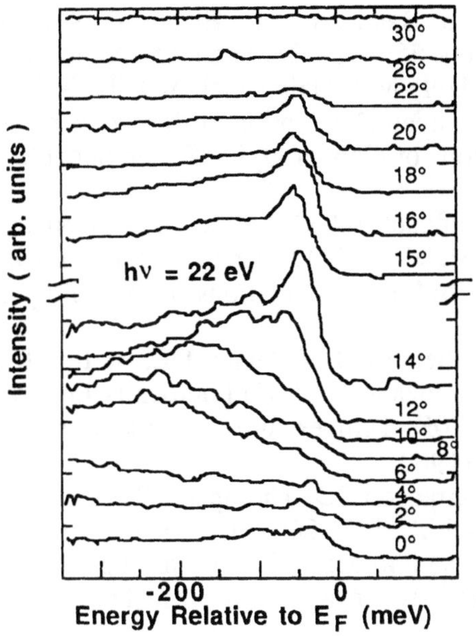

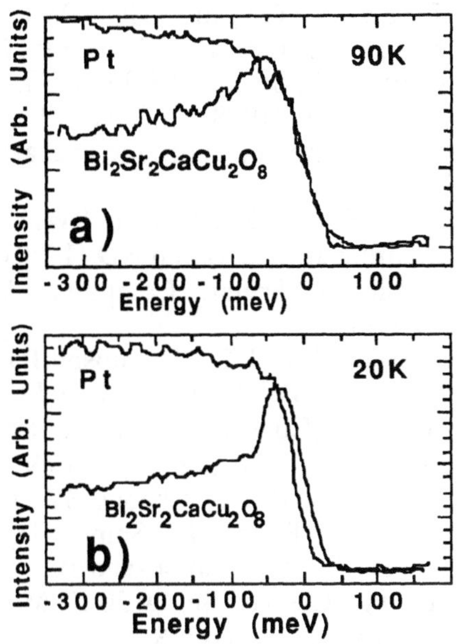

Fig. 3 *Energy dispersive curves for a single crystal of $Bi_2Sr_2CaCu_2O_8$ along the Γ-M direction in the Brillouin zone at T = 20 K and hν = 22 eV. The Fermi energy crossing of the dispersive peak is near 20°. Note the lineshape change near E_F.*

Fig. 4 *Comparison of the Fermi edge of Bi-2212 (at θ=18°) to that of Pt: a) at 90 K; b) at 20K.*

IV) CONCLUSIONS

It is difficult to distinguish between Fermi liquid behavior in the new high-T_c superconductors and behavior expected from the novel new non-Fermi liquid theories. The differences are predicted to be on an energy scale smaller than our experimental resolution. However, while deviations from simple band theory certainly do exist in the form of core and valence band satellites, band narrowing, and rapid photoemission peak broadening away from E_F, there are sufficient agreements for the overall DOS (at least for the 123 structure) that it should be considered a good starting point for the electronic structure. The calculated Fermi surface for both the 123 and 2212 structures is reasonably-well reproduced experimentally; the bands at E_F consist of p-d hybridized orbitals as predicted. A large BCS-like energy gap is observed below T_c.

This work was performed under the auspices of the U.S. Department of Energy. The Ames Lab. acknowledges DOE support under contract # W-7405-ENG82 while Argonne National Lab acknowledges DOE support under contract # W-31-109-ENG-38. The Synchrotron Radiation Center, whose staff we wish to gratefully acknowledge for their assistance, is supported by the NSF under contract #DMR8601349. KV is supported under contract #STC-880954.

1. For a compilation of the various approaches see, (1988) Woods Halley J (ed),"Theories of High Temperature Superconductivity," Addison-Wesley
2. Wendin G (1987) High-Energy Spectroscopy and High-T_c Superconductors: An Overview. J. Phys.(Paris) Colloq. 48: C9-483 ; Wendin G (1989) Spectroscopic Views of High-T_c Superconductors. Physica Scripta,
3. Takahashi T, Maeda F, Katayama-Yoshida H, Okabe Y, Suzuki T, Fujimori A, Hosoya S,

Shamoto S , and Sato M , (1988) Photoemission Study of Single - Crystalline ($La_{1-x}Sr_x)_2CuO_4$. Phys. Rev. B37: 9788-9791.

4. Anderson PW, (1987) Resonating Valence Bond State in La_2CuO_4 and Superconductivity. Science 235: 1196-1198; Anderson PW, Baskaran G, Zou Z, and Hsu T (1987) Resonating-Valence-Bond Theory of Phase Transitions and Superconductivity in La_2CuO_4-Based Compounds. Phys. Rev. Lett. 58: 2790-2793.
5. For a complete list of references on various band structure calculations, see Picket WE (1989) Electronic Structure of the High-Temperature Oxide Superconductors. Rev. Mod. Phys. 61: 433-512.
6. Zaanen J, Sawatsky GA, and Allen JW (1985) Band Gaps and Electronic Structure of Transition Metal Compounds. Phys. Rev. Lett. 55: 418-421: Shen ZX, Allen JW, Yeh JJ, Kang JS, Ellis WP, Spicer WE, Lindau I, Maple MB, Dalichauch YD, Torikachlivi MS, Sun JZ, and Geballe TH (1988) Anderson Hamiltonian Description of the Experimental Electronic Structure and Magnetic Interactions of Copper Oxide Superconductors. Phys. Rev. B36: 8414.
7. Arko AJ, List RS, Fisk Z, Cheong SW, Thompson JD, O'Rourke JA, Olson CG, Yang AB, Pi TW, Schirber JE, and Shinn ND (1988) Metallic, Band-like Behavior of $EuBa_2Cu_3O_{6.7}$ As Seen From UPS Spectra: Comparison With Band Calculations. J. Mag. Mag. Mater. Lett. 75: L1-L6.
8. List RS, Arko AJ, Fisk Z, Thompson JD, Pierce CB, Peterson DE, Bartlett RJ, Shinn ND, Schirber JE, Veal BW, Paulikas AP and Campuzano Jc(1988) Photoemission From Single Crystals of $EuBa_2Cu_3O_{7-x}$ Cleaved Below 20K: Temperature-Dependent Oxygen Loss. Phys. Rev. B 38 (Rapid Comm.) : 11966 -11969.
9. List RS, Arko AJ, Fisk Z, Cheong SW, Conradson SD, Thompson JD, Pierce CB, Peterson DE, Bartlett RJ, O'Rourke JA, Shinn ND, Schirber JE, Olson CG, Yang AB, Pi TW, Veal BW, Paulikas AP, and Campuzano JC, (1989) Photoemission From Single Crystal $EuBa_2Cu_3O_{6.9}$ Cleaved Below 20 K; Metallic-to- Insulating Surface Transformation. AIP Conf. Proc. 182: 283-288.
10. Redinger J, Freeman AJ, Yu J, and Massida S (1987) Local Density Theory of X-Ray and Photoemission From $YBa_2Cu_3O_{7-d}$:The High-T_c Superconductor. Phys. Lett. A124, 469-473.
11. Marksteiner P, Massidda S, Yu J, Freeman AJ, Redinger J (1988) Calculated Photoemission and X-Ray emission Spectra of $Bi_2Sr_2CaCu_2O_8$. Phys. Rev B38: 5098-5101.
12. Takahashi T, Matsuyama H, Katayama-Yoshida H, Okabe Y, Hosoya S, Seki K, Fujimoto H, Sato M, and Inokuchi H (1989) Band Structure of $Bi_2Sr_2CaCu_2O_8$ Studied by Angle-Resolved Photoemission. Phys. Rev. B39 : 6636.
13. Takahashi T, Matsuyama H, Katayama-Yoshida H, Y. Okabe, S. Hosoya, K. Seki, H. Fujimoto, M. Sato, and H. Inokuchi (1988) Evidence From Angle-Resolved Resonant Photoemission for Oxygen-2p Nature of the Fermi Liquid States in $Bi_2CaSr_2Cu_2O_8$. Nature 334: 691-692.
14. Olson CG, Liu R, Yang AB, Lynch DW, Arko AJ, List Rs, Veal BW, Chang YC, Jiang PZ, Paulikas AP (1989) High-Resolution Angle-Resolved Photoemission Studies of High-Temperature Superconductors. Submitted to the Stanford Conference on High-T_c Superconductors, Stanford, CA, July 24-27, 1989.
15. Arko AJ, List RS, Bartlett RJ, Fisk Z, Cheong SW, Thompson JD, Olson CG, Yang AB, Liu R, Gu C, Veal BW, Liu JZ, Paulikas AP, Vandervoort K, Claus H, Campuzano JC, Schirber JE, and Shinn ND (1989),Large Dispersive Photoelectron Fermi Edge and the Electronic Structure of $YBa_2Cu_3O_{6.9}$ Single Crystals Measured at 20 K, Phys. Rev. B 40: 2268-2278
16. Anderson PW, (private comm.).
17. Varma CM, Littlewood PB, Schmitt-Rink S, Abrahams E, and Ruckenstein AE, (Submitted), Phenomenology of the Normal State High-Temperature Superconductors, To be published in Phys. Rev. B.
18. Olson CG, Liu R, Yang AB, Lynch DW, Arko AJ, List Rs, Veal BW, Chang YC, Jiang PZ, Paulikas AP (1989) Superconducting Gap in Bi-Sr-Ca-Cu-O by High-Resolution Angle-Resolved Photoelectron Spectroscopy. Science 245: 731-733.

Layered Cuprates: Structure, Valence State and Superconductivity

C. MICHEL, M. HERVIEW, F. STUDER, and B. RAVEAU

Laboratoire de Cristallographie et Sciences des Matériaux – ISMRa, Bld Maréchal Juin – 14032, Caen Cedex, France

ABSTRACT

The problems of oxygen stoichiometry and valence state of the different metallic elements present in the superconductive cuprates are closely related. They are presented and discussed in connection with the structure and the results from electron microscopy and X-ray absorption studies.

KEYWORDS : Oxygen non stoichiometry, valence state, superstructure, X ray absorption spectroscopy.

INTRODUCTION

The superconductive cuprates, synthesized up to now, are characterized by a bidimensional character of their structures [1, 2 and included references]. This results from the property of copper to take coordinations lower than six and, for most of them, from the ability of the perovskite structure AMO_3 and of the rock salt- type structure A'O to adapt to each other leading to intergrowth phenomena with the general formulation $(AMO_{3-x})_m(A'O)_n$. These oxides can then be represented by the symbol [m,n] where m and n are the thickness of the perovskite and of the rock salt-type slabs, respectively. For m=1, the perovskite slab is built up from a $[CuO_3]_\infty$ single layer of corner-sharing CuO_6 octahedra; for m=2, it is characterized by two $[CuO_{2.5}]_\infty$ of corner-sharing CuO_5 pyramids; and for m>3, (m-2) $[CuO_2]_\infty$ layers of corner-sharing CuO_4 square planar groups are intercalated between two adjacent $[CuO_{2.5}]_\infty$ layers; $YBa_2Cu_3O_7$ is the limit member (n=0, m=∞). In the perovskite slabs, each copper oxygen layer is separated from the adjacent one by a plane of cations which are alkaline or rare earths. The other common characteristic is the mixed valence of copper Cu(II)-Cu(III), keeping in mind that Cu(III) must only be considered as a formal valence. This mixed valency is, with the low dimensionality of the structures, one of the two parameters which govern the superconductivity in these oxides. Since it results from electroneutrality calculations, the mean oxidation state of copper is closely related to the oxygen stoichiometry and to the valence state of the other elements present in the materials, when it can vary. However it is not necessarily representative of the real trivalent copper content which, in fact, depends on the structure by the possibility of copper disproportionation which would involve the formation of Cu(I).

COPPER DISPROPORTIONATION

This phenomenon appears to be very rare and corresponds in fact to a metastable state closely related to the nature of the structure. Among the different layered copper oxides only three compounds exhibit such a behaviour : $YBa_2Cu_3O_{7-\delta}$ [3], $Pb_2Sr_2Ca_{0.5}Y_{0.5}Cu_3O_8$ [4] and $Pb_2Sr_{2-x}La_{1+x}Cu_2O_6$ [5].

The first one corresponds to a progressive transition of the 92 K orthorhombic superconductor $YBa_2Cu_3O_7$ (fig. 1a) , characterized by the mixed valence Cu(II)-Cu(III) in pyramidal and square planar coordination, into the tetragonal semiconducting oxide $YBa_2Cu_3O_6$ (fig. 1b), characterized by divalent copper in a

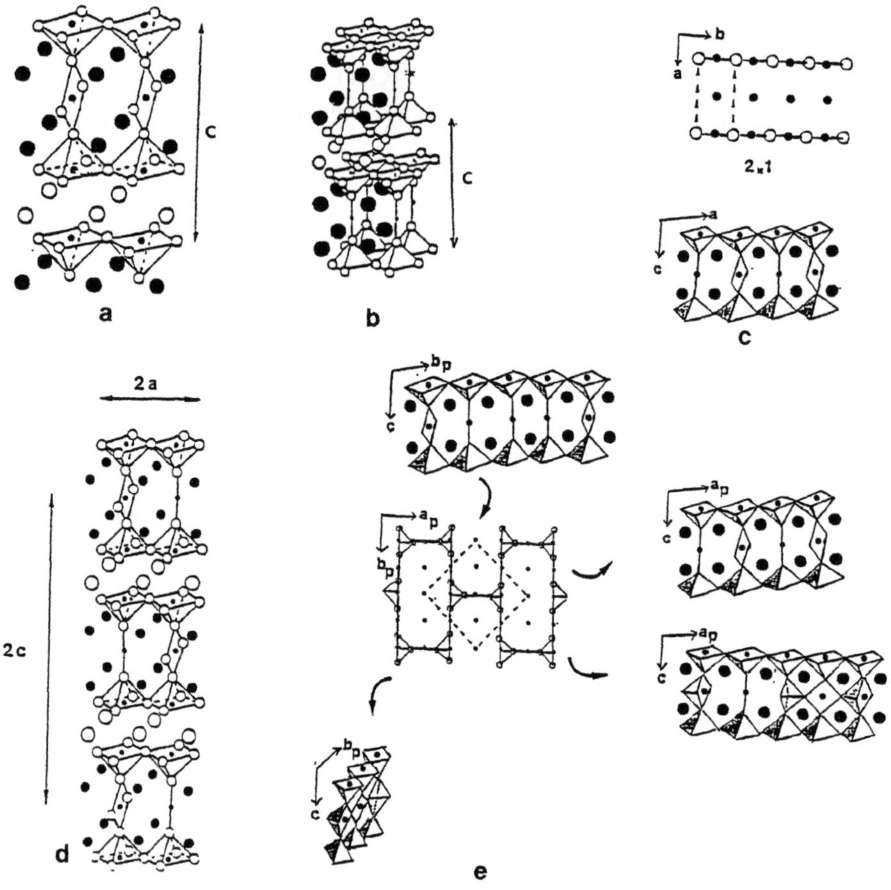

Fig.1 : Structure of the orthorhombic superconductor $YBa_2Cu_3O_7$ (a) and of the tetragonal semiconducting oxide $YBa_2Cu_3O_6$ (b). Models of local superstructures : "2a x a x c" (c), "2a x a x 2c" (d) and "2a√2 x 2a√2 x c" (e).

pyramidal environment and monovalent copper in a twofold coordination. From a structural point of view, the deviation from "O_7" stoichiometry cannot correspond to a simple removing of the oxygen atoms at the level of copper in a square planar coordination which would implie copper in a threefold coordination which has never been observed in oxides. Electron diffraction and high resolution electron microscopy studies have shown, in samples with $0.37 < \delta < 0.45$ [6] the existence of local or extended superstructures : "2a x a x c", "2a x a x 2c", "3a x a x c", "2a√2 x 2a√2 x c", "2a x a√5 x c" and "2a x a√10 x c" which can be related to an inhomogeneous distribution of the oxygen vacancies in the crystals. To explain such superstructures a model of copper disproportionation was proposed : 2Cu(II) ⟶ Cu(I) + Cu(III) [3]. It was supported by X ray absorption measurements [7] which show the presence of monovalent copper even for low δ values, which increases with δ. For instance, respecting the usual coordination for copper (twofold for Cu(I), fourfold, fivefold and sixfold for Cu(II) and Cu(III)), the existence of "2a x a x c" superstructure (fig. 1c) implies that one row of corner sharing CuO_4 groups alternates along a with one row of CuO_2 sticks, whereas the "2a x a x 2c" superstructure (fig. 1d) corresponds to the shifting of one layer of rows out of two of a/2. For these two superstructures the corresponding composition is $YBa_2Cu_3O_{6.5}$ which would involve only divalent copper from electroneutrality calculation; but considering the proposed models, the formula should be written $(YBa_2Cu_3^{II,III}O_7)_{0.5}(YBa_2Cu_2^{II}Cu^{I}O_6)_{0.5}$. The other superstructures can also be described by such models as shown for "2a√2 x 2a√2 x c" in figure 1e. This model corresponds to 6.875 oxygen atoms per formula unit. Thus the oxygen non stoichiometry in the oxides $YBa_2Cu_3O_{7-\delta}$ corresponds to the coexistence in the same crystal of ordered microdomains according to the formula: $(YBa_2Cu_3O_7)_{1-\delta}(YBa_2Cu_3O_6)_\delta$ where the "O_7"

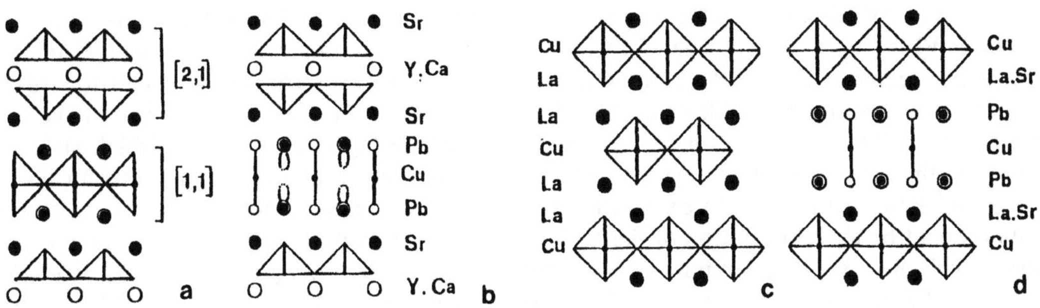

Fig. 2: Structure of $Pb_2Sr_2Ca_{0.5}Y_{0.5}Cu_3O_8$ (a), $Pb_2Sr_2Ca_{0.5}Y_{0.5}Cu_3O_{10}$ (b), La_2CuO_4 (c) and $Pb_2Sr_{2-x}La_xCu_2O_6$.

regions are superconducting and the "O_6" ones are insulating.

The oxides $Pb_2Sr_2Ca_{0.5}Y_{0.5}Cu_3O_8$ (T_c= 46 k) and $Pb_2Sr_{2-x}La_xCu_2O_6$ (T_c= 40 K) whose syntheses require a very poor oxygen atmosphere also exhibit copper disproportionation. The structure of the first one (fig. 2b) derived from the limit structure $Pb_2Sr_2Ca_{0.5}Y_{0.5}Cu_3O_{10}$ (fig. 2a) corresponding to a double intergrowth of [2,1] and [1,1] terms by removing the oxygen atoms in the basal plane of the octrahedra of the perovskite monolayer, leading to the formation of (Cu^IO_2) sticks. In such a structure the theoretical amount of monovalent copper is 33%, in agreement with that deduced from X ray absorption spectroscopy measurements at Cu K edge [8]. In the same way, the structure of the second one derives from that of La_2CuO_4 (m=1, n=1) (fig.2c) by elimination, in one perovskite monolayer out of two, of the oxygen atoms in the basal plane of the octahedra (fig. 2d), leading again Cu(I) in a twofold coordination. The stabilization of these structures may be realized by the $6s^2$ lone pair of Pb(II) which is probably extended towards the oxygen vacancies. Extra oxygen atoms take place in the oxygen vacancies by oxidization, leading to an increase of the oxidation state of copper and/or lead. Since in the same time superconductivity disappears, this suggests that holes are mainly transferred from copper-oxygen layers to lead layers by transformation of a part of Pb(II) into Pb(IV), showing the importance of the nature of the oxidation state of the other elements present in the material with regard to copper.

VALENCE STATES IN BISMUTH CUPRATES

The problem of valence state of copper and its relation with the oxygen stoichiometry is more complex in the case of bismuth, thallium and lead superconducting cuprates since these metals are able to take several oxidation states which can influence the valence of copper. Moreover the oxygen content is not known with accuracy owing to the difficulty of determination either by analysis or structural studies.

The bismuth family is represented by the symbol [m,3]. There is no doubt that mixed valency of copper is present, so the ideal formulation $Bi_2Sr_2Ca_{m-1}Cu_mO_{2m+4}$ cannot be considered as representative if one admits that bismuth is trivalent. Several structural studies [9, 10] have shown that the existence of the incommensurate structure results in a waving of the A'O layers which are not only characterized by modulated displacements of the ions but also by the presence of an excess of oxygen allowing the mixed valency of copper and consequently the appearence of the superconducting properties. Nevertheless the nature of the valence of bismuth in these oxides is so far not clear as it appears from a preliminary X-ray absorption study of these phases [12]. The patterns at Bi LIII edge of the three superconductors (m=1, 2, 3) have been compared with those of bismuth metal, Bi(III) (Bi_2O_3 and $Bi_2SrNb_2O_9$) and Bi(V) ($NaBiO_3$) as standards and with that of the isostructural phase $Bi_2Sr_2CaFe_2O_9$. The patterns of Bi_2O_3 and $Bi_2SrNb_2O_9$ are absolutely identical and characteristic of Bi(III). The comparison of those of the superconductive cuprates and of the non superconductive $Bi_2Sr_2CaFe_2O_9$ with that of $NaBiO_3$ shows that both oxides exhibit small amounts of Bi(V) identified by the prepeak corresponding to 6s empty levels (fig. 3a). Nevertheless the amounts of Bi(V) in the cuprates is smaller than that observed for the isostructural iron compound. but the most interesting feature concerns the shifting of the curves of the superconductive oxides towards lower energy with respect to Bi(III) oxides (fig. 3b). Such a feature is not observed neither in the layered $Bi_2SrNb_2O_9$ (Aurivillius phase) nor in the iron oxide $Bi_2Sr_2CaFe_2O_9$ which seems to exhibit similar Bi-O distances as in the superconductors, due to similar incommensurate structure. This suggests for bismuth a mean oxidation state smaller than three which has never been observed to our knowledge. An alternative explanation would be an hybridization of the 6s-6p of bismuth allowing a narrow band to be built up by overlapping with the oxygen 2p orbitals. In this hypothesis the bismuth oxygen layers would be conductive and would play the role of reservoir of holes for the superconductive copper-oxygen layers. The presence of small amounts of Bi(V) would only be a secondary phenomenon resulting from an excess of oxygen in the Bi-O layers.

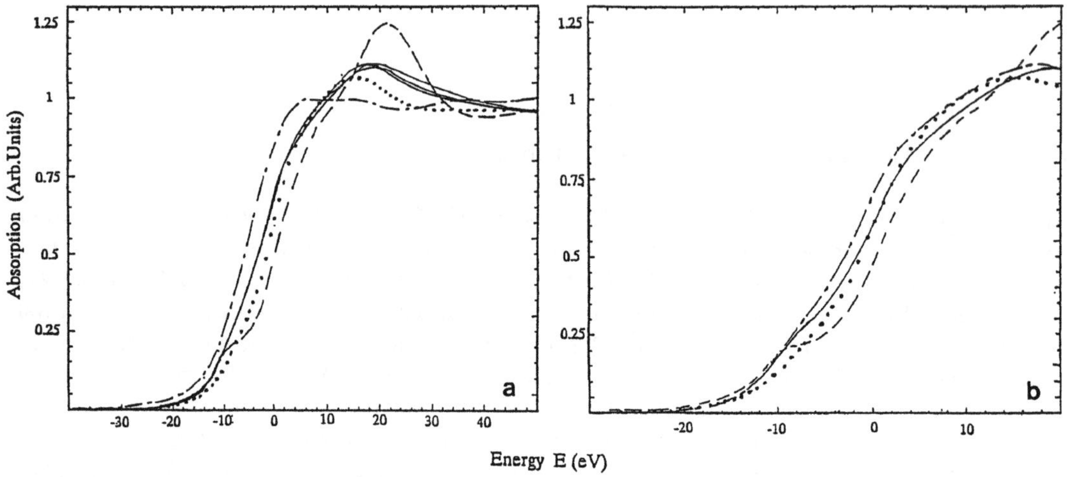

Fig. 3 Bi LIII-edge spectra for : (a) - — Bi metal, ···· Bi_2O_3, — — $NaBiO_3$, —— the three bismuth superconductors; (b) - — Bi metal, ···· Bi_2O_3,— —$NaBiO_3$, —— $Bi_2Sr_2CaFe_2O_9$.

THALLIUM AND LEAD CUPRATES

If the mixed valence of copper is easy to understand in the superconductive thallium cuprates $TlA_2Ca_{m-1}Cu_mO_{2m+3}$ (A=Sr,Ba) ([m,2] series), the same problem as in bismuth cuprates occurs in the series $Tl_2Ba_2Ca_{m-1}Cu_mO_{2m+4}$ ([m,3]) for which no deviation from the ideal oxygen stoichiometry has been observed up to now from the different structural studies. On the other hand, it seems that thallium deficiency (5%-10%) or partial occupation of thallium sites by calcium occur, allowing the mixed valence of copper to appear. Contrary to the bismuth phases, these oxides exhibit many extended defects [12] such as [TlO] and [$CaCuO_2$] intergrowth defects and excess of thallium on the alkaline-earth sites which can also influence drastically the superconducting properties of these materials. But the existence of such extended defects is sufficient to understand other issues such as the oxidation state of thallium. For instance, the observation of satellites with very weak intensities in the electron diffraction patterns of some samples of the [m,3] superconductors is assumed to be correlated with AO layers distortions associated with different structural features such as the presence of monovalent thallium . Moreover, in the TlO_6 octahedra, the four Tl-O distances in the basal plane (2.7-2.9 Å) are close to those observed in univalent thallium oxides, whereas the apical distances (2.0-2.1 Å) are close to those observed in Tl_2O_3. However, isolated Tl(I) cations have not been detected by X-ray absorption study [13]. The LI and LIII edge spectra of some of these oxides have been compared to those of Tl(I) (TlBr and $Tl_2Ta_2O_6$) and Tl(III) (Tl_2O_3) standards. They are closely related to that of Tl_2O_3 (Fig.4a). Since Tl_2O_3 is a very good metallic conductor, one can think that the Tl-O layers in these oxides are also conductive and would play the role of electron acceptor.

The case of lead cuprates has been partly discussed in a previous paragraph, we will just discuss now about the superconductor $Pb_{0.5}Sr_{2.5}Ca_{0.5}Y_{0.5}Cu_2O_{7-\delta}$ [14]. From chemical analysis, the oxygen content was found to be close to 6.75. The superconducting properties ($T_c \approx 50$ K), the resistivity value at room temperature (1.6 10^{-2} Ω cm) and the metallic behaviour above T_c attest of the presence of the mixed valence of copper in this material and consequently of a part of lead as divalent. Its structure (fig. 4b) belongs to the [2,2] type, it consists of double pyramidal copper layers intergrown with double rock salt layers. The rock salt slabs are formed of mixed $[Sr_{0.5}Pb_{0.5}O]_\infty$ layers sandwiched by Sr-O layers, but the real structure appears to be more complex since the preliminary X ray study suggests that the oxygen vacancies are located in the Sr-O layers which seems to be correlated with the presence of the $6s^2$ lone pair of Pb(II). The numerous satellites observed in this phase, which are associated with the layer distortion, are also in agreement with the presence of divalent lead and/or oxygen vacancy. The fact that the resistive transition (fig. 4c) is broad and changes with the experimental conditions (oxygen pressure, annealing) shows that the oxygen distribution in the structure should play a capital role in the superconducting properties.

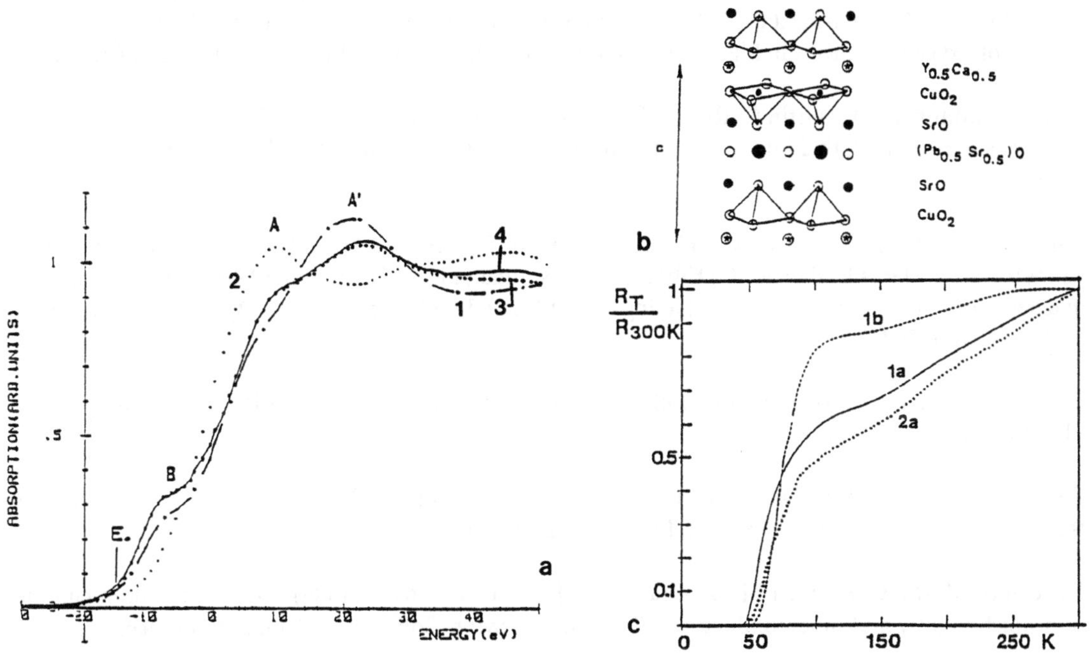

Fig. 4 (a) Tl LIII edge spectra for Tl_2O_3 (1), $Tl_2Ta_2O_6$ (2) and the superconductors $TlBa_2CaCu_2O_7$ (3) and $Tl_2Ba_2CaCu_2O_8$ (4); (b) structure of the superconductor $Pb_{0.5}Sr_{2.5}Y_{0.5}Ca_{0.5}Cu_2O_{7-\delta}$; (c) resistive transition for as synthesized (a) and annealed under oxygen samples(b) (1 and 2 refer to samples with different nominal oxygen compositions).

CONCLUSION

The oxygen non stoichiometry is an important factor for superconductivity in cuprates, but the nature of the rock salt layers appears also to be active. The different X ray absorption studies are in favour of conductive rock salt layers which could play the role of reservoir of holes and could improve the superconducting properties by proximity effects as recently developped by R. Tournier [15].

REFERENCES

1. Raveau B. Michel C. (1989) Crystal chemistry and properties of mixed valence copper oxides. Ann. Rev. Mat. Sci. 19 : 319-339.
2. Yvon K. François M. (1989). Crystal structures of high-T_c oxides. Z. Phys. B, 76 : 413-44.

3. Raveau B. Michel C. Hervieu M. Provost J. (1988). Crystal chemistry of perovskite superconductors. Physica C, 153-155 : 3-8.

4. Cava RJ. Batlogg B. Kajewski JJ. Rupp LW. Schneemeyer LF. Siegrist T. Van Dover RB. Marsh P. Peck WF. Gallagher PK. Glarum SH. Marshall JH. Farrow RC. Waszczak JV. Hull R. Trevor P(1988) Superconductivity near 70 K in a new family of layered copper oxides, Nature, 336 : 211-212.

5. Zandbergen HW. Fu WT. Van Ruitenbeek JM. de Jongh LJ. Van Tendeloo G. Amelinckx S (1989) Superconductivity in $(Pb,Bi)_2Sr_{2-x}La_xCu_2O_{6+\delta}$. Physica C, 159 : 81-86.

6. Hervieu M. Domengès B. Raveau B. Tarascon JM. Post M. Mc Kinnon WR (1989) Order-disorder phenomena in the "60 K" superconductor $YBa_2Cu_3O_{7-\delta}$ ($0.37 < \delta < 0.45$) by means of HREM study. Mat. Lett. 8 : 73-82.

7. Oyanagi H. Ihara H. Matsubara T. Tokumoto M. Matsushita T. Hirabayashi M. Murata K. Terada N. Yao T. Iwasaki H. Kimura Y.(1987) Valence study of orthorhombic and tetragonal $Ba_2YCu_3O_y$: the role of oxygen vacancies in high-T_c superconductors. Jpn. J. Appl. Phys. 26 : L1561-L1564.

8. Studer F. Bourgault D. Martin C. Retoux R. Michel C. Raveau B. Dartyge E. Fontaine A. (1989) Valence states of lead, thallium and copper in superconducting cuprates by XANES. Physica C, 159 : 609-615.

9. Tarascon JM. Miceli PF. Barboux P. Hwang DM. Hull GW. Giroud M. Greene LH. Lepage Y. Mc Kinnon WR. Tselepis E. Pleizier G. Eibschutz M. Neumann DA. Rhyne JJ. (1989) Structure and magnetic properties of non superconducting doped Co and Fe $Bi_2Sr_2Cu_{1-x}M_x O_y$ Phases. Phys. Rev. B; 39 : 11587-11598.

10. Hewat EA. Capponi JJ. Marezio M. (1989) A model for the superstructure of $Bi_2Sr_2CaCu_2O_{8.2}$. Physica C, 157 : 502-508.

11. Retoux R. Studer F. Michel C. Raveau B. Fontaine A. Dartyge E (1989) Valence state for bismuth in the superconducting bismuth cuprates. Phys. Rev. B, in press.

12. Raveau B. Martin C. Hervieu M. Bourgault D. Michel C. Provost J (1989) Layered thallium cuprates non stoichiometry and superconductivity. Proc. Intern. Symp. Solid State Chem., Pardubice.

13. Studer F. Retoux R. Martin C. Michel C. Raveau B. Dartyge E. Fontaine A (1989) Valence state and local environment of thallium and copper ions in some new high-T_c superconductors by XAS. Mod. Phys. Lett. B, 3 : 1085-1096

14. Rouillon T. Provost J. Hervieu M. Groult D. Michel C. Raveau B (1989) Superconductivity up to 100 K in lead cuprates : a new superconductor $Pb_{0.5}Sr_{2.5}Y_{0.5}Ca_{0.5}Cu_2O_{7-\delta}$. Physica C, 159 : 201-209.

15. de Rango P. Giordanengo P. Tournier R. Sulpice A. Chaussy J. Deutscher G. Genicon JL. Lejay P. Retoux R. Raveau B. (1989) The irreversibility line of $Bi_{2-x}Pb_xSr_2Ca_2Cu_3O_{10}$: a possible breakdown of an intrinsic proximity effect, J. Phys., 50 : 2857-2868.

X-ray Absorption Near Edge Studies fo $Nd_{2-x}Ce_xCuO_{4-y}$

H. OYANAGI, H. YAMAGUCHI, Y. YOKOYAMA, T. KATAYAMA, and Y. NISHIHARA
Electrotechnical Laboratory, Umezono, Tsukuba, Ibaraki, 305 Japan

ABSTRACT

The effect of Ce doping and oxygen deficiency on the electron states at the copper sites in $Nd_{2-x}Ce_xCuO_{4-y}$ has been stduied by X-ray absorption near-edge structure on the Cu K- and Ce L_3 edges. The results indicate that (1) the doped carriers are electrons which may occupy the Cu d-holes and (2) the number of carriers initially increases with the increase of Ce concentration for x<0.15 and saturates at x=0.15-0.16 coinciding with x having the maximum T_c value. We find that (3) further doping decreases the electron concentration suggesting the coexistence of holes for x>0.16, while (4) the oxygen vacancies introduced by annealing under reducing conditions provide the extra electrons to the CuO_2 plane.

KEY WORDS: X-ray absorption near-edge structure, electron states, $Nd_{2-x}Ce_xCuO_{4-y}$, synchrotron radiation, valence

INTRODUCTION

The nature of charge carriers in $Nd_{2-x}Ce_xCuO_{4-y}$ [1] is an interesting yet still a controversial problem which remains to be solved. For instance, the recent X-ray absorption near edge study [2] reported the formation of Cu^+ (d^{10} configuration) upon Ce doping, indicating that doped electrons occupy Cu d_{x2-y2} states. On the other hand, the phtoemission study [3] and more recent X-ray absorption study [4] claim that doped electrons take other states near the Fermi level, such as the Cu 4s states. Moreover, the effect of oxygen vacancies is not well understood although there exists the critical electron concentration above which superconductivity is initiated. Second, the negative Hall coefficient decreases as the Ce concentration x increases which even reverses its sign in the higher Ce concentration range where superconductivity dissapears. To understand the phase diagram of electron states, we have studied the effect of Ce doping and oxygen vacancies on electron states in relation with the transport properties.

EXPERIMENTAL

Samples were synthesized from a mixture of CeO_2, Nd_2O_3 and CuO as reported in ref. 1. The mixed powder was sintered at 1100 °C and quenched in air to room temperature. The Ce-doped samples were annealed at 900 °C in a stream of Ar atmosphere for 24 hours while a series of samples to study the effect of oxygen vacancies were prepared by annealing at 1050 °C for 20 hours with various pO_2 (1-10^{-4} atm). X-ray absorption experiments were performed using synchrotron radiation from the 2.5 GeV storage ring at the Photon Factory. The Cu K-edge spectra were obtained in a transmission mode while the Ce L_3-edge data were collected in a fluorescence mode using NaI scintillation detectors. An energy resolution is ca. 2 eV at 9 keV using an encoded Si(111) double crystal monochromator. The energy scale was calibrated by recording Cu metal data with the characteristic feature indicating the Fermi level (8980.3 eV). All spectra were sequentially measured at room temperature to minimize the experimental error in an energy scale.

RESULTS AND DISCUSSION

Figure 1 shows the Ce L3-edge spectra of $Nd_{1.85}Ce_{0.15}CuO_{4-y}$ and two reference compounds, CeO_2 and $CeTiO_3$. The near-edge features of Ce^{4+} compounds can be easily distinguished from Ce^{3+} compounds as they show a characteristic doublet structure due to the final state effects. Thus the valence of doped Ce ions in $Nd_{1.85}Ce_{0.15}CuO_{4-y}$ is 4+.

In Fig. 2, the Ce concentration dependence of the Cu K near-edge structure is shown for $Nd_{2-x}Ce_xCuO_{4-y}$. It is obvious that Ce doping induces the absorption edge shift to lower energy indicating that electron doping directly affects the Cu valence. A systematic change of the near-edge structure can be observed with the increase of the Ce concentration. A bump structure observed at around 8982-8988 eV is due to the 1s-4p*(p) transition.

In order to analyze the effect of Ce doping in more detail, the difference spectra are shown in Fig. 3. Each spectrum is obtained by the normalization and subtraction of the data for Nd_2CuO_4 as a standard, which can thus single out the effect of Ce doping on the Cu valence. Doping induces new features at 8982 eV and 8994 eV which grow with the increase of x. The results have been interpreted as

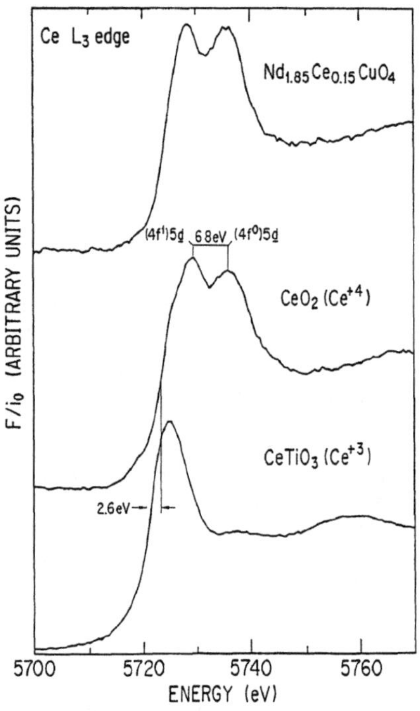

Fig. 1 The Ce L3-edge spectra of $Nd_{1.85}Ce_{0.15}CuO_{4-y}$, CeO_2 and $CeTiO_3$.

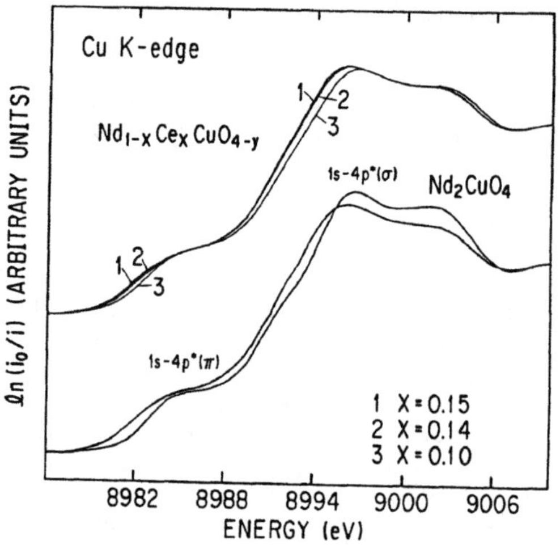

Fig. 2 The Cu K near-edge spectra for $Nd_{2-x}Ce_xCuO_{4-y}$ (x=0.10-0.15).

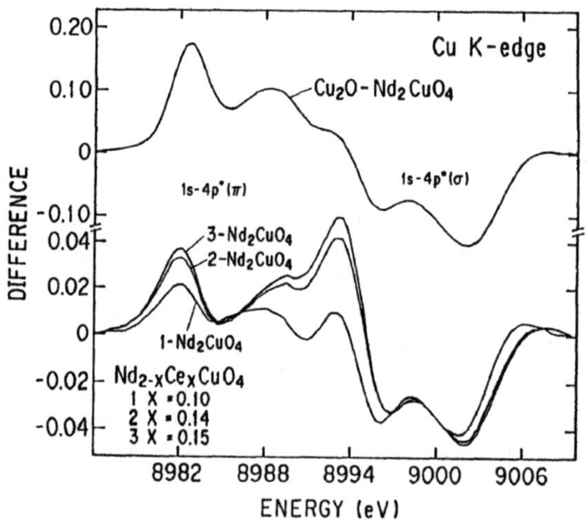

Fig. 3 The difference spectra between the doped and undoped Nd_2CuO_4.

indicating that doped electrons occupy Cu d-holes [2] while the absence of the higher energy peak in the difference spectrum of Cu_2O led to the different interpretation over the location of charge carriers [4].

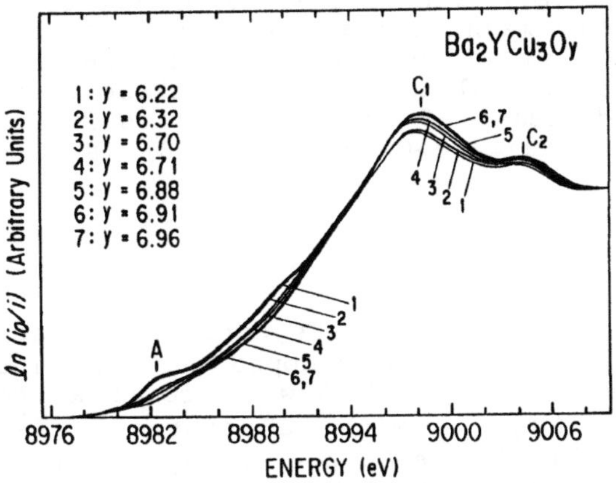

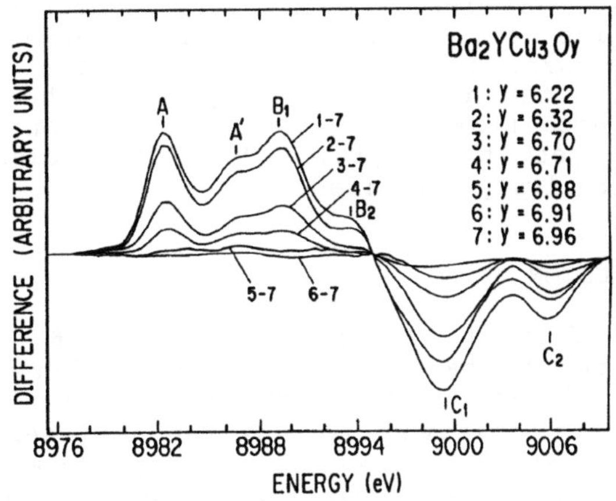

Fig. 4 The Cu K-edge spectra (left) and the difference spectra (right) for $Ba_2YCu_3O_{7-y}$.

In Fig. 4, the raw and difference spectra are shown for $Ba_2YCu_3O_{7-y}$ with various y values where $Ba_2YCu_3O_{6.96}$ is used as a standard to analyze the effect of oxygen deficiency [5]. The new doping-induced states at 8982 eV and 8994 eV in Fig. 3 coincide with the oxygen deficiency-induced features in Fig. 4. Since oxygen vacancies in $Ba_2YCu_3O_{7-y}$ system cause the Cu^+ (d^{10} configuration), the results for Ce doped Nd_2CuO_4 suggest that the doped electrons occupy the Cu d-holes.

Now we analyze the effect of oxygen deficiency in a similar manner. Figure 5 shows the Cu K difference spectra for $Nd_{1.85}Ce_{0.15}CuO_{4-y}$ with various oxygen pressure ($pO_2=1-10^{-3}$ atm). The sample with $pO2=1$ atm shows the semiconductive temperature dependence of resistivity while the samples with $pO_2<10^{-1}$ atm are metallic and show superconductivity. The doping-induced features at 8982 eV and 8994 eV increase in intensity as pO_2 decreases or as oxygen vacancies are introduced. Clearly, the transport properties, in particular, the T_c value strongly correlates with the doping-induced states which are enhanced by oxygen deficiency. These results suggest that there exists the critical electron concentration to initiate superconductivity. The electron concentration provided by Ce doping is insufficient to reach the critical value and the oxygen vacancies provide extra electrons to the CuO_2 plane.

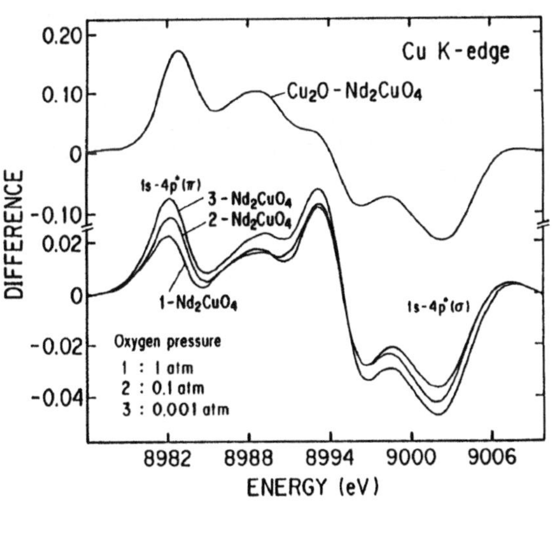

Fig. 5 The difference spectra between the data for $Nd_{2-x}Ce_xCuO_{4-y}$ annealed with $pO_2=1-10^{-3}$ atm and Nd_2CuO_4

Let us assume that the intensity of doping- or oxygen vacancy-induced states measures the number of doped electrons. The electron concentration estimated from the integrated intensity of the two characteristic features in the difference spectra increases with x and saturates at x=0.15-0.16 around which the highest T_c value is obtained [1]. However, it decreases sharply with further doping where the negative Hall coefficient sharply decreases its magnitude and superconductivity disappears. These results suggest that the heavily doped Ce ions (x>0.16) now provide the holes which compensate the electrons. Although whether these holes originate from the oxidation of Ce ions or decrease of oxygen vacancies is not clear yet, the positive Hall coefficients in $Nd_{2-x}Ce_xCuO_{4-y}$ (x>0.18) is thus qualitatively explained.

CONCLUSION

The effect of Ce doping on the Cu K-edge spectra of $Nd_{2-x}Ce_xCuO_{4-y}$ has been analyzed by taking the difference between the doped and undoped samples. The doping-induced states were found to grow at 8982 eV and 8994 eV which coincide the features observed for oxygen-deficient $Ba_2YCu_3O_{7-y}$ with Cu^+ states or d^{10} configuration, suggesting the doped electrons in $Nd_{2-x}Ce_xCuO_{4-y}$ occupy the Cu d-holes. The charge carrier concentration estimated from the intensity of these features increases with the Ce concentration for x<0.15 and saturates at x=0.15-16 coinciding with x having the maximum T_c value. We find that further doping decreases the electron concentration for x>0.16 indicating the presence of holes which compensate the doped electrons. The oxygen vacancies introduced by annealing under reducing conditions provide extra electrons to the CuO_2 plane.

ACKOWLEDGEMENT

The authors appreciate invaluable discussions with J. Kondo and K. Kajimura. This work has been performed as a part of a project (Proposal No. 87-065) approved by the Photon Factory Program Advisory Committee.

REFERENCES

1. Tokura Y. Takagi H. Uchida S. Uchida (1989) A Superconducting copper oxide compound with electrons as the charge carriers. Nature 337: 345-347

2. Tranquada JM. (1989) Nature of the charge carriers in electron-doped copper oxide superconductors. Nature 337: 720-721

3. Fujimori A. Tokura A. Eisaki H. Takagi H. Uchida S. Takayama-Muromachi E. (1989) Electronic structure of the electron-doped superconductor $Nd_{2-x}Ce_xCuO_{4-y}$ studied by phtoemission spectroscopy preprint.

4. Alp EE. Mini SM. Ramanathan M. Dabrowski B. Richards DR. Hinks DG. (1989) Effect of Ce doping on the Cu charge in the electron superconductor $Nd_{2-x}Ce_xCuO_4$. Phys. Rev. B40: 2617-2619

5. Oyanagi H. Ihara H. Matsubara T. Tokumoto M. Matsushita T. Hirabayashi M. Murata K. Terada N. Yao T. Iwasaki H. Kimura Y. (1987) Valence study of orthorhombic and tetragonal $Ba_2YCu_3O_y$: The role of oxygen vacancies in high-Tc superconductivity. Jpn. J. Appl. Phys. 26: L1561-L1564

Photoelectron Spectroscopic Study of $(Y_{1-x}Ca_x)Ba_2Cu_4O_8$

R. ITTI, T. MIYATAKE, K. IKEDA, S. TAJIMA, and N. KOSHIZUKA

Superconductivity Research Laboratory, International Superconductivity Technology Center, 10-13, Shinonome 1-chome, Koto-ku, Tokyo, 135 Japan

ABSTRACT

Photoelectron spectroscopic studies of superconducting $(Y_{1-x}Ca_x)Ba_2Cu_4O_8$ (x=0.0-0.1) were performed. The surface of this material was stable and oxygen degradation from surface seemed to be negligible in vacuum. This is consistent with the fact that, $(Y_{1-x}Ca_x)Ba_2Cu_4O_8$ has an excellent thermal stability in oxygen content up to high temperature in air. In photoelectron spectra , edges at Fermi level were observed even at room temperature (normal state). The result provides another evidence that the existence of Fermi-like edge in normal state seems to be an intrinsic feature and essential for high Tc superconductors.

KEYWORDS: $(Y_{1-x}Ca_x)Ba_2Cu_4O_8$, XPS, UPS, Fermi energy, Y-Ba-Cu-O system

INTRODUCTION

$YBa_2Cu_4O_8$ was first reported as an ordered-defect structure in epitaxial $YBa_2Cu_3O_7$ thin films [1]. The existence of this 80K superconducting $YBa_2Cu_4O_8$ was also discovered as lattice defects in the $YBa_2Cu_3O_7$ by Zandbergen et al.[2]. Later, syntheses of bulk $YBa_2Cu_4O_8$ have been reported by several groups[3-5]. Crystal structure of this compound was studied by neutron [6] and X-ray diffraction [7]. The significant difference between the crystal structure of this compound and the 123 compound is that it contains double Cu-O chain along the b-axis. The oxygen deterioration of $YBa_2Cu_3O_7$ upon heating over room temperature is widely known. However, $YBa_2Cu_4O_8$ has an excellent thermal stability in oxygen content up to high temperature in air. Although the reason for this is not yet clear at the moment, the relation between this property and the existence of the double Cu-O chain in this compound is noteworthy. Recently, preparation of $(Y_{1-x}Ca_x)Ba_2Cu_4O_8$ has been reported by Miyatake et al. [8]. Substitution of Ca for Y up to 10% raises the critical temperature, Tc, of this system towards 90K, which is comparable to that of $YBa_2Cu_3O_7$. This feature and the fact that Ca substituted $(Y_{1-x}Ca_x)Ba_2Cu_4O_8$ also has excellent thermal stability in oxygen content give advantage to $(Y_{1-x}Ca_x)Ba_2Cu_4O_8$ over $YBa_2Cu_3O_7$ and should be considerably important in practical applications.

Regarding the photoelectron spectroscopic study of high Tc superconductor, whether edge structure at the Fermi energy exists or not has been controversial. Most of the earlier works showed the lack of edge structure at the Fermi energy. However, recent experimental results showed that the edge structure can be observed in general, provided that the measurement was performed on well defined, i.e. superconducting phase, sample surfaces.

In this paper, we report on the photoelectron spectroscopic study of $(Y_{1-x}Ca_x)Ba_2Cu_4O_8$, especially on the UPS results. The main purposes are to investigate the surface stability in vacuum and the existence of edge structure at the Fermi energy of $(Y_{1-x}Ca_x)Ba_2Cu_4O_8$.

EXPERIMENTAL

$(Y_{1-x}Ca_x)Ba_2Cu_4O_8$ samples were prepared by solid state reaction method with the oxygen-HIP (Hot Isostatic Pressing) treatment. Sample preparation was described in details elsewhere[8]. High quality polycrystalline $(Y_{1-x}Ca_x)Ba_2Cu_4O_8$ (x=0.0,0.05 and 0.1) samples with no secondary phases obtained by this method were used in this study. Measurements were made on the VG Scientific ESCALAB MkII spectrometer equipped with an Al X-ray source (1486.6 eV) for XPS and a noble gas discharge lamp for UPS.
Pressure in the analysis chamber during the measurements was typically $2x10^{-10}$ Torr.

RESULTS AND DISCUSSION

Figure 1 shows the XPS result for Ca 2p inner core level. Two points should be mentioned. First, the intensity of Ca 2p peaks increased as the content of Ca in the samples increased. Second, the result showed that the binding energy of Ca $2p_{3/2}$ in $(Y_{1-x}Ca_x)Ba_2Cu_4O_8$ was about 345 eV. Considering that the binding energy of Ca 2p core level is about 347 eV for insulating $CaCO_3$ and about 344 eV for Ca metal, these results indicate that no $CaCO_3$ existed in our samples and Ca substituted for Y in $(Y_{1-x}Ca_x)Ba_2Cu_4O_8$ .This is consistent with the X-ray diffraction results.

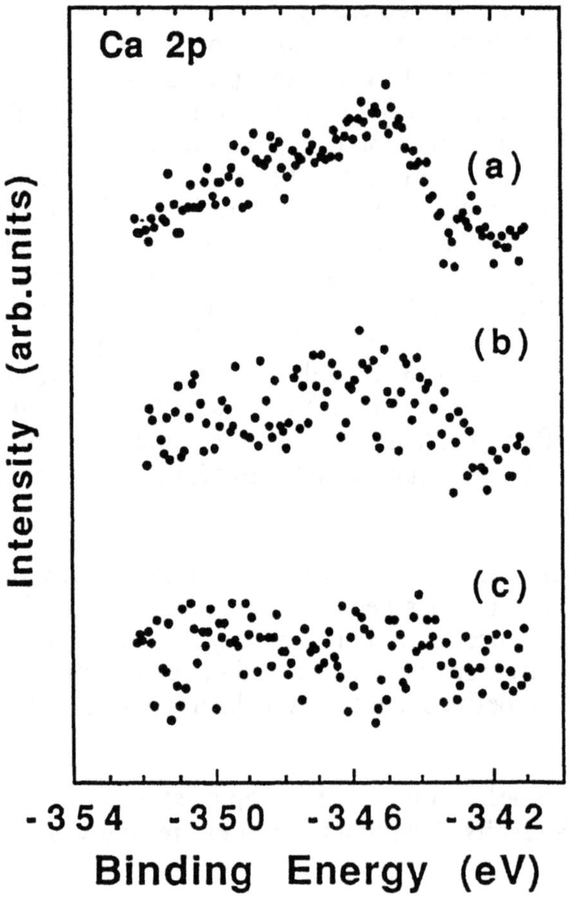

Fig.1. XPS result for Ca 2p inner core level of $(Y_{1-x}Ca_x)Ba_2Cu_4O_8$
(a) x=0.10
(b) x=0.05
(c) x=0 (background)

Figure 2 shows the UPS (He I, 21.2 eV) results in the range of the Fermi energy for $(Y_{1-x}Ca_x)Ba_2Cu_4O_8$ and $YBa_2Cu_3O_7$ measured at room temperature. Each of the sample was scraped by diamond file before measurement. For $(Y_{1-x}Ca_x)Ba_2Cu_4O_8$, edges at Fermi energy can be observed quite clearly for all of the three distinct samples. In the case of $YBa_2Cu_3O_7$ no clear edge could be observed.

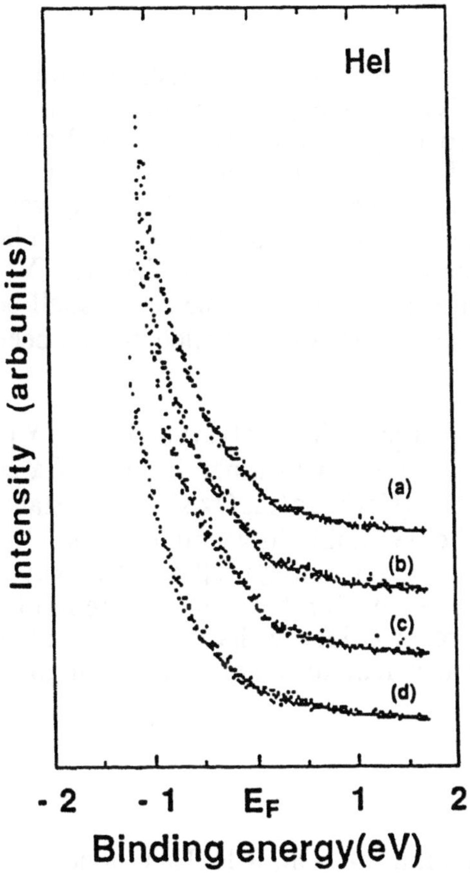

Fig.2. UPS results at room temperature for $(Y_{1-x}Ca_x)Ba_2Cu_4O_8$ (a) x=0.10, (b) x=0.05, (c) x=0.0 and (d) $YBa_2Cu_3O_7$ (for comparison)

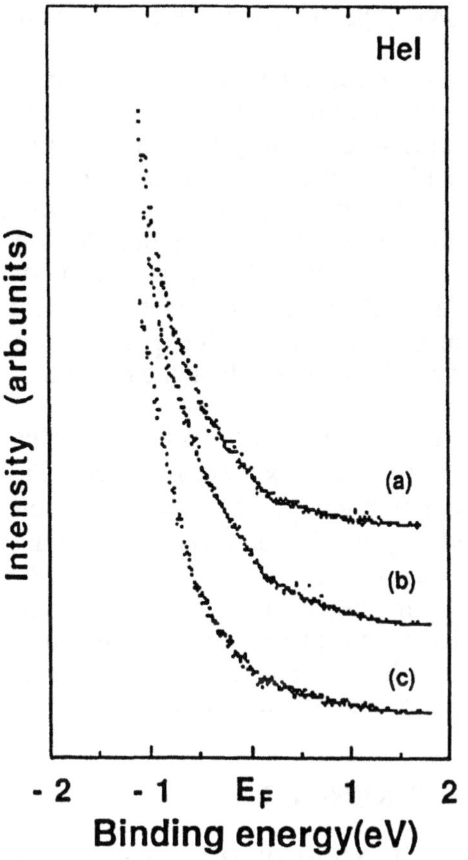

Fig.3. UPS results at room temperature for $(Y_{0.9}Ca_{0.1})Ba_2Cu_4O_8$ (a) after first measurement (b) 2 weeks after (a) in vacuum (c) $YBa_2Cu_3O_7$(for comparison)

For $YBa_2Cu_3O_7$, because of its oxygen content instability in vacuum, oxygen degradation from the surface of $YBa_2Cu_3O_7$ occurs. This causes the surface composition to change from superconducting to insulating composition and thus no edge structure reflecting the existence of the density of states at Fermi energy should be observed in the photoelectron spectrum. Arko et al. argued that clear Fermi edge for $YBa_2Cu_3O_7$ could be observed if the surface was cleaved and measured at temperature as low as 20K[9]. The fact that clear edges were observed for $(Y_{1-x}Ca_x)Ba_2Cu_4O_8$ even at room temperature shows that $(Y_{1-x}Ca_x)Ba_2Cu_4O_8$ was stable in oxygen content even under ultra high vacuum and this is consistent with the fact that, $(Y_{1-x}Ca_x)Ba_2Cu_4O_8$ has an excellent thermal stability in oxygen content up to high temperature in air.

To test the stability of $(Y_{1-x}Ca_x)Ba_2Cu_4O_8$ in vacuum, measurements on the same sample were performed twice. The results are shown in Fig. 3. Spectrum (a) is the result of the first measurement on a well characterized sample, while spectrum (b) was the result remeasured after the sample was left in UHV for 2 weeks at room temperature. Edge structures can be observed in both spectra. This indicates that $(Y_{1-x}Ca_x)Ba_2Cu_4O_8$ surface was stable in vacuum. The result for $YBa_2Cu_3O_7$ at room temperature is also shown (spectrum (c)) for comparison.

Since UPS is a very surface sensitive technique and the fact that polycrystalline samples were used in this study, one has to be careful in analyzing the experimental results. If the sample was not single phase, there might exist secondary phases between crystal grain boundaries. If these secondary phases are metallic, the observed edge at the Fermi energy may be caused by these metallic secondary phases. However, since in this study very high quality single phase samples were used, and no metallic secondary phases can be detected by either X-ray diffraction or XPS, we believe that the observed edges for $(Y_{1-x}Ca_x)Ba_2Cu_4O_8$ are intrinsic for this system and essentially the same result shall be reproduced by the UPS experiment on a single crystal of this system, which has not been yet available generally at the moment.

In conclusion, the first photoelectron spectroscopies for polycrystalline $(Y_{1-x}Ca_x)Ba_2Cu_4O_8$ were reported. Ca substitution for Y was confirmed from the Ca 2p inner core level XPS results. It is concluded that $(Y_{1-x}Ca_x)Ba_2Cu_4O_8$ surface is stable and oxygen degradation from surface seemed to be negligible in vacuum. In the UPS spectra, edges at the Fermi energy were observed for $(Y_{1-x}Ca_x)Ba_2Cu_4O_8$ samples even at room temperature. The results were reproducible, even after the samples were left for a long time in vacuum. This study provides another evidence that the existence of Fermi-like edge in normal state seems to be an intrinsic feature and essential for high Tc superconducting materials.

REFERENCES

1. A.F. Marshall, R.W. Barton, K. Char, A. Kapitulnik, B. Oh, R.H. Hammond, S.S. Laderman, Phys. Rev.B **37**(1988)9353.
2. H.W. Zandbergen, R. Gronski, K. Wang, G. Thomas, Nature **331**(1989)596.
3. J. Karpinski, E. Kaldis, E. Jilek, S. Rusiecki, B.Bucher, Nature **336**(1988)660.
4. D.E. Morris, J.H. Nickel, J.Y.T. Wei, N.G. Asmer, J.S. Scott, U.M. Scheven, C.T. Hultgren, A.G. Markelz, J.E. Post, P.J. Heaney, D.R. Veblen, R.M. Hazen, Phys.Rev.B **39**(1989)7347.
5. R.J. Cava, J.J. Karajewski, W.F. Peck Jr., B. Batlogg, L.W. Rupp Jr., R.M. Fleming, A.C.W.P. James, P. Marsh, Nature **338**(1989)328.
6. P. Fisher, J. Karpinski, E. Kaldis, E. Jilek, S. Rusiecki, Solid State Commun. **69** (1989)531.
7. R.M. Hazen, L.W. Finger, D.E. Morris, Appl.Phys.Lett. **54**(1989)1057.
8. T. Miyatake, S. Gotoh, N. Koshizuka, S. Tanaka, Nature **341**(1989)41.
9. A.J. Arko, R.S. List, R.J. Bartlett, S.-W. Cheong, Z. Fisk, J.D. Thompson, C.G. Olson, A.-B. Yang, R. Liu, C. Gu, B.W. Veal, J.Z. Liu, A.P. Paulikas, K. Vandervoort, H. Claus, J.C. Campuzano, J.E. Schirber, N.D. Shinn, Phys.Rev.B **40**(1989)2268.

X-ray Absorption Studies of Tl-Ba-Ca-Cu-O Superconductors

HIROTAKA YAMAGUCHI[1], HIROYUKI OYANAGI[1], HIDEO IHARA[1], RYOJI SUGISE[2], and TAKEHIKO SHIMOMURA[3]

[1] Electrotechnical Laboratory, Umezono, Tsukuba, Ibaraki, 305 Japan
[2] Ube Industries Ltd., Ube, Yamaguchi, 755 Japan
[3] Unitika R & D Center, Uji, Kyoto, 611 Japan

ABSTRACTS

The local structures and valence states of $TlBa_2Ca_{n-1}Cu_nO_{2n+3}$ (n=2,3,4) and $Tl_2Ba_2Ca_{n-1}Cu_nO_{2n+4}$ (n=2,3) have been investigated by X-ray absorption spectroscopy on the Cu K- and the Tl L_{III}-edges. XANES (X-ray absorption near-edge structure) results on the Cu K-edge have shown that the Cu valence in these systems is nearly 2.2, which hardly depends on the structures. On the other hand, those on the Tl L_{III}-edge suggest the occurrence of mixed valence states of Tl^{+1} and Tl^{+3}. These suggest that Tl ions supply the hole to the Cu-O conductive planes.The Cu-O vibrational properties have been studied by the Cu K-edge EXAFS (extended X-ray absorption fine structure), and no anomalies are found in the mean square relative displacement, $\sigma^2(k)$ in the temperature range from 30 K to room temperature.

KEY WORD: extended X-ray absorption fine structure(EXAFS), X-ray absorption near-edge structure(XANES), local structure, valence states, Tl-Ba-Ca-Cu-O, oxide superconductor

INTRODUCTION

Tl-Ba-Ca-Cu-O superconductors [1,2] consist of stacking Cu-O layers between which Ca layers are sandwiched, and they form perovskite-like layered units. The units are separated by Tl-O monolayers and bilayers in $TlBa_2Ca_{n-1}Cu_nO_{2n+3}$ (Tl_1-system) and $Tl_2Ba_2Ca_{n-1}Cu_nO_{2n+4}$ (Tl_2-system), respectively. The superconducting transition temperature T_c depends on the number of Cu-O layers in the perovskite-like unit and the highest T_c occurs at 122 K in $TlBa_2Ca_3Cu_4O_{11}$ (1234). However, the origin of such a structural dependence of T_c is still unclear. Suzuki *et al.* [3] reported the occurrence of mixed valences of Tl^{+1} and Tl^{+3} and discussed the stability of the layered structure and origin of superconducting carriers.

We have measured the Cu K-edge absorption fine structure to study the valence state of Cu ions and local structure around them in the Tl-Ba-Ca-Cu-O system. Tl L_{III}-near-edge structures have also been presented.

EXPERIMENTAL

The samples were prepared by mixing powders of Tl_2O_3, BaO_2, CaO_2 and CuO. Well mixed powders were cold pressed into pellets and sintered at 890 °C in an oxygen atmosphere for 30-50 minutes [4].

X-ray absorption measurements were performed at the BL-4C beam line at the Photon Factory using synchrotron radiation from the 2.5 GeV storage ring. White X-ray was monochromatized by a sagittaly focusing Si(111) double crystal monochromator. Both Cu K- and Tl L_{III}-edge spectra were measured with powdered samples by transmission mode at 300 K and 30 K. The measurements of the temperature variance of Cu K-EXAFS of 1234 were also carried out between these temperatures.

RESULTS AND DISCUSSION

Figure 1 shows the Cu-K edge absorption spectrum for $Tl_2Ba_2Ca_2Cu_3O_{10}$ (2223). The near-edge spectrum of the 2223 is shown with those of La_2CuO_4, $Ba_2YCu_3O_{6.96}$ and Nd_2CuO_4 in Fig. 2. It is found that the near-edge spectrum includes twin peaks due to the 1s-4p*(π) transition in the pre-edge region and the 1s-4p*(σ) transition around the dominant peak. Pure La_2CuO_4 and Nd_2CuO_4 are typical compounds composed of Cu-O octahedra and planes, respectively while $Ba_2YCu_3O_{6.96}$ is of both of Cu-O pyramids and planes. The near-edge structure of the 2223 is found to be very similar to that of $Ba_2YCuO_{6.96}$, which reflects the similar average coordination environment of the Cu ions comprising the layers and pyramids. The absorption edge of the 2223 is 0.47 eV higher in energy than that of La_2CuO_4. The Cu valence is estimated at 2.2 by using the value of average edge shift of 2.47 eV/valence [5]. Figure 3 shows the near-edge structures of the Tl_1- and the Tl_2-systems of n=2 (the 1212 and the 2212), which shows the edge position of the 1212 is slightly higher in energy than that of the 2212. The tendency for the edge of the Tl_1-system to lie slightly higher in energy than that of the Tl_2-system was also found in the case of n=3. On the other hand, the systematic variations of the near-edge structure with n were not found and no meaningful edge-shifts were found. The nominal valence based on the stoichiometric composition is given by $2+\frac{1}{n}$ and 2

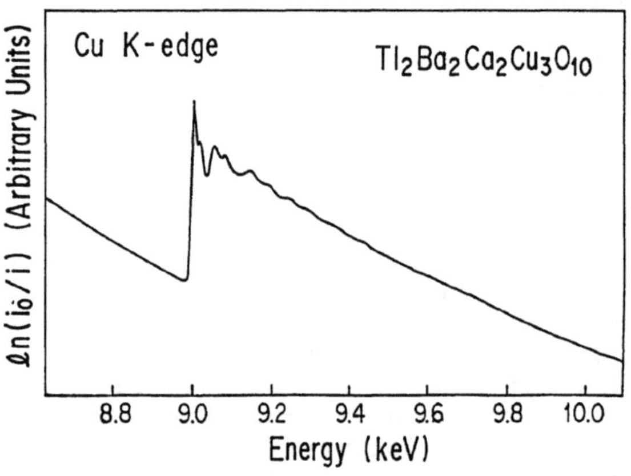

Fig. 1 The Cu K-edge absorption spectrum of $Tl_2Ba_2Ca_2Cu_3O_{10}$.

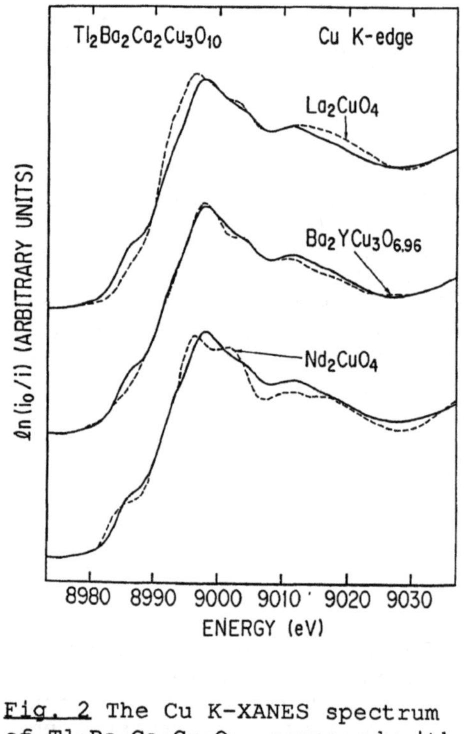

Fig. 2 The Cu K-XANES spectrum of $Tl_2Ba_2Ca_2Cu_3O_{10}$ compared with those of La_2CuO_4, Ba_2YCuO and Nd_2CuO_4(dashed lines).

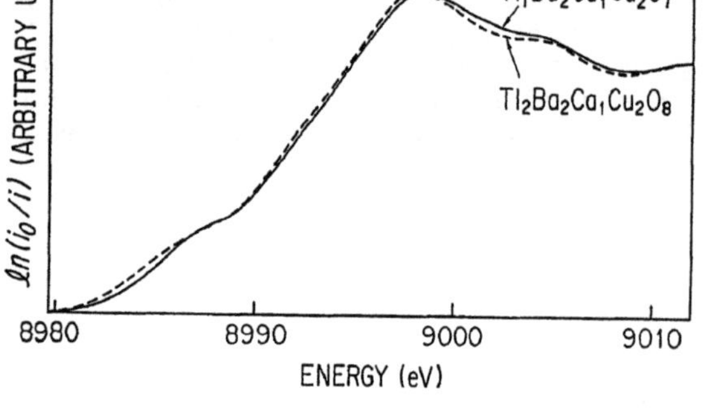

Fig. 3 The Cu K-XANES spectra of $Tl_1Ba_2Ca_1Cu_2O_7$ and $Tl_2Ba_2Ca_1Cu_2O_8$.

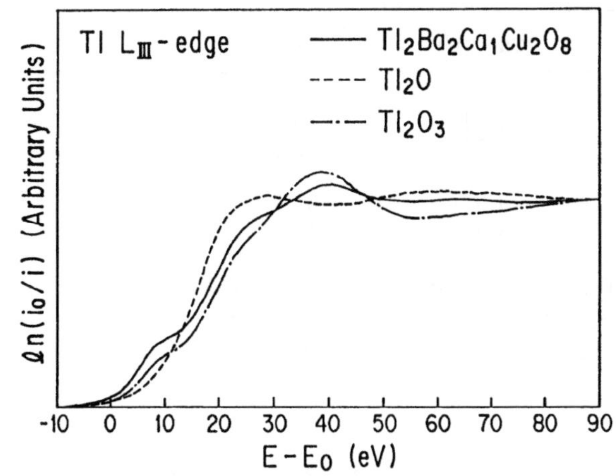

Fig. 4 The Tl L_{III}-XANES spectrum of $Tl_2Ba_2Ca_1Cu_2O_{10}$.

for the Tl_1- and Tl_2-systems, respectively. However, for Tl_1-system, n dependence of the Cu valence was not observed. Higher values of the Cu valence in the Tl_2-system may be ascribed to the occurrence of Tl^{1+}. Suzuki *et al.* [3] reported that the Tl valence in this system was between +3 and +1 by X-ray photoelectron spectroscopy. We have examined it by the Tl L_{III}-edge absorption and a typical result is shown in Fig. 4. The absorption spectra of Tl_2O and Tl_2O_3 have characteristic features due to the difference of the Tl valence (+1 and +3, respectively); the spectrum of Tl_2O (Tl^{+1}) has no fine structure in the pre-edge region and the edge energy is lower than that of Tl_2O_3 (Tl^{+3}) while two peaks are found in the pre-edge region of the spectra of Tl_2O_3. Roughly speaking, one finds that the spectrum of 2212 has both features of Tl^{+1} and Tl^{+3} and the edge position, which is defined by a midpoint of absorption, for 2212 lies between those for Tl^{+1} and Tl^{+3} in energy. This suggests the occurrence of Tl^{+1} and agrees with the results of the photoelectron spectroscopy [3]. However, more detail investigations are required to clarify the valence state of the Tl ions. In addition, for Tl_1-system, the occurrence of Tl^{+1} was observed.

Figure 5 shows the Fourier transform of the Cu-K EXAFS at temperatures of 30 and 300 K. Most dominant peak indicates the contribution of the nearest planer-coordinated oxygen ions while the peak corresponding to the apical

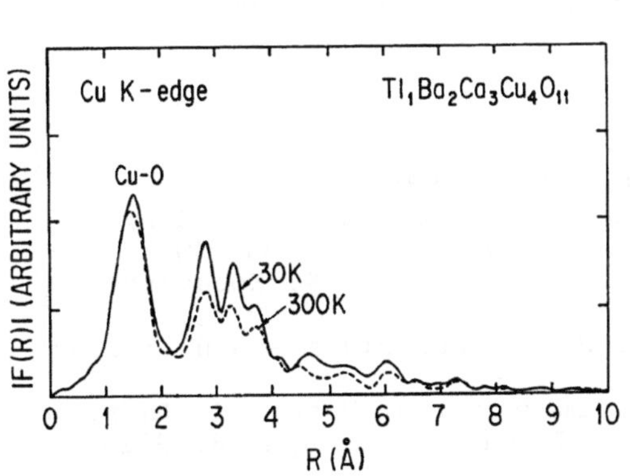

Fig. 5 The Fourier transform of the Cu K-EXAFS.

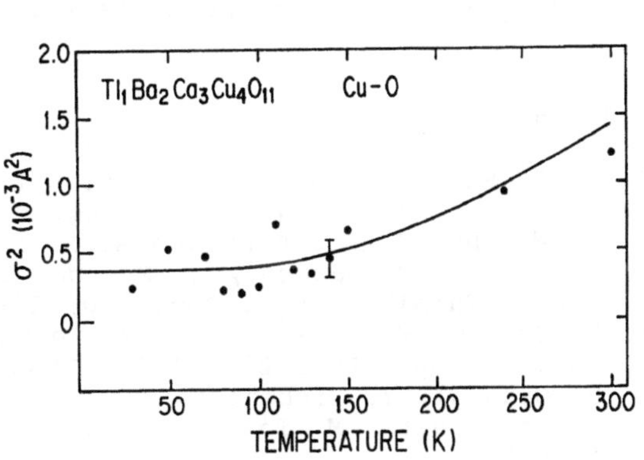

Fig. 6 The mean square relative displacements of $Tl_1Ba_2Ca_3Cu_4O_{11}$. The solid line is a fit of an Einstein model to the data.

Compound	a/2 (Å)	l_{Cu-O} (Å)	σ^2 (10^{-3}Å^2)
1212	1.92_2	1.92_2	0.69
1223	1.92_2	1.91_7	1.49
1234	1.92_1	1.91_6	0.73
2212	1.92_8	1.92_1	0.19
2223	1.91_9	1.91_0	1.89

Table 1 Cu-O bond length (l_{Cu-O}) and mean square relative displacement (σ^2) of Tl-Ba-Ca-Cu-O system. A half of lattice parameter (a/2) is also shown for a reference.

oxygens of the CuO_5 pyramid is hardly found. This is because the apical oxygens are fewer in number and located further than those in plane, or it is likely to be the occurrence of structural or static disorder of the apical oxygens. We performed the Fourier filtering of the contribution of the first-nearest oxygen in $k^3\chi(k)$ between 1.0 and 2.3 Å, and fitted it using single-shell model. The evaluated bond lengths and the mean square relative displacement, $\sigma^2(k)$ are summarized in Table 1. The temperature dependence of the mean square relative displacements of the 1234 is shown in Fig. 6. Anomaly is not found and the temperature dependence is fitted to the Einstein model [6]:

$$\sigma^2(T) \approx (\hbar^2/2\mu k_B\Theta_E)\coth(\Theta_E/2T),$$

where $\hbar$ is Planck's constant, k_B is Boltzmann's constant, μ is the reduced mass and Θ_E is the Einstein temperature. The Einstein temperature obtained from the fitting to the experimental values is 577 K, which is almost the same value as of $La_{2-x}Sr_xCuO_4$, x=0.15 (603 K) [7] and of $Ba_2YCu_3O_{6.98}$ (596 K) [8].

In this paper, we have presented the valence states of Cu and Tl ions and Cu-O vibrational properties by X-ray absorption spectroscopy on the Cu K- and the Tl L_{III}-edges. Recently, Shimakawa *et al.* [9] showed the apparent oxygen-content dependence of T_c in Tl_2-system. Hereafter, to clarify the superconductive properties in this system, the role of oxygen ions should be taken into account.

We would like to thank Prof. Y. Shono of Tohoku University for variable discussions. We also acknowledge his helpful assistance with Dr. Y. Kuwahara of RIKEN. This work has been performed as a part of a project (Proposal No. 87-065) approved by the Photon Factory Program Advisory Committee.

References

1. Z. Z. Sheng and A. M. Hermann: Nature **332**, 55 (1988).
2. L. Gao, Z. J. Huang, R. L. Meng, P. H. Hor, J. Bechtold, Y. Y. Sun, C. W. Chu, Z. Z. Sheng and A. M. Hermann: Nature **332**, 623 (1988).
3. T. Suzuki, M. Nagoshi, Y. Fukuda, Y. Syono, M. Kikuchi, N. Kobayashi and M. Tachiki: Phys. Rev. B**40**, 5184 (1989).
4. R. Sugise, M. Hirabayashi, N. Terada, M. Jo, T. Shimomura and H. Ihara: Jap. J. Appl. Phys **27**, L1709 (1988).
5. H. Oyanagi, Y. Nishihara, K. Murata, H. Yamaguchi, H. Unoki, H. Ihara, T. Matsushita, M. Tokumoto and Y. Kimura: J. Phys. Soc. Jpn **58**, 3324 (1989).
6. E. Sevillano, H. Meuth and J. J. Rehr: Phys. Rev. B**20**, 4908 (1979).
7. J. M. Tranquada, S. M. Heald and A. R. Moodenbaugh: Phys. Rev. B**36**, 8401 (1987).
8. J. B. Boyce, F. Bridges, T. Claeson and M. Nygren: Phys. Rev. B**39**, 6555 (1989).
9. Y. Shimakawa, Y. Kubo, T. Manako and H. Igarashi: To be published in Phys. Rev. B.

Valence State of $Ba_{1-x}K_xBiO_{3-\delta}$ Superconductor Controlled by the Oxygen Content

Kazuhiro Ueki[1], Ayako Tokiwa[1], Masae Kikuchi[1], Teruo Suzuki[2], Masayasu Nagoshi[2], Reiko Suzuki[3], Norio Kobayashi[1], and Yasuhiko Syono[1]

[1] Institute for Materials Research, Tohoku University, Katahira, Sendai, 980 Japan
[2] Steel Research Center, NKK Corporation, Kawasaki, 210 Japan
[3] College of Education, Tohoku University, Kawauchi, Sendai, 980 Japan

ABSTRACT

The oxygen content of the high T_c superconductor $Ba_{1-x}K_xBiO_{3-\delta}$ (x=0.35, 0.4 and 0.5) was systematically varied by annealing at different temperatures in N_2 atmosphere. The lattice parameter, a, decreased with increasing δ. T_c decreased with increasing Bi^{5+} concentration and increasing a. Disappearance of superconductivity seems to be accompanied by semiconductor-metal transition around 50% Bi^{5+} concentration and a= 4.30 A.

KEYWORDS: Ba-K-Bi-O system, high T_c oxides, oxygen content,

INTRODUCTION

Discovery of a new superconductor $Ba_{1-x}K_xBiO_{3-\delta}$ (BKBO) with T_c 30 K, considerably higher than 13 K of $BaPb_xBi_{1-x}O_3$ (BPBO), gave a strong impact on the research of high temperature superconducting oxides[1-4], because it is not based on Cu oxides, but is three dimensional, s-electron superconductor. There is marked similarity between BKBO and BPBO. Both systems become metallic by substituting either K for Ba or Pb for Bi in semiconducting CDW $BaBiO_3$, thereby reducing electron number in the 6s band and contracting Bi-O bond length in the perovskite structure. In the case of BKBO, superconductivity is reported in the cubic, metallic phase for 0.35 < x < 0.5.

Contrary to BPBO, BKBO may be susceptible to oxygen loss due to substitution of monovalent K ions. Such oxygen loss may give rise to substantial change in the electronic and structural properties which affect the superconductivity. However, no systematic investigation has been done for the well characterized specimens. The aim of the present work is to synthesize a series of BKBO specimens under various preparation condition and to find the effect of oxygen vacancy formation on the superconductivity.

EXPERIMENTAL

A series of $Ba_{1-x}K_xBiO_{3-\delta}$ for 0 < x < 0.5 were synthesized from high purity materials of BaO_2, KO_2 and Bi_2O_3[3, 4]. The oxide powders in the proper metal ratios were mixed and ground under dry nitrogen. The pelletized mixture was sintered at 735 - 650°C for 0.3 - 1 hour in N_2 atmosphere and quenched to room temperature. The sintered material was subsequently annealed at 450 - 400 °C in O_2 atmosphere for 1 - 4 hours. This procedure was repeated three times to get good crystalline materials, incorporating oxygen as much as possible, which were used for starting materials.

Table 1 Preparation condition, lattice parameter, and oxygen vacancy content determined by iodometry(IM) and thermogravimetry(TG) of $Ba_{1-x}K_xBiO_{3-\delta}$.

x	TG holding condition in N_2 T(℃)/ t(h)	Lattice parameter a (A)	Oxygen vacancy content δ IM*2	TG*3	Color	Resistance (ohm)*4
0.35	--- / ---*1	4.3015(4)	0.125	0.125*2	purple	500
	378 / 1.0	4.3058(2)	---	0.166	purple	7k
	400 / 1.0	4.3082(4)	---	0.206	brown	10k
	427 / 1.0	4.3109(4)	---	0.227	brown	10k
	467 / 1.0	4.3197(5)	---	0.329	brown	70k
	546 / 1.0	4.3278(7)	---	0.430	brown	100k
0.40	--- / ---*1	4.2964(4)	0.121	0.121*2	blue black	30
	378 / 0.5	4.2973(3)	---	0.143	purple	200
	416 / 1.0	4.3055(4)	0.255	0.251	purple	5k
	431 / 1.0	4.3068(4)	0.288	0.261	brown	25k
	453 / 1.0	4.3148(4)	0.332	0.341	brown	100k
	530 / 1.0	4.3240(7)	0.455	0.472	ocher	50k
0.50	--- / ---*1	4.2833(4)	0.182	0.182*2	blue black	7
	397 / 0.2	4.2865(5)	0.211	0.217	blue black	30
	402 / 0.4	4.2915(4)	0.247	0.240	purple	400
	402 / 1.0	4.2978(4)	0.305	0.309	purple	1k
	418 / 1.0	4.3003(4)	0.356	0.359	purple	5k
	516 / 1.0	4.3228(12)	---	0.535	brown	20k

*1 Initial sample was sintered three times and finally at 450° C for 1 hour in O_2 atmosphere.
*2 Determined by iodometry(IM).
*3 Determined by thermogravimetry(TG).
*4 Resistance between two points with ca.2mm distance on the sample surface.

To determine the oxygen content, thermogravimetric analysis (TGA) was carried out in N_2 atmosphere; Weight change due to oxygen loss during heating with a rate of 5°C/min was measured. To obtain specimens with varied oxygen content, the starting material was heated to appropriate temperature between 378 - 546 °C for 1 hour in N_2 atmosphere and then rapidly quenched to room temperature. Preparation conditions are summarized in Table 1.

The oxygen content of the single phase material was also measured by iodometry. The end point was determined by using potentiometric titration method, because visual judgement was difficult due to deep yellow color of Bi ions. The synthesized material was examined by X-ray powder diffraction (XPD) analysis with Cu Kα radiation. The lattice parameter was accurately determined from measured d-spacings by least squares method.

The superconducting critical temperature (T_c) was determined by measuring the temperature variation of diamagnetic response with an AC susceptometer (Sumitomo Heavy Industry). The lowest temperature available for measurements was 12 K.

RESULTS AND DISCUSSION

BKBO changes its color from golden yellow of $BaBiO_3$ to brown, black and blue black with increasing amount of K doping, corresponding to a transition from semiconductor to metal. Crystal symmetry also changes from monoclinic in the semiconductor phase to cubic in the metallic phase, and unit cell volume monotonically decreases with x. These results are generally in good agreement with the previous results by ATT Bell and Argonne group.

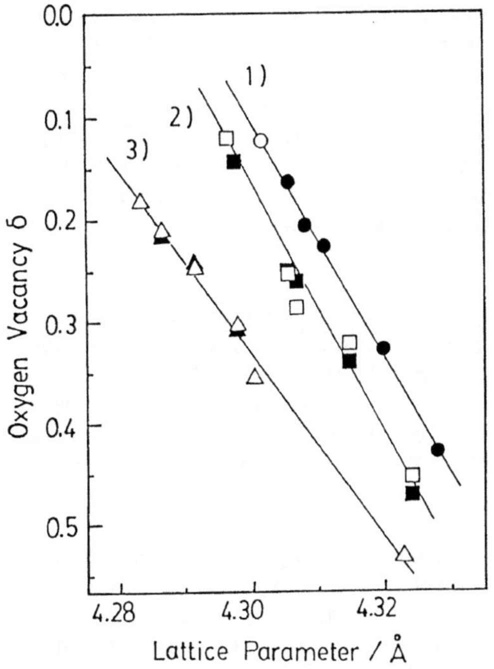

Fig.1 Oxygen vacancy content, δ, determined by TG(open symbol) and iodometry(solid symbol) versus lattice parameter of $Ba_{1-x}K_xBiO_{3-\delta}$. x=0.35(circle); x=0.4(square); x=0.5 (triangle).

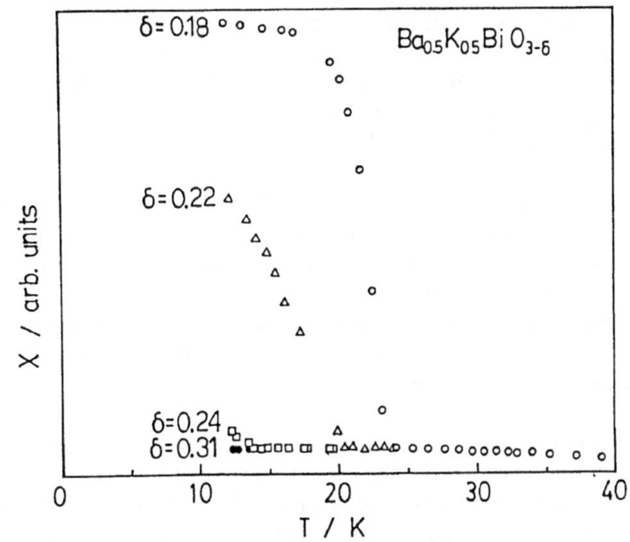

Fig.2 Temperature variation of diamagnetic response of $Ba_{0.5}K_{0.5}BiO_{3-\delta}$ with various oxygen content measured by AC susceptmeter.

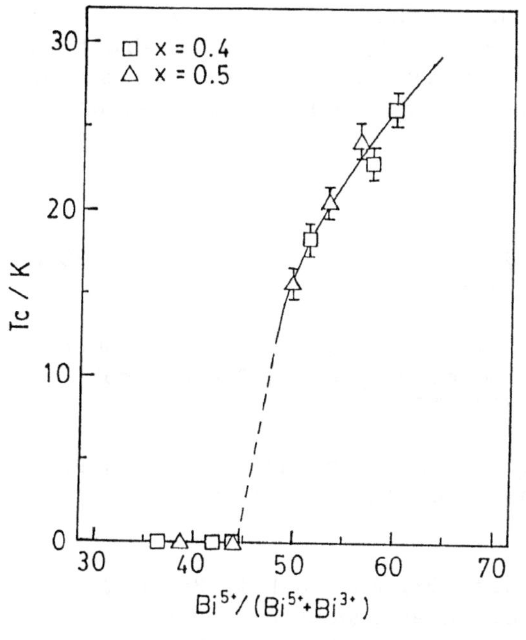

Fig.3 Dependence of Tc on the $Bi^{5+}/(Bi^{3+}+Bi^{5+})$ ratio of $Ba_{1-x}K_xBiO_{3-\delta}$ for x=0.4 and 0.5.

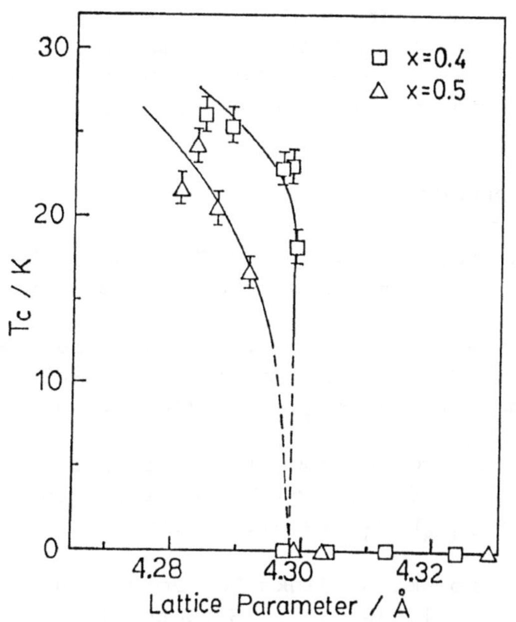

Fig.4 Dependance of Tc on the lattice parameter, a, of $Ba_{1-x}K_xBiO_{3-\delta}$ for x=0.4 and 0.5.

Detailed study on the relation between the amount of oxygen vacancy, lattice parameter and T_c was carried out for the specimens with x = 0.35, 0.4 and 0.5, which were prepared by quenching in N_2 atmosphere from various temperatures. The oxygen content determined by iodometry and thermogravimetry are plotted against measured lattice parameter a (Fig.1) Increasing amount of oxygen vacancy is accompanied by remarkable increase in a, which is twice Bi-O bond length in the cubic perovskite structure. The observed increase in a is explained by the increase in the s electron number (or decrease in the ratio $Bi^{5+}/(Bi^{3+} + Bi^{5+})$.

Temperature dependence of diamagnetic response of $Ba_{0.5}K_{0.5}BiO_{3-\delta}$ measured by AC susceptometer is shown in Fig.2. T_c as well as Meissner fraction decreases with increasing content of oxygen vacancy. T_c is plotted against Bi^{5+} concentration calculated from measured δ in Fig.3. T_c gradually decreases with increasing Bi^{5+} concentration (increasing s electron concentration) and superconductivity seems to disappear suddenly at 50 % Bi^{5+}, irrespective to the K concentration x.

Dependence of T_c on the lattice parameter is shown in Fig.4. T_c decreases with increasing lattice parameter which is a function of oxygen vacancy concentration and suddenly drops below the measurable temperature range of susceptometer. The critical value of the lattice parameter is close to 4.30 A, again irrespective to x. Interestingly semiconductor-metal transition seems to occur when the Bi-O bond length is reduced below a critical distance of 2.15 A, and T_c is apparently controlled by s electron concentration.

ACKNOWLEDGEMENTS

The authors are grateful to Professors Y. Muto and M. Tachiki for helpful discussion and warm encouragements. The work was partly supported by Grant-in-Aid for Scientific Research for Priority Areas "Mechanism of Superconductivity" given by Ministry of Education, Science and Culture, Japan.

REFERENCES

1. Mattheiss LF, Gyorgy EM, Johnson DW Jr (1988) Superconductivity above 20 K in the Ba-K-Bi-O system. Phys. Rev. B37:3745-3746.

2. Cava RJ, Batlogg B, Krajewski JJ, Farrow RC, Rupp LW Jr, White AE, Short KT, Peck WF Jr, Kometani TY (1988) Superconductivity near 30 K without copper: $Ba_{.6}K_{.4}O_3$ perovskite. Nature 332:814-816.

3. Hinks DG, Dabrowski B, Jorgensen JD, Mitchell AW, Richards DR, Shiyou Pei, Donglu Shi (1988) Synthesis, structure and superconductivity in the $Ba_{1-x}K_xBiO_3$ system. Nature 333:836-838.

4. Dabrowski B, Hinks DG, Jorgensen JD, Richards DR, Shiyou Pei, Zheng Y, Mitchell AW (1989) Structural properties of superconducting and non-superconducting $Ba_{1-x}K_xBiO_3$. Preprint of paper presented at 4th Ultrastructure Conference, Tucson, Arizona, Feb. 1989.

5. Batlogg B, Cava RJ, Schneemeyer LF, Espinosa GP (1989) High-T_c superconductivity in bismuthates - How many roads lead to high T_c ? IBM J. Res. Dev. 33:208-214.

Electronic States of Cu_nO_m Clusters by DV-Xα Calculation

TSUYOSHI YAMAGUCHI and NOBUHISA FUJIMA
Faculty of Engineering, Shizuoka University, Hamamatsu, 432 Japan

ABSTRACT

Electronic states of various Cu_nO_m clusters which consist of 1- and 2-dimensional network of superconductor oxides are calculated by the DV-Xα method. In the 1-dimensional cluster along the b-axis, a hole exists on O 2p-orbital.

KEY WORDS: electronic state, Cu_nO_m cluster, DV-Xα method

INTRODUCTION

The high T_c superconductor oxides have the 2-dimensional Cu-O network layers. The superconductivity in the CuO based superconductors is correlated with the electronic holes in a nearly filled valence band. The experiments show that the unoccupied O 2p states exist near the Fermi edges and that this O 2p hole is polarized in the a-b plane, that is, in the Cu-O network plane. Therefore, it is interesting to clarify the electronic structure of the unoccupied O 2p orbital.

To clarify the properties of CuO based superconductor, the first-principle calculation of electronic states of Cu_nO_m clusters has been done.[1,2,3,4] Sarma and Sreedhar[1] calculated electronic structure of square planar CuO_4^{6-} cluster by the spin-polarized MS-Xα method. They showed that the ground state of the CuO_4^{6-} cluster is essentially non-magnetic in spite of odd number of electrons in the system, as found in the high T_c superconductors. This arises from the fact that the unpaired electron resides in a molecular orbital with primarily oxygen 3s character, because the Cu-O distance is short.

Sekine et al.[2] calculated electronic structure of CuO_m (m=4,5,6) clusters by the DV-Xα method. In the calculation they varied the electron deficiency of the cluster. They revealed that the electron was drawn out from the O 2p on the plane site and not from Cu 3d nor from the O at the apex site. The increase in the electron deficiency enhances the hybridization between Cu 3d and O 2p orbitals.

Adachi and Takano[3] calculated electronic states of $Cu_3O_{12}^{17-}$ and $Cu_3O_{10}^{15-}$ clusters by the spin-polarized DV-Xα method. These clusters line up along the c-axis of CuO based superconductor. They showed that the Cu 3d is hybridized with the O 2p orbital and that a hole is located on the O 2p band in the a-b plane.

By using the Hartree-Fock, the full-valence-configuration-interaction, and the complete-active-selfconsistent-field methods, Yamaguchi et al.[4] obtained the electronic structures of deformed CuO_6 octahedron cluster and Cu-O-Cu chain cluster where a single hole or a couple of holes exist. They compared the results obtained by different methods.

To our knowledge, there has been no calculation of electronic states of planar Cu_nO_m clusters in the a-b plane, although the Cu 3d and O 2p orbitals are hybridized and the hole exists in the a-b plane. Then, in this work, we calculate the electronic structure of planar Cu_nO_m clusters by the DV-Xα method to discuss the electronic properties of CuO superconductors.

ELECTRONIC STATES OF Cu_nO_m CLUSTERS

$Cu_4O_{20}^{32-}$ and $Cu_4O_{20}^{28-}$ Clusters

$(La_{1-x}TM_x)_2CuO_{4-\delta}$ superconductor where TM represents a transition-metal atom has the tetragonal K_2NiF_4 structure at high temperature and the orthorhombic structure at low temperature. Roughly, a Cu ion is surrounded by 6 O ions. The Cu-O distance is 1.9 A for the O ion in the a-b plane and 2.4 A for the O ion along the c-axis. The deformed CuO_6 octahedrons consist of a 2-dimensional network.

Figure 1(b) shows the energy levels of a 2-dimensional $Cu_4O_{20}^{32-}$ cluster shown in Fig.1(a). Assuming that the system has a hole per Cu ion, we calculate the electronic structure of $Cu_4O_{20}^{28-}$ cluster as shown in Fig.1(c).

In both clusters, Cu 3d levels are located near HOMO. These Cu 3d levels are less hybridized with O 2p levels. The electron configuration is Cu:$3d^{8.84}4s^{0.07}4p^{0.04}$, O(1):$2p^{6.00}$, O(2):$2p^{5.99}$ and O(3):$2p^{6.00}$ for $Cu_4O_{20}^{32-}$ cluster and Cu:$3d^{8.32}4s^{0.04}4p^{0.03}$, O(1):$2p^{5.76}$, O(2):$2p^{5.90}$ and O(3):$2p^{6.00}$ for $Cu_4O_{20}^{28-}$ cluster.

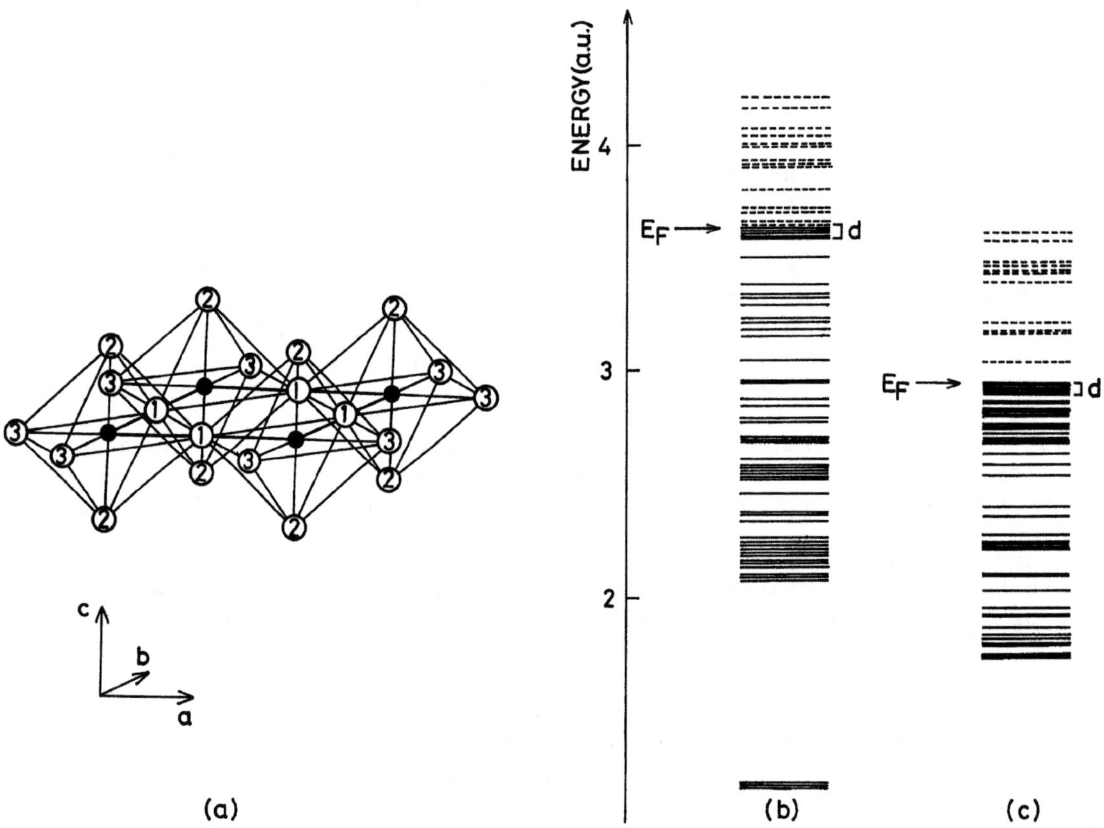

Fig.1 Atomic and electronic structures of $Cu_4O_{20}^{32-}$ and $Cu_4O_{20}^{28-}$ clusters. (a) Atomic structure. Filled and open circles represent Cu and O ions, respectively. Numeral shows the inequivalent ion. (b) Valence energy levels of $Cu_4O_{20}^{32-}$ cluster in the atomic unit. Dashed and full lines represent unoccupied and occupied levels, respectively. E_F shows the highest occupied molecular orbital (HOMO) or the Fermi level. The label d shows energy levels which have much Cu 3d-character by Mulliken charge analysis. (c) Valence energy levels of $Cu_4O_{20}^{28-}$ cluster.

$Cu_4O_{12}^{16-}$ and $Cu_4O_{12}^{12-}$ Clusters

$YBa_2Cu_3O_{7-\delta}$ superconductor has a 2-dimensional network of 3 layers. The top and bottom layers consist of pyramidal CuO_5 clusters in the a-b plane and the middle layer consists of CuO chains along the b-axis. However, the Cu-O distance 2.30 A for the apex O ion of the pyramid is much longer than that 1.93 A for planar O ion. Furthermore, the Cu-O distance 1.93 A for the chain along the b-axis is longer than that 1.84 A between the chain and the apex O ion of the pyramid. Then, we consider two types of Cu-O networks. One is a 2-dimensional network of square CuO_4 clusters in the Nd_2CuO_4 structure, and the other is a 1-dimensional network of square CuO_4 clusters in the Sr_2CuO_3 structure. Figure 2(b) shows the energy levels of a 2-dimensional $Cu_4O_{12}^{16-}$ cluster shown in Fig.2(a). Figure 2(c) shows the energy levels of a $Cu_4O_{12}^{12-}$ cluster where a single hole per Cu ion is located.

Although Cu 3d levels near HOMO are narrow in Fig.2(b), they are relatively broad in Fig.2(c). Cu 3d-levels are less hybridized with O 2p-levels in $Cu_4O_{12}^{16-}$ cluster but much hybridized in $Cu_4O_{12}^{12-}$ cluster. The HOMO has 58% Cu 3d-character, 11% O(1) 2p-character and 29% O(2) 2p-character. The electron configuration is Cu:$3d^{8.97}4s^{0.03}4p^{0.01}$, O(1):$2p^{5.89}$ and O(2):$2p^{6.00}$ for $Cu_4O_{12}^{16-}$ cluster and Cu:$3d^{8.44}4s^{0.11}4p^{0.03}$, O(1):$2p^{5.48}$ and O(2):$2p^{5.97}$ for $Cu_4O_{12}^{12-}$ cluster.

$Cu_4O_{13}^{18-}$ and $Cu_4O_{13}^{14-}$ Clusters

Figure 3(b) shows the energy levels of a 1-dimensional $Cu_4O_{13}^{18-}$ cluster shown in Fig.3(a). Figure 3(c) shows the energy levels of a $Cu_4O_{13}^{14-}$ cluster where a single hole per Cu ion is located.

Cu 3d-levels near HOMO are relatively broad in both Figs.3(b) and (c). Cu 3d-levels are hybridized with O 2p-levels. The HOMO has 65% Cu(1) 3d-character and 13% Cu(2) 3d-character for $Cu_4O_{13}^{18-}$ cluster and 12% Cu(1) 3d-character, 26% Cu(2) 3d-character and 55% O(5) 2p-character for $Cu_4O_{13}^{14-}$ cluster. The electron configuration is Cu(1):$3d^{8.82}4s^{0.05}4p^{0.02}$, Cu(2):$3d^{9.10}4s^{0.15}4p^{0.06}$, O(1):$2p^{6.00}$, O(2):$2p^{5.92}$, O(3):$2p^{6.00}$, O(4):$2p^{5.90}$ and O(5):$2p^{5.97}$ for $Cu_4O_{13}^{18-}$ cluster and Cu(1):$3d^{8.72}4s^{0.04}4p^{0.02}$, Cu(2):$3d^{9.07}4s^{0.12}4p^{0.04}$, O(1): $2p^{5.87}$, O(2):$2p^{5.80}$, O(3):$2p^{5.59}$, O(4):$2p^{5.84}$ and O(5):$2p^{5.18}$ for $Cu_4O_{13}^{14-}$ cluster. A hole exists mainly on O 2p-orbital of $Cu_4O_{13}^{14-}$ cluster.

REFERENCES

1. Sarma DD. Sreedhar K (1988) Electronic structure of square planar CuO_4^{6-} cluster. Z.Phys.B-Condensed Matter 69:529-534
2. Sekine R. Kawai M. Adachi H (1989) Electronic states of the CuO_x (x=4,5,6) model cluster. Physica C 159:161-164
3. Adachi H. Takano M (1989) Electronic states of $Cu_3O_{12}^{17-}$ and $Cu_3O_{10}^{15-}$ clusters. Physica C 159:169-170
4. Yamaguchi K. Takahara Y. Fueno T. Nasu K (1987) Ab initio MO calculation of effective exchange integrals between transition-metal ions via oxygen di-anions: nature of the copper-oxygen bonds and superconductivity. Jpn.J. Appl.Phys. 26:L1362-1364

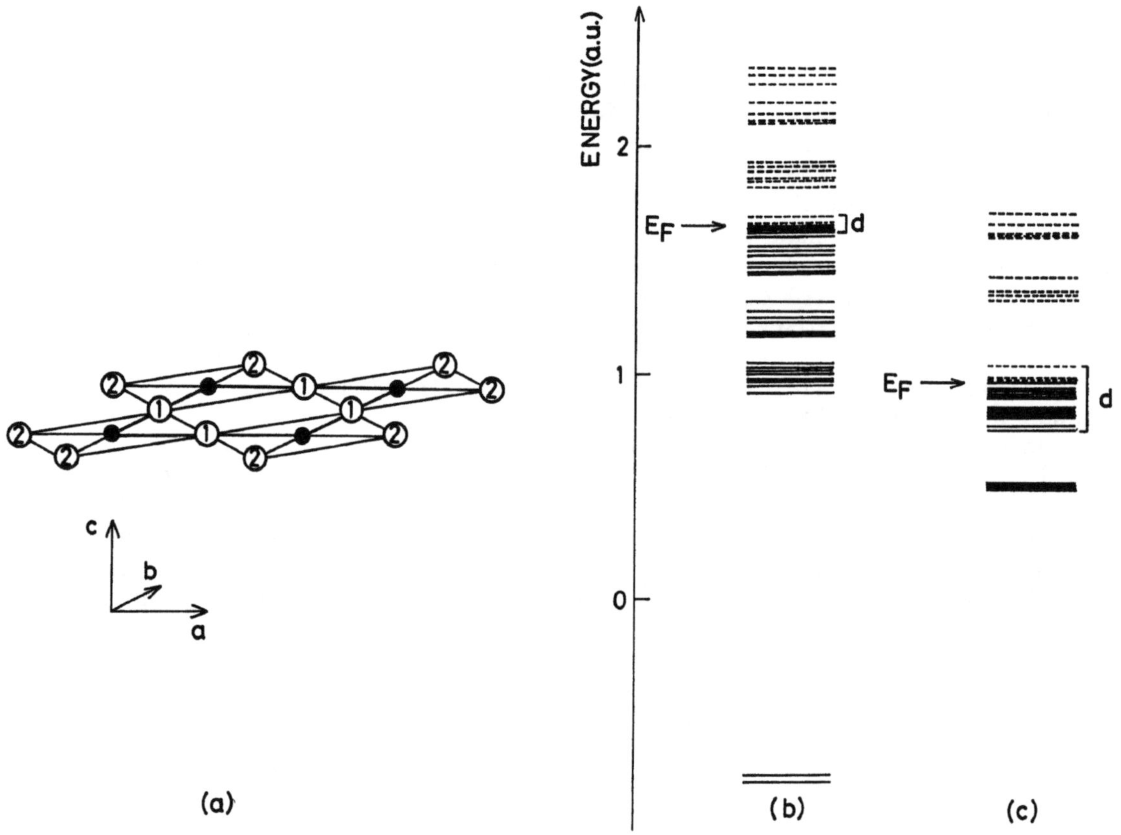

Fig.2. The same as Fig.1 for $Cu_4O_{12}^{16-}$ and $Cu_4O_{12}^{12-}$ clusters.

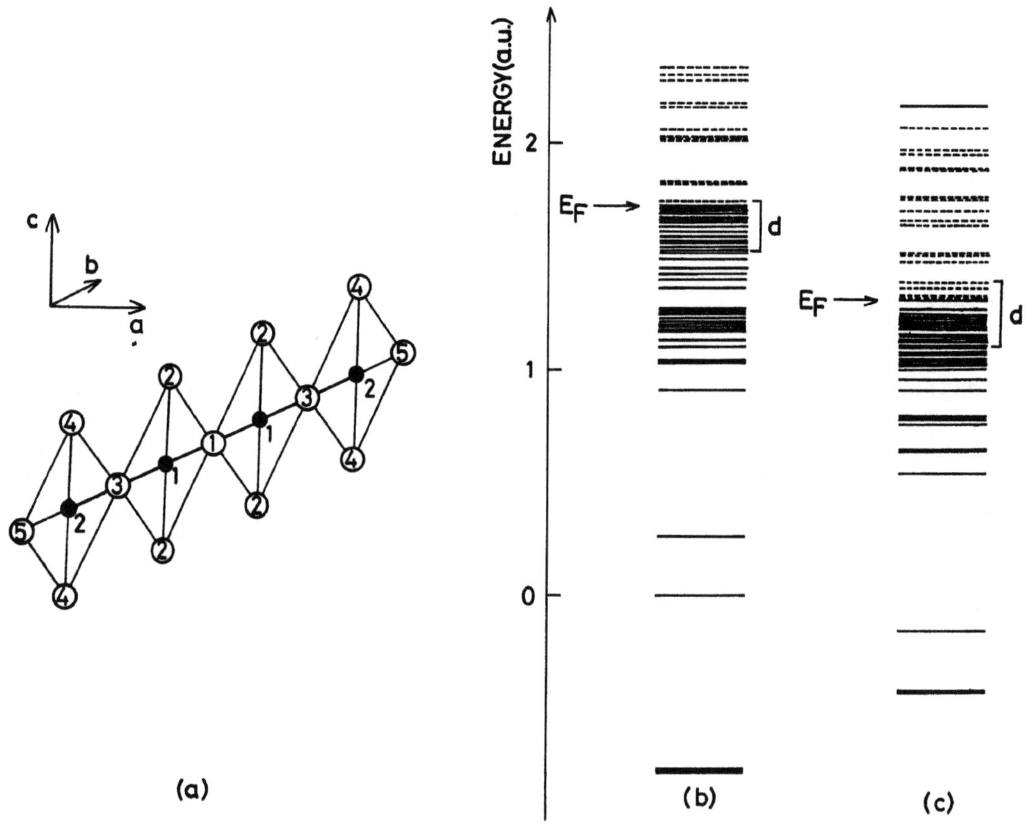

Fig.3. The same as Fig.1 for $Cu_4O_{13}^{18-}$ and $Cu_4O_{13}^{14-}$ clusters.

3.3 Transport and Tunneling Properties

Tunneling and Transort Experiments on Single-Crystal $Nd_{2-x}Ce_xCuO_{4-y}$ and $YBa_2Cu_3O_7$

T.W. Jing[1], Z.Z. Wang[1], T.R. Chien[1], N.P. Ong[1], J.M. Tarascon[2], and E. Wang[2]

[1] Joseph Henry Laboratories of Physics, Princeton University, Princeton, NJ 08544, USA
[2] Bell Communications Research, Red Bank, NJ 07701, USA

Abstract

Several anomalous features are observed in tunneling spectra obtained on cleaved crystals of $Nd_{2-x}Ce_xCuO_{4-y}$. The conductance curve is strongly asymmetric in the bias, and the gap parameter Δ increases as the temperature is warmed towards T_c. In one sample a weak magnetic field completely suppress distinctive structures which appear at energies above the gap. Hall measurements in $Nd_{2-x}Ce_xCuO_{4-y}$ crystals at low temperatures provide strong evidence for an anomalous contribution to the Hall scattering in the normal state. The Andreev reflection spectrum in $YBa_2Cu_3O_7$ crystals is also reported.

Key words: Tunneling gap, Hall effect, $Nd_{2-x}Ce_xCuO_{4-y}$

1. Introduction

We have used tunneling and transport measurements to probe the high-T_c superconductivity in the recently discovered[1] T'-phase system $Nd_{2-x}Ce_xCuO_{4-y}$ ("NCCO"). The energy spectrum shows several anomalous features which cannot be accounted for by conventional theory. We compare our results with those of Ekino and Akimitsu[3] (EA) taken on polycrystalline samples. Results on the Hall effect[4] of NCCO crystals and the Andreev reflection spectrum obtained on $YBa_2Cu_3O_7$ ("YBCO") crystals are also reported.

2. Tunneling spectra of $Nd_{2-x}Ce_xCuO_{4-y}$

Superconducting crystals of NCCO were grown as described by Tarascon et al[5]. To form the junctions for the tunneling experiments, crystals with particularly sharp resistive transitions were cleaved in air at room temperature and a film of Pb (2,500 Å) was evaporated. The typical junction resistance is between 100 and 200 Ω, and only weakly T dependent between 150 and T_c. (The tunneling current is injected parallel to the CuO_2 planes.) Below T_c, the zero-biased resistance $R(0)$ rises approximately linearly with the reduced temperature $t = T_c - T$. Figure 1 shows the conductance curves for Samples 1 and 2 (plotted as the differential conductance $G = dI/dV$ vs. the bias V). Although $R(0)$ remains finite at our lowest T (2 K), there is clear evidence for a (pseudo) gap developing below T_c. If we call the energy at which G intersects the normal state curve V_x, we see that states that are removed from below V_x are inserted above V_x. Figure 1 also shows the distinct asymmetry in the tunneling curves. At any bias, G is higher on the negative branch (NCCO negative) than on the positive, i.e. it is easier to inject holes into NCCO than to inject electrons. Whether this is a manifestation of a fundamental asymmetry between electron and hole excitations, or a signature of exotic quasi-particles, is an interesting question which is being pursued.

Examination of Fig. 1 reveals that the width of the central "well" appears to increase as $T \rightarrow T_c$ from below. This is brought out more clearly in Fig. 2 where we show G at three T, normalized to the normal curve (The value of the ratio is one at large V). Whether we use the intersection point V_x, the

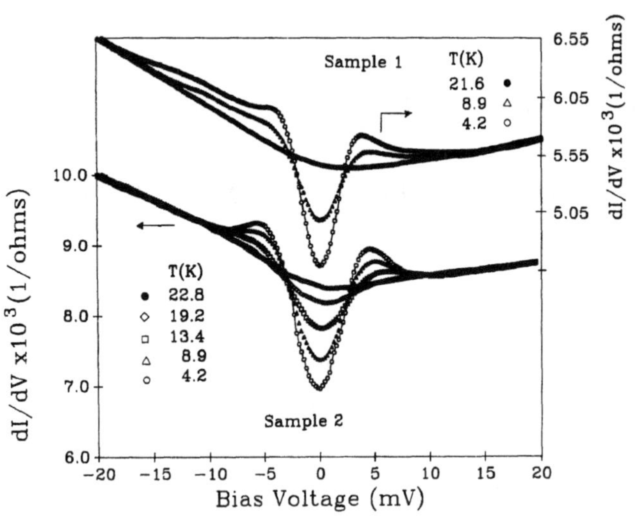

FIG 1 The junction conductance G vs. bias V of $Nd_{2-x}Ce_xCuO_{4-y}$ crystals for Samples 1 and 2 in zero magnetic field. As T falls below T_c, G is progressively enhanced above (suppressed below) the normal-state value for V larger (smaller) than the bias V_x at which G intersects the normal-state curve. (Ref. 2).

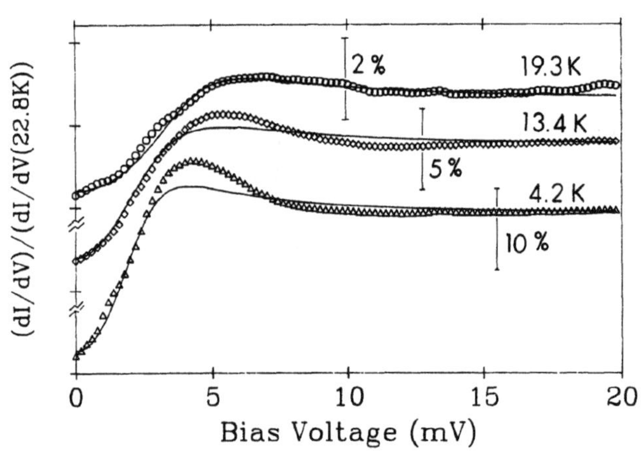

FIG 2 Comparison of the data for G in Sample 2 (normalized to the pretransition curve) with the optimum fit using Eq. 1. The values of G obtained are 1.8, 2.4 and 2.5 mV at 4.2, 13.4, 19.3 K respectively. The deviation from Eq. 1 becomes progressively worse at low T near the peak. The temperature dependence of the "gap" Δ_1 extracted from the fits is shown in Ref. 2.

half-width, or the broad peak as a measure of the gap parameter, we see that the width of the central well tends to increase with T. To get a more quantitative estimate, we have tried to fit the normalized G to an equation suggested by the lifetime smearing model of Dynes, Narayanmurti and Garno[6]

$$(dI/dV)/(dI/dV)_N \quad = \quad A \; Re[\,(E-i\Gamma)/\surd\{(E-i\Gamma)^2 - \Delta^2\}] \;\; + G_0, \qquad (1)$$

where Γ is a parameter proportional to the inverse lifetime, A a scale factor, and G_0 the background conductance. Good fits (solid lines in Fig. 2) can be made close to T_c. However, as T decreases, Eq. 1 cannot be made to accomodate the peak near the gap edge, as is clear from the fit to the 4.2 K curve. Nevertheless, it is interesting to see how Δ determined from Eq. 1 varies with T. Consistent with the behavior of V_x, we find that Δ shows a 25 % increase as T warms towards T_c. This runs counter to our expectation that the gap parameter should decrease as we move towards the normal phase. In contrast to our findings, EA[3] report that Eq. 1 gives a good fit to their curves, and that Δ is T-independent.

At present, we have no explanation for this anomalous variation. We note that this behavior can be discerned in previously published[7] tunneling curves on the 2-plane Bi 2212 system, although the authors did not discuss the trend of the gap. This appears to be an important point for further research. It is possible that the high-T_c order parameter behaves in the conventional way (grows with increasing t),

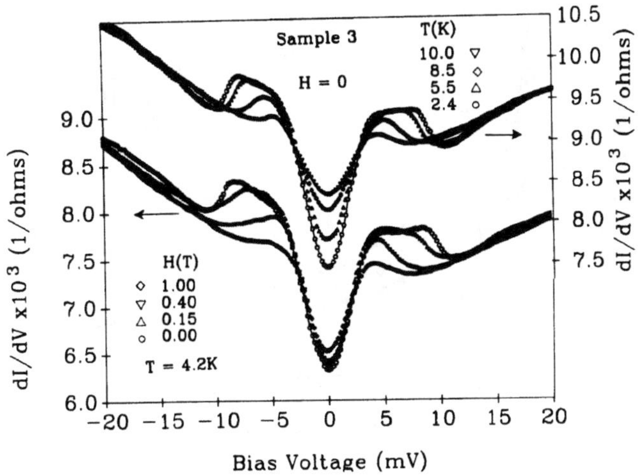

FIG 3 The junction conductance G vs. V of $Nd_{2-x}Ce_xCuO_{4-y}$ (Sample 3). The upper set shows the variation with temperature in zero-field, while the lower set shows the variation with field at 4.2 K. (**H** is applied parallel to CuO_2 planes.) The lower set is taken ~24 h. after the upper set. The curve at $T = 4.2$ K and $H = 0$ shows clearly the very broad range of states ("wings") extending from 10 to 20 mV. The suppression of the "mesa" structures and the wings is roughly similar in both sets. However, **H** has only a weak effect on the central "gap" structure (Ref. 2).

but that $G(V)$ may not be directly related to the density of states in the direct way familiar in low-T_c systems. (This may be the case, for example, if the charge carriers in the high-T_c oxide are not simple electrons, but composite objects[8].)

In a third sample, we observe a tunneling spectrum which is richer in structure (Fig. 3). In this run a field **H** was applied parallel to the CuO_2 planes (and normal to the junction interface). The upper set of curves represents the variation of the spectra with T in zero field, while the lower set shows the variation with field at fixed T (4.2 K). There exist in this sample prominent supra-gap features which we call a mesa structure (for V between 3 and 8 mV) and a wing structure (between 12 and 22 mV). The overall shape of the spectrum recalls the spectrum obtained in YBCO crystals by Gurvitch et al[8]. The effects of raising T and H on these structures are broadly similar. At high T or large H, these structures are suppressed, so that the spectrum resembles that in Sample 2 (Fig. 1). The field has negligible effect on the depth of the central well since H_{c2} for **H** parallel to the a-b plane greatly exceeds 1 T. The existence of these supra-gap structures is intriquing. The relatively weak fields at which they are suppressed suggests some relation to the existence of the superconductivity in Pb. However, it is difficult to reconcile the high energies of these features (20 mV) with the weak gap (1.3 mV) in the Pb spectrum. The field sensitivity also rules out any connection with phonon-induced structures. Futhermore, the fact that superconductivity in NCCO persists when these structures are suppressed by field implies that they are not intrinsic to the high-T_c mechanism. (Note that they are absent in Sample 2 in zero field.)

3. Hall effect of $Nd_{2-x}Ce_xCuO_{4-y}$

The discovery of superconductivity in NCCO has generated great interest, since it appears to be the first high-T_c system in which the carriers are electron-like[1]. However, we regard the question of the sign as open. More experimental work appears necessary before this question can be answered with certainty. The Hall measurements of Takagi et al[9] showed that the Hall coefficient R_H is negative over the range of Ce content $0 < x < 0.17$. R_H becomes positive when x exceeds 0.17. In the data of Takagi et al,

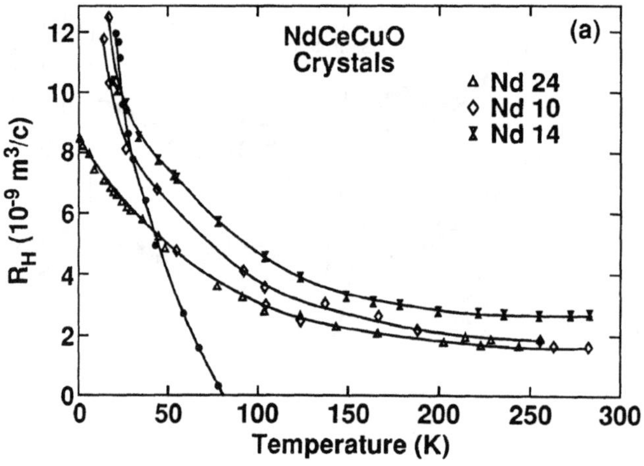

FIG 4 The temperature dependence of the Hall coefficient R_H in four crystals of $Nd_{2-x}Ce_xCuO_{4-y}$ with **H** // **c** and current in-plane. In one sample, R_H is negative above ~80 K. Note that R_H is strongly T dependent down to 1 K (Ref. 3).

the superconducting region appears to fall in the region where R_H is negative. Subsequent work has shown that R_H is actually sensitive to both T and the oxygen content 4-y. Thus, the exact place at which R_H crosses zero depends on both T and y (for fixed x). Rather than determining the exact point in the phase diagram where R_H changes sign, it is more meaningful to study the trend of R_H as x and y are varied. Recent work has shown that, in the superconducting phase, R_H tends towards positive values. However, this question is complicated by the anomalous behavior of the Hall effect in the whole class of high-T_c materials[3]. We have exploited the relatively low H_{c2} (for **H**//**c**) to examine how R_H varies as $T \rightarrow 0$. From such studies, we find good evidence that in addition to the usual Lorentz term, there exists a strongly T dependent contribution to R_H.

Figure 4 shows the variation of R_H with T in four crystals of NCCO. In all samples, the current is in-plane while **H** // **c**. In three crystals R_H is positive at all T below 290 K. In a fourth crystal R_H is initially negative at room T, but changes sign at 80 K to become large and positive below 80 K. (We had erroneously reported at the Stanford meeting that one of the samples has a negative R_H at all T[11]. Subsequent measurement shows that the sign is positive at all T.) In all samples, R_H is observed to increase rapidly to large positive values as $T \rightarrow 0$. It is interesting to ask if R_H continues to diverge below T_c. By using a 15 T field, we have suppressed the superconductivity so that the normal state behavior of R_H becomes observable. In the normal state, ρ_{ab} is almost T independent below 30 K, whereas R_H remains strongly T dependent even at 1 K. From this behavior we infer that the T dependence of R_H is not due to two competing bands of carriers. In the usual two-band model, the T dependence of R_H is caused by the relative variation of the mobilities of the two bands. However, when ρ_{ab} becomes T independent, the mobilities also become T independent, and it is difficult to see how R_H could show the strong T dependence in Fig. 5. In our opinion, the behavior of R_H (and ρ_{ab}) at low T constitutes the strongest evidence to date that there exists an anomalous transverse scattering (in addition to the Lorentz force).

In view of the large moment on the Nd ions and the existence of an antiferromagnetic instability in the undoped Nd system, it is natural to ask if the anomalous scattering is due to magnetic skew scattering[12]. We do not believe that this is likely. First, since R_H is proportional to the magnetic susceptibility in magnetic scattering, we should observe a kink in R_H near 70 K[5]. This is absent in our data. Secondly, the T dependence of R_H is very similar to what is found in YBCO (and, to a smaller extent, in the Bi and Tl systems). We need a more general mechanism. Clayhold et al[13] have explored the correlation between the suppression of T_c in YBCO and LaSrCuO by dopant impurities on the one hand, and the decrease in the slope of dR_H/dT, on the other. It looks increasingly evident that there exists an anomalous skew scattering operating within the CuO_2 planes which is quite *unlike* the conventional magnetic skew scattering observed in ferromagnets or in nonmagnetic hosts doped with transition metal ions[12]. In clean high-T_c samples, this mechanism is very strong and is responsible for the strong T dependence of R_H. However, when impurities are introduced into the CuO_2 planes they destroy the skew scattering mechanism, and apparently suppress as well the mechanism for high-T_c superconductivity. The existence of this skew mechanism may explain the unexpectedly large Hall signal observed in the Tl based 2223 and 2212 compounds[14]. Anderson[9] has recently proposed an enhanced susceptibility model which relates the $1/T_1$ NMR relaxation data to angle-resolved photoemission data. In this model the pair susceptibility diverges as $(\log J/T)^3$ with decreasing T, in contrast to the $\log J/T$ behavior in BCS theory. Anderson hypothesizes that R_H may have a contribution from the pair susceptibility which grows as $(\log J/T)^2$. This may provide an explanation of the anomalous T dependence in R_H. Fits to R_H in single-crystal YBCO are quite good, but less satisfactory in NCCO. These fits will be discussed elsewhere.

4. Andreev reflection spectrum in $YBa_2Cu_3O_7$ crystals

Andreev reflection[15,16] involves the reflection of an electron with change of character (into a hole) at the interface between a normal metal N and a superconductor S. Hoevers et al[17] have reported observing at 1.2 K Andreev reflection off the Ag-*123* interface, using a thin-film sample. Recently, we developed a technique to form gold contacts on YBCO crystals cleaved in high vacuum[18]. We have extended Hoevers et al's work, using our Au-*123* junctions to investigate the T dependence of the Andreev signal[19]. As in Ref. 17, we use a point contact (Au) to inject current into the junction. The inset in Fig. 5 shows the experimental geometry. By tuning the bias V across the point contact, we change the energy of the injected electrons relative to the Fermi energy ε_F in the Au film (which is equal to μ_s in the *123*), and measure the junction differential resistance $R = dV/dI$. (In the figure, note that R is the quantity plotted instead of G.) We identify the abrupt decrease in R at $|V| \sim 10$ mV with the gap at the Au-*123* interface. This value is similar to that obtained by Hoevers et al.

At a finite bias, the electrons are injected "hot" into the Au film. They undergo multiple scattering (mostly elastic at low T) within the Au film before they reach the Au-*123* interface. If Andreev reflection occurs at the interface, the reflected hole retraces the path of the incident electron, undergoing the same scattering events in reversed order. However, if some of these scattering events are inelastic (involving phonon emission or absorption) the phase memory of the electron or hole is disrupted, and the reflected hole fails to return to the point of injection. Therefore, we expect the Andreev signal to be strongly suppressed as T is increased. Figure 6 shows this thermal smearing quite clearly. As T increases from 4.2 K, the gap feature broadens until it is barely resolved at 62 K.

The shape of the Andreev spectrum is interesting. From Figs. 5 and 6, we observe that the spectrum has a rather "flat" bottom, i.e. R (or G) is independent of V for $|V| \leq \Delta$. In the model of Blonder, Tinkham and Klapwijk (BTK)[16], this implies that the strength of the interface barrier (measured by the dimensionless parameter Z) is zero. If Z is non-zero, $G(0)$ is suppressed while its value at the gap, $G(\Delta)$, remains unaffected. The resulting spectrum has a conductance minimum at zero-bias and sharp conductance maxima at $V = \pm\Delta$. The absence of minima in R at $\pm\Delta$ in Figs. 5 and 6 suggests that the interface at which the Andreev reflection occurs has a vanishing (classical) barrier, i.e. the barrier strength Z at the Au-*123* interface is close to zero. Electrons are transmitted through the Au-*123* interface with negligible scattering.

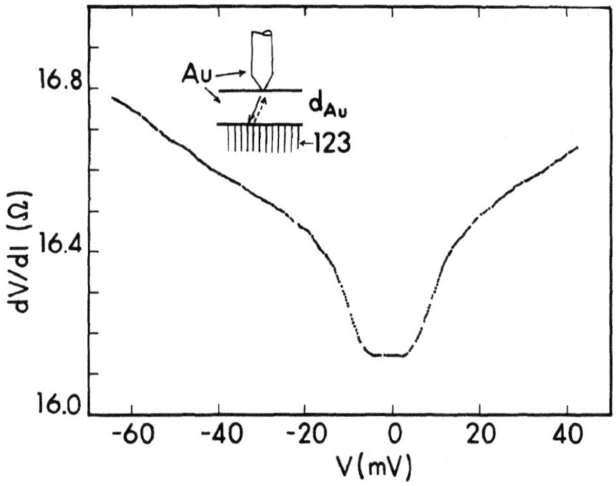

FIG 5 The differential resistance vs. bias of point contact junction showing Andreev reflection spectrum in Au-123 interface. The insert shows the experimental geometry (Ref.19).

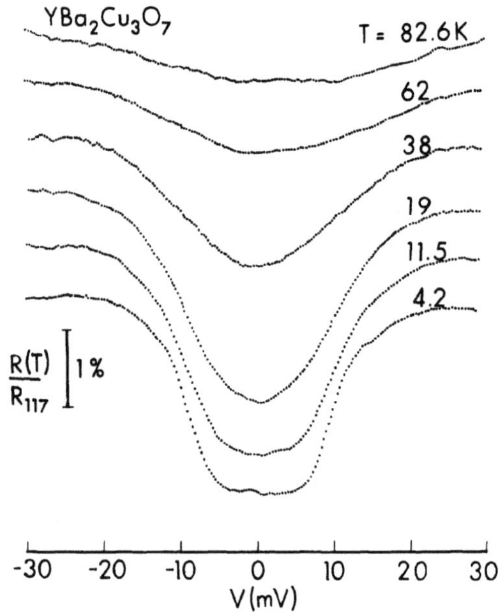

FIG 6 The I-V curves of the point contact junction in Fig. 3 at various *T*. All curves are normalized to the curve at 117 K (Ref. 19).

We have benefitted from discussions with P.W. Anderson, P. M. Chaikin, J. Clayhold, R.C. Dynes, and J. Rowell. The work at Princeton University is supported by the U.S. Office of Naval Research (Contract N00014 -90 -J -1013). The high field experiments were performed at the National Magnet Lab. (Cambridge) which is a facility supported by the U.S. National Science Foundation.

References

1. Y. Tokura, H. Takagi, and S. Uchida, Nature **337**, 345 (1989)
2. T.W. Jing, Z.Z. Wang, N.P. Ong, J.M. Tarascon, and E. Wang, Phys. Rev. B, submitted.
3. Z.Z. Wang, T.R. Chien, N.P. Ong, J.M. Tarascon and E. Wang, to be published.
4. T. Ekino and J. Akimitsu, Phys. Rev. B **40**, 7364 (1989).
5. J.M. Tarascon et al, Phys. Rev. B **40**, 4494 (1989).
6. R.C. Dynes, V. Narayanamurti, and J.P. Garno, Phys. Rev. Lett. **41**, 1509 (1978).
7. Hiroshi Ikuta, Atsutaka Maeda, Kunimitsu Uchinokura and Shoji Tanaka, Jpn. Jnl. Appl Phys. **27**, L1038 (1988); T. Ekino and J. Akimitsu, Phys. Rev. B **40**, 6902 (1989).
8. M. Gurvitch, J.M. Valles Jr., A.M. Cucolo, R.C. Dynes, J.P. Garno, L.F. Schneemeyer, and J.V. Waszczak, Phys. Rev. Lett. **63**, 1008 (1989).
9. P.W. Anderson, to appear in Nature; and private communication.
10. H. Takagi, S. Uchida and Y. Tokura, Phys. Rev. Lett. **62**, 1197 (1989).
11. Z.Z. Wang et al, *Proceedings of Inter. M^2S-HTSC Conference*, Stanford 1989, to appear as a special issue in Physica.
12. A. Fert and P.M. Levy, Phys. Rev. B **36**, 1907 (1987); J.J. Rhyne, Phys. Rev. **172**, 523 (1968).
13. J. Clayhold, N.P. Ong, Z.Z. Wang, J.Tarascon, and P. Barboux, Phys. Rev. B **39**, 7324 (1989).
14. J. Clayhold, N.P. Ong, P.H. Hor, and C.W. Chu, Phys. Rev. B **38**, 9280 (1988); J. Clayhold et al, to be published.
15. A.F. Andreev, Zh. Eksp. Teor. Fiz. **46**, 1823 (1964) [Sov. Phys. JETP **19**, 1228 (1964)].
16. G.E. Blonder, M. Tinkham and T.M. Klapwijk, Phys. Rev. B **25**, 4515 (1982).
17. H.F.C. Hoevers et al, Physica C **152**, 105 (1988).
18. T.W. Jing, Z.Z. Wang and N.P. Ong, Appl. Phys. Lett., Nov. 1st 1989
19. T.W. Jing, N.P. Ong, Z.Z. Wang, and P.W. Anderson, unpublished.

Transport and Magnetic Properties in Tl System Superconductors Showing Large T_c-Variations

YOSHIMI KUBO, YUICHI SHIMAKAWA, TAKASHI MANAKO, and HITOSHI IGARASHI

Fundamental Research Laboratories, NEC Corporation, 1-1, Miyazaki 4-chome, Miyamae-ku, Kawasaki, 213 Japan

ABSTRACT

T_c values for Tl system superconductors $Tl_2Ba_2Ca_{n-1}Cu_nO_{2n+4}$ can be controlled by slightly changing the oxygen content as small as 0.10-0.15 per formula unit, which changes the carrier density in the system. In particular, the n=1 system (Tl2201 phase) is most interesting because it shows a remarkable phase transition from a normal metal to a 85-K superconductor as the carrier density decreases. Transport and magnetic properties of the Tl2201 phase with various T_c values will be reported.

KEY WORDS: Tl system superconductor, transport, magnetic susceptibility

INTRODUCTION

It has been quite mysterious that T_c values for Tl system superconductors showed large variations without significant difference in the crystal structure. We first pointed out that the T_c-variations were caused by very small change in oxygen content which affected the carrier density in the system [1-3]. In a series of compounds $Tl_2Ba_2Ca_{n-1}Cu_nO_{2n+4}$, variations in T_c were 0K-85K, 85K-110K and 116K-110K for n=1,2 and 3, respectively [3]. Here, the former T_c values were for samples fully oxygenated at ~350°C and the latter values were for samples reduced in argon at ~550°C. The argon-annealing slightly decreased the oxygen content by 0.10-0.15 per formula unit for each system. Corresponding decrease in the hole carrier concentration was confirmed by the increase in both Hall coefficient(positive) and normal resistivity [2].

Since both hole concentration and T_c are easily controlled by slightly changing the oxygen content, these systems seem to be good models for the study of superconductivity mechanism. In particular, the n=1 system, which contains a single Cu-O octahedron layer between Tl_2O_2 layers, is most interesting because it shows a remarkable phase transition from a normal metal to a 85-K superconductor as the hole concentration decreases. The fact that superconductivity was degraded if the carrier density exceeded an optimum value was first reported for La-Sr-Cu-O system [4], where a 40-K superconductor changed to a normal metal. We observed a very similar superconductor-normal metal transition in $TlBa_{1+x}La_{1-x}CuO_5$ [5], where the maximum T_c value was also 40K. The n=2 system in the present study also showed significant degradation of superconductivity as the carrier density increased. Some recent works on Bi system superconductors supported the same tendency, as well [6,7]. Therefore, it seems to be well established that there exists an optimum value of carrier density for occurrence of superconductivity, although the absolute value may depend on the particular crystal structure of each system.

Until now, many studies have focused on an insulator-superconductor transition which takes place in the materials with lower carrier density than the optimum carrier density. This was partly due to the fact that samples with lower carrier density have been easily available for most systems. On the contrary, studies of the metal-superconductor transition have been restricted to the La-Sr-Cu-O system, so far [8,9], undoubtedly because it was the only available system. However, if we consider that superconductivity appears in an intermedeate range between an antiferromagnetic insulator and an normal metal, or in other words, between spin liquid description and Fermi liquid description, investigating approach to the superconductivity from the metallic side should be as important as the approach from the insulating side.

The n=1 system of the present study (Tl2201 phase) provides us a novel and superior example of the metal-superconductor transition. In particular, it shows much higher transition temperatures up to 85K which is more than twice of that for La-Sr-Cu-O system. Moreover, the crystal structure may be

clearer and more ordered throughout whole range of carrier density because a small number of oxygen atoms are considered to be incorporated in and released from the Tl-O double layer [3,10], preserving the BaO-CuO_2-BaO layers unchanged. This makes a sharp contrast with the hole doping mechanism by Sr substitution for La in the La-Sr-Cu-O system which is essencially a solid-solution system. One interesting open question as to why the T_c for Tl2201 phase is exceptionally so high in comparison with other systems having Cu-O octahedrons might be related to this difference.
In this paper, some recent results on transport and magnetic properties of Tl2201 phase will be reported.

EXPERIMENTAL

Single phase polycrystalline samples were prepared by a conventional powder method similar to those discribed in [1]. Samples sintered in oxygen atmosphere were then reduced in argon atmosphere at 300-590°C for 5 hours, resulting in T_c increase and weight loss. Following annealing in oxygen atmosphere recovered both T_c and sample weight, confirming the complete reversibility of the process. Resistivity and Hall effect were measured by van der Pauw method using a disk sample of about 10 mm dia.× 0.5 mm thick under a magnetic field of up to 8 T. Meissner effect and normal susceptibility were measured using a Quantum Design MPMS SQUID magnetometer under magnetic fields of 0.01 and 1 T, respectively.

RESULTS AND DISCUSSION

Resistivities and Hall coefficients for samples with various T_c's are shown in Fig.1. All data in Fig.1 were obtained using a particular single sample whose T_c was repeatedly varied by annealing in argon or oxygen. As T_c increased, both normal resistivity and positive Hall coefficient increased, which suggested the decrease in hole carrier density. Hall number per Cu atom ($V_0/R_H e$) showed a T-linear behavior which has been observed often in high-T_c materials. Hall number at 120-130K, where the Hall coefficients showed maximum values, was smoothly decreased from ~1.0 to ~0.5 as T_c increased from ~0K to ~80K. This change in Hall number was much larger than the estimated value of ~0.2 from the decrease in oxygen content (~0.1). Moreover, considering that the present sample was a sintered body of 88% density [2], single crystal samples are expected to show smaller Hall coefficients, making the discrepancy much larger. Such discrepancy means that a simple concept of chemical doping is no longer applicable in a carrier-rich region. Such a tendency has also been observed in La-Sr-Cu-O system [8].

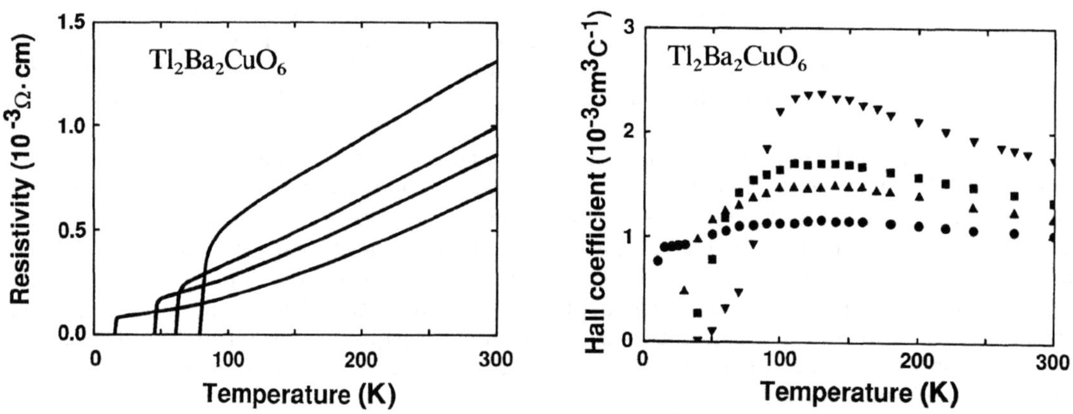

Fig. 1. Temperature dependences of (a) resistivity and (b) Hall coefficient for $Tl_2Ba_2CuO_6$ with various Tc's.

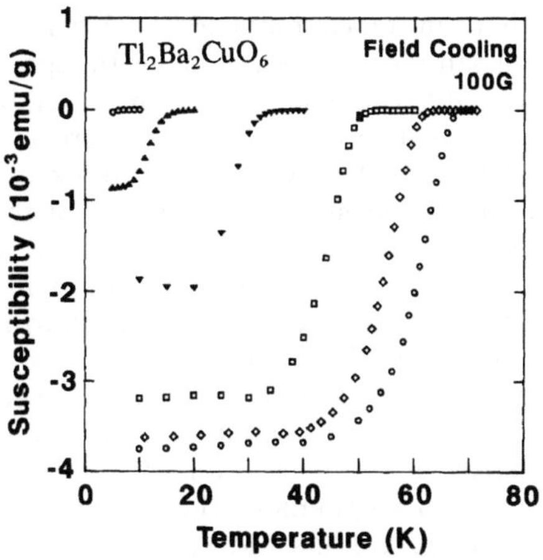

Fig. 2 Meissner effect for $T_2Ba_2CuO_6$ with various T_c's.

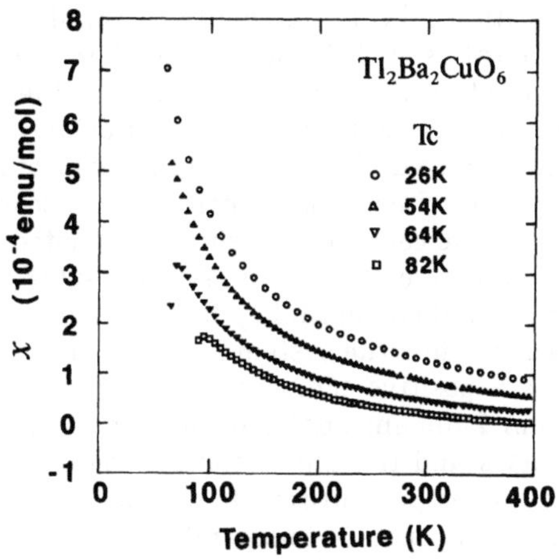

Fig. 3 Temperature dependence of the magnetic susceptibility for $Tl_2Ba_2CuO_6$ with various T_c's.

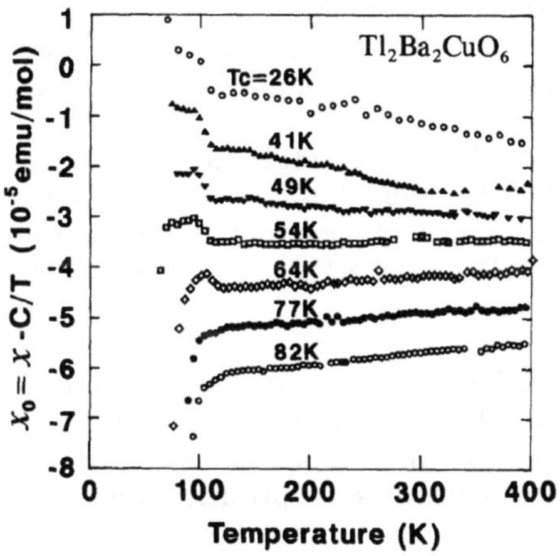

Fig. 4 Temperature dependence of the magnetic susceptibility for $Tl_2Ba_2CuO_6$ with various T_c's after subtracting the Curie term.

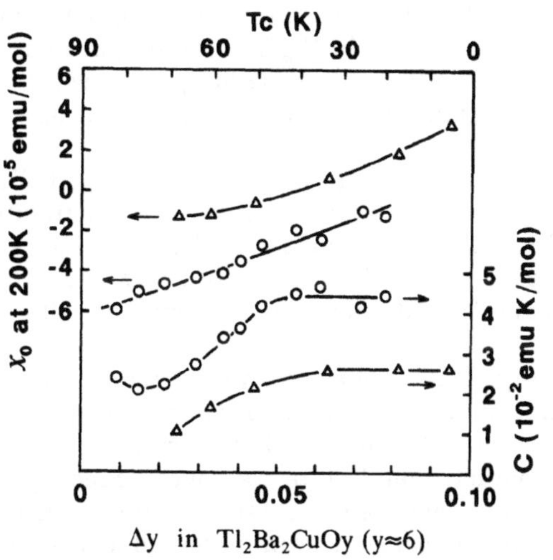

Fig. 5 χ_0 at 200K and the Curie constant for $Tl_2Ba_2CuO_6$ plotted against the relative change in oxygen content Δy. Circles and triangles represent two separately prepared samples.

Meissner effects for samples with various T_c's are shown in Fig.2. The superconducting volume fraction reached to 35-40% for samples with $T_c \geqq 60K$, but it showed a remarkable tendency to decrease toward to 0 as the T_c decreased to 0K. Similar behaviors have been observed in other systems including La, Y, and Bi systems [11]. Thus, it is indeed a universal tendency that the superconducting volume fraction approximately scales to the transition temperature when the system's T_c is changed by doping. This makes the concept of "optimum carrier density" much more significant because it is only at the optimum carrier density where the perfect occurrence of superconductivity is achieved. There still remains a possibility that the optimum carrier density might be related to some parameters of the crystal structure.

Figure 3 shows temperature dependent normal susceptibilities. They all exhibited Curie-like behaviors and were well described as $\chi = \chi_0 + C/T$, where C was the Curie constant and χ_0 was an nearly temperature-independent term. We subtracted the Curie term so as that the temperature dependence of χ_0 became minimum. The obtained χ_0's were almost temperature-independent as shown in Fig.4. C and χ_0 values for samples with various T_c's are plotted against the relative change in oxygen content in Fig.5, where the relative change in oxygen content was estimated from T_c using the linear relationship reported in [3]. In Fig.5, circles and triangles represent two separately prepared samples. Since the Curie constant depends on the sample preparation, it seems to be not intrinsic but due to impurities

like $BaCuO_2$ with local Cu^{2+} moments. Assuming that the Curie term comes from Cu^{2+} with spin 1/2 and g=2.2, the ratios of Cu^{2+}/Cu are about 10% and 6% for both samples, respectively. Although X-ray diffraction patterns for both samples contained no impurity peaks, this may not be unreasonable because of the low scattering factor for $BaCuO_2$. The decreasing Curie constant with decreasing oxygen content is also understood as a behavior of the reduced $BaCuO_2$. However, we cannot exclude the possibility of intrinsic small Curie-like contribution because it is impossible to separate out that from the total Curie constant. It should be noted that Curie-like behaviors have been observed for samples with excessive carriers in La-Sr-Cu-O [8,9] and Bi-Sr-Ca-Cu-O [12]. According to previous works on the normal susceptibility for La [8,9], Y [13] and Bi [12] systems, we can see the following general tendency when the system is doped. For lower carrier densities, the normal susceptibility is small and has a positive temperature coefficient, which is interpreted as a tail of a broad peak originated from 2-dimensional antiferromagnetic correlations between Cu^{2+} spins. As the carrier density increases and the antiferromagnetic interaction is destroyed, the susceptibility gradually increases and the temperature coefficient decreases. At the optimum carrier density, the susceptibility becomes almost temperature-independent for Y and Bi systems while it still has a positive slope for La system. For higher carrier densities, a Curie-like behavior begins to appear in La and Bi systems, which seems to be due to the remaining Cu^{2+} spins released from the antiferromagnetic interactions. Here, we can see a similar tendency in Fig.4, where the susceptibility increases and the temperature coefficient changes from positive to negative as the carrier density increases. (Small jumps at 110-100K seem to be due to a trace of ferromagnetic impurity.) The same tendency was also observed for another sample shown in Fig.5 (triangle), although the values are somewhat larger. Therefore, the normal susceptibility of the Tl2201 phase seems to show essencially the same behavior as those of other systems, and thus it is an important question for future studies whether an intrinsic Curie term appears or not.
Considering the core diamagnetic susceptibility of -1.93×10^{-4} emu/mol, the temperature independent paramagnetic susceptibility is estimated as 1.3-1.7×10^{-4} and 2.0-2.3×10^{-4} emu/mol for T_c=85K and T_c=0K, respectively. These values are also close to the corresponding values (at 200K) for La system which also consists of Cu-O octahedrons; 1.7-1.8×10^{-4} and 2.2-2.4×10^{-4} emu/mol for T_c=40K and T_c=0K (metallic), respectively [8,9].

In summary, Tl2201 phase showed a remarkable phase transition from a normal metal to a 85-K superconductor as a small number of oxygen atoms were removed by 0.10-0.15 per formula unit. The Hall measurement qualitatively confirmed the corresponding decrease in hole carrier density, showing that the normal metal state was over-doped by oxygen. The normal susceptibility was almost temperature-independent after subtracting the Curie term which seemed to be due to impurities. The temperature-independent term increased as the carrier density increased, which was essencially the same behavior as observed in other oxide superconductors. The Meissner volume fraction showed a remarkable decrease as the T_c decreased to 0K. Such a behavior is quite universal in high-T_c materials and suggests that the full appearance of superconductivity is rather restricted to the vicinity of the optimum carrier density.

REFERENCES

1. Y. Shimakawa, Y. Kubo, T. Manako, T. Satoh, S. Iijima, T. Ichihashi and H. Igarashi, Physica C 157,279(1989).
2. Y. Kubo, Y. Shimakawa, T. Manako, T. Satoh, S. Iijima, T. Ichihashi and H. Igarashi, in Proceedings of M^2S-HTSC, Stanford, California, 1989 [Physica C(to be published)].
3. Y. Shimakawa,Y. Kubo,T. Manako and H. Igarashi, Phys. Rev.B (in press).
4. J. B. Torrance et al., Phys. Rev. Lett. 61,1127(1988).
5. T. Manako, Y. Shimakawa, Y. Kubo, T. Satoh and H. Igarashi, Physica C 158,143(1989).
6. R. G. Buckley et al., Physica C 156,629(1988).
7. J. Zhao and M.S. Seehra, Physica C 159,639(1989).
8. H. Takagi et al., Phys. Rev. B 40,2254(1989).
9. J. B. Torrance et al., Phys. Rev. B (in press).
10. J. B. Parise et al., physica C 159,239(1989).
11. Y. Kubo and H. Igarashi, in Mechanisms of High Temperature Superconductivity, eds. H. Kamimura and A. Oshiyama(Springer-Verlag, Berlin,1989)pp.313-321.
12. T. Tamegai et al., Jap. J. Appl. Phys. 28,L112(1989).
13. Y. Nakazawa and M. Ishikawa, Physica C 158,381(1989).

Microwave Resistance of $YBa_2Cu_3O_y$ Ceramics Caused by Weak Links

M. SATO, T. KONAKA, and K. ISHIHARA

NTT Transmission Systems Laboratories, Tokai-mura, Ibaraki, 319-11 Japan

ABSTRACT

Microwave surface resistance Rs of $YBa_2Cu_3O_y$ ceramics with extremely different sized grains (~μm,~10μm, and ~mm) is measured using a TE_{011} cavity at 17GHz. Rs is also measured under a magnetic field (<220 Oe). It is shown that Rs is the sum of Rs_{grain} (caused by superconducting grains), $Rs_{weak\text{-}link}$ (weak links between the grains), and $Rs_{non\text{-}super}$ (non-superconducting phases and surface imperfections). Rs_{grain} is dominant in the large-sized grain sample just below Tc and is thought to be much lower than the measured Rs in lower temperature regions. $Rs_{weak\text{-}link}$ is dominant in the small-sized grain sample.

KEY WORDS: microwave, surface resistance, high Tc superconductor, $YBa_2Cu_3O_y$ ceramics, weak links

INTRODUCTION

According to the BCS theory, high Tc superconducting materials are expected to be suitable for microwave applications because of their large energy gaps, which raise the upper frequency limitation of these applications, and also because of their large critical field, which enables high power applications. Moreover, surface resistance Rs in high Tc superconductors is expected to be lower than in conventional low Tc superconductors.

Ceramic $YBa_2Cu_3O_y$, however, has a surface resistance higher than copper above 10GHz, even at liquid He temperature, so far [1-3]. The energy gap ratio $2\Delta/kTc$ determined from Rs temperature dependence using the Mattise-Bardeen relation [4] is much smaller than the BCS theoretical value [5]. While these disagreements still exist, the reported residual Rs values in ceramic samples seem to follow the quadratic frequency dependence expected by BCS theory [6].

It is questionable whether the measured Rs represents the intrinsic property of high Tc superconductors, especially in sintered ceramic samples. Though existence of grain boundaries has been mentioned to explain a large residual Rs in $YBa_2Cu_3O_y$, clear evidence of the grain boundary influence in surface resistance has not been demonstrated [7-11].

To clarify the influence of grain boundaries in surface resistance in $YBa_2Cu_3O_y$ ceramics, we measured Rs of the three ceramic samples with extremely different grain sizes (~μm,~10μm, and ~mm). The grain size variation was expected to cause the variation in boundary density. We also measured the Rs under magnetic fields to identify the existence of grain boundaries.

We found that the temperature dependence of each sample was quite different and that the magnetic field enhanced the differences. We show that the temperature dependence in the measured Rs comprises three parts: Rs of the intrinsic superconducting grains, Rs of the weak link between grains, and Rs of the non-superconducting phases and surface imperfections.

SAMPLE PREPARATION

Three samples with extremely different sized grains were prepared. To prepare the small-sized grain sample (~μm in diameter), the raw powder was prepared using coprecipitation method. The powder coprecipitated from a nitric solution was dryed at 400°C in air and then calcined. The middle-sized grain sample (~10μm) was prepared by an ordinary mixed powder method. The large-sized grain sample (~mm) was made using the partial melt method described in ref. [12]. Raw powder preparation methods, calcining and sintering temperatures, and grain sizes are shown in Table 1. All heat treatments were carried out in an O_2 flow atmosphere.

Table 1. Sample preparation

Powder	Calcining temperature	Sintering temperature	Grain size
(a) Coprecipitation method	925°C	975°C	~μm
(b) Mixed powder method	950°C	950°C	~10μm
(c) Mixed powder method	940°C	1100°C	~mm

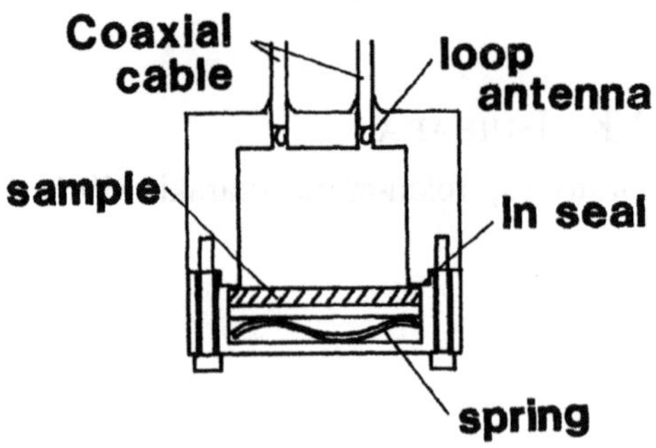

Fig. 1. A 17GHz cylindrical TE_{011} cavity. The sample is the bottom surface of the cavity.

The calcined powders were ground and size-selected by passing through a mesh (#500) for the fine powder, then pressed into disks and sintered. The sintered disks were shaped into an appropriate size for measurement (about 3mm in thickness and 26mm in diameter). One side of each disk was polished to achieve a flat surface with dry polishing sheets and washed in alcohol. Finally, before measurement, the disks were annealed at 950°C in O_2 flow for 5 hours and cooled at a rate of 0.5 deg/min.

Grain sizes were measured from SEM photographs. It was ascertained that there was no remarkable difference in the X-ray diffraction patterns, including grain axis orientation. It is natural to think that electrical property variation in these samples is related to the grain size variation.

MEASUREMENT

Surface resistance was measured by a transmission type cylindrical TE_{011} mode microwave cavity with a resonant frequency of 17GHz [13]. Figure 1 shows the cavity having one end plate used as a sample. The microwave field is exited and received using small loop antennas located on the coaxial cable ends. The cavity is made of gold plated copper. An external magnetic field is applied perpendicularly to the sample surface. Because the microwave electrical field in the TE_{011} mode has only a circular component, electrical contact is not necessary between the sample and the cavity wall.

Surface resistance Rs is determined from the cavity Q value with following relations.

$$Q_0 = \pi f_0 \mu r\{1+(\pi/p_{01})^2(r/h)^2\}/\{R_{cavity}+(R_{cavity}+Rs)(\pi/p_{01})^2(r/h)^3\},$$
$$Q_0 = (1+\beta) Q_L,$$

where Q_0 is the unloaded Q and Q_L is the loaded Q. Q_L is obtained as a ratio of resonant center frequency f_0 and FWHM Δf_0. Rs_{cavity} is the surface resistance of the cavity wall, β is the coupling constant, μ is the magnetic permeability of the vacuum, r is the cavity radius and h is the cavity height. p_{01} is the cavity mode constant; $p_{01}=3.832$.

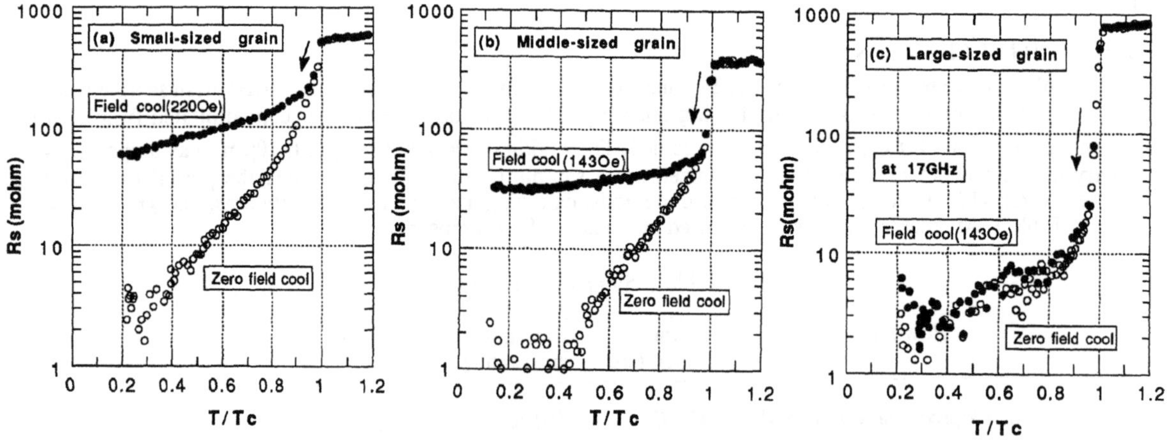

Fig.2. Temperature dependencies of surface resistance at 17GHz in (a) the small-sized grain sample prepared by the coprecipitation method, (b) the middle-sized grain sample made by the ordinary mixed powder method and (c) the large-sized grain sample prepared by the method described in ref. [12]. Open circles represent the zero field cooling data and filled circles represent the cooling data under magnetic fields high enough to brake the weak links.

RESULTS

Zero field cool

Figure 2 shows temperature dependence of Rs in the three samples. Zero field cool surface resistances are plotted with open circles in Fig.2. From room temperature down to Tc, Rs shows a slow linear decrease. Rapid decrease in Rs is observed just below Tc. We can observe that the temperature dependence is different in each sample. For the Rs drops near Tc, in the small grain sample, the drop is about 30% of $Rs_{(Tc)}$; in the middle grain sample, the drop is about 80%; and in the large grain sample, the drop is more than 95%. In the ensuing low temperatures, a relatively slow Rs decrease is observed. In this region, logarithmic Rs is almost proportional to temperature. The gradient increases with decreasing grain size.

Field cool

Weak links are broken by an external magnetic field. We can detect the Rs increase caused by the break of weak links in the sample surface [13]. Therefore, Rs measurement under a magnetic field clarifies the existence of weak links between grains.

The results of Rs measurements in magnetic fields that are high enough to break weak links are shown in Fig.2 with filled circles. The field required to break the weak links is determined using an Rs-H curve of the initial field application as the point where the Rs rapid increase stops [13], that will be mentioned later.

There is no increase in Rs above Tc and in the sharp drop in Rs near Tc under a magnetic field. The largest Rs increment and residual Rs were observed in the small grain sample at low temperatures ($T/Tc<0.8$). The level where the Rs starts to increase was nearly equal to the level where the sharp Rs drop ended. In the large grain sample, however, there was no clear increase.

Field dependence

The field strength needed to break weak links depends on the grain size. In Fig.3, Rs changes under an external field in the initial application are shown at several temperatures. The Rs-H curve in the middle grain sample is a typical pattern for ordinal ceramic samples. The difference in Rs changes separates the curve into two regions. In a low field, Rs increases rapidly with increasing field strength. In the ensuing high field, Rs increases slowly. We define the magnetic field where the rapid Rs increase stops as H_J. It is shown that H_J shifts linearly to a larger value as the temperature decreases, and that H_J is the field strength needed to break weak links in the sample surfaces [13].

The H_J in the small grain sample is larger than that in the middle grain sample and is too large to identify at 25K because of our limited coil current capacity. In field cool Rs data in this sample, the external field was not high enough to rise over H_J below 40K. However, the external field was expected to be strong enough to break weak links because of field trapping. While in the large grain sample, H_J is very small.

DISCUSSION

Because the small grain sample has a large number of links, a strong magnetic field dependence is expected. In fact, the Rs in the small grain sample is influenced most by the magnetic field. Since the Rs increase occurred only in the low temperature region ($T/Tc<0.8$), the linear temperature dependence of logarithmic Rs is caused by weak links.

The influence of weak links in the middle grain sample is less than that in the small grain sample. The Rs in low temperatures, however, is thought to come from weak links. The f^2 law in ceramic $YBa_2Cu_3O_y$ is perhaps caused by the weak links.

In the large grain sample, weak links exert only a small influence relative to the small number of grains. Though the influence of weak links is almost removed, the temperature dependence curve of Rs at low temperatures is too flat to conclude that it is caused by an intrinsic Rs of $YBa_2Cu_3O_y$. It is natural that non-super conducting phases in the sample have this flat temperature dependence. In addition, surface imperfections may raise the residual Rs simultaneously.

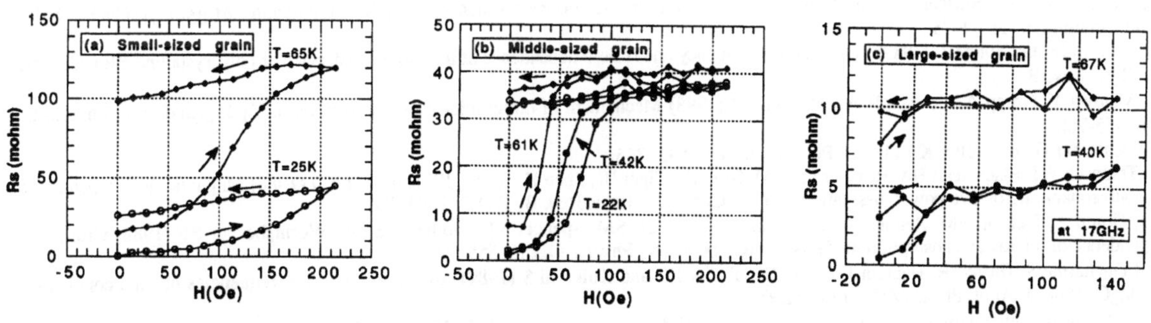

Fig. 3. Magnetic field dependence of surface resistance in the initial cycle for (a) the small-sized grain sample, (b) the middle-sized grain sample and (c) the large-sized grain sample. The field is applied perpendicularly to the sample surface.

The sharp Rs drop is thought to result from intrinsic properties of the superconducting grain. Because Rs increase in a magnetic field is very small near Tc, the Rs drop is not related to the weak links. Besides, near Tc the intrinsic grain Rs is larger than the Rs due to weak links or Rs due to non-superconducting phases. Therefore, Rs in the sharp drop region is expected to show intrinsic grain surface resistance. We can estimate the Rs at liquid N_2 temperature for the large grain sample, which has the largest Rs drop, using a curve fitting method with Tinkham's complex conductivity expression [4]. A very roughly estimated Rs is $5 \times 10^{-3} \Omega$. This estimation suggests that the ceramic $YBa_2Cu_3O_y$ has a lower Rs than copper at 17GHz at liquid N_2 temperature. Moreover, we expect that c-axis oriented grains have one order lower Rs.

The grain size dependence of H_J is estimated to be due to the number of weak links again. In the small grain sample, an external field penetrates into the ceramic sample homogeneously because the paths for flux movement exist as a large number of grain boundaries. Therefore, the local field in the sample is almost equal to the external field. In the large grain sample, the small number of grain boundaries limits the flux paths. Therefore, the fluxes condense in the limited grain boundaries, and the local field around the grain boundary where the weak links exist becomes higher than that in the small grain sample. This difference of the local field near the weak links causes the grain size dependence of H_J.

Finally, according to our results, the measured surface resistance consists of three components:

$$Rs_{measured} = Rs_{grain} + Rs_{weak\text{-}link} + Rs_{non\text{-}super.}$$

The first is the intrinsic Rs of the superconducting grains, Rs_{grain}, which appears in the Rs sharp drop area near Tc. The second is the Rs caused by weak links, $Rs_{weak\text{-}link}$, which is dominant below Tc, except near Tc, and is dominant in small grain samples. The third is the Rs from the non-superconducting phase, $Rs_{non\text{-}super}$, which is found in the large grain sample, is not magnetic dependent, but is metallic temperature dependent.

For microwave applications, we should investigate how to prepare low Rs $YBa_2Cu_3O_y$ in magnetic fields and at liquid N_2 temperature. Our results seem to show a way to achieve this, that is, to prepare weak link-free material. Making a single crystal is one of the final solutions. In this case, it becomes important to remove non-superconducting phases that are usually needed as a flux to grow crystals. Making links strong between grains is another one with improving the process of making ceramics.

CONCLUSIONS

We measured temperature dependence and magnetic field dependence of surface resistance in $YBa_2Cu_3O_y$ ceramics. The following experimental results have been obtained.

1. The Rs drop near Tc is large in the large grain sample.
2. The Rs increase induce by magnetic field is large in the small grain sample.
3. H_J is large in the small grain sample.
4. The Rs change in the low temperature region ($T/Tc<0.8$) is large in the small grain sample.

The pattern variation in the Rs-T curves in ceramic $YBa_2Cu_3O_y$ samples is caused by the variation of weak link density, which becomes larger with decreasing grain size.

REFERENCES

[1] Bohn CL, Delayen JR and Lanagon MT (1989) RF measurements on high-Tc superconductors., Proceedings of workshop on high-Tc superconductivity, Huntsville, AL

[2] Piel H, Hein M, Klein N, Klein U, Michalke A, Muller G and Ponto L (1988) Superconducting perovskites in microwave fields., Physica C, 153-155: 1604-1609

[3] Sridhar S, Shiffman CA and Hamdeh H (1987) Electrodynamic response of $Y_1Ba_2Cu_3O_y$ and $La_{1.85}Sr_{0.15}CuO_{4-\delta}$ in the superconducting state., Phys. Rev. B, 36(4): 2301-2304

[4] Mattis DC and Bardeen J (1958) Theory of the anomalous skin effect in normal and superconducting metals., Phys. Rev., 111(2): 412-417

[5] Kennedy WL and Sridhar S (1988) Low temperature microwave surface resistance of $Y_1Ba_2Cu_3O_y$ and $La_{1.85}Sr_{0.15}CuO_4$., Solid State Commu., 68(1): 71-75

[6] Delayen JR and Bohn CL (1989) Temperature, frequency and rf field dependence of the surface resistance of polycrystalline $YBa_2Cu_3O_{7-x}$., Phys. Rev. B,

[7] Maniwa Y, Grupp A, Hentsch F and Mehring M (1988) Mechanism of direct microwave absorption in $Y_1Ba_2Cu_3O_{7-y}$., Physica C, 156: 755-760

[8] Deutscher G and Müller KA (1987) Phys. Rev. Lett., 59: 1745

[9] Tomasch WJ, Blackstead HA, Ruggiero ST, McGinn PJ, Clem JR, Shen K, Weber JW and Boyne D (1988) Magnetic field dependence of nonresonant microwave power dissipation in $YBa_2Cu_3O_{7-x}$., Phys. Rev. B, 37(16): 9864-9867

[10] Gould A, Jackson EM, Renouard K, Crittenden R, Bhagat SM, Spencer ND, Dolhert LE and Wormsbecher RF (1988) Grain size dependence of microwave absorption in $Y_1Ba_2Cu_3O_7$ powders., Physica C, 156: 555-558

[11] Ducharme S, Durny R, Hautala J, Zheng DJ, Taylor PC and Kulkarni S (1989) Absorption at radio frequencies in superconducting $Y_1Ba_2Cu_3O_y$., J. Appl. Phys., 66(3): 1252-1260

[12] Salama K, Selvamanickam V, Gao L and Sun K (1989) High density in bulk $YBa_2Cu_3O_x$ superconductor., Appl. Phys. Lett., 54(23): 2352-2354

[13] Sato M, Konaka T and Sankawa I (1989) Microwave surface resistance of YBaCuO ceramics in low magnetic fields., Extended abstracts of ISEC'89: 247-250

Hall Effect of a $YBa_2Cu_3O_{7-\delta}$ Epitaxially Grown Thin Film

T.R. NICHOLS[1], K. MURATA[1], I. ITOZAKI[2], and Y. NISHIHARA[1]

[1] Electrotechnical Laboratory, 1-4, Umezono 1-chome, Tsukuba, Ibaraki, 305 Japan
[2] Sumitomo Electric Industries, Itami, Hyogo, 664 Japan

ABSTRACT

We present measurements of the Hall coefficient and resistivity in the ab-plane of an epitaxially grown YBCO thin film. The resistivity is found to be only approximately linear in T and the calculated mobility is found to vary strictly as $T^{-1.5}$. We use the mobility as a parameter with which to compare other published work. The Hall number varies exponentially with 1/T to within 10% of T_c. We find the existence of a small (0.022eV) activated energy gap and find the number of carriers/ unit cell at T=0 K to be 1.1 ± 0.05.

KEYWORDS: Hall Effect, Resistivity, Mobility, Activated Energy Gap

INTRODUCTION

The experimental investigation of the properties of high T_c superconductors has revealed many unusual and interesting characteristic properties in both the superconducting and normal state. The normal state transport properties reveal much about the properties of the charged carriers and collision processes, the interpretation of which is essential in understanding the mechanism of superconductivity. In this work we present measurements of the Hall coefficient and resistivity in the ab-plane of the $YBa_2Cu_3O_{7-\delta}$ compound. Whilst the temperature dependence of the resistivity and Hall coefficient have been determined qualitatively with ceramic samples [1], high quality single crystals and thin films have provided more precise quantitative information and showed the existence of anisotropy between the ab-plane and c-axis [2]. The sample used in this work is an epitaxially grown single crystal film, the ab-plane lying parallel to the MgO substrate. The film thickness is 3100 Å. The RHEED pattern suggests a very flat ab-plane (on the atomic scale) whilst the X-ray diffraction pattern shows no ab component along the c-axis and very little tilt within the film plane.

EXPERIMENTAL

We present a brief description of the experimental details. Further details will be described in a separate paper. Contact to the sample was provided by 50 μm gold wires attached by indium to 0.5 mm square pads arranged in an 8-terminal configuration pattern. The sample was aligned with the magnetic field applied parallel to the c-axis of the sample. The experiment was performed twice for each value of applied dc current in the range 0.03-1mA. The sample resistivity was measured with no applied magnetic field and the experiment repeated with an applied magnetic field. Over the range of current values no difference in the values of the measured Hall resistance was observed. The Hall resistance can in principle be extracted simply by taking the difference in measured value of that obtained in zero field to the measured value obtained in 5 tesla. However we must take account of the possible contribution of magnetic field to the longitudinal resistance. This magnetic contribution to the longitudinal resistance, and hence resistivity, was estimated from the measurements to be $\Delta\rho/\rho = 10^{-4}$. Since the width across the Hall voltage terminals is some 50 times less we estimate the magnetoresistance contribution to the Hall resistance to be less than 0.003% of the measured value. This value compares with an experimental uncertainty of 5% in the measured Hall resistance.

DISCUSSION

The resistivity data, shown in Fig. 1, is of particular interest. A comparison of the absolute magnitude of the resistivity at 100 K and 250 K with the values obtained by other work is presented in Table 1. Whilst there is large variation in resistivity values between all the published values listed in Table 1, the resistivity of our sample is particularly low. The width of the superconducting transition ΔT_c is less than 0.2 K which is further evidence for the quality of this particular sample. We point out that the resistivity is **not** strictly linear in temperature as has been previously believed.

Fig. 1. Resistivity in the ab-plane ρ_{ab} versus temperature T.

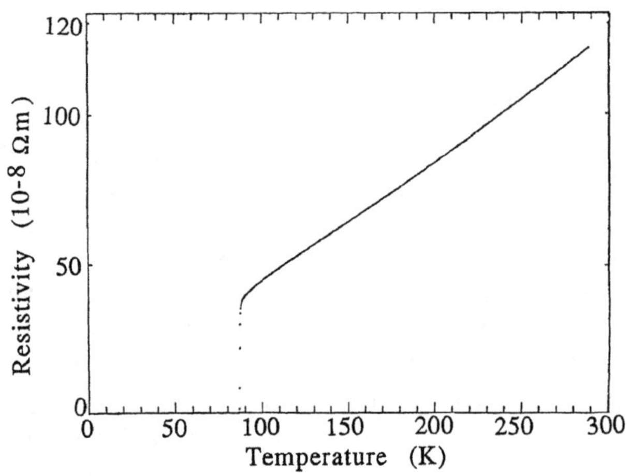

Table 1. Comparison of sample resistivity values. Temperature in kelvin and resistivity in units of 10^{-8} Ωm.

Sample	Resistivity values in the ab-plane 100K	250K	T_c	Refs
Ceramic	320	710	85	1.
Single Crystal	200	520	84	3.
Single Crystal	80	200	92	4.
Thin Film	100	220	87	5.
Epit. Thin Film	400	810	90	6.
Epit. Thin Film	42	107	87	This Work

The positive 'hole like' Hall coefficient R_H (Fig. 2) was found to vary roughly linear with inverse temperature in the range 115 K to 300 K. However the value of R_H is smaller than that obtained on ceramic and other single crystal work [1,4-6], in contrast with the much lower values of resistivity.

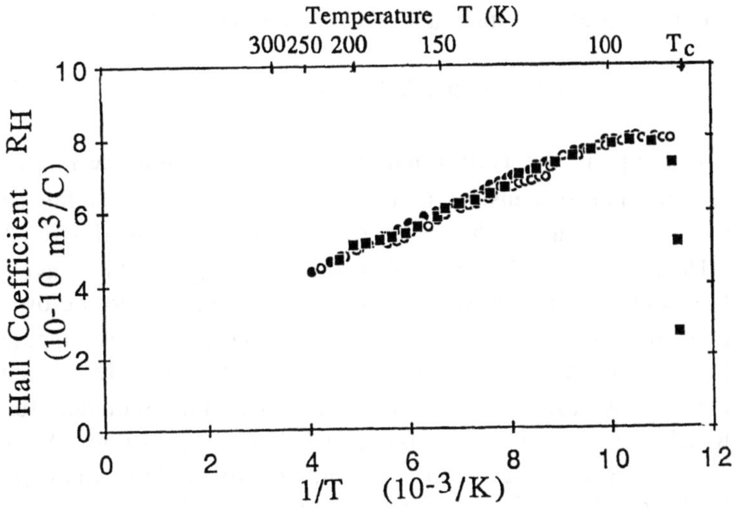

Fig. 2. The Hall Coefficient R_H versus inverse temperature 1/T. Note the deviation from linearity around 120 K.

We use now the values of Hall coefficient R_H and resistivity ρ_{ab} to calculate the mobility μ

$$\mu = R_H / \rho_{ab}. \qquad 1$$

If we take the approach of a resistivity proportional to T and R_H proportional to 1/T we would expect the mobility to be proportional to the square of the inverse temperature. In contrast to the above argument we find the in-plane mobility to depend on $T^{-1.5}$ over the temperature range 300 K to below 100 K (Fig. 3). Presumably this results from

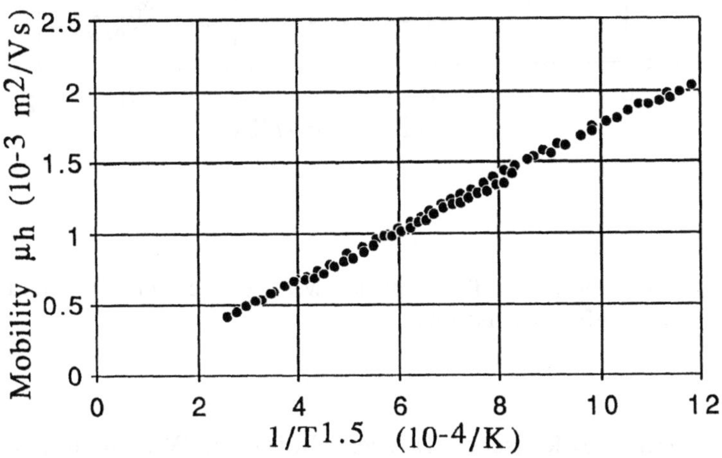

Fig. 3. The in-plane mobility μ plotted versus $T^{-1.5}$. A linear dependence is observed to below 100 K (above 10^{-3} on the x-axis).

the non-linearity of ρ_{ab} and the deviation in linearity of Hall number around 120 K. Moreover the mobility values of other groups [1,4,6] may, within experimental scatter, be fitted to the same power law dependence. Mobility values from Ref 4 are found to be in excellent agreement with the results of this work and to a good approximation may be fitted by the same functional form. On this basis we suggest that the mobility may be used as a convenient scaling parameter with which the data from various works may be compared.

We return now to the problem of a temperature dependent Hall coefficient. Within a free electron-like model we may assume there exists some localized state from which carriers can transfer to the carrier band so that the carrier density can increase with temperature. Such a mechanism suggests the existence of an energy gap between a possible localized state and the carrier band. We find we can fit our data by

$$n_H = \alpha \exp(-\Delta/T) + n_o \qquad 2$$

where n_H is the carrier number density per unit cell (with volume V_o=176 Å^3), n_o is the carrier number density/ unit cell at T=0K and Δ is the activated energy gap. To obtain a fit we have varied the value of n_o and find that above 100 K n_o lies in the range 1.05 to 1.15 holes/unit cell. This fit is shown in Fig. 4. We have also chosen to plot the results of this analysis by chosing values of n_o above and below our fitted value of 1.1 to show that in this range curvature in the data is readily apparent. We make the comment that this is not an unreasonable value for n_o and lends support to the analysis presented here. In addition, from this analysis we find evidence for the existence of an energy gap which we estimate to be 0.022 ± 0.003 eV in value. We note that the value we obtain for an activated energy gap is some 20 times less than the value of 0.5 eV obtained by angle resolved photoemission experiments [7].

Fig. 4. Determination of the activated energy gap Δ and carrier density per unit cell at T=0 K. A value of n_o = 1.1 carriers/unit cell (at T=0K) was used as the best value for a straight line fit. Values of n_o=0 and n_o=1.25 are shown to illustrate the deviation from linearity using different values of n_o. The value of Δ = 0.022 eV.

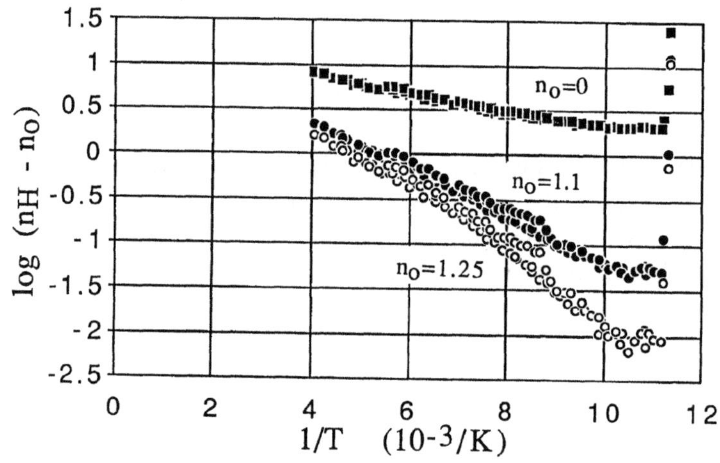

ACKNOWLEDGEMENTS

We would like to acknowledge the kind help of T.Komazaki with the experiment and Dr.K.Yamaji and Dr.N.Fortune for many useful discussions.

REFERENCES

[1] U.Gottwick, R.Held, G.Spard, F.Steglich, H.Rietschel, D.Ewert, B.Renker, W.Bauhofer, S. von Molnar, M.Wilhelm and H.E.Hoenig, Europhys. Lett., **4**, 1183-1188 (1987).
[2] S.W.Tozer, A.W.Kleinsasser, T.Penney, D.Kaiser and F.Holtzberg, Phys.Rev.Lett. **59**, 1768 (1987)
[3] L.Forro, M.Raki, J.Y.Henry and C.Ayache, Solid State Comm. **69**, 1097 (1989).
[4] T.Penney, S. von Molnar, D.Kaiser, F.Holtzberg and A.W.Kleinsasser, Phys.Rev **B38**, 2918 (1988)
[5] A.I.Golovashkin, S.I.Krasnosvobotsev, I.V.Kucherenko and E.V. Pechen', JETP **11**, 27 (1988).
[6] P.Chaudhari, R.T.Collins, P.Frietas, R.J.Gambino, J.R.Kirtley, R.H.Koch, R.B.Laibowitz, F.K.LeGoues, T.R.McGuire, T.Penney, Z.Schlesinger, A.P.Segmuller, S.Foner and E.J.McNiff, Jr., Phys.Rev **B36**, 8903 (1987)
[7] H.Matsuyama, T.Takahashi, H.Katayama-Yoshida, T.Kashiwakura, Y.Okabe and S.Sata, N.Kosugi, A.Yagishita, K.Tanaka, H.Fujimoto and H.Inokuchi, Physica **C160**, 567, (1989).

Hall Effect in Oxide Superconductors Having Fluorite-Type Layers

S. IKEGAWA, M. KOSUGE, T. WADA, A. ICHINOSE, Y. YAEGASHI, K. NAKAO, T. YAMASHITA, T. SAKURAI, and H. YAMAUCHI

Superconductivity Research Laboratory, International Superconductivity Technology Center, 10-13, Shinonome 1-chome, Koto-ku, Tokyo, 135 Japan

ABSTRACT

The Hall effect at temperatures between T_c and 300 K was investigated for the superconducting ceramics of $(Nd,Ce)_2CuO_{4-y}$, $(Ln,Sr)_2CuO_{4-y}$ where Ln=(Nd,Ce) or (La,Sm), and $(Ln,Ce)_4(Ln,Ba,Sr)_4Cu_6O_y$ where Ln=Nd or (Y,La). The sign of the Hall coefficient (R_H) was positive except for $(Nd,Ce)_2CuO_{4-y}$. The magnitude of R_H for all the three compounds exhibits a broad peak around 120-150 K with respect to temperature. It is contrary to the case of the K_2NiF_4- and the 1:2:3-type superconductors, in which R_H increases monotonically with decreasing temperature. The peak structure is not likely due to the magnetic moment of the rare earth elements, but due to the influence of the fluorite layer contained in the three compounds.

KEY WORDS: Hall effect, copper oxide, fluorite, 4f moment

INTRODUCTION

The Hall effect in the high-Tc superconducting oxides has been extensively studied for understanding the normal-state electronic states. The monotonic increase of Hall coefficient(R_H) with decreasing temperature was reported for $(La,Sr)_2CuO_{4-y}$ (T-phase), $YBa_2Cu_3O_{7-y}$ ("123" compounds)[1], Bi-Ca-Sr-Cu-O[2], and $Tl_2Ca_2Ba_2Cu_3O_y$[2], even though all of those compounds showed metallic conduction in the normal state. This feature has been regarded as a characteristic of high-T_c superconducting oxides[3], although the mechanism has not been fully understood.

In the last thirteen months, a new type of oxide superconductors which have fluorite-type layers in their crystal structure has been discovered. It includes $(Nd,Ce)_2CuO_{4-y}$ (T'-phase)[4], $(Nd,Ce,Sr)_2CuO_{4-y}$ (T^*-phase)[5], and $(Nd,Ce)_4(Nd,Ba,Sr)_4Cu_6O_y$ ("446" compounds).[6,7] We have investigated temperature dependence of the Hall effect in the ceramic samples of these superconductors. All the R_H-vs-T curves showed a common feature, which was different from those of the T-phase and the "123" compounds. In this paper, we compare the experimental results with those of the T-phase and the "123" compounds and discuss the origin of this anomalous feature.

EXPERIMENTAL

All the samples were prepared by a solid state reaction. From powder X-ray diffraction, all the samples were of single phase, except for $(La,Sm,Sr)_2CuO_4$, in which a small impurity phase was detected. Detailed preparation conditions and characterizations are given elsewhere.[7-10] The resistivity of the sample was measured by a standard four-probe method. In this paper the superconducting transition temperature T_c is defined as the midpoint of resistive transition. For the Hall measurements, sintered pellets were thinned to plates with a typical dimension of $7\times2\times0.15$ mm^3. The electrical contacts were made using gold paste being followed by a heat treatment, except for the case of T'-phase in which contacts were formed by In solder. The Hall effect was measured using a specially designed automatic measuring unit, in which a superconducting solenoid and a 10nV-resolution digital voltmeter are incorporated. The applied field was ±6 T and the applied current was ±50 mA. Temperature, which was measured by means of calibrated Pt and carbon-glass resistance sensors, was kept constant within ±0.02 K during one set of measurements.

RESULTS AND DISCUSSION

First of all, we show the results for superconducting compounds without fluorite-type layers. Figure 1 shows R_H of $La_{1.85}Sr_{0.15}CuO_{4-y}$ (T-phase) as a function of temperature together with those of the samples in which Cu was partly substituted by Zn.[10] The magnitude of R_H increases linearly with decreasing temperature. Figure 1 also shows R_H for two "123" compounds of $Y(Ba_{1-x}Sr_x)_2Cu_3O_{7-y}$. R_H increases monotonically with decreasing temperature and behaves as $R_H^{-1}=A+BT$, as was pointed out by Cheong et al.[1] Clayhold et al.[3] reported that as the amount of dopant increased, the T_c was reduced and the slope of R_H-vs-T curves was suppressed in Ni- and Co-doped $YBa_2Cu_3O_7$ and Ni-doped $La_{2-x}Sr_xCuO_4$. In Fig. 1, similar behavior is observed in the Zn-doped T-phase and Sr-doped "123" compounds. Thus, the monotonic increase of R_H with decreasing temperature was regarded as a property intrinsic to high-T_c systems.[3] However, another type of R_H-vs-T curves was observed for the recently discovered oxide superconductors.

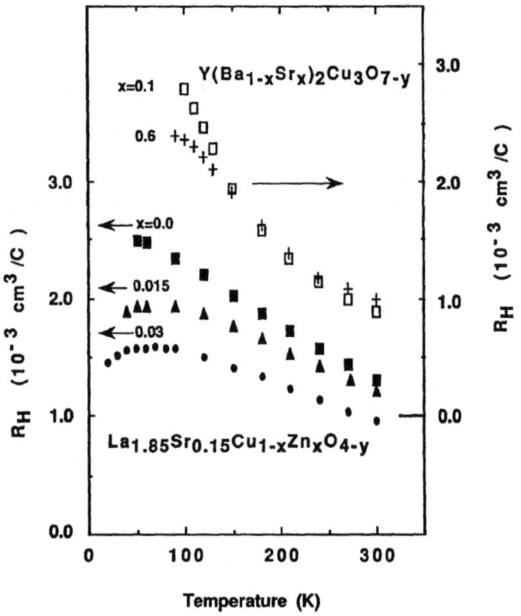

FIG. 1. Temperature dependence of Hall coefficient(R_H) for $La_{1.85}Sr_{0.15}Cu_{1-x}Zn_xO_{4-y}$: x=0($T_c$=38.9K), x=0.015($T_c$=22.7K), x=0.03($T_c$<4K) and $Y(Ba_{1-x}Sr_x)_2Cu_3O_7$: x=0.1(T_c=87.0K), x=0.6(T_c=63.3K).

Figure 2 shows R_H of $Nd_{1.85}Ce_{0.15}CuO_{4-y}$ and $Nd_{1.85}Ce_{0.15}Cu_{0.99}Zn_{0.01}O_{4-y}$ (T'-phase).[10] The R_H's are negative at all the measured temperatures between 30 and 300 K. The R_H-vs-T curves have broad peaks around 130 K. Below 130 K the absolute value of R_H is diminished with decreasing temperature, as Takagi et al. first reported.[4] The R_H of $(Nd_{0.8-x}Ce_{0.2}Sr_x)_2CuO_{4-y}$ (T^*-phase) is shown in Fig. 3. The R_H-vs-T curves have broad peaks at 120-150 K for x=0.0875-0.15, where T_c ranges from below 4 K to 23 K.[9] Figure 3 also shows R_H of $(Nd_{2/3}Ce_{1/3})_4(Nd_{1/3}Ba_{5/12}Sr_{1/4})_4Cu_6O_y$ ("446"). The R_H-vs-T curve also shows a broad peak around 130 K. This behavior is similar to that reported by Tamegai et al.[11] Thus, the R_H-vs-T curves in $(Nd,Ce)_2CuO_{4-y}$, $(Nd,Ce,Sr)_2CuO_{4-y}$ and $(Nd,Ce)_4(Nd,Ba,Sr)_4Cu_6O_y$ commonly possess an extremum in the temperature range of 120-150 K irrespective of the magnitude of T_c.

Why is there a difference between the temperature dependence of R_H in Fig. 1 and that in Figs. 2 and 3? For a direct comparison, we summarize R_H schematically for the former two 'old' compounds (T & "123") and the latter three

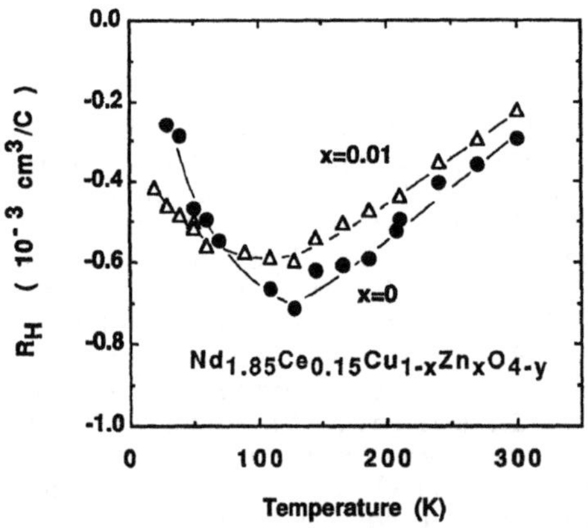

FIG. 2. R_H vs temperature for $Nd_{1.85}Ce_{0.15}CuO_{4-y}$($T_c$=20.1K) and $Nd_{1.85}Ce_{0.15}Cu_{0.99}Zn_{0.01}O_{4-y}$ (T_c=12.5K).

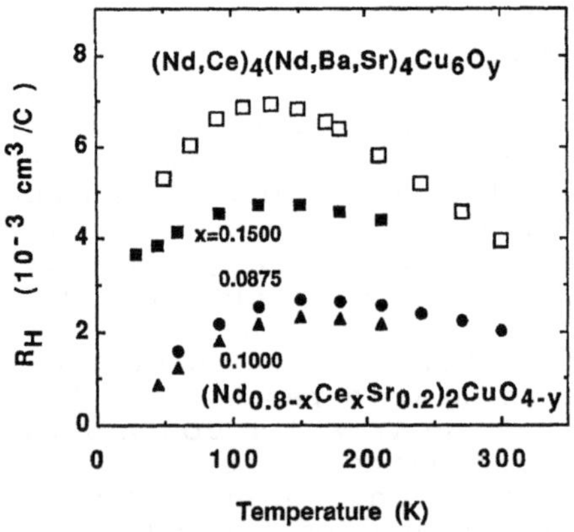

FIG. 3. R_H vs temperature for $(Nd_{0.8-x}Ce_xSr_{0.2})_2CuO_{4-y}$: x=0.0875($T_c$=17.7K), x=0.10($T_c$=23.0K), x=0.15($T_c$<4K), and $(Nd,Ce)_4(Nd,Ba,Sr)_4Cu_6O_y$($T_c$=35.9K).

'new' compounds (T', T^* & "446") in Fig. 4. It seems natural to expect that the broad peak for the 'new' compounds arises from common causes. The following three causes are possible. The first is the 4f moment of Nd atoms. The second is a Ce doping effect. The third is the influence of the fluorite-type layers in their crystal structures. All of these three effects are commonly possible in the 'new' compounds but not in the 'old' compounds. In the following, we examine each of the three causes.

The R_H-vs-T curves for $REBa_2Cu_3O_7$ where RE=Nd or Gd was nearly identical with that of $YBa_2Cu_3O_7$.[12,1] The 4f moment was regarded not to significantly affect transport properties of the normal-state(>90 K) in the "123" compounds. However, the number of rare earth elements per conducting [CuO_2] plane in a unit cell is larger in the 'new' compounds than in "123" compounds. (In T', T^*, "446" and "123", the numbers are 1.85, 1.4, 1, and 0.5, respectively.) Therefore, there is a possibility that the 4f moment of rare earth elements affects the Hall coefficient at low temperatures through certain mechanisms such as the demagnetizing field effect and the skew scattering.[13] In order to investigate the influence of the 4f moment on the normal state conduction, we measured R_H of $(Y_{2/3}Ce_{1/3})_4(La_{1/3}Ba_{1/3}Sr_{1/3})_4Cu_6O_y$, where Nd in the "446" compound given in Fig. 3 is replaced with nonmagnetic La and Y. This compound exhibited superconductivity below 13 K when the sample was annealed in high pressure oxygen gas of 200 atm.[8] The present sample was annealed in 80 atm oxygen gas at 850 °C for 0.5 h, at 600 °C for 5 h, and at 400 °C for 5 h. Consequently, only a small sign of superconductivity was observed at around 9 K by resistivity measurement. The R_H data for $(Y,Ce)_4(La,Ba,Sr)_4Cu_6O_y$ shown in Fig. 5 are similar to those for $(Nd,Ce)_4(Nd,Ba,Sr)_4Cu_6O_y$ given in Fig. 3. Hence, it is unlikely that the 4f moment is the main cause of the broad peak in R_H-vs-T curves.

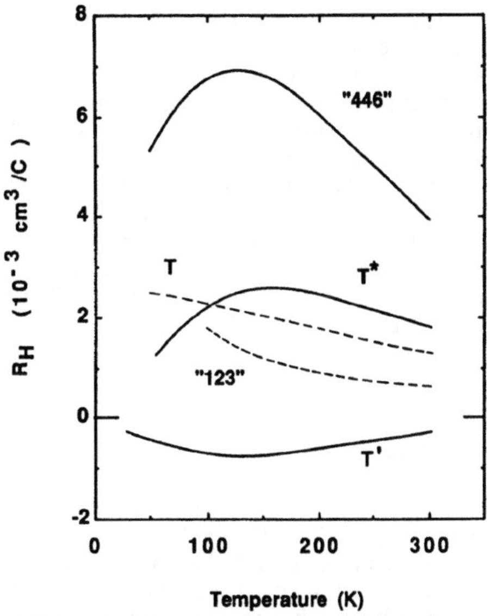

FIG. 4. Typical temperature dependence of R_H for five superconducting ceramics.

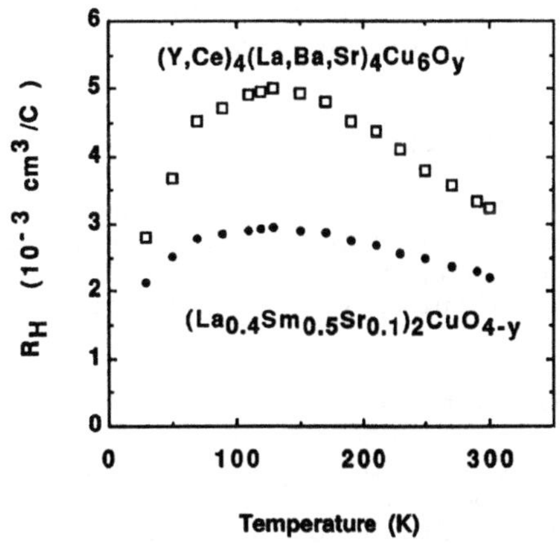

FIG. 5. R_H vs temperature for $(Y,Ce)_4(La,Ba,Sr)_4Cu_6O_y$ and $(La_{0.4}Sm_{0.5}Sr_{0.1})_2CuO_{4-y}$.

In above discussion, we neglected the effect of Ce^{3+}, because the valence state of Ce in T^* and "446" is considered to be +4 ($4f^0$).[14] In order to check the second cause, we investigated R_H of the Ce-free T^*-phase of $(La_{0.4}Sm_{0.5}Sr_{0.1})_2CuO_{4-y}$, which was discovered by Tokura et al.[15] The sample was sintered at 1130 °C and then annealed in oxygen gas of 80 atm as was the case of (La,Y) contained "446" compounds. From X-ray diffraction, it was found that the sample included a T'-phase as an impurity phase to the amount of 8-10 %. Needles to say, this T'-phase did not include Ce. The T_c of the sample was 17.1 K. The resultant curve given in Fig. 5 is parallel to that of $(Nd,Ce,Sr)_2CuO_{4-y}$(Fig. 3). Thus, it may be concluded that the broad peak was not due to the influence of Ce ions.

From the above discussions, the first and second causes are now excluded. The third cause that is the influence of the fluorite-type layers seems to be most plausible. The crystal structures of T'- and T^*-phase are similar to that of the T-phase, however the arrangement of apical oxygens are different in these three phases. The crystal structure of "446" can be described as a combination of an oxygen-disordered "123" structure and the T^*-structure. The most striking difference in the crystal structure of the 'new' compounds from that of the 'old' compounds is the fluorite-type layer sandwiched between two

$[CuO_2]$ planes. The $[CuO_2]$ planes are believed to be responsible for electrical conduction. The fluorite layer probably affect the electronic states of a neighboring $[CuO_2]$ plane or the band structure, and subsequently affect the temperature dependence of R_H. In this event, the degree of anisotropy of R_H, which was observed in T- and "123"-single crystals, is expected to be altered in the 'new' compounds. From the stand point of the temperature dependence of R_H, the electronic transport properties of the superconductors having fluorite layers are somewhat different from those of superconductors not having fluorite layers. This may be related to the fact that T_c is limited below 43 K so far in the compounds having fluorite layers. In fact, it has been pointed out that the thickness of the fluorite layer proportional to the ionic radius of the rare earth elements strongly correlates to the occurrence of superconductivity.[8]

SUMMARY

The temperature dependence of the Hall coefficient for $(Nd,Ce)_2CuO_{4-y}$ (T'-phase), $(Ln,Sr)_2CuO_{4-y}$ where Ln=(Nd,Ce) or (La,Sm) (T^*-phase), and $(Ln,Ce)_4(Ln,Ba,Sr)_4Cu_6O_y$ where Ln=Nd or (Y,La) was measured. The R_H-vs-T curves had a common feature of a broad peak at 120-150 K, which was different from the behavior of the R_H-vs-T curves for the T-phase and the "123" compounds. Therefore, a monotonic increase of R_H with decreasing temperature was not a generic property of the high-T_c oxides. The broad peak was due to neither the skew scattering by the moment of rare earth elements nor an influence of Ce ions. It is most likely caused by the modification of electronic states in $[CuO_2]$ planes due to the fluorite-type layer inserted between two $[CuO_2]$ planes. Nevertheless, it is still open for discussion why the Hall coefficient depends on temperature strongly in cuprate superconductors.

ACKNOWLEDGMENT

We thank Prof. S. Uchida and Dr. H. Takagi of University of Tokyo for helpful advice for the Hall measurements. We also thank Dr. Koshizuka of SRL-ISTEC for helpful discussions. We thank Mr. T. Noguchi and Mr. H. Okubo of Vacuum Metallurgical Co., Ltd. for their help in the development of a new instrument.

REFERENCES

1. S. W. Cheong, S. E. Brown, Z. Fisk, R. S. Kwok, J. D. Thompson, E. Zirngiebl, G. Gruner, D. E. Peterson, G. L. Wells, R. B. Schwarz, and J. R. Cooper, Phys. Rev. B 36, 3913-3916 (1987).
2. J. Clayhold, N. P. Ong, P. H. Hor, and C. W. Chu, Phys. Rev. B 38, 7016-7018 (1988).
3. J. Clayhold, N. P. Ong, Z. Z. Wang, J. M. Tarascon, and P. Barboux, Phys. Rev. B 39, 7324-7327 (1989).
4. H. Takagi, S. Uchida, and Y. Tokura, Phys. Rev. Lett. 62, 1197-1200 (1989).
5. J. Akimitsu, S. Suzuki, M. Watanabe, and H. Sawa, Jpn. J. Appl. Phys. 27, L1859-L1860 (1988).
6. H. Sawa, K. Obara, J. Akimitsu, Y. Matsui, and S. Horiuchi, J. Phys. Soc. Jpn. 58, 2252-2255 (1989).
7. A. Ichinose, T. Wada, Y. Yaegashi, H. Yamauchi, and S. Tanaka, Jpn. J. Appl. Phys. 28, L1765-L1768 (1989).
8. T. Wada, A. Ichinose, Y. Yaegashi, H. Yamauchi, and S. Tanaka, submitted to Jpn. J. Appl. Phys. (1989).
9. M. Kosuge, S. Ikegawa, N. Koshizuka, and S. Tanaka, unpublished.
10. S. Ikegawa, T. Yamashita, K. Sakurai, H. Yamauchi, and S. Tanaka, unpublished.
11. T. Tamegai, Y. Iye, M. Ogata, K. Obara, and J. Akimitsu, Jpn. J. Appl. Phys. 28, L1537-L1540 (1989).
12. K. Takita, H. Akinaga, H. Katoh, and K. Masuda, Jpn. J. Appl. Phys. 27, L607-L609 (1988).
13. J. R. Cullen, J. J. Rhyne, and F. Mancini, J. Appl. Phys. 41, 1178-1179 (1970); A. Fert and A. Hamzic, in The Hall Effect and Its Applications. edited by C. L. Chien and C. R. Westgate (Plenum, New York, 1980), pp 77-98.
14. M. Kosuge and K. Kurusu, Jpn. J. Appl. Phys. 28, L810-L812 (1989).
15. Y. Tokura, H. Takagi, H. Watabe, H. Matsubara, S. Uchida, K. Hiraga, T. Oku, T. Mochiku, and H. Asano, Phys. Rev. B 40, 2568-2571 (1989).

Transport Properties of High-T_c Superconductors

NORIAKI HAMADA[1], SANDRO MASSIDDA[2], JAEJUN YU[2], and ARTHUR J. FREEMAN[2]

[1] NEC Fundamental Research Laboratories, 34 Miyukigaoka, Tsukuba, 305 Japan
[2] Department of Physics, Northwestern University, Evanston, IL 60208, USA

ABSTRACT

Transport properties of $Ba_{1-x}K_xBiO_3$ and $Nd_{2-x}Ce_xCuO_4$ are calculated using the electronic energy band structure obtained with the local-density full-potential linearized augmented-plane-wave method. For $Ba_{1-x}K_xBiO_3$, the calculated Hall coefficient R_H has the correct (negative) sign. For $Nd_{2-x}Ce_xCuO_4$, on the other hand, a positive Hall coefficient for the magnetic field oriented perpendicular to the Cu-O planes contrasts with a negative experimental value. Recent experiments, however, show a change of sign of this Hall coefficient at x=0.18 from negative to positive with increasing x, indicating a trend towards a regime where the conventional band-theoretical discription becomes in better agreement with experiment.

KEY WORDS: transport property, band structure calculation, Boltzmann theory

INTRODUCTION

The discovery[1] of superconductivity in $Nd_{2-x}Ce_xCuO_4$ has provided a new point of discussion towards understanding the physics of the high temperature cuprate superconductors, in which it is obtained by electron doping of the parent compound, Nd_2CuO_4, which is an antiferromagnetic insulator. Superconductivity occurs for a narrow range of x values between 0.14 and 0.18. The electronic structure of $Nd_{2-x}Ce_xCuO_4$, determined previously by the local-density band-structure calculation[2], showed important similarities with the hole-doped superconducting oxides, including a property of a (single) strongly hybridized Cu-O dpσ antibonding band crossing the Fermi level.

A major issue is the understanding of the normal state of these systems, and, in particular, how well a Fermi liquid picture in which the quasiparticle energy band structure is approximated by the eigenvalues obtained within the local density approximation (LDA) works in discribing their normal state properties. Normal state transport properties are an important test. Such calculations have previously been performed for $La_{2-x}M_xCuO_4$ and $YBa_2Cu_3O_7$ by Allen et al.[3,4], and gave important results; these include a prediction of a change of sign in the current carriers at x=0.24 for $La_{2-x}M_xCuO_4$, and the positive and negative signs for the Hall coefficients R_H in $YBa_2Cu_3O_7$ when the magnetic fields is aligned parallel and perpendicular to the c-axis, respectively. The actual x value at which R_H changes sign is larger (x>0.28) in experiments on $La_{2-x}Sr_xCuO_4$, while the latter prediction has been confirmed by single crystal experiments[5]. In this paper, we present the transport properties calculated from the LDA energy band structure for the copper-less $Ba_{1-x}K_xBiO_3$ system and electron-deped $Nd_{2-x}Ce_xCuO_4$ system.

CALCULATIONAL METHOD

We recall the method used for the calculation of the transport properties[3,4]. Since we do not have any information on the relaxation time τ, we will mainly focus our attention on quantities which are independent of τ under an assumption of τ being constant on the Fermi surface. In the presence of an electric and magnetic field, we write the electric current as

$$j_i = \sigma_{ij}E_j + \sigma_{ijk}E_jB_k + \dots , \qquad (1)$$

where i,j,k = x,y, or z, and summations over repeated indices are implied. Within the Bloch-Boltzmann theory, these conductivity tensors are given by

$$\sigma_{ij} = \tau \langle v_i(k)v_j(k)\rangle \qquad (2)$$

$$\sigma_{ijk} = -\tau^2 \langle v_i(k)[v_i(k)v_{jj}(k)-v_{ij}(k)v_j(k)]\rangle \qquad (3)$$

in atomic unit, where $\langle\rangle$ denotes the Fermi surface average, k is the wave vector in the Brillouin zone. $v_i=\partial\varepsilon(k)/\partial k_i$, and $v_{ij}=\partial^2\varepsilon(k)/\partial k_i\partial k_j$. These expressions neglect the anisotropy of the scattering (in τ). Owing to the tetragonal symmetry of the Nd-Ce-Cu-O system, two independent elements exist for σ_{ij}, $\sigma_{xx}=\sigma_{yy}\neq\sigma_{zz}$; the anisotropy ratio σ_{xx}/σ_{zz} does not depend on τ. The only non-zero components of σ_{ijk} have all the three axes defferent, with $\sigma_{ijk}=-\sigma_{jik}$ due to the Onsager relations. The independent components are $\sigma_{yzx}=\sigma_{zxy}\neq\sigma_{xyz}$. For the magnetic field oriented along the c axis, the Hall coefficient is given by

$$R_{xyz} = E_y/j_xB_z = \sigma_{xyz}/\sigma_{xx}\sigma_{yy} \qquad (4)$$

and by cyclic permutation on this expression for $R_{yzx}=R_{zxy}\neq R_{xyz}$. The Hall coefficient does not depend on the relaxation time τ. We have also calculated the electronic plasma energies, defined as

$$\Omega_{pij}^2 = 4\pi N(E_F) \langle v_i(k)v_j(k)\rangle . \qquad (5)$$

The electronic energies $\varepsilon(k)$ for the single band crossing the Fermi energy E_F are obtained within the local density approximation (LDA) from the full-potential linearlized augmented-plane-wave method (FLAPW) calculation of Refs. [6] and [2] for $Ba_{1-x}K_xBiO_3$ and $Nd_{2-x}Ce_xCuO_4$, respectively. The K and Ce doping are treated within a rigid band model.

TRANSPORT PROPERTIES OF $Ba_{1-x}K_xBiO_3$

Figure 1 presents the results for the Hall coefficient of the $Ba_{1-x}K_xBiO_3$ system in its cubic perovskite structure, as a function of K concentration. Our results are expected to reflect the experimental situation only in the region x>0.25, since pure $BaBiO_3$ is an insulator with a monoclinic crystal structure. In the metallic region, the Hall coefficient is seen to be negative, and varies substantially with x. A negative sign of the Hall coefficient is expected from the electron-like Fermi surface of this system, as calculated in Ref. [6]. Experimentally, Sato et al.[7] measured a negative Hall coefficient, $-0.4\times10^{-9}m^3/C$, which is about a half of our result.

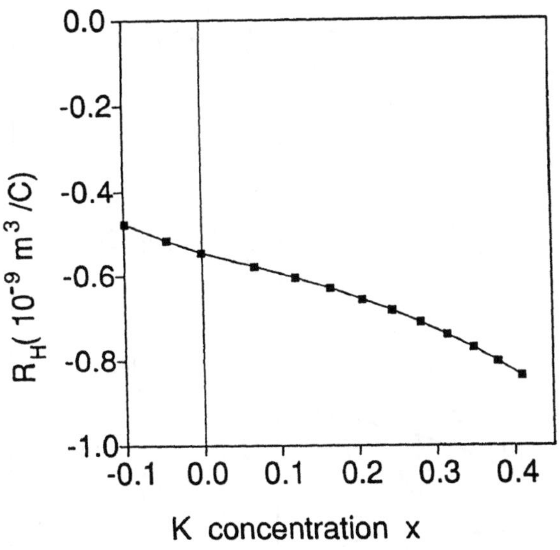

Fig.1. The Hall coefficient for $Ba_{1-x}K_xBiO_3$

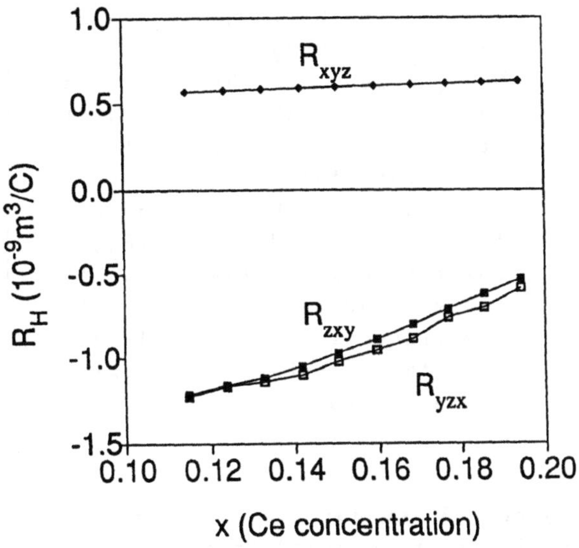

Fig.2 The Hall coefficient for $Nd_{2-x}Ce_xCuO_4$. The discrepancy between R_{zxy} and R_{yzx} is due to the numerical error.

The electronic plasma energy is calculated as Ω_p=4.0eV for x=0.37, which is compared with the experiment of Sato et al.[7]. Their fitting of the reflectivity data with a Drude model gives Ω_p=2.8eV (for the bare plasma energy) with $\varepsilon_\infty=2$. The agreement can be considered to be satisfactory if one considers the large uncertainty in the ε_∞ value.

TRANSPORT PROPERTIES OF $Nd_{2-x}Ce_xCuO_4$

The transport properties are highly anisotropic in this system compared with other high T_c superconductors. We have calculated the electronic plasma energies to be Ω_{pxx}=3.95eV and Ω_{pzz}=0.07eV at x=0.15, and to decrease very little (less than 1%) with increasing x. The anisotropy ratio $\sigma_{xx}/\sigma_{zz} = (\Omega_{pxx}/\Omega_{pzz})^2$ is about 2700. This ratio is much larger than that of $La_{2-x}M_xCuO_4$. Our estimate of the anisotropy ratio can be compared with the experimental anisotropy of the critical field H_{c2}, according to $\sigma_{xx}/\sigma_{zz} \sim (H_{c2}^{\parallel}/H_{c2}^{\perp})^2 \sim 440$ [8].

The calculated Hall coefficients for $Nd_{2-x}Ce_xCuO_4$ as a function of x over a range including the superconducting region are shown in Fig.2. R_{xyz} has a positive sign, as expected from the hole-like two-dimensional projection of the Fermi surface. The two other coefficients ($R_{zxy}=R_{yzx}$) are negative, and have magnitudes comparable to that of R_{xyz}. Measurements by Takagi et al.[9,10] on both polycrystaline samples and single crystals (magnetic field parallel to the c axis) showed a negative Hall coefficient with a change of sign (from negative to positive) for x=0.18. This value seems to coincide with the value at which superconductivity disappears. If we compare our results with experiment, we see that an agreement on the sign of R_{xyz} exists only for x>0.18. Actually, the transition from the insulating (x<0.14) regime, characterized by a negative Hall coefficient, to the metallic region, in which band structure effects should produce a positive R_{xyz} value, is to be clarified from the theoretical point of view.

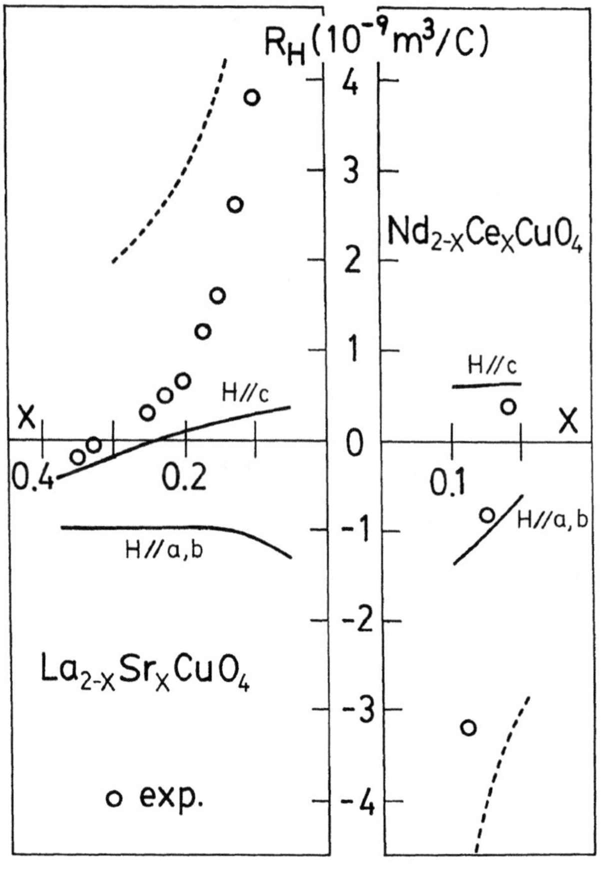

Fig.3. The Hall coefficients for $La_{2-x}Sr_xCuO_4$ and $Nd_{2-x}Ce_xCuO_4$, which are compared with the experimental values[10]. The solid lines are the theoretical ones, denoted by H//c and H//a,b for R_{xyz} and $R_{zxy}=R_{yzx}$, respectively. The dashed lines indicate the Hall coefficients proportional to the reciprocal doping concentration, 1/x.

Figure 3 summarizes the Hall coefficients obtained from the LDA band structure calculation. The Hall coefficients for $La_{2-x}Sr_xCuO_4$ have been calculated by Allen et al.[3]. Those theoretical results are compared with the experimental values taken from the paper of Uchida et al.[10]. The experimental values approach to the theoretical R_{xyz} (H//c) rapidly with increasing x, indicating that the band theory gives the better discription for the quasiparticles on the Fermi surface in the higher concentration region, x>0.18 and x>0.28 for $Nd_{2-x}Ce_xCuO_4$ and $La_{2-x}Sr_xCuO_4$, respectively.

ACKNOWLEDGEMENT - Work at Northwestern University is supported by the National Science Foundation (through the Northwestern University Materials Research Center, Grant No. DMR88-21571 and by a grant of computer time from its Division of Advanced Scientific Computing at the National Center for Supercomputer Applications, University of Illinois, Urbana/Champaign).

REFERENCES

1. Y.Tokura, H.Takagi and S.Uchida, Nature 337, 345 (1989).
2. S.Massidda, N.Hamada, Jaejun Yu and A.J.Freeman, Physica C157, 571 (1989).
3. P.B.Allen, W.E.Pickett and H.Krakauer, Phys.Rev.B36, 3926 (1987)
4. P.B.Allen, W.E.Pickett and H.Krakauer, Phys.Rev.B37, 7482 (1988)
5. S.W.Tozer, A.W.Kleinsasser, T.Penney, D.Kaiser and F.Holtzberg, Phys.Rev. Lett.59, 1768 (1987).
6. N.Hamada, S.Massidda, A.J.Freeman and J.Redinger,Phys.Rev.B40,4442,(1989).
7. H.Sato, S.Tajima, H.Takagi and S.Uchida, Nature 338, 241, (1989); Proceedings of 2nd Intl.Conf. on Materials and Mechanisms of Superconductivity: High-Temp. Superconductors (Stanford, 1989).
8. Y.Hidaka and M.Suzuki, Nature 338, 635 (1989).
9. H.Takagi, S.Uchida and Y.Tokura, Phys.Rev.Lett.62, 1197 (1989).
10. S.Uchida, H.Takagi and Y.Tokura, Proc. of 2nd Intl.Conf. on Materials and Mechanisms of Superconductivity: High-Temp.Superconductors(Stanford,1989).

Unusual Physical Properties of $Bi_2Sr_2Ca_{n-1}Cu_nO_y$ Family Materials

ATSUTAKA MAEDA, ICHIRO TERASAKI, TERUYUKI NAKAHASHI, SEIKI TAKEBAYASHI, MASASHI HASE, and KUNIMITSU UCHINOKURA

Department of Applied Physics, The University of Tokyo, 3-1, Hongo 7-chome, Bunkyo-ku, Tokyo, 113 Japan

ABSTRACT

In $Bi_2Sr_2Ca_{n-1}Cu_nO_y$ family materials including single crystals with various carrier concentrations, the superconducting transition temperature, Hall effect, optical conductivity were investigated as functions of hole concentration, and a simple model describing the electronic states is proposed. However, conduction noise measured remains unusual. Result of surface resistance measurement at 24 GHz in superconducting state is also reported.

KEY WORDS: $Bi_2Sr_2Ca_{n-1}Cu_nO_y$ family materials, phase diagram, electronic-state description, surface resistance, noise measurement

INTRODUCTION

$Bi_2Sr_2Ca_{n-1}Cu_nO_y$ family forms a broad range of materials which have the superconducting transition temperature T_C between 0 K (semiconductor or normal metal) and 110 K with various number of CuO_2 layers in a unit cell[1,2]. The comparison of various properties among these materials will be expected to provide important information on electronic states, mechanism of high-temperature superconductivity, etc. We prepared single-phase polycrystalline samples, and single crystals of these materials doped with various cations. After the sufficient characterization and the determination of the relation between T_C and hole concentration, several properties both in normal state and in superconducting state were investigated. Based on the results, we propose a simple model describing electronic state of these materials.

EXPERIMENTAL

Polycrystals were prepared by ordinary solid state reaction method. Hole concentration was changed by the partial substitution of La for Sr and Bi for Pb in $Bi_2Sr_2CuO_y$ (2201) material, and Y for Ca in $Bi_2Sr_2CaCu_2O_y$ (2212) material. Exact compositions will be published later. Single crystals were prepared by CuO self-flux method. They were characterized by X-ray and electron diffraction method, and Meissner-effect measurement. Chemical composition of single crystals was determined by the comparison of the lattice constants with that of polycrystals. Average valence of Cu in polycrystals was determined by iodometric titration under the assumption of Bi^{3+} and Pb^{2+}. Optical conductivity was obtained by the Kramers-Kronig transformation of the reflectivity data measured by a Fourier spectrometer and by using sychrotron radiation. Surface resistance at 24 GHz was measured by transmission cavity method, details of which will be published elsewhere.

EXPERIMENTAL RESULTS AND DISCUSSION

Figure 1(a) shows the superconducting transition temperature T_C as a function of hole concentration p, which is the deviation of the average Cu valence from 2.0. In 2201 materials, T_C increases with increasing p, and after taking maximum it decreases with further increasing temperature. At p=0.25, the bulk superconductivity almost disappears. In 2212 compounds, a similar behavior was observed. These behaviors are similar to those obtained in $(La,Sr)_2CuO_4$ (214 material) and $LnBa_2Cu_3O_y$ (123 material; Ln = Y and lanthenides)[3,4]. In $Bi_2Sr_2Ca_2Cu_3O_y$ (2223 material), we have not succeeded in changing p. From Fig. 1, it can be said that different curve exists in materials with different n. This suggests that an extra factor other than p

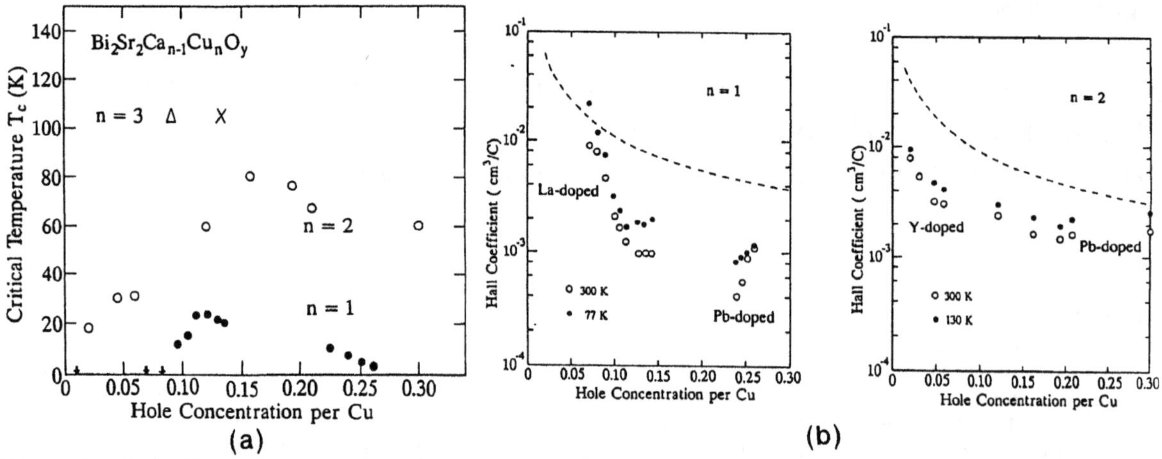

Fig. 1 (a) The superconducting transition temperature T_C as a function of the hole concentration in Bi-2201, 2212 and 2223 materials. Δ and X for the 2223 materials represent the points obtained under the assumption that holes are distributed with equal probability in three and two CuO_2 layers, respectively. (b) Hall coefficients as functions of the hole concentration determined by the titration for 2201 (n=1) and 2212 (n=2) materials. See the text for detail.

affects T_C strongly.

Figure 1(b) shows the Hall coefficients R_H of 2201 and 2212 materials as functions of p. In 2201 materials, R_H strongly deviates from the expected value in simple single-hole-band model (=1/p) even at very small p. On the other hand, in 2212 materials, apart from the difference in absolute magnitude, R_H roughly scales with 1/p. Similar difference in the relation between R_H and p was also found in 214 and 123 compounds[5,6]. This difference also suggests the importance of an extra factor which controls T_C other than p.

Figure 2(a) shows the optical conductivity of three 2212 single crystals with different hole concentration. Three structures were observed; (1) a peak around 1.5 eV clearly seen in semiconducting samples, which can be regarded as a charge transfer (CT) absorption, (2) Drude-like structure which is clearly seen in undoped sample, and (3) another broad peak observed in intermediate sample. Because other structures are absent except for a broad absorption located in high-energy region of CT absorption, we tried a fitting of these data under the assumption of the existence of a Drude term and two Lorentz oscillators in every sample. The obtained results were also shown in Fig. 2(a) as solid curves. The simple fitting was found to reproduce the data nicely. From the obtained parameters, the "band" diagram shown in Fig. 2(b) were obtained. In terms of the picture shown in Fig. 2(b), the Drude term corresponds to the plasma oscillation by doped holes, and the Lorentz oscillators correspond to the CT excitation and the excitation of hole from upper Hubbard band to newly developed state shown by arrows in Fig. 2(b). The fitting shows that the location of this band is very sensitive to the hole concentration. More detailed experiment is now in progress.

A model proposed just above may explain several characteristic features of anomalous normal-state properties in this compound. Several properties, however, remain unusual. One of them is the conduction noise. In ordinary metals, noise can be observable only in thin films. On the other hand, in high-T_C oxides Testa et al. reported that conduction noise was observed even in bulk polycrystalline samples of $YBa_2Cu_3O_y$[7]. We investigated the size effect of the conduction noise in Bi-2212 bulk samples, and found that the noise is generated as bulk effect, as is shown in Fig. 3(a). Figure 3(b) shows the power spectrum of the noise observed in a single crystal of 2212 material, in which so-called 1/f spectrum appeared. The noise is of almost the same orders of magnitude as that in polycrystals, which strongly suggests that the noise has intragrain origin. The magnitude of the noise is 7-8 orders of magnitude larger than that in conventional metals. Although the origin of the extremely large noise is quite puzzling, it may be correlated with unusual scattering mechanism which leads to T-linear temperature dependence of the resistivity characteristic of these materials.

Superconducting state in high-T_C superconductor is anomalous as well as normal state. In an experiment of the substitution of Cu site, we have shown

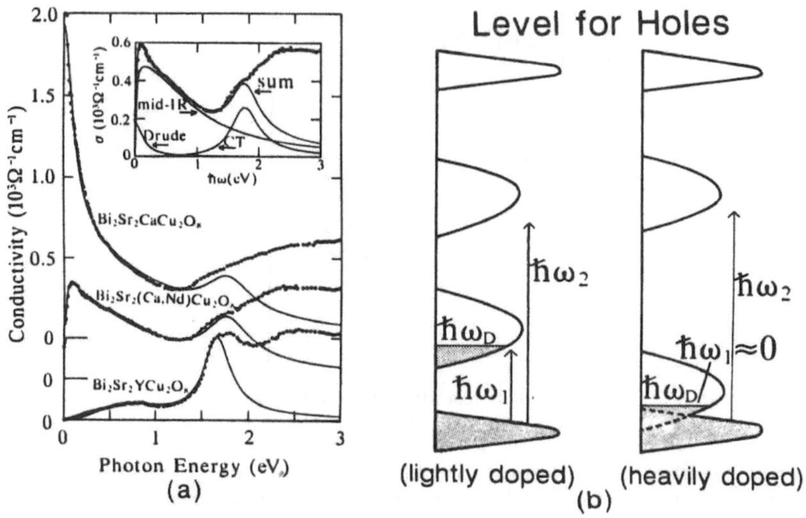

Fig. 2 (a) Optical conductivity obtained for three different single crystals of Bi-2212 samples. See the text for detail. (b) A model for the electronic structure of Bi-2212 system proposed based on the optical conductivity measurement.

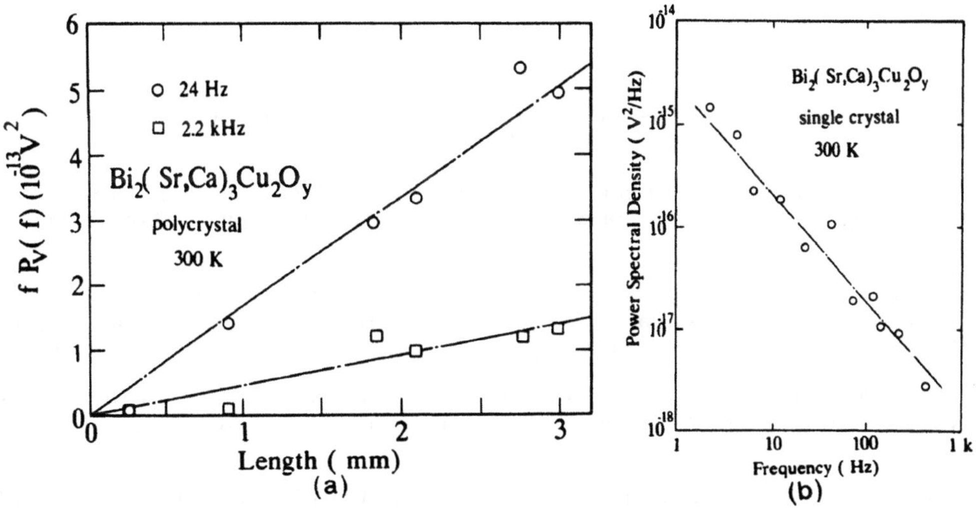

Fig. 3 (a) The dependence of the noise power spectral density on the distance between the electrode. The power spectral density are normalized by the inverse of the frequency. Dashed-solid lines are only guide for eye. (b) The frequency spectrum of the noise measured at 300 K in a single crystal of Bi-2212 material.

that nonmagnetic Zn decreases T_C as well as magnetic Fe, Ni, and Co by almost the same amount[8]. This cannot be understood in terms of the BCS theory, This strongly suggests that the pairing mechanism of high-T_C superconductor is unusual. For further discussion, precise determination of parameters in superconducting state is needed. However, most of them have been obscure. A typical example is the energy gap in quasi-particle excitation spectrum. So far investigated methods for the determination of the gap strongly depend on the condition of surface. Microwave surface resistance measurement is another candidate which is expected to be a better method in the sense that it reflects the bulk properties. Figure 4 shows the temperature dependence of the surface resistance R_S normalized to that at T_C of two different single crystals. Although R_S shows a sharp drop at T_C, the different temperature dependences were observed on the two samples. Sample B, which has the lowest R_S at T_C (1 ohm), shows the sharpest drop among several samples measured. Although these data exclude the possibility of smaller gap than that expected

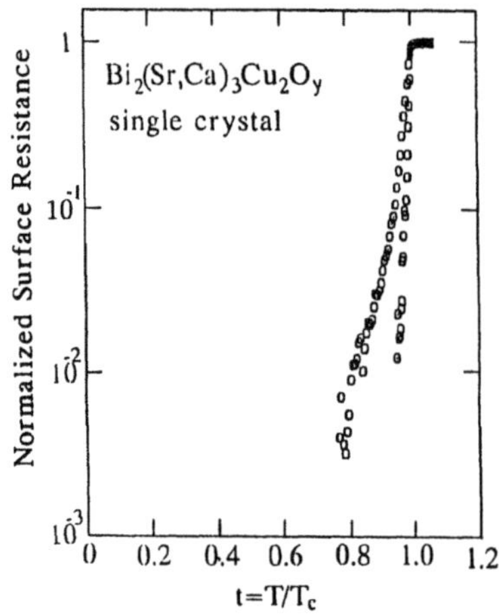

Fig. 4 The temperature dependence of the surface resistance R_S at 24 GHz for two Bi-2212 single crystals. R_S and T is normalized by R_S at T_C and T_C, respectively.

by the BCS theory, further studies with lower-R_S samples and measuring system with higher dynamic range are necessary for discussing the details of the magnitude and the temperature dependence of the energy gap. This is in progress.

REFERENCES

1. Akimitsu J, Yamazaki A., Sawa H., Fujiki H. (1987) Superconductivity in the Bi-Sr-Cu-O system, Jpn. J. Appl. Phys. 26: L2080-L2081.

2. Maeda H., Tanaka Y., Fukutomi M., Asano T. (1987) A New high-T_C oxide superconductor without a rare earth element, Jpn. J. Appl. Phys. 27: L209-L210.

3. Torrance J. B., Tokura Y., Nazzal A. I., Bezinge A., Huang T. C., Parkin S. S. (1988) Anomalous disappearance of high-T_C superconductivity at high hole concentration in metallic $La_{2-x}Sr_xCuO_4$, Phys. Rev. Lett. 61: 1127-1130.

4. Tokura Y., Torrance J. B., Huang T. C., Nazzal A. I. (1988) Broader perspective on the high-temperature superconducting $YBa_2Cu_3O_y$, Phys. Rev. B38: 7156-7159.

5. Takagi H., Ido T., Ishibashi S., Uota M., Uchida S., Tokura Y. (1989) Superconductor-to-nonsuperconductor transition in $(La_{1-x}Sr_x)_2CuO_4$ as investigated by transport and magnetic measurements, Phys. Rev. B40: 2254-2261.

6. Takita K., Akinaga H., Masuda K., Asano H., Takeda Y., Takano M., Nishiyama K., Nagamine K. (1989) Hole concentration dependences of T_C and T_N in $Nd_{1+x}Ba_{2-x}Cu_3O_{7+y}$, In: Masuda K. et al.(ed) Proc. Tsukuba seminor on high-T_C superconductivity, 31 May - 2 June 1989, Tsukuba University, Tsukuba, p.11-20.

7. Testa J. A., Song Y., Chen X. D., Golben J., Lee S., Patton B. R. Gaines J. R. (1988) 1/f-noise-power measurements of copper oxide superconductors in the normal and superconducting states, Phys. Rev. B38: 2922-2925.

8. Maeda A., Yabe T., Takebayashi S., Hase M., Uchinokura K., Study on the substitution of 3d metals for Cu in $Bi_2(Sr_{0.6}Ca_{0.4})_3Cu_2O_y$, Phys. Rev. B, in press.

STM Inestigations of the Grain Surface of $Bi_{0.8}Pb_xSr_1Ca_1Cu_{1.6}O_{8+y}$ Superconductors

CONGPING CHAO and SATOAKI ARAI
Tokyo Denki University, Kanda Nishiki-cho, Chiyoda-ku, Tokyo, 101 Japan

ABSTRACT

Scanning tunneling microscopy was used to investigate the surface topographs of bared crystal grain in Pb-free or Pb-dopped $Bi_{0.8}Pb_xSr_1Ca_1Cu_{1.6}O_{8+y}$, x=0.0~0.5, bulk superconductors. Images with a scanning electron microscopy resolution revealed that the profiles of grain surface was in the form of periodic rugged wavelet with the amplitude of 10 nm and the length of 20 nm. This result was considered to provide direct proof for the percolation processes between randomly oriented grains and grain boundaries in bulk superconductor.

KEY WORDS: scanning tunneling microcsopy, crystal structure, grain surface, Bi-system

INTRODUCTION

Crystal structure analysis has been a great concern of material scientists in many areas. Electron diffraction pattern and transmission or scanning electron microscopy techniques are noted to be powerful instruments in investigating crystal structures, but none of them having the resolution up to single atomic image scale but the symmetricity of atoms in crystal structure. Scanning tunneling microscopy (STM) is the singular scanning probe microscope capable of resolving material surface details down to the atomic level [1]. Although much of the applications are focussed on its ability to image objects on the atomic and molecular scale [2], STM also emerges as a singular device for mapping and measuring three-dimentional surface profiles of objects in the range from 0.01 to 10 μm [3]. This size range is well within the resolution capability of scanning electron microscopy (SEM), but SEM has poor capability for measuring vertical distance.

The first STM observation of superconducting energy gap and spatial variations of properties over microscopic lenghts was reported by de Lozanne [4]. Recent STM studies on high-Tc superconductors have been focussed on local electronic properties to investigate the state density near the Fermi-level, and the surface crystal atomic superstructure to probe the atom incommensurate modulations or missing atoms [5-7]. However we have not as yet been able to obtain a single phase 110K sample since the 110K-phase forms mainly during decomposion. It may be that each particle is surrounded by a lower Tc, or non-superconducting material. The similarity in Meissner and shielding effects and the existance of the resistive foot, for example in our previous work [8], are consistent with this statement. This is also consistent with the observation of extra peaks in the powder x-ray diffraction, as shown in Fig. 3 in our recent work [9], which would preferentially sample the outside of high-Tc part. It is therefore the purpose of this paper to present the investigations of the surface profiles of bared crystal grains in bulk sample of Bi-based Pb-free or Pb-dopped superconductor $Bi_{0.8}Pb_xSr_1Ca_1$-$Cu_{1.6}O_{8+y}$, x=0.0~0.5, using STM in the resolution range of a SEM. Topographically, we find an optical flat crystal surface observed by SEM is really periodic rugged wavelet.

EXPERIMENTALS

High-Tc bulk samples were prepared by solid reaction method. Superconductivity and X-ray diffraction results were described previously [9,10]. The best superconducting sample was obtained in the composition $Bi_{0.8}Pb_xSr_1Ca_1Cu_{1.6}O_{8+y}$ with x=0.2~0.3. From the magnetization, weight, and volume of the sample we caculated a Meissner signal which was about 80% of that expected and therefore evidence for bulk superconductivity in the sample. The value is over-estimated because the demagnetization factor (n) was not taken into account. Complete Meissner susceptibility should lead to $M=-1/(1-n)(1/4\pi)$ times the aplied field ($H_{applied}$) whereas we taken $M=-(1/4\pi)H_{applied}$ for caculation as done in the work of Tarascon et al [11]. Experiment setup for STM had been successfully summarized in several articals [2,3]. Nanoscope II apparatus was used in our STM experiments. Auger Electron Microscopy measurements were performed either before or after STM experiments to ensure a clean sample surface. All the STM pictures were taken under ultrahigh vacuum in liquid Nitrogen temperature.

RESULTS

STM images of ethanel cleaned Bi-based superconductor surface were taken at sample bias of -25.0 mV and stablized tunneling current of 1.5 nA. Scans at the bared crystal surface are showing in Figure (a) with the STM resolusion in the range of a SEM. The usual SEM image of optical flat-like surface profiles for crystal grains, A, B, and C showing in Figure (c), can be seen to be really rugged wavelet. The magnitude of the wavelet, showing in Fig. (b) is about 10 nm with a period of about 20 nm. These sizes are not in the dimension of any atom in Bi-Sr-Ca-Cu-O composition or perovskite cell for low or high-Tc phase, (2212) or (2223), indicating that this character is not crystal structural but crystal surfaces. This character had been previously observed to be general in Bi-based Pb-free or Pb-dopped multiphase system [8,9], $Bi_{0.8}Pb_xSr_1Ca_1Cu_{1.6}O_{8+y}$, whether the superconductivity is the best for x=0.2~0.3, or not for x<0.2 or x>0.3.

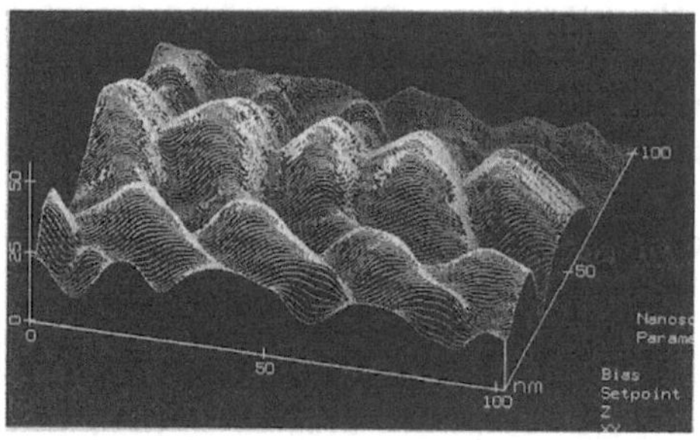

(a)

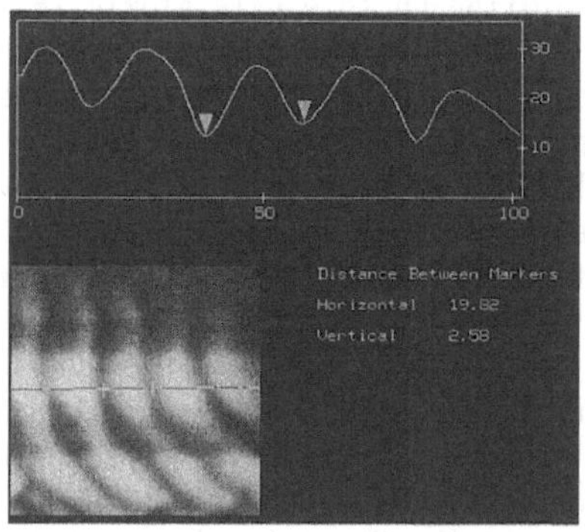

(b)

Fig. a Top view STM image of bared grain surfaces in $Bi_{0.8}Pb_xSr_1Ca_1Cu_{1.6}O_{8+y}$, x=0.0~0.6 bulk samples, showing the rugged wavelet topograph. Sample bias voltage and tunneling current are -25.0 mV and 1.5 nA respectively.

Fig. b Crossection of the wavelet showing in Fig. a. The duration of each peak can be seen to be about 20 nm with the amplitude of 10 nm. Sample bias voltage and tunneling current are the same as the above.

DISCUSSIONS

Grain Surface: High-Tc superconductor crystal surfaces have been recognized to be very important in superconducting mechanism and very different from that inside the crystals in electronic properties. STM analysis revealed that the bulk of the grain for $Y_1Ba_2Cu_3O_{7-y}$ superconductors is semiconductive and that a conducting percolative network of grain and domain boundaries may be responsible for the superconductivity [12]. But photoemission data [13] revealed an abnormally low density of states at the Fermi energy, which indicate that the surfaces of Y-Ba-Cu-O system may be intrisically non-metallic. Moreover, infrared reflectivity

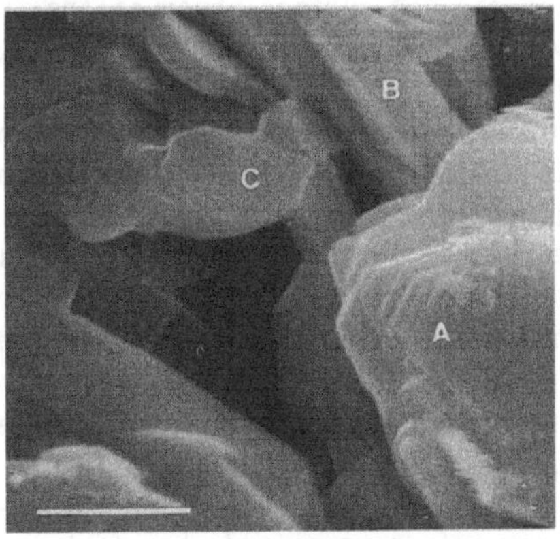

Fig. c Typical SEM photograph of bulk samples $Bi_{0.8}Pb_xSr_1Ca_1Cu_{1.6}O_{8+y}$, x=0.0~0.5 annealed at 850℃ for 36 hrs with A, B, and C representing different bared crystal grains. The scale is 1μm.

measurements on polycrystalline samples reveal much lower energy gaps when compared with the tunneling technuque [14]. This indicates that the superconducting properties of the surface at which the reflection takes place can differ considerably with that of the bulk of the sample. Here it should be pointed out that the averaging nature in photoemission measurements and other techniques would not be ignored. However for Bi-based system, one interesting feature which distingwishes the photoemission spectra from other oxide copper superconductors is the intensity of signals near the Fermi-level [15], it is therefore encouraging that surfaces of Bi-based superconductive oxide clearly are metallic. But the magnetic relaxation for Bi-based system is generally recognized to be larger than Y-Ba-Cu-O samples, and greatly affected by preparation processes, as shown in our previous work [8,9]. This means that the large relaxation of bulk samples is not an intrisic property but due to large flux creep at grain boundaries. The reason why we always find contradictory conclusions on the state density near Fermi-level is our lack of very large single crystals and incompletely understanding the crystal surface properties. The results showing in Figure (a) and (b) are very good explanation for the existance of intergrain phenomena observed for Bi-based superconductors since this nature will result point connection between grains that produce weak-link regions. The percolation processes between randomly oriented grains in bulk samples neecessitates a redistribution of current between the Cu-O planes and grain boundaries. We consider it is a direct proof for the percolative processes in bulk superconductors.

Surface Structure: Structures of Bi-Sr-Ca-Cu-O superconductor are characterized by two adjacent Bi-O planes, about 3.2Å apart, which are weakly held together [11]. It is this weak link in the crystal structure that results in a preferential cleavage between the Bi-O planes. Many studies on structure, including STM, photoemission, and transmission electron microscopy, confirm that the cleaved surface is a Bi-O plane that has a stable structure [16]. The lamellar nature of the perovskite superstructure suggests that the surface structure of Bi-O plane may be similar to that in the bulk. In other words, the most probable cleavage plane is the one that is perpendicular to the c-axis, i. e., the a-b planes. It is most plausible that the Bi-O plane terminates the crystal surfaces in bulk sample. Therefore the observed surface profiles in Figure (b) may be attributed to be in Bi-O plane in the crystal structure. However the size is much beyond the bonding distances of 2.2Å or 2.7Å for Bi-O(2) or Bi-O(3), and the high-Tc (2223) or low-Tc(2212) perovskite cells.

Critical Assessment: There are several points worthy of mentioning. First the large vertical corrugation amplitude in STM studies is an artifact of the imaging process, resulting from either an effect of the surface electronic structure, or a mechanical interaction between the tip of STM and sample surface [7]. But this is no problem for STM to be used in the resolution range of a SEM in our present work. Second the Bi-based superstructure is mostly displacive in nature [17], the presence of Bi vacancies, which can be considered a compositional modulation, is a key feature of this problem [6]. Compressive and explacements along a-axis with the amplitute of about 1.1Å are agreed in two STM experiments [6,7] in atomic range. However the present results showing periodic row of 10 nm in amplitude and 20 nm in length on bared crystal surface in bulk sample, are much different from that of cleaved surface observations in those work. Of course, we can not rule out the possibility of a multiple-tip effect existing in any STM measurements, for example, the filling and emptying of localized electron traps in thin oxide layers on the electrodes [3]. However experiments in references 6 and 7 are performed on cleaved crystal surface and based on the assumption that single crystals of (2212) compounds are most likely to be cleaved along the Bi-O plane without any damadge. That is to say these observations are perhaps not the real crystal surface feature since it is now well established that ion milling induces defects into microstructures in sample [18]. We think it is necessary to clarify the validity of etching technique for sample preparation in atom position determine experiments. The present results identified on bared superconducting crystal surface in bulk samples can be considered to be true observations. We consider the formation are related to the crystal growing processes. Further analysis is in progress. The third consideration is that STM uses a point contact which may have large current densities. This could tend to suppress the superconducting gap [19]. This discrepancy will probably only be resolved when techniques are developed by using tunneling in a planar geometry or with true vacuum tunneling.

CONCLUSIONS

STM technique under ultrahigh vacuum and liquid Nitrogen temperature had been used in the resolusion range of a SEM to investigate the topographs of bared high-Tc grain surface in Pb-free or Pb-dopped $Bi_{0.8}Pb_xSr_1Ca_1Cu_{1.6}O_{8+y}$, $x=0.0\sim0.5$, bulk superconductors. Grain surfaces were found to be characterized by rugged wavelet profiles with the incommensurate period of about 20 nm in length and 10 nm in amplitute. This results was considered to be a direct proof for the percolative processes between randomly oriented grains and grain boundaries in bulk superconductor.

ACKNOWLEDGMENTS

This work was partially supported by a Grant-in-Aid for scientific research from the Center For Research at the University. The authors are grateful to Professors H. Takai and A. Tamaki for valuable discussions, and to Mr. K. Yada of Toyo Techniques Ltd. Co. for his help in STM work.

REFERENCES

1 Binning G & Rohrer H, Rev. Mod. Phys., 59, 615(1987)
2 Golovchenko J A, Science, 232, 48(1986)
3 Hansma P K, Tersoff J, J. Appl. Phys., 61, R1(1987)
4 de Lozanne A L, Elord S A, Quate C F, Phys. Rev. Lett., 54, 2433(1985)
5 Kirtley J R, Feenstra R M, Fein A P, et al, J. Vac. Sci. Technol., A6, 259(1988)
6 Shih C K, et al, Phys. Rev. B, 40, 2682(1989)
7 Kirk M D, Nogarni J, Baski A A, et al, Science, 241, 1673(1988)
8 Chao C P, Arai S, Tamaki A, et al, BHTSC'89, MC-17, Peking.
9 Chao C P, Arai S, Takai H, Physica C, to be publ.
10 Chao C P, Arai S, Hoshikawa K, et al, The Beijing Inter. Conf. on High-Tc Supercond., BHTSC'89, MC-15.
11 Tarascon J M, Le Page Y, Barboux P, et al, Phys. Rev. B, 38, 9382(1988)
12 Garcia N, Vieira S, Baro A M, et al, Z. Phys. B, 70, 9(1988)
13 Egdell R G, Flavell W R, ibid, 74, 279(1989)
14 Gijs M A M, de Vries J W C, Stollman G M, Phys. Rev. B, 37, 9837(1988)
15 Onellion M, Tang M, Chang Y, et al, ibid, 38, 881(1988)
16 for axample, Lindberg P A P, Shen Z X, Wells B O, et al, ibid, 39, 2890(1989)
17 Gao Y, Lee P, Coppens P, et al, Science, 241, 954(1988)
18 Sarikaya M, Thel B L, Aksay I A, Weber W J, Frydych W S, J. Mater. Res., 2, 736(1987)
19 Fein A P, Kirtley J R, Shafer M W, Phys. Rev. B, 37, 9737(1988)

Survey of Superconductivity in a Layered Compound 1T-VSe_2

K. TSUTSUMI[1], Y. ISHIHARA[1], and H. SUZUKI[2]

[1] Department of Physics, Kanazawa University, Kanazawa, 920 Japan
[2] Department of Physics, Tohoku University, Sendai, 980 Japan

ABSTRACT

We have performed a survey of superconductivity in a layered compound 1T-VSe_2 by means of measuring electrical resistance down to 12 mK. Our result is negative. If 1T-VSe_2 undergoes the superconductivity phase transition, the transition temperature is lower than 12 mK. However we observed a superconductivity transition below 1.0 K in 2H-TaS_2 measured together with 1T-VSe_2. The superconductivity of 2H-TaS_2 is very sensitive to the current for the measurement of the electrical resistance. In addition the current dependence of the electrical resistance is clearly observed at 1.4 K and 4.2 K. This dependency is discussed in terms of the distribution of the superconductivity tansition temperature within the specimen.

KEY WORDS : layered transition-metal compound, MX_2, charge-density wave, superconductivity,

INTRODUCTION

The superconductivity phase transition and the charge-density wave (CDW) phase transition in the layered transition-metal dichalcogenides MY_2 (M=V,Nb,Ta;S=S,Se,Te) have been extensively studied [1]. Balseiro and Falicov have insisted that all metallic CDW systems should be superconducting at low temperatures[2].
The crystal structure of MX_2 consists of the stacking of the layer in which the transition-metal atom sheet lies between the chalcogen atom sheets. In each sheet atoms are hexagonally close packed. The chemical bonding within the layer is covalent, while the bonding between layers is of a van der Waals type. As the interactions between layers are weak, many two-dimensional behaviors are observed in MX_2.
In the layer the transition metal is surrounded by the chalcogen atoms in the octahedral or trigonal prismatic coordination. MX_2 can be prepared in a number of polytypes. These polytypes differ from each other in the stacking arrangement of the octahedral or trigonal prismatic coordination layer. The simplest polytype is the 1T structure which has one octahedral coordination layer per unit cell and trigonal symmetry . Designations such as 1T or 2H come from the number of the layers in the unit cell and from overall symmetry.
MX_2 undergoes the CDW phase transition at low temperature and exhibits superconductivity at lower temperature. When CDW phase transition temperature T_{CDW} is higher, the onset temperature T_c of superconductivity is lower. For example, T_{CDW} is 33.5 K and T_c is 7.2 K in 2H-$NbSe_2$. For 2H-TaS_2, T_{CDW} is 78 K and T_c is 0.8 K. However there is no observation of superconductivity in 1T polytype. 1T-VSe_2 undergoes the CDW phase transiton at 110 K. So we expect a very low superconductivity transition temperature. We have performed the electrical resistivity measurement in order to survey superconductivity in the temperature range from room temperature down to 12 mK together with 2H-TaS_2.
Our conclusion is that if 1T-VSe_2 undergoes superconductivity phase transition, the transition temperature is lower than 12 mK. While we found a highly current-sensitive superconductivity below 1.0 K and the current-dependency of the electrical resistance at 1.4 K and 4.2 K in 2H-TaS_2. We report these results and discuss in terms of the distribution of the superconductivity transition temperature within the single crystals of 2H-TaS_2.

EXPERIMENTAL

Single crystals of 1T-VSe_2 were prepared by chemical transport method with iodine as the transport agent. As the starting materials 99.95 % pure vanadium wire and 99.999 % selenium shot were used. Stoichiometric content was reacted in an evacuated qualtz tube with excess selenium . The growth temperature was 620 ° C. The temperature dependence of the electrical resistance was measured in two small specimen with an AC bridge and an ordinary dc four-probe methods. The sharp kink corresponding to the CDW phase transition was observed at 110 K. The residual resistivity ratios of the two specimens are 20 and 23. Fig.1 shows the temperature dependence of the resistivity measured with an ordinary dc four-probe method.
The single crystals of 2H-TaS_2 were synthesized by a coventional vapor phase transport reaction at 900 ° C via the 1T polytype and a subsquent annealing procedure. The residual resistivity ratio of the crystal (sample #1) used for investigating the superconductivity transition is 70. Another crystal (sample #4) exhibited the CDW phase transition at 78 K. This transition temperature is in good agreement with that obtainned in the measurement of heat-capacity [3]. Our result of the temperature dependence measurement of the resistance around 78 K is shown in Fig. 2.
The measurement of the electrical resistance below 1.5 K was performed by using a $^3He/^4He$ dilution refrigerator. We pasted the single crystals on the sapphire substrates by a commercial varnish (GE7031) and attached four gold wires to the specimen with silver paint. The sapphire sabstrares were also pasted to a thin plate of copper by GE7031 varnish. Next we set the plate to the mixing chamber of the dilution refrigerator. The temperature of the mixing chamber was measured by a calibrated carbon glass resistance thermometer and a ^{60}Co nuclear orientation thermometer.

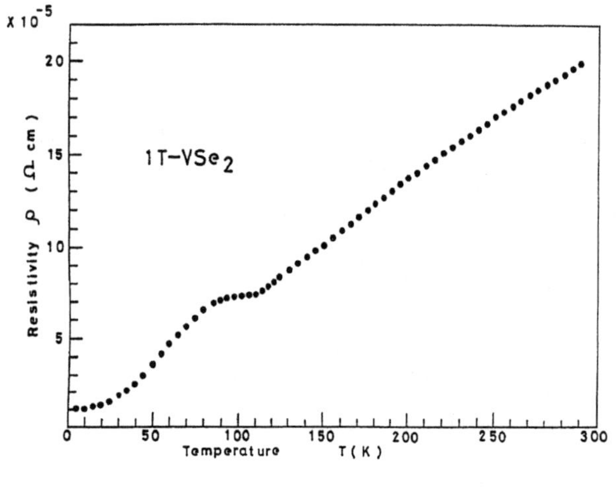

Fig. 1. Temperature dependence of the resistivity in 1T-VSe_2. A sharp kink of the resistivity at 110 K corresponds to the CDW phase transition.

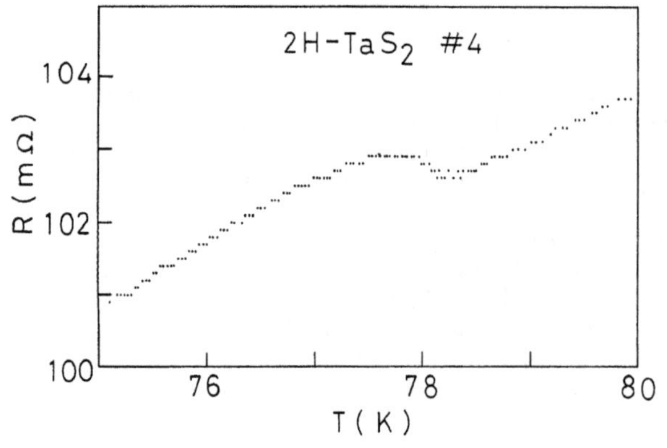

Fig. 2. Temperature dependence of the resistance in 2H-TaS_2 in the vicinity of 78 K at which the CDW phase transition occurs.

RESULT

Fuqure 3 shows the temperature dependence of the electrical resistance on two specimens of 1T-VSe_2 in the temperature range below 1.5 K down to 12 mK. The data presented by closed circles were obtained by using an AC bridge on sample #1. On the other hand those presented by closed triangles were obtained by measuring with an ordinary dc four-probe method on sample #2. Neither of the two samples showed any decrease of the resistance which should be accompanied by superconductivity phase transition down to 12 mK. Our conclusion is that if 1T-VSe_2 undergoes superconductivity phase transition, the transision temperature is lower than 12 mK.

We also measured the superconductivity transition in 2H-TaS_2 together with 1T-VSe_2 with only an ordinary dc four-probe method. The results are shown in Fig. 4. The superconductivity transition observed in the present measurement is very sensitive to the sample current. When the sample cuurent is 100 mA, the transition temperature is lower than 0.5 K. While the transition temperature is higher than 0.9 K, when the sample current is 10 mA.

Next we measured the cuurrent dependency of the electrical resistance at 1.4 K with an ordinary dc four-probe method. The results of this measurement is shown in Fig. 5. One can find a very current-sensitive behavior of the electrical resistance. Furthermore we have performed the measurement of the current-sensitive resistance on other two samples (sample #2 and sample #3) with a dc two-probe method at 4.2 K in the small amount of current regime. We showed the results in Fig. 6. One can clearly find the current-sensitive electrical resistance at 4.2 K too. One should note that the scale of the sample current in Fig. 6 is μA range, while that of the current in Fig. 5 is in mA range.

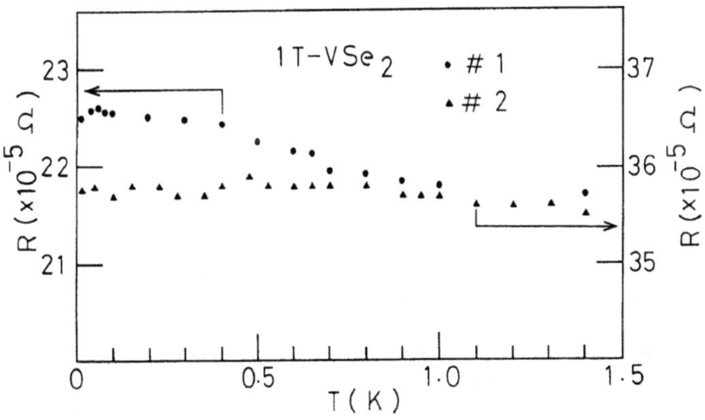

Fig. 3. Temperature dependence of the resistance below 1.5 K in 1T-VSe_2. The closed circles show the results measured by an AC bridge and the closed triangls the results measured with an ordinary dc four probe method.

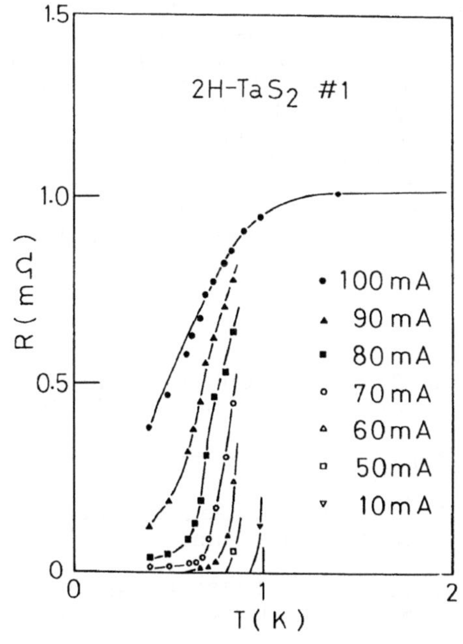

Fig.4. Superconductivity transition in 2H-TaS_2. Note that the transition is broad and highly current-sensitive.

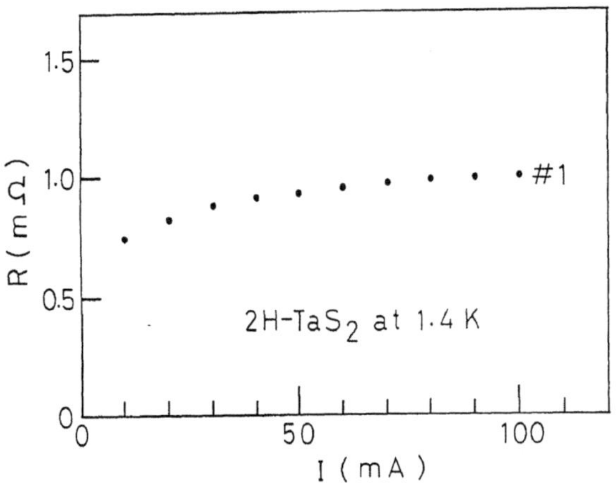

Fig. 5. Current dependence of the electrical resistance in 2H-TaS_2 at 1.4 K. Measurements were performed with an ordinary dc four-probe method.

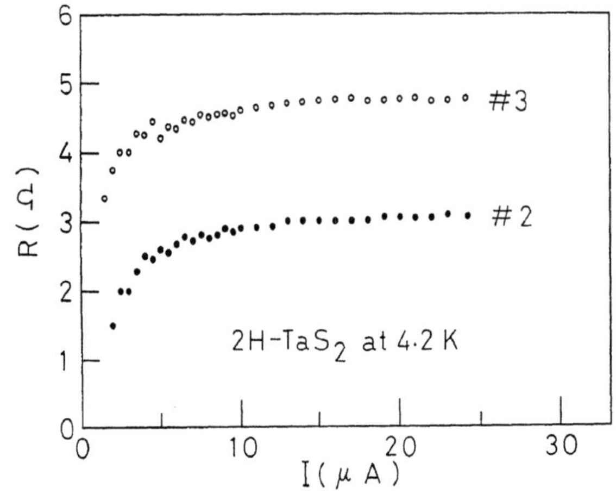

Fig. 6. Current dependence of the electrical resistance in 2H-TaS_2 at 4.2 K. Measurements were performed with a dc two probe method.

DISCUSSION

The broad superconductivity transition and the ditribution function which yields the fraction of the sample having some particular value of the superconductivity transition temperature in 2H-TaS_2 is already reported by the measurements of the specific heat [4]. The distribution function obtained by the specific heat measurement is a bell-shaped curve whose maximum of the ditribution is located very close to 0.6 K. It has a kind of asymmetry showing the existence of the segments whose superconductivity transition temperatures are higher than 1.0 K. So we speculate that the current-sensitive resistance observed in our present study might have some relation to superconductivity. The wrinkling which is likely to be due to the change of the volume of the unit cell during the transformation from the 1T plytype to the 2H plytype is usually present in all the crystals of 2H-TaS_2. It gives rise to a stress distribution throughout the specimen. It is well known that stress and defects produce a substancial variation of the transition temperature in a superconductor. This is seemed to be the origin of the asymmetry distribution function of the transition temperature and then the current-sensitive resistance reported for the first time in this article.

REFERENCES

1. Willson JA. DiSalvo FJ. Mahajan S (1975) Charge-density waves and superlattices in the metallic layered transition metal chalcogenides. Adv Phys 24 : 117-201
2. Balseiro CA. Falicov LM (1979) Superconductivity and charge-density wave. Phys Rev B 20 : 4457-4464
3. Harper JME. Geballe TH. DiSalvo FJ (1977) Thermal properties of layered transition-metal dichalcogenides at charge-density-wave transitions. Phys Rev B 15 : 2943-2951
4. Garoche P. Manuel P. Veyssie JJ. Molinie P (1978) Dynamic Measurements of the Low-Temperature Specific Heat of 2H-TaS_2 Single Crystals. J Low Temp Phys 30 : 323-336

Anomalous Transport Properties of Organic Superconductor κ-(BEDT-TTF)$_2$Cu(NCS)$_2$

H. MORI[1], K. NAKAO[1], I. HIRABAYASHI[1], S. TANAKA[1], K. OSHIMA[2], and G. SAITO[3]

[1] International Superconductivity Technology Center, 10-13, Shinonome 1-chome, Koto-ku Tokyo, 135 Japan

[2] Okayama University, 1-1, Tsushimanaka 1-chome, Okayama, 700 Japan

[3] Institute for Solid State Physics, 22-1, Roppongi 7-chome, Minato-ku, Tokyo, 106 Japan

ABSTRACT

The critical magnetic field (H_{c2}) by a magnetization measurement was investigated for an organic superconductor, κ-(BEDT-TTF-d_8)$_2$Cu(NCS)$_2$ with T_c=11.0 K. The linear temperature dependence of H_{c2} between 11 K and 5 K was obtained, which is consistent to Ginzburg-Landau theory. However, this determined H_{c2} was totally different from the H_{c2} defined by a resistivity recovery with applying field, which shows an anomalous upward curve with decreasing temperature, previously reported. The difference is discussed in terms of flux flow or flux creep.

KEY WORDS: organic superconductor, critical magnetic field(H_{c2}), static magnetization, critical current

INTRODUCTION

After the discovery of the organic superconductor with T_c=10.4 K under an ambient pressure, κ-(BEDT-TTF)$_2$Cu(NCS)$_2$, less difficult experimental technique due to comparatively higher T_c advanced elucidation on physical properties of an organic superconductor.(1) Extensive investigations have afforded some anomalous behaviors of κ-(BEDT-TTF)$_2$Cu(NCS)$_2$: First, the temperature dependence of H_{c2} along the a^*-axis indicated not a saturation curve to a constant at lower temperatures, but an upward curve with decreasing temperatures.(1b,1j) Secondly, ^{1}H-NMR relaxation rate showed an anomalous enhancement around 4 K at H=3.28kG. This enhancement was strongly dependent on to which direction and how large a magnetic field was applied.(2) Thirdly, an anomaly was found in a specific heat measurement. The C/T vs. T^2 plot gave not a linear slope but a slight curve around 4K.(3)

In this report, the origin of upward curve of $H_{c2} /\!/ a^*$, mentioned first, is discussed, compared with newly determined H_{c2} by a magnetization measurement. Finally, we briefly report the first study of transport critical current density for a single crystal of an organic superconductor, κ-(BEDT-TTF)$_2$Cu(NCS)$_2$, by an estimation of current-voltage curve.

EXPERIMENTAL

Black plate crystals, κ-(BEDT-TTF)$_2$Cu(NCS)$_2$, were prepared by an electrochemical oxidation of BEDT-TTF with using a supporting electrolyte of (18-crown-6 ether)KCu(SCN)$_2$ in 1,1,2-trichroloethane for a week to ten days. Especially a magnetization measurement and a critical current density study were carried out with purified sample crystals; KSCN was recrystallized in EtOH twice. CuSCN was purified with recrystallized KSCN three times. 18-crown-6 ether was recrystallized in CH_3CN. Transparent crystals of a crown salt were obtained with their purified starting materials and were used for a supporting electrolyte. As shown in Fig. 1, the most purified sample (a) shows comparatively higher and sharper superconducting transition.

The electrical resistivity was measured by a conventional four-probe technique with gold paste as contacts. The magnetization measurement was carried out with Quantum Design SQUID magnetometer model MPMS. The diamagnetic contribution caused by an Apiezon grease and a straw was subtracted from the observed magnetization signal. The transport critical current density was determined from a current-voltage curve.

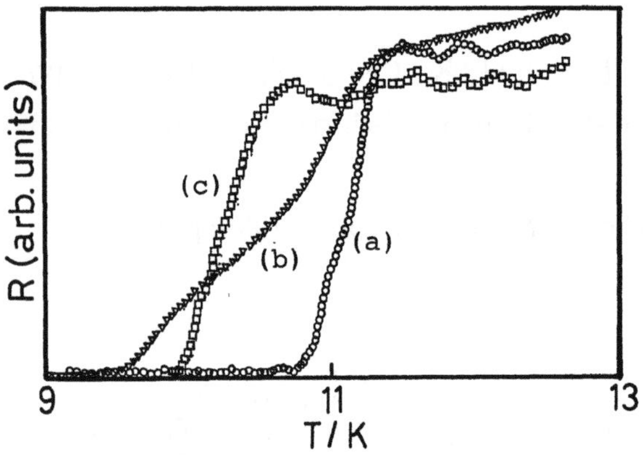

Fig. 1. Superconducting transition behaviors for (a) most purified, (b) purified, and (c) no purified κ-(BEDT-TTF-d_8)$_2$Cu(NCS)$_2$.

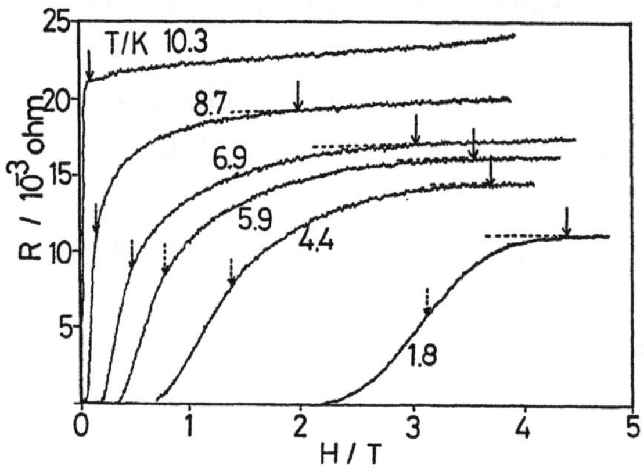

Fig. 2. The normal resistance recovery by the applied magnetic field along the a^*-axis for κ-(BEDT-TTF-d_8)$_2$-Cu(NCS)$_2$.

RESULTS AND DISCUSSION

The superconducting to normal state transitions by applying magnetic field at various temperatures of κ-(BEDT-TTF-d_8)$_2$Cu(NCS)$_2$ are depicted in Fig. 2. At higher temperatures sharper superconducting transition appeared, while a rather broad one was observed below 4.4 K. The transport H_{c2} was defined as the field at a half of a normal resistance. The corresponding result and other data (1b) were plotted as closed squares and circles in Fig. 3. This shows an anomalous upward curve with decreasing temperatures as previously reported (1b, 1j).

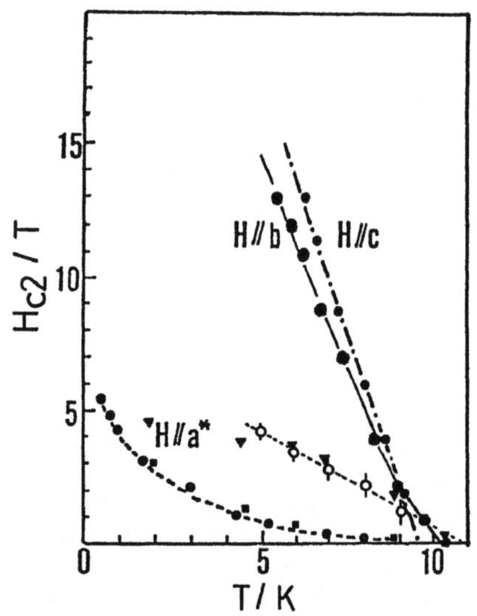

Fig. 3. The temperature dependence of H_{c2} for κ-(BEDT-TTF)$_2$Cu(NCS)$_2$ (closed circles··1/2 of normal resistance)and deuterated samples (closed squares··1/2 of normal resistance, closed triangles··offset of resistance recovery, and open circles··magnetization curve).

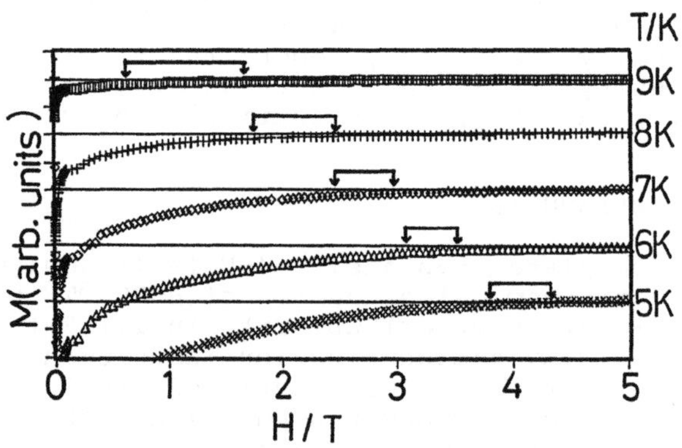

Fig. 4. The magnetization curve of H_{c2} by SQUID for κ-(BEDT-TTF-d_8)$_2$Cu(NCS)$_2$.

Recently we obtained magnetization curves up to 5 T at various temperatures as shown in Fig. 4 when single crystals (13.0 mg) of κ-(BEDT-TTF-d_8)$_2$Cu(NCS)$_2$ were aligned with applying a magnetic field along the a^*-axis. The samples were cooled to a set temperature under the zero field and the magnetization was observed at each magnetic field step. H_{c2} was determined as the applied field when M=0. On account of an experimental error of SQUID magnetometer ($<1\times10^{-5}$emu), H_{c2} was estimated as a range indicated by arrows in Fig. 4. The temperature dependence of these estimated H_{c2} was plotted as open circles in Fig. 3. Between 11 K to 5 K, H_{c2} follows a linear temperature dependence, which agrees well with Ginzburg-Landau theory.

What makes difference between transport upward temperature dependence of H_{c2} and magnetic linear one of H_{c2}? Figure 5 shows a total magnetization curve when a magnetic field was applied in the a^*-axis to aligned single crystals. The critical current at 50 G was estimated from hysteresis curve to be 1330 A/cm^2 (4) which is consistent with 1060 A/cm^2 at 50 G and 4.9 K for randomly oriented crystalline samples reported by K. Nozawa et al.(5) Due to the flux pinning caused by crystal defects, other imperfections, and etc., the hysteresis curve was obtained. The existence of pinning centers is also proved by a Meissner signals. M. Tokumoto et al. reported the Meissner signals that was a half as large as the shielding signals especially when an applied field was perpendicular to the conducting plane.(6) Though there exist flux pinning centers, enough thermal or magnetic energy release flux from pinning which causes flux creep or flux flow. Flux creep affords a broadening in the resistance recovery with field. The transport H_{c2} defined by the field at a half of normal resistance was a underestimated value. The H_{c2} by magnetization curve are reasonably correct. Similar results come from the measurement of the offset of the resistivity recovery as a function of field (arrows in Fig. 2 and closed triangles in Fig.3). This defined H_{c2} is much closer to the H_{c2} by magnetization curve, especially around T_c.

H_{c2} estimation in Y-Ba-Cu-O superconducting system resembles that of κ-(BEDT-TTF)$_2$Cu(NCS)$_2$.(7) Ac susceptibility measurement, the onset of resistance in transport measurements, and disappearance of irreversible magnetization and the corresponding critical current or pinning force determination techniques tend to show a upward curvature of the apparent $H_{c2}(T)$, while dc reversible magnetization measurements similar to our experiment afforded the linear temperature dependence of H_{c2}. Their result indicate a giant flux creep in oxide superconductors.

The flux pinning mechanism can be explained by measurements of a transport critical current density. This time, a current was applied along the b-axis, whereas a magnetic field was along the c-axis for a single crystal. The transport critical current density (J_c) at 5.1 K and 0 T was estimated as 100 A/cm^2, which is one order lower than that obtained by hysteresis loop, mentioned above. The difference might be that the former was crystalline samples and the latter was a single crystal. However, this J_c when Lorentz

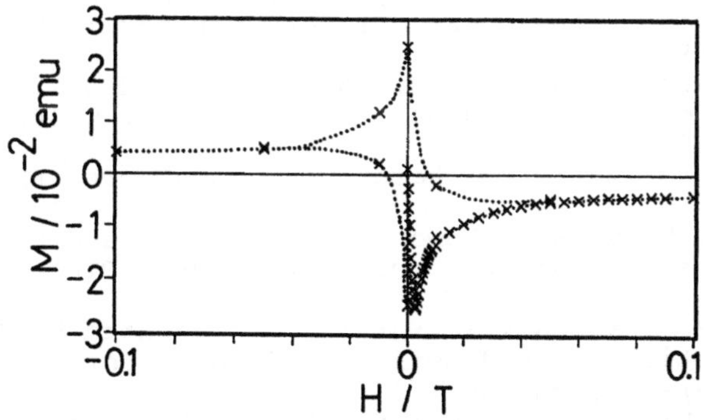

Fig. 5. The magnetization curve for κ-(BEDT-TTF-d_8)$_2$Cu(NCS)$_2$ at 5 K.

Force worked on a flux along the a^*-axis might be higher than J_c when a force was imposed to a flux in the two-dimensional plane, since an insulating plane can be a pinning center. More detail of pinning mechanism of κ-(BEDT-TTF)$_2$Cu(NCS)$_2$ will be presented elsewhere.(8)

In summary, we newly determined H_{c2} by measurements of magnetization curve. The H_{c2} was rather higher and in a linear temperature dependence, compared with that defined by electrical resistivity recovery with field. The anomalous behavior of upward temperature dependence on transport H_{c2} is caused by a flux creep or a flux flow.

REFERENCE

1. (a)Urayama H. Yamochi H. Saito G. Nozawa K. Sugano T. Kinoshita M. Sato S. Oshima K. Kawamoto A. and Tanaka J. (1988) A new ambient pressure organic superconductor based on BEDT-TTF with T_c higher than 10 K (T_c=10.4 K). Chem. Lett. 1988: 55; (b)Oshima K. Urayama H. Yamochi H. and Saito G.(1988) Peculiar critical field behavior in the recently discovered ambient pressure organic superconductor (BEDT-TTF)$_2$Cu(NCS)$_2$ (T_c=10.4 K). J. Phys. Soc. Jpn. 57: 730; (c)Urayama H. Yamochi H. Saito G. Sato S. Kawamoto A. Tanaka J. Mori T. Maruyama Y. and Inokuchi H (1988). Crystal structure of organic superconductor, (BEDT-TTF)$_2$Cu(NCS)$_2$, at 298 K and 104 K. Chem. Lett. 1988: 463; (d)Oshima K. Mori T. Inokuchi H. Urayama H. Yamochi H. Saito G. Shubnikov-de Haas effect and the fermi surface in an ambient-pressure organic superconductor ((bisethylenedithiolo)tetrathiafulvalene)$_2$-Cu(NCS)$_2$. (1988) Phys. Rev. B 37: 938; (e)Urayama H. Yamochi H. Saito G. Sugano T. Kinoshita M. Inabe T. Mori T. Maruyama Y. and Inokuchi H. (1988) Valence state of copper atoms and transport property of an organic superconductor, (BEDT-TTF)$_2$Cu(NCS)$_2$, measured by ESCA, ESR, and thermoelectric power. Chem. Lett. 1988: 1057; (f)Oshima K. Urayama H. Yamochi H. and Saito G. (1988) A new ambient pressure organic superconductor (BEDT-TTF)$_2$Cu(NCS)$_2$ with T_c above 10 K. Physica C 154: 1148; (g)Saito G. Urayama H. Yamochi H. and Oshima K. (1988) Chemical and physical properties of a new ambient pressure organic superconductor with T_c higher than 10 K. Synth. Met. 27: 331; (h)Urayama H. Yamochi H. Saito G. Sato S. Sugano T. Kinoshita M. Kawamoto A. Tanaka J. Inabe T. Mori T. Maruyama Y. and Inokuchi H. (1988) Crystal and electronic structures and physical properties of T_c=10.4 K superconductor, (BEDT-TTF)$_2$Cu(NCS)$_2$. Synth. Met. 27: 393 (i)Oshima K. Mori T. Inokuchi H. Urayama H. Yamochi H. and Saito G. (1988) Fermi surface and pressure effect in (BEDT-TTF)$_2$Cu(NCS)$_2$. Synth. Met. 27: A413; (j)Oshima K. Urayama H. Yamochi H. and Saito G. (1988) Superconducting critical field in (BEDT-TTF)$_2$Cu(NCS)$_2$ Synth. Met. 27: A419; (k)Oshima K. Urayama H. Yamochi H. and Saito G. (1988) Superconductivity and deuteration effect in (BEDT-TTF)$_2$Cu(NCS)$_2$ Synth. Met. 27: A473; (l)Mori H. Tanaka S. Yamochi H. Saito G. and Oshima K. (1989) An ambient pressure organic superconductor κ-(BEDT-TTF-h_8 and -d_8)$_2$Cu(NCS)$_2$ with Tc higher than 10 K. (to be published) The physics and chemistry of organic superconductors. eds. Saito G. and Kagoshima S., Springer-Verlag
2. Takahashi T. Tokiwa T. Kanoda K. Urayama H. Yamochi H. and Saito G. (1988) NMR relaxation in the organic superconductor (BEDT-TTF)$_2$Cu(NCS)$_2$. Synth. Met. 27: A319
3. Katsumoto S. Kobayashi S. Urayama H. Yamochi H. and Saito G. (1988) Low-temperature specific hear of organic superconductor κ-(BEDT-TTF)$_2$Cu(NCS)$_2$. J. Phys. Soc. Jpn. 57; 3672
4. Bean CP. (1962) Magnetization of hard superconductors. Phys. Rev. Lett. 8: 250
5. Nozawa K. Sugano T. Urayama H. Yamochi H. Saito G. and Kinoshita M. (1988) Meissner effect in an organic superconductor (BEDT-TTF)$_2$(Cu(NCS)$_2$). Chem. Lett. 1988: 617
6. Tokumoto M. Anzai H. Takahashi K. Murata K. Kinoshita N. Ishiguro T. (1988) Anisotropy of magnetization and meissner effect in organic superconductor κ-(BEDT-TTF)$_2$Cu(NCS)$_2$. Synth. Met. 27: A305
7. Malozemoff AP. Worthington TK. Yeshurun Y. and Holtzberg F. (1988) Frequency dependence of the ac susceptibility in a Y-Ba-Cu-O crystal: A reinterpretation of H_{c2}. Phys. Rev. B 38: 7203
8. Mori H. Nakao K. Nagaya S. Hirabayashi I. and Tanaka S. (to be submitted)

3.4 ESR and Mossbauer Studies

Superconducting Properties and ESR of Mn-Doped $YBa_2Cu_3O_{7-x}$

MASAHIRO KAISE, MASAGI MIZUNO, CHIZUKO NISHIHARA, HISAKAZU NOZOYE, and HITOSHI SHINDO

Division of Basic Research, National Chemical Laboratory for Industry, Tsukuba, Ibaraki, 305 Japan

ABSTRACT

Materials of partially substituted Cu with manganese in high-Tc $YBa_2Cu_3O_{7-x}$ (YBCO) were prepared, and found that Tc decreases with an increase of the Mn concentration. However, the superconducting behavior to temperature of the materials show that $YBa_2Cu_{3-y}Mn_yO_{7-x}$(YBCMO) is metastable and composition variation occurs between bulk phase and grain boundaries.
The ESR measurements show that neither Cu nor Mn does not exist as isolated divalent cations in the materials. However, all powder samples after manual grinding have shown similar ESR absorption ascribed to Cu^{2+} with orthorhombic g-tensor. They are considered to be a mechano-radical which produced on the surface of the crystallites.

KEY WORDS : $YBa_2Cu_3O_{7-x}$, Mn doping, ESR, Mechano-radical

INTRODUCTION

The characterization of the 90 K high-Tc compounds with defects and impurities are important not only to get more insight into the conduction mechanism[1,2] but also to the technological utility of the material. There are many reports on the study of substitution and mixing effect on the YBCO and related compounds.[3-7] A rather low critical current density (Jc) is ascribed to nonsuperconducting phase or defects on the grain boundaries and voids or mismatch between the grains. The detailed study on the structure, defect morphology, and composition of grain boundaries therefore remains of utmost importance.[11,13] The present paper describes some results obtained with YBCO and manganese doping for copper. The electric conduction measurements and the ESR measurements were performed with the prepared superconducting materials.

EXPERIMENTAL

The reagent grade powders of Y_2O_3, $BaCO_3$, CuO, and crystallites of $Mn(NO_3)_2$ were used as starting materials. The disk-shaped tablets (1-2 g, 10 mm in diameter) of the YBCO samples were prepared by the standard method in air. The Mn-doped YBCMO were obtained with similar method of the YBCO after the addition of proper amount of $Mn(NO_3)_2$ and drying. The electrical resitivities as a function of temperature for the samples of calcined tablets were measured by a d.c. four-probe method with contact probes which were needless any conducting pastes to connect sample and lead wire. The ESR signals were detected at room temperature with a JEOL JES-RE3X X-band and a Varian E-112 Q-band spectrometer systems. The microwave power used was 0.1- 1.0 mW. The magnetic field was calibrated for isotropic 6 h.f. lines of Mn^{2+} doped in MgO powder. The mg order samples were filled in a Suprasil tube.

RESULTS AND DISCUSSION

There are many reports on the effect of foreign ion doping except Mn on the superconductivity of YBCO.[3-7] We prepared samples of the form $YBa_2Cu_{3-y}Mn_yO_{7-x}$ (y=0, 0.02, 0.04, 0.08) because the ionic size and orbital structure of the Mn is close to those of Cu, and could be expected that the Mn occupy the Cu sites if it is substituted in the YBCO. Mn has versatile charged states ; Mn^{3+}, Mn^{2+}, Mn^{+}, Mn^{0}, Mn^{-}, and most of those have unpaired electrons which should produce substantial changes in the superconducting and magnetic properties to be elucidated. The composition of Y, Ba, Cu, and Mn were determined by ICP atomic emission spectorometry. It has turned out that the composition of the Mn-doped YBCO (YBCMO) preserve those of stating reagents except oxygen. This is one of indications to a substitution of the Mn for the sites of Cu ions.

Figure 1 shows the temperature dependence of the electric resitivity for $YBa_2Cu_{3-y}Mn_yO_{7-x}$ (y=0.08). The YBCO without Mn has the Tc of 90 K and a very narrow transition width. (A) is obtained by a twice calcined disk. Substitution by Mn produces a reduction in Tc, and the transition width is broader than those of YBCO. In the normal conducting phase above the transition , the resitivity behavior to temperature is metallic and decreasing monotonously from high to low temperature, which is different from Fe, Co and Ga.[4] Figure 1(B) is obtained from a five time calcined disk. The difference of (B) from (A) is the higher Tc onset and the lower Tc end with small step to the low temperature side. It is considered that the materials have two phases with different Mn density ; the YBCMO is intrinsic metastable and Mn segregate toward grain boundaries by repeated heat treatment in air.

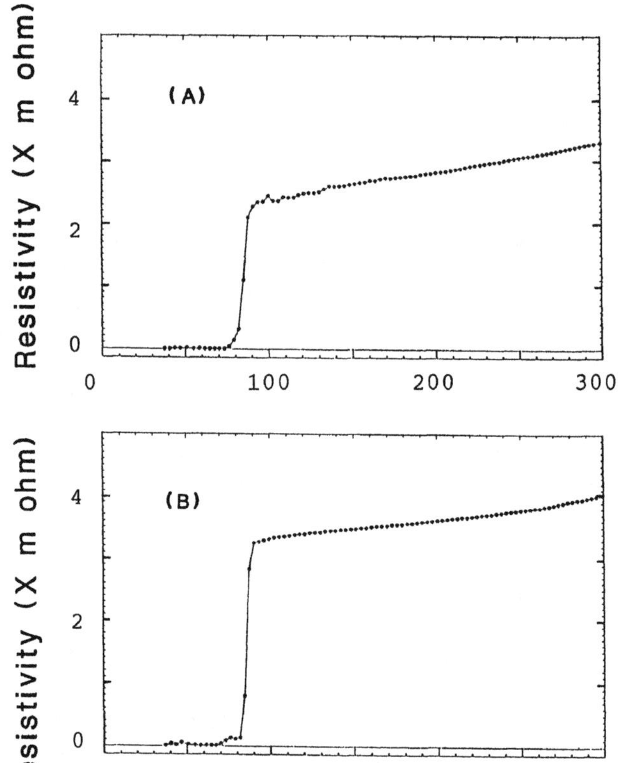

Fig. 1. Resistivity vs. temperature of two Mn-doped $YBa_2Cu_{3(1-y)}Mn_yO_{7-x}$ (y = 0.08): (A) is from a disk of powder which obtained by twice calcining; (B) is from a disk of powder which obtained by five time calcining.

Since the discovery of YBCO and related superconducting ceramic oxides, many ESR studies have been carried out because of its exclusive high sensitivity to paramagnetic species in the materials.[8-10] The Cu^{2+} ions could exist in conductive CuO_2 planes to almost all copper oxides superconductors or as impurities and defects. The ions are one of the most popular paramagnetic species in the ESR study. If we use the grains in mm size from calcined disks, no appreciable ESR absorption was observed for any prepared YBCO and Mn-doped YBCMO. On the other hand, definite ESR absorption were observed from YBCO fine

powder which obtained by mechanical grind of the calcined disks in an agate mortar. There are some reports on the variations of local composition within grain boundaries and chemical instability at high mechanical pressure to the YBCO.[11,12] Figure 2 shows ESR spectra of ground powder (between 200 and 250 mesh) of the superconducting YBCO at room temperature.
Figure 2(A) is a X-band spectrum operating at 9.446 GHz of 0.1 mW microwave.
The spectrum shows a typical powder line shape with rhombic symmetry of ESR tensor parameter for the spin Hamiltonian. Figure 2(B) is a Q-band spectrum operating at 35.07 GHz of 1 mW microwave. It is concluded from the comparison of the X-band and the Q-band ESR spectra that ESR tensor parameter does depend on the intensity of magnetic field and could be assigned to g-tensor.

Obtained g-values are g_x=2.049, g_y=2.122, g_z=2.234 from X-band spectrum, and g_x=2.048, g_y=2.125, g_z=2.240 from Q-band spectrum. These g-values are in fair agreement with each other. There are no resolved structure by h.f. of Cu^{2+}(I=2/3) in both spectra. The fact indicates that the Cu^{2+} derived paramagnetic species do not exist as isolated states in the sample, and an unpaired electron is delocalized around the vicinity of the defects. The h.f. is narrowed less than the individual intrinsic line width of the defects becase of the strong exchange interaction between the Cu^{2+} ions and Cu of Cu-O sheets. The stronger ESR absorption was obtained from the finer powders in the same quantity . This indicates that the species are present on the surface of powder particles. On the other hand, a considerable reduction of the ESR absorption was observed from the heat-treated powders: 520 °C (12 h) in air. The paramagnetic species are ascribed to the defects which were caused by oxygen dispersion from the surface of the crystallites. A similar ESR absorption spectra were observed from the Mn-doped YBCMO, but no detectable Mn derived ESR signal was detected from any prepared samples. The doped Mn does not exist as isolated paramagnetic species in the YBCO at room temperature, though it sensitively influences to the superconductivity.

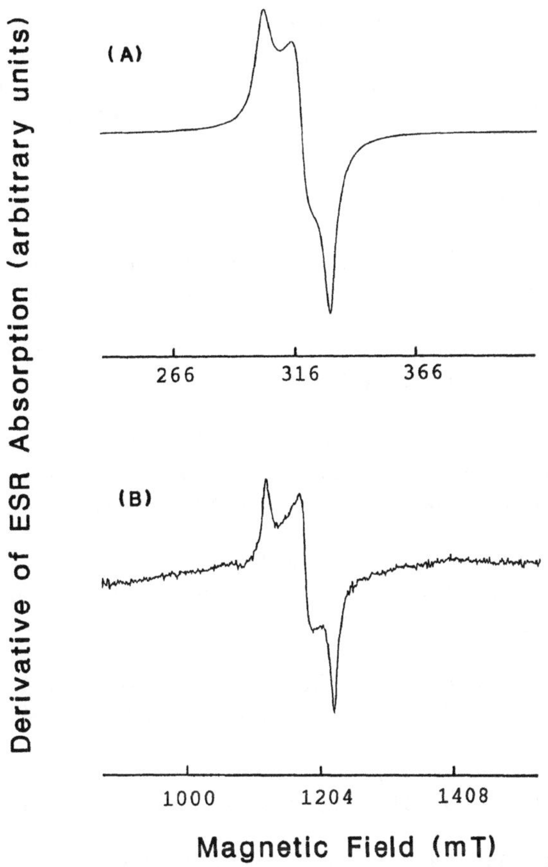

Fig. 2. ESR spectra of a ground powder (200-250 mesh) of the superconducting $YBa_2Cu_3O_{7-x}$ at room temperature: (A)is a X-band ESR spectrum operating at 9.446 GHz of microwave; (B) is a Q-band ESR spectrum operating at 35.07 GHz .

In conclusion, we have demonstrated that the Mn-doped YBCO is intrinsic metastable and Mn segregates toward grain boundaries by repeated calcining processes, which shows two-step resitivity behavior at the Tc of the superconductor. Mn is substituted for Cu site and sensitively decreases Tc, though Mn do not exist as an isolated paramagnetic species at room temperature. This is another fact that copper-oxygen sheet is prerequisite to the superconducting nature of the high-Tc materials.

The Cu^{2+} derived paramagnetic species with orthorhombic g-tensor was discovered for the first time on the surface of the YBCO crystallites. This is one of the mechano-radicals produced by a mechanical force of grinding. The species were identified to oxygen decipated defects in which unpaired electron is delocalized beyond adjacent Cu atoms.

REFERENCES

1. W. A. Little, Experimental Constraints on Theories of High-Transition Temperature Superconductors. Science, 242,1390 (1988).

2. A. W. Sleight, Chemistry of High-Temperature Superconductors. Science, 242, 1519 (1988).

3. Y. Maeno, T. Tomita, M. Kyogoku, S. Awaji, Y. Aoki, K. Hoshino, A. Minami & T. Fujita, Substitution for copper in a high-Tc superconductor $YBa_2Cu_3O_{7-x}$. Nature, 328, 512 (1987).

4. G. Xiao, F. H. Streitz, A. Garvin, Y. W. Du and C. L. Chien, Effect of transition-metal elements on the superconductivity of Y-Ba- Cu- O. Phys. Rev. B, 35, 8782 (1987).

5. T. Miyatake, S. Gotoh, N. Nishizuka & S. Tanaka, Tc increased to 90 K in $YBa_2Cu_4O_8$ by Ca doping. Nature, 341, 41 (1989).

6. N. Imanaka, F. Saito, H. Imai and G. Adachi, Critical Current Characteristics of $YBa_2Cu_3O_{7-x}$-Ag Composite. Japn. J. Appl. Phys., 28, L580 (1989).

7. R. Sonntag, D. Hohlwein, A. Hoser, W. Prandl, W. Schafer, R. Kiemel, S. Kemmler-Sack, S. Losch, M. Schlichenmaier and A. W. Hewat, Structural Changes of the Superconductor $YBa_2Cu_3O_7$ by Cobalt Substitution. A High-resolution Neutron Powder Diffraction Study. Physica C, 159, 141 (1989).

8. D. Shaltiel, H. Bill, P. Ficher, M. Francois, H. Hangemann, M. Peter, Y. Ravisekhar, W. Sadowski, H. J. Scheel, G. Triscone, E. Walker and K. Yvon, Single Crystal ESR Studies on Tetragonal $YBa_2Cu_3O_{6+x}$. Physica C, 158, 424 (1989).

9. A. G. Vedeshwar, Md. Shahbuddin, P. Chand, H. D. Bist, S. K. Agarwal, V. N. Moorthy, C. V. N. Rao and A. V. Narlikar, EPR and Low Magnetic Field Microwave Absorption in Hafnium Doped $YBa_2Cu_3O_y$. Physica C, 158, 385 (1989).

10. Y. Hayashi, M. Fukui, T. Fujita, H. Shibayama, K. Iwahashi and K. Adachi, ESR and Nonresonant Microwave Absorption in a High-Tc Superconductor of the Tl-Ba- Ca- Cu- O System. Japn. J. Appl. Phys., 28, L910 (1989).

11. S. E. Babcock and D. C. Larbalestier, Evidence for local composition variations within $YBa_2Cu_3O_{7-x}$ grain boundaries. Appl. Phys. Lett., 55, 393 (1989).

12. B. C. Hendrix, T. Abe, J. C. Borofka and J. K. Tien, Chemical instability of $YBa_2Cu_3O_{7-x}$ at high mechanical pressures. Appl. Phys. Lett., 55, 313 (1989).

13. R. A. Camps, J. E. Evetts, B. A. Glowacki, S. B. Newcomb, R. E. Somekh & W. M. Stobbs, Microstructure and critical current of superconductivity $YBa_2Cu_3O_{7-x}$. Nature, 329, 229 (1987).

ESR Study on High-T_c Superconducting Oxides

Y. Yamada, K. Sugawara, and Y. Shiohara

Superconductivity Research Laboratory, International Superconductivity Technology Center, 10-13, Shinonome 1-chome, Koto-ku, Tokyo, 135 Japan

ABSTRACT

$YBa_2(Cu_{1-X}Mn_X)_3O_{7-\delta}$ (x=0.03) was investigated by electron spin resonance(ESR). Two types of signals were detected which have different temperature dependences. It seems that one is due to the intrinsic of $YBa_2(Cu_{0.97}Mn_{0.03})_3O_{7-\delta}$, but the other is due to an impurity phase composed with Ba, Mn, Cu and O.

KEY WORDS: oxide superconducter, electron spin resonance

INTRODUCTION

A number of ESR studies of $YBa_2Cu_3O_{7-\delta}$ have been investigated [1-3], and revealed that no signal appears in the sample of δ=0[4,5]. The similar behavior has been found in both La-system[6] and Bi-system[7,8]. No signal was detected from Cu^{2+} ions in the superconducting oxides,which seems to be due to the characteristics of these specimens having high Tc. Since Cu ions are essential components to high Tc, it is important to investigate whether or not partly substituting Cu ions with other metal ions affects the electronic state of surrounding Cu ions. And it is interesting to understand a relationship between Tc and the electronic state of Cu ions. In addition, substituting ions may provide an atomic probe when ESR signals are observed from them.

EXPERIMENTAL

$YBa_2Cu_{3-X}Mn_XO_{7-\delta}$ (x= 0.09) were prepared from the mixture of appropriate amounts of oxides Y_2O_3, CuO, MnO_2, and $BaCO_3$, followed by calcining and firing in air at 920°C for 12hrs., then pulverized. After twice firings, the sample was pressed into pellets and sintered in a tube furnace at 920°C under an oxygen gas flow for 12hrs.

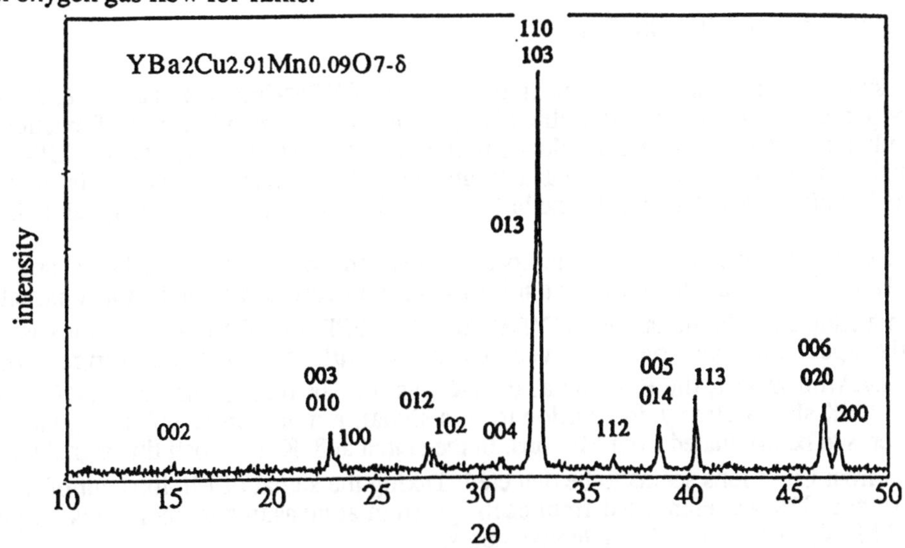

Fig.1 X-ray powder diffraction spectrum of $YBa_2Cu_{2.91}Mn_{0.09}O_{7-\delta}$

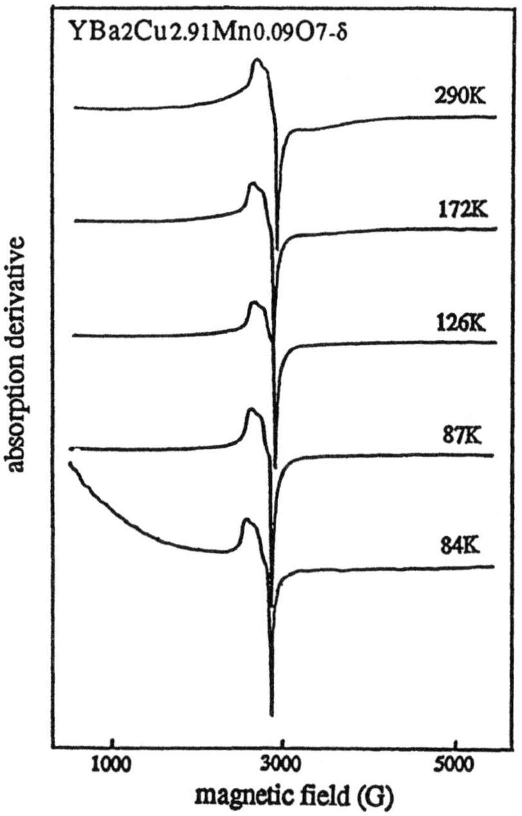

Fig.2 Typical ESR powder spectra of $YBa_2Cu_{2.91}Mn_{0.09}O_{7-\delta}$ at the different temperature

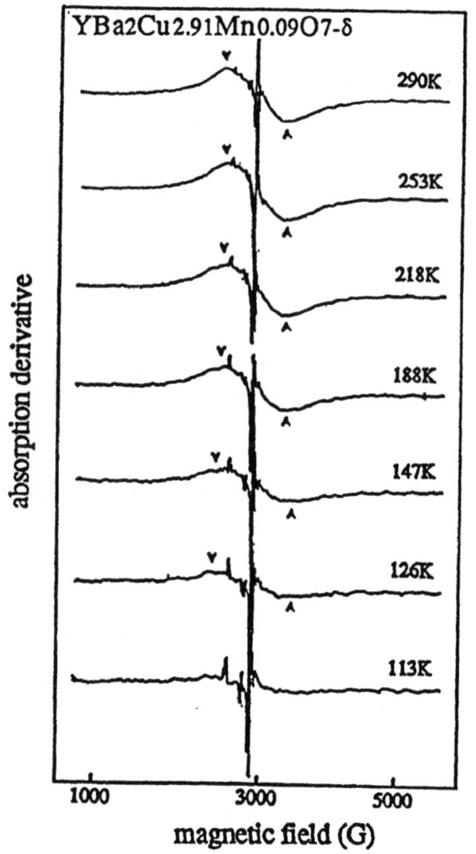

Fig.3 Mn^{2+} signals at different temperatures

ESR measurements were performed on powdered samples by a conventional X band(9.1GHz) ESR equipment, and in the temperature range between Tc and 473K. The signal sensitibity was estimated by comparing with Mn^{2+} signal whose ions dissolved in MgO.

RESULTS AND DISCUSSION

Figure 1 shows the powder X-ray diffraction spectrum of $YBa_2Cu_{2.91}Mn_{0.09}O_{7-\delta}$. No spectrum from impurity phase is seen. An orthorhombic phase pattern is disturbed because of existience of Mn ions. This result suggests that Mn ions are doped in the 123 phase structure.However, EPMA(electron prob microanalizer) analisis indicated that a small amount of inpurity phases exist such as CuO and compounds with Ba, Mn, Cu and O. The influence of these inpurities to ESR signal is described later.

ESR line shapes detected at different temperatures are shown in Fig.2. The line shape at 84K shows a typical field dependent non-resonant microwave absorption which is the characteristic of the superconducting state. In the sample of $YBa_2Cu_3O_{7-\delta}$, ESR signal from Cu^{2+} ions is often observed, and this spectrum can be observed over the full temperature range (10K-473K). In $YBa_2Cu_{2.91}Mn_{0.09}O_{7-\delta}$, the ESR signal at 87K was similar to the signal in $YBa_2Cu_3O_{7-\delta}$, while the signal at 290K shows clearly, in addition to this signal, a much broader line signal. This broad line signal seems to be originated from Mn ions. In the signal at 87K, no broad line signal was seen and only the signal from Cu^{2+} ions seems to be detected. Taking the signal of Mn ions out of these two signals, the spectrum at 87K was eliminated from each spectrum at measured temperatures. Figure 3 shows the remained Mn signals at the different temperatures.

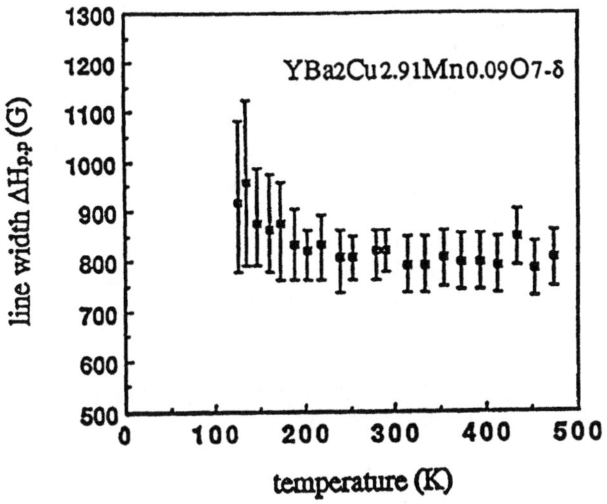

Fig.4 Temperature dependence of line width of Mn^{2+} signals

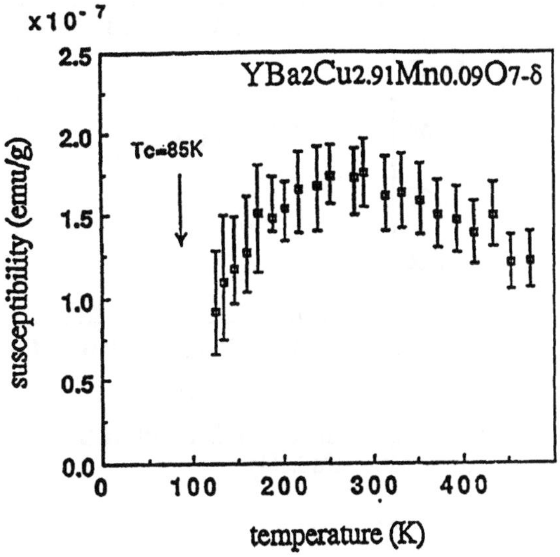

Fig.5 Temperature dependence of susceptibility of Mn^{2+} signals

The line width of this signal is almost independent of temperature, while the signal heights decrease significantly with decreasing temperature. Therefore, the intensity of the signal decreases with decreasing temperature. Temperature dependence of the line width and spin susceptibility deduced from the signal intensity are shown in Figs.4 and 5. The line width is broadened slightly below about 200K. This behavior indicates that the motion of spin fluctuation is slowing down at low temperatures.

To consider the physical meanig of these behaviors, it is important whether these ESR signals from Mn ions are intrinsic of $YBa_2Cu_{2.91}Mn_{0.09}O_{7-\delta}$. No inpurity phase was detected by XRD, however, a small amount of CuO and Ba-Mn-Cu-O were detected by EPMA.We examine whether or not CuO and Ba-Mn-Cu-O provide ESR signals. From CuO, no ESR signal was detected at room temperature. CuO is an antiferromagnetic material whose neel temperature is about 220K which has been determined by its specific heat and nutron diffraction, while, it is 453K evaluated by the susceptibility measurement. This magnetic property is the reason why CuO does not provide any ESR signal.

In order to examine the ESR signal of Ba-Mn-Cu-O, we prepared $BaMn_{0.5}Cu_{0.5}O_y$ by sintering. Intensive signals were detected and Figs.6 and 7 show its line width and signal intensity respctively.Comparing these data with Figs.4 and 5, they are almost consistent. This fact suggests that the signals of Mn ions from the sample of $YBa_2Cu_{2.91}Mn_{0.09}O_{7-\delta}$ were originated possibly from Ba-Mn-Cu-O phase.

It is obvious that superconducting phase includes Mn ions By EPMA analisis, while it is not clear whether the ESR signals from Mn ions in superconducting phase exist. The intensity of the central line of $YBa_2Cu_{3-X}Mn_XO_{7-\delta}$ increases with increasing Mn contents (Fig.8) and X-ray diffraction spectra of $YBa_2Cu_3O_{7-\delta}$ phase varies systematicaly. This suggests that Mn ions affect the structure and electronic state of $YBa_2Cu_3O_{7-\delta}$.

SUMMARYS

In summary, we have shown that ESR spectra of $YBa_2(Cu_{1-x}Mn_x)_3O_{7-\delta}$ consist of more than two signals. One of them is due to Mn ions. The line width of the Mn signal is gradually broadened below 200K because of spin fluctuation. Spin susceptibility shows a peak at about 300K, and decreases with decreasing temperature. Ba-Mn-Cu-O provides the ESR signal, therefore, the ESR spectra of the sample of $YBa_2Cu_{3-X}Mn_XO_{7-\delta}$ includes Ba-Mn-Cu-O spectra. It is not clear the origin of other signals and whether the ESR signals from Mn ions in superconducting phase exist. The line shape variations as a function of Mn content suggests that Mn ions affect the structure and electronic state of $YBa_2Cu_3O_{7-\delta}$.

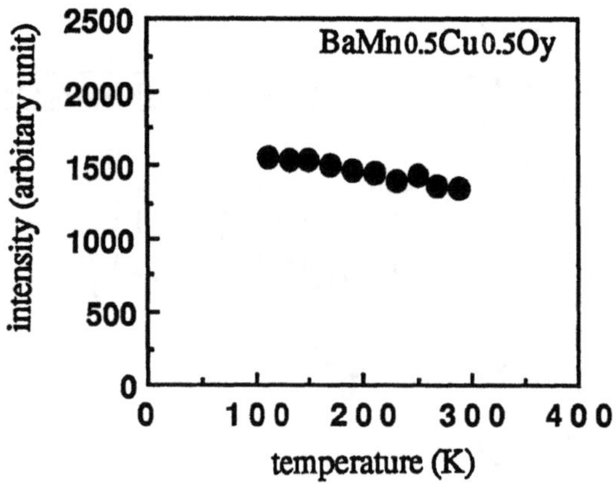

Fig.7 Temperature dependence of intensity of $BaMn_{0.5}Cu_{0.5}O_y$

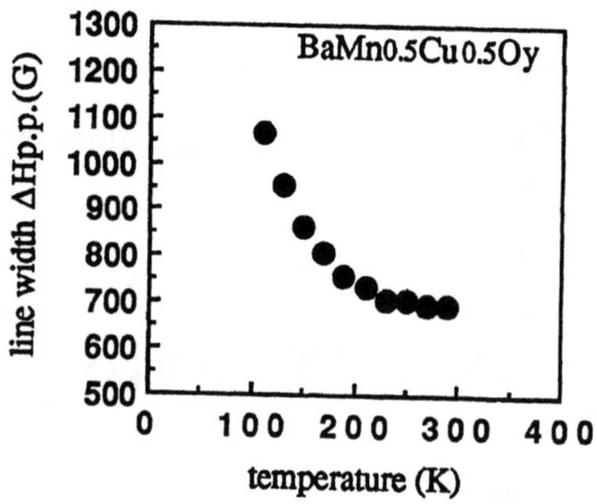

Fig.6 Temperature dependence of line width of $BaMn_{0.5}Cu_{0.5}O_y$

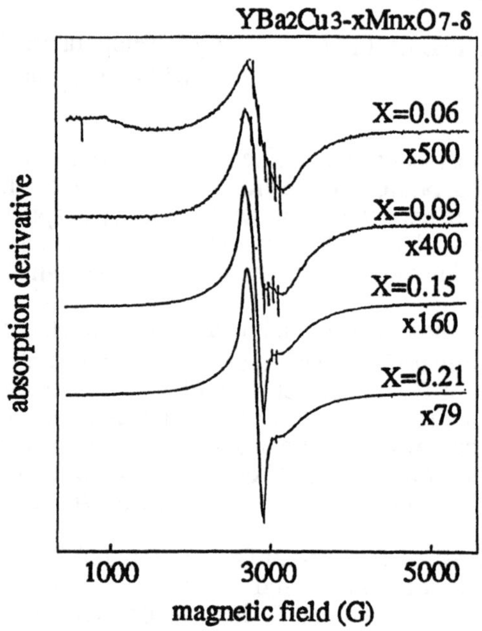

Fig.8 Typical ESR power spectra of $YBa_2Cu_{3-x}Mn_xO_{7-\delta}$ at the different Mn contents

It is nessecery that pure single phase sample or single crystal are prepered for ESR measurement. This work has been supported by the R&D Basic Technology for Future Industries through New Energy and Industrial Technology Development Organization(NEDO).

REFFERENCES

1. Oseroff SB. (1987) Solid State Commun. 64: 241
2. Owens FJ. (1989) Solid State Commun. 70: 173
3. de Mesquita RN. Castilho JH. Barberis GE. Rettori C. Torriani I. Terrile MC. Basso H. and Nascimento OR. (1989) Phys. Rev. B 39: 6694
4. Albino J. de Aguiar O. Menovsky AA. Van Den Berg J. and Brom HB.,(1988) J.Phys.C:Solid State Phys. 21:L237
5. Mehran F. Bornes SE. McGuire TR. Dinger TR. Kaiser Dl. and Holtzberg F. (1988) Solid State Commun.66:299
6. Kikuchi H and Ajiro Y (1988) J.Phys.Soc.Jpn. 57: 2628
7. Owens FJ. and Iqbal Z (1988) Solid State Commun. 68: 523
8. Tagaya K (1989) Jpn. J. Appl. Phys. 28: L566

Interactions among Gd and Cu-2 in $GdBa_2Cu_3O_y$ Compounds

F. NAKAMURA[1], K. SENOH[1], T. TAMURA[1], S. NAKADA[1], H. SHIMIZU[3], Y. OCHIAI[2], and Y. NARAHARA[1]

[1] Institute of Physics, University of Tsukuba, Tsukuba, 305 Japan
[2] Institute of Materials Science, University of Tsukuba, Tsukuba, 305 Japan
[3] Faculty of Science and Technology, Science University of Tokyo, Noda, 278 Japan

ABSTRACT

Electron-spin-resonance (ESR) and dc-susceptibility are measured for $GdBa_2(Cu_{1-x}M_x)_3O_y$ (M= Ni and Co) compounds. Curie-Weiss susceptibility and the frequency dependence of the linewidth are observed. The Gd magnetism can be explained by the Gd dipolar and the exchange interaction in this system. The origin of the exchange interaction is the superexchange interaction via the oxygen on Cu-2 plane. The Gd linewidths largely depend on Ni concentrations. The results indicate that the substitution of Ni makes the superexchange interaction on Cu-2 plane weak.

INTRODUCTION

In $GdBa_2Cu_3O_7$ compounds, superconductivity occurs below 90 K. The antiferromagnetic ordering of Gd ions is observed below about 2.3K.[1] The static susceptibility in the normal phase obeys a Curie-Weiss law. The effective magnetic moments are close to that of Gd^{3+} free ions. The atomic distance in Gd plane is larger than the ionic radius of the free ion. The 4f-electron among Gd plane are considered to be localized. The Gd dipole-dipole interaction can be expected in the Gd plane. The Néel temperature is calculated from the dipole-dipole interaction to be about 1.4K.[2] However, the observed Néel temperature is about 2.3 K. Then it cannot be explained from only the dipolar interaction. Recent neutron-diffraction studies report that the ordering are three-dimensional.[3] We expect that these results should be explained not only with the dipole interaction but also with considering other interactions. Some reports[4] suggest the possibilities of the interactions between Gd ions and Cu-O plane. It is well known that the Cu-O plane takes an important role in the superconductivity. In order to explain such interactions, we try to measure X(9.5GHz), K(23.5GHz) and Q-band(35.2GHz) electron-spin-resonance (ESR) and dc-susueptibility in our compounds. We discuss the susceptibilities, the g shift, and the relaxation time. The discussions will make clear the origin of the Gd magnetism in these compounds and the magnetic behavior of Ni and Co ions.

EXPERIMENT

Our samples are prepared from mixtures of Gd_2O_3, $BaCO_3$, NiO, CoO, and CuO powders. These mixtures are sintered in flowing nitrogen atmosphere at about 850 C for 24h, and are cooled down to room temperature. These pellets are reground and resintered in flowing oxygen atmosphere at 950 C for 24h, and are cooled down to 600 C at the rate of 20 C/h and then down to room temperature at the rate of 10 C/h. A powder X-ray diffraction study is performed. The compounds show a single-phase orthorhombic $GdBa_2Cu_3O_7$ structure. In order to control the oxygen content y, these compounds are annealed at constant high temperatures (from 700 to 400 C) in a flowing nitrogen atmosphere for about 20 h, and cooled down to room temperature at the rate of 20 C/h. For all the compounds, we determine the oxygen content using a modification of standard iodometric titration techniques.[5]

ESR measurements are performed for the compounds. The X-band ESR spectra are recorded using a JEOL JES-FE spectrometer and Varian E-7000 series magnet system, equipped with an Oxford Instruments E9 helium flow-through cryostat in the X-band measurements. The K-band ESR spectrum are recorded using these spectrometer system and MICRO DEVICE MWG-24KR microwave unit. The Q-band ESR spectra are observed using a Varian E-Line EPR System. Dc-susceptibilities are measured using a SQUID magnetometer (SHE VTS model 805) at the range of from 300 to 1.6 K. Spin susceptibilities are determined from the linewidth and intensity of the signals.

RESULTS AND DISCUSSION

In Fig.1, the dc and spin susceptibilities are shown for $GdBa_2(Cu_{1-x}M_x)_3O_y$ (M= Ni and Co) compounds. These susceptibilities obeys a Curie-Weiss law. For x=0 compounds, we determine the Curie constant to be 9.55 emu K/g, and the effective magnetic moments to be 7.54 μ_B. There are no difference between the spin and dc susceptibilities. Therefore, the susceptibility of this compound is roughly determined by magnetism of Gd 4f-electron. The Weiss temperature, θ, is -3±1K. It is well known that an exchange integral J can be estimated from the Weiss temperature. In $GdBa_2Cu_3O_y$ compounds, we estimate J to be 6×10^{-18} erg.

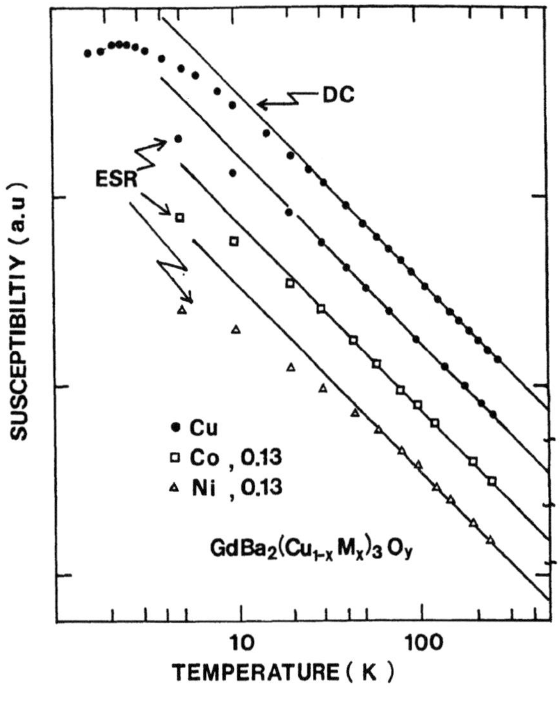

Fig. 1. The dc and spin susceptibility of $GdBa_2(Cu_{1-x}M_x)_3O_y$ (M=Ni and Co) compounds. These obey a Curie-Weiss law. The oxygen contents of non-doped compounds are y=6.2.

ESR spectra for Gd 4f-electron in $GdBa_2Cu_3O_7$ compounds are observed. On the K and Q-band spectra, the shapes are a Lorentzian shape. It is well known that an exchange or a motional narrowing make a line shape Lorentzian form. Since the Gd 4f-electron is localized in this system, it is difficult to expect a motional narrowing. The exchange interaction is expected and is stronger than the dipolar interaction. The linewidths depend on the Zeeman frequencies. The observed peak to peak linewidths ΔH are 2800 Oe in f=9.05 GHz, 2100 Oe in f=23.5 GHz and 1700 Oe in f=35.2 GHz, respectively. The temperature dependence of the linewidth is reported in previous results.[6] The linewidths are determined by spin-spin relaxation time. The linewidths can be determined mainly dipole, ω_d , exchange, ω_{ex}, and Zeeman frequencies, ω_z . Generally, theoretical expression[7] for the frequency dependence in ESR linewidths ΔH is given by

$$\Delta H= \frac{3}{5}\,B\sigma^2\tau + B\sigma^2\{\frac{\tau}{1+\omega_z{}^2\tau^2} + \frac{2}{5}\,\frac{\tau}{1+4\omega_z{}^2\tau^2}\} \qquad (2)$$

where B is constant, σ is natural unit for the frequency associated with the dipolar interaction, τ is correlation time, and ω_z is zeeman frequency. By using the equation (2), the correlation time is estimated to be 1.3×10^{-10}sec in this system. Since the Gd 4f-electron is localized, the τ is determined by the exchange mechanism. This leads to the exchange integral J among Gd spins to be 8.1×10^{-18} erg. This estimation of J is consistent with that from the Weiss temperature. In this system, the exchange field amounts to about 5 kOe. The dipolar interaction is calculated to be 3kOe.[2]

In order to make clear the origin of this exchange field, the ESR measurements are performed for $GdBa_2(Cu_{1-x}M_x)_3O_y$ (M=Ni and Co) compounds. A broadening of the linewidth is observed with increasing x from x=0 to 0.16. The broadening of the Ni doped compounds is larger than that of the Co doped compounds. It is shown in Fig.2. It is well known that Ni ions substitute Cu ions on Cu-2 planes, and Co ions replace Cu ions on Cu-1 chains. This result is obtained that the Gd ions interact with Cu-2 plane. The linewidths also depend on the Zeeman frequencies, ω_z . By use of the equation (2), the exchange fields can be estimated, and are shown in Fig.3. H_{ex} are reduced with increasing the Ni concentration. The origin of the exchange field, H_{ex}, is related to magnetism in Cu-2 plane. The substitution of Ni ions makes this exchange interaction weak. The superexchange interaction among Gd ions via the oxygen on Cu-2 plane can explain the origin. It is well known that the superexchange interaction among Cu ions on Cu-2 plane is related with superconductivity. It is difficult that the Gd magnetism is independent of the superconductivity.

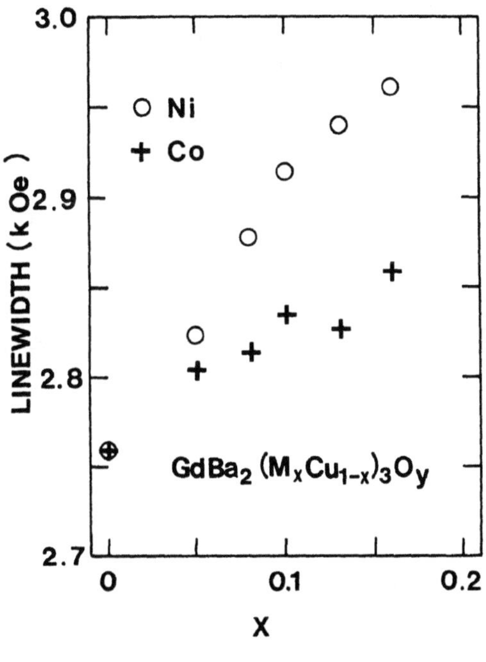

Fig. 2. The Gd linewidths in $GdBa_2(Cu_{1-x}M_x)_3O_y$ (M=Ni and Co) compounds. In Ni doped compounds, a broadening is observed with increasing doped concentration x. In Co doped compounds, the broadening is smaller than that in Ni doped compounds.

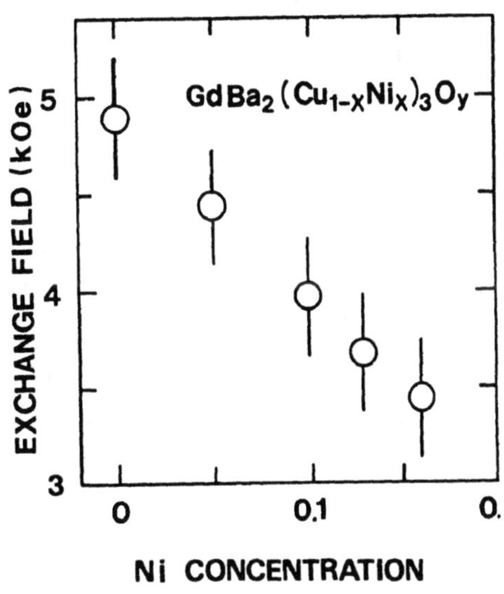

Fig. 3. The exchange fields on Gd ions are reduced with increasing Ni concentration. The origin of these exchange fields are superexchange interaction via the oxygen on Cu-2 plane.

The g shift are observed for the compounds having a different oxygen contents. The oxygen concentration dependence of the g value at room temperature is shown in Fig.4. The g values decrease from about 2.1 to 2.0 with increasing the oxygen contents from 6.0 to 7.0. The difference in the g value becomes 0.08 between y=6.1 and 7.0 compounds. This difference corresponds to a field of 130 Oe. For Gd^{3+} free ions, the g value is expected to be 1.99. The y=7.0 compounds have the g value which is close to the free ions. It is well known that the magnetic and electric behaviors on Cu-2 planes are controlled by the oxygen contents, y. The difference should be related to these behaviors. The origin of this difference can be explained with the internal field due to the magnetic ordering of Cu ions or the weak interaction between Gd ions and carriers on Cu-2 plane. In dc-resistivity measurements near the Néel temperature for y<6.4 compounds, anomaly due to the Gd ordering is observed.[8] It is possible that there is a small interaction between Gd and carriers on Cu-2 plane.

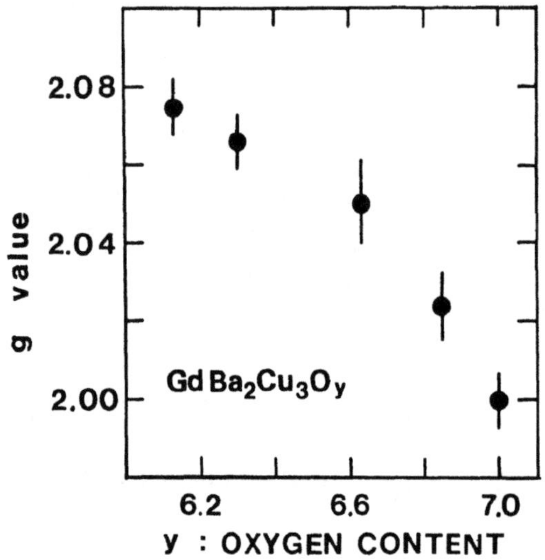

Fig. 4. The oxygen concentration dependence of g value. The difference of g value is 0.08 between y=6.1 and 7.0. This correspond to a field of 130 Oe. It is well known that the carrier concentration in Cu-2 plane is controlled by the oxygen contents. The g value of Gd ions also depend on the contents. The Gd magnetism is related with behaviors of the Cu ions.

In conclusion, from the effective magnetic moment, the atomic distance, and the g value for the compounds, the dipolar interaction is important in order to explain the Gd magnetism in this system. The exchange interaction is estimated from the Zeeman frequency dependence of linewidth. The Gd magnetic interaction can be explained with the Gd dipolar fields of about 3kOe and the exchange fields of about 5kOe. The superexchange interaction via the oxygen on Cu-2 plane is the origin of this interaction from the experimental results for Ni doped $GdBa_2(Cu_{1-x}Ni_x)_3O_y$ compounds.

ACKNOWLEDGMENTS

The authors express their sincere gratitude to Professor Kei Yosida for stimulated discussions. They thank Dr. M Kaise for Q-band ESR measurements. The ESR and susceptibility measurements are performed at Cryogenics Center in University of Tsukuba.

REFERENCES

1. J.O.Willis, Z.Fisk, J.D.Thompson, S.W.Cheong, R.M.Aikin, J.L.Smith, and E.Zirngiebl, J.Magn.Mater. **67**,L139(1987). F.Nakamura, A.Tominaga, and Y.Narahara, Jpn.J.Appl.Phys.lett. **26**,L1734(1987).
2. K.Umeda and K.Yosida (private communication).
3. D.Mck.Paul, H.A.Mook, A.W.Hewat, B.C.Sales, L.A.Boatner, J.R.Thomson, and M.Mostller, Phys.Rev.B **37**,2341(1988). H.A.Mook, D.McK.Paul, B.C.Sales, L.A.Boatner, and L.Cussen, Phys.Rev.B **38**,12008(1988).
4. D.Shaltiel, J.Gnossar, A.Grayevsky, Z.H.Kalman, B.Fisher, and N.Kaplan, Solid State Commun. **63**,987(1987). F.Mehran, S.E.Barnes, C.C.Tsuel, and T.R.McGuire,Phys.Rev.B **36**,7266(1987).
5. A.I.Nazzal, V.Y.Lee, E.M.Engler, R.D.Jacowitz, Y.Tokura, and J.B.Torrance, Physica C **153-155**,1367(1988).
6. F.Nakamura, K.Senoh, T.Tamura, Y.Ochiai, and Y.Narahara, Phys.Rev.B **39**, 12283(1989).
7. R.Kubo, and K.Tomita, J.phys.Soc.Japan,9,316(1954). P.W.Anderson, and P.R.Weiss, Rev.mod.Phys. **25**,269(1953).
8. Y.Ochiai, et.al. Physica C (in press).

ESR, Nonresonant Microwave Absorption and Static Magnetic Susceptibility in Tl-Ba-Ca-Cu-O System

YOSHIKAZU HAYASHI[1], KENGO ADACHI[2], KATSUTOSHI IWAHASHI[2], HIDEO SHIBAYAMA[2], TETSUO FUJITA[2], MINORU FUKUI[3], and SANSHIRO SAKO[4]

[1] Materials Science, University of Osaka Prefecture, Sakai, Osaka, 591 Japan
[2] Department of Physics, Nagoya University, Chikusaku, Nagoya, 464 Japan
[3] Nakanihon Automotive College, Sakahogicho, Kamogun, Gifu, 505 Japan
[4] Department of Physics, Faculty of Education, Mie University, Tsu, Mie, 514 Japan

ABSTRACT

Measurements of ESR of Cu^{2+} ions in high-T_c phase of Tl-Ba-Ca-Cu-O system are made. Two kinds of ESR signals are detected. The one having broad linewidth disappears at T_c indicating a strong correlation to the superconductivity. The number of spins responsible for this resonance is estimated to be an order of 10^{20}/g. A nonresonant microwave absorption near zero magnetic field is strongly enhanced at T_c. Results of measurements of the static magnetization are presented and the relation to the behavior of flux lines is discussed.

KEY WORDS: ESR, microwave absorption, Tl-Ba-Ca-Cu-O system, static magnetic susceptibility

INTRODUCTION

Extensive studies of the magnetic field dependence of the microwave power absorption in high-T_c superconductors have been made [1-5]. Unfortunately however, most of the published results for ESR of Cu^{2+} ions are considered to orignate from impurites contaminating in the sample because detected signals are weak or lack direct correlations to the onset of the superconductivity.

Recently, we have succeeded to detect ESR signals in high T_c superconductors having three sheets of a CuO_2 plane [6-8]. These spectra consist of a broad isotropic line and a relatively sharp anisotropic line. A correlation to the superconductivity has been confirmed by an observation of the disappearance of the broad isotropic line at T_c. One of the purpose of the present paper is to report an estimated number of spins responsible for the ESR and to discuss the origin of this broad line.

On the other hand, the nonresonant microwave absorption at low magnetic field gives useful informations for the behavior of magnetic flux lines. Another purpose of this paper is to present results of static measurements of magnetic susceptibility. The relation to the microwave absorption at low magnetic field region is discussed.

EXPERIMENTAL

Almost single phase samples with a nominal composition of Tl:Ba:Ca:Cu=2.18: 1.91:2.5:3.5 were used in the present experiment. The method of preparation, data for an X-ray analysis and a resistivity measurement are the same as those reported previously [7]. ESR and nonresonant absorption measurements were performed by a conventional 9 GHz spectrometer with 100 kHz field modulation. Several samples containing about 50 to 100 mg of fine powder sealded in a quartz tube were used for ESR measurements. The number of spins was estimated by comparing the intensity of ESR with that of the standard sample [9]. Errors in estimated values of spin numbers were approximately 40 %. A SQUID magnetometer was used for static measurements of the magnetization.

RESULTS AND DISCUSSION

Figure 1 shows typical examples of a temperature dependence of the profile of the microwave response under the magnetic field. The ESR signal is clearly visible at room temperature. The spectrum can be decomposed into two components as shown at the right hand side of Fig.1. A broad isotropic line denoted (B) disappears at T_c. Below T_c, only an anisotropic spectrum labeled (A) can be observable. These characteristic features are similar to those of the case for Bi system [6,8].

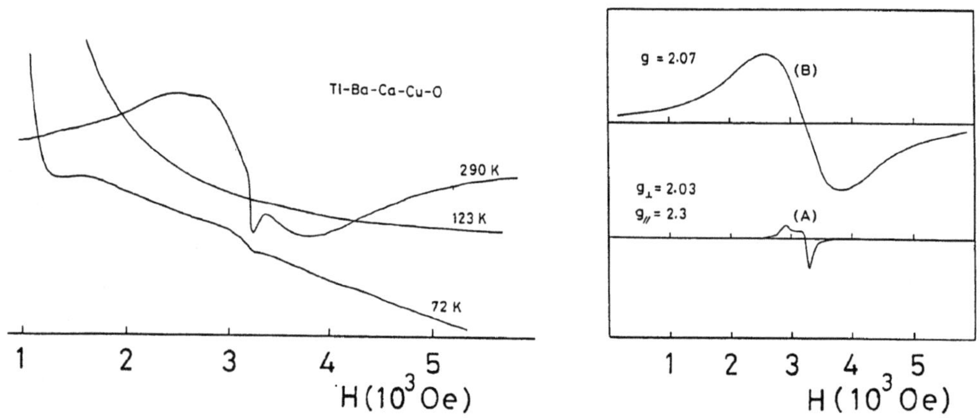

"Fig.1" ESR and nonresonant microwave power absorption (left), and (A) and (B) components of ESR (right).

Although a definite conclusion is not yet obtained, we have tentatively assigned Cu^{2+} ions as the origin of ESR [7]. The intensity of (B) spectrum is more than 100 times stronger than that of (A) spectrum. The estimated value of the number of spins for (B) spectrum is about $1x10^{20}$/g at room temperature which corresponds to about 5 % of the Cu ions contained in the sample. Since the contaminated Cu ions as impurity phases are confirmed to be less than 1 % by an analysis of X-ray diffraction data, the center responsible for (B)spectrum is not considered to be impurities.

Behavior of Holes

Here, a model proposed in the previous paper [6,7] is examined. There are many experimental results showing that holes are doped at oxygen sites and

Cu ions remain divalent [10-12]. According to a theoretical results [13], holes are easily doped at the appex oxygen denoted "O1" in Fig.2. The material may be an insulator in this case. However, holes can also occupy "O2" sites when the number of doped holes increases. Holes doped at "O2" sites can move arround in a CuO_2 plane and the material becomes conductive. In this case, "O1" sites are considered to act as trapping centers of holes and parts of doped holes may always be trapped at "O1" sites.

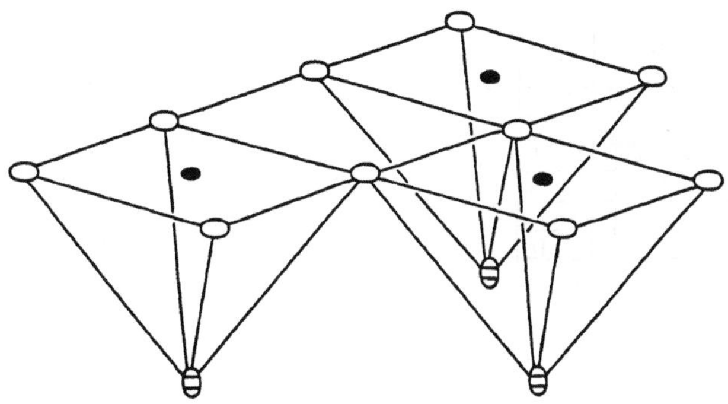

"Fig.2" CuO_2 plane and appex oxygen "O1" in a high-T_c superconductor

In the previous paper [6,7], we speculate that an origin of (B) spectrum is Cu^{2+} ions under or over the appex oxygen which is occupied by a hole. Because one to one correspondence exists between an "O1" site and a Cu ion in a CuO_2 plane, the number of spins responsible for ESR is equal to the number of trapped holes at "O1" sites. We have estimated that the number of spins is about 5 % of Cu ions contained in the sample. Therefore, the same amount of doped holes may be trapped at "O1" sites and do not contribute to conduction.

Nonresonant Microwave Absorption and Static Susceptibility

In Fig.1, two examples of a profile for nonresonant absorption are shown. The origin of these absorption has not yet been fully understood. At first sight, power absorption seems to increase and becomes maximum at zero field. However, the absorption is known to become minimum at zero field [14]. The relation between the Josephson current and the penetration of the magnetic flux into the superconductor are interesting [15].

These effects are also important for static magnetization as shown in Fig.3. The effect of a trapping of the flux are observed in (a) and (b). Curves shown in (c) are taken by decreasing the temperature under the magnetic field of 10 Oe (upper curve) and by increasing the temperature under the magnetic field after cooling to 4.2 K without magnetic field. Flux trapping effect at high temperature is obvious.

In the present case, grains are considered to be coupled each other by the Josephson effect. A hump appearing in (d) may be due to the suppression of the Josephson critical current flowing through grain boundaries similar to that observed in the nonresonant microwave absorption [15].

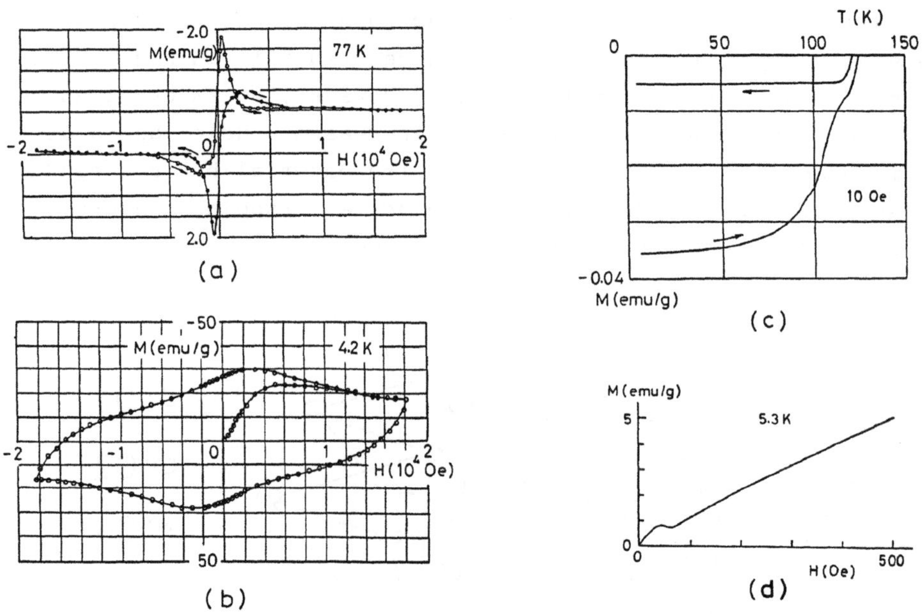

"Fig.3" Characteristics for static magnetization

REFERENCES

1 D.C.Vier,S.B.Oseroff,C.T.Salling,J.F.Smyth,S.Schultz,Y.Dalichaouch,B.W.Lee, M.B.Maple,Z.Fisk and J.D.Thompson.Phys.Rev.B36(1987)8888.

2 W.R.McKinnon,J.R.Morton and G.Pleizier:Solid State Commun.66(1988)1093.

3 F.Mehran and P.W.Anderson:Solid State Commun.71(1989)29

4 F.J.Owens and Z.Iqbal:Solid State Commun.67(1988)523.

5 S.V.Bhat,P.Ganguly,T.V.Ramakrishnan and C.N.R.Rao:J.Phys.C20(1987)L559.

6 Y.Hayashi,M.Fukui,H.Sasakura,S.Minamigawa,T.Fujita and K.Nakahigashi:Jpn.J. Appl.Phys.28(1989)L759.

7 Y.Hayashi,M.Fukui,T.Fujita,H.Shibayama,K.Iwahashi and K.Adachi:Jpn.J.Appl. Phys.28(1989)L910.

8 Y.Hayashi,S.Sako,M.Fukui,T.Fujita,H.Sasakura,S.Minamigawa and K.Nakahigashi:Jpn.J.Appl.Phys.28(1989)L1531.

9 C.P.Poole,Jr.,:Electron Spin Resonance,Interscience,New York,1967.

10 A.Fujimori,E.Takayama-Muromachi and Y.Uchida:Solid State Commun.63(1987)857

11 N.Nucker,J.Fink,J.C.Fuggle,P.J.Darham,W.M.Temmerman:Phys.Rev.B37(1988)5158.

12 N Nucker,H.Romberg,X.X.Xi,J.Finck,G.Gegenheimer,Z.X.Zhao:Phys.Rev. B39(1989)6619.

13 J.Kondo:J.Phys.Soc.Jpn.58(1989)2884.

14 E.J.Pakulis and T.Osada:Phys.Rev.B37(1988)5940.

15 Y.Hayashi,M.Fukui,T.Fujita,H.Shibayama,K.Iwahashi and K.Adachi:Jpn.J.Appl. Phys.28(1989)L1746

^{57}Fe and ^{57}Co Mossbauer Studies of High-T_c Y-Ba-Cu Oxides

S. NASU[1], M. YOSHIDA[1], Y. ODA[1], K. ASAYAMA[1], F.E. FUJITA[1], K. UEDA[2], T. KOHARA[2], T. SHINJO[3], S. KATSUYAMA[4], Y. UEDA[4], and K. KOSUGE[4]

[1] Department of Material Physics, Faculty of Engineering Science, Osaka University, Toyonaka, Osaka, 560 Japan
[2] Himeji Institute of Technology, Himeji, Hyogo, 671-22 Japan
[3] Institute for Chemical Research, Kyoto University, Uji, Kyoto, 611 Japan
[4] Department of Chemistry, Faculty of Science, Kyoto University, Kyoto, 606 Japan

ABSTRACT

Utilizing ^{57}Fe absorption and ^{57}Co emission Mossbauer spectroscopy, the chemical and physical properties of Fe-doped and Co-doped $YBa_2Cu_3O_{7-y}$ oxide superconductors have been investigated. Mossbauer spectra obtained consist of at least three components. The relative intensities of the each components depend largely on the oxygen concentration of the specimens. Fe and Co atoms mainly substitute at Cu1 chain sites, while small portions of Fe and Co atoms occupy the Cu2 plane sites indicating an antiferromagnetic long-range order in oxygen deficient compounds. The direction of Fe magnetic moment at Cu2 sites in oxygen deficient Fe-doped specimen is normal to the c-axis. Preferential occupancy of Fe atoms at Cu1 sites has been analyzed thermodynamically.

KEY WORDS: ^{57}Fe absorption Mossbauer spectroscopy, ^{57}Co emission Mossbauer spectroscopy, high-T_c Y-Ba-Cu oxide, oxygen deficient Y-Ba-Cu oxide, excess free-energy of mixing.

INTRODUCTION

In order to understand the microscopic nature of Fe in Fe-doped high-T_c oxide superconductors, the ^{57}Fe Mossbauer measurements have been performed and reported by several researchers [1-4]. Especially, many research groups in the world have been devoted to clarify the site occupation problems and the magnetic properties of Fe atoms at two different Cu sites in Fe-doped $YBa_2(Cu_{1-x}Fe_x)_3O_{7-y}$. However, it is not yet clear even for the site occupation problems of Fe atoms at two different Cu sites in Fe-doped 1-2-3 compounds.

In this report, we present the results from ^{57}Fe Mossbauer measurements in Fe- and Co-doped 1-2-3 compounds and conclude that the Fe and Co atoms in these compounds substitute mainly at Cu1 chain sites, while some small portion of Fe and Co atoms occupy the Cu2 plane sites indicating an antiferromagnetic long-range order in oxygen deficient compounds. The spin direction of Fe at Cu2 was determined to be normal to the c-axis using an oxygen deficient specimen having a strong texture along the c-axis. The distribution of Fe or Co atoms at two different Cu sites has been determined by using a thermodynamical analysis of Cu-Fe or -Co binary mixture at Cu1 and Cu2 sites.

^{57}Fe TRANSMISSION MOSSBAUER MEASUREMENTS

^{57}Fe transmission Mossbauer measurements were carried out on ^{57}Fe enriched powder specimens prepared as $YBa_2(Cu_{1-x}Fe_x)_3O_{7-y}$ (x=0.018-0.10) by standard solid-state reaction in air. Oxygen deficient compounds are prepared by the quenching from 1173 K into 77 K. Resistivity and ac susceptibility measurements have been performed by similar specimens and reported previously [5]. Room temperature Mossbauer spectra obtained from x=0.08 specimens are shown in Fig. 1. Spectrum A obtained from slowly cooled superconductor, whose zero resistance temperature was 34 K, consists of 3 doublets denoted by D-1, D-2 and D-3 having different quadrupole splitting and isomer shift values. Spectrum B obtained from quenched non-superconductor shows a

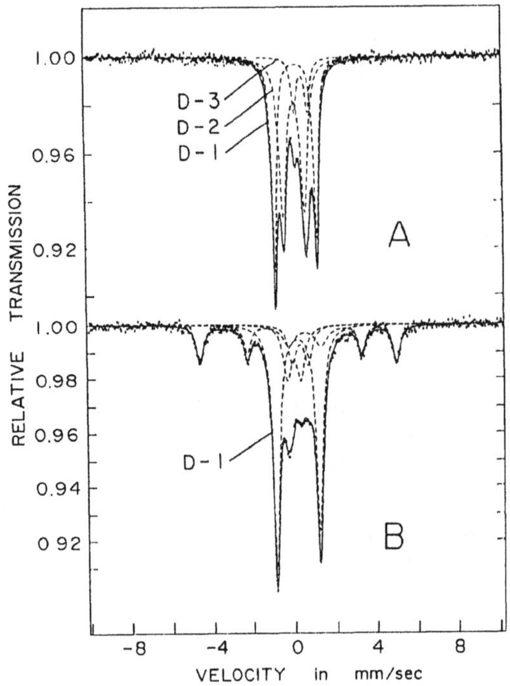

Fig. 1 ^{57}Fe Mossbauer spectra of $YBa_2(Cu_{.92}Fe_{.08})O_{7-y}$ at RT.
A: Superconductor
B: Semiconductor quenched from 1173 K.
Velocity scale is relative to bcc Fe at RT.

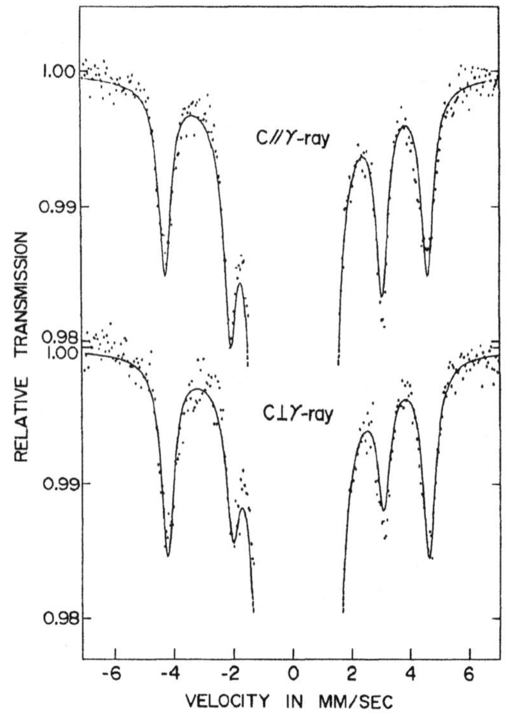

Fig. 2 ^{57}Fe Mossbauer spectra in expanded absorption scale of $YBa_2(Cu_{.94}Fe_{.06})_3O_{7-y}$. Specimen was quenched from 1173 K and has a strong texture along c-axis.

superposition of a magnetically-split 6 line pattern, an intense D-1 component and the other two minor doublets. The intensities of the D-1 and D-2 components depend on the heat-treatments, especially on the oxygen deficiency of the specimen. It implies that these components are attributed to the Fe atoms at Cu1 chain sites having different oxygen coordinations around them as described later in the experimental results from ^{57}Co Mossbauer measurements. We assigned the magnetically-split component of spectrum B and the D-3 component to the Fe atoms at Cu2 plane sites, since the neutron diffraction experiment shows an antiferromagnetic long-range order of CuO_2 plane in oxygen deficient compounds [6]. In order to get further confirmation of the site assignment, the ^{57}Fe Mossbauer measurements using a strong texture along c-axis were performed at room temperature. In Fig. 2, the Mossbauer spectra obtained from the quenched x=0.06 powder specimens are shown in expanded absorption scale in order to show clearly the relative intensities of the each lines in magnetically-split components. These powder specimens were fixed in a glue under the external magnetic field of 1.8 T and have a strong texture of c-axis. Upper figure of Fig. 2 shows the spectrum under the condition in which the Mossbauer gamma-ray propagate along the c-axis. Lower figure is the spectrum obtained from the perpendicular direction to the c-axis. Second and fifth lines are correspond to the m=0 transitions of ^{57}Fe nucleus whose largest intensity could be observed when the gamma-ray propagation direction is to be normal to the direction of the magnetic hyperfine field. Since the second and fifth lines in upper figure have much larger intensities than those of the lower figure, it is shown clearly that the direction of the magnetic moment of the Fe atoms is to be normal to the c-axis. The Fe atoms which show the magnetically-split component observed from oxygen deficient compounds are surely incorporated into the specimen matrix occupying Cu2 plane sites and do not form any impurity phases, while the magnetic moment of Fe is quite large compared to the Cu spin.

EXCESS FREE-ENERGY OF MIXING

The above Mossbauer results indicate that the Fe atoms mainly occupy the Cu1 chain sites and are not distributed randomly to Cu1 and Cu2 sites (if randomly distributed, Cu1:Cu2 must be 1:2). In order to understand the Fe distribution at Cu1 and Cu2 sites mentioned above, we analyzed the thermodynamical occupation probability of Fe at Cu2, p, as a function of temperature T, using an expression for the excess free energy of a Cu-Fe binary mixture having two different sites. The fraction of Fe at Cu2 sites, p, is given by the following equation taking only the configurational entropy term into account:

$$p[1/3 - (1-p)]/[(1-p)(2/3-px)] = \mathrm{Exp}\ [\ -\ \Delta\Omega\ /kT\ ],$$

where x is Fe concentration and $\Delta\Omega$ is the difference in excess enthalpy changes induced by the substitution of Fe at Cu2 and Cu1 sites. Assuming p is constant, the relation between x and T is given by

$$[px/(2/3 - px)]^{p}[[(1-p)x]/[1/3 - (1-p)x]]^{(1-p)} = \mathrm{Exp}\ [\ -\ \Delta H/kT\ ].$$

where ΔH is the fractional change in excess enthalpy.

Figure 3 shows the relationships between p and T as a function of $\Delta\Omega$. Since the p values obtained experimentally are ranged between 0 to 0.3 when the specimens were annealed in air or O_2, the $\Delta\Omega$ value is roughly estimated to be 0.2 eV which predicts a stable oxygen configuration of Fe at Cu1 sites. When the difference in excess enthalpy changes, $\Delta\Omega$, is small, which means the case of the annealing in low oxygen concentration like in N_2, p values should become to be large comparing to the case of annealing in O_2 at the temperature range to be able to take place the Fe atomic diffusion, because the oxygen coordination number around Fe at Cu1 chain sites will be restricted by the low oxygen concentration and the enthalpy change at Cu1 sites becomes to be large. This situation occurred surely by the annealing at 1073 K in N_2. Figure 4 shows the relationships between T and x which indicate physically the Fe solid solubility curve for Cu-Fe binary mixture having two different sites. From these curves, one might expect the precipitation reactions at the temperatures lower than the solid curves shown in Fig. 4.

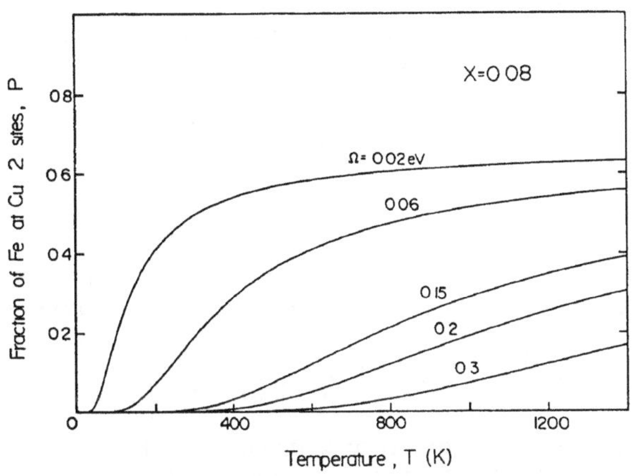

Fig. 3 Equilibrium fraction of Fe atoms, p, occupied Cu2 plane sites for x=0.08 as a function of temperature, T. $\Delta\Omega$ is the difference in excess enthalpy changes induced by the substitution of Fe at Cu2 and Cu1 sites.

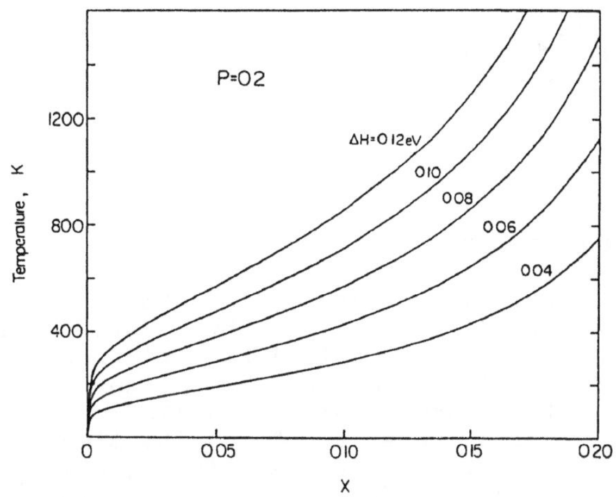

Fig. 4 Solid solubility curves of Fe in Cu-Fe binary mixture having two different sites, assuming fraction of Fe at Cu2 sites, p, equal to 0.2. ΔH is the fractional change in excess enthalpy.

^{57}Co EMISSION MOSSBAUER MEASUREMENTS

In order to understand the behavior of Co atoms in Co-doped 1-2-3 compounds, we performed source experiments using a ^{57}Co-doped $YBa_2(Cu_{.96}Co_{0.04})_3O_{7-y}$ specimen. Figure 4 shows typical room temperature spectra obtained after various heat-treatments. Figure 4(A) is obtained after slow cooling from 1173 K in O_2 and the specimen was superconductor having T_c=75 K. B was obtained after annealing for 20 hours at 1073 K in N_2 and the specimen was semiconductor. C shows a spectrum after subsequent annealing for 20 hours at 673 K in O_2 and the specimen recovered again to be a superconductor having T_c=70 K. All spectra can be decomposed into D-1, -2, -3 and -4 components. When the specimen was annealed in N_2, the magnetically-split component appeared and suggested the Co atoms moved from Cu1 to Cu2 sites because of the decrease in difference of enthalpy change between these two sites as discussed in previous section using Fig. 3. Further annealing in O_2 at 673 K, in which temperature the Co atoms cannot diffuse in lattice, induce the disappearance of the magnetic component and the enhancement of the intensity of D-3 component. It is suggested that the behavior of the Co atoms doped into the 1-2-3 compounds is quite similar to the behavior of the Fe atoms in this compound, while the intensity of D-3 component for the specimen annealed in O_2 is relatively small comparing to the case of Fe atoms.

This work is supported partially by Grand-in-Aid for Scientific Research on Priority Area No.01644004 of the Ministry of Education, Science and Culture.

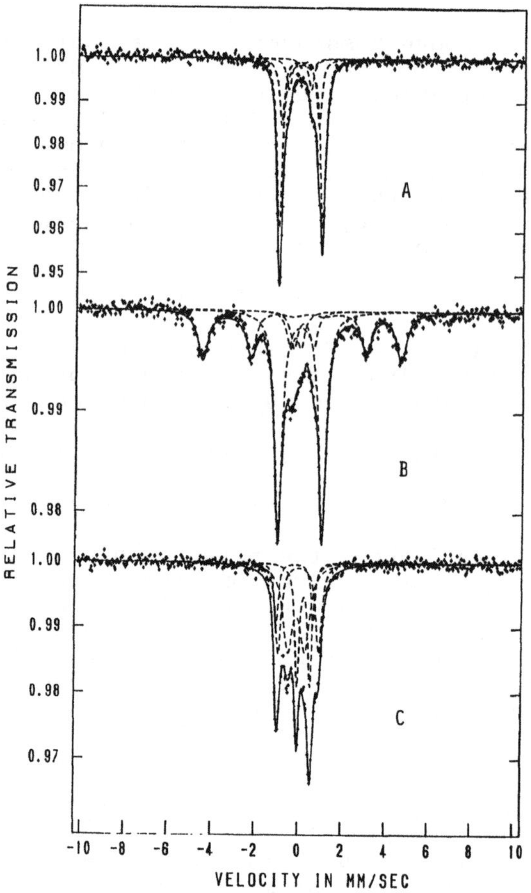

Fig. 5 Typical ^{57}Co emission Mossbauer spectra obtained from $YBa_2(Cu_{.96}Co_{.04})_3O_{7-y}$ at RT. Velocity scale is to bcc Fe at RT.
A:Superconductor annealed in O_2.
B:Semiconductor annealed in N_2.
C:Superconductor annealed subsequently in O_2 using same specimen of B.

REFERENCES

1. Saitovitch EB et al (1988) ^{57}Fe Mossbauer study of the superconductor $YBa_2(Fe_{1-x}Cu_x)_3O_y$. Phys. Rev. B 37: 7967-7970
2. Bottyan L et al (1988) Evidence for Fe^{4+} in $YBa_2(Cu_{1-x}M_x)_3O_{7-y}$ (M=^{57}Fe, ^{57}Co) by absorption and emission Mossbauer spectroscopy. Phys Rev B 38: 11373-11381
3. Brand RA et al (1988) The sign of the EFG in semiconducting and grain-oriented high-T_c $YBa_2(Cu_{1-x}Fe_x)_3O_y$ by Mossbauer spectroscopy. Physica C 156: 539-546
4. Nasu S et al (1989) ^{57}Fe Mossbauer study of high-T_c Y-Ba-Cu oxide. MRS Int'l. Mtg. on Adv. Mat. 6: 635-640
5. Oda Y et al (1987) Superconductivity of $Y_1Ba_2(Cu_{1-x}Fe_x)_3O_y$. J. Phys. Soc. Jpn. 26: L1660-L1663
6. Tranquada JM et al (1988) Neutron-diffraction determination of antiferromagnetic structure of Cu ions in $YBa_2Cu_3O_{6+x}$ with x=0.0 and 0.15. Phys. Rev. Lett. 60: 156-159

Anisotropy of Lattice Vibration in Aligned $Ba_2EuCu_3O_7$ Observed by ^{151}Eu Mössbauer Spectroscopy

T. MURAKI, M. TANIWAKI, and K. SHIRAMINE

Department of Electronic Engineering, Faculty of Engineering, Hokkaido University, Sapporo, 060 Japan

ABSTRACT

The anisotropy of the lattice vibration of europium in $Ba_2EuCu_3O_7$ is studied by Mössbauer spectroscopy of two kinds of aligned samples. The lattice vibration is analyzed by line broadening, line saturation and the temperature dependence of the line area. It is shown that the lattice vibration along c-axis is softer than that in ab-plane. From the temperature dependence of the line area and the direct estimation of the recoilless fraction, the mean square displacements along c-axis and in ab-plane are obtained as a function of temperature.

KEY WORDS: high Tc, anisotropy, Mössbauer effect, lattice vibration, europium

INTRODUCTION

It is well known that the high temperature superconductor $Ba_2LnCu_3O_7$(Ln = lanthanide) has the anisotropic superconductivity and lattice structure. The anisotropy in the lattice structure means that this material has anisotropies in the electronic state and the lattice vibration, which cause the anisotropic superconductivity. For this reason, to clarify the anisotropies of the electronic state and the lattice vibration may be a key to understand the high temperature superconductivity. As little has been known about the anisotropies of them, the present authors have investigated the electronic state and the lattice vibration of lanthanide in $Ba_2LnCu_3O_7$ by ^{151}Eu Mössbauer effect[1,2]. In these studies, we succeeded in the detection of the anisotropy of the lattice vibration using an unaligned sample and an c-axis aligned sample. It was clarified that the Debye temperature of the lattice vibration along c-axis is smaller than that in ab-plane, i.e., the lattice vibration along c-axis is softer than that in ab-plane. In our previous work, the Debye temperatures were estimated by applying Debye approximation to the temperature dependence of the Mössbauer line area which is the function of the recoilless fraction. However Debye temperature may not been estimated correctly by this analysis, in case the lattice vibration is not approximated by the Debye model. In fact, our Mössbauer data showed that the lattice vibration in the superconductor is not expressed correctly by the Debye model. This suggests that it is necessary to analyze the lattice vibration from the direct estimation of the recoilless fraction, f.

In the present work we tried the direct estimation of the recoilless fraction, f, by two kinds of analyzing methods. One method is to obtain f from the line broadening with increasing σn, where σ is Mössbauer cross-section and n is Mössbauer nucleus number per unit area in the specimen. The other is to obtain f from the σn dependence of the line area. Furthermore, the temperature dependence of the mean square displacement of the lattice vibration was deduced from the temperature dependence of the line area and the estimated f value. To detect the anisotropy of the lattice vibration, two kinds of aligned samples were prepared. One is for the measurement of the lattice vibration along c-axis, and the other is for the measurement of the lattice vibration in ab-plane.

EXPERIMENTAL DETAILS

Powder of high temperature superconductor $Ba_2LnCu_3O_7$ was produced by the conventional method and grinded well so that almost all particles were single crystals. The aligned disk samples were produced by mixing the powder in epoxy and curing it in a magnetic field of 0.5 T - 2 T at room temperature. The c-axis in the sample AB were aligned to be parallel to the disk plane. Those in the sample C were aligned to be perpendicular to the disk plane.

Mössbauer measurement was performed using ^{151}Sm gamma ray source at room temperature. Mössbauer spectra of samples AB and C were taken as a function of σn at room temperature. The lattice vibrations in ab-plane and along c-axis were studied by sample AB and sample C, respectively. The temperature dependence of line area was measured at 4.2 K - 320 K, using samples of σn=3. The velocity axis was standardized by the line position of EuF_3 and the Zeeman splitting of natural iron. Spectrum parameters were obtained by approximating the measured spectrum with a single Lorentzian using the non-linear least square methods.

RESULTS AND DISCUSSION

1. Line broadening

The line width of an observed Lorentzian, Γ_e is broaden due to the finite thickness of the absorber. Γ_e is

$$\Gamma_e = \Gamma_S + \Gamma_A + 0.27\,\Gamma\,Ta \qquad (1)$$

where Γ_S and Γ_A are the line widths of the source and the absorber, and Γ is the natural width[3]. Ta is Mössbauer effective absorber thickness, given by

$$Ta = \sigma n f \qquad (2)$$

where f is the recoilless fraction in the absorber. f is given by

$$f = \exp\left(-\frac{\langle u^2\rangle}{\lambda^2}\right) \qquad (3)$$

where $\langle u^2\rangle$ is the mean square displacement and $2\pi\lambda$ is the γ- ray wavelength. The recoilless fraction, f, is obtained from the slope of $\Gamma_e - \sigma n$ relation. The line widths in samples AB and C are

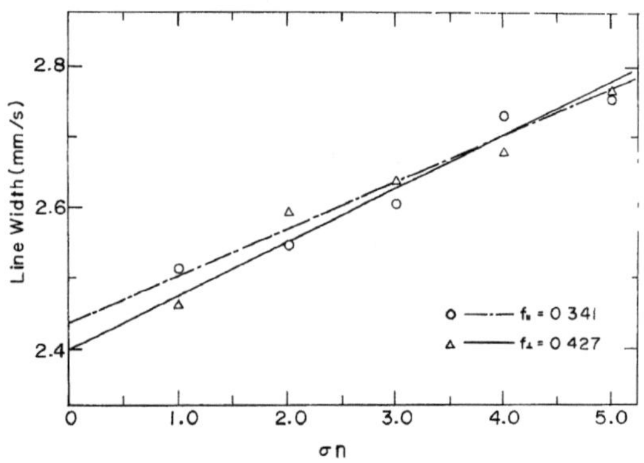

Fig. 1. The f estimation from the line broadening. $f_{\parallel}$ and $f_{\perp}$ represent the recoilless fractions in the case of γ ray//c-axis and γ ray$\perp$c -axis, respectively.

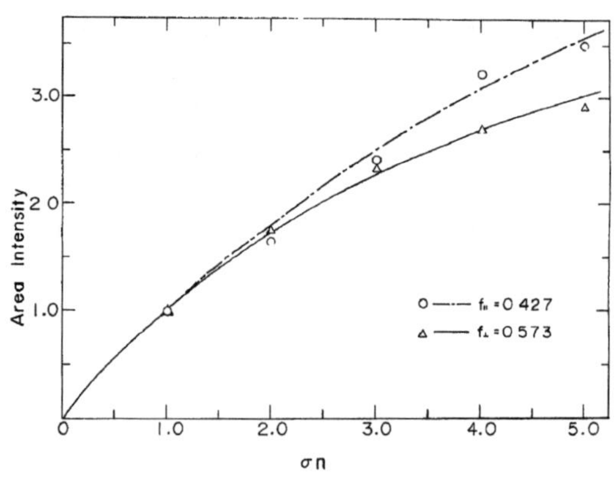

Fig. 2. The f estimation from the line area saturation.

plotted against σn (Fig. 1). The recoilless fraction of europium in sample AB is 0.427 at room temperature and that in sample C is 0.341. The effective Debye temperatures estimated from these recoilless fractions, $\Theta_{D\perp}$(the lattice vibration in ab-plane) and $\Theta_{D\parallel}$(the lattice vibration along c-axis), are 199 K and 177 K, respectively. This result leads again to the previous conclusion that the lattice vibration along c-axis is softer than that in ab-plane, though these values are small comparing to those(230 K and 276 K) reported in ref. 2. Precise estimation of f by this analysis is difficult, because the line broadening is very small and substantial error is induced in spectrum fitting by the non-linear least square method. Especially, this method is meaningless for analyzing of overlapped spectra. The europium Mössbauer spectrum surely has a quadrupole splitting, judging from its line width.

2. Line area saturation

The Mössbauer line area increases linearly at small Ta with increasing Ta($=\sigma nf$), however it reveals a tendency of saturation at large Ta. The line area, A(Ta) is given by

$$A(Ta) = \frac{\pi}{2} \Gamma_e I(Ta) \qquad (4)$$

where I is the observed line intensity, given by

$$I(Ta) = K(1 - \exp(-Ta/2)\, Jo(iTa/2)) \qquad (5)$$

where Jo(x) is the zero-order Bessel function and K is a factor involving S/N and the recoilless fraction of Mössbauer γ-ray source[4]. Since Ta = σnf, A is a function of nf. So, f is obtained as a parameter to fit eq. (4) to A - σn data. The σn dependences of the line areas in the samples AB and C are shown in Fig. 2. The recoilless fractions, $f_\perp$ and $f_\parallel$ best fitting to the data, are 0.573 and 0.427, from which $\Theta_{D\perp}$= 246 K and $\Theta_{D\parallel}$ = 199 K are obtained. This result agrees qualitatively with those already obtained. This analysis gives more precise f than the line broadening analysis, since the line area saturation due to the increase of Ta is large.

3. Temperature dependence of line area

The temperature dependence of the line area for sample AB is shown in Fig. 3, and that for sample C is in Fig. 4. The best fitting Debye temperature is 225 K for sample AB which is smaller than 230 K for sample C. However this will be attributed to the fact that the lattice vibration of europium in this superconductive oxide is not expressed correctly by the Debye model as we have already reported[1,2].

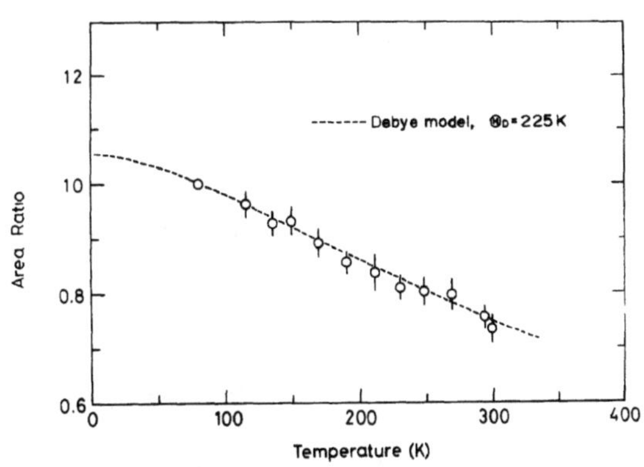

Fig. 3. The temperature dependence of the line area of sample AB. The Debye approximation gives Θ_D=225 K.

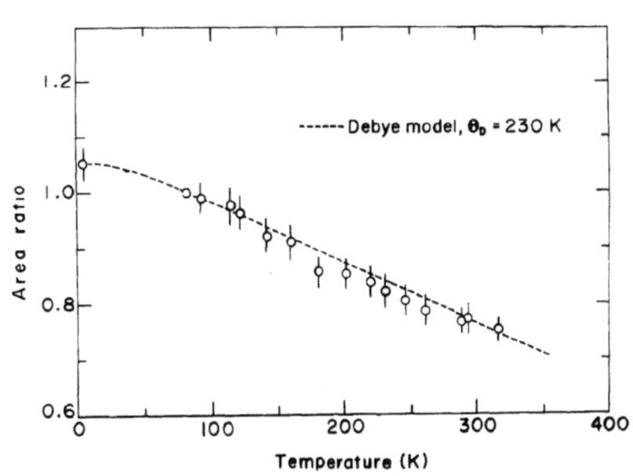

Fig. 4. The temperature dependence of the line area of sample C. The Debye approximation gives Θ_D=230 K.

We can deduce valuable informations on the lattice vibration by combining the T dependence of line area and the direct f estimation described above. Inserting the line area and the value of f at room temperature into eq. (4), K in eq. (5) is obtained. And by using this value, the temperature dependences of f and $\langle u^2 \rangle$ are obtained. Figure. 5. shows the temperature dependence of the mean square displacement along c-axis, $\langle u^2 \rangle_{\parallel}$, and that in ab-plane, $\langle u^2 \rangle_{\perp}$. The both dependences are similar, however, it may be noticeable that below 100 K the mean square displacement in ab-plane is substantially smaller than that along c-axis.

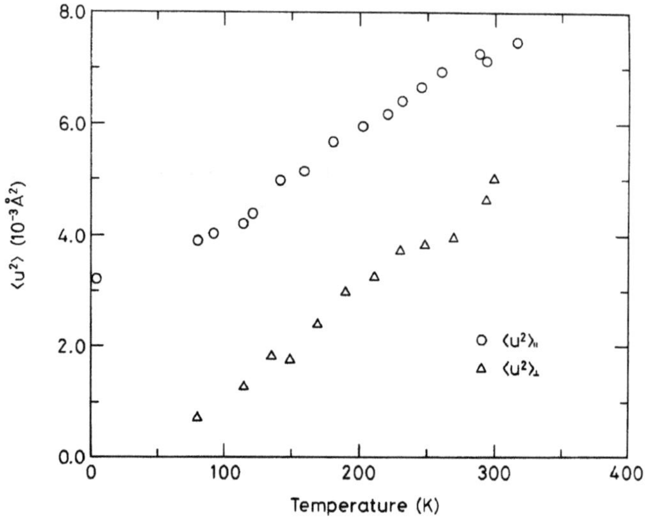

Fig. 5. The temperature dependence of the mean square displacement of europium. $\langle u^2 \rangle_{\parallel}$ and $\langle u^2 \rangle_{\perp}$ represent the displacements along c-axis and in ab-plane.

CONCLUSION

By three kinds of analysis, the anisotropy of the lattice vibration of europium in $Ba_2EuCu_3O_7$ was estimated. The direct estimation of the recoilless fraction showed that the lattice vibration along c-axis is softer than that in ab-plane. Particularly below 100 K, the mean square displacement in ab-plane is substantially smaller than that along c-axis.

REFERENCES

1. Taniwaki M. Sasaki H. (1988) The electronic state and the lattice vibration of europium in the high temperature superconductor $Ba_2EuCu_3O_y$ observed by Mössbauer effect spectroscopy. Pyhsica C 153-155: 1549-1550

2. Taniwaki M. Muraki T. Shiramine K. (1989) Anisotropy of lattice vibration uniaxialy aligned high temperature superconductor $Ba_2EuCu_3O_7$ observed by Mössbauer spectroscopy. Physica C: in press

3. Frauenfelder H. (1963) The Mössbauer effect. W.A. Benjamin, Inc.. New York. p45

4. Margulies S. Ehrman J.R. (1961) Transmission and line broadening of resonanceradiation incident on resonance absorber. Nucl. Inst. Meth. 12: 131-137

3.5 Optical Properties

Optical Study of the Electronic States in the High-T_c Cuprates

S. TAJIMA[1], S. TANAKA[1], T. IDO[2], and S. UCHIDA[2]

[1] Superconductivity Research Laboratory, International Superconductivity Technology Center, 10-13, Shinonome 1-chome, Koto-ku, Tokyo, 135 Japan

[2] Department of Applied Physics, The University of Tokyo, 3-1, Hongo 7-chome, Bunkyo-ku, Tokyo, 113 Japan

ABSTRACT

The optical reflectivity spectra for single crystals of $La_{2-x}Sr_xCuO_4$ were investigated at room temperature over a wide energy range from far-infrared to vacuum ultraviolet and over a wide x-range $0 \leq x \leq 0.34$. With increasing Sr-content, the spectral weight moves from the charge-transfer excitation to the lower energy part which seems to consist of two components, a Drude-like component and a mid-infrared peak. In the heavily doped region both components seems to be combined into a single Drude-like absorption. Based on these observation the electronic state characteristics of the high-T_c cuprates are discussed.

KEY WORDS: optical spectrum, electronic states, $La_{2-x}Sr_xCuO_4$, charge transfer gap, doping effect

INTRODUCTION

Elucidation of the mechanism of high-T_c superconductivity is one of the most attractive thema in the recent physics. The first step to this problem is to study the electronic states in the normal state and to find any abnormal property whose origin may intimately be related to the occurrence of the high-T_c superconductivity. For this purpose optical investigation is one of powerful probes.

The deviation from an usual Drude-type metallic state in the optical spectrum was first reported by Thomas et al. for oxygen deficient $YBa_2Cu_3O_y$ (YBCO) with T_c=50K.[1] Their succeeding study for YBCO with various oxygen content revealed the existence of some absorption in the mid-infrared region. However, the YBCO-system is disadvantageous to the study of electronic states in the Cu-O plane for the following reasons: (i) There are two inequivalent Cu-sites, Cu(1) in the chain and Cu(2) in the plane. This fact makes the carrier distribution complicated and even a nominal hole number cannot be determined definitely. (ii) There are twin-structures in almost all of the single crystals with superconducting composition, which makes it impossible to extract the pure a-(or b-)direction spectrum from the observed spectrum. (iii) It is difficult to realize the heavily doped region where the superconductivity probably dissapeares as in the case of $La_{2-x}Sr_xCuO_4$.[2]

For these reasons we believe that $La_{2-x}Sr_xCuO_4$ (LSCO) is a simpler and better system where Sr-substitution for La introduces carriers in the Cu-O plane.

In this study, using the single crystals of $La_{2-x}Sr_xCuO_4$ with various Sr-contents, the reflectivity spectra were measured at room temperature over a wide energy range from 5 meV to 40eV. Based on the spectral change with doping, a peculiar electronic states inherent to this Cu-O plane emerges.

EXPERIMENTAL

The single crystals of $La_{2-x}Sr_xCuO_4$ were prepared by the CuO-flux method. Sr-composition was determined by the Electron Probe Microanalyser. The temperature dependence of resistivity for each sample is shown in Fig.1. The samples with x=0, 0.02 and 0.06 are semiconductive, resistivity of which scale out of the same figure. With increasing Sr-composition x, the system changes from an insulator to a metal and shows the maximum value of T_c=25K at x=0.15. Although the T_c-value is relatively lower than that of polycrystal with the same composition, the superconducting transition width ΔT_c of 2K is very

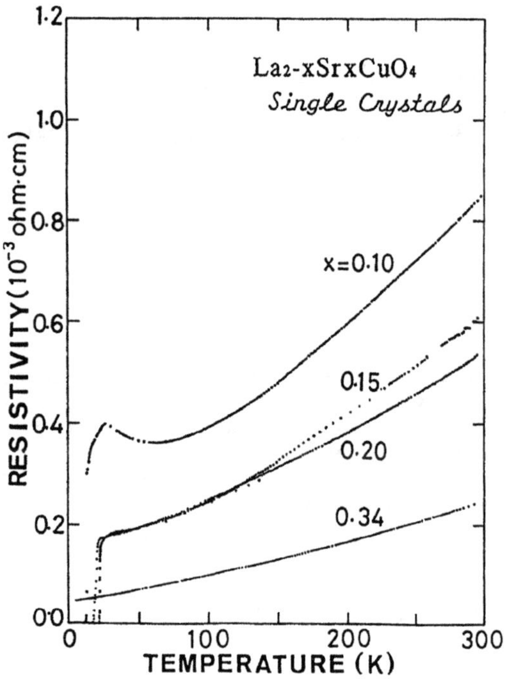

Fig.1 Temperature dependence of resistivity for $La_{2-x}Sr_xCuO_4$ with x=0.10,0.15,0.20 and 0.34.

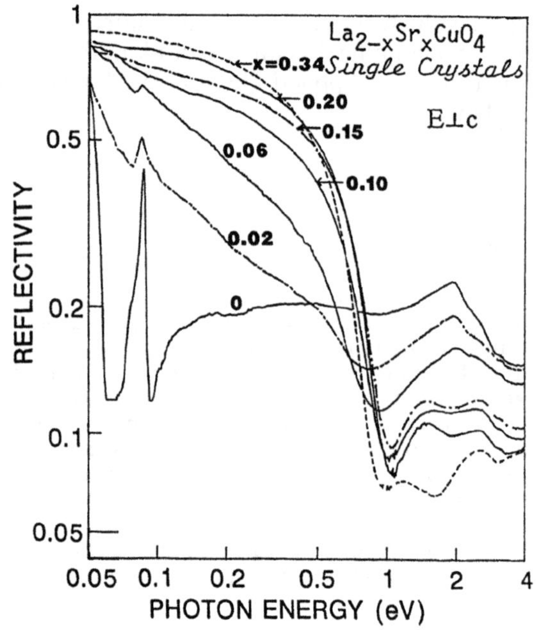

Fig.2 Reflectivity spectra for $La_{2-x}Sr_xCuO_4$ with various x.

narrow. It should be noticed that the heavily doped sample is not superconductive but more conductive than the highest T_c sample. The average sample size is 5×7×1 mm^3. Before optical measurement the sample surface perpendicular to c-axis was polished by Al_2O_3 powder with acetone.

The reflectivity spectrum was measured by Fourier-type spectrometers for 5 meV$\lesssim\hbar\omega\lesssim$1eV and by a Grating-type monochromator for 1eV$\lesssim\hbar\omega\lesssim$40eV. For measurement in the energy region higher than 2eV, was used the beam from the synchrotron orbital radiation ring of Institute for Solid State Physics in the University of Tokyo.

RESULTS AND DISCUSSIONS

The reflectivity spectra for E⊥c are shown in Fig.2 for all the samples. In the spectrum of the parent material La_2CuO_4, in addition to the far-infrared phonon peaks, there is a dominant peak at 2eV which is commonly obseved for all the cuprates with Cu-O plane.[3] Considering that the Coulomb repulsion energy U_d is as large as 6eV and the doped holes have dominantly O 2p character, the origin of this 2eV-peak is supposed to be the charge-transfer (CT) excitation from O 2p to Cu 3d.

When La is substituted by Sr, the reflectivity in the CT-excitation region decreases rapidly and, by contrast, the reflectivity in the lower energy region below 1eV increases which makes a reflectivity edge around 1eV. With increasing Sr-content the low energy reflectivity rises steeper and steeper but the edge position does not change appreciably. Only for the heavily doped sample with x=0.34, this edge shifts to lower energy, which is consistent with the band picture where the electron density decreases with Sr doping. These spectral behaviors do not allow simple interpretation based on a "rigid band" model where holes are introduced into the O 2p valence band and as a consequence the spectral weight is transfered from the excitation across the insulating CT-gap to the intraband excitation associated with free holes. In this case the reflectivity edge due to free carrier should shift with increasing hole concentration.

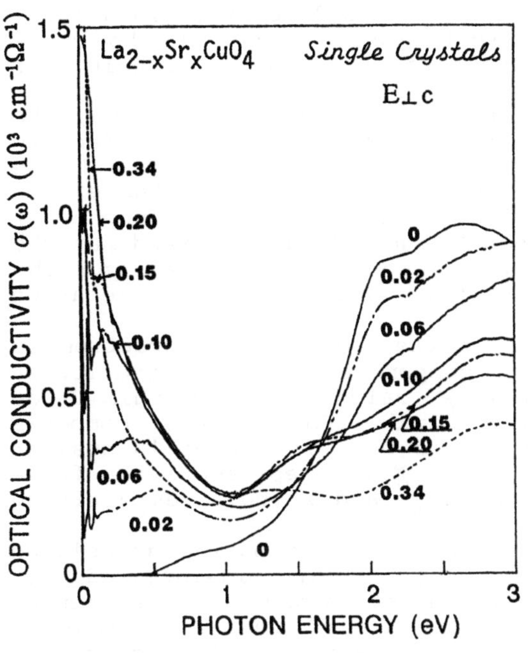

Fig.3 Optical conductivity spectra of $La_{2-x}Sr_xCuO_4$ for various x, which are calculated from the reflectivity spectra in Fig.2 by the Kramers-Kronig analysis.

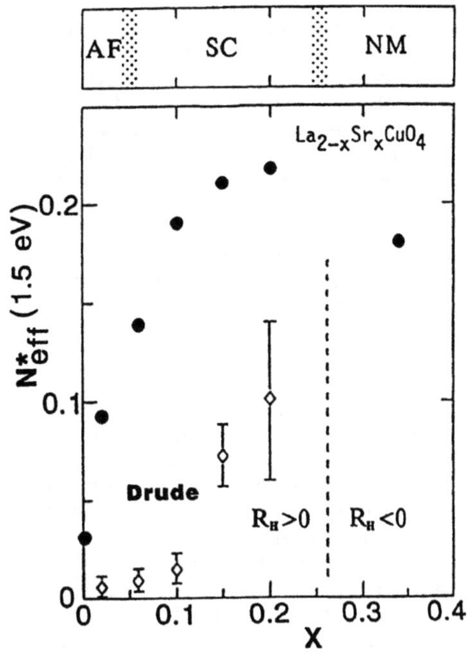

Fig.4 Composition dependence of the effective electron density at 1.5eV and that calculated from the Drude term of the conductivity in eq.(1).

The details of the lower energy spectral behavior can be more clearly seen in the optical conductivity spectra in Fig.3, which were calculated from the reflectivity spectra in Fig.2 by Kramers-Kronig transformation with use of higher energy spectrum up to 40eV and appropriate extrapolation procedures. When holes are lightly doped into the system, a broad absorption peak appears around 0.5eV, whereas the 2eV-CT absorption becomes weakened. As Sr-content increases, this mid-infrared peak grows and its center of gravity moves towards lower energy. In the metallic composition, for example at x=0.10, the low energy spectrum below 1eV seems to consist of two components. One is the above-mentioned mid-infrared absorption and the other is the Drude-like conductivity centered at $\omega=0$. Similar structures are also observed by Thomas et al. in the spectrum of YBCO with low T_c-values. With further Sr-substitution, both conductivity components grow and finally form a single absorption band centered at $\omega=0$.

From these spectral behaviors, a unique electronic structure emerges. The starting material at x=0 is a charge-transfer insulator and at the other end for larger x it seems like a Fermi-liquid state where the sign of Hall coefficient is negative which is expected in the band picture. The question is what electronic state is realized between them.

One of the key point is whether the CT-gap is destroyed by doping. As mentioned above, the optical excitation across the CT-gap seems to be seriously diminished by small amount of doping. Considering the fact that Sr-substitution has almost no effect on the spectrum in the energy region higher than 4eV[4], under the sum rule the oscillator strength of the CT-gap should be transfered mostly to the lower energy part with increasing Sr-content. Figure 4 shows the composition dependence of the oscillator strength below 1.5eV which is proportional to the effective electron (or hole) number N^*_{eff} calculated from the conductivity spectrum by using the following relation

$$N^*_{eff}(\omega)=\frac{2m_0V}{\pi e^2}\int_0^{\omega}\sigma(\omega')\,d\omega' \qquad (1)$$

where m_0 is the free electron mass and V is the volume of a formular unit. As seen in this figure, the oscillator strengh in the lower energy spectrum increases steeply towards heavily doped region only by small doping, in accordance with the rapid diminishing of the CT-excitation. Here it should be noted that the value of N_{eff}^* does not represent the bare number of holes (or electrons) N_{eff}, because it is renormalized by the electron correlation, as described in terms of effective mass m^* as $N_{eff}^* = (m_0/m^*)N_{eff}$. In fact, the values of N_{eff}^* at 3eV (~0.45 indepenent of x) is smaller than the effective electron number contributing to the CT-excitation, which is expected to be approximately 1.0. This discrepancy is one of the evidences that the effective mass is enhanced in the low energy region. The saturation tendency of N_{eff}^* for x>0.1 and subsequent decrease at x=0.34 suggests that already at the superconducting compositions the CT-gap is seriously destroyed and the electronic state gets near a Fermi liquid state, although the correlation remains strong.

An attempt to separate the two components in the lower energy spectrum is made by an estimate of the weight of the Drude component by fitting the lowest energy part of $\sigma(\omega)$. In Fig.4 is shown N_{eff}^D calculated from this Drude conductivity, which is indicated by an open diamond. For small Sr compositions N_{eff}^D is proportional to x but its value is by a factor of 10 smaller than x which coincides with the hole density. If N_{eff}^D represents the free-carrier contribution, the differnece between N_{eff}^D and x indicates a strong enhancement of effective mass.

The difference between N_{eff}^* and N_{eff}^D is the contribution associated with the mid-infrared absorption. As seen in Fig.4, this mid-infrared component grows with increasing x, as N_{eff}^D does. In this sense, it might represent the additional <u>intra</u>band contribution with a different - and/or T-dependence from the Drude-like contribution. This is suggested by a school of theories, based on the t-J model[5]. According to them, the Drude and mid-infrared components are interpreted as coherent and incoherent motion of carriers with respect to spin excitations. However, at the present stage, the origin of the mid-infrared absorption is not clear. For more detailed understanding, it is necessary to study the temperature dependence of the spectrum which is now under way.

SUMMARY

Optical reflectivity spectra were measured for single crystals $La_{2-x}Sr_xCuO_4$ with various x. When carriers are doped into the CT-insulator La_2CuO_4, the spectral weight transfers rapidly from the CT-excitation to the lower energy part contributing to the conduction. This rapid transfer suggests the significant deformation or destruction of the CT-gap at the high-T_c compositions.

REFERENCES

[1] Thomas GA, Orenstein J, Rapkine DH, Capizzi M, Millis AJ, Bhatt RN, Schneemeyer LF, Waszczak JV (1988) Phys.Rev.Lett. 61:1313-1316.
[2] Takagi H, Ido T, Ishibashi S, Uota M, Uchida S (1989) Phys.Rev.B.40: 2254-2261.
[3] Uchida S (1989) International Seminar on High Temperature Superconductivity, Dubna USSR, World Scientific.
[4] Tajima S, Ishii H, Nakahashi T, Takagi H, Uchida S, Seki M, Suga S, Hidaka Y, Suzuki M, Murakami T, Oka K, Unoki H (1989) J.Opt.Soc.Am.B 6:475-482.
[5] For example, Maekawa S, Inoue J, Tohyama T (1989) "Strong Correlation and Superconductivity", Springer Series in Solid State Sciences 89 (Springer-Verlag) eds. Fukuyama H, Maekawa S, Malozemoff AP: p66.

Electronic Structure and Midinfrared Exciton Bands of Cuprate Superconductors

J. TANAKA[1], M. SHIMIZU[1], S. MIYAMOTO[1], C. TANAKA[1], K. KAMIYA[2], and H. OZEKI[2]

[1] Department of Chemistry, Faculty of Science, Nagoya University, Chikusa, Nagoya, 464-01 Japan
[2] Institute for Molecular Science, Okazaki, 444 Japan

ABSTRACT

The midinfrared exciton bands of cuprate superconductors are interpreted by the molecular orbital (MO) calculation of model cuprate ions. Two electronic transitions are found in the CuO_2 (xy) plane at 0.1 - 0.9 eV regions, and they are assigned to the transitions between the highest occupied MO(HOMO) and lowest unoccupied MO(LUMO) energy bands composed of O(2px) and O(2py) anti-bonding π type orbitals. The strong bands are found along the b-axis of $YBa_2Cu_3O_7$ as well as $EuBa_2Cu_3O_7$. They are explained as the transition from the highest occupied MO to the lowest unoccupied MO of the oxidized O-Cu-O-Cu-O chain elongated along the b axis.

KEY WORDS: electronic spectra, infrared spectra, electronic structure, exciton

INTRODUCTION

Electronic structure and spectra of cuprate superconductors are important in view of finding fundamental aspect on properties of free carrier and their interaction with exciton. Exciton bands in the mid-infrared region attracted many attention because of its significance for the mechanism of Cooper pair formation.

In this paper we will summarize our experimental results on the single crystal reflection spectra of several cuprate superconductors, and present the interpretation of spectra based on the molecular orbital calculation of model cuprate ions.

EXPERIMENTAL

We measured reflection spectra of single crystals of $La_{2-x}Sr_xCuO_4$, $Bi_2(Sr,Ca)_3Cu_{8+y}$, $YBa_2Cu_3O_{7-z}$, $EuBa_2Cu_3O_{7-z}$ and $Tl_2Ca_2Ba_2Cu_3O_w$.

We used microspectrophotometers designed for far infrared, mid infrared and near infrared to ultraviolet regions. Kramers-Kronig transformations of reflectances are calculated to obtain the dielectric functions and optical conductivities for 100 - 30000 cm^{-1} region.

We estimated preliminary values of plasma frequencies of free carrier and oscillator strengths of individual bands of these crystals with simulation of reflectances and optical conductivities by using a Drude term and harmonic oscillator model for these crystals. At least two electronic excited states are found in the mid-infrared region of the CuO_2 plane at about 0.05 - 0.15 eV and 0.3 - 0.9 eV, the positions are dependent on the layer structures.[3] Brief reports on these spectra are presented previously.[1-3]

We report here the spectra of YBaCuO. The spectra of YBaCuO and EuBaCuO [1,2] measured with single domain of (001) face show strong anisotropy for the b and a-axes (Fig.1). The anisotropy is dependent on the oxygen content of the crystals.

The spectrum polarized parallel to the c-axis on the (100) and (010) mossaic plane (Fig.2) show no prominent peak, accordingly both the free carrier band and mid-infrared exciton bands observed perpendicular to the c axis are attributed to the electrons on the CuO_2 plane and the O-Cu-O chain parallel to the b axis. This anisotropy was also reported by Bosovic et al.[4]

ASSIGNMENT OF EXCITON BANDS

Two exciton states of the CuO_2 plane in the mid-infrared region are interpreted as the transitions between the highest occupied MO(HOMO) and the lowest unoccupied MO(LUMO) bands, which are composed of O2px and O2py (π type) orbitals. (Fig.3)[5]

The oscillator strengths for these transitions are estimated by the MO caluculation at 0.018 (0.1 eV band) and 0.054(0.3 eV band) ,respectively, the experimental values are larger than the calculated values by a factor of two or three. It implies the mixing of the exciton states with the motion of free carrier, in other words, the exciton couples with conduction electron and borrows intensity from the Drude term.

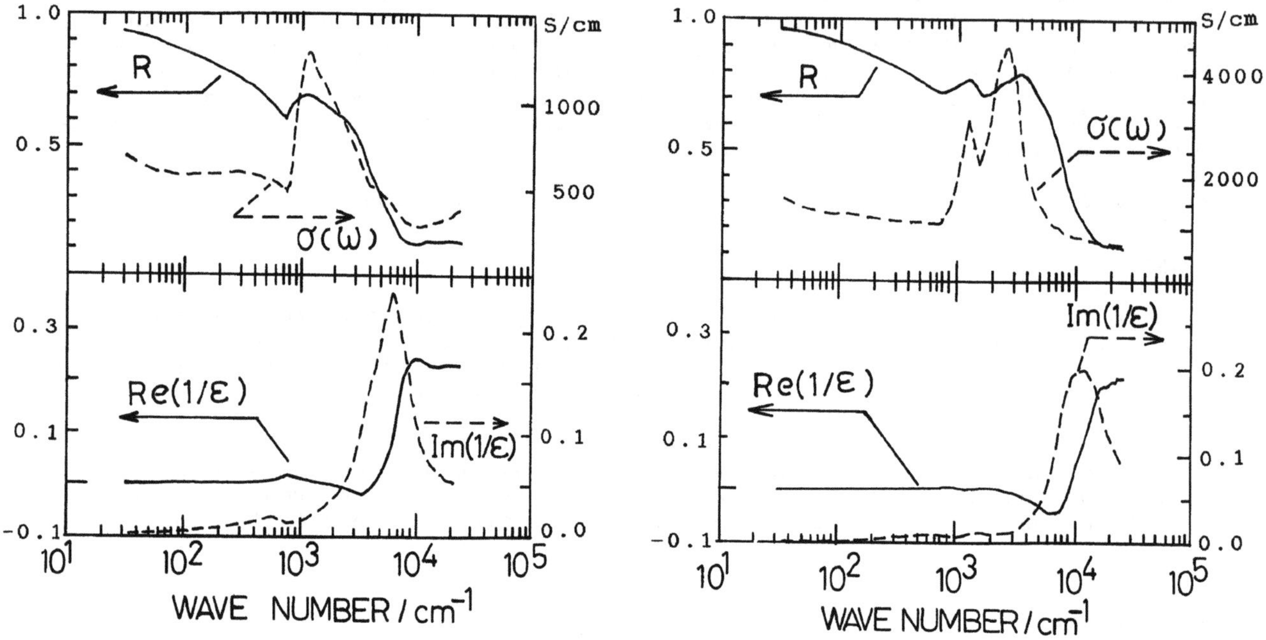

Fig.1 Upper curves: Reflection spectra and optical conductivity spectra of $YBa_2Cu_3O_{7-z}$; light polarized parallel to the a axis (left) and light polarized parallel to the b axis (right). Bottom curves: Real and imaginary parts of inverse dielectric functions.

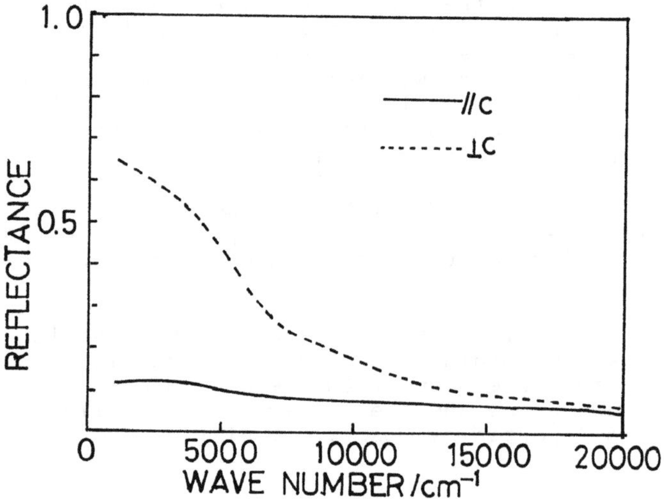

Fig.2

Reflection spectra of $YBa_2Cu_3O_{7-z}$ with light polarized parallel and perpendicular to the c axis.

The Cu-O chain (b-axis spectrum of Fig.1) shows excited state at about 0.3 eV, and the intensity is larger than those of the exciton bands of the CuO_2 plane. The MO calculation on the model compound of the chain is carried out to find a precise assignment for this strong band.

MOLECULAR ORBITALS OF $Cu_2O_7^{10-}$ AND $Cu_2O_7^{9-}$

Cu_2O_7 ion is a model compound for the chain as shown in Fig.4. Ab initio unrestricted (UHF) selfconsistent field (SCF) MO caluclation is used to find correct energy sequences and shapes of orbitals.

The calculations are carried out on the $Cu_2O_7Be_6$ ion, six Be^{2+} ions are placed in the positions of surrounding Cu sites in order to provide approximate electrostatic potentials of surrounding ions but without any conjugation effect.

In $YBa_2Cu_3O_{7-z}$ crystals, the holes are introuduced into the chain and the plane when z approaches to zero. Following chemical substitutional study of Tokura and Torrance [6] , holes are introuduced into the CuO chain first and then they are injected into the CuO_2 plane. In the utmost hole doping, every half oxygen atoms in the chain are oxidized.

In order to study electronic structure of the chain, the calculations are performed on $(Cu_2O_7Be_6)^{2+}$ (singlet) and $(Cu_2O_7Be_6)^{3+}$ (doublet) with a Gauss 82 program. The basis sets are Huzinaga's Gaussian functions; Cu(2D)(5333/53/5) and O(3P)(43/4). We referred the structural data for the geometry of Cu_2O_7 ion.[7]

By the oxidation of $Cu_2O_7^{10-}$ ion, an electron in the MO composed of mostly O 2py orbital, which is not HOMO but 7th highest among 68 MO, is removed and the shape of HOMO does not changed. The HOMO and LUMO of $Cu_2O_7^{9-}$ ion is shown in Fig.4 , where the hole orbital is depicted with dotted line in the fiugre of HOMO. The transition from the HOMO to LUMO is polarized along direction of the O-Cu-O-Cu-O chain, accordingly it is b-axis polarized in agreement with experiment.

The intensity of the b-axis polarized band is estimated by MO caluclation at 0.21, which is same order of magnitude of experimental value. This result gives assignment for strong mid-infrared exciton band in YBaCuO at 0.3 eV to the excitation in the O-Cu-O-Cu-O chain, while 0.1 eV bands clearly found in Fig.1 are assigned to the in-plane CuO_2 band.

Bulaevskii et al.[8] showed in a tunnelling experiment on $EuBa_2Cu_3O_7$ that pairing originates from the electron electron interaction via electronic excitation of $\sim$ 0.1 eV. Combining with this result, the primary important exciton state in cuprate superconductor is the in plane 0.1 eV band. [5]
A direct evidence for the involvement of the chain exciton of 0.3 eV has not been found. Further study will be required to elucidate the interaction of chain exciton with a pair of electrons.

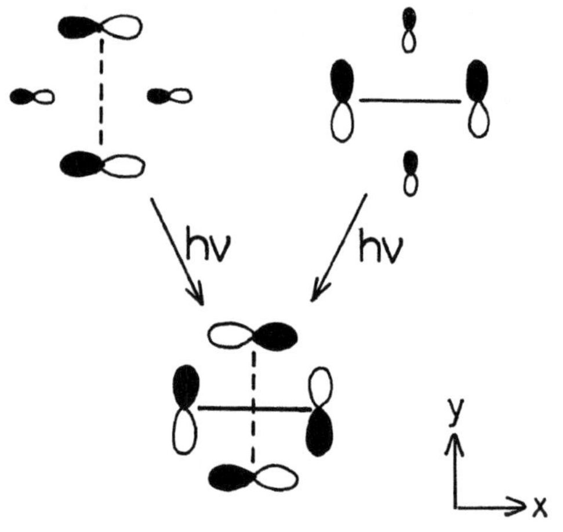

Fig.3

Electronic transitions between 2pπ anti-bonding type MOs composed of O(2px) and O(2py) orbitals.

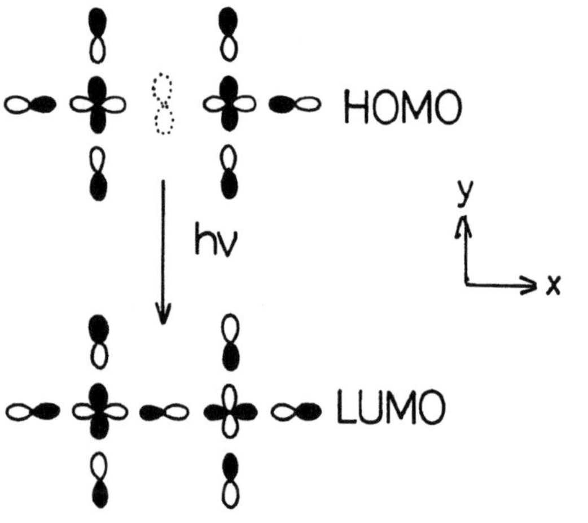

Fig.4

Electronic transition from the HOMO to LUMO of $Cu_2O_7^{9-}$ ion. The orbital with a hole is shown by a dotted line.

REFERENCES

1. Tanaka J, Kamiya K, Tsurumi S (1988) Anisotropy in optical spectra of single crystal of $EuBa_2Cu_3O_{7-y}$. Physica C 153: 653-654

2. Tanaka J, Kamiya K, Shimizu M, Shimada M, Tanaka C, Ozeki H, Adachi K, Iwahashi K, Sato F, Sawada A, Iwata S, Sakuma H, Uchiyama S (1988) Optical spectra and electronic structures of high Tc oxide superconductors. Physica C 153: 1752-1755

3. Tanaka J, Kamiya K, Shimizu M, Tanaka C, Ozeki H, Miyamoto S (1989) Optical spectra and electronic structure of high Tc oxide superconductors Physica C: to be published.

4. Bozovic I, Char K, Yoo S.J.B, Kapitulnik A, Beasley M.R, Geball T.H, Wang Z.Z, Hagen S, Ong N.P, Aspnes D.E, Kelly M.K (1988) Optical anisotropy of $YBa_2Cu_3O_{7-x}$: Phys Rev B38: 5077-5080

5. Tanaka J, Kamiya K, Tanaka C (1989) Electronic structure of the CuO_2 plane of cuprate superconductor. Physica C: to be published.

6. Tokura Y, Torrance J.B, Huang T.C, Nazzal A.I (1988) Broader perspective on the high-temperature superconducting $YBa_2Cu_3O_y$ system: The real role of oxygen content. Phys Rev B38: 7156-7159

7. Izumi F, Asano H, Ishigaki T, Takayama-Muromachi E, Uchida Y, Watanabe N, Nishikawa T (1987) Rietveld refinement of the structure of $Ba_2YCu_3O_{7-x}$ with neutron powder diffraction data. Jap J Appl Phys 26: L649-651

8. Bulaevskii L.N, Dolgov O.V, Kazakov I.P, Maksimovskii S.N, Pittsyn M.O, Stepanov V.A, Vedeneev S.I (1988) A tunnelling study of the oxide superconductors $La_{2-x}Sr_xCuO_{4-y}$ and $EuBa_2Cu_3O_7$. Supercond Sci Technol 1: 205-209.

Optical Properties of $Ln_{2-x}Ce_xCuO_4$ (Ln=Pr, Nd, and Sm) Single Crystals

IZUMI TOMENO, MASASI YOSIDA, KAZUTO IKEDA, TSUTOMU TAKATA, KEISI TAI, and NAOKI KOSHIZUKA

Superconductivity Research Laboratory, International Superconductivity Technology Center, 10-13, Shinonome 1-chome, Koto-ku, Tokyo, 135 Japan

ABSTRACT

Optical properties of $Ln_{2-x}Ce_xCuO_4$ (Ln=Pr, Nd and Sm) single crystals were investigated using both X-ray photoemission and Raman spectroscopies. The Cu 3d core-level shift with the Ce doping in Pr_2CuO_4 is attributed to the increase in Cu^+/Cu^{2+} ratio. For each parent crystal, two-magnon peak was observed at 2800 cm^{-1}. Two-magnon features persist in the Ce-doped crystals at room temperature, although their peaks are suppressed considerably. The broadening of two-magnon Raman spectra in the Ce-doped samples means that the spin fluctuations remain in the CuO_2 layer where spin-half $Cu3d^9$ atoms are diluted with no-spin $Cu3d^{10}$ atoms.

KEY WORDS: $Ln_{2-x}Ce_xCuO_4$, Raman scattering, Two-magnons, X-ray photoemission, Valence band, Core level

INTRODUCTION

The electron-doped superconductors $Ln_{2-x}Ce_xCuO_4$ (Ln=Pr, Nd and Sm) have received much attention because of their unique properties.[1-5] These crystals belong to the tetragonal crystal structure, where the Cu atoms are square-planar coordinated by oxygens. Recent neutron-diffraction studies [3-5] revealed that Nd_2CuO_4 and Pr_2CuO_4 are two-dimensional Heisenberg antiferromagnets with T_N=250-260 K. We are concerned with Cu 3d electronic states and spin fluctuations on the CuO_2 layers in the $Ln_{2-x}Ce_xCuO_4$ systems. First, we investigate the valence band and Cu3d core-levels for both Pr_2CuO_4 and $Pr_{1.86}Ce_{0.14}CuO_4$ single crystals by X-ray photoemission experiments. Next, we determine two-magnon Raman spectra for each parent crystal in order to evaluate the exchange interaction energy J between Cu ions. Then we present the Ce-doping effect on spin fluctuations in Raman spectra, and compare the present results with those for hole-doped cuprate superconductors [6-9].

SAMPLE PREPARATION

Single crystals of $Ln_{2-x}Ce_xCuO_4$ (Ln=Pr, Nd and Sm) were grown from a CuO flux, based on the phase diagram for the Nd_2O_3-CuO system.[10] A mixture in a concentration ratio of $Ln_2O_3:CeO_2:CuO$=(15-7.5x):7.5x:85 (mole %) were heated in alumina crucibles up to 1300 °C, held at this temperature for 2 hours, cooled to 1000 °C at a rate of 5 °C/h, and then furnace-cooled to room temperature. A number of platelike single crystals with large c-plane appeared on the surface of CuO flux, as shown in Fig.1. Typical crystal size was 3x3x0.5 mm. Their planes were occasionally perpendicular to the flux surface. The Ce-doped samples were annealed in Ar atmosphere at 950 °C for 12 hours, and then quenched to room temperature. The Ce concentration x was determined using an energy dispersive x-ray spectroscopy. Magnetization measurements were performed using a SQUID magnetometer. Figure 2 shows that the $Pr_{2-x}Ce_xCuO_4$ samples with x=0.17 are superconducting at T_C=15 K.

RESULTS AND DISCUSSION

X-RAY PHOTOEMMISON

Photoemission experiments on Pr_2CuO_4 and $Pr_{1.86}Ce_{0.14}CuO_4$ single crystals were performed using a VG spectrometer at pressures less than $5x10^{-10}$ Torr. Spectra were taken at room temperature with a typical energy resolution of 0.8 eV using an Al $K\alpha$($h\nu$=1486.6 eV) radiation. All the data were obtained in reference with C 1s at -285.0 eV.

Figure 3 shows the valence band spectra. The two peaks for the parent crystal are related with Pr 4f features overlapped with the Cu-O bands [11]. The Ce doping enhances the intensity around -2.5 eV, while there is no evidence for the increase in the density of states at E_F with the Ce doping.. No difference at E_F is due primarily to the fact that the Pr 4f features are overlapped with the low-density antibonding Cu-O bands. Figure 4 shows the Cu2p core-level results for the two crystals. The subtraction from the spectrum of non-doped crystal demonstrates that the Cu^+ state appears in the Ce-doped crystal. This is in marked contrast to the case of hole-doped superconductors, where the holes are introduced at the oxygen sites.

RAMAN SCATTERING

The Raman measurements were carried out in backscattering configuration at temperatures between 6 and 300K. The spectra were excited with 50-200 mW of argon ion laser at either 4880 or 5145 Å. Scattered light was analyzed with a double-monochromator (Jobin Yvon U1000) The sample was mounted on the cold finger of an Oxford Instruments flow cryostat for the low-temperature experiments.

Figure 5 shows Raman spectra of Nd_2CuO_4 and Pr_2CuO_4 single crystals at room temperature. A broad peak appears at 2800 cm^{-1} in (a,a) configuration for each parent crystal, while Raman intensity in (c,c) configuration for Pr_2CuO_4 is practically frequency independent in the range between 2500 and 4800cm^{-1}. Both 4880 and 5145 Å incident lines give rise to the Raman peak at the same frequency. Furthermore, the broad peak is nearly temperature independent between 6 and 300 K. In view of neutron-diffraction results [3-5], the Raman features are characterized by two-magnon scattering in two-dimentional system. We obtain the exchange interaction value J=1040 cm^{-1} using the conventional relation $\hbar\omega$=2.7J [12]. The present results are consistent with the Nd_2CuO_4 spectra reported by Sugai et al.[13]. The extremely large J value means that the short-range order in $Ln_{2-x}Ce_xCuO_4$ persists at temperatures far above T_N. Similar situation has been found for $YBa_2Cu_3O_{6+x}$ [6-8] and $La_{2-x}Sr_xCuO_4$ [9]. Raman results for the Ce-doped Pr_2CuO_4 are shown in Fig.6. Comparison with the non-doped data shows that the two-magnon scattering persists in the Ce-doped Pr_2CuO_4, although the suppressed peak is shifted to lower frequencies. The spin dilution effect explains the broadening of two-magnon Raman spectra in the Ce-doped samples. In the Ce-doped crystal, the spin system is diluted with zero-spin Cu^+ ions. In the case of $YBa_2Cu_3O_{6+x}$ and $La_{2-x}Sr_xCuO_4$, holes introduced at oxygen sites may disturb the Cu-Cu interaction.

REFERENCES

1.T.Tokura,H.Takagi, and S.Uchida, Nature 337,345 (1989).
2.H.Takagi, S.Uchida, and Y.Tokura, Phys.Rev.Lett.62, 1197 (1989).
3.J.Akimitsu, H.Sawa, T.Kobayashi, H.Fujiki, and Y.Yamada, J.Phys.Soc. Jpn, 2646 (1989).
4.Y.Endo, M.Matsuda, K.Yamada, K.Kakurai, Y.Hidaka, G.Shirane R.J.Birgeneau, Phys.Rev.B40,7023 (1989).
5.D.E.Cox, A.I.Goldman, M.A.Subramanian, J.Gopalakrishnan, and A.W.Sleight, Phys.Rev.B40,6998 (1989).
6.K.B.Lyons, P.A.Fleury, J.P.Remeika, A.S.Cooper, and T.J.Negran, Phys.Rev.B37,2353 (1988).
7.K.B.Lyons, P.A.Fleury,L.F.Schneemeyer, and J.V.Waszczak, Phys.Rev.Lett.60, 732 (1988).
8.K.B.Lyons and P.A.Fleury, J.Appl.Phys.64, 6075 (1988).
9.S.Sugai, S.Shamoto and M.Sato, Phys.Rev.B38, 6436 (1988).
10.K.Oka and H.Unoki, Jpn.J.Appl.Phys. 28,L937 (1989).
11.J.H.Weaver,H.M.MeyerIII,T.J.Wagener,D.M.Hill,Y.Gao,D.Petersn, Z.Fisk, and A.J.Arko, Phys.Rev.B38,4668 (1988).
12.J.B.Parkinson, J.Phys.C 2, 2012 (1969).
13.S. Sugai, T.Kobayasi and J.Akimitsu, Phys.Rev.40,2686(1989).

Figure 1. Single crystals $Pr_{2-x}Ce_xCuO_4$ with CuO flux in alumina crucible.

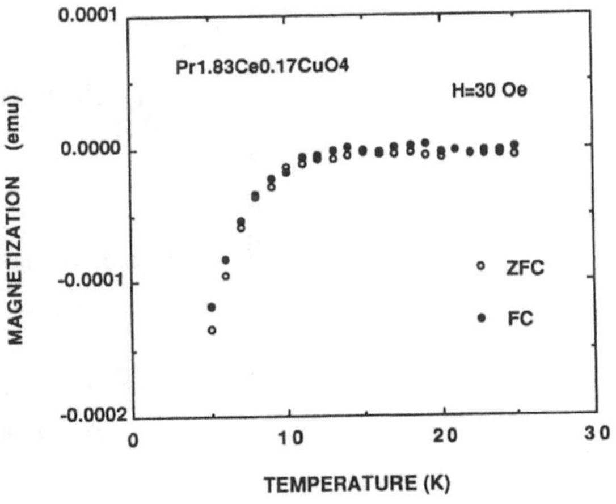

Figure 2. Magnetization M at H=30 Oe as a function of temperature for $Pr_{1.83}Ce_{0.17}CuO_4$ single crystal.

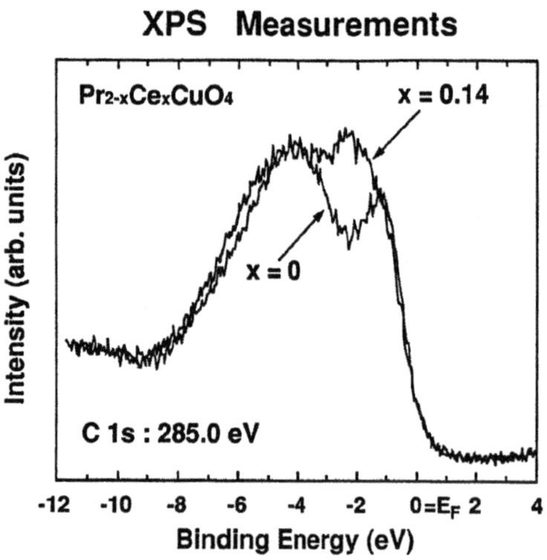

Figure 3 Valence band spectra in Pr_2CuO_4 and $Pr_{1.86}Ce_{0.14}CuO_4$.

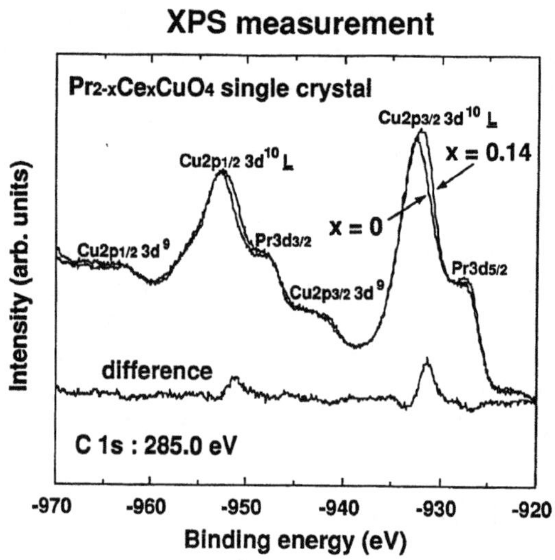

Figure 4. X-ray photoemission spectra for Cu 2p states in Pr_2CuO_4 and $Pr_{1.86}Ce_{0.14}CuO_4$.

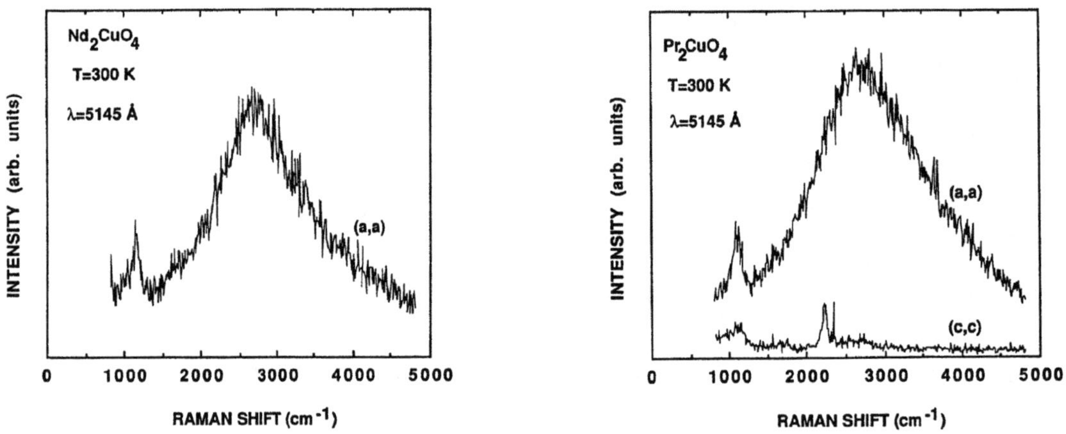

Figure 5. Raman spectra of Nd_2CuO_4 and Pr_2CuO_4 single crystals at room temperature.

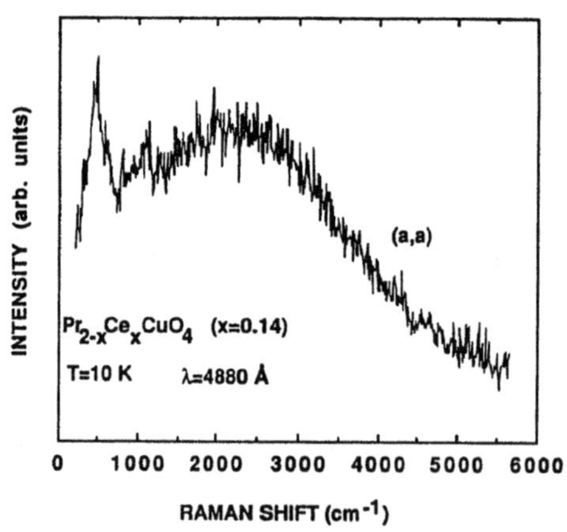

Figure 6. Raman spectrum of $Pr_{1.86}Ce_{0.14}CuO_4$ single crystal.

Raman Scattering Spectra of $NdBa_2Cu_3O_x$ Single Crystals

M. Yoshida, S. Gotoh, T. Takata, and N. Koshizuka

Superconductivity Research Labolatory, International Superconductivity Technology Center, 10-13, Shinonome 1-chome, Koto-ku, Tokyo, 135 Japan

ABSTRACT

Raman scattering from $NdBa_2Cu_3O_x$ single crystals have been studied. Single phonon peaks have been observed at 136, 168, 316, 448, 508 and 580 cm^{-1} among which the 316 and 580 cm^{-1} peaks are strong in the (x,x) configuration while the 448 and 508 cm^{-1} peaks in the (z,z) configuration. These polarization dependence can be explained considering the characteristics of the electronic transitions. The 448 cm^{-1} peak is asscribed to Raman forbidden B_{2u} or B_{3u} modes being different from Ag mode in the previous literature. In the (x,x) configuration 2-magnon peak has been observed at 2400cm^{-1}, while in the (z,z) configuration, up to three phonon scattering of the 508cm^{-1} peaks heve been observed.

KEY WORDS: Raman scattering, phonon, magnon, $NdBa_2Cu_3O_x$, $YBa_2Cu_3O_x$

INTRODUCTION

Intensive studies of Raman spectra of $YBa_2Cu_3O_x$ (YBCO) have been performed to investigate the role of phonons[1] and magnons[2] in the high Tc superconductivity. Strongly anisotropic scattering spectra have been obtained.[3-5] Five Ag mode phonons are identified[4, 5] and the behavior of these modes under the substitution of constituent atoms has been examined.[1] However, the polarization characteristics of these phonons cannot be explained by the group theoretical considerations.[3] Also, the magnon scattering spectra have been obtained only in the ab-plane[2] and the information concerned about the polarization along the c-axis has not been obtained yet. $NdBa_2Cu_3O_x$ (NBCO) is a superconductor with Tc of 90K and the YBCO-type crystal structure.[6] In the present study we have investigated the Raman spectra of NBCO single crystals. Anisotropic scattering spectra due to phonons and magnons have been observed. The polarization dependence of the spectra is explained considering the properties of the electronic states which act as the intermediate states in the light scattering process.

EXPERIMENTAL

Single crystals of NBCO were obtained by heating the sintered NBCO samples at 1150C in air for 24 hours. The crystals were thin platelets with typical dimensions of 1x1x0.1mm^3. They were found to be oriented with the c-axis perpendicular to the large surface of the platelet by the x-ray diffraction. The superconducting transition temperature of 60K in the zero-field cooled magnetic susceptibity was achieved by annealing the crystals at 400C in O_2 environment for 4 days. The Raman scattering measurements were carried out using 5145A light from an argon laser, which was focused onto the ac-plane of the sample with diameter about 0.1mm. Laser power was maintained below 50 mW in order to avoid damaging samples. The scattered light was detected with a Jobin Ybon U-1000 double monochromator and a conventional photon counting detection system.

RESULTS AND DISCUSSION

In Figs.1 (a) and (b) are shown the Raman spectra of NBCO at room temperature in the backscattering geometry in the (z, z) and (x, x) configurations, respectively, where (z, z) ((x, x)) means that the polarizations of both the incident and scattered lights are parallel to the crystallographic c-axis (a-axis). Due to twinning in the ab-plane, we could not distinguish between (x, x) and (y, y) Raman polarization. There is no detectable depolarized (z, x) or (x, z) scattering to within noise. In the (z, z) configuration, peaks are seen at 136, 168, 448 and 508 cm^{-1} and broad structures at around 780 and 1000cm^{-1}. The 508^{-1} peak is shifted to lower energy side in comparison with the 513 cm^{-1} of sintered NBCO sample[7]. This fact implys that our sample has oxygen deficiency[8], which is consistent with the Tc of 60K of our sample. The Raman shift of the 1000cm^{-1} structure is about twice that of the 508cm^{-1} peak. Thus, it is supposed that the 1000cm^{-1} structure is caused by the two phonon scattering of the 508cm^{-1} phonon.

In the (x, x) configuration, peaks are seen at 136, 316 and 580 cm^{-1} and a broad structure at around 1160cm^{-1}. The 1160cm^{-1} structure can be ascribed to the two phonon scattering of the 580cm^{-1} phonon. It should be noticed that the scattering intensity is about ten times weaker in the (x, x) configuration than in the (z, z) configuration. Sharp peaks around 100cm^{-1} are caused by the rotational vibration of oxygen molecules in the air. From the absence of all phonon features under cross polarization, it is concluded that observed peaks have Ag symmetry.

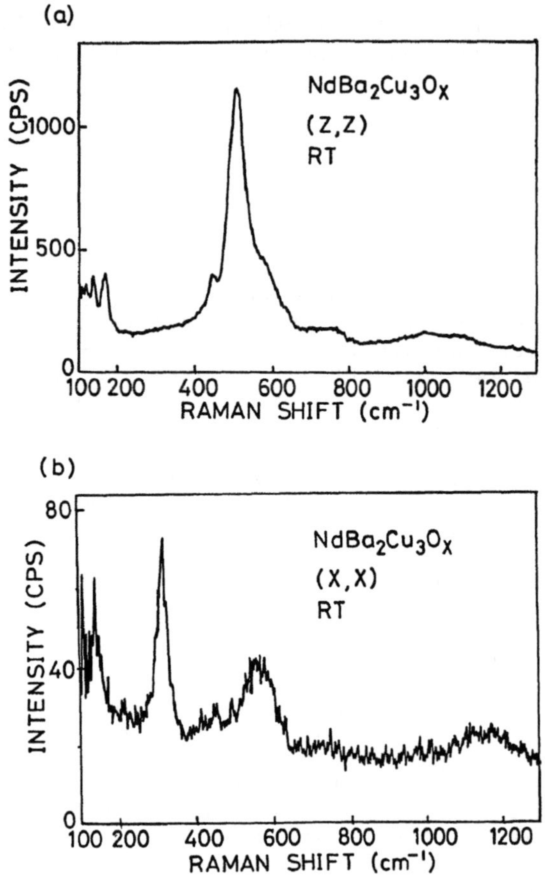

Fig. 1 Raman spectra of $NdBa_2Cu_3O_x$

Anisotropic phonon features have been obtained as shown in Fig. 1. The crystal structure of NBCO is shown in Fig. 2. There are five Ag mode phonons in this structure. In the Raman spectra of YBCO, Ag symmetry peaks are observed at 116, 150, 340, 440 and 500 cm^{-1} [3], where the 340cm^{-1} peak is strong in the (x, x) and the 440 and 500 cm^{-1} peaks are strong in the (z, z) configurations.[3,4] Liu et al[4] gave the assignments of these peaks on the basis of their phonon energy calculation. According to them, the two lower energy peaks are due to the vibration of Ba (116cm^{-1}) and Cu (150cm^{-1}). The three higher energy peaks are due to the vibration of oxygen atoms where the 340cm^{-1} and 440cm^{-1} peaks are assigned to the out-of-phase and in-phase vibration of O(3) and O(4) atoms along the c-axis, respectively, and the 500cm^{-1} peak the vibration of O(2) atom along the c-axis. However, the polarization dependence of these three peaks cannot be explained by the simple group theoretical consideration since scattering by Ag mode phonons is allowed both in (x, x) and (z, z) configurations. To explain the polarization dependence of these lines the characteristics of the electronic transition must be considered, which has not been taken into account in the previous literature.[3-5]

The Raman scattering is the third order perturbation process which consists of (1) virtual creation of electron-hole pair by the absorption of the light, (2) scattering of electron or hole by a phonon and (3) radiative annihilation of the electron-hole pair.[9] The polarization of light has a crucial role in determining the electron and the hole which take part in the Raman scattering. By the band structure calculation of YBCO by Mattheiss and Hamann[10] it has been shown that the valence bands of YBCO are constructed by anti-bonding Cu-O bonds, among which the bands constructed mainly by $2p_x$ of O(3) and $2p_y$ of O(4) orbitals are the highest in energy.[11] The higher conduction bands where electrons are excited by the absorption of light is supposed to be the upper Hubbard band of Cu 3d states. Thus, according to the selection rule for the dipole transition, the radiative transition occurs mainly between $2p_x$ ($2p_y$) of O(3) (O(4)) atom and $3d_{x^2-y^2}$ state of Cu (2) for the x-polarized (y-polarized) light and between $2p_z$ of O(2) and $3d_{3z^2-r^2}$ of Cu(1) for the z-polarized light.

In Figs. 1 (a) and (b) are shown that the $508cm^{-1}$ peak is strong in the (z, z) configuration. The assignment of this peak to the vibration of O(2) atom is consistent with the above consideration since holes on O(2) atom are expected to couple strongly with the vibration of O(2) atom unless the coupling is forbidden by the symmetry condition. The fact that the $508cm^{-1}$ peak is not observed in (x, x) configuration implys that the holes are localized on O(2) cite near perfectly and tight binding picture holds well in this compound.

The $316cm^{-1}$ peak is strong in the (x, x) configuration as shown in Fig. 1. This is in accord with the assignment of this peak to the out-of-phase vibration of O(3) and O(4) atoms. However, the assignment of the $448cm^{-1}$ peak which is strong in the (z, z) configuration to the in-phase viblation of O(3) and O(4) atoms do not agree with the above consideration. It should also be pointed out that the energy difference over $100cm^{-1}$ between the $316cm^{-1}$ and $448cm^{-1}$ peaks is too large in comparison with the expected value of about $20cm^{-1}$ as the difference between the in-phase and out-of-phase vibration of O(3) and O(4) atoms.[4] We suppose that the $448cm^{-1}$ peak is due to the bending vibration of O(2) atom along x or y direction (B_{2u} or B_{3u}) which becomes Raman allowed in the (z, z) configuration through the intra-band scattering of the $2p_z$ hole on O(2) by the Frohlich type interaction.[12] It may be realized that the $316cm^{-1}$ peak has contribution from both the in-phase and out-of-phase vibration of O(3) and O(4) atoms.

In Fig.1 (b) a broad peak has been observed at $580cm^{-1}$ and $1160cm^{-1}$. According to the polarization characteristics, these peaks can be ascribed to the vibration of O(3) and O(4) atoms. We assign these peaks to the stretching vibration of O(3) and O(4) atoms along the a- or b-axis since this mode is expected to have an energy around $580 cm^{-1}$.[4]

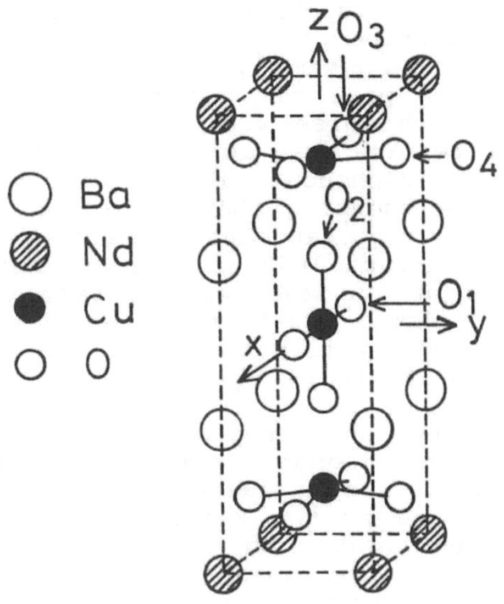

Fig. 2 Crystal structure of $NdBa_2Cu_3O_x$

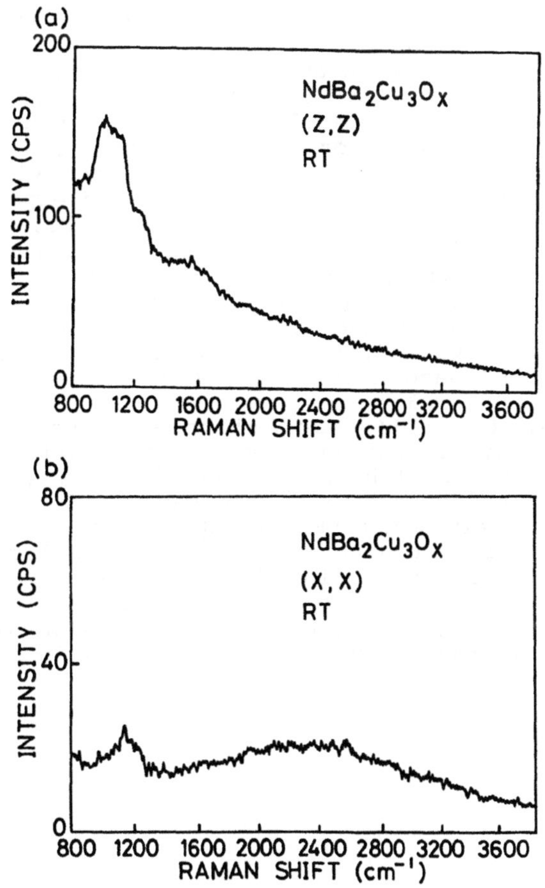

Fig. 3 Raman spectra of $NdBa_2Cu_3O_x$

In Figs. 3 (a) and (b) are shown the Raman spectra of NBCO in the range between 800 and 3800cm^{-1} in the (z, z) and (x, x) configurations, respectively. A broad peak is observed at around 2400cm^{-1} in the (x, x) configuration, which can be ascribed to 2-magnon scattering.[2] The exchange coupling constant is estimated as 890cm^{-1}, which is nearly equal to the value 950cm^{-1} of YBCO.[2] Two-magnon peak has not been observed in the (z, z) configuration (Fig. 3 (a)), which agree with the selection rule[2]. In the (z, z) configuration, structures are seen at around 1000cm^{-1} and 1600cm^{-1}. The 1000cm^{-1} structure is caused by 2-phonon scattering as described before. This structure has a broad width representing the dispersion of the phonon since phonons in the entire Brillouin zone can contribute to the two phonon scattering process. The structure at around 1600cm^{-1} can be ascribed to the three phonon scattering line because of its energy. The fact that higher order scattering lines of the 508cm^{-1} phonon can be observed in the Raman scattering implys strong coupling of this mode with holes, which also suggests the localized nature of the holes on the O(2) site.

In summary, anisotropic scattering spectra of $NdBa_2Cu_3O_x$ single crystals by phonons and magnons have been obtained. The polarization dependence of the spectra can be explained by considering that holes are excited on the O(3) (O(4)) site by the x-polarized (y-polarized) light and O(2) site by the z-polarized light. On the basis of the polarization characteristics, the 448 cm^{-1} peak is assigned to the vibration of O(2) atom being different from the previous assignment to the vibration of O(3) and O(4) atoms.

REFERENCES

1. Feile R. (1989) Lattice vibrations in high-Tc superconductors: optical spectroscopy and lattice dynamics. Physica C 159, 1-32
2. Lyons KB. Fleury PA. Schneemeyer LF. and Waszczak JV. (1988) Spin fluctuations and superconductivity in $Ba_2YCu_3O_x$. Phys. Rev. Letters 22, 732-735
3. Krol DM. Stavola M. Weber M. Schneemeyer LM. Wasczcak JV. Zahurak SM. and Koninski SG. (1987) Raman spectroscopy and normal-mode assignments for $Ba_2MCu_3O_x$ (M=Gd, Y) single crystals. Phys. Rev. B36, 8325-8328
4. Liu R. Thomsen T. Kress W. Cardona M. and Gegenheimer G. (1988) Frequencies, eigenvectors, and single-crystal selection rules of k=0 phonons in $YBa_2Cu_3O_x$: theory and experiment. Phys. Rev. B 37, 7971-7974
5. Cooper SL. Klein MV. Pazol BG. Rece JP. and Ginsberg DM. (1988) Raman scattering from superconducting gap excitations in single-crystal $YBa_2Cu_3O_x$. Phys. Rev. B 37, 5920-5923
6. Hosoya S. Shamoto S. Onoda M. and Sato M. (1987) High-Tc superconductivity in new oxide systems II. Jpn. J. Appl. Phys. 26 L456-457
7. Stavola M. Krol DM. Weber W. Sunshine SA. Jayaraman A. Kouroulis GA. Cava RJ. and Rietman EA. (1987) Cu-O vibrations of $Ba_2YCu_3O_x$. Phys. Rev. B-38 850-853
8. Rosen HJ. Macfarlane RM. Engler EM. Lee UY. and Jacowitz RD. (1988) Systematic Raman study of effects of rare-earth substitution on the lattice modes of high-Tc superconductors. Phys. Rev. B 38, 2460-2465
9. Loudon R. (1964) The Raman effects in crystals. Adv. Phys. 13, 423
10. Mattheiss LF. and Hamann DR. (1987) Electronic Structure of the high Tc superconductor $Ba_2YCu_3O_{6.9}$. Solid State Commun. 63, 395-399
11. Nucker N. Fink J. Fuggle JC. Durham PJ. Temmerman WM. (1988) Evidence for holes on oxygen sites in the high-Tc superconductors $La_{2-x}Sr_xCuO_4$ and $YBa_2Cu_3O_{7-y}$ Phys. Rev. B37 5158-5168
12. Frohlich H. (1954) Electrons in lattice fields. Adv. Phys. 3, 325

Raman Scattering in $(Ca_{0.86}Sr_{0.14})CuO_2$ and Nd_2CuO_4

TAKUYA UZUMAKI, KAZUNORI YAMANAKA, ATSUSHI TANAKA, NOBUO KAMEHARA, and KOICHI NIWA

Fujitsu Laboratories Ltd., Morinosato-Wakamiya, Atsugi, Kanagawa, 243-01 Japan

ABSTRACT

We fabricated single-phased $(Ca_{0.86}Sr_{0.14})CuO_2$ and Nd_2CuO_4 and determined the Raman scattering spectrum. A broad peak centered at about 3000 cm^{-1} was observed in the two systems. These peaks are interpreted as being due to two-magnon scattering which has been observed in La_2CuO_4 and $YBa_2Cu_3O_{6+y}$. The exchange interaction constant was estimated as 1100 cm^{-1} in $(Ca_{0.86}Sr_{0.14})CuO_2$ and as 1000 cm^{-1} in Nd_2CuO_4 at room temperature. The temperature dependence of the Raman spectra was measured. The broad peak shifted to higher frequency as the sample was cooled. We fabricated the quenched samples in order to dope the carriers in $(Ca_{0.86}Sr_{0.14})CuO_2$ and found that the resistivity at room temperature depended on the quench temperature.

KEY WORDS: $(Ca_{0.86}Sr_{0.14})CuO_2$, Nd_2CuO_4, Raman scattering, Two-magnon, Exchange interaction constant

INTRODUCTION

The parent structure of the layered high-Tc superconductors, $(Ca_{0.86}Sr_{0.14})CuO_2$ was reported by T. Siegrist et al.[1] and has infinite stacks of Cu-O sheets in the crystal structure in which Ca or Sr ions separate the Cu-O sheets. Conductivity in this system shows semiconducting behaviour. $A_2B_2Ca_{n-1}Cu_nO_y$(A = Bi, Tl, B = Sr, Ba) superconductors have critical temperature (Tc) above 100 K[2,3] and Tc increases as the number of Cu-O sheets (n) increase from 1 to 3. $(Ca_{0.86}Sr_{0.14})CuO_2$ has no apical oxygen atoms in the Cu-O sheets, in contrast to K_2NiF_4 structure with Cu-O octahedra. Recently the superconductor $Nd_{2-x}Ce_xCuO_{4-y}$ with electrons as charge carriers, was discovered by Y. Tokura et al. [4], and also has no apical oxygen atoms in the Cu-O sheets.

The semiconducting phases of $(La_{2-x}Sr_x)CuO_4$ and $YBa_2Cu_3O_{6+y}$ show antiferromagnetism, and superconductivity as the carrier was doped. K. B. Lyons et al.[5,6] examined the Raman spectrum of these antiferromagnetic phases. The spectra had a broad peak at 2600 cm^{-1} for $YBa_2Cu_3O_{6+y}$ and 3000 cm^{-1} for La_2CuO_4. The broad peaks are interpreted as being due to scattering by spin pairs. The exchange interaction constant J was estimated as 1100 cm^{-1} for La_2CuO_4 and 950 cm^{-1} for $YBa_2Cu_3O_{6+y}$.

We studied Raman scattering in $(Ca_{0.86}Sr_{0.14})CuO_2$ and Nd_2CuO_4. The focus of our Raman scattering observation was the broad peak caused by two-magnon scattering. Temperature dependence of two-magnon frequency was measured in $(Ca_{0.86}Sr_{0.14})CuO_2$. We estimated the exchange interaction constant J and compared in the two systems. We tried to dope the carriers in $(Ca_{0.86}Sr_{0.14})CuO_2$ and examined the quench temperature dependence of the resistivity at room temperature.

EXPERIMENTAL PROCEDURES

The sample of $(Ca_{0.86}Sr_{0.14})CuO_2$ and Nd_2CuO_4 were prepared by the solid-state reaction of appropriate amounts of $CaCO_3$, $SrCO_3$, Nd_2O_3 and CuO. $(Ca_{0.86}Sr_{0.14})CuO_2$ was sintered at 965°C for 6 hours after calcining at 800°C for 5 hours. Nd_2CuO_4 was sintered at 1140°C for 12 hours after calcining at 943°C for 5 hours.

X-ray diffraction patterns were obtained with monochromatic Cu Kα radiation and a diffractometer. Raman spectra were obtained using a photodetection system with an Ar laser (4880 Å and 5145 Å) and a double monochrometer. A 200 mW Ar laser beam was focused to 500 μm using a cylindrical lens. Temperature of the sample was changed using a cryostat from 30 K to 300 K and was measured by a thermocouple (Au-0.07%Fe, Chromel). The sample was attached to the holder by Ag-paste because of heat dissipation.

We fabricated quenched samples in order to dope the carriers in $(Ca_{0.86}Sr_{0.14})CuO_2$. The samples were kept at the quench temperature for 30 minutes after sintering at 965°C for 6 hours and then quenched to 77 K. The resistivity was measured at room temperature.

RESULTS AND DISCUSSION

X-ray diffraction patterns for $(Ca_{0.86}Sr_{0.14})CuO_2$ and Nd_2CuO_4 are shown in Figs. 1, 2. To determine the crystal structure, we calculated the lattice constants of the samples by the least squares method. The diffraction angles of the samples were calibrated by Si powder. The peaks in Figs. 1, 2 have been indexed assuming that the unit cell is tetragonal. The lattice constants were calculated to be a=3.863 Å and c=3.213 Å for $(Ca_{0.86}Sr_{0.14})CuO_2$ and a=3.942 Å and c=12.165 Å for Nd_2CuO_4. The samples consist of single phased.

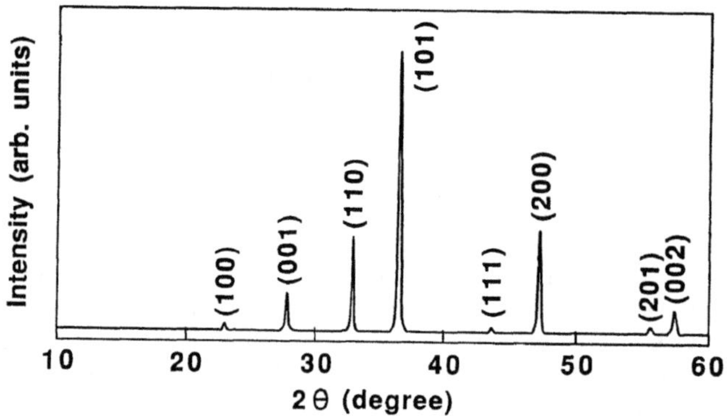

Fig. 1 Powder X-ray diffraction pattern of single-phased $(Ca_{0.86}Sr_{0.14})CuO_2$ sintered at 965°C for 6 hours. The lattice constants were calculated as a = 3.863 Å and c = 3.213 Å.

Raman spectra measured with different wavelengths of Ar laser (4880 Å and 5145 Å) are shown in Fig. 3. The primary feature is the broad peak with a 3000 cm^{-1} frequency shift. The frequency shift of the broad peak was independent of the wavelength of the Ar laser as shown in Fig. 3. Therefore, this peak is not due to luminescence. This peak is interpreted as being due to two-magnon scattering similar to that which has been observed in La_2CuO_4. The fre-

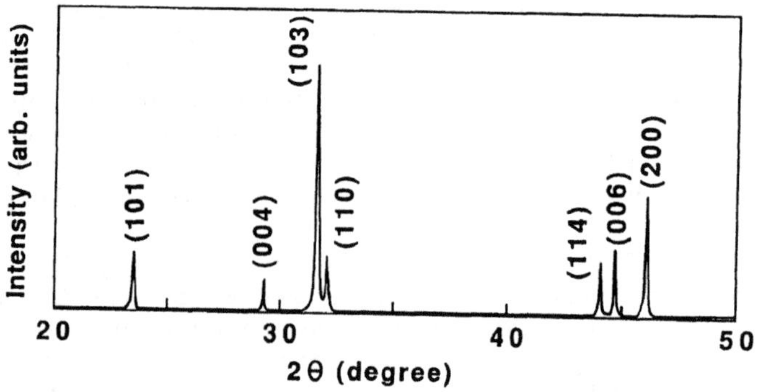

Fig. 2 Powder X-ray diffraction pattern of single-phased Nd_2CuO_4 sintered at 1140°C for 12 hours. The lattice constants were calculated as a = 3.942 Å and c = 12.165 Å.

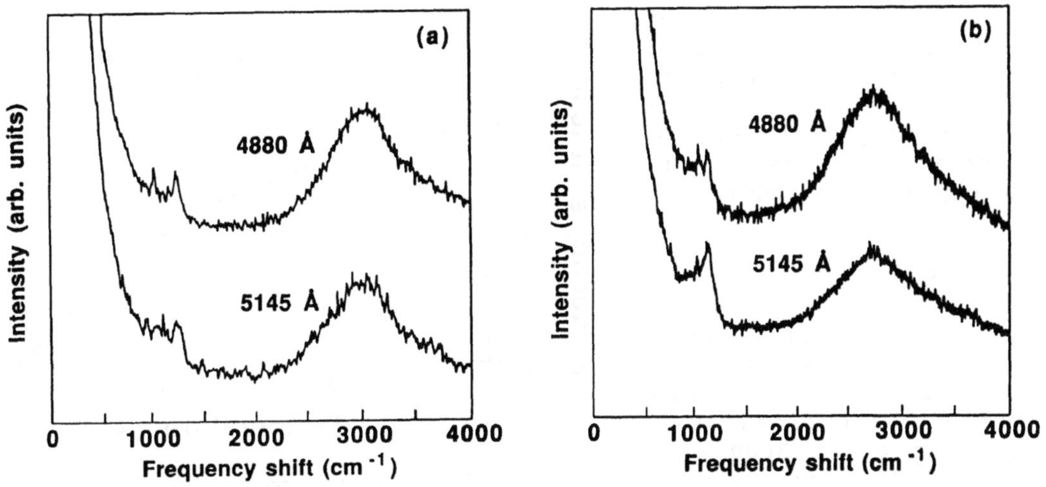

Fig. 3 Raman spectra obtained in (a) $(Ca_{0.86}Sr_{0.14})CuO_2$ and (b) Nd_2CuO_4 at room temperature. The wavelength of Ar laser is indicated. 200 mW Ar laser was used at wavelengths of 4880 and 5145 Å. The broad peaks are independent of the laser wavelength. The frequency shift is 3010 cm^{-1} for $(Ca_{0.86}Sr_{0.14})CuO_2$ and 2750 cm^{-1} for Nd_2CuO_4.

quency shifts were determined by the least squares method assuming that the spectra are Lorentzian type. From the magnon dispersion curve, we estimated the two-magnon energy at the zone boundary as 4 J, where J is the exchange interaction constant. The peak position was calculated from the magnon-magnon interaction as 2.7 J. We then estimated that J is 1100 cm^{-1} in $(Ca_{0.86}Sr_{0.14})CuO_2$ and 1000 cm^{-1} in Nd_2CuO_4 at room temperature. It seems that the difference of J value in the systems is related to Cu-Cu distance(a-axis) in Cu-O sheets.

Figure 4 shows the temperature dependence of the Raman spectra in $(Ca_{0.86}Sr_{0.14})CuO_2$. The broad peak shifted to higher frequency as the sample was cooled. Figure 5 shows the temperature dependence of the two-magnon frequency. The frequency shifts monotonically to higher values at lower temperatures. Two-dimensional anti-ferromagnetic order increases with decreasing temperature. If the electron can be doped in the $(Ca_{0.86}Sr_{0.14})CuO_2$, it will lead to superconductivity.

The resistivity at room temperature for samples quenched at various temperature is shown in Fig. 6. The resistivity decreases from 3810 Ω·cm to 240 Ω·cm. The conductivity of the samples shows semiconducting behaviour. X-ray

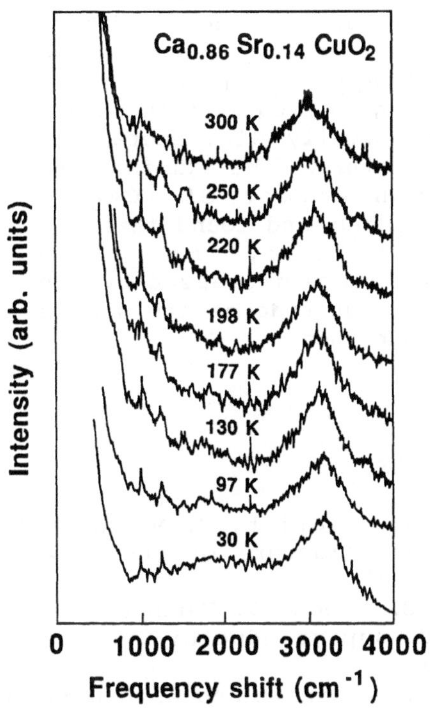

Fig. 4 Raman spectra in $(Ca_{0.86}Sr_{0.14})CuO_2$ between 30 K and 300 K, excited by Ar laser at 4880 Å.

diffraction patterns suggest that the samples consist of the single phase. It seems that the resistivity change is caused by the difference of the carrier concentration due to the oxygen deficiency.

CONCLUSIONS

We fabricated single-phased $(Ca_{0.86}Sr_{0.14})CuO_2$ and Nd_2CuO_4 and studied the Raman scattering. The broad peaks centered at about 3000 cm^{-1} were observed in both systems. This peak is assigned to the two-magnon scattering. The exchange interaction constant J at room temperature was estimated as 1100 cm^{-1} in $(Ca_{0.86}Sr_{0.14})CuO_2$ and as 1000 cm^{-1} in Nd_2CuO_4. It seems that the difference of J values is related to the Cu-Cu separation in Cu-O sheets. Temperature dependence of two-magnon frequency was measured in $(Ca_{0.86}Sr_{0.14})CuO_2$. The two-magnon frequency increased monotonically with decreasing temperature. We examined the quench effect of $(Ca_{0.86}Sr_{0.14})CuO_2$ and found that the resistivity at room temperature decreased by one order magnitude as the sample was quenched.

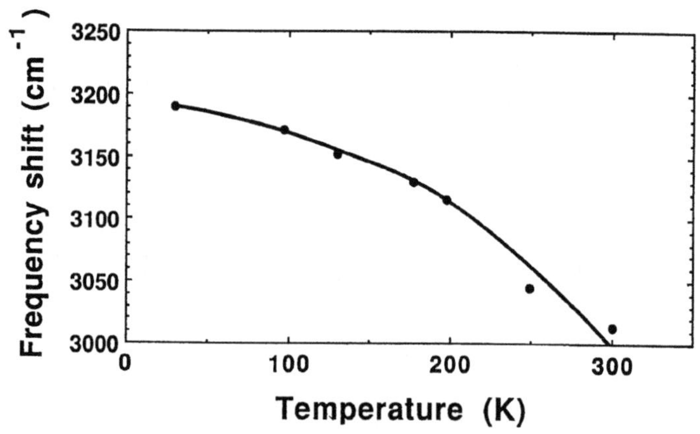

Fig. 5 Temperature dependence of two-magnon frequencies in $(Ca_{0.86}Sr_{0.14})CuO_2$, excited by Ar laser at 4880 Å.

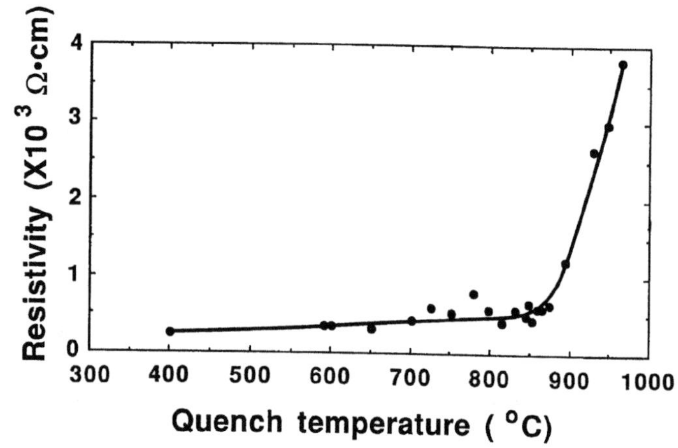

Fig. 6 Quench temperature dependence of resistivity at room temperature for $(Ca_{0.86}Sr_{0.14})CuO_2$. Samples were kept at the quench temperature for 30 minutes and quenched to 77 K.

REFERENCES

[1] T. Siegrist, S. M. Zahurak, D. W. Murphy, and R. S. Roth (1988) The parent structure of the layered high-temperature superconductors. Nature 334, 231

[2] H. Maeda, Y. Tanaka, M. Fukutomi, and T. Asano (1988) A new high-Tc oxide superconductor without a rare earth element. Jpn. J. Appl. Phys. 27, L209

[3] Z. Z. Sheng, and A. M. Herman (1988) Bulk superconductivity at 120 K in the Tl-Ca/Ba-Cu-O system. Nature 332, 138

[4] Y. Tokura, H. Takagi, and S. Uchida (1989) A superconducting copper oxide compound with electrons as the charge carriers. Nature 337, 345

[5] K. B. Lyons, P. A. Fleury, J. P. Remeika, A. S. Cooper, and T. J. Negran (1988) Dynamics of spin fluctuations in lanthanum cuprate. Phys. Rev. B37, 2353

[6] K. B. Lyons, P. A. Fleury, L. F. Schneemeyer, and J. V. Waszczak (1988) Spin fluctuations and superconductivity in $Ba_2YCu_3O_{6+y}$. Phys. Rev. Lett. 60, 732

3.6 Mechanical Properties

Phonon Eohoes in Superconducting Powders of Tl-Ba-Ca-Cu-O and (Bi, Pb)-Sr-Ca-Cu-O

H. NISHIHARA[1], K. HAYASHI[2], M. TAKANO[3], K. KISHIO[4], T. OHTANI[2], K. KAJIMURA[5], Y. OKUDA[6], T. TAMEGAI[7], and K. MOTOYA[8]

[1] Faculty of Science and Technology, Ryukoku University, Seta, Ohtsu, 520-21 Japan
[2] Laboratory for Solid State Chemistry, Okayama University of Science, Okayama, 700 Japan
[3] Institute for Chemical Research, Kyoto University, Uji, Kyoto, 611 Japan
[4] Department of Industrial Chemistry, The University of Tokyo, Hongo, Bunkyo-ku, Tokyo, 113 Japan
[5] Electrotechnical Laboratory, Tsukuba, Ibaraki, 305 Japan
[6] Department of Applied Physics, Tokyo Institute of Technology, Ohokayama, Tokyo, 152 Japan
[7] Institute for Solid State Physics, The University fo Tokyo, Roppongi, Minato-ku, Tokyo, 106 Japan
[8] Department of Physics, Saitama University, Saitama, 338 Japan

ABSTRACT

Radio-frequency phonon echoes in powdered samples have been observed in the superconducting states of new high-T_c oxide systems of Tl-Ba-Ca-Cu-O and (Bi,Pb)-Sr-Ca-Cu-O. The coupling mechanism between rf fields and acoustic modes is found to be different from that in usual phonon echoes. The temperature dependence of the echo intensity and the echo decay rate T_2^{-1} has anomalies at low T_c of the coexisting second superconducting phases. This property of the phonon echoes provides a new method of severe tests for the quality of high-T_c superconducting powders.

KEY WORDS: phonon echo, superconductivity, high-T_c oxide, acoustic oscillation, sample characterization

INTRODUCTION

We reported recently the first observation of the two-pulse phonon echoes in powders of high-T_c superconducting oxide $YBa_2Cu_3O_{7-\delta}$ [1]. The echoes, which are not usual NMR spin echoes but are due to acoustic oscillations of loosely packed powders in a static field and called also radio-frequency (rf) powder echoes or dynamic polarization echoes, are similar to those observed in powder samples of type-II superconductors of V-Ti and Nb-Zr alloys, ferrite powders, piezoelectric powders or even normal metallic powders[2,3]. However, judging from a sudden disappearance of the echoes at T_c of 90 K as well as the angular dependence of the echo intensity for the angle between static and rf fields, we have suggested that the coupling mechanism between rf fields and acoustic modes in $YBa_2Cu_3O_{7-\delta}$ is different from that in usual phonon echoes. We extend the experiment to the new high-T_c oxide systems of Tl-Ba-Ca-Cu-O and (Bi,Pb)-Sr-Ca-Cu-O, and suggest that the coupling mechanism between rf fields and acoustic modes in the new oxides is the same as that in $YBa_2Cu_3O_{7-\delta}$. We also report that the temperature dependence of the echo intensity and the echo decay rate T_2^{-1} has anomalies at low T_c of the coexisting second superconducting phases. This property of the phonon echoes may be used to check the quality of powders of high-T_c oxides.

EXPERIMENTAL RESULTS AND DISCUSSION

Phonon echoes have been observed for powders of new high-T_c oxides of Tl-Ba-Ca-Cu-O and (Bi,Pb)-Sr-Ca-Cu-O sealed in quartz ampoules with helium gas for heat exchange by the use of a conventional phase-coherent NMR apparatus.

Properties of the observed echoes are quite similar to those of phonon echoes reported[2,3] : (i) The amplitude is not strongly frequency dependent. (ii) No echo was observed in weak applied fields of less than 3 kOe, but the amplitude of the echoes increased with increasing applied field strength with a hysteresis in earlier sweeps. (iii) The rf phase of the secondary echo is the same as that of the main echo. (iv) The echoes are of maximum strength when the pulse widths of the two rf pulses are approximately equal, and their amplitude is much stronger than that expected from an NMR signal. (v) A weak stimulated echo is observed following the third rf pulse. (vi) All the echoes disappear when powders are fixed by solidification with liquid nitrogen. (vii) As the pulse interval (τ) is increased, the echo amplitude decays monotonically (exponentially with a time constant T_2). (viii) T_2 becomes longer as the frequency is decreased although τ must be increased because of the longer blocking time of the receiving system. In addition to these properties, the temperature dependence of the echo intensity and that of echo-decay rate T_2^{-1} are similar to those for $YBa_2Cu_3O_{7-\delta}$ in which the echo intensity decreases rapidly as the temperature approaches the superconducting critical temperature T_c and the echo disappears abruptly at T_c even in cases where T_2 are sufficiently long just below T_c. Examples of temperature dependence of echo intensity and echo-decay rate are shown in Fig. 1 (a) and (b) for a sample of $Tl_2Ba_2Ca_2Cu_3O_x$ with T_c of 114 K. These data support the model proposed in our previous paper[1] in which we made following assumptions on the signal enhancement in phonon echoes in the superconducting state. Echoes are formed by the anharmonic acoustic oscillations of powders as in usual phonon echoes because the properties (i)-(viii) are quite similar to those of usual phonon echoes. However, the coupling mechanism between the rf field and the acoustic oscillations is probably different. The lower and upper critical fields in high-T_c oxides have large anisotropies which result in a large anisotropy in the diamagnetic magnetization in the mixed superconducting state[2]. Because of this property, rf fields induce acoustic oscillations of each particle and created acoustic echo oscillations can be detected by the oscillations of diamagnetic magnetizations which induce echo voltages across the rf coil directly. Although this mechanism of the signal enhancement in observation of phonon echoes are effective only in the superconducting state of strongly anisotropic superconductors, a similar mechanism of the enhancement will arise if the pinning of the fluxoids to each particle is sufficiently strong.

In view of the data in Fig.1 (a) and (b) in detail, temperature dependences of echo intensity and echo-decay rate are not simple functions of T as in the cases in $YBa_2Cu_3O_{7-\delta}$ [1], but have anomalies at about 75 K. The echo intensity drops rapidly with increasing temperature toward 75K followed by a gradual decrease and finally the signal disappears at T_c of 114 K. We believe that these behaviors are due to a low-T_c phase with T_c of 75 K in the sample in spite of the good quality of the sample as probed by SQUID magnetization measurement shown in Fig. 1 (c). The mixing of the low-T_c phase can strongly be exaggerated in the intensity measurement at a constant τ if T_2 in high-T_c phase is much shorter than T_2 in the low-T_c phase, as is the case in Fig. 1(b), and/or if the signal enhancement described above is much larger in the low-T_c phase than in the high-T_c phase, i.e., the anisotropy is much larger or the pinning is much stronger. The damping of the acoustic oscillations of each particle characterized by T_2 seems to be strongly affected by a coexistence of different superconducting phases. The divergent character of the echo-decay rate at each T_c in these high-T_c oxides seems to make clear the coexistence. The experiment of phonon echoes may, therefore, provide a new method of severe test for the quality of high-T_c superconducting powder.

Examples of data of temperature dependences of echo intensity and echo-decay rate together with magnetization are shown in Fig. 2 (a), (b) and (c) for an artificially prepared sample of somewhat low grade with the nominal composition of $Bi_2Sr_2CaCu_2O_x$. The amount of high-T_c phase in the powders is estimated as about 4 %, but we could not observe the signal from the high-T_c phase. Figure 2 (b) may suggest an existence of a further low-T_c phase with T_c of 35 K. Examples of data are also shown in Fig. 3 for powders of (Bi,Pb)-Sr-Ca-Cu-O system whose high-T_c phase is stabilized by an addition of Pb [4]. The magnetization data in Fig. 3 (c) shows the sample is mostly consists of the high-T_c phase, but similar anomalies which seem to be due to the coexistence of a low-T_c phase are seen in Fig. 3 (a) and (b) as was observed in the case of $Tl_2Ba_2Ca_2Cu_3O_x$. The sample quality at present probed by our phonon-echo method is in the order of $YBa_2Cu_3O_{7-\delta}$, $Tl_2Ba_2Ca_2Cu_3O_x$, and (Bi,Pb)-Sr-Ca-Cu-O followed by $Bi_2Sr_2CaCu_2O_x$.

In summary, phonon echoes observed in the new high-T_c oxide systems of Tl-Ba-Ca-Cu-O and (Bi,Pb)-Sr-Ca-Cu-O support the model in which the coupling mechanism between rf fields and acoustic modes in high T_c-oxides is different from that in usual phonon echoes. The experiment of phonon echoes provides a new method of severe test for the quality of high-T_c superconducting powders.

REFERENCES

1. Nishihara H, Hayashi K, Okuda Y, Kajimura K (1989) Phonon echoes in powders of high-T_c superconducting $YBa_2Cu_3O_{7-\delta}$, Phys Rev B39: 7351-7353
2. See references cited in ref.1.
3. See for a review, Tsuruoka F, Kajimura K (1980) Dynamic polarization echoes in metallic powders, Phys Rev B 22: 5092-5109
4. Ikeda Y, Takano M, Hiroi Z, Oda K, Kitaguchi H, Takada J, Miura Y, Takeda Y, Yamamoto O, Mazaki H (1988) The high-T_c phase with a new modulation mode in the Bi,Pb-Sr-Ca-Cu-O system, Jpn J Appl Phys 27: L2067-L2070

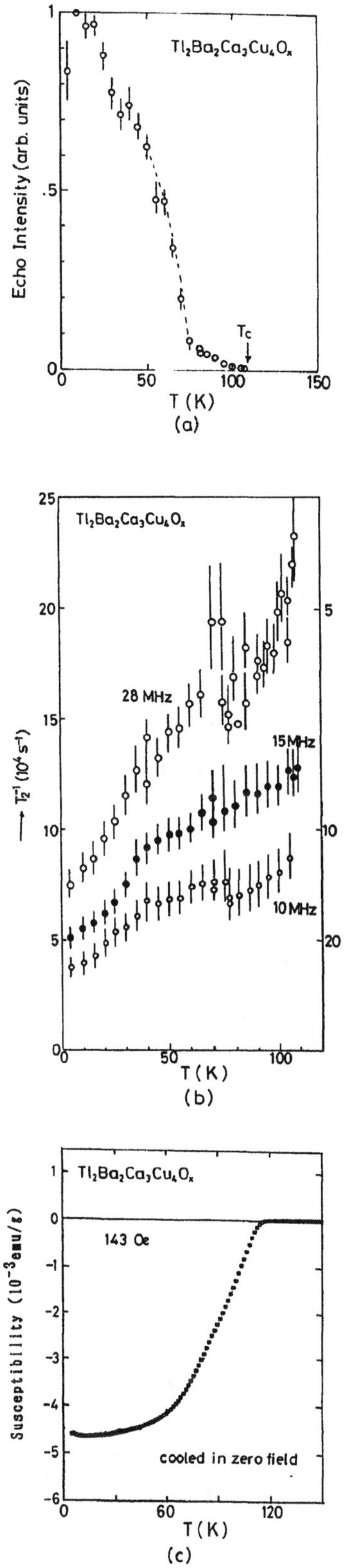

Fig. 1. A typical example of temperature dependence of echo intensity (a), echo-decay rate (b) and susceptibility (c) for a sample of $Tl_2Ba_2Ca_3Cu_4O_x$.

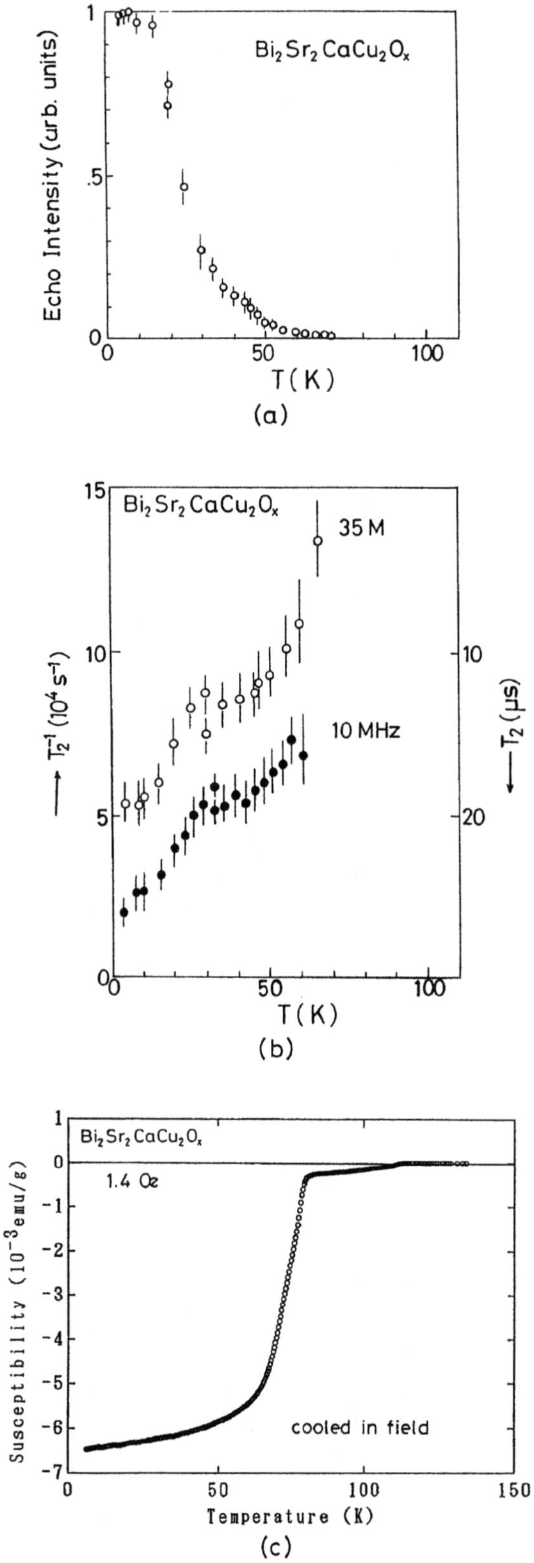

Fig. 2. A typical example of temperature dependence of echo intensity (a), echo-decay rate (b) and susceptibility (c) for a sample of $Bi_2Sr_2CaCu_2O_x$.

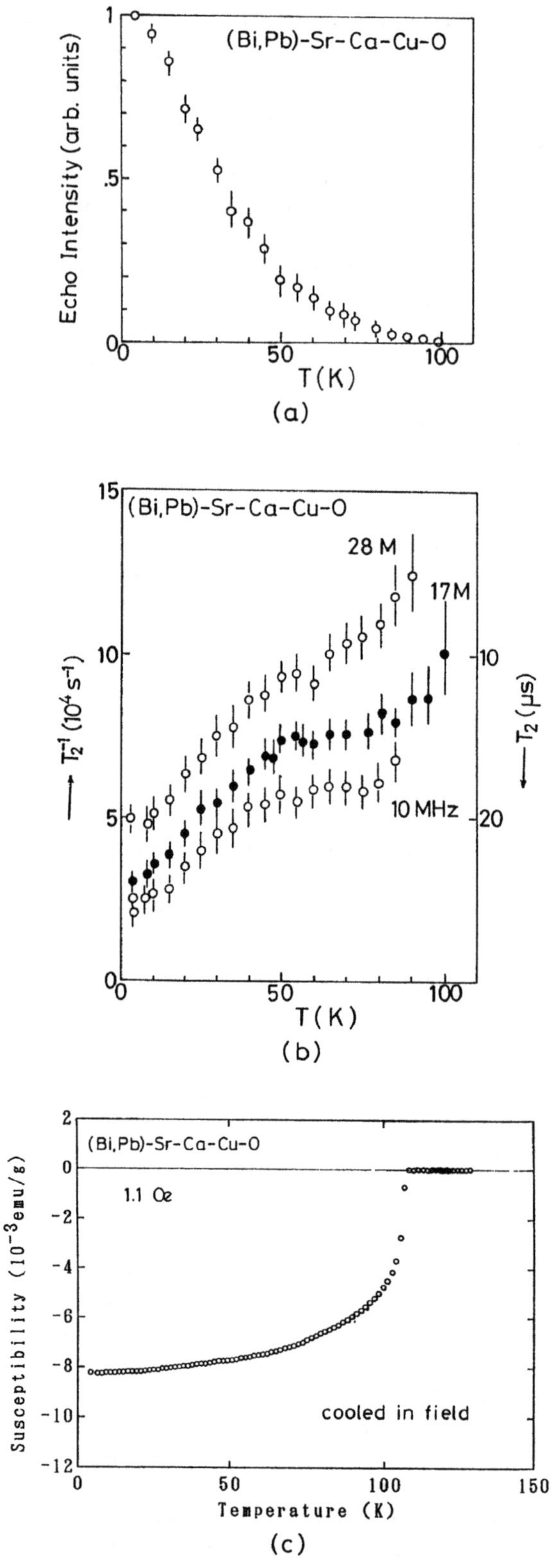

Fig. 3. A typical example of temperature dependence of echo intensity (a), echo-decay rate (b) and susceptibility (c) for a sample of (Bi,Pb)-Sr-Ca-Cu-O.

Ultrasonic, Vibrating Reed and X-Ray Study of the Structural Phase Transition in Single Crystal $La_{2-x}Sr_xCuO_{4-y}$ High-T_c Superconductor

T. LAEGREID[1], WU TING[1], O.-M. NES[1], M. SLASKI[1], E. EIDEM[1], E.J. SAMUELSEN[1], K. FOSSHEIM[1], and Y. HIDAKA[2]

[1] Division of Physics, The Norwegian Institute of Technology, and Division of Applied Physics - SINTEF, N-7034 Trondheim, Norway
[2] Nippon Telegraph and Telephone Corporation, Tokai, Ibaraki, 319-11 Japan

ABSTRACT
Measurements of elastic and structural properties of a single crystal $La_{2-x}Sr_xCuO_{4-y}$ superconductor are presented. The longitudinal ultrasonic measurements were performed at 13MHz. A strong minimum in the sound velocity and a corresponding sharp peak in the sound attenuation is found and associated with a lattice instability. Distinct change in the vibrating reed response is found in the same temperature range. These anomalies in the elastic properties are ascribed to interactions with a soft optical phonon mode leading to a tetragonal to orthorhombic phase transition at $T_0 \simeq 255K$. X–ray intensity measurements support these observations. The interesting possiblity of additional low temperature structural transition(s) is also investigated.

KEY WORDS: high–T_c superconductors, structural phase transition, elastic properties, critical attenuation, low–temperature structural transition

INTRODUCTION

Lattice instabilities and structural phase transitions (SPT) have turned out to be a common occurrence in copper–oxide high–T_c superconductors with perovskite structure[1 2] and are observed even in other superconductors, e.g. A–15 compounds. There are reasons to belive that the electron–phonon interaction contributes, at least partly, in forming the superconducting state in the high–T_c copper–oxides, even if the complete mechanism leading to superconductivity in these materials has not been fully understood yet. Therefore, lattice instabilities are of potential importance, also for the superconducting properties.
In this paper we study the lattice instabilities and SPT in single crystal $La_{2-x}Sr_xCuO_{4-y}$ (LASCO), $x \simeq 0.12$. The elastic properties have been investigated by ultrasonic[3] and vibrating reed methods. X–ray intensity measurements of an orthorhombic superlattice reflection have also been performed. First we discuss the results connected to the well known tetragonal[4] high–temperature (THT, I4/mmm) to orthorhombic medium temperature (OMT, Cmca) phase transition. Further, we discuss the experimental evidence for an SPT at lower temperature, possibly similar to the observations[5 6] in the isostructural compound $La_{2-x}Ba_xCuO_4$ (LABCO).

EXPERIMENTAL

A single crystal of LASCO was grown from a CuO flux and labelled NTT–33. Standard ultrasonic technique using Matec equipment was employed to measure the longitudinal sound velocity (v_s) and attenuation (α_s) of a 7x7x2 mm^3 sample consisting of two bonded plates. v_s and α_s determine the real and imaginary part of the effective elastic constant, respectively. A new Vibrating Reed system similar to that described in Ref. [7] was used to measure the resonance frequency (f_r) of a 7x3x0.3 mm^3 reed. Youngs modulus, E, is given by a combination of elastic constants, and is proportional to f_r^2. The X–ray measurements were performed with a standard 2–axis diffractometer (Siemens).

An iodometric method was used to determine the oxygen content of the crystal, and was found to be close to 4 ($y \simeq 0$). The superconducting transition occured at $T_c \simeq 12K$ ($\Delta T_c \simeq 2K$), found from AC–susceptibility measurements.

THEORY

Coupling between strain and the order parameter at an SPT results in characteristic anomalous temperatur dependences of the corresponding complex elastic constants. These anomalies are reflected in the elastic measurements. A Landau–Ginzburg free energy expansion, including the elastic free energy (F_e) and the strain to order parameter coupling free energy (F_c), has to be constructed to analyze the elastic anomalies in detail[8]. The symmetries of the LASCO–system places the tetragonal to orthorhombic transition in the universality class d=3 XY with cubic anisotropy[9]. This gives the following free energy:

$$F = F_l + F_e + F_c$$

where F_l is the Landau order parameter free energy expansion with cubic anisotropy. F_c is found by applying the Unsöld theorem from group theory, which states that only couplings between strains, e, and powers of the order parameter belonging to the same irreducible representations can exist[10]. The rotation of the CuO_6 ochtahedra which cause the STP is described by the degenerate order parameter $\underline{Q} = (Q_1,Q_2)$. Details of this calculation will be given elsewhere[8]. When e is a nonsymmetry breaking strain, the effect of the strain is to renormalize the transition temperature T_0. This leads to a change ΔC of the corresponding elastic constant C:

$$\Delta C \sim \frac{\partial^2 F}{\partial e^2} \sim t^{-\alpha}$$

where α is the specific heat exponent. When e is a symmetry breaking strain ($T>T_0$):

$$\Delta C \sim \frac{\partial^2 F}{\partial e_m^2} \sim t^{-[\alpha+2(\varphi_m-1)]} \equiv t^{-\mu}$$

where φ_m is the crossover exponent. To estimate μ we use previously measured values of φ in perovskites as a guide. For instance, $\varphi=1.27\pm0.06$ in[11] $SrTiO_3$ and $\varphi=1.31\pm0.07$ in[12] $LaAlO_3$. Taking $\varphi\sim1.25$ and $\alpha\approx0$ then $\mu \approx 0.5-0.6$.

RESULTS AND DISCUSSION

Longitudinal ultrasonic waves with a frequency of 13MHz were propagated along the tetragonal a (or b)–axis probing the elastic constant C_{11}. The corresponding measured sound velocity, v_s, and attenuation, α_s, are shown as a function of temperature in Fig. 1.

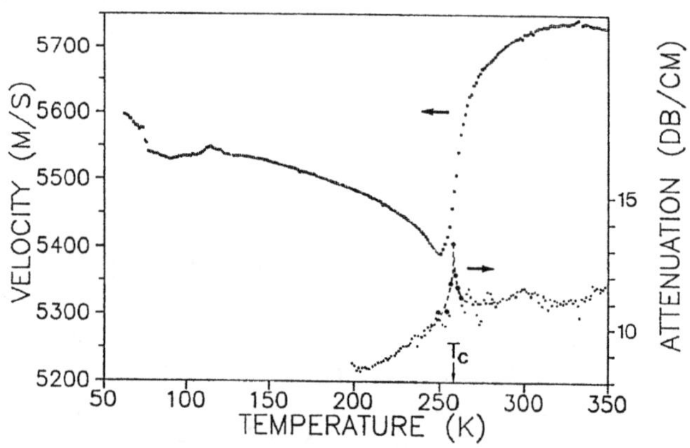

Fig. 1. Longitudinal ultrasound of 13MHz (v_s and α_s) in single crystal $La_{1.88}Sr_{0.12}CuO_4$ propagated along the tetragonal a (or b)–axis. $T_0 \approx 255K$. From Ref. [3].

The Vibrating Reed measurements were performed on a small reed from the same batch of LASCO single crystal. The measured temperature dependence of f_r is shown in Fig. 2. X–ray diffraction measurements were performed on a sample from the same batch of LASCO single crystal. The variation with temperature of the scattering intensity of an orthorhombic superlattice reflection is shown in Fig. 3.

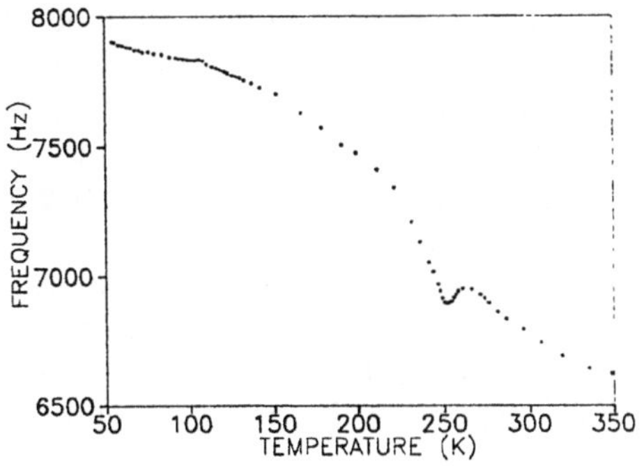

Fig. 2. Vibrating Reed resonance frequency f_r of single crystal $La_{1.88}Sr_{0.12}CuO_4$. $T_0 \approx$ 255K. (Youngs modulus $E \propto f_r^2$.)

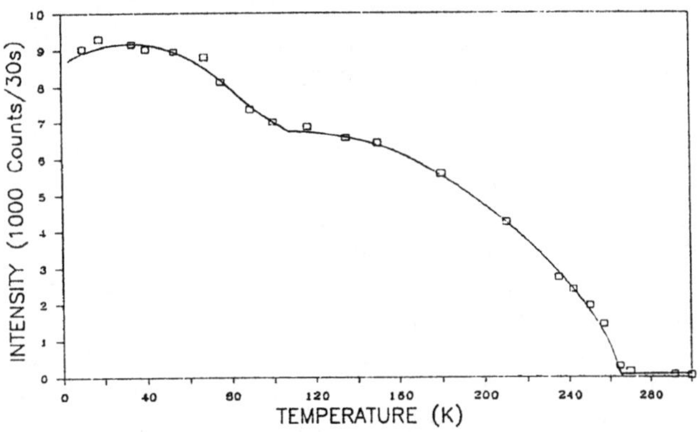

Fig. 3. X–ray intensity of the orthorhombic (1.5, 1.5, –16) superlattice reflection in single crystal $La_{1.88}Sr_{0.12}CuO_4$. The line is drawn as guide to the eye.

We first discuss our results in light of the well known THT (I4/mmm) to OMT (Cmca) SPT. The transition is clearly observed in all three measurements, see Figs. 1–3. The sound velocity, v_s, shows a pronounced minimum at the transition, while the attenuation, α_s, displays a sharp maximum (Fig. 1). The transition is found to take place at $T_0 \approx$ 255K. The velocity minimum is found at a slightly lower temperature (251K), which may indicate additional softening from domain wall motion below the SPT.
Consistent with the ultrasonic observations a minimum in the Vibrating Reed resonans frequency, f_r, is found at about 252K (Fig. 2). Finally, a (1.5, 1.5, –16) orthorhombic superlattice reflection is found to emerge below about 260K (Fig. 3).
From the observed value of $T_0 \approx$ 255K we deduce the Sr–doping[13] to be $x \approx 0.12 \pm 0.02$, which is consistent with the observed superconducting transition temperature $T_c \approx$ 12K.
The sharply peaked α_s is characteristic for critical behaviour, as is the influence on v_s far above the transition. As has been shown in Ref. [3], the observed velocity exponent $\mu \approx 0.5 - 0.6$, consistent with the theoretical prediction for the d = 3 XY universality class. However, T_0 is a strong function of Sr–doping[13] ($dT_0/dx \approx$ –2000K), so even a small spread in x will lead to a substantial smearing of the transition, influenceing the elastic measurements. Below the THT to OMT transition the effective lattice constant C_{11} stays soft, unlike in an ordinary SPT, where the elastic constants recover. This behaviour signals that the lattice remains near instability, similar to what has been observed[1 14] in LABCO. In that material such soft–mode behaviour was confirmed by X–ray and neutron

scattering experiments[5 6 15], in which also a SPT to a tetragonal low temperature structure (TLT, $P4_2/ncm$) was found. It should be noted that domain scattering may also cause softening below T_0.

What experimental evidence can be found for an SPT at low temperature? A bump in the ultrasonic velocity, v_s, is observed at about $T_1 \approx 110K$, clearly departing from the smoothly growing v_s below T_0 (see Fig. 1). We mention that below about 80K the trasonic pulsetrain deteriorated so that the measured v_s can not be trusted below that temperature. Further, a similar bump as in v_s is observed in the Vibrating Reed resonans frequency, f_r, at about $T_1 \approx 108K$ (see Fig. 2). Finally, an extra upturn in the superlattice X–ray intensity is found at $T_1 \approx 110K$ (see Fig. 3). These are three independente measurements, all showing some anomaly at about 110K. We belive this to indicate the existence of a low temperature SPT in $La_{1.88}Sr_{0.12}CuO_4$, similar to the isostructural Ba–compound[5 6]. The actual symmetry change, however, has not been determined in our measurements.
Finally, we comment that a clear downturn has been observed below 30–35K in several X–ray intensity measurements, perhaps indicating even another SPT in this temperature range.

CONCLUSIONS
Elastic measurements have proven to be sensitive probes of SPT in solids. The THT to OMT transition leads to pronounced anomalies in the elastic quantities v_c, α_s and f_r and is easily observed in the X–ray orthorhombic superlattice intensity. Critical fluctuation damping is qualitatively observed in the ultrasonic attenuation data. The critical exponent for v_s, $\mu \approx 0.5 - 0.6$ for $T > T_0 \approx 255K$, is consistent with predictions for the $d = 3$ XY universality class, to which this system is expected to belong.
Ultrasonic, Vibrating Reed and X–ray intensity measurements on samples from the same single crystal $La_{1.88}Sr_{0.12}CuO_4$ batch, all show an anomaly at $T_1 \approx 110K$. We conclude that these three independente measurements support the existence of a low temperature SPT, similar to that observed[5 6 14] in LABCO.
Further measurements on high quality single crystal LASCO with different Sr–doping x is needed for a detailed investigation of the possible low temperature SPT. In particular, X–ray and neutron scattering experiments should be employed to determine the symmetry at low temperature. Such work is presently carried out in our group.

ACKNOWLEDGEMENTS The autors are grateful for stimulating cooperation with Dr. J.D. Axe and other members of the Trondheim Superconductivity Group. This work was supported by NAVF, NORSK HYDRO A/S and STATOIL A/S.

REFERENCES
1. K.Fossheim and T.Lægreid, IBM J.Res.Dev., 33, 365, (1989).
2. T.Lægreid et al., Physica C 153–155, 1096–1099 (1988).
3. Wu Ting et al., Proc. of HTSC–M²S, Stanford, July 1989, Physica C.
4. P.Böni et al., Phys.Rev.B 38, 185 (1988), and refs. therein.
5. J.D.Axe et al., Phys.Rev.Lett. 62, 2751 (1989).
6. T.Suzuki and T.Fujita, Physica C159, 111 (1989).
7. A.K.Raychaudhuri and S.Hunklinger, Z.Phys. B57, 113 (1984).
8. Wu Ting et al., to be published.
9. T.R.Thurston et al., Phys.Rev., B39, 4327, (1989).
10. J.O.Fossum, J.Phys C, 18, 5531, (1985).
11. S.Stokka et al., Phys.Rev.B 25, 4896 (1982)
12. K.A.Müller et al., Multicritical Phenomena (Plenum), 143 (1984).
13. R.J.Birgenau et al., Phys.Rev., B39, 2868, (1989).
14. K.Fossheim et al., Solid State Comm., 63, 531, (1987).
15. J.D.Axe et al., IBM J.Res.Dev., 33, 382, (1989).

Degradation by Mechanical Grinding, and Recovery by Annealing, in the Superconducting Phases of the Bi-Sr-Ca-Cu-O System

TSUNEYUKI KANAI, TOMOICHI KAMO, and SHIN-PEI MATSUDA
Hitachi Research Laboratory, Hitachi, Ibaraki, 317 Japan

ABSTRACT

It has been found that mechanical grinding may degrade phases from supercounducting to non-superconducting; i.e., a sample with 73vol% of the high-Tc phase (105K phase), 18vol% of the low-Tc phase (75K phase) and 8vol% of a non-superconducting phase resulted in a sample with 98vol% of non-superconducting phase after grinding for 62min. X-ray analysis showed that cleavage fracture of the high-Tc phase in the {001} planes occurred more easily than that of the low-Tc phase by mechanical grinding. The {001} fracture gradually changed to a more isotropic fracture on prolonged grinding. It was also found that the degraded sample did not recover easily to the initial conditions compared with the amorphous samples initially free of the high-Tc phase. A process which includes crystal grinding, causing the degradation of the superconducting phases, may not therefore be appropriate for fabricating practical superconducting wires or sheets.

KEY WORDS: high-Tc superconductor, Bi(Pb)-Sr-Ca-Cu-O system, mechanical grinding.

INTRODUCTION

Many studies have reported on the high-Tc superconducting oxides since the discovery of the 40K-class superconductor of the La-Ba(Sr)-Cu-O system[1,2] and 90K-class superconductor of the Y-Ba-Cu-O system[3]. Discoveries of superconductivity in the Bi-Sr-Ca-Cu-O system[4] and Tl-Ba-Ca-Cu-O system[5], with Tc's above 100K, followed.
For some practical applicatoions, these high-Tc supercounducting oxides need to be fabricated into sheets or wires. During the process, the powders are often subjected to mechanical grinding to achieve a desired grain size for a doctor blade method for sheets or a silver-sheathed method for wires.
In this paper, changes of the superconducting properties by mechanical grinding, and subsequent recovery of the crystal structures by annealing, in the Bi(Pb)-Sr-Ca-Cu-O system were investigated.

EXPERIMENTAL

The Bi-Sr-Ca-Cu-O system contains two main superconducting phases; i.e., the high-Tc phase (105K) with the formal composition of $Bi_2Sr_2Ca_2Cu_3O_y$ and the low-Tc phases (75K) formally $Bi_2Sr_2Ca_1Cu_2O_y$. It has been reported that partial substitution of bismuth with Pb[6] or the reduction of the oxygen partial pressure during sintering[7] produced Bi-Sr-Ca-Cu-O ceramics with a large amount of the high-Tc phase.
Samples composed of nearly sigle high-Tc phases were fabricated as follows: Commercial powders of 3N purity Bi_2O_3, PbO, $SrCO_3$, $CaCO_3$ and CuO were mixed and calcined at 800°C for 50h in air, pulverized, pressed into pellets and sintered at 850°C for 210h in Ar-7.6%O_2 atmosphere. The samples obtained were mechanically ground for 2-62min to investigate changes in the supercounducting phases, monitored by X-ray power diffraction analysis using CuK_α radiation. The absolute volume of superconducting phases and non-superconducting phases in the samples was evaluated from the susceptibility change, AC inductive method (100Hz, 0.04mT), of the high-Tc phases and the low-Tc phase, as was described elsewhere[8].

RESULTS AND DISCUSSION

Figure 1 shows X-ray diffraction patterns for various grinding times in air; the main phase of the sample ground 2min is the high-Tc phase which is easily damaged by prolonged grinding time as shown in Fig. 1(b)-(e). As the grinding time increases from 17 to 62min, the intensities of the X-ray lines weaken significantly and the impurity phase of Ca_2PbO_4 become relatively large in it's intensity. The broad peak around 30° indicates the formation of an amorphous phase by mechanical grinding.
Because water vapor or CO_2 in air might be promoting the degradation of the crystals, the grinding atmosphere was changed to Ar atmosphere from air. Fig. 2 shows a X-ray diffraction pattern after mechanical grinding for 32min in Ar atmosphere. Comparison with the X-ray pattern of the sample ground in air for 32min shown in Fig. 1(d), indicates the extent of the degradation of the superconducting phase into amorphous phase is almost the same; the degradation of the superconducting phases is not primarily due to the water or CO_2 in air.
Figure 3 shows the temperature dependence of the AC susceptibility for the mechanically ground samples with time. Zero min in Fig. 3 shows the data from the lightly hand crushed powder. The susceptibility change for both 105K and 75K phases decreases with grinding time, the change for each phase is defined in the inserted figure in Fig. 3. The relationship between the volume fraction of each phase, estimated from the susceptibility change and the mechanical grainding time, is shown in Fig. 4. The initial volume of high-Tc, low-Tc and non-superconducting phases is 73%, 18% and 8%, respectively, the volume of the superconducting phases decreased rapidly, whereas the non-superconaucting phases increased rapidly, with grinding time so that after 62min the sample was predominantly non-superconducting phase.
Figure 5 shows that the relationship between the volume ratio of the high-Tc phase V_H to the low-Tc phase V_L with mechanical grinding time in the early stage. The ratio of the V_H/V_L over the grinding time was not estimated because of the uncertainty from the minute volume fraction of the high-Tc and low-Tc phase. The initial of the high-Tc phase, 4.1, decreases rapidly with mechanical grinding time. After grinding for 7min, it becomes 2.3, almost the same as the sample ground for 17min. The high-Tc phase apparently changes into non-superconducting phase more easily than the low-Tc phase in the early stage, but this tendency lessens with prolonged grinding.
The fracture morphology also changes during mechanical grinding. SEM observation showed that plate like grains of about ten microns for the sample ground for 2min, changed to sub-micron grains after 62min. The fracture morphology change was examined in more detail by X-ray diffraction. Fig. 6 shows the calculated X-ray diffraction intensities for the high-Tc phase using an X-ray pattern generation program. The relative intensity from each Miller index is totally different from the X-ray pattern of the initial state of powders; i.e., the peaks of {001} planes in Fig. 1 (a)-(b) extremely enhanced. The enhancement of the {001} planes is due to the ease of the cleavage fracture along {001} planes; the {001} planes, fractured in a plate-like fashion, tend to align parallel to an X-ray sample holder in a conventional powder preparation. On the other hand, the X-ray intensities of the samples for the prolonged mechanical grinding (Fig. 1(c)-(e) and Fig. 2) are changed gradually to patterns similar to the calculated intensities. This means the cleavage fracture along {001} planes mainly occurs at the initial stage and isotropic fracture, together with amorphous material formation occurs gradually on prolonged grinding.
The layer distance between Bi-Bi atoms is 3.3A which is much larger than the expected values from the ionic radius of Bi and the bonds between the layers are only weak van der Waals force in the Bi-Sr-Ca-Cu-O system[9], whereas the bonds in the basal plane are much stronger. Therefore, the cleavage fracture of {001} planes occurs first. When comparing the degradation of the high-Tc and low-Tc phase, shown in Fig. 5, the easier degradation of the high-Tc phase might be resulted from the difference of the c/a ratio of the crystal structures. Assuming that the same shear stress applied parallel to the a-b plane of the high-Tc or the low-Tc phase, the shear strain produced in the Bi-Bi plane of the high-Tc phase would be higher than that of the low-Tc phase by 1.2, which is the ratio of the c axis of the high-Tc phase to that of the low-Tc phase. Therefore, the {001} fracture of the high-Tc phase might occur more easily than that of the low-Tc phase.
The amorphous powders produced by the mechanical grinding were annealed to examine the potential recovery of superconducting phases. Fig. 7 shows the

X-ray diffraction patterns of the samples, annealed at 850°C for 50h in Ar-7.6%O_2 atmosphere, after grinding at various times. Compared to Fig. 1, the peaks of the high-Tc phase increase on annealing in samples especially after prolonged grinding. The volume fraction estimated from the susceptibility is shown in Fig. 8. The volume of the low-Tc phase, about 20%, is independent of the grinding time. Whereas, the volume fraction of the high-Tc phase seems to be associated with the initial volume of the high-Tc phase. The longer ground sample, which has smaller volume fraction of the high-Tc phase, shows the small fraction of the high-Tc phase. The samples ground for 17-62min show only about 30vol% of the high-Tc phase.
Previously, we reported that the addition of the seed crystals with the high-Tc phase was effective in forming the high-Tc phase in shorter annealing time, but a smaller volume fraction of the high-Tc phase was obtained after annealing[8]. That is, a sample which was seeded with the high-Tc phase by 5- 10vol% in melt-quenched amorphous powders showed about 30vol% of the high-Tc phase after annealing, however, the volume of the amorphous samples free of the high-Tc phase seeding was much higher of about 60vol% in the same annealing condition. The mechanical ground samples are considered to be the same as the samples seeded with the high-Tc phase; the ground sample on prolonged annealing mainly consists of the amorphous material with a small amout of the high-Tc and the low-Tc phase so that the volume fraction of the high-Tc phase after annealing is about 30vol% for both the samples. Therefore, a process which includes crystal grinding, causing non-superconducting phases to form from the superconducting phases, is not appropriate for obtaining high volume fractions of the high-Tc phase. The causes for the smaller volume fraction of the high-Tc phase are not clear at present.

REFERENCES

1. J. G. Bednorz and K. A.Muller:Z. Phys. B64 (1986) 189.
2. S. Uchida, H. Takagi, S. Tanaka and K. Kitazawa: Jpn. J. Appl. Phys. 26 (1987) L1.
3. M. K. Wu, J. R. Ashburn, C. J. Torng, P. H. Hor, R. L. Meng, L. Gao, Z.J. Haung, Y. Q. Wang and C. W. Chu: Phys. Rev. Lett. 58 (1987) 909.
4. H. Maeda, Y. Tanaka, M. Fukutomi and T. Asano: Jpn. J. Appl. Phys. 27 (1988) L209.
5. R. M. Hazen, L. W. Finger, R. J. Angle, C. T. Prewitt, N. L. Ross. C. G. Hadidacos, P. J. Heaney, D. R. Veblen, Z. Z. Sheng, A. El Ali and A. M. Hermann: Phys. Rev. Lett. 60(1988) 1657.
6. M. Takano, J. Takada, K. Oda, H. Kitaguchi, Y. Miura, Y. Ikeda, Y. Tomii and H. Mazaki: Jpn. J. Appl. Phys. 27 (1988) L1476.
7. U. Endo, S. Koyama and T. Kawai: Jpn. J. Appl. Phys. 27 (1988) L1476.
8. T. Kanai, T. Kamo and S. Matsuda: accepted to Jpn. J. Appl. Phys.
9. Y. Syono, M. Kikuchi: Bulletin of the Japan Institute of Metals 27 (1988) 574.

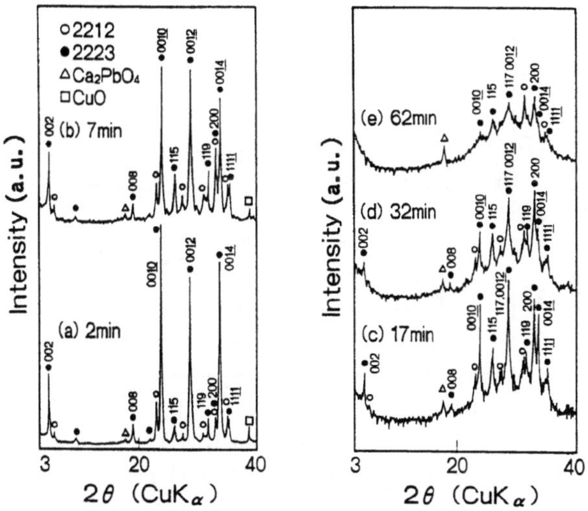

Fig.1 Comparison of X-ray diffraction patterns for various grinding times in air.

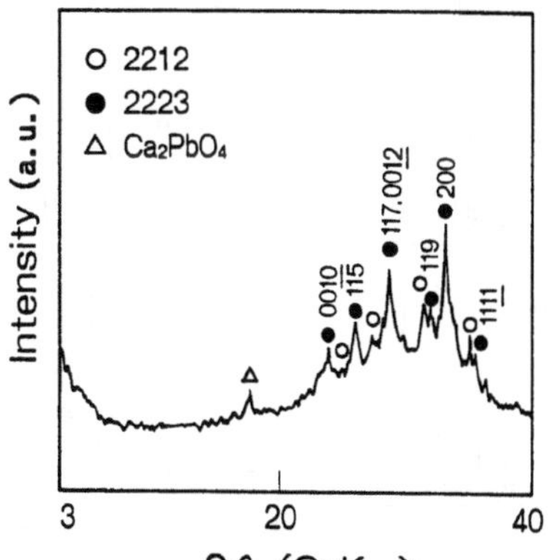

Fig.2 X-ray diffraction pattern after mechanical grinding for 32min in Ar atmosphere.

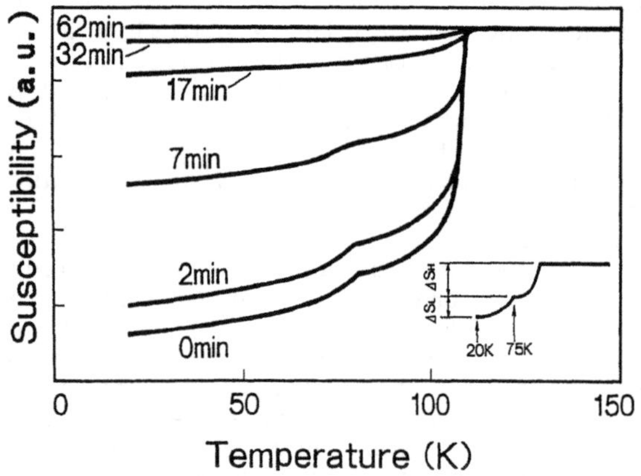

Fig.3 Relationship between susceptibility change and various mechanical grinding time.

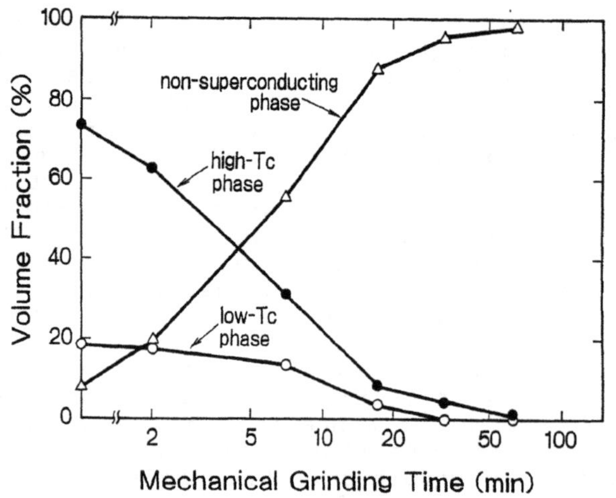

Fig.4 Relationship between volume fraction and mechanical grinding time.

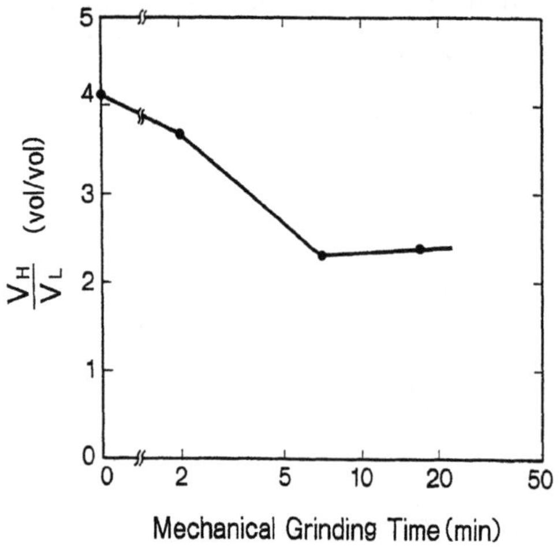

Fig.5 Relationship between volume fraction ratio of the high-Tc phase V_H to the low-Tc phase V_L and mechanical grinding time.

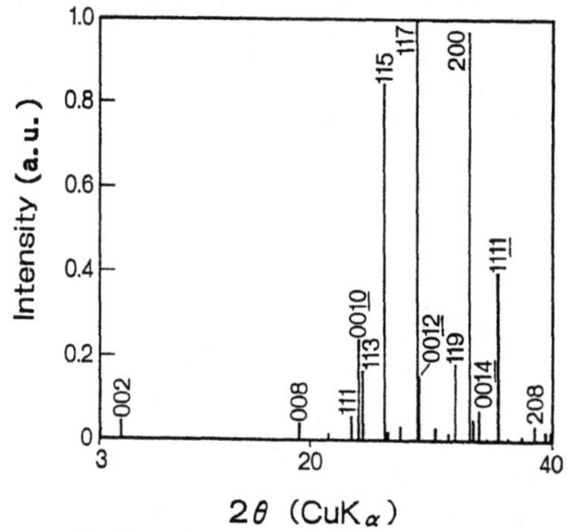

Fig.6 Calculated X-ray diffraction intensity for the high-Tc phase of the $Bi_2Sr_2Ca_2Cu_3O_x$. The lattice parameters of a, b and c are assumed to be 5.4A, 5.4A and 37A, respectively.

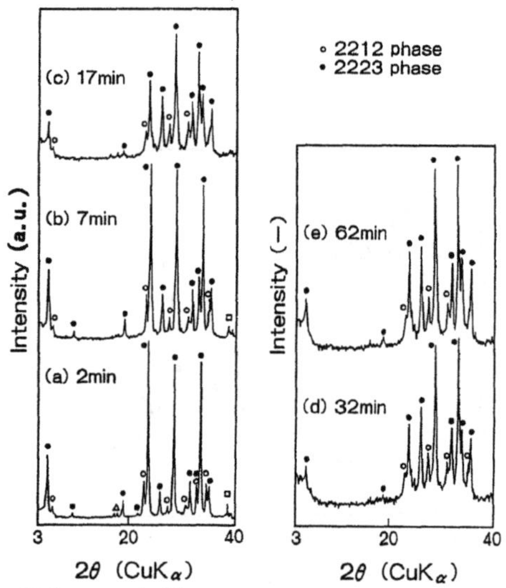

Fig.7 X-ray diffraction patterns annealed at 850℃ for 50h in Ar-7.6%O_2 atmosphere after grinding at various times.

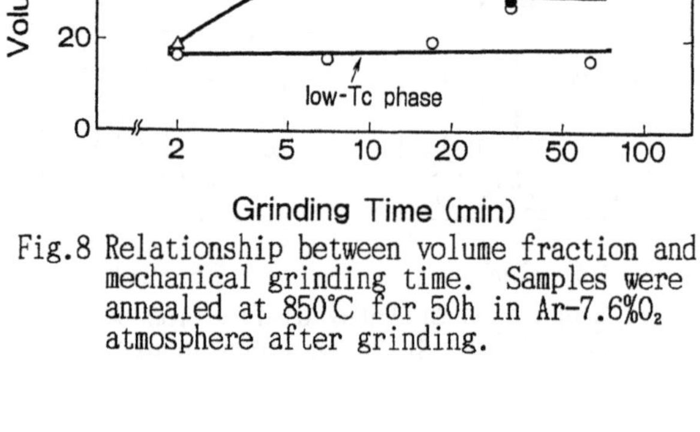

Fig.8 Relationship between volume fraction and mechanical grinding time. Samples were annealed at 850℃ for 50h in Ar-7.6%O_2 atmosphere after grinding.

Electron Microscopic Studies of Shock Loading Effects on High T_c Superconductor: Layered Bi System

MASAE KIKUCHI[1], YASUHIKO SYONO[1], MASAYASU NAGOSHI[2], AYAKO TOKIWA[1], EIJI AOYAGI[1], TERUO SUZUKI[2], KEIJI KUSABA[1], and KIYOTO FUKUOKA[1]

[1] Institute for Materials Research, Tohoku University, Katahira, Aoba-ku, Sendai, 980 Japan
[2] Steel Research Center, NKK Corporation, Kawasaki-ku, Kawasaki, 210 Japan

ABSTRACT

Shock loading effects on Bi-Sr-Ca-Cu-O superconductor were studied for the pressure to 10 GPa by TEM/SEM observations and magnetic measurements. Shock deformation of Bi-compounds showed very unique texture of kink bands which were characteristic deformation structure in mica-like compounds. $Bi_2Sr_2CaCu_2O_8$ shocked to 10 GPa showed considerable degradation of Tc in contrast to other superconducting oxides. Annealing of the shocked specimen at 870 C in air for 10 hours recovered superconductivity with improvement in Tc. The comparison of shocked and subsequently annealed specimen by TEM/SEM observation explained the characteristic change in superconductivity.

KEYWORDS: $Bi_2Sr_2CaCu_2O_8$, shock-loading, kink-band, improvement of Tc, electron microscopy.

INTRODUCTION

Shock compression method has been utilized for the synthesis and densification of high Tc superconductors[1]. The shock process also creates atomic scale defects which are important for pinning the flux and improving Jc. The structure defects induced by shock loading are closely related to the structure and bonding type of materials[2]. Shock deformation of $La_{1.85}Sr_{0.15}CuO_4$ with K_2NiF_4 type structure is due to dislocation with crossed networks. $YBa_2Cu_3O_7$ with triple perovskite structure shows stacking fault and mechanical twinning by shock loading[3].

The deformation structures of Bi-compounds with layered structure are quite different from those in the La- and Y-systems. The shocked specimen of $Bi_2Sr_2CaCu_2O_8$ shows considerable degradation of Tc compared with that of other superconducting oxides and remarkable improvement in Tc is achieved by subsequent annealing. It seems that the characteristic features depend on the mica-like layered structure. Here we reports on the comparison of shocked and subsequently annealed specimen of $Bi_2Sr_2CaCu_2O_8$ by TEM/SEM observation to explain the characteristic changes in superconductivity.

EXPERIMENTAL

The specimen of $Bi_2Sr_2CaCu_2O_8$ was synthesized from Bi_2O_3, $SrCO_3$, $CaCO_3$ and CuO by solid state reaction in air at 870°C for 10 hours. Synthesized specimens were examined to be a single phase by powder X-ray diffraction analysis, and unit cell dimensions measured were in good agreement with the previous work[3]. Shock-loading experiments were carried out by using gun method. A sintered specimen formed as disc of 10 mm in diameter and 1 mm in thickness was encased in a stainless steel container, and impacted by a teflon or Al2024 flyer accelerated to high velocity. The pressure was estimated from the flyer velocity on the basis of impedance matching. The sintered specimen, recovered shocked specimen and subsequently annealed specimen were examined by XRD in the plane normal to the propagation direction of shock-wave. The deformation texture by shock-loading and subsequently annealed specimen was observed by TEM using JEM-200B, JEM-4000EX and by SEM using JSM-5300. For observation by TEM, the recovered disc specimen was cut perpendicularly to shock direction and thinned down by ion milling. Diamagnetic response due to change in mutual inductance was measured by a magnetic susceptometer made by Sumitomo Heavy Industry.

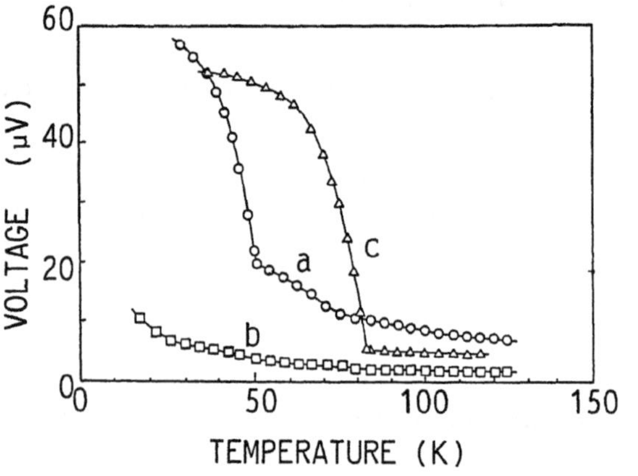

Fig.1 Diamagnetic response vs. temperature of $Bi_2Sr_2CaCu_2O_8$, (a) sintered, (b) shocked to 10 GPa, (c) subsequently annealed specimen.

RESULTS AND DISCUSSION

Figure 1(a),(b) and (c) show diamagnetic responses vs. temperature of $Bi_2Sr_2CaCu_2O_8$ unshocked, shocked to 10 GPa and subsequently annealed at 870°C in air for 10 hours respectively. The superconductivity disappeared in the shocked specimen, but subsequently annealing recovered remarkably the diamagnetic response.

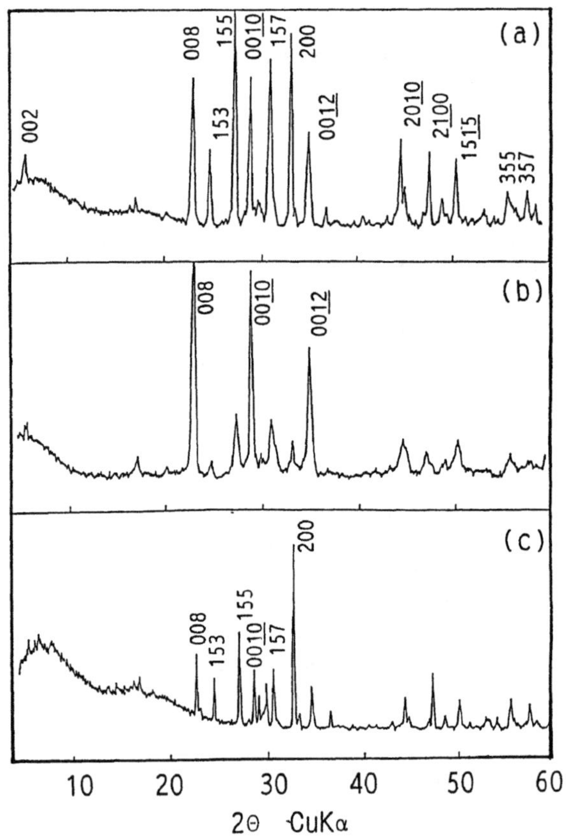

Fig.2 XPD of $Bi_2Sr_2CaCu_2O_8$, (a) sintered, (b) shocked to 10 GPa, (c) subsequently annealed specimens.

Figure 2 shows XRD patterns of pelletized specimen(a), shocked specimen(b) and annealed specimen(c) respectively. The XRD of the shocked specimen(b) exhibits highly ordered [001] orientation with broadening of diffraction lines, while the annealed specimen(c) have sharp and strong diffraction line of 200.

The textures of those three specimens by SEM observation (Fig.3) showed characteristic features: starting material of the sintered specimen(a) reveal a rather porous texture consisting of sharp and platy crystals. The shocked specimen(b) is found to be very compact due to complete filling of voids and breakage of grains by shock loading. The annealed specimen(c) shows a compact homogeneous texture accompanied by crystal growth resulting from the subsequent annealing.

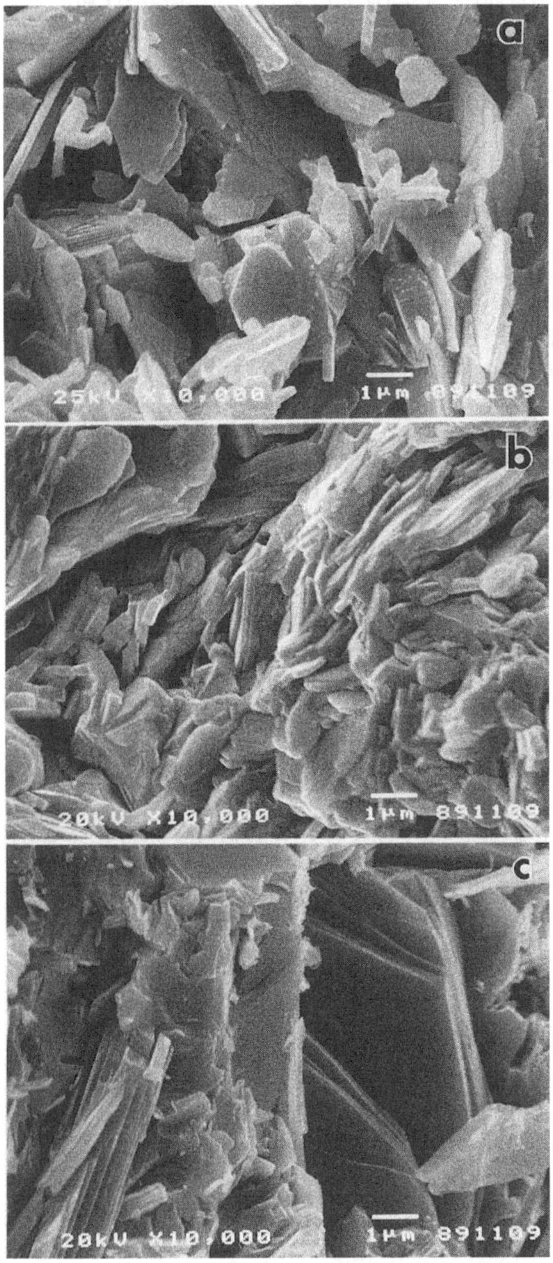

Fig.3 SEM image of $Bi_2Sr_2CaCu_2O_8$, (a) sintered, (b) shocked to 10 GPa, (c) subsequently annealed specimens.

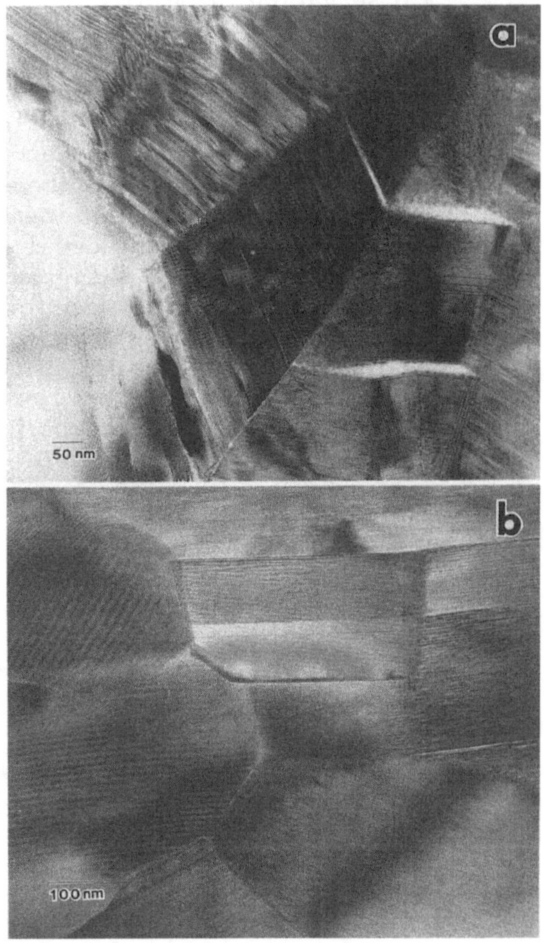

Fig.4 TEM images of shocked to 10 GPa(a) and subsequently annealed specimen(b).

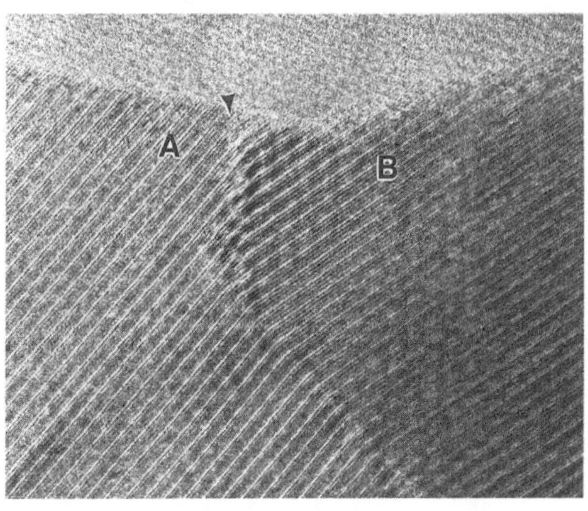

Fig.5 HRTEM of annealed specimen. ➤ An arrow shows grain boundary.

The TEM observation of microstructure in atomic scale of shocked specimen (Fig. 4) demonstrates very unique slip textures between planes and kink bands which are characteristic for deformation structure in mica-like compounds[4]. The kink bands are observed only by taking the incident beam parallel to the layer. A strain and open cracks along the layer are observed everywhere and in those cracks there are some decomposed particles. Annealed specimen after shock loading(Fig.4(b)) shows remarkable change of microstructure; the crystal growth along the plane, disappearance of the strain contrast and recovery of kink bands. The easy development of crystal growth in the layer direction is closely related to the mica-like layered structure; joined boundary between A and B as seen in Fig. 5 to form a single domain, and growth of platy crystals as seen in SEM(Fig.3(c)). While a crack in slipped layers including some deposited particles seems to stick together by SEM level observation, but complete joint in atomic-level is not achieved as shown in high resolution electron microscopy. Such severe deformation due to kinking is probably the origin of easy degradation and superior recovery of superconductivity, if the short coherence length of the oxide superconductors is taken into consideration.

ACKNOWLEDGEMENTS

The authors are grateful to Professors Y. Muto, N. Kobayashi and K. Hiraga of Tohoku University for helpful discussion. The work was partly supported by Grant-in-Aid on Scientific Research for Priority Areas "Mechanism of Superconductivity" given by Ministry of Education, Science and Culture, Japan.

REFERENCES

1. Murr LE, Monson T, Javadpour J, Strasik M, Sudarsan U, Eror NG, Hare AW, Brasher DG, and Butler DJ (1988) Shock-induce microstructures in explosively fabricated superconductors. J.Metals 1:19-23.

2. Syono Y, Nagoshi M, Kikuchi M, Tokiwa A, Aoyagi E, Suzuki T, Kusaba K and Fukuoka K (1990) Shock loading effects on high Tc superconductor Bi-Sr-Ca-Cu-O. Shock Waves in Condensed Matter-1989, edited by Schmidt SC, North-Holland, Amsterdam, in press.

3. Syono Y, Hiraga K, Kobayashi N, Kikuchi M, Kusaba K, Kajitani T, Shindo D. Hosoya S, Tokiwa A, Terada S, and Muto Y(1988) An X-ray diffraction and electron microscopic study of a new high-Tc superconductor based on the Bi-Ca-Sr-Cu-O system. Jpn.J.Appl.Phys.27: L87-L90.

4. Horz F, and Ahrens TJ (1986) Deformation of experimentally shocked biotite. Am.J.Sci.267:1213-1229.

3.7 Magnetic Properties

Why Is Meissner Effect Dependent on Field Intensity and Surface-to-Volume Ratio of Samples in Oxide Superconductors?

K. KITAZAWA[1], O. NAKAMURA[2], T. MATSUSHITA[3], Y. TOMIOKA[1], N. MOTOHIRA[1], M. MURAKAMI[4], and H. TAKEI[5]

[1] Department of Industrial Chemistry, The University of Tokyo, Hongo, Tokyo, 113 Japan
[2] Toa Nenryo Kogyo Ltd., R & D Laboratory, 3-1, Nishitsurugaoka 1-chome, Oimachi, Iruma, Saitama, 354 Japan
[3] Department of Electronics, Kyushu University, 10-1, Hakozaki 6-chome, Higashi-ku, Fukuoka, 812 Japan
[4] Superconductivity Research Laboratory, International Superconductivity Technology Center, 10-13, Shinonome 1-chome, Koto-ku, Tokyo, 135 Japan
[5] Sumitomo Electric Industries, Ltd., R & D Laboratories, 1-3, Shimaya 1-chome, Konohana-ku, Osaka, 554 Japan

ABSTRACT

The Meissner curves were measured systematically under a wide range of magnetic fields for $Ba_2YCu_3O_7$ and $Bi_2Sr_2CaCu_2O_y$ in the forms of single- and poly-crystalline and powder specimens. The Meissner fraction was smaller for single crystals than for powders. The poly-crystalline specimens exhibited the Meissner fraction similar to the single crystals under a low field while they did similar to the powder under a high field. Roughly speaking, the Meissner fraction decreases with the field but in a complex manner. A model has been proposed to systematically explain the observed behaviors based on the temperature dependent pinning force exerted on the flux expulsion.

INTRODUCTION

The Meissner effect is the most essential and unique feature of the superconductivity. This property, therefore, has been used in order to prove the occurrence of superconductivity in the new materials.[1,2] Furthermore, the Meissner fraction f_M, i.e., the ratio of the observed diamagnetic magnetization to the perfect diamagnetism, has been frequently employed as a measure of the quality of the specimen, relating it with the volume fraction of the superconducting phase in the mixture with the possible non-superconducting phase.[3,4] Also, the sharpness of the change in the Meissner curve, i.e., susceptibility vs. T, which is taken on cooling of the specimen under a constant magnetic field (field cooling curve), has been often regarded as indicative of the uniformity of the critical temperature of superconductivity of the specimen especially in the early studies on the high temperature oxide superconductors, because it has often been difficult to obtain the pure and homogeneous phase of the oxide specimens.

However, the criteria based on the Meissner effect could be seriously misleading because of the hysteretic nature of magnetization in the type II superconductor. The flux lines, that are to be expelled due to the Meissner effect, might be trapped in the superconductor by the pinning of flux lines, Krusin-Elbaum, Malozemoff et al. were perhaps the first to have clearly pointed out that the Meissner fraction of the oxide superconductor, $Ba_2YCu_3O_y$, was dependent on the intensity of the magnetic field; it decreased significantly as the field was increased.[5] Furthermore, if the pinning is involved, the ratio of surface area to volume of the specimen should be expected to become an important factor to determine the Meissner fraction because the flux is expelled out of the specimen through the surface. But there have not been any systematic studies performed in this context.

Therefore, in this report, we have systematically investigated the dependences of the Meissner curves on the field and the form of the specimens. Based on the significant dependences observed on both of the factors, a model is proposed to systematically explain the factors to determine the Meissner curve in the oxide superconductors.

EXPERIMENTAL

The magnetic susceptibility was measured in a SQUID susceptometer (Hoxan HSM-2000X) under magnetic fields from about 1000 Oe down to 0.1 Oe. The susceptometer was equipped with a Nb metal tube inside the magnet to shield the external field. The field was calibrated with the aid of a pure Pb metal as the standard specimen each time after changing the field by warming and destroying the superconductivity of the Nb shield. This was done for the field below 10 Oe for which the reading of the current of the superconducting Nb magnet was found to give increasingly inaccurate values due to the hysteresis of the superconducting magnet. The data were taken only when the symmetrical current was observed for the two asymmetrically-wound pick up coils located above and below the central position of the specimen around which it was moved up and down with a stroke of 50 mm.

The specimens employed were the single crystals of $Bi_2Sr_2CaCu_2O_y$ (BSCCO) grown by the traveling solvent floating zone method,[6] sintered bodies of BSCCO[7] and $Ba_2YCu_3O_y$ (BYCO) as prepared by the standard solid state reactions and powders prepared by crashing and grinding the sintered bodies.

Figure 1 shows the Meissner curves obtained in field cooling processes at a speed of 2K/min for a platelet of single crystalline BSCCO under various fields. As seen in the figure, the curves show saturation towards lower temperatures and the Meissner fraction $f_M=-4\pi M/H$ can be obtained from the saturation region. The f_M value sharply decreases with increasing the field as has been reported by Krusin-Elbaum et al.[5]

Figure 2 shows the range of scattering of the f_M values for various specimens, all of which are supposed to be single phased according to EPMA and

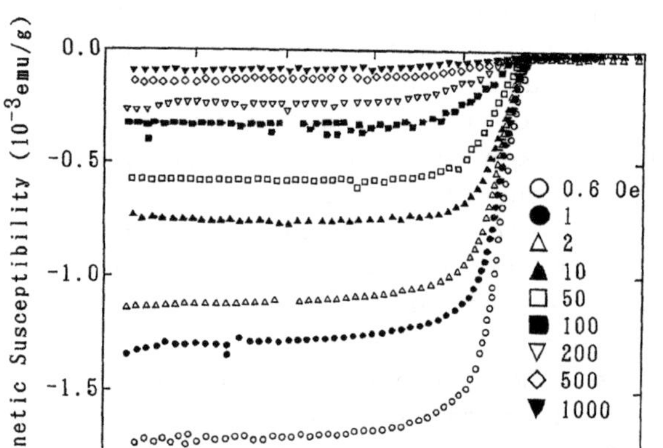

Fig.1. The Meissner curves observed for a single-crystalline platelet of $Bi_2Sr_2CaCu_2O_y$ ($0.1x1.5x4mm^3$) grown by the TSFZ method under different magnetic fields applied along the ab-plane.

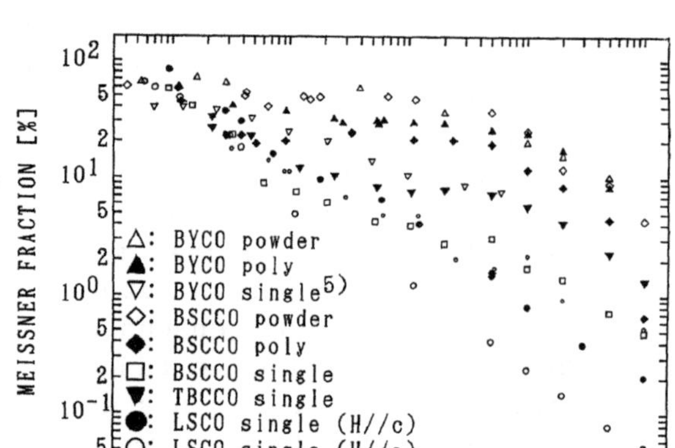

Fig.2. The Meissner fraction f_M observed as a function of the field for various kinds of specimens of $Ba_2YCu_3O_y$, $Bi_2Sr_2CaCu_2O_y$, $(La_{1-x}Sr_x)_2CuO_4$ (x=0.07) and $Tl_2Ba_2CaCu_2O_y$ in various forms.

X-ray observations, and for various magnetic fields. It is noted at first that f_M values are significantly variable depending both on the field and the form of the specimens: single- or poly-crystalline or powder. As the general trends, the data points merge to the value 100% as the field is decreased regardless of the form of the specimens and of the difference in the materials, while they tend to be diverted with increase in the field but roughly speaking they decrease under the highest fields.

Figure 3 compares the f_M values obtained for single crystal,[5] poly-crystalline and powder specimens of BYCO. It is noted in this figure that f_M decreases monotonically with increase in the field for single crystal while that for the powder keeps to be nearly 100% up to the field about 10 Oe and then decreases monotonically under the higher fields. The poly-crystalline specimen on the other while shows quite an interesting behavior. It behaves similar to the single crystal when the field is low, while it does similar to powder specimen when the field is high. Under the intermediate range of field strength, the f_M value of the polycrystalline specimen stays nearly constant,i.e., there is a rather wide transition region for the poly-crystalline specimen from the pseudo-single crystalline behavior to the pseudo-powder behavior. A similar but less systematic observation has been reported by Celani et al.[8] This seems to suggest that grain boundaries in the polycrystalline specimen behave like the bulk when the field is weak but they behave like easy paths of magnetic flux lines when the field is strong.

Figure 4 gives the equivalent comparisons for BSCCO (T_c=80K) among the single-, poly-crystalline and powder specimens. It is obvious that one can obtain essentially the same conclusions from this figure as from Fig.3.

DISCUSSIONS

From the observation above, it is evident that the Meissner fraction f_M is dependent on the effective surface area of the specimen as well as on the intensity of the field H. The effect of the surface area tells us that the flux expulsion can be much less complete in a large specimen than in a small one, indicating the involvement of flux pinning. Because flux lines must move

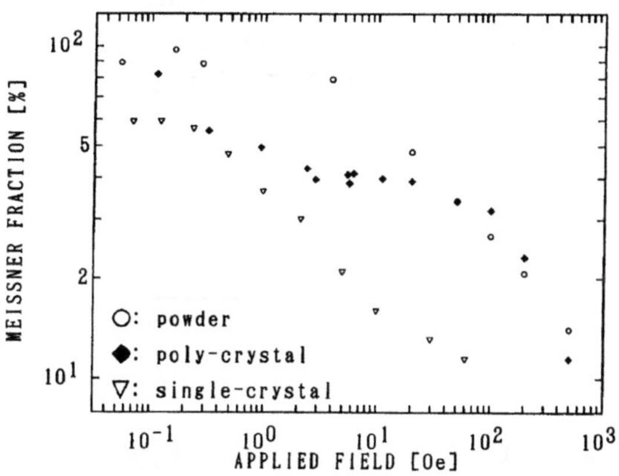

Fig.3. Dependence of the Meissner fraction f_M of single crystal,[5] poly-crystal (grain size 2-20μm) specimens of $Ba_2YCu_3O_y$ on the intensity of magnetic field.

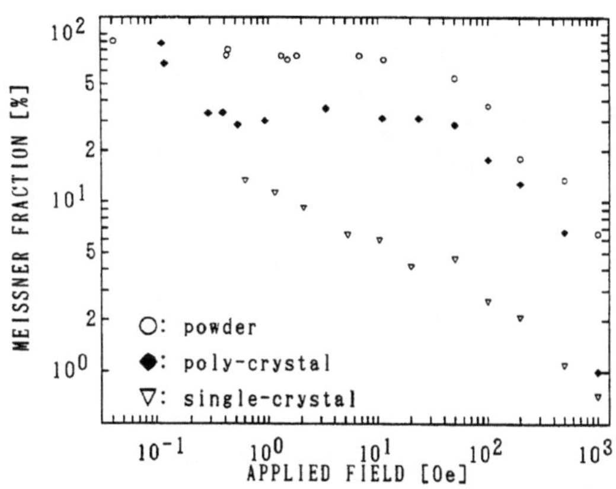

Fig.4. Dependence of Meissner fraction f_M of single-crystal, poly-crystal(grain size 2-50μm) and powder (1-20μm) specimens of $Bi_2Sr_2CaCu_2O_y$ on the intensity of magnetic field.

towards the surface of the specimen prior to the expulsion, there must be a gradient of flux density developed toward the surfaces according to the Bean's critical state model.[9)]

Therefore, we propose a semi-quantitative model to systematically explain the dependence of f_M on the relative surface area and H. For simplicity we take a plate with infinite surface area and a semi-infinite thickness for the sample shape. Fig.5a shows the distribution of the flux density B in the specimen which is a function of H and T.

When T is lowered below the critical temperature T_c under the field H, i.e., T_c for $H = H_{c2}$, B in the specimen starts to decrease because of the set in of the Meissner effect. Near T_c, we assume the motion of the flux lines are reversible, and hence we expect a decreasing but flat distribution of B(H,T) down to T* until the pinning force become strong to start impeding the flux motion.

Let us assume for simplicity, that we lower the temperature step-wise at first from T* to T_1 as shown in Fig.5b. Because of the finite pinning force $F(T_1,H)$, a slope develops near the surface where $B = B(T_1,H)$, the equilibrium lux density. The slope is proportional to $F(T_1,H)$ according to the critical state model. We ignore the flux creep effect for simplicity. What would result then is that the flux expulsion is no more the reversible process but that only the amount of flux lines corresponding to the shaded area can be expelled

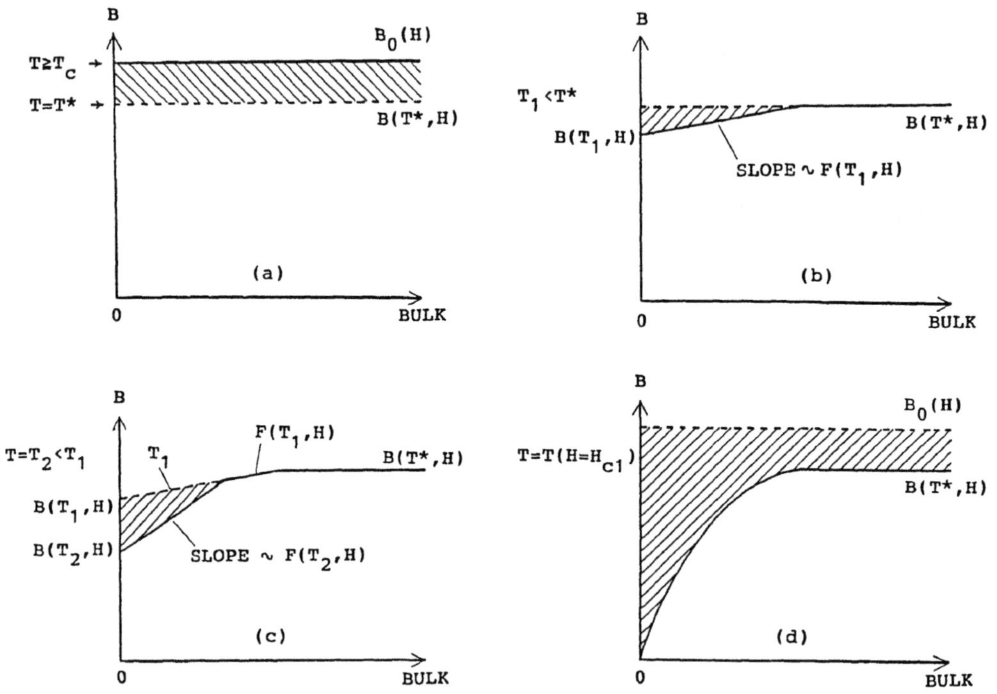

Fig.5. The distribution of magnetic flux density B expected in a type II superconductor of a semi-infinite thickness expected for the proposed model in the field-cooling process when the temperature is lowered down to (a) T^* (see text), (b) then step-wise to T_1, (c) then again step-wise to T_2 ($<T_1$) and (d) for the case when T is lowered continuously from above T_c down below $T=T(H=H_{c1})$. The shaded area stands for the amount of flux lines expelled during each of the cooling processes.

during the decrease in T from T* to T_1. In the absence of the pinning force, the flux density would be a flat line at the level of $B(T_1,H)$

When we further decrease T, again in a step-wise manner from T_1 to T_2, the new resulting distribution of B will be as shown in Fig.5c. The flux lines expelled during this process corresponds to the shaded area. The slope near the surface is determined by the pinning force $F(T_2,H)$ which is larger than $F(T_1,H)$.

Therefore, if we lower the temperature continuously, we expect the final distribution of B to be as shown in Fig.5d when $T(H=H_{c1})$ is attained. The slope dB/dx in the immediate vicinity of the surface is determined by F(T,H) with $T = T(H=H_{c1})$ and B(x=0) becomes zero. The Meissner fraction f_M will then be given by the ratio of the shaded area to the total area below $B_0(H)$ line. The area un-shaded represents the the amount of flux lines which should have been expelled if the pinning force were absent.

Krusin-Elbaum et al.[5] assumed f_M to be determined by the flux expulsion corresponding to the area defined by the two lines $B_0(H)$ and B(T*,H) because the flux lines stop their motion below the irreversibility temperature T*(H). Their model , therefore, is ignoring the flux expulsion occurring below T*(H). The model proposed here gives an additional contribution to f_M as expressed by the shaded area below the B(T*,H) line, S_p.

Now let us consider the case if the sample thickness is finite and equal to L. Then f_M should be related as;

$$f_M = \frac{\{B_0(H)-B(T^*,H))\}L + 2S_p}{B_0(H)\ L} = \left\{ 1 - \frac{B(T^*,H)}{B_0(H)} \right\} + \frac{2S_p}{B_0(H)L}, \qquad 1)$$

considering the flux expulsion from both sides of the plate. The first term has been evaluated by Krusin-Elbaum et al. to be approximately proportional to $H^{-1/3}$. The model proposed by Krusin-Elbaum et al. should be a good approximation if the relative contribution from the second term is small. But the significant dependence of f_M observed on the form of the sample indicates that their model is valid only under limited conditions.

From eq.1, it is obvious that f_M decreases as the size of the specimen increases, or, to be more specific, it increases as the surface to volume ratio of the sample increases. This directly explains why f_M is always smaller for single-crystal than for powder samples. Since the grain boundary in the polycrystalline oxide sample prepared by the usual process is known to give a very weak couple between the adjacent grains, it is understandable why f_M exhibited by polycrystalline samples is close to that of single crystal only under the weakest field. As the field is intensified, the effective size of the sintered bodies decreases gradually and approaches to the saturated minimum effective size presumably equal to the grain size which is close to the particle size of the powder samples in the present experiments. This can explain the reason why f_M values of the sintered bodies are close to those observed for the respective powder samples when the field is strong.

The area S_p is a function of the equilibrium B(T,H) value and the pinning force F(T,H). The quantitative analysis will be reported elsewhere.[10] But it is obvious that the area S_p becomes smaller with increase in F(T,H).

The above arguments lead us to the following predictions; 1) A sintered body of which grain boundaries are improved in terms of the strength of coupling should exhibit a smaller f_M value than for the non-improved one. 2) A sample of stronger pinning force should exhibit a smaller f_M value.

In this context, it is quite interesting to note that we have observed very small f_M values such as 3% under H = 2 Oe for a QMG (Quench and Melt

Growth) specimen of BYCO which has been reported to exhibit a very high J_c value due to the improved coupling at the grain boundaries and due to the strengthened pinning force by the dense precipitation of microscopic BaY_2CuO_5.[11] This value may be compared with the one obtained for a flux grown single-crystalline BYCO; about 50% under a similar field.[5]

Furthermore, it is understood from the present model that the sharpness of the Meissner curve should not be simply related with the uniformity of T_c in the sample because the area S_p gives an additional contribution to f_M which is determined by the dependence of the pinning force on T. If the pinning force grows the more gradually as T is lowered, the broader change in the Meissner curve is expected. We have experienced that the Meissner curves change sharply below T_c for BYCO, but rather gradually for BSCCO and TBCCO in accordance with the larger pinning force for the former than for the latter two materials.

In summary, a model has been proposed based on the assumption that the flux expulsion proceeds reversively down to a certain temperature below T_c but it is impeded by the pinning force the more significantly as the temperature is lowered. The present model, in combination with the model proposed by Krusin-Elbaum et al., can systematically explain the dependences of the Meissner fraction observed for various forms of sized specimens under various field intensities. This model also predicts that the sharpness of the Meissner curve is determined by the way how pinning force develops as the temperature is lowered, hence suggesting that the frequently used criterion to be often misleading to relate the broad change in the Meissner curve as the indication of non-uniformity of the specimen.

REFERENCES

[1] J.G. Bednorz, M. Takashige and K.A. Muller, Europhys. Lett. 3 379 (1987).

[2] S. Uchida, H. Takagi, K. Kitazawa, and S. Tanaka, Jpn. J. Appl. Phys. 26 1 (1987).

[3] S.H. bloom, M.V. Kuric, Y.S. Yao, R.P. Guertin, D. Nichols, C. Jee, A. Kebede, J.E. Crow, T. Mihalisin, G.N. Myer and P. Shulottmann, "High-Temperature Superconductors", Eds, M.B. Brodsky, R.C. Dynes, K. Kitazawa, H.L. Tuller, MRS (1988) p.19

[4] R.B. van Dover, R.J. Cava, B. Batlogg and E.A. Rietman, Phs. Rev. B 35 5337 (1987).

[5] L. Krusin-Elbaun, A.P. Malozemoff, Y. Yeshurun, D.C. Cronemeyer and F.Holtzberg, Physica C153-155, 1469 (1988).

[6] M. Motohira, K. Kuwahara, T. Hasegawa, K. Kishio, K. Kitazawa, Jpn. J. Cer. Soc. Int. Ed. in press.

[7] A. Maeda, T. Yabe, H. Ikuta, Y. Nakayama, T. Wadaura, S. Okuda, T. Itoh, M. Izumi, K. Uchinokura, S. Uchida and S.Tanaka, Jpn. J. Appl. Phys. 27 (1988) L661

[8] A. Celani, R. Messi, N. Sparvieri, S. Pace, A. Saggese, C. Giovannella, L. Fruchter, C. Chappert and I.A. Campbell, J. de Physique, to be published.

[9] C.P. Bean, Phys. Rev. Lett. 8 (1962) 250

[10] K. Kitazawa, T. Matsushita, Y. Tomioka, O. Tamura, I. Tanaka and H. Kojima, Jpn. J. Appl. Phys., to be submitted.

[11] M. Murakami, M. Morita, K. Doi, K. Miyamoto and H. Hamada, Jpn. J. Appl. Phys. 28 (1989) 399.

Resistive State of High Temperature Superconductors in Magnetic Fields

Y. IYE[1], S. NAKAMURA[1], T. TAMEGAI[1], T. TERASHIMA[2], and Y. BANDO[2]

[1] The Institute for Solid State Physics, The University of Tokyo, Roppongi, Minato-ku, Tokyo, 106 Japan

[2] Institute for Chemical Research, Kyoto University, Uji, Kyoto, 611 Japan

Experimental studies of the resistive state of high temperature superconductors in the mixed state have been carried out on thin film samples of $YBa_2Cu_3O_{7-y}$ and $Bi_2Sr_2CaCu_2O_{8+y}$. Precise angular dependence measurements have revealed not only the critical field anisotropy with respect to the a, b, and c-axes but also a feature associated with twin boundaries. Dependence on the angle between the transport current and the magnetic field is investigated as a key test for the model based on the Lorentz-force-driven flux motion. The Hall effect in the resistive transition region shows a series of peculiar behavior including sign reversal.

Key words: resistive state, anisotropy, Lorentz force, Hall effect

1. INTRODUCTION

Broadening of the resistive transition of high temperature superconductors (HTSCs) in magnetic fields is the subject of much current interest. The fact that the broadening is consistently observed even in the best available samples indicates the intrinsic nature of the phenomenon. Since the current carrying ability without energy dissipation at high temperature is the most attractive property of HTSCs as practical material for application, its deterioration in magnetic fields is certainly a matter of great concern. The mechanism of energy dissipation in high temperature superconductors is also an interesting fundamental problem. In this paper, we present our recent experimental findings pertinent to the transport in the mixed state of the HTSCs. Some parts of the work presented here are already published.[1-3]

2. EXPERIMENTAL METHODS

Samples used in the present work are thin films of $Bi_2Sr_2CaCu_2O_{8+y}$ and $YBa_2Cu_3O_{7-y}$. Highly c-axis oriented polycrystalline filmes of $Bi_2Sr_2CaCu_2O_{8+y}$ were prepared at ISSP, Univ. of Tokyo by single-target rf-sputtering method on MgO (100) substrates. Epitaxial films of $YBa_2Cu_3O_{7-y}$ were fabricated at ICR, Kyoto Univ. by activated reactive coevaporation technique on $SrTiO_3$ (100) substrates.[4] The $YBa_2Cu_3O_{7-y}$ films were heavily twinned, consisting of densely interwoven two regions with interchanged a- and b-axes. The average spacing between the twin boundaries is not precisely known. It is presumably much less than a micron, since the twin patterns are not resolved by usual optical microscopy.

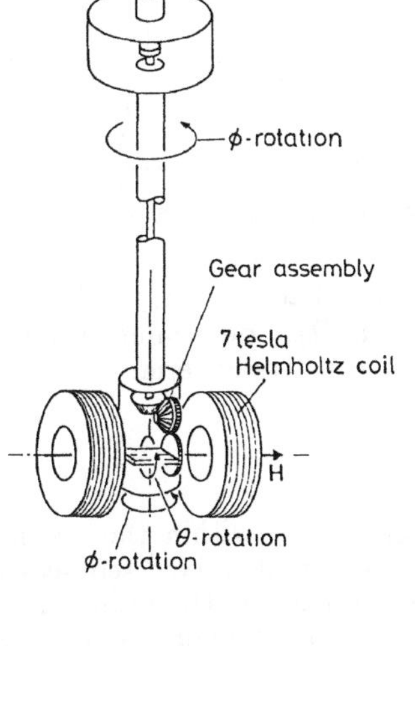

Fig. 1 Schematic of the expeimental setup for precise angular dependence measurements.

All the samples were prepared in a standard Hall bar pattern, Electrical leads were attached with silver paste either directly or on evaporated silver pads. The resistivity and Hall measurements were done by usual d.c. methods. Two sets of superconducting magnet systems were used for the magnetotransport experiments. One system is a 9 *tesla* solenoid combined with a cryostat with a gas thermometer designed for field independent temperature regulation during field sweep. This system was mainly used for $\rho_{xx}(H)$ and $\rho_{xy}(H)$ measurements at fixed T, and for $\rho_{xx}(T)$ measurements at fixed H. Another system schematically illustrated in Fig. 1 utilizes horizontal magnetic fields generated by a 7 *tesla* superconducting Helmholtz coil. A sample holder equipped with a gear assembly is used to rotate samples around a horizontal axis (θ-rotation) with angular reproducibility better than $\sim 1°$. The sample holder itself can be rotated around the vertical axis (ϕ-rotation) by a computer-controlled stepping motor with an angular resolution of 0.0036°. This setup with the double-axis rotation capability enabled us to carry out precise angular dependence measurements. It was powerful in the studies of various types of anisotropy, and also in the automated measurements of the temperature dependent Hall effect.

3. ANISOTROPY IN THE RESISTIVE STATE

3.1 The c-Axis Anisotropy

Figure 2 shows the resistivity of a $Bi_2Sr_2CaCu_2O_{8+y}$ film as a function of field angle with respect to the *c*-axis, at a fixed temperature slightly below T_c for several values of magnetic field. One can derive the angular dependence of the "H_{c2}" by taking resistive criteria like $0.5\rho_n$ (midpoint), $0.1\rho_n$ *etc.* The angular dependence thus determined is shown in Fig.3. Extremely sharp angular dependence near $\vec{H} \parallel ab$-plane manifests a highly two-dimensional nature of superconductivity in this material.[3]

By contrast, similar curves for $YBa_2Cu_3O_{7-y}$ show much less sharp feature. The angular dependence of "H_{c2}" for $YBa_2Cu_3O_{7-y}$ can be fitted by the anisotropic Ginzburg-Landau model (so-called effective mass model) with the anisotropy ratio ~ 5, indicating a fairly three-dimensional character of this material.[5] The large difference in the dimensionality between $YBa_2Cu_3O_{7-y}$ and $Bi_2Sr_2CaCu_2O_{8+y}$ has also been demonstrated by a recent torque magnetometry study.[6]

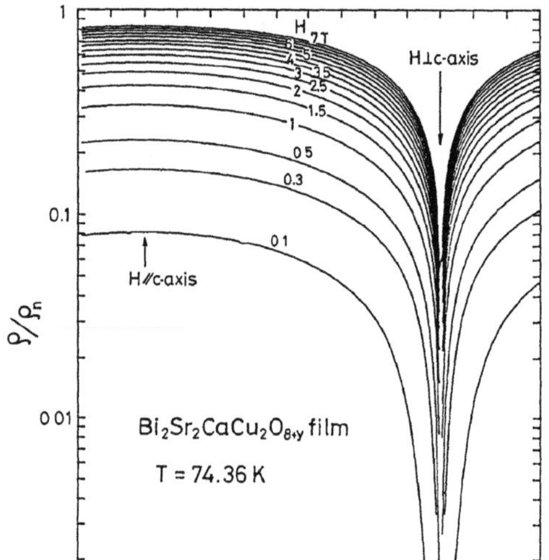

Fig. 2 Angular dependence of the resistivity in $Bi_2Sr_2CaCu_2O_{8+y}$ for different magnetic fields at fixed temperature below T_c. Extremely sharp angular dependence near $\vec{H} \parallel ab$-plane is seen.

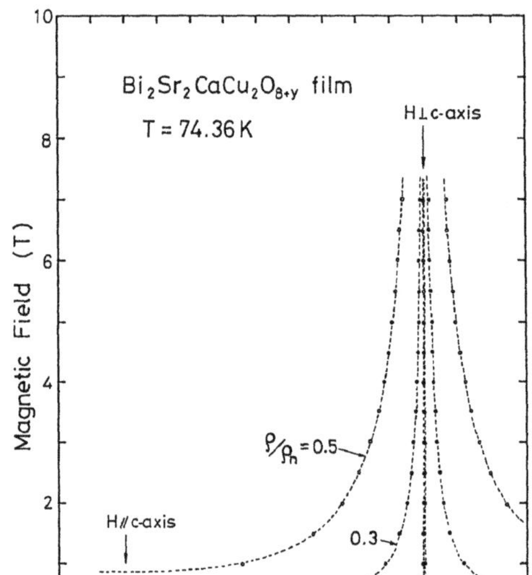

Fig. 3 Angular dependence of "H_{c2}" of $Bi_2Sr_2CaCu_2O_{8+y}$ determined from Fig. 2 by different resistive criteria.

3.2 The ab-Plane Anisotropy

The $YBa_2Cu_3O_{7-y}$ epitaxial film samples allowed us to study the anisotropy within the *ab*-plane. Information on the *ab*-plane anisotropy can be gained, by measuring resistivity at a fixed temperature and field strength, as a function of field angle while keeping the field direction strictly within the *ab*-plane. In reality, however, this is not an easy experimental task, since even a small angular deviation of the field away from the *ab*-plane can totally distort the result.

When an epitaxial $YBa_2Cu_3O_{7-y}$ film with its *ab*-plane aligned horizontally, is rotated around the vertical axis in the setup illustrated in Fig. 1, the angular dependence of resistivity will come from the following factors:

(a) The *c*-axis anisotropy can mix in on account of possible angular misalignment between the *c*-axis and the rotation axis. Such a misalignment-originated angular dependence should be two-fold symmetric, relative to the direction of misalignment.

(b) The intrinsic *ab*-plane anisotropy should be basically two-fold symmetric. In case of heavily twinned crystal, the two-fold symmetry is statistically averaged to yield a four-fold symmetry.

(c) Another four-fold symmetry shifted by 45° from that of (b) may arise in a heavily twinned epitaxial film, from the interaction of fluxes with twin boundaries.

(d) On the basis of usual picture of Lorentz-force-driven flux motion, the dependence on the relative angle between transport current and magnetic field is expected. It should yield a two-fold symmetric structure relative to the transport current direction.

Figure 4 shows the longitudinal and transverse resistivities of the epitaxial $YBa_2Cu_3O_{7-y}$ film as a function of field angle within the *ab*-plane. The Hall-bar pattern of this sample is such that the current direction is along the $a(b)$-axis. There are several noteworthy features in the angular dependence curves. Firstly, the resistivity takes minima both at $\theta = 0°$ and $90°$. This four-fold symmetric feature arises from the "H_{c2}" anisotropy in the *ab*-plane. The "H_{c2}" is presumably higher for $\vec{H} \parallel b$-axis, *i.e.* along the CuO chain direction. The fact that it is clearly observed even after averaged over the twin structure suggests sizable anisotropy between the *a*- and *b*-axes. Although conversion from the resistivity data to critical field anisotropy is not straightforward, the *ab*-plane anisotropy manifested in the present experiment appears to be much larger than the penetration depth anisotropy of 1.15 obtained by a decolation experiment[7]. Secondly, there are another set of resistivity minima at angles shifted by 45° from the $a(b)$-axis. These resistivity minima for field directions aligned to the twin boundary direction may be either due to an enhancement of H_{c2} for thie field direction, or due to strong pinning of flux lines at twin boundaries.

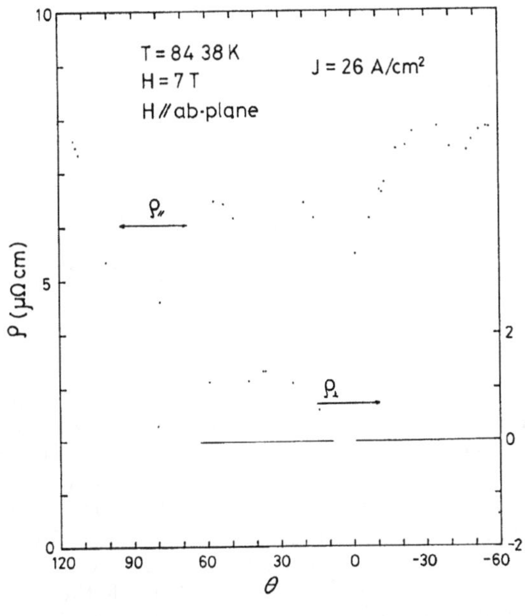

Fig. 4 Angular dependence of longitudinal and transverse resistivities in $YBa_2Cu_3O_{7-y}$ for $\vec{H} \parallel$ *ab*-plane configuration. The resistivity minima occurs when the field direction coin-

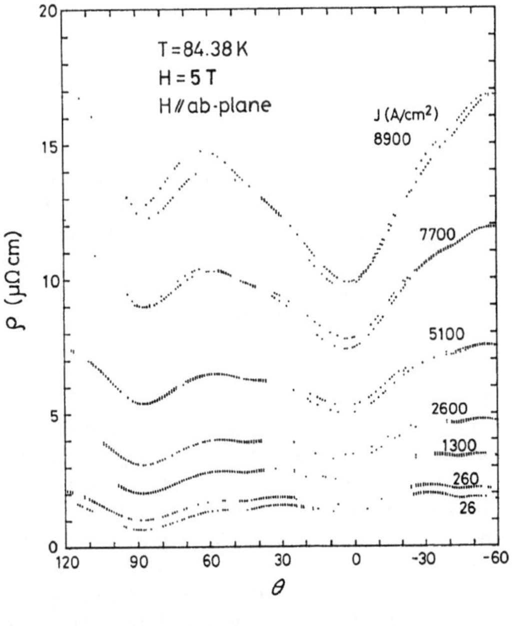

Fig. 5 Angular dependence of resistivity in the same $YBa_2Cu_3O_{7-y}$ sample as Fig. 4 for different transport current densities. The relative height of the minima at $\theta = 0°$ and 90° changes with the current density.

3.3 Dependence on the Current Direction

The energy disspation mechanism in the resistive state under magnetic field is currently discussed in terms of flux creep model.[8] According to the standard theories for flux flow and flux creep, the driving force is the $\vec{J} \times \vec{B}$ Lorentz force. A critical test for the applicability of the picture for the HTSCs is provided by investigating whether or not the phenomenon depends on the relative angle between the field and the transport current. Figure 5 shows the angular dependence of resistivity for the same

$YBa_2Cu_3O_{7-y}$ film as Fig. 4, for different tranport current densities. It is seen that the resistivity minimum at $\phi = 90°$ ($\vec{H} \perp \vec{J}$) increases relative to that at $\phi = 0°$ ($\vec{H} \parallel \vec{J}$), as the current density is increased. This behavior at high current densities is consistent with what one expects for a Lorentz-force-driven process. On the other hand, in the low current density limit, where the *I-V* characteristic is Ohmic, the resistivity minimum at $\phi = 90°$ is lower than that at $\phi = 0°$, contrary to the expected behavior. The relative height of the two minima varied from run to run, and was probably affected by angular misalignment.

In order to differenciate the angular dependence relative to the current direction from that relative to the crystal axes, a second sample was prepared with the Hall-bar pattern intentionally made at an angle $\sim 20°$ from the $a(b)$-axis. The angular dependence can be decomposed to a two-fold symmetric component relative to the current direction ($\phi = 0°$) and a four-fold symmetric part relative to the $a(b)$-axis ($\phi = 20°$). For this sample, the two-fold symmetry component showed up much pronounced than the four-fold symmetry component. Thus, a definite conclusion is yet to be reached for $YBa_2Cu_3O_{7-y}$, with regard to the Lorentz-force-related angular dependence *in the low current limit.*

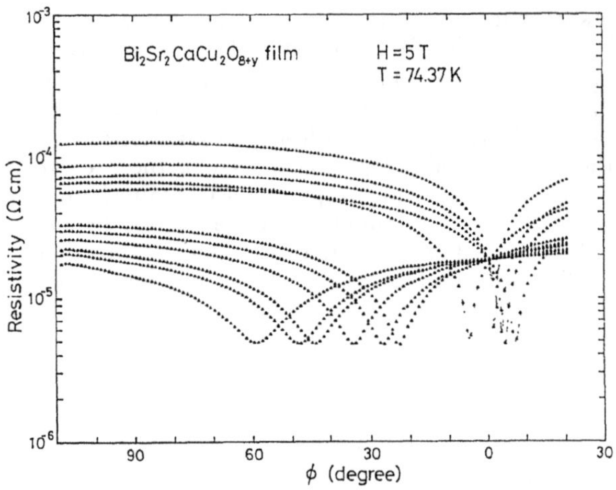

Fig. 6 Angular dependence of the resistivity in $Bi_2Sr_2CaCu_2O_{8+y}$ for angular misalignment varied slightly from each other. The resistivity minima correspond to $\vec{H} \parallel ab$-plane. The values of resistivity minima are independent of the relative angle between $\vec{H}$ and $\vec{J}$.

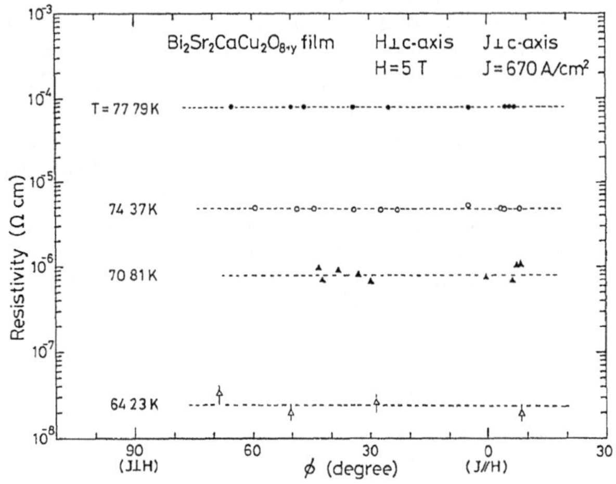

Fig. 7 Resistivity minima plotted against the relative angle between $\vec{H}$ and $\vec{J}$, domonstrating the Lorentz force independence of the resistive state in $Bi_2Sr_2CaCu_2O_{8+y}$.

For the $Bi_2Sr_2CaCu_2O_{8+y}$ system, on the other hand, we have recently demonstrated the complete independence of the relative angle between field and current.[1] Figure 6 shows a series of traces taken with angular misalignment slightly changed for each sweep. The minimum of resistivity in each trace corresponds to the accurate alignment of the field direction to the ab-plane. By comparing the minimum values, therefore, one can determine the dependence on the angle between the field and current strictly under the condition of $\vec{H} \parallel ab$-plane. Plotted in Fig. 7 are the minimum values of resistivity as a function of angle at four different temperatures in the resistive transition region. It appears that the dissipation mechanism in the case of $Bi_2Sr_2CaCu_2O_{8+y}$ is completely independent of whether or not the Lorentz force is at work. This surprising result cannot be easily reconciled with the usual picture of Lorentz-force-driven flux motion. Although it may be argued that possibly meandering current flow can randomize the local direction between field and current, such effect tends to be averaged out for the $\vec{H} \parallel \vec{J}$ configuration. Similar conclusion is also reached for $YBa_2Cu_3O_{7-y}$[9] and $(La_{1-x}Sr_x)_2CuO_4$[10] bulk crystals in the $\vec{H} \parallel \vec{J} \parallel c$-axis configuration.

3.4 Flux Dynamics in a Highly Anisotropic Layered Superconductor

In an attempt to understand the current direction independence observed in $Bi_2Sr_2CaCu_2O_{8+y}$, we consider the following model. In a highly anisotropic layered superconductor with extremely short c-axis coherence length like $Bi_2Sr_2CaCu_2O_{8+y}$, the core of a vortex parallel to the ab-plane is so small in radius that it fits confortably in the interlayer spacings. There is an substantial energy barrier against vortex hopping from one interlayer spacing to the next. On the other hand, vortex meandering within the same interlayer spacing can occur without much energy cost. In such a situation, vortex motion may occur in a two-step process depicted in Fig. 8. First, a short segment of a vortex hops to the next interlayer spacing by thermal activation. A pair of kinks thus created will be driven by the Lorentz force. In the $\vec{H} \perp \vec{J}$ configuration, the kinks will propagate to yield an end result indistinguishable from a vortex hop as a whole. Note that a pair of kinks created by a local hop to the "wrong" direction if forced to shrink. In the $\vec{H} \parallel \vec{J}$ configuration, the Lorentz force drives the pair of kinks towards an spiral instability. The fate of the deformed vortex is not easy to visualize. It will collide with other vortices and will probably go through compricated processes involving vortex cutting and cross-joining.[11]

The observed angular independence might be qualitatively understood, if we postulate that the first of the two steps, *i.e.* the local hopping of a vortex segment is ***the rate-limiting process.*** Quantitatively, however, the microscopic processes involved for the two configurations are so different that it still seems very difficult to get an identical dissipation rate for the two cases. Thus, although there is no doubt that flux dynamics is an essential ingredient of energy dissipation mechanism in the mixed state of HTSCs, further studies are clearly needed to gain a full understanding of the dissipation mechanism in the mixed state of the HTSCs. It is also important to clarify the conceptual connection between the fluctuation-based approach from above T_c and the flux-dynamics-based approach from below T_c.

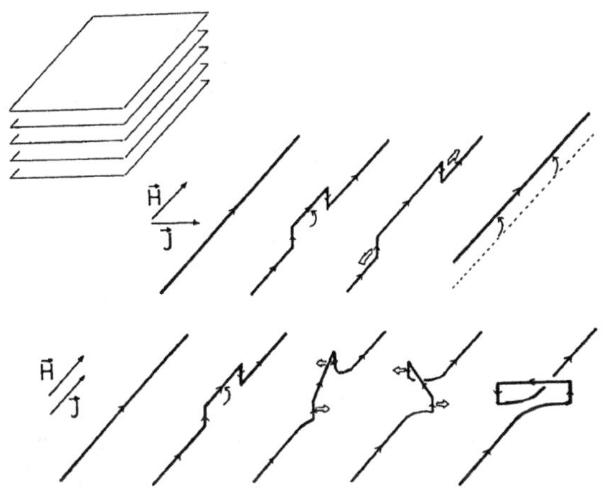

Fig. 8 Schematic illustration of possible vortex motion in a highly anisotropic layered superconductor both in the $\vec{H} \perp \vec{J}$ and $\vec{H} \parallel \vec{J}$ configurations. The process consists of two steps, local hop of a vortex segment followed by propagation of thus created kinks.

4. HALL EFFECT

Figures 9 and 10 show the magnetoresistivity ρ_{xx} and Hall resistivity ρ_{xy} of the $YBa_2Cu_3O_{7-y}$ epitaxial film sample for fixed temperatures near T_c. Figure 11 shows the $\rho_{xx}(H)$ curves of Fig. 8 re-plotted in a logarithmic scale. From these curves, we can define the "$H_{c2}(T)$" curves by taking the $0.5\rho_n$ (midpoints),

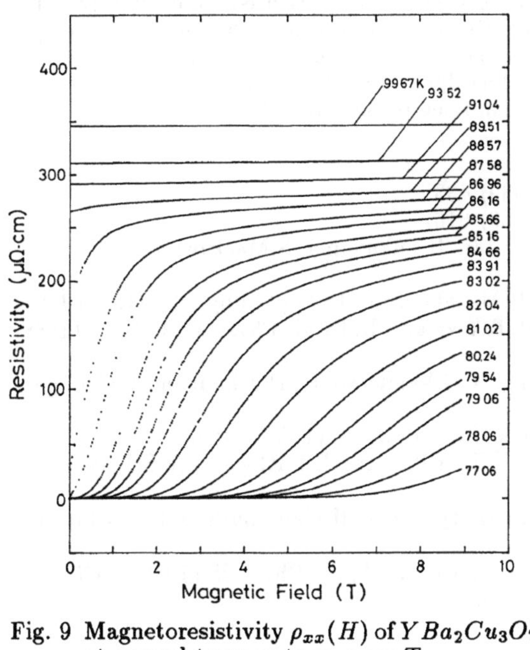

Fig. 9 Magnetoresistivity $\rho_{xx}(H)$ of $YBa_2Cu_3O_{7-y}$ at several temperatures near T_c.

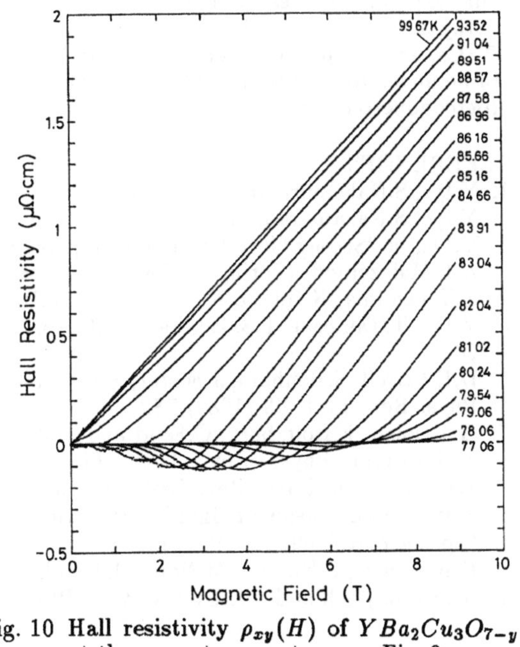

Fig. 10 Hall resistivity $\rho_{xy}(H)$ of $YBa_2Cu_3O_{7-y}$ at the same temperatures as Fig. 9.

$0.1\rho_n$, $10^{-2}\rho_n$, $10^{-3}\rho_n \cdots$. The "$H_{c2}(T)$" curves thus defined are plotted in Fig. 12. It is noted that they all show upward curvature. As for the ρ_{xy} traces, notable feature is that there exists a low field range for each temperature where ρ_{xy} remains vanished while ρ_{xx} is already finite. Another feature noteworthy is the apparent parallel shift of the $\rho_{xy}(H)$ curves. To be quantitative, we define two kinds of characteristic field from the $\rho_{xy}(H)$ curves. One is the threshold field at which ρ_{xy} starts to deviate from zero. The other is the "bias" field characterizing the parallel shift of the ρ_{xy} curves in the high field range. It is defined by the intercept with the H-axis, of the high field part of the ρ_{xy} curves linearly extrapolated back. It is interesting to note that the threshold field and the "bias" field obtained from the ρ_{xy} data show T-linear behavior, in contrast to the upward-curvature of the "$H_{c2}(T)$" curves determined from ρ_{xx}.

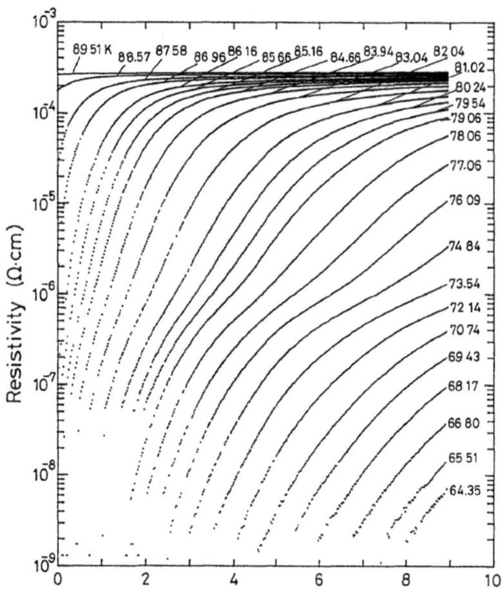

Fig. 11 Traces of magnetoresistivity in Fig. 9 replotted in a logarithmic scale.

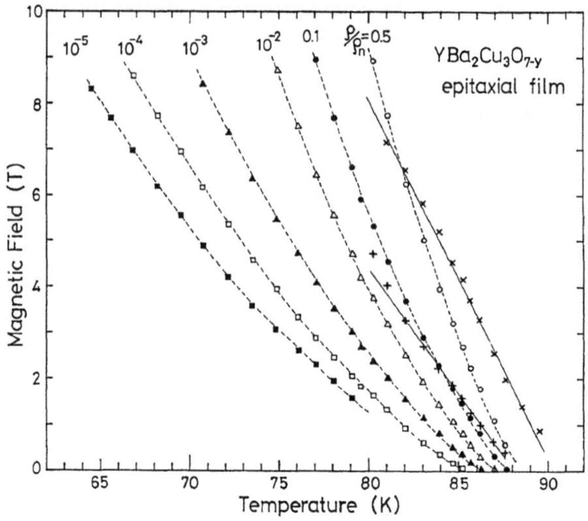

Fig. 12 Temperature dependences of "H_{c2} determined from the ρ_{xx} curves in Fig. 11 by different resistive criteria. The data points connected with solid straight lines represent characteristic field obtained from the ρ_{xy} curves in Fig. 10 $\cdots$ (+ : threshold field, × : "bias" field).

5. CONCLUSION

Some experimental results relevant to the resistive state of HTSCs are presented. Particularly noteworthy are, the current direction independence of the resistive state observed in the highly anisotropic HTSCs, and the peculiar behavior of the Hall effect including the sign reversal. Whether such behavior can be understood in terms of the conventional picture of flux dynamics, or it is possibly a sign of exotic superconductivity in the HTSC materials, is an outstanding issue to be elucidated.

REFERENCES

1. Y.Iye, S.Nakamura and T.Tamegai, Physica C159 (1989) 433.
2. Y.Iye, S.Nakamura and T.Tamegai, Physica C159 (1989) 616.
3. Y.Iye, in "Strong Correlation and Superconductivity", *eds.* H.Fukuyama, S.Maekawa and A.P.Malozemoff, (Springer, Heidelberg 1989) p.213.
4. T.Terashima, K.Iijima, K.Yamamoto, Y.Bando and H.Mazaki, Jpn. J. Appl. Phys. 27 (1988) L91.
5. Y.Iye, T.Tamegai, T.Sakakibara, T.Goto, N.Miura, H.Takeya and H.Takei, Physica C153-155 (1988) 26.
6. D.E.Farrell, S.Bonham, J.Foster, Y.C.Chang, P.Z.Jiang, K.G.Vandervoort, D.J.Lam and V.G.Kogan, Phys. Rev. Lett. 63 (1989) 782.
7. G.J.Dolan, F.Holtzberg, C.Feild and T.R.Dinger, Phys. Rev. Lett. 62 (1989) 2184.
8. M.Tinkham, Phys. Rev. Lett. 60 (1988) 1658: T.T.M.Palstra, B.Batlogg, L.F.Schneemeyer, J.V.Waszczak, Phys. Rev. Lett. 60 (1988) 1662.
9. Y.Iye, "Mechanisms of High Temperature Superconductivity" , *eds.* H.Kamimura and A.Oshiyama, (Springer, Heidelberg, 1989) p.263.
10. K.Kitazawa, S.Kambe, M.Naito, I.Tanaka and H.Kojima, Jpn. J. Appl. Phys. 28 (1989) L555.
11. A.M.Campbell and J.E.Evetts, Adv. Phys. 21 (1972) 199.

Two-Dimensional Phase Fluctuation in High-T_c Superconductor under Magnetic Field

MASASHI BAN[1], TSUNEO ICHIGUCHI[1], TOSHIYUKI ONOGI[2], and TOSHIYUKI AIDA[1]

[1] Central Research Laboratory, Hitachi Ltd., Tokyo, 185 Japan
[2] Advanced Research Laboratory, Hitachi Ltd., Tokyo, 185 Japan

ABSTRACT

Two kinds of power laws, $V \propto I^{n(T,H)}$ and $V \propto H^{m(T,I)}$, are observed in high T_c superconducting $ErBa_2Cu_3O_{7-x}$ films. The Nelson-Kosterlitz jumps and the related new kinks are found in the temperature dependences of the respective exponents n and m, giving direct evidences of the Kosterlitz-Thouless transition. These behaviors are the results of vortex-antivortex pair excitation caused by the phase fluctuation on the each CuO_2 plane. Dynamics of the pair excitation is turned out to be strongly affected by an applied magnetic field. The observed power laws in a magnetic field with an arbitrary direction suggest that the pair excitation in a magnetic field is related to the flux entanglemant state.

INTRODUCTION

High-T_c superconductors have a finite resistance even below T_c, and this is a major obstacle to many promising applications. The origin of the resistive state and its behavior have recently attracted much attention from the viewpoints of physics and application. As the resistance is strongly enhanced in a magnetic field, the resistive state has often been discussed in connection with the thermal behavior of magnetic flux in the mixed state of these type-II superconductors. Many authors have proposed models, for example, giant flux creep[1], flux flow[2], flux lattice melting[3], superconducting glass[4], vortex-antivortex (or fluxon-antifluxon) pair excitation[5-9] and so on. However, none of these models have been established in high-T_c superconductors.

On the other hand, it is well known that high-T_c superconductors have two-dimensional properties owing to the layered structure of the CuO_2 superconducting planes, which interact weakly with each other. The two dimensional feature, as well as the higher temperature condition, will have an important effect on the peculiar resistive state in high-T_c superconductors essentially through enhancement of the phase fluctuation of the order parameter.

The large phase fluctuation can bring about excitation of vortex-antivortex pairs on the CuO_2 planes. The many body interaction of the pairs gives rise to the Kosterlitz-Thouless (KT) transition[10], which is a binding-unbinding transition of the pairs. The KT transition was observed in ultra-thin films of conventional superconductors[11]. The signs of this transition in YBCO[5-8] and in BSCCO[9] were already obtained from the power-law transport properties and from the temperature dependent resistivity. Power laws, suggesting a scaling law, can never be derived from the standard flux flow or flux creep phenomena. It is, however, quite important to provide direct evidence of the phase transition to clarify that it exists. The most characteristic feature, direct evidence of the KT transition, is the universal Nelson-Kosterlitz jump[12] in the power exponent. This indicates a sudden change in the nonlinearity of transport properties at the transition temperature.

In this paper, we report the first observation of the Nelson-Kosterlitz jump and related phenomena in the exponents of both I-V and H-V curves in $ErBa_2Cu_3O_{7-x}$, giving clear evidence of the KT transition[13]. Very recently, the Nelson-Kosterlitz jump was also observed in BSCCO by Artemenko et al. [14] independently of us, but without an external magnetic field. Therefore, the KT transition seems to be a common feature in high-Tc superconductors with two-dimensional layered structures. This is an important subject to be studied further. It is also pointed out in this paper that an applied magnetic field or external vortices, as well as current, play a crucial role in the dynamics of the KT transition. The newly found coexistence of the symmetric power laws against current and magnetic field will be a key point in the total understanding of the KT transition and of the resistive state. Furthermore, preliminary results of transport experiments in magnetic fields with tilted angles to the c-axis are also presented and will be discussed in connection with the KT behavior and the flux entanglement state.

EXPERIMENTAL RESULTS

The measured sample is a 0.7μm-thick film of $ErBa_2Cu_3O_{7-x}$ grown epitaxially on a MgO[100] substrate by conventional rf-magnetron sputtering [15]. The film is a single crystal with its c-axis perpendicular to the surface. The sample is cut mechanically to form a Hall bridge along the a, b-axis. The channel width is 0.2mm and the distance between the voltage probes is about 3mm. The superconducting transition temperature in the absence of an external magnetic field is T_c=87K, where T_c is defined as the mid-point temperature so that $R(T_c)=R_N/2$. The induced voltage drop V is measured along a transport current I ($|I|\leq$300mA) in the presence of an external magnetic field H ($|H|\leq$ 14kOe). The sample is mounted on a copper block whose temperature can be controlled between 65K and 100K with an accuracy of 0.1K. The self-heating effect is carefully verified to be negligible under these experimental conditions.

Typical I-V characteristics for the temperature range 73K$\leq T\leq$86K are shown in Fig. 1, where the magnetic field (H=8kOe) is applied parallel to the c-axis. As the log$I$-log$V$ plots form quite good straight lines in all our temperature and magnetic field regions, the power law

$$V \propto I^{\,n(T,H)}, \qquad (1)$$

is obtained. Here, the exponent n depends on the temperature and the magnetic field. It increases as the temperature and the magnetic field decrease.

The temperature dependence of the exponent $n(T, H)$ in various magnetic fields is shown in Fig. 2. The remarkable feature in this figure is the clear kink at n=3. We should note that the kinks for all the magnetic fields appear at n=3. This peculiar feature corresponds to the Nelson-Kosterlitz jump which describes the abrupt change in the power exponent from 3 to 1 at the KT transition temperature T_{KT}. The observed tendency toward smearing of the universal jump, to produce not a step but a kink, is possibly due to the finite size effect of the two dimensional system and to the inter-plane coupling effect. In this case, T_{KT} is defined as the temperature where n=3. From Fig. 2, it is found that T_{KT} depends on the magnetic field and shifts from 83K to 80K when the magnetic fields vary from 2kOe to 13.5kOe. In the absence of a magnetic field, a clear kink is not observed. It is surprising that the magnetic field reveals the remarkable feature of a universal jump. The crucial role of a magnetic field in the KT transition will be discussed later.

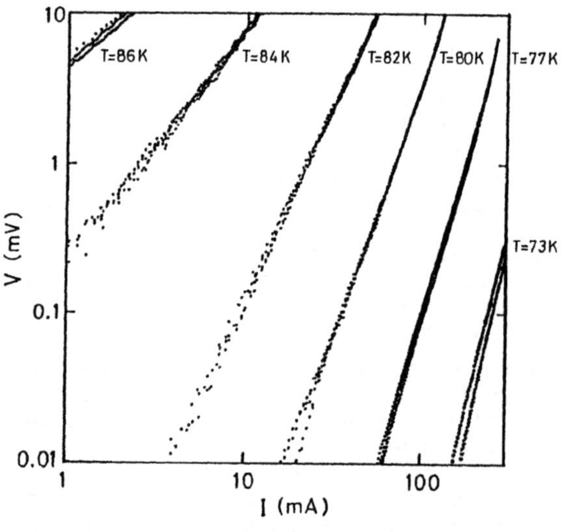

Fig. 1 Plots of logI-logV for various temperatures below T_c in the presence of a magnetic field of H=8.0kOe.

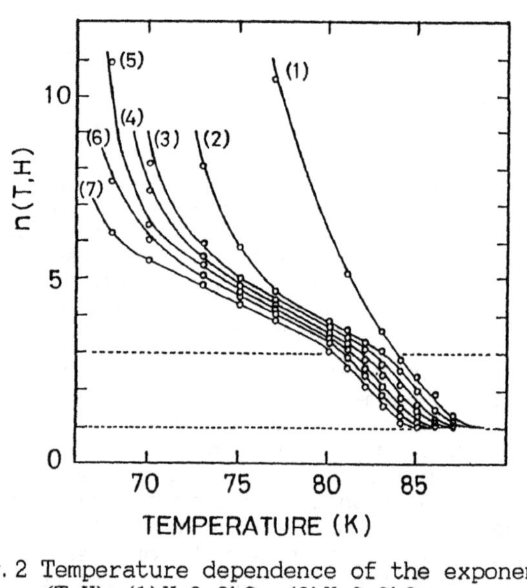

Fig. 2 Temperature dependence of the exponent n(T, H) : (1) H=0.0kOe, (2) H=2.0kOe, (3) H=4.0kOe, (4) H=6.0kOe, (5) H=8.0kOe, (6) H=10.0kOe and (7) H=13.5kOe.

The H-V characteristics for the temperature range 80K≦T≦85K are given in Fig. 3, where it is found that the log$H$-log$V$ plots also become good straight lines. Thus, we have another power law in the form of

$$V \propto H^{m(T,I)}. \qquad (2)$$

Here, the exponent m depends on both temperature and current. This kind of the power law for BSCCO has been obtained in a much smaller magnetic field and without current dependence [9]. The exponent $m(T,I)$ increases as the temperature and the current decrease. Figure 4 shows the temperature dependence of the exponent $m(T,I)$ for various currents. The temperature dependence bends sharply at the temperature (near T_{KT}) where m=2. The bending appears more clearly for the smaller currents. It is interesting to point out that this bending feature in the exponent m must have a close relation to the Nelson-Kosterlitz jump in the exponent n described above, and that the importance of the magnetic field is suggested also by this result.

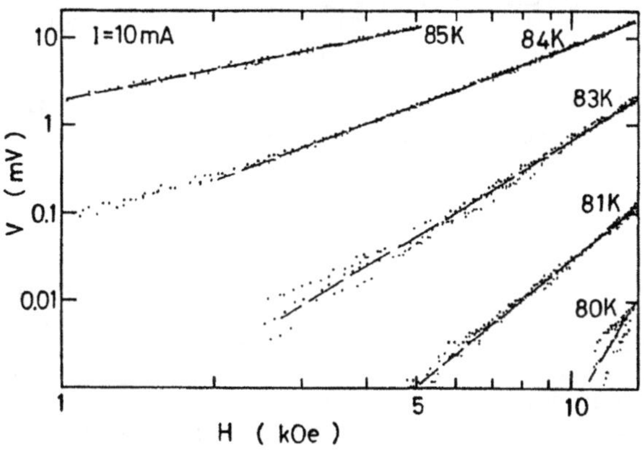

Fig. 3 Plots of logH-logV for various temperatures below T_c at transport current of I=10mA.

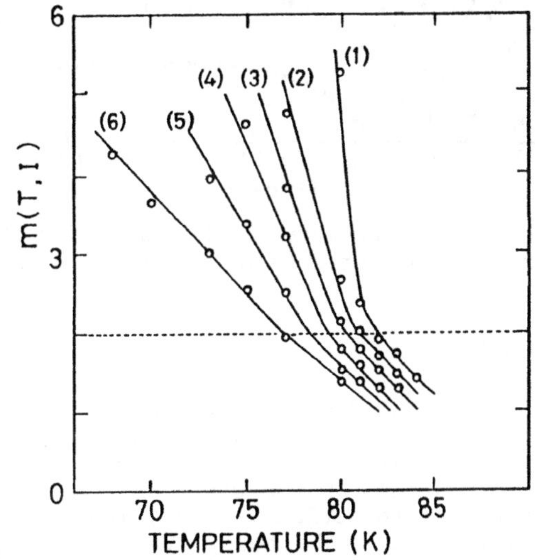

Fig. 4 Temperature dependence of the exponent m(T, I): (1) I=20mA, (2) I=40mA, (3) I=60mA, (4) I=80mA, (5) I=120mA and (6) I=200mA.

Tilting the magnetic field will help to elucidate the effect of the field, (as pointed out in Ref. [16]). The direction of the magnetic field is varied from perpendicular to parallel to the surface in the transverse configuration (i. e. $H \perp I$), and the angle θ is measured from the sample surface (see Fig. 5). The angular dependence of the induced voltage V in our sample is shown in Fig. 6 for various currents. The strength of the magnetic field and the temperature are fixed at 6.0kOe and 77K, respectively. The voltage V varies depending on the angle θ, but the dependence does not seem to have a simple form of sinθ. In Fig. 7, the observed voltages V are plotted against the current I for various angles θ of the magnetic field (H=6.0kOe). It is easily seen that the I-V curves are described by power laws with the same exponent (n≒4.2 for H=6kOe) for an arbitrary angle θ. The exponent depends on the magnetic field strength but not on the angle θ, while the voltage V (or the resistance V/I) is a function of both field strength and angle. Thus, the I-V characteristics can be represented by

$$V=\alpha(T,H,\theta)\cdot I^{n(T,H)}. \qquad (3)$$

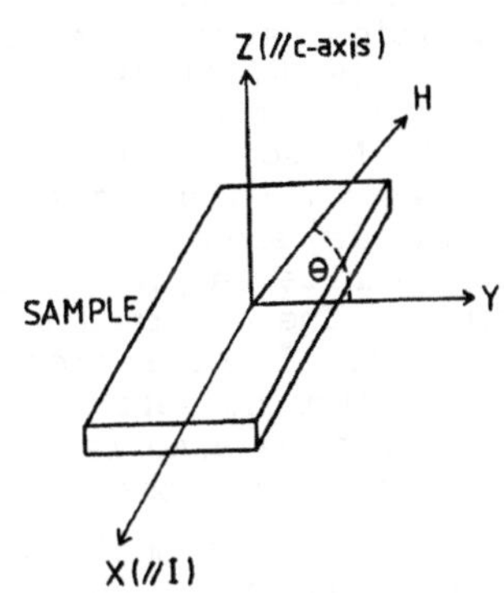

Fig. 5 The directions of current and magnetic field.

This behavior of the angular dependence suggests a close relation with vortex-antivortex pairs excited in the flux entanglement state as discussed later.

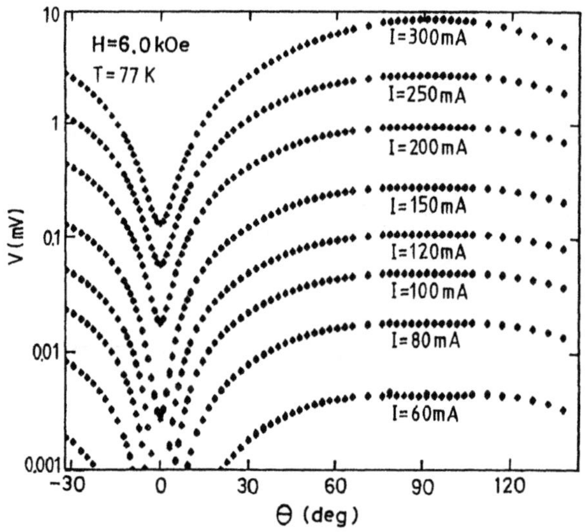

Fig. 6 Angular dependence of the induced voltage for various currents at T=77K and H=6.0kOe.

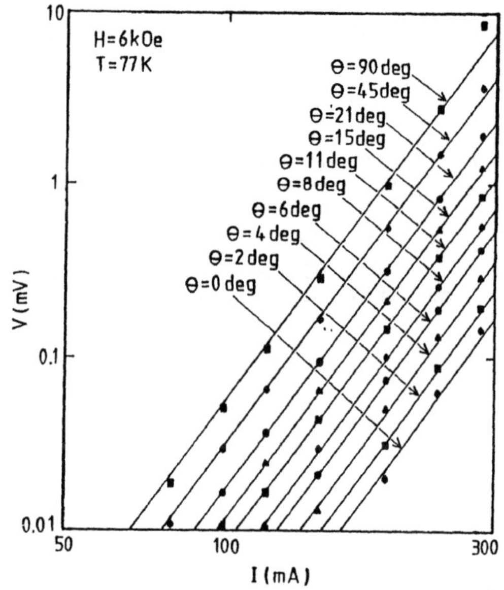

Fig. 7 Plots of logI-logV for various angles at T=77K and H=6.0kOe.

DISCUSSIONS

In a two-dimensional superconductor, the thermal fluctuation of the phase of order parameter will give rise to the topological excitation of vortices and antivortices with the same number. The number of the pairs is settled in the thermal equilibrium through the creation and annihilation of the pairs by the fluctuation. An attractive interaction ($\sim (q^2/\varepsilon)\cdot \log(r/\xi)$) between vortex and antivortex makes the pair into the bound state below T_{KT}. (Here, q, ε and ξ are the topological charge of the vortex, the dielectric constant representing the many body screening effect, and the coherence length in *ab*-plane, respectively.) The observed power law, $V \propto I^n$, is derived from the dissociation process of bound vortex-antivortex pairs by current flow. The vortex and antivortex are suffered from the Lorentz force of $J\Phi_0/c$ in opposite direction (J:current density, Φ_0:flux quantum), and some of the pairs are broken up to be released from binding. The broken pairs contribute to the flux-flow resistance. The ratio of the broken pairs to the bound pairs is given by $\sim (J/J_0)^{n-1}$, ($n=1+q^2/(2\varepsilon(T)k_B T)$), which leads the power law of the I-V characteristics with the exponent n.

The many body effect of the similar systems was treated with renormalization group method[10-12]. The renormalization scaling behaves quite differently, depending on the temperatures of $T>T_{KT}$ or $T \lesssim T_{KT}$, and predicts the existence of a topological phase transition between the binding state and unbinding state of the vortex-antivortex pairs. (i.e. the KT transition). The theory has revealed that $n \gtrsim 3$ for $T \lesssim T_{KT}$ and $n=1$ for $T>T_{KT}$. Thus, the exponent n changes from 3 to 1 discontinuously at T_{KT}, which is called as the Nelson-Kosterlitz jump[12]. Since all pairs are unbound above T_{KT}, their flux flow resistance gives a linear I-V relation (hence $n=1$). The Nelson-Kosterlitz jump feature is smeared by the finite size effect of the two dimensional system, because the finite sample prevents the renormalization scaling from reaching the final fixed points. The observed kinks in the temperature dependence of the exponent at $n=3$ as shown in Fig.2 agree precisely with the feature of smeared universal jump. The inter-plane coupling effect and the critical fluctuation of the superconducting transition might be also responsible for the smearing. However, it should be stressed that we give the first observation of a direct evidence of the KT transition.

Since two-dimensionality is essential for the KT transition, it is interesting to estimate the effective thickness d of the high-Tc superconductor. The following relation has been given[17] for a thin film superconductor in the dirty limit of BCS scheme (where $2\Delta_0/k_B T_c=3.5$ is assumed) and in approximation of $T_c \sim T_{KT}$.

$$T_c/T_{KT}=1+0.173\varepsilon\rho_{ab}/(d\cdot R_c), \tag{4}$$

where $R_c=h/(2\pi e^2)=4.1\text{k}\Omega$, ρ_{ab} ($\fallingdotseq 70\mu\Omega\cdot$cm for our samples) is the resistivity in the normal state near T_c, and $\varepsilon(\sim 1)$ is an effective dielectric constant of vortex-antivortex pairs. As T_c=87K and T_{KT}=84K in the absence of a magnetic field, then we get $d_{H=0}$~8Å which is comparable to or even smaller than the lattice constant(=11.7Å) along c-axis. This guarantees the two dimensionality and means that the closed ring of the vortex lines pierces only one double-plane of CuO_2. Assuming that Eq. (4) is satisfied in a magnetic field, d_H~4Å is obtained from T_{KT}=81K under H=13.5kOe. Therefore, one of the important role of the magnetic field seems to be the enhancement of the two-dimensionality through the suppression of the inter-plane correlation. This may be a reason why the Nelson-Kosterlitz jump becomes clear in the presence of magnetic field.

Another important role of the magnetic field is the enhancement of dissociation of the bound vortex-antivortex pairs. This is clear from decreasing of the transition temperature T_{KT} in the higher magnetic field (see Fig. 2). As the power-law behavior in the H-V curves is observed, the magnetic field has the same effect as the applied current on the vortex-antivortex pairs. The external field introduce the excessive vortices in the system where vortex-antivortex pairs exist. The interaction between the vortices with the same circulation is repulsive, while the interaction between vortex and antivortex is attractive. These repulsive and attractive forces acting on the pairs from the external vortex can break up the pairs, and give the power law behavior in H-V curves in the same manner as current.

We have found that two kinds of power laws, given by Eqs. (1) and (2), are satisfied simultaneously in I-V and H-V curves. The necessary and sufficient condition is expressed by the following relations.

$$\log(V/V_0)=\gamma\cdot\log(I/I_0)\times\log(H/H_0)$$
$$=m(T,I)\,n(T,H)/\gamma, \qquad (5)$$

where $\gamma(<0)$ is a temperature dependent coefficient, and $V_0(>V)$, $H_0(>H)$ and $I_0(>I)$ are the normalization factors. This relation has an unusual form because it involves a multiplication of two logarithmic functions. It is also interesting that the logarithmic voltage is given by the product of two exponents. In this case, the exponents $n(T,H)$ and $m(T,I)$ can be represented by

$$n(T,H)=\gamma\cdot\log(H/H_0), \quad \text{and} \quad m(T,I)=\gamma\cdot\log(I/I_0). \qquad (6)$$

Equation (6) can be examined directly from the experimental data, and turns out to be satisfied well in a wide range of temperatures below Tc. (see ref. [13]). It should be noted that the exponents $n(T,H)$ and $m(T,I)$ are not independent as seen from Eqs. (5) and (6). This shows that the synthetic effect of transport current and external magnetic field is important, and that the significant modification of the renormalization group for the KT transition is necessary under a magnetic field.

In the high T_c superconductor, the vortex lines created by the external magnetic field can wander about rather randomly due to the large thermal fluctuation. As the result, the vortex lines penetrate the sample in a winding shape like spaghetti to minimize the free energy with increasing the entropy. This picture is similar to the flux line (vortex line) entanglement model proposed by Nelson[18]. The configuration R of the vortex line can be divided into two parts, the average $\langle R\rangle$ and the fluctuation ΔR ($\langle\Delta R\rangle$=0), where $\langle R\rangle$ depends on the direction of the magnetic field and ΔR (or $\langle\Delta R^2\rangle$) is determined by thermal fluctuation. Then, the total density of the field-created voritces (including antivortices) in the CuO_2 planes is given by $N=\langle N\rangle+\Delta N$. Here, $\langle N\rangle$ is determined by both strength and direction of the applied magnetic field, and it is described as the averaged configuration $\langle R\rangle$, while ΔN is due to the thermal fluctuation part ΔR. The ΔN is a function of the temperature and the magnetic field strength and is possibly independent of the direction of the magnetic field. The condition $\langle N\rangle\ll\Delta N$ might be achieved in our experimental condition of higher temperature, unless the applied magnetic field is not strong enough. We call it flux spaghetti state in extension of the flux line entanglemant state[18].

The flux spaghetti state, been stirred dynamically by the thermal fluctuation, will assist the excitation of the vortex-antivortex pairs, and will also stimulate the depairing of them. The magnetic field dependence of the exponent $n(T,H)$ is determined by the interaction between the vortex-pairs and the external vortices. Here, the experimental length scale at the fixed temperature would be defined as L_H~$\log(H_{c2}/N\Phi_0)$ ~$\log(H_{c2}/\Delta N\Phi_0)$, which is independent of the field direction. This is the reason why the exponent does not depend on the direction of a magnetic field as observed in Fig. 7. On the other hand, the induced voltage V depends on the average configulation $\langle R\rangle$ of the external vortices in some extent, because the fluctuation term in V will be more or less canceled out. Thus, the voltage V depends on the direction of magnetic field as shown in Fig. 6. The dynamical fluctuation of the flux entanglement state, like a spaghetti in boiling water, will have potential influence on the KT transition. However, the further investigation will be necessary to clarify the relationship between the KT transition and the entangled vortex lines.

CONCLUSION

Two kinds of power laws, $V \propto I^n$ and $V \propto H^m$, have been observed in superconducting $ErBa_2Cu_3O_{7-x}$ films. The Nelson-Kosterlitz jump in the temperature dependence of the exponent $n(T, H)$ has been found clearly in the presence of an external magnetic field. This provides direct evidence of the KT transition associated with vortex-antivortex pairs excited thermally on the CuO_2 planes. Sharp bending in the temperature dependence of the exponent $m(T, I)$ has also been found. This is closely related to the KT transition. Furthermore, the power-law I-V curve has been observed for an arbitrary angle of the magnetic field to the c-axis. In this case, the exponent $n(T, H)$ is independent of the angle, although the induced voltage V is not. All our results show that thermally excited vortex-antivortex pairs caused by the phase fluctuation of order parameter play an essential role in the peculiar transport properties of the resistive state in high T_c superconductors.

ACKNOWLEDGMENTS

The authors thank many co-workers for providing the excellent samples. They also grateful to Dr. Y. Murayama for valuable discussions.

REFERENCES

[1] Y. Yeshurun & A. P. Malozemoff, Phys. Rev. Lett. 60 (1988) 2202: A. P. Malozemoff, K. Worthington, Y. Yeshurun, F. Holtzberg & P. H. Kees, Phys. Rev. B38 (1988) 7202.
[2] R. B. van Dover, L. F. Schneemeyer, E. M. Gyorgy & J. V. Waszczak, Phys. Rev. B39 (1989) 4800.
[3] P. L. Gammel, et al., Phys. Rev. Lett. 59 (1987) 2592, and Phys. Rev. Lett. 61 (1988) 1666.
[4] K. A. Muller, M. Takashige & J. G. Bednorz, Phys. Rev. Lett. 58 (1987) 1143: A. C. Mota, A. Pollini, P. Visani, K. A. Muller & J. G. Bednorz, Phys. Rev. B36 (1987) 401.
[5] M. Sugahara, M. Kojima, N. Yoshikawa, T. Akeyoshi & N. Haneji, Phys. Lett. 125A (1987) 429.
[6] P. C. E. Stamp, L. Forro & C. Ayache, Phys. Rev. B38 (1988) 2847.
[7] N. C. Yeh & C. C. Tsuei, Phys. Rev. B39 (1989) 9078.
[8] T. Onogi, T. Ichiguchi & T. Aida, Solid State Commun. 69 (1989) 991.
[9] S. Martin, A. T. Fiory, R. M. Fleming, G. P. Espinoza & A. C. Cooper, Phys. Rev. Lett. 62 (1989) 677.
[10] J. M. Kosterlitz & D. J. Thouless, J. Phys. C6 (1973) 1181: J. M. Kosterlitz, J. Phys. C7 (1974) 1046: The review is given by P. Minnhagen, Rev. Mod. Phys. 59 (1987) 1001.
[11] A. M. Kadin, K. Epstein & A. M. Goldman, Phys. Rev. B27 (1983) 6691: J. C. Garland & Hu Jong Lee, Phys. Rev. B36 (1987) 3638.
[12] D. R. Nelson & J. M. Kosterlitz, Phys. Rev. Lett. 39 (1977) 1201: B. I. Halperin & D. R. Nelson, J. Low Temp. Phys. 36 (1979) 599.
[13] M. Ban, T. Ichiguchi & T. Onogi, Phys. Rev. B40 (1989) 4419.
[14] S. N. Artemenko, I. G. Gorlova & Yu. I. Latyshev, Phys. Lett. A138 (1989) 428.
[15] T. Aida, T. Fukazawa, A. Tsukamoto, K. Takagi, T. Shimotu & T. Ichiguchi, in *Advances in Superconductivity*, edited by Kitazawa and Ishiguro, (Springer-Verlag, Berlin, 1989), p. 539.
[16] Y. Iye, S. Nakamura & T. Tamegai, Physica C159 (1989) 616.
[17] M. R. Beasley, J. E. Mooij & T. P. Orlando, Phys. Rev. Lett. 42 (1979) 1165.
[18] D. R. Nelson, Phys. Rev. Lett. 60 (1988) 1973: D. R. Nelson & H. S. Seung, Phys. Rev. B39 (1989) 9153.

Superconducting Properties of $(GdCe)_4(LaBaSr)_4Cu_6O_{18.8}$

T. KANEKO, T. WADA, A. ICHINOSE, Y. YAEGASHI, S. IKEGAWA, and H. YAMAUCHI

Superconductivity Research Laboratory, International Superconductivity Technology Center, 10-13, Shinonome 1-chome, Koto-ku, Tokyo, 135 Japan

ABSTRACT

The superconducting properties of a 40 K class superconductor, $(Gd_{0.667}Ce_{0.333})_4(La_{0.334}Ba_{0.333}Sr_{0.333})_4Cu_6O_{18.8}$ (446 phase), were investigated by transport and magnetization measurements. The lower critical field, $H_{c1}(0)$, was obtained to be 110 Oe. The upper critical field, $H_{c2}(0)$, was estimated to be 72.7 T. The superconducting material parameters were derived using these values for $H_{c1}(0)$ and $H_{c2}(0)$: $\xi_{GL}(0)$=21 A, $\lambda_{GL}(0)$= 2680 A, κ_{GL}= 126 and $H_c(0)$= 4.1 kOe. Furthermore the effective magnetic moment per Gd ion was found to be nearly 7.97 μ_B which is the same as that of a Gd ion in the free ionic state.

KEY WORDS: $(Ln,Ce)_4(Ln,Ba)_4Cu_6O_z$ structure, superconducting material parameters, lower critical field, upper critical field, effective magnetic moment

INTRODUCTION

The new-type oxide superconductors with $T_c\sim$ 40 K discovered by Sawa and his coworkers[1] were represented by the chemical formula of $(Ln,Ce)_4(Ln,Ba)_4Cu_6O_z$ (Ln=Nd, Sm and Eu)("446" phase). The crystal structures of these superconductors are all perovskite-related[2] being similar to the triperovskite structure of $YBa_2Cu_3O_z$ [3, 4]. In our previous papers[5, 6], we reported on the syntheses and characterization of $(Nd_{0.500}(Ba_{1-x}Sr_x)_{0.334}Ce_{0.166})_8Cu_6O_z$ and $((Gd_{1-x}La_x)_{0.500}(Ba_{1-y}Sr_y)_{0.334}Ce_{0.166})_8Cu_6O_z$. In this paper we report on the magnetic properties of the superconductor, $(Gd_{0.667}Ce_{0.333})_4(La_{0.334}Ba_{0.333}Sr_{0.333})_4Cu_6O_{18.8}$, which has the highest critical temperature in our series of synthesized 446 superconductors, being measured by transport and magnetization methods.

SAMPLES

The samples were prepared from high pure La_2O_3, Gd_2O_3, CeO_2, $BaCO_3$, $SrCO_3$ and CuO powders. These powders were mixed in ethonol to form mixture with the nominal composition of $(Gd_{0.667}Ce_{0.333})_4(La_{0.334}Ba_{0.333}Sr_{0.333})_4Cu_6O_z$. The mixture was calcined at 1020 ℃ for 15 h in O_2 gas. The calcined powder was molded and sintered at 1030 ℃ for 15 h in O_2 gas. Then the sintered sample was annealed at 600 ℃ for 3 h and then at 400 ℃ for 3 h in a 1000kg/cm^2 mixed gas consisting of 80 % Ar and 20 % O_2. The last high pressure annealing procedure was effective in increasing the O content to yield good superconductors.
The phases present and the lattice parameters were determinded by powder X-ray diffraction using Cu-Kα radiation. The sample was single phase. We analyzed the crystallographic structure of this sample using the Rietveld method for the X-ray diffraction data. It is a tetragonal with the lattice parameters of a = 0.384 nm and c = 2.8375 nm as shown in Fig.1.
In this structure, we confirmed that Gd atoms preferentially occupied the A1 sites and La or Sr atoms occupied the A2 sites. We analyzed the oxygen content "z" of this

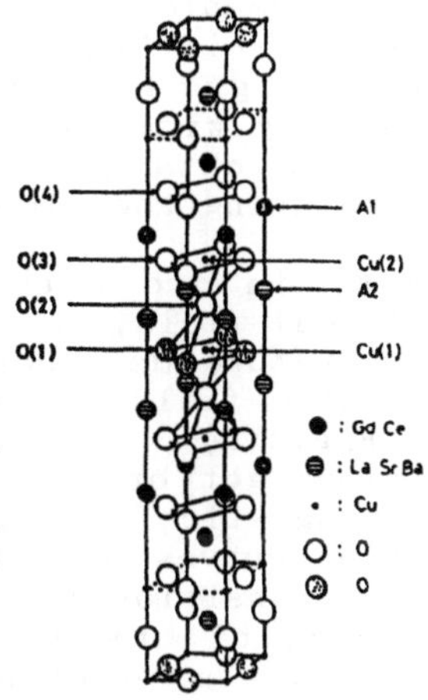

Fig.1 Crystal structure of $(Gd_{0.667}Ce_{0.333})_4(La_{0.334}Ba_{0.333}Sr_{0.333})_4Cu_6O_{18.8}$.

sample using an intert gas fusion nondispersive IR method (Horiba: Model EGMA-2800). The determined value for z was 18.8. It is thus that there are vacancies in the O(1) site defined in Fig.1.

RESULTS and DISCUSSION

The electrical resistivity of the samples was measured by a standard four-probe technique and the dc magnetic susceptibility by a SQUID magnetometer(QUANTUM DESIGN: Model MPMS). The temperature dependence of resistivity of the samples annealed at different level of O_2 partial pressure is shown in Fig.2(a). All the samples show metallic behavior and sharp superconducting transitions. The zero resistance temperature of the samples annealed in O_2 gas of 1, 50, 200 atm were 28, 35 and 39 K respectively. The higher the O_2 partial pressure during annealing was, the higher was T_c of resultant sample. The temperature dependence of dc magnetic susceptibility of the sample with the highest T_c is shown in Fig.2(b). The magnetic field used was 10 Oe. The measurement was done in field cooling. A sharp diamagnetic transition was observed. The superconducting onset temperature was magnetically determined 38 K. The magnitude of diamagnetism at 10 K was about 35 % of a full Meissner effect. The susceptibility saturated at about 20 K. That means there is no other phases with low T_c's contained in this sample.

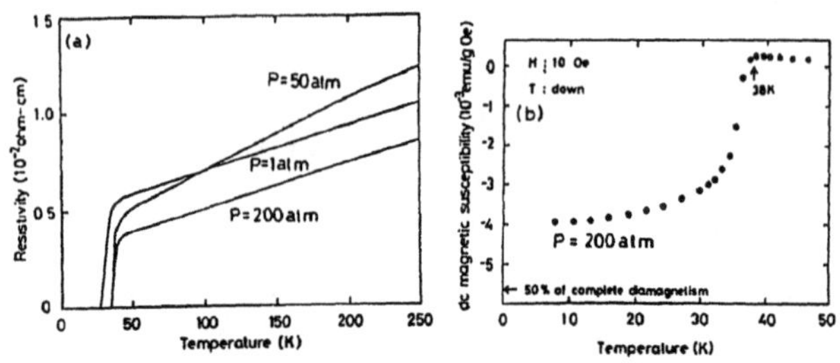

Fig.2 Temperature dependence of (a) the resistance and (b) the dc magnetic susceptibility for $(Gd_{0.667}Ce_{0.333})_4(La_{0.334}Ba_{0.333}Sr_{0.333})_4Cu_6O_z$ ceramics.

The lower critical fields, $H_{c1}(T)$, of this system were determined from the field dependences of magnetization by using the SQUID. The low-field magnetization curves are shown in Fig.3(a). Here, $H_{c1}(T)$ is defined as the field where the magnetization curve, M(H), first deviates from linearity. $H_{c1}(T)$ of the sample is plotted against temperature in Fig.3(b). It is observed in this figure that $H_{c1}(T)$ increases linearly with decreasing temperature. Then we obtain $H_{c1}(0)$ = 110 Oe by extrapolating the linear H_{c1}-vs-T plot to 0 K.

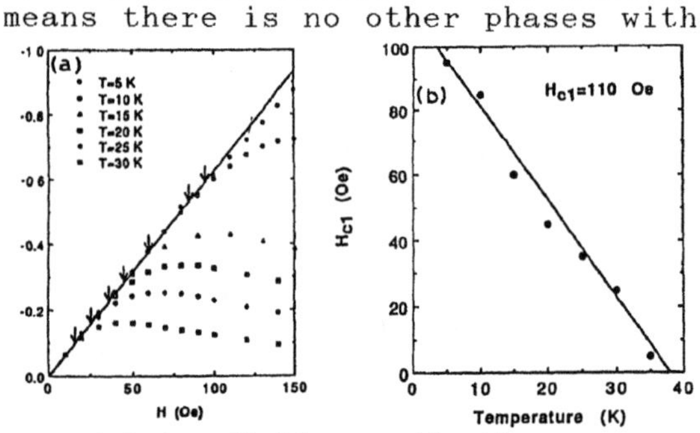

Fig.3 (a) Low-field magnetization curves (b) temperature dependence of $H_{c1}(T)$ for $(Gd_{0.667}Ce_{0.333})_4(La_{0.334}Ba_{0.333}Sr_{0.333})_4Cu_6O_{18.8}$.

Next, in order to estimate the upper critical field, $H_{c2}(0)$, we measured the temperature dependence of resistivity in magnetic fields. The results are shown in Fig.4(a). The superconducting transitions got broaden as the magnitude of the magnetic field increased. The transition curves have two steps. The two-step transition may not be due to the existence of any low T_c phases, because there was no evidence for such phases in the susceptibility data. The upper critical field may be difined as the field at which resistivity is equal to 80 %, 50 % or 20 % of the normal state resistivity at the superconducting onset temperature. The results are shown in Fig.4(b). The H_{c2}-vs-T curves of 50 % and 20 % criteria are not linear. It is likey that the nonlinearity of the two curves is due to the movement of the magnetic flux lines trapped in the sample. In order to check the magnetic

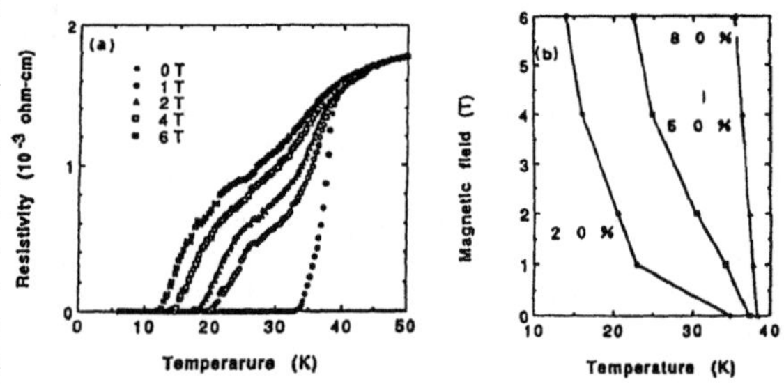

Fig.4 Temperature dependence of (a) the resistivity in magnetic fields and (b) upper critical field, $H_{c2}(T)$, for $(Gd_{0.667}Ce_{0.333})_4(La_{0.333}Ba_{0.333}Sr_{0.333})_4Cu_6O_{18.8}$.

flux line movements, Arrhenius plot of the electrical resistivity in magnetic fields from 1 to 6 T is shown in Fig.5. An approximately linear relationship between log R and 1/T is observed in the low resistivity region ($R < 10^{-4}$ Ω cm) as shown by dashed lines. We suspect that the slope of the curve indicates the thermal activation energy related to the magnetic flux line movement[7]. Therefore it is understood the H_{c2}-vs-T curves of 50 % and 20 % criteria may not represent an intrinsic material property. In the present analysis, we estimate the intrinsic $H_{c2}(0)$ of this "446" compound employing the slope for the 80 % resistivity criterion in the Werthamer-Helfand-Hohenberg formula for dirty type-II superconductor,[8]i.e.,

$$H_{c2}(0)=0.69\cdot T_c\cdot (dH_{c2}/dT_c)_{T_c}.$$

Thus, the upper critical field at 0 K obtained was 72.7 T. This value is very close to 71 T reported for $(LaSr)_2CuO_z$ which have similar $T_c\sim 40$ K [9]. It is also comparable to the critical field limited by the Pauli paramagnetism, H_p, of 70 T,i.e.

$$H_p=1.84\cdot T_c=1.84\cdot 38\ K.$$

Tamegai et al[10] reported that $H_{c2}(0)$ of $(Eu_{0.67}Ce_{0.33})_4(Eu_{0.33}Ba_{0.67})_4Cu_6O_z$ was 21.8 T. This value seems to be significantly lower than the present value, because it was determined by employing the slope for 50 % criterion.

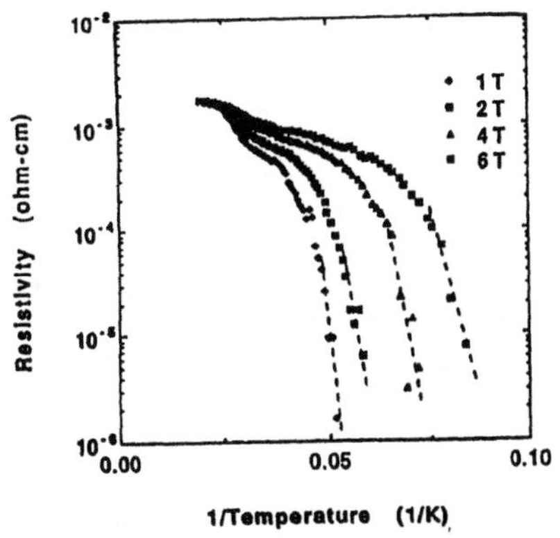

Fig.5 Arrhenius plot of the electric resistivity of the sample for magnetic field from 1 to 6 T. An approximetely linear relationship in the low resistivity region ($R<10^{-4}$ Ω cm) is shown by dashed line.

Table 1 Material parameters for $(Gd_{0.667}Ce_{0.333})_4(La_{0.334}Ba_{0.333}Sr_{0.333})_4Cu_6O_{18.8}$

magnetic T_c (K)	38.0
$[dH_{c2}/dT]_{Tc}$* (kOe/K)	26.7
$H_{c2}(0)$ (kOe)	727
$H_{c1}(0)$ (kOe)	0.11
$H_c(0)$ (kOe)	4.08
$\xi_{GL}(0)$ (Å)	21
$\lambda_{GL}(0)$ (Å)	2680
κ_{GL}	126
γ (mJ/(mol·K^2))	49.5

*:determined from 80 % resistivity criterion.

Now that $H_{c1}(0)$ and $H_{c2}(0)$ have been obtained, it is possible to estimste superconductivity material parameters of this "446" compound such as the GL coherence length, $\xi_{GL}(0)$, penetration depth, $\lambda_{GL}(0)$, GL parameter, κ, thermodynamical critical field, $H_c(0)$, and Sommerfeld parameter, γ. These quantities were calculated with the aid of the following relations[11],i.e,

$$H_{c2}(0)=(\Phi_0/2\pi\,\xi_{GL}(0)^2),\qquad H_{c2}(0)/H_{c1}(0)=2\kappa^2/\ln\kappa,$$

$$\kappa=\lambda_{GL}(0)/\xi_{GL}(0),\qquad H_{c2}(0)/H_c(0)=\sqrt{2}\kappa,\qquad \gamma=0.17(H_c(0)/T_c)^2,$$

The calculated values are given in Table 1. The unusually large value of κ indicates that this oxide is an extreme type-II superconductor.

We also measured the magnetic susceptibility of this sample in normal state by SQUID. The sample was annealed at 700 ℃ for 3 h in N_2 gas.

The temperature dependence of resistivity was semiconductor-like and didn't show superconductivity down to 4.2 K. The inverse susceptibility, χ^{-1}, is plotted in terms of temperature in Fig.6. The magnitude of magnetic field employed was 5000 Oe. Figure 6 shows that there is an approximately linear relationship between χ^{-1} and T. The data obey the following fomula consisting of the Curie-Weiss term and a constant Pauli-like contribution term.

$$\chi(T)=\chi_0+NP_{eff}^2\mu_B^2/(T+\theta),$$

where N is the number of magnetic ions per gram, P_{eff} is the effective number of Bohr magnetons and θ is the Curie Weiss temperature. The χ_0 is nearly zero for this sample.

We assume that Gd ions in this compound have the major contribution to the mag netic susceptibility. Then experimental value of P_{eff} for one Gd ion was derived to be 7.97. This value is in good agreement with the theoretical value of 7.94 for a Gd^{3+} free ion calculated by Hund's rule[12]. Furthermore the experimental value of P_{eff} for a Dy ion was obtained to be 10.58 that is in good agreement with the theoretical value of 10.63[12]. The Curie-Weiss temperatures, θ , were obtained to be 5.9 K and 5.2 K for the samples with Gd and Dy respectively. As a conclusion, the rare earth ion located between the Cu-O pyramids in the "446" compounds commonly have the following properties: (1) the value of of P_{eff} is close to that of the same ions in nearly free ionic state and (2) the positive Curie-Weiss temperature. It should be noted that similar magnetic properties have been reported for "123" compounds having rare earth ions in place of Y.

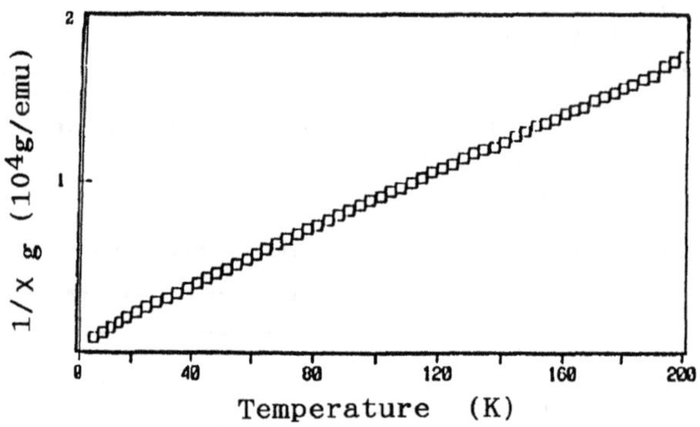

Fig.6 Inverse dc magnetic susceptibility χ^{-1} vs temperature for $(Gd_{0.667}Ce_{0.333})_4(La_{0.334}Ba_{0.333}Sr_{0.333})_4Cu_6O_z$ annealed at 700 ℃ for 3 h in N_2 gas.

REFERENCES

[1] Sawa H, Obara K, Akimitsu J, Matsui Y, Horiuchi S (1989) A New Family of Superconducting Cooper Oxide : $(Ln_{1-x}Ce_x)_2(Ba_{1-y}Ln_y)_2Cu_3O_{10-\delta}$ (Ln=Nd, Sm, Eu). J. Phys. Soc. Jpn. 58: 2252-2255.

[2] Galasso FS (1970) Structure and Properties of Inorganic Solids, Pergamon Press, New York.

[3] Izumi F, Asano H, Ishigaki T, Muromachi ET, Uchida Y, Watanabe N: (1987) Crystal Structure of the Orthorhombic Form of $Ba_2YCu_3O_{7-x}$ at 42 K. Jpn. J. Appl. Phys. 26: L1193-L1196.

[4] Cava RJ, Batlogg B, van Dover RB, Murphy DW, Sunshine S, Sierist T, Remeika JP, Reitman EA, Zahurak S Espinosa GP (1987) Bulk Superconduc tivity at 91 K in Single-Phase Oxigen-Deficient Perovskite $Ba_2YCu_3O_{9-\delta}$. Phys. Rev. Lett. 58: 1676-1679.

[5] Wada T, Ichinose A, Yaegashi Y, Yamauchi H, Tanaka S (in press) Prepara tion of New Oxide Superconductors, $(La, Gd, Ba, Ce)_8Cu_6O_z$ and $(La, Gd,Ba, Sr, Ce)_8Cu_6O_z$. Phys. Rev. B.

[6] Ichinose A, Wada T, Yaegashi Y, Yamauchi H, Tanaka S (1989) Preparation and Superconducting Properties of $[Ln, Ce, (Ba_{1-x}Sr_x)]_8Cu_6O_z$ (Ln=Nd, Sm and Eu). Jpn. J. Appl. Phys. 28: L1765-L1768.

[7] Palstra TTM, Batlogg B, Schneemeyer LF, Waszczak JV (1988) Thermally Ac tivated Dissipation in $Bi_{2.2}Sr_2Ca_{0.8}Cu_2O_{8+\delta}$. Phys. Rev. Lett. 61: 1662 -1665.

[8] Werthamer NR, Helfand E, Hohenberg PC (1965).

[9] Kobayashi N, Sasaoka T, Oh-ishi K, Sasaki T, Kikuchi M, Endo A, Matsuzaki K, Inoue A, Noto K, Syono Y, Saito Y, Matsumoto T, Muto Y (1987) Upper Critical Fields on High Temperature Superconductivity in La-Sr-Cu-O System. Jpn. J. Appl. Phys. 26: L358-L360.

[10]Tamegai T, Iye Y, Ogata M, Obara K, Akimitsu J (1989) Transport Properties of New High-Tc Superconductors $Ln_{2-x+y}Ce_xBa_{2-y}Cu_3O_{10-\delta}$ (Ln=Nd,Eu). Jpn. J. Appl. Phys. 28: L1537-L1540.

[11]Tinkaham M (1975) Introduction to superconductivity, McGraw-Hill, New York

[12]Kittel C (1986) Introduction to Solid State Physics, John Wiley & Sons,Inc, New York

Influence of Transition Metal Doping on Antiferromagnetic Order in the $Pb_2Sr_2(Ca, Y)Cu_3O_8$ System

HIROMI NIU, NOBURU FUKUSHIMA, HIROSHI KUBOTA, and KEN ANDO

Advanced Research Laboratory, Research and Development Center, Toshiba Corporation, Saiwai-ku, Kawasaki, 210 Japan

ABSTRACT

Changes in magnetic properties by Fe or Zn substitution in $Pb_2Sr_2Ca_{0.25}Y_{0.75}Cu_{3-x}TM_xO_8$ superconductor and in related antiferromagnetic insulator $Pb_2Sr_2YCu_{3-x}TM_xO_8$ were examined. Superconductivity was reduced with increasing Fe or Zn content, and Curie paramagnetism was observed. Apparent magnetic moment J per one Zn atom in $Pb_2Sr_2Ca_{0.25}Y_{0.75}Cu_{2.86}Zn_{0.14}O_8$ was 0.33. $Pb_2Sr_2YCu_{2.9}Zn_{0.1}O_8$ substituted Zn in AFI shows Curie paramagnetic character. Apparent J=0.4 for $Pb_2Sr_2YCu_{2.9}Zn_{0.1}O_8$ was derived. 0.66 Cu spins for $Pb_2Sr_2Ca_{0.25}Y_{0.75}Cu_{2.86}Zn_{0.14}O_8$ or 0.8 Cu spins for $Pb_2Sr_2YCu_{2.9}Zn_{0.1}O_8$ surrounding non-magnetic Zn^{2+} show paramagnetic character. Considering these results, disappearance of superconductivity by Fe or Zn dopig is attributed to destruction of fluctuated antiferromagnetically ordered Cu spins, which agrees with model in which localized Cu spins and antiferromagnetic order exist, even in the superconducting state.

KEY WORDS: superconductor, related antiferromagnetic insulator, paramagnetism

INTRODUCTION

Since the discovery of high-Tc superconductors, extensive studies to elucidate superconducting mechanism have been carried out. It is well known that antiferromagnetic insulator phases exist near high-Tc superconductor phases.[1),2),3)] Therefore, it is suggested that magnetism plays an important role in superconductivity of high-Tc superconductor. Recent investigations of magnetism in high-Tc superconductors using neutron-scattering [4),5),6)] and muon-spin-rotation techniques [7)] have indicated the coexistence of the fluctuated antiferromagnetic order and superconductivity. The relation between magnetism and superconductivity, however, has not been clarified, yet.

One way to examine the role of the spins on the Cu site for the superconductivity is the substitution by other transition metals with and without the magnetic moment for Cu, and that has been attempted in the $La_{2-x}Sr_xCuO_4$ [8)] and $YBa_2Cu_3O_{7-d}$ systems [9),10)]. In both systems, it has been reported that the superconductivity was reduced with the increase in the Fe or Zn content substituted for Cu. In the $YBa_2Cu_3O_{7-d}$ system, however, it was found that the disappearance of the superconductivity by the Fe substitution was attributed to the decrease in the hole concentration, as Fe^{3+} was substituted for Cu^{2+} without changing the oxygen content [11),12),13)]. The influence of Fe or Zn substitution on magnetism of the CuO_2 plane is an interesting problem. The experiments of Fe and Zn substitutions in those superconductors are not sufficient to explain the change in magnetism. As a new attempt to examine that change, we studied the Fe and Zn substitutions in the related antiferromagnetic insulator (AFI) $Pb_2Sr_2YCu_3O_8$. In this paper, we report about the changes in magnetization caused by Fe or Zn substitution in the AFI phase, in addition to the change in magnetization caused by these substitutions in $Pb_2Sr_2Ca_{0.25}Y_{0.75}Cu_3O_8$ superconductor .

EXPERIMENTAL

The $Pb_2Sr_2Ca_{0.25}Y_{0.75}Cu_{3-x}TM_xO_8$ (TM=Fe, Zn, $0 \leq x \leq 0.1$, x=0.14) and $Pb_2Sr_2YCu_{3-y}TM_yO_8$ (TM=Fe, Zn, $0 \leq y \leq 0.1$) samples were prepared by an ordinary solid-state reaction using PbO, $SrCO_3$, $CaCO_3$, Y_2O_3, CuO, Fe_2O_3 and ZnO as starting materials. They were mixed and ground, and then calcined at 750 C for 24 hours in air. After they were reground, they were pressed into the rectangular bars, and sintered at 850 C for 24 hours in N_2 atmosphere. X-ray diffraction analysis was performed using Philips PW 1710. The magnetic measurements for the obtained samples were carried out using HOXAN HSM-2000 and Quantum Design MPMS SQUID susceptometers.

RESULTS AND DISCUSSION

All samples obtained were single phase, as seen the X-ray diffraction pattern of $Pb_2Sr_2YCu_{2.9}Zn_{0.1}O_8$ in Fig.1. The slight changes in the lattice parameters of the Fe- or Zn-substituted samples and non-substituted sample were observed. It has been reported that there are two Cu sites in the crystal structure of $Pb_2Sr_2(Ca,Y)Cu_3O_8$ [14]. One Cu site is the CuO_2 plane which is common in a series of high-Tc superconductors. Another Cu site is an only-Cu layer sandwiched by the Pb-O layer. The latter Cu is +1 valent and has two oxygen coordinations. It is considered that Fe and Zn are selectively substituted for Cu in the CuO_2 plane, because Fe and Zn are difficult to coordinate two oxygens [15]. The below results are attributed to the changes in the properties by these substitutions for Cu in the CuO_2 plane. The temperature dependences of the magnetic susceptibility (H=10 gauss) for $Pb_2Sr_2Ca_{0.25}Y_{0.75}Cu_{3-x}Fe_xO_8$ and $Pb_2Sr_2Ca_{0.25}Y_{0.75}Cu_{3-x}Zn_xO_8$ are shown in Fig.2 and Fig.3, respectively. The superconductivity was reduced with the increase in Fe or Zn content. And no superconductivity could be observed at 0.06 mol Fe substitution or at 0.1 mol Zn substitution. After the superconductivity disappeared, the $Pb_2Sr_2Ca_{0.25}Y_{0.75}Cu_{3-x}Fe_xO_8$ (x=0.6, 0.8, 1.0) and $Pb_2Sr_2Ca_{0.25}Y_{0.75}Cu_{2.9}Zn_{0.1}O_8$ samples showed Curie paramagnetic characters. The Fe substitution reduces the superconductivity more quickly than the Zn substitution. The non-magnetic Zn, however, tends to destruct the superconductivity as well as the magnetic Fe.

The Fe- or Zn-substituted sample for the related antiferromagnetic insulator, the $Pb_2Sr_2YCu_{3-y}TM_yO_8$ (TM=Fe, Zn, $0 \leq y \leq 0.1$) systems, showed Curie paramagnetic characters below the room temperature, as shown in Fig.4. The Zn content dependence of the Curie constant is plotted in Fig.5. The paramagnetism of non-substituted sample seems to be attributed to the defect in the CuO_2 plane. The Curie constants were derived from χ value within the limits of 100 K -200 K assuming the following equation.

$$\chi = \chi_0 + C/T \qquad (1)$$

where χ_0 is the temperature-independent term. The χ_0 values are in the range of $\pm 10^{-10}$-10^{-8} and are smaller than the measurement resolution. It is thought that the effects of the Zn substitution on the antiferromagnetic order of the Cu spins are expressed in the C/T term in (1), because the Curie constant increases with increasing Zn content, as shown in Fig.5. It is suggested that one non-magnetic Zn^{2+} ion in the CuO_2 plane creates one spin-vacancy in the antiferromagnetic ordered Cu spins and causes the paramagnetic character of adjacent Cu spins. Figure 6 and Figure 7 show the magnetization (M) vs. the magnetic field (H) at 5 K for the $Pb_2Sr_2YCu_{2.94}Fe_{0.06}O_8$ and $Pb_2Sr_2YCu_{2.9}Zn_{0.1}O_8$ samples, respectively. The Fe- or Zn-substituted content in the related antiferromagnetic insulator $Pb_2Sr_2YCu_3O_8$ are the same content as the samples which caused the disappearance of the superconductivity in the

Fe- or Zn-substituted $Pb_2Sr_2Ca_{0.25}Y_{0.75}Cu_{3-x}TM_xO_8$ system, respectively. The M-H curves of the $Pb_2Sr_2YCu_{2.94}Fe_{0.06}O_8$ and $Pb_2Sr_2YCu_{2.9}Zn_{0.1}O_8$ samples keep linearity to about 2 T, and have no hysteresys. If Fe ions exist as an impurity phase such as Fe_2O_3 in the $Pb_2Sr_2YCu_{2.94}Fe_{0.06}O_8$ sample, the M-H curve will show a distinctive curve [16]. Furthermore, if Fe ions exist as a cluster in the $Pb_2Sr_2YCu_{2.94}Fe_{0.06}O_8$ sample, it will become a superparamagnet and the M-H curve will saturate at about 500 gauss. As the M-H curve of the $Pb_2Sr_2YCu_{2.94}Fe_{0.06}O_8$ sample does not satisfy the above mentioned factors, it is thought that Fe is substituted for Cu in $Pb_2Sr_2YCu_3O_8$. The slopes of the M-H curves for the $Pb_2Sr_2YCu_{2.94}Fe_{0.06}O_8$ and $Pb_2Sr_2YCu_{2.9}Zn_{0.1}O_8$ samples are smaller than the assumed value where all Cu spins in the CuO_2 plane contribute to the paramagnetism. It is suggested that the paramagnetic characters for $Pb_2Sr_2YCu_{2.94}Fe_{0.06}O_8$ and $Pb_2Sr_2YCu_{2.9}Zn_{0.1}O_8$ are attributed not to all Cu spins in the CuO_2 plane, but to a certain portion of the Cu spins in the CuO_2 plane. The apparent magnetic moment J value per one Zn atom is given from

$$M = \frac{n J (J+1) g^2 \mu_B^2 H}{3kT} \quad (2),$$

where M is total magnetic moment in a unit volume, g is Lande factor, and n is the number of particles. Assuming g=2 and $n=[Zn]=2.17*10^{20}$, the apparent J per one Zn atom in the $Pb_2Sr_2YCu_{2.9}Zn_{0.1}O_8$ sample is derived as 0.4. The paramagnetism of non-substituted AFI was subtracted as the background. When one non-magnetic Zn^{2+} ion exists in the CuO_2 plane, one spin-vacancy is created in the antiferromagnetic ordered Cu spins in the CuO_2 plane. It is thought that the antiferromagnetic order of the Cu spins surrounding the spin-vacancy is destructed and that 0.8 Cu spins surrounding one spin-vacancy show the paramagnetic character. The decrease in the Neel temperature with increasing the Zn content is expected, but it was not observed within the measured range.

On the other hand, the magnetic property was examined in regards to the $Pb_2Sr_2Ca_{0.25}Y_{0.75}Cu_{2.86}Zn_{0.14}O_8$ sample which did not show the superconductivity. The M-H curve of the $Pb_2Sr_2Ca_{0.25}Y_{0.75}Cu_{2.86}Zn_{0.14}O_8$ sample is illustrated in Fig.8. The apparent J=0.33 per one Zn atom was derived using the equation (2). It is thought that 0.66 Cu spins surrounding one spin-vacancy, that is created by one non-magnetic Zn^{2+} ion substitution for the antiferromagnetic ordered Cu site, show the paramagnetic character.

CONCLUSION

We examined the changes in the magnetic properties by Fe or Zn substitution both in the $Pb_2Sr_2Ca_{0.25}Y_{0.75}Cu_3O_8$ superconductor and in the related antiferromagnetic insulator $Pb_2Sr_2YCu_3O_8$. The superconductivity was reduced with increasing x in $Pb_2Sr_2Ca_{0.25}Y_{0.75}Cu_{3-x}TM_xO_8$ (TM=Fe, Zn) and the superconductivity disappeared at above x=0.6 for Fe substitution and at x=1.0 for Zn substitution. Then, the paramagnetism was observed . The apparent J value per one Zn atom, derived from the slope of the M-H curve for $Pb_2Sr_2Ca_{0.25}Y_{0.75}Cu_{2.86}Zn_{0.14}O_8$, was 0.33. On the other hand, the samples in the $Pb_2Sr_2YCu_{3-y}TM_yO_8$ (TM=Fe, Zn) systems showed Curie paramagnetic characters and Curie constants changed continuously with increasing Zn content. The apparent J per one Zn atom in the $Pb_2Sr_2YCu_{2.9}Zn_{0.1}O_8$ sample came to 0.4. When one non-magnetic Zn^{2+} ion exists in the CuO_2 plane, one spin-vacancy is created in the antiferromagnetically ordered Cu spins. Thus, the antiferromagnetic order of the Cu spins around the spin-vacancy is destructed. It is thought that 0.8 Cu spins adjacent to one spin-vacancy show the paramagnetic character.

Assuming the model in which the localized Cu spins and the antiferromagnetic order exist even in the superconductor, the paramagnetism in the $Pb_2Sr_2Ca_{0.25}Y_{0.75}Cu_{2.86}Zn_{0.14}O_8$ sample can be explained as the result of the local destruction of the antiferromagnetic order around doped Zn ions.

References
1) Y.Kitaoka, K.Ishida, S.Hiramatsu and K.Asayama: Jpn. J. Appl. Phys. 27 (1988) L734
2) N.Nishida, H.Miyake, D.Shimada, S.Okuma, M.Ishikawa, T.Takabatake, Y.Nakazawa, Y.Kuno, R.Keiel, J.H.Brewer, T.M.Riseman, D.L.Williams, Y.Watanabe, T.Yamazaki, K.Nishiyama, K.Nagamine, E.J.Ansald and E.Torikai: J. Phys.Soc. Jpn. 57 (1988)597
3) N.Nishida, H.Miyatake, S.Okuma, T.Tamegai, Y.Iye, R.Yoshizaki, K.Nishiyama and K.Nagamine: Physica C 156 (1988)625
4) R.J.Birgeneau, Y.Endoh, K.Kakurai, Y.Hidaka, T.Murakami, M.A.Kastner, T.R.Thurston, G.Shirane and K.Yamada: Phys. Rev. B 39 (1989) 2868
5) G.Shirane, R.J.Birgeneau, Y.Endoh, P.Gehring, M.A.Kastner, K.Kitazawa, H.Kojima, I.Tanaka, T.R.Thurston and K.Yamada: Phys. Rev. Lett. 63 (1989) 330
6) T.R.Thurston. R.J.Birgeneau, M.A.Kastner, N.W.Prewer, G.Shirane, Y.Fujii,K.Yamada, Y.Endoh, K.Kakurai, M.Matsuda, Y.Hidaka and T.Murakami: Phys. Rev. B 40 (1989) 4585
7) A.Weidinger, ch.Niederermayer, A.Golnik, R.Simon, E.Recknagel, J.I.Budnick, B.Chamberland and C.Baines: Phys. Rev. Lett. 62 (1989)102
8) G.Xiao, A.Bakhshai, M.Z.Cieplak, Z.Teasanovic and C.L.Chien: Phys. Rev. B 39 (1989) 315
9) J.M.Tarascon. P.Barboux, P.F.Miceli, L.H.Greene, G.W.Hull, M.Eibschutz and S.A.Sunshine: Phys. Rev. B 37 (1988) 7458
10) H.B.Tang, Y.Ren, Y.L.Liu, Q.W.Yan and Z.Zhang: Phys. Rev. B 39 (1989) 12291
11) H.Obara, H.Oyanagi, K.Murata, H.Yamasaki, H.Ihara, M.Tokumoto, Y.Nishida and Y.Kimara:Jpn. J. Appl. Phys. 27 (1988) L603
12) H.Oyanagi, H.Obara, H.Yamaguchi, K.Murata, H.Ihara, T.Matsushita, M.Tokumoto, Y.Nishihara and Y.Kimura: J. Phys. Soc. Jpn. 58 (1989) 2140
13) Y.Nishihara, H.Obara, T.Katayama, H.Oyanagi, K.Murata, Y.Kimura and K.Kajimura: J. Phys. Soc. Jpn. 58 (1989) 2925
14) R.J.Cava, B.Batlogg, J.J.Krajewski, L.W.Rupp, L.F.Schneemeyer, T.Siegrist, R.B.vanDover, P.Marsh, W.F.Peck,Jr, P.K.Gallagher, S.H.Glarum, J.H.Marshall, R.C.Farrow, J.V.Waszczak, R.Hull and P.Trevor: Nature 336 (1988) 211
15) R.D.Shannon : Acta Cryst. A32 (1976) 751
16) A.Tasaki and S.Iida: J. Phys. Soc. Jpn. 18 (1963) 1148

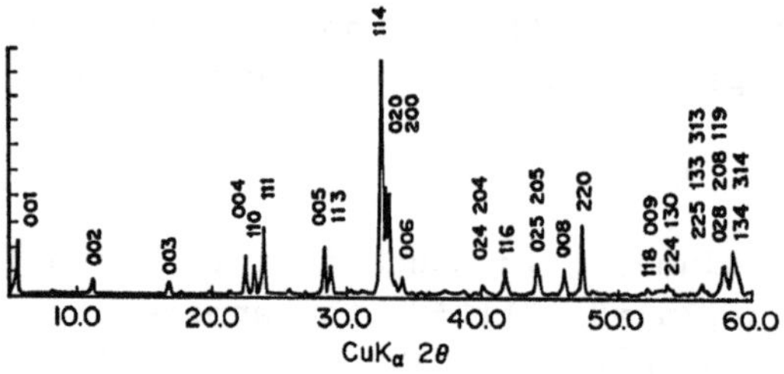

Figure 1 The X-ray diffraction pattern for $Pb_2Sr_2YCu_{2.9}Zn_{0.1}O_8$

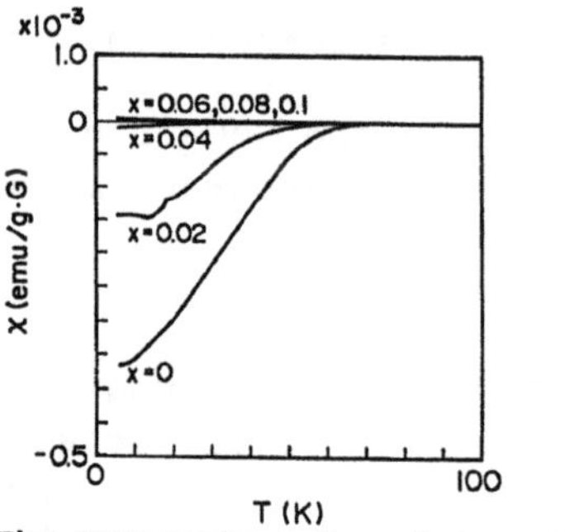

Figure 2 The temperature dependence of the magnetic susceptivility for the $Pb_2Sr_2Ca_{0.25}Y_{0.75}Cu_{3-x}Fe_xO_8$ system

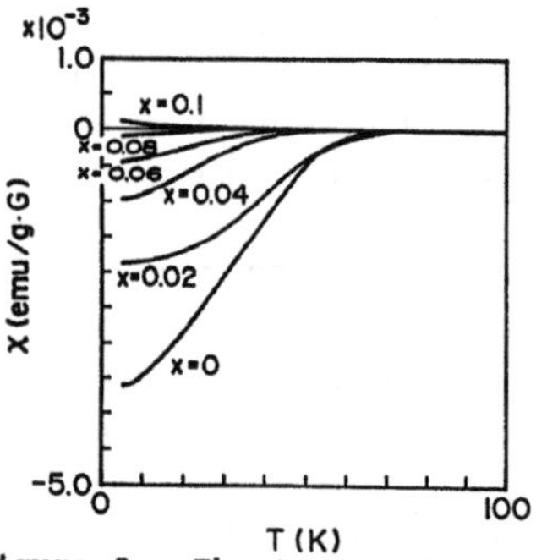

Figure 3 The temperature dependence of the magnetic susceptivility for the $Pb_2Sr_2Ca_{0.25}Y_{0.75}Cu_{3-x}Zn_xO_8$ system

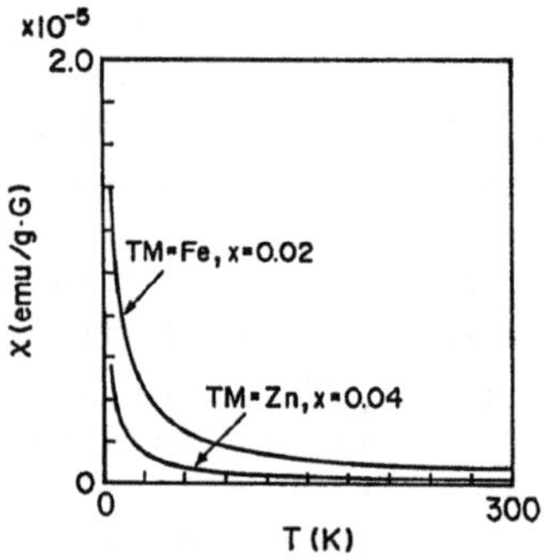

Figure 4 The temperature dependences of the magnetic susceptivility for $Pb_2Sr_2YCu_{2.98}Fe_{0.02}O_8$ and $Pb_2Sr_2YCu_{2.96}Zn_{0.04}O_8$

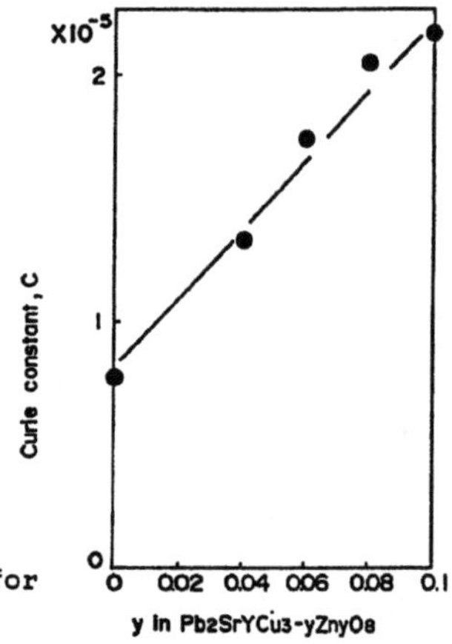

Figure 5 The Zn contents dependence of the Curie constant in the $Pb_2Sr_2YCu_{3-y}Zn_yO_8$ system

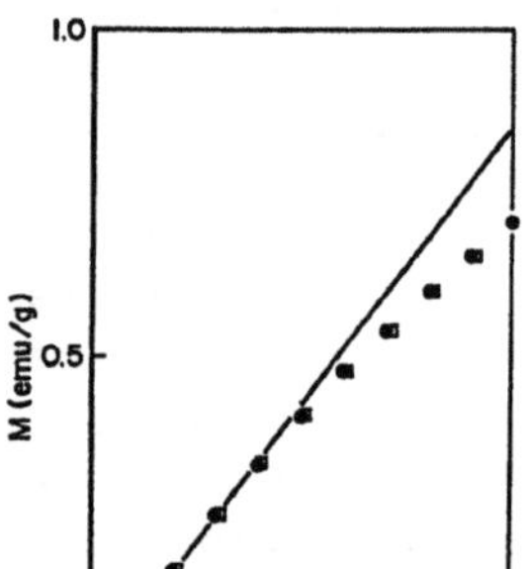

Figure 6 The magnetization (M) vs the magnetic field (H) at 5K for $Pb_2Sr_2YCu_{2.94}Fe_{0.06}O_8$

● increasing the magnetic field

□ decreasing the magnetic field

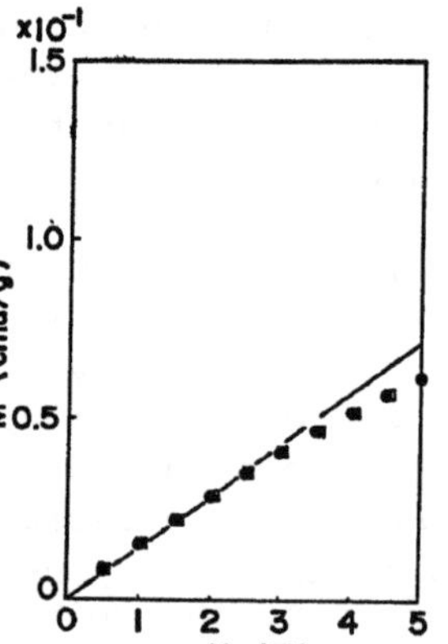

Figure 7 The magnetization (M) vs the magnetic field (H) at 5K for $Pb_2Sr_2YCu_{2.9}Zn_{0.1}O_8$

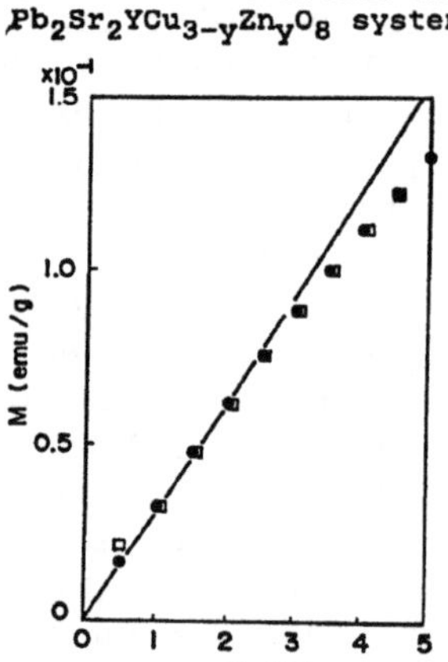

Figure 8 The magnetization (M) vs the magnetic field (H) at 5K for $Pb_2Sr_2Ca_{0.25}Y_{0.75}Cu_{2.86}Zn_{0.14}O_8$

Direct Observation of Superconducting Magnetic Fluxons Using Electron Holography

S. HASEGAWA[1], T. MATSUDA[1], J. ENDO[1], A. TONOMURA[1], and R. AOKI[2]

[1] Advanced Research Laboratory, Hitachi, Ltd., Kokubunji, Tokyo, 185 Japan
[2] Faculty of Engineering, Osaka University Suita, Osaka, 565 Japan

ABSTRACT

The magnetic flux lines of a quantized flux (fluxon) penetrating through a superconducting Pb film were observed directly and individually by the electron holography technique using the Aharonov-Bohm effect. The digital phase analysis at the optical reconstruction stage in electron holography confirms the quantized flux value h/2e, and has potentiality to analyze the fluxon core structre in detail.

KEY WORDS: fluxon, flux line lattice, electron holography
fringe scanning interferometry, Aharonov-Bohm effect

INTRODUCTION

The essential nature of superconductivity has been unveiled through its peculiar magnetic behaviors, the Meissner effect, flux quantization, flux line lattice structures, and so on. They were phenomenologically understood by the Ginzburg-Landau theory, and recognized to be closely related to the fundamental of superconductivity through the Gor'kov theory besed on the BCS theory.

Although a great amount of efforts have been made to clarify the high-T_c superconducting mechanism since its discovery, we have not yet reached a consistent understanding. As in the case of low T_c superconductivity, important information on the high T_c superconducting mechanism is expected to be brought through their magnetic structure analysis.

Dynamical behaviors of fluxons also play an important role in the transport characteristics, limitting the critical current. Especially in the high T_c superconductors, the flux creep, flux line lattice melting, and fluxon-antifluxon pairs are expected to dominatly affect the transport characteristics.

Electron holography, invented by D. Gabor to improve the resolution of electron microscopes, has been realized with an electron microscope equipped with a cold field emission electron gun, which produces highly coherent electron beams, and an electron biprism. Its new applications have been developed, direct observation of magnetic field with high spatial resolution. Magnetic field can be revealed in the form of magneic flux line distribution[1]. We have successfully employed the electron holography technique for the direct observation of superconducting magnetic fluxons. Although this technique has a potentiality of dynamical observation of fluxons, we focus ourselves here on the static observation.

OBSERVATION OF MAGNETIC STRUCTURES IN SUPERCONDUCTORS

Various kinds of experimental techniques have been employed to investigate the magnetic structures in superconductors[2]. We summarize them based on their spatial resolution and the sensitivity for magnetic flux. Figure 1 shows rough estimates of availability of typical experimental methods.

The shadowed area covers the resolution and sensitivity necessary for observing the mixed state in type II superconductors. The Bitter method has been most widely used, including its recent applications to high T_c superconductors[3], only for qualitative discussions. The neutron diffraction method allows the very quantitative analysis on the flux structures, only when the flux is periodically distributed . Recent observations using scanning tunneling microscopy reveal a flux line lattice[4], which, however, does not probe the magnetic flux itself, but the electronic structure surrounding the fluxon at the surface. Electron holography enables one to directly observe the fluxons very quantitatively with high spatial resolution and analyze the individual fluxon core structure, even when the flux is not periodically distributed[5].

ELECTRON HOLOGRAPHIC OBSERVATION

The principle underlying the electron holographic observation of magnetic flux is the Aharonov-Bohm effect[6]. It predicts that the magnetic flux Φ causes the phase shift $\Delta\phi$ beween the electron waves passing through the different sides of the flux (Fig. 2):

$$\Delta\phi = 2\pi \frac{\Phi}{(h/e)}$$

A single flux quantum h/2e, therefore, causes the phase shift of π. Electron holography makes it possible to explicitly measure the phase of electron wave. When we draw the contour phase lines (the interference fringes) at a phase interval of π, they directly correspond to the magnetic flux lines in units of h/2e.

Our superconducting films were prepared by vacuum evaporation of Pb on one side of a tungsten wire of 30 μm-diameter. The critical temperature of the Pb film was 7.2 K and the residual resistance ratio between T=300K and 7.5K was 50~80. The film was cooled down to 4.5K to be superconducting with fluxons under a transverse magentic field of 3G in our holography electron microscope (Fig. 3). We can regard the illuminating electron wave as a plane wave. Transmitting through the sample region, the wavefront is deformed by the magnetic field; the localized field of a fluxon causes abrupt phase change, although the wavefront passing far from the Pb film is inclined smoothly because of a uniform field. By electron biprism action, the transmitted wave is devided into two parts, superimposed, and interfered to each other. Interference fringe is recorded on a hologram.

After developing and fixing the hologram, it was set in a laser interferometer (Fig. 4). The He-Ne laser beam is devided into two beams, and each irradiates the hologram with a different angle. A set of the $\pm$first-order diffracted waves emerges for each illuminating beam. Only the first-order diffracted wave from one beam and the $-$first-order one from the other beam are selected, and are made interfere with each other to form a interference micrograph. The interference fringes in this image are contour phase lines of π interval, which correspond to the magentic flux lines in units of a single fluxon.

Figure 5 shows the interference micrographs thus obtained. The lower black parts are the films and the upper, vacuum. The fringes show magnetic flux lines that fan out into free space after penetrating through the superconductor. In the case of Pb film of 1.0 μm thickness, the penetrating flux is a bundle of several fluxons, which is the type-I behavior. When the film thickness decreases, 0.2 μm, the flux becomes a single fluxon, the type-II behavior. We also observed fluxon-antifluxon pairs. They may have been created when the film was cooled through the Kosterlitz-Thouless regime, just below the T_c[7], and pinned so that the opposite fluxons would have not met to annihilate each other. The polarity of the flux cannot be distinguished with any other methods but electron holography.

The electron wavefront itself can be reconstructed from the hologram using "fringe scanning interferometry"[8]. Stepwise movement of the mirror A in Fig. 4, driven by a piezoelectric transducer(PZT), causes a fringe shift in the ineterference micrograph. Images at four different mirror positions, of which position interval is $\lambda/8$ (λ is the wavelength of the laser beam), were synchronously stored through a TV camera in a computer. The phase value at each pixel on the image was calculated from the brightness values at the corresponding pixel in the four images, and the original electron wavefront is numerically reconstructed.
The wavefront reconstructed from the hologram of 1.0 μm Pb film (Fig. 5(a)) is 3-dimensionally shown in Fig. 6, which is an expected one in Fig. 3. The abrupt phase shifts at the flux exits on the surface are multiples of π, and their multiples are the number of fringes in Fig. 5(a). This precisely means the flux quantization in units of h/2e. From the curvature of the wavefront at the fluxon root, the fluxon core structure can be analyzed in detail. Such analysis and its theoretical simulation are now in progress.
We here used a low T_c superconductor, Pb, to check the availability of our electron holographic method for research on superconductors. We are now carrying out the experiment on high-T_c superconductors, not only in the static, but also dynamical manner.

REFERENCES

1. Tonomura A. (1987) Applications of electron holography. Rev Mod Phys 59:639-669.
2. Huebener RP. (1979) Magnetic flux structures in superconductors. Springer-Verlag.
3. Gammel PL. Bishop DJ. Dolan Gj. Kwo JR. Murray CA. Schneemeyer LF. Waszczak JV. (1987) Observation of hexagonally correlated flux quanta in $YBa_2Cu_3O_7$. Phys Rev Lett 59: 2592-2595.
4. Hess HF. Robinson RB. Dynes RC. Valles JM. Waszczak JV. (1989) Scanning-tunneling-microscope observation of the Abrikosov flux lattice and the density of states near and inside a fluxoid. Phys Rev Lett 62: 214-216.
5. Matsuda T. Hasegawa S. Igarashi M. Kobayashi T. Naito M. Kajiyama H. Endo J. Osakabe N. Tonomura A. Aoki R. (1989) Magnetic field observation of a single flux quantum by electron-holographic interferometry. Phys Rev Lett 62: 2519-2522.
6. Aharonov Y. Bohm D. (1959) Significance of electromagnetic potentials in the quantum theory. Phys Rev 115: 485-491.
7. Kosterlitz JM. Thouless D. (1973) Ordering, metastability and phase transitions in two-dimensional systems. J Phys C6: 1181-1203.
8. Hasegawa S. Kawasaki T. Endo J. Tonomura A. Honda Y. Futamoto M. Yoshida K. Kugiya F. Koizumi M. (1989) Sensitivity-enhanced electron holography and its application to magnetic recording investigations. J Appl Phys 65: 2000-2003.

FIGURE CAPTIONS

Fig. 1. Experimental methods to observe magnetic flux structures in superconductors.
Fig. 2. Phase shift of electron wave caused by magnetic flux.
Fig. 3. Electron wavefront deformation in a holography electron microscope.
Fig. 4. Optical reconstruction system with fringe scanning interferometry.
Fig. 5. Interference micrographs directly showing magneic fluxons.
Fig. 6. Electron wavefront reconstructed from the hologram by fringe scanning interferometry.

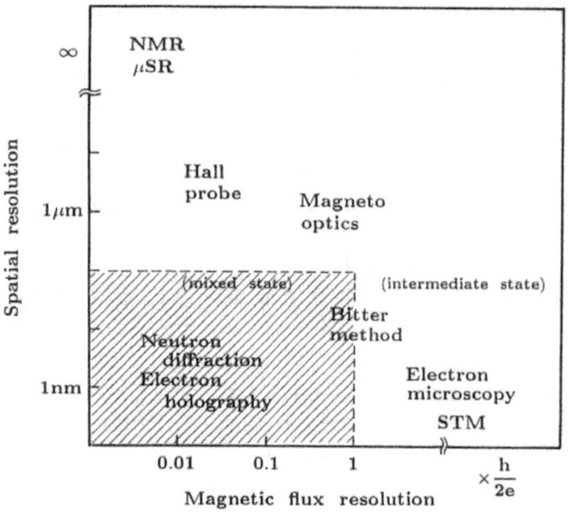

Fig. 1.

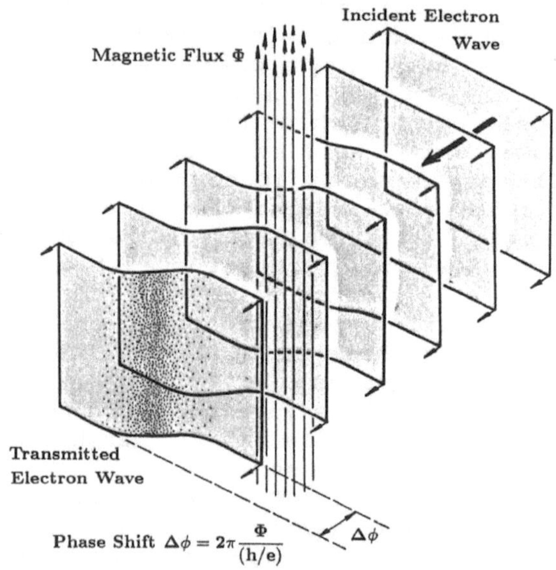

Fig. 2.

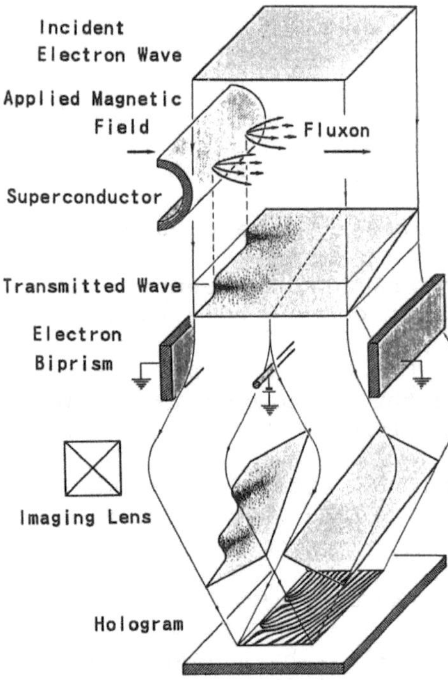

Fig. 3.

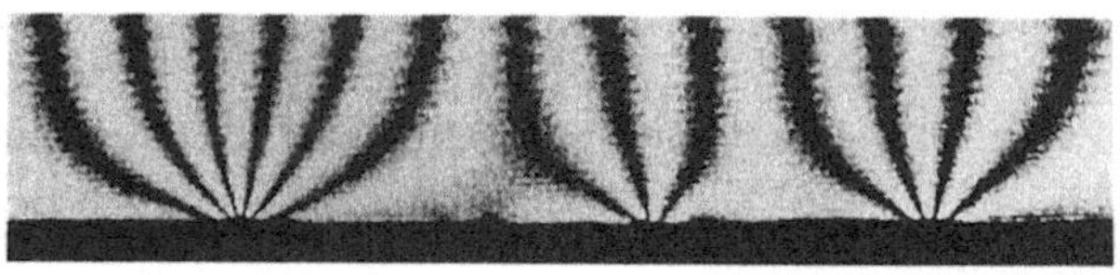

(a) Pb film of 1.0 μm thickness.

(b) Pb film of 0.2 μm thickness.

Fig. 5.

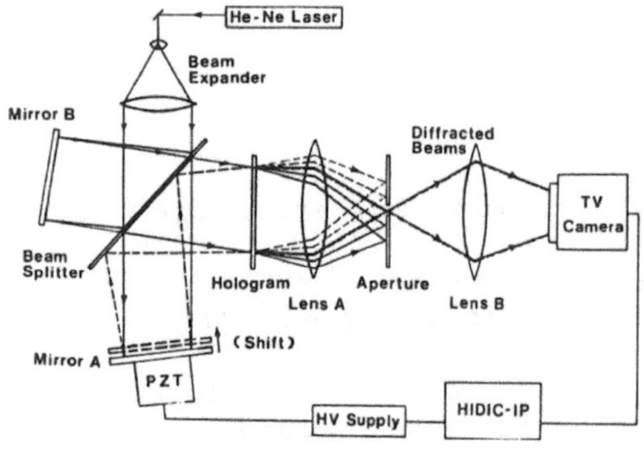

Fig. 4.

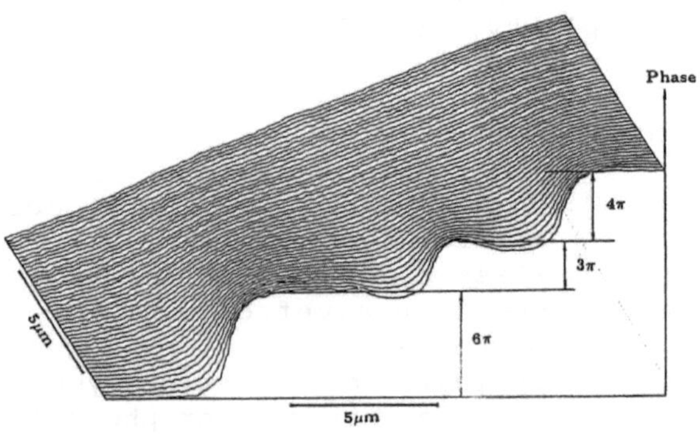

Fig. 6.

3.8 Flux Creep and Flux Motion

Magnetization Relaxation and Resistive Behaviour of High-T_c Superconductors

R. GRIESSEN[1], R.J. WIJNGAARDEN[1], and B. DAM[2]

[1] Natuurkundig Laboratorium, Vrije Universiteit, 1081 HV, Amsterdam, The Netherlands
[2] Natuurkundig Laboratorium, Philips, P.O. Box 80000, 5600 JA Eindhoven, The Netherlands

ABSTRACT

From the analysis of magnetization relaxation data on various types of samples and Monte-Carlo calculations Hagen and Griessen concluded that the existing experimental data can be understood in terms of a model which incorporates a distribution $m(U^*)$ of activation energies for thermally activated flux motion. This model is extended to include flux-flow in order to reproduce the characteristic features of current-voltage curves of high-T_c superconductors. In particular the recent data of Koch et al. and Zeldov et al. can be explained without invoking a continuous phase transition (freezing into a superconducting vortex-glass phase) or a logarithmic current dependence of the activation energy.

INTRODUCTION

Since the discovery of high-T_c superconductivity a wealth of information has been gathered about the resistive transition near T_c in applied magnetic fields. In early experiments the R(T) curves showed a smooth, structureless transition from a normal state value above T_c to essentially zero at lower temperatures. Tinkham [1] first succeeded in describing quantitatively the broadening of R(T) curves measured by Iye et al. [2] in a magnetic field. However, as the quality of single crystals and thin films improved a knee became apparent in the R(T) curves in high magnetic fields. This characteristic feature has recently be interpreted by Hebard and Palstra [3], Batlogg et al. [4] and Malozemoff et al. [5] as a crossover from thermally activated flux-creep at low temperatures to viscous flux-flow just below T_c.
Recently Koch et al. [6] and Zeldov et al. [7] investigated in detail the temperature dependence of the electric field-current density curves (E-j), respectively the resistivity-current density curves (ρ-j) of laser ablated epitaxial films. For a given magnetic field Koch et al. found the E-j-isotherms to exhibit power-law behaviour at a given temperature, T_g, which is identified as a second order phase transition between a normal and a true superconductor (with resistivity identically zero). Further the value of the critical exponents determined from the experimental data are claimed to be indicative of a transition into a vortex-glass superconductor. On the same type of epitaxial films Zeldov et al. [7] observed a power-law behaviour over three decades in ρ-j curves at temperatures between 77 and 81 K and current densities between 10^8 and 10^{10} A/m^2. These data are shown by Zeldov et al. to be consistent with a thermally activated flux-creep model with a current dependent activation energy $U(T,H,j) = U(T,H)\ln(j/j_0)$ where j_0 is the current for which U approaches zero. The parameter j_0 is both field and temperature dependent. We have recently shown [8] that both assumptions i) the existence of a continuous phase transition to a glassy superconductor and ii) a logarithmic current dependence of the activation energy are unnecessary if one takes into account that even good quality epitaxial films of $YBa_2Cu_3O_7$ are probably still fairly disordered systems. Before describing the model we review some results obtained from the analysis of the temperature dependence of giant-flux creep in polycrystalline and single-crystalline $YBa_2Cu_3O_7$.

PACS. 74.60.Ge, 74.70.Vy, 74.75.+t

MAGNETIZATION RELAXATION

On the basis of Monte-Carlo simulations Hagen et al. [9,10] concluded that the decay of the magnetization M(t,T,B) in a critcal state at t = 0 via thermally activated flux-motion was well described by the following relation (even when $U(T,B) \simeq kT$)

$$M(t,T,B) = M(0,T,B)\left[1 - \frac{kT}{U(T,B)} \ln \left(1 + \frac{t}{\tau} \right) \right] \tag{1}$$

where U(T,B) is the temperature and field dependent activation energy barrier separating two pinning regions and τ is related to the attempt frequency for a flux-line to jump from one pinning region to an adjacent one. The value of the magnetization at t = 0, temperature T and magnetic field B is M(0,T,B). Eq. 1 is valid as long as $t < t^*$ with $t^* = \tau[\exp(U(T,B)/kT)-1]$. For $t < \tau$

$$M(t,T,B) = M(0,T,B)\left[1 - \frac{kT}{U(T,B)\tau} t \right] \tag{2}$$

i.e. it decreases linearly with time while for $t \gg \tau$ it exhibits the usual logarithmic time dependence. In the limit of very long times, $t > t^*$, M(t,T,B) decreases exponentially to zero [9,10].
As discussed by Hagen and Griessen [9], the temperature dependence of the decay rate dM/dlnt cannot be explained by means of a single activation energy model. It is necessary to introduce a distribution $m(U^*)$ of activation energies U^* in such a way that $m(U^*)dU^*$ is the fraction of the pinning regions with a pinning energy between U^* and $U^* + dU^*$ at T = 0 and B = 0. The actual activation energy U(T,B) is related to U^* by

$$U(T,B) = b(T,B)U^* \tag{3}$$

as we assume that the temperature and field dependence of the activation energy is independent of its actual strength U^*. By definition b(0,0) = 1. From eq. 3 and the normalization condition

$$\int_0^\infty m(U^*)dU^* = 1 \tag{4}$$

follows that the distribution $\tilde{m}(U;T,B)$ of activation energies U at temperature T and field B is given by

$$\tilde{m}(U;T,B) = m\left(\frac{U}{b(T,B)} \right) \frac{1}{b(T,B)} \tag{5}$$

$$\text{with} \int_0^\infty \tilde{m}(U;T,B)dU = 1 \tag{6}$$

Since b(T,B) vanishes on the B_{c2} (T) line in the T-B-plane the distribution $\tilde{m}(U;T,B)$ is compressed to lower activation energies when the temperature increases (at constant field) or when B approaches $B_{c2}(T)$ at constant temperature T. A family of distributions $\tilde{m}(U;T,B)$ for various values of b, which derive all from the same $m(U^*)$ is indicated in Fig. 1 for the case of a log-normal distribution function

$$m(U^*) = \sqrt{\frac{\gamma}{\pi}} \frac{e^{-\frac{1}{4\gamma}}}{U_p^*} \exp\left[-\gamma \left(\ln\left(\frac{U^*}{U_p^*} \right) \right)^2 \right] \tag{7}$$

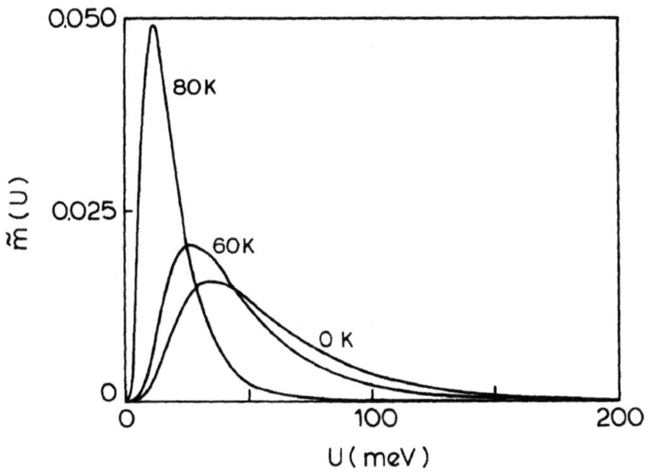

Fig. 1 Log-normal distribution function $\tilde{m}(U)$ of activation energies $U = U^*b(T,B)$ (see eq. 5) for B = 4T and T = 0, 60 and 80 K. All these curves derive from the same $m(U^*)$ distribution where U^* is the activation energy at 0 K and zero field. The parameters for $m(U^*)$ are $U_p^* = 38$ meV and $\gamma = 1.4$

In refs. 10 and 11 it is shown that the distribution function $m(U^*)$ and b(T,B) at fixed B can be unambiguously determined from M(t,T,B)-experimental data by means of an inversion scheme for the integral equation

$$M(t,T,B) = \int_{U_o^*(t,T,B)}^{\infty} m(U^*) \left[1 - \frac{kT}{U^*b(T,B)} \ln\left(1 + \frac{t}{\tau}\right) \right] dU^* \tag{8}$$

The lower integration limit $U_o^*(t,T,B) = \dfrac{kT}{b(T,B)} \ln\left(1 + \dfrac{t}{\tau}\right)$ differs from zero because at the time of observation t all the regions with $U^* < U_o^*$ have fully relaxed (i.e. M(t) tends to zero asymptotically) <u>before</u> the first data point is taken.
For polycrystalline $YBa_2Cu_3O_7$ as well as a single crystal of the same material it is found that $m(U^*)$ resembles strongly a log-normal distribution function (see. Eq. 7) although no a priori analytical form for $m(U^*)$ is introduced in the inversion scheme. Furthermore, for B = 0.1 T it is found that the requirement of a constant characteristic relaxation time τ can only be satisfied for $b(T,B) = 1-\theta^4$ with $\theta = T/T_c$. This suggests that $U(T,B) \propto H_c^2\xi^2$ as $H_c \propto 1-\theta^2$ and $\xi \propto [1+\theta^2/(1-\theta^2)]^{\frac{1}{2}}$. As $b \to 0$, $U_o^* \to \infty$ and this explains why large activation energies (typically of order 1 eV) are determined from experiments near T_c by various authors. On the other hand when $T \to 0$, all relaxation processes contribute to M(t,T,B) and the average activation energy is of the same order of magnitude as the energy U_p^* corresponding to the maximum in $m(U^*)$.

THE PARALLEL RESISTOR MODEL

From the existence of a relatively broad distribution of activation energies Hagen and Griessen conclude that single crystals have still a fair amount of disorder. Further as the width of m(U*) for single crystals and polycrystalline ceramic samples are similar, it is clear that this disorder arises from intragrain rather than intergrain properties.
In the description of magnetization relaxation we assumed that each pinning domain with a given U^* decayed with time independently from the other domains. For the calculation of current-voltage curves such an approximation is not realistic as an overall transport current is imposed to flow through a given sample. In a disordered system the current flow exhibits a complicated pattern and the calculation of E-j-curves is a formidable task. In this work we drastically simplify the problem by assuming that all domains are electrically in parallel, i.e. the voltage drop over each domain is equal to the applied voltage. From calculations and experiments of resistivity measurements on segregated alloys it is known that the main features of the resistive behavior of disordered systems can be understood by means of such a parallel resistor model [12,13]. This is evidently not true for the series resistor model in which the current is forced equally through all domains regardless of the dissipation level in a given domain. In a forthcoming publication [13] the predictions of the parallel resistor model will be compared to calculations on a network of random non-linear resistors.

In the present model each domain is characterized by a temperature and field dependent activation energy U(T,B) which separates adjacent pinning regions in a given domain. In the flux-creep regime the drift velocity of flux-lines is given by the usual expression [14]

$$v_c = v_o e^{-\frac{U(T,B)}{kT}} \sinh\left(\frac{Aj}{kT}\right) \qquad (9)$$

where v_o is a velocity prefactor related to the attempt frequency for flux-line hopping, j is the local current density and Aj is the change in the energy of a flux-line associated with the Lorentz force acting on a vortex [15].
In the flux-flow regime the velocity of flux-lines is limited by a viscous drag, in which the simple Stephen-Bardeen [16] model, is given by

$$v_f = \frac{\rho_n}{B_{c_2}(T)} j \qquad (10)$$

where ρ_n is the normal state resistivity and $B_{c2}(T)$ the upper critical field [17] at temperature T, i.e. $B_{c2}(T) = B_{c2}(0)(1-(T/T_c)^2)$.
A smooth interpolation from the weak driving force (flux-creep) regime to the strong driving force regime (flux-flow) is obtained by assuming that the motion of a flux-line is governed by flux-creep (Eq. 9) over a distance L_c and by flux-flow over a distance L_f. The average velocity is then given by

$$\langle v\rangle = \frac{L_c + L_f}{L_c/v_c + L_f/v_f} \qquad (11)$$

Near T_c in zero field or, in fields near $B_{c2}(T)$ the system is in the flow-regime as the activation energy U vanishes when $B = B_{c2}(T)$. Consequently $L_c(B_{c2}(T)) = 0$. For simplicity we assume that $L_c \propto (B_{c2}(T)-B)$ with $L_c \ll L_f$ and by using Eqs. 9 and 10 we obtain finally that the electric field E set up by the motion of the flux-lines in a magnetic field B is

$$E = \frac{1}{S \exp\left(\frac{U(T,B)}{kT}\right)\left[\sinh\left(\frac{Aj}{kT}\right)\right]^{-1} + \frac{B_{c2}}{Bj\rho_n}} \qquad (12)$$

with $S = (L_c/L_f)v_o^{-1}B^{-1} = S_0(B_{c2}(T)/B-1)$.
Expression (12) has the correct limiting behavior for $j \to 0$ and $j \to \infty$, if $S \exp[U(T,B)/kT] \gg B_{c2}(T)/B\rho_n)$. At low current densities the creep-term dominates while at current densities $j \gg U(T,B)/A$ the sinh-term increases so rapidly that the flow-term becomes the dominant term in the denominator of Eq. 12.
So far we have only considered flux motion within a single domain. In the spirit of our analysis of magnetization relaxation data we consider now a sample consisting of a distribution of domains which are coupled electrically in parallel. For all domains E is then equal to the applied electric field. This implies that the velocity of flux-lines is the same everywhere irrespective of the local values of the activation energies U(T,B). The average current density through the sample is given by

$$\langle j\rangle = \int_{U^*_{min}}^{U^*_{max}} j(U^*,E,T,B)m(U^*)dU^* \qquad (13)$$

where $j(U^*,E,T,B)$ is the current density in a domain with activation energy U^* (at T = 0 and B = 0) obtained by solving Eq. 12 for j. For the distribution $m(U^*)$ we take a log-normal function (see Eq. 7) as this type of function resembles strongly the distribution derived from giant flux-creep experiments by means of the Hagen-Griessen inversion scheme [10,11]. The integration limits in Eq. 13 are determined by the requirement that pinning regions cannot be arbitrarily small. To be physically meaningful, in our model, the size of a pinning region must be larger than the coherence length. This implies that for a filmstrip of typically 20 μm width (as used by Koch et al. and Zeldov et al.) and $\xi \approx 5$ Å, $m(U^*) \gtrsim 10^{-5}$ meV^{-1}. In Eq. 13, U^*_{min} and U^*_{max} are chosen to satisfy the condition $m(U^*_{min}) = m(U^*_{max}) = 10^{-5}$ meV^{-1}.

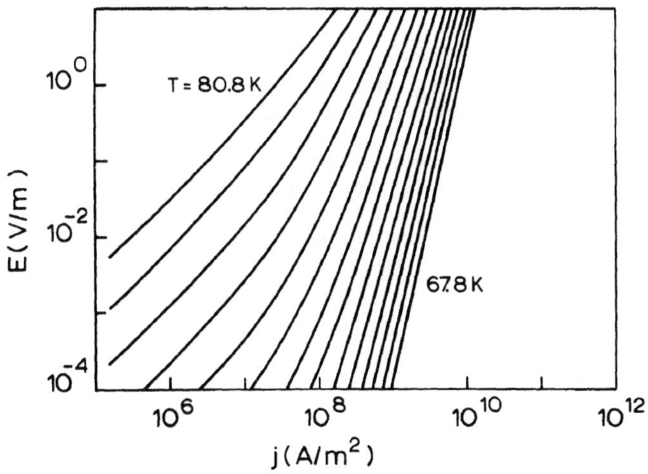

Fig. 2 Electric field versus current density curves calculated by means of eq. 13 with the distribution m(U*) of activation energies indicated in Fig. 7a for temperatures between 67.8 K and 80.8 K at 1 K interval in a magnetic field of 4 T. For comparison with experimental data, see ref. 6.

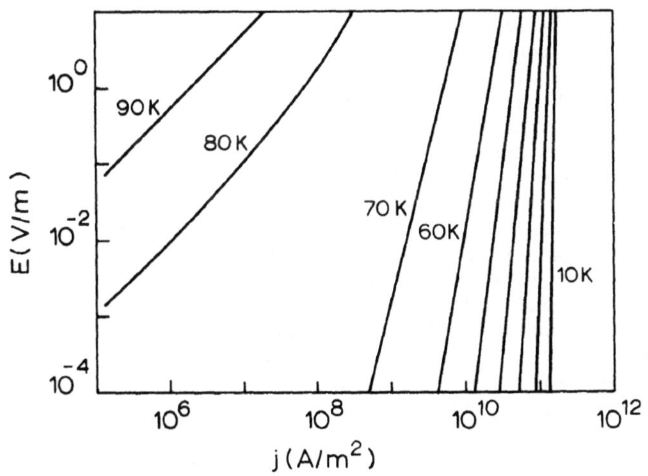

Fig. 3 E-j-isotherms as in Fig. 2 but for temperatures between 10 K and 90 K at 10 K interval.

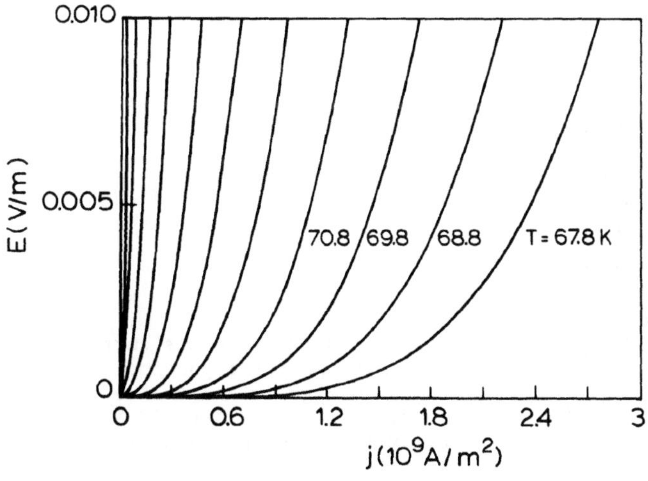

Fig. 4 Linear plot of the E-j isotherms in Fig. 2.

RESULTS

In ref. 8 we showed that the data of Koch et al. [6] and Zeldov [7] could be reproduced by choosing reasonable values for the parameters U_o^*, γ, A and S_0. Here we consider in detail the resistive behavior predicted by our model for a superconductor with $U_p^* = 38$ meV, $\gamma = 1.4$, $A = 3 * 10^{-10}$ meV/(Am^{-2}), $S_0 = 4 * 10^{-7}$ m/V, $B_{c2}(0) = 60$ T in a field B = 4 T.
In Fig. 2 the E-j-isotherms are plotted for temperatures between 67.8 K and 80.8 K at 1 K interval. The main features of these isotherms are the same as those in the data of Koch et al. [6]. In Fig. 3 the E-j-isotherms are indicated for temperatures from 10 K to 90 K at 10 K interval. Below 70 K a power-law corresponding to high powers of the current are evident. The critical current $j_c(0)$ at 0 K may be estimated directly from

$$j_c(0) = \frac{1}{A} \int_{U^*_{min}}^{U^*_{max}} U^* m(U^*) dU^* \cong \frac{U_p^*}{A} \exp\left(\frac{3}{4\gamma}\right) \qquad (14)$$

and is in good agreement with an extrapolation to 0 K of the data in Fig. 3.
For a direct comparison with experimental data we have also plotted the E-j-isotherms linearly in Fig. 4 and in Fig. 5 we have indicated the ρ-j-isotherms where the resistivity ρ is defined as $\rho = E/j$. The curves in Fig. 5 exhibit the same trend as the data of Zeldov et al. In particular a power-law behavior over three decades in ρ is also found around 70 K. Near 85 K ρ is independent of current density as long as $j < 10^9$ A/m^2. Logarithmic plots as those in Fig. 5 emphasize the low resistivity behavior. For comparison with $\rho(T)$ curves in a magnetic field we also calculated the resistivity as a function of temperatures near T_c in a field B = 4 T. The results of this calculation are shown in Fig. 6. It is gratifying to see that a knee is clearly present around 83 K. It can be shown that for T > 83 K all domains are in the flow-regime, even that with $U^* = U^*_{max}$.
The low temperature and low resistivity part of the $\rho(T)$ curves have often been analyzed in the literature in terms of a simple thermally activated process for which $\rho = \rho_0 \exp(-U_{res}/kT)$ where U_{res} is the corresponding activation energy (res: resistivity). With such an analysis various authors (among others Palstra et al. [18]) obtained unrealistically large values for U_{res} and the prefactor ρ_0. Recently, however, Malozemoff et al. [5] and Hebard et al. [3] showed that the temperature dependence of U_{res} has to be included in the analysis of the data. To investigate this aspect of the resistive behavior of high T_c superconductors we plotted the low-current data in Fig. 5 as a function of 1/T and found an activation energy U_{res} at B = 4 T of 857 meV (9943 K). This energy is approximately 23 times larger than the energy U_p^* corresponding to the maximum of the energy distribution $m(U^*)$. This example shows clearly that the large differences between values reported in the literature for "an" activation energy are in some cases due to an incorrect treatment (e.g. neglect of T and B dependence of U_{res}) and in other cases simply to the specificity of the experiments themselves. For example, low temperature magnetization relaxation experiments are sampling all relaxation processes at low temperature and consequently the relevant activation energy is approximately equal to the energy at the maximum of $m(U^*)$, while resistivity measurements near T_c are sensitive mainly to domains with large activation energies. This follows from the current distribution in the sample indicated in Figs. 7b and 7c. As $m(U^*)dU^*$ is the fraction of pinning regions with pinning energy $U^* \epsilon [U^*, U^*+dU^*]$, $m(U^*)j(U^*,E,T,B)dU^*$ represents the current carried by the pinning

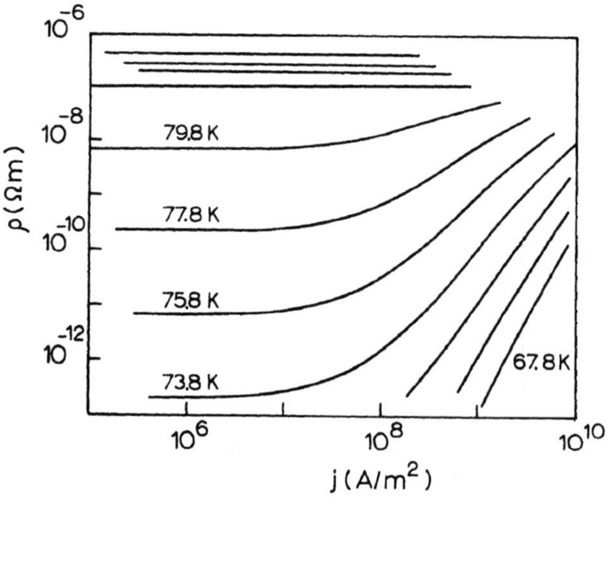

Fig. 5 Resistivity versus current density isotherms corresponding to the E-j-curves in Figs. 2 and 4. The temperature varies from 67.8 K to 87.8 K at 2 K interval. For comparison with similar data, see ref. 7.

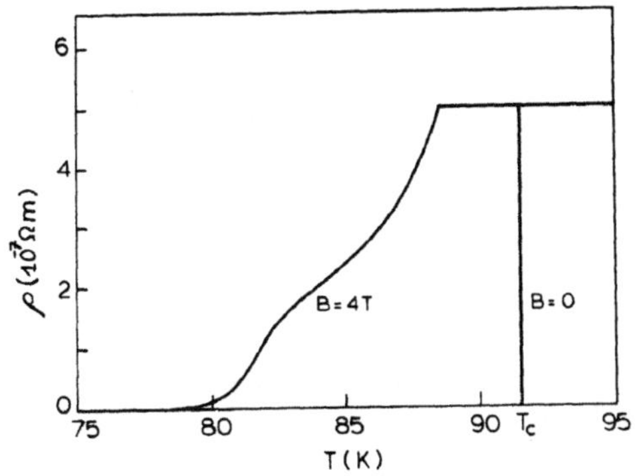

Fig. 6 Resistivity versus temperature curve corresponding to the E-j or ρ-j curves in Figs. 2 to 5 at B = 4 T. These calculated values apply to current densities lower than 10^6 A/m² for which ρ is independent of j (see Fig. 5).

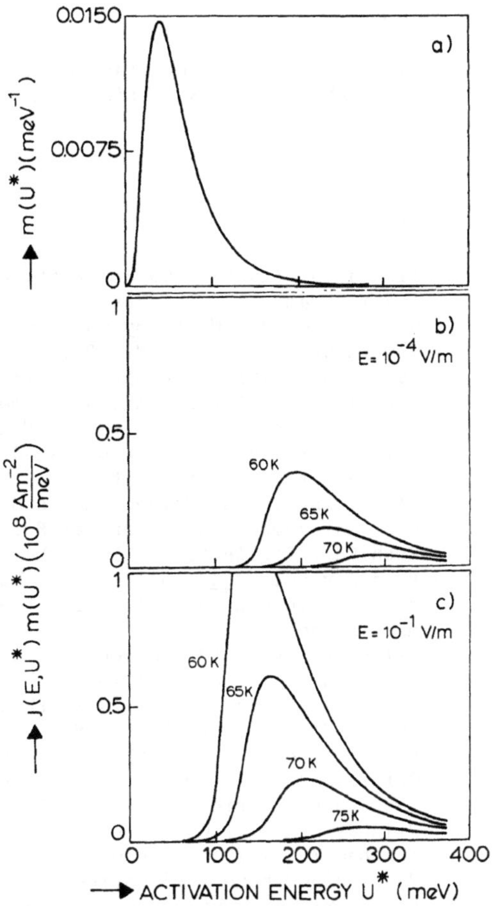

Fig. 7 Contribution of the various domains (characterized by their activation energy U^* at T = 0 and B = 0) to the total current flowing in the sample for an electric field E = 10^{-4} in Fig. 7b and E = 10^{-1} V/m in Fig. 7c. For comparison the distribution function $m(U^*)$ is shown in the top panel (7a).

regions with a certain activation energy. As expected, most of the current is carried by the domains with large U^* so that the maximum of $m(U^*)j(U^*,E)$ is significantly shifted with respect to the peak at $U^* = U_p^*$ in the log-normal distribution function.
The peak in $m(U^*)j(U^*,E)$ depends both on the electric field and the temperature.
The peak position shifts toward higher energies with increasing temperature as a results of the decrease in U(T,B) when $B_{c2}(T)$ approaches B. Then most of the pinning regions have rather low activation energies. In these regions the current density is small so as to avoid excessive dissipation.

CONCLUSIONS

Using the Hagen-Griessen inversion scheme we have determined distributions $m(U^*)$ of activation energies for thermally activated flux-motion from data on the time dependence of the magnetization M(t,T,B). These distributions resemble a log-normal distribution with a maximum at energies of the order of 50 meV at 0.1 T. The existence of a wide distribution in $YBa_2Cu_3O_7$ is subsequently used in a model to describe the resistive behavior of high-T_c superconductors in a magnetic field. This model incorporates both flux-creep and flux-flow and a log-normal distribution of activation energies for the various pinning domains in a sample. All domains are assumed to be electrically in parallel. The main features of recent current-voltage measurements by Koch et al. and resistivity-current curves by Zeldov et al. are reproduced with a log-normal distribution with a maximum at $U_p^* = 38$ meV in a field B = 4T. The model,also leads to a knee in the ρ(T) curves as observed in good quality crystals. Furthermore it is shown that if the ρ(T)-curve at temperatures below the knee are fitted to a simple Arrhenius relation $\rho = \rho_0 \exp(-U_{res}/kT)$ both ρ_0 and U_{res} are large and are not simply related to the basic parameters of the model. The activation energy U_{res} determined from this portion of the ρ(T)-curves is for example 23 times larger than $U_p^* = 38$ meV and almost 3 times larger than the largest activation energy $U_{max}^* = 373$ meV entering the model (this is the value of U^* for the domain with the strongest pinning).
As the parallel coupling of domains is a rather drastic approximation we are presently calculating the E-j-curves of random superconducting arrays. These results will be published elsewhere.

Acknowledgments

We acknowledge interesting discussions with drs. R. Koch, E. Zeldov and T.K. Worthington. We are grateful to the Stichting voor Fundamenteel Onderzoek der Materie (FOM) for its financial support.

REFERENCES

1. Tinkham M. (1988) Phys.Rev.Lett. 61: 1658
2. Iye Y., Tamegai T., Takeya H., Takei H., (1988) in "Superconducting Materials", ed. S. Nakajima and H. Fukuyama, Jpn.Appl.Phys. Series 1 (Tokyo) p. 46
3. Hebard AF., Palstra TTM., Physica C. (1990) Proc.Int.Conf. Materials and Mechanisms of Superconductivity and High Temperature Superconductors: to be published
4. Batlogg B., Palstra TTM., Schneemeyer LF., Waszczak JV., (1989) in "Strong Correlation and Superconductivity"; ed. H. Fukuyama, S.Maekawa and A.P. Malozemoff, Springer Ser. Solid State Sci.: to be published
5. Malozemoff AP., Worthington TK., Zeldov E., Yeh NC., McElfresh MW., Holtzberg F., (1989) in "Strong Correlation and Superconductivity"; eds. H.Fukuyama, S.Maekawa and A.P. Malozemoff, Springer Ser.Solid State Sci.
6. Koch RH., Foglietti V., Gallagher WJ., Koren G., Gupta A., Fisher MPA. (1989) Phys.Rev.Lett. 63: 1511. (Note that the existence of a "true" R=0 superconducting state is in contradiction with the large flux creep observed at low temperatures by many authors).
7. Zeldov E., Amer NM., Koren C., Gupta A., McElfresh MW., Gambino RJ. (1989) preprint
8. Griessen R., to be published
9. Hagen CW., Griessen R. (1989) Phys.Rev.Lett. 62: 2857
10. Hagen CW., Griessen R. (1989) in "Studies of High Temperature Superconductors" Vol.III; ed. A.V. Narlikar, Nova Science Publishers
11. Griessen R., Lensink JG., Hagen CW., to be published
12. Geerken BM., Griessen R., (1983) J.Phys.F: Met.Phys. 13: 963
13. Wijngaarden RJ., Griessen R. to be published
14. Tinkham M. (1975) Introduction to Superconductivity, McGraw Hill Inc.,New York
15. We consider here a situation where the external magnetic field is applied parallel to the c-axis and the current flows in the a-b-plane. There is thus certainly a Lorentz force acting on each flux-line.
16. Bardeen J., Stephen MJ. (1965) Phys.Rev. 140A: 1197
17. Kadowaki K., Roeland LW., Kes PH., Li Jiang-ning, Menovsky A., Huang Ying-Kay., van den Berg J., van der Beek CJ., Menken MJV., de Boer FR. (1989) Physica B155: 136
18. Palstra TTM., Batlogg B., Schneemeyer LF., Waszczak JV., (1988) Phys.Rev.Lett. 60; 1662

Flux Pinning Mechanism and Critical Current Density in High-Temperature Superconductors

T. MATSUSHITA

Department of Electronics, Kyushu University, 10-1, Hakozaki 6-chome, Higashi-ku, Fukuoka, 812 Japan

ABSTRACT

Flux pinning strength and resultant critical current density in high-temperature superconductors are argued by comparing with those in ordinary low-temperature superconductors. Because of smaller thermodynamic critical field and coherence length, the potentiality in flux pinning in Y-Ba-Cu-O at 77.3 K is lower than that in Nb_3Sn at 4.2 K. This means that the effect of thermally activated flux creep is significant when Y-Ba-Cu-O is used at 77.3 K. The irreversibility line, the relaxation of persistent current and the broad resistive transition in Y-Ba-Cu-O are discussed.

KEY WORDS: high-temperature superconductor, flux pinning, critical current density, flux creep

INTRODUCTION

When a transport current is applied to a superconductor in a magnetic field, quantized magnetic flux lines (fluxoids) suffer the Lorentz force. If the fluxoids move due to this force, the electric field is induced. Therefore, flux pinning interactions which prevent the fluxoid motion are necessary. Under the influence of flux pinning interactions, the transport current density can be applied without energy dissipation up to a certain value, the critical current density J_c. This value is given by F_p/B, where F_p is the macroscopic pinning force density and B is the magnetic flux density. For an improvement of the critical current density in a superconductor, the macroscopic pinning force density is required to be increased.

The macroscopic pinning force density is expressed as a function of the elementary pinning force f_p, i.e., the maximum pinning strength of each pinning center and the concentration N_p under given conditions of the temperature and the magnetic field. It is known[1] that, in the regime of strong pinning, F_p is proportional to a product of f_p and N_p. Hence, the flux pinning characteristic in this regime can be approximately estimated from f_p. In addition, f_p is evaluated in terms of the phenomenological quantities such as the thermodynamic critical field and the coherence length. This enables us to compare the potentiality in flux pinning in high-temperature superconductors with that in ordinary low-temperature ones. It is clarified that the potentiality in Y-Ba-Cu-O at 77.3 K is higher than that in Nb-Ti at 4.2 K but is lower than that in Nb_3Sn at 4.2 K. However, if we can introduce favorable structure for flux pinning by metallurgical manipulation, this disadvantageous characteristic can be compensated.

If high-temperature superconductors are used at 77.3 K, a combination of larger thermal energy and smaller pinning potential energy leads to a significant flux creep. One of the important results from the flux creep is a disappearance of the irreversibility at relatively high fields. Our interest is the irreversibility field above which the critical current density is reduced to zero. According to measurement by Yeshurun and Malozemoff[2] on a single crystal Y-Ba-Cu-O, the irreversibility field was only 0.7 T at 77.3 K for the field directed along the c-axis. How can this characteristic field be increased by increasing the pinning strength? Another result from the flux creep is a relaxation of a persistent current. Even if the critical current density takes a nonzero value, the persistent current decays with time[3]. Can this supercurrent persist for sufficiently long period for application? These points will be discussed in this paper.

Recently, a broad resistive transition in high-temperature superconductors under the magnetic field has been agreed and various mechanisms have been proposed[2,4-7]. This phenomenon will also be discussed in this paper from the view of the fluxoid motion caused by the thermally activated flux creep.

FLUX PINNING IN SUPERCONDUCTORS

In a mixed state of a superconductor, a fluxoid lattice is formed. The fluxoid lattice contains two spatially varying structures; the magnetic flux density and the superconducting order parameter. If there is an inhomogeneous region with different superconducting parameters, therefore, the energy varies during a displacement of the fluxoid lattice. This causes the flux pinning interaction. The elementary pinning force is given by a maximum value of the derivative of the energy with respect to the displacement. The variation in the magnetic flux density δB is at most of the order of $B_{c2}/\kappa^2 \sim 2B_{c1}/\ell n\kappa$, where B_{c1} and B_{c2} are the lower and upper critical fields and κ is the Ginzburg-Landau parameter. Since κ in high-temperature superconductors is very large (around 100), δB is negligibly small except for the vicinity of B_{c1}. Hence, the magnetic pinning interaction is not effective in most cases. The spatial structure of the order parameter is, on the other hand, responsible for the flux pinning. This interaction is called condensation energy interaction. The flux pinning in ordinary commercial superconductors originates from this interaction. Typical pinning centers are normal precipitates in Nb-Ti and grain boundaries in Nb_3Sn.

Flux pinning in high-temperature superconductors seems also to be discussed in terms of the phenomenological Ginzburg-Landau theory, since any incompatibility with this theory has not been reported. In this paper, two kinds of pinning centers, i.e., normal precipitates and planar defects such as grain boundaries or twinning planes, are considered as examples, and potentiality for flux pinning in high-temperature superconductors is compared with that in ordinary commercial superconductors.

Since the order parameter in a normal core of the fluxoid is depressed, and hence this region has higher free energy density than in the surrounding region with larger order parameter. When the normal core meets a normal precipitate, the volume of the higher energy region decreases in the superconducting matrix. This causes an attractive pinning interaction by normal precipitates. For a precipitate sufficiently larger than the fluxoid spacing a_f, its elementary pinning force at low fields is given by[8]

$$f_p = \pi B_c{}^2 \xi L^2 / 4\mu_0 a_f, \tag{1}$$

where B_c the thermodynamic critical field and L a size of the precipitates.

Electrons are scattered at planar defects such as grain boundaries and twinning planes. This causes a reduction in the coherence length ξ. The transverse cross sectional area of the normal core, $\pi\xi^2$, decreases at such planar defects. Hence, these defects also work as attractive pinning centers. It is to be noted that such planar defects are effective only when the fluxoids parallel to them cross them[9]. The elementary pinning force of planar defects per unit length of the fluxoid at low fields is given by[10]

$$\hat{f}_p = AB_c{}^2\xi/2\mu_0 \tag{2}$$

where A is a numerical factor of the order of 0.1.

We note that the both elementary pinning forces given by eqs. (1) and (2) are proportional to $B_c{}^2\xi/2\mu_0$. This suggests that a material with a large value of this product has a high potentiality for the flux pinning. In Table 1, Nb_3Sn at 4.2 K, Nb-Ti at 4.2 K and Y-Ba-Cu-O at 77.3 K are compared. It is clarified that the potetiality in Y-Ba-Cu-O at 77.3 K is much higher than that in Nb-Ti at 4.2 K but is about one third of Nb_3Sn at 4.2 K. This may suggest that the critical current density in Y-Ba-Cu-O at 77.3 K can scarcely exceed that in Nb_3Sn at 4.2 K. However, we note that the difference in J_c between Nb_3Sn and Nb-Ti is not as large as the difference in $B_c{}^2\xi/2\mu_0$ in Table 1. Dominant pinning centers in Nb-Ti are normal α-Ti precipitates with a shape of ribbon. Normal precipitates are strong pinning centers and the effective surface area for the pinning is quite large for such a ribbon-shape. These favorable conditions increase the efficiency of pinning in Nb-Ti. If we can introduce a suitable pinning structure in Y-Ba-Cu-O, therefore, the resultant critical current density may exceed that in Nb_3Sn.

FLUX CREEP

When Y-Ba-Cu-O is applied at 77.3 K, slightly weaker pinning force and twenty-fold larger thermal activation lead to much more significant flux creep than in Nb_3Sn at 4.2 K. Here we will quantitatively investigate the effect of flux creep on the irreversible property in high-temperature superconductors.

Yeshurun and Malozemoff[2] reported on Y-Ba-Cu-O single crystal that the irreversibility disappears due to the flux creep above 0.7 T at 77.3 K. Our interest is a quantitative estimate of the irreversibility field in strongly pinned high-temperature superconductors in the future. For this purpose the pinning potential in which flux bundle is trapped is required to be estimated. The pinning potential is formally given by[11]

$$U_0 = \alpha_L d_i{}^2 V/2, \tag{3}$$

where α_L is the Labusch parameter[12], d_i is a radius of the effective pinning potential and V is a volume of the flux bundle. The radius d_i can be measured by the ac inductive method[13] and is called interaction distance. These parameters are generally related to the critical current density as $\alpha_L d_i = J_{c0}B$, where J_{c0} is the ideal critical current density in the absence of thermal agitation. It is empirically known that d_i is proportional to the fluxoid spacing a_f[8,11] and we put $d_i = a_f/\zeta$, where ζ is a constant. We assume that

Table 1 Comparison of superconducting parameters

	Nb_3Sn (4.2 K)	Nb-Ti (4.2 K)	Y-Ba-Cu-O (77.3 K)
B_c [T]	∿0.5	∿0.1	∿0.4
ξ [nm]	∿3.9	∿5.5	∿1.9[a]
$B_c{}^2\xi/2\mu_0$ [arb.u.]	1	0.056	0.31

a: mean value $(\xi_{//}{}^2\xi_\perp)^{1/3}$ is used.

the volume of the flux bundle is determined by the correlation lengths as $V=\ell_{44}\ell_{66}^2$, where ℓ_{44} and ℓ_{66} are the longitudinal and transverse elastic correlation lengths, respectively. These correlation lengths are determined by the elastic moduli of the fluxoid lattice and the Labusch parameter[11]. We are interested only in the regime of strong pinning, where the transverse correlation length in shorter than the cut-off length, a_f. The longitudinal one is given by $(C_{44}/\alpha_L)^{1/2}$ with $C_{44}=B^2/\mu_0$. In this case, the expression of U_0 is

$$U_0 = (8\phi_0{}^7 J_{c0}{}^2/27\sqrt{3}\zeta^6\mu_0 B)^{1/4}, \tag{4}$$

where ϕ_0 is the flux quantum.

In the resistive measurement, J_c is defined by the electric field criterion E_c, e.g., 100 μV/m. The irreversibility line is defined as the condition at which the above J_c reaches zero. Thus, the irreversibility line is given by[2]

$$U_0 = k_B T \ell n(B a_f \nu_0/E_c), \tag{5}$$

where ν_0 is an oscillation frequency of the flux bundle. In the above, the logarithmic term is approximately estimated to be 20[11]. From eqs. (4) and (5) with a suitable magnetic filed dependence of $J_{c0}(B)$, we can estimate the irreversibility field B_i. If we assume J_{c0} of 2×10^{10} A/m^2 at 5 T and 77.3 K by introduction of strong pinning centers[14], B_i at 77.3 K is estimated to be 74 T for the field directed parallel to the a-b plane, where we used $\zeta=6$. If we assume $\zeta\sim4$ observed for melt-processed Y-Ba-Cu-O[15], B_i is expected to be larger than the above estimate. In fact, B_i in CVD processed Y-Ba-Cu-O film with a smaller J_c value is estimated to be 62 T[16]. Thus, it can be concluded that the effect of flux creep is not so severe for getting nonzero J_c at high fields at 77.3 K.

The second interest is a relaxation of the persistent current in strongly pinned high-temperature superconductors at 77.3 K. We suppose an application at 77.3 K and 5 T. If the intrinsic critical current density of $J_{c0}=2\times10^{10}$ A/m^2 is attained under these conditions as assumed previously, U_0 is estimated to be 3.46×10^{-20} J (0.22 eV). The creep rate during one hour after the establishment of the critical state is approximately given by $(k_B T/U_0)\ell n3600$, which amounts to 0.253 at 77.3 K. This means that the persistent current decreases by 25.3% during first one hour. If J_{c0} takes only one quarter of the above value, eq. (4) suggests a degradation of the persistent current by 50.6% during first one hour. Hence, the effect of flux creep is significant for the persistent current. However, the operation current is usually designed to be smaller than the critical current. Hence, the persistent current can be practically kept constant during some period. In the former case of degradation by 25% from the complete critical state during first one hour, if we settle the operation current to be 50% of the critical current, this operation current does not decay for 3600^2sec≃5 months. Thus, we can practically avoid problems caused by relaxation.

RESISTIVE TRANSITION

Recently broadened resistive transition in high-temperature superconductors is observed under the magnetic field[4,5]. A fluctuation was proposed for the origin of the broad resistive transition[5,6]. However, the resistance is initiated by the flux creep at lower temperatures. This means that the resistive transition width between the temperature of appearance of the resistance and T_c is determined by the mechanism of fluxoid motion due to the Lorentz force. This suggests that the resistive transition is mostly determined by the fluxoid motion. In a usual plot of the resistance vs temperature curve in a linear scale, where the region of low resistivity due

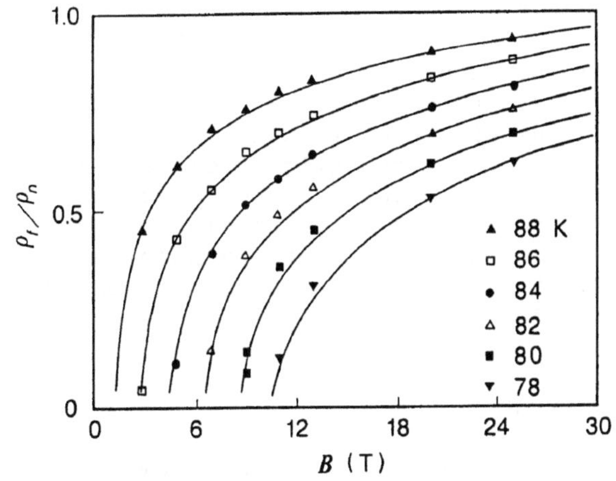

Fig. 1 Reduced flow resistivity in Y-Ba-Cu-O film prepared by chemical vapor deposition for the field along the c-axis (data from Kobayashi[18]).

to the pure flux creep is invisible, the resistive behavior seems to be principally described by the simple flux-flow model:

$$E = \rho_f(J-J_c), \tag{6}$$

where E is the electric field and ρ_f is the flow resistivity. In fact, the broad resistive transition can be explained by this mechanism[7] with the Bardeen-Stephen model[17] for the flow resistivity, $\rho_f=(B/B_{c2})\rho_n$ with ρ_n denoting the normal resistivity. The flux flow model explains the anisotropy in the resistive transition through the anisotropic B_{c2} and the appearance of knee observed in the transition curves. However, there still remains a difference in the shape of the transition curve between the simple model and experiments. If we use the observed flow resistivity[18] shown in Fig. 1, we obtain the resistive transition shown in Fig. 2(a), which is quite similar to the observed result[18] shown in Fig. 2(b). Thus, the flux flow model exactly explains the resistive behavior in high-temperature superconductors. That is, the resistance occurs due to the fluxoid motion. It is to be noted that the flow resistivity shown in Fig. 1 is different from those observed in low-temperature superconductors. Hence, the simple Bardeen-Stephen model does not exactly hold in high-temperature superconductors. This may suggest an anomalous characteristic of the fluxoid lattice or an enhanced effect of fluctuation.

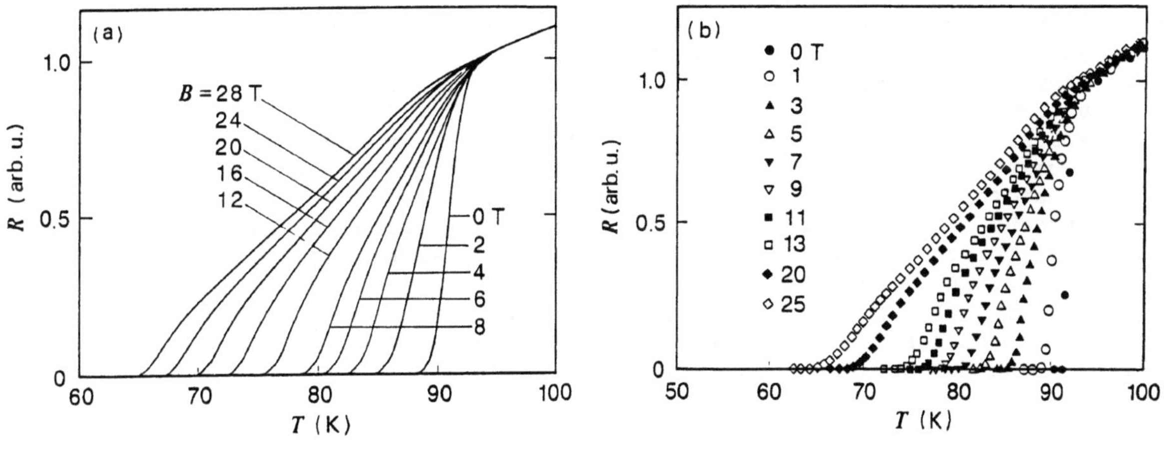

Fig. 2 Resistive transition under the magnetic field along the c-axis: (a) theoretical result from the flux flow model and (b) experimental result on CVD processed Y-Ba-Cu-O thin film[18].

SUMMARY

(1) The potentiality for flux pinning in high-temperature superconductors at 77.3 K is lower than that in Nb_3Sn at 4.2 K. If those are used at 77.3 K, the effect of flux creep is expected to be significant.
(2) The irreversibility field at 77 K is estimated to be sufficiently high for high-temperature superconductors with strong pinning centers.
(3) The relaxation of persistent current is significant at high fields even for the case of strong pinning.
(4) The broad resistive transition is principally explained by the mechanism of flux flow.

REFERENCES

1. Campbell AM, Evetts JE, Dew-Hughes D (1968) Pinning of flux vortices in type II superconductors. Phil. Mag. 18: 313-343
2. Yeshurun Y, Malozemoff AP (1988) Giant flux creep and irreverisibility in an Y-Ba-Cu-O crystal: an alternative to the superconducting-glass model. Phys. Rev. Lett. 60: 2202-2205
3. Yeshurun Y, Malozemoff AP, Holtzberg F (1988) Magnetic relaxation and critical current in an YBaCuO crystal. J. Appl. Phys. 64: 5797-5799
4. Tinkham M (1988) Resistive transition of high-temperature superconductors. Phys. Rev. Lett. 61: 1658-1661.
5. Kitazawa K, Kambe S, Naito M, Tanaka I, Kojima H (1989) Broadening mechanism of resistive transition under magnetic field in single crystalline $(La_{1-x}Sr_x)_2CuO_4$. Jpn. J. Appl. Phys. 28: L555-L556
6. Ikeda R, Ohmi T, Tsuneto T (1989) Renormalized fluctuation theory of resistive transition in high-temperature superconductors under magnetic field. J. Phys. Soc. Jpn. 58: 1377-1386
7. Matsushita T, Ni B (submitted) Broad resistive transition under magnetic field in high-temperature superconductors. Physica C
8. Matsushita T (1981) Pinning force of a nonideal superconductor containing pins with large interaction forces. II. low pin density. Jpn. J. Appl. Phys. 20: 1955-1966
9. DasGupta A, Koch CC, Kroeger DM, Chou YT (1978) Flux pinning by grain boundaries in niobium bicrystals. Phil. Mag. B 38: 367-380
10. Zerweck G (1981) On pinning of superconducting flux lines by grain boundaries. J. Low Temp. Phys. 42: 1-9
11. Matsushita T (submitted) Flux creep and critical currents in oxide superconductors. Physica B
12. Labusch R (1969) Calculation of the critical field gradient in type II superconductors. Crystal Lattice Defects 1: 1-16
13. Campbell AM (1971) The interaction distance between flux lines and pinning centers. J. Phys. C 4: 3186-3198
14. Matsushita T (1988) Strong flux pinning by cracks in films of superconducting oxide. Jpn. J. Appl. Phys. 27: L1712-L1714
15. Keller C, Küpfer H, Meier-Hirmer R, Wiech U, Selvamanickam V, Salama K (submitted) Irreversible behavior of oriented grained $YBa_2Cu_3O_x$. Part I: transport and shielding currents. Cryogenics
16. Watanabe K, Kobayashi N, Yamane H, Kurosawa H, Hirai T, Kawabe H, Muto Y (1989) Critical current criterion in high-T_c superconducting films. Jpn. J. Appl. Phys. 28: L1417-L1420
17. Bardeen J, Stephen MJ (1965) Theory of the motion of vortices in superconductors. Phys. Rev. 140: A1197-A1207
18. Kobayashi N: private communication

Effects of Non-Linearity in the Variation of Pinning Potential with Current-Density on Magnetic Flux Creep in High and Low T_c Superconductors

D.O. WELCH, M. SUENAGA, YOUWEN XU, and A.R. GHOSH

Materials Science Division, Brookhaven National Laboratory, Upton, NY 11973, USA

ABSTRACT

A theoretical analysis is presented of the effects on magnetic flux creep of non-linearity in the relation between the activation energy required for the depinning of magnetic flux and the current density. The non-linearity results in an apparent temperature dependence of the depinning energy as deduced by the conventional methods of analysis of flux-creep data. The discussion is illustrated with data for $YBa_2Cu_3O_7$ and Nb_3Sn.

INTRODUCTION

The time-dependent decrease in the hysteresis in the magnetization of type II superconductors (i.e., flux creep) has been discussed by many investigators using an analysis based on thermally-activated flux flow [1-3]:

$$D = -(\nabla B/|\nabla B|)Bwv_o \exp[-U(B,|\nabla B|)/kT] \qquad (1)$$

where D is the magnetic flux-flow density, B the magnetic induction, v_o the attempt frequency of a pinned flux line, w the distance of flux-bundle motion upon depinning, and U the effective pinning energy. [The time dependence of B is obtained from the continuity equation $\partial B/\partial t = -\nabla\cdot D$.]

The pinning energy U is a function of the magnetic induction B and its gradient ∇B, which is proportional to the current density J, (as well as temperature, because of the variation of the superconducting parameters such as coherence length and thermodynamic critical field with T/T_c). It has customarily been assumed, e.g. [3], that U decreases linearly with $B\nabla B$ due to the Lorentz force acting on the pinned flux, although it was recognized early on by Beasley et al. [1] that non-linear effects can be quite important. Recently Xu et al. [2] have shown that the apparent pinning energy of $YBa_2Cu_3O_7$, deduced on the assumption of linearity, can have a misleading temperature dependence. In this paper we illustrate the effects of non-linearity using the analytical model of Hagen et al. [3] (but without their assumption of linearity) for thermally-activated flux motion, and present flux-creep data on $YBa_2Cu_3O_7$ and Nb_3Sn obtained with a SQUID magnetometer which show these effects.

RESULTS AND DISCUSSION

Hagen et al. [3] have presented a simple analytical model for the flux distribution in a type II superconductor evolving from the critical state under the influence of thermally-activated depinning. The results obtained from this simple model agree very well with those obtained by more elaborate Monte Carlo methods in the case of a linear activation-energy versus force relation, and we extend it here to the case of non-linearity where the thermally-activated depinning energy is given by $U(B,\nabla B) = U_p(B)[1-(\nabla B/\nabla B_{max})]^n$, U_p is the pinning well depth and ∇B_{max} corresponds to the maximum current density which the pinning potential can sustain in the absence of thermal activation. (When n=1, one recovers the linear approximation.) In the general case the time dependent of the magnetization becomes:

$$\frac{t}{\varepsilon^{(1-1/n)}\tau_o} = \int_0^{|\Delta M(t)|\varepsilon^{1/n}/M_o} e^{x^n}\,dx \qquad (2)$$

where $\varepsilon \equiv U_p/kT$, M_o is the initial value of the magnetization before thermally-activated depinning begins, and ΔM is the decrease in magnetization caused by flux creep. For n = 1, Eq. (2) yields the familiar case of ΔM proportional to ℓnt, after a transient period of length τ_o, as discussed by Hagen et al. [3]. This is shown in Fig. 1 together with results for n = 3/2 and 2.

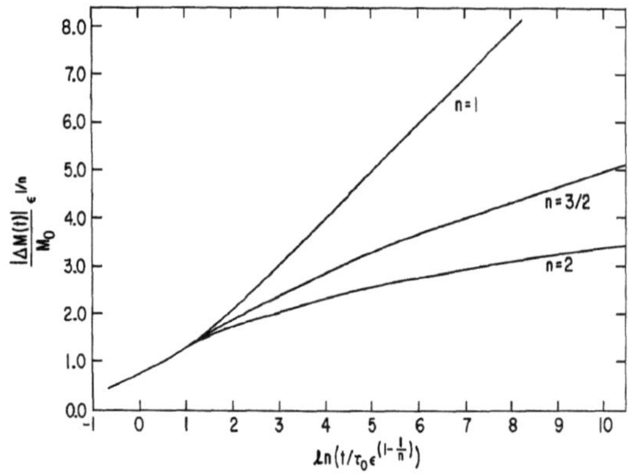

Fig. 1. Time-dependent magnetization for varying degrees of non-linearity.

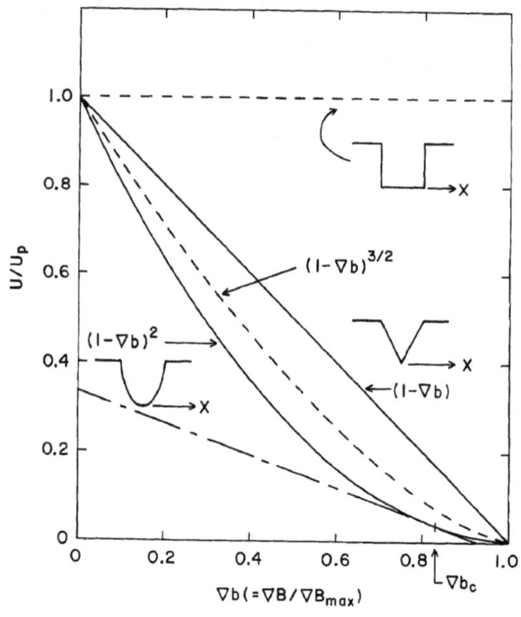

Fig. 2. The dependence on ∇B of the activation energy for depinning for different pinning potential shapes.

The value of the exponent n depends upon the details of the shape of the pinning-energy-versus-displacement curve (the "pinning" well"). Values appropriate to several idealized pinning wells are shown in Fig. 2. Detailed calculations for a wide variety of more realistic well-shapes yield $3/2 \leq n \leq 2$; for potentials with no discontinuities in their slope, $n \rightarrow 3/2$ as $\nabla B \rightarrow \nabla B_{max}$.

As seen in Fig. 1, when $n > 1$ (as is the case for realistic potentials) M(t) does not vary exactly as ℓnt, although it does so approximately over a limited range of ℓnt. This can be seen more clearly by examining the derivative $dM/d\ell nt$, as shown in Fig. 3; for $n > 1$ there is well-defined "dip", i.e. a maximum in $|dM/d\ell nt|$. Experimental data obtained with a SQUID magnetometer for melt-textured, bulk $YBa_2Cu_3O_{7-x}$ [4] are shown in Fig. 4 and indeed show the presence of just such a dip, indicating that n is greater than unity for this material.

The non-linearity of the depinning activation energy, demonstrated above, may account for some of the temperature dependence of the apparent pinning-well depth deduced from flux-creep data by assuming $n = 1$, as discussed in [2]. (Another contributing factor is a spectrum of pinning energies [5].) Because of complications associated with the granular nature of bulk $YBa_2Cu_3O_{7-x}$, it is not fruitful to analyze the data shown in Fig. 4 in more detail. Thus, to avoid such complications, we have examined the flux-creep behavior of the low-temperature superconductor Nb_3Sn, since it is expected to be more well-behaved. The temperature and magnetic-field dependence of the flux-creep rate for multifilamentary Nb_3Sn, as measured with a SQUID magnetometer, is shown in Fig. 5. However, the temperature and magnetic-field dependence of the apparent pinning-well depth obtained from the data, when the conventional assumption is made that $n = 1$, are shown in Fig. 6. Generally accepted theories of pinning interactions in a superconductor such as Nb_3Sn yield a pinning potential which slowly decreases with temperature until T approaches T_c, where it rapidly decreases to zero. The results shown in Fig. 6 do not behave this way. The apparent rise with temperature at low T is a consequence of the assumption of linearity ($n = 1$). It is possible to obtain a consistent fit to the data in Fig. 5, with the assumption that $n = 3/2$, and the resulting pinning-well depth is then found to be essentially temperature-independent at low T, with values of ~0.4 and 0.25 eV at fields of 1 and 2 T, respectively, for temperatures below about 10 K.

The magnetic flux-creep data for both $YBa_2Cu_3O_{7-x}$ and Nb_3Sn are thus seen to show manifestations of non-linearity in the dependence of the activation energy for depinning by thermal fluctuations upon ∇B (i.e., upon current density J). It appears to be essential to take this into account in order to obtain realistic values of the pinning-well depth from the analysis of flux creep data.

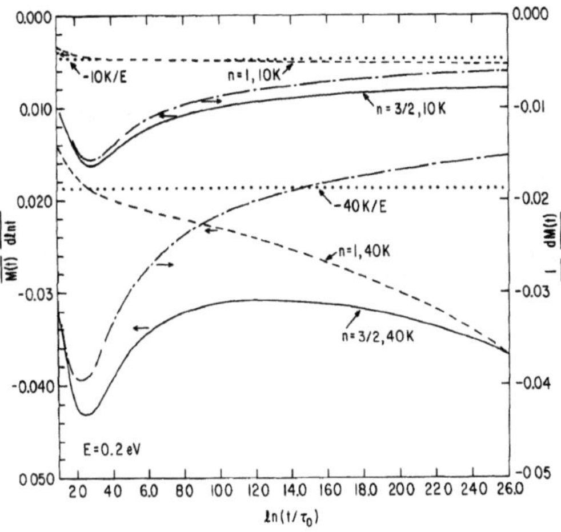

Fig. 3. Calculated values of $dM(t)/d\ell nt$ at two temperatures for n = 1 and n = 3/2. (The pinning energy $U_p \equiv E = 0.2$ eV.) Two different normalizations are used: M(t) [solid lines and dashed lines] and M(t = 0) [dotted lines and dot-dash lines.]

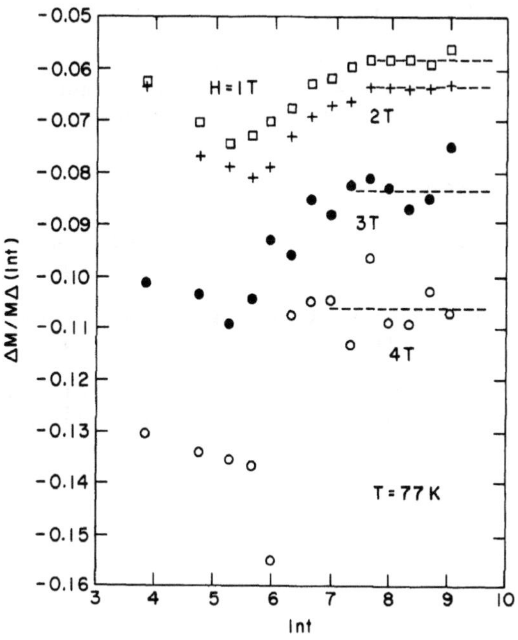

Fig. 4. Experimental data for the slope of M vs ℓnt, normalized by M(t), for melt-textured, bulk $YBa_2Cu_3O_{7-x}$.

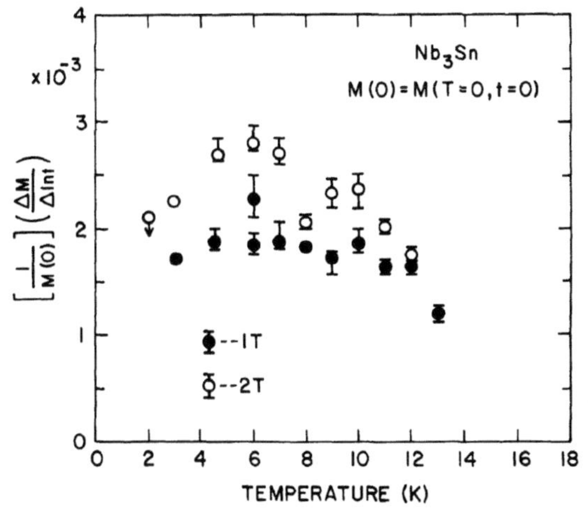

Fig. 5. The magnetic flux creep rate for multifilamentary Nb_3Sn.

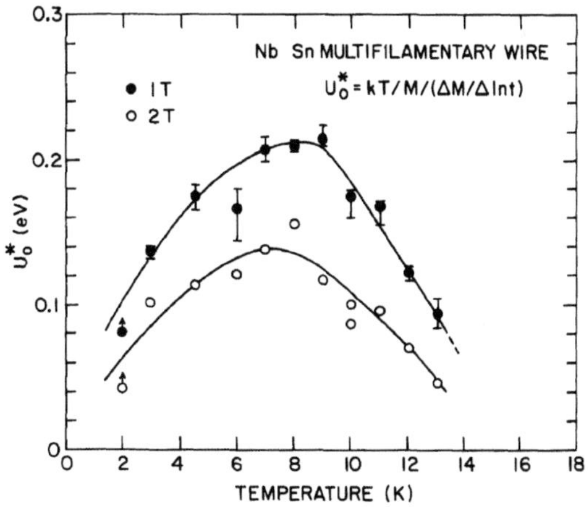

Fig. 6. The temperature dependence of the apparent pinning-well depth obtained from the data of Fig. 5 by assuming n = 1.

ACKNOWLEDGMENTS

This work was performed under the auspices of the U.S. Department of Energy, Division of Materials Science, Office of Basic Energy Sciences under Contract No. DE-ACO-276CH00016. We are grateful to Dr. Murakami for providing us with a sample of melt-textured $YBa_2Cu_3O_{7-x}$.

REFERENCES

1. M. R. Beasley, R. Labusch, and W. W. Webb, Phys. Rev. 181, 682 (1969).
2. Youwen Xu, M. Suenaga, A. R. Moodenbaugh, and D. O. Welch, Phys. Rev. B (to appear 1 Dec. 1989).
3. C. W. Hagen, R. Griessen, and E. Salomons, Physica C 157, 199 (1989).
4. The melt-textured $YBa_2Cu_3O_{7-x}$ was kindly supplied to us by M. Murakami of Nippon Steel.
5. C. W. Hagen and R. Griessen, in High-Temperature Superconductors, A. V. Narlikar, editor, Nova Sci. Publ. Inc., New York, 1989.

Flux Creep of Melt Processed $YBa_2Cu_3O_7$

M. MURAKAMI, H. FUJIMOTO, S. GOTOH, N. KOSHIZUKA, and S. TANAKA
Superconductivity Research Laboratory, International Superconductivity Technology Center,
10-13, Shinonome 1-chome, Koto-ku, Tokyo, 135 Japan

ABSTRACT

We report results of flux creep measurements for YBaCuO crystals prepared by quench and melt growth process. The crystals contain fine Y_2BaCuO_5 precipitates which are expected to work as pinning centers. Relaxation of magnetization was measured at various magnetic fields from 2kOe to 8kOe. It was found that the pinning energy U was almost field independent and approximately 0.3 eV. Although this value is higher than that of single crystal, relaxation of remanent magnetization yields much higher U value. We believe that some extrinsic effects such as fluctuation of magnetic field affect the flux creep data.

KEYWORDS: flux creep, magnetic relaxation, quench and melt growth process, $YBa_2Cu_3O_7$, Y_2BaCuO_5

INTRODUCTION

Oxide superconductors have high critical temperature, and therefore, reasonably high potential for practical applications. For most applications, high Jc values of 10^4 - 10^6 A/cm^2 are required in the presence of large magnetic field. Recent developments of processing[1,2] revealed that such high Jc values are obtainable even in bulk oxide superconductors at 77K. However, it has been also observed by several research groups [3,4] that remarkably large flux creep occurs in both single crystals and high Jc thin films. Since dissipative flux motion can reduce Jc, these observations led to the conclusions that applications of high Tc oxide superconductors are not feasible at 77K.

According to the thermally activated model of flux creep[5], the rate of flux creep, v is described as

$$v = v_0 \exp(-U/kT) \qquad (1)$$

where v_0 is the attempt frequency of the flux bundle to move, U is the depth of the pinning potential, k is Boltzman constant and T is temperature. When we consider the use of oxide superconductors at 77K, thermal energy kT becomes extremely large, leading to large flux creep rate. On the other hand the pinning energy U is also considered to be remarkably small in oxide superconductors due to extremely small coherence length. The pinning energy originates in the condensation energy of the superconducting state. The elementary pinning energy Up per unit length is given by the following relation:

$$Up = (Hc^2/8)\, \xi^2 \qquad (2)$$

where Hc is the thermodynamical critical magnetic field, ξ is Ginzburg-Landau coherence length. Since ξ is extremely small in oxide superconductors, Up is also considered to be very small. Although macroscopic pinning energy U scales with Up, U is not necessarily small. In order to obtain U value, we need to determine macroscopic pinning force Fp. The elementary pinning force fp is given by

$$fp = Up/2\xi \qquad (3)$$

Therefore, the value of fp scales with ξ and hence is not so small compared with Up which scales with ξ^2. Bulk or macroscopic pinning force Fp is obtained by summing up the elementary pinning force in unit volume. In a simple case, Fp becomes

$$Fp = N\ fp \qquad (4)$$

where N is the number of interactions between the pinning center and the fluxoid per unit volume. This relation indicates that Fp can be increased by increasing N or the pinning-site density. Then the pinning energy U is obtained by the following relation:

$$U = Fp\ V\ x \qquad (5)$$

where V is the volume of flux bundle to hop collectively, x is the distance moved by the flux bundle. Therefore it is clear that the pinning energy can be increased by either increasing V or x, though the values of V or x are obtained empirically and closely related to the stiffness of flux line lattice[6]. It is also known that Fp is simply related to Jc through Fp=JcB. Therefore higher Jc implies higher U as far as V and x are constant. Recently the small U values in thin films are ascribed to small V[6,7]. When the interaction distance of flux line lattice l_{44} is longer than that of film thickness d, U is reduced by a factor of d/l_{44} compared to the bulk sample with the same Jc value.

It is also clear that small U values observed in single crystals are not intrinsic values to high Tc oxide superconductors. It has been found that Jc can be increased by the improvement in processing conditions[8] and Jc values exceeding 30000 A/cm^2 were attained in YBaCuO samples prepared by quench and melt growth(QMG) process[9]. In this paper, we report the results of flux creep measurements in QMG processed YBaCuO samples and show that the U value can be improved by the introduction of pinning centers.

EXPERIMENTAL

The YBaCuO samples were prepared by quench and melt growth(QMG) process. Details of the process are described in Ref. 10. The sample with dimensions of 2 x 3 x 5 mm^3 was cut from the QMG processed YBaCuO plate. The c axis was parallel to the long axis of the sample. Magnetic field was applied parallel to the long axis of the sample, therefore, in this situation, magnetic field was parallel to the c axis. It has been reported that flux creep rate is much larger in this direction than $H \perp c$ axis in single crystals[3].

Measurements of magnetic relaxation were performed at a fixed field in the increasing and decreasing field process. Then we determined relaxation of $(|M^+| + |M^-|)/2$. By this treatment we can remove the effects of surface magnetization[11]. All measurements were conducted at 77K.

RESULTS AND DISCUSSION

Figure 1 is magnetic relaxation of the QMG processed YBaCuO sample in magnetic fields from 2 to 8 kOe. When the data are normalized by the initial value M_0 in M/M_0, Figure 2 is obtained. Surprisingly, the results indicate that magnetic field dependence is almost negligible and the pinning energy U is obtained to be 0.3 eV. Considering Fp=JcB equation (5) is reformed as

$$U = Jc\ B\ V\ x \tag{6}$$

Recent analyses[9] indicated that Jc scales with $B^{-1/2}$ in the QMG processed sample. Dew-Hughes[12] suggested that x is the flux-line spacing a_f and $x = a_f = 1.075\ (\phi_0/B)^{1/2}$. In that case U should be field independent. The present experimental analyses are consistent with such hypothesis. But we have also measured the decay of remanent magnetization[13] and found that U values of 0.7 eV or even higher values are obtained in the full remanent state where remanent magnetization is saturated. Average magnetic field inside the sample in that condition was 2kOe. We believe that the difference in the U value is attributed to the extrinsic effects during measurements in flux creep. Two possible sources are considered to affect the data extrinsically. First, fluctuation of magnetic field may affect the data. Second, mechanical oscillation needed to determine magnetization may increase relaxation rate. Recently Griessen et al.[14] suggested that magnetization process also affects results of flux creep measurements strongly, and in some cases, flux creep rate becomes very small even though relaxation is occurring at much larger rate. But in that case, the initial relaxation becomes extremely large and will be observable in measurements. In our experiments we could not observe such fast relaxation. Furthermore, such effects do not affect the decay of remanent magnetization. Therefore, larger U values in the case of relaxation of remanent magnetization indicate that such effects are negligible in the present case.

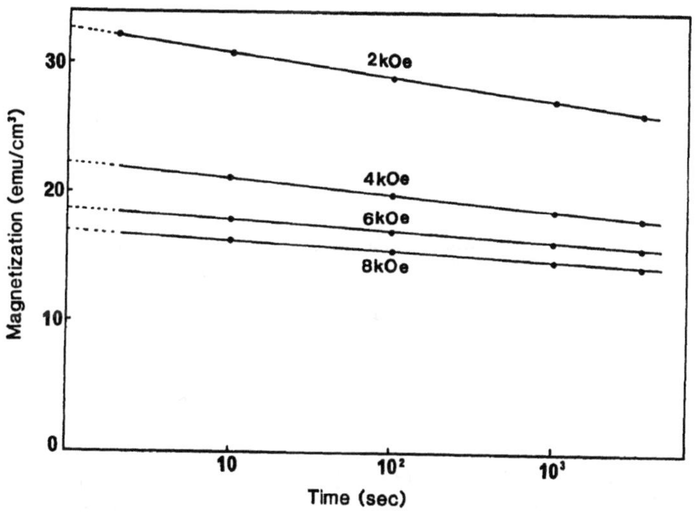

Figure 1. Relaxation of magnetization (average of M^+ and M^-) in various magnetic fields for QMG processed YBaCuO at 77K.

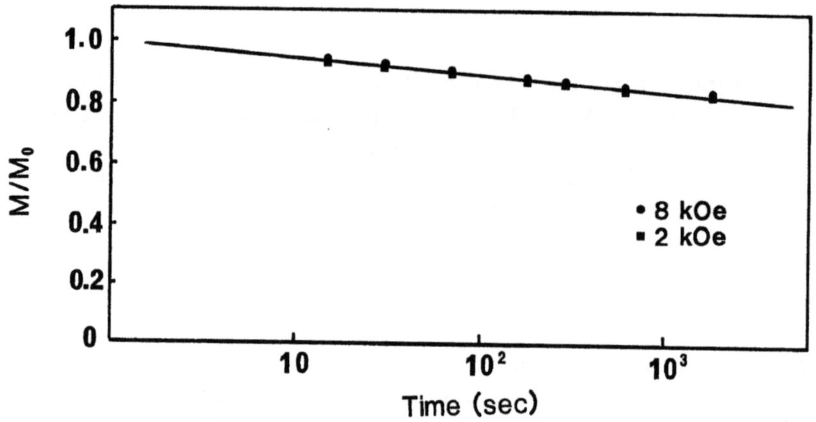

Figure 2. Magnetic relaxation normalized to initial value obtained from Figure 1.

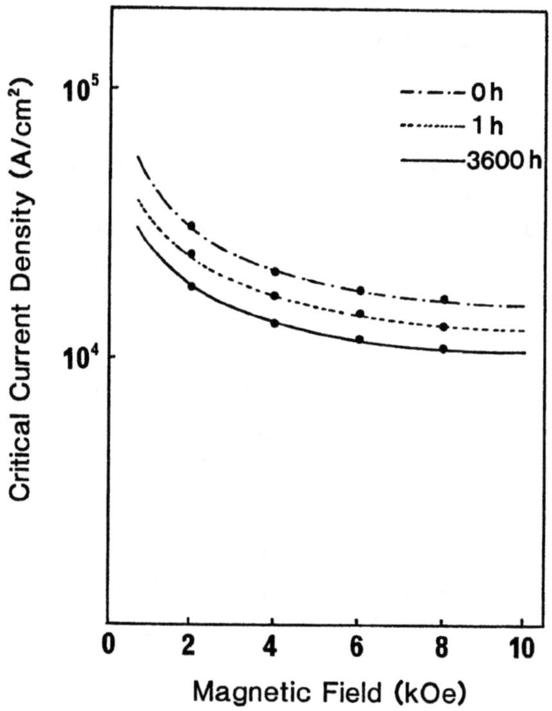

Figure 3. Magnetic field dependence of Jc under the effect of flux creep. Note that a Jc value exceeding 10000 A/cm^2 is obtained even after 3600 h has passed. The data were extrapolated from results flux creep measurements.

Although the present data may include extrinsic effects, we believe we do not overestimate the U values in magnetic fields from our measurements. It is possible to obtain Jc values under the effects of flux creep using these data. Figure 3 indicates how Jc values decay with time. Although the effect of flux creep is not negligible, we can achieve a Jc value exceeding 10000 A/cm^2 after 5 months. It is also notable that the U value of 0.3 eV is higher than those of single crystals, indicating that the pinning energy can be increased by the introduction of pinning centers.

CONCLUSIONS

We measured flux creep of the YBaCuO sample prepared by quench and melt growth process at 77K in the presence of magnetic fields. Flux creep rate was almost field independent and the pinning energy of 0.3 eV was obtained in magnetic fields from 2 to 8kOe. As compared to the data of relaxation of remanent magnetization, we believe that some extrinsic effects affected the data. Even though the data cannot give us any conclusive result on intrinsic flux creep, an important conclusion was deduced from comparison with single crystals that the pinning energy can be increased by introduction of pinning centers.

REFERENCES

[1] S. Jin, T. H. Tiefel, R. C. Sherwood, M. E. Davis, R. B. van Dover, G. W. Kammlott, R. A. Fastnacht, and H. D. Keith: Appl. Phys. Lett. 52 (1988) 2074.

[2] M. Murakami, M. Morita and K. Miyamoto: Progress in High Temperature Superconductivity 15 (World Scientific, Singapore, 1989) p. 95.

[3] Y. Yeshurun and A. P. Malozemoff: Phys. Rev. Lett. 60 (1988) 2202.

[4] S. Hatta, Y. Ichikawa, H. Adachi and K. Wasa: Jpn. J. Appl. Phys. 28 (1989) L422.

[5] P. W. Anderson: Phys. Rev. Lett. 9 (1962) 309.

[6] K. Yamafuji, T. Fujiyoshi, K. Toko and T. Matsushita: Physica C 159 (1989) 743.

[7] G. B. Smith, J. M. Bell, S. W. Filipczuk and C. Andrikidis: Physica C 160 (1989) 333.

[8] M. Murakami, S. Matsuda, K. Sawano, K. Miyamoto, A. Hayashi, M. Morita, K. Doi, H. Teshima, M. Sugiyama, M. Kimura, M. Fujinami, M. Saga, M. Matsuo and H. Hamada:Proc. ISS 89 (Springer-Verlag, Tokyo,1989) 247.

[9] M. Murakami, S. Gotoh, N. Koshizuka and S. Tanaka: Cryogenics (1990) to be published.

[10] M. Murakami, M. Morita, K. Doi and K. Miyamoto: Jpn. J. Appl. Phys. 28 (1989) 1189.

[11] M. R. Beaseley, R. Labusch and W. W. Webb: Phys. Rev. B 181 (1969) 682.

[12] Dew-Hughes: Cryogenics 28 (1988) 674.

[13] M. Murakami, M. Morita and N. Koyama: Jpn. J. Appl. Phys. 28 (1989) L1254.

[14] R. Griessen : Cryogenics (1990) to be published.

Intergranular Flux Creep in Ceramic $YBa_2Cu_3O_x$

N. NAKAMURA, M. ISHIDA, and M. SHIMOTOMAI

New Materials Research Center, High Technology Research Laboratories, Kawasaki Steel Corporation, 1, Kawasaki-cho, Chiba, 260 Japan

ABSTRACT

Magnetic relaxation in crystallographically aligned superconducting ceramics $YBa_2Cu_3O_x$ was measured to study the kinetics and anisotropy of flux penetration into intergranular region. With use of a thermally activated flux motion model, the activation energy of the intergranular flux creep was evaluated as 4.0±1.0 meV and 3.7±1.1 meV for external magnetic fields perpendicular and parallel to the ab-plane, respectively. Implications of the apparent isotropy are discussed.

KEY WORDS: flux creep, $YBa_2Cu_3O_x$, intergranular region

INTRODUCTION

Magnetic relaxation in single crystals of $YBa_2Cu_3O_x$ was observed in fields above the lower critical field Hc_1 and was interpreted in terms of thermally activated motion of flux inside the crystal /1/. The derived activation energy was unusually small and anisotropic, 20 meV for external magnetic fields perpendicular to the ab-plane (H⊥ab) and 150 meV for fields parallel to the ab-plane (H∥ab).

As regards polycrystalline samples of $YBa_2Cu_3O_x$, such magnetic relaxation has been observed even in fields below Hc_1 /2-4/, indicating an occurrence of flux creep at intergranular region. It is the purpose of this paper to investigate intergranular flux motion and its anisotropy in well-aligned polycrystalline samples of $YBa_2Cu_3O_x$.

EXPERIMENTAL

Crystallographycally aligned samples of single-phased $YBa_2Cu_3O_x$ were prepared by a magnetic casting method similar to one described in reference 5. A precursor for $YBa_2Cu_3O_x$ was prepared by spray-drying of a mixture of aqueous solutions of acetic salts of the component metals, the details of which were described elsewhere /6/. The spray-dried precursor was calcined at 900°C and then subjected to jet-milling. Final powder size was in the range of 3 to 5μm. Suspended in isopropanol, the powders were placed in a magnetic field of 6 T. After the liquid evaporation, the dried cake was sintered at 950°C for 12 hr in oxygen gas. The apparent density of the sintered body was 90% of the theoretical value and the alignment factor was estimated as 90% by X-ray diffration. Resistive Tc of the samples was found to be 91 K. Figures 1-(a) and -(b) are SEM micrographs of fractured surfaces of an aligned sample parallel and perpendicular to the ab-plane, respectively. It is easy to see that plate-like crystallites of roughly 12 μm in diameter and 5μm in thickness are stacked like bricks and that grain-boundaries (GB) are classified into ones perpendicular to the ab-plane (GB-I) and ones parallel to the ab-plane (GB-II).

Magnetic measurements were performed on zero-field-cooled samples of rod-shape 1 mm in diameter and 5 mm in length with a vibrating sample magnetometer. Due to the instrumental limitation, magnetic isotherms were taken between 4.2 and 30 K. The Hc_1 value defined as the magnetic field where the initial magnetization curve deviates from linearity, has turned out to be 410 Oe for H$\perp$ab and 230 Oe for H$\|$ab at 4.2 K, decreasing to 125 Oe and 105 Oe at 30 K, respectively. We chose a field of 50 Oe for measurements of magnetic relaxation. It took approximately 30 sec to raise the magnetic field from zero to the desired value.

RESULTS AND DISCUSSION

Figures 2 and 3 show the time dependences of magnetic isotherms for H$\perp$ab and H|ab, respectively. It is important to note that the diamagnetization decays with time logarithmically. Deviations from logarithmic behavior in initial transient stage is ascribed to flux penetration during stepping up of the field. The decay rate increases with temperature, indicating a contribution of a thermally activated process.

By assuming such a process to the dacay of magnetization, the hopping velocity of flux v is given by /7/

$$v = 2\nu_0 L\exp(-U_0/kT)\sinh(JBV_c L/kT), \qquad (1)$$

where ν_0 is the attempt frequency of the flux of volume V_c to hop over an energy barrier U_0 and move a distance L, J the current density and B the magnetic induction. The quantity $W=JBV_cL$ means the work done by Lorentz force. We now assume that J is approximately the same value as the transport critical current density, B the applied field, V_c the cube of the coherence

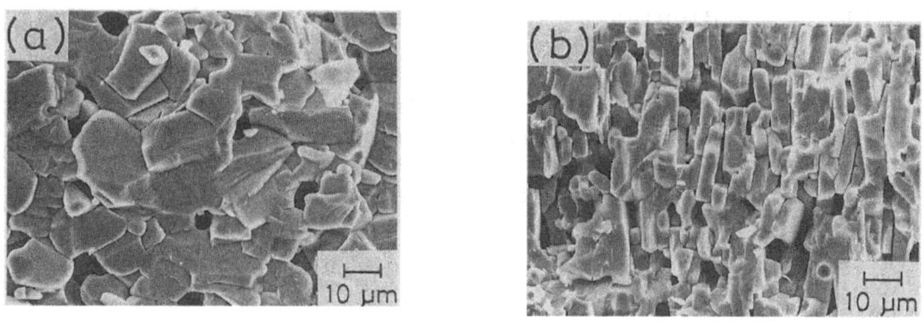

Fig.1 SEM micrographs of the fractured surfaces of an aligned sample; (a) is for the surface parallel to the ab-plane and (b) is for the surface perpendicular to the ab-plane.

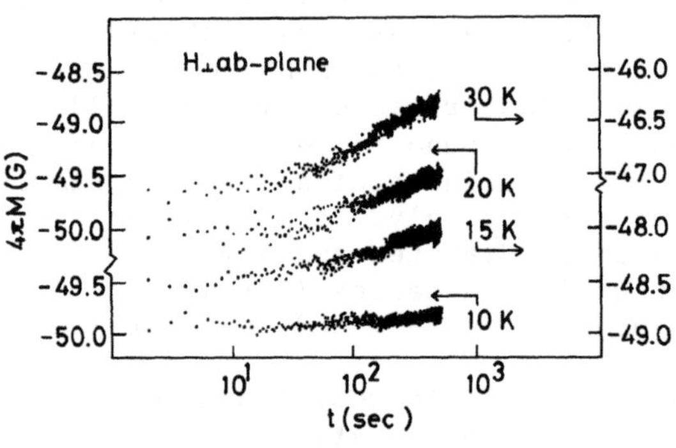

Fig.2 Time-dependent magnetic decay in the aligned sample for H$\perp$ab.

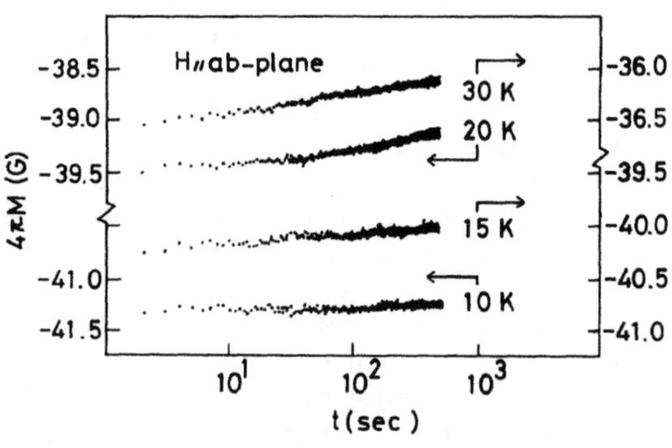

Fig.3 Time-dependent magnetic decay in the aligned sample for H$\|$ab.

length, and L the penetration length at grain boundaries. Assuming the values of $J=10^4$ A/cm^2, B=50 Oe, $V_c=10^{-21}$ cm^3, and $L=10^{-3}$ cm /8/ at measured temperatures, we find W_c of the order of 10^{-20} erg, much smaller than thermal energy kT of about 10^{-15} erg. Now, we may safely replace $\sinh(JBV_cL/kT)$ in eq.(1) by its argument JBV_cL/kT. Substitutung the v of eq.(1) into the equation of flux conservation

$$\partial B/\partial t+\mathrm{div}(vB) = 0, \tag{2}$$

we obtain a logarithmic relaxation rate given as

$$d(4\pi M)/d(\ln t) = C\exp(-U_0/kT)/T, \tag{3}$$

where we have used a relation

$$4\pi M+H = \int BdV/V. \tag{4}$$

During the derivation, we have implicitly assumed the pre-exponential factor $C \propto \nu_0 JB^2 V_c L^2$ is independent of temperature.

We have obtained experimental time-logarithmic relaxation rates $d(4\pi M)/d(\ln t)$ from the data for t≥50 sec by the least squares approximation method. Plots of $\ln(Td(4\pi M)/d(\ln t))$ versus 1/kT are shown in Fig.4 for H⊥ab and for H∥ab. Data at 4.2 K have been omitted here because the relaxation rates were negligibly small. Approximately linear fits are obtained over the temperature range of 10 to 30K. The activation energy U_0, evaluated from the slope, is 4.0±1.0 meV and 3.7±1.1 meV for H⊥ab and H∥ab, respectively. They are apparently isotropic within the experimental error. Furthermore, these values are one to two orders of magnitude smaller than those for intragranular flux motion /1/.

To check the validity of the above results, measurements of magnetic relaxation at a field of 1 kOe, well above Hc_1, for the same aligned sample was carried out. Analyzing data with the model in reference 1, we have obtained values of 19 meV and 72 meV at 4.2 K for H⊥ab and H∥ab, respectively. It certainly shows anisotropic character, although the value for H∥ab is about one half of the single crystal value of 150 meV /1/. Therefore, the apparent isotropy of the intergranular U_0 values has resulted from the characteristic of flux motion in intergranular region.

In fields below Hc_1 flux cannot enter the grains and is accordingly compressed along GB /9/. One possibility interpreting the apparent isotropy is that flux motions along both kinds of GB are intrinsically isotropic.

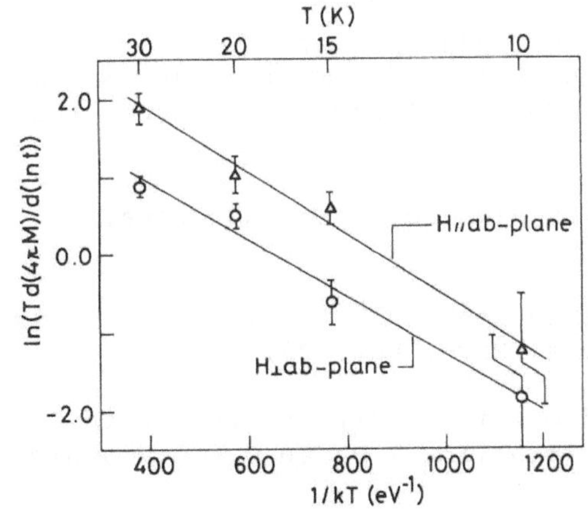

Fig.4 Plots of $\ln(Td(4\pi M)/(d\ln t))$ vs 1/kT for the alined sample.

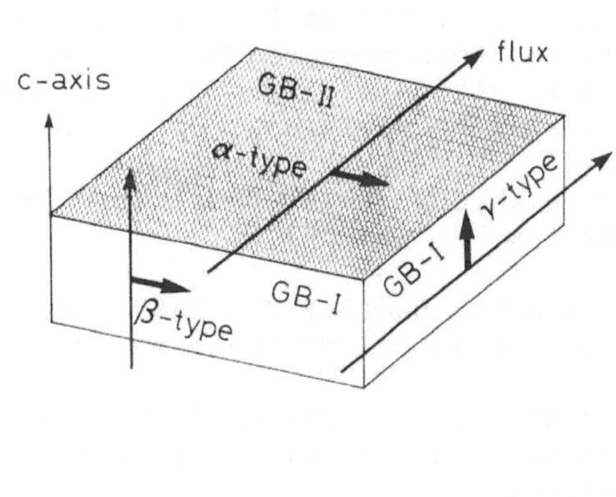

Fig.5 A model illustrating three types (α, β and γ) of flux movement on two kinds of grain boundaries (GB-I and GB-II) in aligned ceramics of $YBa_2Cu_3O_x$.

However, such is unlikely because of reported anisotropy of intergranular properties /8/. It would be reasonable to postulate that an elementary flux movement, encountering the largest U_0 value, acts as a rate-determining activation process (RDAP) for both external field directions. There are three types of flux movements as shown in Fig.5, denoted as α-, β- and γ-types. In H⊥ab case, α- and β-types should taken place. In H∥ab case, all of three types are involved. Among these γ-type is ruled out as RDAP because it occurs only in H∥ab case. Next question is which of α or β is RDAP. It has been observed that critical current density J_c flowing perpendicular to GB-I is larger than GB-II case /8/. As J_c is dominated by flux pinning at GB, the activation energy associated with β-type should be larger than that associated with α-type. This leads to a conclusion that β-type flux movement is RDAP.

In summary, intergranular flux creep in aligned polycrystalline $YBa_2Cu_3O_x$ was investigated. By using a thermally activated flux motion model, the activation energy was evaluated to be 4.0 ± 1.0 meV and 3.7 ± 1.1 meV for H⊥ab and H∥ab, respectively. It is suggested that the rate-determining activation process is flux movement perpendicular to the c-axis on grain boundaries perpendicular to the ab-plane of $YBa_2Cu_3O_x$.

REFERENCES

1. Yeshurun Y, Malozemoff AP, Holtzberg F, Dinger TR (1988) Magnetic relaxation and the lower critical fields in a Y-Ba-Cu-O crystal. Phys. Rev. B38: 11828-11831

2. Wong D, Stamper AK, Stancil DD, Schlesinger TE (1988) Low-field structure in the magnetization of polycrystalline $YBa_2Cu_3O_{7-x}$ and $ErBa_2Cu_3O_{7-x}$. Appl. Phys. Lett. 53: 240-242

3. Tjukanov E, Cline RW, Krahn R, Hayden M, Reynolds MW, Hardy WN, Carolan JF, Thompson RC (1987) Current persistence and magnetic shielding properties of $Y_1Ba_2Cu_3O_x$ tubes. Phys. Rev. B36: 7244-7247

4. Norling P, Svedlindh P, Nordblad P, Lundgren L, Przyslupsky P (1988) Low field magnetic relaxation phenomena in YBaCuO. Physica C153-155: 314-315

5. Ostertag CP, Shull RD, Vaudin MD, Blendell JE, Stearns LC, Fuller ER Jr (1988) Alignment of superconducting grains by magnetic casting. In: Yan MF (ed) Ceramic superconductors II: The American ceramic society, Inc., Westerville, Ohio, pp 332-342

6. Nakamura N, Nakano T, Gotoh S, Shimotomai M (1988) High-Tc oxide superconductors prepared by spray-drying method. In: Mayo WE (ed) Processing and applications of high Tc superconductors: The metallurgical society, Inc., Warrendale, Pennsylvania, pp 17-22

7. Dew-Hughes D (1988) Model for flux creep in high Tc superconductors. Cryogenics 28: 674-677

8. Hylton TL, Beasley MR (1989) Effect of grain boundaries on magnetic field penetration in polycrystalline superconductors. Phys. Rev. B39: 9042-9048

9. Evetts JE, Glowacki BA (1988) Relation of critical current irreversibility to trapped flux and microstructure in polycrystalline $Y_1Ba_2Cu_3O_7$. Cryogenics 28: 641-649

Flux Creep and Temperature Dependence of the Transport Critical Current Density in High-T_c Superconductors

NICK SAVVIDES

Division of Applied Physics, CSIRO, Sydney, 2070 Australia

ABSTRACT

We report measurements of the transport critical current density J_c of thin films and bulk ceramic specimens of high-T_c superconductors as a function of temperature in the range 4.2 K to T_c under self-field conditions. Theoretical models for thermally activated flux creep in homogeneous superconductors, and for current transport across Josephson weak links (S-I-S and S-N-S junctions) are used to analyze the temperature dependence of J_c. We find that flux creep determines the temperature dependence of the critical current density $J_c(T)$ of specimens, and that $J_c(T)$ increases approximately linearly with decreasing temperature for $t \lesssim 0.7$ (where $t = T/T_c$). Fits to the data are obtained using a barrier potential $U(0) = 10kT_c = 65$-90 meV.

KEY WORDS : critical current density, flux pinning, flux creep, activation energy.

INTRODUCTION

Recent studies of magnetic and critical current phenomena in high-T_c superconductors have shown the importance of classical flux-pinning ideas and the Anderson-Kim flux creep model in understanding the electro-magnetic properties of high-T_c superconductors [1-10].

In this paper we report measurements of the transport critical current density J_c in thin films and bulk ceramic specimens of $YBa_2Cu_3O_7$, and in bulk Bi-Pb-Sr-Ca-Cu-O. The pinning well depth or barrier potential U(0), one of the parameters which characterises the flux-creep process, is determined from the temperature dependence of the critical current density, $J_c(T)$, predicted by the flux creep model. We also use expressions for $J_c(T)$ predicted by using S-I-S and S-N-S models, and show that none of our specimens has the S-I-S behaviour, and only one special specimen can be fitted by the S-N-S model.

THEORETICAL MODELS

A. Flux Creep Model for Homogeneous Superconductors

According to the Anderson-Kim flux creep model [10-13] in the absence of an external magnetic field the transport current provides the flux density gradient for flux jump or creep. The critical current density J_c is controlled by flux creep, and the model gives the following expression for the thermally-activated flux creep velocity [9]:

$$v = v_0 \exp(-U/kT) \sinh(J_c \Phi_0 \xi^2/kT). \tag{1}$$

Here U is the depth of the pinning well or barrier potential which pins the flux lines or vortices, Φ_0 is the quantum of flux, and ξ is the coherence length. The parameters U, J_c and ξ are functions of temperature, and we express this dependence in terms of the reduced temperature $t = T/T_c$. Associated with the movement of flux lines is the induced electric field $\underline{E} = \underline{v} x \underline{B}$, where $\underline{B}$ is the magnetic self-field. $\underline{E}$ is parallel to $\underline{J}$ and under self-field conditions $E = v\beta J_c$, where β is a constant that depends on the sample geometry. Equation (1) is then an implicit relation of $J_c(t)$ and can be written as

$$\frac{E}{v_0 \beta J_c(t)} = \exp\{-U(t)/kT\} \sinh\{J_c(t)\Phi_0 \xi^2(t)/kT\}. \tag{2}$$

The pinning potential for the case of a weak magnetic field is given by

$$U(t) = H_c^2(t)\xi^3(t) \tag{3}$$

where H_c is the thermodynamic critical field. Following [9], we assume the temperature dependences of H_c and ξ are given by the Ginzburg-Landau (G-L) theory [13]:

$$H_c(t) \propto (1-t^2), \qquad \xi(t) = \xi(0)\{(1+t^2)/(1-t^2)\}^{\frac{1}{2}} . \tag{4}$$

The temperature dependence of U(t) is then

$$U(t) = U(0)(1+t^2)^{\frac{3}{2}}(1-t^2)^{\frac{1}{2}} . \tag{5}$$

To solve equation (2), we define dimensionless parameters c, s and u_0:

$$c = E\Phi_0\xi^2(0)/v_0\beta kT_c, \qquad s = \Phi_0\xi^2(0)J_c(t)/kT_c, \qquad u_0 = U(0)/kT_c, \tag{6}$$

where T_c is the critical temperature for zero resistivity. Substituting into equation (2), we obtain

$$\frac{c}{s} = \exp\{(-u_0/t)(1+t^2)^{\frac{3}{2}}(1-t^2)^{\frac{1}{2}}\}\sinh\{(s/t)(1+t^2)/(1-t^2)\}. \tag{7}$$

Using fixed values of u_0 we can generate a family of curves for different values of c. Some results of calculations are shown in Fig. 1 as normalized critical current density $J_{CN} = J_c(t)/J_c(0)$ versus reduced temperature t. Only curves for $u_0 = 10$ are shown. We obtain the best fit to the experimental data using this value of u_0.

B. S-I-S and S-N-S Models for Granular Superconductors

A granular superconductor may be represented as an array of identical Josephson-coupled superconducting grains arranged in a cubic lattice with lattice constant a_0.

Superconductor-insulator-superconductor (S-I-S) junctions are formed when the intergranular material is an insulator. If we neglect gap suppression by the supercurrent, then the critical current density due to tunneling between grains is given by the Ambegaokar-Baratoff (A-B) expression [14] for the maximum dc current of a Josephson junction,

$$I_0 = \{\pi\Delta(T)/2eR_n\}\tanh\{\Delta(T)/2kT\} \tag{8}$$

where $\Delta(T)$ is the gap parameter from the BCS theory, e is the electron charge, and R_n is the junction's normal-state tunneling resistance, which is assumed to be temperature independent. Close to T_c the Josephson current I_0 is proportional to $\Delta^2(T)$ and the model predicts $J_c \propto (1-t)$. Experimental results, however, show $(1-t)^{\frac{3}{2}}$ temperature dependence as expected from G-L theory.

This discrepancy has been corrected by Clem [15] who took into account the ability of the supercurrent to suppress the gap parameter. Using G-L theory he derived an expression for the critical current density of S-I-S junctions:

$$J_c(t) = J_0(t)g(\varepsilon). \tag{9}$$

Here $g(\varepsilon)$ is the factor by which gap suppression reduces the net critical current J_c below the A-B value J_0. The relevant dimensionless parameter in the theory is ε_0, and near T_c

$$\varepsilon_0 = 2\xi^2/a_0{}^2. \tag{10}$$

In the weak-coupling limit, $\varepsilon_0 \ll 1$, any intergranular currents are too small to suppress the gap parameter and J_c is simply given by the A-B expression. For $\varepsilon_0 \simeq 1$ the A-B temperature dependence holds only at low temperature, and gives way to the G-L $(1-t)^{\frac{3}{2}}$ behaviour above a cross-over temperature $T_x = T_c(1-0.882\varepsilon_0)$. Experimental results on conventional [15] and high-T_c superconductors [9] yield $\varepsilon_0 = 0.1 - 0.3$. Figure 2 shows normalized critical current versus reduced temperature according to A-B theory (ε_0=0) and Clem's treatment (ε_0=1).

Superconductor-normal metal-superconductor (S-N-S) junctions are formed when the intergranular material is a normal metal. The maximum Josephson current is given by the de Gennes-Werthamer-Clarke theory of the proximity-effect junction [16-18]:

$$I_0(t) \propto (1-t)^2 \exp\{-a_N t^{\frac{1}{2}}/\xi_N(T_c)\}. \tag{11}$$

Here a_N is the thickness of the normal metal barrier and ξ_N is the distance electron pairs penetrate the barrier. Near to T_c equation (11) gives $J_c \propto (1-t)^2$. Although the equation is strictly valid close to T_c, it holds at lower temperatures as well [19]. Figure 2 shows theoretical calculations for $a_N/\xi_N(T_c) = 0.1$, 1, 3, and 10. It will be seen that the temperature dependence of J_c is sufficiently different for S-I-S and S-N-S junctions so that measurements of J_c versus T are quite useful in discriminating between the two models. Also, in practice the maximum Josephson current of a S-I-S junction is considerably smaller than the corresponding S-N-S junction.

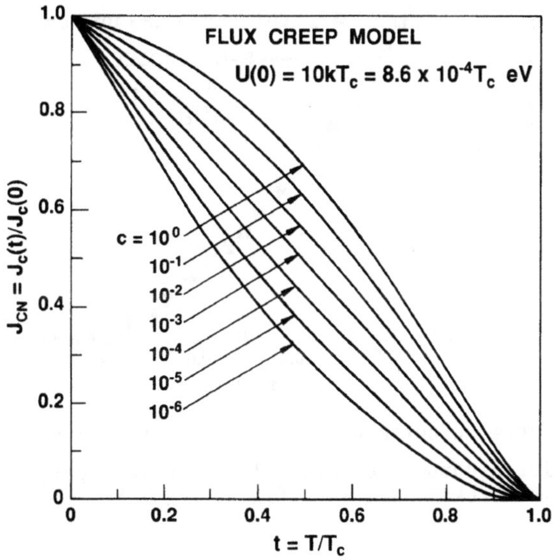

Fig.1 Normalized critical current density versus reduced temperature according to the flux creep model.

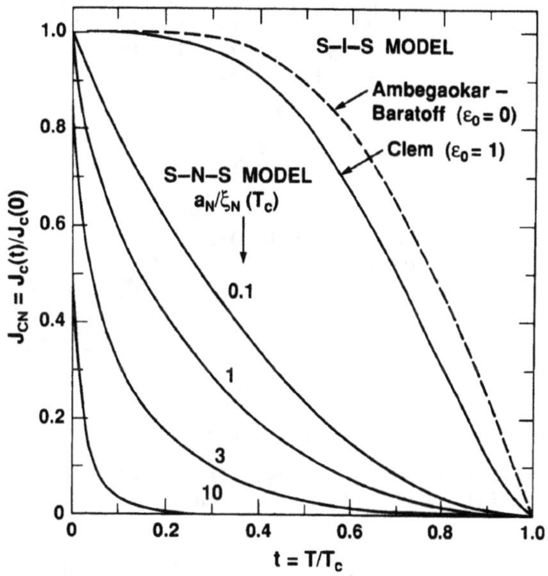

Fig.2 Normalized critical current density versus reduced temperature according to the S-I-S and S-N-S models.

EXPERIMENTAL

Thin film specimens of $YBa_2Cu_3O_7$ were prepared by unbalanced dc magnetron sputtering from a single stoichiometric target [20]. The films were deposited onto (100) yttria-stabilized zirconia and MgO wafers, and post annealed at 850-930°C in flowing oxygen. The preparation conditions were deliberately varied to obtain films of varying quality in the belief that the poorer specimens are more likely to show intrinsic weak-link behaviour. The best films had a high degree of epitaxial order with the c-axis normal to the film plane. Four-terminal dc measurements of critical current were done on films typically 1-2 μm in thickness and 12x3 mm^2 in surface area. It should be noted that doing measurements on large film areas would cause us to underestimate J_c either because the current flow is inhomogeneous or because structural defects in the films, such as microcracks and large voids, would lead to thermal propagation for currents well below the true critical current. Bulk ceramic specimens of $YBa_2Cu_3O_7$ and $(Bi_{1.6}Pb_{0.4})Sr_2Ca_2Cu_3O_{10}$ were prepared by pressing and then sintering powders derived from stir-dried nitrate solutions.

In the case of a homogeneous superconductor, the transition from flux-pinning to the full flux-flow state extends over a range of currents, and thus there is no precise critical state. We used E=1 $\mu V/cm$ as the criterion for flux-creep-controlled critical current J_c, and similarly for determining J_c in granular materials.

RESULTS AND DISCUSSION

We first analyze the temperature dependence of J_c in the framework of the flux creep model. Figure 3 shows plots of the normalized critical current density versus reduced temperature for the investigated $YBa_2Cu_3O_7$ thin films. Thermal activation increases the flux-creep rate thereby reducing J_c. To a first approximation J_c for the best films is linear in t at low temperatures ($t \lesssim 0.7$). Theoretical fits to the data are obtained using a single value $u_0 = U(0)/kT_c = 10$ for the barrier or activation energy, and a very narrow range of values for c. For films with $T_c \simeq 93$ K this corresponds to $U(0) = 80$ meV. These values agree with values obtained by others [1,8-10] using V-I characteristics and transport J_c measurements on epitaxial thin films of $YBa_2Cu_3O_7$.

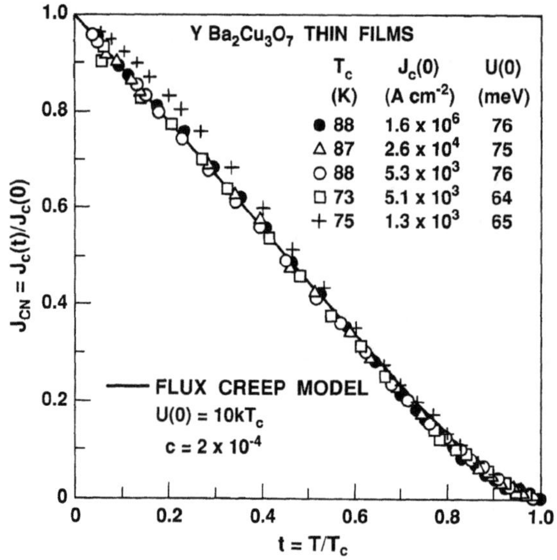

Fig.3 Experimental data of J_{CN} versus t for $YBa_2Cu_3O_7$ thin films. The solid line is a fit by the flux creep model.

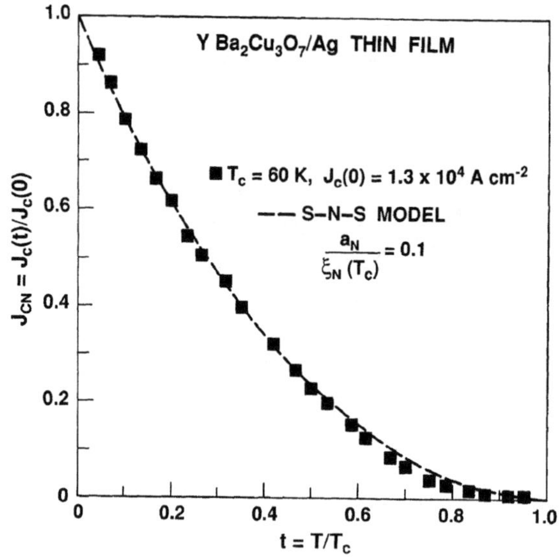

Fig.4 Experimental data of J_{CN} versus t for a composite thin film of $YBa_2Cu_3O_7$/Ag. The broken line is a fit by the S-N-S model.

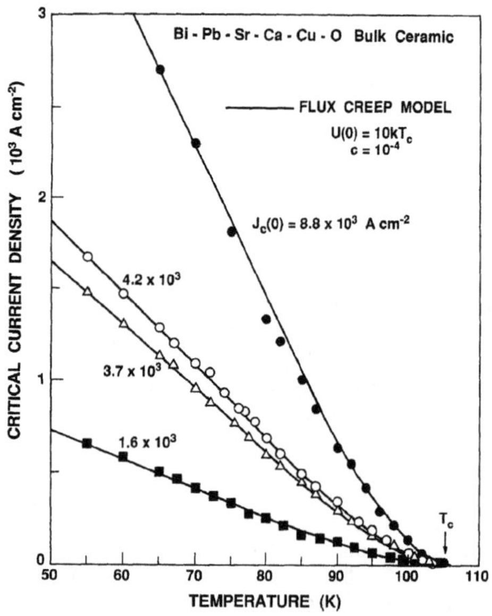

Fig.5 J_{CN} versus t for bulk specimens of $(Bi_{1.6}Pb_{0.4})Sr_2Ca_2Cu_3O_{10}$. The solid lines are fits to the experimental data by the flux creep model using measured values of J_c at 65K.

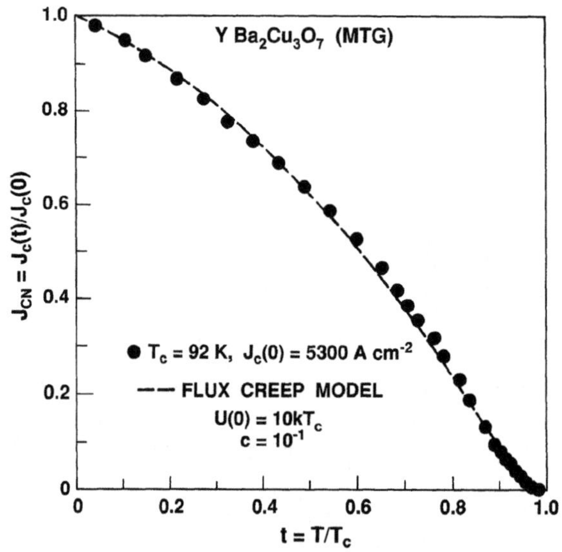

Fig.6 J_{CN} versus t for a melt-texture-growth bulk specimen of $YBa_2Cu_3O_7$. The broken line is a fit by the flux creep model.

Our attempts to produce films showing S-I-S behaviour were unsuccessful. We succeeded, however, in producing films showing S-N-S behaviour by depositing films on a thick Ag buffer layer. During post-annealing the Ag metal layer was consumed, and SEM examination revealed that the silver diffused along grain boundaries to form metal barriers between superconducting crystallites. Figure 4 shows typical results. These films had very low normal-state resistivity, and resistivity ratio $\rho_{300}/\rho_{100} \simeq 4$.

The bulk ceramic specimens examined were rectangular in shape (typically 1x4x20 mm²) and were necked down in the centre to a cross section of about 1x1 mm². All the specimens studied showed flux-creep controlled J_c (Figs. 5 and 6). Interestingly, the data on Bi-Pb-Sr-Ca-Cu-O specimens could be fitted to theory using the same values of u_0 and c as for thin films. For these specimens T_c = 106 K, and so the value u_o = 10 used corresponds to U(0) = 91 meV.

CONCLUSION

The temperature dependence of the transport critical current density $J_c(T)$ in thin films and bulk ceramic specimens of high-T_c superconductors was analyzed in the framework of the flux creep model, and that of S-I-S and S-N-S models. We find that $J_c(T)$ is flux-creep-controlled, and can be fitted in all cases assuming a value for the zero-temperature depth of the pinning well potential or activation energy of $U(0) = 10kT_c = 8.616x10^{-4}T_c$ eV.

ACKNOWLEDGEMENT

The author is pleased to acknowledge the help of D.M. Eagles in the numerical solution of equation (7), and useful discussions with K.-H. Müller. Bulk specimens and powders were kindly supplied by S.X. Dou and E. Vance. This work was supported in part by Department of Industry Technology and Commerce GIRD grant No.15017.

REFERENCES

1. J. Mannhart, P. Chaudhari, D. Dimos, C.C. Tsuei, and T.R. McGuire, Phys. Rev. Lett. 61, 2476 (1988).
2. T.T.M. Palstra, B.Batlogg, L.F. Schneemeyer, and J.V. Waszczak, Phys. Rev. Lett. 61, 1662 (1988).
3. Y. Yeshurun and A.P. Malozemoff, Phys. Rev. Lett. 60, 2202 (1988).
4. M.A.-K. Mohamed, W.A. Miner, J. Jung, J.P. Frank, and S.B. Woods, Phys. Rev. B 37, 5834 (1988).
5. M. Tinkham, Phys. Rev. Lett. 61, 1658 (1988).
6. T.T.M. Palstra, B. Batlogg, R.B. van Dover, L.F. Schneemeyer, and J.V. Waszczak, Appl. Phy. Lett., 54, 763 (1989).
7. B.D. Biggs, M.N. Kunchur, J.J. Lin, and S.J. Poon, Phys. Rev. B 39, 7309 (1989).
8. K. Enpuku, T. Kisu, R. Sako, K. Yoshida, M. Takeo, and K. Yamafuji, Jpn. J. Appl. Phys. 28, L991 (1989).
9. J.W.C. de Vries, G.M. Stollman, and M.A.M. Gijs, Physica C 157, 406 (1989).
10. J.D. Hettinger, A.G. Swanson, W.J. Skocpol, and J.S. Brooks, Phys. Rev. Lett. 62, 2044 (1989).
11. P.W. Anderson, Phys. Rev. Lett. 9, 309 (1962); P.W. Anderson and Y.B. Kim, Rev. Mod. Phys. 36, 39 (1964).
12. M.R. Beasley, R. Labusley, and W.W. Webb, Phys, Rev. 181, 682 (1969).
13. M. Tinkham, Introduction to Superconductivity (Krieger, Florida, 1980).
14. V. Ambegaokar and A. Baratoff, Phys.Rev.Lett. 10, 486 (1963); ibid 11, 104 (1963).
15. J. R. Clem, B. Bumble, S.I. Raider, W.J. Gallagher, and Y.C. Shih, Phys. Rev. B 35, 6637 (1987); J.R. Clem, Physica C 153-155, 50 (1988).
16. P.G. de Gennes, Rev. Mod. Phys. 36, 225 (1964).
17. N.R. Werthamer, Phys. Rev. 132, 2440 (1963).
18. J. Clarke, Proc. Roy. Soc. A 308, 447 (1969).
19. S. Greenspoon and H.J.T. Smith, Can. J. Phys. 49, 1350 (1971).
20. N. Savvides, C. Andrikidis, D.W. Hensley, R. Driver, J.C. Macfarlane, N.X. Tan, and A.J. Bourdillon, Proc. MRS Fall Meeting, Boston, Nov.27-Dec.2 (1989).

Magnetic Properties of Bi- and Tl-Based Single Crystals

T. KOTANI, K. OHKURA, H. TAKEI, and K. TADA

Basic High-Technology Laboratories, Sumitomo Electric Industries, Ltd., 1-3, Shimaya 1-chome, Konohana-ku, Osaka, 554 Japan

ABSTRACT

Crystal growth and magnetic relaxation of $Tl_2Ba_2CaCu_2Ox$, $Tl_2Ba_2Ca_2Cu_3Ox$, $(Tl,Bi,Pb)Sr_2CaCu_2Ox$ and $Bi_2Sr_2CaCu_2Ox$ single crystals were investigated. According to a study on the phase relations of these systems, four kinds of mm-sized single crystals with different transition temperatures were grown by the self-flux method. The single crystals showed a large magnetic relaxation, which was approximately linear with the logarithm of the time. Based on the thermally activated flux creep model, the effective flux-pinning potential energy for each crystal was estimated from these results.

KEY WORDS: Tl-Ba-Ca-Cu-O system, Bi-Sr-Ca-Cu-O system, crystal growth, flux creep, magnetic relaxation

INTRODUCTION

The magnetic properties of high-Tc oxide superconducting materials have been extensively studied for a understanding of these superconductivity and technical applications such as superconducting magnets. Since a large relaxation in the magnetization of the La-Ba-Cu-O system was found [1] , a number of studies have been carried out for an analysis of this behavior. For a basic understanding of this phenomenon, precise analyses based on a sample of single crystal are necessary, because a polycrystal sample, such as a sintered sample, has complex problems due to inhomogeneity, non-orientation and grain boundaries. A recent study on a single crystal of Y-Ba-Cu-O showed that this magnetic relaxation could be explained as thermally activated flux creep in the type II superconductor with low pinning potential energy [2] . However, only a few attempts have been made to examine other oxide superconducting systems with samples of single crystals.

In this study, we have investigated the growth of mm-sized single crystals in the Bi- and Tl-based superconducting systems by the flux method, and then measured the magnetic relaxation in the grown crystals with a DC SQUID magnetometer to estimate the effective flux-pinning potential energy of these materials based on the classical flux creep model.

EXPERIMENTAL

Based on the preliminary study of the phase relations in the Bi and Tl systems, four kinds of single crystals were grown by the slow-cooling method from CuO-rich solutions or stoichiometric melts. Raw materials for the crystal growth of $Bi_2Sr_2CaCu_2Ox$ were prepared by thoroughly mixing the powders of Bi_2O_3, $SrCO_3$, $CaCO_3$ and CuO in several ratios. These materials were melted at 1050°C for 4 hours in aluminum crucibles with lids, and then slowly cooled at a rate of 1.0°C/hour to room temperature. Single crystals were mechanically isolated by crushing these melt-processed materials. Single crystals of $Tl_2Ba_2CaCu_2Ox$, $Tl_2Ba_2Ca_2Cu_3Ox$ and $(Tl,Bi,Pb)Sr_2CaCu_2Ox$ were grown in gold tubes by the self-flux method. The detailed growth conditions for the crystals are described elsewhere [3] . The grown crystals were characterized by X-ray diffraction measurements(XRD) and energy-dispersive X-ray spectroscopy(EDX). The superconducting onset transition temperatures and magnetic relaxation of the crystals were measured with a

DC SQUID magnetometer. The samples for the measurements of the magnetization were typically areas of 1 to 5mm^2 with weights of 0.5 to 1mg. The process of measurement is shown next. The field H=0.1T, parallel to the c-axis of the crystal, was applied after cooling the crystal in a zero field. Thereafter, the applied field was returned to zero and the time dependence of remanent magnetization was measured at different temperatures.

RESULTS AND DISCUSSION

The grown crystals were plate-like in shape with a predominantly grown face of (001) for the Tl system, but bulky for the Bi system. A large single crystal of $Bi_2Sr_2CaCu_2O_x$ with maximum dimensions of 4mm×12mm×2mm was successfully grown from the raw materials with the ratio of Bi:Sr:Ca:Cu= 2:2:1:2. The single crystals in the Tl system and Tl-Bi-Pb system were typically 1mm×1mm wide and about 0.2mm thick. The temperature dependence of DC magnetic susceptibility for these crystals is shown in Fig.1. The Bi-based crystal showed a superconducting onset transition temperature of 88K. The Tl-based superconducting crystals of $Tl_2Ba_2CaCu_2O_x$ and $Tl_2Ba_2Ca_2Cu_3O_x$ have higher transition temperatures, 110K and 116K respectively. In contrast, the $(Tl,Bi,Pb)Sr_2CaCu_2O_x$ crystal with double Cu-O layers exhibited a transition temperature of 85K, close to that of $Bi_2Sr_2CaCu_2O_x$ crystal.

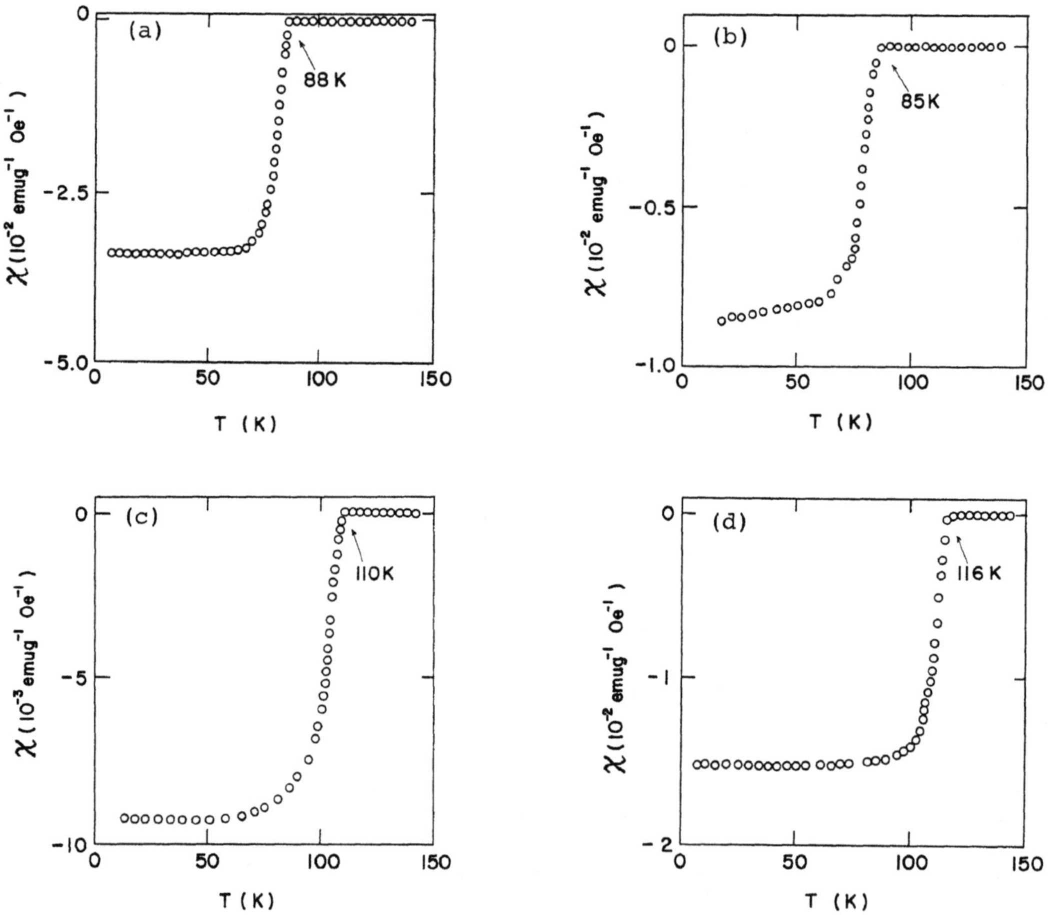

Fig.1. Temperature dependence of DC magnetic susceptibility for single crystals of (a) $Bi_2Sr_2CaCu_2O_x$, (b) $(Tl,Bi,Pb)Sr_2CaCu_2O_x$, (c) $Tl_2Ba_2CaCu_2O_x$ and $Tl_2Ba_2Ca_2Cu_3O_x$.

Lattice constants and analyzed compositions of each crystal are summarized in Table I. The compositions are almost consistent with those of the ideal chemical formulas in each system.

Figure 2 shows the decay of the magnetization in the single crystal of $Bi_2Sr_2CaCu_2Ox$ at different temperatures. The Bi-based crystal showed a large magnetic relaxation with a decrease of up to 20% of the initial value during about 1 hour. The relaxation rate roughly tends to increase with increasing temperature, and showed a peak at 50K. The decay was approximately linear with the logarithm of the time in most cases. This behavior can be explained by the thermally activated flux creep model. However, a deviation from this relationship was observed at short times in the case of increasing temperature. Therefore, the following discussions are based on the regions with a linearly time-logarithmic decay.

Table I. Lattice constants and analyzed compositions of the grown crystals.

Crystal		Lattice constant c	Composition						
System	Phase	(Å)	Tl	Bi	Pb	Sr	Ba	Ca	Cu
Bi	2212	30.9	-	2.1	-	1.9	-	0.8	2
Tl-Bi-Pb	1212	12.2	0.7	0.2	0.3	1.8	-	0.9	2
Tl	2212	29.2	2.0	-	-	-	2	0.7	2.0
Tl	2223	35.8	1.8	-	-	-	2	1.6	2.9

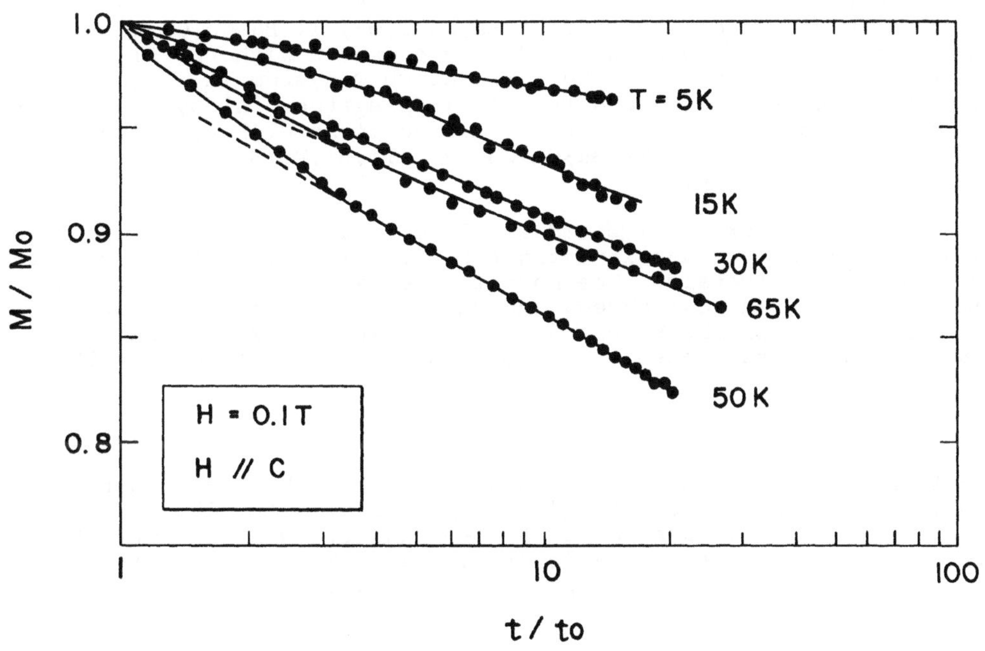

Fig.2. Relaxation in the magnetization of a $Bi_2Sr_2CaCu_2Ox$ single crystal. M_0 is the first measured value which is taken at about 200s after the applied field is returned to zero.

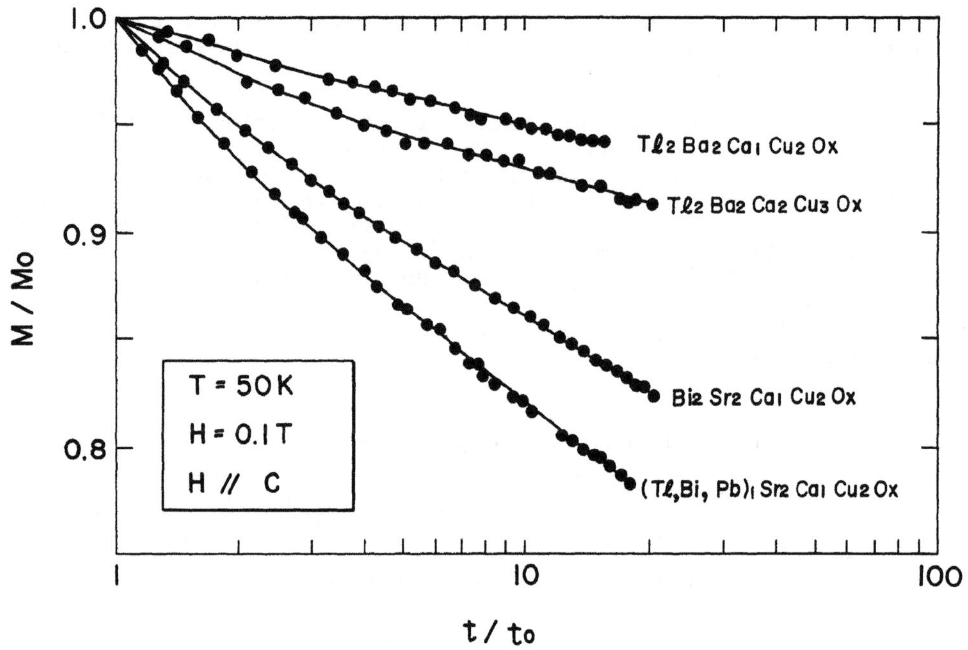

Fig.3. Relaxation in the magnetization for the Bi- and Tl-based single crystals at 50K.

Based on the flux creep model, the relaxation rate is related to the effective flux-pinning potential energy U_0 by the equation of $(1/M_0)(dM/d\ln t)=-kT/U_0$. The effective flux-pinning potential energy of the Bi-based single crystal was estimated to be 0.03-0.14eV at the temperature range in this study.
The relaxation in single crystals of $Tl_2Ba_2CaCu_2O_x$, $Tl_2Ba_2Ca_2Cu_3O_x$ and $(Tl,Bi,Pb)Sr_2CaCu_2O_x$ is shown in Fig.3, as compared with the Bi-based crystal. The relaxation rate of the Tl-Ba-Ca-Cu-O system is roughly half as large as that of the Bi system and the Tl-Bi-Pb system. Also, the effective pinning potential energy at 50K is calculated to be 0.19eV, 0.19eV, 0.06eV and 0.08eV for $Tl_2Ba_2CaCu_2O_x$, $Tl_2Ba_2Ca_2Cu_3O_x$, $(Tl,Bi,Pb)Sr_2CaCu_2O_x$ and $Bi_2Sr_2CaCu_2O_x$, respectively.

In conclusion, we have grown mm-sized single crystals of the Bi and Tl system by the slow-cooling method and investigated the relaxation of the magnetization in the grown crystals. The magnetic relaxation was linear with the logarithm of the time in most cases, and showed a large decrease of up to 20% in the initial value for the Bi-based crystal. The relaxation rates of the Tl system were lower than those of the Bi and Tl-Bi-Pb systems at 50K.

REFERENCES

1. K.A. Müller, M. Takashige and J.G. Bednorz (1987) Phys. Rev. Lett. 58:1143.
2. Y. Yeshurun and A.P. Malozemoff (1988) Phys. Rev. Lett. 60:2202.
3. H. Takei, T. Kotani, T. Kaneko and K. Tada (1988) Superconductivity on single crystals of Tl-Ca-Ba-Cu-O system. In: K. Kitazawa and T. Ishiguro(ed) Advances in Superconductivity, Springer-Verlag, Tokyo, p229.

Distribution of Flux Pinning Energies in Superconducting Thin Films

H. FURUKAWA, K. KAWAGUCHI, and M. NAKAO
Sanyo Tsukuba Research Center, 2-1, Koyadai, Tsukuba, Ibaraki, 305 Japan

ABSTRACT

The flux pinning behavior of Tl-Ca-Ba-Cu-O and Er-Ba-Cu-O thin films have been observed directly. Time-logarithmic magnetic relaxation M(t, T) was observed as well as in a single crystal. Within a thermally activated flux motion (TAFM) model, the distributions ρ (U_0) of activation energies U_0 are determined from data on the time and temperature dependence of M(t, T) for the Tl-Ca-Ba-Cu-O and Er-Ba-Cu-O films. The distribution for Tl-Ca-Ba-Cu-O film extends from 10 meV to well above 400 meV. The distribution for Er-Ba-Cu-O film exhibits a peak around 40 meV and has little dependence on the grain orientation.

KEY WORDS: flux pinning, activation energy, thin film, Tl-Ca-Ba-Cu-O, Er-Ba-Cu-O

INTRODUCTION

Shortly after the discovery of the high-T_c superconductors large flux creep effects in the magnetization have been reported [1]. Especially at low temperature the time dependence of the magnetization followed approximately a logarithmic law. The logarithmic time dependence of the magnetization has been found to be a prominent feature in the recently discovered high-T_c oxide superconductors. Logarithmic decay of magnetization with time has also been observed in conventional superconductors and successfully explained in terms of thermally activated flux motion (TAFM) model. In high-T_c superconductors, many authors pointed out that the magnetic relaxation also rises from TAFM. In this paper we analyze magnetic relaxation data within a TAFM model and show that the distribution of activation energies in oxide superconducting thin films are obtained from the temperature dependence of the magnetic relaxation by an inversion scheme of C.W.Hagen and R.Griessen [2,3].

EXPERIMENT

Four types of thin film samples were investigated in this study. Thin films of $TlCa_2Ba_2Cu_3O_9$ and $Tl_2CaBa_2Cu_2O_8$ were prepared by molecular beam deposition[4]. Each element of the film was evaporated from all metal sources in a conventional MBE apparatus. As deposited, Tl-Ca-Ba-Cu-O thin films are amorphous and need a post-deposition anneal to make the superconducting phases. The $TlCa_2Ba_2Cu_3O_9$ and $Tl_2CaBa_2Cu_2O_8$ films with a preferential orientation along the c-axis perpendicular to the film plane showed a superconducting transition at 110 K and 98 K, respectively. Thin films of Er-Ba-Cu-O were prepared by rf magnetron sputtering. When the films were deposited on MgO substrates at optimized temperature 650 ℃, the $ErBa_2Cu_3O_{7-y}$ thin films showed the superconducting transition at 62 K. The x-ray diffraction pattern showed a preferential orientation along the c-axis perpendicular to the film plane. Er-Ba-Cu-O thin films deposited without heating substrates are amorphous and insulating. After a post-deposition anneal around 900 ℃ for 1 h, the films showed a superconducting transition at 72 K and random orientation was observed in the x-ray diffraction pattern.

Susceptibility measurements were made using a HOXAN superconducting quantum interference device (SQUID) magnetometer. Values of critical current density J_c were determined using the Bean formula: $J_c = 30\ M_r/R$, where M_r is the remanent magnetization in emu/cm^3, R is an average sample radius in cm, and J_c is in A/cm^2. The remanent magnetization was measured as a function of temperature in zero field after cooling in a field (FC) perpendicular to the film surface and removing the applied field. Magnetization versus time measurements were made, also with the HOXAN SQUID magnetometer. The sample was cooled to appropriate temperature ranging from 4.5 K to 90 K, an external field of 100 or 200 Oe was applied perpendicular to the film surface. Subsequently, the magnetization was measured as a function of time. The procedure for acquiring the magnetic relaxation data consisted of cooling in zero field (ZFC) and, after waiting to attain a stable temperature, application of the field. The first magnetization point was taken after the 200 sec required to latch the field. Subsequent magnetization points were taken every 30 sec for the duration of the experiment which was typically 1 h. Between runs the field was removed and the sample was heated to above its transition temperature to expel all remaining flux.

RESULTS AND DISCUSSION

In the Bean critical state model, the critical current density J_c is related to the magnetization. Repeatedly, the remanent magnetization was measured in the Tl-Ca-Ba-Cu-O thin film (1.5 μm) cut in half and the remanent magnetization versus average sample radius was investigated. For the thin film the remanent magnetization M_r linearly decreased with decreasing average sample radius R. This linear dependence indicates that the simple Bean model is applicable in the critical state of thin film, when a field applied perpendicular to the film plane. Figure 1 shows the temperature dependence of J_c for the highly oriented Tl-Ca-Ba-Cu-O and Er-Ba-Cu-O thin films and the randomly oriented Er-Ba-Cu-O thin films. The J_c values at 4.5 K were 5.9×10^4, 2.5×10^4, 6.3×10^6 and 3.3×10^3 A/cm^2 in zero field for the thin films of $TlCa_2Ba_2Cu_3O_9$, $Tl_2CaBa_2Cu_2O_8$, as-deposited $ErBa_2Cu_3O_{7-y}$ and annealed $ErBa_2Cu_3O_{7-y}$, respectively. The randomly oriented $ErBa_2Cu_3O_{7-y}$ film has low J_c's of the order of 10^3 A/cm^2 in zero field, implying the existence of many weak links.

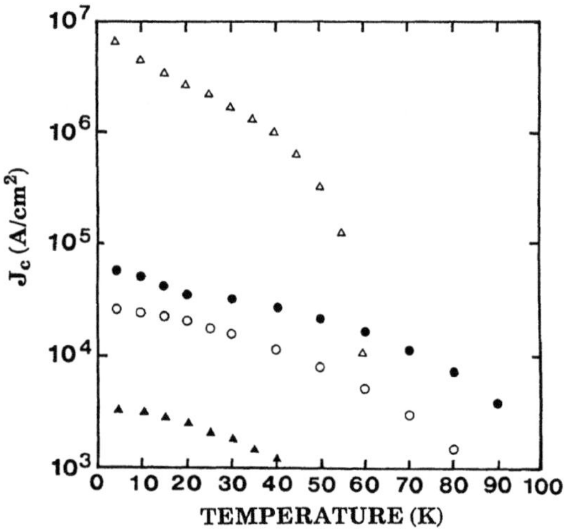

Fig. 1 Temperature dependence of the critical current densities in a randomly oriented $ErBa_2Cu_3O_{7-y}$ thin film(▲), highly oriented $ErBa_2Cu_3O_{7-y}$ (Δ), $TlCa_2Ba_2Cu_3O_9$(●) and $Tl_2CaBa_2Cu_2O_8$(○) thin films.

Magnetization of superconducting thin films was measured as a function of time. For the thin films of both Tl-Ca-Ba-Cu-O and Er-Ba-Cu-O, the time dependence of the magnetization followed a logarithmic law, which is explainable within a TAFM model. Figure 2 illustrates the temperature dependence of logarithmic magnetic relaxation rates dM/dlnt for the highly oriented $TlCa_2Ba_2Cu_3O_9$ and $Tl_2CaBa_2Cu_2O_8$ thin films. The samples were cooled initially in zero field, and a field of 100 Oe was applied. The relaxation rates dM/dlnt change little in the temperature range from 4.5 K to 60 K.

Figure 3 shows the temperature dependence of relaxation rates for the highly and randomly oriented $ErBa_2Cu_3O_{7-y}$ thin films. The relaxation rates exhibit a broad peak around 12 K for both samples and decrease with increasing temperature. The difference of relaxation rates between randomly and highly oriented Er-Ba-Cu-O thin films is attributed to the difference of driving force, which is related to critical current density.

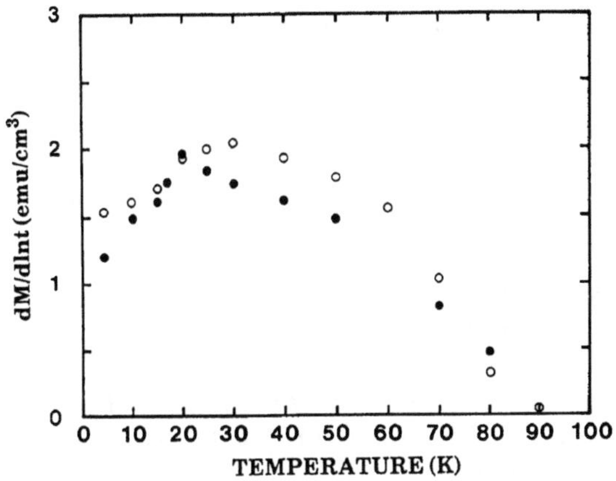

Fig.2 Temperature dependence of the logarithmic relaxation rates in highly oriented $TlCa_2Ba_2Cu_3O_9$ (•) and $Tl_2CaBa_2Cu_2O_8$ (○) thin films.

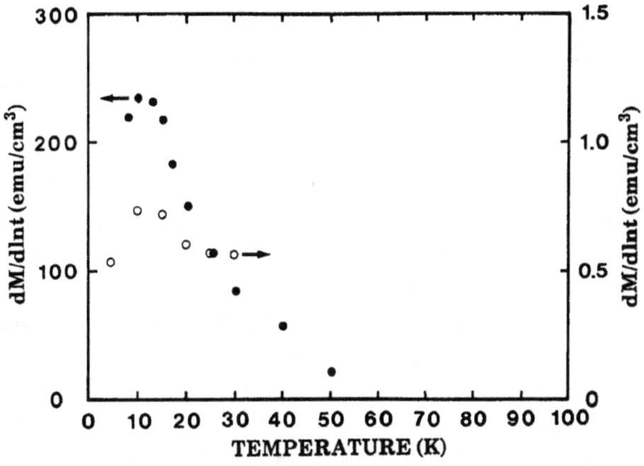

Fig.3 Temperature dependence of the logarithmic relaxation rates in highly oriented $ErBa_2Cu_3O_{7-y}$ (•) and randomly oriented $ErBa_2Cu_3O_{7-y}$ (○) thin films.

The relaxation rate is closely related to thermal activation energy U_0 in the following equation: $dM/d\ln t = M_0 k_B T/U_0$, where M_0 is M_r in the absence of thermal activation. The activation energies estimated by the above equation increased with increasing temperature. The fact that the effective value of U_0 depends seriously on temperature is concerned with a distribution of U_0. The distribution functions $\rho(U_0)$ of activation energies in oxide superconducting thin films are obtained from the temperature dependence of the magnetic relaxation rates by the inversion scheme of C.W.Hagen and R.Griessen [2]. Figures 4 and 5 show the distribution functions $\rho(U_0)$ of activation energies U_0 calculated from the relaxation data for the highly oriented thin films of Tl-Ca-Ba-Cu-O and the highly oriented and randomly oriented thin films of Er-Ba-Cu-O, respectively. The distribution for $TlCa_2Ba_2Cu_3O_9$ phase is quite similar to that for $Tl_2CaBa_2Cu_2O_8$ phase. This result suggests no difference

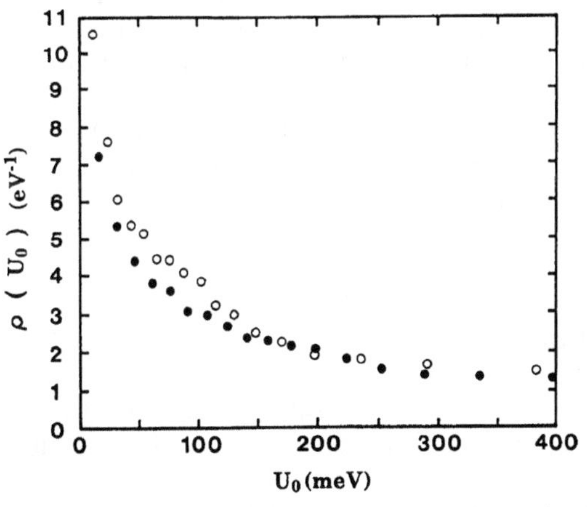

Fig.4 Distribution functions $\rho(U_0)$ of activation energies U_0 for depinning in highly oriented $TlCa_2Ba_2Cu_3O_9$ (•) and $Tl_2CaBa_2Cu_2O_8$ (○) thin films.

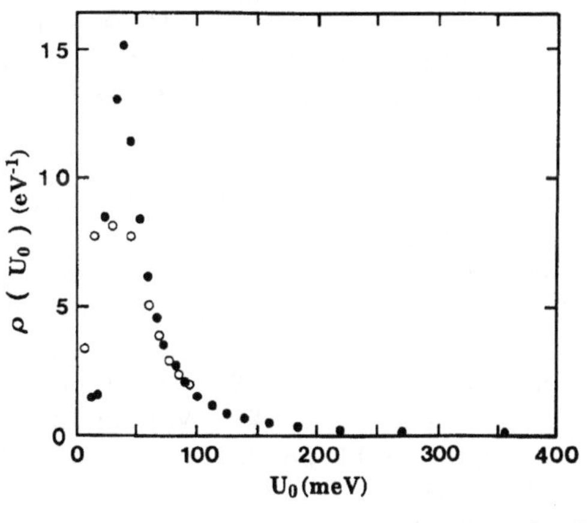

Fig.5 Distribution functions $\rho(U_0)$ of activation energies U_0 for depinning in highly oriented (•) and randomly oriented $ErBa_2Cu_3O_{7-y}$ (○) thin films.

in flux pinning behavior between these phases. The presence of high-energy tail in the distributions of the Tl-Ca-Ba-Cu-O thin films is consistent with the activation energy of 0.33 eV reported by M.E.McHenry *et al* [5] in poly-crystalline Tl 2:2:2:3 oxide superconductor. The distribution $\rho(U_0)$ for the highly oriented thin films of Er-Ba-Cu-O exhibits a peak around 40 meV, as seen in Fig. 5, and is quite similar to the result obtained by C.W.Hagen *et al* [2] in a Y-Ba-Cu-O single crystal, implying the effect of the same flux pinning mechanism. In comparison between the highly oriented and the randomly oriented film of Er-Ba-Cu-O the distribution $\rho(U_0)$ has little dependence on the grain orientation. Note that the distributions $\rho(U_0)$ for the thin films are basically the same as a single crystal, whether the grains have random orientation or highly orientation. If anisotropic pinning centers are effective to flux pinning in Er-Ba-Cu-O, the distribution will depend on the grain orientation. But the obtained distribution functions are similar. As a result of this similarity of the distribution $\rho(U_0)$, in flux pinning, anisotropic pinning centers are not effective in Er-Ba-Cu-O thin films. The fact that the distribution $\rho(U_0)$ resemble between the thin films and a single crystal indicates that flux pinning in grain boundaries makes no significant contribution to the pinning behavior in these films. This suggests that flux creep process is dominated by the flux pinning in the interior of grains, which shows no remarkable anisotropy to flux pinning.

CONCLUSIONS

(1) The linear dependence of the sample dimension versus the remanent magnetization indicates that the critical state in a thin film can be described by the simple Bean model.
(2) The temperature dependence of the relaxation rate can be explained by the distribution of activation energies $\rho(U_0)$.
(3) The pinning centers in Er-Ba-Cu-O might be related to defects in the interior of grains, which give no remarkable anisotropy to flux pinning.

The authors would like to thank M.Nemoto and Y.Matsuta for preparation of sputtered thin films.

REFERENCES

1. Müller KA. Takashige M. Bednorz JG. (1987) Flux trapping and superconductive glass state in La_2CuO_{4-y}:Ba. Phys Rev Lett 58: 1143-1146
2. Hagen CW. Griessen R. (1989) Distribution of activation energies for thermally activated flux motion in high-T_c superconductors: an inversion scheme. Phys Rev Lett 62: 2857-2860
3. Griessen R. Hagen CW. Lensink J. Salomons E. Flipse CFJ. Dam B. (1989) Thermally activated flux motion in high-T_c superconducting and high critical current epitaxial films. 2nd Workshop on High-Temperature Superconducting Electron Devices R&D Association for Future Devices. 7-9 June 1989 in Shikabe, Hokkaido, Japan. pp 273-280
4. Furukawa H. Nakao M. (1989) Growth of highly oriented Tl-Ca-Ba-Cu-O thin films using shutter-controlled MBE techniques. Tsukuba Seminar on High T_c Superconductivity. May 31-June 2 1989. Tsukuba Japan. pp 155-160
5. McHenry ME. Maley MP. Venturini EL. Ginley DL. (1989) Magnetic relaxation in sintered $Tl_2Ca_2Ba_2Cu_3O_x$ and $YBa_2Cu_3O_{7-x}$ superconductors. Phys Rev B 39: 4784-4787
6. Nakao M. Kawaguchi K. Furukawa H. Shikichi K. Matsuta Y. (1989) Comparison of flux pinning in superconducting Tl-Ca-Ba-Cu-O single crystals and thin films. Materials and Mechanisms of Superconductivity High-Temperature Superconductors. 23-28 July 1989 Stanford, California

Relaxation in High Transport Current Bi-Sr-Ca-Cu-O System

HIRONORI MATSUBA, AKIHITO YAHARA, and KUMIKO IMAI

Yokohama R & D Laboratories, The Furukawa Electric Co., Ltd., 4-3, Okano 2-chome, Nishi-ku, Yokohama, 220 Japan

ABSTRACT

The relaxation of a persistent current circulating in a toroid made of $Bi_2Sr_2Ca_1Cu_2O_{8+x}$ superconductor has been measured. The relaxation is described of the form of t^{-a} and is explained by a equation deduced from a power-law characteristic of the I-V property, which is originated from a 2-dimensional coupled Josephson junction array. It is further revealed that the power-law can be applied to the superconductors over very wide ranges of voltages from 0.1mV to 0.1pV.

KEY WORDS:relaxation,persistent current,power-law,flux creep,$Bi_2Sr_2Ca_1Cu_2O_{8+x}$

INTRODUCTION

Much research on high temperature oxide superconductors has revealed that thermally activated flux creep controls the dissipation behavior of the superconductors[1-3]. The bismuth oxide system has the smallest activation energy, implying that 'zero' resistivity does not exist at the temperatures around 77K and that the low activation energy reduces critical current. The most direct proof of zero resistivity of the superconductors is the existence of persistent current. Persistent current measurements of logarithmic decays have been reported with toroids of the Y-Ba-Cu-O system[4-6] and of the Tl-Ca-Ba-Cu-O system[7]. The relaxation behavior of the persistent currents has been explained by the spin-glass-like state[8] or by the flux creep theory[9].

In a bismuth oxide system, the authors have studied the low Tc phase with a transition temperature of 75-85K to improve both the critical transport current and the critical temperature. We have achieved a superconducting transport current greater than 1000 A/cm^2(defined by the electric field of 1μV/cm) at 77K, and 'zero' resistance at temperature exceeding 95K with a $Bi_2Sr_2Ca_1Cu_2O_{8+x}$ composition[10].

In this report, we shall present a persistent current experiment using a toroid made of the bismuth compound with a critical temperature of 95K, and explain the relaxation behavior.

SAMPLE PREPARATION AND EXPERIMENT

The toroidal sample was prepared by means of the solid state reaction method on the basis of the proper heating method. The toroid was made in a ring shape with a rectangular cross section. The size of the toroid was 23mm outer diameter, 8mm inner diameter, and 1.5mm thickness. Fig.1 shows the temperature dependence of the resistivity and susceptibility for a sample made with the concurrent fabrication process. The sample shows 'zero' resistivity at 95K with 1mA current. Fig.2 shows the current-voltage (I-V) curves measured by the four probe method with a pulsive current, increases to 100A within 25 msec. The I-V property follows a power-law[10].

A schematic figure of the experimental assembly is shown in Fig.3. First the sample was cooled in liquid nitrogen with zero magnetic field, and then the external magnetic field B_0 was applied coaxially to the sample for 1.2 sec.

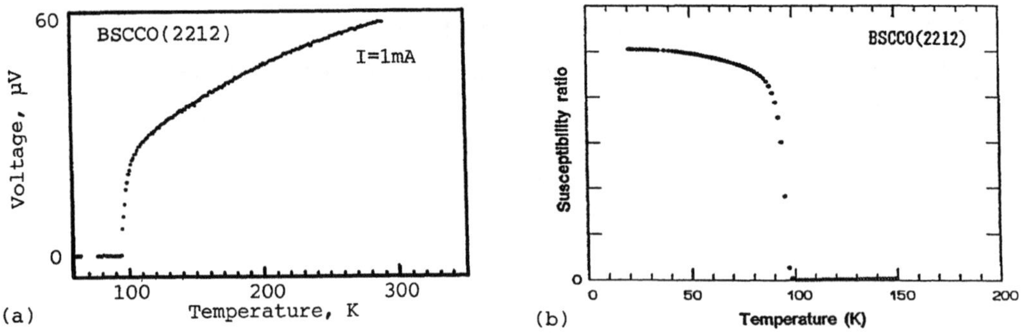

Fig.1 Temperature dependence of (a)resistivity and (b)susceptibility for a BSCCO(2212) sample. 'Zero' resistivity is 95K with 1mA.

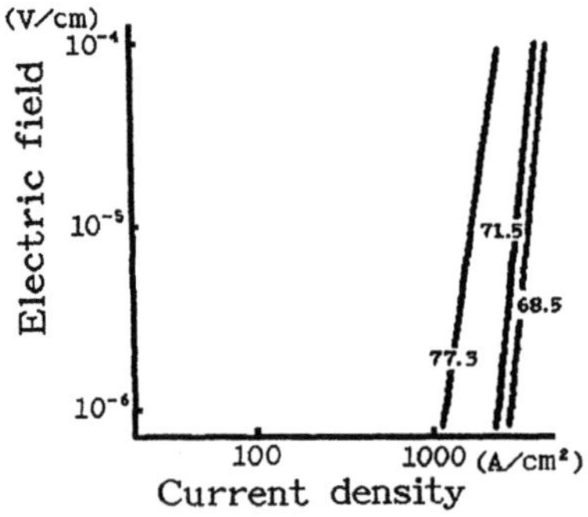

Fig.2 Current-voltage(I-V) curves for BSCCO(2212) samples by the four probe method. Numbers indicate the measured temperatures.

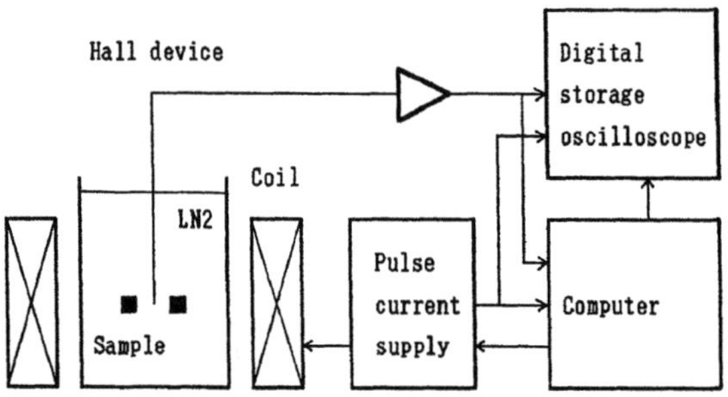

Fig.3 Experimental assembly for the relaxation measurement.

Thereafter, a field B_{it} remained inside the toroid which resulted from the persistent current I_p circulating in the toroid. Relaxation of the remanent field B_{it} was measured by using a Hall device from 10^{-1} to 10^5 seconds.

If the applied field B_0 was greater than a critical value, then the relaxation behaved in similar manner under different B_0s. Fig.4(a) shows the relaxation data plotted according to the flux creep model [B_{it} vs. log(t)], and Fig.4(b) shows the plots based on a power-law model [log(B_{it}) vs. log(t)]. The results fit the power-law model better than the flux creep model.

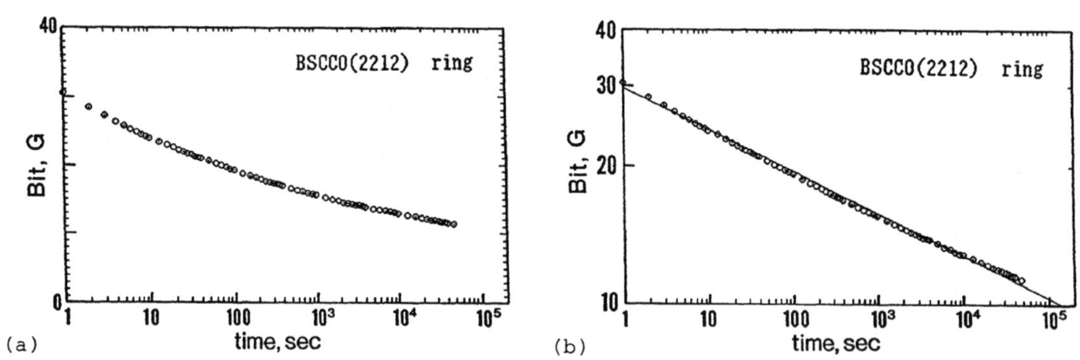

Fig.4 Relaxation of a trapped field in a BSCCO(2212) toroid;(a) plotted according to the flux creep model [B_{it} vs. log(t)], and (b) plotted based on the power-law model [log(B_{it}) vs. log(t)]. Solid line indicates the calculated curve on the basis of the equation(4).

On the other hand, if the applied field B_0 was smaller so that the initial remanent field B_{it} has a smaller value than that in Fig.4, relaxation behavior depended on the initial value of B_{it}. Typical results are shown in Fig.5.

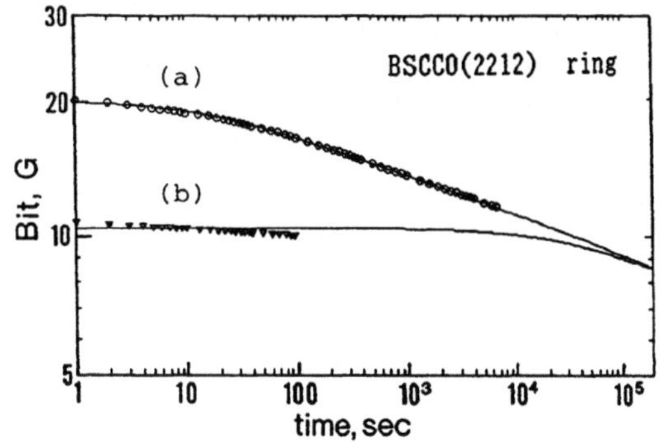

Fig.5 Relaxation of a trapped field [log(B_{it}) vs. log(t)] in a BSCCO(2212) toroid with smaller applied fields. Solid curves are derived from the equation(3).

DISCUSSION

The current-voltage (I-V) property of this bismuth compound follows a power-law, as shown in Fig.2, as;

$$v/v_k=(i/i_k)^n, \qquad ---- (1)$$

where i represents current density flowing in the superconductor and v represents the electric field generated by the current. v_k, i_k, and n are constant independent of i and of v. The decay of the persistent current I_p is caused by the generated voltage V_r induced in the toroid, according to:

$$V_r = - L\frac{dI_p}{dt}, \qquad ---- (2)$$

where L is the self-inductance of the toroid. Then, substituting the relations $V_r=2\pi r*v$, $I_p=S*i$, where r and S represent the mean radius and the cross-section area of the toroid, respectively, and substituting equation(1) into equation(2), we obtain

$$I_p = \frac{I_{p0}}{(1+t/t_0)^{\frac{1}{n-1}}}, \qquad t_0 = \frac{L}{n-1}\cdot\frac{I_{p0}}{V_{r0}}, \qquad ---- (3)$$

where I_{p0} and V_{r0} indicate the initial conditions of I_p and V_r at t=0, respectively.

Taking the case t<<t0, relation(3) is simplified as follows:

$$I_p = I_{p0},$$

and when t>>t0, relation(3) describes a power-law as:

$$I_p = I_{p0}(t/t_0)^{-\frac{1}{n-1}}. \qquad ---- (4)$$

The relation(4) explains the relaxation behavior when the applied field B_0 was greater than a critical value. In the case of Fig.4(b), n=12, t_0=1sec and I_{p0}=34A were used in the calculation. On the other hand, the relation(3) explains the relaxation behavior when the applied field B_0 was smaller value as shown in Fig.5. Solid lines in Fig.5 are calculated from the equation(3); n=12.7, t_0=13sec and I_{p0}=23A are used for the case(a), and n=12.7, t_0=22000sec and I_{p0}=12A for the case(b). These results imply that the relation(3) describes the relaxation behavior of the persistent current under various initial conditions.

The I-V property of the bismuth compound at very low generated voltages can be indirectly deduced from the equation(3). Fig.6 shows the deduced I-V property on the basis of the persistent current measurement shown in Fig.4(b). The I-V property measured by a four probe method is also plotted in Fig.6. Both plots are well linked and show that the I-V property of the bismuth compound follows a power-law over very wide ranges of voltage. The current-voltage exponent n increases, from 11 to 14, with decreasing current, from 34A to 13A, respectively.

It is known that the I-V characteristics of a square array of Josephson junctions exhibit a power-law behavior, and the current-voltage exponent decreases with increasing magnetic field[11]. The temperature dependence of the critical current density of the bulk samples also implies the existence of layered structures coupled by Josephson junctions[10]. The samples were structured with thin layered grains as shown in Fig.7. The results suggest that the dissipation behavior of $Bi_2Sr_2Ca_1Cu_2O_{8+x}$ superconductor is controlled by 2-dimensional coupled Josephson junctions that exist between layered grains.

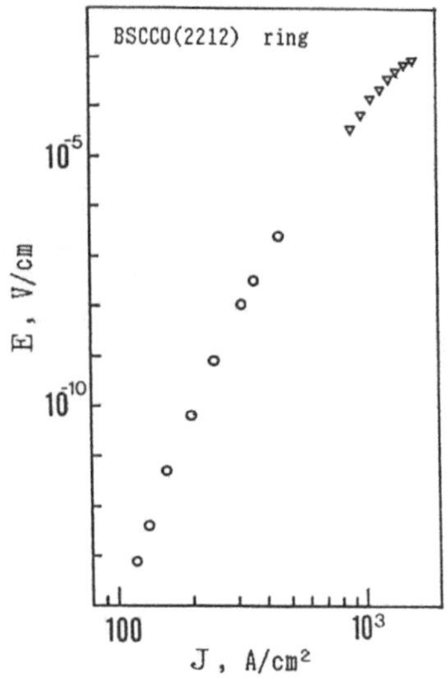

Fig.6 I-V curve of a BSCCO(2212). ▽: from a four probe measurement, O: from a relaxation measurement.

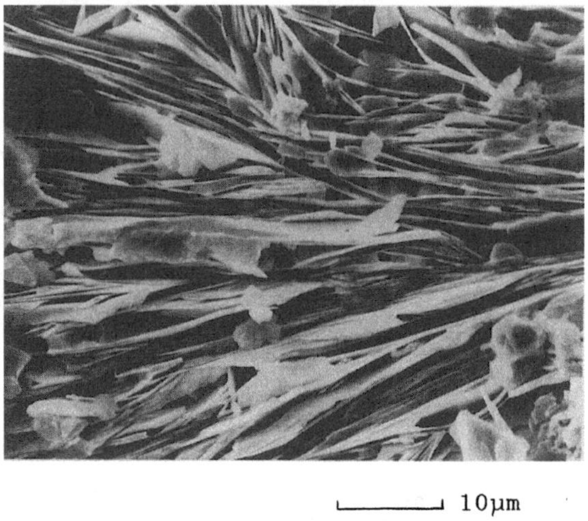

Fig.7 SEM photograph of the fractured surface of the BSCCO(2212) sample with a 'zero' resistivity at 95K and with a critical current of 1000 A/cm^2(1μV/cm).

CONCLUSION

We have measured the relaxation of a persistent current circulating in a toroid made of $Bi_2Sr_2Ca_1Cu_2O_{8+x}$ superconductor. The circulating current decayed of the form of t^{-a} and the relaxation behavior is explained by a power-law characteristic of the I-V property of the superconductor. The results implies that the dissipation behavior of the polycrystalline bulk $Bi_2Sr_2Ca_1Cu_2O_{8+x}$ superconductor is controlled by a square array of Josephson junctions between layered grains. We have further demonstrated that the power-law can be applied to the superconductors over very wide ranges of voltages from 0.1mV to 0.1pV.

REFERENCE

1 Y.Yeshurun and A.P.Malozemoff;Phys.Rev.Lett.**60**(1988)2202

2 Y.Yeshurun,A.P.Malozemoff,T.K.Worthington,R.M.Yandrofski,L.Krusin-Elbaum, F.H.Holtzberg,T.R.Dinger,and G.V.Chandrashekhar;Cryogenics,**29**(1989)258

3 H.Kumakura,K.Togano,E.Yanagisawa,K.Takahashi,M.Nakao,and H.Maeda; Jpn.J.Appl.Phys.**28**(1989)L24

4 W.J.Yeh,L.Chen,F.Xu,B.Bi,and P.Yang;Phys.Rev.B**36**(1987)2414

5 E.Tjukanov,R.W.Cline,R.Krahn,M.Hayden,M.W.Reynolds,W.N.Hardy,J.F.Carolan and R.C.Thompson;Phys.Rev.B**36**(1987)7244

6 O.G.Symko,W.J.Yeh,and D.J.Zheng;J.Appl.Phys.**65**(1989)2142

7 G.S.Grader,E.M.Gyorgy,L.G.Van Uitert,W.H.Grodkiewicz,T.R.Kyle,and M.Eibschutz;Appl.Phys.Lett.**53**(1988)319

8 C.Ebner and D.Stroud;Phys.Rev.B**31**(1985)165

9 P.W.Anderson;Phys.Rev.Lett.**9**(1962)309

10 K.Imai and H.Matsuba;proc Workshop on High Temp.Superconductivity, Huntsville,Alabama,USA,23-25 May 1989(to be published)

11 J.P.Carini;Phys.Rev.B**38**(1988)63

Resistive Behavior of High T_c Superconducting Thin Film in Magnetic Field

T. MATSUURA, S. TANAKA, K. HARADA, H. ITOZAKI, and S. YAZU

Itami Research Laboratories, Sumitomo Electric Industries Ltd., 1-1, Koyakita 1-chome, Itami, Hyogo, 664 Japan

ABSTRACT

We have measured the temperature dependence of resistance for epitaxial and polycrystalline YBCO thin films in a magnetic field. The resistive behavior of the epitaxial thin film could be explained by the flux creep model. The pinning potential estimated from that model had strong anisotoropy and this anisotropy was consistent with that of the coherence length. On the other hand, the resistive behavior of the polycrystalline thin film could not be explained by the flux creep model. It was suggested that the broadening of the resistivity for the polycrystalline thin film is caused by weakly linked grain boundaries.

KEY WORDS: pinning potential, thin film, resistive behavior, anisotropy

INTRODUCTION

The resistive transition of the high Tc superconductors become broad remarkably in a magnetic field. There are several models for this broadening, which are the flux creep model[1,2], the giant fluctuation model[3] and the glassy state model[4]. However, it is not sure which model can explain this behavior well. In this paper , I will report on the analysis of resistive behavior of YBCO thin film based on the flux creep model, in order to make sure whether the flux creep model can be adapted to the resistive behavior of YBCO thin film.

THEORETICAL BACKGROUND

The flux creep model represents that the resistance of superconductors is generated by thermally activated hopping of flux over a pinning barrier. The conventional equation for the resistivity based on that model can be written

$$R=R_0\cdot\exp(-(U-J\cdot B\cdot L\cdot Vc)/kT) \qquad (1)$$

where R_0 is the prefactor, U is the pinning potential, J is the applied current density, B is applied field, Vc is the volume of the flux bundle, L is the dimension parameters of flux bundle and k is the Boltzman constant[5]. $J\cdot B\cdot L\cdot Vc$ means Lorentz force energy. Here we used pinning potential, U which depends on temperature and magnetic field, as proposed by Tinkam[6].

$$U=\alpha Hc^2 la^2=\beta(1-t)^{1.5}/H \qquad (t=T/Tc) \qquad (2)$$

where Hc is the thermodynamic critical field, l is the coherence length along the direction of the field direction, H is the applied field, and a is the flux lattice spacing. α and β includes all numerical factors. Using Eq.(2) and ignoring the Lorentz force energy part of Eq.(1) (because measurement current is much lower than critical current), we obtain

$$R=R_0 \cdot \exp(-\beta/H \cdot k(1-t)^{1.5}/T) \qquad (3)$$

From Eq.(3), log R can be linearly plotted as the function of $(1-t)^{1.5}/T$. The slope of that straight line represents β/Hk.

EXPERIMENTAL PROCEDURE

We prepared two types of YBCO thin films, epitaxial and polycrystalline films. These films were deposited on MgO(001) single crystals by RF magnetron sputtering[7]. Both films were c-axis oriented and the thickness was 600nm. The critical temperature of the epitaxial film was 85K and critical current density at 77.3K was 1.2 x 10^6 A/cm^2 in a zero magnetic field. The critical temperature of the polycrystalline film was 60 K.

The specimens were cut to the size of 0.2cm x 1cm. The dc current density for the measurement was 8 A/cm^2. The magnetic field up to 8 Tesla was applied by a superconducting magnet. After the sample was cooled down to 10k in a zero magnetic field, the magnetic field was increased to the target field(0.3-8.0 Tesla) and the resistance was then measured with increasing temperature. The measurement limit of the system for the resistance was 10 mΩ. The magnetic fields were applied parallel and perpendicular to the a-b plane of the film.

RESULTS and DISCUSSION

The resistance of the epitaxial film with each perpendicular and parallel field up to 8 Tesla is plotted as a function of temperature in Fig.1.

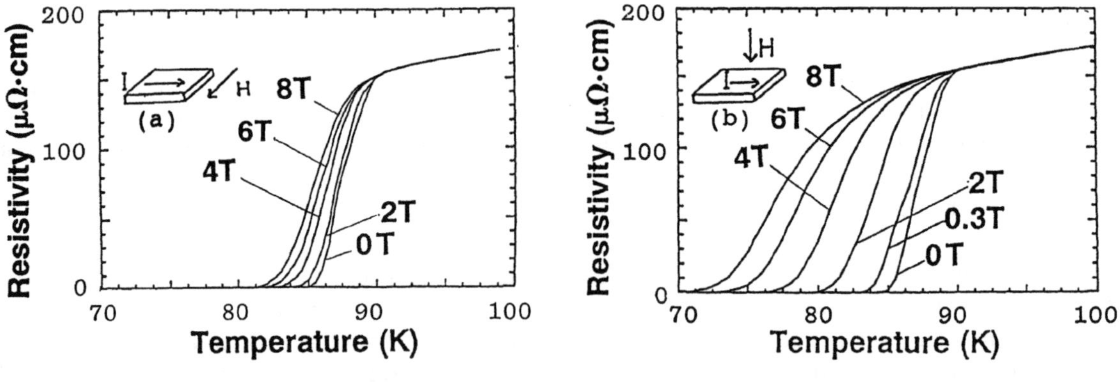

Fig.1 Temperature dependence of resistivity of YBCO epitaxial thin film in various magnetic fields parallel to the a-b plane (a) and perpendicular to the a-b plane (b).

As shown in the figure, the broadening of the resistive transition increases with the increase of the field and the broadening with the perpendicular field is larger than that with the parallel field. In Fig.2, these data are plotted in log R vs $(1-t)^{1.5}/T$. These plotted lines are linear in the resistivity below about 10 μΩ•cm. This linear fitting indicates that the broadening of the resistive transition in the low resistivity is caused by thermally activated flux creep and that pinning potential energy has the temperature dependence shown in Eq.(2). Thus we can calculate the pinning potential in any temperature using the slope of the straight lines which represents β/kH. In Fig.3, the pinning potential for each parallel and perpendicular fields at 77.3K is plotted as a function of 1/H. Plotted points show good fitting with the linear line. The slope of the linear line represents β. The linear fitting indicates that pinning potential has the field dependence shown in Eq.(2). From these fitting we obtain β=31 eV•Tesla in a parallel field and β=5 eV•Tesla in a perpendicular field. This anisotropy of β means that the pinning potential for a parallel field is about six times larger than that for a perpendicular field. The difference of the pinning potential correspond to that of the coherence length, because in Eq.(2), only the coherence length has anisotropy. Therefore the anisotropy of the pinning potential indicates that the coherence length of the a-b plane is about 6 times longer than that of c-axis. This anisotropy is almost the same as that estimated from the temperature dependence of the upper critical field[8]. These results indicate that the flux creep model can be adapted to the resistive behavior in the low resistivity(below 10μΩ•cm) regime for the epitaxial thin film in a magnetic

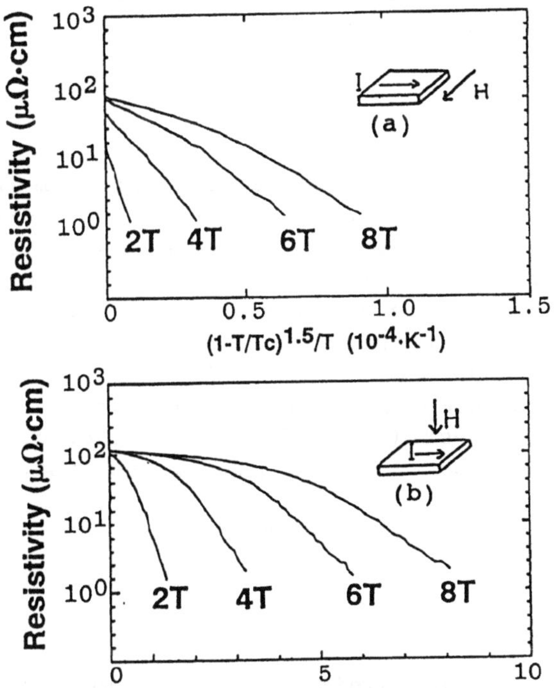

Fig.2 log R vs $(1-T/Tc)^{1.5}$of YBCO epitaxial film in various magnetic fields parallel to the a-b plane (a) and perpendicular to the a-b plane (b).

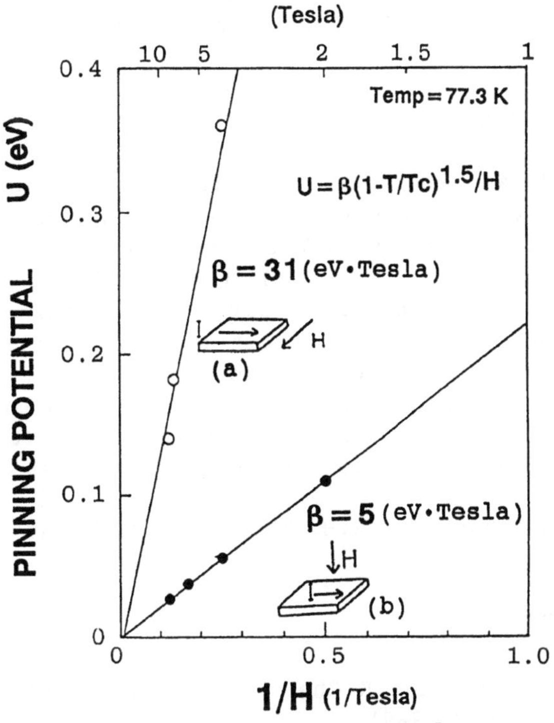

Fig.3 Pinning potential as a function of 1/H for YBCO epitaxial film in fields perpendicular to the a-b plane (closed circle) and parallel to the a-b plane (open circle).

field. On the other hand, the resistance for the polycrystalline thin film was plotted in the same way as the epitaxial film in Fig.4. The broadening of the resistive transition is much larger than that of the epitaxial film and

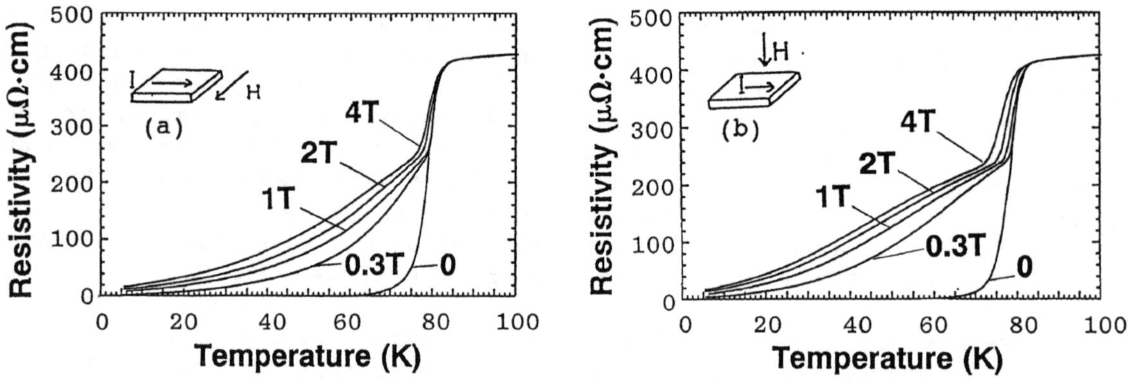

Fig.4 Temperature dependence of resistivity of YBCO polycrystalline thin film in various magnetic fields parallel to the a-b plane (a) and perpendicular to the a-b plane (b).

the sample doesn't show the superconductivity in the magnetic field more than 0.3 Tesla. Furthermore, the resistive behavior have neither a magnetic field dependence nor an anisotropy so much as compared with the epitaxial film. This indicates that the resistive behavior of the polycrystalline thin film is not caused by the flux creep. Judging from the large deterioration of the superconductivity in low fields, it is suggested that the broadening of the resistance is caused by the weakly linked grain boundaries.

SUMMARY

It was indicated that the resistive behavior of YBCO epitaxial thin film in the low resistivity (below 10μΩ•cm) regime in a magnetic field can be explained very well by the flux creep model. The pinning potential estimated based on the model had both the temperature and magnetic field dependence predicted by the theory. The pinning potential for the field parallel to the a-b plane was six times larger than that for perpendicular to the plane. This anisotropy is much consistent with that of the coherence length. On the other hand, the flux creep model couldn't be adapted to the resistive behavior of the polycrystalline thin film. It was suggested that the superconductivity of a polycrystalline film is broken at the weakly linked grain boundaries.

We thank Mr. S.H. Thomsen for the assistance of the measurement.

REFERENCES

1. J.Z.Sun,K.Char,M.R.Hahn,T.H.Geballe and Kapitulnik, Appl.Phys.Lett. 54, 663(1989).
2. T.T.M.Palstra,B.Batlogg,R.B.vanDover,L.F.Schneemeyer and J.V.Waszczak, Appl.Phys.Lett.54,763(1989).
3. K.A.Muller,M.Takashige and J.G.Bednorz,Phys.Rev.Lett.58,1143(1987).
4. K.Kitazawa,S.Kambe and M.Naito,"Strongly correlated electrons and supercon ductivity"Eds.by H.Fukuyama and S.Maekawa to be published by World Scientific.
5. Y.B.Kim,C.F.Hempstead and A.R.Strnad,Phys.Rev.131,2486(1963).
6. M.Tinkam,Phys.Rev.Lett.61,1658(1988).
7. S.Tanaka and H.Itozaki,Jpn.J.App.Phys.27,L622(1988).
8. M.Hikita and M.Suzuki,Phys.Rev.B39,4756(1989).

Intergranular Vortex and Weak-link Structure in the Ag_2O Doped $LaBa_2Cu_3O_{7-y}$

F. MIZUNO, H. MASUDA, and I. HIRABAYASHI

Superconductivity Research Laboratory Nagoya Division, International Superconductivity Technology Center, 4-1, Mutsuno 2-chome, Atsuta-ku, Nagoya, 456 Japan

ABSTRACT

Low-field magnetization hysteresis curves have been investigated in the Ag_2O doped $LaBa_2Cu_3O_{7-y}$ system. An anomaly is observed in the initial magnetization curve at a certain field of H* less than 1 mT. It is neither a lower critical field of the weak link region H_{c1J} nor a critical field for flux entry H_{en}. Sample thickness dependence of the magnetization loop width ΔM and field dependence of the hysteresis loss Q suggest a relationship between this anomaly and the maximum applied field Hs which can be screened at the midplane of samples in the framework of the critical state model. The behavior of the temperature dependence of ΔM at low temperature range agrees with the thermally activated flux creep model.

KEY WORDS: intergranular vortex, weak link, critical current, flux creep

INTRODUCTION

It has been generally accepted that the bulk oxide superconductor can be described as an agglomerate of superconducting grains connected by weak links. The intergranular or intragranular weak links form a complicated network in this system. The critical current is dominated by the properties of weak link network,[1] if the current through the sample and the applied field are less than the critical current and lower critical field of the grain, respectively.

In the present study, we have investigated the low field magnetization hysteresis, the initial magnetization curve, and the temperature dependence of the irreversible magnetization width ΔM at lower temperature range for the Ag_2O doped $LaBa_2Cu_3O_{7-y}$ system which is suitable for the observation of the weak link properties separating from the intragranular properties.

EXPERIMENTAL

$LaBa_2Cu_3O_{7-y}$-Ag composites were prepared in highly reduced atmosphere by coventional solid state reaction method described in ref. [2]. The X-ray diffraction patterns of the samples doped with Ag_2O showed exsistence of the orthorhombic phase of $LaBa_2Cu_3O_{7-y}$ and metallic silver. The superconducting transition temperature was found to be about 90 K. Magnetization measurements were performed using a low-drift SQUID magnetometer. The final shape of the sample for the present study was a thin flat plate (typically 2.0 x 6.0 x 0.05~0.24 mm^3). Transport critical current density was determined from the I-V characteristic curve using 10_3 μV/cm criterion. The density of the sample with Ag_2O 25 wt % was 6.2 g/cm^3 excluding the amount of the metallic silver precipitated at the grainboundary.

RESULTS AND DISCUSSION

Figure 1 shows a typical low-field (0-10 mT) magnetization hysteresis curve which is related to the weak link properties at 4.2 K in the sample $LaBa_2Cu_3O_{7-y}$ with Ag_2O 25 wt %. The hysteresis loop of the magnetization appears below 5.4 mT denoted by H_w* and disappears above it. The hysteresis curve indicates an irreversibility due to the intergranular vortex pinning. The evidence that the low field magnetization hysteresis is attributed to intergrain properties is attained from the powdered sample. Figure 1(b) shows the magnetization curve of the powdered sample with a grain size below 5 μm. Hysteresis is not clearly observed in this powdered sample because the weak

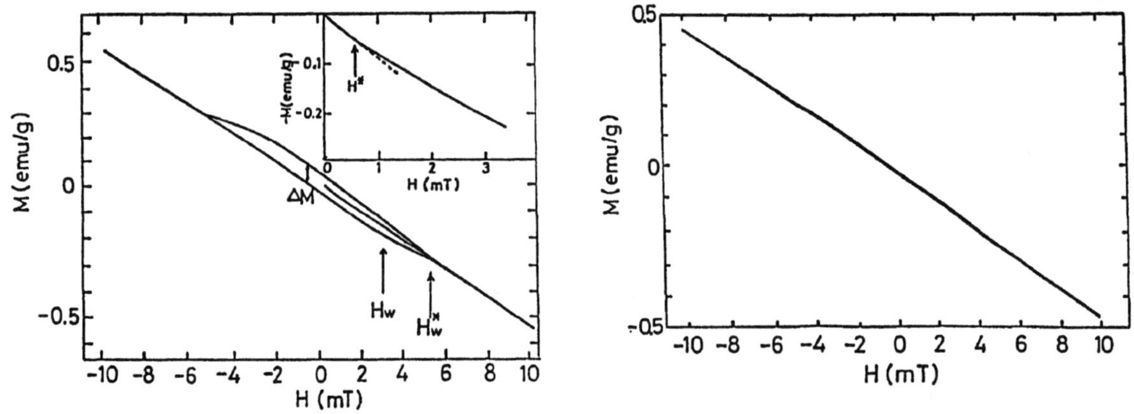

"Fig.1": Magnetization curves at 4.2 K. (a) bulk sample with the thickness 0.08 mm in $LaBa_2Cu_3O_{7-y}$ with Ag_2O 25 wt % $|H|=10$ mT. The inset shows the first deviation of the slope of the initial magnetization curve. (b) powdered sample $LaBa_2Cu_3O_{7-y}$ with Ag_2O 25 wt % $|H|=10$ mT.

link between the grains almost removed from the sample.

A kink anomaly in the low field initial magnetization curve is observed at the field H* about 0.5 mT as shown in the inset of Fig. 1 (a). There are some questions in assigning H* as the lower critical field in the weak link region H_{c1J} in contrast to the intragrain case. This problem is discussed in detail.

As a typical model for granular superconductors, several authors [1,3] modelled the bulk oxide superconductor as a cubic array of identical cells with a lattice parameter a and assumed nearest neighbor cells to be coupled by identical weak links. In the weak external field and at the low temperatures, it is reasonably assumed that the order parameter of each cell is phase-locked and the phase change occurs only at the junction region. For such a granular medium, the penetration of the magnetic field into the medium is estimated by Eq. (1) which is similar to the Josephson penetration depth [1,3].

$$\lambda_J = (c\phi_0/8\pi^2 J_c a)^{1/2}, \qquad (1)$$

where ϕ_0 is the flux quantum, J_c is the junction critical current density. On the analogy of the critical field in type-II superconductor, we obtain the junction lower critical field as [3]

$$H_{c1J} = (\phi_0/4\pi\lambda_J^2)\ln(2\lambda_J/a). \qquad (2)$$

As pointed out by Tinkham et al.[4], if we identify a/2 in the logarithmic term in Eq. (2) with junction coherence length ξ_J, an analogy between Eq. (2) and usual GL results is established. When the applied magnetic field exceeds H_{c1J}, the vortex starts to enter into the weak link regions. This situation is parallel to the mixed state in type-II superconductors. The hysteresis in the magnetization curve is obsereved in the range of $H_{c1J} < |H| < H_w^*$. It is due to the irreversibility induced by the intergranular vortex pinning in the range of $H_{c1J} < |H| < H_w^*$. With the definition of ξ_J, the upper critical field H_{c2J} is naturally defined as a magnitude of the maximum magnetic field which causes the 2π phase change around an entered unit flux without variation of phase within each grain in the cubic array model,[1] i. e.,

$$H_{c2J} = 3\pi\,\phi_0/8a^2 . \qquad (3)$$

If the field H_w at low temperature is identical to H_{c2J} in the Eq. (3), H_{c2J} is estimated to be 3 mT. This value is comparable to that of estimated in the same way from the transport critical current density. Using Eqs. (1) and (3), we can estimate λ_J and a at 4.2 K as 5.32 μm and 0.90 μm, respectively. The estimated value of a is smaller than the grain size (typically 2-3 μm). Using these results and $J_c=1030$ A/cm^2 which is estimated from the magnetization curve, we obtain 0.014 mT for H_{c1J} from Eq. (2). The value is 1/30 times smaller than H*, which indicate that the field H* cannot be regarded as the lower critical field H_{c1J}.

Next, we will discuss whether the field H* is attributable to the surface entry field H_{en} [4]. Considering the repulsive interaction between a vortex line and the surface field and the attractive one of the mirror force between the flux line and its image line, an energy barrier for flux entry develops near the surface in the range $H_{c1} < H < H_{en}$. H_{en} is calculated as a critical magnetic field at which the gradient of the Gibbs free energy normal to the surface becomes zero at the surface. Neglecting the demagnetization effect, H_{en} is given by

$$H_{en}=\phi_0/4\pi\lambda_J\xi_J. \qquad (4)$$

Using above estimation, we obtain 0.09 mT for H_{en} at 4.2 K, which is smaller than the magnitude of H*. Since it is without question to neglect the demagnetization effect in our configuration i.e., thin slub parallel to the field, H_{en} is only dependent of the intrinsic parameters such as λ_J and ξ_J. It is difficult to explain the sample thickness dependence of H* by the critical entry field H_{en} as described later.

Another possibility for H* we considered is the field H_s which is the maximum external field that can be completely screened out at the midplane of the sample. If the Bean's model holds true, the irreversible magnetization width ΔM for the flat thin plate sample should follow the relation, that is

$$\Delta M=J_c(H)D/2c, \qquad (5)$$

where D is the thickness of the sample. The field H_s is also given by

$$H_s=2\pi J_c(H)D/c. \qquad (6)$$

As described below, we will assign H* with H_s. Chaddah et al.[5] discussed the relationship between H_s and the feature of the initial curve of the magnetization hysteresis. According to their results, if $H_s < H_{c1}$, the first minimum in the initial magnetization curve occurs at H_{c1}, whereas if $H_s > H_{c1}$, it occurs not at H_{c1} but at a certain field which is related to H_s. In order to clarify the relationship between H* and H_s, we measured the sample thickness dependence of H* and ΔM which was related to H_s by the formula $\Delta M=(1/4\pi)H_s$. It is easily derived from Eqs. (5) and (6). Figure 2 shows the experimental value of H* and the ΔM at zero field as a function of sample thickness D by lower field (10 mT) sweep. A linear relationship was confirmed between ΔM and D for D < 0.09 mm. It means the Bean's model holds true for the sample of D < 0.09 mm. On the other hand, the ΔM becomes almost constant for

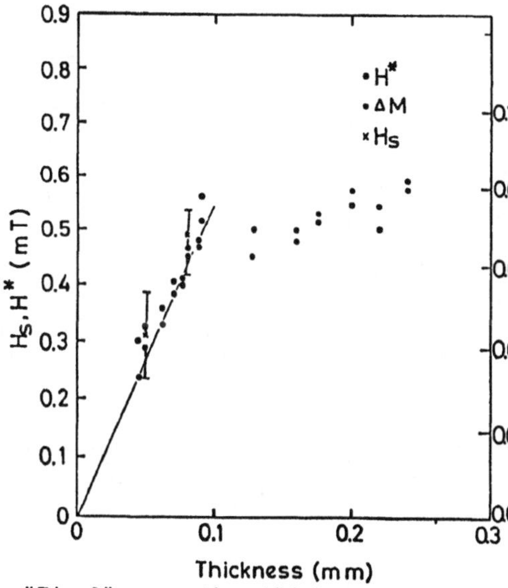

"Fig.2": •: the thickness dependence of the magnetization loop width ΔM at 5 K and 0 mT (a) |H|=5 T, (b) |H|=10 mT. o: the value of H* from the deviation of the virgine curve of the magnetization. χ; the field H_s in the sample with the thickness 0.048 and 0.08 mm

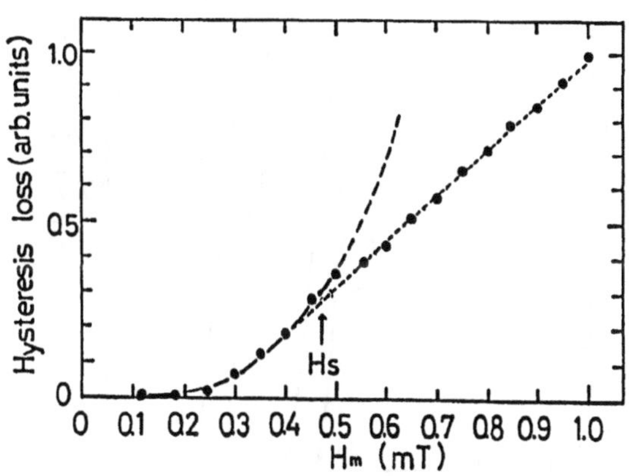

"Fig.3": The field sweep dependence of the hysteresis loss in the sample with thickness 0.08 mm.

the thicker samples than D=0.09 mm. It suggests that the size of crack or macroscopic silver segregation in the thick samples may influence the scale which define the critical state. As a consequence, we can conclude that the Bean's model is only applicable for estimation of intergranular J_c from the ΔM for our samples of D < 0.09 mm. The thickness dependence of the experimental H* is also displayed in Fig. 2. It is well consistent with H_s estimated from intergranular signals.

Figure 3 shows the field sweep dependence of hysteresis loss Q in the low-field hysteresis curve for the samples with D=0.08 mm at 4.2 K. The Q increases as H_m^3 in the range of H_m < 0.5 mT and linearly increases above 0.5 mT. This behavior of the hysteresis loss is consistent with that of the Bean's model and the value of H_s is estimated to be 0.5 mT for this sample. The fields H_s estimated by this way for two different thickness samples agrees with the field H_s estimated from ΔM and the experimental H* as shown in Fig. 2. Therefore, the kink anomaly in the initial magnetization curve is interpreted as the specific field Hs which is related to the irreversible field penetration as been discussed.

Figure 4 shows the magnetic field dependence of ΔM. Reflecting the sharp cut off of ΔM at H_w*, the estimated ΔM sharply decreases to zero. The field value of H_w* is highly temperature dependent even at very low temperatures lower than 20 K. On the other hand, transport J_{cJ} as a function of the applied field for a zero-field-cooled (ZFC) specimen at 4.2 K linearly falls (~1/H) with increasing the applied field. This result is typical of bulk oxide superconductors and have been given a detailed analysis in terms of a Josephson-junction-Airy-current pattern [6]. The magnetic field dependence of ΔM and transport J_{cJ} at 4.2 K show the large discrepancies. The ΔM above H is reduced by a factor ~1/H and then strongly falls down. It may be because a phase randomization occurs more drastically than expected by the theory. It is interesting to note that the experimental H_w*'s at each temperature are different even in rather low temperature apart from the critical temperature. It may be due to a characteristic behavior of the magnetic flux in the junction region.

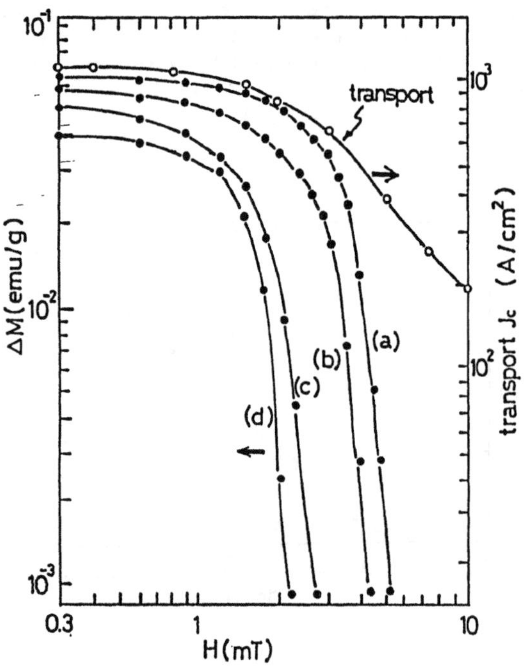

"Fig.4": Magnetic field dependence of ΔM for in $LaBa_2Cu_3O_{7-y}$ with Ag_2O 25 wt % (a) at 4.2 K, (b) at 7 K, (c) at 10 K, and (d) at 15 K.

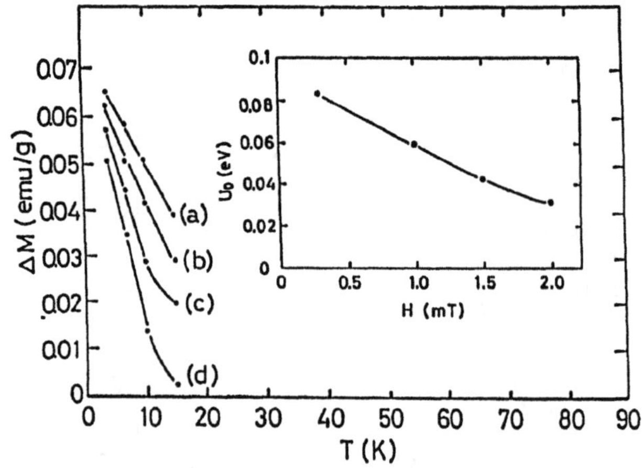

"Fig.5": Temperature dependence of ΔM for bulk sample $LaBa_2Cu_3O_{7-y}$ with Ag_2O 25 wt % (a)H=0.3 mT, (b) H=1.0 mT, and (d) H=2.0 mT

Figure 5 shows the temperature dependence of ΔM. The slope of $\Delta M(T)$ becomes steeper with increasing the magnetic field. The ΔM's fall down following the relationship that $\Delta M \propto 1-\alpha T$ near at zero temperature. The extrapolated point to the zero $\Delta M(T)$ is much lower than the critical temperature and shifts to the lower temperature with increasing the magnetic field. This behavior occurs at below 15 K. When the temperature exceeds above 20 K, the hysteresis curve of the magnetization in the lower magnetic field loses in shape and H_W* fades away from the hysteresis curve. It suggests that the flux start to enter the grains before the intergranular coupling is severely weakened by the magnetic field about at 20 K. On the other hand, in the case of transport J_{cJ}, cut off of J_c slowly continues up to its critical temperature. This characteristic behavior of sharp cutoff in the ΔM at the lower temperature range is explained with a flux creep model.

Thermal activation increases a flux creep rate, thereby it reduces the observed critical current density. As the Bean's model is established below 1.5 mT, we may regard ΔM as J_{cJ}. The $J_{cJ}(\equiv \Delta M)$ is given by

$$J_{cJ}=J_{cJ0}(1-(k_BT/U_0)\ln(t/t_0)), \qquad (7)$$

where J_{cJ0} is the value of J_{cJ} in the absence of thermal activation of flux lines, U_0 is the depth of the potential wells which trap the flux lines, t is a characteristic time for an experiment, and t_0 is a constant which depends on parameters such as the flux lines oscillation frequency and the averaged hopping distance of the flux lines. From the values of t=600 sec and $t_0=10^{-12}$ sec, we determined $\ln(t/t_0)=30$. A value of t_0 is comparable to a typical phonon frequency which determines the flux line hopping. Equation (7) has two consequences. First, the temperature dependence is much stronger than the time dependence. Therefore, by a rough approximation, J_{cJ0} is time independent and J_{cJ} depends linearly on T at low temperatures (where J_{cJ0} is time independent of T), as is evident from Fig. 5. Secondly, from Eq. (7) and the observed slope of the temperature in ΔM versus T, we found U_0=0.083 eV at B=0.3mT. The inset of Fig. 5 shows the magnetic field dependence of U_0. These values are less than 1 order of magnitude smaller than the U_0 value of the grains at 77 K [7].

The authors would like to thank Professor S. Maekawa for his helpful discussions.

REFERENCES

1. Tinkham M, and Lobb CJ, (1989) Physical Properties of the new superconductors. In: Seitz F, Turnbull D, and Ehrenreich E (eds) Solid State Physics. Academic, New York, pp 91-103 (vol. 42)
2. Mizuno F, Masuda H, Hirabayashi I, Tanaka S, (1989) Preparation of a 90 K Supercondueyor ($LaBa_2Cu_3O_{7-y}$) with Ag-Related Additives. Jpn. J. Appl. Phys. 28: 1780-1783
3. Clem JR (1988) Granular and superconducting-gless properties of the high temperature superconductors. Physica C 153-155: 50-55
4. Hüebener RP, (1979) Magnetic flux structures in superconductors. In: Fulde P (ed). Springer Berlin Heidelberg New York. pp82-86 (Springer series in solid-state sciences. vol. 6)
5. Chaddah P, Ravi Kumar G, Grover AK, Radhakrishnamurty C, Subba Rao GV, (1988) Critical stste model and the magnetic behavior of high T_c superconductors. : proceedings of a international conference on critical currents in high-temperature superconductors. 16-19 August 1988, Snowmass Village, Colorado, USA
6. Peterson RL, Ekin JW (1988) Josephson-junction model of critical current in granular $Y_1Ba_2Cu_3O_{7-y}$ superconductors. Phys. Rev. B 37: 9848-9851
7. Matsushita T, Ni B, Murakami M, Morita M, Miyamoto K, Saga M, Matsuda S, Tanino M (1989) Critical current densities in superconducting Y-Ba-Cu-O prepared by the quench and melt growth technique. Jpn. J. Appl. Phys. 28: L1545-1548

Enlargement of Flux Pinning Forces in X-ray Irradiated $Gd_1Ba_2Cu_3O_{7-x}$ Superconducting Thin Films

S. Kohiki[1], S. Hatta[1], K. Setsune[1], K. Wasa[1], Y. Higashi[2], S. Fukushima[2], and Y. Gohshi[2]

[1] Central Research Laboratories, Matsushita Electric Industries Co. Ltd., Moriguchi, Osaka, 570 Japan
[2] Faculty of Engineering, The University of Tokyo, 3-1, Hongo 7-chome, Bunkyo-ku, Tokyo, 113 Japan

ABSTRACT

We have successfully introduced stable pinning centers into $Gd_1Ba_2Cu_3O_{7-x}$ superconducting thin films by x-ray irradiation before the oxygen annealing. A large enhancement of critical current density with the small rate of flux creep was realized. The activation energy, estimated with the flux creep model, increased from 0.1eV to 0.25eV with the x-ray irradiation treatment. The significantly increased magnetization showed both the temperature independency and the small magnetic relaxation.

KEY WORDS: x-ray irradiation, strong pinning center, large critical current density, increase of activation energy, small rate of flux creep

INTRODUCTION

Since large supercurrent density (J_c) induced large flux creep, large time dependence of magnetization has been the most serious problem for practical applications of cuprate superconductors [1,2].

The exact nature of pinning centers of cuprate superconductors has not yet been clarified, but imperfections in the lattice such as lattice defects, twin boundaries, dislocations, grain boundaries, voids, and nonsuperconducting inclusions are believed to pin the fluxoid against the Lorentz forces. Energy beam irradiations were studied as another approach for creating pinning centers. Ion bombardment [3] and neutron irradiation [4,5] appeared not to be effective for obtaining large J_c with small flux creep.

In order to describe the properties of cuprate superconductors, the use of crystalline thin film is very important because crystalline thin films provide more useful information on the physical properties of such an anisotropic two-dimensional system than the bulk ceramics.

We have artificially created strong pinning centers in $Gd_1Ba_2Cu_3O_{7-x}$ superconducting thin films by x-ray irradiation before the annealing in oxygen, and as a result, we obtained large J_c (3.4×10^6 A/cm^2 below 43K) with a small rate of flux creep. The activation energy (U_0) in a flux creep model increased from 0.1eV to 0.25eV due to the x-ray irradiation treatment.

EXPERIMENTAL

$Gd_1Ba_2Cu_3O_{7-x}$ thin films were prepared by rf magnetron sputtering using the Gd-Ba-Cu-O sintered target. Typical sputtering conditions are listed in table I [6]. The ratio of Gd:Ba:Cu was determined to be 1:2:3 by using electron probe microanalysis and inductively coupled plasma optical emission spectroscopy within the experimental uncertainty of 5%.

Table I Sputtering conditions

Target	$Gd_1Ba_2Cu_{4.5}O_x$
Substrate	MgO (100)
Substrate temperature	650°C
Sputtering gas	Ar:O_2=5:1
Gas Pressure	0.4Pa
Rf input power	150W
Film thickness	250nm

1 Permanent address: Matsushita Technoresearch, Inc., Moriguchi, Osaka 570

X-ray irradiation for the as-deposited film was carried out by using an Rh x-ray tube operated at 50kV and 50mA in a vacuum of 10^{-3} Torr for 100 h. Both films x-ray irradiated and unirradiated were annealed simultaneously at 900°C for 2 h and then at 450°C for 2 h in flowing oxygen.

The time dependence of magnetization relaxation was measured at constant temperatures by a SQUID susceptometer, as described elsewhere [2]. Demagnetization was carried out by zero field cooling from the normal state. The external field (H_{ex}) was applied parallel to the c axis of the film because of maximizing the output of diamagnetization.

RESULTS AND DISCUSSION

Figure 1 shows the J_c^M of the x-ray irradiated and unirradiated films derived from the shielding effect with the Bean's model [7]. The diamagnetization M (emu/cc) is related to J_c^M (A/cm^2) as follows;

$$M=RJ_c^M/30,$$

where R (cm) is the effective radius.

Peculiar temperature dependence of diamagnetization in the x-ray irradiated film was observed. The J_c^M at H_{ex}=150 Oe was constant (3.4×10^6 A/cm^2) below 43K and it decreased suddenly above 43K. This characteristic temperature dependence of the J_c^M has not yet been reported anywhere in the field of cuprate superconductors.

The time dependences of the magnetization for the films are shown in fig.2. The relaxation of magnetization for x-ray irradiated film is very slight even at around 70K in comparison with that for unirradiated film which showed the serious logarithmic time dependence of relaxation.

In the flux creep model the time-logarithmic dependence of magnetization is expressed as

$$M/M_0=1-(kT/U_0)\ln(t/t_0),$$

where M/M_0 is the normalized magnetization at t=0. The estimated values of U_0 of the films with and without x-ray irradiation from fig.2 are 0.25eV and 0.1eV, respectively. This increase of U_0 implies that the flux pinning forces in the film were greatly strengthened by the x-ray irradiation treatment.

The J_cs from the I-V characteristics with 100μm patterned films were almost identical to those from the magnetization measurement. The agreement between intragranular and intergranular superconducting current densities suggests that the influences of weak-link at grain boundaries and nonsuperconducting inclusions due to lack of composition uniformity in the films are negligible, and then the grain boundaries and nonsuperconducting inclusions cannot act as strong pinning centers in cuprate superconductors.

This conclusion from the superconducting current density measurements can be supported by the secondary electron microscopic (SEM) observation and x-ray

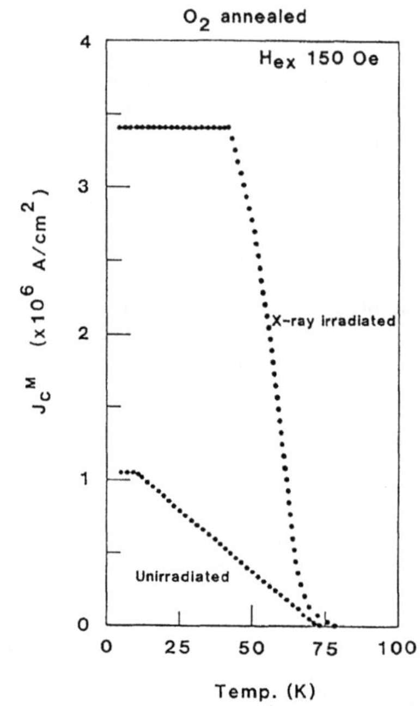

Fig.1 Temperature dependences of J_c^M.

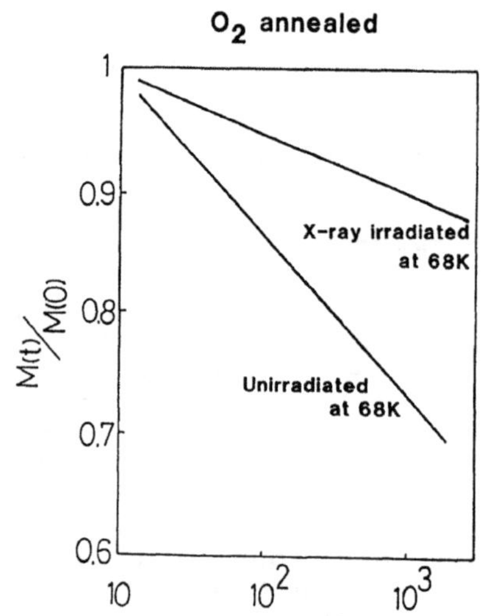

Fig.2 Time dependences of magnetization with a logarithmic time scale.

diffraction of the films. The grain size, approximately 300nm, was invariant and the number of grains decreased with the x-ray irradiation. The number of grain boundaries decreased but both the J_c and the U_o increased for the x-ray irradiated films.

The x-ray diffraction patterns of the films as-deposited, x-ray irradiated without oxygen annealing, oxygen annealed without x-ray irradiation, and x-ray irradiated followed by oxygen annealing are shown in fig.4. The degree of c-axis orientation became higher with the x-ray irradiation treatment as well as seen in the SEM observation. This enhancement of macroscopic periodity of cations ordering with the x-ray irradiation must be due to the release of potential energy of photons caused by impingement into the crystal. The high periodic peaks attributed to (001) reflection of the c plane of the x-ray irradiated film shifted to higher diffraction angle side with the oxygen annealing. The diffraction patterns of the oxygen annealed films both x-ray irradiated and unirradiated are essentially identical. There was no peaks from other crystal structure such as non-superconducting 2115 (green) phase. The absence of other diffraction peaks from the superconducting 123 phase and the agreement of J_c between the magnetization and the electron transport property measurements suggest that the nonsuperconducting inclusions are less than 10% in the films and cannot act as strong pinning sites. Both films had highly oriented structure with c axis perpendicular to the film surface. The apparent change of c-axis lattice parameter due to the x-ray irradiation was not observed after the oxygen annealing. It is reasonable to expect that x-ray irradiation for oxides gives rise to the oxygen defects in the crystal but it is impossible to determine the periodity of the oxygen defects in the crystal directly by x-ray diffraction. It is well known that the electron transport properties reflect the oxygen deficiency (x) of the crystal.

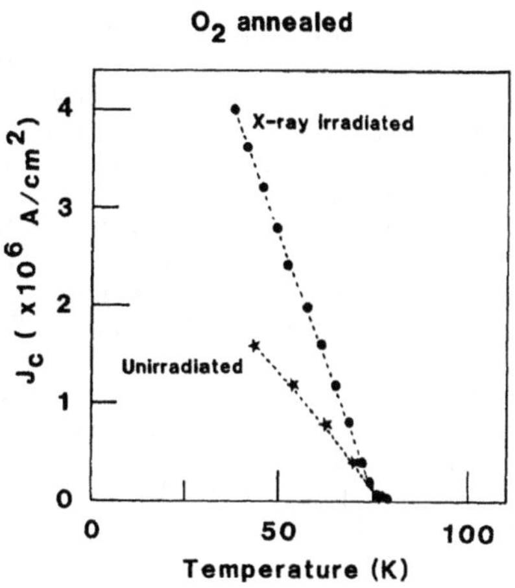

Fig.3 Critical current densities from the I-V measurement vs temperature.

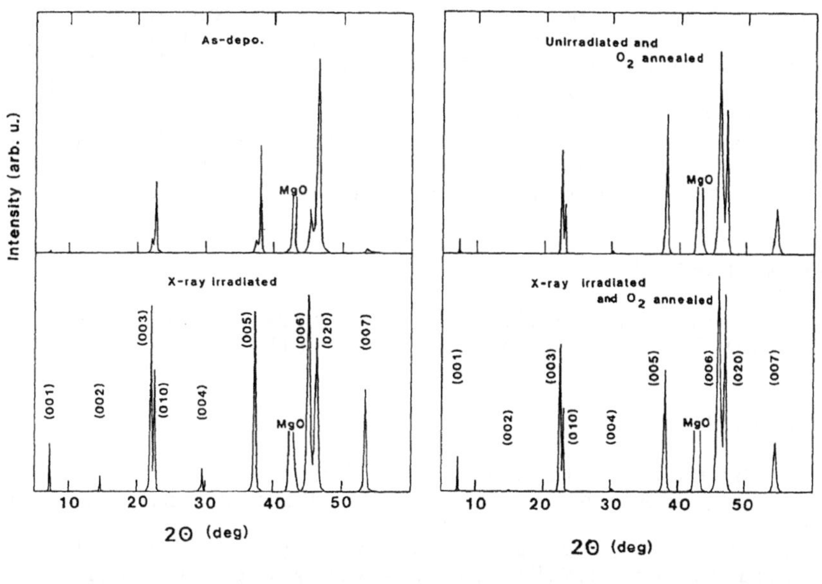

Fig.4 X-ray diffraction patterns with Cu Kα.

For the resistivity measurement, gold electrodes were evaporated on the surface of the films, and the standard four-probe method was used. Figure 5 shows the temperature dependence of the resistivity of the films as-deposited, x-ray irradiated without oxygen annealing, oxygen annealed without x-ray irradiation, and x-ray irradiated followed by oxygen annealing. The resistivity of the as-deposited film, showed two-step superconducting transition, indicated semiconductor like behavior after the x-ray irradiation. This reflects the creation of the oxygen defects in the crystal since the x-rays induce photo-chemical reaction accompanying the oxygen deficiency but they do not take place the composition change of the cations in oxides. The resistivity of the x-ray irradiated film was smaller (1/2) than that of the unirradiated one after the annealing in oxygen. The onset temperature around 90K of the oxygen annealed films was invariant with the x-ray irradiation. The zero-resistance temperature was 78K for the oxygen annealed films with x-ray irradiation and that for such films without irradiation was 74K. Essentially the superconducting transition property does not change by the x-ray irradiation.

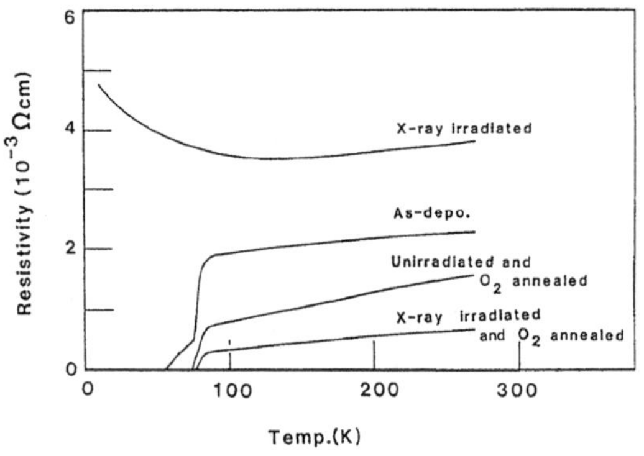

Fig.5 Temperature dependences of the resistivity.

SUMMARY

Here, we reported the great reduction of magnetization relaxation with large J_c by x-ray irradiation. This corresponds to the increase of the activation energy from 0.1eV to 0.25eV. The characteristic temperature dependence of the diamagnetization was obtained in the x-ray irradiated thin films. It is suggested that x-ray irradiation caused the creation of the oxygen vacancies and then the redistribution of both the oxygen defects and the cations in the crystal. This redistribution reflected to both the electron transport properties and x-ray diffraction patterns. The oxygen annealing of the x-ray irradiated crystal stabilizes the flux pinning centers and increases the flux pinning forces. The details of this stabilization mechanism has not yet been clarified. But the x-ray irradiation process followed by the annealing in oxygen provides an effective treatment method for cuprate superconductors in the field of pinning control.

ACKNOWLEDGMENTS

The authors would like to express their thanks Drs. T.Nitta and F.Konishi for their support in this work.

REFERENCES

1 Y.Yeshurun and A.P.Malozemoff, Phys. Rev. Lett. 60, 2202 (1988).
2 S.Hatta, H.Adachi, Y.Ichikawa, S.Hayashi, S.Kohiki and K.Wasa, J. Phys. Soc. Jpn. 58, 4132 (1989).
3 S.Takamura, T.Aruga and T.Hoshiya, Jpn. J. Appl. Phys. 28, L1118 (1989).
4 R.L.Fleischer, H.R.Hart,Jr., K.W.Lay and F.E.Luborsky, Phys. Rev. B40, 2163 (1989).
5 R.B.van Dover, E.M.Gyorgy, L.F.Schneemeyer, J.W.Mitchell, K.V.Rao, R.Puzniak and J.V.aszczak, Nature 342, 55 (1989).
6 T.Kamada, K.Setsune, T.Hirao and K.Wasa, Appl. Phys. Lett. 52, 1726 (1988).
7 C.P.Bean, Phys. Rev. Lett. 8, 250 (1962).

Properties of Bi-Pb-Sr-Ca-Cu-O Bulk Superconductors Prepared by a Hot-Press Method

R. YOSHIZAKI[1], H. IKEDA[1], K. YOSHIKAWA[2], and N. TOMITA[3]

[1] Institute of Applied Physics, Cryogenics Center, University of Tsukuba, Ibaraki, Japan
[2] Research and Development Center, Mitsubishi Heavy Industries, Ltd., Takasago, Japan
[3] Shipyard and Machinery Works, Mitsubishi Heavy Industries, Ltd., Kobe, Japan

ABSTRACT

Synthesis of single-phase $Bi_{2-x}Pb_xSr_2Ca_2Cu_3O_y$ compounds has been carried out a new method, an uniaxial hot-press method. Physical properties of the compounds are studied and compared to those of an intermediate-pressed sample and a normally sintered one. Critical current density was measured at 77 K for each sample, and the best result was obtained for the hot-pressed sample.

KEY WORDS: superconductivity, critical current density, Pb-doped Bi-Sr-Ca-Cu-O superconductor, high-Tc phase

INTRODUCTION

Critical current-density, Jc, of bulk oxide-superconductors measured so far is of the three or four orders of magnitude lower than those for single crystals or thin films [1]. The degradation of Jc is attributed to weak links along grain boundaries for sintered materials. To improve inter-grain linkage, compaction effect on sintered substance was investigated in the Pb-doped Bi-Sr-Ca-Cu-O (hereafter abbreviated to BSCCO) system by employing intermediate pressing process [2]. After this cold work, heat treatment of annealing process was inevitably necessary to restore the connection among grains. Hence, it is expected that heating of a sample during the compaction process (so-called hot-press method) will yield realignment of crystals without or less breaking of grains. Moreover the hot-press method yielded grain growth in an oxide superconductor, La-Sr-Cu-O (LSCO) system [3], but was not succeeded for the BSCCO system [4]. Remind to the strong two dimensionality of the BSCCO structure, we have not adopted hydrostatic pressure as in the case of the LSCO system [3], but employed an uniaxial pressure in the present case of BSCCO.

EXPERIMENTAL

Nominal composition of $Bi_{1.84}Pb_{0.34}Sr_{1.91}Ca_{2.03}Cu_{3.06}O_y$ [5] was prepared from high-purity (99.99%) reagents of Bi_2O_3, PbO, $SrCO_3$, $CaCO_3$ and CuO. The mixed powder was calcined at 810 °C for 12 h in air. Sintering was performed twice under the conditions at 842 °C for 60 h and at 844 °C for 100 h in 100 % nitrogen. The obtained samples (called NP) were used as the starting material of a hot-press method. In the hot press, uniaxial pressure of 300 kg/cm² was applied during the sintering at 822 °C for 2 h in air. We call this sample as HP. We have prepared another samples (called IP) with intermediate pressing process [2] for comparison. In all the nitrogen-flowing procedure mentioned above, gas flow (3 l/min) was stopped during the cool-down process.

Critical current-density was estimated from a standard four probe method for a sample shaped so as to have a narrow current channel (1x0.8 mm^2) between two voltage terminals. Current was swept by a stair-like pattern; each step has 1.5 A height with rising time of 0.1 s and the duration of 0.45 s. Jc is determined at the current when 1 uV/cm appears between the voltage terminals.

RESULTS AND DISCUSSION

The samples were evaluated by means of x-ray powder diffraction, ac magnetic susceptibility and Berman-balance measurements. In the last measurement, we have get the ideal value of density, 6.5, for HP sample, while 6.2 for IP and 6.0 for NP. All samples were confirmed to be composed of a single 2223 phase, and their Tc's determined by magnetic measurement were 106 K.

The effect of the hot-press on the crystal packing was observed with SEM images for fractured sections as shown in Fig. 1 with the reference pictures for the NP and IP samples. Thin flake-like grains are closely packed for HP with their planes perpendicular to the direction of the uniaxial pressure. The stacking of grains for IP or NP is fairly good but not so close as HP. These features are consistent with the result of sample densities mentioned above.

(a) NP (b) IP (c) HP

Fig. 1 Secondary electron microscope images of fractured cross section for the sample before pressing process (NP) (a), the intermediate pressed sample (IP) (b) and hot-pressed sample (HP) (c) under the magnification of 550.

Temperature dependence of dc magnetic susceptibility is shown in Fig. 2 for the samples measured after zero-field cooling (ZFC) (a) and in field cooling (FC) (b). In ZFC process, the diamagnetic profiles for HP and IP samples are quite similar to each other, and the magnitude at 5 K is close to the value estimated from 100 % bulk superconductor. Knickpoints observed on the curves around 100 K suggests that flux flow occurs above that temperature. In contrast, the magnitude of the diamagnetism for NP is extraordinarily large at low temperature. This result indicates that the sample contains so many open spaces among flake-like grains as confirmed on a SEM image, and the shielding current flows peripheral grains of the sample through the inter-grain structures. This effect continues up to breaking of current flow through the structures, which is about 70 K. Then the flux penetrates into grains above 100 K.

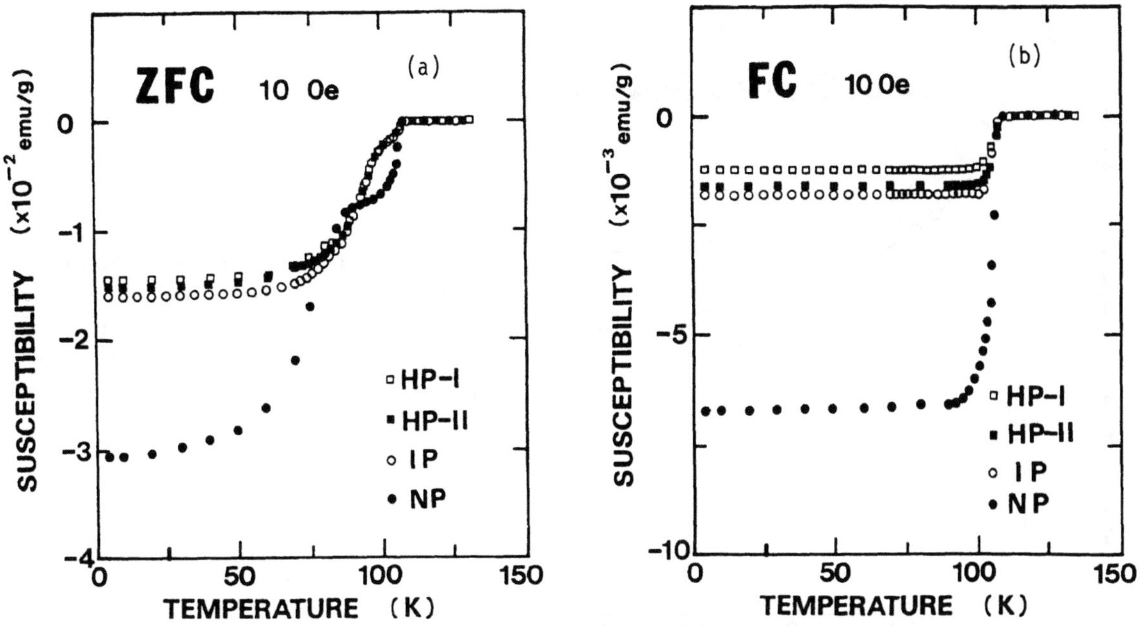

Fig. 2 Temperature dependence of magnetic susceptibility measured in zero-field cooling (ZFC) (a) and field-cooling (FC) process (b).

In FC process, the Meissner effect give us the knowledge about the flux motion within the superconductor. For NP samples, many fluxes are expelled out of the superconductor, judging from the large saturated value of the diamagnetism, as seen in Fig. 2(b). Furthermore the saturating temperature is slightly lower than that for the pressed samples. Then it is expected that pinning forces are weak and the number of pinning center is small in NP sample. This result implies the quality of single crystals composing the grains is fairly good in NP. On the other hand, there are many strong pinning centers in the pressed samples, since the magnitude of the diamagnetism for each pressed sample is much smaller than that of NP. Some strong pinning centers must be induced into the samples due to pressing. It is noted that the number of the pinning centers is almost independent of the cold work (IP) or hot one (HP). Pinning effect to prohibit flux expulsion is the strongest in HP-I sample, and the highest Jc value is expected for this sample.

Jc of HP measured as a function of external magnetic field is shown in Fig. 3. Magnetic field was applied parallel or perpendicular to the direction of current. The observed maximum Jc is about 7,000 A/cm^2 at 77.3 K and H=100 Oe. We cannot get Jc at zero field because of the instrumental limitation. We have obtained $Jc=1x10^3$ A/cm^2 at H=2 kOe. On the other hand, the maximum Jc at 77.3 K for IP sample is 6,250 A/cm^2 at H=0 and $Jc=1.3x10^3$ A/cm^2 at H=2 kOe, and Jc for NP at 77.3 K is much smaller of 650 A/cm^2 at H=0. These values are quite reproducible and the scattering of the highest Jc is less than 1000 A/cm^2 for several samples of a few batches. We can conclude from the results that the critical current density at 77 K for HP is better than that of IP and NP.

Hysteresis curve of the magnetization (M) against the external field was observed for all samples at 5 K and 77.3 K up to 10 kOe. Varying the sample thickness, we get little change in the characteristic features of the hysteresis even for HP. This means that the connection among grains are not sufficiently strong to behave like a single crystal but flux will penetrate into each grain. Details of the features will be discussed elsewhere.

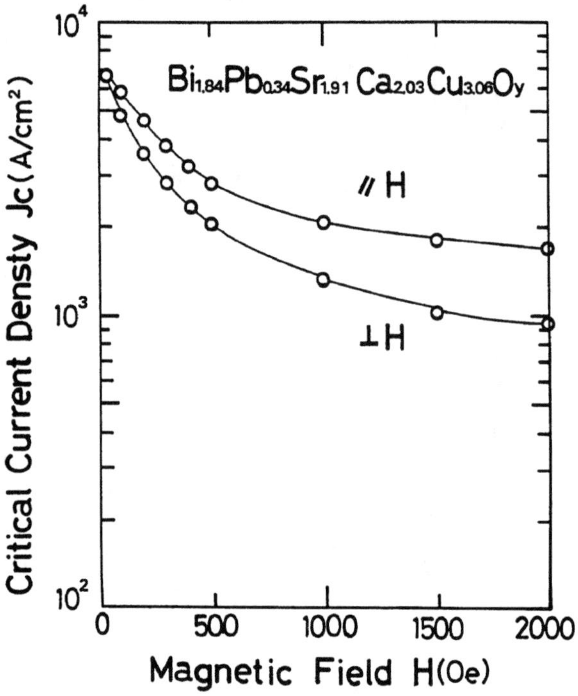

Fig. 3 Jc for the hot-pressed sample measured against the external magnetic field applied parallel (//) and perpendicular (⊥) to the current. The curves are guides for eyes.

In summary we have synthesized Pb-doped Bi-Sr-Ca-Cu-O compound by employing a hot-press technique. The hot-pressed specimen has a density almost equal to the ideal value. Critical current density measured is of the order of 10^4 A/cm^2 at 77.3 K 0 Oe. Magnetic properties of the compound are discussed with the relevance to the critical current density. These profiles of the hot-pressed samples are superior to other samples, without pressing process or with intermediate pressing one, from the point of high Jc bulk materials.

ACKNOWLEDGEMENTS

The authors are in debt to Y. Tanaka and T. Asano for the Jc measuring technique. A part of this work is supported by the Grant-in-Aid for Scientific Research on Priority Areas "Mechanism of Superconductivity" from the Ministry of Education, Science and Culture, Japan.

REFERENCES

1. See for example, Kitazawa K. Ishiguro T. (eds) (1988) Proc. the 1st Intern. Symposium on Superconductivity (ISS'88).
2. Asano T. Tanaka Y. Fukutomi M. Jikihara K. Maeda H. (1989) Properties of Pb-doped Bi-Sr-Ca-Cu-O superconductors prepared by the intermediate pressing process. Jpn. J. Appl. Phys. **28** L595-L597.
3. Iwazumi T. Yoshizaki R. Sawada H. Uwe H. Sakudo T. Matsuura E. (1987) Preparation and properties of $La_{1.85}Sr_{0.15}CuO_4$ single crystal. Jpn. J. Appl. Phys. **26** L386-L387.
4. Murayama N. Torii Y. (to be published) Increasing in Jc of Bi-Pb-Sr-Ca-Cu-O superconductors by hot pressing. Proc. Intern. Conf. on Materials and Mechanisms of Superconductivity High-Temperature Superconductors, Stanford.
5. Koyama S. Endo U. Kawai T. (1988) Preparation of single 110 K phase of the Bi-Pb-Sr-Ca-Cu-O superconductor, Jpn. J. Appl. Phys. **27** L1861-L1863.

Novel Magnetic Transition in X-ray Irradiated GdBaCuO Films

S. HATTA, S. KOHIKI, K. SETSUNE, and K. WASA

Central Research Laboratories, Matsushita Electric Industrial Co. Ltd., Moriguchi, Osaka, 570 Japan

ABSTRACT

A new transition in shielding magnetization was observed in LnBaCuO thin films annealed in oxygen after x ray irradiation. When the external field was applied after zero field cooling, the large absolute value of diamagnetization suddenly dropped to the very small value, similar to the type I superconductor. Once it changed to the small value state, it never recovered to the original state. Such an irreversible transition was thought to be due the steep flux invasion to overcome pinning barriers. These new pinning centers might be produced by oxygen defects with irradiation.

KEYWORDS : x ray irradiation, LnBaCuO thin films, irreversible magnetization transition, flux jump, pinning centers

INTRODUCTION

In order to realize very high critical current density (Jc) under strong magnetic field in high Tc superconductor, it is important how to clarify an origin of the fluxoid pinning, and how to strengthen the pinning forces. However, the fluxoids in the high Tc superconductor is originally considered to be easily mobile along the Cu-O basal plane direction. This poor pinning ability for the high Tc superconductor is thought to be due to the intrinsic superconducting properties such as high Tc, extremely small coherent length, etc. The easily activated fluxoid motion called as flux creep or flux flow would be observed as many physical phenomena like the resistive transition broadening[1-3], magnetic relaxation[4,5], etc. As the origin of pinning center of the oxide superconductor, various lattice defects (twin boundary[6], precipitation[7], variation of oxygen stoichiometry[8], etc) were assumed. However, it is not yet clarified what is the most effective pinning defects. For the purpose of the practical application, many trials were performed to introduce more effective pinning centers. In particular, some investigators used a lot of forms of nuclear radiation such as neutrons[9], protons[10], heavy ions[11], electrons[12], γ rays[13], etc. Most of these results shows that Tc decreases, intragranular critical current density (Jc^m) increases, and the transport critical current density (Jc^t) decreases. Thus, these pinning centers seemed to degrade the superconductivity seriously.

Recently, Kohiki et al [14,15] introduced new pinning centers into LnBaCuO thin films annealed in oxygen after x ray irradiation. This treatment made both Jc^t and Jc^m increase without serious destruction of superconductivity. In particular, magnetic properties was largely changed, compared to the non-irradiated specimens. It seemed that the pinning situation was quite different from the previous results[16].

In this paper, we report magnetic and magneto-resistive properties of the x ray treated LnBaCuO films, and discuss the effect of the new pinning centers on magnetization.

EXPERIMENTAL

Both GdBaCuO and ErBaCuO films were prepared by an rf magnetron sputtering as specimens. The film thickness was about 2500 A. These films were irradiated by using x ray tube operated with 50 kV and 50 mA in 0.001

Torr for 100 h. Then they were postannealed at 900°C for 2 h and then at 450 °C for 2 h in flowing oxygen. Magnetization measurements were carried out using an rf SQUID susceptometer. The resistivity of the film was measured by a standard four probe method under DC magnetic field. up to 1 T.

RESULTS AND DISCUSSION

As reported in another paper [15], after the irradiation, the resistive behavior was semiconductive. However, both Jc^m and Jc^t were increased by the additional postannealing. Strongly correlated with the transport result, the magnetic properties of the treated films were very unusual. Also as shown in another paper [15], the shielding magnetization (zero field cooling and warming to measure it) of the non-irradiated film exhibited almost linear decreasing with increasing temperature. On the contrary, for the irradiated film, the initial level at 4.2 K remained up to much higher temperature, at which this initial level suddenly decayed with showing a discontinuous drop. The calculated Jc^m by the Bean model was significantly increased up to several millions A/cm^2 in the wide temperature range. The magnetic relaxation was much smaller in the initial level region, and was slightly smaller in the decreasing region than that of non-irradiated specimen. The Meissner effect (cooling in the field to measure it)was much smaller than the shielding signal, for both the irradiated and non-irradiated specimens.

The most dramatical results could be observed in the initial magnetization curve, as seen in Fig. 1. When the external field was applied perpendicular to the non-irradiated film plane, the initial magnetization curve usually showed a broad minimum at about 100 Oe. Above this field, an absolute value of diamagnetization was continuously decreased. This shape of the initial magnetization curve was also seen widely in other film specimens [16].

However, for the irradiated GdBaCuO specimen, the very large magnetization above M =10000 emu/cc is discontinuously changed to the very small value around Hex = 100 Oe at 4.2 K. The magnetization in the small value state is not perfectly zero, but is the very small diamagnetization below M = 100 emu/cc. This novel transition is apparently similar to the behavior of the type I superconductor. The transition field (Hp) is not perfectly definite with showing some distribution, but it seems that the transition field exhibits the broad peak at around 50 K, as shown in Fig. 2. The initial susceptibility before the transition is smoothly decreasing with increasing temperature. At about 50 K, the magnetization maintains the very high value up to the transition field. Near the critical region, such a steep transition can not be observed any more. In this region, the initial magnetization curve becomes similar to the non-irradiated specimens with showing the continuous change. The same results are also observed in the irradiated ErBaCuO film.

As shown in Fig. 3, its hysteresis curve is very different from that of the non-irradiated specimen. For the high Tc superconducting films, the

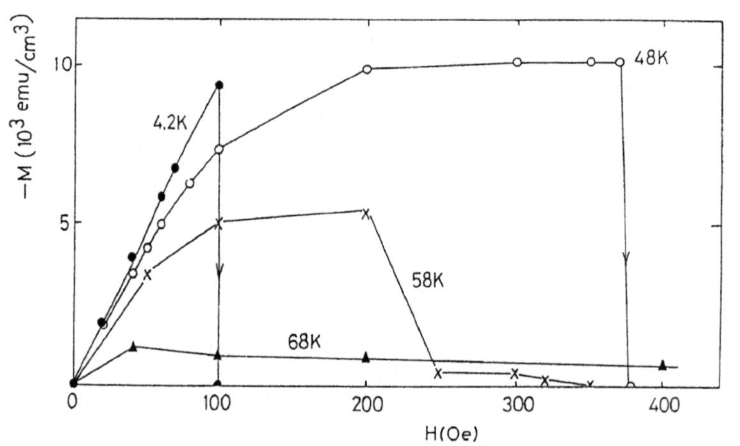

Fig. 1 The initial magnetization curves for x ray irradiated and postannealed GdBaCuO films at constant temperature.

shape of the hysteresis curve is generally an imperfect rhombic curve[16]. But, in the hysteresis curve of the irradiated specimen, once it changes to the small value state, it never recovered to the original state.

In order to observe the effect of the treatment on the transport supercurrent, the resistivity was measured under DC magnetic field up to 1 T. Different from the magnetization results, even if the field is increased perpendicular to the film plane, any anomalous change in the resistivity was not observed at constant temperature. The resistive broadening in the critical region is relatively smaller than that of the non-irradiated specimen. The apparent dHc2/dT is obtained from the mid-points between the onset resistivity and zero. They are 1.1 T/K and 0.66 T/K for the irradiative and non-irradiated GdBaCuO specimens, respectively.

From these results, we will try to explain the magnetization anomalies. We assume the magnetization of superconductor as

$$4\pi M = - 1/V\int(Hex - B)dV$$

, where $B=\phi_0 n(r)$, ϕ_0 is a flux quantum, and $n(r)$ is its number density per unit area. In a shielding state below Hc1, the origin of diamgnetization is surface supercurrent with $n(r)=0$ all over the specimen. Above Hc1, fluxoids begins to invade from the surface into the specimen. In the case of the weak pinning centers, the fluxoids invasion is considered to be smooth with showing the broad minimum in the initial magnetization curve, as shown in Fig. 1. The very rapid decreasing of the shielding magnetization with increasing temperature in the non-irradiated films is also thought to be due to the easy invasion of fluxoids. On the contrary, the Meissner effect generally shows the very small magnetization with the very poor temperature dependence. This is because the shielding magnetization might be canceled by the large number of fluxoids invading into the whole volume just below Tc. If the new pinning centers are created by the treatment, the pinning forces would work to prevent the invasion of fluxoids into the shielded volume. When the pinning forces may be sufficiently strong to fix the invading fluxoids locally in the surface area, the large shielded volume could be held to show the giant diamagnetization until the invading pressure will overcome the pinning force to cause a rush of fluxoids into the specimen. By this model, the anomalies of our results would be well explained. Namely, in a low temperature region, the created pinning forces prevent a rush of fluxoids into the specimen. But, with increasing temperature, finally, the fluxoids may destroy the weakest part of pinning barrier to invade into the specimen. This may be called as a flux jump. It is also very reasonable that the magnetic relaxation is significantly suppressed before the steep invasion. The transition in the initial magnetization curve in Fig. 1 would be explain by the same consideration. In contrast with the magnetic results, the magneto-resistive result was not very exciting. Considering this result, the created pinning forces might be suppress the superconducting fluctuation which could cause the resistive broadening in the critical region.

It is very clear to change the pinning situation by this treatment. However, it is not yet clarified what is the pinning origin due to this anomalies. As described in another paper[15], we could not observe any macroscopic imperfection such as twin boundary, precipitation, etc. An x ray diffraction pattern indicated that any other peak was not observed except the (123) orthorhombic phase. Therefore, we speculate that the origin might be

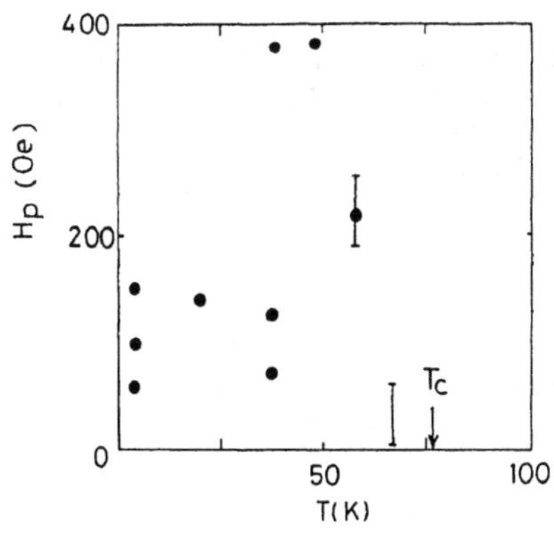

Fig. 2 The transition fields as a function of temperature in the initial magnetization curves for x ray irradiated and postannealed GdBaCuO films.

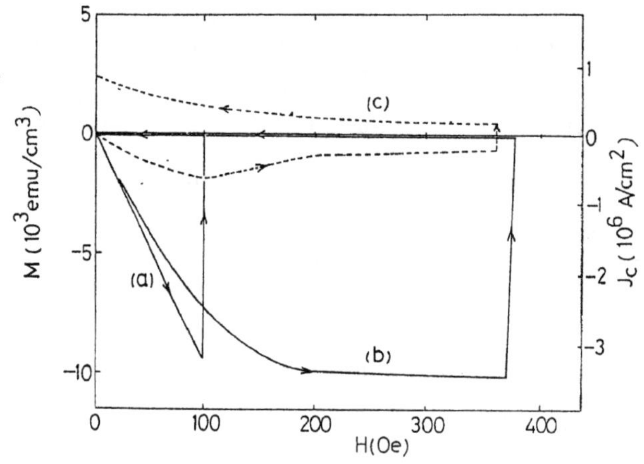

Fig. 3 The hysteresis curves for

(a) x ray irradiated and postannealed GdBaCuO film at 4.2K,

(b) x ray irradiated and postannealed GdBaCuO film at 48 K,

(c) non-irradiated and annealed GdBaCuO film at 4.2 K.

more microscopic defects as small as atomic size, which should be comparable to the very small coherent length. Considering the energy of x ray radiation, it is suggested that the oxygen defect may be most probable for the pinning centers.

REFERENCES

1. J. Z. Sun, K. Char, M. R. Hahn, T. H. Geballe and A. Kapitulnik : Appl. Phys. Lett. 54 (1989) 663
2. T. T. M. Palstra, B. Batlogg, R. B. van Dover, L. F. Schneemeyer and J. V. Waszczak: Appl. Phys. Lett. 54 (1989) 763
3. K. Kitazawa, S. Kambe, M. Naito, I. Tanaka and H. Kojima : Jpn. J. Appl. Phys. 28 (1989) L555
4. Y. Yeshurun and A. P. Malozemoff : Phys. Rev. Lett. 60 (1989) 2202
5. S. Hatta, H. Adachi, Y. Ichikawa, S. Hayashi, S. Kohiki, and K. Wasa : J. Phys. Soc. Jpn. 58 (1989) 4132
6. T. Matsushita : Proc. ISTEC WORKSHOP ON SUPERCONDUCTIVITY (1989) Oiso p31
7. M. Murakami, M. Morita and N. Koyama : Jpn. J. Appl. Phys. 28(1989)L1125
8. M. Tinkham : Proc. ISTEC WORKSHOP ON SUPERCONDUCTIVITY (1989) Oiso p9
9. R. L. Fleischer, H. R. Hart, Jr., K. W. Lay and F. E. Luborsky : Phys. Rev. B 40 (1989) 2163
   R. B. van Dover, E. M. Gyorgy, L. F. Schneemeyer, J. W. Mitchell, K. V. Rao, R. Puzniak and J. V. Waszczak : nature 342 (1989) 55
10. G. C. Xiong, H. C. Li, G. Linker and O. Meyer : Phys. Rev. B 38 (1988) 240
11. S. Takamura, T. Aruga and T. Yoshiya : Jpn. J. Appl. Phys. 28 (1989) L1118
12. W. G. Maisch, G. P. Summers, A. B. Campbell, C. J. Dale, J. C. Ritter, A. R. Knudson, W. T. Elam, H. Herman, J. P. Kirkland, R. A. Neiser and M. S. Osofsky : IEEE Trans. Nucl. Sci. NS-34 (1987) 1782
13. B. B. Boiko, F. P. Korshunov, G. V. Gatalskii, A. I. Akimov, V. I. Gatalskaya, S. E. Demyanov and E. K. Strikuk : Phys. Status Solidi A 107 K139 (1988)
14. S. Kohiki, S. Hatta, K. Setsune and K. Wasa submitted to Appl. Phys. Lett.
15. S. Kohiki, S. Hatta, K. Setsune, K. Wasa, Y. Higashi, S. Fukushima, and Y. Gohshi : to be published in Proc. 2nd ISS (1989) Tsukuba
16. S. Hatta, Y. Ichikawa, H. Adachi, K. Kamada, Y. Ichikawa and K. Wasa : Jpn. J. Appl. Phys. 27 (1988) 1646

Magnetic Shield of Bi-Pb-Sr-Ca-Cu-O High-T_c Superconductors at 77K in a Weak Magnetic Field

T. NAKAYAMA[1], H. OHTA[2], H. TAKAYAMA[1], K. HOSHINO[1], K. SHIGEMATSU[1], E. SUDOH[3], S. YAMAZAKI[1], K. KATOH[1], H. TAKAHARA[1], and M. AONO[2]

[1] Core Technology Laboratories, Mitsui Mining & Smelting Co., Ltd., Ageo, Saitama, 362 Japan
[2] RIKEN (Institute of Physical and Chemical Research), Wako, Saitama, 351-01 Japan
[3] Research & Development Laboratory, Tokyo Super Refractories Co., Ltd., Asamuta, Ohmuta, Fukuoka, 836 Japan

ABSTRACT

Magnetic shield of Bi-Pb-Sr-Ca-Cu-O high-Tc superconductor which is applicable to neuromagnetic measurements was investigated. To measure a magnetic shielding effect of a Bi-Pb-Sr-Ca-Cu-O superconducting vessel, alternating magnetic fields were applied from the outside of the vessel and detected by a pick-up coil placed inside. A cooled FET amplifier at 77K was used to improve S/N ratios of very small voltages induced in the pick-up coil. Field attenuation ratios (magnetic shielding effects) were around 2×10^{-6} when magnetic fields of $1\times10^{-5}\sim1\times10^{-4}$ T at 10~1000Hz were applied to the superconducting vessel at 77K.

KEY WORDS: Bi-Pb-Sr-Ca-Cu-O high-Tc superconductor, magnetic shield, cooled FET amplifier, neuromagnetic measurement

INTRODUCTION

High-Tc superconductors of Bi-Pb-Sr-Ca-Cu-O system have been widely investigated about its structure, phase diagram and superconducting mechanisms. Also, many applications to electronic devices have been proposed and studied. The material, however, have some unsolved problems and difficulties to be used as superconducting magnets, wires and so on. One of the most serious problems is its low critical current density. For example, superconducting magnets requires the superconducting tapes whose critical current densities are at least $10^5\sim10^6$ A/cm^2 at 77K. Unfortunately, it is very difficult to obtain tapes of such high quality by well-known method such as sputtering, MBE, CVD or screen printing.

This paper deals with the magnetic shield of high-Tc oxide superconductors of Bi-Pb-Sr-Ca-Cu-O system. One of the important applications of magnetic shield is measurement of neuromagnetic fields from human brains. The neuromagnetic fields of human brains are $10^{-11}\sim10^{-13}$ T and terrestrial magnetic field is $\sim3\times10^{-5}$ T in average. Second order gradiometers of a SQUID magnetometer can reduce the disturbance of spacially-homogeneous terrestrial magnetic fields by a factor of a thousand or more. Attenuation of the terrestrial magnetic field by a factor of a thousand is not enough for detecting the neuromagnetic fields. Therefore, the reduction of spurious ambient magnetic fields by the magnetic shield of high-Tc superconductor by a factor of hundred thousands is essential to detect neuromagnetic fields. It should be pointed out that a low critical density as $\sim10^2$ A/cm^2 is required to shield weak magnetic fields of $10^{-5}\sim10^{-4}$ T for this application.

EXPERIMENTAL

Although the Bi-Pb-Sr-Ca-Cu-O system has an unknown phase diagram and has difficulties to get a single phase material, we have successfully prepared Bi-Pb-Sr-Ca-Cu-O vessels which have sharp phase transitions from a normal conducting state to a superconducting state at 100K~105K with the following sequence. Homogeneous and fine powders were condensed in ethanol as oxalates of each components. The chemical composition of Bi:Pb:Sr:Ca:Cu was fixed to 0.8:0.2:0.8:1.0:1.4 in this study. The powders were calcined at 500°C for 5 hours to dissociate oxalic acid, and ground in a ball mill. They were again calcined at 800°C for 12 hours, and ground. We then pressed them into disk shapes with a pressure of 30MPa. The disks were annealed at 830°C for 24 hours, and finally ground into fine powders. The 105K phase of Bi system became dominant in the powder through the process described above. Fig.1 shows the superconducting vessel which was made of powders mentioned above and pressed using a cold isostatic press (CIP) with a pressure of 10^2MPa. The vessel had size of 46mm in inner diameter, 67mm in depth, and 2mm in thickness. The Bi-Pb-Sr-Ca-Cu-O superconducting vessel had transition temperature of 103K as shown in Fig.2. Figure 3 shows critical current density of the superconducting vessel at 77K.

Fig.1. A Bi-Pb-Sr-Ca-Cu-O superconducting vessel. Its critical temperature was measured as 103K, and its critical current density was 139 A/cm². It is 46mm in inner diameter, 67mm in depth and 2mm in thickness.

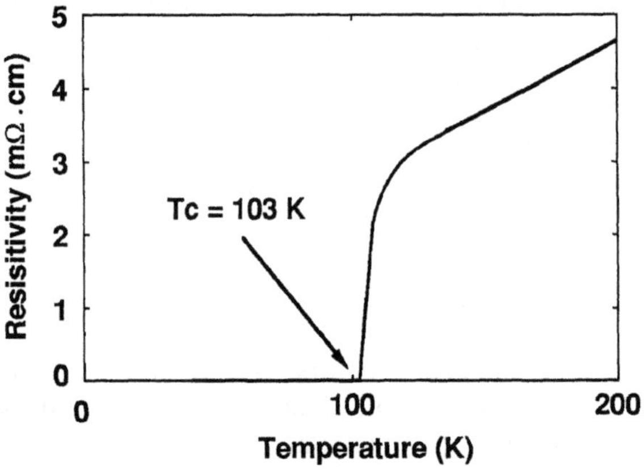

Fig.2. Critical Temperature of the Bi-Pb-Sr-Ca-Cu-O superconducting vessel is shown. Sample was cut from the bottom of the vessel.

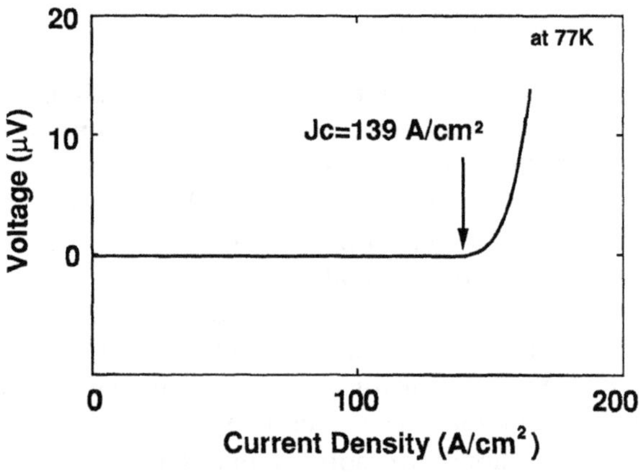

Fig.3. Critical current density of the Bi-Pb-Sr-Ca-Cu-O superconducting vessel. For a magnetic shield against a weak magnetic field, the value of 139 A/cm² is considered to be enough.

Figure 4. shows the system used to measure a magnetic shielding effect. The function generator generates sinusoidal waves, and the coil placed just outside the vessel generates alternating magnetic field. The magnetic field was varied in the range of $0.1\sim1\times10^{-4}$ T and 10~1000Hz. All experiments have been done in an electromagnetic shield room.

It has been experimentally confirmed that magnetic shielding effects are remarkably affected by small cracks in the vessel. We have actually investigated the correlation between shielding effects and cracks using X-ray photographs. Some of cracks were caused by thermal stress when the vessel was rapidly cooled from room temperature down to 77K. And the others were caused by soaking liquid nitrogen into the vessel. Since liquid nitrogen in small pores of the vessel expands rapidly during its changes from the liquid phase to the gas phase, the vessel can not stand such expansion. Alumina powder has ensured a poorer thermal conduction to prevent the vessel from rapid quenching and soaking liquid nitrogen while temperature of the vessel was monitored with a thermocouple of Au•Fe/Cr. Thus we have successfully cooled a thin sample as shown in Fig.1 without making any crack in the vessel.

When a Bi-Pb-Sr-Ca-Cu-O vessel was in the superconducting state, induced voltages in the pick-up coil was extremely small($10^{-6} \sim 10^{-10}$ volts). So we designed and assembled "Cooled FET Amplifier" which operates at 77K and amplifies such small signals described above. The cooled FET amplifier was placed just outside the vessel in order to improve the S/N ratio. Figure 5 shows a circuit diagram of the amplifier. Since all devices used in the amplifier are not guaranteed to operate at 77K, we had to test characteristics of all components at 77K carefully. The noise equivalent voltage of the amplifier was 1.7nV/√Hz at 77K. Because the cooled FET amplifier could be placed near the pick-up coil, the system has become sensitive enough to detect the extremely small signal from the pick-up coil in the superconducting vessel at 77K.

RESULTS AND DISCUSSION

It has proved that the Bi-Pb-Sr-Ca-Cu-O superconducting vessels can shield weak magnetic fields with high efficiency. Figure 6 shows magnetic shielding effects of the vessel in various frequencies. Magnetic shielding effects did not depend on strengths and frequencies of magnetic fields applied to the superconducting vessel from the outside. These tendencies are different from the results obtained with the Y-Ba-Cu-O vessel we have reported previously[1]. The difference seems to come from the process in preparation of the vessel, especially in pressing and annealing of the vessels. We

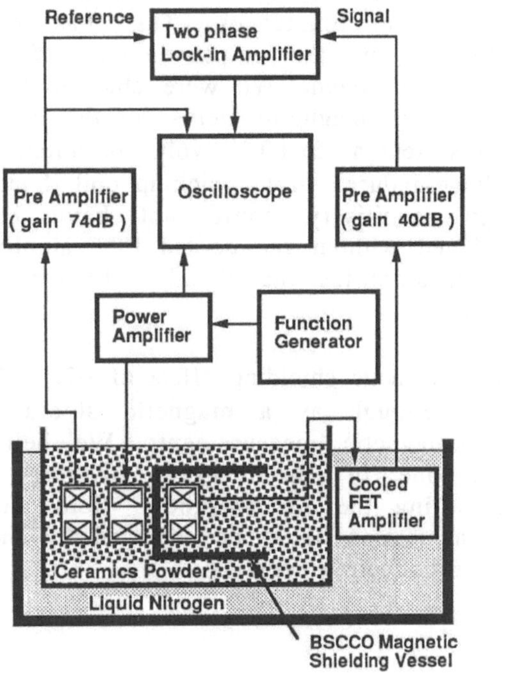

Fig.4. Block diagram of the measuring system for magnetic shielding effect of the vessel. The superconducting vessel is surrounded by ceramics powder in order to prevent it from thermal shock. A cooled FET amplifier is immersed in liquid nitrogen to improve signal to noise ratio.

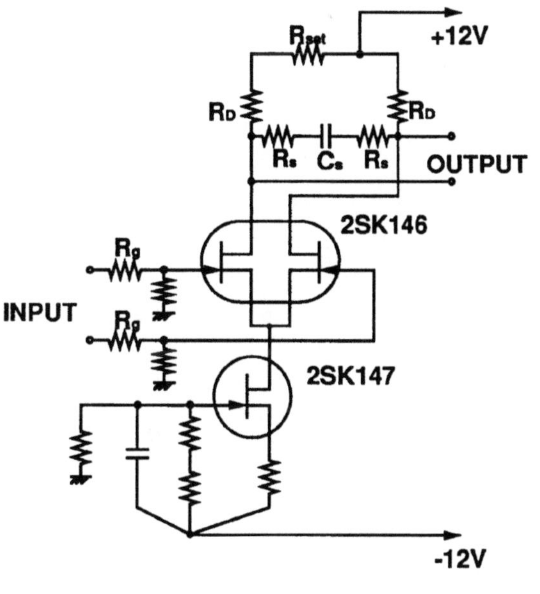

Fig.5. Circuit diagram of a cooled FET amplifier. Devices for this amplifier should be carefully chosen for the purpose of operating at 77K.

must note that this piece of data was taken under daily conditions, i.e. under noisy ambient circumstances. This has been achieved by the improvements of the measuring system. We were able to detect such weak magnetic fields by the simple coil system as 5×10^{-10} volt, the minimum voltage induced in the pick-up coil. Such a high sensitivity could not have been achieved without the cooled FET amplifier which amplifies the signals effectively at 77K.

This magnetic shielding effect of $\sim2\times10^{-6}$ is good enough as a magnetic shield for neuromagnetic measurements. We believe that one of the biggest barriers for measuring the neuromagnetic fields from human brains could be removed by these superconducting vessels.

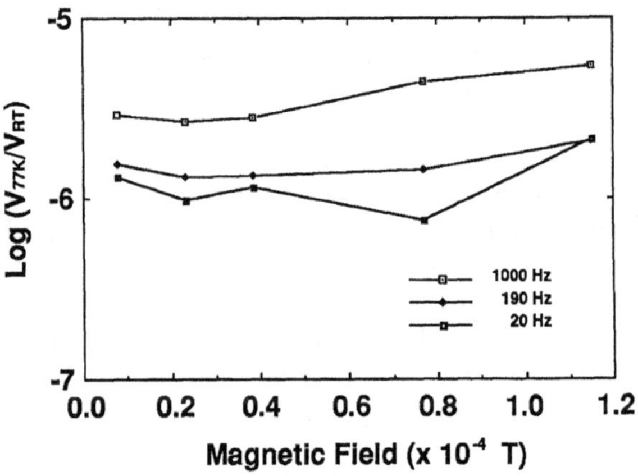

Fig.6. Magnetic shielding effect of the Bi-Pb-Sr-Ca-Cu-O superconducting vessel for weak magnetic fields. Note that terrestrial magnetic field is about 3×10^{-5} T.

REFERENCE

1. Shigematsu K. Ohta H. Hoshino K. Takayama H. Yagishita O. Yamazaki S. Takahara H. Aono M. (1989) Magnetic Shield of High-Tc Oxide Superconductors at 77K. Jpn.J.Appl.Phys 28: L813-815

Evaluation of Magnetic Hysteresis Curves for Various Specimens

Y. Ishikawa, M. Kojima, and S. Yoshizawa
Central Research Laboratory, Dowa Mining Co., Ltd., Tobuki, Hachioji, Tokyo, 192 Japan

ABSTRACT

Hysteresis curves of various specimens were measured using vibrating sample magnetometer (VSM) to estimate the magnetic charactors. Used specimens are powders of different particle size and bulks made from those powders .
The difference of initial magnetization curves and hysteresis curves between powders and bulks were studied. In case of bulks initial magnetization curve has a shoulder below Hc1,irrespective of no shoulder for powders. This shoulder comes from breaking of the weaklinks between grains. Value of Hc1 increased as the particle size of powders decreased from 50 um to 1um. Volume fraction of superconducting phase for powders was relatively proportional to the particle size. By increase of scanning rate of magnetic field, hysteresis loop widened. This result indicates that current induced from the change of magnetic field increases the magnetization.

KEY WORDS: initial magnetization curve, hysteresis curve, induced current transport current, shielding current

Introduction

It is not suitable to determine low critical current density by magnetic hysteresis ΔM used Bean-London model (BLM). The low Jc, for example 10A/cm², by BLM is much higher than one by four-probe method(FPM). Thus, we suppose that magnetization Ma consists of the transport current Mt which means intergranual shielding currents and the shielding current Ms. From the magnetic measurements of bulk specimens by VSM, it is difficult to discriminate between Mt and Ms. There are many grain boundaries contributing to insulators or weaklinks in polycrystalline specimens prepared by the sintering method. In order to calculate exactly the Jc value, when the transport current is much less than the shielding current, it is better to use the grain size as the width of sample D. A factor increasing magnetization in spite of Mt and Ms has been studied in this report. The value of magnetization under the alteration of magnetic fieldcontains a certain contribution by induced-current Mi. This portion of magnetization Mi increased corresponding to the alteration ratio of magnetic field. After all, it is probably mentioned that the magnetization consists of three portion of magnetization attributed from a shielding current, transport current, and induced current.
The magnetization by shielding current and induced current can be estimated by measuring magnetization of pressed powder specimen, because we could neglect transport current of this specimen like insulator.In bulk specimens sinterd pressed-powders, the magnetization value contains the contribution from the transport current in addition to shield current and induced current.
We examined the dependence of magnetization on the size of the powder and the difference of magnetization between powders and bulks, and the magnetization dependence on the alteration ratio of magnetic field.

EXPERIMENTAL

Y_2O_3 ,$BaCO_3$ and CuO powders were mixed with Zr balls in atheton for 10h and calcined at 900 C for 30h in air. The calcined powder was pulverized in an agate mortar for 1hr. The powder was calcined at 950 C in air and pulverized again. After more than 50 um screened ,the size of powder was controled from 10 um to 1 um by ball milling method. We prepared an average size powder 1 um, 2 um , 7 um, 10 um. Furthermore the powders were pressed to 1 ton /cm^2 into pellets of 20 mm in diameter and 2 mm in thickness and then sintered at 950 C in air for 20 h, followed by cooling to room temperature at a rate of 1 C/min.

Measuring magnetization data of powders and bulks on rod-shaped sample cooled in zero field by VSM (TOEI) is two kinds. Firstly we measured initial magnetization curves to estimate volume fraction of superconducting relatively , breaking of weak-coupling and lower critical magnetic field. Secondly we measured magnetic hysteresis to estimate Jc. Transport currents were measured by FPM to compare with Jc ditermined from hysteresis curves.

RESULTS AND DISCUSSION

Initial magnetization curves of the powder and the YBCO bulk at 77 K in low field are shown in Fig 1. the curve of polycrystalline $YBa_2Cu_3O_7$ has a shoulder at about 10 Oe bellow Hc1 while there is no shoulder for powders. This shouder is derived from the breaking of weak-links among crystallites [1].We compared magnetic field breaking weak-links of YBCO with TlBaCaCuO and BiSrCaCuO in Fig 2. It seems that its field increases as critical temperature increases. One of the reason is that ξ (77K) becomes longer as the critical temperature increases according to an equation which describes a relation between ξ and temperature [2]. Lower critical magnetic field Hc1 determined from initial magnetization curves is defferent between powders and bulks. The Hc1 of powders is changed by the powder size as shown in Fig 3. Hc1 decreases to powder size. As shown in Fig 4,volume fraction of powders increases by the increase of powder size, which relationship is caused by the penetration depth [3]. Volume fraction of bulk increased as one of powders increases.The hysteresis loops depending on the powder size are shown in Fig 5. The loops of bulks sintered at 950 C for 20 h in air from different powder size are shown in Fig 6. The magnetization of powders is less than its sintered powders. As is shown in Table 1, Jc for all specimens determined from FPM

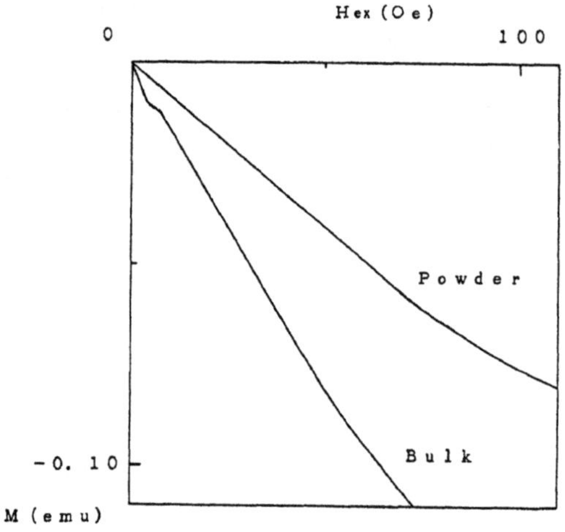

Fig. 1 Initial magnetization curves of the powder and the YBCO bulk at 77K in a low field.

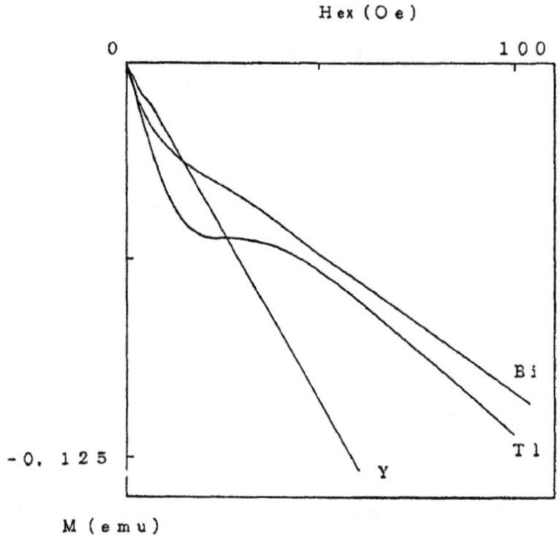

Fig. 2 Initial magnetization curves for TlBaCaCuO, BiPbSrCaCuO as other kinds of superconducting materials to compare the applied field of breaking weaklinks.

is nearly equal. These results indicate that the magnetization of low Jc specimens is mainly contributed to shielding current though transport current of low Jc specimens is unchanged irrespective of the shape of hysteresis loop. Furthermore hysteresis loops aredepending on sintering conditions. As shown in Fig 7, hysteresis loop of the specimen sintered for 40 h at 950 C is larger than one for 20 h, though same powders are used and Jc of its sintered powders by FPM is nearly equal to 30 A/cm². It seems that this is derived from the increase of shielding current by a growth of grains in sintering process.

Table.1
Comparison of some parameters for bulks made from different powder size.

SAMPLE	AVERAGE PARTICLE DIAMETER (um)	DENSITY (g/cm3)	Tc (K)	Jc (A/cm2)
A	1.00	5.85	86.3	6.62
B	2.20	6.00	87.1	8.99
C	7.55	5.65	90.5	18.11
D	10.50	5.40	91.0	28.11

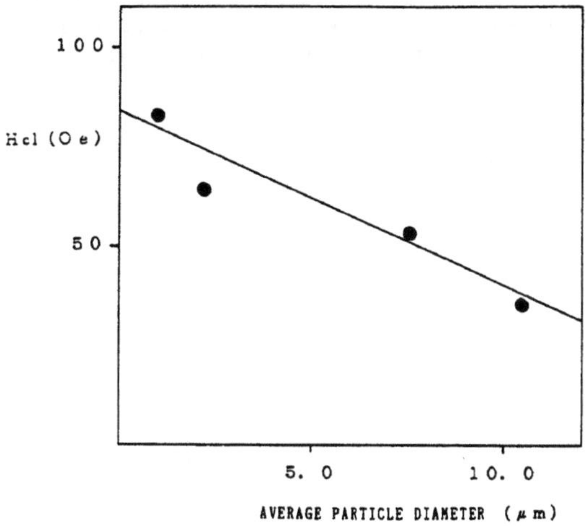

Fig. 3 Hcl of powders is inversely proportional to powder size.

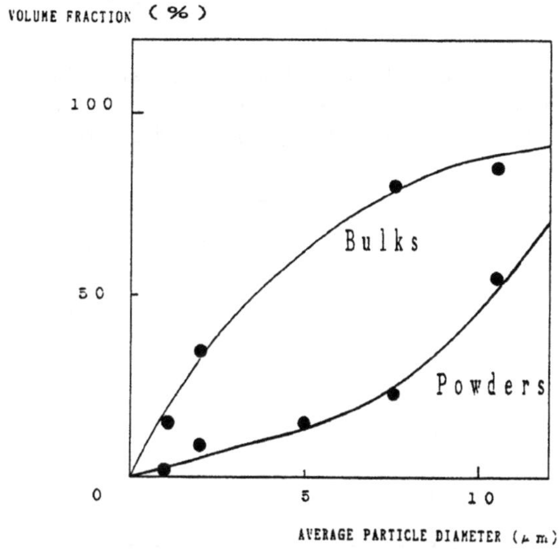

Fig. 4 Volume fraction of powders and bulks is proportional to powder size.

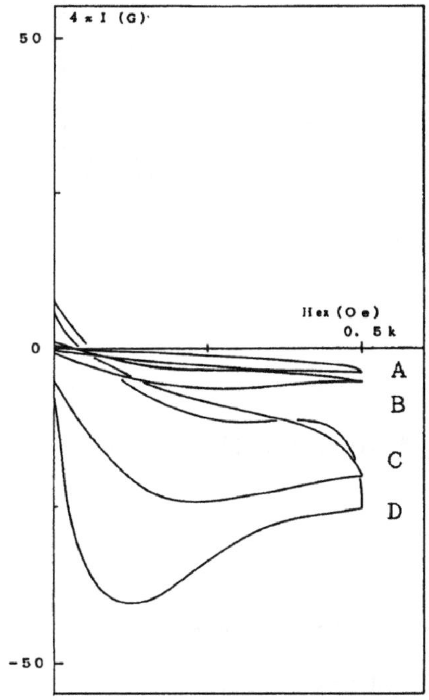

Fig. 5

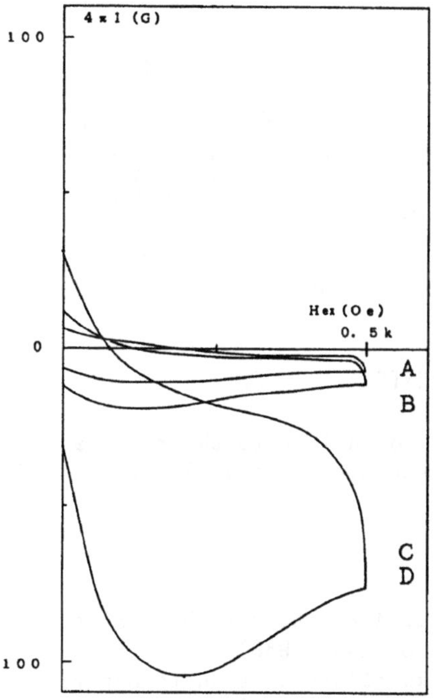

Fig. 6

The hysteresis loop depending on powder size (Fig.5) and bulks of its sintered powders (Fig.6)

Hysteresis loops of 50 um powders size and bulks made from those powders widened with the increase of scanning rate of magnetic field as shown in Fig 8, Fig 9. It is mentioned that this result is caused by induced current inspecimens with changing magnetic field. In changing scanning rate of magnetic field, this hysteresis curves for bulks and powders have almost same variation rate. That is, induced current occures to intragrains and intergrains.

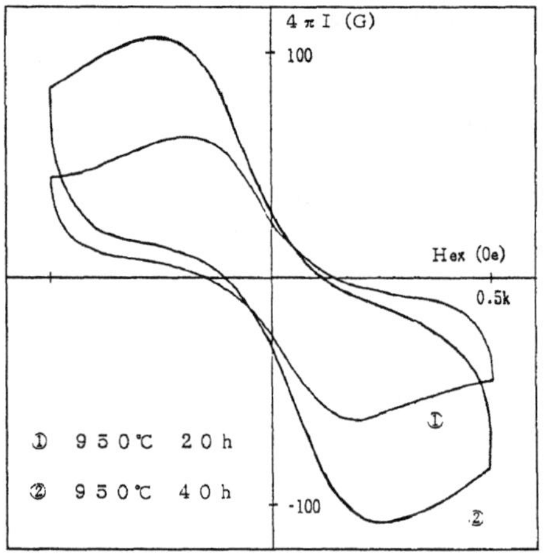

Fig. 7 Hysteresis loop is changed with sintering conditions of powders changed though same powders are used and Jc of its sintered powders by FPM is nearly equal.

Hysteresis loop of 50 um powders size (Fig.8) and bulks of its sintered powders (Fig.9) widened with scanning rate of magnetic field set fast.

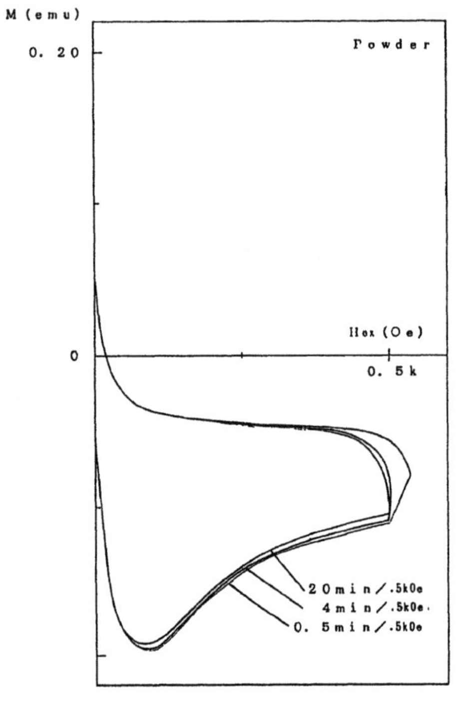

Fig. 8

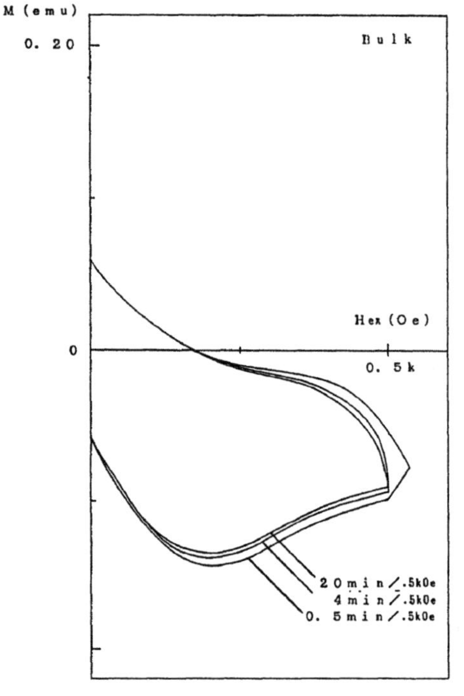

Fig. 9

ACKNOWLEDGMENTS

The authors would like to thank H. Masuda and H. Momiura (TOEI) for his valuable discussions and analysis with VSM-5.

REFERENCES

[1] D. Wang, A.K. Stamper, D.D. Stancil and T.E. Schlesinger:Appl.phy. Lett. 53, 240 (1988)
[2] M. Tinkham:Introduction to Superconductivity (McGraw-hill,1975)
[3] E. Shimizu, D. Ito:Advances in Superconductivity (Nagoya,1988)

Critical Current Density Measurements by Opposite Polarized Magnetic Dipoles

KOICHI SHINTOMI and HIRONORI MATSUBA

Yokohama R & D Laboratories, The furukawa Electric Co., Ltd., 4-3, Okano 2-chome, Nishi-ku, Yokohama, 220 Japan

ABSTRACT

The authors prepared an apparatus to form a magnetic field on the superconductor surface by dipoles consisting of two small opposite coils. It was found that current density could be determined from the penetration of magnetic field into a superconductor, or from the residual magnetic field distribution.

KEY WORDS: superconductor, critical current density, magnetic dipole flux penetration, permanent current

INTRODUCTION

Evaluating a change of the characteristics of superconductors at various points is important for producing and evaluating large superconductors, and several evaluation methods have been examined. For example, J.H.Claassen revealed that the penetration of a magnetic field into a superconductor can be judged by measuring the third harmonic voltage across a coil by applying an alternate magnetic field to a superconductor from the coil [1], while F.Hellman et al, C.Moon et al, and D.B.Marshall et al reported that the superconductive characteristics can be evaluated by measuring the repulsive force produced between a superconductor and a permanent magnet [2] [3] [4]. But it may be difficult to evaluate the characteristic at every measured point, because the range of the magnetic field to be applied to the superconductors changes according to their characteristics. Such these circumstances, we propose an apparatus which can apply a magnetic field within a limited range of superconductors, and thus can evaluate the characteristics of the superconductor at every measured point by monitoring their magnetic behavior.

MEASURING APPARATUS

Figure 1-a shows a pair of two coils which works as a magnetic field source and a magnetic sensor (Hall device), which measures the magnetic field component in the direction perpendicular to a superconductor. The field direction of coil A is opposite to that of coil B, so that the spread of the magnetic field distribution on the superconductor surface is reduced as shown in figure 1-b. The magnetic field on the superconductor surface is parallel to the surface when no magnetic field penetrated. While, when a magnetic field penetrated, a magnetic field component is produced in the direction perpendicular to the superconductor surface as shown in figure 2. As a result, the penetration of magnetic field can be judged by a magnetic sensor mounted at the position shown in the figure.

EXPERIMENTAL

We examined three measuring methods. Method 1 is used to judge the penetration of a magnetic field into a superconductor, method 2 is used to measure the intensity of residual magnetic field, and method 3 is used to examine the distribution of the residual magnetic field.

Method 1: The magnetic field is measured while applying a constant current to coils. The measurement is repeated with increasing the current.

Method 2: The current is turned off after applying a constant current to coils, and the residual magnetic field is measured. The measurement is repeated with increasing the current.

Method 3: The current is turned off after applying a current to coils enough for the magnetic field to penetrate through a superconductor. The residual magnetic field distribution is measured after one minute.

Table 1 shows samples employed for measurement as well as the coil current and measured points in method 3. In addition, we measured defective sample and deviations of characteristics at various points in the same sample.

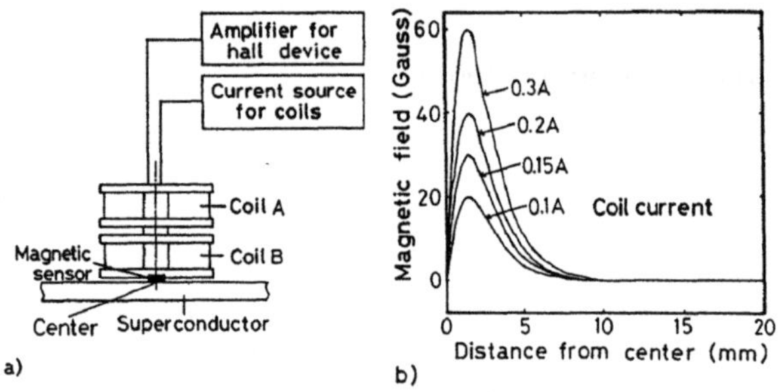

Fig. 1-a Schematic diagram of measuring system
Fig. 1-b Magnetic field formed by pair coils as a function of current.

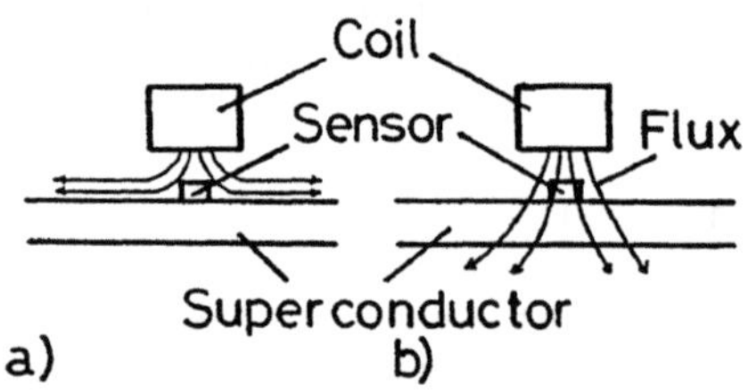

Fig. 2 Magnetic flux distribution. a) In case of weak field b) In case of strong field.

RESULTS AND DISCUSSIONS

Figure 3 shows results of method 1 for sample No. 1 together with the magnetic field on the rear side. The gradient of an increase of the magnetic field changes at a coil current of 0.08A (point A in the figure) and 0.14A (point B in the figure). Also, the magnetic field on the rear side begins increasing with point B. This shows that a magnetic field started penetrating into the superconductor at 0.08A, and then, fed-through to the rear side at 0.14A.

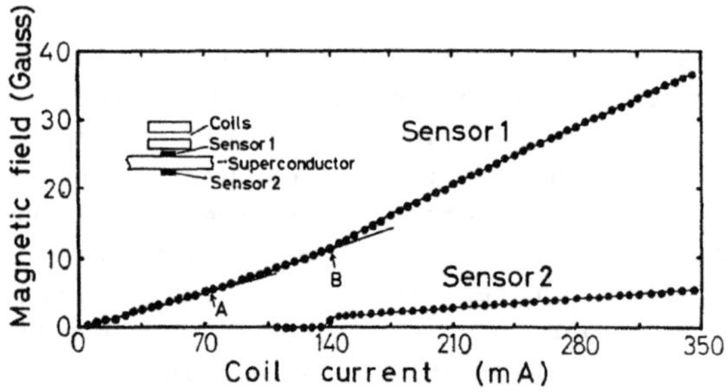

Fig. 3 Magnetic fields behaviors when coil current increased.

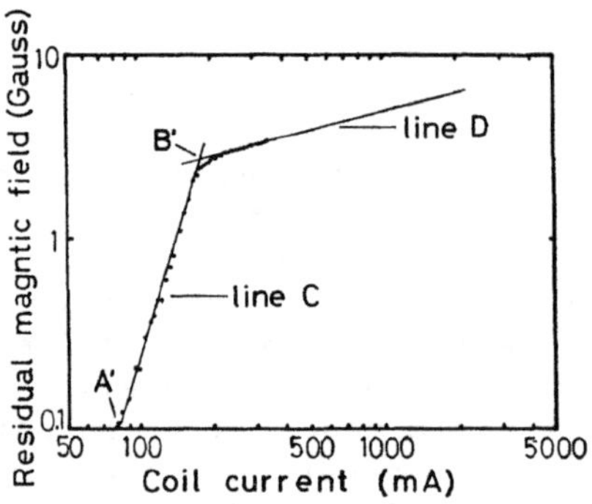

Fig. 4 Residual magnetic field depending on pair coil current.

Figure 4 shows results of method 2 for sample No. 1. From this figure, it is understood that a residual magnetic field started penetrating at a coil current of about 0.08A (point A' in the figure), and this A' point corresponds to point A in figure 3. Also, the residual magnetic field increases in two ways as shown in straight line C and D in the figure, and point B' corresponds to the point B in figure 3. These tendencies were also observed in other samples. Figure 5 shows the result of examining YBCO and BSCCO at point A' of sample having a different Jc. There is little relation between Jc and the coil current at point A'. We examined the relation between Jc and the behavior after the penetration of a magnetic field into a superconductor. Figure 6 shows the relation between Jc and the gradient of the portion corresponding to straight line C in figure 4 about samples having various Jc. From this figure, we can understand that Jc increases correspondingly as the gradient of straight line C decreases.

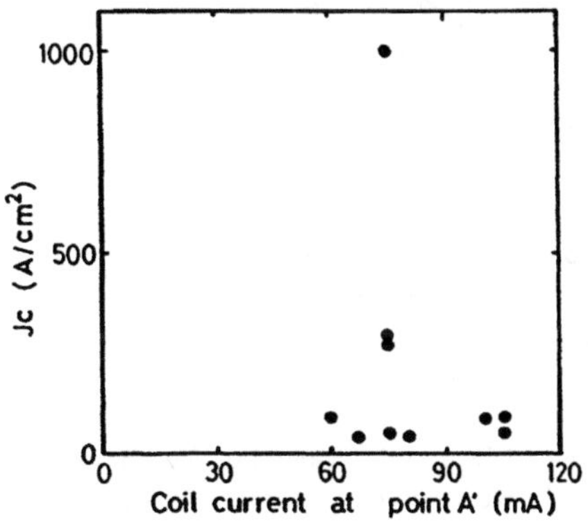

Fig. 5 Relation between Jc and penetration current (point A').

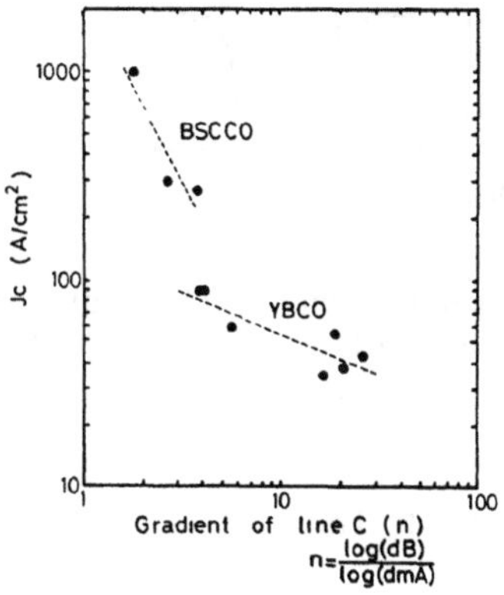

Fig. 6 Relation of Jc with gradient of line C.

Figure 7 shows results of the residual magnetic field distribution by method 3 for sample 1. The applied coil current was 0.4A in range of straight line D in figure 4, because it is necessary that the residual magnetic field remains unchanged, irrespective of the coil current. A residual magnetic field is present in the vicinity of the coil center axis in the same direction as in the applied magnetic field direction. It direction is reversed at positions slightly apart from the coil center axis. This distribution is rotationally symmetric. From these results, a permanent flowing to the residual magnetic field portion is presumable to be cylindrical. We obtained the current density from the residual magnetic field distribution. Assume that thickness of a superconductor is infinite, and the residual magnetic field has only a component which is perpendicular to the superconductor, and the current density j is represented by;

$$j=-1/\mu_0 * dB/dr$$

where, r shows the radial position. The current density is obtainable, assuming that the magnetic field on the superconductor surface is equal to the magnetic field inside the superconductor. But on samples having different Jc, there is little correlation between Jc and the current density obtained from the residual magnetic field distribution. This fact may be caused by the following;

1) Whether this model is suitable or not; 2) The thickness of samples differs from each other. 3) In the method of obtaining the current density from the residual magnetic field distribution, obtained Jc corresponds to a criterion much lower voltage than that in the 4 probe method. These factors must be examined hereafter.

Figure 8 shows the residual magnetic field distribution of sample No. 2. In this figure, the residual magnetic field distribution is concaved at position corresponding to crack positions, and also, the maximum residual magnetic field position is deviated from the center. From these symptoms, we understand that the residual magnetic field distribution is affected by defects and other factors. In case of sample No. 3, the residual magnetic field distribution was found to be disordered, irrespective of the presence of defects in appearance. This may be caused by the presence of defects that cannot be identified in appearance.

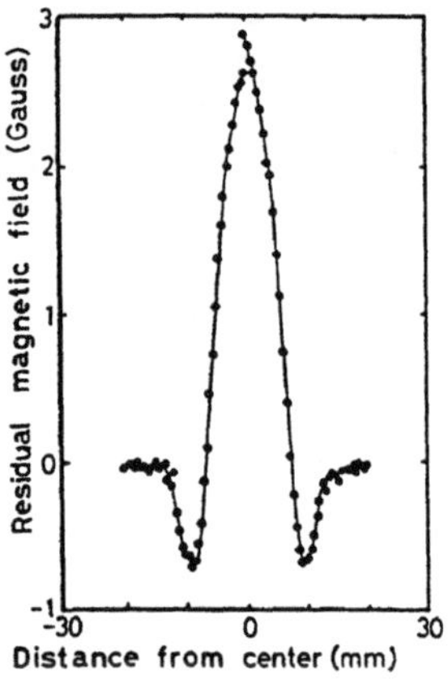

Fig. 7 Residual magnetic field distribution as a function of distance from the center of pair coil.

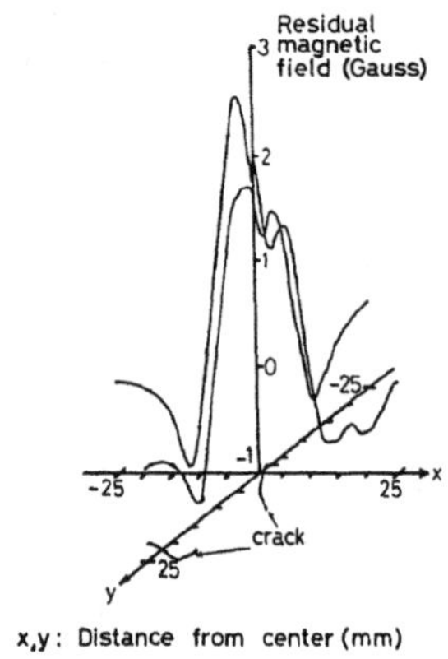

Fig. 8 Residual magnetic field distribution of a sample containing a crack.

Table 2 indicates the deviations of characteristics of sample No. 3 and 4 by measured points. The measurement at each point was executed at positions apart more than 15mm in radius as a measuring area which is presumable from the above measuring results.

Table 1 Characteristics of employed samples, and measured point and coil current in method 3.

Sample No.	Dimensions (mm)	Jc (A/cm^2)	Measuring method No.	Measured point	Coil current(A)
No.1 YBCO	100x100x1.0	90	1,2,3	Center	0.4
No.2 BSSCO	100x100x0.6	300	2,3	by crack	1.0
No.3 YBCO	100x100x5.0	90	2,3	4 points	1.25
No.4 BSCCO	100x100x0.25	1000	2,3	4 points	1.0
No.5 BSCCO	100x100x0.6	270	2,3	Center	0.75
No.6,7,8,9 and 10 YBCO	100x100x0.6	35, 40 50, 60	2	Center	

Table 2 Variation of residual magnetic field in location at the center of pair coil.

Measured point (mm)	Sample No.3 YBCO		Sample No.5 BSCCO	
	Residual field (Gauss)	Standard deviation (Gauss)	Residual field (Gauss)	Standard deviation (Gauss)
X=25 Y=0	9.67	0.119	8.08	0.646
X=0 Y=25	9.44		6.55	
X=-25 Y=0	9.50		7.28	
X=0 Y=-25	9.73		8.12	

CONCLUSIONS

We prepared coils having a small magnetic field distribution range, and examined the behavior of magnetic field penetration and residual magnetic field by applying a magnetic field into the superconductor. It was elucidated that the characteristics of superconductors can be evaluated from the behavior ranging from the penetration of magnetic field into a superconductor to the feed-through of magnetic field into the superconductor, or from residual magnetic field distribution by applying a magnetic field enough to allow the magnetic field to feed through superconductors. In addition, this method is applicable to locate various defects.

REFERENCES

1 J. H. Claassen, IEEE Trans. on Magn., Vol. 25, No. 2, March 1989, 2233-2236

2 F. Hellman, E. M. Gyorgy, D. W. Johnson, Jr., H. M. O'Bayan, and R. C. Sherwood, J. Appl. Phys., 63 (2), 15 January 1988, 447-450

3 F. C. Moon, M. M. Yanoviak, and R. Ware, Appl. Phys. Lett., 52 (18), 2 MAy 1988, 1534-1536

4 D. B. Marshall, R. E. DeWames, P. E. D. Morgan, and J. J. Ratto, Appl. Phys. A, 48, 87-91 (1989)

Nonresonant Microwave Absorption and Critical Current Density in High T_c Superconductors

A. MORIMOTO, M. MAKIDA, A. MOTO, and T. SHIMIZU

Department of Electronics, Faculty of Technology, Kanazawa University, Kanazawa, 920 Japan

ABSTRACT

Nonresonant microwave absorption (NRMA) and ESR measurements were performed for high T_c superconductors, which were prepared by solid phase reaction from powder mixture. The results were compared with the result of the critical current density measurement. The presence of a small amount of BaY_2CuO_5 impurity phase enhances hysteresis in the nonresonant microwave absorption spectrum for superconducting $Ba_2YCu_3O_x$ samples, suggesting that the impurity phase can play a role of pinning centers.

KEYWORDS: nonresonant microwave absorption, ESR, critical current density, $Ba_2YCu_3O_x$, pinning center

INTRODUCTION

High temperature superconductors in bulk form has a relatively low critical current density compared with the low temperature superconductors, though their critical temperature is extremely high. For a wide field of application, it is essential to enhance the critical current density. Weak links between the superconducting grains, which are loosely connected with each other [1], are probably responsible for the low critical current density. Furthermore, a lack of efficient pinning centers may be responsible for it through a mobile flux in the superconducting grains. Thus, in order to increase the critical current density J_c in high T_c superconductors, it seems to be important to reduce the weak links between grains and to introduce the pinning centers in grains. We distinguish two critical current densities for detailed discussion. One is the macroscopic one (intergrain), which is measured by the conventional four probe method. The other is the microscopic one (intragrain), which is usually measured by the magnetization methods [2]. Therefore, it is important to estimate each of J_c separately in order to increase the macroscopic J_c.

Here, we will point out an interesting behavior on nonresonant microwave absorption (NRMA) for the high temperature superconductors. NRMA signal often appears around the zero field region in the ESR measurement below the critical temperature [3]. Although the origin of the NRMA has not be known exactly, it seems to be related to flux slippage around the superconducting loops including weak links [3]. The field dependence of the NRMA has a hysteresis arising from flux trapping in the superconductors. The hysteresis curves for the Y-system is different from that for the Bi-system. The difference is perhaps caused by the difference in the pinning effect between the two systems [4]. What is the pinning centers in the high temperature superconductors? The candidate is a twin boundary or an impurity phase. But the exact identification is still an open problem.

In this work, we investigated the role of the impurity phase in $Ba_2YCu_3O_x$ superconductors. We have already reported that the impurity phase in $Ba_2YCu_3O_x$ is easily detected by the ESR measurement with a high sensitivity [5]. Therefore, combined NRMA and ESR measurements are suitable for clarifying the pinning role of a small amount of the impurity phase. For varying the amount of the impurity phase, in this work, two procedures were employed. First, the amount of the impurity phase is varied by a variation of the sintering temperature (the first series of samples), and second, it is varied by a variation of the amount of the impurity additives such as BaY_2CuO_5 green phase in samples (the second series of samples).

EXPERIMENTAL

For the first series of samples, $Ba_2YCu_3O_x$ superconductors were prepared by the solid-state reaction methods using Y_2O_3, $BaCO_3$ and CuO powders at various sintering temperatures. The mixtures were fired in a furnace at 930 ℃ for 8 hours in dry oxygen. After the firing, collected powder was pressed into pellets. The pellets were sintered at the various temperatures for 12 hours in dry oxygen and slowly cooled. For the second series of samples, various amounts of BaY_2CuO_5 green phase is added into the superconducting $Ba_2YCu_3O_x$. The mixture was thoroughly ground and pressed into a pellet. Then it was sintered at 945 ℃ for 12 hours in dry oxygen, and slowly cooled. For NRMA and ESR measurements the pellet was pulverized and reground into powder, resulting in several micron size.

The amount of nonsuperconducting impurity phase, which is expected to contribute both to the weak links and the pinning centers, was estimated by ESR measurements. ESR measurements were carried out by an X-band Varian E109 system at room temperature. As the ESR intensity changes approximately with $1/T$ law, the ESR spin density probably originates from localized unpaired electrons without a strong interaction [4]. NRMA measurement instead of the magnetization measurement was performed for powder samples in order to characterize the superconducting properties mainly within grains. NRMA measurements were carried out by the same ESR spectrometer at 77K. The microwave power was fixed at 0.1 mW, the modulation frequency is 100 kHz and the modulation amplitude is 0.11 G. The transport critical current measurement was performed by the four probe method in order to obtain the macroscopic J_c. The critical current was defined at an electric field of 1 μV/cm.

RESULTS AND DISCUSSION

(1)Variation of Sintering Temperature

The amount of the impurity phase as a function of the sintering temperature was examined by ESR measurement as shown in Fig. 1. Since Cu ions in the superconducting phase is known to be ESR silent, the ESR signal intensity measured at room temperature should be proportional to the amount of the impurity phase [4]. The ESR spin density is shown by the number of spins per one Cu ion in the sample. The dependence has a minimum around 1000 ℃. X-ray diffraction measurement revealed that above 1000 ℃ $Ba_2YCu_3O_x$ phase is decomposed into BaY_2CuO_5 phase. The increase in the spin density with decreasing the temperature

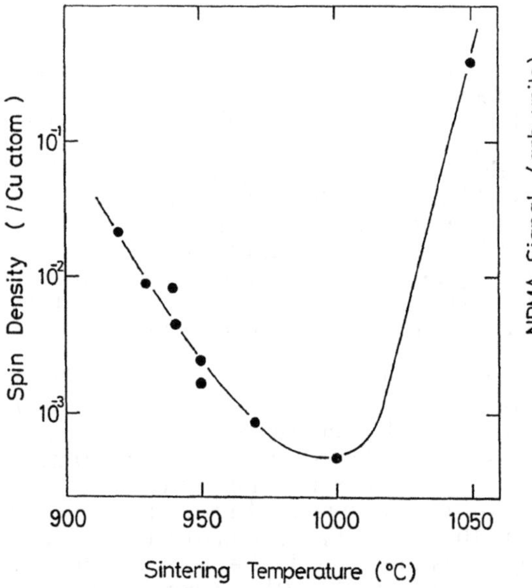

Fig. 1 Density of ESR spins arising from the impurity phase as a function of sintering temperature for the first series of samples in powder form

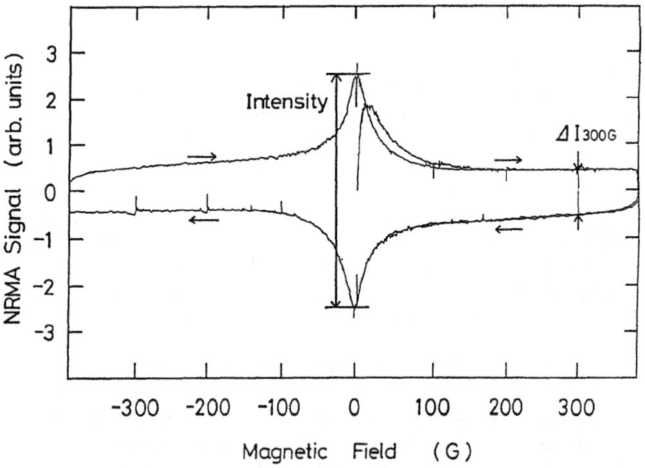

Fig. 2 Magnetic field dependence of nonresonant microwave absorption for powdered sample (NRMA spectrum)

is probably due to a formation of a non-superconducting impurity phase, though the identification of the impurity phase has not yet been done.

A typical NRMA spectrum is shown in Fig. 2. The spectrum is a derivative of the field dependence of the microwave absorption. Here we define the NRMA intensity as a peak-to-peak height, and ΔI_{300G} as an intensity hysteresis at 300 G for quantitative characterization of NRMA.

Figure 3 shows the sintering temperature dependence of J_c and the NRMA intensity. At a glance it seems that both J_c and the NRMA intensity have a correlation to some extent, and show maxima around 950 ℃. In fact, the sample with the largest J_c around 200A/cm^2 also shows the largest NRMA intensity, but the amount of the impurity phase in that sample is not a smallest one as shown in Fig. 1. The former finding suggests that J_c in this level is affected not only by the intergrain properties but also by the intragrain properties. Principally, J_c measured by a transport method should be affected by the macroscopic structure such as grain boundaries or cracks, while the NRMA using the powder form should be affected by the microscopic structure. Therefore, a further detailed experiment is needed for detailed discussion.

Figure 4 shows the NRMA intensity and ΔI_{300G} as a function of the spin density. With increasing the spin density, both the NRMA intensity and ΔI_{300G} increase and then decrease. The decease in a high spin density region is attributed to a decrease in the volume fraction of superconducting phase. The increase in a low spin density region suggests a possibility that the impurity phase acts as pinning centers. However, this suggestion is not so reliable because the variation of the spin density is brought about by the variation of the sintering temperature, accompaning a variation of the grain size *etc.* The sample located at the highest spin density is prepared at 1050 ℃ and partially melted. So, the data for the sample is omitted from the above discussion.

(2) Variation of Amount of Impurity Phase Additives

For confirming the above estimation on pinning effect arising from impurity phase, the second series of samples were prepared at a fixed sintering temperature of 945 ℃ by changing the amount of impurity phase additives. Figure 5 shows the NRMA intensity and J_c as a function of the spin density. J_c decreases with increasing the spin density, while the NRMA intensity increases and then decrease

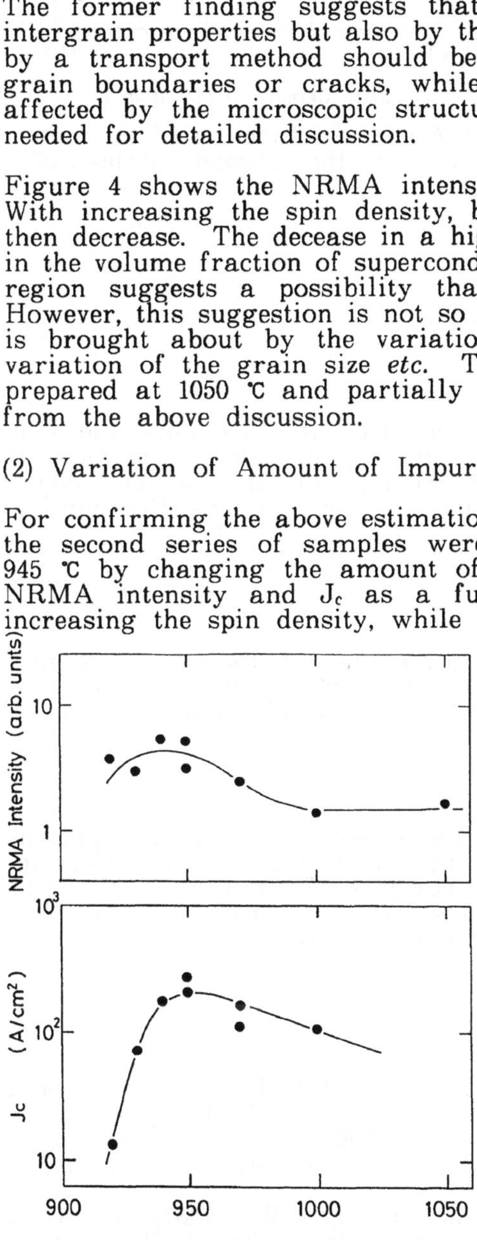

Fig. 3 Sintering temperature dependence of J_c measured by transport method and NRMA intensity for the first series of samples

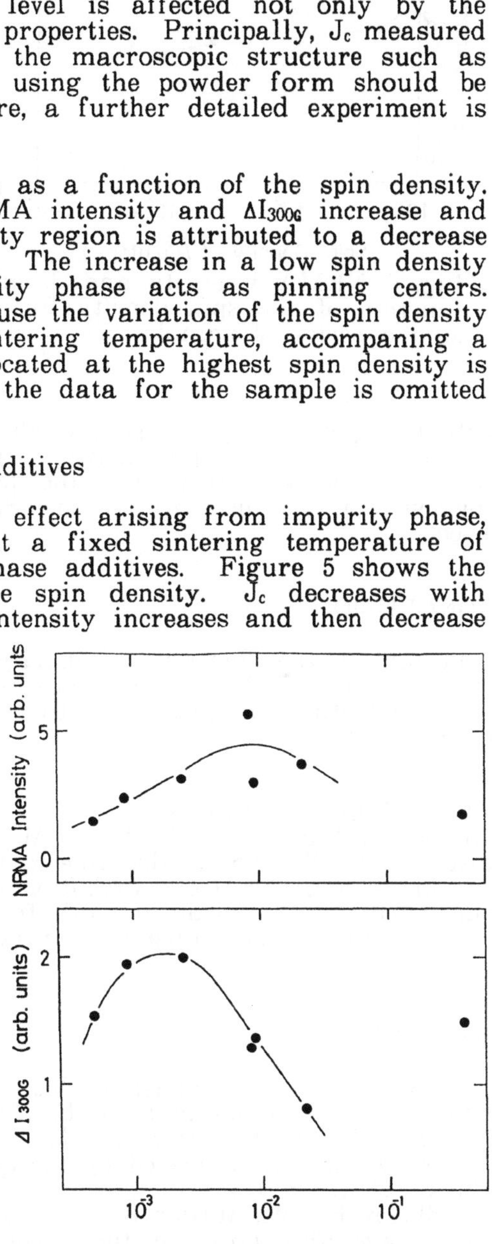

Fig. 4 NRMA intensity and ΔI_{300G} as a function of the spin density for the first series of samples in powder form

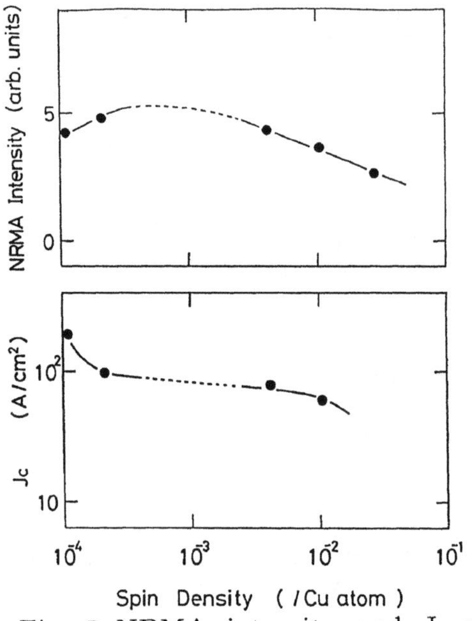

Fig. 5 NRMA intensity and J_c as a function of the spin density for the second series of samples

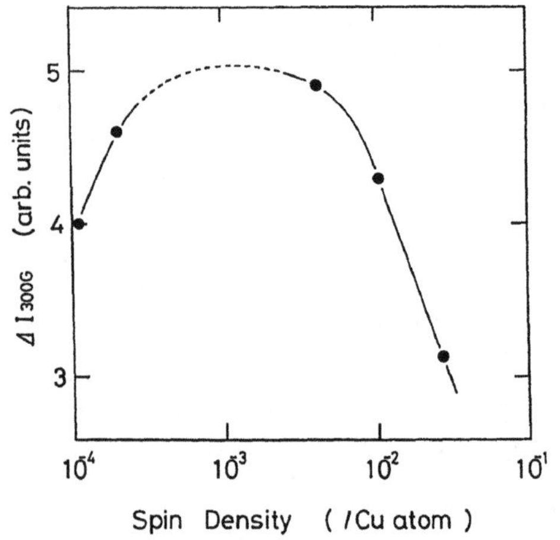

Fig. 6 ΔI_{300G} as a function of the spin density for the second series of samples

with the spin density. ΔI_{300G} also shows a similar increase with increasing the spin density in a small spin density region as shown in Fig. 6. These increases in NRMA signal with increasing the density of spins arising from the impurity phase suggest that the presence of the BaY_2CuO_5 impurity phase enhances the pinning effect.

CONCLUSION

Addition of the BaY_2CuO_5 impurity phase in superconducting $Ba_2YCu_3O_x$ sample enhances hysteresis in the nonresonant microwave absorption spectrum for the superconducting sample, suggesting that the impurity phase can play a role of pinning centers.

ACKNOWLEDGEMENTS

We would like to thank Prof. M. Kumeda of Kanazawa University for helpful discussions. We would also like to thank Dr. T. Ogawa of Murata Mfg. Co, Ltd. for measuring the size distribution of the sample powders. This work was partly supported by a Grant-in-Aid for Scientific Research on Chemistry of New Superconductors of Ministry of Education, Science and Culture.

REFERENCES

1. Muller K. A., Takashige M., and Bednorz J. G. (1987) Flux trapping and superconductive glass state in La_2CuO_{4-y}:Ba. Phys. Rev. Lett. **58** 1143–1146.
2. Murakami M., Morita M., and Koyama N. (1989) Magnetization of a $YBa_2Cu_3O_7$ crystal prepared by the quench and melt growth process. Jpn. J. Appl. Phys. **28** L1125–1127.
3. Blazey K. W., Muller K. A., Bednorz J. G., and Berlinger W. (1987) Low-field microwave absorption in the superconducting copper oxides. Phys. Rev. B **36** 7241–7243.
4. Kohl M., Odehnal M., Plintovic M. Safrata S., and Vacek K. (1988) Non-resonant microwave absorption in low magnetic fields in samples of $Ba_2YCu_3O_{7-x}$ and $BiCaSrCu_2O_{55}$ at 77K. Supercond. Sci. Tech. **1** 326–328.
5. Morimoto A., Maeda T., Moto A., Kumeda M., and Shimizu T. (1988) ESR and X-ray diffraction studies on Ba-Y-Cu-O superconductors. Jpn. J. Appl. Phys. **27** L407–L410.

Magnetization Hysteresis in the Bi(Pb)SrCaCuO Superconducting System

M.L. GREEN, K. NAKATANI, T. HASEGAWA, K. KISHIO, and K. KITAZAWA

University of Tokyo, Department of Industrial Chemistry, 3-1, Hongo 7-chome, Bunkyo-ku, Tokyo, 113 Japan

ABSTRACT

Magnetization hysteresis of polycrystalline Bi(Pb)SrCaCuO samples was measured over a wide temperature range at applied fields up to 9 T using vibrating sample magnetometry (VSM). The irreversibility (ΔM) was found to decrease nearly exponentially with temperature over the lower temperature regime and then more rapidly at higher temperatures, similar to behavior observed for YBaCuO samples [1]. The irreversibility of the Bi(Pb)SrCaCuO samples was also found to depend strongly on the applied field, particularly at higher temperatures, in contrast to YBaCuO, for which ΔM showed very little field dependence. The phase distribution of the polycrystalline Bi(Pb)SrCaCuO samples had a strong effect on the ΔM vs. magnetic field behavior.

KEY WORDS: vibrating sample magnetometry, irreversibility, phase distribution, magnetic flux

INTRODUCTION

In order to use high T_c oxide superconductors in practical applications, the critical current density (J_c) under high magnetic field must be increased. Since the onset of resistance in a superconductor is associated with the motion of magnetic flux lines through the body, the study of macroscopic magnetic properties (particularly under conditions of high field and high temperature) is extremely important in understanding and improving the critical current density. Measurement of magnetization hysteresis is very suitable for such studies, providing a direct measure of the relative difficulty (or ease) with which magnetic flux lines move throughout the sample.

Since the problems of critical current and flux creep are particularly acute in the BiSrCaCuO system, it is necessary to study this system in detail. Also, of particular interest are the differences between the Bi-based system and the YBaCuO system because it is still not clear why there is greater irreversibility and less flux creep in the YBaCuO system (Yeshurun et al. [2]). Thus, in the present study, hysteresis curves were also obtained for polycrystalline YBaCuO samples, and these results are compared with data obtained for Bi-based samples.

EXPERIMENTAL PROCEDURE

Polycrystalline samples were prepared from oxides and carbonates by a conventional solid state reaction method consisting of the calcination of mixed powders and the subsequent sintering of pressed pellets, with intermediate grinding steps. Samples in the BiSrCaCuO system generally consist of several phases: a high T_c (110 K) phase, a lower T_c (80 K) phase, and other phases.

The resulting phase distribution is known to be sensitive to composition and firing conditions; the addition of lead to the BiSrCaCuO system has been found to enhance the formation of the high T_c, 110 K phase [3,4]. Our starting composition had a metals ratio of Bi:Pb:Sr:Ca:Cu as 0.90:0.21:1.00:1.00:1.60. Calcination was done at 800 °C and sintering at 850 °C for 340 h. X-Ray powder diffraction analysis indicated the sample to consist of mostly the high T_c phase with a small amount of the 80 K phase (about 10 %) and some other minor phases. SQUID measurements indicated a Meissner fraction of about 60 percent. Synthesis of the 80 K phase was done using a composition of Bi:Sr:Ca:Cu as 4:3:3:4 and firing at 850 °C for about 90 h. X-Ray diffraction confirmed the product to be predominantly the 80 K phase. The YBaCuO sample was also prepared by a solid state reaction method; details are reported elsewhere (Sawano et al.[5]).

Magnetization measurements were performed using a vibrating sample magnetometer (EG&G Princeton Applied Research, Model 4500) equipped with a 12 T cryostat (Janis Research). A constant field sweep rate of 20 mT/s was used throughout the present study.

RESULTS AND DISCUSSION

Figure 1 shows magnetization curves (up to 9 T) over a wide range of temperatures for the high T_c phase Bi(Pb)SrCaCuO sample. The most striking feature of the data is the extremely strong dependence of the hysteresis on temperature. At the lowest temperature (4.2 K), the hysteresis is substantial, persisting out to the highest fields measured. However, as the temperature is increased, the hysteresis decreases rapidly. For example, at 40 K, after a magnetic field of about 3 T is reached, the curve becomes more or less reversible. At 60 K, reversibility sets in at about 1.5 T. Hysteresis curves of similar shape, showing a fairly similar temperature dependence, were obtained for the 80 K phase sample.

In order to quantify the magnitude of the irreversibility, we consider the quantity ΔM defined as the difference in the magnetization (at some specific value of applied field H) between the increasing and decreasing branches of the magnetization curve. It should be noted that this quantity, ΔM, is related in the familiar Bean model [6] to a pinning current density, J_c. Since J_c is directly proportional to ΔM, it is equivalent to consider either quantity here.

In Fig. 2, ΔM is shown as a function of applied magnetic field, H, for the Bi(Pb)SrCaCuO samples and for the YBaCuO sample. Considering first the data for the high T_c phase (Fig. 2b), there is only a rather small dependence of ΔM on H at 4.2 K. However, at higher temperatures, 20 K and higher, ΔM decreases rapidly with increasing

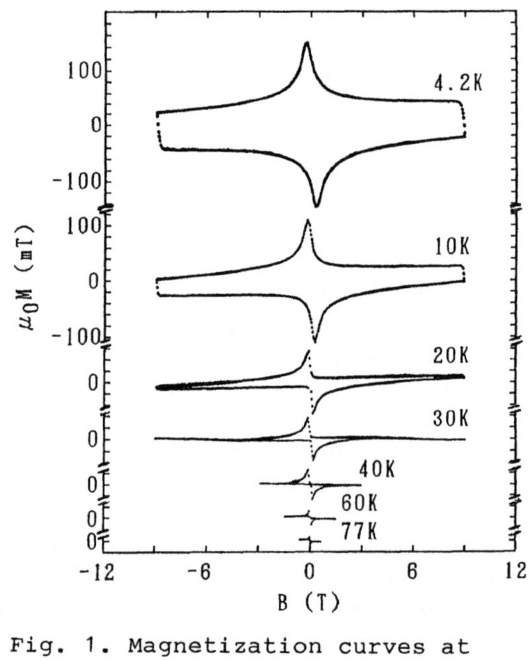

Fig. 1. Magnetization curves at various temperatures for the high T_c phase of Bi(Pb)SrCaCuO.

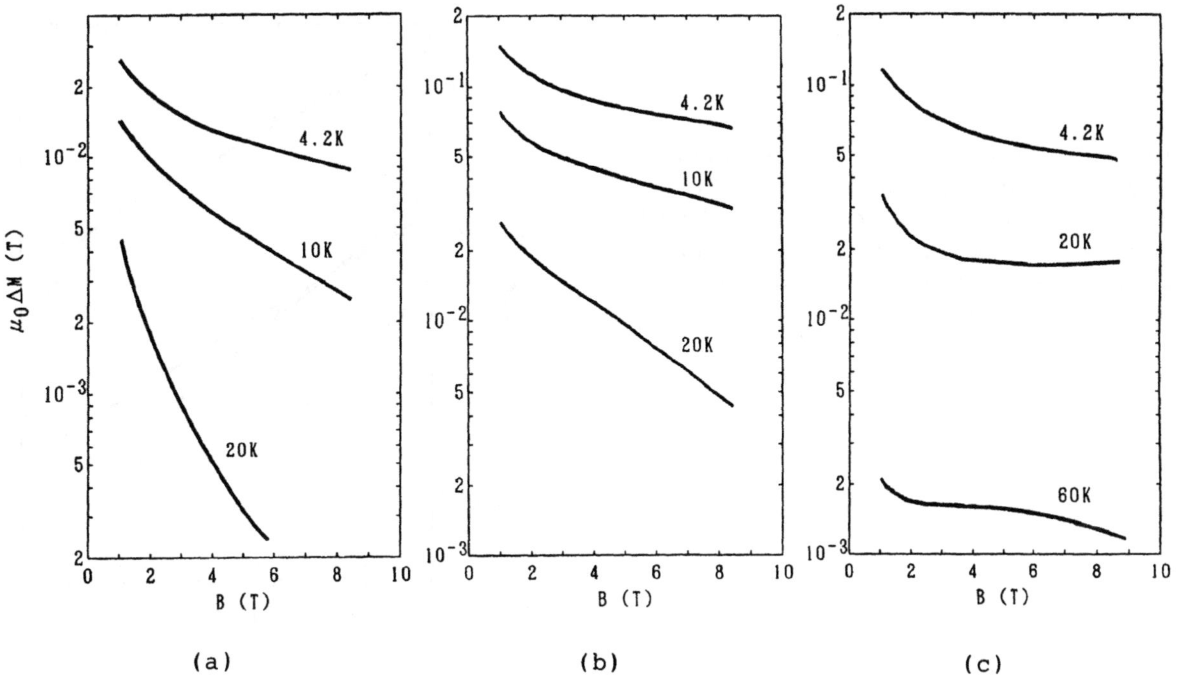

Fig. 2. ΔM vs. magnetic field at various temperatures (semilog scale).
a) 80 K phase of Bi-system; b) high T_c (110 K) phase of Bi-system;
c) YBaCuO.

H. For comparison, the data for the YBaCuO system (Fig. 2c) should be considered. In this case, very little dependence of ΔM on H was observed. For example, at 20 K, almost no change at all in ΔM with H was observed from a few Tesla up to the highest field measured, 9 T. Even at 60 K, very little effect of magnetic field was observed.

The difference between the 80 K phase and the high T_c phase is particularly apparent in the plots of ΔM vs. H in Figs. 2a and 2b. The 80 K phase (Fig. 2a) shows a greater dependence of ΔM on H than does the high T_c phase, at all temperatures. While it is expected that the grain size will affect the magnitude of ΔM, differences in grain size are not expected to affect the dependence of ΔM on variables such as magnetic field or temperature. Therefore, this plot of ΔM vs. H for the high T_c phase and the 80 K phase demonstrates the important effect of phase distribution on magnetic hysteresis. It should be noted that in the present study, the polycrystalline samples have been assumed to consist of randomly oriented grains. However, since the high T_c phase exists in elongated grains, grain shape and orientation may affect the magnetization, even in polycrystalline samples. Thus, it is possible that the phase effects observed here are due to grain shape and orientation changes and not due to any effects inherent to the structures of the phases themselves. Further work is being directed toward understanding these effects.

Next we consider the temperature dependence of ΔM for the various samples. Figure 3 shows a semilog plot of ΔM vs. temperature at a specific value of applied magnetic field. The figure contains data for the high T_c phase, the 80 K phase, and the YBaCuO samples, all taken at 1.5 T. For comparison, the single crystal YBaCuO data (also taken at 1.5 T) from Senoussi et al.[1] is included in the figure. In this discussion, the relative vertical positions of the curves (absolute value of ΔM) will not be considered since this may

Fig. 3. ΔM vs. temperature (semilog scale) at an applied field of 1.5 T. a) 80 K phase; b) high T_c (110 K) phase; c) polycrystalline YBaCuO; d) single crystal YBaCuO (after Senoussi et al.[1]).

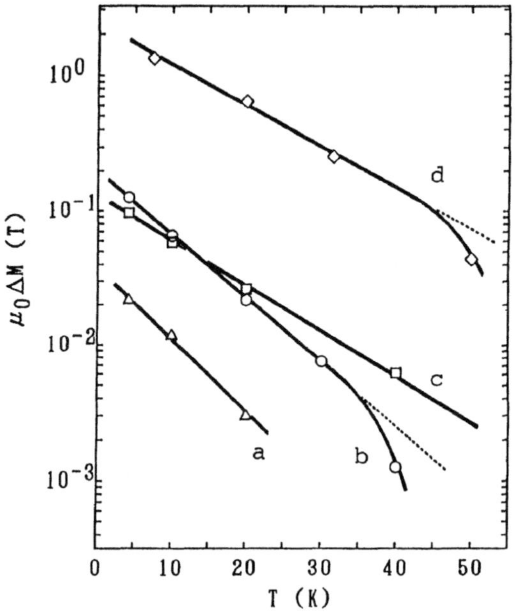

depend on the grain size. Here we consider the dependence of ΔM on temperature. In all cases, the log(ΔM) plots vs. temperature are nearly linear up to a certain temperature range, after which the decrease is even more rapid. For example, the high T_c phase shows linear behavior up to around 30 K, after which it decreases more rapidly. As shown in the figure, the nearly exponential temperature dependence we observed for the Bi-based samples and the YBaCuO sample was also observed for the YBaCuO single crystal studied by Senoussi et al.[1]. Furthermore, although not shown in this figure, as the applied magnetic field is increased, the temperature at which deviation from linearity occurs becomes lower. This was true both for the Bi-based samples in this study and the YBaCuO single crystal.

An interesting feature of the data in Fig. 3 is that the slopes of ΔM vs. temperature are not particularly different, despite the fact that these samples are quite different in terms of composition and microstructure and are expected to exhibit different values for the critical current density. The slopes do change in the following order: the magnitude of the slope is greatest for the 80 K phase, followed by the 110 K phase, followed by the YBaCuO polycrystalline sample. The slopes for the YBaCuO polycrystalline sample and the YBaCuO single crystal are fairly similar. At any rate, while there are differences in slope, they seem to be rather small. In a related study, Morita et al.[7] compared the magnetic hysteresis behavior of melt-processed YBaCuO (QMG process [8]) with that of ordinary sintered polycrystalline YBaCuO. Although these samples have very different microstructures and vastly different critical current characteristics [9], their ΔM-temperature dependences were quite similar.

More significant differences in the samples might become evident at higher temperatures (above 30 to 40 K), where the dependence on temperature becomes greater. However, at higher temperatures, the value of ΔM is quite small for polycrystalline samples, so there is too much uncertainty in the ΔM data in this range to make any conclusions about differences in samples at this stage. Perhaps with samples containing much larger grains it will be possible to study this higher temperature regime.

It is particularly interesting to compare the data for the Bi-based system with data for the YBaCuO system. As discussed above, at higher temperatures the Bi-based samples show reversible behavior once a certain relatively low magnetic field is reached, while for the YBaCuO samples, irreversibility persists at higher temperatures and magnetic fields. For example, at 60 K the Bi-based sample (high T_c phase) becomes reversible above about 1.5 T. On the other hand, at the same temperature, the YBaCuO polycrystalline sample still showed irreversible behavior at the highest fields measured, 9 T. Our YBaCuO data is supported by that of Senoussi et al.[10] on polycrystalline YBaCuO. They also found that irreversibility in the magnetization persisted at 60 K and 3.5 T, the highest field measured.

It is clear that at higher temperatures flux creep is substantially greater in the BiSrCaCuO system than in the YBaCuO system [2]. Gammel et al.[11] studied flux-lattice melting in single crystal YBaCuO and in a single crystal of the Bi 80 K phase. They found possible evidence for flux-lattice melting at around 75 K for the YBaCuO crystal (T_c= 87 K) and at around 30 K for the BiSrCaCuO crystal (T_c= 75 K). The results in the present study are consistent with the suggestion that pronounced changes occur in the BiSrCaCuO system in the region of 30 to 40 K. Moreover, the fact that irreversibility persists to higher temperatures and higher magnetic fields in the YBaCuO system than in the Bi(Pb)SrCaCuO system is also consistent with the Gammel et al. study. It was mentioned above that the temperature at which the log(ΔM) vs. temperature curves begin to deviate from linearity tends to decrease as the applied field is increased. This is consistent with the fact that the possible flux-lattice melting transition was found to occur at lower temperatures as the applied field was increased.

It should be noted, however, that in our experiments on the 80 K phase of the BiSrCaCuO system, some finite irreversibility does persist at temperatures above 30 K. At 40 K for example, reversible behavior does not set in until an applied field of around 1.2 T is reached. Similarly, Kumakura et al.[12], in a study of the high T_c phase, found some irreversibility in the magnetization even at 77 K. Reversible behavior did not set in until a field of about 0.2-0.3 T was reached. Further studies are needed to elucidate whether or not a small, yet finite irreversibility at temperatures of 40 K and higher is compatible with the proposed condition of a vortex-liquid state at temperatures in excess of 30 K.

CONCLUSION

While at very low temperature the Bi-based system shows high irreversibility and only a small magnetic field dependence, as the temperature is increased, the irreversibility decreases rapidly and becomes strongly field dependent. In all cases the irreversibility (ΔM) was found to decrease nearly exponentially with temperature, corresponding to a decrease in J_c of more than an order of magnitude over a temperature range of 4 to 30 K at an applied field of 1.5 T. It was observed that there were only relatively small differences in the slope of the log(ΔM)-temperature relations among the samples, despite differences in composition and microstructure. However, differences in samples became evident as the ΔM-magnetic field dependences were considered, with the Bi-based samples showing a strong dependence on magnetic field at temperatures of 20 K and higher.

ACKNOWLEDGMENTS

We thank K. Sawano and M. Morita of Nippon Steel Co. for helpful discussions. This work was partially supported by a Grant-in-Aid for Scientific Research from the Ministry of Education, Science and Culture of Japan. One of us (M.L.G.) was supported by a fellowship from the Japan Society for the Promotion of Science.

REFERENCES

1. Senoussi S, Oussena M, Collin G, Campbell IA (1988) Exponential H and T decay of the critical current density in $YBa_2Cu_3O_7$ single crystals. Phys Rev B 37: 9792-9795.

2. Yeshurun Y, Malozemoff AP, Worthington TK, Yandrofski RM, Krusin-Elbaum L, Holtzberg FH, Dinger TR, Chandrashekhar GV (1989) Magnetic properties of YBaCuO and BiSrCaCuO crystals: a comparative study of flux creep and irreversibility. Cryogen 29: 258-262.

3. Takano M, Takada J, Oda K, Kitaguchi H, Miura Y, Ikeda Y, Tomii Y, Mazaki H (1988) High-T_c phase promoted and stabilized in the Bi,Pb-Sr-Ca-Cu-O system. Jpn J Appl Phys 27: L1041-L1043.

4. Endo U, Koyama S, Kawai T (1988) Preparation of the high T_c phase of Bi-Sr-Ca-Cu-O superconductor. Jpn J Appl Phys 27: L1476-L1479.

5. Sawano K, Hayashi A, Ando T, Inuzuka T, Kubo H, Processing of superconducting ceramics for high critical current density (submitted for publication).

6. Bean CP (1962) Magnetization of hard superconductors. Phys Rev Lett 8: 250-253.

7. Morita M, Sawano K (in preparation).

8. Murakami M, Morita M, Doi K, Miyamoto K (1989) A new process with the promise of high J_c in oxide superconductors. Jpn J Appl Phys 28: 1189-1194.

9. Murakami M, Morita M, Koyama N (1989) Magnetization of a $YBa_2Cu_3O_7$ crystal prepared by the quench and melt growth process. Jpn J Appl Phys 28: L1125-L1127.

10. Senoussi S, Oussena M, Hadjoudj S (1988) On the critical fields and current densities of $YBa_2Cu_3O_7$ and $La_{1.85}Sr_{0.15}CuO_4$ superconductors. J Appl Phys 63: 4176-4178.

11. Gammel PL, Schneemeyer LF, Waszczak JV, Bishop DJ (1988) Evidence from mechanical measurements for flux-lattice melting in single-crystal $YBa_2Cu_3O_7$ and $Bi_{2.2}Sr_2Ca_{0.8}Cu_2O_8$. Phys Rev Lett 61: 1666-1669.

12. Kumakura H, Togano K, Uehara M, Maeda H, Takahashi K, Yanagisawa E (1988) Magnetic properties of Pb-doped Bi-Sr-Ca-Cu-O superconductors. Jpn J Appl Phys 27: L1514-L1516.

A.C. Susceptibility of YBaCuO Prepared by Quench and Melt Growth Process

S. GOTOH, M. MURAKAMI, H. FUJIMOTO, and N. KOSHIZUKA

Superconductivity Research Laboratory, International Superconductivity Technology Center, 10-13, Shinonome 1-chome, Koto-ku, Tokyo, 135 Japan

ABSTRACT

A.C. susceptibility measurements have been carried out on YBaCuO prepared by the quench and melt growth process. The temperature dependence of the real component, X', and the imaginary component, X", of the susceptibility does not indicate the presence of weak-link regions in the sample. The superconducting transition width of X' is very narrow and only a single peak of X" is observed as differed from the bulk sintered materials. The observed relationship between X' and X" is well explained in terms of Bean's critical state model.

KEY WORDS: A.C. susceptibility, superconductors, weak-link, critical state model, hysteresis loss

INTRODUCTION

The measurement of A.C. susceptibility, X=X'-iX", has been so far used to determine the accurate critical temperatures of the conventional metallic superconductors, and recently used to decide the onset temperatures for diamagnetism of the high-Tc oxide superconductors. In general, in the high-Tc bulk sintered superconductors A.C. susceptibility measurements as functions of temperature show two drops in the real component, X' and two corresponding peaks in the imaginary component, X" when A.C. field amplitude exceeds the threshold value, for example, 1 Oe[1-8]. This behaviour indicates the presence of superconducting regions or grains coupled by weak-links or Josephson type junctions[1,3,9], reflecting low values of transport critical current density, Jc. Extremely short coherence length and low carrier density, which are characteristic of high-Tc oxide superconductors, or oxygen deficiency at grain boundaries seem to be the source of the weak-link behaviour[10]. But it has become clear that low Jc is not intrinsic to them by the fact that very high Jc values are obtained in single crystals[11] and thin films[12].

Recently, large YBaCuO crystals have been fabricated by the quench and melt growth process[13]. They have shown high Jc values that exceeded 1×10^4 A/cm^2 at 77K and 1T. Their D.C. magnetization measurements indicated that Bean's critical state model[14] was well established in such samples[15]. This implied no weak-link region inside the samples. In this paper , we report the A.C. susceptibility of the QMG processed YBaCuO and show that the weak-link behaviour is absent according to the temperature and field dependence of the susceptibility.

EXPERIMENTAL

Samples of $YBa_2Cu_3O_x$ were prepared by the quench and melt growth (QMG) process. Since the details of the process is given in ref. 13, only a brief account is given here. $YBa_2Cu_3O_x$ powders which had been calcined in advance were heated to 1400°C and quenched by using copper hammers. The quenched plates were heated to 1100°C and held for 20 min, and then cooled to 1000°C at the rate of 100°C/h followed by slow cooling at the rate of 5°C/h.

A.C. susceptibility as a function of temperature was measured by an A.C. susceptometer(Lake Shore Cryotronics, Inc. ; Model 7000). Samples for measurements were prepared in form of small plates, in which the c axis was parallel to the surface plane. Magnetic field was applied parallel to the c axis, since the sample contained cracks along the ab plane, which complicated the analysis of experimental results. The frequency of the applied A.C. field was f=125 and

1000 Hz, and the exciting field amplitude, Hm, ranged from 0.05 Oe to 10 Oe. No attempt to shield the earth magnetic field was done. All the values of measurements were corrected by the approximate demagnetizing factors.

RESULTS AND DISCUSSION

Fig. 1 shows the temperature dependence of A.C. susceptibility for various A.C. field amplitudes with f=125 Hz. The onset of diamagnetism (X') and the energy loss peak (X") start at the same temperature, 92.0 K, independently of Hm. The transition width from the normal to superconducting state is very narrow with the perfect diamagnetism at 88K, which is in good agreement with the result in D.C. susceptibility measurements[15] (Fig. 1(a)). The X' curves became slightly broader with increasing Hm values, but no kink or no shoulder in the transition region of the curve appeared, while such shoulder was observed in the X' curve of the bulk sintered sample. For all measured Hm values X" curve exhibits only a single peak as differed from sintered materials[1-8] (Fig. 1(b)). When Hm increased, the X" peak became broad and its height increased and did not saturate in the present experiment. The peak temperatures were almost constant, 91 K, for Hm less than 1 Oe. As Hm exceeds 1 Oe the peak temperature slightly shifts to 90.6 K. As the transition from the normal to superconducting state completed 4π X" reached a constant value of ~ 0.035 almost independently of Hm.

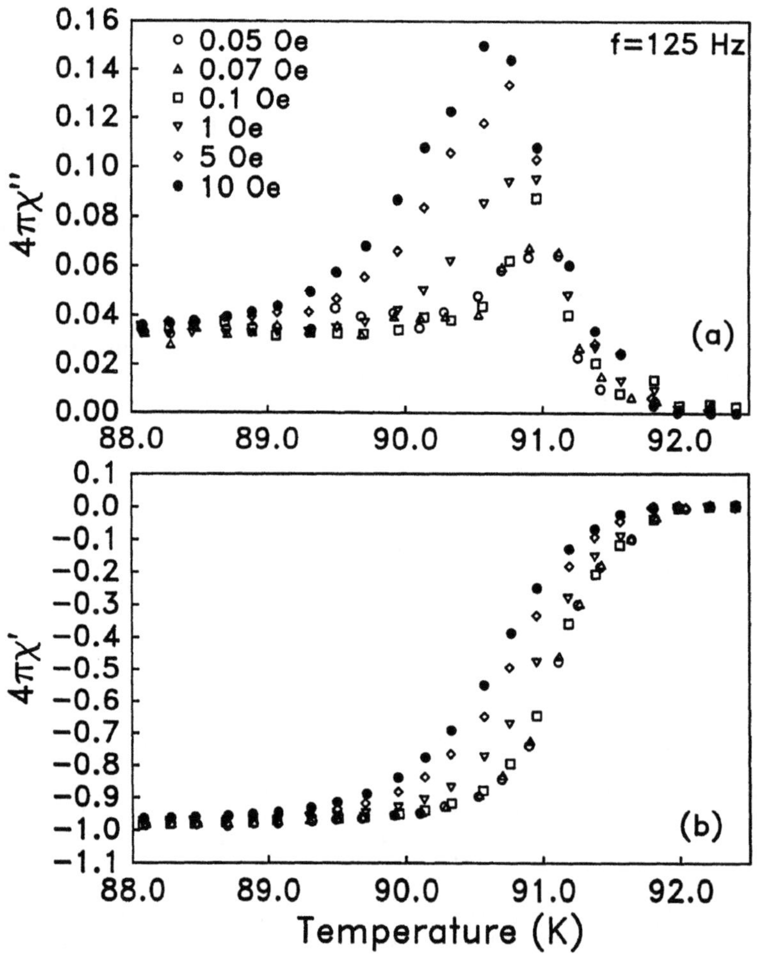

Fig. 1 (a) Imaginary component and (b) real component of the A.C. susceptibility with f=125 Hz

Fig. 2 shows the results with f=1000 Hz. As compared with the curves with f=125 Hz no appreciable change in the X' curves was observed with the increase of the frequency, however, a large frequency dependence was obtained for the X" curves. Though the peak temperature of the X" curve did not shift, the peak height increased and 4π X" reached a larger value of ~ 0.065 when the transition from the normal to superconducting state completed (Fig. 2(b)).

It is well known that in the bulk sintered YBaCuO materials the X' curve shows a two-stage transition and two corresponding X" peaks are observed at a relatively large applied field because of the weakly-coupled grains[1-3,5,7,8,16]. The low temperature peak arises from the penetration of intergranular supercurrents and the high temperature peak originates from the penetration of supercurrents inside the grains. However, the QMG processed YBaCuO does not show such a behaviour. This indicates that YBaCuO prepared by the QMG process has no weak-link region inside the sample. Therefore, there is no need to use the superconducting multiconnected structure model[17,18] for the analysis of susceptibility. The behaviour of the X" curve can be qualitatively understood in terms of Bean's critical state model. The observed X" is related to the magnetic hysteresis loss caused by the irreversible flux motion inside the specimen. Dubots and Cave[6] attributed the X" peaks to the incomplete flux penetration and the complete flux penetration during the transition from the normal to superconducting state. Our results are well explained by this model, which could also explained the D.C. magnetization behaviour[15].

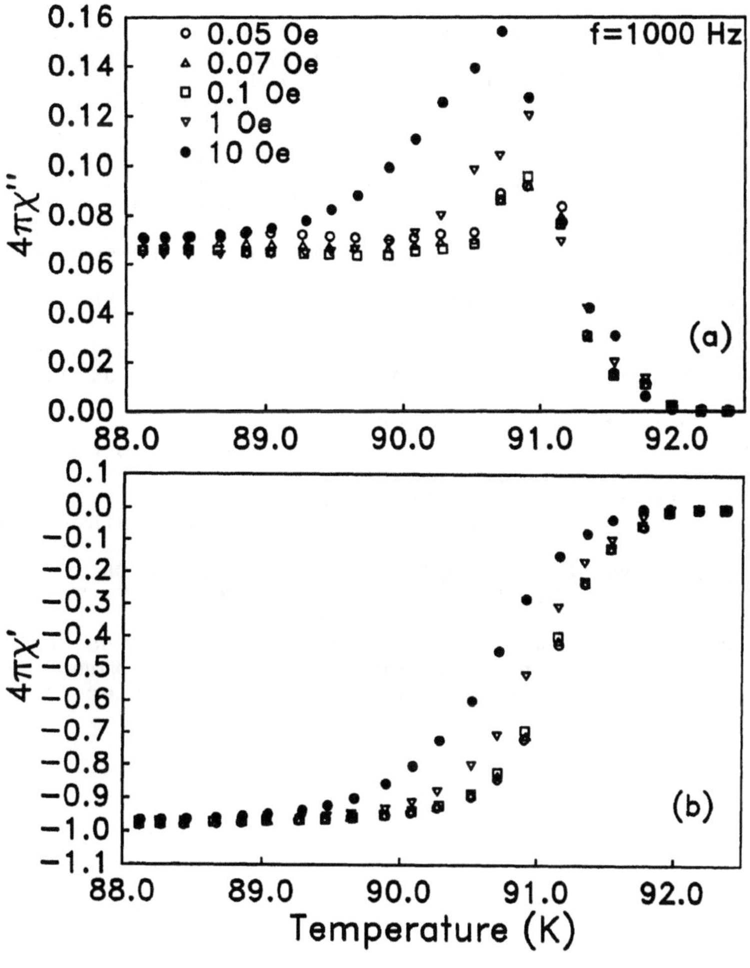

Fig. 2 (a) Imaginary component and (b) real component of the A.C. susceptibility with f=1000 Hz

Some reports on the bulk sintered YBaCuO has shown the frequency independence of the A.C. susceptibility[1,19,20], others has shown the large frequency dependence[8,21,22]. We believe this discrepancy is caused by the sample quality. As for frequency dependence of X" in the completely superconducting state, we have no definite explanation. One possible source is the effects of non-superconducting 211 particles trapped in the QMG processed YBaCuO.

CONCLUSIONS

We have measured the A.C. susceptibility of YBaCuO prepared by the QMG process and have shown that no weakly-coupled region in the sample. The A.C. field amplitude dependence was explained by Bean's critical state model in which most part of energy loss, expressed as X", is considered to be the magnetic hysteresis loss.

REFERENCES

1. R. B. Goldfarb, A. F. Clark, A. I. Braginski and A. J. Panson, Cryogenics 27 (1987) 475.
2. J. Garcia, C. Rillo, F. Lera, J. Bartolomé, R. Navarro, D. H. A. Blank and J. Flokstra, J. Magn. Magn. Mat. 69 (1987) 225.
3. D.-X. Chen, R. B. Goldfarb, J. Nogués and K. V. Rao, J. Appl. Phys. 63 (1988) 980.
4. J. S. Muñoz, A. Sanchez, T. Puig, D.-X. Chen and K. V. Rao, Physica C 153-155 (1988) 1531.
5. H. Küpfer, I. Apfelstedt, R.Flükiger, C. Keller, R. Meier-Hirmer, B. Runtsch, A. Turowski, U. Wiech and T. Wolf, Cryogenics 28 (1988) 650.
6. P. Dubots and J. Cave, Cryogenics 28 (1988) 661.
7. V. Calzona, M. R. Cimberle, C. Ferdeghini, M. Putti and A. S. Siri, Physica C 157 (1989) 425.
8. D.-X. Chen, J. Nogues and K. V. Rao, Cryogenics 29 (1989) 800.
9. D. C. Larbarestier, S. E. Babcock, X. Cai, M. Daeumling, D. P. Hampshire, T. F. Kelly, L. A. Lavanier, P. J. Lee and J. Seuntjens, Physica C 153-155 (1988) 1580.
10. G. Deutcher, Proceedings of ISS'88, Kitazawa and Ishiguro ed., Elsevir, Tokyo, 1989 p.383.
11. T. K. Worthington, W. J. Gallagter and T. R. Dinger, Phys. Rev. Lett. 59 (1987) 1160.
12. S. Tanaka and H. Itozaki, Jpn. J. Appl. Phys. 27 (1988) L662.
13. M. Murakami, M. Morita, K. Doi and K. Miyamoto, Jpn. J. Appl. Phys. 28 (1989) 1125.
14. C. P. Bean, Phys. Rev. Lett. 8 (1962) 250.
15. M. Murakami, S. Gotoh, N. Koshizuka and S. Tanaka, to be published in Proceedings of ICMC, Los Angels, July 1989.
16. K.-H. Müller, Phisica C 159 (1989) 717.
17. Y. Oda, H. Takenaka, H. Nagano and I. Nakada, Solid State Commun. 35 (1980) 887.
18. T. Ishida and H. Mazaki, J. Appl. Phys. 52 (1981) 6798.
19. H. Mazaki, M. Takano, Y. Ikeda, Y. Bando, R. Kanno, Y. Takeda and O. Yamamoto, Jpn. J. Appl. Phys. 26 (1987) L1749.
20. F. Gömöry and P. Lobotka, Solid State Commun. 66 (1988) 645.
21. B. Loegel, A. Mehdaoui and D. Bolmont, Solid State Commun. 70 (1989) 667.
22. C. Giovannella, C. Lucchini, B. Lecuyer and L. Fruchter, IEEE Trans. Mag. 25 (1989) 3521.

Is Oxygen Deficiency at Grain Boundaries the Origin of Weak Links in $YBa_2Cu_3O_y$ Sintered Materials?

KUNIHIKO EGAWA, TOSHIO UMEMURA, MITSUNOBU WAKATA, and KIYOSHI YOSHIZAKI

Mitsubishi Electric Corp., Materials & Electronic Devices Laboratory, 1-57, Miyashimo 1-chome, Sagamihara, Kanagawa, 229 Japan

ABSTRACT

In order to elucidate the relationship between the oxygen deficiency and the weak links in sintered $YBa_2Cu_3O_y$, the inter- and intra-grain critical current densities (J_{cT} and J_{cM}) were measured for the samples with different oxygen contents. It is found that the J_{cT}-B characteristics become worse with decreasing oxygen contents, while, the J_{cM}-B characteristics are almost unchanged. These results suggest that the oxygen deficiency exists at the grain boundaries and that is the origin of weak links in $YBa_2Cu_3O_y$.

KEY WORDS : inter-grain critical current densities, intra-grain critical current densities, oxygen deficiency, weak links, grain boundary

INTRODUCTION

The critical current densities (J_c's) for the sintered samples of $YBa_2Cu_3O_y$ superconductors are much lower and have stronger field dependence than those for the samples like the single-crystal because of the existence of the weak links at the grain boundaries. In our previous studies [1,2,3,4], we have observed a low resistive state above the J_c, so measured the V-I characteristics up to much higher current region and found that the resistivity is decreased with decreasing the temperature. Furthermore, we have found that J_c varies as $[1-(T/T_c)^2]^{\alpha}$, where T is the temperature, T_c is the critical temperature and the index α is about 2. Therefore, it can be estimated that the weak links in $YBa_2Cu_3O_y$ are S-N-S or S-S'-S type.

A candidate of the normal metallic phase in Y-Ba-Cu-O system is an oxygen deficient $YBa_2Cu_3O_y$ phase. The purpose of the present investigation is to discuss the influence of the oxygen deficiency on the weak links in this system. We have prepared $YBa_2Cu_3O_y$ sintered samples with various oxygen contents and measured J_{cT}-B and J_{cM}-B characteristics at 77K.

EXPERIMENTAL

The $YBa_2Cu_3O_y$ powder was prepared by the conventional solid state reaction. The calcination (and grinding) was repeated twice, at 900°C for 20h and then at 930°C for 20h in an oxygen atmosphere. The powder was pressed into pellets at 1000kgf/cm^2. The pellets were sintered at 930°C for 5h in an oxygen atmosphere. In order to obtain the samples having various oxygen contents, they were cooled down to the temperature, T_q (=room temperature, 500 ,600, 700, 800 and 900°C for the samples (a), (b), (c), (d), (e) and (f), respectively) at the rate of 100°C/h, and then quenched into a vessel filled with liquid nitrogen, except the sample (a).

The crystal phases for the sintered samples were determined by the X-ray diffraction (XRD) technique. The grain size for each sample was estimated by scanning electron microscopy (SEM) observation.

The T_c was evaluated by the temperature dependence of the resistivity and the a.c. susceptibility. The inter-grain critical current density (J_{cT}) was determined by the transport current measurements at the magnetic field up to 5T. It has been reported that the most of the width of the magnetic hysteresis curves for sintered $YBa_2Cu_3O_y$ samples at 77K is due to the intra-grain currents at higher field than about 0.05T [3]. Then the intra-grain critical current density (J_{cM}) was estimated by the magnetization measurements as follows :

$$J_{cM}[A/cm^2] = \frac{160\Delta M[emu/cm^3]}{3\pi r[cm]} \quad (1)$$

where ΔM was the width of the magnetic hysteresis curve and r was the mean radius of the grains which were assumed to be spherical.

RESULTS AND DISCUSSION

By the XRD analysis, the samples (a), (b) and (c) were single-phase of $YBa_2Cu_3O_y$ orthorombic. Samples (e) and (f) were tetragonal phase. The sample (d) was mixture phases of them.

With increasing Tq, the oxygen contents of quenched samples were monotonously decreased. For example, the oxygen content of samples for T_q=500, 600, 700, 800 and 900°C are smaller than that of unquenched sample by 0.28, 0.53, 0.76, 0.97 and 1.35wt%, respectively.

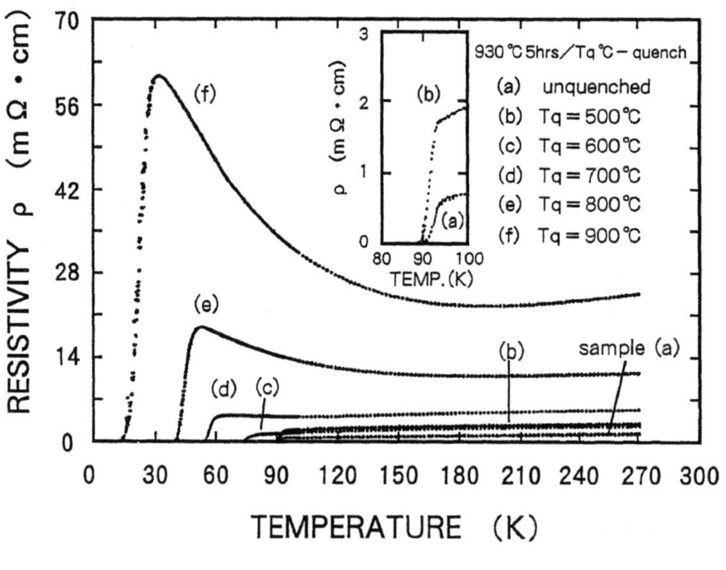

Fig.1 Temperature dependence of the resistivity for the samples quenched from various temperatures (T_q). The inset indicates the transition region for sample (a) and (b) between 80 and 100K.

Figure 1 shows the temperature dependence of the resistivity for all the samples. As shown in this figure the $T_c(\rho=0)$ is decreased and the resistivity is increased with increasing T_q. Especially, the temperature dependence of the resistivity for the samples with T_q above 800°C behaves like semiconductors so as to have tetragonal phase.

According to a.c. susceptibility measurements, the volume fraction of superconducting phase was equivalent for all samples.

The samples (a) and (b) showed T_c higher than the liquid nitrogen temperature (90 and 89K), so the J_c's were evaluated at 77K for these two samples.

Figure 2 shows the magnetic field dependence of the J_{cT}, where the field is applied perpendicular to the current and to the pressed surface of the samples. $J_{cT}(B=0)$'s for the samples (a) and (b) are 193 and 59A/cm^2, respectively. It is clear that the field dependence of J_{cT} for the sample (b) is much stronger than that for the sample (a).

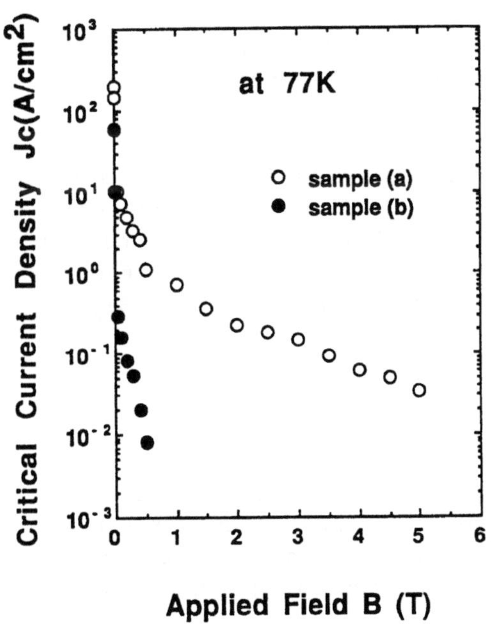

Fig.2 The magnetic field dependence of the J_{cT} for the sample (a) and (b).

In order to compare the intra-grain critical current densities between the samples (a) and (b), the magnetization curves were measured at 77K. By the SEM observations, any difference on the mean grain radius was not found among all the samples and it was about 4 micrometers. The J_{cM}'s for the samples (a) and (b) calculated by using the formula (1) are shown by closed circles in Figure 3 as a function of the applied magnetic field.

The J_{cM}-B characteristics for the samples (a) and (b) are almost the same in spite of the large discrepancy of the J_{cT}-B characteristics which are shown by open circles in Figure 3.

Furthermore, the J_{cM}-B characteristics were almost unchanged with grinding samples unless the particle size was smaller than the grain size.

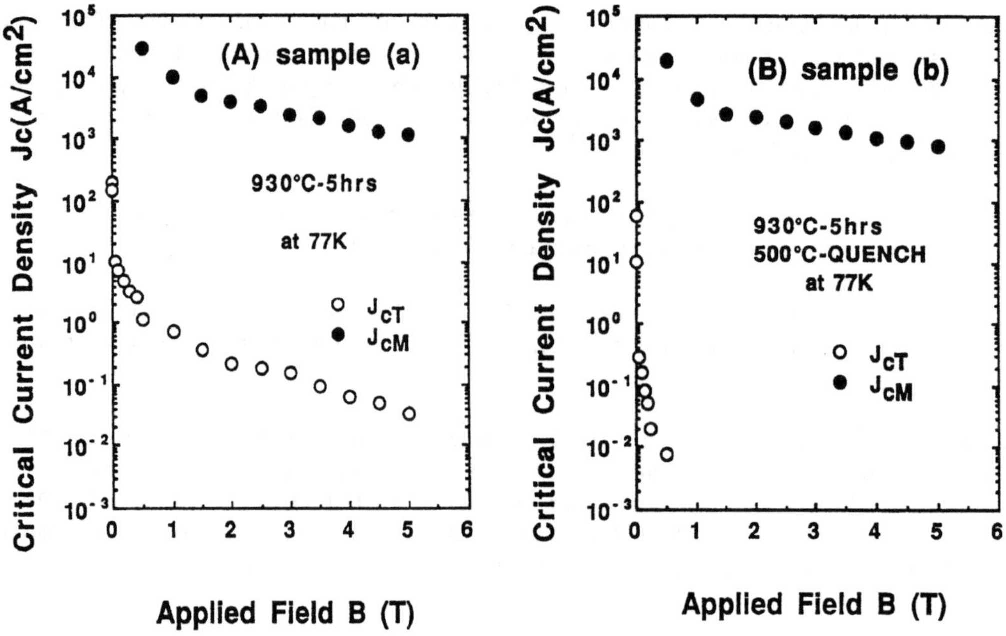

Fig.3 The magnetic field dependence of the J_{CM} for the sample (a) (A) and the sample (b) (B) shown by closed circles. The open circles show the J_{CT}'s.

The above results suggest that the oxygen deficiency occurs mainly at the surface of grain by quenching procedure and it enhances the influence of the weak links on the transport currents. Furthermore, it may be said that the oxygen deficiency occurs easily at grain boundaries even if the sample is not quenched and that is the main origin of the weak links in $YBa_2Cu_3O_y$ sintered materials.

CONCLUSION

In this study we have found that the oxygen deficiency at grain boundaries appeared to be the origin of the weak links in sintered $YBa_2Cu_3O_y$. In order to make the J_c's high enough for the applications, it would be necessary to develop processes to increase the oxygen content at grain boundaries.

REFERENCES

[1] M. Wakata et al: 5th Seminar on Frontier Technology (1988) Shuzenji
[2] K. Shimohata et al: New Developments in Applied Superconductivity (1988) P.238
[3] S. Yokoyama et al: Proc. ISS'88 (1988) P.411
[4] M. Wakata et al: ISTEC Workshop on Superconductivity - Program & Abstracts (1989) P.39

4 Preparation and Properties of Thin Films

4.1 YBCO and Related Films

Hetero Epitaxial Growth Mechanism of Thin Film for High-T_c Superconductors

TAKAHITO TERASHIMA[1], YOSHICHIKA BANDO[1], KENJI IIJIMA[2], KAZUNUKI YAMAMOTO[2], KAZUTO HIRATA[2], KATSUHIKO HAYASHI[2], KOUSEI KAMIGAKI[3], and HIKARU TERAUCHI[3]

[1] Institute for Chemical Research, Kyoto University, Uji, 611 Japan
[2] Research Institute for Production Development, Kyoto, 606 Japan
[3] Department of Physics, Kwansei-Gakuin University, Nishinomiya, 662 Japan

ABSTRACT

Initial stage of the epitaxial growth and growing manner of the $YBa_2Cu_3O_{7-x}$ thin films on the $SrTiO_3(100)$ and MgO(100) were investigated by means of in situ reflection high energy electron diffraction (RHEED) and X-ray diffraction. In situ RHEED observation showed that the formation of the perovskite structure occurred even at initial deposit and the growth manner was layer by layer. X-ray analysis of the 100 Å thick film on the $SrTiO_3(100)$ showed that the orthorhombic distortion disappeared because of the in-plane lattice deformation.

KEY WORDS: in-situ reflection high energy electron diffraction, epitaxial growth, $YBa_2Cu_3O_{7-x}$, thin film, layer-by-layer growth

INTRODUCTION

High quality thin films of high-Tc oxides have a important role in the study of the fundamental physics and device applications. The recent trend in the preparation of high-Tc films has been toward "in-situ" growth of the superconducting phase at relatively low temperatures. The purpose of "in-situ" growth is to attain not only a smooth film surface suitable for fabricating of film devices but also high quality film. We have succeeded in the in-situ growth of superconducting $YBa_2Cu_3O_{7-x}$ (YBCO) thin films by activated reactive evaporation.[1] The investigation of the initial stage of the epitaxial growth and the growth manner in the in-situ growth gives essential informations for the progress of the thin film growth. In-situ reflection high energy electron diffraction (RHEED) observation is one of the most effective methods of investigating growth manner.[2] Besides, the investigation of the crystal structure of the ultra-thin film (<100 Å) by X-ray diffraction would give informations about the effect of the substrate lattice on the growing film.[3]

In this paper we mention about the results for the growth of YBCO by activated reactive evaporation on (100) planes of $SrTiO_3$ and MgO. $SrTiO_3$ has a perovskite structure similar to that of YBCO and a lattice mismatch of less than 2% with tetragonal YBCO. MgO has the NaCl structure and a large mismatch of 9% with YBCO. It is very interesting to study the effects of the magnitude of mismatch on the initial growth manner and on the crystal structure of the resultant film. Generally, epitaxial film can be stabilized by accommodating the mismatch with the substrate by elastic strain up to the critical layer thickness "h_c". Over h_c, the in-plane lattice of epitaxial film changes rapidly to its bulk value by the formation of the mismatch dislocations. In this paper, we report the results of the in-situ RHEED observation during deposition and the crystal structure of the ultra-thin films investigated by X-ray diffraction.

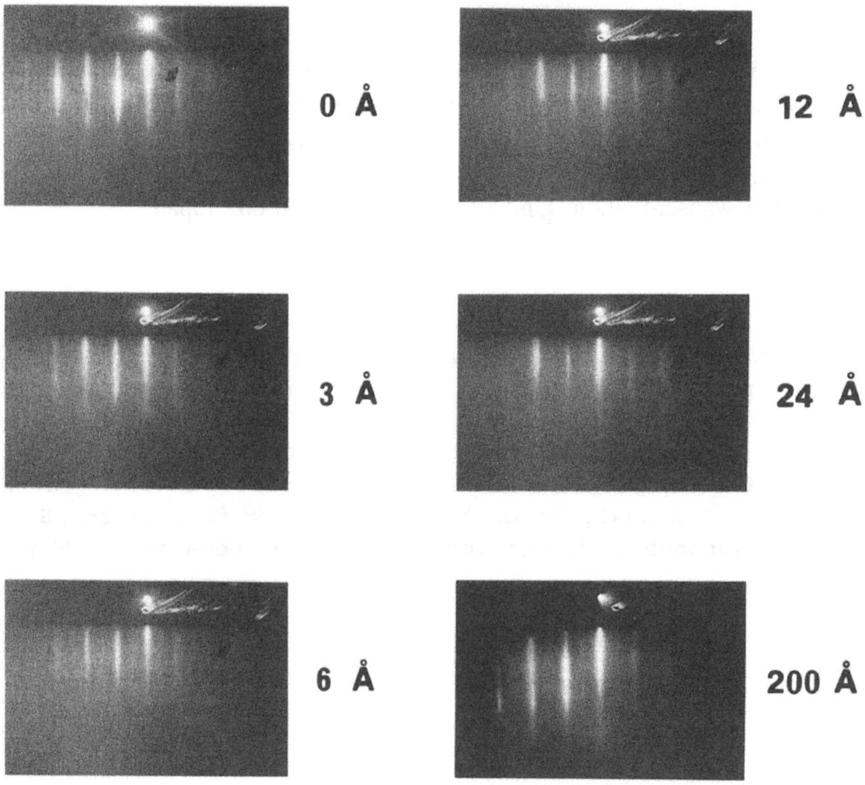

Fig.1 In-situ RHEED patterns observed during the growth of the YBCO film on $SrTiO_3$(100).

EXPERIMENTAL

The epitaxial growth of YBCO was performed by activated reactive evaporation. The detail of this method was described elsewhere [1,2]. We have sputtered the surface of the substrate by Ar^+ ion beam bombardment (acceleration voltage : 500 eV) at 650 ℃ for 1 min before the deposition. After this treatment, the RHEED pattern exhibited very sharp streaks and low background. Subsequently, YBCO was deposited on the substrate at the same temperature.
X-ray diffraction measurement were performed using a conventional double-axis diffractometry. A Cu Kα radiation (power of 55 kV x 250 mA) monochromatized by a pyrolytic graphite crystal was employed. The rectangular divergent- and receiving-double slits were employed, giving the sharp resolution in the tracing reciprocal plane and the wide one in the normal direction.

RESULTS AND DISCUSSION

Figure 1 shows in-situ RHEED patterns of the growing film on the $SrTiO_3$(100), where the electron beam is parallel to the [100] direction of $SrTiO_3$. The pattern at the initial growing stage of 3 Å is the sharp streaks and is the same as that of the substrate. The 3 Å layer corresponds to one or two atomic planes. The sharp streaks at the deposition of one or two atomic planes suggest that the initial deposition of YBCO occurs in the monolayer overgrowth without formation of three dimensional nuclei. The sharp streaks for every

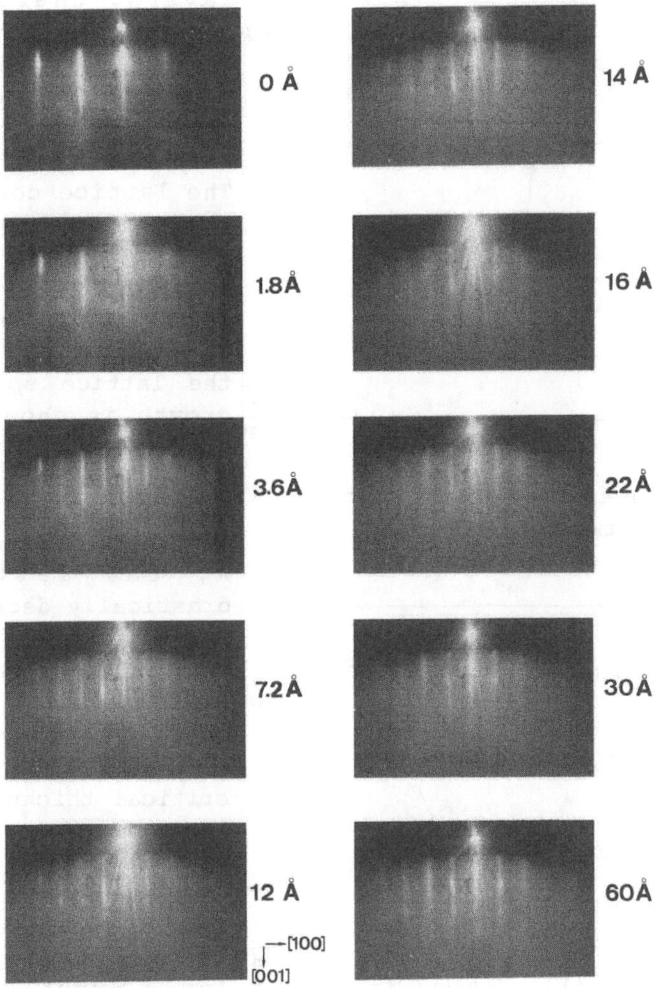

Fig.2 In-situ RHEED patterns observed during the growth of the YBCO film on MgO(100).

stage reveal that the growing surface is atomically smooth and the growth manner is layer by layer.

Extra streaks appeared in the middle of streaks during the growth and remained after the film was cooled to room temperature at the same oxygen pressure. The extra streaks were also observed for incidence of electron beam along the [110] direction of the substrate. The [h00] scanning of X-ray diffraction confirmed the same superstructure along the a-axis as observed in the RHEED.[3] The oxygen pressure as low as 10^{-1} Torr during the growth resulted in the formation of YBCO with a deficient oxygen stoichiometry. After the film was oxidized in the chamber below the growth temperature under the oxygen pressure of 200 Torr, the extra streaks disappeared. We believe that the superstructure originated from an oxygen vacancy ordering in the oxygen deficient YBCO crystal.

In-situ RHEED patterns have also been observed during the growth of the film on MgO(100), as shown in Fig.2. The as-polished MgO substrate was chemically etched by nitric acid aqueous solution followed by Ar^+ ion beam sputtering.

The diffraction pattern of the MgO substrate is distinctly different from that of the perovskite structure. On the pattern of the deposit of two atomic

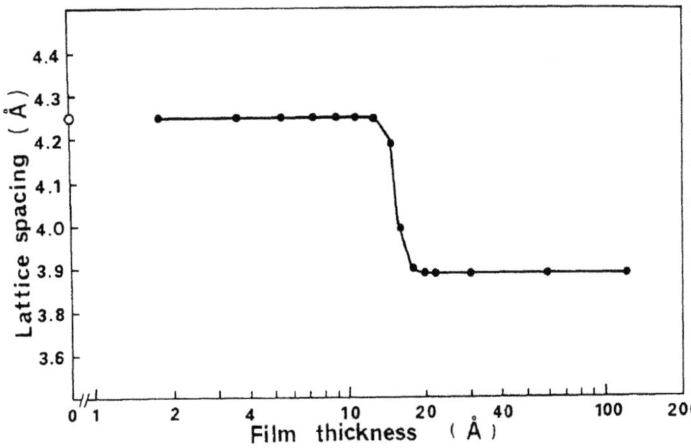

Fig.3 Lattice spacing vs thickness of the YBCO film on the MgO(100) calculated from the distance of the streaks.

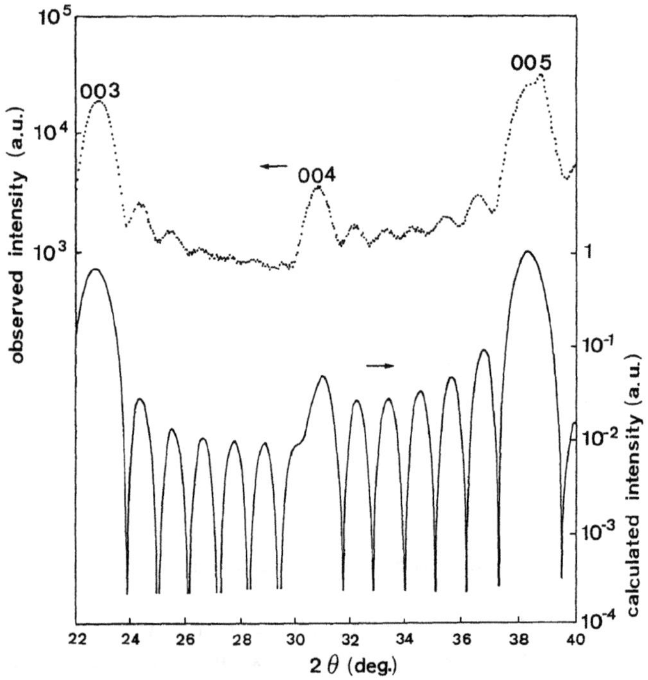

Fig.4 X-ray diffraction profile for the 100 Å thick YBCO film on MgO(100) (dotted line) and calculated one (solid line).

layers, there appeared new streaks revealing the perovskite structure in addition to that of MgO. The streak intensity increased with increase of the thickness.

The lattice constant a_0 of MgO at 650 ℃ can be estimated from the thermal expansion coefficient, to be 4.25 Å, which is the reference for the lattice spacing of YBCO. The change of the lattice spacing during the growth is shown in Fig.3. The film with thickness from 3 Å to 12 Å had the same in-plane lattice spacing as MgO. When the thickness became larger than 12 Å, the lattice spacing drastically decreased and became 3.9 Å at a thickness of 24 Å. This value is nearly the same as 3.88 Å for the calculated bulk value of tetragonal YBCO [4] in the growth condition. The critical thickness h_c of YBCO is about 12Å for the MgO substrate with the large lattice mismatch of 9%. Above h_c, the YBCO film should form misfit dislocations at the interface. The layer-by-layer growth on the MgO was demonstrated by the streak pattern.

Figure 4 shows the X-ray diffraction spectrum scanned along the [001]YBCO direction for a 100 Å thick YBCO film on MgO(100)(dotted line). The thickness of 100 Å means that the film consists of only eight or nine unit cells of YBCO structure (c_0=11.7Å). The oscillation of the Laue function by the finite size effect is clearly shown in the spectrum. This spectrum can be analyzed considering that the film consists of 7 unit cells, using the reported data of the atomic positions by Jorgensen et al. [4] (solid line). The RHEED observation suggested that misfit dislocations should be introduced when the first unit cell was formed for the relaxation of the large misfit (9%), so there may be incoherent layers corresponding to one or two unit cells at the interface. The discrepancy between the deposited thickness and calculated coherent one may be caused by these incoherent layers.

Figure 5 represents the X-ray scattering profiles around the (407) reciprocal point scanned along the [100] direction for the (a) 600 Å thick film and (b)

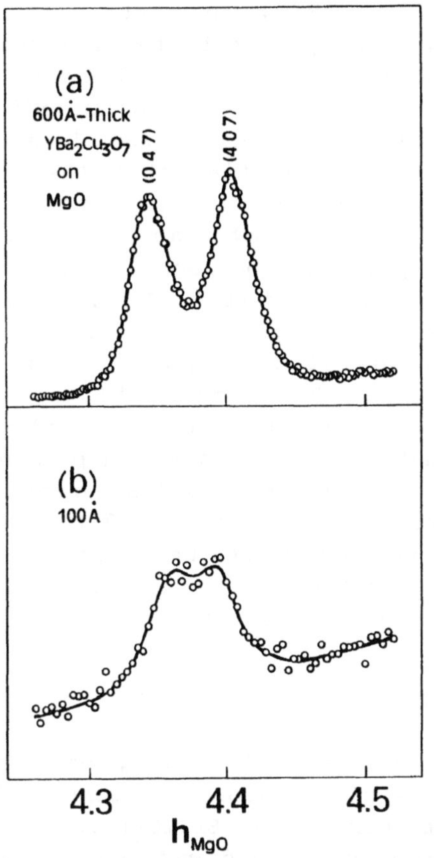

Fig.5 X-ray scattering spectra for (a) 600 Å and (b) 100 Å thick YBCO on MgO(100) scanned along the [100] direction around the (407) reflection.

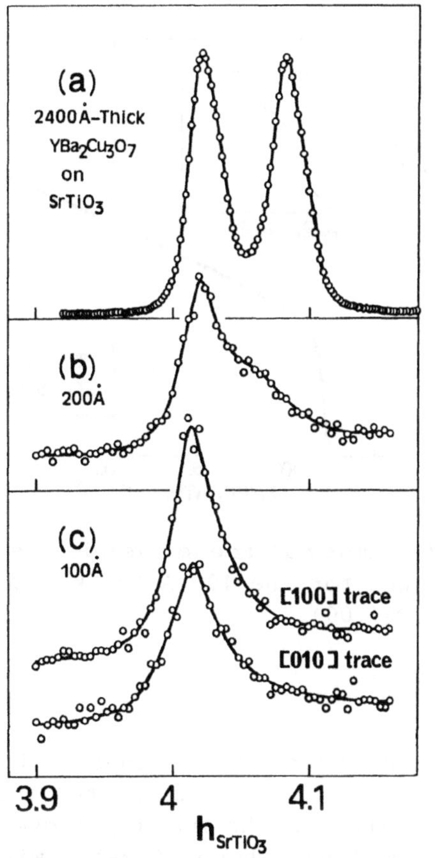

Fig.6 X-ray scattering spectra for (a) 2400 Å and (b) 200 Å thick YBCO on $SrTiO_3$(100) scanned along the [100] direction around the (407) reflection, and (c) 100 Å thick YBCO scanned along the [100] and [010] directions around the (407) and (047) reflections.

100 Å thick film on the MgO(100). The resolutions are estimated by using the MgO and Miller index h is defined by using the MgO lattice. Distinct doublet in Fig.5(a) comes from the twining in the orthorhombic crystal, where left and right peaks correspond to the (047) and (407) reflections. The lattice constants a and b were derived to be 3.827 Å and 3.879 Å from the two peak positions. The values are in good agreement with those of bulk values. The peak splitting due to $a \neq b$ is also seen in the 100 Å crystal in Fig.5(b). From the in situ RHEED observation the critical thickness h_c of the YBCO on MgO was estimated to be 12 Å. As the thickness is enough larger than h_c, the lattice of the 100 Å thick YBCO is assumed to be free from the substrate lattice except the distorted region near the interface.

Figure 6 represents the X-ray scattering profiles around the (407) reciprocal point scanned along the [100] direction for the (a) 2400 Å thick (b) 200 Å thick and (c) 100 Å thick films on the $SrTiO_3$(100). In this case the Miller index h is defined by using the $SrTiO_3$ lattice. In the case of YBCO crystal on $SrTiO_3$, it is expected that the in-plane lattice accommodation would occur until rather large thickness as compared with the case of the crystal on MgO because of small lattice mismatch. It is found that the spectra in Fig.6(c) shows obviously different feature from the Fig.5(b). In

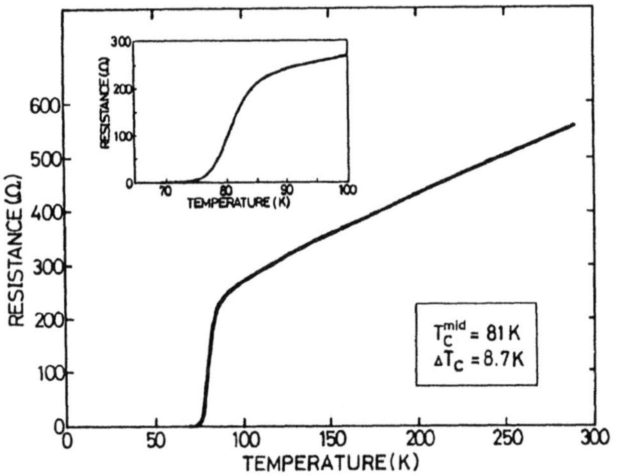

Fig.7 Temperature dependence of a resistance for the 100 Å thick YBCO film on $SrTiO_3(100)$.

100 Å YBCO, in particular, twin splitting is not observed and peak positions in the [100] and [010] traces are the same. In addition, peak widths are almost independent of h, indicating that there exist no orthorhombic distortion. Formation of the orthorhombic crystal can be seen in the spectrum for 200 Å thick YBCO. Figure 7 shows a typical resistivity versus temperature result for a 100 Å thick YBCO thin film on the $SrTiO_3(100)$. The film showed a metallic temperature dependence of the resistivity and superconducting transition at about 80 K. The fact implies that the orthorhombic distortion is not essential for superconductivity in YBCO.

CONCLUSION

We have demonstrated the in-situ RHEED observation during the growth of YBCO film. The RHEED patterns for the initial stage of the growth suggest that YBCO was deposited in the monolayer overgrowth manner on the $SrTiO_3(100)$ and even on the MgO(100), whose mismatch with the YBCO is 9%, and was never deposited as the three-dimensional nuclei. In the deposition on the MgO, the initial deposit of one or two atomic layers in thickness already had the perovskite structure. Its in-plane lattice spacing of the film converted drastically from the bulk value of MgO to that of YBCO when the film thickness exceeded 12 Å. X-ray diffraction study revealed that the structure of the ultra-thin YBCO films strongly depended on the lattice mismatch between the YBCO and substrate crystals. No orthorhombic distortion was observed in 100 Å thick YBCO crystal on $SrTiO_3$, implying that the orthorhombic distortion itself is not essential for superconductivity in YBCO.

REFERENCES

1. Terashima T, Iijima K, Yamamoto K, Bando Y, Mazaki H (1988) Single-Crystal $YBa_2Cu_3O_{7-x}$ Thin Films by Activated Reactive Evaporation. Jpn.J.Appl.Phys. 27: L91-L93

2. Terashima T, Iijima K, Yamamoto K, Hirata K, Bando Y, Takada T (1989) In Situ Reflection High Energy Electron Diffraction Observation During Growth of $YBa_2Cu_3O_{7-x}$ Thin Films by Activated Reactive Evaporation. Jpn.J.Appl.Phys. 28: L987-L990

3. Kamigaki K, Terauchi H, Terashima T, Bando Y, Iijima K, Yamamoto K, Hirata K (1989) X-RAY STUDY ON A 100 Å-THICK $YBa_2Cu_3O_{7-\delta}$ EPITAXIAL FILM : THE RELATIONSHIP BETWEEN THE ORTHORHOMBIC SYMMETRY AND SUPERCONDUCTIVITY. Physica C 159: 505-512

4. Jorgensen JD, Beno MA, Hinks DG, Soderholm L, Volin KJ, Hitterman RL, Grace JD, Schuller IK, Segre CU, Zhang K, Kleefish MS (1987) Oxygen Ordering and the Orthorhombic-Tetragonal Phase Transition in $YBa_2Cu_3O_{7-x}$. Phys.Rev.B 36: 3608

Thin Film of High-T_c Superconductor

HIDEO ITOZAKI

Itami Research Laboratories, Sumitomo Electric Industries, Ltd., 1-1, Koyakita 1-chome, Itami, Hyogo, 664 Japan

ABSTRACT

High Tc and high Jc superconducting thin films can be made by various deposition methods, such as a sputtering, a co-evaporation and a laser ablation. TEM observation shows that these films have many micro twins. Their size became larger in annealing process. These twin boundaries located parallel to the c axis. These defects might pin the magnetic flux. Pinning potential was estimated from temperature dependence of resistivity near transition temperature. The pinning potential was about 0.02-0.4eV. It depends on the magnetic field and has an anisotropy.

KEY WORDS: Thin film, High critical current, Twin, Pinning potential

INTRODUCTION

High Tc superconductor is expected to be used for application of electronics, such as high speed devices or transmission lines for microwaves or high speed signals. In order to use the high Tc superconductor in electronics fields, a thin film should be developed. Here properties and processes of high Tc superconducting thin films were investigated.

HIGH Jc SUPERCONDUCTING THIN FILM

High Tc superconducting thin film have been developed since they were discovered. Their properties have been improved drastically [1-3]. Especially, their critical current density becomes more than three million A/cm^2 at 77.3K for YBCO[1-3], BSCCO [4] and TBCCO [5] superconducting thin films. Their main properties are listed in Table 1.

Table 1 High Tc superconducting thin film

	YBaCuO	BiSrCaCuO	TlBaCaCuO
Tc	90K	105K	115K
Jc(77.3K)	4MA/cm^2	3.4MA/cm^2	3.2MA/cm^2

These critical current are much higher than those of sintered materials, even when some improved process such as a partial melting method was developed [6]. Here we tried to explain how the thin film can get such high critical current density compared with the sintered bulky materials.

The critical current density at certain magnetic field is determined by grain boundaries, pinning potential energy and density of pinning center. It has been observed that the high Jc thin film grew epitaxially on the substrate [7] and it is rather difficult to find grain boundaries in the high Jc thin film [3]. Grain boundaries of sintered materials reduce their critical current density largely. Therefore, the main reason of the high Jc of the thin film is that there exist no grain boundaries which are barriers or weak links for supercurrent. The high Jc thin film still has high Jc in high magnetic field such as more than ten Tesla. In such high magnetic field, magnetic flux passes through the superconductor. If this magnetic flux moves by the Lorentz force caused by the current and the magnetic field, flux movement induces voltage. As the thin film sustains high Jc in high magnetic field, the magnetic flux should be pinned strongly. We investigate the pinning potential energy and density of pinning center for the high Jc YBCO superconducting thin film. Samples were made by an RF magnetron sputtering on a single crystal of MgO. Preparation conditions are described in the previous papers in detail [3].

Pinning Potential Energy of YBCO Thin Film

The pinning potential of the YBCO thin film was evaluated from temperature dependence of resistivity near the critical temperature in various magnetic field. Fig.1 shows that resistivity vs. temperature near critical temperature in magnetic field of 0 to 8 tesla. Two cases are investigated. One is that the magnetic field is parallel to the c-axis, and the other is that the magnetic flux is perpendicular to the c-axis. The geometry of the magnetic field and the thin film was shown in Fig.1. We use the following equations to evaluate the pinning potential energy [8,9]

$$R=R_0\exp(-U/kT) \qquad (1)$$

$$U=\beta(1-T/Tc)^{1.5}/H \qquad (2)$$

R is resistivity, U is pinning potential energy, k is Boltzmann constant, T is temperature, Tc is critical temperature in zero magnetic field, H is applied magnetic field and β is constant.

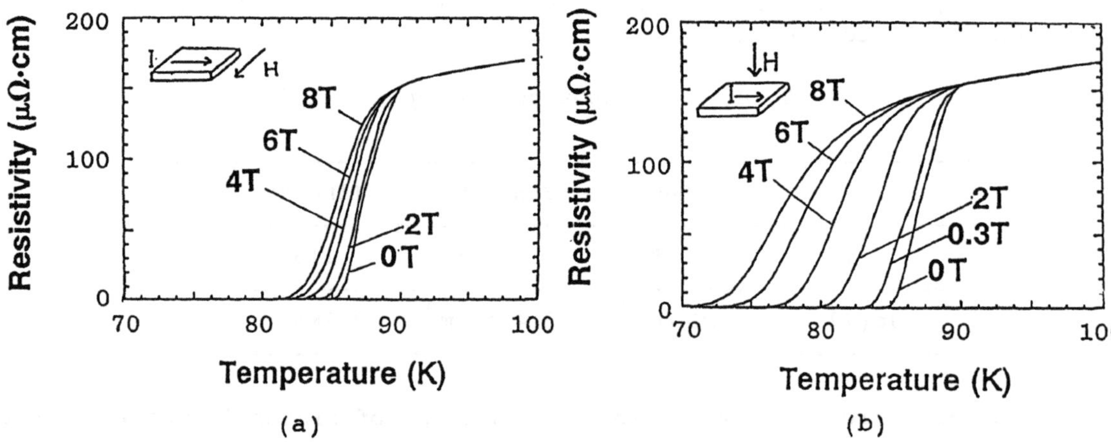

Fig.1 Temperature dependence of resistivity in the various magnetic field. (a) is for magnetic field perpendicular to the c-axis and (b) is for magnetic field parallel to the c-axis.

Fig. 2 shows the pinning potential as a function of inverse of the magnetic field. Data shows that U is proportional to the 1/H. This indicate that we can use the equation (2) to analyze the data of resistivity transition. The pinning potential at 77.3K is from 0.02 to 0.4eV as a function of 1/H, and it depends on the direction of the magnetic field. These potential energy are similar to those of a single crystalline bulk material [10]. Therefore, the pinning potential is not a key factor of high Jc in the magnetic field.

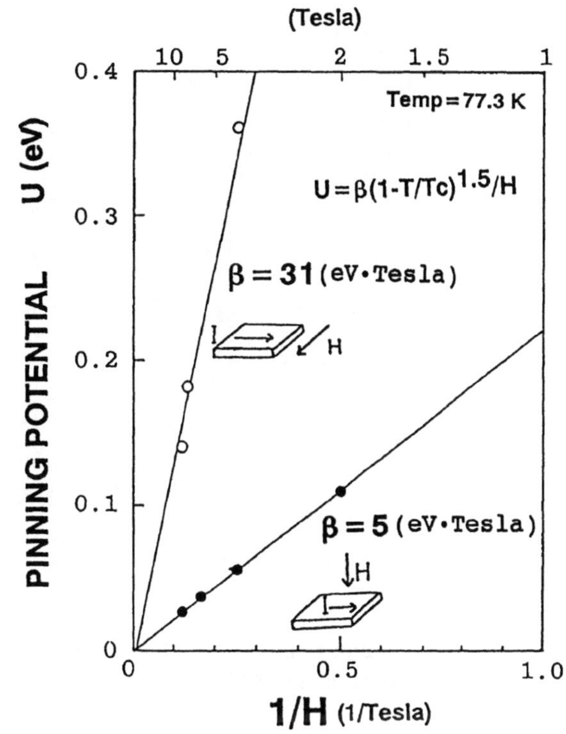

Fig.2 Pinning potential as a function of the magnetic field for YBCO thin film.

Microstructure of YBCO Thin Film

Structure of the YBCO superconducting film has been investigated, by X-ray, RHEED and high resolution SEM [3]. These data indicates that the high Jc film grew epitaxially and was almost a single crystal. This means that the high Jc film has no grain boundaries which reduce the critical current. We have investigated its microstructure by a transmission electron microscopy. Fig.3 shows the micrograph of the as-deposited YBCO thin film. Although electron diffraction is indicating that the film is an epitaxially grown single crystal, there are fine structures in the film. These plate like fine structures were aligned in the direction of [110]. High resolution transmission electron micrograph of this line structures shows that atomic planes continue at these defect lines. These facts indicates that these defect lines are twin boundaries. TEM images has a lot of tiny dark areas where the crystal planes are tilted very slightly by residual stress. Fig.4 show the TEM micrograph of post annealed sample. This photograph shows that the post anneal process makes twin domains larger and rearranged its structure more regularly, and make twin boundaries sharper. We also performed cross sectional observation by TEM. It is found that twin boundaries are almost perpendicular to ab planes. It was already reported that this kind of planer defect might work as a pinning center[11]. They prevent the magnetic flux from moving by the Lorentz force. Although twins are observed in sintered materials or wires, their twin domain size was much larger than that observed in the thin film here. Hence, high Jc of thin film in the magnetic field is explained not by strong pinning potential but by high density of pinning center.

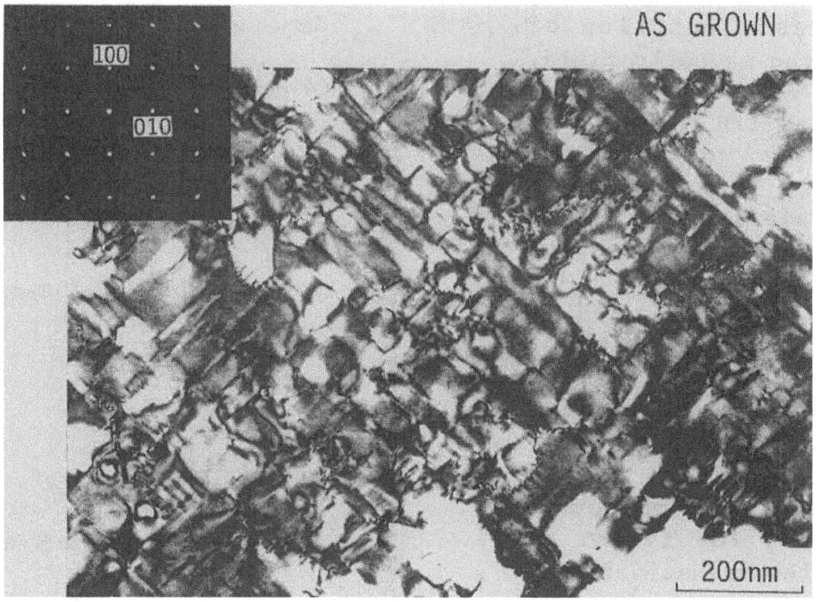

Fig.3 Transmission electron micrograph (TEM) for as-deposited YBCO thin film.

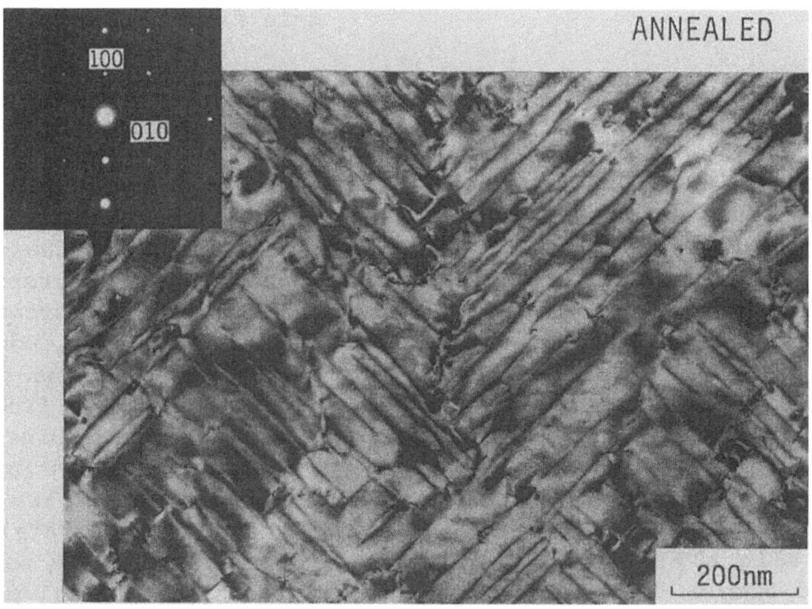

Fig.4 Transmission electron micrograph (TEM) for annealed YBCO thin film.

AS-GROWN SUPERCONDUCTING YBCO THIN FILM

In order to adopt the superconducting thin film to electronics, it is important to reduce process temperature. An as-grown thin film by sputtering deposition has usually low Tc and it should be annealed at more than 900C in oxygen atmosphere[3]. However, recently deposition techniques are improved and they can make an as-grown thin film which has high Tc and high Jc like more than million A/cm^2 at 77.3K without post-annealing [12]. Deposition rate is also important for the production of a superconducting thin film. The deposition rate of sputtering was very slow, but the deposition rate was improved by some deposition methods and they can also make an as-grown high Jc film. These improvement was shown in Fig.5, as a function of Jc and the deposition rate. It shows only reports which has more than a million A/cm^2. A co-evaporation method [13] and a laser ablation method [14,15] are superior in these viewpoints, now. The author also obtained deposition rate of 6 A/sec by the co-evaporation method and got an as-grown film with Jc of 1.75 million A/cm^2 at 77.3K. This film is almost epitaxially grown on the $SrTiO_3$ (100) substrate. The deposition was done with an oxygen RF plasma near the substrate. This kind of active oxygen is effective to obtained the as-grown superconducting thin film, because the deposition method is more or less low pressure process. This sort of the as-grown high speed deposited film are attractive to electronics applications.

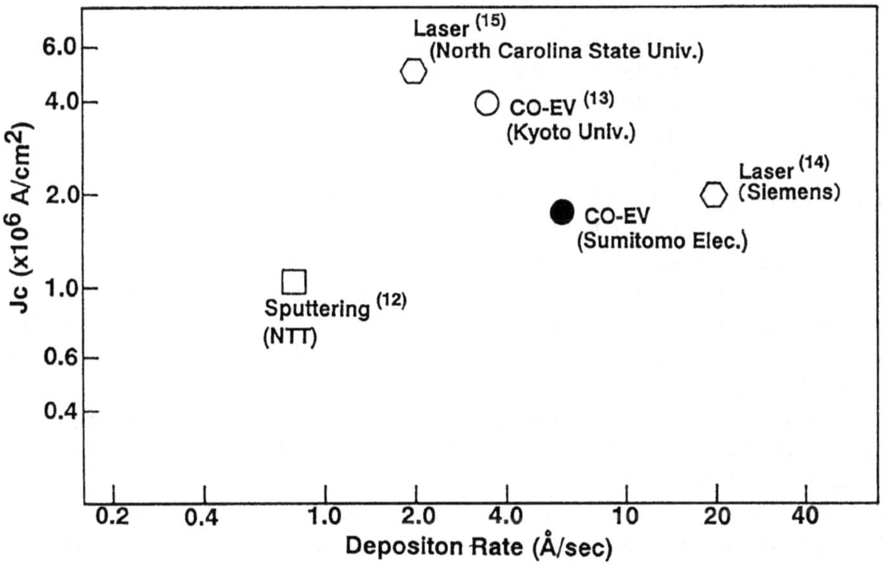

Fig.5 High speed deposition of high Jc as grown thin film.
CO-EV: Co-evaporation

SUMMARY

High Tc superconducting thin films were developed to have high critical current density of more than 3 million A/cm^2 at 77.3K for YBCO, BSCCO or TBCCO. In order to understand the mechanism of such high critical current, pinning potential energy and micro-structure were investigated. While their pinning potential energy was normal, we observed very high density of micro-twins in the YBCO thin film. These high dense twins might work as pinning center of magnetic flux. We should make more direct observation of magnetic flux pinning to realize the high Jc. Fabrication techniques of the superconducting thin film have been developed. The superconducting thin film can be made without post-annealing with high temperature. Deposition rate to obtain high Jc as-grown thin film becomes high like more than 5 A/sec. These data indicates that the high Tc superconducting thin film is hopeful and is applied to the electronics in the near future.

REFERENCES

1. Y.Enomoto,T.Murakami,M.Suzuki and K.Moriwaki,Jpn.J.Appl.Phys.26,L1248 (1987).
2. T.Terashima,K.Iijima,K.Yamamoto,Y.Bando and H.Mazaki,Jpn.J.Appl.Phys.27, L91(1988).
3. S.Tanaka and H.Itozaki,Jpn.J.Appl.Phys.27,L622(1988).
4. H.Itozaki,K.Higaki,K.Harada,S.Tanaka,N.Fujimori and S.Yazu,ed.by K.Kitazawa and T.Ishiguro,Advances in Superconductivity,Nagoya,1988 (Springer-Verlog,Tokyo,1989)p.599.
5. S.Yazu,ed.by K.Kitazawa and T.Ishiguro,Advances in Superconductivity, Nagoya,1988(Springer-Verlog,Tokyo,1988)p.563.
6. S.Jin,T.H.Tiefel,R.C.Sherwood,R.B.van Dover,M.E.Davis,G.W.Kammlott and R.A.Fastnacht,Phys.Rev.B37,7850(1988).
7. S.Tanaka and H.Itozaki,Jpn.J.Appl.Phys.28,L441(1989).
8. Y.B.Kim,C.F.Hempstead and A.R.Strnad,Phys.Rev.131,2486(1963).
9. M.Tinkam,Phys.Rev.Lett.61,1658(1988).
10. A.P.Malozemoff,T.K.Worthington,E.Zeldov,N.C.Yeh,M.W.McElfresh and Holtzberg in Strong Correlation and Superconductivity, ed. H.Fukuyama, S.Maeda and A.P.Malozemoff(Springer,Heidelberg 1989), to be published.
11. L.Ya.Vinnikov,L.A.Gurevich,G.A.Emelchenko and Yu.A.Ossipyan,Solid State Commun.67,421(1988).
12. H.Asano,M.Asahi and O.Michikami,Jpn.J.Appl.Phy.28,L981(1989).
13. T.Terashima,K.Iijima,K.Yamamoto,J.Takada,K.Hirata,H.Mazaki and Y.Bando, J.Crystal Growth,95,617(1989).
14. B.Roas,L.Schultz and G.Endres,Appl.Phys.Lett.53,1557(1988).
15. R.K.Singh,J.Narayan,A.K.Singh and J.Krishnaswamy,Appl.Phys.Lett.54,2271 (1989).

Hetero-Epitaxial Growth of YBaCuO Thin Films

KEN SAKUTA[1], MASAHIRO IYORI[1], UKI KABASAWA[1], MINORU NAKAJIMA[2], and TAKESHI KOBAYASHI[1]

[1] Faculty of Engineering Science, Osaka University, 1-1, Machikaneyama-cho, Toyonaka, Osaka, 560 Japan
[2] Yokohama R & D Laboratories, The Furukawa Electric Co., Ltd., 4-3 Okano 2-chome, Nishiku, Yokohama, 220 Japan

ABSTRACT

We have fabricated the YBaCuO/MgO/YBaCuO double-heterostructure using conventional rf-magnetron sputtering. The formation of both (110)YBaCuO/(100)MgO/(110)YBaCuO on (110)$SrTiO_3$ substrate, and (001)YBaCuO/(100)MgO/(100)YBaCuO and (001)YBaCuO/(100)MgO/(001)YBaCuO on (100)MgO substrate, became feasible. According to the RHEED observations, each layer of the double-heterostructure was grown epitaxially. The cross sectional TEM observation was carried out for characterizations of the heterostructure, bringing a number of informations concerning the interfacial problems. Final discussion is on the electric field effect of Al/MgO/YBaCuO, as a new application of the heterostructure.

KEYWORDS: double-heterostructure, epitaxial growth, rf-magnetron sputtering, MIS structure, $YBa_2Cu_3O_y$

INTRODUCTION

Since a discovery of high-T_c oxide superconducting ceramic by Bednorz and Müller[1], there has been a rush of new high-T_c superconducting materials consisting of a modified perovskite structure[2,3]. Very recently, Wu et al. reported their success in getting the T_c of 160K[4], much higher than the liquid nitrogen boiling temperature.

The birth of high-T_c superconductive materials is necessarily taking us upward from the material processing point of view, and therefore, no one engaged in the electronics field can avoid the epitaxial growth of thin films. As well known, the single crystal technology greatly helps the physical research and practical use of the semiconductors. The epitaxial growth of high-T_c superconductor is very important in view not only of the application but also of the scientific usage, like semiconductor engineering. The later means that the high-T_c superconductivity mechanism of the oxide materials can not be explored without ideal crystal. The oxide superconductor epitaxy should be followed by the hetero-epitaxial growth technique, without which really nice electronic devices like a superconducting transistor[5,6] can never be brought out.

In the present work, we demonstrate the fabrication of the YBaCuO/MgO/YBaCuO double-heterostructure. Though the new technique itself seems sophisticated, it can provide the layered structures with combination of various crystal orientations. The cross sectional TEM observation is done for the film with an ultra-thin MgO intermediate layer. The final discussion is given to the electric field effect on the Al/MgO/YBaCuO MIS diode, which did not work well until the hetero-epitaxy was in hands.

YBaCuO/MgO/YBaCuO EPITAXIAL DOUBLE-HETEROSTRUCTURE

Since our aim is placed on a fabrication of the epitaxial layered structure, the selection of intermediate layer material is important. For instance, the intermediate layer must be grown epitaxially at a temperature as low as possible, it has a lattice constant matched to the lattice constant of YBaCuO, it must not take oxygen away from the superconductor and it has less alloying or inter-diffusion with the adjacent layer. In the superconductor device, the carriers or superconducting particles near the interface between the superconductor and intermediate layer do most work than those inside the films. The coherent length of oxide superconductor is so short that the disorder of a few lattice layer near the interface causes the degradation of electrical properties. A candidate of the intermediate layer, meeting these conditions, is MgO. $SrTiO_3$ is not suitable for intermediate layer material. The reason of this is the diffusion of a number of Sr atoms to superconductor. An MgO is not so hard to grow by the simple sputtering deposition unless the extremely high quality is required for that.

(110)YBaCuO/MgO/(110)YBaCuO Structure

The thin film growth of both $YBa_2Cu_3O_y$ and MgO layers was done with the rf-magnetron sputtering. A target of $Y_1Ba_2Cu_{4.5}O_y$ composition was made by sintering the mixture of Y_2O_3, $BaCO_3$ and CuO. The source material was enough ground down and pressed in a form of the target plate. The MgO growths used the sintered MgO target. The details of the deposition condition are given in Table I.

Under the condition of Table I, the YBaCuO thin film was deposited on (110)$SrTiO_3$ substrate as a bottom layer. Next, MgO 20 nm thick was grown as an intermediate layer. At the final stage, a top layer YBaCuO was successively grown. In the epitaxial growth, crystallinity of each layer must be good, because the layer serves as the substrate of the next layer.

The schematic double-heterostructure we fabricated, is shown in Fig.1. Figure 2 shows the RHEED patterns of each layer. Figure 2(a) is an RHEED pattern of the bottom YBaCuO layer. This pattern shows single- and tri-layered perovskite structures, so this YBaCuO film grew epitaxially and (110) oriented. But this pattern is somewhat spotty, showing that the film

Table I Epitaxial growth conditions

YBaCuO growth	
Discharge gas	$Ar+O_2$ (50%)
Gas pressure	4Pa
Substrate temperature	500-700°C
Target composition	$Y_1Ba_2Cu_{4.5}O_y$ (powder)
Discharge power	50W
Growth rate	4 nm/min
Substrate material	(100)MgO, (110) and (100)$SrTiO_3$
MgO growth	
Discharge gas	Ar
Gas pressure	1Pa
Substrate temperature	300-600°C
Target composition	MgO (sinter)
Substrate material	YBaCuO (as made)

face is not so smooth. Figure 2(b) is the intermediate MgO layer RHEED pattern. This MgO growth was done at 300°C. It looks like the (100) oriented epitaxial MgO film. It is not clear why the (100)MgO grows on the (110)YBaCuO, yet. The crystalline quality of the top YBaCuO layer can be seen in the RHEED pattern (c), where one can find a series of the spots corresponding to the tri-layered perovskite structure. It indicates that at least a long distance ordering of the crystal is possibly obtained. However, the observed ring-like pattern is a sign of the crystalline imperfection.

The crystallinity of the intermediate MgO layer depended on the substrate temperature at the depositing. When the intermediate MgO layer is deposited at the temperatures higher than 550°C, MgO falls in several orientations. This cases complete missing of the long distance ordering of the top YBaCuO crystallinity (Fig.3(c)). We can only find the diffraction patterns from the perovskite structure.

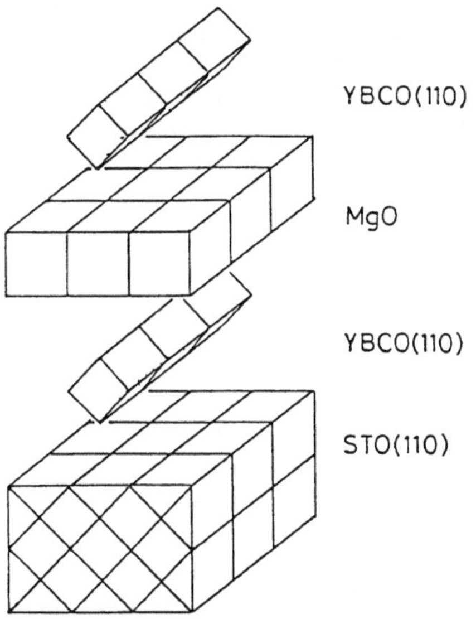

Fig.1 A schematic illustration of the double-heterostructure. (110)YBaCuO/MgO/(110)YBaCuO/(110)SrTiO substrate.

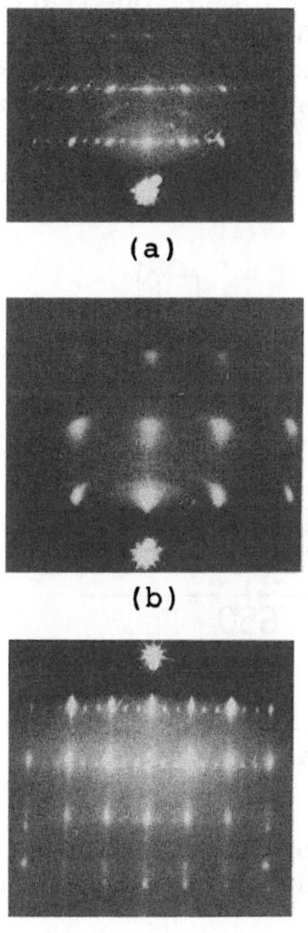

Fig.2 An intermediate MgO layer was grown at 300°C. RHEED patterns of a top YBaCuO (a), intermediate MgO (b) and bottom YBaCuO (c).

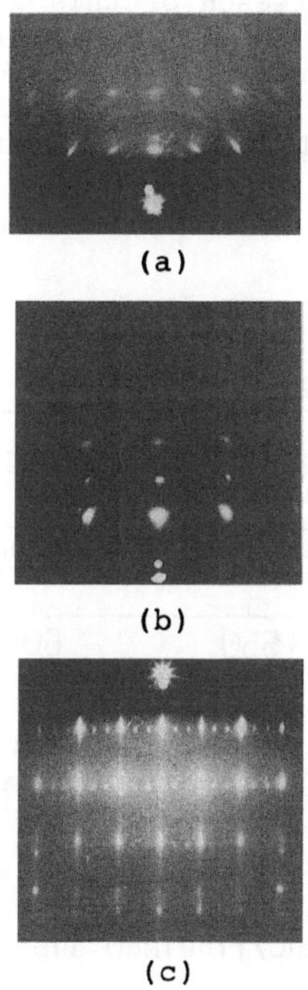

Fig.3 An intermediate MgO layer was grown at 550°C. RHEED patterns of a top YBaCuO (a), intermediate MgO (b) and bottom YBaCuO (c).

(001)YBaCuO/MgO/(100)YBaCuO Structure

As the next step, we tried the fabrication of the asymmetric (001)YBaCuO/MgO/(100)YBaCuO structure, shown in Fig.4. It may be the most interesting structure using the tri-layered perovskite oxide superconductor. This structure is characterized by the vertical facing of the conduction planes, which might allow unexpected interaction of the tunneling carriers.

Prior to the heterostructure formation, it is worth noting the dependence of the growing YBaCuO orientation on the MgO substrate temperature. The results are summarized in Fig.5. In the low temperature region, YBaCuO grows with (110) orientation. At the temperatures between $570^{\circ}C$ to $650^{\circ}C$, the YBaCuO orients (100) or $(\bar{1}00)$. (010)YBaCuO may be possible to grow here. Since the same probability is shared by four growing axes, the film results in the polycrystal. With further increase in the growth temperature, higher than $650^{\circ}C$, the YBaCuO grows (001) oriented. In this case, the film grows as a quasi-single crystal except the twin- and/or subgrain-growth.

On the basis of knowledges mentioned above, the double-heterostructure shown in Fig.5 can be obtained. The RHEED patterns of each layer of this structure are shown in Fig.6. The RHEED pattern (Fig.6(a)) of the bottom YBaCuO layer grown at $600^{\circ}C$ shows no indication of existence of tri-perovskite structure. The reason of this is that, as mentioned above, a polycrystal layer grew at this temperature. Figure 6(b) reveals the single crystal growth of the intermediate MgO layer, though the underlying YBaCuO film has a mixture of (100) and $(\bar{1}00)$ domains. The top YBaCuO layer offers a streak pattern in its RHEED pattern, showing the (001) oriented growth. From these observation, each layer of YBaCuO and MgO can grow epitaxially.

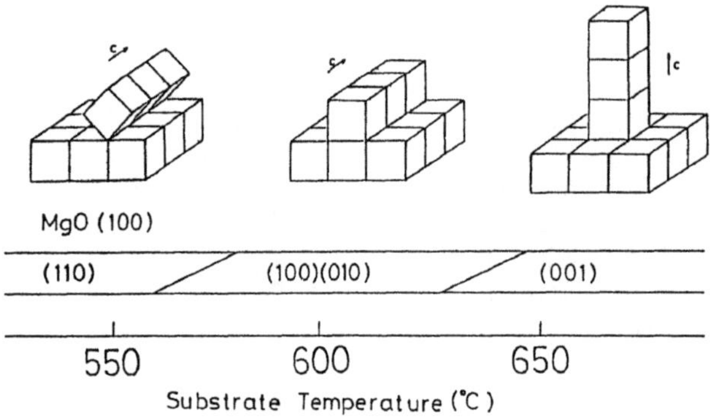

Fig.4 Dependence of the orientation of YBaCuO films on the substrate temperature.

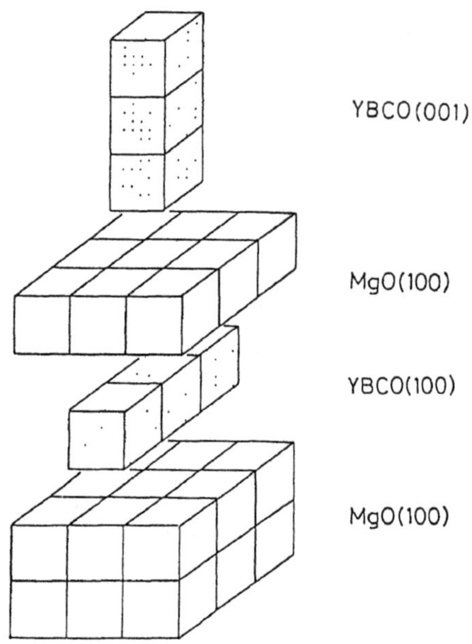

Fig.5 A schematic illustration of the double-heterostructure (001)YBaCuO/(100)MgO/(100) YBaCuO/(100)MgO substrate.

DOUBLE-HETEROSTRUCTURE WITH VERY THIN MgO FILMS

The thickness of the intermediate MgO layer in the above experiments was 20 nm. This is rather thick. For use of this structure to S-I-S Josephson diode, the intermediate layer must be thin as possible. So we fabricated the double-heterostructure with MgO layer as thin as 2 nm, and with the top and bottom YBaCuO layers oriented (001). Though data are omitted here, the RHEED patterns successively taken for each layer indicate no sign of crystal degradation. However, the diffraction pattern from the ultra-thin MgO layer was somewhat weak as compared with those of the above. What happens with it? The cross sectional TEM observation answered this question.

Figure 7 is a TEM lattice image taken for the cross section of the double-heterostructure wafer, where we can see pretty nice epitaxial growth of each

(a)

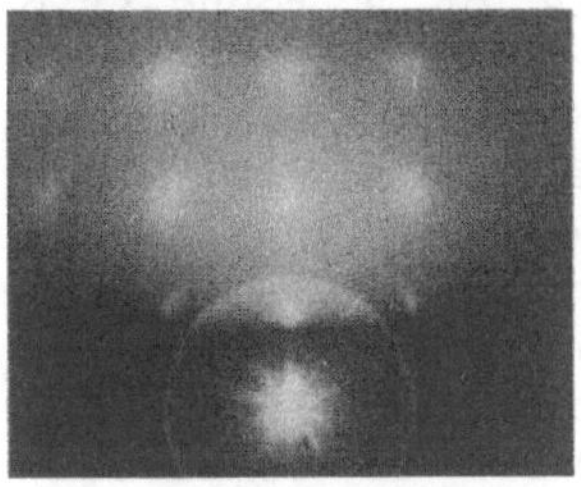

(b)

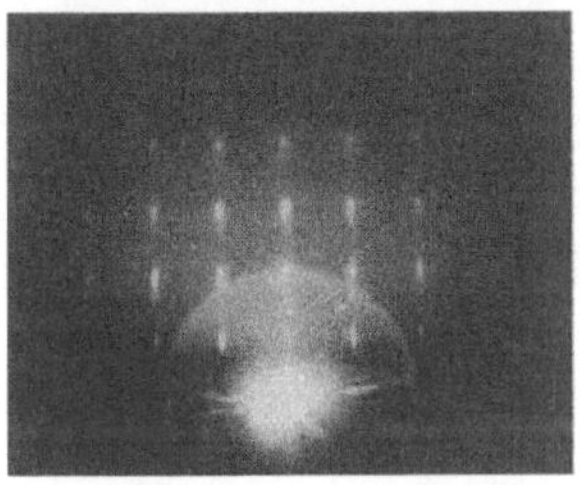

(c)

Fig.6 RHEED patterns of a top YBaCuO (a), intermediate MgO (b) and bottom YBaCuO (c). Bottom and top YBaCuO layers were grown at 600°C and 670°C, respectively.

layer. One point we should notice is a non-uniform MgO growth : It grows like islands rather than a thin film. Our plane view TEM image observation suggests that the coverage of MgO layer is about 50 %. Because of this, the MgO layer became twice thick as our expectation. Anyway, the island growth may be responsible for the rather weak RHEED pattern of the MgO layer. According to the nucleation theory, the island growth is very likely to derive from low adsorption energy of Mg atom (or MgO molecular, its cluster) on the YBaCuO surface and/or a large value of MgO/YBaCuO interface free energy.

FABRICATION OF YBaCuO-MIS STRUCTURE

One of the applications of the heterostructure is superconductor MIS devices. The success of MIS devices is based on whether or not we can prepare the excellent interface, as well known in the case of Si-MOS devise. Heteroepitaxial growth is a key technology for challenging this subject.

The MIS structure we fabricated is schematically drawn in Fig.8. The gate insulator was made by in-situ epitaxial growth of MgO. The gate electrode was made of Al deposition and lift off. The carrier concentration of the YBaCuO film was intentionally reduced down one or two orders of magnitude less than the best film.

The C-V curves were measured at room temperature under the probe signal with 1 MHz and 15 mV amplitude. The round-trip dc bias was swept at a rate of 2 V/s between -10 and 10 V. The raw C-V data were calibrated by taking the series resistance effect into account. A typical example

Fig.7 A cross sectional TEM image of YBaCuO(150nm)/MgO(2nm)/YBaCuO(150nm) structure.

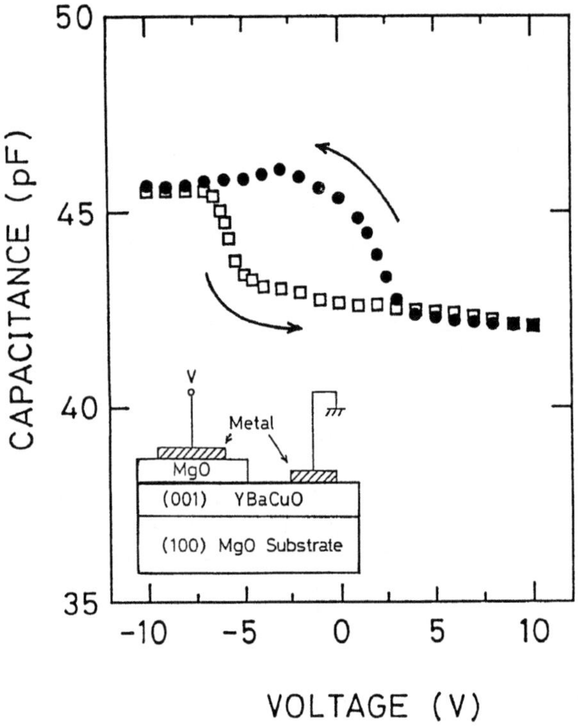

Fig.8 A typical example of the measured C-V curve with a schematic cross section view of the Al/MgO/YBaCuO MIS diode.

of the Al/MgO/YBaCuO MIS C-V curve is given in Fig.8, clearly showing that the MIS diode works well just like a semiconductor MIS device. In this figure, the arrows indicate the direction of the applied voltage change. A shape of this C-V curve is like that of p-type semiconductor MIS diode. A large hysteresis of a round-trip C-V curve in Fig.8 is seen. It is due probably to the ion-drift inside the insulated MgO layer and/or to a slow trap and detrap of free carriers in YBaCuO at the interface region. Preparation of excellent interface and insulating layer will promise to completely overcome the hysteresis.

CONCLUSION

The epitaxial growth of the double-heterostructure consisting of YBaCuO/MgO/YBaCuO was investigated. Some methods were proposed to obtain several combination of the crystal axes differed between the bottom and top YBaCuO layers. The cross sectional TEM diagnosis revealed a non-uniformity in growing ultra-thin (2 nm) MgO as an intermediate layer. Finally, we demontrated an Al/MgO/YBaCuO MIS diode, a possible application of the heterostructure at least at present.

ACKNOWLEDGMENTS

This work was partly supported by Science Research Grant-in-Aid from Ministry of Education, Science and Culture of Japan.

REFERENCES

[1] J. G. Bednorz and K. A. Müller, Z. Phys. B64 (1986) 189.
[2] M. K. Wu, J. R. Ashburn, C. J. Torng, P. H. Hor, R. L. Meng, L. Gao, Z. J. Huang, Y. Q. Wang and C. W. Chu, Phys. Rev. Lett. 58 (1987) 908.
[3] H. Maeda, Y. Tanaka, M. Fukutomi and T. Asano, Jpn. J. Appl. Phys. 27 (1988) L209.
[4] J. M. Liang, R. S. Liu, L. Chang and P. T. Wu, Appl. Phys. Lett. 53(15) (1988) 1434.
[5] T. Kobayashi, H. Sakai and M. Tonouchi, Electron Lett. 22 (1986) 659.
[6] T. Kobayashi, K. Hashimoto, U. Kabasawa and M. Tonouchi, IEEE Trans. Magn. MG-25(2) (1989) 927.

Thin Film, OMCVD Process for High-T_c Superconductivity

HITOSHI ABE, RYODO KAWASAKI, and TAIJI TSURUOKA

Research Laboratory, Oki Electric Industry Co., Ltd., 550-5, Higashiasakawa-cho, Hachioji, Tokyo, 193 Japan

ABSTRACT

This paper describes the potentiality of the OMCVD method for a high-Tc thin film formation technique. At the beginning of 1988, Naval Research Institute, Tohoku University and Oki Electric have independently succeeded in the high-Tc ceramics film formation by the OMCVD method. Y-Ba-Cu-O films could be prepared with a quality of Tc higher than 90K and $Jc=1.9x10^6 A/cm^2$ at 77K, which is comparable to that obtained by sputtering or MBE method. Basic studies on the OMCVD deposition process of high-Tc films have been done. Very recently, by using N_2O as an oxygen source, we have succeeded in radically reducing the temperature of film formation.

KEY WORDS: Thin film High-Tc OMCVD Crystal growth

1. INTRODUCTION

For practical applications of high-Tc superconductive oxides, thin film formation techniques have been developed as the key technology to fabricate various devices. Thin film formation technique can be broadly classified into the chemical vapor deposition (CVD) method and the physical vapor deposition (PVD) method. While the PVD method is a method of forming thin film by vacuum vaporization or sputtering of the material, the CVD method is a method of forming the desired thin film by chemical reaction in gaseous phase or on substrate surface, by supplying a gas source onto the substrate surface. The OMCVD method was applied by H. M. Manasevit to the thin film growth of GaAs in 1968[1]. Because of poor quality of the source materials, it did not become popular in the early period. However, after the GaAs/AlGaAs laser as well as the quantum well laser were produced by the OMCVD method, this method stepped into the limelight. The OMCVD method is characterized as following excellent features;

Abbreviation		R	R′
acac	acetylacetonate	$-CH_3$	$-CH_3$
dpm (thd)	dipivaloymethanate (tetramethyl-heptanedion)	$-C(CH_3)_3$	$-CH_3$
fod	heptafluoro-butanoyl pivaloylmethane	$-C_3F_7$	$-C(CH_3)_3$
tfa	trifluoro-acetylacetonate	$-CF_3$	$-CH_3$
hfa	hexafluoroacetylacetonate	$-CF_3$	$-CF_3$

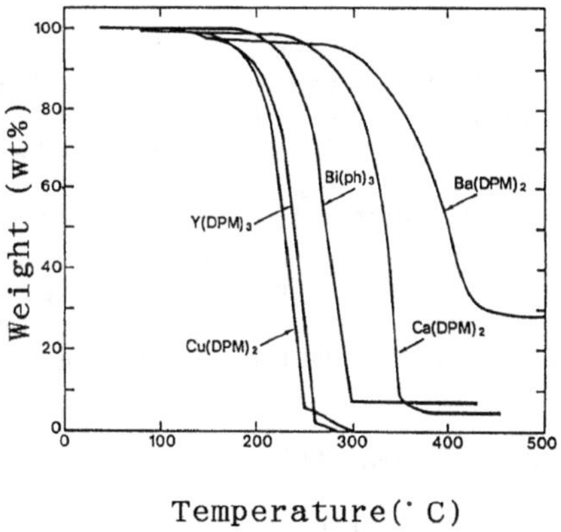

Table 1: β-diketone ligands   Fig.1 Thermogravimetric curves of metal β-diketonetes and triphenylbismuth for High-Tc film formation source materials.

1) Film thickness control about one molecular layer
2) Easy composition control of compound crystals
3) Formation of heteroepitaxial multi-layer of thin films
4) Uniformity of epitaxial layer
5) High throughput

Table 2: Experimental Conditions

Precursor	Y	$Y(DPM)_3$	Bi	$Bi(Ph)_3$
	Ba	$Ba(DPM)_2$	Sr	$Sr(DPM)_2$
	Cu	$Cu(DPM)_2$	Ca	$Ca(DPM)_2$
			Cu	$Cu(DPM)_2$
Bubbler Temperature	Y	120~140℃	Bi	95~125℃
	Ba	220~240℃	Sr	220~260℃
	Cu	110~140℃	Ca	175~220℃
			Cu	110~140℃
Carrier Gas Flow	Y	50	Bi	50~70
Ar (cc/min)	Ba	50	Sr	50
	Cu	50	Ca	50
			Cu	50
O_2, N_2O Flow Rate		400 cc/min		40 ~ 400 cc/min
Gas Pressure		5000 Pa		5000 Pa
Substrate Temperature		600 ~ 800 ℃		600 ~ 800 ℃

2. CHARACTERISTICS OF GAS SOURCES FOR HIGH-Tc FILMS

For formation of a thin film by the OMCVD method, an organometallic gas source intended object is necessary. In taking an example of Y-Ba-Cu-O film, an organometallic chelates containing each element of Y, Ba and Cu becomes necessary. Furthermore, appropriate vapor pressure and thermal stability are required for the organometallic chelate. However, until an oxide high-Tc superconductor was discovered, hardly any research had been carried out on gas source such as ⅡA group elements, rare earth elements and copper. β-diketone coordination complex forms stable complexes in combining with various metals. β-diketone ligands are shown in Table 1 in an arranged form. The degree of volatility in β-diketone coordination complex of metal is in the order of hfa>tfa>fod>dpm(thd)>acac.

TG (thermogravimetry) results of β-diketonates $Y(DPM)_3$, $Cu(DPM)_2$, $Ca(DPM)_2$, and $Ba(DPM)_2$ and triphenylbismuth $Bi(ph)_3$, used for Y-Ba-Cu-O and Bi-Sr-Ca-Cu-O film formation, are measured and shown in Fig.1. Reduction in mass for $Y(DPM)_2$, $Ba(DPM)_2$ seen at around 100°C and 150°C is considered to represent vaporization of water or organic solvents contained as impurities, while much larger reduction in mass at higher temperatures represents vaporization of their chelates. $Y(DPM)_3$, $Cu(DPM)_2$, $Bi(ph)_3$ and $Ca(DPM)_2$ show their mass reduced to practically zero at 250°C, 260°C, 300°C and 340°C respectively. On the contrary, $Ba(DPM)_2$ has a considerable amount of unvaporized mass remaining until 500°C.

3. EXPERIMENTAL

The experimental equipment used here is very similar to the standard OMCVD equipment. The growth chamber is a horizontal cold-wall type[2]. A difference is that the bubbling temperature is much higher than room temperature, so the gas inlet system has to be heated. The substrates are heated on the susceptor by radio frequency induction. The substrate temperature is measured with a chromel-almel thermocouple which is set in the susceptor. The growth chamber is evacuated to less than 10 Pa after the substrate is set, and then gases are introduced.

Experimental conditions for Y-Ba-Cu-O films and Bi-Sr-Ca-Cu-O films are shown in Table 2. Total gas pressure is maintained at 5000Pa during the deposition. After the deposition is complete, the films are then allowed to cool in 1 atm or 5000 Pa oxygen atmosphere by the programmed cooling procedure. The growth rate of films is measured by a microbalance. Substrates used in these experiments are $SrTiO_3$(100) and (110), MgO(100) and Si(100). The electrical resistance of films is measured by the standard four-probe method and the crystal structures are examined by the X-ray diffraction (XRD) method.

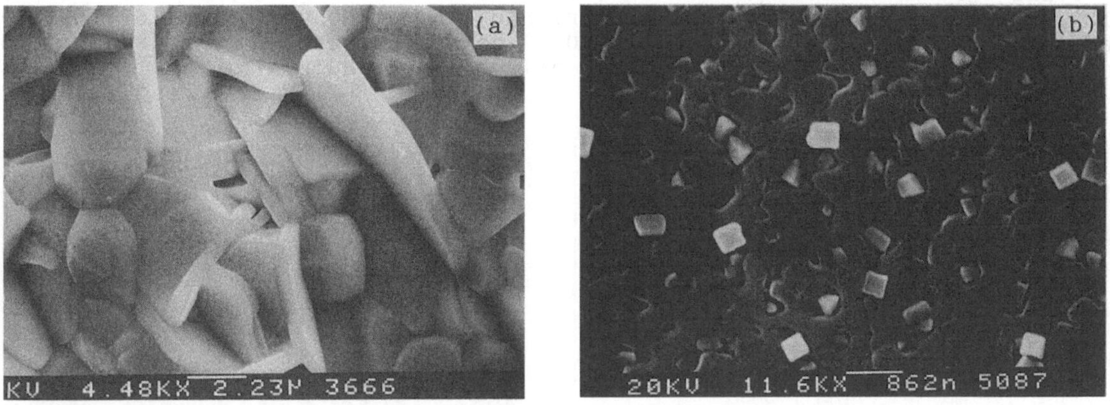

Fig.2 Scanning electron micrographs of Y-Ba-Cu-O films (a) with post annealing and (b) without annealing.

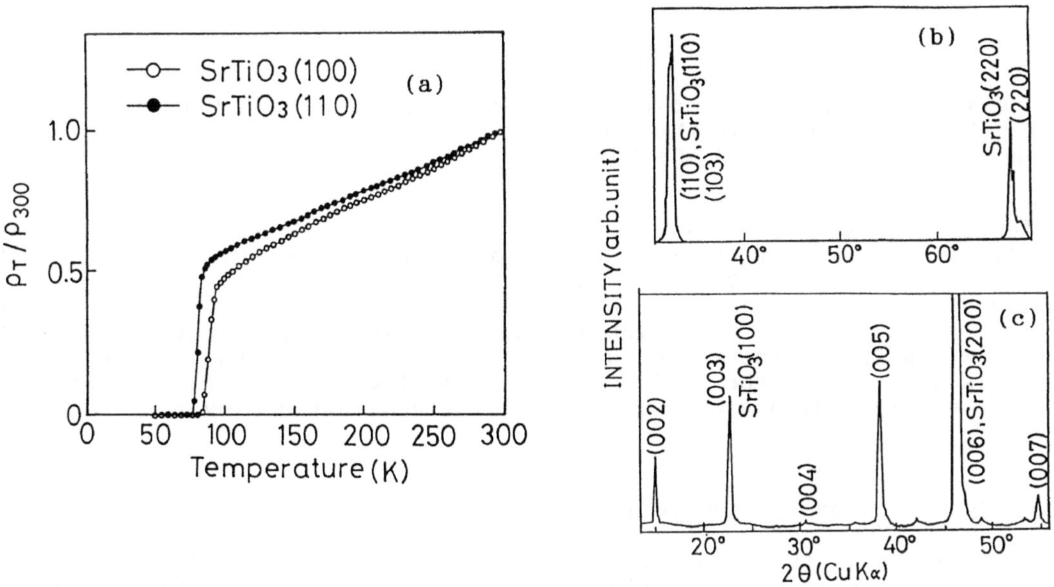

Fig.3 The temperature dependence of normalized resistance (a), X-ray diffraction patterns (b) and (c) for Y-Ba-Cu-O films of 1000A thick grown on $SrTiO_3$(110) and (100) substrates.

4. SUPERCONDUCTING Y-Ba-Cu-O FILMS AND Bi-Sr-Ca-Cu-O FILMS

The smooth surface morphology is inevitably needed for microelectronic applications. Fig.2 shows the scanning electron micrographs of Y-Ba-Cu-O films prepared at the substrate temperature of 800°C ; (a) after post annealing at 950°C and (b) as grown (Tc=88K). These SEM photographs suggested clearly that superconducting oxides films must be prepared without annealing process at the higher temperature than the growth temperature. For superconducting devices, very thin films are required. We have prepared a superconducting film less than a few hundred angstroms by the OMCVD method. Fig. 3(a) shows the temperature dependence of normalized resistance for Y-Ba-Cu-O films of 1000 A thick grown on $SrTiO_3$(100) and $SrTiO_3$(110) substrates at the growth temperature of 800°C. Films on $SrTiO_3$(100) and (110) substrates show Tc=84K and 75K. Some of films prepared on $SrTiO_3$(100) substrate showed the critical current density more than 10^6 A/cm^2. Fig. 3(b) and 3(c) show the XRD patterns of films grown on $SrTiO_3$(110) and (100) substrates. Strong peaks of (110), (013) and (103), a,b-axis preferred orientation, are observed on $SrTiO_3$(110) substrate. On the other hand, only (00n) peaks are observed on $SrTiO_3$(100) substrate, which suggests the highly C-axis orientated crystal.

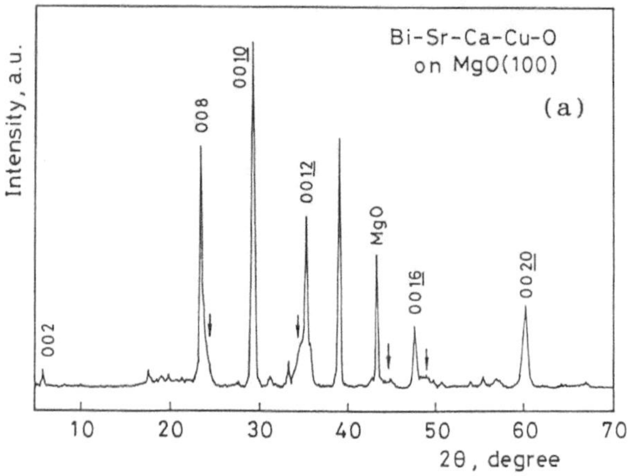

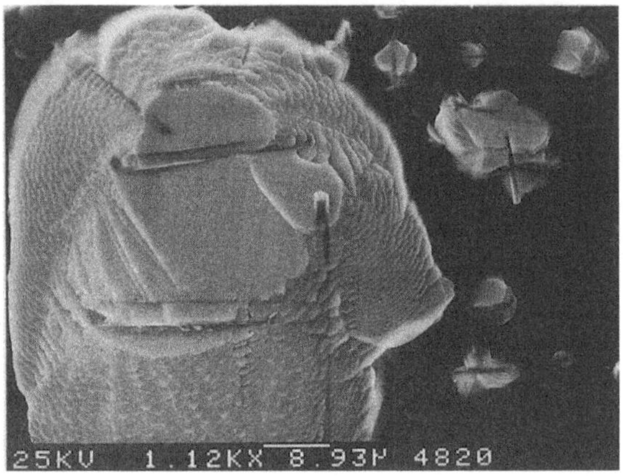

Fig.4 X-ray diffraction pattern of (a) a Bi-Sr-Ca-Cu-O film grown on a MgO(100) substrate at the substrate temperature of 800°C, (b) a SEM photograph of the Bi-Sr-Ca-Cu-O film.

Fig. 4(a) shows X-ray diffraction pattern of a Bi-Sr-Ca-Cu-O film grown on a MgO(100) substrate at the substrate temperature of 800°C. This film shows the zero resistance at 78K. This diffraction pattern indicates the film is composed of nearly single low Tc phase, whose C-axis length is 30 A. Arrows in the figure show peaks of the high Tc phase, whose C-axis length is 37 A. Fig. 4(b) show a SEM photograph of a Bi-Sr-Ca-Cu-O film. A sheet like crystal growth is observed in the Bi-Sr-Ca-Cu-O film. The sheet surface of the film seems to be very flat and as smooth as a single molecular layer.

5. EFFECTS OF N_2O ON THE GROWTH TEMPERATURE

Fig. 5(a) shows the temperature dependence of normalized resistance for Y-Ba-Cu-O films grown on $SrTiO_3$(100) substrates at growth temperatures from 600°C to 800°C by using N_2O gas as an oxygen source. The film grown at a substrate temperature above 650°C showed metallic-like behavior as a function of temperature. Films grown at substrate temperatures of 650°C, 700°C, and 800°C showed a zero resistance at 79K, 84k and 85K, respectively. Films prepared at a substrate temperature of 600°C did not show the zero resistance. In the case of films grown by using O_2 gas shown in Fig. 5(b), these were not as good as those films grown by using N_2O gas. When using O_2 gas, with decreasing growth temperature from 800°C to 700°C, Tc decreased from 88K to 68K. The films grown at 700°C showed a semiconductor-like resistance behavior. Fig. 5(c) shows the resistance change of Y-Ba-Cu-O films deposited on $SrTiO_3$(100), MgO(100) and Si(100) substrate at the growth temperature of 650°C using N_2O as an oxygen source. The temperature of zero resistance were 79K, 65K for films on $SrTiO_3$(100) and MgO(100). Films grown on Si(100), in which the onset of superconducting transition appeared at about 30K, did not show zero resistance above 20K[3].

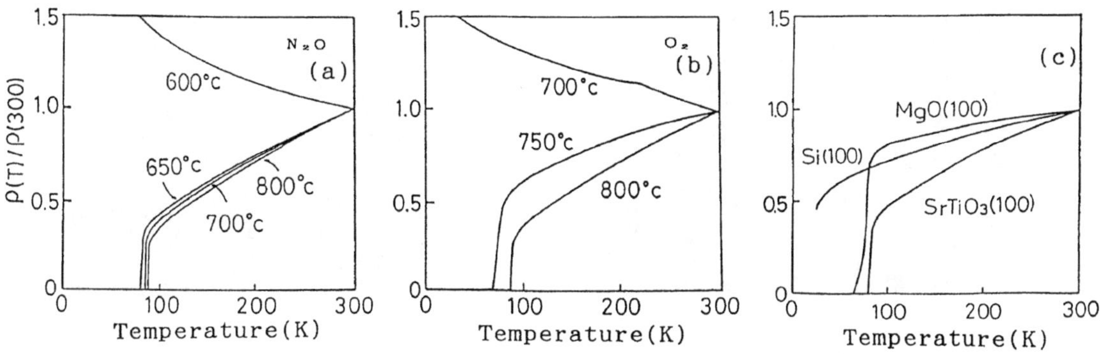

Fig.5 Temperature dependence of normalized resistance for Y-Ba-Cu-O films grown on $SrTiO_3$(100) (a) at growth temperatures from 600°C to 800°C by using N_2O gas as an oxygen source (b) at growth temperature from 700°C to 800°C by using O_2 gas as an oxygen gas source, (c) temperature dependence of resistances for the films grown on $SrTiO_3$(100) , MgO(100) and Si(100) substrates at the growth temperature of 650°C.

6. EFFECTS OF SUBSTRATES ON THE SUPERCONDUCTING PROPERTIES

The effects of substrates on superconducting thin films other than orientation have not been known. The effects of substrates on the superconducting properties of Y-Ba-Cu-O films by forming on $SrTiO_3$(100) and MgO(100) substrates under different growth conditions. The source temperature determines the vapor pressure of the gas source and inlet rate of gas source into the growth chamber. The source temperature was set as follows, Condition 1 ; Y=126°C, Ba=228°C, Cu=125°C Condition 2 ; Y=125, Ba=228°C, Cu=125°C Condition 3 ; Y=124°C, Ba=224°C, Cu=128°C. Other growth conditions are kept constant[4].

Fig. 6 shows the temperature dependence of normalized resistance for Y-Ba-Cu-O films formed on MgO(100) and $SrTiO_3$(100) substrates. The composition ratio of films is examined by EPMA. By this evaluation, films formed under condition 2 show stoichiometric composition. Films of the condition 2 shows metallic-like resistivity changes for MgO and $SrTiO_3$ substrates. Films became Y rich under condition 1 and Cu rich under condition 2. For MgO substrate, temperature dependence of the resistivity changed from metallic-like to semiconductor-like. However, for $SrTiO_3$ substrate that of resistivity was only slightly changed.

SUBJECTS FOR THE FUTURE AND REFERENCES

Research results on OMCVD have been already reported by more than 10 research groups in less than one year from the start. This suggests that many people consider the great potentiality of the OMCVD method. However, a few item of problems to be solved exist. The first item for the future is research and development of the gas source. The largest cause in falling behind other thin film formation techniques consists in a lack of suitable gas source materials for OMCVD, which should have a higher volatility and a simpler coordination complex of metal. The second item is to elucidate the crystal growth mechanism. The current problems, such as improvement of surface morphology, control of grain boundary, reduction in growth temperature and control of stoichiometry. The third item is analysis of the structure and composition in micro-region of deposited thin film. From this analysis, information such as orientation characteristics, grain boundary surface and deposition surface of crystal grain can be obtained in order to prepare single crystal thin film of superconducting oxides.

[1] Manasevit H M (1968) Appl. Phys. Lett. 12: 156
[2] Abe H, Tsuruoka T, Nakamori T (1988) Jpn. J. Appl. Phys. 27: L1473-L1475
[3] Tsuruoka T, Kawasaki R, Abe H (1989) Jpn. J. Appl. Phys. 28: L1800-L1802
[4] Takahashi H, Kawasaki R, Tsuruoka T, Kanamori T (1989) Extended Abstracts of 1989 International Superconductivity Electronics Conference: 78-81

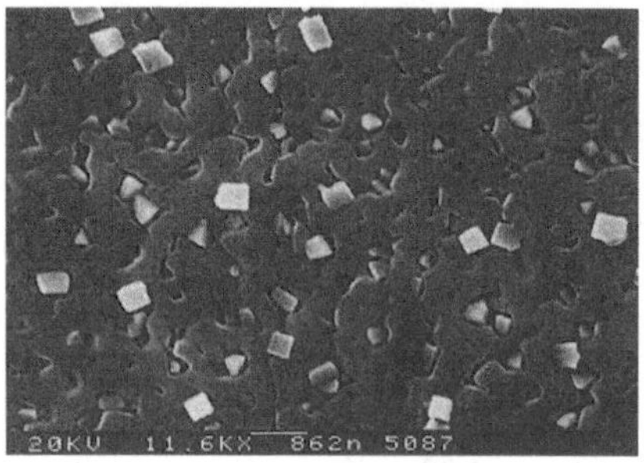

Condition 1

Condition 1

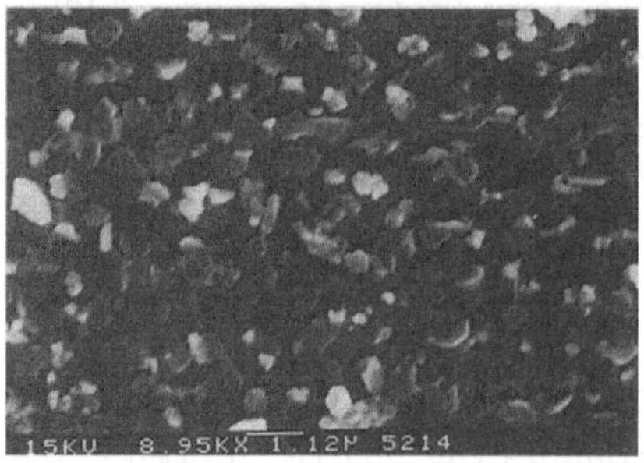

Condition 2

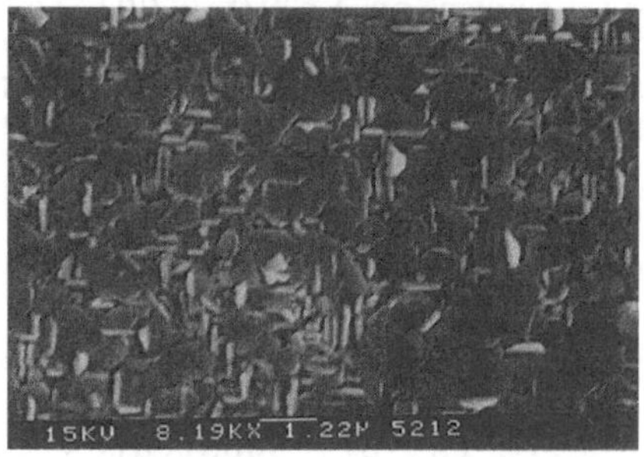

Condition 2

Condition 3

Condition 3

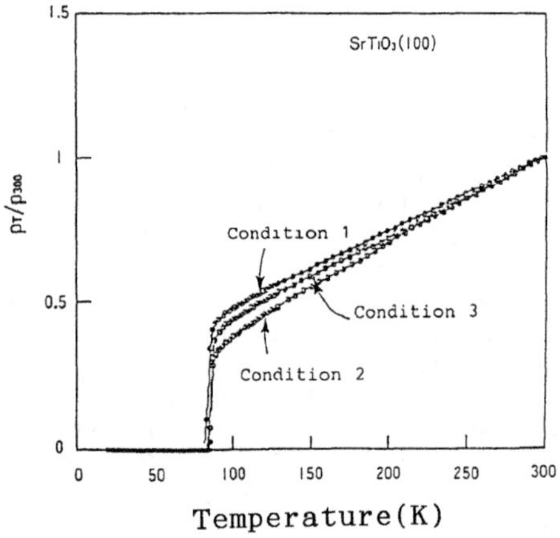

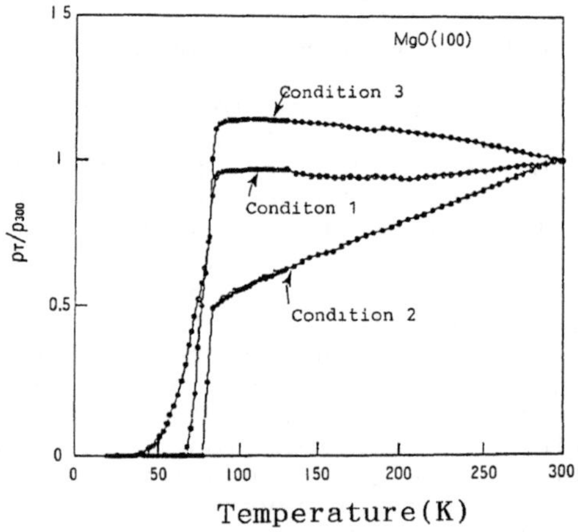

Fig.6 The temperature dependence of normalized resistance and SEM photographs for Y-Ba-Cu-O films formed on $SrTiO_3(100)$ and $MgO(100)$ substrate under different growth conditions 1-3.

High-T_c Superconducting Oxide Films Prepared by CVD

H. YAMANE[1], T. HIRAI[1], H. KUROSAWA[2], A. SUHARA[1], K. WATANABE[1], N. KOBAYASHI[1], H. IWASAKI[1], E. AOYAGI[1], K. HIRAGA[1], and Y. MUTO[1]

[1] Institute for Materials Research, Tohoku University, 1-1, Katahira 2-chome, Aoba-ku, Sendai, 980 Japan
[2] Riken Co., 810 Kumagaya, Kumagaya, 360 Japan

ABSTRACT

YBaCuO superconducting films were prepared on $SrTiO_3$(100) and MgO(100) by chemical vapor deposition using β-diketone metal chelates. The films were composites in which the grains of CuO and a-axis oriented $YBa_2Cu_3O_y$ dispersed in a matrix of c-axis oriented $YBa_2Cu_3O_y$. The films deposited on $SrTiO_3$(100) showed Tc of 87-92 K and Jc above 10^5 A/cm^2 at 77.3 K and 0 T. Tc and Jc of the films on MgO(100) were 80-89 K and below 10^4 A/cm^2 at 77.3 K and 0 T, respectively. Jc of the films on $SrTiO_3$(100) did not depend on the film thickness up to 2.4 μm.

KEYWORDS: YBaCuO film, CVD, β-diketone metal chelates, composites, thickness dependence

INTRODUCTION

Recently, chemical vapor deposition(CVD) has been a representative technique of high-Tc superconducting oxide film preparation. Superconducting transition temperatures(Tc) above liquid nitrogen temperature(77.3 K) were obtained for CVD films of superconducting oxides in the systems of Y-Ba-Cu-O(YBaCuO)[1,2], Bi-Sr-Ca-Cu-O[3], and Tl-Ba-Ca-Cu-O[4]. Critical current density(Jc) at 77.3 K was, however, reported only for CVD-YBaCuO films. We reported Jc of 2.0x10^6 A/cm^2 at 77.3 K and 0 T for a YBaCuO film prepared on a $SrTiO_3$(100) single crystal substrate[5]. Takahashi et al. obtained Jc in excess of 2x10^5 A/cm^2 at 77 K for the films on $SrTiO_3$(100)[6]. Critical current densities in the range of 10^4 A/cm^2 under magnetic fields up to 27 T were also measured at 77.3 K for the CVD-YBaCuO films[7,8]. Schmaderer and Wahl reported Jc of 10^5 A/cm^2 at 77.3 K and 5.5 T[9]. These high-Jc films were prepared on $SrTiO_3$(100) single crystal substrates. The relationship between the chemical composition of these films and superconducting properties has not yet been reported.

High Jc above 10^6 A/cm^2 were measured for thin films prepared by physical vapor deposition(PVD). The thickness of the PVD films having a high Jc was usually less than 1 μm. Sintered bulk samples gave Jc values that were two or three orders of magnitude smaller than for the PVD thin films. It is important to investigate the thickness dependence of Jc so as to consider the large over-all critical current Ic.

This paper describes the chemical composition, structure, and superconducting properties of CVD-YBaCuO films prepared on $SrTiO_3$(100) and MgO(100). The thickness dependence of Jc is reported for the films on $SrTiO_3$(100).

EXPERIMENTAL

A CVD reactor was a vertical hot-wall type. The source materials used were β-diketonates of $Y(thd)_3$, $Ba(thd)_2$, and $Cu(thd)_2$ (where (thd) represents 2,2,6,6-tetramethyl-3,5-heptanedionato). CVD conditions are summarized in Table 1. The films having various chemical compositions were prepared at different vaporizer temperatures in the ranges shown in Table 1. Each evaporated source was introduced into the reactor. After deposition, the films were cooled from the deposition temperature of 850°C to room temperature at a rate of 15°C/min under an atmosphere of 1 atm of oxygen. Substrates used were $SrTiO_3$(100) and MgO(100) single crystals(5x10x1 mm^3). Part of each substrate was masked with Pt foil in order for the film thickness to be measured using a stylus instrument.

The film composition was determined by inductively coupled plasma emission spectroscopy. The microstructure of the film surfaces was observed with a scanning electron microscope. A plane-view sample was prepared by mechanically griding and then ion-milling to electron transparency for transmission electron microscope observation. Microanalysis of the film was carried out by using a electron microprobe analyzer(EPMA) and a scanning Auger electron spectroscope(AES). X-ray diffraction (XRD) patterns of the deposited films were obtained by the standard 2θ-θ method. The resistivity and Jc of the films were measured by a DC four-probe method with Au electrodes sputtered on the film. For the Jc measurement, the sample was immersed directly in the liquid nitrogen. Magnetic fields were applied perpendicular to the current by use of a hybrid magnet at Tohoku University.

Table 1 Deposition Conditions.

Vaporizer Temperature	$Y(thd)_3$: 110-160°C $Ba(thd)_2$: 240-275°C $Cu(thd)_2$: 110-150°C
Deposition Temperature	850°C
Total Gas Pressure	10 Torr
Carrier Gas(Ar) Flow Rate	450 ml/min
O_2 Gas Flow Rate	250 ml/min
Deposition Time	1 h
Substrate	$SrTiO_3$(100) MgO(100)

thd : 2,2,6,6-tetramethyl-3,5-heptanedionato

EXPERIMENTAL RESULTS

Figure 1 shows chemical compositions and Tc(resistivity zero: R=0) of the films on $SrTiO_3$(100)(a) and MgO(100)(b). The values of Tc(R=0) of 87-92 K were measured for the films which were prepared on $SrTiO_3$(100) and in the compositional region from Y:Ba:Cu=1:2:3 to Cu-rich compositions. The films obtained in almost the same compositional region on MgO(100) showed Tc(R=0) above 80 K. The highest value of Tc(R=0) among the films on MgO(100) was 89 K(MG-2).

The values of Jc measured at 77.3 K and 0 T by a voltage criterion of 2 μV/cm were above 10^5 A/cm^2 for the films grown on $SrTiO_3$(100). The region of composition of the films spreads from around Y:Ba:Cu=1:2:3 to Cu-rich compositions as indicated in Fig. 1 with a shadow. The highest Jc value measured resistively among the films on MgO(100) was 8.0×10^3 A/cm^2 at 77 K and 0 T for MG-2. The dimensions of MG-2 for critical-current measurement were approximately 5 mm wide, 1.6 μm thick, and 2 mm long. Jc of the films on $SrTiO_3$(100) with the same dimensions could not be determined due to Joule heating by the contact resistance. Critical currents of the films on $SrTiO_3$(100) were measured through bridges with sizes of 0.3-0.6mm wide, 0.6-2.4 μm thick, and 1-2 mm long, which were prepared by masking during the deposition.

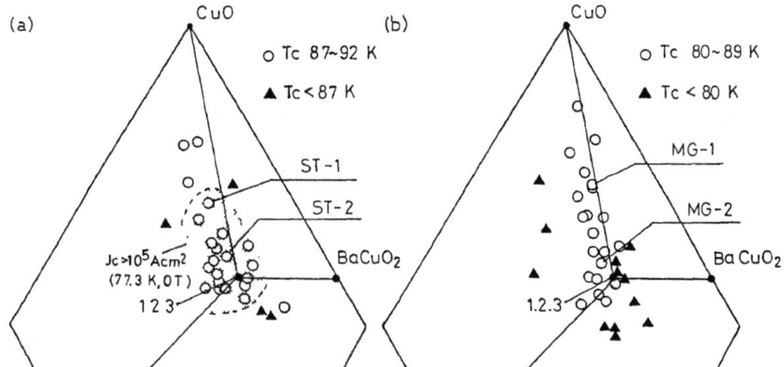

Fig.1 Compositions and Tc(R=0) of CVD-YBaCuO films on $SrTiO_3$(100)(a) and MgO(100)(b).

Precipitates with a size of about 2-3 μm were seen in scanning electron micrographs of the film surface of ST-1 (Fig. 2(a)) and MG-1(b). EPMA and AES analyses revealed that these precipitates were CuO and were grown from the substrates. The size of the precipitates became larger with increasing Cu content. The AES analysis could not detect the elements of the substrates, carbon contained in the source gas, or any other elements of the impurities in the films.

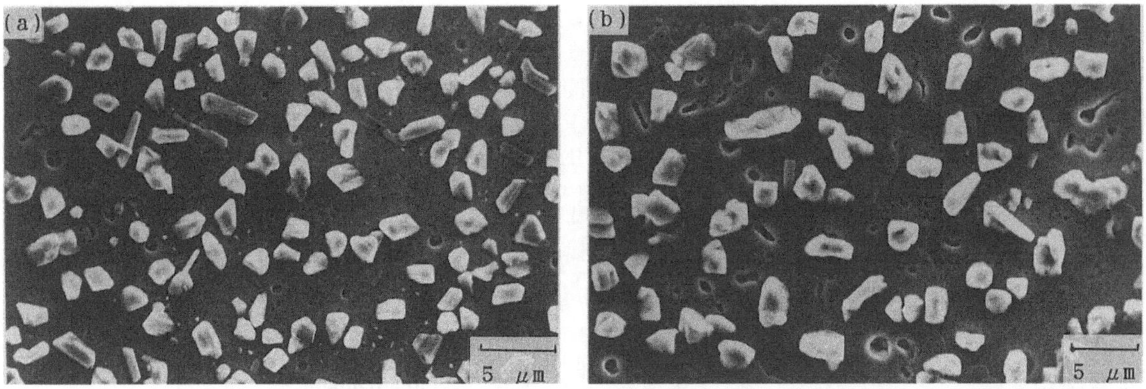

Fig.2 Scanning electron micrographs of CVD-YBaCuO films;(a) ST-1 and (b) MG-1.

The XRD patterns of ST-1 and MG-1 are shown in Fig. 3(a) and (b). Observed high relative intensities of (001) peaks indicate that the c-axis is oriented perpendicular to the substrate. A very small peak of CuO(111) could be revealed under magnification. Figure 4 shows the XRD patterns of ST-2(a) and MG-2(b) whose compositions are near to 1:2:3 as shown in Fig. 1(a) and (b). In addition to the prominent peaks of (001), a small peak of (200) reflection is observed. This means that the films consist mostly of c-axis oriented grains with a few a-axis oriented grains.

A plane-view transmission electron micrograph of the film on $SrTiO_3$(100) is shown in Fig. 5. Grains oriented with their c axis in the plane of the substrate is observed. As indicated by arrows in the micrograph, strips with a length of about 20 nm are bedded in the grains along the layers of $YBa_2Cu_3O_y$. Strain contrasts are observed in the part of the c-axis oriented $YBa_2Cu_3O_y$.

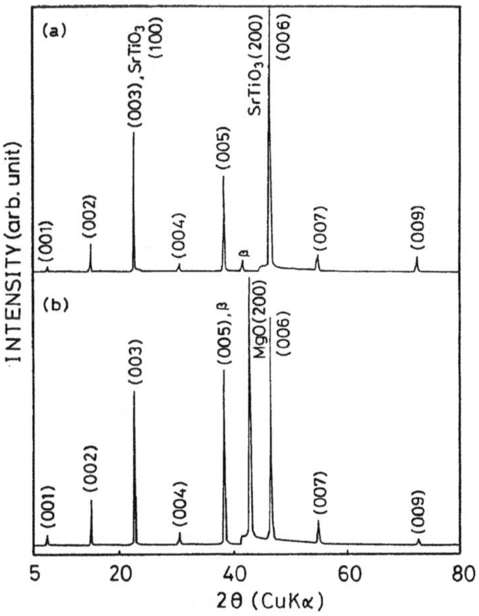

Fig.3 X-ray diffraction patterns of CVD-YBaCuO films; (a) ST-1 and (b) MG-1.

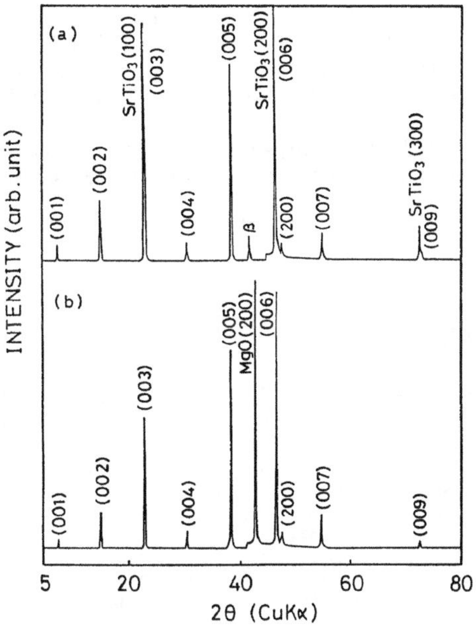

Fig.4 X-ray diffraction patterns of CVD-YBaCuO films; (a) ST-2 and (b) MG-2.

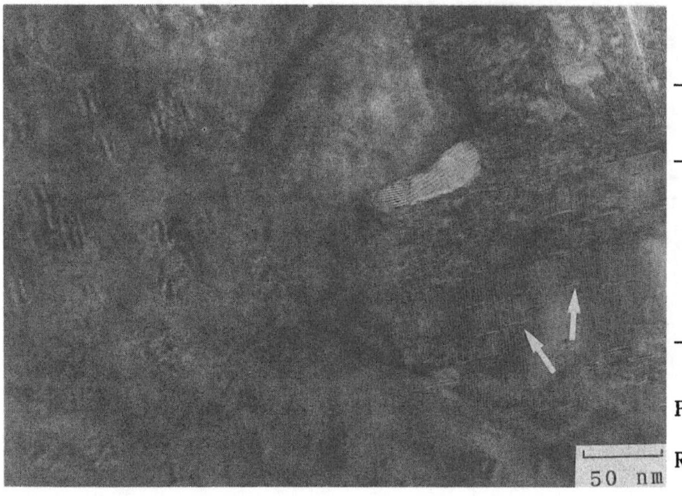

Table 2 XRD data of $YBa_2Cu_3O_y$ in YBaCuO films

substrate	$SrTiO_3(100)$	MgO(100)
C_0 (nm)	1.1680(5)*	1.1694(6)
PIW(007) (2θ)	0.26°(3)	0.25°(2)
RIW(007) (θ)	0.45°(3)	0.64°(13)

* Standard deviations are given in parentheses.
PIW : average integral width of peaks
RIW : average integral width of rocking curves

Fig.5 Plane-view micrograph of a CVD-YBaCuO film on $SrTiO_3(100)$.

Table 2 summarized the averages and standard deviations of the c-axis lengths(c_0) calculated from the (007) peak position, integral width of the (007) peak(PIW), and integral width of the rocking curve for the (007) peak(RIW). The c-axis length of the films on $SrTiO_3(100)$ was smaller than that of the films on MgO(100). The films on $SrTiO_3(100)$ and MgO(100) have the same integral width. The crystallite size evaluated from the integral width is above 200 nm. The films on MgO(100) have a larger integral width of the rocking curve than the films on $SrTiO_3$. This indicates that the degree of the c-axis orientation for the films on MgO(100) is inferior to that for the films on $SrTiO_3(100)$.

Magnetic field dependence of the critical current densities of ST-2 and MG-2 are shown in Fig. 6. A Jc of $4.6x10^5$ A/cm^2 at 77.3 K and 0 T was measured for ST-2. When magnetic fields were applied parallel to the film plane, Jc of ST-2 decreased gradually at B > 3 T and reached $7.3x10^3$ A/cm^2 at 27 T. The Jc of ST-2 in magnetic fields perpendicular to the film plane diminished steeply above 3 T. Similar field direction dependence of Jc was observed for the films on MgO(MG-2), although the value of Jc was two orders of magnitude smaller than that of the films on $SrTiO_3(100)$.

The thickness dependence was investigated for films prepared on $SrTiO_3(100)$ at the fixed evaporation temperatures of the source materials($Y(thd)_3$:130°C,

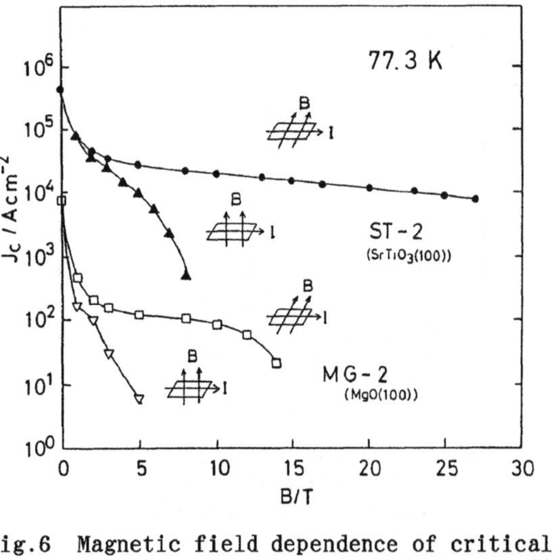

Fig.6 Magnetic field dependence of critical current densities at 77.3 K for CVD-YBaCuO films on $SrTiO_3(100)$(ST-2) and MgO(100)(MG-2).

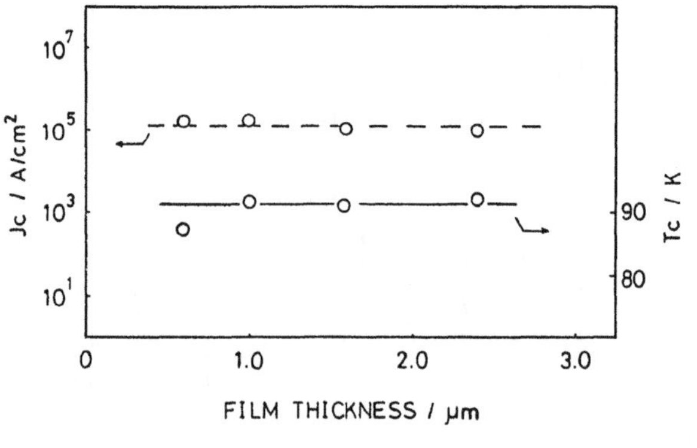

Fig.7 Thickness dependence of Tc and Jc for CVD-YBaCuO films on $SrTiO_3(100)$.

$Ba(thd)_2$:255°C, $Cu(thd)_2$:123°C). The thickness was increased linearly with deposition time. Figure 7 shows the thickness dependence on Tc and Jc. The films having thicknesses above 1.0 μm show Tc(R=0) above 90 K. Jc of the films with thicknesses of 0.6-2.4 μm were 1.0-1.9x10^5 A/cm^2 at 77.3 K and 0 T.

DISCUSSION

YBaCuO sintered bulk samples with Tc above 88 K were obtained in a small compositional region around Y:Ba:Cu=1:2:3[10]. But the CVD-YBaCuO films prepared on $SrTiO_3$ had Tc(R=0) above 87 K, even when their composition was rich in Cu content as shown in Fig. 1(a). Excess amounts of Cu present as island precipitates of the CuO in the films did not affect Tc(R=0). However, it is considered that the presence of CuO precipitates decreases the effective volume of the current path and causes a small compositional region of the high-Jc film as compared to the region of high-Tc film(Fig. 1(a)).

Tc(R=0) of the films on MgO(100) was lower than that of the films on $SrTiO_3$(100) and did not exceed 90 K. According to Komatsu et al., the value of Tc decreased to 70-75 K and c-axis length changed from 1.167 to 1.157 nm when the x of $Y(Ba_{1-x}Mg_x)_2Cu_3O_y$ was 0.025[11]. Yan et al. reported that addition of 5 wt% $SrTiO_3$ decreased the Tc from 91 to 82.5 K and shortened the c-axis length from 1.167 to 1.163 nm[12]. However, we could not detect Mg, Sr, and Ti in the films by AES analysis. Cheung reported that MgO was less reactive to $YBa_2Cu_3O_y$ than $SrTiO_3$[13]. Moreover, the c-axis length of the CVD-YBaCuO films on MgO(100) was larger than the length(1.168 nm) of the films on $SrTiO_3$(100).

It is well-known that the oxygen deficiency of $YBa_2Cu_3O_y$ reduces Tc and expands c-axis length[14]. The Tc and c-axis length in the present study agree with the reported values. The oxygen content of the films on MgO(100) might be lower than that of the films on $SrTiO_3$(100), although the films on both substrates received the same in-situ oxygen treatment after deposition. We suppose that this might be caused by the difference in the degrees of c-axis orientation or in the discrepancies of the thermal expansion coefficient between the film and substrate.

Critical current density of the films on MgO(100) was also two orders of magnitude lower than that of the films on $SrTiO_3$(100). It was reported that the misorientation of the $YBa_2Cu_3O_y$ grains depressed the values of Jc[15]. The integral width of the rocking curve revealed that the degree of orientation of the films on MgO(100) was inferior to that of the films on $SrTiO_3$(100). The reason for this may be ascribed to the large lattice mismatch between $YBa_2Cu_3O_y$ and MgO(8-10%) compared with the mismatch between $YBa_2Cu_3O_y$ and $SrTiO_3$(2%) or to the differences in single crystal perfection of the substrates.

Xi et al. reported high Jc up to 9x10^5 A/cm^2 at 77.3 K and 0 T for the films on MgO(100)[16]. They measured critical current through a 30 μm wide and 300 μm long bridge with a thickness of about 0.3 μm. In our study, Jc of the films on MgO was measured for the samples with dimensions of 0.6-2.0 μm thickness, about 5.0 mm width, and 2 mm length. The values of Jc in the range from 10^4 to 10^5 A/cm^2 were measured through a microbridge(0.2-1.0 μm thick, 1 μm wide, and 1 μm long) made with the CVD-YBaCuO films on MgO(100) by wet chemical etching[17]. Higher Jc measured through the microbridge may be derived from the smaller number of grain boundaries contained in the measured part of the samples.

Various pinning mechanisms for high Tc superconducting oxides have already been proposed. The CVD-YBaCuO films on $SrTiO_3$(100) showed high Jc at high magnetic fields up to 27 T. The strips of nano-meter size, included in the $YBa_2Cu_3O_y$ grains with the c axis parallel to the film plane(Fig. 5), may be a candidate for a flux pinning center. The relationship between the strips and the strain contrasts observed in the c-axis oriented $YBa_2Cu_3O_y$ region is under investigation.

Luborsky et al. prepared the YBaCuO films having the thickness of 0.2-2.1 μm by annealing amorphous sputtering films on $SrTiO_3$(100)[18]. A random

$YBa_2Cu_3O_y$ layer caused by nucleation near the film surface was found on top of the oriented $YBa_2Cu_3O_y$. They showed that the values of Jc decreased from 10^5 A/cm^2 to 10^4 A/cm^2 as the film thickness increased from 0.2 to 2.1 μm. Jc of CVD-YBaCuO films were independent of the film thickness. This can probably be attributed to the formation of the crystalline $YBa_2Cu_3O_y$ from the substrate with epitaxial relation during the CVD process.

SUMMARY

YBaCuO superconducting composite films were prepared by CVD. The films contained CuO precipitates and a small amount of a-axis oriented grains in the c-axis oriented $YBa_2Cu_3O_y$ matrix. Tc and Jc of the films on $SrTiO_3$(100) were superior to those of the films on MgO(100). The films on $SrTiO_3$(100) had Tc(R=0) of 87-92 K and Jc of 10^5 A/cm^2 at 77.3 K and 0 T. Any thickness dependence of Jc was not observed up to the film thickness of 2.4 μm.

REFERENCES

1. Berry AD, Gaskill DK, Holm RT, Cukauskas EJ, Kaplan R, Henry RL (1988) Formation of high Tc superconducting films by organometallic chemical vapor deposition. Appl Phys Lett 52: 1743-1745
2. Yamane H, Kurosawa H, Hirai T (1988) Preparation of $YBa_2Cu_3O_{7-x}$ films by chemical vapor deposition. Chem Lett: 939-940
3. Ihara M, Kimura T, Yamawaki H, Ikeda K (1989) High-Tc BiSrCaCuO superconductor grown by CVD technique. IEEE Trans Magn 25: 2470-2473
4. Richeson DS, Tonge LM, Zhao J, Zhang J, Marcy HO, Marks TJ, Wessels BW, Kannewurf CR (1989) Organometallic chemical vapor deposition routes to high Tc superconducting Tl-Ba-Ca-Cu-O films. Appl Phys Lett 54: 2154-2156
5. Yamane H, Kurosawa H, Hirai T, Watanabe K, Iwasaki H, Kobayashi N, Muto Y (1989) High critical-current density of Y-Ba-Cu-O superconducting films prepared by CVD. Supercond Sci Technol 2: 115-117
6. Takahashi H, Kawasaki R, Tsuruoka T, Kanamori T (1989) Characteristics of Y-Ba-Cu-O thin films formed by OM-CVD. In: Extended Abstracts of 1989 International Superconductivity Electronics Conference (ISEC'89). 12-13 June 1989. Tokyo: 78-81
7. Watanabe K, Yamane H, Kurosawa H, Hirai T, Kobayashi N, Iwasaki H, Noto K, Muto Y (1989) Critical currents at 77.3K under magnetic fields up to 27 T for an Y-Ba-Cu-O film prepared by chemical vapor deposition. Appl Phys Lett 54: 575-577
8. Hirai T, Yamane H, Kurosawa H, Watanabe K, Kobayashi N, Iwasaki H, Muto Y (1989) Preparation of high-Jc Y-Ba-Cu-O films by CVD. In: Extended Abstracts of 1989 International Superconductivity Electronics Conference (ISEC'89). 12-13 June 1989. Tokyo: 425-428
9. Schmaderer F, Wahl G (1989) CVD of superconductive $YBa_2Cu_3O_{7-\delta}$. In: Ducarroir M, Bernard C, Vandenbulcke L (eds) Proceedings of the Seventh European Conference on Chemical Vapour Deposition. 19-23 June 1989. Perpignan: C5-119-129
10. Wadayama Y, Kudo K, Nagata A, Ikeda K, Hanada S, Izumi O (1988) Phase compatibility and superconductivity of Y-Ba-Cu-O compounds. Jpn J Appl Phys 27: L1221-1224
11. Komatsu T, Meguro H, Sato R, Tanaka O, Matusita K, Yamashita T (1988) Effect of Mg addition on superconducting properties of Ba-Y-Cu-O ceramics prepared by the melt quenching method. Jpn J Appl Phys 27: L2063-2066
12. Yan MF, Rhodes WW, Gallagher PK (1988) Dopant effects on the superconductivity of $YBa_2Cu_3O_7$ ceramics. J Appl Phys 63: 821-828
13. Cheung CT, Ruckenstein E (1989) Superconductor-substrate interactions of the Y-Ba-Cu oxide. J Mater Res 4: 1-15
14. Cava RJ, Batlogg B, Chen CH, Rietman EA, Zahurak SM, Werder D (1987) Single-phase 60-K bulk superconductor in annealed $Ba_2YCu_3O_{7-\delta}$ ($0.3<\delta<0.4$) with correlated oxygen vacancies in the Cu-O chains. Phys Rev 36: 5719-5722
15. Dimons D, Chaudhari P, Mannhart J, LeGoues FK (1988) Orientation dependence of grain-boundary critical currents in $YBa_2Cu_3O_{7-\delta}$ bicrystals. Phys Rev Lett 61: 219-222
16. Xi XX, Linker G, Meyer O, Nold E, Obst B, Ratzel F, Smithey R, Strehlau B, Weschenfelder F, Geerk J (1989) Superconducting and structural properties of YBaCuO thin films deposited by inverted cylindrical magnetron sputtering. Z Phys B 74: 13-19
17. Yamashita T, Era M, Noge S, Irie A, Yamane H, Hirai T, Kurosawa H, Matsui T (1989) Transport current in micro-bridges of $YBaCu_3O_{7-\delta}$ thin films prepared by CVD. Jpn J Appl Phys 28: in press
18. Luborsky FE, Kwasnick RF, Borst K, Garbauskas MF Hall EL, Curran MJ (1988) Reproducible sputtering and properties of Y-Ba-Cu-O films of various thicknesses. J Appl Phys 64: 6388-6391

Characterization of Superconducting Oxide Thin Films by X-ray Diffraction

T. Iwata, Y. Enomoto, S. Kubo, K. Moriwaki, and A. Yamaji
NTT Opto-electronics Laboratories, Tokai, Ibaraki, 319-11 Japan

ABSTRACT

Microstructural characteristics of $YBa_2Cu_3O_y$ thin films have an important influence on their physical properties. Structures of $YBa_2Cu_3O_y$ thin films are studied by X-ray diffractometry with a 4-axis goniometer. Semi-quantitative analysis was carried out to evaluate the ratio of (110) and (103) domains. The films in which the RHEED method had detected only the (110) domain revealed a large amount of (103) domain. The results indicate that the films grow in both (110) and (103) directions at the first stage of deposition. To characterize the structure of the films, both the RHEED method and the X-ray diffraction method are needed.

KEY WORDS: film, preferred orientation, X-ray diffraction, 4-axis goniometer.

1. INTRODUCTION

Film properties depend on their crystal structures. Oxide superconductors have a large anisotropy and the properties are different depending on their direction. A $YBa_2Cu_3O_y$ superconductor has a large anisotropy between the *a-b* axis and the *c* axis, so electric conductivity and optical properties vary with its crystal orientation. Anisotropy measurement is very important not only for application but also for understanding the superconducting mechanism of oxide superconductors. However, accurate anisotropy measurement is difficult for $YBa_2Cu_3O_y$, because large single crystals cannot be obtained, and in paticular, the *c* axis is very short. To clarify the anisotropy of this material, a large single crystal is needed. If a good single crystal film with a (110) growth direction is obtained, anisotropy can be measured, because the *c* axis of the film will be long enough. Therefore, the synthesis of single crystal films with a (110) growth direction has been studied.

It is difficult to completly characterize the structure of thin film. Electron diffraction methods such as RHEED give us information on thin film structure, and because RHEED is a good characterization method for epitaxially grown film, many films are evaluated this way. However, the data thus obtained is only from the surface of the samples. The X-ray diffraction method gives information about the internal structure of the film. Conventional X-ray diffraction gives us only information from a plane parallel to the substrate. Therefore, for film with a preferred orientation along the *c* axis, only the (00L) diffraction peaks can be obtained, which means that information from the *a* and *b* axis cannot be obtained. For film with a (110) preferred orientation on a $SrTiO_3$ (110) substrate, the situation is even worse. The peak obtained from the film is very near the substrate peak, furthermore, the obtained peak cannot be separated into (110) and (103) diffraction as shown in Fig. 1. To obtain more information on the films, X-ray diffraction with a tilted sample using a 4-axis goniometer was undertaken.[1] In this study, we characterized the structure of $YBa_2Cu_3O_y$ thin film using X-ray diffractometry with a 4-axis goniometer to obtain information from a plane not parallel to the substrate.

2.EXPERIMENTAL PROCEDURE

Sample films are deposited onto $SrTiO_3$ substrates by rf-sputtering or co-evaporation in the presence of oxygen. The substrate temperature is varied from 450 to 560 °C and rf power is varied from 1.0 to 1.8 kV for rf-sputtering. Details of the synthesis procedure have already been published elsewhere.[2] Co-evaporation is carried out in an oxygen atmosphere of about 10^{-4} Torr. The source materials are metals. During evaporation rf power is introduced to excite the oxygen atoms.

X-ray diffraction is carried out using a 4-axis goniometer with a rotating anode Cu Kα source and a single graphite monochromator. The geometry of the apparatus is as follows. The θ axis and 2θ axis are vertical which is the same as a conventional X-ray diffractometer. Samples can be tilted along the χ axis which is set perpendicularly and samples can also be rotated around ϕ axis which is in plane rotation.

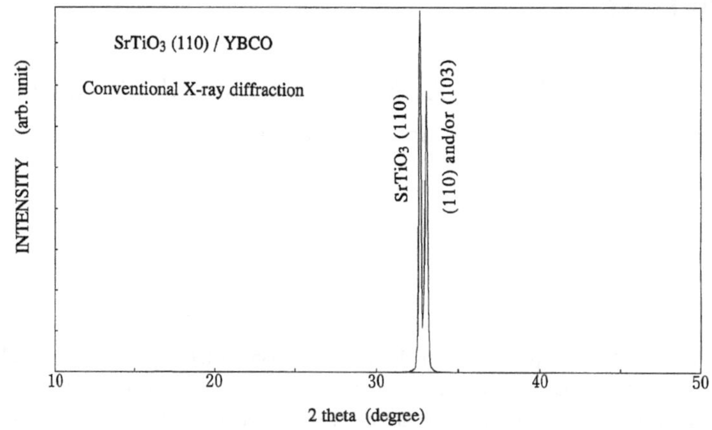

Fig. 1 X-ray diffraction pattern from a $YBa_2Cu_3O_y$ film on a $SrTiO_3$ (110) substrate. The data was taken by conventional X-ray diffractometer.

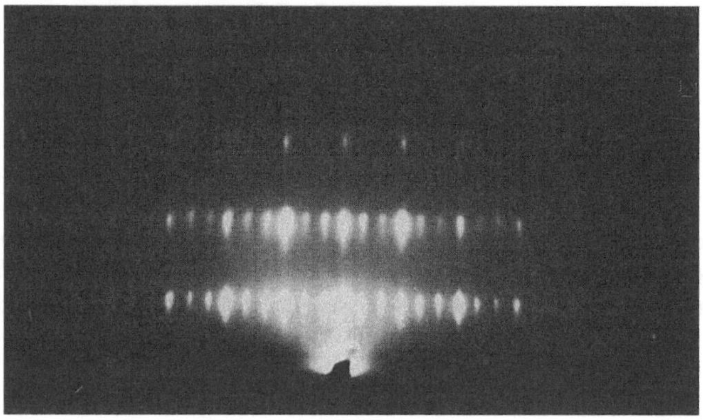

Fig. 2 RHEED pattern from a $YBa_2Cu_3O_y$ film on a $SrTiO_3$ (110) substrate.

3. EXPERIMENTAL RESULTS AND DISCUSSION

3.1 Qualitative analysis

The $YBa_2Cu_3O_y$ films deposited on $SrTiO_3$ (110) substrates have two preferred orientations, (110) and (103), because of lattice matching between the substrate and the film. To identify this preferred orientation, the (102) diffraction peak was measured because it is not masked by the diffraction from the substrate. The tilting angle can be calculated if the diffraction peak to be obtained and a plane parallel to the substrate is assumed. If a sample is tilted to 54° along the χ axis, the (102) diffraction peak from a domain with a (110) preferred orientation appears. And, if the tilting angle is set at 11°, the (102) diffraction peak from the (103) domain can be detected.

Figure 2 shows the RHEED pattern from the sample. This figure shows that the surface of the sample consists of (110) domain. Even when the sample was rotated in plane, signals from the (103) domain were not detected. This result suggests that the sample is (103) domain free.

The sample was measured by X-ray diffractometry with a 4-axis goniometer. Figure 3(a) shows X-ray intensity data, where $\theta/2\theta$ was set to the (102) peak position and χ was set at 54° and the sample was rotated in plane (ϕ) during data sampling. This data shows that the (110) domain really exists. However, the (103) domain was also detected as shown in Fig. 3(b), where χ is set at 11°. These results show that the sample contains the (103) domain, even if no evidence of it is detected at the surface. This may indicate that the (103) domain is easy to grow at the first stage of growth and that the (110) domain grows subsequently.

3.2 Semi-quantitative analysis

To clarify the growth conditions of the (110) domain, many films were made under various conditions. The substrate temperature and applied rf voltage were varied. The films were measured by the above method. The obtained integrated intensity data was corrected with

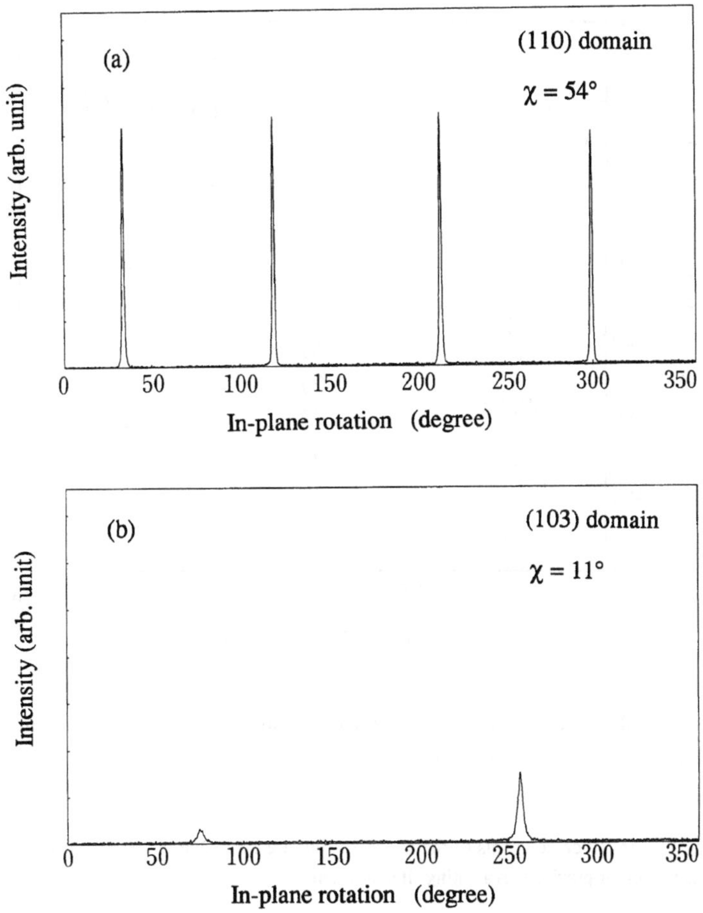

Fig. 3 X-ray intensity data from a $YBa_2Cu_3O_y$ film on a $SrTiO_3$ (110) substrate, where θ/2θ was set to the (102) peak position and χ was set at 54° (a) and 11° (b), respectively. The sample was rotated in plane (ϕ) during data sampling.

regard to their χ dependence by deduction from their difference from ideal geometry. Data obtained from Si powder was used as standard data.

Figure 5 shows the ratio of the corrected intensity of the (103) domain in the film i.e. $I_{(103)} / I_{(103)} + I_{(110)}$. The numbers in this figure indicate the ratio. The ratio decreases if either the substrate temperature or the applied rf voltage decreases. These results show that the (110) domain is easy to grow in both the low substrate temperature region and the low power region. The (110) grown film without the (103) domain has not yet been obtained. Other conditions must be optimized in order to obtain (103) free films.

Even if the films contain two preferred orientation domains, the growth directions of the films are well correlated with the direction of the substrates. These results indicate that two preferred orientation domains do not grow randomly but in a way that is similar to epitaxy.

These results are different from those for *a* axis grown film and *c* axis grown film in $EuBa_2Cu_3O_y$ on a $SrTiO_3$ (100) substrate, where the samples can easily grow with a single domain.[3] The *c* axis grown films are obtained at high substrate temperatures, and the *a* axis grown films are obtained at low substrate temperatures. The result is confirmed by the above method. From the comparison of films on $SrTiO_3$ (110) and (100) substrates, the domains where the *c* axis is parallel to the substrate surface grow easily at low substrate temperatures. We believe that it is harder to grow the (110) domain on a $SrTiO_3$ (110) substrate than the *a* axis domain on a $SrTiO_3$ (100) substrate.

3.3 Lattice constant measurement

The lattice constants of a *c* axis oriented $YBa_2Cu_3O_y$ film made by co-evaporation were measured. The goniometer angle was set to detect the (203) and (023) diffraction peaks. The as-deposited film showed only one peak with a broad peak width. The film was annealed at 900 °C and held at 900 °C for 1 hour and then cooled to 520 °C and held at that temperature for 5 hours and finally slowly cooled to room temperature. The annealed sample showed two peaks under the previously mentioned measuring conditions. This means that the

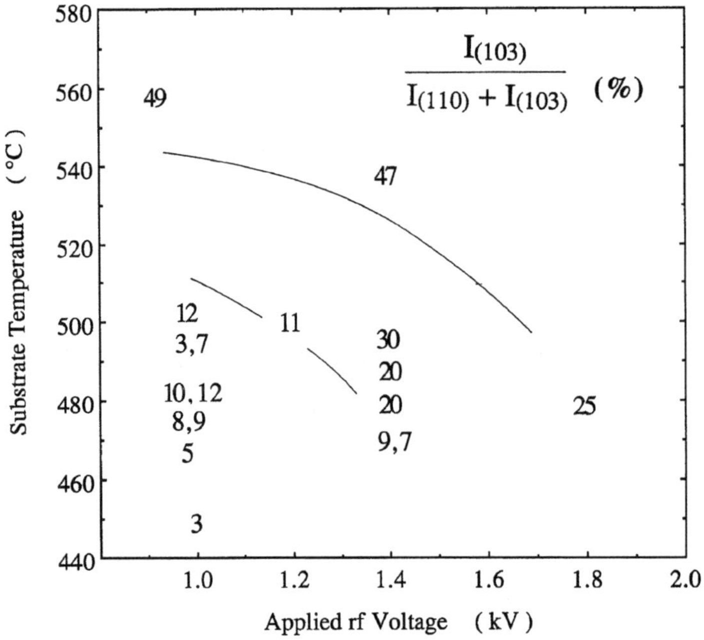

Fig. 4 The ratio of the corrected intensity of the (103) domain in the film i.e. $I_{(103)} / I_{(103)} + I_{(110)}$. The numbers in this figure indicate the ratio.

(203) and (023) peaks separated. The lattice constants of the annealed film are estimated as follows: *a*=3.89, *b*=3.85, *c*=11.70Å. This result shows that the method is useful for precisely evaluating film structure.

4. CONCLUSIONS

The structure of $YBa_2Cu_3O_y$ film is studied by X-ray diffraction using a 4-axis goniometer. The films deposited onto a $SrTiO_3$ (110) substrate have a preferred orientation of (110) at low substrate temperatures and (103) at high substrate temperatures. Even if the sample is characterized as having a good epitaxial like surface by RHEED, a domain grown in another orientation can be seen in the film. Therefore, precise characterization is required for the purpose of anisotropy studies. Lattice constants are evaluated independently by this method. These results indicate that total characterization is needed to precisely evaluate the structure of a film. To this end not only RHEED, but also X-ray diffraction should be employed.

AKNOWLEGEMENTS

We thank H. Asano for his $EuBa_2Cu_3O_y$ films and useful discussions.

REFERENCES

[1]. Sizemore J, Barton R, Marshall A, Bravman J.C, Naito M, Char K (1988) Microstructural characterization of $YBa_2Cu_3O_{7-d}$ thin films on SrTiO3 using four-axis X-ray diffraction. Proceedings of the applied superconductivity conference, Aug. 22, 1988

[2]. Enomoto Y, Murakami T, Suzuki M, Moriwaki K (1987) Largely anisotropic superconducting critical current in epitaxially grown $Ba_2YCu_3O_{7-y}$ thin film. Jpn. J. Appl. Phys. **26** : L1248-L1250

[3]. Asano H, Asahi M, Michikami O (1989) Epitaxy and orientation of $Eu_1Ba_2Cu_3O_{7-y}$ films grown *in situ* by magnetron sputtering. Jpn. J. Appl. Phys. **28** : L981-L983

Preparation of Thin-Film Oxide Superconductor, $Y_1Ba_2Cu_3O_{7-\delta}$ by Facing Target Sputtering Deposition Technique

YASUO TAKAGI, HIDEO YAMADA, and NOBORU KOYAMA

R&D Laboratories-I, Nippon Steel Corporation, 1618 Ida, Nakahara-ku, Kawasaki, 211 Japan

ABSTRACT

Thin-film oxide superconductor, $Y_1Ba_2Cu_3O_{7-\delta}$ was prepared onto single crystal MgO (100) substrates by facing target sputtering(FTS) deposition technique. Throughout the various preparation conditions little deviation of the compositions of the metallic components in the as-prepared films from those of the target: i.e., Y:Ba:Cu=1:2:3 were observed. The films prepared under the conditions, $T_s > 650$ °C and $P_{O_2} > \sim 1 \times 10^{-2}$ Torr have c-axis oriented $Y_1Ba_2Cu_3O_{7-\delta}$ structure, having Tc > 65 K. Under other preparation conditions polycrystalline phase(s) having ca. 1/3 periodicities of c-axis of $Y_1Ba_2Cu_3O_{7-\delta}$ were mixed in the $Y_1Ba_2Cu_3O_{7-\delta}$ phase. The "1/3 periodicity" phase could not be transformed into the $Y_1Ba_2Cu_3O_{7-\delta}$ phase by low temperature annealing at around 400 to 600 °C.

KEY WORDS: facing target sputtering deposition, plasma confinement, composition deviation, 1/3 periodicity, low temperature annealing

INTRODUCTION

Various methods have been applied to prepare thin-film oxide superconductors[1]. Several sputtering deposition methods like magnetron sputtering deposition were widely used. There are, however, several common problems in the sputtering deposition methods. In the sputtering deposition process a plasma is used to sputter targets. In the process, the plasma components not only sputter the target(s) but also reach the substrate(s) and re-sputter or displace the film components selectively. These phenomena cause composition deviations of the film components from those of the target, and also leave various crystalline defects in the films, resulting in amorphization in extreme cases.

In preparing oxide thin-films, in particular, the negative particles: ie.. O and γ-electron sputtered or emitted from the targets, respectively, are accelerated by the sheath potential applied between the targets and the plasma, promoting further damages in the depositing films.

To improve such problems, a separation of the sputtering plasma from the substrate is indispensable. Facing target sputtering(FTS) deposition system which was originally developed by Hoshi et al. is one of the most promising one since the plasma can be confined almost completely in the space between the facing targets[2]. In this study, thin-film oxide superconductor, $Y_1Ba_2Cu_3O_{7-\delta}$ was prepared by FTS under several conditions of substrate temperatures(T_s's) and oxygen partial pressures(P_{O_2}'s).

EXPERIMENTAL

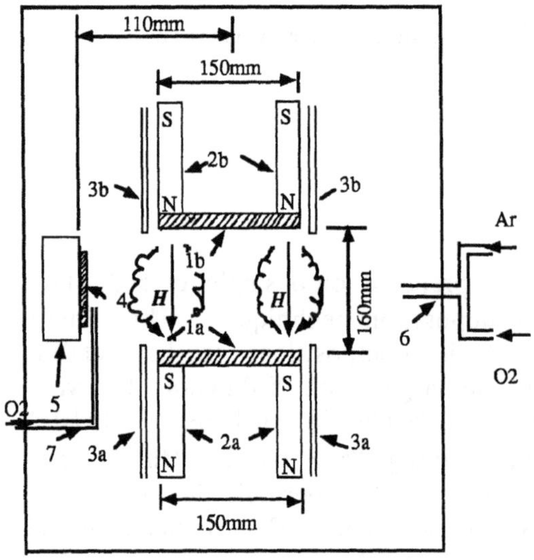

Fig.1 Schematic drawing of the facing target sputtering(FTS) deposition system.
1: sputtering targets, 2: plasma confinement magnets, 3: shield cylinders, 4: a substrate, 5: a substrate heater, 6: a premixed gas line for sputtering, 7: a separate gas line for film oxidation

The FTS system used in the present study is shown in Fig. 1.

In this system two ceramic targets having the same stoichiometric composition, $Y_1Ba_2Cu_3O_{7-\delta}$ are facing to each other.
Two ring-shaped magnets having the opposite poles are mounted behind the targets to generate the magnetic confinement field perpendicular to the target surfaces. The electrons emitted from the targets repeat Larmor gyration along the magnetic field, not only confine the plasma components between the targets but increase the collision frequency between the neutral atoms and electron resulting in a higher ionization efficiency.
Argon gas was used for sputtering; and oxygen, for oxidation by a premixed and/or a separate feed-line toward the substrate. RF(13.65 MHz, 250 W) was applied to the targets.
The deposition conditions, structures, and properties of the as-prepared films were summarized in Table I. The atomic structures of the films were determined by $\theta-2\theta$ x-ray diffraction using Cu Kα radiation. The temperature dependences of the electric resistivity were measured by a conventional DC four-probe method.

Table I. The deposition conditions, structures, and properties of the as-prepared films. A: Film Number, B: Substrate Temperature (°C), C: Sputtering Pressure (Torr), D: O_2 Partial Pressure (Torr), E: Atomic Composition Ratios of (Ba/Y, Cu/Y) by EDS , F: T_{c0} (K), G: $T_{c,\ on\text{-}set}$ (K), H: Crystal Structure: c-(1,2,3) means c-axis oriented (1,2,3) phase, (1,2,3), polycrystalline (1,2,3), and 1/3, the "1/3 periodicity" phase.

A	B	C	D	E	F	G	H
1	700	$5.0x10^{-2}$	$1.6x10^{-2}$	2.0, 3.0	72	96	c-(1,2,3)
2	700	$5.0x10^{-2}$	$1.7x10^{-2}$	-------	66	85	c-(1,2,3)
3	650	$1.2x10^{-2}$	$6.5x10^{-3}$	2.1, 3.1	30	80	(1,2,3),1/3
4	600	$4.6x10^{-3}$	$1.7x10^{-3}$	2.0, 3.0	31	75	(1,2,3),1/3
5	600	$5.0x10^{-2}$	$1.7x10^{-2}$	-------	**	**	(1,2,3),1/3
6	400	$4.6x10^{-3}$	$1.7x10^{-3}$	2.0, 3.1	**	**	1/3

RESULTS AND DISCUSSION

Although the atomic structures and electric properties of the films largely depended on T_s and P_{O2}, the compositions of the metallic components in the as-prepared films were almost identical with those of the target within the accuracy of 2 at. % by EDS analyses over the ranges of the preparation conditions. As the result, no traces of the second phase(s) were observed in the diffraction patterns.

Group I: (1, 2, 3) Structure(Films #1 & #2)

Figure 2 shows the x-ray diffraction pattern from Film #1.

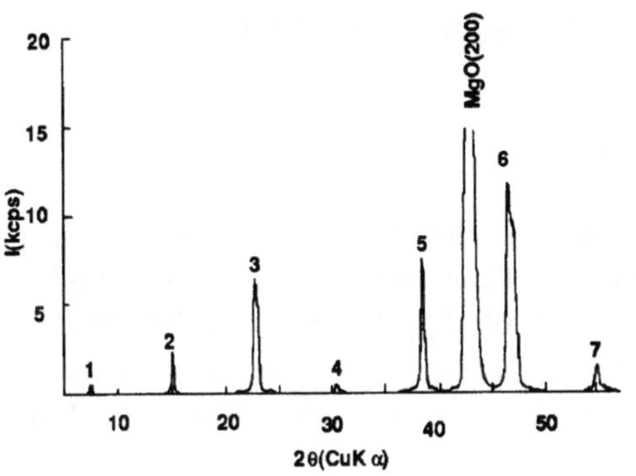

Fig. 2 The X-ray diffraction pattern from the as-prepared film(#1) in Group I. The figures on the peaks show the orders of reflection, i.e. (0, 0, n) of the $Y_1Ba_2Cu_3O_{7-\delta}$ phase.

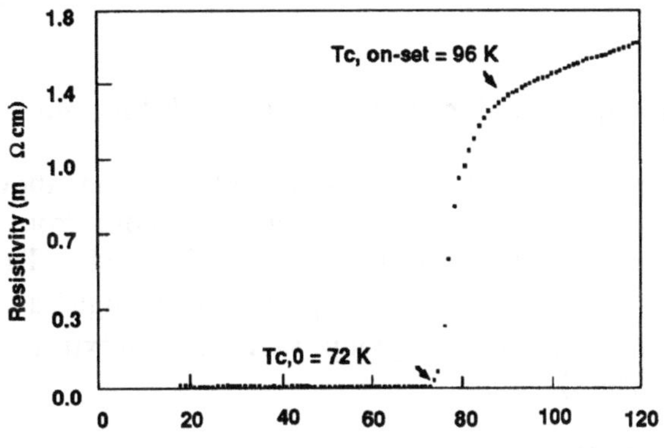

Fig. 3 Temperature dependence of electric resistivity of Film #1 in Group I.

The film consisted of c-axis oriented $Y_1Ba_2Cu_3O_{7-\delta}$ structure,[1] having slightly larger c-axis length(= ca. 11.7 A)[2] than that of the bulk material. The temperature dependence of the electric resistivity of the film is shown in Fig. 3. The value of T_{c0} was about 72 K, and $T_{c,\ on\text{-}set}$, ca. 96 K. These findings suggest the films in the group consisted of a mixture of the ortho-I and -II phases.

---

[1]Here and below the structure will be refereed as (1, 2, 3) phase.
[2]The length of the c-axis was calculated by the higher order limits of the (0, 0, n) reflections.

Group II: "1/3 Periodicity' Phase(Films #5 & #6)

When T_s < 550 °C and $P_{O2,}$ < ~$5x10^{-3}$ Torr, one or two series of reflections having periodicities of about 1/3 of the c-axis of (1, 2, 3) phase were observed in the diffraction patterns(Fig. 4)(see also Table I). Similar phenomenon had been found by Michikami et al.[3]. The films were insulators, and could not be transformed into superconductors by annealing at around 400 ~ 600 °C under O_2 flows of 1 atm.

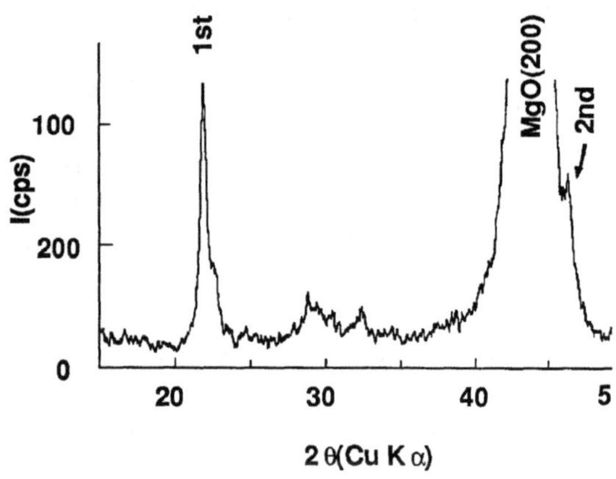

Fig. 4 The X-ray diffraction pattern from the as-prepared films(#6) in Group II.

Group III: Transition Region(Films #3 & #4)

There is a transition region between the above two. In the diffraction patterns of the films in this group the reflections both from polycrystalline the (1,2,3) phase and the "1/3 periodicity" phase(s) were observed. The values of T_{c0} were much lower(ca. 30 K) although $T_{c,\ on\text{-}set}$ kept relatively high(ca. 80 K), resulting in broad transitions. The films in the group will be a mixture of the (1,2,3) and the "1/3 periodicity" phase(s).

CONCLUSIONS

It was revealed that the films prepared by FTS have the following characteristics compared with other sputtering techniques.

(1) The films having the almost same compositions of metallic components as those of the target can be prepared over the relatively wide ranges of preparation conditions.

(2) The deposited films crystallized without post-annealing as low as 400 °C. However, T_s < 550 °C the resulting crystal structure of the films was not superconducting (1, 2, 3) phase but an "1/3 periodicity" insulator phase(s) which have the same metallic composition as that of (1, 2, 3) phase.

REFERENCES

1. Broussard P.R., and Wolf S.A.(1988): *Film growth of high transition temperature superconductors,* J. Cryst. Growth, **91,** 340.

2. Hoshi Y., Naoe M. and Yamanaka S. (1977): *High rate deposition of iron films by sputtering from two facing targets.* Jpn. J. Appl. Phys., **16,** 1715.

3. Michikami O., Asahi M. and Asano H. (1989): *Dependence of superconducting properties on substrate temperature in Y-Ba-Cu-O thin films prepared by magnetron sputtering,* Jpn. J. Appl. Phys., **28,** L448.

Polarized Plasma Annealing of Y-Ba-Cu-O Thin Films

HARUO SHIMADA, MUNEYUKI IMAFUKU, WATARU ITO, SATOSHI ITO,
and SHOICHI MATSUDA
R&D Laboratories-I, Nippon Steel Corporation, 1618 Ida, Nakaharaku, Kawasaki, 211 Japan

ABSTRACT

Superconducting Y-Ba-Cu-O films were prepared by a new annealing method, the *polarized plasma annealing*, of $Ba/Y_2O_3/Cu$ multilayered films formed by sequential electron beam evaporation on MgO(100) substrates.The essential point of this method is as follows. The film was set in the positive column of the O_2 plasma(frequency:50 Hz) during the annealing. The electric potential of the film was controlled independently of the plasma by using a large counter electrode. The effects of electric potential on the oxidation and phase formation of these films during the annealing were studied. It was found that the positive potential is effective for the oxidation and the negative one promotes the c-axis orientation perpendicular to the plane of the film. The combination of negative and positive potentials can successfully form the superconducting phase having Tc(zero) above 77K.

KEYWORDS: polarized plasma annealing, Y-Ba-Cu-O, superconducting film

INTRODUCTION

Some characteristics of Y-Ba-Cu-O oxide superconductor are sensitively affected by the process of high temperature annealing as well as by the atomic stoichiometory. The oxygen content and crystal structure have been controlled by the high temperature annealing under oxygen atmosphere. Some people tried to control the oxygen content of Y-Ba-Cu-O superconductors by plasma oxidation [1,2,3]. In these cases, they were only interested in the control of the oxidation state and didn't try to apply it to the annealing process. Moreover, from a viewpoint of plasma technology, no field-assisted motions of charges were considered by them. On the other hand, there have been many studies on the plasma anodization of semiconductors [4,5]. As the name of "anodization" indicates, the sample itself forms the anode of the apparatus. Higher rates and larger thicknesses can be obtained by this method. In this paper we describe the application of plasma technique, especially the polarized plasma, during the annealing for the preparation of Y-Ba-Cu-O superconductors.

EXPERIMENTAL

We prepared $Ba/Y_2O_3/Cu$ multilayered films onto MgO(100) by sequential electron beam evaporation. The periodicity and the total thickness of the film were 0.1μm and 2μm, respectively. In advance, Ag was deposited on the substrate as to be the electrode of the film. This buffer-layer had an optical flat surface and a thickness of 0.5μm. The composition of the film was determined with EDX analyzer. In this study, the average atomic ratio of the multilayered film was proven to be

Y:Ba:Cu=0.91:2.00:3.01. As we know from the vapor sources, the oxygen content of the film was obviously much less than that of the stoichiometoric Y-Ba-Cu-O.

The equipment used in this study is shown in Fig.1.

Quartz tube chamber was evacuated and then filled back with He and O_2 mixed (1:1) gas to 10 Torr. During the annealing, O_2+He plasma had been generated between the two Pt electrodes by 50 Hz AC oscillator as a power of 4 watts. The sample was installed in the center of the positive column of the plasma. DC bias potential of the sample was controlled independently of the plasma potential by using the Pt counter electrode (CE). Therefore the plasma was "*polarized*" between the sample and CE. So we call it polarized plasma.

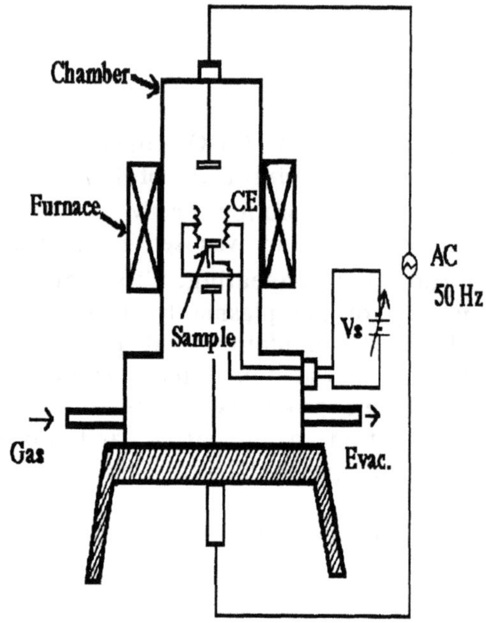

Fig.1. Schematic drawing of the equipment for the polarized plasma annealing.

The gross area of CE was 500 times larger than the sample exposed to the plasma. Therefore, the electric field was effectively imposed near the surface of the sample. We measured it by using double probes, but don't describe the details here. We also monitored the current of the sample during the annealing. It consists of electronic one and ionic one.

Infrared image furnace was used to heat up and cool down the films. The films were annealed at 840°C for five minutes with heating rate of 7°C/min and cooling rate of 1.75°C/min. Without plasma, we couldn't obtain superconducting phase after the treatment of this annealing pattern. In case of the post-annealing treatment of our films in a pure oxygen atmosphere (1 atm), we had to heat up at least to 920°C and hold it at least for 1 hour to obtain superconducting phase. Therefore, the highest temperature and its holding time were relatively lower and shorter than those of the annealing in 1 atm oxygen atmosphere, respectively.

RESULTS AND DISCUSSION

We studied the effects of bias potential on the resistive transition and crystal orientation of Y-Ba-Cu-O films. The bias potential, Vs, was varied from -150V to +300V. We could obtain superconductors in this bias potential range. The current of positive biased sample was more than ten times larger than that of negative biased one due to the large electronic and ionic currents. When the negative potential exceeded -150V, there occurred abnormal large sample current at high temperature region and the film was melted. The film was partially melted even at Vs=-150V. So the limit of negative potential was smaller than -150V in this study. The origin of this abnormal current is not clearly understood by now. As for the positive potential, such a phenomenon was not observed.

Figure 2 shows the Vs dependence of resistive transitions Tc(onset) and Tc(zero). The results of partially melted sample at Vs=-150V were inserted in this figure for reference.

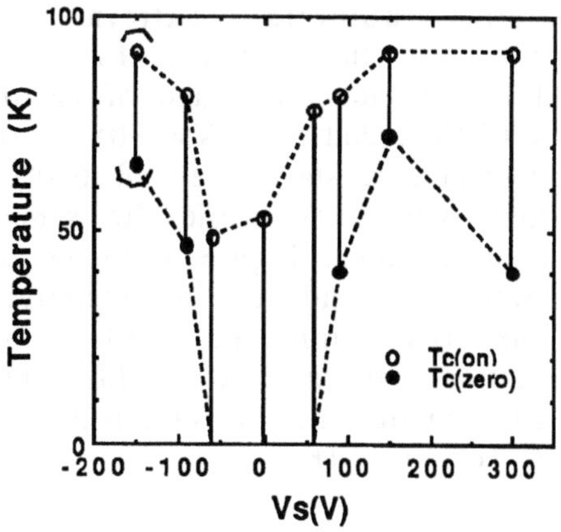

Fig. 2. Vs dependence of the onset and zero transition temperatures. The data at Vs=-150V are inserted with ( ) for reference.

The sample annealed without bias potential (Vs=0) showed the superconducting transition at 52K(Tc(onset)). The situation of zero-bias annealing corresponds to the low pressure plasma oxidation annealing. we can see the effectiveness of plasma by comparing with the result of annealing without plasma.

Tc(onset) gradually increased and Tc(zero) appeared with the increase of Vs. At Vs=150V, Tc(onset) and Tc(zero) were 91K and 72K, respectively. It seems that the oxygen content of the film was increased with the positive bias. we think that the effect of positive bias potential is similar to the mechanism of plasma anodization. Tc(zero) decreased at Vs=300V. In this case, the bias-field was so large that the film was damaged and that the fluctuation of oxygen content or week links tended to be generated.

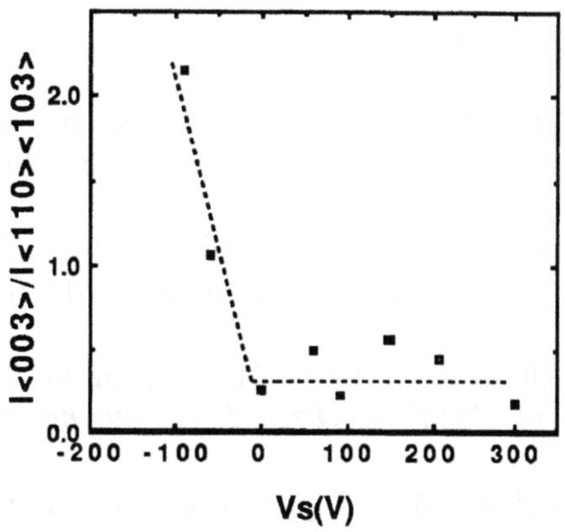

Fig. 3. Vs dependence of the X-ray peak intensity ratio, I<003>/I<110><103>.

The resistive transition was also improved by the negative bias potential. Positive ion of oxygen in the plasma probably plays an important role under this condition. We couldn't utilize, however, the negative one sufficiently because of the abnormal current mentioned above. Broad transition still remained in the negative region of Vs.

We investigated the crystallization behaviors of the films by X-ray diffraction (XRD) measurement. XRD patterns indicated that the films were all in polycrystalline state. The dominant orientations perpendicular to the surface were <001> and <110><103> (We couldn't separate the reflection peak of <110> from that of <103> because the d-values of them are almost the same). We found that the crystal orientation of the film was considerably affected by the sign of Vs. Figure 3 explains the tendency of preferred orientation by the parameter of XRD peak intensity ratio as I<003>/I<110><103>. We can see from this figure that the XRD peak intensity of <003> becomes significantly larger than that of <110><103> at the negative bias potential region. On the other hand, the value of I<003>/I<110><103> at positive bias potential was not so much different from the result at Vs=0. The negative bias potential promotes the c-axis preferred crystal growth.

We examined the combination of positive and negative bias potentials during the annealing. The heat pattern was the same as the previous one. The new bias potential pattern was composed of three steps. 1st step: Vs=+150V at the early stage of heating (below 650°C). 2nd step: Vs=-90V at the high temperature stage (beyond 650°C). 3rd step: Vs=+150V while the sample was cooling down. Figure 4 shows the temperature dependence of resistivity for the film annealed under this combination pattern. We could obtain the superconducting Y-Ba-Cu-O film having Tc(zero)=81K.

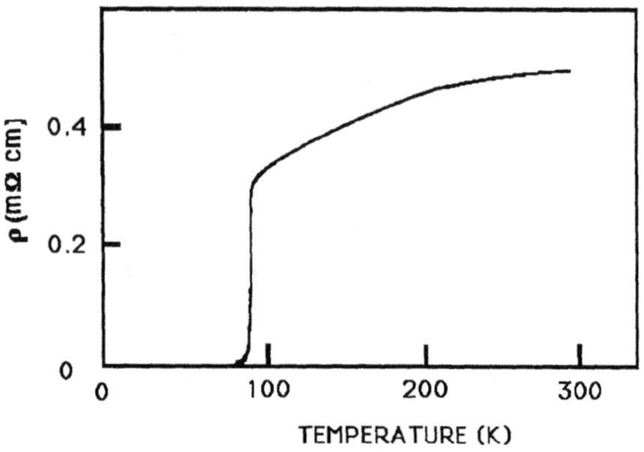

Fig.4. ρ vs. T for the film annealed under bias combination pattern.

In conclusion, we found that the positive bias potential is of advantage to oxidation and that the negative one promotes the preferred-orientation of the Y-Ba-Cu-O films. We established the polarized plasma annealing technique for the preparation of this system.

REFERENCES

1. Bagley BG, Greene LH, Tarascon JM, Hull W (1987) *Plasma oxidation of the high Tc superconducting perovskites.* Appl. Phys. Lett. **51**: 622-624.

2. Tamura H, Yoshida A, Morohashi S, Hasuo S (1988) *Plasma oxidation of* $Ba_2YCu_3O_{7-y}$ *thin films.* Appl. Phys. Lett. **53**: 618-620.

3. Yoshida A, Tamura H, Morohashi S, Hasuo S (1988) *Oxygen diffusion into oxygen-deficient* $Ba_2YCu_3O_{7-y}$ *films during plasma oxidation.* Appl. Phys. Lett. **53**: 811-813.

4. Copeland MA, Rappu R (1971) *Comparative study of plasma anodization of silicon in a column of a dc glow discharge.* Appl. Phys. Lett. **19**: 199-201.

5. Sugano T, Mori Y (1974) *Oxidation of* $GaAs_{1-x}P_x$ *surface by oxygen plasma and properties of oxide film.* J. Electrochem. Soc. **121**: 113-118.

Composition and Deposition Temperature Dependences on T_c for the RF Sputtered Y-Ba-Cu-O Films

N. AKUTSU[1], M. FUKUTOMI[2], K. KATOH[1], H. TAKAHARA[1], Y. TANAKA[2], T. ASANO[2], and H. MAEDA[2]

[1] Corporate R&D Center, Mitsui Mining & Smelting Co., Ltd., Ageo, Saitama, 362 Japan
[2] National Research Institute for Metals, 2-1, Sengen 1-chome, Tsukuba, Ibaraki, 305 Japan

ABSTRACT

High Tc superconducting Y-Ba-Cu-O films were prepared by rf magnetron sputtering with three sintered targets; Y_2O_3, $BaCuO_2$, and CuO. Resputtering by high kinetic energy negative ions makes serious effect on compositional deviation of films. To avoid the effect of negative ions during deposition, MgO (100) substrates were placed outside of the region above the three targets. The films, deposited at 700°C in the mixture of argon and oxygen ($Ar:O_2=4:1$) of 0.28Pa and in-situ post annealed in oxygen atmosphere at the deposition temperature, showed zero resistance at 84K. The c-axis of $Y_1Ba_2Cu_3O_{7-x}$ was well oriented perpendicular to the film plane. The Tc has a good correlation with full width at half maximum (w) of the (007) x-ray diffraction peak.

KEY WORDS: high Tc, sputter, negative ions, resputter, compositional deviation

INTRODUCTION

Sputter deposition is one of the most common thin-film fabrication methods. Various sputtering techniques have been employed to synthesize the high Tc superconducting $Y_1Ba_2Cu_3O_{7-x}$ films. One common technique is to sputter from a sintered single target of $Y_1Ba_2Cu_3O_{7-x}$. In this material, however, the composition of the film which is deposited by using an oxide target had a different composition from that of target, and the deposition rate is very small.[1-2] This is due to the preferential resputtering by negative ions accelerated away from the oxide target. The negative ion effect on sputtering is well documented in the literature.[3] The negative ions have sufficient kinetic energy to resputter deposited films, so that it is difficult to synthesize stoichiometric $Y_1Ba_2Cu_3O_{7-x}$ films. One good way to synthesize stoichiometric $Y_1Ba_2Cu_3O_{7-x}$ films is to place the substrate in the space where the target plane does not face. The grain boundaries and crystal defects may create harmful effect on superconducting properties. The oxygen content of $Y_1Ba_2Cu_3O_{7-x}$ has a large effect on Tc and the c-axis lattice constant.[4] In this study, we fabricated many Y-Ba-Cu-O films with different rf power and deposition temperature by using rf magnetron sputtering method with three targets. The films were deposited on MgO (100) placed in the space where negative ions did not bombard. We measured the X-ray diffraction peak position and the full width at half maximum for the $Y_1Ba_2Cu_3O_{7-x}$ (001) peak. The peak position varies with the c-axis lattice constant. The peak width changes with grain size, defects and distortion. In order to clarify the effect of crystallinity on Tc, we studied the relation between Tc and the peak width of the (001) reflection, and the relation between Tc and peak position.

EXPERIMENTAL

Y-Ba-Cu-O films were prepared on MgO (100) by rf magnetron sputtering with three sintered targets 10cm in diameter Y_2O_3, $BaCuO_2$ and CuO. A schematic diagram of the sputtering system is shown in Fig. 1. The distance from the center of the substrate holder to the center of each target was about 12cm. The rotation speed of the substrate holder was 3rpm. The base pressure of the chamber was 2×10^{-4}Pa. The sputtering gas was argon and oxygen mixture ($Ar:O_2=4:1$). The sputtering gas pressure was 0.28Pa. The films were deposited on MgO (100) at the temperature range from 630°C to 740°C. After deposition, the main valve of the sputtering chamber was closed and the films were annealed introducing oxygen at a rate of 5sccm at deposition temperature for two hours, and then the films were cooled down in oxygen atmosphere to room temperature at a cooling rate of 2°C/min. The volume of the chamber was $0.077m^3$. The composition of the films was determined by inductively coupled plasma emission spectroscopy (ICP). The deposition rate of the film prepared was 2500Å/hour with following conditions: 0.28Pa, 700°C, Y_2O_3(200W), $BaCuO_2$(126W), CuO(64W). Deposition time was 2 hours. The crystal structure and full width at half maximum of the (001) peak were determined from x-ray diffraction using CuKα radiation. The electrical resistance measurements were done by using a four probe method.

RESULTS AND DISCUSSION

Generally, many x-ray diffraction peaks {(00l), l=1, 2, 3, ...} by the c-plane of $Y_1Ba_2Cu_3O_{7-x}$ were detected in well c-axis oriented film. It is, however, difficult to use the (00l) peak except the (007) peak in order to analyze data of many different films, because the (00l) peak {l≠7} often had weak intensity or overlapped with other different (hkl) peaks of $Y_1Ba_2Cu_3O_{7-x}$ and/or different phase. Therefore, in this report, the (007) peak was mainly used for data analyses. Figure 2 shows the deposition temperature dependence of Tc and the full width at half maximum of the (003) peak and (007) peak. Generally, the line shape of x-ray diffraction peaks varies with x-ray diffractmetry system, so that it is necessary to measure a line shape of a standard sample for calibration. α-SiO_2 powder (10-30μm) was used as a standard sample. The α-SiO_2 diffraction peak which is appearing near the (007) peak of $Y_1Ba_2Cu_3O_{7-x}$ was split due to the wave length difference between CuKα1 and CuKα2. The split makes difficulty to determine the full width at half maximum, but the α-SiO_2 (100) peak which is appearing near the (003) peak of $Y_1Ba_2Cu_3O_{7-x}$ was not split. The full width at half maximum of the α-SiO_2 (100) peak was 0.15 degrees. Therefore, in this figure not only the (007) peak but the (003) peak was used. The triangle symboles represent the nearly stoichiometric films which were prepared with following rf power conditions: Y_2O_3(200W), $BaCuO_2$(126W), CuO(64W). The square symbols represent the composition deviated films (5at.% Cu less than stoichiometric composition) which were prepared with following rf power conditions: Y_2O_3(200W), $BaCuO_2$(126W), CuO(64W). The Tc of nearly stoichiometric films which is prepared for a wide temperature range (630-710°C) was higher than 80K, but the Tc of the films which were poor in Cu was not higher than 80K. These high Tc films were obtained without high temperature post annealing. The films prepared at the temperature higher than 730°C included unknown phase. This phase might be $Cu_2Y_2O_5$ or $Ba_2Y_2O_5$. Figure 3 shows the x-ray diffraction pattern of films prepared at 700°C {Y_2O_3(200W), $BaCuO_2$(126W), CuO(64W)}. This was a well c-axis oriented film to

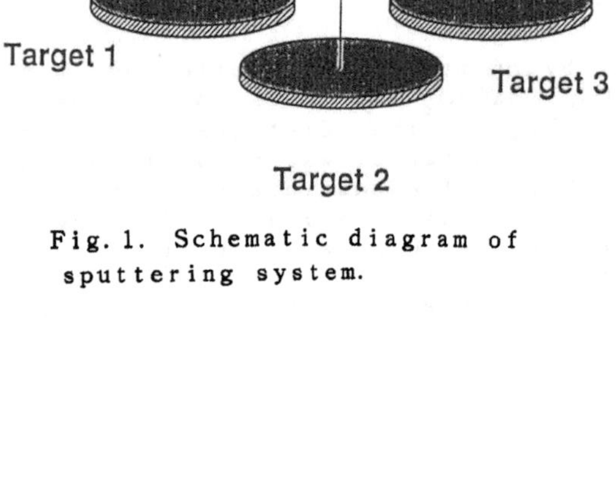

Fig. 1. Schematic diagram of sputtering system.

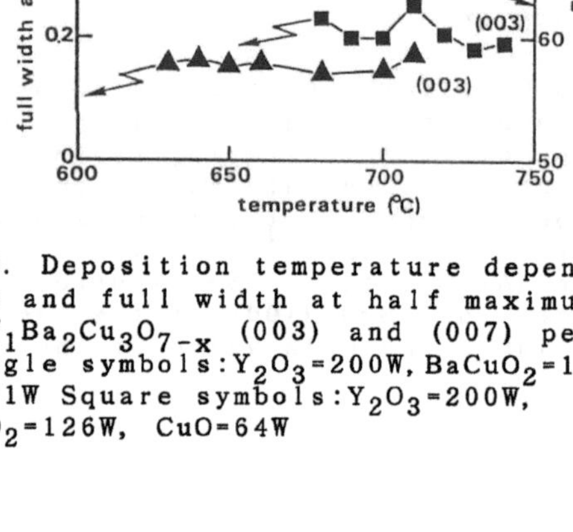

Fig. 2. Deposition temperature dependence of Tc and full width at half maximum for the $Y_1Ba_2Cu_3O_{7-x}$ (003) and (007) peaks. Triangle symbols: Y_2O_3=200W, $BaCuO_2$=126W, CuO=71W Square symbols: Y_2O_3=200W, $BaCuO_2$=126W, CuO=64W

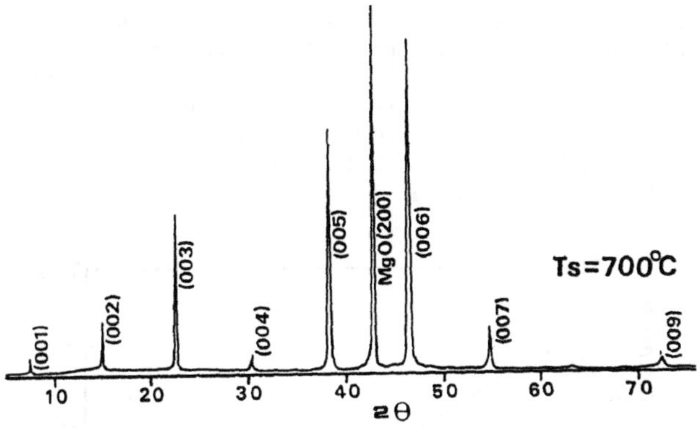

Fig. 3. X-ray diffraction pattern of Y-Ba-Cu-O film.
Deposition temperature=700°C, Y_2O_3=200W $BaCuO_2$=126W, CuO=71W

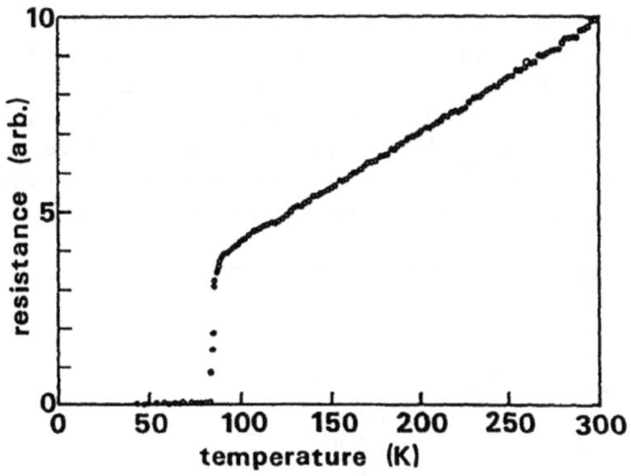

Fig. 4. Temperature dependence of resistance for the Y-Ba-Cu-O film.

the film plane, and only (00l) peaks (l=1, 2, 3, ...) were detected. The Tc of this film was about 84K (Fig. 4). The full width at half maximum of the (003) and (007) peaks had a good correlation with Tc, the width was small when Tc is high, and large when Tc is low. But, films including other phases ($Cu_2Y_2O_5$ and/or $Ba_2Y_2O_5$) had a different tendency. Figure 5 shows the dependence of the full width at half maximum and Tc on the rf power of the CuO target. The rf power of the other targets was fixed. Figure 6 shows the dependence of the full width at half maximum and Tc on the rf power of the Y_2O_3 target. From Fig. 5 and Fig. 6, it is clear that there is the same relation as that shown in Fig. 2 between the Tc and the half value width. The Tc and the full width at half maximum are very sensitive in Cu composition ratio. The film's composition and deposition temperature have an influence on Tc and full width at half maximum. Figure 7 shows the relation between Tc and full width at half maximum. The data was obtained from Fig. 2 , 5 , 6, and others. In Fig. 7, the suffix n represent the not-oriented $Y_1Ba_2Cu_3O_{7-x}$ film. The suffix u represent the composite film consist of $Y_1Ba_2Cu_3O_{7-x}$ and unknown phases. The Tc of the well c-axis oriented films has a good correlation with the full width at half maximum (w) of the (007) peak. This relation is $Tc=89.7\pm1.0-(23.2\pm2.0)\times w$.

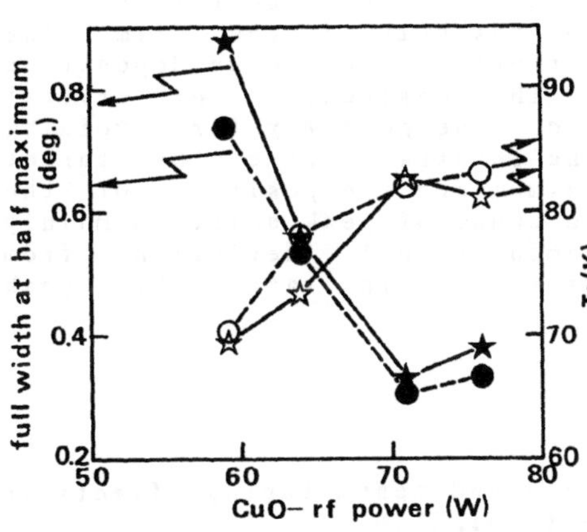

Fig. 5. CuO target rf power dependence of full width at half maximum for the (007) peak and Tc.
Deposition temperature=700°C.
Circle symbols:Y_2O_3=200W, $BaCuO_2$=126W, CuO=xW Star symbols:Y_2O_3=180W, $BaCuO_2$=126W, CuO=xW

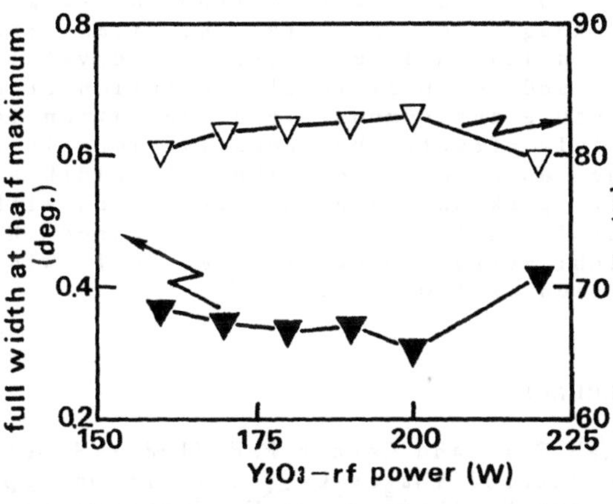

Fig. 6. Y_2O_3 target rf power dependence of full width at half maximum for the (007) peak and Tc.
Deposition temperature=700°C.
Y_2O_3=xW, $BaCuO_2$=126W, CuO=71W

Figure 8 shows a relation between the peak position and Tc and full width at half maximum for the films deposited with various CuO target's power at 700°C and 710°C. The Tc becomes high with increasing peak position, and the full width at half maximum become low. The lattice constant of c-axis for $Y_1Ba_2Cu_3O_{7-x}$ bulk is 11.693Å. The peak position of the (007) from the value is 54.98 degrees. The Tc extrapolated from Fig.8 is about 86K. This transition temperature is a few degrees lower than that of the bulk.

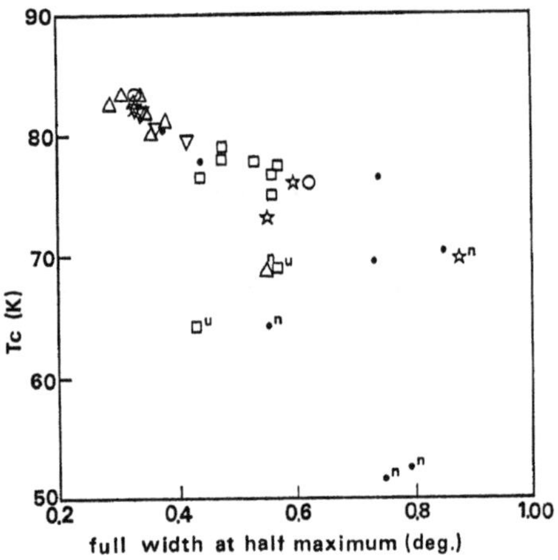

Fig.7. Tc vs full width at half maximum of the (007) peak for the films prepared on various sputtering conditions. All symbols except are same in Fig.2, 5, and 6. Symbol ● indicate the films prepared on the other deposition condition. Suffix "n" indicate not-oriented film and suffix "u" the composite film consist of high Tc Y-Ba-Cu-O crystal and unknown crystal.

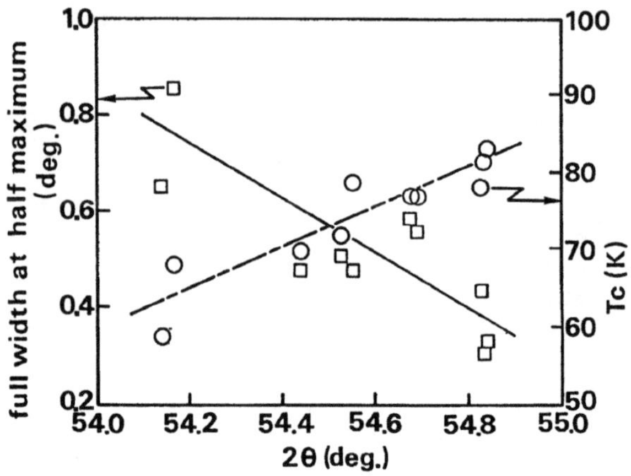

Fig.8 (007) peak position vs Tc and full width at half maximum for the films prepared with various CuO target rf power. Deposition temperature : 700°C and 710°C rf power : Y_2O_3=200w, $BaCuO_2$=126W, CuO=59-76W

CONCLUSIONS

The cause of compositional deviation and small deposition rate of the sputtered films which is prepared by using oxide target might be etching by negative ions from the target. The kinetic energy of ordinary sputtered particle is about 10eV, but negative ions from the target have a kinetic energy about 1keV which is enough to resputter the film. The negative ions bombard the film. The bombard induce a destruction of crystal and resputtering which lower deposition rate and compositional deviation from target. The destruction of crystal structure and compositional deviation lower the Tc. One good way for avoiding harmful effects by negative ions is placing the substrate where the target plane dose not face. The full width at half maximum and the position of the (007) peak have a good correlation with Tc. The cause of peak shift is mainly crystal defect which is caused by the compositional deviation from stoichiometry, because the full width at half maximum is large when peak position is low.

REFERENCES

1) Shah S. I. and Carcia P.F. (1987):Superconductivity and resputtering effects in rf sputtered $YBa_2Cu_3O_{7-x}$ thin films, Appl. Phys. Lett., 51, 2146

2) Terada N., Ihara H., Jo M., Hirabayashi M., Kimura Y., Matsutani K., Hirata K., Ohno E., Sugise R. and Kawashima F. (1988):Sputter synthesis of $Ba_2YCu_3O_y$ as-deposited superconducting thin films from stoichiometric target --- A mechanism of compositional deviation and its control, Jpn. J. Appl. Phys., 27, L639

3) Cuomo J. J., Gambino R. J., Harper J. M. E., Kuptsis J. D. and Webber J. C. (1978): Significance of negative ion formation in sputtering and SIMS analysis, J. Vac. Sci. Technol., 15, 281

4) Ono A. and Ishizawa Y. (1987):Preparation and properties of three types of orthorhombic superconductor $Ba_2YCu_{3-x}O_{7-y}$, Jpn. J. Appl. Phys., 26, L1043

In-situ Preparation of YBCO Thin Films on Single Crystal $LaAlO_3$

Y. HANG[1], D.P. FAN[2], S.U. ZHANG[2], H.C. ZHANG[2], P.H. WU[3], M. QIAN[3], B.X. JIANG[3], Z.J. SUN[3], S.Z. YANG[3], and Z.M. JI[3]

[1] Anhui Institute of Optics and Fine Mechanics, Academia Sinica, Hefei, Anhui, P.R. China
[2] Department of Physics, Nanjing University, and
[3] Department of Information Physics, Nanjing University, Nanjing 210008, P.R. China

ABSTRACT

Superconducting YBCO films are successfully prepared on single crystal $LaAlO_3$ substrates by rf magnetron sputtering. The R-T curve, X-ray diffraction pattern and the critical current density of the film, as well as the dielectric constant and loss tangent of $LaAlO_3$ measured at 36GHz, Show that YBCO/$LaAlO_3$ is a better candidate for electronic applications than those films on other substrates.

KEY WORDS: high Tc superconducting thin film, $LaAlO_3$ substrate, rf magnetron sputtering, in-situ plasma oxidation

INTRODUCTION

Since high Tc superconductor $Y_1Ba_2Cu_3O_{7-\delta}$ was successfully prepared, thin films of this material with Tc exceeding 90K and high critical current density have also been obtained on several single crystal substrates such as $SrTiO_3$, ZrO_2 and MgO. Especially $SrTiO_3$, whose lattice constant matches that of 1-2-3 film to within 2.1%, has been the best choice for growing high quality epitaxial 1-2-3 film. Its high dielecric constant and loss tangent, however,would limit its practical utility, particularly in high-freqency microelectronic applications.

Recently some alternate substrates, such as $NdGaO_3$ and $LaGaO_3$, which have dielectric constabts of about an order of magnitude smaller than that of $SrTiO_3$ and quite low loss at high frequencies, have been employed to fabricate 1-2-3 films resulting in Tc 90K and Jc 10^6 A/cm^2 which are comparable in quality to those on $SrTiO_3$. R.W.Cimon et al [1] have produced EBCO films on single crystal $LaAlO_3$ and measured the loss of these substrates at microwave band.

In this paper we report our results on the fabrication of $LaAlO_3$ substrate, film deposition and the properties of both substrate and film.

EXPERIMENTS

$LaAlO_3$ substrates are single crystal wafers cut from single crystal rod with a diameter of 20mm,which is prepared using Czochralski growing process. Then the wafers are polished finely.

YBCO thin films are grown on single crystal $LaAlO_3$ by rf magnetron sputtering method [2].The sputtering target is sintered $Y_1Ba_2Cu_3O_{7-\delta}$ pellet, 3-4mm thick and 72-80mm in diameter.Sputtering is carried out under the following conditions: substrate temperature 650°C; mixture gas pressure 0.3 Torr with 30% O_2; deposition rate 0.5Å/sec,and anode voltage 1.0KV. Here, film thickness is typically 3000Å. After sputtering, YBCO films are plasma oxidized for 80 minutes which is performed in-situ with anode voltage 300V and anode current

---

1. Anhui Institute of Optics and Fine Mechanics, Academia Sinica.
2. Department of Physics, Nanjing University.

10-20mA. During plasma oxidation the substrates are kept at 500°C, while the pressure of O_2 is increased to 1 Torr. After plasma oxidation, substrates are cooled down to room temperature.

RESULTS AND DISCUSSION

Figure1. shows the R-T curve of the film deposited on $LaAlO_3$, which exhibits zero resistance at 90K. The critical current density is $1.6X10^4$ A/cm^2 at 77K .The X-ray diffraction pattern of the film on $LaAlO_3$ shown in Fig.2, where (00ℓ) peaks are dominating, indicates that the C axis is reasonably well oriented.

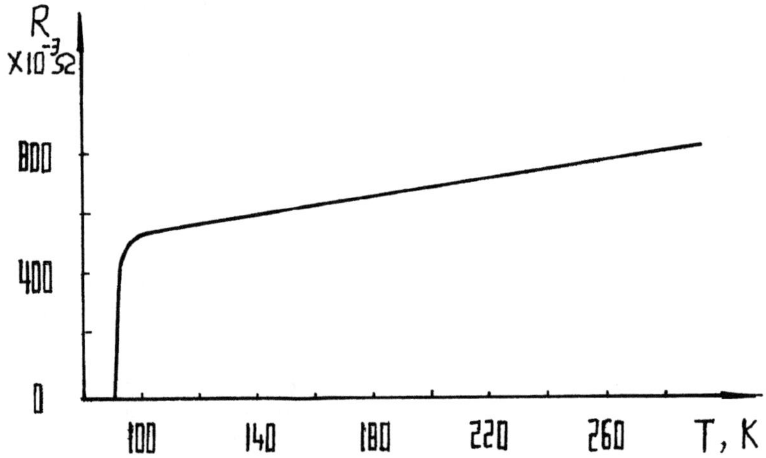

Fig.1. R-T curve of the film YBCO/$LaAlO_3$

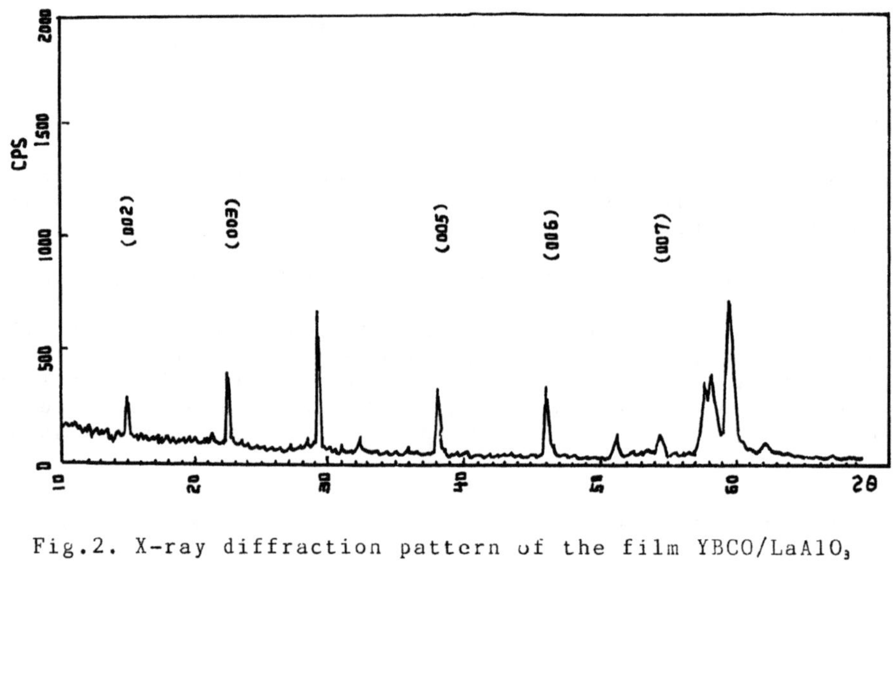

Fig.2. X-ray diffraction pattern of the film YBCO/$LaAlO_3$

In order to evaluate the suitability of $LaAlO_3$ as a substrate for mm wave application, we have measured both the dielectric constant and loss tangent of these substrate at 36GHz. The apparatus [3] used for this purpose consists of a waveguide bridge system with two dielectric antennas. The dielectric constant of $LaAlO_3$ and loss tangent are found to be 20 and 0.05, respectively, at 36 GHz.

Although the loss seems a little too high, which we are doing our best to reduce, our preliminary results show that YBCO/$LaAlO_3$ is a better candidate for electronic application than those films on other substrates.

REFERENCES

1. R.W.Simon, C.E.Platt,et al, Appl.Phys.Lett.53(26), (1988)2677.

2. Z.J.Sun, R.J.Zhang, et al, BHTSC'89.

3. J.Musil, F.zacek, A.Burger and J.Kariovsky, 4th European Microwave Conference 1979, p.66.

In-situ Preparation of Superconducting Y-Ba-Cu-O Films by Sequential Deposition Using 40MHz Magnetron Sputtering

H. TAKAHASHI, N. HOMMA, S. KAWAMOTO, H. KONDO, K. SUZUKI, and T. MORISHITA

Superconductivity Research Laboratory, International Superconductivity Technology Center,
10-13, Shinonome 1-chome, Koto-ku, Tokyo, 135 Japan

ABSTRACT

Superconducting Y-Ba-Cu-O thin films have been prepared by sequential deposition using three-target 40.68MHz magnetron sputtering. This sequential deposition is a very suitable method to control a chemical composition of Y-Ba-Cu-O films, because the deposition time on each target can be precisely adjusted. Films deposited on MgO(100) using $YBa_2Cu_3O_x$, $BaCuO_x$ and Cu targets at 750°C show the c-axis normal to the film and a zero resistance Tc of 78K. Films prepared at 700°C are polycrystalline with a zero resistance Tc of 45.6K.

KEY WORDS: Y-Ba-Cu-O, magnetron sputtering, sequential deposition, 40MHz

INTRODUCTION

In preparation of superconducting Y-Ba-Cu-O thin films, sputtering techniques are very common methods for preparing superconducting oxides films. In the case of single target, it is known that the chemical composition of a film deviates from the target composition, and additionally the film composition is extraordinarily sensitive to deposition conditions. The deficiency of Ba in films is explained by selective resputtering with bombardment of energetic negative oxygen emitted from the target. And the deficiency of Cu is explained by small stacking coefficients of Cu and CuO at higher substrate temperature.[1] These problems are resolved by adopting a multi-target sputtering method.[2-3] The target self-bias voltage reduced by increasing a radio frequency is reported.[4] This low target self-bias is probably effective to reduce the resputtering with bombardment of energetic negative oxygen. We have successfully obtained the stoichiometric $YBa_2Cu_3O_{7-\delta}$ film using three targets and 40.68MHz magnetron sputtering. This sequential deposition is a very suitable method to control a chemical composition of Y-Ba-Cu-O films, because the deposition time on each target can be precisely adjusted. In this report, the effects of the substrate temperature and repetition number of sequential deposition, corresponding to a thickness deposited at each target, on the crystal structure and electric properties of $YBa_2Cu_3O_{7-\delta}$ are presented.

EXPERIMENTAL

$YBa_2Cu_3O_{7-x}$ films were deposited by 40.68MHz magnetron sputtering with $YBa_2Cu_3O_x$, $BaCuO_x$ and Cu targets on MgO(100) substrates. Figure 1 shows the diagram of sequential deposition. RF powers continuously supplied for $YBa_2Cu_3O_x$, $BaCuO_x$ and Cu targets were 80, 80, and 50W, respectively. The ambient gas pressure was 6.5 mTorr (Ar+20%O_2) and the distance from the substrate to the target was fixed at 30mm. Substrates were set on the turning table. After the deposition, films were transfered to the next chamber and cooled down in an oxygen atmosphere to room temperature. The sputtering conditions and the deposition time on each target obtaining the stoichiometry of $YBa_2Cu_3O_{7-\delta}$ are summarized in Tables 1 and 2. Deposition rates of these films deposited at the radio frequency of 40.68MHz were 0.95 ~ 1.03 Å/sec. The chemical

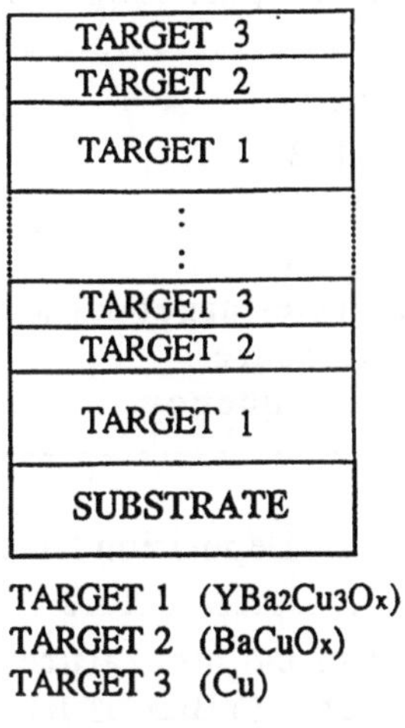

Fig.1. Diagram of sequential deposition using three-target.

composition of films was analyzed by inductively coupled plasma atomic emission spectroscopy (ICP). The crystal structure and surface morphology were characterized by X-ray diffraction (XRD), scanning electron microscopy (SEM) and Auger electron spectroscopy (AES). DC resistivity was measured using a four -lead method.

RESULTS AND DISCUSSION

Figure 2 shows the concentration of Y-Ba-Cu-O films prepared at 700°C under radio frequencies of 13.56MHz and 40.68MHz. The sputtering conditions are shown in Tables 1 and 2. The concentration of Ba in films is increased for the higher radio frequency discharge, 40.68MHz. This is probably caused by reduced selective resputtering with bombardment of negative oxygen emitted from the target. Generally, the concentration of Cu in films has a tendency to decrease with increasing substrate temperatures. However, the stoichiometry of $YBa_2Cu_3O_{7-\delta}$ is obtained only by increasing the deposition time for the Cu target in our system, as shown in Table 2.

Figure 3 shows the XRD patterns and SEM photographs of films prepared at the substrate temperature of 650°C, 700°C, and 750°C. Temperature dependences of resistance normalized by that at room temperature are shown in Fig.4. Films prepared at 650°C show weak (00l) peaks. As is seen in Fig. 3(a) the surface is not so smooth and there exist some particulates on the film. Measurements of AES revealed that particulates had higher Cu and lower Ba concentration than those in $YBa_2Cu_3O_{7-\delta}$ and that Cu atoms deposited from the Cu target did not diffuse over the whole film enough to crystallize the $YBa_2Cu_3O_{7-\delta}$ structure. Films prepared at 700°C show very strong (110), (220) peaks in addition to week (00l) peaks. The preferential growth parallel to MgO<011> is observed in the SEM photograph.(Fig. 3(b)) The resistance vs temperature curve shows a semiconductive behavior. A small resistance still remains just above a zero resistance Tc of 45.6K. Films prepared at 750 °C present a zero resistance Tc of 78K. Flake-like $YBa_2Cu_3O_{7-\delta}$ crystals grow normal to the substrate.(Fig. 3(c))

Figure 5 shows the composition variation determined by ICP in samples deposited with different repetition numbers of the sequential deposition. SEM photographs corresponding to samples in Fig.5 are shown in Fig.6. For all samples, the total deposition time was fixed at 95min and the deposition times were changed for each target, as listed in Table 3. When films were sputtered by five repetitions, Ba is deficient but Cu is in excess comparing with the stoichiometry of $YBa_2Cu_3O_{7-\delta}$. The particulates appeared on films are seemed to mainly contribute this deviation from the stoichiometry. As the repetition is increased to 10 or 20 cycles, the number of particulates decreased. As shown in Figs.6(a) and (b), the texture along MgO<011> directions is clearly seen at the surface for the repetition of 10 while there is rather random orientation for 20 repetitions.

Table 1. Sputtering conditions.

Target 1	$YBa_2Cu_3O_x$ (64mmφ)
Target 2	$BaCuO_x$ (64mmφ)
Target 3	Cu (64mmφ)
Substrate	MgO(100)
Substrate-target distance	30mm
Gas pressure	6.5mTorr(Ar+20%O_2)
Radio frequency	13.56MHz, 40.68MHz

Table 2. Deposition time.

Substrate temperature	650°C	700°C	750°C
Target 1 ($YBa_2Cu_3O_x$)	5min	5min	5min
Target 2 ($BaCuO_x$)	2min	2min	2min
Target 3 (Cu)	2min	2.5min	3min
	(	10 cycle	)
Total deposition time	90min	95min	100min

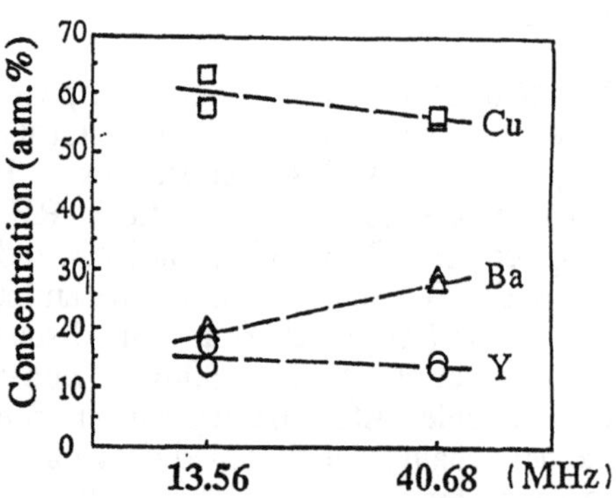

Fig.2. Concentration of Y-Ba-Cu-O films prepared at 700°C under the radio frequency of 13.56MHz and 40.68MHz.

Fig.3. X-ray diffraction patterns and SEM photographs of films prepared at the substrate temperatures : (a) 650°C, (b) 700°C and (c) 750°C.

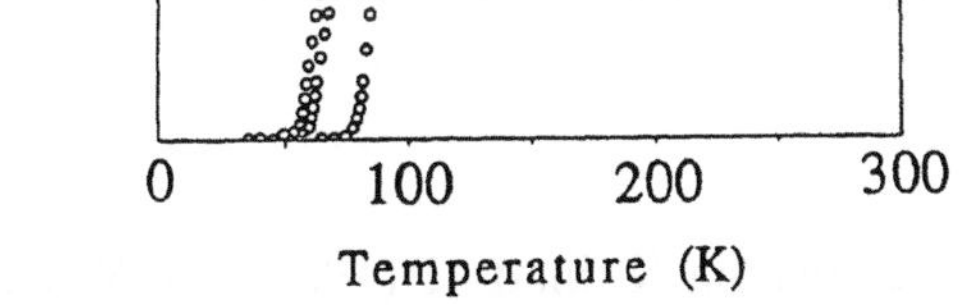

Fig.4. Temperature dependences of relative resistance of films prepared at substrate temperatures : (a) 650°C, (b) 700°C and (c) 750°C.

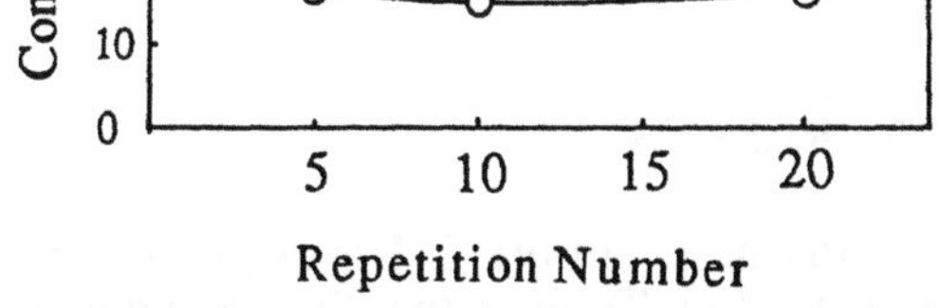

Fig.5. Composition changes with the repetition number of sequential deposition. Total deposition time was fixed at 95min and the deposition times were changed for each target, as listed in Table 3.

Table 3. Repetition number and deposition time.

Substrate temperature			700°C
Repetition number	5	10	20
Target 1	10min	5min	2.5min
Target 2	4min	2min	1min
Target 3	5min	2.5min	1.25min
Total deposition time	95min	95min	95min

Fig.6. SEM photographs of films prepared at the repetition number : (a) 5, (b) 10 and (c) 20.

CONCLUSION

In situ superconducting $YBa_2Cu_3O_{7-\delta}$ films have prepared by three-target sequential sputtering using 40.68MHz and 13.56MHz power sources. The stoichiometry of $YBa_2Cu_3O_{7-\delta}$ has been easily obtained by increasing the deposition time for the Cu target. At a substrate temperature of 750°C, deposition times of 5min on $YBa_2Cu_3O_x$, 2min on $BaCuO_x$ and 3min on Cu target were suited to get quite good films with a zero resistance Tc of 78K. By using three targets, it is rather easy to adjust the composition to the stoichiometry on an average, but there often appears to be particulates with compositions deviated from $YBa_2Cu_3O_{7-\delta}$. The concentration of Ba in films is increased for the higher radio frequency discharge, 40.68MHz. This is probably caused by the reduced selective resputtering with bombardment of negative oxygen emitted from the target at the higher radio frequency discharge.

This work was performed under the management of the R&D of Basic Technology for Future Industries supported by New Energy and Industrial Technology Development Organization.

REFERENCES

1.Ishibashi K, Hiraya K, Kim K, Hosokawa N, Asamaki T, Kawakita K, Inada T (1989) 2nd Work Shop on High-Temperature Superconducting Electron Devices R&D Association for Future Electron Devices Jun 7 -9 in Shikabe Hokkaido Japan : 39-44

2.Chang J, Seo S, Sayama A, Matsui M, Yamamoto K, Harada N (1988) ISS'88, August, Nagoya : 575

3.Kuroda K, Kojima K, Tanioku M, Yokoyama K, Hamanaka K (1989) 2nd Workshop on High-Temperature Superconducting Electron Devices R&D Association for Future Electron Devices Jun 7 -9 in Shikabe Hokkaido Japan : 29-34

4.Ohmi T, Kuwabara H, Shibata T, Kiyota T (1987) ULSI Science and Technology 87-11 : 574-592

Epitaxially Grown Superconducting Y-Ba-Cu-O Films Prepared by Multi-Target Magnetron Sputtering

MASAYUKI SAGOI, YOSHIAKI TERASHIMA, and TADAO MIURA

Advanced Research Laboratory, Research and Development Center, Toshiba Corporation, 1, Komukai Toshiba-cho, Saiwai-ku, Kawasaki, 210 Japan

ABSTRACT

Y-Ba-Cu-O films were prepared using the reactive magnetron sputtering method with three targets. The critical temperature, *Tc*, was significantly dependent on the distance between the targets and the substrate. *Tc*'s of the film on $SrTiO_3$ (110) substrates increased to more than 80K by decreasing the distance, in spite of a very low oxygen partial pressure of 0.3 Pa. The *c*-axis oriented films on the (100) substrates showed an extremely smooth surface, and were found to be single-crystalline from the RHEED observation. In conclusion, it was found that the influence of plasma existing above the targets is essential in order to obtain high-quality films.

KEY WORDS: superconducting film, sputtering, plasma, epitaxial growth

INTRODUCTION

Thin film preparation is the most important subject for the application of high *Tc* superconductors to electronics. Several methods have been tried to obtain epitaxially grown films. Both reactive evaporation [1] and laser ablation [2] have been reported to enable us to make up single-crystalline Y -Ba-Cu-O films. Recently, sputtering [3] has shown successful results in the preparation of high quality Y-Ba-Cu-O films. However, the key parameters for obtaining such excellent films have not yet been clearly described.

We have found that plasma influence is essential for obtaining single-crystalline Y-Ba-Cu-O films using the reactive magnetron sputtering with three sources. In this report, the structural features and superconducting properties of the films are evaluated in relation to several sputtering conditions.

EXPERIMENTAL

Y-Ba-Cu-O films were deposited by reactive magnetron sputtering with Y, Cu and Ba_2CuO_3 sources onto $SrTiO_3$ (100) and (110) substrates. The targets are placed on a large circumference. The substrates rotate within the circumference so as to avoid direct irradiation of negative ions from the targets. The details of the experimental procedure were described in our previous paper.[4]

RESULTS AND DISCUSSION

Figure 1 shows changes in the resistance of YBCO films prepared on the (110) substrates. The substrate-target distance was 90 mm or 150 mm during the deposition. The set temperature of the substrate holder was 680°C. The critical temperature, *Tc*, was significantly dependent on the distance between the targets and the substrate. *Tc*'s of the films on $SrTiO_3$ (110) substrates increased to more than 80K by decreasing the distance, in spite of a very low

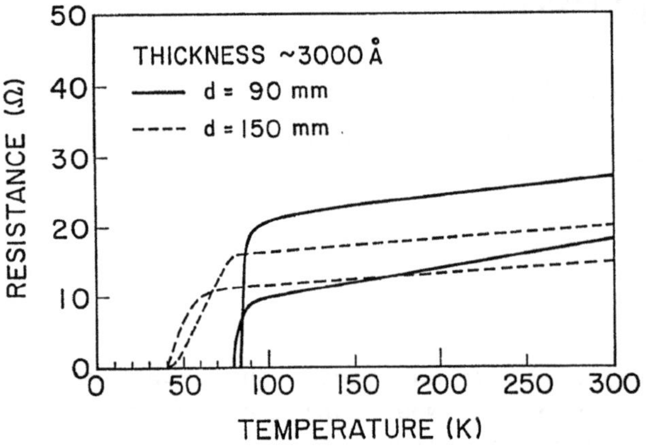

Fig.1 Changes in temperature dependence of resistance for films prepared on $SrTiO_3$ (110) substrate setting the substrate-target distance 90 mm or 150 mm. *Tc* depends on the distance.

oxygen partial pressure of 0.3 Pa. *Tc* of more than 80 K was obtained only for films prepared setting the distance 90 mm. For the 150 mm, the as-grown films exhibited *Tc* below 50 K. This implies that the amount of oxygen taken into the film during the deposition increases by holding the substrates close to the targets. Discharge plasma producing activated oxygen near the targets will prompt film oxidation, as pointed out by Adachi et al [5]. An effective film oxidation is considered to take place only during the sputtering, because the prepared film was cooled down in the low oxygen content atmosphere identical to the sputtering one after the sputtering.

Fig.2 shows the morphology, viewed by a SEM, for the films prepared onto the (110) by setting the distance 90mm or 150mm. Elongated grains are seen in both photographs, but the grain size is quite different. Though the holder temperature for the 150 mm film was set higher than that for the 90mm film by 30°C, the grains of the 90 mm film were larger than the former. In other words, the film prepared at lower temperature appears to grow better than that prepared at higher temperature. A holder-temperature gap between 640 °C and 710 °C brought about little difference in grain size of the films on the (110) in the case of the large substrate-target gap. This indicates that the contribution of the elevated substrate temperature to the grain growth is not so great under the environment without plasma-enhancement. Besides, an excessively high substrate temperature in general lowers *Tc* or makes the film tetragonal.

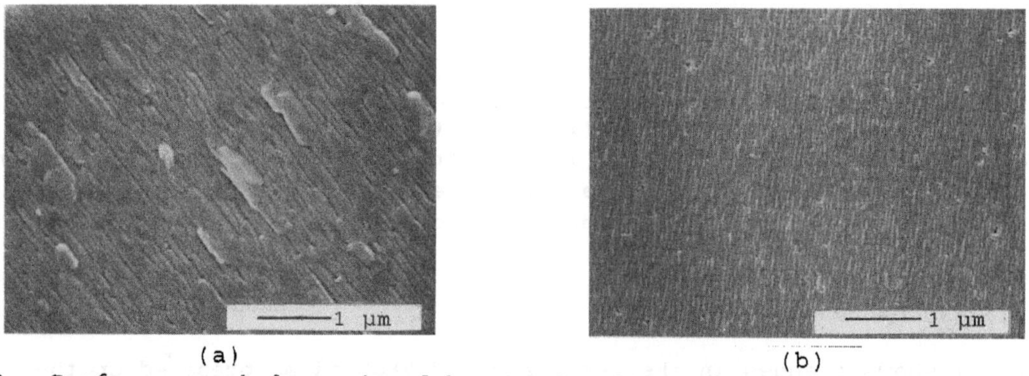

Fig.2 Surface morphology viewed by SEM for films prepared on $SrTiO_3$ (110) substrate setting the distance (a) 90 mm or (b) 150 mm.

Therefore, the large difference of the grain-size, as shown in Fig.2, was obviously caused by the plasma existing just above the targets. Plasma is, therefore, considered to work not only on increasing substrate temperature but also on prompting film to crystallize and to grow effectively.

The films on the (100) show lower *Tc* than those on the (110). The *Tc* value increased up to only about 70 K even when optimizing the substrate-target distance, as shown in Fig.3. The crystalline structure of the films changed from an a-axis orientation to a c-axis one with decreasing the substrate-target distance. As for X-ray diffraction data for the films on the (100), the *a*-spacing of the film prepared at the distance of 150 mm was about 3.85 A, indicating a low *Tc*. The *c*-spacing of the film prepared at 90 mm was 11.70 A-11.71 A. Therefore, the film oxidation is prompted by holding the substrate close to the targets also in this case.

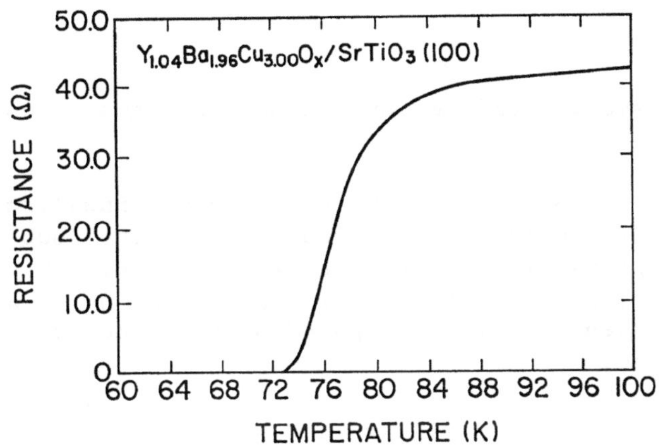

Fig.3 Temperature dependence of resistance near *Tc* for film prepared onto $SrTiO_3$ (100) substrate by setting the substrate-target distance 90 mm.

Fig.4 shows a surface of the film prepared at 90 mm. The scratch on the surface was introduced to get a sharp focus. The surface had no structure, and was extremely smooth, implying that the film was single-crystalline. The higher magnification photograph of the film is also shown. No crystal grains are seen even in the photograph, except for several pores and a dust particle.

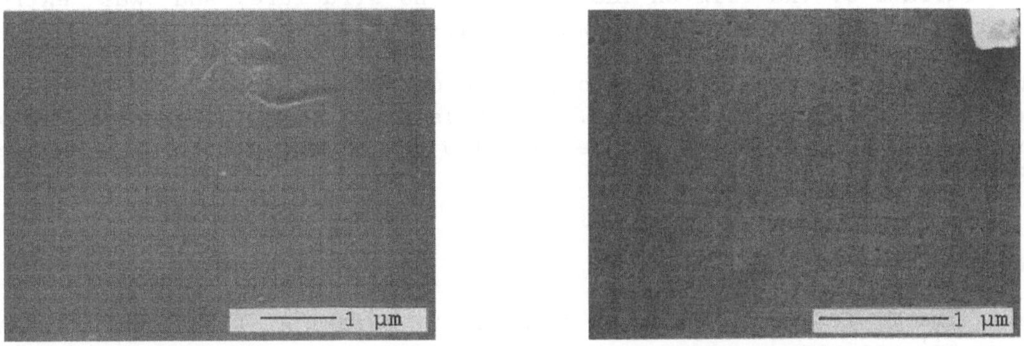

Fig.4 Surface morphology of the film on the (100) prepared by setting the distance 90 mm. No grains are seen even in the higher magnification photo.

This smooth surface is considered to be brought about by the plasma enhancement. RHEED observation was carried out for the film. The pattern was taken using a TEM option for RHEED. The acceleration voltage was 200 kV. The result is shown in Fig.5. Streak-patterns will become easier to observe, while ring-patterns will become harder, as the acceleration voltage is lowered. RHEED observation with the voltage as high as 200 kV reveals crystallinity or surface roughness more severely than that with several 10s kV. The pattern was

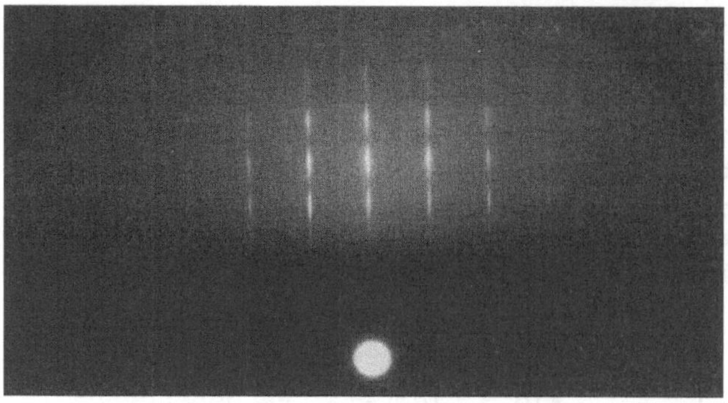

Fig.5 RHEED pattern of the film on the (100). The pattern is composed of streaks without any spots and rings, indicating an epitaxially grown film.

composed of streaks resulting from a smooth surface, and it had no rings, thus corresponding to epitaxial growth.

The films were annealed at 550 °C for 3 hours. Such low temperature annealing was effective only to the *c*-axis oriented films, prepared setting the substrate-target distance 90 mm. It improved the *Tc* value by about 10 K. This enables us to obtain film on the (100) showing *Tc* of more than 80 K. On the other hand, the *Tc* value of *a*-axis oriented films, prepared setting the distance 150 mm, declined by the annealing.

CONCLUSION

An important parameter for obtaining Y-Ba-Cu-O films of high-quality has been found to be the distance between the substrate and the targets, and several results have been obtained as follows.

(1)Tc's of films on $SrTiO_3$ (110) substrates increased to more than 80K by decreasing the substrate-target distance, in spite of a very low oxygen partial pressure of 0.3 Pa.

(2)The surface of the film on the (100) had no structure, and was extremely smooth, implying that the film was single-crystalline.

(3)Plasma influence was taken into Y-Ba-Cu-O film preparation by holding the substrate close to the targets. Plasma is considered to work not only on increasing substrate temperature but also on prompting film to grow and to oxidize effectively.

REFERENCES

1. Terashima T. Iijima K. Yamamoto K. Bando Y. Mazaki H. (1988) Jpn. J. Appl. Phys. 27: L91-L93

2. Wu XD. Inam A. Venkatesan T. Chang CC. Chase EW. Barboux P. Tarascon JM. Wilkens B. (1988) Appl. Phys. Lett. 52: 754-756

3. Asano H. Asahi M. Michikami O. (1989) Jpn. J. Appl. Phys. 28: L981-L983

4. Sagoi M. Terashima Y. Kubo K. Mizutani Y. Miura T. Yoshida J. Mizushima K. (1989) Jpn. J Appl. Phys. 28: L444-L447

5. Adachi H. Hirochi K. Setsune K. Kitabatake M. Wasa K. (1987) Appl. Phys. Lett. 51: 2263-2265

Preparation and Superconductive Properties of $YBa_2Cu_3O_y$ Thin Films by Coevaporation with ECR Ion Source

H. OBARA, S. KOSAKA, M. UMEDA, and Y. KIMURA

Electrotechnical Laboratory, 1-4, Umezono 1-chome, Tsukuba, Ibaraki, 305 Japan

ABSTRACT

Superconducting $YBa_2Cu_3O_y$ films have been prepared by coevaporation with ECR ion source in relatively low pressure ($< 1 \times 10^{-4}$ Torr). In the present work, Y, Ba and Cu metal elements were individually evaporated from three effusion cells on MgO substrates with (100) plane, and O_2 or N_2O gas was introduced to ECR plasma near the substrate. As-deposit films had an orientation with the c-axis perpendicular to the substrates, and showed T_c as high as approximately 50 K. After in-situ annealing at low temperature (500°C), T_c of films increased to 80 K without any changes of crystallinity and morphology. Samples were analyzed by in-situ Auger and X-ray photoelectron spectroscopy (XPS) in a closed system. Further, superconductive properties of samples were measured by resistivity measurement, and magnetization measurement was performed using SQUID magnetometer. Finally, critical current properties of thin films were discussed.

KEY WORDS: oxide superconductor, $YBa_2Cu_3O_y$ thin film, evaporation, ECR ion source, magnetization

INTRODUCTION

A number of studies have been reported on preparation of thin films of high-T_c oxide superconductors. One of the purposes of these researches is to apply the high-T_c superconductors to electronic devices and another is to observe the intrinsic properties of these systems because of good quality of thin films. Many techniques have been successfully used to generate thin films of the high-T_c superconductors with good crystallinity, smooth surface and high critical current density (J_c).

Coevaporation in an ultra-high vacuum (UHV) deposition system, which we used in the present work, is one of the ideal methods for thin-film preparation because high-quality samples can be prepared and this system has the possibilities of the epitaxial growth of films. In UHV systems, the most important subject for thin-film preparation of oxide superconductors is oxidation of samples. To prepare films of oxide superconductor in relatively low pressure ($\leq 1 \times 10^{-4}$ Torr), it is necessary to activate oxygen. Neutral O_2 gas is not so effective for oxidation of samples. For activation of oxygen, ozone [1, 2] rf-plasma [3, 4] and ECR (electron cyclotron resonance) ion source [5, 6] have been found to be useful.

In the present work, we have prepared $YBa_2Cu_3O_y$ thin films by coevaporation in an UHV system and used an ECR ion source to activate oxygen. First, we have deposited Cu-O films and discussed the effects of the ECR ion source. Second, we have prepared $YBa_2Cu_3O_y$ thin films and characterized our samples using X-ray diffraction, resistivity measurement and magnetization measurement. Finally we have discussed the superconducting properties and critical current properties of our films.

EXPERIMENTAL

In the present work, samples were prepared by coevaporation in an UHV system for molecular beam epitaxy (MBE). The background pressure was 1×10^{-10} Torr and the pressure during deposition was less than 1×10^{-4} Torr. Evaporation source were three effusion cells (the maximum temperature is 1800°C) and Y, Ba and Cu metal elements were individually evaporated. The deposition rate of each element was $0.1 \sim 1$ Å/s.

Oxygen or N_2O gas was introduced to the ECR ion source. The distance between the discharge tube of ECR ion source and the sample holder was approximately 10 cm. The ion current density at the sample holder was typically 0.1 mA/cm^2, which depends on the pressure and voltage of the discharge tube and grid of ECR ion source.

The resistive measurement was performed using an usual four terminal technique and magnetization was measured by a SQUID magnetometer.

RESULTS AND DISCUSSION

Figure 1 shows XPS spectra of Cu films deposited on MgO substrates with (100) plane at room temperature. Samples C1 and C2 were deposited using ECR ion source and not using ECR ion source, respectively. Sample C1 shows a satellite structure near the Cu $2p_{3/2}$ peaks which is typical of the Cu^{2+} state. On the other hands, sample

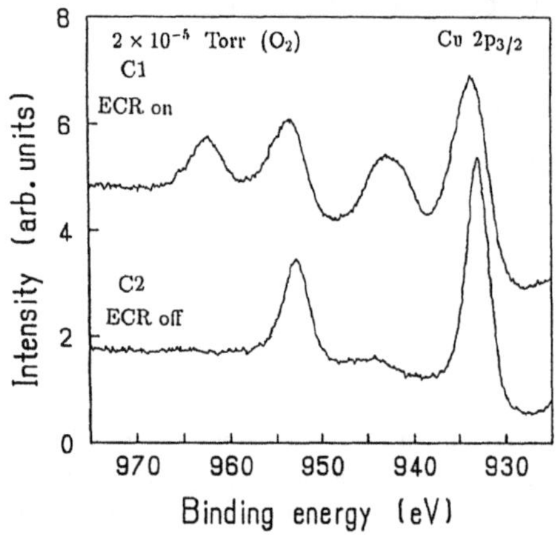

Fig. 1 XPS spectra of Cu films deposited using ECR ion source and not using ECR ion source.

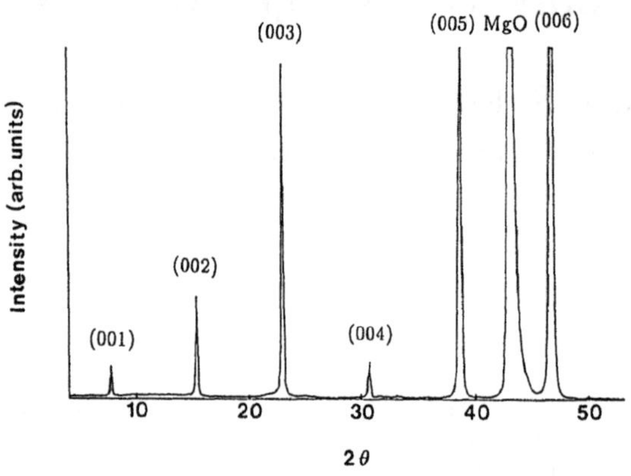

Fig. 2 X-ray diffraction pattern of $YBa_2Cu_3O_y$ film which is deposited on a MgO substrate with (100) plane.

C2 shows no satellite structure and coincides with the spectrum of Cu metal. From these XPS spectra, it is clear that ECR ion source is effective for oxidation of films.

Typical X-ray diffraction pattern of $YBa_2Cu_3O_y$ film is shown in Fig. 2. $YBa_2Cu_3O_y$ films were deposited of MgO substrates with (100) plane and substrates temperature was 650°C. Thickness of films was about 1300 Å. Films are highly c-axis oriented perpendicular to the substrates. The condition of sample preparation, lattice parameters, and the onset of superconducting transition temperature (T_c), which were measured using a SQUID magnetometer, are listed in Table 1. Sample C was transferred in the specimen transfer rod without exposure to the air and annealed at 500°C in another chamber where the pressure of pure oxygen was 500 Torr.

Sample B, which was deposited using N_2O and ECR plasma, has shorter lattice constant c, i.e., more oxygen content y, than sample A which was deposited using O_2. The partial pressure of oxygen was expected to be lower in the N_2O case because N_2, NO and O_2 were mainly observed using quadrupole mass spectroscopy. Therefore the gasses which were produced in ECR plasma of N_2O is more effective than that of O_2 for preparation of oxide superconductor films. In the present work, however, we did not directly detect the ions which were produced in ECR plasma and further study is needed to clarify the effects of ECR plasma. In samples B and C, no N atom was detected using Auger spectroscopy.

Fig. 3 shows the temperature dependences of magnetization and resistivity of sample C. The zero-field-cooled (ZFC) and field-cooled (FC) magnetization was measured for magnetic field applied perpendicular to the films. Zero-resistance temperature of sample C is 79 K, which corresponds to the onset temperature of the transition curve of magnetization. A sharp transition of magnetization indicates that the sample is a homogeneous superconductor and had high-critical-current density, J_c.

Magnetization measurement offers more information about superconductivity of samples than resistivity measurement. If we can consider that samples are uniformly magnetized ellipsoid, magnetization , M, is calculated as follows,

$$M = \frac{\chi H_0}{1 + 4\pi\nu\chi} V \,, \tag{1}$$

where χ is susceptibility, H_0 is applied magnetic field, V is volume and ν is the demagnetizing factor of sample, which is approximately given by,

Table 1 The condition of sample preparation, lattice constant c and T_c of $YBa_2Cu_3O_y$ films. (The substrate temperature was 650 °C.)

Sample	gas	P (Torr)		c (Å)	T_c
A	O_2	2.5×10^{-5}	as-deposited	11.88	
B	N_2O	7.3×10^{-5}	as-deposited	11.73	50
C	N_2O	7.3×10^{-5}	annealed	11.66	80

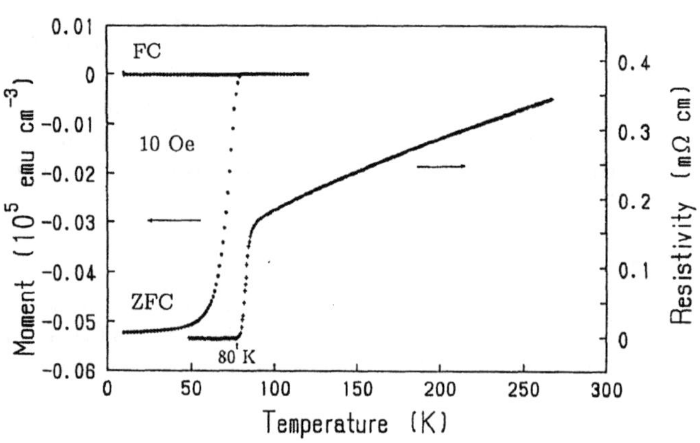

Fig. 3 Temperature dependence of magnetization and resistivity of sample C. ZFC and FC magnetization measurements were performed for field applied perpendicular to the film.

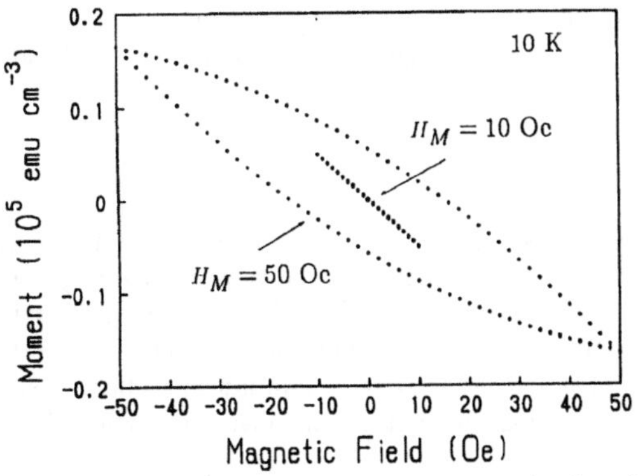

Fig. 4 Magnetic field dependence of magnetization of sample C at 10 K. H_M is the maximum applied magnetic field.

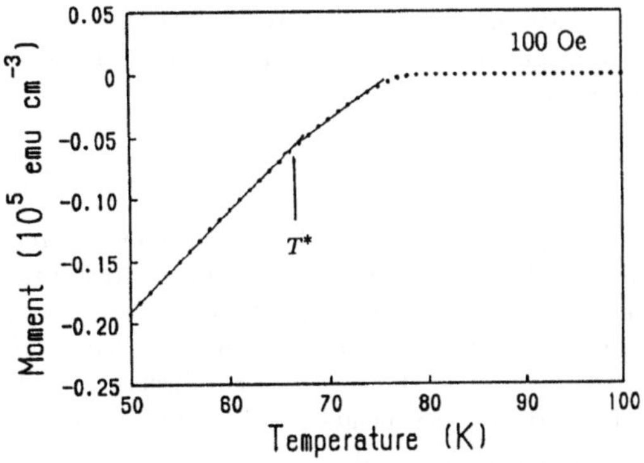

Fig. 5 Temperature dependence of ZFC magnetization of sample C at 100 Oe. Solid lines are guide for the eye.

$$\nu \approx 1 - \frac{\pi}{2}\frac{d}{R} \quad (d \ll R) , \tag{2}$$

where d is thickness and R is radius of sample. Typical value of R is about 3 mm. In the case of thin film, i.e., $\nu \to 1$, which has perfect diamagnetism, $\chi = -1/4\pi$, equation (1) indicates that the apparent susceptibility, M/H_0, gets very large. The magnetization of sample C at low temperature almost corresponded to the perfect diamagnetism.

The magnetic field dependence of magnetization of sample C is shown in Fig. 4. The curvature of magnetization depends on the maximum value of applied field, H_M. When H_M is 10 Oe, the magnetization behaves reversibly and hysteresis behavior was observed when H_M is 50 Oe. Because thin films have weak Meissner effect for field applied perpendicular to films, the trapped flux can not be expelled and magnetization shows hysteresis behavior. So we can consider that sample C can shield the magnetic field at 10 Oe in spite of large demagnetization effect.

Further, critical current density of thin films can be estimated from magnetization because good quality films have no weak link and Bean-critical-state model can be easily applied to this case. Figure 5 shows the temperature dependence of ZFC magnetization of sample C for field applied perpendicular to the film. The temperature dependence of ZFC magnetization changes at temperature denoted T^* and the change of temperature dependence of ZFC magnetization is similar to single crystals [7] except for very low value of apparent lower critical field H_{c1}, which is due to demagnetization effect. T^* can be interpreted as the temperature, at which the flux fronts meet at the center of samples. So the flux penetration picture can be also applied to thin films and a conventional Bean-critical-state model is available.

Figure 6 shows the temperature dependence of the remanent magnetization of sample C for several fields applied perpendicular to the film. In this measurement field was applied well above T_c, the sample was cooled, and field was turned off at low temperature. Above 200 Oe, the value of remanent magnetization saturated and critical current density is calculated from these data using the simple equation of Bean-critical-state model [8],

$$J_c = 30M/R , \tag{3}$$

where M is the remanent magnetization and R is radius of sample. We did not consider the "dB/dH" effect [9] and some corrections are needed to calculate the exact value of J_c. In the present work, however, the temperature dependence of J_c and high J_c ($\sim 10^6$A/cm^2) have been clearly observed. These properties indicate that our sample is homogeneous superconductor [10].

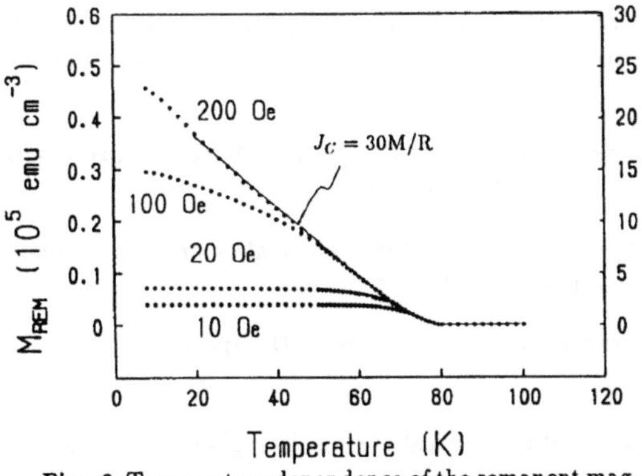

Fig. 6 Temperature dependence of the remanent magnetization of sample C. Criitical current density is calculated using Bean-critical-state model (solid line).

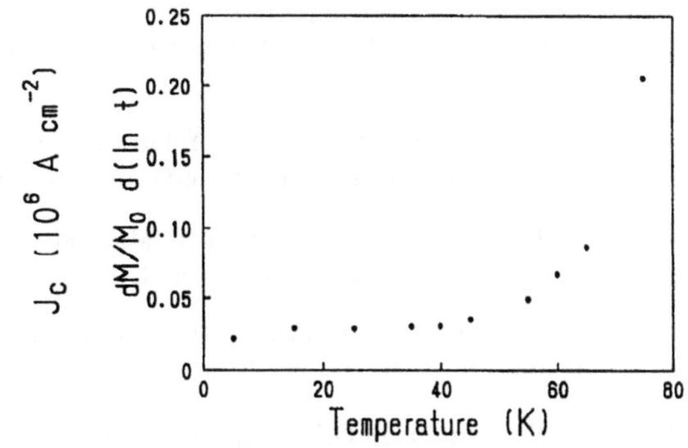

Fig. 7 Temperature dependence of normalized relaxation rate of magnetization.

Finally, we have measured the decay of remanent magnetization of thin films. Samples show an almost linear dependence of magnetic relaxation upon the logarithm of time. Temperature dependence of normalized relaxation rate of magnetization,

$$\frac{dM}{d(\ln t)}\frac{1}{M_0}, \tag{4}$$

where t is time and M_0 is initial value of remanent magnetization, is plotted in Fig. 7. If we adopt the simple flux creep model, the normalized relaxation rate of remanent magnetization reduces to the value, which is given by kT/U_0, where k is the Boltzman constant and U_0 is activation energy of flux motion [11]. Because the temperature dependence of kT/U_0 was not linear, It is expected that U_0 of the thin film is not constant and has some temperature dependence. At the present stage, we can not discuss the details of superconductivity in thin films and further study is needed. From these experimental data of thin films, however, we believe that the intrinsic properties of oxide superconductors can be discussed.

CONCLUSION

Using coevaporation in an UHV system with ECR ion source, we have successfully prepared $YBa_2Cu_3O_y$ thin films at relatively low pressure. The ECR ion source is considered to be effective for oxidation of samples from in-situ XPS measurement. Superconductive properties of our sample were similar to the single crystals and critical current properties were estimated from magnetization measurements. The extensive studies of magnetization were useful method for observation of intrinsic properties of oxide superconductors.

ACKNOWLEDGEMENT

The authors would like to express their thanks to K. Kajimura for his encouragement throughout this research.

REFFERENCES

1. Berkley D.D., Johnson B.R., Anand N., Beauchamp K.M., Conroy L.E., Goldman A.M., Maps J., Mauersberger K., Mecartney M.L., Morton J.,Tuominen M., Zhang Y.J. (1988) *In situ* formation of superconducting $YBa_2Cu_3O_{7-x}$ thin films using pure ozone vapor oxidation. Appl. Phys. Lett. **53**: 1973-1975.

2. Nakayama Y., Ochimizu H., Maeda A., Kawazu A., Uchinokura K., Tanaka S. (1989) *In situ* growth of Bi-Sr-Ca-Cu-O thin films by molecular beam epitaxy technique with pure ozone. Jpn. J. Appl. Phys. **28**: L1217-L1219.

3. Terashima T., Iijima K., Yamamoto K., Bando Y., Mazaki H. (1988) Single-crystal $YBa_2Cu_3O_y$ thin films by activated reactive evaporation. Jpn. J. Appl. Phys. **27**: L91-L93.

4. Akoh H., Matsumoto M., Yamano K., Takada S. (1989) As-grown Y-Ba-Cu-O thin films using reactive coevaporation method. In: 2nd Workshop on High-Temperature Superconducting Electron Devices, 7-9 June 1989, Shikabe, Japan.

5. Noriwaki K., Enomoto Y., Kubo S., Murakami T. (1988) As-deposited superconducting $Ba_2YCu_3O_{7-x}$ films using ECR ion beam oxidation. Jpn. J. Appl. Phys. **27**: L2075-L2077.

6. Aida T., Tsukamoto A., Imagawa K., Fukazawa T., Saito S., Shindo K., Takagi K., Miyauchi K. (1989) Thin film growth of $YBa_2Cu_3O_y$ by ECR oxygen plasma assisted reactive evaporation. Jpn. J. Appl. Phys. **28**: L635-L638.

7. Krushin-Elbaum L., Malozemoff A.P., Yehurun Y., Cronemeyer D.C., Holtzberg F. (1989) Temperature dependence of lower critical fields in Y-Ba-Cu-O crystals. Phys. Rev. **B39**: 2936-2939.

8. Chaudhari P., Koch R.H., Laibowitz R.B., MacGuire T.R., Gambino R.J. (1987) Critical-current measurements in epitaxial films of $YBa_2Cu_3O_y$ compound. Phys. Rev. Lett. **58**: 2684-2686.

9. Campbell A.M., Evetts J.E. (1972) Flux vortices and transport currents in type-II superconductors. Adv. Phys. **21**: 199-429.

10. de Vries J.W.C., Stollman G.M., Gijs M.A.M. (1989) Analysis of the critical current in high-T_c superconducting films. Physica C **157**: 406-414.

11. Yeshurun Y. and Malozemoff A.P. (1988) Giant flux creep and irreversibility in Y-Ba-Cu-O crystal: An alternative to the superconducting-glass model. Phys. Rev. Lett. **60**: 2202-2205.

Preparation of $Y_1Ba_2Cu_3O_{7-\delta}$ Thin Films by MBE Using Metal Chelates

K. Endo, Y. Ikedo[a], S. Hayashida[b], J. Ishiai[c], K. Nakatsuka[d], S. Misawa, and S. Yoshida

Electrotechnical Laboratory, 1-4, Umezono 1-chome, Tsukuba, Ibaraki, 305 Japan

ABSTRACT

We have fabricated YBCO superconducting thin films by MOMBE, for the first time. As source materials, β-diketonate chelates of Y, Ba and Cu were used. The films annealed at 900°C show the superconductivity with Tc(onset) = 80K and Tc(zero resistance) = 74K.

KEYWORDS: MOMBE, YBCO thin film, β-diketonate chelate, oxide superconductor

INTORODUCTION

Metalorganic molecular beam epitaxy (MOMBE) using organometallic compound sources has the advantages of both metalorganic chemical vapor deposition (MOCVD) and molecular beam epitaxy (MBE). In particular, MBE is superior in the higher control of the composition and the film thickness in the order of a monoatomic layer through in-situ monitoring and shutter control. So far, MOCVD has been actively investigated in the semiconductor field and demonstrated to be an effective method for the epitaxial growth of semiconductors such as GaAs and InP.

In order to develop MOMBE technique for the growth of oxide superconductor, we carried out MBE using metal chelates as source materials. In conventional MBE [1], refractory metals, Y and Ba, are used as source materials. So that we are forced to use the electron beam gun for evaporation of these metals, which restricts the oxygen pressure during growth. The use of metal chelate sources in MBE allows a higher oxygen pressure compared with that of metal sources, because the source temperatures of metal chelates are much lower than those of metal. Also, the lower source temperatures make the apparatus simple, where an electron beam gun is not necessary.

In this study, we report MOMBE growth of YBCO using β-diketonate metal chelates as source materials, for the first time.

EXPERIMENTAL

Figure 1 is a schematic drawing of the MOMBE apparatus used in this study. The chamber was evacuated by turbo molecular pump down to less than 10^{-6} Torr. As source materials, β-diketonate chelates of Y, Ba and Cu were used. Each source was loaded into the PBN crucible

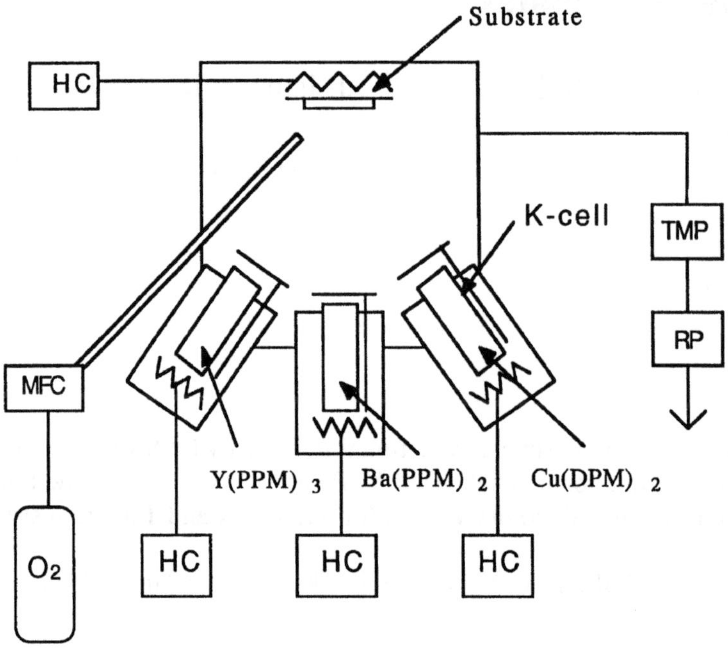

Fig. 1 Schematic drawing of MOMBE apparatus.
TMP: Turbo Molecular Pump, RP: Rotary Pump,
HC: Heater Controller, MFC: Mass Flow Controller

of Knudsen-cell with a shutter and was heated at 120, 160, and 45°C by a tantalum heater, respectively. Oxygen gas was introduced through a nozzle near the substrates. The flow rate of oxygen gas was about 5 sccm. During the deposition, the total pressure in the chamber was 1×10^{-4} Torr. MgO(100) crystals were used as substrates. The substrates were mounted on a molybdenum block holder with indium metal solder.

The compositions and the structures of films were examined by inductively coupled plasma spectroscopy (ICP) and X-ray diffraction (XRD). The incorporation of fluorine and carbon in the film were analyzed by electron probe microanalysis (EPMA). Resistance measurements were carried out by a standard four probe DC method, where Cu wire were attached with Ag paste to Au electrodes evaporated onto the films.

RESULTS AND DISCUSSION

Selection of MO source materials

The candidates of MO source materials for Ba are the β-diketonate metal chelates; $Ba(DPM)_2$ and $Ba(PPM)_2$. Also, $Y(DPM)_3$ and $Y(PPM)_3$ for Y source, and $Cu(DPM)_2$ for Cu. In order to choose suitable source materials, thermogravimetric analysis (TGA) has been carried out. In the TGA curve for $Ba(DPM)_2$, multiple steps of the decrease in weight were observed, which suggests that $Ba(DPM)_2$ contains some impurities. In contrast, in that for $Ba(PPM)_2$, only one step of the decrease was observed. Only half amount of $Ba(DPM)_2$ vaporizes in weight even at 500°C, whereas 90% of $Ba(PPM)_2$ vaporizes at the same temperature. These results lead us to

Tab . 1 Deposition rate (metal μg / $cm^2 \cdot hr$)

Substrate temp.	25 °C (D25)	600°C (D600)	Yield (D25 / D600)
$Y(DPM)_3$	35	2	5.7 %
$Y(PPM)_3$	65	10	15.4 %
$Ba(PPM)_2$	-	22	-
$Cu(DPM)_2$	-	29	-

choose $Ba(PPM)_2$ as a source material for Ba. Both yttrium chelates, $Y(DPM)_3$ and Y $(PPM)_3$, show the TGA curves with single step of the decrease and almost all source materials vaporize around 300°C, although the onset vaporization temperature of $Y(PPM)_3$ is somewhat lower than that of $Y(DPM)_3$. These results suggest that both yttrium chelates can be used as a source material for Y. Also, TGA for $Cu(DPM)_2$ revealed that $Cu(DPM)_2$ is a suitable source material for Cu.

For the further selection of these chelates, the deposition rates of metal oxides at the substrate temperature of 600°C were examined for each source material. The amount of individual metal contained in the deposited film was measured by ICP. The evaporation rate was determined from deposition rate at a substrate temperature of 25°C, since these MO sources deposited on the substrate without decomposition at 25°C. In the case of $Y(PPM)_3$, 15.4% of evaporated metal is contained in the film deposited at 600°C, whereas only 5.7% in the case of Y $(DPM)_3$. The higher yield for $Y(PPM)_3$ suggests that $Y(PPM)_3$ is the better source material for Y, compared with $Y(DPM)_3$.

Preparation of YBCO thin film

On the basis of TGA and the deposition rates for individual metal chelate, $Y(PPM)_3$, Ba $(PPM)_2$ and $Cu(DPM)_2$ were selected as the source materials. Using these chelates, the MOMBE growth of YBCO films was carried out at the substrate temperature of 600°C for 1 hour. The thicknesses of the YBCO films were around 2700Å, and the growth rate was about 45Å/min. The atomic ratio Y:Ba:Cu in the film was determined to be 1:1.38:4.32 by ICP. As-deposited films were amorphous and insulating. The films were set in an infra-red furnace

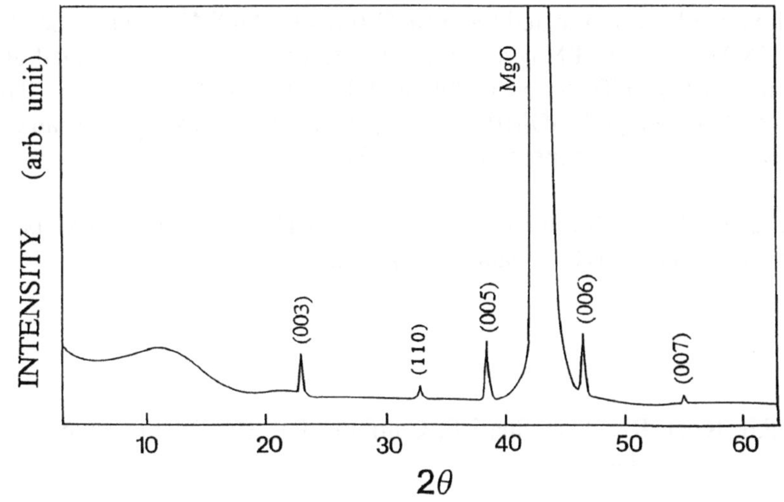

Fig. 2 XRD pattern of an annealed film.

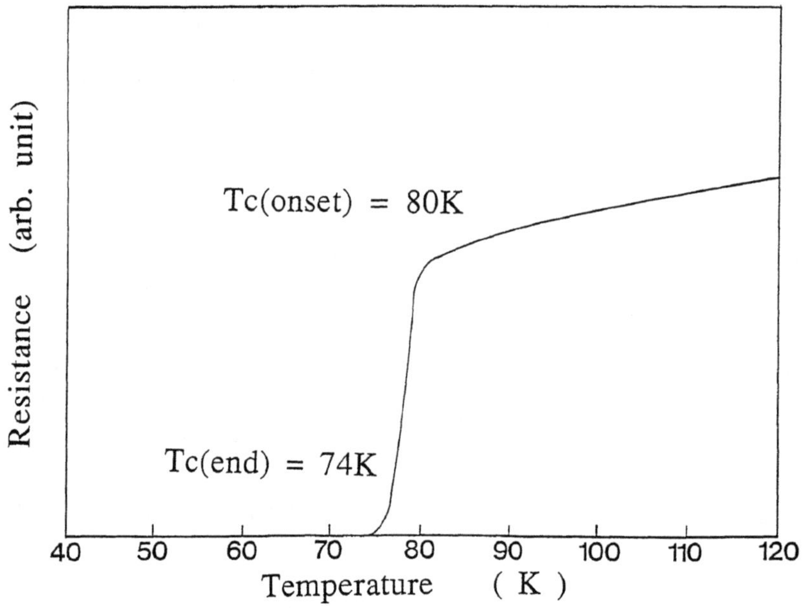

Fig. 3 Temperature dependence of the electrical resistance of an annealed film.

and annealed at 900°C for 1 hour in flowing oxygen. Electron probe microanalysis (EPMA) revealed that the annealed film is free from fluorine and carbon. Figure 2 shows the XRD pattern of an annealed film. All peaks were assigned as those from the (00n) planes of YBCO except YBCO (103) peak. This shows that the film grows along the c-axis preferentially orientated normal to the substrate. Figure 3 shows the temperature dependence of the resistance for the annealed film. The sharp superconducting transition was observed at Tc(onset) = 80K, and Tc(zero resistance) = 74K.

CONCLUSION

For the first time, we have fabricated the YBCO films by MOMBE using metal chelates of Y$(PPM)_3$, $Ba(PPM)_2$ and $Cu(DPM)_2$ as source materials. The films annealed at 900°C show the superconductivity with Tc(onset) = 80K and Tc(zero resistance) = 74K. This result suggests that the newly developed MOMBE technique has the possibility to be an effective method for the growth of the oxide superconducting films.

On leave of absence from [a)]Showa Electric Wire & Cable Co.,LTD., [b)]Nippon Sanso K.K., [c)]Dowa Mining Co.,LTD., and [d)]Iwatani Co.,LTD..

REFERENCE

1. Laibowitz RB, Koch RH, Chaudhari P, Gambino RJ (1987) Thin superconducting oxide films. Phys Rev B35: 8821-8823

Preparation and Properties of Y-Ba-Cu-O Thin Films on Flexible Ysz Substrates

SHIGERU OKUDA[1], NORIKI HAYASHI[1], SATOSHI TAKANO[1], HAJIME HITOTSUYANAGI[1], SEIICHIRO TERAI[2], and KIYOSHI HASEGAWA[2]

[1] Osaka Research Laboratories, Sumitomo Electric Industries, Ltd., Konohana-ku, Osaka, 554 Japan
[2] Research and Development Group, The Kansai Electric Power Co., Kita-ku, Osaka, 530 Japan

ABSTRACT

Y-Ba-Cu-O thin films were prepared on flexible polycrystalline YSZ (Yittria stabilized zirconia) by RF magnetron sputtering. Both as-grown films and post-annealed films had c-axis orientation, and Tc(R=0) of 90.3K and Jc of 1.2 x10^4 A/cm^2 were obtained after annealing at 950°C. Magnetic field dependence of Jc and strain effects by bending were measured. The degradation of Jc by compressive strain was very small compared to the tensile strain. Continuous deposition on a 20-cm long YSZ tape was tried by moving the tape, and resulted in Tc above 64K.

KEY WORDS: RF sputtering, thin film, Y-Ba-Cu-O film, critical current dencity

INTRODUCTION

Many processes are being investigated to apply high Tc ceramic superconductors to superconducting wires for power transmission lines and high field magnets. [1] As high Tc ceramic superconductors show strong anisotropic properties[2], [3], alignment of superconducting grains is very important to obtain high critical current density. High Jc values which exceed 10^6 A/cm^2 have been reported for the thin films on single crystal substrates of MgO[4] and $SrTiO_3$[5], [6], because they are single crystal and have c-axis alignment perpendicular to the substrates. We are developing a thin film process for high Tc superconductors to make superconducting tapes. Flexible YSZ film of 100 ∿ 200 μm thickness was chosen as a substrate material, as the substrate should have toughness and should not react with the thin film. We have prepared c-axis aligned Y-Ba-Cu-O thin films on flexible polycrystalline substrates, and measured their superconducting properties. Since it is possible to use high Tc ceramic superconductors in liquid helium as well as in liquid nitrogen, Jc at 4.2K and 77.3K were measured in high magnetic fields.

EXPERIMENTAL

Flexible polycrystalline YSZ substrates of 100 ∿ 200 μm thickness were used. Sputtering was carried out in an Ar +10% O_2 atmosphere with pressure of 8 x 10^{-2} Torr and the substrates were heated to 700°C. The growth rates were 0.2 ∿ 0.4 A/sec and the film thicknesses were 0.5 ∿1.0 μm. The samples were annealed in O_2 at 850 ∿950°C for 10 min, cooled down at the cooling rate of 10°C/min and annealed at 400°C for 3h subsequently.
Jc at 77.3K and 4.2K were measured resistively in magnetic fields of up to 1T using a normal-conducting magnet, and in magnetic fields up to 23T using a hybrid magnet at Tohoku University. Magnetic fields were applied parallel to the film surface and perpendicular to the current. Jc vs B characteristics of the films on YSZ were compared with those of the films on single-crystalline MgO (100). For MgO substrates, Sputtering was carried out in an Ar + 10% O_2 atmosphere with pressure of 3 x 10^{-2} Torr, and the substrates were heated to 700°C. The films on MgO were annealed in O_2 at 950°C for 30 min, cooled down at the cooling rate of 1°C/min and annealed at 400°C for 5h subsequently.
Strain effects on the superconducting properties were measured by bending the samples. To make a superconducting tape, a 20 cm-long YSZ tape was moved at the rate of 2.4 mm/h in the sputtering chamber, and a Y-Ba-Cu-O film of 1 μm thickness was deposited continuously. Tc and Jc of sections of the tape were measured.

RESULTS AND DISCUSSION

A. Effects of post-annealing

X-ray diffraction patterns showed that both as-grown films and post annealed films had c-axis orientation perpendicular to the substrate surface, and Tc (R=0) of 90.3K was obtained after annealing at 950°C. As-grown Y-Ba-Cu-O thin films on YSZ substrate by RF magnetron sputtering were composed of island-like grains which were partially united with each other. Figure 1 shows SEM micrographs of film surfaces annealed at 850°C, 900°C, and 950°C for 10 min. The degree of the union of grains increased as the annealing temperautre increased from 850°C to 950°C. Jc was improved by the union and the highest Jc was 1.2 x 10^4 A/cm^2 at 77.3K.

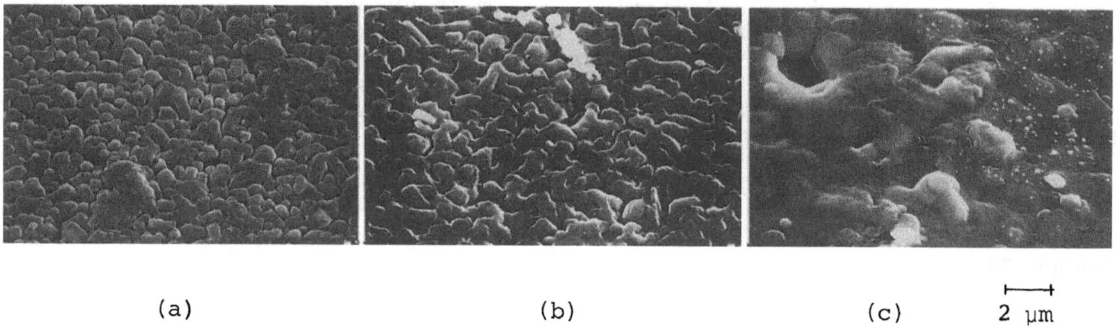

Fig. 1 Scanning electron micrographs of Y-Ba-Cu-O thin films annealed at (a) 850°C, (b) 900°C, (c) 950°C, and Jc were 1,200 A/cm^2, 9,000 A/cm^2, 12,000 A/cm^2 respectively.

The RHEED pattern of this film on YSZ showed a ring pattern, which means it is a poly-crystal. RHEED patterns of the thin films on MgO with Jc above 1 x 10^6 A/cm^2 showed streak patterns, which means they are single crystals. This is the reason for the difference in Jc between thin films on YSZ and on MgO.

B. Magnetic field dependence of Jc

Magnetic field dependence of Jc of Y-Ba-Cu-O thin films on MgO and YSZ at 77.3K is shown in Fig. 2, and Jc at 0T were 1.18 x 10^6 A/cm^2 and 3.50 x 10^3 A/cm^2 respectively. The decrease of Jc of the film on YSZ was larger than that on MgO, as the film on YSZ is polycrystalline. The ratios of Jc(1T)/Jc(0T) of the films on MgO and YSZ were 0.41 and 0.02, respectively.
Figures 3 (a) and (b) show the results of Jc measurement in magnetic fields of up to 23T at 4.2K. Jc(0T) of the film on YSZ at 4.2K was 3.49 x 10^4 A/cm^2, and dropped rapidly when the magnetic field was applied. However, Jc between 5T and 23T was almost constant, and the value was about 1.1 x 10^4 A/cm^2. Jc(0T) of the film on MgO was 2.1 x 10^7 A/cm^2 and Jc between 15T and 23T was also constant, and the value of Jc(15 ∿23T) was about 1.6 x 10^7 A/cm^2. As the decrease of Jc in the magnetic field was very small, it is hoped that high Tc superconductors will be applied to high field use at 4.2K.

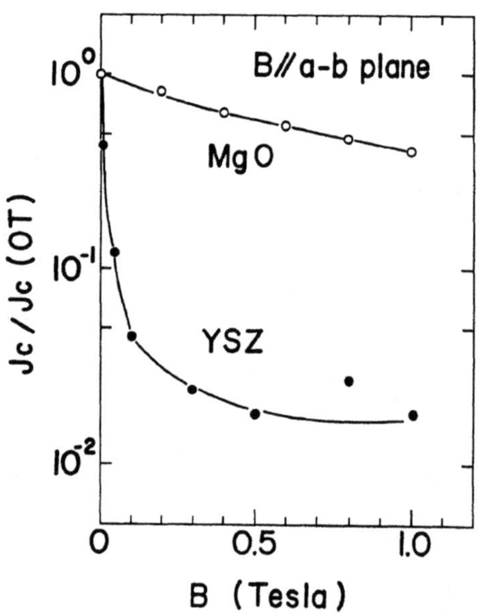

Fig. 2 Magnetic Field dependence of Jc at 77.3K

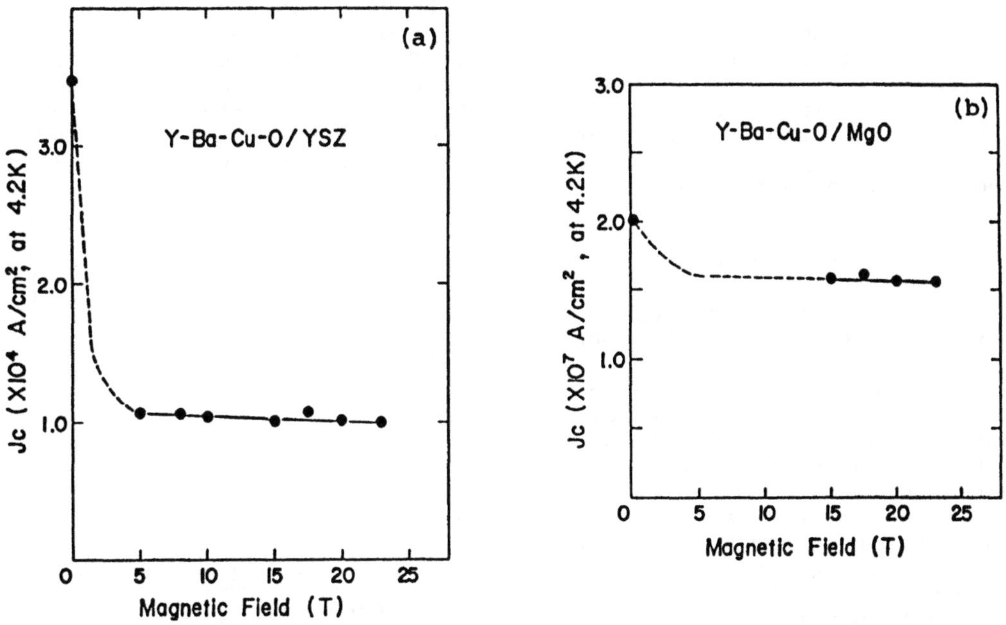

Fig. 3 Magnetic field dependence of Jc at 4.2K.
(a) Y-Ba-Cu-O film on YSZ, (b) Y-Ba-Cu-O film on MgO.

C. Effects of strain

Figures 4 (a) and (b) show the effects of compressive and tensile strain on Jc at 77.3K and 4.2K. Jc at 77.3K and 4.2K did not change under a compressive strain of up to 0.3%. When tensile strain was applied, Jc decreased rapidly and the value of Jc at 0.3% tensile strain was about 30% of the value without the strain.

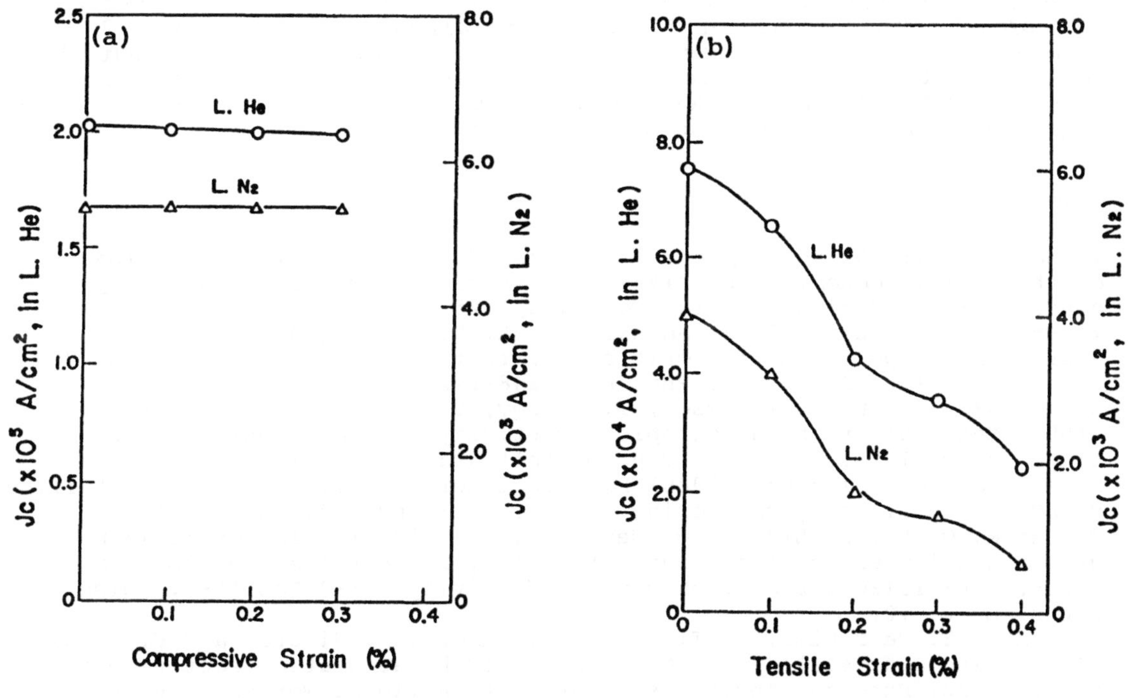

Fig. 4 Effects of strains on Jc at 77.3K and 4.2K
(a) Compressive strain. (b) Tensile strain

Interestingly, Tc(R=0) was raised from 82.2K to 84.6K as the compressive strain increased from 0% to 0.3%. Tc dropped from 84.1K to 82.0K when a tensile strain of 0.4% was applied to the film. The difference in superconductivity between tensile and compressive strain might be explained by the difference of the thermal expansion coefficients of YSZ and Y-Ba-Cu-O thin film. As the thermal expansion coefficients of YSZ and Y-Ba-Cu-O are $1.0 \times 10^{-5}/°C$ and $1.4 \sim 1.7 \times 10^{-5}/°C$, the films on YSZ substrates received tensile stress during cooling after annealing.

D. Continuous deposition on a YSZ tape

To make long superconducting tapes for power applications, a continuous deposition technique is necessary. We tried the deposition of Y-Ba-Cu-O thin film on a 20 cm-long YSZ tape. The length of the deposited film was 16 cm. Fig. 5 shows the results of Tc values which were measured at various sections of the tape and shows that this tape has Tc of above 64K. Parts of the beginning and the end of the deposition have Tc above the liquid nitrogen temperature, but Tc values of the center are below 77.3K. The reason for this is that we couldn't avoid deposition on the tape at the areas where compositions of Y, Ba, and Cu were not ideal in the chamber. An adequate mask between the moving tape and the target will solve this problem. The compositions of the tape measured by ICP were Ba, Cu rich at the center of the tape. The highest Jc value among the parts of the tape was 1.62×10^3 A/cm^2 at 4.2K.

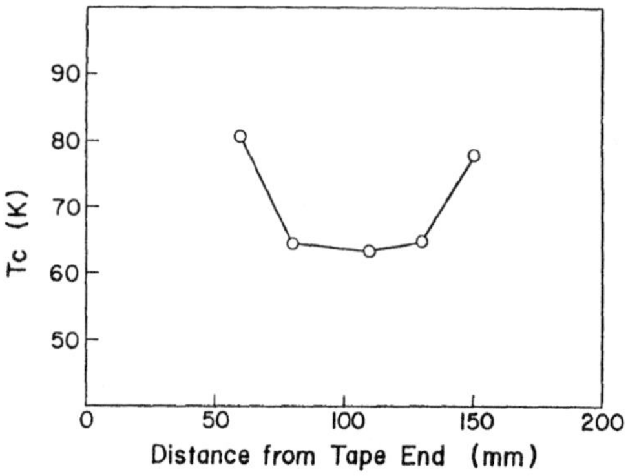

Fig. 5 Tc of Y-Ba-Cu-O thin film on YSZ tape by continuous deposition.

CONCLUSION

The Y-Ba-Cu-O thin film deposited on flexible polycrystalline YSZ showed Jc of 1.2×10^4 A/cm . The decrease of Jc from 5T to 23T was very small at 4.2K. Jc did not decrease under a compressive strain of up to 0.3%. It seems that these results are due to the difference of the thermal expansion coefficients between Y-Ba-Cu-O and YSZ. Continuous deposition was tried on a YSZ tape and a 16 cm-long superconducting tape with Tc above 64K was obtained.

REFERENCES

1. T. Nakahara (1989) Development of superconducting materials and impacts on the social economy. In: Kitazawa, Ishiguro (Eds) Advances in superconductivity. Springer-Verlag Tokyo: pp 725-730.
2. Y. Enomoto, T. Murakami, M. Suzuki, and K. Moriwaki (1987) Largely anisotropic superconducting critical current in epitaxially grown $Ba_2YCu_3O_{7-y}$ thin film. Jpn J Appl Phy 26: L1248-L1250.
3. S. W. Tozer, A. W. Kleinsasser, T. Penney, D. Kaiser, and F. Holtzbert (1987) Measurement of anisotropic resistivity and hall constant for single-crystal $YBa_2Cu_3O_{7-x}$. Phy Rev Lett 59: 1768-1771.
4. S. Tanaka and H. Itozaki (1988) High-Jc superconducting single crystalline HoBaCuO thin films by sputtering. Jpn J Appl Phys 27: L622-L624.
5. J. Geerk, Q. Li, G. Linker, O. Meyer, X.X.Xi (1989) Relation between growth quality and critical current density in sputtered YBaCuO thin films. In: ISTEC workshop on superconductivity. 1-3 Feb 1989. Oiso. Japan. pp 67-70.
6. Y. Bando, T. Terashima, K. Iijima, K. Yamamoto, K. Hirata, and H. Mazaki (1988) Single-crystal $YBa_2Cu_3O_{7-y}$ thin film by activated reactive evaporation. In: Extended abstracts of 5th international workshop on future electron devices, Topical meeting on high-temperature superconducting electron devices. 2-4 Jun 1988. Zao. Japan. pp 11-16.

Preparation of As-Deposited Superconducting YBaCuO Thin Films on Metallic Substrate by Magnetron Sputtering

M. FUKUTOMI[1], Y. TANAKA[1], T. ASANO[1], H. MAEDA[1], N. AKUTSU[2], K. HOSHINO[2], and H. TAKAHARA[2]

[1] National Research Institute for Metals, Tsukuba Laboratories, Tsukuba, Ibaraki, 305 Japan
[2] Central Research Laboratory, Mitsui Mining & Smelting Co., Ltd., Ageo, Saitama, 362 Japan

ABSTRACT

Superconducting YBaCuO thin films were prepared on metallic substrate in an attempt to fabricate a superconducting tape. Deposition was made by co-sputtering from three oxide targets, CuO, Y_2O_3, and $BaCuO_2$. MgO and BaF_2 thin films were found to be an effective buffer layer. All films with good superconducting properties were prepared outside the region of head-on negative ion flux from each target. The highest zero resistance temperature obtained so far of the YBaCuO thin films on metallic tape was 84.0 K in its as grown state, and a critical current density was $4.8 \times 10^3 A/cm^2$ at zero field and 77.4 K.

KEY WORDS: YBaCuO thin film, metallic substrate, three-target magnetron sputtering, buffer layer, negative ion bombardment

INTRODUCTION

There have been many reports of YBaCuO thin films deposited on single crystal substrates, such as MgO, $SrTiO_3$, and yttria-stabilized zirconia. Most films, however, have been prepared mainly for the electronic applications. Thin film technologies of high Tc superconductors have many other potential applications, including the fabrication of conductors for high field superconducting magnets, electromagnetic shields, etc. It is worthwhile, therefore, to develop superconducting films on various substrate materials suitable for the purpose intended.

An object of this study is to determine the feasibility of preparing YBaCuO films on metallic substrates in an attempt to fabricate a superconducting tape [1]. Deposition method employed was a three-target rf magnetron sputtering. Studies were focused on the selection of buffer layers to prevent the adverse reaction between the superconducting film and the metallic substrate. Particular attention was given to the preferential resputtering effect due to the bombardment of negative ions during sputtering. The substrate-cathode arrangement was described to achieve stoichiometric film composition.

PREPARATION OF BUFFER LAYER

Buffer layers examined in this study include MgO, BaF_2, CaF_2 and LiF, while Hastelloy-X was used as substrate. Buffer layers were magnetron-sputtered onto the substrate under an argon pressure of 20 mTorr using a sintered compound target of each material. There was no intentional substrate heating. The as-sputtered films were always amorphous. However, MgO buffer layers with (200) preferred grain orientation were easily obtained by post-deposition annealing at 1000 °C for 5 min in air. On the other hand, X-ray diffraction pattern of the fluoride film gave no evidence of preferred orientation or texture. We will discuss later a difference in superconducting properties of the films grown on a crystallized and non-crystallized buffer layer.

DEPOSITION OF YBaCuO THIN FILMS AND RESULTS

Arrangement Of Targets And Substrate To Reduce A Preferential Resputtering Effect

A schematic view of the cathode and the substrate is shown in Fig.1. In our sputtering system, the three targets installed in a triangular arrangement were designed to be inclined by 30 degrees in such a way that each target pointed at the substrate holder [2]. According to our previous work [3], a marked compositional change was observed in the film prepared using inclined target. It should be noted that films once formed using the two targets (Y_2O_3 and CuO) were etched away when an rf power was applied to the $BaCuO_2$ target. From these experimental results, we concluded that resputtering due to oxygen anion, mainly produced from barium oxide, is responsible for a drastic film compositional change. This result agrees well with that by Kester et al [4], who proposed a method to predict negative ion resputtering in thin films. In the present study, therefore, all the films were deposited using non-inclined target arrangement. As shown in Fig.1, the substrate was placed outside the region of head-on negative ion flux from each target. This geometrical arrangement prevented the detrimental negative oxygen ion bombardment of the growing films [5,6].

Film Preparation And Electrical Measurement

The targets (100 mm in diameter) used were a sintered Y_2O_3, $BaCuO_2$ and CuO. Films were prepared by simultaneous cosputtering and their composition was adjusted to achieve a correct stoichiometry by regulating the power supplied to each target. The substrate used was a Hastelloy-X tape ($20x3x0.1\ mm^3$) precoated with a buffer layer. They were mounted on a heated substrate holder which was rotated at about 6 rpm during the run. Typical sputtering conditions were as follows: sputtering pressure (Ar-50%O_2); 2 mTorr, substrate holder temperature; 600 - 720°C, film thickness;0.7-1.5 μm(deposition rates; 0.5-3 Å/s). Pure argon gas was flowed over the target and pure oxygen was introduced near the growing films separately. After deposition, the system was back-filled with oxygen to an atmospheric pressure. The film temperature was decreased down to 500°C, held at that temperature for 1 h and finally slowly cooled to room temperature. The composition of the film was determined by energy-dispersive X-ray microanalysis (EDAX) using $Y_1Ba_2Cu_3O_x$ single crystal as the standard.
Measurements of the temperature dependence of the resistance and transport critical current for tape samples were performed using a standard dc four-probe method. Silver paste was used to form contacts. The temperature was measured by Au7%Fe-Chromel thermocouples. No special patterning of the films was conducted for Jc measurements.

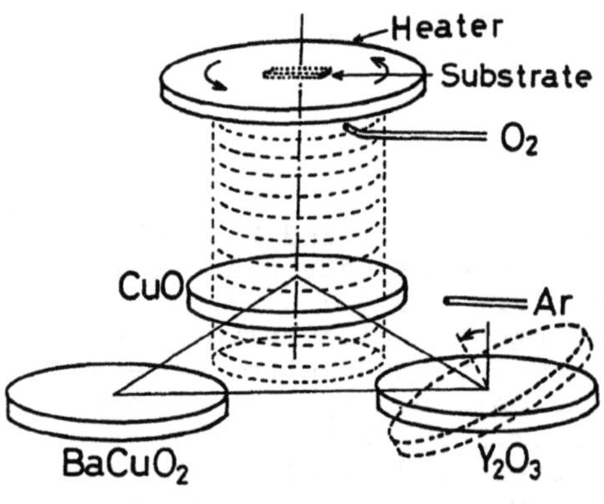

Fig.1 Schematic view of three-target rf magnetron sputtering system

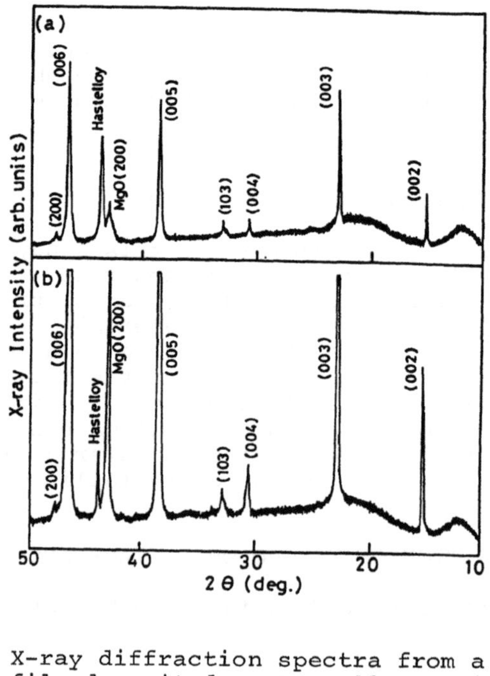

Fig.2 X-ray diffraction spectra from a YBaCuO film deposited on Hastelloy-X with an amorphous (a) and pre-crystallized (b) MgO buffer layer

Preferential Orientation And Scanning Electron Microscope (SEM) Observation Of The YBaCuO Films

Figure 2 shows the X-ray diffraction pattern of YBaCuO thin films deposited on Hastelloy-X with an amorphous (a) and a (200) preferentially oriented MgO buffer layer (b). Both films are highly oriented with the c-axis perpendicular to the substrate. The diffraction pattern includes very weak peaks from (n00) planes indicating that the films contained a small amount of grains with a-axis perpendicular to the surface also. From the film pattern (a), we can see that the crystallization of amorphous MgO buffer layer proceeded with (200) preferred orientation during the deposition of YBaCuO films. The crystallization of the amorphous MgO buffer layer may help the YBaCuO orthorhombic phase grow with c-axis perpendicular to the substrate. Preparation of YBaCuO films on Hastelloy-X with a BaF_2, CaF_2 and LiF buffer layer was also studied, but only a randomly oriented polycrystalline film has been obtained so far. Figure 3 is a typical scanning electron micrograph of a surface of the YBaCuO films. The films had a smooth and shiny surface. Note the morphology which consists of grains of about 0.5 μm in size with no discernible defect such as crack and exfoliation. The cross-sectional view indicated that grains have a columnar morphology.

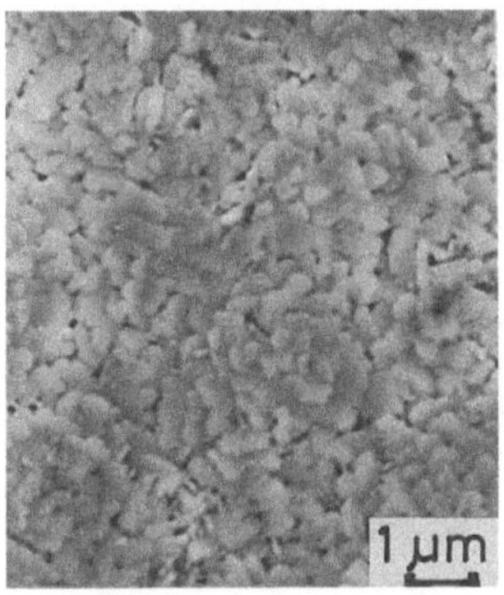

Fig.3 SEM micrograph of the surface of YBaCuO thin film prepared on Hastelloy-X

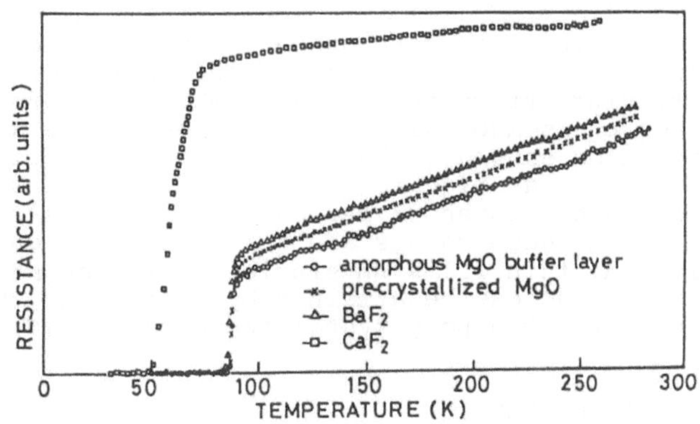

Fig.4 Resistance as a function of temperature for YBaCuO thin film on Hastelloy-X with different buffer layers

Superconducting Properties Of YBaCuO Thin Films Grown On The Metallic Substrate

Curves of the resistance versus temperature of the films are shown in Fig.4. In this figure the best result of YBaCuO films obtained using each buffer layer is shown. Of all the buffer layers examined in the present study, the best film was obtained on amorphous MgO buffer layer and had a zero resistance temperature of as high as 84 K in its as grown state. The resistivity of the film was measured and found to be $0.5x10^3$ μΩ·cm at 300 K. The film composition determined by EDAX was Y:Ba:Cu = 1.0:1.9:2.8. As far as the transition curve is concerned, the YBaCuO films using both pre-crystallized MgO and BaF_2 buffer layer also gave almost the same results as is shown in Fig.4. Films grown on a Hastelloy-X with CaF_2 buffer layer exhibited rather poor superconducting properties with zero resistance temperature around 50 K, while those on LiF buffer layer showed no superconductivity due to severe substrate reactions.

Figure 5 shows temperature dependence of critical current density of YBaCuO films grown on the metallic substrate. For the best films, which were prepared on an amorphous MgO buffer layer, its critical current density was $5.7x10^4$ A/cm^2 at 4.2 K and $4.8x10^3$ A/cm^2 at 77.4 K. In the present study, an amorphous MgO film was found to be the most promising buffer layer, followed by a pre-crystallized MgO with (200) preferred orientation and an amorphous BaF_2 layer in this order.

Fig.5 Critical current densities as a function of temperature for YBaCuO thin film on Hastelloy-X with a MgO buffer layer

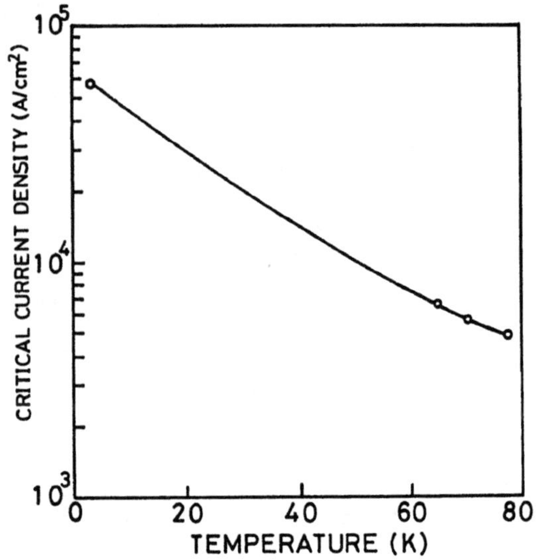

CONCLUSION

YBaCuO thin films were prepared on metallic substrate using a three-target cosputtering. Particular attention was given to the preferential re-sputtering effect due to the bombardment of negative ions during sputtering. An amorphous MgO thin film was found to be a promising buffer layer. The best films showed a sharp resistive transition with zero resistance temperature as high as 84 K. The zero field critical current density measured by transport method was 5.7×10^4 A/cm^2 at 4.2 K and 4.8×10^3 A/cm^2 at 77.4 K. These values are still very low compared with those reported for epitaxial thin films on single crystal substrates. The point is how high values of Jc can be attained by optimizing the deposition conditions in highly-oriented polycrystalline superconducting films.

ACKNOWLEDGEMENT

Authors would like to thank Mr.J.Saitoh of Mitsuba Electric MFG. Co.,LTD. for his support and helpful discussions.

REFERENCES

1.Fukutomi M,Machida J,Tanaka Y,Asano T,Wada H,Maeda H(1989) Effect of various buffer layers on the properties of high-Tc superconducting films deposited onto metallic substrate.MRS Int'l.Mtg. on Adv. Mats. vol 16, 863-868

2.Char K,Kent AD,Kapitulnik A,Beasley MR,Geballe TH(1987) Reactive magnetron sputtering of thin-film superconductor $YBa_2Cu_3O_7$. Appl.Phys.Lett.51 :1370-1372

3.Akutsu N,Fukutomi M,Katoh K,Takahara H,Tanaka Y,Asano T,Maeda H (submitted) Study of high-Tc Y-Ba-Cu-O films by the three target magnetron sputtering. Jpn.J.Appl.Phys.

4.Kester DJ,Messier R(1986) Predicting negative ion resputtering in thin films. J.Vac.Sci.Technol.A4(3):496-499

5.Terada N,Ihara H,Jo M,Hirabayashi M,Kimura Y,Matsutani K,Hirata K,Ohno E, Sugise R,Kawashima F(1988) Sputter synthesis of $Ba_2Y_1Cu_3O$ as-deposited superconducting thin films from stoichiometric target-- a mechanism of compositional deviation and its control. Jpn.J.Appl.Phys.27:L639-L642

6.Sandstrom RL,Gallagher WJ,Dinger TR,Koch RH,Laibowitz RB,Kleinsasser AW, Gambino RJ,Bumble B,Chisholm MF(1988) Reliable single-target sputtering process for high-temperature superconducting films and devices. Appl.Phys. Lett.53:444-446

Formation Method of Ybco Thin Films with Large Critical Current Ic

H. KAJIKAWA, Y. FUKUMOTO, S. HAYASHI, R. OGAWA, and Y. KAWATE

Superconducting & Cryogenic Technology Center, Kobe Steel, Ltd., 5-5, Takatsukadai 1-chome, Nishi-ku, Kobe, 673-02 Japan

ABSTRACT

The effect of metallic composition on the critical current (density) of YBCO thin films was investigated. Large critical current Ic of more than 1 A at 60 K were obtained in the films with the Cu-poor composition of $Y_{l\%}Ba_{m\%}Cu_{n\%}Ox$ (l=16-19, m=37-41, n=41-45, l+m+n=100). These films were almost free from the insulating second phase $BaCuO_2$ and comprised of the densely packed mixture of a-axis and c-axis oriented grains forming the macroscopic current path.

KEY WORDS: RF-diode sputtering method, two-step heat treatment, $YBa_2Cu_3O_{7-y}$, $SrTiO_3(100)$, critical current density Jc.

INTRODUCTION

We have investigated systematically the effect of non-stoichiometry in metallic composition and heat treatment temperature on the crystal orientations of sputtered YBCO films on $SrTiO_3(100)$ substrates [1], [2], [3]. It was found that the crystal orientations were dependent mainly on the yttrium content, and that the c-axis oriented grains, the mixture of c-axis and a-axis oriented grains and the a-axis oriented grains became predominant in turn as the yttrium content increased. It was also found that the yttrium content classifying the types of the crystal orientations decreased as the heat treatment temperature for the crystallization was raised. And it was suggested that large Ic would be obtained in the films comprised of the mixture of both a-axis and c-axis oriented grains [2],[3].

In this work, we investigate the procedures to obtain the superconducting YBCO films with large critical current Ic by using the RF-diode sputtering method followed by the two step heat-treatment.

EXPERIMENTAL

There are two major methods to make superconducting YBCO films by using the sputtering techniques. One method is to make in-situ crystallized films at high substrate temperature. The other method is to deposit amorphous films at substrate temperature of lower than 350°C and to crystallize them by the post-heat treatment. In this work, the latter one was used noting that the film composition is well adjustable at low substrate temperature. The sputtering conditions are shown in Table 1. The surface composition of the target varies depending on the sputtering time. Therefore, the film composition can be adjusted by changing the pre-sputtering time. In order to form the orthorhombic $YBa_2Cu_3O_{7-y}$ phase, a two-step heat-treatment was made in an oxygen atmosphere as shown in Fig.1. The heating and cooling rates were carefully kept constant for each film in order to obtain the good reproducibility of the results.

The metallic compositions of the films were measured by electron probe micro-analysis calibrated using the powder of sintered $YBa_2Cu_3O_{7-y}$ as a standard.

The crystal structure was identified by X-ray diffraction measurement, and the surface and cross-sectional morphologies were observed by scanning electron microscopy (SEM).

The superconducting properties of Tc and Ic were measured by the conventional d.c. four-point-probe method without external magnetic field. The strips used for the measurement were about 2 mm wide and 10 mm long. The electrodes at intervals of 1 mm were formed on the surface of the specimen by baked silver paste. The contact resistance was below 100 micro-ohm cm^2.

RESULTS AND DISCUSSION

Figure 2 shows the crystal orientations of the films having the metallic composition of $Y_{l\%}Ba_{m\%}Cu_{n\%}$ (l=10-30, m=30-50, n=40-60, l+m+n=100) under the heat treatment at 880 °C. The critical current densities (Jc) of the films of more than 1000 A/cm^2 at 70 K or at 77 K are denoted in the same figure. The ones at 77 K are marked by asterisk. The definition of the film types are shown in Table 2 [2]. The ac-type films were distributed in the region where the yttrium content is 16-19%, corresponding to our previous work [2].

The large critical current densities were obtained in the ac-type films with the metallic composition of $Y_{l\%}Ba_{m\%}Cu_{n\%}Ox$ (l=16-19, m=37-41, n=41-45, l+m+n=100). The large critical current density (Jc) might be related to the content of the insulating second phase $BaCuO_2$ and the film morphology. The ratios of the maximum X-ray diffraction peak for $BaCuO_2$ to that for $YBa_2Cu_3O_{7-y}$ were plotted on the 3-phase diagram as shown in Fig.3. The maximum peak for $BaCuO_2$ appeared mainly at $2\theta=29.3°$ and that for $YBa_2Cu_3O_{7-y}$ was I(005). Supposing that these ratios reflect the content of $BaCuO_2$ in the films, the region of the large Jc were projected roughly to the region of less $BaCuO_2$. The surface and cross-sectional morphologies of an typical ac-type film were shown in Fig 4. It is observed that the crystal grains were less than 0.8 micron in diameter and were closely packed.

The critical current Ic of more than 1 A at 60 K could be obtained in the ac-type films. Figure 5 shows the temperature dependence of the critical current of the two ac-type films having Tc of 84.6 K and 77.6 K. With decreasing the temperature, each curve shows abrupt increase in Ic below each transition temperature and each Ic reached 3 A around 20 K. The abrupt increase in Ic below Tc means that the films are comprised of strongly coupled grains. The large Ic suggests the possibility of the films for practical applications.

CONCLUSIONS

The effect of metallic composition on the critical current (density) of the YBCO thin films was investigated using the RF-diode sputtering method and the two-step heat treatment method. Large critical current Ic of more than 1 A at

60 K were obtained in the films with the Cu-poor composition of $Y_{l\%}Ba_{m\%}Cu_{n\%}Ox$ (l=16-19, m=37-41, n=41-45, l+m+n=100), deposited in the amorphous state at RF power of 300 W and at a temperature of <350°C, and then crystallized at 880°C. In these films, the insulating second phase $BaCuO_2$ was reduced and the mixture of a-axis and c-axis oriented grains were closely packed to form the macroscopic current path.

REFERENCES

1. Kajikawa H, Hase T, Okuda M, Nishimura K, Kawate Y (1988) Effect of heat treatment conditions on the crystal orientations and morphology of Y-Ba-Cu-O superconducting films. Extended Abstract of 5th Workshop on FED: 95-100
2. Kajikawa H, Hase T, Okuda M, Nishimura K, Kawate Y (1988) Effect of non-stoichiometry in metallic composition and heat treatment temperature on the superconducting properties of sputtered Y-Ba-Cu-O films. Progress in High Temperature Superconductivity, Vol.15, World Scientific: 226-231
3. Hase T, Kajikawa H, Kannan H, Kawate Y (1989) Reduction of second phase $BaCuO_2$ in Y-Ba-Cu-O polycrystalline superconducting thin films. Extended Abstract of ISEC'89: 70-73

Table 1. Sputtering conditions.

Target	$Y_1Ba_{5.7}Cu_{6.0}O_x$
Substrate	$SrTiO_3$ (100)
Sputtering gas	$0.5>O_2/(Ar+O_2)>0.2$
Gas pressure	~1 Pa
RF power	300 W
Substrate temperature	< 350°C
Deposition rate x time	0.35-0.45 µm/h x 2 h

Table 2. Definition of film types.

Ⓡ ; r-type film with $D_1>3.3$
Ⓐ ; a-type film with $D_1<3.3$, $D_2<0.01$
(ac) ; ac-type film with $D_1<3.3$, $0.01<D_2<0.17$
Ⓒ ; c-type film with $D_1<3.3$, $D_2>0.17$

D_1=Max.[I(110),I(103)]/Max.[I(200),I(002)]
D_2=I(002)/I(200)
D_1=3.3 and D_2=0.17 for sintered bulk samples.

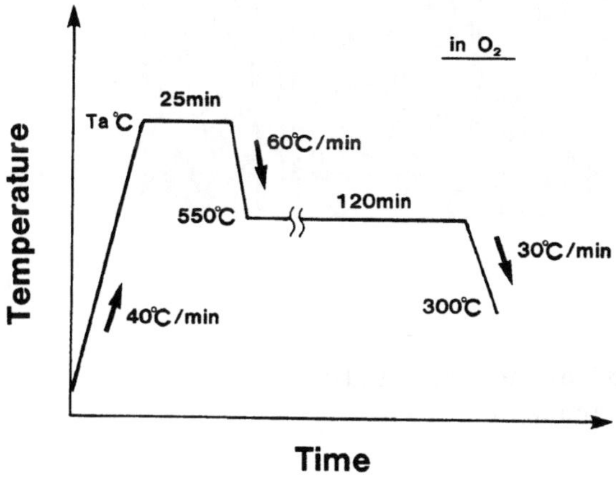

Fig.1. Heat treatment process Ta=880 C.

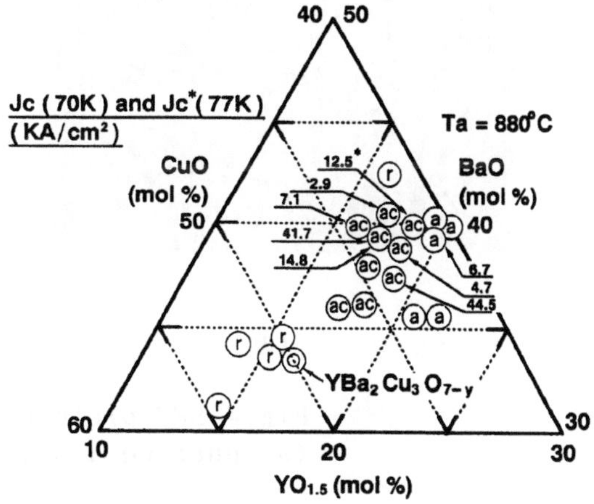

Fig.2. Dependence of crystal orientations and Jc on film composition.

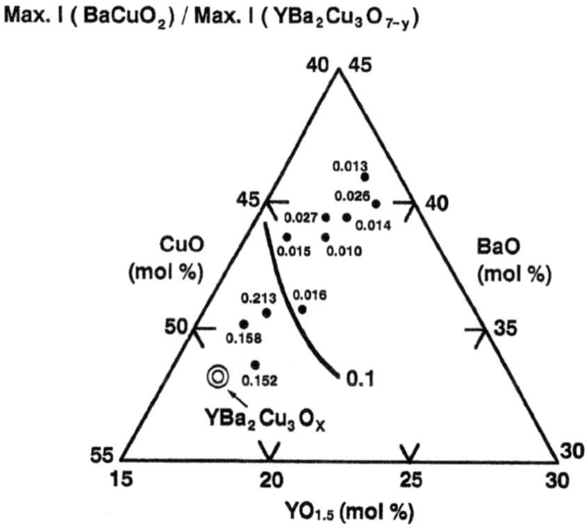

Fig.3. Ratio of maximum peak for $BaCuO_2$ to that for $YBa_2Cu_3O_{7-y}$.

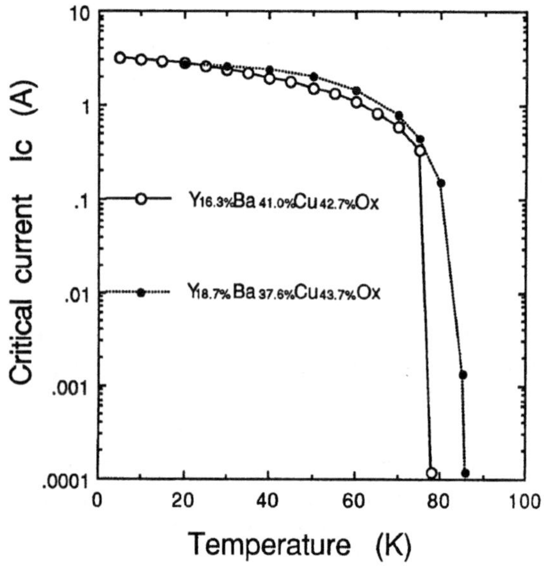

Fig.5. Dependence of critical current on the temperature.

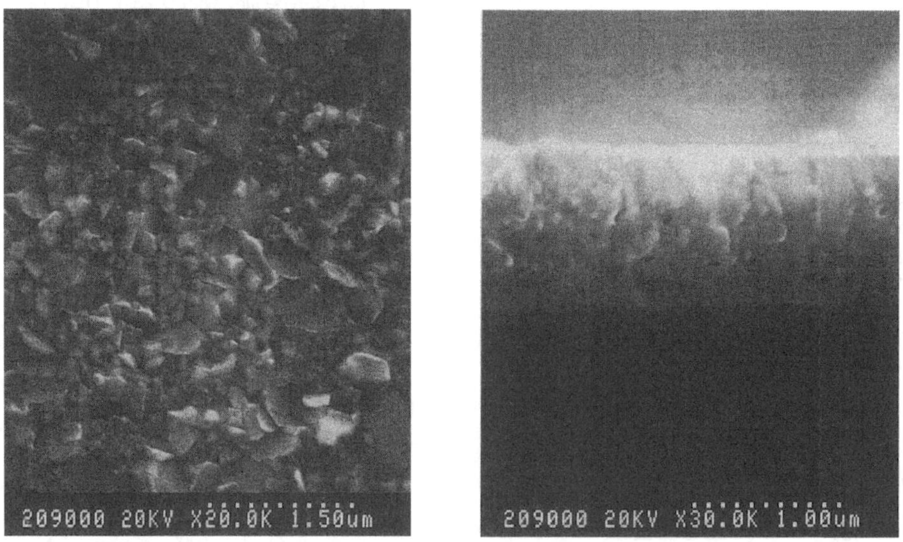

(a) (b)

Fig.4 SEM pictures of an ac-type film (a) surface and (b) cross-section.

Preparation of Superconducting Ti Doped Y-Ba-Cu-O Films on Metallic Substrates with a Thin Buffer Layer

JOJI SHINOHARA[1], KAZUTOSHI INOUE[2], MICHIAKI NOZAWA[2], and SHUNJI IDO[2]

[1] Research Institute, Ishikawajima-Harima Heavy Industries Co., Ltd., 1-15, Toyosu 3-chome, Koto-ku, Tokyo, 135 Japan

[2] Department of Electrical Engineering, Faculty of Engineering, Saitama University, Shimo-Ookubo 255, Urawa 338 Japan

ABSTRACT

Ti-doped Y-Ba-Cu-O superconducting films were prepared on single crystals and metal substrates by RF magnetron sputtering method. Post-growth annealing was carried out in an oxygen atmosphere. Ni base superalloy Inco-718 was used successfully as the substrate for high-Tc superconducting thin films with a thin buffer layer of MgO. The films exhibit sharp transition profile of the resistivity. The high-Tc superconducting phase of the thin film are higher stable than Ti-free sample. A superconducting transition was obtained Tc_{onset}=83K. The film composition was estimated by inductively coupled plasma emission spectrometry (ICP). The composition of Ti-Y-Ba-Cu film was 0.02-0.31 :1:2.3:3.4.

KEY WORDS: Y-Ba-Cu-O, high-Tc superconductor, rf magnetron sputtering, buffer layer, thin film

INTRODUCTION

After discovery of the 90 K class superconductivity in the mixed oxide system Y-Ba-Cu-O, a considerable research effort has been carried out on the superconducting thin films. It has been difficult to form superconducting films on metal and Si substrates because of inter diffusion between superconducting films and the substrates [1],[2]. For the application of superconducting thin film on electric device, electromagnetic shields, etc., however, it is necessary to develop the technique which can make the superconducting films on Si and metallic substrates.

It is well known that the superconducting properties of Y-Ba-Cu-O compounds often depend on fabrication process and heat treatment conditions. For preparation of the superconducting bulk materials, it was confirmed that a Ti-doping for Y-Ba-Cu-O was effective to improve the properties of the superconductor $Y_1Ba_2Cu_3O_{7-X}$. This approach was also useful for preparation of the superconducting thin films.

Ti-doped Y-Ba-Cu-O has some advantage of its properties, as follows.

(1) $Y_1Ba_2Cu_3O_{7-X}$;1-2-3 phase is more stable

(2) a critical superconducting transition temperature, Tc, becomes steeply

(3) an annealing process can be shortened

In this paper we report the Ti-doped Y-Ba-Cu-O films and the preparation of superconducting films on metallic substrates with buffer layer. The preparation of buffer layers which has controlled the crystalline has been described in proceeding paper[3].

EXPERIMENTAL

Figure 1 shows the temperature dependence of the resistivity of Ti-doped Y-Ba-Cu-O bulk materials. The $Ti_1Y_1Ba_3Cu_3O_X$ superconductor was made from Y_2O_3, $BaCO_3$, CuO and Ti_2O_3 powders. The mixed powder was ground, isostatically pressed, and reacted at 950°C for 1 hour in 1 atm atmosphere, and quenched after cooling down to 500°C. $Y_1Ba_2Cu_3O_X$ was also made by same the manner and the heat treatments. On an ordinary preparation process of $Y_1Ba_2Cu_3O_X$, it is necessary to add annealing at 500°C for 8-12 hours in oxygen after sintered process.
Films have been deposited in rf magnetron sputtering system which has a 6 inch target electrode. The sputtering target was a sintered Y-Ba-Cu-O compound which composition was $Y_1Ba_{2.5}Cu_{3.5}O_X$. The typical sputtering conditions are shown in Table 1.

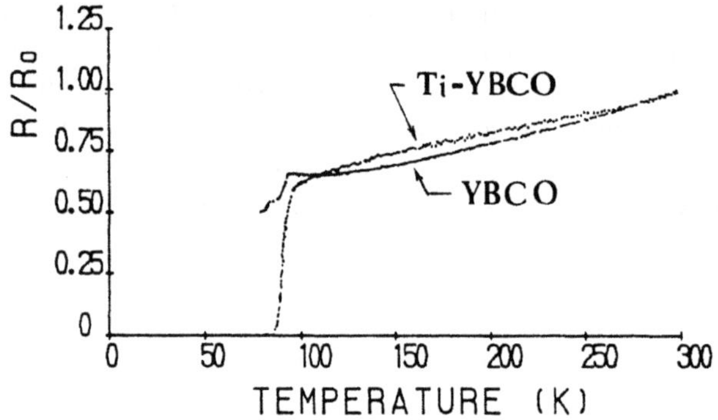

Fig.1 Temperature dependence of resistivity for Ti-doped and non-doped YBaCuO which are quenched at 500°C.

Table1 Sputtering conditions

Target compound	$Y_1Ba_{2.5}Cu_{3.5}O_x$
Sputtering gas	Ar
Gas pressure	0.67Pa
Substrate temperature	100°C
Film growth rate	0.1~0.28µm/hour
Sputtering power	200W

Ti-Y-Ba-Cu-O films have been deposited onto MgO(100) crystal and nickel based alloy, Inco 718, with buffer layer. In order to obtain a constant composition of sputtered Y-Ba-Cu-O film, the substrate temperature was kept low temperature (at 100°C) and the films was heat treated after the deposition. As a result of preliminary sputtering test, Ti-dopants were used pellets of $BaTiO_3$. The amount of Ti-dopants in Y-Ba-Cu-O film was controlled by the numbers of $BaTiO_3$ pellets.
The composition of Ti-Y-Ba-Cu-O films were analyzed by induced coupled plasma emission spectroscopy (ICP). The temperature dependence of resistivity was measured by a conventional four-probe technique.

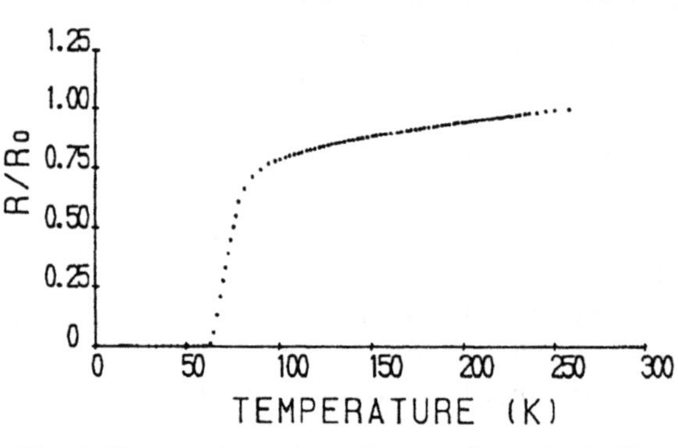

Fig.2 Temperature dependence of resistivity for YBaCuO film on MgO (100) single crystal substrate.

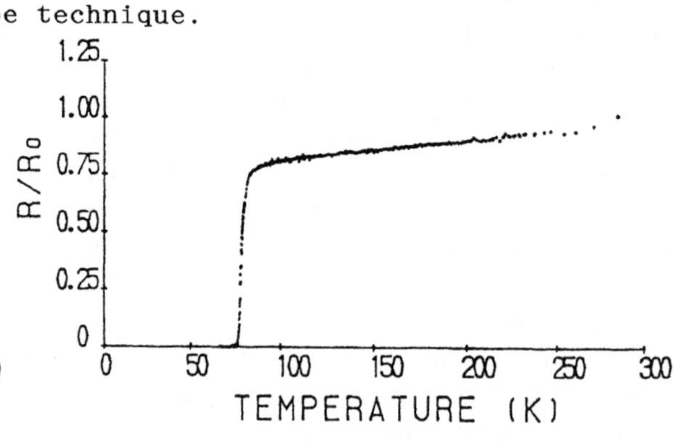

Fig.3 Temperature dependence of resistivity for Ti-doped YBaCuO film on MgO (100) single crystal substrate.

RESULTS & DISCUSSION

Ti-Y-Ba-Cu-O on MgO single crystal substrate

Figure 2 shows the dependence of electric resistance on temperature for the Y-Ba-Cu-O film on MgO(100) single crystal substrate. The film was sintered at 930°C for 1 hour and annealed at 550°C for 2 hours. A superconducting characteristics with a Tc_{onset} 93 K and a Tc_{end} around 60 K was observed. The composition of film was Y:Ba:Cu = 1:2.0:3.5 and the room temperature resistivity of film was 3.6 mohm·cm.
Figure 3 shows the dependence of electric resistance on temperature for Ti-Y-Ba-Cu-O film on MgO(100) single crystal substrate. The film was sintered at 930°C for 1 hour and cooled down without annealing. The composition of the film was Ti:Y:Ba:Cu= 0.08:1:2.3:3.4. It was considered that there was no effect of the variation in Ti concentration, although the ratio of Ti/Y was changed in the region of 0.02-0.31. The resistivity of Ti doped film dropped steeply at superconducting transition point. The room temperature resistivity of film was obtained 0.2 mohm·cm and this value was comparable to that of a single crystal. From these results, it is found that a Ti-doping was effective to improve the superconducting characteristics (of Y-Ba-Cu-O).

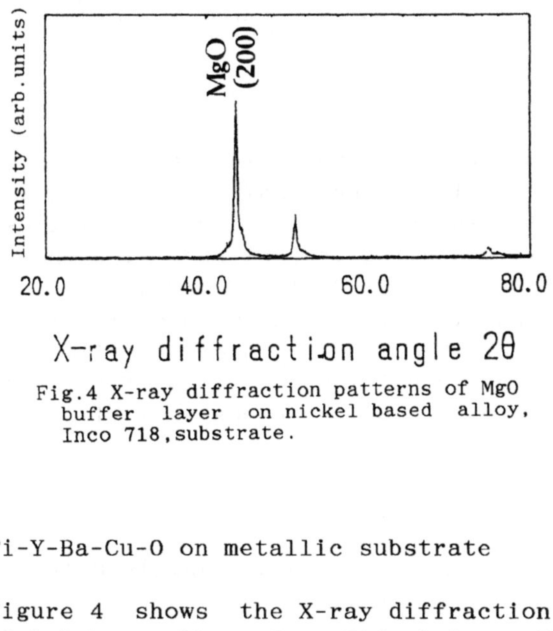

Fig.4 X-ray diffraction patterns of MgO buffer layer on nickel based alloy, Inco 718, substrate.

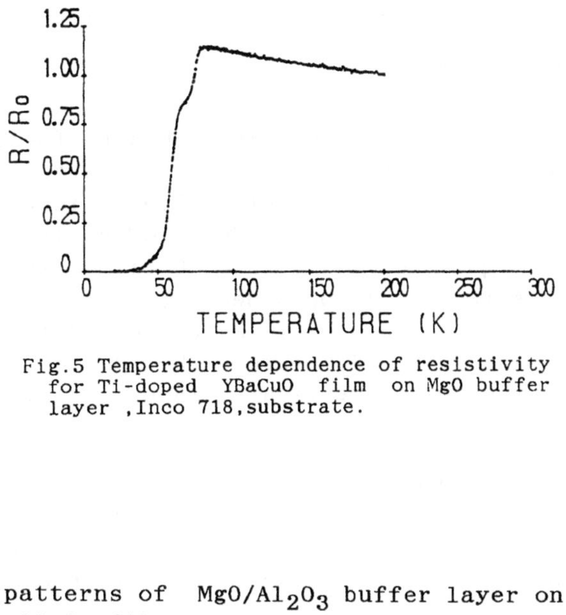

Fig.5 Temperature dependence of resistivity for Ti-doped YBaCuO film on MgO buffer layer, Inco 718, substrate.

Ti-Y-Ba-Cu-O on metallic substrate

Figure 4 shows the X-ray diffraction patterns of MgO/Al_2O_3 buffer layer on nickel base alloy, Inco 718 plate. The Al_2O_3 film was used for diffusion barrier from metallic substrate. The MgO buffer layer was sputtered at 400°C and the film thickness was 1 µm. A sharp peak of MgO(200) is observed and other small peaks are that of nickel base alloy, Inco 718 substrate.
Figure 5 shows the temperature dependence of resistivity on Ti-Y-Ba-Cu-O film. This sample was made and heat treated at the same time as that of Figure 3. The resistivity of this film rose slightly as the temperature fall in spite of its low resistivity which was about 0.2 mohm·cm at room temperature. The superconducting characteristics with a Tc_{onset} 79 K and a Tc_{end} around 30 K was observed.
Figure 6 shows the X-ray diffraction patterns of Ti-Y-Ba-Cu-O films on MgO (100) single crystal and metallic substrate. The diffraction peaks of Ti-Y-Ba-Cu-O on MgO (100) single crystal correspond to 1-2-3 phase of Y-Ba-Cu-O, showing highly oriented growth of the c-axis perpendicular to substrate.

Similar tendency of the crystal orientation has been observed in the film on metallic substrate (b), although their intensities were relatively small. Figure 7 shows AES(Auger Electron Spectroscopy) analysis of the Ti-Y-Ba-Cu-O on $MgO-Al_2O_3$/Inco 718 substrate. In Figure 7, the diffusion of Ni from the metallic substrate was observed at a depth of a third in Al_2O_3 layer, and Cu from Ti-Y-Ba-Cu-O was detected in MgO layer.

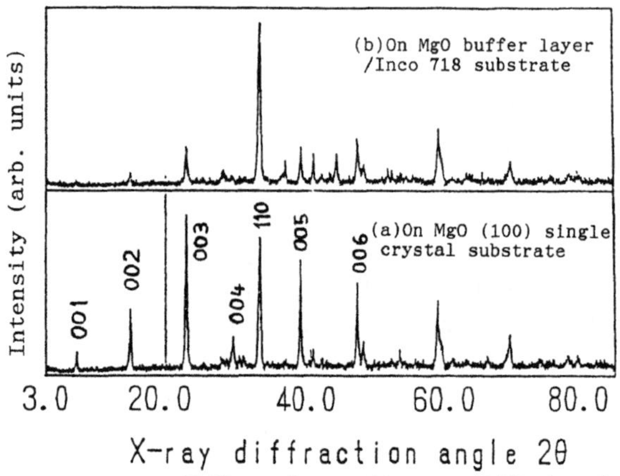

Fig.6 X-ray diffraction patterns of Ti-doped YBaCuO film on (a)MgO single crystal and (b)MgO buffef layer,Inco 718,substrate.

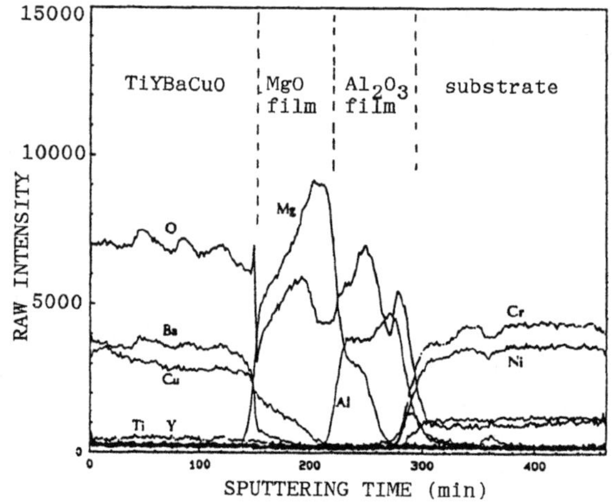

Fig.7 AES depth profile of YBaCuO film on MgO buffer layer/Inco 718 substrate. (After heat treatment of 930°C, 1hr)

CONCLUSION

Ti-doped Y-Ba-Cu-O thin film were formed on MgO(100) substrates and metallic substrates by rf magnetron sputtering. It is found that a Ti-doping to Y-Ba-Cu-O film is effective to improve the superconducting characteristics, the temperature dependence of resistivity, annealing process and 1-2-3 phase of Y-Ba-Cu-O. Ti-Y-Ba-Cu-O film composition was 0.02-0.31:1:2.3:3.4.

Although we obtained the superconducting film of Ti-Y-Ba-Cu-O on metallic substrate and the film exhibited Tc_{onset} of 70 K, the Tc_{end} of the film was low as 30 K. To improve the superconducting transition and the buffer layer, further experiments are planed to investigate both of these possibilities.

Acknowledgment

The authors would like to thank Professor J. Akimitsu(Aoyama-gakuin Univ.), Dr. M. Ayabe and Miss T. Fukuoka for useful discussions.

REFERENCE

1. H. Nasu, H. Myouren, Y. Ibara, S. Makida, Y. Nishiyama, T. Kato, T.Imura, and Y.Osaka (1988) Formation of High-Tc superconducting $BiSrCaCu_2O_x$ Films on ZrO_2/Si(100). Jpn. J. Appl. Phys. 27: L634-L635
2. H. Myoren, Y. Nishiyama, H. Fukumoto, H. Nasu and Y. Osaka (1989) As-Grown Preparation of super conducting Epitaxitial $Ba_2YCu_3O_x$ Thin Films Sputtered on Epitaxially Grown ZrO_2/Si(100). Jpn. J. Appl. Phys. 28 : 351-355
3. J.Shinohara, Y.Ikegami and T.Kawamoto (1988) Formation of Crystallized Buffer Layer for High-Tc Superconducting Thin Film. Proceedings of the 1st international. Symposium on Superconductivity, Aug. : 641-645

Oriented $YBa_2Cu_3O_{7-x}$ Superconductive Thin Film Growth on Metallic Substrate by ICB Deposition Method

H. YOSHINO, M. YAMAZAKI, T.D. THANH, T. YAMASHITA, and K. ANDO

Advanced Research Laboratory, Research and Development Center, Toshiba Corporation, 1, Komukai, Toshiba-cho, Saiwai-ku, Kawasaki, 210 Japan

ABSTRACT

An oriented $YBa_2Cu_3O_{7-x}$ superconducting thin film on silver substrate was prepared by the ICB(Ionized Cluster Beam) deposition method with oxygen plasma. Silver, nickel, nichrome and stainless steel alloys were investigated as candidate substrates without any buffer layer. The deposition was done in rf glow discharged oxygen plasma at a substrate temperature of about 650C. Only the film on silver substrate showed high orientation of (001) plane of $YBa_2Cu_3O_{7-x}$. The zero resistance Tc was 76K and the critical current density was $2x10^5 A/cm^2$ at 4.2K in zero magnetic field. The acceleration for the cluster is effective for the improvement of its orientation and crystallization.

KEY WORDS: highly oriented thin film, metallic substrate, ICB deposition method, oxygen plasma, without buffer layer

INTRODUCTION

There have been many reports of vapor codeposition techniques for growing high Tc superconducting films. These films exhibit excellent properties such as high Jc more than 10^6 A/cm^2 at 77K [1]. Almost all films were deposited on insulating substrates such as $SrTiO_3$ and MgO for the application of electronics devices. These substrates have a lattice parameter nearly equal to that of $YBa_2Cu_3O_{7-x}$, and hardly react with $YBa_2Cu_3O_{7-x}$. Therefore, it is easy to prepare $YBa_2Cu_3O_{7-x}$ thin film having the c-axis which is perpendicular to its substrate.

On the other hand, efforts to deposit $YBa_2Cu_3O_{7-x}$ on the metallic substrate have been made for the application of power electronics [2]. Usually buffer layers like the MgO are used to prevent the reaction between $YBa_2Cu_3O_{7-x}$ and the substrates. But a buffer layer is not desirable from the view point of stability of superconductivity. When it becomes possible to deposit $YBa_2Cu_3O_{7-x}$ film on metallic substrate without any buffer layer and deposit films with the c-axis perpendicular to the metallic substrate, superconducting wire with high Jc may be realized.

In this study we tried to prepare the oriented $YBa_2Cu_3O_{7-x}$ superconducting films on metallic substrates without buffer layer using the ICB deposition method. The ICB method has the advantage of increaseing the kinetic energy of ionized particles compared with the conventional deposition method[3]. Therefore, an improvement of the crystallization of $YBa_2Cu_3O_{7-x}$ can be expected.

EXPERIMENTAL

The schematic diagram of the ICB deposition system is shown in Fig.1. Source materials of Y,Ba and Cu are vaporized through the nozzle of refractory crucibles into a vacuum chamber. Each cluster of Y,Ba and Cu which is an atomic aggregate cluster is formed by the adiabatic expansion process. The size of the clusters is determined by the shape of the nozzle, the vapor pressure of the inner side of the crucible and the pressure in the vacuum chamber. Under the conditions in this study, it was supposed that each cluster has atoms of between a few hundreds and one thousand. Each cluster is ionized positively by an electron bombardment. The film deposition rates and composition are controlled by monitoring the deposition rates of Y,Ba and Cu individually, using crystal sensors near the substrate. The chemical composition of $YBa_2Cu_3O_{7-x}$ was controlled within three atomic percent of stoichiometry. The substrate is heated by lamp heater and the temperature measuredby thermo couple is about 650 C. It is important to supply oxygen ions into high Tc superconducting film in order to prepare a high-grade superconductor by the vapor codeposition method. Recently, ECR activated oxygen plasma [4], ozone [5], N_2O [6]and reactive oxygen [8] are used as oxygen sources. In this study we supply reactive oxygen generated by rf glow discharged oxygen plasma in a quartz tube which has an orifice of about 2 mm in diameter. The distance between the orifice and the substrate is about 3 cm. The base pressure of the system was 10^{-7} Torr and the pressure during deposition was maintained at $2x10^{-4}$ Torr. The estimated pressure near the substrate was the order of 10^{-3}Torr. The deposition rate was approximately 0.1nm/s. After deposition of the 400nm thick films, they were coold slowly in oxygen plasma. $SrTiO_3$ substrates were used to clarify the effect of the parameter of the ICB method. Ag, Au, Ni, NiCr alloy and stainless steel deposited by Ag were investigated as metallic substrates. Prepared films were analyzed for the degree of crystallization and orientation by X-ray diffraction. Surface morphology was also observed by SEM.

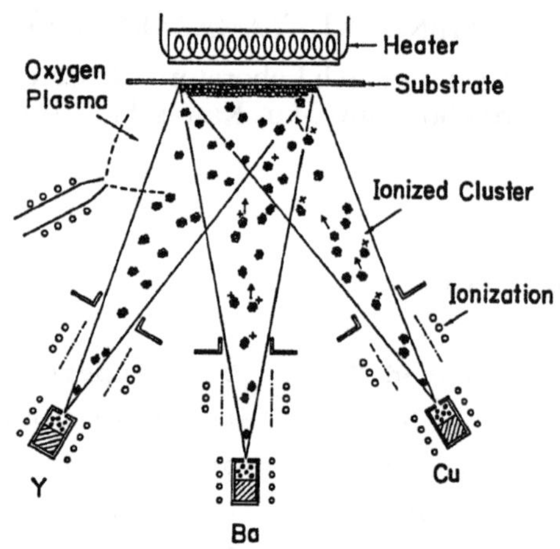

Fig. 1 Schematic diagram of ICB deposition method.

RESULTS AND DISCUSSION

The X-ray diffraction patterns of the films which were deposited on $SrTiO_3$ (100) substrates are shown in Fig.2. When the acceleration voltage, ionization current and rf power are zero, the diffraction pattern shows a polycrystalline state. When supplying accelerate voltage and rf power, the c-axis orients perpendicular to the film. And the length of the c lattice parameter decreases by supplying rf power. So, it is said that the acceleration of the ionized clusters and the oxygen plasma are effective to improve the degree of crystallization and oxygen contents. Zero resistance Tc is 82K when the rf power is 50W. Figure 3 shows the X-ray diffraction patterns which were deposited on various metallic substrates, under the conditions of accelerate voltage being 0.2kv and rf power being 50W. Diffraction peaks of $YBa_2Cu_3O_{7-x}$ deposited on Ni, NiCr and stainless steel coated by Ag are weak. On the other hand, $YBa_2Cu_3O_{7-x}$ films deposited on an Ag substrate show strong orientation of the c-axis perpendicular to the substrate. And the a-axis is perpendicular when deposited on Au substrate. It is supposed that Ni and NiCr are oxidized at the first stage of deposition by oxygen plasma and react with $YBa_2Cu_3O_{7-x}$. But Ag and Au do not react with $YBa_2Cu_3O_{7-x}$ and high orientation obtained.

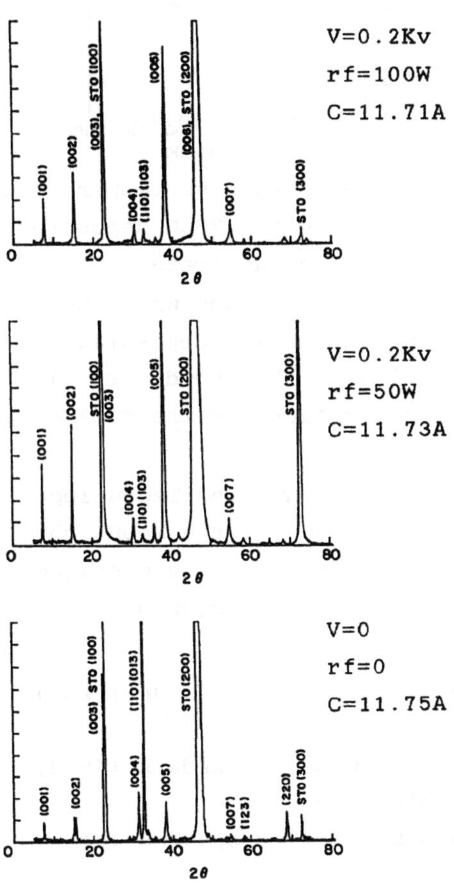

Fig. 2 Effect of the acceleration voltage and rf power for the orientation of YBCO deposited on $SrTiO_3$(100).

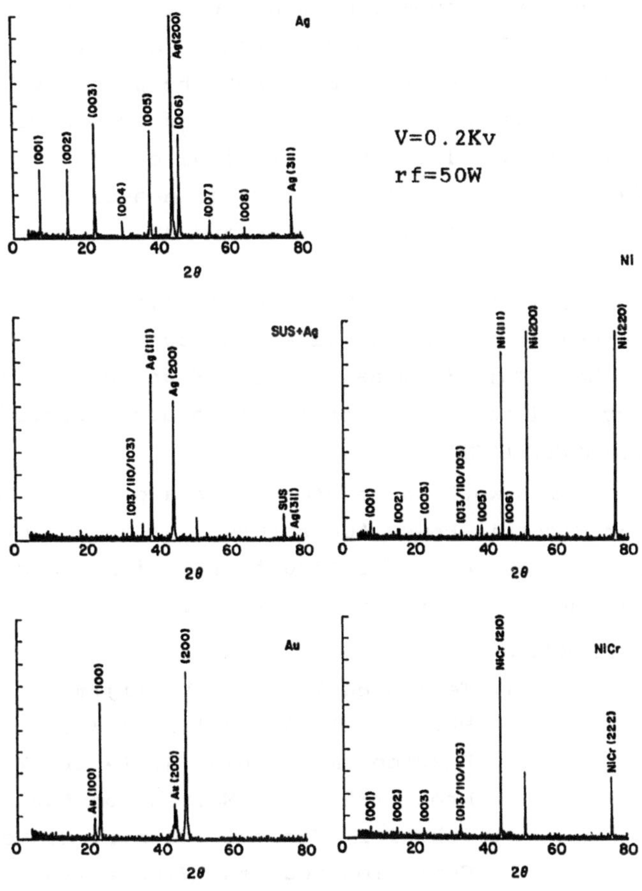

Fig. 3 X-ray diffraction patterns of YBCO deposited on various substrates.

Figure 4 shows the dependence of the crystal orientation of $YBa_2Cu_3O_{7-x}$ on that of Ag substrates. It seems that the degree of the c-axis orientation of $YBa_2Cu_3O_{7-x}$ is high in case of an Ag (200) substrate. This result is concerned with the matching of the lattice parameter of Ag substrates and $YBa_2Cu_3O_{7-x}$. The morphology of the surface of $YBa_2Cu_3O_{7-x}$ film deposited on Ag substrate was observed by SEM. The grain size was about 0.5um, and the flat planes arrayed parallel to the substrate. No microcracks and voids were observed.

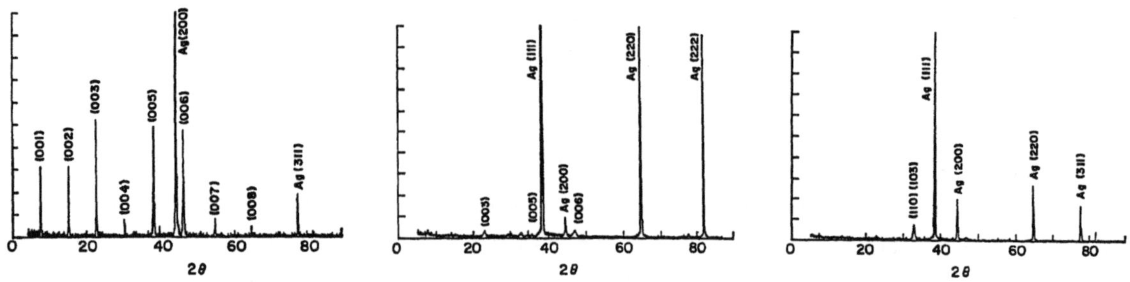

Fig. 4 X-ray diffraction patterns of YBCO deposited on Ag substrate which has various orientation.

A curve of the resistance versus temperature of the film deposited on Ag substrate is shown in Fig.5. Zero resistance Tc is 76K and the the normal state show the property of Ag substrate. This composit superconductor has a flexibility and a critical current density of $2x10^5$ A/cm^2 which is measured under the condition of 4.2K and 0T.

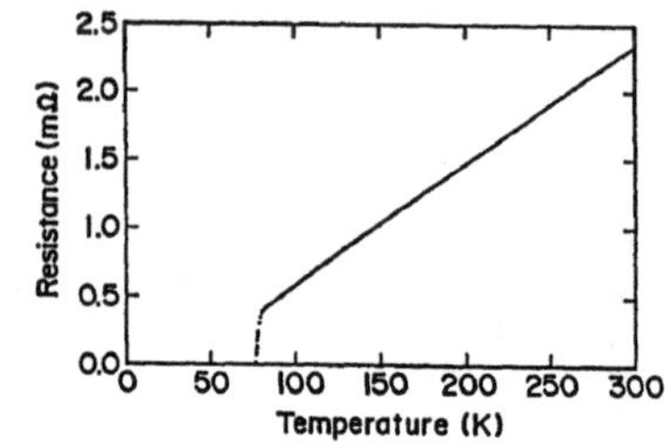

Fig. 5 Resistance vs temperature curve for YBCO deposited on Ag substrate.

In conclusion, we have prepared highly oriented $YBa_2Cu_3O_{7-x}$ films which were deposited on Ag substrates without a buffer layer using the ICB deposition method. This film has Tc of 76K and Jc of $2x10^5A/cm^2$ at 4.2K. The degree of crystallization and orientation are improved by acceleration for the clusters.

ACKNOWLEDGMENT

This work was performed as a part of "R&D on Superconducting Technology for Electric Power Apparatuses" as a subject of Super-GM under the Moonlight Project of Agency of Industrial Science and Technology, MITI, being consigned by New Energy and Industrial Technology Development Organization(NEDO).

REFERENCES

1. Terashima T, Bando Y, Iijima K, Yamamoto K, Hirata K (1988) Appl. Phys. Lett. 53 (22) 2232-2234
2. Fukutomi M, Machida J, Tanaka Y, Asano T, Wada H, Maeda H (1988) Proc. MRS. Int. Meeting on Advanced Materials, Tokyo
3. Takagi T, Yamada I, Kunori M, Kobiyama S (1972) Proc. 2nd Int. Conf. Ion Sources, Vienna 790
4. Moriwaki K, Enomoto Y, Kubo S, Murakami T (1988) Jpn. J. Appl. Phys. 27 L2075-2077
5. Berkley DD,Johnson BR, Anand N, Beauchamp KM, Conroy LE, Goldman AM, Maps J, Tuominen M, Zhang YJ (1988) Appl. Phys. Lett. 53 1973-1975
6. Yamanishi K, Yasunaga S, Imada K, Sato K, Hashimoto Y (1989) Mat. Res. Soc. Symp. Proc. vol 99 343-346
7. Tabata H, Kawai T, Kanai M, Murata O, Kawai S (1989) Jpn. J. Appl. Phys. 28 L823-826
8. Kwo J, Hong M, Trevor DJ, Fleming RM, White AE, Farrow RC, Kortan AR, Short KT (1989) Appl. Phys. Lett. 53 2683-2685

Complex Susceptibility in Single-Crystal $YBa_2Cu_3O_{7-x}$ Thin Films

HIROSHI YASUOKA[1], HIROMASA MAZAKI[1], KAZUNUKI YAMAMOTO[2], KAZUTO HIRATA[2], KENJI IIJIMA[2], KATSUHIKO HAYASHI[2], TAKAHITO TERASHIMA[3], and YOSHICHIKA BANDO[3]

[1] Department of Mathematics and Physics, The National Defense Academy, Yokosuka, 239 Japan
[2] Research Institute for Production Development, Kyoto, Japan
[3] Institute for Chemical Research, Kyoto University, Uji, Japan

ABSTRACT

Higher-harmonic susceptibilities were measured for single-crystal $YBa_2Cu_3O_{7-X}$ thin films located in ac magnetic field perpendicular to the film plane. We found that the higher-harmonic complex susceptibilities have nonzero values in the region of superconducting transition in terms of fundamental susceptibility. The nonzero higher harmonic susceptibilities indicate that the magnetization responses nonlinearly. This nonlinear response and the nonzero imaginary part of the fundamental susceptibility suggest that the magnetization curve traces hysteresis loop. These must be caused by the appearance of the obstruction of vortex motion.

KEY WORDS: oxide superconductor, single-crystal thin film
complex susceptibility, $YBa_2Cu_3O_{7-x}$

INTRCDUCTION

The measurement of complex susceptibility is one of the useful method for characterizing the superconducting properties, the crystallinity and uniformity of the samples. For the sintered samples, the bulk mode due to Meissner effect of intragrains and the weak-link mode due to shield effect of intergrains are observed in the complex susceptibility transition.[1] The weak-link mode depends on the amplitude of the applied ac magnetic field.[3] On the contrary, we have observed a single mode transition and peculiar dependence on the magnetic field amplitude in the complex susceptibility of the single-crystal $YBa_2Cu_3O_{7-x}$(YBCO) thin films.[2] The measurement of the higher-harmonic complex susceptibilities gives us detailed informations on the non-linear magnetic response.[3],[4],[5],[6]

EXPERIMENTAL

Single-crystal thin films were prepared by activated reactive evaporation technique.[7] The (001) plane of YBCO was epitaxially grown on the (100) surface of single-crystal $SrTiO_3$ substrate. The X-ray diffraction patterns of the films show that they consists of the single-phase YBCO whose (001) plane is parallel to the surface plane of the substrate. The results of reflection high-energy electron diffraction show that the films are well crystallized even at the top surface of the film. From the measurement of critical current density, being in the order of 10^6 A/cm^2, we believe the films do not contain the grain boundaries which act as the weak-link junction. The film thickness of 100-6000A was used.

The fundamental and the higher-harmonic complex susceptibility were measured by the Hartshorn bridge.[3],[4] The ac magnetic field of frequency f_1 was applied perpendicular to the film plane, where f_1 is 13 to 132 Hz and the amplitude of the magnetic field is 1 to 2 Oe. The fundamental complex susceptibility was probed by a lock-in amplifier for the reference signal of frequency f_1. The n-th harmonic complex susceptibility was probed for the reference signals of frequency $n*f_1$, where the reference signals were synchronized with the ac magnetic field signal of frequency f_1. The sample temperature was lowered quite slowly, typically less than 0.2 deg/min. A null adjustment of the bridge was made for frequency f_1 at the sample temperature of 100K. Phase adjustment of the lock-in amplifier was made so as to give variation only to the in-phase signal against the change in the bridge inductance.

RESULTS AND DISCUSSION

The fundamental, third, 5th and 7th harmonic complex susceptibilities were measured. In Fig. 1(a), the temperature dependence of the fundamental susceptibility are shown with the resistance measured by the dc four-probe method. The third, 5th and 7th harmonic susceptibility are shown in Fig. 1(b),(c) and (d), respectively. From these results, we notice the following aspects.

(a) The curve of $-\chi_1'$ grows rapidly and smoothly, and χ_1'' forms a single peak. This indicates that the sample has satisfactory quality for susceptibility measurement.

(b) Each higher-harmonic complex susceptibility has nonzero value in the temperature region where χ_1'' forms a single peak. The temperature dependence of the higher-harmonic components becomes complicated, as n increases.

(c) The onset temperature of each component is the same, 89.9 K. And this onset temperature is the same or slightly below the end point of resistive transition.

These aspects suggest that the behavior of χ_1 cannot be explained by the Maxwell-Strongin model.[8] According to this model, it is expected that the onset and the end point of χ_1 should coincide with those of resistive transition, and the higher-harmonic components do not exist. The nonzero higher-harmonic susceptibility indicate that the magnetization responses nonlinearly. This nonlinear response and the nonzero imaginary part of the fundamental susceptibility mean that the magnetization curve traces hysteresis loop in the region of the susceptibility transition.

Meanwhile, we should take into account of the demagnetization effect for thin films. The internal magnetic field is enhanced by the factor of R/d for the films of $k \gg 1$ and $d/\lambda \ll 1$ below the transition temperature, where k is the Ginzburg-Landau parameter, d is the film thickness, λ is the London penetration depth and R is the sample radius.[9] In our case, R/d is estimated to be $10^{-4} \sim 10^{-5}$. Consequently, the internal magnetic field should exceed the lower critical field H_{c1}, and films become in the vortex state. The existence of the higher-harmonic components implies that the flux sweeping in and out of the specimen is the nonlinear process. This process must be caused by the obstruction of the flux motion due to the bulk pinning and surface barrier.[10],[11] We applied the modified critical state model which involves the above two factors.[12] The temperature dependences of the higher-harmonic susceptibilities are explained by the modified critical state model on the whole.

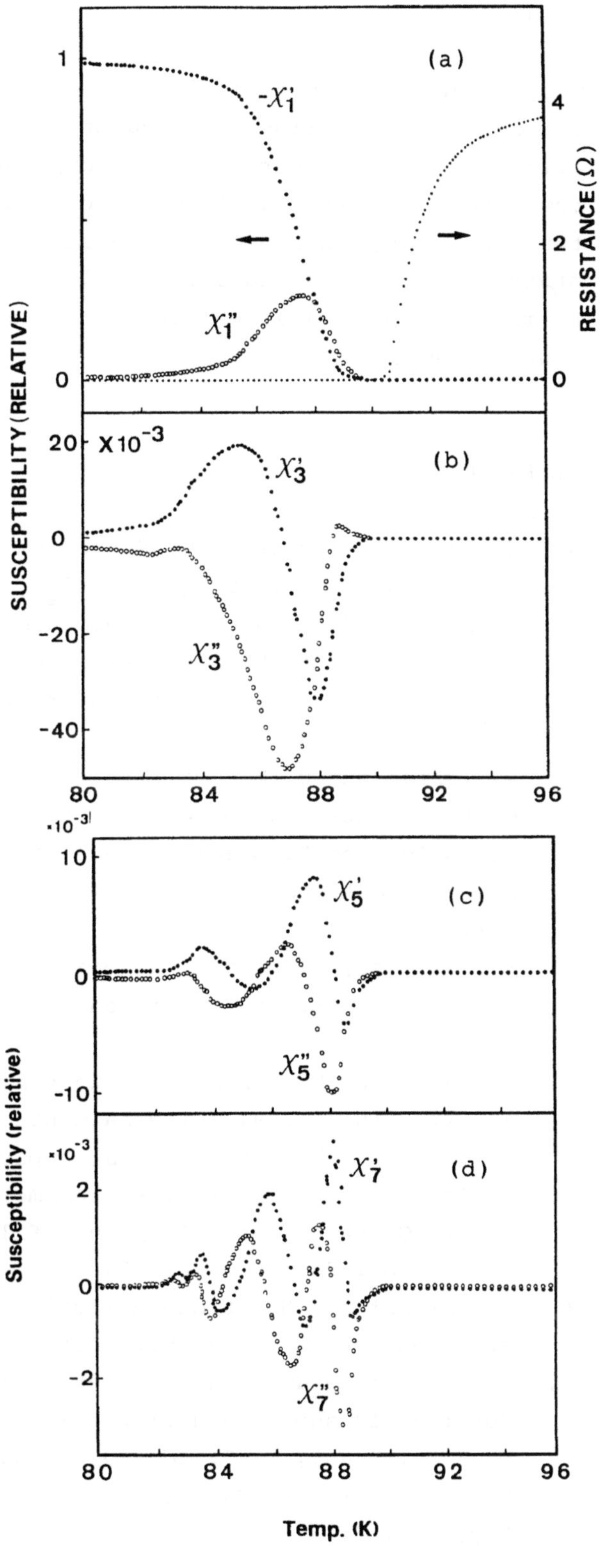

Fig. 1 The temperature dependences of χ_n (n=1,3,5,7) and resistance for a 1000A YBCO (001) thin film.

SUMMARY

The higher-harmonic susceptibilities were studied on the single crystal $YBa_2Cu_3O_{7-x}$ thin films. The magnetic field is applied perpendicular to the film plane. The higher-harmonic components were observed in the region of the susceptibility transition. The existence of these components suggests that the magnetization curve traces hysteresis loop. The modified critical state model can qualitatively explains the observed behavior of the higher-harmonic susceptibilities. The vortex motion must be obstructed by the bulk pinning and the surface barrier.

REFERENCES

1. H. Mazaki, M. Takano, R. Kanno and Y. Takeda (1987) Complex susceptibility of $La_{1.9}Sr_{0.1}CuO_{4-\delta}$. J. J. Appl. Phys. 26: L780-L782
2. H. Mazaki, H. Yasuoka, T. Terashima, Y. Bando, K. Yamamoto, K. Hirata and K. Iijima (1989) Complex susceptibility of single crystal $YBa_2Cu_3O_y$. Extended abstracts of 40th International Society of Electrochemistry Meeting 1: 94-95
3. T. Ishida and H. Mazaki (1981) Superconducting transition of multiconnected Josephson network. J. Appl. Phys., 11: 6798-6805
4. T. Ishida and H. Mazaki (1982) Higher-harmonic susceptibilities of a multiconnected Josephson network in a superconducting transition. Phys. Lett. 87A: 373-375
5. K. Yamamoto, K. Hirata, H. Mazaki, H. Yasuoka, K, Iijima, T. Terashima and Y. Bando (To be published) Ac susceptibility in single-crystal $YBa_2Cu_3O_{7-x}$ II.
6. K. Yamamoto, H. Mazaki, H. Yasuoka, K. Hirata, T. Terashima, K. Iijima, and Y. Bando (1989) Third-harmonic susceptibility in single-crystal $YBa_2Cu_3O_{7-x}$ thin films. J. J. Appl. Phys. 28: L1568-L1570
7. T. Terashima, K. Iijima, K. Yamamoto, Y. Bando and H. Mazaki (1988) Single-Crystal $YBa_2Cu_3O_{7-x}$ thin films by activated reactive evaporation. J. J. Appl. Phys. 27: L91-L93
8. E. Maxwell and M. Strongin (1963) Filamentary structure in superconductors. Phys. Rev. Lett. 10 : 212-215
9. J. Pearl (Plenum, New York 1965) Distinctive properties of quantized vortices in superconducting films. Low Temperature Physics-LT9, edited by J. G. Daunt, D. O. Edward, F. J. Milford and M. Yagub : 566-570
10. C. P. Bean (1964) Magnetization of high-field superconductors. Rev. Mod. Phys. 36: 31-39
11. C. P. Bean and J. D. Livingston (1964) Surface barrier in type-II superconductors. Phys. Rev. Lett. 12: 14-16
12. W. I. Dunn and P. Hlawiczka (1968) Generalized critical-state model of type II superconductors. Brit. J. Appl. Phys. 1: 1469-1476

As Grown Y-Ba-Cu-O Thin Films with Fine Grains and the Fabrication of the Superconducting Lines

E. OHNO, M. NAGATA, H. SHINTAKU, H. NOJIMA, and M. KOBA
Sharp Corporation, Central Research Laboratories, 2613-1, Ichinomoto-cho, Tenri, Nara, 632 Japan

ABSTRACT

Superconducting thin films of $Y_1Ba_2Cu_3O_{7-x}$ have been prepared by off-axis rf magnetron sputtering using the single target with the stoichiometric composition. We obtained the films with fine grains 0.2μm in average size and the smooth surface. The c-axis of the specimen was highly oriented normal to MgO(100) plane in the wide area. The superconducting lines having 2μm width and 500μm length were fabricated from the films of 200nm thickness by chemical wet etching. The superconducting line showed a zero resistance at 78K, which was only 2K lower than that of the film before being etched.

KEY WORDS: thin film, rf magnetron sputtering, superconducting line, chemical wet etching

INTRODUCTION

The fabrication of superconducting electronic devices requires the high quality thin films with high-T_c $Y_1Ba_2Cu_3O_{7-x}$. The preparation of Y-Ba-Cu-O thin films by single-target rf magnetron sputtering has been studied intensively. However, it is suggested that negative oxygen ions and high energy secondary electrons are directly incident on the substrate in the usual single-target sputtering geometry[1]. The film composition is deviated from the stoichiometric one by selective resputtering of high sputtering yield constituents. So far the off-axis sputtering geometry have been investigated to avoid the resputtering damage from the negative ion and the secondary electron[2,3]. The substrate is placed on the side of the target in the off-axis sputtering geometry. Using the off-axis sputtering deposition, high quality Y-Ba-Cu-O thin films have been obtained.
In this paper, the substrates were mounted normal to the target as shown in Fig. 1. The effect of substrate location on the composition and c-lattice parameter of the film is presented. We fabricated the superconducting line having 2μm width from the high quality film by chemical wet etching. The superconducting transition temperature of the line was 78K, which was degraded by only 2K.

EXPERIMENTAL

The off-axis sputtering geometry used in our experiment is shown in Fig. 1. The distance between the target and the substrate is represented by the parameters x and y as shown in Fig. 1. We used the stoichiometric target $Y_1Ba_2Cu_3O_7$-x in a typical run. In some cases, the target of composition $Y_1Ba_{2.2}Cu_{3.3}O_{7-x}$ was used. The targets with 10cm in diameter were prepared from sintered powders. The sputtering atmosphere consisted of, in most cases, 20mTorr Ar and 20mTorr O_2. Pure oxygen was spouted to the substrate as shown in Fig. 1. The flow rate was about 2cc/min. The rf power was 100W. The substrate temperature was held at 700-750°C during film growth and measured by an infrared radiation thermometer. The films were grown on MgO(100) substrates. After the deposition, the film was cooled below 200°C for about 10 minutes. Film thickness was about 200nm. The compositions of the films were measured with an electron microprobe. The results were verified in some cases by inductively coupled plasma spectroscopy. The superconducting lines were fabricated by chemical etching. The PFR-3003B resist with 1μm thickness was spin coated on as grown film and prebaked at 90°C. The resist was irradiated with UV light by the contact printing method and developed. After the postbake, the film was etched by Feliox-115[4]. The resist was removed by dissolved in acetone and ethyl alcohol. Finally, the oxygen plasma ashing was used to make a clean surface. The resistance of the line was measured by the standard four prove method.

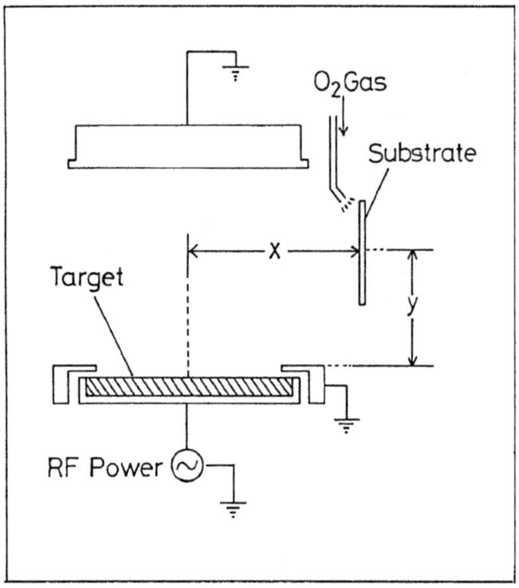

Fig. 1. Sputtering system geometry.

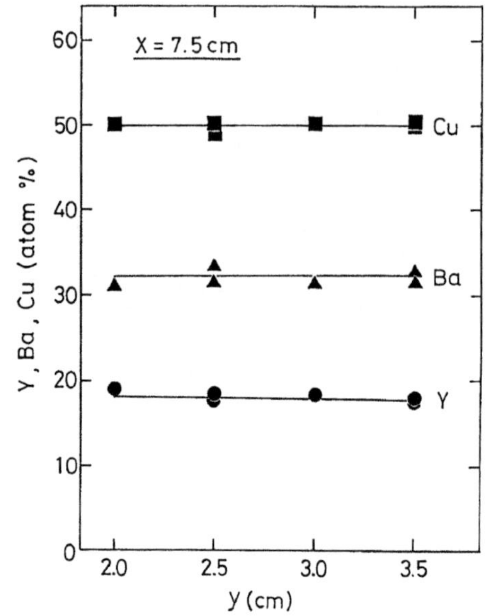

Fig. 2. Compositions of films vs. the parameter y. y is defined in Fig. 1.

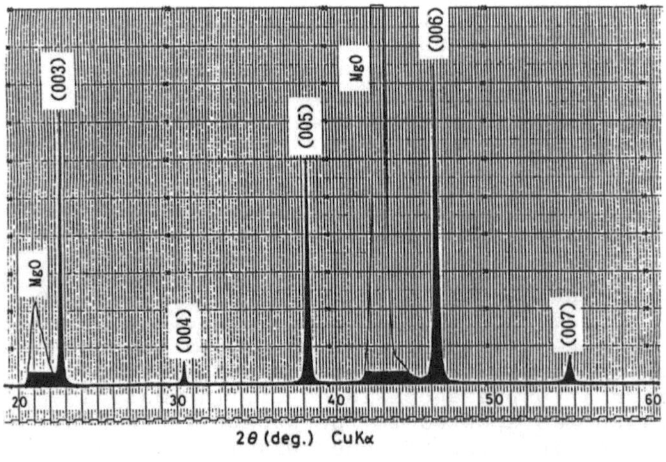

Fig. 3. X-ray diffraction pattern of as grown film using CuKα source.

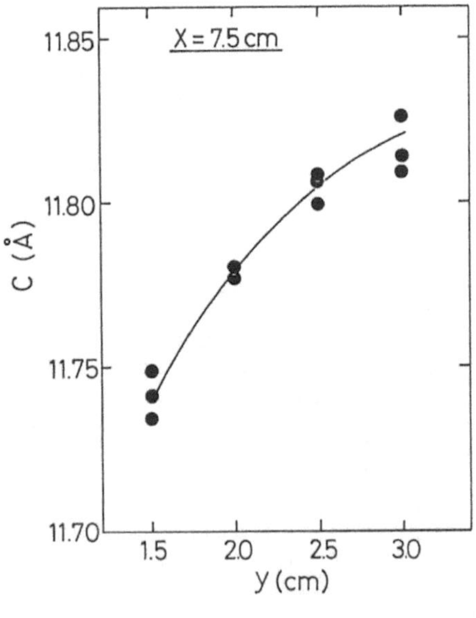

Fig. 4. c-lattice parameters vs. the parameter y. y is defined in Fig. 1.

(a)

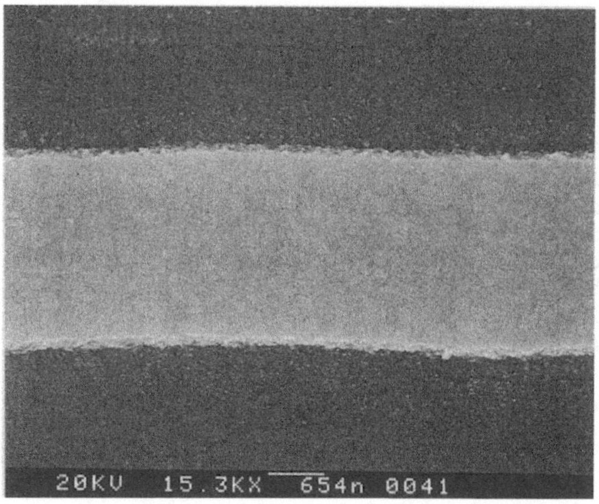

(b)

Fig. 5. Scanning electron micrographs of as grown films. (a)as grown film with 200nm thickness. (b)superconducting line with 2µm width.

RESULTS AND DISCUSSION

Figure 2 shows the dependence of the film composition on the location of the substrate. Y, Ba and Cu concentrations of the film are plotted against the parameter y. The parameter x was fixed at 7.5cm. The films with the almost 1-2-3 stoichiometric composition were obtained over the wide area y from 2.0 to 3.5cm. We confirmed that the compositions of the film were independent of y from 1.5 to 4.0cm. The c-axis of the film prepared in this area was highly oriented normal to MgO(100) plane. Figure 3 shows a typical x-ray diffraction scan of the films. The c-lattice parameter, however, depended on the location of the substrate as shown in Fig. 4. The c-lattice parameter increased with increasing the parameter y. The superconducting transition temperature T_{ce} of the film decreased monotonously with increasing the parameter y. We obtained the film with T_{ce}=80K at y=1.5cm. The density of the oxygen radical in the plasma seems to decrease with increasing the distance from the target. Thus the oxygen content of the film probably depends on the location of the substrate to cause the y dependence of the superconducting properties. Figure 5(a) shows the SEM image of the film. The average grain size was 0.2µm. We infered from the SEM image of the film that the surface of the specimen was very smooth.

The superconducting lines having 2µm width and 500µm length were fabricated by the chemical etching described above. Figure 5(b) shows the SEM image for a 2µm wide line. After removing the resist by organic liquid, there remained carbonic contaminations on the surface of the film. Thus we used the oxygen plasma ashing to clean the surface of the etched film. Figure 6 shows the spectra of Auger electron spectroscopy for the patterned film before and after removing the carbonic contaminations by O_2 plasma ashing for 10 minutes. The ashing power of O_2 plasma was 300W. The spectrum of carbon disappeared after the O_2 plasma ashing. We also confirmed from the Auger depth profile that the compositions of the film were uniform. Figure 7 shows the temperature dependence of the resistance of the film before and after being etched. The transition temperature at zero resistance of the superconducting line was 78K, which was only 2K lower than that of the film before being etched.

In summary, the high quality Y-Ba-Cu-O thin films were prepared by the off-axis magnetron sputtering. The superconducting line was fabricated by chemical wet etching. The transition temperature of the line was degraded by only 2K during the patterning process.

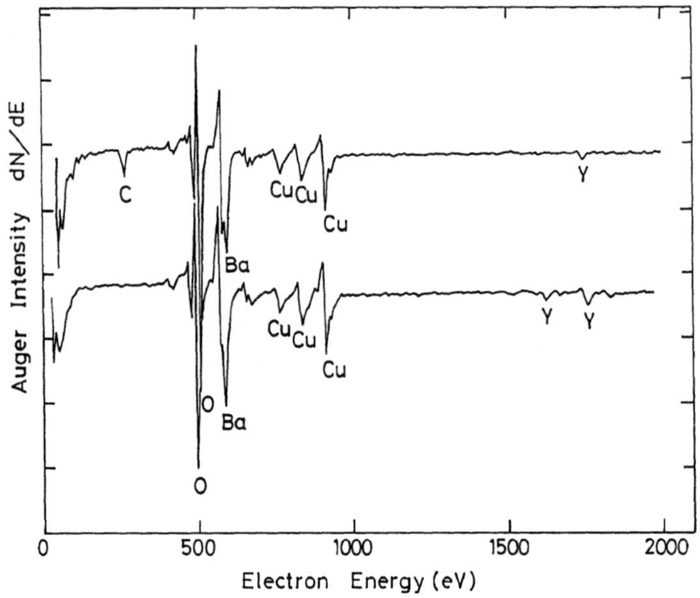

Fig. 6. Spectra of Auger electron spectroscopy for the patterned film before and after O_2 plasma ashing. The upper and lower lines show the spectra for the film before and after plasma ashing, respectively.

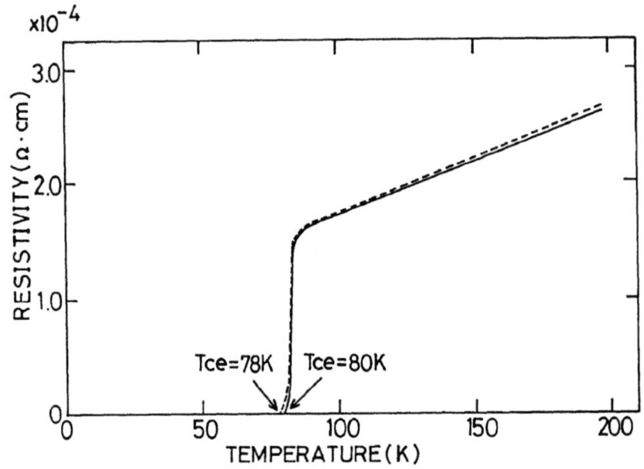

Fig. 7. Resistive transitions for the film before and after being etched. The solid and dashed lines represent the resistivity of the film before and after being etched, respectively.

REFERENCES

1. Sandstrom RL. Gallagher WJ. Dinger TR. Koch RH. Laibowitz RB. Kleinsasser AW. Gambino RJ. Bumble B. Chisholm MF (1988) Reliable single-target sputtering process for high-temperature superconducting films and devices. Appl. Phys. Lett. 53:444-446

2. Terada N. Ihara H. Jo M. Hirabayashi M. Kimura Y. Matsutani K. Hirata K. Ohno E. Sugise R. Kawashima F (1988) Sputter Synthesis of $Ba_2Y_1Cu_3O_y$ As--deposited Superconducting Thin Films from Stoichiometric Target. Jpn. J. Appl. Phys. 27:L639-L642

3. Eom CB. Sun JZ. Yamamoto K. Marshall AF. Luther KE. Geballe TH (1989) In situ grown $YBa_2Cu_3O_{7-d}$ thin films from single-target magnetron sputtering. Appl. Phys. Lett. 55:595-597

4. Tonouchi M. Sakaguchi Y. Kobayashi T. (1988) Chemical Etching of High-Tc Superconducting Films in Feliox-115 Solution. Jpn. J. Appl. Phys. 27:L98 -L100

Properties of Sputter-Deposited $YBa_2Cu_3O_{7-x}$ Thin Films

MICHIHITO MUROI, YUKO OKAMURA, TAKESHI SUZUKI, KOICHI TSUDA, MEGUMI NAGANO, and KAZUO MUKAE

Fuji Electric Corporate Research and Development Ltd., 2-1 Nagasaka 2-chome, Yokosuka, Kanagawa, 240-01 Japan

ABSTRACT

Properties of sputter-deposited $YBa_2Cu_3O_{7-x}$ thin films are discussed in terms of the kinetic energy of the depositing atoms. When the sputtered atoms were thermalized before reaching the substrate, the superconducting phase was formed only at the substrate temperature above 650°C. In contrast, when the sputtered atoms traveled straight to the substrate and arrived there with most of their initial energies, the superconducting phase was formed even at 500°C due to the enhanced migration of the deposited atoms on the substrate. The Tc of the as-deposited film increased with the increase of the substrate temperature and reached 83K at the substrate temperature of 650°C.

KEYWORDS: oxide superconductor, $YBa_2Cu_3O_{7-x}$, thin film, sputtering, low temperature growth

INTRODUCTION

The discovery of oxide superconductors with unprecedentedly high critical temperatures (Tc) including $YBa_2Cu_3O_{7-x}$ (YBCO) [1,2] has touched off world-wide research activity in this field. A major focus of this effort has been the fabrication of high-quality thin films of these materials for both scientific and technological purposes. Among the various deposition techniques which have been attempted, sputtering using a single oxide target is one of the simplest methods for preparing thin films of oxide superconductors such as YBCO.

The sputtering system is so flexible that various arrangements of the target and the substrate are possible and the operating gas pressure can be chosen over a wide range. This flexibility allows us to control the energy of the atoms arriving at the substrate by adjusting the system geometry and the operating gas pressure. In this paper, we discuss properties of the sputter-deposited YBCO thin films in terms of the kinetic energy of the depositing atoms.

EXPERIMENTAL

We have deposited YBCO thin films by rf-magnetron sputtering. We tried two kinds of depositions, namely, (a)diffusion process, and (b)line-of-sight process. The deposition conditions for both processes are listed in Table I. In the diffusion process, high-energy particles are interrupted by a metal mask placed between the target and substrate, and the film is formed mostly by the thermalized atoms with low energies below 0.1eV (we call it "diffusion process" because the atoms are transported to the substrate mainly by thermal diffusion in this process). In the line-of-sight process, almost all the sputtered atoms travel straight to the substrates without colliding with the gas atoms, arriving there with most of their initial energies of about several eV on average. After a deposition, 1atm oxygen was introduced into the chamber to prevent the release of oxygen from the film and it was cooled down to 200°C in about 20 minutes before being removed from the chamber.

Crystal structure of the films was analyzed by X-ray diffractometer (XRD) using CuKα radiation, and the film compositions by inductively coupled plasma emission spectroscopy (ICP). Resistivity-temperature characteristics (R-T characteristics) were measured using a conventional dc four-probe method.

Table I. Deposition conditions of YBCO thin films.

	(a) Diffusion Process	(b) Line-of-sight Process
Target	$YBa_2Cu_6O_y$	$YBa_2Cu_4O_y$
Substrate	MgO(100) $SrTiO_3$(100)	MgO(100)
Configuration	See Fig.1(a)	See Fig.1(b)
Substrate temperature	600-700°C	450-650°C
Operating gas	Ar(50%)+O2(50%)	<--------
Gas pressure	0.15-10Pa	0.5Pa
Input rf-power	300W	<--------
Deposition time	2h	1h
Deposition rate	0.3-0.4Å/s	2Å/s

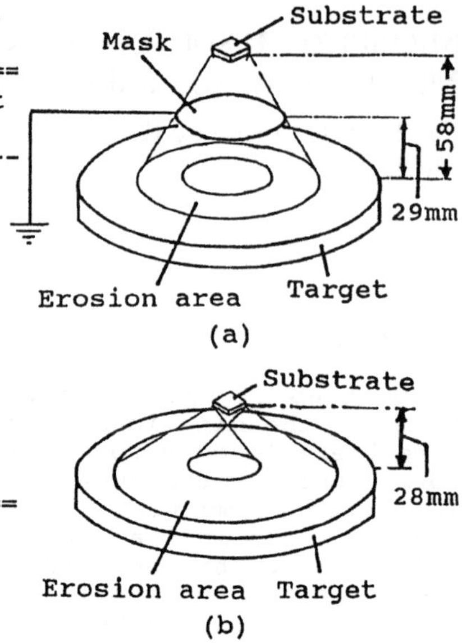

Fig.1. Configuration of the system.
(a) diffusion process
(b) line-of-sight process

RESULTS AND DISCUSSION

a. Diffusion Process

With the conditions listed in Table I(a), no film was formed at the gas pressure below 0.5Pa as can be seen in Fig.2, due to the long mean free path of the sputtered atoms. The film was deposited above 1Pa, because the thermalized atoms formed the "diffusion front" from which they were transported toward the substrate by thermal diffusion [3]. The use of the mask naturally decreased the deposition rate and it was 0.3-0.4Å/s with the gas pressure between 2Pa and 10Pa.

According to XRD, the films deposited at 650°C on both MgO and $SrTiO_3$ had the 123-structure with the c-axis normal to the substrate. At a lower substrate temperature of 600°C, even the crystallization did not occur and the crystal structure of the film was amorphous. We show the variations of the film composition and the lattice constant c_0 with the gas pressure, in Fig.3 and Fig.4, respectively. When the mask was not used, the contents of Ba and Cu decreased and the c_0 increased with the decrease of the gas pressure due to the heavier bombardment of the growing film by high-energy charged particles at the lower gas pressure [4]. In contrast, the composition and the

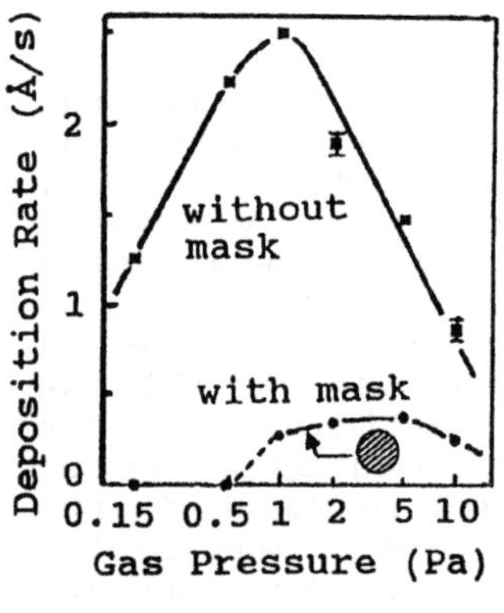

Fig.2. Variation of the deposition rate with the gas pressure.

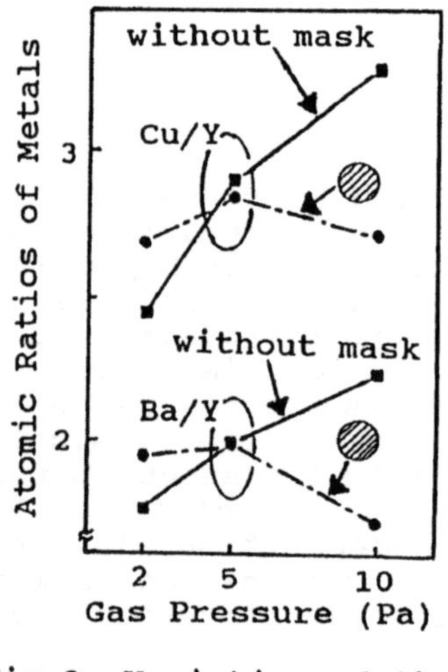

Fig.3. Variation of the composition with the gas pressure.

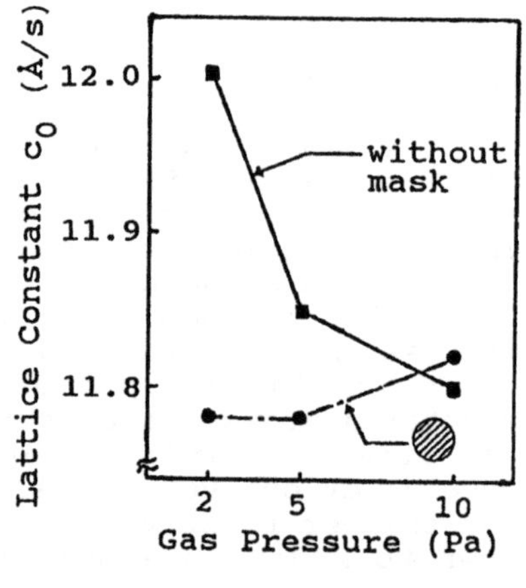

Fig.4. Variation of the lattice constant c_0 with the gas pressure.

c_0 of the film deposited with the mask were considerably constant regardless of the gas pressure, suggesting that the mask effectively prevented the bombardment by high-energy charged particles [3]. The Tc's of the films deposited at 650°C were almost the same (55-65K) regardless of the gas pressure and the substrate. However, an apparent resistance drop was observed at about 80K only for the film on $SrTiO_3$ as can be seen in Fig.5. This fact suggests that the properties of the film are strongly affected by the substrate and that the choice of the substrate is crucial for the fabrication of high-quality YBCO films with this process. The deposition at 700°C increased the ratio of the high-Tc phase, but the Tc was only 70K as shown in Fig.5.

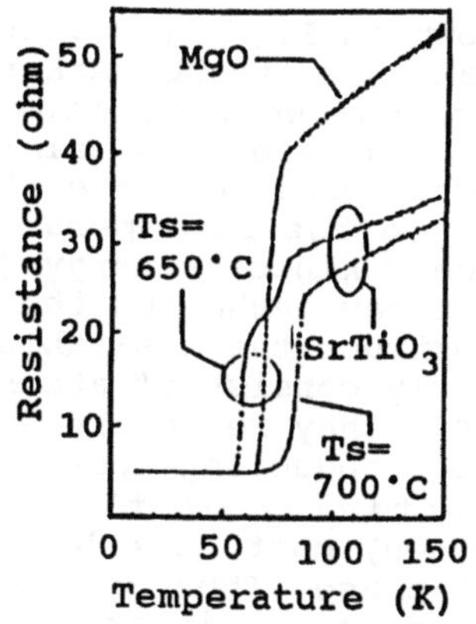

Fig.5.
R-T characteristics.

b. Line-of-sight Process

As shown in Fig.2, no film was formed below 0.5Pa in the deposition with the mask, guaranteeing the line-of-sight process with the conditions listed in Table I(b) (gas pressure: 0.5Pa, target-substrate distance: 28mm).

Figure 6 shows the variation of the film composition with the substrate temperature. Little deviation from the target composition and the constancy of the Ba/Y ratio suggest that the film is almost free from the bombardment by high-energy particles, otherwise the film should suffer a serious deficiency in Ba and Cu due to the resputtering effect at such a low gas pressure [4]. The bombardment must have been prevented by the configuration shown in Fig.1(b) in which the substrate is located as close as possible to the target to avoid the strong flux of charged particles.

In Fig.7 are shown the R-T characteristics of the films deposited at various substrate temperatures. It can be seen that the superconducting phase was formed even at 500°C. The Tc of the as-deposited film increased with the increase of the substrate temperature and reached 83K at 650°C. Since the increase of the Tc accompanied the decrease of the c_0, we attribute the change of the Tc to the change of the oxygen content in the Cu(1) plane. The higher the substrate temperature is, the more easily the Cu is oxidized, resulting in the smaller c_0 and higher Tc.

XRD revealed that the films deposited at above 500°C had the 123-structure. Figure 8 shows the variation of the peak intensity with the substrate temperature. It can be seen that the intensity is almost constant above 525°C. Moreover, no other peak corresponding to the orientations other than

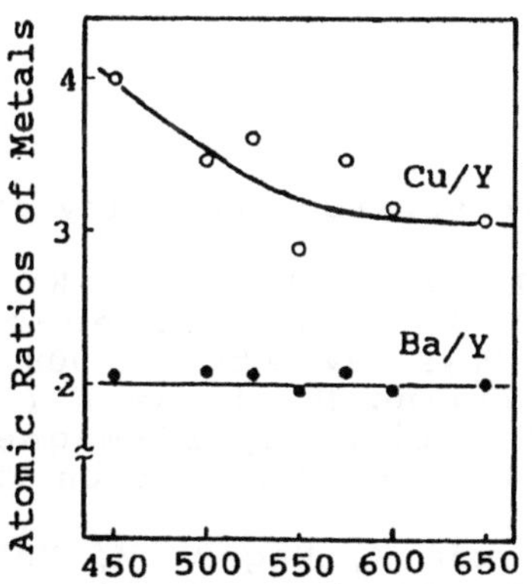

Fig.6. Variation of the film composition with the substrate temperature.

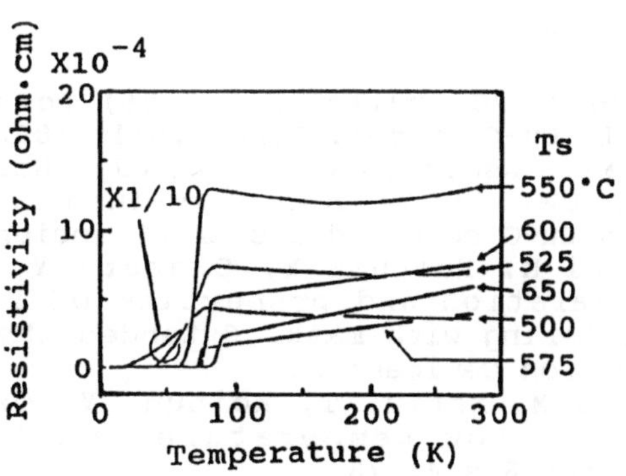

Fig.7. R-T characteristics of the films deposited at various substrate temperatures.

shown in Fig.8 was detected, and the intensity of the background was quite weak so as to be weaker than the peak intensity by three orders of magnitude, showing the excellent crystallinity. We believe that the following two factors made it possible to achieve excellent crystallinity in spite of the high deposition rate (2Å/s), even at the lower substrate temperatures. First, due to the low operating gas pressure, each of the depositing atoms has higher kinetic energy (probably several eV on average). This kinetic energy must have enhanced the migration of the deposited atoms on the substrate to construct the desirable crystal structure. Second, the depositing particles mostly consist of atoms as they were when they left the target and it is not until they reach the substrate that they react with one another to form the superconducting phase, again due to the low operating gas pressure. The high reactivity of the atoms (both metals and oxygen) must have helped the formation of the 123-phase. Thus sputtering with low operating gas pressure is very effective in fabricating high-quality YBCO thin films.

The films deposited at the substrate temperature above 575°C show the preferred orientation with the c-axis normal to the substrate, while those deposited at 525-550°C with the c-axis parallel to the substrate. The change from one orientation to the other occurs in a very narrow temperature range (note the log scale in the vertical axis of Fig.8). This change in the orientation well corresponds to the change in the R-T characteristic. The resistivity above Tc suddenly increases and the temperature dependence of resistivity in the normal state changes from metallic to semiconducting when the substrate temperature decreases from 575°C to 550°C as can be seen in Fig.7. Below 525°C, (110) and/or (013) orientation begins to appear and it is dominant for the film deposited at 500°C. The mechanism which determines the preferred orientation is discussed elsewhere [5].

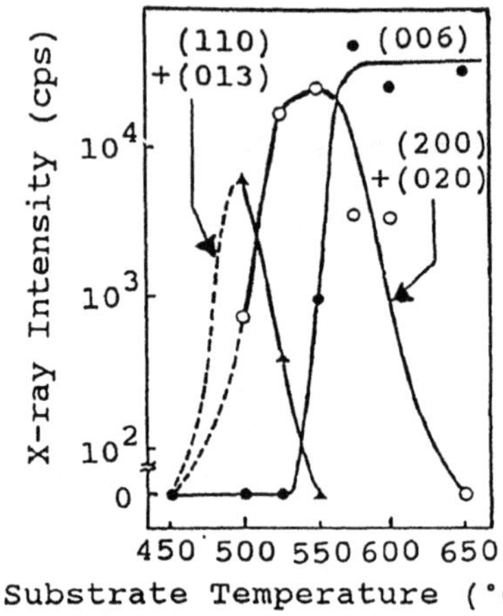

Fig.8. Variation of the XRD peak intensity with the substrate temperature.

CONCLUSION

We have studied properties of sputter-deposited YBCO thin films in terms of the kinetic energy of the depositing atoms. When the film was formed by the thermalized atoms, the superconducting phase was obtained only at substrate temperature above 650°C. When the film was deposited by the atoms with higher kinetic energies, on the other hand, the superconducting phase was formed even at a low substrate temperature of 500°C, and the Tc of the as-grown film reached 83K at the substrate temperature of 650°C.

REFERENCES

1. Bednorz JG, Müller KA (1986) Possible high Tc superconductivity in the Ba-La-Cu-O system. Z.Phys.B64: 189.
2. Wu MK, Ashburn JR, Torng CJ, Hor PH, Meng RL, Gao L, Huang ZJ, Wang YQ, Chu CW (1987) Superconductivity at 93K in a new mixed-phase Y-Ba-Cu-O compound system at ambient pressure. Phys.Rev.Lett.58: 908.
3. Muroi M, Koinuma Y, Okamura Y, Tsuda K, Nagano M, Mukae K (1989) Preparation and properties of Y-Ba-Cu-O thin films by rf-magnetron sputtering with mask. Extended Abstracts of 6th Int. Workshop on Future Electron Devices: 23.
4. Muroi M, Matsui T, Koinuma Y, Okamura Y, Tsuda K, Nagano M, Mukae K (1989) Low temperature synthesis of $YBa_2Cu_3O_{7-x}$ thin films. J.Mater.Res.4: 781.
5. Muroi M, Okamura Y, Suzuki T, Tsuda K, Nagano M, Mukae K (submitted) Sputter deposition of $YBa_2Cu_3O_{7-x}$ thin films with low gas pressure. Jpn.J.Appl.Phys.

As-Deposited Superconducting Bi-Sr-Ca-Cu-O Thin Films Prepared by RF Magnetron Sputtering

KAZUSHIGE OHBAYASHI[1], SHIGERU SUZUKI[2], YOSHIAKI TAKAI[2], and HISAO HAYAKAWA[2]

[1] NTK Technical Ceramics Division, NGK Spark Plug Co., Ltd., 14-18, Takatsuji-cho, Mizuho-ku, Nagoya, 467 Japan

[2] Department of Electrical Engineering, Nagoya University, Furo-cho, Chikusa-ku, Nagoya, 464 Japan

ABSTRACT

As-deposited superconducting thin films of the Bi-Sr-Ca-Cu-O system were prepared on MgO(100) substrates by a three targets rf magnetron sputtering with a regulated shuttering technique. Shuttering was employed in order to control the chemical composition of the films and to help form the desired crystal structure with c-axis in the growth direction. X-ray diffraction patterns from the as-grown films deposited at the substrate temperatures in the range of 580-660°C, showed the low-Tc phase and all the peaks indicated can be indexed as (002n) to a 30Å unit cell. The superconducting transition temperature of the film deposited at 660°C was 70K.

KEY WORDS: Bi-Sr-Ca-Cu-O, thin film, sputtering, multitarget, as-grown

INTRODUCTION

Since a perovskite-related high temperature oxide superconductor of Bi-Sr-Ca-Cu-O system was discovered [1], many efforts have been made to clarify the structures of these materials. The compositions of these oxides are represented by $Bi_2Sr_2Ca_{n-1}Cu_nO_x$ (n=1,2 and 3) [2], the number of $Cu-O_2$ layers with Ca cations between double Bi-O layers determined the superconducting properties of these materials. It was found that the low-Tc phase (Tc=80K) was composed of two $Cu-O_2$ layers and high-Tc phase (Tc=110K) was composed of three layers [3],[4]. But it is difficult to make a single phase material in this system. Matsui et al. reported intergrowth of the modulated structure which contained four and five $Cu-O_2$ layers in ceramics [5]. For the purpose of controlling the number of $Cu-O_2$ layers, a thin film process is a promising technique. These oxide thin films have been prepared by various methods, such as sputtering [6]-[8], molecular beam epitaxy [9], and coevaporation [10]. These methods produced highly oriented films without heat treatment. In this report as-deposited superconducting thin films of Bi-Sr-Ca-Cu-O were prepared by multi-target magnetron sputtering. In order to control the c-dimension of the films, shuttering apparatus were installed between targets and a substrate. It made possible to control the c-dimension of these films. Using this system we obtained a number of different layerings of Bi-Sr-Ca-Cu-O single phase films. Some of the films grown in this way showed superconducting transition.

EXPERIMENTAL

A schematic representation of the system used in our experiments is shown in Fig.1. Shutters were given to the individual targets. Operations of each shutter, open sequence were programed in advance. Controlling these shuttering operations and regulating rf input power of each target, the deposition sequence of atomic layers and their thickness were controlled in the growth direction. The compositions of the targets are Bi_2O_3(4N, conventional powder), $Bi_{2.0}Sr_{1.6}Cu_{0.9}O_x$(calcined powder) and $Ca_{1.0}Cu_{1.0}O_x$(sintered). Films were grown on MgO(100) substrates at a variety of substrate temperature in the range of 580-660°C. The substrate was fixed tightly to the substrate table which turns round at 24 r.p.m., so that the substrate passes through periodically just above the center of each target. The pressure of Ar and O_2 ($Ar:O_2$=4:1) mixture gas was maintained at 1Pa with a cryopump. After deposition oxygen gas was introduced to the chamber approximately up to 1×10^5Pa, and the substrate was cooled to 300°C in about 20 minutes. The resistivity of the films was measured as a function of temperature by a standard four probe method using pressed indium contact at a current of 10μA from 15K to 280K. Crystal structure of films was examined by an x-ray diffraction method with Cu-Ka radiation.

RESULTS AND DISCUSSION

In order to realize the in-situ growth of Bi-Sr-Ca-Cu-O superconducting films, shuttering equipment was designed. Detailed sputtering conditions are listed in Table 1 for getting 2201, 2212 and 2223 phases.

The films of Bi-Sr-Ca-Cu-O grown under 2212 shuttering operation showed to have a low-Tc phase and all the peaks indicated in the x-ray diffraction can be indexed as (002n) of a 30Å unit cell except for the diffraction peaks from the substrate. Typically it has a broad resistive transition starting at approximately 75K and reaching zero around 25K. On the other hand, the films grown with all the sources deposited simultaneously by opening all shutters, were insulating. The thickness of these films was about 1500Å and the substrate temperature was kept at 580°C. Figure 2 shows the x-ray diffraction patterns of these films. When using shuttering, as grown c-axis oriented superconducting films were obtained at relatively low temperature of 580°C. This is probably because, layer by layer deposition using shuttering could fabricate directly or make it easy to grow the perovskite-related structure.

During sputtering, the substrate temperature was kept constant at various different temperatures in the range of 580-660°C. The input power of each target and the opening time of each shutter are the same as above. The films deposited at 580°C and 640°C showed superconducting transition temperatures of 22K and 43K respectively. The film deposited at 660°C showed a higher zero-resistivity temperature of 70K. These films deposited at 580°C and 640°C had smooth surface and no segregation of the constituent element.

All the films described above were cooled in oxygen gas which was introduced into the chamber after deposition. This cooling process, however, was found to influence the superconducting properties, so that we prepared several films by sputtering identically as above but the substrates were cooled in various atmospheres. The superconducting transition temperature and c-dimension

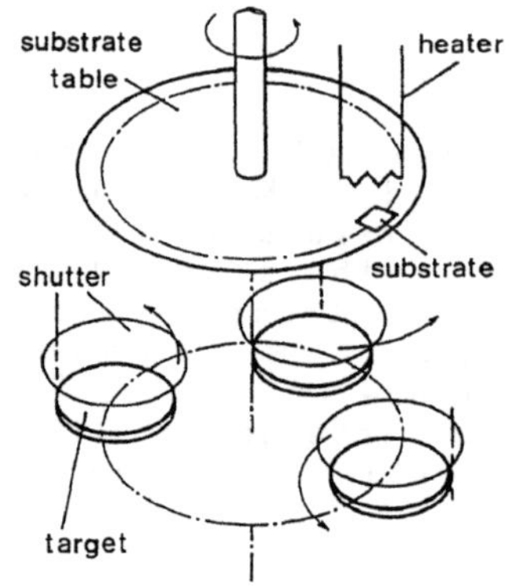

Fig.1 Schematic view of multitarget sputtering system.

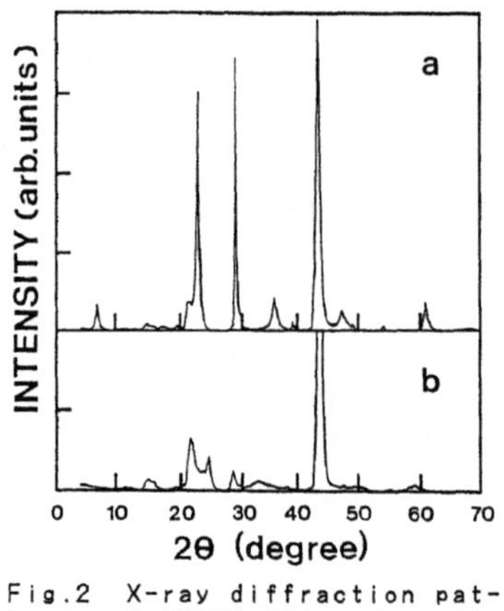

Fig.2 X-ray diffraction patterns of BSCCO films deposited (a) with shuttering and (b) without shuttering.

Table 1. Sputtering conditions.

	Target	2201 phase	2212 phase	2223 phase
Input Power	Bi_2O_3	20 W	20 W	20 W
	$Bi_{2.0}Sr_{1.6}Cu_{0.9}O_x$	50 W	50 W	50 W
	$Ca_{1.0}Cu_{1.0}O_x$	-	65 W	65 W
Shutter opening time	Bi_2O_3	60 s	60 s	60 s
	$Bi_{2.0}Sr_{1.6}Cu_{0.9}O_x$	60 s	60 s	60 s
	$Ca_{1.0}Cu_{1.0}O_x$	-	60 s	110 s
Shuttering order	Bi_2O_3	1	2n-1	2n-1
	$Bi_{2.0}Sr_{1.6}Cu_{0.9}O_x$	1	2n-1	2n-1
	$Ca_{1.0}Cu_{1.0}O_x$	-	2n	2n

were measured as a function of the oxygen partial pressure at the cooling step. These data are shown in Fig.3 and Fig.4. The film which was cooled from 660°C to 300°C in 20 minutes under vacuum (P_{O2}:1x10^{-4}Pa) had an on set temperature at approximately 55K but did not show a superconducting transition above 12K. Whereas the film cooled at 0.2Pa oxygen partial pressure (Ar:O_2=4:1) had a transition at 70K and had zero temperature of 15K. Further, the films cooled in pure oxygen gas at 200Pa and atmospheric pressure (1x10^5Pa), had zero temperature at 35K and 70K respectively. Thus the superconducting transition temperature depended upon the oxygen partial pressure at the cooling step. The x-ray analysis identified these films as low-Tc phase and showed c-axis orientation. Figure 4 shows that the c-dimension calculated from (002n)

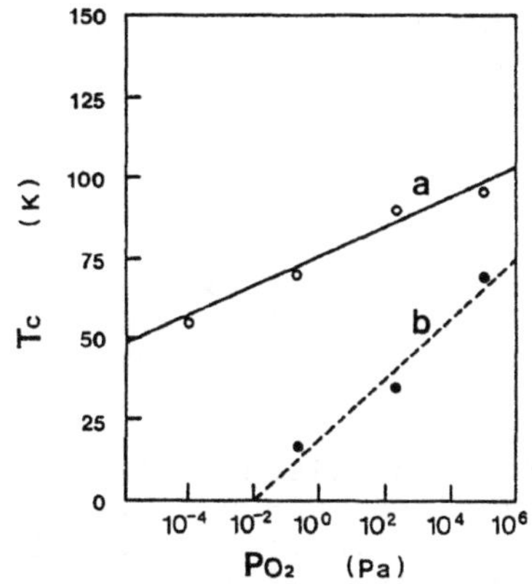

Fig.3 Superconducting transition temperature vs O_2 partial pressure at cooling step. Cooled from 660°C to 300°C in 20 minutes (a) Tc on and (b) Tc zero.

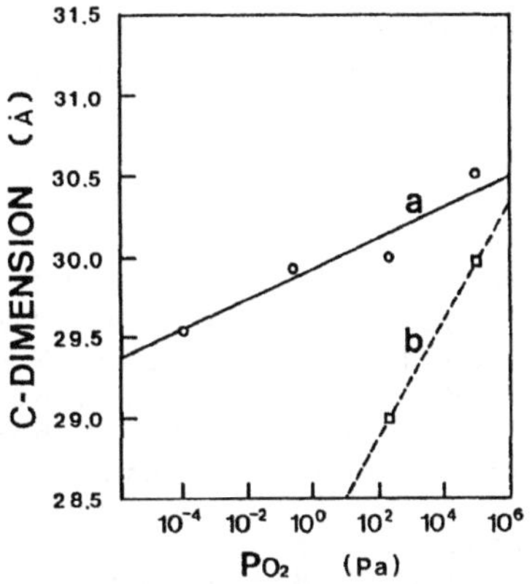

Fig.4 C-dimension vs O_2 partial pressure at cooling step. Cooled from 660°C to 300°C (a) in 20 minutes and (b) in 80 minutes.

peaks is influenced by the oxygen partial pressure at the cooling step. The c-dimension increased with increasing the oxygen gas partial pressure and also increasing the cooling rate. The c-dimension of the film which was cooled under 1x10^{-4}Pa oxygen partial pressure was shorter by about 1Å than that cooled under 1x10^5Pa within experimental accuracy. Such behaviors may indicate insufficient oxygen content in the film. Thus the basic layer structure is formed in the sputtering process. However the superconducting property is influenced by the easily disordered oxygen deficiency. This result suggests that the controlling of the oxygen quantity in Bi-Sr-Ca-Cu-O system is very important for relevant processes.

Several films with a different number of Cu-O_2 layers were obtained by controlling the shutter system. Each film was composed of 50 cyclic units of $Bi_2Sr_2Ca_{n-1}Cu_nO_x$ unit layer. Figure 5 shows a θ-2θ scan of films corresponding to the 2201, 2212 and 2223 structures. C-dimensions calculated from these (002n) peaks were 24.3Å, 29.7Å and 35.7Å respectively. Although it was difficult to obtain single phase film by annealing after deposition [11], shuttering operation controlling the number of the Cu-O_2 layers, easily gave the c-axis oriented single phase films along with a desired structure. Detailed sputtering conditions are listed in Table 1. The resistive superconducting transitions of these films are shown in Fig.6. The 2201 film was semiconducting, the 2212 and 2223 films were superconducting. As for the x-ray diffraction pattern of the 2223 film, all the peaks indicated can be indexed to a high-Tc

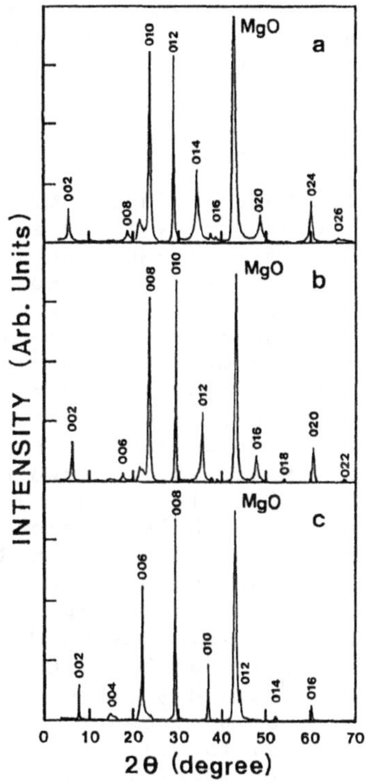

Fig.5 X-ray diffraction patterns of BSCCO films deposited at 660°C. The peaks were indexed as (002n) to (a) 35.7Å, (b) 29.7Å and (c) 24.3Å unit cell.

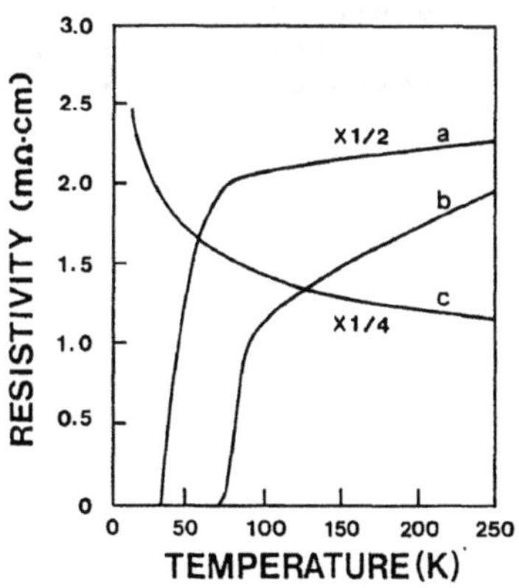

Fig.6 Resistive transitions of BSCCO films of (a) 2223 phase, (b) 2212 phase and (c) 2201 phase.

36Å unit cell. Nevertheless the film has a superconducting transition at 70K, which is substantially lower than the ideal value 110K. The composition of the 2223 films defined by EPMA was $Bi_{2.0}Sr_{1.8}Ca_{1.9}Cu_{3.2}O_x$. Thus high-Tc phase films were obtained, the chemical composition were nearly 2223 and the x-ray diffraction patterns indicated 36Å unit cell, but the superconducting transitions were different from that of the ideal 110K high-Tc phase. We have little knowledge on this subject.

CONCLUSION

We have succeeded in the growth of high temperature superconducting films with their $Cu-O_2$ layers controlled during sputtering by shuttering technique. Furthermore we obtained as-grown superconducting films with zero resistance temperature at 70K and the x-ray diffraction patterns showed highly c-axis orientated single phase.

REFERENCES

1 H.Maeda,Y.Tanaka,M.Fukutomi and T.Asano (1988) A new high-Tc oxide superconductor without a rare earth element. Jpn.J.Appl.Phys.27: L209-L210

2 E.Takayama-Muromachi,Y.Uchida,A.Ono,F.Izumi,M.Onoda,Y.Matsui,K.Kosuda, S.Takekawa and K.Kato (1988) Identification of superconducting phase in the Bi-Ca-Sr-Cu-O system. Jpn.J.Appl.Phys.27: L365-L368

3 Y.Bando,T.Kijima,Y.Kitami,J.Tanaka,F.Izumi and M.Yokoyama (1988) Structure and composition analysis of high-Tc superconducting Bi-Sr-Ca-Cu-O oxide by high resolution analytical electron microscopy.Jpn.J.Appl.Phys.27:L358-L360

4 E.Takayama-Muromachi,Y.Uchida,Y.Matsui,M.Onoda and K.Kato (1988) On the 110 K superconductor in the Bi-Ca-Sr-Cu-O system: Jpn.J.Appl.Phys.27: L556-L558

5 Y.Matsui,S.Takekawa,H.Nozaki,A.Umezono,E.Takayama-Muromachi and S.Horiguchi (1988) High resolution electron microscopy of intergrowth and modulated in 110K high-Tc superconductor $Bi_2(Sr,Ca)_4Cu_3O_y$:Jpn.J.Appl.Phys.27:L1241-L1244

6 H.Adachi,S.Kohiki,K.Setsune,T.Mitsuyu and K.Wasa (1988) Formation of superconducting Bi-Sr-Ca-Cu-O thin films with controlled c-axis lattice spacings by multitarget sputtering. Jpn.J.Appl.Phys.27: L1883-L1886

7 K.Kojima,K.Kuroda,M.Tanioku and K.Hamanaka (1989) As-grown superconductivity of BiSrCaCuO thin films prepared by magnetron sputtering with three targets:$Bi_{2+a}(SrCa)_2Cu_3O_x$, $Bi_2(SrCa)_{2+b}Cu_3O_x$ and $Bi_2(SrCa)_2Cu_{3+c}O_x$ (a=b=1,C =1.5).Jpn.J.Appl.Phys.28:L643-L645

8 J.Fujita,T.Tatsumi,T.Yoshitake and H.Igarashi (1989) Epitaxial film growth of artificial (Bi-O)/(Sr-Ca-Cu-O) layered structure. Appl.Phys.Lett.54:2364 -2366

9 Y.Nakayama,H.Ochimizu,A.Maeda,A.Kawazu,K.Uchinokura and S.Tanaka (1989) In situ growth of Bi-Sr-Ca-Cu-O thin films by molecular beam epitaxy technique with pure ozone: Jpn.J.Appl.Phys.28 L1217-L1219

10 T.Satoh,T.Yoshitake,S.Miura,J.Fujita,Y.Kubo and H.Igarashi (1989) As-grown superconducting Bi-Sr-Ca-Cu-O thin films by coevaporation. Appl.Phys.Lett. 55: 702-704

11 K.Ohbayashi,T.Ushida,T.Tunooka,K.Ohya and H.Banno (1988) Properties of superconducting Bi-Sr-Ca-Cu-O thin films prepared by controlling sputtering conditions and annealing atmospheres. Proceedings of 1st International Symposium on Superconductivity,August 28-31,1988,Nagoya: 551-556

Formation Mechanism of As-Deposited Epitaxial $YBa_2Cu_3O_x(x = 6 - 7)$ Thin Films in Laser Deposition

M. OHKUBO and T. KACHI

Toyota Central Research and Development Labs., Inc., Nagakute-cho, Aichi, 480-11 Japan

ABSTRACT

It has been revealed that the oxygen content x of as-deposited thin films by laser deposition depends strongly on oxygen pressure (Po_2) during a rapid cooling following deposition at 730°C. The x–Po_2 characteristic fits with that in the thermal equilibrium at 500°C. This temperature of 500°C agrees with that at which activation energy for oxygen diffusion begins to decrease. Therefore, it is concluded that the thin films are in thermal equilibrium above about 500°C during the rapid cooling because of small activation energy, and x at about 500°C is retained down to room temperature.

KEY WORDS: $YBa_2Cu_3O_x$, thin film, laser deposition, oxygen content

INTRODUCTION

It is well recognized that the structure and electronic properties of $YBa_2Cu_3O_x$ depend strongly on oxygen content (x). As x increases from 6 to 7, superconducting transition temperature (Tc) varies from 0 to 90 K. It has been usually reported that $YBa_2Cu_3O_x$ phases are superconductive in x>6.4. The mobile oxygens are located on the Cu(1) planes. In x near 7, the mobile oxygens order along b-axis, and form Cu(1)O linear chains. This results in an orthorhombic lattice.

By laser deposition, one can synthesize single-crystalline $YBa_2Cu_3O_x$ thin films, which have the crystal structure with correct positions of cations without post-annealing. However, it has been reported that some as-deposited thin films are non-superconductive and tetragonal, while some films are superconductive and orthorhombic. It is obvious that this is caused by a difference in x. Therefore, in order to obtain as-deposited superconducting thin films without post-annealing, it is important to know the mechanism determining oxygen content. In this paper, we describe an in-diffusion mechanism of oxygen in a period of cooling following laser deposition.

EXPERIMENTAL

The target of a $YBa_2Cu_3O_x$ polycrystalline pellet was irradiated by pulsed beams from an ArF excimer laser. The $SrTiO_3${100} was used as substrate for epitaxial growth. The deposition was done at a substrate temperature of 730±10°C, and at an oxygen pressure of 13 Pa, which is slightly higher than a critical decomposition pressure of 10 Pa.[1] After deposition, oxygen pressure (Po_2) was controlled at 0.013–27000 Pa and then substrate was rapidly cooled. The substrate temperature as a function of time in the cooling at 2700 Pa is shown in Fig.1. The times for cooling from 730 to 400°C were between 2 and 2.5 min. The thin films were not treated by any post-annealing. The detail of sample preparation was reported in previous paper[2].

X-ray diffraction (XRD), Rutherford backscattering spectroscopy including channeling (RBS), and transmission electron microscopy (TEM) were applied to calculate x and to study crystalline quality of the thin films. Oxygen content x was calculated from a relation between *c*-lattice parameter and *x* (Fig.2).[3],[4] The solid line in Fig.2 shows a linear least-square fitted one of c=12.736-0.1501x.

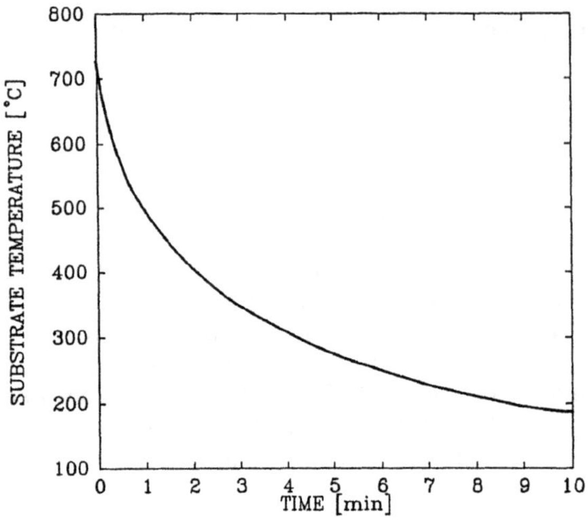

Fig.1. Cooling process after deposition.

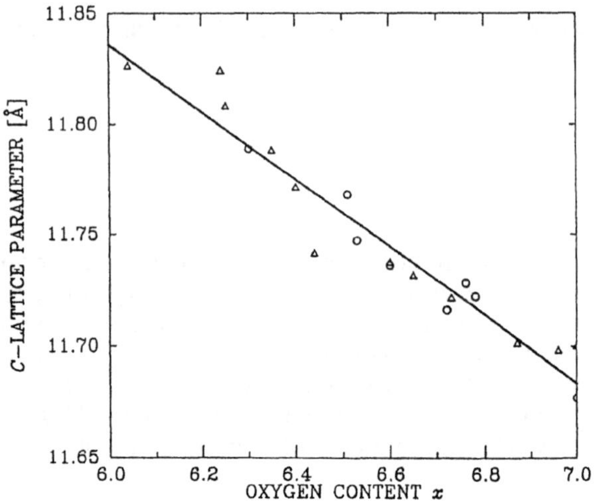

Fig.2. Relation between *c*-lattice parameter and oxygen content *x*; o[3], Δ[4].

For resistivity measurement, samples were cut with widths of about 1 mm and lengths of about 10 mm, and four gold electrodes were formed by vacuum evaporation. Resistivity was measured down to 12 K by a DC four prove method.

RESULTS AND DISCUSSION

From XRD, it has been shown that the thin films cooled at Po_2 above 0.13 Pa are single-phasic, while other thin films cooled at low pressures have secondary phases. The observed impurity diffraction lines with intensities less than 2% of that from (005) of $YBa_2Cu_3O_x$ were considered to be from decomposed products due to Po_2 considerably lower than the critical decomposition pressure.[1] Quality of the crystalline structure was evaluated by ratio of axial channeling yield to random one for barium near surface [χ_{min}(Ba)] in RBS spectra. The χ_{min}(Ba) values ranged from 9 to 30%. TEM and ion-channeling results showed that a relation between $YBa_2Cu_3O_x$ and $SrTiO_3$\{100\} crystals was (001)$YBa_2Cu_3O_x$ ‖ (001)$SrTiO_3$ and [100]$YBa_2Cu_3O_x$ ‖ <100>$SrTiO_3$.[5]

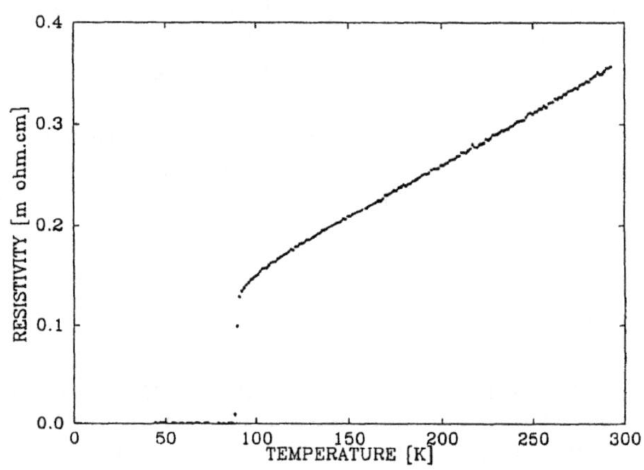

Fig.3. Temperature dependence of resistivity ρ_{ab} of the thin film cooled at 2700 Pa.

Figure 3 shows typical temperature dependence of resistivity in the *ab* plane (ρ_{ab}) for the thin film cooled at 2700 Pa. This film has a $Tc(\rho=0)$=88 K, a transition width of 1.4 K, and a Jc(77 K) of 5.5 x 10^5 A/cm^2. In normal state, the relation between ρ_{ab} and T is completely linear.

Figure 4 shows *c*-lattice parameter as a function of Po_2 during cooling process. The *c* decreases monotonously with increasing Po_2. These *c* values are converted to oxygen contents, and plotted against Po_2 with the thermal equilibrium x at 450, 500, 550°C in Fig.5.[6] It is obvious that the data in this work fits the x–Po_2 curve in the thermal equilibrium at 500°C. It has been revealed by *in situ* resistance measurement that the tetragonal phase is formed during laser deposition.[7] It can be therefore concluded that tetragonal $YBa_2Cu_3O_{\approx 6}$ phase grows epitaxially on $SrTiO_3$ during deposition, and in-diffusion of oxygens occurs to x in the thermal equilibrium at about 500°C during the rapid cooling. This incorporation of oxygens results in a formation of superconducting thin films in as-deposited state.

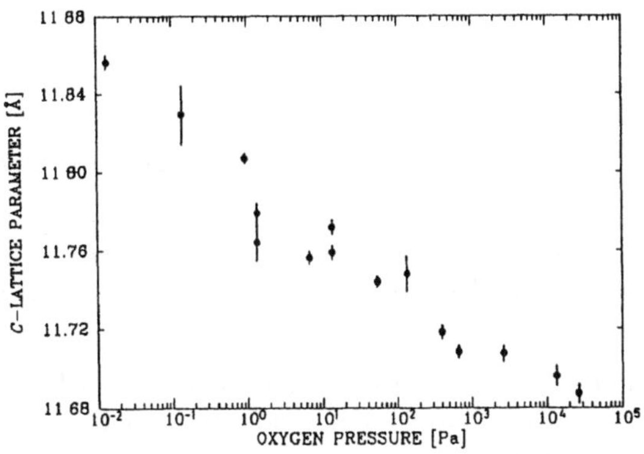

Fig.4. Dependence of c–lattice parameter on oxygen pressure (Po_2) during cooling.

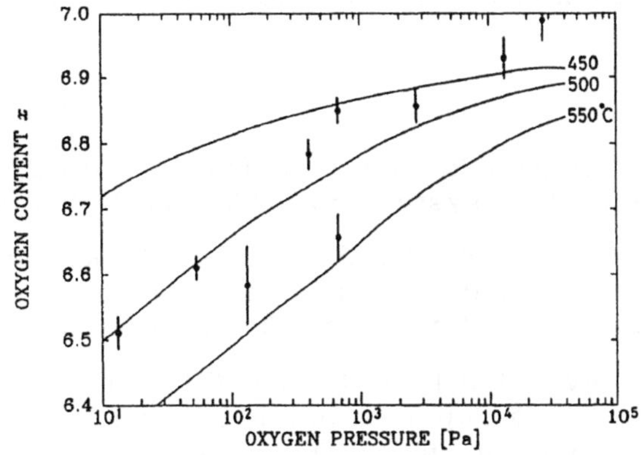

Fig.5. Oxygen content x as a function of oxygen pressure Po_2. The solid lines show the thermal equilibrium x.[6]

The oxygen diffusivity in $YBa_2Cu_3O_x$ is in controversy in the literature.[8] This can be caused by uncertainties of experiments using sintered samples with porosity and granularity. However, it is known that the in-diffusion is much faster than the out-diffusion.[9] Recently, data of internal friction peaks have been applied to determine the diffusivity within a one-dimensional random walk diffusion mechanism in the Cu(1) plane.[8] In this model, the oxygen diffusivity is expressed by $D = D_0 \exp(-H_B/kT)$, where $H_B = H_A + \Delta E$, H_B is potential barrier height for oxygen jump from oxygen site along b–axis (O_B) to vacancy site along a–axis (O_A), H_A is that for oxygen jump from O_A to O_B, and ΔE is potential energy difference between two oxygen sites. It has been observed that both of $D_0 = 3.5 \times 10^{-4}$ cm^2/s and $H_A = 1.03$ eV are independent on temperature, while ΔE is almost constant at 0.23 eV in a range from 400 to 500°C and decreases to 0 eV at 670°C. The 670°C corresponds to that for orthorhombic-tetragonal transition temperature. In other word, activation energy for oxygen diffusion decreases above 500°C. This temperature agrees with that at which x of the thin films are determined, as shown in Fig.4. Therefore, the formation mechanism of the as-deposited superconducting thin films can be explained as follows. During the laser deposition, x in the thin films is near 6. After the deposition, the thin films are in thermal equilibrium above 500°C in spite of the rapid cooling, because H_A is small. On the other hand, below 500°C, the oxygen diffusion almost stops and the x at about 500°C retains to room temperature.

Above diffusion of oxygens is within the Cu(1) plane. On the other hand, since the ab–plane of the thin films is parallel to the substrate surface, we must consider the oxygen diffusion along the c–axis. The diffusivity along the c–axis should be very small, although the mechanism has not been understood. However, in the thin films, there are stacking faults or some lattice defects.[5] It is presumable that such defects act as sources to supply oxygens along the c–axis.

CONCLUSION

We have shown that the oxygen content x in the as-deposited epitaxial $YBa_2Cu_3O_x$ thin films depends strongly on the oxygen pressure during the rapid cooling. The x increases from 6 to 7 as the oxygen pressure increases from 0.013 to 27000 Pa. In the rapid cooling, the thin films have x in the thermal equilibrium at about 500°C. This has been understood by the in-diffusion mechanism of oxygens during the rapid cooling.

ACKNOWLEDGEMENT - We are grateful to T.Noritake, T.Ishiguro, T.Hioki, and Y.Taga for experimental support and encouragement.

References

1. Bormann R, Nölting J (1989) Stability limits of the perovskite structure in the Y-Ba-Cu-O system. *Appl Phys Lett* 54: 2148-2150

2. Ohkubo M, Kachi T, Hioki T, Kawamoto J (1989) Oxygen content control for as-deposited $YBa_2Cu_3O_x$ thin films by oxygen pressure during rapid cooling following laser deposition. *Appl Phys Lett* 55: 899-901

3. Cava RJ, Batlogg B, Chen CH, Rietman EA, Zahurak SM, Werder D (1987) Single-phase 60-K bulk superconductor in annealed $Ba_2YCu_3O_{7-\delta}$($0.3<\delta<0.4$) with correlated oxygen vacancies in the Cu-O chains. *Phys Rev* B36: 5719-5722

4. Tranquada JM, Heald SM, Moodenbaugh AR, Xu Y (1988) Mixed valency, hole concentration, and *Tc* in $YBa_2Cu_3O_{6+x}$. *Phys Rev* B38: 8893-8899

5. Ohkubo M, Kachi T, Noritake T (1989) Ion-channeling analysis for epitaxial $YBa_2Cu_3O_x$ superconducting thin films by laser deposition. *Physica* C160: 480-488

6. Kishio K, Shimoyama J, Hasegawa T, Kitazawa K, Fueki K (1987) Determination of oxygen nonstoichiometry in a high-*Tc* superconductor $Ba_2YCu_3O_{7-\delta}$. *Jpn J Appl Phys* 26: L1228-L1230

7. Ying QY, Kim HS, Shaw DT, Kwok HS (1989) Nature of *in situ* superconducting film formation. *Appl Phys Lett* 55: 1041-1043

8. Xie XM, Chen TG, Wu ZL (1989) Oxygen diffusion in the superconducting $YBa_2Cu_3O_{7-x}$. *Phys Rev* B40: 4549-4556

9. Tu K N, Park SI, Tsuei CC (1987) Diffusion of oxygen in superconducting $YBa_2Cu_3O_{7-x}$ oxide upon annealing in helium and oxygen ambients. *Appl Phys Lett* 51: 2158-2160

Preparation of High T_c Superconducting Compound by Laser Ablation

K. ONABE, N. SADAKATA, and O. KOHNO

Materials Research Laboratory, Fujikura Ltd., 5-1, Kiba 1-chome, Koto-ku, Tokyo, 135 Japan

ABSTRACT

High T_c superconducting YBCO compounds have been prepared by Ar-F excimer laser ablation on $SrTiO_3$ and on metal substrates. Critical temperature (T_c) of the specimen formed on $SrTiO_3$ (100) substrate was 88K and critical current densities (J_c) at 77K over 2 mm x 2 mm squared area at 0.5, 1.0 and 8.0T were 2.3×10^4, 1.5×10^4 and 2.0×10^3 A/cm^2, respectively. YBCO on the metal substrate, Hastelloy C-276 with $SrTiO_3$ thin buffer layer prepared by this method showed high T_c of 83K and J_c of 500 A/cm^2 at 77K, 0T. The excellent J_c vs. magnetic field property is attractive for the application of the compound as practical conductors.

KEY WORDS: YBCO, excimer laser ablation, $SrTiO_3$, Hastelloy C-276, high J_c in magnetic field

INTRODUCTION

Since the discovery of the new high T_c superconductors[1)], a number of studies have been conducted to investigate their properties, and establish their fabrication process for practical applications. Results have been reported using chemical vapor deposition method[2)], rf-magnetron sputtering[3)], ion beam sputtering, laser ablation[4-8)] and other deposition methods. In particular, the excimer laser ablation method has a great advantage in view of long wire fabrication because of its rapid formation rate and controllability of composition. In this paper, we report the results of the preparation of YBCO superconducting compounds by excimer laser ablation on single crystalline $SrTiO_3$ and metal substrate. Investigation of electrical and magnetic properties, X-ray diffraction, surface observation by SEM and grain alignment are discussed.

EXPERIMENTAL

Our experimental setup for the laser ablation consists of Ar-F excimer laser (193 nm) and a stainless steel high-vacuum chamber (Fig. 1). The vacuum chamber contained a YBCO sintered target and a substrate holder to be heated and shifted to X-Y direction. The YBCO targets used in the experiments were prepared with standard ceramics techniques and their composition of Y :Ba :Cu chosen as 0.9 :1.8 :3.0. The pulsed laser beam was introduced into the chamber through two reflection mirrors, a condensing lens, and a quartz glass window. Incident angle of the beam relative to the surface of the target was approximately 45 degree. The pulsed Ar-F excimer laser was operated at 1 to 2 J/cm^2·shot of the laser energy density and 2 to 20 Hz of the repetition rate. Substrates used in the experiments were Hastelloy C-276 with approximately 0.5 micrometer $SrTiO_3$ thin layer. $SrTiO_3$(100) single crystalline substrates were also used for the experiments. The YBCO compounds of 1 to 2 micrometer thickness were formed at 600 to 700°C of the substrate temperature and 9.0×10^{-4} torr of oxygen pressure, and then cooled down to room temperature at the rate of 3 degree per minute in 1 atm of oxygen. Crystal structure of the as-deposited samples were studied by X-ray diffractometry XRD) and resistivities were measured by the standard four probe technique after etching a pattern. We also measured magnetic field dependence of critical current density (J_c) for some samples. Observation of the surface morphology by scanning electron microscope (SEM) was also carried out.

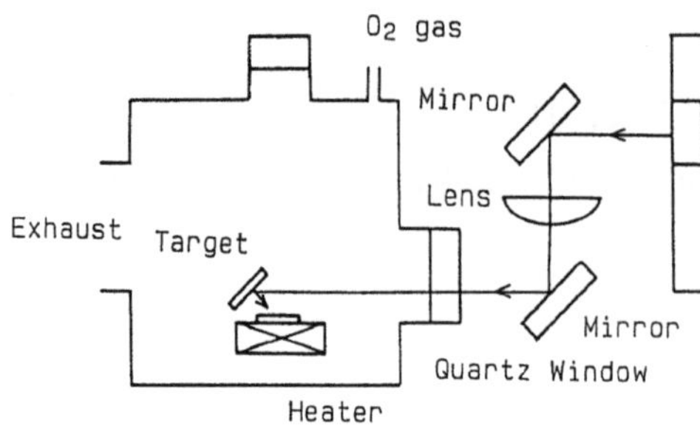

Fig. 1 Schematic diagram of the experimental setup for laser ablation.

RESULTS AND DISCUSSION

Samples of 1 to 2 micrometer thickness was prepared under following conditions;1 to 2 J/cm^2·shot of the laser energy density, 2 to 20 Hz of the repetition rate and 30 to 60 minutes of deposition duration. It was found that the deposition rate can be attained 10 micrometer per hour under a condition. Table 1 shows the analyzed chemical composition by atomic absorption spectrophotometry of the samples prepared at 680 to 720°C and oxigen pressure of 9.0 x 10^{-4} torr from the Y :Ba :Cu = 0.9 :1.8 :3.0 target on $SrTiO_3$(100) substrate. It was found that Y :Ba :Cu = 0.9 :1.8 :3.0 was optimal target composition ratio to obtain stoichiometry in the samples deposited under this condition, and stoichiometric composition, Y :Ba :Cu = 1 :2 :3 was almost attained in this temperature range. Figure 2 and Table 2 show the X-ray diffraction patterns and change of normalized (00n) peak intensity by energy density in the sample deposited at 680°C. Relative (00n) peak intensities tended to increase with decreasing laser energy density in the range of 1.1 to 1.6 J/cm^2·shot. Figure 3 shows the temperature dependence of resistance of the sample of approximately 1.0 micrometer thickness prepared using the $Y_{0.9}Ba_{1.8}Cu_{3.0}O_x$ target at 2 Hz and 1.3 J/cm^2·shot. Superconducting onset and zero-resistance temperatures were 92K and 88K, respectively. Critical current densities (J_c) of the sample under magnetic field at 0.5, 1.0 and 8.0T at 77K were 2.3 x 10^4 , 1.5 x 10^4 and 2.0 x 10^3 A/cm^2, respectively, as shown in figure 4 over 2 mm x 2 mm squared area. The critical currents were determined with 1 microvolt per centimeter criterion. Magnetic field was applied normal to the sample surface and the current was pararell to the surface in the measurements.

Table 1 Analyzed chemical compositions of YBCO compounds refered to Cu = 3.0.

	Composition ratio(Cu=3.0)		
Temp(℃)	Y	Ba	Cu
680	1.0	2.0	3.0
700	0.9	1.9	3.0
720	1.0	1.9	3.0

Table 2 Normalized (00n) peak intensity dependence of laser energy density.

Laser energy density (J/cm^2·shot)	Σ (00n) Peak Int. / Σ Total Peak Int.
1.6	0.23
1.3	0.67
1.1	0.92

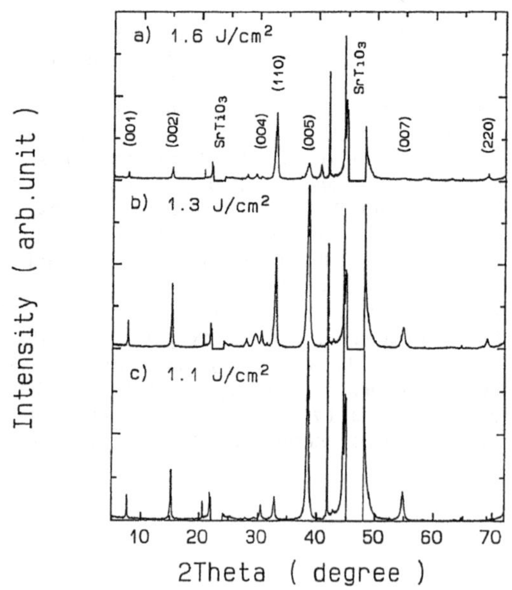

Fig. 2 X-ray diffraction patterns of YBCO compounds deposited at a)1.6, b)1.3, c)1.1 J/cm^2·shot.

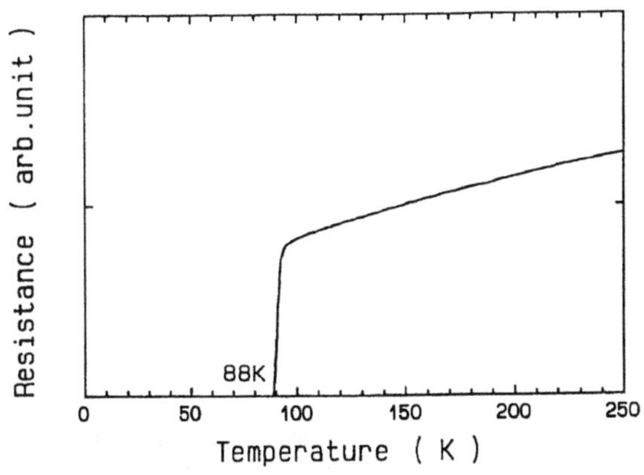

Fig. 3 Temperature dependence of resistivity for the sample by laser ablation at 1.3 J/cm²·shot.

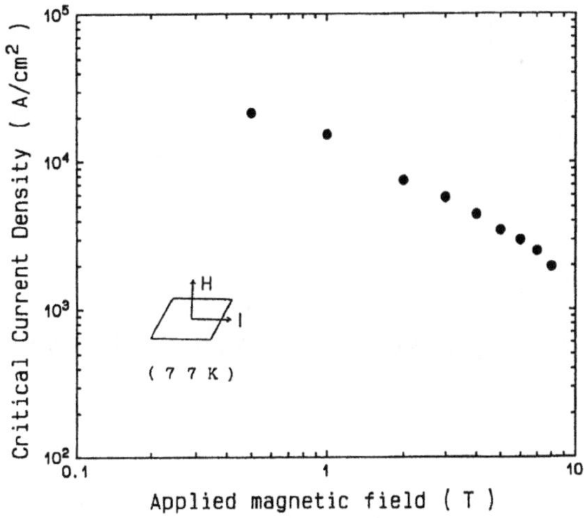

Fig. 4 Magnetic field dependence of critical current density for the sample by laser ablation at 1.3 J/cm²·shot.

An SEM image of this sample is shown in figure 5. Small rectangular grains typically as large as 0.3 x 2 micrometer squared are connected closely and have a mesh-like structure as shown in the figure.

Figure 6 and Figure 7 show an X-ray diffraction pattern and temperature dependence of resistance of the sample formed on Hastelloy C-276 substrate with $SrTiO_3$ thin buffer layer. The resistance shows onset near 92K and zero-resistance was attained at 83K. Critical current density of this sample at 77K, 0T was 500 A/cm². The diffraction pattern revealed the sample consisted mostly of YBCO and no severe contamination from the metal substrate was not observed, where $SrTiO_3$ buffer layer peak is within the background of YBCO. It is clear that the $SrTiO_3$ buffer layer had an effect to reduced the intrusion of elements from the metal substrate. The fact that high J_c was obtained in the sample formed on metal substrate although the c-axis alignment was not clear suggests a further enhancement of J_c would be achieved by controlling the texture of the poly-crystalline compound.

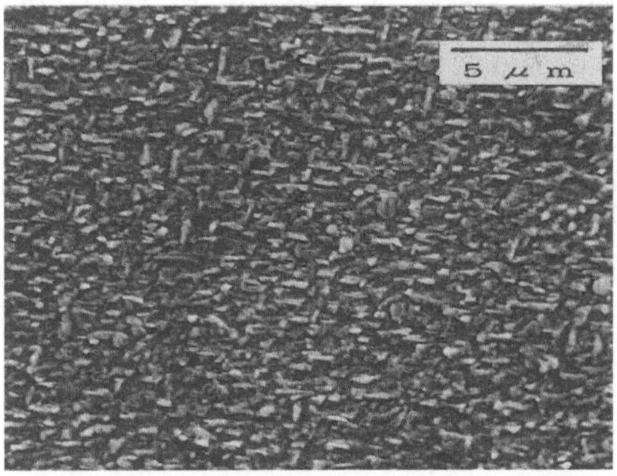

Fig. 5 SEM image of the surface of YBCO compound on $SrTiO_3$(100). X-ray diffraction pattern of this sample shows the weak c-axis grain alignment.

CONCLUSION

We studied the formation of excimer laser ablation and the laser energy dependence of the grain alignment. We prepared YBCO compounds on both $SrTiO_3$ and metal substrate, and measured their superconducting properties. High T_c with 88K and J_c at 77K in the magnetic field of 2.3 x 10^4 A/cm²(0.5T), 1.5 x 10^4, A/cm²(1.0T) and 2.0 x 10^3 A/cm²(8.0T) have been realized despite of weak c-axis grain alignment on single crystalline $SrTiO_3$(100) substrate. The sample prepared on metal substrate, Hastelloy C-276, realized T_c with 83K and J_c with 500 A/cm²(0T) at 77 K. The excellent J_c properties encourages practical conductor application by laser ablation technique.

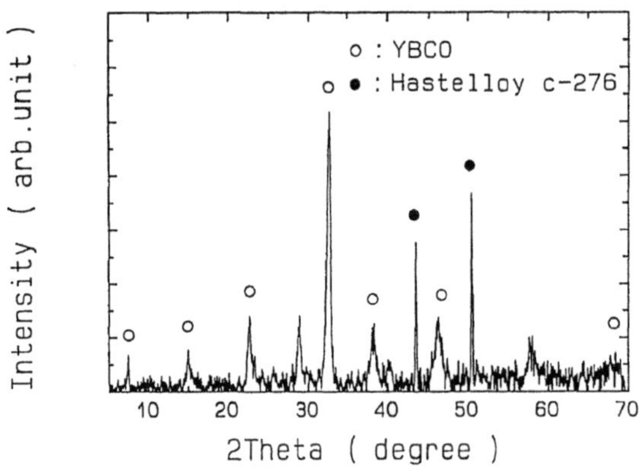

Fig. 6 X-ray diffraction pattern of YBCO compound on Hastelloy C-276 with $SrTiO_3$ thin buffer layer.

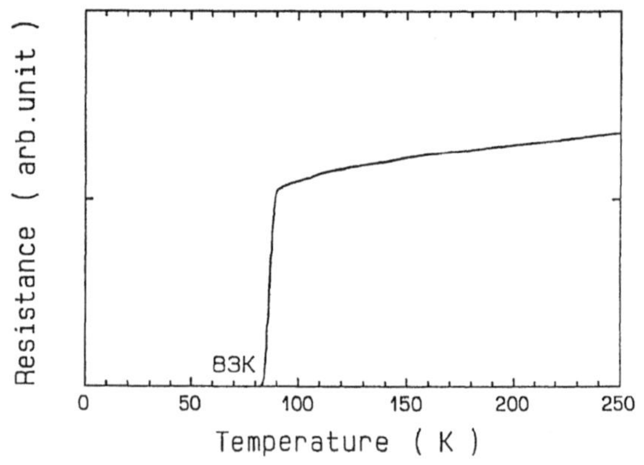

Fig. 7 Temperature dependence of resistivity for the sample on Hastelloy C-276 with $SrTiO_3$ thin buffer layer.

ACKNOWLEDGMENT

This work was performed as a part of " R & D on Superconducting Technology for Electric Power Apparatuses " as a subject of Engineering Research Association for Superconductive Generation Equipment and Materials (Super-GM) under the Moonlight Project of Agency of Industrial Science and Technology, MITI, being consigned by New Energy and Industrial Technology Development Organization (NEDO).

REFERENCES

1) J. Bednorz, K. Muller, Z. Physik B64, 189 (1986).
2) K. Watanabe, N. Kobayashi, H. Yamane, H. Kurosawa, T. Toshio, H. Kawabe and Y. Muto, Jpn. J. Appl. 28, L1417 (1989).
3) H. Itozaki, S. Tanaka, K. Higaki, K. Harada, N. Fujimoto, and S. Yazu, Proc. MRS Int. Meeting on Adv. Mat. Vol. 6, (1989).
4) D.Dijkamp, T. Venkatesan, X.D. Wu, S.A. Shaheen, N. Jisrawi, Y.H. Min Lee, W.L. Mclea and M. Croft, Appl. Phys. Lett. 51, 619 (1987).
5) T. Kawai, M. Kanai, and M. Kawai, Prc. MRS Int Meeting on Adv. Mat. Vol. 6, (1989).
6) O. Eryu, K. Murakami, K. Takita, K. masuda, H. Uwe, H. Kudo, and T. Sakudo, Jpn. J. Appl. Phys. 27, L628 (1988).
7) B. Roas, L. Schultz, and G. Endres, Appl. Phys. Lett. 53, 1557 (1988).
8) U. Sudarsan, N.W. Cody, M.J. Bozack, and R. Solanki, J. Mater. Res. 3, 825 (1988).

Preparation of Y-Ba-Cu-O Superconducting Films by Excimer Laser Ablation

NORIYUKI YOSHIDA[1], MARI KUBOTA, SATOSHI TAKANO[1], KEN-ICHI SATO[1], HAJIME HITOTSUYANAGI[1], MAUMI KAWASHIMA[1], TSUKUSHI HARA[2], KIYOSHI OKANIWA[2], and TAKAHIKO YAMAMOTO[2]

[1] Osaka Research Laboratories, Sumitomo Electric Industries, 1-3, Shimaya 1-chome, Konohana-ku, Osaka, 554 Japan
[2] Engineering Research Center, Tokyo Electric Power Company, 4-1, Nishi-Tsutsujigaoka 2-chome, Chofu, Tokyo, 182 Japan

ABSTRACT

Y-Ba-Cu-O films were deposited on MgO substrates by KrF excimer laser ablation of a stoichiometric $Y_1Ba_2Cu_3O_x$ target. The substrate temperature was kept at 700 ~800 ℃. The best superconducting properties of as-deposited films were end point of transition temperature Tc(R=0) of 89K and critical current density Jc(77.3K) of 1.9×10^6A/cm^2. The microscopic structures of the films were studied using cross-section transmission electron microscopy. Different images were clearly observed between the high Jc and the low Jc film. We also tried high speed deposition using a high repetition rate of laser pulses. At the deposition rate of 300Å/min , a film with Jc(77.3K) of 2.3×10^5A/cm^2 was obtained. Up to the rate of 700 Å/min, Tc(R=0) was as high as 80K. Even at the rate of 2200 Å/min , films were found to have the almost perfect c-axis orientation normal to the substrate surface.

KEY WORDS:high Tc superconducting film, laser ablation, high speed deposition

INTRODUCTION

Recently the laser ablation technique has been used to deposit high quality superconducting films.[1)-5)] This technique makes it possible to form high Jc films without high temperature(800 ~900 ℃) post annealing. An outstanding feature of this technique when a pulsed laser is used is 'pulsed deposition', which means that the formation of the film occurs soon after the ablated fragments arrive at the substrate,as reported by Koren et al.[6)] This suggests the possibility of high speed deposition using a high repetition rate of laser pulses. In this paper, we describe the superconducting properties and the film structures of as-deposited films with relatively low deposition rates. We also report on the relation between superconductivity and the deposition rate up to 2200Å/min .

EXPERIMENTAL

The experimental schematic is shown in Fig.1. A stoichiometric $Y_1Ba_2Cu_3O_x$ target was mounted in a vacuum system and irradiated with a KrF excimer laser (λ=248nm) at 45 ° angle of incidence. The laser energy

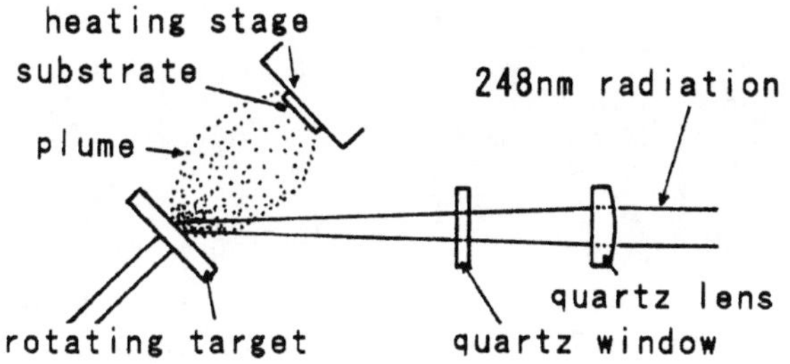

Fig.1.The schematic of laser ablation deposition system for Y-Ba-Cu-O superconducting films. The target was rotated to reduce local cratering. The laser was irradiated on the target through a quartz lens and a quartz window.

densities on the target were between 1.5~2.5J/cm^2. The laser was operated at 1Hz. For higher deposition rates, repetition rates up to 20Hz were also used. A MgO(100) single crystal substrate was mechanically clamped onto a heating stage parallel to the target surface,normal to the laser spot. During deposition the substrate temperature was kept at 700~800 ℃. The oxygen pressure in the system was kept constant between 100 ~200mTorr using a mass flow controller and a variable control valve. No attempt was made to introduce more oxygen gas into the system during the cooling process after deposition. Thicknesses of the films were between 300~500nm.

X-ray diffraction patterns of the as-deposited films were measured on a θ-2θ diffractometer with Cu-K α radiation. The microstructure of the films was examined using a scanning electron microscope(SEM) and a cross-section transmission electron microscope(TEM). The conventional four-probe technique was used to measure the resistance versus temperature curve and the Jc(77.3K). To measure Jc, constriction (typical length of 100 μm and width of 200 ~300 μm) was produced using a sapphire scribe.

RESULTS AND DISCUSSION

Fig.2 shows the temperature versus resistance curve of as-deposited film with the best values of Tc and Jc. The end point of transition was 89K and the transition width ΔT(10~90%) was 3K. Jc(77.3K) was 1.9 $\times 10^6$ A/cm^2. This film was obtained at the conditions of 780 ℃ substrate temperature

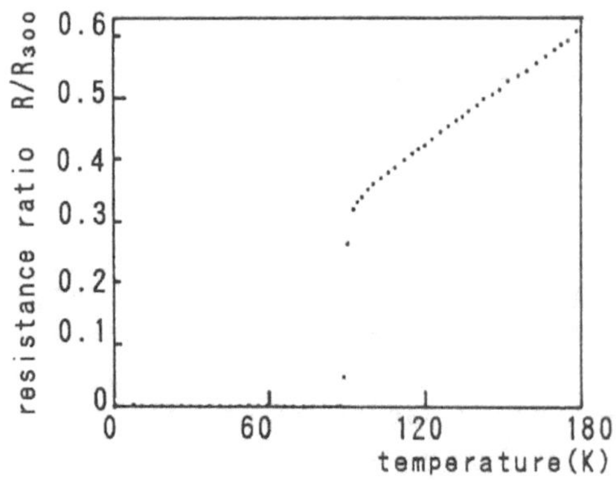

Fig.2. The temperature vs. resistance curve of as-deposited Y-Ba-Cu-O film on MgO(100) by laser ablation. Jc(77.3K) is 1.9$\times 10^6$ A/cm^2.

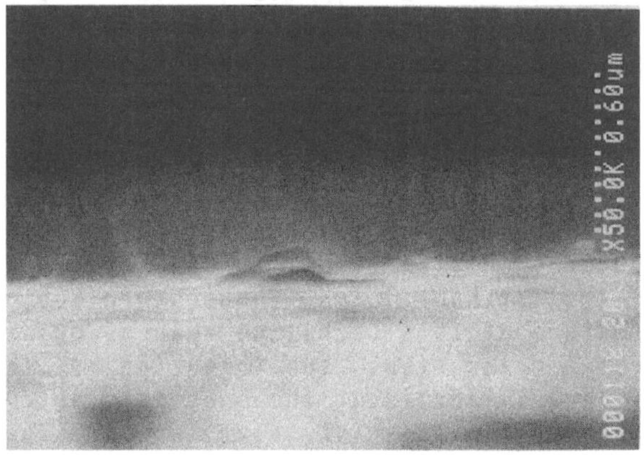

0.6 μm

Fig.4. SEM photo of the cleaved edge of a film with Jc(77.3K) of 2.2$\times 10^4$ A/cm^2.

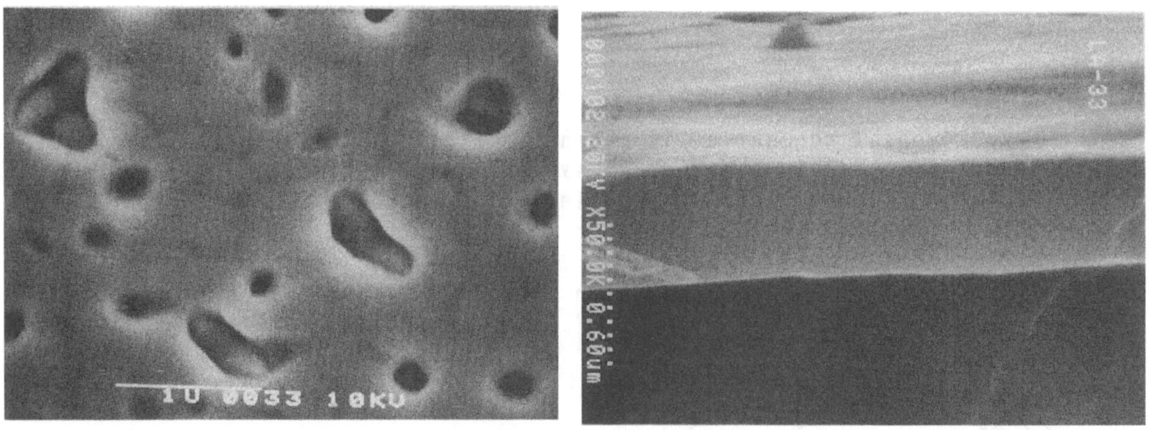

(a) 1μm (b) 0.6 μm

Fig.3. SEM photos of the Y-Ba-Cu-O film with Jc(77.3K) of 1.9 $\times 10^6$A/cm^2.
(a) SEM photo of the surface;(b) SEM photo of the cleaved edge

and 100mTorr oxygen pressure during deposition. The laser was operated at 1Hz. The deposition rate was about 50Å/min. The film was found to have almost perfect c-axis orientation from the X-ray diffraction measurement. The lattice parameter c_0 calculated from the Bragg angle of (00<u>10</u>) reflection was 11.69 Å, which is very close to the ideal c_0 of 11.68 Å.

Fig.3 shows the SEM photos of the surface and the cleaved edge. Holes with typical depths of several tens of nm for this 330nm thick film can be seen in the surface. The film was found to be entirely dense and smooth. Neither clear grain boundaries nor columnar structures were observed. A cleaved edge SEM photo of another film with a relatively lower Jc of 2.2×10^4A/cm^2 is shown in Fig.4. Columnar structures were observed in the cross-section of this film. By comparing this with Fig.3(b), the critical current density seems to be strongly related to the microstructure of the film, especially grain boundaries.

We also tried high speed depositions using higher repetition rates of laser pulses up to 20Hz. The other deposition conditions were the same as those of the above-mentioned film with Jc of 1.9×10^6A/cm^2. The relations between superconducting properties of as-grown films and deposition rates are plotted in Fig.5. The deposition rates in this figure correspond to the repetition rates of 1, 5, 10, and 20Hz, respectively. Although Tc(R=0) values were gradually degraded as the deposition rates were increased, they remained as high as 80K up to the rate of 700 Å/min. Jc(77.3K) at the rate of 300Å/min has the relatively high value of 2.3×10^5A/cm^2. Above the rate of 500 Å/min, Jc(77.3K)'s result in rather low values, less than 1×10^2A/cm^2.

Fig.6 is a comparison between the cross-section TEM photos of the 1Hz and the 5Hz deposition films. Both reveal the clear stripe patterns which are interpreted as c lattice spacing parallel to the substrate surfaces. In Fig.6(a) of the 1Hz deposition, the c lattice spacings were very uniform and no disordered structures were observed. On the other hand, in Fig.6(b) of the 5Hz deposition, disordered and partly unclear lattice spacings were recognized. We believe that the degradation of Jc with the increase in the repetition rate is partly because the disordered structures become dense at the higher repetition rate of deposition.

Fig.7 is an X-ray diffraction pattern of as-deposited Y-Ba-Cu-O film formed at the deposition rate of 2200 Å/min. The film had the c-axis strong orientation normal to the substrate surface. The lattice parameter calculated from the

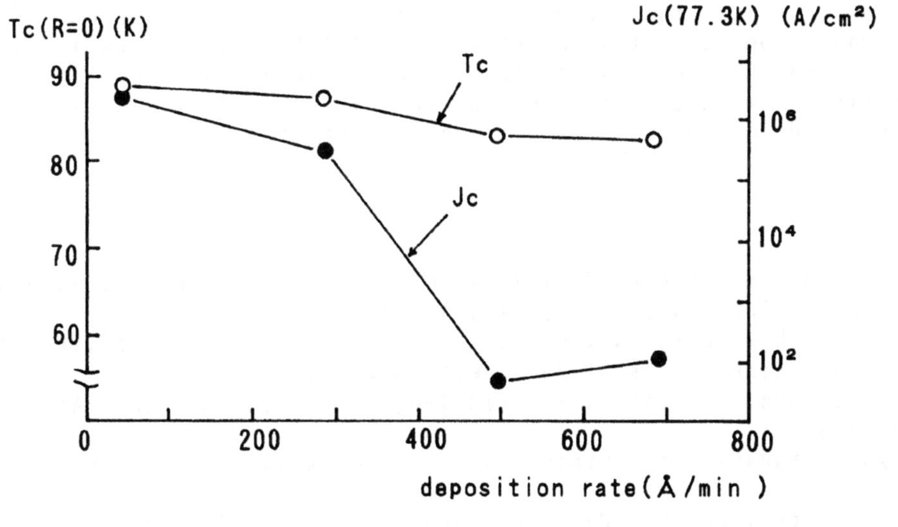

Fig.5. The dependence of Tc(R=0) and Jc(77.3K) of as-grown Y-Ba-Cu-O films on the deposition rate.

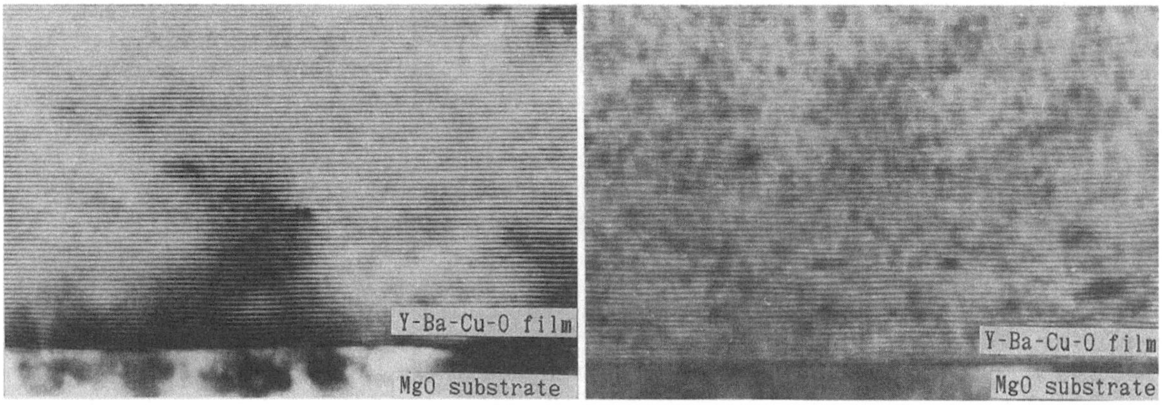

(a) (b)

Fig.6.The cross section TEM photos of the as-deposited Y-Ba-Cu-O films on MgO(100) by laser ablation. (a) is for the 1Hz deposition ; (b) for the 5Hz deposition.

Bragg angle of (00<u>10</u>) reflection was 11.95 Å ,which is distinctly larger than those corresponding to superconducting Y-Ba-Cu-O. This film was semiconducting with no superconducting transition. A method for more oxidation will be primarily necessary for the film to exhibit superconductivity at such a high speed of deposition.

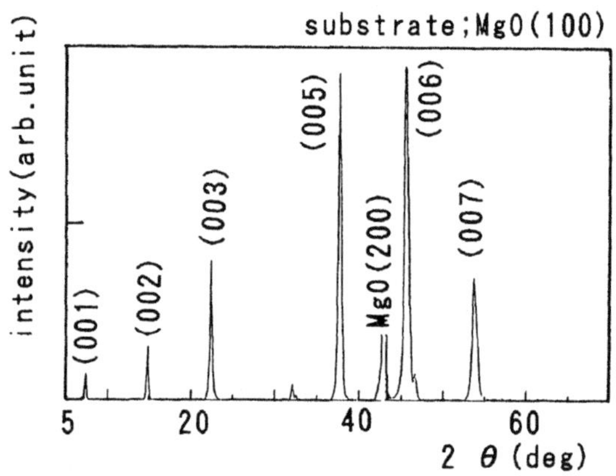

Fig.7. X-ray diffraction pattern of as-deposited Y-Ba-Cu-O film formed at the deposition rate of 2200Å/min.

CONCLUSION

We have successfully deposited high Jc superconducting films on MgO single crystal substrates by excimer laser ablation technique. Our trials of high speed depositions have demonstrated that at the deposition rates of several hundreds Å/min , the films have good superconducting properties. SEM,TEM observations and X-ray diffraction measurements have shown three possible origins for the degradation of superconductivity with an increase in deposition rate ; namely, grain boundaries ,disordered lattice structure , and oxygen deficiency.

REFERENCES

1) X.D.Wu, A.Inam, M.S.Hegde, B.Wilkens, C.C.Chang, D.M.Hwang, L.Nazar, and T.Venkatesan, Appl.Phys.Lett.**54**,754(1989)

2) Q.Y.Ying, H.S.Kim, D.T.Shaw, and H.S.Kwok, Appl.Phys.Lett.**55**,1041(1989)

3) B.Roas, and L.Schultz, Appl.Phys.Lett.**53**,1557(1988)

4) G.Koren, A.Gupta, G.A.Giess, A.Segmüller, and R.B.Laibowitz, Appl.Phys.Lett.**54**,1054(1989)

5) R.K.Singh, J.Narayan, A.K.Singh, and J.Krishnaswamy, Appl.Phys.Lett.**54**,2271 (1989)

6) G.Koren, A.Gupta, and R.J.Baseman, Appl.Phys.Lett.**54**,1920(1989)

Effect of N_2O on Preparation of $Ba_2YCu_3O_x$ Films by Excimer Laser Ablation with Laser Irradiation on Growing Surface

TOSHIHARU MINAMIKAWA[1], YASUTO YONEZAWA[1], SHIGERU OTSUBO[2], TOSHIHIRO MAEDA[2], AKIHARU MORIMOTO[2], and TATSUO SHIMIZU[2]

[1] Industrial Research Institute of Ishikawa, Kanazawa, 920-01 Japan
[2] Department of Electronics, Faculty of Technology, Kanazawa University, Kanazawa, 920 Japan

ABSTRACT

Superconducting Ba-Y-Cu-O films were prepared with an N_2O ambient by the excimer laser ablation method. It was revealed that N_2O ambient enhances the c-axis orientation for the film on crystalline Si substrate, compared with O_2 ambient. Moreover, it was suggested that the laser irradiation onto growing film surface with N_2O ambient improves superconducting properties for the films on MgO substrate with a relatively low substrate temperature.

KEY WORDS : preparation of $Ba_2YCu_3O_x$ film, excimer laser ablation
laser irradiation on growing film surface, N_2O ambient

INTRODUCTION

There are many methods to prepare Ba-Y-Cu-O superconducting films; sputtering[1], laser ablation[2],chemical vapor deposition[3] etc. Each method has been attempted to prepare the superconducting film at lower substrate temperature to reduce the chemical reaction between the film and the substrate. Among these methods the laser ablation method has been found to be promising. The merits are (1) a small difference between the target composition and the film composition, (2) reactive deposition with active gas, such as O_2, (3) contamination-free deposition due to the absence of energy source within the chamber, etc.

So far, we had reported a low temperature preparation technique for $Ba_2YCu_3O_x$ superconducting films using the ArF excimer laser ablation with an O_2 ambient[4,5,6]. We had revealed that a laser irradiation onto the growing film surface improves the crystallinity, superconducting properties, and surface morphology for films on various substrates at a relatively low substrate temperature. The ArF excimer laser irradiation onto the growing film surface enhances a formation of oxygen radicals, activating the film surface atoms and gives the thermal energy only to the film surface. On the other hand, Kanai et al.[7] reported that N_2O gas has a high ability of oxidation. So, the use of the N_2O gas is interesting for our low temperature preparetion technique. In this study, the effect of N_2O instead of O_2 on the preparation of superconducting films was examined in a similar technique.

EXPERIMENTAL

The film deposition was carried out by the excimer laser (SHIBUYA SQL2240, 193nm wavelength, 10ns pulse duration, 5Hz repetition rate) ablation of a target material in a vacuum chamber with N_2O ambient. The pressure is 13-133Pa. The deposition was performed with and without the ArF excimer laser irradiation onto the growing film surface. During the deposition with excimer laser irradiation the excimer laser beam was split into two beams for ablating the target and for irradiating the growing film surface on the substrate. The beam energy density for ablation was roughly 2 or 5Jcm^{-2} and that irradiation was varied over 0-90mJcm^{-2} by moving the position of the lens. The target was the pellet of $Ba_2YCu_3O_x$, which was prepared by the solid-state reaction of the powder. Two kinds of substrates, (100)MgO and (100)Si were used. Nominal substrate temperature was varied over 550-700°C. After the ablation, the films were cooled to the nominal temperature of 490°C with a rate of 3°C/min, kept for 2 hours at that temperature, and then cooled to room temperature at a rate of 1°C/min. During the cooling, the chamber was filled with N_2O up to 1.3kPa in order to increase the oxygen content in the film. The film thicknesses were 0.8-2μm.

For investigating the structure of the films, X-ray diffraction measurements were carried out with CuKα at room temperature. The temperature dependence of the resistivity was measured by the conventional four-probe technique using a cryostat equipped with He refrigerator (Osaka-sanso cryo-mini D). The surface morphology for films was inspected by scanning electron microscope (Akashi DS-130C).

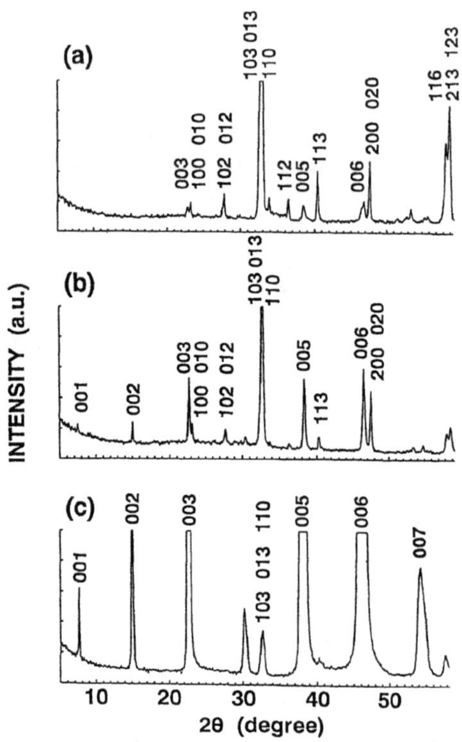

Fig.1 XRD Spectra for the films prepared with the ambient of 27P O_2 at 670°C (a), 27Pa N_2O at 670°C (b), 53Pa N_2O at 600°C (c)

RESULTS AND DISCUSSION

Firstly, we investigate the difference in the film property among the films. Some were prepared with an O_2 ambient and others were prepared with an N_2O ambient. Both of them were prepared on (100)Si substrate without laser irradiation onto the growing film surface (i.e irradiation energy density =0mJcm^{-2}). Figure 1 shows the XRD spectra for the films prepared with the ambient of 27Pa O_2 at a nominal substrate temperature of 670°C (a), 27Pa N_2O at a nominal substrate temperature of 670°C (b), 53Pa N_2O at a nominal substrate temperature of 600°C (c), respectively. The preparation conditions of (a) and (b) are the same except for the ambient gas. The N_2O ambient is found to enhance c-axis orientation of the films compared with the O_2 ambient. Moreover, in spite of the lower nominal substrate temperature, increasing the pressure of N_2O enhances c-axis orientation, as shown in Fig 1(c).

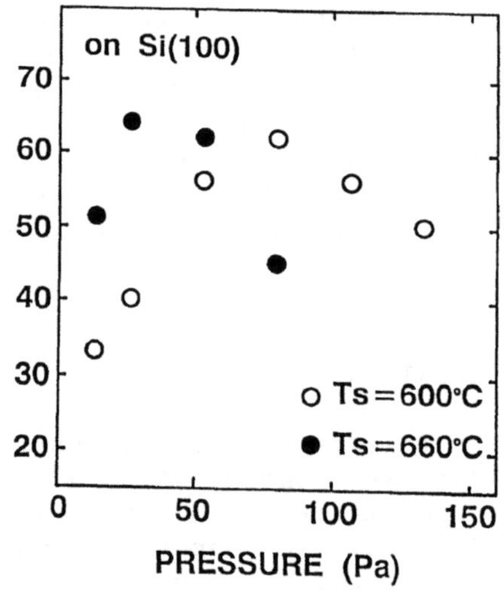

Fig.2 Pressure dependence of T_{c0} for films prepared with an N_2O ambient without irradiation.

Figure 2 shows pressure dependences of the zero resistance temperature T_{c0} for the films prepared on (100)Si substrate with

(a) (b) (c)

Fig.3 SEM photographs for the films prepared with an N_2O ambient of 13Pa (a), an N_2O ambient of 53Pa (b) and an N_2O ambient of 133Pa (c).

an N_2O ambient without laser irradiation onto the growing film surface. Closed circles and open circles indicate the data for the films prepared at nominal substrate temperatures of 660°C and 600°C, respectively. The optimum condition shifts toward higher pressure side with lower substrate temperature. This suggests that oxygen atoms play an important role for enhancing the crystallization of the $Ba_2YCu_3O_x$ film. Figure 3 shows SEM photographs for the films prepared with an N_2O ambient of 13Pa (a), an N_2O ambient of 53Pa (b) and an N_2O ambient of 133Pa (c). Their nominal substrate temperature was 600°C. The length of white bars on the photographs corresponds to 500nm. We can see that the grain sizes become large as the ambient pressure is increased. The N_2O ambient gas is effective for crystallization of the films.

Figure 4 shows the temperature of zero resistance as a function of the laser energy density for irradiation. Films on (100)MgO substrate were prepared with a nominal substrate temperature of 580°C, an N_2O ambient of 27Pa and an irradiation onto the growing film surface. These data were rather scattered because of a little uncontrollability of the irradiated area. However, the irradiation with energy densities from 20 to 60mJcm^{-2} is found to enhance the zero resistance temperature. But the irradiation with energy densities above 60mJcm^{-2} reduces it. One of the reasons is that the irradiation with the excess laser energy density re-ablates the film, resulting in a compositional deviation from the stoichiometry.

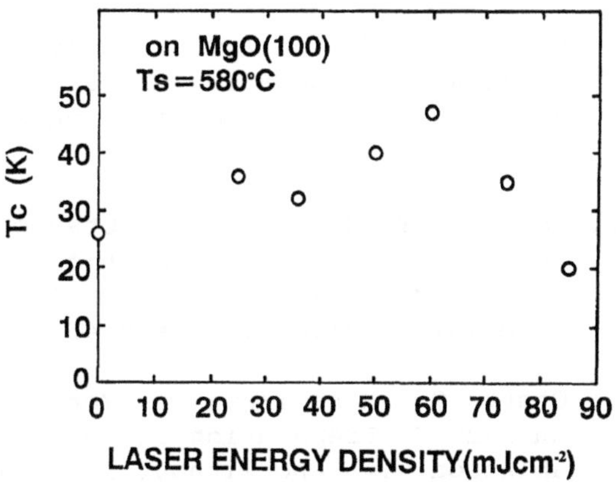

Fig.4 Dependence of Tc_0 on laser energy density for irradiation

CONCLUSIONS

(1)Compared with an O_2 ambient, an N_2O ambient enhances the c-axis orientation of the films on the (100)Si and improves the crystallinity of the films prepared at lower substrate temperature.
(2)The optimum pressure for preparing the high Tc films shifts toward higher pressure side at a lower substrate temperature.

(3)A laser irradiation onto growing film surface on MgO with N_2O ambient also tends to improve suprconducting property.
These results suggest that the oxygen atoms play an important role to reduce the substrate temperature.

ACKNOWLEDGEMENTS

We would like to thank Mr. S.Shinmura and Mr. K.Segawa of the Industrial Research Institute of Ishikawa Prefecture for helping with our experiment. We would also like to thank Prof. M.Kumeda of Kanazawa University for his many helpful discussions. We are grateful to Shibuya Kogyo Co., Ltd. for supplying the excimer laser system. This work was partly supported by a Grant-in-Aid for Scientific Research on Chemistry of New Superconductors from the Ministry of Education, Science and Culture.

REFFERENCES

1. Enomoto Y. Murakami T. Suzuki M. and Moriwaki K. Largely anisotropic superconducting critical current in epitaxially grown $Ba_2YCu_3O_{7-Y}$ thin. films. Jpn. J. Appl. Phys. 26, 1987:1248

2. Dijkkamp D. Venkatesan T. Wu X.D. Shaheen N. Jisrawi N. Min-Lee Y.H. McLean W.L. and Croft M. (1987) Preparation of Y-Ba-Cu oxide superconductor film Using pulsed laser evaporation from high Tc bulk material. Appl. Phys. Lett. 51(8),24 Aug 1987:619-621

3. Yamane H. Kurosawa H. Iwasaki H. Matsumoto H. Hirai T. Kobayashi N. and Muto Y. Tc of c-axis-oriented Y-Ba-Cu-O films by CVD. Jpn. J. Appl. Phys. 27 1988:1275

4. Minamikawa T. Yonezawa Y. Otsubo S. Maeda T. Moto A. Morimoto A. and Shimizu T. Preparation of $Ba_2YCu_3O_x$ superconducting films by laser evaporation and rapid laser annealing. Jpn. J. Appl. Phys.Lett. 27 1988: 619-621

5. Yonezawa Y. Minamikawa T. Otsubo S. Maeda T. Moto A. Morimoto A. and Shimizu T. Preparation of high Tc oxide superconducting films by laser annealing. Extended Abstract 1988 Int. Conf. Solid State Devices & Materials, Tokyo, 1988:435-438

6. Otsubo S. Minamikawa T. Yonezawa Y. Maeda T. Moto A. Morimoto A. and Shimizu T. Preparation of Ba-Y-Cu-O superconducting films by laser ablation with and without laser irradiation on growing surface. Jpn. J. Appl. Phys. 27 1988:2442-2444

7. Kanai M. Kawai T. Kawai S. and Tabata H. Low-temperature formation of multilayered Bi(Pb)-Sr-Ca-Cu-O thin films by successive deposition using laser ablation. Appl. Phys. Lett., 54 1989:1802

Effect of Oxygen Pressure During Laser Deposition on Crystal Orientation in $YBa_2Cu_3O_{7-\delta}$ Films

H. IZUMI, K. OHATA, T. HASE, K. SUZUKI, and T. MORISHITA

Superconductivity Research Laboratory, International Superconductivity Technology Center, 10-13, Shinonome 1-chome, Koto-ku, Tokyo, 135 Japan

ABSTRACT

The effects of the oxygen partial pressure on the crystal orientation have been investigated for the $YBa_2Cu_3O_{7-\delta}$ films deposited on $SrTiO_3$ (100) at 700 °C with an ArF excimer laser. The films deposited in pressures of less than 10 mTorr have grown with the c-axis normal to substrate. By contrast, the films prepared in the pressure region between 20 mTorr and 1Torr have oriented the a-axis perpendicular to the film. The critical current density Jc of the c-axis-oriented film is large, $1.0*10^5$ A/cm^2 at 77 K, compared to that of the a-axis-oriented one of $1.8*10^4$ A/cm^2 at 40 K. It is expected that this difference in the superconducting properties is caused by anisotropy of the material itself.

KEY WORDS : laser ablation, oxygen pressure effect, orientation, anisotropy

INTRODUCTION

Excimer laser ablation is a particularly promising method for the preparation of as grown highly qualified superconducting ceramics thin films[1-3], although several techniques for deposition of high-Tc superconducting films have been reported[4-6]. This method has the advantage of being carried out under higher oxygen pressures enabling to provide enough oxygen to films during deposition, particularly for the preparation of $YBa_2Cu_3O_{7-\delta}$ superconducting films. To date, the preparation of the $YBa_2Cu_3O_{7-\delta}$ films using laser ablation technique has been quite a few reported but oxygen partial pressures have been limited within a narrow range around 0.1 Torr. Thus, the effect of the oxygen partial pressure on the formation of superconducting films is still not clear. To confirm the effect of the oxygen partial pressure on the crystal orientation of the resulting films, we have investigated in a wide oxygen partial pressure range from 10^{-5} to 5 Torr. We also discuss anisotropy of the superconducting properties related to the preferential crystalline orientation.

EXPERIMENTAL

The experimental arrangement is shown in Fig.1. ArF excimer laser pulses of 193 nm wave length and 10 ns duration were used for ablation with a repetition rate of 5 Hz. They were introduced to the vacuum chamber through the quartz window. The laser beam was focused on the target down to an area of 1x3 mm^2, to produce a fluence about 1J/cm^2. The target was a 30-mm-diam sintered cylinder of a high-purity $YBa_2Cu_3O_{7-\delta}$ and the laser beam was allowed to impinge on the side of the target. By rotating and forwarding this target along the axis, the fresh surface was always exposed to the laser beam. Mirror polished $SrTiO_3$ (100) substrates held at 700 °C were used. The substrate heater was separated from the processing chamber by the quartz window and the substrate was heated by radiation. This arrangement enables the stable operation even in a high oxygen pressure. Substrates were mounted at 50 mm above the target. Deposition was carried out in oxygen partial pressures ranged 10^{-5} to 5 Torr. The deposited films were cooled in an oxygen pressure of 300 Torr. The crystallinity of the resulting films was evaluated by a X-ray diffraction (XRD) measurement, and a detailed characterization of the structure of these films was carried out by a small-angle XRD technique. The critical temperature and the critical current density were determined by a resistive measurement using a DC four-probe method.

RESULTS AND DISCUSSION

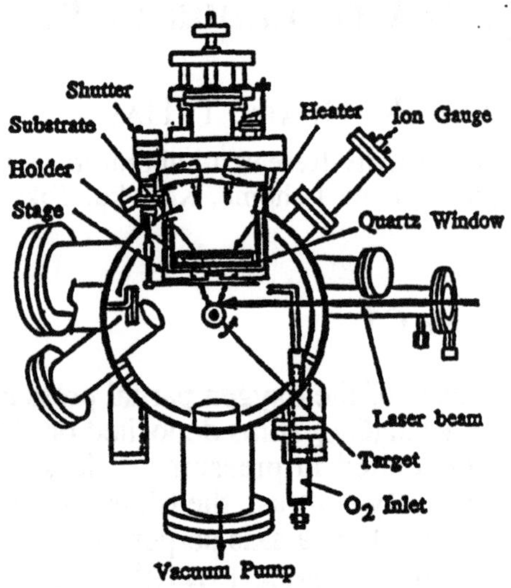

FIG.1. Schematic diagram of the laser ablation apparatus.

The influence of the deposition pressure on the X-ray intensity ratio of the (002)-peak to the (200)-peak is shown in Fig.2. As is seen in Fig.2, the films deposited in pressures of less than 10mTorr have preferentially grown with the c-axis normal to the film as reported by many researchers[1-6]. While the films deposited in the pressure range between 20 mTorr and 1Torr are a-axis oriented. The excimer laser ablation technique is accompanied with a flash plasma, what is called *laser plume*, originated by laser beam. The expansion of this plume gradually became smaller with increasing deposition pressures. The laser plume covered the whole surface of substrate in the deposition pressure range of less than 10mTorr, but the top of the plume did not reach the substrate in higher pressures. It is most likely that the plasma in the laser plume affects crystallization, and this accelerates the preferential growth of the c-axis normal to the substrate. Considering the mean free path of the species (Y, Ba, Cu, and their oxide molecules) blown up from the target by laser ablation, the deposition pressure of less than 1mTorr should be needed to keep these species away from the scattering effect on the way to the substrate from the target separated by 50 mm. In addition, the deposition rate of films was always about 1 A/sec (0.2 A/shot) in the deposition pressure range of less than 0.1 Torr, and it was rapidly decreased to 0.4 A/sec or less in the range of more than 0.5 Torr. Both of the mean free path and the deposition rate did not change at pressures of around 10 mTorr, so one should attribute the change in the preferential orientation to a loss in the kinetic energy of the deposition species by the scattering, and not to the deposition rate.

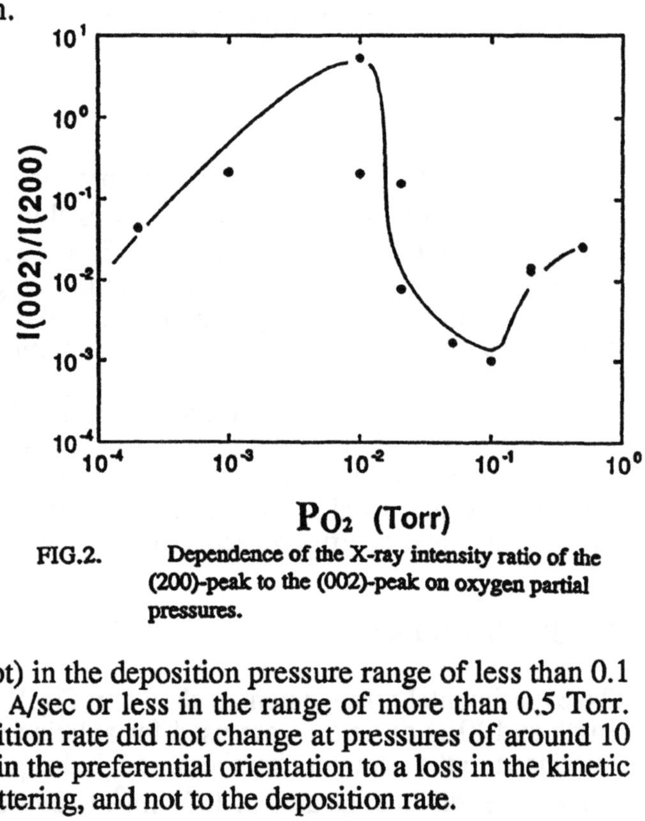

FIG.2. Dependence of the X-ray intensity ratio of the (200)-peak to the (002)-peak on oxygen partial pressures.

Small-angle XRD experiments were carried out for a characterization of the quality of these films in detail. Figure 3 (a) shows the result for the c-axis oriented $YBa_2Cu_3O_{7-\delta}$ films for a fixed incidence angle (13.8°) and detector angle (68.8°), corresponding to (108)-peak, with rotating azimuth angle ω between -60 and 150 degrees. The azimuth angle ω was measured as angles from the <100> direction of the substrate. It should be noted that the a and b axes of the films are in good alignment with the crystal axes of the $SrTiO_3$ substrate. The result from a fixed incidence (14.7°) and detector angle (47.8°), corresponding to the (201)-peak for the a-axis oriented film is shown in Fig.3 (b) showing that the c-axis of $YBa_2Cu_3O_{7-\delta}$ is parallel to the [100] directions of the $SrTiO_3$ substrate.

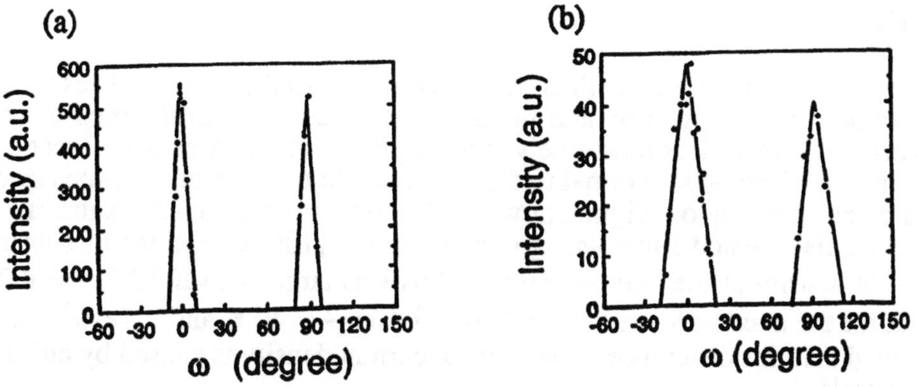

FIG.3. X-ray diffraction intensity vs azimuth angle ω for (a) the c-axis oriented $YBa_2Cu_3O_{7-\delta}$ film using fixed angle of incidence at 13.8° and detector angle at 68.8° corresponding to the (108)-peak, and (b) the a-axis oriented film using incidence and detector angles of 14.7° and 47.8° corresponding to the (201)-peak.

It is expected that $YBa_2Cu_3O_{7-\delta}$ films with less oxygen deficiency can be prepared in a higher oxygen pressure. However, the crystallinity of the resulting films in the pressures of more than 2 Torr was not satisfactory and no films were obtained in 5 Torr, because the laser plume was extremely small in these conditions and was not reached the substrates. To realize a high quality film in such high pressure conditions, it is necessary to decrease the distance between the substrate and the target so that the plume reaches the substrate, and/or to use another technique to improve the crystallinity of films.

Superconducting properties are quite different between the a-axis oriented $YBa_2Cu_3O_{7-\delta}$ films and the c-axis oriented ones. Figure 4 shows results of the resistive measurements for both the c-axis-oriented and the a-axis-oriented films using a DC four-probe method. The resistivities at the onset temperature of the superconducting transition for the films with the c-axis and the a-axis normal to the substrate are quite different each other, 0.4mΩ*cm and 3.5mΩ*cm, respectively. The corresponding zero-resistance temperature Tc is almost same, about 85 K with the transition widths within 2 K. These values are reflected in the gradients of the resistivity versus temperature curves in the normal conduction region. The slope for the c-axis-oriented film indicates metallic behavior, being extrapolated to the origin. By contrast, the resistivity for the a-axis-oriented film slightly decreases with decreasing temperature until Tc. The critical current density Jc measurement was made by taking the 10 μV/cm criterion. Jc of the c-axis-oriented films is considerably large, $1.0*10^5$ A/cm^2 at 77 K, compared to that of the a-axis-oriented films being $1.8*10^4$ A/cm^2 at 40 K and $2.4*10^2$ A/cm^2 at 50 K. It is expected from the results in Fig.3 that the superconducting behavior for the c-axis-oriented films reflects the transportation characteristics in the CuO_2 plane of $YBa_2Cu_3O_{7-\delta}$, whereas that for the a-axis oriented ones indicates the characteristics averaged over properties along of the b and c axes. Further investigation should be needed but it is expected that anisotropy on superconducting properties of $YBa_2Cu_3O_{7-\delta}$ is the origin of this difference.

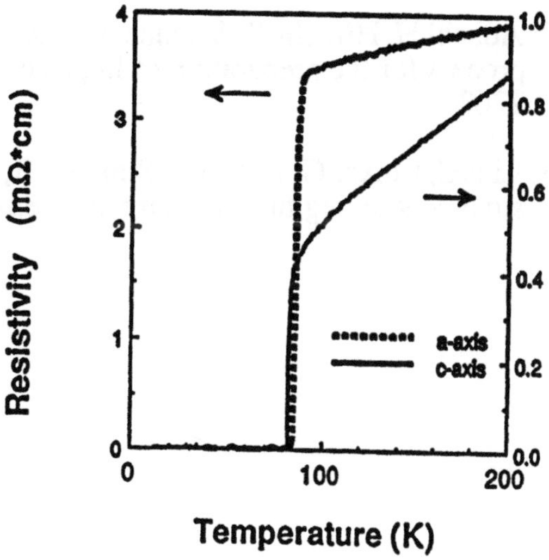

FIG.4. Resistivity vs. temperature curve of the c-axis oriented $YBa_2Cu_3O_{7-\delta}$ film and the a-axis oriented film on $SrTiO_3$ (100).

CONCLUSION

We have prepared the c-axis-oriented films in the oxygen partial pressure of less than 10 mTorr during excimer laser deposition and the a-axis oriented films in the pressure range between 20 mTorr and 1 Torr. It is most likely that the plasma in the laser plume accelerates the preferential growth of the c-axis normal to the substrate. In the case of the c-axis oriented films, the a and b axes are in good alignment with [100] directions in $SrTiO_3$ substrate, and the c-axis of the a-axis-oriented films was grown along the [100] directions of substrates. The critical current density of the c-axis-oriented films was large, $1.0*10^5$ A/ cm^2 at 77 K, compared to that for the a-axis-oriented films, $1.8*10^4$ A/cm^2 at 40 K and $2.4*10^2$ A/cm^2 at 50 K. It is expected that this difference in the critical current density is caused by anisotropy of $YBa_2Cu_3O_{7-\delta}$ itself.

This work was performed under the management of the R & D of Basic Technology for Future Industries supported by New Energy and Industrial Technology Development Organization (NEDO).

REFERENCES

1) Inam A, Hedge MS, Wu XD, Venkatesan T, England P, Miceli PF, Chase E.W, Chang CC, Tarascon JM, Watchman JB (1988) As-deposited high Tc and Jc supreconducting thin films made at low temperature. Appl. Phys. Lett.,**53**: 908

2) Witanachchi S, Kwok HS, Wang XW, Shaw DT (1988) Deposition of superconducting Y-Ba-Cu-O films at 400 °C without post-annealing. Appl. Phys. Lett. **53**: 234

3) Koren G, Gupta A, Giess FA, Segmuller A, Laibowitzand RB (1989) Epitaxial films of $YBa_2Cu_3O_{7-\delta}$ on $NdGaO_3$, $LaGaO_3$, and $SrTiO_3$ substrates deposited by laser ablation. Appl. Phys. Lett.,**54**: 1054

4) Terashima T, Iijima K, Yamamoto K, Bando Y, Mazaki H (1988) Single-crystal $YBa_2Cu_3O_{7-x}$ thin films by activated reactive evaporation. Jpn. J. Appl.Phys. **27**: L91

5) Adachi H, Hirochi K, Setsune K, Kitabatake M, Wasa K (1987) Low-temperature process for the preparation of high Tc superconducting thin films. Appl. Phys. Lett. **51**: 2263

6) Li HC, Linker G, Ratzel F, Smithey R, Geerk J (1988) In situ preparation of Y-Ba-Cu-O superconducting thin films by magnetron sputtering. Appl. Phys. Lett. **52**: 1098

Characterization of Oxide Superconducting Films Prepared by Pyrolysis of Organic Acid Salts

M. FUJIOKA[1], T. SEKI[1], T. OHHASHI[1], K. YAMAGUCHI[2], and S. SAWADA[1]

[1] Nippon Mining Co., Ltd. Isohara Plant, Usuba, Hanakawa-cho, Kitaibaraki, Ibaraki, 319-15 Japan
[2] Nikko CS Chemical Co., Ltd., Hitachi Miyata-cho, Hitachi, Ibaraki, 317 Japan

ABSTRACT

Bi-Sr-Ca-Cu-Pb-O system superconducting film was prepared on MgO substrates by a spray pyrolysis method using organic acid salts of constituting elements as raw materials. A study on the influence of various heat treatments showed that the film subjected to a rapid heating and quench (RHQ) treatment exhibited improved Tc, as high as 103 K, than films without such practice. Further survey showed that the superior property by RHQ treatment was brought by the improvement in film morphology, and also crystal orientation.

KEY WORDS

Bi-Sr-Ca-Cu-Pb-O system, spray pyrolysis method, organic acid salt, effect of heat treatment

INTRODUCTION

Since the discovery of high Tc superconductivity in perovskite oxides, unprecedented research efforts were accumulated worldwide in many fields including synthesis of new higher Tc material, characterization of material, development of practical applications and so on. In order to utilize this material in engineering applications, it must be formed into certain shapes such as films or wire etc. From the nature of material, film is believed to be the most important form, and has potential to be used in electrical devices, magnetic shields, and so on. Many film preparation techniques were studied such as sputtering, CVD, and printing for instance, but among them, less attention is given to spray pyrolysis. This technique is, however, as important as others, since it is simple, less costly, and easy to be applied on large area. This paper presents the film preparation technique by the spray pyrolysis of organic salts and the resultant properties of films obtained in Bi-Sr-Ca-Cu-Pb-O system.

EXPERIMENTAL PROCEDURE

Octylic acid salts of Bi, Sr, Ca, Cu and naphtenic acid salts of Pb solved in either toluen or xylen were mixed in proportion, and then coated on sintered MgO substrates by spin coating. The coated substrates were then annealed for 10 minutes at 500 °C to decompose organic salts, which was repeated until the desired thickness was reached. The effect of RHQ treatment, which is rapidly heating to 1030

°C and holding for 30 to 60 seconds followed by quenching to room temperature, was studied after the annealing for 150 hours at 840 °C. Electric resistivity was measured by a conventional four probe method. Crystal structure of the oxide was determined with Rigaku RINT-1100 X-ray diffractometer. Film morphology was observed under a scanning electron microscope.

RESULTS AND DISCUSSION

The temperature dependence of electrical resistivity of the films which received (a) RHQ treatment for 60 seconds at 1030 °C and annealing for 150 hours at 840 °C, (b) annealing only and (c) RHQ treatment only is shown in Fig. 1. Although each film shows superconducting properties, the film treated with both RHQ treatment and annealing exhibited the highest Tc at 103 K.
X-ray diffraction study in Fig. 2 revealed that as decomposed film contains unreacted phases such as Cu_2O, $SrCO_3$ and Bi_2O_3, whereas the as RHQ'd material, without annealing, consists of 2212 phase and the unreacted phases. Annealing after RHQ treatment promotes the reaction of elemental phases, and only 2212 and 2223 phases could be detected in the RHQ'd then annealed sample. These two phases only could also be detected in the film that received annealing without the RHQ treatment, however, the peak height of $(002)_{2223}$ plane is smaller than that of the RHQ'd then annealed sample.
SEM observation of the each film are presented in Fig. 3. The surface of the as decomposed sample is characterized by the numerous cracks and rough texture. Annealing without the RHQ treatment seems to improve the surface texture by the promotion of the new phases, however, the cracks are still visible. As RHQ'd sample exhibits rather smooth surface with few cracks, and the following annealing results in the growth of the superconducting phases which can be seen in acicular relief on the film surface.

SUMMARY

High Tc oxide superconducting film prepared by the spray pyrolysis of organic salts was characterized by the temperature vs. electrical resistivity measurement, X-ray diffraction study, and SEM observation. Followings are the summary of the results.

(a) Usefulness of the spray pyrolysis of organic salts in obtaining good superconducting film is demonstrated.
(b) Superior properties, especially high Tc, can be achieved by the RHQ treatment followed by conventional annealing.
(c) This improvement stems from the drastic change in film surface morphology, namely the great reduction in number of cracks and surface texture.
(d) RHQ treatment improves the crystal orientation of film in such manner that larger proportion of the grains show $(002)_{2223}$ plane parallel to substrates. Thus superconducting links are enhanced between adjacent grains.

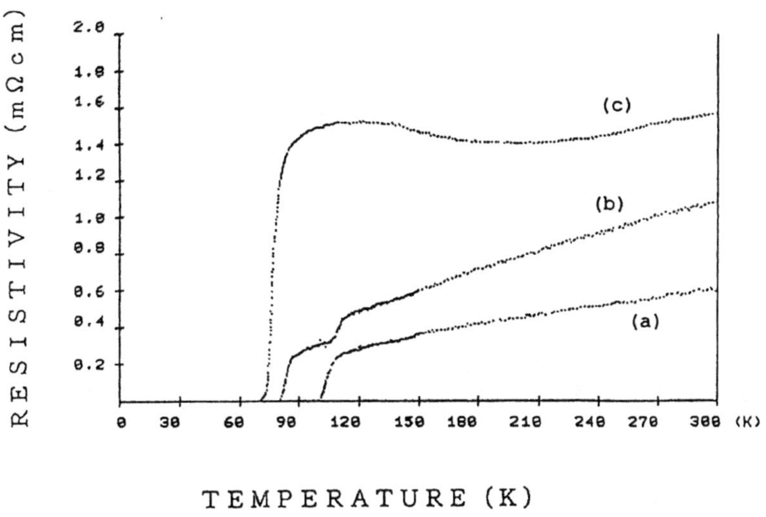

Fig.1 Temperature dependence of resistivity of $Bi_2Sr_2Ca_2Cu_3Pb_{0.4}O_x$ films which were obtained by (a) treated for 60sec at 1030°C followed by annealing for 150hr at 840°C (b) annealed for 150hr at 840°C (c) treated for 60sec at 1030°C

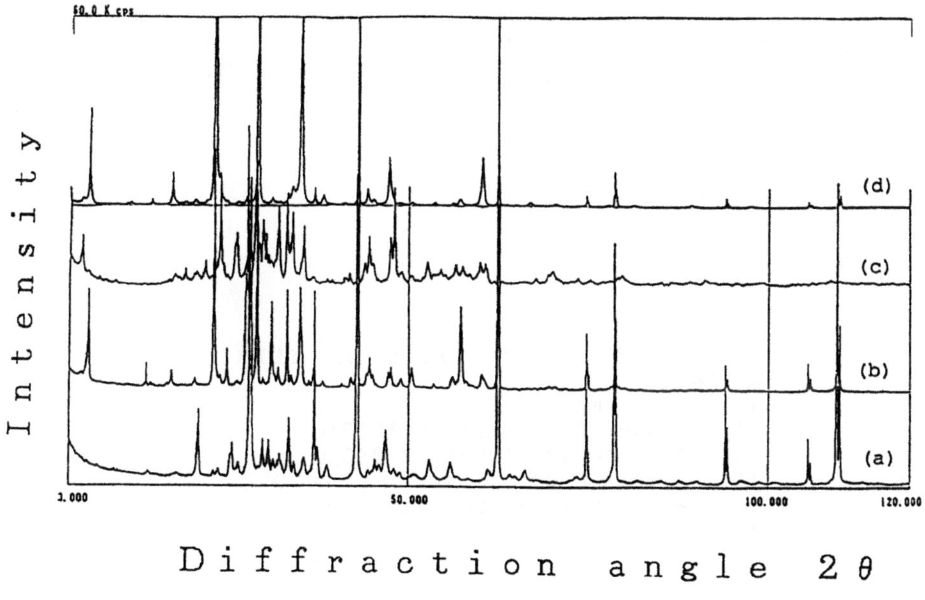

Fig.2 X-ray diffraction patterns of $Bi_2Sr_2Ca_2Cu_3Pb_{0.4}O_x$ films
(a) As decomposed
(b) treated for 60sec at 1030°C
(c) treated for 60sec at 1030°C followed by annealing for 150hr at 840°C
(d) annealed for 150hr at 840°C

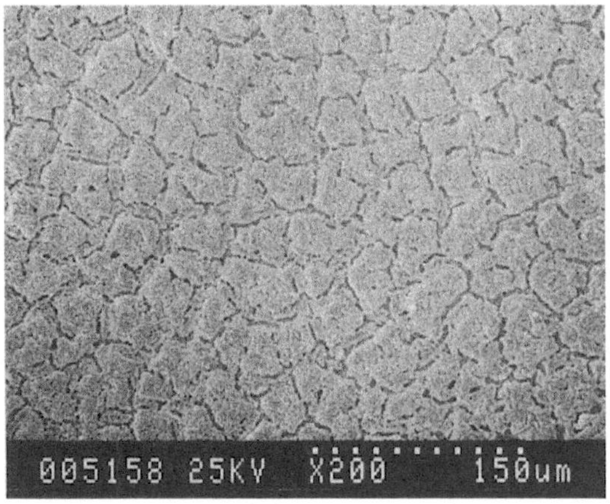

(a)

(c)

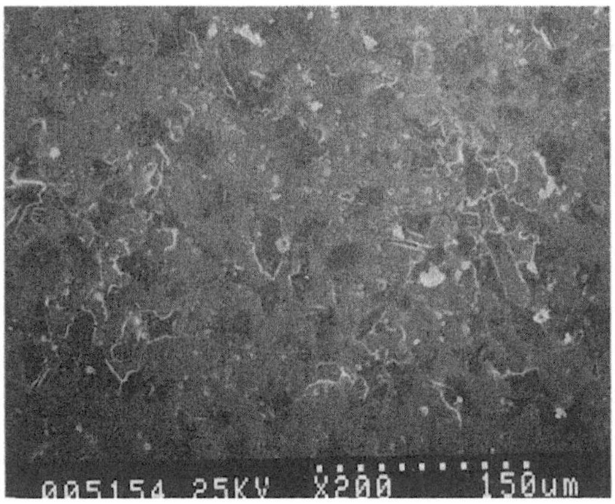

(b)

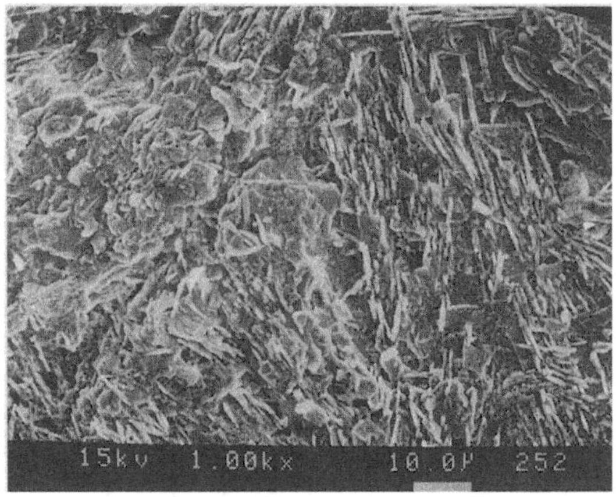

(d)

Fig.3 SEM observations of film surface
(a) As decomposed
(b) treated for 60sec at 1030°C
(c) annealed for 150hr at 840°C
(d) treated for 60sec at 1030°C followed by annealing for 150hr at 840°C

Observations on Boundary Layer Between Oxide Superconducting Films and Various Substrates

S. FUCHINO, K. AGATSUMA, T. OHARA, K. KAIHO, and H. TATEISHI
Electrotechnical Laboratory, 1-4, Umezono 1-chome, Tsukuba, Ibaraki, 305 Japan

ABSTRACT

In this paper, the annealing effect on the growth of both the superconducting oxide and the intermediate products is investigated by using an energy dispersive X-ray (EDX) analysis and a scanning electron microscopy (SEM) examination to get the basic information on the suitable substrate for Y-Ba-Cu-O (YBCO) or Bi-Sr-Ca-Cu-O (BSCCO) film. The results show that the heat treatment yields Ba compounds between YBCO and the substrates and Sr and Bi compounds between BSCCO and the substrates. Since the produced boundary layer of the intermediate product is the thinnest, MgO substrate can be recommended for YBCO film, and $SrTiO_3$ substrate for BSCCO film.

KEYWORDS: oxide superconducting film, substrate, intermediate product

INTRODUCTION

Since the discovery of the new high Tc superconductor[1], various kinds of film growth technique have been reported on fabrication of an oxide superconducting film, especially YBCO and BSCCO above liquid nitrogen temperature. All of the techniques require the so-called post-growth heat treatment or a heating of the substrate during the deposition in order to grow an adequate polycrystal which can realize the superconducting state in the oxide. Such a heating process, however, yields some intermediate products at a boundary layer between the starting materials and the substrate, and the products result in degradation of the superconducting property[2].

In this paper, the annealing effect on the growth of YBCO or BSCCO oxide superconducting film and the substrate is investigated by using EDX analysis and SEM examination. In order to enhance the annealing effect, relatively thin YBCO and BSCCO films are fabricated with the starting materials pasted on substrates , and used rather than sputtered thin films for the analysis. MgO, Al_2O_3, $SrTiO_3$, 5% or 8% Y stabilized ZrO_2 (YTZ or YSZ) and so on are chosen as a substrate.

PREPARATION OF THE SAMPLES

YBCO films were prepared on various substrates by pasting $Y_1Ba_2Cu_3O_{7-x}$ powder with methanol. This powder was produced by calcination at 900°C for 15 hours in flowing oxygen gas and grinding of Y, Ba and Cu oxalate fine powder with about 0.3 μm in diameter by coprecipitation method. Thereafter pasted films were annealed again at 900°C for 15 hours in an oxygen ambient. Tc onset temperature of a bulk sample pelleted by pressing the calcinated and ground powder was about 90 K and the endpoint was about 80 K.

BSCCO films were also prepared in almost same way as YBCO films by annealed at 830°C for 12 hours in an oxygen atmosphere.

RESULTS AND DISCUSSION

YBCO film on various substrates

Figure 1 shows a EDX point analysis of boundary layer between YBCO film and MgO substrate. This analysis suggests no intermediate product on the boundary layer. A little bit of Cu can be observed, however, MgO substrate is seemed to be excellent for YBCO.

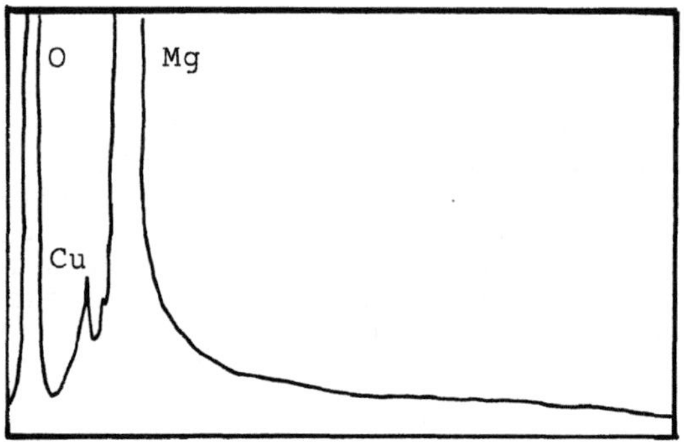

Fig.1 EDX analysis of the boundary layer between YBCO film and MgO substrate

A SEM picture of the cross section and a EDX distribution analysis of each element show the existence of the very thin boundary layer between YBCO and $SrTiO_3$ substrate[3]. The boundary layer of the intermediate products on $SrTiO_3$ is revealed to consist of of Zr, Y and Ba by means of a EDX analysis as shown in Figure 2.

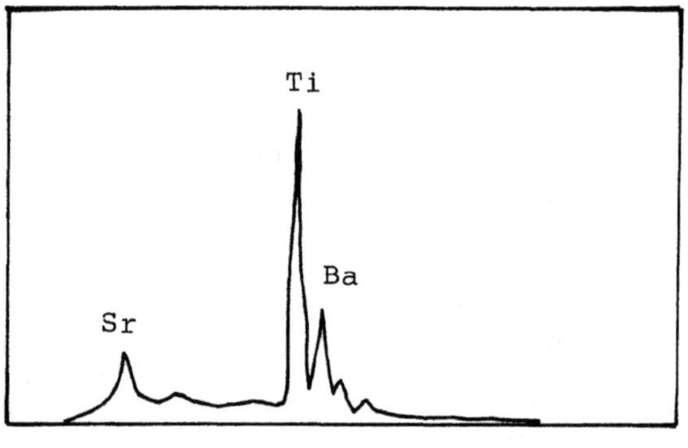

Fig.2 EDX analysis of the boundary layer between YBCO film and $SrTiO_3$

A boundary layer can be clearly observed in a SEM picture of the cross section of a YBCO pasted film on Al_2O_3(99.5%)[3]. A EDX point analysis suggests the boundary layer consists of Ba and Al as shown in Figure 3.

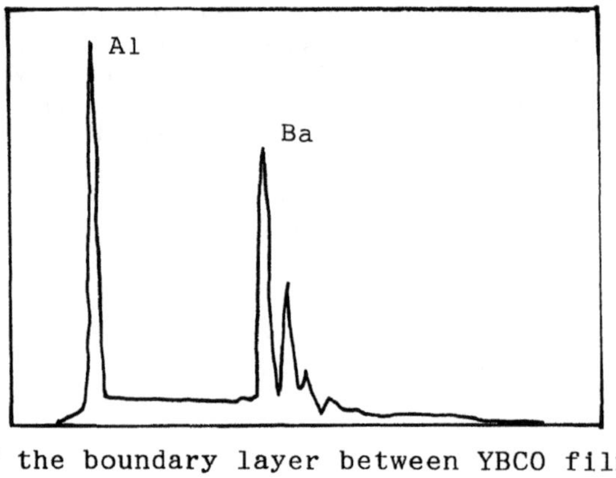

Fig.3 EDX analysis of the boundary layer between YBCO film and Al_2O_3

Figure 4 shows a EDX point analysis of the boundary layer of a YBCO film on α-Al_2O_3 substrate. This also suggests the boundary layer consists of Ba and Al.

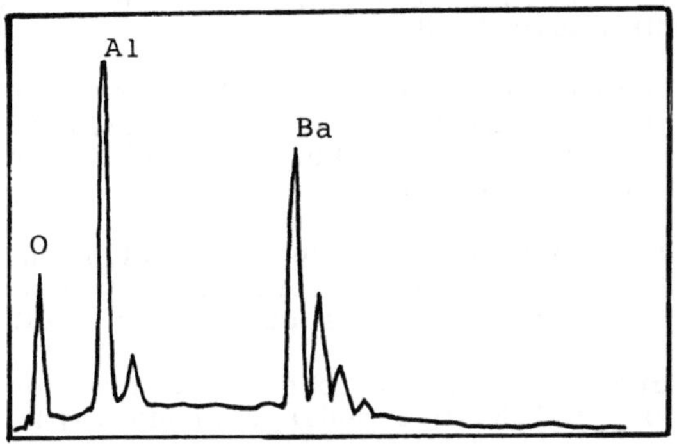

Fig.4 EDX analysis of the boundary layer between YBCO and α-Al_2O_3

A boundary layer of a thin intermediate products can be seen in a SEM picture of the cross section of YBCO film on YSZ substrate[3]. This boundary layer is revealed to be confirmed to consist of Zr and Ba by means of a EDX analysis as shown in Figure 5.

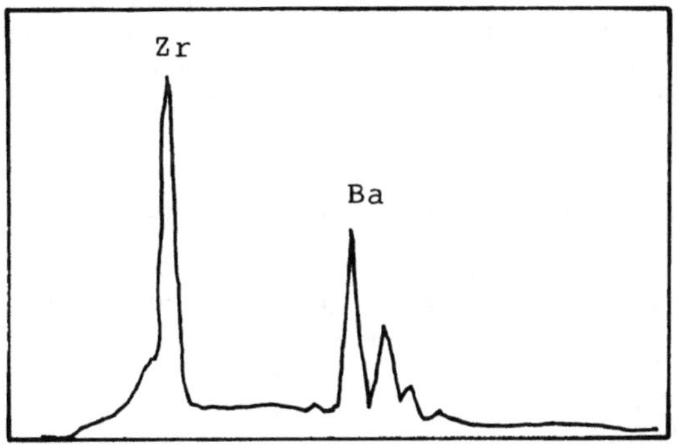

Fig. 5 EDX analysis of the boundary layer between YBCO film and YSZ substrate

BSCCO films on various substrates

A SEM picture of the cross section and a EDX distribution analysis of each element of a BSCCO pasted film on MgO (100) substrate shows a thin boundary layer clearly[3] . The EDX analysis suggests the boundary layer consists of Bi, Sr, Ca and Mg shown in figure 6.

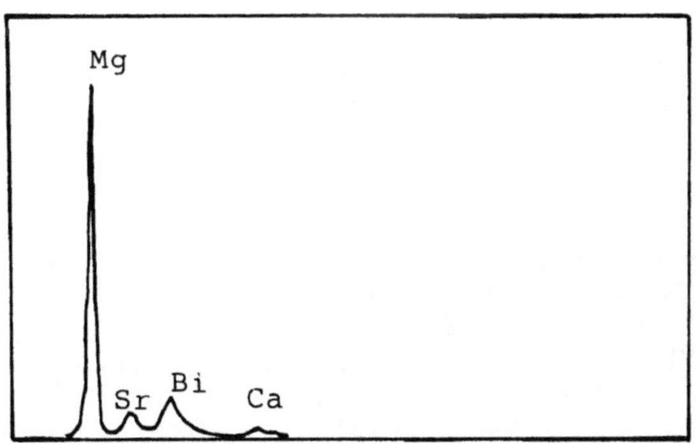

Fig.6 EDX analysis of the boundary layer between BSCCO film and MgO substrate

A SEM picture of the cross section and a EDX distribution analysis of each element including EDX line analysis along the specified line of BSCCO film on YTZ substrate show the existence of a boundary layer between BSCCO film and YTZ substrate[3]. This boundary layer is revealed to consist of Zr, Sr, Ca and Bi by means of a EDX analysis[3].

A SEM picture of the cross section and a EDX distribution analysis of each element including EDX line analysis along the specified line of BSCCO film on $SrTiO_3$ substrate are not able to show the existence of a boundary layer between BSCCO and $SrTiO_3$ substrate[3].

A SEM picture of the cross section of the BSCCO film on YSZ substrate shows a boundary layer of a relatively fat intermediate products[4]. This suggests that YSZ is not a good candidate for a substrate of BSCCO.

A SEM picture of the cross section of the BSCCO film on SiO_2 substrate shows also a boundary layer of thick intermediate products[4].

CONCLUSIONS

The annealing effect on the growth of YBCO or BSCCO oxide superconducting film on the various substrates is investigated by using a SEM examination and a EDX analysis. Table 1 shows the results of the investigation. Thus the basic information on the suitable substrate for YBCO or BSCCO superconducting film is experimentally collected. Consequently, it is shown by the investigation that the heat treatment yields Ba compounds between YBCO and the substrates and Sr and Bi compounds between BSCCO and the substrates. Since the produced boundary layer of the intermediate product is the thinnest, MgO substrate can be recommended for YBCO film, and $SrTiO_3$ substrate for BSCCO film.

Table 1 Choice of substrate materials.

SUBSTRATE	YBCO	BSCCO	COMMENT
MgO	◎	⊙	
$SrTiO_3$	⊙	◎	
Al_2O_3	×	—	
α-Al_2O_3	△	—	
YSZ	○	×	Cheap
YTZ	—	○	Cheap
SiO_2	×	×	

◎ Excellent ⊙ Good
○ Passable △ Bad
× Fail — Not examined

ACKNOWLEDGMENTS

We wish to acknowledge to Mr. T. Aoshima in TOTO CO. Ltd. and Mr. N. Sadakata in Fujikura Ltd. to help SEM and EDX examinations.

REFERENCES

1. Bednorz JG, Muller K (1986) Possible high Tc superconductivity in Ba-La-Cu-O system. Z Phys B 64:189-193
2. Agatsuma K, Ohara T, Kaiho K, Tateishi H (1988) Observation of boundary layer of $YBa_2Cu_3O_{7-x}$ films pasted on Al_2O_3, $SrTiO_3$ and YSZ substrates Physica C 153-155:814-815
3. Agatsuma K, Ohara T, Kaiho K, Tateishi H (1989) Boundary layers of oxide superconductors films pasted on Al_2O_3, MgO, $SrTiO_3$ and YSZ. IEEE Trans Mag vol 25 no 2:2487-2490
4. Agatsuma K, Ohara T, Kaiho K, Tateishi H (1988) Boundary layers between oxide superconductor films and substrates. Tech Rep IEE Jpn SA-88-48

Preparation of Ag-YBCO Composite Fine Powder by Spray-Pyrolysis

K. NISHIO, T. SAKAI, N. OGAWA, and I. HIRABAYASHI

ISTEC Superconductivity Research Laboratory, Nagoya Division, 4-1, Mutsuno 2-chome, Atsuta-ku, Nagoya, 456 Japan

ABSTRACT

Fine powders of the Ag-YBCO composite have been prepared by thermal decomposition of the nitrate salts of Ag, Y, Ba, and Cu in the presence of oxygen. The powder particles were spherical and their average diameters were between 0.1 μm and 1 μm dependent on the concentration of the nitrates solutions. The X-ray diffraction patterns of the powders are the mixture of the silver metal and YBCO. The results of SEM-EDX and ESCA indicate that the silver metal is uniformly distributed in the powder particles. Magnetic susceptibility measurements show that the powders become superconducting state at 90K. There is no degradation of superconducting properties caused by silver metal blending with YBCO.

KEY WORDS: thermal decomposition process, fine powder production, Ag-YBCO composite powder.

INTRODUCTION

High-purity chemically homogeneous fine powders are necessary for the fabrication of superconduting wires, tapes, and magnetic shields. The most common method for superconducting powder production: mechanical grinding is not easy to obtain submicron particles having a sharp particle size distribution, and often occurs contamination from the grinding media. Other techniques such as sol-gel method, coprecipitation method, and freeze drying do not directly produce superconducting particles. These particles must usually be dried and heated to remove residual carbon and finally calcined at high temperatures before the material becomes superconducting. On the other hand, the thermal decomposition of the aerosol particles produces high purity fine superconducting powders in a simple continuous process[1]. Furthermore, an introduction of flux pinnig centers are important to enhance the potential for applications in the magnetic field. It has been reported that Ag-YBCO composite samples exibit very strong magnetic flux pinning which demonstrates magnetic suspension[2-3]. Besides we can expect improvement of the grain boundary and weak-link properties as another effect of Ag addition. Therefore, we have explored the thermal decomposition of the aerosol particles for the production of the Ag-$YBa_2Cu_3O_{7-x}$ composite powders.

EXPERIMENTAL APPARATUS

The experimental apparatus is shown in Figure 1. The system consists of an ultrasonic nebulizer, a reaction furnace and a powder collector. An aqueous solution was prepared by mixing silver nitrate salts and the powder precursor consisted of the nitrate salts of Y, Ba, and Cu in a 1:2:3 molecular ratio in ultra-pure water. The prepared solution was vaporized with an ultrasonic nebulizer, which was operated at 1.7 MHz. The genatrated vapor was led to the reactor tube (87mm in inner diameter and 1200mm in length), with carrier gas (oxygen), and then it was thermally decomposed at the appropriate temperature. The flow rate of the carrier gas was 5 l/min. The powders were collected using the tetrafluoroethylene filter. The crystalline structure and the shape of the powder particles were examined using X-ray diffraction (CuKα) and a scanning microscope (SEM), respectively. The inclusion form of silver in the powders was examined using X-ray microanalyzer (XMA) and ESCA. The temperature dependence of the magnetization and hysteresis were measured by the DC-SQUID magnetometer.

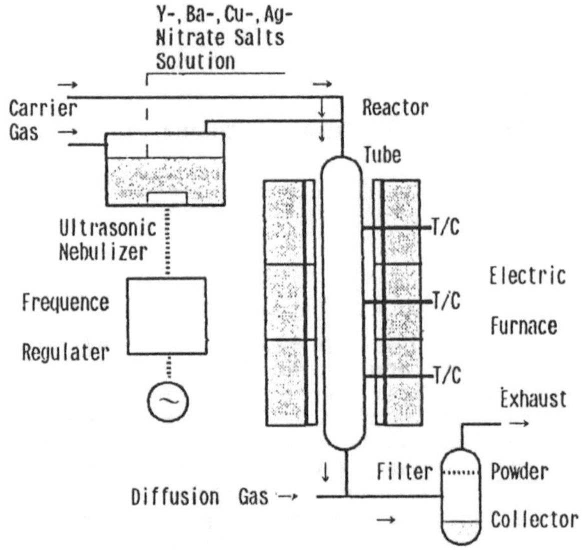

Figure 1. Schematic diagram of the experimental system.

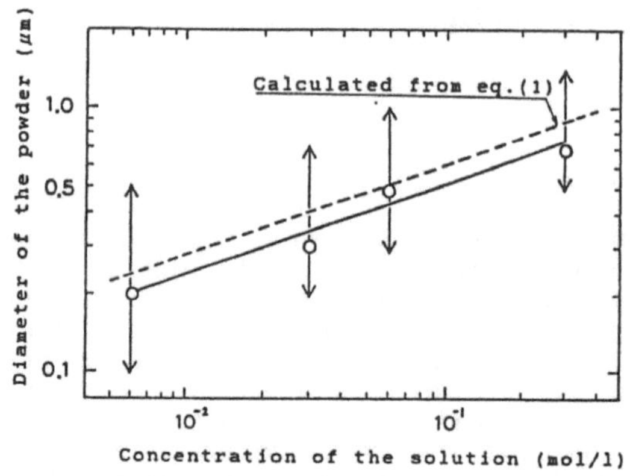

Figure 2. concentration of the solution and particle size obtained at 900°C.

RESULTS AND DISCUSSION

To make clear the condition of the preparation of the fine superconducting powders, we examined (1) the relation between the diameter of the final powders and concentration of the starting solutions, and (2) the optimal thermal decomposition temperatures for producing superconducting powders.

The diameter of the powder is given by

$$d = (m \cdot M/G)^{1/3} \cdot D, \qquad (1)$$

where m, M, G, and D are the concentration of the solution, the molecular weight of the powder (g/mol), the specific weight of the powder (g/cm^3), and the diameter of the particle in the mist (cm), respectively. Concerning about the diameter of the particle in the mist, R.J.Lang reported that the diameter of the particle is given by an empirical formula:

$$D = 0.34 \cdot (8\pi \cdot T/\rho \cdot F^2)^{1/3}, \qquad (2)[4]$$

where T, ρ and F are the surface tention of the particle in the mist (dyn/cm^2), the liquid density of the solution (g/cm^3), and the exciting ultrasonic frequency (Hz), respectively. The diameter of the powders obtained at various concentrations of the solutions are shown in Figure 2. The experimental result shows that the diameters of the obtained powders are proportional to the cube root of the concentration of the starting solution. They agree well with the values calculated by eqs. (1) and (2).

To get the optimal thermal decomposition temperature, we have examined crystalline phases of the powders thermally decomposed at verious temperatures using X-ray diffraction (CuK_α). The X-ray diffraction patterns of the powders obtained at 900°C, 950°C, 1000°C, and 1050°C are as shown in Figure 3. In the patterns of the powders obtained at 900°C, 950°C, and 1000°C, most of the peaks can be assigned to the orthorhombic phase of YBCO. Moreover it is noted that the crystallinity of these powders improves with an increasing thermal decomposition temperature. However, when the temperature exceeds 1050°C, Y_2BaCuO_5 phase appears. The above results show that the $YBa_2Cu_3O_{7-x}$ powders in an orthorhombic form can be synthesized in the temperature range from 950°C to 1000°C, using oxygen as a carrier gas.

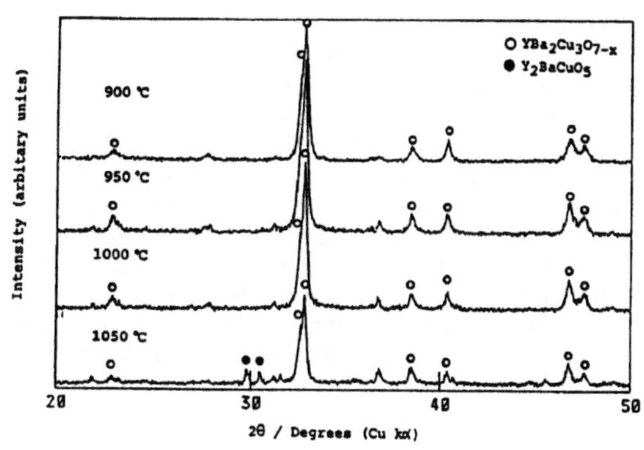

Figure 3. X-ray diffraction patterns of the powders obtained at various decomposition temperatures.

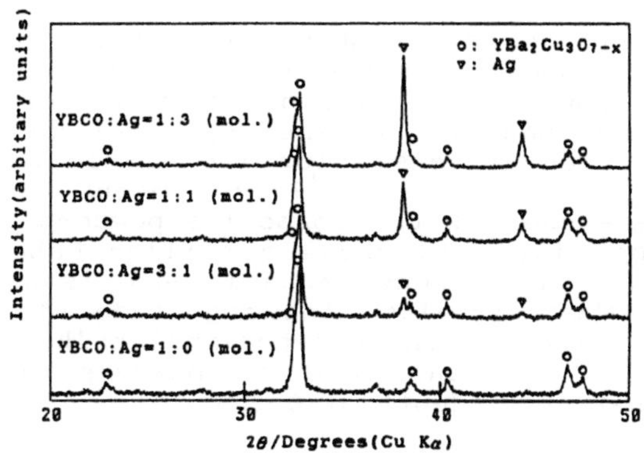

Figure 4. X-ray diffraction patterns of the Ag-YBCO composite powders obtained at 950°C

We have prepared the Ag-YBCO composite powders under the conditions determined by the above experiments. The X-ray diffraction patterns of the powders obtained at 950°C using the solution contained various content of silver nitrate salt are shown in Figure 4. In those patterns, most of the peaks can be assigned $YBa_2Cu_3O_{7-x}$ phase with and silver metal phase. In these patterns, the intensity at $2\theta=38°$ for silver metal phase is proportional to the content of silver nitrate salts in the starting solution.

We have examined the distribution of silver metal in the obtained powders using SEM·XMA and ESCA. The typical XMA result of the obtained composite powders is shown in Figure 5 with the SEM photograph. Comparing the composition measured by XMA from the total view area of the SEM photograph with that from the local part "A", we find the intensity ratios of silver metal to YBCO are almost the same. It means that there is not segregation of silver metal in the particle size order. The ESCA result shows that the surface layer within the escape length of photoelectron consists of both silver metal and YBCO. It means that silver does not capsule YBCO and vice versa. On the basis of these results, we expect that silver metal is uniformly distributed in the powder particle.

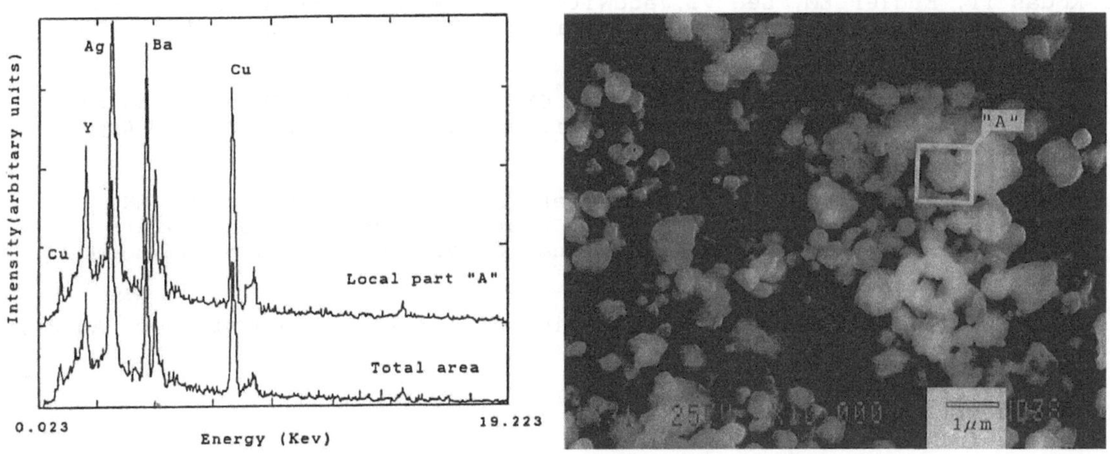

Figure 5. The typical SEM-XMA result of the obtained Ag-YBCO composite powders

Finaly we have examined the superconductivity of the composite powders using the magnetic mesurements. The temperature dependences of the magnetization of the powders with various contents of silver metal are shown in Figure 6. The silver addition to YBCO does not decrease the critical temperature. Meissner fraction increases for the sample with doped silver metal. But there is an optimal value of silver content for Meissner fraction. The magnetization hystereses of the composite powders are shown in Figure 7. The magnetization loop width: ΔM shows a similar behavior to Meissner fraction, i.e., silver addition enhances ΔM, and there is an optimal value of silver content for ΔM. In above results, the magnitude of the magnetization is considerably lower than that of the conventional bulk samples of the same volume. It is because that the dimension of the powders is comparable orders of the superconducting penetration depth of 0.2μm.

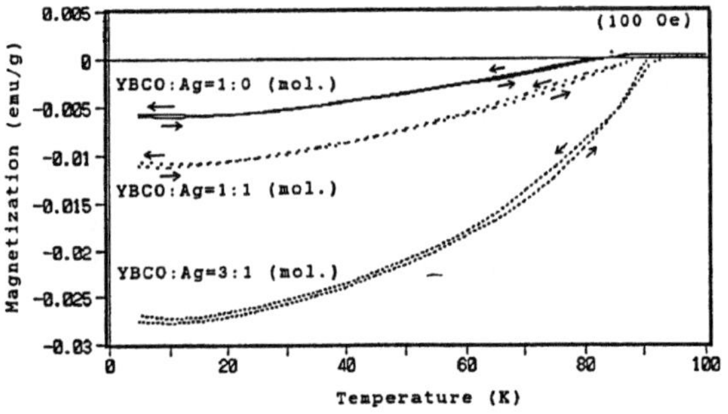

Figure 6. The temperature dependences of the magnetization of the obtained Ag-YBCO composite powders.

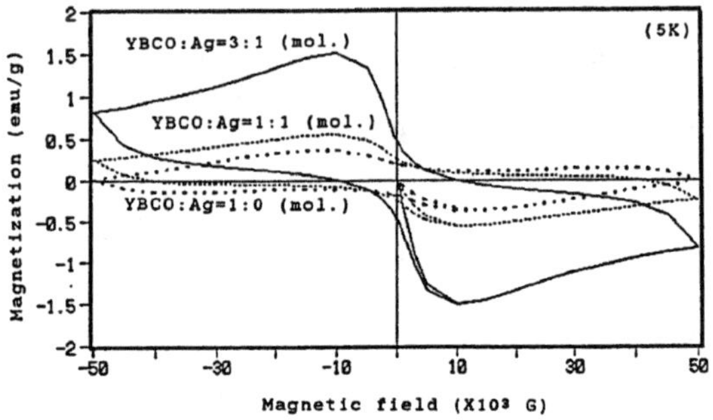

Figure 7. The magnetization hystereses of the magnetization of the obtained Ag-YBCO composite powders.

CONCLUSION

In summary, we have produced Ag-YBCO composite powders by thermal decomposition of the aerosol particles. The addition to YBCO powders does not degrade the superconductivity, and can enhance Meissner fraction and the magnetization loop width at the optimal content of the silver.

REFERENCES

1. Kodas TT, Engler EM, Lee VY,Jacowits R, Baum TH, Roche K, Parkin SSP, (1988) Aerosol flow reactor production of $YBa_2Cu_3O_7$ powder. Appl. Phys. lett.52,1622.

2. Peters PN, Sisk CR, Urban EW, Huang CY, Wu MK, (1988) Observation of enhanced properties in samples of silver oxide doped $YBa_2Cu_3O_X$. Appl. Phys.Lett.52,2066.

3. Huang CY, Shapira Y, Mcniff EJ, Peters Jr.,PN, Schwartz BB, Wu MK, Shull RD, Chiang CK,(1988) Magnetic hysterisis of high-temperature $YBa_2Cu_3O_X$-AgO superconductors:explanation of magnetic suspension. Mod. Phys.Lett,B,Vol 2,7.

4. Lang RJ, (1962) Ultrasonic atomization of liquids. J.Acoust. Soc. Am,34,6.

Preparation of $YBa_2Cu_3O_{7-y}$ Superconducting Films by the Organic Transport-Chemical Vapor Deposition

S. MATSUNO, F. UCHIKAWA, and K. YOSHIZAKI

Materials & Electronic Devices Laboratory, Mitsubishi Electric Corporation, 1-57, Miyashimo 1-chome, Sagamihara, Kanagawa, 229 Japan

ABSTRACT

High Tc $YBa_2Cu_3O_{7-y}$ superconducting films were prepared by metalorganic chemical vapor deposition using a new source transport technique to increase deposition rates. Ar and organic vapor were used as carrier gases. Metastable addition products formed could be transported smoothly to the reactor. This method could give a major decrease of the sublimation temperature of Ba source. Using this method, deposition rates of the films were able to be controled in the condition of 0.5~20μm/hr. The superconducting zero-resistance temperature and the critical current density at 77K were 88K and $1.4 \times 10^6 A/cm^2$ for the thin film respectively.

KEY WORDS: superconducting oxide film, Y-Ba-Cu-O, CVD, addition products, high-Jc

INTRODUCTION

The recent discovery of superconducting materials such as La-Ba-Cu-O [1], Y-Ba-Cu-O [2], and Bi-Sr-Ca-Cu-O [3] have generated much interest in developing their technological applications. Particularly the high Tc superconducting films are expected to use for a tape-shaped superconducting wire and new electronic devices. The superconducting films have been synthesized by sputtering [4], ICB (ionized cluster beam) deposition [5], electron-beam evaporation [6], laser ablasion [7], CVD (chemical vapor deposition) [8], Sol-Gel [9] and spray pyrolysis [10]. These methods have some advantages for preparing high Tc superconducting films. For example, CVD process is more suitable for better control of composition and crystal structure. Also CVD method has generally high deposition rates, compatibility with variously shaped substrates and flexibility to large scale processing. But those advantages have not been realized for preparing high Tc superconducting films because of the problem caused by the source materials, such a poor volatility and easiness of decomposition for especially Ba source material.

In this paper, we report the successful preparation of high Tc $YBa_2Cu_3O_{7-y}$ superconducting films by the Organic Transport-CVD (OT-CVD) as a new method. Not only stable volatility of the source materials but also the high speed synthesis was achieved by use of this method.

EXPERIMENTAL

Figure 1 shows a schematic diagram of the OT-CVD system used in this study. This system is a conventional hot-wall type CVD except a source transport technique. $Y(DPM:C_{11}H_{19}O_2)_3$, $Ba(DPM)_2$ and $Cu(DPM)_2$ were adopted as source materials. Those were heated at 130~150°C, 180~200°C and 130~150°C respectively in source cylinders which were made of stainless steel. Argon was used as the carrier gas at flow rates of 3~20CCM for Y and Cu. For the Ba source, argon and THF (tetrahydrofuran) as a organic vapor were mixed as the carrier gases. The THF vapor was generated by a bubbling method. Flow rate of the organic vapor was 10~100CCM. The vaporization of THF was carried out at -5~10°C and the flow rates of argon bubbling gas was 5~10CCM. Flow rate of oxygen as a reaction gas was 200CCM. Three kinds of metal source gases with a carrier gas and oxygen were mixed into the reactor with a resistive heating stage for the deposition. The temperature of the susceptor was set at 850°C and the reactor pressure was kept at 10 Torr during the deposition. A $SrTiO_3$(100) single crystal was used as the substrate. After the deposition time of 20~30 minutes, the as-deposited films were cooled to room temperature under oxygen flow of 1atm. The cooling rate was about 15°C/min.

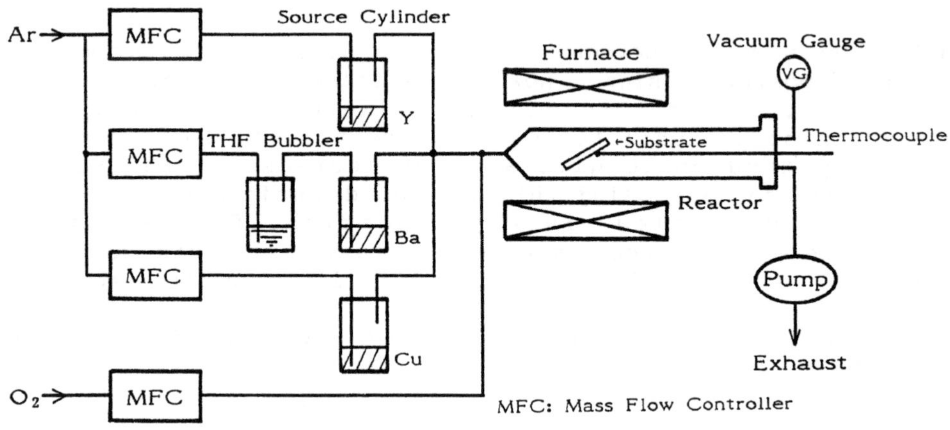

Fig. 1. Schematic diagram of the organic transport-CVD system for $YBa_2Cu_3O_{7-y}$ superconducting films.

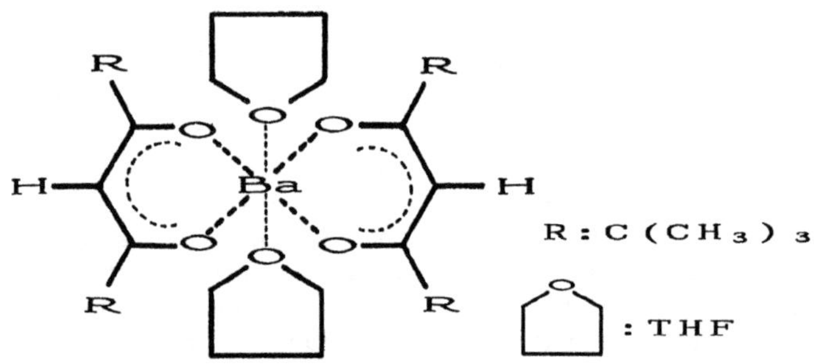

Fig. 2. The proposed structure of $Ba(DPM)_2$-THF addition products.

The crystal structure of the films was analyzed by X-ray diffractometer (Cu Kα) and the surface morphology was examined by scanning electron microscopy (SEM). The film thickness was measured by SEM and a needle-probe thickness gauge. The resistivity and critical current of the as-deposited films were measured by a standard four probe method.

RESULTS AND DISCUSSION

In this method, metastable addition products formed between the THF vapor and source material in the source cylinder were transported smoothly to the reactor. Figure 2 shows the proposed structure of $Ba(DPM)_2$-THF addition products. THF molecules are thought to be added to the central Ba atom by some coordination bonds.

Figure 3 shows SEM photograph of the diagonal view for the as-deposited thick film on $SrTiO_3(100)$ substrate (film surface and cross-section of the film and substrate). As shown in this figure, the film has smooth surface. The film thickness was 5.5μm and the deposition rate was achieved as high as 20μm/hr. In the case of thin film the thickness was 0.2μm and the deposition rate was 0.5μm/hr. The surface of thin film was smoother than that of the thick film.

The X-ray diffraction patterns of the thick film (a) and the thin film (b) are shown in Figure 4. The enhanced intensities of the (00ℓ) reflections for both films indicate that the c-axes are preferentially oriented perpendicular to the substrate surface. It is thought that only $YBa_2Cu_3O_{7-y}$ structure should be formed.

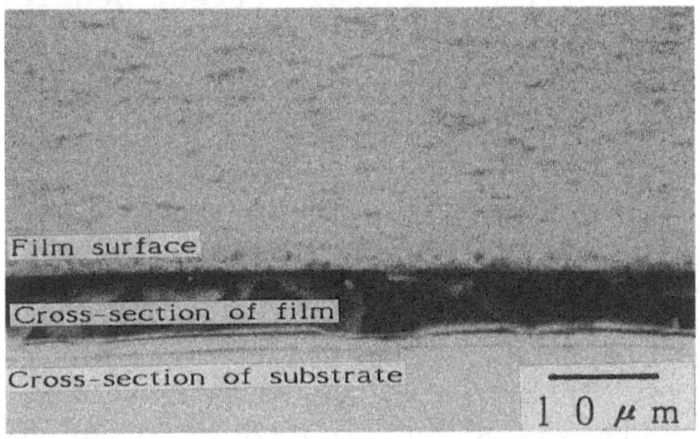

Fig. 3. The SEM photograph of diagonal view for the as-deposited thick film on $SrTiO_3(100)$ substrate.

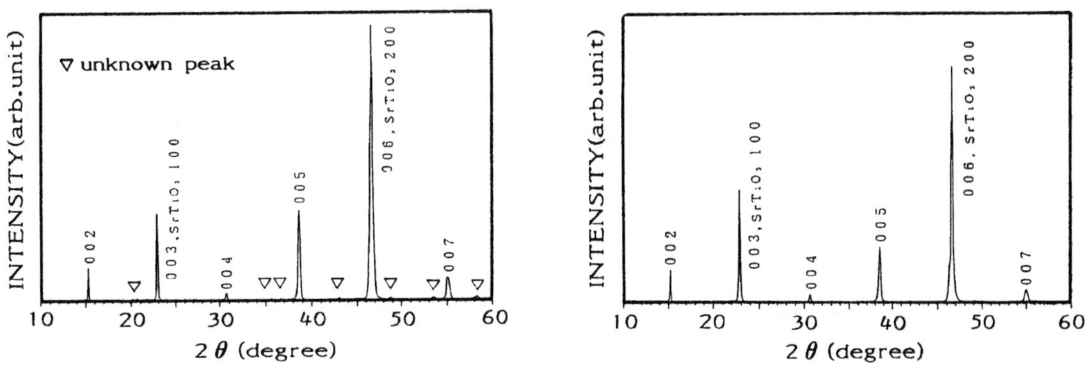

Fig. 4. The X-ray diffraction patterns of the thick film (a) and thin film (b) on $SrTiO_3(100)$ substrate.

The zero-resistance temperature Tc and the critical current density Jc at 77K for the thick film were 84K and $1.0x10^3 A/cm^2$ respectively. Figure 5 shows temperature dependence of the resistivity for the thin and thick films. The thin film in Figure 5 showed a sharp transition with zero-resistance at 88K. Figure 6 shows the temperature dependence of the Jc for the thin film. As shown in figure 6, the Jc was $1.4x10^6 A/cm^2$ at 77K. This value is a top-level data for thin films of Y-system superconductor. The Jc was degraded up to about 1/50 at a magnetic field of 5T which was applied in the direction of perpendicular to the c-axis, and more degraded in the direction of parallel. The Jc under magnetic fields can be improved by introducing pinning centers into the films.

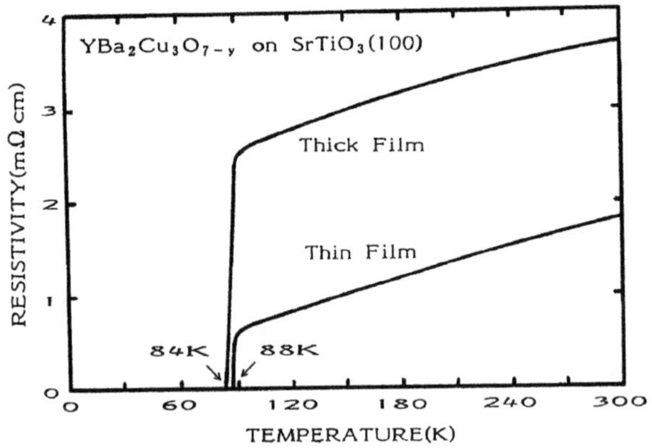

Fig. 5. Temperature dependence of the resistivity for the thin and thick films.

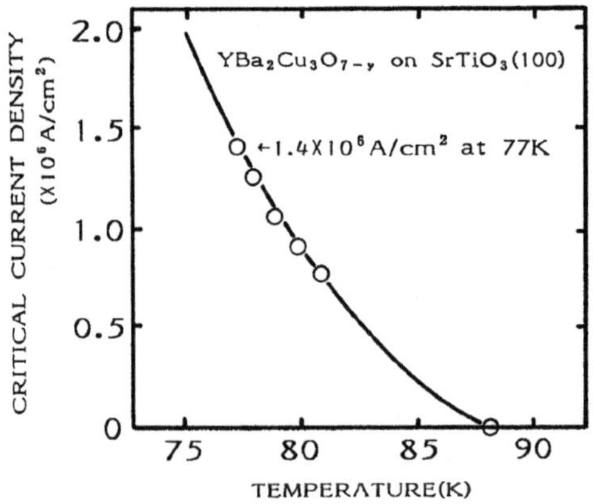

Fig. 6. Temperature dependence of the critical current density for the thin film.

CONCLUSIONS

High Tc $YBa_2Cu_3O_{7-y}$ superconducting films were prepared by metalorganic chemical vapor deposition method using a new source transport technique (Organic Transport-Chemical Vapor Deposition) on $SrTiO_3$(100) single crystal substrate. Using this method, Ba source was evaporated smoothly at the much lower heating temperature than conventional methods. Deposition rates of the films were achieved up to 20μm/hr. The thin film having Tc of 88K showed a high Jc value of $1.4x10^6 A/cm^2$. The thick film was shown to have a Tc of 84K and Jc of $1.0x10^3 A/cm^2$.

ACKNOWLEDGMENT

This work was performed as a part of " R&D on Superconducting technology for Electric Power Apparatuses " as a subject of Super-GM under the Moonlight Project of Agency of Industrial Science and Technology, MITI, being consigned by New Energy and Industrial Technology Development Organization (NEDO).

REFERENCES

[1] J. G. Bednorz and K. A. Müller, Z. Phys. B64, 189 (1986).
[2] M. K. Wu, J. R. Ashburn, C. J. Torng, P. H. Hor, R. L. Meng, L. Gao, Z. J. Huang, Y. Q. Wang, and C. W. Chu, Phys. Rev. Lett. 58, 908 (1987).
[3] H. Maeda, Y. Tanaka, M. Fukutomi, and T. Asano, Jpn. J. Appl. Phys. 27, L209 (1988).
[4] M. Hong, S. H. Liou, J. Kwo, and B. Davidson, Appl. Phys. Lett. 51, 694 (1987).
[5] K. Yamanishi, S. Yasunaga, K. Imada, K. Sato, and Y. Hashimoto, Mat. Res. Soc. Symp. Proc. 99, 343 (1988).
[6] R. B. Labowitz, R. H. Koch, P. Chaudhari, and R. J. Gambino, Phys. Rev. B35, 8821 (1987).
[7] X. D. Wu, D. Dijkkamp, S. B. Ogale, A. Inam, E. W. Chase, P. E. Miccli, C. C. Chang, J. M. Tarascon, and T.Venkatesan, Appl. Phys. Lett. 51, 861 (1987).
[8] H. Yamane, H. Masumoto, T. Hirai, H. Iwasaki, K. Watanabe, N. Kobayashi, Y. Muto, and H. Kurosawa, Appl. Phys. Lett. 53, 1548 (1988).
[9] F. Uchikawa and J. D. Mackenzie, J. Mater. Res. 4, 787 (1989).
[10] S. Matsuno, K. Egawa, K. Yoshizaki, and H. Watarai, MRS Int'l. Mtg. on Adv. Mats. 6, 857 (1989).

4.2 BSCCO and Related Films

In-situ Preparation of Bi-Sr-Ca-Cu-O Films by Coevaporation Assisted by Energy Controlled ECR Oxygen Ion Beam

M. Kamei, I. Yoshida, H. Teshima, M. Nemoto, K. Suzuki, and T. Morishita

Superconductivity Research Laboratory, International Superconductivity Technology Center, 10-13, Shinonome 1-chome, Koto-ku, Tokyo, 135 Japan

ABSTRACT

Both substrate precleaning and in-situ preparation of Bi-Sr-Ca-Cu-O films by coevaporation have been performed using an ECR oxygen plasma. Before deposition the substrates have been cleaned by the high energy oxygen ion beam. The kinetic energy of ions has been controlled by changing a bias voltage of the drift tube. During coevaporation the bias voltage has been kept low so as not to damage films. The 2212 and 2201 mixed phase has been grown in-situ.

KEY WORDS: ECR oxygen plasma, energy controlled ion beam
in-situ Bi-Sr-Ca-Cu-O preparation, substrate precleaning

INTRODUCTION

Since the discovery of the high Tc oxide superconductors, thin films of these materials have been prepared using various methods [1][2][3]. Coevaporation is one of the promising methods and its advantage is the independent control of metal fluxes[3]. Coevaporation must be operated in so low oxygen partial pressure as to protect evaporators. Therefore oxygen activation is necessary for in-situ crystallization of copper oxide superconducting films. The ECR oxygen plasma has been noticed as an oxygen source enable to oxidize metals in a low oxygen partial pressure. Thin films of Y-Ba-Cu-O have been successfully prepared in-situ by coevaporation using an ECR oxygen plasma [4]. However in-situ preparation of Bi-Sr-Ca-Cu-O system in an ECR oxygen plasma has still been a challenging problem. It is also expected that the ECR plasma accelerated by a rather high voltage is available for cleaning of substrates prior to deposition.

EXPERIMENTAL

The system is shown schematically in Fig. 1a. Bi_2O_3 and Cu were evaporated from two independent electron beam guns. The evaporation rates of Bi_2O_3 and Cu were monitored by two quartz-crystal thickness sensors. Sr and Ca were evaporated from two K-cells and their flux densities were monitored by one of electron impact emission spectroscopy (EIES) sensors [5].

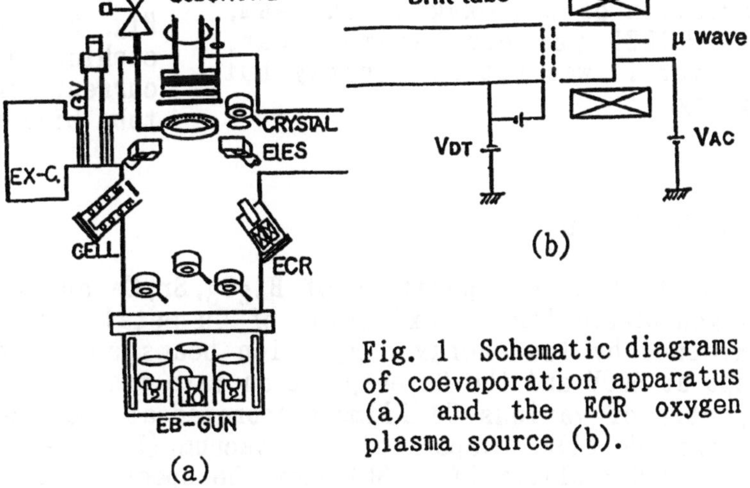

Fig. 1 Schematic diagrams of coevaporation apparatus (a) and the ECR oxygen plasma source (b).

Oxygen ions generated in the ECR plasma chamber are extracted by the drift tube bias voltage (Vdt in Fig. 1b) and introduced to the substrate. The distance between substrates and the plasma chamber is 51 cm, which is too far for oxygen plasma to reach the substrate without the drift tube. Typical conditions of generating oxygen ion beams are shown in Table 1. Vdt was changed between 250 and 1000 V. Vdt's higher than 650V were used for ion beam etching and lower bias voltages were used in order to enhance oxidation of constituent elements during coevaporation. Ion current density was measured by the probe placed between the drift tube and the substrate. MgO (100) single crystals were used as substrates. Substrate temperature was changed between 700 and 800C. The surface structure of films and substrates were monitored in-situ by the reflection high energy electron diffraction (RHEED). The structure of films were investigated by x-ray diffraction (XRD). The film compositions were determined by inductively coupled plasma spectroscopy (ICP). The degree of oxidation of Cu were examined by x-ray photoemission spectroscopy (XPS).

Table 1 The conditions of oxygen ion beam generation

microwave power	70-100W	
drift tube bias voltage	oxidation 250-650V	etching 650-1000V
oxygen partial pressure	$1x10^{-5}$	$2x10^{-4}$torr

RESULTS AND DISCUSSION

1 Substrate Precleaning

Vdt dependence of etching rate for MgO (100) substrates is shown in Fig. 2. Surface structures of MgO(100) substrates were examined before and after ion beam irradiation by RHEED. The RHEED pattern of substrates heated at 500C in a vacuum showed a halo pattern before the irradiation. After 15 minute irradiation of the oxygen ion beam, the RHEED pattern became clear, however it was not so streaky but spotty.

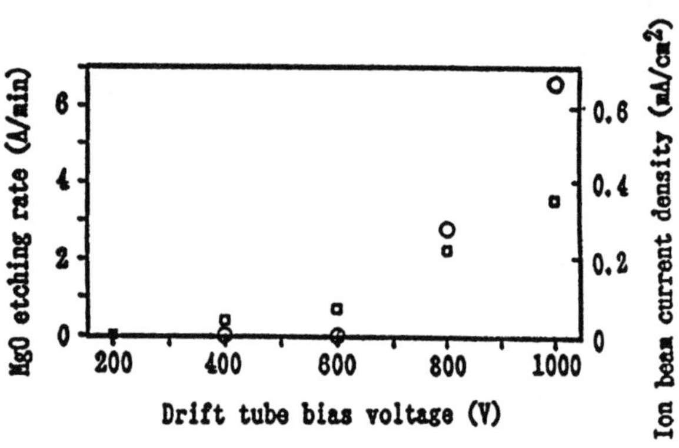

Fig. 2 Dependences of MgO (100) etching rate (circle) and ion beam current density (square) on drift tube bias voltages.

2 Cu Oxidation

Prior to the coevaporation of Bi_2O_3, Sr, Ca and Cu, the efficiency of an ECR oxygen plasma for the oxidation of Cu metal was investigated. Cu vapor was oxidized by low energy oxygen ion beams so as not to damage films. Vdt was kept at 500V and the Cu evaporation rate was 0.45 A/sec. X-ray photoemission spectra of various Cu films evaporated at room temperature are shown in Fig. 3; the Cu films evaporated in a vacuum (Fig. 3a) and in $1.0x10^{-4}$ Torr oxygen without ECR plasma (Fig. 3b) show the spectrum of metallic Cu. But the Cu film evaporated with oxygen plasma shows another spectrum (Fig. 3c) where there are satellite peaks due to Cu^{2+}. This means that oxygen gas of $1.0x10^{-4}$Torr is insufficient for the oxidation of Cu deposited at 0.45A/sec. The ECR oxygen plasma, however, is effective for formation of cupreous oxides.

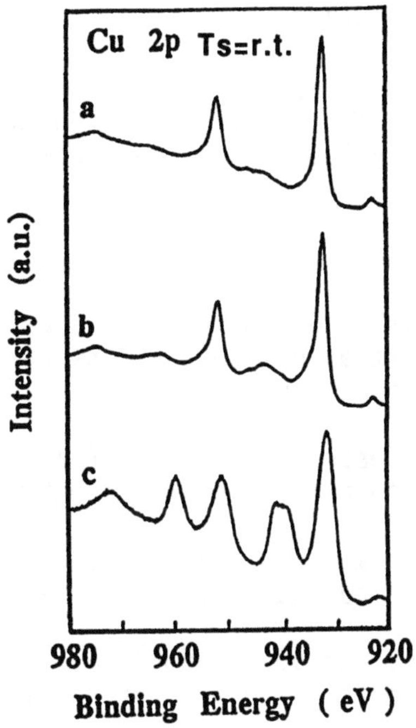

Fig. 3 X-ray photoemission spectra of Cu films evaporated in a vacuum (a) ,1x10^{-4} Torr oxygen gas (b) and with ECR oxygen plasma (c).

3 Bi-Sr-Ca-Cu-O Thin Film Preparation

Substrates were cleaned before coevaporation by the method described above. Vdt was decreased to 500V. Coevaporation from Bi_2O_3, Sr, Ca and Cu sources was assisted by this low energy oxygen ion beam in an oxygen partial pressure of 1x10^{-4}Torr in order to enhance oxidation. Sr and Ca are easily oxidized compared with Cu. Bi_2O_3 was used instead of Bi metal because the sticking coefficient of Bi metal on substrates is too small at substrate temperatures between 700 and 800C. Typical deposition rate and total film thickness were 0.7A/sec and 2500A, respectively. The substrate temperature was changed between 700 and 800C. The single 2201 phase film was obtained in-situ at a substrate temperature of 730C. The film was highly oriented with the c axis normal to the substrate (Fig. 4). The atomic composition of this film was determined to be Bi:Sr:Ca:Cu=1.63:1.56:0:3 by ICP. The film with the composition of 2.82:1.67:1.93:3 presents the XRD pattern coexisting the 2212 phase and some other unidentified phases (Fig. 5).

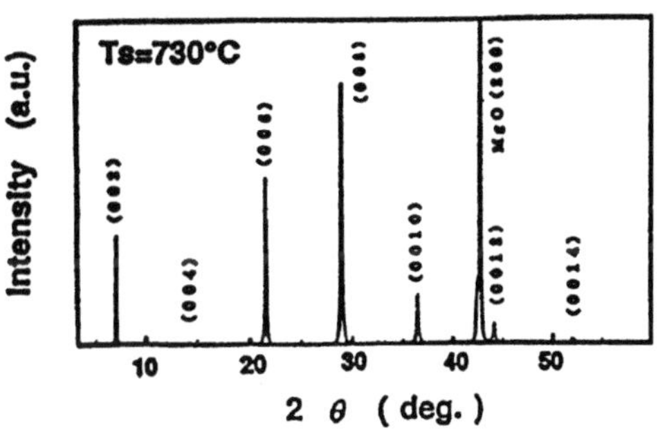

Fig. 4 X-ray diffraction pattern of a single phase $Bi_2Sr_2CuO_x$ thin film.

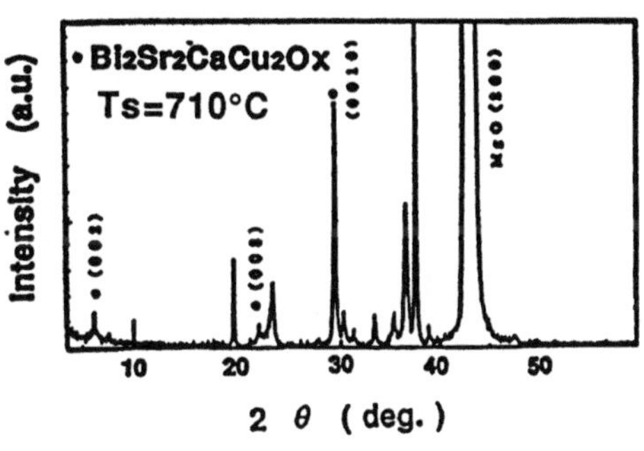

Fig. 5 X-ray diffraction pattern of a mixed phase $Bi_2Sr_2CaCu_2O$ thin film.

CONCLUSION

Oxygen ion beam generated by the ECR plasma source was able to use for substrate cleaning and in-situ preparation of $Bi_2Sr_2CuO_x$ films by adjusting a drift tube bias voltage. An oxygen partial pressure of 1x10^{-4}Torr, even though assisted by an ECR plasma, is not enough to crystallize the 2212 structure.

ACKNOWLEDGMENTS

The authors are grateful to Dr. R.Itti for XPS measurements and helpful discussion. This work was performed under the management of the R&D of basic technology for future industries supported by NEDO.

REFERENCES

1. Michikami O., Asahi M. and Asano H. (1989) Dependence of superconducting properties on substrate temperature in Y-Ba-Cu-O thin films prepared by magnetron sputtering Jpn. J. Appl. Phys. 28-3:L448-L451

2. Venkatesan T., Wu X.D. et.al. (1989) High-temperature superconductivity in ultrathin films of $Y_1Ba_2Cu_3O_{7-x}$ Appl. Phys. Lett. 54-6 :581-583

3. Terashima T. et.al. (1988) Single crystal $YBa_2Cu_3O_{7-x}$ thin films by activated reactive evaporation Jpn. J. Appl. Phys. 27-1:L91-L93

4. Aida T. et.al. (1989) Thin film growth of $YBa_2Cu_3O_{7-x}$ by ECR oxygen plasma assisted reactive evaporation Jpn. J. Appl. Phys. 28-4:L635-L638

5. Gogol C.A. and Cipro C. (1985) Proc. 1st. Int. Symp. Silicon Molecular Epitaxy 85-7:415

Oxide Superconductor BSCCO Films Prepared by the Rapid Melting and Resolidification Process of Ceramic Powder Using CO_2 Laser Beam

K. AGATSUMA, F. UCHIYAMA, K. TSUKAMOTO, T. OHARA, and T. YANAGISAWA
Electrotechnical Laboratory, 1-4, Umezono 1-chome, Tsukuba, Ibaraki, 305 Japan

ABSTRACT

A rapid melting and resolidification process of ceramic powder using CO_2 laser beam has been developed. A high power CO_2 laser beam is applied to an extremely small quantities (a few micro grams per second) of BSCCO fine powder which is dropped from a micro feeder and deposited onto a substrate. This process has a remarkable advantage of rapid growth of the thickness of these ceramic superconductors to make a superconducting tape rather than other techniques, for examples, sputtering or chemical vapor deposit process. We have successfully made BSCCO films on ZrO_2 (YSZ; 5% Y partially stabilized ZrO_2) substrate by this rapid melting and resolidification process. These BSCCO films have about 85 K superconducting onset temperature.

KEY WORDS: rapid melting and resolidification process, CO_2 laser, BSCCO film, micro feeder, high Tc superconductor

INTRODUCTION

Since the discovery of the new high Tc superconductor[1] there has been considerable progress in the production of thin films of these new materials, especially Y-Ba-Cu-O(YBCO)[2], Bi-Sr-Ca-Cu-O(BSCCO)[3] and Tl-Ba-Ca-Cu-O(TBCCO)[4]. Thin film growth techniques of new high Tc oxide superconductors are really important especially for the electronic device applications. Also the production techniques of new high Tc superconducting wires or tapes are intensely desired to be developed for energy applications. We have been developing a new micro feeder which can supply an extremely small quantities of ceramic powder and also a rapid melting and resolidification process of ceramic powder using CO_2 laser beam which applies to form the new high Tc superconducting films of YBCO or BSCCO. In this process a high power CO_2 laser beam is applied to an extremely small quantities of oxide ceramic superconducting fine powder which is dropped from a micro feeder and deposited onto a substrate.

RAPID MELTING AND RESOLIDIFICATION PROCESS OF CERAMIC POWER BY CO_2 LASER BEAM

Distinctive Features of This Process

This rapid melting and resolidification process of ceramic powder using laser beam has several distinctive features as follows;(1) Rapid growth of the thickness of these ceramic superconductors, (2) Substrate is not damaged because a laser beam does not irradiate the substrate directly, (3) It is able to be applied for coating either materials of low or high melting point and also the mixture of them, (4)There are many choices of the independent irradiation experimental parameters, for instance the power density of laser beam, temperature of substrate, flow rate of the powder and the atmospheres etc. We have successfully made BSCCO films on ZrO_2 (YSZ; 5% Y partially stabilized ZrO_2) substrate by this rapid melting and resolidification process using laser beam. These films have about 85 K superconducting onset temperature. The investigations of these ceramic superconducting filmes by energy dispersive X-ray analysis (EDX), wave length dispersive X-ray analysis (WLDX) and scanning electron microscopy (SEM) examinations are shown in this papaer.

Schematic Diagram of Apparatus

The schematic diagram of this rapid melting and resolidification process of ceramic powder by CO_2 laser beam is shown in Figure 1. High power density CO_2 laser beam (1) through a focusing lens (2) irradiates above the substrate (3) which has been preheated by a heater (7). An extremely small quantity of oxide superconducting fine powder (5) is dropped from a special designed micro feeder system (4) into the melting zone (6) by the laser beam to be rapidly melted, and deposited onto the substrate. The ambient of a deposition is at the option.

Micro Feeder

A special designed micro feeder to supply an extremely small quantity of powder material has been developed as shown in Figure 2. The neck (3) of the hopper (2) is vibrated by a piezo-electric oscillator(6) and diaphragm (7). An extremely small quantity (for instance a few micro grams per second) of fine powder is continually fed from a guide tube (4) through an orifice (5) controlled quantitatively by an amplitude and or frequency of an oscillator (9). A typical data of the micro feeder is shown in Figure 3.

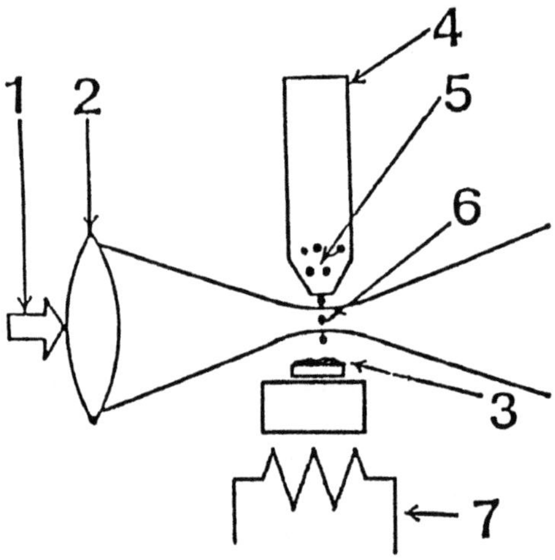

Fig.1 Schematic diagram of rapid melting and resolidification process by laser. (1)lasser beam, (2)condensing lens, (3)substrate, (4)micro feeder, (5)deposit material, (6)melting zone, (7)heater

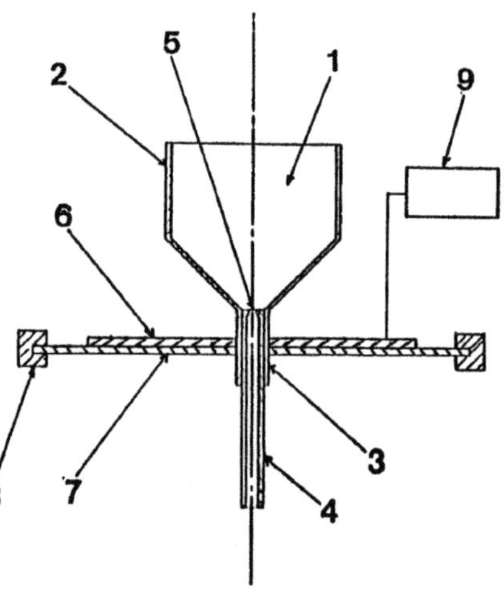

Fig.2 Micro feeder. (1)fine powder, (2)hopper, (3)neck, (4)guide tube, (5)orifice, (6)piezoelectric oscillator, (7)diaphram, (8)suport, (9)oscillator

EXPERIMENTS AND RESULTS

Preparation of Powder

YBCO and BSCCO oxide superconducting ceramic powder are provided for these laser irradiation experiments. YBCO powder is prepared by milling a fired bulk sample made from mixtures of Y, Ba and Cu oxalate fine powder (about 0.3 micro meter) for composition of Y_1 Ba_2 Cu_3 at 900°C 15 hours in flowing oxygen gas. Also BSCCO powder is made as same by milling sintered mixtures of Bi, Sr, Ca and Cu oxalate fine one for composition Bi_1 Sr_1 Ca_1 Cu_2 at 830°C 12 hours in oxygen gas ambient.

Experiments

Experiments have been done as shown schematically in Figure 1. High power CO_2 laser beam of 80 mm diameter is focused in about 2 mm diameter at the melting zone through the focusing lens. Maximum power of this CO_2 laser beam is about 10 Kw. In these experiments the operating power of the laser beam was changed about from 1.5 Kw to 7.0 Kw. The substrate is mounted onto a stage that can be preheated and moved by stepper motors. YSZ(8% Y stabilized ZrO_2, and 5% Y partially stabilized ZrO_2) and SrTiO3 were tested as the substrate. The typical preheated temperature was about 670°C . Both YBCO and BSCCO powder were tested. During laser irradiating, melted material is deposited onto the substrate. A BSCCO film made by this

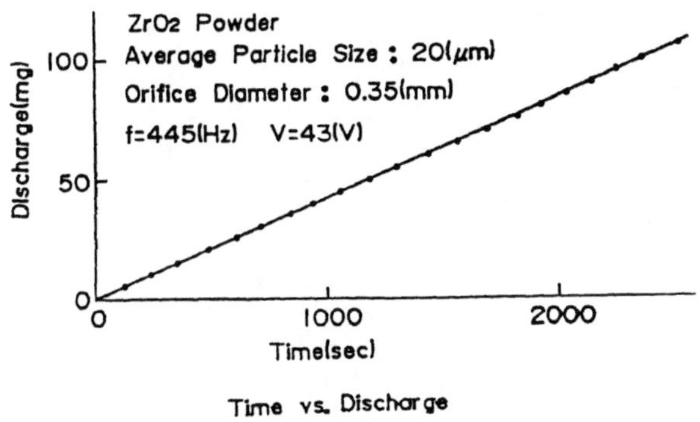

Fig.3 A typical characteristics of the micro feeder.

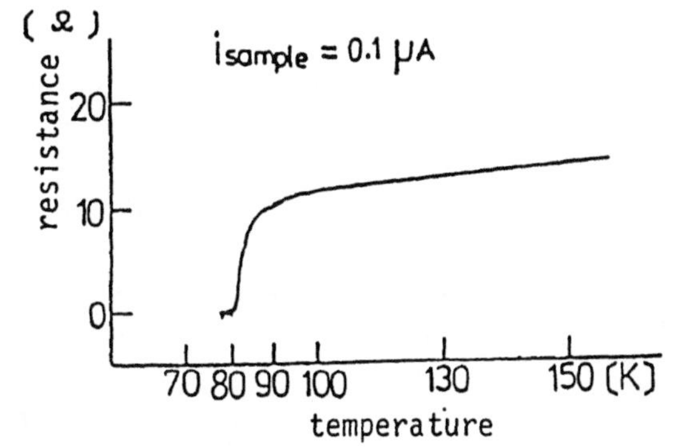

Fig.4 Temperature dependence of the resistance of BSCCO film on YSZ substrate made by this process.

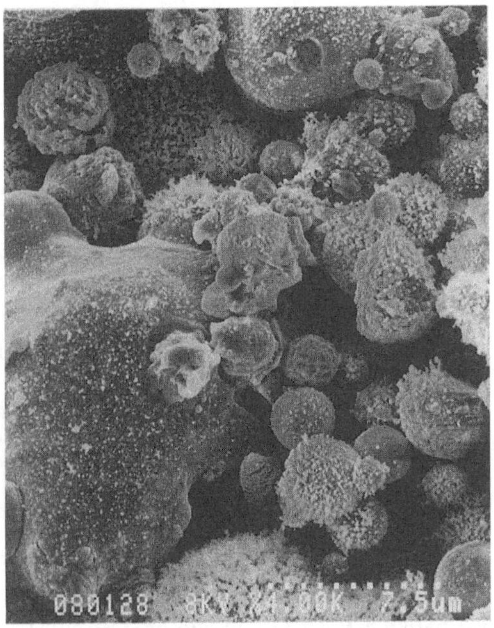

Fig.5 SEM picture of a surface of BSCCO film on YSZ substrate before annealed.

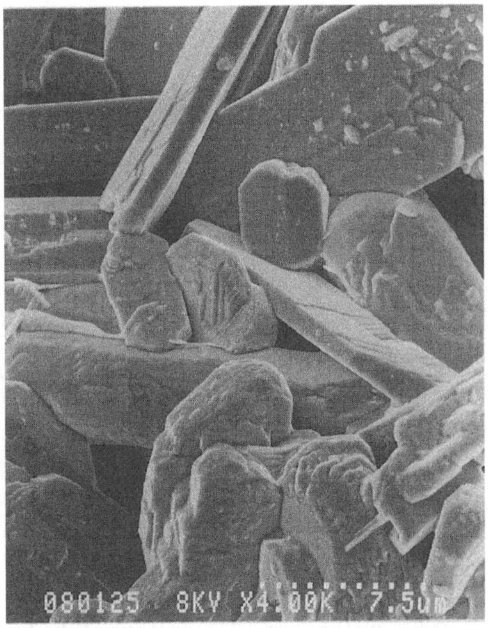

Fig.6 SEM picture of a surface of BSCCO film on YSZ substrate after annealed in oxygen atomsphere.

process on YSZ substrate shows superconductivity after annealed in oxygen atmosphere. A typical temperature dependence of resistance of this BSCCO film on YSZ substrate is shown in Figure 4. The superconducting onset temperature is about 85 K and zero resistance temperature is about 80 K. A YBCO film made by this process does not show the superconductivity at 77 K and also at 4.2 K.

FILM SYNTHESIS

Figure 5 shows a SEM picture of a surface of the BSCCO film as deposited without annealing. Figure 6 shows a SEM picture of a surface of another one after annealed in flowing oxygen gas. A drastic change of a crystal orientation is observed. The investigations of these filmes by EDX,WLDX and SEM examinations before and after annealed have been performed. Figure 7 shows a SEM picture and WLDX picture of the distribution of elements. The boundary layer of new products between BSCCO film and the YSZ substrate can be observed in Figure 7. The boundary layer is revealed to be an oxide compound of Bi, Sr, Ca, Cu and Zr.

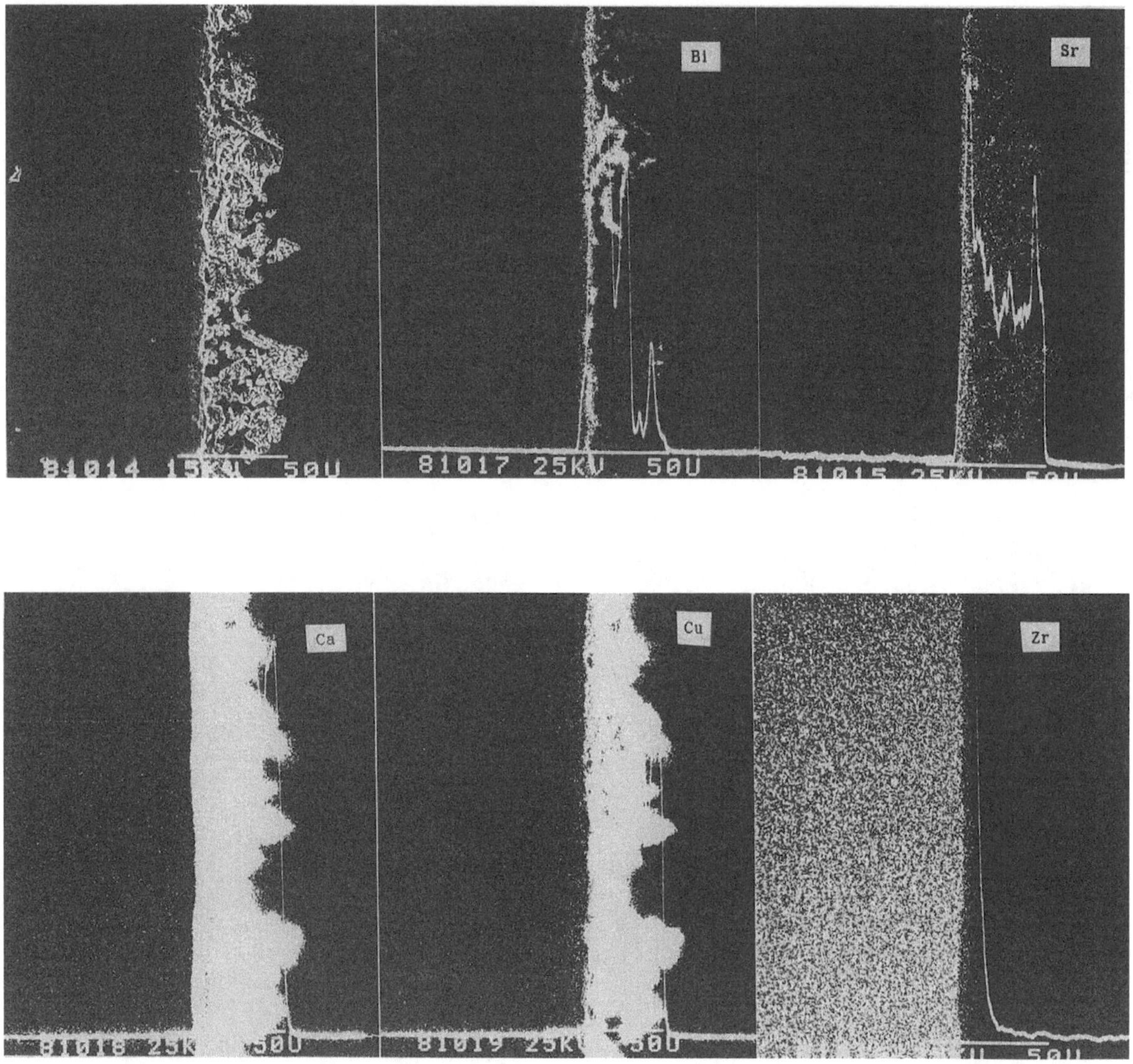

Fig.7 SEM picture and EDX analysis of the boundary layer between BSCCO film and on YSZ substrate after oxygen ambient annealing.

ACKNOWLEDGEMENT

We wish to acknowledge to Mr. T. Aoshima in TOTO Co. Ltd. to help SEM, EDX and WLDX examinations and to Mr. Y. Nobue of a graduate student from Waseda university to assist our experiments and type our manuscript. Also we wish to acknowledge to Dr. Onishi and the colleagues of our laboratory who made the useful arguments with us.

REFERENCE

1. J.G.Bedonortz and K.Muller, "Possible high Tc superconductivity in Ba-La-Cu-O system.", Z.Phys. ,B64. 18, 1986
2. C.W.Chu, P.H.Hor, R.I.Meng, I.Gao, Z.J.Haung and Y.Q.Wang, Phys.Rev.Lett., vol.58, pp.405, 1987
3. H.Maeda, Y.Fukutomi and T.Asano, "A new high Tc oxide superconductor without rare earth elements.", Jpn.J.Appl.phys., 27, L209, 1988
4. P.Claudtdhari, R.H.Koch, R.B.Laibowitz, T.R.Maguire and R.J.Gambinon, Phys.Rev.Lett.vol.58, pp.2684, 1987

UV Light Irradiation Effects on $Bi_2Sr_2CaCu_2O_x$ Superconducting Thin Films

A. Enokihara[1], S. Kohiki[1,2], K. Setsune[1], and K. Wasa[1]

[1] Central Research Laboratories, Matsushita Electric Industrial Co., Ltd., Motiguchi, Osaka, 570 Japan
[2] Permanent address: Matsushita Technoresearch Inc., Moriguchi, Osaka, 570 Japan

ABSTRACT

$Bi_2Sr_2CaCu_2O_x$ superconducting thin films prepared by rf magnetron sputtering were irradiated with ultra-violet (UV) light from a low pressure mercury lamp in helium gas of about 500 Pa. The superconducting properties were gradually degraded with irradiation time and were almost retrieved to the initial state by oxygen annealing at 700 °C. No remarkable change in crystallinity caused by the irradiation was observed from the X-ray diffraction (XRD) analysis. The UV light irradiation must induce oxygen vacancies in the Cu-O planes, which can be filled by the oxygen annealing.

KEYWORDS: irradiation effect, UV light, oxygen vacancy, $Bi_2Sr_2CaCu_2O_x$, thin film

I. INTRODUCTION

Irradiation effects on oxide superconductors have been investigated to examine physical properties of the materials and to control their superconducting characteristics by using several kinds of particles, such as ions[1-3], electrons[4] and photons[5-7]. Irradiation of heavy particles, ions or electrons, dominantly induces mechanical damages, such as amorphization and change of chemical composition of the materials[1,3]. In contrast to that, the photon irradiation primarily induces the photo-chemical reactions in the superconductors so that the change of the superconducting properties with the irradiation should strongly depend on the chemical state of the superconductors[6]. For the cuprate superconductors, the excitation energy of electrons in the Cu 3d-O 2p bonding orbital to the continuous level is regarded as about a few electron-volt, which corresponds to the photon energy of UV light. If the UV light irradiation can change the chemical state of the Cu-O bond by exciting those electrons, the superconducting properties will be controlled by the irradiation since the Cu-O bonding state is closely related with those properties. Moreover, UV light of such energy can be emitted by a conventional mercury lamp and a usual photolithographic technique is also available for selective irradiation of the sample surface with high resolution. Therefore, it is worthwhile to study the UV light irradiation effects on the oxide superconductors not only for study of the physical characteristics of these materials but also for development of a new fabrication process for high-T_c superconducting devices.

In this work, $Bi_2Sr_2CaCu_2O_x$ superconducting thin films were irradiated with UV light and the effects on electrical properties and crystallinity were investigated.

II. EXPERIMENT

$Bi_2Sr_2CaCu_2O_x$ superconducting thin films were prepared on (100) MgO single-crystal substrates by means of rf magnetron sputtering. The sputtering target was 100 mm in diameter and its composition was (Bi : Sr : Ca : Cu =) 2.1 : 1 : 0.5 : 1. Distance between the substrate and the target was set at 25 mm. Input rf power was 150 W. Typical substrate temperature was 580 °C during the deposition. Atmospheric gas pressure of 0.5 Pa was composed of 20 % argon and 80 % oxygen. After the deposition for 10 minutes, in situ annealing at 650 °C

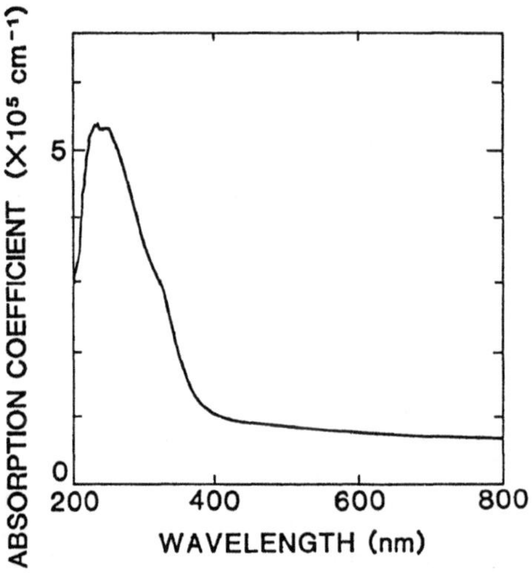

Fig.1. Optical absorption spectrum of the $Bi_2Sr_2CaCu_2O_x$ superconducting thin film.

was performed for 15 minutes in oxygen gas of about 100 Pa. The films were 110 nm thick.

We observed an optical absorption spectrum of the film as shown in Fig.1. The curve of the spectrum has a peak at about 250 nm wavelength. The peak must be due to the absorption by electrons in the Cu-O bonding orbital because energy for the excitation of these electrons to the continuous level is a few electron-volt, which corresponds to the photon energy of around 250 nm wavelength. Therefore, the Cu-O bonding state is expected to be changed by irradiated with UV light.

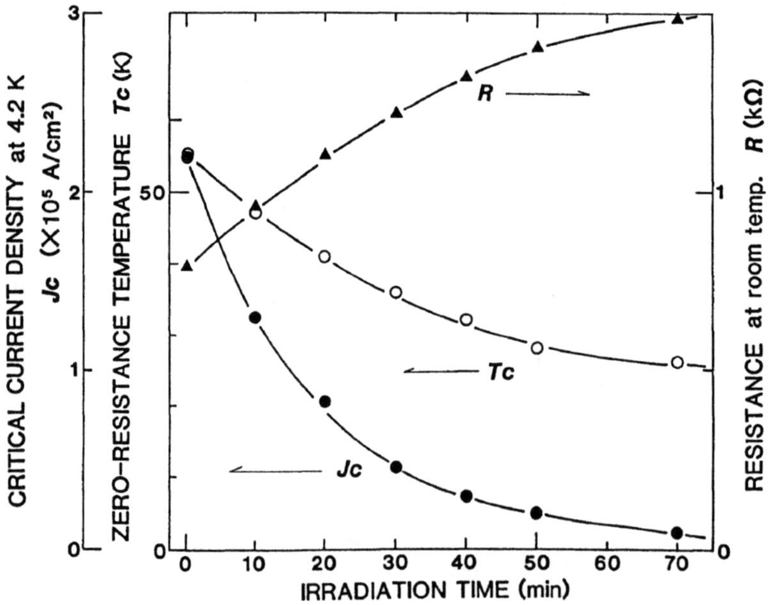

Fig.2. Zero-resistance temperature (T_c), critical current density at 4.2 K (J_c) and resistance at room temperature (R) of the $Bi_2Sr_2CaCu_2O_x$ thin film as functions of UV light irradiation time.

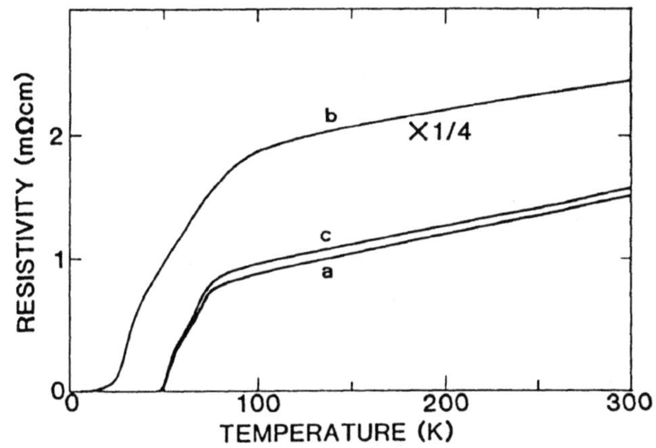

Fig.3. Temperature dependence of resistivity of the $Bi_2Sr_2CaCu_2O_x$ thin film before irradiation (a), after UV light irradiation for 3 hours (b) and after oxygen annealing at 700 °C for 2 hours (c).

UV light of that wavelength can be emitted by a usual mercury lamp. In these experiments, a low pressure mercury lamp (VUV-163 manufactured by ORC) of 160 W was employed for irradiation of the $Bi_2Sr_2CaCu_2O_x$ films. The film sample was set in flowing He gas of about 500 Pa to eliminate the creation of ozone around the sample since ozone gas may oxidize the film material to induce unexpected influence. Radiation power at a sample holder was measured to be about 180 W/cm^2 with an optical power-meter. The sample holder was cooled at about 10 °C. The film was patterned to a Dayem-bridge shape with minimum width of 10 um by Ar ion-beam etching in order to measure the critical current accurately. Figure 2 shows the changes of zero-resistance temperature (T_c), critical current density at 4.2 K (J_c) and resistance at room temperature (R) of the film as functions of UV-light irradiation time. The J_c was calculated by dividing the critical current by the minimum cross section area of the film. It is seen that the superconducting properties were damaged with the UV light irradiation. The J_c was exponentially decreased with the irradiation time and, after irradiation for 70 minutes, became less than a twentieth of that before irradiation. The decrease of the T_c was not so remarkable compared with the J_c. The absorption coefficient of the film material is considerably large at 250 nm wavelength as seen from Fig.1 so that intensity of the incident UV light should be quickly attenuated as an exponential function of the depth in the film. Therefore, the radiation damage gradually extended from the surface into the film and the effective thickness for the superconducting current might be exponentially decreased.

The $Bi_2Sr_2CaCu_2O_x$ film was irradiated with the UV light for 3 hours and then was annealed at 700 °C for 2 hours in flowing oxygen gas of 1 atm to examine an effect of oxygen diffusion. It has been confirmed that the annealing at such temperature has no effect on crystallinity from the XRD analysis and only results in oxygen doping into the $Bi_2Sr_2CaCu_2O_x$ film material. Figure 3 shows the temperature dependence of resistivity of the film before irradiation, after irradiation and after oxygen annealing. The zero-resistance state disappeared after the irradiation and the resistivity at 273 K was increased to more than eight times that before irradiation. After the annealing, the T_c and the resistivity were almost revived to those before the irradiation as seen from the figure.

Figure 4 shows the XRD patterns for the film before and after UV light irradiation for 5 hours. It is seen that the sample presented c-axis orientation normal to the film surface and that no change in the crystallinity was induced by the irradiation.

From the results, it is found that the UV light irradiation damages the superconductivity in spite of no change in the crystallinity and that the oxygen annealing improved the damaged superconductivity almost to the state before irradiation.

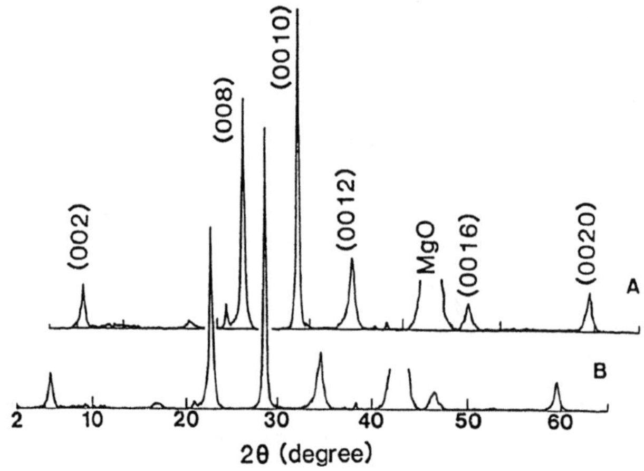

Fig.4. X-ray diffraction (XRD) patterns of the $Bi_2Sr_2CaCu_2O_x$ thin film before irradiation (A) and after UV light irradiation for 5 hours (B).

III. DISCUSSION

Cu-O bonds in these oxide superconductors are known to be covalent because of the strong Cu 3d - O 2p hybridization [8]. When the UV light excites electrons in the Cu-O bonds by the photo-chemical reaction in the materials, oxygen vacancies may be induced in the Cu-O planes since the oxygen atoms would remove from their original sites and the excited electrons might be relaxed to locate on the Cu sites. These oxygen vacancies could cause the damage of the superconductivity, and be filled by the oxygen annealing so that the superconductivity was revived. On the other hand, 'cations are primarily fixed because UV light photons cannot induce the ejection of cations. This is reason no change in the crystal structure was observed from the XRD analysis. Therefore, UV light irradiation must selectively produce the oxygen vacancies, which can be filled by oxygen annealing.

ACKNOWLEDGMENTS

The authors wish to thank Dr. T. Nitta for his continuous encouragement.

REFERENCES

1. G.J.Clark, F.K.LeGoues, A.D.Marwick, R.B.Laibowitz and R.Koch (1987) Ion beam amorphization of $YBa_2Cu_3O_x$. Appl.Phys.Lett. 51: 1462-1464.
2. A.E.White, K.T.Short, R.C.Dynes, A.F.J.levi, M.Anzlowar, K.W.Baldwin, P.A.Polakos, T.A.Fulton and L.N.Dunkleberger (1988) Controllable reduction of critical currents in $YBa_2Cu_3O_{7-x}$ films. Appl.Phys.Lett. 53: 1010-1012.
3. A.Enokihara, H.Higashino, S.Kohiki, K.Setsune and K.Wasa (1989) Effects of Ar ion-beam etching on Gd-Ba-Cu-O superconducting thin films. Jpn.J.Appl.Phys. 28: L452-L455.
4. M.Nastasi, D.M.Parkin, T.G.Zocco, J.Koike and P.R.Okamato (1988) Electron irradiation induced amorphization in $YBa_2Cu_3O_7$ and $GdBa_2Cu_3O_7$ superconductors. Appl.Phys.Lett. 53: 1326-1328.
5. J.Bohandy, J.Suter, B.F.Kim, K.Moorjani and F.J.Adrian (1987) Gamma radiation resistance of high T_c superconductor $YBa_2Cu_3O_{7-x}$. Appl.Phys.Lett. 51: 2161-2163.
6. A.Enokihara, S.Kohiki, K.Setsune, K.Wasa, Y.Higashi, S.Fukushima and Y.Gohshi (1989) X-ray irradiation effects on $ErBa_2Cu_3O_{7-x}$ superconducting thin films. to be published in Physica C.
7. H.Tamura, A.Yoshida, S.Morohashi and S.Hasuo (1988) Ozone-UV irradiation effects on $Ba_2YCu_3O_{7-x}$ thin films. Appl.Phys.Lett. 52: 2183-2185.
8. S.Kohiki, S.Hayashi, H.Adachi, S.Hatta, K.Setsune and K.Wasa (1989) Electron spectroscopy of $Nd_{2-x}Ce_xCuO_{4-y}$ (x=0, 0.15, and 0.23) thin films. to be published in J.Phys.Soc.Jpn. 58.

Fabrication of Bi-Sr-Ca-Cu-O/Ferromagnet Layered Thin Films

T. MATSUSHIMA[1], Y. ICHIKAWA[2], H. ADACHI[2], S. HATTA[2], K. SETSUNE[2], and K. WASA[2]

[1] Semiconductor Research Laboratory, Matsushita Electric Works Ltd., Kadoma, Osaka, 571 Japan
[2] Central Research Laboratories, Matsushita Electric Industrial Co. Ltd., Yagumo-Nakamachi 3-15, Moriguchi, Osaka, 570 Japan

ABSTRACT

Thin films with layered structure of Bi-Sr-Ca-Cu-O superconductor and Ni-ferrite ($NiFe_2O_4$) have been prepared by RF magnetron sputtering. The $NiFe_2O_4$(900Å)/$Bi_2Sr_2CaCu_2O_y$(2100Å) layered films showed an onset superconducting temperature T_c of 87K and a zero resistance temperature T_c of 60K. From magnetization measurements, it was revealed that the observed magnetization of the layered film is a resultant of the diamagnetization of $Bi_2Sr_2CaCu_2O_y$ and the ferromagnetization of $NiFe_2O_4$ as similar to the magnetization of ferrimagnetic materials.

KEY WORDS: Bi-Sr-Ca-Cu-O, Ni-ferrite, layered structure, diamagnetization, ferromagnetization.

INTRODUCTION

The study of the interaction between superconductivity and magnetism has attracted much attention. In previous studies of metallic-superconductor/ferromagnetic-material superlattices such as V/Ni[1], V/Fe[2], Al/EuO[3] and Bi-Pb/EuS[4], various attractive phenomena of superconductivity have been reported: the dimensional cross over on the upper critical field[1,2], the suppression of T_c[2], the spin-polarized electron tunneling[3] and the critical current enhancement.[4] The proximity effect is supposed as one of reasons for these phenomena, because Ginzuburg-Landau coherence lengths of these superconductors are comparable with the modulation wavelengh of superlattices. On the other hand, short coherence lengths of high-T_c copper oxide superconductors have been reported. Thus a new novel behavior on superconductivities is expected for oxide-superconductor/magnetic-material layered films. In addition these layered films may open the way for new applications. The high-T_c superconductor on a magnetic substrate $YbFeO_3$ has been reported without magnetic properties.[5] In order to study the interaction between the superconductivity of Bi-Sr-Ca-Cu-O system and magnetism, we have prepared the thin films with layered structure of $Bi_2Sr_2CaCu_2O_y$ and $NiFe_2O_4$. The $NiFe_2O_4$ compound is a ferromagnetic insulator with Curie temperature of 858 K and with spinel structure. The layered thin films have been prepared by RF magnetron sputtering. The most important point on the preparation of the layered films is the prevention of the interdiffusion of atoms at the interface. Hence it is desirable to lower the deposition temperature. We reported earlier on the low-temperature preparation of Bi-Sr-Ca-Cu-O films.[6] In this study, superconducting $Bi_2Sr_2CaCu_2O_y$ films have been prepared by our low-temperature process and preliminary results were reported on structure, superconducting and magnetic properties of layered films.

EXPERIMENTAL PROCEDURE

The layered films have been prepared by RF magnetron sputtering on (100) MgO single-crystals. At the first, a Bi-Sr-Ca-Cu-O film was deposited on MgO and then the films were took out from the sputtering chamber. Subsequently a Ni-Fe-O film was directly deposited on the Bi-Sr-Ca-Cu-O film. During the deposition, the substrate temperature was kept at about 650°C. Film thicknesses of Bi-Sr-Ca-Cu-O and Ni-Fe-O were 2100Å and 900Å, respectively.

Detailed sputtering conditions are listed in Table I. The crystal analysis was carried out by means of an X-ray diffractometer using a Cu target. The resistivity of the layered films was measured by a standard DC four-probe method using evaporated Au electrodes. Magnetization measurements were carried out using a vibrating sample magnetometer in fields up to 16 kOe at room temperature and using an RF-SQUID susceptometer in fields up to 7 kOe at temperatures from 4.2 K to 100K.

Table I. Sputtering conditions.

Target	Bi-Sr-Ca-Cu-O and Ni-Fe-O
Substrate	MgO (100)
Substrate Temperature	650°C
Sputtering Gas	Ar/O_2=1/4 (BSCCO) and 5/1 (NiFeO)
RF Input Power	100 W (BSCCO) and 80 W (NiFeO)
Growth Rate	70 Å/min (BSCCO) and 15 Å/min (NiFeO)

RESULTS AND DISCUSSIONS

Figure 1 shows magnetization curves of the layered film at room temperature. The external magnetic field was applied perpendicular to the film plane. As shown in Fig.1, the layered film showed a ferromagnetic property. The saturation magnetization was about 290 emu/cm^3; this value corresponds to that of bulk material. It is well known that the crystal structure of $NiFe_2O_4$ is spinel type and a lattice constant a is 8.34 Å. This value is twice as large as that of MgO. In Fig. 2, the X-ray diffraction pattern of the layered film is shown. The diffraction pattern showed the high degree of orientation with c-axis perpendicular to the film plane and these peaks were assignable to (00l) lines of the $Bi_2Sr_2CaCu_2O_y$ phase. No extra peak was observed except for peaks of $Bi_2Sr_2CaCu_2O_y$ and MgO. Judging from results of the crystal analysis and the magnetization measurement, it seems that Ni-Fe-O grew with c-axis perpendicular to $Bi_2Sr_2CaCu_2O_y$ film. The observation of no extra peak on the X-ray diffraction pattern is explained by the overlap of (400) line for $NiFe_2O_4$ with (200)line for MgO, because the lattice constant a of $NiFe_2O_4$ is twice as large as that of MgO.

A characteristic state of the interface between $Bi_2Sr_2CaCu_2O_y$ and $NiFe_2O_4$ was investigated by Auger electron spectroscopy. Metallic elements were almost constant inside the film and considerable inter-diffusion was not observed at interfaces.

Figure 3 shows the temperature dependence of the resistivities of as-grown films. The film of $Bi_2Sr_2CaCu_2O_y$ without $NiFe_2O_4$ showed an onset T_c of 80 K and a zero resistivity

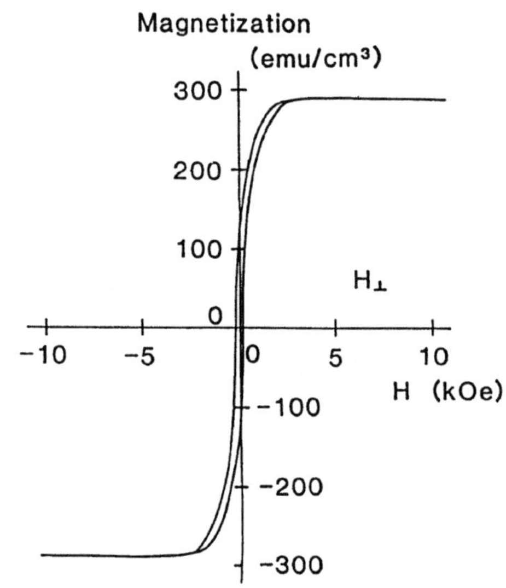

Fig. 1. Magnetization curve of the layered film at room temperature.

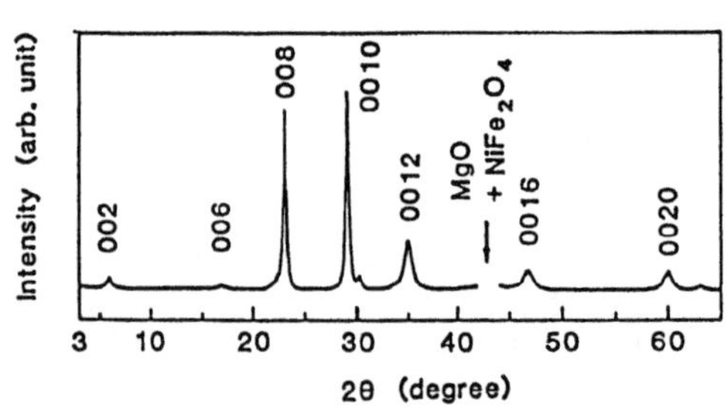

Fig. 2. X-ray diffraction pattern of the layered film.

T_c of 68 K. On the layered film, striking suppression of T_c is not observed; the layered film showed slightly higher onset T_c of 87 K and slightly lower zero resistivity T_c of 60 K. The layered films had higher resistivity than that of $Bi_2Sr_2CaCu_2O_y$ film at 250K. An origin of this result is caused by slight interdiffusion of atoms at the interface which was not observed by our Auger analysis.

Magnetization curves of the $Bi_2Sr_2CaCu_2O_y$ film and the layered film at 4.2 K were shown in Fig. 4. The external magnetic field was applied perpendicular to the c-plane after cooling the sample to 4.2 K. The magnetization of the layered film exhibited the negative value in fields up to 150 Oe and it exhibited the positive value over one.

The temperature dependence of the magnetization in magnetic field of 100 Oe for the films are shown in Fig. 5. Magnetization measurements were carried out after cooling the sample to 4.2 K without any external fields. On the layered film, the magnetization in the filed of 100 Oe exhibited the negative value at 4.2 K and it increased rapidly with increasing temperature. At 9 K, the sign of the magnetization changed from negative to positive and the magnetization saturated at about 20 K. A similar temperature dependence of the magnetization is observed on a ferrimagnetic material[7] in which the sign of the magnetization dose not change but the magnetization exhibits a maximum value after exhibiting a minimum value with increasing temperature. The magnetization of the ferrimagnetic material is a resultant of sublattice magnetizations having different values and an anti-parallel configuration. By similar analogy, when the magnetization of the layered film is found by the adding the diamagnetization of $Bi_2Sr_2CaCu_2O_y$ and the ferromagnetization of $NiFe_2O_4$, the magnetic properties of the layered films are explained well. On the magnetization curve of the layered film, as shown in Fig.4, the magnetization of $NiFe_2O_4$ increases rapidly with increasing the field, while that of $Bi_2Sr_2CaCu_2O_y$ increases gradually. Thus the sign of the magnetization changed from negative to positive at 150 Oe. The temperature dependence of the magnetization of the layered film is also explained by similar way.

From our experimental results, it was revealed that the interaction between superconductivity of $Bi_2Sr_2CaCu_2O_y$ and magnetism is not observed in our layered film with the thick(2100Å) $Bi_2Sr_2CaCu_2O_y$ film.

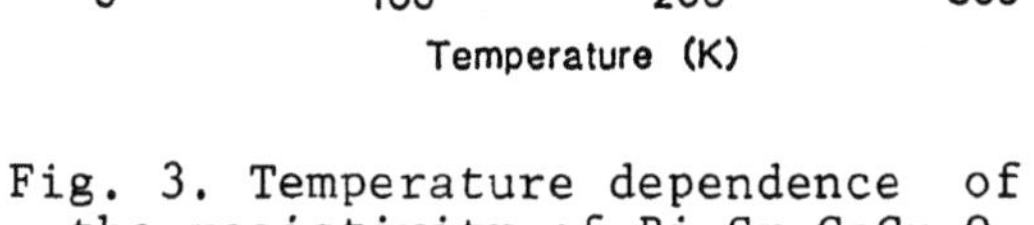

Fig. 3. Temperature dependence of the resistivity of $Bi_2Sr_2CaCu_2O_y$ film and the layered film.

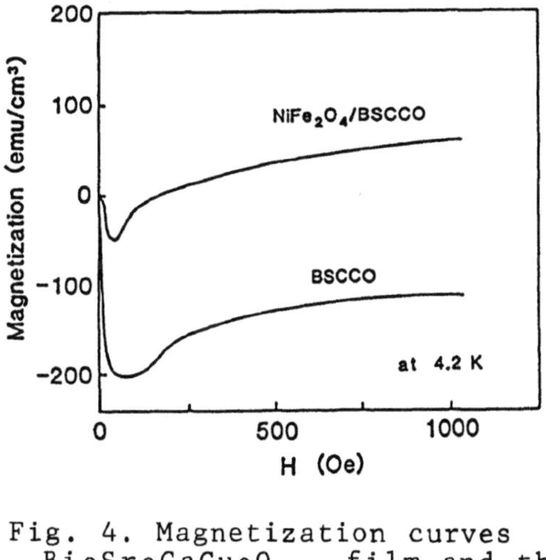

Fig. 4. Magnetization curves of $Bi_2Sr_2CaCu_2O_y$ film and the layered film at 4.2 K.

SUMMARY

We report for the first time of the preparation of thin films with layered structure of $Bi_2Sr_2CaCu_2O_y$ and $NiFe_2O_4$. The layered film showed the onset T_c of 87 K and the zero resistivity T_c of 60K. Striking suppression of T_c is not observed. The magnetization of layered films is the resultant of the diamagnetization of $Bi_2Sr_2CaCu_2O_y$ and the ferromagnetization of $NiFe_2O_4$. The interaction between superconductivity and magnetism is not found in our layered films. Further study is desired to reveal the interaction between superconductivity and magnetism on thinner oxide-superconducting films.

We would like to thank Dr. T. Nitta for his support of this work and to thank M. Shoji and K. Kakite of Matsushita Electric Works, for their encouragement throughout this work.

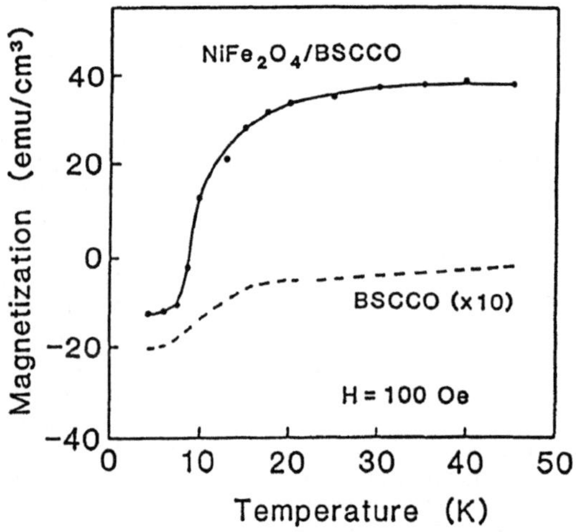

Fig. 5. Temperature dependence of the magnetization of $Bi_2Sr_2CaCu_2O_y$ film and the layered film.

REFERENCES

[1] H. Homma, C. S. Chun, G. G. Zheng, and Ivan K. Schuller, Phys. Rev. B33, (1986) 3562.
[2] H. K. Wong and J. B. Ketterson, J. Low. Temp. Phys. 63, (1986) 139.
[3] P. M. Tedrow and J. E. Tkaczyk, Phys. Rev. Lett. 21, (1986) 1746.
[4] R. Aoki, J. Yamashita, Y. Takabatake and H.Hamada, Proc. MRS Int'l. Mtg. on Adv. Mats. 10 (1989) Multilayers 33.
[5] R. Ramesh, A. Inam, W. A. Bonner, P. England, B. J. Wilkens, B. J. Meagcher, L. Nazar, X. D. Wu, M. S. Hegde, C. C. Chang, T. Venkatesan, and H. Padamsee, Appl. Phys. Lett. 55, (1989) 1138.
[6] T. Matsushima, K. Hirochi, H. Adachi, K. Setsune, and K. Wasa, Jpn. J. Appl. Phys. 28, (1989) L97.
[7] E. W. Gorter, Philips Research Rept. 9, (1954) 295, 321 and 403.

Bi Based Oxide Superconducting Thin Film Prepared by CVD Method

KEN-ICHI SAIKUSA, TADASHI SUGIHARA, and TAKUO TAKESHITA

Central Research Institute, Mitsubishi Metal Corporation, 1-297, Kitabukuro, Omiya, Saitama, 330 Japan

ABSTRACT

BiSrCaCuO superconducting thin films were prepared on MgO (100) substrates by metal organic chemical vapor deposition (MOCVD) using $Bi(ph)_3$, Sr-, Ca-, $Cu(DPM)_2$, and effects of deposition parameters (substrate temperatures (Ts), oxygen flow rates (FO_2) and cooling rates (C.R)) were investigated with regards to superconducting properties of films. A film as prepared at Ts=800°C, in FO_2=200ccm, and with C.R=15°C/min possesses the (2212) structure. This film shows a micro mosaic structure with c-axis orientation perpendicular to the substrate, and has zero-resistance temperature of 72K.

INTRODUCTION

Since the high-Tc superconducting oxide was discovered, many studies have been made on high-Tc superconductors in bulk form or in thin film form concerning with their preparation. In particular, the thin film technology for preparation of these superconducting oxides is needed for their application of microelectronic devices, and thin films have been prepared by many methods such as magnetron sputtering [1], pulse laser evaporation [2], electron beam evaporation [3], and so on. Although studies of preparation of superconducting oxide films by chemical vapor deposition have been reported [4], the superconducting properties of these CVD films are not as good as those of films prepared by physical vapor deposition methods. Therefore, it is important to study the formation mechanism of CVD thin films in order to obtain good quality superconducting thin films.

In this paper, we report the preparation of BiSrCaCuO thin films on MgO (100) substrates by MOCVD method using $Bi(ph)_3$, Sr-, Ca-, $Cu(DPM)_2$ as source gases, and describe the effects of preparation parameters on superconducting properties of the films. Substrate temperature, oxygen partial pressure and cooling rate were investigated as experimental parameters which are considered to affect the deposition mechanism.

EXPERIMENTAL

Figure 1 shows a schematic diagram of MOCVD apparatus used in this work. The source materials were kept in vaporizers which were made by glass, and were heated up to temperatures at which desired vapor pressures of these gases were maintaind. Ar gas as a carrier and O_2 gas as a reactant were separately introduced into the quartz reactor. The single crystal MgO (100) substrate was placed on a quartz susceptor in the reactor. The substrate temperature was maintained with infrared image furnace. The gas line was kept at 260°C in order to prevent the condensation of source gases.

$Bi(ph)_3$, $Sr(DPM)_2$, $Ca(DPM)_2$, and $Cu(DPM)_2$ were heated to 115, 230, 210, and 123°C, respectively. Substrate temperatures of 750, 800, and 850°C, and oxygen gas flow rates of 100, 200, and 300ccm, and cooling rates of 15, 45, and 135°C/min in 1 atm oxygen atmosphere were used. Ar used as the carrier gas was passed through each heated vaporizer at a rate of 50 ccm. The total gas pressure was kept at 15 torr during the film growth, and the growth rate of BiSrCaCuO films could be maintained in the range 0.5~1.0 μm/h. The film thicknesses of all samples were kept about 1μm in this work. Superconducting properties and microstructures of the films were evaluated by X-ray diffractometry, EPMA, SEM, and conventional DC four probe resistivity measurement method. Typical deposition conditions are summarized in Table-1.

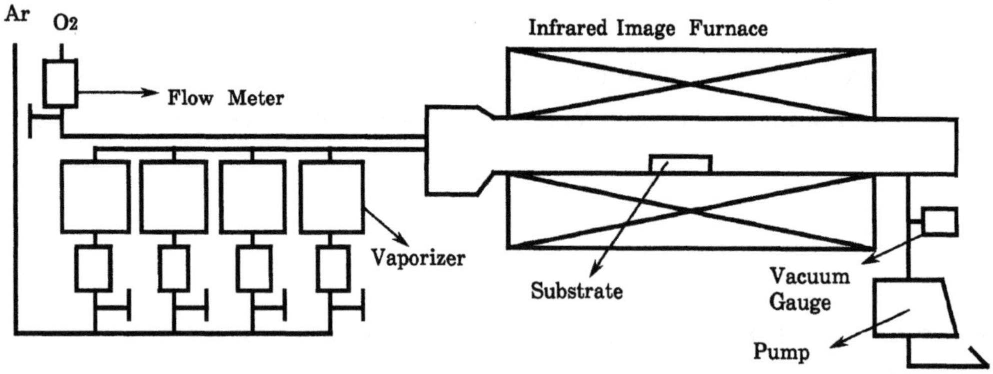

Fig. 1. A schematic diagram of MOCVD apparatus.

Table - 1. Deposition conditions of BiSrCaCuO films.

	$Bi(ph)_3$	$Sr(DPM)_2$	$Ca(DPM)_2$	$Cu(DPM)_2$
Vaporizer Temp. (°C)	115	230	210	123
Substrate Temp. (°C)	750	800	850	
O_2 Gas Flow Rates (ccm)	100	200	300	
Cooling Rates (°C/min)	15	45	135	
Ar Gas Flow Rates (ccm)		50		
Gas Line Temp. (°C)		260		
Total Pressure (Torr)		15		
Period (Hr)		1		
Substrates	MgO (100) 10 × 10 × 0.5 mm			

RESULTS AND DISCUSSION

Figure 2 shows the X-ray diffraction patterns of BiSrCaCuO thin films deposited at various substrate temperatures (FO_2=200ccm, C.R=15°C/min). The X-ray diffraction patterns of the films deposited at substrate temperature of 750°C and 800°C could be assigned mainly to (00ℓ) plane of the low-Tc (2212) phase with c=30.6A, and they also show good preferred orientation of c-axis perpendicular to the substrate surface plane.

In contrast, the films deposited at the substrate temperature of 850°C could be assigned to (00ℓ) plane of the low-Tc (2212) phase and to those of the semiconductor (2201) phase with c=24.4A. This indicates that the (2201) phase was formed by partial melting in films prepared at 850°C. Figure 3 shows the X-ray diffraction patterns of BiSrCaCuO films deposited at various oxygen flow rates (Ts=800°C, C.R=15°C/min). The film formed at 800°C with FO_2=100ccm shows the random crystalline orientation in contrast to the films with FO_2=200ccm and 300ccm.

Figure 4 shows the temperature dependence of the resistivity of BiSrCaCuO films deposited at various substrate temperatures and oxygen flow rates (C.R=15°C/min). In all samples, the resistivities decrease with decreasing temperature showing the metallic behavior, and drop abruptly at around 90K. The zero-resistance temperatures obtained were 44K, 72K, 30K, for the films deposited at 750°C, 800°C, and 850°C, respectively.

The resistance of a film deposited with FO_2=200ccm begins to drop at around 100K, and reaches to zero at 72K. Tc of the film produced with FO_2=300ccm is 69K. However, the film produced with FO_2=100ccm shows no superconductivity in the measuring condition, ie., I=1mA (corresponding to 10^{-3} A/cm^2).

From these results, it is seen that the oxygen flow rate is one of the important deposition parameters for growing good superconducting film of a well oriented structure and with good interconnection paths between grains.

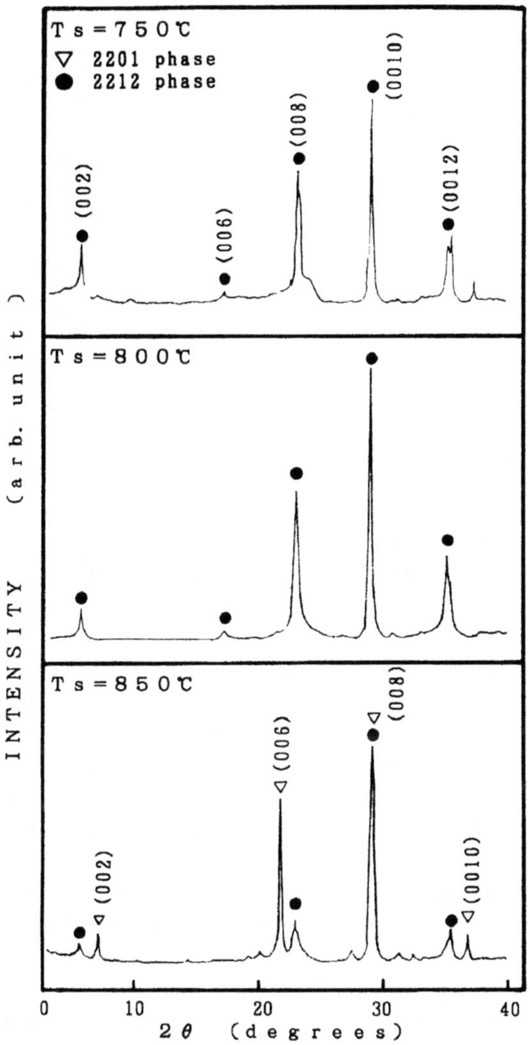

Fig. 2. X-ray diffraction patterns of BiSrCaCuO films deposited at various substrate temperatures (Ts).

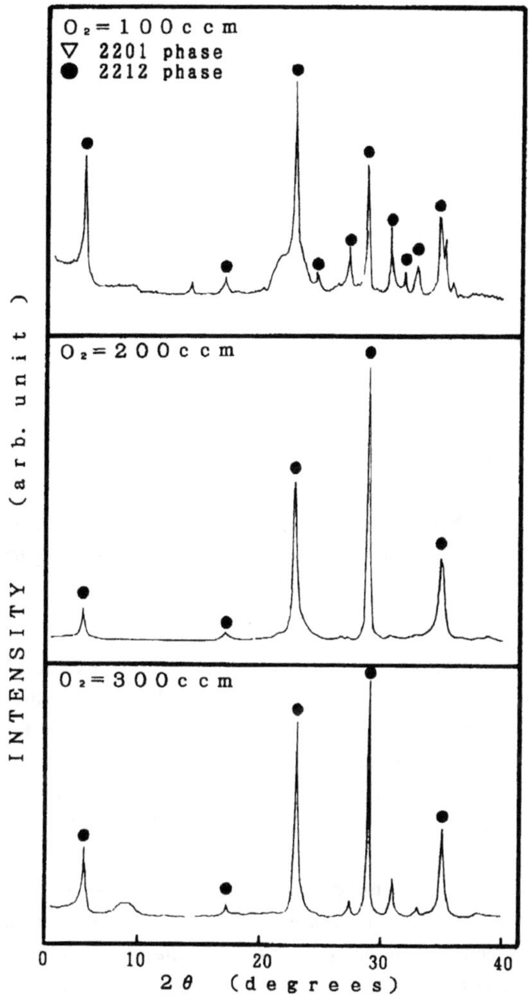

Fig. 3. X-ray diffraction patterns of BiSrCaCuO films deposited at various oxygen flow rates (FO2).

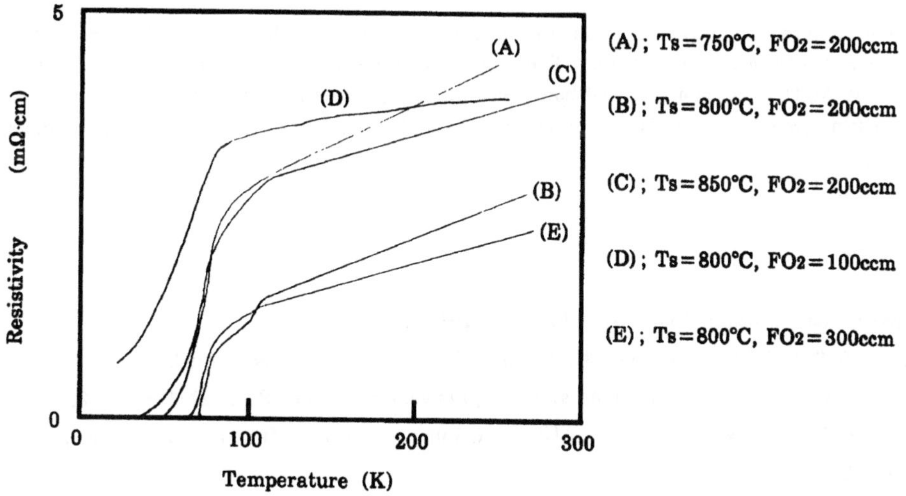

Fig. 4. Temperature dependence of the resistivity of BiSrCaCuO films deposited at various substrate temperatures (Ts) and oxygen flow rates (FO2).

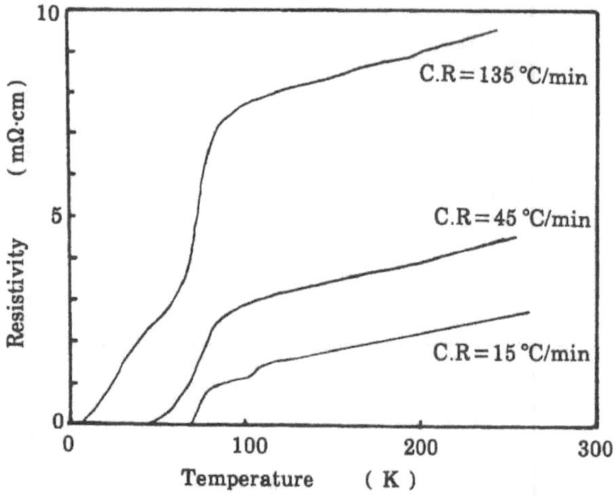

Fig. 5. Temperature dependence of the resistivity of the BiSrCaCuO films deposited at various cooling rates (C.R).

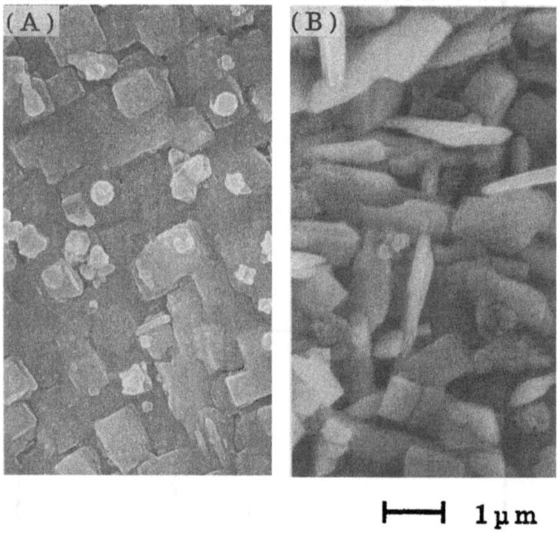

Fig. 6. SEM photographs of the films produced with C.R=15°C/min (A), and C.R=135°C/min (B).

In addition to those results, we also examined the effects of sample cooling rates on superconducting properties of films. Figure 5 shows the temperature dependence of the resistivity of BiSrCaCuO films deposited at various cooling rates (C.R). All samples were deposited at 800°C in FO_2=200ccm. The resistivity curves of these films produced with C.R=15°C/min have a shoulder around 100K indicating the formation of the (2223) phase in some amount. The resistance of all samples decreases with decreasing temperature and the zero-resistance temperature is 72K for C.R=15°C/min, 46K for C.R=45°C/min, and 6K for C.R=135°C/min, respectively.

The slow cooling rates produce films with well behaved metallic resistivity than the rapid one, and this result indicates that the slow cooling after film deposition is effective for grain growth and strengthening interconnections between grains as the post-annealing treatment in oxygen atmosphere does. This is also seen clearly by SEM photographs as shown in Figure 6. In summary, high quality (2212) phase films were obtained at the substrate temperature of 800°C and with oxygen flow rates above 200 ccm. It is also shown that the cooling rate after deposition is the important parameter to obtain high quality films.

CONCLUSION

We have prepared the BiSrCaCuO superconducting thin films by MOCVD method using $Bi(ph)_3$, $Sr(DPM)_2$, $Ca(DPM)_2$, and $Cu(DPM)_2$ as reactants. The superconducting thin films with the zero-resistance of 72K were formed on the MgO (100) substrate without post-annealing process, and it is found that the oxygen partial pressure on film growth process is very important as well as substrate temperature in producing good quality CVD thin films.

REFERENCES

[1] S.Tanaka, and H.Itozaki: Jpn. J. Appl. Phys. 27 (1988) L 622.

[2] M.Kanai, T.Kawai, M.Kawai and S.Kawai: Jpn. J. Appl. Phys. 27 (1988) L 1293.

[3] K.Kuroda, M. Mukaido, M.Yamamoto and S.Miyazawa: Jpn. J. Appl. Phys. 27 (1988) L 625.

[4] H.Yamane, H.Kurosawa, H.Iwasaki, T.Hirai, N.Kobayashi, and Y.Muto: Jpn. J. Appl. Phys. 28 (1989) L 127.

A 110-K Phase BiSrCaCuO Thin Film Grown by Halide CVD

M. IHARA, H. YAMAWAKI, T. KIMURA, and H. NAKAO
Fujitsu Laboratories Ltd., Atsugi, 10-1, Morinosato-Wakamiya, Atsugi, 243-01 Japan

ABSTRACT

We established the conditions for growing a 110-K phase non-doped BiSrCaCuO thin film on an MgO (100) substrate. Films were grown by chemical vapor deposition (CVD) using metal-halide sources and oxygen gas in an open tube reactor. We found that the X-ray diffraction intensity of a 110-K phase BiSrCaCuO layer strongly depends on the concentration of the $BiCl_3$ vapor in the reactor tube. The X-ray diffraction pattern of a BiSrCaCuO layer grown at an optimum concentration of $BiCl_3$ shows that the film contained only 110-K phase crystals. Resistivity ρ at 300 K was ρ = 1 mΩ·cm to 3 mΩ·cm. Critical temperature Tc was Tc = 89 K to 97 K. Critical current density was Jc = 2 x 10^5 A/cm^2 at 10 K for a BiSrCaCuO layer 20 µm in width.

This paper discusses the halide-CVD system and the characteristics of a 110-K phase BiSrCaCuO layer on an MgO substrate.

INTRODUCTION

Many researchers have been investigating thin film growth of high-Tc superconducting materials, and have used several methods to fabricate electronic devices which operate at liquid-nitrogen temperature. Films have been produced by RF diode sputtering, EB sputtering, ICB deposition, laser ablation deposition, MBE, and CVD. We developed a CVD technique for growing high-quality superconducting BiSrCaCuO layers on MgO substrates based on the $BiCl_3$-SrI_2-CaI_2-CuI-O_2 system [1-3].

We fabricated 110-K phase BiSrCaCuO superconducting as-grown film by halaide-CVD. This paper discuses the relationship between film characteristics and growth conditions.

GROWTH SYSTEM AND CONDITIONS

The BiSrCaCuO growth apparatus shown in Fig. 1 consists of a quartz reactor tube, two source chambers, a substrate holder, a horizontal furnace with five separate heating zones, a gas flow control system, and $BiCl_3$, CuI, CaI_2, and SrI_2 boats. The MgO (100) substrate was 30 mm x 30 mm. Before growth, substrates were etched with HCl (90) + H_2O_2 (10) or similar compounds. Halide-source temperatures were T_{BiCl3} = 150°C-170°C, T_{CuI} = 450°C-470°C, T_{CaI2} = 800°C-850°C and T_{SrI2} = 800°C-850°C. MgO substrate temperatures were T_{sub} = 750°C-850°C and the oxygen gas flow rate was 2.5 liters/min. The He carrier gas flow rate into the source chambers and reactor tube was 30 liters/min. The oxygen concentration in the He carrier gas was about 7.7 % and the H_2O concentration was about 500 ppm. The growth rate and layer composition of the BiSrCaCuO superconductor were controlled by individual furnace heating for the halide sources, and oxygen concentration and humidity in the reactor tube. After film growth, the substrates were cooled rapidly at about 50C/min. The layer thickness was measured by a surface profiler after chemically etching part of the grown layer. The growth rate of the BiSrCaCuO layers was about 0.5 nm/min to 1 nm/min. The layer composition was $Bi_2Sr_{1.5-3}Ca_{1-1.5}Cu_{3-4.5}O_x$ by fluorescence X-ray analysis (FXA).

RESULTS AND DISCUSSION

Figure 2 shows (a) a photomicrograph of the as-grown surface of a 0.20-µm-thick BiSrCaCuO layer on an MgO substrate, and (b) a SEM image of the same sample. The layer surface was good and domain of the BiSrCaCuO crystal was large. The domain size was about 100 µm to 1 mm wide. The X-ray diffraction intensities of the same sample are shown in Fig. 3. The X-ray diffraction pattern for a layer grown with the $BiCl_3$ source at 165°C shows all peaks in relation to the C-axis of the higher peak for 80-K phase BiSrCaCuO crystal and the extra peaks for 110-K and 20-K phase crystal. Growth conditions for the sample were T_{sub} = 775°C, T_{CuI} = 460°C, T_{CaI2} = 825°C, and T_{SrI2} = 850°C at a growth rate of 1 nm/min. Figure 4 shows the X-ray diffraction pattern for a layer grown with the $BiCl_3$ source at 160°C. The main peak for 110-K phase and the extra peaks for 80-K and 20-K phases are smaller than in Fig. 3. The temperature dependence of the electrical resistance of the layers is shown in Fig. 5. The curves are normalized to the resistance at room temperature (300 K). The resistance of sample A (T_{BiCl3} = 165°C) decreases abruptly below 105 K and has a tail from 100 K to 80 K. The critical temperature (zero-resistance) for sample A was 81 K. The resistance of sample B (T_{BiCl3} =160°C) decreases abruptly below 110 K. The critical temperature for sample B was 95 K.

Figure 6 shows the X-ray diffraction pattern for a layer grown with the $BiCl_3$ source at 155°C. Only 110-K phase BiSrCaCuO crystal is present. 80-K and 20-K phase BiSrCaCuO diffraction patterns could not be detected even at high magnification. The temperature dependence of the electrical resistance of the layers is shown in Fig. 7. The layers were grown under the same conditions using a batch of three MgO substrates. The critical temperature for the first sample (upper stream) was 97 K. That for the second sample (center) was 93 K. That for the third sample (lower stream) was 89 K. The layer thickness after 300 min for the first sample was 0.25 µm. That for the second sample was 0.20 µm. That for the third sample was 0.15 µm. The critical current densities for the first, second, and third samples were 3×10^3 A/cm^2 to 2×10^5 A/cm^2 at 10 K for a width of 20 µm and a length of 50 µm formd by photolithograph. The X-ray diffraction pattern for a layer grown with the $BiCl_3$ source at 150°C shows two kinds of peaks for 110-K and 80-K phase BiSrCaCuO crystal. The diffraction intensity of the layer was smaller than that of the other samples, indicated in Fig. 3, 4, and 6. Therefore, the highest quality in a 110-K phase BiSrCaCuO layer on a MgO substrate is obtained at T_{BiCl3} = 155°C ± 2°C.

CONCLUSION

We developed a technique for growing a 110-K phase non-doped BiSrCaCuO thin film on an MgO (100) substrate by halide CVD. We found that the diffraction intensity of a 110-K phase BiSrCaCuO layer strongly depends on the concentration of $BiCl_3$ vapor in the reactor tube. The X-ray diffraction pattern of the BiSrCaCuO layer shows that the film containd only 110-K phase crystal. The critical temperatures for layers were between 89 K and 97 K. The critical current densities for layers were 3×10^3 A/cm^2 to 2×10^5 A/cm^2 at 10 K.

REFERENCES

[1] M. Ihara, T. Kimura, H. Yamawaki, and K. Ikeda: ASC-1988-San Francisco: IEEE Transactions on Magnetics MAG-25, 2, 2470 (1989).

[2] T. Kimura, M. Ihara, H. Yamawaki, K. Ikeda, and M. Ozeki: Proc. Advances in Superconductivity, ISS 88-Nagoya, 495 (1988).

[3] M. Ihara, T. Kimura, H. Yamawaki, and O. Ueda: Ex. Abstracts of FED HiTcSc-ED Workshop, June, 1989, Hokkaido, Japan, 135 and 309.

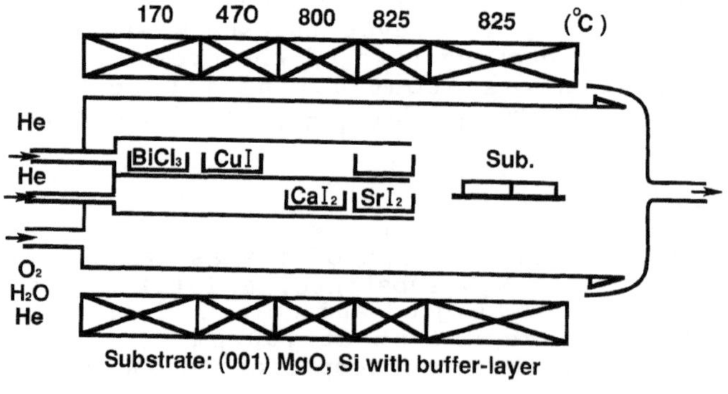

Fig. 1. BiSrCaCuO halide CVD system.

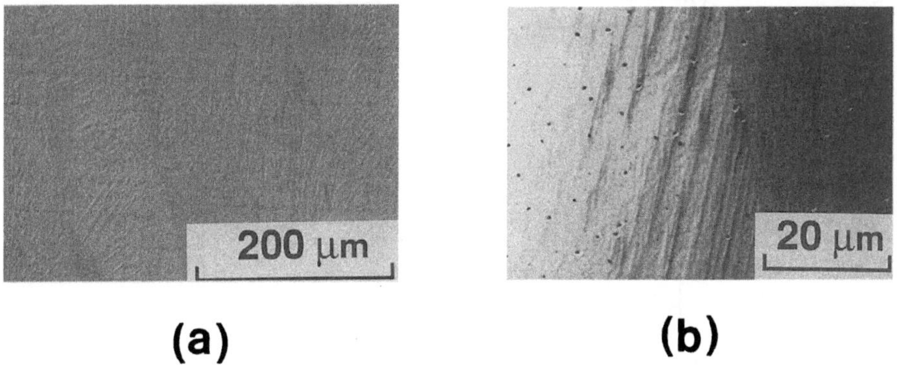

Fig. 2. (a) photomicrograph of an as-grown surface of a BiSrCaCuO layer on a MgO substrate. (b) SEM image of the same sample.

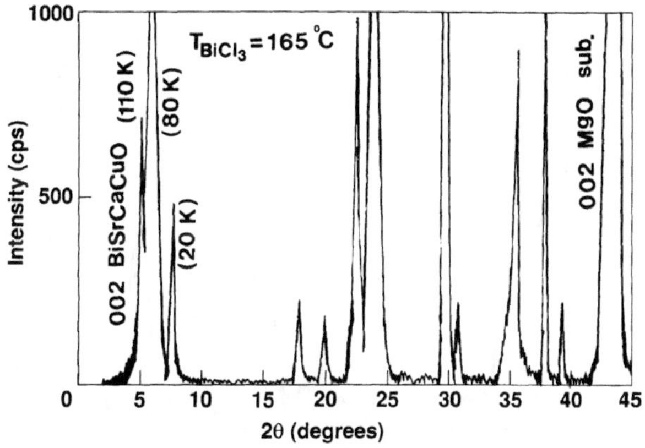

Fig. 3. X-ray diffraction pattern for a layer grown with the $BiCl_3$ source at 165°C.

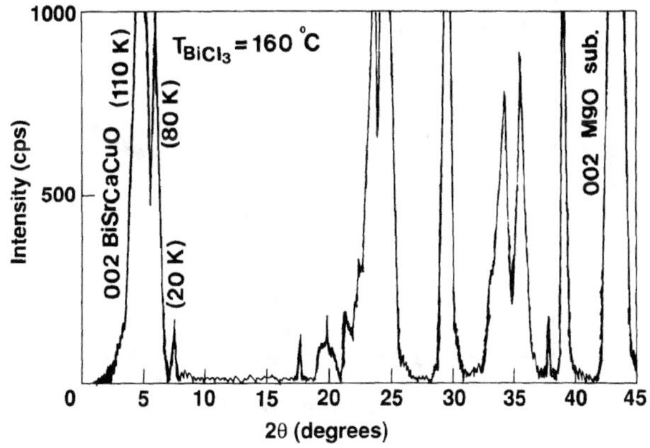

Fig. 4. X-ray diffraction pattern for a layer grown with the $BiCl_3$ source at 160°C.

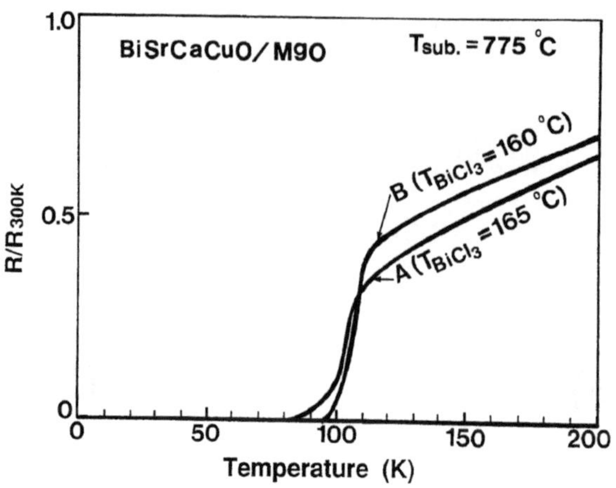

Fig. 5. Temperature dependence of the electrical resistance of the films. Sample A was grown with the $BiCl_3$ source at 165°C. Sample B was grown with the $BiCl_3$ source at 160°C.

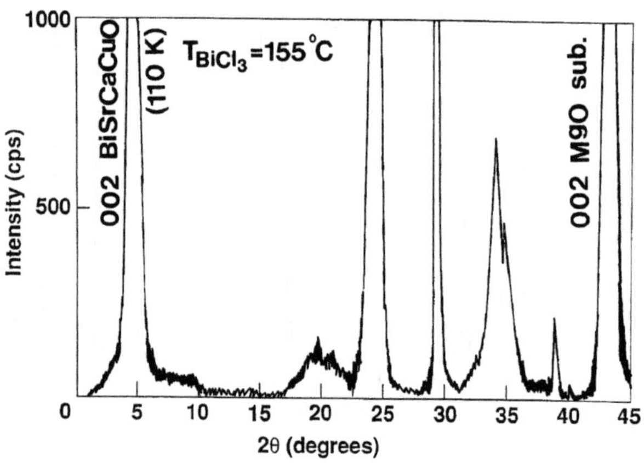

Fig. 6. X-ray diffraction pattern for a layer grown with the $BiCl_3$ source at 155°C.

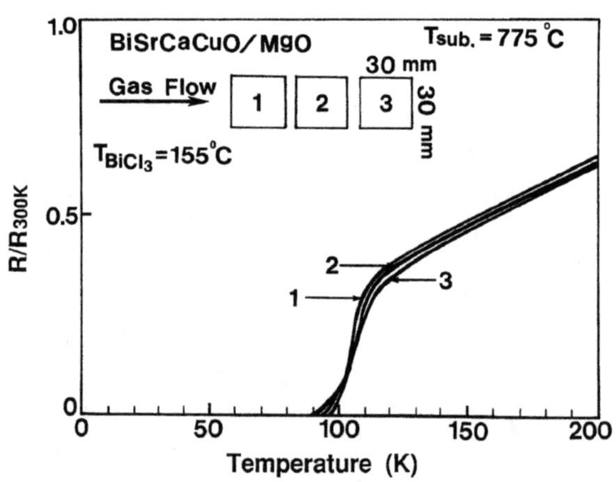

Fig. 7. Temperature dependence of the electrical resistance of the films. The layers were grown with the $BiCl_3$ source at 155°C on a batch of three MgO substrates.

Thin Film Fabrication of BSCCO Superconductors Using MOCVD

T. SUGIMOTO, S. YUHYA, Y. YAMADA, H. HAYASHI, K. KIKUCHI, M. YOSHIDA, K. SUGAWARA, and Y. SHIOHARA

Superconductivity Research Labolatory, International Superconductivity Technology Center, 10-13, Shinonome 1-chome, Koto-ku, Tokyo, 135 Japan

ABSTRACT

Bi-Sr-Ca-Cu-O (BSCCO) superconducting thin films were fabricated by metalorganic chemical vapor depositon (MOCVD) using triphenylbismuth, $Sr(DPM)_2$, $Ca(DPM)_2$, and $Cu(DPM)_2$ (DPM= dipivaloylmethanate) for metalorganic (MO) sources. We have constructed an MOCVD apparatus succeeding in high reproducibility of films with improving line-heatings for the uniformity of MO transport at relatively low evaporation temperatures for the MO sources. X-ray diffraction data indicated that the films prepared at 820-880 °C consisted predominantly of the high-Tc phase and had preferencial c-axis orientation perpendicular to the (100) single-crystal MgO substrate surface. The resistance versus temperature for these films reveals an onset of superconductivity at about 110 K and zero-resistance by 69-77 K. We have also investigated the relationship between the substrate temperature and the deposition rate of components for a BSCCO thin film.

KEY WORDS: Bi-Sr-Ca-Cu-O system, oxide superconductor, thin film, high-Tc phase, MOCVD

INTRODUCTION

Much attention has been paid to high-Tc superconductors without rare earth elements since the discovery of the BSCCO system [1]. The system is superior to the Ln-Ba-Cu-O system (Ln=rare-earth) in not only Tc but its chemical stability. Unfortunately, this system is multi-phased with $Bi_2Sr_2Ca_2Cu_3O_x$ (high-Tc phase), $Bi_2Sr_2CaCu_2O_x$ (low-Tc phase), and $Bi_2Sr_2CuO_x$ (semiconducting or insulating phase) and it is considerably difficult to isolate the high-Tc phase. The BSCCO thin films have been prepared using physical vapor deposition such as sputtering [2], electron beam evaporation [3], and laser ablation [4]. In addition, metalorganic chemical vapor deposition (MOCVD) [5,6] and metal halide chemical vapor deposition (CVD) [7] have been reported recently. MOCVD is a promising thin film fabrication process because of its precise controllability of film composition, wide range controling capability of deposition rates, and compatibility with variously shaped substrates. However, the precise control of film composition has not yet realized because the stability of metalorganic (MO) sources is very weak or unclear. We have constructed an MOCVD apparatus succeeding in high reproducibility of film composition with improving line-heatings for the uniformity of MO sources transport at relatively low vaporization temperatures of the MO sources. In this paper, we report the fabrication of high-Tc BSCCO films by the MOCVD process and the characterization of these thin films. In addition, we report for the first time the relationship between the substrate temperature and the deposition rate of a BSCCO thin film.

EXPERIMENTAL

Figure 1 shows the MOCVD apparatus used in this study. It consists of a gas-handling system, four evaporators, a horizontal cold wall type reactor, and an evacuation system. All gases are mixed prior to entering the reactor. The MO source evaporators are placed into accurate temparature-controled ovens and the whole MO lines are also placed into another oven maintained at 200 °C in order to improve the uniformity of line-heatings and prevent MO vapors from condensation and/or decomposition. Only strontium transport tube and line were wrapped with heating tapes and maintained at 230 °C. Volatile ß-diketonate complexes of Cu, Sr, and Ca with ligand of dipivalylmethanate (DPM) were employed. For volatile bismuth source, ß-diketonates, alkoxides, or trialkyl- and triaryl compounds were considered. Bismuth alkoxides and trialkylbismuth were difficult in handling

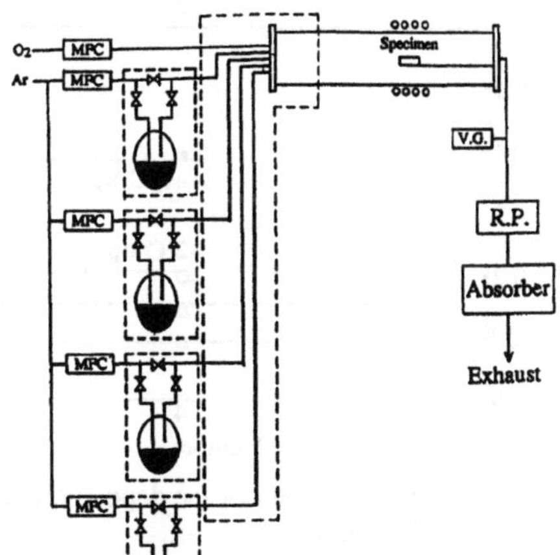

Fig. 1 Apparatus for MOCVD system.
The regions of dashed line are placed into ovens.

because of strong moisture-sensitivity and/or air-sensitivity. Bismuth ß-diketonates such as $Bi(DPM)_3$ was found to decompose gradually at relatively low temperature and the residue was likely bismuth oxide because the residue of 25 wt% left without evaporation in the thermogravimetric mesurement, as shown in fig. 2. On the contrary, triphenylbismuth almost evaporated and no detectable residue was observed. Consequently, we chose triphenylbismuth which was volatile and stable. The depositon of BSCCO films was carried out on (100) single crystal MgO substrates (10x10x0.5 mm) degreased with acetone prior to placing in the reactor. After the deposition process, the substrate was cooled to 500 °C at 10 °C/min under an atmosphere of O_2 gas flow. Typical deposition rates selected were 0.1-0.5 µm/h, and typical film thicknesses were 0.2-1.0 µm. The detail deposition conditions were listed in table 1. The as-deposited films were characterized by four-probe resistance mesurements, X-ray diffraction (XRD) using Cu-Kα radiation. The avarage composition of films was analyzed by inductively coupled plasma chemical analysis (ICP).

Table 1. Depodition condition of BSCCO thin films

MO source	$Bi(ph)_3$	$Sr(DPM)_2$	$Ca(DPM)_2$	$Cu(DPM)_2$
Evaporator Temperature (°C)	80-90	205-215	185-200	95-105
Flow rate of Ar gas (sccm)	50	50	50	50
Substrate temperature (°C)	700-940			
Gas pressure(torr)	10			
Flow rate of O_2 (sccm)	400			
depodition time (h)	2			

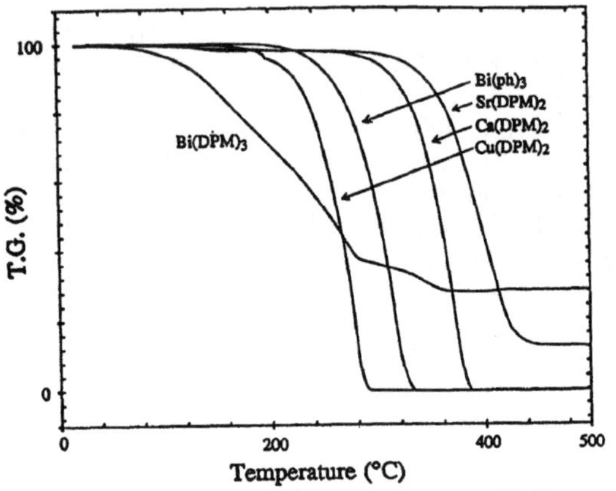

Fig. 2 Thermogravimetric measurement of MO sources.

RESULTS AND DISCUSSION

The reproducibility of the deposition rates is shown in fig. 3. The every deposition was carried out under the same condition and deposition rates were estimated by ICP analysis. The evaporator temperatures of Sr and Ca sources were 215 °C and 180 °C, respectively. Those of Bi and Cu sources were respectively 85 °C and 100 °C for the first three runs and then increased the evaporator temperatures to 87 °C and 105 °C, respectively. The deposition rates of Bi and Cu increased 1.3 times and wice, respectively. The deposition rates of Sr and Ca changed within the range of 10 %, while those of Bi and Cu changed within the range of 5 %. Accordingly, the deposition rates of Bi and Cu are found to be more stable than those of Sr and Ca. Figure 4 shows the relationship between the substrate temperature and the deposition rates of the components for a BSCCO thin film.

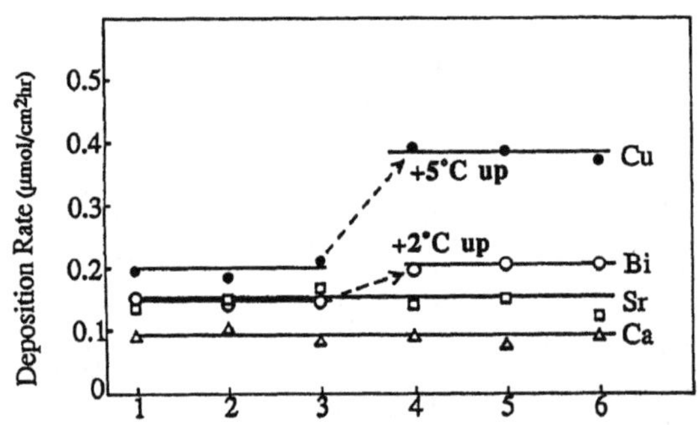

Fig. 3 Reproducibility of the deposition rates of BSCCO thin films.

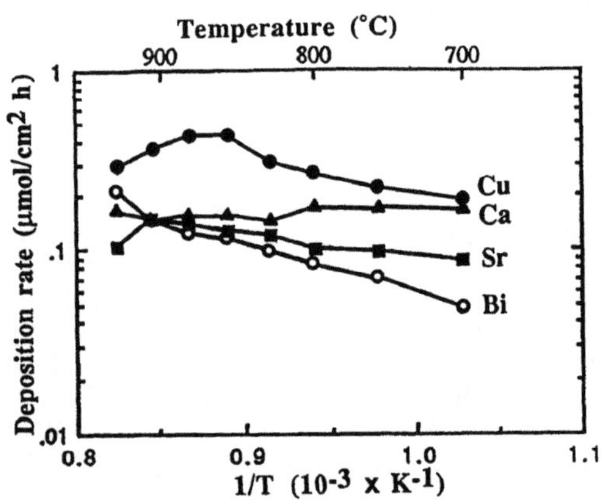

Fig. 4 Relationship between the substrate temperature and the deposition rates of BSCCO thin films.

Evaporator temperatures of the Bi, Sr, Ca, and Cu sources were 85°C, 215°C, 190°C, and 105°C, respectively. Ar gas carrier flow rates of the Bi, Sr, Ca, and Cu sources were 50 sccm for each. As can be seen in fig. 4, the substrate temperature dependence of the deposition rate for each component was different each other. This result indicates that the composition of a BSCCO thin film clearly depends on the substrate temperature even when the evaporator temperatures of MO sources are constant. The depodition rate of Bi increased proportionally as the substrate temperature increased. This suggests that triphenylbismuth decomposed or oxidized on the surface of the substrate rather than in the gas phase and this is consistent with the fact of the low thermal decomposition rate of triphenylbismuth [6]. The deposition rate of Ca tended to be slightly dependent on the substrate temperature, which is similar to the case of the MOCVD for compound semiconductors such as GaAs [8]. The deposition rates of Sr and Cu together increased as the substrate temperature increased in the range of the substrate temperatures lower than 880 °C, but decreased with the substrate temperature in the range of the substrate temperatures higher than 880 °C. This suggests that $Cu(DPM)_2$ and $Sr(DPM)_2$ decomposed and/or oxidized in the gas phase at the substrate temperature higher than 880 °C before the MO gases reach the substrate.

BSCCO thin films with composition of closely Bi:Sr:Ca:Cu=2:2:2:3 were prepared at different substrate temperatures. The evaporator temperatures of the MO sources were determined based on the results of the substrate temperature dependence of the film composition (see fig. 5). Figure 6 shows the X-ray diffraction (XRD) pattern of the BSCCO thin films. Diffraction lines for 2θ were calibrated based on MgO (200), 2θ=42.895° of the MgO substrate. The XRD pattern of the sample at Ts=720 °C (Ts is the substrate temperature) reveals the presence of the semiconducting phase (c=24.4 A) with weak intensities of the (00*l*) reflections, CuO, and the unknown pluses. XRD petterns of the samples at Ts=750 °C and 780 °C reveal the presence of the low-Tc superconducting $Bi_2Sr_2CaCu_2O$ phase (c=30.6 A) with weak intensities of the (00*l*) reflections, CuO, CaO, and the unknown pluses. XRD patterns of the samples at Ts=820 °C, 850 °C, and 880 °C indicate the presence of the high-Tc superconducting $Bi_2Sr_2Ca_2Cu_3O_x$ phase (c=37 A), the low-Tc superconducting $Bi_2Sr_2CaCu_2O_x$ phase, CuO, and CaO. Both superconducting phases have preferential c-axis orientations perpendicular to the (100) MgO surface and the intensities of diffraction peaks related to the high-Tc phase are higher than those related to the low-Tc phase. The broad peaks of the high-Tc phase suggest the presence of many intergrowth defects similar to a BSCCO bulk oxides [9]. The intensities of diffraction peaks related to the low Tc-phase are much higher than those of the high-Tc phase at Ts=910 °C, oppositely the XRD patterns at Ts=820 °C, 850 °C, and 880 °C. This result suggests that the high-Tc phase can be formed more preferably at 820-880 °C of the substrate temperature. The oxygen partial pressure in the system was about 0.01 atm in our experiment, and the high-Tc phase in a bulk BSCCO superconducting oxides appears preferably between 810 °C and 830 °C in the similar range of the oxygen partial pressure [10]. It should be noted that the surface temperature of the substrate is about 50 °C lower than the measured substrate temperature. It means that the substrate temperature for the formation of the high-Tc phase is similar to that of a bulk BSCCO superconducting oxides.

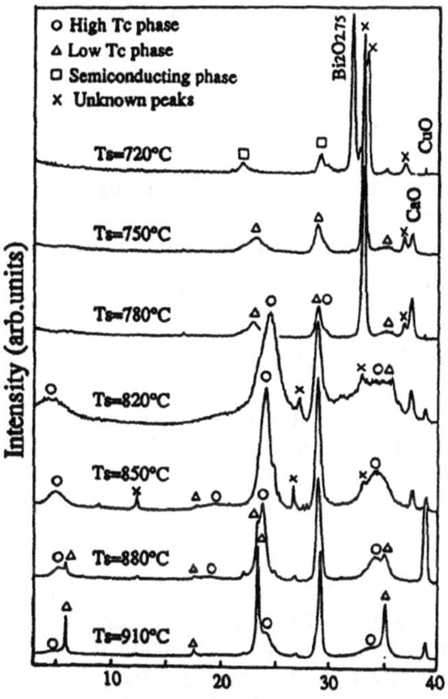

Fig. 5 X-ray diffraction patterns of BSCCO thin films deposited at Ts=720-910 °C (Ts is the substrate temperature).

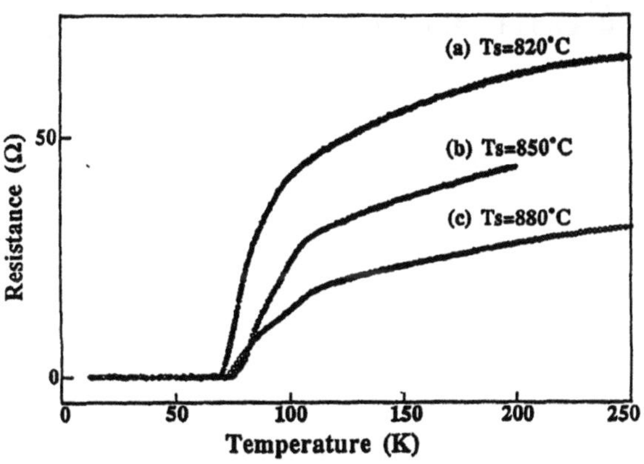

Fig. 6 Resistance versus temperature of BSCCO thin films deposited at Ts=820, 850, and 880 °C, respectively. The onset temperatures of all samples were obtained around 110 K and the zero-resistivity temperatures were 69, 73, 77 K for the films deposited at Ts=820, 850, and 880 K, respectively.

The resistance versus temperature of the samples, which is shown in fig. 6, was measured in a hellium-sealed cryostat. Electrical contacts were made to the films by four indium-soldered contacts arranged in a standard in-line geometry. For the temperature of the zero-resistance, avaraging of the data obtained by cooling and heating the samples was adopted to minimize the error of the temperature measurements due to the geometrical difference between the samples and the thermocouple. Figure 7 illustrates the resistances versus temperature for the BSCCO thin films prepared at 820 °C, 850 °C, and 880 °C of the substrate temperatures. The resistances of the samples prepared at 820 °C, 850 °C, and 880 °C clearly decreased around 110 K, but the temperatures of the zero-resistance were 69 K, 73 K, and 77 K, respectively, in spite of the high-Tc phase preferably appearing in the XRD patterns as shown in fig. 6. This result suggests that the crystal property of these films is poor and the high-Tc phase does not electrically link due to presence of many intergrowth defects.

CONCLUSION

We fabricated BSCCO thin films contained the high-Tc phase on a (100) MgO single crystal substrate, at the substrate temperatures between 820 °C and 880 °C, by the MOCVD process without postannealing. It was found that the composition of the BSCCO thin films depended on the substrate temperature even though the evaporation temperatures of MO source were constant.

REFERENCES

[1] Maeda H. Tanaka Y. Fukutomi M. Asano T. (1988) Jpn.J.Appl.Phys. 27 : L209.
[2] for example: Ichikawa Y. Adachi H. Hiroshi K. Setsune K. Hatta S. Wasa K. (1988) Phys.Rev.B. 38:765
[3] for example: Yoshitake T. Satoh T. Kubo Y. Igarashi H. (1988) Jpn.J.Appl.Phys. 27 : L1089.
[4] for example: Guarnieri CR. Roy RA. Seanger KL. Shivashankar SA. Yee DS. Cuomo JJ. (1988) Phys.Lett. 53 :532
[5] Berry AD. Holm RT. Cukauskas EJ. Fatemi M. Gaskill DK. Kaplan R. Fox WB. (1988) J.Cryst.Growth 92:344 ; Yamane H. Kurosawa H. Hirai T. (1988) Chem.Lett.:1515 ; Yamane H. Kurosawa H. Hirai T. Iwasaki H. Kobayashi N. Muto Y. (1988) Jpn.J.Appl.Phys. 27:L1495 ; Yamane H. Kurosawa H. Iwasaki H. Hirai T. Kobayashi N. Muto Y. (1989) Jpn.J.Appl.Phys. 28:L827 ; Zhang J. Zhao J. Marcy HO. Tonge LM Wessels TW. Marks TJ. Kannewurf CR. (1989) Appl.Phys.Lett. 54:1166.
[6] Kobayashi K. Ichikawa S. Okada G. (1989) Chem.Lett.:1415.
[7] Ihara M. Kimura T. Ex. abstracts of FEX HiTcHc-ED workshop, June 1988. Miyagi-Zao, Japan :137 Kimura T. Ihara M. Yamawaki H. Ikeda K. Ozeki M.(1989) Proc.1st.Int.Symp.Superconductor (ISS'88). Nagoya,Japan. Spring-verlag, Tokyo. pp495.
[8] Reep DH. Ghandhi SK. (1983) J.Electrochem.Soc. 130 :675.
[9] Adachi S. Hirano H. Takahashi Y. Inoue O. Kawasima S. (1989) Jpn.J.Appl.Phys. 28 :L209.
[10] Endo U. Koyama S. Kawai T. (1989) Jpn.J.Appl.Phys 27 :L1476.

In-situ Growth of Bi-Sr-Ca-Cu-O Films with High-T_c Superconducting Phase by MOCVD

K. ENDO, S. HAYASHIDA[a)], K. NAKATSUKA[b)], J. ISHIAI[c)], Y. IKEDO[d)], S. MISAWA, and S. YOSHIDA

Electrotechnical laboratory, 1-4, Umezono 1-chome, Tsukuba, Ibaraki, 305 Japan

ABSTRACT

The effect of *in-situ* Pb-doping in MOCVD on the structures and superconductivity of Bi-Sr-Ca-Cu-O films has been studied for the first time. We have obtained the films with the high-Tc single phase by Pb-doping in the high oxygen pressure (≈50 Torr) region, where only the films with the low-Tc phase are grown without Pb-doping. On the contrary, Pb-doping resulted in the formation of the mixed phase in the low oxygen pressure (≈25 Torr) region, where the high-Tc single phase appears without Pb-doping.

KEY WORD: Bi-Sr-Ca-Cu-O thin films, MOCVD, *in-situ* Pb-doping, high-Tc phase, oxide superconductor

INTRODUCTION

Since the discovery of superconductivity in the Bi-Sr-Ca-Cu-O system [1], many studies have been made on obtaining the high-Tc single phase. Bi-Sr-Ca-Cu-O system has at least two superconducting phases; the high-Tc phase ($Bi_2Sr_2Ca_2Cu_3O_x$) with Tc around 110K and the low-Tc phase ($Bi_2Sr_2Ca_1Cu_2O_y$) with Tc around 80K. It has been reported that Pb-doping enhances the high-Tc phase formation in the sintering of bulk materials [2] and in the annealing of sputtered films [3].

In the previous study [4], we have succeeded in fabricating the Bi-Sr-Ca-Cu-O films with the high-Tc single phase on MgO(100) substrates without postannealing by metalorganic chemical vapor deposition (MOCVD) using triphenyl bismuth and b-diketonates of Sr, Ca and Cu as source materials , for the first time. However, Tc(zero resistance) of the films was lower than 100K.

In this study, we have carried out *in-situ* Pb-doping in the growth of Bi-Sr-Ca-Cu-O films by MOCVD and have studied its effect on the crystal structures and superconductivity of the films, for the first time.

EXPERIMENTAL

Figure 1 is a schematic drawing of the MOCVD apparatus used in this study. The source materials were triphenyl bismuth $Bi(C_6H_5)_3$, bis(-2,2,6,6-tetramethyl-3,5-heptanediono)-strontium Sr $(DPM)_2$, -calcium $Ca(DPM)_2$ and -copper $Cu(DPM)_2$. Bis(dipivaloylmethanato)-lead $Pb(DPM)_2$ was selected as the MO source material for Pb-doping. Each source material was loaded into individual vaporizer, which was put in an oil-bath. The temperatures of the vaporizers of $Bi(C_6H_5)_3$, $Sr(DPM)_2$, $Ca(DPM)_2$, $Cu(DPM)_2$ and $Pb(DPM)_2$ were 115°C, 215°C, 200°C, 120°C and 135-145°C, respectively. These source materials were carried into the quartz reactor through a mixing tube with argon gas. The gas line and mixing tube were heated up to the temperatures higher than those of the vaporizers to prevent condensation of source materials. Oxygen gas was separately in-

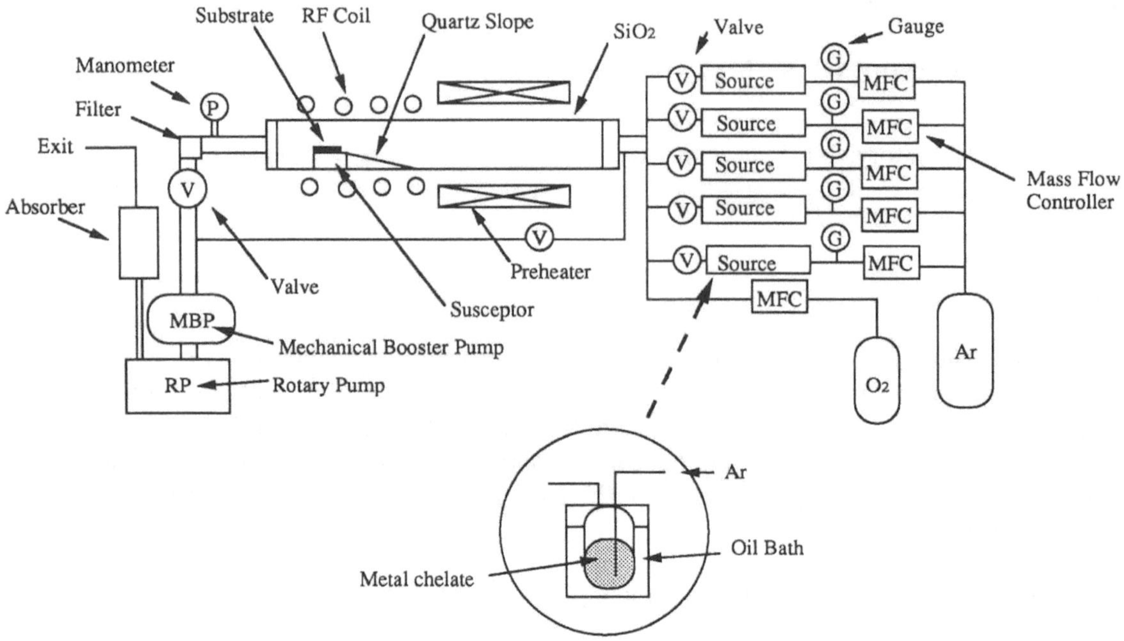

Fig.1 Schematic drawing of MOCVD apparatus.

phases; high-Tc phase, low-Tc phase and so-called 24Å phase ($Bi_2Sr_2Cu_1O_z$). In the figures, H (00n), L(00n) and L^*(00n) correspond to the indices of the high-Tc, the low-Tc and 24 Å phases, respectively. All the films grew along the c-axis oriented normal to the substrate surface. Figure 3 shows the temperature dependences of the electrical resistance of samples A, B and C. All the samples show the abrupt decrease of resistance at around 80-90K. Sample A does not show zero resistance. Samples B and C show zero resistance at around 40K.

At the low oxygen pressure(≈25 Torr), we grew two types of the films with and without Pb-doping, denoted samples E and D, respectively. The atomic ratio Pb/Bi in sample E is 0.06. Figures 4(a) and (b) show the XRD patterns of samples D and E, respectively. It is found that sample E is a mixture of the low-Tc and high-Tc phases, though the films grown without Pb-doping consist of the high-Tc single phase like sample D. In the figures, all the peaks were indexed as H(00n) and L(00n), which shows the preferred orientation of c-axis normal to the substrate surface. Samples D and E show zero resistance at around 67K and 33K, respectively, as shown in Fig. 5. For sample D, in particular, a superconducting transition occurred clearly at Tc(onset) of 110K. However, there were no effect that Pb-doping brings about the rise of Tc(zero resistance).

Figure 6 shows the SEM photograph of the surface of the sample E. The surface is very smooth, though small precipitates are seen on the surface. Electron probe microanalysis (EPMA) revealed that these precipitates are copper oxide.

Thus, Pb-doping has a strong effect on the condition of the oxygen partial pressure for the formation of the high-Tc and low-Tc phases. This result implies that Pb-doping affects the kinetics for the high-Tc phase formation, where oxygen plays an important role, rather than the stability of the crystal structure of the high-Tc phase. To make clear the role of Pb-doping in Bi-Sr-Ca-Cu-O system, we should have the knowledge on the formation mechanism of the high-Tc and low-Tc phases.

phases; high-Tc phase, low-Tc phase and so-called 24Å phase ($Bi_2Sr_2Cu_1O_z$). In the figures, H(00n), L(00n) and L^*(00n) correspond to the indices of the high-Tc, the low-Tc and 24 Å phases, respectively. All the films grew along the c-axis oriented normal to the substrate surface. Figure 3 shows the temperature dependences of the electrical resistance of samples A, B and C. All the samples show the abrupt decrease of resistance at around 80-90K. Sample A does not show zero resistance. Samples B and C show zero resistance at around 40K.

At the low oxygen pressure(≈25 Torr), we grew two types of the films with and without Pb-doping, denoted samples E and D, respectively. The atomic ratio Pb/Bi in sample E is 0.06. Figures 4(a) and (b) show the XRD patterns of samples D and E, respectively. It is found that sample E is a mixture of the low-Tc and high-Tc phases, though the films grown without Pb-doping consist of the high-Tc single phase like sample D. In the figures, all the peaks were indexed as H(00n) and L(00n), which shows the preferred orientation of c-axis normal to the substrate surface. Samples D and E show zero resistance at around 67K and 33K, respectively, as shown in Fig. 5. For sample D, in particular, a superconducting transition occurred clearly at Tc(onset) of 110K. However, there were no effect that Pb-doping brings about the rise of Tc (zero resistance).

Figure 6 shows the SEM photograph of the surface of the sample E. The surface is very smooth, though small precipitates are seen on the surface. Electron probe microanalysis (EPMA) revealed that these precipitates are copper oxide.

Thus, Pb-doping has a strong effect on the condition of the oxygen partial pressure for the formation of the high-Tc and low-Tc phases. This result implies that Pb-doping affects the kinetics for the high-Tc phase formation, where oxygen plays an important role, rather than the stability of the crystal structure of the high-Tc phase. To make clear the role of Pb-doping in Bi-Sr-Ca-Cu-O system, we should have the knowledge on the formation mechanism of the high-Tc and low-Tc phases.

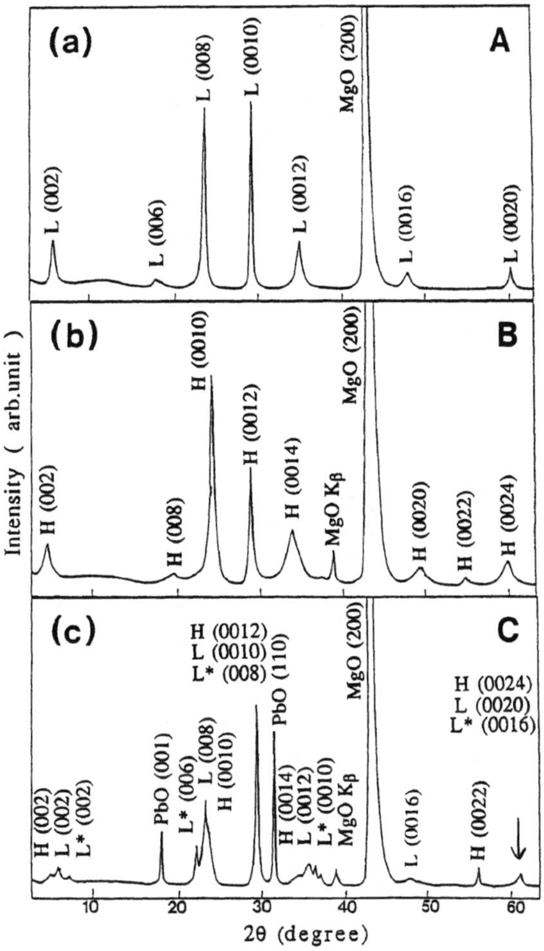

Fig.2 XRD patterns of the films (a) without Pb-doping (sample A), (b) lightly doped with Pb (sample B), and (c) heavily doped with Pb (sample C). H, L and L* denote the high-Tc, the low-Tc and so called 24Å phases, respectively.

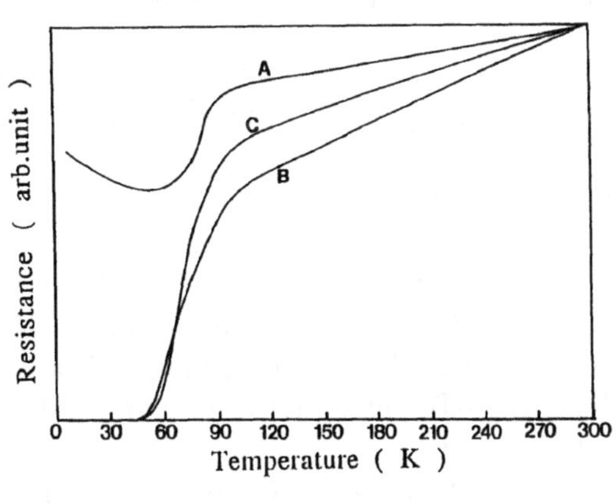

Fig.3 Temperature dependences of the electrical resistance of samples A, B and C.

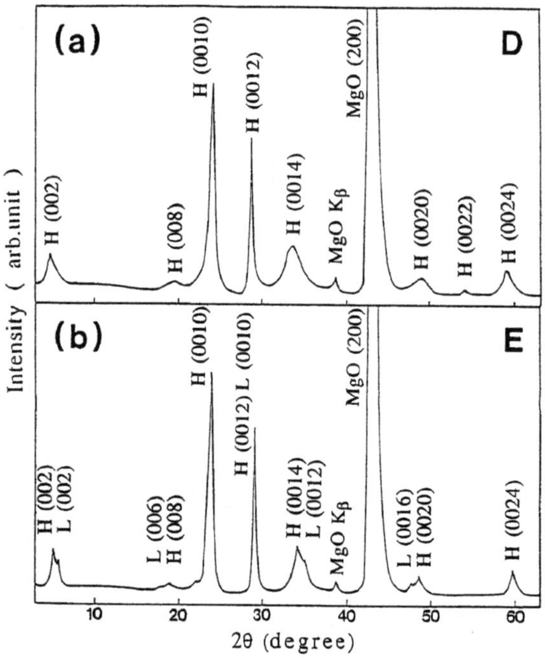

Fig.4 XRD patterns of the films (a) without and (b) with Pb-doping (denoted sample D and E, respectively).

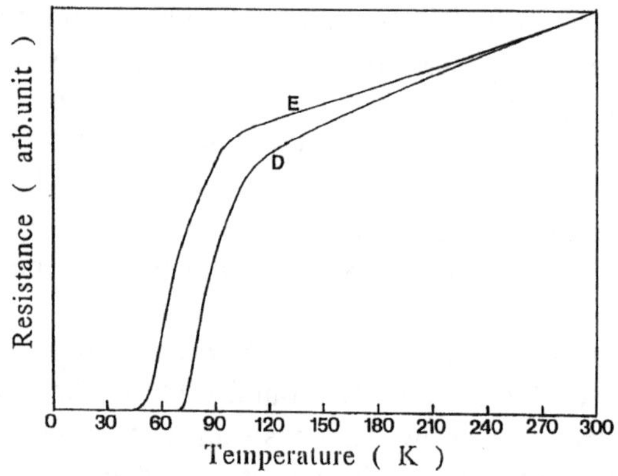

Fig.5 Temperature dependences of the electrical resistance of samples D and E.

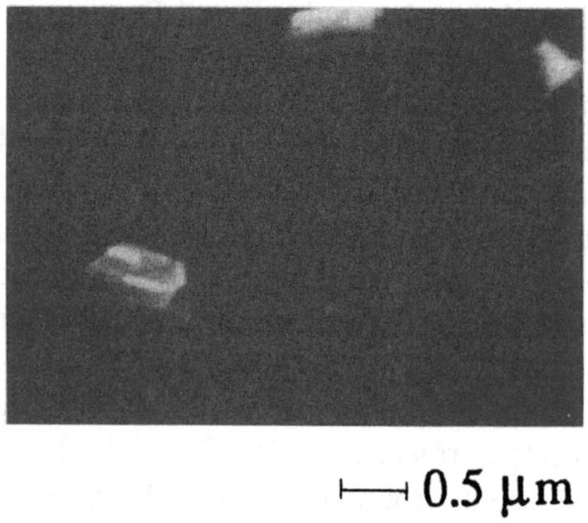

Fig.6 SEM photograph of the surface of an as-deposited film.

CONCLUSION

We have studied the effect of *in-situ* Pb-doping in MOCVD on the crystal structures and superconducting properties of Bi-Sr-Ca-Cu-O films. In the high oxygen pressure region (≈50 Torr), the films without Pb-doping consist of the low-Tc single phase. Lightly doping of Pb (Pb/Bi ≈0.03) brings about the high-Tc single phase, though Tc(zero resistance) = 42K. The films heavily doped with Pb (Pb/Bi ≈0.5) are mixed phase. On the other hand, in the low oxygen pressure region (≈25 Torr), Pb-doping causes the growth of the film with a mixed phase, whereas the high-Tc single phase is formed without Pb-doping.

On leave of absence from [a)]Nippon Sanso K.K., [b)]Iwatani Co.,LTD. [c)]Dowa Mining Co.,LTD. and [d)]Showa Electric Wire &Cable Co.,LTD..

REFERENCES

1. Maeda H. Tanaka Y. Fukutomi M. Asano T (1988) A new high-Tc oxide superconductor without a rare earth element. Jpn J Appl Phys 27: L209-L210

2. For instance, Takano M. Takada J. Oda K. Kitaguchi H. Miura Y. Ikeda Y. Tomii Y. Mazaki H. (1988) High-Tc phase promoted and stabilized in the Bi, Pb-Sr-Ca-Cu-O system. Jpn J Appl Phys 27: L1041-L1043

3. For instance, Dew SK. Osborne NR. Mulhern PJ. Parsons RR. (1988) Effects and loss of lead in doped Bi-Sr-Ca-Cu-O films. Appl Phys Lett 54: 1929-1931

4. Endo K. Hayashida S. Ishiai J. Matsuki Y. Ikedo Y. Misawa S. Yoshida S. (submitted) As-deposited Bi-Sr-Ca-Cu-O films with high Tc superconducting phase by MOCVD. Jpn J Appl Phys

Synthesis of Bi-Pb-Sr-Ca-Cu-O Thin Films by PbO Vapor Annealing

HIROTOSHI NAGATA[1], EUNGI MIN[1], SHOICHI HASHIGUCHI[1], KYOICHI SHIBUYA[1], AKIHIRO TAKANO[2], and HIDEOMI KOINUMA[2]

[1] Sumitomo Cement Co. Ltd., Central Research Laboratory, 585, Toyotomi-cho, Funabashi, Chiba, 274 Japan

[2] Tokyo Institute of Technology, 4259 Nagatsuta, Midori-ku, Yokohama, 227 Japan

ABSTRACT

Bi-Sr-Ca-Cu-O thin films were annealed in PbO atmosphere to get $Bi_2Sr_2Ca_2Cu_3O_x$ (2223), high Tc phase. By this annealing method, (2223) was successfully achieved. However, excess of PbO vapors causes decomposition of the superconductive phases.

KEY WORDS: Bi-Sr-Ca-Cu-O, film, PbO vapor, chemical reaction

INTRODUCTION

In Bi-Sr-Ca-Cu-O system, Pb ions act as effective dopant to yield $Bi_2Sr_2Ca_2Cu_3O_x$ (2223). However, thin film synthesis of this material with sputtering and post-deposition annealing has difficulty for the film doping with proper amounts of Pb ions, as Pb and PbO volatilize from the film during the synthesis process at elevated temperatures.[1,2] To overcome this difficulty, thin films were doped during annealing in the PbO atmospheres above 800°C.

Actual Pb amount of doped bulk ceramics was measured with an analytic transmission electron microscope.[3,4] This was estimated to be from 0.1 to 0.2 using ratio Pb/(Bi+Pb) with assumption of Pb substitution into Bi sites. We tried similar measurements on the Pb doped film to clear proper amount of Pb.

EXPERIMENTAL

Films of thickness 400nm, with no Pb contents, were deposited on <100>MgO, <100> and <110>$SrTiO_3$ substrates at ambient temperatures of about 200°C by plasma controlled magnetron sputtering method.[5,6] The film deposition was designed to iterate alternatingly a Bi-Sr-O rich layer and a Ca-Cu-O rich layer each 1nm thick.[7,8] Film composition was measured to be Bi:Sr:Ca:Cu=2.4:2.0:1.8:2.8 by ICP. Deposited films were annealed with 1.0g of PbO powder in a capped alumina boat at temperatures between 820 and 865°C for various durations with subsequent slow cooling in the furnace. Observations of the Pb doped annealed films were carried out with an Auger electron spectroscope (AES) and an electron probe micro-analyzer (EPMA) both having spatial resolutions of about 1μm.

RESULTS AND DISCUSSION

Films on MgO Substrates

Although some films annealed without PbO at 880°C showed clear drops of electrical resistivity around 110K, yet X-ray diffractions (XRD) of these films indicated no clear existence of (2223).[6] Use of PbO vapor during the annealing improved film crystallization and (2223) formation. Figure 1 exhibits the typical XRD patterns of films annealed without and with PbO vapor. Annealing at 820°C for 1h without PbO was observed to be insufficient to form even $Bi_2Sr_2Ca_1Cu_2O_y$ (2212). PbO vapors, however, made (2212) dominant in the film annealed at 820°C, as shown in Fig.1(b). Pb diffusion through the film after PbO vapor annealing was confirmed by AES depth measurements. Further annealing with PbO vapors at 860°C for 8h yielded (2223) among other superconductive phases, as shown in Fig.1(c). This film showed zero-resistivity at 103K at a current density of $0.28A/cm^2$, but zero-resistivity temperature decreased gradually when current was increased.

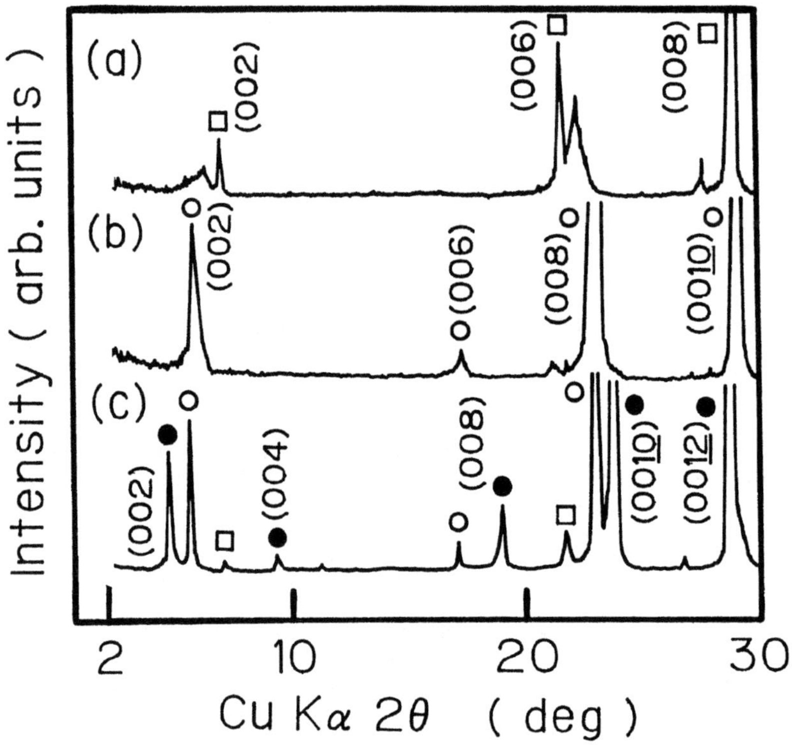

Fig.1 XRD patterns of films annealed (a): at 820°C for 1h without PbO, (b): with PbO, and (c): at 860°C for 8h with PbO. Closed circles are for (2223), open circles are for (2212) and open squares are for (2201).

In our experiments, (2223) could not be singled out by PbO vapor annealing, even though longer durations, up to 16h at temperatures between 820 and 865 °C were used. Longer annealing with PbO made film surface very rough and decreased the superconductive phases. This might be due to chemical reaction between Bi-Sr-Ca-Cu-O and excess PbO vapors. The similar experiment on PbO vapor annealing of bulk ceramics showed that Ca_2PbO_4 is more stable than superconductive phases at the annealing temperature, furthermore Pb ions diffuse into the bulk from excess PbO atmosphere to combine with Ca ions of (2223) and (2212). These reactions should also decompose superconductive phases in the film.

Figure 2 shows a back-scattered electron image of the film mentioned in Fig.1(c) and its composition as determined by EPMA. Brighter regions in Fig.2 might correspond to (2223) and (2212). At most measured points, atomic ratio of Pb to (Bi+Pb) were between 0.08 and 0.12. These values were close to that measured for bulk ceramics, 0.1[3] and 0.19[4].

To prevent decompositions of superconductive phases and to form (2223) single phase, accurate control of PbO vapors along with proper amounts of the dopant during the annealing will be necessary.

Films on $SrTiO_3$ Substrate

(2223) formation on $SrTiO_3$ by PbO vapor annealing did not succeed. This might be due to chemical reactions between the film and the $SrTiO_3$ substrate rather than PbO vapors. Contrary to MgO, in case of $SrTiO_3$ even (2212) could not be achieved well.

Depth profiles of secondary ion mass spectroscopy (SIMS) on the film annealed at 850 °C for 1h without PbO are shown in Fig.3. Large amounts of Ca from the film diffused into the substrate. Because of unhomogeneity of the film within the analyzed area, we could not judge difference of profiles between films on <100> and <110> substrates, and Ti diffusion onto film surface. XRD measurements on the annealed films showed formation of $Ca_4Ti_3O_{10}$, $Bi_2Ti_2O_7$ and Bi_2-Sr_2CuO_z (2201), as shown in Fig.4. These reactions robbed Ca ions preferentially from Bi-Sr-Ca-Cu-O films as well as the reaction with PbO, and ultimately decomposed superconductive phases into (2201).

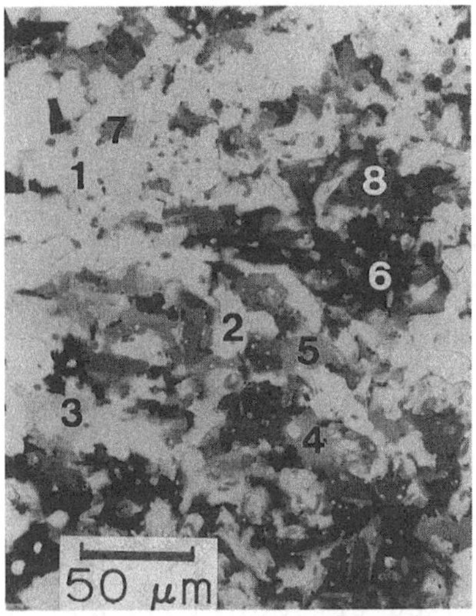

no.	($BiO_{1.5}$	PbO)	SrO	CaO	CuO
1	(0.91	0.09)	1.13	1.08	2.53
2	(0.88	0.12)	1.89	1.46	4.00
3	(0.92	0.07)	0.88	0.79	1.24
4	(0.90	0.10)	0.84	0.75	2.45
5	(0.90	0.10)	0.82	0.79	2.27
6	(0.93	0.07)	3.50	4.80	22.30
7	(0.94	0.06)	0.60	0.60	1.69
8	(0.91	0.09)	0.42	0.19	0.94

Fig.2 Compositional distribution image of the film annealed at 860 °C for 8h with PbO and results of point analysis on some grains by EPMA.

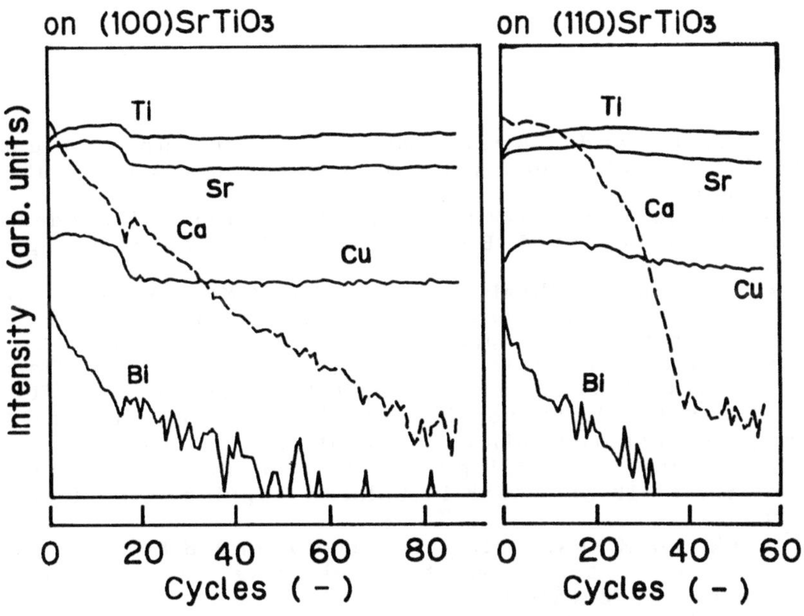

Fig.3 SIMS depth profiles of films on <100> and <110>$SrTiO_3$ after annealing at 850 °C for 1h without PbO. SIMS measurements was done with O_2^+ ion beam 8μm in diameter at 8kV and 1μA detecting $^{48}Ti^+$, $^{87}Sr^+$, $^{63}Cu^+$, $^{40}Ca^+$ and $^{209}Bi^+$ ions.

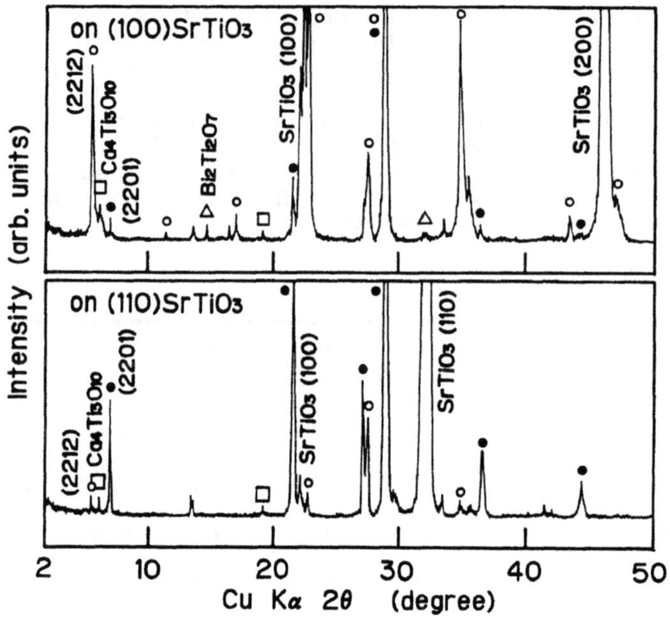

Fig.4 XRD patterns of films on <100> and <110>$SrTiO_3$ after annealing at 850 °C for 1h without PbO.

CONCLUSION

Annealing in PbO vapors at 860 °C for 8h improved crystallization and formation of (2223) in the film on MgO substrate. Amount of doped Pb were measured to be about Pb/(Bi+Pb)=0.1 by EPMA, if Pb substituted Bi sites. However, as annealing time was prolonged, superconductive phase decreased because of preferential chemical reactions between Ca and excess Pb ions into Ca_2PbO_4 and (2201). Similar reactions resulting into decomposition of superconductive phases occurred between the film and $SrTiO_3$ substrate during annealing. In this annealing method, precise control of PbO vapor amounts will be necessary to single out (2223).

REFERENCES

1. H.Masumoto, T.Goto, and T.Hirai (1989) As-grown superconducting Bi(-Pb)-Sr-Ca-Cu-O films by electron cyclotron resonance plasma sputtering. Appl. Phys. Lett. 55: 498-500
2. A.Tanaka, N.Kamehara, and K.Niwa (1989) Composition dependence of high Tc phase formation in Pb-doped Bi-Sr-Ca-Cu-O thin films. Appl. Phys. Lett. 55: 1252-1254
3. H.Nobumasa, T.Arima, K.Shimizu, Y.Otsuka, Y.Murata, and T.Kawai (1989) Observation of the high-Tc phase and determination of the Pb position in a Bi-Sr-Ca-Cu oxide superconductor. Jpn. J. Appl. Phys. 28: L57-L59
4. R.Ramesh, G.Thomas, S.M.Green, Y.Mei, and H.L.Luo (1988) Microstructure of Pb-modified Bi-Sr-Ca-Cu-O superconductor. Appl. Phys. Lett. 53: 1759-1761
5. H.Koinuma, H.Nagata, A.Takano, M.Kawasaki, and M.Yoshimoto (1989) Bi-Sr-Ca-Cu-O compositional modulation and multilayered film synthesis by plasma controlled sputtering. Proceedings of Conf. on Sci. and Tech. of Thin Film Supercondoctors. 14-18 Nov. 1988 in Colorado Spring, CO. Plenum Press, New York. pp205-213
6. H.Nagata, A.Takano, M.Kawasaki, M.Yoshimoto, and H.Koinuma (1989) Preparation of Bi-Sr-Ca-Cu-O thin films by sputtering under a variable magnetic field. J. Am. Ceram. Soc. 72: 680-683
7. A.Takano, M.Yoshimoto, H.Koinuma, and H.Nagata (1989) Annealing in a PbO atmosphere for high Tc superconductivity of Bi-Sr-Ca-Cu-O films. Appl. Phys. Lett. 55: 798-800
8. H.Nagata, S.Hashiguchi, and E.Min (in press 1989) Doping of Pb ions into Bi-Sr-Ca-Cu-O thin films from PbO vapor. J. Am. Ceram. Soc. 72

In-situ Formation of Bi-System Thin Films Formed by RF Magnetron Sputtering from Three Pb-doped Targets

KEN'ICHI KURODA[1], KAZUYOSHI KOJIMA[1], MASAMI TANIOKU[1], KAZUO YOKOYAMA[2], and KOICHI HAMANAKA[1]

Central Research Laboratory[1], and Materials and Electronic Device Laboratory[2], Mitsubishi Electric Corporation, Tsukaguchi, Amagasaki, Hyogo, 661 Japan

ABSTRACT

Superconducting Bi system thin films have been formed on MgO(100) substrates by RF magnetron sputtering from three Pb-doped targets: $Bi_{2.4}Pb_{0.6}Sr_2Ca_2Cu_3O_x$, $Bi_{1.6}Pb_{0.4}Sr_3Ca_3Cu_3O_x$ and $Bi_{1.6}Pb_{0.4}Sr_2Ca_2Cu_{4.5}O_x$. The as-grown film formed at 660°C did not contain Pb; the film showed a resistivity drop at 108 K and a zero resistivity at 72 K. These values are nearly equal to those of films formed from Pb-nondoped targets.
In situ annealing process at the same substrste temperature and gas pressure as the sputtering conditions improved the superconducting properties; a resistivity drop occurred at 115K and zero resistivity was obtained at 83K. The J_c value of the film was 4×10^5 A/cm^2 at 77 K and 3×10^7 A/cm^2 at 20 K.

KEY WORDS: Bi(Pb)SrCaCuO, superconductor, thin films, sputtering, as-grown superconductivity

INTRODUCTION

It is essential to obtain the superconductivity in as-grown state for applications in microelectronics, however, there have been few reports [1-4] that have obtained as-grown superconductivity of Bi system thin films. Moreover, in these reports, T_c values of the films were rather low compared with those of bulk samples or annealed films [5]. In our previous papers [6,7], we also reported the results of superconducting properties for as-grown BiSrCaCuO thin films. The films had onset temperature above 100 K, however, zero resistivity temperature was 75 K. The main reason why a zero resistivity temperature as high as that of bulk samples cannot be obtained lies in difficulty in forming a pure high-T_c phase of 110 K in the as-grown state. As is well known [8], the addition of Pb enhances the formation of the high-T_c phase. Thus, we attempted to form as-grown Bi-system thin films using Pb-doped targets. Furthermore, we carried out in situ annealing process to improve the superconductivity of the as-grown films, since it has been reported [3,4] that T_c of the film with the low-T_c phase increases due to low-temperature annealing.

EXPERIMENTAL

The Bi(Pb)SrCaCuO thin films were grown on MgO(100) substrates by RF magnetron sputtering from three targets: $Bi_{2.4}Pb_{0.6}Sr_2Ca_2Cu_3O_x$, $Bi_{1.6}Pb_{0.4}Sr_3Ca_3Cu_3O_x$ and $Bi_{1.6}Pb_{0.4}Sr_2Ca_2Cu_{4.5}O_x$. Each target had a 50 % excess of Bi+Pb, SrCa and Cu compared with the composition of the high-T_c phase (2223), and had a Pb/Bi value of 0.2. Figure 1 is a schematic illustration of the sputtering system with the multitarget. As shown in Fig.1, the three targets were simultaneously discharged and the substrate was rotated alternately over each target. The substrate was kept directly over each target for 10-90 seconds. The total of the three staying times was typically fixed at 100 s. It was possible to precisely control the chemical composition of the film by adjusting the staying time over each target. The substrates were heated at 540°C-670°C during the growth. The sputtering gas was pure O_2 and the sputtering pressure was 300 mTorr. Typical sputtering time was 100 min; an approximately 100nm-thick film was deposited. Within 1 min after deposition, films, except for a few samples, were quickly cooled down to room temperature by introducing air into the chamber. A few samples, after deposition, were kept at 660°C in O_2 atmosphere of 300 mTorr for 5 h without breaking the vacuum.

The temperature dependence of resistivity was measured by the AC four-probe method. After resistivity measurement, the surface morphology was observed by scanning electron microscopy (SEM),and then, X-Ray Diffraction (XRD) spectra were measured to study structural properties. Afterwards, the chemical composition of metallic elements in the film was determined by inductively coupled plasma emission spectroscopy (ICP).

RESULTS AND DISCUSSION

First, Bi(Pb)SrCaCuO thin films were formed by adjusting the staying time on each target under the above-mentioned sputtering conditions; the substrate temperature was fixed at 660°C. Zero resistivity temperature T_c(zero) of as-grown films was highest when the staying time of the substrate was 80 s on the BiPb-rich target, 10 s on the SrCa-rich target and 10 s on the Cu-rich target. Figure 2 shows the temperature dependence of normalized resistivity of the best superconducting as-grown film. As shown in the figure, resistivity of the film dropped at 108 K and zero resistivity was obtained at 72 K. The superconducting properties were nearly equal to those of films prepared by Pb-nondoped targets [6,7]. From ICP analysis, It was found that the film composition was Bi:Pb:Sr:Ca:Cu=2.0:0.0:1.8:1.8:2.6. That is, no trace of Pb was detected in the film formed at 660°C.
Figure 3 shows the surface morphology of the as-grown film shown in Fig.2. There is a flat surface similar to those of the films by Pb-nondoped targets [6]. XRD measurement showed that the film has a c-axis length of 3.4nm. We speculate [7] that the film has ordered intergrowth structure of the low-T_c phase and high-T_c phase.
Second, as-grown films were formed by varying the substrate temperature (T_{sub}); the staying times over each target were fixed at the values mentioned above. Figure 4 shows T_{sub} dependence of T_c; we defined the onset temperature T_c(onset) as the temperature of resistivity dropping. As shown in Fig.4, both T_c(onset) and T_c(zero) become lower a decrease in T_{sub} below 650°C, and films do not show superconductivity below 600°C. From XRD measurement, the c-axis lattice constant also becomes shorter with a decrease in T_{sub} below 650°C. As-grown films formed at 640°C-670°C are considered to have the ordered intergrowth phase with a c-axis length of 3.4nm and films at 600-620°C have the low-T_c phase with a c-axis length of 3.1nm. A film formed at 580°C has both phases with c-axis length of 3.1nm and 2.4nm. These experimental results correspond to those in Fig.4; for example, the films with the 3.4nm phase have T_c(onset) above 100 K and the films with the 3.1nm phase have T_c(onset) below 100 K. From these experimental results, it was concluded that T_{sub} above 640°C is necessary to form the superconducting phase containing the high-T_c phase.
The T_{sub} dependence of the composition was examined by ICP analysis. The value of 2Sr/Bi ranged from 1.75 to 1.80 and that of 2Cu/Bi ranged from 2.45 to 2.65, when T_{sub} was varied from 540°C to 670°C; however, the 2Ca/Bi value varied widely from 1.55 to 1.95 due to measurement errors [9] of the ICP analysis. Pb was detected in the film when the T_{sub} was lower than 600°C.
Lastly, a few films were in situ annealed to improve the superconductivity of the as-grown films. A film, which had been formed at 580°C, was kept at 660°C in O_2 atomosphere of 300 mTorr for 5 h after deposition. This annealed film showed T_c(onset) of 90 K and T_c(zero) of 60 K and had a low-T_c phase, while the as-grown film formed under the same sputtering conditions did not show superconductivity. The annealed film also contained Pb, though Pb decreased in quantity by 60 % compared with the as-grown film. The experimental result indicates that the high-T_c phase was not formed under this in situ annealing conditions in spite of the existence of Pb.
Figure 5 shows the temperature dependence of resistivity of the film formed at 660°C and annealed under the same condition as the above sample; that is, after deposition, the film was kept at the same substrate temperature and gas pressure as the sputtering conditions. The film has T_c(onset) of 115 K and T_c(zero) of 83 K. The critical current density of the film was 4 x $10^5 A/cm^2$ at 77 K and 3 x $10^7 A/cm^2$ at 20 K. It is concluded from these results that the in situ annealing process improves superconductivity. However, XRD patterns of the film showed the annealed film to have a 3.4nm phase, not distinguishably different from those of the as-grown film. That is, this process does not clearly enhance the formation of the high-T_c phase.
Figure 6 is an SEM micrograph of the surface of the in situ annealed film

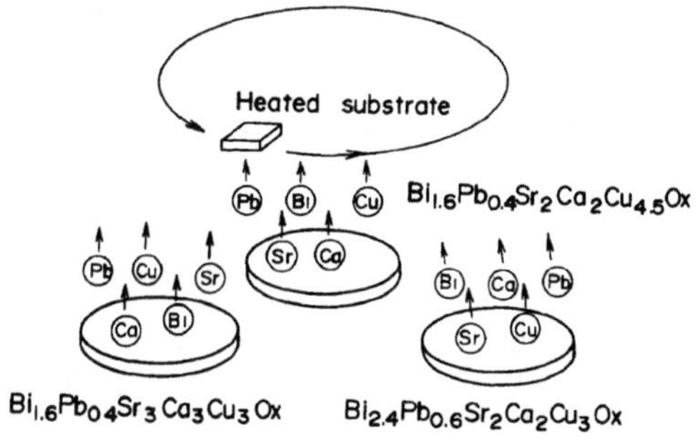

Fig.1. Schematic illustration of the sputtering system.

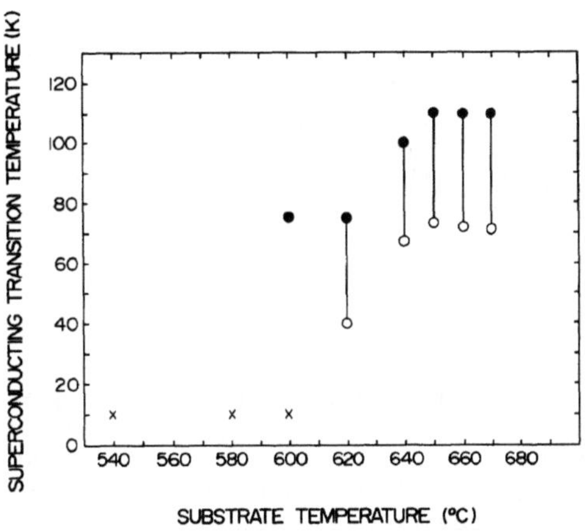

Fig.4. Relationship between T_c and substrate temperature. Closed and open circle indicate onset temperature and zero resistivity temperature, respectively. Cross indicates that film is non-superconductive down to 10 K.

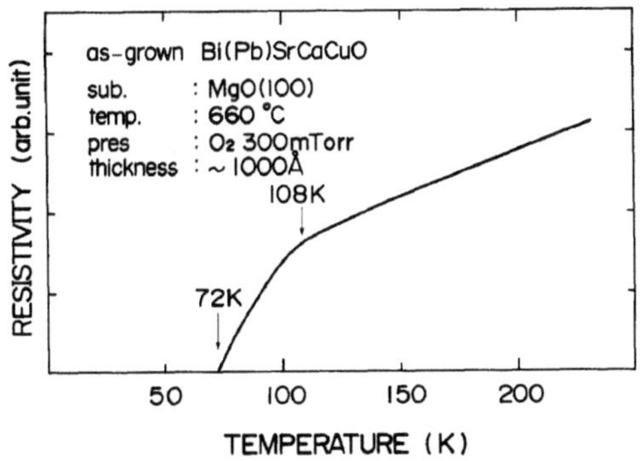

Fig.2. Resistivity vs. temperature curve of as-grown Bi system film with optimum composition.

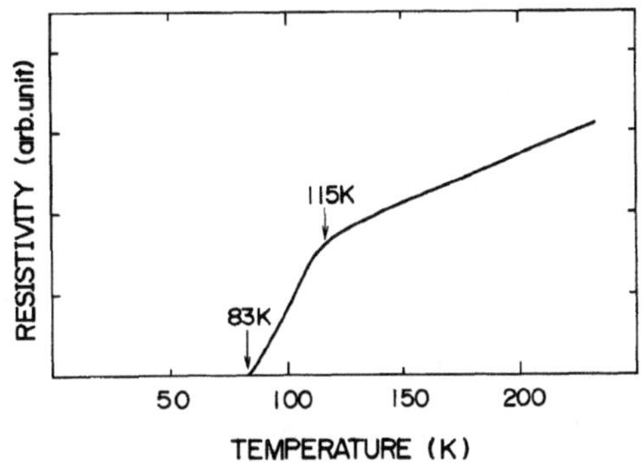

Fig.5. Resistivity vs. temperature curve of in situ annealed film; the film was kept at same substrate temperature and gas pressure as sputtering condition for 5h after deposition.

Fig.3. SEM micrograph of as-grown film surface. Speck of dust was taken at corner of picture to show sharpness of focusing.

Fig.6. SEM micrograph of surface of in situ annealed film shown in Fig.5.

shown in Fig.5. The surface of the film is not as smooth as that of the as-grown film shown in Fig.3. From the ICP analysis of the in situ annealed film, it was found that Bi,Sr,Ca and Cu respectively decreased in quantity compared with the as-grown film by 30 %, 15 %, 10 % and 20 %. The metal elements in the film were probably re-evaporated under the in-situ annealing. The decrease in metal elements is considered one reason for the degradation of the surface morphology.

CONCLUSIONS

To form a pure high-T_c phase, as-grown Bi system thin films were prepared on the MgO (100) substrates by RF magnetron sputtering from three Pb-doped targets. Nevertheless, T_c values of the films were nearly equal to those of films from Pb-nondoped targets. The reason why the improvement of the superconductivity cannot be obtained is that the films from Pb-doped targets do not contain Pb elements at the substrate temperature above 600°C. From these results, a decrease in the crystallization temperature is considered necessary for forming a pure high-T_c phase because of preserving Pb. Moreover, we carried out in situ annealing of as-grown films. As a result, it was found that the superconductivity of the films was fairly improved by this process; a T_c(onset) at 115 K, T_c(zero) at 83 K and J_c of $4x10^5$ A/cm^2 at 77 K were obtained.

REFERENCES

1. M.Fukutomi, J.Machida, Y.Tanaka, T.Asano, T.Yamamoto and H.Maeda (1988) New technique for preparation of BiSrCaCuO thin films with T_c of 100K and above. Jpn J Appl Phys 27: L1484-1486

2. H.Asano, M.Asahi, K.Katoh and O.Michikami (1988) Low-temperature growth of high-T_c Bi-Sr-Ca-Cu-O films by magnetron sputtering. Jpn J Appl Phys 27: L1487-1488

3. S.Miura, T.Yoshitake, T.Manako, Y.Miyasaka, N.Shohata and T.Satoh (1989) Low-temperature annealing effect of as-grown Bi-Sr-Ca-Cu-O thin films on superconducting property and the c-axis lattice constant. Appl Phys Lett 55: 1360-1362

4. T.Satoh, T.Yoshitake, S.Miura, J.Fujita, Y.Kubo and H.Igarashi (1989) As-grown superconducting Bi-Sr-Ca-Cu-O thin films by coevaporation. Appl Phys Lett 55: 702-704

5. A.Tanaka, N.Kamehara and K.Niwa (1989) Composition dependence of high-T_c phase formation in Pb-doped Bi-Sr-Ca-Cu-O thin films. Appl Phys Lett 55: 1252-1254

6. K.Kojima, K.Kuroda, M.Tanioku and K.Hamanaka (1989) As-grown superconductivity of BiSrCaCuO thin films prepared by magnetron sputtering with three targets. Jpn J Appl Phys 28: L643-645

7. K.Kuroda, K.Kojima, M.Tanioku, K.Yokoyama and K.Hamanaka (1989) Surface morphology and crystal structures of as-grown BiSrCaCuO thin films prepared by magnetron sputtering from three targets. Jpn J Appl Phys 28: 1586-1592

8. M.Takano, J.Takada, K.Oda, H.Kitaguchi, Y.Miura, Y.Ikeda, Y.Tomii and H.Mazaki (1989) High-T_c phase promoted and stabilized in Bi, Pb-Sr-Ca-Cu-O system. Jpn J Appl Phys 27: L1041-1043

A New Fabrication Method of Monolithic Lateral S-I-S Structure from Amorphous Bi-O/Sr-Ca-Cu-O Layer

T. Usuki, I. Yasui, Y. Yoshisato, and S. Nakano

Functional Materials Research Center, Sanyo Electric Co., Ltd., 18-13, Hashiridani 1-chome, Hirakata, Osaka, 573 Japan

ABSTRACT

A superconducting Bi-Sr-Ca-Cu-O (BSCCO) area and a nonsuperconducting Sr-Ca-Cu-O (SCCO) area can be obtained on the same surface of the substrate by controlling compositions for the first time. Patterned Bi-O films were deposited by an electron beam evaporation method on as-sputtered SCCO films. Annealed at 870°C, Bi-O films were diffused into SCCO films to be crystallized to the BSCCO. The BSCCO area, with a 150 nm thickness, indicates that the highest zero resistance temperature is 109 K. The SCCO area shows semiconductor-like temperature dependence of resistivity.

Key words: BiSrCaCuO high-Tc phase, SrCaCuO, Bi content, annealing, lateral isolation

INTRODUCTION

New high-Tc oxide superconductors [1-3] ,whose Tc's are higher than the temperature of liquid nitrogen, have had a great impact on superconductive devices. The electronic devices of superconductors, especially functional devices, need junction structures of superconductors (S), semiconductors (SE), and insulators (I). Monolithic devices are required for junction structures not only for vertical direction but also for lateral direction as in many semiconductor devices .

The electrical characteristics are changed from those of superconductors to insulator or semiconductor with little modification of the same compositions themselves in oxide superconductors. Several kinds of junction structures of S-I-S, S-N-S and S-SE are proposed using these characteristics. There are some experimental reports on ion implantation in thin films [4] and local oxidation of oxygen deficient films of the Y-system [5] and so on. However, adequate characteristics of the junctions have not been achieved. The lateral S-I-S structure using different phases, such as the superconducting and semiconducting phases, is effective, because the coherent lengths of the high-Tc superconductor along the a-axis and b-axis are longer than those along the c-axis.

This paper reports on a new fabrication method of monolithic lateral S-I-S structure to maintain c-axis orientation. This method uses the crystallization process of oxide films of the Bi-system [6]. Local isolation between superconductors and nonsuperconductors is achieved by Bi compensation for Sr-Ca-Cu-O films during crystallization. Our method is characterized by the isolation of the same elements as superconductors.

EXPERIMENTAL

Sr-Ca-Cu-O (SCCO) films were prepared on (100) MgO single-crystal substrates by magnetron sputtering. The target is $SrCaCu_2O_x$ compounds. We sputtered at lower temperatures to examine crystal growth after annealing. Sputtering conditions are shown in Table 1. As-sputtered films were amorphous and had high resistivities.

Bi-O film was deposited on as-sputtered SCCO films on MgO substrates by electron beam evaporation. The source material is Bi_2O_3. Deposition rate is uniformly 10 Å/s by the feed back system of a quartz thickness monitor. Bi-O films were patterned by a lift-off process with photoresist or by a metal mask deposition method.

Table 1 Sputtering condition for SCCO films

target	$Sr_1Ca_1Cu_2O_x$ ceramics ϕ100
power	150W
gas	$Ar:O_2=1:1$
gas pressure	~0.4 Pa
substrate	MgO(100)
substrate temp.	280 °C
thickness	100-600nm

Figure 1 shows the schematic diagram of the lateral-isolation method for BSCCO films. Typical thickness of the SCCO film and the Bi-O film is 100 nm and 70 nm, respectively. SCCO films and SCCO/Bi-O films were annealed for 3 hours at 870 °C in air.

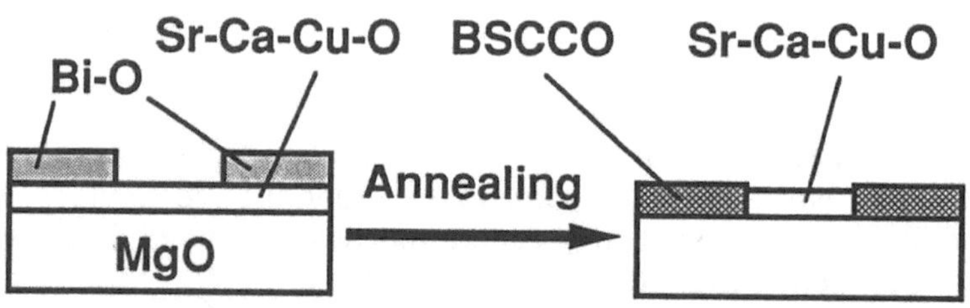

Fig.1 Schematic diagram of the fabrication method for lateral S-I-S structures.

The composition of the films was determined using ICP analysis. Depth profile of the films was measured by XPS. Surfaces of the films were observed by SEM. Film structures were determined and identified by X-ray diffraction measurements using CuKα radiation. Resistivity measurements as a function of temperature for the films were made at a constant current in a standard four-prove configuration.

RESULTS AND DISCUSSION

According to ICP analysis, the composition of SCCO films and BSCCO films were Sr:Ca:Cu=2.0:1.6:3.8 and Bi:Sr:Ca:Cu=2.3:2.1:1.2:3.3, respectively. Bi composition in the BSCCO films was nearly the same as values calculated by the thickness of Bi-O films. A change of composition through heat treatment was scarcely seen in either film. XPS analysis shows that as-deposited Bi-O films were mainly composed of Bi-oxide, and contained slight Bi-metal. Depth profile of the films composition by XPS analysis shows that Bi-O films were diffused uniformly into SCCO films.

Both as-sputtered SCCO films and SCCO/Bi-O films were almost amorphous phase, according to X-ray diffraction analysis. The films were annealed to be crystallized. Figure 2 and Fig.3 show the X-ray diffraction patterns of BSCCO films. Figure 2 is for the sample which shows patterns of Bi-system low-Tc phase. Figure 3 is for the sample of two phase, low-Tc and high-Tc. The patterns and peak heights of Fig.3 show that BSCCO films were largely composed of the high-Tc phase.

In Fig.4(a) and Fig.4(b) are the SEM photographs of the surface of the SCCO films and the BSCCO films, respectively. The surface of SCCO films seems to have multi-phase with granular-like grains. On the other hand, BSCCO films seem to grow to large plate-like grains. The grain size is larger than 6 x 6 μm^2 in BSCCO films. BSCCO films seem to partially melt during annealing, because Bi rich layers play an important role, like self-flux. It has been reported that the volume fraction of high-Tc phase in the BSCCO can be increased by partially substituting Pb for Bi [7]. The exact role of such reactions in the absence of Pb require further investigation of various compositions and thicknesses of films, and annealing atmospheres.

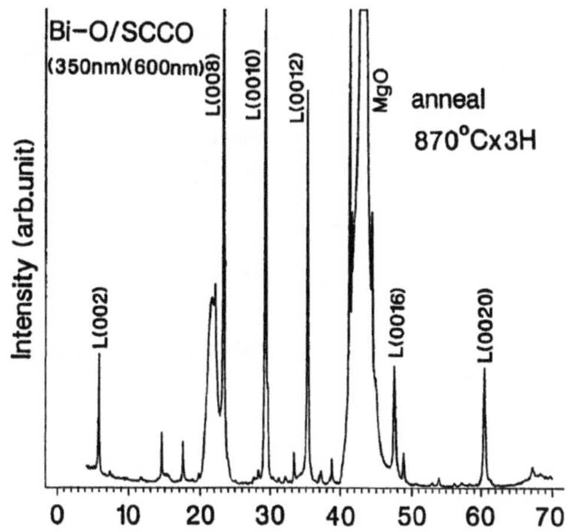

Fig.2 X-ray diffraction patterns of BSCCO films which were made from Bi-O films (350 nm) and SCCO films (600 nm).

Temperature dependence of resistivity for the SCCO area and the BSCCO area are shown in Fig.5 and Fig.6, respectively. The resistivity of the SCCO area shows nonsuperconducting and semiconductor-like temperature dependence. Figure 6 indicates that the zero resistance temperature of BSCCO films is 109 K. In spite of the absence of Pb, high-Tc at 109 K was shown for the sample whose thickness was 150 nm.

A lateral BSCCO/SCCO/BSCCO structure was obtained by using patterned Bi-O films. We have been trying to obtain SIS or SNS characteristics by preparing fine patterned Bi-O films and the adequate annealing conditions.

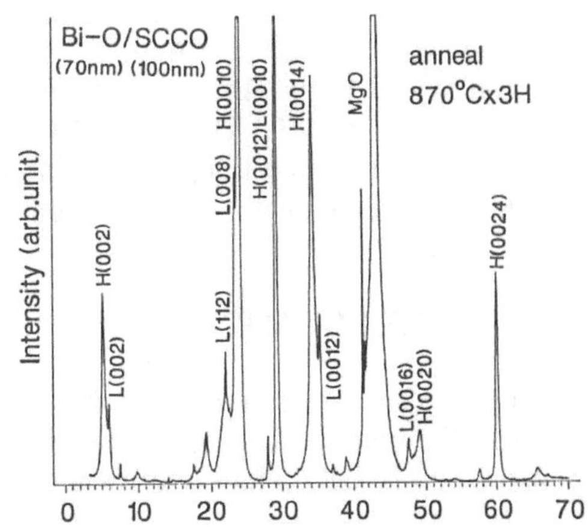

Fig.3 X-ray diffraction patterns of BSCCO films which were made from Bi-O films (70 nm) and SCCO films (100 nm).

(a) SCCO films

(b) BSCCO films.

Fig.4 SEM photographs of the surfaces of the films

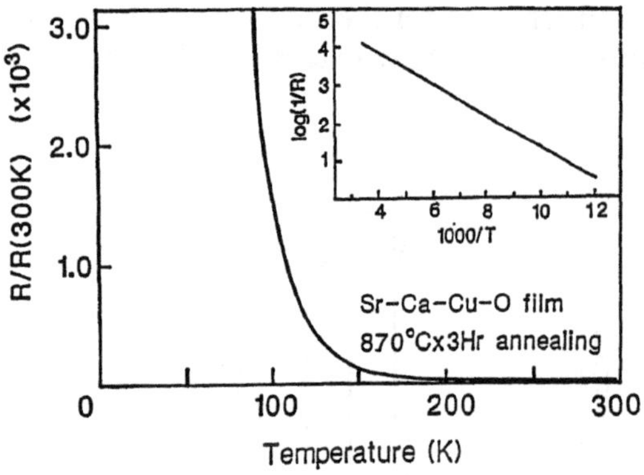

Fig.5 Temperature dependence of resistivity for SCCO films.

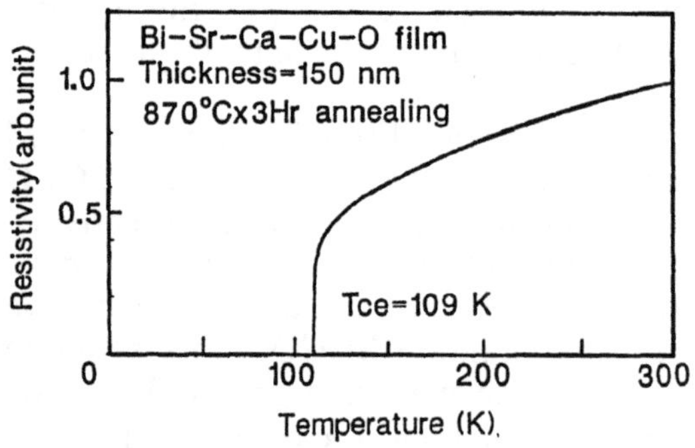

Fig.6 Temperature dependence of resistivity for BSCCO films.

CONCLUSION

We reported on a new fabrication method of monolithic lateral S-I-S structures to maintain c-axis orientation. A superconducting BSCCO area and a nonsuperconducting SCCO area can be obtained on the same surface of the substrate by controlling compositions. Patterned Bi-O films were deposited by an electron beam evaporation method on as-sputtered SCCO films. Annealed at 870°C, Bi-O films were diffused into SCCO films to be crystallized to the BSCCO. The BSCCO area, with a 150 nm thickness, indicates that the highest zero resistance temperature is 109 K, in spite of the absence of Pb. The SCCO area shows semiconductor-like temperature dependence of resistivity.

ACKNOWLEDGEMENTS

This work was performed under the management of the R & D Association for Future Electron Devices as a part of the R & D of Basic Technology for Future Industries sponsored by NEDO (New Energy and Industrial Technology Development Organization).

REFERENCES

[1] M. K. Wu, J. R. Ashburn, C. J. Trong, P. H. Hor, R. L. Meng, L. Gao, Z. J. Huang, Y. Q. Meng and C. W. Chu : Phys. Rev. Lett., **58**, 908 (1987)
[2] H. Maeda, Y. Tanaka, M. Fukutomi and T. Asano : Jpn. J. Appl. Phys., **27**, L209 (1988)
[3] Z. Z. Sheng and A. M. Herman : Nature, **332**, 55 (1988)
[4] R. Yuasa, M. Nakao, A. Chayahara and M. Sato : Ext. Abs. of 6th Int. Ws. on Future Electron Devices, 145 (1989)
[5] A. Yoshida, H. Tamura, K. Gotoh, N. Fujimaki and S. Hasuo : Proc. of ISEC'89, Tokyo, 516 (1989)
[6] T. Usuki, I. Yasui, H. Suzuki, Y. Chikano, F. Konishi, Y. Yoshisato and S. Nakano : IEICE Technical Report SCE **89-16**, 29 (1989)
[7] M. Takano, J. Takada, K. Oda, H. Kitaguchi, Y. Miura, Y. Ikeda, Y. Tomii and H. Mazaki : Jpn. J. Appl. Phys., **27**, L1041 (1988)

Growth and Property of High-T_c Phase in Bi(Pb)-Sr-Ca-Cu-O Sputtered Films

K. MAEDA[1], T. KITAMURA[1], H. KOBAYASHI[1], T. HASEGAWA[1], T. SHIONO[1], M. KATO[2], H. YAMAMOTO[2], and M. TANAKA[2]

[1] Chemical Material R & D Department, Showa Electric Wire & Cable Co., Ltd., 2-1, Odasakae, Kawasaki-ku, Kawasaki, Kanagawa, 210 Japan

[2] College of Science & Technology, Nihon University, 7-24, Narashinodai, Funabashi, Chiba, 274 Japan

ABSTRACT

The purpose of this work is to investigate annealing process or conditions for Bi(Pb)-Sr-Ca-Cu-O sputtered films in order to promote the growth of the high-Tc phase. The composition of the specimen film was adjusted to the stoichiometry of the high-Tc phase (the atomic ratio of Bi:Pb:Sr:Ca:Cu = 1:0.2:1:1:1.5)by changing the composition of the target and sputtering conditions.

A new method, in which a MgO plate was put on the film surface, was developed to keep the concentration of Pb in the film during the anneal. The optimum annealing temperature was 830 ~ 850℃ . A volume fraction of the high-Tc phase was attained above 70% in superconducting phases with good reproducibility. A preferential growth of c-axis perpendicular to the film was observed.

As a result, Tc of zero resistance of 105K and a critical current density in excess of 3×10^4 A/cm² were achieved.

1. INTRODUCTION

A new superconducting Bi-Sr-Ca-Cu-O system[1)] was discovered after the discovery of Y-Ba-Cu-O system[2)] , of which critical temperature Tc is above the boiling temperature of liquid nitrogen. This system was found to have three superconducting phases in the formula Bi Sr Ca_{n-1} Cu_n Ox:Tc=10K for n=1(semiconducting phase), 85K for n=2(low-Tc phase), 110K for n=3(high-Tc phase) . [3)] So it is very difficult to get a single high-Tc phase . Takano et al.[4)] reported that a small amount of Pb of about 10% to the Bi concentration effectively stabilize the high-Tc phase, though strict conditions for sintering are needed. Furthermore, Kawai et al. [5)] suggested that the high-Tc phase can be obtained with comparatively ease at low oxygen partial pressure.

It is, however, difficult in a thin film process to utilize efficiently the addition of Pb, because Pb is evapolated easily. A method with annealing of a film in a closed Pt crucible together with a sintered bulk which supply enough constituent atoms in vapor[6)], and a method with increasing Pb contentent in a film were reported in order to compensate Pb evaporated. [7)] Surely these methods make the high-Tc phase stable, but they need to prepare sintered bulks and much time annealing. The purposes of this work are to newly develop a simple heat treatment technique peculiar to thin films and to investigate the properties of the films obtained.

2. EXPERIMENTAL METHOD

A facing targets sputtering was adopted to prepare specimen films. The targets materials were Bi_2O_3, PbO, $SrCO_3$, $CaCO_3$ and CuO powder(>3N). The powder of the mixture of Bi:Pb:Sr:Ca:Cu = 1.2: 0.2:1.0:1.0:1.6(mol ratio)was calcined at 800 ℃ for 24 hrs in air. The disk type of targets were made by sintering of the calcined powder at 860℃ for 8 hrs in air. Sputtering conditions were shown in Table.1. A schematic foundation of specimen films during an anneal is shown in

Fig.1. The film surface was simply covered by a MgO plate to suppress a change of the film composition during anneal. This method will be named as "hermetic anneal". The anneal was done in a furnace heated by infrared rays. The temperature was raised in the rate of 60℃/min and cooled down in the rate of 5℃/min. The atmosphere was a mixture gas of oxygen and nitrogen. The partial pressure of oxygen was 1/13 atm. The film composition was measured by ICP. The structure of the film was analized by XRD method. The observation of morphology and the ultimate analysis of the film were done by SEM and EDX, respectively. Tc and critical current density Jc were measured by a four probe method using electrodes of Ag films. The thickness of the film was measured by DEKTAK 3030.

Table1. Sputtering Conditions

substrate	MgO(100)
rf power	1.5 W/cm²
sputtered gas	$Ar:O_2$ =1:1
gas pressure	3 mTorr
substrate temperature	400 ℃
film thickness	about 4000Å
deposition time	6 hrs

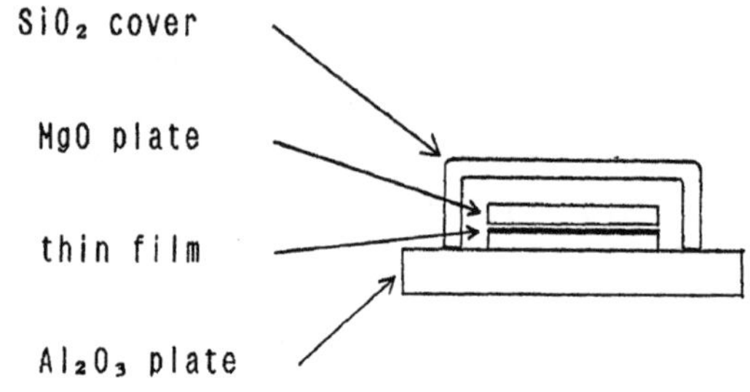

Fig.1 Schematic foundation of specimen films during an anneal.

3. RESULTS AND DISCUSSION

3.1 Composition and crystal structure

The growth of the high-Tc phase is enhanced by addition of Pb and the partial replacement of Pb into the Bi-site is observed.[8),9)] The concentration of Pb in films before and after a conventional anneal decreased to approximately zero at about 800℃, while the hermetic anneal resulted in no change of the Pb content even after 860℃ anneal. Typical XRD patterns of the films by the hermetic anneal are shown in Fig.2. The low-Tc phase is a dominant superconducting phase in the film annealed for 15min. As annealing time increased, the x-ray intensity of the high-Tc phase become larger than that of the low-Tc phase. In all these films, a preferential growth of c-axis perpendicular to the film plane was observed. Since the a-b plane with superior superconductivity is pararelled to the substrate, the film has advantage to get a high Jc. In Fig.3, the volume fraction of the high-Tc phase (high-Tc ratio) estimated from the XRD peak ratio is shown as a function of annealing time. The high-Tc ratio was defined by the ratio of the x-ray (002)peak intensity of the high-Tc phase to that of all superconducting phases. At 830 ~ 850℃, the value of the high-Tc ratio increased with increasing annealing time. The maximum value of the high-Tc ratio exceeded 70%. At 860 ℃ the high-Tc ratio was attained to the maximum for 30min and deteriorated for 45min. The result indicates that the optimum annealing temperature is quite narrow, about several tens degree. In the temperature of 860℃, the degradation of the high-Tc phase overcomes the rate of synthesis because of occurrence of partial melt. It is concluded that the hermetic anneal contributed effectively to the growth of the high-Tc phase because Pb was sufficiently diffused in the thin film.

3.2 Microstructure

Figure 4 shows the SEM photographs of the film surfaces before and after annealing. The surface morphology of the annealed film was like a pile of scales, though the surface of the as-deposited film was very smooth. From the ultimate analysis of the film annealed for 30min, Ca-Cu oxide like a needle at(A) and Sr-Ca-Cu oxide like a needle at(B) were observed. In the case of annealing time of 45min, Ca-Cu oxide in the film decreased. This result shows that the high-Tc phase is synthesized by the reaction between the low-Tc phase and Ca-Cu oxide.

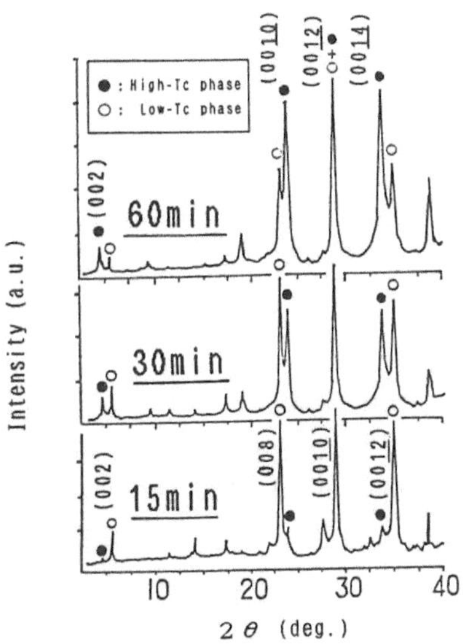

Fig.2. Change of x-ray diffraction patterns of films by annealing time at annealing temperature of 840 ℃.

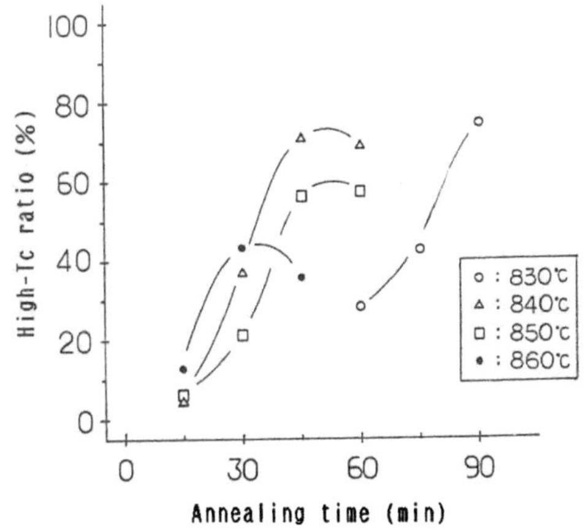

Fig.3. Relationship between annealing condition and volume fraction of high-Tc phase.

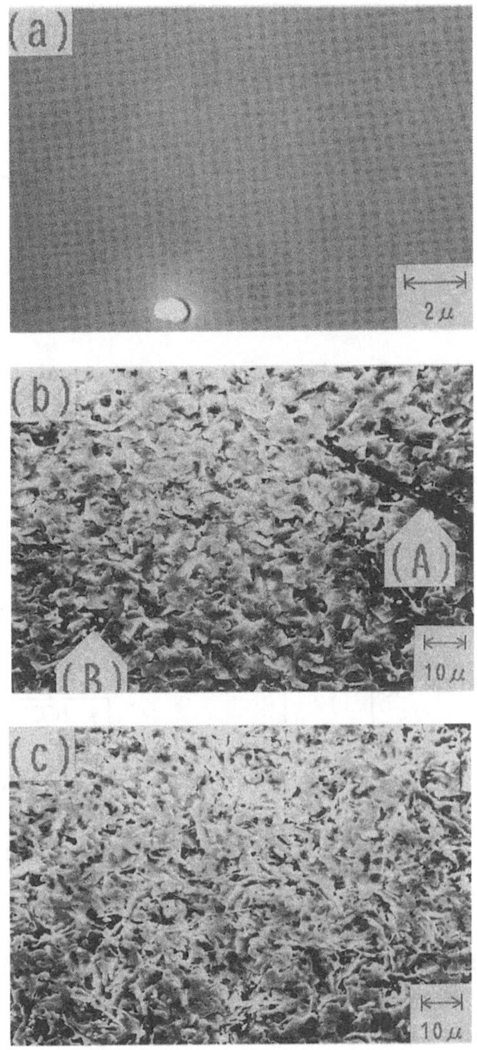

Fig.4. SEM photographs of specimen films before and after anneal;(a)before anneal,(b) after anneal at 840 ℃ for 30 min,(c) at 840 ℃ for 45 min.

3.3 Electric conduction

Figure 5 shows the typical temperature dependences of resistivity of the annealed films. The film annealed for 15min revealed Tc of zero resistance Tc(end) of 67K accompanying with a shoulder of resistivity. As increasing annealing time,Tc(end) rose up 102K in the film annealed for 30min and 103K for 60min. The maximum Tc(end) attained up 105K in the film annealed at 830 ℃ for 90min. The relationship between Tc(end) and the high-Tc ratio is shown in Fig.6. When the high-Tc ratio was above about 20%,Tc(end) exceeded 100K. The optimum result was obtained by the annealing condition;above 60min at 830 ℃ and above 30min at 840℃ and 850 ℃. This result corresponds to the result of the high-Tc ratio shown in Fig.3.
The value of Jc was determined by the current density at which a significant voltage appeared, like as the case shown in Fig.7. Figure 8 shows the relationship between the high-Tc ratio and observed Jc's. The observed Jc increased exponentially as increasing the high-Tc ratio, probably because the superconducting percolation abruptly increased. The maximum Jc of 3.1×10^4 A/cm² was obtained. Hereafter ,the improvement of Jc will be expected by increasing of the high-Tc ratio and decreasing of non-superconducting phases.

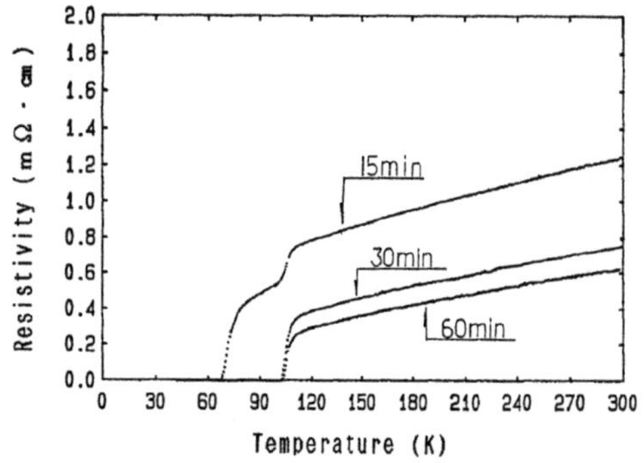

Fig.5. Resistivity vs. temperature of films annealed at 840℃.

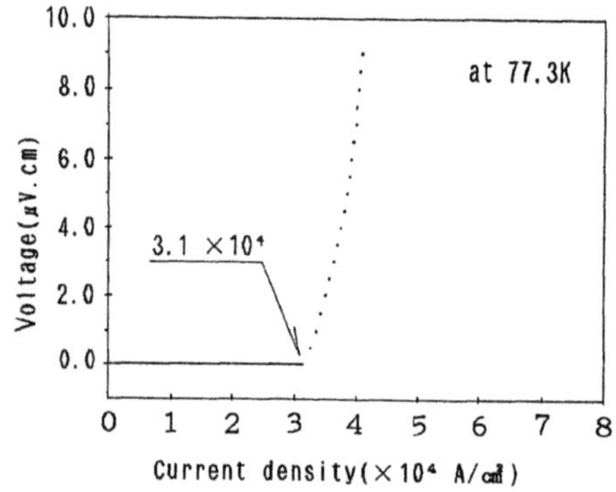

Fig.7. J-V characteristic in the film annealed at 850℃ for 60min.

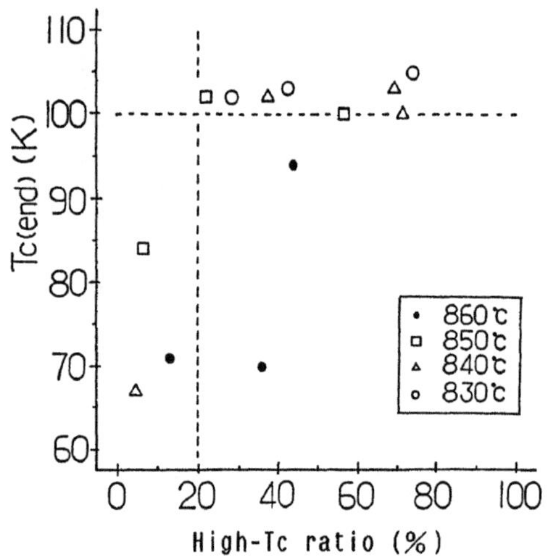

Fig.6. Relationship between Tc(end) and volume fraction of high-Tc phase

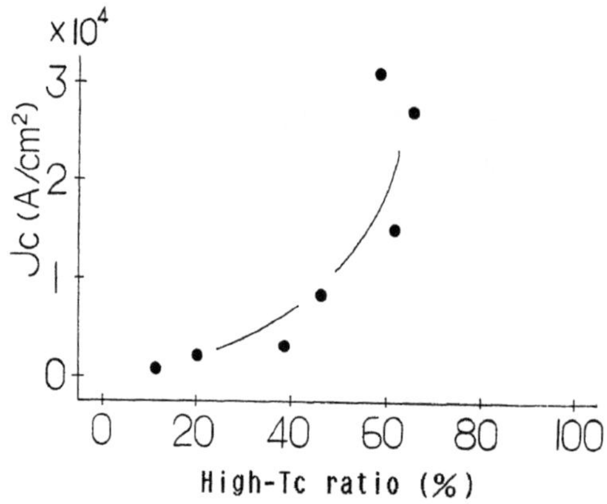

Fig.8. Relationship between volume fraction of high-Tc phase and observed Jc.

4. SUMMARY

The hermetic anneal technique was demonstrated to obtain the high volume fraction of the high-Tc phase in the Bi(Pb)-Sr-Ca-Cu-O sputtered film. The content of Pb in the annealed film was not changed even after a high temperature anneal. The optimum annealing condition was about 90min at 830℃ in atmosphere of 1/13 oxygen pressure. The films prepared were poly-crystal with the microstructure like a pile of scales. The non-superconducting, Ca-Cu oxide reduced as increasing annealing time. By the newly developed hermetic anneal, Tc(end) above 100K and the high-Tc ratio of about 70% were obtained with good reproducibility. At this point the highest Tc(end) and Jc were attained up 105K and 3.1 $\times 10^4$ A/cm², respectively.

REFERENCES

1) M.K.Wu et al, Phys.Rev.Lett.58,908(1987)
2) H.Maeda et al, Jpn.J.Appl.Phys.Vol 27 (1988)
3) J.M.Tarascon et al, Phys.Rev.Lett.38,8885(1988)
4) M.Takano et al, Jpn.J.Appl.Phys.Vol 27, L1041 (1988)
5) T.Kawai et al, Jpn.J.Appl.Phys.Vol 27, L1476 (1988)
6) K.Ogawa et al, Jpn.J.Appl.Phys.Vol 28, L967 (1989)
7) A.Tanaka et al, Appl.Phys.Lett.54,1362(1989)
8) T.Kawai et al, Jpn.J.Appl.Phys.Vol 28, L187 (1989)
9) T.Kawai et al, Jpn.J.Appl.Phys.Vol 28, L190 (1989)

Focused Ion Beam Processes for Bi-Ca-Sr-Cu-O Superconducting Thin Films

S. FUJIWARA, R. YUASA, H. KUWAHARA, M. NAKAO, and S. SUZUKI
Tsukuba Research Center, Sanyo Electric Co., Ltd., 2-1, Koyadai, Tsukuba, Ibaraki, 305 Japan

ABSTRACT

Focused ion beam (FIB) processes have been used for the patterning of Bi-Ca-Sr-Cu-O high-Tc superconducting thin films. Maskless sputter etching was carried out using Bi^+ FIB to minimize the effect of impurity. V-shaped fine grooves with the 0.2 μm width near the surface of the film and the depth of 2 μm were obtained by the FIB processes. Bridge structures with the width of 0.7 μm were fabricated with little degradation in Tc. Planer S-N-S Josephson junctions were made by depositing Au into the grooves. Superconducting current flowing across the junctions was clearly observed.

KEY WORDS: focused ion beam, superconducting thin film, bismuth, patterning, S-N-S junction

INTRODUCTION

Since the discoveries of high-Tc superconductors, many studies have been made on superconducting mechanisms and application to electronic devices. Microfabrication techniques of high-Tc superconducting thin films including both wet or dry processes[1,2] have been also studied.

Focused ion beams (FIB) obtained by use of a liquid metal ion source have been expected as new tools for microfabrication of semiconductor devices with a submicron dimension. Also FIB must be a useful technique for patterning of high-Tc superconducting thin films. Several authors have reported patterning processes for superconducting thin films using maskless sputter etching with Au^+ or Ga^+ FIB [2-4]. Submicron features were obtained by the FIB processes, and devices such as DC-SQUID's were fabricated successfully. In general, the FIB processes offer less degradation of the film properties than the other processes, because they do not involve wet processes. It is thought, however, the doped atoms act as impurities if the ions used are not the compositional elements of the films.

In this paper, we demonstrate FIB processes to pattern Bi-Ca-Sr-Cu-O (BCSCO) thin films. For the first time, the Bi^+, which is a component of the BCSCO films, is used to make patterns in the films. These processes avoid the problem of impurities. Therefore, it is expected that the films are patterned with little degradation of the film properties. In fact, bridge structures and S-N-S Josephson junctions[5,6] are fabricated by the FIB processes.

EXPERIMENTAL

Thin films of Bi-Ca-Sr-Cu-O were prepared on (100) MgO substrates by rf magnetron sputtering from a single target[7]. The nominal composition of the target was $Bi_3Ca_3Sr_2Cu_6O_x$. Sputtering was carried out at room temperature in a pure Ar atmosphere of 30 mTorr. The thickness of the films used in this study was about 1 μm. Coarse patterns were made by a lift-off method. Metal mask with windows of 100 μm wide and 5 mm long was mounted on the substrates during the deposition. Each stripe had bonding pads on both ends. The films were annealed at 800 ℃ for 1 h in a mixture of O_2 and N_2 gases before patterning by FIB.

A FIB apparatus of the JIBL-100A (JEOL Ltd.) was used in this study; Bi^+ ions were emitted from a liquid metal ion source and accelerated up to 100 keV. We employed a Bi liquid metal ion source with a NiCr wire needle, which however had a short life time of less than 50 hours. As reported by Komuro[8], Ni^+ ions are observed in the ion beam emitted from a Bi ion source with a similar needle. These facts indicate that a reaction between Bi and the needle causes a loss and a dullness of the needle apex. To suppress the reaction, In-Bi alloy (In : Bi = 78 : 22(at %)) was used because of a very low melting point of 78 ℃. As a result, long time and quite stable operations were obtained using this In-Bi alloy source. A typical probe current of Bi^+ ions was 20 pA and a beam diameter defined by a full width at half maximum in the current density distribution was about 0.1 μm. Maskless sputter etching was carried out by the repetitive line scanning of the FIB with the scan length of 100 μm in an ambience of $<1\times10^{-8}$ Torr.

For the fabrication of planer S-N-S junctions, the Au film was deposited into the grooves with the thickness of about 1 μm. A standard four-probe technique was employed for electrical measurements.

RESULTS AND DISCUSSION

Figure 1 shows a SEM micrograph of a groove formed by maskless sputter etching with the Bi^+ FIB, where the line dose was 5.2×10^{13} ions/cm. This groove was formed in an as-deposited BCSCO film with a thickness of 4.5 μm to see the shapes of the patterns clearly. A V-shaped groove with a width of about 0.2 μm near the surface of the film and a depth of more than 2 μm was obtained in Fig.1. As seen in Fig.1, there is little redeposition of sputtered atoms in the grooves. On the other hand, a large quantity of redeposited material was clearly observed in the grooves when it was formed by a single line scanning, where the FIB was scanned slowly so that the line dose is equal to the repetitive scanning case. The redeposited material was decreased by a repetitive scanning with a faster speed. In fact, a vacant groove was cut by more than 10 times repetitive line scanning.

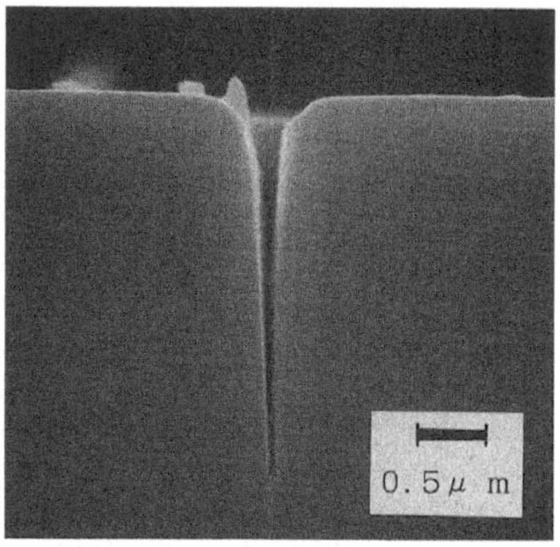

Fig.1 A SEM micrograph of a groove formed by Bi^+ FIB

Bridge structures were fabricated to examine the degradation of the film properties caused by the FIB processes. Bridges with widths of 0.5 and 0.7 μm were formed by the Bi^+ FIB. Figure 2 shows a SEM micrograph of the 0.7 μm bridge, where the grooves were formed with a line dose of 7×10^{13} ions/cm. Resistance vs temperature curves of these bridges and the virgin film are shown in Fig.3. It is confirmed that the electrical conductivity of the bridge was isolated by the groove. The resistance was more than the order of GΩ at room temperature. The bridge of 0.7 μm width exhibited little change in Tc as compared with the virgin film. The 0.5 μm bridge, however, showed a little degradation in Tc,onset.

The degradation is probably due to the FIB irradiation damage of the film, because the Bi^+ ions do not act as impurities. It is well known that the crystal structure of the film is easily destroyed by the multiple scattering

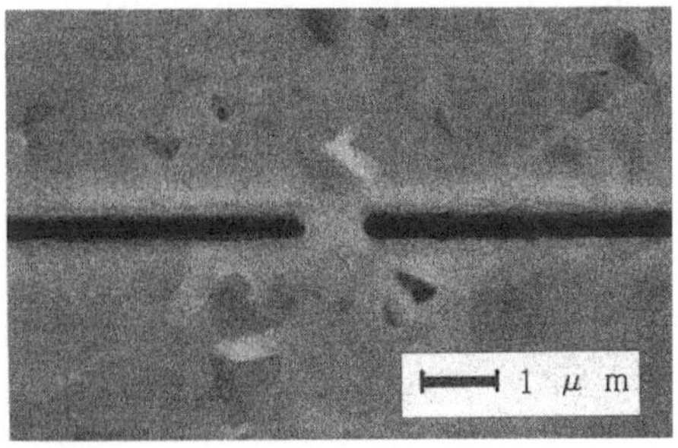

Fig.2 A SEM micrograph of a bridge structure with a width of 0.7 μm

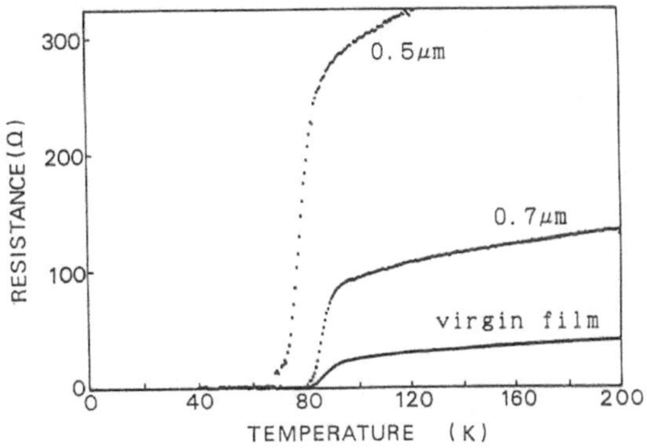

Fig.3 Resistance vs temperature curves of the fabricated bridges and the virgin film

between irradiated ions and component atoms. That is, the damaged layers are, more or less, formed even in the sidewalls of the grooves. The width of such layers is estimated to be about 0.1 μm from the penetration depth of the 100 keV Bi^+ in the BCSCO film. Thus, the formation of damaged layers is a serious problem when a structure of a lower submicron size is fabricated. In our samples, the effective channel width of superconducting currents is narrower than the real width of the bridge. It is expected, however, that the 0.5 μm bridge have an effective channel of about 0.3 μm width. If there are grain boundaries in the channel, superconducting current is extremely reduced in such a narrow channel of a lower submicron size. Accordingly, the FIB process is not a prime reason for the degradation in the 0.5 μm bridge. FIB process using Bi^+ ions is effective for patterning down to 0.7 μm.

In order to demonstrate the feasibility of the FIB process, we attempted to make planer S-N-S Josephson junctions. The procedures are shown in Fig.4. After the deposition of the Au film, the sample was annealed for 30 min at 750 ℃ in O_2 and N_2 gas mixture. It is a point of this procedure that the heat treatment is made after the fabrication of the patterns. If the other ion beams are used to etch the films, the ions will diffuse into the film and degrade the junctions during the heat treatment. We expect that the interface between BCSCO and Au film will be improved using Bi^+ ions.

Figure 5 shows the I-V characteristics of the S-N-S junctions at 15 K before and after the heat treatment. Before the heat treatment, there exist a finite resistance of 1.4 Ω which implies a normal-state interface between the BCSCO and the Au film. After the anneal, the sample has no resistance within experimental accuracy. From these results, it is confirmed that the heat treatment improves the interface between the BCSCO and the Au film. It was found that superconducting current flows across the junctions, indicating that the the BCSCO banks were superconductively connected through the Au film in terms of the proximity effect. A critical current of the junction was about 1 mA at 15 K and the resistance was about 40 mΩ at 10 μV, so the IcR product of about 40 μV was obtained.

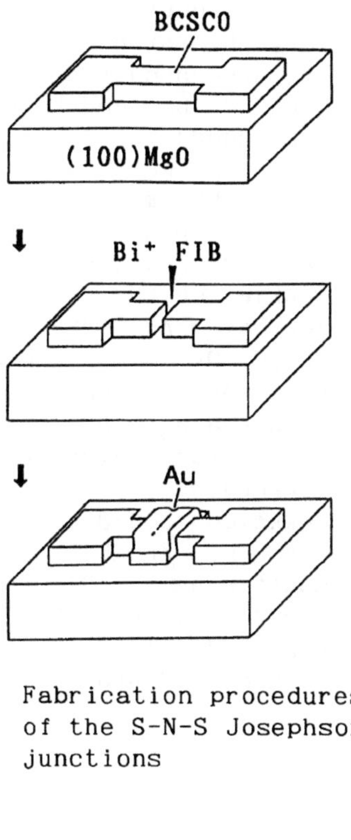

Fig.4 Fabrication procedures of the S-N-S Josephson junctions

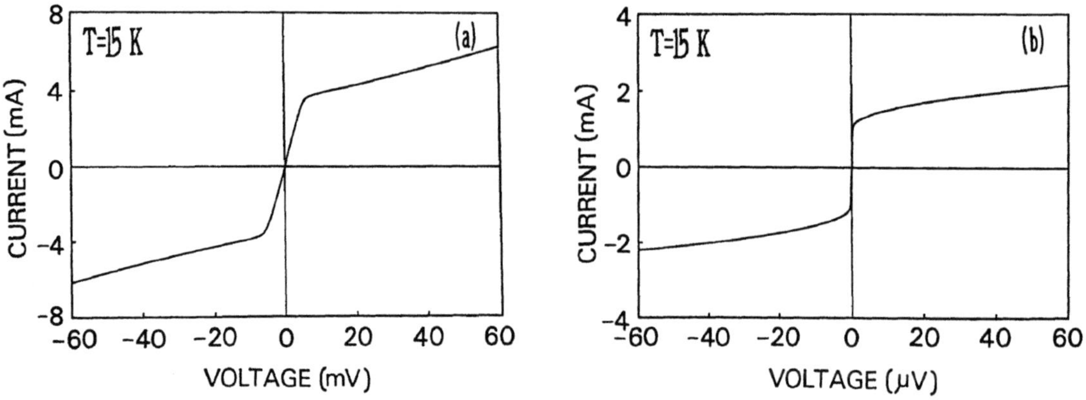

Fig.5 I-V characteristics of the S-N-S junctions at 15 K (a) before the heat treatment, (b) after the heat treatment

CONCLUSIONS

We have demonstrated FIB processes with Bi^+ ions for patterning Bi-Ca-Sr-Cu-O thin films. V-shaped fine grooves with a width of about 0.2 μm near the surface of the film and a depth of more than 2 μm were obtained. Bridge structures were fabricated to examine the degradation of the film properties caused by the FIB processes. The bridge with a width of 0.7 μm exhibited little degradation in Tc. The Au film was deposited into the grooves to make planer S-N-S Josephson junctions. We observed superconducting current flowing across the junctions. It is a remarkable advantage of this process that the heat treatment is made after the fabrication of the patterns, because the irradiated Bi atoms cause no degradation of the film properties.

REFERENCES

1. Tonouchi M, Sakaguchi Y, and Kobayashi T (1988) Chemical etching of high-Tc superconducting films in Feliox-115 Solution. Jpn J Appl Phys 27: L98-L100
2. Tsuge H, Matsui S, and Wada Y (1989) Microfabrication technology for high-Tc superconducting films. Extended Abs of ISEC'89, 12-13 June 1989: PR-1
3. Harriott LR, and Polakos PA (1989) Focused ion beam patterning of high Tc superconductor films. Proc SPIE 1089: 2-10
4. Yuasa R, Nakao M, Fujiwara S, Kaneda K, Suzuki S, and Mizukami A (1988) DC-SQUID's made from high-Tc superconducting oxide thin films using focused ion beam processes. Extended Abs of 5th Int Workshop on Future Electron Devices, 2-4 June 1988, Miyagi-Zao: 225-229
5. Mankiewich PM, Schwartz DB, Howard RE, Jackel LD, Straughn BL, Burkhardt EG, Dayem AH (1988) Fabrication and characterization of an $YBa_2Cu_3O_7$/Au/$YBa_2Cu_3O_7$ S-N-S Microbridge. Extended Abs of 5th Int Workshop on Future Electron Devices, 2-4 June 1988, Miyagi-Zao: 157-160
6. Hayakawa H (1989) Possible applications of proximity effect to high temperature superconductor devices. Extended Abs of 2nd Workshop on High-Temperature Superconducting Electron Devices, 7-9 June 1989, Shikabe-Hokkaido: 353-355
7. Nakao M, Kuwahara H, Yuasa R, Mukaida H, and Mizukami A (1988) Magnetron sputtering of Bi-Ca-Sr-Cu-O thin films with superconductivity above 80 K. Jpn J Appl Phys 27: L378-L380
8. Komuro M (1983) Liquid metal ion sources. Proc Int Ion Engineering Congress, Kyoto: 337-348

Synthesis and Magnetic Properties of Pb-doped Bi-Sr-Ca-Cu-O Films

ATSUSHI TANAKA, KAZUNORI YAMANAKA, JASON CRAIN, TAKUYA UZUMAKI, NOBUO KAMEHARA, and KOICHI NIWA

Fujitsu Laboratories Ltd., Atsugi Inorganic Materials Laboratory, 10-1, Morinosato-Wakamiya, Atsugi, 243-01 Japan

ABSTRACT

We synthesized Pb-doped Bi-Sr-Ca-Cu-O thin films by rf magnetron sputtering. We attempted to reduce Pb concentration to 0.8 the amount of Bi and deposited slightly Cu-rich films. After sintering for only one hour, nearly single-phase high-Tc thin films were synthesized. We studied their magnetic properties. Flux creep was observed to obey logarithmic time decay of magnetization. Pinning potential for flux motion perpendicular to c-axis was estimated to be 0.028 eV at 10 K, which is much smaller than that for 1-2-3 compounds and Bi-layered bulk samples. Flux creep was observed to depend on sintering conditions. Faster heating rate during sintering led to slower flux creep rate.

KEY WORDS: Bi-Pb-Sr-Ca-Cu-O, thin film, high Tc phase, flux creep, magnetic relaxation

INTRODUCTION

The Bi-layered superconductor is characterized by its short coherence length[1], ξ_c which is much shorter than the lattice constant of c-axis and comparable to the distance between CuO planes. In the Bi-layered system, magnetic flux behaves like a disk, not a tube[2], because each CuO plane interacts weakly and flux moves easily under the influence of Lorentz forces or thermally activated processes. Flux creep and low Jc(critical current) in a magnetic field cause severe problems for potential practical applications. We synthesized nearly single-phase high-Tc thin films and studied their physics properties focusing on magnetic behavior.

EXPERIMENTAL

Thin films were prepared by rf magnetron sputtering by the same method mentioned in our previous report[3]. The films were deposited on MgO (100) single crystals at 400°C. We used three targets; Bi-Sr-Ca-Cu-O (BSCCO) with the composition of Bi:Sr:Ca:Cu=3:2:2:3, PbO and CuO targets. PbO and CuO layers were stacked repeatedly on BSCCO layers to precisely control compositions. The film compositions were determined by inductively coupled plasma analysis (ICP).
As-deposited films were amorphous. After deposition they were sintered in air for 1 hour at about 850°C. The films were examined using x-ray diffraction and a scanning electron microscope. The electrical resistivity was measured with the four-point probe method. Magnetic measurements were performed by SQUID system. The temperature dependence of magnetization, hysteresis curve and remnant moment time decay were measured. For nearly single phase high-Tc samples, the relationship between sintering process and flux creep rate was measured.

RESULTS AND DISCUSSION

We investigated the high-Tc phase formation dependence on the Pb and Cu compositions. Figure 1 shows the composition dependence of the high-Tc phase formation. The peak ratio was calculated from the intensity of the high-Tc phase (00<u>14</u>) and low-Tc phase (00<u>12</u>). Although heavily Pb-doped thin films form the high-Tc phase rapidly, they are multiphased and have poor morphology. We attempted to minimize the amount of Pb needed to enhance high-Tc phase formation. The amount of the high-Tc phase formed depends on the Cu content. A slightly Cu-rich composition relative to 2:2:2:3 is effective for promoting high-Tc phase growth. The sintering temperature also greatly affects the high-Tc

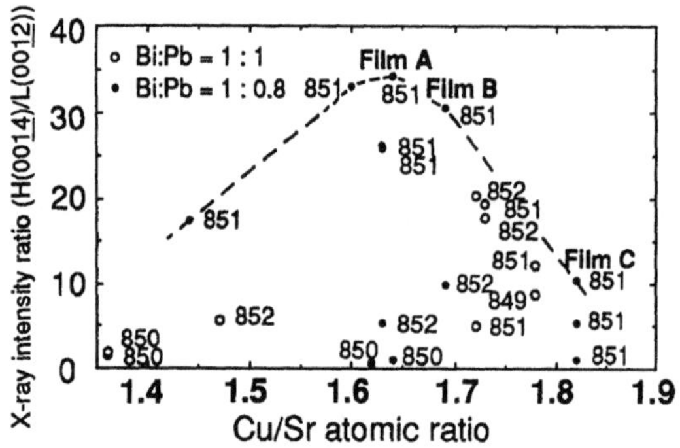

Fig. 1. X-ray intensity ratio of high-Tc phase (0014) and low-Tc phase (0012). The numbers inside the graph indicate sintering temperature. The high-Tc phase formation strongly depends on the Cu content and sintering temperature. Slightly Cu-rich films sintered at 851°C gave the best results.

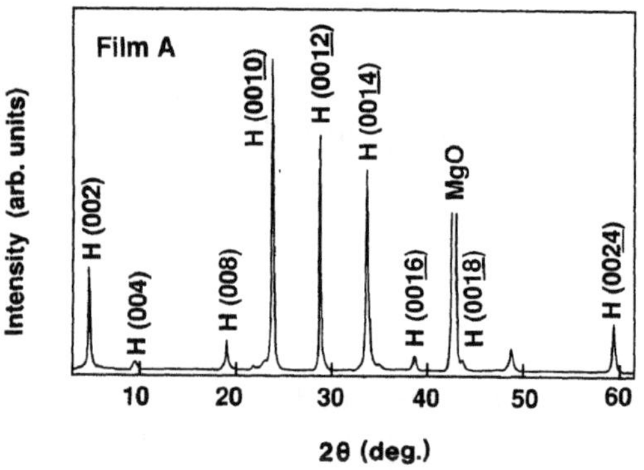

Fig.2. X-ray diffraction pattern for Film A. There are strong (00n) peaks from the high-Tc phase. This implies the film has a high c-axis orientation perpendicular to the substrate. As-deposited film composition was Bi:Pb:Ca:Sr:Cu = 1.02 : 0.80 : 1.00 : 0.99 : 1.64.

phase formation. The low-Tc phase forms below 848°C and above 853°C. For Bi:Pb:Sr:Ca:Cu = 1.00:0.80:1.00:0.99:1.64, nearly single-phase high-Tc films were obtained after one hour of sintering at 851°C in air. During sintering, most of Pb evaporated and the sintered film contained 1/4 less Pb than the as-deposited film. The composition of the sintered film is close to the stoichiometric value of 2:2:2:3. Figure 2 is an X-ray diffraction pattern of film A. The peaks from the low-Tc phase are barely detectable but the (00n) peaks from the high-Tc phase are strong. This implies that film has high c-axis orientation perpendicular to the substrate. Thin plate-like crystals are observed by an SEM to be stacked on each other with a c-axis orientation. At this level of Pb doping, the surface morphology of the film improves relative to more heavily Pb-doped samples. The onset temperature was 118 K and Tc endpoint was 107 K.

Magnetic properties were measured in a SQUID system. The magnetic field was applied perpendicular to the substrate and parallel to the c-axis of the films. Figure 3 shows the Meissner signal dependence on temperature for mixed- and single-phase films. The external magnetic field was about 4 Gauss. In mixed-phase films, the transition has a kink corresponding to the low-Tc phase transition. For the high-Tc phase film, the transition has tail above 60 K. This implies that, in thin films, the superconducting state is easily broken by a weak external magnetic field.

The hysteresis curve shown in Fig. 4 was measured for a high-Tc phase film at 10 K. At just below 20 Gauss, the magnetization has a peak. This means that H_{c1} (lower critical field) is very low even at 10 K. At around 70 Gauss, a minimum in the magnetization was observed. At present, we do not understand the mechanism of this singular behavior, but similar phenomena was observed in the low-Tc phase of bulk samples[4]. J_c was estimated from the difference in magnetization (ΔM, in Gauss, the magnetization difference per unit volume) between increasing and decreasing fields.

According to Bean's model J_c in A/cm^2 is estimated using the following equation:

$$J_c(\mathrm{B}) = 30\ \Delta M(\mathrm{B})/r$$

where, r is sample size in cm. Since $\Delta M(\mathrm{B})$ has strong dependence on magnetic field, Bean's model is not well suited for our samples and the estimated J_c gives a lower limit[4]. J_c decreased as the magnetic field increased.

The pinning potential U_o was estimated from the relaxation of the remnant magnetization. After cooling in approximately 5 gauss to 10 K, a 1 kGauss magnetic field was applied parallel to the c-axis of the films and then removed. Figure 5 shows the relaxation. The magnetization decreased logarithmically in time, and the decay was well described by the flux creep model[5,6]. According to this model the remnant magnetization decays according to the law:

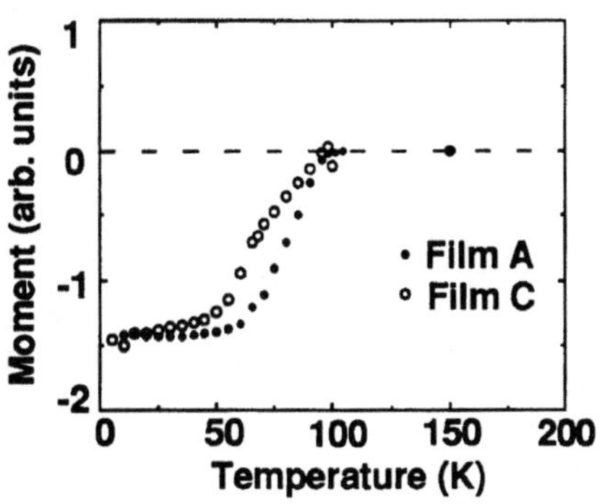

Fig. 3. Meissner signal vs temperature for mixed- (FilmC) and single-phase film (FIlm A).

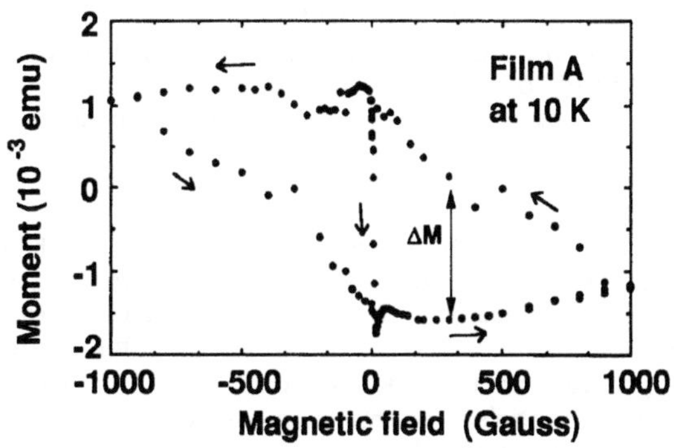

Fig. 4. Hysteresis curve at 10 K for Film A. At around 70 Gauss, a minimum in the magnetization was observed.

$$M(t)=Mo\{1-kT/Uo\ \ln(t/to)\}$$

where, *Mo* is a magnetization at *to*. The pinning potential *Uo* at 10 K was calculated to be 0.028 eV from our measurements. This value is much smaller than that of Y-Ba-Cu-O[7]. For powders of low-Tc phase single crystals of Bi-based materials, *Uo* was estimated to be 0.13 eV[8], which is about 5 times larger than our estimation. Our samples have high c-axis orientation, and the *Uo* value reported here is expected to be relevant for flux motion perpendicular to the c-axis of the film. Transport experiments in magnetic field[9] indicate that the pinning potential is highly anisotropic and depends on the direction of the magnetic field. It is found that flux flows much more easily if it penetrates perpendicular to the a-b plane. This means that the flux parallel to the c-axis feels less pinning potential than that in the a-b plane. This is consistent with our estimation which is smaller than that of the bulk samples with a random orientation.

We have also made some preliminary investigations of the relationship between the flux flow rate and sintering conditions. The decay rate is determined by intra-grain and inter-grain flux creep rate. Inter-grain coupling strongly depends on the sintering process. Film A and Film D were prepared under almost same conditions except the heating rate during the sintering process was different. It was 1°C/min for Film A and 2°C/min for Film D. Film D contained

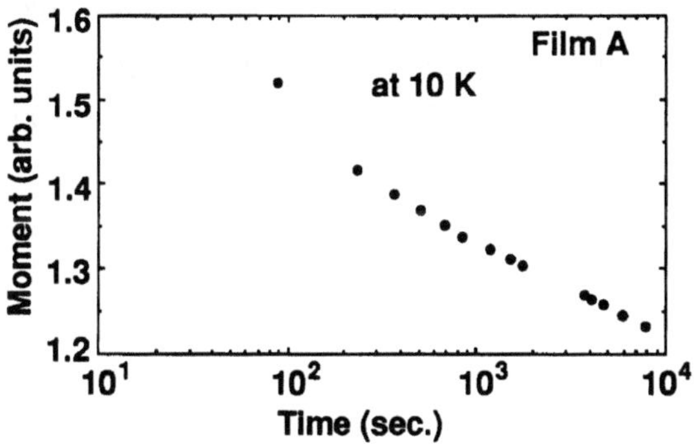

Fig. 5. Relaxiation of the remnant magnetization. The magnetization decreased logalithmically in time. Using the decay rate, the pinning potential was estimated to be 0.028 eV.

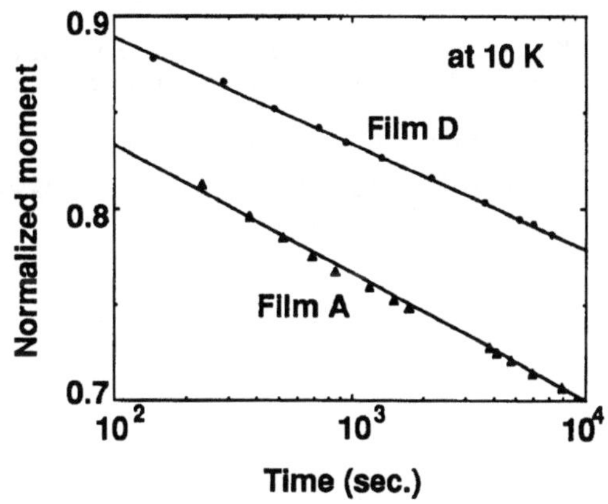

Fig. 6. Relationship between flux flow rate and sintering conditions. Magnetization is normalized by Mo(t=1 sec.). These two samples were prepared under almost same conditions except the heating rate during the sintering process was different. It was 1°C/min for Film A and 2°C/min for Film D. The flux creep rate of Film D was reduced by 20% compared to that for Film A.

larger amounts of Pb just after it reached 850°C. Figure 6 shows the decay of magnetization normalized to its value at *t*=1 sec. The flux creep rate of Film D was reduced by about 20% compared to that for Film A. This suggests that, for Film D, intergrain coupling are stronger than that for Film A.

CONCLUSIONS

We synthesized Pb-doped Bi-Sr-Ca-Cu-O high-Tc films and studied their magnetic properties. We reduced Pb concentration to 0.8 the amount of Bi to improve surface morphology and to obtain single phase material. Slightly Cu-rich composition is effective to form the high-Tc phase. Nearly high-Tc phase films were synthesized after 1 hour of sintering at 851°C. Sintering temperature range for high-Tc phase formation is very narrow.

Critical current density, *Jc* exhibited strong dependence on magnetic field. Flux creep was observed by logarithmic time decay of magnetization. Pinning potential parallel to the c-axis was estimated to be 0.028 eV which was much smaller than that for 1-2-3 compounds. This value is about one fifth of that for bulk samples with random orientation. Flux creep rate decreased by changing sintering condition. Faster heating rate during sintering leads strong coupling of grains and could improve *Jc* in magnetic field.

REFERENCES

1 Y. Hidaka, M. Oda, M. Suzuki, Y. Maeda, Y. Enomoto, and T. Murakami (1988) Large anisotropy of the upper critical magnetic field in single crystal Bi-(Sr,Ca)-Cu-O. Jap. J. Appl. Phys. 27, L538-L541

2 Peter H. Kes (1989) Flux pinning in thin films and single crystals of high-Tc oxides. In: 9th workshop on new superconducting materials symposium, 20 Oct. 1989, Tokyo.

3 A. Tanaka, N. Kamehara and K. Niwa (1989) Composition dependence of high Tc phase formation in Pb-doped Bi-Sr-Ca-Cu-O thin films. Appl. Phys. Lett. 55, 1252-1254

4 R. B. van Dover, L. F. Schneemeyer, E. M. Gyorgy, and J. V. Waszczak (1988) Critical current densities in single-crystal $Bi_{2.2}Sr_2Ca_{0.8}Cu_2O_{8+\delta}$. Appl. Phys. Lett. 52, 1910-1912

5 M. R. Beasley, R. Labush and W. W. Webb (1969) Flux creep in type-II superconductors. Phys. Rev. 181, 682-700

6 A. M. Campbell and J. E. Evetts, Adv. Phys. 21, 299 (1972).

7 T. Matsushita, S. Funaba, Y. Nagamatsu, B. Ni, K. Funaki and K. Yamafuji (1989) Flux creep in sintered superconducting Y-Ba-Cu-O. Jap. J. Appl. Phys. 28, L1508-1510

8 H. Kumakura, K. Togano, E. Yanagisawa, K. Takahashi, M. Nakao, and H. Maeda (1989) Magnetic relaxation in high-Tc oxide superconductors. Jap. J. Appl. Phys. 28, L24-L26

9 T.T.Palstra, B.Batlogg, L.F. Schneemeyer, R.B.van Dover, and J.V.Waszczak (1988) Angular dependence of the upper critical field of $Bi_{2.2}Sr_2Ca_{0.8}Cu_2O_{8+\delta}$. Phys. Rev. B 38, 5102-5105

Tailored Thin Films of Superconducting Bi-Sr-Ca-Cu Oxide Prepared by the Incorporation of Exotic Atoms: Superconductivity and the Distance Between CuO_2 Layers

HITOSHI TABATA[1], OSAMU MURATA[1], TOMOJI KAWAI[2], and SHICHIO KAWAI[2]

[1] Technical Institute, Kawasaki Heavy Industries Ltd., Kawasakicho, Akashi, Hyogo, 673 Japan
[2] The Institute of Scientific and Industrial Research, Osaka University, Mihogaoka, Ibaraki, Osaka, 567 Japan

ABSTRACT

The layer-by-layer successive deposition method utilizing laser ablation has been applied to the controlling the structures and the compositions of Bi-Sr-Ca-Cu-O superconducting films. With this method, we achieve the formation of thin films having good morphology by controlling the deposition conditions such as energy density of laser ablation, substrate temperature, and atmosphere. Furthermore the structures and the superconductivity can be controlled by the substitution of various ions for Ca or Sr in the atomic scale.

KEY words: Bi-Sr-Ca-Cu-O superconductor, laser ablation, layer-by-layer site selective substitution

INTRODUCTION

In the cuprate superconductors, the achievement of high quality, smooth and epitaxial thin film is important not only for physical purpose but also for the applications to superconducting devices. The layer-by-layer successive deposition is one of the most effective methods to control the structures[1)] and to substitute exotic ions for Ca or Sr[2)]. In order to control the structure and compositions in the atomic level and to achieve the artificial formations of these layered oxide superconductors, the improvement of morphologies and epitaxial growth of these films by means of layer-by-layer method is essential. In this paper, we report the synthetic conditions to achieve good morphology of $Bi_2Sr_2Ca_1Cu_2Oy$ superconducting thin films using the laser ablation method, and the effect of replacement of Ca or Sr by the various ions having different radii and valence state.

EXPERIMENTAL

The films were prepared by the layer-by-layer successive deposition method using a pulsed ArF excimer laser(193nm). The laser beam was sequentially focused on the multi-targets to form a film on a substrate placed at the opposite side in the presence of N_2O and/or O_2 gas atmosphere. Typical experimental conditions were : energy density for ablation after focusing = 0.5 J/cm^2 , and repetition rate = 10 Hz. The substrate used was a MgO(100) single crystal heated at 580 - 600°C. For the deposition, sintered disks of $Bi_7Pb_3O_y$, $Sr_1Cu_1O_y$, $Ca_1Cu_1O_y$ were used as targets for the standard BSCCO superconducting films. The targets such as MCuOy(M=Li, Na, K, Mg, Ba, Y, La and Nd) were used for substitution at Ca or Sr site.

(I)Deposition condition

There are several factors to obtain smooth and epitaxial thin films in the laser ablation method. It is important to keep these factors suitably. So we have examined these conditions as follows, (1) the laser intensity (0.2 to 3 J/cm^2), (2)substrate temperature (580 to 700 ℃) and (3) gas pressure (10^{-3} to 10^{-1} Torr) and a kind of gases (O_2 and/or N_2O).

(II)Substitution of various ions(+1,+2 and +3) for Ca or Sr

Under the suitable conditions which from former experiments(I), we tried to control the structure and the superconductivity of the Bi-Sr-Ca-Cu-O superconductors by incorporation of exotic ions.

Namely partial substitution of mono-(Li, Na and K), di-(Mg, Sr and Ba), and tri-valent(Y, La and Nd) ions for Ca were examined at first. In this case, the deposition was carried out successively from the targets, BiPbO-SrCuO-CaCuO-MCuO-CaCuO-SrCuO-BiPbO. This consists of one cycle which was repeated. Typical irradiation periods for successive deposition are 25s for $BiPbO_y$, 15s for $SrCuO_y$, 10s for MCuOy and 28s for $CaCuO_y$.

The second, partial substitution of the ions for Sr has also been performed as for the Ca site. Typical irradiation periods for the successive deposition are 25s for BiPbO, 5s for SrCuO or MCuO and 65s for CaCuO.

RESULTS AND DISCUSSION

(I) When the laser density was more than 2-3 J/cm^2, we observed a lot of clusters and particles on the surface of the films. For less than 0.2 J/cm^2, on the other hand, no ablation occurrs on the target. Accordingly the films were deposited under the condition of about 0.3-1 J/cm^2. In this condition, the films show mirror like morphology.

The film deposited from Bi_7Pb_3Oy target had a very smooth mirror surface. On the films deposited from SrCuOy and CaCuOy, on the other hand, the particles were observed on their surface in the presence of O_2 atmosphere. These particles were proved with EDX analysis that their compositions were Cu-rich, that is, Cu or CuO. One possible explanation for the particle formation is that the oxygen in the SrCuOy and CaCuOy target was released out and metallic copper or CuO was formed with the laser ablation. Then the Cu or CuO would be ablated as particles and deposited on the substrate surface. The morphologies of the films did not change so much during the changes of the substrate temperature(from 580 to 700 °C). Accordingly, the films made under O_2 gas have particles in any conditions(Fig.1(b)).

The film which was deposited in the presence of O_2+N_2O(1:1, total pressure is 0.1 Torr), on the other hand, had mirror like smooth surface(Fig.1(a)). The existence of particles on the surface was not observed. Introduction of N_2O gas was quite effective to get rid of particle deposition.

(a)

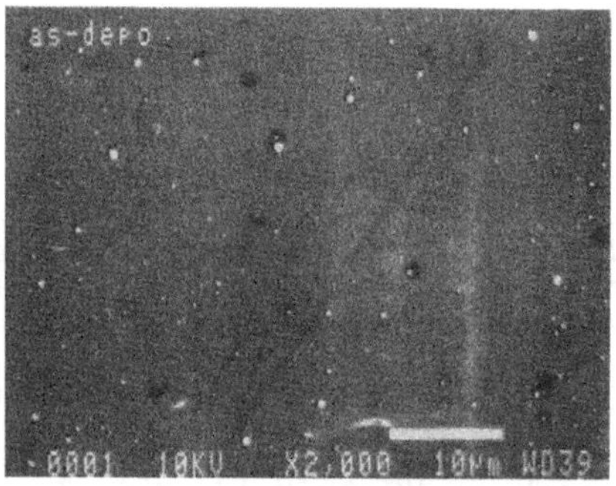

(b)

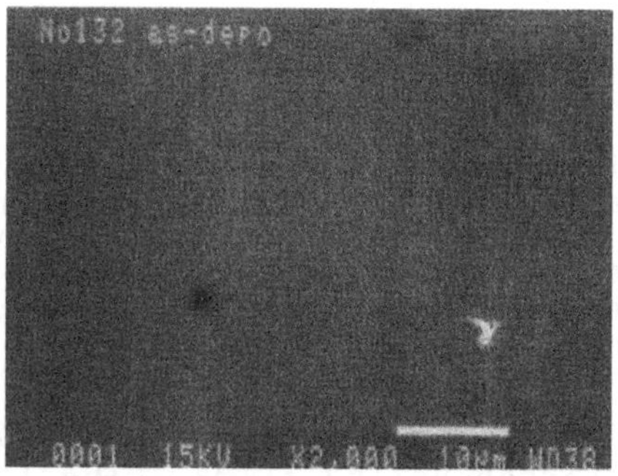

Fig.1. SEM images of the films which was prepared layer by layer successive deposition in the presence of (a) O_2 and (b) O_2+N_2O gas atmosphere.

(II) Partial substitution of +2 ions for Ca by about 30%(atomic) shows broad X-ray diffraction patterns of double $Cu-O_2$ layered structure. All films were heated at 800°C to keep oxygen incorporation enough. No apparent impurity peaks have been observed, indicating that these atoms were incorporated into the BSCCO crystal structure.

The lattice parameter c is found to be dependent on the ionic radii of the incorporated ions(Fig. 2). When small Mg ion is incorporated at the Ca site, the c axis shrink. On the other hand, with the incorporation of larger ions such as Sr and Ba, the lattice constant c increases. This is because the Ca ions contact closely with the neighboring oxygen ions.[3,4] It would be reasonably understood that the distance between CuO_2 planes increases by inserting the large ions and decreases with small ions(Fig.2(a)).

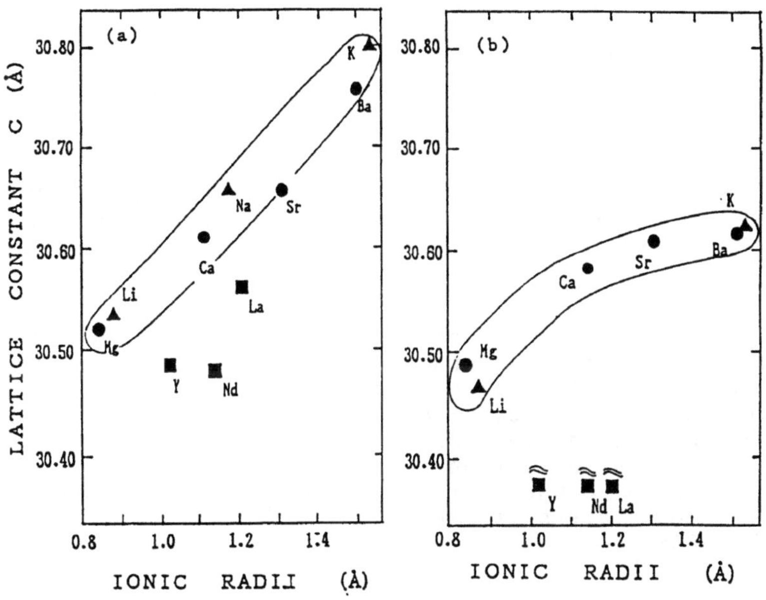

Fig.2. (a) Lattice constant c versus ionic radii of the ions incorporated at the Ca site in the BSCCO films, and (b) at the Sr site.

When the Sr site is substituted by the Mg, Ca and Ba ions, on the other hand, the lattice constant c decreases only in case of Mg, but not increases much even by the substitution by the large ion like Ba. This behavior is expected by the reason that there is a room between Sr and above oxygen, since oxygen around Sr has close contact with Bi[4]. The changes of the lattice parameter c by the incorporation of +1 ions, Li and Na, and +3 ions, Y, Nd and La have also been examined(Fig.2(b)). The +1 ions show similar behavior to the +2 alkaline earth ions, that is, the larger ions leads to the larger lattice constant. The +3 ions, on the other hand, show distinct shrinking of the c axis. This is presumably because highly positive valence works to attract negative oxygen ions. The R-T curves for the incorporation of +2 ions at the Ca site are shown in Fig.3(a). It is clearly observed that the larger the substituted ion is, the higher the Tc_{mid} is. The highest Tc is observed for the Ba substitution, Tc_{onset} to be 90K and Tc_{mid} to be 83K. Accordingly the correlations are observed between the Tc and ionic radii of the substituted ions, in other word, Tc and lattice constant c shown in Fig.3(b). For the substitution of Ba at the Sr site, we have shown the R-T curve in the Fig.2(a) to be semiconductive. This is consistent with the smaller expansion of lattice constant c for the Sr site substitution.

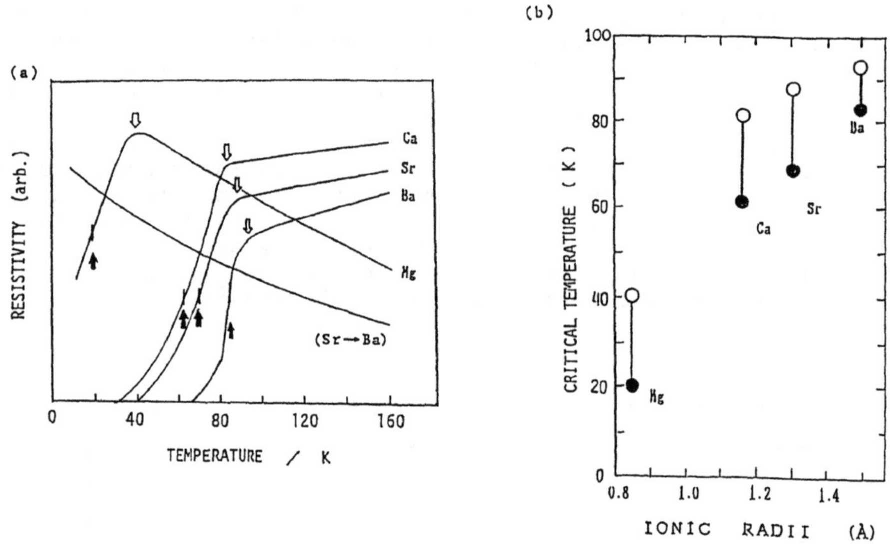

Fig.3. (a) Temperature-resistivity curves for the BSCCO films which contain various +2 ions (b) Tc_{onset}(○) and Tc_{mid}(●) versus ionic radii of the ions incorporated at the Ca site.

In this manner, the site-selective partial substitution of exotic atoms at the Sr and the Ca site shows drastic difference. The Ca site substitution make c axis longer, while that at the Sr site does not elongate the c axis. Accordingly, the spacing between double CuO_2 planes can be changed artificially by this technique, and we found that this spacing between CuO_2 planes has an important effect on the Tc value. It has been reported that there is a hopping of Cooper pairs between CuO_2 layers and this kind of interaction between layers is important for the Tc value.[5] The present experimental results indicate that the distance between the layers is really important factor of the determining Tc value.

CONCLUSION

1. The films show smooth surface by the deposition under the suitable condition, especially in N_2O and O_2 mixed atmosphere with weaker laser intensity.
2. With the layer-by-layer successive deposition method utilizing laser ablation, site selective partial substitution of +1, +2 and +3 ions for the Ca and Sr site has been performed.
3. The incorporation of smaller +2 ion, Mg, at the Ca site makes the c axis shrink, while that of larger Sr and Ba ions makes the c axis longer. On the other hand, the incorporation of even Ba at the Sr site does not elongate the c axis so much.
4. The Tc value has correlation with the expansion of c axis due to the incorporation of exotic atoms. The larger lattice parameter c produces higher Tc.

REFERENCES

[1] M.Kanai, T.Kawai, S.Kawai and H.Tabata: Appl. Phys. Lett., 54, 1802(1989).
[2] H.Tabata, T.Kawai, M.Kanai, O.Murata and S.Kawai, Jpn. J. Appl.Phys., 28, L823(1989).
[3] M.A.Subramanian, C.C.Torardi, J.C.Calabrese, J.Gopalakrishnan, K.Morrissey, T.R.Askew, R.B.Flippen, U.Chowdhry and A.W.sleight, Science, 26, 1015(1988).
[4] K.Imai, I.Nakai, T.Kawashima, S.Sueno and A. Ono, Jpn.J.Appl.Phys., 27, L1661(1988).
[5] J.M.Wheatley, T.C.Hsu and P.W. Anderson, nature, 333, 121(1988).

Preparation of High T_c Bi-System Film with Magnetron Sputtering and Physical Properties

M. SUZUKI*, H. NAKANO**, D. ABUKAY, and L. RINDERER

Institute of Experimental Physics, University of Lausanne, CH-1015 Lausanne-Dorigny, Switzerland

ABSTRACT

High T_C Bi-system film were deposited on the (100) face of MgO single crystal with rf magnetron sputtering method using the single target Bi-(Pb)-Sr-Ca-Cu-O and then annealed. From the temperature dependence of resistivity and the electron microscope photograph, it was found that the critical temperature T_C of Bi-system film depends on the annealing condition and the film thickness, and that the grain size grows with doping Pb into Bi-Sr-Ca-Cu-O. Further, from the currernt-voltage (*I-V*) characteristic of Bi-system films under microwave radiation and magnetic field and tunneling spectroscopy, the physical properties of these films were investigated. The first Shapiro step was observed in the *V-dI/dV* characteristic under the microwave radiation (50GHz). The one electron tunneling spectrum at 38K gave the superconducting gap 2Δ=60meV.

KEY WORDS: High T_C, Bi, thin film, Josephson effect, tunneling spectroscopy

INTRODUCTION

Since the High T_C ceramic superconductor, Ba-La-Cu-O system had been discovered by Bednorz and Muller [1] at 1986, many High T_C ceramic superconductors have been developed. These oxide superconductors have been very attractive in both sides of applied and basic physics, because they have not only a high critical temperature but also extraordinary properties which are different from predictions of the BCS theory [2]. The High T_C Bi-system has the second highest T_C of High T_C oxide ceramics and the surface of this system is relatively stable. These properties are favorable in a fundamental physics as well as an electronic engineering field. We have developed a formation method of high quality Bi-system film and studied physical properties of it.

EXPERIMENTAL RESULTS AND DISCUSSIONS

Film preparation:

The High T_C Bi-system film was deposited onto the (100) face of single crystal MgO substrate by the rf magnetron sputtering method with a $Bi_2PbSr_2Ca_2Cu_4O_x$ target prepared by usual ceramic techniques without heating substrate, where the sputtering power, environmental gas and gas pressure were respectively 100W, pure Ar and 0.05mbar. The resistivity of the samples was measured through mechanically patterned bridges (length: ≈1.5mm, width: ≈0.5mm) by the 4-probe technique in a closed-cycle refrigerator.

Annealing effect of the electrical resistivities - temperature (ρ-T) characteristic:

All the samples, as deposited, were found in amorphous phase. The temperature dependences of the electrical resistivities after the first annealing are shown in Fig. 1(a), where S7-7 (860°C,0.25h) means that the annealing at 860°C for 0.25hour was applied for the sample of S7-7. The ρ-T characteristics of film annealed at 850°C and 860°C are semiconductive, and the one at 840°C is metallic. But we couldn't get sharp transition with only one annealing. After the second annealing at 840°C for 2h, the samples with first annealing of 850-860°C become a rather good superconductor with sharp transition of 80K phase as shown in Fig.1(b). These results suggest that the superconducting phase is not stable above 840°C, but first annealing at temperature above 840°C is necessary to crystal growth. The third annealing for the sample S7-8 confirms this fact as shown in Fig. 1 (c).

Film thickness dependence of ρ-T characteristic:

Figure 2 shows that the thin films have higher proportions of 110K phase in the ρ-T characteristic than the thick one in the region of film thickness between 0.5mm and 3mm.The first annealing conditions of samples S8-3 and S9-3 are respectively 850°C for 1h and 855°C for 20 minutes, and second annealing conditions of them are equally 840°C for 2h.

* On leave from Saga University, Saga, 840 Japan

** On leave from Nagoya University, Nagoya, 464-01 Japan

Lead doping effect:

The electron micro probe analysis has shown that there remained no lead in films after heat treatments [3]. Even if the annealing temperature was the same value, the grain sizes in the films deposited by a target without lead ($BiSrCaCu_2O_x$) were much smaller and their surface morphology was less rough than the one prepared by the target with lead as shown in Fig. 3. Because of the small grain sizes, the film sputtered from the target without lead does not show the sharp transition.

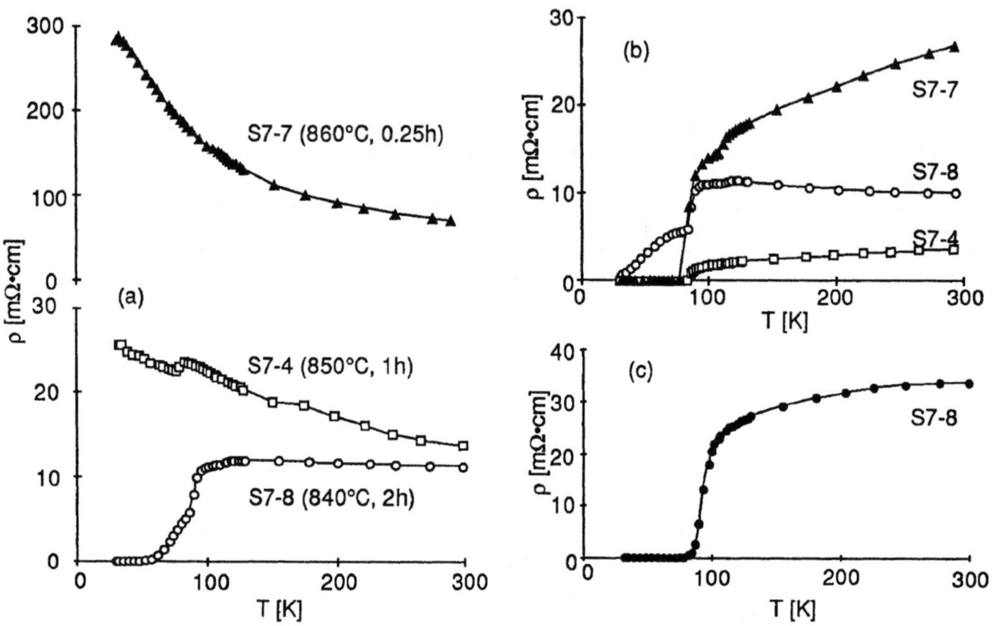

Fig.1 Temperature dependences of the electrical resistivities after the first annealing (a), after second annealing at 840°C for 2h (b) and after third annealing (870°C for 1h, and then, 840°C for 2h) for S7-8 (c).

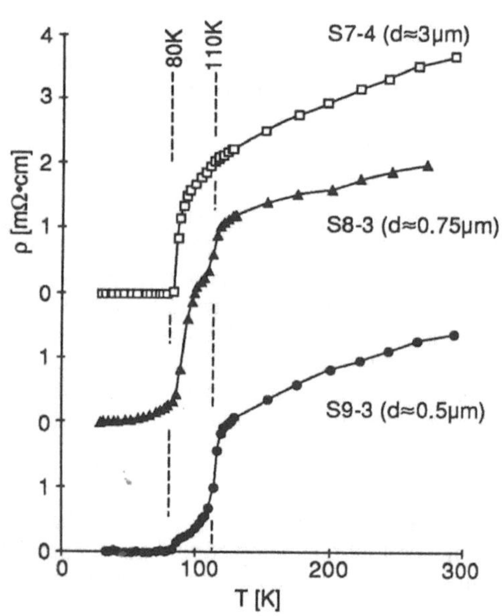

Fig.2 Temperature dependences of the electrical resistivities with various thickness *d*.

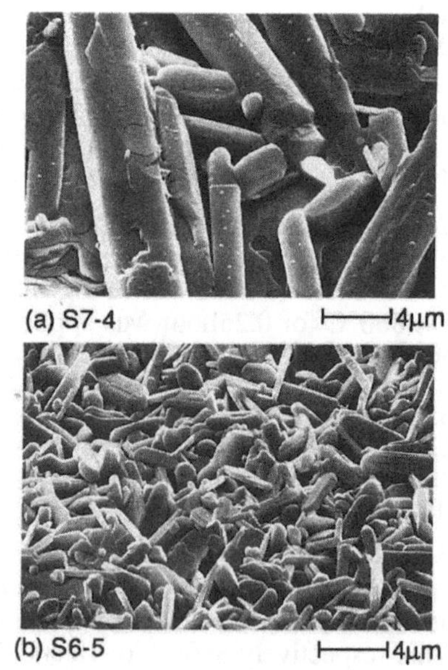

Fig.3 SEM pictures of films sputtered by the targets with Pb (a), and without Pb (b). (S6-5 was annealed at 850°C for 2h)

Microwave radiation and magnetic field effects:

Figure 4 shows the *V-dI/dV* characteristics of S7-8 after the third annealing with various power of microwave radiation. The peak appears at about 100mV which corresponds to the first step of Shapiro steps for a single Josephson junction given by $hf/2e$ ($f \approx$ 50GHz). As increasing the microwave radiation power, the peak is increasing, and then, decreasing. This phenomena also corresponds to the Shapiro step. These results show the existence of Josephson junctions made naturally in the grain boundaries. This film might have a large number of Josephson weak links with series and parallel connections, but the step structure was slightly observed as like as Y-Ba-Cu-O films [4]. Magnetic field dependence of voltage at various currents were measured and there are some structures reappearably at the lower voltage region [5]. It seems to be corresponding to the loop structures of Josephson weak links. The effects of microwave radiation and magnetic field were observed below 50K and became clear at low temperature. One of the possibility, the junction type is Superconductor-Normal-Superconductor (SNS) junctions, because the normal coherence length of SNS junction increases as decreasing temperature.

One electron tunneling spectroscopy:

The Pb-SiO-High T_C Bi-system junctions were fabricated on the High T_C Bi-system film with two phases of T_C=80 and 110K by depositing SiO and then Pb, where the 80K phase is dominant in volume. The *V-dI/dV* characteristic at 38K has two structures near ±30mV as shown in Fig.5. The lower curve is the *V-dI/dV* characteristic of the junction fabricated on High T_C Bi-system ceramic and has clear gap structures. From a comparison of the upper with the lower one, it may be concluded that the two structures in the upper one are superconducting gap structures. Thereby, a value of superconducting gap of high T_C Bi-system film is estimated as 2Δ=60meV [6].

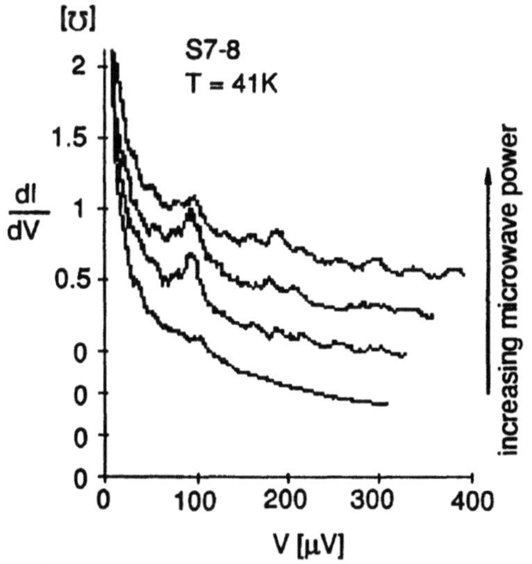

Fig.4 Microwave radiation dependence of *V-dI/dV* characteristics.

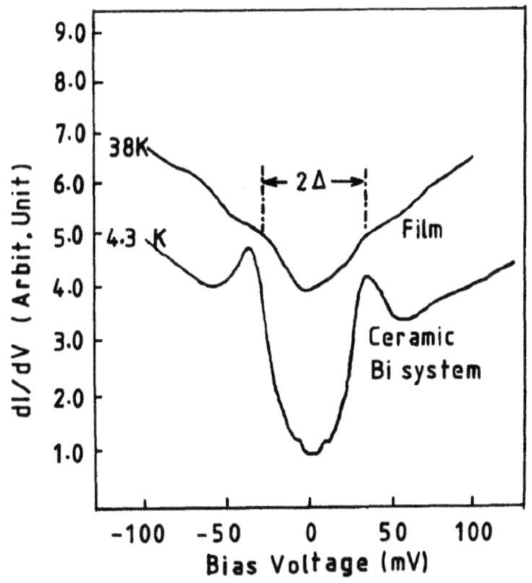

Fig.5 The *V-dI/dV* characteristics for the junctions fabricated on High T_C Bi-system film and on High T_C Bi-system ceramic.

CONCLUSIONS

It is very important in a fundamental physics and an electronic engineering to make high quality superconducting film. In this paper, the High T_C Bi-system film was formed on the (100) face of MgO single crystal with the magnetron sputtering method. The annealing effect and the film thickness dependence of T_C were investigated, further, the physical properties of the High T_C Bi-system were studied by the Josephson effect and the one electron tunneling spectroscopy. The results are as follows; when the first annealing is followed by the second annealing at 840°C for 2h, the first annealing process at temperature from 850°C to 870°C gives a transition in ρ-T characteristics sharper than the one at 840°C. In the region of film thickness between 0.5μm and 3μm, a proportion of 110K phase increases with decreasing thickness. The dope of Pb into Bi-Sr-Ca-Cu-O makes the grain sizes large with the same annealing temperature. The Josephson effect was found in the *V-dI/dV* characteristics under the microwave radiation. The effects of microwave radiation and magnetic field indicate that the Josephson junctions are naturally made at the grain boundaries. From the results of one electron tunneling spectrum, the*V-dI/dV* characteristic at 38K gave superconducting gap 2Δ=60meV.

ACKNOWKEDGMENTS

The authors thank Prof. Aomine and Mr. Soares for their valuable suggestions and encouragement. They also thank Mr. Burri and Mr. Nicoud for their assists of experiments and are indebted to the Fonds National Suisse de la Recherche Scientifique.

REFERENCES

1. Bednorz J G, and Muller K A (1986) Possible high T_c superconductivity in the Ba-La-Cu-O system. Z Phys B 64: 189-193
2. The American Physical Society (1987) High-Temperature Superconductivity. American Physical Society, New York
3. Abukay D, Nakano H, Suzuki M, Burri G, Rinderer L (submitted) A study of the effects of annealing on thin films of Bi-(Pb)-Sr-Ca-Cu-O. Helv Phys Acta
4. Gallop J C, Radcliffe W J, Langham C D, Sobolewski R, Kula W, Gierlowski P (1989) Josephson effects in a YBaCuO thin film. Suprercond Sci Technol 2: 1-4
5. Nakano H, Suzuki M, Abukay D, Nicoud S, Rinderer L (submitted) I-V characteristics in Bi-(Pb)-Sr-Ca-Cu-O films. Helv Phys Acta
6. Suzuki M, Abukay D, Nakano H, Rinderer L (submitted) Tunneling study of High T_c Bi-system film. Helv Phys Acta

4.3 Other Superconducting Films

Preparation and Characteristics of $Nd_{2-x}Ce_xCuO_{4-y}$ Thin Films

S. HAYASHI, K. HIROCHI, H. ADACHI, S. KOHIKI, S. HATTA, K. SETSUNE, T. HIRAO, and K. WASA

Central Research Laboratories, Matsushita Electric Industrial Co., Ltd., 3-15, Yagumo-Nakamachi, Moriguchi, Osaka, 570 Japan

ABSTRACT

Characteristics of the electron-doped Nd-Ce-Cu-O system were studied using highly c-axis oriented thin films. $Nd_{2-x}Ce_xCuO_{4-y}$ films with various Ce concentrations (x) were prepared on $SrTiO_3$(100) substrates by sputtering and post-annealing. After the reducing treatment, the x=0.15 film exhibited a sharp superconducting transition with zero resistivity at 22K, in consistent with diamagnetic measurement. Transport and superconducting properties were sensitive to the Ce concentration and the reduction process. Optical spectra showed that the Nd_2CuO_4 is a semiconductor with an energy gap of 1.3 eV, and is transformed into a metal where the Drude-like reflection and near-infrared absorption are induced by the Ce doping.

KEY WORDS: Nd-Ce-Cu-O, thin films, Hall effect, diamagnetism, optical spectra

INTRODUCTION

The discovery of new cuprate superconductors $Ln_{2-x}Ce_xCuO_{4-y}$ (Ln=Nd,Sm and Pr)[1,2] has significant implications for understanding of the high-T_c superconductivity. In the previous cuprate superconductors, the charge carriers in the normal state are holes, and oxidation process is important to obtain excellent superconducting properties. In contrast, in these new materials, the charge carriers are electrons added by substituting Ce^{4+} ions for Ln^{3+} ones, and reduction process is indispensable to obtain superconductivity.

For scientific and technological reasons, the study of thin films is of interest. Especially, epitaxial films are of great use for fundamental research [3]. Thin film samples are suitable for the reduction due to their large ratio of surface. Here, we report the preparation of highly c-axis oriented $Nd_{2-x}Ce_xCuO_{4-y}$ films and their transport, diamagnetic and optical properties[4-6].

THIN FILM PREPARATION

$Nd_{2-x}Ce_xCuO_{4-y}$ thin films were prepared by an rf magnetron sputtering method. Typical sputtering conditions are listed in Table I. Sputtering targets were fabricated by sintering a mixture of Nd_2O_3, CeO_2 and CuO at 1050 °C in air for 5 hours with various Nd/Ce ratios. Perovskite single crystal substrates of $SrTiO_3$ were selected for the epitaxial growth. The Ce concentrations x of deposited films were almost equal to those of the targets.

Table I. Sputtering conditions.

Target	$Nd_{2-x}Ce_xCu_{1.5}O_y$ (x= 0 - 0.18)
Substrate	$SrTiO_3$ (100)
Substrate Temperature	500 - 600 °C
Sputtering Gas	Ar
Gas Pressure	0.4 Pa
Rf Input Power	160 W
Growth Rate	10 - 12 nm/min

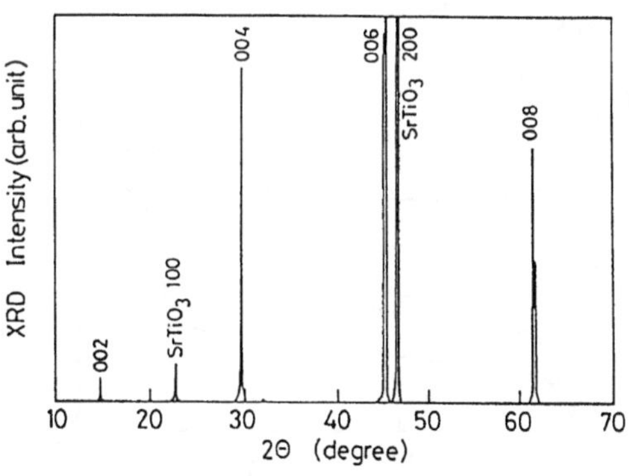

Fig.1 X-ray diffraction pattern for a $Nd_{1.85}Ce_{0.15}CuO_{4-y}$ film before reduction.

As-deposited films were poor in crystallinity. In order to improve crystallinity, the films were post-annealed at 1100°C for 2 hours in air, followed by quenching. The films showed a highly oriented structure with the c-axis normal to the substrate, as shown in Fig.1 for the x=0.15 film. The c-axis lattice constant decreased from 12.15 A to 12.03 A as x increased from 0 to 0.18.

The post-annealed films were not superconducting. Superconductivity was induced by the reducing treatment. This treatment was done by annealing the films in a vacuum (10^{-2} -10^{-6} Torr). The annealing time was adjusted to give the highest T_c. It seems to depend on the annealing temperature and on x. The change of c-axis lattice constant after reduction was less than 0.01A.

Transport properties

Figure 2 shows the temperature dependence of the resistivity(ρ) for the x=0.15 film before and after reduction (800 °C, 1 hour). Semconductor-like behavior ($d\rho/dT<0$) was observed below 100 K before reduction. After reduction, the film exhibited metallic temperature dependence and a sharp superconducting transition with zero resistivity at 22 K. The resistivity at room temperature is much lower than that of bulk ceramics[1,2], reflecting high quality of the film.

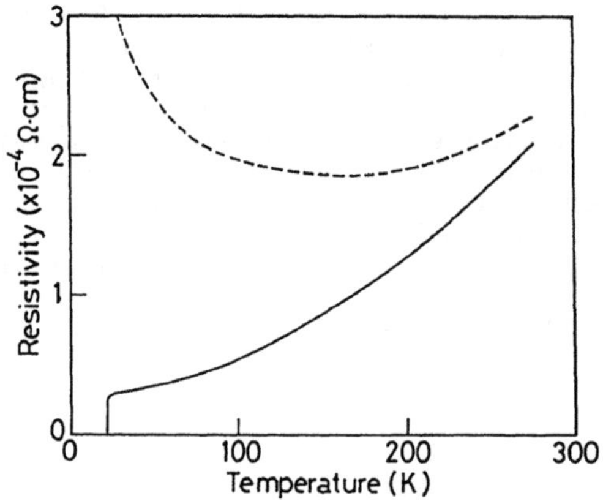

Fig.2 Temperature dependence of the resistivity for a $Nd_{1.85}Ce_{0.15}CuO_{4-y}$ film. The broken and solid lines are before and after reduction, respectively.

Figure 3 shows the temperature dependence of the resistivity for the films with various Ce concentrations x $(0<x0.18)$ after reduction. Due to optimum reducing treatments, the $d\rho/dT$ increased for all x and metallic behavior ($d\rho/dT>0$) was obseved for x > 0.10. Superconductivity was obseved for $0.14<x<0.18$. With increasing x from 0.14 to 0.18, T_c tended to lower, similarly to bulk samples[1,2].

Measurement of the Hall effect was done by the Van der Pauw method. The results were shown in Fig.4. The Hall coefficient R_H was negative throughout the measurement, indicating that the charge carriers are electrons in the normal state. The temperature dependence of R_H for the x=0.15 film is

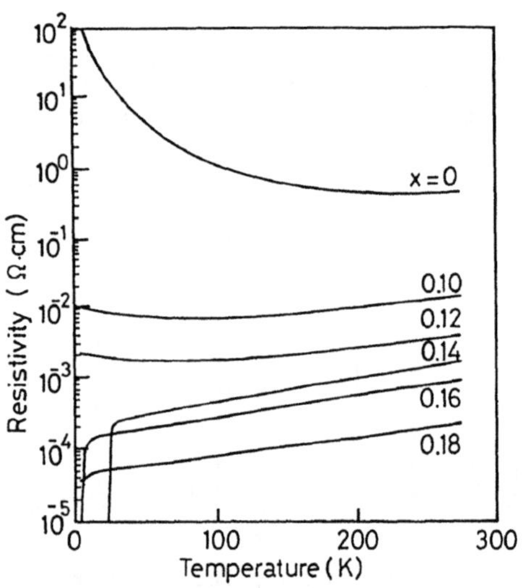

Fig.3 Temperature dependence of the resistivity for $Nd_{2-x}Ce_xCuO_{4-y}$ films after reduction.

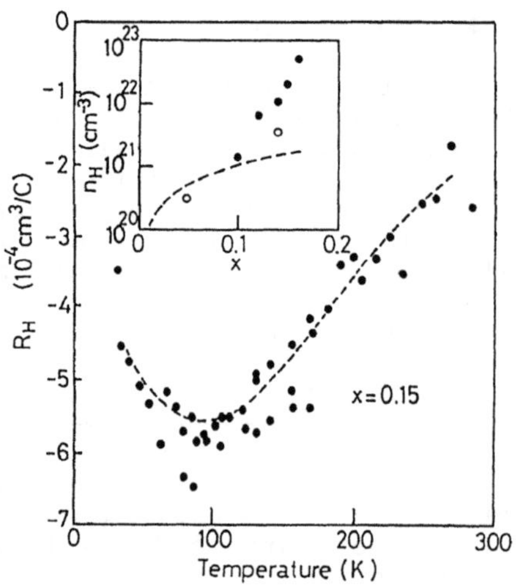

Fig.4 Temperature dependence of the Hall coefficient for a $Nd_{2-x}Ce_xCuO_{4-y}$ film (x=0.15) after reduction. The inset shows the Hall carrier density at room temperature for various x (o, before reduction;•, after reduction).

anomalous, but very similar to that of $La_{1.85}Sr_{0.15}CuO_4$ single crystal film [3]. It is difficult to understand the temperature dependence only by the effects of phonon scattering and carrier localization.

The inset shows the Hall carrier density n_H at room temperature for various x, which are derived from the relationship $R_H=-1/n_He$. The broken line indicates the Ce concentration in cm^{-3}. As x exceeds 0.1, data points deviate from this line. This may suggest the emergence of the metallic state above x>0.10. The increase of n_H was observed in the x=0.14 film by the reduction, which may be attributed to the oxygen vacancies with y=0.03.

DIAMAGNETIC PROPERTIES

The diamagnetism of the films was measured by an rf SQUID susceptometer under a magnetic field of 10 Oe. Figure 5 shows the temperature dependence of the diamagnetization for the films with various x after the reduction. The diamagnetization was measured on warming the film after zero field cooling (the shielding effect). Before reduction, diamagnetism was not observed. After reduction, in the x=0.15 film, the diamagnetization was observed below 25 K. It rapidly decreased at around 20 K with increasing the temperature.

According to Bean's formula, the diamagnetization at 4.2 K for the x=0.15 film corresponds to a critical current density of $7x10^5$ A/cm^2, which is consistent with the result of electrical measurement. The diamagnetization is highly dependent on x. The value at 4.2 K remarkably decreased as x deviates from 0.15.

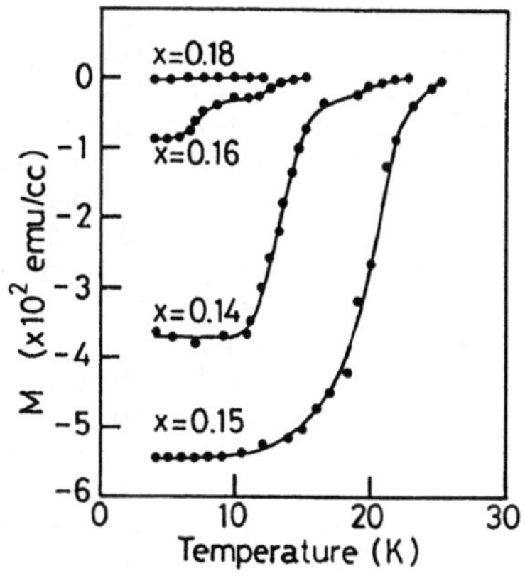

Fig.5 Temperature dependence of the diamagnetization for the $Nd_{1.85}Ce_{0.15}CuO_{4-y}$ films with various x after reduction.

OPTICAL PROPERTIES

Optical measurements have been carried out to investigate the change of the electronic state by the Ce doping. Reflectivity and absorption spectra were measured from 0.4 to 2.5 μm. The polarization of the incident light was normal to the c-axis. The spectra were measured before and after reduction, but no appreciable change was observed except for the x=0 film. The change of the spectra by the reduction seems to be due to the electron doping effect originating from oxygen vacancies.

Figure 6 shows the reflectivity spectra. For the Nd_2CuO_4 film(x=0), some features were observed, but no sign of plasma reflection. With increasing the Ce concentration x up to 0.12, a Drude-like plasma reflection appeared. At higher x of 0.14<x<0.16, the reflection became much clearer with a plasma edge at around 1.0 μm. The solid lines in this figure are the Drude fits. The plasma frequency ωp was estimated to be 1.1 eV, and was almost constant over the range of 0.14<x<0.18. The ωp value is about 0.2 eV larger than that of La-Sr-Cu-O system [3], but its constant behavior against dopant concentration is common.

Figure 7 shows the absorption spectra. A solid line for x=0 shows the absorption spectrum of the Nd_2CuO_4 film before reduction. The intense absorption below 1 μm seems to originate from the interband transition with

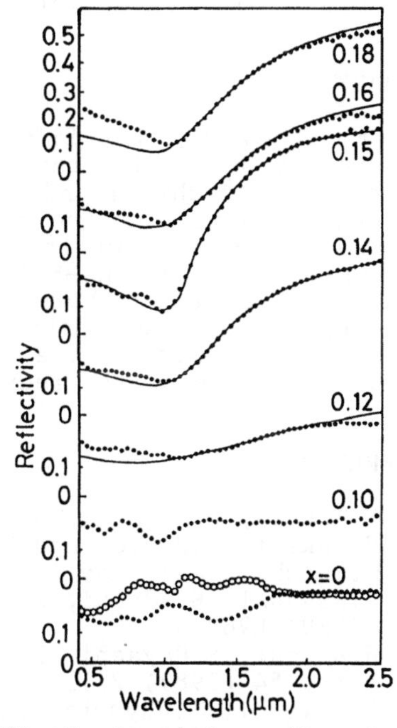

Fig.6 Optical reflectivity spectra for the $Nd_{2-x}Ce_xCuO_{4-y}$ films after reduction. Open circles for x=0 are before reduction.

an energy gap of 1.3eV. According to the Mott-Hubbard energy scheme, it may be interpreted as a charge transfer excitation between the O 2p states (valence band) and the upper Hubbard band (conduction band). Suzuki [3] reported an energy gap of 2.0 eV for La_2CuO_4 single crystal film, and Tajima et al.[7] observed charge transfer excitation at 2.0 and 1.5 eV for La_2CuO_4 and Nd_2CuO_4 single crystals, respectively. These gap energies may be inherent in different Cu-O configurations, octahedra for La_2CuO_4 and squares with no apical oxygen for Nd_2CuO_4.

As x increases from 0 to 0.12, the interband absorption decreased losing intensity at the lower energy side, and a broad absorption appeared in the near-infrared region. In the superconducting compositional region, $0.14<x<0.18$, the spectra were dominated by the near-infrared absorption and the interband absorption was completely suppressed at x=0.18.

These features reflect the change of the electronic state by the electron doping into the conduction band, which evolves from the semiconducting state to the metallic state. The integrated oscillator strength in this spectral region was enhanced in the superconducting compositional region ($0.14<x<0.18$). This may suggest the development of density states and its correlation with T_c. However, the optical spectra in this region are rather insensitive to the reduction whereas transport and superconducting properties are highly sensitive to it. The effect of reduction may be a key to elucidate the superconducting mechanism. Due to the photoemission study [8], the mean Cu valence decreases from 2+ for x=0 to about 1+ for x=0.15 with Ce doping, and the reduction eliminates the remaining 2+ species. It is important to distinguish superconductivity-correlated behavior from simple doping dependence and oxygen stoichiometry dependence.

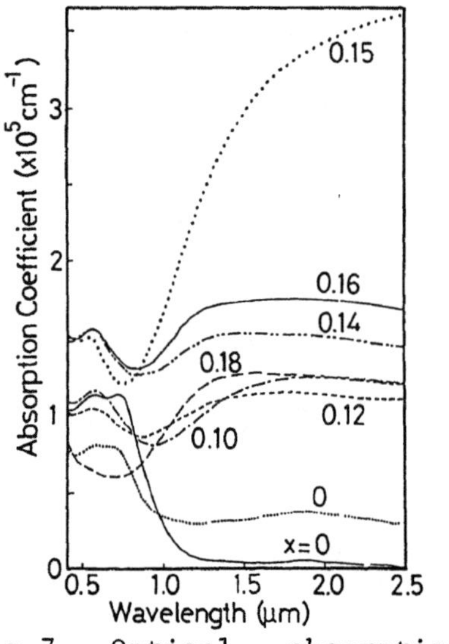

Fig.7 Optical absorption spectra for $Nd_{2-x}Ce_xCuO_{4-y}$ films after reduction. Solid line for x=0 is before reduction.

SUMMARY

$Nd_{2-x}Ce_xCuO_4$ thin films with excellent superconducting properties were prepared by sputtering and post-annealing. The resistivity at room temperature as low as 10^{-4} ohm cm and its temperature dependence are metallic, reflecting the high quality of the films. Superconductivity of the Nd-Ce-Cu-O system emerges in the transitional region from a semiconductor to a metal, analogously to the La-S-Cu-O system. Transport, diamagnetic and optical properties revealed here, almost in consistent with previous studies on polycrystalline ceramics or single crystal, more clearly show the Ce concentration dependence and the effect of reduction. It is important to clarify the role of the reduction in the emergence of superconductivity.

REFERENCES

[1] Y.Tokura, H.Takagi and S.Uchida, Nature **337** (1989) 345.
[2] H.Takagi, S.Uchida and Y.Tokura, Phys.Rev.Lett.**158** (1989)178.
[3] M.Suzuki, Phys. Rev. B **39** (1989) 2312.
[4] S.Hayashi, H.Adachi, K.Setsune, T.Hirao and K.Wasa, Jpn. J. Appl. Phys. **26** (1989) L962.
[5] H.Adachi, S.Hayashi, K.Setsune, S.Hatta, T.Mitsuyu and K.Wasa, Appl. Phys. Lett. **54** (1989) 2713.
[6] K.Hirochi, S.Hayashi, H.Adachi, T.Mitsuyu, T.Hirao, K.Setsune and K.Wasa, Physica C, **160** (1989) 273.
[7] S.Tajima, H.Ishii, T.Nakahashi, S.Uchida, M.Seki, S.Suga, Y.Hidaka, M.Suzuki, T.Murakami, K.Oka and H.Unoki, J. Opt. Soc. Am. B **6** (1989) 475.
[8] S.Kohiki, J.Kawai, S.Hayashi, H.Adachi, S.Hatta, K.Setsune and K.Wasa, submitted, Phys.Rev.B.

Superconducting Properties of Artificially Superstructured Films Composed of Nitride Materials

MITSUGU SOHMA, KENJI KAWAGUCHI, and SHIGEMITSU SHIN

National Chemical Laboratory for Industry, 1-1, Higashi, Tsukuba, Ibaraki, 305 Japan

ABSTRACT

Artificially superstructured films (ASF's) [niobium nitride (NbN)/iron nitride (FeN)] and [molybdenum nitride (MoN)/iron nitride (FeN)] have been prepared by an alternate reactive deposition method. The MoN/FeN ASF's consist of γ-Mo_2N layers and amorphous like iron nitride layers. The nitrogen composition of iron nitride layers is extremely poor judged from the Moessbauer measurement of MoN/FeN ASF's. An ASF MoN(20nm)/FeN(10nm) showed ferromagnetism at room temperature and superconducting transition at 4.0K. It is suggested that ferromagnetism and superconductivity coexist below Tc. Niobium nitride layers of NbN/FeN ASF's have a cubic structure similar to the δ-NbN_x phase.

KEY WORDS: artificially superstructured film, superconductivity, ferromagnetism, MoN, FeN

INTRODUCTION

There has been many reports in artificially superstructured films (ASF's) composed of superconducting materials[1]. It is interesting to study physical properties of ASF's with the combination of ferromagnetic and superconducting materials such as magnetic proximity effect, dimensional crossover and so on [2]. Furthermore, it will be possible to develop new devices utilizing the interaction between either properties. Most of those ASF's which have been studied consist of metals and alloys, however a lot of superconducting materials with high critical temperature (Tc) are ceramics, e.g. nitrides (NbN, MoN), oxides ($(La,Ba)_2CuO_4$, $Ba_2YCu_3O_7$) and chalcogenides (chevrel compounds). In this paper, we present the study of ASF's of [niobium nitride (NbN)/iron nitride (FeN)] and [molybdenum nitride (MoN)/iron nitride (FeN)] prepared for various conditions. NbN and MoN have relatively high Tc and have a simple structure (B1-structure) [3-6]. FeN is a typical ferromagnetic compound and a promising magnetic recording material [7,8]. The combination of NbN and FeN was examined at first, but reproducible superconducting transition was not obtained. Instead of NbN, MoN was attempted. Superconducting transition of the MoN/FeN ASF's was observed reproducibly, though the Tc was low.

EXPERIMENTAL PROCEDURE

Samples were prepared by an alternate reactive deposition method. The deposition chamber was first evacuated to the base pressure about 10^{-7} Torr. The deposition was performed in the atmosphere of nitrogen or ammonia. The atmospheric pressure was examined in the range of 10^{-5} - 10^{-3} Torr. At need, atmospheric gas was activated by R.F. plasma. In general, nitride films are prepared in nitrogen atmosphere, and bulk nitride materials are often synthesized in ammonia atmosphere. We examined both atmosphere. Metal ingots (Nb, Mo and Fe) were evaporated by electron beam guns. The substrates employed were cleaved MgO single crystal plates and polyimido films (UPILEX, trade name by Ube Industries, Ltd.,). Samples were deposited at the various substrate

temperature (Ts) from 200° to 600°C. Very low deposition rate 0.02 - 0.05nm/sec was adopted for the purpose of enough nitrogen composition of the samples and precise thickness control. The thickness of the sample was monitored by the method of quartz crystal oscillation. But precise calibration of the absolute film thickness has not been done yet. The film thickness described in this paper is tentative value. NbN single layered films (SLF's) and MoN SLF's were also studied to compare with ASF's.
Structural properties of ASF's were characterized by means of X-ray diffraction (XRD) using a standard powder diffractmeter. Electric resistance was measured by a conventional four-prove method. The measuring current density was less than 1.0 A/cm^2. We define Tc as a midpoint of residual resistance. Magnetic properties of the samples were measured by Moessbauer spectroscopy.

RESULT AND DISCUSSION

▪ NbN/FeN ASF's

Figure 1 shows three XRD patterns of NbN/FeN ASF's deposited on MgO substrates in nitrogen atmosphere (10^{-3} Torr). The three ASF's have the same composition of NbN(4.0nm)/FeN(0.8nm). No superstructure reflections and no nitride diffraction peaks were observed for the sample (Ts=600°C) in Fig.1(a). When nitrogen atmosphere was activated by R.F. plasma, a δ-NbN_x (200) peak was seen in the middle angle region (Fig.1(b)). For the sample (Ts=300°C) (Fig.1(c)), the δ-NbN_x (200) peak and the superstructure peaks were observed in the middle angle and low angle region, respectively. Higher order low angle peaks indicate that the ASF has well defined artificial periodicity. Such ASF's did not show superconducting transition.
Reproducible superconducting transition was not observed in NbN SLF's prepared under various conditions, either. Each XRD pattern of NbN SLF's deposited on a MgO substrate and a UPILEX substrate is shown in Fig.2 (a),(b), respectively. The XRD pattern in Fig.2(a) indicates that NbN film was epitaxially grown with the direction of <100> perpendicular to the film plane. The result of the NbN SLF on a UPILEX substrate in Fig.2(b) shows a polycrystal pattern. Both the samples of Fig.2(a) and (b) have almost the same lattice parameter judged from (200) peaks. The lattice parameter of the films calculated from

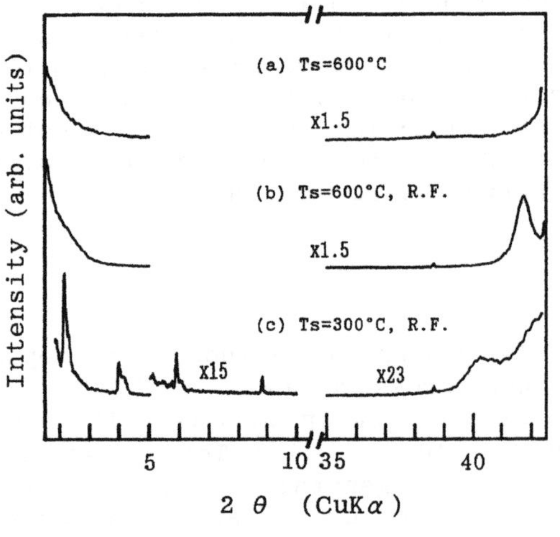

Fig.1 XRD patterns of [NbN(4.0nm)/FeN(0.8nm)] ASF's prepared in nitrogen atmosphere (10^{-3} Torr) (a) Ts=600°C, (b) Ts=600°C, in R.F. plasma, (c) Ts=300°C, in R.F plasma.

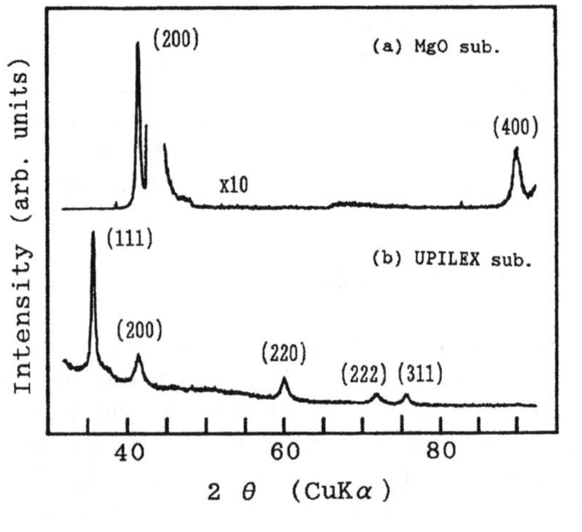

Fig.2 XRD patterns of NbN SLF's deposited in ammonia atmosphere (10^{-5} Torr) at Ts=200°C. The samples were deposited (a) on a MgO substrate, (b) on a UPILEX substrate.

the (200) peaks are mostly within the range of 0.432 - 0.438nm. On the other hand, the reported value of δ-NbN_x (cubic) is within the range of 0.438 - 0.4395nm (0.86 <x< 1.06) [9]. The lattice parameter of c-axis of γ-NbN_x (tetragonal) is within the range of 0.4310 - 0.4335nm (0.67 <x< 0.78) [9]. Our NbN SLF's prepared in this study is considered to have B1-structure with somewhat small lattice parameter which has not been reported yet. Nitrogen composition of NbN_x SLF's is supposed to be x<0.86. Probably, such undesirable superconducting behavior is due to the poor nitrogen composition.

■ MoN/FeN ASF's

There are three well known superconducting molybdenum nitrides, e.i. B1-MoN (Tc=5 - 14K), δ-MoN (12K) and γ-Mo_2N (5K). Superconducting molybdenum nitrides are obtained over wide nitrogen compositional range (30 - 50 atomic %). We tried to prepare MoN/FeN ASF's instead of NbN/FeN ASF's. Typical XRD patterns of MoN/FeN ASF's are shown in Fig.3. The samples were deposited on a UPILEX substrate (Fig.3(a)) and a MgO substrate (Fig.3(b)) simultaneously under the condition (Ts=300°C, the sample composition MoN(20nm)/FeN(10nm), 10^{-4} Torr ammonia atmosphere, deposition rate 0.02nm/sec). The MoN layers of the ASF's on both a UPILEX substrate and a MgO substrate have a γ-Mo_2N structure as shown in Fig.3(a),(b). The sample on a UPILEX substrate showed a polycrystal XRD pattern in Fig.3(a). While the preferred orientation of (100) plane were observed for the sample on a MgO substrate shown in Fig.3(b). No diffraction peaks derived from FeN layers were observed in both samples. The information about artificial superstructure was obtained from the low angle diffractions (Fig.3(b)). We could not confirm the periodicity for the sample on a UPILEX substrate due to experimental difficulties of low angle XRD. Electric resistance measurement of the ASF's on a UPILEX substrate shown in Fig.4 indicates superconducting transition at 4.0K. The XRD result in Fig.3(a) and very sharp superconducting transition in Fig.4 represent the formation of single phase γ-Mo_2N. Moessbauer spectroscopy measurements were performed using a conventional absorption method. The sample used on a UPILEX substrate was measured at room temperature. The spectrum in Fig.5 shows a major magnetic sextet and a small doublet components. The sextet peak is considered to be attributed to α-Fe. The spectrum suggests that major part of the FeN layers is an iron metal rather than an iron nitride. The magnetic properties of the doublet component are now investigated. The Moessbauer result implies that interdiffusion at interfaces is limited. It is believed that a few atomic % magnetic impurities collapse the superconductivity. A compositional mixing is negligible judged from both the superconductivity and magnetism of the ASF's. A superconducting dimensional crossover phenomenon has been reported for the ASF

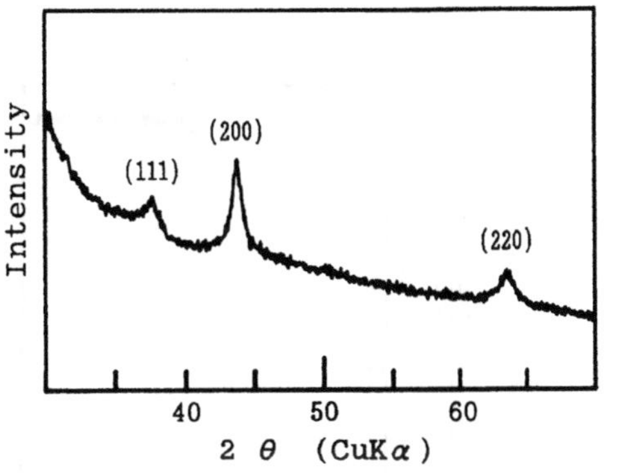

Fig.3(a) XRD pattern of [MoN(20nm)/FeN(10nm)] ASF deposited at 400°C on a UPILEX substrate in ammonia atmosphere (10^{-4} Torr).

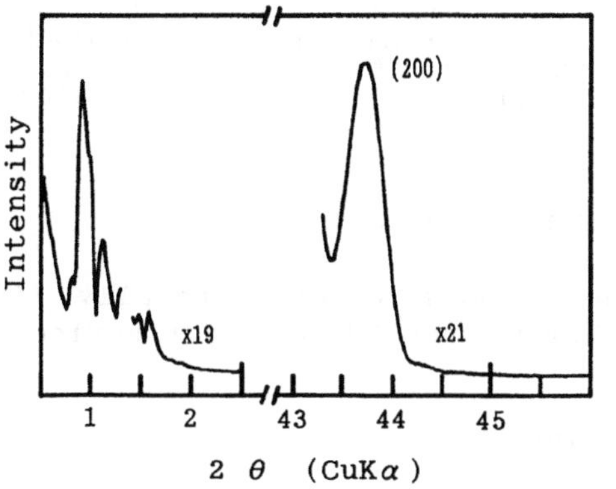

Fig.3(b) XRD pattern of [MoN(20nm)/FeN(10nm)] ASF deposited at 400°C on a MgO substrate in ammonia atmosphere (10^{-4} Torr).

with the combinations of non-superconducting and superconducting materials [1,10]. In the case of an ASF with a magnetic material (Fe) and a superconducting material (V), a dimensional crossover phenomenon is observed [2]. It is considered that superconducting electron pairs diffuse through the non-superconducting layers in the temperature region where three dimensional superconductivity appears. It was proposed that both the superconductivity and the ferromagnetism coexist in the Fe layers, however detailed magnetic properties have not been reported. The FeN layers in our MoN/FeN ASF are ferromagnetic at room temperature and it is reasonable to consider that those layers are ferromagnetic at lower temperature than Tc. Thus the coexistence of ferromagnetism and superconductivity is also expected. Advanced studies of ferromagnetic and superconducting properties are now in progress.

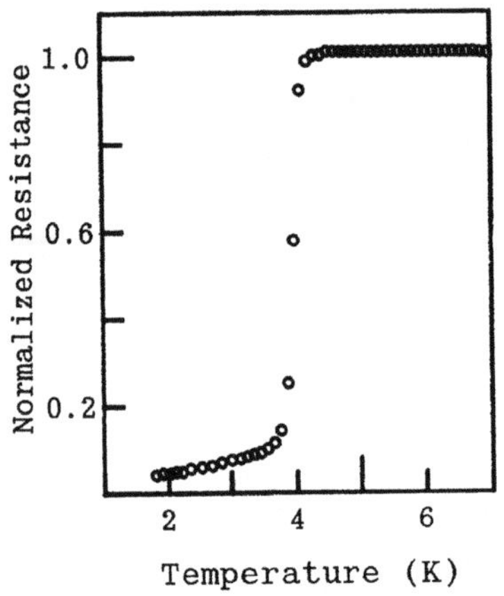

Fig.4 The temperature dependence of the resistance normalized by the resistance at 7.0K. The sample is the same in Fig.3(a).

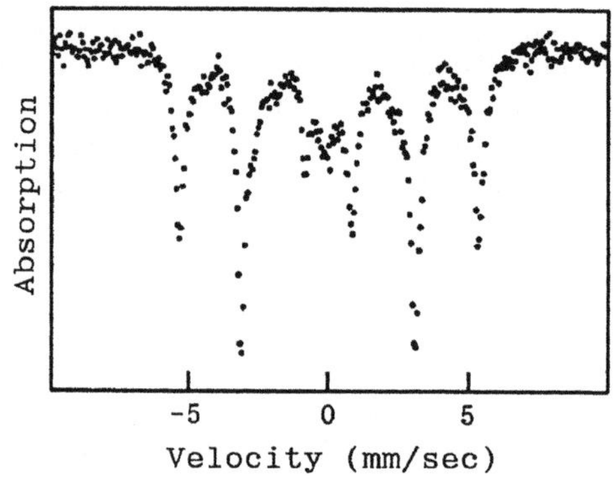

Fig.5 ^{57}Fe Moessbauer spectrum of [MoN(20nm)/FeN(10nm)] ASF at room temperature. The sample is the same in Fig.3(a).

ACKNOWLEDGEMENT

The authors would especially like to thank K.Fukuda for helpful comments, K.Nomura and Y.Oosawa for advice of XRD and resistivity measurement.

REFERENCE

1. T.Shinjo, T.Takada (eds), Metallic Superlattices, (Elsevier, Netherlands, 1987)
2. H.K.Wong, B.Y.Jin, H.Q.Yang, J.E.Hilliald and J.B.Ketterson, Superlattices and Microsructure, 1, 259 (1985)
3. E.F.Skelton, M.R.Skokan, E.Cukauskas, J.Appl.Cryst, 14, 51 (1981)
4. K.Kawaguchi, S.Shin, MRS Symposium Proc., 103, 281 (1988)
5. H.Ihara, K.Senzaki, Y.Kimura, M.hirabayashiand, N.Terada and H.Kezuka, Adv.Cryog.Mat., 32, 603 (1986)
6. Y.H.Shi, B.R.Zhao, L.Li and J.R.Lio, Phys.Rev.B, 38, 4488, (1988)
7. M.Kume, T.Tsujioka, K.Matsuura and Y.Abe, IEEE Trans.Mag., MAG-23, 3633 (1987)
8. C.Chang, J.M.Sivertsen and J.H.Judy, IEEE Trans.Mag., MAG-23, 3636, (1987)
9. L.E.Toth, Transition Metal Nitrides, (Academic Press, New York, 1971)
10. I.Banerjee, Q.S.Yang, C.M.Falco and I.K.Sculler, Phys.Rev.B, 28, 5037, (1983)

5 Applications

5.1 Electronic Use

High Temperature Superconducting Junctions

ANTONIO BARONE*

Dipartimento Scienze Fisiche, Università di Napoli, P. Tecchio 80, Napoli, Italy and INFN, Napoli

ABSTRACT

Some aspects of the Physics of high-temperature superconducting junctions are discussed. More recent results are outlined which may cast some light on the whole topic indicating, in spite of the severe material constraints, an encouraging trend.

1. INTRODUCTION

Since the discovery of high-Tc superconductors [1] an enormous effort has been devoted to the realization of junctions both for the understanding of the underlying physics and for the perspectives of stimulating applications.
Among the high Tc proposed devices there are some already realized which are not clearly competitive, as well as some absolutely competitive which however don't fall completely within today's technological possibilities.

Referring to the SQUID, which remains the basic device for many applications, we observe in fact a variety of d.c.(as well as r.f.) SQUIDs, made by weak link structures, whose performances are not yet fully satisfying, whereas technical progress in this field legitimate the expectation of good quality devices in a near future.[2]
For almost all devices in the small scale high-T_c superconductivity applications the crucial point lies in the lack of a good Josephson tunnel junction. This situation is roughly represented by the sketch of Fig. 1.

I shall try to give a brief account of what is the situation to date in this respect, namely what has been done, what appears to be the most severe material constraints to make reliable Josephson junctions and which projections, can be made for the future. Of course in so doing, I shall refer to a number of results which represent a very small fraction of the vast literature available.

In the next Section I shall discuss those aspects which pertain to the tunneling spectroscopy of high Tc superconductive junction dealing thereby mainly with physical aspects.

Device considerations concerning Josephson structures will be the subject of Section 3. This division is somewhat fictitious since it is hard to make a distinction between what can be learned about material properties, by working on presently available junction structures, and what can be inferred from the present knowledge of material properties in order to fabricate more reliable high-T_c junctions.

2. PHYSICAL ASPECTS

In studying "conventional" low-T_c superconducting Josephson junctions [3] we were used over a period of 25 years to perform suitable inspections of current-voltage (I-V) characteristics to infer essential aspects of the superconductor forming the electrodes as well as of the specific junction properties. Time by

*Work partially supported by the CNR under the Progetto Finalizzato "Superconducting and Cryogenic Technologies"

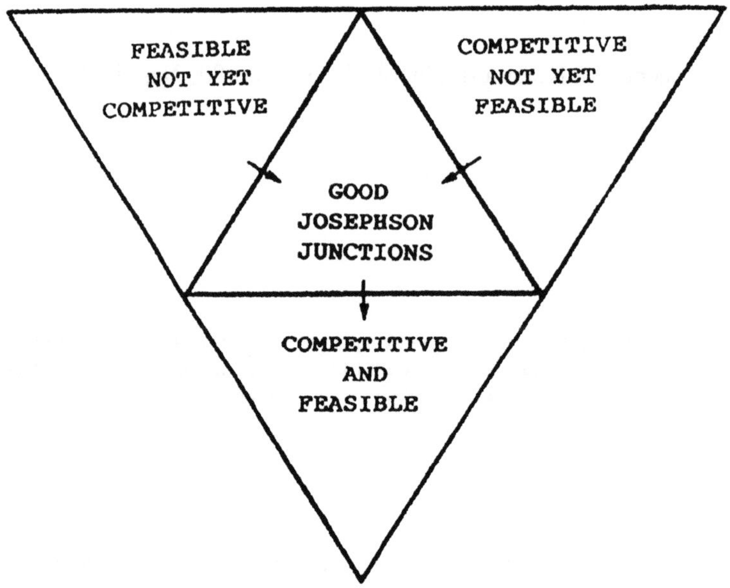

Fig. 1 - Proposed devices needing good reliable Josephson tunnel junctions.

time, depending on the specific task, the analysis was extended to first (dI/dV vs.V) or higher order derivatives. Presently for high-T_c materials, as we know, the situation is reversed, a fantastically large zoology of dI/dV vs.V curves is reported in the literature to investigate a variety of junction structures. There is no doubt that this is imposed by the lack of information which can be extracted by the I-V characteristics of the samples. However, if these show too little, conversely, there is a risk that dI/dV curves show too much.

Two of the main features on which the attention of the investigators has been addressed in this kind of measurements are the size of the energy gap (if any) and the background conductance.

2.1 - Gap Structure.

Since the discovery of high-T_c materials a great attention was concentrated on tunneling spectroscopy studies [4] in order to observe the energy gap and estimate its value. The obvious reason is that an evaluation of the ratio $\gamma = 2\Delta(0)/k_BT_c$ would greatly help in deciding about the underlying mechanism of superconductivity. Indeed depending on the values of γ one can decide on whether we are dealing with "traditional" BCS superconductors ($\gamma \sim 3.5$) or strong coupling superconductors or whatever else.

Earlier measurements were performed by using point contact structures, namely a point (often made by conventional superconductor) properly pressed on a sinterized pellet of high-T_c material or a bulk junction in which the junction was realized by bulk material overlayed by a conventional superconductor counterelectrode · The second step was that of the use of crystals. In this case it was possible to investigate both in the direction of the c-axis or on the a, b plane to study anisotropy In this case however, the necessary pressure of the point on the crystal could very likely induce a damage on the sample so that, as a final result, information on the crystal structure is lost. A different interesting approach has been proposed by Tsai [5]. A further possibility is that of planar structures such as a crystal overlayed by a film and finally the sandwich (film-film) archetype junction. Break-junction configuration was also proposed showing interesting results as well as other techniques.

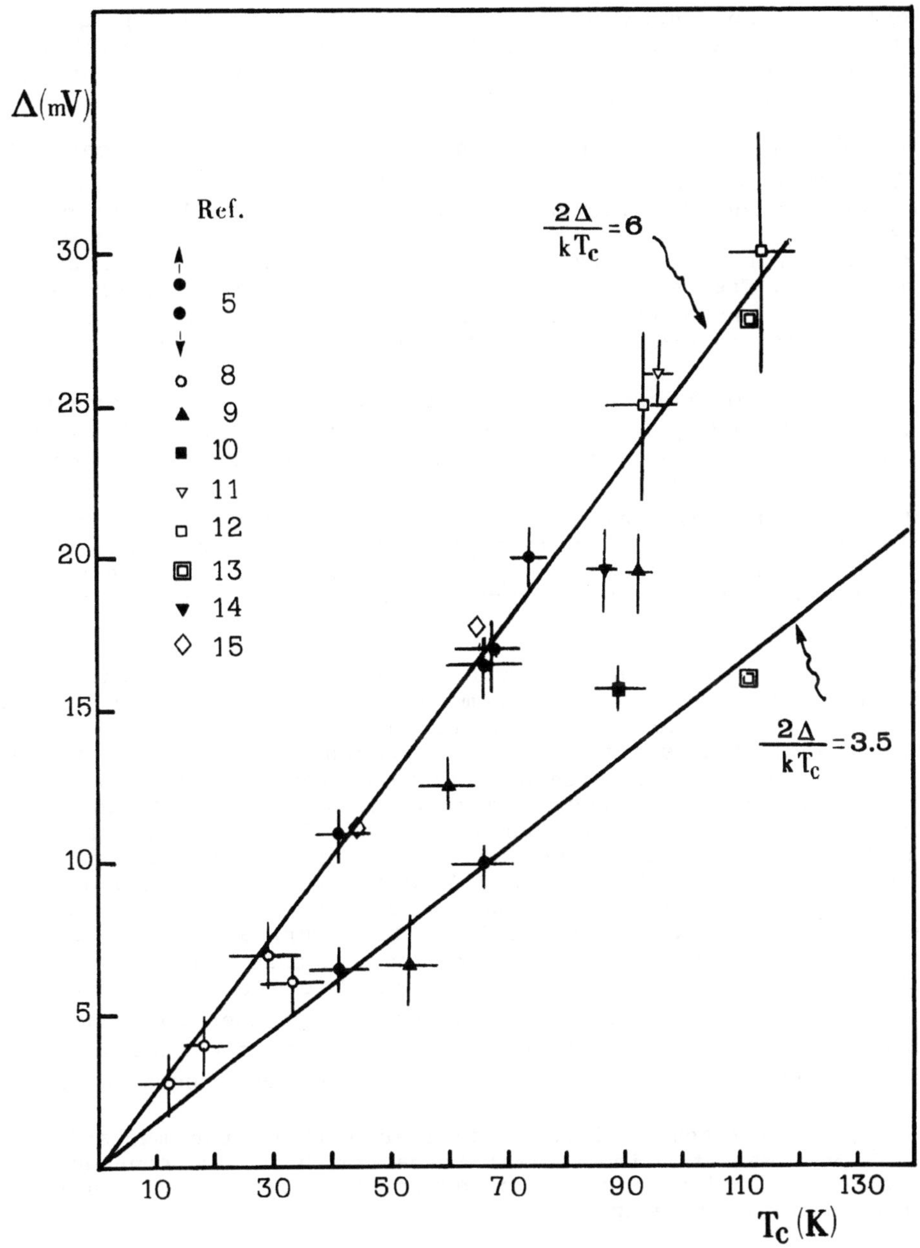

Fig. 2 - Energy gap vs. critical temperatures for a variety of junctions structures and high-T_c materials.

Without going into details of the single structures, let us here confine ourselves to a few general considerations. First we can observe that the problems of surface degradation coupled to the small coherence length did not allow so far to make a tunnel junction of the quality obtainable with low-T_c superconductors and pose serious doubts as to whether such a goal will ever be reached. It is interesting to observe that, due to such material constraints, the diagnostic power of a point contact structure in tunneling spectroscopy can be in some cases greater than that of planar junctions. Indeed the former type of devices the point going through the material can explore a more representative region of the superconductor (provided of course a careful "reading" of the measurements to exclude multiple grain-grain junction effects etc.) whereas nice film sandwich type junction may suffer more of surface and interface degradation.

For the problem of the identification and measure of the energy gap, lack of space cannot allow to make here an analytical discussion of the various results reported in the literature so that I shall just point out some general, though perhaps controversial, indications on the whole subject.

As mentioned above, there are experimental results which indicate BCS-like values for γ, others giving significantly different values of this parameter and some which even exclude the existence of a gap. From the largest part of the results however, it tends to emerge some consistent picture (see the excellent paper by Kirtley [6]), which among other things confirms some points just outlined in Ref.[7]. The term gap is here intended in a less strict sense, that is, as an energy around E_F where the density of states is drastically reduced, though non zero. This is simply reflected in a conductance peak.

In Fig. 2 a variety of experimental results Δ vs.T_c are reported [5,8-14] which refer to different junction structures and to various high-T_c materials. The data fall essentially within the two lines corresponding to values of the parameters $\gamma = 2\Delta/k_BT$ ranging roughly within 3.5 and 6. Indeed in the analogous representation reported by Tsai et al.[5] the values obtained by these Authors corresponding to the case of tunneling perpendicular and parallel to the a, b plane, were essentially grouped along these two lines. It is not surprising therefore that measurements "averaging" somewhat these two situations can be "distributed" within the two lines.

There are of course different reasons for being quite careful in drawing conclusions about the gap values. Some of these are related, as outlined above, to the peculiar experimental situation which is encountered especially in connection with the short coherence length of high-T_c materials. In reference [7] several aspects which contribute to the difficulty underlying the diagnostic use of tunneling spectroscopy, were indicated. Among others, one can call the attention to the numerous broadening effects on the gap structure. Dynes et al.[15] discussed a broadening of BCS density of state caused by the shortening of quasiparticle lifetime and Zhao et al.[16] reported on a lifetime broadening in connection to strong coupling effects. Both approaches apply to some extent, to high-T_c junctions results. Moreover, a classic effect of gap smearing is that caused by magnetic impurities. In this case, as we know, part of the pairs do not "belong to the condensate" but are characterized by a smaller binding energy. In this case the quantity Δ does no longer represent both the number of Cooper pairs in the condensate and the binding energy. (In Fig. 3a we shall denote these two quantities by Δ and ω respectively).

Indeed it should be remembered that whatever will be the mechanism of pair breaking (not only magnetic impurities) it can produce such consequences (e.g. proximity effects). Anisotropy is also expected to lead to alterations of the usual feature as for example in Fig. 3b [17].

The occurrence of small coherence lengths in high-Tc superconductors permits also the detection of multigap structure [18]. All these circumstances can in principle have a rôle in the variety of data available in the literature.

Another feature which has attracted the attention of the investigators is the peculiar background conductance at high voltage values. RVB approach has been considered to explain this circumstance [19] as well as charging effects (e.g.[20]). Although this behavior remains a debatable point, it should be remembered in any case that at this voltage level the conditions which in the simplest "golden rule" approach, allow to treat the tunneling probability as constant, do no longer hold [6].

In recent papers by Huang et al.[11,13] an unexpected decreasing conductance with voltage has been observed in point contacts employing a variety of high-Tc materials show that such a behavior occurs as soon as the normal junctions resistance decreases: i.e. for junctions of high barrier transparency. They conclude that this "anomaly" follows from the peaking of normal state conductance at E_F and proposed that it could be ascribed to a strongly energy dependent normal-state dos.

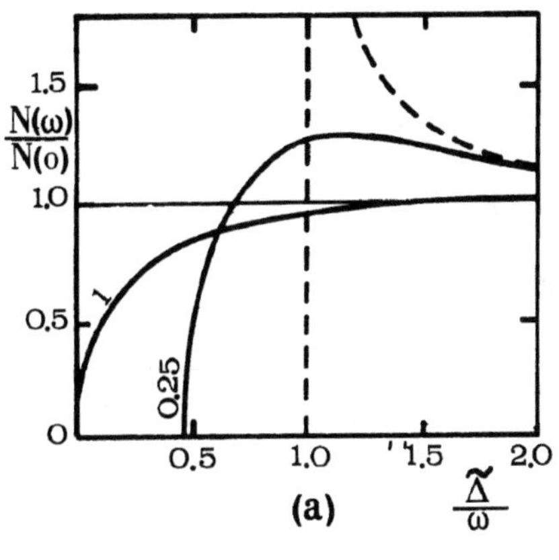

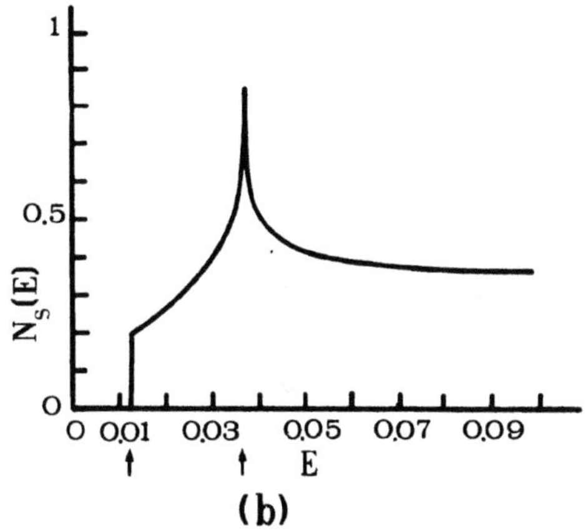

Fig. 3 - (a) Tunneling density of states for different values of depairing parameter; (b) Density of states for anysotropic gap (Ref.[17])

In conclusion we can say that small values of E_F and v_F (the latter violating the Migdal condition $v_F \gg v_{sound}$) as well as the marked anysotropy and the short ξ length scale which characterizes high-T_c materials, all contribute to the puzzling aspects of the tunneling spectroscopy in high-T_c materials.

3. - DEVICES CONSIDERATIONS

The obvious answer to the question of whether low-T_c or high-T_c superconductors should be considered for the applications is that "conventional" superconductor devices operating at very low temperatures optimize performances, whereas high-Tc ones allow easier conditions for the required cryogenic environment. Indeed low temperature operation is mandatory whenever more stringent low noise limits have to be reached. Conversely the need of less severe cryogenic requirements can dominate in several situations at the cost of lower performances.

Less obvious, though interesting, are the possibilities of using high-T_c superconductors devices at low temperature and low-T_c superconductors at relatively high temperature when properly proximized by new oxide superconductors.

In the former case the advantage arises from the higher attainable values of the parameter $\alpha(T)=T_c/T$ (where T is the operating temperature) which represents in fact in several situations an actual figure of merit. The latter possibility, which is discussed by V. Kresin at this conference, represents a unique chance of combining good features of conventional superconductors with a higher T_c realized by proximity effect with the new superconductors. To put it concisely, considering a bilayer high-T_c/low-T_c superconductors, it would be possible to get, for instance, Nb with a significant higher working T_c temperature. To this end, unfortunately the extremely short coherence length of oxide superconductors is a severe obstacle.

Indeed it should be stressed that the extent of the proximization in each material is driven by its own coherence length, that is, provided a proper thickness ratio between the two superconducting layers, the short coherence length of the oxide material would not be a problem by itself. The ξ length represents a severe obstacle in that it "competes" with the size of interface nonuniformities.

In Fig.4 is reported a qualitative sketch which indicates, in a proximity bilayer composed of a high-T_c superconductor (HTS) and a conventional one (e.g. Nb), the change of the critical temperature of the system by varying the

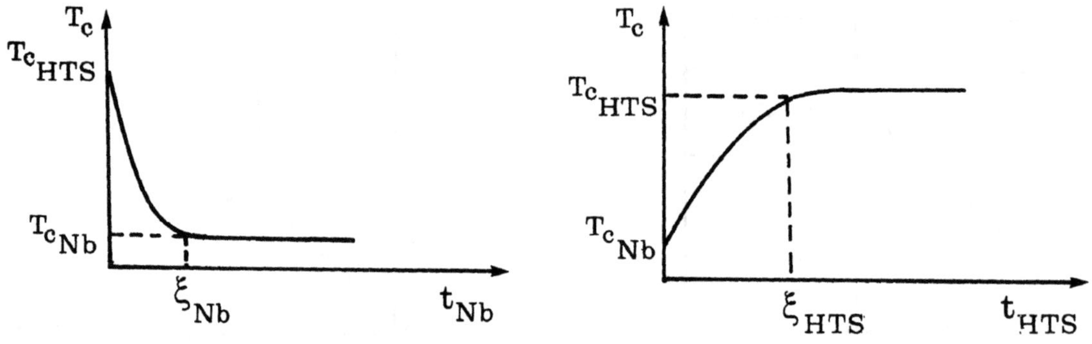

Fig. 4 - High-T_c/Low-T_c bilayer-Crical current vs.layers thickness

relative thicknesses of the layers t_{Nb} and t_{HTS}. The saturation of T_c occurs at values of t which gives a measure of the "effective" coherence length.

On the proximity effect is based the success of high-T_c in S-N-S junctions (e.g. see Ref.[21]).

The actual possibility of realizing a "good" tunnel junction is fixed by the required quality of the interface which is set by the short coherence length. This characteristic length scale, when compared to the usually observed surface-interface degradation, represents indeed the main constraint for making layered film structures.

Probably the answer to these problems will come when such layered structures will be made by a fully reliable process of sequential epitaxial growth in-situ with a thickness control on the sub single atomic layer scale and stoichiometry control within 1%.

In the meantime a number of Josephson junctions have been realized which are not good enough nor reliable. Indeed, in the available data, zero-voltage current and energy gap very seldom appear to coexist. Finally, as far as the characterization of the Josephson junction is concerned, attention should be paid on the dependence of the Josephson current on the applied magnetic field Ha [3] considering that anisotropy implies a tensorial form for the junction equation [24] which results in a quite peculiar I_j(Ha) pattern.

The trend is however encouraging and the success in heteroepitaxial structures such as that realized at Bellcore [22] and by Poppe et al.[23] (as well as results presented at this Conference) contribute significantly in the direction of High-T_c superconducting electronics.

4. REFERENCES

1. Bednorz JG. Müller KA (1986) Z. Phys. 64, 189

2. Carke J. Koch RH (1988) Science 242, 217

3. Barone A. Paternò G.(1988) Physics and Applications of the Josephson Effect" John Wiley 1982.

4. For general collections of papers on the various junction structures the reader is referred to "Novel Superconductivity" S.A.Wolf and K.V. kresin Plenum Press (1987)

5. Tsai JS. et al. (1988) Physica C153-155, 1385;and (1989) Physica C157, 537

6. Kirtly JR. (1989) Int. J. Mod. Physics B.

7. Barone A. (1988) Physica C153, 1712

8. Fein AP. Kirtley JR. Shafer MW. (1988) Phys. Rev. B37, 9738

9. Volodin AP. Khaikin MS. (1987) Sov. Phys. JETP Lett. 46

10. Gurvitch M. et al. (1989) Phys. Rev. Lett.63, 1008

11. Qiang Huang, Zasadzinsky JF. Gray KE. Liu JZ. and Claus H. (preprint)

12. Qiang Huang, Zasadzinsky J.F. Gray KE. Bukowsky EO Ginsberg DM. (preprint)

13. Takeuchi I.et al. (1989) Physica C158, 83

14. Barone A. Di Chiara A. Puluso G. Scotti di Uccio U. Cucolo AM. Vaglio R. Matacotta FC. Olzi E. (1987) Physical Review B36 235

15. Dynes RC. et al. (1984) Phys. REV. Lett. 53, 2427

16. Zhao Shipand et al. (1988) Solid State Communication 67, 1179

17. Schneider T. Frick M. - IBM Japan Intern. Symp. on Strong Correlation and Superconductivity - Springer-Verlag (in press)

18. Kresin VS. Wolf SA. (1988) Journ. Superconductivity 1, 143

19. Flensberg K. Hedegard P. Brix M. (1988) Phys. Rev. B38, 841

20. Barone A. et al. (1988) NATO ASI Superconductive Electronics (H.Weinstock and M. Nisenoff Eds., Springer Verlag) pag. 431 1989.

21. Akoh H. Shinoki F. Takahashi M. Takada S. (1988) Japan. J. Appl. Phys.27, L519

22. Rogers CT. Inam A. Hedge MS. Duta B. Wu XD. Venkatesan T. (preprint)

23. Poppe U. Prieto P. Schubert J. Soltner H. Urban K. Buchal Ch. (preprint)

24. Mints RG. (1989) Modern Phs. Lett. B 3, 51

Proximity Effect and High T_c Superconductivity

VLADIMIR Z. KRESIN

Materials and Chemical Sciences Division, Lawrence Berkeley Laboratory, 1, Cyclotron Road, Berkeley, CA 94720, USA

Abstract

Proximity system containing HT films (HT is a high T_c superconductor) allows one to take advantage of the properties of both HT and conventional materials. Josephson junctions with HT and Sc films (Sc is a semiconductor) display a strong field effect. The problems of the critical current and the microwave properties of proximity systems are discussed.

Key words: proximity effect, field effect, critical current, impedance

I. Introduction

Proximity systems containing HT films are promising from the point of view of various applications. This problem has been discussed by the author [1]. Recent progress in thin film preparation has resulted in observations of the proximity effect in S_h–N system [2-6] (S_h is a HT film, N is a normal film [2-6]; S_h = Y–Ba–Cu–O, N = Ag, Au, Pt). The proximity effect allows one to induce the superconducting state in a normal film and to take advantage of superconducting properties and the properties of the N film simultaneously. The high values of T_c and the energy gap Δ_h make use of HT films very attractive.

In this paper we describe some proximity systems containing HT films. The structure of the paper is as follows. Section II is concerned with the properties of S_h–N–S_γ Josephson junctions, where S_γ is a HT or conventional (S_c) film. Section III addresses the problem of the field effect and the possibility of making a three-terminal device. The systems S_h–N–I–S_γ are described in Sec. IV (I is an insulator). The possibility of obtaining high critical currents with the help of the proximity effect is discussed in Sec. V. In Sec. VI we discuss the microwave properties of the proximity systems.

II. S_α^h–N_β–S_γ Josephson Junction

Evaluation of the Josephson current for an S_α^h–N_β–S_γ system can be carried out with the use of the method developed by the author [7]. The latter paper contains an analysis of the field effect for a junction containing a semiconductor barrier, however, the method developed there can be used for any S-N-S junction.

The flow of nondissipative Josephson current through an S-N-S junction differs from the usual case of an S-I-S system because the barrier N contains its own carrier system. The proximity effect plays a key role. A complex state is induced in the N film; it is determined by two superconductors with different phases. The thickness L_N is usually large, and it is necessary to take into account the space dependence of the order parameter. The evaluation is based on a special diagrammatic technique [7] developed in coordinate space.

Consider first the "clean" case, when $\zeta_N << l$ (l is the mean free path and ζ_N is the coherence length in the N film; $\zeta_N = \hbar V_F/\pi T$ [8]). The amplitude of the Josephson current is described by the equation:

$$j_m = AT \sum_{n \geq 0} F_{h;n}^{\gamma}\, \omega_n^{-1/2} \exp(-2\omega_n m_\beta L/p_{F;\beta}) \tag{1}$$

Here $\omega_n = (2n+1)\pi T$; $F^{\gamma}_{h;n} = \Delta_h\Delta_\gamma(\omega_n^2 + \Delta_h^2)^{-1/2}(\omega_n^2 + \Delta_\gamma^2)^{-1/2}$; m_β, $p_{F;\beta}$ and L are the mass, the Fermi momentum, and the thickness of the N_β film, respectively; $A = \tilde{\gamma}\, p_{F;\beta}(m_\beta)^3$, $\tilde{\gamma} = \pi e m_\beta^{-1} S^2_{\beta\alpha}$; $S_{\beta\alpha}$ is the effective vertex describing the proximity effect [7], $S_{\beta\alpha}\sim v_\alpha$, v_α is the density of states in the HT film; Δ_h and Δ_γ are the energy gaps (for simplicity, we are not considering the effects of strong coupling; the generalization is straightforward, see Ref. 7). For an S_h–N–S_h system with the same HT superconductor on both sides (e.g., HT ≡ Y–Ba–Cu–O, see Refs. 3 and 5), we obtain $F^{h}_{h;n} = \Delta_h^2(\omega_n^2+\Delta_h^2)^{-1}$.

In the temperature region where $\zeta_n << L \equiv L_N$, one can only retain the term n = 0 and we obtain

$$j_m = A\, F^{\gamma}_{h;0};\ \exp(-2\pi T m_\beta L/p_{F;\beta}) \tag{2}$$

Then $\Delta_h >> \pi T$ and therefore $F^{\gamma}_{h;0} = \Delta_\gamma^2[(\pi T)^2+\Delta_\gamma^2]^{-1/2}$. For the S_h–N–S_h junction, $F \cong 1$. Therefore, junction S_h–N–S_h is characterized by a large value of j_m. If $\zeta_N > L$, then one should use the more general expression (1), and the Josephson current is no longer described by a simple exponential dependence (2).

Consider now the "dirty" case ($l << \zeta_N$) which corresponds to the region $\hbar V_F/2nl > T > T_o = \hbar V_F l/6\pi k_B L^2$. Then the amplitude of the current is described by the expression (see Refs. 7 and 9):

$$j_m = \tilde{\gamma}(m_\beta^3/4p_{F;\beta})(\pi\tau T)^{-1/2}\exp(-L/\zeta_{N;D}) \tag{3}$$

where $\zeta_{N;D} = (\zeta_N l/3)$ and τ is the transport time between collisions. The system S_h–N–S_c has been studied [2,4,6] (S_h ≡ Y–Ba–Cu–O, N ≡ Au, and S_c ≡ Nb [2] or Pb [4,6]). In this case $T < T_{c,conv.}$ ($T \simeq$ 4.2K in Ref. 2). For Au $V_F \simeq 1.4 \times 10^8$ sm sec^{-1}, $l \simeq 6 \times 10^2$ Å [2], and $\zeta_N > l$ at T < 4.2K. Therefore, the situation studied [2] corresponds to the "dirty" limit. The expression (3) can be used only for relatively thick Au films $L > 10^3$ Å). For smaller thickness, one should use the more general expression

$$j_m = \tilde{\gamma}\pi T \sum_{k=-\infty}^{\infty}(-1)^k \sum_{n\geq 0}\frac{\Delta_h\Delta_c}{(\omega_n^2+\Delta_h^2)^{1/2}(\omega_n^2+\Delta_\gamma^2)^{1/2}}\left(2|\omega_n|+\frac{1}{3}(\pi k/L)^2 V_F^2\tau_{tr}\right)^{-1} \tag{4}$$

The situation is different for S_h–N–S_h systems studied in [3,5]. The temperature range is broader, and for T > 15K one can satisfy the criterion $\zeta_N < l$ ("clean" case).

III. Field Effect and Three-Terminal Device

In this section we are going to discuss an important application of HT and the proximity effect, namely, the opportunity to constructing a three-terminal device. The field effect in conventional superconductors has been discussed by the author [7] in connection with a remarkable experiment [10]. In [1] the problem was discussed for HT films. Recent progress makes the use of the field effect realistic. Consider the S_h–N–Sc–N–S_h system (Fig. 1), here S_h is a HT film, N is a thin normal film (e.g., Ag, Au, Nb, or NbN) and Sc ≡ InAS. InAS contains a natural inversion layer. The N film provides a contact between S_h and Sc; therefore, the thickness L_N should be small. The theoretical analysis can be carried out in a way, similar to [7]. An external field will affect the carrier concentration in the inversion layer, which allows one to control the magnitude of the Josephson current.

Note that the use of HT films noticeably improves the parameters of the field effect. The main reason is the possibility of increase in the temperature. This decreases ζ_N and it is possible to enter the regime of the "clean" limit.

In this case the dependence of j_m on the carrier concentration in InAs is strong ($\propto \exp(\rho/N_s^{1/2})$; N_s is a surface carrier concentration, $\rho = (2\pi)^{1/2}\, T m_\beta L$. The field effect is weaker in a "dirty" case. For example, if $T \simeq 2$K then $l/\zeta_N = 0.15$ ("dirty" limit, see Ref. 7). However, if $T \gtrsim 20$K (this is realistic if a HT film is used), then we are dealing with the "clean" case and the field effect becomes stronger.

IV. S-N-I-S Junction

Consider now a different type of the Josephson junction, namely, the S_α^h–N_β–I–S_γ system. In this case, the Josephson current flows between the β and γ films; the superconducting state in the β film is caused by the proximity contact (S_α–N_β). Let us focus on the case $L_\beta << \zeta_N$. In this case the order parameter Δ_N is uniform across the β film, and one can use the McMillan tunneling model [11]. This type of Josephson junction has been studied in detail by the author. [12]

The amplitude of the Josephson current is described by the expression

$$i_m = \frac{r2\pi T}{eR} \sum_{n>0} \frac{\Delta_h \Delta_\gamma}{[X_n^2(1+\alpha t\tilde{k})^2 + \Delta_n^2]^{1/2} (X_n^2 + r^2)^{1/2}} \qquad (5)$$

Here $r = \Delta_\gamma(0)/\Delta_h(0)$, $X_n = \omega_n/\Delta_h(0)$, $\alpha = \Delta_h(0)/\pi T_c$, $\tilde{k} = (X_n^2+1)^{1/2}\alpha$, $t = \pi T_c/\Gamma$.

The quantity Γ has been introduced by McMillan [11]; it describes the quality of the contact. It also depends on the thickness: $\Gamma \propto L_\beta^{-1}$.

An increase in T leads to a decrease in the current, and the slope of this decrease depends parametrically on the thickness L_β. As a result, one can observe strong deviations from the dependence [13] for the S_α–I–S_β junction. According to Ref. 13, the dependence $j_m \sim (T_c-T)$ can be observed only in the region near $T_c(T/T_c < 0.9)$. According to Ref. 12, j_m is proportional to (T_c-T) down to $T/T_c \simeq 0.65$. We think that the presence of this linear dependence throughout such a wide temperature range is a direct manifestation of the proximity effect.

The proximity effect is also manifested in the dependence of j_m on the thickness L_β (at fixed temperature). An increase in L_β leads to a decrease in the current; the slope of the curve $j_m(L_\beta)$ depends parametrically on the ratio r. For the S_h–N_β–I–S_h system, r = 1. If S_γ is a conventional superconductor, such as Pb, Nb, or NbN, then r << 1. For example, if $S_h \equiv$ Y-Ba-Cu-O and $S_\gamma \equiv$ NbN, then $r \simeq 1/6$. According to the author's paper [12], the Josephson current in the S_h–N_β–I–S_h system decreases with increasing L_β much faster than in the S_h–N_β–I–S_c junction (S_c is a conventional superconductor). This means that use of S_h–N–I–S_c junctions is more favorable for obtaining large values of j_m.

V. Proximity effect and critical current

Consider the proximity system S_h–S_β–S_h (see Fig. 1), (S_β is a conventional superconductor). The most interesting case is when S_β is an A-15 or B1 compound; in the case where S_β is characterized (in the isolated state) by high values of j_c, H_{c2}, etc. The critical temperature, T_c, of the system of interest will be lower than T_c^h but, which is most important, higher than T_c^β (the critical temperature of an isolated β film). Hence, as a consequence of the proximity effect, we have obtained and A-15 or B1 superconductor with a high T_c.

If the system is placed in an external magnetic field, then one can observe vortex penetration (up to $H = H_{c2}$). In the region $T_c^\beta < T < T_c$, the system can be treated as an S-N-S proximity contact.

It is important that new high T_c materials are characterized by high values of H_{c2}. As a result, the critical field, H_{c2}, for the proximity system of interest, S_h–S_β–S_h, will be larger than H_{c2}^β (the critical field of S_β in the isolated state; see analysis of the S-N-S system in Ref. 14).

Consider now the flow of transport current through the system; assume that the current is perpendicular to the magnetic field. We are interested in the current density in the β film. As is known, the critical value of the current, j_c, is determined by the pinning force, F_p, and can be evaluated from the expression $j_cB = F_p$. The pinning force F_p is proportional to H_{c2}^n (usually $n \simeq 2$).

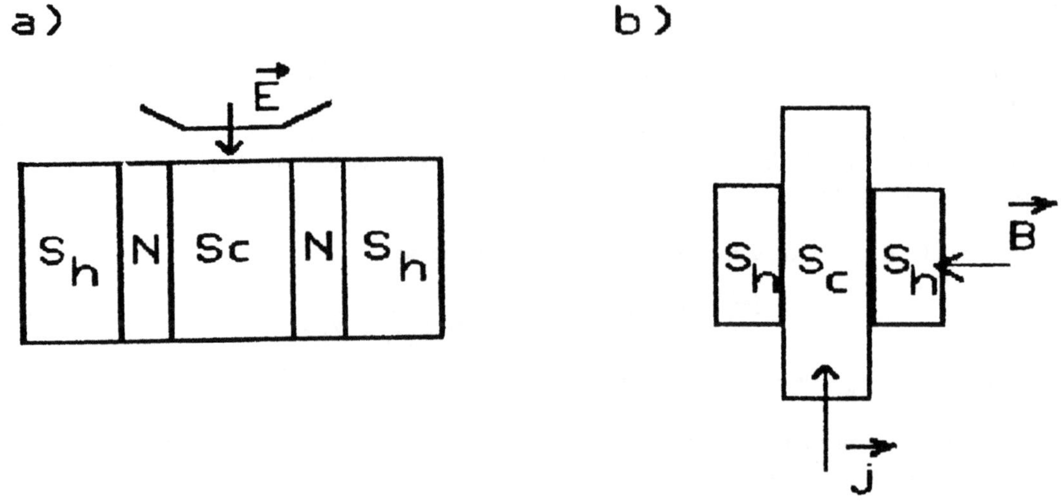

Fig. 1 a) S_h–N–Sc–N–S_h junction; Sc = InAS and
b) S_h–S_c–S_h proximity system; S_c = {A-15, B1}

Since $H_{c2} > H_{c2}^{\beta}$ (see above), the pinning force can be increased by a proximity contact with a high T_c material. As a results, one can obtain an effective increase of j_c of the β film. However, most importantly, because of the proximity effect, the conventional β film is able to carry a high critical current at temperatures $T > T_c^{\beta}$. It would be interesting to study the properties of the proximity system S_h–S_β–S_h, where S_β is an Nb, NbN, A-15, or another conventional superconducting film.

VI. Microwave Properties of the Proximity System

Consider the system S_h–N. Static screening by the proximity system has been studied by the author [15]. In this section we generalize the method [15] to the case of an alternating field and focus on the situation when S_h is a high T_c film.

Because of the short coherence length ζ_h, a high T_c superconductor is described by local electrodynamics. The situation might be different in the N film. Indeed, in the low temperature region ζ_N is large, so that $\zeta_N >> L_N$ and $\zeta_N >> \delta$ (δ is a penetration depth).

The kernel $Q(\vec{q},\omega_o)$ defined by the relation $\vec{j}_\beta(\vec{q},\omega) = -\,Q_\beta(\vec{q},\omega)\,\vec{A}(\vec{q},\omega)$, can be written in the form [16]:

$$\Delta_\beta(\vec{q},\omega_o) = \frac{3\pi^2 ne^2}{4m_\beta V_F^\beta q}\, T \sum_n \left\{ 1 + \frac{\Delta_\beta^2(\omega_n) - \omega_n\tilde{\omega}_n}{[(\omega_n^2+\Delta_\beta^2(\omega_n))(\tilde{\omega}_n+\Delta_\beta^2(\omega_n))]^{1/2}} \right\}$$

Here $\tilde{\omega}_n = \omega_n - \omega_o$, $\omega_n = (2n+1)\pi T$, $\omega_o = -\,i\omega$, ω is the frequency of the applied field. The order parameter $\Delta_\beta(\omega_n)$ can be evaluated using the theory of the proximity effect [11,12] and equal to

$$\Delta_\beta \simeq \Gamma\Delta_h(\Gamma+\Delta_h)^{-1} \tag{6}$$

Therefore, the kernel $Q(\vec{q},\omega)$ and the impedance can be evaluated within the framework of the theory [17,18], but with a gap, given by Eq. (6).

The function $\tilde{Q} \equiv \mathrm{Im}\, Q$, which determines the absorption of radiation, is given by the equation (cf., Ref. 19)

$$\tilde{Q} = \frac{3\pi}{b}\,\frac{ne^2}{m_\beta c}\,\frac{\omega}{V_F^\beta q}\,\frac{\Delta_\beta}{T}\,ch^{-2}\frac{\Delta_\beta}{2T}\,ln\,\frac{\Delta_\beta}{\omega} \tag{7}$$

We assume that $\omega << \Delta_\beta$; the impedance can be calculated from the dependence (see, e.g., Ref. 16): $Z \propto i\omega(\Delta Q)^{-1/3}$. In the low temperature region $Z \propto \exp(-\Delta_\beta/T)$. Because of the proximity effect, Δ_β depends strongly on the thickness of the N film. Indeed, $\Gamma \propto L_\beta^{-1}$. If L_β is small, then $\Delta_\beta \simeq \Delta_h$ (see Eq. (6)). For large L_N, we obtain $\Delta_\beta \simeq \Gamma$. In addition, doping of the HT film strongly affect the value of $\Gamma \propto \nu_\alpha$ and Δ_h (ν_α is the density of states in α film). Therefore, by changing the thickness L_N or by doping, one can vary the impedance by an large amount.

Conclusion

The proximity effect leads to a number of unusual properties. In the presence of a high T_c film one can observe strong field effect, which is important from the point of view of constructing a three-terminal device (see Sec. III). The proximity sandwich S_h–S_{A-15}–S_h can be used to obtain high values of the critical current (see Sec. V). Cavities containing proximity systems will display interesting microwave properties.

This work was supported by the U. S. Office of Naval Research under Contract No. N00014-89-F-0077 and carried out at the Lawrence Berkeley Laboratory under Contract No. DE-AC03-76SF00098.

References

(1) V. Z. Kresin, Solid State Sci., 1, 221 (1987); in *Novel Superconductivity,* ed. by S. Wolf and V. Kresin, p. 309, Plenum, New York (1987); Cryogenics 28, 409 (1988); 29, 1096 (1989).

(2) H. Akoh, *et al.*, IEEE Mags. Trans. 25, 795 (1989); Jpn. J. Appl. Lett. 27, L519 (1989); '89 ISEC Procs., Tokyo, Japan.

(3) D. Schwartz;, *et al.*, IEEE Mags. Trans. 25, 1298 (1989).

(4) J. Yoshida, *et al.*, in *Science and Technology of Thin Films Superconductors,* ed. by R. McConnell and S. Wolf, p. 459, Plenum, New York, (1989).

(5) J. Moreland, *et al.*, Appl. Phys. Lett. 54, 1477 (1989).

(6) L. Greene, *et al.*, The Int. Conf. Procs., Stanford, California, 1989, Physica C (in press).

(7) V. Z. Kresin, Phys. Rev. B34, 7587 (1986).

(8) J. Clarke, R. Soc. Procs. London, Ser. A308, 447 (1969).

(9) J. Seto and T. Van Duzer, 13th Int. Conf. on Low Temp. Phys. Procs., eds. O'Sullivan, K. Timmerhaus, and E. Hamel, p. 328, No. 3, Plenum, New York (1972).

(10) H. Takayanagi and Kawakami, Phys. Rev. Lett. 54, 2449 (1985).

(11) W. McMillan, Phys. Rev. 175, 537 (1968).

(12) V. Z. Kresin, Phys. Rev. B28, 1254 (1983).

(13) A. Ambegaokar and A. Barafoff, Phys. Rev. Lett. 10, 486 (1963).

(14) K. Biagi, *et al.*, Phys. Rev. B32, 7165 (1985).

(15) V. Z. Kresin, Phys. Rev. B32, 145 (1985).

(16) A Abrikosov, L. Gor'kov, and I. Dzyaloshinskii, *Methods of Quantum Field Theory in Statistical Physics,* (Prentice-Hall, Englewood Cliff, 1963).

(17) D. Matis and J. Bardeen, Phys. Rev. 111, 412 (1958).

(18) A. Abrikosov, L. Gor'kov, and I. Khalatnikov, Sov. Phys. JETP 10, 132 (1960); I. Khalatnikov and A. Abrikosov, Adv. Phys. 8, 879 (1976).

(19) E. Lifshitz and L. Pitaevskii, *Physical Kinetics,* Pergamon Press, Oxford (1981).

Proximity Effect in the System of Y-Ba-Cu-O/Au/Nb Films

HIROSHI AKOH
Electrotechnical Laboratory, 1-4, Umezono 1-chome, Tsukuba, Ibaraki, 305 Japan

ABSTRACT

SNS Josephson junctions consisting of Y-Ba-Cu-O/Au/Nb thin-film sandwiches with use of the proximity effect have been described. The ac Josephson effect is confirmed by the observation of Shapiro steps in the junctions under microwave radiation. The magnetic field dependence of the critical current shows the Fraunhofer pattern with the self-field effect, due to the dc Josephson effect. *SNS* junctions using epitaxial Y-Ba-Cu-O films with *c*-axis orientation have no Josephson current, while the junctions with *a-b* plane orientated Y-Ba-Cu-O films behave as Josephson junctions. This anisotropic behavior reveals that the proximity effect is strongly affected by the anisotropic coherence length of Y-Ba-Cu-O films.

KEY WORDS: *SNS* Josephson Junction, Y-Ba-Cu-O/Au/Nb films sandwich, Shapiro step, Fraunhofer pattern, anisotropic proximity effect

INTRODUCTION

Josephson junctions using high-T_c oxide superconductors [1,2] are greatly attractive for device applications because of not only high operating temperature, but also high gap voltage. Recently, many efforts have been made to fabricate Josephson junctions, especially with the film-layered structure [3-7]. Fabrication of this sandwich-type junction requires the superior film qualities of a high-T_c, a high-J_c, smooth surfaces, and so forth. Moreover, further complications concerning the surface superconductivity of the films and interfacial properties between the barrier and films are encountered for the sandwich-type devices. For instance, there exists non-superconducting layer on the surface of high-T_c superconductors [8]. In addition, direct connections between films of Y-Ba-Cu-O and other superconducting metals such as Pb and Nb, frequently show the semiconducting or insulating behaviors. In the interface between the films of oxide superconductors and metal superconductors, the surface of the metal film may be oxidized by taking oxygen atoms from the oxide superconducting film, while the surface of the oxide superconducting film may result in an oxygen-poor phase, consequently producing a thick semiconducting or insulating surface layer.

The proximity effect between high-T_c superconductors and normal metals is very important to understand the surface properties of high-T_c superconductors, since this effect reflects electronic properties of high-T_c superconductors, such as short and anisotropic coherence lengths, surface superconductivity, and so forth [9-11]. An *SNS* (superconductor-normal metal-superconductor) junction is one of the typical examples to investigate the proximity effect, since the junction clearly shows the Josephson effect which gives us the information of the contact between high-T_c superconductors and normal metals [4,6,7,12,13]. This junction also has a potential application as high-T_c superconducting devices. *SNS* junctions [14] are more easily fabricated, compared to *SIS* junctions. Since the high-T_c superconducting film has a short coherence length, an insulating barrier is required to have the same order of thickness as the coherence length. In the case of *SNS* junctions, junction properties are dominantly governed by the coherence length of normal layer in the junction. For instance Au has the coherence length of ~100 nm. Therefore, the Au barrier can be thicker than the insulating barrier in *SIS* junctions.

In this paper, *SNS* Josephson junctions due to the proximity effect, which consist of Y-Ba-Cu-O/Au/Nb film sandwiches are presented. The junctions reveal the Shapiro steps due to the ac Josephson effect under microwave radiation. The magnetic field dependence of the critical current shows the Fraunhofer pattern due to the dc Josephson effect. Moreover, the relationship between the proximity effect and crystalline orientation of Y-Ba-Cu-O films is demonstrated in this *SNS* system using epitaxial Y-Ba-Cu-O films.

FABRICATION OF *SNS* JOSEPHSON JUNCTIONS

The full procedure for fabricating *SNS* junctions consisting of Y-Ba-Cu-O/Au/Nb film sandwich is schematically shown in Fig. 1. Y-Ba-Cu-O films were fabricated by using an rf diode sputtering method and by a reactive coevaporation method (Fig. 1(a)). For sputtered Y-Ba-Cu-O films on $SrTiO_3$ substrates [15], post-deposition annealing was carried out in flowing O_2 at 710°C for 10 h to achieve the superconducting films. In spite of the annealing process, the films showed the smooth surface with small grains of less than 100 nm. The x-ray diffraction (XRD) analysis showed that sputtered films had polycrystalline structure and no preferred orientation . The sputtered films had the thickness of 700-800 nm and T_c of 60-70 K. By the reactive coevaporation method with the rf-plasma cooling process in the low oxygen pressure [16], as-grown superconducting films were obtained at the substrate temperature of 650°C. The XRD analysis revealed that the films on (100) MgO and (110)$SrTiO_3$ substrates had *c*-axis orientation and *a-b* plane orientation, respectively. RHEED and RBS studies showed that all of the coevaporated films were epitaxially grown. The coevaporated films had the thickness of 100 nm and T_c of 70-80 K.

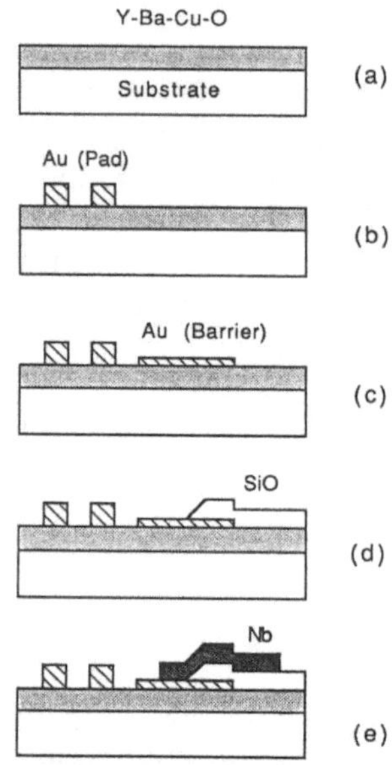

Fig. 1. Fabrication process. (a) Y-Ba-Cu-O film base electrode, (b)Au contact pads, (c) Au barrier, (d) SiO insulating layer, (e) Nb counter electrode.

Second, we formed contact pads of an Au film with the thickness of 30-100nm (Fig. 1(b)). To achieve a good contact on the Y-Ba-Cu-O film for the measurements, the plasma oxidation was carried out with an O_2 pressure of 13.3-133 Pa and an rf power of 10 W for 10-60 min before evaporating the Au film. This surface improvement due to the O_2 plasma oxidation is very important, since the surface layer of the Y-Ba-Cu-O film is contaminated and/or has other non-superconducting phases. Third, a thin film of Au as a barrier of *SNS* junctions was evaporated after the same O_2 plasma treatment to improve the surface of the Y-Ba-Cu-O film (Fig. 1(c)). Au was chosen as a barrier in this junction to avoid the diffusion of oxygen atoms from the oxide superconducting film into the Nb film. Au is also hardly diffused into the Nb film at room temperature, in contrast to the combination of Au and Pb films. Next, an SiO film (320 nm) was deposited to insulate the Y-Ba-Cu-O film from a counterelectrode of the Nb film (Fig. 1(d)). Finally, the junction was accomplished by sputtering deposition of an Nb film (220 nm) (Fig.1(e)). Patterning of Au, SiO, and Nb films were achieved by using a metal mask. Junction area was about 0.8×0.8 mm^2.

SNS JUNCTIONS WITH POLYCRYSTALLINE Y-Ba-Cu-O FILMS

An *SNS* junction with the Au barrier of 30 nm and the sputtered Y-Ba-Cu-O film with the polycrystalline structure had dc Josephson current of 27 mA at 4.2 K [4]. This fact indicates that the superconducting current flows across the junction, indicating that the Y-Ba-Cu-O film is superconductively connected to the Nb film through the Au barrier due to the proximity effect. The normal resistance R_n is 0.48 mΩ, so that the $I_c R_n$ product is 13 μV. In order to verify the junction as a Josephson junction, a microwave was applied to the junction. Figure 2 shows the *I-V* characteristic of the junction under the microwave radiation with the frequency f of 9.216 GHz at 4.2 K. The critical current I_c is decreased to 18 mA under the microwave radiation. As is seen in the figure, the first harmonic step is observed at the voltage of $hf/2e$ due to the ac Josephson effect, where h is Planck's constant and e is the electronic charge. In addition, a subharmonic step appears at the voltage of $(1/2)(hf/2e)$.

Figure 3 shows the temperature dependence of the critical current I_c of the junction. I_c is proportional to $(1-T/T_{cj})$ near T_{cj}, where T_{cj} is the transition temperature of the junction. This linear temperature dependence of I_c suggests that the coherence length ξ_n of Au film is larger than the thickness of Au barrier [17].

Figure 4 shows the magnetic field dependence of the critical current I_c for the junction at 4.2 K, which exhibits the Fraunhofer pattern as observed in the Josephson junction. The magnetic field was applied parallel to the junction. This magnetic field dependence of I_c shows the behavior of the self-

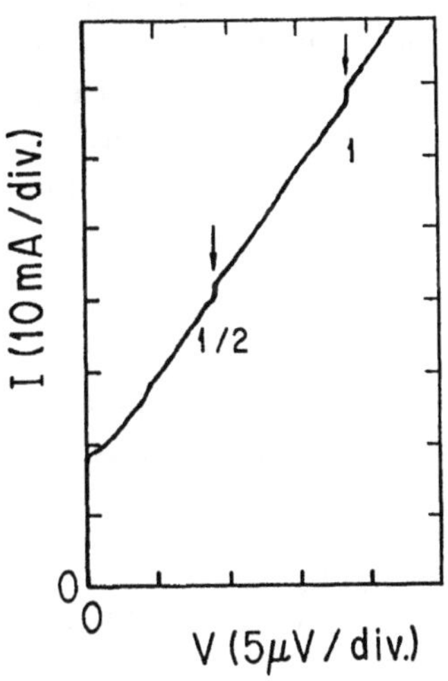

Fig. 2. *I-V* characteristic of *SNS* junction with polycrystalline Y-Ba-Cu-O film and a 30-nm Au barrier under a 9.216-GHz microwave radiation at 4.2 K.

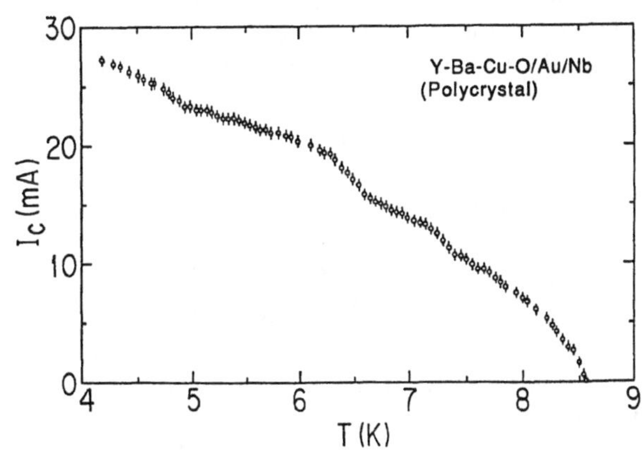

Fig. 3. Temperature dependence of critical current I_c for *SNS* junction with polycrystalline Y-Ba-Cu-O film and a 30-nm Au barrier.

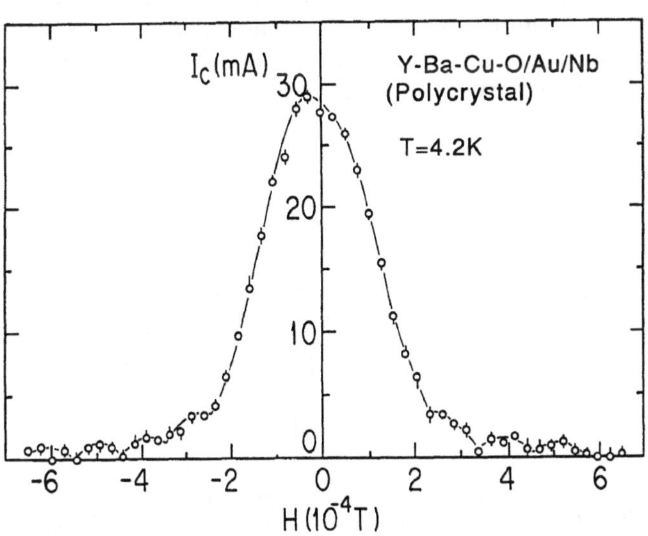

Fig. 4. Magnetic field dependence of critical current I_c for *SNS* junction with polycrystalline Y-Ba-Cu-O film and a 30-nm Au barrier at 4.2 K.

field limited junction [18], which means that the junction size is larger than the Josephson penetration depth. Self-field effect also may produce a shift in the maxima in the magnetic field pattern. It is emphasized that the Fraunhofer pattern occurs definitely in the junction of Y-Ba-Cu-O/Au/Nb sandwich structure, *not in the grain boundaries of the Y-Ba-Cu-O film*, since the critical current of the patterned Y-Ba-Cu-O film with even a 100-nm width had no changes when the magnetic field was increased up to 65×10^{-4} T.

We fabricated different junctions with the thicknesses of the Au barrier of 50 and 100 nm. All the junctions were found to exhibit the Josephson effect [6].

SNS JUNCTIONS WITH EPITAXIAL Y-Ba-Cu-O FILMS

In the system of *SNS* Josephson junctions, the crystalline orientation of the high-T_c superconducting film is very important, since the high-T_c superconductors have very short and anisotropic coherence lengths. If the Y-Ba-Cu-O film has a *c*-axis orientation in the Y-Ba-Cu-O/Au/Nb structure, the coherence length along the *c*-axis is so short (~a few tenth nm), so that the thickness of Au barrier may be required to be less than a few nm. It may be difficult to obtain the very thin and uniform Au film on Y-Ba-Cu-O films. The Y-Ba-Cu-O film with *a-b* plane orientation, on the other hand, has a longer coherence length compared with that of the film with *c*-axis orientation. Thus, in order to confirm the relationship between the proximity effect and the crystalline orientation of the high-T_c superconductors, we have fabricated *SNS* junctions using epitaxial Y-Ba-Cu-O films with *c*-axis and *a-b* plane orientations, which have been made by a reactive coevaporation method.

C-axis Orientated Junction

Epitaxial Y-Ba-Cu-O films with *c*-axis orientation were fabricated on (100)MgO substrates. By using these films, 14 numbers of *SNS* junctions were fabricated, changing the thickness of Au barrier in the range from 5 nm to 50nm. All the junctions had *no Josephson current*, but resistive *I-V* characteristics. This fact suggests that the proximity effect does not reach the Au film in the interface between Y-Ba-Cu-O and Au films, since the coherence length of *c*-axis oriented Y-Ba-Cu-O films is very short, resulting in no Josephson coupling between Y-Ba-Cu-O and Nb films.

A-b Plane Orientated Junction

A-b plane orientated Y-Ba-Cu-O films with epitaxial growth were fabricated on (110) $SrTiO_3$ substrates. Y-Ba-Cu-O films had (103) orientation. Two different *SNS* junctions with different directions of (103) orientated Y-Ba-Cu-O films were fabricated as schematically shown in Fig. 5. One is the junction *J1* using the film with the Cu-O plane perpendicular to the direction of junction (Fig. 5(a)). Another is the junction *J2* using the film with the Cu-O plane along the direction of junction (Fig. 5(b)). The thickness of Au barrier was 15nm.

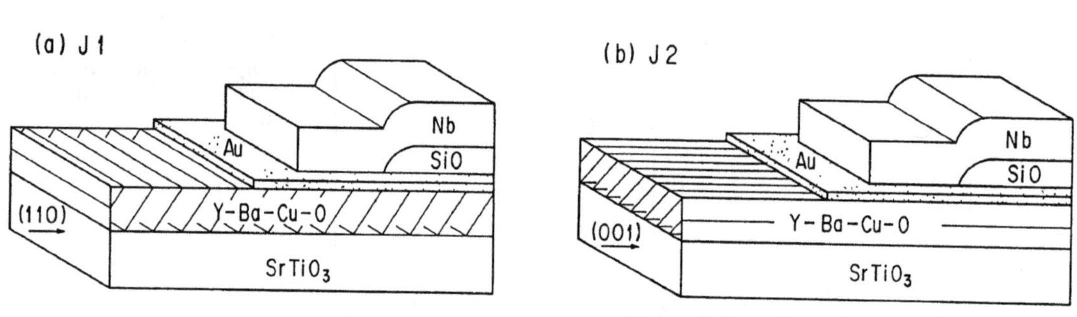

Fig. 5. Schematic configuration of *SNS* junctions with *a-b* plane orientated Y-Ba-Cu-O films. (a) junction *J1* with the Cu-O plane perpendicular to the direction of junction, (b) junction *J2* with the Cu-O plane along the direction of junction.

In order to confirm the direction of the film-orientation, the temperature dependence of resistivity was measured. Figure 6 shows the temperature dependence of resistivity for the films of junctions *J1* and *J2*. As seen in the figure, the film for the junction *J1* has a higher resistance, compared to that of the film for the junction *J2*. In the film for *J1*, the current flows perpendicular to the Cu-O plane, resulting in the higher resistance. The values of T_C are 70 K and 76K for the Y-Ba-Cu-O films of junctions *J1* and *J2*, respectively.

Figure 7 shows *I-V* characteristics of the junctions *J1* and *J2* with and without the microwave radiation at 4.2 K. Dc Josephson currents of 2.9 mA and 2.7 mA are observed for the junctions *J1* and *J2*, respectively. Both of the junctions have Shapiro steps due to the ac Josephson effect by microwave radiation with the frequencies of 8.52 GHz and 8.31 GHz for the junctions *J1* and *J2*, respectively. The fact confirms that both of the junctions are Josephson junctions.

The magnetic field dependences of critical current I_c for the junctions *J1* and *J2* are shown in Fig. 8. A large difference is seen in the magnetic field dependence of critical current between *J1* and *J2*. The junction *J1* shows the Fraunhofer pattern with the self-field effect. The junction *J2*, on the other hand, has a weak magnetic field dependence of I_c. The origin of this large difference is not yet solved clearly. It may reflect the difference of the Josephson penetration depth λ_J which is related to the magnetic penetration depth λ_L of the Y-Ba-Cu-O film [19]. In the junction *J1*, the Y-Ba-Cu-O film has an anisotropic λ_L with a larger value for the external magnetic field parallel to the junction.

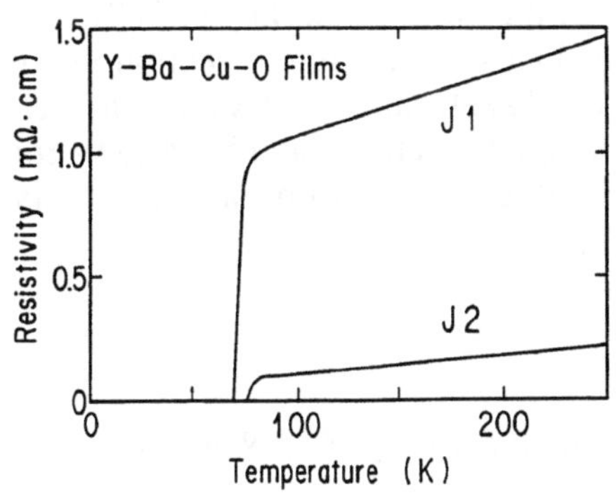

Fig. 6. Temperature dependence of resistivity for *a-b* plane orientated Y-Ba-Cu-O films of *SNS* junctions *J1* and *J2*.

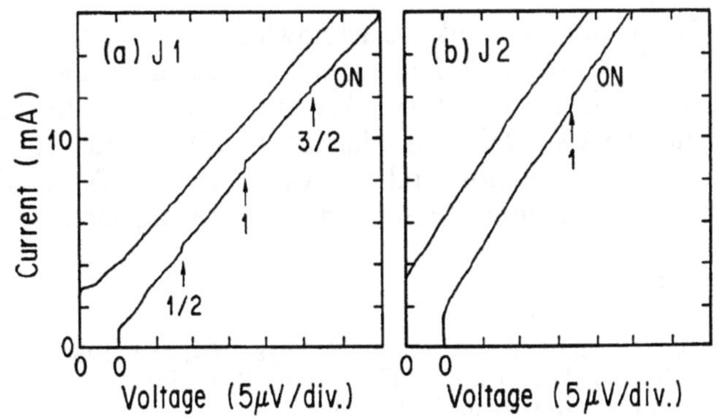

Fig. 7. *I-V* characteristics of *SNS* junctions with *a-b* plane orientated Y-Ba-Cu-O film. (a)junction *J1* with and without microwave radiation of 8.52GHz, (b)junction *J2* with and without microwave radiation of 8.31GHz.

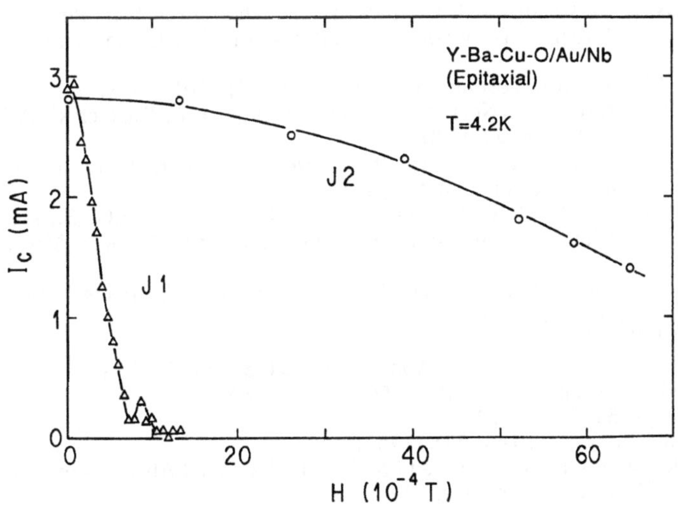

Fig. 8. Magnetic field dependence of critical current I_c for *SNS* junctions *J1* and *J2*.

In the junction *J2*, however, the Y-Ba-Cu-O film has a small λ_L for the external magnetic field. Thus, λ_J of *J1* is smaller than that of *J2*, taking account of the relationship of $\lambda_J \propto (\lambda_L)^{-1/2}$. Therefore, the critical field H_{c0} to reduce the critical current to zero, which is proportional to λ_J has a large value for the junction *J2*, compared to that for the junction *J1*. In addition, the tilt of the Cu-O plane may affect the effective magnetic penetration depth of Y-Ba-Cu-O film.

CONCLUSION

In summary, sandwich-type *SNS* Josephson junctions which consist of Y-Ba-Cu-O/Au/Nb layered films have been described. The junctions exhibit Shapiro steps under microwave radiation, according to the ac Josephson effect. Magnetic field dependence of the critical current shows a Fraunhofer pattern with the self-field effect, indicating that the junction size is larger than the Josephson penetration depth. From *SNS* junctions using the epitaxial films with *c*-axis and *a-b* plane orientations, it is found that the Josephson effect and the proximity effect depend strongly on the orientation of Y-Ba-Cu-O films, due to the anisotropic coherence length and the magnetic penetration depth of Y-Ba-Cu-O films.

ACKNOWLEDGEMENTS

The author would like to acknowledge helpful discussions with Drs. C. Camerlingo of CNR, F. Shinoki, S. Kojiro, S. Takada, and the members of the superconductivity electronics section in the Electrotechnical Laboratory. He also would like to thank M. Takahashi of ULVAC Ltd., K. Yamano of SANYO Co., M. Matsumoto of Kyocera Co., M. Aoyagi, H. Tanoue, and S. Misawa for the sample fabrication and analysis. Dr. T. Tsurushima and Dr. K. Kajimura are greatly appreciated for continuous support and encouragement.

REFERENCES

[1] J.G. Bednorz and K.A. Müller, Z. Phys. B**64**, 189 (1986).
[2] M.K. Wu, J.R. Ashburn, C.J. Torng, P.H. Hor, R.L. Meng, L.Gao, Z.J. Huang, Y.Q. Wang, and C.W. Chu, Phys. Rev. Lett. **58**, 908 (1987).
[3] A. Nakayama, A. Inoue, K. Takeuchi, and Y. Okabe, Jpn. J. Appl. Phys. **26**, L2055 (1987).
[4] H. Akoh, F. Shinoki, M. Takahashi, and S. Takada, Jpn. J. Appl. Phys. **27**, L519 (1988).
[5] Y. Katoh, H. Asano, M. Asahi, and O. Michikami, Extended Abstracts of 5th International Workshop on Future Electron Devices, Miyagi-Zao, p. 261, 1988.
[6] H. Akoh, F. Shinoki, M. Takahashi, and S. Takada, IEEE Trans. Magn. **25**, 795 (1989).
[7] L.H. Greene, J.B. Barner, W.L. Feldmann, L.A. Farrow, P.F. Miceli, R. Ramesh, B.J. Wilkens, B.G. Bagley, J.M. Tarascon, J.H. Wernick, M. Giroud, and J.M. Rowell, to be published in Proceedings of International Conference on Materials and Mechanisms of Superconductivity High-Temperature Superconductors, Stanford, 1989.
[8] R.S. List, A.J. Arko, Z. Fisk, S-W. Cheog, S.D. Conradson, J.D. Thompson, C.B. Pierce, D.E. Peterson, R.J. Bartlett, N.D. Shinn, J.E. Schirber, B.W. Veal, A.P. Paulikas, and J.C. Campuzano, Phys. Rev. B**38**, 11966 (1988).
[9] K. Mizushima, M. Sagoi, T. Miura, and J. Yoshida, Appl.Phys. Lett. **52**, 1101 (1988).
[10] J.W. Ekin, T.M. Larson, N.F. Bergren, A.J. Nelson, A.B. Swartzlander, L.L. Kazmerski, A.J. Panson, and B.A. Blankenship, Appl. Phys. Lett. **52**, 1819 (1988).
[11] H. Akoh, K. Kasahara, K. Yamano, and S. Takada, Extended Abstracts of 1989 International Superconductivity Electronics Conference, Tokyo, p. 525, 1989.
[12] P.M. Mankiewich, D.B. Schwarz, R.E. Howard, L.D. Jackel, B.L. Straughn, E.G. Burkhardt, and A.H. Dayem, Extended Abstracts of 5th International Workshop on Future Electron Devices, Miyagi-Zao, p. 157, 1988.
[13] T. Hatano, A. Fujimaki, Y. Nakasha, Y. Takai, and H. Hayakawa, Extended Abstracts of 1989 International Superconductivity Electronics Conference, Tokyo, p. 233, 1989.
[14] J. Clarke, Proc. Roy. Soc. A**308**, 447 (1969).
[15] H. Akoh, F. Shinoki, M. Takahashi, and S. Takada, Appl. Phys. Lett. **52**, 1732 (1988).
[16] M. Matsumoto, H. Akoh, and S. Takada, J. Appl. Phys. **66**, 3907 (1989).
[17] K.K. Likharev, Rev. Mon. Phys. **51**, 101 (1979).
[18] C.S. Owen and D.J. Scalapino, Phys. Rev. **164**, 538 (1967).
[19] K. Yamano, H. Akoh, M. Matsumoto, K. Kasahara, and S. Takada, Extended Abstracts of 1989 International Superconductivity Electronics Conference, Tokyo, p. 529, 1989.

Three-Terminal Devices of High-T_c Superconductors

HIDETAKA HIGASHINO, KENTARO SETSUNE, and KIYOTAKA WASA

Matsushita Electric Industrial Co., Ltd., Central Research Laboratories, 3-15, Moriguchi, Osaka, 570 Japan

ABSTRACT

Variable critical-current-type Josephson junctions using the thermal effect (VCJJ), and current-injection-type three-terminal devices are described. The VCJJ using a Bi-Sr-Ca-Cu-O (BSCCO) thin film exhibited almost linear decrease in the critical current with increasing a heating current, and a current-gain of 1.6 was obtained. Rf-induced current-step heights in the device can be controlled by the heating current. The current-injection-type device using the BSCCO film and aluminum gate electrodes exhibited a differential current-gain as high as 7.9.

KEY WORDS: critical current, heater, Shapiro step, current injection, tunnel barrier

INTRODUCTION

The discovery of superconducting materials which exhibit high critical temperatures (Tc) such as La-Ba-Cu-O by Bednorz and Müller [1] and Y-Ba-Cu-O (YBCO) by Wu et al. [2] brought an epoch to the study on the high-Tc superconductors and their applications. The new high-Tc oxide superconductors of Bi-Sr-Ca-Cu-O (BSCCO) discovered by Maeda et al. [3] attracted much attention because of their higher Tc over 110 K and of their stability. The high-Tc superconductors have unique characteristics compared with the low-Tc materials. The high-Tc over 77 K and the large energy gap in the order of 10 mV are help for easy maintenance devices and intrinsic high-speed operations. The low density of state ranging from 10^{21} to 10^{22} cm^{-3} suggests a possibility for a realization of field-sensitive devices or carrier-sensitive ones. However, it is still unknown that whether the anisotropy and the nature of the oxide, which requires a high preparation temperature of 400 - 700 °C, take sides with us or do not. The short coherent length as a few Å - some nm [4] is lying as an obstacle in the device fabrication. Most of the reported high-Tc devices such as dc SQUIDs [5-7] or Josephson junctions [8], except for a proximity-type superconductor/normal-metal/superconductor (S/N/S) junction with a 1 µm bridge-length [9], avoid this problem by using grain boundary Josephson junctions (GBJJ) whose characteristics are difficult to be controlled or poor in reproducibility, . However, even in a S/N/S junction, there lies a difficulty that the S/N interface must be perfect in the order of the coherence length. A variable critical-current-type Josephson junction (VCJJ) [10,11] is one of the three-terminal devices which give us some solutions in these difficulties at present.

Another reported high-Tc three-terminal devices are of current-injection-type [12,13] and of field-effect-type [14,15] in this stage. In the former, Kobayashi et al. have developed the superconducting current switching transistor (SCST) using YBCO films and has obtained a differential current-gain of 5 - 7 [12]. Also in our group, the same kind of device using a $Gd_1Ba_2Cu_3O_y$ (GBCO) film has been developed, but any current-gain was not obtained. The reason seems to lie on an interface instability between an aluminum injection gate and the GBCO film, because 1-2-3 system superconductors have the nature of the surface instability due to the oxygen deficiency. On the other hand, the Bi-based superconductors are more stable and are promising to fabricate tunnel junctions.

In this paper, the characteristics of the VCJJ [10,11] and the current-injection-type three-terminal device using a BSCCO film [13] are described.

CHARACTERISTICS OF THE VCJJ

A configuration of the VCJJ is shown in Figure 1. The superconducting BSCCO thin film has a constricted bridge-shape of 10 μm width and in length of 10 μm. Over the bridge, a Ta heater is deposited separated by a Ta_2O_5 buffer layer. The heater has an arched shape with 0.5 μm height and 2 μm length. The basic idea of the VCJJ is as follows: With putting a heating current, the heater heats the film through the buffer layer. The film temperature under the heater except for the arch goes up over the Tc. The superconducting current flows only an area of the film remaining below the Tc. Increasing the heating current, narrowing a width of the area. In this simple structure, a very narrow superconducting bridge can be realized without employing a nano-fabrication technology and the characteristics can be trimmed by the heating current. This configuration can be applied not only to the BSCCO film but also to another superconductors, e.g. the YBCO or even a Nb.

BSCCO thin films were prepared on a MgO (100) single crystal using a conventional rf sputtering technique [16]. The sputtering and subsequent annealing conditions are listed in Table 1. The prepared BSCCO film has a composition of $Bi_2Sr_2Ca_3Cu_{4.5}$ and a c-axis oriented mixed phase of (2212) and (2223) phases. The thickness of the film is 320 nm. A formation of the bridge was carried out by an Ar ion milling. The detailed fabrication process of the VCJJ is described in [11]. The thickness of the buffer layer was about 150 nm. The 240 nm-thick Ta heater was fabricated by a liftoff technique.

Table 1. Sputtering and annealing conditions

Target	Bi:Sr:Ca:Cu=1.5:1:1.5:2
Substrate	(100) plane of MgO
Sputtering gas	Ar/O_2=1.5
Gas pressure	0.5 Pa
Rf power	50 W
Substrate temperature	670 - 680 °C
Film thickness	320 nm
Growth rate	4 nm/min
Annealing	845 °C x 10 hours
Annealing gas	O_2 gas flow

The VCJJ exhibited the onset temperature (Tc_{on}) of 115 K and zero-resistivity temperature (Tc_0) of 77 K. These values and the resistance of the bridge were hardly changed through the process. The temperature dependence of the critical current (Ic) is shown in Fig. 2. The Ic decreases linearly with increasing the temperature up to 50 K and is approximately proportional to $(1-T/Tc_0)^a$ over 50 K where the values of 'a' are 2.2 and 2.3, before and after liftoff, respectively. The Ic at 4.2 K was 49 mA which corresponds to a 5% degraded value from that before liftoff, and does to a critical current density (Jc) of $1.5x10^6$ A/cm^2. This slight degradation is presumed to be due to the thermal stress. Fig. 3 shows a voltage-current (V-I) characteristics measured at 4.2 K with a parameter of heating current (Ih). The device exhibited similar characteristics as a bi-directional depression-type transistor with a zero saturation-voltage. The Ih dependence of the Ic at 4.2 K is shown in Fig. 4. The Ic decreases almost linearly with increasing the Ih. Over the Ih of 40 mA, the superconductivity was disappeared. A current-gain defined by a ratio of the Ic_0 (at Ih=0 mA) to the Ih_0 (at Ic=0 mA) was obtained as 1.2. The maximum differential current-gain is obtained as 1.6. It is confirmed that the VCJJ can operate as a current amplifier if high-Jc superconducting films are employed, although the heating power was rather high in the order of 100 mW. Fig. 5 shows a photograph of Shapiro steps observed in the VCJJ using a Er-Ba-Cu-O film where the Ih was 27 mA and an rf frequency was 6.2 GHz [10]. Current steps were appeared with a voltage interval of hf/2e. These Shapiro steps could be observed with the Ih over 20 mA and up to 55mA which was the Ic-cutoff value, and the rf response curve could be trimmed by varying the Ih. This result suggests us that the observed steps are due to a GBJJ which is left by thermally cutting back another parallel GBJJs, otherwise, that they are due to the thermally constricted narrow bridge. In fact, this thermal narrowing effect was observed even in a Nb-VCJJ.

CHARACTERISTICS OF THE CURRENT-INJECTION-TYPE DEVICE

Figure 6 shows a configuration of the current-injection-type three-terminal device with dual gates. A Dayem-type bridge of a BSCCO film is fabricated on a MgO substrate. Both of a minimum width and a length of the bridge are 10 μm and a thickness of the film was 320 nm. To examine the operation mechanism, two separated aluminum gate-electrodes are directly deposited on the bridge. The electrode width is 10 µm and a gap space between two gates is 2 µm. The gate current is injected to and/or extracted from the BSCCO film through the two gate junctions, resulting in a decrease of the superconducting current. The fabrication process is similar as that of the VCJJ and the details are described in [13]. The gate aluminum electrodes are formed by the liftoff technique.

The bridge shows the Tc_{on} of 115 K and the Tc_0 of 69 K and they were not changed through the process. However, the bridge resistance became as much as 34 % higher after the gate formation. This increase suggests a possibility of a grain boundary diffusion by the aluminum atoms. The Ic before the gate formation was approximately proportional to $(1-T/Tc_0)^{2.1}$ and was 35 mA at 4.2 K which corresponds to the Jc of 1.1×10^6 A/cm^2. After the gate formation, it was decreased down to 2.5 mA and the Jc was 7.8×10^4 A/cm^2 which corresponds to only 7.1 % of the value before formation. This degradation seems to support a possibility of the grain boundary diffusion by the aluminum not only of the chemical reaction at the junction interface. The V-I characteristics of the gate junctions are measured with two probe method at 4.2K. Fig. 7 and Fig. 8 show the data in the cases where a current flows to a superconductor from a G1 gate and from a G2 gate, respectively. In Fig. 8, two kinks are clearly observed at ±18.7 mV which correspond to an energy gap voltage of the BSCCO film. From the results, it is proved that a normal-metal/insulator/ superconductor (N/I/S) tunnel junction is naturally formed by a chemical reaction at the interface between the aluminum gate and the BSCCO film. Fig. 9 shows the V-I characteristic when a current flows from the G1 gate to the G2 gate through the BSCCO. A solid line shows a measured curve. With a simple thought, it is considered that this curve must coincide with a summation of that in Fig. 7 and of the polarity inverted one in Fig. 8, which is indicated with a dotted line in Fig. 9. The measured resistances are a little higher than the synthesized vales. It is unknown at present whether this discrepancy is caused by a resistance increase in each junction due to the difference of the gate-current distribution, or by that in bonding pads. The V-I characteristic of the device with a parameter of the gate current (I_{g-g}) at 4.2 K is shown in Fig. 10. With increasing the I_{g-g}, the Ic decreases and the curve closes to the voltage axis. Furthermore, a differential resistance at the same voltage becomes smaller comparing with that in the case of zero I_{g-g}. Fig. 11 shows the I_{g-g} dependence of the Ic. The Ic was measured on the condition that an 1 uV voltage was appeared between the two banks. The curve is almost symmetric against the vertical axis at I_{g-g}=0 mA. A slight asymmetry of the curve must be caused by a difference in two junction characteristics and by that in the kinds of the injected carriers, electrons or holes. The Ic of 2.5 mA at I_{g-g}=0 mA was suppressed down to 1.0 mA with an injection of the I_{g-g} as 0.3 mA which corresponds to the gate current-density of 750 A/cm^2. The maximum differential current-gain which is defined by a slope of the curve was obtained as 7.9 near the zero I_{g-g}. The operation mechanism is not so clear at present. The non-equilibrium superconductivity [17], the thermal effect, and the field effect [14,15,18] can be considered. With putting the I_{g-g} of 300 µA, the voltage of about 3 V was appeared between two gate electrodes, and it may produce a inversion layer as a few nm at the interface of the BSCCO. However, the thickness of the BSCCO film was 320 nm and therefore the field effect may not be dominant in this device configuration. With consideration of the fact that the measurement was carried out at 4.2 K while the Tc_0 was 69 K and the input power to the junction near I_{g-g} of 50 µA was under 100 µW, and of the fact that the aluminum gates work as heat sinks, it seems to be difficult to explain the mechanism as the simple heat effect. As Kobayashi mentioned concerning with the SCST [12], it is most plausible to explain the current-gain of this device as due to the non-equilibrium superconductivity.

CONCLUSIONS

Two types of three-terminal devices using high-Tc superconductors are described.

Variable critical-current-type Josephson junctions using Bi-Sr-Ca-Cu-O thin films exhibited a current amplifying operation with a current-gain of 1.6. It was demonstrated on the devices using an Er-Ba-Cu-O thin film or a Nb film that current-step heights in the rf-induced V-I curve were controlled by varying the heating current.

The current-injection-type device using the BSCCO film with aluminum gate electrodes exhibited a differential current-gain as high as 7.9. It is proved that the aluminum metal which is deposited on the BSCCO film forms a good N/I/S tunnel junction at the interface and that a combination of the BSCCO and the aluminum is promising in this configuration.

The introduced devices are preliminary ones. The goal of the three-terminal devices seems to be still far at present. Further investigations on the high-Tc materials and devices, and the development of the epitaxial oxide-thin-film technology are expected to bring us a fruitful future.

ACKNOWLEDGMENT

The authors would like to express their appreciation to the members of superconducting science division for their useful discussions and to Dr. T. Nitta of central research laboratories for his encouragement at Matsushita.

REFERENCES

1. J.G.Bednoz and K.A.Müller, Z.Phys., **B64**, 189 (1986).
2. M.K.Wu, J.R.Ashburn, C.J.Torng, P.H.Hor, R.L.Meng, L.Gao, Z.J.Huang, Y.Q.Wang and C.W.Chu, Phys.Rev.Lett., **58**, 908 (1987).
3. H.Maeda, Y.Tanaka, M.Fukutomi and T.Asano, Jpn.J.Appl.Phys., **27**, L209 (1988).
4. R.J.Cava, B.Batlogg, R.B.Van Dover, D.W.Murphy, S.Sunshine, T.Siegrist, J.P.Remeika, E.A.Rietman, S.Zahurak and G.P.Espinose, Phys.Rev.Lett., **58**, 1676 (1987).
5. R.H.Koch, C.P.Umbach, G.J.Clark, P.Chaudhari and R.B.Laibowitz, Appl.Phys. Lett., **51**, 200 (1987).
6. H.Nakane, Y.Tarutani, T.Nishino, H.Yamada and U.Kawabe, Jpn.J.Appl.Phys., **26**, L1925 (1987).
7. B.Haüser, M.Diegel and H.Rogalla, Appl.Phys.Lett., **52**, 844 (1987).
8. H.Tanabe, S.Kita, Y.Yoshizako, M.Tonouchi and T.Kobayashi, Japn.J.Appl. Phys., **26**, L1961 (1987).
9. P.M.Mankiewich, D.B.Schwartz, R.E.Howard, L.D.Jackel, B.L Straughn, E.G.Burkhardt and A.H.Dayem, SPIE **948** High-Tc Super-conductivity: Thin Films and Devices 37 (1988).
10. H.Higashino, A.Enokihara, K.Mizuno, T.Mitsuyu, K.Setsune and K.Wasa, Ext.Abst. of FED HiTcSc-ED Workshop, 267 (1988).
11. H.Higashino, K.Setsune and K.Wasa, Ext.Abst. of ISEC'89, 229 (1989).
12. T.Kobayashi, K.Hashimoto, U.Kabasawa and M.Tonouchi, IEEE Trans. on Magnetics, **25**, 927 (1989).
13. H.Higashino, A.Enokihara, K.Mizuno, K.Setsune and K.Wasa, Ext.Abst. of FED 2nd Workshop on HiTcSc-ED, 257 (1889).
14. D.F.Moore, Ext.Abst. of FED 2nd Workshop on HiTcSc-ED, 281 (1889).
15. S.Morohashi, H.Suzuki, K.Gotoh, N.Fujimaki and S.Hasuo, Proc. of Japn.Sympo.Appl.Phys.(in Japanese), 29a-Q-3 (1989).
16. H.Adachi, Y.Ichikawa, K.Setsune, S.Hatta, K.Hirochi and K.Wasa, Jpn.J.Appl.Phys., **27**, L643 (1988).
17. C.S.OWen and D.J.Scalapino, Phys. Rev. Lett., **28**, 1559 (1972).
18. A.F.Hebard, A.T.Fiory and R.H.Eick, IEEE Trans. on Magnetics, **MAG-23**, 1279 (1987).

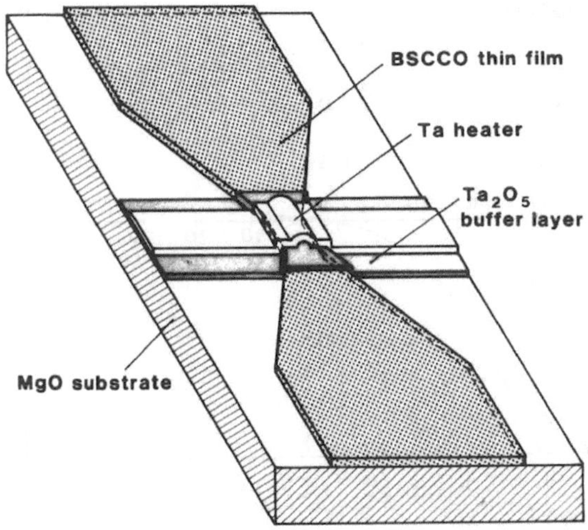

Fig.1. Configuration of a variable critical - current - type Josephson junction (VCJJ).

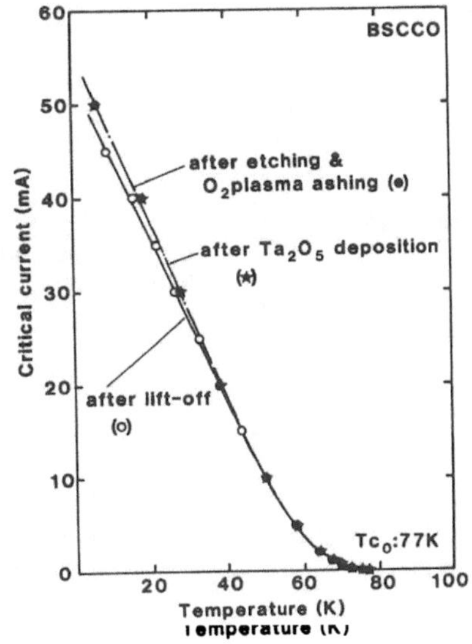

Fig.2. Temperature vs. critical current.

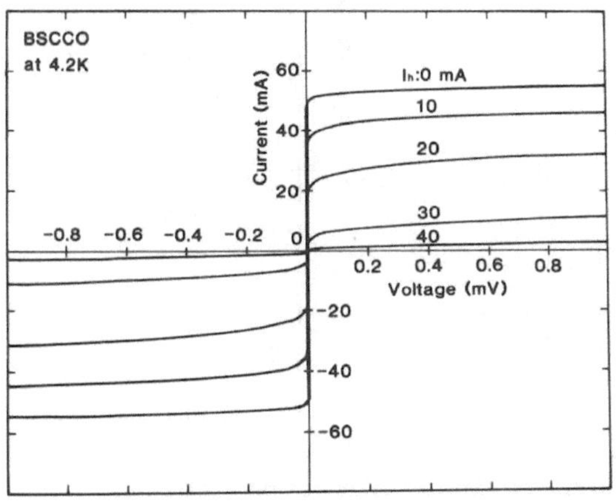

Fig.3. V-I characteristics of the VCJJ with a parameter of heating current.

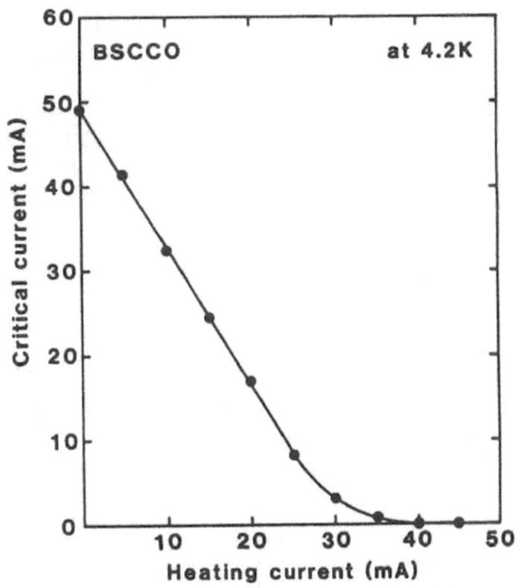

Fig.4. Heating current vs. critical current.

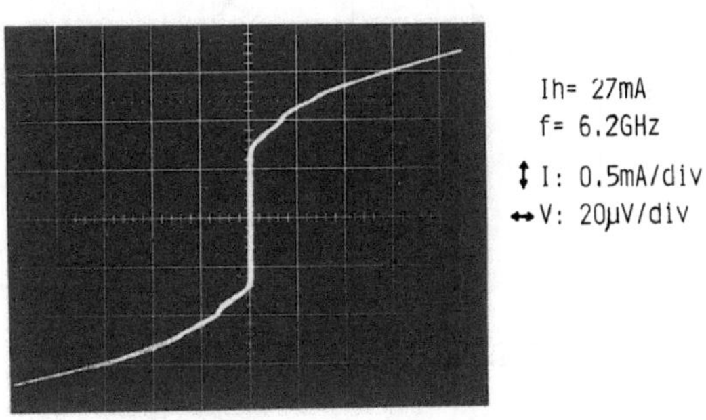

Fig.5. The Shapiro steps observed on the EBCO-VCJJ.

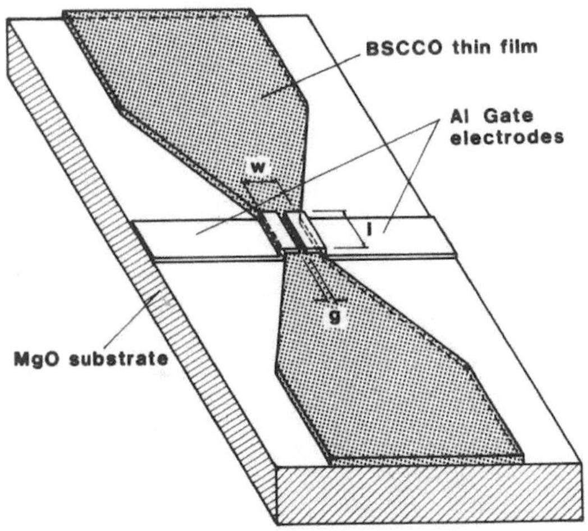

Fig.6. Configuration of the current-injection-type three terminal device.

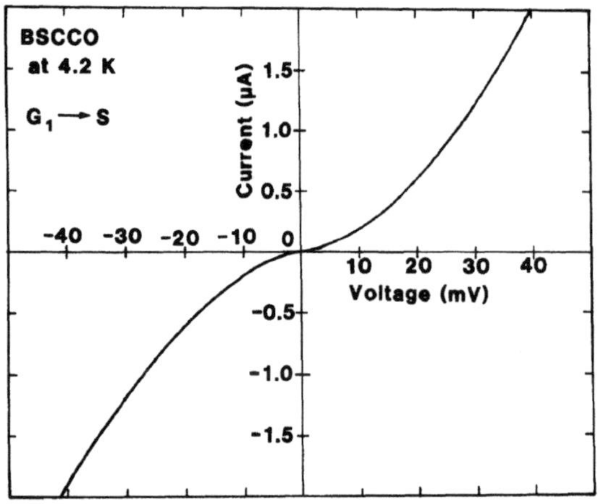

Fig.7. V-I characteristic of G 1 gate junction:G_1 -> BSCCO.

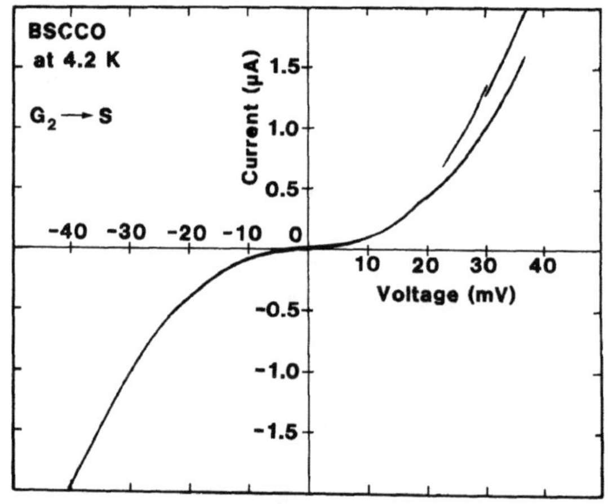

Fig.8. V-I characteristic of G 2 gate junction:G_2 -> BSCCO.

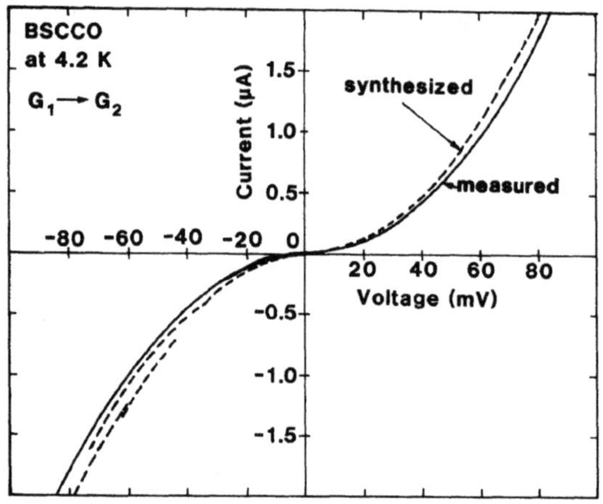

Fig.9. V-I characteristic of series gate junctions:G_1 -> G_2

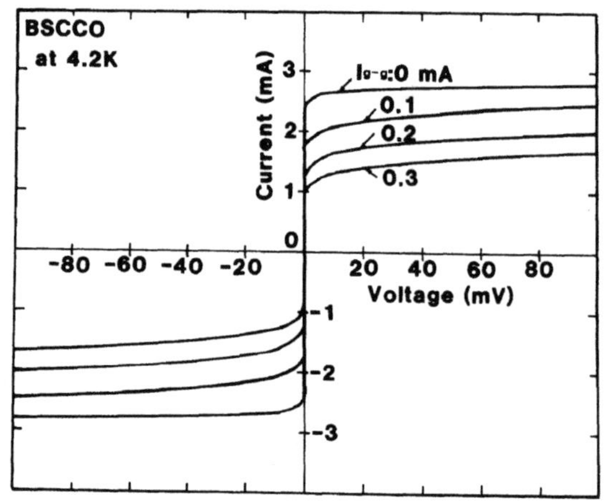

Fig.10. V-I characteristic of the current-injection-type device.

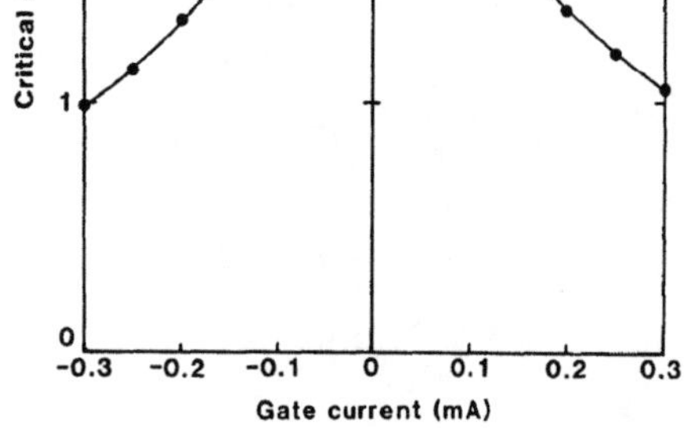

Fig.11. Gate current vs. critical current:G_1 -> G_2

Fabrication of Dc-SQUID with As-Grown YBaCuO Thin Film by Focused Ion Beam

M. TANIOKU[1], K. KURODA[1], K. KOJIMA[1], K. HAMANAKA[1], Y.H. HISAOKA[2], A. SHUHARA[2], and H. MURAKAMI[2]

Central Research Laboratory[1], Manufacturing Development Laboratory[2], Mitsubishi Electric Corporation, Tsukaguchi, Amagasaki, Hyogo, 661 Japan

ABSTRACT

We have fabricated dc superconducting quantum interference devices (dc-SQUIDs) from an as-grown YBaCuO thin film of 50 nm thickness which was deposited on a MgO(100) substrate by RF magnetron sputtering. The film is highly textured with the c-axis perpendicular to the substrate surface. Microbridges of SQUIDs were formed by an implantation technique with a Focused Ion Beam (FIB). The scale of microbridges could be reduced down to 0.5 μm or less. SQUIDs indicated both current steps induced by a microwave and a periodic response to an applied magnetic field.

KEY WORDS: as-grown thin film, focused ion beam, dc-SQUID

INTRODUCTION

Superconducting quantum interference devices (SQUIDs) fabricated with thin films of high temperature superconductors have been reported by a number of groups.[1)-3)] They used the grain boundaries of polycrystalline films as weak links, observed hysteresis, aperiodicity, and low frequency noise greater than that of Nb material SQUIDs; no reproduciblity will be expected from grain-boundary Josephson junctions produced spontaneously. One should use single crystal thin films and fabricate deliberate Josephson elements, though both of them are hard to develop. As for noise reduction, SQUIDs will also require single crystal thin films or at least well oriented films, because considerable flux noise has been observed in polycrystalline loops of high-T_c superconductors.[4)]

We used as-grown, highly-oriented, smooth surface thin films,[5),6)] to fabricate Constant Thickness Bridges (CTB). Recently submicrometer constrictions from post-annealed films have reportedly been formed by FIB etching.[7)] We constricted the effective width of superconducting pass and formed extremely narrow bridges by using FIB implantations. The SQUIDs with such microbridges showed a periodic response to an applied magnetic field. In this paper the properties of these SQUIDs will be described in detail.

EXPERIMENTAL

Preparation of YBaCuO thin film

Films were grown on MgO(100) substrates by RF magnetron sputtering from three targets. Each target was made of a calcined Y-Ba-Cu oxide with one of cation elements in 50 % excess of the superconducting composition $Y_1Ba_2Cu_3O_x$, *i.e.*, $Y_{1.5}Ba_2Cu_3\ O_x$, $Y_1Ba_3Cu_3O_x$, and $Y_1Ba_2Cu_{4.5}O_x$. The three targets were simultaneously discharged, and the substrate was alternately positioned over each target. The chemical composition of the film was precisely controlled by adjusting the staying time over each target. The substrate was heated at 620 °C in the atmosphere of O_2 at 300 mTorr. The growth rate obtained was 0.5 nm/min. The film thickness was controlled to be 50 nm. After deposition, the film was cooled to below 200 °C within 30 min in the oxygen gas atmosphere.

Microbridge fabrication requires a smooth surface, since radiation damage can not be controlled in a film with a rough surface. By varying the film composition, the surface presented a variety of morphology, and a very smooth surface was obtained at the stoichiometric composition corresponding to staying times over the three targets of 5, 25, and 70 s respectively. The superconducting transition

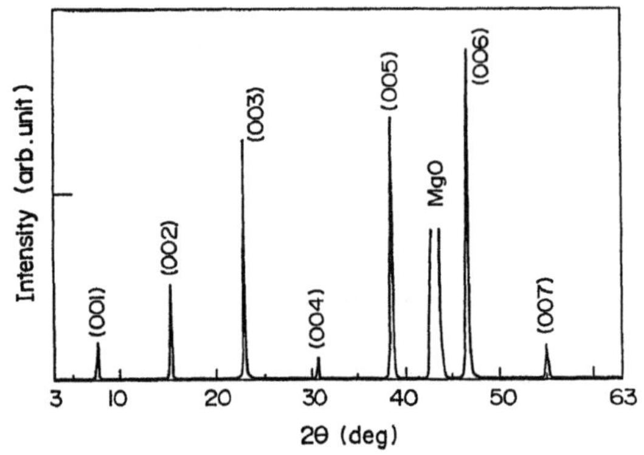

Fig.1 X-ray diffraction pattern for as-grown thin film.

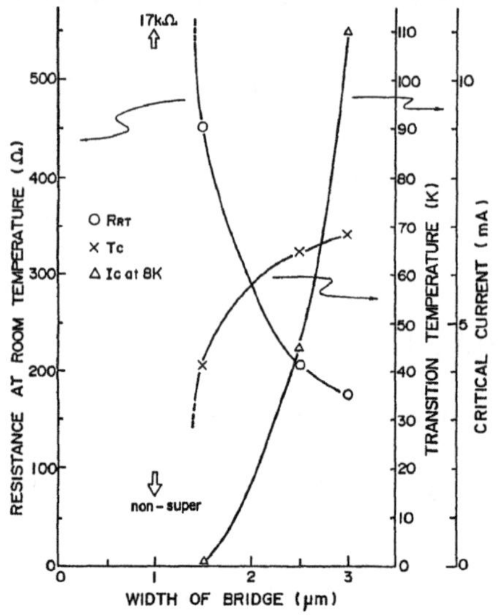

Fig.2 Resistance at room temperature, T_c, and I_c at 8 K as a function of microbridge width of SQUID.

temperature (T_C) was 74K, which was not the highest attainable temperature. The highest superconductive temperature of 81 K was achieved with a rough-surface film containing an excess of Cu. Figure 1 shows the X-ray diffraction pattern for the as-grown film; the film is highly oriented with the c-axis perpendicular to the substrate plane. The c-axis parameter was found to be 11.67 Å which is smaller than those of other sputtered films [8),9)] and equal to those of bulk samples.[10)]

Fabrication of SQUIDs

We made four SQUIDs with microbridges of different widths (1, 1.5, 2.5, and 3.0 μm) in an as-grown film. Microbridges were patterned by the focused Ga^+ ion beam implantation. The FIB at 80 kV with a spot size of 0.2 μm was scanned onto the film directly to draw a straight line at a rate of 6 ms/μm, *i.e.*, with a dose of 10^{16} ions/cm^2 and was blanked for a certain time to form a space in the line. The irradiated line was damaged to be a near insulator and the space became a microbridge. In a subsequent step the SQUIDs pattern with each loop area of 40 μm × 10 μm was defined by photolithography and chemical etching process.

Figure 2 shows resistances at room temperature, T_cs, and I_cs at 8 K of four SQUIDs measured by the four probe method. Monotonous dependence of them as a function of the microbridge width proves good homogeneity of the film and good controllability of the FIB. The microbridge was degraded as observed in the figure by the decrease in T_c with the constriction of the designed width and was finally turned into non-superconducting below 1 μm width. Then, in the case of the designed width of 1.5 μm, the effective width for the superconducting transition was estimated to be below 0.5 μm. Such degradation and reduction mainly owe to the crystal damage caused by a small number of ions distributed far from the beam spot. They are not avoidable due to striving for a CTB Josephson junction using a fairly thin film such as this one of 50 nm thickness.

RESULTS AND DISCUSSION

The performance characteristics of SQUIDs were measured in a He atmosphere magnetically shielded by the enclosed Pb sheet cooled under liquid He. Figure 3 (a) shows the I-V curve at 8 K for the SQUID with designed microbridge width of 1.5 μm. The dynamic resistance was 2 Ω. Induced current steps were observed with microwave irradiation of 15 GHz as shown in Fig. 3 (b). The value of Rn was estimated to be about 5 Ω from the slope of the I-V curve. The Shapiro steps were also observed for SQUIDs with bridge widths of 2.5 and 3.0 μm.

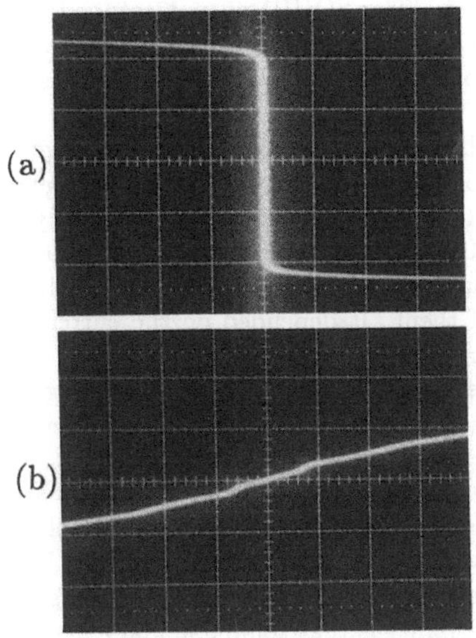

Fig.3 Current-voltage characteristics at 8 K for SQUID with 1.5 μm-width microbridges (a) without microwave irradiation, (b) with irradiation of 15 GHz.

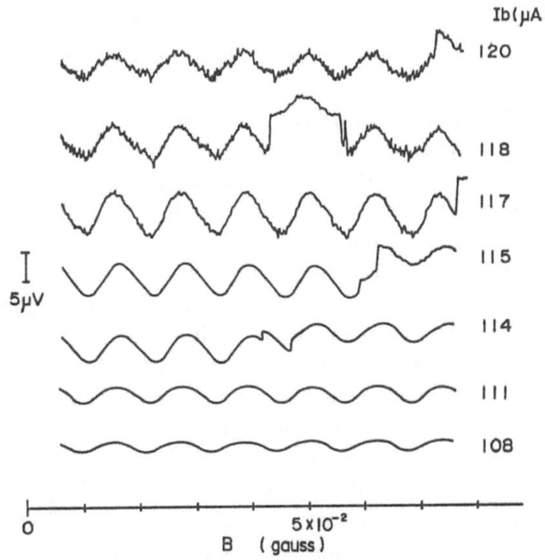

Fig.4 Flux-voltage characteristics at 8 K for SQUID with 1.5 μm-width microbridges biased at different currents. Magnetic flux was biased at 0.01 Hz.

Figure 4 shows the flux-voltage characteristics of the SQUID with 1.5 μm-width microbridges operated at 8 K for the different currents biased. The magnetic flux was biased by a coiled Cu wire with a frequency of 0.01 Hz. The SQUID clearly showed a periodic response to the magnetic flux with the maximum transfer function $dV/d\phi$ of 19 $\mu V/\phi_0$. The modulation depth of the critical current by the applied flux was very small compared with one predicted from the value of parameter $2LI_c/\phi_0$ which is 2 using geometrical inductance. It means equivalently that the inductance of the SQUID was larger than the geometrical one. The period at 8 K for the SQUID was about three times smaller than that calculated from the loop size. This discrepancy arises from a "flux focusing" effect [11] due to magnetic screening by superconducting pads. In the figure, considerable noise is noticeable in the flux-voltage curve in accordance with the increasing bias current. Since the sweep of flux was extremely slow, such noise appeared under the influence of low frequency noise. In fact, it was not observed at higher frequencies.

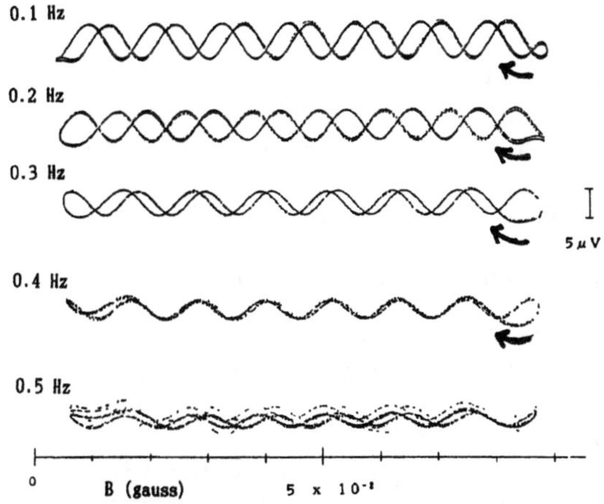

Fig.5 Hysteresis in the flux-voltage characteristics of the SQUID.

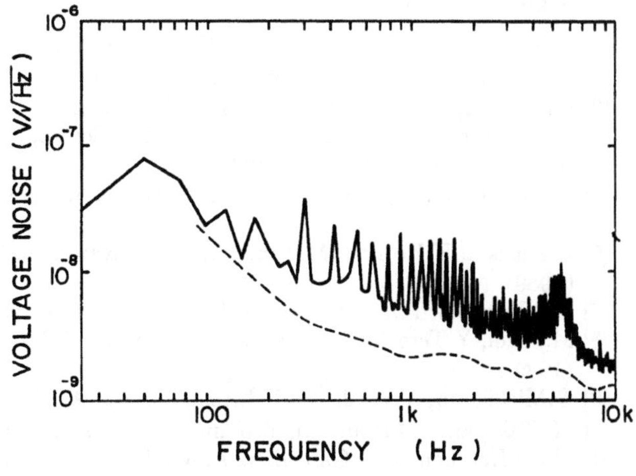

Fig.6 Spectral densities of voltage noise at 8 K of the SQUID. The dashed line is electronic noise level.

Hysteresis of the F−V curve was observed as shown in Fig. 5. The SQUID output responded behind the current in the flux modulation coil with a lag time of less than 1.5 s estimated from the figure. The recorder traced the same curve with repetition of the forward and backward sweep, that is, no irregularity existed.

Its voltage noise was measured. The current was biased with a battery and resistor, and the voltage across the SQUID was measured through a preamplifier. Figure 6 plots spectral densities of the voltage noise at 8 K of the SQUID biased at the highest modulation point. The voltage noise was completely limited by electronic noise. The flux noise level was about $3 \times 10^{-4}\ \phi_0/\mathrm{Hz}^{1/2}$ at 100 Hz. This value is rather large compared with that of low noise SQUIDs,[1] because the transfer parameter is small. However, the low voltage noise level was almost equal to those of them. It depended on the current biased. The voltage noise level rose slightly as the bias current increased.

SQUIDs with designed bridge widths of 2.5 and 3.0 μm were also measured on the flux-voltage characteristics. They seemed to respond to the applied flux at their suitable temperatures, but their modulated voltage values were extremely small at levels below electronic noise.

All the results of four SQUIDs were given in Table I. A coherence length of YBaCuO [12] requires a nanoconstriction for a Josephson element. Nevertheless, Josephson effects were observed independently of the bridge width of the SQUID. Both D/ξ_c and Le/ξ_{ab} were estimated to be 100 assuming that D (thickness) was 50 nm, Le (effective length) 200 nm, ξ_c 0.5 nm, and ξ_{ab} 20 nm. This value is so large that microbridges hardly work as Josephson elements. Another possibility is that some boundary weak links might have led to Josephson effects. In a micrograph taken by a field emission SEM, boundary-like morphology was seen, though it is not yet identified.

Table I Results of SQUIDs

Width (μm)	T_c (K)	I_c at 8 K (mA)	Shapiro steps	SQUID operation
1	×	–	–	–
1.5	41	0.1	○	○
2.5	64	4.5	○	△
3	68	11	○	△

CONCLUSIONS

We have demonstrated and studied YBaCuO SQUIDs fabricated from as-grown thin films. The FIB implantation technique was successfully employed to pattern microbridges of SQUIDs. Shapiro steps were observed for the SQUIDs independently of the width of their microbridges. The SQUID with the narrowest superconducting microbridges showed the periodic response appreciably. It showed hysteresis, but no irregularity in the flux-voltage characteristics, and its voltage noise was completely limited by the electronic noise.

REFERENCES

[1] R.H.Koch, W.J.Gallagher,and B.Bumble, Appl.Phys.Lett.54 (10),951 (1989).
[2] B.Hauser, M.Diegel,and H.Rogalla, Appl.Phys.Lett.52 (10),844 (1988).
[3] I.Takeuchi, J.S.Tsai, H.Tsuge,and N.Matsukura, Jpn.J.Appl.Phys.27, 2265 (1988).
[4] M.J.Ferrari, M.Johnson, F.C.Wellstood, J.Clarke, P.A.Rosenthal, R.H.Hammond, and M.R.Beasley, Appl.Phys.Lett.53 (8),695 (1988).
[5] K.Kuroda, K.Kojima, M.Tanioku, K.Yokoyama, and K.Hamanaka, Jpn.J.Appl.Phys.28, 1586 (1989).
[6] K.Kuroda, K.Kojima, M.Tanioku, K.Yokoyama, and K.Hamanaka, Jpn.J.Appl.Phys.28, 1797 (1989).
[7] L.R.Harriott, P.A.Palakos, and C.E.Rice, Appl.Phys.Lett.55 (5),495 (1989).
[8] M.Sagoi, Y.Terashima, K.Kubo, Y.Mizutani, T.Miura, and J.Yoshida, Jpn.J.Appl.Phys.28, L444 (1989).
[9] O.Michikami, M.Asahi, and H.Asano, Jpn.J.Appl.Phys.28, L448 (1989).
[10] E.Takayama-Muromachi, Y.Uchida, Y.Matsui, and K.Kato, Jpn.J.Appl.Phys.26, L619 (1987).
[11] M.B.Ketchen, W.J.Gallagher, A.W.Kleinsasser, S.Murphy, and J.R.Clem, SQUID'85, edited by H.D.Hahlbohm and H.Lubbig (Walter de Gruyter, Berlin,1985).
[12] T.Sakakibara, T.Goto, Y.Iye, N.Takeya, and H.Takei, Jpn.J.Appl.Phys.26 ,L1892 (1987).

Microwave Propagation on High-J_c YBCO Transmission Lines

K. HIGAKI, S. TANAKA, H. ITOZAKI, and S. YAZU

Itami Research Laboratories, Sumitomo Electric Industries, Ltd., 1-1, Koyakita 1-chome, Itami, Hyogo, 664 Japan

ABSTRACT

We have prepared YBCO superconducting transmission lines on the substrates of a MgO single crystal and have studied the propagation of microwave signals from 200MHz to 20GHz on these lines. The transmission lines are co-planar type and are 0.5μm thick, 30-100μm wide, and 12mm long. Their critical current densities which were directly measured through the whole line (12mm long) were more than $10^5 A/cm^2$ at 77.3K. The transmission properties of the YBCO superconducting co-planar lines were measured by a network analyzer. The results were compared with those of Al transmission lines. The transmission losses on the YBCO co-planar line (30μm wide) were less than 0.5dB for frequencies up to 9GHz. They are lower than those on the equivalent Al line on a MgO single crystal at 77.3K.

KEY WORDS: superconducting transmission line, microwave propagation, low attenuation, superconducting interconnection

INTRODUCTION

Electronics applications of high Tc superconducting materials have been studied by many researchers and several significant results have been reported on SQUID or Josephson junction [1-3]. Recently, it has been demonstrated that YBCO has lower surface resistance than normal metals such as copper [4,5]. Therefore, a microwave application of superconductor such as a transmission line or microwave elements attracts a great deal of attention and some experimental studies have been reported [6-8]. On the other hand, attenuation and dispersion of microwave signals at chip-to-chip or board-to-board normal metal interconnections of CPU should be made even smaller because a operating speed has been getting higher at the computer devices. It is consequently expected that high Tc superconducting materials which are estimated to have lower attenuation and lower dispersion than normal metals will be adopted to these interconnections. We have prepared YBCO co-planar transmission lines and have studied the microwave propagation on these lines.

EXPERIMENTAL

Preparation of Co-planar Transmission Line

The YBCO co-planar lines were prepared with etching high-Jc YBCO thin films on the substrates of a 1mm thick MgO single crystal. Preparation conditions of these thin films are described in the previous papers [9,10]. The transmission lines are 100μm and 30μm wide which were patterned by chemical wet etch-

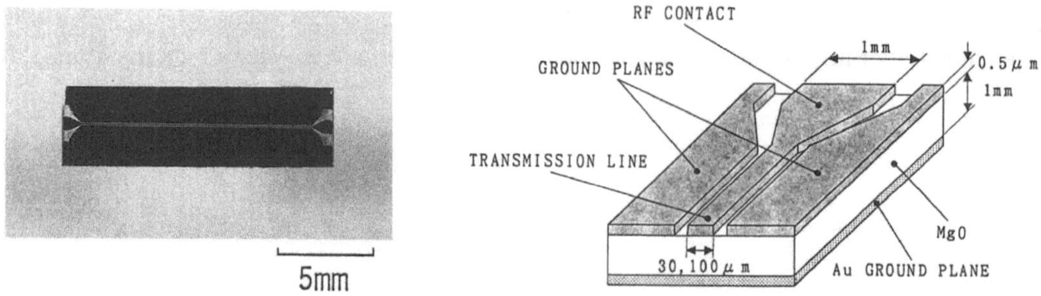

Fig.1 Photograph and cross-sectional diagram of the co-planar line.

ing and Ar ion milling respectively, with the length of 12mm and the thickness of 0.5μm. They are spaced with the prescribed width from the YBCO ground planes on both sides and the Au ground plane is deposited on the back side of the substrate to adjust the characteristic impedance to 50Ω. Both ends of the lines are patterned 1mm wide in order to contact the line with the connectors, as shown in Fig.1. The equivalent Al co-planar lines were also prepared in the same way so as to compare the transmission properties with the YBCO lines.

Measurement of Electrical Property

The temperature dependence of the resistivities and the critical current densities of the lines were directly measured through the whole line (12mm long) by a four-probe method in a cryostat. The transmission and reflection properties of microwave signals from 200MHz to 20GHz on these lines were measured by a network analyzer at room temperature and 77.3K in liquid nitrogen. Figure 2 shows the schematic diagram of the propagation measurement. Both connectors were pressed on the ends of the line inserting a indium sheet. Microwaves were fed to the co-planar line through the semi-rigid cable. The transmitted signals as well as the reflected signals were also fed to the network analyzer through each semi-rigid cable.

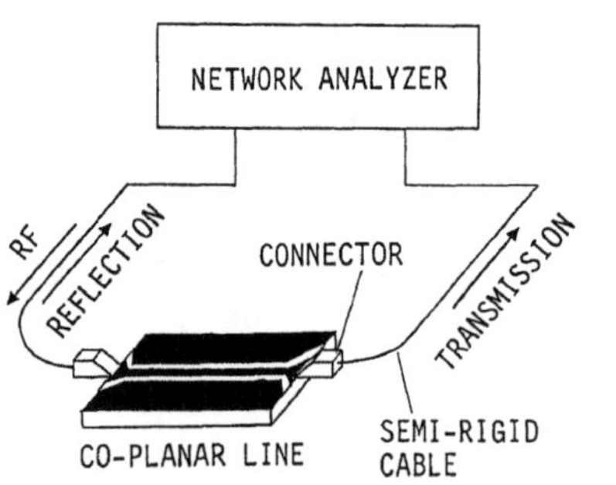

Fig.2 Schematic diagram for propagation measurement

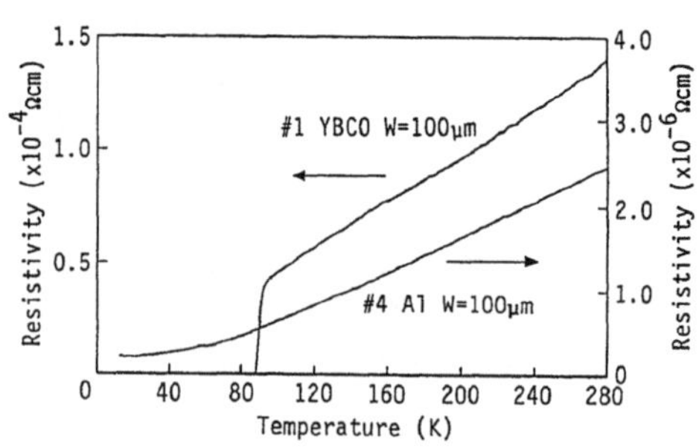

Fig.3 Temperature dependence of resistivities for YBCO and Al co-planar lines.

RESULTS AND DISCUSSION

The temperature dependence of the resistivities for YBCO and Al co-planar lines measured through the whole line (100μm wide, 12mm long) is shown in Fig.3. Resistivities of Al line were found to be about 2.7μΩcm at room temperature and about 0.43μΩcm at 77.3 K, which are considerably close values to

Table 1 Direct current properties of YBCO and Al co-planar lines.

SAMPLE No.	MATERIAL	LINE WIDTH,μm	Tc(R=0),K	Jc(77.3K),A/cm^2	RESISTANCE,Ω R.T.	77.3K
#1	YBCO	100	87	4.3x10^5	340	0
#2	YBCO	30	85	3.4x10^5	1643	0
#3	YBCO	100	83	4.3x10^4	380	0
#4	Al	100	-	-	7.5	1.2
#5	Al	30	-	-	23	3.7

those of 99.99%-Al. Table 1 shows the DC properties of the co-planar lines used for the study of propagation. The critical temperatures Tc(R=0) are about 85K and the critical current densities Jc(77.3K) are more than 10^5A/cm^2 for both YBCO lines with the width of 100μm and 30μm. Low-Jc YBCO line ($4.3x10^4$A/cm^2) was also prepared in order to study Jc dependence of the transmission properties.

Figure 4 shows the frequency dependence of the transmission loss on the YBCO (#2) and the Al(#5) co-planar lines with the line with W=30μm at room temperature and 77.3K. Peaks of the transmission loss were mainly existed over frequencies more than 10GHz. It is also found that the reflected power increased correspondingly at these points. It is thus presumed that they were caused by resonance which arose between discontinuous points such as contacts of the connectors and the line. The transmission losses of the Al line at room temperature were considerably high around 4dB for the frequency of 4GHz. But these losses of the same line became lower at 77.3K, because the resistance of the line became about one sixth lower at 77.3K than that at room temperature as shown in table 1. However, there still existed the transmission losses about 0.5dB even around MHz range. They are supposed to have been attributed to the resistance of the line which remained 3.7Ω at 77.3 K. The transmission losses of the 100μm wide Al line (#4) were correspondingly lower than those of 30μm wide Al line. The transmission losses of the YBCO line at room temperature were more than 25dB because the resistance of the line was very high. As the line was cooled down to 77.3 K and then became superconducting, they were remarkably getting lower as shown in Fig.4. These losses were less than 0.5dB for the frequencies up to 9GHz and less than 4dB for the frequencies up to 18GHz. They were lower than those of the equivalent Al line at 77.3K. These results indicate that our high-Jc YBCO thin films have lower surface resistance than Al and that YBCO could be a superior interconnection to normal metal in case that resistance of normal metal line is considerably high.

We also investigated the Jc dependence of the transmission loss on the YBCO co-planar line. Figure 5 shows the frequency dependence of the transmission losses for a high-Jc line (#1) and a low-Jc line (#3) with the line width of 100μm at 77.3K. Though the transmission losses of the low-Jc line was almost 0dB at the lowest frequencies around 200MHz, they rose as increase of frequency and became obviously larger than those of high-Jc line. Thus, excellent transmission properties in

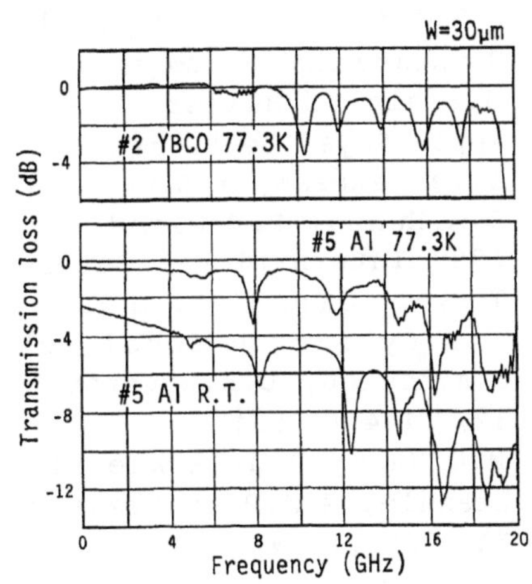

Fig.4 Frequency dependence of transmission loss on YBCO and Al co-planar lines with the line width of 30μm.

the high-Jc line are not obtained in the line whose Jc is about $4x10^4 A/cm^2$. High frequency signals must be scattered by grain boundaries which exist only in the low-Jc line. These results are consistent with the report that the surface resistance for polycrystalline YBCO is ten times larger than that of YBCO single crystal [4].

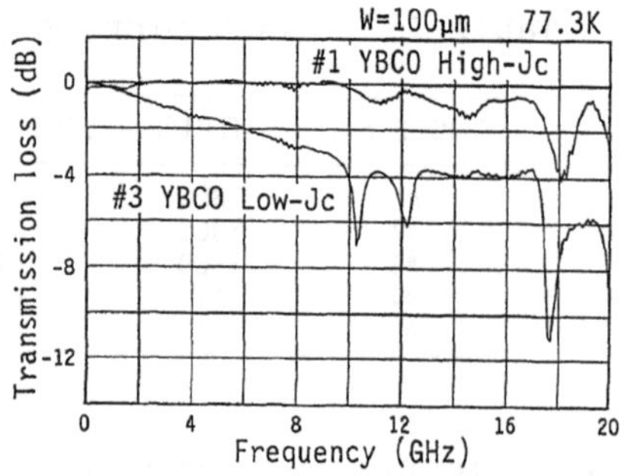

Fig.5 Frequency dependence of transmission loss on high-Jc and low-Jc YBCO co-planar lines with the line width of 100μm at 77.3K.

SUMMARY

Microwave propagation study on YBCO superconducting co-planar line was performed. The excellent transmission properties were obtained at 77.3K with the lines whose Jc measured through the whole line (12mm long) were more than $10^5 A/cm^2$. The transmission losses of such a high-Jc YBCO line with the line width of 30μm were less than 0.5dB for the frequencies up to 9GHz and those with the line width of 100μm were less than 0.3dB for the frequencies up to 10GHz. They were lower than those of the equivalent Al lines. However , the low-Jc YBCO line ($4.3x10^4 A/cm^2$) had obviously larger transmission losses than the high-Jc line. It is thus suggested that only high-Jc superconducting transmission lines could take the place of normal metal lines whose resistance is considerably high.

We are grateful to Mr. H. Yoshino for the measurement of microwave propagation.

REFERENCES

1. C.T.Rogers, A.Inam, M.S.Hedge, B.Dutta, X.D.Wu and T.Venkatesan, to be published.
2. Y.Higashino, T.Umezawa, T.Takahashi, K.Mizobuchi and S.Naito, Extended Abstracts of 1989 International Superconductivity Electronics Conference (ISEC'89), Tokyo, June 1989, p.218.
3. S.Tanaka and H.Itozaki, submitted to Appl. Phys. Lett.
4. N.Klein, G.Müller, H.Piel, B.Roas, L.Schults, U.Klein and M.Peiniger, Appl. Phys. Lett. 54, 757 (1989).
5. T.Venkatesan, X.D.Wu, A.Inam, C.T.Rogers, L.Nazar B.Wilkens, C.C.Chang, D.M.Hwang, P.England, M.S.Hedge and H.Padamsee, Extended Abstracts of 1989 International Superconductivity Electronics Conference (ISEC'89), Tokyo, June 1989, p.1.
6. D.R.Dykaar, R.Sobolewski, J.M.Chwalek, J.F.Whitaker, T.Y.Hsiang, G.A.Mourou, D.K.Lathrop, S.E.Russec and R.A.Buhrman, Appl. Phys. Lett. 52, 1444 (1988).
7. M.C.Nuss, P.M.Mankiewich, R.E.Howard, B.L.Straughn, T.E.Harvey, C.D. Brandle, G.W.Berkstresser, K.W.Goossen and P.R.Smith, Appl. Phys. Lett. 54 2265 (1989).
8. M.Morisue, J.asahina, K.Yoneda and K.Araki, Extended Abstracts of 1989 International Superconductivity Electronics Conference (ISEC'89), Tokyo, June 1989, p.251.
9. S.Tanaka and H.Itozaki, Jpn. J. Appl. Phy. 27, L662 (1988).
10. H.Itozaki, S.Tanaka, K.Higaki and S.Yazu, Physica C 153-155, 1155 (1989).

Tapered Tube Lenses for Intense Electron Beams (Supertrons)

HIDENORI MATSUZAWA, YOSHIHARU ISHIBASHI, TOMOAKI OSADA, and TETSUYA AKITSU
Faculty of Engineering, Yamanashi University, 3-11, Takeda 4-chome, Kofu, 400 Japan

ABSTRACT

Supertrons, which were previously proposed and demonstrated to function as lenses for intense electron beams (310 keV, 1∿3 kA, a few nanosecond pulse widths), were experimentally evaluated in focusing ability for a tapered tube. The tube was made of Bi-compound and was 20 and 10 mm at the inlet and the exit, respectively, with an axial length of 31 mm. The electron beam focused was 1.7 kA and 2 mm in diam at the exit of the tube. The focusing ability of the tapered tube was better than those of straight tubes which were previously reported.

KEY WORDS: lenses, Supertrons, intense electron beams

INTRODUCTION

Soon after high-Tc yttrium compounds were reported, we conceived an idea of lenses (Supertrons) to focus and transport intense electron beams (relativistic electron beams, REB). The principle of the lenses is as follows: When REBs are injected into small apertures of superconducting tubes, self magnetic field of the REBs cannot penetrate the tubes owing to the Meissner effect. The magnetic field is thus confined and compressed into a narrow space between the tubes and the REBs. The enhanced magnetic field will then focus the REBs.

In previous papers [1∿3], focusing ability of Y- and Bi-compound Supertrons was successfully demonstrated using straight tubes with inner diams of 20, 10, and 5 mm. For the 10- and 5-mm inner-diam tubes, REBs (310 keV, 1∿3 kA, a few nanosecond duration time) hit the corners where funnel-type inlets were connected to the straight tubes. In the present paper, we used a tapered Bi-compound tube instead of straight ones to avoid sputtering of superconductors due to REB bombardment. Thinner REBs were achieved with the tapered tube.

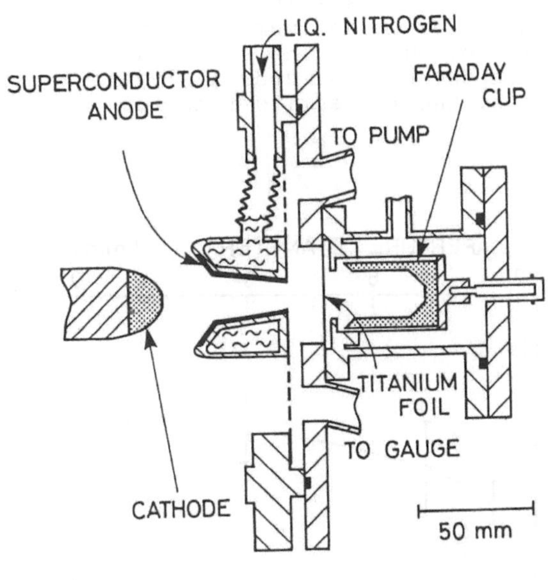

Fig. 1. Experimental apparatus.

EXPERIMENTAL APPARATUS

Figure 1 shows experimental apparatus used. High-voltage pulses (310 kV, 40 ns) were applied to the diode which consisted of a carbon-made cathode and a superconductor anode. The diode was immersed in an atmosphere of neon gases of the order of 0.1 Torr to generate higher electron current beams [4] and to charge-neutralize space charges for efficient focusing and transport of REBs. The REBs focused were detected with a Faraday cup which was set downstream a titanium foil of 20 μm thickness. The Faraday cup was in a vacuum of pressure less than 10^{-4} Torr and collected electron beams of more than 60 keV.

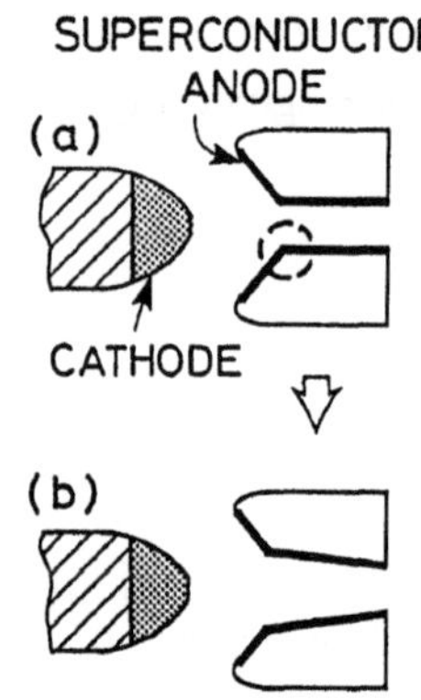

Fig. 2. Anode shapes. (a) previous anode [1∿3] and (b) present one.

As mentioned above, anodes were previously composed of a funnel-type inlet and straight tubes (Fig. 2a). The corners were damaged with REBs when the straight tubes were narrower than 20 mm in diam. The anode in the present paper was thus improved as in Fig. 2b. The Bi-compound anode (Tc = 99 K) was cemented on a copper-made heat sink with electrically conducting epoxy resin. Carbon-made apertures (2 to 11 mm in diam and 2 mm thick) were placed at the exit of the anode to observe radial profiles of focused REBs.

The anode had only a narrow single inlet for liquid nitrogen. So, a feeder was devised for liquid nitrogen to be poured with little loss [5]. To form gently tapered tubes, we pressed calcined Bi-compound powders with a jig whose surface was chromed to be more smooth and hardened [6]. Figure 3 shows the jig used. One of the key points of the jig is that the outer vessels, 1, are composed of two parts, not of a single one.

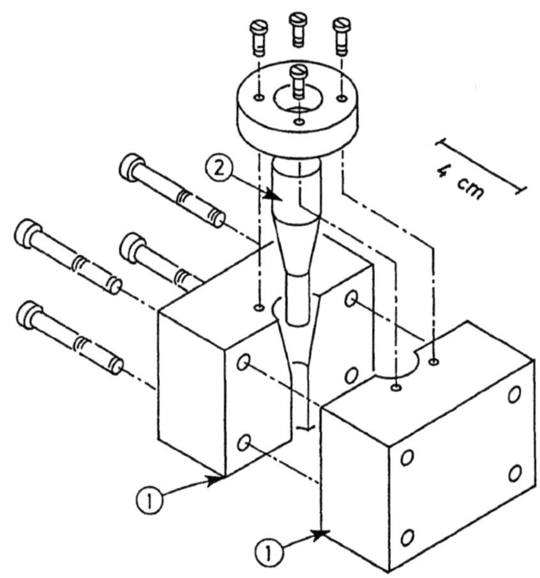

Fig. 3. Jig for tapered tubes.

EXPERIMENTAL RESULTS

Figure 4 shows REBs collected with the Faraday cup as a function of apertures for an optimum pressure of 0.15-Torr neon. The error bars indicate the highest and the lowest values at the respective time for

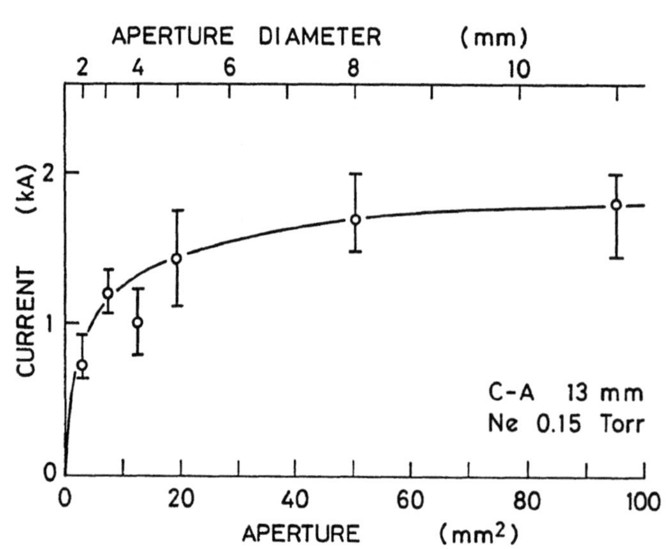

Fig. 4. REBs collected as function of apertures.

fifteen shots. The point for the 4-mm-diam aperture deviated from an approximated curve. This is because the aperture was a little eccentric in regard to the exit of the anode. After differentiating the curve in Fig. 4 with respect to diam, we had a radial profile of current density of focused REBs, which is shown in Fig. 5a with open circles. The REBs focused had a 2-mm diam (full width at half maximum). Triangles in Fig. 5a are for a tapered copper-made anode with an 8-mm diam exit, for 0.15-Torr neon. The superconducting anode certainly focused the REBs much sharply.

For comparison, open circles in Fig. 5b [7] shows radial profiles for straight Bi-compound (105 K) anodes with a 10-mm inner diam. The filled circles in Fig. 5b are for a straight Bi-compound anode of poor quality (98 K). From Fig. 5, the tapered anode focused higher REBs to thinner beams than the straight anodes even though the tapered anode had a lower Tc (99 K) than those of the straight ones. The current density on the beam axis was as high as 400 A/mm^2 or more. The value is about two orders of magnitude higher than those of electron beams for gyrotrons (high-power microwave tubes).

To visualize spatial distribution of focused REBs, we took open-shutter photographs [2, 3] through a prism for a single shot of REBs. Figure 6 shows one of the examples which indicates contours of intensity of fluorescence emitted from neon atoms. The contours show the diam of focused REBs to be about 2 mm, corresponding to that obtained above with apertures.

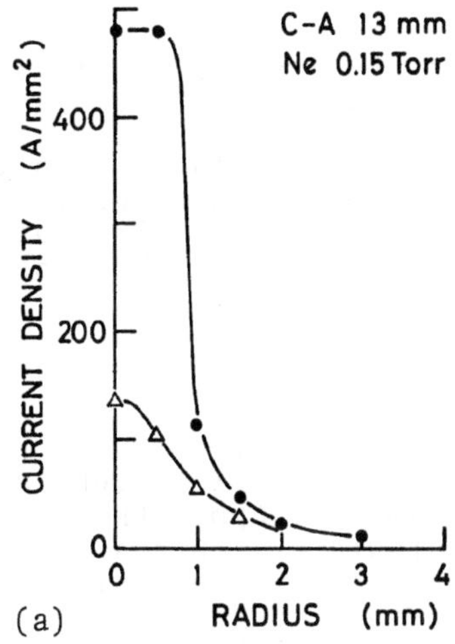

(a)

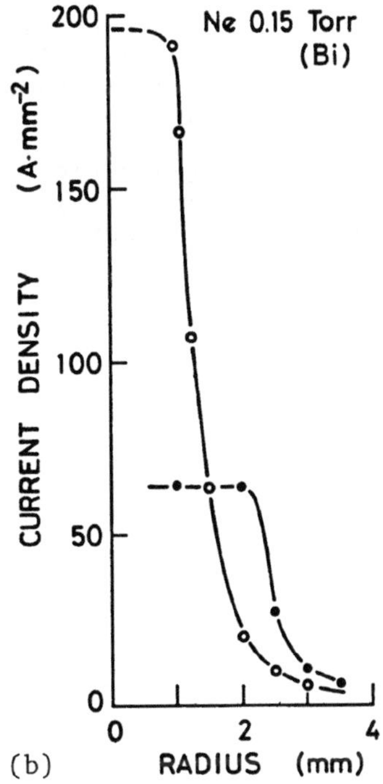

(b)

Fig. 5. Radial profiles of focused REBs. (a) tapered superconducting anode (●) and copper-made one (Δ), and straight superconducting anode (o) [7].

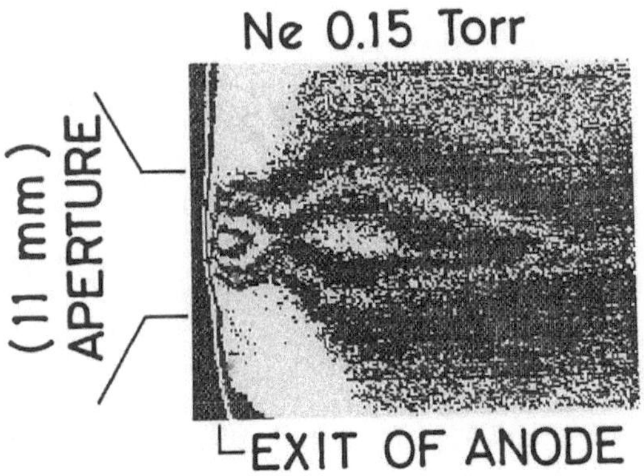

Fig. 6. Spatial distribution of fluorescence emitted from neon atoms for tapered anode.

CONCLUSIONS

A tapered tube anode focused REBs to thinner beams than straight tube ones. These results suggest that Supertrons will find many applications by modifying their shapes and dimensions.

The authors would like to thank M. Ohishi, M. Kawamura, Y. Kobayashi, A. Tohyama, T. Nagakura, H. Marumo, N. Morimoto, T. Yokozawa, and H. Wada for their help in carrying out the experiment, and Y. Chino for his cooperative help in preparing the experimental apparatus.

REFERENCES

1. Matsuzawa H. Ohmori O. Yamazaki H. Ueno J. Furumizu A. Saito A. Takahashi T. Akitsu T (1989) High-Tc superconducting lenses for relativistic electron beams. J. Appl. Phys. 65: 2596-2603
2. Matsuzawa H. Osada T. Ishibashi Y. Irikura K. Okamoto K. Mochizuki A. Wada H. Akitsu T (1989) Focusing of intense electron beams with novel high-Tc superconducting lenses. Jpn. J. Appl. Phys. 28: L717-L719
3. Evans R (1989) High-temperature superconductor lens. Nature 340: 191
4. Matsuzawa H. Akitsu T (1988) High-pressure operation of a beam diode for relativistic electron beams. J. Appl. Phys. 63: 4388-4391
5. Ishibashi Y. Irikura K. Matsuzawa H (1989) An effective liquid-nitrogen feeder for a narrow single inlet. J. Vac. Sci. Technol. A 7: 2818
6. Osada T. Tohyama A. Chino Y. Matsuzawa H (1989) Jig for tapered high-temperature superconductor tubes used as electron beam lenses (Supertrons). Rev. Sci. Instrum. 60: No. 12 to be published
7. Matsuzawa H. Ishibashi Y. Irikura K. Osada T. Okamoto K. Mochizuki A. Wada H. Akitsu T (1989) Supertrons——High-Tc superconducting lenses for intense electron beams. In: Extended Abstracts of 1989 International superconductivity electronics conference (ISEC '89) DE2-3. 12-13 June 1989. Tokyo pp521-524

Low Noise Operation of Novel Magnetic Sensor Using Ceramic High T_c Superconductor Film

H. SHINTAKU, H. NOJIMA, M. NAGATA, E. OHNO, and M. KOBA
Sharp Corporation, Central Research Laboratories, 2613-1, Ichinomoto-cho, Tenri, Nara, 632 Japan

ABSTRACT

The low noise operation of a magnetic sensor employing magnetoresistive effect of ceramic high Tc superconductor films is studied. The sensor was fabricated using $Y_1Ba_2Cu_3O_{7-x}$ superconductor film prepared by a spray pyrolysis method. By applying a modulation of a.c. magnetic field, the external magnetic field was detected with lock-in amplifier. A magnetic field resolution of 2×10^{-5} gauss/$(Hz)^{1/2}$ at 1 Hz is obtained, which is about 1 order less than that obtained directly from the voltage-magnetic field transfer function (dV/dB) and the voltage noise.

KEY WORDS: magnetic sensor, low noise operation, magnetoresistive effect, voltage noise, field modulation

INTRODUCTION

We have already reported a novel magnetic sensor made of high Tc ceramic superconductor [1] [2]. A principle of operation of the magnetic sensor is based on the magnetoresistive effect in the ceramic superconductor. This sensor has following particular features: 1) it undergoes a very abrupt change in the resistance at a low magnetic field, leading to a great potential of a very high sensitivity; 2) it can be used for analogue as well as digital operations; 3) it has a very simple structure; 4) it can be very easily operated.

In this paper, we report the low noise operation of the magnetic sensor and the magnetic field resolution obtained with this method. We compare the experimental results with those obtained directly from the voltage-magnetic field transfer function (dV/dB) and the voltage noise, and discuss the sensitivity of the sensor.

EXPERIMENTAL

A. DEVICE FABRICATION

Y-Ba-Cu-O ceramic superconductor films were prepared by a spray pyrolysis method [3], which is a convenient technique to obtain a superconducting film. The sample investigated in this paper had a composition of $Y_1Ba_2Cu_3O_{7-x}$ with c-axis oriented perpendicular to yttrium stabilized ZrO_2 substrate surface. The thickness of the film thus obtained was about 5 μm. The SEM observation showed that the film consisted of small grains ranging from 2 μm to 5 μm in diameter. The rectangular film of 5mm x 10 mm x 5 μm was patterned to a meander shape mechanically to increase the magnetoresistance. Current and voltage electrodes were formed by Ti evaporation onto the film. The contacts between the Ti electrodes and the lead wires were made by silver paste. The zero resistance temperature Tc of the element was 81 K. The element was mounted in a package with dry N_2 gas to prevent from degradation by a reaction with atmospheric moisture. We confirmed that the properties of the sensor were stable for more than 1 year. The schematic structure of the element is shown in Fig.1. The photograph of the sensor is shown in Fig.2.

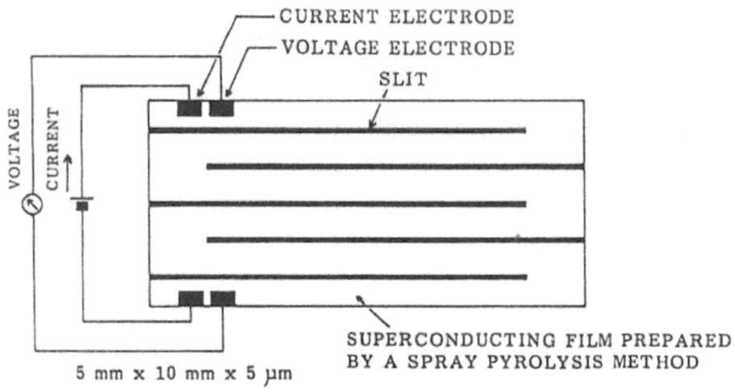

Fig.1. Schematic structure of the element.

Fig.2. Photograph of the sensor.

B. MEASUREMENT SYSTEM

A schematic diagram of the measurement system is shown in Fig.3. A d.c. transport current through the element was fed by a constant current source. With applying an a.c. magnetic field generated by a solenoidal coil surrounding the element, a modulated voltage signal of the element was passed through a low-noise differential amplifier and detected with lock-in amplifier. To generate the a.c. modulation field, a sinusoidal a.c. current

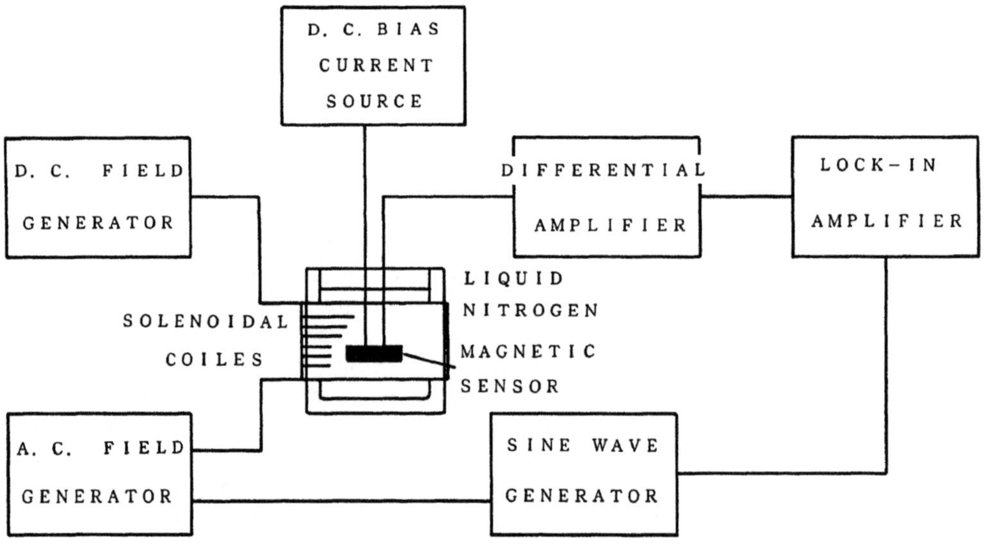

Fig.3. Schematic diagram of the measurement system.

was applied to the solenoidal coil. In this experiment, a frequency and a magnitude of the modulation field was 710 Hz and 360 mgauss(rms), respectively. A high-permeability shield was used to reduce the Earth's field to less than 2×10^{-3} Oe. In addition, in some cases, d.c. magnetic field generator was utilized to cancel the Earth's field. The measurements were performed at 77K (while immersed in liquid nitrogen).

RESULTS AND DISCUSSION

Figure 4 shows the fundamental magnetoresistive characteristics of the element with a d.c. transport current as a parameter at 77 K. As shown in this figure, the resistance of the element increases very steeply with the increase of the magnetic field. These magnetoresistive characteristics form the basis of the magnetic sensor.

The output signal of lock-in amplifier as a function of the external magnetic field is shown in Fig.5. The signal represents a first derivative of the voltage across the electrodes of the element with respect to the magnetic field, dV/dB. The magnetic field resolution obtained with this measurement system as a function of the frequency of the external magnetic field is shown in Fig.6. As shown, the field resolution of $10^{-4}\sim10^{-5}$ gauss/(Hz)$^{1/2}$ was obtained in the frequency range from 0.1 Hz to 10 Hz. For comparison, the magnetic field resolution obtained directly from the voltage-magnetic field transfer function dV/dB and the voltage noise is shown in this figure. It is found that the field resolution at low frequency is greatly improved with this present operation.

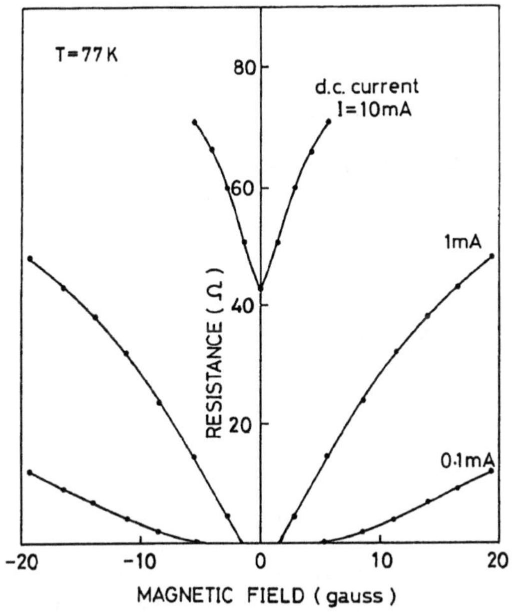

Fig.4. Fundamental magnetoresistive characteristics of the element with a d.c. transport current as a parameter.

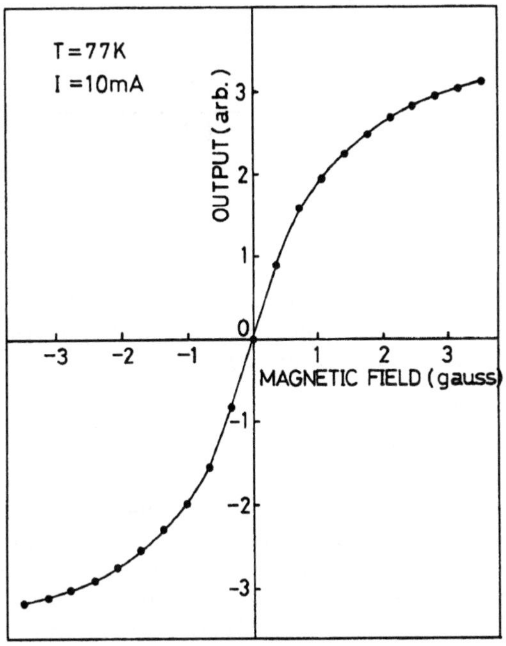

Fig.5. Output signal of lock-in amplifier in the measurement system shown in Fig.3. The d.c. transport current through the element is 10mA.

The improvement of the magnetic field resolution with this method can be attributed to an unconventional voltage noise characteristics. The voltage noise, which is considered due to a motion of magnetic flux in a high Tc superconductor film[4], showed $f^{-1/2}$-like frequency spectrum and was 2~3 orders larger than thermal noise[5]. The noise increases with the increase of the external magnetic field, has a peak, and then decreases. The peak almost corresponds with a peak of dV/dB[5]. When the sensitivity is obtained directly from dV/dB and the voltage noise, this characteristic is not favorable for a reduction of the magnetic field resolution. In this measurement system, on the other hand, the output signal represents dV/dB and thus the magnetic field resolution is obtained from d^2V/dB^2 and the voltage noise. The peak of d^2V/dB^2 does not correspond with that of the voltage noise. As a result, by adjusting the operation conditions, the magnetic field resolution can be improved with this method.

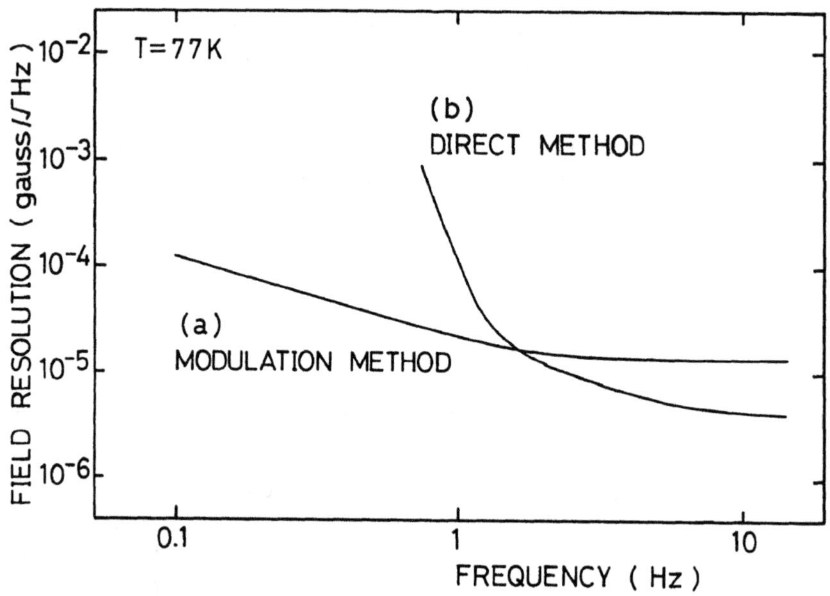

Fig.6. Magnetic field resolution obtained with the modulation method (a) and with the direct method (b) as a function of the frequency of the external magnetic field.

CONCLUSION

Low noise operation of the magnetic sensor employing magnetoresistive effect has been demonstrated. A magnetic field resolution of about 10^{-4}~10^{-5} gauss/(Hz)$^{1/2}$ at 0.1~10 Hz is obtained with the field modulation method. The sensitivity at low frequency is greatly improved as compared with that obtained directly from dV/dB and the voltage noise. The improvement can be explained by considering the voltage noise characteristics. Since the operation conditions are not fully optimized yet, the realization of a much higher sensitivity can be expected.

ACKNOWLEDGEMENTS

The authors wish to thank I.Fujimoto, S.Kataoka, S.Misaka, N.Hashizume and S.Tsuchimoto for valuable discussions and for continuous encouragement throughout this work.

REFERENCES

1. Nojima H, Tsuchimoto S, Kataoka S (1988) Galvanomagnetic effect of an Y-Ba-Cu-O ceramic superconductor and its application to magnetic sensors. Jpn.J. Appl.Phys.27:746-750
2. Nojima H, Kataoka S, Tsuchimoto S, Nagata M, Kita R, Shintaku H, Ohno E, Hashizume N (1988) Improvement in sensitivity of novel magnetic sensor using Y-Ba-Cu-O ceramic superconductor film. IEDM Tech. Digest:892-893
3. Kawai M, Kawai T, Masuhira H, Takahashi M (1987) Formation of Y-Ba-Cu-O superconducting film by a spray pyrolysis method. Jpn.J.Appl.Phys.26: L1740 -L1742
4. Ferrari MJ, Johnson M, Wellstood FC, Clarke J, Rosenthal PA, Hammond RH, Beasley MR (1988) Magnetic flux noise in thin-film rings of $YBa_2Cu_3O_{7-\delta}$. Appl.Phys.Lett.53:695-697
5. Nojima H, Shintaku H, Nagata M, Ohno E, Koba M (1989) Dissipative properties of ceramic high Tc superconductor film. in Proceedings of the 6th International Workshop on Future Electron Devices-High Temperature Superconducting Devices(R. and D. Assoc. for Future Electron Devices, Tokyo, 1989):375-378

Application of High-T_c Superconducting Thick Film to Superconducting Interconnection and Contact

K. HATANAKA[1], M. KAITOU[2], Y. MATSUZAKI[2], M. TOYOKAWA[3], S. KASHU[2], and C. HAYASHI[1]

[1] ULVAC JAPAN, Ltd., 2500 Hagisono, Chigasaki, 253 Japan
[2] Vacuum Metallurgical Co., Ltd., 516 Yokota, Sanbu, 289-12 Japan
[3] Tohoku Vacuum Metallurgical Co., Ltd., Ichikawa, Hachinohe, 031 Japan

ABSTRACT

Line shaped superconducting thick films, 0.8 mm width and 10 μm thick, have been formed on MgO by gas deposition of fine(<1 μm) powder. Y-Ba-Cu-O(YBCO), Ag doped YBCO, Bi-Sr-Ca-Cu-O(BSCCO), Bi-Pb-Sr-Ca-Cu-O (BPSCCO) and layered Pb/BSCCO films are reproducibly fabricated and all films show superconductivity above 70 K. BPSCCO films, which were sintered at 770 °C, exhibit zero resistance at 105 K and the critical current density at 77 K in H=0 is 600 A/cm². The contact resistivity between base electrode film and overlaid interconnect film are 1.5 m Ωcm² for YBCO/YBCO, 5.8 m Ωcm² for BSCCO/BSCCO and 0.52 m Ωcm² for BPSCCO/BPSCCO at 120 K. The critical contact current density is 120 A/cm² for BPSCCO/BPSCCO contact at 77 K.

KEY WORDS: superconducting interconnection, superconducting contact, gas deposition, ultrafine powder

INTRODUCTION

Recently, there has been increasing interests in application of high-T_c superconductor(HTSC) to interconnection[1,2]. This is partly because zero resistance interconnection is essential for low noise or high speed operations in thin film superconducting devices such as superconducting quantum interference device(SQUID), and partly because of the possibility of operating high speed digital semiconductor integrated circuits in liquid nitrogen with superconducting wiring of HTSC. We have developed the gas deposition system which can write lines of HTSC films on a substrate in serial fashion. Using this system a HTSC interconnect film, as well as perfect superconducting contacts at both ends, is easily obtained, which in other thin film process is very difficult to produce. In this article, we first explain the preparation of the samples in detail and then discuss the superconducting characteristics of the films and contacts. Some applications of our system to HTSC device are also presented.

EXPERIMENTAL

We have previously shown that the evaporation of ultrafine powder(<100 nm) of Y, Ba, Cu and simultaneous mixing and gas deposition process made c-axis oriented YBCO films on (100)MgO substrate[3]. Although this in-situ generation and deposition process has the advantage of mixing the constituents on an ultrafine powder scale, there is difficulty in controlling composition of HTSC. We have recently modified the process and developed a new deposition system in which film growth starts from HTSC composition fine (<1 μm) powder. The schematic diagram of this gas deposition system is shown in Fig. 1. Fine powder of HTSC is stored in a reservoir which plays similar role of a bubbler of a chemical vapor deposition(CVD) system. The starting fine powder is obtained by standard ceramic powder processing. Grinding and heating several times produce good start-up fine powder. Fine powder is transported to the evacuated deposition chamber by the gas flow (O_2 gas for YBCO and air for BSCCO) and sprayed onto a heated (100)MgO crystal substrate.

For preparing Ag doped YBCO films, two powder reservoirs - one for YBCO and the other for commercial Ag ultrafine powder(70 nm) - are used. Mixing of the powder is achieved by joining the flow tubes from each reservoir at just before the deposition chamber. This system makes lower normal resistivity -

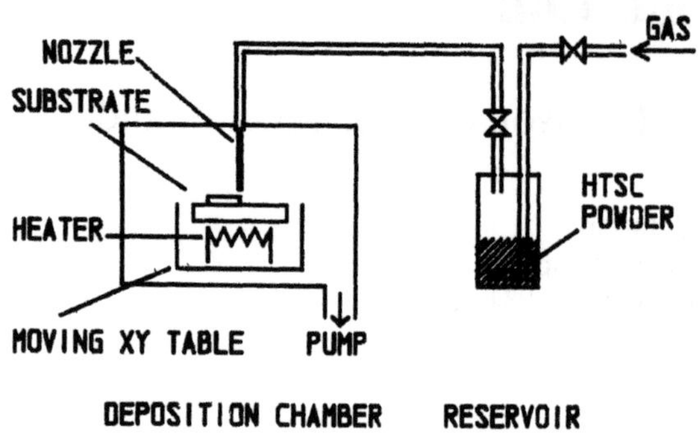

TABLE 1. **Preparation conditions of samples**

Film	Gas	T_{SUB} (°C)	Heat treatment (°C x h)
YBCO	O_2	550	930 x 2
Ag-YBCO	O_2	500	900 x 2
BSCCO	air	400	830 x 10-30
Pb/BSCCO	air	400	>830 x 10-30
BPSCCO	air	400	770 x 10-30

Fig. 1. Gas deposition system

and hence having large thermal stability - films. It also provides ideal interconnecting film for superconductor/normal metal contacts[4].

For BSCCO film it is well known that Pb substitution is vary effective to obtain high T_c phase. We tried Pb substitution by solid diffusion method[5] at first. In this method, BSCCO film is deposited first and then Pb film is overlaid. Subsequently, we found that by mixing Pb powder with start-up BSCCO fine powder(about 50wt% Pb to 50wt.% Bi) high-T_c BPSCCO films were quite easily and reproducibly obtained.

After depositing films, heat treatments are applied. For YBCO samples, post-deposition heat treatment are carried out at 900 to 930°C for 2 h and the condition is not optimized. Layered Pb/BSCCO films are sintered in an oven at above 830°C and Pb substitution is achieved. Since superconducting properties of BPSCCO films strongly depend on heat treatment, we sought the optimum heat treatment condition for BPSCCO film carefully[6]. The highest critical temperature film is obtained for the post-deposition heat treatment at 770°C for 10 to 30 h, which, as we know, is the lowest heat treatment temperature reported. The reason for this low temperature sintering condition, we believe, is the starting powder size, which from transmission electron microscope(TEM) observations and Brunauer-Emmett-Teller(BET) surface area measurements is confirmed to be smaller than 1 μm. Note also that in the gas deposition system only smaller(< 1 μm) particles are transported by gas flow. Typical conditions for preparing films are summarized in Table 1.

Some of the film samples are utilized as a base electrodes of contact sample. An overlaid film is deposited on these films in the same way and makes cross shaped contacts. Although post-deposition heat treatment is not necessarily if sufficient high (more than 650°C for YBCO) substrate heating is applied, we found that post-depostion heat treatment at the same condition as the base electrode film results in good contacts. In this article we only discuss contact samples fabricated in this way.

The value of T_c and J_c was determined directly by the dc resistivity method in a four-probe arrangement. Critical current is defined as a current at which 1 μV is appeared across 1 cm length for films, or between underlaid and overlaid films for contacts.

RESULTS AND DISCUSSIONS

Figure 2 shows the resistive transition and the critical current of film samples. The critical temperature of YBCO films are in 70 K range. This is probably due to oxygen deficiency. We also found that the powder size of YBCO is larger than BSCCO, and this may have an effect on the deposited film properties. A BPSCCO film shows the maximum critical temperature of 105 K, and the critical current density is 600 A/cm² at 77 K. Although all films shows superconductivity above 70 K, only BPSCCO films are applicable as an interconnection in liquid nitrogen.

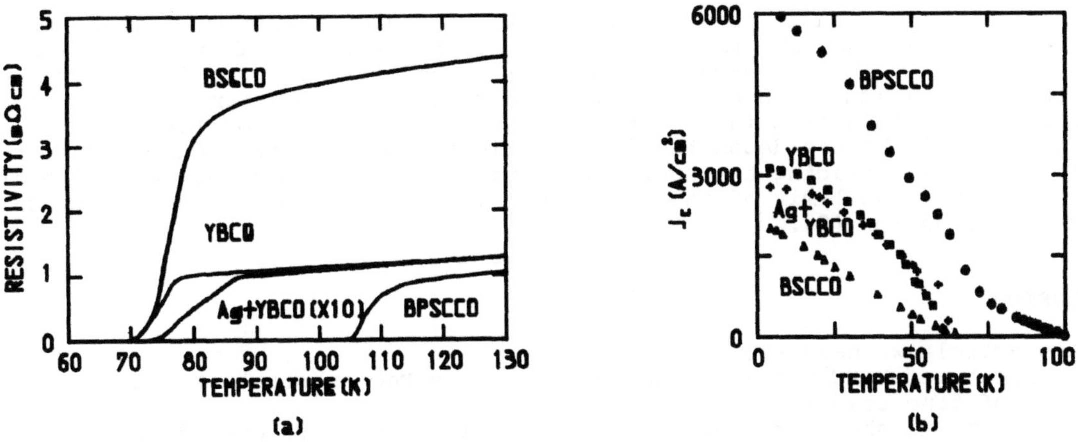

Fig. 2. Temperature dependence of the (a) resistivities and (b) critical current of typical films.

Figure 3 shows the resistive transition and the critical current at contacts. YBCO/YBCO contact become superconducting at 70 K. Comparing with BSCCO/BSCCO contact, which also shows superconductivity at about 70 K, we notice that YBCO/YBCO contact exceeds in both contact resistivity at 120 K and critical current at 4.2 K. This is probably due to the crystal structure of BSCCO film. Since our films are c-axis oriented[3,6], and at contact current must flow in c-axis direction, BSCCO film which has large two dimensionality must exhibit higher contact resistivity. The highest contact critical temperature 99K is obtained by the high T_c phase BPSCCO/BPSCCO sample. The critical contact current density at 77K in zero magnetic field is 120 A/cm². Contact resistivity and hence contact critical current density may be improved by O_2 plasma cleaning or etching at the underlaid film surface, or Ag powder doping.

We have also tried to fabricate a hetero contact of BSCCO/YBCO and found that this contact show very large contact resistivity(6 kΩ cm² at 120 K) and semiconductor-like temperature dependence. In this contact, underlaid part of YBCO film is deteriorated.

Since the gas deposition is a direct-write process requiring neither mask nor sophisticated etching equipments, this system has many advantages[7] in: (1) Chip to chip or chip to bonding pad wiring for superconducting devices. (2) On-chip wiring of superconductor or semiconductor device where main junctions or gates were formed by thin film lithography process. On-chip wiring is used only for minor modification of functions of the chip. (3) Repair of pattern-

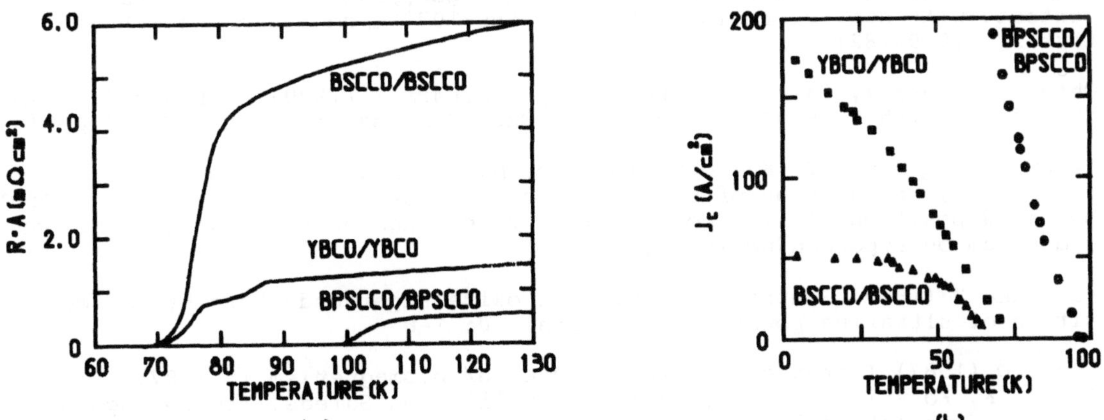

Fig. 3. Temperature dependence of the (a) resistivities and (b) critical current of the contacts.

ing of thin film device. (4) Fabrication of simple magnetic sensors which utilize weak coupling of grain boundaries. (5) Forming HTSC pick-up coil in the vacuum space of the dewar and leading signal to 4.2 K or 77 K operating SQUID[8]. (6) Trimming and tuning of the planer gradiometer in which coil balance is most important in environmental magnetic noise reduction.

Line width of the gas deposited film is limited by the inner diameter of the nozzle used. By replacing the nozzle with finer one, lines of 0.1 mm width are easily written. The limit of the line width is estimated by aerodynamical calculations to be 3 μm[9].

CONCLUSION

In this article we have shown that:

(1) A variety of HTSC films can be fabricated on MgO substrate by gas deposition of fine powder.

(2) BPSCCO films show zero resistance critical temperature exceeding 100 K. Due to smaller start-up powder size, sintering temperature is reduced to 770 °C.

(3) We have succeeded in fabricating the superconducting contact of HTSC. The critical contact current density at 77 K is 120 A/cm² for BPSCCO/BPSCCO.

ACKNOWLEDGMENT

This research was supported by High Technology Consortium which is promoted by Research Development Corporation of Japan. We wish to thank M. Aizawa and H. Ohtsu, for their generous assistance. We also wish to thank T. Ohtsuka, H. Naruse and Y. Yamakawa for their encouragements throughout this work.

REFERENCES

1. Dykaar DR, Sobolewski R, Chwalek JM, Whitaker JF, Hsiang TY, Mourou GA (1988) High-frequency characterization of thin film Y-Ba-Cu oxide superconducting transmission lines. Appl Phys Lett 52: 1444-1446

2. Hsiang TY, Whitaker JF, Sobolewski R, Martinet S, Golob LP (1989) High-frequency characterization of superconducting transmission structures from picosecond transient measurements. In: Extended abstracts of ISEC'89 pp 510-515

3. Hatanaka K, Kaitou M, Umehara M, Kashu S, Hayashi C (1989) Preparation of superconducting thick films of Y-Ba-Cu-O by gas deposition of ultrafine powder. In: Kitazawa, Ishiguro(eds) Advances in Superconductivity. Springer-Verlag. Tokyo. pp 341-345

4. Ekin JW, Larson TM, Bergren NF, Nelson AJ, Swartzlander AB, Kazmerski LL, Panson AJ, Blankenship BA (1988) High T_c superconductor/noble-metal contacts with surface resistivities in the 10^{-10} Ω cm range. Appl Phys Lett 52: 1819-1821

5. Ohta H, Iimura Y, Shinada K, Fujino S, Sugihara T (1989) Bi-Pb-Sr-Ca-Cu-O thin films with T_c(R=0) of 110 K. In: Extended abstracts of ISEC'89 p.93

6. Kashu S, Matsuzaki Y, Kaito M, Toyokawa M, Hatanaka K, Hayashi C (submitted) Preparation of superconducting thick films of Bi-Pb-Sr-Ca-Cu-O by gas deposition of fine powder. In: Proceedings of the 2nd International symposium on superconductivity.

7. Hatanaka K (1989) Synthesis of high-T_c oxide superconductors by gas deposition of ultrafine powder. VMC Journal 10: 1-8

8. Cohen D (1987) Introduction: An overview of biomagnetism in 1987. In: Atsumi K, Kotani M, Ueno S, katila T, Williamson SJ(eds) Proceedings of the 6th International Conference on Biomagnetism, Tokyo Denki University Press, pp 2-9

9. Naruse H (1987) private communication

Gapless Characteristics of Superconductivity in Surface Layer of HTSC

N. YOSHIKAWA, T. MURAKAMI, and M. SUGAHARA

Faculty of Engineering, Yokohama National University, Hodogaya, Yokohama, 240 Japan

ABSTRACT

We make theoretical study of the degradation of superconducting properties in the surface region of the high-T_c oxide superconductor due to the pair breaking in the proximity effect assuming the existence of a thin nonsuperconducting surface layer. It is found that the existence of the normal surface layer has strong influence on the tunneling characteristics and the absorption characteristics of the electromagnetic wave.

KEY WORDS: high-T_c superconductor, proximity effect, tunnel spectroscopy, electromagnetic absorption

INTRODUCTION

The superconductivity of the surface layer of the high-T_c oxide superconductor (HTSC) tends to be deteriorated by nonideal stoichiometry and/or imperfect morphology. Non-superconducting $BaCuO_2$ surface layer of 1nm thickness is found in Y-Ba-Cu-O[1]. The existence of nonmetallic Bi-O surface layer is also suggested by the STM observation of the Bi-Sr-Ca-Cu-O surface[2]. These deteriorated surfaces are believed to be the main cause of the nonideal superconductivity found in the tunneling spectroscopy and the electromagnetic absorption. In this report we represent theoretical calculations of the density of states, quasiparticle tunneling characteristics and the absorption of the electromagnetic wave in the surface layer of the HTSC based on the pair-breaking model in the proximity effect. It is supposed that the surface layer itself is normal, and affects the superconductivity of the interface region. Besides we assume that ideal HTSC has BCS-like energy gap structure.

On the SN interface where steep orderparameter variation takes place spatially, the electron pairs suffer pair-breaking perturbation. As the result the gap structure broadens and the orderparameter is depressed. The effect of the orderparameter depression by this mechanism may be especially serious in HTSC because of their short coherence length and of the surface deterioration.

GAP DEPRESSION BY PROXIMITY EFFECT

The theory of the gapless superconductivity in superconductor doped with magnetic impurities is well established[3]. Fulde and Maki[4] showed that the pair breaking similar to the magnetic impurity effect is caused by the spatial variation of the order parameter in SN boundaries. This similarity permits us to treat the degradation of superconductivity in the proximity region after the theory of the magnetic impurity effect. In the dirty superconductor ($\ell/\xi_0 \ll 1$, ℓ: mean free path, ξ_0: intrinsic coherence length) the transition temperature T_c of the SN interface layer is obtained from

$$\ln(T_c/T_{c0}) = \psi(1/2) - \psi(1/2 + \alpha/2\pi k_B T_c) \qquad (1)$$

where $\psi(x)$ is di-gamma function, α is the strength of pair-breaking perturbation and T_{c0} is the unperturbed transition temperature[4]. The dependence of T_c obtained from Eq.(1) is shown in Fig.1, where $\Delta_0(0)$ is the unperturbed order parameter at T=0. In the temperature region slightly below T_c, the pair-breaking strength α is found to be

$$\alpha_S = \hbar \ell_S v_F k_S^2/6 \text{ or } \alpha_N = \hbar \ell_N v_F k_N^2/6, \qquad (2)$$

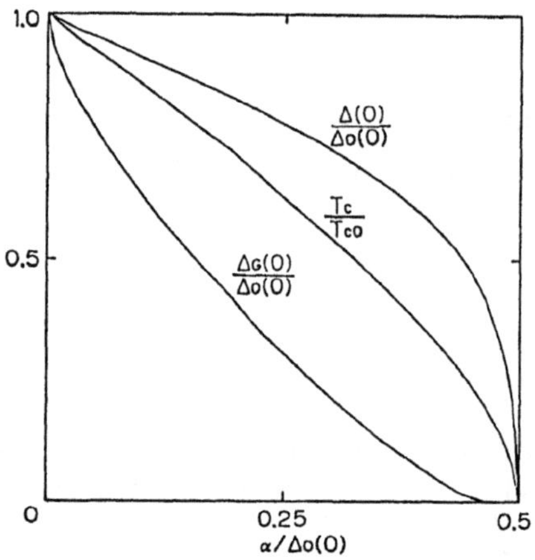

Fig.1 The critical temperature T_c, the orderparameter $\Delta(0)$, and the excitation edge $\Delta_G(0)$ at T=0 versus pair-breaking strength α. (After Skalski et al. [5]).

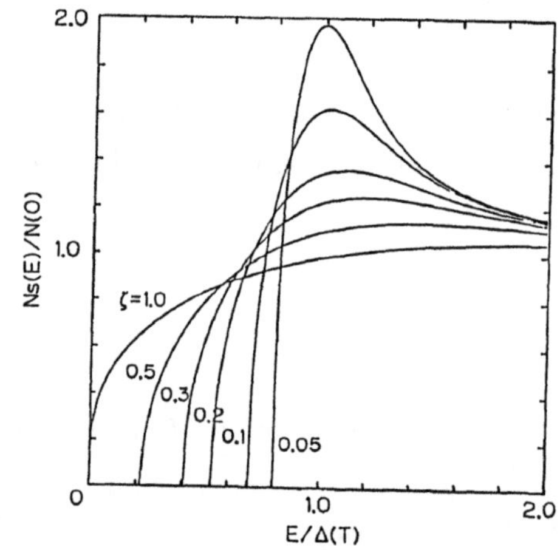

Fig.2 The density of states as a function of the reduced energy for several values of the reduced pair-breaking strength $\zeta=\alpha/\Delta(T,\alpha)$.

using the orderparameter expressions

$$\Delta_S(z) \propto \cos[k_S z+\phi_S] \quad \text{(superconductor side)} \qquad (3)$$
$$\Delta_N(z) \propto \cosh[k_N z+\phi_N] \quad \text{(normal side)},$$

where v_F is the Fermi velocity and ϕ is constant[4]. Skalski et al.[5] gave the density of states $N_S(E)$ under the pair-breaking perturbation by

$$N_S(E)=N(0)\sinh 2x(\cosh 2x-\cos 2y)^{-1}, \qquad (4)$$

where E is the excitation energy and N(0) is the density of states in the normal state at the Fermi surface. The variables x and y are real and determined by

$$\sin^3 2y+(\varepsilon^2+\zeta^2-1)\sin 2y-2\varepsilon\zeta=0 \qquad (5)$$
$$\cosh x\cdot\cos 2y=\varepsilon\cos y-\zeta\sin y, \qquad (6)$$

where $\varepsilon=E/\Delta(T,\alpha)$, the normalized pair-breaking strength $\zeta=\alpha/\Delta(T,\alpha)$. $\Delta(T,\alpha)$ is orderparameter under the pair-breaking α at finite temperature T. In Fig.2 is shown $N_S(E)$ calculated from Eqs.(4)-(6). With the increase of $\zeta=\alpha/\Delta$, the gap structure broadens and the minimum excitation edge Δ_G is depressed. When $\zeta=1$, $N_S(E)$ fall in a gapless state. The dependences of $\Delta(0,\alpha)$, $\Delta_G(0,\alpha)$ and T_c are inconsistent with the BCS relation $\Delta(0)=\Delta_G(0)=1.76k_BT_c$ as shown in Fig.1.

TUNNEL SPECTROSCOPY

The SNIN' structure in Fig.3 is the model of the HTSC/N' tunnel junction, where a thin normal layer (N) (or weakly superconducting layer whose critical temperature $T_{cN} \ll T$) exists between a thick HTSC (S) and a insulating tunnel barrier (I) and the normal electrode (N'). It is assumed that in the S region (z>0) the orderparameter suffers steep reduction near the SN interface over a distance of the order of the GL coherence length ξ_{GL}. If we use the de Gennes[6]-Werthamer[7] expression for the boundary condition of SN interface, we get following orderparameter variation.

$$\Delta_S(Z)=aV_S\cos k_S(x-\xi_{GL})/\cos k_S\xi_{GL}, \qquad (0<x<\xi_{GL}) \qquad (7)$$
$$\Delta_N(Z)=aV_N\cosh k_N(x+d_N)/\cosh k_Nd_N, \qquad (-d_N<x<0) \qquad (8)$$
$$k_S\tan(k_S\xi_{GL})=k_N\tanh(k_Nd_N), \qquad (9)$$

where we chose $\ell_S=\ell_N$, $N_S=N_N$ (N: the density of states). V_S,V_N are the inter-electron potential and a is the constant. Supposing $T_{cN}\to 0$ (so that putting $T_{c0}\to 0$ in eq.(1)), we get $k_N\cong(\hbar\ell_Nv_F/6\pi k_BT_c)^{-1/2}=\xi_N^{-1}$. Thus, the pair-breaking strength in the N region is given from eqs.(2) by

$$\zeta_N=\alpha_N/\Delta(T,\alpha)=\pi k_BT_c/\Delta. \qquad (10)$$

With $d_N \ll \xi_N$, and assuming the validity of above expressions when $k_Nd_N \ll 1$, we find from eq.(9)

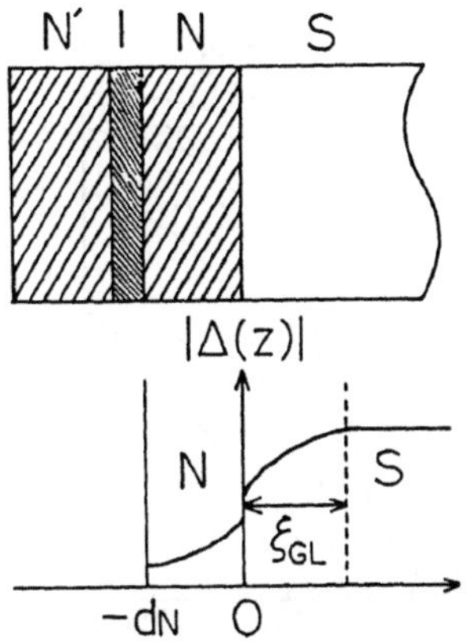

Fig.3 The spatical variation of the order parameter in the SNIN' tunnel junction.

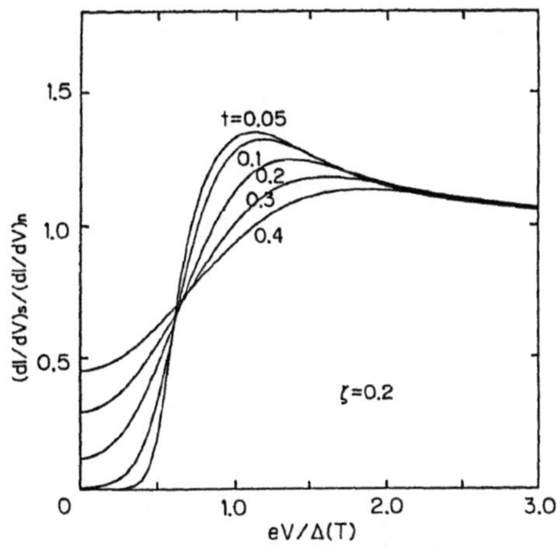

Fig.4 The tunneling conductance as a function of the voltage for several values of reduced temperatures $t=k_BT/\Delta(T,\alpha)$ at $\zeta=\alpha/\Delta(T,\alpha)=0.2$.

$$k_S \cong (d_N/\xi_{GL}\xi_N^2)^{1/2} \tag{11}$$

for $k_S\xi_{GL}<<1$. Thus, ζ in the S region near the SN interface is given from eqs.(2) and (11) by

$$\zeta_S=\alpha_S/\Delta(T,\alpha)=(\pi k_BT_c/\Delta)(d_N/\xi_{GL}). \tag{12}$$

In the tunneling experiments we must consider the "proving length" [6] over which we can observe the state density of the interface region. Reference 6 suggests that the "proving length" is very large at low temperature in the N region (perhaps of the order of the diffusion length for the inelastic scattering time), while in the S region the length is about $\xi_S=(\hbar\ell_Sv_F/6\pi k_BT_c)^{1/2}$. We assume that the effective ζ observed in tunneling measurement is the one averaged over the SN interface and get

$$\zeta\cong(d_N\zeta_N+\xi_S\zeta_S)/(d_N+\xi_S). \tag{13}$$

When $d_N<<\xi_S$, the effective ζ of SNIN' system is simply given by eq.(12). In eq.(12) $\pi k_BT_c/\Delta$ depends on α and takes the value from 0.71 to 1.78. Since ξ_{GL} of HTSC is very small, the effect of N layer may be serious in HTSC.

The tunnel conductance of a SN tunnel junction is given by

$$\frac{(dI/dV)_S}{(dI/dV)_N}=\int_{-\infty}^{\infty}\frac{N_S(E)}{N(0)}\left(-\frac{\partial f(E+eV)}{\partial(eV)}\right)dE\,, \tag{14}$$

where $(dI/dV)_S$ and $(dI/dV)_N$ are respectively the differential tunnel conductance in the superconducting and in the normal state, and f(E) is the Fermi function. In Fig.4 is shown the voltage dependences of the tunnel conductance in the presence of the pair breaking perturbation at finite temperatures, where the density of states in Fig.2 is used. In comparison with the BCS result, remarkable broadening of the gap is seen. The tunnel spectroscopy for HTSC shows very broadened conductance variation in comparison with BCS case. In Fig.5 are shown the SN tunneling experiment on the single-crystal Y-Ba-Cu-O sample[8] and our calculation (ζ=0.1) with fitting. Fairly good agreement is found except slight discrepancy near V=0. We can fit our calculation also to the data of Bi-Sr-Ca-Cu-O, La-Sr-Cu-O, etc. using ζ=0.1∿0.5.

ELECTROMAGNETIC ABSORPTION

In the extreme anomalous limit, where the penetration depth of the field is smaller than the superconducting coherence length, the absorption of the electromagnetic wave in the presence of the pair-breaking perturbation can be computed using eq.(6.23) of reference 5.

$$\begin{aligned}\sigma_{S1}/\sigma_N=&(2/\hbar\omega)\int_{\Delta_G}^{\infty}dE[n(E)n(E+\hbar\omega)(1+m(E)m(E+\hbar\omega))(f(E)-f(E+\hbar\omega))]\\&+(1/\hbar\omega)\int_{\Delta_G-\hbar\omega}^{-\Delta_G}dE[n(E)n(E+\hbar\omega)(1+m(E)m(E+\hbar\omega))(1-2f(E+\hbar\omega))]\end{aligned} \tag{15}$$

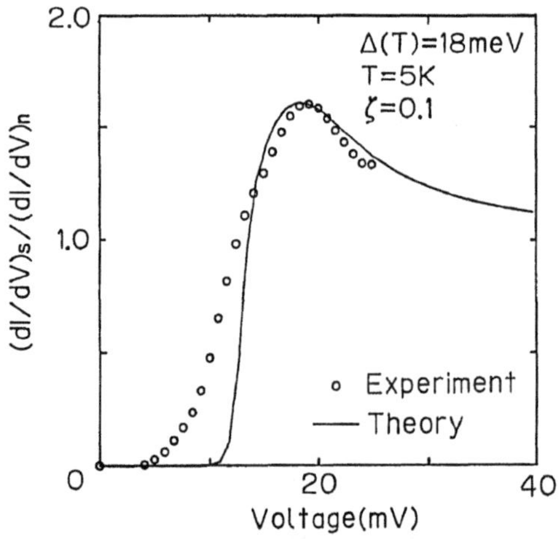

Fig.5 The comparison of the theoretical tunneling conductance with the experimental data reported by Kirtley et al. [8].

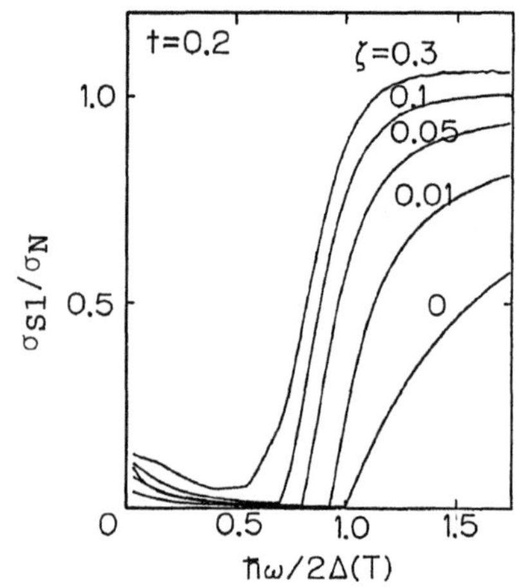

Fig.6 the real part of the complex conductivity for several values of pair-breaking strength $\zeta=\alpha/\Delta(T,\alpha)$ at $t=k_BT/\Delta(T,\alpha)=0.2$.

where $n(E)=\sinh 2x(\cosh 2x-\cos 2y)^{-1}$, $m(E)=\cos y/2\cosh x$, and x, y are obtained solving the eqs.(5) and (6). σ_{S1} is the real part of the complex conductivity for the superconducting state and σ_N is the normal state conductivity. The second term of eq.(15) appears only for $\hbar\omega>2\Delta_G$. The first integral of eq.(16) represents the effect of the thermally excited normal electron, while the second integral represents the contribution of photon-excited quasiparticles. When $\zeta=0$, eq.(15) reduces to the Mattis-Bardeen result[9]. The calculation of eq.(15) is shown in Fig.6, where we found the absorption edge considerably smaller than $2\Delta(T,\alpha)$ when $\zeta\neq 0$.

CONCLUSION

We calculate the density of states, the quasiparticle tunneling characteristics, and the electromagnetic absorption of HTSC having the normal surface layer. The steep orderparameter gradient on the SN interface causes strong pair-breaking perturbation, which result in the modification of BCS state density. It was found that even the existence of very thin normal surface layer can seriously affect the tunneling characteristics and the electromagnetic absorption in the HTSC whose coherence length is very short.

REFERENCES

1. Chang CC, Hegde MS, Wu XD, Dutta B, Inam A, Venkatesan T, Wilkens BJ, Wahtman JB (1989) Surface layers on superconducting Y-Ba-Cu-O films strudied with X-ray photoelectron spectroscopy. Appl Phys Lett 55: 1680-1682
2. Shich CK, Feenstra RM, Kirtley JR, Chandrashekhar GV (1989) Surface structural and electronic properties of cleaved single crystal of $Bi_{2.15}Sr_{1.7}CaCu_2O_{8+\delta}$ compounds. Phys Rev B40: 2682-2685
3. Abrikosov AA, Gor'kov LP (1960) Contribution to the theory of superconducting alloys with paramagnetic impurities. Zh. Eksp. Teor. Fiz 39: 1781-1796
4. Fulde P, Maki K (1965) Gapless superconductivity induced by metallic contacts. Phys Rev Lett 15: 675-677
5. Skalski S, Betbeder-Matibet O, Weiss PR (1964) Properties of superconducting alloys containing paramagnetic impurities. Phys Rev 136: A1500-A1518
6. Deutscher G, de Genns (1969) Proximity effect. In: Parks RD (ed) Superconductivity. vol 2 chap 17 Marcel Dekker, New York
7. Werthamer NR (1963) Theory of the superconducting transition temperature and energy gap function of superposed metal films. Phys Rev 132: 2440-2445
8. Kirtly JR, Collins RT, Schlesinger Z, Gallagher WJ, Sandstrom RL, Dinger TR, Chance DA (1987) Tunneling and infrared measurements of the energy gap in the high-critical-temperature superconductor Y-Ba-Cu-O. Phys Rev B35: 8846-8849
9. Mattis DC, Bardeen J (1958) theory of the anomalous skin effect in normal and superconducting metals. Phys Rev 111: 412-417

Josephson Effect in Epitaxial $Ba_2YCu_3O_x$ Thin Films on ZrO_2/Si

HIROAKI MYOREN[1], YUKIO NISHIYAMA[1], NAOKAZU MIYAMOTO[1], YUKIO OSAKA[1], and TOSHIHIKO HAMASAKI[2]

[1] Department of Electrical Engineering, Faculty of Engineering, Hiroshima University, Saijo, Higashi-Hiroshima, 724 Japan

[2] ULSI Research Center, Toshiba Corporation, 1, Komukai Toshiba-cho, Saiwai-ku, Kawasaki, 210 Japan

ABSTRACT

Using the silicon substrates patterned with the trench, the superconducting microbridge junctions have been fabricated in as-grown $Ba_2YCu_3O_x$ thin films by rf magnetron sputtering. The microbridge junctions with constrictions as small as submicron dimensions were obtained. These microbridge junctions behaved as Josephson junctions and microwave-induced steps were observed. The observed characteristics seem to be consistent with those expected for a homogeneous wide superconducting bridge (constriction size >> coherence length).

KEY WORDS: Josephson effect, wide superconducting bridges, $Ba_2YCu_3O_x$ thin films on Si, VLSI technology based on Si

INTRODUCTION

For microelectronic application of the high-Tc superconducting oxide, it is highly desirable to fabricate thin films of the new materials on Si. In previous papers[1], we reported on the successful as-grown preparation of the epitaxially grown $Ba_2YCu_3O_x$ thin films on Si substrate with epitaxially grown ZrO_2 as a buffer layer and the Tc(end) of these films exceeded liquid nitrogen temperature. For realizing the applications such as microbridge type Josephson junction, it is also necessary to develop appropriate microfabrication technique. Fortunately, the established VLSI technology based on Si have the appropriate microfabrication technique and the obtained linewidth as small as submicron dimensions may be possible.

This paper demonstrates a simple, novel fabrication method of high-Tc superconducting $Ba_2YCu_3O_x$ microbridge junctions on silicon substrates patterned with a 3μm-deep trench. This paper also reports Josephson effect in the wide superconducting bridges made by these $Ba_2YCu_3O_x$ thin films on Si.

FABRICATION OF MICROBRIDGE JUNCTION

The main steps in the fabrication process of superconducting microbridge junctions are described in Fig. 1. The Si substrate was patterned with a 3μm-deep trench by the established VLSI technology (a). Then, a YSZ (9mol% Y_2O_3) buffer layer was deposited on the patterned Si(100) substrate by electron beam evaporation (b), and finally, the $Ba_2YCu_3O_x$ thin film was deposited by rf magnetron sputtering (c). The buffer layer and the film deposition processes are described in detail elsewhere[1]. The crystalline quality of the YSZ (9mol% Y_2O_3) buffer layer was estimated by Rutherford backscattering channeling (2MeV He^+ ions). The resulting χ_{min} was 0.04[2]. The thicknesses of the buffer layer and the $Ba_2YCu_3O_x$ films were typically about 100nm and 500nm, respectively. Figure 2 shows a scanning electron micrograph of an

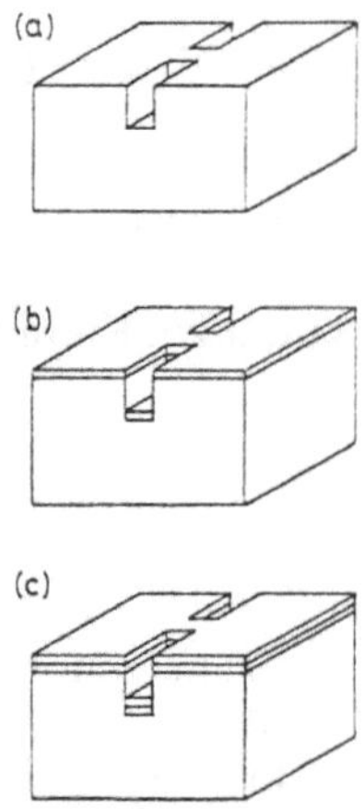

Fig. 1. Main steps in the fabrication process of the microbridge junction on a Si substrate patterned with the trench.

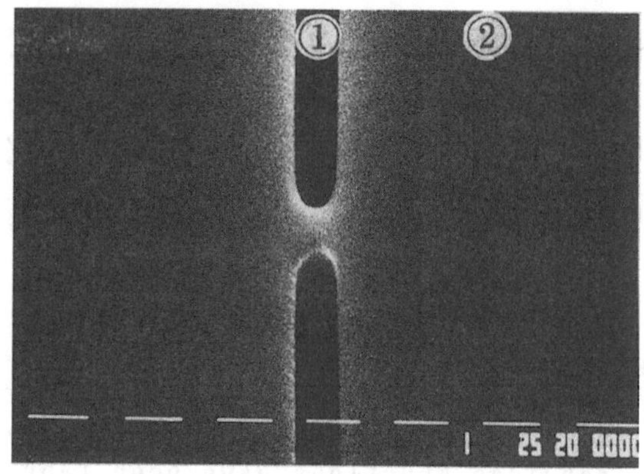

Fig. 2. SEM photograph of a fabricated microbridge junction on a Si substrate patterned with the trench. The bridge dimensions are W=0.8 μm and L=0.8 μm. (①:the trench fabricated by VLSI technology, ②:$Ba_2YCu_3O_x$ thin film)

obtained superconducting microbridge junction on a Si substrate patterned with the trench. The width W and length L of the microbridge in this micrograph are both about 0.8 μm. Resistivity between devices separated by the trench was several MΩ, and the isolation between devices by the trench was almost complete.

I-V CHARACTERISTICS AND MICROWAVE RESPONSE OF MICROBRIDGE JUNCTION

The I-V characteristics of the microbridge junctions were measured by the conventional four-probe method with dc current. The samples were cooled down in a cryostat with a mechanical compressor (Cryo Mini model D310). Microwaves with frequencies of 11.4GHz were generated by a reflex klystron and fed to the sample in the cryostat via an attenuator, waveguide and coaxial cable.

Figure 3 shows the I-V characteristic of the microbridges whose constriction sizes were (a) L=0.8 μm and W=8 μm and (b) L=2 μm and W=8 μm. Generally the nearly linear I-V characteristics ($V \propto (I-Ic)$), as shown in Fig. 3(a), were observed, but they sometimes consisted of nearly linear

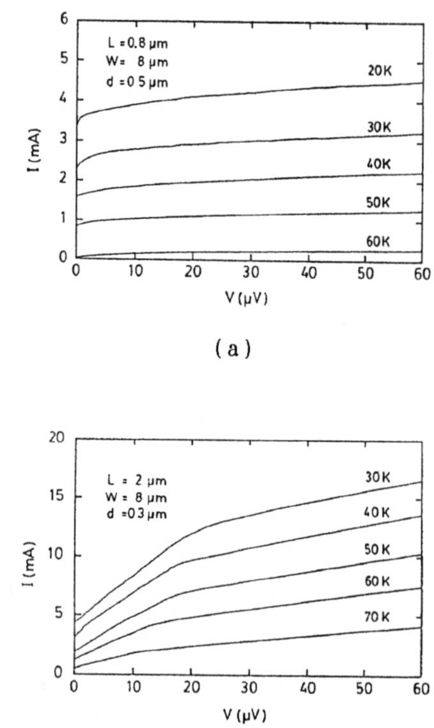

Fig. 3. I-V characteristics of the microbridge made of $Ba_2YCu_3O_x$ thin films on Si. The bridge dimensions are (a) W=8 μm, L=0.8 μm and (b) W=8 μm, L=2 μm. d is the thickness of the superconducting film.

sections with an overall parabolic behavior, shown in Fig. 3(b), i.e., $V \propto (I-Ic)$ at low currents and $V \propto (I-Ic)^2$ at high currents. This latter feature was also observed by Kuriki et al. for the microbridges of Nb_3Ge[3]. Aslamazov and Larkin[4] showed that for currents $I \leq 2Ic$, $V \propto (I-Ic)$ and for currents $I \geq 2Ic$, $V \propto (I-Ic)^2$, which agree qualitatively with our results.

Figure 4 shows the temperature dependence of the critical current of the $Ba_2YCu_3O_x$ microbridge shown in Fig. 3(a). In this case, we find that the law $Ic \propto (1-t)$ with t being the reduced temperature, applies for most of the bridges, and the critical current densities of these bridges were greater than $10^5 A/cm^2$ at 15K. This large value of critical current density shows that the I-V characteristics of these microbridges are not determined by grain boundaries of the film, because the Ic of the grain-boundary current for $Ba_2YCu_3O_x$ was observed to be of the order of $10^3 A/cm^2$[5]. It has been shown that the wide bridges produced by Nb_3Ge obey a temperature law $Ic \propto (1-t)^{2.5}$[6]. This type of variation was commonly observed in bridges of A15 materials (Nb_3Sn[7], V_3Si[8], Nb_3Ge[9]) and attributed to inhomogeneities in the election mean free path or film thickness[10]. Since high-Tc A15 materials are not in the equilibrium phase but in the metastable phase, the films may contain various kinds of defects or disorders. Grain boundaries may also act as inhomogeneities. It is interesting that wide bridges made by Nb obey the temperature law $Ic \propto (1-t)$[11].

Figure 5 shows the microwave response of the $Ba_2YCu_3O_x$ wide bridge shown in Fig. 3(a) at 15K. (The Ic decreases with increase of microwave power.) A step structure appears in the I-V characteristic, and the step separation in voltage satisfies the relation $V_0=fh/2e(\sim 23.6\mu V)$. No subharmonic steps were observed. The lack of sharpness in the step structure is partly due to the noises of the mechanical compressor.

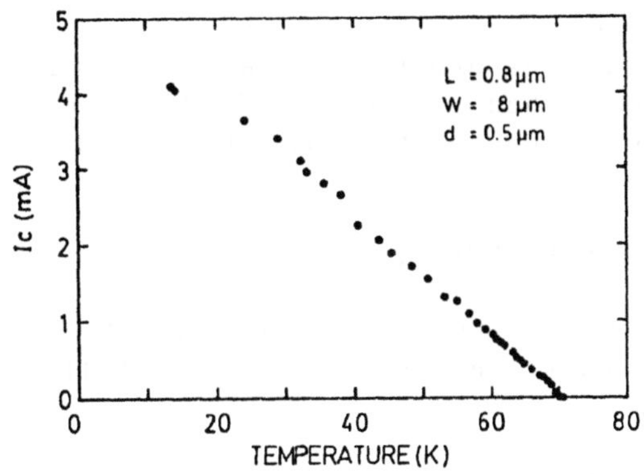

Fig. 4. The temperature dependence of the critical current of the microbridge shown in Fig. 3(a).

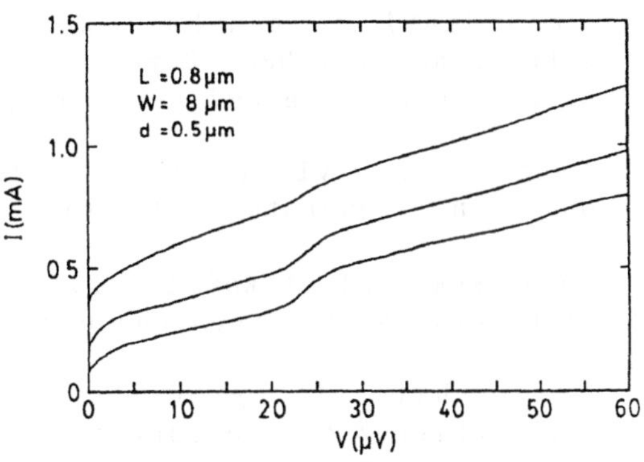

Fig. 5. Microwave response of the microbridge made of $Ba_2YCu_3O_x$ thin film on Si shown in Fig. 3(a).

CONCLUSION

Using silicon substrates patterned with the trench, the superconducting microbridge junctions have been fabricated in as-grown $Ba_2YCu_3O_x$ thin films by rf magnetron sputtering. The microbridge junctions with constrictions of as small as submicron dimensions were obtained. The critical current densities of these bridges were greater than $10^5 A/cm^2$ at 15K, and this shows that the electrical characteristics of these microbridges are not determined by grain boundaries of the film. These microbridge junctions behaved as Josephson junctions and were observed to have microwave-induced steps. The temperature dependence of the critical current and the observed I-V characteristics seem to be consistent with those of a homogeneous wide superconducting bridge.

ACKNOWLEDGEMENT

We would like to thank Mr. M. Iwase of the ULSI Research Center, Toshiba Corporation, for his help with the Si microbridge pattern fabrication.

REFERENCES

1. Myoren H, Nishiyama Y, Fukumoto H, Nasu H, Osaka Y (1989) As-Grown Preparation of Superconducting Epitaxial $Ba_2YCu_3O_x$ Thin Films Sputtered on Epitaxially Grown ZrO_2/Si(100). Jpn J Appl Phys 28: 351-355

2. Fukumoto H, Imura T, Osaka Y, Nishiyama F (1989) Evaluation of Crystalline Quality of Heteroepitaxial Yttria-Stabilized Zirconia Films on Silicon by Means of Ion Beam Channeling. J Appl Phys 66: 616-619

3. Kuriki S, Yoshida A, Konishi H (1983) Vortex Behavior in Nb_3Ge Microbridges. J Low Temp Phys 51: 149-163

4. Aslamazov LG, Larkin AI (1975) Josephson Effect in Wide Superconducting Bridges. Sov Phys JETP 41: 381-386

5. Chaundhari P, Mannhart J, Dimos D, Tsuei CC, Chi J, Oprysko MM, Scheuermann M (1988) Direct Measurement of the Superconducting Properties of Single Grain Boundaries in $Y_1Ba_2Cu_3O_{7-\delta}$. Phys Rev Lett 60: 1653-1656

6. Rogalla H, David B, Rühl J (1984) Thin Film Nb_3Ge dc-SQUID with High Operating Temperature. J Appl Phys 55: 3441-3443

7. Golovashkin AI, Lykov AN (1978) Investigation of Bridge Junctions Made of High-Temperature Superconductor Nb_3Sn. Sov Phys JETP 47: 110-115

8. Golovashkin AI, Lykov AN, Prishchepa SL (1979) Investigation of the Applicability of the Vortex Model for Bridge Junctions of Superconductors with the A-15 Lattice. Sov Phys JETP 46: 668-673

9. Hikita M, Nakamura K, Kubo S, Igarashi M, Kakuchi M, Kogure O (1982) Fabrication and I-V Characteristics of High-Tc Nb_3Ge Microbridges. Jpn J Appl Phys 21: L10-L12

10. Larkin AI, Ovchinnikov YuN (1974) Electrodynamics of Inhomogeneous Type-II Superconductors. Sov Phys JETP 38: 854-858

11. Kuriki S, Goto K (1981) Regime of Vortex flow in Short Nb Variable-Thickness Bridges. J Appl Phys 52: 5257-5261

Josephson Junction Using Layered Bi-Based Oxide Thin films

KOICHI MIZUNO, HIDETAKA HIGASHINO, KENTARO SETSUNE, and KIYOTAKA WASA
Central Research Labolatories, Matsushita Electric Industrial Co. Ltd., Moriguchi, Osaka, 570 Japan

ABSTRACT

Sandwich type S/N/S Josephson junction using Bi-Sr-Ca-Cu-O high-Tc oxide thin film was successfully fabricated. Both counter and base electrodes were constructed by superconducting "2212" phase. "2201" phase of Bi system was chosen as normal conducting layer. The sandwich structure was obtained by in-situ sputter deposition process. The junction area of $20X40\mu m^2$ was restricted by photo-lithography technique and Ar ion milling. The junction exhibited clear ac Josephson effect under rf irradiation. This is , in our knowledge, the first report of the sandwich type Josephson junction using Bi-Sr-Ca-Cu-O thin film.

KEY WORDS: Bi-Sr-Ca-Cu-O, S/N/S, thin film, Josephson junction

INTRODUCTION

A sandwich type Josephson junction has much advantage for practical application because of its potential for reproducibility and stability of characteristics and integration. This type junctions using conventional metal superconducting thin films, such as Nb/AlOx/Nb junction, have been realized by sharp transitions of each materials at the interface between each layer. Similar sharp profile of interface will be required for high-Tc oxide superconducting junctions.

Several preliminary attempts for high-Tc oxide superconductor junctions have been done [1-5]. C.T.Rogers et.al. [5] have proposed the S/N/S structure using $Y_1Ba_2Cu_3O_x$ film for the both superconducting electrodes and $Pr_1Ba_2Cu_3O_x$ for the barrier layer. They sequentially deposited all these films in one chamber on the heated substrate and suggested josephson characteristics.

Recently we successfully prepared Bi-based superconducting thin films by magnetron sputtering. The crystal phase of those films were well controlled by using multitarget sputtering system. It was confirmed that as-deposited low Tc phase of Bi-Sr-Ca-Cu-O thin films showed the superconducting transition at above 60K. These techniques were applied to our Josephson junction constructed by high-Tc Bi-based oxide superconductor.

In this paper, we report the fabrication method and characteristics of sandwich type Josephson junction on which the Bi-based oxide were used for each superconducting electrode and barrier layer.

EXPERIMENTAL

So called "2212" phase of Bi-Sr-Ca-Cu-O (low Tc phase:BSCCO) films are used for both counter and base superconducting electrodes. The c-axis oriented single phase was easily obtained for "2212" BSCCO. Its superconducting transition temperature is about 60 K. "2201" phase Bi-Sr-Cu-O (BSCO) was chosen for barrier layer. A c-axis oriented single phase "2201" BSCO were prepared more easily than BSCCO. The sintered BSCO, $Bi_2Sr_2Cu_1O_x$, was reported as superconductor which transition temperature is 8 K [6]. But our deposited films of "2201" phase did not exhibit superconducting transition even at 4.2 K. Its electrical conductivity is about 100 ohm•cm in the direction of c axis at the temperature of 4.2 K. Using this "2201" BSCO film as barrier layer, we expected S/N/S type junction in BSCCO/BSCO/BSCCO structure.

A multitarget rf magnetron sputtering system which has four targets was used for the film preparation. Rf power was supplied to four target in turn and those were sputtered one by one. On each sputtering deposition, adequate presputtering were done. The superconducting Bi-Sr-Ca-Cu-O (BSCCO) and Bi-Sr-Cu-O (BSCO) for barrier layer were deposited from different target. These targets were made of mixed oxide powder and composition ratios are arranged for the film of "2212" or "2201" stoichiometry. The thin films of BSCCO, BSCO and BSCCO were successively prepared on the heated MgO(100) single crystal substrate. The substrate temperature during the deposition was kept at 650°C. These sputtering conditions will be reported elswhere.

After cooling the substrate to the room temperature, the Pt metal film was deposited in the same chamber without breaking the vacuum. This Pt layer played a roll of passivation layer during the fabrication process and realized good ohmic contact to the junction. In this way, our multi-layered structure was fabricated by fully in-situ deposition.

Using the deposited multi-layered film, the junction in area of $20X40\mu m^2$ was fabricated by photo-lithography and Ar ion milling. This process was as same as the process called SNIP(Self-aligned Niobium Insulating Process)[7]. Figure 1 shows the fabrication process schematically. In this study, the typical thickness of BSCCO(base electrode), BSCO, and BSCCO (counter electrode) are 300nm, from 40 to 100nm, and 200nm, respectively. At the first step of fabrication, the junction area was patterned using negative type photo-resist and by Ar ion milling technique (fig.1(b)). By controlling the milling time, the top BSCO/BSCCO layer could be successfully etched off. In this step the surface of BSCCO film of base electrode was physically bombarded by Ar atom, but less degradation of superconducting property was occurred on the film[8]. After a CaF_2 film was evapolated as an insulator layer (fig.1(c)), the photo-resist and CaF_2 film on the junction were removed by lift-off method. Then second Pt film was deposited over the substrate (fig.1(d)). Before this deposition, oxygen plasma cleaning was done to ashed off the contamination of polymer on the Pt film and this second Pt film was contacted to the clean surface of the top Pt layer of junction area. At the last steps of the fabrication, the Pt film was patterned by Ar ion milling technique as contact electrode (fig.1(e)). Through this process, the top Pt layer of multi layered film acted as passivation layer.

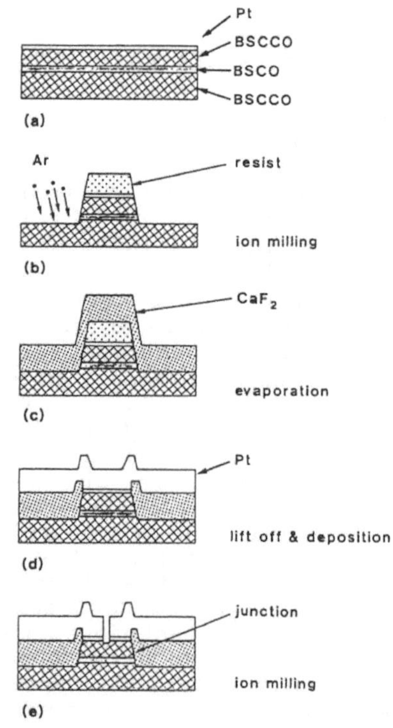

Fig.1. Fabrication process of BSCCO/BSCO/BSCCO Josephson junction.

RESULTS AND DISCUSSION

The X-ray diffraction (XRD) pattern of one of our multi-layered film of BSCCO/BSCO/BSCCO was shown in fig.2. In fig.2(a), one peak caused by Bi_2O_3 (impurity phase) was identified. The other peaks can be assigned to (00n) peak of "2212" phase BSCCO (indicated by the character of "L") or "2201" phase BSCO (indicated by the character of "S"). We etched off the top BSCCO film or the top BSCO/BSCCO film by means of Ar ion milling technique to ensure the crystallinity of each layer. The XRD patterns resulted from these etching are also shown in fig.2(b) and fig.2(c). Figure 2(b) shows the XRD pattern of BSCCO(base electrode)/BSCO film. figure 2(c) shows that of only BSCCO

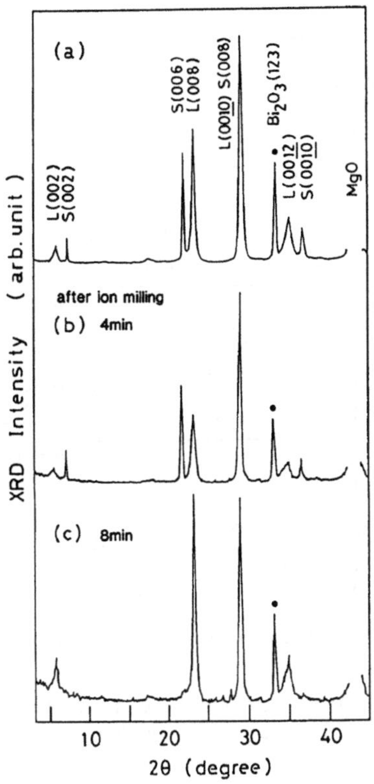

Fig.2. The X-ray diffraction pattern(XRD) of BSCCO/BSCO/BSCCO multi-layered film; (a) BSCCO/BSCO/BSCCO; (b) BSCCO/BSCO ; (c) BSCCO. Each sample was obtained from the same BSCCO/BSCO/BSCCO.

film and no peaks concerning to "2201" phase. From this fact, the crystal phase of base electrode is confirmed to be almost single "2212".

The interdiffusion between each layer or the depth profile of atom concentration was examined by Auger electron spectroscopy (AES) measurement. The depth profile is shown in figure 3. The sputtering target without Ca was used to deposit BSCO. However the signal from Ca atom was detected even in the barrier layer. The transition profile at each interfaces are fairly sharp. These results from XRD and AES analyses are summarized as follows. 1) The base electrode BSCCO has surely "2212" phase structure. 2) The barrier layer may have "2201" BSCO phase structure and the counter electrode also may have "2212" BSCCO phase structure. 3) Sharp interfaces are realized, however , a little amounts of Ca atom diffused into barrier layer during the sputtering deposition of barrier layer or counter electrode.

According to the fabrication process mentioned above, the junction structure was fabricated using such

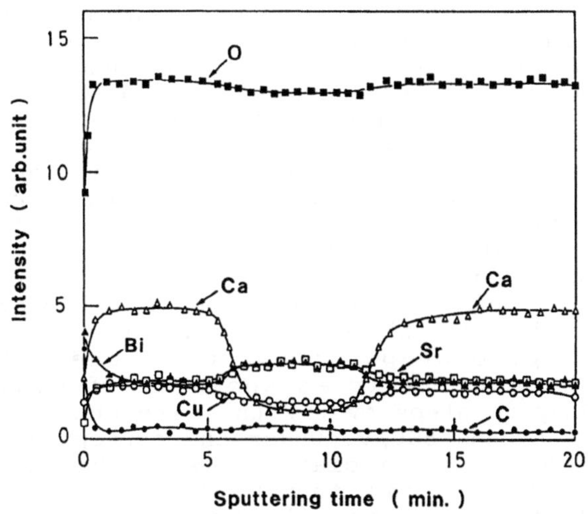

Fig.3. The Auger electron spectroscopy depth profile of BSCCO/BSCO/BSCCO multi-layered film.

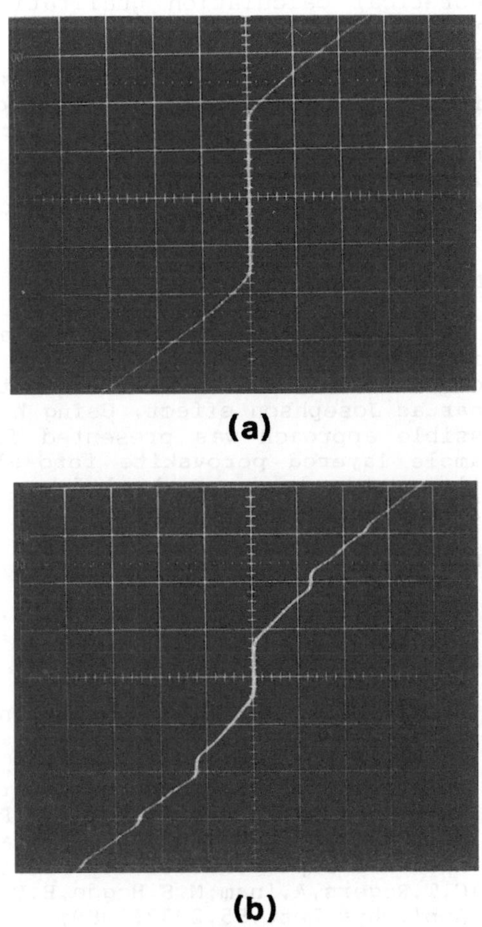

Fig.4. Current-voltage curves for BSCCO/BSCO(60nm)/BSCCO junction at 4.2K (0.5mA/div.vert. and 20 μV/div. horiz.); (a) without rf radiation; (b) under 12.4GHz radiation.

layered film. The transition temperature to zero resistance for BSCCO film was 62 K, and our junctions became superconductive state at around 50 K. Figure 4 shows the I-V characteristics of the junction in which BSCO (barrier layer) thickness was 60 nm. In this figure, S/N/S like behavior was observed at 4.2 k. The Ic·Rn product of the order of 0.1mV was obtained.

Under the irradiation of rf power at higher than 12 GHz , the clear Shapiro steps appeared in the I-V characteristic as shown in fig.4. Shapiro steps over 20th were observed at around 500μV. Under the appropriate level of the radiation power, the step width of the first Shapiro step decreased. As like this the step width changes according to applying power level. These power level dependences of step width are shown in fig.5. The dashed lines in fig.5 are calculated from RSJ model and fitted to each step width dependence. The experimental data do not fit to the theoretical calculation qualitatively. Further analysis must be needed in this result.

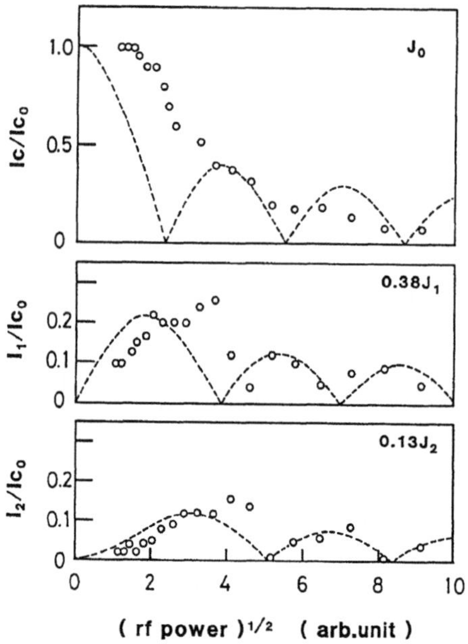

Fig.5. Dependence of Shapiro step width on induced rf power.

Such S/N/S like characteristic is also observed in the point contact structure. Our junction might be clarified as some pin holes and operate as point contact type Josephson junction. However, it is concluded that our junction is S/N/S type junction according to the following reasons; 1) the barrier layer was rather thick, 2) clear AES depth profile was obtained and 3) these junction characteristic was obtained reproducibly.

CONCLUSION

Preliminary results were obtained for sandwich type Josephson junction which is constructed by Bi-based oxide thin films. The junction fabricated from BSCCO/BSCO/BSCCO film exhibited S/N/S type I-V characteristic and also clear ac Josephson effect. Using high-Tc Bi-based oxide superconductor, one of possible approach was presented for S/I/S Josephson tunnel junctions. For example layered perovskite fero-electric materials such as Bismuth-Titanate will be able to be applied for tunneling barrier. We expect much attention will be paid for such approach.

ACKNOWLEDGMENT

The authors wish to thank Dr.T.Nitta for his continuous encouragements.

REFERENCES

(1)H.Akoh,F.Shinoki,M.Takahashi, and S.Takada,Jpn.J.Appl.Phys.27,L519(1988)
(2)A.Nakayama,A.Inoue,K.Takeuchi, and Y.Okabe,Jpn.J.Appl.Phys.26,L2055(1987)
(3)T.Shiota,K.Takechi,Y.Takai, and H.Hayakawa,Proceedings of the 1st International Symposium on superconductivity (ISS'88),Nagoya,755(1988)
(4)T.Venkatesan,X.D.Wu,A.Inam,C.T.Rogers,L.Nazar,B.Wilkens, C.C.Chang, D.M.Hwang,P.England,M.S.Hedge, and H.Padamsee, Extended Abstracts of International Superconductivity Electronics Conference (ISEC'89),Tokyo,1(1989)
(5)C.T.Rogers,A.inam,M.S.Hegde,B.Dutta,X.D.Wu,and T.Venkatesan, Appl.Phys.Lett.55,2032(1989)
(6)J.Akimitsu,A.Yamazaki,H.Sawa, and H.Fujiki,Jpn.J.Appl.Phys.26,L2080(1987)
(7)A.Shoji,F.Shionoki,S.Kosaka,M.Aoyagi, and H.Hayakawa, Appl.Phys.Lett.41,1097(1982)
(8)A.Enokihara,Private communications.

Preliminary Study of YBCO/Au/YBCO Josephson Junction

S. TANAKA, H. ITOZAKI, and S. YAZU

Itami Research Laboratories, Sumitomo Electric Industries, Ltd., 1-1, Koyakita 1-chome, Itami, Hyogo, 664 Japan

ABSTRACT

We have grown a YBCO/Au/YBCO tri-layered structure using RF magnetron sputtering for YBCO and high vacuum evaporation for Au. X-ray Diffraction Analysis revealed a strong c-axis orientation of the YBCO thin film even on the Au barrier. We tried to fabricate an SNS junction with an area of 90 um x 90 um. After applying microwave radiation of 8.96 GHz, the critical current was suppressed. Also, weak induced voltage steps in the junction were observed. These steps suggest that an ac-Josephson effect might occur in this device.

KEY WORDS: microwave irradiation, thin film, Josephson junction, YBCO, SNS

INTRODUCTION

The interest of many scientists is focused on the fabrication of Josephson Junctions using high Tc superconducting thin films. Recently, results using Tunnel Junctions combining low Tc superconducting materials with high Tc superconducting materials have been reported. An example is the Nb/AlO_x/YBCO junction [1,2]. Further more, characteristics of YBCO/$Pr_1Ba_2Cu_3O_x$/YBCO junctions by an in-situ growth technique have been reported by C. T. Rogers [3]. Fabrication of a junction using high Tc superconducting materials without an in-situ growth technique was thought impossible because of the deterioration of the superconducting bottom layer's surface. Therefore there have been a few reports on fabricating junctions using high Tc superconducting materials through ex-situ techniques [4]. In this paper, we will discuss preliminary results describing the characteristics of a YBCO/Au/YBCO junction fabricated through ex-situ techniques.

EXPERIMENTAL

A base layer of an epitaxial YBCO thin film (400 nm) was deposited on a MgO single crystalline substrate by an RF magnetron sputtering techniques with a single target. The preparation conditions have been previously reported in detail [5,6]. The specimen was then taken out of the sputtering chamber and evaporated with an Au barrier layer (60 nm) in a vacuum evaporation chamber. Before the Au evaporation, the evaporation chamber was evacuated to $1.3x10^{-6}$ Pa by an oil free turbo-pumping system. The specimen was not intentionally heated during the evaporation. The thickness was controlled by a quartz thickness monitor (INFICON IC-6000). This two-layered specimen was then set in the sputtering chamber again and deposited with an another YBCO layer (400nm). After formation of the top YBCO layer, an Ag/Au electrode was deposited on the top YBCO layer. This layer was deposited without heating the specimens. The Ag/Au layer works not only as an electrode but also as a preventing layer for reactions between the top YBCO layer and photoresist used to form the junctions. To reduce the contact resistance between the electrode and the YBCO, the specimens were then heat treated at 400C in the air for 30 min. We employed a photolithographic technique to form the Junction. The processing steps are shown in Fig.1. (a)--initially, an area of contact electrodes was protected with a positive type photoresist (OFPR-800). The exposed portion of the Ag/Au layer was then removed by an Ar ion beam with an acceleration voltage of 300 V and a current density of 0.5 mA/cm2. (b)--The junction area of 90 um x 90 um was then defined by a photoresist mask. The wafer was exposed to the Ar ion beam to etch both the YBCO layers and Au barrier. The etching was stopped in the middle of the bottom YBCO layer. (c)--The insulation for the side wall of the junction was then added. A negative type photoresist (OMR-83) was temporarily used as an insulator, even at temperatures lower than 4.2 K. (d)--Ti/Au leads and wire bonding pad were deposited.

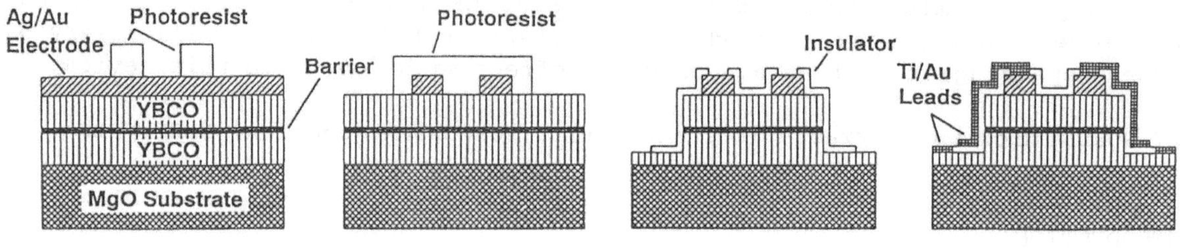

Fig.1. Processing steps of SNS junctions. (a) Formation of electrodes. (b) Definition of the junction area. (c) Ion beam etching and device isolation. (d) Formation of Ti/Au leads.

RESULT AND DISCUSSION

X-Ray Analysis

Figure 2 shows the X-ray diffraction pattern for a three layered structure which consists of the YBCO layer, the Au layer, and the top YBCO layer. The X-ray diffraction pattern indicates that both the top and bottom YBCO layers have a (001) plane parallel to the surface, and that no other aligned peaks are included. Also, the Au barrier is aligned with the (111) plane.

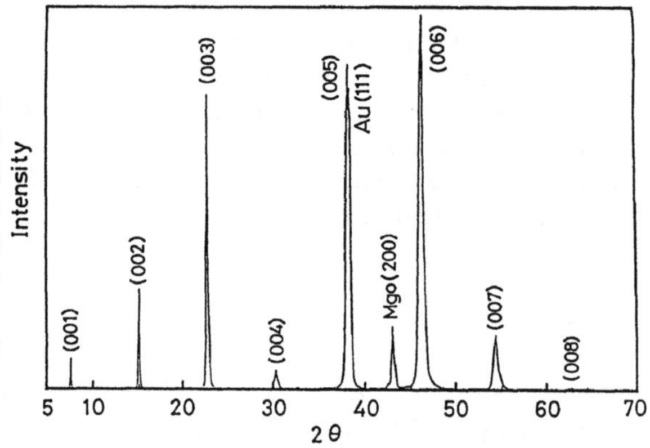

Fig.2. X-ray diffraction pattern for three layered structure. Both the top and bottom YBCO layers have (001) planes parallel to the surface. Au barrier is aligned with (111) planes.

Electrical Properties

Electrical measurements were made in a magnetic shielded cryostat with temperature control. The temperature dependence of the resistance of the junction is shown in Fig.3. The resistance decreases linearly and slightly drops at about 75 K. The resistance continues to decrease until 35 K and becomes zero at 25K. We think that the first drop at 75 K represents the transition of the top YBCO layer and the second one at 35 K represents the transition of the bottom YBCO layer. It seems that the transition temperature of the bottom superconducting layer is lower than that of the top superconducting layer because of an oxygen deficiency by ionic damage during the Ar ion beam etching process. Results indicate that superconducting transport occurs between the top and bottom layer via an Au barrier apart from the possibility of a micro-short. Microwave irradiation was provided by placing an antenna near the device inside the cryostat. The frequency of the microwave irradiation was 8.96 GHz. I-V curves of a junction with and without microwave irradiation are shown in Fig.4. The critical current Ic of the junction was decreased from 0.31 mA to 0.18 mA at 18 K when the irradiation was applied. However, no clear constant voltage steps on the curves were observed. Thus the differential resistance was measured to enhanced the steps. Figure 5 shows the differential resistance dV/dI as a function of voltage at 18 K when the junction was irradiated with microwaves. There are small bumps at the expected interval

$$V=n(h/2e)f,$$

where h is Plank's constant, e is electron unit charge, f is the frequency of irradiating microwaves and n=1,2,3 --. Arrows in Fig.5 indicate the expected voltage points. Subharmonic steps in the middle of the interval (indicated by broken arrows) are also observed. The results suggest that an ac-Josephson effect might occur in this device. These steps are very weak compared with the scale of the decrement of the Ic when the microwave are applied. We think that the Ic is mostly suppressed by the bolometric effect and that a small Josephson current flows through the junction. The irradiated microwaves might then interfere with this small current and cause weak voltage steps. Since a top YBCO layer might have many grain boundaries which provide weak links, there is a possibility that these weak links respond to the microwave irradiation. We measured I-V characteristics of several samples with different thickness of the Au barrier (up to 100 nm). When the thickness of the Au

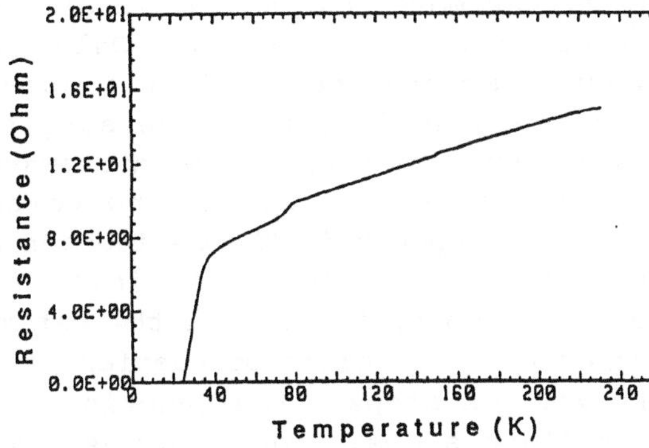

Fig.3. Temperature dependence of device resistance.

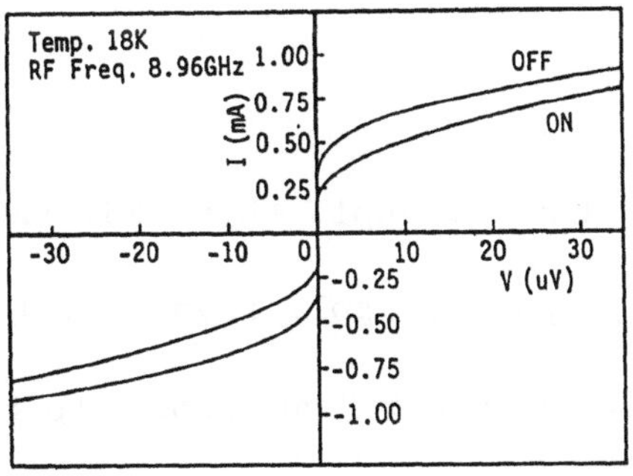

Fig.4. I-V characteristics of the device. Critical current Ic is suppressed under microwave irradiation.

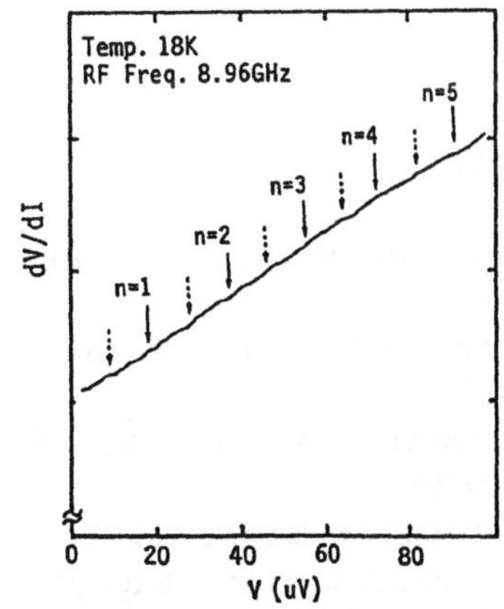

Fig.5. Differential resistance dV/dI as a function of voltage.

barrier was more than 80 nm, Ic of the devices and constant voltage steps were not observed even at 4.2 K. Only the bolometric change in the shape of the I-V curve around the origin was observed. With an Au barrier of 20 nm to 60 nm in the junction, both the suppression of Ic and the occurrence of weak constant voltage steps under microwave irradiation were observed. In cases of thickness less than 20 nm, no voltage steps were observed. This lack of microwave response in the device can be explained by a strong coupling of both the YBCO layers. An upper limit of about 60 nm, in which Ic was observed, shows good consistency with the estimated value of coherence length in the Au barrier [7]. If grain boundaries of the top YBCO layer work as weak links and make voltage steps, the junction with even the thick Au barrier should show the voltage steps. None, however, were observed. Therefore, voltage steps in Fig.5 come not from grain boundaries but from the junction. However, the possibility of a micro-short between the top and bottom YBCO layers still remains. We think that there exists two types of paths for a critical current. One is an SNS junction or a micro-short, which can respond to the microwaves. The other is a large short path, which provides a large amount of Ic. The large leakage current passing through the short path may conceal the tiny modulation occurring in the junction. The magnetic field dependence of the current was measured. It was expected that the Ic of the Josephson junction decreases and is modulated as the magnetic field increases. However, the Ic didn't decrease under a magnetic field which ranges from 0 to 1×10^{-3} Tesla.

CONCLUSIONS

Preliminary study of the fabrication of SNS junctions using all high Tc superconducting materials was performed. We employed not in-situ techniques, but ex-situ techniques. X-ray diffraction analysis revealed a strong c-axis orientation of the YBCO thin film even on the Au barrier. Constant voltage steps on the I-V curves were observed when the device was irradiated with microwaves, but they were weak. It suggests that the AC Josephson effect may occur in the device. We think not only a Josephson current but also a short current might flow through the device. However, these results encourage us to develop future Josephson devices.

REFERENCES

1. A.Nakayama,A.Inoue, K.Takeuchi and Y.Okabe, Jpn. J. Appl. Phys. 26, L2055 (1987).
2. H.Akoh, F.Shinoki, M.Takahashi and S.Takada, Jpn. J. Appl. Phys. 27, L519 (1988).
3. C.T.Rogers, A.Inam, M.S.Hegde, B.Dutta, X.D.Wu, and T.Venkatesan, to be published.
4. T.Shiota, K.Takechi, Y.Takai and H.Hayakawa, Proc., 1st Int. Sympo. on Superconductivity, Nagoya, Japan, August 28-31, 1988, 755
5. S.Tanaka and H.Itozaki, Jpn. J. Appl. Phys. **27**, L662 (1988).
6. S.Tanaka and H.Itozaki, Jpn. J. Appl. Phys. **28**, L441 (1989).
7. K.K.Likharev, Rev. Modern Phys. 51, 101 (1979).

Fabrication of YBCO/Barrier/YBCO Structure Junctions

TOSHIYUKI MATSUI[1], GENSOH MATSUBARA[2], AKIYOSHI NAKAYAMA[3], NORITADA SATOH[1], KAZUO MUKAE[1], and YOICHI OKABE[2]

[1] Basic Research Laboratory, Fuji Electric Corporate Research and Development, 2-1, Nagasaka 2-chome, Yokosuka, 240-01 Japan
[2] Department of Electronic Engineering, Faculty of Engineering, University of Tokyo, 3-1, Hongo 7-chome, Bunkyo-ku, Tokyo, 113 Japan
[3] Department of Electrical Engineering, Faculty of Engineering, Kanagawa University, 27-1, Rokkakubashi 3-chome, Kanagawa-ku, Yokohama, 221 Japan

ABSTRACT

YBCO/barrier/YBCO junctions were fabricated with the rf magnetron sputtering system. The barrier morphology of these junctions was studied. Results indicated that the barriers, such as $SrTiO_3$, MgO and Y_2O_3, formed island structures on the YBCO.

KEY WORDS: YBCO/barrier/YBCO junction, morphology of barrier layer, island structure

INTRODUCTION

Since the discovery of oxide superconductors such as Y-Ba-Cu-O (YBCO), Bi-Sr-Ca-Cu-O (BSCCO) and Tl-Sr-Ca-Cu-O (TSCCO), many experimental results about the applications of these materials have been reported. According to the BCS theory, the superconducting gap parameter Δ in weak coupling state is proportional to the superconducting transition temperature (Tc), and it is likely that these high-Tc materials have large gap parameter. Furthermore, the intrinsic switching delay τ of a Josephson logic gate is quantum mechanically limited by the equation $\tau > \Phi_0/2\Delta$, where Φ_0 is the flux quantum. Hence these high Tc materials are quite interesting as materials of Josephson junctions. In the fabrication of the tunnel structure junctions using these high-Tc materials, it was reported that oxide insulators such as AlO_x [1], MgO [2,3] and Y_2O_3 [4] were used for the tunnel barrier. However, only in a few reports, Josephson effects was obtained, the effective area of the junction being much smaller than the junction area [1,2]. It seems that this phenomenon is caused by the morphology of the barrier layer.

In this paper, we will report the results of the estimation of the barrier morphology for YBCO/barrier/YBCO junctions.

FABRICATION OF YBCO/BARRIER/YBCO JUNCTIONS

Both YBCO electrodes and barrier layer were formed by using an rf magnetron sputtering system. Sputtering conditions are summarized in Table I. In those conditions, the obtained YBCO films were (110) oriented and critical temperatures were 60K. On the other hand, we used $SrTiO_3$ and MgO as the barrier materials, because their lattice constants almost match with that of YBCO. And we also used Y_2O_3 to avoid a reaction between YBCO and the barrier.

Table I Sputtering conditions

	Lower YBCO	Barrie materials			Upper YBCO
		$SrTiO_3$	MgO	Y_2O_3	
Rf power	300w	300w	300w	300w	300w
Pressure	10Pa	10Pa	10Pa	10Pa	10Pa
Gas	Ar+50%O_2	Ar+50%O_2	Ar+50%O_2	Ar+50%O_2	Ar+50%O_2
Substrate temp.	620°C	470°C	500°C	600°C	600°C
Target	$Y_1Ba_{2.6}Cu_6O_x$	$SrTiO_3$	MgO	Y_2O_3	$Y_1Ba_{2.6}Cu_6O_x$

Figure 1 shows the schematic drawing of the junction. The patterning of the junctions were performed by the metal mask process. The junction area was $500x500\mu m^2$. The thickness of the lower YBCO was 0.5μm and the upper one was 0.73+t μm (t:the thickness of the barrier layer).

MORPHOLOGY OF BARRIER LAYER

Figure 2 shows the temperature dependence of resistance of the lower and the upper YBCO electrodes. Both electrodes shows superconductivity around 60K. It suggests that there are no damages during the junction fabrication process.

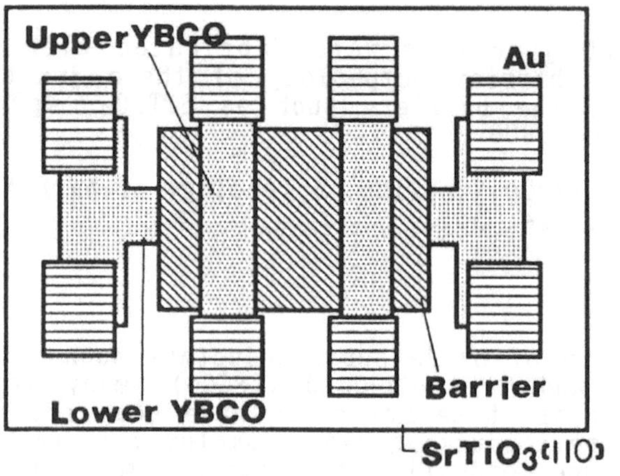

Fig. 1 The schematic drawing of the YBCO/barrier/YBCO junction.

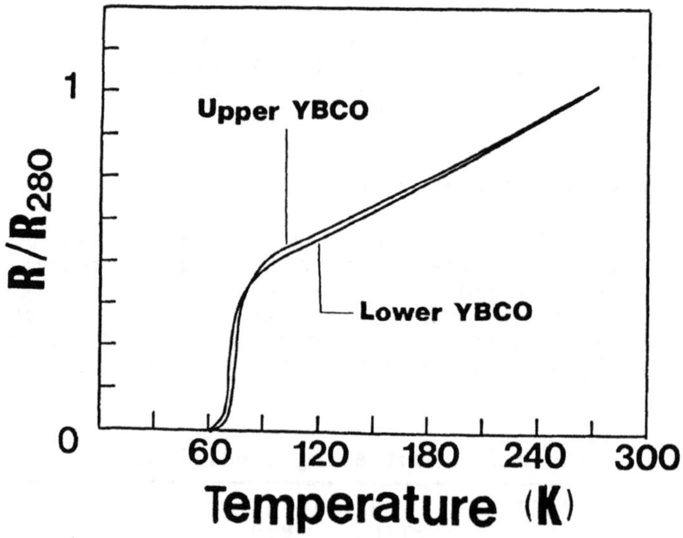

Fig. 2 The temperature dependence of the resistance of the lower and the upper YBCO electrodes. Both YBCO electrodes show superconductivity at 60K.

In order to obtain the information about the junction, the critical current across the junctions was measured. Figure 3 shows the temperature dependence of the critical current of the YBCO/$SrTiO_3$/YBCO junctions for the various barrier thickness. Although the thickness of the barrier layer was as much as 0.3μm, the superconducting current flowing across the junction was detected. It suggests that the barrier layer, $SrTiO_3$ layer, is not uniform and many parts of the junction form YBCO/YBCO junctions, that is to say "micro shorts". To investigate the morphology of the barrier layer, the scanning electron microscope (SEM) was used. Figure 4 shows the surface morphology of the $SrTiO_3$ barriers. The lower YBCO had a smooth surface as seen in Fig. 4(a). When $SrTiO_3$ was deposited for the barrier layer on this YBCO, it grew into crystal grains which were (110) oriented and formed the island structure as seen in Fig. 4(b). This is probably due to the morphology and the crystal orientation of the lower YBCO, because the $SrTiO_3$ film deposited on the other substrate such as MgO or $SrTiO_3$ single-crystal was flat and uniform in this sputtering condition. Therefore, when the upper YBCO was formed on the barrier layer in the next stage, many parts of the upper one were deposited on the lower one directly. This forms YBCO/YBCO weak links. And with the increasing of the barrier thickness, these YBCO/YBCO regions decrease as seen in Fig. 4(c). Hence, increasing of the barrier thickness means the decrease of the current paths through the junction. Therefore, on the I-V curve, where $SrTiO_3$ thickness was 0.3μm, the periodic steps caused by the vortex motion, which is typical in the micro bridge structure [5], were obtained as shown in Fig. 5.

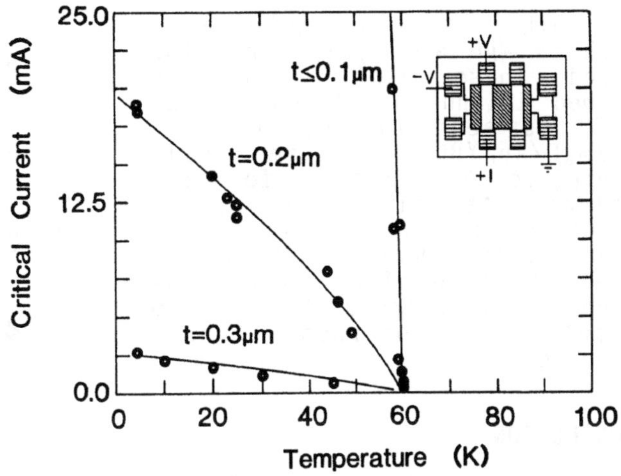

Fig. 3 The temperature dependence of the critical current of the YBCO/barrier/YBCO junctions for the various barrier thickness. t is the thickness of the barrier layer.

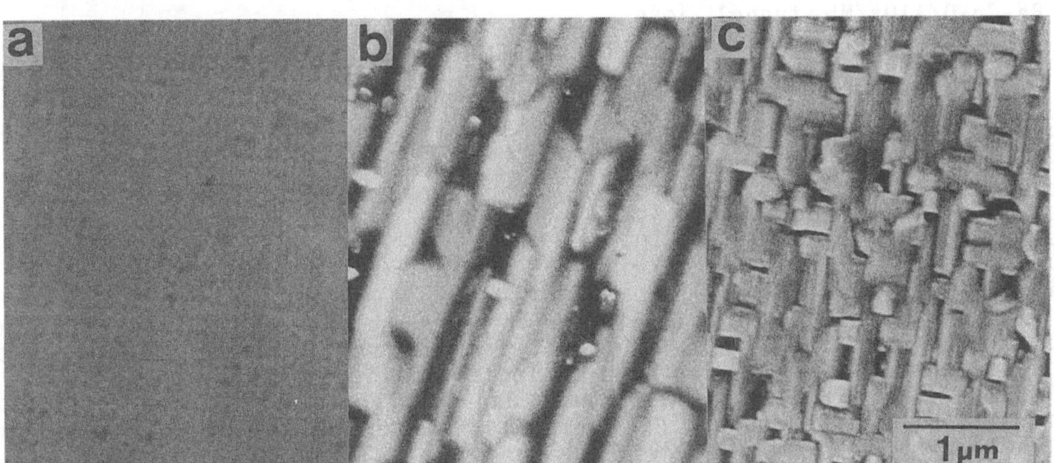

Fig. 4 The surface morphology of the $SrTiO_3$ barriers for the various thickness. (a):The surface of YBCO. $SrTiO_3$ was deposited on YBCO for (b):0.02μm, and (c):0.1μm.

Other barrier materials, such as MgO and Y_2O_3, were also tried. These materials similarly formed island structures on the YBCO as seen in Fig. 6. In Nb/MgO/Au/YBCO system [2], therefore, MgO seems to form the island structure, causing a reduction in the effective area for the Josephson junction. As the interface of Nb/YBCO forms high resistive layer, this system results in Josephson tunnel junctions.

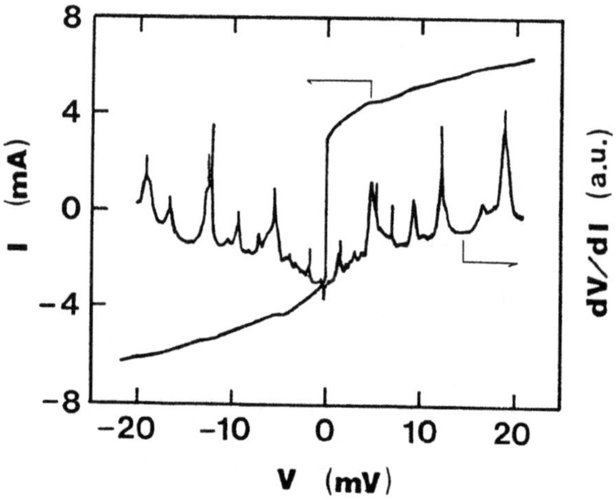

Fig. 5 dV/dI-V and I-V characteristics of YBCO/$SrTiO_3$/YBCO junction, where the thickness of $SrTiO_3$ was 0.3μm. This measurement was carried out at 4.5K.

CONCLUSION

In conclusion, we fabricated junctions of YBCO/barrier/YBCO structure and estimated the morphology of the barrier layers. Results showed that the oxide insulators, such as $SrTiO_3$, MgO and Y_2O_3, formed island structures on the YBCO film.

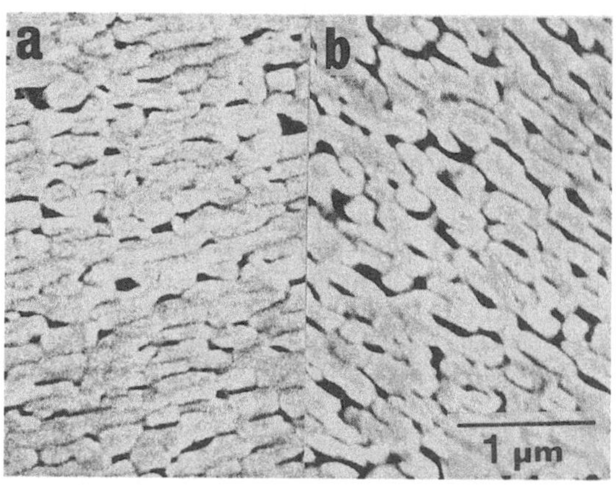

Fig. 6 The surface morphology of the barrier layers deposited on YBCO films. (a):MgO (0.1μm), (b):Y_2O_3(0.1μm)

ACKNOWLEDGMENT

The authors would like to thank G. S. Wong, H.Ide and K.Sugita for their help.

The project is partially supported from Special Grant in Aid of Ministry of Education and Hoso Bunka Fund.

REFERENCES

1. A.Nakayama, A.Inoue, K.Takeuchi and Y.Okabe (1987) Y-Ba-Cu-O/AlOx/Nb tunnel Josephson junctions. Jpn. J.Appl. Phys. 26:L2055
2. A.Nakayama, T.Matsui and Y.Okabe (1989) Fabrication of YBCO/Nb Josephson tunnel junctions. Proc. 6th Int. 2nd Workshop on Future Electron Devices-. Research and Development Association for Future Electron Devices.Tokyo. p199
3. T.Kobayashi and M.Iyori (1989) Interface problems of HTSC SIS junction. Proc. 6th Int. 2nd Workshop on Future Electron Devices-. Research and Development Association for Future Electron Devices. Tokyo. p415
4. M.G.Blamire, G.W.Morris, R.E.Somekh and J.E.Evetts (1987) Fabrication and properties of superconducting device structures in $YBa_2Cu_3O_{7-x}$thin films. J.Phys. D:Appl. Phys. 20:1330
5. M.A.M.Gijs, D.Terpstra and H.Rogalla (1989) Vortex transport in $Y_1Ba_2Cu_3O_{7-x}$ thin film microbridge. Extended Abstracts of Int. Superconductivity Electronics Conference. p243

Detection of 6 keV X-rays by Using Large-Size Nb-Based Tunnel Junctions

K. Ishibashi[1], K. Takeno[1], K. Mori[1], T. Sakae[1], Y. Matsumoto[1], A. Katase[1], S. Takada[2], H. Nakagawa[2], M. Aoyagi[2], H. Akoh[2], and S. Kohjiro[2]

[1] Department of Nuclear Engineering, Kyushu University, Hakozaki, Higashi-ku, Fukuoka, 812 Japan
[2] Electrotechnical Laboratory, Umezono, Tsukuba, 305 Japan

ABSTRACT

Nb-based tunnel junctions may be applicable to a high resolution x-ray measurement. Large-size tunnel junctions with a Nb/Al/AlOx/Nb structure have been fabricated to study their performance in x-ray detection. The lithographic technique was employed to make the junctions. The niobium layer being sensitive to x rays had a large size of 100 μm□ width × 1.2 μm thickness. The junction detected 5.9 keV x rays with an energy resolution of 1 keV. The detector characteristics such as the pulse height spectrum and the signal to noise ratio were well understood.

KEY WORDS: niobium, superconducting, tunnel junction, x ray, quasiparticle

INTRODUCTION

The structure of superconductor/insulator/superconductor is known as a typical superconducting tunnel junction. The energy gap of the superconductor is a few meV, and is about three orders of magnitude lower than that of semiconductor. When an x ray deposits its energy in the superconductor, a great number of quasiparticles (free electrons) are created. If the quasiparticles are all collected to produce an electronic signal through the tunnel junction, it may produce the energy resolution which is more than an order of magnitude better than that of semiconductor detectors.

D. Twerenbold and A. Zehnder[1] have recently measured 5.9 keV x rays using a tunnel junction. The junction was fabricated by evaporation of tin, and was used at 0.3 K in the measurement. They demonstrated that the tunnel junction showed the energy resolution which was comparable to that of silicon detector. As is well known, however, the tin junction has a property where the performance readily deteriorates through both the preservation at room temperature and the thermal cycle from room to cryogenic temperature. The junction, therefore, is not considered to be suitable for practical application. Unlike the case of tin, the tunnel junctions made of niobium are free from this property, and accordingly they may be applicable to practical radiation measurements. The transition temperature of niobium is 9.2 K, being about three times as high as that of tin. When junctions are used at an operating temperature, such values as the thermal leakage current and the life time of quasiparticles are principally determined by the ratio of the operating to the transition temperature. Hence, the use of a Nb-based tunnel junction may produce good characteristics at an operating temperature that is higher, by roughly three times, than that of tin junction.

Fabrication technique of Nb-based tunnel junctions has recently been advanced to develop Josephson logical circuits for digital computers. The technique employs the lithographic method[2]. The Nb-based tunnel junctions for radiation detection have been made by an European[3] and two Japanese[4,5] groups. Reference 3 described the detection of x rays by Nb/Al/AlOx/Al/Nb junctions of 10~50 μm□, and reference 4 measured them by Nb-based junctions of 20 μm□. The tunnel junctions which detected x rays were usually composed of x-ray sensitive layers with a thickness of about 0.2 μm. When the efficiency of x-ray detection is concerned, however, the layers are required to be as thick as possible. We attempt to make large-size Nb/Al/AlOx/Nb junctions for detecting radiations. A simple geometric configuration of junctions are chosen to make clear their characteristics as a radiation detector.

Nb/Al/AlOx/Nb TUNNEL JUNCTION

The tunnel junctions with the Nb/Al/AlOx/Nb structure were fabricated in the same way as in reference 5. The picture of the junctions on a chip is presented in Fig. 1. There are many squares with different sizes in the central region of the chip. These are the upper niobium layers of the junctions, and have the size of 30 ~ 1000 μm□. These layers were located on the common lower niobium layer, which was rectangular with a size as large as 2×4 mm. The thickness of the lower and the upper layers was 0.2 and 1.2 μm, respectively. A 1.5nm-thick aluminum-oxide layer acted as the tunnel barrier between these layers. The upper layers had current leads with a line width of 2.5 μm.

For the junction in reference 1, the lower layer had the same size as the upper one. The configuration in Fig. 1 then is different from that of this reference. Since the multiple tunneling (back tunneling) of quasiparticles[1] is expected to produce an insignificant effect for the junctions in Fig. 1, the present configuration is suitable for analyzing the basic characteristics of the quasiparticle and phonon behavior.

CURRENT-VOLTAGE CHARACTERISTICS

Figure 2 shows the leakage current of the junctions as a function of temperature. The size of the junction was 100 μm□, and the oxidation time for barrier formation was 12 hrs. The current was measured with a bias voltage of 1.0 mV. The Josephson current was suppressed during the measurement by applying a magnetic field of 200 G parallel to the tunnel barrier. The marks of square show the results for the junction which was made at an initial stage of this work. The experimental leakage current decreased with temperature, but the reduction was gradually saturated at 2.5 K. The leakage current at the low

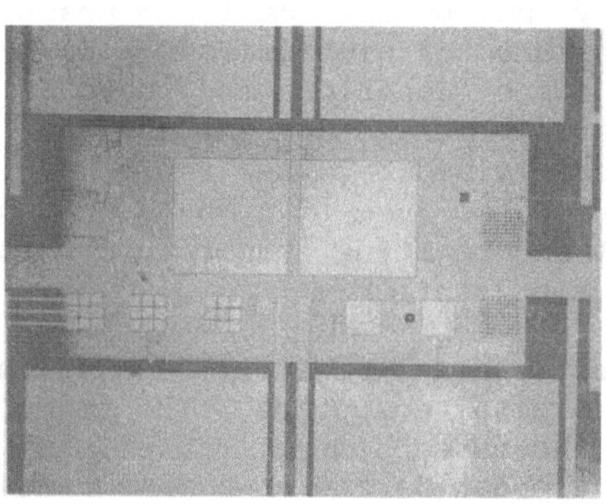

Fig. 1 Picture of the tunnel junctions on a chip.

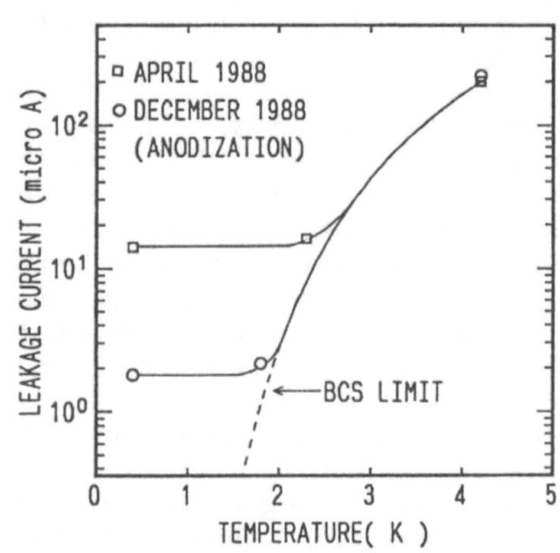

Fig. 2 Leakage current as a function of temperature.

temperature was too large to use the junction as a radiation detector. This current was considered to be originated from some kind of imperfection in the edge region of the junction[5]. Hence, the junction edge was oxidized by adopting an anodization process[6]. The marks of circle show the leakage current of the junction that was made using this process. The current decreased until the temperature reached about 2 K. The leakage current shown by a dashed curve indicates the BCS limit. This suggests that the imperfection in the edge region was reduced, but still remained to some extent.

MEASUREMENT OF X RAYS

A ^{55}Fe source (5.9 keV x ray) was used for the measurement. The experiment was made at 0.4 K. The 1.2 μm thick upper layers were sensitive to the x rays. Figure 3 shows the pulse height spectrum obtained by the 100 μm□ junction. The energy resolution (FWHM) is 1.0 keV. The x ray peak is well separated from the electronic noise. So far, the layer thickness of 1.2 μm is the largest among the junctions that have been reported to detect x rays. It was demonstrated that such a thick layer successfully produced the x ray peak. The x ray signals by Nb-based tunnel junctions have so far been reported for junction size up to 50 μm□ . The area of the present junction then is 4 times as large as the maximum value.

As seen in Fig. 3, there is only one peak in the spectrum. Unlike this, clear double peaks were reported[1] for 5.9 keV x rays. In reference 1, the lower and upper layers had the same area. The paper concluded that the two layers produced signals with different pulse heights, and accordingly produced double peaks in the pulse height spectrum. On the other hand, it was observed[7] that the quasiparticles readily diffuse over a distance beyond 100 μm when the temperature is 0.4 K. In the present work, as described in the preceding section, the lower layer has the quite large size of 2×4 mm. Therefore, if an x ray deposits its energy in the lower layer, the generated quasiparticles easily diffuse out of the tunneling region. For this reason, such an energy deposition is considered to result in a pulse height spectrum like background events. The double peak may thus be avoidable by using the lower layer much wider than the upper.

For the junctions of 30 μm□ , x rays also produced the signal pulses. The pulse height was about 10 % smaller than that of 100 μm□ junctions. This shows that phonons or quasiparticles escaped more greatly from the 30 μm□ than from the 100 μm□ junctions.

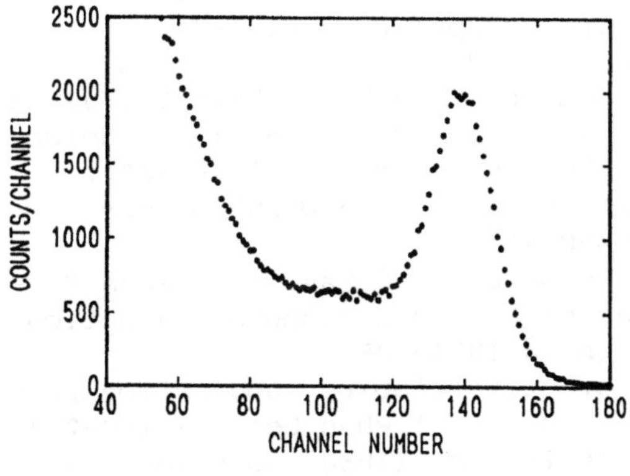

Fig. 3 Pulse height spectrum for 5.9 keV x rays.

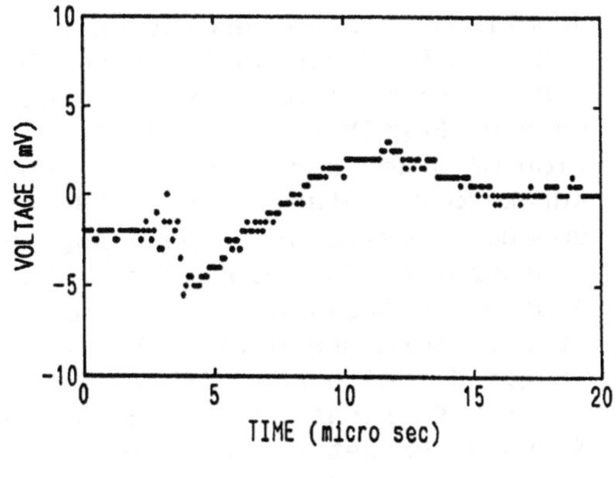

Fig. 4 Output signal from a pre-amplifier for incidence of x rays.

DISCUSSION ON THE ENERGY RESOLUTION

The output signal from the charge-sensitive preamplifier is shown in Fig. 4 for the 100 μm□ junction. The pulse had a negative polarity. An appreciable undershoot appeared because of the circuit employed in the preamplifier. The output signal was analyzed by using an equivalent circuit. The time constant for the quasiparticle tunneling was derived from the normal resistance of the junction, and was estimated as 10 μs. Another time constant was required for representing the quasiparticle relaxation. This time constant indicates the quasiparticle decay for other reasons than the tunneling. The relaxation time of 0.30 μs was found to reproduce the experimental output signal. This short relaxation time is deduced to be caused mainly by the leakage of phonons from the superconducting layer. When either the quasiparticle tunneling rate is increased or the phonon escape rate is reduced, the pulse height is raised. Since the tunneling rate is proportional to the critical current density, the junction of larger current density will make a signal of higher pulse height.

The use of the experimental pulse height enabled us to estimate the energy that was required to produce an electron from the detector. The energy per electron was found to be 100 meV, and it was much lower than the value of 3.8 eV for silicon detector. The junction had a resistance of 0.5 kΩ in the subgap region. This value is several orders of magnitude lower than the operational resistance of usual semiconductor detectors. The low impedance of junction induced a considerable amount of series noise of FET, which was used in the head of preamplifier. Thus, the series noise exceeded the FWHM of 0.8 keV. When the resistance is raised by an order of magnitude, the series noise will be reduced to the level that is determined by the junction capacitance alone.

CONCLUSION

Nb/Al/AlOx/Nb tunnel junctions with a special configuration were fabricated for radiation detection. The junctions had the upper layer as thick as 1.2 μm. The junction of 100 μm□ detected 5.9 keV x rays with a resolution of 1.0 keV. The characteristics such as the pulse height, the single peak in the spectrum and the signal to noise ratio were understood for the junction. When either the phonon escape rate is reduced or the quasiparticle tunneling rate is increased, the energy resolution is expected to be improved.

The authors acknowledge Dr. Y. Akiyama of Electrotechnical Laboratory for his kindness in official procedures required for fabricating the junctions.

REFERENCES

1. Twerenbold D. Zehnder A. (1987) Superconducting Sn/Sn-oxide/Sn tunneling junctions as high-resolution x ray detectors. J Appl Phys 61:1-7
2. Nakagawa H. Kurosawa I. Takada S. Hayakawa H. (1987) Josephson 4-bit digital circuit made by Nb/Al-oxide/Nb junction. IEEE Trans Magn 23:739-742
3. Gare P. Engelhardt R. Peacock A. Twerenbold D. Lumley J. (1989) The detection of 6 keV x rays with Nb junctions. IEEE Trans Magn 25:1351-1353
4. Kurakado M. Matsumura A. (1989) X-ray detection with Nb/AlOx/Nb superconductor detectors. Jpn J Appl Phys 28:L459-461
5. Ishibashi K. Takeno K. Oae Y. Sakae T. Matsumoto Y. Katase A. Takada S. Akoh H. Nakagawa H. (1989) Characteristics of Nb-based Josephson junction at a temperature below 1 K. IEEE Trans Magn 25:1354-1357
6. Kroger H. Smith L.N. Jillie D.W. (1981) Selective niobium anodization process for fabricating Josephson tunnel junctions. Appl Phys Lett 39:280-282
7. Kraus H. Probst F. Freilizshn F. von Peterreins Th. (1988) Location selective irradiation of superconducting tunnel junctions with low energy x-rays. In: Borone A. (Ed) Superconductive Particle detectors. World Scientific. Singapore. pp 66-82

5.2 Energy Systems

Application of Superconductivity for Power Systems

Y. AIYAMA

Electrotechnical Laboratory, 1-4, Umezono 1-chome, Tsukuba, Ibaraki, 305 Japan

ABSTRACT

The application of the superconductor(SC) technologies to the electric power supply system is mainly described with aiming to the final goal to realize the totally SC-rized power system which will have the highest efficiency, power density, stability, availability and reliability. On a synchronous generator with SC field windings, the first target of the MITI/AIST/ML Project, some problems, specific to the applications for giant systems, are discussed.

KEY WORDS: superconductor(SC) application, electric power system, SC generator, critical current density, power system stability

INTRODUCTION

On behalf of the Project team of MITI/AIST(Min. of Internat. Trade & Industry/ Agency of Industrial Science & Technology), NEDO(New Energy & Industrial Technology Development Organization), Super-GM(Engineering Research Association for SC Generation Equipment & Materials), ETL and the others for "R&D of SC Technology for Application to Power Apparatuses", the author, who chairs the NEDO Project Committee, reviews the purpose and problems of the Project[1,2].

Among the power applications of superconductor(SC) technologies in various fields such as electric power supply system, industrial power, industrial process, transportation, chemical analysis, medical diagnostics, scientific research and so on, the ones for the electric power supply system are discussed. For the applications in this field, the following specific features of the system should be considered, those are peculiar to applications to giant systems with some analogy to the biological evolution.

a) That, at present, there have been already constructed the gigantic electric power supply systems in many countries of the world. This obliges us to abandon our plan to construct a new independent SC supply system as an unfeasible plan, because its construction necessitates a huge initial investment and the switchover to the new system will confuse a great many consumers and make their burden too heavy. We must begin with replacing an existing apparatus in the conventional system by an SC one. In this meaning, the word "superconductorization(SC-rization)" is used in this text.

b) That, in conventional power supply systems, the apparatuses and the system structures have been designed on condition of finite electrical resistivity of the conductors. For example, the higher voltages and the lower currents of the power transmission lines have been introduced for the less transmission losses. We can not introduce an SC transmission line of lower voltage as a part of the conventional high voltage system, simply because SC's prefer higher currents. The first SC-rized apparatus must have the same rated specifications as the conventional one in the system. In the stage when the SC apparatuses will be predominant in a sub-system, we will be able to adopt the design which is optimized as an SC apparatus.

c) That the electric power systems have their histories of 100 years and the apparatuses have been advanced almost to their technological limits. The energy conversion efficiency of a generator has reached to 98.7% and a power transformer, 99.7%. Even if their SC-rization could reduce these losses by half, the incentives for their SC-rization cannot be very high only with the loss reduction. The new functions or the advanced performances of the SC apparatuses in the system are desired and promissing.

These features may likely reduce the innovative nature of the SC technology but, considering that the past steady efforts have realized the present advanced electric power system, we can well expect the realization, by a steady way, of the future totally SC-rized power system, the supply system with the least loss and the highest stability, that is, with the lowest rate and the highest reliability.

PRESENT STATUS OF ELECTRIC POWER SYSTEM

Table 1 Energy Consumption in the World(1986)

Continent/Coutry	Population 10^6	Primary Energy Cons. Oil equiv. 10^6 ton	Primary Energy Cons. Per head kg	Electricity Cons* Elec.E. 10^9 kWh	Electricity Cons* Per head kWh
World total	4,917	6,525	1,327	9,962	2,026
Africa(Rep.S.Africa)	572	170	297	(122)	(3,202)
N.America (U.S.A.)	406	1,896	4,667	(2,583)	(10,756)
S.America (Brazil)	274	194	708	(202)	(1,454)
Asia (P.R.China)	2,880	1,388	482	(444)	(422)
Europe (F.R.Germany)	493	1,533	3,108	(406)	(6,690)
Soviet Union	281	1,258	4,472	1,599	5,684
Oceania (Australia)	25	87	3,472	(128)	(8,029)
Japan	121	308	2,538	672	5,533

*Figures in parentheses correspond to the data of the designated countries.

As shown in Table 1, in 1986, electricity of 10PWh(10^{13}kWh) was consumed in the world. Japan, who had 2.5% of the world population, consumed 6.7% of the total electricity, and, as obvious in this example, the regional differences of electricity use are very large. On the one hand, in North America, Europe and Japan, the gigantic electric power supply systems have been constructed and 5∿15MWh/year per head are consumed, but, on the other hand, the greater part of the Nepal people (population 17.1 millions) can not been graced with power utility (electricity consumption 400GWh/year; 23kWh/year per head).

An electric power supply system is a complex system composed of power generation, transmission, transformation, distribution apparatuses and other auxiliaries. In Japan, there are nine main electric power utility companies and the others, those supply 87% of total electricity. The nine companies have their own power systems respectively and cover their monopolistic supply areas of the whole of Japan except Okinawa, being provided with the interconnections to interchange electricity between companies.

In each system, the highest voltage lines(500 or 275kV; 1MV in plan) form its arterial lines and the lines of lower voltages cover its local and distribution networks. Around a few large cities (Tokyo et al.), 500kV loop lines surround the dense consumer districts for the stable power supply.

In 1987, the electric power plants of 163GW(figure only with the electric utilities) produced 640TWh electricity and the simply averaged utilization factor was 45%(nuclear 76%, thermal 41%, hydraulic 24%). The average total loss rate[(1-saled/generated electricity)x100] of the nine companies was 9.6%. Some details of this total loss can be estimated as shown in Fig.1, in which the turbine outputs correspond to 100. The parts which can be reduced by SC-rization of generators, transformers and transmission lines may be about 4%.

After the oil crises(1973 & 1979), the total end energy consumption in Japan had been almost unchanged with annual increasing rate of -0.1%(1973∿83) and the electric energy consumption had been increased with lower annual increasing rate of average 2.8% in the same period instead of over 10% before 1973. From around 1983, however, the energy consumption has begun to increase again with higher rates contrary to the every demand forecast made before by various organizations including the authorities. Especially from the spring of 1987, coincident with the activities for high critical temperature superconductors,

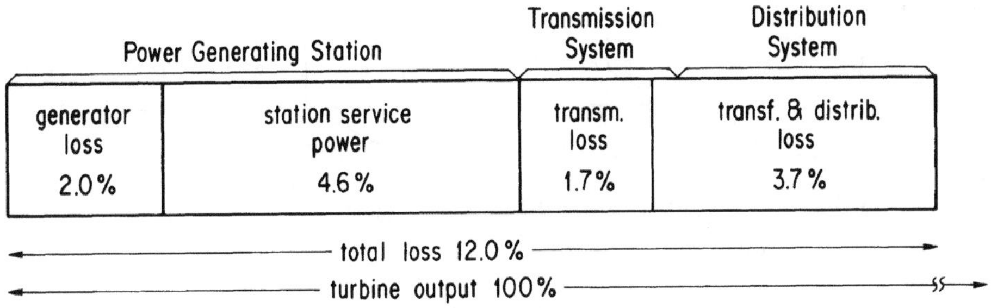

Fig.1 Details of Total Loss in Electric Power Supply System
(9 electric power companies in Japan; estimated by the author)

the annual increasing rate of energy consumption over 5% has continued. Moreover, the so-called "shift to electricity", the increment of the ratio of electricity to the total energy demand, has been going on constantly.

For these demands, the electric power companies must augment their supply capacities by constructing power stations and by expanding and improving their networks, those derive the difficulties of securing plant sites and transmission routes. Technically, the following problems are already present now and will be more serious multiplicatively in the future.

a) To improve the operating efficiencies and availabilities of the apparatuses in the systems. E.g.; decrease by one per cent of the total loss can eliminate a construction plan of an 1 GW power generating station.

b) To improve the stabilities of power systems with long-distance transmission lines. E.g.; increase by 40 per cent of the stability of a system can increase its transmission capacity by 40 per cent, because the capacity of a long-distance transmission line is determined not by its thermal threshold but by the system stability.

c) To make more compact, or to increase the power density of, the apparatuses in the systems. E.g.; the construction of a new culvert in the centre of a large city, where the existing culverts are full of power cables and have no space for new ones, will be prohibitively costly and a cable with higher power density is necessary to increase the transmission capacity of the circuit.

d) To enlarge the rated unit capacities of the apparatuses in the systems, which are limited by the mechanical conditions in their operation or by the weight restraints for their transports, to have the benefit of their scale effects in their operational and voluminal efficiencies.

e) To ensure the voltage stabilities at receiving ends to guarantee the supply of high-quality electric power.

f) To provide energy storage devices to meet the decreasing utilization factors of the system apparatuses. For this measure, the hydroelectric power stations of pumped storage type are now provided but henceforth they can hardly find out their adequate sites in Japan.

There are some measures to solve these problems individually even in the conventional technology scheme. Owing to the mutually contradicting natures of these problems, however, the system has been faced with the difficulty to settle the problems totally. E.g., the forced-cooled conductor windings in a synchronous generator can increase its output coefficient and make its rated large capacity possible but, on the other hand, this strengthens its copper machine character, i.e. the share of its electric loading becomes much larger than that of its magnetic loading and this deteriorates the system stability, necessitating the other stabilizing auxiliaries in the system. Moreover, the forced-cooling is a measure against energy conservation. The SC technology can be considered to be very potent exclusively to settle the problems wholly.

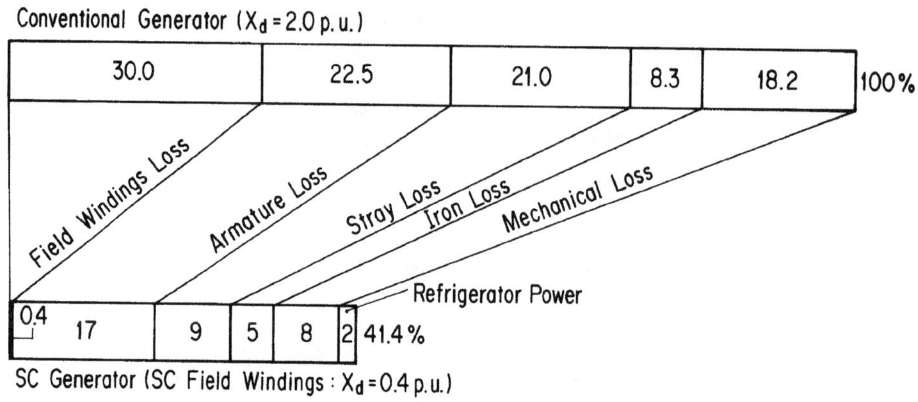

Fig.2 Comparison of Loss for SC vs. Conventional Generator: 2-poles; 1,200MVA

SC APPLICATIONS FOR ELECTRIC POWER APPARATUSES

A natural concept of SC application to the power system apparatuses is to replace the conductors by SC ones to eliminate or reduce their copper(Ohmic) losses. The effect will be larger for longer conductors, e.g. ones for coil windings or transmission lines. A example is shown in Fig.2.

A problem of copper conductors is their low current density due to the finite resistivity(<8A/mm^2, even with good cooling) and, for generating magnetic flux, usually they make use of iron-cores. Superconductors, having higher critical current density(J_c=$10^2\sim10^3$A/mm^2) and higher critical fields, can produce magnetic flux density far higher than the saturation magnetization B_s of ferromagnetic materials, without cores. It is to be noted that their high J_c is the indispensable condition for this potential and that a certain stabilization method of the conductors is necessary corresponding to the "critical" nature of their characteristics. The lower current density results in a so large winding volume as to be impracticable and the unstabilized conductors result in a premature "quenching" of the windings. To develop a generator with SC field windings, it is now critical whether the SC coils up to 25MJ can be made with their high current density($J_{overall}$=$1\sim2\times10^2$A/mm^2) as well as their sufficient stability in the high-speed rotation.

The problem of SC's, in addition to their J_c's and their stability, is their loss caused by the magnetic flux change in them. By the progress till now in SC technologies, the AC loss of the SC of fine-multi-filamentary twisted structure can be reduced to the sufficient extent to be operated in several to several tens Hz. Much R&D efforts, however, is necessary to have the SC's for the power frequency use(50,60Hz). This status restricts technically the SC applications in the near future to the ones to be used in DC mode in principle; SC field windings of the synchronous generator, the DC power transmission cable, the small SMES(<1MWh), etc..

After the three years studies(1985∿8) on the technical and economical feasibility of generators with SC field windings, the Project has been started in FY1988 in our full conviction, which will establish all of the component technologies necessary for the 200MW pilot generator, the technological demonstration plant scheduled in the Project Phase II(1996∿).

The technical feasibility of SC/AC apparatuses can not be foreseen definitely now due to the uncertainty of AC SC's and so the scheduled R&D programme for, e.g., power transformers can not be planned. The R&D of AC SC's and oxide SC's are included in the Project without a concrete target. On the other hand, the technical possibility of SC cables has been demonstrated by several research groups but the economical feasibility of these can be foreseen only with the capacity over several GVA (even with the 80K SC, the break-even point may be 3GVA). These scales are far beyond those of conventional cables and more efforts will be necessary to reduce the SC cable cost. The feasibility study of these power apparatuses including SMES is now under way by ISTEC under contract with ANRE(Agency of Natural Resources & Energy) of MITI.

SC GENERATOR IN ELECTRIC POWER SYSTEM

The effects of SC-rization of the field windings of a generator to the generator itself and to the electric power system are as follows:

a) To reduce the generating loss by about 60% resulted from disappearance of field winding loss and reduction of mechanical and stray losses due to its size reduction as shown in Fig.2.. This will reduce its annual operating expense by 600M¥/year in an 1GW class generator.

b) To reduce the manufacturing cost by about 10% resulted from its size and weight reduction(40%) even with costly SC conductors. The complex structure of an SC rotor, however, has possibility to cancel out the above cost reduction and, in the case of quick response excitation type, the high cost of the exciter will certainly overcome it.

c) To increase the stability of long-distance transmission lines connected to the generator, which has a much lower synchronous reactance ∿0.3 p.u. compared to 1.5 of a conventional one, resulting from the increase its magnetic loading. This stability increase will save the transmission facilities by 10 to 60% and this effect will be larger economically than the one by the loss reduction a). Its low reactance will also serve to increase voltage stability at receiving ends.

d) To increase the fault current of the system owing to its low impedance, necessitating some current-limiting devices to protect the system components. SC current limiters are now under study.

The R&D problems for SC generators are as follows and now vigorous efforts are being concentrated on them by the Project team.

i) Multi-layered cylindrical rotors with enough mechanical strength against the strong centrifugal and electro-magnetic forces and with enough thermal insulation compatible with torque transmission. They must be free of the temperature differences between its R.T. outer and LHe temp. inner parts. The materials tests and the model tests for field winding support, torque tube, dampers, LHe transfer coupling, etc. are now going on.

j) High current density SC windings which can be operated stably on the condition of high-speed rotation and electric and mechanical disturbances. The excitation control should be adopted to them and two types of SC generators, slow & quick response excitation type, are planned. The SC conductors for them must be of high current density, be of low AC losses, be stable in the operating conditions and not quench in ordinary system failure conditions.

k) Air-gap armature windings with sufficient strength and low AC loss.

l) A highly reliable He liquefier and cooling system for SC windings in the rotor which can be operated continuously for 10,000 hours or more with the least maintenance works. Two R&D items of enhancing the reliability of liquefier system, based on conventional & advanced system, are now going on.

m) The load-test of a new large generator can not be made without connecting it to existing power system, for lack of a proper load. Being connected with the system in active service, however, such a failure test can not be made as to affect the power supply of that system. A new conventional generator used to be tested by the test standards, which will be insufficient for a new type of generator. The tests of the model machines, equivalent to a 70MW output, are planned to be made independently by G-M method, in which the loading synchronous motor will be directly coupled to the driving motor with varying phase difference angles(load angles), and to include the tests by severities such as sudden short circuit.

This situation is common to the apparatuses of giant systems. The computer simulation experiments will be used as much as possible to cover the deficiencies due to the hardware constraints.

ELECTRIC POWER APPARATUSES WITH HIGH T_C OXIDE SUPERCONDUCTOR

In the case, when oxide SC's in long conductor form have their SC characteristics (J_C, AC loss etc.) at 80K & 5T comparable to those of NbTi SC at 4K & 5T and can be applicable to the field windings of a generator, the following merits superior to the generator at LHe temperature are foreseen:

u) The thermal insulation structure of its multi-layered rotor can be simplified through eliminating its radiation shield, resulting in its size and weight reduction which may be 5% or less.

v) The necessary power for its coolant liquefier can be reduced to a twentieth or less, resulting in its efficiency increase by 0.03% or less.

w) With the simplified rotor structure and the simplified liquefier system, it will be easier to make the generator system reliable, that is very important for the apparatuses in electric power system.

It is to be noted that, the high current density in the windings(150A/mm^2) is essential to these merits. E.g., if it is half(75A/mm^2), the rotor will be 30∿40% heavier and its efficiency lower. That is, the merits of SC-rizing will be almost cancelled out.

The fact that the merits of higher operating temperature in SC generators are rather small comes from that the burden of the generator for the liquefier is not so large owing to its thickset shape. In an SC transmission cable, of which the burden for the liquefier is larger owing to its slim or slender shape, the greater merits of higher operating temperature can be expected. As already mentioned, however, the feasibility of SC cables can be foreseen only above too large scale.

The author recommends that oxide SC's should be operated at 20∿30K from the view points of their high and stable critical current and the stabilization method of the conductors.

CONCLUDING REMARKS

Only for an SC generator with SC field windings, the feasibility can be foreseen at present and the Project team is making their strenuous efforts to establish its feasibility. The R&D's of AC/SC, high H_C/SC and high T_C/SC are expected to make feasible other SC electric apparatuses or to lighten the constraints for their feasibility.

To enlarge their feasible regions, it is very effective to SC-rize two kinds of neighbouring apparatuses in the system. In such arrangement, the current leads connecting the two can be kept at low temperature throughout their whole length without an intermediate warm portion which causes large heat leak and also the common use of a coolant liquefier can be possible. These are expected to reduce the system cost and to increase the efficiency.

This concept leads us developmentally to the idea of totally SC-rized electric power system which will have the highest efficiency, power density, stability, availability and reliability, provided that the system and apparatus designs are optimized as SC ones. Remarking in addition, in this system it is necessary that the auxiliaries such as switches, converters, current limiters, etc. are SC-rized or are operated at low temperatures.

REFERENCES

1. Uyeda K. (1988) Current situation of R&D on superconductive power application technologies carried out by Super-GM. Proc ISS'88

2. Tanaka T. et al (1989) Current situation of R&D on superconducting generators carried out by Super-GM. Proc MT11

Power System Control Experiments Using 1MJ SMES

HIDEKI FUJITA

Chubu Electric Power Co., Inc., Electric Power Research & Development Center, 20-1, Kitasekiyama, Ohdaka-cho, Midori-ku, Nagoya, 459 Japan

ABSTRACT

Superconducting Magnetic Energy Storage (SMES) system is considered to be useful in electrical power system because of high efficiency and quick response. This is an attractive technology for load leveling and system stabilization in electrical utility. Chubu Electric Power Co., Inc. and Hitachi Ltd. have been studying SMES since 1988 and developed 1MJ SMES composed of pulsed superconducting magnet (1000A, 2H). We tested its basic characteristics in this August and power system control effect in this September. In this paper, I report on this system and several results of power system control experiments connecting this system to the simulated power system which contains generators, transmission lines and fluctuating load.

KEY WORDS: SMES, NbTi, GTO-thyristor, power system, fluctuating load

INTRODUCTION

Large quantity storage of electric power is difficult. It can only done by electric batterry cell or capacitors, which are too small for improving utility demand curve. Most popular method of large quantity energy storage, which is really used, is pumping-up power plant generator only. But efficiency of energy storage is 65-75 % or so. By this reason, concerning power system operation, it is desired to develop higher efficiency storage system. The reason of low efficiency of pumping-up power plant is that it transfers electric energy to the another type of energy. (namely the head of water level) And there are many needs in future power system shown in Fig.1. It is said that superconducting magnetic energy storage (SMES) system will be a new and flexible device for operation of power systems with load leveling capability, dynamic operation advantages, high efficiency, and high reliability. The rapid, response time for changes in magnitude and direction of energy transfers will make an SMES unit extremely valuable in improving the dynamic system operation. SMES can provide load following with instantaneous response to sudden load changes or loss of generation; damping of low frequency, sub-synchronous or transient oscillations; and permit more economical operation of system generation. It is a new power system component

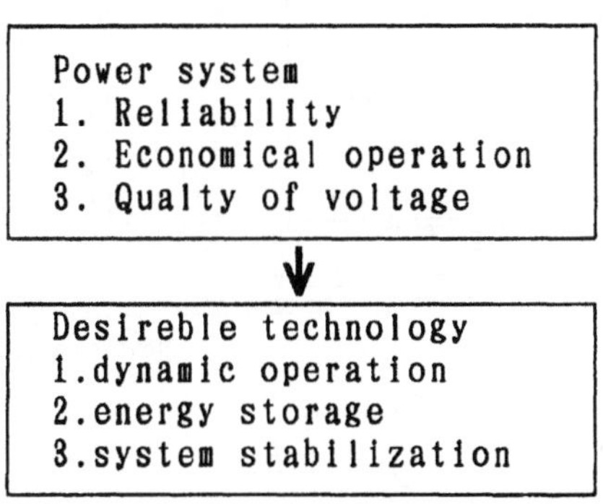

Fig.1 Needs to power system

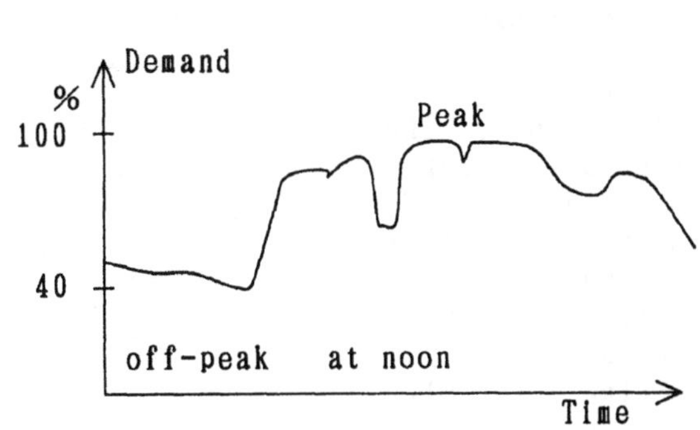

Fig.2 Typical one day demand curve

with great potential as an operational tool. Chubu Electric Power Co., Inc. and Hitachi Ltd. have been studying SMES since 1988. We studied not only by paper work but also by manufacturing little size but real SMES system. And in order to understand the SMES effects from utility side, we planned various kinds of system stabilization tests. They are currently underway using simulated power transmission systems with fluctuating load.

PRINCIPLE OF ENERGY STORAGE

Fig.3 shows three conditions. First stage is charging stage. In this stage, SW1 is opened and SW2 is closed. DC-output voltage of Power Conversion System (P.C.S), equal to the magnet-voltage, is plus to earth. In this condition, the current that flows utility power source to magnet become larger and larger. Namely, in this condition, stored energy in magnet is growing. Next stage is energy holding stage. In this stage, SW1 is closed and SW2 is opened. Magnet-voltage is zero. In this condition, the current that flows in the magnet flows through the same loop circuit without power dissipation by its superconductivity. The third stage is discharging stage. In this stage, SW1 is opened and SW2 is closed as the First stage. But, in this case, DC-output voltage of P.C.S. is minus to earth. In this condition, the current that flows in the magnet decreases. Namely this is discharging stage. Active power that was stored in the magnet is poured to the utility source from SMES. When we consider the usage of SMES in power system, efficiency and rapidness between charging and discharging stage are two important items. When we use power system stabilization effect, only First stage and Third stage are used. Second stage is used only by load leveling by permanent current mode.

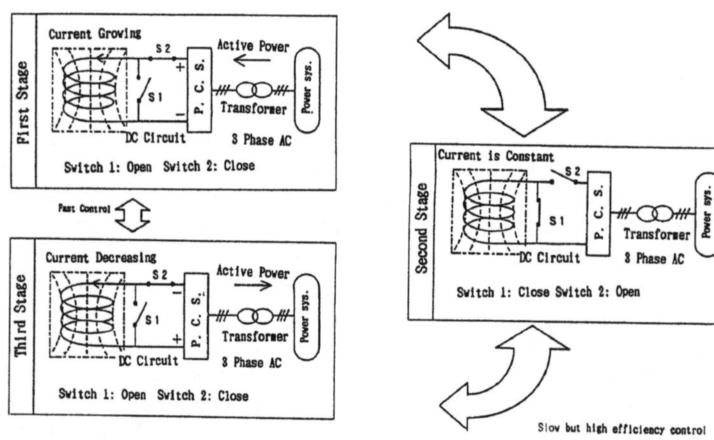

Fig.3 The principle of energy storage

SPECIFICATION OF SMES

This system consists of a superconducting coil, cryostat, GTO thyristor converter, main circuit system and control system. Fig.4 shows the circuit of SMES. The superconducting magnet is solenoid type. The conductor consists of a 7 μm super fine NbTi wire combined with a copper supporting material, and it assures stability to pulse application.

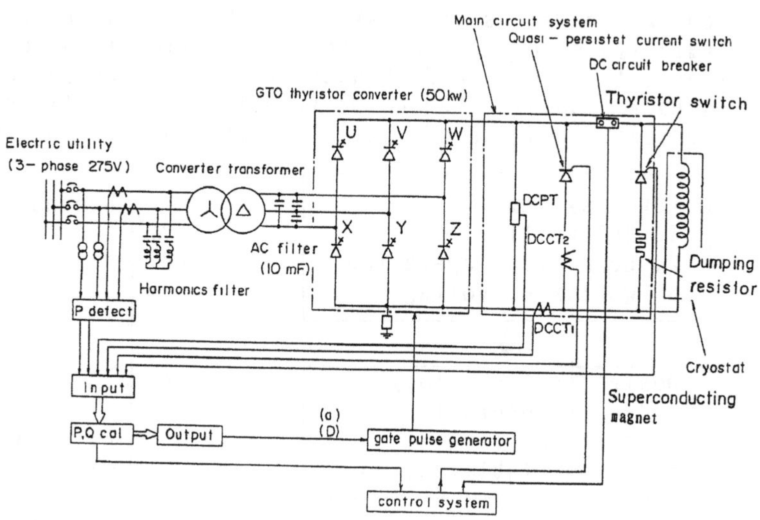

Fig.4 The circuit of SMES

Fig.5 shows cross-sectional view of the superconductor. The coil inductance is 2H for 1MJ energy storage at a rated current of 1,000A. 32 pieces of doble pancake type coils designed with 2880 total turns are used. Fig.6 to Fig.9 show constructed SMES system appearance. Table 1 shows conductor parameters and table 2 shows the specification of SMES. The cryostat is made with a non-magnetic material SUS304, which uses vacuum heat insulating layers and using a liquid N2 container , and the coil is dipped into liquid He at 4.2 degree temperature. The 1000A DC-current connectors are located on the top of the cryostat to minimize heat transfer from outside. The GTO converter consists of a single bridge circuit of rated voltage 50V, rated current 1,000A combined with a quasi-persistent current switch, a DC-circuit breaker and a dumping resistor at the DC-output side. The control system consists of a 32 bit high speed CPU and an AD/DA converter, and uses a universal system which allows programming at site to comply with various tests.

Table 1 Conductor parameters

Cable height	8.45mm
Cable width	2.1mm
Number of strands	13
Strand diameter	1.18mm
Number of filaments	78000
Filaments diameters	7μm
Cu/CuNi/NbTi ratio	3.6/1.3/1.0
RRR of stabilizer copper	80
Cable twist pitch	94
Critical current in cable	4550A at 5T

Table 2 Specifications

Magnet	Type	solenoid
	Rated current	1,000A
	Inductance	2.0H
	Storage energy	1 MJ
	material	NbTi
Cryostat	Height	2.00 m
	Outer radius	1.00 m
GTO thyristor	Rated current	1,000A
	Rated voltage	50 V
	Rectify phases	6 phases
	Dumping resistor	0.25 ohm
Control system	32 bit cpu	
	Output interval	1 ms

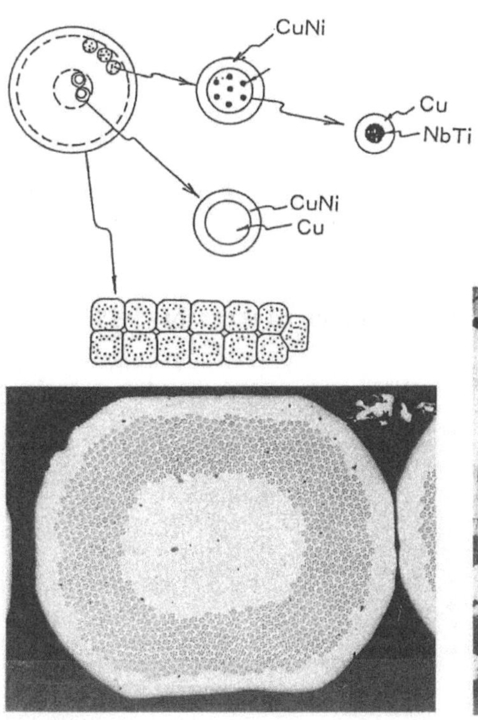

Fig.5 Superconductor

Fig.6 Completed coil

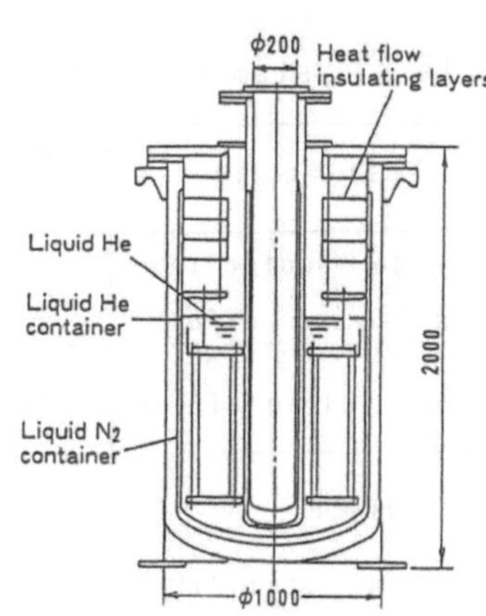

Fig.7 Cryostat

CONTROL FUNCTION

Fig.10 shows output/input P-Q control circle-area of P.C.S. Radius W0 is determined by stored energy, namely DC-current flowing in magnet. We can control P and Q independently by different target signal. As shown in Fig.11, this SMES can control P by three method, namely time-scheduled ,line-P feed back and voltage-frequency feed back and Q only by voltage-magnitude. Table 3 shows samples of system control experiments.

Fig.8 GTO Thyrister converter

Fig.9 Control system

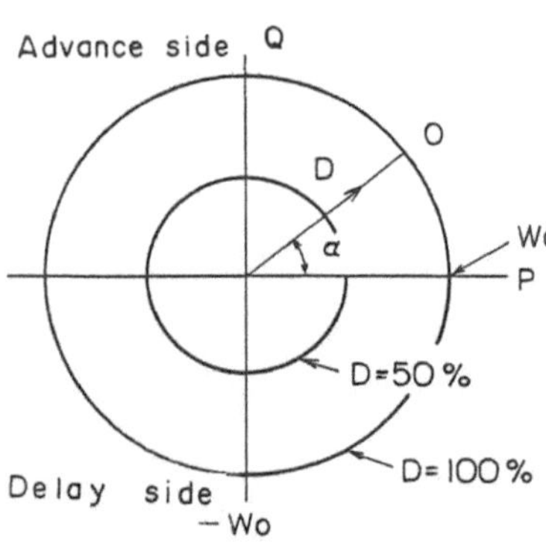

Fig.10 P-Q control area

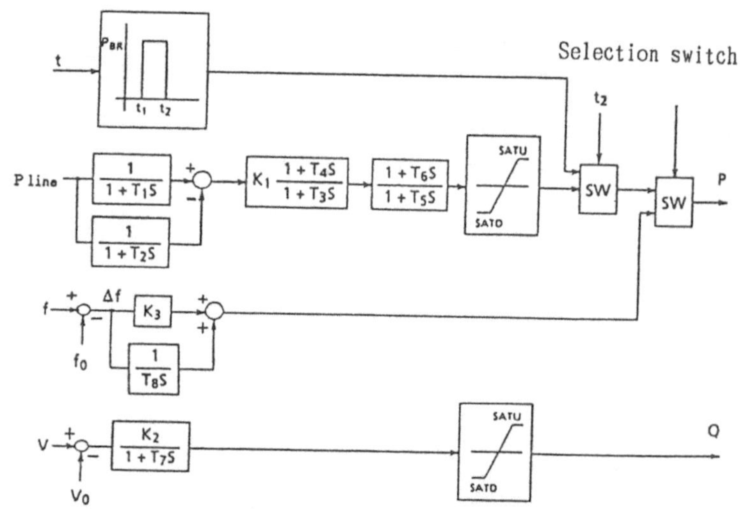

Fig.11 Block diagram of SMES

Table 3 System control experiments

Test name	SMES control signal	SMES output
Voltage fluctuation stabilizing effect	ΔV	Q
Load leveling effect	ΔV, ΔP	P, Q
AFC effect	Δf	P
Transient stability improving effect	ΔV, ΔP	P, Q

Fig.12 Simulated generators

POWER SYSTEM CONTROL EXPERIMENTS AND RESULT

This SMES has been connected to the simulated transmission systems, which is a miniture size power system model composed by real generators (6kVA ×2, 4.42kVA×1, 4kVA×1; Fig12) and transformer and transmission line models used real coil, which are simulated 150 ~300km, 275kV transmission lines. And they are operated at 1/1000 of actual electric utility voltage (1/100 for current). Fig.13 shows the change of operation mode to the condition of P-Q decoupling test. In order to operate SMES for P-Q control preparing mode, We must store energy in magnet at first step. This mode is called excitation mode.

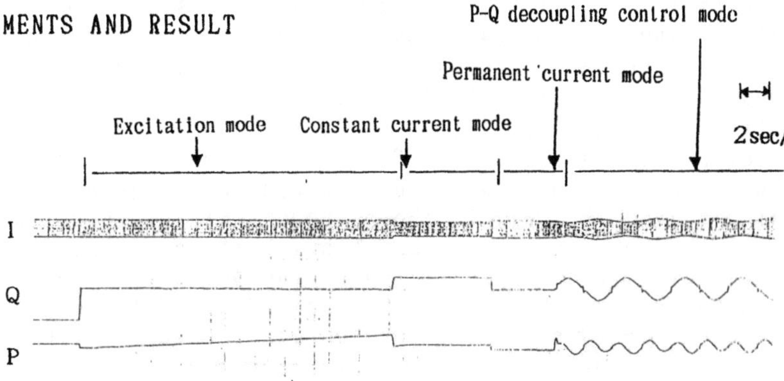

Fig.13 P-Q decoupling confirming test

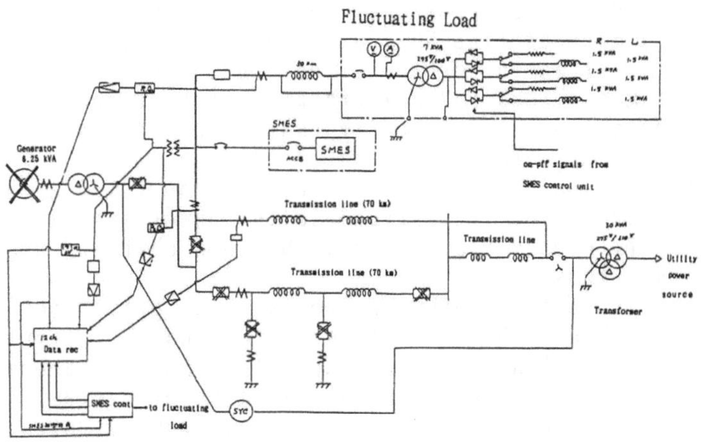

Fig.14 The circuit of test

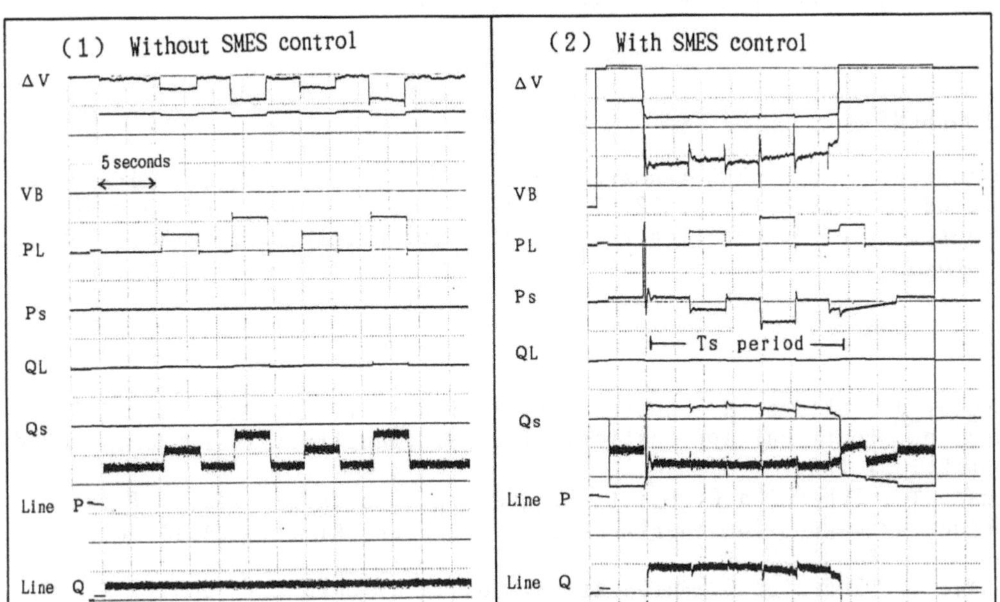

Fig.15 The results of load fluctuation stabilizing effect

In this mode, constant DC-output voltage is charged on magnet, and the current grows larger. When the current becomes target value, DC-output voltage becomes nearly zero and the current is controlled to the constant value. This is "constant current mode". Then quasi-persistent current switch acts. (This is "persistent current mode".) And then, it becomes P-Q control mode. From Fig.13, we can understand the decoupling characteristic between P and Q, where the frequency of target waveform of Q is twice that of P. Fig.14 shows the circuit of load fluctuation test. Fig.15 shows the measured waveforms of bus voltage and active power that flows in transmission line. this Fig shows the period,Ts, which is the period when SMES is under P-Q control mode. By the effect of the P-Q control of SMES, there are clear differences between the case of (a) and (b).

CONCLUSION

1. Chubu Electric Power Co., Inc. and Hitachi Ltd. have manufactured 1MJ SMES using 7 μm super fine NbTi wire combined with a copper supporting material.
2. For confirmation of the effect in the electric utility and also for extracting problems for commercialization of such a system, this system is connected to the simmulated transmission lines and some experiments of power system control were carried out.
3. SMES for power system use would be consisted a large superconductive energy storage coil interfaced to the power system through a reversible AC/DC P.C.S. Efficiency and rapidness between charging and discharging are two important items.
4. The manufactured SMES has not so high efficiency as expected for energy-storage. This is because of laboratory use. The ratio of thyristor loss is large compare with its rated capacity. But rapidness response of SMES was confirmed to be enough to stabilizing power system.

We are now prepareing the next tests of power system control effects which is more detailed than the first one.

We confirmed SMES technology using low-temperature superconductors. But if the high-temperature superconductors which are now studying become useful, the SMES which we are researching becomes more realistic equipments in power system.

ACKNOWLEDGEMENT

The development of our SMES research is due to the past results which have been studied in universities, CRIEP and other laboratories. And this report is made as a result of cooperation reserch of Chubu Electric Power Co., Inc. and Hitachi Ltd.

REFERENCES

1. R.W Boom, J.J.Skiles, R.L.Kustom. SMES for utility applications, EPRI 1988 Conference on Electrical Applications of Superconductivity, Orlando, Florida, September 21-23, 1988.
2. Tominaga T., Takashiba O, Fujita H, Asano K. Design and tests of the superconducting magnet for energy storage, Tsukuba, MT-11, August 28 to September 1, 1989.
3. Tominaga T, Takashiba O, Shibazaki M, Fujita H, Goto M. Tsuchiya K. Hasegawa H, Ishigaki Y, Toita H. Power control experiments using a PWM GTO thyristor. San Diego, IEEE/IAS,October 1-5, 1989.
4. Takashiba O, Fujita H, Shibazaki M, Goto M, Ishigaki Y, Toita H. System control with SMES, Fukuoka, IEEJ, July 19-20, 1989.

Development of Superconducting Linear Induction Motor

O. TSUKAMOTO[1], Y. TANAKA[2], K. OISHI[2], T. KATAOKA[3], Y. YONEYAMA[3], T. TAKAO[1], and S. TORII[1]

[1] Faculty of Engineering, Yokohama National University, Yokohama, Japan
[2] Furukawa Electric Co., Yokohama, Japan
[3] Nippon Steel Co., Kitakyushu, Japan

ABSTRACT

We have developed superconducting linear induction motor using ultra fine filament NbTi AC superconductor. The superconducting primary windings, three phase and three coils, are 222mm long and 90mm wide and put in liquid helium. The secondary conductor is a metal plate placed in the room temperature region. Generally, AC superconductors are highly unstable and easily quenched by a frictional heating caused by a wire motion of only a few μm. To prevent the windings from premature quench caused by conductor motions, they are epoxy-impregnated. The SLIM was tested by applying 3-phase 50Hz AC current and the maximum input was 5.5kVA at 360Arms. Measured thrust was 5.3gf at 310Arms with a copper plate as the secondary conductor.

KEY WORDS: superconducting linear induction motor, AC superconducting wire, epoxy-impregnation

INTRODUCTION

By the development of NbTi ultra-fine filamentary wires for 50Hz/60Hz use, new aspects of superconducting applications have been created, and a superconducting linear motor (SLIM) has become feasible. It is difficult for a conventional linear motor with iron cores and copper-wire windings to obtain a large thrust in a wide gap, because of flux saturation of the iron cores. Windings of very high ampere tunes can be realized by using superconducting wires and the SLIM with the superconducting primary windings can easily produce a large thrust in a wide gap. Therefore, by using a SLIM, a linear electromagnetic drive system can be introduced and give a great impact to many fields where it has been considered very difficult for a conventional system to be applied. A typical example of those fields is a steel making process.[1] A SLIM can be applied to a tension controller of steel strips, a stirrer of melting steel, a controller for winding motions of steel strips, etc., and is expected to drastically improve productivity and quality of the products.

We are developing a SLIM for steel making processes. The most important technology to realize SLIM is the stabilization of AC superconducting windings. The AC ultra-fine filamentary wires use highly resistive metals, usually cupro-nickel, as the matrices to reduce losses caused by the coupling of the NbTi filaments. Therefore, AC superconducting wires are highly unstable because of high Joule heating dissipation at the normal zone and easily quenched by a frictional heating due to an abrupt wire motion of only a few μm. The AC superconducting wires are also twisted in a very short pitch, typically 1mm, to reduce the coupling losses, and the diameter of the wire is less than about 0.2mm, that is, the current capacity of the wire is around ten amperes. In a practical SLIM, a current capacity of several hundreds amperes is required to the conductor, therefore, tens of AC superconducting strand wire are bundled. For the SLIM winding to be stable, the conductor should be fixed so tightly that any strand in the windings does not move. It is well known that epoxy-impregnation is an effective technique to fix wires in a DC superconducting coil, but, for a AC winding, the epoxy-impregnation has been considered to be impractical, because AC losses in the winding is hard to be cooled due to the low thermal conductivity of the epoxy. However, we have demonstrated that AC losses can be cooled by thermal conduction through the epoxy, provided that thickness of the impregnated winding is less than 1mm and that AC winding can be effectively stabilized.[2][3] Applying this epoxy-impregnation technique to the SLIM windings, we have developed 5kVA class SLIM.

CONFIGURATION OF SLIM

A schematic diagram and a photograph of the 5kVA class SLIM is shown in Fig.1. The cross section of the AC superconducting cable for the SLIM is shown in Fig.2. The first stage cable is made by bundling 7 of AC superconducting strands and the final cable is made by bundling 7 of the first stage cables. The specifications of the strand are shown in the table of Fig.2. NbTi filaments are embedded in the CuNi matrix and, to improve the stability of the strand, a copper section segmented by CuNi barriers, is placed at the core of the strand.
The primary windings consist of 3 saddle-shape coils, coil Ⅰ, Ⅱ and Ⅲ, as are shown in Fig.1. In this project, we considered it most important to demonstrate the feasibility of a SLIM as a hardware, and we designed the windings to ensure the stability and for the liquid helium to easily circulate around the winding layers, so that the three coils have different sizes and number of turns, as is shown in Fig.1. GFRP (glass fiber reinforced plastic) plates with teeth and slots are placed on the coil frame and the superconducting cables are placed in the slots. After the cables were wound, the whole coil was vacuum-impregnated with epoxy.

EXPERIMENTS

The experimental arrangement is shown in Fig.3. The superconducting primary windings were put in the inner glass dewar containing liquid helium. The outer dewar containing liquid nitrogen was for thermal shield. Three kinds of 1mm thick reaction plate, aluminum, copper and stainless steel (SUS304), were tested. The gap between the reaction plate placed in the room temperature region and the surfaces of the primary windings was about 90mm.
3 Phase/50Hz AC currents were supplied to the SLIM. The voltage of the power supply was adjusted by a three-phase auto-transformer. The current vs. voltage lines of the coils are shown in Fig.4. Due to the impedance unbalance of the coils, the coils have different excitation characteristics. Quenches occurred around 360Arms of the coil Ⅱ current. From observed transient voltages across the coils (Fig.5), a quench was always initiated by the coil Ⅱ which had the highest current and about 400ms later, a quench was induced in the coil Ⅲ, but no quench occurred in the coil Ⅰ. The quench current was lower than the value calculated from the short-sample critical current of the strand. We consider that, in the windings, there was a part where the strands were not fixed tightly and that the premature quenches were caused by strand motions. The input to the SLIM reached a maximum value 5.47kVA at the quench. Thrust vs. coil Ⅱ current is shown in Fig.6(a) for different reaction plate materials and in Fig.6(b) for a copper reaction plate with gaps of 90mm and 94mm. Measured values of the thrust were small, because the gap between the reaction plate and winding surfaces was large compared to the size of the SLIM due to the thick walls of the glass dewars.

CONCLUSION

It has been demonstrated that a SLIM was feasible as a hardware. The developed 5kVA SLIM had a rather small thrust. To increase the thrust, it is necessary,

i) to decrease the gap between the reaction plate and the superconducting primary winding by making the walls of the cryostats thinner,

ii) to increase the size and ampere turns of the primary windings.

In the next step of this project, we are planning to develope 30kVA SLIM which will have higher thrust and have specially designed cryostat with thinner wall. We are also continuing the effort to develop the technique to increase the stability of large-scale AC superconducting windings.

REFERENCE

1. T. Kataoka, T. Nasada and Y. Yoneyama: " Applicaions of the Electromagnetic Force to Steel Making Processes", Applied Electromagnetics in Materials, p.197, October (1988)
2. O. Tsukamoto, H. Kobayashi and S. Akita: " Stability of Epoxy-Impregnated AC Superconducting Winding", IEEE Trans. on Magn., p.1170-1173, Vol.24, No.2, March (1988)
3. O. Tsukamoto, M. Yamamoto, T. Ishigohka, Y. Tanaka, T. Takao and S. Torii: " Development of Large-Current Capacity Epoxy-Inpregnated 50Hz Superconducting Coil", p.724-728, Proc. of the ICEC 12 (1988)

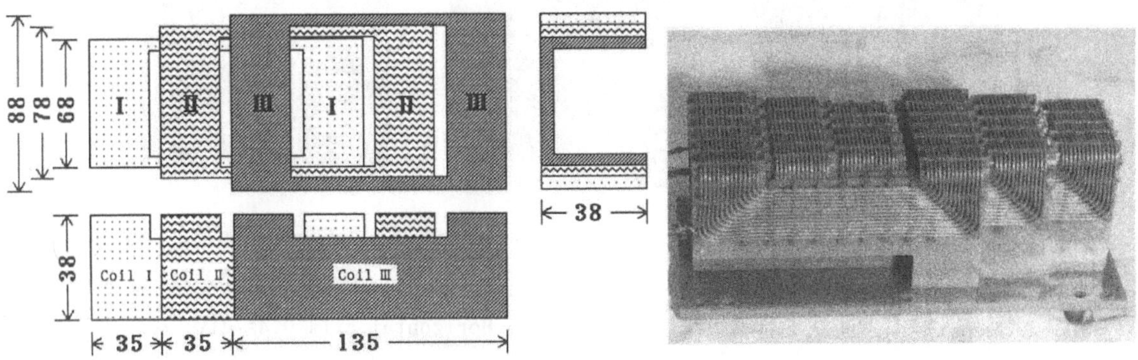

a) Dimensions of Coil

b) Photograph of primary windings

Fig.1 Configuration of 5kVA class SLIM
(Coil I, II 28turns, Coil III 26turns)

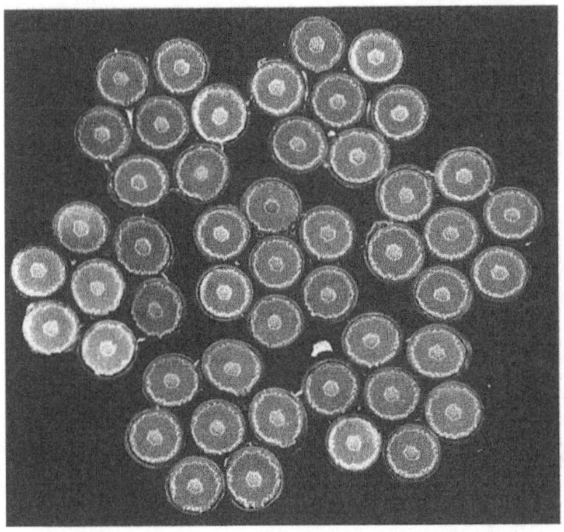

a) Cross-section of cable

Diameter of the strand	**ϕ 0.12mm(Bare)** **ϕ 0.14mm(With insulation)**
Cu / CuNi / NbTi	**0.23 / 2.53 / 1**
Twist pitch	**3.5mm**
Filament diameter	**0.5μm**
Number of filament	**15,300**
Critical current	**16.9A (at 1T)**

b) Specifications of strand

Fig.2 Cross-section of AC superconducting cable and specifications of strand

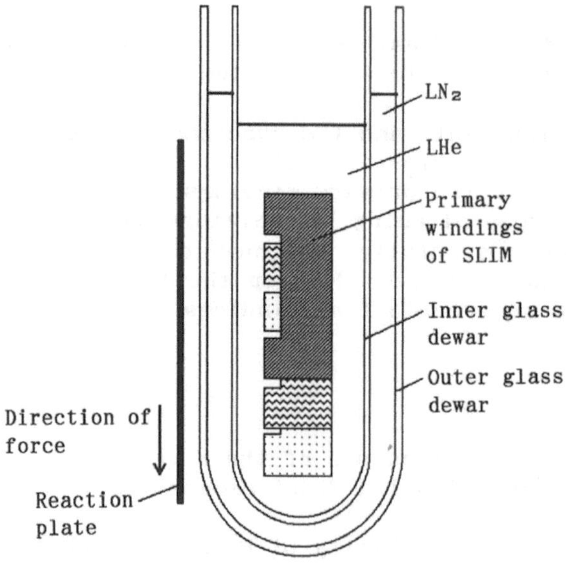

Fig.3 Experimental arrangement of 5kVA SLIM

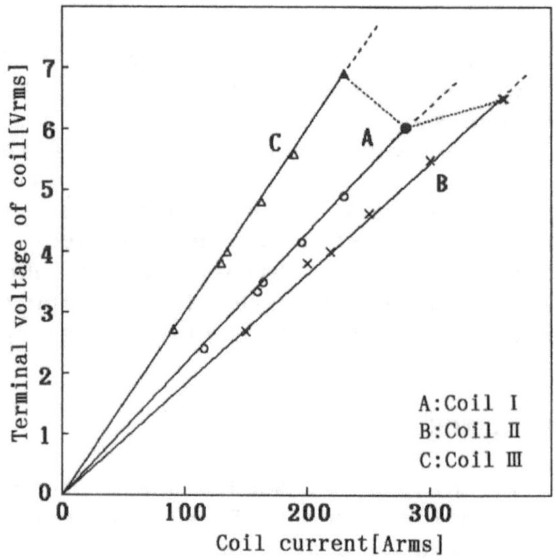

Fig.4 Current vs. Voltage lines of coils

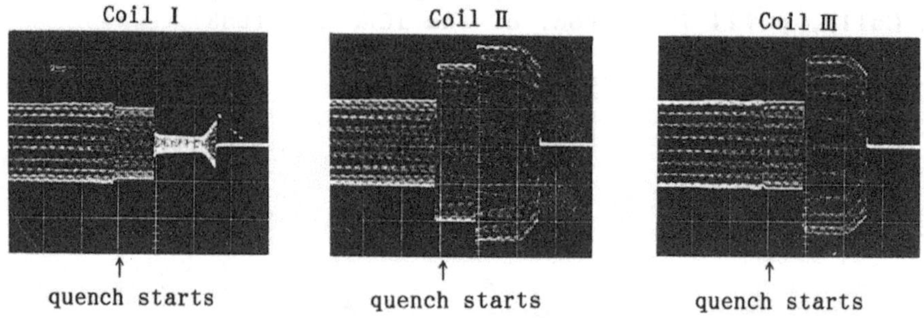

Fig.5 Transient voltage signals of coils at quench
(Vertical axis 8V/div, Horizontal axis 0.4s/div)

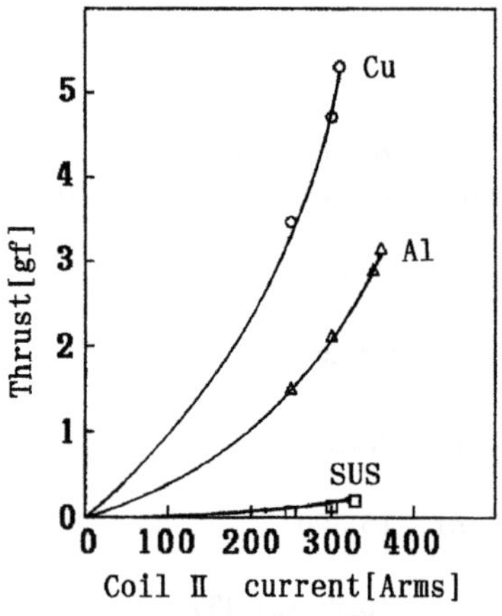

a) Various reaction plate materials(gap=90mm)

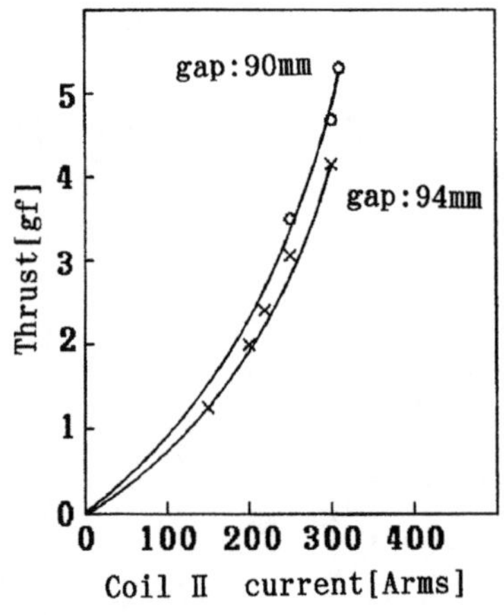

b) Different gaps for copper reaction plate

Fig.6 Coil current vs. Thrust curves
(Thicknesses of the reaction plate are 1mm)

High-T_c Oxide Superconducting Magnet with an Iron Core

T. USHIJIMA, S. YOKOYAMA, K. SHIMOHATA, and T. YAMADA

Central Research Laboratory, Mitsubishi Electric Corp., 1-1, Tsukaguchi-Honmachi 8-chome, Amagasaki, Hyogo, 661 Japan

ABSTRACT

High-Tc oxide superconducting magnet with an iron core, which has an air gap of 1mm, was developed. The magnet has an iron core, an induction coil and 11 pieces of $YBa_2Cu_3O_{7-y}$ ring (OD25mm, ID18mm, t5–7mm). The rings were sintered by 970 °C–10 hours in O_2 gas. The critical current measured by induction method was about 30 A at 77 K, B=0 T. The sum of the critical current of each ring was 310 A. The persistent currents of $YBa_2Cu_3O_{7-y}$ rings were induced by induction coil. The magnetic field of 2 kGauss was generated in the air gap. The magnetomotive force of the rings estimated by the magnetic field was 200 A. The leakage magnetic field on inner periphery of the ring calculated by two dimensional analysis was 25 Gauss. The sum of the critical current of each ring measured by induction method was 200 A under the external magnetic field of 25 Gauss. The decay of the magnetic field was measured and the flux creep was observed. The magnetic field generated by the currents of 56%Ic scarcely decayed.

KEY WORDS: superconducting magnet, $YBa_2Cu_3O_{7-y}$ ring, flux creep, two dimensional analysis

INTRODUCTION

Since the discovery of high Tc oxide superconductors, a great deal of activity has been made to apply these oxides to various technical applications.[1][2][3] We also have been engaged in research on power applications of superconductors. [4][5]

A superconducting magnet is able to generate high and stable magnetic field. High Tc oxide superconducting magnets are possible to be used in liquid nitrogen. With the present techniques, the critical current densities of bulk high Tc oxide superconductors are lower than those of metal superconductors under magnetic fields. It is possible to decrease magnetic fields, which are applied to superconducting coils, by the adoption of high permeability materials.

In this paper, we report some experimental results and a magnetic field analysis about the $YBa_2Cu_3O_{7-y}$ superconducting magnet with an iron core. One of the advantages of the magnet is that the magnetic fields can be changed.

EXPERIMENTAL

Figure 1 shows the apparatus of the superconducting magnet. The magnet has an iron core with an air gap of 1mm, a copper coil, and $YBa_2Cu_3O_{7-y}$ rings. A material of the iron core is the pure iron. Magnetic fields of the air gap were measured by a hall sensor. The magnetic fields generated by the Cu coil, which was located in the ring region without $YBa_2Cu_3O_{7-y}$ rings, were shown in Figure 2. The magnetic field of 2 kGauss was generated as the magnetomotive force of the Cu coil was 200 A. The residual magnetization was 80 Gauss.

The rings were made by shaping $YBa_2Cu_3O_{7-y}$ powder with a press (1 ton/cm^2) into rings (OD25mm, ID18mm, t5–7mm). The sintering condition was 970 °C–10hours in O_2 gas.

The critical current of a ring was estimated by the magnetization. Figure 3 shows the measurement system of the magnetization of the ring. The magnetization of the ring was measured by the magnetic field at the center of the ring. The typical value of the critical current and the critical current density of one ring was 30 A and 150 A/cm^2.

Figure 4 shows the magnetic field dependence of the critical current densities of the rings at 77K. The error bar shows the scattering of the critical current density of each ring. The critical current densities sintered at 930 °C varied widely compared with those sintered at 970 °C. The critical current densities of the rings sintered at 970°C were higher than those sintered at 930 °C. 11 pieces of the formers were used in the magnet. Excitation procedure of the rings is expressed as follows using the experimental result as is shown in figure 5. The magnet is put into liquid nitrogen, then the rings change in superconductive state. The currents are induced in the rings as the Cu coil is excited in the first step①in the figure. The magnet is excited till the current more than the critical currents of rings in the second step②. When the current of the Cu coil is reduced slowly in the third step③ the persistent currents corresponding to the critical currents are induced into the rings, and the magnetic field is generated in the air gap.

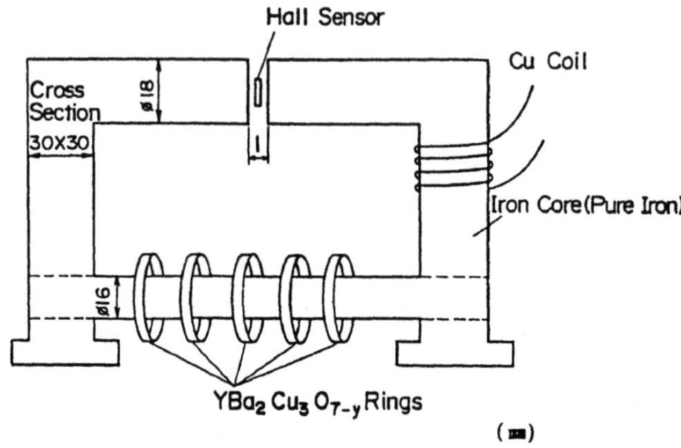

Fig.1 Apparatus of the superconducting magnet.

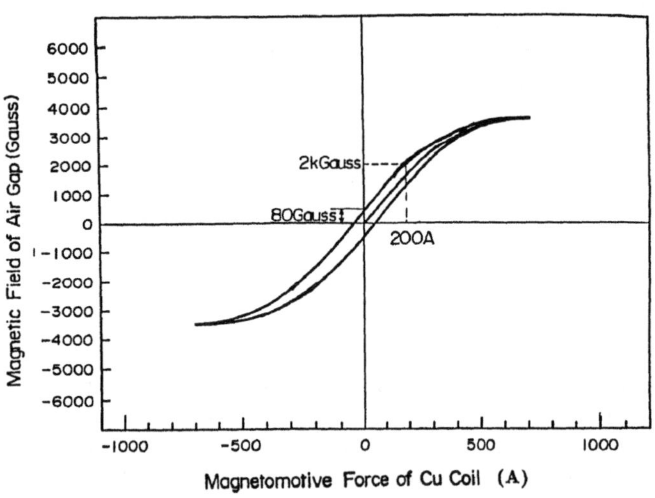

Fig.2 Magnetomotive force of the Cu coil dependence of the magnetic field. The coil is located in the ring region.

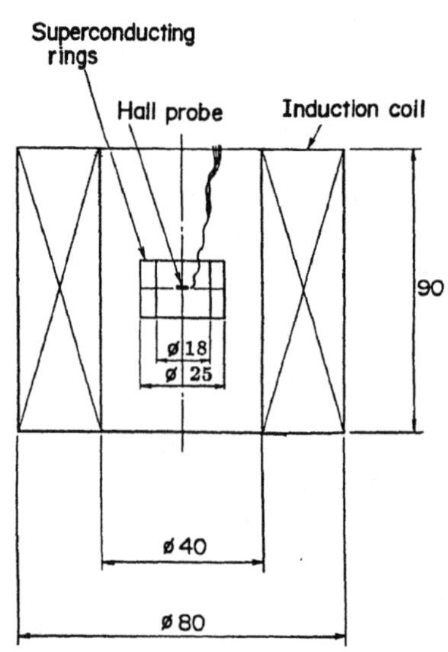

Fig.3 Measurement system of the critaical current Ic of the rings by induction method.

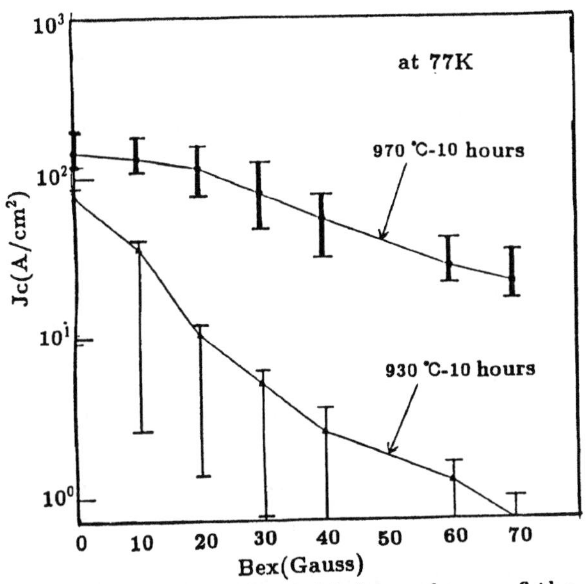

Fig.4 External field dependence of the Jc of the rings.

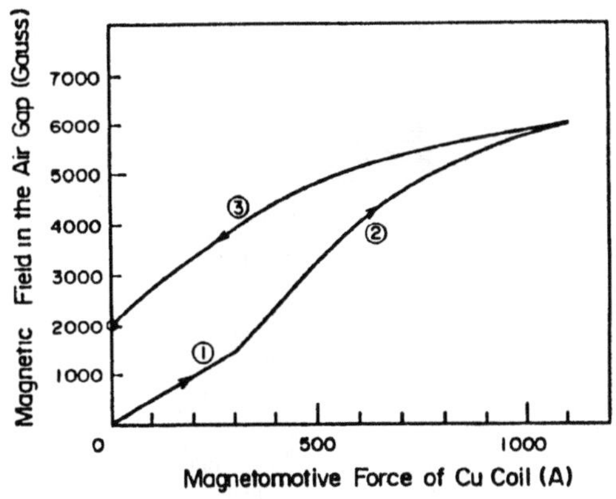

Fig.5 Magnetic fields generated by the persistent currents of the rings.

RESULTS AND DISCUSSION

The magnetic field generated by the persistent current of the 11 pieces of ring was 2 kGauss as is shown in figure 5. From figure 2, the magnetomotive force of the rings estimated by the magnetic fields was 200 A. Figure 6 shows the external magnetic field dependence of the sum of the critical current of each ring at 77 K. The sum of the critical current at B=0T, which was measured one by one, was 310 A. The value is higher than the value estimated from figure 2. This difference suggests that the leakage magnetic field of the iron core in the ring region reduces the critical current of the each ring.

Two dimensional field analysis was carried out to obtain the leakage magnetic field of the magnet. The leakage magnetic field on inner periphery of the rings was 25 Gauss, as the magnetic field in the air gap was 2 kGauss. The magnetic flux distribution was shown in figure 7. The magnetic field of 25 Gauss reduced the critical currents of rings. The sum of the critical current of each ring at B=25 Gauss was 200 A as is shown in figure 6. The value agreeded with the magnetomotive force estimated from figure 2.

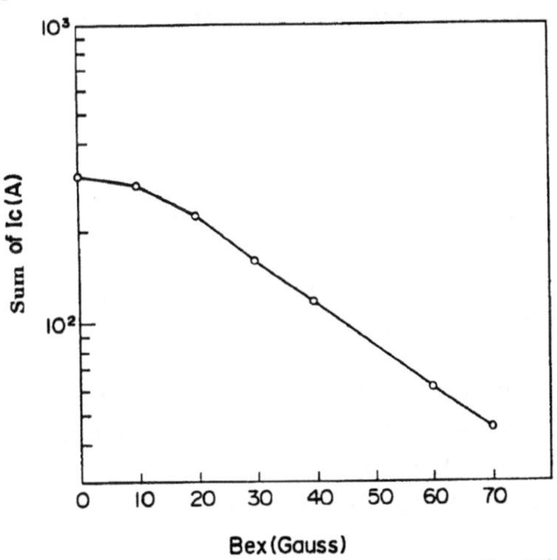

Fig.6 External field dependence of the sum of the critical current of each ring. The critical current of each ring was measured by induction method.

The time dependence of the magnetic field in the case of the initial magnetic field of 2 kGauss was measured. The magnetic field decayed in proportion to $\log t$. It is suggested that the decay is due to the flux creep of rings. In each condition of 79% , 66% and 56% of the sum of the critical cuurent, the time dependence of the magnetic field was measured. The results are shown in figure 8. The decay in proportion to $\log t$ was observed in all condition. The decay of the magnetic field was lower as the currents in the rings were lower. In operation of 56% of the sum of critical current, the decay of the magnetic field was scarcely observed.

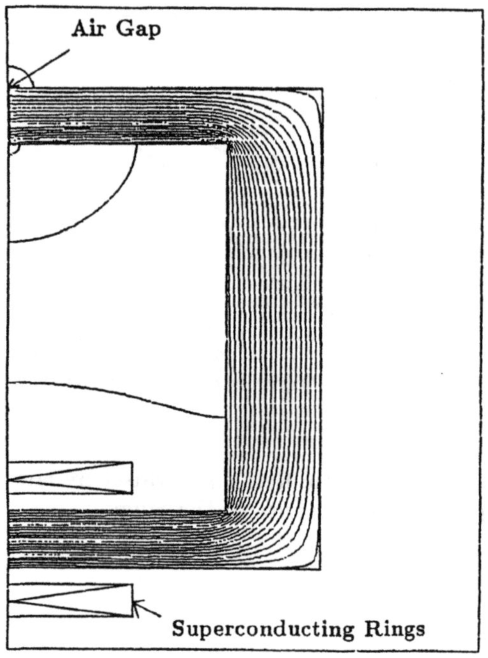

Fig.7 Magnetic flux distribution obtained by two dimensional field analysis in the case that the magnetic field of the air gap is 2 kGauss.

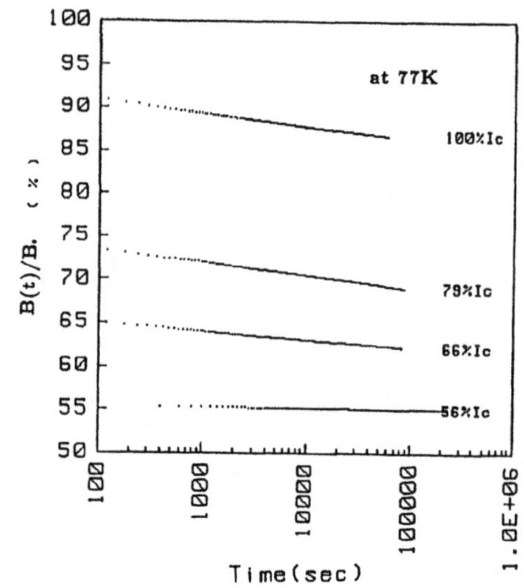

Fig.8 Time dependence of the magnetic fields normalized to 2 kGauss.

CONCLUSION

The $YBa_2Cu_3O_{7-y}$ superconducting magnet with an iron core was developed. The magnet can be used in liquid nitrogen. One of the advantages of the magnet is that the magnetic field can be changed.
The magnetic field of 2 kGauss was obtained. The leakage magnetic field of 25 Gauss was present on inner periphery of the rings.
The time dependence of the decay of the magnetic field was measured and the flux creep was observed. In operation of 56% of the sum of the critical current, the magnetic field scarcely decayed.

REFERENCES

[1] S.L.Wipf and H.L.Laquer, IEEE Transaction On Magnetics, Vol. 25, No. 2, March 1989, pp1877–1880
[2] W.J.Gallagher, R.H.Koch, R.L.sandstrom, R.B.Laibowitz, A.W.Kleinsasser, B.Bumbale, and M.F.Chisholm, Proceedings of the 1st International Symposium on Superconductivity(ISS'88), August 28-31, 1988, Nagoya, pp679–682.
[3] A.Takeoka, A.Ishikawa, M.Suzuki, Y.Kishi, and Y.Kuwano, Proceedings of the 1st International Symposium on Superconductivity(ISS'88), August 28-31, 1988, Nagoya, pp695–698.
[4] T.Yamada, S.Yamamoto, T.Matsuda, M.Morita, M.Takeuchi, and Y.Shimada, Proceedings of the 1st International Symposium on Superconductivity(ISS'88), August 28-31, 1988, Nagoya, pp55–58.
[5] S.Yokoyama, M.Morita, T.Yamada, H.Higuma, Proceedings of the 1st International Symposium on Superconductivity(ISS'88), August 28-31, 1988, Nagoya, pp411–415.

Study of Vertical Transportation System Using Superconducting Linear Motor

H. NAGANO[1], M. KINUGASA[2], T. TOKIZAWA[2], K. HAYAKAWA[3], and K. SASAKI[3]

[1] University of Toyama, Toyama, Japan
[2] Power Reactor and Nuclear Fuel Development Corporation, Tokyo, Japan
[3] Mitsubishi Heavy Industries, Ltd., Kobe, Japan

ABSTRACT

A basic experimental equipment of linear motor using superconducting magnet was constructed and a series of experiments was carried out with it, in order to investigate whether a linear motor method could be used for a vertical transportation equipment or not.

As a result of the experiment, it was found that 30 kgf of maximum thrust force could be obtained, and the capsule could keep on floating and remain stationary.

And, it was proved that measured values by experiment provided good agreement with prospected ones by calculation.

KEY WORDS : Vertical Transportation System, linear motor, Superconducting Magnet, Normal Conducting armature Magnet

INTRODUCTION

A basic investigation and examination about vertical transportation system using superconducting linear motor were conducted.

The program has as its goal the development of enough verified and effective transportation system, in order to serve the utilization of underground space deeper than several hundred meters.

Furthermore, it intends to serve certain development as a part of application fields of superconductor technology.

In this basic experiment, a design goal of thrust force was previously settled to be about 30 kgf, and so experimental equipment was constructed, and then experiment required was successfully carried on, resulting in capsule sustaining in air without mechanical contact.

SPECIFICATION OF EXPERIMENTAL EQUIPMENT

Conceptional drawing and general specification of the experimental equipment are shown in Fig.-1 and Table-1 respectly.

The experimental equipment is principally composed of capsule (cryostat), superconducting magnet, normal conducting armature magnets, capsule guide, superconducting magnet exciting power source and normal conducting armature magnet exciting power source.

(1) Capsule (Cryostat)

Capsule (Cryostat) is 250mm-outside diameter, 200mm-inside diameter, 600mm-depth and 17kgf-weight FRP made vessel.

It is double wall vessel, composed of vacuum vessel outside and liquid helium vessel inside.

And supercunducting magnet will be refrigerated, soaking in liquid helium.

Superconducting magnet is hanged with 3 hanger jigs from upperside frange of capsule.

Thus, the thrust force induced on superconducting magnet will be transmitted to capsule through these jigs.

(2) Superconducting magnet (SCM)

Superconducting magnet is 190mm-outside diameter, 130mm-inside diameter and 76mm-height solenoid type, and NbTi is used as coil material.

It is rated to 70A/mm^2-current density, 25A-current and 50000AT-magneto motive force.

Its weight is 5.5 kgf.

And also, permanent current switch using superconductor is arranged.

(3) Normal conducting armature magnet (NCM)

Normal conducting armature magnet is 400mm-outside diameter, 270mm-inside diameter and 70mm-height solenoid type, and copper is used as coil material.

It is rated to 10A-current and 250V-voltage.

4 magnets with same specification are vertically placed one upon another.

Then, horizontal magnetic field will be induced there when matualy opposite directional currents are sent into alternately placed each pair of magnets respectly.

(4) Superconducting magnet exciting power source

This specification is shown in Table-1.

(5) Normal conducting armature magnet exciting power source

This specification is shown in Table-1.

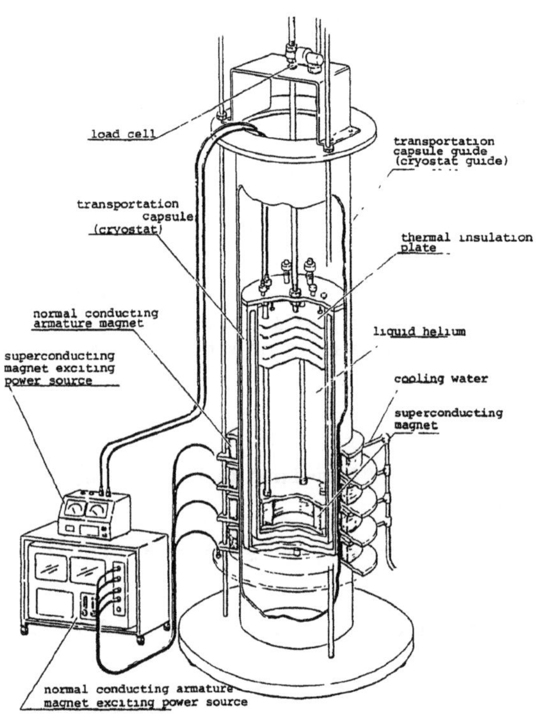

Figure-1 Conceptional drawing of experimental equipment

Table-1 Specification of experimental equipment

No.	Item	Specification
1	cryostat	material FRP size outside diameter 250 mm inside diameter 200 mm height 660 mm depth 600 mm L.He volume 10 ℓ weight 17 kgf
2	superconducting magnet (SCM)	material NbTi size outside diameter 190 mm inside diameter 130 mm length 76 mm current 25 A (50000 AT) current density 70 A/mm^2 weight 5.5 kgf thrust force 27 kgf with permanent current switch
3	normal conducting armature magnet (NCM)	size outside diameter 400 mm inside diameter 270 mm length 45 mm current 10 A (5000 AT)
4	cryostat guide	size outside diameter 270 mm inside diameter 260 mm length 1600 mm
5	superconducting magnet exciting power source	output current 0 ∿ 50 A output voltage 0 ∿ 5 V
6	normal conducting armature magnet exciting power source	output current 10 A output voltage 250 V

TEST RESULTS

After the superconducting magnet on capsule side and the normal conducting armature magnets on outside were assembled, rated current was sent to normal conducting armature magnets.

Then, thrust force induced on superconducting magnet which was in permanent current mode state, was measured by means of load cell.

These test conditions are shown in Table-2.

Table-2 Test condition

	Test 1	Test 2	Test 3
NCM coil	1 piece	2 pieces	4 pieces
NCM current	10 A	10 A	10A
SCM current	25 A	25 A	25 A
Test result	Fig-2	Fig-3	Fig-4

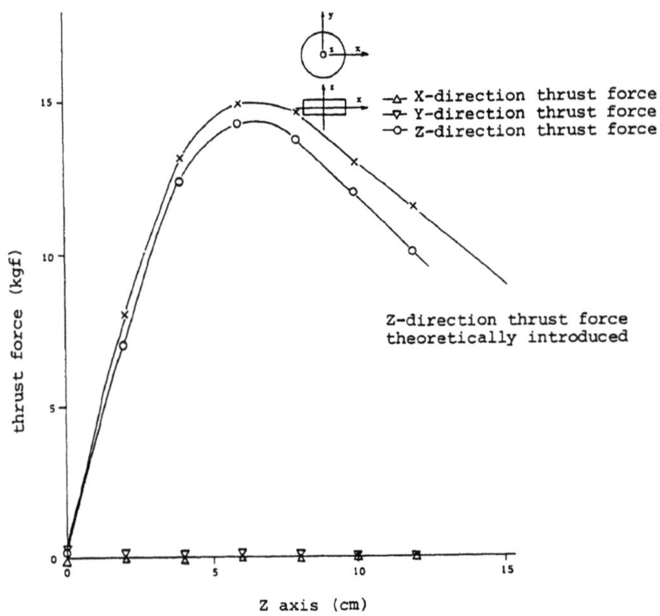

Figure-2 Measured thrust force induced on capsule

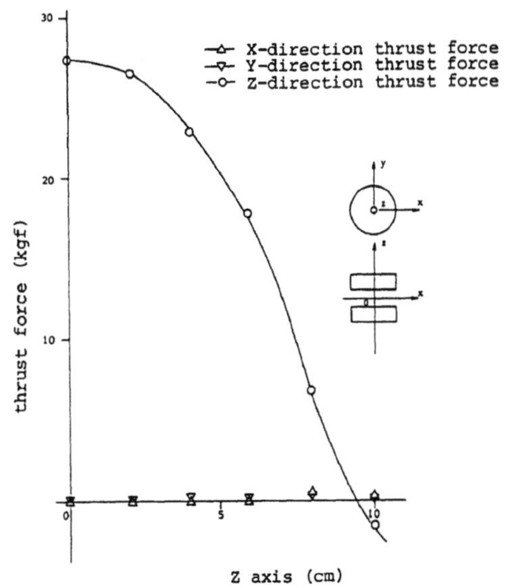

Figure-3 Measured thrust force induced on capsule

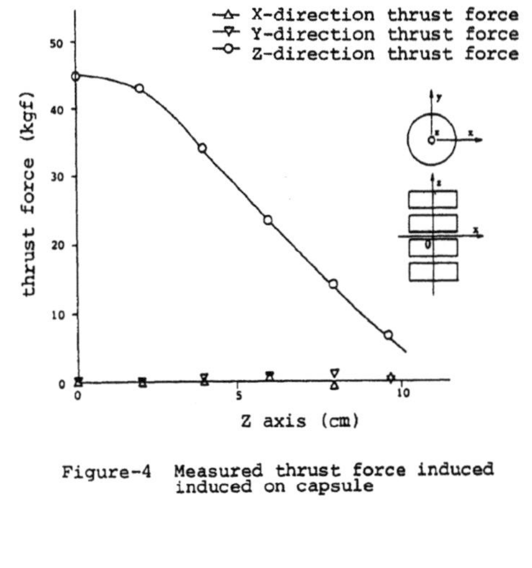

Figure-4 Measured thrust force induced induced on capsule

CONCLUSION

When electric current was sent to two normal conducting armature magnets, capsule (cryostat) began floating and remained stationary at certain level.

This level was depending on the balance position between capsule weight and thrust force induced on magnets, as position control was not applied to this experimental equipment.

But when sine wave type and step type current command was sent to the magnet remotely, the capsule began reciprocating motion among the capsule guide following the command.

The capsule could smoothly run up and down among the capsule guide without pushing the capsule guide wall by any eccentric force.

As result, it was cleared that we could make the capsule vertically run as we like if the capsule position was detected by some detectors and sine curve type currents synchronized to the capsule position were sent to the magnets.

There might be many problems and items to be developed and verified after this.

But it might be concluded that massive and heavy vertical transportation system to underground space deeper than several hundred meters is quite feasible.

Feasibility Study of Compact High-T_c Superconducting Cables by Bean Model for Urban Power System

TSUKUSHI HARA, KIYOSHI OKANIWA, and TAKAHIKO YAMAMOTO

Engineering Research Center, Tokyo Electric Power Company, 4-1, Nishitsutsujigaoka 2-chome, Chofu, Tokyo, 182 Japan

ABSTRACT

Since high-Tc superconductors have very small lower-critical fields Hc_1, a high-Tc superconducting cable of practical size for power transmission should be so designed that its self-field exceeds Hc_1. A compact high-Tc superconducting cable has been designed using a newly-developed method based on the Bean model so that it is installed in existing, cable ducts (150-mm diameter). The results indicate that a compact high-Tc superconducting cable system can increase the power transmission capacity without the need to construct new cable ducts or tunnels. This could cut power transmission costs to less than those of normal-conductor cable systems.

Keywords: high-Tc superconducting cable, power system, Bean model, cable design, power transmission cost,

INTRODUCTION

High-Tc superconductors have very small lower-critical fields Hc_1. Therefore, if a surface-current technique (as in metal superconductors) is used for conceptual designs of the cables, then cable outer diameters become greater than inner diameters of existing ducts. If compact large-capacity cables by new design technique can be fitted into existing cable ducts and tunnels without new construction, costs can be reduced to a large extent.

This paper aims to establish a future model system, assuming that the demand for electric power doubles and based on the present-day underground transmission cable systems in large cities. Furthermore, to apply this model system, a conceptual design for a compact high-Tc superconducting cable was implemented using the Bean model. Hysteresis curves were measured for yttrium silver-sheathed wires, and the validity of the Bean model was studied. A method for designing a compact high-Tc superconducting cable using the Bean model was developed. This design method was used to determine the optimum system voltage of a high-Tc superconducting cable that would fit in ducts with inner diameters of 150 mm. Conceptual designs for cables with 1000 MVA/cct and 600 MVA/cct capacities were studied, and their required critical current density (Jc) was determined.

In addition, analysis of a high-Tc superconducting cable voltage stabilization, short-circuit current, and stability was performed, along with cost evaluation and comparative studies for normal-conductor cable systems.

FUTURE MODEL SYSTEM

The model of the present-day power system, which forms the basis for the present study, is shown in Fig. 1 (a). Based on this system, a model of the future system is shown in Fig. 1 (b) and (c). Normal-conductor cable systems with internal cooling have an outside diameter of about 150 mm or more and require new construction of more cable tunnels and ducts below substations B_1 and B_2 to meet twice the present-day power demand. On the other hand, if a high-Tc superconducting cable with twice the power transmission capacity of present-day cable systems could be fitted into existing ducts and cable tunnels, construction costs would be reduced significantly. Furthermore, a high-Tc superconducting cable has a greatly reduced power loss compared to normal-conductor cables and can carry very large current so that low-voltage power transmission becomes possible. This would permit uniform voltage distribution at low levels (e.g., 66 kV), which would achieve significant cost reduction by eliminating primary substations.

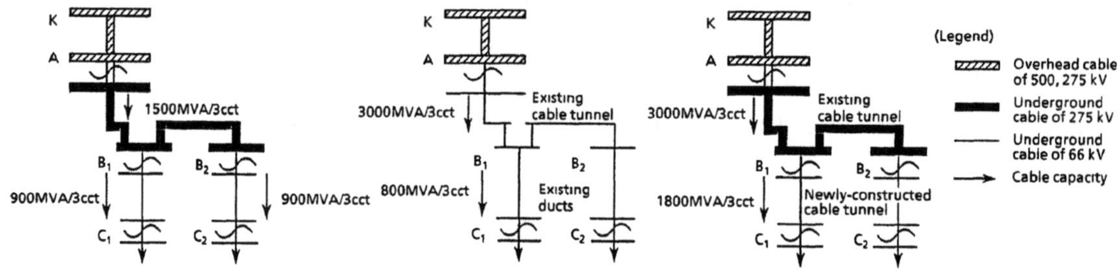

(a) Current model system (b) High-Tc superconducting cable system (The superconductor is used in sections A-B-C.) (c) Normal-conductor cable system

Fig. 1 Model Systems

DESIGN OF COMPACT HIGH-TC SUPERCONDUCTING CABLE BY USING BEAN MODEL

A. Surface Current Density and Conductor Diameter

In conventional superconducting cable designs, the surface current density concept has been used. For a superconducting cable cooled with liquid helium, the surface current density was designed so that the magnetic self field at the conductor surface did not exceed the superconductor lower critical field Hc_1. When the field applied to the conductor is equal to the lower critical field, the surface current density is given by Eq. (1).

$$Is\,(A/cm) = \frac{5}{2\sqrt{2}\,\pi} Hcl\,(0e) \tag{1}$$

As a result, the superconducting tape is in a perfect Meissner state and the conductor AC losses are negligibly small. This was justified by the necessity to minimize heat generation from the conductor AC losses. In a cable design based on surface current, the minimum conductor diameter D is derived from the power transmission current I, as shown in Eq. (2).

$$D = \frac{I}{\pi \cdot Is} \tag{2}$$

At equal power transmission currents, the greater the surface current density, the smaller the conductor diameter can be made. However, as is clear from Eq. (1), the surface current density is restricted by a physical value that is inherent to the superconductor called the lower critical field. High-Tc superconductors have smaller values of Hc_1 than those of metal superconductors.[1] For this reason, high-Tc superconducting cable materials in the region of Hc_1 to Hc_2 should be used for cable. However, if this is done, conductor losses due to the penetration of field will occur. Consequently, qualitative evaluation of AC losses in high-Tc superconducting cable materials is a new technical challenge.

B. AC Loss Evaluation with Bean Model

AC losses in high-Tc superconducting wire include hysteresis loss and viscous-resistance loss in the superconducting layer, and coupling loss and eddy-current loss in stabilizers. Because cables have a self-field of 50/60 Hz operating frequency mainly at the superconducting layer, it is assumed that hysteresis loss is the major conductor loss in cables, as is the case with metal superconductors. Based on this assumption, magnetization curves of Y-Ba-Cu-O silver-sheathed wire were measured by the DC magnetization technique. The hysteresis loss Wh is derived from the magnetization curve, as shown in Eq. (3).

$$Wh = \oint M dHs \ (J/m3 \cdot cycle) \tag{3}$$

Hs: surface magnetic field
M : magnetization of superconductor

Assuming that the maximum value of Hs is Hm, the hysteresis loss ($J/m^3 \cdot$cycle) in the superconducting plate derived from the Beam model is given by the following equation which depends on Jc.

$$Hm \leqq Hp \quad Wh = \frac{2\,\mu o}{3} \cdot \frac{Hm^3}{Jc \cdot d}$$
$$Hm \geqq Hp \quad Wh = 2d\mu oJcHm\left(1 - \frac{2}{3} \cdot \frac{Hp}{Hm}\right) \tag{4}$$

Where, Hp denotes the magnitude of the surface field observed when the quantum flux penetrates to the center of the superconducting layer. For comparison, Fig. 2 show the hysteresis losses observed from the magnetization curves and the values calculated using the Bean model based on Jc obtained from Hp. The values, both observed and calculated, show

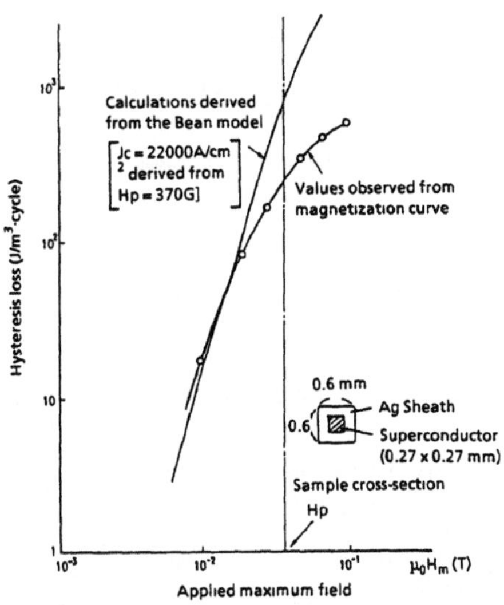

Fig. 2 Hysteresis Loss in Y type Silver-Sheathed Wire Material (77K)

similar field dependency where Hm<Hp with a reasonable agreement in the order of magnitude. However, where Hm>Hp, the calculated values are one or two orders of magnitude greater. This suggests that because Jc of the wire material decreases with increasing field when H_m>Hp.

In this cable design, assuming that only the self-field is applied to the wire material, Hm<Hp is always true for a conductor around which one layer of superconducting tape is wound helically. Consequently, the Bean model seems applicable for approximation of hysteresis loss in high-Tc superconductors. Using the Bean model, the hysteresis loss Wc (W/m) in an entire cylindrical cable conductor is given by Eq. (5).[2)]

$$Wc = \frac{2\sqrt{2}\mu o}{3} \cdot \frac{I^3}{Jc \cdot Pe^2} \cdot f \qquad (5)$$

Where Pe is the conductor perimeter and f is the frequency.

C. Conceptual Design of Compact High-Tc Superconducting Cable and Required Jc

Fig. 3 shows the structure of the high-Tc superconducting cable under review. To reduce the cable size, a three-core construction was used with three cable cores enclosed in

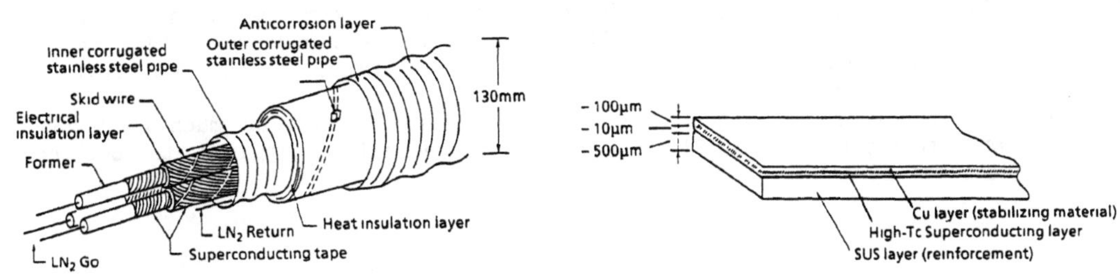

(a) AC Flexible High-Tc Superconducting Cable Concept (b) Structure of High-Tc Superconducting Tape

Fig. 3 Structure of High-Tc Superconducting Cable

a heat-insulating tube as shown in Fig. 3 (a). A former in each cable core formed an cooling channel for liquid nitrogen and a superconductor was wound around its outer layer to conduct electric current. To make the cable flexible, the superconducting layer was sandwiched between a stabilizing copper layer and a reinforcing stainless-steel layer in the form of a tape as shown in Fig. 3 (b). The heat-insulation tube enclosed the three cable cores and a coolant return channel and consisted of stainless-steel corrugated pipes and heat-insulation material (superinsulation, etc.).

Table 1 shows the basic requirements for the conceptual design discussed in this paper. Fig. 4 illustrates the high-Tc superconducting cable with the cooling system.

The variation in coolant pressure loss P (atm/section) was investigated while the value of Jc for high-Tc superconductor of 66 kV, 1000 MVA was changed. The results are shown in Table 2. In this case, it is distinct that Jc= $1\times10^6 A/cm^2$ approx. (with conductor surface flux density of 0.15 T) is sufficient to design a 1000 MVA 66 kV high-Tc superconducting cable of 130-mm diameter.

In addition, the voltage stabilization, short-circuit current and stability of a high-Tc superconducting cable were analyzed with respect to future model systems and a comparative evaluation study relative to normal-conductor cable systems was carried out. The results indicated that high-Tc superconducting cable systems will not pose any problem in actual operation.

Table 1 Basic Requirements

Item	Requirements
Insulation	• Semi-synthetic paper (OPPL) impregnated with liquid nitrogen (5.9 mm thick), $\tan\delta = 0.1\%$, $\varepsilon = 2.8$[3]
Heat insulation	• Superinsulation (SI) (10 mm thick) • $W_T = 2\pi K\Delta T/(\ln(Do/Di))$ (W/m) where W_T: heat inleak, K: thermal conductivity of SI ΔT: temperature difference, Do, Di: outside and inside diameters of heat-insulating tube
Conductor loss (hysteresis loss)	• See Eq. (5)
Pressure loss	• $P = 32\,FM^2L/(\pi^2D^5\rho)$ (N/m^2) where P: pressure loss, F: coefficient of friction, M: coolant mass flow rate L: cooling distance, D: coolant tube diameter, ρ: coolant density

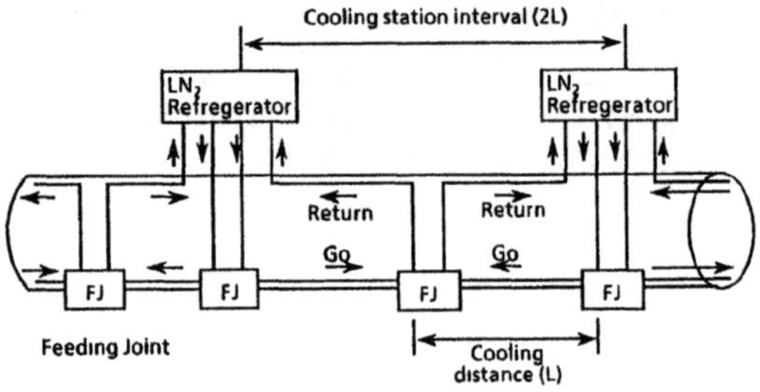

Fig.4 High-Tc superconducting cable cooling system

Table 2 Conceptual Design of 66kV 1000MVA High-Tc Superconducting Cable

Coolant approach pipe diameter	[mm]		19.4	
Conductor outside diameter	[mm]		24.1	
Shield conductor inside diameter	[mm]		35 9	
Core outside diameter	[mm]		38.1	
Heat-insulation tube inside diameter	[mm]		87 7	
Heat-insulation tube outside diameter	[mm]		130 0	
Critical-current density	[A/cm^2]	10^5	10^6	10^7
Conductor loss	[W/m]	30.2	3 0	0.3
Dielectric loss	[W/m]	0.5	0.5	0.5
Thermal penetration	[W/m]	0 7	0 7	0 7
Coolant temperature rise	[K]	15 0	15.0	15 0
Coolant pressure loss	[atm]	367 3	10.9	1.9

COST CONSIDERATIONS

Costs were studied for the model systems shown in Fig. 1 (b) and (c). With due regard to not only the construction cost but also the refrigerators, refrigerator operating power and power loss, power transmission costs were compared on an annual cost. If the high-Tc superconducting wire costs as much as Nb_3Sn, the high-Tc superconducting cable between B and C permits use of existing ducts and eliminates the cost of cable tunnel construction, the high-Tc superconductor cable generates a 15% cost reduction over the entire cable system compared with normal conductor cable. In addition, with high-Tc superconducting cable, the need for primary substations is eliminated by the uniform voltage, so that the total cost of substations is be reduced by 28%.

Assuming that the high-Tc superconducting wire material costs as much as Nb_3Sn, the high-Tc superconductor achieves a 21% overall cost reduction compared to the normal-conductor in the model system.

SUMMARY

A conceptual design of a compact high-Tc superconducting cable was developed using the Bean model. Based on the results, required values of Jc for accommodating superconducting cable in existing ducts (150-mm diameter) for present-day power systems were clarified. The costs were evaluated and the applicability was studied.

(1) Using the Bean model, we developed a conceptual design for a high-Tc superconducting cable by which the pressure loss of liquid nitrogen coolant is below 10 atm and the cable outside diameter is minimized. This design confirms that the optimal cable system voltage is 66 kV so that the cable can fit into the existing 150-mm ducts.

(2) Conceptual designs for high-Tc superconducting cables of 1000 MVA and 600 MVA (with 130-mm cable overall outside diameter) showed that the required value of Jc is 1×10^6 A/cm^2 (0.15 T) and 2×10^5 A/cm^2 (0.09 T), respectively.

(3) Assuming that the cost of high-Tc superconducting wire is about the same as Nb_3Sn, the high-Tc superconducting systems effects a 21% cost reduction (including construction cost, refrigerator operating power, and power loss) over a normal conductor.

REFERENCES

1) Murayama, N. et al., JJAP Vol. 27 No. 9 L1629-1630 Sep. 1988
2) Bean, C.P., Phys. Rev. Letters 8 p250 1962
3) Fukuda, T. et al., Development of VHV OPPL Insulated Oil-filled Cable, FURUKAWA REVIEW No. 6 1988

Design Study of a 100 kWh SMES

T. NAKANO and K. HAYAKAWA

Mitsubishi Heavy Industries, Ltd., 1-1, Wadasaki-cho, 1-chome, Hyogo-ku, Kobe, 652 Japan

ABSTRACT

SMES system will be much more suitable on the points of efficiency and quick responsibility, comparing with other power storage systems.

This report is concerned to the result of conceptual design of the toroidal SMES as a prototype plant, in which magnetic field and electro-magnetic force analysis, superconductor, coil structure, vessel structure, thermal insulating support, cooling system etc. are also discussed.

This study was performed by Mitsubishi Heavy Industries, Ltd. under the contract with Kansai Electric Power Co. Inc.

KEY WORDS : Superconducting Magnetic Energy Storage, conceptual design, toroidal SMES

INTRODUCTION

The conceptual design of the commercial based SMES (5 GMh toroidal type SMES) which might be the goal was performed, and its general specification was decided. (1)

A development scenario to realize the practical SMES will be following :

1. 100 kWh class prototype plant
2. 1-10 MWh class original type plant
3. several hundred MWh class demo plant
4. 1-5 GWh class commercial based plant

According these steps, elementary technical subjects might be overcome one after another.

Then, the basic specification of 100 kWh toroidal SMES was determined, based on the conception of future SMES, and also after due consideration on verification of elementary technology.

CONCEPTUAL DESIGN

The toroidal type of SMES will be more effective to reduce leakage magnetic field by confining magnetic field in the coil than the solenoid type of SMES. On the other hand, it is necessary to reinforce the coil structure, as load supporting by the bed rock can not be effectively used in this case. In case of a 100 kWH SMES, coils can be installed shallow under the ground because the centripetal force will not be so much. The basic specification is described in Tab.-1 and the basic structual feature is shown in Fig.-1.

Table - 1 The basic specification

stored energy	100 KWh (3.6×10^8 J)
superconductor	Nb_3Sn/Cu
stabilizing material	pure aluminium
reinforcing material	aluminium alloy
major radius of coil	10 m
minor radius of coil	0.5 m
center magnetic field	5 T
rated current	25 kA
critical current	50 kA
number of coils	36 units
number of turns	144 turns per coil

1. Conductor

Figure-2 shows a cross section of coil. Nb_3Sn is used as a superconductor, and pure aluminium as a fully stabilizing material. Aluminium alloy is used as a reinforcing material.

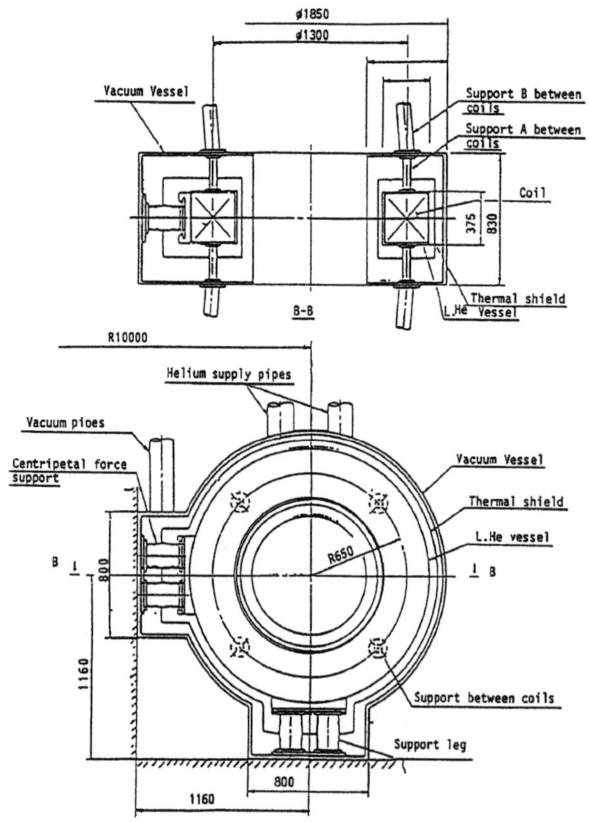

Figure-1 Conception of Main Structure

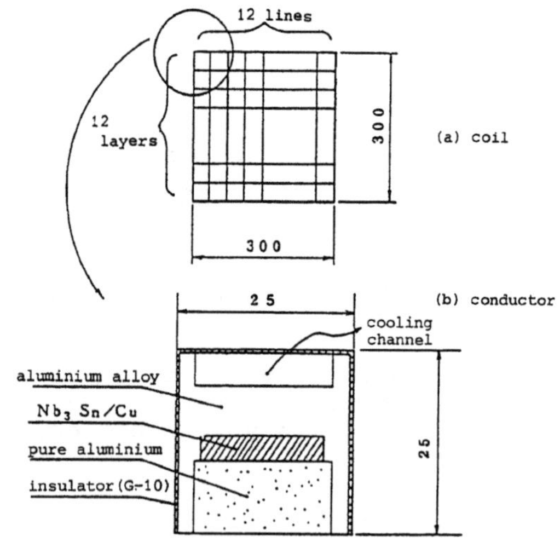

Figure-2 Cross Section of Coil

2 Result of calculation

We have performed calculation of distribution of magnetic field, electromagnetic force and leakage magnetic field, according to this specification. Results of calculation are shown in Fig.-3, and Tab.-2.

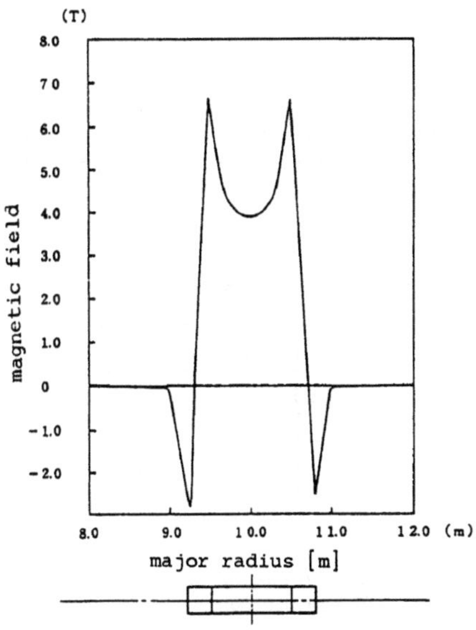

Figure-3 Distribution of Magnetic Field

Table-2 Electromagnetic force on coil (per single coil)

Centripetal force	F_R	1.97×10^5 [N]
hoop force	F_r	2.76×10^7 [N]

3 Vessel and support structure

i) Helium vessel

The feature of configuration of this structure will be that superconducting coils themselves bear the hoop force induced on them, and vessel bear the centripetal force as the resultant force.

This may be called fitting up seat structure.

Aluminium alloy will be selected as the most likely candidate of vessel material.

Concerning with eddy current, generated heat on the coil will not result in any difficulty in the case of 30 minutes charging and discharging pattern, so any measures will not be necessary to take.

ii) Vacuum vessel

The basic design of each part of structures of vacuum vessel was performed, depending on the parameter survey technique as well as the helium vessel.

Though aluminium alloy should be principally used as vacuum vessel material like helium vessel, stainless steel will be rather suitable in this case on the point of vessel design and manufacture.

iii) Support structure

Three kinds of support structures such as support legs, centripetal force support and support between coils (A,B) were investigated, putting special consideration to each part.

Concerning with structure's material, GFRP (Glass Fiber Reinforced Plastic), will be applied to the insulating parts and aluminium alloy to non-insulating parts respectively.

And concerning with support legs and centripetal force support structure, sliding bogies were considered, so that they could bear their own weights or centripetal force, and also could meet heat shrinkage of the structures.

4 Thermal load

Table-3 shows thermal load of the cooling system.

As A.C. loss will be small enough under the one day operating condition of each 30 minutes charging and discharging cycle, this system (SMES) may be estimated to be quite feasible.

Table-3 Thermal load of the cooling system

		4.2 K	80 K
heat penetration	support between coils	55.9 W	196 W
	support leg	61.2 W	336 W
	radiation	1.5 W	590 W
	thermal conduction of gas	16.9 W	49 W
	transfer line	59.3 W	750 W
	current lead	8.4 W + 3.6 g/s	
A.C. loss 1)	wire hysteresis	0.0030W(0.036wh/cycle)	-
	strand cuppling	0.016 W(0.19 wh/cycle)	-
	eddy current on stabilizing material	0.0032W(0.038wh/cycle)	-
	eddy current on L.He vessel	3.7 W(0.97wh/cycle)	-
elastic hysteresis		0.078 W	-
total		207.0 W +4.2 KGHe 3.6 g/s	1921 W

notice 1) average value when 1 cycle/1 day charging and discharging

CONCLUSION

The conceptual design of 100 kWh class toroidal type SMES as the first stage of development of commercial based SMES was performed.

As the results, the basic specification of 100 kWh class SMES prototype which might contribute to verify the technical subjects to be develcped with the MWh class basic plant was decided.

REFERENCE

(1) T. Nakano, K. Hayakawa, M. Shimizu
Design study of a 5 GWh SMES. Proc. of 23rd IECEC Vol. 2'88

(2) M. Shimizu

Study of superconducting magnet energy storage system (Part 2).
Kansai Electric Power Co. Inc. Technical Report No. 42'89

Quench Protection Studies in 20 MWh Solenoidal SMES Magnet

T. ISHIHARA[1], T. DOI[1], T. SHINTOMI[2], and T. TANAKA[3]

Research Association of Superconducting Magnetic Energy Storage, IHI[1], KEK[2], and CRIEPI[3], Tokyo-Chuo Bldg. (IHI), 6-2, Marunouchi 1-chome, Chiyoda-ku, Tokyo, 100 Japan

ABSTRACT

We performed a quench analysis on a 20 MWh solenoid SMES magnet. There are several protection alternatives available, however, of these we have attached primary importance to the "self-protection" method.[1]

KEY WORDS: energy storage, superconducting magnet, quench, solenoid

QUENCHING ASSUMPTION

We assumed the following cases as a case of quenching occuring on a conductor which had been designed according to the fully stabilizing concept.

[1] Degradation of a superconducting cable (case 1)
If the cable is distorted during coil assembly, a quench would occur in the initial current charge.

[2] Degradation of vacuum (case 2)
In the case of the vacuum vessel, if there is a leak at the welded sections of the helium container or the liquid N_2 cooling pipe, the pressure in the vacuum vessel would rise and in the worst instance the liquid helium level would decline and a quench would occur.

It seems there are a number of cases other than the above; quench has occured, for example, due to refrigerator malfunction. The critical temperature of Nb_3Sn, however, at its maximum field of 7.5 T is about 15 K and quench does not seem to occur when at most one or two turns exposed. Moreover, it can be assumes that quench would occur as a result of the wire motion caused by the Lorentz force, however, this could be confirmed by the test plant and we have thus excluded this case from our quench assumption at this time.

QUENCH ANALYSIS

We investigated each of the above mentioned quench factors.

1) Basic parameters
The followings are the basic parameters of the 20 MWh SMES coil that was analyzed.

[1]	β-value (length/diameter)	:	0.084
[2]	Radius	:	30 m
[3]	Inductance	:	3.6 H
[4]	Current	:	200 kA
[5]	Maximum field	:	7.5 T
[6]	Conductor cross section	:	-Fig. 1
[7]	Copper ratio	:	25%
[8]	High purity aluminum	:	Grade - 5N, RRR - 3000

ANALYZING METHOD

We carried out the investigation using the theory introduced by M.N. Wilson[2], and simulated each cases using computer program by giving the propagation velocity as a initial value.

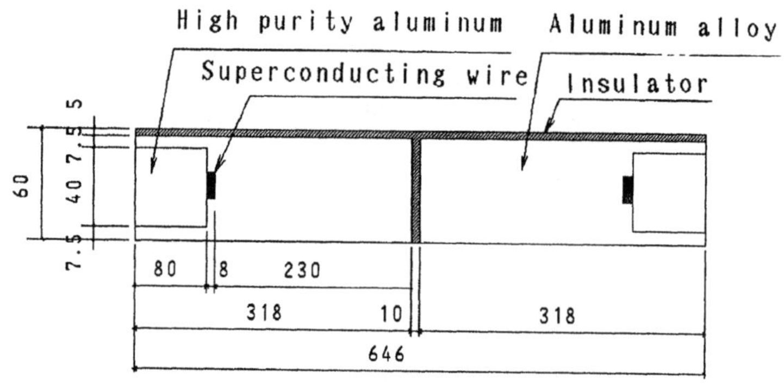

Fig. 1 Coil cross section

QUENCH WITH DEGRADATION OF SUPERCONDUCTING WIRE (CASE 1)

1) Discussion

First we examined conditions of quench propagation and carried out our analysis under these conditions. The value of θ_c, that is enough lowered to propagate the normal zone, is given by the heat balance equation[2], assuming that the propagation velocity is constant.

$$\theta_c < 4.321 \text{ K}$$

The quench may propagate, when the superconducting wire of the coil has deteriorated and some sections of the wire have an θ_c of 4.3 K. The quench does not propagate in the remainder of the wire and thus the deteriorate section may become a hot spot.

$$\nu = 12.8 \text{ mm/sec}$$

The velocity is very slow. We simulated the quench using this value as the initial propagation velocity and obtained the results as shown in Fig. 2. The coil temperature reached 900 K although the coil current did not drop very much. Consequently, the coil cannot be left to stand when quench has occured.

2) Solution

We investigated a method to increase the propagation velocity by transferring coolant into the dump tank. When the conductor does not come into contact with the coolant, the propagation velocity is calculated as follows.

$$\nu_{ad} = 521 \text{ mm/sec}$$

Figure 3 shows the simulation at this initial velocity. The following values were determined as a result.

Maximum coil temperature : 309 K
Maximum coil voltage : 1599 V

These values may be acceptable, however, since the coil temperature rises at the rate shown in Fig. 2 until the coolant is dumped on it, we had to take this effect into account. When the coolant is dumped on it 120 seconds after quenching, the following coil temperature was obtained.

$$110 \text{ K} + 309 \text{ K} = 419 \text{ K}$$

this value may be unacceptable.
To allow the energy to be uniformly dissipated, we assumed a model

whereby 20 points of the coil were heated up then force quenched. For a quench propagation velocity the same as above, we calculated the simulated quench using a 1/20 model, that is, the results for which the following characteristics were obatined.

Maximum coil temperature	:	230 K
Maximum coil voltage	:	94 v

To the above value we need to add the margin of temperature rise upon dumping all the coolant, whereupon the maximum temperature is about 340 K.

QUENCH INDUCED BY VACUUM DEGRADATION (CASE 2)

[1] When vacuum degradation occurs, heat leakage increases. The incremental rate depends on the amount of vacuum leakage, the operating condition of the vacuum pump appratus and the performance of the vacuum pumps.

[2] When the incremental heat leakage ratio is comparatively low, and the transient is slow, the normal energy discharge, magnet warm-up and repair processes are possible.

[3] Extending the above, we see that the quench induced by vacuum degradation occurs in the case of rapid transient, i.e. vacuum failure. But, quench does not occur while the coil is in the liquid helium. After the liqid helium has been completely lost and the coil is heated above the critical temperature, quench then occurs. The helium vessel is then warmed up.

[4] The use of a forced quench may be considered for purposes of vessel protection. In this case the dumping of the liquid helium is done as rapidly as possible and the quench is induced by warming up the coil with heaters.

[5] The characteristics of the forced quench induced by coil warm-up using heaters following the dumping of the liquid helium are the same as in Fig. 4.

SUMMARY INVESTIGATION OF THE DUMP TANK

As shown in above section, when quench causing factors exist, the liquid helium must be separated from the vessel. Consequently, a summary investigation of the dump tank was carried out.

1) Configuration of the dump tank

The configuration of the dump tank is as shown in Fig. 5. The data on the dump tank configuration design are as follows:

[1]	Volume of liquid helium	300 m^3
[2]	Configuration radius of dump tank	32.1 m
[3]	Dump tank temperature	4.2 K
[4]	Natural pressure of dump tank	1 kg/cm^2

From this data, the diameter of the dump tank is determined to be 1.4 m. However, dump tank must be partially cut away due to the interference of the axial support rods of the coil. The resulting diameter would then be about 2 m.

2) Piping of the dump tank

Several design values have been evaluated based upon a 100 sec. dump time for the liquid helium and 1.8 kg/cm^2 for the helium vessel pressure. The evalation indicated that the piping should consists of more than 10 pipes with an inside diameter of 140 mm. Thermal expansion bellows are also necessary. These should, however, be reviewed along with the dump tank temperature and the helium gas vent pipes after coming to a precise decision on the configuration of the coil axial support rods, vacuum vessel, thermal shield and so on.

ACKNOWLEDGEMENT

The authers are extremely thankful for the help they have received during the preparation of this paper to the members of Research Association of Superconducting Magnetic Energy Storage.

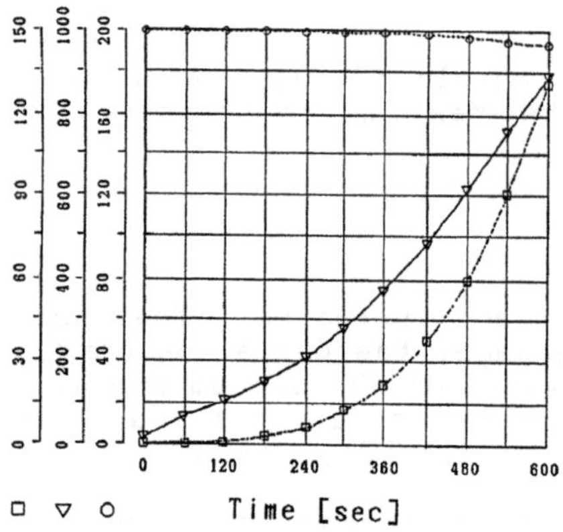

Fig. 2 Quench analysis results (θ_o=4.2 K, v=12.8 mm/sec, L=3.6 H)

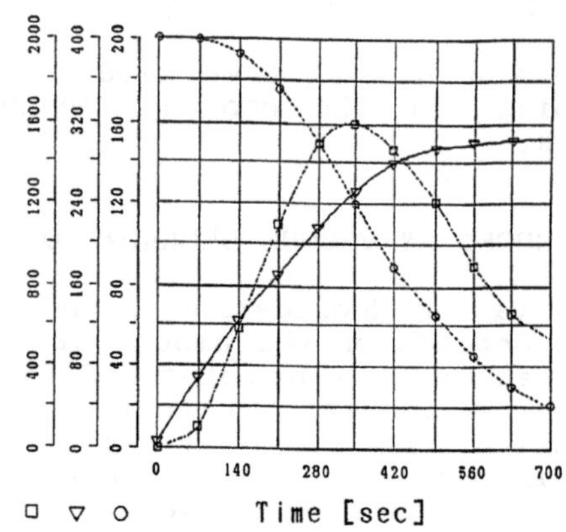

Fig. 3 Quench analysis results (θ_o=4.2 K, v=521 mm/sec, L=3.6 H)

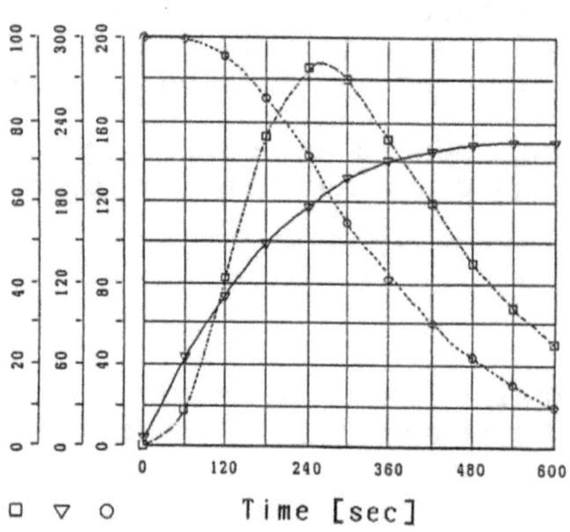

Fig. 4 Quench analysis results (θ_o=4.2 K, v=521 mm/sec, L=0.18 H)

□ Coil Voltage [V]

▽ Coil Temperature [K]

○ Coil Current [A]

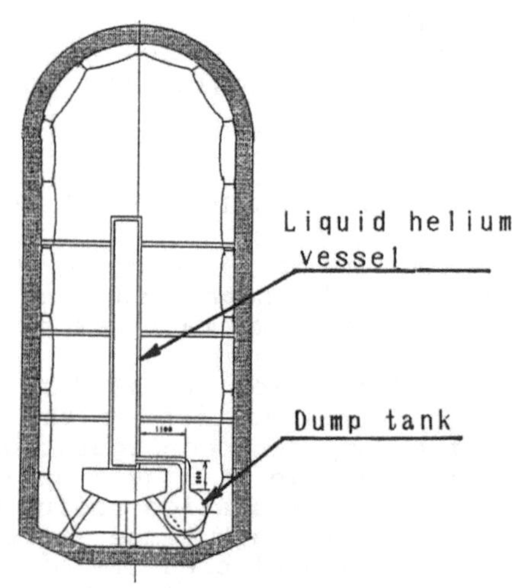

Fig. 5 Configuration of dump tank

REFERENCES

1. Research Association of Superconducting Magnetic Energy Storate (1988) FY 1987 Research Report p126 (Japanese)

2. Martin N. Wilson (1983) Superconducting Magnets. Oxford University Press, New York. pp 200 - 208

Basic Concept of Superconductor Test Items for Application to Power Apparatuses in Super-GM

K. UYEDA, K. TAKAHASHI, T. SAITOH, S. HIROSE, M. SUNADA, and H. HATAKEYAMA

Engineering Research Association for Superconductive Generation Equipment and Materials (Super-GM), Umeda UN Bldg., 14-10, Nishitenma 5-chome, Kita-ku, Osaka, 530 Japan

ABSTRACT

Super-GM conducts the research and development of superconductivity technology for application to power apparatuses such as superconducting generators. In the research of metal based superconductors to be applied to superconducting generators, the requirements for use of the superconductors as field windinds were clarified and the test items and methods for evaluating their properties were examined. With regard to the research of oxide based superconducting materials aiming at the application to power apparatuses, techniques for wire fabrication are being studied using verious synthesis processes. Examination will be also required on test items and methods for oxide based superconducting materials.

KEY WORDS : power application, superconducting gnerator, metal based superconductor oxide based superconducting materials, wire fabrication

INTRODUCTION

Super-GM started the research and development of superconductivity technology for application to power apparatuses from FY 1988 as an 8-year project under the Moonlight Project of Agency of Industrial Science and Technology, Ministry of International Trade and Industry, being consigned by New Energy and Industrial Technology Development Organization (NEDO).

The development of superconductors for field windings of generators is being conducted by manufacturers in charge of developing generators and manufacturers in charge of developing conductors in mutual cooperation under the promotion by Super-GM. In this paper we describe the results of studies on the requirements for superconducting field windings and the test items of their properties which are respectively very important in developing superconductors.

For oxide based superconducting materials, R&D of wire fabrication is being conducted aiming at application to power apparatuses. As the first step, various methods for synthesizing oxide based superconducting materials including solid phase processes, liquid phase processes and vapor phase processes are being examined, and the improvement of superconductivity properties such as improvement of current density is being pursued from the viewpoints of homogenization and compactness intensification of material, control of grain size and grain boundaries and control of crystal orientation. In this paper we also describe the present states in the research of oxide based superconducting materials and in the examination of evaluation items for wire fabrication in Super-GM.

REQUIREMENTS FOR FIELD WINDINGS OF SUPERCONDUCTING GENERATORS AND TEST ITEMS OF SUPERCONDUCTORS

Super-GM worked out the conceptual designs of 200,000kW class superconducting generators in FY 1987, and in this study, clarified the electric and mechanical properties of the rotor of superconducting generator [1]. Especially coutious and challenging develpment is required on the superconducting field winding which is a major component for deciding the performance of superconducting generator.

(1) Main Requirements for the field windings of superconducting generators

In the transient phase such as power system fault, the current of field winding changes rapidly and it causes AC loss. It is very important in the design of field winding to sufficiently examine how and what faults affect the field winding. A three-phase sudden short-circuit fault in a transmission line nearest to a power station is considered as the most severe fault. The field winding must be stably operated without quenching, even when AC loss is caused by rapid current rise during a short-circuit fault and by the slow current variation after removal of fault.

As for the integrity of the field winding in fault operation phase, in addition to the above, sufficient and careful examination is required also on the effects of the short-circuit at generator terminals, unbalanced load operation, load shedding, etc. on the field winding.

(2) Test items of superconductors

The R&D items of superconductors for field windings of generators and test items required for verifying them are summarized in Table 1. Almost all of test methods for the test items are not standardized and require long-term study because of containing many R&D elements respectively. According to the progress in the development of conductors and generators, standardization will be achieved one after another.

Test items can be considered in the following classification:

Table 1 R&D Items and Test Items for Superconducting Field Windings (Metal Based Superconductor)

R&D ITEMS	TEST ITEMS
Higher Critical Current Density	·Critical current (Ic) ·Temperature dependency of Ic ·Variation of Ic relating to cabling ·Variation of Ic under strain ·Current distribution in strand and cable
Higher Stability	·Resistivity (ρ) of stabilizing metal ·ρ of superconducting composite ·Temperature dependency of ρ (RRR) ·Effect of magnetic field on ρ ·Heat transfar coefficient to LHe ·Heat capacity of a composite ·Heat conductivity of a composite ·Minimum Propagating Current ·Minimum Propagating Zone ·Minimum Recovery Current ·Minimum Propagating Energy ·Quench current of model coil ·Quench Propagation Velocity ·Simulation of wire movement ·Temperature (Ic) margin ·Effect of impregration
Reduction in AC losses	·Total loss measurment (thermal and electrical) ·Dependency on frequency of changing field ·Dependency on amplitude of changing field ·Evaluation of transverse resistivity (ρ) ·Effect of trnsport current ·Effect of stress on strand coupling ·Proximity effect ·Effect of biasing DC field ·Temperature dependency of loss
Higher Mechanical Strength	·Stress-strain property (tension and compression) ·Bending property ·Fatigue property ·Evaluation of rigidity

Table 2 Test Items Being Standardized in Super-GM (Metal Based Superconductor)

Classifications		Test Items
Simple property	Geometrical properties	Dimensions
	Electrical properties	Resistivity(RRR)
		Strand critical current
		Conductor critical current
		AC losses
	Mechanical properties	Tensile properties
		Compressive properties
Complex property		Resistivity under strain
		Critical current under strain
		Stability
		Rigidity

- Properties which can be evaluated as independent conductors in separation from the service environments of field windings of generators
- Properties which cannot be examined without considering the service environments of field windings of generators

The former as simple properties and the latter as complex properties to be examined urgently are listed in Table 2.

PRESENT STATES OF RESEARCH ON OXIDE BASED SUPERCONDUCTING MATERIALS, AND STUDY ITEMS CONCERNING WIRE FABRICATION TECHNIQUES

As the first step in the research of wire fabrication from oxide based superconducting materials, various synthesizing methods are being studied. The processes for synthesizing oxide based superconducting materials include solid phase processes, liquid phase processes and vapor phase processes. For R&D, the basic research for wire fabrication from oxide based materials is classified into three major categories:

- Study on the homogenization and compactness intensification of crystral structure
- Study on the control of grain size and grain boundaries
- Study on crystal orientation

and furthermore subclassified into seven synthesis process subjects. Table 3 lists research subjects and main data of superconductivity properties.

As the second step, research will be promoted on the enhancement of critical current density and wire fabrication, and finally, the applicability of oxide based superconducting materials to power apparatuses will be evaluated.

Moreover, to compare the measured results of properties of oxide based superconducting materials, methods for measuring the critical current density and the critical temperature were examined [2]. A draft of study items with wire fabrication taken into account is shown in Table 4. In future, in order to evaluate the various methods for synthesizing oxide based superconducting materials, it is required to study the methods for measuring superconductivity properties and to examine the evaluation items from the viewpoint of wire fabrication.

Table 3 Research Subjects and Main Results of Oxide Based Superconducting Materials in Super-GM

Reserch subjects		Main results
Homogenization Compactness	Improving homogenization and compactness in wire structure	By submicron particle sintered bulk method: Tc=89K, Jc=1000A/cm^2
	Improving homogenization and compactness in large area	By plasma spraying method: Tc=91K, Jc=750-1050A/cm^2
Controlling grain and grain boundary	Study of non-equilibrium reaction technology	By cluster ion beam method: Tc=82K, Jc (on testing)
Controlling crystal orientation	Crystal orientation control by mechanical stress, etc.	By laser evaporation method: Tc=88K, Jc=2.3x10^4A/cm^2 (77K, 0.5T)
	Study of single crystal wire processing by CVD, etc.	By thermal decomposition CVD method: Tc=84K, Jc=1000A/cm^2
	Study of single crystal wire processing by melt growth	By melt processing (laser pedestal etc.): Tc=106K, Jc=1560A/cm^2 (Bi system)
	Study on process synthesizing fine crystallites and highly oriented ceramic materials	By laser beam irradiation: Tc=91-92K (onset), Jc=73A/cm^2

Table 4 Evaluation Items for Wire Processing of Oxide Based Superconducting Materials (Draft)

Classification	Evaluation and test itemes
Characteristics of materials	- Uniformity of chemical composition - Synthesized ratio - Critical temperature - Critical current density - Magnetic field dependence of critical current density etc.
Characteristics of wires	- Dimension (cross ection, length) - Processing (synthesizing) ratio - Critical current density and critical current - Environment resistance - Mechanical characteristic (tensile stress, flexibility) - Characteristic of conductivity on stress (tension, compression bending, repeated load) - Heat stability etc.

CONCLUSION

The requirements for the superconductors used as field windings of generators were studied, and test items were clarifield. At present, for simple properties, a test code is being prepared. Henceforth, the properties of developed superconductors will be quantitatively evaluated by applying the test code, and in order to develop conductors with more excellent properties, the results will be fed back to conductor design.

As future research of oxide based superconducting materials for application to power apparatuses, further optimization will be pursued in the seven synthesis processes and research and development will be conducted on wire fabrication techniques utilizing features of the synthesis processes. Futhermore, based on the basic concept in the metal based superconductors, test items and methods for evaluating the wire fabrication techniques in the project will be examined.

ACKNOWLEDGEMENT

This research has been carried out as a part of "R&D on Superconducting Technology for Electric Power Apparatuses" under the Moonlight Project of Agency of Industrial Science and Technology, Ministry of International Trade and Industry, being consigned by New Energy and Industrial Technology Development Organization (NEDO).

The authors would like to thank the members of the working group in charge of preparing the test code for superconductors in Super-GM.

REFERENCES

1. "Feasibility study on superconducting machinery and materials technology to electrical power generation", prepared by Super-GM for Moonlight Project of AIST, MITI (March 1988)
2. Kimura K. et al IEEE Trans. Mag. 25, 2032 (1989)

Author Index

Key Word Index

MIX
Papier aus verantwortungsvollen Quellen
Paper from responsible sources
FSC® C105338

If you have any concerns about our products,
you can contact us on
ProductSafety@springernature.com

In case Publisher is established outside the EU,
the EU authorized representative is:
Springer Nature Customer Service Center GmbH
Europaplatz 3, 69115 Heidelberg, Germany

Printed by Libri Plureos GmbH
in Hamburg, Germany